4-D Framework of Continental Crust

Edited by

Robert D. Hatcher Jr.
Department of Earth & Planetary Sciences
and
Science Alliance Center of Excellence
University of Tennessee
Knoxville, Tennessee 37996-1410
USA

Marvin P. Carlson
Nebraska Geological Survey
102 NH UNW NEB
Lincoln, Nebraska 68588-0517
USA

John H. McBride
Brigham Young University
Department of Geological Sciences
P.O. Box 24606
Provo, Utah 84602
USA

José R. Martínez Catalán
Departamento de Geología
Universidad de Salamanca
37008 Salamanca
Spain

THE
GEOLOGICAL
SOCIETY
OF AMERICA®

Memoir 200

3300 Penrose Place, P.O. Box 9140 ▪ Boulder, Colorado 80301-9140 USA

2007

Published by The Geological Society of America, Inc.
3300 Penrose Place, P.O. Box 9140, Boulder, Colorado 80301-9140, USA
www.geosociety.org

Printed in U.S.A.

GSA Books Science Editors: Marion E. Bickford and Donald I. Siegel

Library of Congress Cataloging-in-Publication Data

4-D framework of continental crust / edited by Robert D. Hatcher ... [et al.].
p. cm. (Memoir ; 200)
Includes bibliographical references and index.
ISBN 978-0-8137-1200-0
1. Continental crust. 2. Plate tectonics. 3. Geology, Structural. I. Hatcher, Robert D., 1940–. Memoir (Geological Society of America) ; 200

QE511.A13 2007
551.1'3--dc22

2007020532

10 9 8 7 6 5 4 3 2 1

Contents

Preface

This volume contains a series of papers related to aspects of the evolution of continental crust through time. Most of the papers were presented as keynote, invited, and volunteered papers at the 17th International Basement Tectonics Conference in Oak Ridge, Tennessee, 27 June–1 July 2004. Several additional papers were submitted independent of the conference. This conference was one of a series conducted under the auspices of the International Basement Tectonics Association, the first of which was held in Salt Lake City, Utah, in 1975 and organized by S. Parker Gay Jr. In addition, the 2004 conference was cosponsored by the Geological Society of America, the American Geological Institute, the American Geophysical Union, the University of Tennessee–Knoxville, and the Oak Ridge National Laboratory. Partial financial support for the conference, primarily for key international participants, some key U.S. participants, and students, was provided by the National Science Foundation. Partial support for field trips was provided by the U.S. Geological Survey. Additional financial support was provided by the University of Tennessee Science Alliance Center of Excellence.

Papers in the volume are broadly grouped into general papers, papers discussing accretionary orogens, structural geology papers, papers on Precambrian crust-forming processes, and papers on Paleozoic crust-forming processes. All make important contributions toward our understanding of the plate tectonics and possibly pre–plate tectonics (Hamilton) processes that form and modify continental crust through time.

All papers were reviewed by qualified specialists in the different topics of papers submitted for this volume. They were revised with considerable improvement of all manuscripts before being accepted by the editors for publication. We acknowledge the following for their conscientious and helpful reviews; their reviews largely determine the quality of papers published in the book: John N. Aleinikoff, J. Lawford Anderson, David Barbeau Jr., Mervin J. Bartholomew, Wouter Bleeker, Jean-Pierre Burg, Peter A. Cawood, Kent C. Condie, John O. Costello, John F. Dewey, William R. Dickinson, William M. Dunne, Bernard W. Evans, Rodger T. Faill, Peter Geiser, M. Charles Gilbert, Warren B. Hamilton, Thomas B. Hanley, Michael W. Higgins, David M. Hirsch, Raymond V. Ingersoll, Suzanne M. Kay, Theodore C. Labotka, Edward G. Lidiak, Harry Y. McSween, Kevin L. Mickus, Brent V. Miller, Eric W. Mountjoy, R. Damian Nance, Charles M. Onasch, Annie Rassios, Steven Schamel, Donald T. Secor Jr., Frank S. Spear, Mark G. Steltenpohl, James F. Tull, W. Randall Van Schmus, Martin J. Van Krannendonk, Cees R. van Staal, and Donald U. Wise. Some reviewers kindly reviewed more than one paper, and the editors, in addition to serving as "associate editors," reviewed a few papers. A few reviewers requested to remain anonymous, and are not listed. We also appreciate the guidance of former GSA Books Editor Abhijit Basu during the completion of this volume.

Robert D. Hatcher Jr.
Marvin P. Carlson
John H. McBride
José R. Martínez Catalán

The Geological Society of America
Memoir 200
2007

The secular evolution of plate tectonics and the continental crust: An outline

John F. Dewey
Department of Geology, University of California, Davis, California 95616, USA
and
University College, High Street, Oxford, UK

ABSTRACT

The main conclusion of this paper is that some form of plate tectonics started at ca. 3.0 Ga or possibly as early as 3.1 Ga., and that, since then, plate tectonics has steadily become dominant over plumes as a mechanism of heat loss. Modern plate tectonics started at ca. 0.6 Ga. The volume history of the continental crust is one of fast Early Archean growth to generate a, probably, globally continuous crust, then a growth, probably exceeded or balanced, long-term, by crustal return to the mantle reservoir.

Keywords: secular tectonic evolution of Earth, inversion, plate tectonics.

INTRODUCTION

This short essay is not intended as a thorough, scholarly, fully referenced work. Rather it is a distillation, a précis, and a personal statement of my views about Earth's evolution based upon seeing, recording, and thinking about rocks and their relationships around the globe for the past fifty years. Like Warren Hamilton (2003), with whose views about the tectonic evolution of Earth, in principle, I largely agree, the paper is not fully referenced with all those papers that are important and deserve reference in a thorough work; I apologize to all those who believe that their relevant work is not referenced. My views about Earth's tectonic evolution have changed and oscillated as I have been influenced by the views and experiences of many geologists. I have now returned, in principle, to my view of the early 1970s, which espouses secular evolution from an Archean, pre-plate-tectonic, permobile Earth to a Phanerozoic plate-tectonic Earth (Burke et al., 1976b).

Of the planets in the solar system, only Earth has extant plate tectonics. Past and present crustal mobility and deformation, indicative of convective processes, is evident in other planets and moons, but only on Earth do plates form a lithospheric thermal boundary conduction layer that is broken into a number of torsionally rigid spherical caps or shells, whose relative motion can be described by an angular rate around a pole of rotation. One-plate (continuous lithosphere) planets (Solomon, 1978) are characterized by early expansion, rifting, and volcanism, followed by contraction and thrusting. In today's Earth, plate tectonics is driven by slab-pull and ridge-push; the generation and conductive cooling of the lithosphere is the principal engine of global heat loss. The other mechanism of heat loss is by the formation of large igneous provinces, the surface expression of mantle plumes, commonly, though not exclusively, associated with mega-dike swarms and continental breakup. These two expressions of convection differ critically in that horizontal motions and slab subduction form narrow linear/arcuate zones to characterize plate tectonics, whereas vertical mass transfer typifies plumes and large igneous provinces. Secular tectonic change and evolution are certain consequences of Earth's thermal evolution; as radiogenic heat production declined, plates have cooled, thickened and strengthened, and flat-slab subduction is likely to have diminished in favor of steeper-slab subduction. Geologic

Dewey, J.F., 2007, The secular evolution of plate tectonics and the continental crust: An outline, *in* Hatcher, R.D., Jr., Carlson, M.P., McBride, J.H., and Martínez Catalán, J.R., eds., 4-D Framework of Continental Crust: Geological Society of America Memoir 200, p. 1–7, doi: 10.1130/2007.1200(01). For permission to copy, contact editing@geosociety.org.

data indicate, clearly, that there have been associated secular changes in rock associations, facies, and structures. Also, there have been episodic/periodic changes related to the assembly, fragmentation, and dispersal of continents with Wilson cycles of oceanic opening and closing. A key question in evaluating the rock assemblages that record, partially, the history of Earth, is the relative importance of plates and plumes with time. Was there a time before which plate tectonics did not operate and when tectonic processes were plume-dominated with globally pervasive convective overturn up to the shallowest levels of the crust?

Although many, if not most, of the processes that make and modify the continental crust (volume 5.25×10^9 km^3, mass 1.44×10^{23} g) are known in a general way (Dewey and Windley, 1981), their mechanisms (e.g., origin of arc volcanism, mechanisms of ultrahigh-pressure [UHP] exhumation, secular evolution) and, especially, rates, such as volume of continental crust, through time) are poorly or imperfectly understood. Although it is difficult to interpret the terrane collage of surface geology, our knowledge of the deep crust is limited to xenoliths, seismic data, and rare direct observation (e.g., Kapuskasing, Ronda, Finero, Beni Bouchera, Ivrea; Dewey, 1986). Whereas the principal engine of global heat loss is the accretion, cooling, subsidence, and subduction of the younger-than-160 m.y., dense and torsionally strong, relatively simple, oceanic lithosphere, the buoyancy and weakness of the exceedingly complicated continental crust allows its growth but prevents, at least in the extant plate-tectonic regime, its wholesale recycling back into the mantle, hence the preservation of rocks from almost 4 Ga.

Addition (transfer of mass from the mantle to the continents) occurs by subduction zone magmatism (major), mafic under- and overplating, and the incorporation of fragments of oceanic crust and seamount/plateaux (minor). Subtraction (transfer of mass from the continents to the mantle) may occur by delamination of a mafic lower crust, and subduction erosion (Dewey, 1980), and does not include shallow recycling, which is merely the transfer of tectonically decreted material from shallow to deeper, crustal subduction levels. Delamination of a mafic lower crust allows the trend toward a more silicic bulk composition. Growth/diminution are, respectively, the net gain/loss of continental crust, and equal addition minus subtraction. Accretion/decretion are, respectively, area increase/decrease by the addition or redistribution of mass at the edges of continents. Enlargement/reduction are, respectively, increase/decrease in continental area by stretching/shortening and accretion/decretion. Differentiation is the development of an upper potassic/silicic minimum-melt crust rich in radioactive and large-ion lithophile elements, and a refractory lower crust. The putative secular trend toward an increasing potash/soda ratio from general Archean values <1.0 to Proterozoic-Phanerozoic values >1.0 would indicate progressive growth, differentiation, and cratonization. My estimate, based upon a long-term and extensive global study of continental geology, is that crust was generated, during the Archean and Archeo-Proterozoic, at a growth rate of $\sim 11.17 \times 10^{15}$ g/yr, some six times faster than the Proterozoic-Phanerozoic rate of 1.64×10^{15} g/yr (Dewey and Windley, 1981).

The Archean nucleic cratons are a tiny remnant of the globally continuous Archean crust. The Armstrong (1968, 1981, 1991) and Fyfe (1978) growth curves probably are close to reality but modified as follows: During the Hadean, a global basaltic/komatiitic crust was recycled into the mantle from ca. 4.0 to 3.85 Ga. During the Archean, from ca. 3.85 to 3.0 Ga, a new global crust was formed, which has been recycled, largely, back into the mantle reservoir. Since the Late Archean, there has been a rough balance between growth of the continental crust, in arcs and by mafic underplating, and its return to the mantle reservoir. The mechanisms of return include the deep subduction and delamination of eclogitized crust in collision zones (Ryan and Dewey, 1997) and tectonic erosion in subduction zones (Dewey, 1980; Scholl and von Huene, this volume). Subtraction of continental crust is well documented in the India-Asia collision zone where there is a substantial volume deficit (Dewey et al., 1989). However, not all subducted, eclogitized, crust is returned to the mantle reservoir; there are several exhumed high-pressure/low-temperature coesite-eclogite terrains (Krabbendam and Dewey, 1998). What determines whether such crust is returned to the mantle or exhumed may be related to its density/hydration history. There are other views of the volume, area, growth, and destruction of continental crust with time (e.g., Hurley and Rand, 1969; Veizer and Jansen, 1979), which will not be discussed and evaluated because this short paper is concerned, principally, with tectonic styles and mechanisms with time; I recognize, however, that the issues are related intimately.

EARTH EVOLUTION

There is little doubt that a secular evolution of tectonic style has occurred through Earth's history (Dewey and Spall, 1975; Hamilton, 2003). Also, there has been diminishing bolide frequency and size. These secular trends were punctuated by episodic and, possibly, periodic events. The assembly, fragmentation, and dispersal of at least three supercontinents, during the Proterozoic-Phanerozoic, has led to a plate-tectonically driven episodic/periodic "cyclic" repetition of structural, sedimentary, igneous, metamorphic, and climatic events (Dewey, 1988a), of which the Wilson cycle of oceanic opening and closing is a consequence. During the assembly of a supercontinent, continental collisions cause mountain building, far-field inversion of rifts, continental shortening, lowering of sea level, a high clastic/carbonate ratio, rise in $^{87}Sr/^{86}Sr$, drop in pCO_2, and low benthic diversity. Continental fragmentation is accompanied or caused by subduction rollback in circum-Pangea extensional arcs, alkaline magmatism, the extensional collapse of intra-Pangea orogens, and the rapid exhumation of mid and lower crustal rocks. Maximum continental dispersal is recorded by Andean-type orogeny, the obduction of supra-subduction-zone ophiolites, blueschists, carbonate shelves, deep-marine and platform euxinites, few clastics, low $^{87}Sr/^{86}Sr$, rise in pCO_2, and high benthic diversity.

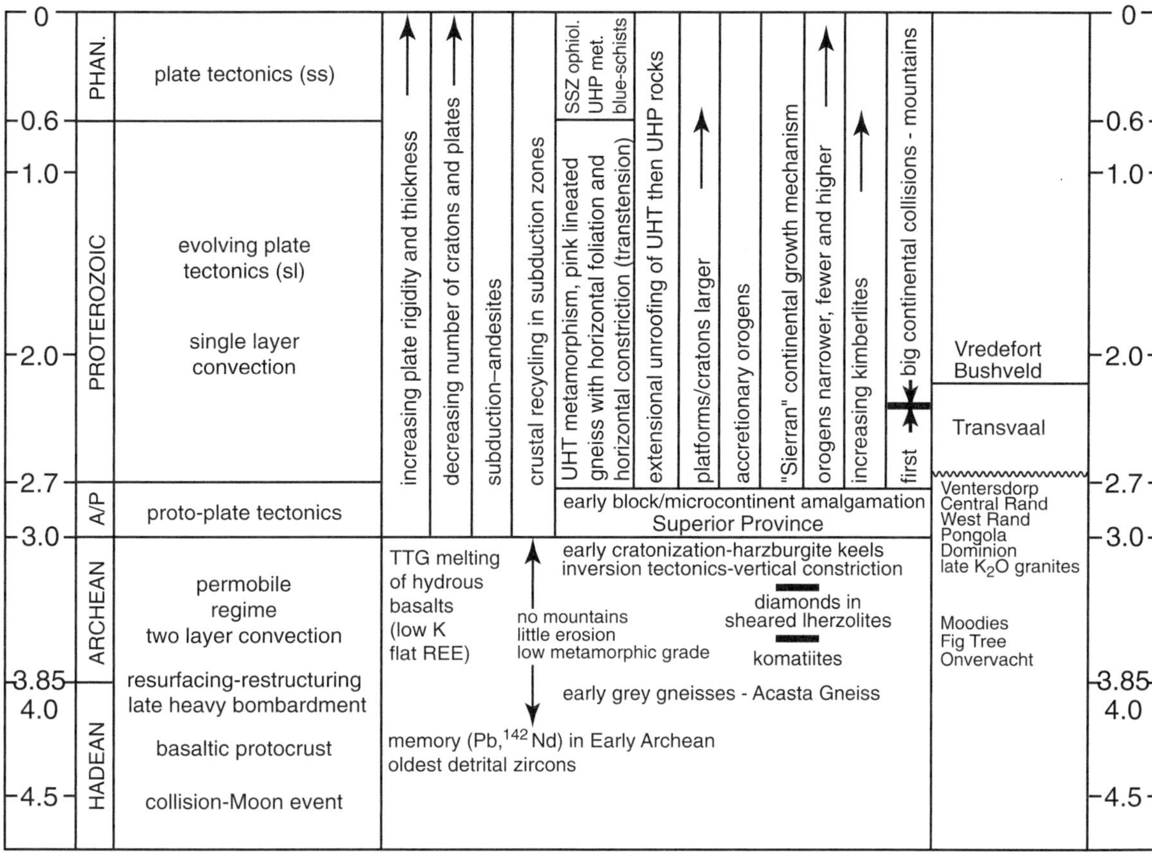

Figure 1. Suggested tectonic evolution of Earth. A/P—Archeo-Proterozoic (new term); REEs—rare earth elements; SSZ—supra-subduction-zone; TTG—tonalite-trondhjemite-granodiorite; UHP—ultrahigh-pressure; UHT—ultrahigh-temperature.

We should place eon and other boundaries at important events and changes in Earth history rather than at arbitrary numbers and unconformities: 3.85 Ga for the Hadean/Archean boundary, 3.0 Ga, perhaps a little earlier at 3.1 Ga, for the Archean/Archeo-Proterozoic (new term) boundary, 2.7 Ga for the Archeo-Proterozoic/Proterozoic boundary, and 0.6 Ga (base of Ediacaran) for the Proterozoic/Phanerozoic boundary. My view of Earth's evolution, summarized in simple tabular form in Figure 1, is similar to that of Burke and Dewey (1972, 1973; especially see Fig. 1 of 1973) and Hamilton (2003) but with temporal differences outlined below.

The Hadean, from 4.7 to 3.85 Ga, was an eon of a global basaltic proto-crust, which was wholly recycled before and during the intense late bombardment, from which only a few 4.4–4.3 Ga zircons, from presumed felsics, remain. Some fragmentary gray gneisses, in which a 4.36–4.1 Ga Pb and ^{147}Nd and Sr "memory" is preserved, could be late Hadean remnants. The Acasta Gneiss (Bowring et al., 1989) and Isua supracrustals (Moorbath et al., 1973) may be the oldest preserved rocks on Earth but not generated by a plate-tectonic process. The Hadean-Archean transition may have been a time, during the late heavy bombardment, of complete crustal/lithospheric turnover such as the process that may characterize the periodic (500 m.y.) evolution of Venus (Phillips and Hansen, 1998).

Arguments about post-Hadean secular tectonic evolution persist (Eriksson et al., 2004), especially processes responsible for Archean evolution. Kroner (1981) argued that rock assemblages and paleomagnetic data indicate that some form of plate tectonics must have operated back to, at least, the Late Archean. One view is that vertical diapiric inversion generated the "greenstone keel/gregarious batholith" pattern so typical of the Archean crust (MacGregor, 1951; Talbot, 1968; Burke and Dewey, 1973; Burke et al., 1976b; Hamilton, 2003). The opposing view (Burke and Dewey, 1972) is that Archean crust was generated by subduction processes some six times faster than modern processes. The geology of the Barberton Mountain Land (Anhaeusser, 1983) appears in some ways similar to that of western Newfoundland and western Ireland during the Ordovician (Van Staal et al., 1998) and that of the Sierra Nevada foothills during the early Mesozoic, where the growth and collisional amalgamation of primitive oceanic arcs was followed by the intrusion of tonalite batholiths from migrated and/or flipped subduction zones. In the Barberton Mountain Land, the Onvervacht could be interpreted as an Archean oceanic crust, the Fig Tree as an accretionary complex, the Moodies as an episutural sequence, the tonalite plutons as crust-forming arc intrusions, and the late adamellite sheets as a cratonizing event. Alternatively, the Onvervacht could be seen as primitive, highly magnesian lava sequences, on an older tonalite-trondhjemite-granodiorite (TTG) continental crust, representing a period of massive mantle partial melting, followed by diapiric inversion. In this gravity-driven model, the two-layer high-density komatiite, low-density TTG package inverts with broad, rounded-profile TTGs rising diapirically and sharp "greenstone keels" sinking. The latter is my preferred model because, especially, it explains the structure of Early Archean rocks with vertical constrictional fabrics in the cores of greenstone keels to the oblate marginal fabrics of ballooning plutons. If the TTGs were oceanic island arcs, it is strange that they lasted 300 m.y., that komatiite sequences rest, unconformably, upon older continental crust (Bickle et al., 1975), and that they have a blobby aspect with no obvious sutures. Andesites are rare in the Archean, suggesting the absence of slab subduction.

During the Archean, from 3.85 to ca. 3.0 Ga (the permobile regime; Burke et al., 1976b), the first continental crust (gray gneisses) formed as the earliest stage of mantle partial melting. The TTGs probably formed by the partial melting of Hadean hydrous basalts (amphibolites) that were "inverted" into the mantle during the late Hadean–Early Archean resurfacing and restructuring. This was followed by the supracontinental eruption of massive thick komatiite sequences, to leave a refractory, depleted harzburgite mantle keel. Data from the Belingwe Greenstone Belt (Bickle et al., 1975) suggest that the komatiite sequences accumulated in deepening basins upon a weak hot TTG crust with thick sequences in basin centers and thin on basin margins. This led to gravitational instability between the heavy komatiitic upper crust and the TTG lower crust, and their inevitable turnover. During the turnover, diapirically rising and ballooning TTGs developed oblate marginal fabrics that transitioned through plane strain into vertical constrictional fabrics in the cores of the greenstone triangles. This was a pre-plate-tectonic period of Earth history when no obvious sign or result of plate-tectonic processes can be discerned in the rock record. We do not know whether this phase generated a continuous global crust that has been mainly recycled into the mantle reservoir or whether the extant Archean nuclei represent microcontinental survivors, between which plate tectonics turned over an Archean oceanic lithosphere; my preference is for the former. I now incline to the Fyfe (1978) view that the Archean nuclear pre–3.0 Ga cratons are remnants of a once continuous global crust and depleted harzburgite upper mantle lithosphere, most of which was returned to the mantle reservoir by inversion in the Archeo-Proterozoic and/or the well-documented process of subduction erosion (Dewey, 1980) during the Proterozoic-Phanerozoic.

Plate tectonics has been the engine of convective/conductive global, radiogenic heat loss during the Proterozoic and Phanerozoic, but the pre–3.0 Ga Earth was tectonically quite different, when heat loss was effected by pervasive convective inversion, as on Venus (Hansen, 2005), by episodic/periodic convective turnover, wholesale in the late Hadean and shallow diapiric during the Archean. Most Early Archean terrains are low-grade, suggesting little erosion of a never very thick crust. Erosion rates were low and there were no mountains. Tectonics was blobby and nonlinear. The only suggestion of some form of subduction, perhaps an ultradeep inversion process, is the existence of Archean diamonds in sheared lherzolites that were erupted in Kaapvaal Craton Cretaceous kimberlites, suggesting the subduction or inversion of surficial carbon into the mantle. By 3.0 Ga, the oldest, extant, cratonic lithosphere was developed with a greenstone/TTG crust, and with a mantle consisting of depleted granular harzburgite above and sheared lherzolite below with diamonds at ~220 km. Clearly, little has happened to the remaining Archean lithosphere since it formed because Archean diamonds stayed in their stability field until their Cretaceous rapid eruption in kimberlites. The late high-K_2O granite sheets (3.2–3.1 Ga) seal this early cratonization, upon which cratons the earliest rift and mobile cratonic shelf sequences (Dominion <3.074 Ga, Pongola 2.985–2.837 Ga, West and Central Rand 2.985–2.764 Ga) were deposited during the latest Archean and Archeo-Proterozoic.

The Archeo-Proterozoic, from ca. 3.0 to 2.7 Ga, during which the Superior province was developed and assembled, was a period of the first widespread andesites, indicating subduction, and geological assemblages and relationships that suggest the first clear evidence of some form of a plate-tectonic-like convective overturn of a lithosphere, including greater linearity and terrane assembly (Burke et al., 1981). Island arc complexes are suggested by petrology and geochemistry (Anhaeusser et al., 1983; Burke and Dewey, 1972; Mueller et al., 2002; Wyman et al., 2002). These terrains have a lenticular rather than the Archean, blobby, "gregarious batholith", triangular greenstone keel form. Possibly extensive flat-slab subduction of thin hot lithosphere was responsible for the early recycling of much of the Archean lithosphere. Missing from the Archeo-Proterozoic are many of the clear plate-tectonic arrangements of facies and assemblages that characterize the Phanerozoic, such as clear arc–forearc–accretionary complex triplets, linear/arcuate orogens, blueschists, ophiolites, low-angle thrusts, and extensional detachments. Perhaps this was a result of early proto-plate-tectonic terrane assembly of arcs and microcontinents operating in a higher-heat-flow, more mobile regime with more smaller and weaker plates. Metamorphic grades remained generally low; high-pressure and -temperature metamorphic rocks are rare, with the exception of eclogites in the Baltic Shield (Volodichev et al., 2004), suggesting a not-very-thick crust, little erosion, and no mountains. Burke and Dewey (1973) suggested that the Archeo-Proterozoic style was the result of weak collisions between small microcontinents and arcs. I see the Archeo-Proterozoic as the transitional period from Archean inversion tectonics to plate tectonics. As radiogenic heat production declined, the thermal/instantaneous lithosphere thickened, during the evolving plate-tectonic regime, to narrow plate boundary zones and enlarge cratons. As remnant Archean and Archeo-Proterozoic microcontinents collided and amalgamated, plate tectonics began and evolved.

The Proterozoic, from 2.7 to ca. 0.6 Ga, has most of the basic hallmarks of the Wilson cycle throughout, and records clear evidence of plate-tectonic processes operating in linear/

sublinear/arcuate belts of widely varying widths, and with most plate-tectonic indicator rock assemblages and structures, including sutures, rift and passive continental margin sequences, and linear/arcuate orogens, such as the Coronation-Wopmay (Hoffman et al., 1974), Labrador, Cape Smith, Grenville, Dahomey, Gariep, Damara, and Mozambique (Dewey and Burke, 1973; Burke and Dewey, 1972, 1973). The Early Proterozoic Trans-Hudson (Lewry and Stauffer, 1990) contains island arc volcanics that are geochemically identical to those of modern oceanic arcs (Wyman et al., 2002) including boninites (Wyman, 1999), and assemblages and relationships that indicate plate tectonics (Lucas et al., 1996). The reported sheeted complex of alleged ophiolite origin (Scott et al., 1992) is more problematic; there is no clear sequential relationship with a full ophiolite sequence; local sheeted dike complexes can be generated in the floors of rifts. Moller et al. (1995) recorded mafic eclogites in the 2.0 Ga Usugarian Belt of Tanzania, which were rapidly exhumed at Phanerozoic rates. The Svecofennian orogen (1.96–1.75 Ga) consists of a collage of terranes accreted to the Karelian Craton, which must have involved horizontal tectonics (Korja and Heikkinen, 2005; Korsman et al., 1999). The first major, transpressive, continental collisions occurred at ca. 2.0 Ga (Limpopo, Ubendides). The 2.1 Ga Birrimian (Abouchami et al., 1990) and 2.45–1.9 Ga Pechenga-Varzuga Belt (Sharkov and Smolkin, 1997) record oceanic and arc terrains. During the Early Proterozoic, the first large cratons were stabilized, as witnessed by the Transvaal sequence. However, these early cratons were more mobile vertically than later cratons in their very thick "platform" sequences; the Transvaal sequence is over 10 km thick. The Proterozoic, uniquely, saw ultrahigh-temperature (UHT) metamorphism, with opx/sapphirine/osumilite/sillimanite assemblages. These were generated in orogenic regions of widespread crustal melting above thick mafic underplates. The crustal partial melts formed granite sheets commonly emplaced in late orogenic transtensional regimes with lineated and subhorizontally foliated, constrictional, pink, potassic gneisses, rapidly extensionally exhumed beneath gently dipping detachments, very similar to Phanerozoic detachments (Dewey, 1988b; Dewey et al., 1993), as in the 1 Ga Namaqua Belt (Dewey et al., 2006). Paleomagnetic data from Proterozoic rocks are not only compatible with but demand the relative motion of continental blocks now welded by Proterozoic sutures (Burke et al., 1976a; Meert and Torsvik, 2003; Pesonen et al., 2003).

Throughout the Phanerozoic, plate-tectonic regimes, facies, assemblages, and structures show little variation. Blueschists and the generation and obduction of supra-subduction-zone ophiolites occurred mainly during the Ordovician and Cretaceous periods of continental dispersal and high sea level. Ultrahigh-pressure (UHP) metamorphic assemblages are, uniquely, Phanerozoic, contrasted to ultrahigh-temperature (UHT) metamorphic assemblages, which are, uniquely, Proterozoic. This must reflect the thermal evolution of Earth with declining heat production and flow with a thickening lithosphere and narrower, steeper, and colder subduction zones. The origin and evolution of higher plants and metazoan life was likely related to the development of well-defined continental margins and their upwelling zones.

COROLLARIES

Eclogites may play an important role in secular crustal evolution. They allow the subduction of continental crust into the mantle, overthickening of continental crust (Dewey et al., 1993), and cause long-term weakening of the continental lithosphere that is one, or a partial, explanation of the Wilson cycle of constant oceanic opening and closing along roughly the same lines (Ryan and Dewey, 1997). Their role as "sinkers" and in crustal delamination is likely to have been increasingly important, especially during the Phanerozoic, as the lithosphere enjoyed secular thickening.

There are many different types of continental crust with different strength (sailboard) profiles that generate many different tectonic responses to shortening and extension; each has a distinctive profile of strength distribution and décollement/detachment/thrust horizons, which generate quite different patterns of thrust stacking and extensional detachments. The secular evolution and thickening of the lithosphere is likely to have profoundly influenced detachment levels, nappe thickness, and structural style. As the lithosphere thickened and heat flow declined, the lithosphere became more difficult to deform in, progressively, narrower zones; décollements, detachments, and thrusts that are rooted in basement went ever deeper.

CONCLUSIONS

Plate tectonics, as operating in modern Earth, is a Phanerozoic process; plumes play a minor role in heat and mass transfer. Conversely, early Earth was dominated by plume mass and heat transfer from a hotter asthenosphere until ca. 3.0 Ga. The Archean (3.85–3.0) Ga saw the growth of a, possibly, globe-covering sialic crust, which, largely, has been recycled back into the mantle reservoir. During the Archeo-Proterozoic (new term), proto–plate tectonics was an early form of heat loss by the creation and destruction of small plates with weak arc and microcontinent collisions. During the Proterozoic, plate tectonics evolved to its Phanerozoic form as the lithosphere thickened. A difficult problem in continental tectonics is that, although plate tectonics generates the cross-sectional templates for plate boundary zones, most plate boundary zones become complex terrane collages that are not easy to decipher and reconstruct; orogenic belts commonly achieve a final linearity by collisional smearing, which rearranges elements of a long and complicated oceanic plate-tectonic history (Van Staal et al., 1998). We are victims of the inversion problem; we can forward model the most complex systems but cannot satisfactorily and uniquely backward model even the simplest systems. Kinematic and numerical forward models are useful in giving us ideas and showing us how complicated, or sometimes simple, is plate boundary evolution, but they can never uniquely describe the real world. They are useful chimeras that, like concepts of stress, paraphrase reality. Research in the earth sciences is moving away, too fast and too far from the reality of the rocks. Ultimately, the current, blinkered, obsession with process must be tempered by the historical perspective of the record preserved in the rocks. Rocks and their relationships are complicated and, commonly, difficult to interpret, but that is scarcely an excuse for not looking at rocks and substituting modeling.

ACKNOWLEDGMENTS

I am grateful to Carl Anhaeusser, Kevin Burke, Warren Hamilton, Alfred Kroener, and Celâl Şengör for many lively discussions about the evolution of Earth.

REFERENCES CITED

Abouchami, W., Boher, M., and Albarede, F., 1990, A major 2.1 Ga old event of mafic magmatism in West Africa: An early stage of crustal accretion: Journal of Geophysical Research, v. 95, p. 17,605–17,629.

Anhaeusser, C.R., editor, 1983, Contributions to the geology of the Barberton Mountain Land: Geological Society of South Africa Special Publication 9, 223 p.

Armstrong, R.L., 1968, A model for the evolution of strontium and lead isotopes in a dynamic Earth: Reviews of Geophysics, v. 6, p. 175–199.

Armstrong, R.L., 1981, Radiogenic isotopes: The case for crustal recycling on a near-steady-state no-continental growth Earth: Royal Society of London Philosophical Transactions, series A, v. 301, p. 443–472.

Armstrong, R.L., 1991, The persistent myth of crustal growth: Australian Journal of Earth Sciences, v. 38, p. 613–630.

Bickle, M.J., Martin, A., and Nisbet, E.G., 1975, Basaltic and peridotitic komatiites and stromatolites above a basal unconformity in the Belingwe Greenstone Belt, Rhodesia: Earth and Planetary Science Letters, v. 27, p. 155–162, doi: 10.1016/0012-821X(75)90024-2.

Bowring, S.A., Williams, I.S., and Compston, W., 1989, 3.96 Ga gneiss from the Slave province, Northwestern Territories, Canada: Geology, v. 17, p. 971–975, doi: 10.1130/0091-7613(1989)017<0971:GGFTSP>2.3.CO;2.

Burke, K.C.A., and Dewey, J.F., 1972, Orogeny in Africa, *in* Dessauvagie, T.F., and Whiteman, A.J., eds., African geology: Ibadan, Nigeria, University of Ibadan, p. 583–608.

Burke, K.C.A., and Dewey, J.F., 1973, An outline of Precambrian plate development, *in* Tarling, D.H., and Runcorn, S.K., eds., Continental drift, seafloor spreading and plate tectonics, volume 2: New York, Academic Press, p. 1035–1045.

Burke, K.C.A., Dewey, J.F., and Kidd, W.S.F., 1976a, Precambrian paleomagnetic results compatible with contemporary operation of the Wilson cycle: Tectonophysics, v. 33, p. 287–299, doi: 10.1016/0040-1951(76)90149-9.

Burke, K.C.A., Dewey, J.F., and Kidd, W.S.F., 1976b, Dominance of horizontal movements, arc and microcontinental collision during the later permobile regime: NATO Symposium (Newcastle, England, 1975), *in* Windley, B.F., ed., Early history of the Earth: London, Wiley, p. 113–129.

Burke, K.C.A., Delano, L., Dewey, J.F., Edelstein, A., Kidd, S.F., Nelson, K.D., Şengör, A.M.C., and Stroup, J., 1981, Rifts and sutures of the world: Greenbelt, Maryland, NASA/Goddard Space Flight Center, Contractors Report, 238 p.

Dewey, J.F., 1980, Episodicity, sequence and style at convergent plate boundaries, *in* Strangway, D.W., ed., The continental crust and its mineral deposits: Geological Association of Canada Special Paper 20, p. 553–573.

Dewey, J.F., 1986, Diversity in the lower continental crust: Geological Society [London] Special Publication 24, p. 71–78.

Dewey, J.F., 1988a, Lithospheric stress, deformation and tectonic cycles: The disruption of Pangea and the closure of Tethys: Geological Society [London] Special Publication 37, p. 23–40.

Dewey, J.F., 1988b, The extensional collapse of orogens: Tectonics, v. 7, p. 1123–1139.

Dewey, J.F., 2003, Ophiolites and lost oceans: Rifts, ridges, arcs, and/or scrapings, *in* Dilek, Y., and Newcomb, S., Ophiolite concept and the evolution of geological thought: Geological Society of America Special Paper 373, p. 153–158.

Dewey, J.F., and Burke, K.C.A., 1973, Tibetan, Variscan and Precambrian basement reactivation: Products of continental collision: Journal of Geology, v. 81, p. 683–692.

Dewey, J.F., and Spall, H., 1975, Pre-Mesozoic plate tectonics: Geology, v. 3, p. 422–426, doi: 10.1130/0091-7613(1975)3<422:PPTHFB>2.0.CO;2.

Dewey, J.F., and Windley, B.F., 1981, Growth and differentiation of the continental crust: Royal Society of London Philosophical Transactions, ser. A, v. 301, p. 189–206.

Dewey, J.F., Cande, S.C., and Pitman, W.C., III, 1989, Tectonic evolution of the India/Eurasia collision zone: Eclogae Geologicae Helvetiae, v. 82, p. 717–734.

Dewey, J.F., Robb, L., and Van Schalkwyk, L., 2006, Did Bushmanland extensionally unroof Namaqualand?: Precambrian Research, v. 150, p. 173 182.

Dewey, J.F., Ryan, P.D., and Andersen, T.B., 1993, Orogenic uplift and collapse, crustal thickness, fabrics and phase changes: The role of eclogites: Geological Society [London] Special Publication 76, p. 325–343.

Eriksson, P., Altermann, W., Nelson, D., Mueller, W., and Catuneanu, O., 2004, The Precambrian Earth: Tempos and events: Amsterdam, Netherlands, Elsevier, 941 p.

Fyfe, W.S., 1978, The evolution of the Earth's crust: Modern plate tectonics to ancient hot-spot tectonics: Chemical Geology, v. 23, p. 89–114, doi: 10.1016/0009-2541(78)90068-2.

Hamilton, W.B., 2003, An alternative Earth: GSA Today, v. 13, n. 11, p. 4–12, doi: 10.1130/1052-5173(2003)013<0004:AAE>2.0.CO;2.

Hansen, V.L., 2005, Venus's shield terrain: Geological Society of America Bulletin, v. 117, p. 808–822, doi: 10.1130/B256060.1.

Hoffman, P., Dewey, J.F., and Burke, K.C.A., 1974, Aulacogens and their genetic relation to geosynclines with a Proterozoic example from Great Slave Lake, Canada, *in* Dott, R.H., Jr., and Shaver, R.H., eds., Modern and ancient geosynclinal sedimentation: Society of Economic Paleontologists and Mineralogists Special Publication 19, p. 38–55.

Hurley, P.M., and Rand, J.R., 1969, Pre-drift continental nuclei: Science, v. 164, p. 1229–1232, doi: 10.1126/science.164.3885.1229.

Korja, A., and Heikkinen, P., 2005, The accretionary Svecofennian orogen: Insights from the BABEL profiles: Precambrian Research, v. 136, p. 241–268, doi: 10.1016/j.precamres.2004.10.007.

Korsman, L.R., Korja, A., Pajunen, M., Virransalo, P., and GGT/SVEKA Working Group, 1999, The GGT/SVECA transect: Structure and evolution of the continental crust in the Palaeoproterozoic Svecofennian orogen in Finland: International Geology Review, v. 41, p. 287–333.

Krabbendam, M., and Dewey, J.F., 1998, Exhumation of UHP rocks by transtension in the Western Gneiss Region, Scandinavian Caledonides: Geological Society [London] Special Publication 135, p. 159–181.

Kroner, A., 1981, Precambrian plate tectonics, *in* Kroner, A., ed., Precambrian plate tectonics: Amsterdam, Elsevier, p. 56–90.

Lewry, J.F., and Stauffer, M.R., editors, 1990, The Early Proterozoic Trans-Hudson orogen of North America: Geological Association of Canada Special Paper 37, 505 p.

Lucas, S.B., Stern, R.A., Syme, E.C., Reilly, B.A., and Thomas, D.J., 1996, Intraoceanic tectonics and the development of continental crust: 1.92–1.84 Ga evolution of the Flinn Flon Belt: Canada: Geological Society of America Bulletin, v. 108, p. 602–629, doi: 10.1130/0016-7606(1996)108<0602:ITATDO>2.3.CO;2.

MacGregor, A.M., 1951, Some milestones in the history of S Rhodesia: Transactions of the Geological Society of South Africa, v. 54, p. 207–251.

Meert, J.G., and Torsvik, T.H., 2003, The making and unmaking of a supercontinent: Rodinia revisited: Tectonophysics, v. 375, p. 261–288, doi: 10.1016/S0040-1951(03)00342-1.

Moller, A., Appel, P., Mezger, K., and Schenk, V., 1995, Evidence for a 2 Ga subduction zone: Eclogites in the Usugarian Belt of Tanzania: Geology, v. 23, p. 1067–1070, doi: 10.1130/0091-7613(1995)023<1067:EFAGSZ>2.3.CO;2.

Moorbath, S.E., O'Nions, R.K., and Pankhurst, R.J., 1973, The evolution of early Precambrian rocks at Isua, West Greenland: Nature, v. 245, p. 138–189, doi: 10.1038/245138a0.

Mueller, W.U., Marquis, R., and Thurston, P., editors, 2002, Evolution of the Archean Abitibi Greenstone Belt and adjacent terranes: New insights from geochronology, geochemistry, structure, and facies analysis: Precambrian Research, v. 115, 374 p.

Pesonen, L.J., Elming, S.A., Mertanen, S., Pisarevsky, S.A., D'Agrella-Filho, M.S., Meert, J., Schmidt, P.W., Abrahamsen, N., and Bylund, G., 2003, Paleomagnetic configuration of continents during the Proterozoic: Tectonophysics, v. 375, p. 289–324, doi: 10.1016/S0040-1951(03)00343-3.

Phillips, R.J., and Hansen, V.L., 1998, Geological evolution of Venus: Rises, plains, plumes, and plateaus: Science, v. 279, p. 1492–1497.

Ryan, P.D., and Dewey, J.F., 1997, Continental eclogites and the Wilson cycle: Geological Society [London] Journal, v. 154, p. 437–442.

Scholl, D.W., and von Huene, R., 2007, this volume, Crustal recycling at modern subduction zones applied to the past—Issues of growth and preservation of continental basement crust, mantle geochemistry, and supercontinent reconstruction, *in* Hatcher, R.D., Jr., et al., eds., 4-D framework of continental crust: Geological Society of America Memoir 200, doi: 10.1130/2007.1200(02).

Scott, D.J., Helmstaedt, H., and Bickle, M.J., 1992, Purtuniq ophiolite, Cape Smith Belt, northern Quebec, Canada: A reconstructed section of Early Proterozoic oceanic crust: Geology, v. 20, p. 173–176, doi: 10.1130/0091-7613(1992)020<0173:POCSBN>2.3.CO;2.

Sharkov, E.V., and Smolkin, V.F., 1997, The Early Proterozoic Pechenga-Varzuga Belt: A case of Precambrian back-arc spreading: Precambrian Research, v. 82, p. 133–151, doi: 10.1016/S0301-9268(96)00041-1.

Solomon, S.C., 1978, On volcanism and thermal tectonics on one-plate planets: Geophysical Research Letters, v. 5, p. 461–464.

Talbot, C.J., 1968, Thermal convection in the Archean crust: Nature, v. 220, p. 552–556, doi: 10.1038/220552a0.

Van Staal, C.R., Dewey, J.F., MacNiocall, C., and McKerrow, W.S., 1998, The Cambrian-Silurian tectonic evolution of the northern Appalachian and British Caledonides: History of a complex southwest Pacific-type segment of Iapetus: Geological Society [London] Special Publication 143, p. 199–242.

Veizer, J., and Jansen, S.L., 1979, Basement and sedimentary recycling and continental evolution: Journal of Geology, v. 87 p. 341–356.

Volodichev, O.I., Slabunov, A.I., Bibikova, E.V., Konilov, A.N., and Kuzenko, T.I., 2004, Archean eclogites in the Belomorian mobile belt, Baltic Shield: Petrology, v. 12, p. 540–560.

Wyman, D.A., 1999, Paleoproterozoic boninites in an ophiolite-like setting, Trans-Hudson orogen, Canada: Geology, v. 27, p. 455–458, doi: 10.1130/0091-7613(1999)027<0455:PBIAOL>2.3.CO;2.

Wyman, D.A., Kerrich, R., and Polat, A., 2002, Assembly of Archean cratonic mantle lithosphere and crust: Plume-arc interaction in the Abitibi-Wawa subduction-accretion complex: Precambrian Research, v. 115, p. 37–62, doi: 10.1016/S0301-9268(02)00005-0.

Manuscript Accepted by the Society 1 February 2007

Printed in the USA

The Geological Society of America
Memoir 200
2007

Crustal recycling at modern subduction zones applied to the past—Issues of growth and preservation of continental basement crust, mantle geochemistry, and supercontinent reconstruction

David W. Scholl*
Department of Geophysics, Stanford University, Stanford, California 94305, USA

Roland von Huene†
Department of Geology, University of California, Davis, California 95616, USA

ABSTRACT

Geophysical and geological observations document that beneath the submerged forearc, processes of sediment subduction and subduction erosion move large volumes of material toward the mantle. The conveying system is the subduction channel separating the upper plate from the underthrusting ocean plate. Globally, the *zero-porosity* or solid-volume rate at which continental debris is shuttled toward the mantle is estimated to be ~2.5 km^3/yr. To deliver this volume, the average thickness of the subduction channel is ~1.0 km. Some deeply subducted material is returned to the surface of Earth as a component of arc magma or as tracks of high-*P*/*T* crustal underplates. But over long periods of time (>50 m.y.), most of the removed material is evidently recycled to the mantle.

Applying Cenozoic recycling rates to the past astonishingly implies that since 2.5 Ga a volume of continental crust equal to the standing inventory of ~6 × 10^9 km^3 has been removed from the surface of Earth. This minimum estimate does not include crustal material recycled at continental collision zones nor reliable estimates of recycling where large accretionary bodies form.

The volume of demolished crust is so large that recycling must have been a major factor determining the areal pattern and age distribution of continental crust. The small areal exposure of Archean rock is thus probably more a consequence of long-term crustal survival than the volume originally produced. Reconstruction of older supercontinents is made difficult if not unachievable by the progressive truncation of continental edges effected by subduction zone recycling, in particular by subduction erosion.

Keywords: subduction erosion, tectonic erosion, crustal recycling, crustal truncation, sediment subduction, subduction channel.

*dscholl@pangea.stanford.edu
†rhuene@mindspring.com

Scholl, D.W., and von Huene, R., 2007, Crustal recycling at modern subduction zones applied to the past—Issues of growth and preservation of continental basement crust, mantle geochemistry, and supercontinent reconstruction, *in* Hatcher, R.D., Jr., Carlson, M.P., McBride, J.H., and Martínez Catalán, J.R., eds., 4-D Framework of Continental Crust: Geological Society of America Memoir 200, p. 9–32, doi: 10.1130/2007.1200(02). For permission to copy, contact editing@geosociety.org.

INTRODUCTION

During the past three decades, offshore geophysical, geological, and submersible investigations documented that large volumes of rock and sediment are missing from the upper plate of ocean margin subduction zones. Concurrently, coastal studies at convergent margins, in particular along western South America, revealed that the structural and lithofacies fabric of exposed cratonic basement rock is truncated at the relatively nearby (50–70 km) trench axis (Miller, 1970a, 1970b; Rutland, 1971; Moran-Zenteno et al., 1996). Examples of tectonically buried truncation scars of Mesozoic age attributed to subduction erosion have been described for coastal California (Page, 1970) and southern Alaska (Clift et al., 2005). Further inboard, the trace element geochemical signatures of arc magmatic rocks established that ocean floor sediment is being recycled in young lavas (Morris et al., 1990; Plank and Langmuir, 1993), and also forearc crustal debris (Kay et al., 2005).

We interpret this spectrum of offshore and onshore observations to mean that

1. lower plate ocean floor sediment is transported landward beneath the submerged forearc; and
2. processes attending the underthrusting oceanic plate actively thin and truncate the crust of the overlying forearc (von Huene and Scholl, 1991; Scholl and von Huene, 2003; Clift and Vannucchi, 2004).

Missing masses of crustal material are transported toward the mantle via the subduction channel that separates the upper and lower plates (Cloos and Shreve, 1988a, 1988b). Well-collected and advanced-processed seismic reflection data image the subduction channel (e.g., Ranero and von Huene, 2000; Park et al., 2002; Collot et al., 2002; Calahorrano et al., 2005).

Two kindred tectonic processes, *sediment subduction* and *subduction erosion* introduce sediment and eroded forearc debris to the subduction channel (Figs. 1 and 2). If these processes recycle volumetrically large amounts of continental material to the mantle, then crustal losses at subduction zones must be importantly involved in creating the existing volume, geometry, and age distribution of continental basement rock.

In this paper we explore the implications of the general hypothesis that, since at least 2.5 Ga, the present is the key to the past and that the observed rate at which continental crust has been recycled at late Mesozoic and Cenozoic convergent margins can be applied to older subduction zones. Although debated (see Hamilton, 2003; Van Kranendonk and Cassidy, 2004), we accept the view that ocean margin subduction has been under way for at least the past 2.5 billion years (i.e., since the beginning of the Proterozoic), seismic images of which have been recorded worldwide (for documentation see van der Velden and Cook, 2005). Applying estimated long-term (i.e., >10–20 m.y.) rates of crustal recycling to this length of time allows us to explore general questions concerning

1. the growth of continental crust through at least the second half of Earth's history;
2. linked implications for the evolution of mantle geochemistry; and
3. the availability of exposed crustal basement to constrain and guide supercontinent reconstructions.

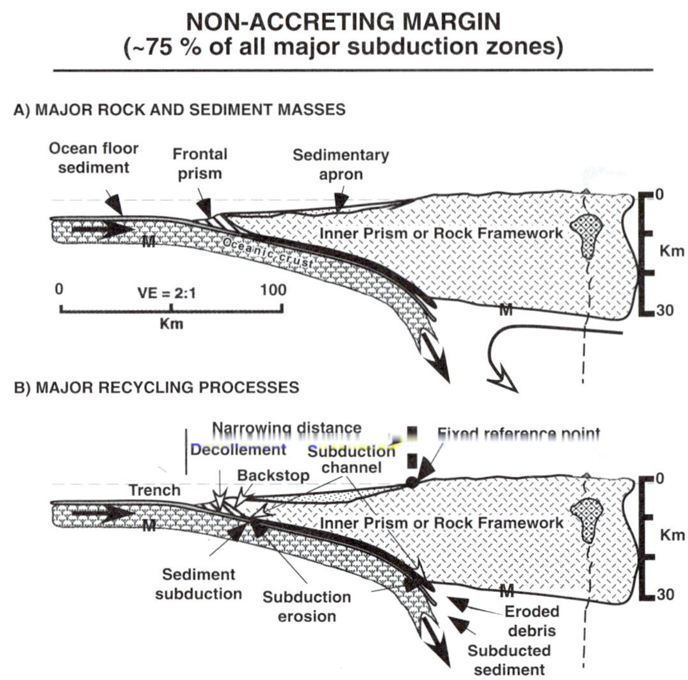

Figure 1. Major rock and sedimentary masses of nonaccreting margins and recycling process operating at them. Nonaccreting margins are characterized by the close approach of the margin's rock framework to the trench and small or lacking older prisms of accreted lower plate sediment (see Fig. 2) behind the actively deforming material of the frontal prism. Material of the frontal prism can be offscraped ocean floor sediment or largely constructed of masses of slope debris. With time, nonaccreting margins narrow with respect to a reference point on the margin; i.e., the trench advances landward toward the reference. M—Moho.

We address these wonderments by first describing evidence from modern ocean margins testifying to missing masses of rock and sediment. Estimates are then presented of rates of sediment subduction and forearc crustal erosion at margins where seismic imaging and dated rock and sediment samples allow us to decipher a long (>10–20 m.y.) history of margin evolution. Recycling rates determined for these best-controlled margins are then applied to the global length of margins that have evolved similarly.

Summing up subduction zone losses provides an estimate of the global rate (in km^3/yr, or Armstrong units, where 1 km^3/yr = 1 AU; Robert Kay, 2004, personal commun.) at which continental crust has been removed from modern ocean margins during the past 150–15 m.y. Applying our estimated averaged flux of recycled continental material (2.5 km^3/yr) to the past 2.5 billion years implies that a genuinely vast quantity of continental material, a mass roughly equal to the standing volume (~7×10^9 km^3; Reymer and Schubert, 1984), has been tectonically erased from the surface of Earth. We argue that the bulk of this material has been recycled to the mantle.

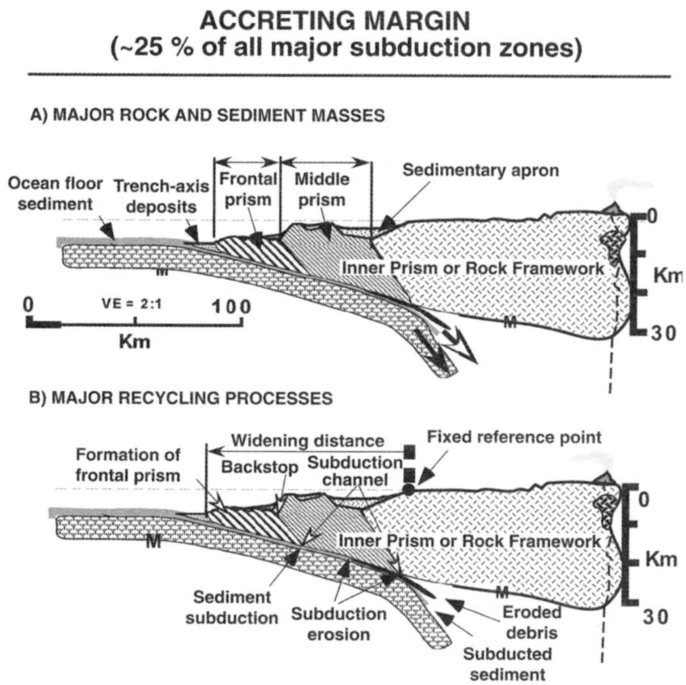

Figure 2. Major rock and sedimentary masses of accreting margins and recycling process operating at them. Accreting margins are characterized by a rock framework that is deeply buried by a wide body of accreted material consisting of a little-deforming, older accreted prism and, at the base of the slope, the frontal prism of actively deforming (shortening) sediment scraped off the subducting lower plate. With time, accreting margins widen with respect to a reference point on the margin; i.e., the trench retreats seaward from the reference. M—Moho.

Not included in our volume estimate of subduction zone recycling of continental crust is (1) crust injected into the mantle at subduction zones where continental collision is under way (e.g., India and southern Eurasia, Australia and Indonesia), (2) a good estimate of recycling occurring at the accretion-dominated subduction zones, and (3) crustal losses that might occur inland of the coastal area, e.g., by delamination or ablative processes (Pope and Willett, 1998).

We thus regard our estimate of crustal recycling at ocean margin subduction zones as a conservative assessment of what the true global rate might be.

CONFIGURATION AND TYPES OF CONVERGENT MARGINS

Rock and Sediment Bodies of Convergent Margins

Geophysical and ocean drilling data demonstrate that four lithostratigraphic units form the bulk of the construct of convergent margins. Looking seaward and upsection these are

1. the *inner prism* or rock framework (always present) of older, well-lithified core rock;

2. the *middle prism* (not always present—commonly referred to as an accreted prism of older material tectonically added to the margin) lies landward or behind the accretionary construct of a frontal prism (see below). The middle prism, unlike the frontal prism, is not the most actively deforming mass but nonetheless is regarded as a constitutive or continuum part of a long-term accretionary process;

3. the *frontal prism* (always present) constructed of actively deforming material of accreted oceanic (lower plate) or downslope-displaced forearc (upper plate) material, disintegrating inner prism basement, or a combination of these sources; and

4. the *sedimentary apron* (always present) of margin-mantling continental shelf and slope deposits including accumulations of mass wasting debris.

Forearc Inner Prism or Rock Framework

The margin's interior rock framework consists of a core of igneous or metamorphic basement rock and younger but well-lithified and deformed sedimentary sequences. Igneous and metamorphic basement rock and infolded or faulted sedimentary units are commonly exposed along the coast and can be traced seaward geophysically and by dredging, drilling, and submersible/ROV observations. Forearc basement is lithologically as diverse as Paleozoic and older igneous and metamorphic cratonic rock (i.e., Peru and Chile) and the magmatic massifs of Cenozoic arcs (i.e., Aleutian and Tonga Ridges).

Many lithified sedimentary sequences included within the rock framework are fossil accretionary prisms that accumulated seaward of the margin's older basement core, for example the Mesozoic to middle Tertiary accretionary bodies of Alaska (Plafker et al., 1994) and Japan (Taira et al., 1982). These fossil bodies are grouped with the margin's inner prism of framework rock because most of them are significantly older or are separated by a significant hiatus in growth, from an ongoing cycle of building an accretionary frontal prism.

Middle Prism

Roughly 25% of the global length of ocean margin subduction zones is bordered by large, widening bodies (50–100 km wide) of ocean floor sediment more or less continuously scraped off the underthrusting ocean plate and added tectonically and progressively to the margin's older inner rock framework. These subduction zones are described below as *accreting margins* (Fig. 2; Table 1).

At accreting margins, growth of the wedge has generally been ongoing over a long period of time (>10–20 m.y.). Large accreted wedges are constructed of a middle prism of little-deforming older accreted material—the middle prism—that is attached landward to the front of the rock framework of the forearc's inner prism and seaward to a frontal prism (see below) of actively accreting and deforming material (Fig. 2). Large and growing wedges (>40 km in width) form

TABLE 1. TECTONIC TYPES OF CONVERGENT OCEAN MARGINS

Region and trench sector	Accreting margin: Length of sector (km)	Nonaccreting margin: Length of sector (km)	Orthogonal convergence (km/m.y.)
Pacific Ocean region			
Northeastern Alaska (east of Kodiak Seamount)	500		60
Southwestern Alaska (Kodiak Seamount to Unimak Pass)		1100	65
Aleutian (Unimak Pass west to Near Islands)		1800	45
Cascadia (Vancouver to central Oregon)	1000		35
Southern Oregon		200	35
Mexico (southeast of Gulf of California)		1100	70
Middle America (Guatemala southeast to Panama)		1400	70
Colombia		750	65
Ecuador		500	65
Peru		2000	85
Northern Chile (south to 33°S)		1500	85
Central Chile (33°–45°S)		1800	80
Southern Chile (45°S to western Tierra del Fuego)	1100		20
Hikurangi	400		30
Kermadec		1300	70
Tonga		1300	170
Vanuatu (New Hebrides)		1500	100
Solomon (San Cristobal)		1500	110
New Britain		900	110
New Guinea		1100	20
Philippine		1200	90
West Luzon (Manila)	1000		90
Northern Ryukyu		1000	60
Southern Ryukyu	500		30
Nankai	700		30
Yap-Palau		600	20
Mariana		2400	90
Izu-Bonin		1200	100
Japan		800	100
Kuril-Kamchatka		2000	90
Indian Ocean region			
Puysegur (southern New Zealand)		500	20
Makran	500		35
Andaman	1500		10
Sumatra (Sunda)	1700		50
West Java	500		75
East Java		1200	75
Atlantic and Mediterranean region			
Antilles	800		20
South Sandwich		600	20
Aegean	1300		25
Sum	11,000	31,250	
Length of all major subduction zones (km)	42,250		
Rounded percent nonaccreting (~31,000 km)	74%		
Rounded percent accreting (~11,000 km)	26%		

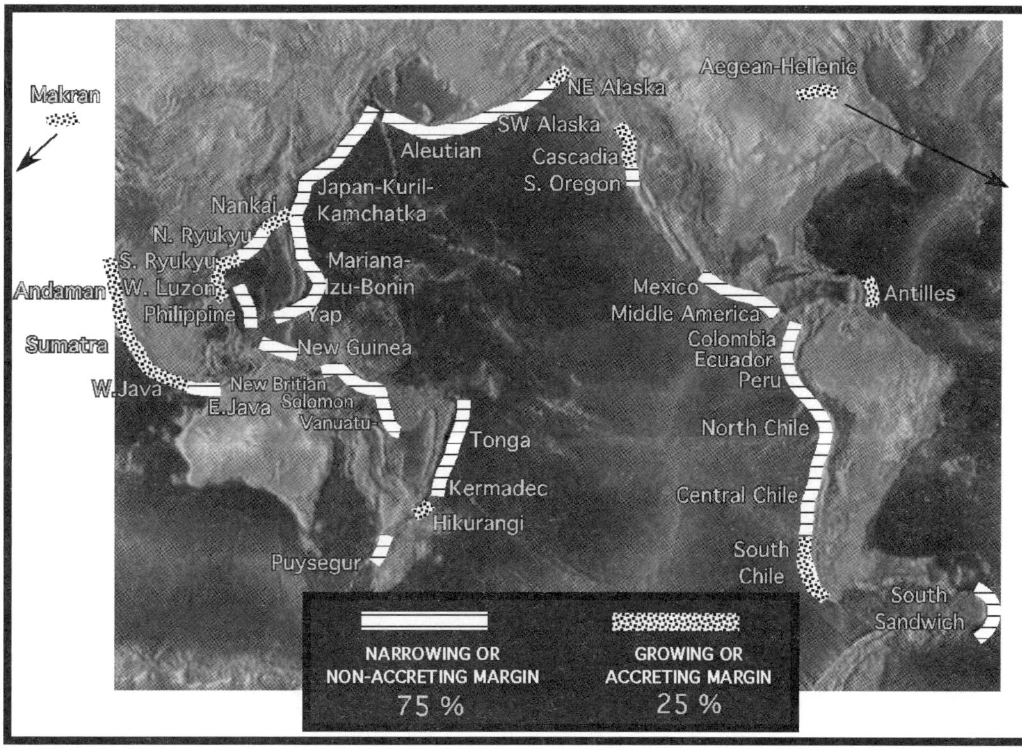

Figure 3. Pacific-centered distribution of major ocean margin subduction zones and their tectonic type—accreting or nonaccreting. Not shown are the Makran accreting margin in the NW corner of the Indian Ocean and the accreting Aegean margin of the eastern Mediterranean Sea.

most commonly at margins bordered by thickly sedimented (>1.5–2.0 km) trenches and where the orthogonal underthrusting rate is less than ~40–50 km/m.y. (e.g., the northern Cascadia, Makran, southern Lesser Antilles, and Sumatra margins; Fig. 3; Table 1).

Frontal Prism

A frontal prism forms at the base of the landward trench slope of all convergent margins (von Huene et al., 2004) (Figs. 1 and 2). Frontal prisms are seismically imaged as a landward-thickening body of clastic and fragmental material exhibiting evidence of active thrusting and shortening (see Fig. 4). The frontal prism functions tectonically to overpressure pore water in the clastic sedimentary sequence entering the subduction zone, thus creating a low-stress surface or horizon of slip commonly referred to as the interplate boundary (Figs. 1 and 2). Below this surface, sediment can be subducted to mantle depths.

Material of the frontal prism accumulates tectonically by the offscraping (i.e., accreting) of lower plate sediment, by the in situ tectonizing of unaccreted depositional and mass wasting debris shed from the margin or produced by the physical disintegration of the seaward terminus of its inner rock framework. In size, the frontal prism ranges from <10 km wide and <1 km thick to a maximum of ~30–40 km wide and thicknesses

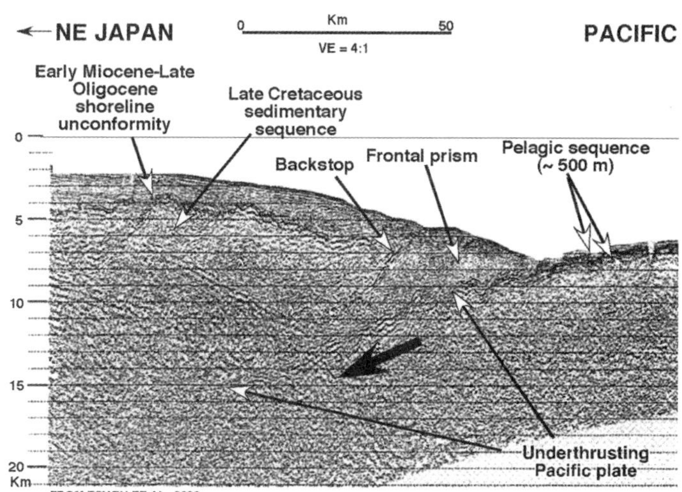

Figure 4. Seismic section off NE Japan showing a frontal prism that is potentially a mass of pelagic deposits accreted to the margin's Late Cretaceous rock framework and also a deeply submerged early Miocene–latest Oligocene shoreline unconformably overlying a Late Cretaceous sedimentary sequence that extends seaward within ~30 km of the Japan Trench. Subsidence of the shoreline documents crustal thinning effected by subduction erosion. The small size (~65 km^3 solid volume) of the frontal prism, roughly 5% of the pelagic section that entered the subduction zone during the past 65 m.y., documents sediment subduction.

of many kilometers. Recycling and accreting processes evidently stabilize the width of frontal prisms of actively deforming material to <30–40 km (von Huene and Klaeschen, 1999; von Huene and Scholl, 2003). The larger frontal prisms border trenches with sedimentary fills exceeding 1.5 km. These frontal prisms appear to be dominantly accreted sediment, and virtually all drilling investigations of frontal prisms have been where they have formed. However, most convergent margins (~53% or ~24,000 km; von Huene and Scholl, 1991) (Table 2) are bordered by trenches underlain by sedimentary sections thinner than ~0.75 km and frontal prisms less than 15 km wide. They appear to be dominantly constructed of upper plate or nonaccretionary material. But little is factually known about the composition and material origin of the smaller-width frontal prisms, the only one of which fully penetrated by drilling lies at the base of the Costa Rica sector of the Middle America Trench (Kimura et al., 1997; Vannucchi et al., 2001).

Long sectors of trenches filled with sediment thicker than ~1.5 km and where the orthogonal convergence rates exceeds ~60 km/m.y. (e.g., the Aleutian and central Chile margins; von Huene and Scholl, 1991) (Table 2) are bordered by a frontal wedge of dominantly lower plate sediment that attains a stable width of 25–40 km. These prisms are either directly or closely attached to the forearc's inner rock framework and appear to reflect the special condition of enhanced rates of glacial-age delivery of turbiditic sediment to high-latitude trenches (McCarthy and Scholl, 1985; Bangs and Cande, 1997).

Landward of the frontal prism, older material of the margin displays much reduced or virtually no evidence of active shortening. The frontal prism is thus defined mechanically rather than by the material undergoing compiling and shortening. This definition is similar to that of Byrne et al. (1993), who describe the landward limit of the deforming frontal prism as the margin's backstop. At a widening or *accreting margin* (see below), the mechanical backstop occurs at the front of the stored, and little-deforming, material of the middle prism (Kopp and Kukowski, 2003). At a narrowing or *nonaccreting margin* (see below), the potential "backstop" is the seaward fragmented and disintegrated projection of the margin's inner prism of older framework rock (von Huene et al., 2004). Thus with respect to sediment accretion the term "backstop" is not germane to the ~53% of all margins at which subduction accretion or offscraping effectively does not occur.

Sedimentary Apron

The submerged forearc is underlain by an upper sequence of deposits, the shelf-slope sedimentary apron. It can include substantial volumes of mass wasting debris (von Huene and Ranero, 2003). Along most margins, the apron extends from the shoreline to near the base of the landward trench slope. DSDP (Deep Sea Drilling Project) and ODP (Ocean Drilling Program) drilling at the NE Japan, Tonga, Peru, Costa Rica, Guatemala, and the upper part of central Mexico margins, and dredge sampling of the southern slope of the Puerto Rico Trench, revealed that the older beds of the slope apron are shallow-water, shoreline, or nonmarine deposits and that the surface of the underlying framework rock is an eroded, wave-based unconformity that has subsided 3–5 km. Across the width of the submerged forearc, the thickness of the aprons ranges from a few meters to as much as 4–5 km beneath forearc basins (Clift and Vannucchi, 2004; Wells et al., 2003). At the base of the slope, the apron continues seaward across all but the youngest of the thrust packets of the frontal prism (Fig. 1; see for example Adam et al., 2004; Bangs et al., 2004).

PROCESSES THAT STORE AND RECYCLE MATERIAL

Three mechanical and hydrogeological processes control the storage, bypassing, and removal of material at convergent ocean margins. These are, respectively, *subduction accretion*, *sediment subduction*, and *subduction erosion* (Scholl et al., 1980; von Huene and Scholl, 1991; Clift and Vannucchi, 2004). The kindred processes of sediment subduction and subduction erosion are the principal ones causing crustal recycling. Subduction erosion tectonically truncates and destroys the seaward edges of continents and island arcs.

Subduction Accretion

Subduction accretion describes the tectonic addition of lower or ocean plate sediment and scraps of igneous crust to the margin's makeup of rock and sediment. Subduction accretion thickens the margin construct and causes it to expand or widen seaward. In general, where sedimentary sections thicker than ~1.5 km enter the subduction zone, accretion typically builds a frontal prism of structurally offscraped sediment. Accretionary growth continues where accumulating material is transferred to an older, only slowly deforming, middle prism inboard of the actively deforming frontal prism (Fig. 2).

Sediment Subduction

Sediment subduction describes the passage of lower plate sediment beneath and landward of the frontal prism. Landward transport is effected by the entrance of ocean floor or slope sediment into the subduction channel that separates the upper and lower plates (Cloos and Shreve, 1988a, 1988b). Sediment subduction allows ocean deposits to bypass the frontal prism and thus avoid involvement in accretionary processes except possibly for crustal underplating in the coastal region.

Subduction Erosion

Subduction erosion describes the transfer of sediment and rock from the upper plate to the subduction channel, within which material moves with the lower plate toward the mantle. Subduction erosion is enhanced, and at some margins greatly so, by the passage of bathymetric relief (e.g., seamounts, ridges, horst and graben topography) down the subduction zone (Ballance et al., 1989; von Huene and Scholl, 1991; Bourgois et al., 1996; Ranero and von Huene, 2000; Clift and Vannucchi, 2004; Hampel et al., 2004; von Huene et al., 2004).

Frontal erosion and truncation of the toe of the margin is linked to the mechanical disruption of the material of the lower slope and the sweeping of some or most of this debris into the inlet of the subduction channel (Cloos and Shreve, 1988a, 1988b; von Huene and Ranero, 2003). Farther landward, beneath the lower and middle forearc, basal subduction erosion is presumed to be effected by hydrofracturing processes that detach material from the base of the upper plate and transfer the tectonically eroded debris to the subduction channel. A detachment process of large lens-shaped masses has been imaged along the Costa Rica sector of the Middle America forearc (Ranero and von Huene, 2000). Frontal and basal subduction erosion work to truncate and thin the rock and sediment bodies of the submerged forearc, processes that cause the margin to subside and narrow in width. A significant amount of basal erosion may be effected at the times of large and great subduction zone earthquakes (Wells et al., 2003).

The Coeval Storage and Removal Workings of Subduction Zone Processes

In our view, and also that of Clift and Vannucchi (2004), a process of storage, bypassing, or erosion of material along a sector of convergent margins does not exclude the simultaneous construction or destruction effects of other processes. This yin and yang tectonic view allows that the forward building of an accreted mass or wedge can take place as sediment is subducted beneath and landward of it. Cotectonically, frontal and basal subduction erosion, in particular as seamounts plow through the margin (see Bangs et al., 2006), can strip material from the frontal and middle prisms as well as the inner prism of the rock framework. Evidence supporting his view is explored by Wells et al. (2003) and Scholl et al. (2003).

The tectonic setting along a sector of convergent margin can change rapidly and alter the way it evolves. For example, the entrance of major physiographic relief into the subduction zone can convert a margin sector dominated by subduction accretion to one dominated by subduction erosion (Bourgois et al., 1996; Fruehn et al., 1999). After subduction of the relief, the evolution of the margin may return to one characterized by accretionary growth (von Huene and Lallemand, 1990). The opposite switch can occur as readily. This has been hypothesized to have happened along the subduction-erosion-dominated Aleutian Ridge (arc) when, in the late Cenozoic, collision-driven mountain building in the Gulf of Alaska began to supply large volumes of synorogenic sediment to the Aleutian Trench (Scholl et al., 1987). The consequence was the construction of a 30–40 km wide frontal prism of mostly accreted trench turbiditic sediment (McCarthy and Scholl, 1985; Ryan and Scholl, 1989).

RECYCLING AT ACCRETING AND NONACCRETING SUBDUCTION ZONES

To explore the long-term implications that recycling poses for the evolution of Earth's storehouse of basement rock, we adopt the margin-type defining criteria of Clift and Vannucchi (2004) and group modern convergent margins into two tectonically contrasting modes, *accreting* (widening) and *nonaccreting* (erosionally narrowing) (Figs. 1, 2, and 3; Table 1). This classification attempts to identify the process that during the late Cenozoic dominantly determined the volumetric growth or wasting of the margins.

The global length of the principal ocean margin subduction zones is presently ~42,000 km, an estimate that does not include short sectors of Philippine and Indonesian trenches nor where continental collision is occurring, i.e., the Banda arc, India, and sectors of the western Mediterranean (see Stern, 2002). Accreting and nonaccreting ocean margins are estimated, respectively, to occur along ~25% (11,000 km) and ~75% (31,000 km) of the global span of major subduction zones (Figs. 1, 2, and 3; Table 1).

Accreting margins are those in which, during a protracted period of time (>5–10 m.y.), the trench axis migrated seaward—retreated—away from a fixed point on the margin. Accreting margins characteristically have middle prisms that are constitutively or genetically linked to an actively deforming accretionary frontal prism. These margins widen with time by the progressive accretionary transfer of lower plate sediment (mostly) and igneous crust to the inner rock framework of the upper plate (Fig. 2). Accreting margins are commonly blessed with a long-lived, well-sedimented trench axis and an orthogonal convergence rate most commonly slower than ~40–50 km/m.y. (Table 1). A typical example is the northern Cascadia margin of North America where the incoming sedimentary sequence is sufficiently thick to mask the geomorphic expression of the trench and the seaward march of thrust-accreted packages of trench sediment is impressive and well imaged (Fig. 3) (Adam et al., 2004). Other exemplary accretionary subduction zones are the Makran of southern Pakistan (not shown on Fig. 3; Fruehn et al., 1997; Kopp et al., 2000) and the Sunda margin off Sumatra (Fig. 3) (Kopp et al., 2001; Kopp and Kukowski, 2003). The subduction zones of these margins are overwhelmingly nourished with sediment delivered west and east, respectively, from the collision-elevated and glaciated Himalaya drainages.

Nonaccreting margins are those in which, during a protracted period of time (>5–10 m.y.), the trench axis is judged to have migrated landward—advanced—toward a margin reference point. The typical migration rate or rate of forearc truncation is 2–3 km/m.y. (Table 3). Landward advance is based on the occurrence of shallow-water or nonmarine deposits overlying the seaward extent of framework rock of the inner prism. A characteristic attribute of nonaccreting margins is that the bedrock framework extends offshore to the back of the frontal prism and thus within a few tens of kilometers of the trench floor (Fig. 1). A typical non-

TABLE 2. OCEANIC SEDIMENT STORED AND RECYCLED (SUBDUCTED) AT ACCRETING (i.e., LONG-TERM WIDENING) AND NONACCRETING (i.e., LONG-TERM NARROWING) SUBDUCTION ZONES

Region and trench sector	Accreting margin (km)	Nonaccreting margin (km)	Orthogonal convergence (km/m.y.)	Porous trench sediment column (km)	Solid trench sediment column (km)	Accreting margin: accreted sediment at 70% of incoming section (km^3/m.y.)	Accreting margin: subducted sediment at 30% of incoming section (km^3/m.y.)	Nonaccreting margin: subducted sediment at 100% of incoming section (km^3/m.y.)
PACIFIC OCEAN								
Northeast Alaska (east of Kodiak Seamount)	500		60	2.0	1.3	27,300	11,700	
Southwestern Alaska (Kodiak Seamount to Unimak Pass)		1100	65	0.8	0.4			
Aleutian (Unimak Pass west to Near Islands)		1800	45	1.5	1.0			81,000
Cascadia (Vancouver south to central Oregon)	1000		35	2.0	1.3	31,850	13,650	
Southern Oregon		200	35	2.0	1.3			9100
Mexico (southeast of Gulf of California)		1100	70	0.6	0.3			23,100
Middle America (Guatemala southeast to Panama)		1400	70	0.5	0.3			24,500
Colombia		750	65	2.5	1.0			46,313
Ecuador		500	65	1.0	0.6			19,500
Peru		2000	85	0.7	0.4			68,000
Northern Chile (south to 33°S)		1500	85	0.3	0.2			19,125
Central Chile (33°–45°S)		1800	80	2.0	1.3			187,200
Southern Chile (45° S to western Tierra del Fuego)	1100		20	3.0	2.0	30,800	13,200	
Hikurangi	400		30	2.0	1.3	10,920	4680	
Kermadec		1300	70	0.4	0.2			21,840
Tonga		1300	170	0.4	0.2			53,040
Vanuatu (New Hebrides)		1500	100	0.4	0.2			36,000
San Cristobal (Solomons)		1500	110	0.4	0.2			39,600
New Britain		900	110	1.0	0.6			59,400
New Guinea		1100	20	1.5	1.0			22,000
Philippine		1200	90	0.4	0.2			25,920
West Luzon (Manila)	1000		90	1.5	1.0	59,850	25,650	
Southern Ryukyu	500		30	1.0	0.6	6300	2700	
Northern Ryukyu		1000	60	0.5	0.3			15,000
Nankai	700		30	1.5	1.0	13,965	5985	
Yap-Palau		600	20	0.4	0.2			2880
Mariana		2400	90	0.4	0.2			51,840
Izu-Bonin		1200	100	0.4	0.2			28,800
Japan		800	100	0.6	0.3			24,000
Kuril-Kamchatka		2000	90	0.6	0.3			54,000
INDIAN OCEAN								
Puysegur (southern New Zealand)		500	20	0.4	0.2			2400
Makran	500		35	7.0	4.5	55,125	23,625	
Andaman	1500		10	2.0	1.3	13,650	5850	
Sumatra (Sunda)	1700		50	1.5	1.0	59,500	25,500	
West Java	500		75	1.6	1.0	26,250	11,250	
East Java		1200	75	0.6	0.3			27,000
ATLANTIC AND MEDITERRANEAN								
Antilles	800		20	2.5	1.7	19,040	8160	
South Sandwich		600	20	0.4	0.2			2880
Aegean	1300		25	4.0	3.0	68,250	29,250	
Sums						395,500	169,500	944,438
Sums (km)	11,000	31,250				At accreting subduction zones, total sediment accreted is ~400,000 km^3/m.y.	At accreting subduction zones, total sediment subducted is ~170,000 km^3/m.y.	At nonaccreting subduction zones, total sediment subducted is ~940,000 km^3/m.y.
Total major convergent margin trench length (km)			42,250	Using rounded numbers				
Percent accreting margin	26%			25%				
Percent nonaccreting margin	74%			75%				

(*continued*)

TABLE 2. *(continued)*

Region and trench sector	Percent of global length of all subduction zones (out of 42,250 km)	Percent of global length of accreting subduction zones (out of 11,000 km)	Percent of global length of nonaccreting subduction zones (out of 31,250 km)	Fraction of global length of orthogonal subduction (km/m.y.)	Fraction of orthogonal convergence at nonaccreting margins (km/m.y.)	Fraction of orthogonal convergence at accreting margins (km/m.y.)
PACIFIC OCEAN						
Northeast Alaska (east of Kodiak Seamount)	0.14	4.55		0.1		2.7
Southwestern Alaska (Kodiak Seamount to Unimak Pass)	2.60		3.52	1.7	2.3	
Aleutian (Unimak Pass west to Near Islands)	4.26		5.76	1.9	2.6	
Cascadia (Vancouver south to central Oregon)	2.37	9.09		0.8		3.2
Southern Oregon	0.47		0.64	0.2	0.2	
Mexico (southeast of Gulf of California)	2.60		3.52	1.8	2.5	
Middle America (Guatemala southeast to Panama)	3.31		4.48	2.3	3.1	
Colombia	1.78		2.40	1.2	1.6	
Ecuador	1.18		1.60	0.8	1.0	
Peru	4.73		6.40	4.0	5.4	
Northern Chile (south to 33° S)	3.55		4.80	3.0	4.1	
Central Chile (33°–45° S)	4.26		5.76	3.4	4.6	
Southern Chile (45° S to western Tierra del Fuego)	2.60	10.00		0.5		2.0
Hikurangi	0.95	3.64		0.3		1.1
Kermadec	3.08		4.16	2.2	2.9	
Tonga	3.08		4.16	5.2	7.1	
Vanuatu (New Hebrides)	3.55		4.80	3.6	4.8	
San Cristobal (Solomons)	3.55		4.80	3.9	5.3	
New Britain	2.13		2.88	2.3	3.2	
New Guinea	2.60		3.52	0.5	0.7	
Philippine	2.84		3.84	2.6	3.5	
West Luzon (Manila)	2.37	9.09		2.1		8.2
Southern Ryukyu	1.18	4.55		0.4		1.4
Northern Ryukyu	2.37		3.20	1.4	1.9	
Nankai	1.66	6.36		0.5		1.9
Yap-Palau	1.42		1.92	0.3	0.4	
Mariana	5.68		7.68	5.1	6.9	
Izu-Bonin	2.84		3.84	2.8	3.8	
Japan	1.89		2.56	1.9	2.6	
Kuril-Kamchatka	4.73		6.40	4.3	5.8	
INDIAN OCEAN						
Puysegur (southern New Zealand)	1.18		1.60	0.2	0.3	
Makran	1.18	4.55		0.4		1.6
Andaman	3.55	13.64		0.4		1.4
Sumatra (Sunda)	4.02	15.45		2.0		7.7
West Java	1.18	4.55		1.58		3.4
East Java	2.84		3.84	3.79	2.9	
ATLANTIC AND MEDITER-RANEAN						
Antilles	1.89	7.27		0.4		1.5
South Sandwich	1.42		1.92	0.3	0.4	
Aegean	3.08	11.82		0.8		3.0
Sums	100.00	100.00	100.00	70.8	79.8	36.2
Sums (km)				Global average orthogonal convergence at all margins is ~70 km/m.y.	Global average orthognal convergence at nonaccreting margins is ~80 km/m.y.	Global average orthogonal convergence at accreting margins is ~35 km/m.y.
Total major convergent margin trench length (km)						
Percent accreting margin						
Percent nonaccreting margin						

TABLE 3. LONG-TERM RATES OF SUBDUCTION EROSION AND FOREARC TRUNCATION AT NONACCRETING MARGINS

Subduction zone sector	Basis for reconstructing missing forearc width	Missing forearc width (km)	Shoreline crustal thickness (km)	Erosion time (m.y.)	Eroded porous volume based on self-similar model (km^3)	Eroded solid volume at 5% of porous volume (km^3)	Truncation or trench advance rate (km/m.y.)	Rate of tectonic erosion of submerged forearc (km^3/m.y./km)	Data references
			CONTINENTAL CRATONIC BASEMENT						
NE Japan	Subsidence	60	25	20	1500	1275	3.0	64	1
Mexico	Coastal arc plutons								
	P. Vallarata (~105° W)	80	25	75	2000	1700	1.1	23	
	Manzanillo (~104° W)	80	25	60	2000	1700	1.3	28	
	Acapulco (~100° W)	120	25	45	3000	2550	2.7	57	
	Average Mexico						**1.7**	**36**	6
Peru	Subsidence	165	25	50	4125	3506	3.3	70	5
Chile	North Chile arc migration	300	25	160	7500	6375	1.9	40	3
	North Chile coastal plutons	250	25	190	6250	5313	1.3	28	4
	Central Chile arc migration	85	25	20	2125	1806	4.3	90	2
	Average Chile						**2.[illegible]**	**57**	
			MESOZOIC ARC OR MAFIC BASEMENT						
Costa Rica	Costa Rica subsidence	60	13	16.5	780	663	3.6	40	7
Guatemala	Guatemala subsidence	20	13	19	260	221	1.[illegible]	12	8
Guatemala	Guatemala subsidence	45	13	50	585	497	0.9	10	8A
	Average Middle America						**1.9**	**21**	
			CENOZOIC ARC BASEMENT						
Tonga	Subsidence	75	13	40	975	829	1.9	21	9
South Sandwich	Arc migration	70	13	15	910	774	4.7	52	10
						AVERAGE RATES			
						Truncation	2.3		
						Erosion		42	
							Range = 0.9 to 4.7 km/m.y.	Range = 10 to 90 km^3/m.y./km	

Note: 1—von Huene and Cullotta (1989), von Huene and Lallemand (1990), von Huene et al. (1994); 2—Kay et al. (2005); 3—Rutland (1971), Levi (1973), Naranjo et al. (1986), von Huene and Scholl (1991), Andriessen and Reutter (1991), Scheuber (1994); 4—Naranjo et al. (1986); 5—von Huene and Lallemand (1990), Clift et al. (2003); 6—Moran-Zenteno et al. (1996), Moran-Zenteno et al. (1999), Schaaf et al. (1995); 7—Vannucchi et al. (2001, 2003); 8—Vannucchi et al. (2004); 8A—Based on identification of shallow-water foraminifera assemblages of early Eocene age overlying basement at Site 569, DSDP Leg 84 (von Huene et al., 1985); 9—Clift and MacLeod (1999); 10—Vanneste and Larter (2002).

TABLE 4. GLOBAL ESTIMATE OF RECYCLED LOWER PLATE (OCEANIC) SEDIMENT, ERODED UPPER PLATE MATERIAL, AND AVERAGE THICKNESS OF SUBDUCTION CHANNEL

		Globally (km^3/yr)	Average for km of margin (km^3/m.y.)	
Subducted sediment				
Subducted toward mantle at nonaccreting margins		0.9	30.2	
Subducted toward mantle at accreting margins		0.2	15.4	
Globally subducted toward mantle		1.1	26.4	
Subduction erosion				
Subduction erosion @ 40 km^3/m.y./km at nonaccreting margins		1.3	42.0	
Subduction erosion @ 12 km^3/m.y./km at accreting margins		0.1	13.0	
Global sum subducted toward mantle		1.4	34.4	
Total subducted material				**Percent**
At nonaccreting margins (~31,000 km)				
	Subducted sediment	0.9	30.2	**41.8**
	Subduction erosion	1.3	42.0	**58.2**
	Global sum subducted toward mantle	2.2	72.2	**100.0**
At accreting margins (~11,000 km)				
	Subducted sediment	0.2	15.4	**54.2**
	Subduction erosion	0.1	13.0	**45.8**
	Global sum subducted toward mantle	0.3	28.4	**100.0**
At all margins (~42,000 km)				
	Subducted sediment	1.1	26.4	**43.4**
	Subduction erosion	1.4	34.4	**56.6**
	Global sum subducted toward mantle	2.5	60.8	**100.0**
Average thickness of subduction channel				**km**
Nonaccreting margins (~31,000 km) at orthogonal convergence of 80 km/m.y.				**0.9**
Accreting margins (~11,000 km) at orthogonal convergence of 35 km/m.y.				**0.8**
For all margins (~42,000 km) at orthogonal convergence of 70 km/m.y.				**0.9**

accreting subduction zone is the Tonga margin (Fig. 3) (Scholl and Vallier, 1985; Ballance et al., 1989; Clift and MacLeod, 1999). Other examples are the subduction zones bordering northern Japan (von Huene, and Cullotta, 1989; Tsuru et al., 2000) and the connecting Kuril-Kamchatka margin to the north (Fig. 3) (Klaeschen et al., 1994), the Chile (Laursen et al., 2002; von Huene and Ranero, 2003; Sallares and Ranero, 2005), Peru (Clift et al., 2003; Sosson et al., 1994), Ecuador and Colombia (Collot et al., 2002) margins of South America, and evidently the length of the Middle America margin including Mexico (Vannucchi et al., 2001, 2003, 2004; Moran-Zenteno et al., 1996).

Volumetrically, crustal recycling rates at the two margin types would appear to differ, although this is not known to be the situation because virtually all studies of accreting margins have focused on sediment storage and fluid expulsion processes rather than companion recycling losses. Nonetheless we treat accreting (presence of a constitutive middle prism) and nonaccreting margins (close approach of the framework rock of the inner prism to the trench axis) separately because where large accretionary bodies have accumulated, our estimates of recycling volumes are least based on constraining observations (Tables 1, 2, and 4).

Nonaccreting and accreting margins as grouped in this paper are similar but not the same as the margin types defined by von Huene and Scholl (1991) and Clift and Vannucchi (2004) (Fig. 3). The principal difference is that we place in the nonaccreting category margins bordered by accretionary frontal prisms that lack or virtually lack an older or middle prism of accreted material and thus are structurally directly attached to the forearc's inner rock framework. As noted, these accretionary frontal prisms are geologically young bodies that reflect the offscraping of thick (>1.5 km), late Cenozoic accumulations of glacial-age trench turbidite deposits.

EVIDENCE AND VOLUMES OF SUBDUCTED SEDIMENT

Physical Evidence

The principal geophysical evidence attesting to sediment subduction is the small width of frontal prisms fringing most convergent margins, in particular nonaccreting margins. At these margins, the volume of the seismically imaged frontal prism is much less than the potential volume if all sediment entering the subduction zone had been scraped off. An example of this observation is the seismic section displayed on Figure 4 for the NE margin of Honshu, Japan (von Huene et al., 1994; Tsuru et al., 2000). The frontal prism is ~30 km wide and lies seaward of the margin's inner rock framework of Late Cretaceous age, which was reached at Site 439, DSDP Leg 56 (von Huene et al., 1982).

The solid volume (zero porosity) of the prism is ~65 km^3. The solid thickness of the column of pelagic sediment entering the subduction zone with the underthrusting Pacific plate is ~0.25 km. At an average convergence rate during the past 65 m.y. of 80 km/m.y. (presently ~90 km/m.y.), the potential volume of an off-scraped mass of pelagic sediment alone is 1300 km^3, of which the existing prism, presuming that it is only constructed of lower plate material, represents 5%. The missing volume of pelagic sediment, including any additional quantities of slope debris or trench-axis turbidite deposits swept into the subduction zone, has evidently been subducted beneath the Cretaceous and older rock framework of the inner prism of the Japanese margin. Similar observations have been made for many other subduction zones, for example those of Middle America margins (Vannucchi et al., 2001, 2004), the South Sandwich Trench (Vanneste and Larter, 2002), and the thickly sediment-flooded trenches typified by the central Chile (Bangs and Cande, 1997; Behrmann and Kopf, 2001) and the Alaska-Aleutian Trench system (von Huene and Scholl, 1991).

Efficiency and Volume of Sediment Subduction

At Nonaccreting Margins

Because nonaccreting margins lack accreted prisms of any significant width, we view their frontal prisms as ephemeral geologic structures that over a relatively short period of geologic time (~5 m.y.) are themselves recycled and replaced with younger material. For the purpose of this paper and estimating the global rate of sediment subduction over very long periods of geologic time (>50 m.y.), we thus assign a sediment-recycling rate of 100% to all nonaccreting margins whether tipped by a frontal prism of dominantly accreted sediment or tectonized slope debris. This estimate is only slightly higher than that, 95%, presented by von Huene and Scholl (1991) for all margins lacking large, growing accretionary bodies.

Table 2 lists the volume of ocean floor sediment orthogonally subducted along each major sector of ocean margin subduction zones. The data compiled in this table are an updating of a similar table presented in von Huene and Scholl (1991). The differences between these tables reflect the incorporation of observations made during the past 15 years and a subduction zone classification scheme that better distinguishes sectors with a similar convergence rate, involving the same plates, and with a similar input thickness of trench floor sediment. Globally, at major nonaccreting margins, which total ~31,000 km in length, the amount of ocean sediment that is subducted landward of the frontal prism is ~0.9 km^3/yr. In a million years, the globally averaged volume of sediment subducted beneath each kilometer of nonaccreting margin is estimated to be ~30 km^3 (see Table 4).

At Accreting Margins

Sediment subduction at accreting margins (~25% of all; Tables 1 and 2) is poorly understood, let alone quantitatively constrained by mass balance calculations. The principal troubling issue is that most of the major accreting margins, e.g., Makran, Sumatra, Luzon, Cascadia, Lesser Antilles, and Aegean, have been accumulating lower plate sediment for tens of millions of years. Reconstructing the average thickness of lower plate sediment entering the subduction zone over this length of time is extremely difficult to do. In addition, good seismic images of the mass and geometry of the older accreted body are commonly unavailable or murky.

For accreting margins with an inner rock framework of arc or mafic volcanic sequences, for example the Lesser Antilles (White and Dupré, 1986) and Cascadia (Church, 1976) subduction zones, a signal of continental contamination in young erupted volcanic rock implies that subducted sediment shed from a mature continental provenance reaches depths where arc magmatism begins (100–125 km). We estimated in von Huene and Scholl (1991) at least 10% sediment subduction at these and other accreting margins. At the frontal prisms of modern accreting margins, the ratio of the frontally accreted section to that passing landward beneath the prism (i.e., below the plate boundary or décollement) is roughly 60% frontally accreting to 40% underthrusting (see, for example, Kopp et al., 2000). We use this partitioning ratio as implying that sediment bypassing of an accreting margin is at least 30% and, potentially, much higher. We are aware that at the Nankai margin of SW Japan, underthrusting sediment is thought to be underplated toward the back of the frontal prism and thus does not continue onward toward the mantle within the subduction channel (Park et al., 2002; Bangs et al., 2004). Thus the question of crustal recycling at convergent margins remains a frontier of subduction zone research where large accretionary prisms have accumulated (Bangs et al., 2006).

However, because the aggregate length of accreting margins (~11,000 km) does not exceed ~25% of the global length of convergent ocean margins (~42,000 km), the global volume of subducted sediment presented on Tables 2 and 4 is somewhat insensitive to current uncertainties about sediment storage and bypassing partitioning at accreting margins. The subordinate percentage of accreting margins of Cenozoic age, however, may not apply to the Cretaceous when massive accretionary wedges widely presumably formed at convergent margins and in particular around the northern perimeter of the Pacific Basin—e.g., the Franciscan Complex of California, the Kodiak Formation of southern Alaska, and the Shimanto Complex of western Japan. Potentially, these large bodies are not preserved frontal and middle prism but rather exhumed, crustal underplates that accumulated subcrustally beneath the inner prism and well inboard of the submerged forearc (Jacobson et al., 1996; Grove et al., 2003).

Table 2 summaries the orthogonal input of sediment to the major sectors of accreting margins and applies a subduction efficiency factor of 30% (i.e., 70% accreted to the frontal prism and 30% bypassed). Accordingly, at these margins the volume of ocean floor sediment subducted landward of the frontal prism is estimated to be ~0.2 km³/yr (solid). For a typical kilometer-long sector of accreting margin, the amount subducted in a million years is ~15 km³ (see Table 4).

Global Mass of Subducted Sediment

The combined mass of sediment subducted landward of the frontal prism of modern accreting and nonaccreting margins is globally estimated to be ~1.0 km³/yr (i.e., 1 Armstrong unit or AU). Using a different approach, which involved a modeling concept to quantify sediment subduction, Clift and Vannucchi (2004) estimate that the efficiency of sediment subduction at convergent margins globally averages ~83%. Our global average is ~74%. If we had estimated sediment subduction at accreting margins as high as 40%, the global average efficiency of sediment subduction would increase to ~80%.

EVIDENCE AND VOLUMES OF ERODED FOREARC MATERIAL

Physical Evidence

Subduction erosion is a form of tectonic erosion that reduces the rock and sediment mass of the submerged forearc. Evidence and indicators most commonly cited for subduction erosion, which are discussed in von Huene and Scholl (1991) and Clift and Vannucchi (2004), are

1. outer forearc subsidence >~3–4 km during the past 5–50 m.y.;
2. seaward and downward tilting of the forearc attending subsidence;
3. substantial (>30–50 km) and sustained (>15 m.y.) inboard migration of the arc magmatic (volcanic) front;
4. seaward truncation of the forearc's igneous and metamorphic basement rock; and
5. coastal and offshore occurrence of Mesozoic and early Tertiary arc volcanic and plutonic rocks missing much of their original forearc width.

Three of these observations—subsidence, arc migration, and coastal arc exposure—can best quantify the amount of rock framework removed from the submerged forearc during a long period of geologic time (Fig. 5, Table 3).

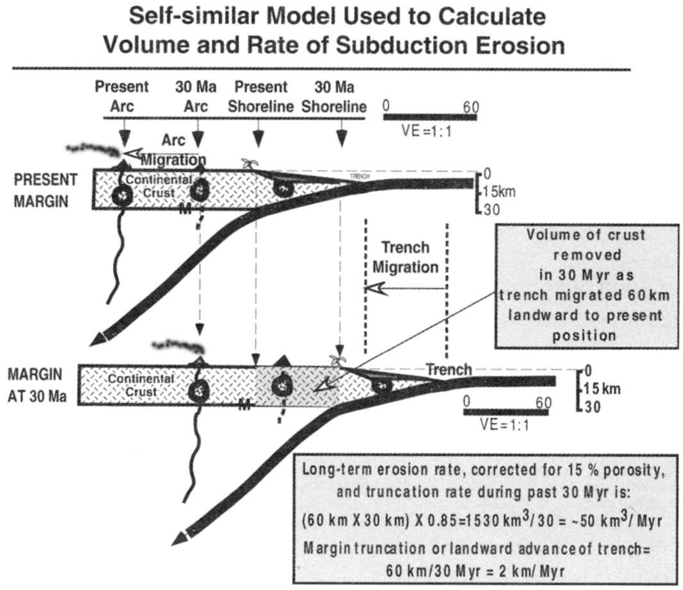

Figure 5. Self-similar model for estimating the volume of material removed from the submerged forearc by subduction erosion. The premise of the model is that as the trench migrates landward the volumetric shape and dimensional relations of the submerged margin stay the same. In the self-similar model for each kilometer of margin, the volume of the demolished forearc is calculated as equivalent to the product of shoreline crustal thickness and the width of forearc truncation (trench advance). Estimates of tectonic erosion of the submerged forearc can be based on recovering shallow-water, shoreline, or nonmarine deposits overlying forearc framework rock within a few tens of kilometers of the trench floor and at a depth of 4–5 km. As shown in this diagram, crustal thinning to explain subsidence is at least 20 km. Landward advance of the trench is shown accompanied by a landward migration of the arc. Occurrence of arc magmatic bodies beneath the submerged forearc is also evidence of margin truncation because arc magmatism occurs 150–200 km inboard of the trench. As shown here, the trench is only about a distance of 70 km from a submerged arc pluton. Thus from 80 to 130 km of its old forearc is missing. This circumstance is typical of the western margin of South America as well as Mexico.

Crustal Thinning and Subsidence

The best documented evidence for deep subsidence of the middle and outer forearc comes from DSDP and ODP drilling, in particular off NE Japan (Legs 56 and 57; von Huene and Cullotta, 1989; von Huene and Lallemand, 1990; von Huene et al., 1994), Tonga (Leg 135; Clift and MacLeod, 1999), Peru (Leg 112; von Huene and Lallemand, 1990; Clift et al., 2003), Costa Rica (Leg 170; Meschede et al., 1999; Vannucchi et al., 2001, 2003), Guatemala (Legs 67 and 84; Vannucchi et al., 2004), central Mexico (Leg 66; Moore et al., 1982; Vannucchi et al., 2004), by submersible observations off northern Mexico (Mercier de Lepinay et al., 1997) and Peru (Sosson et al., 1994), and by dredging the landward trench slope north of Puerto Rico (Grindlay et al., 2005) (Fig. 3).

The principal evidence documenting subsidence is the recovery within a few tens of kilometers of the trench axis of deeply subsided (3–5 km) nearshore, shoreline, or nonmarine deposits resting unconformably on the rock framework of the forearc's inner prism. Crustal thicknesses in coastal areas are typically 15–35 km; thus recovery near the trench of coastal deposits overlying crust only 5–10 km thick requires crustal thinning of 10–25 km (Fig. 5). Because evidence is not observed that thinning is accomplished by structural lengthening of the margin in response to extensional faulting, margin thinning, subsidence, and narrowing is attributed to subduction-effected tectonic erosion (Scholl et al., 1980; von Huene and Scholl, 1991; Ranero and von Huene, 2000; Clift and Vannucchi, 2004; Sallares and Ranero, 2005).

The seismic image of a margin collapsed and truncated by subduction erosion is displayed on Figure 4 for the lower trench slope off the NE margin of Honshu, Japan (Tsuru et al., 2000). DSDP drilling at nearby Site 439 established that the margin's rock framework is constructed of Late Cretaceous sedimentary rock terminated upward by a wave-base-cut unconformity overlain by fossiliferous shoreline deposits of late Oligocene–early Miocene age (von Huene et al., 1982). Seismic data further reveal that the Late Cretaceous rock framework of the inner prism and overlying wave-based unconformity is truncated to seaward at the back of the frontal prism. Near the trench, the crust lying beneath the unconformity and the top of the underthrusting Pacific plate is ~6–8 km thick. But when the wave-based unconformity was carved, the crustal thickness was ~30 km; thus at least 20 km of the crust underlying the late Oligocene coastal area of Japan has been destroyed during the past ~30 m.y.

Landward Arc Migration

Landward arc migration is documented evidence for subduction erosion where migration is sustained over long periods of geologic time and other data (i.e., arc geochemistry and plate reconstructions) do not identify long-term and continuous shallowing of the descent angle of the underthrusting slab that is generally thought to cause landward arc-front migration. But the implications of Chilean observations is that flat-slab subduction does not cause inward arc-front migration but instead a landward broadening or smearing out of the footprint of arc magmatism (Abbruzzi, 1996; Kay et al., 2005). With respect to subduction erosion, the fundamental posit is that arc migration matches landward trench advance (toward a fix reference on the forearc) in response to tectonic-erosion-caused truncation and thinning of the forearc (Fig. 5).

Arc migration evidence for subduction erosion is perhaps best displayed along the western South America margin from the Early Jurassic to the present, in particular for northern and central Chile, 20°–33° S (Rutland, 1971; Levi, 1973; Ziegler et al., 1981; Andriessen and Reutter, 1991; Scheuber, 1994), and since the early Miocene for south-central Chile, 33°–36° S (Kay et al., 2005). Cenozoic arc migration has also been recognized for the South Sandwich arc (Vanneste and Larter, 2002) and also the Aleutian Ridge (Kay et al., 1990; Scholl et al., 2003).

Coastal Arc Plutons

Arc magmatic eruptive and intrusive bodies are typically emplaced ~150–250 km inboard of the trench and 100–125 km from the coast. The occurrence of coastal and submerged arc magmatic bodies, in particular of Mesozoic and early Tertiary age, signals that much or even most of the original width of their forearcs is missing. This realization arises because the width of the offshore or submerged forearc seaward of the coastal arc is commonly less than 50% of that of the arc's original forearc. The truncated forearc is a manifestation of subduction erosion if strike-slip faulting, rifting, or subduction of a spreading ridge cannot explain the coastal setting of arc plutonic bodies. Coastal arc plutons and linked eruptives have thus been cited as documenting subduction erosion for the Mexico (Moran-Zenteno et al., 1996; Mercier de Lepinay et al., 1997), Panama (Lissinna et al., 2002), and Chile margins (Rutland, 1971; Naranjo et al., 1986).

Estimating Volume of Eroded Material

At Sectors of Ocean Margin

The rate of subduction erosion along a sector of convergent margin is estimated by reconstructing the volumetric size of a former extant margin and dividing this volume by the time it took to reduce the margin to the present one (von Huene and Lallemand, 1990). For the purposes of this paper, we adopt a self-similar modeling approach as shown on Figure 5. The simplification of the self-similar method states that, unless evidence exists to the contrary, the present margin's volumetric shape and dimensions are similar to what they were in the past (see Vannucchi et al., 2003). We also adopt the simplification that at most continental margins the underthrusting slab passes beneath the Moho underlying the continental or island-arc crust of the upper plate in the vicinity of the coastline (Ruff and Tichelaar, 1996). Also supporting this notion is the circumstance that the recovery in the outer forearc of nearshore or nonmarine deposits resting on framework rock of the inner prism implies that if subduction erosion is responsible for the required crustal thinning, the underthrusting plate was in contact with the base of the coastal-zone crust. A supporting observation is that Wadati-Benioff zone great thrust earthquakes nucleate beneath the inner zone of the submerged forearc. Presumably this requires interplate contact.

Adopting the self-similar geometry and dimension model and a coastal location of the seaward edge of the upper plate's Moho allows a simple geometric calculation to estimate the volume of tectonically eroded upper plate crust between the trench and the vicinity of the shoreline. Calculating the missing volume requires estimating the crustal thickness in the shoreline area and evidence to reconstruct the former seaward position of the trench axis. For each km of margin, the volume of the missing margin, as shown diagrammatically on Figure 5, is the product of the landward advance of the trench (i.e., distance separating the reconstructed trench position and the present one) and the present coastal thickness of upper plate crust.

The estimated volumes do not include erosive removal of the upper plate's inboard crust by ablative process, for example that proposed attending Andean orogenesis (Pope and Willett, 1998). Similarly, lower crust caught up in foreland lithospheric delamination is not included in our accounting of recycled continental crust summarized on Table 4. Also not included is crustal tectonic erosion effected where continental masses collide at subduction zones, the evidence for which is the exposure of ultrahigh-pressure (UHP) metamorphic rocks and the likelihood that not all UHP material produced at continental subduction zones is returned to the upper plate by exhumation processes (Hacker et al., 2006).

Only long-term rates of subduction erosion can be meaningfully applied to approximate crustal recycling over long periods of geologic time. Hence we do not include on Table 3 short-term or sector-specific rates that reflect enhanced crustal removal where high bathymetric relief enters the subduction zone. Short-term, sector-specific rates can be many times to an order of magnitude higher than long-term rates. Examples of well-constrained but enhanced rates of crustal destruction are where the Chile spreading ridge enters the subduction zone of south-central Chile (Bourgois al., 1996), where the Louisville Ridge enters the Tonga subduction zone (Ballance et al., 1989), and where the Bahama Platform underthrusts the north-facing slope of the Puerto Rico Trench (Grindlay et al., 2005).

Although many margin sectors exhibit evidence of long-term subduction erosion, estimates of volume loss are commonly made unreliable by equivocal observations, for example the Mariana and Izu margins of the western Pacific with respect to evidence documenting subsidence (Clift and Vannucchi, 2004), or inadequate reconstructive information, for example the Aleutian margin (Scholl et al., 2003).

The estimates of long-term (>~10 m.y.) rates of subduction erosion listed on Table 3 are thus only for those margin sectors where we feel sufficient long-term reconstructive evidence, in particular reliable age data, has been acquired. Eight trench sectors are identified on this table, NE Japan, Tonga, South Sandwich, Chile, Peru, Costa Rica, Guatemala, and central Mexico. Different erosion-assessing methodologies within a sector are averaged to avoid introducing a bias from one sector. We presume long-term rates sum the combined effects of slower background crustal removal punctuated by episodes of rapid thinning. The rates shown on this table are based on the self-similar model, and, as a consequence, the value shown may be different from published rates (see references in Table 3). They also different from the rates tabulated by Clift and Vannucchi (2004) because our self-similar model used a crustal thickness typical of the shoreline area rather than the thicker crust underlying the more inboard volcanic arc.

Global Volume of Eroded Crust

Constraining information bearing on estimating rates of subduction erosion is only available from nonaccreting margins. Converting the crustal volume destroyed to its equivalent solid volume requires correcting for void spaces. Conversion from rock and sediment volume to solid material was estimated by assuming that the void space of the margin's volume of inner prism framework rock averages 10%, and that of the shelf-slope sedimentary apron 30%.

For nonaccreting margins (Fig. 1), the average proportion of their volume that is apron sediment is estimated at ~20%. Thus the average solid volume of the submerged nonaccreting margin is ~85% of the geophysically measured volume. This value is integrated into the calculations on Table 3 to estimate the long-term volume of material eroded from eight margin sectors (expressed as km^3/m.y./km) and transported toward the mantle in the subduction channel. For each kilometer of nonaccreting subduction zone, the average long-term rate is estimated to be ~42 km^3/m.y., which we round to ~40 km^3/m.y. This rate is higher than our earlier estimate of ~35 km^3/m.y./km (von Huene and Scholl, 1991). Applied to the present ~31,000 km of major nonaccreting margins, an erosion rate of ~40 km^3/m.y. for each kilometer of subduction zone estimates the global annual rate of eroded material at 1.3 km^3 (i.e., 1.3 AU).

At accreting margins, information constraining or documenting long-term rates of subduction erosion is lacking or at best equivocal. But we know of no reason to suppose that subduction erosion does not cause crustal thinning at margins where subduction accretion is widening the submerged forearc. Wells et al. (2003) argue that submerged forearc basins are the loci of interplate seismic asperities, i.e., where most slip occurs during large subduction zone earthquakes (see also Song and Simons, 2003). This happens because the crust is thinnest here, and thus the effective stress on the interplate boundary is least beneath forearc basins. They link basin formation to a focused region of enhanced basal subduction erosion resulting from rapid, subbasin coseismic slipping. High moment release here in turn hydrofractures the base of the forearc crust, transfers upper plate material to the subduction channel, and hastens the transport of recycling debris toward the mantle.

Both the Cascadia and Nankai accreting margins have large forearc basins underlain by a middle prism and more inboard framework rock of the inner prism. These basins, accordingly, can be ascribed to crustal thinning and subsidence effected by earthquake-enhanced rates of basal subduction erosion that, in positive feedback, works to sustain the localizing of high moment release beneath forearc basins. For the northern Cascadia margin, Calvert et al. (2003) and Calvert (2004) also recognize reflection and refraction evidence consistent with subduction erosion and the transport of crustal material into the mantle wedge.

We thus presume that basal subduction erosion at an accreting margin is concentrated beneath the margin's inner rock prism and perhaps the older accreted material of the middle prism (Fig. 2; see also Wells et al., 2003, their Fig. 19). Based on the limited width of framework rock beneath accreting margins, we assign a subduction erosion rate of only 30% of that typical of nonaccreting margins (~40 km^3/m.y./km). The reduced rate, 12 km^3/m.y./km, could be higher if seamounts and bathymetric relief plowing through the margin also flush significant parts of the frontal prism and older accreted prism into the subduction channel (Fig. 2) (Bangs et al., 2006). At the present global length of accreting margins, ~11,000 km, the estimated annual volume of eroded forearc material is thus a mere 0.1 km^3, an estimate regarded as a minimum evaluation.

RECYCLED CONTINENTAL CRUSTAL VOLUME

Global Volume Moved toward Mantle

Globally, the volume of continental material in transport toward the mantle in the subduction channel is the combined mass of subducted sediment and eroded forearc material (Figs. 1 and 2). The volume of subducted sediment ingested at both nonaccreting and accreting margins, as stated earlier, is estimated in a rounded figure to be ~1.0 km^3/yr or 1 AU (Table 4).

At nonaccreting margins, the quantity of eroded material inserted into the subduction channel is estimated to be 1.3 AU. Adding to this volume our estimate of that eroded at accreting margins (~0.1 AU) increases the global movement of eroded material to 1.4 AU or km^3/yr (Table 4).

Combining this flux of eroded material with that of subducted sediment means that a volume of ~2.5 AU of continental debris is annually conveyed beneath the forearc toward the mantle. At a global averaged orthogonal subduction rate of ~70 km/m.y. (Table 2), delivery of an annual volume of ~2.5 km^3 at ~42,000 km of convergent margins means that the average thickness, rounded off, of sediment and rock in the subduction channel is ~1.0 km (Table 4). Along many sectors of margin, the apparent geophysically imaged subduction channel is this thick or thicker (Park et al., 2002; Ranero and von Huene, 2000; Collot et al., 2002; von Huene and Ranero, 2003; Calvert, 2004).

Subsurface Fate of Subducted Continental Debris

Three subsurface fates await continental debris transported by subduction toward the mantle (Fig. 6):

1. underplating of the upper plate in the coastal area;
2. involvement in arc melting; and
3. mixing into a path of deep mantle circulation.

Underplating and arc magma assimilation follow short return paths to the upper plate. Underplating is, effectively, just a repositioning of crustal material eroded from the surface of Earth and the submerged forearc to the base of the coastal crustal column below which, except where flat-slab subduction occurs, the underthrusting slab commonly sinks into the mantle.

Underplating is a viable process and has been applied to the nonaccreting Chilean and Costa Rica margins to account for coastal uplift coeval with subsidence of the offshore forearc (Hartley et al., 2000; Gardner et al., 2001, respectively). The hallmarking evidence of underplated material is a history of high-*P*/*T* metamorphism. An example of an underplate is the Orocopia-Pelona Schist of southern California and Arizona. The schist is interpreted as an exposed mass of Cretaceous and early Tertiary sediment underplated beneath the backstopping basement rock of the inner prism of the Franciscan margin (Jacobson et al., 1996; Grove et al., 2003). In effect, the Orocopia-Pelona Schist is an exhumed subduction channel.

The rounded rate at which subducted sediment and eroded debris is transported toward the coastal region (i.e., at ~70 km^3/m.y./km at nonaccreting margins and ~30 km^3/m.y./km at accreting margins; Table 4) is so prodigious that if at the global average rate of ~60 km^3/m.y./km all subducted material underplated the coastal region, the entire volume of the forearc (~4000 km^3/km) from arc to trench would be replaced in less than 70 m.y. with underplated material that passed through a high-*P*/*T* metamorphic environment. But around the rim of the Pacific, exposed tracks of high-pressure crustal underplates are volumetrically small compared to the volume of continental material missing. Most coastal regions, for example virtually the Pacific length of Mexico and South America, are underlain by high-*T*/*P* metamorphic sequences and intruding arc igneous masses (see for example Rutland, 1971; Moran-Zenteno et al., 1999; Thomson et al., 2001; Ducea et al., 2004; Kay et al., 2005). We conclude that the bulk of underplated material is only temporarily stored beneath the coastal area except where flat-slab subduction has, far inboard of the coastal region, replaced the lower crust of the margin with an underplate of subducted sediment (Jacobson et al., 1996; Grove et al., 2003). Evidently, as subduction erosion progressively narrows and thins the margin's inner rock prism, the coastal underplate is flushed into the mantle. It is also likely that the "accretionary prisms" of Mesozoic and early Tertiary age widely exposed in Pacific-rim fold belts are exhumed crustal underplate rather than preserved frontal or middle prisms. Preservation of underplates evidently requires a cessation of underthrusting, a significant width of accretionary growth, or flat-slab subduction.

It is known that subducted sediment and fluid are geologically quickly recycled to the upper plate in arc magmas (Morris et al., 1990; Plank and Langmuir, 1993). But probably no more than a few volumetric percent of arc magmas are subducted crustal components (see overview discussion by Stern, 2002). The small percentage, however, is adequate to identify infiltration of the arc melting process by crustal debris injected to the 100–150 km depth typical of arc magma generation (Takagi, 2004; Kay et al., 2005). It is both interesting and important that Kay et al. (2005) reports for central Chile that subduction erosion pulses the flow of crustal debris to depths of arc magma genesis, a phenomenon linked tectonically to an inboard jump in the arc magmatic front.

Evidently, with respect to the fate of subducted sediment and eroded debris, the combined volume of material directly returned to the upper plate via underplating and arc magmatism is much smaller than that transported over a long time toward the mantle. We reason that over the long term (tens of millions of years) more than 90% of subducted continental material is recycled to deeper mantle circulation.

Thus, in summary, our estimate of the solid volume of upper plate material missing from the submerged forearc and that

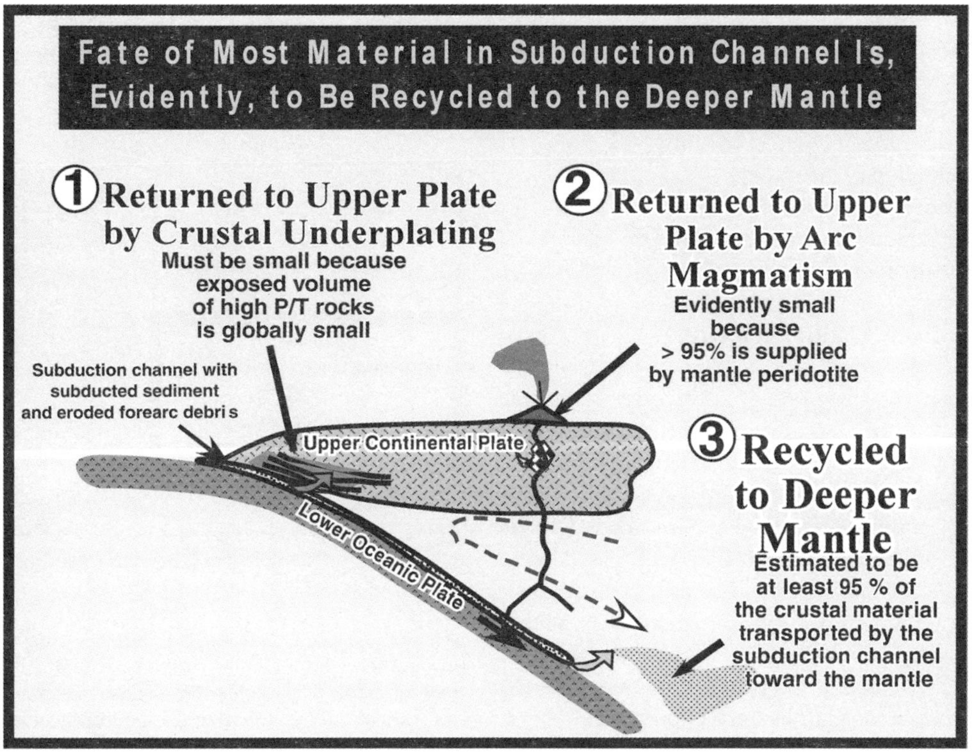

Figure 6. Subducted sediment and eroded forearc debris is moved toward the mantle in the ~1 km thick subduction channel separating the upper and lower plates. So little of the volume of subducted material (~60 km^3/m.y. per kilometer of convergent margin) is directly returned to the upper plate via underplated material or arc magmatism that most (~95%) of the subducted volume presumably inters into deep mantle circulation.

entered mantle circulation is ~2.5 AU (km^3/yr), of which ~55% is tectonically eroded debris and ~45% subducted sediment (Table 4). At nonaccreting margins alone, the recycling volume is ~2.2 AU. In rounded figures, for each kilometer of convergent margin the average rate of mantle recycling is ~60 km^3/m.y.

It is difficult to assign an error bar to these estimates. The biggest uncertainty concerns the rate of sediment subduction and subduction erosion at accreting margins. Fortunately, accreting margins make up no more than 25% of all margins, and subduction at them is slow (~35 km/m.y.; see Table 2). If we assign a ±50% error to our estimate of the global rate of sediment subduction and subduction erosion at accreting margins, a corresponding error of ±0.2 AU can be linked to the global estimated rate of 2.5 AU.

EXPLORING IMPLICATIONS OF RECYCLING OVER A PROTRACTED PERIOD OF GEOLOGIC TIME

Mass of Recycled Continental Crust

In addressing the geologic consequences of crustal recycling, we explore the presumption that the long-term rate of recycling at contemporary convergent margins can be applied as far back in time as the modern process of subduction has been under way. Carlson et al. (2000) recognize evidence that subduction was under way at least by 3 Ga; Condie (2005, p. 113) supposes as long ago as 3.5–4.0 Ga, but Hamilton (2003) reads the rock record as documenting plate tectonics and subduction as beginning only ca. 2.0 Ga, and Stern (2005) identifies the Neoproterozoic (ca. 1000–550 Ma) as the beginning of modern subduction systems. We take the position of van der Velden and Cook (2005) that seismic images of fossil subduction zones are prima facie evidence documenting that subduction processes have been under way since at least 2.7 Ga. We are thus willing to exploratively apply estimates of ocean margin crustal recycling at Cenozoic and Mesozoic subduction zones to the past 2.5 billion years, or since the opening of the Proterozoic. If subduction processes were under way for a significant length of time before 2.5 Ga, then most likely plate convergence was globally faster than currently, and the recycling rate at ocean margins would have been commensurately higher as well, in particular if an Archean subduction zone were ingesting an abundance of high ocean floor relief (i.e., large igneous provinces and hotspot-generated aseismic ridges) that significantly enhances recycling rates (see conclusions of Clift and Vannucchi, 2004).

We explore below a number of implications of crustal recycling at ocean margins with respect to their perceived first-order impact on the evolution of continental basement and Earth's mantle. These implications address

1. volumetric destruction of continental crust;
2. crustal contamination of the mantle;
3. growth of continental crust through time;
4. preservation of continental crust through time; and
5. reconstructing supercontinents.

Volumetric Destruction of Continental Crust

Applying an annual recycling rate of 2.5 AU (km^3/yr) to the past 2.5 billion years assumes that during this time the average rate of recycling along each kilometer of ocean margins was ~60 km^3/m.y. and the average global length of convergent margins was ~42,000 km (Tables 1 and 4). Accordingly, since the end of the Archean, on the order of 6×10^9 km^3 of continental crustal material has been recycled to the mantle (Fig. 7). Earth's existing volume of continental crust is ~7×10^9 km^3 (Reymer and Schubert, 1984). Thus since the beginning of the Proterozoic a volume of continental crust roughly equal to the existing volume has been recycled to the mantle at ocean margin subduction zones.

This amount of destroyed crust is so large that ocean margin recycling must have been a major factor determining the areal pattern and volumetric growth of continental crust. If subduction took place well before 2.5 Ga, then the volume of continental crust that has been removed and recycled to the mantle is greater, probably much greater, than our estimate of 6×10^9 km^3. As emphasized above, this estimate does not consider crustal losses at continental collision zones (see below) or inboard of the coastal region by ablative or delamination processes.

The ^{40}Ar geochemical model of Coltice et al. (2000) implies that as little 30% of Earth's standing volume has been recycled to the mantle, the remaining volume having been mechanically underplated (sequestered) beneath the continents. According to our figures, during the past 2.5 billion years this mass of the underplate would be on the order 2×10^9 km^3 (30% of 6×10^9 km^3), which is adequate to form the lower crust of the extant expanse of continental areas. We suppose instead that because the lower crust is mafic material (see Fountain et al., 1992), and because and the global exposure of deeply underplated crustal material, which would be concentrated at modern and ancient convergent margins, is areally small, the bulk of recycled material makes its way to asthenospheric depths (see also Helffrich and Wood, 2001).

Crustal Contamination of the Mantle

During the past 2.5 billion years, the mass of continental crust recycled to the mantle at ocean margins is a small fraction (~0.6%) of its bulk (~920×10^9 km^3; Fig. 7). But this fraction is sufficiently large to impart a distinct continental aroma to the mantle's trace element composition.

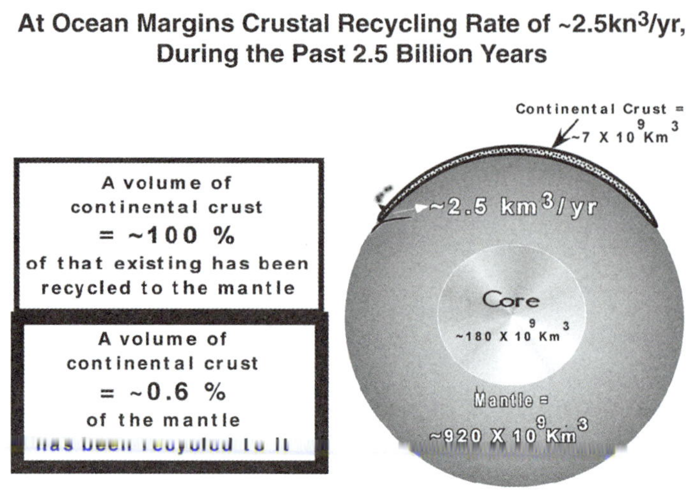

Figure 7. If long-term rates of crustal recycling at modern margins can be applied to extended periods of geologic time, then since 2.5 Ga a volume of continental crust equal to the standing volume of continental crust, ~7×10^9 km^3, has been recycled to the mantle. This volume is trivial (~0.6%) with respect to that of the mantle, yet adequate to provide the whole mantle with a geochemical aroma of recycled continental crust. These calculations assume that subduction of oceanic lithosphere at continental and island-arc convergent margins has been under way since at least the beginning of the Proterozoic (i.e., since 2.5 Ga; Van der Velden and Cook, 2005).

Based on geochemical modeling, the notion that recycling of crustal material takes place at any significant rate is contentious, an issue vigorously recounted by Armstrong (1991) and reviewed recently by Sylvester (2000) and Condie (2005).

Nonetheless, we note, and comment more about below, that our findings applied to at least the past 2.5 billion years are in step with the geochemical model for the evolution of continental crust advocated by Armstrong (1968, 1981, 1991). His model calls for a prominent recycling component on an Earth that has maintained a quasi-steady-state volume of continental crust during the past ~4 billion years. The volume of recycled continental material we estimate is consistent with the likely abundance of clots and blobs of crust-dirtied heterogeneities described by Helffrich and Wood (2001) for a top-to-bottom convecting mantle and early recycling implied by Hf isotopes in detrital grains of Archean zircon (Harrison et al., 2005).

Growth of Continental Crust

Models of continental growth (i.e., the net accumulation—additions minus losses—of mantle-derived igneous rock to form the massifs of island arcs and continents) differ greatly. Condie (2005) reviews most of these models and the underlying assumptions and principles involved. During the Phanerozoic, the rate at which arc magmatism added juvenile continental crust to the surface of Earth has commonly been estimated at ~1 AU (km^3/yr), which corresponds to a delivery rate of ~30 km^3/m.y. for each kilometer of subduction zone (Crisp, 1984; Reymer and Schubert, 1984).

Based on modern geophysical imaging of arc structures, the flux of arc magma to construct the Aleutian Ridge during the past ~45 m.y. has been much higher, at least 65 km^3/m.y. per kilometer of arc (Holbrook et al., 1999; Lizarralde et al., 2002). Similar conclusions have been reached that high magmatic productivity (~60–180 km^3/m.y./km) is characteristic of western Pacific arc massifs (Cosca et al., 1998; Dimalanta et al., 2002; Stern et al., 2003; Arculus, 2004). For the Aleutian Ridge or arc mass, which is basically intact across its width, if crustal losses caused by surface and subduction erosion are factored in, and also the likelihood that the arc massif does not include an eruptively engulfed slab of oceanic crust, then the rate of arc productivity increases to at least 150 km^3/m.y. per kilometer of arc, if not higher (Jicha et al., 2006).

If this rate is applied to the global length of major convergent margins (~42,000 km), the annual addition of juvenile arc crust increases to at least 5 AU (Fig. 8). This rate is by a factor of 2 higher than can be achieved by recycling at modern ocean margins. We earlier (von Huene and Scholl, 1991) noted the similarity of crustal production and recycling at subduction zones, but only at lower estimated rates of arc magmatic growth. If, in fact, the greater part of crustal addition takes place at subduction zones (e.g., Crisp, 1984; Reymer and Schubert, 1984), and the new and higher estimates of arc productivity are correct and applicable to the past, then our recycling estimates (Table 4) at ocean margin subduction zones are not equal to the task of maintaining a quasi-steady-state volume of continental crust during at least the past 2.5 billion years. If Armstrong's steady-state model is accepted (Armstrong (1991), then the long-term rate of crustal recycling at continental collision zones, where ultrahigh-pressure metamorphic rocks occur (Hacker et al., 2006), has been a match (i.e., at 2–3 AU) to that occurring at ocean margin subduction zones. If significantly less, then since at least the Archean, arc magmatism has been steadily increasing Earth's standing inventory of continental crust (Scholl and von Huene, 2006).

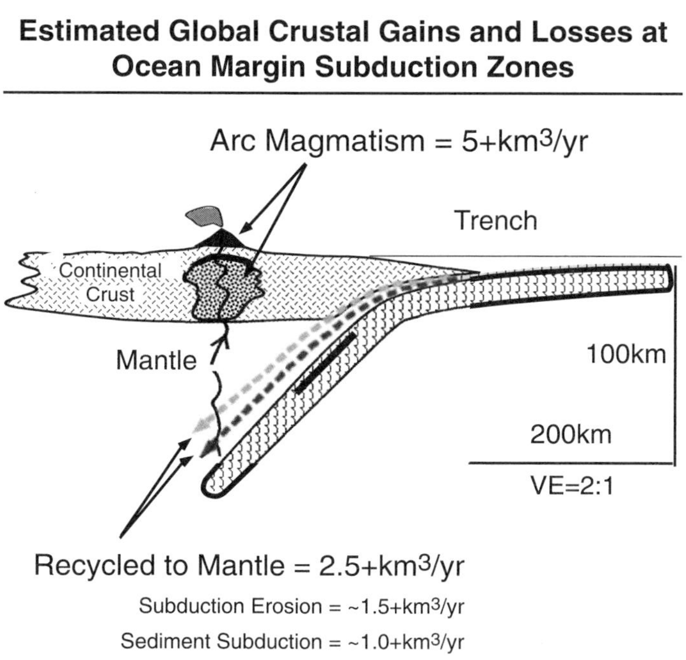

Figure 8. At ocean margin subduction zones, the annual global addition of new arc crust is similar to that lost by recycling. The long-term estimate of arc magmatism is based on new data compiled by Cosca et al. (1998), Holbrook et al. (1999), Lizarralde et al. (2002), Dimalanta et al. (2002), Stern et al. (2003), Arculus (2004), and Jicha et al. (2006). Based on these observations from western Pacific (Izu-Bonin-Mariana) arcs and the Aleutian arc, magmatic productivity since the middle Eocene has averaged at least 150 km^3/m.y. per kilometer of arc, which is equivalent to a global output volume of ~5 km^3/yr.

Preservation of Continental Crust

Figure 9 from Condie (2005), reveals that areas of Archean crust constitute only a small fraction (~5%–10%) of Earth's surface, and that of this surface less than 1% was formed before 3.5 Ga. Crustal recycling at the rates and volumes we list on Table 4 and show graphically on Figures 7 and 8 imply that in the past 2.5 billion years a volume of continental crust at least equal to that existing has been demolished at ocean margin subduction zones. But more significant to the implications of the limited exposure of older crust is the rate of crustal truncation—the landward migration of the trench axis toward a fixed reference—that occurs at the ocean margin subduction zones.

Along continental and island-arc edges, the average rate of truncation of the inner prism is ~2.5 km/m.y. (range = 0.9–4.5 km/m.y.; Table 3). This rate is tectonically leisurely when compared to the rate of orthogonal convergence (~80 km/m.y.), but it is adequate to destroy the width of an orogenic belt within a few hundred million years. Hence the relatively small amount of exposed Archean crust is likely to be more a consequence of rock preservation than the volume originally produced (see discussions by Armstrong, 1991; Sylvester, 2000; Condie, 2005).

Figure 9 reveals that no area of Archean crust older than 3.5 Ga occurs along a convergent margin and, with the exception of far eastern central Asia, no Archean crust is exposed anywhere near a subduction zone. Seemingly, preservation of old cratonic rock requires sequestering within a continental mass that has grown or expanded seaward by magmatism, rifting, or terrane accretion. These tectonic processes can shield older basement units from coming into contact with a subduction zone where frontal and basal subduction erosion would work to destroy them. For example, the two eastern Asian cratonic masses lie inside the Sea of Japan and Okhotsk Sea, both of which formed by back-arc spreading in the middle Tertiary that tectonically pushed the Kuril and Japan subduction zones seaward of the Asian mainland and away from the Archean cratons.

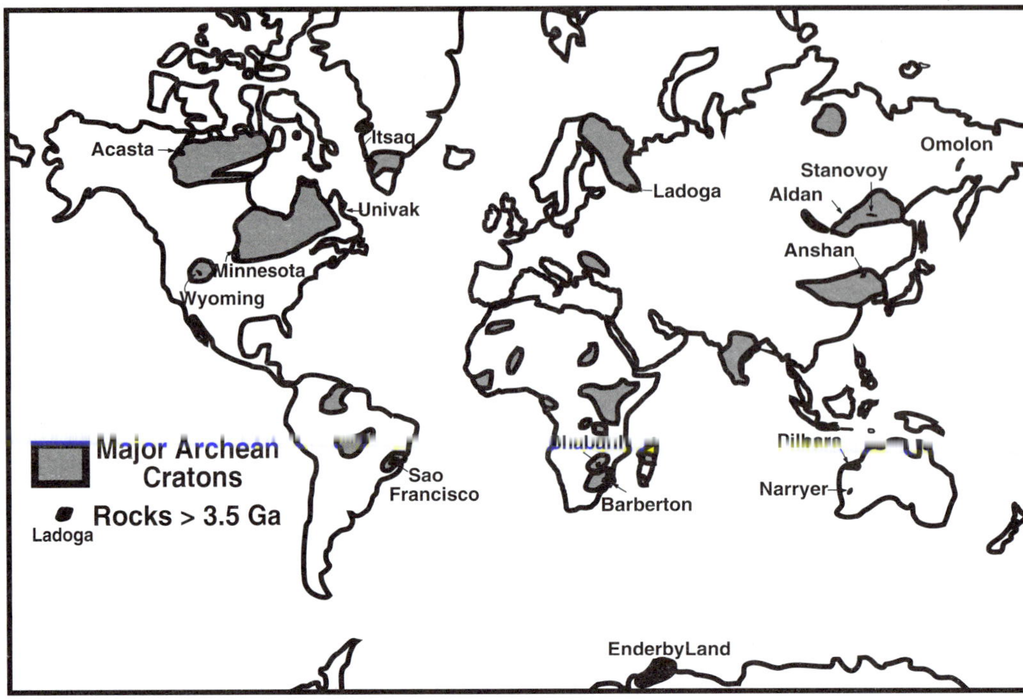

Figure 9. Distribution of major Archean cratons. Adapted, with permission, from Condie (2005, see his Fig. 8.2). No Archean cratonic mass is near a modern subduction zone, and the areal exposure of crust older than 3.5 Ga is globally tiny. These relations are consistent with the notion that Earth's inventory of Archean crust is more a consequence of what has been preserved of a formerly much greater volume than the volume of Archean crust originally produced.

Reconstructing Supercontinents

Reconstructing supercontinents, for example Pangea (fully assembled 450–300 Ma; Condie, 2005), Gondwana (600–500 Ma), and Rodinia (1100–1000 Ma), requires rounding up and replacing terranes of crust globally dispersed by the breakup of a nearly hemisphere-size mass of continental material. Reassembly of Pangea is relatively straightforward because it fragmented only geologically recently, and backtracking and refitting major continental fragments is guided by the magnetic and physiographic fabric of the ocean basins.

Ocean floor crustal fabric, however, is not available to put older supercontinents back together. Reconstructive techniques of geologic nature can be called upon to do this, for example by identifying now-unconnected basement masses that can be refitted by virtue of common or complimentary geometric, structural, and tectonic patterns or depositional settings and provenances of included or overlying sedimentary masses. Based on geologic evidence alone, reconstructive views can differ significantly, for example, in the Late Proterozoic, the contrasting SWEAT (Moores, 1991) and AUSWUS (Karlstrom et al., 1999; Burrett and Berry, 2000) models to reassemble Rodinia's western and southwestern Laurentian margin and, to the relative east, the apparently less controversial refitting of parts of South America to the North American craton (Thomas, 1991; Thomas and Astini, 1999; Thomas et al., 2004; Thomas, 2006).

Subduction zone recycling predicts that with increasing age of supercontinent reconstruction the assembly process is going to be challenged by a progressive and permanent loss of material to rebuild the supercontinent or even to establish that it once existed. For western South America, consider the hunt by Mpodozis and Kay (1992) for the vanished western Chilean Equis or X terrane. Crustal destruction is a consequence of ocean margin truncation caused by frontal and basal subduction erosion. Working in tandem, these two crust-destroying actions can trim a convergent ocean margin at a long-term rate of 2–3 km/m.y. (Table 3). Although slow (~1/30 of the convergence rate), over a long period of time a wide strip of continental edge can be removed (Fig. 10). If truncation also occurs at continental collision zones, then the global tectonic erasing of continental edge is faster than just that estimated for ocean margins alone.

It seems likely that many examples of modern erosionally dominated convergent margins, for example that of much of Middle America and western South America, must have existed in the past. Hence some of the recognized continental truncations, including those identified along the western side of the North American Cordillera, may well be tectonically buried scars originally or terminally sculpted by subduction erosion. Candidate

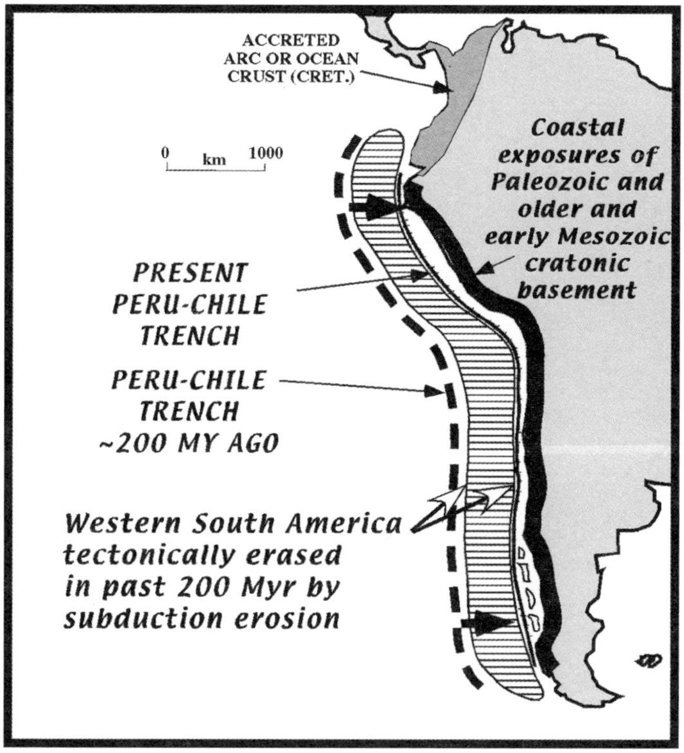

Figure 10. At a truncation rate of 2.5 km/m.y., since the breakup of Pangea, subduction erosion along the Peru-Chile Trench has destroyed a strip of western South America 500–600 km wide. If western South America was extracted in the Cambrian by continental rifting from the southeastern side of Laurentia (Thomas et al., 2004; Thomas, 2006), then, if not protected from the effects of the underthrusting of Pacific lithosphere, South America's original rift margin may have been recycled to the mantle.

scars are the Sur-Nacimiento fault of western California (marks the contact of the Franciscan accretionary complex and its Jurassic-Cretaceous arc basement; Page, 1970), sectors of the Border Ranges fault of south-central Alaska (marks the regional contact of the Jurassic Talkeetna arc and its accretionary prism; Clift et al., 2005), and at least the northwestern or California part of the much-debated Mojave-Sonora megashear of western North America (Anderson et al., 2005).

ACKNOWLEDGMENTS

We are extremely grateful to John McBride and Suzanne Mahlburg Kay for their insightful guidance in helping us craft this paper. Sharpening and shaping of the recycling ideas we explore and describe also benefited hugely from challenging questions from many colleagues, in particular from Paola Vannucchi, Peter Clift, Cesar Ranero, and, although they might not have realized it, Sue and Bob Kay, Steve Kirby, Bill Normark, Tracy Vallier, Peter Kelemen, Simon Klemperer, Norm Sleep, Bob Stern, George Thompson, Kent Condie, and Gene Yogodzinski.

REFERENCES CITED

Abbruzzi, 1996, Magmatic evidence for Neogene lithospheric evolution of the central Andean "flat slab" between 30° and 32°S: Tectonophysics, v. 259, p. 15–28.

Adam, J., Klaeschen, D., Kukowski, N., and Flueh, E., 2004, Upward delamination of Cascadia Basin sediment infill with landward frontal accretion thrusting caused by rapid glacial-age material flux: Tectonics, v. 23, p. TC3009, doi: 10.1029/2002TC001475.

Anderson, T.H., Nourse, J.A., McKee, J.W., and Steiner, M.B., editors, 2005, The Mojave-Sonora megashear hypotheses: Geological Society of America Special Paper 393, 712 p.

Andriessen, P.A.M., and Reutter, K.J., 1991, K-Ar and fission track mineral age determinations of igneous rocks related to multiple magmatic arc systems along the 23°S latitude of Chile and NW Argentina, *in* Reutter, K.J., et al., eds., Tectonics of the southern central Andes: Berlin, Springer-Verlag, 333 p.

Arculus, R.J., 2004, Evolution of arc magmas and their volatiles, *in* Sparks, R.S.J., and Hawkesworth, C.J., eds., The state of the planet: Washington, D.C., American Geophysical Union, p. 95–108.

Armstrong, R.L., 1968, A model for the evolution of strontium and lead isotopes in a dynamic Earth: Reviews of Geophysics, v. 6, p. 175–199.

Armstrong, R.L., 1981, Radiogenic isotopes: The case for crustal recycling on a near-steady-state no-continental growth Earth: Royal Society of London Philosophical Transactions, ser. A, v. 301, p. 443–472.

Armstrong, R.L., 1991, The persistent myth of crustal growth: Australian Journal of Earth Sciences, v. 38, p. 613–630.

Ballance, P.F., Scholl, D.W., Vallier, T.L., Stevenson, A.J., Ryan, H.F., and Herzer, R.H., 1989, Subduction of a Late Cretaceous seamount of the Louisville chain at the Tonga Trench: A model of normal and accelerated tectonic erosion: Tectonics, v. 8, p. 953–962.

Bangs, N.L., and Cande, S.C., 1997, The episodic development of a convergent margin inferred from structures and processes along the southern Chile margin: Tectonics, v. 16, p. 489–503, doi: 10.1029/97TC00494.

Bangs, N.L., Shipley, T.H., Gulick, P.S., Moore, G.F., Kuromoto, S., and Nakamura, Y., 2004, Evolution of the Nankai Trough décollement from the trench into the seismogenic zone: Inferences from three-dimensional seismic reflection imaging: Geology, v. 32, p. 273–276, doi: 10.1130/G20211.2.

Bangs, N.L., Gulick, P.S., and Shipley, T.H., 2006, Seamount subduction erosion in the Nankai Trough and potential impact on the seismogenic zone: Geology, v. 34, p. 701–704, doi: 10.1130/G22451.1.

Behrmann, J.H., and Kopf, A., 2001, Balance of tectonically accreted and subducted sediment at the Chile triple junction: Geologische Rundschau, v. 90, p. 753–768.

Bourgois, J.H., Martin, H., Lagabrielle, Y., Le Moigne, J., and Frutos, J., 1996, Subduction erosion related to spreading-ridge subduction: Taitao Peninsula (Chile margin triple junction area): Geology, v. 24, p. 723–726, doi:.

Burrett, C., and Berry, R., 2000, Proterozoic Australia–western United States (AUSWUS) fit between Laurentia and Australia: Geology, v. 28, p. 103–106, doi: 10.1130/0091-7613(2000)28<103:PAUSAF>2.0.CO;2.

Byrne, D.E., Wang, W-h., and Davis, D.M., 1993, Mechanical role of backstops in the growth of forearcs: Tectonics, v. 12, p. 123–144.

Calahorrano, A., Sallares, V., Collot, J.-Y., Sage, F., and Ranero, C., 2005, Physical Properties of the subduction channel off the Gulf of Guayaquil (Ecuador) from seismic reflection data: Eos (Transactions, American Geophysical Union), v. 86, Fall Meeting Supplement, Abstract T53B-1424.

Calvert, A.J., 2004, Seismic reflection imaging of two megathrust shear zones in the northern Cascadia subduction zone: Nature, v. 428, p. 163–166, doi: 10.1038/nature02372.

Calvert, A.J., Fisher, M.A., Mamachandran, K., and Trehu, A.M., 2003, Possible emplacement of crustal rocks into the forearc mantle of the Cascadia subduction zone: Geophysical Research Letters, v. 30, 2196, doi: 10.1029/2003GL018541.

Carlson, R.W., Boyd, F.R., Shirley, S.B., Janney, P.E., Groves, T.L., Bowring, S.A., Schmitz, M.D., Damn, J.C., Bell, D.R., Gurney, J., Richardson, S.H., Tredoux, M., Menzies, A.H., Pearson, D.G., Hart, R.J., Wilson, A.H., and Moser, D., 2000, Continental growth, preservation, and modification in southern Africa: GSA Today, v. 10, no. 2, p. 1–6.

Church, S.E., 1976, The Cascade mountains revisited: A re-evaluation in light of new lead isotopic data: Earth and Planetary Science Letters, v. 29, p. 175–188, doi: 10.1016/0012-821X(76)90037-6.

Clift, P.D., and MacLeod, C.J., 1999, Slow rates of tectonic erosion estimated from the subsidence and tilting of the Tonga forearc basin: Geology, v. 27, p. 411–414, doi: 10.1130/0091-7613(1999)027<0411:SROSEE>2.3.CO;2.

Clift, P.D., and Vannucchi, P., 2004, Controls on tectonic accretion versus erosion in subduction zones: Implications for the origin and recycling of the continental crust, Reviews of Geophysics, v. 42, RG2001, doi: 10.1029/2003RG000127.

Clift, P.D., Pecher, I., Kukowski, N., and Hampel, A., 2003, Tectonic erosion of the Peruvian forearc, Lima Basin, by subduction and Nazca Ridge collision: Tectonics, v. 22, p. 1023, doi: 10.1029/2002TC001386.

Clift, P.D., Pavlis, T., DeBari, S.M., Draut, A.E., Rioux, M., and Kelemen, P.B., 2005, Subduction erosion of the Jurassic Talkeetna-Bonanza arc and the Mesozoic accretionary tectonics of western North America: Geology, v. 33, p. 881–884, doi: 10.1130/G21822.1.

Cloos, M., and Shreve, R.L., 1988a, Subduction-channel model of prism accretion, mélange formation, sediment subduction, and subduction erosion at convergent plate margins, 1: Background and description: Pure and Applied Geophysics, v. 128, p. 456–500.

Cloos, M., and Shreve, R.L., 1988b, Subduction-channel model of prism accretion, mélange formation, sediment subduction, and subduction erosion at convergent plate margins, 2: Implications and discussion: Pure and Applied Geophysics, v. 128, p. 501–545, doi: 10.1007/BF00874549.

Collot, J.-Y., Charvis, P., Gutscher, M.-A., and Operto, S., 2002, Exploring the Ecuador-Colombia active margin and interplate seismogenic zone: Eos (Transactions, American Geophysical Union), v. 83, p. 185–190, doi: 10.1029/2002EO000120.

Coltice, N., Albarede, F., and Gillet, P., 2000, ^{40}K-^{40}Ar constraints on recycling continental crust into the mantle: Science, v. 288, no. 5467, p. 845–847, doi: 10.1126/science.288.5467.845.

Condie, K.C., 2005, Earth as an evolving planetary system: Amsterdam, Elsevier Academic Press, 447 p.

Cosca, M.A., Arculus, R.J., Pearce, J.A., and Mitchell, J.G., 1998, ^{40}Ar/^{39}Ar and K-Ar geochronological age constraints for the inception and early evolution of the Izu-Bonin-Mariana arc system: Island Arc, v. 7, p. 579–595.

Crisp, J.A., 1984, Rates of magma emplacement and volcanic output: Journal of Volcanology and Geothermal Research, v. 20, p. 177–211, doi: 10.1016/0377-0273(84)90039-8.

Dimalanta, C., Taira, A., Yumul, G.P., Jr., Tokuyama, H., and Mochizuki, K., 2002, New rates of western Pacific island arc magmatism from seismic and gravity data: Earth and Planetary Science Letters, v. 202, p. 105–115, doi: 10.1016/S0012-821X(02)00761-6.

Ducea, M.N., Gehrels, G.E., Shoemaker, S., Ruiz, J., and Valencia, V.A., 2004, Geologic evolution of the Xolapa Complex, southern Mexico: Evidence from U-Pb zircon geochronology: Geological Society of America Bulletin, v. 116, p. 1016–1025, doi: 10.1130/B25467.1.

Fountain, D.M., Arculus, R., and Kay, R.W., editors, 1992, Continental lower crust: Amsterdam, Elsevier, 486 p.

Fruehn, J., White, R.S., and Minshull, T.M., 1997, Internal deformation and compaction of the Makran accretionary wedge: Terra Nova, v. 9, p. 101–103.

Fruehn, J., von Huene, R., and Fisher, M.A., 1999, Accretion in the wake of terrane collision: The Neogene accretionary wedge off Kenai Peninsula, Alaska: Tectonics, v. 18, p. 263–277.

Gardner, T., Marshall, J., Merritts, D., Bhavani, B., Burgette, R., Burton, E., Cooke, J., Kehrwald, N., Protti, M., Fisher, D., and Sak, P., 2001, Holocene forearc block rotation in response to seamount subduction, southeastern Peninsula de Nicoya, Costa Rica: Geology, v. 29, p. 151–154, doi: 10.1130/0091-7613(2001)029<0151:HFBRIR>2.0.CO;2.

Grindlay, N., Mann, P., Dolan, J., and van Gestel, J.-P., 2005, Neotectonics and subsidence of the northern Puerto Rico–Virgin Islands margin in response to the oblique subduction of high-standing ridges, *in* Mann, P., and Prentice, C., eds., Active tectonics and seismic hazards of Puerto Rico, the Virgin Islands, and offshore areas: Geological Society of America Special Paper 385, p. 31–60.

Grove, M., Jacobson, C.E., Barth, A.P., and Vucic, A., 2003, Temporal and spatial trends of Late Cretaceous–early Tertiary underplating of Pelona and related schist beneath southern California and southwestern Arizona, *in* Johnson, S.E., et al., eds., Tectonic evolution of northwestern México and the southwestern USA: Geological Society of America Special Paper 374, p. 381–406.

Hacker, B.R., McClelland, W.C., and Liou, J.G., editors, 2006, Ultrahigh-pressure metamorphism: Deep continental subduction: Geological Society of America Special Paper 403, 206 p.

Hamilton, W., 2003, An alternative Earth: GSA Today, v. 13, no. 11, p. 4–12, doi: 10.1130/1052-5173(2003)013<0004:AAE>2.0.CO;2.

Hampel, A., Kukowski, N., Bialas, J., Huebscher, C., and Heinbockel, R., 2004, Ridge subduction at an erosive margin: The collision zone of the Nazca Ridge in southern Peru, Journal of Geophysical Research, v. 109, B02101, doi:10.1029/2003JB002593.

Harrison, T.M., Blichert-Toft, J., Muller, W., Albarede, F., Holden, P., and Mojzsis, S.J., 2005, Heterogeneous Hadean hafnium: Evidence of continental crust at 4.4 to 4.5 Ga: Science, v. 310, p. 1947–1950, doi: 10.1126/science.1117926.

Hartley, A.J., May, G., Chong, G., Turner, P., Kape, S.J., and Jolley, E.J., 2000, Development of a continental forearc: A Cenozoic example from the central Andes, northern Chile: Geology, v. 28, p. 331–334, doi: 10.1130/0091-7613(2000)28<331:DOACFA>2.0.CO;2.

Helffrich, G.R., and Wood, B.J., 2001, The Earth's mantle: Nature, v. 412, p. 501–507, doi: 10.1038/35087500.

Holbrook, W.S., Lizarralde, D., McGeary, S., Bangs, N.L., and Diebold, J.B., 1999, Structure and composition of the Aleutian island arc and implications for continental crustal growth: Geology, v. 27, p. 31–34.

Jacobson, C.E., Oyarzabal, F.R., and Haxel, G.B., 1996, Subduction and exhumation of the Pelona-Orocopia-Rand Schists, southern California: Geology, v. 24, p. 547–550, doi: 10.1130/0091-7613(1996)024<0547:SAEOTP>2.3.CO;2.

Jicha, B.R., Scholl, D.W., Singer, B.S., Yogodzinski, G.M., and Kay, S.M., 2006, Revised age of Aleutian island arc formation implies high rate of magma production: Geology, v. 34, p. 661–664, doi: 10.1130/G22433.1.

Karlstrom, K.E., Harlan, S.S., Williams, M.L., McLelland, J., Geissman, J.W., and Åhäll, K.-I., 1999, Refining Rodinia: Geologic evidence for the Australia–western U.S. connection in the Proterozoic: GSA Today, v. 9, no. 10, p. 2–7.

Kay, S.M., Kay, R.W., Citron, G.P., and Perfit, M., 1990, Calc-alkaline plutonism in the intra-oceanic Aleutian arc, Alaska, *in* Kay, S.M., and Rapel, C.W., eds., Plutonism from Antarctica to Alaska: Geological Society of America Special Paper 241, p. 233–255.

Kay, S.M., Godoy, E., and Andrew, K., 2005, Episodic arc migration, crustal thickening, subduction erosion, and magmatism in the south-central Andes: Geological Society of America Bulletin, v. 117, p. 67–88, doi: 10.1130/B25431.1.

Kimura, G., Silver, E., Blum, P., et al., 1997, Proceedings of the Ocean Drilling Program, Initial Reports, Volume 170: College Station, Texas, Ocean Drilling Program, http://www-odp.tamu.edu/publications/170_SR/front.htm.

Klaeschen, D., Belykh, I., Gnibidenko, H., Patrikeyev, S., and von Huene, R., 1994, Structure of the Kuril Trench from seismic reflection records: Journal of Geophysical Research, v. 99, no. B12, p. 24,173–24,188, doi: 10.1029/94JB01186.

Kopp, H., and Kukowski, N., 2003, Backstop geometry and accretionary mechanics of the Sunda margin: Tectonics, v. 22, p. 1072, doi: 10.1029/2002TC001420.

Kopp, H., Fruehn, J., Flueh, E.R., Reichert, C., Nukowski, N., Bialas, J., and Klaeschen, D., 2000, Structure of the Makran subduction zone from wide-angle and reflection seismic data: Tectonophysics, v. 329, p. 171–191, doi: 10.1016/S0040-1951(00)00195-5.

Kopp, H., Flueh, E.R., Klaeschen, D., Bialas, J., and Reichert, C., 2001, Crustal structure of the central Sunda margin at the onset of oblique subduction: Geophysical Journal International, v. 147, p. 449–474, doi: 10.1046/j.0956-540x.2001.01547.x.

Laursen, J., Scholl, D.W., and von Huene, R., 2002, Neotectonic deformation of the central Chile margin: Deepwater forearc basin formation in response to hot spot ridge and seamount subduction: Tectonics, v. 21, p. 1038, doi: 10.1029/2001TC901023.

Levi D., B., 1973, Eastward shift of Mesozoic and early Tertiary volcanic centers in the Coast Range of central Chile: Geological Society of America Bulletin, v. 84, p. 3901–3910, doi: 10.1130/0016-7606(1973)84<3901:ESOMAE>2.0.CO;2.

Lissinna, B., Hoernle, K., and van der Bogaard, P., 2002, Northern migration of arc volcanism in western Panama: Evidence for subduction erosion (?): Eos (Transactions, American Geophysical Union), v. 87, p. F1463–F1464.

Lizarralde, D., Holbrook, W.S., McGeary, S., Bangs, N.L., and Diebold, J.B., 2002, Crustal construction of a volcanic arc, wide-angle seismic results from the western Alaska Peninsula: Journal of Geophysical Research, v. 107, no. B8, doi: 10.1029/2001JB000230.

McCarthy, J., and Scholl, D.W., 1985, Mechanisms of subduction accretion along the central Aleutian Trench: Geological Society of America Bulletin, v. 96, p. 691–701.

Mercier de Lepinay, B., Michaud, M., Calmus, T., Bourgois, J., Poupeau, G., Saint-Marc, P., and the NAUTIMATE Team, 1997, Large Neogene subsidence event along the Middle America Trench off Mexico (18–19° N): Evidence from submersible observations: Geology, v. 25, p. 387–390, doi: 10.1130/0091-7613(1997)025<0387:LNSEAT>2.3.CO;2.

Meschede, M., Zweigel, P., and Kiefer, E., 1999, Subsidence and extension at a convergent plate margin: Evidence for subduction erosion off Costa Rica: Terra Nova, v. 11, p. 112–117, doi: 10.1046/j.1365-3121.1999.00234.x.

Miller, H., 1970a, Das Problem des hypothetischen "Pazifischen Kontinentes" gesehen von der chilenischen Pazifikkuste: Geologische Rundschau, v. 59, p. 927–938, doi: 10.1007/BF02042277.

Miller, H., 1970b, Vergleichende Studies an pramesozoischen Gesteinen Chiles unter besonderer Berucksichtigung ihrer Kleintektonik: Geotektonische Forschungen, v. 36, p. 1–64.

Moore, J.C., Watkins, J.S., and Shipley, T.H., 1982, Summary of accretionary processes, Deep Sea Drilling Project Leg 66: Offscraping, underplating, and deformation of the slope apron: Deep Sea Drilling Project, Initial Reports, Volume 66: Washington, D.C., U.S. Government Printing Office, p. 825–836.

Moores, E.M., 1991, Southwest U.S.–East Antarctic (SWEAT) connection: A hypothesis: Geology, v. 19, p. 425–428, doi: 10.1130/0091-7613(1991)019<0425:SUSEAS>2.3.CO;2.

Moran-Zenteno, D.J., Corona, C.P., and Tolson, G., 1996, Uplift and subduction erosion in southwestern Mexico since the Oligocene: Pluton geobarometry constraints: Earth and Planetary Science Letters, v. 141, p. 51–65, doi: 10.1016/0012-821X(96)00067-2.

Moran-Zenteno, D.J., Tolson, G., Martinez-Serrano, R.G., Martiny, B., Schaaf, P., Silva-Romo, G., Macias-Romo, C., Alba-Aldave, L., Hernandez-Bernal, M.S., and Solis-Pichardo, G.N., 1999, Tertiary arc magmatism of the Sierra Madre del Sur, Mexico, and its transition to the volcanic activity of the Trans-Mexican volcanic belt: Journal of South American Earth Sciences, v. 12, p. 513–535, doi: 10.1016/S0895-9811(99)00036-X.

Morris, J.D., Leeman, W.P., and Tera, F., 1990, The subducted component in island arc lavas: Constraints from Be isotopes and B-Be systematics: Nature, v. 344, p. 31–36.

Mpodozis, C., and Kay, S.M., 1992, Late Paleozoic to Triassic evolution of the Gondwana margin: Evidence from Chilean Cordilleran batholiths (28°–31°S): Geological Society of America Bulletin, v. 104, p. 999–1014, doi: 10.1130/0016-7606(1992)104<0999:LPTTEO>2.3.CO;2.

Naranjo, J.A., Puig, A., and Suarez, M., 1986, A note on Lower Jurassic magmatism in the Coastal Cordillera of Atacama, Chile: Geological Magazine, v. 123, p. 699–702.

Page, B.M., 1970, Sur-Nacimiento fault zone of California: Continental margin tectonics: Geological Society of America Bulletin, v. 81, p. 667–690, doi: 10.1130/0016-7606(1970)81[667:SFZOCC]2.0.CO;2.

Park, J.-O., Tsuru, T., Kodaira, S., Cummins, P.R., and Kaneda, Y., 2002, Splay fault branching along the Nankai subduction zone: Science, v. 297, p. 1157–1160, doi: 10.1126/science.1074111.

Plafker, G., Moore, J.C., and Winkler, G.R., 1994, Geology of the southern Alaska margin, *in* Plafker, G., and Berg, H.C., eds., The geology of Alaska: Geological Society of America, Geology of North America, v. G-1, p. 389–449.

Plank, T., and Langmuir, C.H., 1993, Tracing trace elements from sediment input to volcanic output at subduction zones: Nature, v. 362, p. 739–742, doi: 10.1038/362739a0.

Pope, D.C., and Willett, S.D., 1998, Thermal-mechanical model for crustal thickening in the central Andes driven by ablative subduction: Geology, v. 26, p. 511–514, doi: 10.1130/0091-7613(1998)026<0511:TMMFCT>2.3.CO;2.

Ranero, C.R., and von Huene, R., 2000, Subduction erosion along the Middle America convergent margin: Nature, v. 404, p. 748–752, doi: 10.1038/35008046.

Reymer, A., and Schubert, G., 1984, Phanerozoic addition rates to the continental crust and crustal growth: Tectonics, v. 3, p. 63–77.

Ruff, L.J., and Tichelaar, B.W., 1996, What controls the seismogenic plate interface in subduction zones?, *in* Bebout, G.E., et al., eds., Subduction top to bottom: American Geophysical Union Geophysical Monograph 96, p. 105–111.

Rutland, R.W.R., 1971, Andean orogeny and ocean floor spreading: Nature, v. 233, p. 252–255, doi: 10.1038/233252a0.

Ryan, H.F., and Scholl, D.W., 1989, The evolution of forearc structures along an oblique convergent margin, central Aleutian arc: Tectonics, v. 8, p. 497–516.

Sallares, V., and Ranero, C.R., 2005, Structure and tectonics of the erosional convergent margin off Antofagasta, north Chile (23°30′): Journal of Geophysical Research, v. 110, no. B6, B06101, doi: 10.1029/2004JB003418.

Schaaf, P., Moran-Zenteno, D., and del Sol Hernandez-Bernal, M., 1995, Paleogene continental margin truncation in the southwestern Mexico: Geochronological evidence: Tectonics, v. 14, p. 1339–1350, doi: 10.1029/95TC01928.

Scheuber, E., 1994, Tektonische Entwicklumg des nordchilenischen aktiven Kontinentalrandes: Der Einflua von Plattenkonvergenz und Rheologie, 131 p., Geotektonische Forschungen, v. 81: Stuttgart, Germany, E. Schweizerbart'sche.

Scholl, D.W., and Vallier, T.L., editors, 1985, Geology and offshore resources of Pacific island arcs—Tonga region: Earth Science Series, v. 2: Houston, Texas, Circum-Pacific Council for Energy and Mineral Resources, 475 p.

Scholl, D.W., and von Huene, R., 2003, Evidence for, and amounts of, continental crustal material recycled to the mantle at Cenozoic subduction zones: Geological Society of America Abstracts with Programs, v. 35, no. 6, p. 139.

Scholl, D.W., and von Huene, R., 2006, Observation-based rates of creating and recycling solid material at convergent margins—Impact on the long-term preservation and growth of continental crust: Geological Society of America Abstracts with Programs, v. 38, no. 7 (in press).

Scholl, D.W., von Huene, R., Vallier, T.L., and Howell, D.G., 1980, Sedimentary masses and concepts about tectonic processes at underthrust ocean margins: Geology, v. 8, p. 564–568, doi: 10.1130/0091-7613(1980)8<564:SMACAT>2.0.CO;2.

Scholl, D.W., Vallier, T.L., and Stevenson, A.J., 1987, Geologic evolution and petroleum geology of the Aleutian Ridge, *in* Scholl, D.W., et al., eds., Geology and resource potential of the continental margin of western North America and adjacent ocean basins—Beaufort Sea to Baja California, Earth Science Series, v. 6: Houston, Texas, Circum-Pacific Council for Energy and Mineral Resources, p. 124–155.

Scholl, D.W., von Huene, R., and Ryan, H.F., 2003, Forming a deep-water forearc basin by subduction erosion—Exploring this idea for the prominent Aleutian forearc basin: Geophysical Research Abstracts, v. 5, abstract EAE03-A-11794.

Song, T.-R.A., and Simons, M., 2003, Large trench-parallel gravity variations predict seismogenic behavior in subduction zones: Science, v. 301, p. 630–633, doi: 10.1126/science.1085557.

Sosson, M., Bourgois, J., and de Lepinay, B.M., 1994, SeaBeam and deep-sea submersible Nautile surveys in the Chiclayo Canyon off Peru (7°S): Subsidence and subduction erosion of an Andean-type convergent margin since Pliocene times: Marine Geology, v. 118, p. 237–256, doi: 10.1016/0025-3227(94)90086-8.

Stern, R.J., 2002, Subduction zones: Reviews of Geophysics, v. 40, no. 4, p. 1012, doi: 10.1029/2001RG000108.

Stern, R.J., 2005, Evidence from ophiolites, blueschists, and ultrahigh-pressure metamorphic terranes that the modern episode of subduction tectonics began in Neoproterozoic time: Geology, v. 33, p. 557–560, doi: 10.1130/G21365.1.

Stern, R.J., Fouch, M.J., and Klemperer, S., 2003, An overview of the Izu-Bonin-Mariana, in Eiler, J.M., ed., Inside the subduction factory: American Geophysical Union Geophysical Monograph 138, p. 175–222.

Sylvester, P.J., 2000, Continental formation, growth and recycling, *in* Sylvester, P.J., ed., Continent formation, growth and recycling: Tectonophysics, v. 322, no. 1/2 (special issue), p. vii–viii.

Taira, A., Okada, H., Whitaker, J.H.McD., and Smith, A.J., 1982, The Shimanto Belt of Japan: Cretaceous–lower Miocene active-margin sedimentation, *in* Leggett, J.K., ed., Trench-forearc geology: Geological Society [London] Special Publication 10, p. 5–26.

Takagi, T., 2004, Origin of magnetite- and ilmenite-series granitic rocks in the Japan arc: American Journal of Science, v. 304, p. 169–202, doi: 10.2475/ajs.304.2.169.

Thomas, W.A., 1991, The Appalachian-Ouachita rifted margin of southeastern North America: Geological Society of America Bulletin, v. 103, p. 415–431, doi: 10.1130/0016-7606(1991)103<0415:TAORMO>2.3.CO;2.

Thomas, W.A., 2006, Tectonic inheritance at a continental margin: GSA Today, v. 16, no. 2, doi: 10.1130/1052-5173(2006)016[4:TIAACM]2.0.CO;2.

Thomas, W.A., and Astini, A., 1999, Simple-shear conjugate rift margins of the Argentine Precordillera and the Ouachita embayment of Laurentia: Geological Society of America Bulletin, v. 111, p. 1069–1079, doi: 10.1130/0016-7606(1999)111<1069:SSCRMO>2.3.CO;2.

Thomas, W.A., Astini, R.A., Muller, P.A., Gehrels, G.E., and Wooden, J.L., 2004, Transfer of the Argentine Precordillera terrane from Laurentia: Constraints from detrital-zircon geochronology: Geology, v. 32, p. 965–968, doi: 10.1011.30/G20727.1.

Thomson, S.N., Herve, F., and Stockhert, B., 2001, Mesozoic-Cenozoic denudation history of the Patagonian Andes (southern Chile) and its correlation to different subduction processes: Tectonics, v. 20, p. 693–711, doi: 10.1029/2001TC900013.

Tsuru, T., Park, J.-O., Jin-Oh, N., Takahashi, N., Kodaira, S., Kido, Y., Yukari, K., Kaneda, Y., Kono, Y., 2000, Tectonic features of the Japan Trench convergent margins off Sanriku, northeastern Japan, revealed by multichannel seismic reflection data: Journal of Geophysical Research, v. 105, no. B7, p. 16,403–16,413.

van der Velden, A.J., and Cook, F.A., 2005, Relict subduction zones in Canada: Journal of Geophysical Research, v. 110, p. B08403, doi: 10.1029/2004JB003333.

Van Kranendonk, M.J., and Cassidy, K., 2004, An alternative Earth: Comment: GSA Today, v. 14, no. 3, p. 14.

Vanneste, L.E., and Larter, R.D., 2002, Sediment subduction, subduction erosion, and strain regime in the northern South Sandwich forearc: Journal of Geophysical Research, v. 107, no. B7, doi: 10.1029/2001JB000396.

Vannucchi, P., Scholl, D.W., and Meschede, M., 2001, Tectonic erosion and consequent collapse of the Pacific margin of Costa Rica: Combined implications from ODP Leg 170, seismic offshore data, and regional geology of the Nicoya Peninsula: Tectonics, v. 20, p. 649–668, doi: 10.1029/2000TC001223.

Vannucchi, P., Ranero, C.R., Galeotti, S., Straub, S.M., Scholl, D.W., and McDougall-Ried, K., 2003, Fast rates of subduction erosion along the Costa Rica Pacific margin: Implications for nonsteady rates of crustal recycling at subduction zones: Journal of Geophysical Research, v. 108, no. B11, p. 2511, doi: 10.1029/2002JB002207.

Vannucchi, P., Galeotti, D.S., Clift, P.D., Ranero, C.R., and von Huene, R., 2004, Long-term subduction erosion along the Middle America Trench offshore Guatemala: Geology, v. 32, p. 617–620, doi: 10.1130/G20422.1.

von Huene, R., and Cullotta, R., 1989, Tectonic erosion at the front of the Japan convergent margin: Tectonophysics, v. 160, p. 75–90, doi: 10.1016/0040-1951(89)90385-5.

von Huene, R., and Klaeschen, D., 1999, Opposing gradients of permanent strain in the aseismic zone and elastic strain across the seismogenic zone of the Kodiak shelf and slope, Alaska: Tectonics, v. 18, p. 248–262, doi: 10.1029/1998TC900022.

von Huene, R., and Lallemand, S., 1990, Tectonic erosion along the Japan and Peru convergent margins: Geological Society of America Bulletin, v. 102, p. 704–720, doi: 10.1130/0016-7606(1990)102<0704:TEATJA>2.3.CO;2.

von Huene, R., and Ranero, C.R., 2003, Subduction erosion and basal friction along the sediment-starve convergent margin off Antofagasta, Chile: Journal of Geophysical Research, v. 108, no. B2, p. 2079, doi: 10.1029/2001JB001569.

von Huene, R., and Scholl, D.W., 1991, Observations at convergent margins concerning sediment subduction, subduction erosion, and the growth of continental crust: Reviews of Geophysics, v. 29, p. 279–316.

von Huene, R., and Scholl, D.W., 2003, Accretion and erosion at convergent margins, current conceptual models and present observations: Geological Society of America Abstracts with Programs, v. 35, no. 6, p. 139.

von Huene, R., Langseth, M., Nasu, N., and Okada, H., 1982, A summary of Cenozoic tectonic history along the IPOD Japan Trench transect: Geological Society of America Bulletin, v. 93, p. 829–846, doi: 10.1130/0016-7606(1982)93<829:ASOCTH>2.0.CO;2.

von Huene, R., Aubouin, J., Arnott, R.J., Baltuck, M., Bourgois, J., Helm, R., Ogawa, Y., Kvenvolden, K.A., McDonald, T.J., Taylor, E., McDougall, K., Filewicz, M.V., Winsborough, B., and Lienert, B., 1985, Proceedings of the Deep Sea Drilling Project, Initial Reports, Volume 84: Washington, D.C., U.S. Government Printing Office, 967 p.

von Huene, R., Klaeschen, D., Cropp, B., and Miller, J., 1994, Tectonic structure across the accretionary and erosional parts of the Japan Trench margin: Journal of Geophysical Research, v. 90, p. 22,349–22,361, doi: 10.1029/94JB01198.

von Huene, R., Ranero, C.R., and Vannucchi, P., 2004, Generic model of subduction erosion: Geology, v. 32, p. 913–916, doi: 10.1130/G20563.1.

Wells, R.E., Blakely, R.J., Sugiyama, Y., Scholl, D.W., and Dinterman, P.A., 2003, Basin-centered asperities in great subduction zone earthquakes—A link between sip, subsidence and subduction erosion?: Journal of Geophysical Research, v. 108, no. B10, p. 2507, doi: 10.1029/2002JB002072.

White, W.M., and Dupré, B., 1986, Sediment subduction and magma genesis in the Lesser Antilles: Isotopic and trace element constraints: Journal of Geophysical Research, v. 99, p. 5927–5941.

Ziegler, A.M., Barrett, S.F., and Scotese, C.R., 1981, Paleoclimate, sedimentation, and continental accretion: Royal Society of London Philosophical Transactions, ser. A, v. 301, p. 253–264.

Manuscript Accepted by the Society 1 February 2007

Printed in the USA

The Geological Society of America
Memoir 200
2007

Crustal recycling in the Appalachian foreland

William A. Thomas*
Thomas P. Becker*
Department of Geological Sciences, University of Kentucky, Lexington, Kentucky 40506-0053, USA

ABSTRACT

Continental crust is recycled into orogenic forelands by the distinct but interrelated processes of tectonic imbrication and sedimentary dispersal. Tectonic loading by the orogen drives flexural subsidence of a foreland basin, and the orogen provides a source of sedimentary detritus to fill the basin. Detrital zircons in Pennsylvanian-age sandstones in the Appalachian (Alleghanian) foreland basin reflect an Alleghanian orogenic source of recycled and primary detritus from Grenville-age basement rocks and Iapetan synrift rocks, which also yield pre-Grenville recycled craton-derived detrital zircons. Minor contributions are from Taconic- and Acadian-age plutons and accreted Gondwanan terranes. Ages of the detrital zircons show that most of the syn-orogenic clastic-wedge sediment was recycled from older continental crustal rocks. Cratonward thrusting of crystalline thrust sheets over the foreland recycles continental crustal rocks from the continental margin and thickens the continental crust. Along the Alleghanian foreland, the Blue Ridge thrust sheet of crystalline basement rocks and Piedmont thrust sheets of metasedimentary rocks represent imbrication and thickening of rocks of continental crustal composition. Both tectonic imbrication in foreland thrust sheets and sediment dispersal into the foreland basin recycle and thicken continental crust.

Keywords: foreland basin, clastic wedge, basement massif, continental crust, recycling.

INTRODUCTION

Two distinct but interrelated processes, tectonic imbrication and sedimentary dispersal, systematically transfer allochthons and sedimentary detritus, respectively, into orogenic forelands. These transfers move materials of continental crustal composition from the continental margin onto the edge of the craton in the orogenic foreland, potentially thickening the continental crust. Plate collisions at continental margins drive orogenic thrust imbrication, translating some upper crustal components toward the orogenic foreland. Orogenic thickening of the continental crust imposes a tectonic load on the elastic lithosphere, leading to flexural subsidence of a foreland basin. Chemical weathering of the uplifted tectonic load and sedimentary dispersal into the foreland basin include petrochemical differentiation that modifies the composition to that of continental crustal rocks. This article uses examples from the Appalachian orogen to consider implications for tectonic and sedimentary recycling of continental crust in orogenic forelands.

*Current address, Becker: ExxonMobil Exploration Company, 233 Benmar, Houston, Texas 77060, USA; e-mail, Thomas: geowat@uky.edu.

Thomas, W.A., and Becker, T.P., 2007, Crustal recycling in the Appalachian foreland, *in* Hatcher, R.D., Jr., Carlson, M.P., McBride, J.H., and Martínez Catalán, J.R., eds., 4-D Framework of Continental Crust: Geological Society of America Memoir 200, p. 33–44, doi: 10.1130/2007.1200(03). For permission to copy, contact editing@geosociety.org.

OROGENIC BELTS AND FORELANDS

General interpretations of the geometry and kinematics of orogenic belts provide the framework to consider recycling of continental crust into an orogenic foreland. A continental-margin orogenic belt along a peripheral foreland basin on a subducting lower plate incorporates an accretionary prism and elements of the colliding forearc and arc complex (e.g., Miall, 1995). In addition to ocean-floor and deepwater-sedimentary imbricates, other imbricates from the downgoing plate may be incorporated into the accretionary prism, including thrust sheets of basement and sedimentary cover from the preorogenic rifted continental margin (Thomas et al., 2004a). In a continent-continent collision orogen, the orogenic belt may include crustal components from either or both continents. In a retroarc thrust belt behind a continental-margin arc on an upper plate, shallow thrust detachments are restricted to the sedimentary cover and basement of the continental plate in the backarc (e.g., Jordan, 1995). Melting in the core of a magmatic arc generates "new" juvenile crust, in the sense of synorogenic to early postorogenic crystallization of magmas from a homogeneous mantle source. In contrast, the thrust sheets in an accretionary prism, in a continent-continent collision orogen, or in a retroarc thrust belt may include "old" crust that was generated in previous orogenic cycles and mechanically recycled into the new orogen by tectonic imbrication and translation. An oblique collision, incorporating transpressional to strike-slip faults in the orogenic hinterland, may produce uplifts of crustal rocks along the orogen (e.g., Sylvester, 1988).

Thrust imbrication and thickening of the upper continental crust imposes a tectonic load that drives flexural subsidence of a foreland basin (e.g., Jordan, 1981; Quinlan and Beaumont, 1984). The topographic relief between the tectonically thickened and uplifted crust in the orogenic belt and the subsiding foreland basin sets the stage for weathering, erosion, and dispersal of sedimentary detritus from the orogenic edifice (the tectonic load) into the foreland basin. Foreland basins are filled to varying proportions by synorogenic clastic wedges, and the petrography of the clastic-wedge sediment in an orogenic foreland basin provides a record of the tectonic composition of the orogenic belt (e.g., Dickinson et al., 1983; Ingersoll, 1988). Surficial weathering processes act on the uplifted tectonic load to varying degrees, depending upon several variables (e.g., Johnsson, 1993); and the net effect is to modify the composition from that of the sediment provenance toward generally higher concentrations of silica. Sedimentary processes redistribute the detritus from the tectonic load, filling the foreland basin with compositionally modified sediment, and increasing the thickness of buoyant continental crust. The sedimentary fill of a foreland basin is not necessarily cratonized into "crystalline basement rocks"; however, the sedimentary thickness contributes to the total volume of materials of crustal composition.

RECYCLING OF CRUSTAL ROCKS INTO THE ALLEGHANIAN CLASTIC-WEDGE SEDIMENT IN THE APPALACHIAN FORELAND

Provenance of Clastic-Wedge Sediment

An Alleghanian synorogenic clastic wedge fills the Appalachian foreland basin with Upper Mississippian to Lower Permian strata (Fig. 1) (summary in Thomas, 1977; Thomas and Schenk, 1988). Petrography of the sedimentary fill of a foreland basin commonly is used to characterize the sediment as recycled orogenic components (e.g., Dickinson et al., 1983); however, the petrographic data do not document the ages of crystallization of the provenance rocks. Analyses of detrital zircons from Pennsylvanian-age sandstones in the Alleghanian clastic-wedge deposits of the Appalachian foreland basin can be used to characterize the crystallization ages of the source terranes (Thomas et al., 2004a; Becker, 2005; Becker et al., 2005) and to evaluate the role of crustal recycling in the Alleghanian foreland.

Detrital-Zircon Ages from Alleghanian Sandstones

Samples from the most proximal preserved lower Pennsylvanian sandstones at seven sites in the Alleghanian Appalachian foreland basin have been analyzed for U-Pb ages of detrital zircons (Fig. 1) (Thomas et al., 2004a; Becker, 2005; Becker

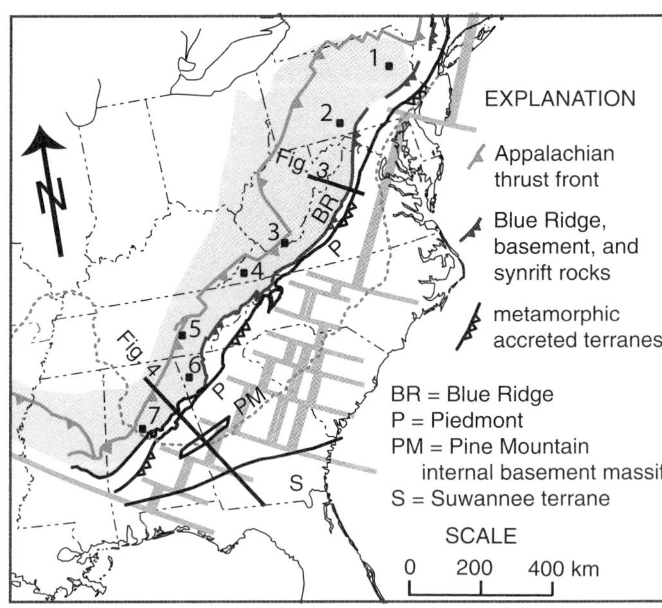

Figure 1. Outline map of Alleghanian orogen, showing location of sandstones sampled for detrital-zircon analysis (numbers correspond to sample numbers in Fig. 2) and lines of cross sections of Figures 3 and 4. Light gray lines show trace of Iapetan rifted margin of Laurentian crust (modified from Thomas, 1977). Gray shading shows approximate extent of preserved sedimentary fill of Alleghanian foreland basin. Dotted line shows limit of Mesozoic-Cenozoic coastal plains.

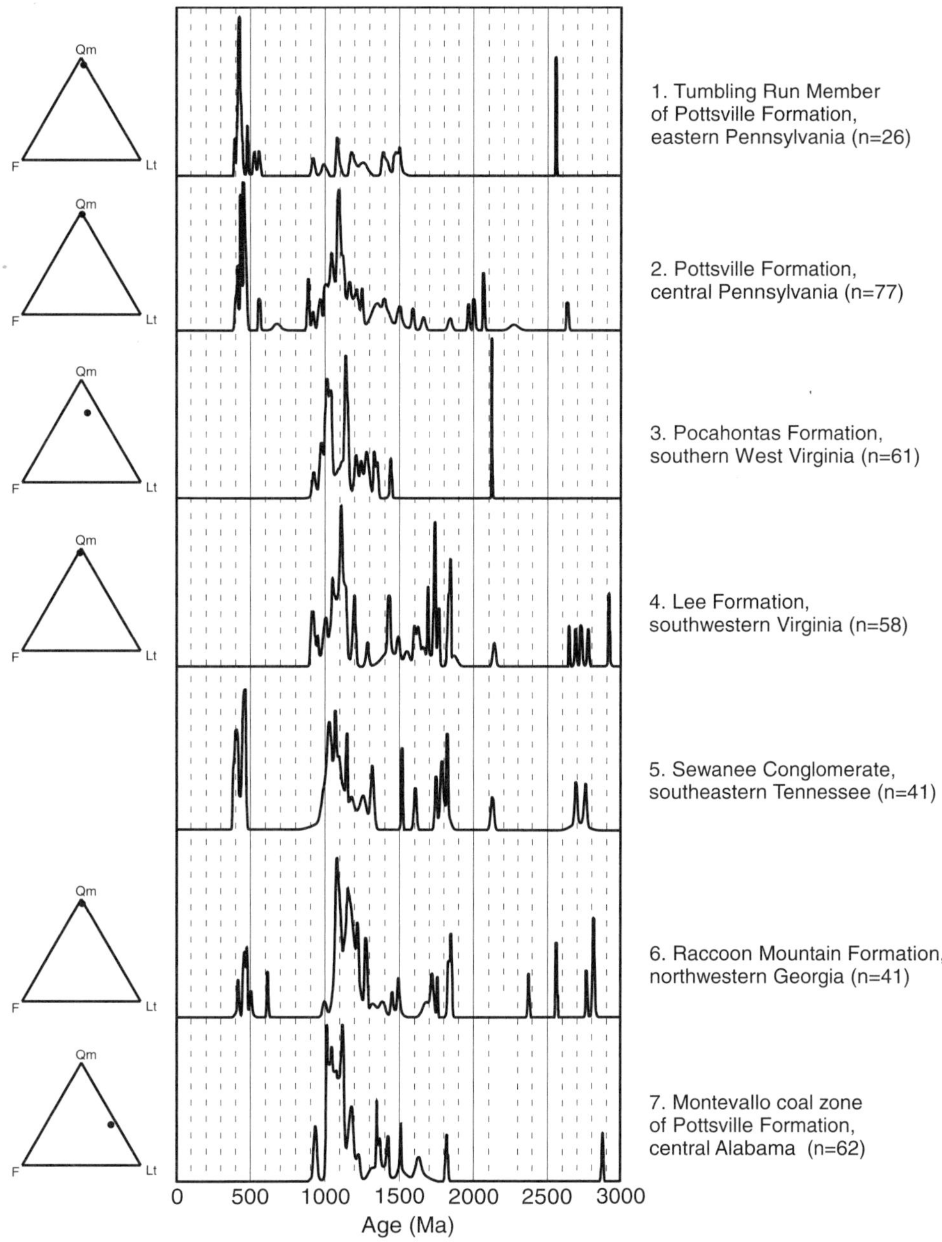

Figure 2. Age-probability plots of detrital-zircon analyses from seven Pennsylvanian sandstone samples (sample locations shown in Fig. 1) and QmFLt plots (from Becker, 2005; Becker et al., 2005) (Qm—monocrystalline quartz grains; F—monocrystalline feldspar grains; Lt—polycrystalline lithic grains, including polycrystalline quartz).

et al., 2005). Taken together, the detrital-zircon populations of each of the samples cluster into several distinctive age ranges (Fig. 2). By far the most dominant age group is ca. 1.0–1.4 Ga, corresponding in age to the Grenville basement beneath the Appalachian allochthon and in Appalachian basement massifs. Zircons older than Grenville comprise several distinctive groups, including 1.4–1.5 Ga, 1.6–1.9 Ga, and 2.5–2.9 Ga. These older zircons correspond in age to crustal elements of the Laurentian craton, including the Granite-Rhyolite province (1.3–1.5 Ga); the Central Plains, Yavapai, and Mazatzal orogens (1.6–1.8 Ga); the Trans-Hudson orogen (1.8–1.9 Ga); and the Superior province (2.5–2.8 Ga) (Hoffman, 1989; Van Schmus et al., 1993). Zircons younger than Grenville are relatively rare and generally cluster at 0.55–0.70 Ga and 0.38–0.48 Ga. The older group (0.55–0.70 Ga) corresponds in age to Iapetan synrift igneous rocks along the pre-Appalachian rifted continental margin (e.g., Su et al., 1994; Aleinikoff et al., 1995; Hogan and Gilbert, 1998; Thomas et al., 2000; Cawood and Nemchin, 2001), and the younger group (0.38–0.48 Ga) spans the range of ages of crystallization of igneous rocks associated with the Ordovician-Silurian Taconic orogeny (Drake et al., 1989) and the Devonian-Mississippian Acadian orogeny (Osberg et al., 1989). Rare zircons have ages of 2.0–2.2 Ga, a crystallization age that is very rare in Laurentian crust (Hoffman, 1989), but that is common in the Trans-Amazonian/Eburnian cratons of western Gondwana (Cordani et al., 2003). Zircon ages of ca. 530–650 Ma correspond in age to the Pan-African/Brasiliano orogens of western Gondwana (Cordani et al., 2003), but that age is not distinct from the age of Iapetan synrift rocks of the Laurentian margin. Significantly, none of the detrital zircons has an age that corresponds to synorogenic Alleghanian crystallization, despite Alleghanian ash beds (tonsteins in coal beds; Lyons et al., 1992) that document synorogenic volcanism.

Implications for Provenance of Pennsylvanian Sandstones in the Alleghanian Clastic Wedge

The dominance of Grenville-age detrital zircons in Pennsylvanian sandstones indicates the importance of Appalachian Grenville-age basement rocks as the source of sediment. In addition to potential primary sources of detrital zircons from basement uplifts of Grenville rocks, Iapetan synrift, pre-Appalachian passive-margin, and Taconic and Acadian synorogenic sedimentary rocks all contain Grenville-age detrital zircons, providing multiple potential sources for sedimentary recycling of Grenville-derived zircons (Thomas et al., 2004a; Becker et al., 2005). This supply of sediment is consistent with lower-plate components in an accretionary prism, with a retroarc thrust belt, or with uplifts along transpressional faults in an oblique collision orogen. Regardless of exact tectonic setting or pattern of recycling, the Grenville-age detrital zircons indicate substantial recycling of Laurentian Grenville-age crust as sedimentary detritus from the orogen into the Alleghanian foreland basin.

The various ages of pre-Grenville zircons indicate ultimate sedimentary provenances within most of the eastern part of the Laurentian craton. First-cycle sediment dispersal from craton provinces to the cratonward (distal) part of the foreland basin is unlikely to account for the older detrital-zircon populations because (1) most of the craton (with the possible exception of part of the Superior province) was covered by Paleozoic sedimentary strata (e.g., Sloss, 1988) and protected from erosion by Early Ordovician time, and (2) the older detrital zircons are in sandstones along the most proximal part of the foreland basin (Becker, 2005; Becker et al., 2005). The proximal location of the sandstone samples in the context of depositional processes within the basin indicates a source within the orogen; however, except for local enclaves within Grenville-age rocks, the present orogen includes no crystalline rocks of pre-Grenville ages. Synrift sandstones of late Precambrian age in Newfoundland (Cawood and Nemchin, 2001) and late synrift sandstones of Early Cambrian age in Alabama (Thomas et al., 2004b) contain detrital zircons with ages that correspond to the full range of ages of Laurentian cratonic provinces, suggesting dispersal of detritus through extensive drainage systems from the central craton (Superior and all younger provinces) to fault blocks along the Iapetan rifted continental margin. In contrast, other synrift sandstones in Newfoundland (Cawood and Nemchin, 2001) and synrift sandstones along the Blue Ridge (Bream, 2002; Bream et al., 2004) contain detrital zircons generally no older than Grenville age, suggesting restricted drainage into local basins along parts of the rift system. The diverse population of detrital zircons in the synrift sediment accumulations, along with the incorporation of the synrift deposits in the orogenic belt, provides a source for the pre-Grenville zircons in the Alleghanian foreland synorogenic sandstones (Thomas et al., 2004a). These zircons represent at least two stages of crustal recycling: first, from the craton to the synrift deposits, and subsequently, from the orogenically uplifted synrift rocks into the foreland basin. A third (or more), intermediate episode(s) of recycling into the Taconic and/or Acadian forelands, followed by deformation and final recycling into the Alleghanian foreland basin, is (are) suggested by ages of detrital zircons in Taconic and Acadian synorogenic sandstones (e.g., Gray and Zeitler, 1997; Cawood and Nemchin, 2001; McLennan et al., 2001; Thomas et al., 2004a).

In addition to the Grenville-age and older detrital zircons, the synrift sandstones in Newfoundland contain some rift-age detrital zircons (0.57–0.76 Ga), indicating a provenance of synrift igneous rocks, which are exposed in several places along the orogen (Cawood and Nemchin, 2001). The few rift-age detrital zircons in the Alleghanian sandstones are compatible with synorogenic exposure of the synrift rocks or recycling from postrift sandstones.

The rare Taconic-age and Acadian-age detrital zircons (Fig. 2) in Alleghanian sandstones show that crystalline rocks from those orogens were uplifted and exposed during the Alleghanian orogeny. The small total area of Taconic and Acadian plutons along the present mountain belt is consistent with the minor component of zircons of those ages, and further suggests relatively minor generation of "new" crust during those events.

The rare zircons of Trans-Amazonian/Eburnian age (2.0–2.2 Ga) suggest that accreted Gondwanan terranes were incorporated into the Alleghanian orogen. The paucity of these components, however, suggests limited sediment dispersal from Gondwanan terranes to the foreland basin. The zircons with ages of 0.55–0.70 Ga may reflect contributions from accreted terranes of the Gondwanan Pan-African/Brasiliano orogens; however, these cannot be distinguished uniquely on the basis of age from the Iapetan synrift rocks of the Laurentian margin.

Variation among Alleghanian Sandstones

Most of the sandstone samples are quartzose, suggesting recycled detritus from older sedimentary deposits (Becker, 2005; Becker et al., 2005). The more quartzose sandstones (samples 1,

2, 4, 5, and 6; Fig. 2) have a wide range of zircon ages, showing recycled contributions from many components of Laurentian crust. In combination, the sandstone composition and detrital-zircon population are consistent with recycling from older sandstones along the Laurentian margin, including Iapetan synrift, pre-Appalachian passive-margin, and Taconic and Acadian synorogenic sandstones.

The more feldspathic and lithic sandstones (samples 3 and 7; Fig. 2) have a more restricted range of detrital-zircon ages, generally more nearly restricted to Grenville age (Becker, 2005; Becker et al., 2005). This coincidence of sandstone composition and detrital-zircon populations suggests first-cycle sediment from Grenville basement rocks in the orogen.

Two of the samples are conglomerate (sample 1 is a quartz-pebble conglomerate; sample 7 is a polymictic lithic conglomerate). The clasts were processed along with the sand-sized matrix, and zircons from both fractions were analyzed together (Becker, 2005). The detrital-zircon ages from the conglomerates are similar to those of the other sandstones along the Alleghanian foreland (Fig. 2), indicating that the provenance of the larger clasts was not different from that of the sand. The clast size indicates that these are proximal deposits.

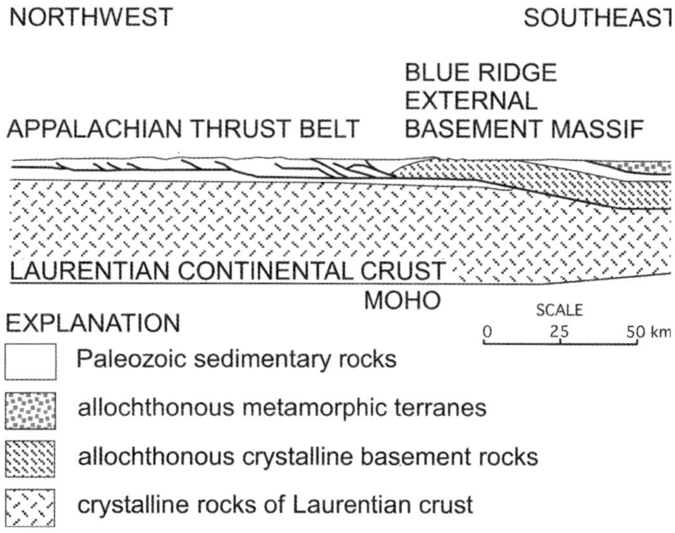

Figure 3. Structural cross section of Alleghanian orogen, showing the Blue Ridge thrust sheet of crystalline basement rocks (constructed by compilation from Harris, 1979; Pratt et al., 1988; Costain et al., 1989; Hatcher et al., 1989). Metasedimentary and metavolcanic cover on the Blue Ridge external basement massif is shown with no pattern. Location of cross section shown in Figure 1.

TECTONIC IMBRICATION OF CRUSTAL ROCKS

Alleghanian Orogeny

Interpretations of a complex tectonic history for the Alleghanian orogeny include an early (Pennsylvanian-age) phase of transpressional deformation (Gates et al., 1986, 1988; Wortman et al., 1998; Hatcher, 2001) followed by a later phase (Permian-age) of cratonward (northwest-vergent) thrusting (Miller and Kent, 1988; Hatcher, 1989; Stamatakos et al., 1996). The Pennsylvanian-age sandstones in the Alleghanian clastic wedge were deposited contemporaneously with early Alleghanian transpression, indicating that the detrital-zircon populations reflect the composition of transpressional tectonic uplifts (Becker et al., 2005). The late Alleghanian compression is recorded in large-scale foreland thrusting, and parts of the thrust belt include thrust sheets of crystalline basement rocks (external basement massifs) above a subhorizontal regional décollement within the cover strata (e.g., the Blue Ridge in Virginia; Fig. 1). In other settings, the thrust belt includes medium- to high-grade metasedimentary rocks (e.g., the Appalachian Piedmont in Alabama; Fig. 1). The metamorphic thrust sheets, both crystalline basement and metasedimentary rocks, constitute thickening and redistribution of continental crustal components by imbrication.

Blue Ridge in Virginia

Precambrian orthogneisses and paragneisses of the basement complex comprise the core of a large-scale anticlinorium, an external basement massif, in the Blue Ridge in Virginia (Figs. 1 and 3). Seismic reflection profiles and structural geometry show that the Blue Ridge anticlinorium is in the hanging wall above a subhorizontal décollement, which extends into the foreland as the basal detachment in the thin-skinned Appalachian sedimentary thrust belt (Harris, 1979; Costain et al., 1989). The Blue Ridge anticlinorium marks the most external position of Precambrian (Grenville) basement rocks in the Appalachian thrust belt, and thrust sheets farther cratonward in the foreland are exclusively sedimentary rocks detached from a level in the lower part of the Paleozoic cover succession. The anticlinorium is a ramp anticline (fault-bend fold) associated with a frontal ramp at which the detachment rises from a flat within the crystalline basement rocks to a flat in the lower part of the Paleozoic sedimentary cover succession (Fig. 3). The amplitude of the frontal ramp corresponds to the amplitude of the ramp anticline, both indicating a vertical thickness of basement rocks of ~6 km in the allochthonous Blue Ridge anticlinorium (Pratt et al., 1988; Costain et al., 1989). The lower Paleozoic sedimentary cover rocks on the limbs of the anticlinorium, as well as the Grenville age and composition of the basement rocks, indicate a palinspastic location near the Iapetan rifted margin of Laurentian continental crust (Thomas, 1991). Cratonward thrusting on the Appalachian décollement translated the basement rocks in the Blue Ridge ramp anticline over a full thickness of autochthonous Laurentian continental crust beneath the Appalachian foreland (Fig. 3). Alleghanian thrusting of the Blue Ridge thickened the crust by imbrication, which resulted in recycling the continental crust from beneath the sedimentary cover near the pre-Appalachian rifted margin into the orogenic edifice along the Appalachian foreland.

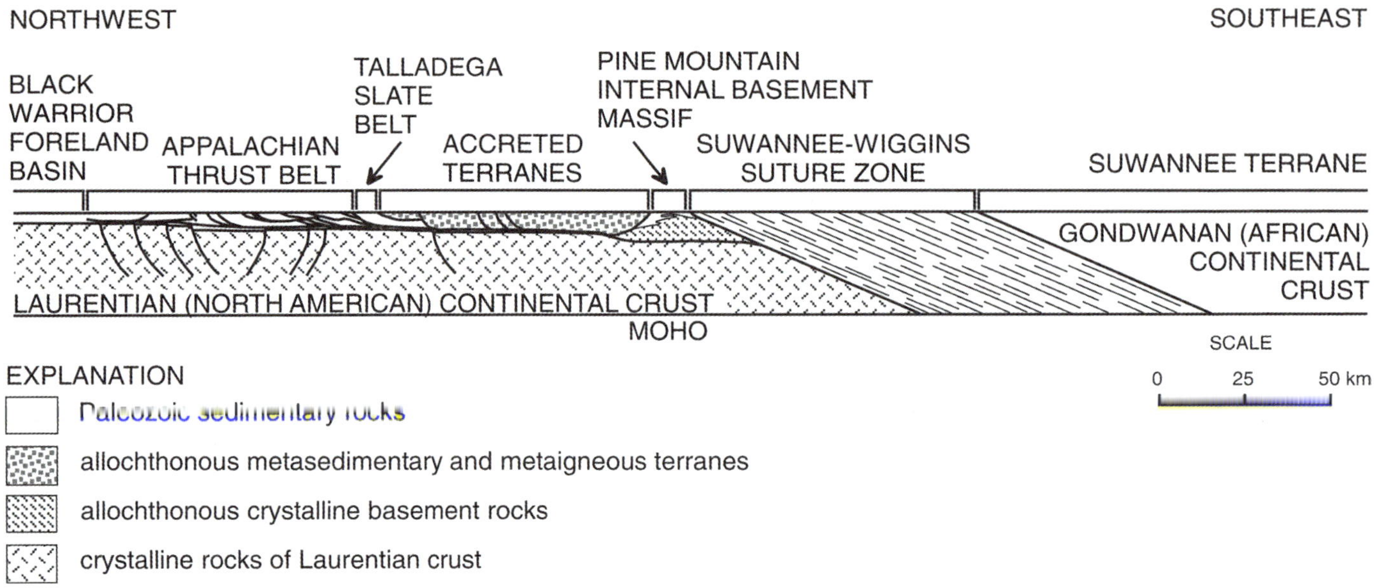

Figure 4. Structural cross section of Alleghanian orogen, showing accreted metamorphic terranes of the Appalachian Piedmont in Alabama (modified from Neathery and Thomas, 1983; Buffler and Thomas, 1994; Thomas, 2004). Metasedimentary cover on the Pine Mountain internal basement massif and metasedimentary succession in the Talladega slate belt are shown with no pattern. Location of cross section shown in Figure 1.

Appalachian Piedmont in Alabama

In contrast to the Precambrian crystalline basement rocks in the Blue Ridge external massif in Virginia, thrust sheets of metasedimentary rocks and less common metaigneous (mafic) rocks comprise the Appalachian Piedmont in Alabama, with the exception of the Pine Mountain internal basement massif (Figs. 1 and 4) (Neathery and Thomas, 1983). Seismic reflection profiles (Consortium for Continental Reflection Profiling [COCORP] Georgia lines 23, 24, and 15) show that the top of autochthonous Precambrian crystalline basement rocks of Laurentian crust dips gently southeastward beneath the Appalachian foreland to the Iapetan rifted margin of Laurentia (Nelson et al., 1987); late synrift basement faults break the regional homocline (Fig. 4) (Thomas, 1991). Thin-skinned Appalachian foreland thrust sheets are detached near the base of the Paleozoic sedimentary cover succession, leaving a thin layer of Lower Cambrian rocks, which have distinctive seismic reflections, above the crystalline basement rocks in the autochthon (Thomas and Bayona, 2005). At the same stratigraphic level of footwall detachment and low-angle southeastward dip, the Appalachian décollement extends beneath the accreted metasedimentary terranes of the Appalachian Piedmont. At the margin of Laurentian crust, the Pine Mountain internal basement massif contains an internally deformed Precambrian basement and infolded metasedimentary cover (Sears and Cook, 1984; Hooper and Hatcher, 1988; Steltenpohl, 1988).

Palinspastic reconstruction of the thin-skinned sedimentary thrust belt in Alabama shows that the thrust sheets of sedimentary cover rocks originally extended over the entire expanse of shallow Laurentian crust that is now beneath the metasedimentary thrust sheets of the Piedmont (Thomas and Bayona, 2005). Specifically, the trailing edge of the sedimentary thrust sheets is palinspastically restored approximately to the present position of the Pine Mountain internal basement massif along the Iapetan margin of Laurentian crust (Fig. 4). Kinematically, emplacement of the metasedimentary thrust sheets drove Alleghanian thrusting, which stripped the sedimentary cover from basement and imbricated the cover strata into the thin-skinned thrust belt. The allochthonous metasedimentary thrust sheets emplaced a tectonic load on Laurentian continental crust. Accretion of the metasedimentary rocks thickened the continental crust by addition of a thickness of at least 6–8 km above the top of the autochthonous basement (Fig. 4). The metasedimentary thrust sheets range in composition from metagraywackes to metaquartzarenites and include metapelites (Neathery, 1975; Thomas and Neathery, 1980; Steltenpohl et al., 1990), suggesting that the protoliths represent sedimentary detritus from continental crust or an arc-continent orogen. Mafic rocks in some of the thrust sheets probably represent oceanic crustal components (Steltenpohl et al., 1990), and the proximity of these rocks to the metasedimentary rocks suggests a lower continental slope and rise as the provenance. The metasedimentary rocks indicate substantial sedimentary recycling of continental crust into sedimentary deposits that subsequently were metamorphosed; Alleghanian thrusting imbricated the metasedimentary rocks to further recycle these crustal components.

CONCLUSIONS

Detrital-zircon populations in the synorogenic clastic-wedge deposits of the Appalachian (Alleghanian) foreland basin record multiple cycles of recycling of continental crust. The dominance of Grenville-age detrital zircons demonstrates the importance of crystalline basement rocks of the Laurentian margin, including primary Grenville sources and recycling from post-Grenville, pre-Alleghanian sedimentary provenances. Detrital zircons of pre-Grenville ages reflect sediment transport from the continental interior to synrift deposits along the rifted continental margin, followed by tectonic uplift of the synrift deposits along with Grenville-age basement to provide a source for the variety of ages of zircons in the foreland basin. Less commonly, Taconic and Acadian synorogenic zircons, as well as some probable Gondwanan zircons from accreted terranes, were recycled into the Alleghanian foreland basin. A lack of Alleghanian synorogenic zircons indicates no significant contribution from synorogenic juvenile crust. The foreland-basin fill was recycled from pre-Alleghanian continental crust, and the thickness of the fill adds to the volume of rocks of continental crustal composition.

Foreland thrusting thickens the crust by imbrication of crystalline thrust sheets above full-thickness continental crust in the orogenic foreland. The Alleghanian foreland includes examples of crystalline basement thrust sheets, as well as metasedimentary thrust sheets. Alleghanian foreland thrusting postdates deposition of the youngest preserved strata in the clastic wedge in the Appalachian foreland basin; however, flexural subsidence of the foreland basin and detrital zircons in the clastic wedge are consistent with orogenic uplifts of upper crustal, crystalline basement rocks within the Alleghanian orogen. Flexural subsidence of a synorogenic foreland basin is a response to an orogenic tectonic load, which provides a source for basin-filling detritus.

ACKNOWLEDGMENTS

We thank Eric Mountjoy and Ray Ingersoll for helpful reviews of the manuscript.

REFERENCES CITED

Aleinikoff, J.N., Zartman, R.E., Walter, M., Rankin, D.W., Lyttle, P.T., and Burton, W.C., 1995, U-Pb age of metarhyolites of the Catoctin and Mount Rogers Formations, central and southern Appalachians: Evidence for two pulses of Iapetan rifting: American Journal of Science, v. 295, p. 428–454.

Becker, T.P., 2005, Paleogeographic and tectonic implications of the late Paleozoic Alleghanian orogen developed from isotopic sedimentary provenance proxies from the Appalachian foreland basin [Ph.D. thesis]: Lexington, University of Kentucky, 214 p.

Becker, T.P., Thomas, W.A., Samson, S.D., and Gehrels, G.E., 2005, Detrital zircon evidence of Laurentian crustal dominance in the lower Pennsylvanian deposits of the Alleghanian clastic wedge in eastern North America: Sedimentary Geology, v. 182, p. 59–86, doi: 10.1016/j.sedgeo.2005.07.014.

Bream, B.R., 2002, The southern Appalachian Inner Piedmont: New perspectives based on recent detailed geologic mapping, Nd isotopic evidence, and zircon geochronology, *in* Hatcher, R.D., Jr., and Bream, B.R., eds., Inner Piedmont Geology in the South Mountains—Blue Ridge Foothills and the Southwestern Brushy Mountains, Central-Western North Carolina: North Carolina Geological Survey, Carolina Geological Society, Annual Field Trip Guidebook, p. 45–63.

Bream, B.R., Hatcher, R.D., Jr., Miller, C.F., and Fullagar, P.D., 2004, Detrital zircon ages and Nd isotopic data from the southern Appalachian crystalline core, Georgia, South Carolina, North Carolina, and Tennessee: New provenance constraints for part of the Laurentian margin, *in* Tollo, R.P., Corriveau, L., McLelland, J., and Bartholomew, M.J., eds., Proterozoic Tectonic Evolution of the Grenville Orogen in North America: Geological Society of America Memoir 197, p. 459–475.

Buffler, R.T., and Thomas, W.A., 1994, Crustal structure and evolution of the southeastern margin of North America and the Gulf of Mexico basin, *in* Speed, R.C., ed., Phanerozoic Evolution of North American Continent-Ocean Transitions: Geological Society of America, The Geology of North America, Continent-Ocean Transect Volume, p. 219–264.

Cawood, P.A., and Nemchin, A.A., 2001, Paleogeographic development of the east Laurentian margin: Constraints from U-Pb dating of detrital zircons in the Newfoundland Appalachians: Geological Society of America Bulletin, v. 113, p. 1234–1246, doi: 10.1130/0016-7606(2001)113<1234:PDOTEL>2.0.CO;2.

Cordani, U.G., Brito-Neves, B.B., and D'Agrella-Filho, M.S., 2003, From Rodinia to Gondwana: A review of the available evidence from South America: Gondwana Research, v. 6, p. 275–283, doi: 10.1016/S1342-937X(05)70976-X.

Costain, J.K., Hatcher, R.D., Jr., Çoruh, C., Pratt, T.L., Taylor, S.R., Litehiser, J.J., and Zietz, I., 1989, Geophysical characteristics of the Appalachian crust, *in* Hatcher, R.D., Jr., Thomas, W.A., and Viele, G.W., eds., The Appalachian-Ouachita Orogen in the United States: Geological Society of America, The Geology of North America, v. F-2, p. 385–416.

Dickinson, W.R., Beard, L.S., Brakenridge, G.R., Erjavec, J.L., Ferguson, R.C., Inman, K.F., Knepp, R.A., Lindberg, F.A., and Ryberg, P.T., 1983, Provenance of North American Phanerozoic sandstones in relation to tectonic setting: Geological Society of America Bulletin, v. 94, p. 222–235, doi: 10.1130/0016-7606(1983)94<222:PONAPS>2.0.CO;2.

Drake, A.A., Jr., Sinha, A.K., Laird, J., and Guy, R.E., 1989, The Taconic orogen, *in* Hatcher, R.D., Jr., Thomas, W.A., and Viele, G.W., eds., The Appalachian-Ouachita Orogen in the United States: Geological Society of America, The Geology of North America, v. F-2, p. 101–177.

Gates, A.E., Simpson, C., and Glover, L., III, 1986, Appalachian Carboniferous dextral strike-slip faults: An example from Brookneal, Virginia: Tectonics, v. 5, p. 119–133.

Gates, A.E., Speer, J.A., and Pratt, T.L., 1988, The Alleghanian southern Appalachian Piedmont: A transpressional model: Tectonics, v. 7, p. 1307–1324.

Gray, M.B., and Zeitler, P.K., 1997, Comparison of clastic wedge provenance in the Appalachian foreland using U/Pb ages of detrital zircons: Tectonics, v. 16, p. 151–160, doi: 10.1029/96TC02911.

Harris, L.D., 1979, Similarities between the thick-skinned Blue Ridge anticlinorium and the thin-skinned Powell Valley anticline: Geological Society of America Bulletin, v. 90, p. 525–539, doi: 10.1130/0016-7606(1979)90<525:SBTTBR>2.0.CO;2.

Hatcher, R.D., Jr., 1989, Tectonic synthesis of the U.S. Appalachians, *in* Hatcher, R.D., Jr., Thomas, W.A., and Viele, G.W., eds., The Appalachian-Ouachita Orogen in the United States: Geological Society of America, The Geology of North America, v. F-2, p. 511–535.

Hatcher, R.D., Jr., 2001, Rheological partitioning during multiple reactivation of the Paleozoic Brevard fault zone, southern Appalachians, USA, *in* Holdsworth, R.E., Strachan, R.A., Magloughlin, J.F., and Knipe, R.J., eds., The Nature and Tectonic Significance of Fault Zone Weakening: Geological Society [London] Special Publication 186, p. 257–271.

Hatcher, R.D., Jr., Osberg, P.H., Drake, A.A., Jr., Robinson, P., and Thomas, W.A., 1989, Tectonic map of the U.S. Appalachians, *in* Hatcher, R.D., Jr., Thomas, W.A., and Viele, G.W., eds., The Appalachian-Ouachita Orogen in the United States: Geological Society of America, The Geology of North America, v. F-2, plate 1.

Hoffman, P.F., 1989, Precambrian geology and tectonic history of North America, *in* Bally, A.W., and Palmer, A.R., eds., Geology of North America—An Overview: Geological Society of America, The Geology of North America, v. A, p. 447–512.

Hogan, J.P., and Gilbert, M.C., 1998, The Southern Oklahoma aulacogen: A Cambrian analog for mid-Proterozoic AMCG (anorthosite-mangerite-charnockite-granite) complexes?, *in* Hogan, J.P., and Gilbert, M.C., eds., Proceedings of the Twelfth International Conference on Basement Tectonics, Central North America and Other Regions: Dordrecht/Boston, Kluwer, p. 39–78.

Hooper, R.J., and Hatcher, R.D., Jr., 1988, Pine Mountain terrane, a complex window in the Georgia and Alabama Piedmont; evidence from the eastern termination: Geology, v. 16, p. 307–310, doi: 10.1130/0091-7613(1988)016<0307:PMTACW>2.3.CO;2.

Ingersoll, R.V., 1988, Tectonics of sedimentary basins: Geological Society of America Bulletin, v. 100, p. 1704–1719, doi: 10.1130/0016-7606(1988)100<1704:TOSB>2.3.CO;2.

Johnsson, M.J., 1993, The system controlling the composition of clastic sediments, *in* Johnsson, M.J., and Basu, A., eds., Processes Controlling the Composition of Clastic Sediments: Geological Society of America Special Paper 284, p. 1–19.

Jordan, T.E., 1981, Thrust loads and foreland basin evolution, Cretaceous, western United States: American Association of Petroleum Geologists Bulletin, v. 65, p. 2506–2520.

Jordan, T.E., 1995, Retroarc foreland and related basins, *in* Busby, C.J., and Ingersoll, R.V., eds., Tectonics of Sedimentary Basins: Cambridge, Massachusetts, Blackwell Science, p. 331–363.

Lyons, P.C., Outerbridge, W.F., Triplehorn, D.M., Evans, H.T., Jr., Congdon, R.D., Caprio, M., Hess, J.C., and Nash, W.P., 1992, An Appalachian isochron: A kaolinized Carboniferous air-fall volcanic-ash deposit: Geological Society of America Bulletin, v. 104, p. 1515–1527, doi: 10.1130/0016-7606(1992)104<1515:AAIAKC>2.3.CO;2.

McLennan, S.M., Bock, B., Compston, W., Hemming, S.R., and McDaniel, D.K., 2001, Detrital zircon geochronology of Taconian and Acadian foreland sedimentary rocks in New England: Journal of Sedimentary Research, v. 71, p. 305–317.

Miall, A.D., 1995, Collision-related foreland basins, *in* Busby, C.J., and Ingersoll, R.V., eds., Tectonics of Sedimentary Basins: Cambridge, Massachusetts, Blackwell Science, p. 393–424.

Miller, J.D., and Kent, D.V., 1988, Regional trends in the timing of Alleghanian remagnetization in the Appalachians: Geology, v. 16, p. 588–591, doi: 10.1130/0091-7613(1988)016<0588:RTITTO>2.3.CO;2.

Neathery, T.L., 1975, Rock units in the high-rank belt of the northern Alabama Piedmont, *in* Neathery, T.L., and Tull, J.F., eds., Geologic Profiles of the Northern Alabama Piedmont: Alabama Geological Society Guidebook, 13th Annual Field Trip, p. 9–47.

Neathery, T.L., and Thomas, W.A., 1983, Geodynamics transect of the Appalachian orogen in Alabama, *in* Rast, N., and Delany, F.M., eds., Profiles of Orogenic Belts: American Geophysical Union Geodynamics Series, v. 10, p. 301–307.

Nelson, K.D., Arnow, J.A., Giguere, M., and Schamel, S., 1987, Normal-fault boundary of an Appalachian basement massif?: Results of COCORP profiling across the Pine Mountain belt in western Georgia: Geology, v. 15, p. 832–836, doi: 10.1130/0091-7613(1987)15<832:NBOAAB>2.0.CO;2.

Osberg, P.H., Tull, J.F., Robinson, P., Hon, R., and Butler, J.R., 1989, The Acadian orogeny, *in* Hatcher, R.D., Jr., Thomas, W.A., and Viele, G.W., eds., The Appalachian-Ouachita Orogen in the United States: Geological Society of America, The Geology of North America, v. F-2, p. 179–232.

Pratt, T.L., Çoruh, C., Costain, J.K., and Glover, L., III, 1988, A geophysical study of the Earth's crust in central Virginia: Implications for Appalachian crustal structure: Journal of Geophysical Research, v. 93, no. B6, p. 6649–6667.

Quinlan, G.M., and Beaumont, C., 1984, Appalachian thrusting, lithospheric flexure, and the Paleozoic stratigraphy of the eastern interior of North America: Canadian Journal of Earth Sciences, v. 21, p. 973–996.

Sears, J.W., and Cook, R.B., Jr., 1984, An overview of the Grenville basement complex of the Pine Mountain window, Alabama and Georgia, *in* Bartholomew, M.J., ed., The Grenville Event in the Appalachians and Related Topics: Geological Society of America Special Paper 194, p. 281–287.

Sloss, L.L., 1988, Tectonic evolution of the craton in Phanerozoic time, *in* Sloss, L.L., ed., Sedimentary Cover—North American Craton: U.S.: Geological Society of America, The Geology of North America, v. D-2, p. 25–51.

Stamatakos, J., Hirt, A.M., and Lowrie, W., 1996, The age and timing of folding in the central Appalachians from paleomagnetic results: Geological Society of America Bulletin, v. 108, p. 815–829, doi: 10.1130/0016-7606(1996)108<0815:TAATOF>2.3.CO;2.

Steltenpohl, M.G., 1988, Geology of the Pine Mountain imbricate zone, Lee County, Alabama: Alabama Geological Survey Circular 136, 15 p.

Steltenpohl, M.G., Neilson, M.J., Bittner, E.I., Colberg, M.R., and Cook, R.B., 1990, Geology of the Alabama Inner Piedmont terrane: Alabama Geological Survey Bulletin 139, 80 p.

Su, Q., Goldberg, S.A., and Fullagar, P.D., 1994, Precise U-Pb zircon ages of Neoproterozoic plutons in the southern Appalachian Blue Ridge and their implications for the initial rifting of Laurentia: Precambrian Research, v. 68, p. 81–95, doi: 10.1016/0301-9268(94)90066-3.

Sylvester, A.G., 1988, Strike-slip faults: Geological Society of America Bulletin, v. 100, p. 1666–1703, doi: 10.1130/0016-7606(1988)100<1666:SSF>2.3.CO;2.

Thomas, W.A., 1977, Evolution of Appalachian-Ouachita salients and recesses from reentrants and promontories in the continental margin: American Journal of Science, v. 277, p. 1233–1278.

Thomas, W.A., 1991, The Appalachian-Ouachita rifted margin of southeastern North America: Geological Society of America Bulletin, v. 103, p. 415–431, doi: 10.1130/0016-7606(1991)103<0415:TAORMO>2.3.CO;2.

Thomas, W.A., 2004, Genetic relationship of rift-stage crustal structure, terrane accretion, and foreland tectonics along the southern Appalachian-Ouachita orogen: Journal of Geodynamics, v. 37, p. 549–563, doi: 10.1016/j.jog.2004.02.020.

Thomas, W.A., and Bayona, G., 2005, The Appalachian thrust belt in Alabama and Georgia: Thrust-belt structure, basement structure, and palinspastic reconstruction: Alabama Geological Survey Monograph 16, 48 p., 2 plates.

Thomas, W.A., and Neathery, T.L., 1980, Tectonic framework of the Appalachian orogen in Alabama, *in* Frey, R.W., ed., Excursions in Southeastern Geology, Volume 2: Geological Society of America 1980 Annual Meeting, Atlanta, Georgia: Falls Church, Virginia, American Geological Institute, p. 465–526.

Thomas, W.A., and Schenk, P.E., 1988, Late Palaeozoic sedimentation along the Appalachian orogen, *in* Harris, A.L., and Fettes, D.J., eds., The Caledonian-Appalachian Orogen: Geological Society [London] Special Publication 38, p. 515–530.

Thomas, W.A., Tucker, R.D., and Astini, R.A., 2000, Rifting of the Argentine Precordillera from southern Laurentia: Palinspastic restoration of basement provinces: Geological Society of America Abstracts with Programs, v. 32, no. 7, p. A-505.

Thomas, W.A., Becker, T.P., Samson, S.D., and Hamilton, M.A., 2004a, Detrital zircon evidence of a recycled orogenic foreland provenance for Alleghanian clastic-wedge sandstones: Journal of Geology, v. 112, p. 23–37, doi: 10.1086/379690.

Thomas, W.A., Astini, R.A., Mueller, P.A., Gehrels, G.E., and Wooden, J.L., 2004b, Transfer of the Argentine Precordillera terrane from Laurentia: Constraints from detrital-zircon geochronology: Geology, v. 32, p. 965–968, doi: 10.1130/G20727.1.

Van Schmus, W.R., Bickford, M.E., Anderson, J.L., Bender, E.E., Anderson, R.R., Bauer, P.W., Robertson, J.M., Bowring, S.A., Condie, K.C., Denison, R.E., Gilbert, M.C., Grambling, J.A., Mawer, C.K., Shearér, C.K., Hinze, W.J., Karlstrom, K.E., Kisvarsanyi, E.B., Lidiak, E.G., Reed, J.C., Jr., Sims, P.K., Tweto, O., Silver, L.T., Treves, S.B., Williams, M.L., and Wooden, J.L., 1993, Transcontinental Proterozoic provinces, *in* Reed, J.C., Jr., Bickford, M.E., Houston, R.S., Link, P.K., Rankin, D.W., Sims, P.K., and Van Schmus, W.R., eds., Precambrian: Conterminous U.S.: Geological Society of America, The Geology of North America, v. C-2, p. 171–334.

Wortman, G.L., Samson, S.D., and Hibbard, J.P., 1998, Precise timing constraints on the kinematic development of the Hyco shear zone: Implications for the central Piedmont shear zone, southern Appalachian orogen: American Journal of Science, v. 298, p. 108–130.

Manuscript Accepted by the Society 1 February 2007

Printed in the USA

The Geological Society of America
Memoir 200
2007

Diagnostic features and processes in the construction and evolution of Oman-, Zagros-, Himalayan-, Karakoram-, and Tibetan-type orogenic belts

Michael P. Searle
Department of Earth Sciences, Oxford University, Parks Road, Oxford OX1 3PR, UK

ABSTRACT

The closing of the Tethys Ocean and continent-continent collision along the Alpine-Himalayan chain ultimately produced large Himalayan-type mountain belts and large plateaux, such as Tibet. Earlier stages in the collision process, however, can be seen in the Oman Mountains of eastern Arabia and the Zagros Mountains of SW Iran. In Oman, a large, intact ophiolite was emplaced onto a Mesozoic passive continental margin, largely by thin-skinned thrust processes, prior to continental collision. The ophiolite and a granulite-amphibolite-greenschist facies inverted metamorphic sole were formed in a subduction zone setting during the early stages of emplacement. Eclogites were formed by the attempted subduction of the continental margin, and its rapid expulsion back up the same subduction zone, during later stages of the orogeny. The early stages of continental collision are best seen in the Zagros Mountains where thick-skinned thrusting and simple folding has resulted in a relatively small amount of crustal shortening (50–70 km) with almost no metamorphic or magmatic consequences. Burial metamorphism may be occurring presently at deep levels of the internal zone and the Turkish-Iranian Plateau where the crust is thicker, but this remains unexposed at the surface.

The collision of the Indian plate with Asia since ca. 50 Ma resulted in formation of the Himalaya along the north margin of India, and the Karakoram–Hindu Kush Mountains along the south Asian margin. Together with renewed uplift and crustal thickening of the Tibetan Plateau, this was arguably the largest continental collision in the last 450 m.y. of Earth history. The Himalayan-type orogeny involved large amounts of crustal shortening (~500–1000 km), early ultrahigh-pressure (UHP) coesite-eclogite facies metamorphism, peak Barrovian facies kyanite and sillimanite metamorphism, and mid-crustal anatexis resulting in garnet, tourmaline, muscovite-bearing migmatites, and leucogranites. Processes involved in the construction of the Tibetan Plateau include crustal shortening and doubling the thickness of the crust to 65–90 km. High-pressure (HP) eclogite and high-temperature/high-pressure (HT-HP) granulite metamorphism may be occurring at depth today in the lower crust beneath Tibet. Widespread ultrapotassic volcanism across Tibet indicates the presence of a hot subcontinental mantle, which was progressively shifted northwards as the cold,

Searle, M.P., 2007, Diagnostic features and processes in the construction and evolution of Oman- Zagros-, Himalayan-, Karakoram-, and Tibetan-type orogenic belts, *in* Hatcher, R.D., Jr., Carlson, M.P., McBride, J.H., and Martínez Catalán, J.R., eds., 4-D Framework of Continental Crust: Geological Society of America Memoir 200, p. 41–61, doi: 10.1130/2007.1200(04).

Indian lithosphere underthrust southern Tibet. Whereas Tibet shows mainly upper crustal sedimentary and volcanic rocks at the present surface, the Karakoram Range, along strike to the west, shows mostly deep crustal high-grade metamorphic rocks, multiple granite intrusions, and over 60 m.y. of high-temperature metamorphism. This paper reviews the salient geological features of Oman-, Zagros-, Himalayan-, Tibetan-, and Karakoram-type orogenic belts. These features can be used in studies of older orogenic belts to give indications of their tectonic origins.

Keywords: orogenesis, metamorphism, magmatism, tectonics.

INTRODUCTION

The Alpine-Himalayan mountain chains of the Middle East and Asia formed largely as a result of the closing of the Tethys Ocean that once separated the Gondwana continents (Africa, Arabia, India) from the southern continental margins of Europe and Asia. Along these mountain ranges, different stages of the continental collision process are best exposed in different mountain ranges, depending mainly on their tectonic position. The earliest aspects of the collision process are best preserved in the Oman Mountains of eastern Arabia, a region where the Arabian plate has not yet collided with the Asian plate. The Gulf of Oman is a relic piece of Tethyan oceanic crust, subducting northwards beneath the Makran accretionary prism. For this reason, Oman contains the best-preserved obducted ophiolite known, and the tectonics of Oman can be used as an analog for other ophiolite belts that have suffered subsequent continental collision overprints. In Oman, a giant, intact ophiolite thrust sheet, consisting of at least 10 km of upper mantle and the entire oceanic crustal sequence, has been emplaced southwestwards onto the passive continental margin of Arabia during the Late Cretaceous (e.g., Lippard et al., 1986).

The earliest stages of a continent-continent collision are best seen in the Zagros fold-and-thrust belt of SW Iran (Fig. 1). Here, the Arabian plate collided with central Iran, probably during the early Miocene, and the early effects of collision

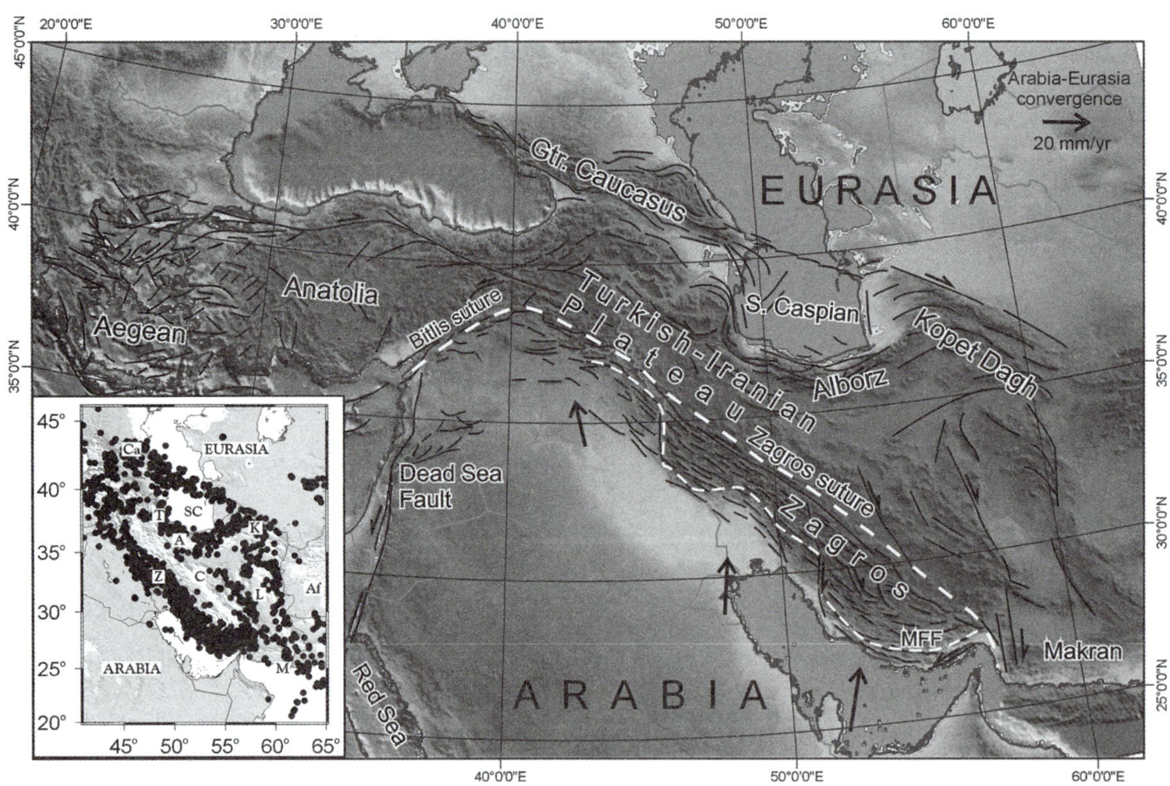

Figure 1. Topographical map showing the Arabia–central Iran collision zone along the Zagros Mountains, after Blanc et al. (2003). MFF—Main Frontal Fault. Inset shows the seismicity of Iran during the period 1964–1998, from Jackson et al. (2002) using the catalog of Engdahl et al. (1998). A—Alborz; Af—Afghanistan; C—central Iran, largely aseimic; Ca—Greater Caucasus; K—Kopet Dagh; L—Lut block; M—Makran; T—Talysh; SC—South Caspian; Z—Zagros.

are well preserved (e.g., Allen et al., 2004). Remnant ophiolites are preserved along the Zagros suture zone (e.g., at Neyriz and Kermanshah) but have suffered later collision-related deformation that has largely wiped out evidence of their earlier obduction onto the Arabian plate. It is not certain whether these ophiolites were large-scale Oman-type ophiolites obducted onto the Zagros passive margin, or whether they were remnant pieces of ocean crust and upper mantle trapped during the continental collision process. Large-scale folding of the shelf sequence across the Zagros Range shows only 50–70 km of NE-SW upper crustal shortening (Blanc et al., 2003).

The later stages of a continent-continent collision are best seen along the Himalaya and Tibetan Plateau. The Indian plate was originally attached to southern Africa, Madagascar, the Seychelles, and Antarctica as part of the Gondwana supercontinent during the early Mesozoic. India-Madagascar broke from the African plate and rifted away at ca. 160 Ma (Reeves and de Wit, 2000). By marine magnetic anomaly 34 (83 Ma) India/Seychelles had separated from Madagascar and begun its rapid northward drift across the Indian Ocean as the mid-Indian Ocean ridges began a period of rapid ocean floor spreading. At magnetic anomaly 22 (ca. 53 Ma), India made first contact with the Asian crust, roughly at equatorial latitude, and drift rates decreased rapidly from ~25–20 cm/yr to ~5 cm/yr (Besse and Coutillot, 1988; Klootwijk et al., 1992). Since 50 Ma, India has indented northwards into Asia, and some 2000 km of crustal shortening has been taken up within the Himalaya-Tibet region since the early Eocene (Dewey et al., 1989). GPS (global positioning system) data show that the Indian plate is continuing to move north relative to the stable Siberian platform (Bilham et al., 1997). Older mountain ranges along the northern margin of Tibet (Pamir, Tien Shan, Kun Lun, Altyn Tagh) have been reactivated and uplifted further as a result of the Indian plate collision and indentation. The effects of the India-Asia collision may extend even farther north, as far as the Altay mountain ranges of Mongolia.

Several Precambrian and Paleozoic orogenic belts are frequently described informally as being ancient Himalayan-type or Tibetan-type orogens, where much of the upper structural levels have been eroded away. The India-Asia collision zone, and the structural and thermal evolution of the Himalaya, Karakoram, and Tibet may form a present-day analog for several older orogenic belts in the geological record, where much of the upper crust has been eroded away, notably for example the Scottish Caledonides, the Trans-Hudson orogen in northern Canada (St. Onge et al., 2006), and the Limpopo belt of South Africa–Zimbabwe. In this paper, the major geological points of four main orogenic belts are described: (1) precollisional ophiolite obduction in the Oman Mountains, (2) early continent-continent collision in the Zagros Mountains, (3) mature continental collision along the Indian plate Himalaya, and (4) long-lived continental collision along the southern margin of the Asian plate in the Tibetan Plateau and Karakoram–Hindu Kush Ranges. The salient geological points of each mountain range are discussed, in particular the amounts and time spans of crustal shortening, thickening, regional metamorphism, magmatism, and volcanism. Neotectonics, erosion, uplift, and exhumation rates are discussed as well as the climatic consequences of mountain uplift. The following sections are not meant to be a comprehensive review but rather a succinct summary of the main points of each orogenic belt.

OMAN OROGEN (PRE–CONTINENTAL COLLISION)

The mountains in eastern Arabia are renown for having the world's largest, best-exposed, and best-studied ophiolite complex (Fig. 2). The Semail ophiolite was obducted from northeast to southwest onto the Mesozoic passive continental margin of Arabia during the Late Cretaceous (see Lippard et al., 1986; Robertson et al., 1990; Searle and Cox, 1999, for reviews). Oman has not undergone any continent-continent collision, so the early, precollisional stages of a typical Himalayan-type orogeny can be well studied here. The salient points concerning the geology of the Oman Mountains are the following:

1. The Semail ophiolite complex, covering an area of 550 km × 150 km, is an intact thrust sheet of Tethyan oceanic crust ~5 km thick, including the Moho transition zone, gabbros, sheeted dikes, pillow lavas, and pelagic sediments (radiolarian cherts, metalliferous Fe- and Mn-rich umbers, etc.) overlying a 12–15 km thickness of upper mantle peridotites including harzburgite, dunite with chromitite bodies, and less common wehrlite (Fig. 3).

2. The ophiolite is the structurally highest of a series of thrust sheets emplaced ~450 km onto a passive continental margin (Arabia) comprising a 3–5 km thickness of Permian to Cenomanian (middle Cretaceous) shelf carbonates.

3. The Semail ophiolite is a "supra-subduction zone" ophiolite formed above a NE-dipping subduction zone (Fig. 3), dipping away from the passive continental margin (Lippard et al., 1986; Searle and Cox, 1999). U-Pb ages from zircons in plagiogranites (95 Ma; Tilton et al., 1981; Warren et al., 2005) are the same age as ^{40}Ar-^{39}Ar and K-Ar hornblende ages from amphibolites in the metamorphic sole. Upper pillow lavas have immature island-arc geochemical signatures. It is not a slice of MORB (mid-oceanic-ridge basalt) crust as claimed by some (e.g., Nicolas et al., 1988), and obduction cannot have been initiated at the spreading center (Boudier et al., 1988) because of the synchroneity of ophiolite lavas and subduction zone amphibolites and the fact that metamorphic sole protoliths are not ophiolite volcanic rocks.

4. Along the base of the ultramafic mantle sequence, the inverted metamorphic sole includes garnet + clinopyroxene amphibolites metamorphosed at 800–840 °C and 10–12 kbar (50–60 km depth) with underlying greenschist facies metasediments and metavolcanics beneath. The metamorphic sole reaches high-temperature (HT) or ultrahigh-temperature (UHT) granulite facies in the United Arab Emirates, with the heat source being the harzburgite mantle sequence (Searle and Cox, 2002).

5. The entire obduction process lasted ~21 m.y. from initial formation of the ophiolite crustal sequence and amphibolite facies metamorphic sole at ca. 95 Ma to final foreland thrusting,

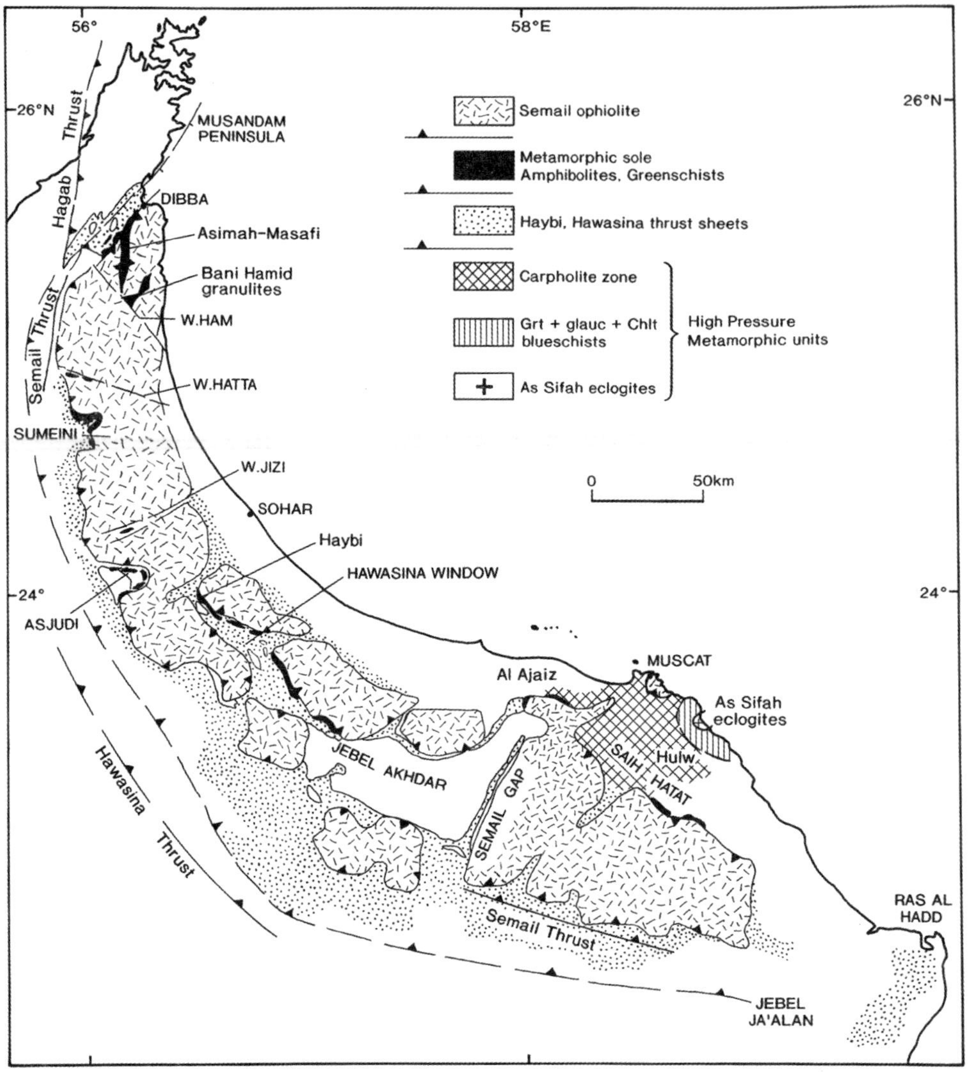

Figure 2. Map showing the geology of the Oman Mountains, after Lippard et al. (1986) and Searle and Cox (1999). Grt—garnet; Glauc—Glaucophane; Chlt—Chloritoid.

which ended at ca. 74 Ma with the deposition of neo-autochthonous shallow marine limestones over all allochthonous units.

6. Thrusting propagated from deep, subduction zone thrusts during formation of the inverted metamorphic sole to shallower-level, thin-skinned thrusts toward the continent, with time. Over 400 km of crustal shortening (from ophiolite to stable foreland) and SW-directed emplacement is shown from structural cross-sectional reconstruction.

7. HP eclogites comprising garnet + omphacite + phengite + quartz metabasites ± chloritoid metapelites formed at 540 °C and 20 kbar (74–80 km depth) in the mid-crustal levels of the leading edge of the Arabian "basement." HP blueschists and carpholite-bearing metasediments occur at structurally higher levels along the NE margin beneath the ophiolite (Searle et al., 1994, 2003).

8. Timing of HP eclogite facies metamorphism (79 Ma; Warren et al., 2003) shows that the thinned leading margin of the Arabian plate basement was dragged down the subduction zone during the later stages of obduction, and then returned to the surface very rapidly beneath an extensional shear zone along the upper margin of the subduction zone (Searle et al., 2004).

9. A flexural foreland basin developed in front of the emplacing thrust sheets as the Arabian margin flexed down beneath the load of the ophiolite and underlying thrust sheets. The youngest thrust tip lines extend into the Late Cretaceous foreland basin sediments southwest of the mountain front, and show that final thrusting ended during the early Maastrichtian (ca. 74 Ma).

10. The northern part of the Oman Mountains in the Musandam Peninsula shows the first effects of the continent-continent collision of Arabia and central Iran. Middle Tertiary thrusts have cut up through the Late Cretaceous thin-skinned thrust stack, placing deeper levels of the Permian-Mesozoic shelf and pre-Permian basement over the foreland basin.

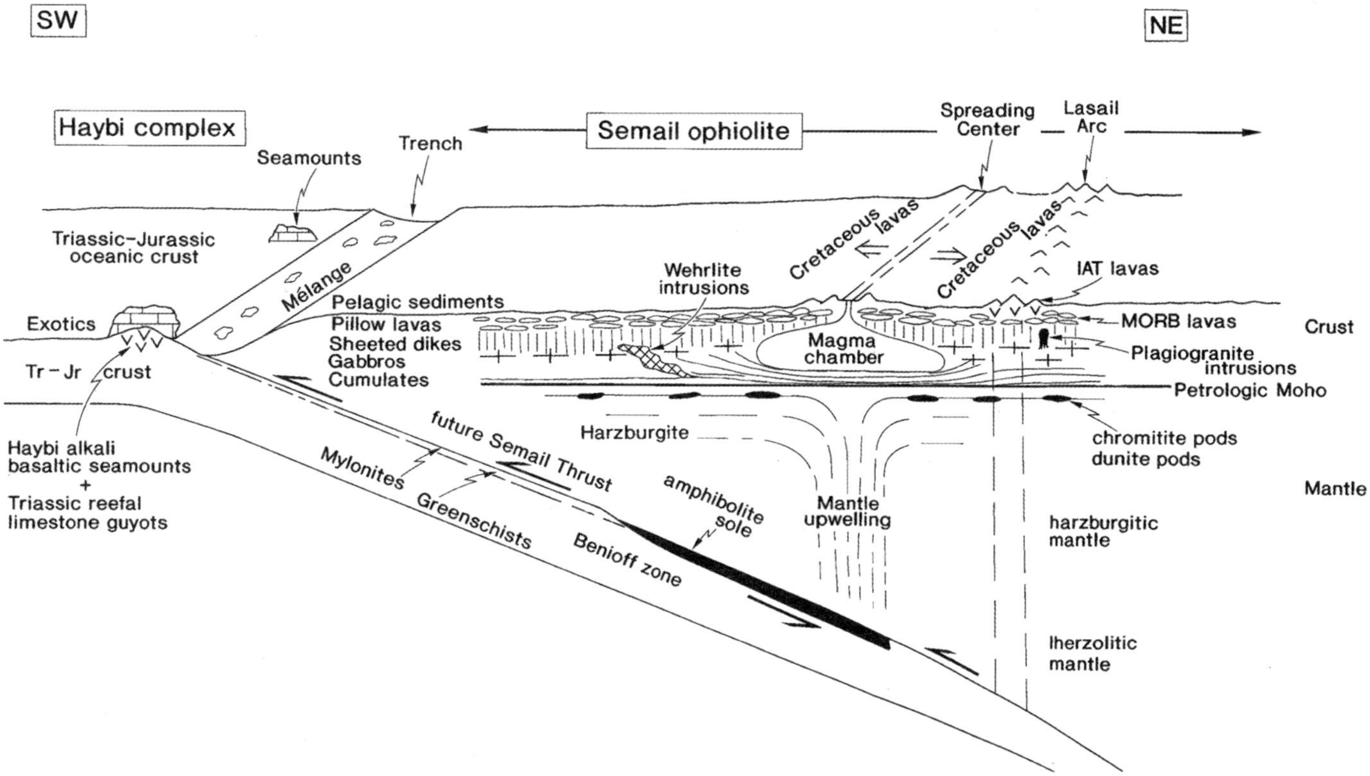

Figure 3. Palinspastic reconstruction of the Semail ophiolite in Oman, showing the upper mantle and crustal sequence of the ophiolite, and amphibolites and greenschists of the metamorphic sole, after Searle and Cox (1999). Early mantle upwelling fabrics in the peridotites are cut by later emplacement-related fabrics along the base of the ophiolite. The lower volcanic sequence has mid-oceanic-ridge basalt (MORB)–like geochemical signatures, while the upper Lasail arc lavas have island-arc tholeiite (IAT) geochemical signatures.

ZAGROS OROGEN (EARLY CONTINENT-CONTINENT COLLISION)

The NE passive continental margin of the Arabian plate collided with the central Iran plate (Sanandaj-Sirjan terrane) probably in the Oligocene in the far northwest (Iraqi Kurdistan and western Zagros, Iran) and during the early Miocene in the southeast (e.g., Blanc et al., 2003; Allen et al., 2004) (Fig. 1). The Zagros fold-and-thrust belt extends for >1800 km from Kurdistan through NE Iraq and SW Iran as far as the Strait of Hormuz and the northern tip of the Oman Mountains (Searle, 1988; Regard et al., 2004). Extensive geological mapping of the Zagros Mountains has been carried out by the Geological Survey of Iran and the National Iranian Oil Company, and much of our knowledge of this orogen results from these works. The Zagros Mountains show the early effects of a continent-continent collision zone. The salient points can be summarized as follows:

1. The Zagros fold-and-thrust belt covers an area of ~1800 km long and 250–350 km wide along the NE margin of the Arabian plate. It forms the link between the strike-slip plate margin of the Dead Sea transform fault in western Arabia and the remnant oceanic crust in the Gulf of Oman to the east.

2. The Zagros suture is marked by a series of highly dismembered ophiolite complexes (e.g., Neyriz, Kermanshah), Mesozoic deep-sea sediments, oceanic basaltic rocks, and younger "molasse-type" sediments (Wells, 1969). The Arabia–central Iran collision occurred in the early Miocene (16–23 Ma; Allen et al., 2004).

3. The structure of the Zagros Range is dominated by very large-scale concentric and periclinal folds (Fig. 4) that affect the entire ~10-km-thick Phanerozoic sedimentary sequence (Colman-Sadd, 1978).

4. Crustal shortening estimates from restoration of balanced cross sections are in the region of 50–70 km across the entire Zagros Range from the suture zone to the Arabian Gulf foreland (Blanc et al., 2003). Comparable amounts of extension occurred along the Red Sea–Gulf of Aden spreading centers, along the western margin of the Arabian plate (Le Pichon and Gaulier, 1988).

5. Several evaporite and shale layers form detachment horizons within the Zagros fold-and-thrust belt, but the major basal detachment lies along the Precambrian Hormuz salt at ~10 km depth. Salt diapirs are common in the Zagros Range where Proterozoic salt domes have risen up to 10 km through the overlying Phanerozoic cover (Kent, 1970).

Figure 4. Aerial photograph of the Zagros fold-and-thrust belt, SW Iran, showing en echelon, doubly plunging periclinal folds. View looking north from above the Gulf of Oman coast.

6. Balancing constraints suggest that the Precambrian basement beneath the basal detachment must have underthrust the Turkish-Iranian Plateau to the north by ~50 km. The basement may have been subducted to depths of 45–50 km beneath the Iranian Plateau but may not have been significantly reactivated, staying within the granulite stability field.

7. There are no regional metamorphic rocks or intrusive granitoids exposed in the Zagros Range (Geological Survey of Iran maps).

HIMALAYAN OROGEN (MATURE CONTINENT-CONTINENT COLLISION)

The collision of the Indian plate with Asia during the early Eocene (ca. 50 Ma; Zhu et al., 2005) resulted in the formation of the Himalaya along the northern margin of India and also much of the uplift of the Tibetan Plateau (Fig. 5). The Indus–Yarlung Tsangpo suture zone marks the zone of collision between India and Asia and contains remnant ophiolites, deep-sea Tethyan sedimentary rocks and Mesozoic ocean island alkali volcanics,

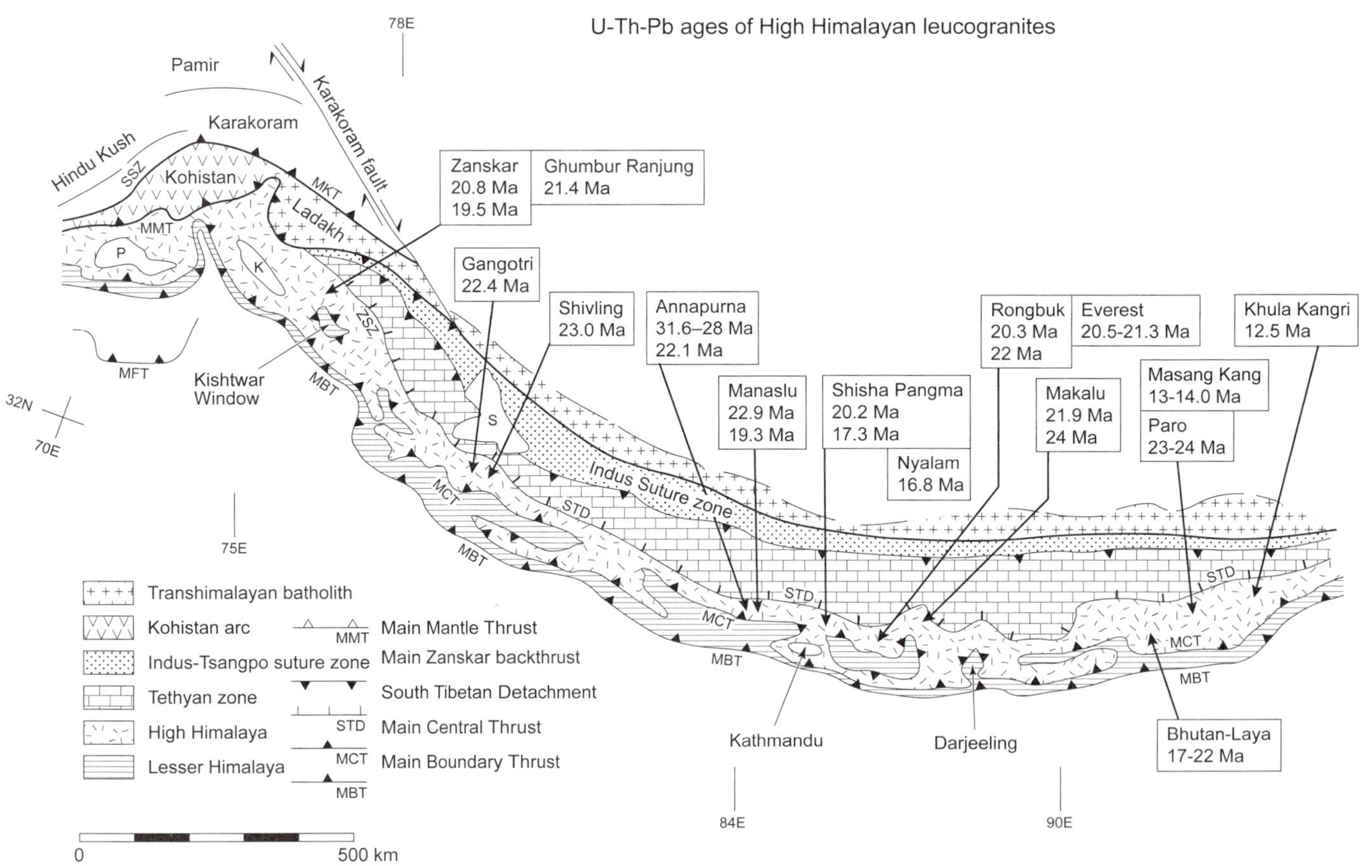

Figure 5. Geological map of the Himalaya, showing the major fault systems and U-Pb ages of the main Himalayan leucogranites. K— Kashmir; MBT—Main Boundary Thrust; MCT—Main Central Thrust; MFT— Main Frontal Thrust; MKT— Main Karakoram Thrust; MMT—Main Mantle Thrust; P— Peshawar basin; S— Sutlej basin; SSZ—supra-subduction zone; STD—South Tibetan Detachment; ZSZ—Zanskar shear zone.

a remnant Cretaceous island arc (Dras arc) and related volcaniclastic sediments. The northern margin of India is a complete Cambrian to early Eocene shelf sedimentary sequence, including the Permian and Mesozoic shelf carbonate sequence, similar to that exposed in Oman and the Zagros Mountains (Gaetani and Garzanti, 1991). These Tethyan shelf carbonates show complex, postcollisional folding and thrusting (Searle et al., 1988; Corfield and Searle, 2000). The Greater Himalayan sequence is a series of mostly high-grade metamorphic rocks, migmatites, and leucogranites that form most of the highest peaks of the Himalaya (Fig. 6). This cross section across the central Himalaya of Nepal–southern Tibet illustrates the mid-crustal channel of the Greater Himalaya flowing south from beneath southern Tibet (Beaumont et al., 2001; Vannay and Grasemann, 2001; Grujic et al., 2002; Searle et al., 2003; Searle and Szulc, 2005). The lower crust is the underthrust Precambrian shield of the Indian plate lower crust. The Lesser Himalaya is composed mainly of low-grade or unmetamorphosed Proterozoic and Paleozoic cover rocks forming the southern thrust sheets of the Himalaya. Lesser Himalayan rocks have been thrust south over the Tertiary Siwalik molasse foreland basin, formed by flexural down-warping of the Indian plate beneath the Himalayan load.

The following features are characteristic of the Himalayan orogenic system:

1. *Morphology*: The Himalaya is ~2500 km long and stretches from the Afghanistan-Pakistan border in the west along the Ladakh, Zanskar, Spiti, Himachal, and Garhwal Ranges of northern India, along the Nepal, Sikkim, Bhutan, and Arunachal Pradesh regions to western Yunnan and northern Burma. The Nanga Parbat syntaxis in the far northwest and the Namche Barwa syntaxis in the far northeast mark the corners of the indenting Indian plate.

2. *Ophiolites*: Ophiolites were obducted onto the passive continental margin of the Indian plate prior to the continent-continent collision. These ophiolite thrust sheets now form the structurally highest unit along the northern Himalayan Tethyan zone and hence are at high altitude. They are the first units to erode and thus are not preserved along much of the Himalaya.

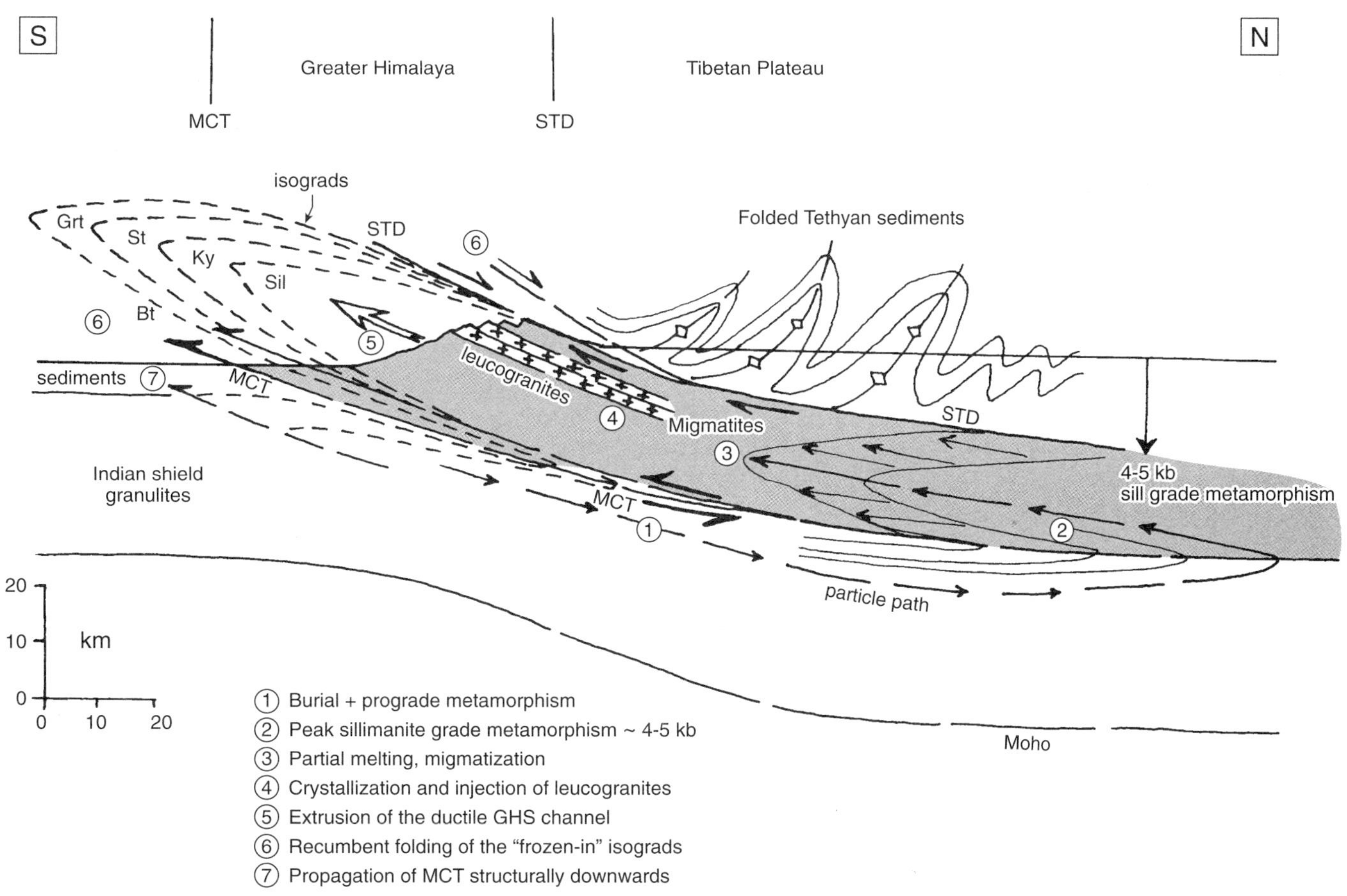

Figure 6. Cross section of the Nepal–southern Tibet Himalaya showing the mid-crustal layer of high-grade gneisses, migmatites, and leucogranites (shaded) separating the brittlely deforming sedimentary rocks of the upper crust from the Indian Shield granulites of the lower crust, after Searle and Szulc (2005). GHS—Greater Himalayan sequence; MCT—Main Central Thrust; STD—South Tibetan Detachment; Bt—biotite; Grt—garnet; Ky—kyanite; Sil—sillimanite; St—staurolite.

The Spontang ophiolite of Ladakh (Corfield et al., 2001) and the Amlang la–Jungbwa–Purang ophiolite sheet of southern Tibet (Gansser, 1964) are examples.

3. The Indus–Yarlung Tsangpo suture zone, marking the Indian plate–Asian margin collision, is marked by a discontinuous line of Tethyan ophiolites that mostly show supra-subduction zone affinity, deep-sea sedimentary rocks, rare blueschists, mélanges, and postcollision conglomerates and molasse sediments.

4. The Himalaya results from the collision of a passive continental margin (Indian plate) with an active Andean-type continental margin (Lhasa block) characterized by the Trans-Himalayan calc-alkaline batholith and its associated calc-alkaline volcanics. The southern margin of the Asian plate (Lhasa block) may have had a crustal thickness and elevation similar to the present-day Andes, prior to the Indian plate collision (England and Searle, 1986).

5. *Structure*: South-verging thrusting is present across most of the Himalaya with large-scale thrusts propagating structurally down-section with time, from the 50–40 Ma deformation along the Indus–Yarlung Tsangpo suture zone to the middle Miocene deformation in the Greater Himalaya to the younger Tertiary thrusting in the Lesser Himalaya. Several phases of thrusting and folding occur in each zone with late Tertiary north-vergent backthrusting along the Indus–Yarlung Tsangpo suture zone and northern Himalaya.

Total crustal shortening across the Himalaya could be as much as 700–1000 km. Some 150–300 km of shortening has been documented across the folded and thrust Tethyan Himalaya in Ladakh (Corfield and Searle, 2000). Minimum amount of shortening across the Lesser Himalaya is ~650 km (DeCelles et al., 2001). Amounts of shortening in the high-grade metamorphic rocks of the Greater Himalaya are impossible to ascertain due to the high degree of ductile strain, internal partial melting, and crustal anatexis.

The base of the Greater Himalayan sequence metamorphic zone is the Main Central Thrust zone—a crustal-scale low-angle ductile shear zone and thrust fault characterized by a zone of inverted and telescoped metamorphic isograds (Fig. 6). The Main Central Thrust ductile shear zone propagates down-section with time to a brittle thrust fault (Stephenson et al., 2000, 2001). The Main Central Thrust acted as the lower boundary of the southward-extruding layer of middle crust and was moving synchronously with the South Tibetan Detachment low-angle normal fault at the top of the channel (Searle and Rex, 1989; Hodges et al., 1992, 1993, 1996; Hodges, 2000; Law et al., 2006).

The South Tibetan Detachment system of low-angle normal faults bound the extruding mid-crustal channel of high-grade metamorphic rocks (Fig. 7). The South Tibetan Detachment is characterized by a right-way-up but telescoped series of metamorphic isograds from sillimanite + K-feldspar up to garnet-biotite. The shear

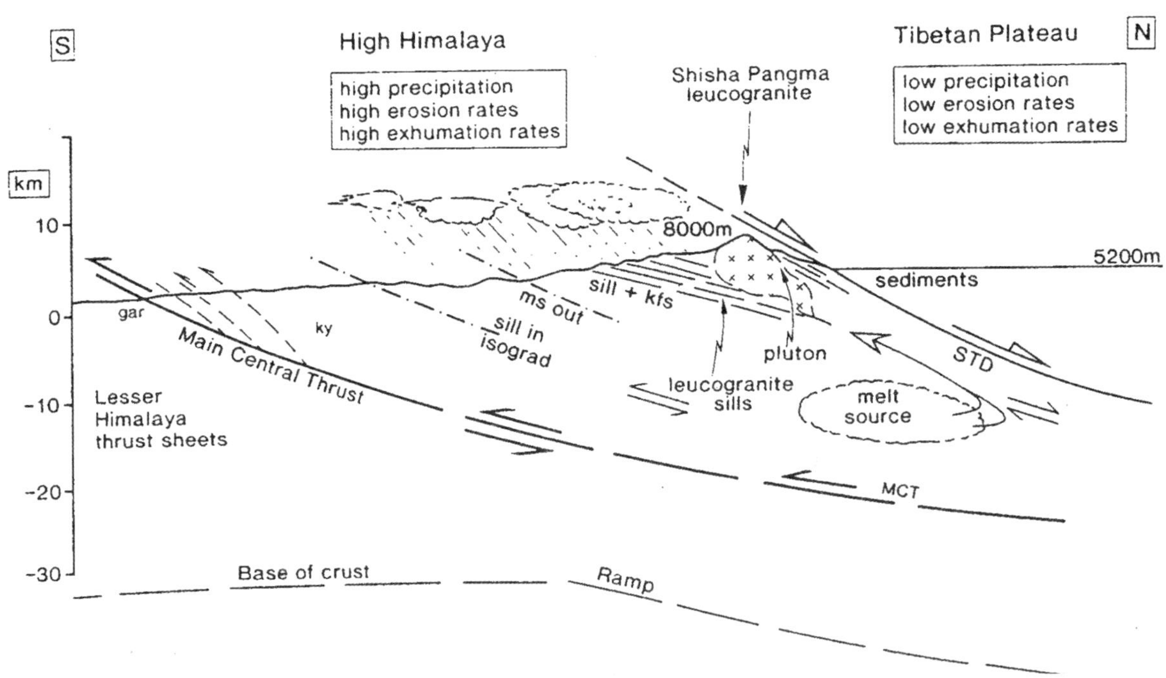

Figure 7. Schematic cross section of the Langtang–Shisha Pangma Himalaya in Nepal and southern Tibet showing major structural features, after Searle et al. (1997). Inverted metamorphic isograds from sillimanite (sill) to kyanite (ky) and garnet (gar) grade are present along the Main Central Thrust (MCT) zone. The South Tibetan Detachment (STD) normal fault bounds the top of the Greater Himalayan sequence. The Himalaya, characterized by steep slopes, high exhumation rates, high erosion rates, and a wet monsoon climate, forms a topographic-climatic barrier to the rain shadow of Tibet, which is high, cold, and arid. kfs—K-feldspar; ms—muscovite.

zone propagates up-section with time from a ductile shear zone to a low-angle brittle normal fault. Geological offsets along the South Tibetan Detachment may be 100 km or more along the Everest profile (Searle et al., 2003; Law et al., 2004).

6. *Metamorphism*: The earliest metamorphism in the Himalaya is recorded by coesite-bearing eclogites along the northern margin of the Indian plate in the Kaghan area of northern Pakistan (O'Brien et al., 2001) and in the Tso Morari dome, Ladakh (Leech et al., 2005). These rocks have basaltic protoliths within late Paleozoic–early Mesozoic carbonates that were subducted to depths of ~100 km at 27–29 kbar and 700–750 °C. Timing of peak UHP metamorphism is almost the same as the timing of initial continent-continent collision, so it is debated as to whether these eclogites resulted from the final stage of the ophiolite obduction process (as for example in Oman; Searle et al., 2004), or the initial stage of the continent-continent collision process (Leech et al., 2005). U-Pb dating of allanite and zircon from the Kaghan eclogites in Pakistan gave an age of 46.4 ± 0.1 Ma interpreted as dating peak UHP metamorphism (Parrish et al., 2006).

The main Himalayan event along the Greater Himalaya is a regional Barrovian facies kyanite-sillimanite grade medium-temperature medium-pressure metamorphism. Early kyanite grade metamorphism peaked at 550–680 °C and 10–12 kbar, corresponding to depths of 36–45 km (Searle et al., 1992b, 1999a; Walker et al., 2001). Timing is constrained as ca. 35–32 Ma from U-Pb dating of monazites and Sm-Nd garnet ages (Vance and Harris, 1999; Walker et al., 1999; Simpson et al., 2000).

Sillimanite grade metamorphism peaked at 650–770 °C and 3.7–4.5 kbar, corresponding to depths of 14–18 km (Searle et al., 1992b, 1999a, 2003). Timing of peak sillimanite grade metamorphism from U-Pb dating of monazites growing in equilibrium with garnet + sillimanite ± cordierite was between 30 and 16.5 Ma (Noble and Searle, 1995; Simpson et al., 2000). Sillimanite metamorphism was concomitant with partial melting that produced widespread migmatites within the Greater Himalayan sequence, and the anatectic event resulted in crystallization of the Himalayan leucogranites (Searle et al., 1999b).

The youngest metamorphism is recorded in the cores of the Nanga Parbat and Namche Barwa syntaxes, where Plio-Pleistocene sillimanite + cordierite + K-feldspar grade gneisses show partial melting and formation of garnet + tourmaline leucogranites <5 m.y. old, and cordierite-bearing melt seams that crystallized from a 5 kbar melt <1 m.y. ago (Zeitler et al., 1993; Bowring et al., 2004; Crowley et al., 2005). Exhumation rates could be as high as 1 cm/yr. Protoliths may be as old as Proterozoic, or even Archean, and there is evidence of several phases of recycling of old crust to depth, remetamorphism, and melting.

7. *Magmatism*: Himalayan leucogranites contain quartz, K-feldspar, and plagioclase with varying amounts of tourmaline, garnet, muscovite, and biotite, and minor amounts of sillimanite, cordierite, apatite, etc. The leucogranites are entirely derived from melted crustal pelitic rocks, probably Proterozoic shales, and result from dehydration melting of muscovite in the presence of some fluids (Harris and Massey, 1994; Prince et al., 2001). The leucogranites were emplaced as sills along the top of the Greater Himalayan sequence channel and beneath the passive roof fault zone of the South Tibetan Detachment normal fault (e.g., Searle et al., 1993, 1997, 2003).

8. *Sedimentation*: Flexure of the Indian plate as it underthrust the Himalaya resulted in the development of a foreland basin to the south that has accumulated debris eroded from the rising Himalaya since the middle Tertiary. The Siwalik molasse basin has a linear depocenter immediately south of the active Main Boundary Thrust. Sediments are transported along river systems into the foreland basin, and the Ganges and Brahmaputra Rivers drain the basin out to the Bay of Bengal. Sediments eroded from the Himalaya have been deposited in the Bengal Fan as far south as offshore Sri Lanka, 1800 km south of the Himalaya (Curray, 2005).

Most Himalayan rivers developed as a result of the rising Himalaya, and many have cut back through the crest of the high mountains by headwall erosion. Some rivers, however, must be antecedent to the rise of the Himalaya. These rivers include the Indus, which drains western Tibet and the western Himalaya, the Yarlung Tsangpo–Brahmaputra River, which drains the northern and central Himalaya, and possibly some of the larger rivers such as the Sutlej and Arun Rivers, which rise on the Tibetan Plateau and cut right through the high topography of the Himalaya.

9. *Active deformation*: GPS rates of convergence today are ~5 mm/yr (Bilham et al., 1997). Convergence has been taken up mainly by movement along the active Main Boundary Thrust, or Main Himalayan Thrust, along which the Indian plate now underthrusts the Himalaya. Most large earthquakes occur along the Main Himalayan Thrust, which forms the active plate boundary today (Bilham et al., 1997).

10. *Climate*: The Greater Himalaya forms a major topographic and climatic divide, separating the wet, forested southern slopes of the Himalaya affected by heavy summer monsoon rains from the dry, barren northern slopes of the Himalaya and Tibetan Plateau (Fig. 8). The highest erosion and exhumation rates occur along the Greater Himalaya, where slopes are steepest and rainfall is highest. Tectonic factors created the Himalaya, and climatic factors such as monsoonal rainfall resulted from tectonic uplift.

TIBETAN PLATEAU

The Tibetan Plateau is the largest elevated plateau on Earth, at an average elevation of just over 5 km. It has low relief and low erosion rates and therefore most of the surface shows upper crustal sedimentary and volcanic rocks. Deep crustal metamorphic rocks are rarely, if ever, exposed. The Tibetan Plateau is made of a series of crustal terranes, the Kun Lun, Songpan Ganzi, Qiangtang, and Lhasa blocks that have successively been accreted to the Asian plate (Fig. 9). The Tarim and Tsaidam basins are stable, undeforming blocks of Precambrian crust, overlain by younger sediments. Suture zones are progressively younger from north to south from the Kun Lun to the Jinsha, Bangong-Nujiang, and finally the Indus–Yarlung Tsangpo sutures. The structure of

Figure 8. View of the Nepal–southern Tibet Himalaya looking south toward Everest (center of photo) from above the Tibetan Plateau. The Tibetan Plateau is dissected by rivers that cut through the Himalaya draining south into Nepal. The snow-capped peaks of the Himalaya mark the geomorphic divide between the Tibetan Plateau (foreground) and the High Himalaya (photo courtesy of National Aeronautics and Space Administration).

the Tibetan lower crust is only known from seismic reflection and refraction profiling carried out by the INDEPTH group (Fig. 10, top) (Zhao et al., 1993; Nelson et al., 1996). Seismic experiments combined with heat-flow data and magnetotelluric data (Wei et al., 2001) show that a mid-crustal layer of high conductivity probably represents a layer of partial melt with pockets of leucogranitic liquid forming today (Fig. 10, bottom). This mid-crustal zone beneath southern Tibet has been traced southward to link with the Greater Himalayan sequence metamorphic rocks and leucogranites of the High Himalaya (Searle, 1999, 2003; Searle et al., 2003).

The following key factors are characteristic of the Tibetan Plateau orogenic belt:

1. *Morphology*: The Tibetan Plateau is the largest area (3500 km × 1500 km) of high elevation (5023 m above sea level) and thick crust (65–85 km) in the world. It has low relief (<1–2 km) and is extremely flat in the internally drained interior of the plateau. The plateau lies in the rain shadow of the northward-flowing Indian monsoon system, is extremely arid, and is under permafrost for more than six months per year.

2. *Structure*: Approximately 1000 km of north-south convergence has been absorbed within the Tibetan Plateau. Around

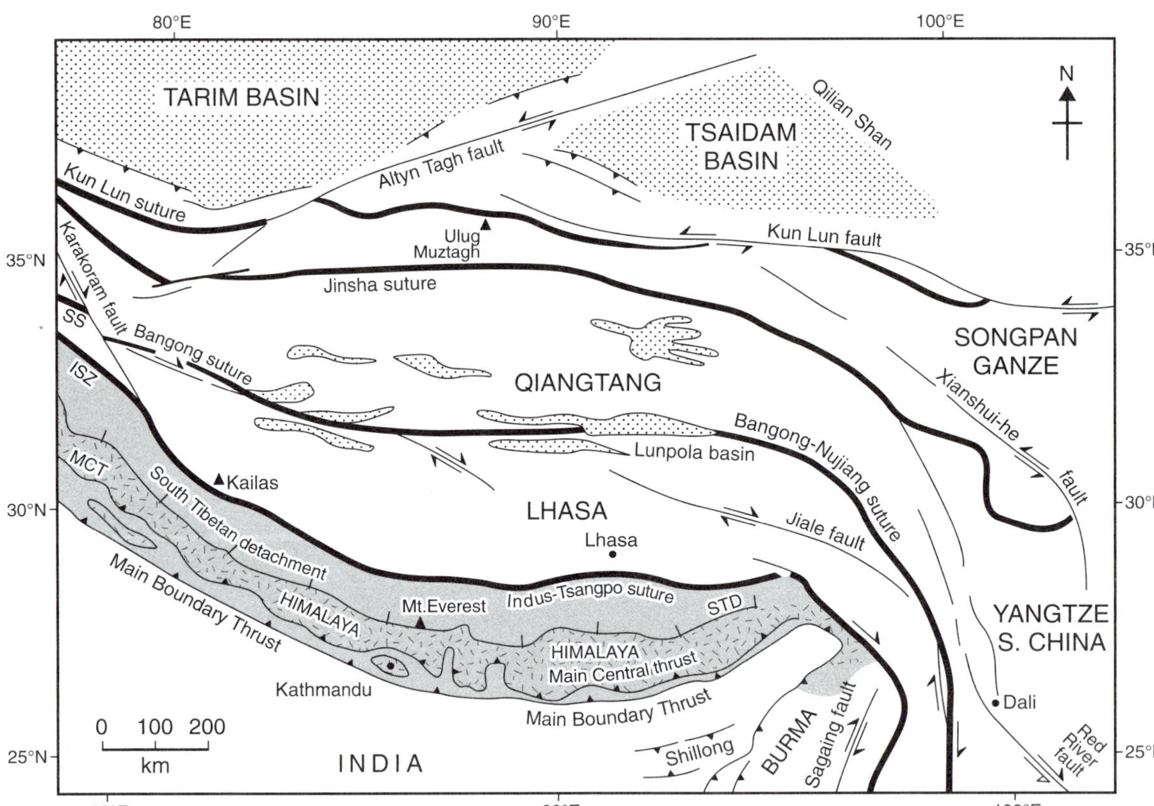

Figure 9. Map of the Tibetan Plateau showing approximate boundaries of the major terranes, suture zones, and strike-slip faults. Shaded area in the south is the Indian plate Himalaya. ISZ—Indus suture zone; MCT—Main Central Thrust; SS—Shyok suture; STD—South Tibetan Detachment.

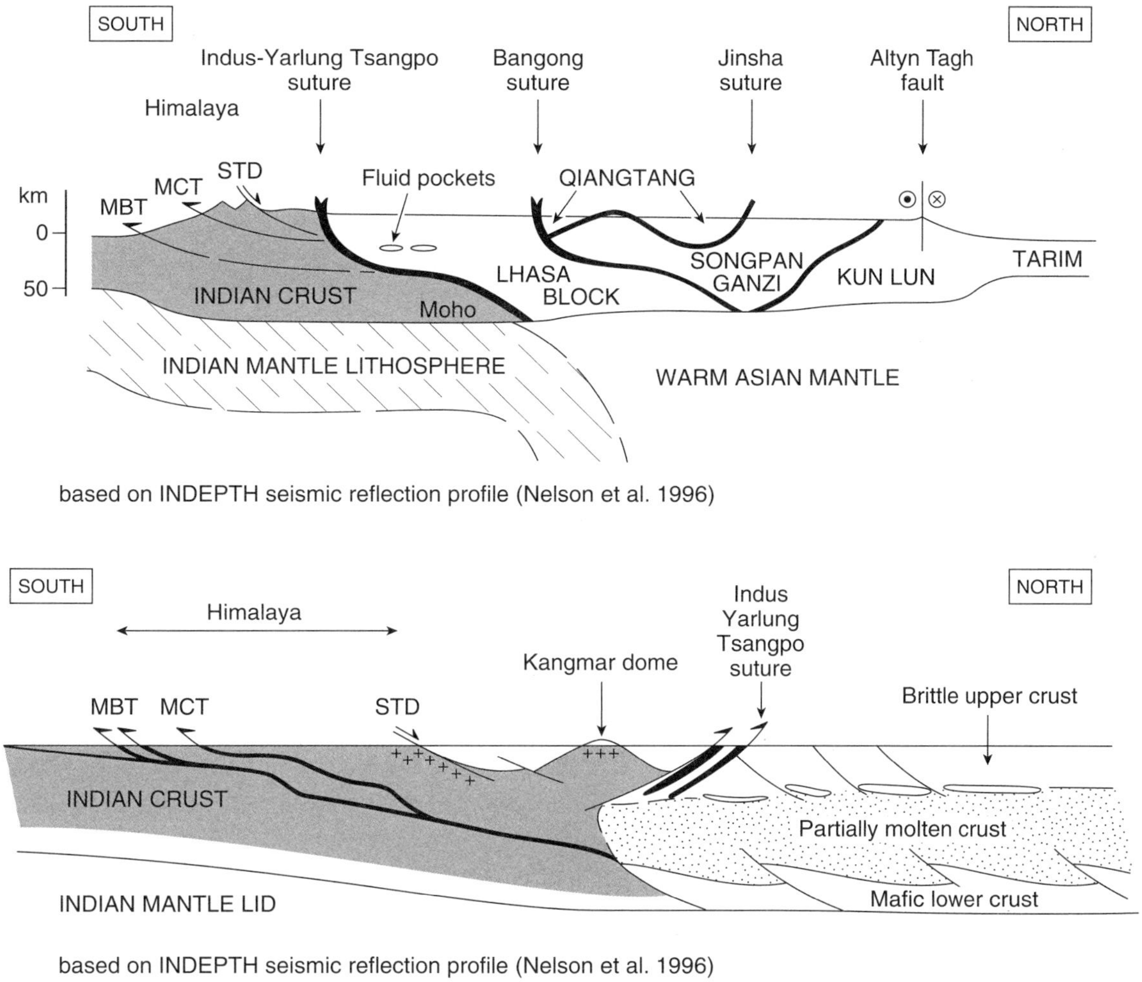

Figure 10. Crustal structure of Tibet interpreted from the INDEPTH seismic reflection and refraction studies (Nelson et al., 1996). Top diagram shows the major sutures and terranes with Indian crust shaded. Bottom diagram shows the mid-crustal layer of partially molten rocks forming today beneath southern Tibet. MBT—Main Boundary Thrust; MCT—Main Central Thrust; STD—South Tibetan Detachment.

200–300 km of underthrusting of lower Indian crust must have occurred beneath the southern margin of Tibet to balance the upper crustal shortening in the Himalaya. These rocks would be the Archean granulites of the Indian Shield. Continental subduction to 60–80 km depth beneath Tibet would impose a Tertiary HP eclogite or HT-HP granulite facies overprint on these Precambrian granulites.

Strike-slip faults bounding the high plateau suggest eastward extrusion of thickened crust out of the way of the Indian indenter (e.g., Molnar and Tapponnier, 1975; Peltzer and Tapponnier, 1988; Armijo et al., 1989). Detailed studies of the dextral Karakoram fault, however, reveal a total finite offset of only 120 km maximum since 16 Ma (Searle et al., 1998; Phillips et al., 2004), not 1000 km as suggested by Peltzer and Tapponnier (1988).

North-south-aligned graben systems show that the upper crust of southern and central Tibet is presently undergoing east-west extension. There are about eight major graben systems that extend from the northern boundary of the Greater Himalaya north as far as northern Tibet. Amounts of east-west extension are low (<5–10 km in each graben), and the throw is largest along the Thakkhola graben north of the Annapurna Range (~5 km).

3. *Seismicity*: Earthquakes in Tibet are restricted to the brittle upper crust where they show east-west extension and some strike-slip motion, and the lowermost crust at 60–80 km depth where the Indian Shield granulites still retain some rigidity (Jackson et al., 2004). Earthquakes in the central and high plateau region mainly show east-west extension, whereas earthquakes around the margins of the plateau show thrust or strike-slip faulting (Molnar and Lyon-Caen, 1989).

4. *Metamorphism*: Metamorphic rocks are rarely exposed in Tibet. The Nyenchen Tanngla Range across central Tibet exposes amphibolite facies mid-crustal rocks, uplifted along the footwall

of low-angle normal faults, similar to core complexes, but in a wholly compressional environment.

It seems likely that HT eclogite and/or HP granulite facies metamorphism is occurring today at depths of 60–80 km beneath the southern part of the Tibetan Plateau at least. Hot and dry deep crustal xenoliths have been recovered from young shoshonitic volcanics in central Tibet (Hacker et al., 2000). Temperatures of 800–900 °C may result in biotite dehydration melting reactions occurring at present.

The south Tibetan crust is rheologically layered. Temperatures up to muscovite dehydration melting (750–700 °C) occur at shallow depths in southern Tibet where "bright spots" on the INDEPTH seismic lines suggest pockets of melt forming today at depths of only 15–18 km (Nelson et al., 1996; Wei et al., 2001). The mid-crustal channel of high-grade metamorphic and migmatitic rocks separate a brittlely deforming upper crust from a rigid lower crust (Indian Shield granulites).

5. *Volcanism*: Shoshonitic or ultrapotassic volcanic rocks occur across the entire Tibetan Plateau (Figs. 11 and 12). Ages of these postcollisional volcanics span 50–30 Ma in the central plateau Qiangtang terrane, 26–10 Ma in the southern plateau Lhasa terrane, and 15–0 Ma along the KunLun terrane in the far north of Tibet (Chung et al., 2005). The spread of age data does not support the model of sudden convective removal of the lithospheric root of the plateau (England and Houseman, 1988). Rather, hot Asian mantle was progressively pushed northward as Indian lower crust and cold mantle was underthrust from the south (e.g., Owens and Zandt, 1997). Adakitic volcanics indicate an eclogitic lower crust source and rare xenoliths of both eclogite and high-pressure granulite lower crust are occasionally enclosed in the ultrapotassic dikes (Hacker et al., 2000; Ducea et al., 2003).

6. *Geomorphology*: The Tibetan Plateau shows a central area of internal drainage (north-central Tibet) that corresponds to the flattest and driest regions. Southern Tibet is externally drained via the Indus River to the west, the Sutlej River to the southwest, and the Yarlung Tsangpo–Brahmaputra River to the east. Southern Tibet has 2–3 km of vertical relief, although the average elevation is similar to the rest of the plateau.

7. *Tectonics*: The Tibetan Plateau has suffered ~50% crustal shortening and doubling of crustal thickness. Some crustal thickening of southern Tibet (Lhasa block) may have occurred before the India-Asia collision. The southern margin of the Lhasa block is characterized by a >2000 km long calc-alkaline batholith (Gangdese-Ladakh-Kohistan granites) and associated extrusive volcanic rocks. It is estimated that up to 30%–40% of the crustal shortening may have occurred before the India-Asia collision in an Andean-type tectonic setting, possibly analogous to the Bolivian Altiplano today (England and Searle, 1986).

Several end-member models have been proposed to explain the crustal shortening, double-thickness crust, and high topography of Tibet (Fig. 13). Model A is the wholesale underthrusting of the plateau by the Indian plate as originally proposed by Argand (1924). Model B is the homogeneous shortening model of Dewey and Burke (1973). Model C (not illustrated here) is the "continental escape" model whereby Tibet is moving east, out of the way of the Indian plate indenter, along giant strike-slip faults (Karakoram, Altyn Tagh and Kun Lun faults), as proposed by Molnar and Tapponnier (1975), Peltzer and Tapponnier (1988), and Armijo et al. (1989). Eastward motion of Tibet by lateral extrusion (Peltzer and Tapponnier, 1988) fails to account for thick crust or high topography of the plateau. Model D is the subduction of Asian lithosphere southward beneath Tibet as proposed by Willett and Beaumont (1994). Southward underthrusting of Tarim lithosphere is also not supported by the lack of significant crustal shortening in the Kun Lun (compared to the Himalaya, for example). Model E is the injection of Indian lower crust into Tibetan crust as proposed by Zhao and Morgan (1985).

KARAKORAM OROGEN

The Karakoram Range along the borders of Pakistan and India with Tajikistan and SW China is the westward geological extension of the Qiangtang block of south-central Tibet. The Karakoram Ranges has the most extreme relief on Earth with many of the world's 7 and 8 km high mountains, deeply eroded gorges, and extensive continental glaciers. Whereas the Tibetan Plateau exposes mainly sedimentary and volcanic rocks and rarely lower crustal metamorphic rocks, the Karakoram Ranges expose mainly lower crustal high-grade gneisses and granitoid batholiths (Fig. 14) (Searle, 1991). Structures and metamorphic-magmatic processes seen in the Karakoram Range probably correlate to similar processes occurring at depth beneath the south and central part of the Tibetan Plateau. Exhumation-cooling rates, as well as erosion rates, are extremely rapid in the southern and central Karakoram Range. Surprisingly, seismicity is limited, especially compared to the Hindu Kush and Pamir Ranges to the west (Pegler and Das, 1998; Searle et al., 2003). This may be because the geothermal gradient is high and most of the crust is deforming ductilely.

The following main points are characteristic of the Karakoram orogen:

1. *Morphology*: Whereas the Tibetan Plateau is high and flat (average 5023 m elevation with <1 km of relief), the Karakoram Ranges have extremely high mountains, including four of the highest 8000 m peaks (K2, Broad Peak, Hidden Peak, Gasherbrum III), hundreds of peaks above 6 km high, and deeply incised valleys. The average elevation is slightly higher than the Tibetan Plateau, and the crustal thickness reaches 90 km beneath the central Karakoram Range (Rai et al., 2006).

2. *Seismicity*: Earthquakes are extremely common along the Hindu Kush seismic zone, the deepest continental seismic zone known (Pegler and Das, 1998; Searle et al., 2001). These earthquakes reflect subduction of the northern, thinned continental crust of the Indian plate beneath the Tadjik basin. The south-dipping Pamir seismic zone merges with the north-dipping Hindu Kush seismic zone at depth. In comparison, the Karakoram Range to the east shows little seismicity, probably reflecting higher heat flow and hotter, more ductilely deforming crust.

3. *Structure*: The structure of the Karakoram Ranges consist of three main zones: a sedimentary upper crust along the northern ranges (Shaksgam–Aghil–Teram Kangri Ranges), an axial batholith along the middle ranges (Baltoro and Muztagh Ranges), and an uplifted deep crustal high-grade metamorphic terrane along the southern Karakoram Range (Searle, 1991; Searle and Tirrul, 1991). High-grade metamorphic "core complexes" form granite-gneiss domes along the southern Karakoram Range (e.g., Dassu dome; Searle, 1991). These have extensional fabrics around the tops but have formed in a wholly compressional tectonic environment.

4. *Metamorphism*: Multiple episodes of high-grade regional Barrovian metamorphism occur along the southern Karakoram Range, spanning ~65 m.y. (Fraser et al., 2001). U-Pb monazite ages from peak sillimanite grade gneisses are both pre–India-Asia collision (65 Ma) and postcollisional (peaking at 45, 30, 25, and 4 Ma; Fraser et al., 2001). Regional staurolite grade metamorphism occurred at 14 Ma in higher structural levels of the Hunza Valley (Fraser et al., 2001). The youngest metamorphic phase is a HT sillimanite grade partial melting event in the core of some metamorphic core complexes along the southern Karakoram Range (Dassu gneiss, 5.4 Ma; Fraser et al., 2001), a metamorphism that is probably still ongoing today at depth.

5. *Magmatism*: The Karakoram batholith includes (1) a series of precollisional diorite-granodiorite-tonalite granitoids subsequently deformed and metamorphosed to amphibolite facies orthogneisses (e.g., K2 gneiss, Muztagh Tower gneiss, Hushe gneiss; Searle et al., 1989, 1990), (2) a massive precollisional I-type biotite and hornblende diorite, granodiorite, and granite batholith, dated as middle Cretaceous (Fraser et al., 2001), and (3) a postcollision crustal-melt, S-type pluton, the Baltoro plutonic unit, comprising biotite monzogranite to garnet two-mica leucogranite, dated at 20 Ma (Searle et al., 1989, 1992; Parrish and Tirrul, 1989).

6. *Volcanism*: There is no known Tertiary or active volcanism in the Karakoram Ranges. A few intrusive dikes of mantle-derived

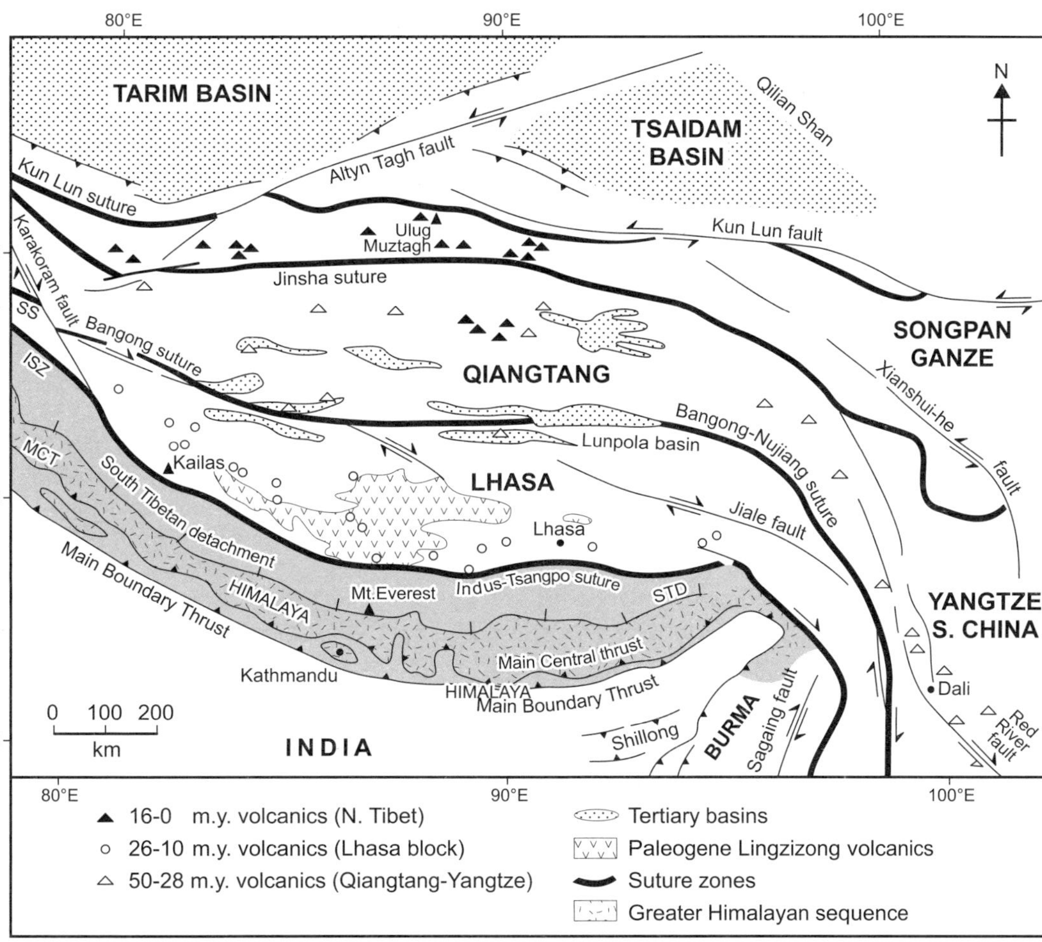

Figure 11. Map of the Tibetan Plateau showing the distribution of volcanic rocks, together with the range of ages, after Chung et al. (2005). The youngest ultrapotassic volcanics occur along the Kun Lun Range of northern Tibet. The Lhasa block of southern Tibet shows ages older than ca. 10 Ma, reflecting the subduction of cold Indian lithosphere beneath the Lhasa block by this time. ISZ—Indus suture zone; MCT—Main Central Thrust; SS—Shyok suture; STD—South Tibetan Detachment.

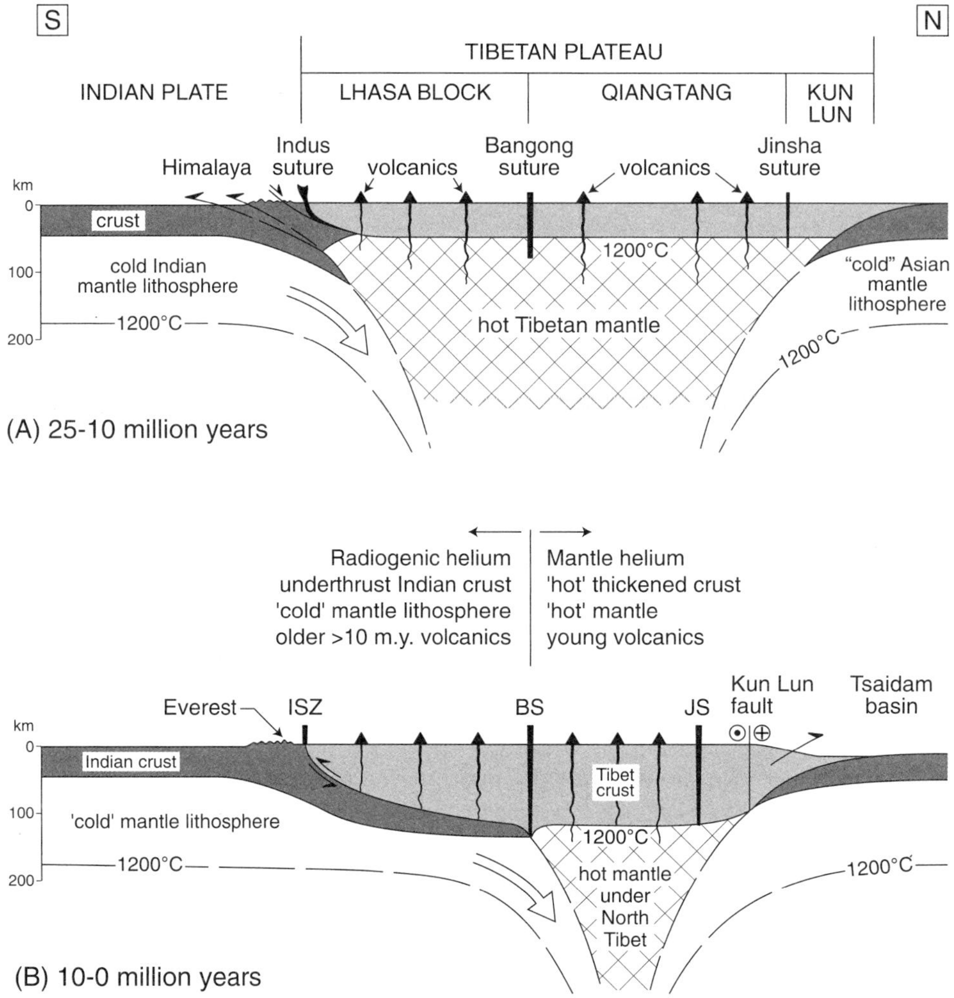

Figure 12. Two sections across the Tibetan Plateau showing how the age distribution of shoshonitic and ultrapotassic volcanic rocks of Tibet could be explained. Hot, Asian mantle is required to produce these shoshonitic volcanics. During the last 10 m.y., underthrusting beneath southern Tibet of cold Indian plate crust and cold mantle lithosphere "shut off" the source for the volcanics; hence all the ultrapotassic volcanics in southern Tibet are older than 10 Ma.

lamprophyres (biotite-bearing vogesites and minettes) intrude the Hushe and northern Karakoram regions. Their Ar-Ar ages are 24–21 Ma, similar to the ages from the Baltoro crustal-melt leucogranites, indicating that limited upper mantle melting occurred simultaneously with extensive lower crustal melting during the early Miocene (Searle et al., 1989, 1992).

7. *Geomorphology*: The Karakoram Range shows some of the highest exhumation, erosion, and uplift rates known, up to 1 cm/yr. The greatest uplift is centered along two main belts: (1) the Miocene Baltoro granite batholith in the Latok, Muztagh, Trango, and Masherbrum Ranges, and (2) the transpressionally uplifted K2–Gasherbrum–Teram Kangri Ranges, parallel to the Karakoram fault. Many glaciers and river valleys cut across the high Karakoram Ranges, including the Hunza Valley, Baltoro Glacier, and Braldu River, and must have initiated prior to the major uplift during the late Tertiary and Quaternary.

8. *Climate*: The Karakoram Ranges are generally arid, lying to the north and in the rain shadow of the Indian monsoon. The high exhumation, erosion, and uplift rates do not appear to be related directly to climate. The highest exhumation rates recorded anywhere are in the Nanga Parbat massif (~1 cm/yr; Zeitler et al., 1993, 2001; Bowring et al., 2003), and this peak lies in a semi-arid region of northern Pakistan, implying that rapid topographic uplift, coupled with rapid exhumation and cooling of rocks, has no link with regions of high precipitation, either today or in the recent past. Uplift must be a result of tectonic factors that caused crustal shortening and thickening, and climate is a secondary factor resulting from topographic uplift.

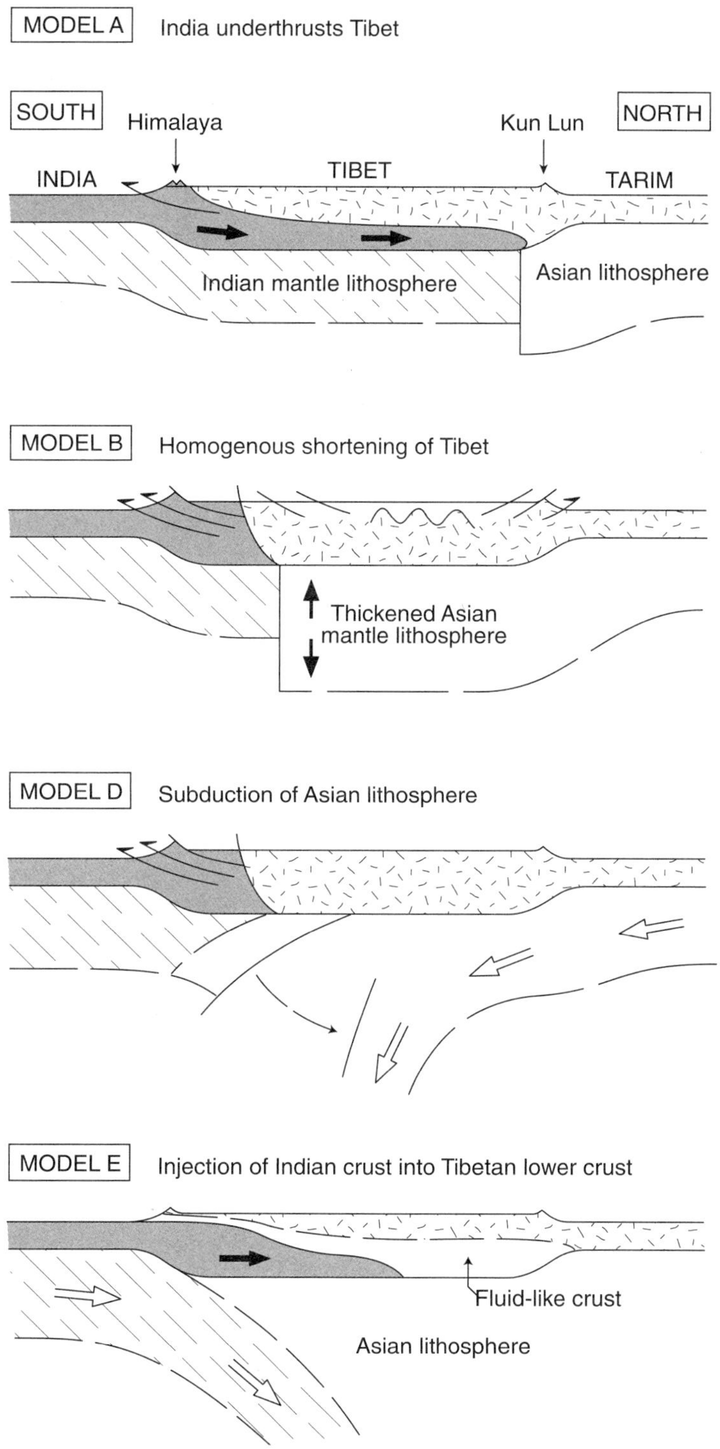

Figure 13. Four models for the structure of Tibet described in the text. Model A is the model of wholescale underthrusting of Tibet by India, after Argand (1924). Model B is the homogeneous thickened crust of Tibet, after Dewey and Burke (1973). Model C (not shown here) is the lateral extrusion of Tibetan crust bounded by the major strike-slip fault systems of the Karakoram, Altyn Tagh, and Kun Lun faults. Model D is the southward underthrusting of Asian plate (Tarim) lithosphere beneath the Tibetan Plateau, after Willett and Beaumont (1994). Model E is the injection of lower Indian crust into southern Tibet, after Zhao and Morgan (1985).

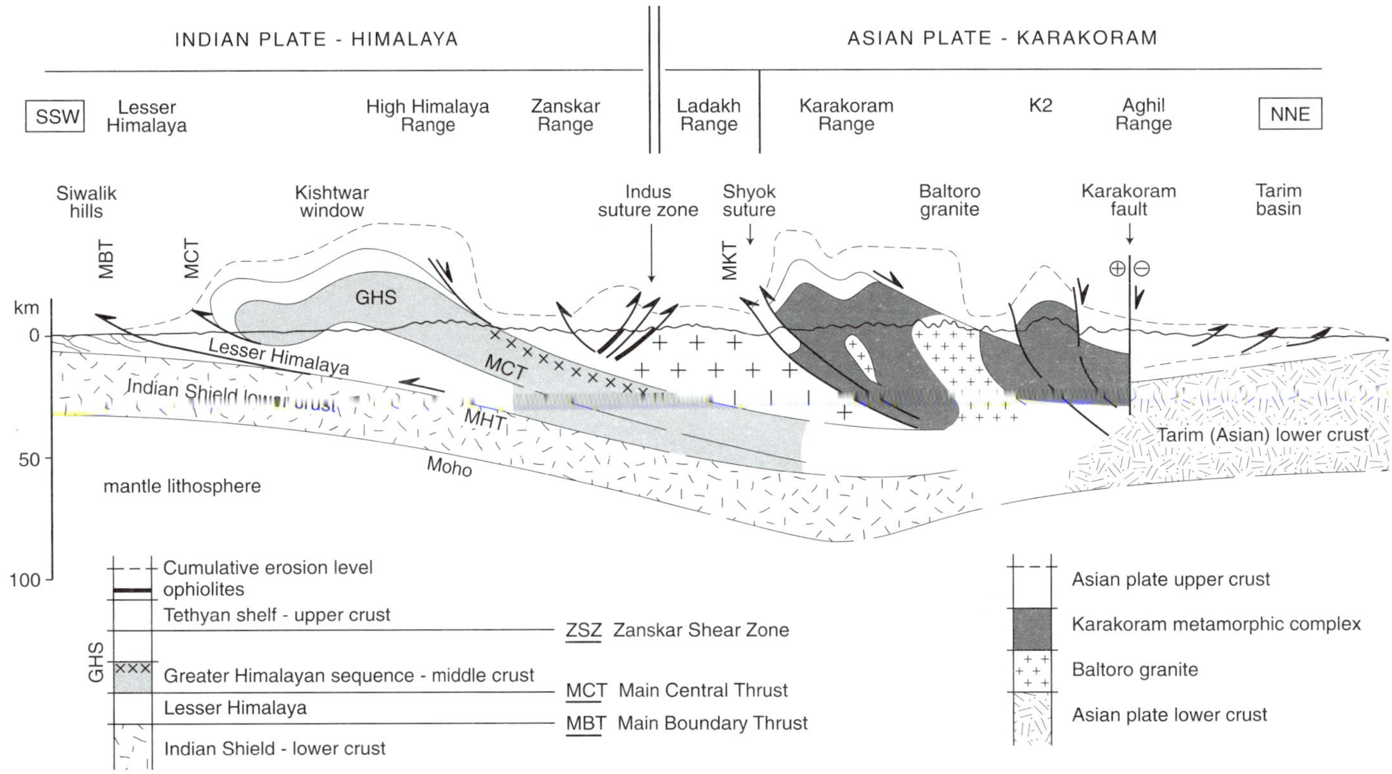

Figure 14. Cross section of the western Himalaya (Kishtwar-Zanskar region, NW India) and Karakoram Ranges (northern Pakistan). Dark shading represents the high-grade metamorphic rocks and crustal-melt granites (Baltoro plutonic unit) of the Asian plate. Lighter shading represents the mid-crustal Greater Himalayan sequence of the Indian plate. Precambrian basement of the Indian plate and Tarim plate is shown in dashed patterns. Crustal thickness beneath the Karakoram Range is ~75–90 km. The Archean granulites and Proterozoic cover of the basement has probably been metamorphosed to eclogite facies or HP granulite facies during the Tertiary. MHT—Main Himalayan Thrust; MKT—Main Karakoram thrust.

EVOLUTION OF THE OMAN-ZAGROS-HIMALAYA-TIBET OROGENIC BELT

This paper has laid out the major geological characteristics of Oman-, Zagros-, Himalayan-, Tibetan-, and Karakoram-type orogenic belts from the type examples. These factors can be used in comparative studies of ancient orogenic belts to determine whether those orogens have common characteristics with the presently active India-Asia collision belt. We can also use the geology of these mountain ranges to infer an evolutionary sequence of orogenesis. A classic Tethyan continental collision orogeny can be simplified into four stages:

1. Ophiolite Obduction Stage (e.g., Oman)

The process of ophiolite obduction onto a passive continental margin has to occur prior to continent-continent collision for obvious reasons. In Oman the entire orogeny, from initial ophiolite crust formation to its final emplacement onto the Arabian continental margin, lasted ~21 m.y. (from 95 to 74 Ma). Initial formation of the ophiolite crustal sequence above a subduction zone dipping away from the passive margin (Fig. 2) was practically synchronous with subduction of oceanic basalts and sediments to 10–12 kbar (37–42 km depth) forming the HT amphibolite metamorphic sole beneath the ophiolite mantle sequence (Searle and Cox, 1999, 2002). Thrusting propagated inboard with time and evolved from deep subduction-related thrusting to a shallow, thin-skinned, foreland fold-and-thrust belt. An ~400–500 km width of oceanic crust to passive continental margin was shortened during the emplacement of the ophiolite and the underlying distal to proximal oceanic sediments. During the later stages of orogeny, the leading edge of the continental margin was subducted to depths of >80 km (20–22 kbar) to form eclogite facies metamorphic rocks in the deeper structural levels (Searle et al., 2004). The entire process occurred beneath sea level and prior to continental collision. The ophiolites preserved along the north Indian plate margin in the Zanskar Himalaya probably had an origin similar to the Oman ophiolite. Some 20 m.y. of early Oman-type orogeny could therefore have occurred along the north Indian margin in the Himalaya, before the collision of India with Asia during the early Eocene.

2. Early Continent-Continent Collision Stage (e.g., Zagros, Iran)

The Zagros Mountains of SW Iran represent an early stage of the collision of two continental plates. The suture zone bounding the Arabian plate in the Zagros Range and the Sanandaj-Sirjan zone of the central Iran microplate shows a few remnant ophiolites (e.g., Neyriz, Kermanshah), but they have been disrupted by faulting and nowhere preserve a complete sequence of oceanic crust and upper mantle as seen in Oman. The Neyriz-Kermanshah ophiolites may have had a spatial and temporal evolution similar to the Oman ophiolite, or may have simply been "suture zone ophiolite" trapped during the collision process, but most of this history has been destroyed by the subsequent collision of the two continental plates, Arabia and central Iran. The Zagros simply folded belt shows a limited amount of crustal shortening on the order of 50–70 km (Blanc et al., 2003). Phanerozoic upper crustal, passive margin sediments of the Arabian plate have been folded by relatively simple periclinal folding above a horizontal detachment lying along the Neoproterozoic Hormuz salt horizon. In places the salt has risen diapirically more than 10–12 km to the surface, forming the famous Zagros salt domes. A similar amount of shortening in the Arabian plate lower crust can only be balanced by underthrusting the Precambrian lower crust northeast beneath central Iran. The Turkish-Iranian Plateau may have an origin similar to the early stages of the Tibetan Plateau, although there are major differences between the two today.

3. Mature Continent-Continent Collision Stage (e.g., Himalaya)

The Himalaya has undergone at least 50 m.y. of postcollision deformation since the last marine sediments were deposited within the Tethyan suture zone. This Himalayan deformation virtually wiped out the earlier Oman-type ophiolite obduction orogenic effects, which are only locally preserved along the base of the Spontang and Purang ophiolites in Ladakh and southern Tibet. In the Himalaya, the earlier Zagros-type simple folding and thrusting has also been completely superimposed by the later, main Himalayan deformation and metamorphism. If erosion levels were taken down only 1 km along the northern Himalaya, all trace of ophiolites obducted onto the passive margin of the Indian plate would be wiped out. If entire orogenies, such as an Oman-type orogeny, can be wiped out by subsequent deformation and erosion along the Himalaya, the most active and ongoing collision known today, then what chance do we have of deciphering the early history of orogenic belts in the Paleozoic or Precambrian? The Himalaya is characterized by hundreds of kilometers of crustal shortening, regional kyanite-sillimanite grade metamorphism, crustal melting, extreme erosion, high and steep topography, thick crust, and active seismicity.

4. Long-Lived Continent-Continent Collision Orogeny (e.g., Tibet, Karakoram)

The Tibetan Plateau and Karakoram Ranges are characteristic of a long-lived orogeny involving multiple collisional events of island arcs or continental plates, a large area of very thick crust, regional long-lived, deep-level metamorphism with up to 60 m.y. of HT sillimanite grade metamorphic "events" (or one long-lasting event), multiple granitoid intrusions, high topography, and high heat flow. At the deepest structural levels, eclogite or HP granulite metamorphism may be occurring on a large scale now. Synchronous with large-scale lower crustal melting, upper mantle melting, indicated by lamprophyric dikes in western Tibet and the northern Karakoram Range, may add a significant input of heat from the mantle into the lower crust. This extra heat might explain why the Baltoro granites are so much larger than the Himalayan crustal-melt leucogranites.

SUMMARY: HIMALAYA-KARAKORAM-TIBET TECTONIC EVOLUTION

The crustal evolution of the Indian plate along the Himalaya involves at least 500 km of upper crustal shortening by folding, thrusting, and crustal thickening. Three main stages of metamorphism include (1) early, deep subduction of thinned continental crust to UHP eclogite depths (~680 °C; 28 kbar; 100 km depth; 49–46 Ma), (2) regional kyanite grade metamorphism (550–680 °C; 10–12 kbar; 36–45 km depth; 35–32 Ma), and (3) widespread, regional sillimanite ± cordierite grade metamorphism (650–770 °C; 3.7–4.5 kbar; 14–18 km depth; 30–16.5 Ma) associated with partial melting and leucogranite formation (U-Th-Pb monazite crystallization ages ranging from ca. 25 to 16 Ma). The very shallow depths of HT metamorphism and melting are consistent with the model of middle Miocene ductile flow of an Indian plate mid-crustal channel, southward from beneath southern Tibet to the Greater Himalaya (Law et al., 2006). This zone of channel flow is bounded by crustal-scale shear zones, the Main Central Thrust, with its zone of inverted metamorphic isograds along the base, and the South Tibetan Detachment system of low-angle normal faults along the top. Restoration of the Greater Himalayan sequence reveals that the protoliths are mainly Proterozoic black shales and gneisses, ca. 500–450 Ma orthogneisses, and Paleozoic–early Mesozoic sediments and Permian volcanics of the northern margin of the Indian plate. A final phase (4) of extremely young sillimanite + cordierite + K-feldspar grade metamorphism, migmatization, and melting occurs in the core of the Nanga Parbat syntaxis of Pakistan, where U-Pb monazite and xenotime ages are as young as 1.1 and 0.77 ± 0.07 Ma (Bowring et al., 2004; Crowley et al., 2005), indicating that 10–15 km of material has been eroded off the summit region of Nanga Parbat in less than one million years (>1 cm/yr erosion rates). This high-grade metamorphism occurs in the middle to upper crust; the lower crust appears to be old (Archean-Proterozoic) dry granulites of the underthrusting Indian Shield.

The Asian side of the collision zone includes the vast Tibetan Plateau north of the Indus–Yarlung Tsangpo suture zone and its extensions to the west, the Karakoram Ranges and the Hindu Raj–Hindu Kush and Pamir Ranges. The geology of the Tibetan Plateau (5 ± 1.5 km altitude, low relief) contains mainly sedimentary and volcanic rocks, rarely deep crustal metamorphic rocks, whereas the Karakoram Range (8.8–2 km altitude, high relief) exposes extensive lower crustal, high-grade metamorphic rocks and both pre- and postcollisional granites. U-Th-Pb dating of accessory minerals from all phases of the Karakoram and Hindu Kush granites has revealed a complex thermal history. Precollisional granites (K2 gneiss, Muztagh Tower gneiss, Hushe gneiss, etc.) are Andean-type hornblende + biotite granite-granodiorite-diorite plutons related to subduction of Tethyan oceanic crust northwards beneath the southern margin of Asia. Postcollisional granites include the huge Baltoro biotite monzogranite to garnet + biotite + muscovite leucogranites, formed at 25–17 Ma by crustal melting of lower crustal gneisses. Lamprophyre dikes, intruded simultaneously with the Baltoro granites, indicate synchronous upper mantle and lower crustal melting during the middle Miocene. Staurolite, kyanite, and sillimanite grade metamorphism is sporadic and semicontinuous from 70 to 0 Ma along the southern Karakoram Range, following collision of the Kohistan arc during the Late Cretaceous and collision of the Indian plate during and after Earliest Eocene. The youngest metamorphism is Pliocene (5–4 Ma) and occurs in the cores of sillimanite grade gneiss domes in the southern Karakoram Range, HT metamorphism that is probably still ongoing today at depth.

CONCLUSIONS

It is hoped that the above criteria used to distinguish the present-day Himalayan, Tibetan, and Karakoram orogenies may be of some use to workers in older and less well-preserved orogenic belts worldwide. It may be appropriate that some old orogenic belts could be described as remnants of a Himalayan- or Tibetan-type orogenic belt. Possible examples could be the Scottish Caledonides (Strachan et al., 2002; Oliver, 2002), the Trans-Hudson belt in northern Canada (St.Onge et al., 2006) and the Limpopo belt of South Africa–Zimbabwe (Shackleton, 1986; de Wit, 1998). These orogens may have had up to 20–30 km of material eroded from the upper structural levels, so the chances of being able to ascertain exactly what the upper crustal levels looked like are, at best, slim. However, if the Tibetan Plateau or Karakoram–Hindu Kush Ranges were eroded down to sea level today, they would probably reveal extensive areas of granulite or amphibolite grade metamorphic "basement" with possible older precollisional "I-type" granitoids and postcollisional "S-type" crustal-melt granites. Himalayan-type leucogranitic crustal melts are extremely rare in Mesozoic and Tertiary mountain belts such as the Alps or the Rockies, but are common in many Precambrian terranes.

Similarly, it is useful to do a reverse correlation. In Tibet today the lower crust is never exposed and the middle crust exposed only very rarely. We could use older, more deeply eroded mountain ranges to gain some idea of the composition and structure of the lower crust of Tibet. Is it possible, for example, that the lower Tibetan crust might look similar to the Archean crust of Greenland or the Proterozoic reworked crust of the Western gneiss region, Norway? The Himalaya has been tilted northwards as India flexes down beneath the Tibetan Plateau, so we do have an idea of the geology of the upper (Tethyan or northern Himalaya) and middle (Greater Himalaya) crust. There is, however, no lower crust exposed anywhere in the Himalaya, so we have to infer the composition and structure of the Himalayan lower crust. The principle of uniformitarianism can be useful to interpret older orogenies, but we can also work it the other way around for the deeper, unexposed parts of the Himalaya-Karakoram-Tibetan orogenic belts. Maybe the basement of the Limpopo belt or the Trans-Hudson orogen can help us to infer composition and structure of the Tibetan lower crust?

Older orogens that show close linkage between UHP eclogite facies rocks and lower crustal UHT granulite facies rocks, such as the Western Gneiss Region of Norway or the Dabie Shan–Sulu belt of China, may also have an analogy with the Tibetan- or Karakoram-type orogeny. Both felsic and mafic granulites and mafic eclogite xenoliths from the lower crust have been found in young ultrapotassic volcanic rocks in Tibet and the Pamir Range (Hacker et al., 2000; Ducea et al., 2003). Successive episodes of metamorphism, dehydration, and melting (muscovite, biotite, and amphibole dehydration) would have created a granulitic residue in the lower crust. The lower crustal "basement" beneath southern Tibet today could well be Archean granulites of the Indian Shield, metamorphosed at ultrahigh temperatures of 800–1000 °C during the middle or late Tertiary Tibetan crustal thickening. The enigmatic potassic volcanism of the Tibetan Plateau may be a consequence of these dehydration melting events.

Recycling of basement rocks is known to occur at all scales. Perhaps the most amazing examples are the Quaternary HT rocks of the Nanga Parbat massif in Pakistan. Himalayan "basement" rocks of Nanga Parbat have a history that includes Archean granulite facies metamorphism, middle Tertiary Himalayan amphibolite facies overprinting, with partial melting producing garnet and tourmaline two-mica leucogranites, preceding the final Pliocene-Quaternary HT sillimanite + cordierite + K-feldspar metamorphism resulting in cordierite-bearing leucogranites less than one million years old (Zeitler et al., 1993, 2001; Bowring et al., 2004; Crowley et al., 2005).

ACKNOWLEDGMENTS

This work was funded by Natural Environment Research Council (NERC) grant NER/K/S/2000/951. I am grateful to Rick Law, Dave Waters, Randall Parrish, and Steve Noble for many years of excellent collaboration and discussion, Maarten deWit for discussions on the Limpopo belt of southern Africa, and Marc St.Onge for discussions about the Trans-Hudson orogen. The manuscript was significantly improved by comments from two unnamed referees and especially from Bob Hatcher, who is also thanked for organizing a thought-provoking conference in Oak Ridge, Tennessee.

REFERENCES CITED

Allen, M., Jackson, J., and Walker, R., 2004, Late Cenozoic reorganization of the Arabia-Eurasia collision and the comparison of short-term and long-term deformation rates: Tectonics, v. 23, p. TC2008, doi: 10.1029/2003TC001530.

Argand, E., 1924, La tectonique de l'Asie: Proceedings of the 13th International Geological Congress, v. 7, p. 171–372.

Armijo, R., Tapponnier, P., and Han, T., 1989, Late Cenozoic right-lateral strike-slip faulting in southern Tibet: Journal of Geophysical Research, v. 94, p. 2787–2838.

Beaumont, C., Jamieson, R.A., Nguyen, M.H., and Lee, B., 2001, Himalayan tectonics explained by extrusion of a low-viscosity crustal channel coupled to focused surface denudation: Nature, v. 414, p. 738–742, doi: 10.1038/414738a.

Besse, J., and Coutillot, V., 1988, Palaeogeographic maps of the continents bordering the Indian Ocean since the Early Jurassic: Journal of Geophysical Research, v. 93, p. 11,791–11,808.

Bilham, R., Larson, K., Freymueller, J., and Project Idylhim members, 1997, GPS measurements of present-day convergence across the Nepal Himalaya: Nature, v. 386, p. 61–64, doi: 10.1038/386061a0.

Blanc, E., Allen, M.B., Inger, S., and Hassani, H., 2003, Structural styles in the Zagros simple folded zone, Iran: Geological Society [London] Journal, v. 160, p. 401–412.

Boudier, F., Ceuleneer, G., and Nicolas, A., 1988, Shear zones, thrusts, and related magmatism in the Oman ophiolite: Initiation of thrusting on an oceanic ridge: Tectonophysics, v. 151, p. 275–296, doi: 10.1016/0040-1951(88)90249-1.

Bowring, S.A., Searle, M.P., Waters, D.J., Hodges, K.V., Schmitz, M., and Crowley, J., 2004, ID-TIMS geochronology of ca. 1 Ma leucogranites from the core of Nanga Parbat: American Geophysical Union fall meeting, San Francisco, 2003, Abstract V22E-08.

Chung, S.-L., Chu, M.-F., Zhang, Y., Xie, Y., Lo, C.-H., Lee, T.-Y., Lan, C.-Y., Li, X., Zhang, Q., and Wang, Y., 2005, Tibetan tectonic evolution inferred from spatial and temporal variations in post-collisional magmatism: Earth and Planetary Science Letters, v. 68, p. 173–196.

Colman-Sadd, S.P., 1978, Fold development in Zagros simply folded belt, southwest Iran: American Association of Petroleum Geologists Bulletin, v. 62, p. 984–1003.

Corfield, R.I., and Searle, M.P., 2000, Crustal shortening estimates across the north Indian continental margin, Ladakh, NW Himalaya, *in* Khan, M.A., Treloar, P.J., Searle, M.P., and Jan, M.Q., eds., Tectonics of the Nanga Parbat syntaxis and the western Himalaya: Geological Society [London] Special Publication 170, p. 395–410.

Corfield, R.I., Searle, M.P., and Pedersen, R.B., 2001, Tectonic setting, origin and obduction history of the Spontang ophiolite, Ladakh Himalaya, NW India: Journal of Geology, v. 109, p. 715–736, doi: 10.1086/323191.

Crowley, J.L., Bowring, S.A., and Searle, M.P., 2005, U-Th-Pb systematics of monazite, xenotime, and zircon from Pleistocene leucogranites at Nanga Parbat (Pakistan Himalaya): Geochimica Cosmochimica Acta, v. 69, Abstract A8

Curray, J.R., 2005, Tectonics and history of the Andaman Sea region: Journal of Asian Earth Sciences, v. 25, p. 187 232, doi: 10.1016/j.jseaes.2004.09.001.

DeCelles, P.G., Robinson, D.M., Quade, J., Ojha, T.P., Garzione, C.N., Copeland, P., and Upreti, B.N., 2001, Stratigraphy, structure and tectonic evolution of the Himalayan fold-thrust belt in western Nepal: Tectonics, v. 20, p. 487 509, doi: 10.1029/2000TC001226.

Dewey, J.F., and Burke, K.C.A., 1973, Tibetan, Variscan and Precambrian basement reactivation: Products of continental collision: Journal of Geology, v. 81, p. 683–692.

Dewey, J.F., Cande, S.C., and Pitman, W.C., III, 1989, Tectonic evolution and the India/Eurasia collision zone: Eclogae Geologicae Helvetiae, v. 82, p. 717–734.

de Wit, M.J., 1998, On Archean granites, greenstones, cratons and tectonic: Does the evidence demand a verdict?: Precambrian Research, v. 91, p. 181–226, doi: 10.1016/S0301-9268(98)00043-6.

de Wit, M., Jeffery, M., Bergh, H., and Nicolaysen, L., 1988, Geological map of sectors of Gondwana reconstructed to their deposition ca. 150 Ma: American Association of Petroleum Geologists and University of Witswatersrand, scale 1:10,000,000.

Ducea, M.N., Lutkov, V., Minaev, V.T., Hacker, B., Ratschbacher, L., Luffi, P., Schwab, M., Gehrels, G.E., McWilliams, M., Vervoort, J., and Metcalf, J., 2003, Building the Pamirs: The view from the underside: Geology, v. 31, p. 849–852, doi: 10.1130/G19707.1.

Engdahl, E.R., van der Hilst, R., and Buland, R., 1998, Global teleseismic earthquake relocation with improved travel times and procedures for depth determination: Bulletin of the Seismological Society of America, v. 88, p. 722–743.

England, P., and Houseman, G., 1988, The mechanics of the Tibetan Plateau: Royal Society of London Philosophical Transactions, ser. A, v. 326, p. 301 319.

England, P.C., and Searle, M.P., 1986, The Cretaceous-Tertiary deformation of the Lhasa block, and its implication for the crustal thickening of Tibet: Tectonics, v. 5, p. 1–14.

Fraser, J.E., Searle, M.P., Parrish, R.R., and Noble, S.R., 2001, Chronology of deformation, metamorphism and magmatism in the southern Karakoram Mountains: Geological Society of America Bulletin, v. 113, p. 1443–1455, doi: 10.1130/0016-7606(2001)113<1443:CODMAM>2.0.CO;2.

Gaetani, M., and Garzanti, E., 1991, Muticyclic history of the north India continental margin (northwestern Himalaya): American Association of Petroleum Geologists Bulletin, v. 75, p. 1427–1446.

Gansser, A., 1964, Geology of the Himalaya: Chichester, UK, John Wiley and Sons, 289 p.

Grujic, D., Hollister, L., and Parrish, R.R., 2002, Himalayan metamorphic sequence as an orogenic channel: Insight from Bhutan: Earth and Planetary Science Letters, v. 198, p. 177–191, doi: 10.1016/S0012-821X(02)00482-X.

Hacker, B.R., Gnos, E., Ratschbacher, L., Grove, M., McWilliams, M., Sobolev, S.V., Wan, J., and Zhenhan, W., 2000, Hot and dry deep crustal xenoliths from Tibet: Science, v. 287, p. 2463–2466, doi: 10.1126/science.287.5462.2463.

Harris, N., and Massey, J., 1994, Decompression and anatexis of Himalayan metapelites: Tectonics, v. 13, p. 1537–1546, doi: 10.1029/94TC01611.

Hodges, K.V., 2000, Tectonics of the Himalaya and southern Tibet from two perspectives: Geological Society of America Bulletin, v. 112, p. 324–350, doi: 10.1130/0016-7606(2000)112<0324:TOTHAS>2.3.CO;2.

Hodges, K.V., Parrish, R.R., Housch, T.B., Lux, D.R., Burchfiel, B.C., Royden, L.H., and Chen, Z., 1992, Simultaneous Miocene extension and shortening in the Himalayan orogen: Science, v. 258, p. 1466–1470, doi: 10.1126/science.258.5087.1466.

Hodges, K.V., Burchfiel, B.C., Royden, L.H., Chen, Z., and Liu, Y., 1993, The metamorphic signature of contemporaneous extension and shortening in the central Himalayan orogen: Data from the Nyalam transect, southern Tibet: Journal of Metamorphic Geology, v. 11, p. 721–737.

Hodges, K.V., Parrish, R.R., and Searle, M.P., 1996, Tectonic evolution of the central Annapurna Range: Nepalese Himalaya: Tectonics, v. 15, p. 1264–1291, doi: 10.1029/96TC01791.

Jackson, J.A., Priestley, K., Allen, M., and Berberian, M., 2002, Active tectonics of the South Caspian Basin: Geophysical Journal International, v. 148, p. 214–245, doi: 10.1046/j.1365-246X.2002.01588.x.

Jackson, J.A., Austreim, H., McKenzie, D., and Preistley, K., 2004, Metastability, mechanical strength, and the support of mountain belts: Geology, v. 32, p. 625–628, doi: 10.1130/G20397.1.

Kent, P.E., 1970, The salt plugs of the Persian Gulf region: Leicester Literary and Philosophical Society, v. LXIV, p. 56–88.

Klootwijk, C.T., Gee, J.S., Peirce, J.W., Smith, G.M., and McFadden, P.L., 1992, An early India-Asia contact: Palaeomagnetic constraints from Ninety East ridge, ODP leg 121: Geology, v. 20, p. 395–398, doi: 10.1130/0091-7613(1992)020<0395:AEIACP>2.3.CO;2.

Law, R.D., Searle, M.P., and Simpson, R.L., 2004, Strain, deformation temperatures and vorticity of flow at the top of the Greater Himalayan Slab, Everest massif, Tibet: Geological Society [London] Journal, v. 161, p. 305–320.

Law, R.D., Searle, M.P., and Godin, L., editors, 2006, Channel Flow, Ductile Extrusion and Exhumation in Continental Collision Zones: Geological Society [London] Special Publication 268, 630 p.

Leech, M.L., Singh, S., Jain, A.K., Klemperer, S., and Manickavasagam, R.M., 2005, The onset of India-Asia continental collision: Early, steep subduction required by the timing of UHP metamorphism in the western Himalaya: Earth and Planetary Science Letters, v. 234, p. 83–97, doi: 10.1016/j.epsl.2005.02.038.

Le Pichon, X., and Gaulier, J.-M., 1988, The rotation of Arabia and the Levant fault system: Tectonophysics, v. 153, p. 271–294, doi: 10.1016/0040-1951(88)90020-0.

Lippard, S.J., Shelton, A.W., and Gass, I.G., 1986, The ophiolite of northern Oman: Geological Society [London] Memoir 11, 178 p.

Molnar, P., and Lyon-Caen, H., 1989, Fault plane solutions of earthquakes and active tectonics of the Tibetan Plateau and its margins: Geophysical Journal International, v. 99, p. 123–153.

Molnar, P., and Tapponnier, P., 1975, Cenozoic tectonics of Asia: Effects of continental collision: Science, v. 189, p. 419 426, doi: 10.1126/science.189.4201.419.

Nelson, K.D., Zhao, W., Brown, L.D., Kuo, J., Che, J., Liu, X., Klemperer, S.L., Makovsky, Y., Meissner, R., Mechie, J., Kind, R., Wenzel, F., Ni, J., Nabelek, J., Leshou, C., Tan, H., Wei, W., Jones, A.G., Booker, J., Unsworth, M., Kidd, W.S.F., Hauck, M., Alsdorf, D., Ross, A., Cogan, M., Wu, C., Sandvol, E., and Edwards, M., 1996, Partially molten middle crust beneath southern Tibet: Synthesis of Project INDEPTH results: Science, v. 274, p. 1684–1688, doi: 10.1126/science.274.5293.1684.

Nicolas, A., Ceuleneer, G., Boudier, F., and Misseri, M., 1988, Structural mapping in the Oman ophiolites: Mantle diapirism along an oceanic ridge: Tectonophysics, v. 151, p. 27–56, doi: 10.1016/0040-1951(88)90239-9.

Noble, S.R., and Searle, M.P., 1995, Age of crustal melting and leucogranite formation from U-Pb monazite dating in the western Himalaya: Geology, v. 23, p. 1135–1138, doi: 10.1130/0091-7613(1995)023<1135:AOCMAL>2.3.CO;2.

O'Brien, P.J., Zotov, N., Law, R.D., Khan, M.A., and Jan, M.Q., 2001, Coesite in Himalayan eclogite and implications for models of India-Asia collision: Geology, v. 29, p. 435–438, doi: 10.1130/0091-7613(2001)029<0435:CIHEAI>2.0.CO;2.

Oliver, G.J.H., 2002, Chronology and terrane assembly, new and old controversies, *in* Trewin, N.H., ed., Geology of Scotland: Geological Society [London], p. 201–211.

Owens, T.J., and Zandt, G., 1997, Implications of crustal property variations for models of Tibetan Plateau evolution: Nature, v. 387, p. 37–43, doi: 10.1038/387037a0.

Parrish, R.R., and Tirrul, R., 1989, U-Pb age of the Baltoro granite, northwest Himalaya, and implications for zircon inheritance and monazite U-Pb systematics: Geology, v. 17, p. 1076–1079, doi: 10.1130/0091-7613(1989)017<1076:UPAOTB>2.3.CO;2.

Parrish, R.R., Gough, S.J., Searle, M.P., and Waters, D.J., 2006, Plate velocity exhumation of ultrahigh-pressure eclogites in the Pakistan Himalaya: Geology, v. 34, p. 989–992, doi: 10.1130/G22796A.1.

Pegler, G., and Das, S., 1998, An enhanced image of the Pamir–Hindu Kush seismic zone from relocated earthquake hypocenters: Geophysical Journal International, v. 134, p. 573–595, doi: 10.1046/j.1365-246x.1998.00582.x.

Peltzer, G., and Tapponnier, P., 1988, Formation and evolution of strike-slip faults, rifts and basins during the India-Asia collision: An experimental approach: Journal of Geophysical Research, v. 93, p. 15,085–15,117.

Phillips, R.J., Parrish, R.R., and Searle, M.P., 2004, Age constraints on ductile deformation and long-term slip rates along the Karakoram fault zone, Ladakh: Earth and Planetary Science Letters, v. 226, p. 305–319, doi: 10.1016/j.epsl.2004.07.037.

Prince, C., Harris, N., and Vance, D., 2001, Fluid-enhanced melting during prograde metamorphism: Geological Society [London] Journal, v. 158, p. 233–241.

Rai, S.S., Preistley, K., Gaur, V.K., Mitra, S., Singh, M.P., and Searle, M.P., 2006, Configuration of the Indian Moho beneath the NW Himalaya and Ladakh: Geophysical Research Letters, v. 33, p. L15308, doi: 10.1029/2006GL026076.

Reeves, C., and de Wit, M., 2000, Making ends meet in Gondwana: retracing the transforms of the Indian Ocean and reconnecting continental shear zones: Terra Nova, v. 12, p. 272–280.

Regard, V., Bellier, O., Thomas, J.-C., Abbassi, M.R., Mercier, J., Shabanian, E., Feghhi, K., and Soleymani, S., 2004, Accommodation of Arabia-Eurasia convergence in the Zagros-Makran transfer zone, SE Iran: A transition between collision and subduction through a young deforming system: Tectonics, v. 23, p. TC4007, doi: 10.1029/2003TC001599.

Robertson, A.H.F., Searle, M.P., and Ries, A., 1990, The geology and tectonics of the Oman region: Geological Society [London] Special Publication 49, 845 p.

Searle, M.P., 1988, Structure of the Musandam culmination (Oman and United Arab Emirates) and the Straits of Hormuz syntaxis: Geological Society [London] Journal, v. 145, p. 831–845.

Searle, M.P., 1991, Geology and tectonics of the Karakoram Mountains: Chichester, John Wiley and Sons, 358 p.

Searle, M.P., 1999, Extensional and compressional faults in the Everest-Lhotse massif, Khumbu Himalaya, Nepal: Geological Society [London] Journal, v. 156, p. 227–240.

Searle, M.P., 2003, Geological map of the Mount Everest massif, Nepal and South Tibet Himalaya: Oxford University, scale 1:100,000.

Searle, M.P., and Cox, J.S., 1999, Tectonic setting, origin and obduction of the Oman ophiolite: Geological Society of America Bulletin, v. 111, p. 104–122, doi: 10.1130/0016-7606(1999)111<0104:TSOAOO>2.3.CO;2.

Searle, M.P., and Cox, J.S., 2002, Subduction zone metamorphism during formation and emplacement of the Semail ophiolite in the Oman mountains: Geological Magazine, v. 139, p. 241–255, doi: 10.1017/S0016756802006532.

Searle, M.P., and Rex, A.J., 1989, Thermal model for the Zanskar Himalaya: Journal of Metamorphic Geology, v. 7, p. 127–134.

Searle, M.P., and Szulc, A.G., 2005, Channel flow and ductile extrusion of the High Himalayan slab: The Kangchenjunga-Darjeeling profile, Sikkim Himalaya: Journal of Asian Earth Sciences, v. 25, p. 173–185, doi: 10.1016/j.jseaes.2004.03.004.

Searle, M.P., and Tirrul, R., 1991, Structural and thermal evolution of the Karakoram crust: Geological Society [London] Journal, v. 148, p. 65–82.

Searle, M.P., Cooper, D.J.W., and Rex, A.J., 1988, Collision tectonics of the Ladakh-Zanskar Himalaya: Royal Society of London Philosophical Transactions, ser. A, v. 326, p. 117–150.

Searle, M.P., Rex, A.J., Tirrul, R., Rex, D.C., and Barnicoat, A., 1989, Metamorphic, magmatic and tectonic evolution of the central Karakoram in the Biafo-Baltoro-Hushe regions of north Pakistan, *in* Malinconico, L.L., Jr., and Lillie, R.J., eds., Tectonics of the western Himalayas: Geological Society of America Special Paper 232, p. 47–74.

Searle, M.P., Parrish, R.R., Tirrul, R., and Rex, D.C., 1990, Age of crystallization and cooling of the K2 gneiss in the Baltoro Karakoram: Geological Society [London] Journal, v. 147, p. 603–606.

Searle, M.P., Crawford, M.B., and Rex, A.J., 1992a, Field relations, geochemistry, origin and emplacement of the Baltoro granite, central Karakoram: Transactions of the Royal Society of Edinburgh, Earth Sciences, v. 83, p. 519–538.

Searle, M.P., Waters, D.J., Rex, D.C., and Wilson, R.N., 1992b, Pressure, temperature and time constraints on Himalayan metamorphism from eastern Kashmir and western Zanskar: Geological Society [London] Journal, v. 149, p. 753–773.

Searle, M.P., Metcalfe, R.P., Rex, A.J., and Norry, M.J., 1993, Field relations, petrogenesis and emplacement of the Bhagirathi leucogranite, Garhwal Himalaya, *in* Treloar, P.J., and Searle, M.P., eds., Himalayan tectonics: Geological Society [London] Special Publication 74, p. 429–444.

Searle, M.P., Waters, D.J., Martin, H.N., and Rex, D.C., 1994, Structure and metamorphism of blueschist-eclogite facies rocks from northeastern Oman: Geological Society [London] Journal, v. 151, p. 555–576.

Searle, M.P., Parrish, R.R., Hodges, K.V., Hurford, A., Ayres, M.W., and Whitehouse, M.J., 1997, Shisha Pangma leucogranite, South Tibetan Himalaya: Field relations, geochemistry, age, origin and emplacement: Journal of Geology, v. 105, p. 295–317.

Searle, M.P., Weinberg, R.F., and Dunlap, W.J., 1998, Transpressional tectonics along the Karakoram fault zone in northern Ladakh: Constraints on tectonic extrusion, *in* Holdworth, R.E., et al., eds., Continental transpressional and transtensional tectonics: Geological Society [London] Special Publication 135, p. 307–326.

Searle, M.P., Waters, D.J., Dransfield, M.W., Stephenson, B.J., Walker, C.B., Walker, J.D., and Rex, D.C., 1999a, Thermal and mechanical models for the structural evolution of Zanskar High Himalaya, *in* MacNiocaill, C., and Ryan, P.D., eds., Continental tectonics: Geological Society [London] Special Publication 164, p. 139–156.

Searle, M.P., Noble, S.R., Hurford, A.J., and Rex, D.C., 1999b, Age of crustal melting, emplacement and exhumation history of the Shivling leucogranite, Garhwal Himalaya: Geological Magazine, v. 136, p. 513–525, doi: 10.1017/S0016756899002885.

Searle, M.P., Hacker, B.R., and Bilham, R., 2001, The Hindu Kush seismic zone as a paradigm for the creation of ultrahigh-pressure diamond- and coesite-bearing continental rocks: Journal of Geology, v. 109, p. 143–153, doi: 10.1086/319244.

Searle, M.P., Simpson, R.L., Law, R.D., Parrish, R.R., and Waters, D.J., 2003, The structural geometry, metamorphic and magmatic evolution of the Everest massif, High Himalaya of Nepal–South Tibet: Geological Society [London] Journal, v. 160, p. 345–366.

Searle, M.P., Warren, C.J., Waters, D.J., and Parrish, R.R., 2004, Structural evolution, metamorphism and restoration of the Arabian continental margin, Saih Hatat region, Oman Mountains: Journal of Structural Geology, v. 26, p. 451–473, doi: 10.1016/j.jsg.2003.08.005.

Shackleton, R.M., 1986, Precambrian collision tectonics in Africa: Geological Society [London] Special Publication 43, p. 331–335.

Simpson, R.L., Parrish, R.R., Searle, M.P., and Waters, D.J., 2000, Two episodes of monazite crystallization during metamorphism and crustal melting in the Everest region of the Nepalese Himalaya: Geology, v. 28, p. 403–406, doi: 10.1130/0091-7613(2000)28<403:TEOMCD>2.0.CO;2.

Stephenson, B.J., Waters, D.J., and Searle, M.P., 2000, Inverted metamorphism and the Main Central Thrust: Field relations and thermobarometric constraints from the Kishtwar window, NW Indian Himalaya: Journal of Metamorphic Geology, v. 18, p. 571–590, doi: 10.1046/j.1525-1314.2000.00277.x.

Stephenson, B.J., Searle, M.P., and Waters, D.J., 2001, Structure of the Main Central Thrust zone and extrusion of the High Himalayan crustal wedge, Kishtwar-Zanskar Himalaya: Geological Society [London] Journal, v. 158, p. 637–652.

St.Onge, M.R., Searle, M.P., and Wodicka, N., 2006, Trans-Hudson orogen of North America and Himalaya-Karakoram-Tibetan orogen of Asia: Structural and thermal characteristics of the lower and upper plates: Tectonics, v. 25, p. TC4006, doi: 10.1029/2005TC001907.

Strachan, R.A., Smith, M., Harris, A.L., and Fettes, D.J., 2002, The Northern Highland and Grampian terranes, *in* Trewin, N.H., ed., Geology of Scotland: Geological Society [London] Memoir, p. 81–147.

Tilton, G.R., Hopson, C.A., and Wright, J.E., 1981, Uranium-lead isotopic ages of the Semail ophiolite, Oman, with application to Tethyan ocean ridge tectonics: Journal of Geophysical Research, v. 86, p. 2763–2775.

Vance, D., and Harris, N.B.W., 1999, Timing of prograde metamorphism in the Zanskar Himalaya: Geology, v. 27, p. 395–398, doi: 10.1130/0091-7613(1999)027<0395:TOPMIT>2.3.CO;2.

Vannay, J.-C., and Grasemann, B., 2001, Himalayan inverted metamorphism and syn-convergence extension as a consequence of a general shear extrusion: Geological Magazine, v. 138, p. 253–276, doi: 10.1017/S0016756801005313.

Walker, C.B., Searle, M.P., and Waters, D.J., 2001, An integrated tectonothermal model for the evolution of the High Himalaya in western Zanskar with constraints from thermobarometry and metamorphic modeling: Tectonics, v. 20, p. 810–833, doi: 10.1029/2000TC001249.

Walker, J.D., Martin, M.W., Bowring, S.A., Searle, M.P., Waters, D.J., and Hodges, K.V., 1999, Metamorphism, melting and extension: Age constraints from the High Himalayan slab of southeastern Zanskar and northwestern Lahoul: Journal of Geology, v. 107, p. 473–495, doi: 10.1086/314360.

Warren, C.J., Parrish, R.R., Searle, M.P., and Waters, D.J., 2003, Dating the subduction of the Arabian continental margin beneath the Semail ophiolite, Oman: Geology, v. 31, p. 889–892, doi: 10.1130/G19666.1.

Warren, C.J., Parrish, R.R., Waters, D.J., and Searle, M.P., 2005, Dating the geologic history of Oman's Semail ophiolite: insights from U-Pb geochronology: Contributions to Mineralogy and Petrology, v. 150, p. 403–422, doi: 10.1007/s00410-005-0028-5.

Wei, W., Unsworth, M., Jones, A., Booker, J., Tan, H., Nelson, D, Chen, L., Li, S., Solon, K., Bedrosian, P., Jin, S., Deng, M., Ledo, J., Kay, D., and Roberts, B., 2001, Detection of widespread fluids in the Tibetan crust by magnetotelluric studies: Science, v. 292, p. 716–718, doi: 10.1126/science.1010580.

Wells, A.J., 1969, The Crush zone of the Iranian Zagros Mountains, and its implications: Geological Magazine, v. 106, p. 385–394.

Willett, S., and Beaumont, C., 1994, Subduction of Asian lithospheric mantle beneath Tibet inferred from models of continental collision: Nature, v. 369, p. 642–645, doi: 10.1038/369642a0.

Zeitler, P.K., Chamberlain, C.P., and Smith, H.A., 1993, Synchronous anatexis, metamorphism and rapid denudation at Nanga Parbat (Pakistan Himalaya): Geology, v. 21, p. 347–350, doi: 10.1130/0091-7613(1993)021<0347:SAMARD>2.3.CO;2.

Zeitler, P.K., Koons, P.O., Bishop, M.P., Chamberlain, C.P., Craw, D., Edwards, M.A., Hamidullah, S., Jan, M.Q., Khan, M.A., Khattak, M.U.K., Kidd, W.S.F., Mackie, R.L., Meltzer, A.S., Park, S.K., Pecher, A., Poage, M.A., Sarker, G., Schneider, D.A., Leonardo Seeber, L., and Shroder, J.F., 2001, Crustal working at Nanga Parbat, Pakistan: Metamorphic consequences of thermal-mechanical coupling facilitated by erosion: Tectonics, v. 20, p. 712 728, doi: 10.1029/2000TC001243.

Zhao, W., and Morgan, W.J., 1985, Uplift of the Tibetan Plateau: Tectonics, v. 4, p. 359–369.

Zhao, W., Nelson, K.D., and Team Project INDEPTH, 1993, Deep seismic reflection evidence for continental underthrusting beneath southern Tibet: Nature, v. 366, p. 557–559, doi: 10.1038/366557a0.

Zhu, B., Kidd, W.S.F., Rowley, D.B., Currie, B.S., and Shafique, N., 2005, Age of initiation of the India-Asia collision in the east-central Himalaya: Journal of Geology, v. 113, p. 265–285, doi: 10.1086/428805.

Manuscript Accepted by the Society 1 February 2007

Printed in the USA

The Geological Society of America
Memoir 200
2007

Continental growth and recycling by accretion of deformed turbidite fans and remnant ocean basins: Examples from Neoproterozoic and Phanerozoic orogens

David R. Gray*
School of Earth Sciences, University of Melbourne, Melbourne 3010, Victoria, Australia

David A. Foster*
School of Geological Sciences, University of Florida, Gainesville, Florida 32611-2120, USA

Roland Maas*
School of Earth Sciences, University of Melbourne, Melbourne 3010, Victoria, Australia

Catherine V. Spaggiari*
Geological Survey of Western Australia, Perth 6004, Western Australia, Australia

Robert T. Gregory*
Department of Geological Sciences, Southern Methodist University, Dallas, Texas 75275, USA

Ben Goscombe*
School of Earth and Environmental Sciences, University of Adelaide, Adelaide 5005, South Australia, Australia

K.H. (Charlie) Hoffmann*
Namibian Geological Survey, Private Bag 13297, Windhoek, Namibia

ABSTRACT

Convergent margin tectonic settings involving accretion of large turbidite fans represent important sites of growth and regeneration of continental crust. The newly accreted continental crust consists of an upper crustal layer of recycled crustal detritus (turbidites) underlain by a lower crustal layer of tectonically imbricated oceanic crust, and/or rifted and thinned continental crust, along with underplated magmatic materials; the new lower crust represents additions to continental crustal volume differentiated from the mantle. This two-tiered crust is of average continental crustal

*Gray: drgray@unimelb.edu.au; Foster: dfoster@geology.ufl.edu; Maas: maasr@unimelb.edu.au; Spaggiari: Catherine.Spaggiari@doir.wa.gov.au; Gregory: bgregory@mail.smu.edu; Goscombe: ben.goscombe@adelaide.edu.au; Hoffmann: khhoffmann@mme.gov.na.

Gray, D.R., Foster, D.A., Maas, R., Spaggiari, C.V., Gregory, R.T., Goscombe, B., and Hoffmann, K.H., 2007, Continental growth and recycling by accretion of deformed turbidite fans and remnant ocean basins: Examples from Neoproterozoic and Phanerozoic orogens, *in* Hatcher, R.D., Jr., Carlson, M.P., McBride, J.H., and Martínez Catalán, J.R., eds., 4-D Framework of Continental Crust: Geological Society of America Memoir 200, p. 63–92, doi: 10.1130/2007.1200(05). For permission to copy, contact editing@geosociety.org.

thickness and is isostatically balanced near sea level, resulting in remarkable stability. The Paleozoic Tasman orogen of eastern Australia is the archetypal example of this style of orogeny, representing continental growth rates of cubic kilometers per year of material that does not return to the mantle by oceanic plate-tectonic recycling. The Neoproterozoic Pan-African Damara orogen of SW Africa is a similar orogen, whereas the Mesozoic Rangitatan orogen or Rakaia wedge of New Zealand illustrates the transition of the convergent margin from a Lachlan-type to more recognizable "ring of fire"-type orogen. These orogens illustrate continental growth from the shortening of deep marine successions and their oceanic crustal basement involving subduction-accretion. The spatial and temporal variations in deformation, metamorphism, and magmatism across these orogens illustrate how large volumes of monotonous turbidites and their relict oceanic basement eventually become stable continental crust. The timing of deformation and metamorphism reflect the crustal thickening phase, whereas the posttectonic granitoids and surficial volcanic deposits give the timing of cratonization. The turbidites represent fertile sources for crustal melting and are the main sources for the S-type granites.

Keywords: continental growth, crustal recycling, turbidites, convergent margins.

INTRODUCTION

Continental growth from marine successions due to subduction-accretion processes involves accretion of large volumes of continental detritus (submarine sediment fans, e.g., Ingersoll et al., 2003) and fragments of juvenile oceanic lithosphere (upper crust, e.g., Şengör and Natal'in, 1996), as well as new addition to the nascent continental crust due to magmatic underplating (lower crust, e.g., Albarède, 1998). Turbidite fan accretion represents an apparent continental growth mechanism that was active and important throughout Earth history with examples in the Phanerozoic of central Asia (Şengör and Natal'in, 1996), Proterozoic Yavapai-Mazatzal provinces of SW Laurentia (Karlstrom et al., 2000), and possibly even Archean granite-greenstone belts (Percival et al., 2004). It has a crustal addition rate comparable to the accretion rate for magmatic arcs (e.g., 13–18 km^3/km/m.y., Smith, 1981, or ~1–1.5 km^3/yr, e.g., Armstrong, 1981) and may dominate in certain parts of Earth history, for example during the Paleozoic along the Pacific margin of East Gondwana (20,000 km long Tasmanide belt, e.g., Foster and Gray, 2000; see also Cawood, 2005).

Turbidite-dominated orogens are a major location for sedimentary recycling (as opposed to recycling of sedimentary rocks into the mantle) of continental crust in the form of detritus, but may also involve continental growth through new crustal addition from the mantle (e.g., Ketilidian terrain of South Greenland; Patchett and Bridgwater, 1984). Turbidite fan systems on oceanic crust or within oceanic backarc basins form the basis for new continental crust consisting of a mixture of basaltic extractions from the asthenospheric mantle along with recycled continental detritus. These orogens tend to develop with a mafic lower crust and felsic upper crust. It is of course likely that some of the detritus is sourced from new crust (e.g., island arcs, large igneous provinces) as well as old continents. Turbidite-dominated orogens throughout the world, and throughout time, are variable mixtures of recycled continental and mantle-derived end members that depend on the nature of the lower crust and source of the turbidite.

Much of our understanding of accretion has come from the study of modern or recent accretionary prisms (e.g., Lundberg and Reed, 1991; Von Huene and Scholl, 1991). In this paper, we investigate some of the parameters of accretion by analyzing ancient orogens, particularly accretionary-style orogens dominated by turbidites. It is based on the premise that the deformed turbidites preserved in the final crustal orogen architecture must reflect the original tectonic settings of the submarine fan and/or wedge, as well as the overall tectonic process. Their metamorphic character also reflects the original tectonic setting and evolutionary history.

Controls on turbidite fan deformation include the plate-tectonic setting, the tectonic position, on either the overriding slab (hanging wall) or downgoing slab (footwall), the degree of coupling between overriding and subducting slabs, the original thickness of the fan, the residence time of the fan on the seafloor, the degree of lithification (diagenesis/metamorphism), the level in the accretionary wedge, and the fluid availability. The deformation style of turbidites includes mélange or broken formation (e.g., parts of the central and eastern Lachlan orogen of eastern Australia, the Pahau terrane of New Zealand), chevron folding (e.g., the western Lachlan orogen, the Ugab domain of the Northern Zone of the Damara orogen), and fold nappes and accompanying schistosity (e.g., Otago Schist belt of the Rangitatan orogen of New Zealand and the Kuiseb Schist of central Namibia). The strain state is variable, with (1) subvertical plane strain or flattening with mostly coaxial increments (e.g., western Lachlan orogen), or (2) dominantly constrictional strain involving noncoaxial deformation (e.g., Otago Schist belt of New Zealand).

This paper investigates the large-scale architecture of crustal accretion with consideration of the position of turbidites relative to tectonic elements in the archetypal Lachlan (eastern Australia), Damara (Namibia), and Rangitatan (New Zealand) orogens (Fig. 1), as well as the spatial and temporal dynamics of accretion, and geodynamic scenarios of crustal evolution. The paper addresses how and why turbidites of large submarine fan systems are accreted, in what environments and by what processes. The paper aims to (1) characterize the nature of turbidite-dominated orogens, (2) use instructive examples from the Lachlan and Damara orogens and Otago Schist belt of New Zealand to outline structural styles and separate evolutionary histories, (3) demonstrate that there is no simple template, and (4) use regional-scale to mesoscale structure to delineate tectonic process.

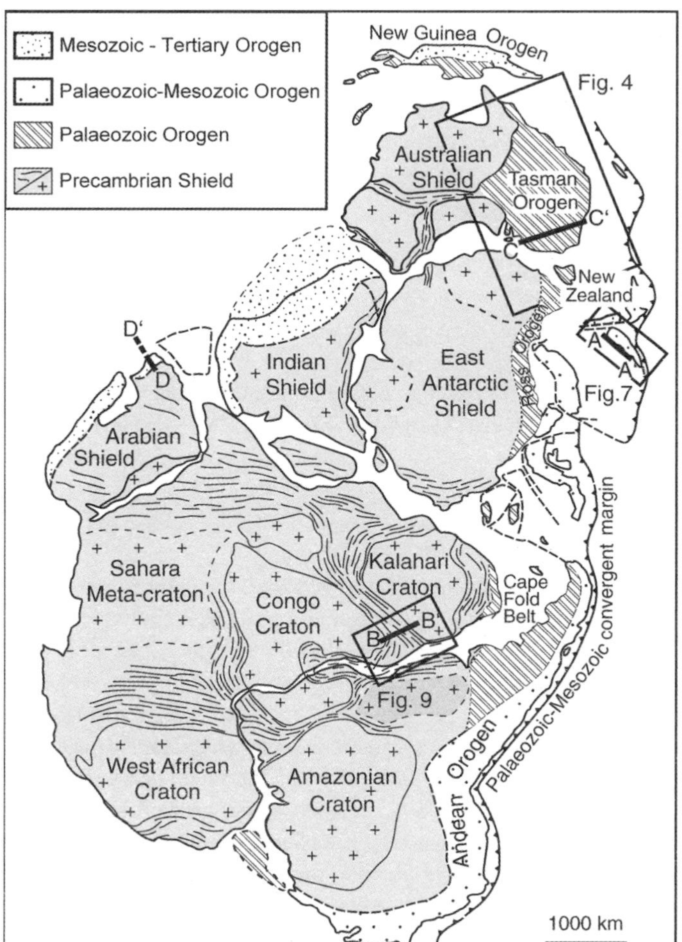

Figure 1. Map of the Gondwana supercontinent showing the distribution of Precambrian shield areas and orogenic belts of Neoproterozoic, Paleozoic, Paleozoic-Mesozoic and Mesozoic-Tertiary ages. The locations of Figures 4, 7, and 9 are shown as well as the schematic crustal profiles A–A′, B–B′, C–C′, and D–D′ in Figure 2. Outline based on IGCP (International Geological Correlation Programme) Gondwana map.

OROGEN ARCHITECTURE AND POSITIONS OF TURBIDITES

Figure 1 shows examples of turbidite-dominated orogens from the Gondwana supercontinent, including the Neoproterozoic Damara belt of the Pan-African orogenic system, the Paleozoic Lachlan orogen of the composite Tasman orogen of eastern Australia, and the Mesozoic Rangitatan orogen of New Zealand. Profiles A–A′, B–B′, and C–C′ (Fig. 2) represent, respectively, (1) structurally thickened turbidites making up most of the Rangitatan orogen of New Zealand apart from the Median batholith arc and forearc sequences inboard of the Dun Mountain ophiolite (section A–A′, Fig. 2), (2) deformed and metamorphosed turbidites occupying the central part of the Inland Branch of the Damara orogen (section B–B′, Fig. 2), and (3) predominantly chevron-folded turbidites across the Tasman orogen, in particular the western and central parts of the Lachlan orogen (WLO and CLO in section C–C′, Fig. 2).

The Damara orogen of Namibia (section B–B′, Fig. 2) is a classic divergent orogen between cratonic nuclei with thrusting of the platform carbonate succession or passive margin sequences back over the bounding cratons to the north and south, respectively. The south-vergent part, however, dominates, defined by the generally homoclinally dipping Southern (Khomas) Zone schist fabrics and Southern Margin Zone thin-skinned thrust system (section B–B′, Fig. 2). The other dominant element is the granite-dominated, high-*T*/low-*P* metamorphic Central Zone that contains domal culminations of basement (e.g., Puhan, 1983). This magmatism, as well as the structural asymmetry, may reflect a subduction interface dipping beneath attenuated Congo craton to form an Andean/Cordilleran-style margin between ca. 560 and 500 Ma (e.g., Barnes and Sawyer, 1980; Miller, 1983a, 1983b; Kasch, 1983a), although Kröner (1982) has argued for ensialic orogeny without a Wilson cycle ocean closure (a common theme for these types of orogens, e.g., Chappell and White, 2001, for the Lachlan).

The Lachlan orogen of the composite Tasman orogen (section A–A′, Fig. 2) includes the distinct western (WLO), central (CLO), and eastern (ELO) parts. It is made up of an attenuated older arc (Ordovician age), a shear zone–bounded high-*T*/low-*P* metamorphic complex with similarities to the Chugach metamorphic complex of the Kodiak accretionary system of Alaska (cf. Hudson and Plafker, 1982), structurally thickened quartz-rich submarine fans where thickening is due to thrust shortening and chevron folding during closure of a marginal backarc basin, and inverted Silurian-Devonian extensional basins within the older arc (e.g., Gray and Foster, 2004a, 2004b). Evidence of subduction is provided by serpentinite-matrix mélanges containing Franciscan-like blueschist blocks or knockers along some of the major fault zones (see Spaggiari et al., 2002a, 2002b, 2004a).

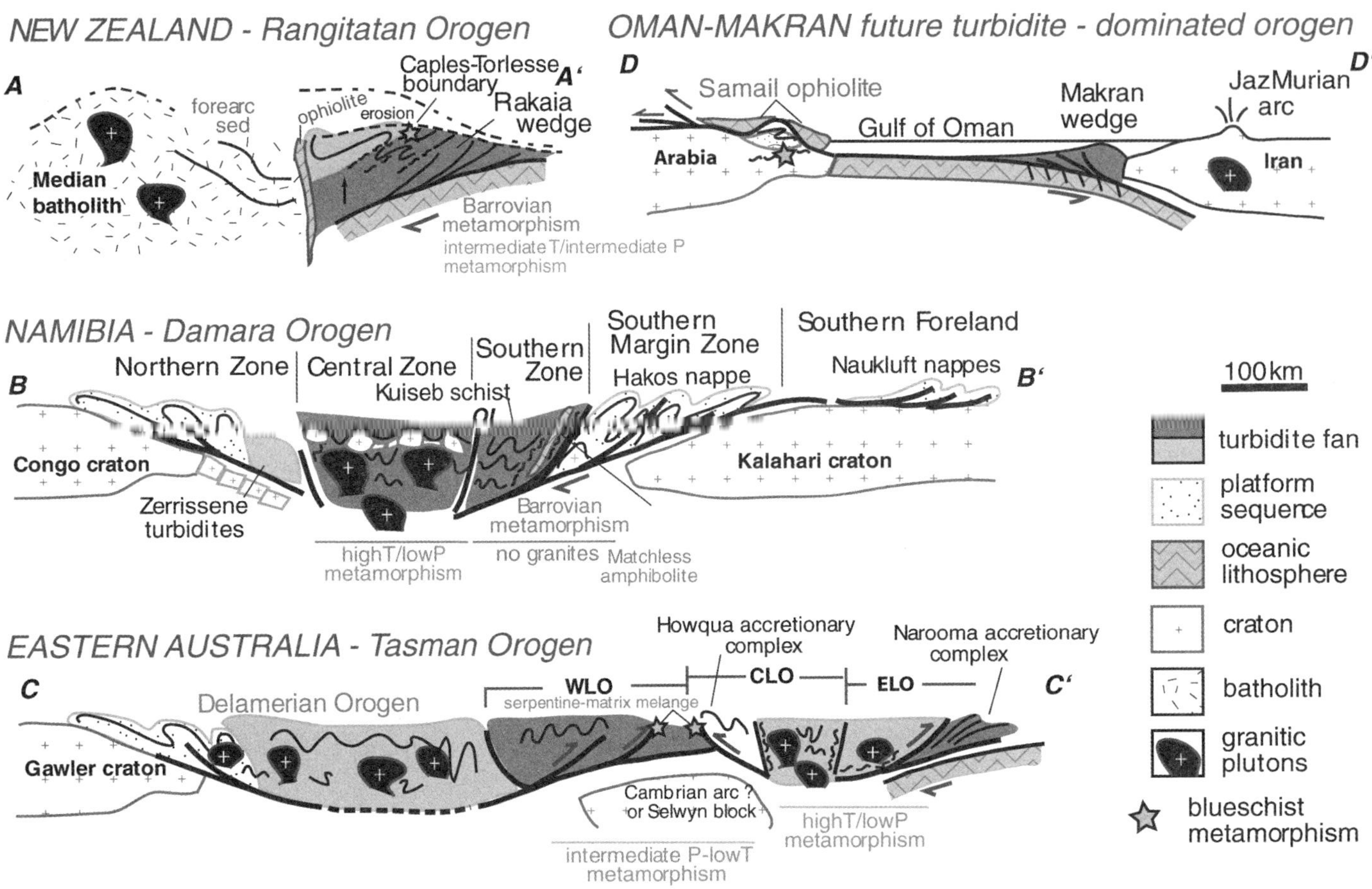

Figure 2. Schematic structural profiles across the various orogens showing the locations of turbidites with respect to tectonic and other geologic elements. Locations of the profiles are shown in Figure 1. CLO—central Lachlan orogen; ELO—eastern Lachlan orogen; OL—Okahandja Lineament; WLO—western Lachlan orogen.

The Rangitatan orogen of New Zealand (section C–C′, Fig. 2) is made up of a structurally thickened sediment wedge (Caples terrane) abutting arc-forearc sequences (Median batholith) and a deformed turbidite fan system (Torlesse of the Rakaia wedge) separated by steeply dipping, fault-bounded, suture-like ophiolite belt (Mortimer, 1993, 2004; Gray and Foster, 2004c). Structural vergence in the schist suggests NE thrusting of the trench volcaniclastic sedimentary sequence over the subducted quartz-rich sedimentary fan (Mortimer, 1993) that lay outboard probably on the subducted slab ocean floor (Coombs et al., 1976).

Evidence of subduction occurs in mafic greenschist-chert sequences close to the trench–submarine fan boundary (the Caples-Torlesse boundary) where relicts of blueschist metamorphism are provided by cores of riebeckite and crossite mantled by actinolite (Yardley, 1982). Eventual closure of the Gulf of Oman (section D–D′, Fig. 2) would result in a structurally thickened Makran wedge juxtaposed against the Jaz-Murian arc of Iran, a scenario similar to that shown in the section across the Rakaia wedge of New Zealand (section A–A′, Fig. 2).

EASTERN GONDWANIDES: EXAMPLE OF PHANEROZOIC CONTINENTAL ACCRETION

The East Gondwana margin from ca. 550 Ma to ca. 90 Ma is a classic accretionary margin providing a spectacular example of Phanerozoic continental accretion in a convergent margin setting (e.g., Foster and Gray, 2000; Cawood, 2005). It shows eastward younging in accretion of oceanic domains, as well as accreted submarine fans, fragments of ocean crust, and arc and forearc basin elements from the Tasman and Ross orogens through to New Zealand (Fig. 3). The Tasman and Ross orogens represent at least 5% of the volume of Gondwanan continental crust (ruled pattern, Fig. 1) and with the South American Tasmanides constitute ~5% of the current continental crustal area. Using a 100 m.y. timescale (e.g., Foster and Gray, 2000), this corresponds to ~3–4 km^3/yr of apparent continental growth, with 1.5 km^3/yr of new mantle-derived material composing the lower crust.

In the eastern Australian part, or Tasman orogenic system, accretion occurred in a stepwise fashion with an eastward

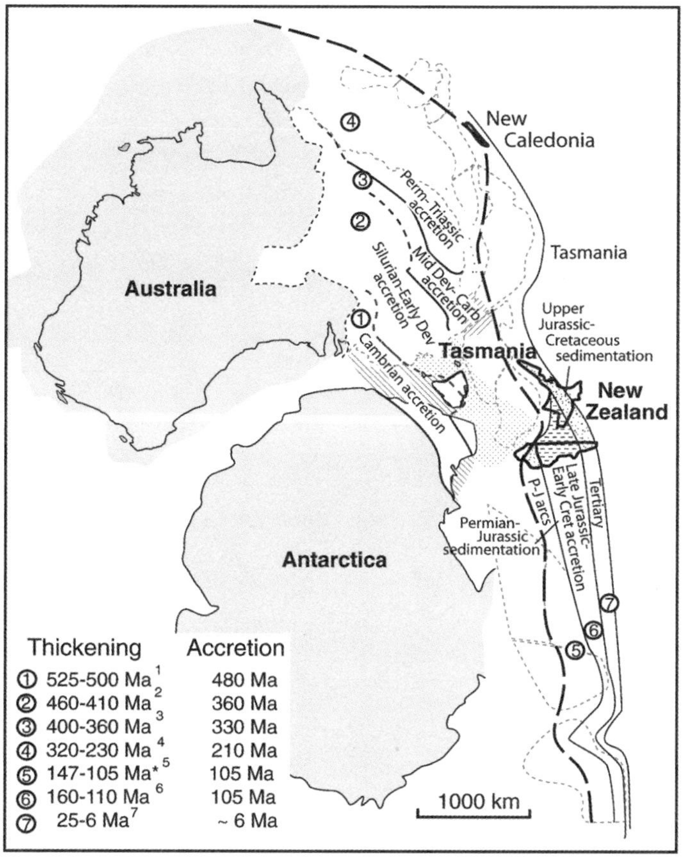

Figure 3. Map showing the succession and timing of accretion of former oceanic successions along the former eastern margin of Gondwana (Australia and Antarctica segment only). Map modified from Veevers (2000a, Fig. 124 therein) and Bradshaw (1989). Structural thickening is the period of deformation and metamorphism of the sedimentary wedge. The timing of accretion is given by the ages of the posttectonic granites. Data sources for the structural thickening and final accretion of the circled-numbered units: 1—Foden et al. (1999) and Turner et al. (1992); 2—Foster et al. (1999) and Foster and Gray (2000); 3—Foster et al. (1998) and Foster and Gray (2000); 4—Murray et al. (1987) and Powell (1984); 5—Muir et al. (1988) and Mortimer et al. (1999a, 1999b); 6—Gray and Foster (2004c); 7—Sporli (1978).

younging from the Cambrian through the Triassic, reflected by peak deformations of Early Ordovician, Early Silurian to Middle Devonian, and Permian-Triassic age in the respective orogenic belts (Foster and Gray, 2000; Gray and Foster, 2004a). This involved deformation of Neoproterozoic platform and rift basinal sequences (Delamerian orogen), a Cambrian-Ordovician composite turbidite submarine fan overlying Cambrian oceanic crust (Lachlan orogen), and Carboniferous-Permian arc, forearc, trench, subduction complex, and foreland basin sequences (New England orogen) (Gray and Foster, 2004a; Cawood, 2005).

Outboard of the Tasman orogen, the continental landmass of New Zealand records continuous sedimentation and accretionary prism accretion (Rakaia and Pahau terranes; Bradshaw, 1989) in the hanging wall to a long-lived subduction system along the margin of the Gondwana supercontinent from Permian to Late Cretaceous times (Fig. 3) (Bradshaw, 1989; Mortimer, 2004).

The Lachlan Orogen of Eastern Australia

The Paleozoic Lachlan orogen part of the Tasman orogenic system (Figs. 1 and 4A) is a composite accretionary orogen that formed along the eastern margin of Gondwana during late Neoproterozoic through Paleozoic times (Fig. 3) (Gray and Foster, 2004a). It is dominated by voluminous and extensive, mostly Cambrian to Ordovician turbidites of low metamorphic grade (Fig. 4B) that formed a large submarine fan system in the Late Cambrian to Ordovician that was comparable in size to the present-day Bengal fan (Gray et al., 1991; Fergusson and Coney, 1992a). The fan accumulated on Middle to Late Cambrian backarc/forearc crust, consisting of predominantly low-K, mid-oceanic-ridge basalts (MORBs) to arc-tholeiite basalts, dolerites and gabbros, high-Mg/low-Ti boninites, ultramafics, and calc-alkaline arc rocks (Crawford and Keays, 1978, 1987), in the Cambrian through Late Ordovician from ca. 505 Ma to ca. 460 Ma (Fergusson and Coney, 1992b; Foster et al., 2005). Closure of the backarc basin took place during plate convergence in an oceanic setting along the eastern margin of Gondwana from ca. 450 Ma to 340 Ma, with accretion of structurally thickened submarine fans, accretionary complexes, former volcanic arcs and oceanic crust, as well as the Tasmania microcontinent (Fergusson and Coney, 1992b; Gray, 1997; Foster et al., 1999; Gray and Foster, 2004b).

The Lachlan orogen comprises three thrust belts that constitute the western, central, and eastern parts, respectively. The western Lachlan orogen (WLO) consists largely of an east-vergent thrust system with alternating zones of northwest- and north-trending structures (Fig. 4A), whereas the central Lachlan orogen (CLO) is dominated by northwest-trending structures and consists of a southwest-vergent thrust belt (Howqua accretionary complex in Fig. 4A) linked to a fault-bounded, high-*T*/low-*P* metamorphic complex (WOMB in Fig. 4A). The eastern Lachlan orogen (ELO) is dominated by a north-south structural grain and east-directed thrust faults. In the south and in the easternmost part, an east-vergent thrust system (Yalmy-Bungonia thrust belt in Fig. 4A) overrides an older, subduction-related accretionary complex (Narooma accretionary complex in Fig. 4A).

An oceanic setting involving oceanic underthrusting (subduction) for the WLO and CLO for parts of their history (Gray and Foster, 2004a) is suggested by the presence of dismembered ophiolite slivers along some major fault zones (Spaggiari et al., 2003a, 2004a), the lower-*T*, intermediate-*P* metamorphic conditions preserved in metasandstone-slate sequences of the WLO and external part of the CLO (Offler et al., 1998; Spaggiari et al., 2003b), the presence of broken formation in the CLO and ELO (Fergusson, 1987; Miller and Gray, 1997; Watson and Gray, 2001), and the presence of serpentinite-matrix mélange incorporating blueschist knockers similar to those in the Franciscan Complex of California (Spaggiari et al., 2002a, 2002b, 2003a).

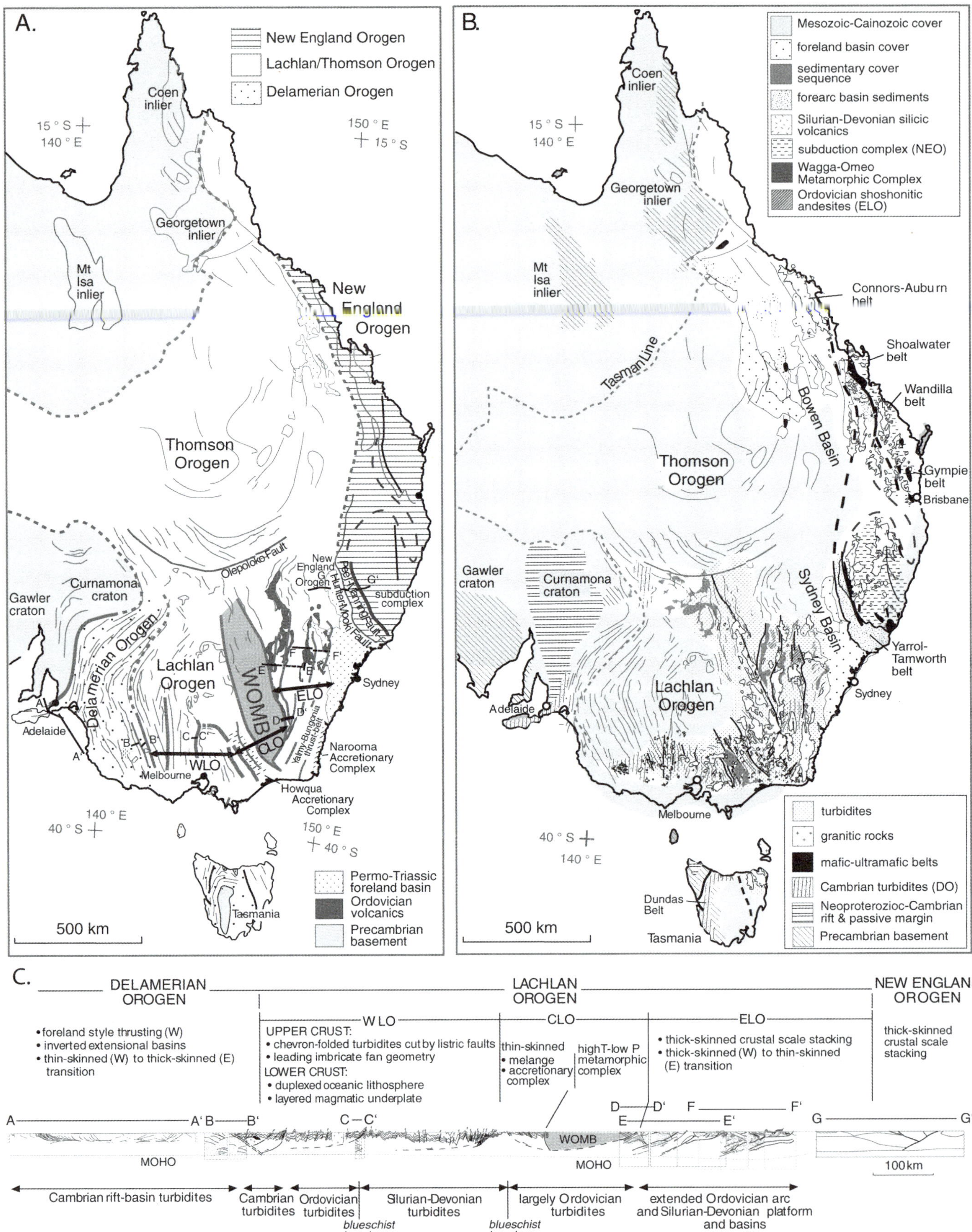
A.
New England Orogen
Lachlan/Thomson Orogen
Delamerian Orogen
Coen inlier
15 ° S
140 ° E
150 ° E
15 ° S
Georgetown inlier
Mt Isa inlier
New England Orogen
Thomson Orogen
Olepoloko Fault
New England Orogen
Peel-Manning Fault
Hunter-Mooki Fault
subduction complex
Gawler craton
Curnamona craton
Delamerian Orogen
Lachlan Orogen
WOMB
ELO
CLO
WLO
Yalmy-Bungonia thrust belt
Sydney
Adelaide
Melbourne
Narooma Accretionary Complex
Howqua Accretionary Complex
140 ° E
40 ° S
150 ° E
40 ° S
Tasmania
500 km
Permo-Triassic foreland basin
Ordovician volcanics
Precambrian basement
B.
Mesozoic-Cainozoic cover
foreland basin cover
sedimentary cover sequence
forearc basin sediments
Silurian-Devonian silicic volcanics
subduction complex (NEO)
Wagga-Omeo Metamorphic Complex
Ordovician shoshonitic andesites (ELO)
Coen inlier
15 ° S
140 ° E
Georgetown inlier
Mt Isa inlier
Connors-Auburn belt
Tasman Line
Shoalwater belt
Wandilla belt
Bowen Basin
Thomson Orogen
Gympie belt
Brisbane
Gawler craton
Curnamona craton
Sydney Basin
Yarrol-Tamworth belt
Adelaide
Lachlan Orogen
Sydney
Melbourne
40 ° S
140 ° E
500 km
Dundas Belt
Tasmania
turbidites
granitic rocks
mafic-ultramafic belts
Cambrian turbidites (DO)
Neoproterozioc-Cambrian rift & passive margin
Precambrian basement
C.
DELAMERIAN OROGEN
LACHLAN OROGEN
NEW ENGLAND OROGEN
• foreland style thrusting (W)
• inverted extensional basins
• thin-skinned (W) to thick-skinned (E) transition
WLO
UPPER CRUST:
• chevron-folded turbidites cut by listric faults
• leading imbricate fan geometry
LOWER CRUST:
• duplexed oceanic lithosphere
• layered magmatic underplate
CLO
thin-skinned
• melange
• accretionary complex
highT-low P metamorphic complex
ELO
• thick-skinned crustal scale stacking
• thick-skinned (W) to thin-skinned (E) transition
thick-skinned crustal scale stacking
A A' B B' C C' D D' E E' F F' G G'
WOMB
MOHO
MOHO
100km
Cambrian rift-basin turbidites
Cambrian turbidites
Ordovician turbidites
Silurian-Devonian turbidites
largely Ordovician turbidites
extended Ordovician arc and Silurian-Devonian platform and basins
blueschist knockers
blueschist knockers

Figure 4. (A) Map of eastern Australia showing tectonic subdivisions and the major structural elements (modified from Gray et al., 2006). The locations of deep crustal seismic profiles A–A′ through G–G′ used to construct the orogen-scale structural profile in C are shown. WLO—western Lachlan orogen; CLO—central Lachlan orogen; ELO—eastern Lachlan orogen; WOMB—Wagga-Omeo metamorphic belt. (B) Map of eastern Australia showing the major lithofacies and tectonostratigraphic elements. The turbidites are shown as a stippled pattern. (modified from Gray and Foster, 2004a). DO—Delamerian orogen. (C) Structural profile across eastern Australia at 37.5° south latitude (modified from Gray et al., 2006). MOHO— Mohorovičić discontinuity or base of crust.

Lachlan Orogen Metamorphism

Turbidite sequences of the Lachlan orogen are generally of low metamorphic grade represented by greenschist (epizonal) and subgreenschist (anchizonal) conditions (Fig. 5A). Most of the turbidites are within the chlorite zone with localized development of biotite in contact aureoles around granites. High-*T*/low-*P* metamorphism is reflected by localized migmatites and K-feldspar-cordierite-sillimanite gneisses in the Wagga-Omeo Complex and Cooma, Cambalong, Jerangle, and Kuark belts of the eastern metamorphic belt (Fig. 5A). Peak metamorphic conditions in the Lachlan orogen were 700 °C and 3–4 kbar in the Wagga-Omeo metamorphic belt (Morand, 1990) and eastern metamorphic complexes (e.g., Cooma Complex). These belts are intimately associated with S-type granitic bodies and anatectic migmatites, and represent deeper parts of the orogenic pile exhumed during crustal-scale extension and voluminous granitic magmatism (e.g., Collins and Hobbs, 2001).

Turbidites from the low-grade parts of the Lachlan orogen show intermediate-*P* metamorphism based on b_0 measurements of phengitic micas (see Fig. 11 of Gray and Foster, 2004b).

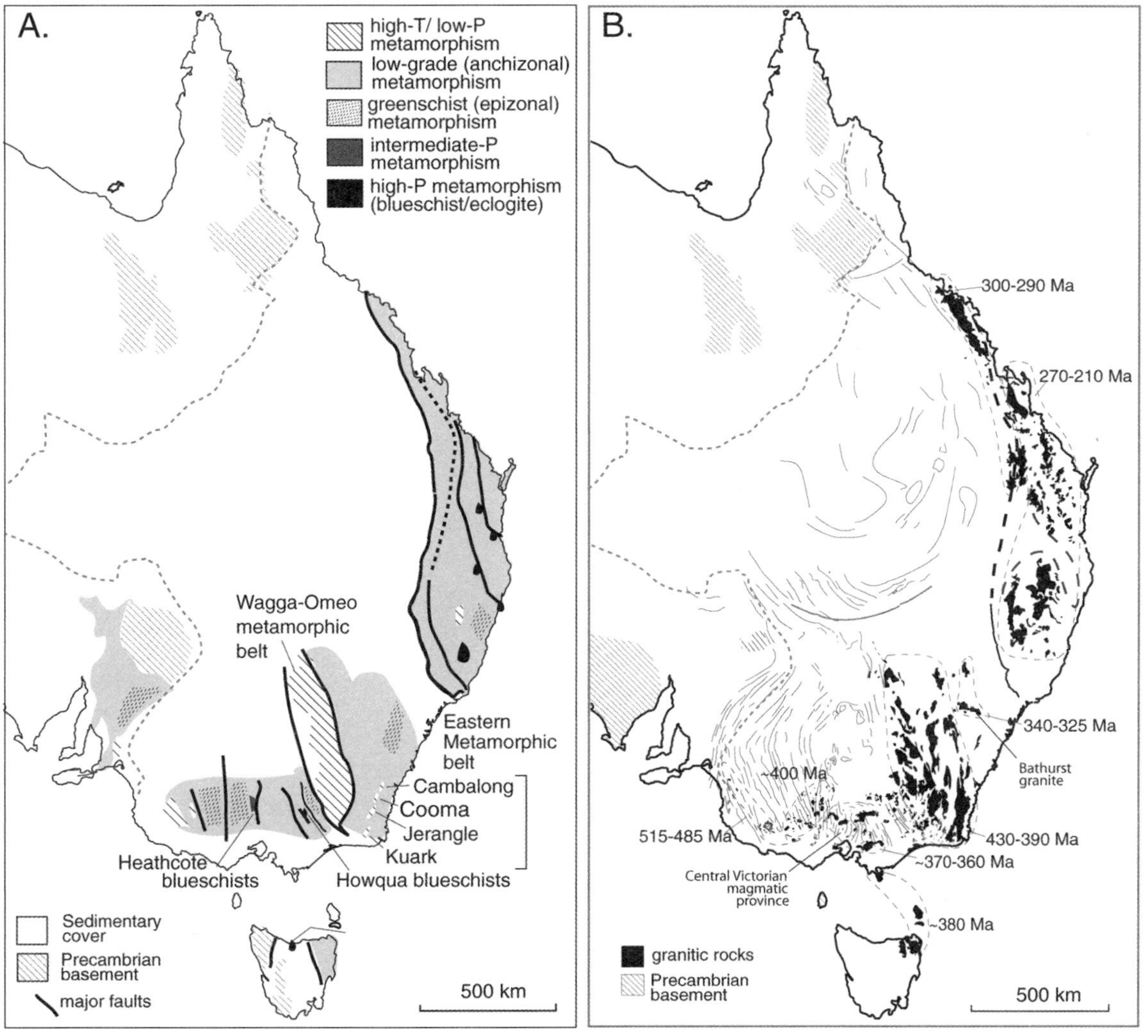

Figure 5. (A) Map of eastern Australia showing the distribution of metamorphic rocks. (B) Map of eastern Australia showing the distribution and shape pattern of granitoids, as well as aeromagnetic trend lines and positions of basement (modified from Gray and Foster, 2004a, 2004b).

Intermediate-*P* metamorphism has been inferred from coexisting chlorite-actinolite from metabasites (Offler et al., 1998), as well as low-*T*/intermediate-*P* blueschist metamorphism recorded by metabasite blueschist blocks (knockers) in serpentinite/talc-matrix mélanges within some major fault zones (Spaggiari et al., 2002a, 2002b). Amphibole compositions (winchite and glaucophane) give estimated pressures of ~5–6.5 kbar (winchite in the WLO) and ~6–7 kbar (winchite and glaucophane in the CLO) with temperatures <450 °C (Spaggiari et al., 2002a, 2002b). Metabasite blueschists from the Franciscan Complex of California and the Sanbagawa belt of Japan yield similar *P-T* estimates. This low-*T*/intermediate-*P* metamorphism occurred at 450–440 Ma during the regional deformation of the Lachlan orogen (Spaggiari et al., 2002a, 2002b, 2003a).

Lachlan Orogen Magmatism

Silurian-Devonian granites are a major component of present outcrop (~20% overall, up to 36% in the eastern and central Lachlan orogen), forming either major batholiths with N-S and NNE orientations roughly parallel to the structural grain, or single plutons (reviews by White and Chappell, 1983; Chappell et al., 1988; Chappell and White, 1992). With few exceptions, the granites are contact-aureole type, i.e., plutons formed at ~2 kbar pressure or less within low-grade upper crustal rocks. Most are posttectonic and unmetamorphosed, although the oldest intrusions (Wagga-Omeo belt, Kosciusko batholith) carry significant foliation and were emplaced syn- to late-kinematically and at mid-crustal depths (e.g., Hine et al., 1978; Keay et al., 1999). Volcanic sequences, mostly dacite-rhyolite, are also widespread (15% of area in the eastern Lachlan orogen, significant posttectonic caldera complexes in the western Lachlan orogen) and associated with shallow granitic plutons. This includes basaltic/andesitic remnants of the dismembered Ordovician Macquarie arc in the eastern Lachlan orogen (e.g., Glen et al., 1998).

K-Ar and Rb-Sr mica, Ar-Ar mica/hornblende, and sensitive high-resolution ion microprobe (SHRIMP) U-Pb zircon ages indicate a total age range of ca. 430 to 370 Ma (excluding the ≥440 Ma Macquarie arc), with two broad west-to-east younging trends in the eastern (430–370 Ma) and western (410–370 Ma) Lachlan orogen, respectively. Carboniferous (ca. 320 Ma) granites form a 100 km wide strip at the eastern edge of the Lachlan orogen near Bathurst; these are probably transitional to the Permian-Triassic granites of the outboard New England Fold Belt (Chappell et al., 1988).

S-type granites make up roughly half of all granite outcrop in the Lachlan orogen and are concentrated in a NNW-trending belt along the center of the orogen. Granodiorite and granite dominate, and coeval/cogenetic S-type volcanics are locally abundant (e.g., Wyborn et al., 1981). Compositions range from Mg-Fe-rich, cordierite-bearing granodiorites (e.g., the classical SE Australian S types of the Bullenbalong Suite; Hine et al., 1978; White and Chappell, 1988) to highly fractionated granite (e.g., Koetong Supersuite; Price et al., 1983). Felsic S-type granites (410–370 Ma) also occur in the western Lachlan orogen; several of the youngest (370 Ma) intrusions here are associated with extensive caldera complexes (Phillips et al., 1981; Clemens and Wall, 1984; Rossiter, 2003).

I-type granites form a broad belt along the eastern seaboard (e.g., the 8620 km^2, 419–370 Ma Bega batholith) but are also abundant throughout the rest of the orogen. I-type volcanics are locally preserved in grabens (Wyborn and Chappell, 1986). Medium- to high-K granodiorites and granites dominate, with lesser tonalite and rare gabbro-diorite (<1%). Surprisingly, trace element patterns in average I and S types are similar and have the Sr-depleted, Y-undepleted (low Sr/Y, i.e., magma sources with residual plagioclase but no garnet) signature found in Australian Proterozoic granites (Wyborn et al., 1992), and in post-Archean upper crust in general. On most Pearce-type trace element discrimination plots, the Lachlan granites plot in the volcanic arc field.

A-type granites are rare and are thought to be derived by high-temperature melting of crustal rocks depleted by earlier I-type magma extraction (Collins et al., 1982; King et al., 1997). Sr-Nd isotope ratios ($^{87}Sr/^{86}Sr$ = 0.704 to 0.720, ε_{Nd} = +4 to −11, e.g., McCulloch and Chappell, 1982; Maas and Nicholls, 2002) define a continuous hyperbolic trend, typical of trends observed in many Phanerozoic granite provinces. I-type granites generally have lower $^{87}Sr/^{86}Sr$ and higher ε_{Nd} than S types, which trend toward the more evolved ($^{87}Sr/^{86}Sr$ = 0.715 to 0.730, ε_{Nd} = −8 to −12; Adams et al., 2005) compositions of the Paleozoic turbidites. However, much isotopic overlap exists between the two granite types. Oxygen isotopes in the I-type (7.9‰–10‰) and S-type (9.2‰–12‰) granites of the eastern Lachlan orogen show correlations with Sr and Nd isotope ratios consistent with mixing of high-$\delta^{18}O$ crustal and low-$\delta^{18}O$ mantle-like components (McCulloch and Chappell, 1982).

Inherited zircon is ubiquitous in S-type granites of the eastern and central Lachlan orogen and has the characteristic pattern (major peaks at ca. 500 and 1000 Ma, with minor peaks to 3.6 Ga; Williams, 1992; Keay et al., 1999; Maas et al., 2001) found in early Paleozoic turbidites of the East Gondwana margin (Veevers, 2000b). By contrast, only weak, ca. 500 Ma zircon inheritance has been reported in S-type granites of the western Lachlan orogen (Elburg, 1996; Bierlein et al., 2001a). The presence of turbidite-derived detrital zircons in many of the S-type granites supports magma sources at least partly within the early Paleozoic metaturbidites, rather than within a hypothetical (but not exposed) Proterozoic basement. Zircon inheritance in I-type granites is typically weak but of the same pattern as in the S-type granites (e.g., Chen and Williams, 1991; Williams, 1992). This indicates a variable sedimentary component in their magma sources.

Debate about the origin of the Lachlan granites pits models that involve melting of distinct crustal protoliths (Chappell and White, 1992, 2001) against models emphasizing large-scale hybridism of mantle and crustal magmas (Gray, 1984, 1990; Collins, 1996, 1998; Keay et al., 1997). The granite compositions are ambiguous in terms of geodynamic setting, probably due to the crustal (sedimentary) influence even on I types. However, mafic dikes, intrusions, and lavas across the orogen carry inferred backarc to arc mantle signatures (Bierlein et al., 2001b; Collins, 2002). This suggests that at least some of the granites, e.g., those in the margin-parallel batholiths of the eastern and central Lachlan orogen, formed within arc/backarc tectonic settings, albeit with much sedimentary (turbidite) input. In contrast, 370 Ma postorogenic magmatism in the western Lachlan orogen and elsewhere is probably related to collision (e.g., Soesoo et al., 1997; Soesoo and Nicholls, 1999).

Lachlan Orogen Tectonic Evolution

Extensive turbidite fan deposition took place in a marginal basin off the Gondwana margin between ca. 490 Ma and 470 Ma, with large turbidite fans spreading out onto recently formed oceanic crust of a developing marginal basin (Gray and Foster, 2004b; Foster et al., 2005). Turbidite deposition occurred at the same time as postorogenic magmatism, cooling, and erosional exhumation inboard in the Delamerian orogen from ca. 490 to 480 Ma (Fig. 5B) (Turner et al., 1992, 1996). Outboard subduction-related arc volcanism was initiated at ca. 485 Ma in the oceanic plate some thousands of kilometers away (based on retrodeformation of the WLO and CLO; Fergusson and Coney, 1992b), leading to the development of the Macquarie volcanic arc complex (Glen et al., 1998).

At ca. 460 Ma, inboard of the arc associated with the older, long-lived subduction zone, subduction has been inferred along both sides of a marginal backarc basin (Gray et al., 1997; Soesoo et al., 1997; Foster and Gray, 2000). Closure of the Lachlan marginal backarc basin has been suggested to have occurred by Woodlark basin–style, doubly divergent subduction (Gray et al., 1997; Soesoo et al., 1997). In this scenario, multiple oceanic thrust systems operated in both the eastern and western parts of the basin, i.e., the ELO and WLO, respectively (Fig. 6A), causing duplexing of oceanic crust, and imbrication and chevron folding in the overlying turbidite wedge (e.g., Spaggiari et al., 2004a, 2004b). Structural thickening of the turbidites in the western thrust system led to erosional exhumation and sediment output into the prograding marginal basin (Melbourne "trough") (Fig. 6D) (Foster et al., 1998; Foster and Gray, 2000). At the same time, west-directed collisional shortening took place in northeastern Tasmania, and deformation of the mineralized Dundas belt (Fig. 4B) was occurring in western Tasmania (Bierlein et al., 2005).

Widespread postorogenic magmatism in the western part of the WLO at ca. 400 Ma (Fig. 5B) was followed by final closure of the marginal basin (Melbourne "trough") at ca. 390 Ma (Fig. 6F). The collision between the WLO accretionary-style thrust belt and the CLO accretionary complex and magmatic arc was accompanied by localized strong to intense north-south folding and regional, meridional crenulation cleavage development in the CLO (Morand and Gray, 1991), and caused reactivation of shear zones in the Delamerian orogen and Precambrian basement massifs (Mount Painter, Broken Hill, and Tyennan of Tasmania) (Gray et al., 2000). This structural thickening and amalgamation of the WLO and CLO led to cratonization of the inner Lachlan (Gray, 1997; Foster and Gray, 2000). Outboard in the ELO, syndeformational ca. 440–435 Ma magmatism and high-T metamorphic belts (e.g., Cooma Complex of Fig. 5A) formed during intermittent east-directed thrusting (Fergusson and VandenBerg, 1990) and periods of extension-related volcanism, particularly at ca. 420 Ma, as part of a Cordilleran or Andean-type margin (Fig. 6F) (Zen, 1995).

Postorogenic magmatism in the eastern part of the WLO (central Victorian magmatic province) occurred at ca. 370–360 Ma (Fig. 5B), while east-directed thrusting in the ELO caused inversion of former extensional basin faults and was followed by postorogenic magmatism (Fig. 5B). Cratonization of the Lachlan orogen was completed by ca. 330 Ma (Powell, 1984).

The Rakaia Wedge (Rangitatan Orogen), New Zealand

The Rangitatan orogen is a collage of accreted arc and forearc assemblages and a turbidite-dominated sediment wedge (Rakaia wedge) that formed as part of Jurassic through Cretaceous accretion along the Mesozoic margin of Gondwana (Coombs et al., 1976; Bradshaw, 1989; Mortimer, 1993). Structurally the sediment wedge is made up of two distinct parts (Fig. 7A): (1) the deeper-level Otago Schist belt characterized by schistosity and transposed layering at the mesoscale and by shear zones and apparent recumbent isoclinal fold nappes at the regional scale (Mortimer, 1993; Gray and Foster, 2004c), and (2) the chevron-folded, poorly cleaved younger sediments of the Pahau terrane that make up the bulk of the succession in the central and northeastern parts of South Island (Bradshaw, 1989).

Deep crustal seismic profiling (Mortimer et al., 2003) has shown that the fault-bounded Permian Dun Mountain ophiolite and Maitai terrane mélange (Cawood, 1987) define a steeply north-dipping interface between the gently folded, interleaved arc-forearc sequences (Median batholith; Brook Street and Murihiku terranes) to the south and the deformed Torlesse composite-terrane submarine fan sediments to the north (Fig. 7B). The Torlesse composite terrane includes a crustal section composed of structurally thickened (~20 km thickness) Permian to Triassic-Jurassic sediments of the Rakaia terrane, structurally overlain by an ~10 km thick wedge of trench sediments (Caples terrane) immediately adjacent to the Livingstone fault (Norris and Craw, 1987). In the immediate hanging wall to the Livingstone fault, intensely deformed monotonous quartzo-feldspathic schist with minor intercalated micaceous schist, greenschist, and metachert (Otago Schist) occupies a domal culmination with a maximum subsurface width of ~220 km and ~20 km of structural relief (Mortimer, 1993) (Fig. 7B). Northwards beyond the Waihemo fault, there is a transition into the tectonically imbricated and weakly metamorphosed Permian-Triassic graywacke sequence of the Rakaia (Older Torlesse) terrane.

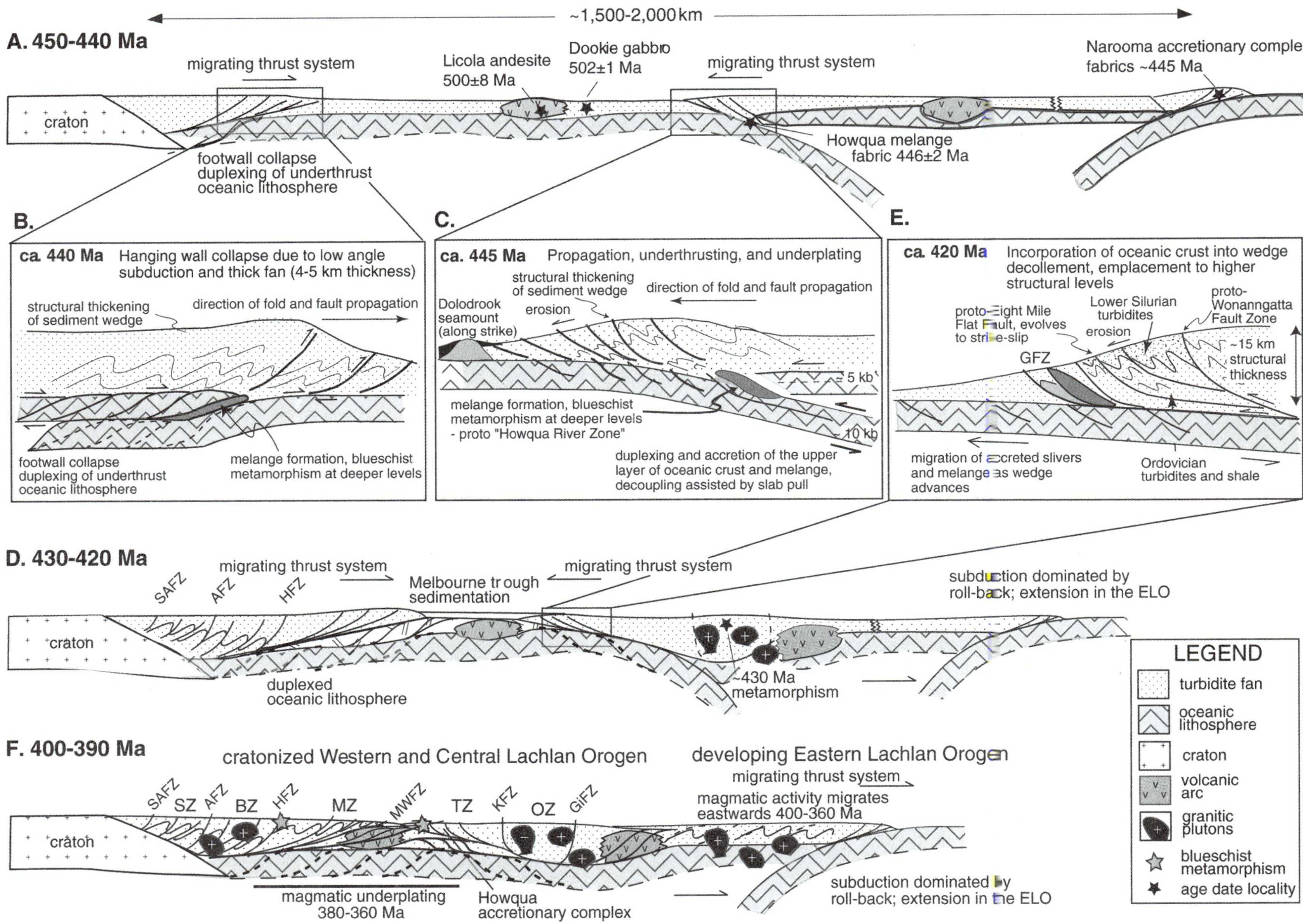

Figure 6. Tectonic evolutionary diagram illustrating the sequential stages in crustal growth for eastern Australia (modified from Spaggiari et al., 2004a, 2004b). AFZ—Avoca fault zone; BZ—Bendigo Zone; ELO—eastern Lachlan orogen; GFZ—Governor fault zone; GiFZ—Gilmore fault zone; HFZ—Heathcote fault zone; KFZ—Kiewa fault zone; MWFZ—Mt Wellington fault zone; MZ—Melbourne Zone; OZ—Omeo Zone; SAFZ—Stawell-Ararat fault zone; SZ—Stawell zone; TZ—Tabberabbera zone

South of the Livingstone fault the crustal section is composed of an ~10–15 km thick succession of Murihiku terrane forearc sediments (Triassic to Jurassic volcanogenic sandstone, siltstone, and tuff) overlying an ~10 km thick arc sequence of Brook Street terrane volcanic rocks (layered ultramafic-gabbro sequences, diorites, and volcaniclastic sediments that are the roots of a Permian intra-oceanic arc), intruded to the south by the Median batholith (see below).

Rakaia Wedge Metamorphism

The metamorphosed part of the Rakaia wedge (Otago Schist) is an ~150 km wide, elongate, NW-trending metamorphic belt (Fig. 7A) that occupies a domal culmination defined by a broad warp in flat-lying schistosity (Fig. 7B). It has lower-metamorphic-grade units (Caples on the south flank and Rakaia on the north flank) structurally overlying the high-grade schist core consisting of garnet-biotite-albite greenschist facies schist (Mortimer, 1993, 2000). Prehnite-pumpellyite facies on the northern and southern flanks (Torlesse and Caples terranes respectively) grade to biotite grade greenschist facies, and then to epidote-amphibolite facies in the broad central portion (Mortimer, 2000).

Mineral parageneses indicate peak *P-T* conditions of 450 °C and 8–10 kbar (Mortimer, 2000), suggesting burial to depths of ~20–30 km and placing the Otago Schist in a moderate- to high-*P/T* metamorphic series (Barrovian-type). Relict blueschist assemblages occur in mafic greenschist-chert sequences close to the Caples-Torlesse boundary (Yardley, 1982). Riebeckite and crossite relicts occur mantled by actinolite, indicating an earlier high-pressure intermediate-type metamorphism that is overprinted by the Barrovian moderate- to high-*P/T* metamorphism.

Rakaia Wedge Magmatism

The Rakaia wedge shows no evidence of magmatism, but subduction-related magmatism was long-lived in the overriding plate, typified by the Median batholith (Mortimer et al., 1999a). Granitic complexes exposed in South Island and recovered from submarine extensions range in age from Late Devonian to Late Cretaceous (Tulloch, 1983, 1988; Wandres and Bradshaw, 2005). The oldest granites were emplaced into Paleozoic metasediments of the Western province. This includes the voluminous Late Devonian (ca. 375 Ma) mixed S- and I-type Karamea batholith, the 375 Ma mafic Riwaka Complex, and Carboniferous (360–300 Ma) I-, A-, and S-type plutons (Cooper and Tulloch, 1992; Muir et al., 1996a, 1996b, 1998). The ca. 375 Ma granites are related to a magmatic episode that is also recognized in SE Australia and West Antarctica.

Fragments of a long-lived magmatic arc (arc volcanics, volcanogenic sediments, and granitoids) are preserved in the Mesozoic Median Tectonic Zone, between the Paleozoic Gondwana margin of the Western province and the accreted Eastern province (Bradshaw, 1993). The granitoids, representing semicontinuous (ca. 230–125 Ma) calc-alkaline, metaluminous, low- to medium-K, I-type magmatism, are exposed in the northwest and south of South Island and have been assigned to the 10,200 km^2 Median batholith (Kimbrough et al., 1993, 1994; Muir et al., 1998; Mortimer et al., 1999a). As in other Mesozoic Andean arcs, mafic granitoids (gabbros, diorites) are common; A-type granites are rare. Although there is little evidence of crustal involvement (no S-type granites, almost no zircon inheritance, $^{87}Sr/^{86}Sr \approx 0.7037$, $\varepsilon_{Nd} \approx +4$; Muir et al., 1998; Mortimer et al., 1999b), the Median batholith arc is thought to have been emplaced at and into the Paleozoic Gondwana margin of the Western province, in response to outboard accretion of the Eastern province (e.g., Muir et al., 1998; Daczko et al., 2001). However, the geometry of the margin and the locus of subduction responsible for the Mesozoic arc are still unclear.

Final arc-margin collision is marked by a chemically distinct suite of stitching granites, the 124–111 Ma Separation Point Suite (in the north of South Island) and their likely lower crustal equivalents, the 126–119 Ma metadiorites and metagabbros of the Western Fiordland Orthogneiss (e.g., McCulloch et al., 1987; Muir et al., 1995, 1998). These I-type suites, while covering a wide range in SiO_2, are distinctly sodic in character and have high to very high Sr/Y ratios, thus resembling Archean TTG (tonalite-trondhjemite-granodiorite) or modern adakite suites (Muir et al., 1995). Tulloch and Kimbrough (2003) drew attention to the spatial and age pattern of early, outboard low-Sr/Y and later, inboard high-Sr/Y granitic magmatism, represented by the Median batholith and Separation Point–Western Fiordland Orthogneiss, respectively.

This margin-normal arrangement, mirrored to a surprising extent at other Mesozoic active margins (e.g., Petford and Atherton, 1996), is thought to reflect melting within metabasaltic subarc crust and mantle in a progressively thickening arc. The youngest and most inboard granites related to this "low-Sr/Y, high-Sr/Y" association, the ca. 110 Ma Rahu Suite, is represented by mildly "high-Sr/Y", I-type and transitional I-S-type plutons in the Hohonu and Paparoa batholiths in NW South Island, intruding Western province Paleozoic metasediments. Zircon inheritance (500–1000 Ma), evolved Sr-Nd isotopic signature ($^{87}Sr/^{86}Sr = 0.7062$ to 0.7085, $\varepsilon_{Nd} \approx -5$), and elevated $\delta^{18}O$ (Tulloch, 1988; Waight et al., 1998a) indicate that Rahu Suite granites may represent blends of Separation Point–type arc crust- and Paleozoic margin-derived magmas (Tulloch and Kimbrough, 2003). Waight et al. (1998a, 1998c) suggest that Rahu Suite magmatism occurred in a continental margin undergoing a switch from convergence to extension. Advanced extension of basement is marked by ca. 82 Ma mafic dikes and A-type granites in the Western province (Hohonu Dike Swarm, Buller Gorge Dikes, French Creek Suite) that are coeval with the opening of the Tasman Sea between SE Australia and New Zealand (Waight et al., 1998b).

In contrast to the Lachlan and Damara orogens, granites do not occur (or are not presently exposed) in the accreted turbidite terranes of New Zealand's Eastern province. All presently

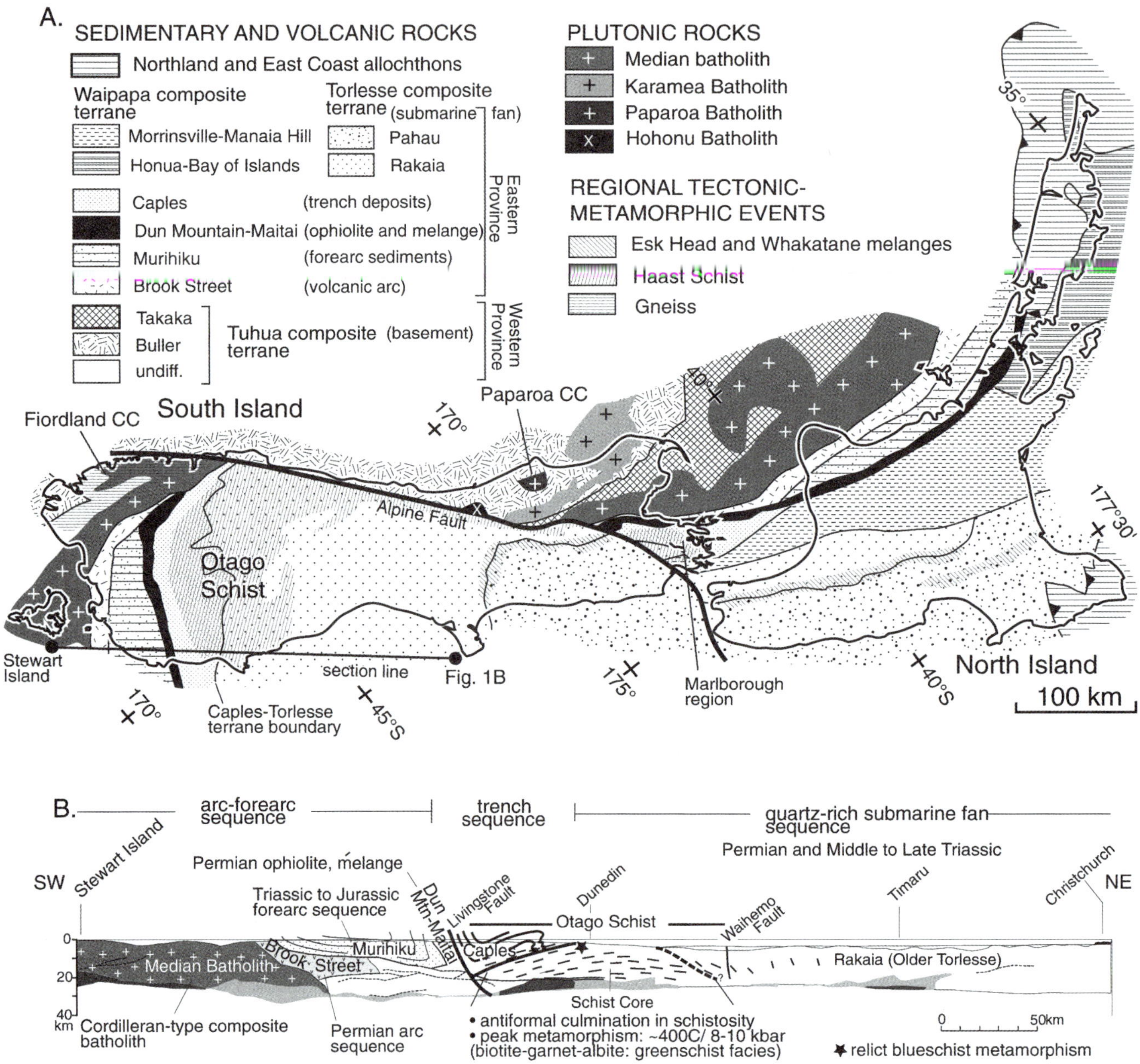

Figure 7 (*on this and following page*). (A) Geologic map of New Zealand showing the key tectonic elements, the main terranes and provinces, the distribution of plutonic rocks, and Torlesse terrane components. CC—core complex. (Map modified from tectonostratigraphic terrane map of the New Zealand Institute of Geological and Nuclear Sciences [IGNS] Qmap Series.) (B) Geological profile showing the crustal architecture of the southern part of South Island (based on seismic reflection profile from Mortimer et al., 2003, Fig. 8 therein).

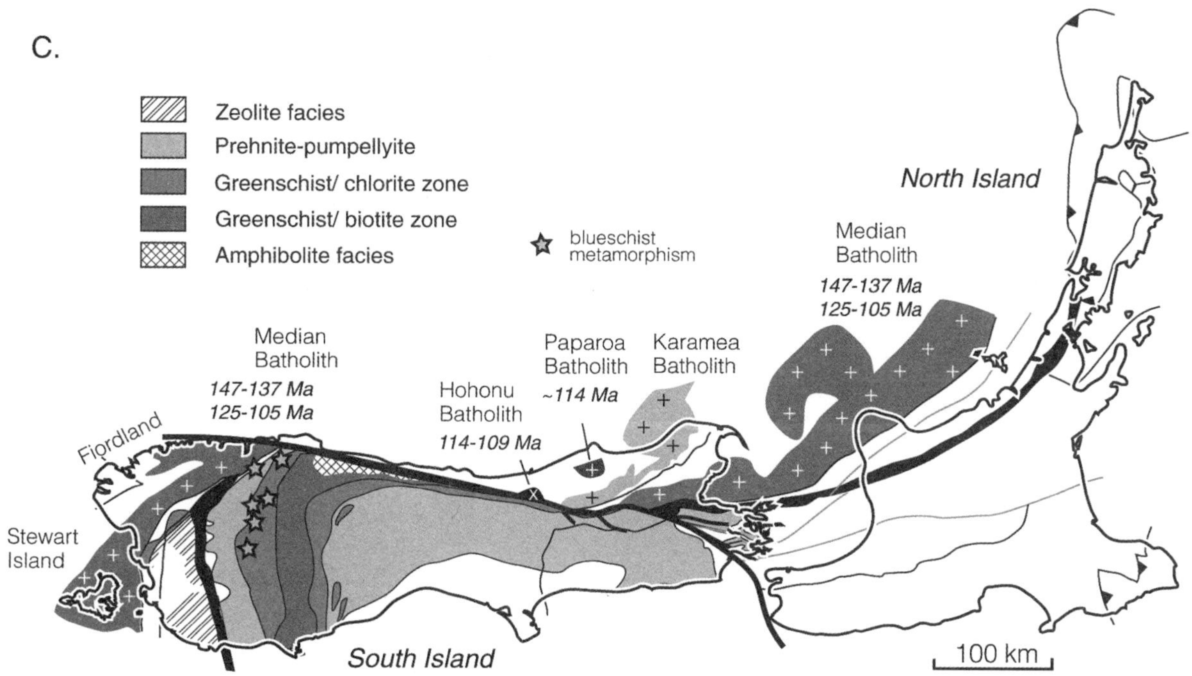

Figure 7 *(continued)*. (C) Metamorphic map of New Zealand also showing the locations and crystallization ages of the various granitoids. Metamorphic data from Yardley (1982), Coombs and Cox (1991), and Mortimer (2000). Granite data from Muir et al. (1988), Tulloch and Kimbrough (1989), Kimbrough et al. (1994), Waight et al. (1998a, 1998b), and Mortimer et al. (1999).

known granites coeval with the Rangitatan orogeny occur within the Mesozoic arc (Median Tectonic Zone, or Tutoko Complex; Wandres and Bradshaw, 2005) or immediately inboard within the Paleozoic Gondwana margin of the Western province. Granite compositions overall are more mafic and almost exclusively I-type (although true granites and A types do occur), zircon inheritance is absent or much weaker (compared to Lachlan orogen granites), and crustally dominated Sr-Nd-O isotopic compositions are rare. This reflects the absence of Proterozoic basement and suggests that equivalents of older oceanic arc terranes of the Eastern province (Permian Murihiku and Brook Street terranes), rather than the crustally derived turbidites accreted farther outboard, became involved in Jurassic–Early Cretaceous granite magma production.

Rakaia Wedge Tectonic Evolution

The generally accepted tectonic evolutionary model for New Zealand (e.g., Coombs et al., 1976; Bradshaw, 1989; Roser and Cooper, 1990) requires convergence of the Permian-Jurassic Torlesse fan with arc–forearc–trench slope sequences (Brook Street, Maitai, Murihiku, and Caples terranes) due to subduction of Permian oceanic crust beneath the Gondwana margin (Fig. 8). Coupling related to this subduction caused structural thickening of the Torlesse turbidite fan in the Late Jurassic (ca. 150–140 Ma), with Early Cretaceous (130–120 Ma) structural interleaving of Caples and deformed Torlesse sediment resulting in final accretion of the turbidite fan to the margin (Bradshaw, 1989; Mortimer, 1993; Gray and Foster, 2004c). This coincided with collision between the Median Tectonic Zone (Median batholith) and the Western province and was marked by the sudden onset of Separation Point–type magmatism at 125–118 Ma (Waight et al., 1998c; Kimbrough et al., 1994). Erosional exhumation and core complex development occurred in the late Early Cretaceous (110–90 Ma) as a precursor to Tasman Sea opening (Ireland and Gibson, 1998; Spell et al., 2000). Thermal decay from Tasman Sea spreading commenced at ca. 84 Ma and was well under way at 70 Ma (Johnson and Veevers, 1984).

NEOPROTEROZOIC CONTINENTAL GROWTH (THE DAMARA BELT OF THE DAMARA OROGEN, NAMIBIA): RECYCLED TURBIDITES INVOLVING A WILSON CYCLE?

The Neoproterozoic Damara orogen is part of the Pan-African orogenic system (e.g., Coward, 1981, 1983) and within Namibia consists of the Damara belt, or Inland Branch, and coastal branches including the Kaoko and Gariep belts in the north and south respectively (Fig. 9A) (e.g., Porada et al., 1983; Prave, 1996). Turbidites constitute a major component of the Damara belt or Inland Branch (Martin and Porada, 1977; Hoffmann, 1983; Martin, 1983; Miller, 1983a, 1983b; Kukla et al., 1988) and represent a Neoproterozoic submarine fan system

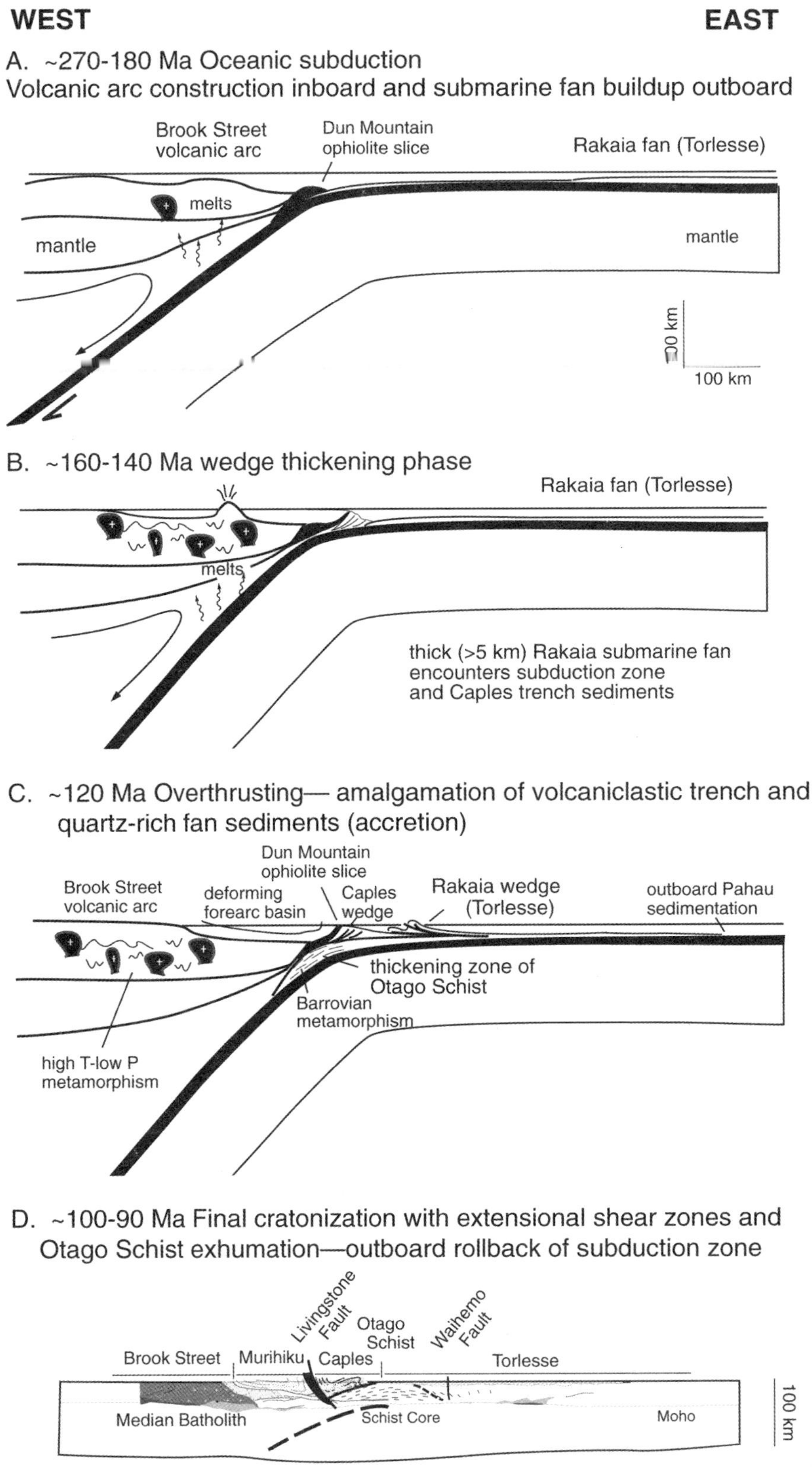

Figure 8. Tectonic evolutionary diagrams illustrating the sequential stages in crustal growth for New Zealand (modified from Coombs et al., 1976). Timing for the schist wedge based on Ar-Ar data from Gray and Foster (2004c).

and/or accretionary complex that formed in part of an ocean basin (Khomas trough) between the Congo and Kalahari cratons (Miller et al., 1983a; Kukla et al., 1988; Kukla and Stanistreet, 1991).

Relict oceanic lithosphere is possibly preserved by the thin (200–300 m thick), but laterally extensive (over 350 km in length) (Fig. 9B), shear zone–hosted Matchless amphibolite (Barnes, 1983; Killick, 2000). This unit consists of intensely deformed basalt, pillow basalt, chert, and gabbro, with tholeiitic geochemistry (Schmidt and Wedepohl, 1983), as well as dispersed serpentinite bodies (boudins). The basalt chemistry has been used to argue for an ensialic origin with basalts formed on thin crust (Schmidt and Wedepohl, 1983), although the dispersed ultramafic bodies have Alpine chemical affinities more typical of oceanic lithosphere (Barnes, 1983).

The flanks of the former basin now consist of craton-verging thrust systems within shelf carbonates (e.g., Hakos and Naukluft nappes in the Southern Margin Zone) (Korn and Martin, 1959; Coward, 1983; Miller, 1983b). The Northern Zone underwent metamorphism from ca. 540 to 530 Ma (Goscombe et al., 2004). The basinal Khomas turbidite sequence is metamorphosed to form the Kuiseb Schist now subdivided into two major structural/metamorphic zones, referred to as the Central and Southern Zones (Fig. 9B). The granite-dominated Central Zone (Fig. 10A), metamorphosed under high-*T*/low-*P* conditions between 538 and 505 Ma, is floored by an attenuated Congo-like craton, with dome and basin fold interference (e.g., Jacob et al., 1983; Coward, 1983; Miller, 1983b, p. 459; Kröner, 1984; Kisters et al., 2004), and possible extensional structures (e.g., Oliver, 1994). The Central Zone has two phases of metamorphism at 538–516 Ma and 511–505 Ma with granitic magmatism between ca. 560 Ma and ca. 470 Ma (Jacob et al., 2000; Jung and Mezger, 2003a, 2003b).

The thicker basinal sequence has undergone intermediate-*T*/intermediate-*P* (Barrovian series) metamorphism and is now part of the ~100 km wide Southern or Khomas Zone (Kasch, 1983a; Kukla, 1993). This zone consists of homoclinally north-dipping Kuiseb Schist with transposed foliation and schistosity (Hälbich, 1977; Miller, 1983b, p. 461–466; Kukla et al., 1988; Kukla and Stanistreet, 1991), which represents a major shear zone interface transitional into the basement-cored fold-and-thrust nappes of the Southern Margin Zone (Fig. 9B) (Miller, 1983b, p. 463–468; Coward, 1983). Based on U-Pb ages for metamorphic monazite, Kukla (1993) suggests an age of 525–515 Ma for peak metamorphism in the Kuiseb Schist; a lower limit is given by the intrusion of the 505 ± 4 Ma posttectonic Donkerhuk granite (Kukla, 1993).

The western Northern Zone (Ugab domain) is influenced by transition between basinal turbidite facies and slope-to-platform carbonates (Miller et al., 1983b; Swart, 1992; Passchier et al., 2002). It also represents the junction between Kaoko and Inland Branches, and is intruded by granites of the Central Zone (Coward, 1983; Maloof, 2000; Passchier et al., 2002; Goscombe et al., 2004).

Damara Belt Metamorphism

The Damara belt has a distinct metamorphic zonation with belts of intermediate-*T*/intermediate-*P* (Barrovian-style) metamorphism bounding the Central Zone (Fig. 10B), which has undergone moderate-*P*/high-*T* metamorphism with peak temperatures ~650 °C and pressures ~4.3 kbar (Kasch, 1983a). The Central Zone has garnet-cordierite granulites (e.g., Puhan, 1983; Masberg, 2000) and clockwise *P-T* paths and shows a high thermal gradient (30–50 °C/km) (e.g., Jung, 2000). It contains A-type, posttectonic granites that originate from mid-crustal, low-pressure melting (McDermott et al., 2000).

Intermediate-*T*/intermediate-*P* (Barrovian-style) metamorphism occurs within the Southern and Northern Zones. The Southern Zone metamorphic conditions are moderate-*P* and -*T*, with peak conditions of ~600 °C and ~10 kbar but showing a decrease in pressure northwards to ~4 kbar toward the Okahandja Lineament (Kasch, 1983a). The Southern Zone has a low thermal gradient (18–22 °C/km), with syntectonic garnet and posttectonic staurolite in metapelites defining a clockwise *P-T* path. The Northern Zone shows high *P* and moderate *T* with a clockwise *P-T* path; peak conditions are estimated at 635 °C and 8–7 kbar and an average thermal gradient of 21 °C/km (Goscombe et al., 2004).

Damara Belt Magmatism

As in the Lachlan orogen, granitic rocks are a major component of the Damara orogen (Inland Branch), composing ~74,000 km^2, or roughly 25% of outcrop (see review by Miller, 1983a). Most of the granitic magmatism was concentrated in the high-temperature Central Zone, with minor activity in the Northern Zone; the very large posttectonic Donkerhuk granite (5000 km^2) represents the only significant granitic complex in the lower-temperature Southern Zone. Granites dominate the compositional spectrum, with an areal ratio of granite to granodiorite-tonalite to diorite-gabbro of 96:2:2. Characterization of Damara granites is hampered by the structural and lithological complexity of syn- and late- to posttectonic intrusions, the high metamorphic grade of the deeply eroded orogen, and the lack of reliable high-precision geochronology. The pre-1990 granite age framework, based largely on Rb-Sr whole-rock isochrons, suggests a period of 190 m.y., from 650 to 460 Ma, for the Pan-African granitic magmatism. More recent studies (U-Pb monazite, titanite, zircon, allanite; Sm-Nd garnet) indicate an age range for granitic magmatism from 560 to 470 Ma (e.g., Jacob et al., 2000; de Kock et al., 2000; Jung and Mezger, 2003a), similar to the age range of regional metamorphism.

McDermott et al. (1996) used geochemistry to divide the granites into three groups irrespective of timing relative to deformation. Group 1 comprises the widespread, mostly peraluminous (S-type) leucogranites (including, for example, the large posttectonic Donkerhuk batholith) and Salem-type granites (a generic term for a diverse group of porphyritic, biotite-rich granites) and

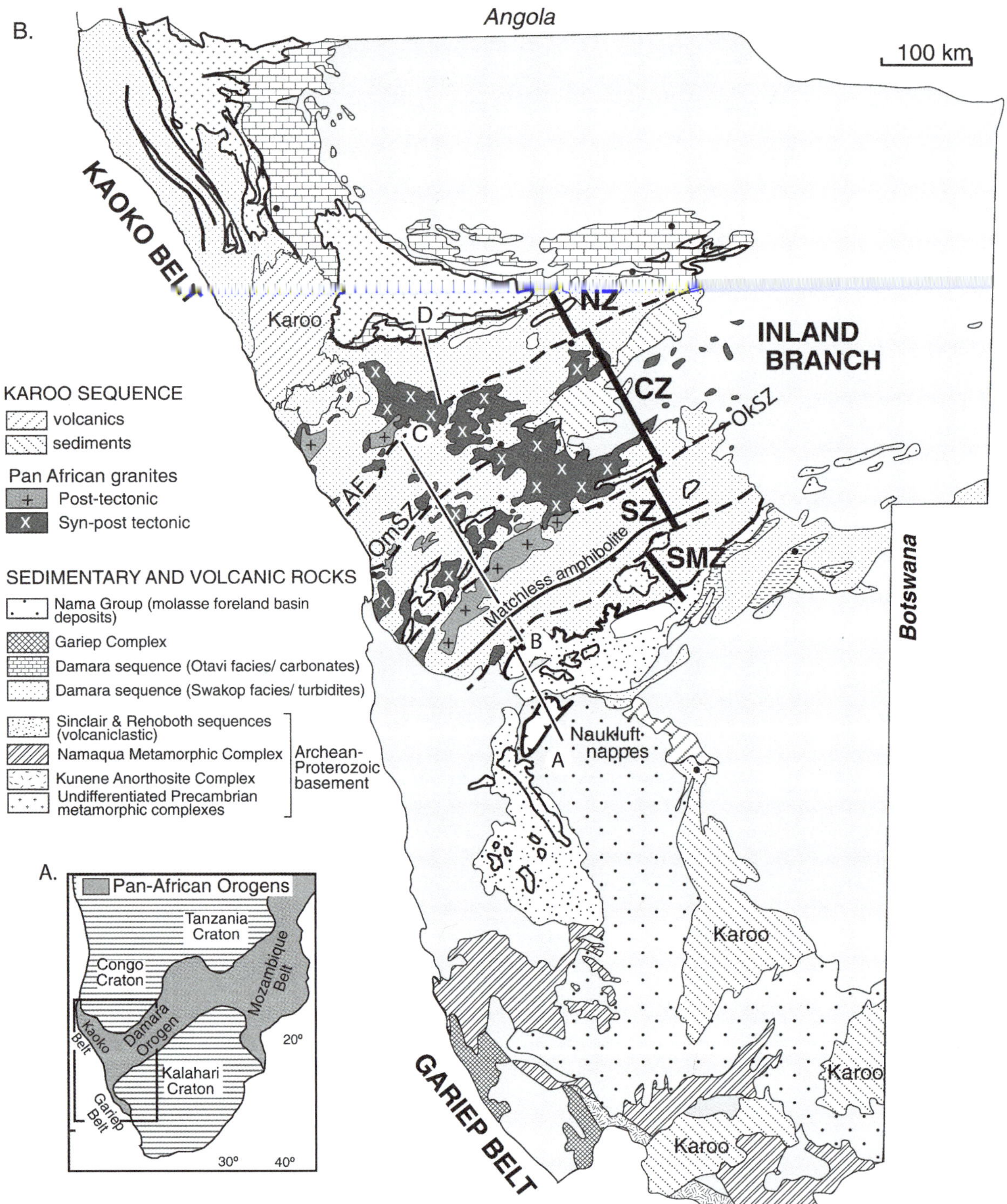

Figure 9 *(on this and following page)*. (A) Simplified map of southern Africa showing the Pan-African orogens and the location of the map shown in B. (B) Geologic map of the Damara orogen, Namibia, showing the key tectonic elements, the major structural zones, the distribution of plutonic rocks and Kuiseb Schist (former Khomas trough turbidites), and the location of the profile A–D in C. (Map modified from geological map of Namibia.) NZ—Northern Zone; CZ—Central Zone; SZ—Southern Zone; SMZ—Southern Margin Zone; AF—Autseib fault; OmSZ—Omaruru shear zone; OkSZ—Okahandja shear zone.

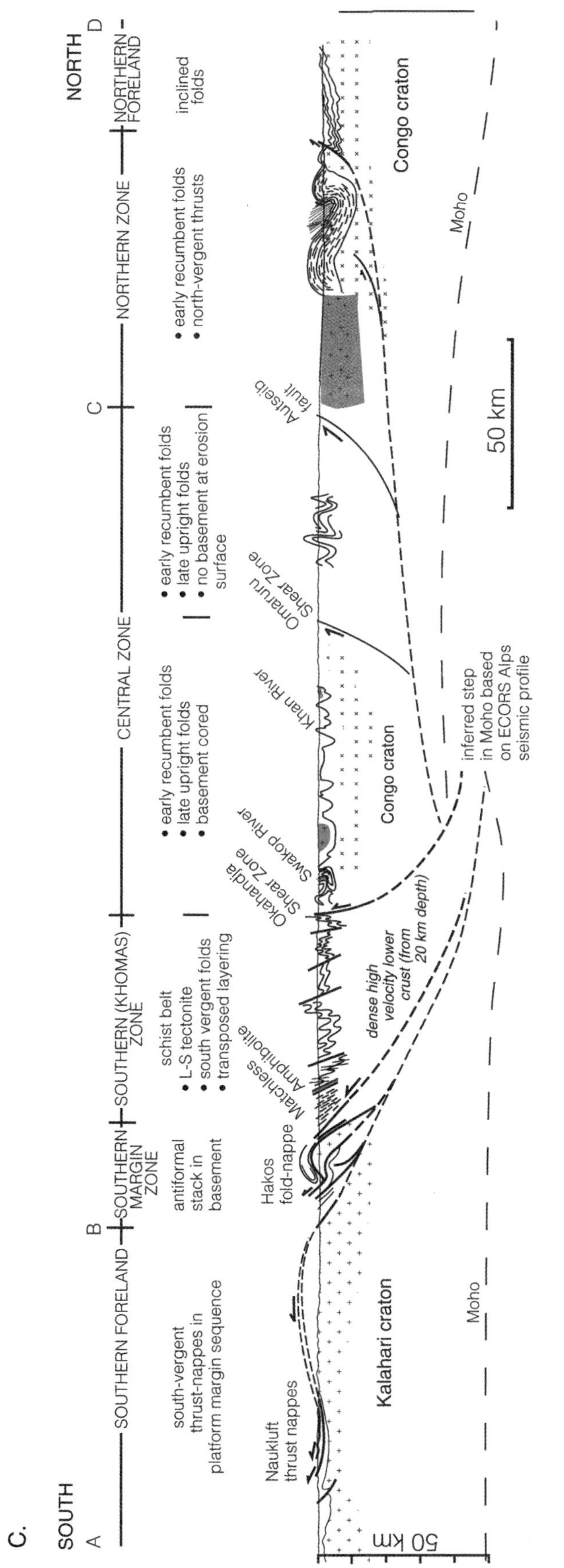

the volumetrically minor but economically important alaskites (e.g., Rössing; Basson and Greenway, 2004). Mildly peraluminous, Nb-Zr-Y (high field strength elements, or HFSE) -rich A-type granites with within-plate signatures make up group 2, while group 3 encompasses the more mafic, metaluminous and partly calc-alkaline quartz diorites and granodiorites.

Group 1 granites, volumetrically by far the most important, show an exceptionally wide range in Sr-Nd isotope ratios ($^{87}Sr/^{86}Sr$ = 0.708 to 0.740, $\varepsilon_{Nd} \approx -2$ to –19; Haack et al., 1982; McDermott et al., 1996; Jung et al., 2001, 2003), reflecting involvement of negative ε_{Nd} Archean-Mesoproterozoic basement gneisses in some complexes (e.g., Jung et al., 2003). Many of the S-type granites, however, have ε_{Nd} near –5, similar to Neoproterozoic (Swakop Group) turbidites (McDermott and Hawkesworth, 1990; Jung et al., 2001). $\delta^{18}O$ values are typically high (11‰–15‰, e.g., Jung et al., 2000, 2003), similar to both Damara metasediments and pre-Damara basement. McDermott et al. (1996) suggest a quasi-secular decrease in ε_{Nd} with decreasing granite age, perhaps due to a retreat of the melting zone into the lower crust after the metamorphic peak (<500 Ma), but this appears to be an oversimplification: The Rb-Sr ages used are unreliable, and large ranges exist even within single complexes (e.g., ~9 ε_{Nd} units for posttectonic Donkerhuk granite).

Chemical and isotopic data for the S-type granites indicate overwhelmingly crustal sources within the Damara metasediments and/or pre-Damara basement highs, with melt generation by dehydration melting at mid-crustal levels (pressures >6 kb). The distinct chemistry of the alaskites suggests they represent rapidly withdrawn (disequilibrium) melts generated by muscovite dehydration melting of pelites (McDermott et al., 1996).

Group 2 granites, with A-type affinities, show a range of ages (525 Ma syntectonic to ca. 486 Ma posttectonic; Jung et al., 1998a, 2000; McDermott et al., 2000), similar to the S-type granites. Sr-Nd isotope signatures are less evolved ($^{87}Sr/^{86}Sr$ = 0.7034 to 0.709, ε_{Nd} = 0 to –6, as low as –16 for the Sorris-Sorris A type), but $\delta^{18}O$ is still high (10‰–13‰). Generation of hot (>800 °C) A-type magmas is thought to involve remelting of a deep-seated (pressures of 8–10 kb) metatonalite source followed by extensive differentiation in some cases. Mixed (mantle + crust) isotopic signatures may be a feature of the metaigneous lower crustal source rocks.

Like group 2, group 3 quartz diorites and granodiorites are quite rare (2%) but have received attention because they may help establish the tectonic setting of the Pan-African granitoids. Group 3 diorites are metaluminous, carry hornblende and relict clinopyroxene and are clearly I-type. However, isotopic signatures vary conspicuously ($^{87}Sr/^{86}Sr \approx 0.704$ to 0.713, ε_{Nd} = 0 to –20, $\delta^{18}O$

Figure 9 *(continued).* (C) Geological profile showing the crustal architecture of the Damara belt (Inland Branch) of the Damara orogen (modified from profiles on Damara orogen 1:500,000 map sheets; Miller and Grote, 1988). ECORS—Étude Continentale et Oceanique par Reflexion et Refraction Sismique.

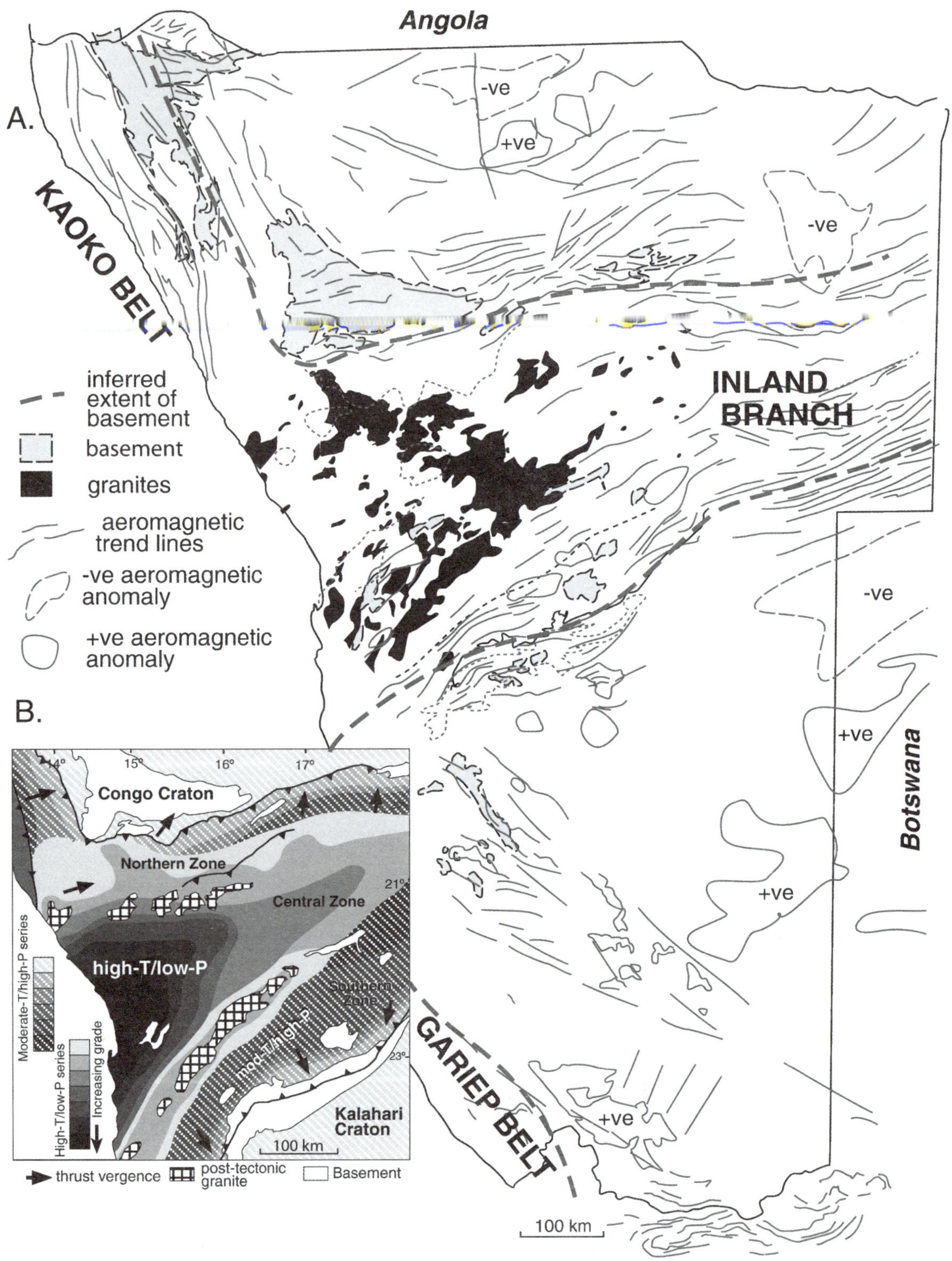

Figure 10. Map of Namibia showing aeromagnetic trend lines and positions of basement (based on interpretation of map 1 from Corner, 2000) and the distribution and shape pattern of granitoids (based on Hawkesworth et al., 1983. +ve—magnetic high, commonly indicating granitoids and/or basement; -ve—magnetic low, commonly indicating thick nonmagnetic sedimentary accumulations.

= 7‰ to 13‰; Haack et al., 1982; Hawkesworth et al., 1981; Miller, 1983a; McDermott et al., 1996; Jung et al., 2002; van de Flierdt et al., 2003) and clearly imply a range of magma sources. Titanite and zircon U-Pb ages near 540 Ma suggest the diorites are consistently among the oldest Pan-African intrusives, coeval with the onset of high-grade metamorphism (Jung and Mezger, 2003a). A distinct suite of ca. 540 Ma syenites (with more limited Sr-Nd isotopic signature; Jung et al., 1998b) was also formed at that time. Petrogenetic models for at least some dioritic complexes (Goas, Okongava in the Central Zone, Bandombaai in the southern Kaoko belt, with mild to high Sr/Y ratios and depletion in heavy rare earth elements) involve remelting of high-K basalts in the deep crust. Occurrence of very low ε_{Nd}, high $\delta^{18}O$, and strongly retarded Pb isotope ratios suggest involvement of Archean-Proterozoic granulitic basement, presumably via deep crustal assimilation. Other diorites, with more primitive Sr-Nd-O isotope signatures (Hawkesworth et al., 1981) appear to lack this assimilation component.

The dominance of S-type granites and the absence of clearly mantle-like (positive ε_{Nd}, low $\delta^{18}O$) isotopic signatures even in mafic I-type granitoids indicate the importance of crustal recycling rather than growth during the Damara orogeny (McDermott and Hawkesworth, 1990; Jung et al., 2003). Clearly, tectonic processes, thermal conditions (including the well-known high heat production in Damara metasediments; Haack et al., 1983), and high fertility combined to produce massive mid/lower crustal melting. The issue of the tectonic setting for Damara Pan-African magmatism is far from settled. Magmatic arc models (e.g., Kasch, 1983a) have been questioned because of the lack of margin-parallel linear batholiths and margin-normal geochemical gradients, the preponderance of felsic S-type granites, and the scarcity of mafic granitoids with mantle-like isotopic signatures (e.g., Jung et al., 2003). With respect to the last point, it should be kept in mind that pervasive crustal melting would set up a density filter preventing further ascent of mafic magmas, and would greatly increase the chance of magma hybridization in the crust. The small volume and crustal isotopic signatures of the rare I-type granitoids thus does not rule out greater mantle contributions at depth.

Damara Tectonic Evolution

Rifting between the Congo and Kalahari cratons has been inferred to develop along three northeast-trending intracontinental rifts between ca. 840 Ma and ca. 730 Ma (Miller, 1983b), although dating of Northern Zone rift-related rhyolites has provided ages 746 ± 2 Ma and 747 ± 2 Ma (Hoffman et al., 1996). Synrifting silicic volcanism in the Central Zone took place at ca. 750 Ma (de Kock et al., 2000). Eventual development of an oceanic spreading center and growth of oceanic crust at ca. 700 Ma in the southern part led to development of the Khomas ocean basin (Miller, 1983b). Like the Lachlan orogen, there are conflicting tectonic hypotheses for closure of the Khomas basin and the nature of the substrate to the turbidites (e.g., Porada, 1979, 1983; Martin, 1983).

The Khomas basin has been depicted either as an intracontinental basin between the Congo and Kalahari cratons with turbidites sitting on attenuated continental lithosphere (Porada, 1979, 1983; Martin and Porada, 1977; Hawkesworth et al., 1983), or as an ocean basin floored in part by oceanic lithosphere that closed by subduction beneath the leading edge of the Congo craton (Fig. 11) to form an Andean-type margin with development of voluminous magmatism (e.g., Barnes and Sawyer, 1980; Miller, 1983b; Kasch, 1983b; de Kock, 1992), not unlike the Silurian-Devonian of the eastern Lachlan orogen (e.g., Zen, 1995). Basement-cored domal culminations in the Central Zone (Jacob et al., 1983; Puhan, 1983; Kröner, 1984; Oliver, 1994; Kisters et al., 2004) and basement-type isotopic signatures in some granites require the presence of attenuated Congo continental crust in this part of the Khomas ocean basin, probably not unlike the present-day Japan Sea, which has a complex distribution of attenuated continental crust, rifted crust, and oceanic crust (Tamaki, 1995).

Closure of the former Khomas ocean basin involved high-angle convergence with overthrusting at the margins to give a doubly vergent orogen (Fig. 9C). Apart from the laterally continuous, shear zone–hosted Matchless amphibolite there is no classic suture. The distinct metamorphic zonation of the Inland Branch, with Barrovian metamorphism on the orogen flanks, reflects structural thickening at the craton margins, while Andean- or Cordilleran-style high-*T*/low-*P* metamorphic conditions in the Central Zone and accretionary prism-like features of the Southern Zone Kuiseb Schist (Kukla and Stanistreet, 1991) reflect subduction beneath this zone (cf. Maloof, 2000). Arguments against subduction-related closure of the Khomas ocean basin have been based on granite geochemistry (Hawkesworth et al., 1983), as well as the lack of blueschists and eclogites or typical subduction zone low-*T*/high-*P* metamorphism (Kröner, 1982); but see Kasch (1983a) for a discussion.

PROCESSES OF ACCRETION IN TURBIDITE-DOMINATED OROGENS

Crustal accretion for the Lachlan, Rangitatan, and Damara orogens as described in this paper involves deformation of turbidite-dominated submarine fans and the underlying oceanic basement, resulting in marked (>50%) shortening and crustal thickening. In the deformed state, the turbidites now occur either as belts of thrust-imbricated, upright chevron-folded, low-grade turbidites or as zones of higher-metamorphic-grade, inclined, homoclinally dipping schistosity and transposition layering.

Turbidite deformation resulted from convergent margin tectonism driven by closure of marginal basins (e.g., western Lachlan orogen), accretion of intraoceanic plateau (e.g., Chatham Rise of the Rangitatan orogen), and Wilson cycle closure of a small ocean basin by convergence between cratonic nuclei (e.g., Damara belt of the Damara orogen). The turbidites are the common link, but each

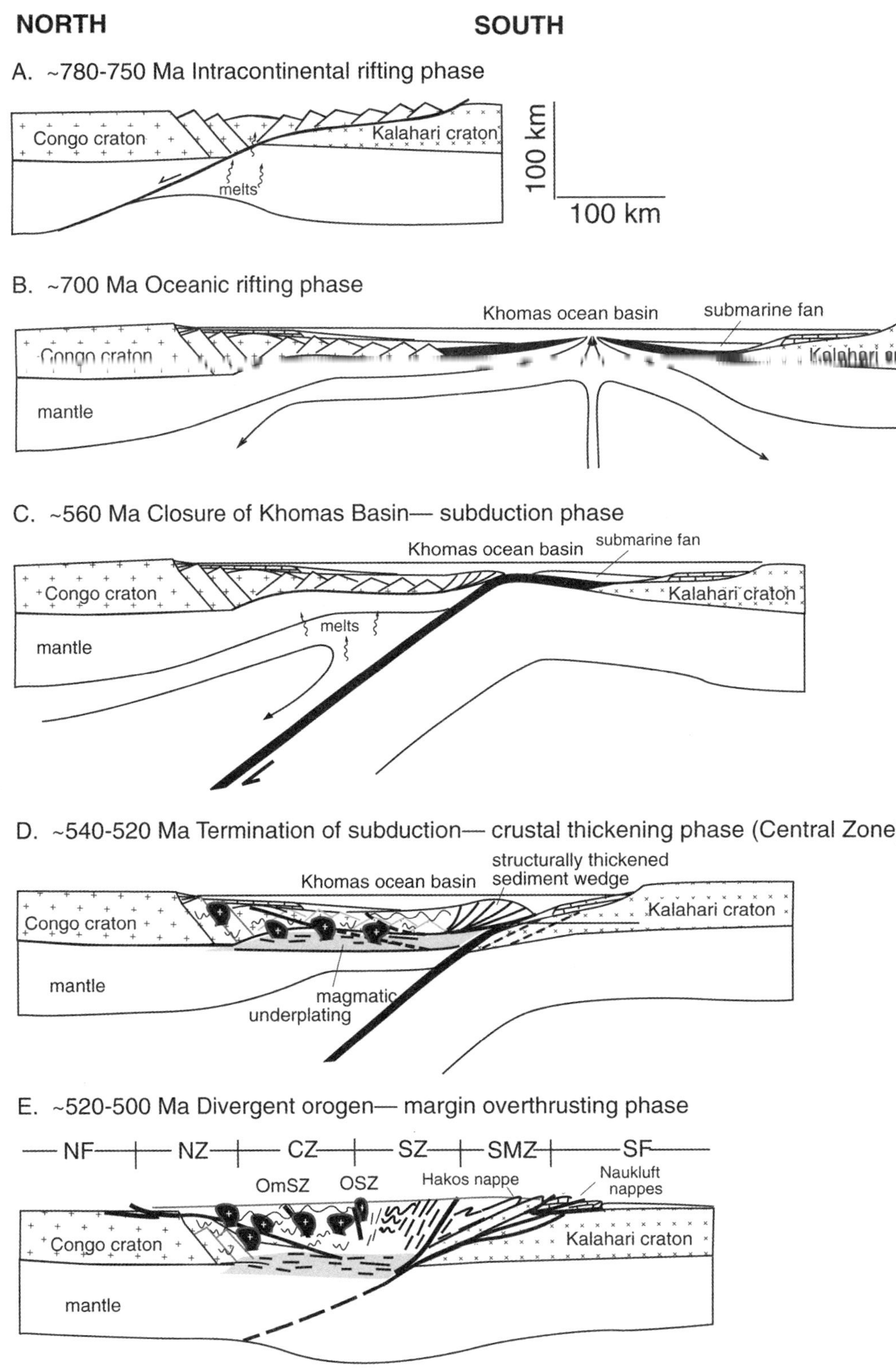

Figure 11. Tectonic evolution diagrams for the Damara belt (Inland Branch) of the Damara orogen highlighting asymmetric rifting between the Congo and Kalahari cratons (modified from Barnes and Sawyer, 1980). CZ—Central Zone; NF—Northern foreland; NZ—Northern Zone; OmSZ—Omaruru shear zone; OSZ—Okahandja shear zone; SF—Southern foreland; SMZ—Southern Margin Zone; SZ—Southern Zone.

orogen has its individual style or crustal architecture as a function of the tectonic setting and the position of the turbidites in this setting. The Lachlan, Rangitatan, and Damara orogens show that parameters such as the thickness, structural style, metamorphic character, and tectonic position of the turbidites, or former submarine fan, as well as the age and nature of the underlying substrate, are all important variables for crustal evolution (Table 1; Fig. 12). Proximity to large continental landmasses with active uplift and relatively rapid erosion is required to produce the large turbidite fans that characterize these orogens and others like them (e.g., Ingersoll et al., 2003).

Positions of Turbidites in Ancient Orogens

The tectonic setting and position of the turbidite fans affect the final architecture and characteristics of the deformed turbidite in each of the three orogens. In most orogens, tectonic reconstructions are based on recognition and interpretations of key geologic or tectonic elements that make up the orogen, as well as the structural architecture and tectonic vergence defined by fault or shear zone dips, the distribution of rock types, and the temporal and spatial distributions of metamorphism and igneous activity (e.g., Gray and Foster, 2004a, 2004b, for the Lachlan orogen; Mortimer, 2004, for the Rakaia wedge of New Zealand). The thickness of the turbidite successions and the degree of shortening (>50%) and thickening (up to 300%) supports an oceanic depositional setting so that the final geometry is not the result of ensialic deformation as inferred from consideration of granite petrogenesis as the determinative factor. The Lachlan orogen has an extensive volume of turbidites, extending over the 750 km width, thrust systems of mixed vergence, and an included high-*T*/low-*P* metamorphic complex. How this belt evolved and in what setting has been debated, but there are certain characteristics of the Lachlan orogen that are important for any tectonic reconstruction (see Gray and Foster, 2004b, p. 792–793). These include

1. three simultaneously operating oceanic thrust systems in different parts of a marginal ocean basin that was behind a long-lived, outboard subduction system along the eastern margin of Gondwana;
2. blueschist blocks in serpentinite-matrix mélange along major faults in the WLO and CLO;
3. mélange and broken formation along faults within the frontal fault system of the CLO (the Howqua accretionary complex);
4. discordant posttectonic granites in the WLO and large elongate composite granitoid bodies in the CLO (northwest-trending) and ELO (north-trending); and
5. shear zone–bounded high-*T*/low-*P* metamorphic complex with regional aureole-style metamorphism related to significant high-level S-type granites.

These features indicate that Lachlan orogen accretion was by thickening and imbrication of extensive and thick submarine fans, including the underlying oceanic lithosphere, in a "Woodlark basin style" doubly divergent subduction system (e.g., Gray et al., 1997; Soesoo et al., 1997).

TABLE 1. TURBIDITE-DOMINATED OROGEN PARAMETERS

	WESTERN LACHLAN	KUISEB/DAMARA	UGAB/DAMARA	TORLESSE
PRECONVERGENCE CHARACTERISTICS				
FAN sedimentation	Cambrian-Devonian 505–460 Ma	Neoproterozoic ca. 750–700 Ma	Neoproterozoic ca. 750–700 Ma	Permian-Triassic 280–200 Ma
THICKNESS	4–6 km	At least 5 km?	~2 km	At least 5 km?
RESIDENCE TIME	~40 m.y.	~140 m.y.	~140 m.y.	~50–60 m.y.
BASEMENT	Oceanic crust ca. 500–510 Ma	Oceanic crust? ca. 600 Ma	Attenuated Congo craton?	Oceanic crust ca. 280 Ma
BASIN WIDTH	~1150 km restored width	~400 km (min) deformed width	~100 km (min) deformed width	~150 km (min) deformed width
CONVERGENCE CHARACTERISTICS				
SUBDUCTION				
Arc	None recognizable	Central Zone?	None recognizable	130–120 Ma
Mica fabrics—wedge	460–440 (420) Ma	560–500 Ma	None reliable	150–140 Ma
COLLISION or CONVERGENCE	ca. 420–400 Ma	550–500 Ma		130–120 Ma
WEDGE STYLE	Chevron folding Leading imbricate fan thrust system	Schist belt	Chevron folding	Schist belt
PLATE ENVIRONMENT	Backarc basin (Late Ordovician–Silurian)	Forearc (FW-HW)	Wrench boundary	Forearc (FW)
CRUSTAL GENESIS				
CRATONIZATION Age of posttectonic granites	ca. 360 Ma	ca. 505 Ma	ca. 530 Ma (post-D2/pre-D3)	—
EXHUMATION Isostatic rebound	Tabberabberan unconf. ca. 370 Ma	Nama foreland basin ca. 510–500 Ma inception		Albian unconf. ca. 110–90 Ma

Note: FW—sediment accumulation on footwall or downgoing slab of subduction zone; HW—sediment accumulation in hanging wall or overriding slab of subduction zone.

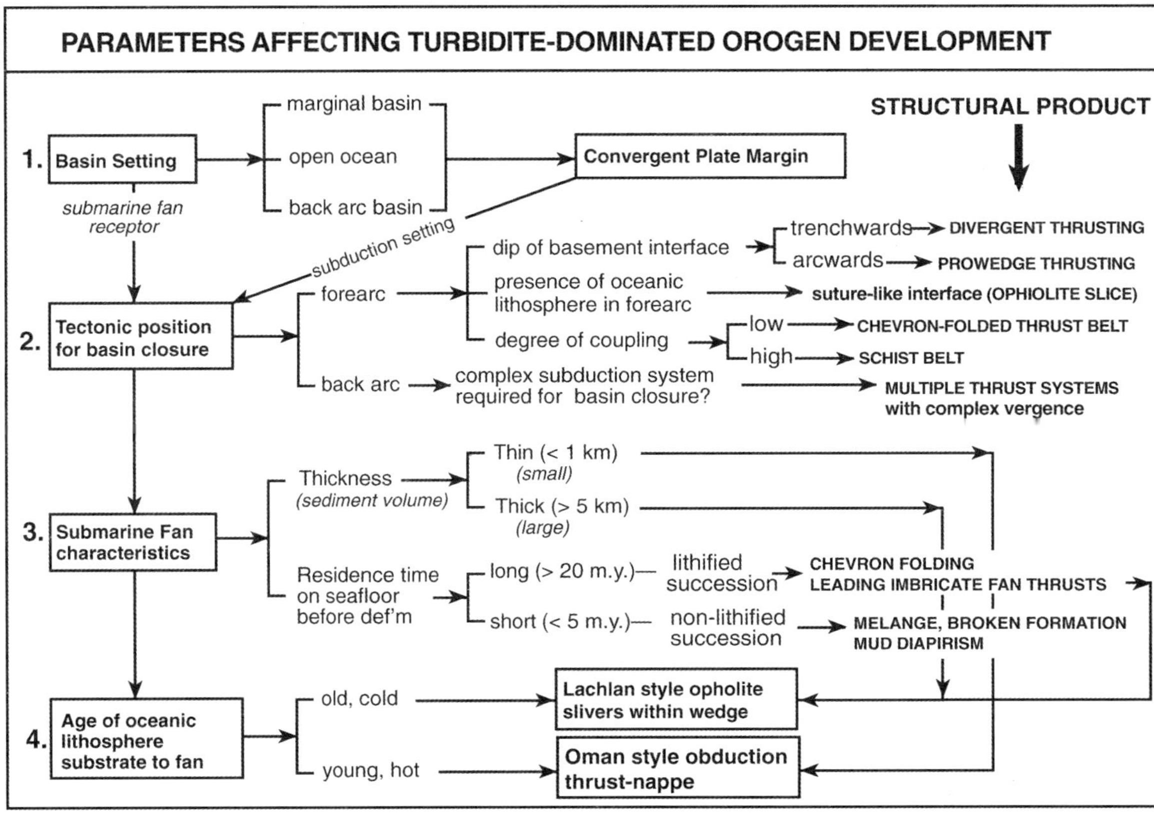

Figure 12. "Flow chart" for accretionary orogen development showing the main parameters 1–4, and links between parameters, that affect the form, structural geometry, crustal architecture, and development of turbidite-dominated accretionary-style orogens. The structural product defines the structural style and geometry of the resultant orogen. Key factors are the initial position of the basin receptor for the turbidites, the thickness and volume of turbidites, the residence time of the turbidite fan on the seafloor prior to deformation, and the age of the underlying oceanic seafloor crust.

In this scenario, subduction zones, essentially zones of underthrusting, were initiated on the opposing sides of a former Cambrian-Ordovician backarc basin (Fig. 6). Low-angle underthrusting in the western Lachlan orogen (WLO, Fig. 4A) resulted in marked shortening (~65%–70%) by chevron folding, cleavage development, and thrust imbrication of the turbidites (e.g., Gray and Willman, 1991; Gray and Foster, 1998) but without development of a magmatic arc. However, subduction along the eastern side of the marginal basin, represented by the Howqua accretionary complex (Fig. 4A), produced large elongate composite batholiths that define a shear zone–bounded, northwest-trending magmatic arc, as well as the associated high-*T*/low-*P* metamorphism (WOMB of Fig. 4A).

Although there is now a general acceptance that the Lachlan turbidite fans developed on backarc basin lithosphere (e.g., Crawford and Keays, 1978, 1987; Foster et al., 2005), the Lachlan substrate is analogous to the Philippine Sea or Japan Sea (see Gray and Foster, 2004b). Each analog has different consequences with possible microcontinental ribbons (e.g., Tasmania and/or the Selwyn block; Cayley et al., 2002) in the former, and relict arcs (e.g., Licola arc; Spaggiari et al., 2003a, 2003b) in the latter providing significant rigidity contrasts during crustal shortening and thickening.

For the Damara orogen, despite the controversy over the tectonic setting of the Central Zone, the accretionary prism-like nature of the Southern Zone schists (Kukla and Stanistreet, 1991) as well as their strong similarities to those of the Otago Schist belt in New Zealand (see photos in Gray and Foster, 2004c) suggest that this orogen is another example of continental growth by subduction-accretion of turbidites (cf. Kukla and Stanistreet, 1991). Damara orogen evolution, however, involves a full Wilson cycle with the opening and closing of a small ocean basin partly floored by attenuated continental crust of the Congo craton and partly underlain by ca. 700 Ma oceanic lithosphere formed during opening of the Khomas ocean basin (e.g., Prave, 1996).

The Rangitatan orogen of New Zealand provides the least complicated setting, representing part of the Mesozoic convergent margin of Gondwana facing a large ocean basin (cf. Busby, 2004), the Rangitatan orogen evolved in a forearc position typical of the modern-day Aleutian arc-subduction complex ("ring of fire"-type arc). In contrast, the Lachlan orogen involves closure of a backarc basin related to possible multiple subduction systems (e.g., Spaggiari et al., 2004a), while the Damara belt of the Damara orogen involves a single subduction system but with final continent-continent collision (e.g., Barnes and Sawyer, 1980).

Parameters Affecting Accretionary Orogen Development: Fan Thickness and Timing

Apart from their tectonic positions, other parameters that affect accretionary orogen development include the thickness of the fan, the depositional age of the fan, the timing of fan shortening and thickening, and the age of the oceanic lithosphere relative to basin closure (Fig. 12). These can be measured or determined for the individual orogens, whereas parameters such as the residence time on the seafloor are derivative of the former (see Table 1).

For the Lachlan orogen, the age of backarc basin crust is 505–495 Ma (e.g., Spaggiari et al., 2003a; Foster et al., 2005), the age of the developing submarine fan system is from 490 to 460 Ma, and the deformation that caused the backarc basin closure was from ca. 450 Ma to 420 Ma (i.e., some 50 m.y. after oceanic lithosphere formation and some 30–20 m.y. after submarine fan deposition). These conditions favor chevron folding, without the significant stratal disruption and mélange or broken formation typical of shallow levels of modern accretionary complexes. Chevron folding requires lithification or precompaction of the sediment fan and therefore time for burial and dewatering on the seafloor before deformation (e.g., closed-system vein formation during shortening; Gray et al., 1991). Metamorphism due to sediment loading is known to occur in the modern-day Bengal fan (Curray et al., 1982).

In this accretionary process, oceanic lithosphere is preserved as fault-bounded slices derived from "peeling" of oceanic lithosphere in the subduction zone (e.g., Kimura and Ludden, 1995). The temperature, and therefore age, of the oceanic lithosphere is important. The western Lachlan orogen shows that thick, lithified or partially lithified turbidite fans (4–5 km thicknesses) sitting on old, cold oceanic lithosphere produce dominantly chevron-folded, thrust-interleaved packages that incorporate fault-bounded, duplexed slivers of the upper parts of the backarc basin oceanic lithosphere (i.e., Cordilleran-style ophiolitic fragments). This process is inferred to occur by low-angle underthrusting with the turbidites predominantly deforming in the overriding plate (see Gray and Foster, 1998; Spaggiari et al., 2004a).

In the case of the Rakaia wedge of New Zealand, the age of the underlying oceanic lithosphere as inferred from the Dun Mountain ophiolite is ca. 280 Ma (Kimbrough and et al., 1992) with Permian to Triassic submarine fan sedimentation (ca. 280 to 200 Ma; Bradshaw, 1989) and eventual wedge thickening at 160–150 Ma (Gray and Foster, 2004c) (i.e., some 120 m.y. after oceanic lithosphere formation and some 40 m.y. after fan deposition).

In this respect, plate convergence for significant parts of the Lachlan and the Rakaia wedge of New Zealand involves old, cold oceanic lithosphere and thick turbidite fans (4–5 km thicknesses). The deformation of the turbidites occurs either in the hanging wall above a shallow-dipping subduction zone or on the downgoing slab (e.g., Spaggiari et al., 2004a, for the Lachlan orogen; Coombs et al., 1976, for the Rakaia wedge deformation). We speculate that the deformation and metamorphism of the Kuiseb Schist of the Damara Inland Branch also occurred on the downgoing slab (cf. Barnes and Sawyer, 1980).

Both the Rangitatan (Otago Schist) and Damara orogens (Southern Zone Kuiseb Schist) show evidence for coupling between the overriding margin and the subducting slab, with the thick turbidite fan sitting on the subducting plate, producing an intensely deformed, highly thickened wedge that undergoes Barrovian-style metamorphism. The result is intensely deformed schist with transposition layering in the former turbidites. In both the Rakaia wedge and Southern or Khomas Zone of the Damara, the wedge underwent significant structural thickening by shear-related, noncoaxial deformation at the subduction interface (e.g., Mortimer, 1993; Gray and Foster, 2004b) to produce transposition layering or schistosity and a pronounced rodding lineation (i.e., S-L tectonites). These are the dominant fabrics produced in this subduction interface environment that have been previously attributed to underplating in a coaxial flattening strain environment (e.g., Lundberg and Reed, 1991; Von Huene and Scholl, 1991).

In both these cases, arc magmatism occurs in the overriding plate but not in the intensely coupled zones. In New Zealand, magmatism occurs inboard of the ophiolite "suture" with a complete lack of granitoids in the thickened wedge (Fig. 7C), presumably a reflection of underlying thin crust and deformation directly above the subducted slab rather than over areas of the asthenospheric wedge. Both arc regions, inboard of the subduction complex schistose rocks, contain granulites in the lower parts of the magmatic arcs: Fiordland in New Zealand (Claypool et al., 2001) and the Central Zone of the Damara (Puhan, 1983).

In the case of the Damara, the presence of oceanic lithosphere, and with it the role of subduction, have been disputed (e.g., Martin, 1983). The MORB character of the Matchless amphibolite and the association of chert, pillow basalt, and gabbros is suggestive of remnants of the upper parts of oceanic crust (e.g., Killick, 2000); similar lithological packages are present in the Cambrian ophiolite fragments of the Lachlan orogen (e.g., Spaggiari et al., 2004a). The unusual elements of the Matchless amphibolite are its length (~350 km), its uniform lateral continuity, and style of deformation. The extent and fault-bounded and/or shear zone–bounded nature of the Matchless amphibolite, however, has certain similarities to the Dun Mountain ophiolite of New Zealand that extends for some 600 km, but separates the arc-forearc sequences from the deformed fan turbidites (Fig. 7), rather than occurring within the deformed sediment wedge (Fig. 9C).

The Template of Accretion

Turbidite-dominated orogens tend to (1) have greater width than the classic Alpine orogen typical of continent-continent collision, and (2) be dominated by turbidites instead of batholithic rocks of the Andean orogen typical of modern Pacific-type subduction. The Lachlan, Damara, and Rangitatan orogens have ~750, ~450, and ~400 km widths, respectively. Despite providing a disconnected history of the tectonic process of accretion through time from the Neoproterozoic to Holocene, the Damara, Lachlan, and Rangitatan orogens clearly show that there is no simple template for the accretionary process. This is reflected in the crustal-scale architecture of each orogen, as well as the temporal and spatial variations in the development of deformation fabrics, metamorphism, and magmatism. The Lachlan orogen is made up of three thrust belts with contrasting vergence, whereas the Damara orogen shows divergent thrust systems, and the Rangitatan orogen one thrust system and a steeply dipping backstop behind the sediment wedge.

Within the orogen architecture, the positions of metamorphic rocks and the spatial variations in different types of metamorphism delineate or help reconstruct the tectonic setting. To a first-order approximation, the low-*T*/high-*P* metamorphic rocks (blueschists and eclogites) define the subduction channel, the moderate-*T*, moderate- to high-*P* (Barrovian-style) metamorphic rocks define the regions of structural thickening above the subduction interface, and the high-*T*/low-*P* regional aureole metamorphic rocks intruded by large, elongate composite granitoids define the magmatic arc.

The Rakaia wedge part of the Rangitatan orogen has relict blueschist metamorphism preserved in intercalated metabasites near the interface between the volcaniclastic trench sediment and the deformed quartz-rich fan (Yardley, 1982), whereas the wedge proper shows Barrovian metamorphism with garnet-oligoclase assemblages in the Otago Schist core (Mortimer, 2000).

Similarly, the schistose part of the Damara orogen shows intermediate-*T*/intermediate-*P* (Barrovian-style) metamorphism (Kasch, 1983b), but relict blueschist metamorphism has not been recognized. In the Damara, the schistose belt dips toward, but verges away from, the Central Zone belt of large, irregularly shaped, syntectonic granitoids and the high-*T* metamorphism (Puhan, 1983).

The Lachlan orogen shows background low-grade, intermediate-*P* metamorphism (Gray and Foster, 2004b), but has two belts of high-*T* regional aureole-style metamorphism associated with northwest-trending (Wagga-Omeo belt of the central Lachlan orogen) and north-trending (eastern Lachlan orogen), large, elongate, composite granitoids. In the western and central Lachlan orogen, blueschist metamorphism is preserved in knockers within serpentinite-matrix mélange of the major faults zones (Spaggiari et al., 2002a, 2002b).

Turbidite-dominated orogens therefore develop in convergent margin settings involving subduction of oceanic crust containing large submarine turbidite fans, but have no specific template or orogen design. They have special character reflected by the large volumes of structurally thickened and metamorphosed turbidites and relicts of oceanic crust as fault-bounded slivers.

The Role of Magmatic Underplating: Chemical Maturation of Recycled Crust

Granitic magmatism plays an important role in the chemical maturation of these structurally thickened and accreted turbidite fan systems. Turbidite-dominated orogens tend to be characterized by large volumes of syn- to posttectonic granite, such as those in all zones of the Lachlan orogen, and in the Central Zone of the Damara orogen.

Turbidites, in particular the more feldspathic (graywacke) and pelitic lithologies, represent fertile sources for crustal melting (Clemens and Vielzeuf, 1987; Thompson, 1996). There is clear isotopic evidence in both the Lachlan and Damara orogens for derivation of most of the S-type granites from the local metaturbidites. Additional source components are sometimes required to explain the chemical/isotopic constraints, e.g., contributions from underlying Cambrian oceanic crust and synmagmatic mantle-derived melts (Lachlan orogen; Keay et al., 1997; Maas et al., 2001), or from old crystalline basement (Damara granitoids with very low ε_{Nd}; Jung et al., 2003).

The large volumes of granite imply significant crustal melting, with a requirement of mafic magma underplating at the base of the crust. Thermal conditions related to tectonic setting as well as a high fertility and, in the case of the Damara, high HFSE contents of the crust, combine to produce the observed massive mid/lower crustal melting in turbidite-dominated orogens.

CONCLUSIONS

Ancient orogens dominated by turbiditic interbedded sandstone and mudstone sequences, such as the Neoproterozoic Damara Inland Branch (Namibia), the Paleozoic Lachlan (eastern Australia), and the Mesozoic Rangitatan (New Zealand) orogens, show that marked shortening and thickening of the former turbidite fan sequence is a requisite for accretion, or incorporation into eventual crust of continental character and thickness. This process occurs along former convergent margins by subduction-related closure of former backarc basins (e.g., Lachlan orogen) and small ocean basins (e.g., Damara orogen), and commonly involves the underlying oceanic lithosphere.

While it is clear that there is no simple template for accretion, the final architecture of turbidite-dominated, accretionary-type orogens is variable and dependent on their previous history and tectonic setting. The marked shortening and thickening of the crust and the involvement of predominantly oceanic basement are key factors in recognizing ancient subduction-accretion processes in such orogens. The presence of the lower crustal layer of imbricated oceanic crust coupled with the overlying thickened sedimentary succession yields a density structure that is remarkably stable and isostatically balanced near sea level. Magmatic underplating and marked mid/lower crustal melting, resulting in significant volumes of granites, are also an important part of the later stages of this continental growth and surficial crustal recycling through the accretion of deformed turbidite fans.

ACKNOWLEDGMENTS

This review and write-up was undertaken as part of an Australian Research Council (ARC) Australian Professorial Fellowship (DP0210178 awarded to Gray). Research on the Tasmanides of eastern Australia has been undertaken over a period of 20 years, supported by ARC grants E8315666, E8315675, A38615754, A38715383, A38930784 (awarded to Gray), and A38615754 (Durney, Gray, Gregory); additional funding is from the Australian Geodynamic Cooperative Research Centre (to both Gray and Foster) and National Science Foundation (NSF) grant EAR-0073638 (to Foster). Research on New Zealand has been supported by ARC grants A39927139 and A39030706 (awarded to Gray) and a Monash University Small Grant (awarded to Gray). Research on Namibia has been funded by ARC Large Grant A00103456 (awarded to Gray) and NSF grant EAR-0440188 (awarded to Foster). We acknowledge discussions with (1) Chris Fergusson, Vince Morand, Nick Woodward, Clive Willman, John Miller, Chris Wilson, Thomas Flöttmann, and Ron Berry on Tasmanides geology; (2) Simon Cox, Richard Norris, Dave Craw, Nick Mortimer, Moses Turnbull, and Alan Cooper on New Zealand geology; and (3) Cees Passchier, Rudolph Trouw, and Thomas Becker on Namibian geology.

REFERENCES CITED

Adams, C.J., Pankhurst, R.J., Maas, R., and Millar, I.L., 2005, Nd and Sr isotopic signatures of metasedimentary terranes around the South Pacific margin, and implications for their provenance, *in* Vaughan, A.P.M., et al., eds., Terrane processes at the margins of Gondwana: Geological Society [London] Special Publication 246, p. 113–142.

Albarède, F., 1998, The growth of continental crust: Tectonophysics, v. 296, p. 1–14, doi: 10.1016/S0040-1951(98)00133-4.

Armstrong, R.L., 1981, The persistent myth of crustal growth: Australian Journal of Earth Sciences, v. 38, p. 613–630.

Barnes, S., 1983, Pan-African serpentinites in central southwest Africa/Namibia and the chemical classification of serpentinites, *in* Miller, R.G., ed., Evolution of the Damara orogen of southwest Africa/Namibia: Geological Society of South Africa Special Publication 11, p. 147–155.

Barnes, S., and Sawyer, E., 1980, An alternative model for the Damara mobile belt: Ocean crust subduction and continental convergence: Precambrian Research, v. 13, p. 297–336, doi: 10.1016/0301-9268(80)90048-0.

Basson, I.J., and Greenway, G., 2004, The Rössing uranium deposit: A product of late-kinematic localization of uriniferous granites in the Central Zone of the Damara orogen, Namibia: Journal of African Earth Sciences, v. 38, p. 413–435, doi: 10.1016/j.jafrearsci.2004.04.004.

Bierlein, F.P., Arne, D.C., Keay, S.M., and McNaughton, N.J., 2001a, Timing relationships between felsic magmatism and mineralisation in the central Victorian gold province, southeast Australia: Australian Journal of Earth Sciences, v. 48, p. 883–899, doi: 10.1046/j.1440-0952.2001.00907.x.

Bierlein, F.P., Hughes, M., Dunphy, J., McKnight, S., Reynolds, P., and Waldron, H., 2001b, Tectonic and economic implications of trace element, $^{40}Ar/^{39}Ar$ and Sm-Nd data from mafic dykes associated with orogenic gold mineralisation in central Victoria, Australia: Lithos, v. 58, p. 1–31, doi: 10.1016/S0024-4937(01)00050-0.

Bierlein, F.P., Foster, D.A., Gray, D.R., and Davidson, G.J., 2005, Timing of orogenic gold mineralization in northeast Tasmania—Implications for the tectonic and metallogenic evolution of Palaeozoic SE Australia: Mineralium Deposita, v. 39, p. 890–903, doi: 10.1007/s00126-004-0458-4.

Bradshaw, J.D., 1989, Cretaceous geotectonic patterns in the New Zealand region: Tectonics, v. 8, p. 803–820.

Bradshaw, J.D., 1993, A review of the Median Tectonic Zone: Terrane boundaries and terrane amalgamation near the Median Tectonic Line: New Zealand Journal of Geology and Geophysics, v. 36, p. 117–125.

Busby, C., 2004, Continental growth at convergent margins facing large ocean basins: A case study from Mesozoic convergent margin basins of Baja California, Mexico: Tectonophysics, v. 392, p. 241–277, doi: 10.1016/j.tecto.2004.04.017.

Cawood, P.A., 1987, Terra Australis orogen: Rodinia breakup and development of the Pacific and Iapetus margins of Gondwana during the Neoproterozoic and Paleozoic: Earth-Science Reviews, v. 69, p. 249–279, doi: 10.1016/j.earscirev.2004.09.001.

Cawood, P.A., 2005, Stratigraphic and structural relations of strata enclosing the Dun Mountain ophiolite belt in the Arthurton-Clinton region, Southland, New Zealand: New Zealand Journal of Geology and Geophysics, v. 30, p. 19–36.

Cayley, R.A., Taylor, D.H., VandenBerg, A.H.M., and Moore, D.H., 2002, Proterozoic–early Palaeozoic rocks and the Tyennan orogeny in central Victoria: The Selwyn block and its tectonic implications: Australian Journal of Earth Sciences, v. 49, p. 225–254, doi: 10.1046/j.1440-0952.2002.00921.x.

Chappell, B.W., and White, A.J.R., 1992, I- and S-type granites in the Lachlan Fold Belt: Transactions of the Royal Society of Edinburgh, v. 83, p. 1–26.

Chappell, B.W., and White, A.J.R., 2001, Two contrasting granite types: 25 years later: Australian Journal of Earth Sciences, v. 48, p. 489–499, doi: 10.1046/j.1440-0952.2001.00882.x.

Chappell, B.W., White, A.J.R., and Hine, R., 1988, Granite provinces and basement terranes in the Lachlan Fold Belt, southeastern Australia: Australian Journal of Earth Sciences, v. 35, p. 505–521.

Chen, Y.D., and Williams, I.S., 1991, Zircon inheritance in mafic inclusions from Bega batholith granites, southeastern Australia: An ion microprobe study: Journal of Geophysical Research, v. 95, p. 17,787–17,796.

Claypool, A.L., Klepeis, K.A., Clarke, G.L., Daczko, N., and Dockrill, B., 2001, The evolution and exhumation of Early Cretaceous lower crustal granulites during changes in plate boundary dynamics, Fiordland, New Zealand: Geological Society of America Abstracts with Programs, v. 33, no. 6, p. A-265.

Clemens, J.D., and Vielzeuf, D., 1987, Constraints on melting and magma production in the crust: Earth and Planetary Science Letters, v. 86, p. 287–306, doi: 10.1016/0012-821X(87)90227-5.

Clemens, J.D., and Wall, V.J., 1984, Origin and evolution of a peraluminous silicic ignimbrite suite: The Violet Town Volcanics: Contributions to Mineralogy and Petrology, v. 88, p. 354–371, doi: 10.1007/BF00376761.

Collins, W.J., 1996, Lachlan Fold Belt granitoids: Products of three-component mixing: Transactions of the Royal Society of Edinburgh, Earth Sciences, v. 87, p. 171–182.

Collins, W.J., 1998, Evaluation of petrogenetic models for Lachlan Fold Belt granitoids: Implications for crustal architecture and tectonic models: Australian Journal of Earth Sciences, v. 45, p. 483–500.

Collins, W.J., 2002, Nature of extensional accretionary orogens: Tectonics, v. 21, p. 1024–1036, doi: 10.1029/2000TC001272.

Collins, W.J., and Hobbs, B.E., 2001, What caused the Early Silurian change from mafic to silicic magmatism in the eastern Lachlan Fold Belt?: Australian Journal of Earth Sciences, v. 48, p. 25–41, doi: 10.1046/j.1440-0952.2001.00837.x.

Collins, W.J., Beams, S.D., White, A.J.R., and Chappell, B.W., 1982, Nature and origin of A-type granites with particular reference to southeastern Australia: Contributions to Mineralogy and Petrology, v. 80, p. 189–200, doi: 10.1007/BF00374895.

Coombs, D.S., and Cox, S.C., 1991, Low- and very low-grade metamorphism in southern New Zealand: Geological Society of New Zealand Miscellaneous Publication 58, 87 p.

Coombs, D.S., Landis, C.A., Norris, R.J., Sinton, J.M., Borns, D.J., and Craw, D., 1976, The Dun Mountain ophiolite belt, New Zealand, its tectonic setting, constitution, and origin, with special reference to the southern portion: American Journal of Science, v. 276, p. 561–603.

Cooper, R.A., and Tulloch, A.J., 1992, Early Paleozoic terranes in New Zealand and their relationship to the Lachlan Fold Belt: Tectonophysics, v. 214, p. 129–144, doi: 10.1016/0040-1951(92)90193-A.

Corner, B., 2000, Crustal framework of Namibia derived from magnetic and gravity data, *in* Miller, R.McG., ed., Henno Martin Commemorative Volume: Communications of the Geological Survey of Namibia, v. 12, p. 13–19.

Coward, M.P., 1981, The junction between Pan-African mobile belts in Namibia: Its structural history: Tectonophysics, v. 76, p. 59–73, doi: 10.1016/0040-1951(81)90253-5.

Coward, M.P., 1983, The tectonic history of the Damaran belt, in Miller, R.G., ed., Evolution of the Damara orogen of southwest Africa/Namibia: Geological Society of South Africa Special Publication 11, p. 409–421.

Crawford, A.J., and Keays, R.R., 1978, Cambrian greenstone belts in Victoria: Marginal sea crust slices in the Lachlan Fold Belt of southeastern Australia: Earth and Planetary Science Letters, v. 41, p. 197–208, doi: 10.1016/0012-821X(78)90010-9.

Crawford, A.J., and Keays, R.R., 1987, Petrogenesis of Victorian Cambrian tholeiites and implications for the origin of associated boninites: Journal of Petrology, v. 28, p. 1075–1109.

Curray, J.R., Emmel, F.J., Moore, D.G., and Raitt, R.W., 1982, Structure, tectonics, and geological history of the northeastern Indian Ocean, in Nairn, A.E.M., ed., The ocean basins and margins, New York, Plenum Press, v. 5, p. 399–450.

Daczko, N.R., Klepeis, K.A., and Clarke, G.L., 2001, Evidence of Early Cretaceous collisional-style orogenesis in northern Fiordland, New Zealand, and its effect on the evolution of the lower crust: Journal of Structural Geology, v. 23, p. 693–713, doi: 10.1016/S0191-8141(00)00130-9.

de Kock, G.S., 1992, Forearc basin evolution in the Pan-African Damara belt, central Namibia: The Hureb Formation of the Khomas Zone: Precambrian Research, v. 57, p. 169–194, doi: 10.1016/0301-9268(92)90001-5.

de Kock, G.S., Eglington, B., Armstrong, R.A., Hermer, R.E., and Walraven, F., 2000, U-Pb and Pb-Pb ages of the Naauwpoort Rhyolite, Kawakeup Leptite and Okongava Diorite: Implications for the onset of rifting and of orogenesis in the Damara belt, Namibia, *in* Miller, R.McG., ed., Henno Martin Commemorative Volume: Communications of the Geological Survey of Namibia, v. 12, p. 81–88.

Elburg, M., 1996, U-Pb ages and morphologies of zircon in microgranitoid enclaves and peraluminous host granite: Evidence for magma mingling: Contributions to Mineralogy and Petrology, v. 123, p. 177–189, doi: 10.1007/s004100050149.

Fergusson, C.L., 1987, Early Palaeozoic backarc deformation in the Lachlan Fold Belt, southeastern Australia: Implications for terrane translations in eastern Gondwanaland, *in* Leitch, E.C., and Scheibner, E., eds., Terrane accretion and orogenic belts: American Geophysical Union Geodynamics Series, v. 19, p. 39–56.

Fergusson, C.L., and Coney, P.J., 1992a, Implications of a Bengal Fan–type deposit in the Paleozoic Lachlan Fold Belt of southeastern Australia: Geology, v. 20, p. 1047–1049, doi: 10.1130/0091-7613(1992)020<1047:IOABFT>2.3.CO;2.

Fergusson, C.L., and Coney, P.J., 1992b, Convergence and intraplate deformation in the Lachlan Fold Belt of southeastern Australia: Tectonophysics, v. 214, p. 417–439, doi: 10.1016/0040-1951(92)90208-N.

Fergusson, C.L., and VandenBerg, A.H.M., 1990, Middle Paleozoic thrusting in the eastern Lachlan Fold Belt, southeastern Australia: Journal of Structural Geology, v. 12, p. 577–589, doi: 10.1016/0191-8141(90)90075-A.

Foden, J., Sandiford, M., Doherty-Page, J., and Williams, I., 1999, Geochemistry and geochronology of the Rathjan Gneiss: Implications for the early tectonic evolution of the Delamerian orogen: Australian Journal of Earth Sciences, v. 46, p. 377–389, doi: 10.1046/j.1440-0952.1999.00712.x.

Foster, D.A., and Gray, D.R., 2000, The structure and evolution of the Lachlan Fold Belt (orogen) of eastern Australia: Annual Review of Earth and Planetary Sciences, v. 28, p. 47–80, doi: 10.1146/annurev.earth.28.1.47.

Foster, D.A., Gray, D.R., Kwak, T.A.P., and Bucher, M., 1998, Chronology and tectonic framework of turbidite-hosted gold deposits in the western Lachlan Fold Belt, Victoria: $^{40}Ar/^{39}Ar$ results: Ore Geology Reviews, v. 13, p. 229–250, doi: 10.1016/S0169-1368(97)00020-6.

Foster, D.A., Gray, D.R., and Bucher, M., 1999, Chronology of deformation within the turbidite-dominated Lachlan orogen: Implications for the tectonic evolution of eastern Australia and Gondwana: Tectonics, v. 18, p. 452–485, doi: 10.1029/1998TC900031.

Foster, D.A., Gray, D.R., and Spaggiari, C.V., 2005, Timing of subduction and exhumation along the Cambrian East Gondwana margin, and the formation of Paleozoic backarc basins: Geological Society of America Bulletin, v. 117, p. 105–116, doi: 10.1130/B25481.1.

Glen, R.A., Walshe, J.L., Barron, L.M., and Watkins, J.J., 1998, Ordovician convergent margin volcanism and tectonism in the Lachlan sector of East Gondwana: Geology, v. 26, p. 751–754, doi: 10.1130/0091-7613(1998)026<0751:OCMVAT>2.3.CO;2.

Goscombe, B., Gray, D.R., and Hand, M., 2004, Variation in metamorphic style along the northern margin of the Damara orogen, Namibia: Journal of Petrology, v. 45, p. 1261–1295, doi: 10.1093/petrology/egh013.

Gray, C.M., 1984, An isotopic mixing model for the origin of granitic rocks in southeastern Australia: Earth and Planetary Science Letters, v. 70, p. 47–60, doi: 10.1016/0012-821X(84)90208-5.

Gray, C.M., 1990, A strontium isotopic traverse across the granitic rocks of southeastern Australia: Petrogenetic and tectonic implications: Australian Journal of Earth Sciences, v. 37, p. 331–349.

Gray, D.R., 1997, Tectonics of the southeastern Australian Lachlan Fold Belt: Structural and thermal aspects, *in* Burg, J-P., and Ford, M., eds., Orogeny through time: Geological Society [London] Special Publication 121, p. 149–177.

Gray, D.R., and Foster, D.A., 1998, Character and kinematics of faults within the turbidite dominated Lachlan orogen: Implications for tectonic evolution of eastern Australia: Journal of Structural Geology, v. 20, p. 1691–1720, doi: 10.1016/S0191-8141(98)00089-3.

Gray, D.R., and Foster, D.A., 2004a, Regional geology, Tasman orogen, Australia: Encyclopedia of geology, volume 1: Amsterdam, Elsevier, p. 237-252.

Gray, D.R., and Foster, D.A., 2004b, Tectonic evolution of the Lachlan orogen, southeast Australia: Historical review, data synthesis and modern perspectives: Australian Journal of Earth Sciences, v. 51, p. 773–817, doi: 10.1111/j.1400-0952.2004.01092.x.

Gray, D.R., and Foster, D.A., 2004c, $^{40}Ar/^{39}Ar$ thermochronologic constraints on deformation, metamorphism and cooling/exhumation of a Mesozoic accretionary wedge, Otago Schist, New Zealand: Tectonophysics, v. 385, p. 181–210, doi: 10.1016/j.tecto.2004.05.001.

Gray, D.R., and Willman, C.E., 1991, Thrust-related strain gradients and thrusting mechanisms in a chevron-folded sequence, southeastern Australia: Journal of Structural Geology, v. 13, p. 691–710, doi: 10.1016/0191-8141(91)90031-D.

Gray, D.R., Gregory, R.T., and Durney, D.W., 1991, Rock-buffered fluid-rock interaction in deformed quartz-rich turbidite sequences, eastern Australia: Journal of Geophysical Research, v. 96, p. 19,681–19,704.

Gray, D.R., Foster, D.A., and Bucher, M., 1997, Recognition and definition of orogenic events in the Lachlan Fold Belt: Australian Journal of Earth Sciences, v. 44, p. 489–581.

Gray, D.R., Foster, D.A., Offler, R., Spaggiari, C.V., and Hartley, M., 2000, Eastern Australia crustal evolution: Timing and temporal links between Delamerian and Lachlan orogens, including Tasmania: Geological Society of Australia Abstracts, v. 59, p. 189.

Gray, D.R., Foster, D.A., Korsch, R.J., and Spaggiari, C.V., 2006, Structural style and crustal architecture of the Tasmanides of eastern Australia, example of a composite accretionary orogen, *in* Butler, R., and Mazzoli, S., eds., Styles of Continental Contraction: Geological Society of America Special Publication 414, p. 119–132.

Haack, U., Hoefs, J., and Gohn, E., 1982, Constraints on the origin of Damaran granites by Rb/Sr and $\delta^{18}O$ data: Contributions to Mineralogy and Petrology, v. 79, p. 279–289, doi: 10.1007/BF00371519.

Haack, U., Gohn, E., and Hartmann, O., 1983, Radiogenic heat generation in Damaran rocks: Geological Society of South Africa Special Publication 11, p. 225–232.

Hälbich, I.W., 1977, Structure and tectonics along the southern margin of the Damara mobile belt, southwest Africa: Annale University of Stellenbosch, Series A1 (Geologie), no. 2, p. 149–247.

Hawkesworth, C.J., Kramers, J.D., and Miller, R.M.G., 1981, Old model Nd ages in Namibian Pan-African rocks: Nature, v. 289, p. 278–282, doi: 10.1038/289278a0.

Hawkesworth, C.J., Gledhill, A.R., Roddick, J.C., Miller, R.McG., and Kröner, A., 1983, Rb-Sr and $^{40}Ar/^{39}Ar$ studies bearing on models for the thermal evolution of the Damara Belt, Namibia, *in* Miller, R.McG., ed., Evolution of the Damara Orogen of South West Africa/Namibia: Special Publication of the Geological Society of South Africa, v. 11, p. 323–338.

Hine, R., Williams, I.S., Chappell, B.W., and White, A.J.R., 1978, Contrasts between I- and S-type granitoids of the Kosciusko batholith: Journal of the Geological Society of Australia, v. 25, p. 219–234.

Hoffmann, K.-H., 1983, Lithostratigraphy and facies of the Swakop Group of the southern Damara belt, SWA/Namibia, *in* Miller, R.G., ed., Evolution of the Damara orogen of southwest Africa/Namibia: Geological Society of South Africa Special Publication 11, p. 43–63.

Hoffman, P.F., Hawkins, D.P., Isachsen, C.E., and Bowring, S.A., 1996, Precise U-Pb zircon ages for early Damaran magmatism in the Summas Mountains and Welwitschia inlier, northern Damara belt, Namibia: Communications of the Geological Survey of Namibia, v. 11, p. 47–52.

Hudson, T., and Plafker, G., 1982, Palaeogene metamorphism of an accretionary flysch terrane, eastern Gulf of Alaska: Geological Society of America Bulletin, v. 93, p. 1280–1290, doi: 10.1130/0016-7606(1982)93<1280:PMOAAF>2.0.CO;2.

Ingersoll, R.V., Dickinson, W.R., and Graham, S.A., 2003, Remnant-ocean submarine fans: Largest sedimentary systems on Earth, *in* Chan, M.A., and Archer, A.W., eds., Extreme depositional environments: Mega end members in geologic time: Geological Society of America Special Paper 370, p. 191–208.

Ireland, T.R., and Gibson, G.M., 1998, SHRIMP monazite and zircon geochronology of high-grade metamorphism in New Zealand: Journal of Metamorphic Geology, v. 16, p. 149–167, doi: 10.1111/j.1525-1314.1998.00112.x.

Jacob, R.E., Snowdon, P.A., and Bunting, F.J.L., 1983, Geology and structural development of the Tumas basement dome and its cover rocks, *in* Miller, R.G., ed., Evolution of the Damara orogen of southwest Africa/Namibia: Geological Society of South Africa Special Publication 11, p. 157–172.

Jacob, R.E., Moore, J.M., and Armstrong, R.A., 2000, Zircon and titanite age determinations from igneous rocks in the Karibib district, Namibia: Implications for Navachab vein-style gold mineralization: Communications of the Geological Survey of Namibia, v. 12, p. 157–166.

Johnson, B.D., and Veevers, J.J., 1984, Oceanic palaeomagnetism, in Veevers, J.J., ed., Phanerozoic Earth history of Australia: Oxford Monographs on Geology and Geophysics, v. 2, p. 17–38.

Jung, S., 2000, High-temperature, mid-pressure clockwise P-T paths and melting in the development of regional migmatites: The role of crustal thickening and repeated plutonism: Geological Journal, v. 35, p. 345–359, doi: 10.1002/gj.866.

Jung, S., and Mezger, K., 2003a, U-Pb garnet chronometry in high-grade rocks: Case studies from the central Damara orogen (Namibia) and implications for the interpretation of Sm-Nd garnet ages and the role of high-U-Th inclusions: Contributions to Mineralogy and Petrology, v. 146, p. 382–396, doi: 10.1007/s00410-003-0506-6.

Jung, S., and Mezger, K., 2003b, Petrology of basement-dominated terranes, I: Regional metamorphic T-t path from U-Pb monazite and Sm-Nd garnet geochronology (central Damara orogen, Namibia): Chemical Geology, v. 198, p. 223–247, doi: 10.1016/S0009-2541(03)00037-8.

Jung, S., Mezger, K., and Hoernes, S., 1998a, Petrology and geochemistry of syn- to post-collisional metaluminous A-type granites—A major and trace element and Nd-Sr-Pb-O isotope study from the Proterozoic Damara belt, Namibia: Lithos, v. 45, p. 147–175, doi: 10.1016/S0024-4937(98)00030-9.

Jung, S., Mezger, K., and Hoernes, S., 1998b, Geochemical and isotopic studies of syenites from the Proterozoic Damara belt (Namibia): Implications for the origin of syenites: Mineralogical Magazine, v. 62, p. 729–730, doi: 10.1180/minmag.1998.62A.2.51.

Jung, S., Hoernes, S., and Mezger, K., 2000, Geochronology and petrogenesis of Pan-African syn-tectonic S-type and post-tectonic A-type granite (Namibia)—Products of melting of crustal sources, fractional crystallization and wall rock entrainment: Lithos, v. 50, p. 259–287, doi: 10.1016/S0024-4937(99)00059-6.

Jung, S., Mezger, K., and Hoernes, S., 2001, Trace element and isotopic (Sr, Nd, Pb, O) arguments for a mid-crustal origin of Pan-African garnet-bearing S-type granites from the Damara orogen (Namibia): Precambrian Research, v. 110, p. 325–355, doi: 10.1016/S0301-9268(01)00175-9.

Jung, S., Hoernes, S., and Mezger, K., 2002, Synorogenic melting of mafic lower crust: Constraints from geochronology, petrology and Sr, Nd, Pb and O isotope geochemistry of quartz diorites (Damara orogen, Namibia): Contributions to Mineralogy and Petrology, v. 143, p. 551–566.

Jung, S., Mezger, K., and Hoernes, S., 2003, Petrology of basement-dominated terranes, II: Contrasting isotopic (Sr, Nd, Pb, O) signatures of basement-derived granites and constraints on the source region of granite (Damara orogen, Namibia): Chemical Geology, v. 199, p. 1–28.

Karlstrom, K.E., Ahall, K.I., Williams, M.L., McLelland, J., Harlan, S.S., and Geissman, J.W., 2000, Long-lived (1.8–1.0 Ga) convergent orogen in southeastern Laurentia: Its extensions to Australia and Baltica and implications for refining Rodinia: Geological Society of America Abstracts with Programs, v. 32, no. 7, p. 317.

Kasch, K.W., 1983a, Continental collision, suture progradation and thermal relaxation: A plate-tectonic model for the Damara orogen in central Namibia, *in* Miller, R.G., ed., Evolution of the Damara orogen of southwest Africa/Namibia: Geological Society of South Africa Special Publication 11, p. 423–429.

Kasch, K.W., 1983b, Regional P-T variations in the Damara orogen with particular reference to early high-pressure metamorphism along the southern margin, *in* Miller, R.G., ed., Evolution of the Damara orogen of southwest Africa/Namibia: Geological Society of South Africa Special Publication 11, p. 243–253.

Keay, S., Collins, W.J., and McCulloch, M.T., 1997, A three-component Sr-Nd isotopic mixing model for granitoid genesis, Lachlan fold belt, eastern Australia: Geology, v. 25, p. 307–310, doi: 10.1130/0091-7613(1997)025<0307:ATCSNI>2.3.CO;2.

Keay, S., Steele, D., and Compston, W., 1999, Identifying granite sources by SHRIMP U-Pb zircon geochronology: An application to the Lachlan Fold Belt: Contributions to Mineralogy and Petrology, v. 137, p. 323–341, doi: 10.1007/s004100050553.

Killick, A.M., 2000, The Matchless belt and associated sulphide mineral deposits, Damara orogen, Namibia, *in* Miller, R.McG., ed., Henno Martin Commemorative Volume: Communications of the Geological Survey of Namibia, v. 12, p. 73–80.

Kimbrough, D.L., Mattison, J.M., Coombs, D.S., Landis, C.A., and Johnston, M.R., 1992, Uranium-lead ages for the Dun Mountain ophiolite belt and Brook Street terrane, South Island, New Zealand: Geological Society of America Bulletin, v. 104, p. 429–443, doi: 10.1130/0016-7606(1992)104<0429:ULAFTD>2.3.CO;2.

Kimbrough, D.L., Tulloch, A.J., Geary, E., Coombs, D.S., and Mattinson, J.M., 1993, Isotopic ages from the Nelson region of the South Island, New Zealand: Crustal structure and definition of the Median Tectonic Zone: Tectonophysics, v. 225, p. 433–448, doi: 10.1016/0040-1951(93)90308-7.

Kimbrough, D.L., Tulloch, A.J., Coombs, D.S., Landis, C.A., Johnston, M.R., and Mattison, J.M., 1994, Uranium-lead ages from the Median Tectonic Zone, South Island, New Zealand: New Zealand Journal of Geology and Geophysics, v. 37, p. 393–419.

Kimura, G., and Ludden, J., 1995, Peeling of oceanic crust in subduction zones: Geology, v. 23, p. 217–220, doi: 10.1130/0091-7613(1995)023<0217:POCISZ>2.3.CO;2.

King, P.L., White, A.J.R., Chappell, B.W., and Allen, C.M., 1997, Characterization and origin of aluminous A-type granites from the Lachlan Fold Belt, southeastern Australia: Journal of Petrology, v. 38, p. 371–391, doi: 10.1093/petrology/38.3.371.

Kisters, A.F.M., Jordaan, L.S., and Neumaier, K., 2004, Thrust-related dome structures in the Karibib district and the origin of orthogonal fabric domains in the south Central Zone of the Pan-African Damara belt, Namibia: Precambrian Research, v. 133, p. 283–303, doi: 10.1016/j.precamres.2004.05.001.

Korn, H., and Martin, H., 1959, Gravity tectonics in the Naukluft Mountains of southwest Africa: Geological Society of America Bulletin, v. 70, p. 1047–1078, doi: 10.1130/0016-7606(1959)70[1047:GTITNM]2.0.CO;2.

Kröner, A., 1982, Rb-Sr geochronology and tectonic evolution of the Pan-African belt of Namibia, southwestern Africa: American Journal of Science, v. 282, p. 1471–1507.

Kröner, A., 1984, Dome structures and basement reactivation in the Pan-African Damara belt of Namibia, *in* Kröner, A., and Greiling, R.O., eds., Precambrian tectonics illustrated: Stuttgart, Nagele und Obermiller, p. 191–206.

Kukla, C., 1993, Strontium isotope heterogeneities in amphibolite facies, banded metasediments—A case study from the Late Proterozoic Kuiseb Formation of the southern Damara orogen, central Namibia: Geological Survey of Namibia Memoir 15, 139 p.

Kukla, P.A., and Stanistreet, I.G., 1991, Record of the Damaran Khomas Hochland accretionary prism in central Namibia: Refutation of an "ensialic" origin of a Late Proterozoic orogenic belt: Geology, v. 19, p. 473–476, doi: 10.1130/0091-7613(1991)019<0473:ROTDKH>2.3.CO;2.

Kukla, P.A., Opitz, C., Stanistreet, I.G., and Charlesworth, E.G., 1988, New aspects of the sedimentology and structure of the Kuiseb Formation in the western Khomas trough, Damara orogen, southwest Africa/Namibia: Communications of the Geological Survey of Namibia, v. 4, p. 33–42.

Lundberg, N., and Reed, D.L., 1991, Continental margin tectonics: Forearc process: Review of Geophysics, v. 29, p. 794–806.

Maas, R., and Nicholls, I.A., 2002, Nd-Sr isotopic evidence for the origin of Devonian granitic and felsic volcanic rocks of the western Lachlan Fold Belt: Geological Society of Australia Abstracts, v. 67, p. 136.

Maas, R., Nicholls, I.A., Greig, A., and Nemchin, A., 2001, U-Pb zircon studies of mid-crustal metasedimentary enclaves from the S-type Deddick Granodiorite, Lachlan Fold Belt, SE Australia: Journal of Petrology, v. 42, p. 1429–1448, doi: 10.1093/petrology/42.8.1429.

Maloof, A.C., 2000, Superposed folding at the junction of the inland and coastal belts, Damara orogen, Namibia, *in* Miller, R.McG., ed., Henno Martin Commemorative Volume: Communications of the Geological Survey of Namibia, v. 12, p. 89–98.

Martin, H., 1983, Overview of the geosynclinal, structural and metamorphic development of the intracontinental branch of the Damara orogen, *in* Martin, H., and Eder, F.W., eds., Intracontinental fold belts: Berlin, Springer-Verlag, p. 473–502.

Martin, H., and Porada, H., 1977, The intracratonic branch of the Damara orogen in southwest Africa, I: Discussion of geodynamic models: Precambrian Research, v. 5, p. 311–338, doi: 10.1016/0301-9268(77)90039-0.

Masberg, P., 2000, Garnet growth in medium pressure granulite facies metapelites from the central Damara orogen: Igneous versus metamorphic history, *in* Miller, R.McG., ed., Henno Martin Commemorative Volume: Communications of the Geological Survey of Namibia, v. 12, p. 115–124.

McCulloch, M.T., and Chappell, B.W., 1982, Nd isotopic characteristics of S- and I-type granites: Earth and Planetary Science Letters, v. 58, p. 51–64, doi: 10.1016/0012-821X(82)90102-9.

McCulloch, M.T., Bradshaw, J.Y., and Taylor, S.R., 1987, Sm-Nd and Rb-Sr isotopic and geochemical systematics in Phanerozoic granulites from Fiordland, southwest New Zealand: Contributions to Mineralogy and Petrology, v. 97, p. 183–195, doi: 10.1007/BF00371238.

McDermott, F., and Hawkesworth, C.J., 1990, Intracrustal recycling and upper-mantle evolution: A case study from the Pan-African Damara mobile belt, central Namibia: Chemical Geology, v. 83, p. 263–280, doi: 10.1016/0009-2541(90)90284-E.

McDermott, F., Harris, N.B.W., and Hawkesworth, C.J., 1996, Geochemical constraints on crustal anatexis: A case study from the Pan-African granitoids of Namibia: Contributions to Mineralogy and Petrology, v. 123, p. 406–423, doi: 10.1007/s004100050165.

McDermott, F., Harris, N.B.W., and Hawkesworth, C.J., 2000, Geochemical constraints on the petrogenesis of Pan-African A-type granites in the Damara belt, Namibia, *in* Miller, R.McG., ed., Henno Martin Commemorative Volume: Communications of the Geological Survey of Namibia, v. 12, p. 139–148.

Miller, J.McL., and Gray, D.R., 1997, Subduction-related deformation and the Narooma anticlinorium, eastern Lachlan orogen: Australian Journal of Earth Sciences, v. 44, p. 237–251.

Miller, R.McG., 1983a, The Pan-African Damara orogen of South West Namibia/Africa, *in* Miller, R.McG., ed., Evolution of the Damara Orogen of southwest Africa/Namibia: Special Publication of the Geological Society of Africa, v. 11, p. 431–515.

Miller, R.McG., 1983b, Evolution of the Damara orogen of southwest Africa/Namibia: Geological Society of South Africa Special Publication 11, 515 p.

Miller, R.McG., and Grote, W., 1988, Geological Map of the Damara Orogen, Namibia: Geological Survey of Namibia, Windhoek, scale 1:500,000, 2 sheets.

Miller, R.McG., Barnes, S.J., and Balkwill, G., 1983a, Possible active margin deposits within the southern Damara Orogen: The Kuiseb Formation between Okahandja and Windhoek, *in* Miller, R.McG., ed., Evolution of the Damara Orogen of southwest Africa/Namibia: Special Publication of the Geological Society of Africa, v. 11, p. 73–88.

Miller, R.McG., Freyer, E.E., and Hälbich, I.W., 1983b, Turbidite succession equivalent to the entire Swakop Group, *in* Miller, R.McG., ed., Evolution of the Damara Orogen of southwest Africa/Namibia: Special Publication of the Geological Society of Africa, v. 11, p. 65–71.

Morand, V.J., 1990, Low-pressure regional metamorphism in the Omeo metamorphic complex, Victoria, Australia: Journal of Metamorphic Geology, v. 8, p. 1–12.

Morand, V.J., and Gray, D.R., 1991, Major fault zones related to the Omeo metamorphic complex, northeastern Victoria: Australian Journal of Earth Sciences, v. 38, p. 203–221.

Mortimer, N., 1993, Jurassic tectonic history of the Otago Schist, New Zealand: Tectonics, v. 12, p. 237–244.

Mortimer, N., 2000, Metamorphic discontinuities in orogenic belts: Example of garnet-biotite-albite zone in the Otago Schist, New Zealand: International Journal of Earth Sciences, v. 89, p. 295–306, doi: 10.1007/s005310000086.

Mortimer, N., 2004, New Zealand's geological foundations: Gondwana Research, v. 7, p. 261–272, doi: 10.1016/S1342-937X(05)70324-5.

Mortimer, N., Tulloch, A.J., Spark, R.N., Walker, N.W., Ladley, E., Allibone, A., and Kimbrough, D.L., 1999a, Overview of the Median batholith, New Zealand: A new interpretation of the geology of the Median Tectonic Zone and adjacent rocks: Journal of African Earth Sciences, v. 29, p. 257–268, doi: 10.1016/S0899-5362(99)00095-0.

Mortimer, N., Gans, P., Calvert, A., and Walker, N., 1999b, Geology and thermochronometry of the east edge of the Median batholith (Median Tectonic Zone): A new perspective on Permian to Cretaceous crustal growth of New Zealand: Island Arc, v. 8, p. 404–425, doi: 10.1046/j.1440-1738.1999.00249.x.

Mortimer, N., Davey, F.J., Melhush, A., Yu, J., and Godfre, N.J., 2003, Geological interpretation of a deep crustal seismic reflection profile across the eastern province and Median batholith, New Zealand: Crustal architecture of an extended Phanerozoic convergent orogen: New Zealand Journal of Geology and Geophysics, v. 45, p. 349–363.

Muir, R.J., Ireland, T.R., Weaver, S.D., Bradshaw, J.D., Evans, J.A., Eby, G.N., and Shelley, D., 1988, Geochronology and geochemistry of a Mesozoic magmatic arc system, Fiordland, New Zealand: Geological Society [London] Journal, v. 155, p. 1037–1053.

Muir, R.J., Weaver, S.D., Bradshaw, J.D., Eby, G.N., and Evans, J.A., 1995, The Cretaceous Separation Point batholith, New Zealand: Granitoid magmas formed by melting of mafic lithosphere: Geological Society [London] Journal, v. 152, p. 689–701.

Muir, R.J., Ireland, T.R., Weaver, S.D., and Bradshaw, J.D., 1996a, Ion microprobe dating of Paleozoic granitoids: Devonian magmatism in New Zealand and correlations with Australia and Antarctica: Chemical Geology, v. 127, p. 191–210, doi: 10.1016/0009-2541(95)00092-5.

Muir, R.J., Weaver, S.D., Bradshaw, J.D., Eby, G.N., Evans, J.A., and Ireland, T.R., 1996b, Geochemistry of the Karamea batholith, New Zealand, and comparisons with the Lachlan Fold Belt granites of southeastern Australia: Lithos, v. 39, p. 1–20.

Muir, R.J., Ireland, T.R., Weaver, S.D., Bradshaw, J.D., Evans, J.A., Eby, G.N., and Shelley, D., 1998, Geochronology and geochemistry of a Mesozoic magmatic arc system, Fiordland, New Zealand: Geological Society [London] Journal, v. 155, p. 1037–1053.

Murray, C.G., Fergusson, C.L., Flood, P.G., Whitaker, W.G., and Korsch, R.J., 1987, Plate-tectonic model for the Carboniferous evolution of the New England Fold Belt: Australian Journal of Earth Sciences, v. 34, p. 213–236.

Norris, R.J., and Craw, D., 1987, Aspiring terrane: An oceanic assemblage from New Zealand and its implications for Mesozoic terrane accretion in the southwest Pacific, *in* Leitch, E.C., and Scheibner, E., eds., Terrane accretion and orogenic belts: American Geophysical Union Geodynamics Series, v. 19, p. 169–177.

Offler, R., McKnight, S., and Morand, V., 1998, Tectonothermal history of the western Lachlan Fold Belt, Australia: Insights from white mica studies: Journal of Metamorphic Geology, v. 16, p. 531–540, doi: 10.1111/j.1525-1314.1998.00153.x.

Oliver, G.J.H., 1994, Mid-crustal detachment and domes in the Central Zone of the Damaran orogen, Namibia: Journal of African Earth Sciences, v. 19, p. 331–344, doi: 10.1016/0899-5362(94)90018-3.

Passchier, C.W., Trouw, R.A., Ribeiro, A., and Paciullo, F.V.P., 2002, Tectonic evolution of the southern Kaoko belt, Namibia: Journal of African Earth Sciences, v. 35, p. 61–75, doi: 10.1016/S0899-5362(02)00030-1.

Patchett, P.J., and Bridgwater, D., 1984, Origin of continental crust of 1.9–1.7 Ga age defined by Nd isotopes in the Ketilidian terrain of South Greenland: Contributions to Mineralogy and Petrology, v. 87, p. 311–318, doi: 10.1007/BF00381287.

Percival, J.A., Bleeker, W., Cook, E.A., Rivers, T., Ross, G., and van Staal, C., 2004, PanLITHOPROBE Workshop IV: Intra-orogen correlations and comparative orogenic anatomy: Geoscience Canada, v. 31, p. 23–39.

Petford, N., and Atherton, M., 1996, Na-rich partial melts from newly underplated basaltic crust: The Cordillera Blanca batholith, Peru: Journal of Petrology, v. 37, p. 1491–1521, doi: 10.1093/petrology/37.6.1491.

Phillips, G.N., Wall, V.J., and Clemens, J.C., 1981, Petrology of the Strathbogie batholith—A cordierite-bearing granite: Canadian Mineralogist, v. 19, p. 47–63.

Porada, H., 1979, The Damara-Ribeira orogen of the Pan-African Brasiliano cycle in Namibia (southwest Africa) and Brazil as interpreted in terms of continental collision: Tectonophysics, v. 57, p. 237–265, doi: 10.1016/0040-1951(79)90150-1.

Porada, H., 1983, Geosynclinal model for the geosynclinal development of the Damara orogen, Namibia, southwest Africa, *in* Martin, H., and Eder, F.W., eds., Intracontinental fold belts: Berlin, Springer-Verlag, p. 503–541.

Porada, H., Ahrendt, H., Behr, J., and Weber, K., 1983, The join of the coastal and intracontinental branches of the Damara orogen, Namibia, southwest Africa, *in* Martin, H., and Eder, F.W., eds., Intracontinental fold belts: Berlin, Springer-Verlag, p. 901–912.

Powell, C.McA., 1984, Ordovician to Carboniferous, in Veevers, J.J., ed., Phanerozoic Earth history of Australia: Oxford Monographs on Geology and Geophysics, v. 2, p. 290–340.

Prave, A.R., 1996, Tale of three cratons: Tectonostratigraphic anatomy of the Damara orogen in northwestern Namibia and the assembly of Gondwana: Geology, v. 24, p. 1115–1118, doi: 10.1130/0091-7613(1996)024<1115:TOTCTA>2.3.CO;2.

Price, R.C., Brown, W.M., and Woolard, C.A., 1983, The geology, geochemistry and origin of Late Silurian, high-Si igneous rocks of the upper Murray Valley, NE Victoria: Journal of the Geological Society of Australia, v. 30, p. 443–459.

Puhan, D., 1983, Temperature and pressure of metamorphism in the central Damara orogen, *in* Miller, R.G., ed., Evolution of the Damara orogen of southwest Africa/Namibia: Geological Society of South Africa Special Publication 11, p. 219–223.

Roser, B.P., and Cooper, A.F., 1990, Geochemistry and terrane affiliation of Haast Schist from the western Southern Alps, New Zealand: New Zealand Journal of Geology and Geophysics, v. 94, p. 635–650.

Rossiter, A.G., 2003, Granitic rocks of the Lachlan Fold Belt in Victoria, *in* Birch, W.D., ed., Geology of Victoria: Geological Society of Australia Special Publication 23, p. 217–237.

Schmidt, A., and Wedepohl, K.H., 1983, Chemical composition and genetic relations of the Matchless amphibolite (Damara orogenic belt), *in* Miller, R.G., ed., Evolution of the Damara orogen of southwest Africa/Namibia: Geological Society of South Africa Special Publication 11, p. 139–145.

Şengör, C.A.M., and Natal'in, B.A., 1996, Turkic-type orogeny and its role in the making of the continental crust: Annual Review of Earth and Planetary Sciences, v. 24, p. 263–337, doi: 10.1146/annurev.earth.24.1.263.

Smith, A.G., 1981, Subduction and coeval thrust belts, with particular reference to North America, *in* McClay, K.R., and Price, N.J., eds., Thrust and nappe tectonics: Geological Society [London] Special Publication 9, p. 111–124.

Soesoo, A., and Nicholls, I.A., 1999, Mafic rocks spatially associated with Devonian felsic intrusions of the southern Lachlan Fold Belt: A possible mantle contribution to crustal evolution processes: Australian Journal of Earth Sciences, v. 46, p. 725–734, doi: 10.1046/j.1440-0952.1999.00740.x.

Soesoo, A., Bons, P.D., Gray, D.R., and Foster, D.A., 1997, Divergent double subduction: Tectonic and petrologic consequences: Geology, v. 25, p. 755–758, doi: 10.1130/0091-7613(1997)025<0755:DDSTAP>2.3.CO;2.

Spaggiari, C.V., Gray, D.R., and Foster, D.A., 2002a, Blueschist metamorphism during accretion in the Lachlan orogen, southeastern Australia: Journal of Metamorphic Geology, v. 20, p. 711–726, doi: 10.1046/j.1525-1314.2002.00405.x.

Spaggiari, C.V., Gray, D.R., Foster, D.A., and Fanning, C.M., 2002b, Occurrence and significance of blueschists in the southern Lachlan orogen: Australian Journal of Earth Sciences, v. 49, p. 255–269, doi: 10.1046/j.1440-0952.2002.00915.x.

Spaggiari, C.V., Gray, D.R., and Foster, D.A., 2003a, Tethyan and Cordilleran-type ophiolites of eastern Australia: Implications for the evolution of the Tasmanides, *in* Dilek, Y., and Robinson, P.T., eds., Ophiolites in Earth's history: Geological Society [London] Special Publication 218, p. 517–539.

Spaggiari, C.V., Gray, D.R., Foster, D.A., and McKnight, S., 2003b, Evolution of the boundary between the western and central Lachlan orogen: Implications for Tasmanide tectonics: Australian Journal of Earth Sciences, v. 50, p. 725–749, doi: 10.1111/j.1440-0952.2003.01022.x.

Spaggiari, C.V., Gray, D.R., and Foster, D.A., 2004a, Ophiolite accretion in the Lachlan orogen, southeastern Australia: Journal of Structural Geology, v. 26, p. 87–112, doi: 10.1016/S0191-8141(03)00084-1.

Spaggiari, C.V., Gray, D.R., and Foster, D.A., 2004b, Lachlan orogen subduction-accretion systematics revisited: Australian Journal of Earth Sciences, v. 51, p. 549–553, doi: 10.1111/j.1400-0952.2004.01073.x.

Spell, T.L., McDougall, I., and Tulloch, A.J., 2000, Thermochronologic constraints on the breakup of the Pacific Gondwana margin: The Paparoa metamorphic core complex, South Island, New Zealand: Tectonics, v. 19, p. 433–451, doi: 10.1029/1999TC900046.

Sporli, K.B., 1978, Mesozoic tectonics, North Island, New Zealand: Geological Society of America Bulletin, v. 89, p. 415–425, doi: 10.1130/0016-7606(1978)89<415:MTNINZ>2.0.CO;2.

Swart, R., 1992, The sedimentology of the Zerrissene turbidite system, Damara orogen, Namibia: Geological Survey of Namibia Memoir 13, 54 p.

Tamaki, K., 1995, Opening of the Japan Sea, in Taylor, B., ed., Backarc basins—Tectonics and magmatism: New York, Plenum Press, p. 407–420.

Thompson, A.B., 1996, Fertility of crustal rocks during anatexis: 3rd Hutton Symposium: Transactions of the Royal Society of Edinburgh, v. 87, p. 1–10.

Tulloch, A.J., 1983, Granitoid rocks of New Zealand—A brief review: Geological Society of America Memoir 159, p. 5–20.

Tulloch, A.J., 1988, Batholiths, plutons and suites: Nomenclature for granitoid rocks of Westland-Nelson, New Zealand: New Zealand Journal of Geology and Geophysics, v. 31, p. 505–509.

Tulloch, A.J., and Kimbrough, D.L., 1989, The Paparoa metamorphic core complex, New Zealand: Cretaceous extension associated with fragmentation of the Pacific margin of Gondwana: Tectonics, v. 8, p. 1217–1234.

Tulloch, A.J., and Kimbrough, D.L., 2003, Paired plutonic belts in convergent margins and the development of high-Sr/Y magmatism: Peninsular Ranges batholith of Baja-California and Median batholith of New Zealand, *in* Johnson, S.E., Paterson, S.R., Fletcher, J.M., Girty, G.H.., Kimbrough, D.L., and Martín-Barajas, A., eds., Tectonic evolution of northwestern México and the southwestern USA: Geological Society of America Special Paper 374, p. 275–295.

Turner, S.P., Kelley, S.P., and Morrison, R.S., 1992, Derivation of some A-type magmas by fractionation of basaltic magma: An example from the Padthaway Ridge, South Australia: Lithos, v. 28, p. 151–179, doi: 10.1016/0024-4937(92)90029-X.

Turner, S.P., Foden, J., Kelley, S.P., VandenBerg, A.H.M., Sandiford, M., and Flöttmann, T., 1996, Source of the Lachlan Fold Belt flysch linked to convective removal of the lithospheric mantle and rapid exhumation of the Delamerian-Ross fold belt: Geology, v. 24, p. 941–944, doi: 10.1130/0091-7613(1996)024<0941:SOTLFB>2.3.CO;2.

van de Flierdt, T., Hoernes, S., Jung, S., Masberg, P., Hoffer, E., Schaltegger, U., and Friedrichsen, H., 2003, Lower crustal melting and the role of open-system processes in the generation of syn-orogenic quartz diorite-granite-leucogranite associations: Constraints from Sr-Nd-O isotopes from the Bandombaai Complex, Namibia: Lithos, v. 67, p. 205–226, doi: 10.1016/S0024-4937(03)00016-1.

Veevers, J.J., 2000a, Billion-year Earth history of Australia and neighbours in Gondwanaland: Sydney, Gemoc Press, 388 p.

Veevers, J.J., 2000b, Antarctic Beardmore-Ross and Mirny provenances saturate Palaeozoic-Mesozoic East Gondwanaland with 0.6–0.5 Ga zircons, *in* Veevers, J.J., ed., Billion-year Earth history of Australia and neighbors in Gondwanaland: Sydney, Gemoc Press, p. 110–130.

Von Huene, R., and Scholl, D.W., 1991, Observations at convergent margins concerning sediment subduction, subduction erosion, and the growth of continental crust: Reviews of Geophysics, v. 29, p. 279–316.

Waight, T.E., Weaver, S.D., Muir, R.J., Maas, R., and Eby, G.N., 1998a, The Hohonu batholith of North Westland, New Zealand: Granitoid compositions controlled by source H_2O contents and generated during tectonic transition: Contributions to Mineralogy and Petrology, v. 130, p. 225–239, doi: 10.1007/s004100050362.

Waight, T.E., Weaver, S.D., Maas, R., and Eby, G.N., 1998b, Bimodal Late Cretaceous alkaline magmatism and the opening of the Tasman Sea: West Coast, South Island, New Zealand: Australian Journal of Earth Sciences, v. 45, p. 823–835.

Waight, T.E., Weaver, S.D., and Muir, R.J., 1998c, Mid-Cretaceous granitic magmatism dating the transition from subduction to extension in southern New Zealand: A chemical and tectonic synthesis: Lithos, v. 45, p. 469–482, doi: 10.1016/S0024-4937(98)00045-0.

Wandres, A.M., and Bradshaw, J.D., 2005, New Zealand tectonostratigraphy and implications from conglomeratic rocks for the configuration of the SW Pacific margin of Gondwana, *in* Vaughan, A.P.M., et al., eds., Terrane processes at the margins of Gondwana: Geological Society [London] Special Publication 246, p. 179–216.

Watson, J.M., and Gray, D.R., 2001, Character, extent and significance of broken formation for the Tabberabbera Zone, central Lachlan orogen: Australian Journal of Earth Sciences, v. 48, p. 943–954, doi: 10.1046/j.1440-0952.2001.00911.x.

White, A.J.R., and Chappell, B.W., 1983, Granitoid types and their distribution in the Lachlan Fold Belt, SE Australia: Geological Society of America Memoir 159, p. 21–34.

White, A.J.R., and Chappell, B.W., 1988, Some supracrustal (S-type) granites of the Lachlan Fold Belt: Transactions of the Royal Society of Edinburgh, v. 79, p. 169–182.

Williams, I.S., 1992, Some observations on the use of zircon U-Pb geochronology in the study of granitic rocks: Transactions of the Royal Society of Edinburgh, v. 83, p. 447–458.

Wyborn, D., and Chappell, B.W., 1986, The petrogenetic significance of chemically related plutonic and volcanic rock units: Geological Magazine, v. 123, p. 619–628.

Wyborn, D., Chappell, B.W., and Johnston, R.M., 1981, Three S-type volcanic suites from the Lachlan Fold Belt, southeast Australia: Journal of Geophysical Research, v. 86, p. 10,335–10,348.

Wyborn, L.A.I., Wyborn, D., Warren, R.G., and Drummond, B.J., 1992, Proterozoic granite types in Australia: Implications for lower crust composition, structure and evolution: Transactions of the Royal Society of Edinburgh, Earth Sciences, v. 83, p. 201–209.

Yardley, B.W.D., 1982, The early metamorphic history of the Haast Schists and related rocks of New Zealand: Contributions to Mineralogy and Petrology, v. 81, p. 317–327, doi: 10.1007/BF00371686.

Zen, E.-A., 1995, Crustal magma generation and low-pressure high-temperature regional metamorphism in an extensional environment: Possible application to the Lachlan Fold Belt, Australia: American Journal of Science, v. 295, p. 851–874.

Manuscript Accepted by the Society 1 February 2007

Printed in the USA

The Geological Society of America
Memoir 200
2007

Pure shear dominated high-strain zones in basement terranes

Christopher M. Bailey
Lina E. Polvi
Adam M. Forte
Department of Geology, College of William & Mary, Box 8795, Williamsburg, Virginia 23187, USA

ABSTRACT

High-strain zones are common in basement terranes, and understanding their tectonic significance requires quantitative knowledge of deformation kinematics. We report on strained rocks from different tectonic settings that record pure shear dominated ($W_m \leq 0.4$) deformations. Mylonitic rocks derived from Mesoproterozoic basement granitoids are exposed in the Lawhorne Mill high-strain zone in the Virginia Blue Ridge. Chemical and mineralogical differences between the leucogranitoid protolith and mylonite are consistent with ~50% volume loss during deformation. Minimum finite strains in *XZ* sections range from 4:1 to 7:1, and three-dimensional strains plot in the field of apparent flattening; however, with volume loss these rocks likely experienced bulk plane strain. The R_s/Θ and quartz *c*-axis vorticity gauges yield W_m values of 0.0–0.6. Fabric asymmetries normal to *both* foliation and lineation are consistent with modest triclinic deformation symmetry. Mylonitic rocks from the Lawhorne Mill high-strain zone record a pure shear dominated deformation that produced ~70% contraction across the zone with only minimal displacement parallel to the zone (<0.5 km).

Pure shear dominated high-strain zones occur in a variety of mid-crustal settings. Ultramylonites from metamorphic core complexes in Arizona record very low vorticity values ($W_m < 0.4$). Well-foliated, steeply dipping, upper amphibolite facies rocks from the Coast shear zone in British Columbia are characterized by orthorhombic fabrics formed during pure shear dominated deformation that accommodated crustal contraction. These zones differ from simple and general shear zones because displacement across these zones is minimal relative to the overall finite strain. However, zone-normal shortening and zone-parallel stretching are significant in pure shear dominated zones. Steeply dipping zones formed in contractional settings serve to effectively shorten and thicken the crust across basement massifs, whereas gently dipping zones formed in extensional settings thin the crust.

Keywords: mylonites, vorticity, monoclinic deformation, triclinic deformation.

Bailey, C.M., Polvi, L.E., and Forte, A.M., 2007, Pure shear dominated high-strain zones in basement terranes, *in* Hatcher, R.D., Jr., Carlson, M.P., McBride, J.H., and Martínez Catalán, J.R., eds., 4-D Framework of Continental Crust: Geological Society of America Memoir 200, p. 93–108, doi: 10.1130/2007.1200(06). For permission to copy, contact editing@geosociety.org.

INTRODUCTION

High-strain zones form an important element of the crustal architecture in mid- and lower-crustal basement terranes. These zones can be many kilometers thick, extend for hundreds to thousands of kilometers, and commonly demarcate significant crustal boundaries. Understanding the kinematics in these zones is a prerequisite for unraveling the tectonic history of basement terranes. In order to accurately characterize displacement and flow across high-strain zones, quantitative measures of strain symmetry, volume change, three-dimensional finite strain, and vorticity are required. In the past decade structural geologists have developed more techniques with which to quantify finite deformation parameters, and quantitative inquiries are yielding surprising results.

Traditionally, strain in shear zones has been assumed to follow a monoclinic deformation path in which the overall tectonic transport direction is parallel to the lineation with asymmetric structures developed on lineation-parallel faces (Fig. 1). Transpressional deformations can produce a lineation normal to the tectonic transport direction (rolling lineations) with fabric asymmetries being well developed on lineation-normal faces (Fig. 1) (Sanderson and Marchini, 1984; Tikoff and Greene, 1997). More recently, triclinic deformation symmetries have been recognized in naturally deformed rocks (Jiang and Williams, 1998; Lin et al., 1998; Bailey et al., 2002). Triclinic deformation produces a complex relationship between fabric elements and the overall transport direction, with asymmetries developed on both lineation-parallel and lineation-normal faces (Fig. 1).

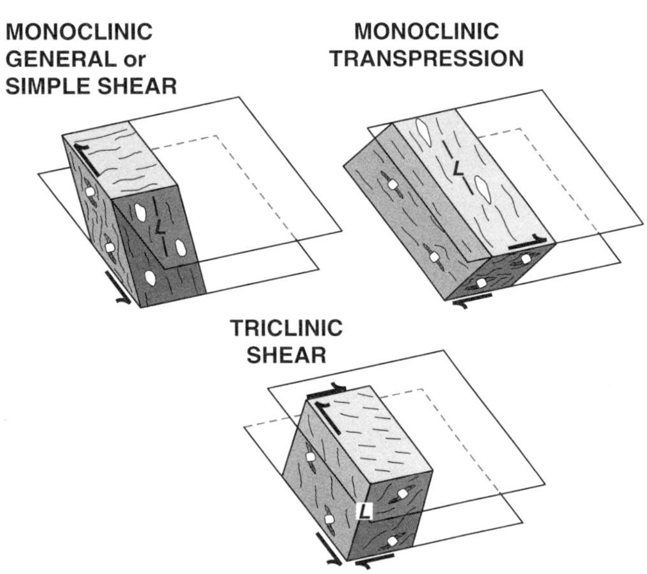

Figure 1. Homogeneous monoclinic simple/general shear, monoclinic transpression, and triclinic shear. In monoclinic simple/general shear the lineation tracks the overall displacement vector, whereas in monoclinic transpression the lineation is a rolling lineation and normal to the overall displacement vector. L is the elongation lineation.

High-strain zones were long considered to be dominated by simple shear deformation (Ramsay and Graham, 1970). Simple shear deformation models became popular because, for parallel-sided zones, compatibility is preserved between the wall rock and deformation zone during simple shear. However, many recent studies have quantified nonsimple shear deformation in mylonitic zones from different tectonic settings (Simpson and De Paor, 1997; Grasemann et al., 1999; Xypolias and Koukouvelas, 2001; Bailey and Eyster, 2003; Law et al., 2004). Quantifying the rotational component of strain is important because the differences between simple, general, and pure shear deformations produce distinct flow paths with significantly different kinematic consequences.

In this work we present quantitative data from a contractional mylonite zone developed in granitic basement rocks of the central Appalachian Blue Ridge, Virginia. These rocks followed a weakly triclinic, pure shear dominated deformation path in which the total displacement parallel to the zone is modest, but the overall zone-normal shortening is high. Integrated over the entire width of the Blue Ridge basement complex, these strains may have shortened the Blue Ridge crust by 40%. We also report evidence for pure shear dominated deformations in other tectonic settings and highlight the kinematic and tectonic implications of pure shear dominated deformations.

LAWHORNE MILL HIGH-STRAIN ZONE, VIRGINIA BLUE RIDGE

Geologic Setting

In Virginia, the Blue Ridge province forms the thick-skinned, hinterland edge of the Appalachian fold-and-thrust belt (Fig. 2). Mesoproterozoic basement rocks, predominantly granites and gneisses formed during the long-lived Grenvillian orogeny (1.2–1.0 Ga), are exposed in the core of the Blue Ridge anticlinorium (Fig. 2). A suite of Neoproterozoic felsic plutons intruded the basement complex and were exhumed during Iapetan rifting. From the late Neoproterozoic to the early Paleozoic, the Blue Ridge was buried under >10 km of rifted-related and passive margin sedimentary rocks. The Blue Ridge anticlinorium developed as a regional-scale structure that formed when the basement complex and its associated cover rocks were displaced northwestward over a major thrust ramp during collisional tectonism in the late Paleozoic (Harris, 1979; Mitra and Elliott, 1980; Evans, 1989). Blue Ridge basement units are cut by an anastomosing network of high-strain zones that experienced thrust displacement and accommodated crustal shortening in the Paleozoic (Bartholomew et al., 1981; Bailey and Simpson, 1993; Bailey, 1994).

The Lawhorne Mill high-strain zone (Evans, 1984; Bartholomew et al., 1991; Bailey and Simpson, 1993) is a heterogeneous belt of mylonitic rock ~400 m wide that extends for ~15 km parallel to the regional trend of the Blue Ridge in central Virginia (Fig. 2). Bailey and Simpson (1993) noted that, in contrast to other Blue Ridge high-strain zones, the Lawhorne

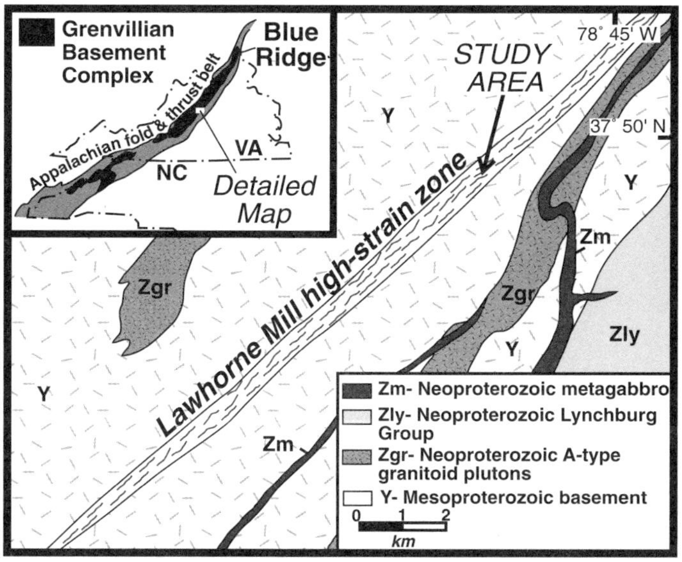

Figure 2. Generalized map of the Blue Ridge with geologic map of the Lawhorne Mill high-strain zone in central Virginia.

Mill high-strain zone preserves dominantly coaxial fabrics and suggested these fabrics developed during Neoproterozoic extensional deformation and that the zone was later reactivated in the Paleozoic. The mineral assemblage and mylonitic microstructures are consistent with greenschist facies deformation (400–450 °C) (Bailey and Simpson, 1993; Polvi, 2003). A $^{40}Ar/^{39}Ar$ date from muscovite in the Lawhorne Mill high-strain zone yields a plateau age of ca. 355 Ma, consistent with pre-Alleghanian deformation in the middle to early Paleozoic (Polvi, 2003).

Study Site

Our data were collected at a well-exposed, subvertical outcrop face of the Lawhorne Mill high-strain zone exposed along Virginia Route 617 and the Rockfish River. The outcrop is ~10 m high and >150 m in length (Fig. 3). Mylonitic rocks in the Lawhorne Mill high-strain zone were derived primarily from Mesoproterozoic granitoid with minor volumes produced from Neoproterozoic (?) mafic intrusives. The high-strain zone boundaries are well defined and sharp. The northwestern boundary of the high-strain zone

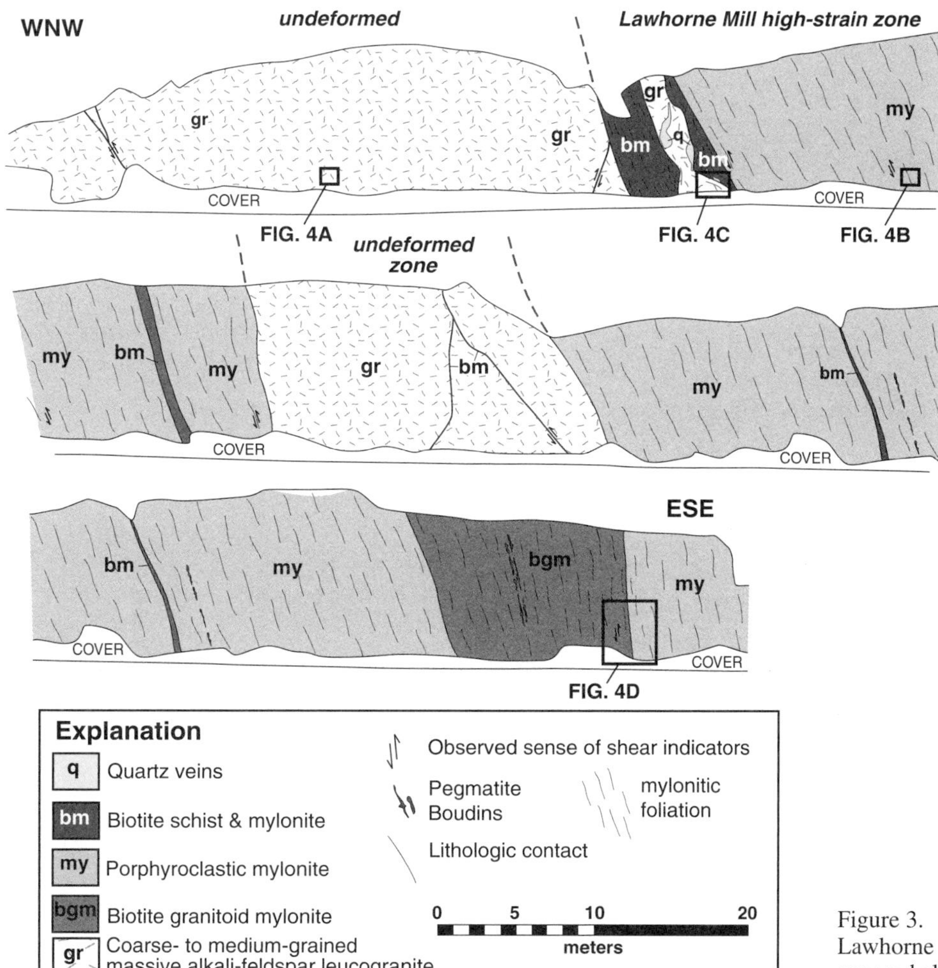

Figure 3. Cross section of the Lawhorne Mill high-strain zone exposed along Virginia Route 617.

is localized along a lithologic contact (Fig. 3). Undeformed domains occur within the zone (Fig. 3). The dominant rock in the Lawhorne Mill high-strain zone is a porphyroclastic mylonite derived from medium- to coarse-grained alkali feldspar granite (Figs. 4A and 4B). The alkali feldspar granite is leucocratic and dominated by perthitic feldspar and quartz with minor plagioclase and biotite. Tabular bodies (20 cm to 3 m thick) of biotite-rich mylonitic schist occur throughout the high-strain zone; these mafic mylonitic rocks are derived from hornblende-bearing metagabbro dikes. Coarse-grained alkali feldspar granite and pegmatite bodies occur as boudins surrounded by less competent rock (Figs. 3 and 4C). Foliated quartz veins are common in boudin necks and along boudin margins. A biotite-bearing granitoid mylonite occurs as an ~10 m wide concordant body exposed at the eastern part of the outcrop (Figs. 3 and 4D).

Mylonitic foliation is defined by aligned micas, elongate quartz lenses, and trains of feldspar fragments. Foliation strikes northeast (~055°) and typically dips steeply to the southeast (Fig. 5A). Lineations, defined by both aligned micas and elongate quartz grains, plunge approximately downdip (Fig. 5A). Kinematic indicators on

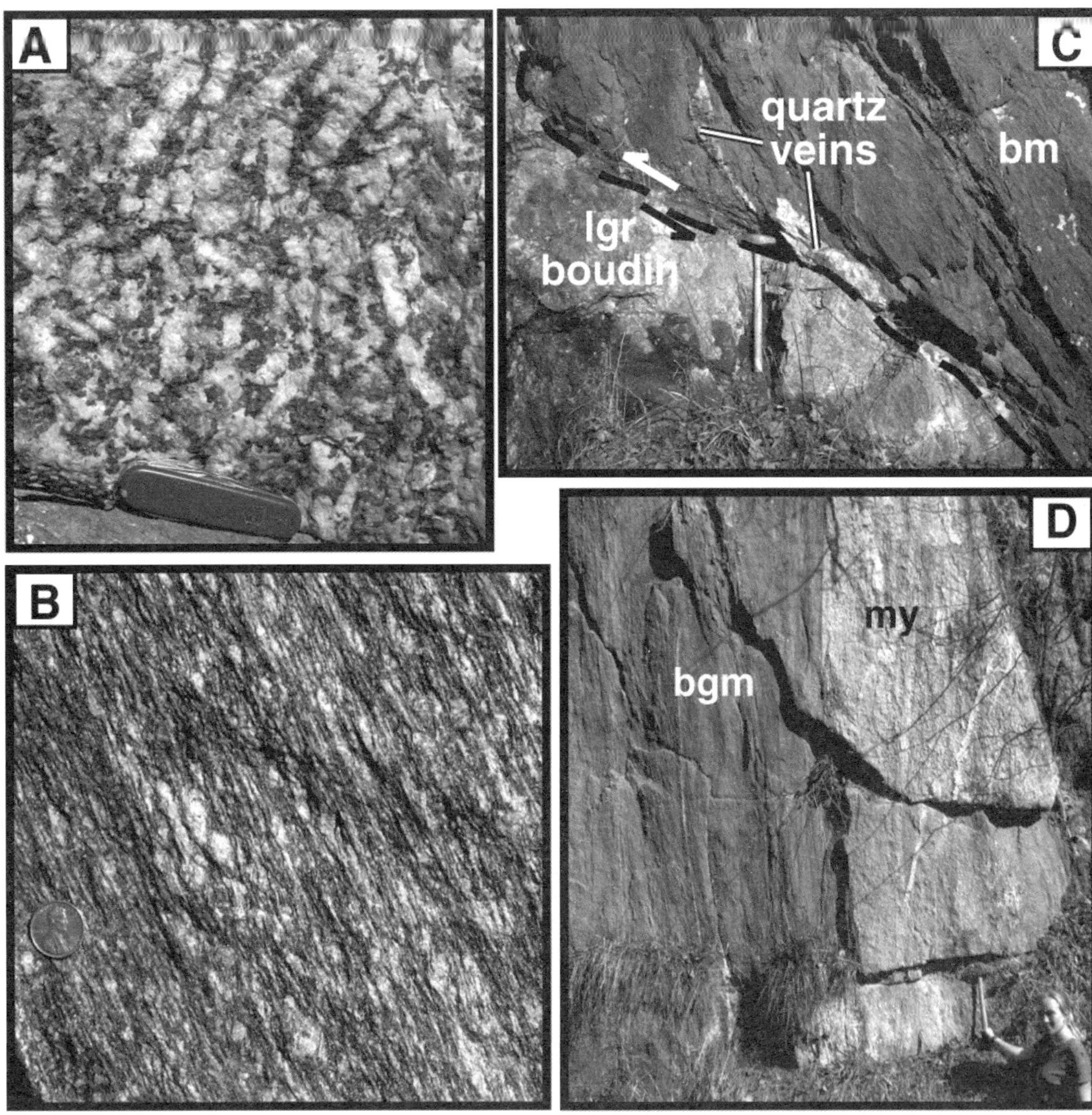

Figure 4. Field photographs, locations referenced in Figure 3. A: Undeformed megacrystic alkali feldspar granite. Knife is ~10 cm in length. B: Porphyroclastic mylonite derived from megacrystic alkali feldspar granite. Coin is 1.5 cm in diameter. C: Alkali feldspar granitoid (lgr) boudin in contact with biotite-rich schistose mylonite (bm). Note deflected quartz veins along boudin margin. The deflection of the quartz vein is consistent with top-to-the-northwest shear. Hammer is ~40 cm in length. D: Subvertical contact between biotite-bearing granitoid mylonite (bgm) and porphyroclastic mylonite (my).

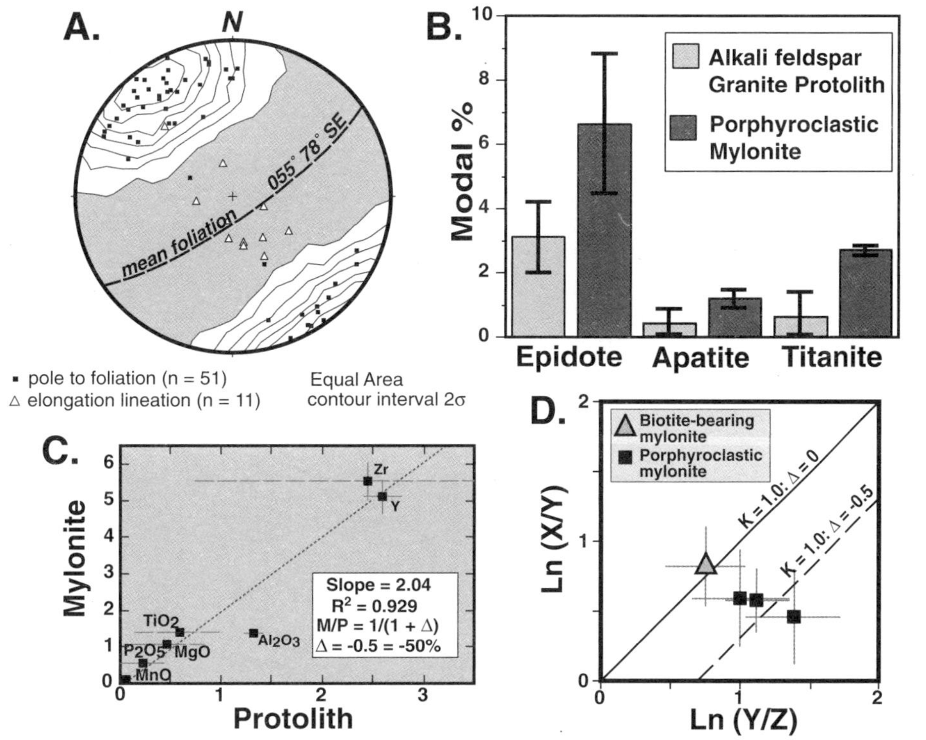

Figure 5. A: Stereogram of Lawhorne Mill foliation and lineation data. B: Modal abundance of epidote, apatite, and titanite in alkali feldspar granite protolith (n = 3) and porphyroclastic mylonite (n = 3). Bars represent maximum and minimum abundance of the phase. C: Isocon plot of immobile elements and oxides. Gray crosses represent maximum and minimum values (n = 3 for both protolith and mylonite). D: Logarithmic Flinn diagram for mylonitic rocks based on R_f/ϕ analysis of quartz grain shapes. Gray crosses represent maximum and minimum values for sectional strain ratios based on measurements from three orthogonal sections.

faces normal to foliation and parallel to lineation have a weak but consistent top-to-the-northwest asymmetry (reverse sense of shear). These asymmetric structures include poorly developed S-C fabrics, mantled porphyroclasts, and stair-stepping feldspar fragments (Fig. 6A). At a few locations quartz veins are asymmetrically deflected along lithologic boundaries (Fig. 4C). Fabric asymmetries *also* occur on foliation-normal/lineation-normal faces and indicate a sinistral shear sense when viewed in the direction of lineation plunge (Fig. 6B).

Mineralogy, Chemistry, and Volume Change

Mylonitic rocks in the Lawhorne Mill high-strain zone exhibit significant mineralogical differences relative to their protolith. Porphyroclastic mylonites are enriched in muscovite, biotite, epidote, apatite, titanite, and zircon (Table 1; Fig. 5B). The dominant feldspar in the granitic protolith is ribbon to rod perthite, which is extensively saussuritized and sericitized in the mylonite. Microtextures are consistent with the formation of muscovite and fine-grained quartz from the reaction of perthite in the presence of a fluid phase. Apatite, titanite, and zircon display similar microstructures in both the granite and mylonite and do not appear to be have been chemically reactive during deformation. Chlorite and calcite are minor phases that appear as posttectonic overgrowths. Ovoid apatite grains are mantled by rims of fine-grained epidote in both the deformed and undeformed rocks, consistent with posttectonic growth (Fig. 6C). Biotite-rich schistose mylonites are dominated by biotite, epidote, and albite with minor quartz, titanite, and apatite.

Geochemical analyses of three samples of the alkali feldspar granite protolith and three samples of the porphyroclastic mylonite indicate that significant compositional changes occurred during deformation. Major and trace element compositions were determined by X-ray fluorescence (performed at ALS Chemex) (Table 1). The alkali feldspar granite protolith is characterized by 72 ± 5% SiO_2, while the porphyroclastic mylonite has 65 ± 1% SiO_2. The ratios of conserved elements (Y/TiO_2, TiO_2/P_2O_5, Zr/TiO_2) are similar for the alkali feldspar granite and porphyroclastic mylonite, consistent with the interpretation that the mylonites were derived from the alkali feldspar granite (Table 1). Mylonitic rocks exhibit approximately a twofold increase in CaO, Fe_2O_3, MgO, P_2O_5, TiO_2, Ba, Nb, Zr, and Y (Table 1). Mylonitic rocks are somewhat depleted in K_2O and Rb with respect to the protolith granitoids (Table 1). The biotite-rich schistose mylonites

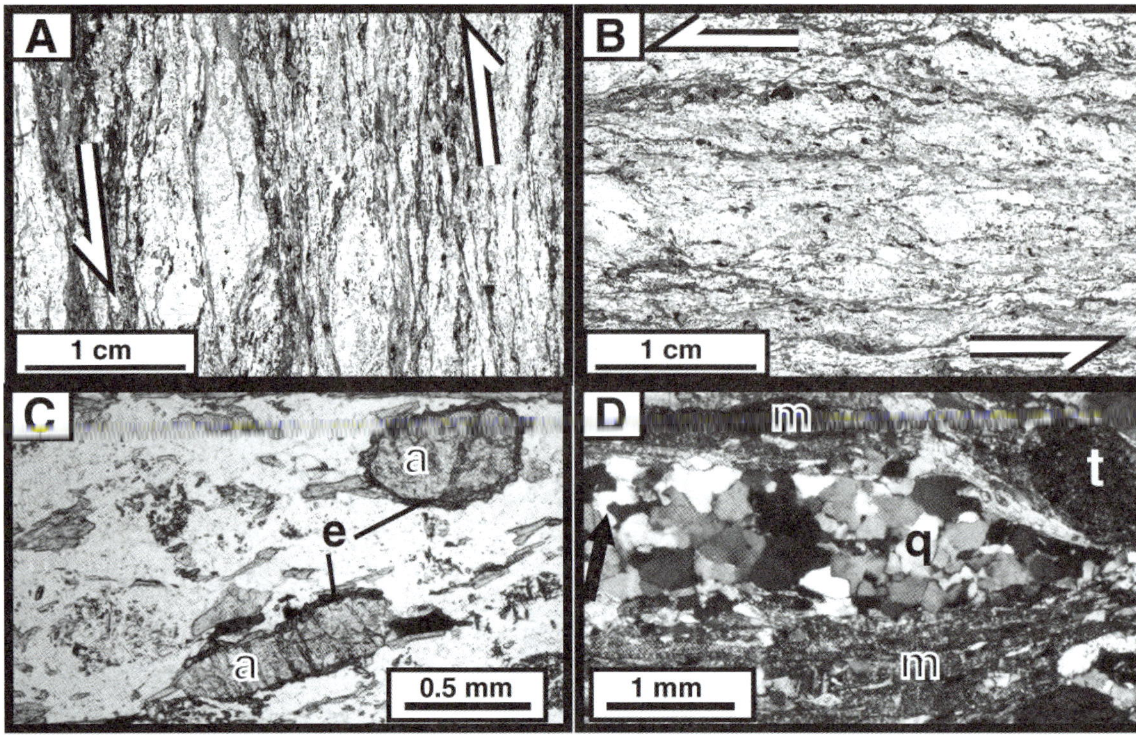

Figure 6. Micrographs. A: Weak asymmetry (top-to-the-northwest sense of shear) in porphyroclastic mylonite. Lineation-parallel/foliation-normal section (*XZ*). Plane-polarized light. B: Weak asymmetry (sinistral sense of shear) in porphyroclastic mylonite. Lineation- and foliation-normal section (*YZ*) viewed in the direction of lineation plunge (~120°). Plane-polarized light. C: Apatite (a) porphyroclasts in mylonite mantled with epidote (e). Plane-polarized light. D: Recrystallized quartz lens (q) in porphyroclastic mylonite. Note the distinct lens boundary and the bulging grain boundaries (arrow) in recrystallized grain. m—fine-grained muscovite and quartz matrix; t—titanite. Cross-polarized light.

are characterized by low SiO_2 values (49%–54%) and >11% Fe_2O_3 (Polvi, 2003).

The modal and chemical differences between the mylonites and their protolith granitoid are consistent with volume change during deformation. O'Hara (1990) demonstrated that bulk-rock volume change (Δ) can be quantified by the relationship

$$M/P = 1/(1 + \Delta),$$

where *M* and *P* refer to the concentration of chemical species in the mylonite and protolith; the slope of the best-fit line for immobile elements on an isocon diagram (Grant, 1985) is used as the *M/P* term. For Lawhorne Mill mylonites and protolith, the slope of the best-fit line for Al_2O_3, MgO, MnO, P_2O_5, TiO_2, Y, and Zr is ~2, corresponding to a volume loss of ~50% (Fig. 5C). Volume losses of up to 50% have been reported for a number of other greenschist facies high-strain zones in similar terranes (O' Hara, 1990; Bailey et al., 1994; Bailey, 1995). Chemical evidence indicates that SiO_2 and K_2O were preferentially removed during mylonization (Table 1). Syndeformational quartz veins occur in the Lawhorne Mill high-strain zone and are well developed in boudin necks and along boudin margins (Fig. 4C), although volumetrically these veins do not compose ~50% of the high-strain zone. A significant quantity of soluble elements was transported out of the high-strain zone.

Strain Analysis

Three-dimensional strain was estimated from quartz grain shapes using the R_f/ϕ method with a hyperbolic stereogram (De Paor, 1988). In the undeformed granitic protolith, quartz forms irregular blebs with no grain shape preferred orientation. During deformation, originally monocrystalline quartz blebs were transformed into polycrystalline lenses and ribbons (Fig. 6D). Relict cores of original quartz grains are rare. New grains (50–300 μm) are characterized by straight extinction with both straight and curved grain boundaries, consistent with both subgrain rotation and grain boundary migration recrystallization. In spite of recrystallization, quartz aggregates form discrete microstructural entities bounded by other phases (Fig. 6D). Quartz grain shapes are not perfect strain markers (e.g., nonelliptical shape, degree of recrystallization), but they have been successfully utilized as a strain marker in a number of greenschist facies high-strain zones (Mawer, 1983; O'Hara, 1990; Bailey et al., 1994; Fletcher and Bartley, 1994). Quartz grain shapes were measured on three mutually perpendicular thin sections for each sample (foliation-parallel,

TABLE 1. MAJOR ELEMENT, TRACE ELEMENT, AND MODAL MINERALOGY FOR ALKALI FELDSPAR GRANITES AND PORPHYROCLASTIC MYLONITES FROM THE LAWHORNE MILL HIGH-STRAIN ZONE

	Alkali feldspar granite		Porphyroclastic mylonite	
Major (%)	Mean (n = 3)	σ	Mean (n = 3)	σ
Al_2O_3	13.3	1.04	13.84	0.30
CaO	1.84	0.80	3.08	0.11
Cr_2O_3	0.02	0.01	0.01	0.00
Fe_2O_3	2.88	2.00	5.78	0.16
K_2O	5.18	0.25	3.14	0.43
MgO	0.47	0.38	1.06	0.10
MnO	0.06	0.02	0.10	0.01
Na_2O	2.80	0.22	3.25	0.47
P_2O_5	0.24	0.22	0.57	0.03
SiO_2	71.5	5.02	65.1	0.75
TiO_2	0.59	0.47	1.40	0.02
LOI	0.57	0.19	1.12	0.12
Trace (ppm)				
Ba	870	425	1753	310
Rb	206	11	99	19
Sr	194	66	313	99
Nb	11	10	27	1
Zr	245	175	555	58
Y	26	17	51	5
Y/TiO_2	52.2	22.6	36.8	3.8
TiO_2/P_2O_5	2.87	0.67	2.44	0.09
Zr/TiO_2	464	108	397	39
Zr/Al_2O_3	17.6	11.3	40.1	4.3
Modal mineralogy	Average (%)	σ	Average (%)	σ
Quartz	35.0	5.2	32.1	3.4
Plagioclase	3.3	1.7	23.1	0.9
Perthite	48.0	11.4	4.1	0.8
Biotite	6.3	7.3	14.9	2.6
Muscovite	3.0	1.7	14.4	3.0
Epidote	3.1	1.2	6.6	2.2
Apatite	0.4	0.4	1.2	0.3
Sphene	0.6	0.8	2.7	0.1
Zircon	0.1	0.0	0.1	0.1
Opaque	0.0	0.0	0.0	0.1
Chlorite	0.1	0.1	0.2	0.2
Garnet	0.0	0.0	0.0	0.0
Calcite	0.0	0.0	0.1	0.1

LOI—loss on ignition

foliation-normal/lineation-parallel, and foliation- and lineation-normal), and 20–30 quartz grains were measured per section. Sectional aspect ratios were greatest on foliation-normal/lineation-parallel sections (*XZ*) and range from 4.4 to 6.3. Stretch estimates, measured from boudinaged pegmatites (subparallel to foliation) at the outcrop scale, range from 1.5 to 2.4 and are similar to the principal stretches derived from quartz strain ratios.

Three-dimensional strain data for porphyroclastic mylonites plot in the field of apparent flattening on a logarithmic Flinn diagram (K = 0.3–0.6) (Fig. 5D). A sample of biotite-bearing granitoid mylonite plots near the isovolumetric plane strain line (Fig. 5D). As noted above, mineralogical and chemical changes between the alkali feldspar granite protolith and mylonite are consistent with ~50% volume loss during deformation. When the plane strain line is adjusted for 50% volume loss, the porphyroclastic mylonite data plot on the plane strain line (K = 1, Δ = –0.5) (Fig. 5D). Macrostructures and microstructures in *YZ* sections reveal no evidence for bulk extension parallel to *Y* (e.g., boudinage), an observation consistent with bulk plane strain.

Vorticity Analysis

We used three techniques to quantify the rotational strain component in Lawhorne Mill mylonites. These include the R_s/Θ method (Tikoff and Fossen, 1995; Bailey and Eyster, 2003), the quartz *c*-axis method (Wallis, 1992, 1995), and the stable position of rigid clasts method (Passchier, 1987; Wallis et al., 1993). The R_s/Θ method utilizes the angle between the long axis of the finite strain ellipsoid (*X*) and the high-strain zone boundary to determine the vorticity. For the Lawhorne Mill high-strain zone, the Θ value is <10° and, when combined with the strain data, yields W_m values of 0.1–0.6 (Fig. 7; Table 2). Although strain measurements based on quartz grain shape fabrics have an inherent uncertainty, for the finite deformation to have approximated simple shear, strain estimates would have to be in error by a factor of 4–5. The R_s/Θ data are consistent with a general shear to pure shear dominated deformation.

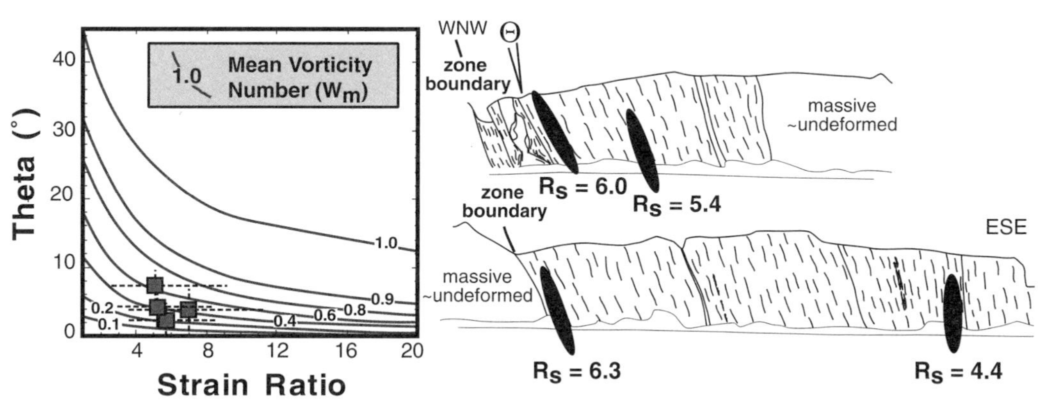

Figure 7. R_s/Θ diagram with sectional strain (*XZ*) data plotted on Lawhorne Mill high-strain zone cross section. Θ is the angle between the zone boundary and foliation. Note the low Θ values (<10°). Dashed gray crosses represent maximum and minimum values for sectional strain ratios based on measurements from three orthogonal sections and uncertainty in Θ.

TABLE 2. STRAIN AND VORTICITY DATA FOR LAWHORNE MILL MYLONITES

Sample	Strain data		Vorticity data		
	XY	*YZ*	R_s/Θ	*c* axes	Clasts
LM9	1.67 ± 0.47	4.17 ± 1.18	0.4 (0.1–0.5)	0.1	0.89–0.93
LM10	1.83 ± 0.38	3.13 ± 0.66	0.2 (0.0–0.3)	0.3	0.82–0.89
LM14	1.90 ± 0.57	2.88 ± 0.87	0.6 (0.4–0.8)	0.3	0.81–0.87
LM15	2.37 ± 0.61	2.20 ± 0.57	0.4 (0.1–0.6)	0.0	0.82–0.92

Note: Strain ratios based on sectional analyses, ± values derived from combining sectional data from 3-planes. Vorticity values in parentheses are minimum and maximum values.

The quartz *c*-axis technique is based on the relationship between β and the *XZ* strain ratio, where β is the acute angle between the foliation and the inferred flow plane as defined by the normal to the central fabric element on a quartz c-axis fabric diagram (Wallis, 1992, 1995). Quartz fabrics for Lawhorne Mill mylonites display moderately developed type I and type II cross-girdle patterns with an orthorhombic symmetry (Fig. 8). The fabric skeleton, visually derived from crests on the contoured plots, yield β values of nearly 0° and always less than 5°. For the finite strains recorded in these mylonites, such low β values yield W_m values of <0.3, in essence pure shear dominated strain (Fig. 8; Table 2).

The stable position of rigid clasts is based on the assumption that for a given vorticity, clasts that exceed the critical aspect ratio (R_c) rotate into a stable orientation, whereas clasts with an aspect ratio below the critical value continuously rotate and hence display no preferred alignment. Clasts having aspect ratios >R_c lie with their long axes subparallel to foliation, but grains with aspect ratios <R_c display a wide range of long-axis orientations. Graphically, R_c defines the boundary between the clasts in a stable position and those with a wide range of orientations (Fig. 9). The critical aspect ratio is a function of W_m, where

$$W_m = (R_c^2 - 1)/(R_c^2 + 1)$$

(Passchier, 1987; Wallis et al., 1993). Apatite grains from Lawhorne Mill mylonites lack microstructural evidence for plastic deformation and appear to have rigidly rotated during deformation (Fig. 6C). Aspect ratios for individual apatite grains range from 1 to 7. The break between clasts in a stable orientation and those that continuously rotated is poorly defined but appears to occur at aspect ratios between 3 and 4, yielding W_m values of 0.8–0.9 (Fig. 9; Table 2). Apatite grains were also measured on sections normal to foliation and lineation (*YZ* section); R_c ranges from 2.2 to 3.5 with W_m values of 0.7–0.8 (Fig. 9; Table 2). These data may be consistent with a component of rotational strain normal to the elongation lineation.

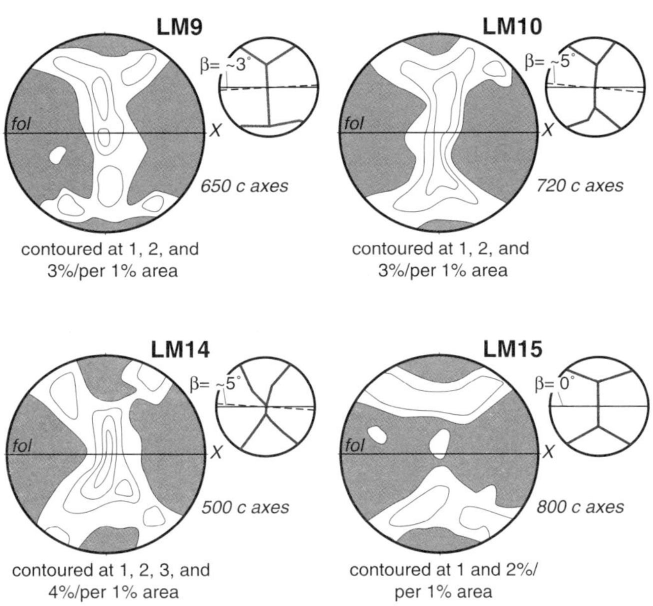

Figure 8. Optically measured quartz *c*-axis fabrics; lower-hemisphere equal-area projections; foliation oriented east-west and vertical; mineral lineation (*X*) horizontal. For each sample, *c* axes were measured on three orthogonal sections and data combined to form a single fabric diagram.

The R_s/Θ and quartz *c*-axis vorticity gauges yield similar results with W_m values of <0.6 and all but one analysis ≤0.4. Collectively, these vorticity indicators, as well as the qualitative indicators (sense of shear), are consistent with pure shear dominated deformation. In contrast, the stable orientation of rigid clasts method yielded much higher W_m values (0.8–0.9) than the other two methods. Lawhorne Mill mylonites experienced "relatively" low strains (R_{XZ} < 7), and as such the finite deformation is not great enough to have rotated all apatite grains into a stable orientation. Many grains with large aspect ratios might have achieved a stable position, subparallel to the foliation, had the deformation been greater (R_{XZ} > 20). Thus, the critical aspect ratio, as graphically determined, is too large, and consequently the W_m values overestimate the simple shear component. This technique has been applied in a number of geologic settings where the overall finite strain is low (<10), and we are concerned that W_m values obtained in these studies are of limited value.

Kinematic Model for the Lawhorne Mill High-Strain Zone

Quantitative kinematic analysis of the Lawhorne Mill high-strain zone indicates (1) ~50% volume loss for mylonites derived from the leucocratic alkali feldspar granite, (2) bulk plane strain, (3) low mean vorticity values (W_m < 0.4), and (4) *XZ* strain ratios of 4:1 to 7:1. We used these values to model the deformation

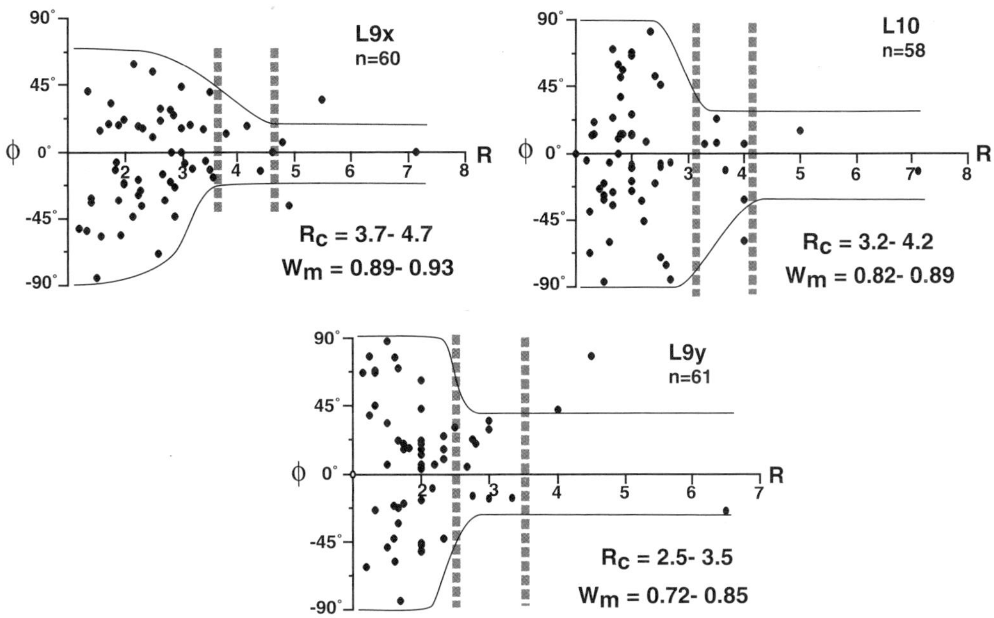

Figure 9. Orientation plots of apatite porphyroclasts. ϕ is the angle between the long axis of apatite grain and the foliation. R is the aspect ratio of apatite clast. Thick dashed lines delineate the transition to clasts in a stable orientation. R_c is the critical aspect ratio. $W_m = (R_c^2 - 1)/(R_c^2 + 1)$ (Passchier, 1987; Wallis et al., 1993).

kinematics across the Lawhorne Mill high-strain zone (Fig. 10). Our model assumes homogeneous ductile deformation with a monoclinic deformation symmetry. The kinematic consequence of a pure shear dominated deformation combined with volume loss produces dramatic shortening (~78%) across the deformation zone and elongation parallel to the zone boundary (Fig. 10). Without volume loss, total shortening across the zone would be considerably less (~30%–40%). The current width (thickness) of the zone is ~400 m; prior to deformation, the rocks that would become the mylonite zone were ~2000 m wide (assuming the zone formed in a subvertical position). The overall shear strain (γ) for this finite deformation is 0.96; integration of shear strain across the width of the Lawhorne Mill high-strain zone yields a displacement of ~400 m. Total shortening across the zone was five times greater than displacement parallel to the zone. If grain boundary sliding or discrete brittle faulting were important, the kinematic model would underestimate the overall displacement. Granitic mylonites at Lawhorne Mill are characterized by a single penetrative foliation, and generally lack extensive mica-rich domains where grain boundary sliding might have been an important deformation mechanism.

The weak asymmetries on both lineation-parallel and -normal sections are consistent with a weakly triclinic deformation path. These asymmetries are inconsistent with the Lawhorne Mill mylonites having experienced monoclinic transpression (Fig. 1), because the asymmetries are best developed on lineation-parallel faces and the component of strike-parallel shear is small. The kinematic consequences of triclinic deformation are a matter of debate,

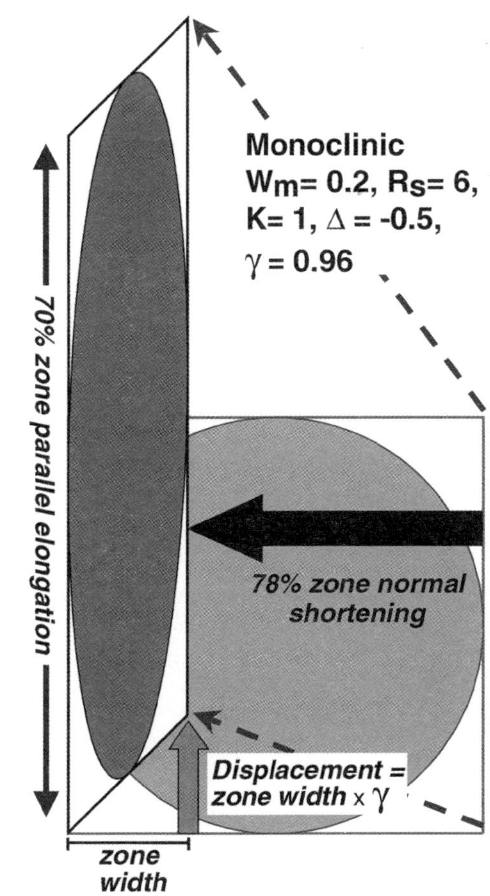

Figure 10. Kinematic model for the Lawhorne Mill high-strain zone.

but the overall transport direction is likely oblique to the fabric elements (not parallel or normal to the lineation). Our data indicate low sectional vorticities on both lineation-parallel and -normal faces, such that, although the overall deformation is weakly triclinic, it does not greatly affect either the total displacement or direction. Two superposed monoclinic deformations could produce an apparent triclinic deformation symmetry, but mylonitic rocks from the Lawhorne Mill high-strain zone display no evidence (at both the macro- and microscale) for two discrete deformation events. The simplest interpretation is that of a single pure shear dominated progressive deformation with a weak triclinic symmetry.

Deformation in nonsimple shear zones leads to compatibility problems with rigid wall rocks because material is typically shortened normal to the zone and stretched parallel to zone boundaries. Compatibility issues have been used as a rationale for the simple shear zone paradigm (e.g., Ramsay and Huber, 1983). However, in transpressional zones material escapes either vertically toward Earth's surface or downwards. Bulk compatibility in general shear zones may be achieved by wall-rock deformation, slip across zone boundaries, stretching both normal and parallel to the shear direction, sectional area changes (volume change), and deformation along curved, non-parallel-sided boundaries (Simpson and De Paor, 1997; Bhattacharyya and Hudleston, 2001). Although individual zones deviate from plane strain and simple shear, arrays of anastomosing high-strain zones can produce bulk strain compatibility (Hudleston, 1999).

Strain compatibility in the Lawhorne Mill high-strain zone was likely achieved by a combination of discrete slip along its boundaries, volume change, and possibly nonparallel sides of the zone. The western high-strain zone boundary forms a discrete fault surface with downdip slickenfibers (Fig. 3). Mineralogical and chemical evidence requires volume loss during mylonite formation; the loss of volume produces a reduction in cross-sectional area that minimizes (but does not eliminate) compatibility differences between deformed rocks in the zone and undeformed wall rocks (Fig. 10). Mylonitic rocks at Lawhorne Mill are well exposed, but the nature of the outcrop precludes a detailed three-dimensional understanding of the zone geometry. Many centimeter- to meter-scale Blue Ridge high-strain zones are non-parallel-sided and commonly change thickness along strike (and/or in the dip direction). From outcrop scale to map scale, Blue Ridge high-strain zones are anastomosing and generally lack parallel sides (Mitra, 1979; Bailey and Simpson, 1993); this geometry and three-dimensional connectivity provides a mechanism with which compatibility is maintained with wall rocks.

Regional Considerations

The Lawhorne Mill high-strain zone is part of a family of Paleozoic mylonite zones that cut Grenvillian basement in the Virginia Blue Ridge (Mitra, 1979; Bartholomew et al., 1981; Bailey and Simpson, 1993). These high-strain zones developed under greenschist facies conditions and accommodated northwest directed contraction. The Lawhorne Mill high-strain zone differs from other Blue Ridge high-strain zones in two significant ways: (1) It records a pure shear dominated deformation rather than a strongly noncoaxial general shear deformation, and (2) its subvertical orientation contrasts many other Blue Ridge mylonite zones that typically dip 35° to 60° southeast. The larger Blue Ridge high-strain zones (e.g., the Rockfish Valley zone) have been inferred to separate distinct Grenvillian basement massifs (Bartholomew et al., 1981; Sinha and Bartholomew, 1984; Hughes et al., 2004). These models require significant displacement across high-strain zones; however, the pure shear dominated nature of the Lawhorne Mill high-strain zone is incompatible with large displacements. Although mylonitic rocks are common in the Blue Ridge basement complex, the general shear to pure shear dominated nature of deformation in these zones did not result in large displacements (>5 km).

We propose that the Lawhorne high-strain zone developed as a steeply dipping structure early in the Paleozoic deformation sequence (pre–355 Ma, Neoacadian to possibly Taconian) and that deformation propagated northwestward across the basement complex toward the foreland during the Alleghanian orogeny. Blue Ridge mylonite zones are truncated by late Paleozoic brittle thrusts related to the later emplacement of the basement and cover sequence over a major thrust ramp during formation of the anticlinorium. Deformation and volume loss were facilitated by significant fluid flow and strain softening that enabled the mechanically stiff basement complex to shorten by development of mylonite zones and a regionally pervasive retrogressive foliation. Rather than folding, Blue Ridge basement granitoids responded to the applied stresses by shortening across high-strain zones. The exposed width (from northwest to southeast) of the basement complex across the Blue Ridge anticlinorium in central Virginia is 25–30 km, and mylonitic rocks compose between 3 and 5 km of that width. Shortening values across the Lawhorne Mill high-strain zone are similar to other Blue Ridge high-strain zones (Mitra, 1979; Berquist and Bailey, 2001; Bailey et al., 2002); thus total shortening across all Blue Ridge mylonite zones is likely 3–10 km (minimum value—50% shortening across 3 km; maximum value—67% shortening across 5 km). Away from high-strain zones, much of the Blue Ridge basement complex is penetratively foliated (although much less so than in the high-strain zones). Collectively, Blue Ridge high-strain zones account for 10%–40% crustal shortening across the province.

PURE SHEAR DOMINATED DEFORMATION IN OTHER TECTONIC SETTINGS

Cordilleran Ultramylonites

Cordilleran metamorphic core complexes represent mid-crustal blocks that were tectonically exhumed during large-magnitude Tertiary extension (Davis and Coney, 1979; Crittenden et al., 1980). The basic architecture of a core complex includes mid-crustal rocks in the lower plate, a gently to moderately dipping detachment fault, and an upper plate with allochthonous cover rocks. Curviplanar mylonite zones in the lower plate of core complexes typically pass downward into lower-strain rocks. In the 1980s and early 1990s the contribution of coaxial versus noncoaxial strain in these mylonites was widely debated (Compton, 1980; Davis, 1983; Miller et al., 1983; Davis et al., 1986; Naruk, 1987; Lee et al., 1987; Wells and Allmendinger, 1990). More recent work demonstrates that mylonitic rocks in a number of Arizona core complexes developed under general shear conditions (W_m = 0.6–0.9) (Force, 1997; Bailey and Eyster, 2003; Forte and Bailey, 2007).

Ultramylonites form a common, but volumetrically small (<1%), component of many Cordilleran core complex high-strain zones. Ultramylonites typically occur in discrete, narrow zones (~0.5 mm to ~1 or 2 m thick) that sharply truncate preexisting markers (Fig. 11A). Some ultramylonite zones are developed along lithologic contacts between very coarse-grained pegmatite sills and their granitoid host rocks, whereas other zones develop in more homogeneous granitoids. The very fine-grained matrix consists of white mica and recrystallized quartz with isolated plagioclase, K-feldspar, epidote, and tourmaline porphyroclasts (Fig. 11B). Microtextures indicate that the matrix developed from the reaction of feldspar and quartz in the presence of H_2O-rich fluids. These rocks underwent dramatic grain size reduction and strain softening. We estimated the vorticity of flow in ultramylonites using the porphyroclast hyperbolic distribution method (Simpson and De Paor, 1993). For a general shear deformation, porphyroclasts rotate both forward and backward with respect to the overall flow (Fig. 11B). The field of backward-rotated porphyroclasts, as plotted on a hyperbolic stereonet, defines the interlimb angle of the flow apophyses, and W_m is the cosine of that angle (Fig. 11C) (Bobyarchick, 1986; Simpson and De Paor, 1993). Vorticity analysis of ultramylonites (n = 11) from three Arizona core complexes (Pinaleño, Santa Catalina, and South Mountains) yield low vorticity values consistent with a pure shear dominated deformation. The methods and results are discussed more thoroughly in Bailey and Eyster (2003) and Forte and Bailey (2005). A consequence of the strain softening associated with ultramylonite formation may be the development of pure shear dominated strain because the mechanically weak ultramylonite shortens more readily normal to foliation relative to surrounding protomylonitic and mylonitic granitoids with a framework of interlocking feldspars.

These pure shear dominated ultramylonite zones are dramatic examples of positive stretching faults (e.g., Means, 1989) in which there is significant material extension parallel to the zone boundary. Furthermore, most Arizona ultramylonites formed with original dips of 30° or less such that the pure shear dominated nature of these high strains accommodate crustal thinning during extension. At a regional scale it appears that core complex mylonite zones are the product of bulk general shear, but narrow ultramylonite zones may play a role in both material elongation and crustal thinning across these complexes.

The Coast Shear Zone, Prince Rupert–Skeena River Area, British Columbia

The Coast shear zone forms a major, but enigmatic, structure that extends for at least 800 km in British Columbia and southeastern Alaska (Fig. 12). Irving et al. (1996) and Hollister and Andronicos (1997) postulated that the Coast shear zone might have accommodated large-scale dextral transpressive motion (hundreds to ~1000 km) that translated the Insular superterrane northward along the North American margin in the Late Cretaceous to

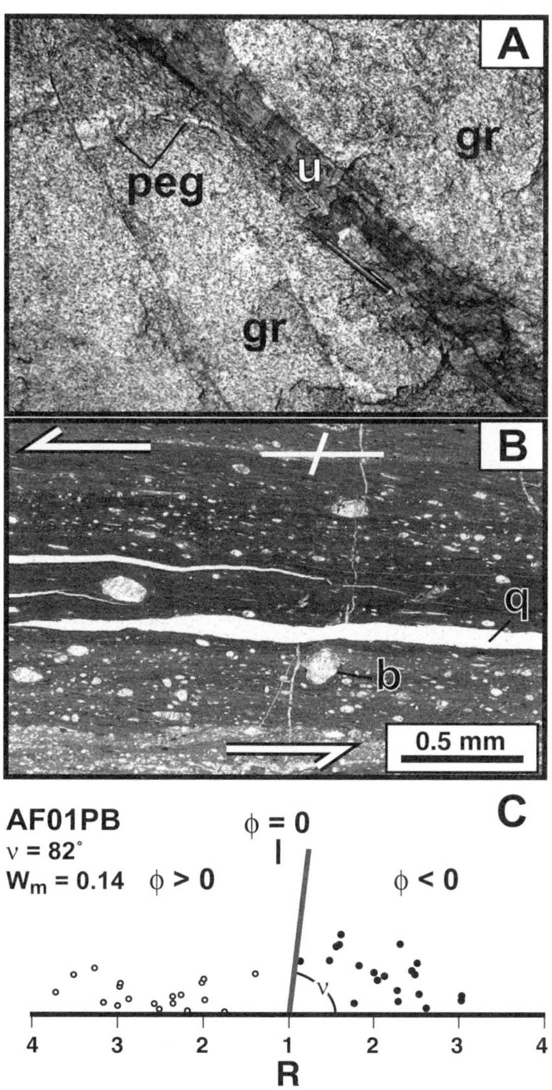

Figure 11. A: Narrow ultramylonite zone (u) cutting protomylonitic granitoid (gr) and pegmatite dike (peg) (Pinaleño Mountains). B: Micrograph of ultramylonite with backward-rotated porphyroclasts (b) and quartz ribbons (q) (Santa Catalina Mountains). Plane-polarized light. C: Porphyroclastic hyperbolic distribution method plot for sample AF-01PB. Solid dots—backward-rotated porphyroclasts; open dots—forward-rotated porphyroclasts. W_m = cos ν (Santa Catalina Mountains).

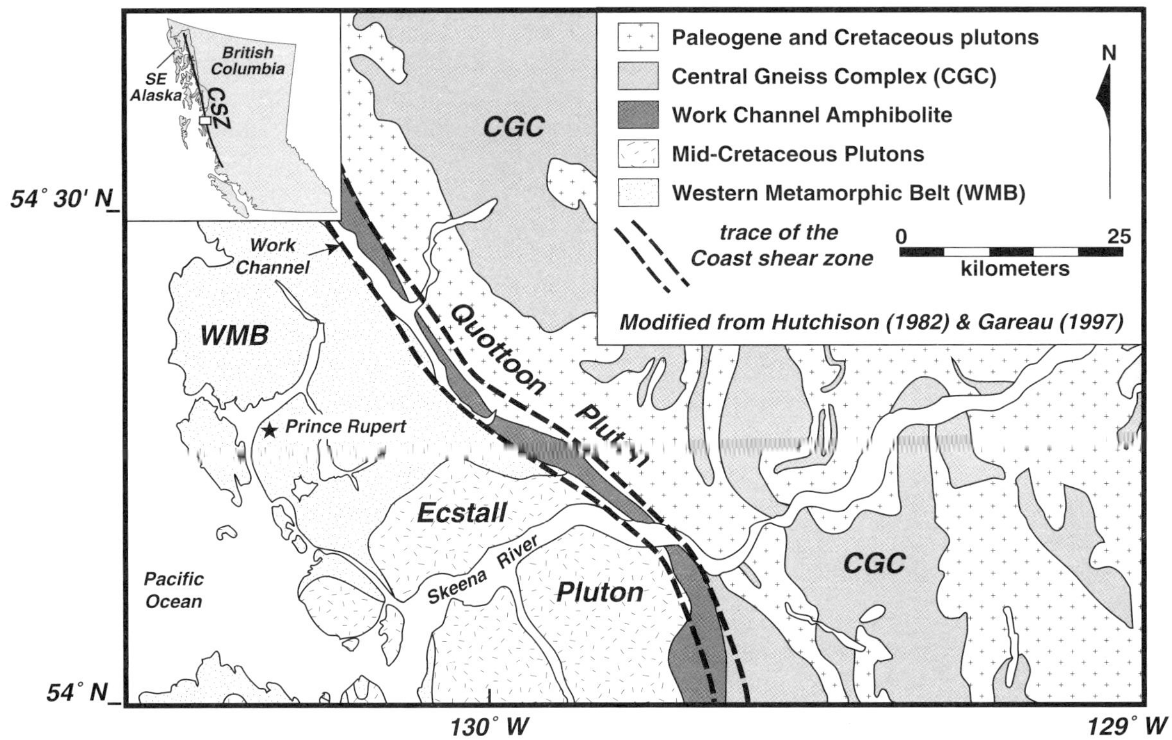

Figure 12. Generalized geologic map of the Prince Rupert and Skeena River area, British Columbia. CSZ—Coast shear zone.

Paleogene. Structural and kinematic studies of ductile deformation in the Coast shear zone have yielded complex and in many cases contradictory interpretations. In southeastern Alaska, Stowell and Hooper (1990) reported east-side-up displacement followed by dextral shear. Klepeis et al. (1998) reported that the Coast shear zone experienced east-side-up shearing between 65 and 57 Ma that was followed by west-side-up shearing between 57 and 55 Ma. Andronicos et al. (1999) also described dip-slip motion on the Coast shear zone as well as an earlier stage of dextral slip. In the Prince Rupert area, Crawford et al. (2000) reported west-side-up shearing.

In the Prince Rupert–Skeena River area, the Coast shear zone forms a 2–5 km thick zone of well-foliated, steeply dipping, upper amphibolite facies rocks that separates the western metamorphic belt from the central gneiss complex (Fig. 12). The Coast shear zone generally coincides with a well-expressed topographic lineament underlain predominantly by amphibolite (Work Channel amphibolite of Hutchison, 1982). West of the Coast shear zone, dioritic rocks of the Ecstall pluton intruded the western metamorphic belt at ca. 90 Ma and were emplaced at depths of 25–30 km (Zen, 1988; Crawford et al., 2000). The ca. 60 Ma tonalitic Quottoon pluton intruded the central gneiss belt along the eastern margin of the Coast shear zone (Gehrels et al., 1991). The Coast shear zone also experienced a number of distinct brittle deformation events into the middle Tertiary (Davidson et al., 2003).

Along the Skeena River, the Coast shear zone is not distinguished by an obvious strain gradient, as rocks to both the east and west are well foliated. Rather, the Coast shear zone is a steep zone where the moderately dipping foliations to the east and west of the zone become subvertical. Foliation in the Coast shear zone strikes northwest, is subvertical, and downdip lineations are common (Fig. 13A). Rocks in the Coast shear zone are tectonites, but not mylonites, as they display no evidence of significant grain size reduction (Fig. 13B). Microstructural and geothermometric data are consistent with upper amphibolite facies conditions during deformation (Roth, 2001; Metcalf and Davidson, 1997). Kinematic indicators are overwhelmingly symmetric on both lineation-parallel and lineation-normal faces. Folded dikes and other marker layers form tight upright folds that lack characteristic vergence (Fig. 13C). Narrow (<10 cm) zones with well-developed asymmetric fabrics cut the dominant steeply dipping fabric and are interpreted to be localized melt filled shear zones (Fig. 13D) (Davidson et al., 1998). These narrow deformation zones occur as conjugate sets that are kinematically compatible with the steep foliation (formed during northeast-southwest directed contraction). Tectonites in the Coast shear zone from the Skeena River area are characterized by orthorhombic fabrics that record northeast-southwest directed contraction and subvertical elongation. Although earlier deformation in the Coast shear zone may record significant dip-slip or even strike-parallel movement, the "strain memory" preserved in these rocks is that of a pure shear

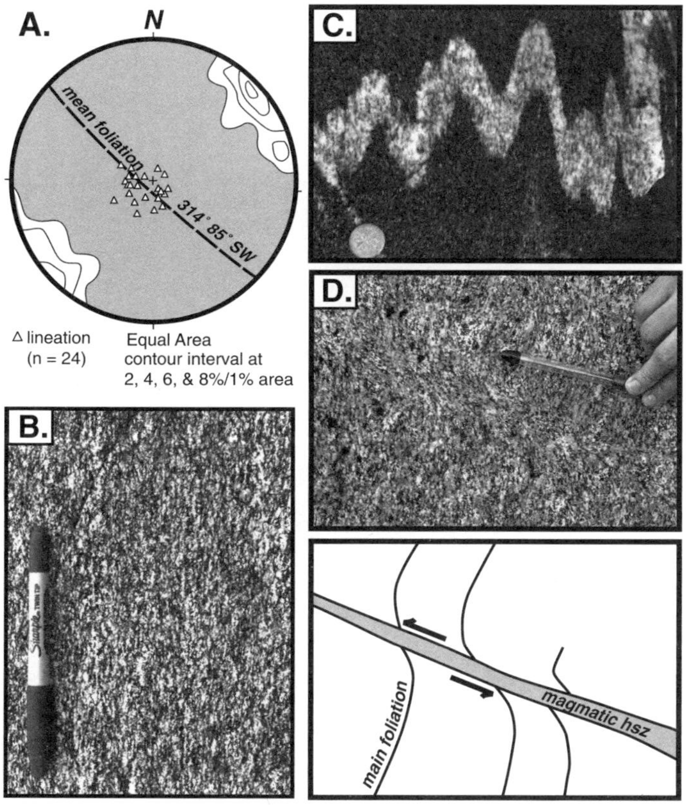

Figure 13. (A) Stereogram of foliation and lineation data for the Coast shear zone. (B) Subvertical foliation in tonalitic rocks near eastern edge of the Coast shear zone. (C) Tight, upright folds in aplite dike surrounded by foliated amphibolite. (D) Steeply dipping foliation in tonalite cut by gently dipping top-to-the-west magmatic shear zone. hsz—high strain zone.

dominated deformation. The fabrics preserved in the Coast shear zone formed in response to crustal contraction normal to the zone, but the total displacement of one side of the shear zone relative to the other is modest. Given the pure shear dominated nature of the strain, the term "shear zone" may not be the appropriate moniker for this structure.

KINEMATIC AND TECTONIC IMPLICATIONS OF PURE SHEAR DOMINATED DEFORMATIONS

General shear forms the continuum between two-dimensional simple shear ($W_m = 1$) and pure shear ($W_m = 0$). Because W_m is a function of the interlimb angle between the flow eigenvectors, the 0–1 scale is not linear in regard to the contribution of simple versus pure shearing (Simpson and De Paor, 1993). An equal contribution of simple and pure shear occurs at a W_m value of 0.81. We define a pure shear dominated deformation as one with a W_m value of ≤0.4, which corresponds to contribution of at least 75% pure shear. Zones dominated by pure shear differ from simple shear and general shear dominated zones because displacement across these zones is minimal relative to the overall finite strain. Displacement across a simple shear zone is typically at least two times greater than displacement across a pure shear dominated zone ($W_m \leq 0.4$) with the same sectional strains (Fig. 14). Although displacement across pure shear dominated zones is relatively low, the amount of zone-normal shortening and zone-parallel stretching is large. Pure shear dominated high-strain zones and their surrounding wall-rock domains are effectively positive (material-lengthening) stretching faults.

For a given strain ratio, a subhorizontal simple shear zone produces significant offset but no thinning or material elongation parallel to the flow plane (Fig. 15). In contrast, a subhorizontal pure shear dominated zone leads to crustal thinning and material elongation. Gently dipping zones formed in extensional settings are effective at thinning the crust. Conversely, steeply dipping zones formed in contractional settings serve to shorten the crust, and the subvertical elongation may facilitate the extrusion of material toward the surface of the orogen (Fig. 15). The extrusion of mid- and deep-crustal materials plays a significant role in the geological, as well as geomorphic, evolution of orogens (Axen et al., 1998; Hodges et al., 2001; Treloar and Blenkinsop, 1995), and pure shear dominated deformation provides a mechanism with which extrusion is facilitated.

Nonsimple shear zones were long viewed with skepticism because of compatibility issues between the deforming zone and its walls, but evidence of nonsimple shear has steadily accumulated over the past decade. Material movement in the core of an orogen is complex, and mid-crustal deformation clearly spans a wide range of vorticity values and three-dimensional strain. Pure shear dominated deformation zones are but one mechanism through which complex kinematics occur. Geologists concerned

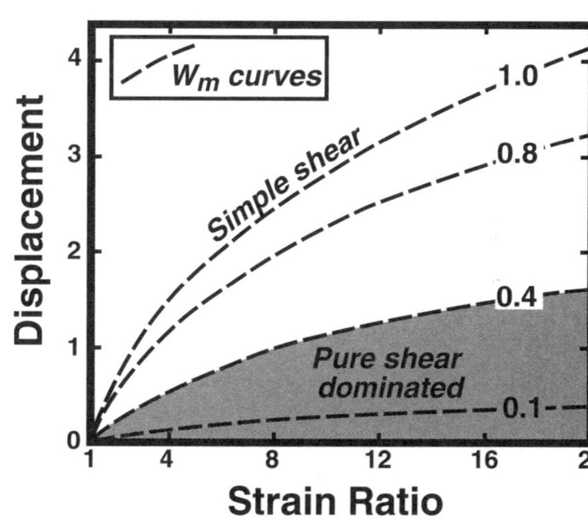

Figure 14. Displacement versus strain ratio for two-dimensional deformations with different W_m values. Displacement is a dimensionless term related to original unit volume of deformed material.

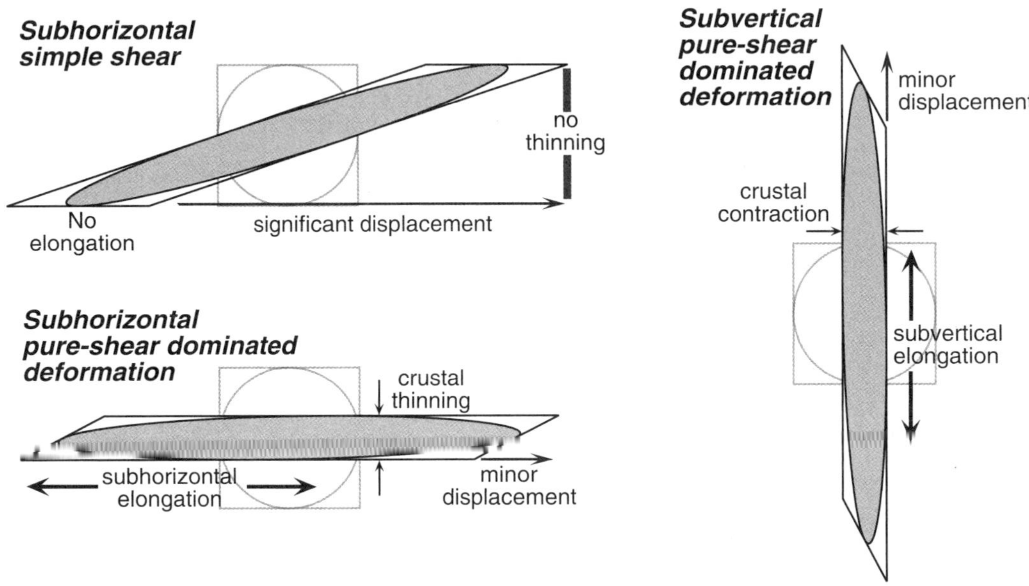

Figure 15. Diagrammatic illustration of the kinematic differences between homogeneous simple shear and pure shear dominated deformations in subhorizontal and subvertical orientations. Strain ellipse and parallelogram derived from unit square and circle. All deformations are plane strain.

with unraveling the tectonic history of basement terranes must not only understand age relations, chemical affinities, and the *P/T* history, but also endeavor to quantify the rotational strain component in order to adequately characterize the flow kinematics.

ACKNOWLEDGMENTS

We thank Bob Hatcher for inviting us to participate in the International Basement Tectonics Association meeting in Oak Ridge, Tennessee, and soliciting our contribution to this volume. Perceptive and thorough reviews were provided by Charles Onasch, Bob Hatcher, and Don Secor. Field assistance at various study sites was provided by the 2002 William and Mary field methods class, Eleanor Fahrney, Jon Spencer, Ken Davis, Carl Tape, John Singleton, Anu Gupta, and Cam Davidson. Financial support for these endeavors came from a William and Mary faculty research grant, William and Mary student research grants, and the Keck Geology Consortium.

REFERENCES CITED

Andronicos, C.L., Hollister, L.S., Davidson, C., and Chardon, D., 1999, Kinematics and tectonic significance of transpressive structures within the Coast plutonic complex, British Columbia: Journal of Structural Geology, v. 21, p. 229–243, doi: 10.1016/S0191-8141(98)00117-5.

Axen, G.J., Selverstone, J., Byrne, T., and Fletcher, J.M., 1998, If the strong crust leads will the weak crust follow?: GSA Today, v. 8, no. 12, p. 1–8.

Bailey, C.M., 1994, Temporal, kinematic, and strain analysis of granitic tectonites from the central Appalachians [Ph.D. thesis]: Baltimore, Maryland, Johns Hopkins University, 254 p.

Bailey, C.M., 1995, Heterogeneous strain in granitic mylonites from the Rockfish Valley fault zone, central Virginia: Journal of Geodynamics, v. 19, p. 177–194, doi: 10.1016/0264-3707(94)00013-L.

Bailey, C.M., and Eyster, E.L., 2003, General shear deformation in the Pinaleño Mountains metamorphic core complex, Arizona: Journal of Structural Geology, v. 25, p. 1883–1892, doi: 10.1016/S0191-8141(03)00044-0.

Bailey, C.M., and Simpson, C., 1993, Extensional and contractional deformation in the Blue Ridge province, Virginia: Geological Society of America Bulletin, v. 105, p. 411–422, doi: 10.1130/0016-7606(1993)105<0411:EACDIT>2.3.CO;2.

Bailey, C.M., Simpson, C., and De Paor, D.G., 1994, Volume loss and tectonic flattening strain in granitic mylonites from the Blue Ridge province, central Appalachians: Journal of Structural Geology, v. 16, p. 1403–1416, doi: 10.1016/0191-8141(94)90005-1.

Bailey, C.M., Bobyarchick, A.R., and Jiang, D., 2002, Kinematics and vorticity of high-strain zones: Virginia Blue Ridge and Piedmont: Geological Society of America Field Forum Report, 26 p., http://www.geosociety.org/fieldForums/02rpt1.htm.

Bartholomew, M.J., Gathright, T.M., II, and Henika, W.S., 1981, A tectonic model for the Blue Ridge in central Virginia: American Journal of Science, v. 281, p. 1164–1183.

Bartholomew, M.J., Lewis, S.E., Hughes, S.S., Badger, R.L., and Sinha, A.K., 1991, Tectonic history of the Blue Ridge basement and its cover, central Virginia, *in* Schultz, A., and Compton-Gooding, E., eds., Geologic evolution of the eastern United States, NE-SE: Geological Society of America Field Trip Guidebook, Virginia Museum of Natural History, Guidebook 2, p. 57–90.

Berquist, P.J., and Bailey, C.M., 2001, Displacement across Paleozoic high-strain zones in the Blue Ridge province, Madison County, Virginia: The Pedlar and Lovingston massifs reconsidered: Geological Society of America Abstracts with Programs, v. 32, no. 2, p. 4.

Bhattacharyya, P., and Hudleston, P., 2001, Strain in ductile shear zones in the Caledonides of northern Sweden: A three-dimensional puzzle: Journal of Structural Geology, v. 23, p. 1549–1565, doi: 10.1016/S0191-8141(01)00020-7.

Bobyarchick, A.R., 1986, The eigenvalues of steady-state flow in Mohr space: Tectonophysics, v. 122, p. 35–51, doi: 10.1016/0040-1951(86)90157-5.

Compton, R.R., 1980, Fabrics and strains in quartzites of a metamorphic core complex, Raft River Mountains, Utah, *in* Crittenden, M.D., Jr., et al., eds.,

Cordilleran metamorphic core complexes: Geological Society of America Memoir 153, p. 385–398.

Crawford, M.L., Crawford, W.A., and Gehrels, G.E., 2000, Terrane assembly and structural relationships in the eastern Prince Rupert quadrangle, British Columbia, *in* Stowell, H.H., and McClelland, W.C., eds., Tectonics of the Coast Mountains, southeastern Alaska and British Columbia: Geological Society of America Special Paper 343, p. 1–21.

Crittenden, M.D., Jr., Coney, P.J., and Davis, G.H., editors, 1980, Cordilleran metamorphic core complexes: Geological Society of America Memoir 153, 490 p.

Davidson, C., Chardon, D., and Hollister, L.S., 1998, Pluton emplacement during east-west stretching at the northern termination of a gneiss dome in the central gneiss complex, British Columbia: Eos (Transactions, American Geophysical Union), v. 79, p. 802.

Davidson, C., Davis, K.J., Bailey, C.M., Tape, C.H., Singleton, J., and Singer, B., 2003, Age, origin, and significance of brittle faulting and pseudotachylyte along the Coast shear zone, Prince Rupert, British Columbia: Geology, v. 31, p. 43–46, doi: 10.1130/0091-7613(2003)031<0043:AOASOB>2.0.CO;2.

Davis, G.A., Lister, G.S., and Reynolds, S.J., 1986, Structural evolution of the Whipple and South Mountains shear zones, southwestern United States: Geology, v. 14, p. 7–10, doi: 10.1130/0091-7613(1986)14<7:SEOTWA>2.0.CO;2.

Davis, G.H., 1983, Shear-zone model for the origin of metamorphic core complexes: Geology, v. 11, p. 342–347, doi: 10.1130/0091-7613(1983)11<342:SMFTOO>2.0.CO;2.

Davis, G.H., and Coney, P.J., 1979, Geologic development of Cordilleran metamorphic core complexes: Geology, v. 7, p. 120–124, doi: 10.1130/0091-7613(1979)7<120:GDOTCM>2.0.CO;2.

De Paor, D.G., 1988, R_f/ϕ strain analysis using an orientation net: Journal of Structural Geology, v. 10, p. 323–333, doi: 10.1016/0191-8141(88)90011-9.

Evans, M.A., 1989, Structural geometry and evolution of foreland thrust systems, northern Virginia: Geological Society of America Bulletin, v. 101, p. 339–354, doi: 10.1130/0016-7606(1989)101<0339:TSGAEO>2.3.CO;2.

Evans, N.H., 1984, Latest Precambrian to Ordovician metamorphism and orogenesis in the Blue Ridge and Western Piedmont, Virginia Appalachians [Ph.D. thesis]: Blacksburg, Virginia, Virginia Polytechnic Institute and State University, 314 p.

Fletcher, J.M., and Bartley, J.M., 1994, Constrictional strain in a non-coaxial shear zone: Implications for fold and rock fabric development, central Mojave metamorphic core complex, California: Journal of Structural Geology, v. 16, p. 555–570, doi: 10.1016/0191-8141(94)90097-3.

Force, E.R., 1997, Geology and mineral resources of the Santa Catalina Mountains, southeastern Arizona: Center for Mineral Resources, Monographs in Mineral Resource Science, no. 1, 134 p.

Forte, A.M., and Bailey, C.M., 2007, Testing the utility of the porphyroclast hyperbolic distribution method of kinematic vorticity analysis: Journal of Structural Geology, v. 29, p. 983-1001, doi: 10.1016/j.jsg2007.01.006.

Gareau, S.A., 1997, Geology of the Scotia-Quaal metamorphic belt, Coast Plutonic Complex, British Columbia: Geological Survey of Canada Map 1868A, scale 1:100,000.

Gehrels, G., McClelland, W.C., Samson, S.D., Patchett, P.J., and Brew, D.A., 1991, U-Pb geochronology of the Late Cretaceous and early Tertiary plutons in the northern Coast Mountains batholith: Canadian Journal of Earth Sciences, v. 28, p. 899–911.

Grant, J.A., 1985, The isocon diagram—A simple solution to Gresens equation for metasomatic alteration: Economic Geology and the Bulletin of the Society of Economic Geologists, v. 81, p. 1976–1982.

Grasemann, B., Fritz, H., and Vannay, J.-C., 1999, Quantitative kinematic flow analysis from the Main Central Thrust zone (NW Himalaya, India): Implications for a decelerating strain path and the extrusion of orogenic wedges: Journal of Structural Geology, v. 21, p. 837–853, doi: 10.1016/S0191-8141(99)00077-2.

Harris, L.D., 1979, Similarities between the thick-skinned Blue Ridge anticlinorium and the thin-skinned Powell Valley anticline: Geological Society of America Bulletin, v. 90, p. 525–539, doi: 10.1130/0016-7606(1979)90<525:SBTTBR>2.0.CO;2.

Hodges, K.V., Hurtado, J.M., and Whipple, K.X., 2001, Southward extrusion of Tibetan crust and its effect on Himalayan tectonics: Tectonics, v. 20, p. 799–809, doi: 10.1029/2001TC001281.

Hollister, L.S., and Andronicos, C.L., 1997, A candidate for the Baja British Columbia fault system in the Coast plutonic complex: GSA Today, v. 7, no. 11, p. 1–7.

Hudleston, P.J., 1999, Strain compatibility and shear zones: Is there a problem?: Journal of Structural Geology, v. 21, p. 923–932, doi: 10.1016/S0191-8141(99)00060-7.

Hughes, S.S., Lewis, S.E., Bartholomew, M.J., Sinha, A.K., and Herz, N., 2004, Geology and geochemistry of granitic and charnockitic rocks in the central Lovingston massif of the Grenvillian Blue Ridge terrane: Geological Society of America Memoir, v. 197, p. 549–569.

Hutchison, W.W., 1982, Geology of the Prince Rupert–Skeena map area, British Columbia: Geological Survey of Canada Memoir 394, 116 p.

Irving, E., Wynne, P.J., Thorkleson, D.J., and Schiarizza, P., 1996, Large (1000 to 4000 km) northward movements of tectonic domains in the northern Cordillera, 83 to 45 Ma: Journal of Geophysical Research, v. 101, p. 17,901–17,916, doi: 10.1029/96JB01181.

Jiang, D., and Williams, P.F., 1998, High-strain zones: A unified model: Journal of Structural Geology, v. 20, p. 1105–1120, doi: 10.1016/S0191-8141(98)00025-X.

Klepeis, K.A., Crawford, M.L., and Gehrels, G., 1998, Structural history of the crustal-scale Coast shear north of Portland Canal, southeast Alaska and British Columbia: Journal of Structural Geology, v. 20, p. 883–904, doi: 10.1016/S0191-8141(98)00020-0.

Law, R.D., Searle, M.P., and Simpson, R.L., 2004, Strain, deformation temperatures and vorticity of flow at the top of the Greater Himalayan slab, Everest massif, Tibet: Geological Society [London] Journal, v. 161, p. 305–320.

Lee, J., Miller, E.L., and Sutter, J.F., 1987, Ductile strain and metamorphism in an extensional tectonic setting: A case study from the northern Snake Range, Nevada, USA, *in* Coward, M.P., et al., eds., Continental extensional tectonics: Geological Society [London] Special Publication 28, p. 267–298.

Lin, S., Jiang, D., and Williams, P.F., 1998, Transpression (or transtension) zones of triclinic symmetry: Natural example and theoretical modelling, *in* Holdsworth, R.E., et al., eds., Continental transpressional and transtensional tectonics: Geological Society [London] Special Publication 135, p. 41–57.

Mawer, C.K., 1983, State of strain in a quartzite mylonite, central Australia: Journal of Structural Geology, v. 5, p. 401–409, doi: 10.1016/0191-8141(83)90026-3.

Means, W.D., 1989, Stretching faults: Geology, v. 17, p. 893–896, doi: 10.1130/0091-7613(1989)017<0893:SF>2.3.CO;2.

Metcalf, J., and Davidson, C., 1997, Quantitative fabric analysis of the Quottoon pluton: Implications for the late deformation history of the Coast shear zone: Geological Society of America Abstracts with Programs, v. 29, no. 6, p. A-277.

Miller, E.L., Gans, P.B., and Garing, J., 1983, The Snake Range décollement: An exhumed mid-Tertiary brittle-ductile transition: Tectonics, v. 2, p. 239–263.

Mitra, G., 1979, Ductile deformation zones in Blue Ridge basement and estimation of finite strains: Geological Society of America Bulletin, v. 90, p. 935–951, doi: 10.1130/0016-7606(1979)90<935:DDZIBR>2.0.CO;2.

Mitra, G., and Elliott, D., 1980, Deformation of the basement in the Blue Ridge and the development of the South Mountain cleavage, *in* Wones, D.R., ed., Proceedings "Caledonides in the USA" IGCP Project 27: Caledonide Orogen: Virginia Polytechnic Institute and State University, Department of Geological Sciences Memoir 2, p. 307–312.

Naruk, S.J., 1987, Displacement calculations across a metamorphic core complex mylonite zone: Pinaleño Mountains, southeastern Arizona: Geology, v. 15, p. 656–660, doi: 10.1130/0091-7613(1987)15<656:DCAAMC>2.0.CO;2.

O'Hara, K., 1990, State of strain in mylonites from the southern Appalachians, western Blue Ridge province: The role of volume loss: Journal of Structural Geology, v. 12, p. 419–430, doi: 10.1016/0191-8141(90)90031-S.

Passchier, C.W., 1987, Stable positions of rigid objects in non-coaxial flow: A study in vorticity analysis: Journal of Structural Geology, v. 9, p. 679–690, doi: 10.1016/0191-8141(87)90152-0.

Polvi, L.E., 2003, Structural analysis of the Lawhorne Mill high-strain zone Blue Ridge province, Nelson County, Virginia: [B.S. thesis]: Williamsburg, Virginia, College of William and Mary, 56 p.

Ramsay, J.G., and Graham, R.H., 1970, Strain variation in shear belts: Canadian Journal of Earth Sciences, v. 7, p. 786–813.

Ramsay, J.G., and Huber, M.I., 1983, The techniques of modern structural geology: London, Academic Press, 307 p.

Roth, D., 2001, Three-dimensional foliation analysis of the Quottoon pluton as a proxy for characterizing the effects of the Coast steep zone, British Columbia: Keck Geology Consortium, 14th Annual Undergraduate Research Symposium, p. 93–97.

Sanderson, D.J., and Marchini, W.R.D., 1984, Transpression: Journal of Structural Geology, v. 6, p. 449–458, doi: 10.1016/0191-8141(84)90058-0.

Simpson, C., and De Paor, D.G., 1993, Strain and kinematic analysis in general shear zones: Journal of Structural Geology, v. 15, p. 1–20, doi: 10.1016/0191-8141(93)90075-L.

Simpson, C., and De Paor, D.G., 1997, Practical analysis of general shear zones using the porphyroclast hyperbolic distribution method: An example from the Scandinavian Caledonides, *in* Sengupta, S., ed., Evolution of geological structures in micro- to macroscales: London, Chapman & Hall, p. 169–184.

Sinha, A.K., and Bartholomew, M.J., 1984, Evolution of the Grenville terrane in the central Virginia Appalachians, *in* Bartholomew, M.J., et al., eds., The Grenville event in the Appalachians and related topics: Geological Society of America Special Paper 194, p. 175–186.

Stowell, H.H., and Hooper, R.J., 1990, Structural development of the western metamorphic belt adjacent to the Coast plutonic complex, southeastern Alaska: Tectonics, v. 9, p. 391–407.

Tikoff, B., and Fossen, H., 1995, The limitations of three-dimensional kinematic vorticity analysis: Journal of Structural Geology, v. 17, p. 1771–1784, doi: 10.1016/0191-8141(95)00069-P.

Tikoff, B., and Greene, D., 1997, Stretching lineations in transpressional shear zones: Journal of Structural Geology, v. 19, p. 29–40, doi: 10.1016/S0191-8141(96)00056-9.

Treloar, P.J., and Blenkinsop, T.G., 1995, Archaean deformation patterns in Zimbabwe: True indicators of Tibetan-style crustal extrusion or not?: Geological Society [London] Special Publication 95, p. 87–107.

Wallis, S.R., 1992, Vorticity analysis in a metachert from the Sanbagawa belt, SW Japan: Journal of Structural Geology, v. 14, p. 271–280, doi: 10.1016/0191-8141(92)90085-B.

Wallis, S.R., 1995, Vorticity analysis and recognition of ductile extension in the Sanbagawa belt, SW Japan: Journal of Structural Geology, v. 17, p. 1077–1093, doi: 10.1016/0191-8141(95)00005-X.

Wallis, S.R., Platt, J.P., and Knott, S.D., 1993, Recognition of synconvergence extension in accretionary wedges with examples from the Calabrian Arc and the Eastern Alps: American Journal of Science, v. 282, p. 463–495.

Wells, M.L., and Allmendinger, R.W., 1990, An early history of pure shear in the upper plate of the Raft River metamorphic core complex: Black Pine Mountains, southern Idaho: Journal of Structural Geology, v. 12, p. 851–867, doi: 10.1016/0191-8141(90)90059-8.

Xypolias, P., and Koukouvelas, I.K., 2001, Kinematic vorticity and strain rate patterns associated with ductile extrusion in the Chelmos shear zone (External Hellenides, Greece): Tectonophysics, v. 338, p. 59–77, doi: 10.1016/S0040-1951(01)00125-1.

Zen, E., 1988, Tectonic significance of high-pressure plutonic rocks in the western Cordillera of North America, *in* Ernst, W.G., ed., Metamorphism and crustal evolution of the western United States (Rubey Volume VII): Englewood Cliffs, New Jersey, Prentice Hall, p. 41–67.

Manuscript Accepted by the Society 1 February 2007

Printed in the USA

The Geological Society of America
Memoir 200
2007

Scale independence of décollement thrusting

John H. McBride*
Department of Geological Sciences, Brigham Young University, P.O. Box 24606, Provo, Utah 84602, USA

Andre J.M. Pugin†
Illinois State Geological Survey, 615 East Peabody Drive, Champaign, Illinois 61820, USA

Robert D. Hatcher Jr.
Department of Earth and Planetary Sciences and Science Alliance Center of Excellence, University of Tennessee, Knoxville, Tennessee 37996-1410, USA

ABSTRACT

Orogen-scale décollements (detachment surfaces) are an enduring subject of investigation by geoscientists. Uncertainties remain as to how crustal convergence processes maintain the stresses necessary for development of low-angle fault surfaces above which huge slabs of rock are transported horizontally for tens to hundreds of kilometers. Seismic reflection profiles from the southern Appalachian crystalline core and several foreland fold-and-thrust belts provide useful comparisons with high-resolution shallow-penetration seismic reflection profiles acquired over the frontal zone of the Michigan lobe of the Wisconsinan ice sheet northwest of Chicago, Illinois. These profiles provide images of subhorizontal and overlapping dipping reflections that reveal a ramp-and-flat thrust system developed in poorly consolidated glacial till. The system is rooted in a master décollement at the top of bedrock. These 2–3 km long images contain analogs of images observed in seismic reflection profiles from orogenic belts, except that the scale of observation in the profiles in glacial materials is two orders of magnitude less. Whereas the décollement beneath the ice lobe thrust belt lies ~70 m below thrusted anticlines having wavelengths of tens of meters driven by an advancing ice sheet, seismic images from overthrust terranes are related to lithospheric convergence that produces décollements traceable for thousands of kilometers at depths ranging from a few to over 10 km. Dual vergence or reversals in vergence (retrocharriage) that developed over abrupt changes in depth to the décollement can be observed at all scales. The strikingly similar images, despite the contrast in scale and driving mechanism, suggest a scale- and driving mechanism–inde-

*john_mcbride@byu.edu
†Current address: Geological Survey of Canada, 601 Booth Street, Ottawa, Ontario, Canada K1A 0E8.

McBride, J.H., Pugin, A.J.M., and Hatcher, R.D., Jr., 2007, Scale independence of décollement thrusting, *in* Hatcher, R.D., Jr., Carlson, M.P., McBride, J.H., and Martínez Catalán, J.R., eds., 4-D Framework of Continental Crust: Geological Society of America Memoir 200, p. 109–126, doi: 10.1130/2007.1200(07). For permission to copy, contact editing@geosociety.org.

pendent behavior for décollement thrust systems. All these systems initially had the mechanical properties needed to produce very similar geometries with a compressional driving mechanism directed subparallel to Earth's surface. Subduction-related accretionary complexes also produce thrust systems with similar geometries in semi- to unconsolidated materials.

Keywords: seismic reflection, glacial sediments, décollements, faulting.

INTRODUCTION

Chapple (1978) listed the attributes of foreland fold-and-thrust belts as "thin-skinned; contain listric thrusts that rise from a basal weak zone of shale, evaporite, or coal just above basement; they ramp steeply across strong units to higher detachments, or to the surface; comprise a continentward-thinning wedge of the undeformed sequence; and maintain, thicken, and shorten the wedge into the deformed state." Cook and Varsek (1994) summarize observations from deep seismic reflection profiles on which décollements have been interpreted from the base of the sedimentary section, the middle crust, near the Moho discontinuity, or from within the uppermost mantle. Hatcher (2004) suggested that crystalline slab thrust sheets are stronger, form detachments along the ductile-brittle transition in the crust, and may ramp into the basal detachments of foreland fold-and-thrust belts, where the ductile-brittle transition descends in cooler cratonic crust. The recognition of regional décollements has become an integral part of interpretations of thrust belts where relatively shallow deformation appears to be detached at depth from deeper undeformed rocks (Rodgers, 1949). Regional seismic reflection recording in the southern Canadian Cordillera and the southern Appalachians confirmed the existence of large-scale décollements undercutting the forelands and orogenic cores of mountain belts (Bally et al., 1966; Cook et al., 1979; Cook and Varsek, 1994). The scales for which décollements have been described and for which geologists are most familiar are either at the level of an orogen (tens to hundreds of kilometers), as typically represented in cross sections, or at the level of isolated occurrences in outcrop (tens of meters), and in drill cores and hand specimens (Fig. 1) (Cook and Varsek, 1994). A disparity thus appears to exist between large- and small-scale observations. This is especially true for representations of décollements in seismic reflection profiles, which have usually provided information only at regional scales.

In this paper we (1) present the first-observed case in the United States, known to us, of a small-scale (hundreds of meters) décollement thrust system driven by surging of ice from a continental glacier and seismically imaged within glacial till, and (2) integrate these new observations with other expressions of décollement thrust systems at various scales, spanning orders of magnitude to demonstrate the scale independence of décollement development. Our interpretation of the glacially driven thrust system is constrained by previous detailed geologic mapping, by nearby quarry observations, and by a dedicated borehole along the key profile. The new availability of detailed high-resolution images of a décollement system provides fresh constraints on outstanding questions such as the role of preexisting structures, how and why thrusts ramp up from the décollement, and the origin of décollement reflectivity.

GLACIAL GEOLOGICAL SETTING OF STUDY AREA

As part of a program for testing new geophysical techniques for imaging shallow strata in glaciated terranes (Pugin et al., 2004), the Illinois State Geological Survey acquired several high-resolution seismic P-wave reflection profiles in the Wadsworth quadrangle in northern Lake County, Illinois, northwest of Chicago (Fig. 2). The survey area is located among glacial moraines and till fields associated with the Wisconsinan glaciation, which occurred approximately from 55 to 10 ka (Grimley and Curry, 2002) (Fig. 2). Northern Illinois was repeatedly invaded by continental glaciers beginning ca. 2.4 Ma. The Wisconsinan glaciation represents the advance of the last major ice sheet into Illinois, while the Laurentide ice sheet covered much of Canada and the northern United States. The study area is mantled by 33–100 m of poorly consolidated glacial drift. According to Curry (2005), the Wisconsinan glacial ice usually advanced in a westerly and southwesterly direction into Illinois from the Lake Michigan basin. In some places in and near the study area, sublobes of ice flowed in a more westerly direction due to constriction by bedrock topography or older moraines (Grimley and Curry, 2002) (see also Hansel and Johnson, 1996; Curry et al., 1999) (Fig. 2). Subglacial streams associated with this flow drained the meltwater out of the basin with a flow direction identical to ice flow. The resulting Wisconsinan glacial stratigraphy is composed of the Wadsworth Till, underlain by the Tiskilwa Till, which rests on Silurian and Ordovician bedrock. Till lithologies typically consist of diamicton, loam, and sand with beds or small channels of sand and gravel (Grimley and Curry, 2002). Illinois State Geological Survey mapping has revealed that drift thicknesses reach 60–90 m directly beneath the seismic survey profiles (Fig. 3). Further description of glacial stratigraphy can be found in Willman and Frye (1970). The study area is located specifically over the Tinley Moraine where the Wadsworth Till occurs at or near the surface.

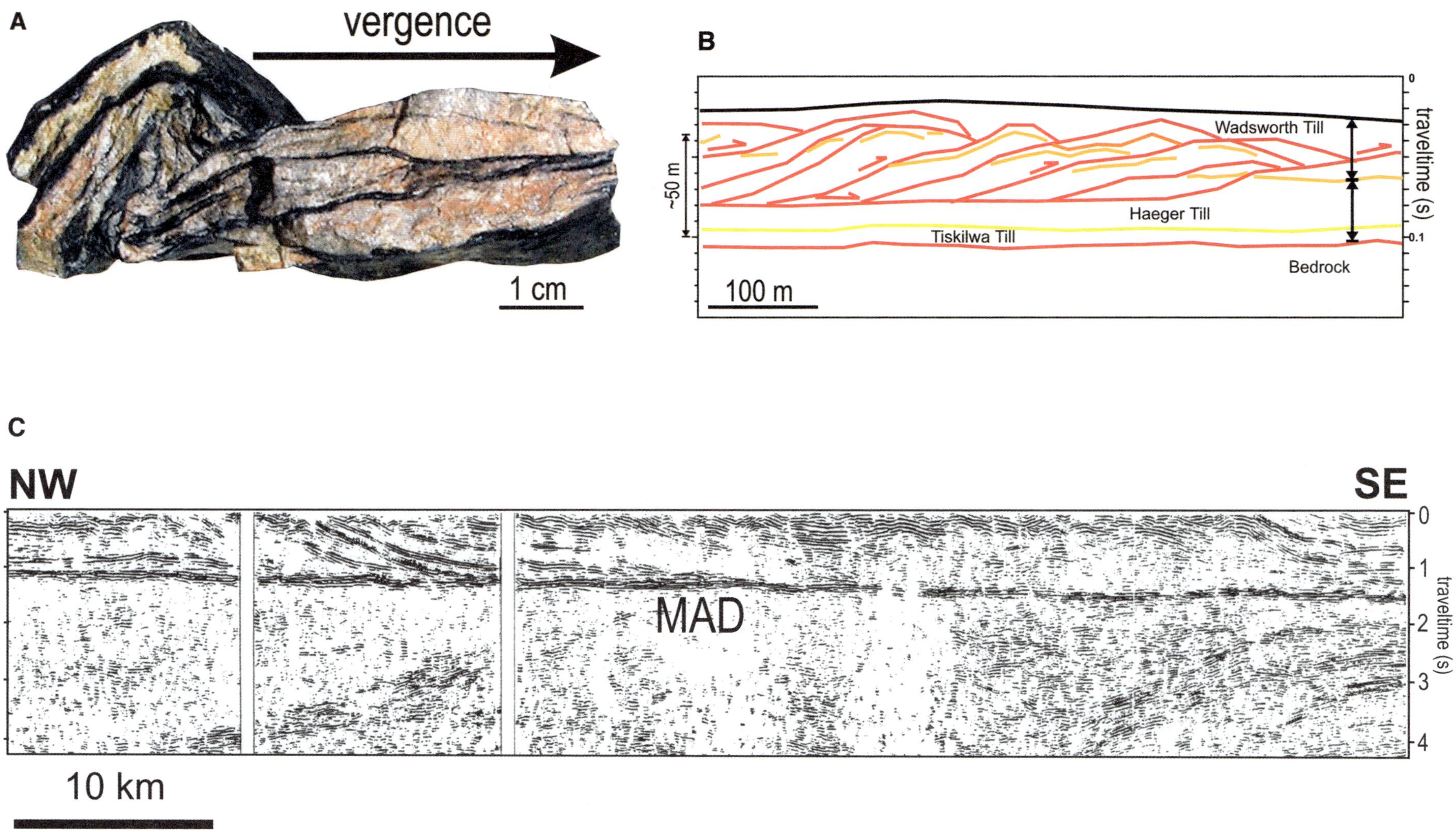

Figure 1. Illustrating the scale independence of décollements. (A) Small mesoscale décollement in thinly layered sandstone and carbonaceous shale from near Dunlap, Tennessee (photo by Robert D. Hatcher Jr.). The detachment cannot readily be identified, except to say that the fault(s) propagated along the carbonaceous shale layers. (B) Sketch of glaciogenic thrust system discussed in detail in the text. (C) Seismic reflection profile across part of the Tennessee Valley and Ridge province (reproduced from Hopkins, 1995). MAD—master Appalachian décollement.

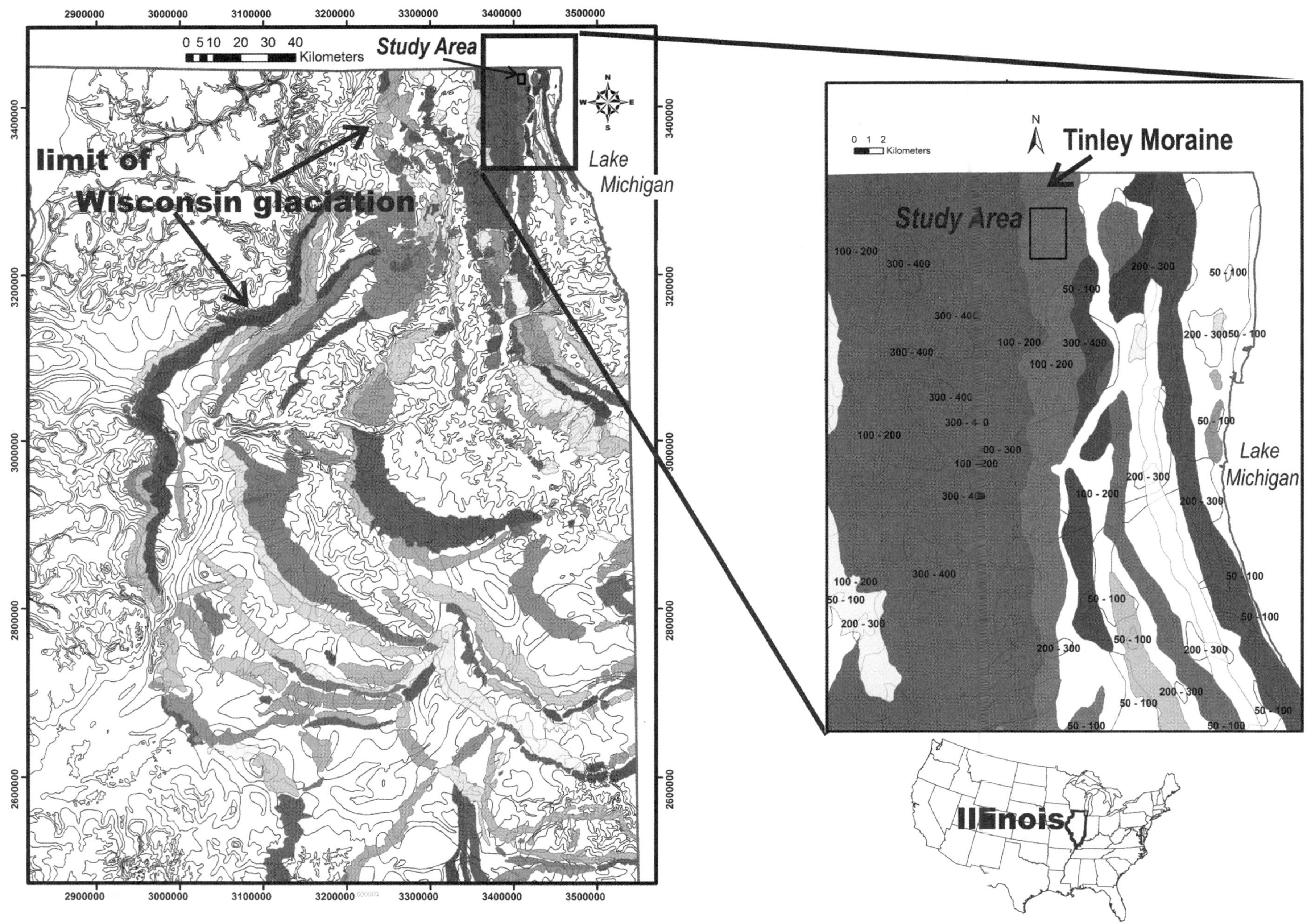

Figure 2. Location map of northeastern Illinois (USA) showing glacial drift thickness contours (thin solid lines) without labeling in order to provide the structure of the pattern. Gray shades indicate different Wisconsinan till units. The inset map shows the region surrounding the study area with the range of Wisconsinan till thickness given (in feet) and the individual Wisconsinan moraines. Source: Illinois State Geological Survey (http://www.isgs.uiuc.edu/nsdihome/webdocs/browse.html—files are drifthk and moraines).

HIGH-RESOLUTION SEISMIC REFLECTION PROFILES IN NORTHERN ILLINOIS

Two orthogonal high-resolution P-wave reflection profiles, totaling 6.44 km, were acquired along local roads in the Wadsworth quadrangle (Fig. 3). A typical acquisition system using a 48-channel seismograph and an accelerated weight-dropper seismic source with 3.05 m geophone and source intervals was employed to collect 24-fold stacked common depth point profiles. A borehole near one of the profiles was used to guide the stratigraphic interpretation of P-wave profiles and to convert them from traveltime to depth. The north-south profile is considered a dip profile ("Wadsworth-North" and "Wadsworth-South" profiles), while the east-west profile ("Kelley" profile) provides control as a strike section. Because the direction of ice movement in the study area is only approximately known, the locations of the profiles relative to dip and strike are approximate. Seismic profiles are shown as two-way traveltime or depth-converted sections.

The high-resolution seismic profiles successfully image stratigraphic and structural details within the Wisconsinan tills (Fig. 4A). Multiple seismic facies on top of bedrock are associated with the local till units, comprising the low-reflectivity upper Wadsworth Till, the lower Wadsworth Till with chaotic and diffractive reflectivity, and the upper part of the Tiskilwa Till with high reflectivity (Fig. 4A). The Tiskilwa Till was deposited within depressions interpreted as filling an east-west buried bedrock valley. A distinct series of mostly north-dipping (within the plane of the section), overlapping reflections appears on the north-south profile within the Wadsworth Till (Fig. 4A). The dipping reflections ramp into south-vergent thrust-cored antiforms higher in the section. As the dipping reflections deepen, dip decreases and reflectors asymptotically approach the base of the lower Wadsworth Till (see alternative interpretation in Fig. 4B). This character is particularly well expressed in a plot of the phase attribute of the signal (Fig. 5). The dipping and antiformal series appears only on the north-south profile and correlates with a fabric of mostly subhorizontal reflections on the east-west profile (Fig. 4C). Neither the southern continuation of the Wadsworth-North profile nor the Kelley profile reveal a fabric of disturbed, dipping reflections such as those observed in the northern part of the Wadsworth-North profile, emphasizing the unique nature of the dipping zone.

The north-south profile describes a ramp-and-flat style of thrusting in which asymmetric anticlines are strongly imbricated into a passive roof complex (Fig. 4A). The observed deformation is interpreted as being related to southwest-directed flow of glacial ice in the Michigan lobe. This structural style implies a local compressive stress that acted on the tills shortly after they were deposited, while they were poorly lithified and water saturated. In the northern half of the profile, the reflections are broken into normal faults, indicating an abrupt change to an extensional mode. Deformation structures appear mostly restricted to the Wadsworth Till. Analogous deformation structures were documented in outcrop several kilometers to the south, where a thrust appears in the Wadsworth Till, defined by a 23° north-dipping sand layer (Fig. 6). The upper part of the outcrop is tilted at the same angle, suggesting that the entire upper part of the section has been thrusted. The deformation observed in the exposed till could represent the uppermost parts of thrust faults such as those observed in the seismic profiles.

The Wadsworth Till section has been qualitatively reconstructed, without any attempt at balancing, to suggest a possible deformation scenario (Fig. 7). Balancing the section was deemed not practicable due to a lack of knowledge about the precise direction of glacial ice movement along the profile, and a lack of ability to assess out-of-section motion in the profile. The reconstructed section is therefore only speculative. This exercise suggests that the center of the section evolved from compression to tension, possibly because (1) the extended part of the thrust system was dewatered; (2) motion of the glacial ice indenter diminished, and the frontal part of the system continued to move; (3) the glacier overrode the segment that exhibits normal faulting, and the ice load triggered dewatering of the till, a less likely possibility; or (4) the sag formed over stagnant ice or tunnel valley (Pugin et al., 1999) or channel beneath the ice. The latter have been documented as products of Wisconsinan glaciation in the Midwest United States and are produced by the flow of subglacial water (Clayton et al., 1999). A change of the ice flow direction from east to south may have triggered abandonment of the channel and flow of ice into the void. This change in flow direction would have caused the ice to push (or compress) the till south of the channel and pull apart the till to the north of the channel (Fig. 7). Décollements and related structures in compressional orogens are typically interpreted as products of contraction only; our glaciotectonic analog shows that décollement systems can develop from a complex interplay of both contractional and extensional effects not unlike the structures so well documented in offshore West Africa (Corredor et al., 2005).

DISCUSSION

Detailed seismic images of a décollement and related structures generated by an ice "indenter" provide an unusual opportunity to probe the limits of décollement formation in general. Cook and Varsek (1994) have also discussed the scale invariance of décollement systems by pointing out many examples of deformation from décollement systems that involve the deep crust and upper mantle. Allmendinger et al. (1987) proposed a consistent set of deformation patterns, which could be applied to orogenic belts in general, observed from deep seismic reflection profiles in which large-scale deformation structures are linked by a master décollement. Our high-resolution seismic profiles that image outcrop-scale décollements beneath a continental glacial foreland include a scale model and furnish structures of opportunity for modeling décollement systems beneath contractional mountain belts. Although this is the first documentation of glaciotectonic seismic images from North America known to us, analogous images have been produced for glacial features for the Devensian (Weichselian)

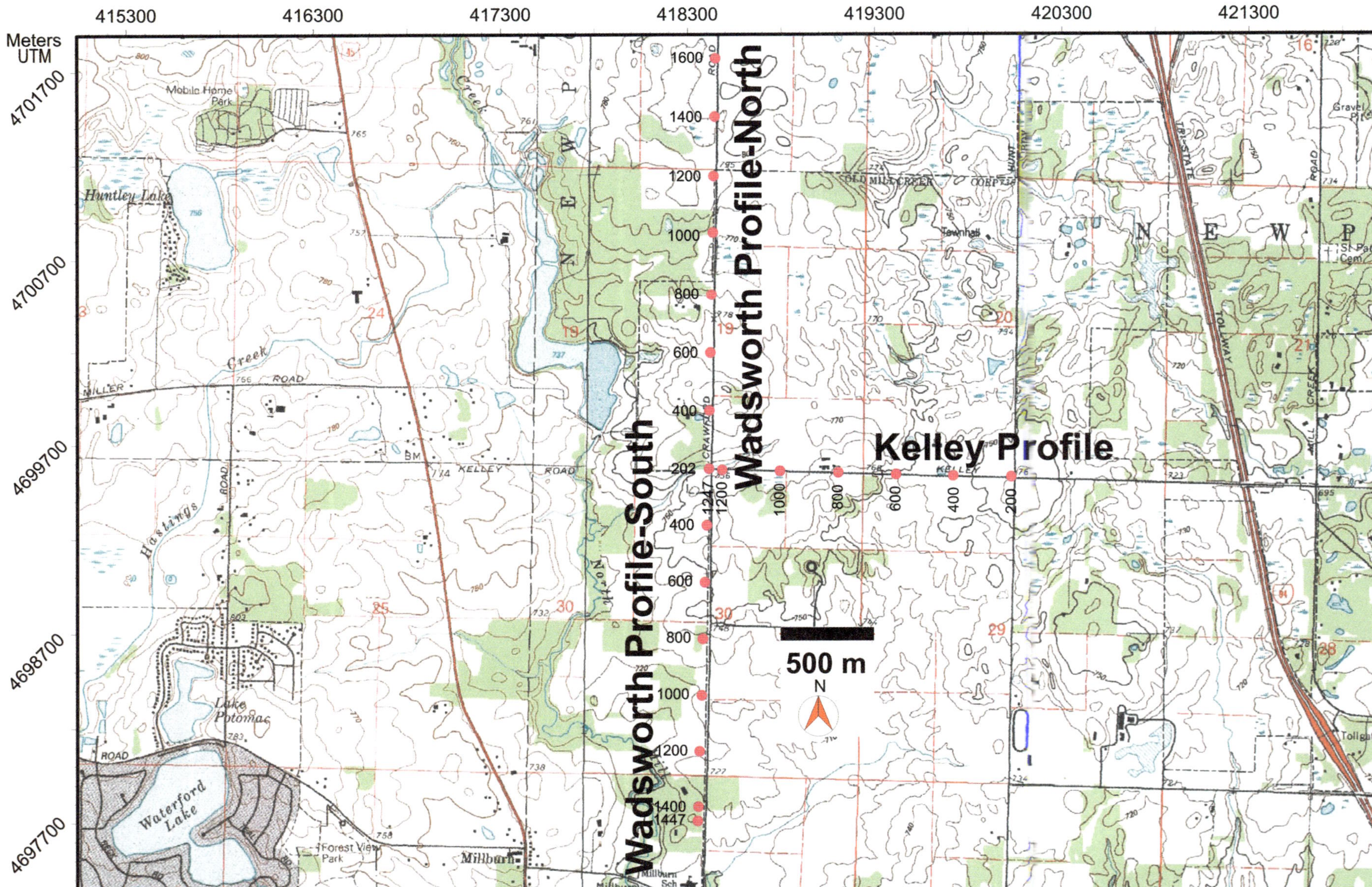

Figure 3. Seismic line location map (with shotpoint numbers) superimposed on the U.S. Geological Survey (USGS) 7.5 Wadsworth (Illinois, Wisconsin) quadrangle map.

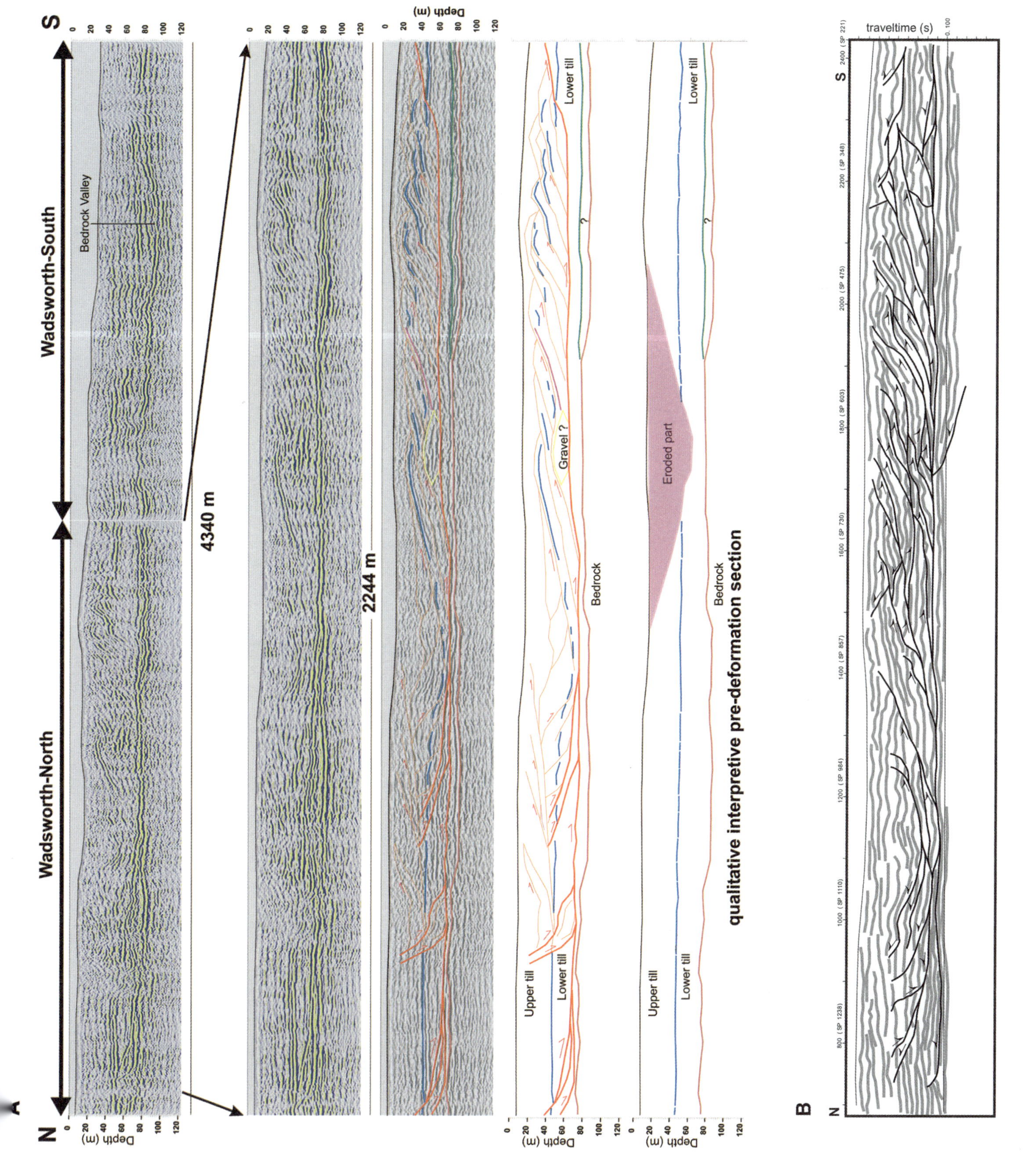

re 4 *(on this and following page)*. (A) Wadsworth-South seismic profile and Wadsworth-North seismic profile with interpretation. (B) Alternative interretation of Wadsworth-North profile, in which the relation of a subhorizontal décollement to thrust faults offsetting dipping reflectors is emphasized.

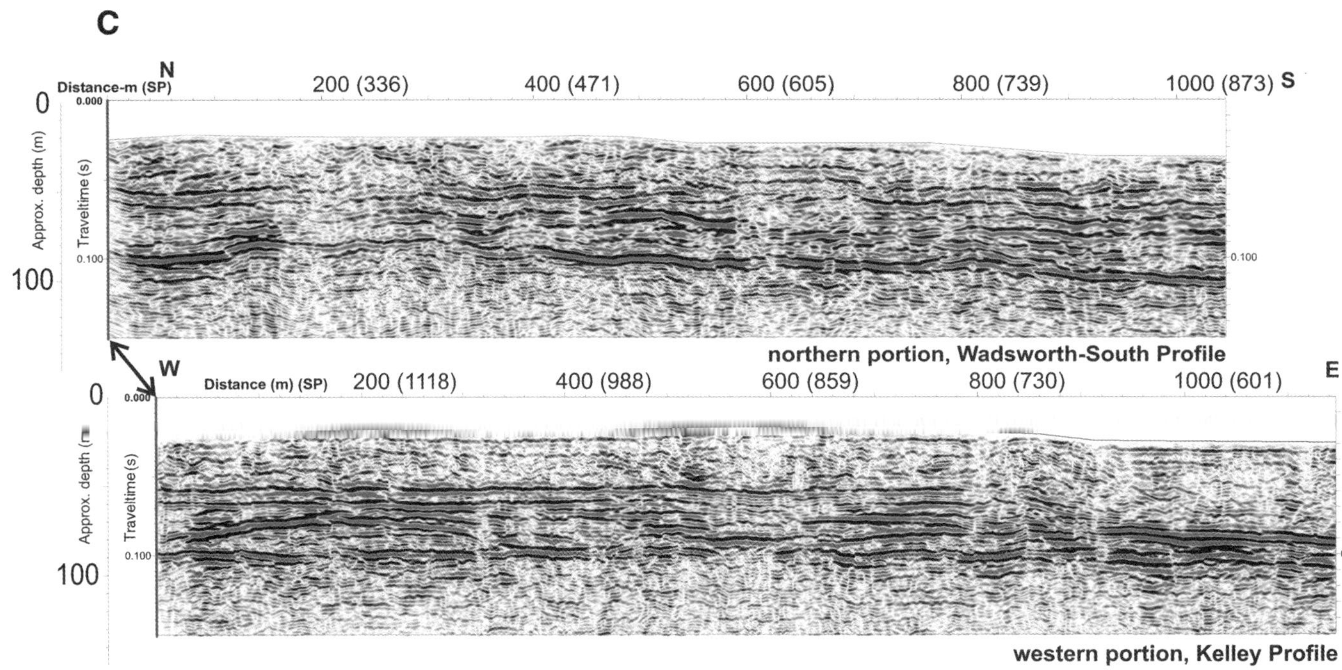

Figure 4 (*continued*). (C) Portions of two seismic profiles that intersect the Wadsworth-North seismic profile at its southern end (locations shown in Fig. 3).

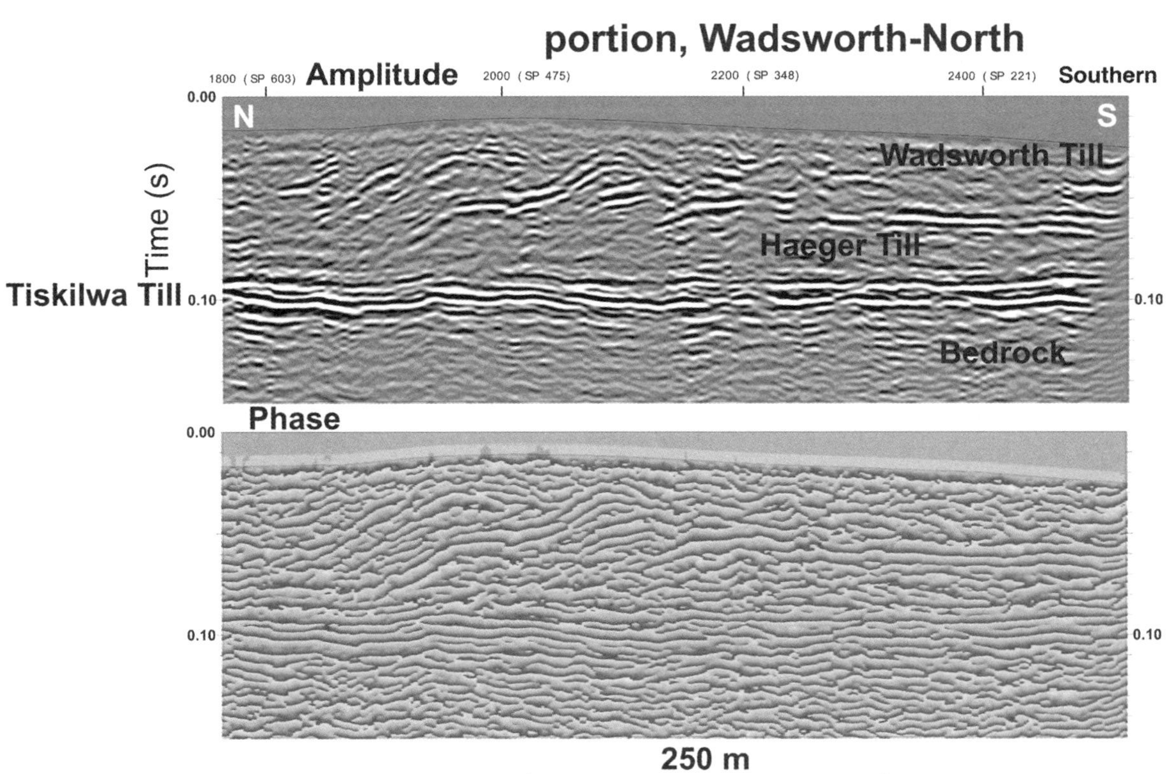

Figure 5. Close-up of Wadsworth-North seismic profile shown as amplitude and instantaneous phase attribute displays.

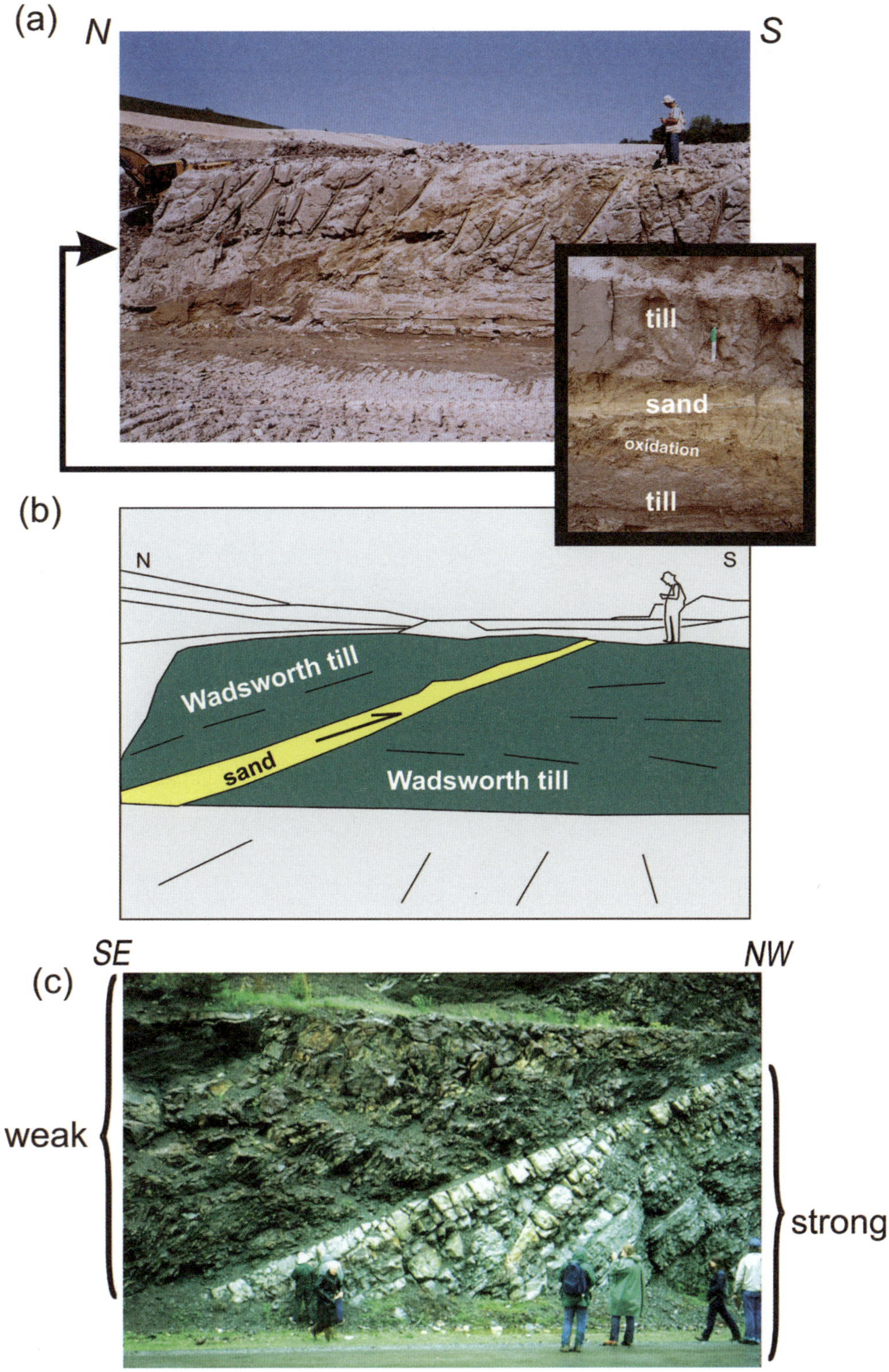

Figure 6. (A) Outcrop showing ice-related thrust structures in a till pit (area is located in USGS Grays Lake 7.5 quadrangle, Illinois, immediately to the southwest of the study area's Wadsworth quadrangle). To the left of the outcrop is another outcrop oriented east-west (perpendicular to that shown); this is where the close-up photo was taken. (B) Interpretation of top photo. Most of the shear deformation is within the sand, and the décollement occurs mainly at the base of the sand layer. (C) Copper Creek fault on highway I-75, north of Knoxville, Tennessee (Rome Formation [Lower Cambrian] over Middle Ordovician limestone with a small fault in the footwall limestone). Weak and strong refer to relative mechanical strength. Photo by Robert D. Hatcher Jr.

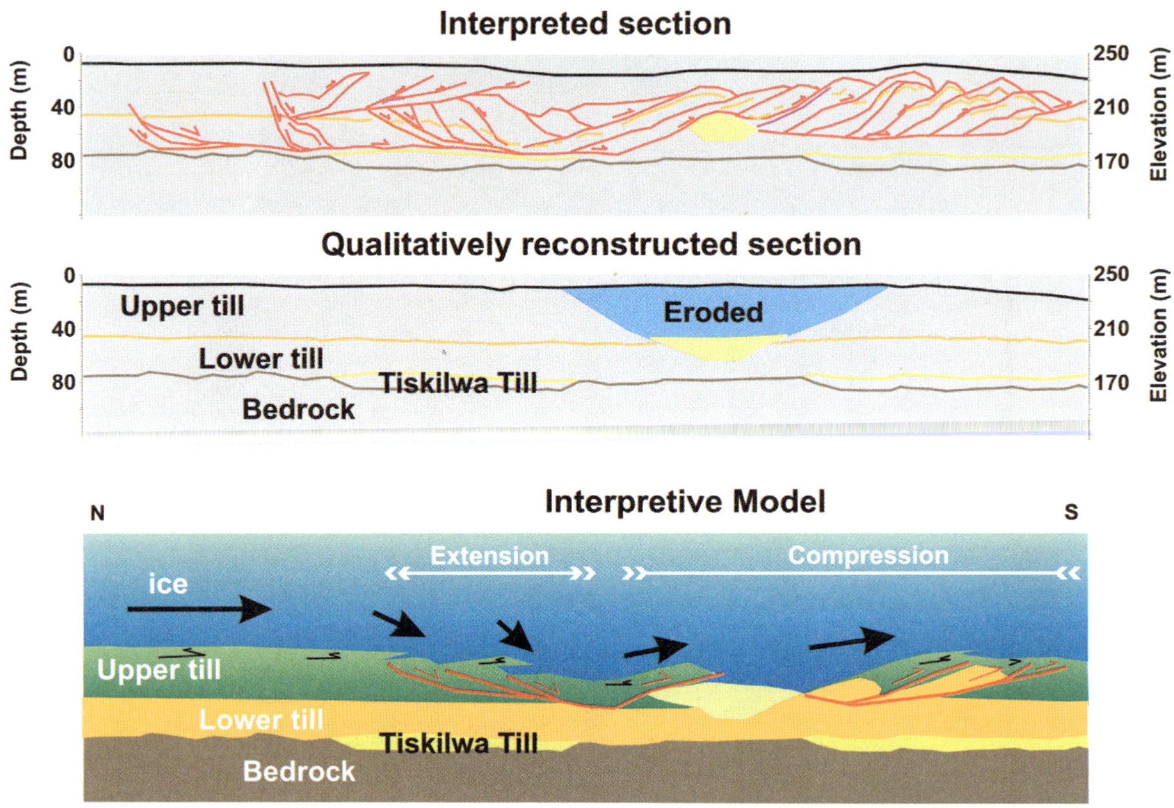

Figure 7. Wadsworth-North seismic section interpreted and qualitatively reconstructed (but not balanced). A possible model is given to explain the dynamics of the deformation structures.

ice margin in Wales and confirmed along a beach exposure (Harris et al., 1997; McCarroll and Rijsdijk, 2003). In order to generalize the interpretation of our glaciotectonic results from Illinois and to better understand processes related to thrusting, we briefly discuss data from seismic profiles acquired in décollement thrust settings in various environments, including results from new data reprocessing and from previously unpublished profiles.

Décollement Surface Characteristics

Where the décollement rests at or just above the top of bedrock, as in the northern part of the Wadsworth-North profile (Fig. 4A), it consists of a relatively smooth, continuous, and seismically coherent surface. Near the southern end of this profile, however, the tills within the upper plate are interrupted by a series of stacked, imbricated asymmetric anticlines beneath which the décollement reflection signature is more irregular and less coherent. Just north of the southern end of the Wadsworth-North profile (Fig. 4A), the décollement ramps upward into the upper till to form a higher décollement and a passive roof complex. Farther south, the décollement is not present along the profile (Fig. 4C), and the reflection pattern at or just above the bedrock is less coherent. From these observations, we suggest that the propagation of a décollement imparts a smoothing effect to the reflectivity pattern. The abrupt upward ramping of asymmetric thrusted anticlines may be controlled by a preexisting structural or lithologic anomaly (e.g., a structurally negative or weak area). The location of abrupt décollement ramps has been observed for crustal-scale systems to be controlled by areas of postorogenic or synorogenic extension at or near the location of crustal thinning on the outboard edge of a craton (Cook and Varsek, 1994; see also Allmendinger et al., 1987; McBride et al., 2005). In such cases, the accommodation space required for restoration of rocks above the décollement can be found at the transition from thick cratonic crust to thinned crust (Cook and Varsek, 1994).

Our glaciotectonic examples (Fig. 4) confirm the importance of structurally negative or weak areas at or below the décollement level in governing ramp and other complex geometries of a fold-and-thrust belt. Tulaczyk (1999) has shown that the granulometry (i.e., grain-size relationships) of the tills beneath the ice influences the degree and depth of ice-related deformation of the till. Coarse, clast-rich tills furnish stronger coupling between the ice and the till and thicker subglacial shearing compared to finer-grained tills (Tulaczyk, 1999). Such lithologic variations could influence the propagation of deformation along and above the décollement. Structural influences on the shape of the décollement could include buttressing by the footwall of a fault, pinchout of a weak unit, or a change in water saturation. The angular relationships of the basal décollement, and the enveloping surface on top of our glaciotectonic thrust system (Fig. 8), are similar to those described by Davis et al. (1983) in the accretionary complex in Taiwan. Similar structures have been observed in other accretionary complexes (e.g., Behrmann et al., 1988), and in gravity-driven collapse of continental margin sediments, as in offshore Nigeria (Corredor et al., 2005).

Thousands of kilometers of published seismic reflection profiles from different orogens (e.g., Bally, 1983) reveal master detachments beneath foreland fold-and-thrust belts. A shallow crustal (usually 6–12 km depth [2–4 s traveltime]) subhorizontal reflection package has been observed beneath the exposed core of the southern Appalachian orogen, which is generally interpreted to track the position of the master décollement (Clark et al., 1978; Cook et al., 1979; Lampshire et al., 1994) or the basal foreland stratigraphy immediately beneath it (Costain et al., 1989; Hatcher, 1989; Hatcher et al., 1989). This reflection package can be traced from beneath the Appalachian foreland (Appalachian Plateau and Valley and Ridge province) at least as far oceanward as the westward edge of the Atlantic Coastal Plain (Harris and Bayer, 1979; Cook et al., 1981; Lampshire et al., 1994).

In order to make detailed comparisons between our Illinois examples and those from the Appalachian orogen, we begin by discussing two examples from the Appalachian foreland of Tennessee, USA: (1) a portion of a long regional industry profile from Hopkins (1995) (Fig. 1C) and (2) two proprietary industry profiles that were incorporated into this study (Fig. 9). These comparisons demonstrate that behavior of the décollement on the 1–2 km long glaciotectonic profiles is very similar to that observed on regional industry and deep seismic profiles across the interior of the southern Appalachian orogen. For example, directly beneath areas where steep thrusts are rooted in continentward-vergent anticlines, the emergence of thrusts upward from the décollement (i.e., where the downdip portion of dipping reflectors meet or approach the décollement) seems to correspond to a loss of coherency and continuity of the décollement reflection pattern. On deep seismic Appalachian profiles in general, irregularities in the structure and depth of the décollement have been ascribed to variations in the thickness of the shelf sedimentary sequence that underlies the décollement (Cook et al., 1983; Lillie, 1984; Lampshire et al., 1994) but may also be related to duplication of the section. This behavior is suggested by the glaciotectonic profiles (Fig. 4) in that original bedrock surface undulations are interpreted to affect the shape of the décollement and the location of emergent thrusts propagating higher into the section.

Hanging-wall cutoffs are not abundantly preserved in the southern Appalachian foreland fold-and-thrust belt (e.g., Fig. 1C), as in many younger foreland fold-and-thrust belts, e.g., the Helvetic Alps (Ramsay, 1981), Sulaiman lobe (western Himalayas; Jadoon et al., 1992), and Canadian Rockies and foothills (Price and Mountjoy, 1970; Thompson, 1981). Hanging-wall cutoffs are preserved, however, in the Wadsworth, Illinois, seismic section (Figs. 4, 5, and 7). An industry section from the Appalachian foreland fold-and-thrust belt buried beneath the Coastal Plain in western Alabama (Fig. 10) also contains hanging-wall cutoffs and an indication of localization of the principal ramp above several preexisting basement normal faults.

Many investigators have interpreted the subhorizontal reflection sequence at the décollement level beneath the Appalachians as originating from relatively undeformed Lower and Middle Cambrian continental shelf strata (Rome Formation, Conasauga Group, and lateral equivalents) (Clark et al., 1978; Cook et al., 1979; Lampshire et al., 1994) (Fig. 6C). Then again, some have argued that a mylonitic zone directly associated with the décollement or other large-scale shear fabrics involving local ductile deformation, or mineralization along a faulted zone may contribute to the reflectivity, in addition to or instead of flat-lying, well-stratified sediments (Iverson and Smithson, 1982; Hurich et al., 1985). Jones and Nur (1984) also considered anisotropy and the degree of lamination within mylonitized fault zones to be capable of producing strong reflectivity. From outcrop-scale studies, Law and Snyder (1997) have shown that the effects of changes in anisotropy and bulk seismic velocity contrasts associated with mylonites can combine to enhance fault-zone reflectivity. Realistically, the reflectivity is probably only related to a shale-sandstone or -dolostone contact in the Rome Formation (Lower Cambrian)—this can be seen on surface exposures where reflectors can be traced to the surface (e.g., Fig. 1C). Localized dewatering of the basal shales would locally produce strain hardening (Chapple, 1978) and create undulations or irregularities in the décollement analogous to variations in the inherent strength of the tills beneath glacial ice.

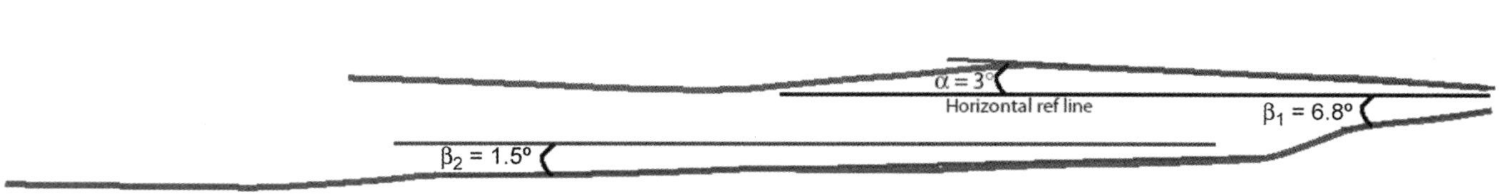

Figure 8. Angular relationships derived from the glaciogenic structure imaged in Wadsworth-North seismic reflection profile. α—dip of décollements. β—dip of "surface" slope (enveloping surface of antiforms).

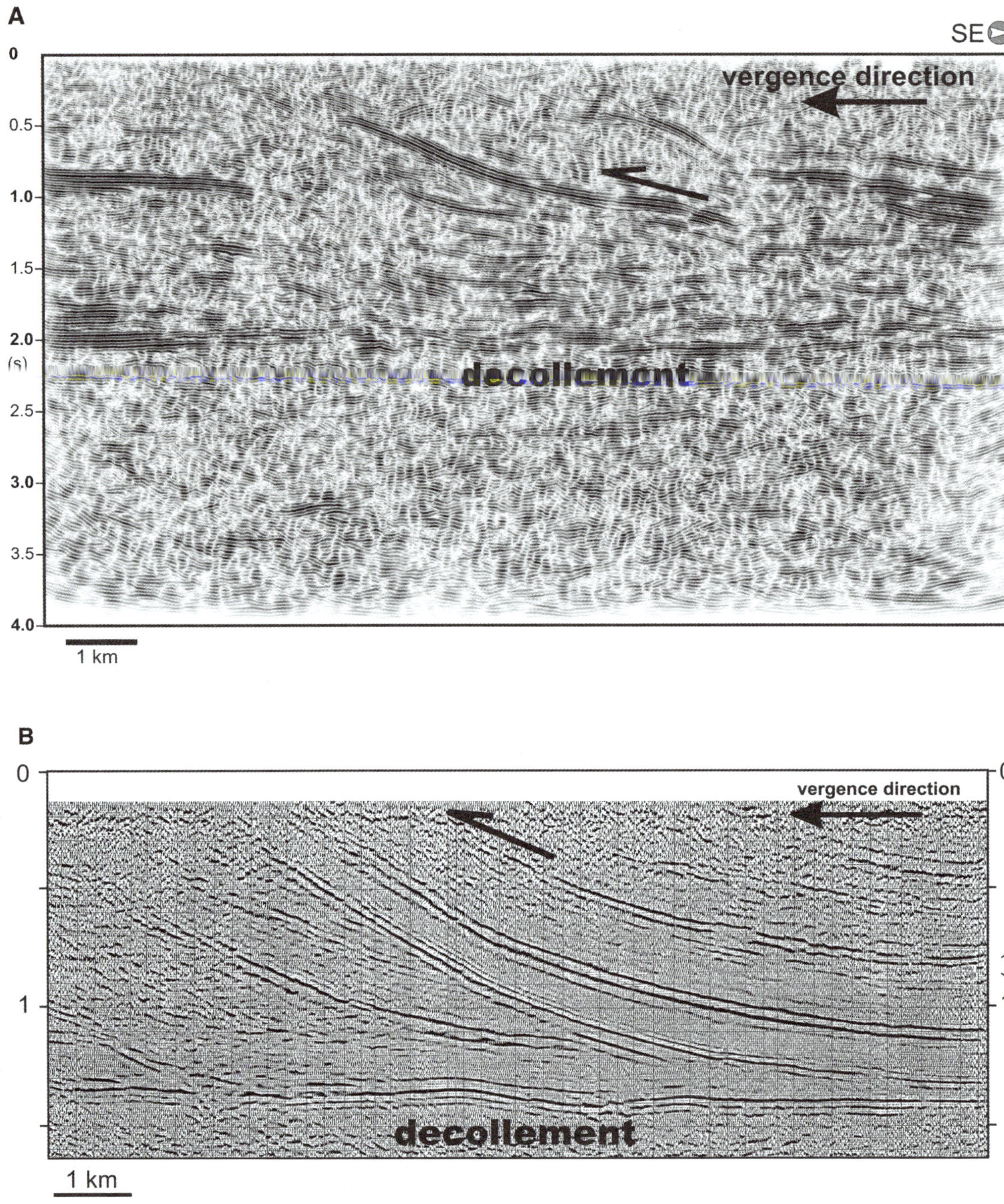

Figure 9. Portions of two regional industry seismic reflection profiles that provide geologic dip sections across the Valley and Ridge province of Tennessee (USA). (A) This profile was reprocessed for this study. Reprocessing included scanning and vectorizing paper copy followed by noise reduction and phase-shift constant-velocity (2500 m/s) migration. This profile demonstrates the relation between décollement reflectivity, planar thrust elements above it, and antiformal reflection patterns above the planar thrusts. See text for more explanation. (B) This migrated industry profile, located southwest of Oak Ridge, Tennessee, shows a high-quality example of the Whiteoak Mountain thrust and a subsidiary footwall thrust repeating the Rome detachment twice (strong reflectors at the base of the Rome Formation and at the top of the overlying Conasauga Group)—above the master décollement that propagated through the basal Rome Formation (strong reflectors) resting on undeformed basement.

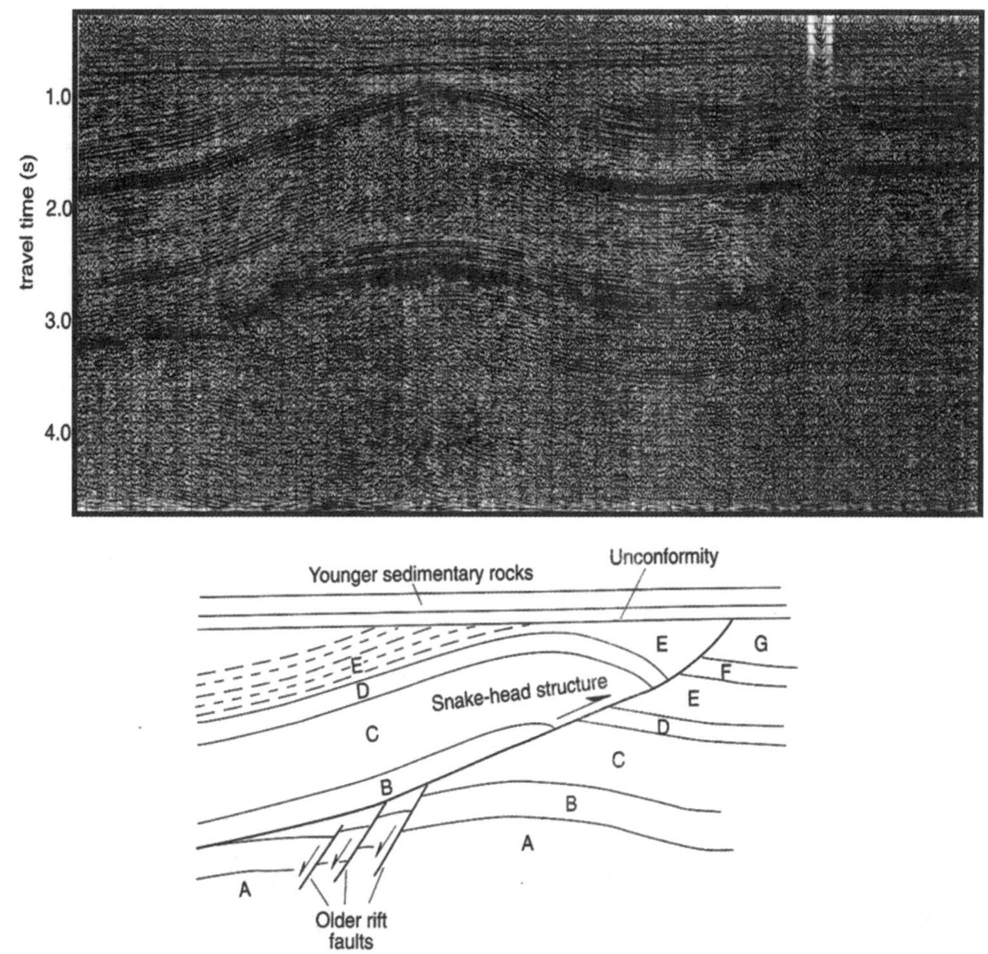

Figure 10. Industry seismic reflection profile of "snake-head" (hanging-wall anticline) structure from the Appalachians beneath the Alabama Coastal Plain. (From Hatcher, 1995, his Fig. 11-34.)

Structure Related to the Décollement and Scale Invariance

The glaciotectonic seismic sections clearly reveal listric surfaces that rise from the décollement and roll over, forming an imbricate stack of south-vergent asymmetric anticlines (passive roof duplex) as the detachment flattens into the higher décollement (Fig. 4). The enveloping surface of the antiforms is gently undulating and can be traced downdip to the basal décollement where the surfaces merge asymptotically (i.e., the dipping reflectors "sole into" the horizontal décollement), outlining near-ideal foreland fold-and-thrust belt geometry (Chapple, 1978). A clear lateral truncation of this pattern of antiformal and dipping reflections appears near the south end of the Wadsworth-North profile (Fig. 4A). Farther south, along the Wadsworth-South profile (Fig. 4C), the reflectivity displays a pattern of subhorizontal to undulating surfaces without formation of anticlines or an obvious décollement, as seen along the Wadsworth-North profile (Fig. 4A). The east-west Kelley profile (Fig. 4C) shows a typical pattern of horizontal, almost "railroad track" reflectivity as would be expected from a strike section. An analogous pattern of juxtaposed antiformal and horizontal reflectors is present in reflection profiles from the southern Appalachian foreland fold-and-thrust belt, e.g., in industry seismic profiles from the Tennessee Valley and Ridge province (Fig. 9), and the Coastal Plain in Alabama (Fig. 10). Interpretation of these Appalachian profiles as thrusts beheading asymmetric anticlines is consistent with thrust development in the Illinois glaciotectonic analog profiles. The general scale invariance is emphasized further by outcrop observations of thrust faults (Fig. 6) that can be interpreted as representing the dipping planar elements appearing beneath anticlinal reflections on seismic profiles in both orogenic and glacial till settings, whether in mechanically strong or weak materials.

Subduction-related accretionary complexes, e.g., those in the Nankai Trough (Shipboard Scientific Party, 2001) or over the Barbados Ridge (e.g., Behrmann et al., 1988; Wallace et al., 2003), also produce thrust systems with similar geometries in semi- to unconsolidated materials (Fig. 11). In these cases, the presence of a weak interval, which may be overpressured, provides an opportunity for propagation of a décollement (Wallace et al., 2003). Peacock (2003) found that the high porosity in the décollement zone beneath the Barbados accretionary complex is sustained by overpressured shale. Fault and décollement geometries in the glaciotectonic example appear similar to sections from water-saturated accretionary prisms, which show plastic or viscous behavior in the deforming wedges. Such deformation appears non-Coulomb, but similar materials in Taiwan are the basis for development of critical-wedge theory (Davis et al., 1983; Dahlen, 1990) and its subsequent application to foreland fold-and-thrust belts. The glaciotectonic fold belt imaged in the Wadsworth, Illinois, seismic section (Figs. 4 and 5), exhibits near-ideal Coulomb behavior (Fig. 8).

During the past two decades, geologists have attempted to explain the appearance of divergent (i.e., dipping in opposite directions) dipping reflector patterns in contractional orogenic belts such as the Appalachians of the United States and Canada as related to an underlying subduction zone that creates a singular point between two mechanically decoupled layers (Phinney, 1986; Beaumont et al., 1994). The diverging pattern of dipping reflections on our glaciotectonic example (Fig. 4) can be explained in a manner not requiring a decoupling mechanism, but by a local asperity in the form of a tunnel valley or other erosional feature.

The scale invariance of thrust structures rising from a décollement is aptly demonstrated further by deep seismic reflection profiles, drill cores and hand specimens, and outcrop data (Figs. 1 and 12C). For example, deep seismic reflection profiles over the Shetland platform north of mainland Scotland provide images of large-scale (tens of kilometers) asymmetrical anticlines beneath the buried Caledonian (Silurian) fold belt that are undercut by downward-soling thrusts (Fig. 12A) (McBride and England, 1999). In this case the interpreted décollement level is within the lower crust and is manifested as a reflective lower crustal layer in profiles orthogonal (see Fig. 9 in McBride and England, 1999) to that shown in Figure 12A. Farther southwest in the West Orkney basin north of mainland Scotland, the DRUM deep seismic reflection line revealed a possible crustal ramp and duplex beneath the shallower Caledonian frontal (Moine) thrust zone, with the lower detachment along the Moho (Butler and Coward, 1984; Snyder 1990) (Fig. 12B).

Hanging-wall cutoffs in the glaciotectonic thrust system occur at the tips of the thrust faults that beheaded several anticlines. Analogous examples of crustal-scale antiform-décollement patterns have been described from compressional orogens (Blundell, 1993; Arnow, 1986; McBride and Nelson, 1991). At the other extreme, anticlines with downward-soling thrusts in their cores have been described at scales of a few centimeters from outcrop and drill cores. For example, Talbot and Von Brunn (1987) presented a restored section for a hand sample interpreted from a micromélange formed by hydraulic overpressures from an advancing glacier that remobilized mudstone layers from the Carboniferous Dwyka Formation, northern Natal, South Africa. The micromélange is extruded along a décollement. Soft-sediment deformation can in general develop décollement structures at the scale of centimeters that are indistinguishable from orogen-scale structures (Figs. 1A and 12C).

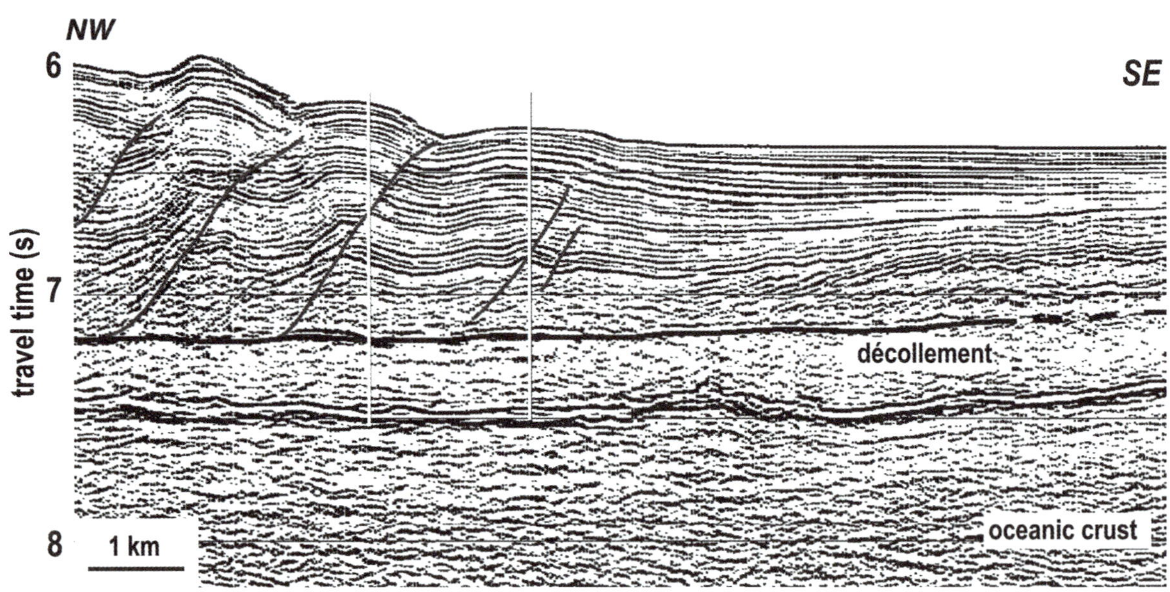

Figure 11. Seismic reflection profile through the Muroto Transect reference from Site 1173 and prism toe sites (Sites 1174 and 808) in the area of the Ocean Drilling Program Leg 190. From Shipboard Scientific Party (2001).

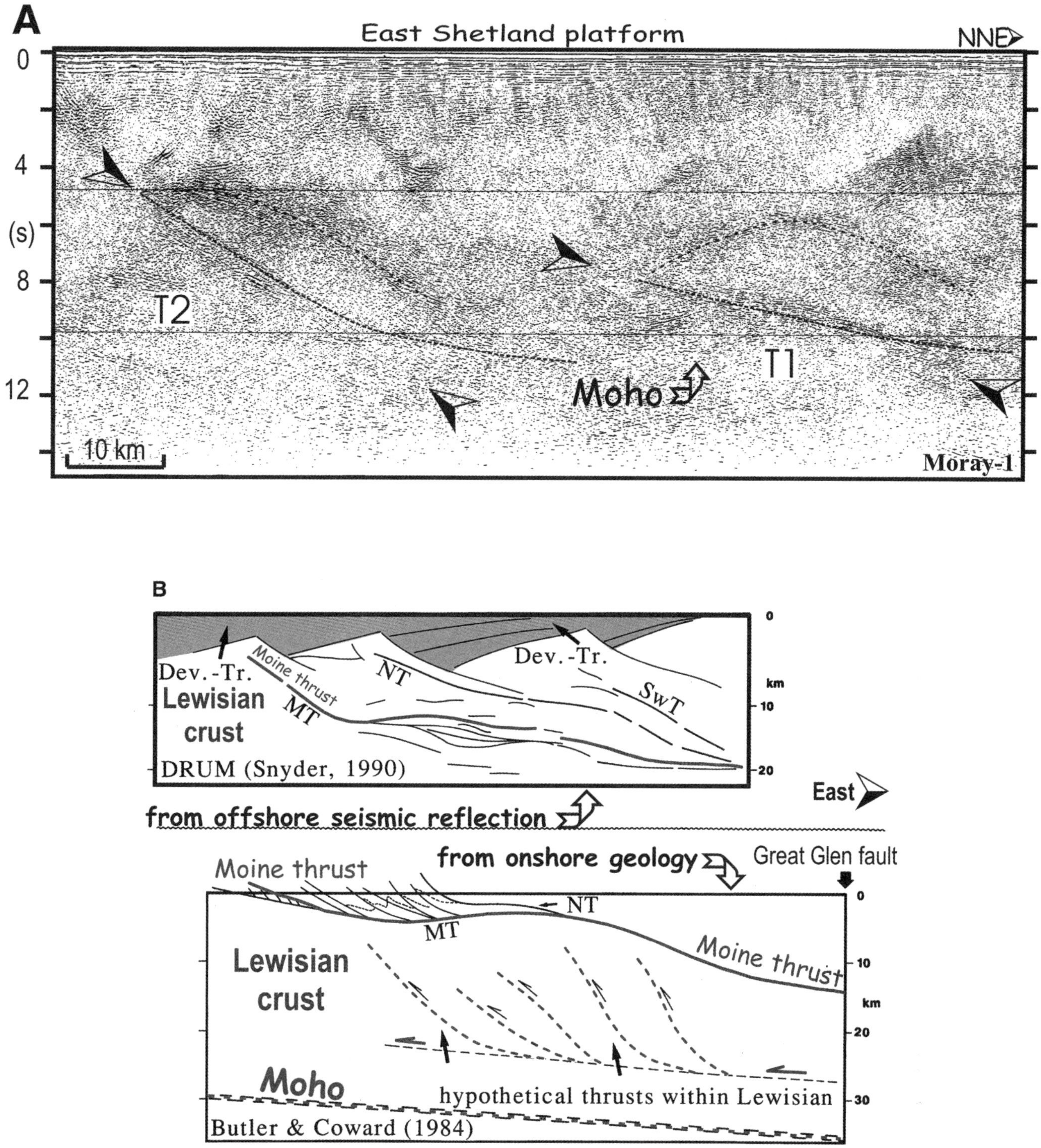

Figure 12 *(on this and following page)*. (A) Excerpt of the British Institutions Reflection Profiling Syndicate (BIRPS) Moray-1 offshore deep seismic reflection profile over the East Shetland platform, northeast of mainland Scotland and southeast of Shetland (McBride and England, 1999). The arrows point to northeast-dipping (in the plane of the section) reflector zones T1 and T2. More detail and information on the seismic sections are available in McBride and England (1999). (B) Two interpretations for the Moine nappes. Top, interpretation of the BIRPS DRUM line in the West Orkney basin, offshore north of the Scottish mainland. Bottom, geological interpretation for Northwest Highlands of Scotland. Dev.-Tr.—Devonian-Triassic; MT—Moine thrust; NT—Naver thrust; SwT—Swordly thrust. Modified from McBride and England (1994).

C

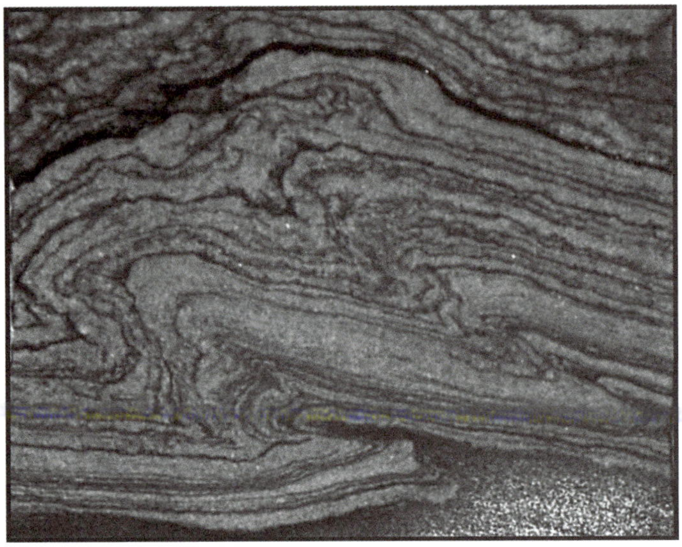

Figure 12 *(continued)*. (C) Close-up view of a 3-inch-diameter cut core from the Ferron Sandstone, Late Cretaceous, south-central Utah, showing small-scale thrusted anticline and décollement. Core provided by the Utah Geological Survey; photo by John H. McBride.

CONCLUSIONS

1. We present the first example, known to us, of high-resolution seismic reflection imaging of a glaciotectonic fold-and-thrust belt associated with Wisconsinan glaciation in the United States. Our interpretation is also delimited by geologic mapping, exposure in a nearby quarry, and a dedicated borehole along the key profile.

2. Complexities in the décollement and related fault structures are explained by irregularities in the underlying bedrock or facies changes or variations in water saturation in the glacial tills.

3. The key profile reproduces many of the structural features expected for a décollement system in a compressional orogen, including ramp-and-flat thrusting, imbricated asymmetric anticlines cored by thrust faults, vergence reversal, and downward soling of listric fault surfaces into the décollement.

4. Integration of observations from a glaciotectonic regime and hand specimen/core thrust systems with examples from major fold-and-thrust belts suggests the processes that produce décollement systems are scale independent over several orders of magnitude.

5. Décollement-generating processes operate independently of material properties, in the sense that décollements can form over a range of physical and geologic conditions. Contrasting material properties do, however, yield different thrust geometries and possibly kinematics. Observations are supported by examples of crustal décollements in crystalline rocks as well as décollements in water-saturated unconsolidated sediments.

6. Glaciotectonic décollement systems provide examples of a complex mechanical process, and additional opportunities should arise to investigate these systems to model deformation processes. Future study could focus on locating and resolving the three-dimensional structure of glaciotectonic décollement systems in areas such as the Midwest United States, where data acquisition would be relatively easy.

ACKNOWLEDGMENTS

We acknowledge the support of this research by Landmark Graphics via the Landmark University Grant Program at Brigham Young University. The authors also express appreciation to Seismic Micro-technology\ (Kingdom Suite™), who kindly provided a university grant of their visualization and mapping software to Brigham Young University and to the Illinois State Geological Survey. The authors gratefully acknowledge reviews by Peter Geiser and Don Wise, which greatly improved the final version of the paper. Recent research in the Appalachian foreland fold-and-thrust belt by R.D.H. has been supported by U.S. Department of Energy (National Energy Technology Laboratory) contract DE-FC26-02NT15341, as well as the University of Tennessee Science Alliance Center of Excellence, ConocoPhillips, Coal Creek Mining and Manufacturing Company, and Tengasco.

REFERENCES CITED

Allmendinger, R., Nelson, K.D., Potter, C., Barazangi, M., Brown, L., and Oliver, J., 1987, Deep seismic characteristics of the continental crust: Geology, v. 15, p. 304–310, doi: 10.1130/0091-7613(1987)15<304:DSRCOT>2.0.CO;2.

Arnow, J.A., 1986, Deep structure from continental collision: COCORP deep seismic reflection profiles across the southern Appalachians [M.S. thesis]: Ithaca, New York, Cornell University, 82 p.

Bally, A.W., 1983, Seismic expression of structural styles: Tectonics of compressional provinces/strike-slip tectonics: American Association of Petroleum Geologists Studies in Geology 15, v. 3.

Bally, A.W., Gordy, P., and Stewart, G., 1966, Structure, seismic data, and orogenic evolution of the southern Canadian Rockies: Bulletin of Canadian Petroleum Geology, v. 14, p. 337–381.

Beaumont, C., Fullsack, P., and Hamilton, J., 1994, Styles of crustal deformation in compressional orogens caused by subduction of the underlying lithosphere: Tectonophysics, v. 232, p. 119–132, doi: 10.1016/0040-1951(94)90079-5.

Behrmann, J.H., Brown, K., Moore, J.C., Mascle, A., Taylor, E., Alvarez, F., Andreieff, P., Barnes, R., Beck, C., Blanc, G., Clark, M., Dolan, J., Fisher, A., Gieskes, J., Hounslow, M., McLellan, P., Moran, K., Ogawa, Y., Sakai, T., Schoonmaker, J., Vroluk, P., Wilkens, R., and Williams, C., 1988, Evolution of structures in the Barbados accretionary prism: Insights from Leg 110 of the Ocean Drilling Program: Journal of Structural Geology, v. 10, p. 577–591, doi: 10.1016/0191-8141(88)90025-9.

Blundell, D.J., 1993, Deep structure of the Anglo-Brabant massif revealed by seismic profiling: Geological Magazine, v. 130, p. 563–567.

Butler, R.W.H., and Coward, M.P., 1984, Geological constraints, structural evolution, and deep geology of the northwest Scottish Caledonides: Tectonics, v. 3, p. 347–365.

Chapple, W.M., 1978, Mechanics of thin-skinned fold-and-thrust belts: Geological Society of America Bulletin, v. 89, p. 1189–1198, doi: 10.1130/0016-7606(1978)89<1189:MOTFB>2.0.CO;2.

Clark, H.B., Costain, J.K., and Glover, L., III, 1978, Structural and seismic reflection studies of the Brevard ductile deformation zone near Rosman, North Carolina: American Journal of Science, v. 278, p. 419–441.

Clayton, L., Attig, J.W., and Mickelson, D.M., 1999, Tunnel channels formed in Wisconsin during the last glaciation, *in* Mickelson, D.M., and Attig, J.W., eds., Glacial processes past and present: Geological Society of America Special Paper 337, p. 69–82.

Cook, F.A., and Varsek, J.L., 1994, Orogen-scale décollements: Reviews of Geophysics, v. 32, p. 37–60, doi: 10.1029/93RG02515.

Cook, F.A., Albaugh, D.S., Brown, L.D., Oliver, J.E., Kaufman, S., and Hatcher, R.D., Jr., 1979, Thin-skinned tectonics in the crystalline southern Appalachians: COCORP seismic-reflection profiling in the Blue Ridge and Piedmont: Geology, v. 7, p. 563–567, doi: 10.1130/0091-7613(1979)7<563:TTITCS>2.0.CO;2.

Cook, F.A., Brown, L.D., Kaufman, S., Oliver, J.E., and Petersen, T.A., 1981, COCORP seismic profiling of the Appalachian orogen beneath the Coastal Plain of Georgia: Geological Society of America Bulletin, v. 92, p. 738–748, doi: 10.1130/0016-7606(1981)92<738:CSPOTA>2.0.CO;2.

Cook, F.A., Brown, L.D., Kaufman, S., and Oliver, J.E., 1983, COCORP seismic reflection profiling of the southern Appalachians: American Association of Petroleum Geologists Studies in Geology, no. 14, 61 p.

Corredor, F., Shaw, J.H., and Bilotti, F., 2005, Structural styles in the deep-water fold-and-thrust belts of the Niger Delta: American Association of Petroleum Geologists Bulletin, v. 89, p. 753–780, doi: 10.1306/02170504074.

Costain, J.K., Hatcher, R.D., Jr., Çoruh, C., Pratt, T.L., Taylor, S.R., Litehiser, J.J., and Zietz, I., 1989, Geophysical characteristics of the Appalachian crust, *in* Hatcher, R.D., Jr., et al., eds., The Appalachian-Ouachita orogen in the United States: Geological Society of America, Geology of North America, v. F-2, p. 385–416.

Curry, B.B., Grimley, D.A., and Stravers, J.A., 1999, Quaternary geology, geomorphology, and climatic history of Kane County, Illinois: Illinois State Geological Survey, Guidebook 28, 44 p.

Curry, B. B., 2005, Crystal Lake Quadrangle, McHenry and Kane Counties, Illinois: Illinois State Geological Survey, Geologic Quadrangle Map IGQ Crystal Lake–SG, scale 1:24,000, 1 sheet.

Dahlen, F.A., 1990, Critical taper model of fold-and-thrust belts and accretionary wedges: Annual Review of Earth and Planetary Sciences, v. 18, p. 55–99.

Davis, D., Suppe, J., and Dahlen, F.A., 1983, Mechanics of fold-and-thrust belts and accretionary wedges: Journal of Geophysical Research, v. 88, p. 1153–1172.

Grimley, D.A., and Curry, B.B., 2002, Surficial geology map of the Geneva 7.5-minute quadrangle, Kane County, Illinois: Illinois State Geological Survey, Geological Quadrangle Map IGQ Geneva-SG, scale 1:24,000, 1 sheet.

Hansel, A.K., and Johnson, W.H., 1996, Wedron and Mason Groups: Lithostratigraphic reclassification of deposits of the Wisconsin Episode, Lake Michigan lobe: Illinois State Geological Survey, Bulletin 104, 116 p.

Harris, C., Williams, G., Brabham, P., Eaton, G., and McCarroll, D., 1997, Glaciotectonized Quaternary sediments at Dinas Dinlle, Gwynedd, North Wales, and their bearing on the style of deglaciation in the eastern Irish Sea: Quaternary Science Reviews, v. 16, p. 109–127, doi: 10.1016/S0277-3791(96)00050-9.

Harris, L.D., and Bayer, K.C., 1979, Sequential development of the Appalachian orogen above a master décollement: A hypothesis: Geology, v. 7, p. 568–572.

Hatcher, R.D., Jr., 1989, Tectonic synthesis of the U.S. Appalachians, *in* Hatcher, R.D., Jr., et al., eds., The Appalachian-Ouachita orogen in the United States: Geological Society of America, Geology of North America, v. F-2, p. 511–535.

Hatcher, R.D., Jr., 1995, Structural geology: Principles, concepts, and problems (2nd edition): New York, Macmillan Publishing Company, 525 p.

Hatcher, R.D., Jr., 2004, Properties of thrusts and the upper bounds for the size of thrust sheets, *in* McClay, K.R., ed., Thrust tectonics and hydrocarbon systems: American Association of Petroleum Geologists Memoir 82, p. 18–29.

Hatcher, R.D., Jr., Thomas, W.A., Geiser, P.A., Snoke, A.W., Mosher, S., and Wiltschko, D.V., 1989, Alleghanian orogen, *in* Hatcher, R.D., Jr., et al., eds., The Appalachian-Ouachita orogen in the United States: Geological Society of America, Geology of North America, v. F-2, p. 233–318.

Hopkins, D.L., 1995, The New York-Alabama magnetic lineament: Its reflection character and relationship to the Grenville Front [Ph.D. thesis]: Blacksburg, Virginia, Virginia Polytechnic Institute and State University, 176 p.

Hurich, C.A., Smithson, S.B., Fountain, D.M., and Humphreys, M.C., 1985, Seismic evidence of mylonite reflectivity and deep structure in the Kettle dome metamorphic core complex, Washington: Geology, v. 13, p. 577–580, doi: 10.1130/0091-7613(1985)13<577:SEOMRA>2.0.CO;2.

Iverson, W.P., and Smithson, S.B., 1982, Master décollement root zone beneath the southern Appalachians and crustal balance: Geology, v. 10, p. 241–245, doi: 10.1130/0091-7613(1982)10<241:MDRZBT>2.0.CO;2.

Jadoon, I.A.K., Lawrence, R.D., and Lillie, R.J., 1992, Balanced and retrodeformed geological cross section from the frontal Sulaiman lobe: Duplex development in thick strata along the western margin of the Indian plate, *in* McClay, K.R., ed., Thrust tectonics: London, Chapman and Hall, p. 343–356.

Jones, T.D., and Nur, A., 1984, The nature of seismic reflections from deep crustal fault zones: Journal of Geophysical Research, B, Solid Earth and Planets, v. 89, p. 3153–3171.

Lampshire, L.D., Çoruh, C., and Costain, J.K., 1994, Crustal structures and the eastern extent of lower Paleozoic shelf strata within the central Appalachians: A seismic reflection interpretation: Geological Society of America Bulletin, v. 106, p. 1–18, doi: 10.1130/0016-7606(1994)106<0001:CSATEE>2.3.CO;2.

Law, A., and Snyder, D.B., 1997, Reflections from a mylonitized zone in central Sweden: Journal of Geophysical Research, B, Solid Earth and Planets, v. 102, p. 8411–8425, doi: 10.1029/96JB03269.

Lillie, R.J., 1984, Tectonic implications of subthrust structures revealed by seismic profiling of Appalachian/Ouachita orogenic belt: Tectonics, v. 3, p. 619–646.

McBride, J.H., and England, R.W., 1994, Deep seismic reflection structure of the Caledonian orogenic front west of Shetland: Geological Society [London] Journal, v. 151, p. 9–16.

McBride, J.H., and England, R.W., 1999, Window into the Caledonian orogen: Structure of the crust beneath the East Shetland platform, United Kingdom: Geological Society of America Bulletin, v. 111, p. 1030–1041, doi: 10.1130/0016-7606(1999)111<1030:WITCOS>2.3.CO;2.

McBride, J.H., and Nelson, K.D., 1991, Deep seismic reflection constraints on Palaeozoic crustal structure and definition of the Moho in the buried southern Appalachian orogen, *in* Meissner, R., et al., eds., Continental lithosphere: Deep seismic reflections: American Geophysical Union Geodynamics Series, v. 22, p. 9–20.

McBride, J.H., Hatcher, R.D., Jr., Stephenson, W.J., and Hooper, R.J., 2005, Integrating seismic reflection and geological data and interpretations across an internal basement massif: The southern Appalachian Pine Mountain window, USA: Geological Society of America Bulletin, v. 117, p. 669–686, doi: 10.1130/B25313.1.

McCarroll, D., and Rijsdijk, K.F., 2003, Deformation styles as a key for interpreting glacial depositional environments: Journal of Quaternary Science, v. 18, p. 473–489, doi: 10.1002/jqs.780.

Peacock, S., 2003, Thin layer tuned reflections from an anisotropic décollement zone: Journal of Geophysical Research, v. 108, no. B11, p. 2518, doi: 10.1029/2001JB000195.

Phinney, R.A., 1986, A seismic cross section of the New England Appalachians: The orogen exposed, *in* Barazangi, M., and Brown, L.D., eds., Reflection seismology: The continental crust: American Geophysical Union Geodynamics Series, v. 14, p. 157–172.

Price, R.A., and Mountjoy, E.W., 1970, Geologic structure of the Canadian Rocky Mountains between Bow and Athabaska Rivers—A progress report: Geological Association of Canada Special Paper 6, p. 7–25.

Pugin, A., Pullan, S.E., and Sharpe, D.R., 1999, Seismic facies and regional architecture of the Oak Ridges Moraine area, southern Ontario: Canadian Journal of Earth Sciences, v. 36, p. 409–432, doi: 10.1139/cjes-36-3-409.

Pugin, A.J.M., Larson, T.H., Sargent, S.L., McBride, J.H., and Bexfield, C.E., 2004, Near-surface mapping using SH-wave and P-wave seismic landstreamer data acquisition in Illinois, U.S.: Leading Edge, v. 23, p. 677–682, doi: 10.1190/1.1776740.

Ramsay, J.G., 1981, Tectonics of the Helvetic Alps, *in* McClay, K.R., and Price, N.J., eds., Thrust and nappe tectonics: Geological Society [London] Special Publication 9, p. 293–309.

Rodgers, J., 1949, Evolution of thought on structure of middle and southern Appalachians: American Association of Petroleum Geologists Bulletin, v. 33, p. 1643–1654.

Shipboard Scientific Party, 2001, Leg 190 summary, *in* Moore, G.F., et al., eds., Proceedings of the Ocean Drilling Program, Initial Reports, Volume 190: College Station, Texas, Ocean Drilling Program, p. 1–87.

Snyder, D.B., 1990, The Moine thrust in the BIRPS data set: Geological Society [London] Journal, v. 147, p. 81–86.

Talbot, C.J., and Von Brunn, V., 1987, Intrusive and extrusive (micro) mélange couplets as distal effects of tidal pumping by a marine ice sheet: Geological Magazine, v. 124, p. 513–525.

Thompson, R.I., 1981, The nature and significance of large "blind" thrusts within the northern Rocky Mountains of Canada, *in* McClay, K.R., and Price, N.J., eds., Thrust and nappe tectonics: Geological Society [London] Special Publication 9, p. 449–462.

Tulaczyk, S., 1999, Ice sliding over weak, fine-grained tills: Dependence of ice-till interactions on till granulometry, *in* Mickelson, D.M., and Attig,

J.W., eds., Glacial processes past and present: Geological Society of America Special Paper 337, p. 159–177.

Wallace, G., Moore, J.C., and DiLeonardo, C.G., 2003, Controls on localization and densification of a modern décollement: Northern Barbados accretionary prism: Geological Society of America Bulletin, v. 115, p. 288–297, doi: 10.1130/0016-7606(2003)115<0288:COLADO>2.0.CO;2.

Willman, H.B., and Frye, J.C., 1970, Pleistocene stratigraphy of Illinois: Illinois State Geological Survey, Bulletin 94, 204 p.

Manuscript Accepted by the Society 1 February 2007

Printed in the USA

The Geological Society of America
Memoir 200
2007

The Southern Oklahoma and Dniepr-Donets aulacogens: A comparative analysis

G. Randy Keller*
School of Geology and Geophysics, University of Oklahoma, Norman, Oklahoma 73071, USA

Randell A. Stephenson*
Faculty of Earth and Life Sciences, Vrije Universiteit, De Boelelaan 1085, 1081 HV, Amsterdam, Netherlands

ABSTRACT

The classic failed continental rift or aulacogen is one that intersects a rifted continental margin at a high angle. Based on recent geological and geophysical studies, we have revisited a classic analogy that was drawn between two major intracratonic rifts, the Southern Oklahoma aulacogen in the southern portion of Laurentia and the Dniepr-Donets aulacogen in the southern portion of Baltica. The Southern Oklahoma aulacogen, also known as the Wichita aulacogen, consists of a linear alignment of extensively inverted rift structures that begins at the rifted margin of Laurentia in northeast Texas and extends northwestward at least to southern Colorado. Deep seismic profiles have revealed the upper crustal structure of this feature, and gravity data provide a regional context for interpreting these results. Velocities low enough to indicate the presence of sedimentary rocks extend to a depth of ~15 km, and the deepest of these sedimentary layers has been interpreted as rift fill. In addition, the main inversion structure of the Southern Oklahoma aulacogen (Wichita uplift) is underlain by very high-velocity and dense mafic material even at upper crustal depths of ~5 km. The Dniepr-Donets aulacogen has been cited as a type example of an aulacogen and is clearly a "failed rift" in the sense that it did not itself lead to continental breakup and ocean crust formation. The main feature of the aulacogen is a Late Devonian rift basin overlain by a substantial (but variable) postrift sedimentary sequence that records several extensional or transtensional events and at least one moderate compressional reactivation. Recent deep seismic reflection and refraction surveys resolve the geometry of the sedimentary succession in the Donets segment of the basin, indicating an asymmetric form with a steeper basement surface in the south than in the north and a total sedimentary thickness of ~20 km. A thick (>10 km) high-velocity (>6.9 km/s) lower crustal body lies beneath the rift basin itself and is offset slightly to the north compared to the main basin depocenter. The Moho displays only slight topography around a depth of 40 km although, based on older deep seismic data, it

*Keller: grkeller@ou.edu; Stephenson: randell.stephenson@falw.vu.nl.

Keller, G.R., and Stephenson, R.A., 2007, The Southern Oklahoma and Dniepr-Donets aulacogens: A comparative analysis, *in* Hatcher, R.D., Jr., Carlson, M.P., McBride, J.H., and Martínez Catalán, J.R., eds., 4-D Framework of Continental Crust: Geological Society of America Memoir 200, p. 127–143, doi: 10.1130/2007.1200(08). For permission to copy, contact editing@geosociety.org.

shallows somewhat under the rift axis in the Dniepr segment to the northwest. Thus, major differences between these two major rifts are the nature of the magmatic modification of the crust and degree of inversion. Both the age of initial rifting and subsequent inversion in the Dniepr-Donets aulacogen are redefined compared to what was thought at the time the original analogy was made with the Southern Oklahoma aulacogen.

Keywords: rifting, continental tectonics, basin structure, geophysics, tectonic inversion.

INTRODUCTION

Rifting is one of the fundamental processes that shape the continents and ocean basins. In terms of continental tectonics, rifts that do not succeed in breaking a continent apart are sometimes referred to as having "failed." The classic failed rift or aulacogen is one that intersects a rifted continental margin at a high angle as part of a triple junction (e.g., Hoffman et al., 1974). These failed rifts usually are the sites of postrift sedimentation that contain important records of continental evolution and prolific petroleum resources. Studies over the past 15 years have shown that the modification of the lithosphere (most evidence is actually for the crust) in failed rifts takes on many forms, is highly variable, and is often very substantial. Although semantic arguments are seldom productive, these results call into question any perception that these intracratonic rifts are not major lithospheric-scale structures. In this paper, we will revisit a classic analogy that was drawn between two major intracratonic rifts, the Southern Oklahoma aulacogen in the southern portion of Laurentia and the Dniepr-Donets aulacogen in the southern portion of Baltica (Shatsky, 1946; Hoffman et al., 1974) (Fig. 1), reviewing a variety of modern geological and geophysical data. Although the term "aulacogen" was originally defined to be a failed rift that experienced subsequent compressional tectonism and structural inversion, it has become a term that is applied to any failed rift. Since failed rifts represent fundamental structural flaws in the lithosphere, we will adopt this usage of the term. Because virtually every failed rift has experienced some reactivation, this distinction has little meaning anyway.

SOUTHERN OKLAHOMA AULACOGEN

The Southern Oklahoma aulacogen, also known as the Wichita aulacogen (Walper, 1977), consists of a linear alignment of extensively inverted rift structures that begins at the rifted margin of Laurentia in northeast Texas (Figs. 1A and 2A). This rifted margin formed as Neoproterozoic supercontinents (Rodinia, Pannotia; e.g., Moores, 1991; Dalziel, 1991, 1997; Hoffman, 1991), broke up in the Early Cambrian, and forms the structural framework for the late Paleozoic Ouachita orogenic belt (e.g., Keller et al., 1989; Thomas, 1991). Gravity anomalies in the Southern Oklahoma aulacogen region (Fig. 3) clearly show that, as expected, the rift structures do not extend southeast of this margin and allow us to follow the margin from the Appalachian orogen across the southern United States to west Texas and northern Mexico (Kruger and Keller, 1986; Keller et al., 1989). Another important consideration is that this northwest-trending feature cuts across the northeast-trending structural grain due to the Mesoproterozoic formation of the crust (e.g., Karlstrom and Bowring, 1988).

The main structures commonly associated with the Southern Oklahoma aulacogen are the Amarillo-Wichita uplift and the Anadarko basin (Fig. 2A). Because of extensive petroleum exploration, the postrift history of this portion of the aulacogen is well known and summarized by Ham et al. (1964), Ham and Wilson (1967), and Perry (1989). However, many other significant structures are involved, and the rifting has been interpreted to extend as far to the northwest as the Paradox basin and Uncompahgre uplift in Utah (Larson et al., 1985) (Fig. 1A). The general sequence of events is (1) rifting and magmatism in the Early to Middle Cambrian, (2) middle Paleozoic postrift subsidence, (3) late Paleozoic deformation, and (4) Mesozoic and Cenozoic stability (e.g., Gilbert, 1983).

Besides the postrift Anadarko basin, the extensive exposures of Cambrian bimodal igneous rocks exposed on the Wichita uplift constitute the major direct geologic evidence for rifting (e.g., Hanson and Al-Shaieb, 1980; Powell et al., 1980; Gilbert, 1983; Lambert et al., 1988; McConnell and Gilbert, 1990). These rocks include a layered mafic intrusion (the Glen Mountains layered

Figure 1. Regional tectonic index maps for the Southern Oklahoma and Dniepr-Donets aulacogens. (A) Southern Oklahoma aulacogen region. EGR—eastern granite-rhyolite province; SGR—southern granite-rhyolite province; CB—Cheyenne belt; MCR—Midcontinent rift system; PMIC—Pecos mafic intrusive complex, which underlies the Central Basin Platform; AGM—Abilene gravity minimum; FM—Franklin Mountains; TU—Tusas uplift; WM—Wet Mountains; UU—Uncompahgre uplift; CU—Cimarron uplift; SGU—Sierra Grande uplift; VH—Van Horn area; LU—Llano uplift. The approximate location of the Ouachita thrust front and the early Paleozoic continental margin are denoted by the orange and green dashed lines respectively (modified from Barnes et al., 1999). Dashed black lines indicate province boundaries. (B) Dniepr-Donets aulacogen region. Map of major tectonic features in Europe. Modified from a map distributed by the EUROPROBE project. TESZ—Trans-European suture zone, which separates Precambrian lithosphere (east; shield areas in pink) from lithosphere accreted to western Europe during Phanerozoic orogenesis.

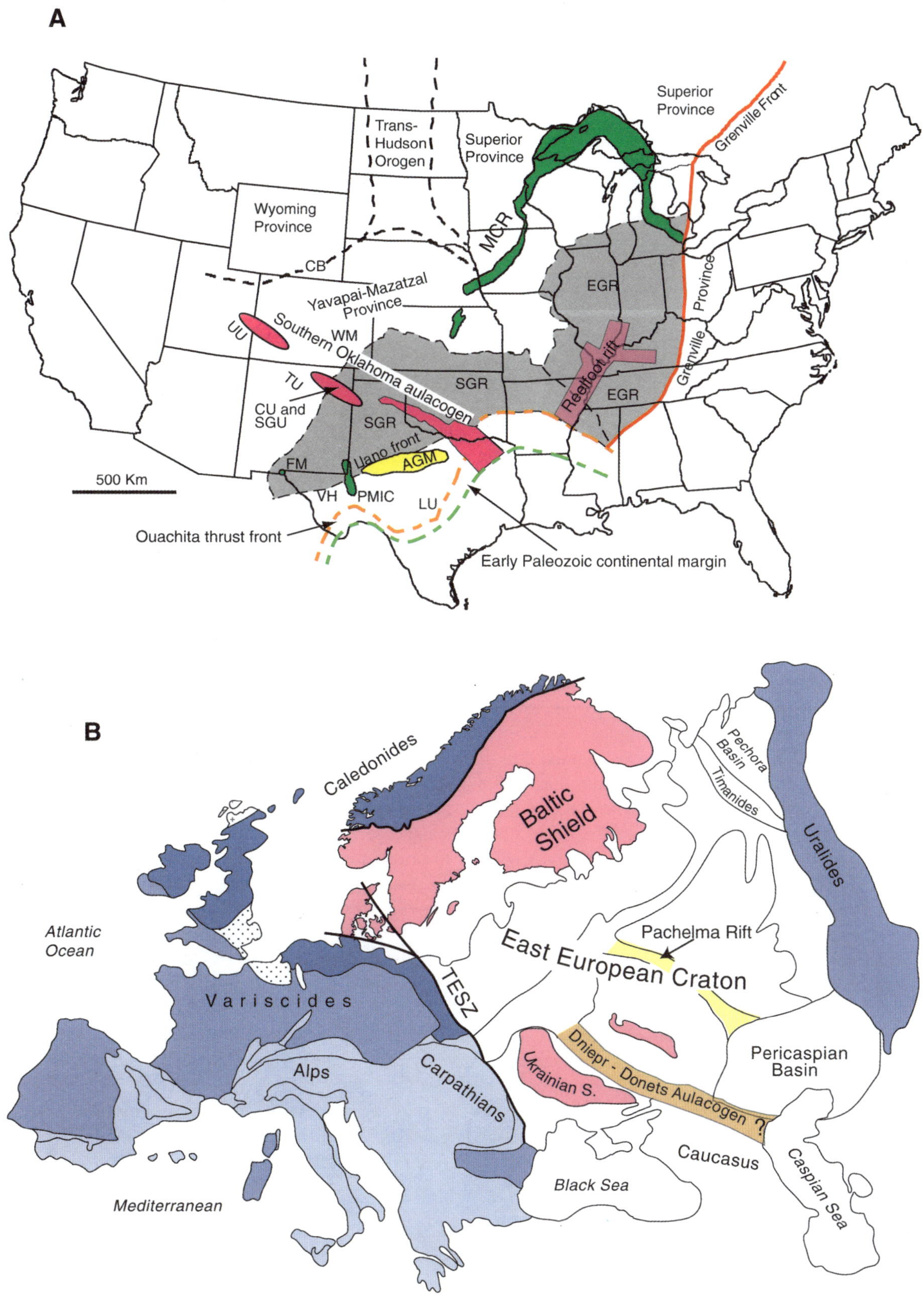
A
Trans-
Hudson
Orogen
Superior
Province
Superior
Province
Grenville Front
Wyoming
Province
MCR
CB
Yavapai-Mazatzal
Province
EGR
Province
UU
Southern Oklahoma aulacogen
WM
Reelfoot rift
Grenville
TU
SGR
EGR
CU and
SGU
SGR
Llano front
AGM
FM
500 Km
VH
PMIC
LU
Ouachita thrust front
Early Paleozoic continental margin
B
Caledonides
Baltic
Shield
Pechora
Basin
Timanides
Uralides
Atlantic
Ocean
Pachelma Rift
East European Craton
TESZ
Variscides
Pericaspian
Basin
Alps
Carpathians
Ukrainian S.
Dniepr - Donets Aulacogen
?
Caucasus
Caspian Sea
Black Sea
Mediterranean

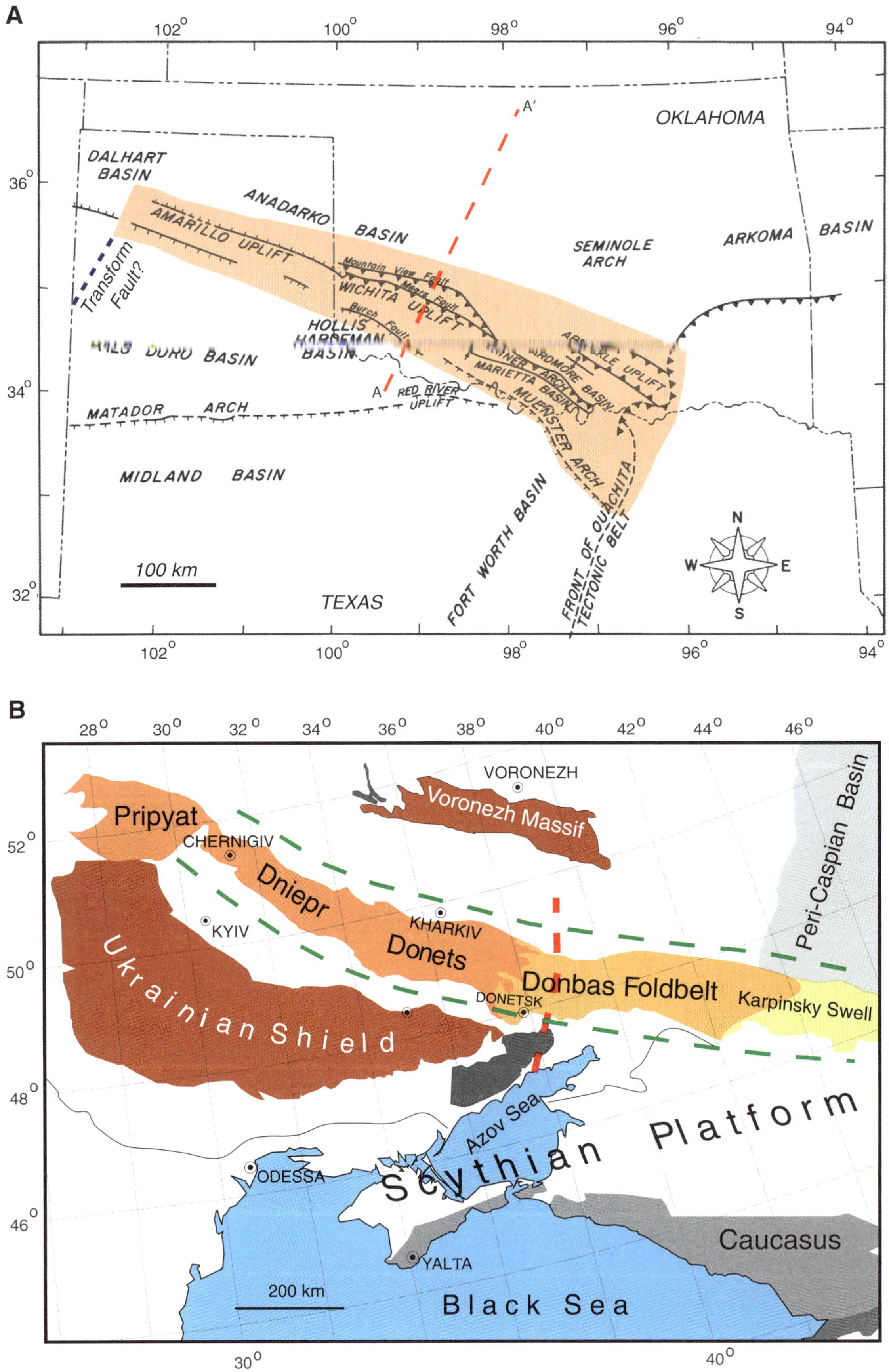
A
OKLAHOMA
DALHART BASIN
ANADARKO BASIN
AMARILLO UPLIFT
Transform Fault?
Mountain View Fault
Meers Fault
WICHITA UPLIFT
Burch Fault
HOLLIS HARDEMAN BASIN
PALO DURO BASIN
SEMINOLE ARCH
ARKOMA BASIN
ARBUCKLE UPLIFT
ARDMORE BASIN
WICHITA-MARIETTA BASIN
MARIETTA BASIN
MUENSTER ARCH
RED RIVER UPLIFT
MATADOR ARCH
MIDLAND BASIN
FORT WORTH BASIN
FRONT OF OUACHITA TECTONIC BELT
TEXAS
100 km
A
A'
N
S
E
W
B
VORONEZH
Voronezh Massif
Pripyat
CHERNIGIV
Dniepr
KYIV
KHARKIV
Donets
Ukrainian Shield
Donbas Foldbelt
DONETSK
Karpinsky Swell
Peri-Caspian Basin
Azov Sea
Scythian Platform
ODESSA
YALTA
Caucasus
200 km
Black Sea

Figure 2. (A) Major tectonic features associated with the Southern Oklahoma aulacogen. A–A′ shows the extent of the integrated gravity model shown in Figure 5. A seismic refraction experiment (Chang et al., 1989) covers the central portion of this profile (100–300 km) and a series of COCORP deep reflection profiles cover the southern portion of it (Brewer et al., 1981, 1983). (B) Major tectonic features associated with the Dniepr-Donets aulacogen (modified from Stephenson et al., 2001). The red dashed line shows the location of the DOBRE reflection and refraction experiments. The green dashed lines delimit the approximate extent of the Dniepr-Donets aulacogen.

complex), basalts, granites, and up to 40,000 km^3 of silicic volcanics (the Carlton Rhyolite Group, 525 Ma). Interestingly, no normal faults that are clearly rift-related have been mapped due to the limited exposures and extensive reactivation. As discussed below, this region is underlain by a massive mafic intrusive feature that is manifested by both high velocities and high densities (e.g., Keller and Baldridge, 1995).

Between the Amarillo uplift and the Uncompahgre uplift (Figs. 1A and 2A), the presence of rift structures is subject to question. However, there are northwest-trending late Paleozoic Ancestral Rocky Mountains structures in this area (Sierra Grande, Tusas, and Cimarron uplifts; Fig. 1A) that could represent inverted rift structures. In addition, there are strong gravity and magnetic signatures along this northwest trend (Suleiman and Keller, 1985; Purucker, 1988; Casillas, 2004) (Fig. 3). West of the prominent and relatively well-studied Amarillo-Wichita uplift portion of the aulacogen, these geophysical anomalies appear to continue across eastern New Mexico while experiencing a large offset that could represent a transform fault (Fig. 3). East of the Tusas uplift in north-central New Mexico (Fig. 1A),

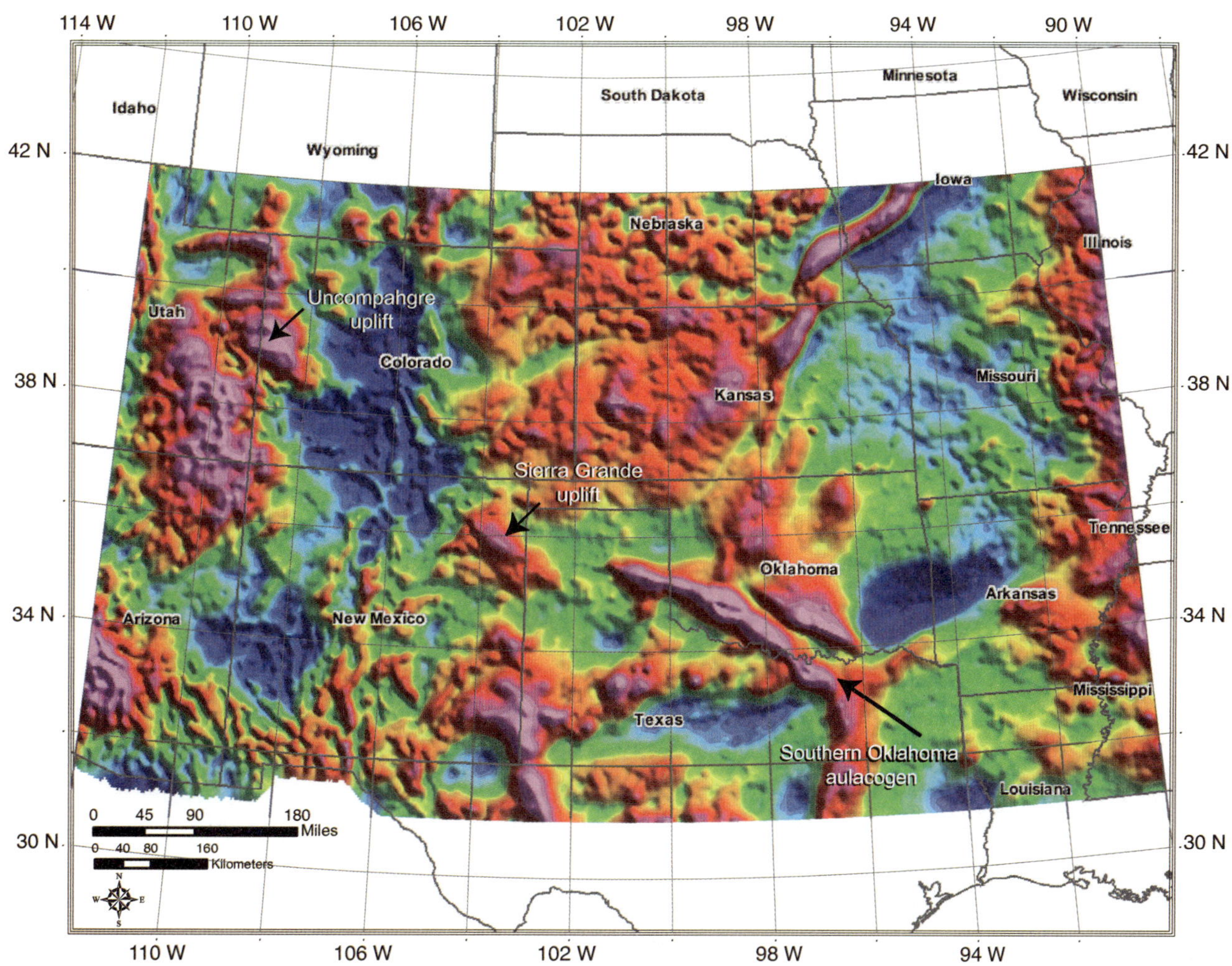

Figure 3. Residual gravity anomaly map produced from complete Bouguer gravity values derived from the U.S. national database (http://paces.geo.utep.edu) that were gridded. A third-order polynomial was fitted to this grid and then subtracted from the grid of complete Bouguer anomaly values to produce the residual values shown here. The colors indicate anomaly values with red indicating highs and blue indicating lows.

younger Rio Grande rift and Laramide features obscure the geophysical signatures of the Paleozoic structures. However, farther to the northwest, Baars and Stevenson (1986) found evidence for late Paleozoic horst blocks along the western edge of the San Juan Mountains that trend northwest and correlate with structures farther west in the Uncompahgre uplift–Paradox basin area (Fig. 1A). Jenkins and Keller (1989) and Casillas (2004) conducted integrated analyses of drilling and gravity data that demonstrated the presence of a mafic mass in the upper crust under the Uncompahgre uplift, and Cambrian igneous rocks do outcrop in this uplift (Larson et al., 1985). The DEEP PROBE seismic profile crossed the eastern portion of the Uncompahgre uplift, and integrated analysis of these data with gravity and drilling data confirmed that there is a high velocity/high density body in this location (Snelson et al., 1998).

The relative roles of thrusting and strike slip during the late Paleozoic inversion have been the subject of considerable debate, but north-northeast-directed compression was the dominant stress regime. On the other hand, there are many northwest-trending faults present that have a significant left-lateral, strike-slip component of movement (e.g., McConnell, 1989; Granath, 1989). In some places, this oblique slip involves approximately equal amounts of strike-slip and dip-slip movement (Granath, 1989). Budnick (1986) suggested that the magnitude of the strike-slip movement was as much as 100 km, but the detailed structural and stratigraphic analysis of McConnell (1989) and Granath (1989) show that the amount of strike-slip offset was unlikely to be greater than 25 km. Additional evidence for the lack of large lateral displacements is the distinctive three-armed gravity anomaly that occurs at the intersection of the Southern Oklahoma aulacogen and the Paleozoic continental margin that shows no evidence of a large offset along a northwest-trending fault (Fig. 3).

The Ouachita orogeny is generally accepted as having played the dominant role in producing the compressive forces that are responsible for the structural inversion of the Cambrian rift (Kluth and Coney, 1981; Kluth, 1986). However, Ye et al. (1996) argue that compression from the southwest rather than the south-southeast produced the compression. In either case, the orogenic activity along the continental margin appears inadequate to produce the extensive deformation in the foreland that is associated with the Ancestral Rocky Mountains (Fig. 1A). For example, the best-constrained geophysical model of lithospheric structure across the Ouachita orogenic belt (Mickus and Keller, 1992) lies just east of the intersection of the Southern Oklahoma aulacogen and the Paleozoic margin (Fig. 2A). This model shows that the Paleozoic passive margin lies largely undisturbed beneath allochthonous Ouachita facies rocks. This general model of a preserved passive margin appears to be valid all along this margin as it extends from Mississippi across Arkansas, Oklahoma, and Texas before entering Mexico (Keller et al., 1989). The fact that this margin is largely intact argues against an energetic continent-continent collision that would produce intraplate deformation as extensive as that of the Ancestral Rocky Mountains orogeny.

The Wichita uplift–Anadarko basin area has been the target of two relatively detailed deep crustal seismic experiments. In addition, intensive petroleum exploration efforts for decades have involved the collection of many thousands of kilometers of seismic reflection data in the region. These data have been used in many integrated analyses (e.g., Arbenz, 1989; Granath, 1989), but very few actual seismic record sections have been released for publication. Although these data have had a major impact on studies of late Paleozoic deformation, they have not yet significantly impacted studies of the Cambrian rifting.

The first study of deep crustal structure was a long seismic refraction line that crossed Oklahoma from southwest to northeast (Mitchell and Landisman, 1970). This profile provides a low resolution velocity model because there were only two shot and the recording locations were ~10 km apart. However, this profile and others in adjacent states provide a regional picture of a crust ~45 km thick with a high-velocity layer ($Vp \approx 7.3$ km/s) at is base (Braile, 1989).

The first detailed deep seismic data gathered were several profiles recorded by the Consortium for Continental Reflection Profiling (COCORP) in southwestern Oklahoma and adjacent areas of north Texas (Brewer et al., 1981, 1983). These data showed that the northern thrust front of the Wichita uplift dips southwestward at an angle of ~40° and extends to mid-crustal depths. In addition, the area south of this uplift is underlain by a zone of strong, layered reflectivity, which was within the Precambrian basement. This zone of reflectivity could represent sedimentary rocks or layered volcanics (Pratt et al., 1992).

The University of Texas, Dallas, and the University of Texas, El Paso, conducted a wide-angle reflection/refraction experiment that crossed the same area as the COCORP surveys (Chang et al., 1989; Suleiman, 1995) (Fig. 2A). The purpose of this survey was to delineate deep structures and derive a velocity model along the profile. This massive data set was investigated with a variety of imaging and tomographic techniques (Chang et al., 1989) as well as detailed ray-trace modeling (Hamilton, 1989; Suleiman, 1995). The velocity model shown in Figure 4 is a synthesis of these modeling efforts and provides the first seismically constrained interpretation of the deepest portion of the Anadarko basin. Velocities low enough to indicate the presence of sedimentary rocks (<6.0 km/s; Christensen and Mooney, 1995) extend to a depth of ~15 km, and the deepest of these sedimentary layers has been interpreted as rift fill (Suleiman, 1995). In addition, the Wichita uplift is underlain by very high-velocity material even at upper crustal depths. In fact, velocities as high as 7.0 km/s are found at a depth of only 8 km (Fig. 4).

The velocity model in Figure 4 was used as a constraint in the construction of a regional gravity model (Fig. 5) that integrates the various seismic and geologic data available in the area of the gravity profile (Fig. 2A). This model provides a clear picture of the massive modification of the middle and upper crust that is the combined result of rifting followed by extensive inversion. The Cambrian igneous rocks that are exposed on the crest of the Wichita uplift lie at depths of as much as 15 km under the

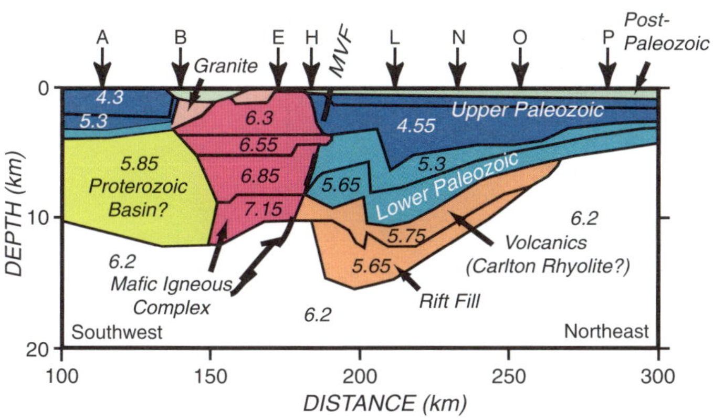

Figure 4. Seismic velocity model derived from the University of Texas, Dallas, and University of Texas, El Paso, seismic experiment. Numbers indicate P-wave velocities in km/s. Letters indicate the position of the shotpoints. MVF—Mountain View fault zone.

adjacent Anadarko basin. Thus, the frontal thrust of the Wichita uplift represents an inversion structure with a remarkable ~15 km of vertical offset. South of the Wichita uplift, the most straightforward way to model the gravity data is by interpreting the layered, intrabasement reflections shown in the COCORP data to represent Precambrian sedimentary rocks that have a relatively low density (2.6 g cm^{-3}) that suggests they are not pervasively metamorphosed. Another surprising result is that the entire 125 mGal gravity high associated with the Wichita uplift is easily modeled as being due to a mafic mass that is entirely contained in the upper plate of the frontal thrust of the Wichita uplift that was imaged by COCORP data (Brewer et al., 1983; Chang et al., 1989).

The changes in structural style along the Southern Oklahoma aulacogen are substantial. For example, the northern boundary fault zone of the Wichita uplift has a moderate dip (~40°) in southwestern Oklahoma (Brewer et al., 1983; Chang et al., 1989), but ~200 km to the northwest along the Amarillo uplift (Fig. 2A), drilling data show that the analogous structure is essentially vertical (e.g., Goldstein, 1984). To the southeast of the Wichita uplift, the Anadarko basin narrows and merges into the Ardmore basin, which is a very deep (~10 km) but narrow structure (e.g., Ham and Wilson, 1967; Granath, 1989; Perry, 1989). Instead of the gradual transition into the cratonal interior that the Anadarko basin experiences to the north, the Ardmore basin is bounded on the north by the Arbuckle uplift, which produces a major outcrop of the Precambrian basement (e.g., Denison et al., 1984).

Many of the deepest petroleum exploration wells ever drilled are located in the Anadarko basin region (e.g., Rowland, 1974), so there are extensive subsurface data available for basin studies. From the standpoint of basin structure, the Anadarko basin shows a vertical variation from a broad postrift sag basin during its early development in the Cambro-Ordovician (Feinstein, 1981) to a narrow trough in the Late Mississippian to Early Pennsylvanian and back to a broad sag basin that existed until the end of the Paleozoic (Garner and Turcotte, 1984). Garner and Turcotte (1984) state that this late Paleozoic evolution required some process besides thermal subsidence and loading due to the structural inversion. Their analysis of elastic flexure of the lithosphere suggested that crustal thinning accompanied by faulting occurred. However, the newer seismic and gravity models (Figs. 4 and 5) indicate that the upper crustal load is very substantial and can all be placed in the hanging wall of the thrust system. Nonetheless, late Paleozoic tectonism is too complex to be explained by simple models of compression due to the Ouachita orogeny. This fact was appreciated by early workers who recognized that at least two pulses (Wichita and Arbuckle) of deformation could be documented in the Arbuckle uplift–Ardmore basin region (Ham and Wilson, 1967).

In summary, we are left with a picture of the southeastern portion of the Southern Oklahoma aulacogen in which pervasive magmatic modification of the upper crust occurred in the Cambrian; cooling and subsidence followed that was in turn followed by a huge (up to 15 km) amount of structural inversion in the late Paleozoic. The scale of this magmatism and deformation is impressive, and the result is that the 15-km-deep Anadarko basin is floored by igneous and basement rocks that outcrop in adjacent uplifts. To the northwest, Mesozoic and Cenozoic tectonism has obscured the aulacogen structures, but evidence for substantial Cambrian mafic magmatism and late Paleozoic structural uplifts is strong. Together, the evidence documents the presence of a major intracratonic structure that is ~1500 km long.

DNIEPR-DONETS AULACOGEN

The Dniepr-Donets aulacogen lies in the southern part of the Russian (or East European) Platform (Figs. 1B and 2B), which was part of Baltica until its early Paleozoic consolidation with Laurentia and was regarded by Shatsky (1946) as being a type example of an aulacogen. The main feature in the aulacogen is the Dniepr-Donets basin, which contains late Paleozoic and younger sediments. Because of its tectonic significance and petroleum resources, the Dniepr-Donets basin has been the target of numerous geological and geophysical studies. Thus, much of the data available is from this basin, and we will be specific when referring to it. Clearly being a "failed rift" in the sense that the Dniepr-Donets aulacogen did not itself lead to continental breakup and ocean crust formation, there remains to this day disagreement on whether it represents the failed third arm of a triple-rift system, the other arms of which ultimately formed a passive continental margin (e.g., Saintot et al., 2003; Stephenson et al., 2006). Indeed, subsequent to Shatsky's work and until quite recently, some authors (e.g., Chekunov et al., 1992; Volozh et al., 1999) have regarded the aulacogen to be part of a very long, singular intracontinental rift system, including the Karpinsky swell (on the northern margin of the northern Caucasus domain; Fig. 2B), that crosses the Caspian Sea and continues into western Asia. Such a model appears at least in part to be untenable within a plate-tectonic framework.

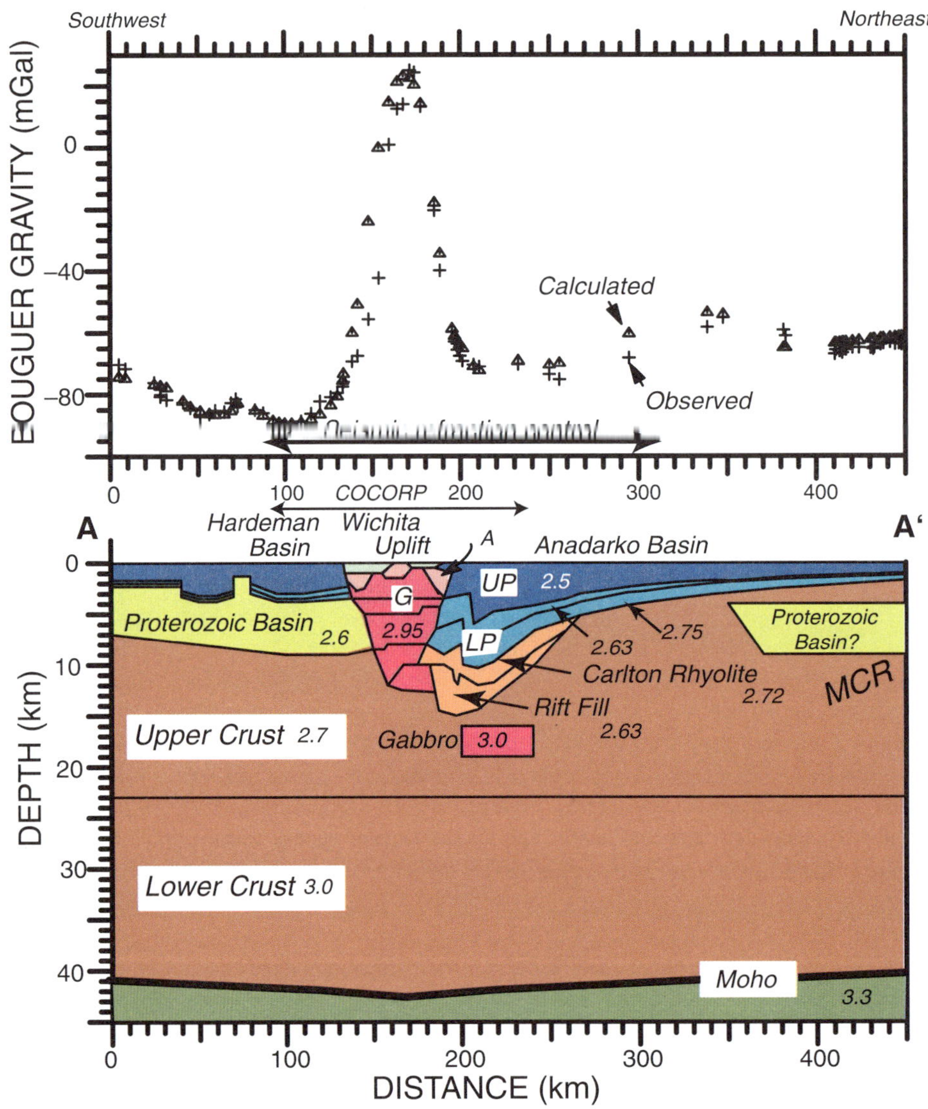

Figure 5. Regional gravity model derived by integrating gravity, seismic, aeromagnetic, surface mapping, and drilling data. The central portion of this profile is constrained by the seismic refraction velocity model (Fig. 4) and COCORP deep seismic reflection data. Numbers are densities in g cm^{-3}. A—felsic intrusions and volcanics that outcrop in the Wichita Mountains (density 2.7); G—gabbroic intrusions that outcrop in the Wichita Mountains and form its core; the densities of the five gabbroic bodies under the Wichita uplift are, from top to bottom, 2.75, 2.8, 2.95, 3.05, and 3.10; UP—upper Paleozoic strata; LP—lower Paleozoic strata; MCR—the southwest portion of the Neoproterozoic Midcontinent rift, which extends into Oklahoma (Robbins and Keller, 1992). If the low-density body shown at the northeastern end of the profile is a basin associated with the MCR, it would in fact extend upward to meet the lower Paleozoic body. Modified from Keller and Baldridge (1995).

Related to the question of whether the Dniepr-Donets aulacogen is a failed arm of a triple rift junction is the age of its initial stage of rifting. Shatsky (1946) and later writers considered the aulacogen to have been initiated as one of a set of Neoproterozoic (Riphean) rifts crosscutting the East European craton, some of which (including the Dniepr-Donets basin and the closely related Pericaspian basin; Fig. 1B) were thought to have been very long-lived and strongly reactivated in late Paleozoic times. For example, Borodulin (1974), Sollogub et al. (1977), and Chekunov et al. (1992) all argued that a Neoproterozoic (Riphean) graben, of much smaller width than the Dniepr-Donets aulacogen underlies the Devonian succession. Its presence was postulated mainly on a seismically inferred, 8–11 km thick layer with velocities in the range 5.6–6.2 km/s (e.g., Chekunov et al., 1992; Ilchenko, 1996). Scattered Neoproterozoic strata are exposed within the Ukrainian shield (e.g., Chekunov et al., 1992), and Neoproterozoic sediments of continental affinity reportedly underlie Devonian sediments in the northwesternmost part of the Dniepr-Donets aulacogen in the vicinity of the western Pripyat trough (Fig. 2B) (Garetsky, 1979) and in its farthest southeastern extent, near the northeast end of the Azov Sea, but nowhere else (e.g., Stephenson et al., 2001). However, Stovba et al. (1996) clearly showed from regional seismic reflection profiling that what was thought to be Neoproterozoic is, in fact, part of the late Paleozoic (Middle Devonian and younger) platformal and rift-related sedimentary sequence. Thus, the pericratonic rifting that formed the Dniepr-Donets basin is convincingly not older than late Paleozoic. Extensional tectonism prevailed until the early Visean (Middle Mississippian) but was punctuated by a period of quiescence during the Tournaisian (Early Mississippian), and extensional reactivations also took place in the middle of the Serpukhovian (Late Mississippian) and in latest Carboniferous–earliest Permian times (Stovba et al., 1996; Stovba and Stephenson, 1999, 2003; Stephenson et al., 2001).

In a pioneering plate-tectonic synthesis of former Soviet Union geology, Zonenshain et al. (1990) represented the Dniepr-Donets aulacogen as the failed arm of a Late Devonian rift system that led to the development of subsequently closed, small ocean basins along the southern margin of the East European craton (see also Kostyuchenko et al., 2004; Saintot et al., 2003). In recent paleogeographic and paleotectonic reconstructions, Nikishin et al. (1996) appear to hold to this view, though it is not explicitly expressed as such. A critical issue in this respect would seem to be the tectonic origins and age of the oldest sediments in the Pericaspian basin (Fig. 2B). Some authors (e.g., Volozh et al., 2003) argue that these are at least as old as early Paleozoic, based primarily on the presence of sediments of this age on the northern, pericratonic platformal margin of the Pericaspian basin. However, others suggest that definitive evidence of the age of these very deeply buried and inaccessible sediments simply does not exist and that a tectonic scenario in which the basin is due to Late Devonian rifting (contemporaneously with the Dniepr-Donets aulacogen, therefore) fully and more satisfactorily fits available observations than other scenarios (e.g., Brunet et al., 1999). At best, this remains an open question. A more comprehensive review of these issues can be found in Stephenson et al. (2006).

In any case, the main feature of the Dniepr-Donets aulacogen is a Late Devonian rift basin (Dniepr-Donets basin) overlain by a substantial (but variable) postrift sedimentary sequence (Fig. 6) that records several extensional or transtensional and at least one compressional reactivation (e.g., Stovba and Stephenson, 1999, 2003; Stephenson et al., 2001). The width of the rift zone (Figs. 2B and 6) varies from 60 to 70 km in the Dniepr segment of the Dniepr-Donets basin to 140–160 km in the southeastern (Donets) segment. The Pripyat trough, which lies along strike to the northwest of the Dniepr-Donets basin (Fig. 2B), is shallower and broader (up to 150 km) than the adjacent Dniepr segment of the aulacogen. Although tectonically part of the same rift system as the Dniepr-Donets basin, the Pripyat trough is, in terms of commonly used terminology, not part of the Dniepr-Donets aulacogen sensu stricto. Sediment thicknesses (comprising Late Devonian synrift and Carboniferous-Paleogene postrift successions) increase from only ~2 km in the Pripyat trough to ~18 km in the Donets segment of the basin (Stovba et al., 1996; Stovba and Stephenson, 1999).

Devonian rifting was accompanied by major magmatic activity (Wilson and Lyashkevich, 1996) and probably by the uplift of the Ukrainian shield and the Voronezh massif, which together form a large-radius dome that is transected by the Dniepr-Donets aulacogen (e.g., Stephenson et al., 1993) (Fig. 2B). Rifting crosscuts the structural grain of the underlying craton as determined by Shchipansky and Bogdanova (1996), with major boundaries within the basement thought to provide conduits for magma expulsion during rifting. Synrift volcanic and intrusive rocks, consisting of a variety of alkali basalts and their differentiates and associated pyroclastics, attain thicknesses of up to 3000 m (Lyashkevich, 1987; Wilson and Lyashkevich, 1996). In the northwest part of the Dniepr-Donets basin, adjacent to the juncture with the Pripyat trough, magmatic rocks can locally form as much as 50% of the synrift succession. Late Devonian volcanic rocks occur on the margins of the basin and are found in cap rocks of salt diapirs along the main axial inversion structure of the southeasternmost Dniepr-Donets basin (Garkalenko et al., 1971).

Potential field anomalies indicate that mafic rocks have also intruded the crust beneath the sedimentary basin. Magnetic anomalies in the Dniepr segment, indicative of both supracrustal and crustal depth sources, correlate well with igneous bodies known to lie within the sedimentary sequence (Orlyuk and Pashkevich, 1994; Stephenson et al., 2001). A distinct linear gravity high is associated with the Dniepr-Donets aulacogen (Fig. 7), and in the northwest near Chernigiv (Fig. 2B), a positive Bouguer anomaly of >90 mGal near the rift axis is among the largest gravity anomalies within the East European craton. Gravity modeling has long suggested the presence of a high-density lower crustal body along the rift axis interpreted as an "axial dike" related to rifting (Starostenko et al., 1986). Yegorova et al. (1999, 2004) show that this feature extends the entire length of the Dniepr-Donets basin to its culmination at the Karpinsky swell. The most recent results (Yegorova et al., 2004) show that this trend takes a 120° bend to the northeast parallel to the recognizable margin of the East European craton at this culmination, providing some evidence that this could have been the triple-rift junction locus (e.g., Stephenson et al., 2006).

The geochemical signature of rift-related magma (Wilson and Lyashkevich, 1996) plus the sheer volume of magma suggest that the origin of the Dniepr-Donets aulacogen is mantle plume/hotspot related (e.g., Gavrish, 1989; Chekunov, 1994). However, there is no clear indication that the intensity of magmatism increases to the southeast in the direction of the supposed triple junction, although the depth of the basin increases significantly in this direction. Various subsidence modeling studies have led to the common conclusion that thinning of the mantle lithosphere

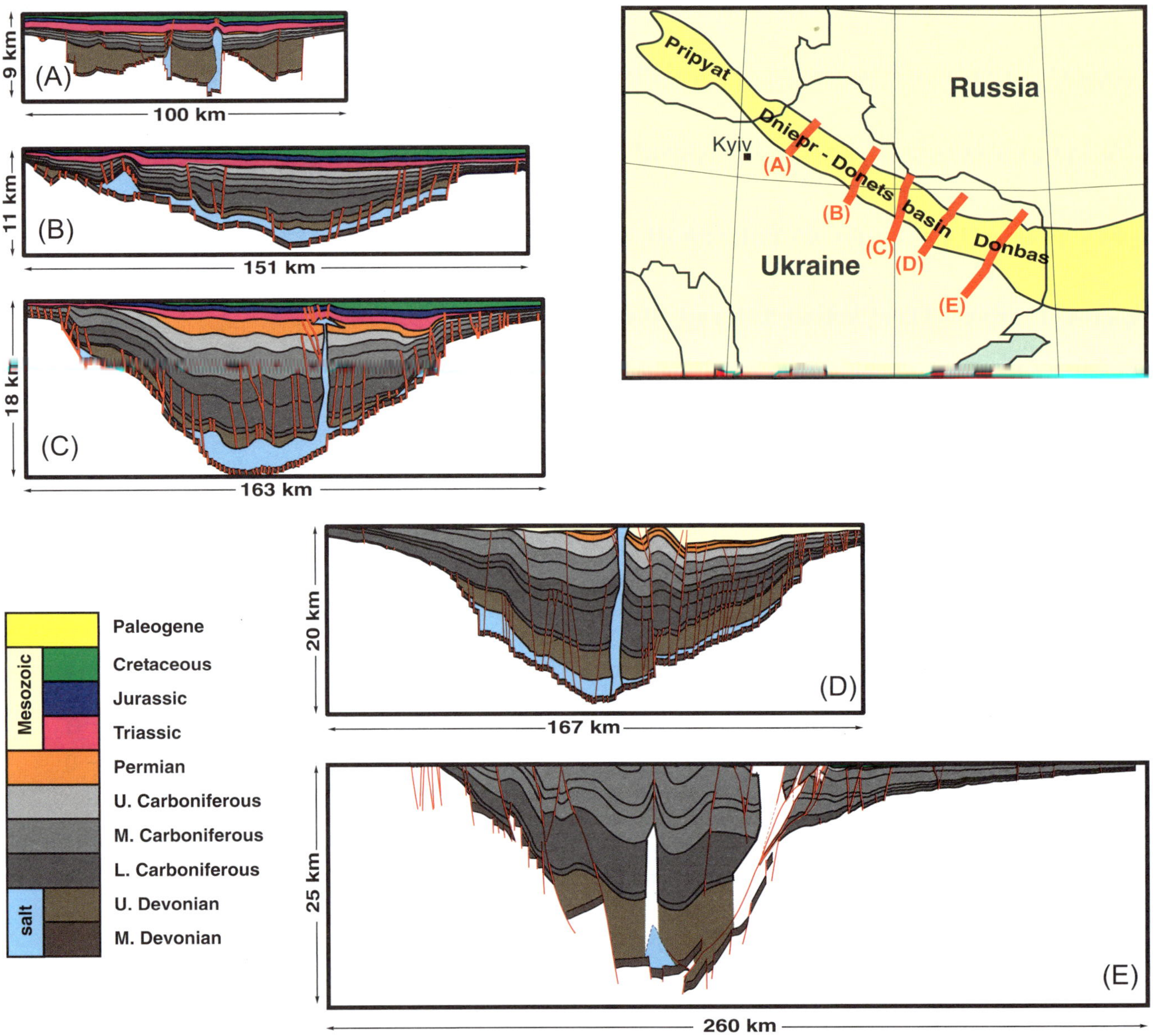

Figure 6. Geologic cross sections through the Dniepr-Donets basin (A–D) and Donbas fold belt (E, from DOBREflection), based on depth-converted versions of interpreted regional seismic reflection profiles (from Stovba et al., 1996, 2003; Stovba and Stephenson, 1999). Locations are shown in the inset. Light blue areas represent salt bodies; Devonian sedimentary rocks are brown; Carboniferous strata are various shades of gray; Permian strata are orange; Mesozoic strata are pink, blue, and green as shown in the stratigraphic legend.

during rifting was greater than crustal thinning (e.g., van Wees et al., 1996; Starostenko et al., 1999; Poplawskii et al., 2001), which is also indirect evidence of a role for a mantle plume/hotspot.

An Early Permian unconformity is evident throughout much of the Dniepr-Donets basin, as it is throughout much of the remainder of the East European craton (e.g., Stephenson et al., 2001). Especially on the southern margin of the Dniepr-Donets basin, this unconformity shows angular discordance with underlying strata. Much of the Upper Carboniferous and younger basin succession has been removed in the Donbas fold belt, which also displays mild folding, thrusting, and reverse faults (e.g., Stovba and Stephenson, 1999). Elsewhere in the basin, especially in the area adjacent to the Donbas fold belt, local folds, domes, and salt diapirs define linear trends that cor-

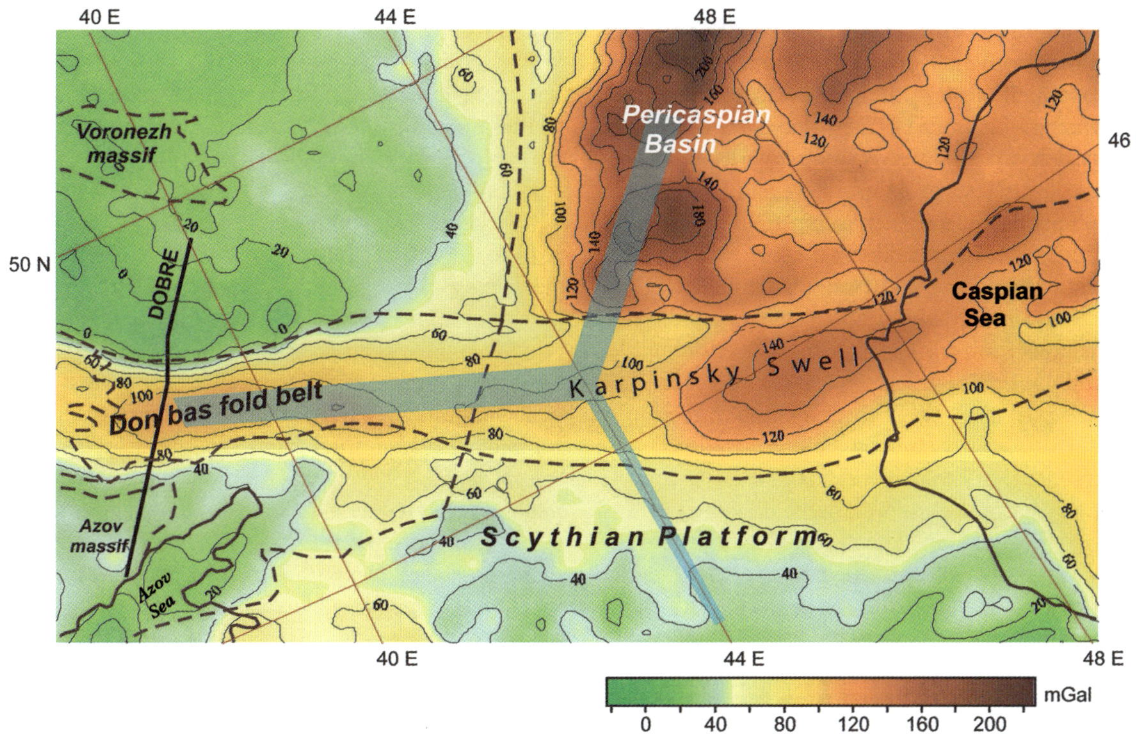

Figure 7. Residual gravity map of the Dniepr-Donets aulacogen–Pericaspian basin area (gravity effect of sedimentary layers removed; Yegorova et al., 1999) with hypothesized triple-junction rift axes (modified from Stephenson et al., 2006; gravity map based on Yegorova et al., 2004). The location of the DOBREfraction profile, crossing the Donbas fold belt, is shown for reference.

respond to the trends of the main folds of the Donbas fold belt. Detailed studies of salt tectonics within the Dniepr-Donets basin document that the latest Carboniferous–Early Permian was a time of active halokinesis in a transtensional setting (Stovba and Stephenson, 1999, 2003). The coal deposits within the deformed Carboniferous rocks of the Donbas fold belt were first studied in the late nineteenth century and are associated with the well-known coal measures of central and western Europe that are found in deformed Variscan foreland basins. This association is the basis for the long-held belief that the Donbas fold belt was part a Variscan-Uralian orogenic foreland along the southern margin of the East European craton. However, recent geologic and seismic studies show that inversion in the Dniepr-Donets basin mainly took place in the Late Cretaceous related to the Eo-Alpine Europe-Africa collision (Stovba and Stephenson, 1999; Saintot et al., 2003; Maystrenko et al., 2003). There is some evidence for compressional deformation during the Late Triassic (Stovba and Stephenson, 1999; Saintot et al., 2003), which is contemporaneous with major orogenic activity on the adjacent Karpinsky swell (Fig. 2B) (Sobornov, 1995).

Abundant hydrocarbon exploration–driven seismic reflection data in the Dniepr-Donets basin demonstrate that the Late Carboniferous–Early Permian tectonic setting of the Dniepr-Donets basin (and by implication the Donbas fold belt) was one of transtension rather than compression as previously thought. In contrast to the northern margin of the Dniepr-Donets basin where the Permian unconformity is manifested as a hiatal surface, this unconformity is angular to the south. This angularity can be explained by differential thermal subsidence increasing toward the basin axis related to rifting. Although some local uplifts and folds of Early Permian age can be ascribed to salt diapirism, there remains abundant evidence that significant uplift affected the southern part of the Dniepr-Donets basin at this time (Stovba and Stephenson, 1999). Modeling of organic maturation in the Donbas fold belt indicates that the Late Carboniferous–Early Permian was also a time of anomalously high heat flow (Shymanovsky et al., 2004).

Deep seismic sounding (e.g., Chekunov et al., 1992; Ilchenko, 1996) and more recent wide-angle reflection/refraction seismic studies (DOBREfraction'99 Working Group, 2003) show that the amount of crustal thinning beneath the Dniepr-Donets basin increases to the southeast, concurrently with increasing sedimentary fill thickness, which is consistent with the predictions of subsidence modeling studies (van Wees et al., 1996; Stovba et al., 2003). Older (deep seismic sounding) studies that suggested a deepening Moho beneath the compressionally deformed Donbas fold belt segment of the Dniepr-Donets basin (e.g., Chekunov et al., 1992; Pavlenkova, 1995; Ilchenko and Stepanenko, 1998) are not confirmed by the more modern wide-angle reflection/refraction data.

The most recent profile is the result of the DOBRE (Donbas Refraction and Reflection) project (DOBREfraction'99 Working Group, 2003), which crossed the inverted Donbas fold belt segment of the Dniepr-Donets basin. The velocity model derived from this project is shown in Figure 8 along with the geometry of the crustal-scale thrust and backthrust revealed by the complementary deep seismic reflection profiling (DOBREflection; Maystrenko et al., 2003). The maximum thickness of the sedimentary basin is interpreted to be ~20 km, in keeping with previous estimates based on the older deep seismic sounding data and with extrapolations from regional seismic reflection profiles crossing the Dniepr-Donets basin northwest of the Donbas fold belt (Stovba and Stephenson, 1999). Such a depth to the top of crystalline basement is also in keeping with interpretations of the DOBREflection profile (Maystrenko et al., 2003).

The sedimentary basin itself is well defined, with velocities from ~2.0–2.4 up to 5.8–5.9 km/s. The lowest velocities are confined to a layer of 2–3 km thickness on the northern margin of the Donbas fold belt, corresponding to a Late Cretaceous sedimentary depocenter that may have formed as a response to inversion processes (e.g., Hansen and Nielsen, 2003; Nielsen et al., 2005). The main crustal layer has velocities ~5.8–6.8 km/s and thins significantly beneath the main sedimentary depocenter. In turn, a high-velocity lower crustal layer, ~6.9–7.2 km/s, thickens significantly in the same part of the profile. The shape of the sedimentary basin is asymmetric with the steepest crystalline basement surface on the southwestern margin of the basin; whereas, the asymmetry of the high-velocity layer displays its steepest upper surface beneath the northeastern margin of the basin. The high-velocity layer has been interpreted as a zone of crustal underplating or "rift pillow" (DOBREfraction'99 Working Group, 2003). The asymmetry may be evidence of some degree of so-called "simple shear" rather than "pure shear" during rifting (e.g., Buck et al., 1988). In contrast to earlier studies (e.g., Chekunov et al., 1992; Stephenson et al., 1993), the Moho depth is more or less constant across the Donbas fold belt (40 ± 2 km). The lower crustal high-velocity layer has presumably incorporated (and been thickened by) mantle-derived magmas during the rifting process. A minimum of 10% but as much as 50% of the high-velocity layer comprises mantle material (e.g., Thybo et al., 2000), implying somewhat more than 50% thinning of the prerift crystalline crust beneath the basin and a maximum crustal "stretching factor" of as much as 2.25–2.5 (DOBREfraction'99 Working Group, 2003), a figure that compares favorably with what has been inferred from subsidence modeling in the vicinity of the DOBREfraction profile (Stovba et al., 2003).

DISCUSSION AND CONCLUSIONS

In order to aid our comparison of Southern Oklahoma and Dniepr-Donets aulacogens, we summarize their key characteristics in Table 1. Although they both have been studied in considerable detail, motivated by both scientific and natural resource exploration interests, it is important to remember that their very nature has resulted in most of the rock record of their early evolution and many key structures being buried. In fact, the postrift basins were the main initial evidence used to infer the existence of these features. Drilling and geophysical data are abundant in both rifts, but conflicting interpretations and gaps in our understanding remain.

A key consideration in the analysis of any major structure is tectonic setting. In the case of both of these rifts, they can be classified as pericratonic because they lie within but near the margins of cratons. It is interesting to note that although preexisting structures have been postulated to play a role in the location of these rifts, they both cut across the deep-seated structural grains of the cratons they traverse. The Southern Oklahoma aulacogen formed in the Cambrian during an episode of continental breakup (e.g., Dalziel, 1997) and at a time when several other intracontinental rifts were forming nearby (e.g., Keller et al., 1983). This is also true of the Dniepr-Donets aulacogen, with rifting (or at least major crustal extension) in the adjacent Pericaspian basin occurring at the same time. The very clear three-armed pattern of highs observed in the gravity anomalies at the southeastern termination of the Southern Oklahoma aulacogen (Fig. 3) and the clear connection to deep seismic control that delineates the rifted margin of Laurentia leave little doubt that the Southern Oklahoma aulacogen is linked to a rift-rift-rift triple junction. On the other hand, the nature of the southeastern portion of the Dniepr-Donets aulacogen is more ambiguous (e.g., Saintot et al., 2006). Stephenson et al. (2006) suggest that the Dniepr-Donets basin opened to the southeast into a deep basin, possibly having oceanic lithospheric affinity, in the area where it adjoins the southern Pericaspian basin and that the rift axes of these basins may constitute two arms of a rift-rift-rift triple junction involved in possible continental breakup in late Paleozoic times (Fig. 7).

As manifested in the upper crust, the aulacogens are of similar dimensions spatially. The length of each feature can be debated based on different interpretations of the upper crustal features delineated geophysically, but we can state with some confidence that they are both at least 1000 km long. Because of the massive inversion in the Southern Oklahoma aulacogen and extensive postrift subsidence in both features, assigning widths is more subjective. Although the rift zones were probably narrower, the aulacogens are both ~150 km wide, using the deep portions of the basins as a reference. The basins in both cases are extremely deep (Southern Oklahoma aulacogen, 15 km; Dniepr-Donets aulacogen, 20 km) with the deepest sections being beyond drillable depths at this time, even though a well in the Anadarko basin was the world's deepest (~9.5 km) for many years (Rowland, 1974).

If one looks at the seismic velocity models in Figures 4 and 8, there are both differences and similarities in the deep structure of these features. The lower crustal structure of the Southern Oklahoma aulacogen region is not known as well as for the Dniepr-Donets aulacogen, so the existence of a local zone of underplating near the Moho as is found in the Dniepr-Donets aulacogen is uncertain. However, given the extensive magmatic modification

TABLE 1. COMPARISON OF KEY CHARACTERISTICS

Characteristic	Southern Oklahoma aulacogen	Dniepr-Donets aulacogen
Age of rifting	Early Cambrian	Late Devonian
Dimensions (maximum)	1500 km × 150 km × 15 km Ouachita to Uncompahgre	1100 km × 150 km × 20 km Dniepr-Donets to Donbas
Volume of volcanics	100,000 km^3 in the intrusive complex; 40,000 km^3 of Carlton Rhyolite	Significant
Nature of magmatic activity	Extensive intrusion into the upper crust initially	Varies axially
	(high-velocity lower crust—a regional feature?)	SE—extensive underplating (high-velocity lower crust)
		NW—upper crustal intrusion more important
β factor[†]	1.2	1.1 to ~2.0
Strike-slip movement	15–20 km	Minor
Regional setting		
Prerift	Pericratonic	Pericratonic
	Along zone of weakness?	Across structural grain
Synrift	Plume-generated triple junction	Plume-generated triple junction?
Postrift	Tectonic stability until reactivation	Visean reactivation
		Salt tectonics
Inversion		
Preinversion	Tectonic stability	Permian uplift
Syninversion	Ouachita orogeny?	Minor in Late Triassic
	NW and/or NE compression	Late Cretaceous–Paleogene
Postinversion	Tectonic stability	Tectonic stability
Vertical movement	15 km vertical	~5 km vertical
Shortening	15 km horizontal	10–12 km horizontal

[†]β (or stretching) factor

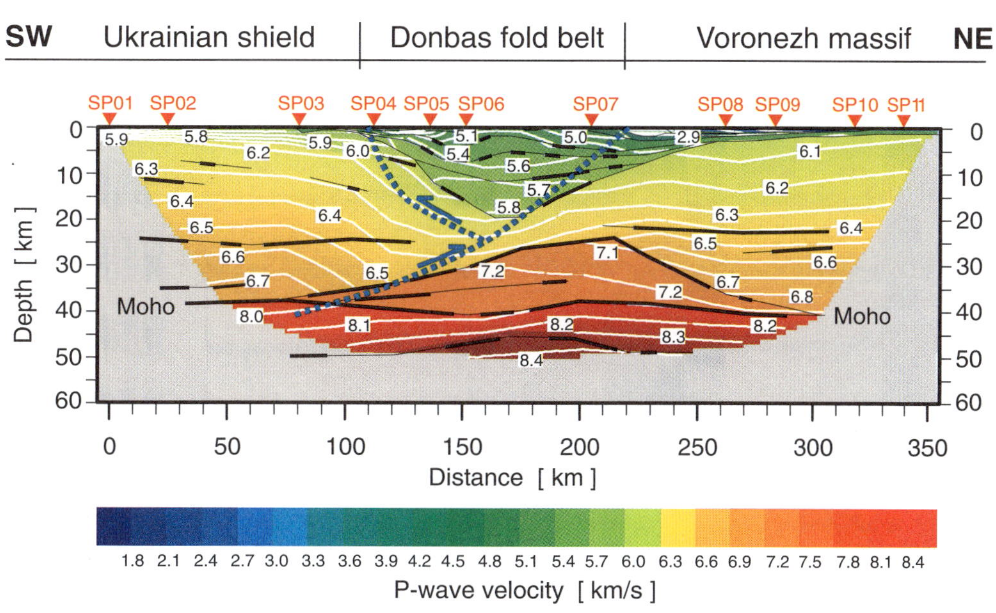

Figure 8. P-wave velocity model along the DOBREfraction profile (modified from DOBREfraction'99 Working Group, 2003), located in Figure 7. Black lines indicate where strong reflections constrain the interfaces. Shotpoint locations are indicated by inverted triangles. The blue dashed lines show the positions of the crustal-scale thrust and backthrust involved in the inversion of the Donbas fold belt as revealed by the DOBREflection profile (e.g., Maystrenko et al., 2003).

in the upper and middle crust at the time of Southern Oklahoma aulacogen rifting, such a layer is likely to have formed as the residuum expected due to the extensive magmatism. In addition, such a layer underlies the nearby Reelfoot rift (Fig. 1A) of the same age, although it experienced much less synrift magmatism, postrift subsidence, and inversion than the Southern Oklahoma aulacogen (e.g., Braile et al., 1986). The magmatic modification in the crust is very different in these features. In the Southern Oklahoma aulacogen, a huge mafic intrusive mass is well mapped in the upper 15 km of the crust via outcrops, drilling, and geophysical data. This feature produces a gravity high that exceeds 100 mGal in places (Robbins and Keller, 1992), but its geometry today is due to massive inversion, indicating that this igneous complex was buried in the middle Paleozoic to a depth of ~5 km before being exhumed by the late Paleozoic inversion. In the Dniepr-Donets aulacogen, small mafic intrusions are present, but there is considerable volcanic material in the synrift strata in the Dniepr-Donets basin, perhaps even more than in the Anadarko basin of the Southern Oklahoma aulacogen.

As discussed by Gilbert (1992), a logical choice for the axis of the rifting in the Southern Oklahoma aulacogen is a graben whose fill is the mafic volcanic units that are exposed today. The boundaries of this graben are not exposed but could be manifested as the steeply dipping boundaries of the igneous complex shown in Figures 4 and 5. In the strongly magmatic portions of the East African rift (Kenya and Ethiopia), there is a strong tendency for the youngest volcanic activity to be concentrated in the axial portions of the rift valleys. However, the area of the rift valley crust that is strongly intruded is significantly narrower than the rift valley (e.g., Simiyu and Keller, 2001; Keranen et al., 2004). The extent of the Carlton Rhyolite interpreted in Figures 4 and 5 is speculative under the deep Anadarko basin, but Gilbert (1983) argued that distribution of the Carlton Rhyolite indicates that this unit was deposited in and overfilled a trough. Thus, it is possible that the rift valley was wider than the exposed igneous complex and that the deep units in the Anadarko basin are some combination of rift valley fill and Carlton Rhyolite. If this is the case, the highly asymmetric structure apparent in these figures is due to more than just the extensive late Paleozoic inversion and northeast-directed thrusting because there is no significant amount of either rift fill or Carlton Rhyolite south of the Wichita uplift. This high degree of asymmetry is in contrast to the minor asymmetry observed in the Dniepr-Donets basin. These observations attest to the complex history of the Anadarko basin, which includes postrift subsidence (early Paleozoic), cratonal basin development (middle Paleozoic), and foreland basin (late Paleozoic) phases (e.g., Feinstein, 1981; Goldstein, 1984; Gilbert, 1992; Garner and Turcotte, 1984). Since we have less control on the deep structure of the Anadarko basin, the estimated amount of extension during rifting (Table 1) is questionable but is probably nowhere as great as in the Donets segment of the Dniepr-Donets aulacogen based on the attenuation of the original crust under the Dniepr-Donets basin.

For both the Southern Oklahoma aulacogen and Dniepr-Donets aulacogen, the ages of their formation (Early Cambrian and Late Devonian respectively) and inversion (Carboniferous and Late Cretaceous–Paleogene respectively) are now well established. In the case of the Dniepr-Donets aulacogen, both of these ages are revised (to be younger) compared to what was thought at the time Shatsky (1946) proposed the similarity of the Dniepr-Donets aulacogen to the Southern Oklahoma aulacogen. However, the general evolution of these structures remains analogous, comprising rifting followed much later by inversion. However, in the Dniepr-Donets aulacogen, there were episodes of postrift, but preinversion, uplift and minor inversion; whereas, inversion in the Southern Oklahoma aulacogen was a relatively short-lived event that may have involved two pulses of deformation. The scale of the inversion in the Southern Oklahoma aulacogen is huge, involving as much as 15 km of vertical displacement on the main thrust system and extensive overthrusting of basement blocks over the Paleozoic strata. The inversion in the Dniepr-Donets aulacogen is not this large, but coupled with salt tectonics locally, the amount of postrift uplift and deformation can also be impressive. It is clear that we still have much to learn about the structure, evolution, and dynamics of these two large intraplate rift zones.

ACKNOWLEDGMENTS

We would like to thank the many colleagues who worked on the DOBRE project. IRIS/PASSCAL is supported by the National Science Foundation (NSF) and provided the majority of the instrumentation used for the DOBREfraction experiment, and most of these instruments were provided through grants to the University of Texas, El Paso (State of Texas Higher Education Coordinating Board, NSF/MRI [National Science Foundation/Research Instrumentation], and Department of Defense). The participation of the University of Texas, El Paso, group was funded by the National Science Foundation Office of International Programs. The GEON (GEOscience Network) project and the Netherlands Centre for Integrated Solid Earth Sciences (ISES) facilitated this synthesis.

REFERENCES CITED

Arbenz, J.K., 1989, Ouachita thrust belt and Arkoma basin, *in* Hatcher, R.D., Jr., et al., eds., The Appalachian-Ouachita orogen in the United States: Geological Society of America, Geology of North America, v. F-2, p. 621–634.

Baars, D.L., and Stevenson, G.M., 1986, The Paradox: A pull-apart basin of Pennsylvanian age, *in* Peterson, J.A., ed., Paleotectonics and sedimentation in the Rocky Mountains region, United States: American Association of Petroleum Geologists Memoir 41, p. 518–519.

Barnes, C.G., Shannon, W.M., and Kargi, H., 1999, Diverse Mesoproterozoic basaltic magmatism in west Texas: Rocky Mountain Geology, v. 34, p. 223–243.

Borodulin, M.I., 1974, New approach of regional seismic investigations techniques and geological referencing of interfaces in the Dniepr basin: Geophysical Collection, v. 60, p. 9–17.

Braile, L.W., 1989, Crustal structure of the continental interior, in Pakiser, L.C., and Mooney, W.D., eds., Geophysical framework of the continental United States: Geological Society of America Memoir 172, p. 285–315.

Braile, L.W., Hinze, W.J., Sexton, J.L., Keller, G.R., and Lidiak, E.G., 1986, Tectonic development of the New Madrid seismic zone: Tectonophysics, v. 131, p. 1–21, doi: 10.1016/0040-1951(86)90265-9.

Brewer, J.A., Brown, L.D., Steiner, D., Oliver, J.E., and Kaufman, S., 1981, Proterozoic basin in the southern Midcontinent of the United States revealed by COCORP deep seismic profiling: Geology, v. 9, p. 569–575, doi: 10.1130/0091-7613(1981)9<569:PBITSM>2.0.CO;2.

Brewer, J.A., Good, R., Oliver, J.E., Brown, L.D., Kaufman, S., and Denison, R.E., 1983, COCORP profiling across the Southern Oklahoma aulacogen: Overthrusting of the Wichita Mountains and compression within the Anadarko basin: Geology, v. 11, p. 109–114, doi: 10.1130/0091-7613(1983)11<109:CPATSO>2.0.CO;2.

Brunet, M.F., Volozh, Y.A., Antipov, M.P., and Lobkovsky, L.I., 1999, The geodynamic evolution of the Precaspian basin (Kazakhstan) along a north-south section: Tectonophysics, v. 313, p. 85–106, doi: 10.1016/S0040-1951(99)00191-2.

Buck, W.R., Martinez, F., Steckler, M.S., and Cochran, J.R., 1988, Thermal consequences of lithospheric extension: Pure and simple: Tectonics, v. 7, p. 213–234.

Budnick, R.T., 1986, Left-lateral intraplate deformation along the Ancestral Rocky Mountains: Implications for late Paleozoic plate motions, *in* Johnson, B., and Balley, A.W., eds., Intraplate deformation: Characteristics, processes and causes: Tectonophysics, v. 132, p. 195–214.

Casillas, H., 2004, An integrated geophysical study of the Uncompahgre uplift, Colorado and Utah [M.S. thesis]: El Paso, University of Texas, 151 p.

Chang, W.F., McMechan, G.A., and Keller, G.R., 1989, Wavefield processing of data from a large-aperture seismic experiment in southwestern Oklahoma: Journal of Geophysical Research, v. 94, p. 1803–1816.

Chekunov, A.V., 1994, The geodynamics of the Dniepr-Donets rift syneclise: Geophysical Journal, v. 16, p. 3–13.

Chekunov, A.V., Gavrish, V.K., Kutas, R.I., and Ryabchun, L.I., 1992, Dnieper-Donets paleorift, *in* Ziegler, P.A., ed., Geodynamics of rifting, volume I: Case history studies on rifts, Europe and Asia: Tectonophysics, v. 208, p. 257–272.

Christensen, N.I., and Mooney, W.D., 1995, Seismic velocity structure and composition of the continental crust: A global view: Journal of Geophysical Research, v. 100, p. 9761–9788, doi: 10.1029/95JB00259.

Dalziel, I.W.D., 1991, Pacific margins of Laurentia and East Antarctica–Australia as a conjugate rift pair: Evidence and implications for an Eocambrian supercontinent: Geology, v. 19, p. 598–601, doi: 10.1130/0091-7613(1991)019<0598:PMOLAE>2.3.CO;2.

Dalziel, I.W.D., 1997, Neoproterozoic-Paleozoic geography and tectonics: Review, hypothesis, environmental speculation: Geological Society of America Bulletin, v. 109, p. 16–42, doi: 10.1130/0016-7606(1997)109<0016:ONPGAT>2.3.CO;2.

Denison, R.E., Lidiak, E.G., Bickford, M.E., and Kisvarsanyi, E.B., 1984, Geology and geochronology of Precambrian rocks in the Central Interior region of the United States: U.S. Geological Survey Professional Paper 1241-C, 20 p.

DOBREfraction'99 Working Group, 2003, "DOBREfraction'99"—Velocity model of the crust and upper mantle beneath the Donbas foldbelt (east Ukraine): Tectonophysics, v. 371, p. 81–110.

Feinstein, S., 1981, Subsidence and thermal history of southern Oklahoma aulacogen: Implications for petroleum exploration: American Association of Petroleum Geologists Bulletin, v. 65, p. 2521–2533.

Garetsky, R.G., editor, 1979, Tectonics of Pripyat trough: Minsk, Nauka I Technika, 222 p.

Garkalenko, I.A., Borodulin, M.I., and Mikhalev, A.K., 1971, About the transition zone between the Dniepr-Donets basin and Donets folded structure: Geological Journal, v. 31, p. 92–98.

Garner, D.L., and Turcotte, D.L., 1984, The thermal and mechanical evolution of the Anadarko basin: Tectonophysics, v. 107, p. 1–24, doi: 10.1016/0040-1951(84)90026-X.

Gavrish, V.K., editor, 1989, Geology and hydrocarbon potential of the Dniepr-Donets depression: Depth, structure, and geotectonic evolution: Kiev, Naukaova Dumka, 208 p.

Gilbert, M.C., 1983, Timing and chemistry of igneous events associated with the Southern Oklahoma aulacogen, *in* Morgan, P., and Baker, B.H., eds., Processes of continental rifting: Tectonophysics, v. 94, p. 439–455.

Gilbert, M.C., 1992, Speculations on the origin of the Anadarko basin, *in* Mason, R., ed., Basement tectonics, 7: Dordrecht, Netherlands, Kluwer Academic Publishers, p. 195–208.

Goldstein, A.G., 1984, Tectonic controls on Paleozoic subsidence in the south-central United States: Journal of Geology, v. 92, p. 217–222.

Granath, J.W., 1989, Structural evolution of the Ardmore basin, Oklahoma: Progressive deformation in the foreland of the Ouachita collision: Tectonics, v. 8, p. 1015–1036.

Ham, W.E., and Wilson, J.L., 1967, Paleozoic epeirogeny and orogeny in the central United States: American Journal of Science, v. 265, p. 332–407.

Ham, W.E., Denison, R.E., and Merritt, C.A., 1964, Basement rocks and structural evolution of southern Oklahoma: Oklahoma Geological Survey Bulletin 95, 302 p.

Hamilton, L.S., 1989, Structure of the Wichita uplift, southern Oklahoma, from a wide-angle seismic experiment [M.S. thesis]: El Paso, University of Texas, 69 p.

Hansen, D.L., and Nielsen, S.B., 2003, Why rifts invert in compression: Tectonophysics, v. 373, p. 5–24, doi: 10.1016/S0040-1951(03)00280-4.

Hanson, R.E., and Al-Shaieb, Z., 1980, Voluminous subalkaline silicic magmas related to intracontinental rifting in the Southern Oklahoma aulacogen: Geology, v. 8, p. 180–184, doi: 10.1130/0091-7613(1980)8<180:VSSMRT>2.0.CO;2.

Hoffman, P., 1991, Did the breakout of Laurentia turn Gondwanaland inside-out?: Science, v. 252, p. 1409–1412, doi: 10.1126/science.252.5011.1409.

Hoffman, P., Dewey, J.F., and Burke, K., 1974, Aulacogens and their genetic relations to geosynclines, with a Proterozoic example from Great Slave Lake, Canada, *in* Dott, R.H., Jr., and Shaver, R.H., eds., Modern and ancient geosynclinal sedimentation: Society of Economic Paleontologists and Mineralogists Special Publication 19, p. 38–55.

Ilchenko, T., 1996, The Dniepr-Donets rift: Deep structure and evolution from DSS profiling: Tectonophysics, v. 268, p. 83–98, doi: 10.1016/S0040-1951(96)00221-1.

Ilchenko, T.V., and Stepanenko, V.M., 1998, A velocity model of the Earth's crust and upper mantle of Donbas and its geological interpretation: Geophysical Journal, v. 20, p. 95–105.

Jenkins, R.D., and Keller, G.R., 1989, Interpretation of basement structures and geophysical anomalies in the southeastern Colorado Plateau: New Mexico Geological Society, 40th Field Conference Guidebook, p. 135–142.

Karlstrom, K.E., and Bowring, S.A., 1988, Early Proterozoic assembly of tectonostratigraphic terranes in southwestern North America: Journal of Geology, v. 96, p. 561–576.

Keller, G.R., and Baldridge, W.S., 1995, The Southern Oklahoma aulacogen, *in* Olsen, K.H., ed., Continental rifts: Evolution, structure, tectonics: Amsterdam, Elsevier, p. 427–435.

Keller, G.R., Lidiak, E.G., Hinze, W.J., and Braile, L.W., 1983, The role of rifting in the tectonic development of the midcontinent, U.S.A., *in* Morgan, P., and Baker, B.H., eds., Processes of continental rifting: Tectonophysics, v. 94, p. 391–412.

Keller, G.R., Kruger, J.M., Smith, K.J., and Voight, W.M., 1989, The Ouachita system: A geophysical overview, *in* Hatcher, R.D., Jr., et al., eds., The Appalachian-Ouachita orogen in the United States: Geological Society of America, Geology of North America, v. F-2, p. 689–694.

Keranen, K., Klemperer, S.L., Gloaguen, R., and the EAGLE Working Group, 2004, Imaging a proto-ridge axis in the Main Ethiopian rift: Geology, v. 32, p. 949–952, doi: 10.1130/G20737.1.

Kluth, C.F., 1986, Plate tectonics of the Ancestral Rocky Mountains: American Association of Petroleum Geologists Memoir 41, p. 353–369.

Kluth, C.F., and Coney, P.J., 1981, Plate tectonics of the Ancestral Rocky Mountains: Geology, v. 9, p. 10–15, doi: 10.1130/0091-7613(1981)9<10:PTOTAR>2.0.CO;2.

Kostyuchenko, S.L., Kostyuchenko, S., Morozov, A.F., Stephenson, R.A., Solodilov, L.N., Vedrentsev, A.G., Popolitov, K.E., Aleshina, A.F., Vishnevskaya, V.S., and Yegorova, T.P., 2004, The evolution of the southern margin of the East European craton based on seismic and potential field data: Tectonophysics, v. 381, p. 101–118, doi: 10.1016/j.tecto.2002.08.002.

Kruger, J.M., and Keller, G.R., 1986, Interpretation of crustal structure from regional gravity anomalies, Ouachita Mountains area and adjacent Gulf coastal plain: American Association of Petroleum Geologists Bulletin, v. 70, p. 667–689.

Lambert, D.D., Unruh, D.M., and Gilbert, M.C., 1988, Rb-Sr and Sm-Nd isotopic study of the Glen Mountains layered complex: Initiation of rifting within the southern Oklahoma aulacogen: Geology, v. 16, p. 13–17, doi: 10.1130/0091-7613(1988)016<0013:RSASNI>2.3.CO;2.

Larson, E.E., Patterson, P.E., Curtis, G., Drake, R., and Mutschler, F.E., 1985, Petrologic, paleomagnetic, and structural evidence of a Paleozoic rift system in Oklahoma, New Mexico, Colorado, and Utah: Geological Society of America Bulletin, v. 96, p. 1364–1372, doi: 10.1130/0016-7606(1985)96<1364:PPASEO>2.0.CO;2.

Lyashkevich, Z.M., 1987, Magmatism of the Pripyat-Dniepr-Donets paleorift: Kiev, Naukova Dumka, 176 p. (in Russian).

Maystrenko, Yu., Stovba, S., Stephenson, R.A., Bayer, U., Menyoli, E., Gajewski, D., Huebscher, C., Rabbel, W., Saintot, A., Starostenko, V., Thybo, H., and Tolkunov, A., 2003, Crustal-scale pop-up structure in cratonic lithosphere: DOBRE deep seismic reflection study of the Donbas fold belt, Ukraine: Geology, v. 31, p. 733–736, doi: 10.1130/G19329.1.

McConnell, D.A., 1989, Determination of offset across the northern margin of the Wichita uplift, southwest Oklahoma: Geological Society of America Bulletin, v. 101, p. 1317–1332, doi: 10.1130/0016-7606(1989)101<1317:DOOATN>2.3.CO;2.

McConnell, D.A., and Gilbert, M.C., 1990, Cambrian extensional tectonics and magmatism within the Southern Oklahoma aulacogen: Tectonophysics, v. 174, p. 147–157, doi: 10.1016/0040-1951(90)90388-O.

Mickus, K.L., and Keller, G.R., 1992, Lithospheric structure of the south-central United States: Tectonophysics, v. 20, p. 335–338.

Mitchell, B.J., and Landisman, M., 1970, Interpretation of a crustal section across Oklahoma: Geological Society of America Bulletin, v. 81, p. 2647–2656, doi: 10.1130/0016-7606(1970)81[2647:IOACSA]2.0.CO;2.

Moores, E.M., 1991, Southwest U.S.–East Antarctic (SWEAT) connection: A hypothesis: Geology, v. 19, p. 425–428, doi: 10.1130/0091-7613(1991)019<0425:SUSEAS>2.3.CO;2.

Nielsen, S.B., Thomsen, E., Hansen, D.L., and Clausen, O.R., 2005, Plate-wide stress relaxation explains European Palaeocene basin inversions: Nature, v. 435, p. 195–198, doi: 10.1038/nature03599.

Nikishin, A.M., Ziegler, P.A., Stephenson, R.A., Cloetingh, S.A.P.L., Furne, A.V., Fokin, P.A., Ershov, A.V., Bolotov, S.N., and Korotaev, M.V., Alekseev, A.S., Gorbachev, V.I., Shipilov, E.V., Lankreijer, A., Bembinova, E.Yu., and Shalimov, I.V., 1996, Late Precambrian to Triassic history of the East European craton: Dynamics of sedimentary basin evolution: Tectonophysics, v. 268, p. 23–63.

Orlyuk, M.I., and Pashkevich, I.K., 1994, Evaluation of consolidated crust magmatism of the Dniepr-Donets aulacogen: Doklady of the Academy of Sciences of Ukraine, v. 5, p. 125–128.

Pavlenkova, N.I., 1995, Double Moho in the Dnieper-Donets basin: Comptes Rendues de l'Académie des Sciences de Paris, v. 321 (IIa), p. 85–93.

Perry, W.J., Jr., 1989, Tectonic evolution of the Anadarko basin region, Oklahoma: U.S. Geological Survey Bulletin 1866, 19 p.

Poplawskii, K.N., Podladchikov, Yu., and Stephenson, R.A., 2001, Two-dimensional inverse modeling of sedimentary basin subsidence: Journal of Geophysical Research, v. 106, p. 6657–6672, doi: 10.1029/1999JB900211.

Powell, B.N., Gilbert, M.C., and Fisher, J.F., 1980, Lithostratigraphic classification of basement rocks of the Wichita province, Oklahoma: Summary: Geological Society of America Bulletin, v. 91, p. 509–514, doi: 10.1130/0016-7606(1980)91<509:LCOBRO>2.0.CO;2.

Pratt, T., Hauser, E.C., and Nelson, D., 1992, Widespread buried Precambrian layered sequences in the U.S. Mid-Continent: Evidence for large Proterozoic depositional basins: American Association of Petroleum Geologists Bulletin, v. 17, p. 1384–1401.

Purucker, M., 1988, Petrologic, paleomagnetic, and structural evidence of a Paleozoic rift system in Oklahoma, New Mexico, Colorado, and Utah: Discussion: Geological Society of America Bulletin, v. 100, p. 1846–1847, doi: 10.1130/0016-7606(1988)100<1846:PPASEO>2.3.CO;2.

Robbins, S.L., and Keller, G.R., 1992, Complete Bouguer and isostatic residual gravity maps of the Anadarko basin, Wichita Mountains, and surrounding areas, Oklahoma, Kansas, Texas and Colorado: U.S. Geological Survey Bulletin 1866-G, 11 p.

Rowland, T.L., 1974, Lone Star 1 Rodgers unit captures world record depth: Oklahoma Geology Notes, v. 34, p. 185–189.

Saintot, A., Stephenson, R.A., Brem, A., Stovba, S., and Privalov, V., 2003, Paleostress field reconstruction and revised tectonic history of the Donbas fold-and-thrust belt (Ukraine and Russia): Tectonics, v. 22, p. 1059, doi: 10.1029/2002TC001366.

Saintot, A., Stephenson, R.A., Stovba, S.M., Brunet, M.-F., Yegorova, T., and Starostenko, V., 2006, The evolution of the southern margin of Eastern Europe (Eastern European and Scythian platforms) from the latest Precambrian–early Palaeozoic to the Early Cretaceous, *in* Gee, D.G., and Stephenson, R.A., eds., European lithosphere dynamics: Geological Society [London] Memoir 32, p. 481–505.

Shatsky, N.S., 1946, The great Donets basin and the Wichita system: Comparative tectonics of ancient platforms: Akademiya Nauk SSSR Izvestiya, Seriya Geologicheskaya, v. 6, p. 57–90.

Shchipansky, A.A., and Bogdanova, S.A., 1996, The Sarmatian crustal segment: Precambrian correlation between the Voronezh massif and the Ukrainian shield across the Dniepr-Donets aulacogen: Tectonophysics, v. 268, p. 109–126, doi: 10.1016/S0040-1951(96)00227-2.

Shymanovsky, V.A., Sachsenhofer, R.F., Izart, A., and Li, Y., 2004, Numerical modelling of the thermal evolution of the northwestern Dniepr-Donets basin (Ukraine): Tectonophysics, v. 381, p. 61–79.

Simiyu, S., and Keller, G.R., 2001, An integrated geophysical analysis of the upper crust of the southern Kenya rift: Geophysical Journal International, v. 147, p. 543–561, doi: 10.1046/j.0956-540x.2001.01542.x.

Snelson, C.M., Henstock, T.J., Keller, G.R., Miller, K.C., and Levander, A., 1998, Crustal and uppermost mantle structure along the Deep Probe seismic profile: Rocky Mountain Geology, v. 33, p. 181–198.

Sobornov, K., 1995, Structural evolution of the Karpinsky swell, Russia: Comptes Rendues de l'Académie des Sciences de Paris, v. 321 (IIa), p. 161–169.

Sollogub, V.B., Borodulin, M.I., and Chekunov, A.V., 1977, Deep structure of Donbas and adjacent territories: Geophysical Journal, v. 37, p. 23–31.

Starostenko, V.I., Kozlenko, V.G., Oganesyan, S.M., Shen, E.L., Oganesyan, M.G., Yegorva, T.P., and Dyadura, G.V., 1986, 3-D distribution of the density in the crust of the Dniepr graben: Geophysical Journal, v. 8, p. 3–19.

Starostenko, V.I., Danilenko, V.A., Vengrovitch, D.B., Kutas, R.I., Stovba, S.M., Stephenson, R.A., and Kharitonov, O.M., 1999, A new geodynamical-thermal model of rift evolution, with application to the Dnieper-Donets basin, Ukraine: Tectonophysics, v. 313, p. 29–40, doi: 10.1016/S0040-1951(99)00188-2.

Stephenson, R.A., Chekunov, A., and Ilchenko, T., Kaluzhna, L., Baranova, Ye., Starostenko, V.I., Krasovskiy, S., Kozlenko, V., Kuprienko, P., Gordienko, V., Kutas, R., Pashkevich, I., Orluk, M., Kivshik, M.K., Stovba, S.M., Turchanenko, M., Savchenko, V.I., Krivchenkov, B.S., Narkiewicz, M., Dadlez, R., Pokorski, J., and Guterch, A., 1993, Continental rift development in Precambrian and Phanerozoic Europe: EUROPROBE and the Dnieper-Donets rift and Polish trough basins: Sedimentary Geology, v. 86, p. 159–175.

Stephenson, R.A., Stovba, S.M., and Starostenko, V.I., 2001, Pripyat-Dniepr-Donets basin: Implications for rift geodynamics and northern Peri-Tethyan tectonic history, *in* Ziegler, P.A., et al., eds., Peri-Tethyan rift/wrench basins and passive margins: Peri-Tethys Memoir 6, Mémoires du Muséum National d'Histoire Naturelle, v. 186, p. 369–406.

Stephenson, R.A., Yegorova, T., Brunet, M.-F., Stovba, S., Wilson, M., Starostenko, V., Saintot, A., and Kusznir, N., 2006, Late Palaeozoic intra- and pericratonic basins on the East European craton and its margins, *in* Gee, D.G., and Stephenson, R.A., eds., European lithosphere dynamics: Geological Society [London] Memoir 32, p. 463–479.

Stovba, S.M., and Stephenson, R.A., 1999, The Donbas Foldbelt: Its relationships with the uninverted Donets segment of the Dniepr-Donets basin, Ukraine: Tectonophysics, v. 313, p. 59–83, doi: 10.1016/S0040-1951(99)00190-0.

Stovba, S.M., and Stephenson, R.A., 2003, Style and timing of salt tectonics in the Dniepr-Donets basin (Ukraine): Implications for triggering and driving mechanisms of salt movement in sedimentary basins: Marine and Petroleum Geology, v. 19, p. 1160–1189.

Stovba, S.M., Stephenson, R.A., and Kivshik, M., 1996, Structural features and evolution of the Dniepr-Donets basin, Ukraine, from regional seismic reflection profiles: Tectonophysics, v. 268, p. 127–147, doi: 10.1016/S0040-1951(96)00222-3.

Stovba, S.M., Maystrenko, Yu.P., Stephenson, R.A., and Kusznir, N.J., 2003, The formation of the southeastern part of the Dniepr-Donets basin: 2-D forward and reverse syn-rift and post-rift modeling: Sedimentary Geology, v. 156, p. 11–33, doi: 10.1016/S0037-0738(02)00280-4.

Suleiman, A.S., 1995, Geophysics of the rifts associated with the Anadarko basin (North America), and the Sirte basin, North Africa [Ph.D. thesis]: El Paso, University of Texas, 165 p.

Suleiman, A.S., and Keller, G.R., 1985, A geophysical study of basement structure in northeastern New Mexico: New Mexico Geological Society, 36th Field Conference Guidebook, p. 153–159.

Thomas, W.A., 1991, The Appalachian-Ouachita rifted margin of southeastern North America: Geological Society of America Bulletin, v. 103, p. 415–431, doi: 10.1130/0016-7606(1991)103<0415:TAORMO>2.3.CO;2.

Thybo, H., Maguire, P., Birt, C., and Perchuc, E., 2000, Seismic reflectivity of the lower crust and magmatic underplating beneath the Kenya rift: Geophysical Research Letters, v. 27, p. 2745–2749, doi: 10.1029/1999GL011294.

van Wees, J.-D., Stephenson, R.A., Stovba, S.M., and Shymanovsky, V.A., 1996, Tectonic variation in the Dniepr-Donets basin from automated modelling of backstripped subsidence curves: Tectonophysics, v. 268, p. 257–280, doi: 10.1016/S0040-1951(96)00233-8.

Volozh, Yu.A., Antipov, M.P., Leonov, Yu.G., Morozov, A.F., and Yurov, Yu.G., 1999, Setting of the Karpinsky swell: Geotectonics, v. 1, p. 28–43.

Volozh, Yu.A., Antipov, M.P., Brunet, M.-F., Garagash, I.A., Lobkovskii, L.I., and Cadet, J.-P., 2003, Pre-Mesozoic geodynamics of the Precaspian basin (Kazakhstan): Sedimentary Geology, v. 156, p. 35–58, doi: 10.1016/S0037-0738(02)00281-6.

Walper, J., 1977, Paleozoic tectonics of the southern margin of North America: Transactions Gulf Coast Association of Geological Societies, v. 27, p. 230–241.

Wilson, B.M., and Lyashkevich, Z.M., 1996, Magmatism and the geodynamics of rifting of the Pripyat-Dnieper-Donets rift, East European Platform: Tectonophysics, v. 268, p. 65–81, doi: 10.1016/S0040-1951(96)00234-X.

Ye, H., Royden, L., Burchfiel, C., and Schuepbach, M., 1996, Late Paleozoic deformation of interior North America: The greater Ancestral Rocky Mountains: American Association of Petroleum Geologists Bulletin, v. 80, p. 1397–1432.

Yegorova, T.P., Stephenson, R.A., Kozlenko, V.G., Starostenko, V.I., and Legostaeva, O.V., 1999, 3-D gravity analysis of the Dniepr-Donets basin and Donbas foldbelt, Ukraine: Tectonophysics, v. 313, p. 41–58, doi: 10.1016/S0040-1951(99)00189-4.

Yegorova, T.P., Stephenson, R.A., Kostyuchenko, S.L., Baranova, E.P., Starostenko, V.I., and Popolitov, K.E., 2004, Structure of the lithosphere below the southern margin of the East European craton (Ukraine and Russia) from gravity and seismic data: Tectonophysics, v. 381, p. 81–100, doi: 10.1016/j.tecto.2002.08.003.

Zonenshain, L.P., Kumin, M.I., and Natapov, L.M., 1990, Geology of the USSR: A plate-tectonic synthesis: American Geophysical Union Geodynamics Series, v. 21, p. 1–242.

Manuscript Accepted by the Society 1 February 2007

Printed in the USA

The Geological Society of America
Memoir 200
2007

Accretionary orogens in space and time

Kent C. Condie
Department of Earth and Environmental Science, New Mexico Institute of Mining and Technology, Socorro, New Mexico 87801, USA

ABSTRACT

Accretionary orogens form along continental margins where oceanic lithosphere is subducted. They are primary sites of juvenile continental crust production and have been active on Earth since the earliest Archean. Orogen lifetimes expressed as accretion intervals range from 50 to over 300 m.y. The short duration of Late Archean accretionary orogens (<70 m.y.) may reflect the short duration of a global mantle plume event at 2.7 and 2.5 Ga. Although there is no simple relationship between the onset or duration of accretionary orogens and the supercontinent cycle, many post-Archean orogens terminate with continent-continent collisions during supercontinent assembly.

Average terrane lifespan is typically 100–200 m.y. in post–1 Ga orogens, 50–100 m.y. in pre–1 Ga Proterozoic orogens, and 70–700 m.y. in Archean orogens. Accretionary orogens can be grouped into two end members: *simple orogens* containing chiefly juvenile terranes with lifespans of <100 m.y., and *complex orogens* with both juvenile accreted components and exotic microcratons, with terrane lifespans of ≥100 m.y. Terrane lifespan is controlled by (1) terrane tectonic setting, (2) complexity of precollisional terrane history, (3) availability of continental crust on Earth, and (4) plate history of ocean basins adjacent to accretionary orogens.

Average accretion rates in accretionary orogens are 70 to 150 km^3/km/m.y. in Phanerozoic orogens and 100 to 200 km^3/km/m.y. in Precambrian orogens. Some orogens at 2.7 Ga have unusually high accretion rates greater than 300 km^3/km/m.y., which may reflect a global mantle plume event. Production rates of juvenile crust in accretionary orogens are typically 10%–30% lower than total accretion rates, but can be up to 50% lower in Phanerozoic orogens.

Keywords: accretionary orogen, orogen, terrane, juvenile crust.

INTRODUCTION

Accretionary orogens are major sites of continental growth and mineralization and include Archean granite-greenstone terranes, many Proterozoic orogens (e.g., the Birimian of West Africa, Svecofennian of Finland and Sweden, Yavapai in the southwestern United States, and the Arabian-Nubian shield), Neoproterozoic-Paleozoic orogens in central Asia and eastern North America, as well as Neoproterozoic to Holocene orogens of the circum-Pacific and Caribbean. Accretionary orogens form at sites of subduction of oceanic lithosphere and consist of accretionary wedges containing material accreted from the downgoing plate and

Condie, K.C., 2007, Accretionary orogens in space and time, *in* Hatcher, R.D., Jr., Carlson, M.P., McBride, J.H., and Martínez Catalán, J.R., eds., 4-D Framework of Continental Crust: Geological Society of America Memoir 200, p. 145–158, doi: 10.1130/2007.1200(09). For permission to copy, contact editing@geosociety.org.

eroded from the upper plate, island arcs, ophiolites, fragments of oceanic plateaus, microcratons, syn- to postaccretion granitic and metamorphic rocks, exhumed high-pressure metamorphic rocks, and syn- to postaccretion clastic sedimentary basins.

The major site of juvenile continental crust production is along convergent plate margins in accretionary orogens. Juvenile crust is produced in arc systems from magmas derived primarily from the mantle wedge, and it also may be added to continents by collision and accretion of oceanic arcs, oceanic plateaus, and oceanic crust (ophiolites).

Accretionary orogens have much in common, yet there are many differences, some of which appear to change with geologic time. Although much is known about a few accretionary orogens, such as the American Cordillera, Archean greenstone terranes of the Superior province, and terranes of the Appalachian orogen, little effort has gone into comparing and contrasting accretionary orogens. In this paper, accretionary orogens of various ages are compared and discussed in terms of terrane lifespan, the supercontinent cycle, and rates of accretion and juvenile crust production.

ACCRETIONARY INTERVALS AND TERRANE LIFESPAN

Two important characteristics of accretionary orogens are (1) the orogen accretion interval and (2) terrane lifespan. In this study, *accretion interval* is defined as the difference between the oldest and youngest age of terrane accretion to a continent in a given orogen. For orogens with long strike lengths, such as the Altaids in the Phanerozoic and the Yavapai-Penokean-Svecofennian in the Paleoproterozoic, accretion interval may be diachronous along the orogen. Some terranes are rifted from continental margins, which provides a time marker for the age of these terranes. However, rifting events have proved impractical to define terrane lifespan because (1) many, if not most, terranes did not begin life by rifting, and (2) the timing of rifting is often unknown. For this reason, *terrane lifespan* is defined as the time interval between the age of the oldest rocks in a terrane (excluding basement) and the time that a terrane collides with and is accreted to a continent. In the case of composite terranes (two or more terranes that collide before collision with a continent), lifespan is the time interval between the oldest terrane and the final collision of the composite terrane with a continent.

Orogen accretion intervals vary widely from ~50 m.y. to over 300 m.y., with the shortest intervals (<100 m.y.) characteristic of the Archean (Fig. 1). Of the Archean orogens studied, only the Yilgarn has an accretion interval in excess of 70 m.y. Phanerozoic and Proterozoic orogens typically have accretion intervals between ~100 and 300 m.y. The Cordilleran, Andean, and Japan orogens are still active, and hence their accretion intervals are minimum values only.

Within a given orogen, terrane lifespan is extremely variable, ranging between 20% and 60% for one standard deviation of the mean (Fig. 1). The shortest lifespans occur in the Late Archean and Paleoproterozoic where mean lifespan is generally less than

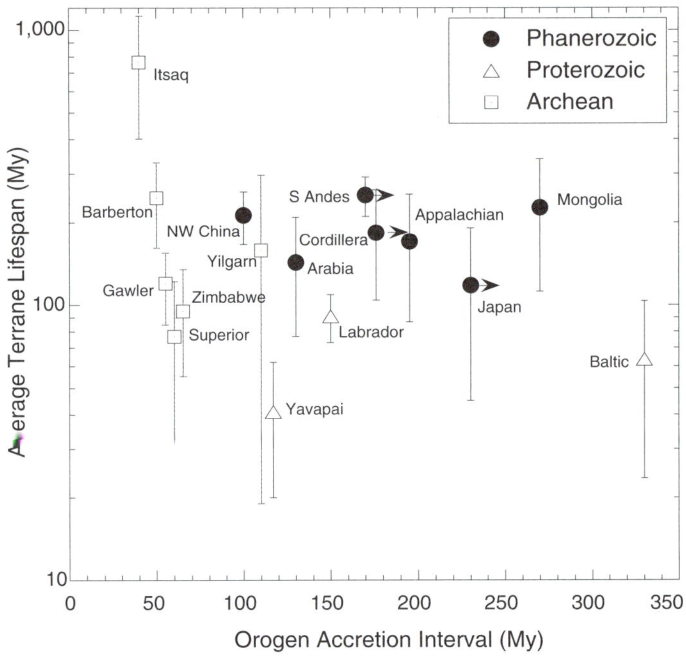

Figure 1. Relationship of average terrane lifespan to orogen accretion interval. Horizontal arrows indicate currently active orogens, and accretion intervals plotted are lower limits only. Data from Table 1 and references given in text.

100 m.y. The longest terrane lifespans are found in the Itsaq orogen in SW Greenland, with an average ~760 m.y. for six terranes, and in the Pilbara craton in Western Australia, with two large terranes with lifespans of 530 and 280 m.y., respectively (Table 1). Average terrane lifespan in Phanerozoic orogens generally falls between 100 and 250 m.y. It is clear from Figure 1 that there is no relation between average terrane (or composite terrane) lifespan and orogen accretion interval.

These results suggest that two end-member scenarios exist for accretionary orogens, which we will refer to as simple and complex orogens. A *simple orogen* is one that is characterized by rapid accretion of juvenile oceanic terranes (arcs, oceanic crust, oceanic plateaus) where average terrane lifespan is <100 m.y. (Fig. 2). In contrast, a *complex orogen* comprises not only juvenile accreted components but also exotic microcratons, and average terrane lifespan is ≥100 m.y. Extensional accretionary orogens (such as the Lachlan orogen in Australia) are a special case of simple orogens, which are characterized by poorly developed foreland basins, little or no crystalline basement, and an extended history of turbidite deposition in backarc basins (Collins, 2002). Unlike most simple orogens, complex orogens may undergo extensive rifting and transcurrent faulting during the accretion cycle, which creates new terranes and moves them laterally along the orogen. Of course, both simple and complex scenarios may occur along strike in the same orogen, but one or the other type often dominates. There also seems to be a time dependence of orogen type: Simple orogens are most common in the Proterozoic, whereas complex orogens dominate

TABLE 1. ACCRETIONARY OROGENS AND TERRANE AND SUPERTERRANE CHARACTERISTICS

Terrane	Oldest rocks (Ma)	Accretion age (Ma)	Lifespan (m.y.)
Cordilleran	AI = 250 – 55 = 195 m.y.		
Alexander	540	185	355
Angayucham	390	100	290
Bridge River	411	246	165
Broad Pass	350	210	140
Cache Creek	380	225	155
Chilliwack	390	220	170
Chugach	150	100	50
Chulitna	390	230	160
Dillinger	420	125	295
Gravina	105	55	50
Kahiltna	270	190	80
Kandik River	635	185	450
Koyukuk	340	200	140
McKinley	400	220	180
McLaren	280	200	80
Mystic	450	125	325
Pacific Rim	130	80	50
Peninsular	390	230	160
Quesnelia	235	140	95
Slide Mountain	395	190	205
Stikine	410	250	160
Susitna	310	200	110
Taku	350	250	100
West Fork	150	100	50
Windy	385	135	250
Wrangellia	370	220	150
Mean	347	177	170
Median	383	195	158
1σ	121	57	103
Japan	AI = 300 – 70 = 230 m.y.		
Akiyoshi	340	300	40
Chichibu	245	140	105
Gosaisho	180	110	70
Hitachi-Takanuki	300	250	50
Kanoashi	295	200	95
Matsugataira-Motai	400	200	200
Mino	330	175	155
Miyamori-Hayachine	450	240	210
Renge	480	250	230
Sanbagawa	300	100	200
Sangun	300	250	50
N Shimanto	145	90	55
S Shimanto	90	70	20
Sorachi-Yezo	140	100	40
Southern Kitakami	440	240	200
Tanba	325	160	165
Mean	298	180	118
Median	300	168	100
1σ	108	77	73
NW China	AI = 360 – 250 = 110 m.y.		
Altai	520	265	255
Balkhash	540	285	255
Bogda	415	260	155
East Junggar	450	260	190

(continued)

TABLE 1. (*continued*)

Terrane	Oldest rocks (Ma)	Accretion age (Ma)	Lifespan (m.y.)
NW China (cont.)	AI = 360 – 250 = 110 m.y.		
Irtysh-Zaysan	420	275	145
Tarbgatay-Chingiz	550	360	190
Tarim	540	250	290
Tian Shan	520	290	230
Turpan	420	260	160
West Junggar	500	280	220
Yili	540	300	240
Mean	492	280	212
Median	520	275	220
1σ	55	30	47
Southern Andes	AI = 600 – 430 = 170 m.y.		
Chilenia	700	430	270
Cordoba	800	600	200
Cuyania	750	460	290
Famatina	690	475	215
Pampia	800	520	280
Precordillera	515	455	60
Mean	709	490	219
Median	750	475	270
1σ	53	66	41
Arabian-Nubian	AI = 760 – 630 = 130 m.y.		
Afif	800	655	145
Ar Rayn	850	650	200
Asir	870	720	150
Hijaz	850	750	100
Jeddah	800	760	40
Midyan	850	630	220
Mean	837	694	143
Median	850	688	148
1σ	29	56	66
Labrador	AI = 1800 – 1650 = 150 m.y.		
Lower Aillik	1880	1795	85
Upper Aillik	1860	1800	60
Groswater Bay	1760	1660	100
Hawke River	1740	1650	90
Lac Joseph	1750	1665	85
Lake Melville	1740	1650	90
Mealy Mountains	1780	1650	130
Molson Lake	1751	1660	91
Wilson Lake	1750	1660	90
Mean	1779	1688	91
Median	1751	1660	90
1σ	53	62	18
Superior	AI = 2700 – 2640 = 60 m.y.		
Abitibi	2820	2675	145
SW Abitibi	2830	2700	130
W Abitibi	2810	2700	110
Batchawana	2730	2677	53
English River	2750	2680	70
Hemlo-Schreiber	2723	2690	33
Lake Abitibi	2806	2689	117
Manitou-Wadge	2690	2680	10

(continued)

TABLE 1. (*continued*)

Terrane	Oldest rocks (Ma)	Accretion age (Ma)	Lifespan (m.y.)
Superior (cont.)	AI = 2700 – 2640 = 60 m.y.		
Michipicoten	2730	2675	55
Minto	2750	2700	50
Shebandowan	2720	2680	40
Timmons	2735	2690	45
Uchi	2780	2700	80
Vermillion	2720	2680	40
Wabigoon	2775	2700	75
Wawa-Quetico	2722	2640	82
Winnipeg River	2880	2700	180
Mean	2763	2686	77
Median	2750	2689	70
1σ	51	15	45
Zimbabwe	AI = 2710 – 2645 = 65 m.y.		
Harare-Shamva	2715	2665	50
S Zimbabwe	2830	2668	162
Zimb 1	2810	2750	60
Zimb 2	2823	2702	121
Zimb 3	2831	2697	134
Zimb 4	2720	2645	75
Zimb 5	2730	2665	65
Mean	2780	2685	95
Median	2810	2668	75
1σ	55	35	43
Pilbara			
East Pilbara	3530	3000	530
West Pilbara	3280	3000	280
Mean	3405	3000	405
Appalachians	AI = 500 – 280 = 220 m.y.		
Arequipa	600	470	130
Avalonia	780	440	340
Brompton	670	450	220
Carolina	670	360	310
Cartoogechaye	675	460	215
Cat Square	400	360	40
Central Maine	500	390	110
Chopawamsic	550	350	200
Cowrock	675	460	215
Ellsworth	550	300	250
Gaffney	620	450	170
Goochland	650	360	290
Juliette	560	445	115
Maritimes 1	590	460	130
Maritimes 2	610	480	130
Maritimes 3	620	470	150
Meguma	500	350	150
Oaxaquia/Chortis	520	280	240
Potomac	570	450	120
Shelburne Falls	500	450	50
Suwannee	400	280	120
Tugaloo	565	360	205
Yucatan	650	280	370
Mean	584	398	182
Median	590	440	170
1σ	90	69	84

(*continued*)

TABLE 1. (*continued*)

Terrane	Oldest rocks (Ma)	Accretion age (Ma)	Lifespan (m.y.)
Mongolia	AI = 520 – 250 = 270 m.y.		
Altai	500	250	250
Atasbogd	440	350	90
Baaran	400	320	80
Bayangol	760	420	340
Bayanhongor	570	450	120
Baydrag	700	490	210
Baytag	400	310	90
Darhad	750	410	340
Dzhida	680	430	250
Edren	420	300	120
Ereendavaa	790	360	430
Gargan	850	600	250
Gobi Altai	510	330	180
Gurvansayhan	450	320	130
Haraa	520	420	100
Hovd	530	380	150
Hug	750	480	270
Idermeg	700	370	330
Lake	570	420	150
Mandalovoo	480	310	170
Tarvagatay	800	450	350
Tsagaan Uul	770	280	490
Zag	500	340	160
Zavhan	850	520	330
Zoolen	480	300	180
Mean	607	380	226
Median	570	370	180
1σ	152	75	116
Svecofennian-Gothian	AI = 1910 – 1580 = 330 m.y.		
Amal	1610	1580	30
Amhorian	1590	1575	15
Belarus-Baltic	1890	1830	60
NW Cent Belarus	1995	1850	145
East Latvian	1850	1830	20
Horred	1640	1580	60
Kiruna	2000	1870	130
Orebro	1850	1830	20
Osmarian	1657	1575	82
Osnitsk-Mikashev	2000	1880	120
Skaldo	1900	1850	50
Skelleftea	1950	1910	40
Sveco	1920	1840	80
Tamper	1930	1900	30
West Lithuanian	1890	1820	70
Mean	1845	1781	63
Median	1890	1830	60
1σ	146	130	41
Yavapai	AI = 1760 – 1650 = 110 m.y.		
Alder	1760	1700	60
Ash Fk	1780	1700	80
Bagdad	1740	1700	40
Big Bug	1760	1700	60
DosCab-Pinal	1680	1660	20
Dubois	1760	1730	30
Grn Gulch	1765	1700	65

(*continued*)

TABLE 1. (*continued*)

Terrane	Oldest rocks (Ma)	Accretion age (Ma)	Lifespan (m.y.)
Yavapai (cont.)	AI = 1760 – 1650 = 110 m.y.		
Grn Mountain	1785	1760	25
Pecos	1720	1700	20
Pedernal	1690	1670	20
Taos	1750	1720	30
Mean	1745	1704	41
Median	1760	1700	30
1σ	34	27	22
Yilgarn	AI = 2750 – 2640 = 110 m.y.		
S Cross	2807	2700	107
E Goldfields	2690	2640	50
Kalgoorlie	2700	2650	50
Murchison	3000	2680	320
Narryer	3100	2750	350
Yeelirrie	2720	2650	70
Mean	2836	2678	158
Median	2764	2665	89
1σ	174	42	139
Gawler craton	AI = 2480 – 2420 = 60 m.y.		
1	2560	2450	110
2	2540	2470	70
3	2545	2430	115
4	2565	2420	145
5	2640	2480	160
Mean	2570	2450	120
Median	2560	2450	115
1σ	40	25	35
Barberton	AI = 3260 – 3210 = 50 m.y.		
Ancient Gneiss	3640	3230	410
Kaap Valley	3306	3210	96
Songimvelo	3504	3260	244
Steynsdorp	3550	3260	290
Umuduha	3413	3230	183
Mean	3483	3238	245
Median	3459	3245	214
1σ	108	24	84
Itsaq	AI = 2740 – 2700 = 40 m.y.		
Aasivik	3600	2720	880
Akia	3220	2700	520
Faeringehavn	3860	2720	1140
Isukasia	3820	2730	1090
Tasiusarsuaq	2920	2730	190
Tri Bodre	3500	2740	760
Mean	3487	2723	763
Median	3550	2725	820
1σ	362	14	361

Note: References given in the text. AI—orogen accretion interval; Cent—Central; Dos Cab-Pinal—Dos Cabazas-Pinal; Fk—Fork; Grn—Green.

during the Phanerozoic (Japan is an exception). The Archean has both types of orogens.

A SURVEY OF ACCRETIONARY OROGENS

One of the most well-known Phanerozoic accretionary orogens is the American Cordillera, and it is perhaps the type example of a complex orogen. Although some of the Cordilleran terranes appear to have traveled great distances before accretion to Laurentia, it is probable that the majority of these terranes formed along or near the west coast of Laurentia (Plafker et al., 1989; Plafker and Berg, 1994). Some appear to have been composite terranes at the time of accretion (Monger et al., 1982). Almost all tectonic settings are represented in Cordilleran terranes, with continental-margin arcs and backarcs dominating (Condie and Chomiak, 1996). Many terranes contain several rock packages separated by unconformities indicating complex histories involving several tectonic settings prior to final accretion to the continent. Terranes were accreted over a long time interval beginning ca. 250 Ma in central Nevada and continuing to 55 Ma when the Gravina terrane was accreted to Alaska ("Cordilleran" in Table 1). Some terranes, such as Alexander, Peninsular, and Wrangellia, are composite terranes, having undergone one or more collisions with constituent blocks before finally colliding with Laurentia (Monger et al., 1982; Plafker et al., 1989; Condie and Chomiak, 1996). Terrane lifespan is quite variable, ranging from ~450 m.y. for Kandik River to 50 m.y. for Chugach, Gravina, and Pacific Rim terranes (Condie and Chomiak, 1996). The average terrane lifespan is ~170 m.y. (Table 1). As long as the Cordilleran orogen opens into the Pacific basin, future terrane accretion is possible.

Although the Appalachian orogen is a classic example of a collisional orogen, prior to the terminal continent-continent collision in the Late Permian, this orogen was a complex accretionary orogen. Terrane collisions occurred between ca. 500 and 280 Ma, and the terminal collision with Baltica–West Africa occurred at ca. 250 Ma. Terranes are very diverse in character, ranging from microcratons (rifted fragments of continent with cratonic sedimentary successions) to both oceanic and continental-margin arcs, ophiolites, and continental rifts (Keppie, 1989). Some microcratons include older Precambrian basement with ages between 1 and 2 Ga (Murphy et al., 1999, 2004; Keppie et al., 2003). Many terranes were rifted from the northern margin of Gondwana, beginning in the Neoproterozoic and Cambrian, and fall into two categories (Murphy et al., 2004): (1) terranes that have 1.3–1.0 Ga basement and formed along the northern Gondwana margin by 650 Ma, and (2) terranes that formed along the West African margin and show recycling of 2–3 Ga crust. Later rifted from Gondwana, these terranes were caught up in Appalachian, Caledonide, and Variscan accretionary orogens and now reside in North America and western Europe. Appalachian terrane lifespan ranges from as little as 50 m.y. (for instance the Shelburne Falls terrane, a North American continental-margin arc; Karabinos et

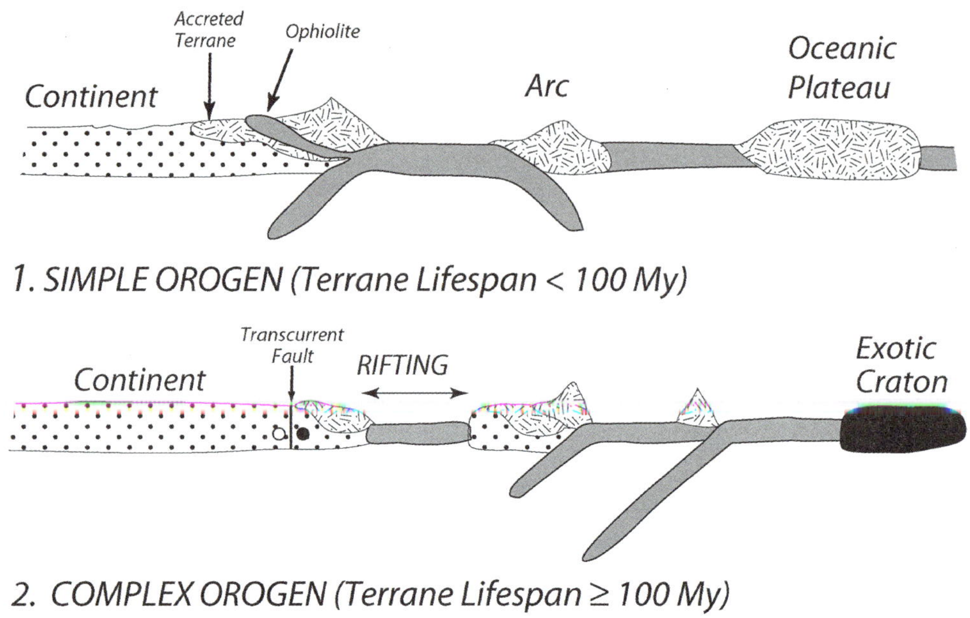

Figure 2. Diagrammatic sections of simple and complex accretionary orogens.

al., 1998) to more than 300 m.y. in some superterranes (Yucatan, Avalonia, and Carolina) ("Appalachians" in Table 1).

Another large and long-lived accretionary orogen in the Phanerozoic is the Altaids in eastern Asia, which has features of both simple and complex orogens (Xiao et al., 2004). Recent work in Mongolia shows the existence of at least 44 accreted terranes in this region (Şengör and Natal'in, 1996; Badarch et al., 2002). Reconnaissance mapping and description of stratigraphic sections in each terrane indicate numerous tectonic settings, including microcratons, island arcs, continental-margin arcs, forearc/backarc regimes, accretionary complexes, and ophiolites. The arcs, ophiolites, and accretionary complexes appear to have accreted around a small Precambrian craton in the Hangay region of northern Mongolia. The oldest rocks in any of the Mongolian terranes (excluding basement) appear to be Neoproterozoic ("Mongolia" in Table 1). Like most other accretionary orogens, the timing and spatial distribution of accretion is complex and does not fall into a simple order outwards from the Precambrian craton. The final accretion ages are chiefly Cambro-Ordovician, Devonian, and Late Permian, with an average accretion age of ca. 380 Ma (Table 1). Although there is a scarcity of Nd isotopic data from this region, inferred tectonic settings indicate that the Mongolian terranes represent a mixture of juvenile and reworked crust. Based on 25 terranes for which there are data, terrane lifespan ranges from 80 to at least 490 m.y., with an average of ~226 m.y. (Table 1).

The Altaids, NW China, and adjacent areas in Siberia comprise at least 11 large terranes accreted to Siberia chiefly in the late Paleozoic ("NW China" in Table 1) (Huang et al., 1987; Coleman, 1989). Most or all of these terranes are composite and include two or more tectonic settings. Rock associations and structure indicate the presence of microcratons, island and continental-margin arcs, and ophiolites. The Tarim and Siberia cratons, south and north of the NW China terranes, respectively, have distinct polar wander paths until the end of the Permian when they collided to form the intervening Altaids orogen. Ages of the oldest rocks in these terranes (excluding basement) are not well known but appear to fall near the Cambrian-Neoproterozoic boundary. Accretion ages for 11 of the terranes range from ca. 360 to 250 Ma, with most falling in the Late Permian. Terrane lifespan ranges only from ~290 to 145 m.y., with an average of ~212 m.y. (Table 1). Although the Altaids orogen came to an end with the collision of the Tarim–North China craton with Asia on the south in the Late Permian, terrane and microcraton accretion continued during the Mesozoic and Tertiary, adding China, Tibet, SE Asia and India. In fact, it is continuing today with the collision of Australia with SE Asia in Indonesia.

The Japanese islands have grown oceanward by ~400 km since the Devonian, chiefly by growth of accretionary complexes (Isozaki, 1997; Maruyama, 1997). The Oki and Southern Kitakami belts in western Japan were originally part of the Yangtze craton and were rifted away ca. 750–700 Ma when Rodinia began to break up. Circa 500 Ma, a subduction zone developed along the eastern margin of these terranes, rifting subsided, and accretionary growth began along the Pacific margin. Unlike many Phanerozoic accretionary orogens, Japan comprises chiefly accreted juvenile oceanic terranes, and for this reason is classified as a simple orogen. Except for the fragments of oceanic crust accreted to the growing continental margin, most terranes in Japan are nearly autochthonous (Isozaki et al., 1990). They have grown chiefly by addition of oceanic crust and trench sediments scraped off into the accretionary prism. Hitachi-Takanuki and related terranes in northern Honshu, however, are allochthonous terranes that collided with Honshu during the collision of the Yangtze and North China cratons ca. 250 Ma. Accretion ages of Japanese terranes range from ca. 70 to 300 Ma, with an average accretion age of 180 Ma ("Japan" in Table 1). Terrane lifespan is highly variable, ranging from 20 m.y. or less to over 200 m.y., with an average of 118 m.y. Today, Japan is still an active, open-ended accretionary orogen.

Recent studies in the Andes indicate that as much as 40% of the growth in this orogen since the Neoproterozoic is due to terrane collision and accretion (Cordani et al., 2000; Vujovich et al., 2004). One of the first terranes identified with Laurentian affinities is the Precordillera terrane, rifted from SW Laurentia during the Cambrian and accreted to South America ca. 455 Ma (Thomas and Astini, 1996). Although terranes and terrane boundaries are not as yet well established, at least six major terranes compose the southern Andes in Argentina and Chile (Ramos, 2004), and at least two of these terranes (Precordillera and Cuyania) were derived from Laurentia. The six Andean terranes were accreted to the Rio de La Plata craton between 600 and 430 Ma and have an average terrane lifespan of ~219 m.y. ("Southern Andes" in Table 1).

The best-known example of a Neoproterozoic accretionary orogen is the Arabian-Nubian orogen, which formed by terrane collisions chiefly between 760 and 630 Ma (Stoeser and Camp, 1985; Abdelsalam and Stern, 1996; Stern, 1994). Terranes are largely juvenile in character with island arcs and ophiolites dominating, and thus, the Arabian-Nubian orogen is generally classified as a simple orogen. Based on six terranes for which adequate data are available ("Arabian-Nubian" in Table 1), terrane lifespan ranges from over 200 m.y. to <50 m.y., with an average of ~143 m.y.

Perhaps the largest accretionary orogen of all time is the Yavapai-Penokean-Svecofennian orogen, which formed along one margin of a Paleoproterozoic supercontinent. This orogen, with a strike length >10,000 km, may have extended from south-central Australia, along the southern and eastern coasts of Laurentia, across southern Baltica, and perhaps into western Amazonia (Karlstrom et al., 2001; Geraldes et al., 2001). In addition to the Yavapai, Penokean, and Svecofennian events, it includes events such as the Labradorian, Mazatzal, and Gothian. Terrane accretion in this orogen began ca. 1900 Ma and continued until the orogen shut down during Grenvillian continent-continent collisions ca. 1100 Ma. Chiefly between 1650 and 1350 Ma, large numbers of A-type granitic plutons were intruded into this orogen. Although any given segment of the orogen preserves a history of terrane collision and accretion of no more than 350 m.y., the orogen's total lifetime approaches 800 m.y. ("Svecofennian-Gothian" in Table 1). Accretion intervals for various segments of the orogen should be considered minima, because the younger parts of the orogen (1.5–1.2 Ga) either are poorly exposed or were rifted away during the breakup of Rodinia 700 Ma. To characterize this immense orogen, we will examine three sections across the orogen: Baltica, NE Canada, and SW United States. Results from Scandinavia and Russia indicate that the Baltic shield grew westward for over 300 m.y. (1910–1580 Ma; Table 1) by terrane accretion (Park, 1985; Skiold, 1988; Gorbatschev and Bogdanova, 1993; Ahall et al., 1995; Bogdanova et al., 2001). Accreted terranes are chiefly continental-margin arcs and associated backarc basins, with a few occurrences of ophiolite and remnants of oceanic plateaus. The last terranes for which we have isotopic ages accreted to Baltica in SW Sweden and adjacent parts of Norway ca. 1580 Ma during the Gothian event (Table 1). The history of the Baltic segment of the orogen between 1580 and 1100 Ma (the terminal Grenvillian collision) remains largely unknown. Paleoproterozoic terrane lifespan in the Baltic shield ranges from 15 to 145 m.y., with an average lifespan of ~63 m.y.

The Labradorian orogen in NE Canada is also part of the Yavapai-Penokean-Svecofennian orogen, and this part of the orogen was active for over 150 m.y., between ca. 1800 and 1650 Ma ("Labrador" in Table 1) (Gower, 1996; Ahall and Gower, 1997; Gower and Krogh, 2002). The Labradorian orogen includes chiefly backarc basins and other arc-related terranes accreted between 1800 and 1650 Ma, and it is largely a simple orogen. Little evidence is preserved for the time period between 1650 and ca. 1500 Ma. Perhaps the plate boundary changed from convergent to divergent (rifting) or transcurrent during this time interval. Between 1500 and 1100 Ma, tectonic settings in the Labradorian orogen ranged from backarc rifts to passive continental margins (Gower, 1996). The orogen terminated with the collision of Baltica and North America as part of the formation of Rodinia ca. 1100 Ma.

The Yavapai-Mazatzal (hereafter, Yavapai) orogen in the southwestern United States formed by terrane accretion chiefly between 1760 and 1650 Ma with an average accretion age of ca. 1704 Ma ("Yavapai" in Table 1) (Condie, 1992; Condie and Chomiak, 1996; Karlstrom et al., 2001). From outcrop, 11 terranes have been recognized in the Yavapai orogen, which include four island arc terranes and seven continental-margin arc and associated backarc basin terranes. Remnants of an accreted oceanic plateau have also been described from west-central Arizona (Condie et al., 2002). Terrane lifespans fall into two groups: four terranes in west-central Arizona with lifespans of 60–80 m.y. and the remaining seven terranes with very short lifespans of 20–40 m.y. (Table 1). Again this segment of the orogen is an example of a simple orogen. Although it may be possible to sort out the post–1650 Ma history of the Yavapai orogen with detrital zircon ages in younger sediments, at present we have little evidence (other than A-type granites) for the tectonic setting of this region between 1630 and 1300 Ma when sedimentary basins developed on older crust (Karlstrom et al., 2001).

Although most investigators agree that Archean granite-greenstone provinces are accretionary orogens that have grown by plate-tectonic processes (Langford and Morin, 1976; Smithies and Champion, 2000), some have challenged this paradigm (Hamilton, 1998; Bleeker, 2002). However, it would seem to me that rather than abandon plate tectonics, which so well explains so much observational data in the Archean, it is better to look for relatively minor changes in tectonic regime to explain some of the differences we see in the Archean (Condie and Benn, 2006). No other Archean crustal province has received such intense study as the Superior province in Canada. This province seems to be the type example of progressive growth of an Archean craton by largely juvenile terrane collisions (Kimura et al., 1993; Mueller et al., 1996; Corfu and Stott, 1998; Percival et al., 2004). Rock associations and geochemistry suggest that collision and accretion of both arcs and oceanic plateaus contributed to the growth of the Superior province (Kimura et al., 1993; Tomlinson and Condie, 2001; Polat and Kerrich, 2001). Growth is concentrated around the southern and northern margins of a 3.0–2.8 Ga nucleus, with most collisions and accretion concentrated between 2700 and 2640 Ma ("Superior" in Table 1). Unlike most Phanerozoic accretionary orogens, terrane lifespan is relatively short, averaging ~77 m.y.

The Yilgarn province in Western Australia formed at about the same time as the Superior province and is another example of an Archean accretionary orogen. Although the exact number of terranes is disputed, there are at least eight terranes, which appear to have collided and accreted over ~110 m.y. at 2700 Ma (Myers, 1993; Griffin et al., 2004) ("Yilgarn" in Table 1). Although many Yilgarn terranes are remnants of oceanic arcs and backarcs, some, such as the Murchison and Narryer terranes, are microcratons containing rocks up to 3.1 Ga. Thus the Yilgarn, unlike the Superior, is an example of a complex orogen. Terrane lifespan is extremely variable, ranging from as low as 50 m.y. for Kalgoorlie and Eastern Goldfields to over 300 m.y. for the Murchison and Narryer microcratons.

In contrast to the Yilgarn craton, the Gawler craton in southern Australia is a relatively short-lived simple accretionary orogen (60 m.y.) ("Gawler craton" in Table 1). The oldest rocks are metasedimentary ca. 2640 Ma, and most igneous activity is concentrated ca. 2550 Ma (Betts et al., 2002; Swain et al., 2004). Both arc- and plume-like volcanic successions have been described from Gawler terranes. Terrane lifespan ranges from ~70 to over 150 m.y.

The Zimbabwe craton is another Late Archean simple accretionary orogen that formed at ca. 2700 Ma. Isotopic ages are consistent with westward crustal growth between 2710 and 2645 Ma (Wilson et al., 1995; Jelsma et al., 1996) ("Zimbabwe" in Table 1). Terranes appear to be chiefly continental-margin arcs with a few remnants of oceanic plateaus. Average terrane lifespan for the Zimbabwe craton is 95 m.y.

There are a few remains of Early Archean accretionary terranes, one of the oldest of which is the Faeringehavn terrane at 3860 Ma in SW Greenland. This terrane amalgamated with five other pre–3000 Ma terranes at ca. 2720 Ma to form the Itsaq orogen (Friend et al., 1988; Nutman et al., 2000, 2002, 2004). Perhaps the most unusual feature of the Itsaq orogen is the large range in terrane lifespan, from 190 m.y. to over 1000 m.y. ("Itsaq" in Table 1). These terranes have undergone long, complex histories before finally becoming part of a Late Archean accretionary orogen. If the six preserved terranes are representative, the relatively short accretionary interval of this orogen (40 m.y.) is also striking.

At least five terranes are recognized in the Barberton region of the Kaapvaal craton in southern Africa, four of which comprise arc-like or oceanic plateau volcanic successions that range in age from 3.3 to 3.6 Ga (Lowe, 1999) ("Barberton" in Table 1). The Ancient Gneiss terrane comprises tonalitic gneisses and other deformed granitoids ca. 3640 Ma. All of these terranes collided and amalgamated in a relatively short time span of 50 m.y. at ca. 3250 Ma. As with the Itsaq orogen, the range in terrane lifespan in the Barberton orogen is significant, varying from over 400 m.y. to less than 100 m.y.

Two terranes have been recognized in the Pilbara craton in Western Australia, each of which had a long complex history before colliding ca. 3000 Ma (Van Kranendonk et al., 2004; Hickman, 2004) ("Pilbara" in Table 1). The East Pilbara has rocks as old as 3530 Ma and a lifespan of ~530 m.y., whereas the West Pilbara has rocks only as old as 3280 Ma and a lifespan of 280 m.y. Because there are only two Pilbara terranes for which we have precise geochronology, accretion interval cannot be estimated for the craton, although it was likely <50 m.y.

ACCRETIONARY OROGENS AND THE SUPERCONTINENT CYCLE

Although it would seem that initiation and cessation of construction of accretionary orogens should be tied to the supercontinent cycle, there are no simple correlations that apply to all orogens, as shown in Figure 3. Most accretionary orogens begin as passive continental margins along which subduction and/or transcurrent faulting is later initiated. Passive margins are remnants of the breakup phase of a supercontinent. For instance, when Rodinia fragmented some 700 Ma, it left passive margins along what are now the east and west coasts of Laurentia and the west coast of South America and around Siberia. As the Iapetus Ocean opened 600 Ma, passive margins formed on the continents facing the Iapetus basin (eastern North and South America, western Africa, and western Europe). Although some of these margins evolved into convergent boundaries that became sites of accretionary orogens, the time intervals between rifting of the supercontinent and onset of subduction and terrane accretion are extremely variable. For instance, it took 400–450 m.y. after rifting before the first terranes were accreted in the Cordilleran and Japan orogens, whereas terranes began to accrete in the Andes and Appalachians after only 120–150 m.y. (Table 1). The Yavapai-Penokean-Svecofennian orogen followed the breakup of one or more Archean supercontinents by 100–200 m.y., and its onset coincides with the formation of a new supercontinent at 1.9 Ga. Unlike most Phanerozoic accretionary orogens, the onset of activity in Late Archean accretionary orogens coincides with the formation of one or more supercontinents (Fig. 3). It is also noteworthy that both of the major periods of Late Archean crustal formation (2.7 and 2.5 Ga) and the onset of the Yavapai-Penokean-Svecofennian orogen correlate with possible

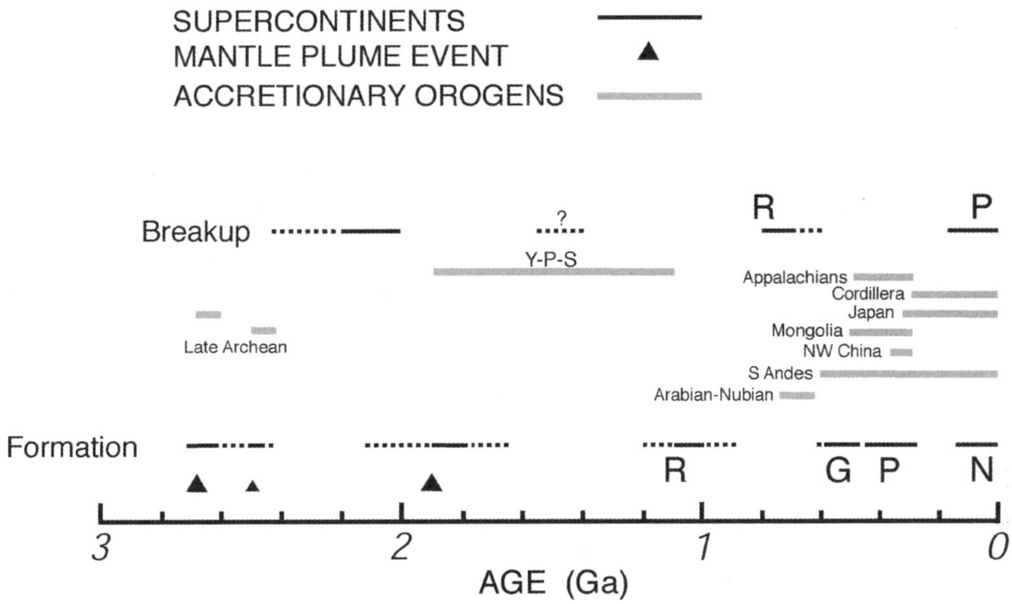

Figure 3. Accretionary orogens in relation to the supercontinent cycle and to possible global mantle plume events. R—Rodinia; P—Pangea; G—Gondwana; N—possible new supercontinent; Y-P-S—Yavapai-Penokean-Svecofennian orogen.

global mantle plume events (Fig. 3). It is possible that the earliest Archean accretionary orogens (e.g., Pilbara and Barberton) formed along the margins of oceanic plateaus, which rapidly evolved into continental crust.

Most accretionary orogens terminate with continent-continent collision. For instance, the Appalachians and the Altaids turned off during major collisions with large continents (Baltica–West Africa with Laurentia in the case of the Appalachians; Tarim–North China with Asia in the case of Altaids) ca. 250 Ma, during the completion of Pangea (Fig. 3). Likewise, the Yavapai-Penokean-Svecofennian orogen ended with the Grenvillian collisions ca. 1100 Ma. Beginning life at ca. 1900 Ma, this orogen was active for at least 800 m.y., perhaps the longest-lived accretionary orogen in Earth history. The Arabian-Nubian orogen is unique among the post-Archean orogens: Its total lifespan correlates with the breakup of the supercontinent Rodinia (Fig. 3).

The termination of Late Archean accretionary orogens is problematic, and the relatively short duration of these orogens (Fig. 1) appears to be unique to the Archean, a feature related perhaps to a global mantle plume event (Condie, 1998). As mentioned above, the Archean orogens correlate with formation of supercontinents and with mantle plume events. For those orogens that formed at or near 2.7 or 2.5 Ga, possible global mantle plume events may have been responsible for both their beginning and ending. A major problem that needs to be addressed is what happened on Archean cratons between the cessation of major accretion at 2.7 and 2.5 Ga and supercontinent breakup, some 200–300 m.y. later?

WHAT CONTROLS TERRANE LIFESPAN?

Several factors contribute to terrane lifespan. First is the tectonic setting or settings of a terrane. Some terranes, such as continental-margin arcs, form very close to a convergent boundary in or near an accretionary orogen, and hence may be short-lived before colliding with and suturing to a continent. As an example, most Yavapai terranes are remnants of continental-margin arcs or associated backarc basins (Condie, 1992), and they have among the shortest of all terrane lifespans (many <50 m.y.) (Fig. 1; Table 1). Not all continental-margin arcs have short lifespans, however, as exemplified by the Koyukuk terrane in the Cordillera with a lifespan of 140 m.y.

Another factor affecting terrane lifespan is the complexity of terrane history. Many terranes, as exemplified by most of those in the Appalachian and Cordilleran orogens have long, complex multitectonic setting histories, and many are superterranes (Plafker et al., 1989; Keppie, 1989; Condie and Chomiak, 1996; Hatcher, 2002; Murphy et al., 2004). Although many of these terranes have a long lifespan, periods of volcanism and sedimentation are separated by significant unconformities (Plafker and Berg, 1994; Plafker et al., 1989; McClelland et al., 1992; Murphy et al., 1999, 2004). For instance, Wrangellia, Chugach, and Peninsular terranes have major age gaps separating supracrustal successions. Although more difficult to recognize because of deformation and lack of fossil control, unconformities also are undoubtedly present in Precambrian terranes. In some cases, however, the total amount of time represented by unconformities in Precambrian terranes is considerably less than in Cordilleran and Appalachian terranes (for instance in the Yavapai orogen; Condie, 1992). In Cordilleran and Appalachian terranes, tectonic settings often differ between unconformity-bounded or fault-bounded rock packages. For instance, in the Alexander terrane in western Canada, the tectonic setting changed from an island arc in the late Paleozoic to a rift in the Triassic and back to an arc in the Cretaceous (Gehrels and Saleeby, 1987). In most terranes in the Yavapai-Penokean-Svecofennian orogen, however, tectonic setting appears to have remained the same throughout the short terrane lifespan (Condie, 1986, 1992). Some terranes, such as Avalonia in the Appalachians, began as a rifted Pan-African continental margin upon which several arcs were built prior to accretion to Laurentia (Murphy et al., 1999). Both Avalonia and Alexander have lifespans in excess of 300 m.y. (Table 1).

Another factor that may be important in controlling terrane lifespan is the amount of continental crust on the surface of Earth. For instance, the long terrane lifespans of Itsaq and Pilbara terranes may reflect the scarcity of continental crust to which they could accrete. And finally, the plate history in the ocean basin adjacent to an accretionary orogen is important in controlling terrane lifespan. For instance, in the case of Japan, from the Mesozoic onwards, ocean floor has converged on the Japanese islands, forming terranes composed of oceanic crust and trench sediment that are scraped off during subduction (Isozaki, 1997). Although occasional oceanic plateaus and islands were also accreted, no microcratons appear to have collided with Japan.

ACCRETION RATES IN ACCRETIONARY OROGENS

It is probable that accretionary orogens are the sites of most continental growth (Taylor and McLennan, 1985; Reymer and Schubert, 1986). Such growth occurs directly by addition of new crust in continental-margin arcs or by collision and accretion of juvenile terranes such as oceanic arcs, ophiolites, and oceanic plateaus (Fig. 2). In addition, repositioning of continental crust can occur by rifting blocks from older cratons and reaccretion of these at other locations in the same orogen or in different orogens. Such repositioning, of course, does not necessarily involve production of new crust, but in some instances new crust can form on microcratons by arc magmatism, before or after amalgamation with a continent.

Nd isotopes are useful in identifying juvenile crustal components in accretionary orogens. A plot of ε_{Nd} in various rocks (excluding basement) from accretionary orogens ranging in age from ca. 3000 Ma to 100 Ma reveals a wide range of ε_{Nd} values from positive to negative (Fig. 4). Most of the values, however, are positive (76%), indicating a large input of juvenile crust into accretionary orogens. Most accretionary orogens contain a mixture of terranes with juvenile and recycled components, with the ratio of these components changing with time and along strike in an orogen. The northern Appalachian and Variscan orogens are unusual in having many rocks with negative $\varepsilon_{Nd}(t)$, reflecting significant contamination of both igneous and sedimentary components with older crust (Murphy et al., 2004). Probably

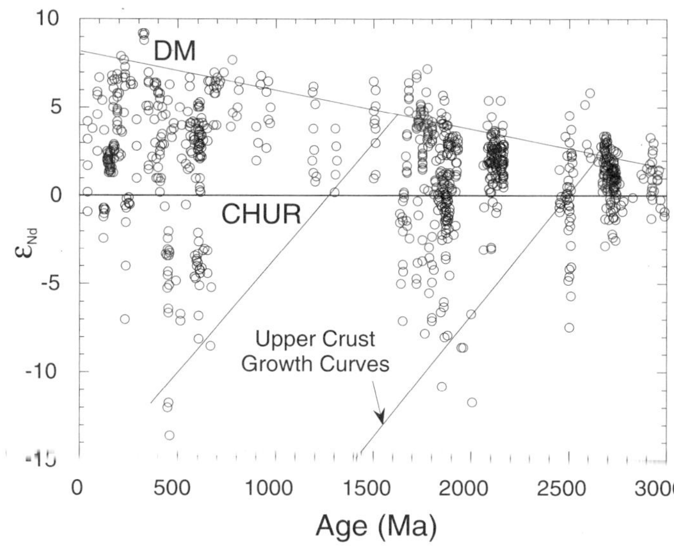

Figure 4. Summary of $\varepsilon_{Nd}(t)$ in rocks from accretionary orogens. Data from many published sources. DM—depleted mantle growth curve; CHUR—chondritic uniform reservoir.

50% of the terranes accreted in these orogens contain recycled older crust. In contrast, in the southern Appalachians (Tugaloo and Carolina terranes) and in the American Cordillera, most terranes have positive $\varepsilon_{Nd}(t)$, indicating a strong juvenile crustal input (Patchett and Gehrels, 1998; Bream et al., 2004). Remnants of oceanic arcs, ophiolites and oceanic plateaus are widespread in the Cordillera. In the Altaids in Mongolia, granitoids emplaced outside the Precambrian basement have juvenile $\varepsilon_{Nd}(t)$ values, whereas those emplaced within the Precambrian basement have negative $\varepsilon_{Nd}(t)$, suggesting contamination by the basement (Jahn et al., 2004). This is also true for the Paleoproterozoic terranes that were accreted or formed close to Archean cratons, such as in the Penokean of Laurentia and the Svecofennian of the Baltic shield (Bennett and DePaolo, 1987; Andersen and Sundvoll, 1995). Even at 2700 Ma, when a large volume of juvenile crust formed, some negative ε_{Nd} data indicate reworking of older crust (Fig. 4). In the western Superior province in Canada only ~50% of the crust was produced at 2700 Ma; the remainder formed at earlier times (Henry et al., 2000).

The distribution of $\varepsilon_{Nd}(t)$ in Figure 4 also clearly shows the episodic nature of juvenile crust production (Condie, 1998). Strong peaks, for instance, are evident at 100–200, 550–600, 1800–1900, 2100, 2500, and 2700 Ma, with notable gaps around 1000 Ma and 2400–2200 Ma. It is still not clear if these gaps reflect times of low crustal production or lack of preservation.

It is possible to estimate both the total accretion rate and the rate of juvenile crust production in accretionary orogens. Orogen widths are measured from geologic maps, and rates are calculated using accretion intervals (Table 1), average crustal thickness, and Nd isotopic data to estimate juvenile input. We assume an original terrane crustal thickness of 40 km, 10 km of which is now removed by erosion. In estimating the total amount of accreted material, the "thin-skinned effect" has been taken into account where seismic data are available, as for instance in the American Cordillera (Monger and Price, 2002; Snyder et al., 2002) and in the northern Appalachians (Clowes et al., 1992). In these orogens, seismic reflection data and Os isotopic compositions of mantle xenoliths (Peslier et al., 2000) suggest that terranes were thrust over Precambrian crust during accretion, and thus, the crust added cannot be assumed to extend to the Moho. In most instances, only 30%–70% of the crustal volume is newly accreted crust, and in our calculations we assume only 50% represents newly accreted material in the Cordilleran and Appalachian orogens. However, since more juvenile crust may be present in the Appalachians in offshore areas, the calculated juvenile crust production rate for the Appalachians should be considered a minimum. In other orogens, where there is no seismic evidence for or against older crust at depth, we assume 80%–100% for the fraction of accreted crust, using Nd isotope data in younger granitoids as a guide.

Results (Fig. 5) compare favorably with those published by Reymer and Schubert (1986) for four accretionary orogens. Both total accretion rate and juvenile crust production rate are higher in the Precambrian than afterwards, averaging between 100 and 200 km^3/km/m.y. (Labrador and Barberton being the exceptions) (Fig. 5). In contrast, Phanerozoic rates are typically between 70 and 150 km^3/km/m.y., NW China being the exception. It is noteworthy that crustal production (or at least preservation) does not parallel the cooling of Earth's mantle, which if it did, should show very high crustal production rates in the Early Archean. Accretion rates are highest in the Yilgarn and Superior provinces in the Late Archean where they exceed 300 km^3/km/m.y. (Fig. 5). Juvenile crust production rates also are high in the Superior province, which may reflect a global mantle plume event at this time (Condie, 1998). It is important to note that the Yilgarn, Zimbabwe, and Itsaq orogens, which have the same accretion age as the Superior province (2700 Ma), do not show this high rate of juvenile crust production.

The lowest juvenile crust production rates occur in some Phanerozoic orogens, such as Japan and the southern Andes, and in the Archean Barberton orogen, all of which have rates <80 km^3/km/m.y. (Fig. 5). The unusually low accretion and juvenile crust production rates in Japan may reflect, in part, the near absence of accreted continental-margin arcs and microcratons. A large number of older microcratons occur in the Andes. In other cases, such as Barberton, the calculated accretion interval may not apply to the entire craton, since the Kaapvaal craton continued to grow until ca. 2700 Ma (which increases the accretion interval from 50 to 280 m.y. and the volume of new crust by an order of magnitude) (Moser et al., 2001).

And finally, it is noteworthy that all juvenile crust production rates in accretionary orogens of all ages are greater than 20–40 km^3/km/m.y., the growth rate of young oceanic arcs (Reymer and Schubert, 1986). This strongly supports the conclusion of Reymer and Schubert (1986) that accretion of oceanic arcs alone

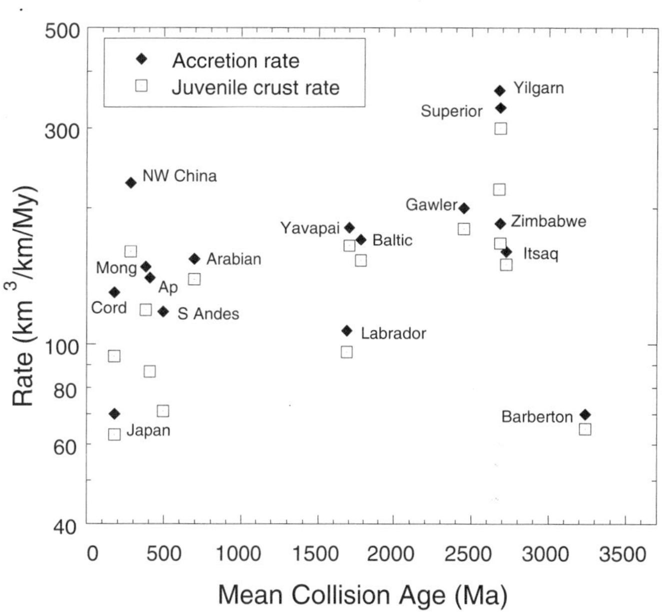

Figure 5. Total accretion rate and juvenile crust production rate in accretionary orogens. Listing of rates, with accretion rate followed by juvenile crust rate (both in km^3/km/m.y.): Cordillera (Cord) (130, 94); Appalachians (Ap) (140, 87); Japan (70, 63); NW China (227, 160); Mongolia (Mong) (148, 119); S Andes (118, 71); Arabian-Nubian shield (Arabian) (154, 139); Labrador (107, 96); Baltic shield (170, 153); Yavapai (181, 165); Superior (333, 300); Zimbabwe (185, 167); Yilgarn (364, 220); Gawler (200, 180); Barberton (70, 65); Itsaq (160, 150).

cannot account for production of juvenile crust in accretionary orogens. It is probable that a large contribution to juvenile crust production occurs in continental-margin arcs. An unknown, but possibly significant, volume of juvenile crust may also be added by accretion of oceanic plateaus (Condie, 1997).

CONCLUSIONS

1. Accretionary orogens are primary sites of juvenile continental crust production and have been active on Earth since the earliest Archean.

2. Orogen accretion intervals (lifetimes) range from ~50 m.y. to over 300 m.y., with Archean intervals less than 100 m.y.

3. Average terrane lifespan in accretionary orogens is typically 100–200 m.y. in post–1 Ga orogens, 50–100 m.y. in pre–1 Ga Proterozoic orogens, and 70–700 m.y. in Archean orogens.

4. Accretionary orogens can be classified into two end members: A *simple orogen* is one that is characterized by rapid accretion of juvenile oceanic terranes with average terrane lifespan <100 m.y. A *complex orogen* comprises not only juvenile accreted components but also exotic microcratons and has an average terrane lifespan ≥100 m.y. Complex orogens may undergo extensive rifting and transcurrent faulting, which creates new terranes and moves them laterally along the orogen.

5. There is no simple relationship between onset and cessation of accretionary orogens and the supercontinent cycle, and many post-Archean orogens terminate with continent-continent collisions during supercontinent assembly. The short duration of Late Archean accretionary orogens (<70 m.y.) may reflect the short duration of global mantle plume events at 2.7 and 2.5 Ga.

6. Terrane lifespan is controlled by (1) terrane tectonic setting, (2) complexity of precollisional terrane history, (3) availability of continental crust on Earth, and (4) plate history of ocean basins adjacent to accretionary orogens.

7. Average accretion rates in accretionary orogens are 70–150 km^3/km/m.y. in Phanerozoic orogens and 100–200 km^3/km/m.y. in Precambrian orogens. Some orogens at 2.7 Ga have unusually high accretion rates greater than 300 km^3/km/m.y., possibly related to a global mantle plume event.

8. Production rates of juvenile crust in accretionary orogens are typically 10%–30% lower than total accretion rates, but can be up to 50% lower in Phanerozoic orogens.

ACKNOWLEDGMENTS

This paper was substantially improved from reviews by Bob Hatcher and Marvin Carlson. Hatcher also provided the author with age estimates of several terranes in the southern Appalachians as compiled in Table 1.

REFERENCES CITED

Abdelsalam, M.G., and Stern, R.J., 1996, Sutures and shear zones in the Arabian-Nubian shield: Journal of African Earth Sciences, v. 23, p. 289–310, doi: 10.1016/S0899-5362(97)00003-1.

Ahall, K.-I., and Gower, C.F., 1997, The Gothian and Labradorian orogens: Variations in accretionary tectonism along a late Paleoproterozoic Laurentia-Baltica margin: GFF, v. 119, p. 181–191.

Ahall, K.-I., Persson, P.-O., and Skiold, T., 1995, Westward accretion of the Baltic shield: Implications from the 1.6 Ga Amal-Horred belt, SW Sweden: Precambrian Research, v. 70, p. 235–251, doi: 10.1016/0301-9268(94)00043-Q.

Andersen, T., and Sundvoll, B., 1995, Nd isotope systematics of the mantle beneath the Baltic shield: Evidence for depleted mantle evolution since the Archean: Lithos, v. 35, p. 235–243, doi: 10.1016/0024-4937(94)00053-5.

Badarch, G., Cunningham, W.D., and Windley, B.F., 2002, A new terrane subdivision for Mongolia: Implications for the Phanerozoic crustal growth of central Asia: Journal of Asian Earth Sciences, v. 21, p. 87–110, doi: 10.1016/S1367-9120(02)00017-2.

Bennett, V.C., and DePaolo, D.J., 1987, Proterozoic crustal history of the western United States as determined by Nd isotopic mapping: Geological Society of America Bulletin, v. 99, p. 674–685, doi: 10.1130/0016-7606(1987)99<674:PCHOTW>2.0.CO;2.

Betts, P.G., Giles, D., Lister, G.S., and Frick, L.R., 2002, Evolution of the Australian lithosphere: Australian Journal of Earth Sciences, v. 49, p. 661–695, doi: 10.1046/j.1440-0952.2002.00948.x.

Bleeker, W., 2002, Archean tectonics: A review, with illustrations from the Slave craton: Geological Society [London] Special Publication 199, p. 151–181.

Bogdanova, S.V., Page, L.M., Skridlaite, G., and Taran, L.N., 2001, Proterozoic tectonothermal history in the western part of the East European craton: $^{40}Ar/^{39}Ar$ geochronological constraints: Tectonophysics, v. 339, p. 39–66, doi: 10.1016/S0040-1951(01)00033-6.

Bream, B.R., Hatcher, R.D., Jr., Miller, C.F., and Fullagar, P.D., 2004, Detrital zircon ages and Nd isotopic data from the southern Appalachian crystalline core, Georgia, South Carolina, North Carolina, and Tennessee: New provenance constraints for part of the Laurentian margin: Geological Society of America Memoir 197, p. 459–475.

Clowes, R.M., Cook, F.A., Green, A.G., Keen, C.E., Ludden, J.N., Percival, J.A., Quinlan, G.M., and West, G.F., 1992, Lithoprobe: New perspectives on crustal evolution: Canadian Journal of Earth Sciences, v. 29, p. 1813–1864.

Coleman, R.G., 1989, Continental growth of northwest China: Tectonics, v. 8, p. 621–635.

Collins, W.J., 2002, Nature of extensional accretionary orogens: Tectonics, v. 21, no. 4, doi: 10.1029/2000TC001272.

Condie, K.C., 1986, Geochemistry and tectonic setting of Early Proterozoic supracrustal rocks in the southwestern United States: Journal of Geology, v. 94, p. 845–864.

Condie, K.C., 1992, Proterozoic terranes and continental accretion in southwestern North America, *in* Condie, K.C., ed., Proterozoic crustal evolution: Amsterdam, Elsevier, p. 447–480.

Condie, K.C., 1997, Contrasting sources for upper and lower continental crust: The greenstone connection: Journal of Geology, v. 105, p. 729–736.

Condie, K.C., 1998, Episodic continental growth and supercontinents: A mantle avalanche connection?: Earth and Planetary Science Letters, v. 163, p. 97–108, doi: 10.1016/S0012-821X(98)00178-2.

Condie, K.C., and Benn, K., 2006, Archean geodynamics: Similar to or different from modern geodynamics?: American Geophysical Union Geophysical, Monograph 164, p. 47–59.

Condie, K.C., and Chomiak, B., 1996, Continental accretion: Contrasting Mesozoic and Early Proterozoic tectonic regimes in North America: Tectonophysics, v. 265, p. 101–126, doi: 10.1016/S0040-1951(96)00148-5.

Condie, K.C., Frey, B.A., and Kerrich, R., 2002, The 1.75 Ga Iron King Volcanics in west-central Arizona: A remnant of an accreted oceanic plateau derived from a mantle plume with a deep depleted component: Lithos, v. 64, p. 49–62, doi: 10.1016/S0024-4937(02)00158-5.

Cordani, U.G., Milani, E.J., Thomaz Filho, A., and Campos, D.A., 2000, Tectonic evolution of South America: Abstract Volume of the 31st International Geological Congress, Rio de Janeiro, Brazil, August 16–17, 2000, 854 p.

Corfu, F., and Stott, G.M., 1998, Shebandowan greenstone belt, western Superior province: U-Pb ages, tectonic implications and correlations: Geological Society of America Bulletin, v. 110, p. 1467–1484, doi: 10.1130/0016-7606(1998)110<1467:SGBWSP>2.3.CO;2.

Friend, C.R.L., Nutman, A.P., and McGregor, V.R., 1988, Late Archean terrane accretion in the Godthab region, southern West Greenland: Nature, v. 335, p. 535–538, doi: 10.1038/335535a0.

Gehrels, G.E., and Saleeby, J.B., 1987, Geologic framework, tectonic evolution, and displacement history of the Alexander terrane: Tectonics, v. 6, p. 151–173.

Geraldes, M.C., Van Schmus, W.R., Condie, K.C., Bell, S., Teixeira, W., and Babinski, M., 2001, Proterozoic geologic evolution of the SW part of the Amazonian craton in Mato Grosso state, Brazil: Precambrian Research, v. 111, p. 91–128, doi: 10.1016/S0301-9268(01)00158-9.

Gorbatschev, R., and Bogdanova, S., 1993, Frontiers in the Baltic shield: Precambrian Research, v. 64, p. 3–22, doi: 10.1016/0301-9268(93)90066-B.

Gower, C.F., 1996, The evolution of the Grenville province in eastern Labrador, Canada: Geological Society [London] Special Publication 112, p. 197–218.

Gower, C.F., and Krogh, T.E., 2002, A U-Pb geochronological review of the Proterozoic history of the eastern Grenville province: Canadian Journal of Earth Sciences, v. 39, p. 795–829, doi: 10.1139/e01-090.

Griffin, W.L., Belousova, E.A., Shee, S.R., Pearson, N.J., and O'Reilly, S.Y., 2004, Archean crustal evolution in the northern Yilgarn craton: U-Pb and Hf-isotope evidence from detrital zircons: Precambrian Research, v. 131, p. 231–282, doi: 10.1016/j.precamres.2003.12.011.

Hamilton, W.B., 1998, Archean magmatism and deformation were not products of plate tectonics: Precambrian Research, v. 91, p. 143–179, doi: 10.1016/S0301-9268(98)00042-4.

Hatcher, R.D., Jr., 2002, Alleghanian (Appalachian) orogeny, a product of zipper tectonics: Rotational transpressive continent-continent collision and closing of ancient oceans along irregular margins, *in* Martínez Catalán, J.R., Hatcher, R.D., Jr., Arenas, R., and Díaz García, F., eds., Variscan-Appalachian dynamics: The building of the late Paleozoic basement: Geological Society of America Special Paper 364, p. 199–208.

Henry, P., Stevenson, R.K., Larbi, Y., and Gariepy, C., 2000, Nd isotopic evidence for Early to Late Archean crustal growth in the western Superior province, Ontario, Canada: Tectonophysics, v. 322, p. 135–151, doi: 10.1016/S0040-1951(00)00060-3.

Hickman, A.H., 2004, Two contrasting granite-greenstone terranes in the Pilbara craton, Australia: Evidence for vertical and horizontal tectonic regimes prior to 2900 Ma: Precambrian Research, v. 131, p. 153–172, doi: 10.1016/j.precamres.2003.12.009.

Huang, T.K., Ren, J., Jiang, D., Zhang, Z., and Quin, D., 1987, Tectonic evolution of China: New York, Springer-Verlag, 203 p.

Isozaki, Y., 1997, Contrasting two types of orogen in Permo-Triassic Japan: Accretionary versus collisional: Island Arc, v. 6, p. 2–24, doi: 10.1111/j.1440-1738.1997.tb00038.x.

Isozaki, Y., Maruyama, S., and Furuoka, F., 1990, Accreted materials in Japan: Tectonophysics, v. 181, p. 179–205, doi: 10.1016/0040-1951(90)90016-2.

Jahn, B.M., Capdevila, R., Liu, D., Vernon, A., and Badarch, G., 2004, Sources of Phanerozoic granitoids in the transect Bayanhongor–Ulaan Baatar, Mongolia: Geochemical and Nd isotopic evidence, and implications for Phanerozoic crustal growth: Journal of Asian Earth Sciences, v. 23, p. 629–653, doi: 10.1016/S1367-9120(03)00125-1.

Jelsma, H.A., Vinyu, M.L., Valbracht, P.J., Davies, G.R., Wijbrans, J.R., and Verdurmen, E.A.T., 1996, Constraints on Archean crustal evolution of the Zimbabwe craton: A U-Pb zircon, Sm-Nd and Pb-Pb whole-rock study: Contributions to Mineralogy and Petrology, v. 124, p. 55–70, doi: 10.1007/s004100050173.

Karabinos, P., Samson, S.D., Hepburn, J.C., and Stoll, H.M., 1998, Taconian orogeny in the New England Appalachians: Collision between Laurentian and the Shelburne Falls arc: Geology, v. 26, p. 215–218, doi: 10.1130/0091-7613(1998)026<0215:TOITNE>2.3.CO;2.

Karlstrom, K.E., Harlan, S.S., Williams, M.L., McLelland, J., Geissman, J.W., and Ahall, K.-I., 2001, Long-lived (1.8–1.0 Ga) orogen in southern Laurentia, its extensions to Australia and Baltica, and implications for refining Rodinia: Precambrian Research, v. 111, p. 5–30, doi: 10.1016/S0301-9268(01)00154-1.

Keppie, J.D., 1989, Northern Appalachian terranes and their accretionary history, *in* Dallmeyer, R.D., ed., Terranes in the Circum-Atlantic Paleozoic orogens: Geological Society of America Special Paper 230, p. 159–183.

Keppie, J.D., Nance, R.D., Murphy, J.B., and Dostal, J., 2003, Tethyan, Mediterranean, and Pacific analogues for the Neoproterozoic-Paleozoic birth and development of peri-Gondwanan terranes and their transfer to Laurentia and Laurussia: Tectonophysics, v. 365, p. 195–219, doi: 10.1016/S0040-1951(03)00037-4.

Kimura, G., Ludden, J.N., Desrockers, J.-P., and Hori, R., 1993, A model of ocean-crust accretion for the Superior province, Canada: Lithos, v. 30, p. 337–355, doi: 10.1016/0024-4937(93)90044-D.

Langford, F.F., and Morin, J.A., 1976, The development of the Superior province of NW Ontario by merging island arcs: American Journal of Science, v. 276, p. 1023–1034.

Lowe, D.R., 1999, Geologic evolution of the Barberton Greenstone Belt and Vicinity, *in* Lowe, D.R., and Byerly, G.R., eds., Geologic Evolution of the Barberton Greenstone Belt, South Africa: Geological Society of America Special Paper 329, p. 287–312.

Maruyama, S., 1997, Pacific-type orogeny revisited: Miyashiro-type orogeny proposed: Island Arc, v. 6, p. 91–120, doi: 10.1111/j.1440-1738.1997.tb00042.x.

McClelland, W.C., Gehrels, G.E., and Saleeby, J.B., 1992, Upper Jurassic–Lower Cretaceous basinal strata along the Cordilleran margin: Implications for the accretionary history of the Alexander-Wrangellia-Peninsular terrane: Tectonics, v. 11, p. 823–835.

Monger, J., and Price, R., 2002, The Canadian Cordillera: Geology and tectonic evolution: Canadian Society of Exploration Geophysicists Recorder, v. 27, p. 17–36.

Monger, J., Price, R.A., and Tempelman-Kluit, D.J., 1982, Tectonic accretion and the origin of the two major metamorphic and plutonic welts in the Canadian Cordillera: Geology, v. 10, p. 70–75, doi: 10.1130/0091-7613(1982)10<70:TAATOO>2.0.CO;2.

Moser, D.E., Flowers, R.M., and Hart, R.J., 2001, Birth of the Kaapvaal tectosphere 3.08 billion years ago: Science, v. 291, p. 465–468, doi: 10.1126/science.291.5503.465.

Mueller, W.U., Daigneault, R., Mortensen, J.K., and Chown, E.H., 1996, Archean terrane docking: Upper crust collision tectonics, Abitibi greenstone belt, Quebec, Canada: Tectonophysics, v. 265, no. 1–2, p. 127–150, doi: 10.1016/S0040-1951(96)00149-7.

Murphy, J.B., Keppie, J.D., Dostal, J., and Nance, R.D., 1999, Neoproterozoic–early Paleozoic evolution of Avalonia, *in* Ramos, V.A., and Keppie, J.D., eds., Laurentia-Gondwana connections before Pangea: Geological Society of America Special Paper 336, p. 267–276.

Murphy, J.B., Pisarevsky, S.A., Nance, R.D., and Keppie, J.D., 2004, Neoproterozoic–early Paleozoic evolution of peri-Gondwanan terranes: Implications for Laurentia-Gondwana connections: International Journal of Earth Sciences, v. 93, p. 659–682, doi: 10.1007/s00531-004-0412-9.

Myers, J.S., 1993, Precambrian history of the West Australian craton and adjacent orogens: Annual Review of Earth and Planetary Sciences, v. 21, p. 453–485.

Nutman, A.P., Bennett, V.C., Friend, C.R.L., and McGregor, V.R., 2000, The Early Archean Itsaq gneiss complex of southern West Greenland: The importance of field observations in interpreting age and isotopic constraints for early terrestrial evolution: Geochimica et Cosmochimica Acta, v. 64, p. 3035–3060, doi: 10.1016/S0016-7037(99)00431-7.

Nutman, A.P., Friend, C.R.L., and Bennett, V.C., 2002, Evidence for 3650–3600 Ma assembly of the northern end of the Itsaq gneiss complex, Greenland: Implication for Early Archean tectonics: Tectonics, v. 21, no. 1, doi: 10.1029/2000TC001203.

Nutman, A.P., Friend, C.R.L., Barker, S.L.L., and McGregor, V.R., 2004, Inventory and assessment of Paleoarchean gneiss terrains and detrital zircons in southern West Greenland: Precambrian Research, v. 135, p. 281–314, doi: 10.1016/j.precamres.2004.09.002.

Park, A.F., 1985, Accretion tectonism in the Proterozoic Svecokarelides of the Baltic shield: Geology, v. 13, p. 725–729, doi: 10.1130/0091-7613(1985)13<725:ATITPS>2.0.CO;2.

Patchett, P.J., and Gehrels, G.E., 1998, Continental influence of Canadian Cordilleran terranes from Nd isotopic study, and significance for crustal growth processes: Journal of Geology, v. 106, p. 269–280.

Percival, J.A., McNicoll, V., Brown, J.L., and Whalen, J.B., 2004, Convergent margin tectonics, central Wabigoon subprovince, Superior province, Canada: Precambrian Research, v. 132, p. 213–244, doi: 10.1016/j.precamres.2003.12.016.

Peslier, A.H., Reisberg, L., Ludden, J., and Francis, D., 2000, Os isotopic systematics in mantle xenolith: Age constraints on the Canadian Cordillera lithosphere: Chemical Geology, v. 166, p. 85–101, doi: 10.1016/S0009-2541(99)00187-4.

Plafker, G., and Berg, H.C., editors, 1994, The geology of Alaska: Boulder, Colorado, Geological Society of America, Decade of North American Geology, vol. G-1, 1056 p.

Plafker, G., Nokleberg, W.J., and Lull, J.S., 1989, Bedrock geology and tectonic evolution of the Wrangellia, Peninsular, and Chugach terranes along the Trans-Alaska crustal transect in the Chugach Mountains and southern Copper River basin: Alaska: Journal of Geophysical Research, B, Solid Earth and Planets, v. 94, p. 4255–4295.

Polat, A., and Kerrich, R., 2001, Geodynamic processes, continental growth, and mantle evolution recorded in Late Archean greenstone belts of the southern Superior province, Canada: Precambrian Research, v. 112, p. 5–25, doi: 10.1016/S0301-9268(01)00168-1.

Ramos, V.A., 2004, Cuyania, an exotic block to Gondwana: Review of a historical success and the present problems: Gondwana Research, v. 7, p. 1009–1026, doi: 10.1016/S1342-937X(05)71081-9.

Reymer, A., and Schubert, G., 1986, Rapid growth of some major segments of continental crust: Geology, v. 14, p. 299–302, doi: 10.1130/0091-7613(1986)14<299:RGOSMS>2.0.CO;2.

Şengör, A.M.C., and Natal'in, B.A., 1996, Paleotectonics of Asia: Fragments of synthesis, *in* Yin, A., and Harrison, T.M., eds., The tectonic evolution of Asia: Cambridge, UK, Cambridge University Press, p. 486–640.

Skiold, T., 1988, Implications of new U-Pb zircon chronology to Early Proterozoic crustal accretion in northern Sweden: Precambrian Research, v. 38, p. 147–164, doi: 10.1016/0301-9268(88)90089-7.

Smithies, R.H., and Champion, D.C., 2000, The Archaean high-Mg diorite suite: Links to tonalite-trondhjemite-granodiorite (TTG) magmatism and tectonic implications: Journal of Petrology, v. 41, p. 1653–1671, doi: 10.1093/petrology/41.12.1653.

Snyder, D.B., Clowes, R.M., Cook, F.A., Erdmer, P., Evenchick, C.C., van der Velden, A.J., and Hall, K.W., 2002, Proterozoic prism arrests suspect terranes: Insights into the ancient Cordilleran margin from seismic reflection data: GSA Today, v. 12, no. 10, p. 4–10, doi: 10.1130/1052-5173(2002)012<0004:PPASTI>2.0.CO;2.

Stern, R.J., 1994, Arc assembly and continental collision in the Neoproterozoic East African orogen: Annual Review of Earth and Planetary Sciences, v. 22, p. 319–351.

Stoeser, D.B., and Camp, V.E., 1985, Pan-African microplate accretion of the Arabian shield: Geological Society of America Bulletin, v. 96, p. 817–826, doi: 10.1130/0016-7606(1985)96<817:PMAOTA>2.0.CO;2.

Swain, G.M., Woodhouse, A., Hand, M., Barovich, K.M., Schwarz, M.P., and Fanning, C.M., 2004, Late Archean tectonothermal history of the Gawler craton: Geochemical and isotopic constraints: Geological Society of America Abstracts with Programs, v. 36, no. 5, p. 340.

Taylor, S.R., and McLennan, S.M., 1985, The continental crust: Its composition and evolution: Oxford, UK, Blackwell, 312 p.

Thomas, W.A., and Astini, R.A., 1996, The Argentine Precordillera: A traveler from the Ouachita embayment of North American Laurentia: Science, v. 273, p. 752–757, doi: 10.1126/science.273.5276.752.

Tomlinson, K.Y., and Condie, K.C., 2001, Archean mantle plumes: Evidence from greenstone belt geochemistry: Geological Society of America Memoir 352, p. 341–357.

Van Kranendonk, M.J., Collins, W.J., Hickman, A.H., and Pawley, M.J., 2004, Critical tests of vertical vs. horizontal tectonic models for the Archean East Pilbara granite-greenstone terrane, Pilbara craton, Western Australia: Precambrian Research, v. 131, p. 173–211, doi: 10.1016/j.precamres.2003.12.015.

Vujovich, G.I., Fernandes, L.A.D., and Ramos, V.A., 2004, Cuyania: An exotic block to Gondwana: Gondwana Research, v. 7, p. 1005–1007, doi: 10.1016/S1342-937X(05)71080-7.

Wilson, J.E., Nesbitt, R.W., and Fanning, C.M., 1995, Zircon geochronology of Archean felsic sequences in the Zimbabwe craton: A revision of greenstone stratigraphy and model for crustal growth: Geological Society [London] Special Publication 95, p. 109–126.

Xiao, W., Windley, B.F., Badarch, G., Sun, S., Li, J., Qin, K., and Wang, Z., 2004, Paleozoic accretionary and convergent tectonics of the southern Altaids: Implications for the growth of central Asia: Geological Society [London] Journal, v. 161, p. 339–342.

Manuscript Accepted by the Society 1 February 2007

The Geological Society of America
Memoir 200
2007

Petrotectonics, climate, crustal thickness, and evolution of geologically young orogenic belts

W.G. Ernst
Department of Geological and Environmental Sciences, Stanford University, Stanford, California 94305-2115, USA

ABSTRACT

New and recycled intermediate to felsic crust in island arcs and continental margins is generated by magmatism surmounting convergent plate junctions. Paired Pacific-type orogens develop adjacent sites of long-lived subduction of oceanic lithosphere. They consist of (1) an outboard accretionary prism deposited in and continentward from the oceanic trench, the filling of which is derived mainly from (2) an inboard calc-alkaline volcanic-plutonic arc. The trench assemblage includes arc-sourced clastic mélanges, minor but widespread deepwater cherts and carbonates, and tectonically disaggregated ophiolites. These relatively incompetent sections recrystallize under high-pressure (HP) conditions and decouple from the descending oceanic plate at depths of ~15–50 km. A massive, slightly older to coeval andesitic-granodioritic arc dominates the landward belt where new sialic crust is added from I-type magmas, and preexisting continental materials are recycled as S-type melts; high temperature (HT) characterizes local-regional metamorphism of the wall rocks. In contrast, Alpine-type orogens result from the underflow of an oceanic plate that transports island arcs, microcontinents, and/or continental salients into the subduction zone. Reflecting lithospheric coherence, the sialic crust may be carried down as much as 90–140 km, becomes thermally softened, and decouples from the sinking plate. Metamorphism of deeply subducted parts of Alpine belts ranges from high pressure to ultrahigh pressure (UHP), and is not paired with a HT calc-alkaline arc. In both types of subduction complex, outboard thrust faults dip landward and fold vergence is seaward, reflecting the polarity of underflow (and rollback) of the oceanic lithosphere. Antithetic thrusting typifies some contractional continental realms. Propelled by buoyancy, allochthonous nappes conduct relatively low-density HP and UHP sections to midcrustal levels. At convergent syntaxes (plate-margin cusps) of overthickened arcs, tectonic aneurysms may produce domical uplifts, further exhuming UHP units surfaceward.

The regional crustal thickness of an active mountain belt partly reflects climate as well as the main convergent-transform plate processes producing the orogen. The volcanogenic Chilean Cordillera parallels the seaward convergent plate junction and the eastward-sinking Nazca plate. The highest ranges and thickest continental crust (~70 km) occur in the compressed north-central Andes at ~25°S, a region of extremely low rain- and snowfall. Aridity and low erosion rates help account for the high-standing calc-alkaline volcanic-plutonic contractional arc and the elevated, internally drained

Ernst, W.G., 2007, Petrotectonics, climate, crustal thickness, and evolution of geologically young orogenic belts, *in* Hatcher, R.D., Jr., Carlson, M.P., McBride, J.H., and Martínez Catalán, J.R., eds., 4-D Framework of Continental Crust: Geological Society of America Memoir 200, p. 159–179, doi: 10.1130/2007.1200(10).

Altiplano downwind. Thus erosional degradation is weakly developed. A very different climate typifies the orogen at ~45°S, where abundant moisture-laden westerlies bring abundant precipitation to the Chilean margin. The rugged mountain belt is of lower elevation compared with the north-central Andes and is supported by a crust of only moderate thickness (35–40 km). Vigorous erosion supplies voluminous detritus to the offshore Chile-Peru Trench as well as eastward, so a plateau is lacking in the lee of the arc. The Himalayas and Tibetan Plateau, the Sierra Nevada and Colorado Plateau, the Japanese island arc, and New Zealand exhibit somewhat similar relationships between crustal thickness and precipitation-linked erosion.

Keywords: precipitation and erosion, continental crust, young mountain belts, convergent plate junctions, Pacific versus Alpine orogens.

INTRODUCTION

Although the geology of mountain belts and their key relationships to continental evolution have been intensively studied for two centuries, our understanding of them is still growing. Most orogens form near the active edges of continents and/or fringing island arcs. All result from the consumption of lithospheric plates, involving underflow of oceanic crust and transport, arrival, and descent of terrigenous debris, oceanic spreading centers and plateaus, island arcs, far-traveled terranes, microcontinents, and salients of continental crust. Long-continued subduction promotes the building of a calc-alkaline volcanic-plutonic arc on the stable, nonsubducted plate. No two orogens are the same. In fact, most mountain chains exhibit significant petrotectonic contrasts along their extents. Many orogens contain crustal rocks reflecting HP or HT recrystallization. A few contain mineralogic relics of UHP metamorphism, but because of intense retrograde reaction, other compressional mountain belts that may have been subjected to similar *P-T* conditions fail to retain any evidence of this metamorphic history. The interpretation of subduction-zone complexes for which critical evidence of deep burial is exceedingly fragmentary—or totally obliterated—is a serious challenge for petrotectonicians.

The architecture and thickness of the continental crust in young, active mountain belts is the result of competing constructional and destructional processes. The former involve calc-alkaline magmatic additions from the mantle; plate-tectonic convergence, transform motion, and/or divergence; terrane suturing; accretionary prism growth; subduction offloading-underplating; and anatexis of fusible, recycled lower crust. The latter include subduction-induced subcrustal erosion, delamination-foundering of the mafic lower crust, mass wastage, and surface erosion. Old continental shields have reached petrotectonic equilibrium due to long-term isolation from upper mantle flow and the establishment of an unperturbed geothermal gradient in the lithosphere. However, the architecture and crustal thickness of young mountain belts dynamically evolve on a regional basis because, although regional isostatic accommodation is closely approximated, the interplay between constructional and destructional forces is influenced to varying degrees by many factors—among them, the erosion rate.

This review tries to assess the orogenic process from a general petrotectonic viewpoint, concentrating on the architectures and rock assemblages of Phanerozoic mountain chains. Although Precambrian analogs probably have resulted from comparable lithospheric plate motions and petrologic processes, with systematic lithotectonic contrasts being related to the higher geothermal gradients of the early Earth, the ancient rock record is much less clear. Accordingly, I concentrate on geologically young mountain building in this synthesis. Two main end members are distinguished (e.g., Bally, 1981), but it is clear that a continuum of intermediate types exists between continent collisional and Circumpacific contractional orogens.

CONTRASTS BETWEEN ANDEAN ARC SUBDUCTION ZONES AND CONTINENTAL COLLISIONAL SUBDUCTION ZONES

Convergent Plate Junctions

Pacific-type convergent boundaries evolve where thousands of kilometers of oceanic lithosphere are consumed without the arrival of significant masses of continental crust at the subduction zone. This plate-tectonic realm produces three subparallel belts: an outboard, largely metasedimentary accretionary trench complex, a medial, longitudinal forearc basin, and an inboard, voluminous, coeval calc-alkaline volcanic-plutonic arc. Seaward, the long, narrow depositional prism consists of a low-temperature, low-heat-flow belt in which folds verge oceanward and outboard thrust faults are defined by the subduction zone and dip beneath the stable, hanging-wall plate, whereas landward, the broad magmatic arc is characterized by open folding and a high-temperature, high-thermal-flux regime (Miyashiro, 1961, 1967; Ernst et al., 1970; Dickinson, 1972). The former is deposited exclusively on the downgoing oceanic crust (e.g., Franciscan, Aleutian, and Chile-Peru subduction systems); the latter is constructed on preexisting basement of the nonsubducted continental margin or island arc and older oceanic crust (e.g., Sierran, Indonesian, and Andean arcs).

Alpine-type convergent boundaries form where large- or small-scale subduction of an ocean plate results in the insertion of a continental promontory, microcontinent, or island arc beneath the stable, nonsubducted, continental crust-capped plate. A salient of the downgoing sialic terrane may descend to great depth, dragged down with the surrounding oceanic lithosphere, because the overall density of the largely oceanic plate exceeds that of the underlying asthenosphere. On subduction, the thickness of the sialic crust is increased by accretion, underplating, shortening, and amalgamation due to continental collision. Collisional orogens exhibit a distinct lack of coeval calc-alkaline igneous activity. Typical examples include the Urals, Alps, and Himalayas (Hamilton, 1970; Dal Piaz et al., 1972; Molnar and Tapponnier, 1975; Molnar et al., 1987; Burchfiel et al., 1989). Of course, it is possible for continental lithosphere to descend beneath young, hot—thus less dense—oceanic crust-capped lithosphere (e.g., Oman and Sulawesi; Searle et al., 2004), but such cases are uncommon because most (old) oceanic plates are denser than the underlying asthenosphere, whereas continental plates are relatively buoyant.

Along the length of a convergent plate junction, underflow may range by degrees between purely Pacific type and Alpine type. The Indonesian archipelago represents a modern plate-tectonic regime transitional between these orogenic end-member types (Hamilton, 1979; Maruyama et al., 1996). This complex suture zone is the product of Pacific and Australian lithospheric underflow beneath the Eurasian and Philippine Sea plates, and of initial stages of subduction suturing of the Australian continental crust. The junction between the Arabian Peninsula and Iran represents another intermediate example (Boudier et al., 1985, 1988; Gnos et al., 1997; Miller et al., 1998; Searle et al., 2004). Although we here consider principally the downdip component of underflow, because of the curvilinear nature of convergent plate junctions, a large component of arc-parallel slip is present in virtually all orogenic belts (e.g., Fitch, 1972; Howell, 1985; Koons et al., 2003).

Subduction Zones

The dynamic evolution of convergent plate junctions has been studied by many workers, employing both numerical simulations (Shreve and Cloos, 1986; Cloos and Shreve, 1988a, 1988b; Beaumont et al., 1996, 1999; Stüwe and Barr, 1998) and experimental scale-model studies (Tapponnier et al., 1982; Chemenda et al., 1995, 1996). The architecture of the crust is a composite function of many input parameters, including the physical states of the converging lithospheric plates (thermal gradients, buoyancy, spatial configurations, ductilities, densities); rates of convergence; amount and nature of sedimentary loading; extents of frictional dissipation; degrees of hydration; fluid expulsion and flow (channeled or pervasive); advective heat transport; kinetics of subsolidus reactions and of partial fusion; extents of coeval tectonic contraction, transpression, or extension; crustal delamination-foundering; subjacent (subduction-induced) erosion; and surficial degradation. In spite of these complexities, computational and scale-model experiments replicate the major features of most convergent plate junctions. During and after underflow, the ascent of subducted lithotectonic packets from great depths to midcrustal levels also is a function of the physical characteristics of the rock units for both Pacific Rim accretionary complexes and collisional orogenic belts.

The temperature structures of convergent plate regimes have been investigated quite extensively (e.g., Oxburgh and Turcotte, 1971; Turcotte and Oxburgh, 1972; Anderson et al., 1978; Thompson and Ridley, 1987; van den Beukel and Wortel, 1988; Peacock, 1990, 1992; Hacker et al., 2003). Rocks are poor conductors, so a rapidly sinking lithospheric slab remains cool at upper mantle depths relative to its surroundings; in contrast, the landward volcanic-plutonic arc occupies a broad thermal high above the nonsubducted mantle wedge and overlying crust, due to advective heat transport. Computations account for paired Circumpacific outboard HP blueschist-eclogite, and inboard HT andalusite-sillimanite metamorphic belts. Geologic relationships and prograde mineral parageneses for such terranes were emphasized by Miyashiro (1961, 1967) prior to the advent of plate-tectonic theory. And, as now recognized for the Japanese island arc and many other Pacific-type paired metamorphic belts, the times of accretion of clastic sedimentary debris and offscraped, far-traveled, preexisting crustal fragments generally decrease monotonically seaward (Banno and Sakai, 1989; Ernst, 1992; Nishimura, 1998).

Calc-Alkaline Volcanic-Plutonic Arcs

Landward calc-alkaline arcs have dominated the post-Paleozoic Circumpacific realm. Massive Andean arcs apparently only form where vast tracts of oceanic lithosphere return to the deep upper mantle, generating a well-developed magma plumbing system in the process (Ernst, 1971, 1999). Bulk-rock isotopic signatures show that, although partially melted quartzofeldspathic and pelitic crust yields recycled, peraluminous, S-type granites and contributes to the growth of mature island-arc and continental-margin magmas (Castro et al., 1999), the metaluminous, I-type calc-alkaline suite contains an important igneous addition to the continental crust from the mantle wedge (hanging wall) and/or the subducted, eclogitized oceanic crust (DePaolo, 1981; Hawkesworth et al., 1993; Clift et al., 2001). Some chemically intermediate volcanic and plutonic rocks thus represent primary additions to the sialic crust, whereas other rock types represent mostly recycled materials. Seaward, deposits of immature, volcaniclastic sediments derived from the arc form fringing accretionary aprons (Dickinson, 1971, 1976); these detrital sediments may overwhelm the forearc basin, spilling into the seaward trench and onto the approaching oceanic plate. Igneous rocks in the outboard, Pacific-type subduction complex and accretionary prism are exclusively tectonized ophiolites and other fragments of oceanic lithosphere (Coleman, 1977); they do not add significantly to the volume of continental crust.

Granitoids and related volcanic rocks are much less common in Alpine-type complexes. Unlike Pacific convergent margins, plutons are not set back on the nonsubducted plate; instead, peraluminous (S-type) leucogranites locally invade the root zone of the imbricate stack of nappes. Ophiolitic rocks occur in many compressional orogens, but tectonic slices of oceanic lithosphere are conspicuously missing from some contractional mountain belts (e.g., Liou et al., 1996). The relative rarity of the ophiolite suite in such collisional complexes may reflect the underflow of only small ocean basins, allowing the sampling of fewer asperities in the oceanic lithosphere (Cloos, 1993), or the lack of sutured backarc basins. In addition, fragments of the oceanic lithosphere tectonically stranded during early stages of growth of accretionary prisms may be preferentially lost due to continued accretionary underplating and consequent uplift-coupled erosion. The subduction of modest tracts of oceanic lithosphere evidently fails to generate sufficient arc magmas and the mantle-wedge plumbing system required to construct a massive calc-alkaline arc in Alpine-type orogens (Ernst, 1971, 1999).

Of course, if long-continued subduction of oceanic lithosphere precedes continental collision, an early-stage Andean arc may be constructed on the stable, nonsubducted plate. For the Himalayan-Tibetan system, the southern margin of Asia was indeed the site of the enormous calc-alkaline Transhimalayan batholith, generated during the consumption of Mesozoic Tethys at 50–110 Ma (Honegger et al., 1982; Searle, 1991; Copeland et al., 1995), prior to the Eocene arrival and progressive northward subduction of the Indian subcontinent. Extinction of I-type magmatic activity accompanied this collision, and Cenozoic leucogranites in the Himalayas represent exclusively anatectic S-type melts (LeFort et al., 1987; Searle et al., 2003).

Metamorphic Belts

Coeval, paired metamorphic belts characterize Pacific-type convergent plate junctions. The outboard subduction complex consists of relatively poorly consolidated units recrystallized at low temperatures and high pressures. Reflecting the subduction channel setting directly below the hanging-wall lithosphere (which acts as a stress guide), folds verge oceanward; landward inclined shear zones, foliation, lineation, and thrust faults are typical. Recovered depths of metamorphism range from 15 to 50 km (Bailey et al., 1964; Ernst, 1971; Aoya, 2001). Landward, voluminous calc-alkaline granitoids invade the preexisting basement and an overlying comagmatic volcanic cover series. Both volcanic and basement rocks show the effects of high-*T*, low-*P* (5–25 km) recrystallization. Metamorphic mineral zonations reflect the additional heat supplied by rising magmas and consist of a series of superimposed contact aureoles surrounding individual plutons (Barton et al., 1988). Hornfels and open, cylindrical folding typify these upper crustal metamorphic terranes (Ernst, 1992).

Alpine collisional complexes consist of preexisting continental massifs, as well as penetratively deformed, superjacent autochthonous and allochthonous units. Metamorphic belts are not paired (Frey et al., 1974). HP subsolidus recrystallization is common, with mineral parageneses comparable to those of Circumpacific subduction complexes. Some contractional orogens include tectonic slices that retain scattered mineralogic relics of coesite and/or microdiamond, chiefly as inclusions in strong, unreactive container phases such as zircon and garnet. These tiny inclusions must have crystallized at ultrahigh pressures and somewhat higher temperatures, but prograde geothermal gradients appear to be similar to *P-T* trajectories of Pacific-type subduction complexes (Chopin, 1984; Smith, 1984; Sobolev and Shatsky, 1990; Coleman and Wang, 1995). Such physical conditions attend the entrance of a segment of old, cold continental crust, well bonded to the spatially associated oceanic lithosphere, into a subduction zone and descent of this complex to depths of 90–140 km; in the process, some sections pass through the brittle-ductile transition and may decouple from the downgoing plate.

HP, HT, and UHP metamorphic conditions are compared in Figure 1, an experimentally determined petrogenetic grid appropriate for rocks of roughly basaltic bulk composition. Contrasts in metamorphic-facies assemblages, and the attendant prograde *P-T* gradients generated in subduction zones versus those produced in the vicinity of calc-alkaline volcanic-plutonic arcs, are clear.

AQUEOUS FLUIDS IN CONVERGENT PLATE JUNCTIONS

Cycling of H_2O in Subduction Zones

Descent of oceanic lithosphere involves the subduction of entrained, water-rich sediment resting on a basaltic-gabbroic-peridotitic basement. Rocks are either offloaded into the accretionary prism at shallow depth, as is much of the weak, low-density sedimentary debris, or returned to the upper mantle, as is most of the strong, dense oceanic crust and mantle lithosphere. Where microcontinental fragments or salients are entrained in the downgoing oceanic plate, sialic crust may be carried to great depth beneath the nonsubducted mantle wedge. Pressure within the subducting slab rises proportionally with depth, but temperature only gradually increases. H_2O may be added to relatively dry portions of the downgoing plate, but is driven off the more volatile-rich parts (Peacock, 1995; Bebout, 1996). Pre-eclogitic stages of these hydrous reactions were treated quantitatively by Frey (1987) and Ernst (1990). Attending deeper subduction of crustal rock types, the remaining H_2O is sequestered exclusively in relatively refractory silicate phases (Schreyer et al., 1987; Poli and Schmidt, 1997; Ernst et al., 1998; Okamoto and Maruyama, 1999). During exhumation, volatiles gain access to the ascending low-density masses, or diffuse away from them, depending on the composition, *P-T* trajectory, grain size, permeability, and extent of shearing of the various units. Some expelled fluids rise into and through the overlying mantle wedge, whereas the rest migrate back up the subduction channel (Cloos, 1984; Reck, 1987; Vrolijk, 1987; Vrolijk et al., 1988).

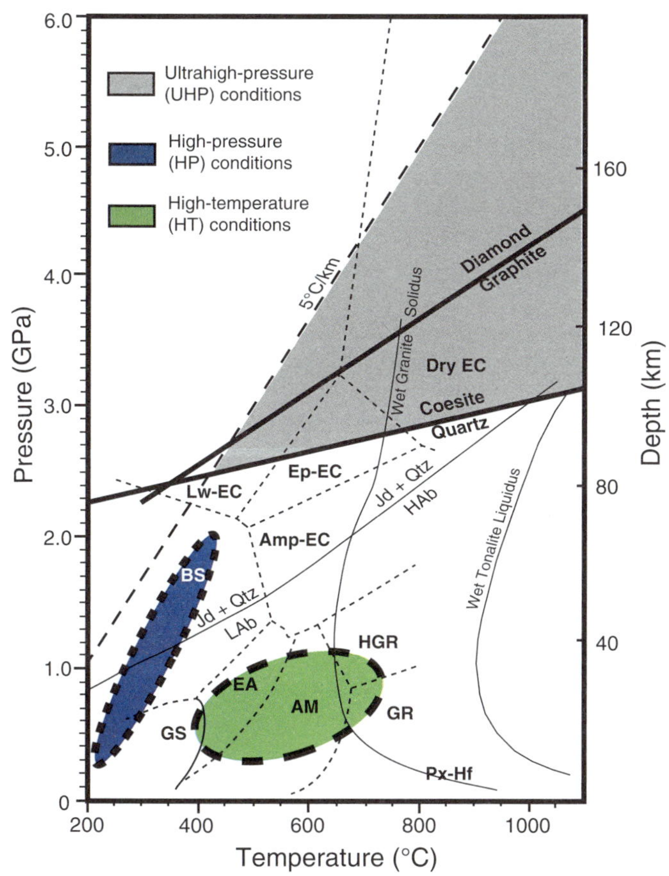

Figure 1. Petrogenetic grid for metabasaltic bulk-rock compositions, modified after Liou et al. (1998) and Okamoto and Maruyama (1999). *P-T* realms for paired Pacific-type outboard (HP) and inboard (HT), and for Alpine-type (HP-UHP) metamorphism are indicated. An extremely low subduction-zone geothermal gradient of 5 °C/km is also shown. Citations to the experimental phase equilibria were provided by Liou et al. (1998). Mineral abbreviations are Jd—jadeite; Qtz—quartz; LAb—low albite; HAb—high albite. Metamorphic-facies abbreviations are AM—amphibolite; Amp-EC—amphibolite-eclogite; BS—blueschist; EA—epidote-amphibolite; EC—eclogite; Ep-EC—epidote-eclogite; GR—sillimanite-granulite; GS—greenschist; HGR—kyanite-granulite; Lw-EC—lawsonite-eclogite; Px-Hf—pyroxene-hornfels. Feebly recrystallized rocks of the zeolite, prehnite, pumpellyite-actinolite, and prehnite-pumpellyite subfacies (not illustrated) lie at lower temperatures and pressures than GS and BS.

H_2O and Generation of Calc-Alkaline Arc Magmas

During subduction, recrystallizing low-grade rock types lose most of their volatiles at modest depths. Deeper than ~50 km, clinoamphibole-bearing metabasalt, metagabbro, and serpentinized peridotite of the oceanic lithosphere appear to be the most significant H_2O-bearing rock types (Liu et al., 1996). Devolatilization of Ca-amphiboles and serpentine minerals should begin at ~70–80 km depth under equilibrium conditions; however, at low temperatures, reactions are sluggish on the timescale of plate descent, so fluid expulsion probably reaches a maximum at somewhat greater depths. Thus, long-continued descent of thousands of kilometers of Pacific-type oceanic lithosphere undoubtedly provides the greatest proportion of deep-seated aqueous fluid rising through the subduction channel, as well as diffusing into the overlying mantle wedge. H_2O-induced partial melting of the refertilized mantle hanging wall and/or the eclogitic oceanic crust results in generation of the calc-alkaline suite making up the massive volcanic-plutonic arc set back 100–300 km from the convergent plate junction (Drummond and Defant, 1990; Kushiro, 1990; Hawkesworth et al., 1993; Morris, 1995; Clift et al., 2001; Hickey-Vargas et al., 2002). Concentrations of incompatible, large-ion-lithophile and light rare earth elements, $^{87}Sr/^{86}Sr$ initial values, and εNd ratios in volatile-bearing arc volcanic and plutonic rocks argue for the partial fusion of ultramafic mantle and/or mafic crustal garnet-bearing protoliths (fertile garnet lherzolite and/or basaltic eclogite, respectively), fluxed by an aqueous fluid in the high-*P*, low-*T* subduction-zone environment.

On the other hand, consumption of a relatively small oceanic basin prior to continental subduction expels only minor amounts of volatiles from the descending oceanic lithosphere, explaining the general lack of H_2O-mediated generation of calc-alkaline arc rocks in many Alpine orogens. The fluid flow under UHP conditions is diminished by the arrival, profound descent, and suturing of continental crust. This is a consequence of the fact that the main hydrous phases present in metamorphosed quartzofeldspathic and pelitic rock types—biotite and white mica—remain stable at great depth, and do not devolatilize at temperatures below ~800 °C, even at a depth of 140 km. Accordingly, continental collision does not generate a massive, I-type calc-alkaline arc, although the partial melting of overthickened continental crust produces modest amounts of peraluminous, S-type granitoids as recycled sialic materials (Ernst, 1999).

H_2O and Kinetics of Metamorphic Reactions

Disequilibrium assemblages and textures produced by prograde subduction-zone HP and UHP metamorphism (e.g., Frey et al., 1974; Rubie, 1990; Hacker, 1996) indicate the need for quantitative data on reaction rates. Presence or absence of aqueous fluid largely determines the kinetics governing mineralogic transformations and the extent of reaction progress (Austrheim, 1990, 1998; Ernst et al., 1998; Leech, 2001; Mosenfelder et al., 2005). The catalytic effect of H_2O is clear (e.g., Rubie, 1986; Mosenfelder and Bohlen, 1997). However, the dependence of reaction rates and transformation mechanisms on fluid chemistry and concentrations in systems with very low fluid:rock ratios is not well constrained.

The ease with which aqueous fluids gain access to, or are expelled from, subducting lithosphere and ascending segments of low-density crustal material is a function of rock compositions, permeabilities, deformation, grain sizes, and availability of volatiles. These factors in turn reflect the past geologic histories and dynamics of lithotectonic units in the subduction zone. Based on computational thermal modeling (Peacock, 1992,

1995; Ernst and Peacock, 1996; Hacker et al., 2003) and laboratory phase-equilibrium studies (Vielzeuf and Holloway, 1988; Schreyer, 1995; Massone, 1995; Liu et al., 1996; Patiño Douce and McCarthy, 1998), the paragenetic evolution of the principal rock types in a subduction zone is reasonably well understood. However, the rates of both prograde and retrograde reactions are almost unknown. Lacking aqueous fluid, low-pressure phase assemblages can persist under prograde HP-UHP conditions (Austrheim, 1990; Zhang and Liou, 1997), and scattered HP-UHP relics can survive as inclusions in strong, unreactive container minerals during retrograde decompression (Hirajima et al., 1993; Liou and Zhang, 1996). For example, in a portion of the Dabie-Sulu belt, closed-system stable isotope exchange took place attending Triassic UHP metamorphism, but preserved the bulk-rock oxygen and H-D isotopic anomalies established during a prior Neoproterozoic near-surface alteration event (Rumble and Yui, 1998; Cosca et al., 2005). Impermeable rocks, characterized by coarse grain sizes, hence small surface:volume ratios, can resist complete recrystallization on both prograde and retrograde *P-T* paths.

EXHUMATION OF SUBDUCTION COMPLEXES

Tectonic Relationships

Most outboard Pacific- and Alpine-type HP and UHP terranes consist of an imbricate stack of tabular sheets (Ernst et al., 1997). Examples of exhumed nappes from the western Alps, northern Kazakhstan, and coastal Norway that were metamorphosed under UHP conditions are illustrated in Figure 2. These remarkably thin tectonic slices have survived exhumation-induced decompression and recrystallization while retaining scattered relics of their HP-UHP subduction-zone histories. Ascent to shallow crustal levels occurs in response to one or more of several possible processes: tectonic extrusion (Maruyama et al., 1994, 1996), buoyant body-force propulsion (Ernst, 1970, 1988; England and Holland, 1979; Hacker, 1996; Hacker et al., 2000), corner flow blocked by a hanging-wall backstop (Cowan and Silling, 1978; Cloos and Shreve, 1988a, 1988b; Cloos, 1993), and/or underplating combined with extensional uplift and erosional collapse (Platt, 1986, 1987, 1993; Ring and Brandon, 1994, 1999). Some relatively old, sinking oceanic plates and submarine trenches appear to be retreating oceanward (i.e., rolling backwards) more rapidly than the encroaching nonsubducted lithosphere (Molnar and Atwater, 1978; Seno, 1985; Busby-Spera et al., 1990; Hamilton, 1995), so for these occurrences, compression and extrusion of subducted continental flakes in the jaws of a convergent junction fail to account for their exhumation. Although faulting associated with extensional collapse and erosional unroofing help to uncover deeply buried terranes, this mechanism cannot explain well-documented, widely occurring major pressure discontinuities across boundaries between subducted and nonsubducted lithologic units (Ernst et al., 1970; Suppe, 1972; Henry, 1990; Michard et al., 1995; Kaneko et al., 2003). However, coupled with erosion and mass wastage, buoyancy provides a reasonable ascent mechanism for the exhumation of low-density crustal slices rising from profound subduction depths. Numerical simulations (Beaumont et al., 1996, 1999), laboratory scale models (Chemenda et al., 1995, 1996), and geologic field relationships such as illustrated in Figure 2 are all compatible with this process.

Buoyancy-Driven Exhumation

Attending subduction of a weak, juicy graywacke and shaley mélange, partial devolatilization and increasing ductility promote the decoupling of slices of quartzofeldspathic material from the downgoing oceanic plate at relatively shallow depths, followed by ascent. For a promontory of continental crust bonded to the sinking mantle lithosphere, the introduction of increasing amounts of light material into the subduction zone reduces the negative buoyancy of the oceanic slab; achievement of neutral buoyancy in turn may result in rupture of the high-density oceanic lithosphere at intermediate upper mantle depths, a realm where the plate is in extension (Isacks et al., 1968), and its continued descent. Break-off of the dense oceanic slab (Sacks and Secor, 1990; von Blanckenburg and Davies, 1995) increases the net buoyancy of the updip sialic subduction complex, and allows low-density sheets separating from the sinking oceanic lithosphere to migrate back up the subduction channel (van den Beukel, 1992; Davies and von Blanckenburg, 1998). Exhumation also may be enhanced as the quartzofeldspathic crustal material gradually warms in the upper mantle and passes through the brittle-ductile transition (Stöckhert and Renner, 1998).

Previous metamorphic studies (e.g., Ernst, 1971; Chopin, 1987) have demonstrated that deep subduction of chiefly low-density sialic material is responsible for the formation of outboard Pacific- and Alpine-type metamorphic belts. Underflow of dominantly oceanic crust-capped lithosphere results in transformation of basaltic crust into yet denser eclogitic crust; such plates continue to sink into the deep upper mantle because tectonic forces are insufficient to allow wholesale return surfaceward. But how have subducted sialic masses returned to shallow crustal levels while preserving relics of HP and UHP phase assemblages? The two-way transport of terranes along subduction zones has been demonstrated (Ernst, 1970; Suppe, 1972; Willett et al., 1993). Although transitional types are common, two end-member convergent plate junctions, both of which may involve large components of arc-parallel slip, are emphasized here: subduction of a weak, Pacific-type graywacke and shale complex carried down on oceanic lithosphere; and Alpine-type suturing of a relatively strong continent, microcontinent, or island arc descending beneath a nonsubducted lithospheric plate. The former decouples at 15–50 km depths, and gives rise to HP trench mélanges and accretionary prisms (Miyashiro, 1967). The latter involves underflow of peninsulas or salients of old continental crust thoroughly embedded in dense oceanic lithosphere (Yin and Nie, 1993; Ernst and Liou, 1995). The sialic section descends relatively rapidly

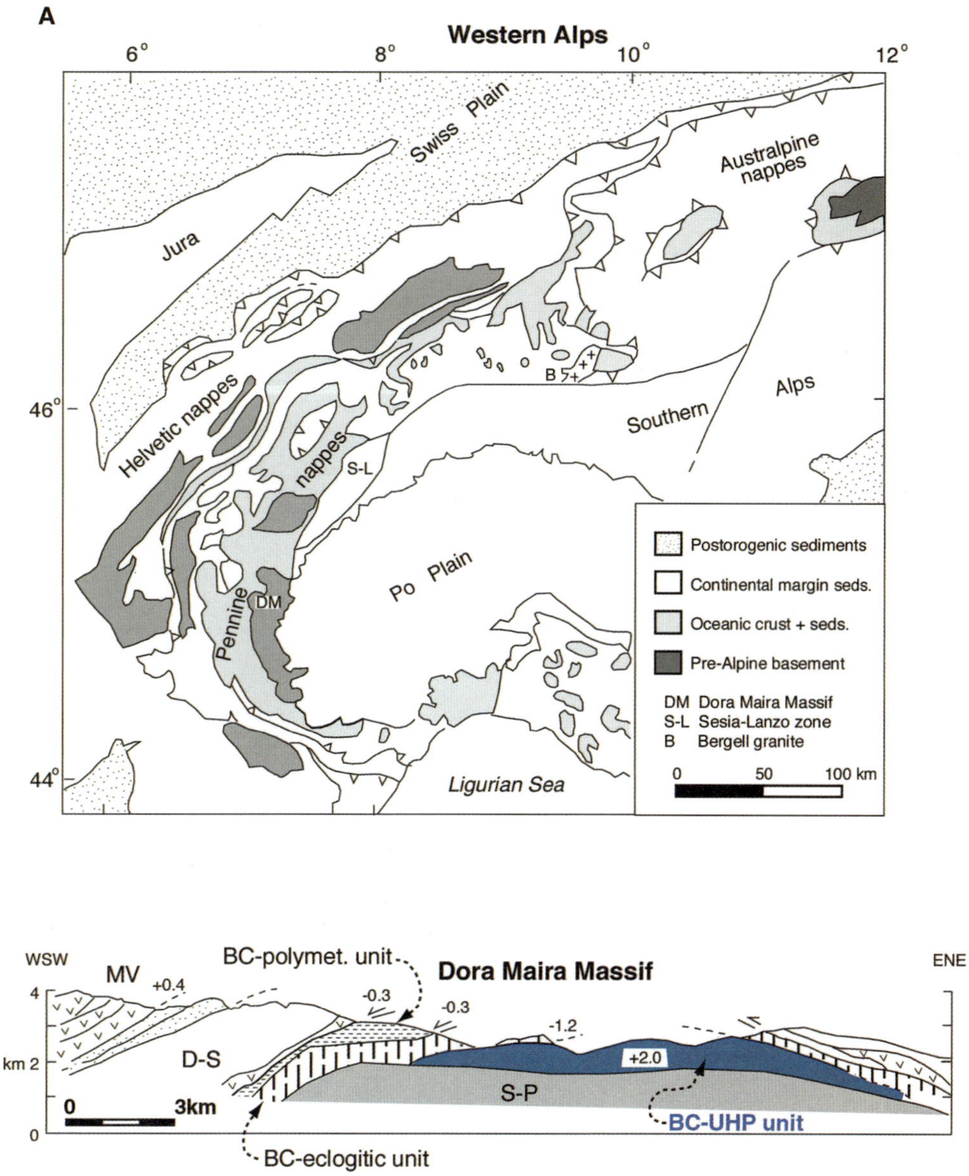

Figure 2. Simplified cross sections of HP-UHP allochthons in the western Alps, including a transect through the Dora Maira massif (A) (after Henry, 1990; Michard et al., 1995), the Kokchetav massif of northern Kazakhstan (B) (after Ishikawa et al., 2000; Ota et al., 2000; Maruyama and Parkinson, 2000), and a well-mapped local area in the Western Gneiss Region of coastal Norway (C) (Terry et al., 2000a, 2000b). In the lower panel of (A), numbers in the diagrammatic cross section for the southern part of the Dora Maira massif (DM) indicate the upward change in recorded pressure (GPa) relative to the underlying unit. Abbreviations of lithotectonic units are BC—basement complex (polymet means polymetamorphic); D-S—Dronero-Sampeyre; MV—Monviso; S-P—Sanfront-Pinerolo. In B, unit II is the diamond- and coesite-bearing sheet of paragneisses. NF—normal fault; RF—reverse fault. In (C) (Mørk, 1985), the structurally highest nappe is the UHP unit. Although each of these terranes include UHP slabs, HP-UHP nappes and décollements also characterize other exhumed subduction-zone complexes, such as the Franciscan Complex of the California Coast Ranges, the Maksyutov Complex of the South Ural Mountains, the Sanbagawa belt of SW Japan, the western Himalayan syntaxis (Nanga Parbat), and the eastern Himalayan syntaxis (Namche Barwa). *(Continued.)*

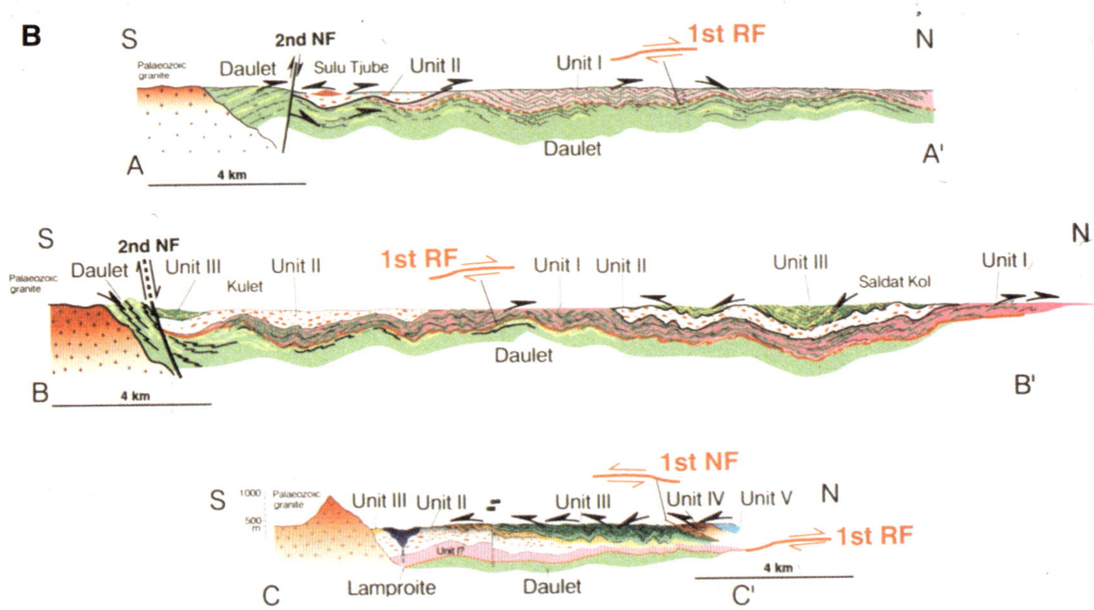

Figure 2 *(continued)*. (B) (after Ishikawa et al., 2000; Ota et al., 2000; Maruyama and Parkinson, 2000), and a well-mapped local area in the Western Gneiss Region of coastal Norway

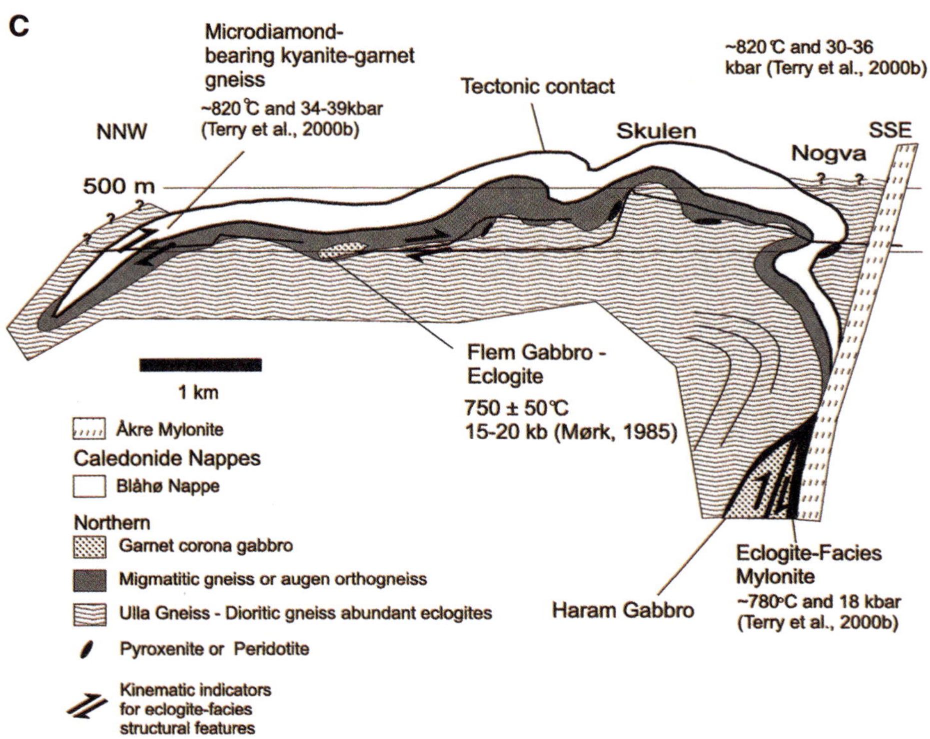

Figure 2 *(continued)*. (C) (Terry et al., 2000a, 2000b).

and decouples at depths as great as 90–140 km, generating the characteristic HP-UHP prograde mineralogy of collisional complexes (Peacock, 1995; Ernst et al., 1997).

But are rising subduction complexes sufficiently buoyant to overcome the traction of the oceanic plate carrying them downward? Ambient densities of unaltered oceanic crust, ~3.0 g cm^{-3}, continental material, ~2.7 g cm^{-3}, and anhydrous mantle, ~3.2 g cm^{-3}, increase with elevated pressure, reflecting the progressive transformation of relatively open, low-coordination silicates to more compact, high-coordination structural analogs. Stable HP and UHP mineralogic assemblages and computed ambient rock densities appropriate for burial depths of ~100 km (Ernst et al., 1997) are metabasaltic eclogite, ~3.6 g cm^{-3}, eclogitic quartzofeldspathic gneiss, ~3.0 g cm^{-3}, and garnet peridotite, ~3.3 g cm^{-3}. Even fully transformed to a dense assemblage, K-feldspar + jadeite + coesite-bearing granitic gneiss remains lighter than garnet or spinel lherzolite, whereas metabasaltic eclogite is considerably denser than the mantle. Because sialic crustal rock types contain muscovite and biotite (phases stable to ~800 °C at subduction depths of at least 140 km) as the principal H_2O carriers, such rock types are unlikely to transform rapidly—or totally—to the UHP mineralogic configuration. In contrast, the main H_2O-bearing phase in mafic rocks is hornblende, a mineral that devolatilizes at any temperature where depths exceed ~80 km, so metabasaltic rocks are far more likely to recrystallize to the stable prograde HP or UHP eclogitic assemblage.

Thus, at any reasonable subduction depth, continental crust transformed completely (or perhaps only incipiently) to HP-UHP phase assemblages remains buoyant relative to the surrounding mantle and tends to rise to midcrustal levels, whereas eclogitized oceanic crust is even more negatively buoyant than basaltic crust-capped lithosphere and continues to sink. This relationship explains why exhumed HP-UHP terranes worldwide consist of 85%–90% low-density quartzofeldspathic and shaley materials, and contain only small proportions of mafic and anhydrous ultramafic rock types. It also helps to account for the volumetrically minor presence of ophiolitic complexes, in contrast to the wealth of thoroughly serpentinized (low-density) masses in many exhumed subduction complexes.

The Thin-Slab Model

The poor thermal conductivities of rocks that are responsible for the formation of high-*P*, low-*T* prograde metamorphic conditions during subduction tend to keep these lithologic units relatively warm on rapid decompression. Rocks cool slowly, and deeply buried, buoyant masses cannot decompress without passing through more normal, low-pressure (in some cases, higher-temperature) crustal environments. Consequently, surviving HP-UHP complexes exhibit the ubiquitous mineralogic overprinting characteristic of granulite-, amphibolite-, and greenschist-facies conditions. In the presence of an aqueous fluid, adiabatic ascent would be expected to result in relatively high-temperature annealing and thorough obliteration of preexisting HP and UHP mineral assemblages. Conversely, lack of catalytic H_2O combined with rapid ascent would substantially decrease the rate of, and time available for, retrogression. Even so, unless heat is efficiently withdrawn from decompressing complexes as they are exhumed, mineralogic evidence of their former HP and/or UHP configurations would not be retained.

The rare preservation of HP and UHP relics in a rising slice of a deeply buried complex is favored by coeval extensional faulting against the overlying, cooler hanging wall combined with continued subduction-induced refrigeration tectonically beneath it (Hacker and Peacock, 1995). Thus allochthons of relatively thin aspect ratio, such as those shown in Figure 2, would lose heat along both upper and lower surfaces by thermal conduction during ascent. Some subduction complexes rise rapidly and virtually adiabatically, whereas others appear to have nearly retraced the prograde *P-T* trajectory during slower decompression (Rubie, 1984; Ernst, 1988; Ernst and Peacock, 1996). Schematic relationships illustrated in Figure 3 apply to the exhumation of both HP and UHP terranes. Forces acting on subducted material are as follows: (1) Descent of a low-density sheet occurs in cases where shear forces caused by underflow (F_s) are greater than the combined influences of the effective buoyancy, $F_b \cos \theta$, and frictional resistance along the upper wall of the subduction channel (F_r). Here, $F_s > F_b \cos \theta + F_r$. (2) Ascent of a slice of the low-density material takes place where the effective buoyancy exceeds the combined effects of shearing along its base and resistance to movement along its upper surface. In this case, $F_b \cos \theta > F_s + F_r$. The hanging wall acts as a tectonic guide, so rising crustal materials are accreted oceanward from the inboard volcanic-plutonic arc. Note that, as the angle of subduction $(90 - \theta)$ decreases, so does the importance of buoyancy for both underflow and exhumation.

For HP-UHP mineral assemblages to be partly preserved on exhumation, the rising slab must overcome frictional resistance to sliding, so must be sufficiently thick for buoyancy-driven ascent; yet it must be thin enough that heat is efficiently removed by thermal conduction across the sheet boundaries—the upper normal and lower reverse faults (Fig. 3). Such shear senses are required by observed structural relations and derived cross sections, for example, in the Franciscan Complex (Ernst, 1970; Suppe, 1972; Platt, 1986; Jayko et al., 1987), the western Alps (Henry, 1990; Compagnoni et al., 1995; Michard et al., 1995), the Sanbagawa belt (Kawachi, 1968; Ernst et al., 1970; Banno and Sakai, 1989), the Kokchetav massif (Kaneko et al., 2000; Ishikawa et al., 2000; Ota et al., 2000; Maruyama and Parkinson, 2000), the Western Gneiss Region of SW Norway (Terry et al., 2000a, 2000b), the Himalayas (Burchfiel et al., 1989; Searle, 1996; Searle et al., 2001; Kaneko et al., 2003), and the Dabie-Sulu belt (Hacker et al., 1995, 1996, 2000; Webb et al., 1999). These and other HP and UHP tectonic regimes are typical of convergent plate junctions worldwide.

Most thoroughly recrystallized and back-reacted UHP complexes, although less dense than relatively anhydrous mantle, become more-or-less neutrally buoyant at mid levels of the continental crust. Further ascent of such slabs to the surface requires

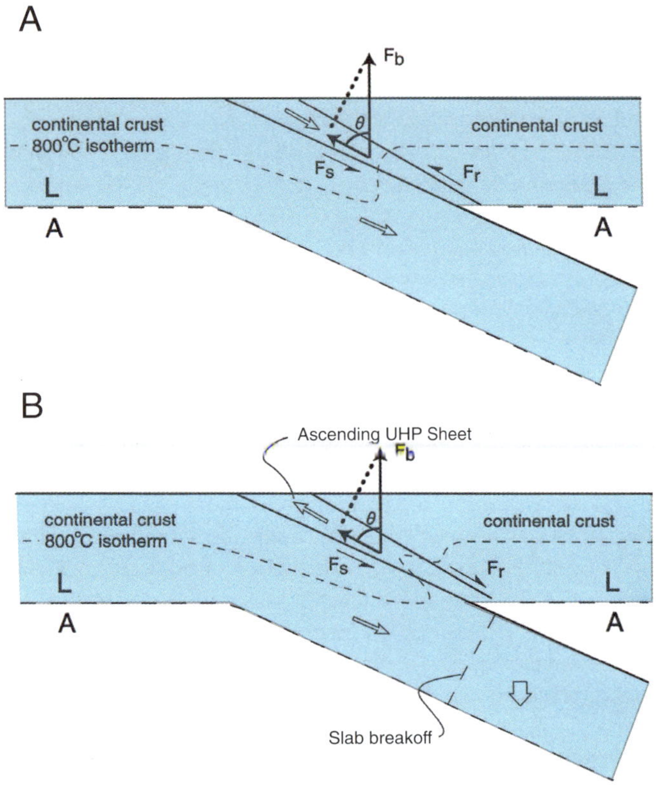

Figure 3. Diagram illustrating forces and approximate thermal structure during subduction of continental crust-capped lithosphere, assuming coherence of the downgoing plate prior to slab breakoff (Ernst et al., 1997): deep burial of a subducted prism of low-density continental crust (A), and later decompression and cooling of a rising slice of buoyant HP-UHP material (B). On ascent of the thin HP-UHP sheet (thickness exaggerated for clarity), cooling of its upper margin takes place where it is juxtaposed against the lower-temperature hanging-wall plate; heat is also withdrawn along the lower margin of the sheet where it is in contact with the lower-temperature, subducting-refrigerating lithospheric plate. Exhumation of low-density slices requires erosive denudation or gravitational collapse driven by the buoyant root at depth. The resolution of forces acting on the low-density slab in stages (A) and (B) is discussed in the text. A—asthenosphere; L—lithosphere (shaded).

contractional tectonism (Maruyama et al., 1994), or some form of low-density crustal underplating (Platt, 1986, 1993), in either case combined with isostatically compensated regional exhumation and erosion. A decrease in bulk density of the continental lithosphere after slab breakoff results in a shallowing of the subduction angle and may be partly responsible for late doming recognized in some exhumed convergent plate junctions (Ernst et al., 1997). Another unloading mechanism involves the antithetic faulting typical of some compressional orogens, in which double vergence is produced during terminal stages of the buoyant ascent of low-density crust, its collapse, and erosional removal (e.g., Dal Piaz et al., 1972; Platt, 1987; Ring and Brandon, 1994, 1999).

Tectonic Aneurysms

Rapid, domical exhumation of sialic crust seems to be occurring along certain collisional boundaries where curvilinear arc segments that intersect at large angles evidently have produced excess crustal mass. These diapiric uplifts have been termed "tectonic aneurysms" (Zeitler et al., 2001). The eastern and western syntaxes of the Himalayan-Tibetan suture zone display marked plate-tectonic discontinuities and attendant rapid exhumation (Burg et al., 1997; Zeitler et al., 2001; Koons et al., 2002). Other cusp-like intersections include compressional plate junctions of the Ryukyu arc and NE Taiwan, the western Aleutians and Kamchatka Peninsula, the eastern Mediterranean Greece-Turkey complex, and possibly the collision of the Indonesian archipelago with Australia. Extremely rapid uplift and erosion in NE Taiwan (Hartshorn et al., 2003; Dadson et al., 2003, 2004) may reflect the presence of a crustal diapir at depth. I speculate that the areal bulge of southern Kamchatka Peninsula reflects decompression partial melting of a tectonic aneurysm, which in turn may partly account for the extensive, volcanic-plutonic-dominated crust. The oroclinal bend of the western Alps and the offset of the Dabie-Sulu orogen (modified by postsubduction left-lateral slip on the Tan Lu fault) also might represent preexisting, now largely deformed, pre-Cenozoic syntaxes.

Many of these arc intersections (plate-margin cusps) are characterized by the domical exposure of UHP mineral assemblages (Maruyama et al., 1996), indicating that the diapiric rise of unusually deep-seated, thick sections of low-density, ductile crust has taken place. Although not directly related to the earlier subduction-zone metamorphism, the ascent from middle to upper crustal levels may be a response to the massive accumulation and heating of deeply buried continental material. UHP minerals and/or phase assemblages have been reported from some of these locales, including the western Himalayan syntaxis (O'Brien et al., 2001; Kaneko et al., 2003), the eastern Himalayan syntaxis (Zhong and Ding, 1996; Ding and Zhong, 1999; Ding et al., 2001), Greece (Mposkos and Kostopoulos, 2001), Indonesia (Kadarusman and Parkinson, 2000; Kadarusman et al., 2004), the western Alps (Chopin, 1984; Schertl et al., 1991; Compagnoni et al., 1995), and the Dabie-Sulu belt (Wang et al., 1989; Liou et al., 1996; Zhang et al., 2003). All of these areas bear evidence of earlier nappe emplacement, so the exhumation of some UHP terranes may reflect the operation in varying degrees of both earlier slab thrusting and later diapirism.

Due to rapid ascent from great depth at moderately high temperatures, the crucial requirement for preservation of HP-UHP relict assemblages in even fragmentary fashion is that the heat must be conducted away effectively—which in turn requires decompressing rock masses to be characterized by large surface-to-volume ratios. This means that, for surviving UHP complexes, transport to midcrustal levels in décollement-type structures within cool, subduction-zone environments evidently occurred first, allowing substantial quenching of the UHP assemblages.

This tectonic regurgitation is followed by domical ascent, in some cases as tectonic aneurysms, toward Earth's surface.

CLIMATIC PATTERNS AND CRUSTAL THICKNESSES IN ACTIVE MOUNTAIN BELTS

General Relationships

The thickness and structure of sialic crust in young orogenic belts is the evolving product of competing constructional and destructional processes. The former involve primary calc-alkaline magmatic additions derived from a descending oceanic slab and/or stable lithosphere; plate-tectonic convergence (contraction), transform motion, and/or divergence (extension); terrane docking; accretionary prism growth; subduction offloading-underplating; and anatexis of fusible, recycled lower crust and subducted sialic materials. The latter include subduction-induced crustal removal, delamination and foundering of the mafic lower crust, mass wastage, and surface erosion. Ancient cratons have attained thermal and structural stasis throughout the lithosphere, reflecting tectonic quiescence, due to long-term isolation from upper mantle flow, and the establishment of a steady-state geothermal gradient. In contrast, geologically youthful contractional mountain belts commonly possess thicker crust than do old continental shields because, although regional gravitative equilibrium is approached, the interplay between constructional and destructional forces is influenced to varying degrees by many factors—among them, the erosion rate. Of course, accelerated surface degradation never anticipates but must lag behind tectonically increased crustal thickness.

Reflecting regional isostatic compensation, the mean elevation and crustal thickness of a growing mountain belt is a time-integrated consequence of the petrotectonically accumulated mass per unit area, but is modified by surficial removal (Burbank, 1992; Horton, 1999; Willett, 1999; Willett and Brandon, 2002; Hilley et al., 2004; Hilley and Strecker, 2004). The regional erosion rate in modern mountain belts may be roughly correlated with precipitation (Strecker et al., 2003; but see also Burbank et al., 2003). This is a gross oversimplification of a much more complex set of variables, including chemical and physical weathering, nature and integrity of the geologic substrate (e.g., resistance of the bedrock to erosion), proximity to local erosional base levels, etc. Other factors being equal, greater relief and steeper topographic gradients promote more rapid degradation for a given precipitation rate. But as snow- and rainfall decrease, contraction and uplift may allow blockage of external drainage systems and result in elevated, internally draining plateaus (Burbank et al., 1996; Whipple and Tucker, 1999; Sobel et al., 2003). The regional climate itself is a complex multivariant function of Earth's near-surface thermal structure, and wind and ocean-surface circulation patterns; the latter two are influenced by latitude, the positions of obstructing landmasses, and the Coriolis effect.

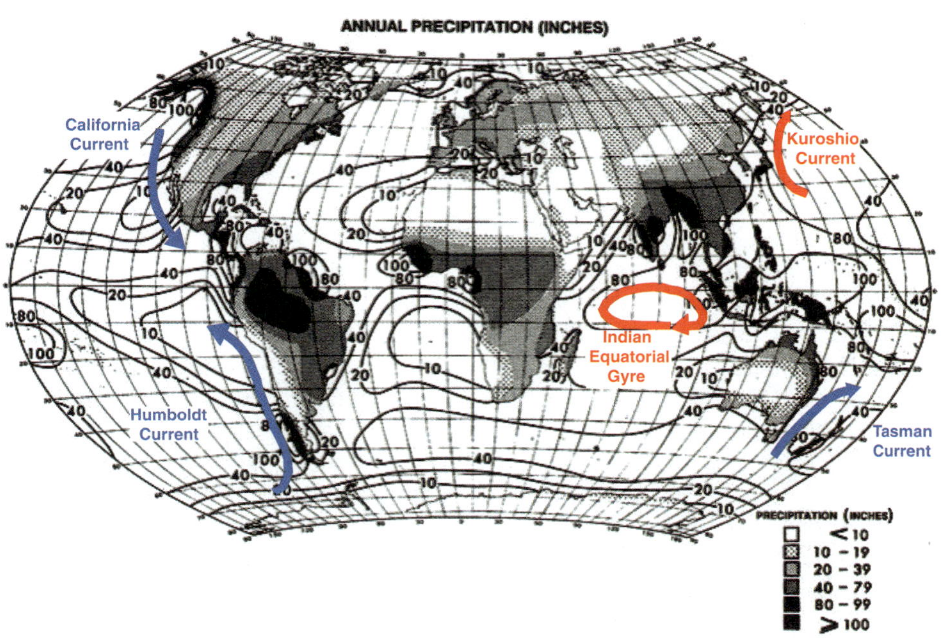

Figure 4. Average annual precipitation in inches around the globe (modified after McKittrick, 1991, their Fig. 3). The cool Tasman, California, and Humboldt Currents, and warm Indian Equatorial Gyre and Kuroshio Current are indicated schematically. Note marked N-S variation of rain- and snowfall along the Sierra Nevada, Chilean Andes, Himalayas and Tibetan Plateau, and the island chains of Japan and New Zealand.

Annual global precipitation is illustrated in Figure 4. Rain- and snowfall are sensitive functions of wind patterns, near-surface ocean currents, oceanic upwelling, and land topography. So how do climatic and related erosional processes impact the primary constructional features of an orogen and its crustal thickness?

The Andes

An illuminating example of a continental-margin mountain belt is the contractional, volcanogenic Chilean Andes, which parallels the outboard Chile-Peru convergent plate junction and the eastward-descending Nazca oceanic crust-capped plate. General motion plate vectors, topographic, and climatic relationships are presented in Figure 5. The highest mountains and most intensely shortened, thickest continental crust (55–75 km) occur in the north-central Andes at 20–25°S (James, 1971; Zeil, 1979; Isacks, 1988; Sheffels, 1990; Wigger et al., 1994; Beck et al., 1996; Baby et al., 1997). This is the atmospheric realm of high-pressure descending, heating Hadley cells, with cold water upwelling and the north-flowing Humboldt Current present directly offshore (e.g., see Hoke et al., 2004). In this maritime environment, westerly winds are relatively cold, thus low in H_2O content; they warm as they pass over the Atacama Desert, and yearly precipitation is only ~0.2–0.3 m in the High Andes (Montgomery et al., 2001). Due to the orographic effect, humidity falls again, with annual precipitation rates ~0.3 m as air currents continue eastward beyond the continental divide and descend over the thick sialic crust of the Altiplano (Mooney et al., 1998; McQuarrie, 2002). Low precipitation, hence a low erosion rate, partially accounts for the elevated calc-alkaline magmatic arc and the internally draining high plateau downwind (Vandervoort et al., 1995; Sobel et al., 2003). Because stream flow and erosion are poorly developed along the north-central Andean margin, insignificant amounts of sediment are supplied to the outboard Chile-Peru Trench (Bangs and Candie, 1997; Lamb and Davis, 2003), so little terrigenous debris is carried down with the subducting plate (von Huene and Scholl, 1991). A sediment-starved subduction channel should enhance subjacent erosion (von Huene and Ranero, 2003) and would tend to thin the crust. Accordingly, the regional thickness of the Andean crust at this latitude cannot reflect substantial subcrustal removal or the underplating and partial melting of sediments, but instead must be a consequence of descending-lithosphere-generated calc-alkaline volcanism-plutonism and tectonic shortening.

The climate is quite different around 45°S. Here the temperature contrast between the Humboldt Current and the land is reversed, with the continent being cooler than the nearby ocean. Low-pressure Ferrel cells in the atmosphere rise at this latitude, bringing moisture-laden winds at ground level to the cool land surface. Consequently, the rugged, fjord-rich but lower-elevation cordillera is richly endowed with ~2–4 m of annual precipitation, producing snowfields and glaciated valleys (Montgomery et al., 2001; Lamb and Davis, 2003); the alpine landscape is supported and accommodated by a continental crust 35–40 km thick

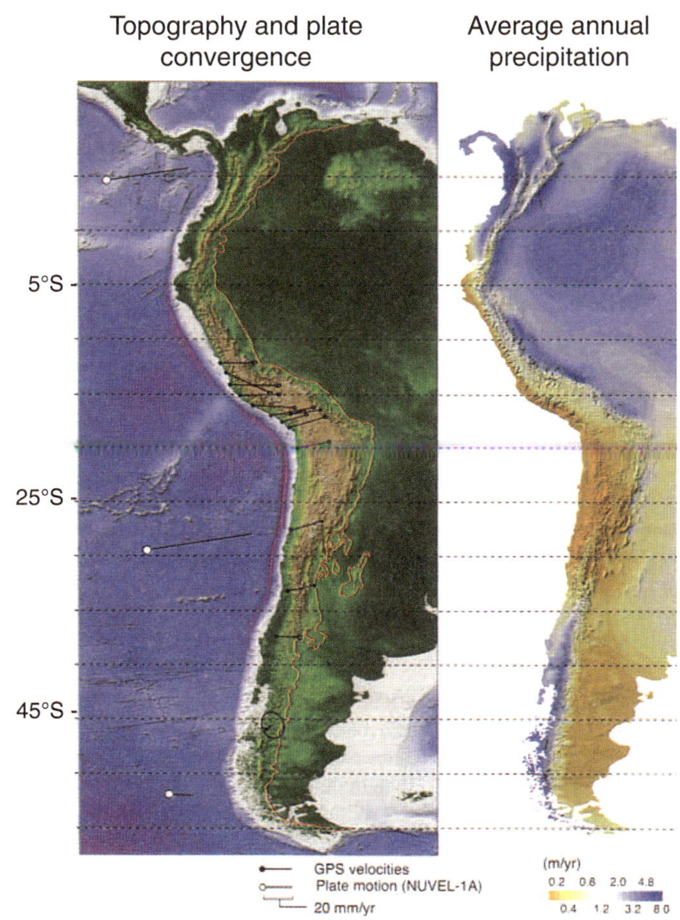

Figure 5. Topography, average annual precipitation in meters, and relative plate motion vectors in the Andes (after Montgomery et al., 2001, their Fig. 1). Note that the mean elevation of the mountain belt at 25°S exceeds 3 km, reflecting thick crust, whereas ~20° farther south, the regional elevation is somewhat less than 1 km, indicating thin crust.

(Hildreth and Moorbath, 1988; Russo and Silver, 1996; Bohm et al., 2002; Ramos et al., 2002; Mooney et al., 2002). Reflecting moderate precipitation downwind, stream drainage on the east is external, and a high plateau is lacking. In the southern Andes, abundant precipitation, active stream flow, and vigorous erosion provide voluminous clastic debris to the offshore Chile-Peru Trench; vast amounts of quartzofeldspathic material are being subducted (Lamb and Davis, 2003). In spite of underflow of this sedimentary mass, the sialic crust is only about half as thick as that at 25°S, demonstrating that intense erosional decapitation (a reflection of climatic patterns), rather than sediment supply and anatexis, is responsible for the moderate crustal thickness.

Although segmented into slabs dipping at different angles, the physical-chemical nature of the Nazca oceanic plate is roughly comparable in both north-central and southern Andes, so the primary calc-alkaline, subduction-zone magmas rising into the continental crust should be similar in both parts of the volcanogenic orogen. Complications to this simple picture reflect time-dependent subcrustal erosion and sialic contributions to arc magmatism, as well as backarc contraction and crustal thickening (Stern, 1991; Kay et al., 1999, 2005). A substantial sediment cushion in the southern Chile-Peru subduction zone should inhibit subjacent erosion and instead favor accretionary offloading, but such processes have not generated a massive Andean arc at 45°S (von Huene and Scholl, 1991; Scholl and von Huene, 2004). In marked contrast, the sediment-starved trench at 25°S (von Huene and Ranero, 2003) favors subjacent crustal removal along the nearly empty subduction channel, but this sector of the crust is almost twice as thick as that farther south.

The Himalayas

Continental collision and intracrustal thrusting accounts for the great regional thickness (65–80 km) of the Himalayan-Tibetan continental crust (Brandon and Romanowicz, 1986; Molnar, 1988; Holt and Wallace, 1990; Houseman and England, 1996; Mooney et al., 1998; Hodges, 2000). The elevation of this orogen dominantly reflects compressional tectonic processes (e.g., Vannay et al., 2004). The topography on the southern side of the orogen is somewhat modified by vigorous erosion attending the monsoonal system. Hot, H_2O-saturated air derived from the tropical Indian Ocean rises and cools as it encounters the Himalayas, resulting in annual precipitation rates of ~3–10 m (Finlayson et al., 2002). On the north, the internally draining Tibetan Plateau lies in the rain shadow of the Himalayas, and thus owes its elevation and great crustal thickness in part to aridity (yearly precipitation considerably less than 0.5 m) and attendant low erosion rates. Elevation also may reflect the presence of a relatively hot mantle substrate. Annual precipitation for the Himalayas and the Tibetan Plateau is shown in Figure 6. Because precipitation isopleths nearly parallel the WNW orientation of the mountain chain (see Fig. 4), this orogen lacks a simple axial gradient in erosion.

The Sierra Nevada

The Sierra Nevada Range of eastern California lies east of the south-flowing California Current. The latter, coupled with dominant westerly wind patterns, brings abundant moisture to the Pacific Northwest and to northern California; farther south, the land is considerably warmer than the ocean, resulting in progressively lower rainfall. Eastward from arid southern California, the Sierra Nevada achieves its maximum regional elevation in the Mount Whitney area. The yearly average precipitation in California is presented in Figure 7. The regional crustal thickness of the Sierra Nevada more-or-less reflects the relationship described above. The crustal mass of the parched southern Sierra Nevada (42–55? km) exceeds that of the snowy northern Sierra Nevada (~35 km), according to sparse available integrated geological and geophysical data (Fuis et al., 1987; Mooney and Weaver, 1989; Fliedner et al., 1996). And what about the Colorado Plateau, located far downwind from the southern Sierra Nevada (and in its rain shadow)? Neogene extensional tectonics have stretched and thinned the continental crust in the Basin and Range province, but it seems possible that in addition to petrotectonic constructional processes, limited erosion linked to an arid climate may

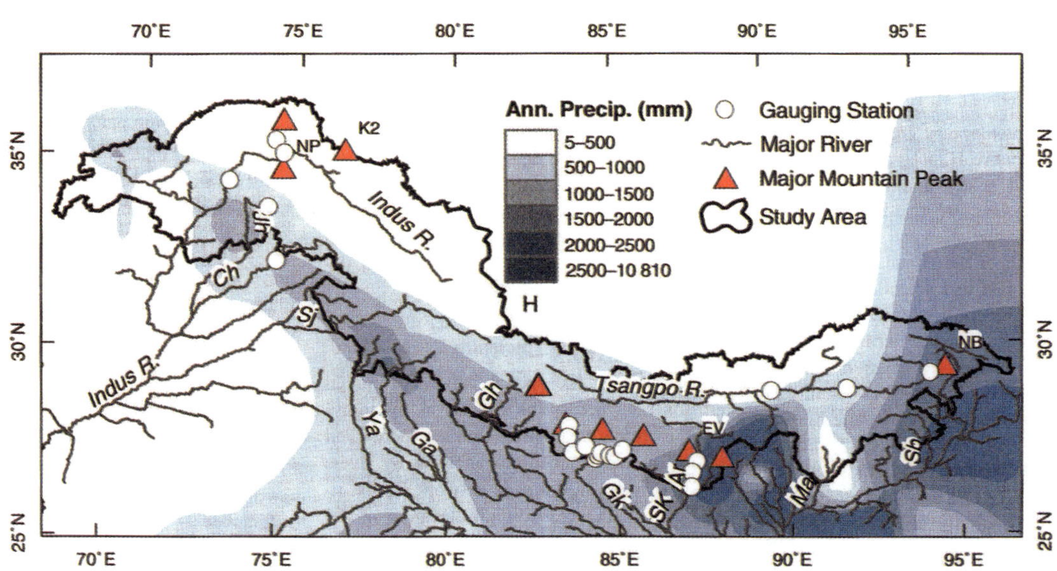

Figure 6. Average annual precipitation in mm, major peaks, drainage network, and gauging stations in the Himalayas and Tibetan Plateau (after Finlayson et al., 2002, their Fig. 1). Note rain shadow north of the mountain belt. (Tributary streams are labeled but names are not of consequence in the summary here.)

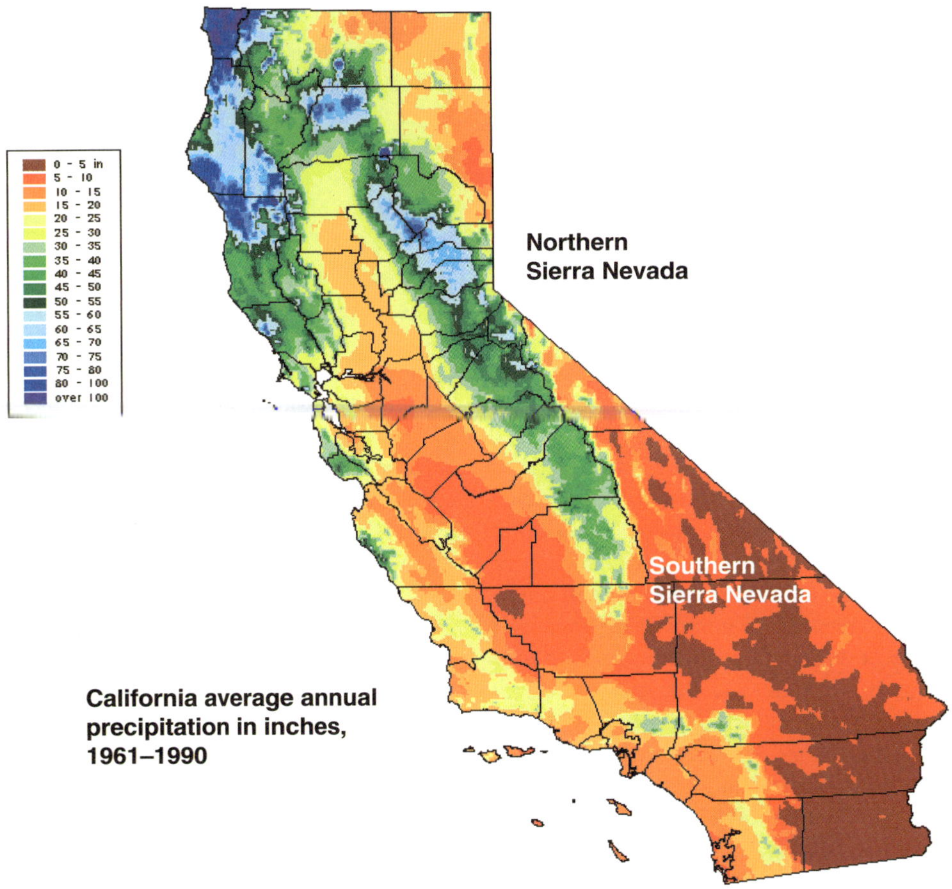

Figure 7. Average annual precipitation throughout California in inches (modified after Oregon Climate Service, 1995). The orographic effect of the Sierra Nevada Range is evident. The Colorado Plateau lies well downwind from the high-elevation southern Sierra Nevada and is separated from it by the extensional Basin and Range province.

have influenced the regional thickness of the crust in the Colorado Plateau as well as the southern Sierra Nevada.

Japan and New Zealand

Two other well-studied, active orogenic belts, Japan and New Zealand, exhibit strongly zoned rainfall patterns (Fig. 4). Annual precipitation in western Chugoku exceeds that in northern Honshu (SW and NE Japan, respectively), reflecting humid winds that blow off the semitropical, north-flowing Kuroshio Current. However, the plate-tectonic environment of the Japanese arc is complex, with the old, cool Pacific lithosphere descending beneath northern Honshu, and the young, warm Philippine Sea plate sinking under western Chugoku (Peacock and Wang, 1999). The Japan Trench receives modest volumes of sediment compared with the substantial clastic load entering the Nankai Trough. Nevertheless, NE Japan appears to have a continental crustal thickness of ~40 km (Hasegawa et al., 1994; Nakajima et al., 2001; Igarashi et al., 2001); in contrast the sialic crust of SW Japan is only ~30–35 km thick (Zhao and Hasegawa, 1993; Kodaira et al., 2000; Murakoshi et al., 2003; Shiomi et al., 2004). Thus, neither a cold lithosphere, sediment offloading, nor subcrustal erosion can account for the thicker crust of NE Japan, whereas lower precipitation and erosion rates can help explain this relationship.

The mountains of New Zealand are even more complicated, with opposite-facing subduction zones linked by a major dextral transform, the Alpine fault. Along the west coast, the north-flowing Tasman Current dumps legendary amounts of H_2O on the southwestern, Fiordland sector of South Island. The land farther north is warmer because westerly winds coupled with the relatively cool oceanic current provide less abundant rainfall to northern North Island. Although a local mountain root is evident, regional crustal thicknesses in southern South Island appear to slightly thicken northward (~25–35 km), with NE North Island slightly thicker still (35–40 km), as documented by seismic studies (e.g., Eberhart-Phillips and Reyners, 1997; Reyners et al., 1999; Eberhart-Phillips and Bannister, 2002). Again, precipitation (erosion) and crustal thickness appear to be inversely correlated.

Active Orogens and Precipitation Summary

For convergent plate-tectonic regimes in temperate latitudes, mean elevations of modern, roughly N-S-oriented mountain chains and regional thicknesses of isostatically compensated continental crust to some degree are inversely related to annual rain- and snowfall and attendant erosional decapitation. This overly simple inverse correlation between crustal thickness and precipitation for the Andes, Himalayas, Sierra Nevada, Japan, and New Zealand is shown in Table 1. Note that, in contrast to the other described examples, the intense ongoing tectonic contraction of the Himalayas far outweighs the ability of erosion to cut down the mountains; moreover, their WNW trend means that the range lacks an axial (latitudinal) precipitation gradient. In any case, due to the orographic effect on the monsoonal winds, the compressional Tibetan Plateau appears to be similar in origin to other rain-shadow plateaus such as the Altiplano and the Colorado Plateau. In addition, precipitation-linked surface erosion has been shown by Pysklywec (2006) to influence the deep-seated lithospheric response to continental collision.

TABLE 1. CORRELATION BETWEEN YEARLY PRECIPITATION AND CRUSTAL THICKNESS OF SOME ACTIVE OROGENS

Mountain belt	Crustal thickness (km)	Precipitation (m)
Andes 25°S	~70	0.2–0.3
Andes 45°S	~35	~3
Himalayas	65–80	3–10
Tibetan Plateau	~70	<0.5
N Sierra Nevada	~40	1.5–2.0
S Sierra Nevada	42–55?	0.7–1.5
NE Japan	~40	~1.5
SW Japan	~30–35	~2.3
North Island, New Zealand	~35–40	~1.1
South Island, New Zealand	~25–35	>2

CONCLUSIONS

Silicic and intermediate crust is generated by calc-alkaline magmatism in island arcs and continental margins sited along convergent plate junctions. Paired Pacific-type orogenic belts develop within and landward from zones of long-lived subduction of oceanic crust-capped lithosphere. They consist of an outboard, low-heat-flow accretionary prism deposited in and continentward from the trench, and an inboard, high-heat-flow volcanic-plutonic arc. The trench assemblage consists dominantly of clastic, arc-derived mélanges, minor but widespread deepwater cherts and/or carbonates, and ubiquitous disaggregated oceanic crust. These relatively incompetent sections recrystallize under HP conditions and decouple from the descending oceanic plate at depths of 15–50 km. A massive, contemporaneous calc-alkaline arc dominates the landward belt where new continental crust is added; high temperatures characterize the associated contact-regional metamorphism. Alpine-type orogenic belts form during underflow of an oceanic plate that transports continental crust into the subduction zone. Reflecting structural integrity, some Alpine-type microcontinents, island arcs, or continental salients are carried down as much as 90–140 km, become ductile, and low-density slices decouple from the sinking lithosphere. Metamorphism of deeply subducted parts of the orogen ranges from high pressure to ultrahigh pressure, and is not paired with a HT belt; calc-alkaline arc magmatism is lacking. In HP-UHP Alpine and outboard HP Pacific metamorphic terranes, thrust faults chiefly dip landward and fold vergence is seaward, reflecting similar convergent plate-tectonic settings; inboard, antithetic thrusting characterizes some contractional realms. High pressures and low temperatures attest to dynamic subduction-zone conditions. Low-density, buoyant crust is recycled in such tectonic realms but does not represent new additions to the continental crust.

In compressional orogens, exhumation of subducted material as thin-aspect-ratio sheets 1–5 km thick, combined with normal faulting above and subduction-zone thrusting below, allows conductive cooling and partial preservation of decompressing HP and UHP metamorphic complexes as they ascend along convergent plate junctions. Worldwide, resurrected subduction-zone terranes consist of only small masses of high-density mafic blueschist-eclogite and/or anhydrous peridotitic lenses surrounded by voluminous, low-density quartzofeldspathic and serpentinitic material, hence the bulk terrane density is less than that of unaltered, dynamically displaced mantle. The return to midcrustal levels is largely buoyancy-driven. Scattered preservation of HP-UHP relict phases reflects rise combined with mineralogic armoring and lack of catalytic aqueous fluid; partial preservation is only possible where thin sheets are exhumed to midcrustal levels and quenched. Some formerly deeply buried sections may be domically uplifted later toward the Earth's surface by tectonic aneurysms (generally associated with convergent plate-tectonic cusps). Exposure of rising HP and UHP metamorphic belts reflects surface erosion and gravitational collapse of the subduction complex.

Reflecting erosional intensity, the regional thickness of the crust in a relatively young contractional mountain belt is a function of climatic patterns as well as petrotectonic processes. For a particular convergent plate-tectonic environment, the mass of isostatically compensated continental crust to some extent is inversely correlated with precipitation and consequent erosional decapitation. In addition, tectonic shortening (crustal thickening) and rapid uplift in extremely arid realms partly defeats stream erosion and promotes formation of internally draining elevated plateaus leeward of high mountain ranges.

ACKNOWLEDGMENTS

This study, inspired by the Seventeenth International Basement Tectonics Conference organized and run by Bob Hatcher, was supported by Stanford University. Suzanne Kay, Walter Mooney, and Mike Searle provided me with important, critical feedback. Frank Spear and Ted Labotka wrote constructive reviews that improved this paper. I thank these researchers and Stanford for help and support.

REFERENCES CITED

Anderson, R.N., DeLong, S.E., and Schwarz, W.M., 1978, Thermal model for subduction with dehydration in the downgoing slab: Journal of Geology, v. 80, p. 731–739.

Aoya, M., 2001, P-T-D path of eclogite from the Sanbagawa belt deduced from combination of petrological and microstructural analyses: Journal of Petrology, v. 42, p. 1225–1248, doi: 10.1093/petrology/42.7.1225.

Austrheim, H., 1990, The granulite-eclogite facies transition: A comparison of experimental work and a natural occurrence in the Bergen Arcs, western Norway: Lithos, v. 25, p. 163–169, doi: 10.1016/0024-4937(90)90012-P.

Austrheim, H., 1998, The influence of fluid and deformation on metamorphism of the deep crust and consequences for geodynamics of collision zones, *in* Hacker, B.R., and Liou, J.G., eds., When continents collide: Geodynamics and geochemistry of ultrahigh-pressure rocks: Dordrecht, Kluwer Academic Publishers, p. 297–323.

Baby, P., Rochat, P., Mascle, G., and Hérail, G., 1997, Neogene shortening contribution to crustal thickening in the backarc of the central Andes: Geology, v. 25, p. 883–886, doi: 10.1130/0091-7613(1997)025<0883:NSCTCT>2.3.CO;2.

Bailey, E.H., Irwin, W.P., and Jones, D.L., 1964, Franciscan and related rocks, and their significance in the geology of western California: California Division of Mines and Geology Bulletin, v. 183, 177 p.

Bally, A.W., 1981, Thoughts on the tectonics of folded belts, in McClay, K.R., and Price, N.J., eds., Thrust and nappe tectonics: Geological Society [London] Special Publication 9, p. 13–32.

Bangs, N.L., and Candie, S.C., 1997, Episodic development of a convergent margin inferred from structures and processes along the southern Chile margin: Tectonics, v. 16, p. 489–503, doi: 10.1029/97TC00494.

Banno, S., and Sakai, C., 1989, Geology and metamorphic evolution of the Sanbagawa belt: Geological Society [London] Special Publication 43, p. 519–535.

Barton, M.D., Battles, D.A., Bebout, G.E., Capo, R.C., Christensen, J.N., Davis, S.R., Hanson, R.B., Michelson, C.J., and Trim, H.E., 1988, Mesozoic contact metamorphism in the western United States, *in* Ernst, W.G., ed., Metamorphic and crustal evolution of the western United States: Englewood Cliffs, New Jersey, Prentice-Hall, p. 110–178.

Beaumont, C., Ellis, S., Hamilton, J., and Fullsack, P., 1996, Mechanical model for subduction-collision tectonics of Alpine-type compressional orogens: Geology, v. 24, p. 675–678, doi: 10.1130/0091-7613(1996)024<0675:MMFSCT>2.3.CO;2.

Beaumont, C., Ellis, S., and Pfiffner, A., 1999, Dynamics of sediment subduction-accretion at convergent margins: Short-term modes, long-term deformation, and tectonic implications: Journal of Geophysical Research, v. 104, p. 17,573–17,602, doi: 10.1029/1999JB900136.

Bebout, G.E., 1996, Volatile transfer and recycling at convergent margins: Mass balance and insights from high P-T metamorphic rocks: American Geophysical Union Geophysical Monograph 96, p. 179–193.

Beck, S.L., Zandt, G., Myers, S.C., Wallace, T.C., Silver, P.G., and Drake, L., 1996, Crustal-thickness variations in the central Andes: Geology, v. 24, p. 407–410, doi: 10.1130/0091-7613(1996)024<0407:CTVITC>2.3.CO;2.

Bohm, M., Lüth, S., Echtler, H., Asch, G., Bataille, K., Bruhn, C., Rietbrock, A., and Wigger, P., 2002, The southern Andes between 36° and 40°S latitude: Seismicity and average seismic velocities: Tectonophysics, v. 356, p. 275–289, doi: 10.1016/S0040-1951(02)00399-2.

Boudier, F., Bouchez, J.L., Nicolas, A., Cannat, M., Ceuleneer, G., Miseri, M., and Montigny, R., 1985, Kinematics of oceanic thrusting in the Oman ophiolites: Model of plate convergence: Earth and Planetary Science Letters, v. 75, p. 215–222, doi: 10.1016/0012-821X(85)90103-7.

Boudier, F., Ceuleneer, G., and Nicolas, A., 1988, Shear zones, thrusts and related magmatism in the Oman ophiolite: Initiation of thrusting on an oceanic ridge: Tectonophysics, v. 151, p. 275–296, doi: 10.1016/0040-1951(88)90249-1.

Brandon, C., and Romanowicz, B., 1986, A "no-lid" zone in the central Chang-Tang platform of Tibet: Evidence from pure path phase velocity measurements of long period Rayleigh waves: Journal of Geophysical Research, v. 91, p. 6547–6564.

Burbank, D.W., 1992, Causes of recent Himalayan uplift deduced from the deposited patterns in the Ganges basin: Nature, v. 357, p. 680–683, doi: 10.1038/357680a0.

Burbank, D.W., Meigs, A., and Brozovic, N., 1996, Interactions of growing folds and coeval depositional systems: Basin Research, v. 8, p. 199–223, doi: 10.1046/j.1365-2117.1996.00181.x.

Burbank, D.W., Blithe, A.E., Putkonen, J., Pratt-Sitaula, B., Gabet, E., Oskin, M., Barros, A., and Ojha, T.P., 2003, Decoupling of erosion and precipitation in the Himalayas: Nature, v. 426, p. 652–655.

Burchfiel, B.C., Deng, Q., Molnar, P., Royden, L., Wang, Y., and Zhang, W., 1989, Intracrustal detachment within zones of continental deformation: Geology, v. 17, p. 748–752, doi: 10.1130/0091-7613(1989)017<0448:IDWZOC>2.3.CO;2.

Burg, J.P., Davy, P., Nievergelt, P., Oberli, F., Seward, D., Diao, Z., and Meier, M., 1997, Exhumation during crustal folding in the Namche Barwa syntaxis: Terra Nova, v. 9, p. 53–56, doi: 10.1111/j.1365-3121.1997.tb00001.x.

Busby-Spera, C.J., Mattinson, J.M., Riggs, N.R., and Schermer, E.R., 1990, The Triassic-Jurassic magmatic arc in the Mojave-Sonoran Deserts and the Sierran-Klamath region, *in* Harwood, D.S., and Miller, M.M., eds., Paleozoic and early Mesozoic paleogeographic relations: Sierra Nevada, Klamath Mountains, and related terranes: Geological Society of America Special Paper 255, p. 93–114.

Castro, A., Fernandez, C., and Vigneresse, J.L., editors, 1999, Understanding granites: Integrating new and classical techniques: Geological Society [London] Journal, v. 168, 278 p.

Chemenda, A.I., Mattauer, M., Malavieille, J., and Bokun, A.N., 1995, A mechanism for syn-collisional rock exhumation and associated normal faulting: Results from physical modeling: Earth and Planetary Science Letters, v. 132, p. 225–232, doi: 10.1016/0012-821X(95)00042-B.

Chemenda, A.I., Mattauer, M., and Bokun, A.N., 1996, Continental subduction and a new mechanism for exhumation of high-pressure metamorphic rocks: New modeling and field data from Oman: Earth and Planetary Science Letters, v. 143, p. 173–182, doi: 10.1016/0012-821X(96)00123-9.

Chopin, C., 1984, Coesite and pure pyrope in high-grade blueschists of the western Alps: A first record and some consequences: Contributions to Mineralogy and Petrology, v. 86, p. 107–118, doi: 10.1007/BF00381838.

Chopin, C., 1987, Very-high-pressure metamorphism in the western Alps: Implications for subduction of continental crust: Royal Society of London Philosophical Transactions, ser. A, v. 1557, p. 183–197.

Clift, P.D., Rose, E.F., Shimizu, N., Layne, G.D., Draut, A.E., and Regelous, M., 2001, Tracing the evolving flux from the subducting plate in the Tonga-Kermadec arc system using boron in volcanic glass: Geochimica et Cosmochimica Acta, v. 65, p. 3347–3364, doi: 10.1016/S0016-7037(01)00670-6.

Cloos, M., 1984, Landward-dipping reflectors in accretionary wedges: Active dewatering conduits?: Geology, v. 12, p. 519–522, doi: 10.1130/0091-7613(1984)12<519:LRIAWA>2.0.CO;2.

Cloos, M., 1993, Lithospheric buoyancy and collisional orogenesis: Subduction of oceanic plateaus, continental margins, island arcs, spreading ridges, and seamounts: Geological Society of America Bulletin, v. 105, p. 715–737, doi: 10.1130/0016-7606(1993)105<0715:LBACOS>2.3.CO;2.

Cloos, M., and Shreve, R.L., 1988a, Subduction-channel model of prism accretion, mélange formation and subduction erosion at convergent plate margins, 1: Background and description, *in* Ruff, L., and Kanamori, H., eds., Subduction zones: Pure and applied geophysics: Basel, Switzerland, Birkhauser, v. 128, p. 455–500.

Cloos, M., and Shreve, R.L., 1988b, Subduction-channel model of prism accretion, mélange formation and subduction erosion at convergent plate margins, 2: Implications and discussion, *in* Ruff, L., and Kanamori, H., eds., Subduction zones: Pure and applied geophysics: Basel, Switzerland, Birkhauser, v. 128, p. 501–545.

Coleman, R.G., 1977, Ophiolites, ancient oceanic lithosphere?: New York, Springer-Verlag, 229 p.

Coleman, R.G., and Wang, X., 1995, Overview of the geology and tectonics of UHPM, in Coleman, R.G., and Wang, X., eds., Ultrahigh pressure metamorphism: New York, Cambridge University Press, p. 1–32.

Compagnoni, R., Hirajima, T., and Chopin, C., 1995, UHPM metamorphic rocks in the western Alps, *in* Coleman, R.G., and Wang, X., eds., Ultrahigh pressure metamorphism: New York, Cambridge University Press, p. 206–243.

Copeland, P., Harrison, T.M., Yun, P., Kidd, W.S.F., Roden, M., and Zhang, Y., 1995, Thermal evolution of the Gangdese batholith, southern Tibet: A history of episodic unroofing: Tectonics, v. 14, p. 223–236, doi: 10.1029/94TC01676.

Cosca, M.A., Giorgis, D., Rumble, D., and Liou, J.G., 2005, Limiting effects of UHP metamorphism on length scales of oxygen, hydrogen, and argon isotope exchange: An example from the Qinglongshan UHP eclogites, Sulu terrain, China: International Geology Review, v. 47, p. 716–749.

Cowan, D.S., and Silling, R.M., 1978, A dynamic, scaled model of accretion at trenches and its implications for the tectonic evolution of subduction complexes: Journal of Geophysical Research, v. 83, p. 5389–5396.

Dadson, S.J., Hovius, N., Chen, H., Dade, W.B., Hsieh, M.L., Willet, S.D., Hu, J.C., Horng, M.J., Chen, M.C., Stark, C.P., Lague, D., and Lin, J.C., 2003, Links between erosion, runoff variability and seismicity in the Taiwan orogen: Nature, v. 426, p. 648–651, doi: 10.1038/nature02150.

Dadson, S.J., Hovius, N., Chen, H., Dade, W.B., Lin, J.C., Hsu, M.L., Lin, C.W., Horng, M.J., Chen, T.C., Milliman, J., and Stark, C.P., 2004, Earthquake-triggered increase in sediment delivery from an active mountain belt: Geology, v. 32, p. 733–736, doi: 10.1130/G20639.1.

Dal Piaz, G.V., Hunziker, J.C., and Martinotti, G., 1972, La zona Sesia-Lanzo e l'evoluzione tettonico-metamorfica delle Alpi nordoccidentali interne: Memoir of the Geological Society of Italy, v. 11, p. 433–460.

Davies, J.H., and von Blanckenburg, F., 1998, Thermal controls on slab breakoff and the rise of high-pressure rocks during continental collisions, *in* Hacker, B.R., and Liou, J.G., eds., When continents collide: Geodynamics and geochemistry of ultrahigh-pressure rocks: Dordrecht, Kluwer Academic Publishers, p. 97–115.

DePaolo, D.J., 1981, A neodymium and strontium isotopic study of Mesozoic calc-alkaline granite batholiths of the Sierra Nevada and Peninsular Ranges, California: Journal of Geophysical Research, v. 86, p. 10,470–10,488.

Dickinson, W.R., 1971, Clastic sedimentary sequences deposited in shelf, slope and trench settings between magmatic arcs and associated trenches: Pacific Geology, v. 3, p. 1–14.

Dickinson, W.R., 1972, Evidence for plate-tectonic regimes in the rock record: American Journal of Science, v. 272, p. 551–576.

Dickinson, W.R., 1976, Sedimentary basins developed during evolution of Mesozoic-Cenozoic arc-trench systems in western North America: Canadian Journal of Earth Sciences, v. 13, p. 1268–1289.

Ding, L., and Zhong, D., 1999, Metamorphic characteristics and geotectonic implications of the high-pressure granulites from Namche Barwa, eastern Tibet: Chinese Science Bulletin, v. 42, p. 491–505.

Ding, L., Zhong, D., Yin, A., Kapp, P., and Harrison, T.M., 2001, Cenozoic structural and metamorphic evolution of the eastern Himalayan syntaxis (Namche Barwa): Earth and Planetary Science Letters, v. 192, p. 423–438, doi: 10.1016/S0012-821X(01)00463-0.

Drummond, M.S., and Defant, M.J., 1990, A model for trondhjemite-tonalite-dacite genesis and crustal growth via slab melting: Archean to modern comparisons: Journal of Geophysical Research, v. 95, p. 21,503–21,521.

Eberhart-Phillips, D., and Bannister, 2002, Three-dimensional crustal structure in the Southern Alps region of New Zealand from inversion of local earthquake and active source data: Journal of Geophysical Research, v. 107, no. B10, 2262, doi: 10.1029/2001JB000567.

Eberhart-Phillips, D., and Reyners, M., 1997, Continental subduction and three-dimensional crustal structure: The northern South Island, New Zealand: Journal of Geophysical Research, v. 102, p. 11,843–11,861, doi: 10.1029/96JB03555.

England, P.L., and Holland, T.J.B., 1979, Archimedes and the Tauern eclogites: The role of buoyancy in the preservation of exotic eclogitic blocks: Earth and Planetary Science Letters, v. 44, p. 287–294, doi: 10.1016/0012-821X(79)90177-8.

Ernst, W.G., 1970, Tectonic contact between the Franciscan mélange and the Great Valley sequence: Crustal expression of a late Mesozoic Benioff zone: Journal of Geophysical Research, v. 75, p. 886–901.

Ernst, W.G., 1971, Metamorphic zonations on presumably subducted lithospheric plates from Japan, California and the Alps: Contributions to Mineralogy and Petrology, v. 34, p. 43–59, doi: 10.1007/BF00376030.

Ernst, W.G., 1988, Tectonic histories of subduction zones inferred from retrograde blueschist P-T paths: Geology, v. 16, p. 1081–1084, doi: 10.1130/0091-7613(1988)016<1081:THOSZI>2.3.CO;2.

Ernst, W.G., 1990, Thermobarometric and fluid-expulsion history of subduction zones: Journal of Geophysical Research, v. 95, p. 9047–9053.

Ernst, W.G., 1992, Metamorphism of the western Cordillera, and its relationship to tectonics, *in* Burchfiel, B.C., et al., eds., The Cordilleran orogen: Conterminous United States: Geological Society of America, Geology of North America, v. G-3, p. 515–538.

Ernst, W.G., 1999, Hornblende, the continent maker—Evolution of H_2O during circum-Pacific subduction versus continental collision: Geology, v. 27, p. 675–678, doi: 10.1130/0091-7613(1999)027<0675:HTCMEO>2.3.CO;2.

Ernst, W.G., and Liou, J.G., 1995, Contrasting plate-tectonic styles of the Qinling-Dabie-Sulu and Franciscan metamorphic belts: Geology, v. 23, p. 353–356, doi: 10.1130/0091-7613(1995)023<0353:CPTSOT>2.3.CO;2.

Ernst, W.G., and Peacock, S., 1996, A thermotectonic model for preservation of ultrahigh-pressure mineralogic relics in metamorphosed continental crust: American Geophysical Union Geophysical Monograph 96, p. 171–178.

Ernst, W.G., Seki, Y., Onuki, H., and Gilbert, M.C., 1970, Comparative study of low-grade metamorphism in the California Coast Ranges and the Outer Metamorphic Belt of Japan: Geological Society of America Memoir 124, 276 p.

Ernst, W.G., Maruyama, S., and Wallis, S., 1997, Buoyancy-driven, rapid exhumation of ultrahigh-pressure metamorphosed continental crust: Proceedings of the National Academy of Sciences, v. 94, p. 9532–9537.

Ernst, W.G., Mosenfelder, J.L., Leech, M.L., and Liu, J., 1998, H_2O recycling during continental collision: Phase-equilibrium and kinetic considerations, *in* Hacker, B.R., and Liou, J.G., eds., When continents collide: Geodynamics and geochemistry of ultrahigh-pressure rocks: Dordrecht, Kluwer Academic Publishers, p. 275–295.

Finlayson, D.P., Montgomery, D.R., and Hallet, B., 2002, Spatial coincidence of rapid inferred erosion with young metamorphic massifs in the Himalayas: Geology, v. 30, p. 219–222, doi: 10.1130/0091-7613(2002)030<0219:SCORIE>2.0.CO;2.

Fitch, T.Y., 1972, Plate convergence, transform faults, and internal deformation adjacent to southeast Asia and the western Pacific: Journal of Geophysical Research, v. 77, p. 4432–4460.

Fliedner, M.M., Rupert, S., and Southern Sierra Nevada Continental Dynamics Working Group, 1996, Three-dimensional crustal structure of the southern Sierra Nevada from seismic fan profiles and gravity modeling: Geology, v. 24, p. 367–370, doi: 10.1130/0091-7613(1996)024<0367:TDCSOT>2.3.CO;2.

Frey, M., editor, 1987, Low temperature metamorphism: Glasgow, Blackie Publishing, 351 p.

Frey, M., Hunziker, J.C., Frank, W., Bocquet, J., Dal Piaz, G.V., Jäger, E., and Niggli, E., 1974, Alpine metamorphism of the Alps: Schweizerische Mineralogisches und Petrografisches Mittelungen, v. 54, p. 248–290.

Fuis, G.S., Zucca, J.J., Mooney, W.D., and Milkereit, B., 1987, A geological interpretation of seismic refraction results in northeastern California: Geological Society of America Bulletin, v. 98, p. 53–65, doi: 10.1130/0016-7606(1987)98<53:AGIOSR>2.0.CO;2.

Gnos, E., Immenhauser, A., and Peters, T., 1997, Late Cretaceous–early Tertiary convergence between the Indian and Arabian plates recorded in ophiolites and related sediments: Tectonophysics, v. 271, p. 1–19, doi: 10.1016/S0040-1951(96)00249-1.

Hacker, B.R., 1996, Eclogite formation and the rheology, buoyancy, seismicity, and H2O content of oceanic crust: American Geophysical Union Geophysical Monograph 96, p. 337–346.

Hacker, B.R., and Peacock, S.M., 1995, Creation, preservation and exhumation of UHPM rocks, *in* Coleman, R.G., and Wang, X., eds., Ultrahigh pressure metamorphism: New York, Cambridge University Press, p. 159–181.

Hacker, B.R., Ratschbacher, L., Webb, L., and Dong, S., 1995, What brought them up? Exhuming the Dabie Shan ultrahigh-pressure rocks: Geology, v. 23, p. 743–746, doi: 10.1130/0091-7613(1995)023<0743:WBTUEO>2.3.CO;2.

Hacker, B.R., Wang, X., Eide, E.A., and Ratschbacher, L., 1996, The Qinling-Dabie ultrahigh-pressure collisional orogen, *in* Harrison, T.M., and Yin,

A., eds., The tectonic development of Asia: New York, Cambridge University Press, p. 345–370.

Hacker, B.R., Ratschbacher, L., Webb, L., McWilliams, M.O., Ireland, T., Calvert, A., Dong, S., Wenk, H.-R., and Chateigner, D., 2000, Exhumation of ultrahigh-pressure continental crust in east-central China: Late Triassic–Early Jurassic tectonic unroofing: Journal of Geophysical Research, v. 105, p. 13,339–13,364, doi: 10.1029/2000JB900039.

Hacker, B.R., Abers, G.A., and Peacock, S.M., 2003, Subduction factory, 1: Theoretical mineralogy, density, seismic wave speeds, and H2O content: Journal of Geophysical Research, v. 108, no. B1, 10.1029-2001JB001127.

Hamilton, W., 1970, The Uralides and the motion of the Russian and Siberian platforms: Geological Society of America Bulletin, v. 81, p. 2553–2576, doi: 10.1130/0016-7606(1970)81[2553:TUATMO]2.0.CO;2.

Hamilton, W., 1979, Tectonics of the Indonesian region: U.S. Geological Survey Professional Paper 1078, 345 p.

Hamilton, W., 1995, Subduction systems and magmatism, in Smellie, J.L., ed., Volcanism associated with extension at consuming plate margins: Geological Society [London] Special Publication 81, p. 3–28

Hartshorn, K., Hovius, N., Dade, W.B., and Slingerland, R.L., 2003, Climate-driven bedrock incision in an active mountain belt: Science, v. 297, p. 2036–2038, doi: 10.1126/science.1075078.

Hasegawa, A., Horiuchi, S., and Umino, N., 1994, Seismic structure of the northeastern Japan convergent margin: Journal of Geophysical Research, v. 99, p. 22,295–22,311, doi: 10.1029/93JB02797.

Hawkesworth, C.J., Gallagher, K., Hergt, J.M., and McDermott, F., 1993, Mantle and slab contributions in an arc magma: Annual Review of Earth and Planetary Sciences, v. 21, p. 175–204.

Henry, C., 1990, L'unite a coesite du massif Dora-Maira dans son cadre petrologique et structural (Alpes occidentales, Italie): Université de Paris VI, 453 p.

Hickey-Vargas, R., Sun, M., Lopez-Escobar, L., Moreno-Roa, H., Reagan, M.K., Morris, J.D., and Ryan, J.G., 2002, Multiple subduction components in the mantle wedge: Evidence from eruptive centers in the Central Southern volcanic zone, Chile: Geology, v. 30, p. 199–202, doi: 10.1130/0091-7613(2002)030<0199:MSCITM>2.0.CO;2.

Hildreth, W., and Moorbath, S., 1988, Crustal contributions to arc magmatism in the Andes of central Chile: Contributions to Mineralogy and Petrology, v. 98, p. 455–489, doi: 10.1007/BF00372365.

Hilley, G.E., and Strecker, M.R., 2004, Steady-state erosion of critical Coulomb wedges with applications to Taiwan and the Himalaya: Journal of Geophysical Research, v. 109, B01411, doi: 10.1029-2002JB002284.

Hilley, G.E., Strecker, M.R., and Ramos, V.A., 2004, Growth and erosion of fold-and-thrust belts with an application to the Aconcagua fold-and-thrust belt, Argentina: Journal of Geophysical Research, v. 109, B01410, doi: 10.1029-2002JB002282.

Hirajima, T., Wallis, S., Zhai, R., and Ye, K., 1993, Eclogitized metagranitoid from the Su-Lu ultrahigh-pressure (UHP) province, eastern China: Proceedings of the Japan Academy, v. 69, section B, p. 249–254.

Hodges, K.V., 2000, Tectonics of the Himalaya and southern Tibet from two perspectives: Geological Society of America Bulletin, v. 112, p. 324–350, doi: 10.1130/0016-7606(2000)112<0324:TOTHAS>2.3.CO;2.

Hoke, G.A., Isacks, B.L., Jordan, T.E., and Yu, J.S., 2004, Groundwater-sapping origin for the giant quebradas of northern Chile: Geology, v. 32, p. 605–606, doi: 10.1130/G20601.1.

Holt, W.E., and Wallace, T.C., 1990, Crustal thickness and upper mantle velocities in the Tibetan Plateau region from the inversion of regional Pnl wave forms: Evidence for a thick upper mantle lid beneath southern Tibet: Journal of Geophysical Research, v. 91, p. 12,499–12,525.

Honegger, K., Dietrich, V., Frank, W., Gansser, A., Thöni, M., and Trommsdorf, V., 1982, Magmatism and metamorphism in the Ladakh Himalayas (the Indus-Tsangpo suture zone): Earth and Planetary Science Letters, v. 60, p. 253–292, doi: 10.1016/0012-821X(82)90007-3.

Horton, B.K., 1999, Erosional control on the geometry and kinematics of thrust belt development in the central Andes: Tectonics, v. 18, p. 1292–1304, doi: 10.1029/1999TC900051.

Houseman, G., and England, P., 1996, A lithospheric-thickening model for the Indo-Asian collision, *in* Yin, A., and Harrison, T.M., eds., The tectonic evolution of Asia: New York, Cambridge University Press, p. 3–17.

Howell, D.G., editor, 1985, Tectonostratigraphic terranes of the circum-Pacific region: Houston, Texas, Circum-Pacific Council for Energy and Mineral Resources, 581 p.

Igarashi, T., Matsuzawa, T., Umino, N., and Hasegawa, A., 2001, Spatial distribution of focal mechanisms for interplate and intraplate earthquakes associated with the subducting Pacific plate beneath the northeastern Japan arc: A triple-planes deep seismic zone: Journal of Geophysical Research, v. 106, p. 2177–2191, doi: 10.1029/2000JB900386.

Isacks, B., 1988, Uplift of the central Andean Plateau and bending of the Bolivian orocline: Journal of Geophysical Research, v. 93, p. 3211–3231.

Isacks, B., Oliver, J., and Sykes, L.R., 1968, Seismology and the new global tectonics: Journal of Geophysical Research, v. 73, p. 5855–5899.

Ishikawa, M., Kaneko, Y., and Yamamoto, H., 2000, Subhorizontal boundary between ultrahigh-pressure and low-pressure metamorphic units in the Sulu-Tjube area of the Kokchetav massif, Kazakhstan: Island Arc, v. 9, p. 317–328, doi: 10.1046/j.1440-1738.2000.00281.x.

James, D.E., 1971, Plate-tectonic model for the evolution of the central Andes: Geological Society of America Bulletin, v. 82, p. 3325–3346, doi: 10.1130/0016-7606(1971)82[3325:PTMFTE]2.0.CO;2.

Jayko, A.S., Blake, M.C., and Harms, T., 1987, Attenuation of the Coast Range ophiolite by extensional faulting, and nature of the Coast Range "thrust," California: Tectonics, v. 6, p. 475–488.

Kadarusman, A., and Parkinson, C.D., 2000, Petrology and P-T evolution of garnet peridotites from central Sulawesi, Indonesia: Journal of Metamorphic Geology, v. 18, p. 193–209, doi: 10.1046/j.1525-1314.2000.00238.x.

Kadarusman, A., Miyashita, S., Parkinson, C.D., and Ishikawa, A., 2004, Petrology, geochemistry, and paleogeographic reconstruction of the East Sulawesi Ophiolite, Indonesia: Tectonics, v. 392, p. 55–83.

Kaneko, Y., Maruyama, S., Terabayashi, M., Yamamoto, H., Ishikawa, M., Anma, R., Parkinson, C.D., Ota, T., Nakajima, Y., Katayama, I., Yamamoto, J., and Yamauchi, K., 2000, Geology of the Kokchetav ultrahigh-pressure–high-pressure metamorphic belt, northeastern Kazakhstan: Island Arc, v. 9, p. 264–283, doi: 10.1046/j.1440-1738.2000.00278.x.

Kaneko, Y., Katayama, I., Yamamoto, H., Misawa, K., Ishikawa, M., Rehman, H.U., Kausar, A.B., and Shirashi, K., 2003, Timing of Himalayan ultrahigh-pressure metamorphism: Sinking rate and subduction angle of the Indian continental crust beneath Asia: Journal of Metamorphic Geology, v. 21, p. 589–599, doi: 10.1046/j.1525-1314.2003.00466.x.

Kawachi, Y., 1968, Large-scale overturned structure in the Sanbagawa metamorphic zone in central Shikoku, Japan: Geological Society of Japan Journal, v. 74, p. 607–616.

Kay, S.M., Mpodozis, C., and Coira, B., 1999, Magmatism, tectonism, and mineral deposits of the central Andes (22°–33°S), *in* Skinner, B.F., ed., Geology and ore deposits of the central Andes: Society of Economic Geologists Special Publication 7, p. 27–59.

Kay, S.M., Godoy, E., and Kurtz, A., 2005, Episodic arc migration, crustal thickening, subduction erosion, and magmatism in the south-central Andes: Geological Society of America Bulletin, v. 117, p. 67–88, doi: 10.1130/B25431.1.

Kodaira, S., Takahashi, N., Park, J.O., Mochizuki, M., Shinohara, M., and Kimura, S., 2000, Western Nankai Trough seismogenic zone: Results from a wide-angle ocean bottom seismic survey: Journal of Geophysical Research, v. 105, p. 5887–5905, doi: 10.1029/1999JB900394.

Koons, P.O., Zeitler, P.K., Chamberlain, C.P., Craw, D., and Meltzer, A.S., 2002, Mechanical links between erosion and metamorphism in Nanga Parbat, Pakistan Himalaya: American Journal of Science, v. 302, p. 749–773, doi: 10.2475/ajs.302.9.749.

Koons, P.O., Norris, R.J., Craw, D., and Cooper, A.F., 2003, Influence of exhumation on the structural evolution of transpression plate boundaries: An example from the Southern Alps, New Zealand: Geology, v. 31, p. 3–6, doi: 10.1130/0091-7613(2003)031<0003:IOEOTS>2.0.CO;2.

Kushiro, I., 1990, Partial melting of mantle wedge and evolution of island arc crust: Journal of Geophysical Research, v. 95, p. 15,929–15,939.

Lamb, S., and Davis, P., 2003, Cenozoic climate change as a possible cause for the rise of the Andes: Nature, v. 425, p. 792–797, doi: 10.1038/nature02049.

Leech, M.L., 2001, Arrested orogenic development: Eclogitization, delamination, and tectonic collapse: Earth and Planetary Science Letters, v. 185, p. 149–159, doi: 10.1016/S0012-821X(00)00374-5.

LeFort, P., Cuney, M., Deniel, C., France-Lanord, C., Sheppard, S.M.F., Upreti, B.N., and Vidal, P., 1987, Crustal generation of Himalayan leucogranites: Tectonophysics, v. 134, p. 39–57, doi: 10.1016/0040-1951(87)90248-4.

Liou, J.G., and Zhang, R.Y., 1996, Occurrences of intergranular coesite in ultrahigh-P rocks from the Sulu region, eastern China: Implications for lack of fluid during exhumation: American Mineralogist, v. 81, p. 1217–1221.

Liou, J.G., Zhang, R.Y., Wang, X., Eide, E.A., Ernst, W.G., and Maruyama, S., 1996, Metamorphism and tectonics of high-pressure and ultrahigh-pressure belts in the Dabie-Sulu region, China, *in* Yin, A., and Harrison, T.M., eds., The tectonic evolution of Asia: New York, Cambridge University Press, p. 300–344.

Liou, J.G., Zhang, R.Y., Ernst, W.G., Rumble, D., III, and Maruyama, S., 1998, High pressure minerals from deeply subducted metamorphic rocks: Reviews in Mineralogy, v. 37, p. 33–96.

Liu, J., Bohlen, S.R., and Ernst, W.G., 1996, Stability of hydrous phases in subducting oceanic crust: Earth and Planetary Science Letters, v. 143, p. 161–171, doi: 10.1016/0012-821X(96)00130-6.

Maruyama, S., and Parkinson, C.D., 2000, Overview of the geology, petrology and tectonic framework of the HP-UHP metamorphic belt of the Kokchetav massif, Kazakhstan: Island Arc, v. 9, p. 439–455, doi: 10.1046/j.1440-1738.2000.00288.x.

Maruyama, S., Liou, J.G., and Zhang, R., 1994, Tectonic evolution of the ultrahigh-pressure (UHP) and high-pressure (HP) metamorphic belts from central China: Island Arc, v. 3, p. 112–121, doi: 10.1111/j.1440-1738.1994.tb00099.x.

Maruyama, S., Liou, J.G., and Terabayashi, M., 1996, Blueschists and eclogites of the world, and their exhumation: International Geology Review, v. 38, p. 485–594.

Massone, H.J., 1995, Experimental and petrogenetic study of UHPM, in Coleman, R.G., and Wang, X., eds., Ultrahigh pressure metamorphism: New York, Cambridge University Press, p. 33–95.

McKittrick, D.A., editor, 1991, Climates of the world: Asheville, North Carolina, National Oceanographic and Atmospheric Administration, 33 p.

McQuarrie, N., 2002, The kinematic history of the central Andean fold-thrust belt, Bolivia: Implications for building a high plateau: Geological Society of America Bulletin, v. 114, p. 950–963, doi: 10.1130/0016-7606(2002)114<0950:TKHOTC>2.0.CO;2.

Michard, A., Henry, C., and Chopin, C., 1995, Structures in UHPM rocks: A case study from the Alps, *in* Coleman, R.G., and Wang, X., eds., Ultrahigh pressure metamorphism: New York, Cambridge University Press, p. 132–158.

Miller, J.M., Gray, D.R., and Gregory, R.T., 1998, Exhumation of high-pressure rocks in northeastern Oman: Geology, v. 26, p. 235–238, doi: 10.1130/0091-7613(1998)026<0235:EOHPRI>2.3.CO;2.

Miyashiro, A., 1961, Evolution of metamorphic belts: Journal of Petrology, v. 2, p. 277–311.

Miyashiro, A., 1967, Orogeny, regional metamorphism, and magmatism in the Japanese islands: Meddelser fra Dansk Geologisk Forening, v. 17, p. 390–446.

Molnar, P., 1988, A review of geophysical constraints on the deep structure of the Tibetan Plateau, the Himalaya and the Karakoram, and their tectonic implications: Royal Society of London Philosophical Transactions, ser. A, v. 326, p. 33–88.

Molnar, P., and Atwater, T., 1978, Interarc spreading and Cordilleran tectonics as alternates related to the age of subducted oceanic lithosphere: Earth and Planetary Science Letters, v. 41, p. 330–340, doi: 10.1016/0012-821X(78)90187-5.

Molnar, P., and Tapponnier, P., 1975, Cenozoic tectonics of Asia: Effects of a continental collision: Science, v. 189, p. 419–426, doi: 10.1126/science.189.4201.419.

Molnar, P., Burchfiel, B.C., Liang, K., and Zhao, Z., 1987, Geomorphic evidence for active faulting in the Altyn Tagh and northern Tibet and quantitative estimates of its contribution to the convergence of India and Eurasia: Geology, v. 15, p. 249–253, doi: 10.1130/0091-7613(1987)15<249:GEFAFI>2.0.CO;2.

Montgomery, D.R., Balco, G., and Willett, S.D., 2001, Climate, tectonics, and the morphology of the Andes: Geology, v. 29, p. 579–582, doi: 10.1130/0091-7613(2001)029<0579:CTATMO>2.0.CO;2.

Mooney, W.D., and Weaver, C.S., 1989, Regional crustal structure and tectonics of the Pacific coastal states, California, Oregon, and Washington, *in* Pakiser, L.C., and Mooney, W.D., eds., Geophysical framework of the continental United States: Geological Society of America Memoir 172, p. 129–161.

Mooney, W.D., Laske, G., and Masters, T.G., 1998, CRUST 5.1: A global crustal model at 5° × 5°: Journal of Geophysical Research, v. 103, p. 727–747, doi: 10.1029/97JB02122.

Mooney, W.D., Prodehl, C., and Pavlenkova, N.I., 2002, Seismic velocity structure of the continental lithosphere from controlled source data, *in* Lee, W.H.K., et al., eds., International handbook of earthquake and engineering seismology: International Association of Seismology and Physics of the Earth's Interior, Committee on Education, v. 81A, p. 887–910.

Mørk, M.B.E., 1985, A gabbro to eclogite transition on Flemsøy, Sunnmøre, western Norway: Chemical Geology, v. 50, p. 283–310.

Morris, P.A., 1995, Slab melting as an example of Quaternary volcanism and aseismicity in southwest Japan: Geology, v. 23, p. 395–398, doi: 10.1130/0091-7613(1995)023<0395:SMAAEO>2.3.CO;2.

Mosenfelder, J.L., and Bohlen, S.R., 1997, Kinetics of the coesite to quartz transformation: Earth and Planetary Science Letters, v. 153, p. 133–147, doi: 10.1016/S0012-821X(97)00159-3.

Mosenfelder, J.L., Schertl, H.P., Smyth, J.R., and Liou, J.G., 2005, Factors in the preservation of coesite: The importance of fluid infiltration: American Mineralogist, v. 90, p. 779–789, doi: 10.2138/am.2005.1687.

Mposkos, E.D., and Kostopoulos, D.K., 2001, Diamond, former coesite and supersilicic garnet in metasedimentary rocks from the Greek Rhodope: A new ultrahigh-pressure metamorphic province established: Earth and Planetary Science Letters, v. 192, p. 497–506, doi: 10.1016/S0012-821X(01)00478-2.

Murakoshi, T., Takenaka, H., Suzuki, S., Shimizu, H., and Uehira, K., 2003, Receiver function analysis of the crust and uppermost mantle structure beneath Kyushu, Japan: Eos (Transactions, American Geophysical Union), v. 84, one page.

Nakajima, J., Matsuzawa, T., and Hasegawa, A., 2001, Three-dimensional structure of Vp, Vs, and Vp-Vs beneath northeastern Japan: Implications for arc magmatism and fluids: Journal of Geophysical Research, v. 106, p. 21,843–21,857, doi: 10.1029/2000JB000008.

Nishimura, Y., 1998, Geotectonic subdivision and areal extent of the Sangun belt, Inner Zone of southwestern Japan: Journal of Metamorphic Geology, v. 16, p. 129–140, doi: 10.1111/j.1525-1314.1998.00059.x.

O'Brien, P.J., Zotov, N., Law, R., Khan, M.A., and Jan, M.Q., 2001, Coesite in Himalayan eclogite and implications for models of India-Asia collision: Geology, v. 29, p. 435–438, doi: 10.1130/0091-7613(2001)029<0435:CIHEAI>2.0.CO;2.

Okamoto, K., and Maruyama, S., 1999, The high pressure synthesis of lawsonite in the MORB + H_2O system: American Mineralogist, v. 84, p. 362–373.

Oregon Climate Service, 1995, Oregon State University, Corvallis, Oregon, http://www.ocs.oregonstate.edu/prism/index.phtml (March 2004).

Ota, T., Terabayashi, M., Parkinson, C.D., and Masago, H., 2000, Thermobarometric structure of the Kokchetav ultrahigh-pressure–high-pressure massif deduced from a north-south transect in the Kulet and Saldat-Kol regions, northern Kazakhstan: Island Arc, v. 9, p. 328–357, doi: 10.1046/j.1440-1738.2000.00282.x.

Oxburgh, E.R., and Turcotte, D.L., 1971, Origin of paired metamorphic belts and crustal dilation in island arc regions: Journal of Geophysical Research, v. 76, p. 1315–1327.

Patiño Douce, A.E., and McCarthy, T.C., 1998, Melting of crustal rocks during continental collision and subduction, *in* Hacker, B.R., and Liou, J.G., eds., When continents collide: Geodynamics and geochemistry of ultrahigh-pressure rocks: Dordrecht, Kluwer Academic Publishers, p. 27–55.

Peacock, S.M., 1990, Fluid processes in subduction zones: Science, v. 248, p. 329–337, doi: 10.1126/science.248.4953.329.

Peacock, S.M., 1992, Blueschist-facies metamorphism, shear heating, and P-T-t paths in subduction shear zones: Journal of Geophysical Research, v. 97, p. 17,693–17,707.

Peacock, S.M., 1995, Ultrahigh-pressure metamorphic rocks and the thermal evolution of continent collision belts: Island Arc, v. 4, p. 376–383, doi: 10.1111/j.1440-1738.1995.tb00157.x.

Peacock, S.M., and Wang, K., 1999, Seismic consequences of warm versus cool subduction metamorphism: Examples from southwest and northeast Japan: Science, v. 286, p. 937–939, doi: 10.1126/science.286.5441.937.

Platt, J.P., 1986, Dynamics of orogenic wedges and the uplift of high-pressure metamorphic rocks: Geological Society of America Bulletin, v. 97, p. 1037–1053, doi: 10.1130/0016-7606(1986)97<1037:DOOWAT>2.0.CO;2.

Platt, J.P., 1987, The uplift of high-pressure, low-temperature metamorphic rocks: Royal Society of London Philosophical Transactions, ser. A, v. 321, p. 87–103.

Platt, J.P., 1993, Exhumation of high-pressure rocks: A review of concepts and processes: Terra Nova, v. 5, p. 119–133.

Poli, S., and Schmidt, M.W., 1997, The high-pressure stability of hydrous phases in orogenic belts: An experimental approach on eclogite-forming processes: Tectonophysics, v. 273, p. 169–184, doi: 10.1016/S0040-1951(96)00293-4.

Pysklywec, R.N., 2006, Surface erosion control on the evolution of the deep lithosphere: Geology, v. 34, p. 225–228, doi:10.1130/G21963.1.

Ramos, V.A., Cristallini, E.O., and Pérez, D.J., 2002, The Pampean flat slab of the central Andes: Journal of South American Earth Sciences, v. 15, p. 59–78, doi: 10.1016/S0895-9811(02)00006-8.

Reck, B.H., 1987, Implications of measured thermal gradients for water movement through the northeast Japan accretionary prism: Journal of Geophysical Research, v. 92, p. 3683–3690.

Reyners, M., Eberhart-Phillips, D., and Stuart, G., 1999, A three-dimensional image of shallow subduction: Crustal structure of the Raukumara Peninsula, New Zealand: Geophysical Journal International, v. 137, p. 873–890, doi: 10.1046/j.1365-246x.1999.00842.x.

Ring, U., and Brandon, M.T., 1994, Kinematic data for the Coast Range fault and implications for the exhumation of the Franciscan subduction complex: Geology, v. 22, p. 735–738, doi: 10.1130/0091-7613(1994)022<0735:KDFTCR>2.3.CO;2.

Ring, U., and Brandon, M.T., 1999, Ductile deformation and mass loss in the Franciscan subduction complex—Implications for exhumation processes in accretionary wedges: Geological Society [London] Special Publication 154, p. 180–203.

Rubie, D.C., 1984, A thermal-tectonic model for high-pressure metamorphism in the Sesia zone, western Alps: Journal of Geology, v. 92, p. 21–36.

Rubie, D.C., 1986, The catalysis of mineral reactions by water and restrictions on the presence of aqueous fluid during metamorphism: Mineralogical Magazine, v. 50, p. 399–415, doi: 10.1180/minmag.1986.050.357.05.

Rubie, D.C., 1990, Role of kinetics in the formation and preservation of eclogites, in Carswell, D.A., ed., Eclogite facies rocks: New York, Chapman and Hall, p. 111–140.

Rumble, D., and Yui, T.F., 1998, The Qinglongshan oxygen and hydrogen anomaly near Donghai in Jiangsu province, China: Geochimica et Cosmochimica Acta, v. 62, p. 3307–3322, doi: 10.1016/S0016-7037(98)00239-7.

Russo, R.M., and Silver, P.G., 1996, Cordillera formation, mantle dynamics, and the Wilson cycle: Geology, v. 24, p. 511–514, doi: 10.1130/0091-7613(1996)024<0511:CFMDAT>2.3.CO;2.

Sacks, P.E., and Secor, D.T., Jr., 1990, Delaminations in collisional orogens: Geology, v. 18, p. 999–1002, doi: 10.1130/0091-7613(1990)018<0999:DICO>2.3.CO;2.

Schertl, H.P., Schreyer, W., and Chopin, C., 1991, The pyrope-coesite rocks and their country rocks at Paragi, Dora Maira massif, western Alps: Detailed petrography, mineral chemistry, and P-T path: Contributions to Mineralogy and Petrology, v. 108, p. 1–21, doi: 10.1007/BF00307322.

Scholl, D.W., and von Huene, R., 2004, Exploring the implications for continental basement tectonics if estimated rates of crustal removal (recycling) at Cenozoic subduction zones are applied to Phanerozoic and Precambrian convergent ocean margins: 17th International Basement Tectonics Conference, Oak Ridge, Tennessee, June 27–July 1, 2004, Program with Abstracts, p. 33–36.

Schreyer, W., 1995, Ultradeep metamorphic rocks: The retrospective viewpoint: Journal of Geophysical Research, v. 100, p. 8353–8366, doi: 10.1029/94JB02912.

Schreyer, W., Massone, H.J., and Chopin, C., 1987, Continental crust subducted to depths near 100 km: Implications for magma and fluid genesis in collision zones, *in* Mysen, B.O., ed., Magmatic processes: Physicochemical principles: University Park, Pennsylvania, Geochemical Society Special Publication 1, p. 155–163.

Searle, M.P., 1991, Geology and tectonics of the Karakorum Mountains: Chichester, UK, John Wiley and Sons, 358 p.

Searle, M.P., 1996, Cooling history, erosion, exhumation, and kinetics of the Himalaya-Karakorum-Tibet orogenic belt, *in* Yin, A., and Harrison, T.M., eds., The tectonic evolution of Asia: New York, Cambridge University Press, p. 110–137.

Searle, M.P., Hacker, B.R., and Bilham, R., 2001, The Hindu Kush seismic zone as a paradigm for the creation of ultrahigh-pressure diamond- and coesite-bearing continental rocks: Journal of Geology, v. 109, p. 143–153, doi: 10.1086/319244.

Searle, M.P., Simpson, R.L., Law, R.D., Parrish, R.R., and Waters, D.J., 2003, The structural geometry, metamorphic and magmatic evolution of the Everest massif, High Himalaya of Nepal–South Tibet: Geological Society [London] Journal, v. 160, p. 345–366.

Searle, M.P., Warren, C.J., Waters, D.J., and Parrish, R.R., 2004, Structural evolution, metamorphism and restoration of the Arabian continental margin, Saih Hatat region, Oman Mountains: Journal of Structural Geology, v. 26, p. 451–473, doi: 10.1016/j.jsg.2003.08.005.

Seno, T., 1985, Age of subducting lithosphere and backarc basin formation in the western Pacific since the middle Tertiary, in Nasu, N., et al., eds., Formation of active ocean margins: Tokyo, Terrapub, p. 469–481.

Sheffels, B.M., 1990, Lower bound on the amount of crustal shortening in the central Bolivian Andes: Geology, v. 18, p. 812–815, doi: 10.1130/0091-7613(1990)018<0812:LBOTAO>2.3.CO;2.

Shiomi, K., Sato, H., Obara, K., and Ohtake, M., 2004, Configuration of subducting Philippine Sea plate beneath southwest Japan revealed from receiver function analysis based on the multivariate autoregressive model: Journal of Geophysical Research, v. 109, B04308, doi: 10.1029-2003JB002774.

Shreve, R.L., and Cloos, M., 1986, Dynamics of sediment subduction, mélange formation, and prism accretion: Journal of Geophysical Research, v. 91, p. 10,229–10,245.

Smith, D.C., 1984, Coesite in clinopyroxene in the Caledonides and its implications for geodynamics: Nature, v. 310, p. 641–644, doi: 10.1038/310641a0.

Sobel, E.R., Hilley, G.E., and Strecker, M.R., 2003, Formation of internally drained contractional basins by aridity-limited bedrock incision: Journal of Geophysical Research, v. 108, no. B7, 2344, doi: 10.1029-2002JB001883.

Sobolev, N.V., and Shatsky, V.S., 1990, Diamond inclusions in garnets from metamorphic rocks: Nature, v. 343, p. 742–746, doi: 10.1038/343742a0.

Stern, C.R., 1991, Role of subduction erosion in the generation of Andean magmas: Geology, v. 19, p. 78–81, doi: 10.1130/0091-7613(1991)019<0078:ROSEIT>2.3.CO;2.

Stöckhert, B., and Renner, J., 1998, Rheology of crustal rocks at ultrahigh pressure, *in* Hacker, B.R., and Liou, J.G., eds., When continents collide: Geodynamics and geochemistry of ultrahigh-pressure rocks: Dordrecht, Kluwer Academic Publishers, p. 57–95.

Strecker, M.R., Hilley, G.E., Arrowsmith, J.R., and Coutand, I., 2003, Differential structural and geomorphic mountain-front evolution in an active continental collision zone: The northwest Pamir, southern Kyrgyzstan: Geological Society of America Bulletin, v. 115, p. 166–181, doi: 10.1130/0016-7606(2003)115<0166:DSAGMF>2.0.CO;2.

Stüwe, K., and Barr, T.D., 1998, On uplift and exhumation during convergence: Tectonics, v. 17, p. 80–88, doi: 10.1029/97TC02557.

Suppe, J., 1972, Interrelationships of high-pressure metamorphism, deformation, and sedimentation in Franciscan tectonics, U.S.A.: 24th International Geological Congress, Montreal, Reports, Section 3, p. 552–559.

Tapponnier, P., Peltzer, G., Le Dain, A.Y., and Armijo, R., 1982, Propagating extrusion tectonics in Asia: New insights from simple experiments with plasticine: Geology, v. 10, p. 611–617, doi: 10.1130/0091-7613(1982)10<611:PETIAN>2.0.CO;2.

Terry, M.P., Robinson, P., and Krogh Ravna, E.J., 2000a, Kyanite eclogite thermobarometry and evidence of thrusting of UHP over HP metamorphic rocks, Nordoyane, Western Gneiss Region, Norway: American Mineralogist, v. 85, p. 1637–1650.

Terry, M.P., Robinson, P., Hamilton, M.A., and Jercinovic, M.J., 2000b, Monazite geochronology of UHP and HP metamorphism, deformation, and exhumation, Nordoyane, Western Gneiss Region, Norway: American Mineralogist, v. 85, p. 1651–1664.

Thompson, A.B., and Ridley, J.R., 1987, Pressure-temperature-time (P-T-t) histories of orogenic belts: Royal Society of London Philosophical Transactions, ser. A, v. 321, p. 27–45.

Turcotte, D.L., and Oxburgh, E.R., 1972, Mantle convection and the new global tectonics: Annual Review of Fluid Mechanics, v. 4, p. 33–68, doi: 10.1146/annurev.fl.04.010172.000341.

van den Beukel, J., 1992, Some thermomechanical aspects of the subduction of continental lithosphere: Tectonics, v. 11, p. 316–329.

van den Beukel, J., and Wortel, R., 1988, Thermo-mechanical modeling of arc-trench regions: Tectonophysics, v. 154, p. 177–193, doi: 10.1016/0040-1951(88)90101-1.

Vandervoort, D.S., Jordan, T.E., Zeitler, P.K., and Alonso, R.N., 1995, Chronology of internal drainage development and uplift, southern Puna Plateau, Argentine central Andes: Geology, v. 23, p. 145–148, doi: 10.1130/0091-7613(1995)023<0145:COIDDA>2.3.CO;2.

Vannay, J.C., Graesmann, B., Rahn, M., Frank, W., Carter, A., Baudraz, V., and Cosca, M., 2004, Miocene to Holocene exhumation of metamorphic crustal wedges in the NW Himalaya: Evidence for tectonic extrusion coupled to fluvial erosion: Tectonics, v. 23, TC1014, doi: 10.1029-2002TC001429.

Vielzeuf, D., and Holloway, J.R., 1988, Experimental determination of the fluid-absent melting relations in the pelitic system: Consequences for crustal differentiation: Contributions to Mineralogy and Petrology, v. 98, p. 257–276, doi: 10.1007/BF00375178.

von Blanckenburg, F., and Davies, J.H., 1995, Slab breakoff: A model for syncollision magmatism and tectonics in the Alps: Tectonics, v. 14, p. 120–131, doi: 10.1029/94TC02051.

von Huene, R., and Ranero, C.R., 2003, Subduction erosion and basal friction along the sediment-starved convergent margin off Antofagasta, Chile: Journal of Geophysical Research, v. 108, no. B2, p. 2079, doi: 10.1029/2001JB001569.

von Huene, R., and Scholl, D.W., 1991, Observations at convergent margins concerning sediment subduction, subduction erosion, and the growth of continental crust: Reviews of Geophysics, v. 29, p. 279–316.

Vrolijk, P.J., 1987, Tectonically driven fluid flow in the Kodiak accretionary complex, Alaska: Geology, v. 15, p. 466–469, doi: 10.1130/0091-7613(1987)15<466:TDFFIT>2.0.CO;2.

Vrolijk, P.J., Meyers, G., and Moore, J.C., 1988, Warm fluid migration along tectonic mélanges in the Kodiak accretionary complex, Alaska: Journal of Geophysical Research, v. 93, p. 10,313–10,324.

Wang, X., Liou, J.G., and Mao, H.K., 1989, Coesite-bearing eclogite from the Dabie Mountains in central China: Geology, v. 17, p. 1085–1088, doi: 10.1130/0091-7613(1989)017<1085:CBEFTD>2.3.CO;2.

Webb, L.E., Hacker, B.R., Ratschbacher, L., McWilliams, M.O., and Dong, S., 1999, ^{40}Ar-^{39}Ar thermochronologic constraints on deformation and cooling history of high- and ultrahigh-pressure rocks in the Qinling-Dabie orogen: Tectonics, v. 18, p. 621–638, doi: 10.1029/1999TC900012.

Whipple, K.X., and Tucker, G.E., 1999, Dynamics of the stream-power river incision model: Implications for height limits of mountain ranges, landscape response timescales, and research needs: Journal of Geophysical Research, v. 104, p. 17,661–17,674, doi: 10.1029/1999JB900120.

Wigger, P.J., Schmitz, M., Araneda, M., Asch, G., Baldzuhn, S., Giese, P., Heinsohn, W.-D., Martinez, E., Ricaldi, E., Röwer, P., and Viramonte, J., 1994, Variation in the crustal structure of the southern central Andes deduced from seismic refraction investigations, *in* Reutter, K.J., et al., eds., Tectonics of the southern central Andes: Berlin, Springer-Verlag, p. 23–48.

Wigger, P., M. Schmitz, M. Araneda, G. Asch, S. Baldzuhn, P. Giese, W.-D., Heinsohn, E. Martinez, E. Ricaldi, P. Röwer, and J. Viramonte, Variation of the crustal structure of the southern Central Andes deduced from seismic refraction investigations, in: Tectonics of the Southern Central Andes (Eds.: Reutter, Scheuber, Wigger), Springer Verlag, 219- 231, 1994.

Willett, S.D., 1999, Orogeny and orography: The effects of erosion on the structure of mountain belts: Journal of Geophysical Research, v. 104, p. 28,957–28,981, doi: 10.1029/1999JB900248.

Willett, S.D., and Brandon, M.T., 2002, On steady states in mountain belts: Geology, v. 30, p. 175–178, doi: 10.1130/0091-7613(2002)030<0175:OSSIMB>2.0.CO;2.

Willett, S.D., Beaumont, C., and Fullsack, P., 1993, Mechanical model for the tectonics of doubly vergent compressional orogens: Geology, v. 21, p. 371–374, doi: 10.1130/0091-7613(1993)021<0371:MMFTTO>2.3.CO;2.

Yin, A., and Nie, S., 1993, An indentation model for the north and south China collision and the development of the Tan-lu and Honam fault systems, eastern Asia: Tectonics, v. 12, p. 801–813.

Zeil, W., 1979, The Andes: A geological review: Berlin, Gebruder Borntraeger, 260 p.

Zeitler, P.K., Koons, P.O., Bishop, M.P., Chamberlain, C.P., Craw, D., Edwards, M.A., Hamidullah, S., Jan, M.Q., Khan, M.A., Khattak, U.K., Kidd, W.S.F., Mackie, R.L., Meltzer, A.S., Park, S.K., Pecher, A., Poage, M.A., Sarker, G., Schneider, D.A., Seeber, L., and Shroder, J.F., 2001, Crustal reworking at Nanga Parbat, Pakistan: Metamorphic consequences of thermal-mechanical coupling facilitated by erosion: Tectonics, v. 20, p. 712–728, doi: 10.1029/2000TC001243.

Zhang, R.Y., and Liou, J.G., 1997, Partial transformation of gabbro to coesite-bearing eclogite from Yangkou, the Sulu terrane, eastern China: Journal of Metamorphic Geology, v. 15, p. 183–202, doi: 10.1111/j.1525-1314.1997.00012.x.

Zhang, R.Y., Liou, J.G., Yang, J., and Ye, K., 2003, Ultrahigh-pressure metamorphism in the forbidden zone: The Xugou garnet peridotite, Sulu terrane, eastern China: Journal of Metamorphic Geology, v. 21, p. 539–550, doi: 10.1046/j.1525-1314.2003.00462.x.

Zhao, D., and Hasegawa, A., 1993, P wave tomographic imaging of the crust and upper mantle beneath the Japanese islands: Journal of Geophysical Research, v. 98, p. 4333–4353.

Zhong, D., and Ding, L., 1996, Discovery of high-pressure basic granulites in Namjagbarwa area, Tibet, China: Chinese Science Bulletin, v. 41, p. 87–88.

MANUSCRIPT ACCEPTED BY THE SOCIETY 1 FEBRUARY 2007

Printed in the USA

The Geological Society of America
Memoir 200
2007

Accretionary growth and crust formation in the Central Asian Orogenic Belt and comparison with the Arabian-Nubian shield

A. Kröner*
Institut für Geowissenschaften, Universität Mainz, 55099 Mainz, Germany, and Tectonics Special Research Centre, University of Western Australia, 35 Stirling Highway, Crawley, WA 6009, Australia

B.F. Windley*
Department of Geology, University of Leicester, Leicester LE1 7RH, UK

G. Badarch†
O. Tomurtogoo*
Institute of Geology and Mineral Resources, Mongolian Academy of Sciences, Ulaanbaatar 210351, Mongolia

E. Hegner*
Department für Geowissenschaften, Universität München, Theresienstrasse 41, 80333 München, Germany

B.M. Jahn*
Institute of Earth Sciences, Academia Sinica, P.O. Box 1-55, Nankang, Taipei 11529, Taiwan

S. Gruschka
Institut für Geowissenschaften, Universität Mainz, 55099 Mainz, Germany

E.V. Khain
Geological Institute, Russian Academy of Sciences, Pyzhevski per. 7, 109017 Moscow, Russia

A. Demoux*
Institut für Geowissenschaften, Universität Mainz, 55099 Mainz, Germany

M.T.D. Wingate*
Tectonics Special Research Centre, University of Western Australia, 35 Stirling Highway, Crawley, WA 6009, Australia

*Kröner: kroener@mail.uni-mainz.de; Windley: brian.windley@btinternet.com; Tomurtogoo: igmr@magicnet.mn; Hegner: hegner@lmu.de; Jahn: jahn@earth.sinica.edu.tw; Khain: khain@ginras.ru; Demoux: demoux@uni-mainz.de; Wingate: mwingate@tsrc.uwa.edu.au.

†Deceased.

Kröner, A., Windley, B.F., Badarch, G., Tomurtogoo, O., Hegner, E., Jahn, B.M., Gruschka, S., Khain, E.V., Demoux, A., and Wingate, M.T.D., 2007, Accretionary growth and crust formation in the Central Asian Orogenic Belt and comparison with the Arabian-Nubian shield, *in* Hatcher, R.D., Jr., Carlson, M.P., McBride, J.H., and Martínez Catalán, J.R., eds., 4-D Framework of Continental Crust: Geological Society of America Memoir 200, p. 181–209, doi: 10.1130/2007.1200(11). For permission to copy, contact editing@geosociety.org.

ABSTRACT

The Central Asian Orogenic Belt is one of the largest accretionary terrains on Earth and records a ca. 800 Ma history of arc and microcontinent accretion, from south to north, during evolution and closure of the southwest Pacific-type Paleo-Asian ocean in the period ca. 1020 to ca. 325 Ma. We contest the evolutionary model for the belt proposed by previous authors in terms of a single, long island arc.

Accretion of ophiolites, arcs, and Precambrian microcontinents took place in southern Siberia in late Neoproterozoic to Cambrian times. Ultrahigh-pressure subduction and metamorphism occurred in the Cambrian at Kokchetav, Kazakhstan, and high-pressure metamorphism took place in the Gorny Altai, together with arcward accretion of a seamount. In the Chinese Altai, Precambrian microcontinents and island arcs collided into the accreting margin.

Overall the Central Asian Orogenic Belt records the formation of small forearc and backarc ocean basins that probably evolved between island arcs and microcontinents and were closed during continuous accretion between the Neoproterozoic and Paleozoic. During this time the southward-growing southern margin of the Siberian craton always faced an open ocean. Final closure of the Paleo-Asian ocean probably occurred in the late Permian when the North China craton was attached to the orogenic belt.

Large volumes of felsic volcanic rocks and the presence of Precambrian zircon xenocrysts as well as ancient detrital zircons in arc-derived sediments suggest substantial reworking of old crust despite seemingly primitive Nd isotopic characteristics. Similar characteristics in arc terranes of the Arabian-Nubian shield in Saudi Arabia suggest that previously proposed anomalously high crust-formation rates in both the Central Asian Orogenic Belt and Arabian-Nubian shield require revision.

Keywords: accretion, Arabian-Nubian shield, central Asia, Kazakhstan, Mongolia, zircon geochronology.

INTRODUCTION

The Central Asian Orogenic Belt (or Altaids) extends from northeastern Asia to the Ural Mountains and is one of the largest accretionary terrains on Earth (Zonenshain et al., 1990; Şengör et al., 1993; Mossakovsky et al., 1993; Yakubchuk et al., 2001; Yakubchuk, 2002) (Fig. 1). There have been many studies on the origin and evolution of this belt since Şengör et al. (1993) proposed an innovative but highly speculative model, deriving the entire orogen from one single, giant intra-oceanic island arc, the Kipchak arc, from the early Cambrian (ca. 540 Ma) to the Permian (ca. 260 Ma). The alternative, and more widely accepted, interpretation explains the Central Asian Orogenic Belt in terms of southwest Pacific-style accretion of arcs and microcontinents and final collision between the Siberian and North China cratons (e.g., Mossakovsky and Dergunov, 1985; Coleman, 1989; Mossakovsky et al., 1993; Abdulin et al., 1995; Fedorovsky et al., 1995; Ruzhentsev and Mossakovskiy, 1996; Filippova et al., 2001; Buslov et al, 2001; Badarch et al., 2002; Dobretsov et al., 2003; Kheraskova et al., 2003; Xiao et al., 2003, 2004a, 2004b). Whichever model is correct, the Central Asian Orogenic Belt represents a site of major crustal growth in the Phanerozoic (Şengör et al., 1993; Jahn et al., 2000, 2004; Jahn, 2004), comparable, in many respects, to the Neoproterozoic Arabian-Nubian shield (Reymer and Schubert, 1987; Johnson and Woldehaimanot, 2003).

Much new information has become available since the Şengör et al. (1993) synthesis was published, particularly as a result of IGCP (International Geological Correlation Programme) Project 473 (for summary see Jahn et al., 2004) and the international project "Atlas of lithological, paleogeographic, structural, palinspastic, and geoecologic maps of central Eurasia" of the Geological Surveys of central Asian countries, Russia, and China (Fedorenko and Militenko, 2002; Filippova et al., 2001). However, there is still a paucity of geochemical data, precise zircon ages, and detailed structural work for many of the terranes proposed in Kazakhstan, Mongolia, northwestern China, and southern Siberia. Nevertheless, a crude age zonation, from north to south, is now apparent in the Central Asian Orogenic Belt (Kröner et al., 2004) that we interpret to have resulted from successive ocean basin closure and accretion of arcs and microcontinents in the long-lived Paleo-Asian ocean (Zonenshain et al., 1990; Khain et al., 2002, 2003). The oldest ophiolitic rocks were generated ca. 1020–1035 Ma near the present margin of the Siberian craton (Khain et al., 2002), followed by the ca. 800 Ma

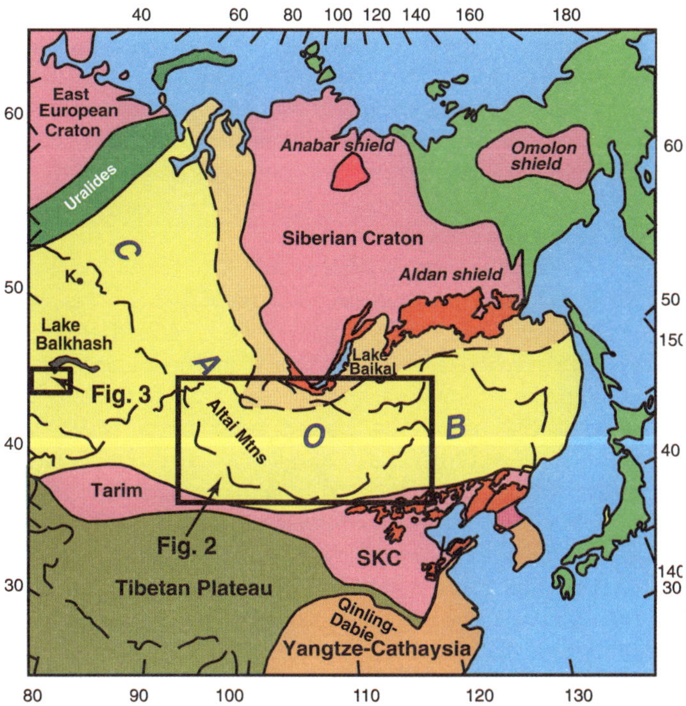

Figure 1. Simplified tectonic divisions of central Asia. The Central Asian Orogenic Belt (CAOB), also known as the "Altaid tectonic collage" (Şengör et al., 1993), is situated between the Siberian craton in the north and the Sino-Korean and Tarim cratons in the south. Red areas are exposed Archean to Paleoproterozoic rocks. Yellow-brown area surrounding Siberian craton is late Meso- to Neoproterozoic part of the Central Asian Orogenic Belt. Brown area is Paleo- to Neoproterozoic Yangtze-Cathaysia craton. Green pattern, including the Japanese islands, represents Pacific fold belts. K—Kokchetav (in northern Kazakhstan); SKC—Sino-Korean craton. Long broken lines are political boundaries; short broken line is approximate border between "Baikalian" and Phanerozoic domains of the Central Asian Orogenic Belt. The approximate locations of Figures 2 and 3 are indicated. Modified from Jahn (2004).

Shishkhid arc and ophiolite complex in northernmost Mongolia (Konnikov et al., 1994; Kuzmichev et al., 2005) and the large ca. 665 Ma Bayankhongor ophiolite in western Mongolia. Farther south, a broad belt of 570 Ma ophiolites extends from southern Tuva (Agardagh Tes-Chem; Pfänder and Kröner, 2004) into Mongolia via Dariv and Khantaishir (Khain et al., 2003; Matsumoto and Tomurtogoo, 2003).

Early Paleozoic arc formation dominates much of Kazakhstan (Heinhorst et al., 2000; Degtyarev and Ryazantsev, 2007; Kröner et al., 2007) and Mongolia (Ruzhentsev and Burashnikov, 1996; Badarch et al., 2002) and, as documented below, was followed by Carboniferous ophiolite formation south of Ulaanbaatar (Tomurtogoo et al., 2005), Silurian to Permian arc formation in southern Mongolia (Lamb and Badarch, 1997; Badarch et al., 2002), and finally, late Permian suturing and accretion of the southern Central Asian Orogenic Belt to the North China craton (Xiao et al., 2003). This indicates that this belt is a long-lived accretionary orogen, details of which still have to be unraveled.

Şengör et al. (1993) proposed that an area of more than five million square kilometers of juvenile crust was generated in central Asia in the Paleozoic, making this the largest Phanerozoic area of crustal growth, comparable, in crust-production rates, to the Neoproterozoic Arabian-Nubian shield (Reymer and Schubert, 1987). However, one striking aspect of the Central Asian Orogenic Belt in comparison to the Arabian-Nubian shield is the large volume of felsic volcanic rocks, particularly in Mongolia, and the relative paucity of true andesites. Furthermore, there is increasing evidence from precise zircon geochronology that Precambrian basement is more widespread in the belt than previously assumed (e.g., Yarmolyuk et al., 2005; Kozakov et al., 2007). In this contribution we particularly examine this aspect and summarize our field relationships and precise zircon ages from Kazakhstan and Mongolia in order to test which of the tectonic models proposed so far is most compatible with the available age and isotopic data. Bykadorov et al. (2003) and Windley et al. (2007) have already advanced important arguments why the Central Asian Orogenic Belt is unlikely to have evolved from a single, long-lived arc, and we discuss some of this evidence and further arguments against the single-arc hypothesis after presentation of our field and age data.

SOUTHERN SIBERIA, NORTHERN AND WESTERN MONGOLIA

The earliest history of ocean opening, probably linked to initiation of the Paleo-Asian ocean, is recorded by the 1020 Ma forearc Dunzhugur ophiolite in East Sayan, southern Siberia, that, together with a volcanic arc and blueschist accretionary wedge, was thrust onto the margin of the Siberian craton (Khain et al., 2002), probably during an accretion-obduction event prior to 790 Ma (Kuzmichev et al., 2001). Farther northeast at Lake Baikal, the ca. 1035 Ma Nurudukan arc and ophiolite suite records a similar tectonic setting (Khain et al., 2002; Dobretsov et al., 2003).

The next younger event, farther south, is exemplified by the evolution of the large but tectonically complex ca. 920–630 Ma Baikal-Muva-Dzhida-Yenisei arc and ophiolite terrane (Khain et al., 1997; Sklyarov et al., 2003), including the 800 Ma Shishkhid ophiolite complex of northern Mongolia that formed in an extensional island-arc environment (Kuzmichev et al., 2005). Kuzmichev et al. (2005) have recently identified a 600 km-long ~750 Ma accretionary prism, known as Oka belt, in southern Siberia and adjacent northern Mongolia and considered this to be an ancient analog of the Tertiary Shimanto belt in Japan.

The largest ophiolite complex of the Central Asian Orogenic Belt at Bayankhongor, western Mongolia (Buchan et al., 2001), has now been dated at 665 ± 15 Ma (sensitive high-resolution ion microprobe [SHRIMP] zircon age of anorthosite from a layered gabbro sequence; Kovach et al., 2005) and lies in the same tectonic position as Shishkhid on the margin of the Tuva-Mongolia block. Kovach et al. (2005) also pointed out that the

Bayankhongor ophiolite has a geochemical signature comparable to that of the Ontong Java Plateau; this may be the first oceanic plateau to be defined in the Central Asian Orogenic Belt, and the comparison is consistent with the fact that this 300 km long ophiolite is far larger than any other ophiolite in the orogenic belt.

This was followed, still farther south, by a broad belt of late Neoproterozoic ophiolites and arc terranes extending from Tuva in southern Siberia (Tes-Chem ophiolite, 570 Ma; Pfänder and Kröner, 2004) to western Mongolia and including the well-exposed and complete ophiolite sections of Dariv and Khantaishir (both ca. 570 Ma; Khain et al., 2003; Matsumoto and Tomurtogoo, 2003).

The northernmost part of the Central Asian Orogenic Belt adjacent to the Siberian craton was strongly deformed and metamorphosed after having been accreted to the Siberian margin, and significant crustal thickening leading to extensive granulite formation occurred at ca. 475 Ma (Donskaya et al., 2000) during a transpressive collisional event. This is particularly well exemplified by the Olkhon terrane fringing the craton along the northern shore of Lake Baikal (Fedorovsky, 1997, 2004), where a Neoproterozoic passive margin sequence and an early Paleozoic island-arc assemblage (previously interpreted as Paleoproterozoic basement) were thrust over Paleoproterozoic rocks of the craton in the early Paleozoic (Fedorovsky, 1997; Fedorovsky et al., 2005). The marginal region of the Siberian craton also shows that orogenic episodes, previously separated into a Neoproterozoic "Baikalian orogeny" (yellow-brown area on Fig. 1) and an early Paleozoic "Caledonian orogeny" are, in fact, part of a continuous sequence of accretion and collision events in the northern Central Asian Orogenic Belt (Fedorovsky et al., 2005).

Similarly, the Tuva-Mongolian Massif, previously interpreted as a Precambrian microcontinent because of the occurrence of high-grade rocks (Mossakovsky et al., 1993; Berzin and Dobretsov, 1994), has now been shown to consist of discrete Neoproterozoic tectonic domains, juxtaposed by thrusting prior to 497 Ma and subjected to high-grade metamorphism between 497 and 489 Ma (Kozakov et al., 2003, 2005; Yarmolyuk et al., 1999; Salnikova et al., 2001).

South of Lake Baikal and extending into northernmost Mongolia is the Dzhida terrane where late Neoproterozoic to early Paleozoic ophiolite, island arc, and backarc basin complexes are interpreted to have accreted in the Paleo-Asian ocean and onto the southward-growing Central Asian Orogenic Belt (Gordienko and Mikhaltsov, 2001; Gordienko and Filimonov, 2005). This terrane also contains a boninite suite (Almukhamedov et al., 2002) and a large allochthonous seamount (Gordienko and Filimonov, 2005).

These results imply that a large part of the northern Central Asian Orogenic Belt, including southern Siberia and northern Mongolia, had accreted to the Siberian craton through the Neoproterozoic until the end of the Ordovician during what were previously considered discrete and different orogenic events, named "Baikalian" (Neoproterozoic) and "Caledonian" (early Paleozoic) in the Russian literature (Zonenshain et al., 1990; Gordienko, 2006). Windley et al. (2004, 2007) suggested that this accreted and stabilized terrane extended as far south as the Main Mongolian Lineament (Yarmolyuk, 1983; Windley et al., 2004), a major tectonic line to the south of which lay an open ocean, the subduction of which gave rise to Silurian, Devonian, and Carboniferous island arcs and accretionary complexes (Fig. 2).

SOUTHERN KAZAKHSTAN

In Kazakhstan, several microcontinental fragments with Precambrian granitic gneisses have been identified (Abdulin et al., 1995) (Table 1), variously considered to be derived by Neoproterozoic rifting from the Siberian margin (Berzin and Dobretsov, 1994) or the East Gondwana margin (Mossakovsky et al., 1993; Kheraskova et al., 2003). The latter view is based on (1) similarities in the late Neoproterozoic and early Paleozoic stable margin sequences between Kazakhstan, Australia, China, and Tarim (Eganov and Sovetov, 1979) and (2) the fact that Kazakhstan has consistently drifted northwards from at least the Early Ordovician through the Permian as indicated by paleomagnetic data (Bazhenov et al., 2003; Collins et al., 2003; Alexyutin et al., 2005). Some investigators consider this rifting event to be related to breakup of the Rodinia supercontinent (e.g., Kheraskova et al., 2003), but the origin of continental fragments in the Central Asian Orogenic Belt remains uncertain. We have dated zircons from one of these crystalline complexes exposed in the Chu-Yili microcontinent (Nikitin and Nikitina, 2000), in the Russian literature also known as the "Zheltau Massif" or "Anrakhai-Altynemel microcontinent" (Avdeyev et al., 1990; Abdulin et al., 1995), and also from the accretionary terranes in traverses southwest of Lake Balkhash (Figs. 3 and 4). For this we used SHRIMP and the Pb-Pb evaporation technique, complemented by Sm-Nd whole-rock data. We have specifically investigated the field relationships in the Koyandisai-Uzunbulak area, the Kendyktas Mountains, the Sulu River area, and at the southwest end of Lake Balkhash near the village of Chiganak (Fig. 3).

The basement complex in the Anrakhai Mountains consists of strongly deformed and partly migmatized granitoid gneisses, granites, strongly foliated amphibolites and rare crystalline schists of likely sedimentary origin. The zircons from a well-foliated granite-gneiss exposed in the Koyandisai River in the Uzunbulak area (Fig. 3) were dated on the Perth SHRIMP II and are long-prismatic with rounded terminations and show oscillatory zoning under cathodoluminescence and in backscattered images (Fig. 5A), typical of magmatic growth. Four grains were analyzed (sample KZ 19, Table 2), and the discordant data yielded an upper concordia intercept at 2791 ± 24 Ma and a lower intercept at 1661 ± 43 Ma (Fig. 6A). The older age is interpreted to reflect the time of gneiss protolith emplacement, whereas the lower intercept age probably reflects Pb loss. However, there is no recorded event at this time in the basement of Kazakhstan, and the zircons do not reveal younger overgrowths (Fig. 5A), so we are hesitant to consider this "age" to signify a major thermal

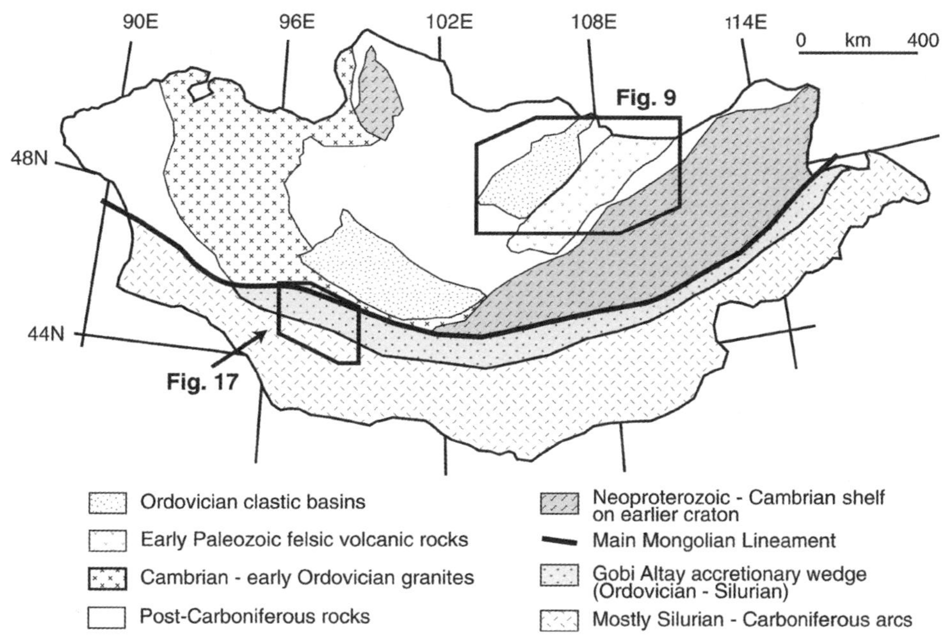

Figure 2. Generalized map of Mongolia showing main tectonic units (from Windley and Xiao, 2004). Locations of Figures 9 and 17 are indicated.

event. In any case, the above age documents Archean basement in the Anrakhai gneiss terrain.

A sample of well-foliated granite-gneiss in the Serektas River (KZ 12) that is intrusive into a sequence of fine-grained felsic gneisses of probably sedimentary origin contains long-prismatic, euhedral zircons, and evaporation of four grains produced a mean ^{207}Pb/^{206}Pb age of 1789.1 ± 0.6 Ma (Table 3; Fig. 7A). Another sample of fine-grained felsic gneiss (KZ 18) from the Uzunbulak River northwest of Kopa (Fig. 3) contains long-prismatic zircons with rounded terminations (Fig. 5B) of which four grains yielded a mean ^{207}Pb/^{206}Pb age of 2187.1 ± 0.5 Ma, whereas one xenocrystic grain is as old as 2431.0 ± 1.0 Ma (Table 3; Figs. 7B and 7C). The mean Nd crustal residence age of this sample is 2.3 Ga (Table 4).

Extensive rifting in the Anrakhai basement is documented by Neoproterozoic bimodal volcanic sequences dominated by dacitic to rhyolitic lavas and tuffs and locally known as the Kopa Formation. We dated single zircons from a coarse-grained, porphyritic and sheared rhyolitic rock (KZ 1) collected in the Kendyktas Mountains (Fig. 3), using the evaporation technique. The grains are long-prismatic with slight rounding at their terminations and well-developed oscillatory zoning defined by low- and high-U domains (Fig. 5C). The mean ^{207}Pb/^{206}Pb age is 775.9 ± 0.8 Ma (Table 3; Fig. 7D), and we interpret this to reflect the time of basement rifting and felsic volcanism. This corresponds to the widespread end-Riphean rifting event documented in central Kazakhstan and the Russian Platform (Kheraskova et al., 2003). The mean Nd crustal residence age of this sample is 1.87 Ga (Table 4) and shows that the source of felsic volcanism is of crustal origin and reflects melting of Paleoproterozoic basement.

Red gneissic granite in the hills northwest of Aschisu (Fig. 3) may be related to the above rifting event since six single zircons of one sample (KZ 22) yielded a mean ^{207}Pb/^{206}Pb age of 741.5 ± 0.7 Ma (Table 3; Fig. 7E). Since the gneiss is overlain by Paleozoic supracrustal rocks, its deformation records a late Neoproterozoic event in southern Kazakhstan. The overlying strata consist of coarse clastic sediments, including conglomerates and interlayered felsic volcanic rocks. Zircons of a strongly deformed metadacite (KZ 20) collected in the northwestern part of the Aschisu River (Fig. 3) vary from perfectly euhedral with only slight rounding at their terminations and well-developed oscillatory zoning (Fig. 5D) to well-rounded with older cores (Fig. 5E). Five grains were analyzed on SHRIMP and are highly variable in their isotopic composition (Table 2). The youngest, euhedral grain is concordant (Figs. 5D and 6B) and has a ^{206}Pb/^{238}U age of 534 ± 7 Ma. Three additional grains define a discordia line whose upper concordia intercept age of 1365 ± 24 Ma is determined by one concordant analysis (Fig. 6B), and the lower intercept is at 503 ± 178 Ma, compatible with the youngest concordant grain. The core of one additional well-rounded grain (Fig. 5E) has a minimum ^{207}Pb/^{206}Pb age of 2441 ± 9 Ma (Table 2; Fig. 6B, inset). We interpret the 1365 and 2441 Ma ages to reflect zircon inheritance from the basement complex, whereas the early Cambrian age of 534 Ma probably represents the time of dacite formation. The interlayered volcanic and sedimentary rocks here

TABLE 1. ISOTOPIC AGE DATA USED IN STRATIGRAPHIC CHART OF THE EARLY PRECAMBRIAN OF KAZAKHSTAN AND KYRGYZSTAN

No.	Composition (rock, mineral)	Sampling area	Formation	Method	Age (Ma)	Laboratory
1	Biotite-amphibole crystalline schist, metamorphogenic zircon first generation	Middle Tian Shan, Sarydzhaz area, Kuilyu River	Kuilyu suite	Pb-Pb isochron	2600 ± 50	Institute of Geology, Bishkek, Kyrgyzstan
2	as above	as above	as above	U-Pb discordia	2600	as above
3	as above	as above	as above	$^{207}Pb/^{206}Pb$	2570 ± 50	Kaz IMS, Alma-Ata, Kazakhstan
4	Biotite gneiss, metamorphogenic zircon first generation	as above	as above	as above	2570 ± 30	Institute of Geology, Bishkek, Kyrgyzstan
5	Biotite-amphibole crystalline schist	as above	as above	as above	1840	as above
6	as above	as above	as above	as above	1900	as above
7	Quartz-garnet-muscovite schist; metamorphic zircons	Kokchetav area, eastern shore of Lake Chelkar	Berlykskaya suite, Zerendinskaya series	U-Pb discordia	>2000	KazIMS, Alma-Ata, Kazakhstan
8	as above	as above	as above	$^{207}Pb/^{206}Pb$	1950	as above
9	Granitized gneiss	Kokchetav massif	Zerendinskaya series, altered due to granitization	as above	1200 ± 75	GEOKHI, Moscow, Russia
10	as above	as above	as above	as above	1050 ± 65	as above
11	as above	as above	as above	as above	1185 ± 80	as above
12	as above	as above	as above	U-Pb discordia	1200	MGU, Moscow, Russia
13	Garnet-micaceous schists and muscovite schists	Aktyuz-Boordy area, Malyi Kemin River	Aktyuz complex, granitization, mylonitization	Pb-Pb isochron	1140 ± 60	KazIMS, Alma-Ata, Kazakhstan
14	as above	as above	as above	U-Pb discordia	1140 ± 60	as above
15	as above	as above	as above	Pb-Pb isochron	1230 ± 50	as above
16	as above	as above	as above	U-Pb discordia	1230 ± 50	as above
17	Garnet-micaceous schists, syntectonic gneiss-granites, metamorphic zircon	Makbal area, western part of the Kyrghyz Ridge, Makbal River	Kirgiz series	as above	2165 ± 100	Institute of Geology, Bishkek,, Kyrgyzstan
18	as above	as above	as above	as above	2010 ± 100	Kaz IMS, Alma-Ata, Kazakhstan
19	Quartzites and schists, detrital and metamorphic zircons (?)	Makbal area, western part of Kyrghyz Ridge, Makbal River	Kirgiz series	U-Pb discordia	2028 ± 11	as above
20	Quartz-garnet-muscovite schist, detrital and metamorphic zircons	Makbal area, northern slope of Kyrghyz Ridge, Mamai-Kaindy River	as above	$^{207}Pb/^{206}Pb$	1870	as above
21	as above	as above	as above	Pb-Pb isochron	1920 ± 50	as above
22	Quartz- eclogite rock, garnet	Makbal area, west of the Kyrghyz Ridge, Tyuekarn Pass	Tyuekaryn suite, Kirgiz series	K-Ar	1688 ± 75	as above
23	as above	as above	as above	as above	1678 ± 75	as above
24	Crystalline schists, metamorphic zircons, later (?) metamorphic stages	as above	Sarychabyn suite	U-Pb discordia	1388 ± 11	as above
25	Granites, magmatic zircons	Karsakpai area, South Ulutau, west of Kar-sakpai village, north of Maityube hill	Zhaunkarskiy Complex, North-Sarysu Massif	Pb-Pb isochron	2230	IGG, UNC

(*continued*)

TABLE 1. ISOTOPIC AGE DATA USED IN STRATIGRAPHIC CHART OF THE EARLY PRECAMBRIAN OF KAZAKHSTAN AND KYRGYZSTAN (*continued*)

No.	Composition (rock, mineral)	Sampling area	Formation	Method	Age (Ma)	Laboratory
26	Granite-gneiss; magmatic zircons	Junggar region	Pioner Massif	U-Pb-Th	1715 ± 120	MGU, Moscow, Russia
27	as above	as above	Granite-gneiss complex	U-Pb discordia	1800	as above
28	as above	as above	Pioner Massif	Pb-Pb isochron (microprobe)	1950	IGG, UNC
29	Nepheline-syenite, magmatic zircons	as above	Karsakpai Massif, intrusive into the surrounding gneisses	U-Pb discordia	1675 ± 110	MGU, Moscow, Russia
30	as above	as above	as above	Pb-Pb isochron (microprobe)	1380	IGG, UNC
31	Sericite-feldspar schist, detrital zircons	West Sarysai-Teniz region, upper reaches of Kirei River	Opar suite	$^{207}Pb/^{206}Pb$ U-Pb-Th	3240 ± 480	MGU, Moscow, Russia
32	as above	as above	as above	Pb-Pb microprobe	3270	IGG, UNC
33	Quartz-biotite schist, detrital and metamorphic zircon	Chu area	Ogiztau suite	$^{207}Pb/^{206}Pb$ U-Pb-Th	1750 ± 50	Kaz IMS, Alma-Ata, Kazakhstan
34	Quartz-garnet-micaceous schist, metamorphic zircon, later metamorphic phases	Kokchetav area, Lake Chelkar	Berlykskaya suite	U-Pb discordia	>2000	as above
35	Pegmatite, muscovite	Ishkeolmes area	Granite-gneiss in Stepnogorsk series	K-Ar	1880 ± 60	IGEM, Moscow, Russia
36	Porphyroids, magmatic zircon	Atasu-Mointy area, upper reaches of Atasu River	Urkendeuskaya suite	Pb-Pb isochron	1850 ± 30	MGU, Moscow, Russia
37	as above	as above	as above	U-Pb discordia	1850 ± 60	as above
38	Granite-gneiss and granites, magmatic zircon	Karakamys area	Localized in Karakamys suite	as above	1900	Kaz IMS, Alma-Ata, Kazakhstan
39	Quartz-biotite crystalline schist, metamorpic zircon (late generation)	as above	Karakamys suite	Pb-Pb isochron	1750	as above
40	Quartz-biotite crystalline schist, metamorphogenic zircon (late generation)	as above	as above	$^{207}Pb/^{206}Pb$ U-Pb-Th	1755	as above
41	Biotite gneiss, metamorphic zircon (late generation)	Karakamys area, out-skirts of Anrakhai Mountains, middle reaches of Koyandsai River	Uzunbulak suite, Anrakhai series	U-Pb discordia	1797 ± 8	as above
42	Gneiss-granites and granitized crystalline schists, metamorphogenic zircon	Karakamys area outskirts of Anrakhai Mountains	Product of granitization of the Anrakhai series	as above	1800 ± 50	as above
43	as above	as above	as above	Pb-Pb isochron	1620 ± 70	as above
44	Granite-gneiss	Kendyktas, Sarybulak and Kerbulak blocks	Sarybulak suite	U-Pb discordia	1977 ± 14	as above
45	as above	as above	as above	$^{207}Pb/^{206}Pb$	1817	as above

Note: From Resolution of 3rd Kazakhstan stratigraphic conference on the Precambrian and Phanerozoic (1991, p. 18–21).

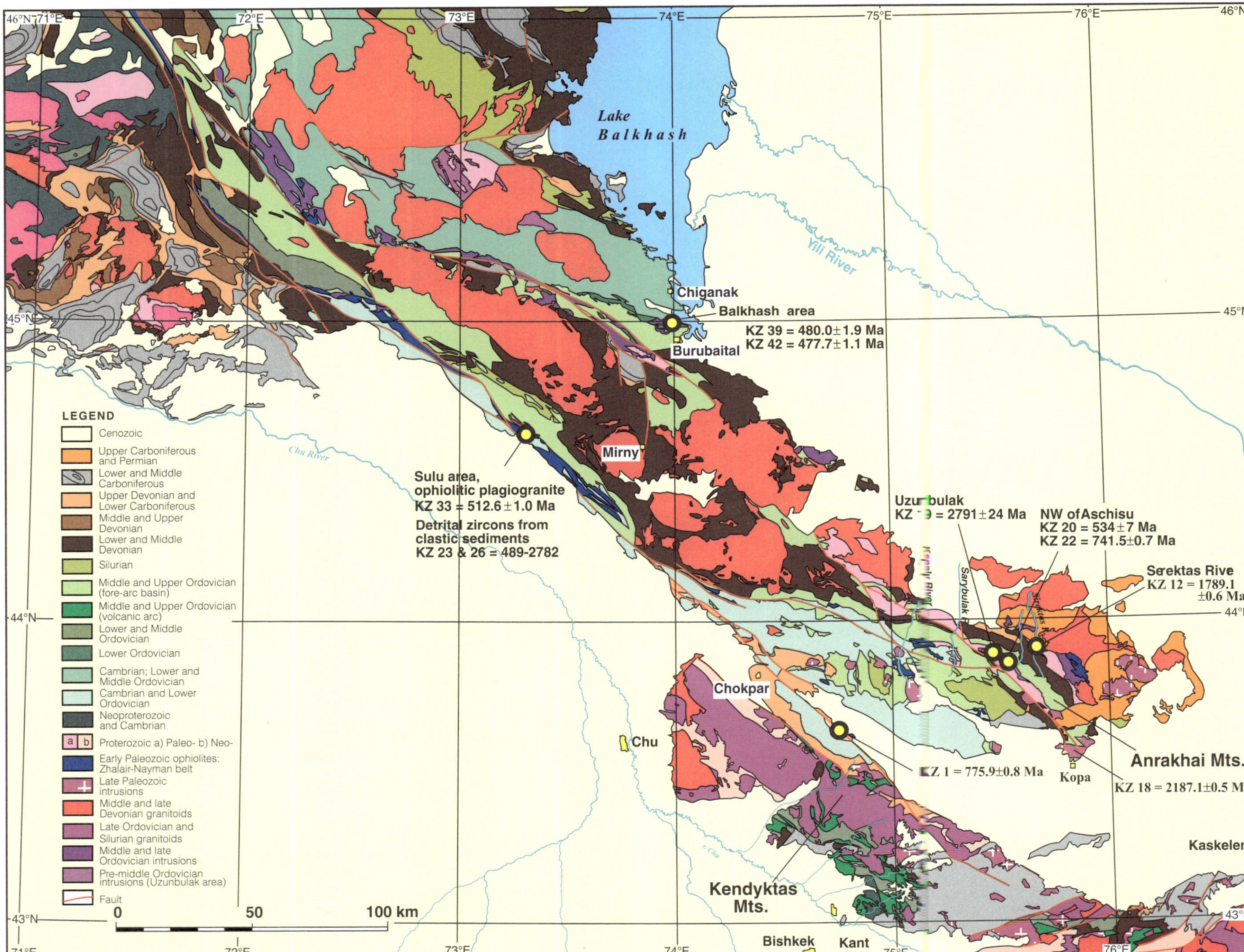

Figure 3. Sample location map and zircon ages of southern Kazakhstan. Base map is from Chakabaev et al. (1979).

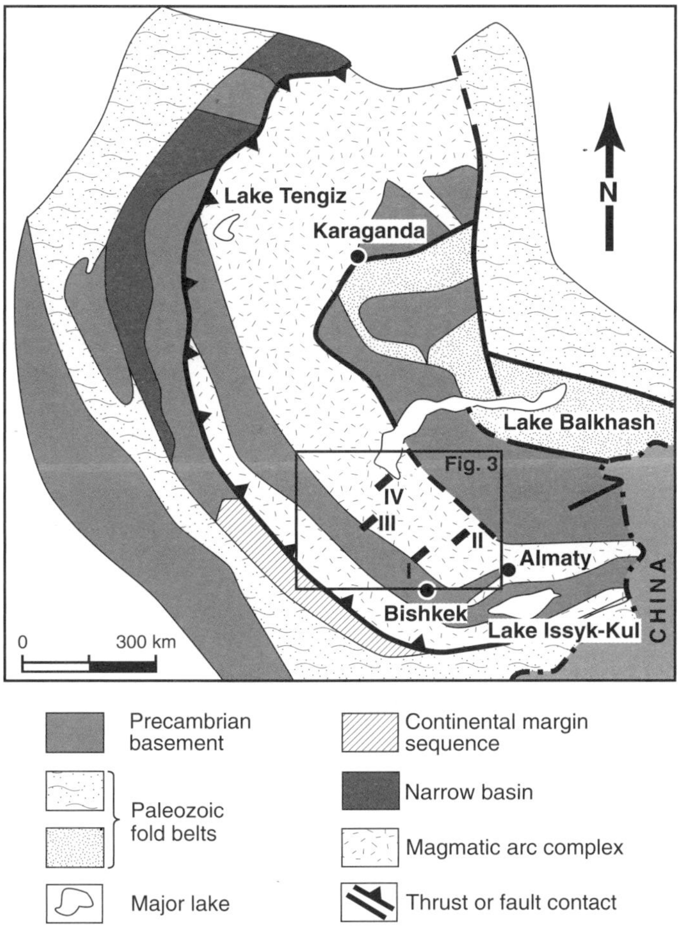

Figure 4. Geological map of southern Kazakhstan (simplified from Chakabaev et al., 1979) showing major tectonic units and working traverses (I–IV). Location of Figure 3 is indicated.

may represent an overlap sequence of a continental margin arc on the edge of the Anrakhai microcontinent.

There are many ophiolites in southern and central Kazakhstan that range in age from Vendian to Ordovician (Yakubchuk, 1994; Avdeev et al., 1995). The Sulu River area near the town of Mirny exposes a sequence of strongly deformed clastic sediments consisting of medium-grained gray to reddish sandstone (Djambul sequence) and laminated sandstone-siltstone (Solushai sequence). These rocks rest directly, with a well-preserved sedimentary contact, on fragmented basaltic pillow lava and red jaspilite. The basalts, in turn, are underlain by isotropic gabbro cut by diabase dikes and containing small pods and lenses of plagiogranite. There is no doubt that this succession constitutes part of an ophiolite sequence, and Avdeev (1984) interpreted the association of ophiolitic rocks with clastic sediments, which is typical of several ophiolite occurrences in southern Kazakhstan, as reflecting formation in a relatively narrow ocean basin. We envision a situation such as in the Red Sea or the Gulf of California since ocean floor sediments in larger oceans usually consist of fine-grained pelagic, cherty, and jaspilitic material (Anonymous, 1972).

We dated zircons from one of the above plagiogranite pods (KZ 33, Figs. 3 and 4), and four short-prismatic, euhedral grains yielded a mean $^{207}Pb/^{206}Pb$ age of 512.6 ± 1.0 Ma (Table 3; Fig. 7F), which we interpret to reflect ocean crust formation in the middle Cambrian. Detrital zircons from the clastic sediments resting on ocean floor basalts are highly variable in age. The youngest, long-prismatic, and euhedral grains (Fig. 5F) from a sample of Djambul sandstone directly resting on basalt (KZ 23; Fig. 3) have a mean early Ordovician $^{207}Pb/^{206}Pb$ evaporation age of 489.6 ± 1.0 Ma (Table 3; Fig. 7G), whereas two rounded detrital grains (Fig. 5G) have Precambrian ages of 2037.6 ± 4.3 and 2782.1 ± 2.1 Ma (Figs. 7H and 7I), probably reflecting input from the nearby Anrakhai microcontinent. Three rounded, detrital zircons from a sample of the Solusai siltstone (KZ 26, Fig. 3) higher up in the sedimentary sequence yielded virtually identical isotopic ratios with a mean $^{207}Pb/^{206}Pb$ age of 2230.6 ± 0.4 Ma (Table 3; Fig. 8A) and also document sedimentary input from the nearby basement.

The antiquity of most detrital zircons proves derivation of the above sediments from an ancient crustal source and supports the view of Avdeev (1984) that the oceanic domains between the microcontinents and early Paleozoic arc complexes in southern Kazakhstan were relatively narrow, probably marginal basins.

Finally, we dated zircons from a well-exposed, early Paleozoic arc complex from the western end of Lake Balkhash, exposed in roadcuts and a railway cut near the village of Chiganak (Fig. 3). A metadacite (KZ 42) with oscillatory-zoned zircons (Fig. 5H) and a granodiorite (KZ 39) with similar zircons (Fig. 5I) from this terrane provided almost identical mean $^{207}Pb/^{206}Pb$ zircon evaporation ages of 477.7 ± 1.1 and 480.0 ± 1.0 Ma respectively and, surprisingly, contain zircon xenocrysts as old as 2288 Ma (Table 3; Figs. 8B–8F). The Nd mean crustal residence age of the metadacite is 1.11 Ga (Table 4), and we consider it unlikely for these rocks to have formed in an intra-oceanic environment, but favor an Andean- or Japan-type setting. This view is compatible with the model of Heinhorst et al. (2000) who, on the basis of geochemistry and Nd isotopic systematics, suggested the central and northern Kazakhstan crust to have formed behind oceanward-drifting continental slivers.

In summary, our data do not support a purely juvenile origin for the early Paleozoic rocks in the accretionary belt of southern Kazakhstan. Furthermore, evidence for early Ordovician to late Carboniferous island-arc magmatism in central and northern Kazakhstan (Zonenshain et al., 1990; Heinhorst et al., 2000; Filippova et al., 2001; Kröner et al., 2007) argue against a stable Kazakhstan block as early as late Ordovician, as inferred by Bykadorov et al. (2003), but support continuous accretion until at least the Carboniferous.

Avdeev et al. (1995) reported ophiolite ages of 680 and 570 Ma from the Altai Mountains and elsewhere in Kazakhstan, and this is complemented by our Middle Cambrian plagiogranite

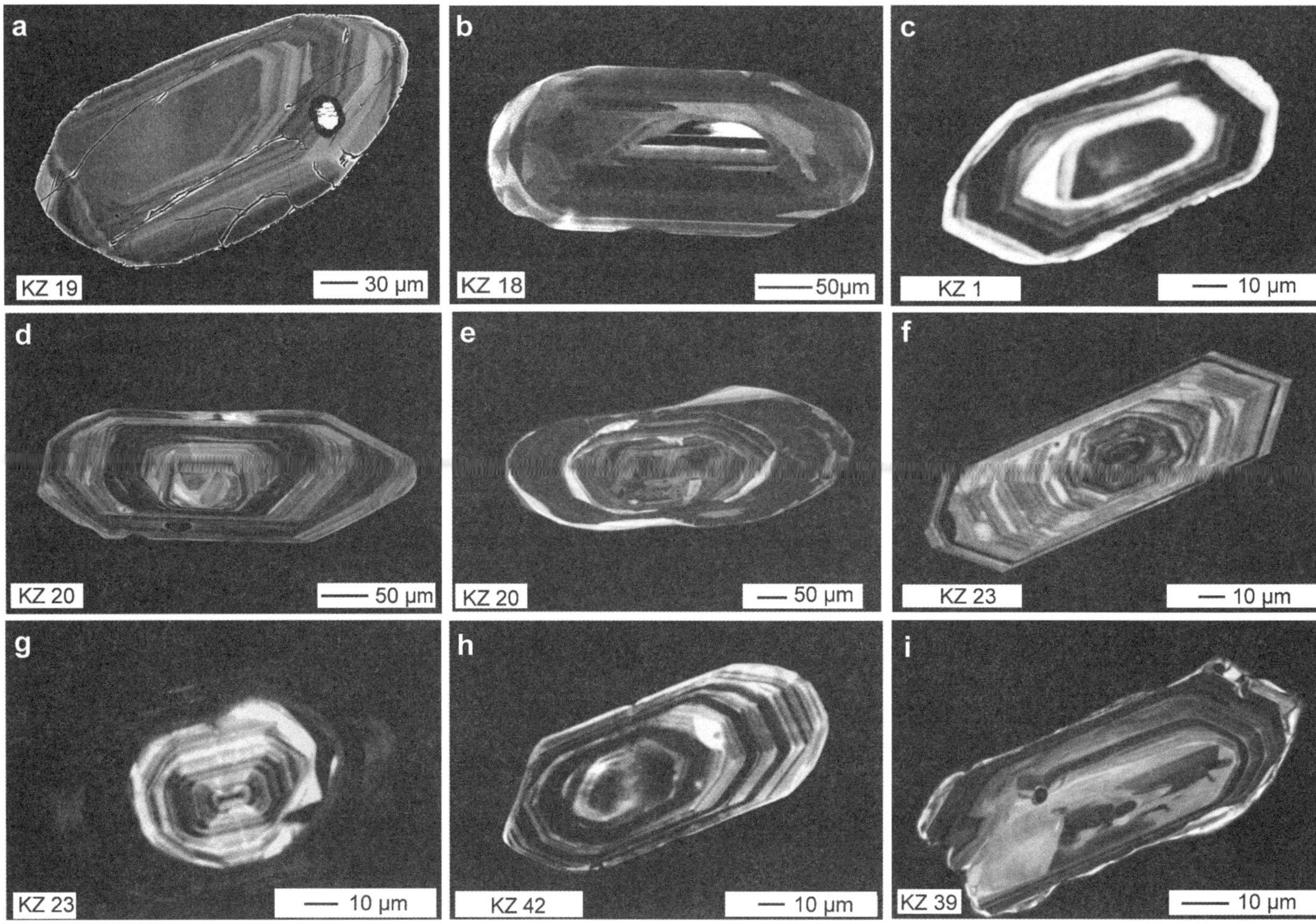

Figure 5. Cathodoluminescence (CL) and backscattered (BS) images of zircons from southern Kazakhstan dated in this study. (A) BS image of zircon from sample KZ 19. White spot is burn-mark of SHRIMP ion beam. (B) CL zircon image of sample KZ 18. (C) CL zircon image of sample KZ 1; note dark (high-U) and light (low-U) zoning. (D) CL image of zircon with well-developed oscillatory zoning. (E) CL image of zircon from sample KZ 20 with older core and overgrowth. (F) CL zircon image of sample KZ 23 showing euhedral shape and oscillatory zoning. (G) CL image of well-rounded, detrital zircon from sample KZ 23. (H) CL image of oscillatory-zoned zircon from sample KZ 42. (I) CL image of oscillatory-zoned zircon from sample KZ 39.

TABLE 2. SENSITIVE HIGH-RESOLUTION ION MICROPROBE (SHRIMP) ANALYTICAL DATA FOR SPOT ANALYSES OF MAGMATIC ZIRCONS FROM GRANITOID GNEISSES OF SOUTHERN KAZAKHSTAN*

Sample no.	U (ppm)	Th (ppm)	$^{206}Pb/^{204}Pb$	$^{208}Pb/^{206}Pb$	$^{207}Pb/^{206}Pb$	$^{206}Pb/^{238}U$	$^{207}Pb/^{235}U$	$^{206}Pb/^{238}U$ age (Ma)	$^{207}Pb/^{235}U$ age (Ma)	$^{207}Pb/^{206}Pb$ age (Ma)
KZ 19-1[†]	197	95	26,643	0.1333 ± 13	0.1615 ± 9	0.4148 ± 48	9.24 ± 13	2237 ± 22	2363 ± 12	2471 ± 10
KZ 19-2	233	93	11,797	0.1078 ± 13	0.1851 ± 9	0.4926 ± 58	12.58 ± 17	2582 ± 25	2648 ± 13	2699 ± 9
KZ 19-3	538	155	11,122	0.0832 ± 8	0.1408 ± 6	0.3621 ± 40	7.03 ± 9	1992 ± 19	2115 ± 11	2237 ± 7
KZ 19-4	135	55	9180	0.1154 ± 18	0.1786 ± 13	0.4708 ± 58	11.59 ± 18	2487 ± 25	2572 ± 14	2640 ± 12
KZ 20-1	228	82	4967	0.1145 ± 13	0.0843 ± 11	0.2009 ± 23	2.34 ± 4	1180 ± 12	1223 ± 13	1301 ± 26
KZ 20-2	164	101	4688	0.1792 ± 31	0.0871 ± 14	0.2371 ± 27	2.85 ± 6	1372 ± 15	1369 ± 16	1364 ± 31
KZ 20-3	249	132	19,303	0.1455 ± 12	0.1586 ± 8	0.4376 ± 50	9.57 ± 13	2340 ± 23	2394 ± 12	2441 ± 9
KZ 20-4	187	106	3824	0.1617 ± 27	0.0861 ± 12	0.2171 ± 25	2.58 ± 5	1267 ± 13	1294 ± 14	1340 ± 28
KZ 20-5	103	64	299	0.2354 ± 186	0.0598 ± 79	0.0863 ± 11	0.71 ± 10	534 ± 7	546 ± 57	596 ± 294

*All analyses were performed on the Perth Consortium SHRIMP II. For analytical details such as instrumental conditions, data reduction procedure, and error assessment see Kröner et al. (2003) and references cited therein. All errors are 1σ.

[†]1 is spot 1 on grain 1, 2 is spot 1 on grain 2, etc.

TABLE 3. SINGLE-ZIRCON EVAPORATION DATA FOR SAMPLES FROM SOUTHERN KAZAKHSTAN

Sample no.	Zircon color and morphology	Grain no.	Mass scan[1]	Evaporation temperature (°C)	Mean ^{207}Pb/^{206}Pb ratio[2] and 2σ error	^{207}Pb/^{206}Pb age and 2σ error (Ma)
KZ 1	pink to clear, long-	1	97	1594	0.065048 ± 46	775.9 ± 1.5
	prismatic, ends	2	174	1601	0.065047 ± 32	775.8 ± 1.0
	rounded	3	131	1599	0.065047 ± 39	775.9 ± 1.2
		4	66	1598	0.065051 ± 68	776.0 ± 2.2
mean of four grains		1–4	468		0.065048 ± 21	***775.9 ± 0.8**
KZ 12	clear to light brown,	1	105	1595	0.109337 ± 72	1788.4 ± 1.2
	long-prismatic,	2	131	1597	0.109406 ± 49	1789.5 ± 0.8
	idiomorphic	3	131	1599	0.109433 ± 66	1790.0 ± 1.1
		4	86	1598	0.109299 ± 82	1787.8 ± 1.4
mean of four grains		1–4	453		0.109378 ± 33	**1789.1 ± 0.6**
		5	122	1598	0.110887 ± 69	1814.0 ± 1.1
KZ 18	clear to light gray-	1	129	1596	0.136825 ± 63	2187.2 ± 0.8
	brown, long-prismatic,	2	129	1598	0.136805 ± 60	2187.1 ± 0.8
	ends slightly rounded	3	108	1597	0.136816 ± 37	2187.3 ± 0.5
		4	74	1598	0.136772 ± 106	2186.7 ± 1.3
mean of four grains		1–4	440		0.136802 ± 37	**2187.1 ± 0.5**
		5	128	1598	0.157690 ± 90	2431.0 ± 1.0
KZ 22	yellowish, transparent,	1	85	1598	0.039883 ± 41	741.2 ± 1.4
	long-prismatic,	2	144	1597	0.063967 ± 43	740.8 ± 1.4
	idiomorphic	3	122	1598	0.064003 ± 60	741.7 ± 2.0
		4	206	1599	0.060049 ± 19	741.8 ± 0.6
		5	132	1597	0.063997 ± 28	741.5 ± 0.9
		6	60	1597	0.063994 ± 99	741.4 ± 3.3
mean of six grains		1–6	749		0.063995 ± 17	***741.5 ± 0.7**
KZ 23	yellowish to clear, long-	1	109	1596	0.056950 ± 36	489.6 ± 1.4
	prismatic, idiomorphic	2	84	1597	0.056947 ± 42	489.5 ± 1.6
		3	84	1597	0.056954 ± 27	489.7 ± 1.1
mean of three grains		1–3	277		0.056950 ± 20	***489.6 ± 1.0**
	red, well rounded	4	43	1595	0.125623 ± 307	2037.6 ± 4.3
		5	86	1598	0.194678 ± 244	2782.1 ± 2.1
KZ 26	brownish, long-	1	88	1598	0.140263 ± 29	2230.5 ± 0.4
	prismatic, idiomorphic	2	61	1599	0.140597 ± 33	2234.6 ± 0.4
		3		1597	0.140131 ± 27	2228.8 ± 0.3
mean of three grains		1–3	277		0.140275 ± 33	**2230.6 ± 0.4**
		62	1596		0.157142 ± 51	2425.1 ± 0.6
KZ 33	yellowish to dark red,	1	102	1595	0.057555 ± 42	512.9 ± 1.6
	short-prismatic,	2	88	1598	0.057551 ± 24	512.7 ± 0.9
	idiomorphic	3	165	1592	0.057552 ± 18	512.8 ± 0.7
		4	132	1596	0.057536 ± 27	512.1 ± 1.0
mean of four grains		1–4	487		0.057548 ± 13	***512.6 ± 1.0**
KZ 39	clear, long-prismatic,	1	129	1597	0.056691 ± 25	479.5 ± 1.0
	idiomorphic	2	109	1598	0.056709 ± 46	480.2 ± 1.8
		3	215	1598	0.056706 ± 22	480.2 ± 0.9
mean of three grains		1–3	453		0.056702 ± 17	***480.0 ± 1.0**
		4	198	1596	0.058483 ± 27	547.9 ± 1.0
		5	44	1597	0.145035 ± 434	2288.2 ± 5.2
KZ 42	clear to yellowish,	1	84	1600	0.056627 ± 58	477.0 ± 2.3
	transparent,	2	91	1596	0.056655 ± 29	478.1 ± 1.1
	idiomorphic	3	116	1598	0.056652 ± 41	478.0 ± 1.6
		4	108	1596	0.056648 ± 29	477.9 ± 1.1
		5	146	1597	0.056638 ± 13	477.5 ± 0.5
mean of five grains		1–5	545		0.056644 ± 15	***477.7 ± 1.1**
		6	99	1598	0.067173 ± 63	843.1 ± 2.0

[1]Number of ^{207}Pb/^{206}Pb ratios evaluated for age assessment.

[2]Observed mean ratio corrected for nonradiogenic Pb where necessary. Errors based on uncertainties in counting statistics.

*Error based on reproducibility of internal standard. For analytical procedures and error assessment see Kröner and Hegner (1998) and references cited therein.

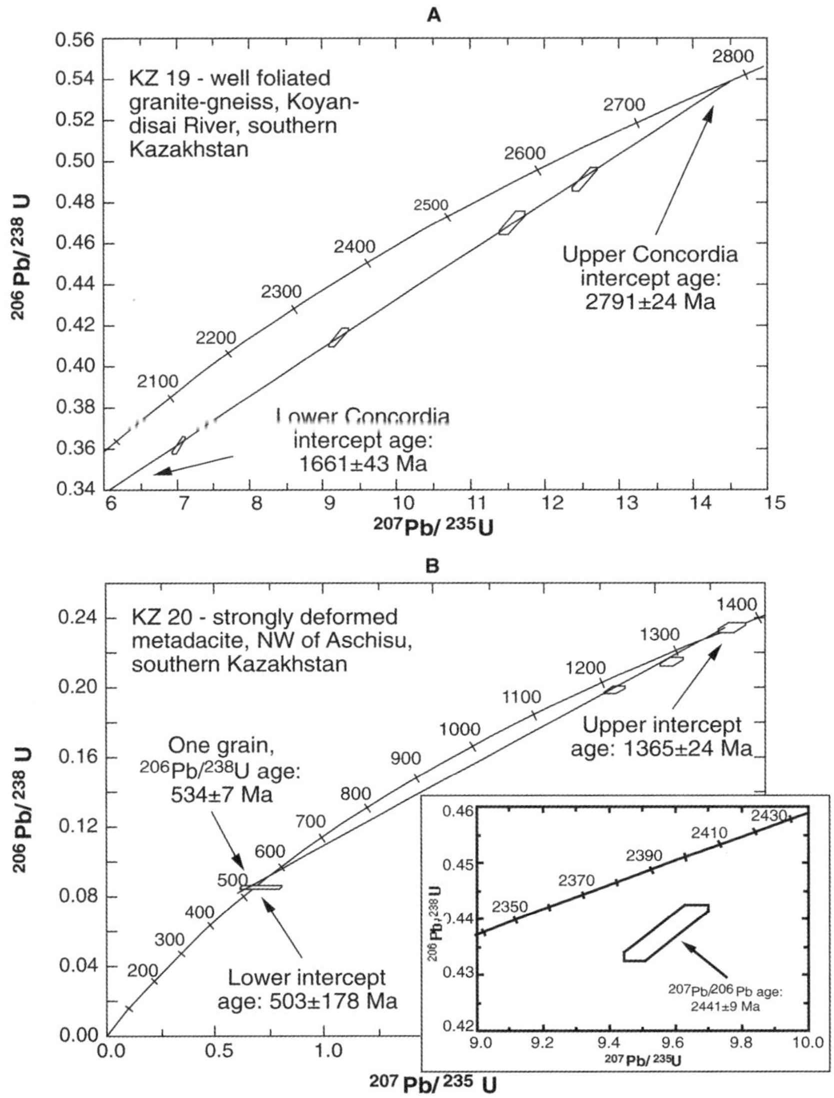

Figure 6. Concordia diagrams for SHRIMP zircon analyses of rocks from southern Kazakhstan. Data boxes for each analysis are defined by standard errors in $^{207}Pb/^{235}U$, $^{206}Pb/^{238}U$, and $^{207}Pb/^{206}Pb$. A: Well-foliated granite-gneiss sample KZ 19, Koyandisai River. B: Strongly deformed metadacite sample KZ 20, NW of Aschisu.

age of 512 Ma presented above. These and further ages reported from Mongolia by Khain et al. (2003) demonstrate that Siberia and the East European craton could not have had a common margin during the late Neoproterozoic and early Paleozoic, because oceanic crust of the Paleo-Asian ocean separated the various cratonic blocks and microcontinental fragments. This is further supported by paleomagnetic results (compare data for Siberia by Smethurst et al. [1998a] and Pisarevsky and Natapov [2003] with data for Baltica by Smethurst et al. [1998b], Popov et al. [2002], and Nawrocki et al. [2004]). The best available paleomagnetic and paleofaunal data (e.g., Hartz and Torsvik, 2002; Murphy et al., 2004; Meert and Lieberman, 2004) indicate that in the early Cambrian, when the Kipchak arc was supposed to have formed (Şengör et al., 1993), Baltica and Siberia were separated by ~30° of paleolatitude by a wide ocean commonly called the Aegir Sea.

NORTHERN AND CENTRAL MONGOLIA

Late Neoproterozoic to early Paleozoic stable continental margin (carbonates) and arc formations dominate much of northern and central Mongolia (Badarch et al., 2002), overlain

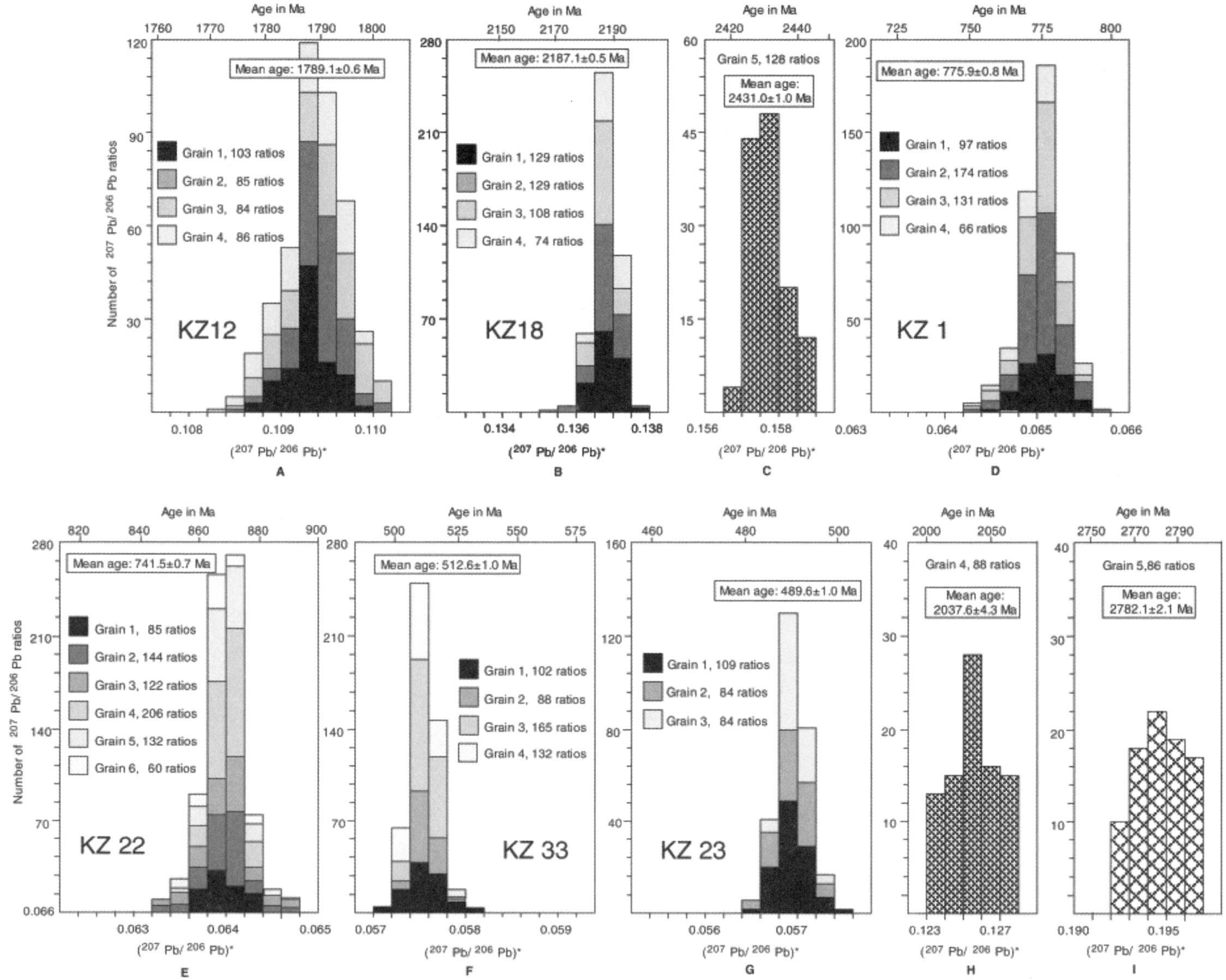

Figure 7. Histograms showing distribution of radiogenic Pb isotope ratios derived from evaporation of single zircons from Precambrian and Paleozoic rocks in southern Kazakhstan. (A) Spectrum for four grains from red granite-gneiss sample KZ 12, Serektas River, NW of Kopa, integrated from 453 ratios. (B) Spectrum for four grains from felsic gneiss sample KZ 18, Uzunbulak River, integrated from 440 ratios. (C) Spectrum for xenocrystic grain from same sample. (D) Spectrum for four grains from porphyritic metarhyolite sample KZ 1, collected near Yeshlilikurda and integrated from 468 ratios. (E) Spectrum for six grains from gneissic granite sample KZ 22, hills NW of Aschisu, integrated from 749 ratios. (F) Spectrum for four grains from ophiolitic plagiogranite sample KZ 33, east of Kompakty River, Mirny area, integrated from 487 ratios. (G) Spectrum for three detrital grains from Djambul sandstone sample KZ 23, Kompakty River near Mirny, integrated from 277 ratios. (H–I) Spectra for Precambrian detrital grains reflecting input from ancient basement.

TABLE 4. Sm-Nd ISOTOPIC COMPOSITION OF IGNEOUS ROCKS FROM SOUTHERN KAZAKHSTAN

Sample no.	Rock type	Age (Ma)	Sm (ppm)	Nd (ppm)	$^{147}Sm/^{144}Nd$	$^{143}Nd/^{144}Nd$	±2σ	eNd(t)	T_{DM} (Ma)
KZ1	Metarhyolite	776	6.01	33.65	0.1079	0.511862	10	–6.4	1.87
KZ18	Felsic gneiss	2190	4.57	25.03	0.1104	0.511594	12	2.1	2.30
KZ33	Plagiogranite	513	5.61	24.33	0.1394	0.512771	14	6.2	0.81
KZ42	Metadacite	478	3.60	24.24	0.0898	0.512263	14	–0.9	1.11

Note: For analytical procedures see Lackschewitz et al. (2003). Isotopic analyses were performed in the Max-Planck-Institut für Chemie, Mainz. T_{DM}—Nd model age assuming a depleted mantle reservoir (DePaolo, 1981).

and intruded by voluminous Permian felsic volcanic rocks and granites. We have investigated a north-south traverse from the Tarvagatay terrane in the north to the Haraa terrane around and west of Ulaanbaatar as well as volcanic sequences in the area around Mörön, in the Idermeg terrane southeast of Ulaanbaatar (Fig. 9). Most of the arc-related rocks of northern and central Mongolia consist of Ordovician to Silurian volcanic and volcaniclastic sequences, intruded by a variety of granitoids, and these rocks were generated in the relatively short time period from 460 to 417 Ma (A. Kröner, B.F. Windley, G. Badarch, unpubl. data) (Table 5). However, there are also Carboniferous arc sequences, particularly around Ulaanbaatar (Popeko, 2002). One striking aspect of these rocks is a predominance of felsic compositions, mainly dacites and rhyolites and their sedimentary derivatives (Fig. 10), and we found these rocks in great abundance north of and around Ulaanbaatar as well as south of Ulaanbaatar and in the Mörön area (Fig. 9). Although these rocks are chemically arc-related, andesites and basalts are relatively rare (Figs. 10 and 11). Murphy and MacDonald (1993) have shown that discrimination diagrams developed for volcanic rocks, in particular HFS (high field strength) elements, are useful to reconstruct the provenance of chemically immature sediments such as those exposed in northern and central Mongolia. We have plotted some of our trace element data for arc-related sedimentary rocks in the Nb-Zr-Y discrimination diagram of Meschede (1986), and there is a well-developed grouping in the volcanic arc field (Fig. 12).

We consider it unlikely that these large volumes of predominantly felsic rocks were generated entirely from juvenile sources since some samples contain Precambrian xenocrystic zircons, and Nd mean crustal residence ages for some of these felsic rocks are between 600 and 1300 Ma, suggesting that at least some older material was involved in their generation. In fact, there is a relatively large crustal block with Archaean to Paleoproterozoic rocks in south-central Mongolia, known as the Baidrag terrane (Kozakov et al., 1997; Badarch et al., 2002). A comparison of Nd isotope data for Central Asian Orogenic Belt rocks from Tuva, Transbaikalia, and Mongolia shows that Transbaikalia and northern Mongolia (Fig. 13) reflect the input of older crust, perhaps derived from the Siberian craton, whereas southern Mongolia appears to be largely, but not exclusively, juvenile, supporting the data of Jahn et al. (2004) and Helo et al. (2006).

By the end of the Ordovician, the northern part of the orogenic belt had probably amalgamated to create a new continental margin (the Main Mongolian Lineament of Windley et al., 2004) (Fig. 2). Evidence of stabilization by this time of the northern region is provided by Eocambrian to Cambrian shelf carbonates, Ordovician clastic basins, and relatively little-deformed, extensive Ordovician ash-fall tuffs, dacites, and rhyolites (Figs. 2 and 10). The Main Mongolian Lineament appears to be a major tectonic boundary separating crustal provinces with different isotopic characteristics. This is in agreement with the conclusion of Kovalenko et al. (2004) that the Neoproterozoic to early Paleozoic area to the north of the lineament (mean age of 570–475 Ma) has a Nd mean crustal residence age of 1.1–0.55 Ga, whereas the late Paleozoic area to the south (mean age of 420–320 Ma) has a T_{DM} of 0.8–0.5 Ga.

The above scenario of a stabilized accreted margin in Ordovician times in northern and central Mongolia is a perfect setting to generate large volumes of felsic volcanic rocks, and we quote from Bryan et al. (2002): "The largest volume silicic igneous provinces occur along accreted continental margins… and ultimately reflect large-scale crustal melting processes in response to lithospheric extension and high thermal input from underlying hot mantle. Partial melting of hydrous, mafic-intermediate composition (amphibolite) crust is critical in generating the large volumes of predominantly I-type silicic igneous melt. In these cases, subduction as much as hundreds of millions of years prior to the emplacement of the silicic igneous province seems crucial in producing a hydrous lower crustal source receptive to melting."

We have reconnaissance-dated zircons from several felsic volcanic samples and clastic sediments, complemented by Sm-Nd whole-rock isotopic systematics, and these data are summarized in Table 5. Specific examples to illustrate our conclusion that older material was involved in the generation of many of these rocks are given below.

Stable depositional conditions in northern Mongolia are documented by a carbonate-quartzite sequence (Darhan Formation) of presumed late Neoproterozoic to earliest Paleozoic age that is exposed in the hills south of the city of Darhan (Byamba, 1996). We have dated detrital zircons from a quartzite exposed in a small abandoned quarry, using SHRIMP and the evaporation method.

Three quartzite samples were taken, NM 16, M 02/115, and M 04/146, and the zircon populations are very heterogeneous, containing transparent, perfectly euhedral grains with no trace

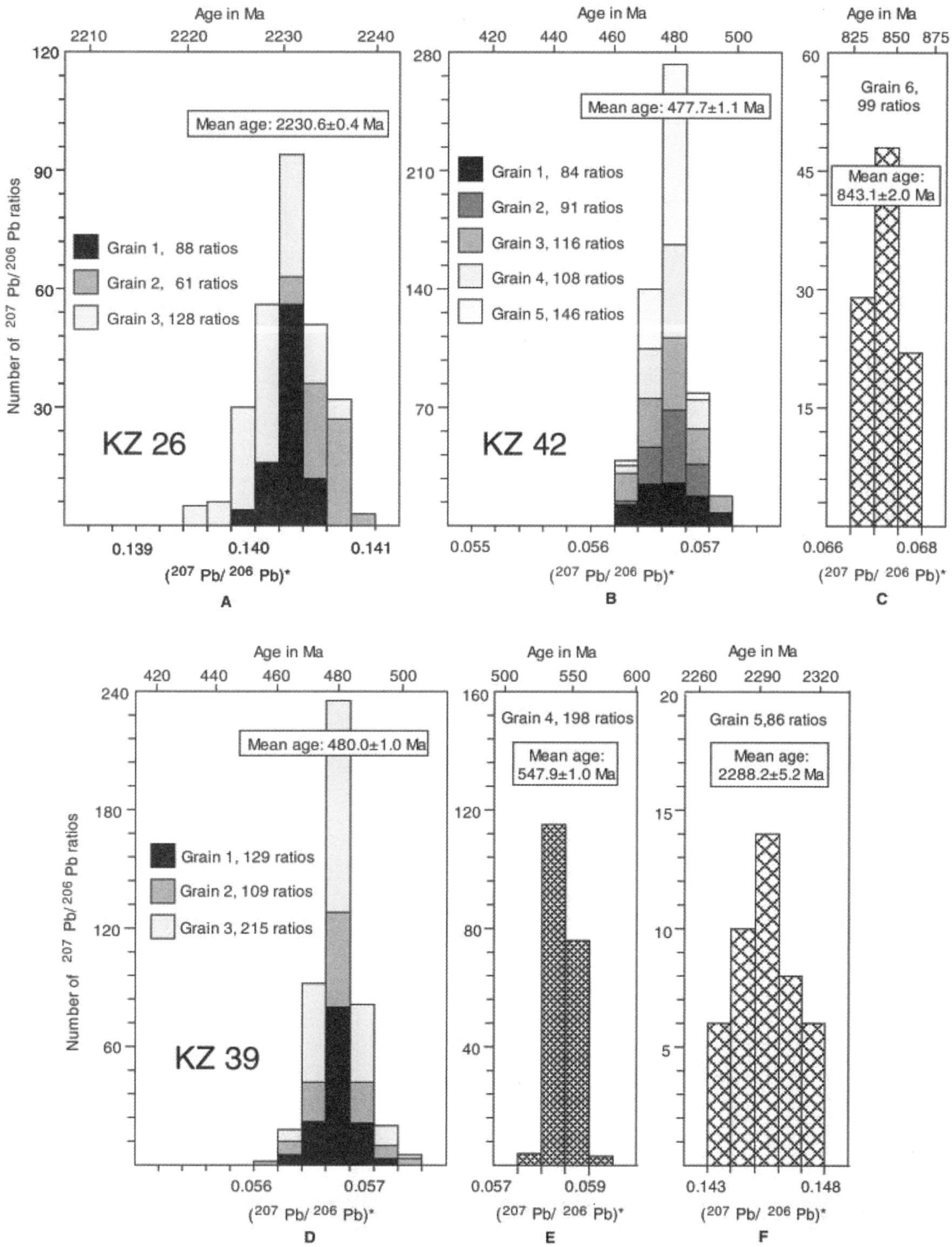

Figure 8. Histograms showing distribution of radiogenic Pb isotope ratios derived from evaporation of single zircons from Paleozoic rocks in southern Kazakhstan. (A) Spectrum for three detrital grains from laminated siltstone sample KZ 26, Kompakty River, Mirny area, integrated from 277 ratios. (B) Spectrum for five grains from metadacite sample KZ 42, north of Chiganak near Lake Balkhash, integrated from 545 ratios. (C) Spectrum for xenocrystic grain from same sample. (D) Spectrum for three grains from arc-related granodiorite sample KZ 39, Chiganak near Lake Balkhash, integrated from 453 ratios. (E–F) Spectra for xenocrystic grains.

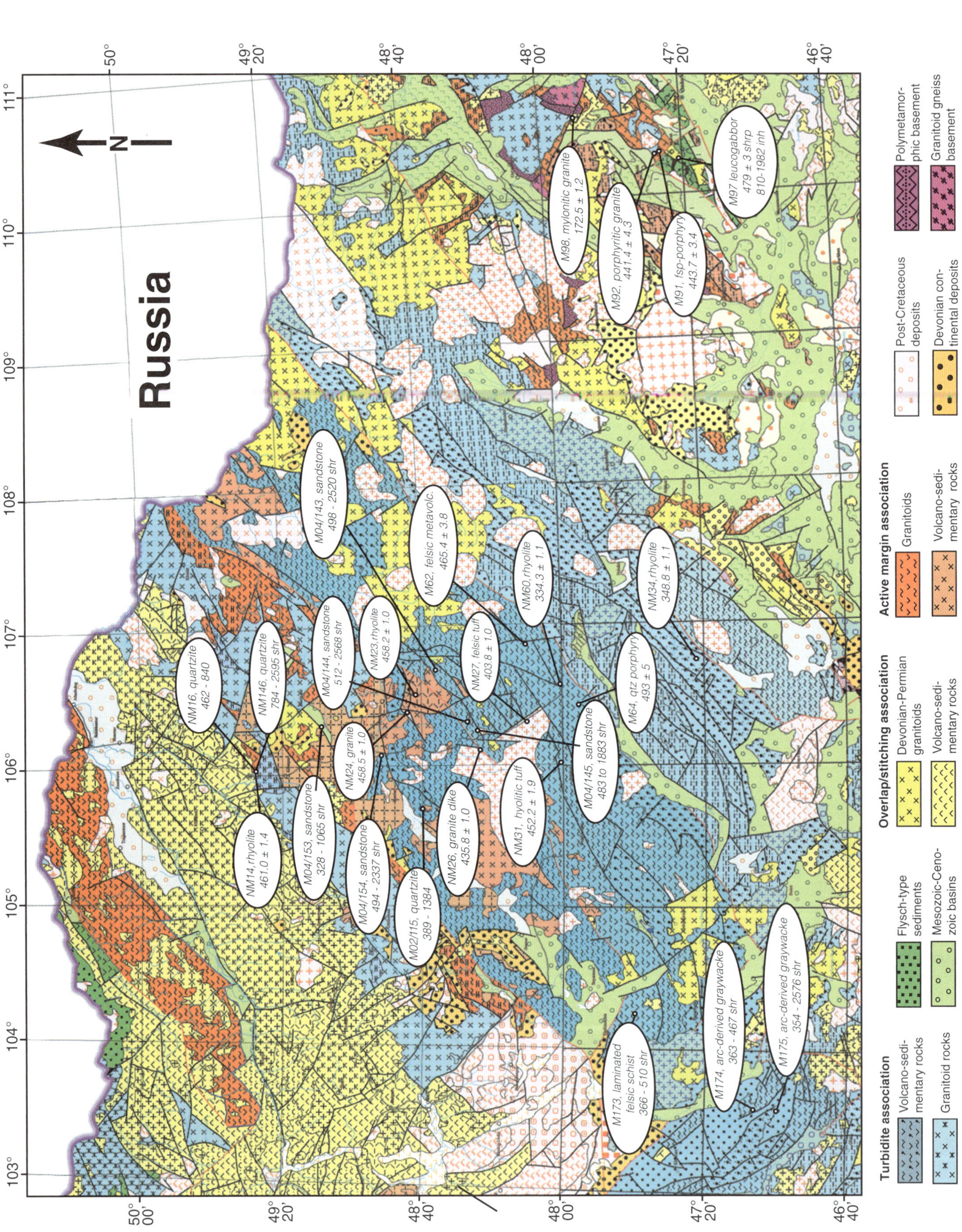

Figure 9. Geological map of part of northern and central Mongolia showing location of dated samples. Ages in Ma. shr—SHRIMP results (other data by zircon evaporation); inh.—ages for inherited zircons. Simplified from Tomurtogoo (1998).

TABLE 5. SUMMARY OF ZIRCON AGES FOR VOLCANIC, GRANITOID, AND SEDIMENTARY ROCKS FROM NORTHERN AND CENTRAL MONGOLIA[†]

Sample no.	Rock type and location	Zircon age (Ma)	Dating method[§]
NM 14	Rhyolite, S of Darhan, northern Mongolia	461.0 ± 1.4	Evap.
NM 16	Detrital zircons from quartzite N or Darham, northern Mongolia	462–840	Evap.
NM 23	Rhyolite, NE of Ulaanbaatar	458.2 ± 1.0	Evap.
NM 24	Granite intrusive into arc sequence, NE of Ulaanbaatar	458.5 ± 1.0	Evap.
NM 26	Granite dyke, intrusive into arc sequence, NE of Ulaanbaatar	435.8 ± 1.0	Evap.
NM 27	Felsic tuff, N of Ulaanbaatar	403.8 ± 1.0	Evap.
NM 31	Rhyolitic tuff, N of Ulaanbaatar	452.2 ± 1.9	Evap.
NM 34	Rhyolite from Carboniferous volcanic sequence, S of Ulaanbaatar	348.8 ± 1.1	Evap.
NM 60	Rhyolite, northeastern outscirts of Ulaanbaatar	334.3 ± 1.1	Evap.
M 01/64	Quartz porphyry, Haraa arc sequence W of Ulaanbaatar	493 ± 5	SHRIMP
M 01/91	Feldspar porphyry, arc sequence of Mörön area, east-central Mongolia	443.7 ± 3.4	SHRIMP
M 01/92	Granodiorite, Buteel Range, northern Mongolia	441.4 ± 4.3	SHRIMP
M 01/97	Leucogabbro from Haraa arc sequence, Mörön River, Herlen terrane	479 ± 3	SHRIMP
M 02/115	Detrital zircons from quartzite S of Darhan, northern Mongolia	389–1384	Evap
M 02/134	Migmatitic orthogneiss, Tseel terrane, southern Mongolia	360.5 ± 1.1	Evap.
M 04/143	Detrital zircons from sandstone along Haraa River NE of Darhan	498–2520	SHRIMP
M 04/144	Detrital zircons from Ordovician volcanic sandstone, Haraa terrane	512–2568	SHRIMP
M 04/145	Detrital zircons from Ordovician volcanic sandsatone, Haraa terrane	483–1883	SHRIMP
M 04/146	Detrital zircons from Eocambrian quartzite S of Darhan, northern Mongolia	784–2595	SHRIMP
M 04/153	Detrital zircons from early Carboniferous sandstone, northern Mongolia	328–1065	SHRIMP
M 04/154	Detrital zircons from Ordovician sandstone N of Baruun	494–2337	SHRIMP

[†]Data of A Kröner, B.F. Windley and G. Badarch, personal communication.
[§]SHRIMP (sensitive high-resolution ion microprobe)—U/Pb single grain age with 2σ error; Evap.—single grain evaporation $^{207}Pb/^{206}Pb$ age with 2σ (mean) error.

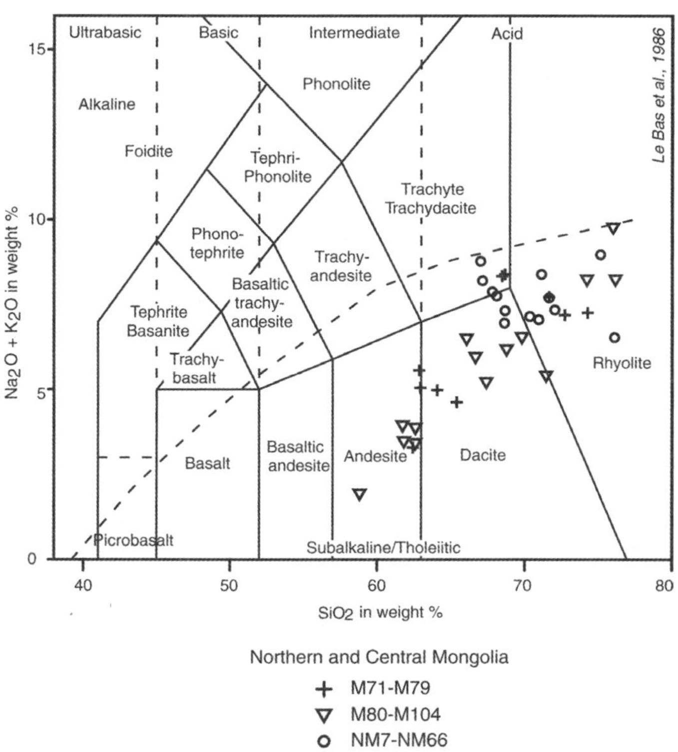

Figure 10. Le Bas et al. (1986) plot of volcanic rocks from northern and central Mongolia. Unpubl. data of A. Kröner, B.F. Windley, and G. Badarch.

of sedimentary transport as well as mostly red to red-brown, strongly abraded, well-rounded to spherical grains. As expected, the euhedral zircons were the youngest, and evaporation of two grains from sample M 02/115 provided a mean $^{207}Pb/^{206}Pb$ age of 389.3 ± 1.4 Ma (Table 5; Fig. 14F), whereas the youngest grain in sample NM 16 has a $^{207}Pb/^{206}Pb$ age of 462.2 ± 1.2 Ma (Table 5; Fig. 14A). It follows from these data that the depositional age of the quartzite must be younger than 389 Ma, presumably middle Devonian. The remaining detrital grains display various degrees of rounding, and the $^{207}Pb/^{206}Pb$ ages range from 812 to 1384 Ma (Table 5; Fig. 14), suggesting detrital input into the Darhan sediments from ancient continental sources. SHRIMP dating of further detrital grains from sample M 04/146 produced ages varying between 784 and 2595 Ma (Table 5; Fig. 15D).

The area along the Haraa River northeast of Darhan is composed of basaltic, andesitic, and dacitic metavolcanic rocks of presumed early Paleozoic age, interlayered with volcanic-derived sandstone, siltstone, and shale. Our sample M 04/143 (location in Fig. 9) is a greenish sandstone, and the detrital zircons are almost exclusively well rounded. SHRIMP (non-representative) reconnaissance dating revealed a large variation in ages with the youngest, idiomorphic grain at 498 ± 4 Ma and the remaining grains varying between 1693 and 2520 Ma (Table 5; Fig. 15A). An Ordovician depositional age is therefore likely for this sandstone, compatible with the volcanic rocks of this region (Table 5). Sample M 04/144 is a green sandstone of presumed Cambro-Ordovician age from a volcano-sedimentary sequence similar to M 04/143 but farther to the southwest

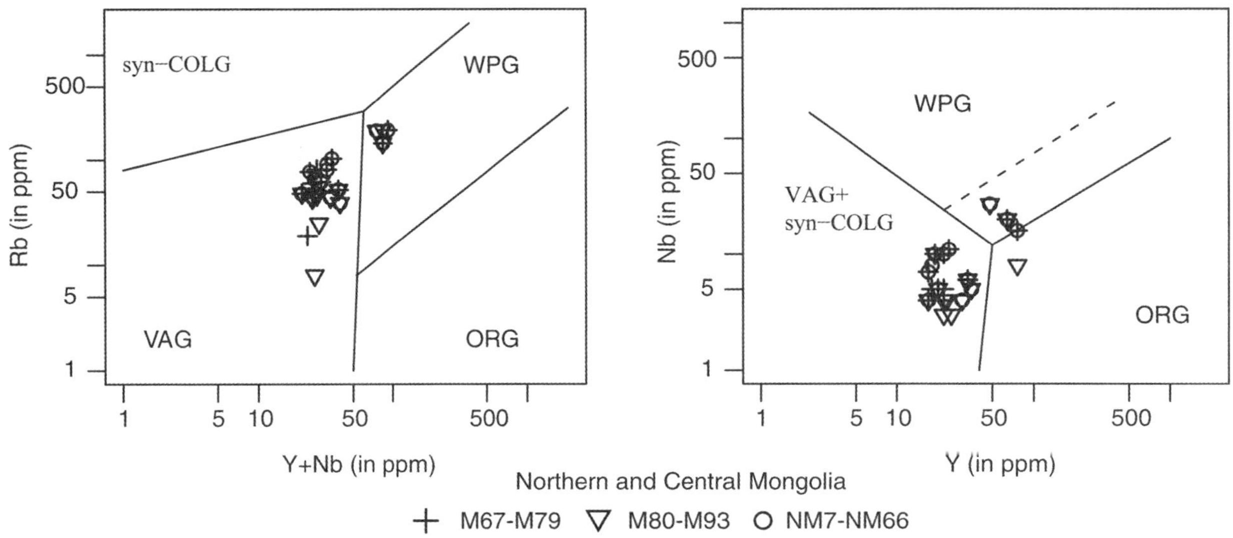

Figure 11. Pearce et al. (1984) plots for northern and central Mongolian rocks. Data as in Figure 10. VAG—volcanic arc granite; ORG—organic granite; WPG—within-plate granite; syn-COLG—syncollisional granite.

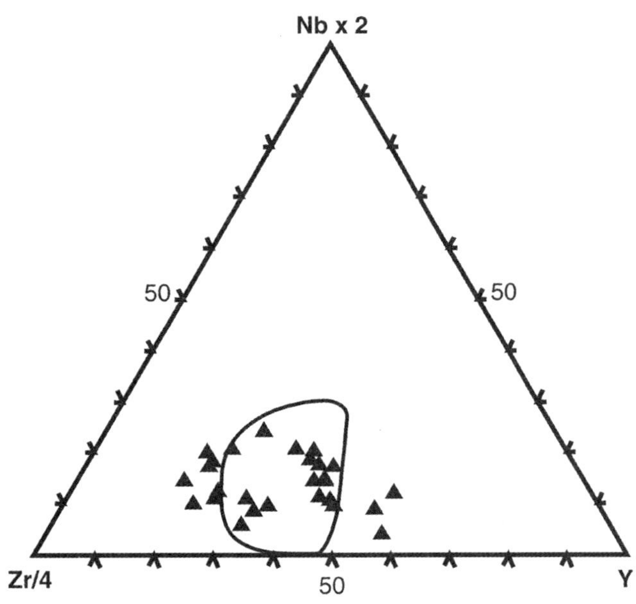

Figure 12. Nb-Zr-Y discrimination plot of Meschede (1986) showing position of northern and central Mongolian arc volcanics and arc-derived sediments (unpublished data of A. Kröner, B.F. Windley and G. Badarch). Black line is field of arc-derived rocks from Murphy and MacDonald (1993).

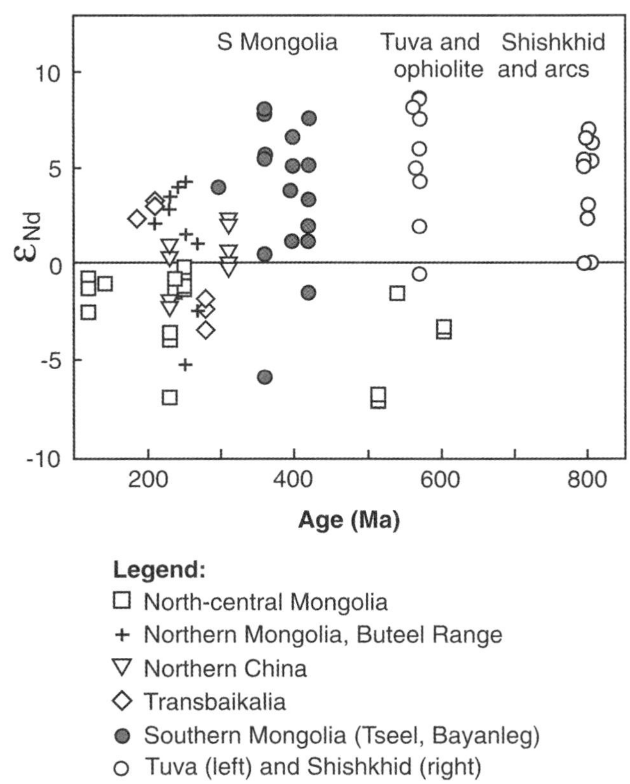

Figure 13. Nd evolution diagram for metaigneous and metasedimentary rocks of the Central Asian Orogenic Belt. Data sources: Shishkhid ophiolite and island arc, northern Mongolia—Kuzmichev et al. (2005); Agardagh Tes-Chem ophiolite and arc, Tuva, Siberia—Pfänder and Kröner (2004); northern and central Mongolia—unpublished data of A. Kröner, Jahn et al. (2004); northernmost China—Chen et al. (2000); Transbaikalia—Litvinovsky et al. (2002); southern Mongolia—Helo et al. (2006).

(Fig. 9). The detrital zircon population is also similar, and two euhedral grains yielded concordant ages of 512 ± 48 and 556 ± 4 Ma, whereas the well-rounded remaining grains range in age from 790 to 2568 Ma (Table 5; Fig. 15B). This sample is therefore also compatible with an early Paleozoic depositional age and contains a record of substantial Precambrian continental input. Lastly, volcanic sandstone sample M 04/145 from the same general sequence was collected from a roadcut still farther west near the village of Bornuur (Fig. 8). As before, idiomorphic zircons are rare, and most grains are little to well rounded. SHRIMP dating produced one young, concordant, age of 483 ± 4 Ma, whereas the remaining analyses range in age between 597 and 1883 Ma (Table 5; Fig. 15C). Again these data are compatible with a maximum Ordovician depositional age and substantial Precambrian detrital input.

We also analyzed detrital zircons from a volcanic-derived and fossiliferous early Carboniferous sandstone northeast of Ulaanbaatar (sample M 04/153, Fig. 9). This sample contains numerous clear, euhedral zircons showing virtually no sedimentary transport, and the youngest concordant age of 328 ± 3 Ma (Table 5; Fig. 16A) suggests deposition in the Carboniferous, compatible with the fossil evidence. The remaining grains vary in age between 367 and 1065 Ma (Table 5; Fig. 16A), and here the input of ancient continental material is less pronounced than in the previous samples.

Our last sandstone sample from the early Paleozoic volcano-sedimentary arc assemblage of northern Mongolia comes from an alternating sandstone-shale sequence exposed in a roadcut along the asphalt road just north of Bayangol, 153 km north of Ulaanbaatar (M 04/154, Fig. 9). SHRIMP dating produced concordant ages ranging between 494 and 2337 Ma (Table 5; Fig. 16B), again revealing input of substantial Precambrian material and compatible with an Ordovician depositional age.

Input of ancient material is recorded not only in volcanic-derived arc sediments but also in igneous rocks of the arc sequences. A massive quartz porphyry from a sequence of strongly deformed felsic volcanic rocks west of Ulaanbaatar (sample M 01/64) yielded a mean SHRIMP $^{206}Pb/^{238}U$ age of 493 ± 5 Ma, based on the analysis of five idiomorphic grains, whereas one euhedral zircon has a concordant age of 888 ± 6 Ma (Table 5; Fig. 16C) thus documenting at least some inheritance from the source terrain of the porphyry. A similar example of crustal contamination is provided by a fresh leucogabbro sample (M 01/97) from a sequence of calc-alkaline volcanic rocks exposed on the Mörön River in the Herlen terrane (Fig. 9). The zircons are clear and idiomorphic, and eight concordant SHRIMP analyses provided a mean $^{206}Pb/^{238}U$ age of 479 ± 3 Ma, whereas four additional analyses are discordant but compatible with the above age (Table 5; Fig. 16D). However, three zircons in this sample are xenocrystic and provided ages of 810, 814, and 1982 Ma (Fig. 16D) thus again documenting involvement of older crust in the generation of the gabbro.

The above examples provide strong evidence for the conclusion that early Paleozoic arc magmatism in northern and central Mongolia was not entirely intra-oceanic as previously assumed. This is also supported by whole-rock Sm-Nd isotopic systematics shown in Figure 13.

SOUTHERN MONGOLIA

Ocean crust formation still occurred south of the Main Mongolian Lineament in the early Carboniferous as indicated by a zircon age of 325 Ma for the Adaatsag ophiolite in south-central Mongolia (Tomurtogoo et al., 2005) that we ascribe to opening of the Mongol-Okhotsk ocean. Early to late Carboniferous magmatic activity in the Zam Bilgikh and Transaltai Ranges in the Gobi Desert of southern Mongolia is documented by zircon ages for a variety of granitoid rocks (K. Schulmann and A. Kröner, unpubl. data).

Several terranes of Mongolia containing metamorphic rocks and previously interpreted as Precambrian microcontinents have since been shown to be much younger, and a good example is the Tuva-Mongolian Massif of southern Siberia and northwestern Mongolia (Salnikova et al., 2001; Kozakov et al., 2005). The Tseel terrane in southern Mongolia (Fig. 17) is another example and consists of low-grade Early Devonian arc volcanics in the north dated at ca. 397 Ma (Kröner, Badarch and Windley, unpubl. data) (Fig. 17), which grade into successively higher-grade assemblages to the south with amphibolite-facies migmatitic gneisses (deformed arc granitoids), one of which (M 02/134, Fig. 17) was dated at 360.5 ± 1.1 Ma (Table 3; Fig. 18). Rare granulite-facies rocks may correlate with a similar high-grade assemblage in the Tsogt block farther west for which Kozakov et al. (2002) obtained a metamorphic age of 365 Ma. We interpret these high-grade rocks as representing the root zone of an arc system.

From southern Mongolia, Lamb and Badarch (1997) described a variety of Devonian and Carboniferous volcanic arcs ranging from submerged island arcs, mature oceanic arcs, and continental margin arcs. Helo et al. (2006) presented geochemical and Nd isotopic data for metaigneous and metavolcaniclastic rocks from a variety of ca. 470–290 Ma island-arc assemblages in the Gurvan Sayan and Zoolen Ranges, in order to define their tectonic environments and their mode of crustal generation. These arc terranes formed during the last phases of the evolution of the Central Asian Orogenic Belt. The whole-rock geochemical data are consistent with an origin in juvenile intra-oceanic arc/forearc and backarc settings. However, the remarkably wide range of initial ε_{Nd} values of ~+8 to –6 indicates the presence of juvenile crust as well as much older material with mean crustal residence ages up to 1.5 Ga, and some Nd isotopic data indicate derivation of the host volcanic rocks from a variably rejuvenated Paleoproterozoic crustal domain (Helo et al., 2006). Our unpublished detrital zircon ages from arc-derived sediments in southern Mongolia (A. Kröner, M.T.D. Wingate, A. Demoux, G. Badarch) support this view (Fig. 17), and all these data convincingly demonstrate that the crust of the Central Asian Orogenic Belt in southern

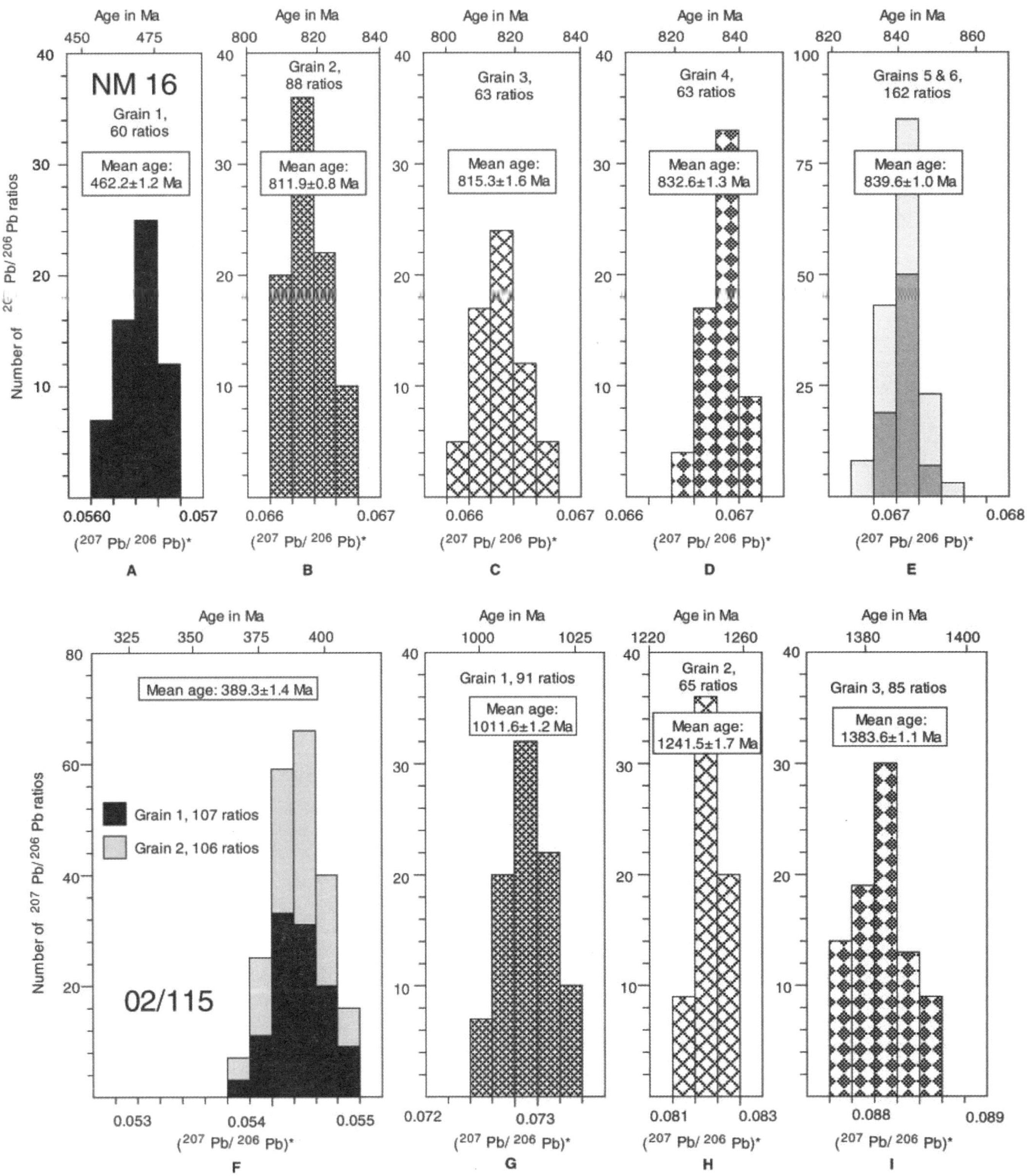

Figure 14. Histograms showing distribution of radiogenic Pb isotope ratios derived from evaporation of detrital zircons from two quartzite samples, quarry south of Darhan, northern Mongolia. (A–E) Zircons from sample NM 16. (F–I) Zircons from sample M 02/115. The spectra have been integrated from ratios as shown for each diagram. Mean ages are given with 2σ (mean) errors.

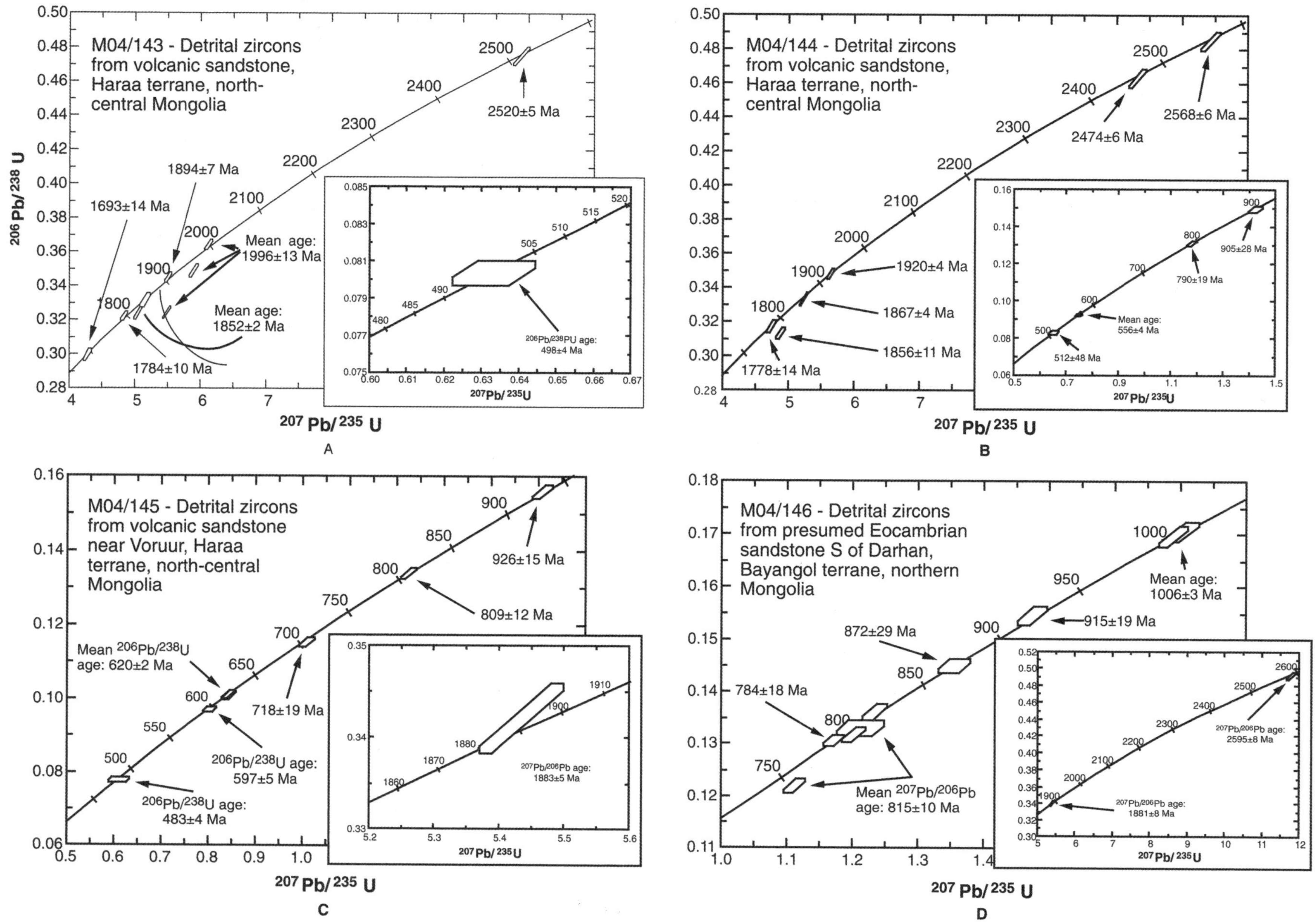

Figure 15. Concordia diagrams showing SHRIMP analyses of detrital zircons from Paleozoic sediments in northern and central Mongolia. Data boxes for each analysis are defined by standard errors in $^{207}Pb/^{235}U$, $^{206}Pb/^{238}U$, and $^{207}Pb/^{206}Pb$. (A) Volcanic sandstone sample M 04/143, Haraa terrane. (B) Volcanic sandstone sample M 04/144, Haraa terrane. (C) Volcanic sandstone sample M 04/145, Haraa terrane. (D) Quartzite sample M 04/146, Bayangol terrane.

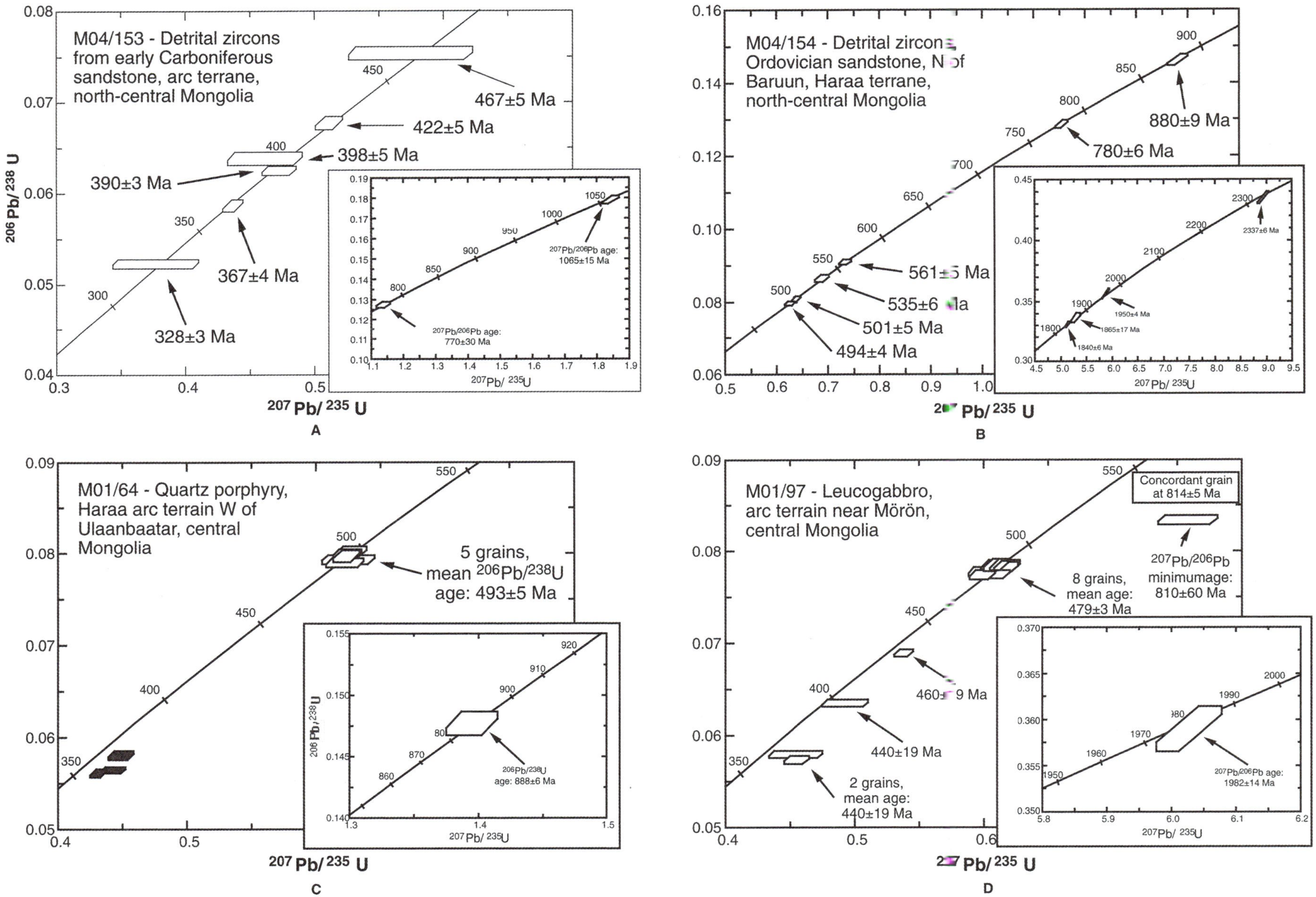

Figure 16. Concordia diagrams showing SHRIMP analyses of detrital zircons from Paleozoic arc-related sediments and magmatic rocks in northern and central Mongolia. Data boxes as in Figure 15. A: Sandstone sample M 04/153, Haraa terrane. B: Sandstone sample M 04/154, Haraa terrane. C: Quartz porphyry sample M 01/64, Haraa terrane. D: Leucogabbro sample M 01/97, Herlen terrane.

Mongolia has an isotopically highly heterogeneous composition. Helo et al. (2006) speculated that a juvenile arc and backarc system was tectonically juxtaposed with a rejuvenated Paleoproterozoic microcontinent, a situation similar to that envisaged for Kazakhstan by Bykadorov et al. (2003). There appear to be only few areas in the orogenic belt where $\varepsilon_{Nd}(t)$ values are consistently positive and as high as +8, and the Junggar terrane in the Xinjiang province of northwestern China is one of them (Chen and Jahn, 2004).

ARABIAN-NUBIAN SHIELD

The Arabian-Nubian shield is a wide Neoproterozoic accretionary orogen at the northern end of the East African orogen (Kröner and Stern, 2005). It consists of continental-margin as well as intra-oceanic arc terranes and a few continental fragments containing Archean and Paleoproterozoic rocks. Magmatic terranes began to form at ca. 870 Ma, and terrane amalgamation was completed by ca. 600 Ma (for recent summary see Johnson and Woldehaimanot, 2003, and references therein). Reymer and Schubert (1987) calculated extraordinarily high crust-production rates for the shield, assuming that most of the arc-related rocks were juvenile, as suggested by low $^{87}Sr/^{86}Sr$ initial isotopic ratios and positive $\varepsilon_{Nd}(t)$ values. These rates are comparable, or even higher, than those proposed for the Central Asian Orogenic Belt by Şengör et al. (1993).

Recent SHRIMP dating of felsic metavolcanic and granitoid rocks from the northern Arabian arc sequences (Fig. 19) (Kennedy et al., 2004, 2005; Hargrove et al., 2006) reveals zircon age patterns similar to the Mongolian felsic rocks. The majority of zircons in these samples are idiomorphic and constitute concordant clusters, interpreted to reflect the age of magmatism (Kennedy et al., 2004, 2005) (Fig. 20). In addition, a variable number of grains, in some cases characterized by variously rounded morphologies, exhibit substantially older ages (Fig. 20), in one case up to 2750 Ma, and these are interpreted as xenocrysts derived from an older basement (Kennedy et al., 2004, 2005; Hargrove et al., 2006). All these rocks reflect various proportions of crustal input and are therefore unlikely to be entirely juvenile. Hargrove et al. (2006) were puzzled by the fact that their trace element and Nd isotopic signatures are primitive and speculated that large volumes of juvenile arc magmas could perhaps melt enough older crust to inherit xenocrystic zircons, yet retain their original arc and isotopic characteristics. Whatever the cause of zircon inheritance in the Arabian-Nubian shield rocks, it would appear that there was more old crust around than previously thought, and previous estimates for crust-production rates are almost certainly too high.

DISCUSSION AND CONCLUSIONS

The above zircon ages show that there appear to be remarkable similarities between the Central Asian Orogenic Belt and the Arabian-Nubian shield, but the origin of ancient xenocrysts in seemingly juvenile arc-derived rocks remains enigmatic. We point out that the term "juvenile crust" cannot be defined precisely on the basis of Nd isotopic systematics alone. Even rocks with positive $\varepsilon_{Nd}(t)$ values may contain ancient crustal material, as shown for some of our samples from Kazakhstan and Mongolia and as demonstrated for the shield by Kennedy et al. (2004, 2005) and Hargrove et al. (2006). It is the proportion of mantle-derived material versus material resulting from crustal recycling that counts, and such estimates contain a large degree of uncertainty (e.g., Jahn et al., 2004). The high crust-production rates suggested for the orogenic belt and the shield were based on the simplistic assumption that most arc-derived rocks were juvenile, whereas more realistic estimates imply significantly lower growth rates.

The Central Asian Orogenic Belt and the Arabian-Nubian shield have frequently been compared to the present southwest Pacific where ancient crustal fragments derived from Australia occur together with juvenile crust produced by subduction-related magmatism, seafloor spreading, and ocean plateau formation (Hall, 2002). This region, once accreted to Asia and caught up in the collision between Asia and Australia in ~50 m.y. from now, will probably look rather similar to the orogenic belt.

There have been some misconceptions about the role of granitic rocks (sensu lato) in the Central Asian Orogenic Belt. Şengör et al. (1993) considered that the subduction-accretion complexes of the Altaids were intruded by vast (granitic) plutons of mainly magmatic arc origin, and that half of these complexes, including the granitic rocks, were of juvenile origin. However, they failed to distinguish the relatively few granitic rocks in mature island arcs from the vast majority that can be broadly termed "crustal-melt granites." For example, granitic rocks occupy more than 40% of the present surface area of the Chinese Altai, and most of these are posttectonic and garnet-bearing (Windley et al., 2002). From their Sm-Nd isotopic study of granitic rocks in several parts of the orogenic belt, Jahn et al. (2000) concluded that the majority are of juvenile origin, "implying a massive addition of new continental crust." However, these crustal-melt granitic rocks are only a proxy for the juvenile arcs that are the main component of the Central Asian Orogenic Belt. The granites are surrounded by, and presumably underlain by, juvenile arc-dominated rocks from which granitic material was derived by partial melting.

There are two different scenarios to explain the generation of silicic magmas. The first is where felsic magmas are derived from a basaltic reservoir by fractional crystallization, or assimilation combined with fractional crystallization (AFC). This generally produces relatively small magma batches, such as those described from seafloor eruptions (Fiske et al., 2001), because to generate large volumes of felsic magmas, unreasonably large amounts of basalt must first be crystallized (Riley et al., 2001). The second scenario is a continental margin where basaltic underplating provides heat for partial melting of crustal rocks, and this is considered more appropriate for large-volume felsic magma bodies such as the Sierra Madre Occidental in Mexico, the Choiyoi province of the Chilean Andes, and the large rhyolite

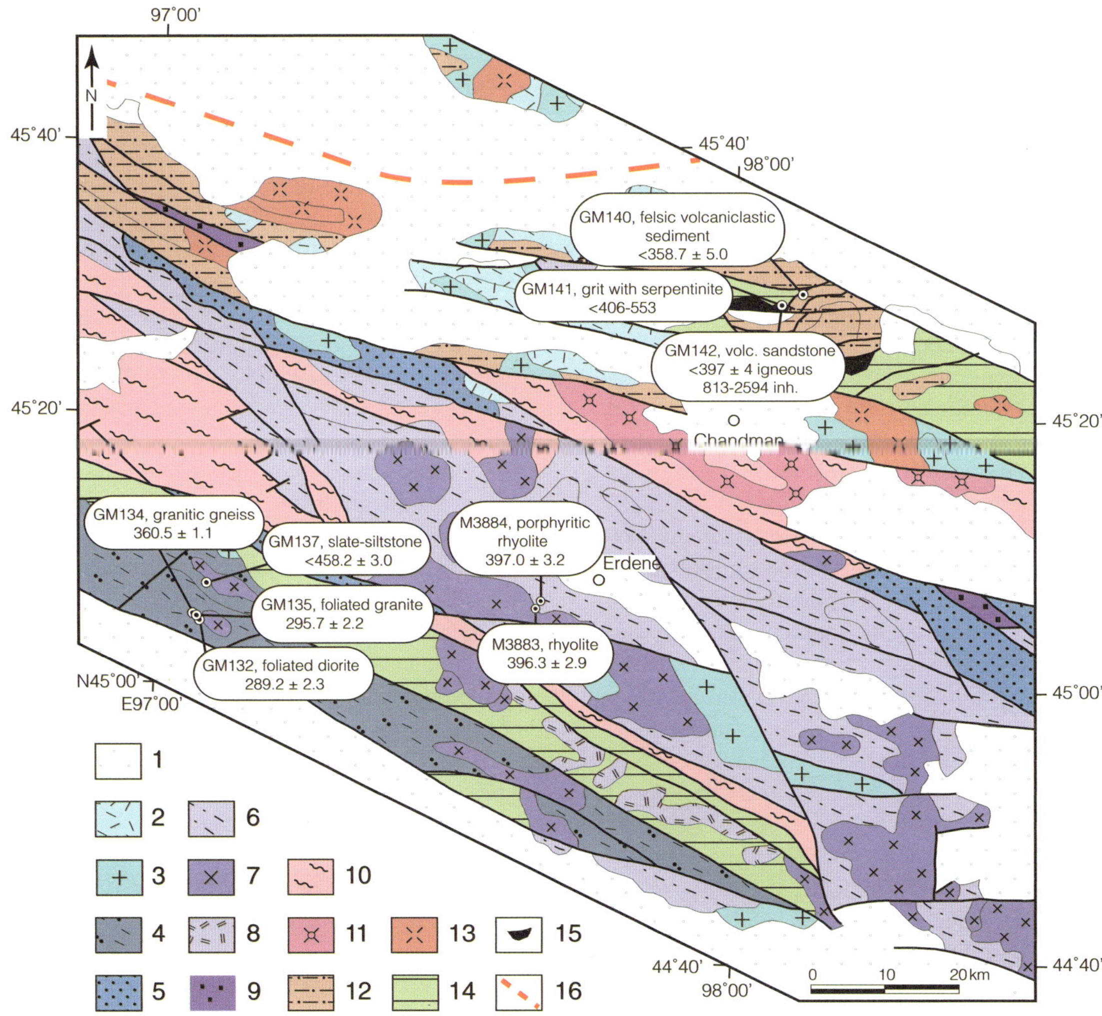

Figure 17. Geological map of the Tseel terrane of southern Mongolia, showing location of dated samples. Ages in Ma. shr—SHRIMP results (other data by zircon evaporation); inh.—ages for inherited zircons. Simplified from Tomurtogoo (1998), sheet L-47. 1—Mesozoic-Quaternary cover; 2—Permian volcanic rocks; 3—Permian granite; 4—Late Carboniferous–Early Permian metavolcano-sedimentary rocks; 5—Early Carboniferous metasediments; 6—Devonian volcano-sedimentary sequence; 7—Middle Devonian granite; 8—Early Devonian volcanic rocks; 9—Silurian volcanogenic rocks; 10—Ordovician metavolcanic rocks; 11—Ordovician granodiorite; 12—early Cambrian volcanogenic schist; 13—Cambrian quartz diorite; 14—Neoproterozoic metasediments; 15—ultramafic mélange; 16—Main Mongolian Lineament.

province in the Antarctic Peninsula and Patagonia (Riley et al., 2001, and references therein).

The latter setting also seems applicable to the Central Asian Orogenic Belt of central and northern Mongolia that was stabilized into a continental margin by early Ordovician time. We suggest that the large volume of predominantly felsic magmatic rocks generated in Ordovician and Silurian times resulted from mixing between a fractionated mafic underplate and partial melts of the previously stabilized lower continental margin crust that consisted predominantly of Neoproterozoic to Cambrian rocks. Therefore, many silicic rocks show apparent juvenile Nd isotopic signatures and model ages up to ca. 1100 Ma (Jahn et al., 2004; Kovalenko et al., 2004; our data for northern Mongolia; see Table 4). The Mongolian felsic rocks have trace element characteristics of subduction-related magmas that are probably inherited from their source. Riley et al. (2001) observed that "water-assisted" generation of partial melts will occur across a broad region adjacent to an evolving arc, and the crust overlying the subduction zone will, over a period of time, become extensively intruded by hydrous, mantle-derived magmas. A large proportion of such magmas will

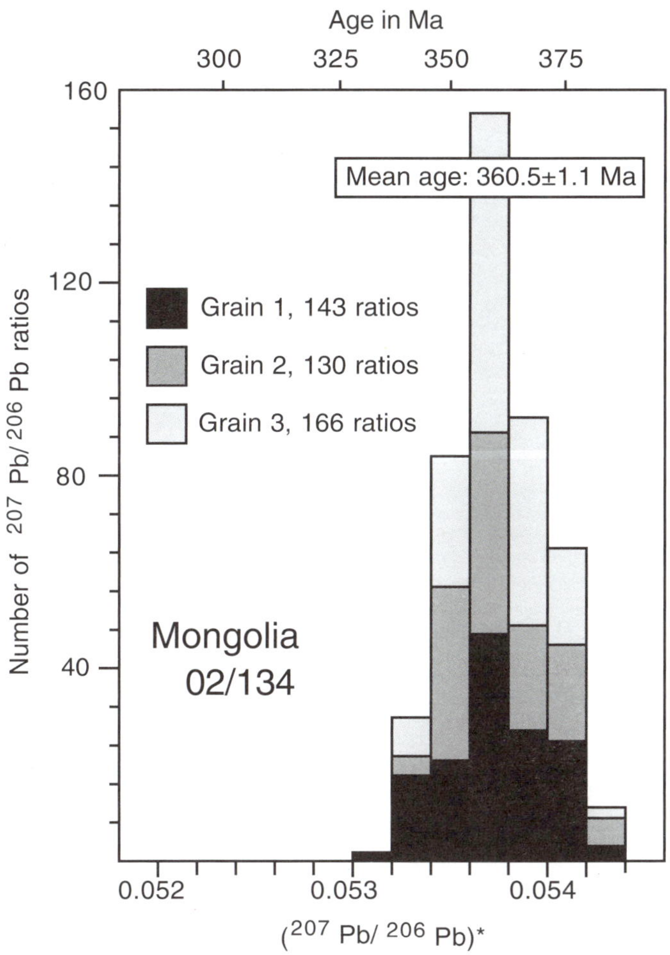

Figure 18. Histogram showing distribution of radiogenic Pb isotope ratios derived from evaporation of three zircon grains from granitic gneiss sample M 02/134, Tseel terrane, southern Mongolia. The spectrum plotted has been integrated from 439 ratios.

Figure 19. Terrane map of the Arabian part of the Arabian-Nubian shield showing major tectonic units (from Johnson, 1998). Magmatic rocks displaying variable degrees of crustal contamination as reported by Kennedy et al. (2004, 2005) occur in the Hijaz, Afif, Ad Dawadimi, and Asir terranes. Considerable inheritance is also described from the Jiddah and Hijaz terranes by Hargrove et al. (2006). f.z.—fracture zone.

collect in the lower crust, generating a hydrous, amphibole-rich mafic layer. We follow this model and suggest that such conditions were initiated in the Early Ordovician, or possibly earlier, through north-dipping subduction of oceanic lithosphere of the evolving Paleo-Asian ocean underneath the relatively young continental margin growing southwards away from the Siberian craton.

We thus agree with Şengör and Natal'in (1996) and Jahn et al. (2004) that juvenile crust formation in the Central Asian Orogenic Belt was considerable, but there was also a significant amount of remelting of older, Archean to earliest Paleozoic crust, to produce the large Ordovician to Permian felsic volcanic province observed in Mongolia and Kazakhstan. Therefore, crust production was spread over a longer time-span than previously estimated, and crust-production rates were not as high as assumed by Şengör and Natal'in (1996). As a corollary, the available ages and geochemical data for the orogenic belt make it unlikely that this vast accretionary belt was derived from one single, giant arc, and we favor a southwest Pacific-type scenario for its origin.

ACKNOWLEDGMENTS

Most of this paper was written while the senior author held a Gledden Visiting Fellowship at the University of Western Australia, and discussions with members of the Tectonics Special Research Centre (TSRC) are appreciated. This is part of a collaborative study with the Mongolian Academy of Sciences, funded by the Volkswagen Foundation of Germany Grant I/76399. Our work in Kazakhstan was funded by the German Science Foundation (DFG). We thank Dmitry Alexeiev (Moscow) for providing Figure 3, Table 1, and information on Kazakhstan geology. Peter Johnson (Jeddah) made available an Open-File Report of the Saudi Arabian Geological Survey and Figure 19, and Ulysses Hargrove (Dallas) provided Figure 20B. Kent Condie and Bob Hatcher helped with constructive reviews. Many of the zircon analyses were carried out on the sensitive high-resolution ion microprobe mass spectrometer in Perth (SHRIMP II), operated by a consortium consisting of Curtin University of Technology,

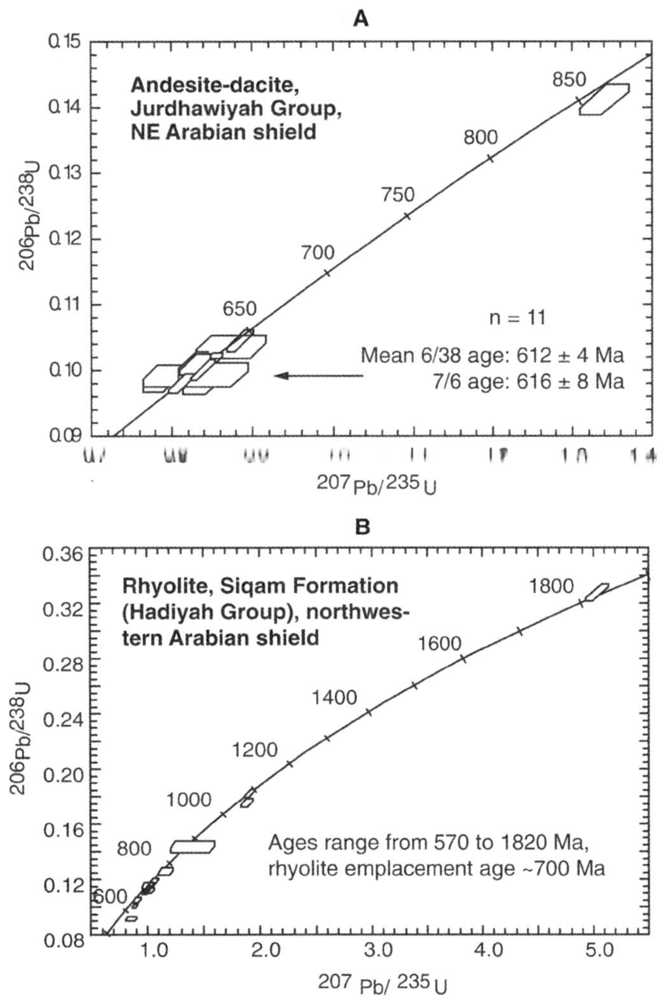

Figure 20. Concordia diagrams showing SHRIMP zircon analyses from Neoproterozoic felsic volcanic rocks of the northern Arabian shield. (A) Age pattern for an andesite-dacite sample, reproduced from Figure 2 of Kennedy et al. (2004). (B) Age pattern for a rhyolite sample, reproduced from Figure 27 of Kennedy et al. (2004).

the Geological Survey of Western Australia, and the University of Western Australia with the support of the Australian Research Council. This is TSRC publication number 364, publication number 50 of the Geocycles Cluster of Mainz University, and a contribution to the ERAS-Project of the International Lithosphere Program and to IGCP Project 480.

REFERENCES CITED

Abdulin, A.A., Avdeev, A.V., and Seitov, N.S., 1995, On the study of microcontinents and ophiolite zones in Kazakhstan: ESRI Occasional Publications, New Series No. 12B, University of South Carolina, USA, p. 97–112.

Alexyutin, M.V., Bachtadse, V., Alexeiev, D.V., and Nikitina, O.I., 2005, Paleomagnetism of Ordovician and Silurian rocks from the Chu-Yili and Kendyktas Mountains, South Kazakhstan: Geophysical Journal International, v. 161, p. 1–12.

Almukhamedov, A.I., Gordienko, I.V., Kuzmin, M.I., Tomurtogoo, O., and Tomurkhuu, D., 2002, Boninites of the Dzidha Caledonide zone, northern Mongolia: Transactions of the Russian Academy of Sciences, v. 377, p. 526–529.

Anonymous, 1972, Penrose field conference on ophiolites: Geotimes, v. 17, p. 24–25.

Avdeev, A.V., 1984, Ophiolite zones and the geologic history of Kazakhstan from the mobilist standpoint: International Geological Review, v. 26, p. 995–1005.

Avdeev, A.V., Khalilov, V.A., Seliverstov, V.A., Ilin, L.I., and Tsygankov, A.I., 1995, New data on geochronology of metamorphic and magmatic formations of South Kazakhstan: Geology and Exploration in Kazakhstan, v. 4, p. 2–6.

Badarch, G., Cunningham, W.D., and Windley, B.F., 2002, A new terrane subdivision for Mongolia: implications for the Phanerozoic crustal growth of Central Asia: Journal of Asian Earth Sciences, v. 21, p. 87–104.

Bazhenov, M.L., Collins, A.Q., Degtyarev, K.E., Levashova, N.M., Mikolaichuk, A.V., Pavlov, V.E., and Van der Voo, R., 2003, Paleozoic northward drift of the North Tien Shan (Central Asia) as revealed by Ordovician and Carboniferous paleomagnetism: Tectonophysics, v. 366, p. 113–141, doi: 10.1016/S0040-1951(03)00075-1.

Berzin, N.A., and Dobretsov, N.L., 1994, Geodynamic evolution of southern Siberia in late Precambrian–early Paleozoic time, *in* Coleman, R.G., ed., Reconstruction of the Paleo-Asian Ocean: Proceedings of the 29th International Geological Congress, part B: Utrecht, VSP Publishers, p. 53–70.

Bryan, S.E., Riley, T.R., Jerram, D.A., Stephens, C.J., and Leat, P.T., 2002, Silicic volcanism: An undervalued component of large igneous provinces and volcanic rifted margins, *in* Menzies, M.A., et al., eds., Volcanic rifted margins: Geological Society of America Special Paper 362, p. 99–120.

Buchan, C., Cunningham, D., Windley, B.F., and Tomurhuu, D., 2001, Structural and lithological characteristics of the Bayankhongor ophiolite zone, central Mongolia: Geological Society [London] Journal, v. 158, p. 445–460.

Buslov, M.M., Saphonova, I.Yu., Watanabe, T., Obut, O., Fujiwara, Y., Iwata, K., Semakov, N.N., Sugai, Y., Smirnova, L.V., and Kazansky, A.Yu., 2001, Evolution of the Paleo-Asian ocean (Altay-Sayan region, Central Asia) and collision of possible Gondwana-derived terranes with the southern marginal part of the Siberian continent: Geosciences Journal, v. 5, p. 203–224.

Byamba, J., 1996, Tectonics of old structures and phosphates of Mongolia: Joint Soviet-Mongolian Scientific Research Geological Expedition, Transactions, v. 57, 179 p.

Bykadorov, V.A., Bush, V.A., Fedorenko, O.A., Filippova, I.B., Miletenko, N.V., Puchkov, V.N., Smirnov, A.V., Uzhkenov, B.S., and Velozh, Y.A., 2003, Ordovician-Permian Palaeogeography of central Eurasia: Development of Paleozoic petroleum-bearing basins: Journal of Petroleum Geology, v. 26, p. 325–350, doi: 10.1111/j.1747-5457.2003.tb00033.x.

Chakabaev, S.E., chief editor, 1979, Geological map of Kazakh SSR, Southern Kazakhstan series: Leningrad, USSR, Aerogeologiya Map Enterprise, scale 1:500,000.

Chen, B., and Jahn, B.-M., 2004, Genesis of post-collisional granitoids and basement nature of the Junggar terrane, NW China: Nd-Sr isotope and trace element evidence: Journal of Asian Earth Sciences, v. 23, p. 691–703, doi: 10.1016/S1367-9120(03)00118-4.

Chen, B., Jahn, B.-M., Wilde, S., and Xu, B., 2000, Two contrasting Paleozoic magmatic belts in northern Inner Mongolia, China: Petrogenesis and tectonic implications: Tectonophysics, v. 328, p. 157–182, doi: 10.1016/S0040-1951(00)00182-7.

Coleman, R.G., 1989, Continental growth of northwest China: Tectonics, v. 8, p. 621–635.

Collins, A.Q., Degtyarev, K.E., Levashova, N.M., Bazhenov, M.L., and Van der Voo, R., 2003, Early Paleozoic paleomagnetism of east Kazakhstan: Implications for paleolatitudinal drift of tectonic elements within the Ural-Mongol belt: Tectonophysics, v. 377, p. 229–247, doi: 10.1016/j.tecto.2003.09.003.

Degtyarev, K.E., and Ryazantsev, A.V., 2007, Cambrian arc-continent collision in the Paleozoides of Kazakhstan: Geotectonics, v. 41, p. 63–86 (translated from Russian).

DePaolo, D.J., 1981, A Nd and Sr isotopic study of Mesozoic calc-alkaline granitic batholiths of the Sierra Nevada and Peninsular Ranges, California: Journal of Geophysical Research, v. 86, p. 10370–10488.

Dobretsov, N.L., Buslov, M.M., and Vernikovsky, V.A., 2003, Neoproterozoic to Early Ordovician evolution of the Paleo-Asian ocean: Implications to the break-up of Rodinia: Gondwana Research, v. 6, p. 143–159, doi: 10.1016/S1342-937X(05)70966-7.

Donskaya, T.V., Sklyarov, E.V., Gladkochub, D.P., Mazukabzov, A.M., Sal'nikova, E.B., Kovach, V.P., Yakovleva, S.Z., and Berezhnaya, N.G., 2000, The Baikal collisional metamorphic belt: Transactions (Doklady) of the Russian Academy of Sciences, Earth Science Section, v. 372, p. 1075–1079.

Eganov, E.A., and Sovetov, Yu.K., 1979, Karatau—A model for phosphorite deposition: Novosibirsk, Nauka Publishing House, Transactions of the Institute of Geology and Geophysics, Siberian Branch, USSR Academy of Sciences, v. 427, 190 p. (in Russian with English abstract).

Fedorenko, O.A., and Militenko, K., editors, 2002, Atlas of lithological, paleogeographic, structural, palinspastic, and geoecologic maps of central Eurasia: Almaty, Yuggeo.

Fedorovsky, V.S., 1997, Domic tectogenesis in the Caledonides collisional system in the western Near Baikal area: Geotectonics, v. 6, p. 56–71.

Fedorovsky, V.S., 2004, Geological map of the southwestern part of Olkhon region: Vernadsky State Geological Museum, Russian Academy of Sciences, Moscow.

Fedorovsky, V.S., Khain, E.V., Vladimirov, A.G., Kargopolov, S.A., Gibsher, A.S., and Izokh, A.E., 1995, Tectonics, metamorphism, and magmatism of collisional zones of the Central Asian Caledonides: Geotectonics, v. 29, p. 193–212.

Fedorovsky, V.S., Donskaya, T.V., Gladkochub, D.P., Khromykh, V.P., Mazukabzov, A.M., Mekhonoshin, A.S., Sklyarov, E.V., Sukhorukov, V.P., Vladimirov, A.G., Volkova, N.I., and Yudin, D.S., 2005, The Ol'khon collision system (Baikal region), *in* Sklyarov, E.V., ed., Structural and tectonic correlation across the central Asia orogenic collage, northeastern segment: Guidebook and abstract volume of the Siberian workshop (International Geological Correlation Programme 480), Irkutsk-Ulan Ude, Russia: Institute of the Earth's Crust, Russian Academy of Sciences, Siberian Branch, Irkutsk, p. 6–97.

Filippova, I.B., Bush, V.A., and Didenko, A.N., 2001, Middle Paleozoic subduction belts: The leading factor in the formation of the Central Asian fold-and-thrust belt: Russian Journal of Earth Sciences, v. 3, p. 405–426.

Fiske, R.S., Naka, J., Iisasa, K., Yuiasa, M., and Klaus, A., 2001, Submarine silicic caldera at the front of the Izu-Bonin arc, Japan: Voluminous seafloor eruptions of rhyolite pumice: Geological Society of America Bulletin, v. 113, p. 813–824, doi: 10.1130/0016-7606(2001)113<0813:SSCATF>2.0.CO;2.

Gordienko, I.V., 2006, Geodynamic evolution of late Baikalides and Paleozoids in the folded periphery of the Siberian craton: Russian Geology and Geophysics, v. 47, p. 53–70.

Gordienko, I.V., and Filimonov, A.V., 2005, The Dzidha zone of the Paleo-Asian Ocean: Main stages of geodynamic evolution (Vendian–early Paleozoic oceanic, island-arc and back-arc basin complexes), *in* Sklyarov, E.V., ed., Structural and tectonic correlation across the central Asia orogenic collage, northeastern segment: Guidebook and abstract volume of the Siberian workshop (International Geological Correlation Programme 480), Irkutsk-Ulan Ude, Russia: Institute of the Earth's Crust, Russian Academy of Sciences, Siberian Branch, Irkutsk, p. 98–164.

Gordienko, I.V., and Mikhaltsov, N.E., 2001, Location of the Vendian–early Cambrian ophiolitic and island-arc complexes in the Dzhida Caledonite zone in the Paleo-Asian ocean structures according to paleomagnetic data: Doklady Earth Sciences, v. 379, p. 508–513.

Hall, R., 2002, Cenozoic geological and plate tectonic evolution of SE Asia and the SW Pacific: Computer-based reconstructions, model and animations: Journal of Asian Earth Sciences, v. 20, p. 353–431, doi: 10.1016/S1367-9120(01)00069-4.

Hargrove, U.S., Stern, R.J., Kimura, J.I., Manton, W.I., and Johnson, P., 2006, How juvenile is the Arabian-Nubian Shield? Evidence from Nd isotopes and pre-Neoproterozoic inherited zircon: Earth and Planetary Science Letters, v. 252, p. 308–326.

Hartz, E.H., and Torsvik, T.H., 2002, Baltica upside down: A new plate tectonic model for Rodinia and the Iapetus Ocean: Geology, v. 30, p. 255–258, doi: 10.1130/0091-7613(2002)030<0255:BUDANP>2.0.CO;2.

Heinhorst, J., Lehmann, B., Ermolov, P., Serykh, V., and Zhuratin, S., 2000, Paleozoic crustal growth and metallogeny of central Asia: Evidence from magmatic-hydrothermal ore systems of central Kazakhstan: Tectonophysics, v. 328, p. 69–87, doi: 10.1016/S0040-1951(00)00178-5.

Helo, C., Hegner, E., Kröner, A., Badarch, G., Tomurtogoo, O., Windley, B.F., and Dulski, P., 2006, Geochemical signature of Paleozoic accretionary complexes of the Central Asian Orogenic Belt in South Mongolia: Constraints on arc environments and crustal growth: Chemical Geology, v. 227, p. 236–257.

Jahn, B.M., 2004, The Central Asian Orogenic Belt and growth of the continental crust in the Phanerozoic, *in* Malpas, J., et al., eds., Aspects of the tectonic evolution of China: Geological Society [London] Special Publication 226, p. 73–100.

Jahn, B.M., Wu, F., and Chen, B., 2000, Granitoids of the Central Asian Orogenic Belt and continental growth in the Phanerozoic: Transactions of the Royal Society of Edinburgh, Earth Sciences, v. 91, p. 181–193.

Jahn, B.M., Capdevila, R., Liu, D.Y., Vernon, A., and Badarch, G., 2004, Sources of Phanerozoic granitoids in the transect Bayanhongor–Ulaan Baatar, Mongolia: Geochemical and Nd isotopic evidence, and implications for Phanerozoic crustal growth: Journal of Asian Earth Sciences, v. 23, p. 629–653, doi: 10.1016/S1367-9120(03)00125-1.

Johnson, P.R., 1998, Tectonic map of Saudi Arabia and adjacent areas: Saudi Arabian Deputy Ministry for Mineral Resources, Technical Report USGS-TR-98-3, scale 1:4,000,000.

Johnson, P.R., and Woldehaimanot, B., 2003, Development of the Arabian-Nubian Shield: Perspectives on accretion and deformation in the northern East African orogen and the assembly of Gondwana, *in* Yoshida, M., et al., eds., Proterozoic East Gondwana: supercontinent assembly and breakup: Geological Society [London] Special Publication 206, p. 289–325.

Kennedy, A., Johnson, P.R., and Kattan, F.H., 2004, SHRIMP geochronology in the northern Arabian Shield, part I: Data acquisition: Saudi Arabian Geological Survey Open-File Report SGS-OF-2004-11, 28 p.

Kennedy, A., Johnson, P.R., and Kattan, F.H., 2005, SHRIMP geochronology in the northern Arabian Shield, part II: Data acquisition, 2004: Saudi Arabian Geological Survey Open-File Report SGS-OF-2005-10, 44 p.

Khain, E.V., Bibikova, E.V., Kröner, A., Zhuralev, D.Z., Sklyarov, E.V., Fedotova, A.A., and Kravbchenko-Berezhnoy, I.R., 2002, The most ancient ophiolite of the Central Asian fold belt: U-Pb and Pb-Pb zircon ages for the Dunzhugur complex, Eastern Sayan, Siberia, and geodynamic implications: Earth and Planetary Science Letters, v. 199, p. 311–325, doi: 10.1016/S0012-821X(02)00587-3.

Khain, E.V., Bibikova, E.V., Salnikova, E.B., Kröner, A., Gibsher, A.S., Didenko, A.N., Degtyarev, K.E., and Fedotova, A.A., 2003, The Palaeo-Asian ocean in the Neoproterozoic and early Paleozoic: New geochronologic data and palaeotectonic reconstructions: Precambrian Research, v. 122, p. 329–358, doi: 10.1016/S0301-9268(02)00218-8.

Kheraskova, T.N., Didenko, A.N., Bush, V.A., and Volozh, Yu.A., 2003, The Vendian–early Paleozoic history of the continental margin of eastern Paleogondwana, Paleoasian ocean, and Central Asian foldbelt: Russian Journal of Earth Sciences, v. 5, p. 165–184.

Konnikov, E.G., Gibsher, A.S., Izokh, A.E., Sklyarov, E.V., and Khain, E.V., 1994, Late Proterozoic evolution of the northern segment of the Paleoasian ocean: New radiological, geological, and geochemical data: Russian Geology and Geophysics, v. 35, p. 131–145.

Kovach, V.P., Jian, P., Yarmolyuk, V.V., Kozakov, I.K., Kovalenko, V.I., Liu, D.Y., and Terent'eva, L.B., 2005, Magmatism and geodynamics of early stages of the Paleoasian ocean formation: Geochronological and geochemical data on ophiolites of the Bayan-Khongor zone: Doklady Earth Sciences, v. 404, p. 1072–1077.

Kovalenko, V.I., Yarmolyuk, V.V., Kovach, V.P., Kotov, A.B., Kozakov, I.K., Salnikova, E.B., and Larin, A.M., 2004, Isotope provinces, mechanism of generation and sources of the continental crust in the Central Asian mobile belt: Geological and isotopic evidence: Journal of Asian Earth Sciences, v. 23, p. 605–627, doi: 10.1016/S1367-9120(03)00130-5.

Kozakov, I.K., Kotov, A.B., Kovach, V.P., and Salnikova, E.B., 1997, Crustal growth in the geologic evolution of the Baidarik Block, central Mongolia: Evidence from Sm-Nd isotopic systematics: Petrology, v. 5, p. 201–207 (translated from Russian).

Kozakov, I.K., Glebovistsky, V.A., Bibikova, E.V., Azimov, P.Y., and Kirnozova, T.I., 2002, Hercynian granulites of Mongolian and Gobian Altai: Geodynamic setting and formation conditions: Doklady Earth Sciences, v. 386, p. 781–785 (translated from Russian).

Kozakov, I.K., Kovach, V.P., Yarmolyuk, V.V., Kotov, A.B., Salnikova, E.B., and Zagornaya, N.Yu., 2003, Crust-forming processes in the geologic development of the Tuva-Mongolian Massif: Sm-Nd isotopic and geochemical data for granitoids: Petrology, v. 11, p. 444–463 (translated from Russian).

Kozakov, I.K., Sal'nikova, E.B., Natman, A., Kovach, V.P., Kotov, A.B., Podkovyrov, V.N., and Plotkina, Yu.V., 2005, Metasedimentary complexes of the Tuva-Mongolian Massif: Age, provenances, and tectonic position: Stratigraphy and Geological Correlation, v. 13, p. 1–20 (translated from Russian).

Kozakov, I.K., Sal'nikova, E.B., Wang, T., Didenko, A.N., Plotkina, Yu.V., and Podkovyrov, V.N., 2007, Early Precambrian crystalline complexes of the Central Asian microcontinent: Age, sources, tectonic position: Stratigraphy and Geological Correlation, v. 15, p. 121–140 (translated from Russian).

Kröner, A., and Hegner, E., 1998, Geochemistry, single-zircon ages and Sm-Nd systematics of granitoid rocks from the Góry Sowie (Owl) Mts., Polish West Sudetes: Evidence for early Palaeozoic arc-related plutonism: Geological Society [London] Journal, v. 155, p. 711–724.

Kröner, A., and Stern, R.J., 2005, Pan-African orogeny: Encyclopedia of geology, v. 1: Amsterdam, Elsevier, p. 1–12.

Kröner, A., Muhongo, S., Hegner, E., and Wingate, M., 2003, Single-zircon geochronology and Nd isotopic systematics of Proterozoic high-grade rocks from the Mozambique belt of southern Tanzania (Masasi area): Implications for Gondwana assembly: Geological Society [London] Journal, v. 160, p. 745–757.

Kröner, A., Windley, B.F., Khain, E.V., Jahn, B.M., Hegner, E., Badarch, G., Tomurtogoo, O., Kuzmichev, A., and Liu, D.Y., 2004, Accretionary growth and crust formation in the Central Asian Orogenic Belt during the Neoproterozoic and Paleozoic and comparison with the present southwest Pacific: Abstract volume, 17th International Basement Tectonics Association Conference "4-D framework of the continental crust—Integrating crustal processes through time," Oak Ridge, Tennessee, p. 77–78.

Kröner, A., Hegner, E., Lehmann, B., Heinhorst, J., Wingate, M., Liu, D.Y., and Ermelov, P., 2007, Palaeozoic arc magmatism in the Central Asian Orogenic Belt of Kazakhstan: SHRIMP zircon ages and whole-rock isotopic systematics: Journal of Asian Earth Sciences (in press).

Kuzmichev, A.B., Bibikova, E.V., and Zhuralev, D.Z., 2001, Neoproterozoic (~800 Ma) orogeny in the Tuva-Mongolia Massif (Siberia): Island arc–continent collision at the northeast Rodinia margin: Precambrian Research, v. 110, p. 109–126, doi: 10.1016/S0301-9268(01)00183-8.

Kuzmichev, A., Sklyarov, E., Postnikov, A., and Bibikova, E., 2007, The Oka belt (southern Siberia and northern Mongolia): A Neoproterozoic analog of the Japanese Shimanto belt? Island Arc, doi:10.1111/j.1440-1738.2007.00568.x.

Kuzmichev, A., Kröner, A., Liu, D.Y., and Wan, Y.S., 2005, The Shishkhid ophiolite, northern Mongolia: A key to the reconstruction of a late Neoproterozoic island-arc system in central Asia: Precambrian Research, v. 138, p. 125–150, doi: 10.1016/j.precamres.2005.04.002.

Lackschewitz, K.S., Mertz, D.F., Devey, C.W., and Garbe-Schönberg, C.-D., 2003, Late Cenozoic volcanism in the western Woodlark Basin area, SW Pacific: The sources of marine volcanic ash layers based on their elemental and Sr-Nd isotopic compositions: Bulletin of Volcanology, v. 65, p. 182–200.

Lamb, M.A., and Badarch, G., 1997, Paleozoic sedimentary basins and volcanic-arc systems of southern Mongolia: New stratigraphic and sedimentologic constraints: International Geological Review, v. 39, p. 542–576.

Le Bas, M.J., Le Maitre, R.W., Streckeisen, A., and Zanettin, B., 1986, A chemical classification of volcanic rocks based on the total alkali-silica diagram: Journal of Petrology, v. 27, p. 745–750.

Litvinovsky, B.A., Jahn, B.-M., Zanvilevich, A.N., Saunders, A., Poulain, S., Kuzmin, D.V., Reichow, M.K., and Titov, A.V., 2002, Petrogenesis of syenite-granite suites from Bryansky Complex (Transbaikalia, Russia): Implications for the origin of A-type granitoid magmas: Chemical Geology, v. 189, p. 105–133, doi: 10.1016/S0009-2541(02)00142-0.

Matsumoto, I., and Tomurtogoo, O., 2003, Petrological characteristics of the Hantaishir ophiolite complex, Altai region, Mongolia: Coexistence of podiform chromitite and boninite: Gondwana Research, v. 6, p. 161–169, doi: 10.1016/S1342-937X(05)70967-9.

Meert, J.G., and Lieberman, B.S., 2004, A palaeomagnetic and palaeobiogeographical perspective on latest Neoproterozoic and early Cambrian tectonic events: Geological Society [London] Journal, v. 161, p. 477–487.

Meschede, M., 1986, A model of discrimination between different types of mid-ocean ridge basalts and continental tholeiites with the Nd-Zr-Y diagram: Chemical Geology, v. 56, p. 207–218, doi: 10.1016/0009-2541(86)90004-5.

Mossakovsky, A.A., and Dergunov, A.B., 1985, The Caledonides of Kazakhstan, Siberia, and Mongolia: A review of structure, development history, and palaeotectonic environments, *in* Gee, D.G., and Sturt, B.A., eds., The Caledonide orogen—Scandinavia and related areas: Chichester, Wiley and Sons, p. 1201–1215.

Mossakovsky, A.A., Ruzhentsev, S.V., Samygin, S.G., and Kheraskova, T.N., 1993, Central Asian fold belt: Geodynamic evolution and formation history: Geotectonics, v. 27, p. 445–473.

Murphy, J.B., and MacDonald, D.L., 1993, Geochemistry and tectonic discrimination of Late Proterozoic arc-related volocaniclastic turbidite sequences, Antigonish Highlands, Nova Scotia: Canadian Journal of Earth Sciences, v. 30, p. 2273–2282.

Murphy, J.B., Pisarevsky, S.A., Nance, R.D., and Keppie, J.D., 2004, Neoproterozoic–early Paleozoic evolution of peri-Gondwanan terranes: Implications for Laurentia-Gondwana connections: International Journal of Earth Sciences, v. 93, p. 659–682, doi: 10.1007/s00531-004-0412-9.

Nawrocki, J., Zylinska, A., Bula, Z., Brabowski, J., Krzywiec, P., and Poprawa, P., 2004, Early Cambrian location and affinities of the Brunovistulian terrane (central Europe) in the light of palaeomagnetic data: Geological Society [London] Journal, v. 161, p. 513–522.

Nikitin, I.F., and Nikitina, O.I., 2000, To the geodynamics of Kazakhstan in early Paleozoic (hypothetical reconstructions), *in* Bespaev, K.A., ed., Geodynamics and minerageny of Kazakhstan: VAC Publishing House, Almaty, p. 64–72.

Pearce, J.A., Harris, N.B.W., and Tindle, A.G., 1984, Trace element discrimination diagrams for the tectonic interpretation of granitic rocks: Journal of Petrology, v. 25, p. 956–983.

Pfänder, J.A., and Kröner, A., 2004, Tectono-magmatic evolution, age and emplacement of the Agardagh Tes-Chem ophiolite in Tuva, Central Asia: Crustal growth by island arc accretion, *in* Kusky T., ed., Precambrian ophiolites and related rocks: Elsevier Science, Amsterdam, p. 207–221.

Pisarevsky, S.A., and Natapov, L.M., 2003, Siberia and Rodinia: Tectonophysics, v. 375, p. 221–245, doi: 10.1016/j.tecto.2003.06.001.

Popeko, L.I., 2002, Carboniferous of the Mongol-Okhotsk orogenic belt: Vladivostok, Russia, Dalnauka Publishing House, 124 p. (in Russian).

Popov, V., Iosifidi, A., and Khramov, A., 2002, Paleomagnetism of Upper Vendian sediments from the Winter Coast, White Sea region, Russia: Implications for the paleogeography of Baltica during Neoproterozoic times: Journal of Geophysical Research, v. 107, no. B11, 2315, doi: 10.1029/2001JB001607.

Resolution of 3rd Kazakhstan stratigraphic conference on the Precambrian and Phanerozoic, 1991, Part 1: Precambrian and Paleozoic, Alma-Ata, 148 p.

Reymer, A.P.S., and Schubert, G., 1987, Phanerozoic and Precambrian crustal growth, *in* Kröner, A., ed., Proterozoic lithospheric evolution: American Geophysical Union Geodynamics Series, v. 17, p. 1–9.

Riley, T.R., Leat, P.T., Pankhurst, R.J., and Harris, C., 2001, Origins of large-volume rhyolitic volcanism in the Antarctic Peninsula and Patagonia by crustal melting: Journal of Petrology, v. 42, p. 1043–1065, doi: 10.1093/petrology/42.6.1043.

Ruzhentsev, S.V., and Burashnikov, V.V., 1996, Tectonics of the western Mongolian Salairides: Geotectonics, v. 29, p. 379–394.

Ruzhentsev, S.V., and Mossakovskiy, A.A., 1996, Geodynamics and tectonic evolution of the Central Asian Paleozoic structures as the result of the interaction between the Pacific and Indo-Atlantic segments of the Earth: Geotectonics, v. 29, p. 294–311.

Salnikova, E.B., Kozakov, I.K., Kotov, A.B., Kröner, A., Todt, W., Bibikova, E.V., Nutman, A., Yakopvleva, S.Z., and Kovach, V.P., 2001, Age of Paleozoic granites and metamorphism in the Tuvino-Mongolian Massif of the Central Asian mobile belt: Loss of a Precambrian microcontinent: Precambrian Research, v. 110, p. 143–164, doi: 10.1016/S0301-9268(01)00185-1.

Şengör, A.M.C., and Natal'in, B.A., 1996, Paleotectonics of Asia: Fragments of a synthesis, *in* Yin, A., and Harrison, T.M., eds., The tectonic evolution of Asia: Cambridge, UK, Cambridge University Press, p. 486–640.

Şengör, A.M.C., Natal'in, B.A., and Burtman, V.S., 1993, Evolution of the Altaid tectonic collage and Paleozoic crustal growth in Eurasia: Nature, v. 364, p. 299–307, doi: 10.1038/364299a0.

Sklyarov E.V., Gladkochub D.P., Mazukabzov A.M., Donskaya T.V., and Stanevich A.M., 2003, Geological complexes in the margin of the Siberian Craton as indicators of the evolution of a Neoproterozoic Supercontinent: Russian Journal of Earth Sciences, v. 5, p. 347–359.

Smethurst, M.A., Khramov, A.N., and Torsvik, T.H., 1998a, The Neoproterozoic and Palaeozoic palaeomagnetic data for the Siberian Platform: From Rodinia to Pangea: Earth-Science Reviews, v. 43, p. 1–24, doi: 10.1016/S0012-8252(97)00019-6.

Smethurst, M.A., Khramov, A.N., and Pisarevsky, S., 1998b, Palaeomagnetism of the Lower Ordovician Orthoceras Limestone, St. Petersburg, and a revised drift history for Baltica in the early Palaeozoic: Geophysical Journal International, v. 133, p. 44–56, doi: 10.1046/j.1365-246X.1998.1331463.x.

Stern, R.A., 1997, The GSC sensitive high-resolution ion microprobe (SHRIMP): Analytical techniques of zircon U-Th-Pb age determinations and performance evaluation, *in* Radiogenic age and isotopic studies: Geological Survey of Canada Report 10-1997-F, p. 1–31.

Tomurtogoo, O., editor, 1998, Geological map of Mongolia: Mineral Resources Authority of Mongolia, Mongolian Academy of Sciences, Ulaanbaatar, scale 1:100,000, CD-ROM (with English summary).

Tomurtogoo, O., Windley, B.F., Kröner, A., and Liu, D.Y., 2005, Zircon age and occurrence of the Adaatsag ophiolite and Muron shear zone, central Mongolia: Constraints on the evolution of the Mongol-Okhotsk ocean, suture and orogen: Geological Society [London] Journal, v. 162, p. 125–134, doi: 10.1144/0016-764903-146.

Windley, B.F., Kröner, A., Guo, J., Qu, G., Li, Y., and Zhang, C., 2002, Neoproterozoic to Paleozoic geology of the Altai orogen, Chinese Central Asia: New zircon age data and tectonic evolution: Journal of Geology, v. 110, p. 719–739, doi: 10.1086/342866.

Windley, B.F., Kröner, A., and Xiao, W., 2004, The tectonic evolution of central Asia, *in* Jahn, B.M., ed., International workshop on accretionary orogens and continental growth [abs.]: Taipei, Taiwan, National Taiwan University, p. 91–93.

Windley, B.F., Alexeiev, D., Xiao, W.J., Kröner, A., and Badarch, G., 2007, Tectonic models for accretion of the Central Asian Orogenic Belt. Journal of the Geological Society of London, v. 164, p. 31–47.

Xiao, W.J., Windley, B.F., Hao, J., and Zhai, M.G., 2003, Accretion leading to collision and the Permian Solonker suture, Inner Mongolia, China: Termination of the central Asian orogenic belt: Tectonics, v. 22, p. 1069, doi: 10.1029/2002TC001484.

Xiao, W., Windley, B.F., Badarch, G., Sun, S., Li, J., Qin, K., and Wang, Z., 2004a, Palaeozoic accretionary and convergent tectonics of the southern Altaids: Implications for the growth of Central Asia: Geological Society [London] Journal, v. 161, p. 339–342.

Xiao, W.J., Zhang, L.C., Qin, K.Z., Sun, S., and Li, J.L., 2004b, Paleozoic accretionary and collisional tectonics of the eastern Tianshan (China): Implications for the continental growth of central Asia: American Journal of Science, v. 305, p.

Yakubchuk, A., 1994, Polychronous ophiolite belts of central Kazakhstan and their evolution, *in* Ishiwatari, A., et al., eds., Circum-Pacific ophiolites: Proceedings of the 29th International Geological Congress, part D: Utrecht, VSP Publishers, p. 235–254.

Yakubchuk, A., 2002, The Baikalide-Altaid, Transbaikal-Mongolian and North Pacific orogenic collages: Similarity and diversity of structural patterns ands metallogenic zoning, *in* Blundell, D.J., et al., eds., The timing and location of major ore deposits in an evolving orogen: Geological Society [London] Special Publication 204, p. 273–297.

Yakubchuk, A., Seltmann, R., Shatov, V., and Cole, A., 2001, The Altaids: Tectonic evolution and metallogeny: Society of Economic Geologists (SEG) Newsletter, v. 46, p. 7–15.

Yarmolyuk, V.V., 1983, Carboniferous volcanism of Mongolia, *in* Luchitsky, I.V., ed., Continental volcanism of Mongolia: Moscow, Nauka, Joint Soviet-Mongolian Scientific Research Geological Expedition, Transactions, v. 36, p. 67–85.

Yarmolyuk, V.V., Kovalenko, V.I., Kovach, V.P., Budnikov, S.V., Kozakov, I.K., Kotov, A.B., and Salnikova, E.B., 1999, Nd-isotopic systematics of western Transbaikalian crustal protoliths: Implications for Riphean crust formation in central Asia: Geotectonics, v. 33, p. 271–286.

Yarmolyuk, V.V., Kovalenko, V.I., Sal'nikova, E.B., Kozakov, I.K., Kotov, A.B., Kovach, V.P., Vladykin, N.V., and Yakovleva, S.Z., 2005, U-Pb age of syn- and postmetamorphic granitoids of South Mongolia: Evidence fort he presence of Grenvillides in the Central Asian foldbelt. Doklady Earth Sciences, v. 404, p. 986–990 (translated from Russian).

Zonenshain, L.P., Kuzmin, M.I., and Natapov, L.M., 1990, Plate tectonics of the USSR: A plate-tectonic synthesis: American Geophysical Union Geodynamics Series, v. 21, 242 p.

Manuscript Accepted by the Society 1 February 2007

Printed in the USA

The Geological Society of America
Memoir 200
2007

Mid-crustal Late Cretaceous metamorphism in the Nason terrane, Cascades crystalline core, Washington, USA: Implications for tectonic models

Harold H. Stowell*
Gerrit R. Bulman
Department of Geological Sciences, University of Alabama, Tuscaloosa, Alabama 35487-0338, USA

Carlos A. Zuluaga
Departamento de Geosciencias, Universidad Nacional de Colombia, Bogotá, Colombia

Douglas K. Tinkham
Department of Earth Sciences, Laurentian University, Sudbury, Ontario P3E 2C6, Canada

Robert B. Miller
Department of Geology, San Jose State University, San Jose, California 95192-0102, USA

Ellen Stein
Department of Geological Sciences, University of Alabama, Tuscaloosa, Alabama 35487-0338, USA

ABSTRACT

The Cordilleran Coast Plutonic Complex comprises the roots of a middle Cretaceous to Paleogene magmatic arc and orogenic belt that extends from the Yukon Territory to Washington State. Exceptional rock exposure and mineral preservation have made the Cascades crystalline core, the southernmost exposure of the Coast Plutonic Complex, a laboratory for understanding mid-crustal processes in contractional magmatic arcs. Perhaps surprisingly, after decades of study, fundamental tectonic models for the Late Cretaceous evolution of the core remain in question. This study evaluates tectonic models using phase equilibrium modeling, thermobarometry, and high-precision geochronology to constrain present crustal structure and the magnitudes, rates, and lateral preservation of Late Cretaceous regional metamorphism across the Nason terrane, Wenatchee block, Cascades crystalline core.

Garnet Sm-Nd ages of 88–86 Ma restrict the last regional metamorphism to no less than 3 m.y. after emplacement of the 96–91 Ma Mount Stuart batholith. These ages and petrologic data reflecting only negligible to moderate pressure increases (0–3.6 kbar) during garnet growth indicate a time lag between a fairly rapid pressure

*hstowell@geo.ua.edu

Stowell, H.H., Bulman, G.R., Zuluaga, C.A., Tinkham, D.K., Miller, R.B., and Stein, E., 2007, Mid-crustal Late Cretaceous metamorphism in the Nason terrane, Cascades crystalline core, Washington, USA: Implications for tectonic models, *in* Hatcher, R.D., Jr., Carlson, M.P., McBride, J.H., and Martínez Catalán, J.R., eds., 4-D Framework of Continental Crust: Geological Society of America Memoir 200, p. 211–232, doi: 10.1130/2007.1200(12). For permission to copy, contact editing@geosociety.org.

increase (up to 0.5 kbar/m.y.) subsequent to low-pressure contact metamorphism associated with emplacement of the Mount Stuart batholith and temperature increases during initial garnet growth. This thermal relaxation signature supports a thrust-loading model for post–Mount Stuart regional metamorphism. The lateral extent and magnitude of regional metamorphism across the Nason terrane and Mount Stuart domain offer additional support for a tectonic loading model for regional metamorphism. Peak recorded pressures decrease from >8 kbar in the northeast of the terrane to 5 kbar southwest of the Mount Stuart batholith, compatible with loading by a tapered thrust sheet, followed by exhumation and shortening after peak metamorphism. The lack of structural evidence for thrusting and steepening of the paleobarometric gradient across the Nason terrane north of the batholith suggest that strain-partitioned folding was dominant at the exposed crustal level during and after the last regional metamorphism. Thus, a tectonic model compatible with metamorphic *P-T* data may include a decoupling of upper-crustal and mid-crustal shortening accommodation mechanisms during Late Cretaceous regional metamorphism.

Keywords: metamorphic P-T paths, tectonics, Chiwaukum Schist, Cascades.

INTRODUCTION

The Cascades crystalline core (Cascades core) of north-central Washington State, USA, is the southernmost exposure of the Cordilleran Coast Plutonic Complex (Fig. 1). Unlike many of the rocks in the northern Coast Mountains, rocks within the Cascades core, and particularly the southern part of the core, contain well-preserved evidence for the Late Cretaceous history of the orogen.

The southern part of the Cascades core includes pelitic metamorphic rocks known as the Chiwaukum Schist and migmatitic schist and gneiss known as the Nason Ridge Migmatitic Gneiss (Fig. 2A). Three metamorphic events have been identified. The earliest metamorphic minerals (M_1) predate intrusion of Late Cretaceous plutons. Andalusite and associated minerals, common near Late Cretaceous plutons, have a contact metamorphic origin and are defined as M_2 (Evans and Berti, 1986; Evans and Davidson, 1999). Later regional metamorphic minerals (M_3), including sillimanite and kyanite, postdate these plutons. Thermobarometric data indicate a pressure gradient from <5 kbar in the south to >8 kbar in the north (Brown and Walker, 1993; Evans and Davidson, 1999; this study, Fig. 2). Mineral textures and compositions in the schist indicate a pressure increase during metamorphism. Evidence for this pressure increase includes kyanite replacing andalusite along the margins of ca. 96–90 Ma plutons that intrude the Chiwaukum Schist and Nason Ridge Migmatitic Gneiss (Evans and Berti, 1986; Brown and Walker, 1993; Paterson et al., 1994) and grossular-rich rims on garnet crystals, which have been used to construct pressure-temperature (*P-T*) paths with pronounced pressure increases (Brown and Walker, 1993; Whitney et al., 1999). Magnitudes of 4–6 kbar have been interpreted for the pressure increase during loading (Brown and Walker, 1993; Whitney et al., 1999), but no systematic geographic variation in the magnitude or timing of loading has been published. Tectonic models have proposed that pressure increases resulted from intra-arc shortening (e.g., Rubin et al., 1990) or thrust loading during collision of the Insular and Intermontane superterranes (e.g., Monger et al., 1982; Whitney and McGroder, 1989; McGroder, 1991). Miller et al. (2003) showed that loading postdates emplacement of most middle Cretaceous plutons, suggesting that intra-arc loading occurred after terrane amalgamation because these plutons stitch terrane boundaries (Tabor et al., 1987a, 1987a). In contrast to thrust loading, Brown and Walker (1993) proposed a magma loading model in which pressure increased due to movement of magma into the upper crust, and that the resulting plutons were subsequently eroded from above the Chiwaukum Schist.

This paper presents new (*P-T*) paths for the Nason terrane—a lithotectonic terrane composed of Tonga Formation, Chiwaukum Schist, and Nason Ridge Migmatitic Gneiss situated between the Mount Stuart batholith and the White River shear zone (Figs. 2A and 2B). These data indicate that pressure increases during garnet growth varied from 0 to as much as 3.6 kbar, and that total pressure increases for metamorphism (M_2 plus M_3) were likely substantially greater in the southwestern part of the Nason terrane. M_3 *P-T* paths presented here are based on garnet zoning and on the *P-T* stability of peak mineral assemblages. In contrast to several earlier studies, observed garnet compositional zoning lacks near-rim grossular increases and, thus, does not require a large pressure increase during garnet growth. Finally, *P-T* paths for metamorphism and geochronologic data for metamorphism and intrusion are used to evaluate models for Late Cretaceous regional metamorphism.

REGIONAL SETTING

The Cordilleran Coast Plutonic Complex comprises the roots of a >1500 km long middle Cretaceous to Paleogene magmatic arc and orogenic belt that extends from Washington to the Yukon. This arc separates the Intermontane superterrane on the east from

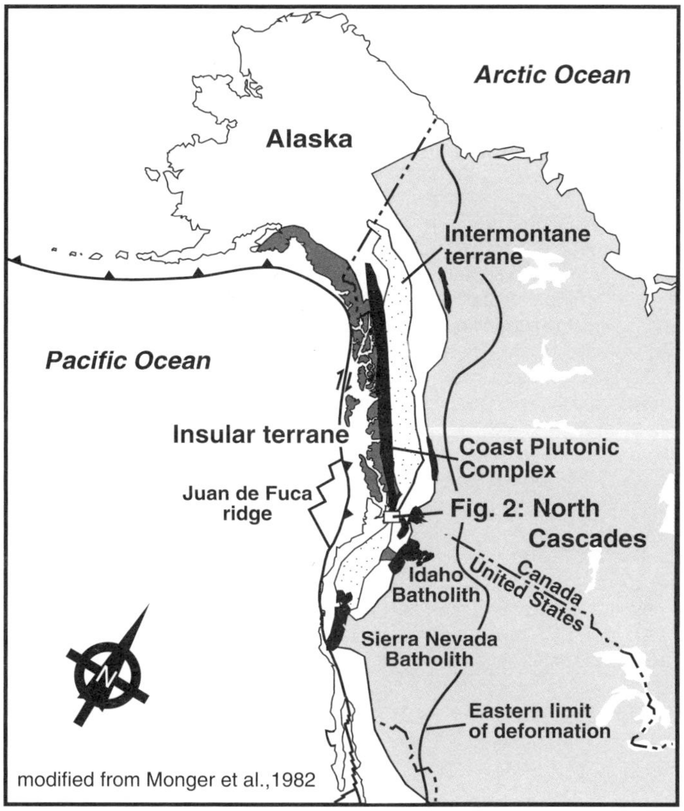

Figure 1. Tectonic map of the northern Cordillera. Note the position of the North Cascades at the southern terminus of the Coast Mountains orogen. The location of Figures 2A and 2B are indicated by the box.

the Insular superterrane on the west (Monger et al., 1982). The North Cascades crystalline core is the southernmost exposure of the Coast Plutonic Complex in Washington (Fig. 1).

The Cascades core experienced a period of major contractional orogenesis in the middle Cretaceous, resulting in a prominent northwest-southeast-trending structural and lithologic fabric. Northeast-southwest-directed shortening resulted in thrusting of the Ingalls Ophiolite Complex over the Chiwaukum Schist on the southwest-dipping Windy Pass thrust (Miller, 1985; Paterson et al., 1994), followed by southwest-directed thrusting, shearing, and southwest-vergent folding (Paterson and Miller, 1998; Miller and Paterson, 2001; Miller et al., 2006). This shortening and subsequent extension resulted in development of several prominent shear zones, including the White River shear zone, a major terrane-bounding reverse shear zone that separates the Napeequa Complex from the Nason terrane to the southwest (Fig. 2). Structural levels in the Cascades core are, from lowest to highest, (1) Swakane Gneiss and adjacent Napeequa Complex (Chelan Mountains terrane), above the White River shear zone, (2) Chiwaukum Schist and Nason Ridge Migmatitic Gneiss (Nason terrane), and (3) the Ingalls Ophiolite Complex above the Windy Pass thrust. The Swakane Gneiss is a thick and homogenous quartzo-feldspathic unit with uncertain protolith. The adjacent Napeequa Complex is a sequence of amphibolite and chert inferred to represent an oceanic assemblage. The lowermost structural level in the Nason terrane is composed of the 93 Ma muscovite-rich trondhjemite of the Wenatchee Ridge Gneiss, which cores a northwest-trending antiform. The gneiss contains a well-developed subsolidus fabric that is transposed into late upright folds. Therefore, significant strain occurred after 93 Ma and prior to folding. The Wenatchee Ridge Gneiss is overlain by the kyanite-rich Nason Ridge Migmatitic Gneiss, which, in turn, is flanked by the pelitic Chiwaukum Schist to the northeast and southwest.

The southern Wenatchee block experienced a complicated polymetamorphic history. Three metamorphic events, M_1 to M_3, have been identified (Plummer, 1980; Evans and Berti, 1986; Paterson et al., 1994; Evans and Davidson, 1999) (Table 1) according to the regional distribution of andalusite, sillimanite, kyanite, and staurolite in pelitic schist (Paterson et al., 1994; Evans and Davidson, 1999; Tinkham, 2002). The earliest metamorphic minerals (M_1) predate intrusion of Late Cretaceous plutons, including the Mount Stuart batholith. Andalusite, common near several Late Cretaceous plutons (e.g., the Mount Stuart batholith), is interpreted to have a contact metamorphic origin and defines the low-pressure M_2 event (Evans and Berti, 1986; Evans and Davidson, 1999), as do cordierite-bearing assemblages found locally between the two main plutons of the Mount Stuart batholith (Whitney et al., 1999). Metamorphic rocks north of the Mount Stuart plutons in the Chiwaukum Schist and Nason Ridge Migmatitic Gneiss contain sillimanite or kyanite and do not contain andalusite or cordierite. Petrographic, thermobarometric (e.g., Evans and Berti, 1986; Evans and Davidson, 1999), and geochronologic data (e.g., Stowell and Tinkham, 2003) indicate that these mineral assemblages postdate intrusion of the Mount Stuart batholith and define a medium- to high-pressure metamorphic event (M_3). Previous thermobarometric studies (e.g., Brown and Walker, 1993; Paterson et al., 1994; Evans and Davidson, 1999) indicate that pressures recorded during M_3 metamorphism increase from the southwest to the northeast across the Nason terrane; however, the magnitude of the M_3 lateral pressure gradient, particularly across the Mount Stuart domain, has been controversial. Evans and Davidson (1999) identified evidence for cryptic higher-pressure M_3 regional metamorphism from veinlets in a few locations in the Mount Stuart domain. These data necessitate a revision of the Brown and Walker (1993) lateral pressure gradient, which relies on data from low-pressure assemblages likely associated with earlier M_2 contact metamorphism. A revised baric gradient is proposed here that incorporates new data and those from Evans and Davidson (1999) (Fig. 2B).

The Wenatchee block northeast of the White River shear zone and the southern Chelan block (Fig. 2A) experienced higher-grade metamorphism than rocks to the southwest (Sawyko, 1994; Whitney et al., 1999). Presumably as a result of higher temperatures, little is known about the prograde metamorphic path for these rocks. Peak metamorphic conditions have been estimated

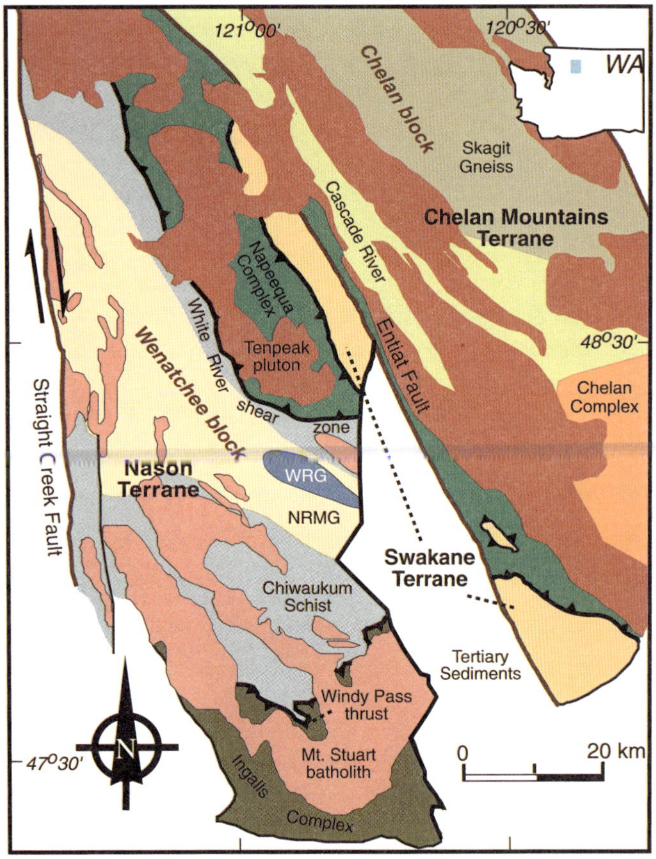

A

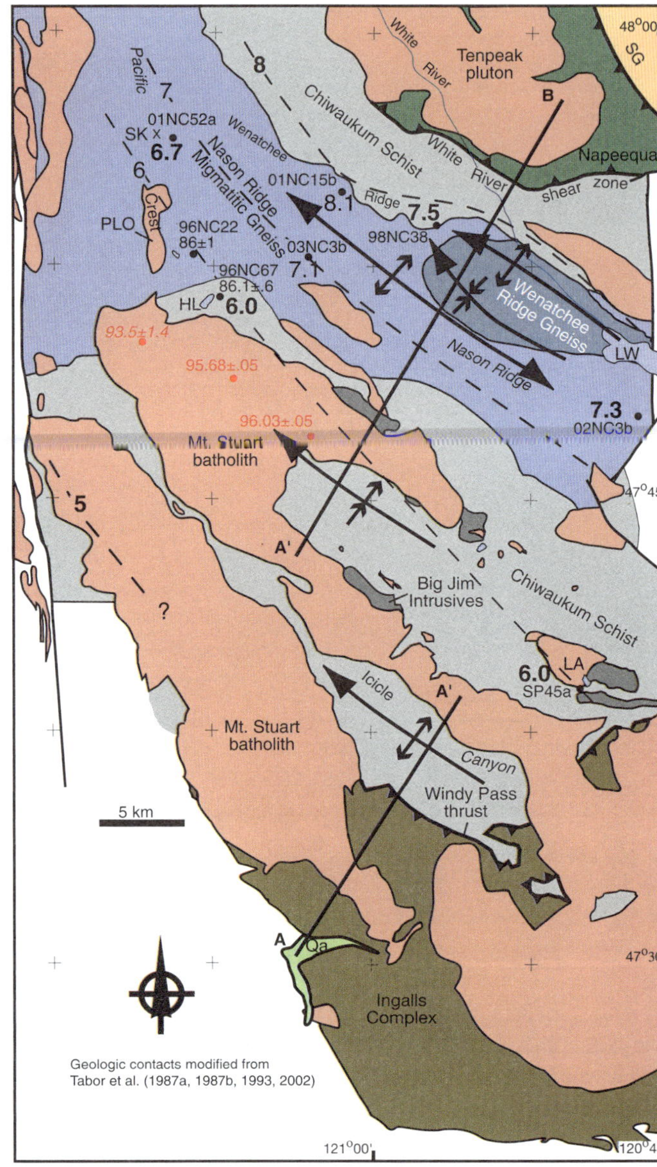

B

Figure 2. Geologic maps of the North Cascades crystalline core. (A) Generalized map of the Wenatchee and Chelan blocks. NRMG—Nason Ridge Migmatitic Gneiss; WRG—Wenatchee Ridge Gneiss. (B) Geologic map of the Wenatchee block. *P-T* paths for the samples shown on the map are presented in Figure 4. Cross sections A–A′ and A′–B are presented in Figure 6. Ages in black are garnet Sm-Nd (Ma) from Stowell and Tinkham (2003). Ages in red are U-Pb zircon from Matzel (2004) and Tinkham (2002)—in italics. Geologic contacts are modified from those in Tabor et al. (1982, 1987a, 1993, 2002). Metamorphic isobars (in kbar) are for M_3 metamorphism and are from Evans and Davidson (1999), Brown and Walker (1993), and data herein. HL—Heather Lake; LA—Lake Augusta; LW—Lake Wenatchee; PLO—Pear Lake Orthogneiss; Qa—Quaternary alluvium; SG—Swakane Gneiss; SK—Skykomish Peak.

at ~640–750 °C and up to ~12 kbar in the Swakane Gneiss and adjacent Napeequa Complex (Valley et al., 2003).

P-T-t PATHS FOR METAMORPHISM

P-T paths are presented for rocks from four regions between the Mount Stuart batholith and the White River shear zone: the northern edge of the Mount Stuart batholith, Nason Ridge, Pacific Crest, and Wenatchee Ridge areas (Fig. 2B; Table 2). *P-T* paths constructed for metamorphism are based primarily on compositional zonation preserved in garnet porphyroblasts and thermodynamic prediction of mineral stability and compositions. A summary of relevant mineral composition is presented in Appendix 1 tables. Observed garnet compositions are used with the compositions of plagioclase, biotite, and muscovite in the matrix and included within garnet in order to constrain pressure and temperature. The portion of a *P-T* path recorded during garnet growth is used independently for individual samples to interpret tectonic history. *P-T* paths are determined from *P-T* phase equilibrium diagrams constructed for the specific bulk-rock compositions of each sample (Appendix 2) (*P-T* pseudosections) using the internally consistent thermodynamic database of Holland and Powell (1998) with the silicate melts model extension (Holland and Powell, 2001; White et al., 2001) and version 3.2 of the program THERMOCALC (Powell and Holland, 1988; Powell et al., 1998). For a more complete discussion of the methodology see Stowell and Tinkham (2003) and Appendix 3 (this study). The *P-T* paths for garnet growth are constructed from two points that are derived, in part, from core and rim garnet compositions; therefore, finite *P-T* paths are constructed as vectors. Although the incremental *P-T* path may be more complicated, simple concentric garnet zoning in most samples (Fig. 3A) suggests that *P-T* paths are correspondingly simple. Plagioclase compositions in equilibrium with garnet are critical for pressure determinations and can be predicted by the pseudosection models. The observed compositions are summarized in Fig. 3B, and model predictions are discussed below. For a few samples, additional textural and/or mineral compositional data are available, and a larger portion of the metamorphic and tectonic history can be interpreted. *P-T* paths for some of the samples are integrated with garnet Sm-Nd ages from Stowell and Tinkham (2003) in order to constrain the timing of garnet growth and to construct quantitative pressure-temperature-time (*P-T-t*) paths.

Garnet almost certainly grew during all of the three metamorphic events recognized in the Wenatchee block (Table 1). Therefore, sampling techniques are critical for fully characterizing the events. Garnet samples analyzed for age determinations (Stowell and Tinkham, 2003) are nearly inclusion-free and subhedral to euhedral because these are the best samples for Sm and Nd isotope analysis. Therefore, one might expect a bias toward late (M_3) garnet. However, sampling for construction of *P-T* paths, which were not dated, had no such bias. Grains that are full of inclusions (e.g., 98NC38) provide better constraints on the *P-T* path because inclusions and adjacent garnet provide *P-T* points.

The Northern Edge of the Mount Stuart Batholith

The northern edge of the Mount Stuart batholith includes rocks from the Chiwaukum Schist within and adjacent to the contact metamorphic aureole of the batholith. The dominantly pelitic rocks within the aureole commonly contain andalusite and/or pseudomorphs after andalusite. Preintrusive M_1, contact M_2, and postintrusive M_3 metamorphism have been identified in these rocks (Kaneda, 1980; Paterson et al., 1994; Evans and Davidson, 1999; Stowell and Tinkham, 2003; Bulman, 2005).

Igneous and metamorphic ages indicate that M_3 metamorphism postdates intrusion of the tonalite plutons in the batholith. U-Pb zircon ages for the Mount Stuart batholith range from ca. 96 to 91 Ma (Matzel, 2004). Garnet Sm-Nd ages of the Chiwaukum Schist near the northwest margin of the batholith range from 88 to 86 Ma (Tinkham, 2002; Stowell and Tinkham, 2003). These ages clearly indicate that regional metamorphic heating and garnet growth postdate intrusion of Mount Stuart tonalite and associated contact metamorphism.

TABLE 1. METAMORPHIC EVENTS IN THE WENATCHEE BLOCK, NORTH CASCADES, WASHINGTON

Metamorphic and tectonic events	Metamorphic facies	Pressure (kbar)	Time (Ma)
M_1 regional metamorphism*	to amphibolite?	3–6?	103–143[†]
M_2 contact metamorphism	to amphibolite andalusite zone	<2–4	93–96
M_3 regional metamorphism	amphibolite	6–~9	86–88

*Regional extent, and temperatures and pressures, are poorly known. However, M_1 amphibole was described by Paterson et al. (1994).

[†]Age range is based on garnet Sm-Nd chronology (Magloughlin, 1993; Chace et al., 1998).

TABLE 2. SAMPLES FROM THE SOUTHERN NASON TERRANE USED TO CALCULATE *P-T* PATHS FOR METAMORPHISM

Sample	Latitude (NAD 27)	Longitude	Peak mineral assemblage (+ H_2O)	Rock type	*P* increase (kbar)
96NC67	47.8615	121.1050	Grt Bt Sil Pl Qtz	CS	≤1.2/2.1*
SP45a	47.6533	120.8380	Grt Bt Pl Qtz	GD	none
02NC3b	47.8096	120.8553	Grt Bt St Ky Ms Pl Qtz	NRMG	≤0.5
01NC52a	47.9562	121.1580	Grt Bt Sil Pl Qtz Liq	NRMG	n.a.
01NC15b	47.9184	121.0420	Grt Bt Ky Pl Qtz	NRMG	1.1–2.6
98NC38	47.9020	120.9600	Grt Bt St Ky Ms Pl Qtz	CS	≤0.5

Notes: P-T increases were determined from garnet zoning, *P-T* pseudosections, peak mineral assemblages, and mineral rim thermobarometry (see text). CS—Chiwaukum Schist; GD—granodiorite dike; NRMG—Nason Ridge Migmatitic Gneiss; n.a.—not applicable.

*Pressure increase during garnet growth / during And to Sil zone metamorphism.

Sample 96NC67

This sample of pelitic schist, collected near Heather Lake ~2 km from the Mount Stuart batholith (Fig. 2B), contains a peak mineral assemblage of garnet + biotite + sillimanite + plagioclase + quartz. Garnet porphyroblasts range up to 15 mm in diameter and have concentric zonation with spessartine decreasing from 10.4% to 1.6%, almandine increasing from 69.6% to 77.6%, and grossular ranging from 4.3% to 5.3% (Fig. 3A). Stowell and Tinkham (2003) presented a *P-T-t* path for this sample. Composition isopleths for garnet cores and rim thermobarometry indicate that garnet growth began at ~595 °C at 4.8 kbar and ended at ~670 °C at 4.9–6.7 kbar (Fig. 4A). Initial garnet growth was in the field of staurolite stability as indicated by the staurolite inclusions near the core of garnet crystals (Stowell and Tinkham, 2003) and supported by intersection of core garnet isopleths within the staurolite field on the *P-T* pseudosection. The pseudosection predicts plagioclase compositions ranging from an anorthite content of 31% during garnet core growth to 30% during garnet rim growth. These compositions are indistinguishable from those determined by electron probe microanalysis (EPMA) (Fig. 3B). The *P-T* path is compatible with a pressure increase of up to 1.2 kbar (disregarding uncertainties). However, given the uncertainties in *P* and *T* estimates for core and rim growth, the path does not require any change in pressure during garnet growth.

A *P-T* path that includes both ca. 93 Ma contact M_2 and later regional M_3 metamorphism can be constructed by considering the above data and M_2 mineralogy around the Mount Stuart batholith. Abundant pseudomorphs of kyanite ± garnet ± staurolite ± muscovite after andalusite are found in rocks closer to the Mount Stuart batholith and within 2 km of sample 96NC67. Evans and Berti (1986), Evans and Davidson (1999), and Paterson et al. (1994) summarize the distribution and nature of these replacement textures. Pseudosection models for Chiwaukum Schist samples indicate small andalusite-bearing *P-T* fields at temperatures above the aluminum silicate triple point (e.g., 96NC67; Fig. 4A). Assuming that the andalusite-bearing *P-T* fields for rocks containing the pseudomorphs are similar, loading was at least 2.1 kbar using the Pattison (2001) triple point at 4.4 kbar and 550 °C, or at least 3 kbar using the Holdaway (1971) triple point at 3.8 kbar and 507 °C.

Sm-Nd data for garnet in sample 96NC67 yielded an age of 86.1 ± 0.6 Ma (Stowell and Tinkham, 2003). Garnet core and rim segments define a single isochron within an uncertainty of 0.6 m.y., indicating rapid growth of the 12 mm crystals. A second garnet sample from ~3 km northwest (96NC22a; Fig. 2B) yielded an age of 86.0 ± 1.0 Ma. These precise ages, combined with a 93.5 ± 1.4 Ma U-Pb zircon age for the adjacent Mount Stuart batholith (Tinkham, 2002), are compatible with rapid M_3 garnet growth 5.1–9.9 m.y. after intrusion of the pluton.

The rate of crustal loading can be calculated from emplacement ages for the Mount Stuart batholith, garnet growth ages, and the total pressure increase estimated above. Utilizing 93.5 Ma (U-Pb zircon ~7 km from sample 96NC67; Tinkham, 2002) for the age of Mount Stuart batholith emplacement and andalusite growth, and the *P-T-t* path, the rate of loading would be ~0.7–1.4 km/m.y. Matzel (2004) reported a precise 96 Ma age for a sample of the Mount Stuart batholith ~10 km south of sample 96NC67. Using this age, loading rates could have been as low as 0.6 km/m.y. During garnet growth, loading rates are constrained to have been between 0 and 2.1 km/m.y.

Sample SP45a

This sample of metamorphosed granodiorite dike adjacent to intrusions in the Mount Stuart batholith near Lake Augusta (Fig. 2B) contains a peak mineral assemblage of garnet + biotite + plagioclase + quartz (Stowell and Stein, 2005). Garnet porphyroblasts range up to 3 mm in diameter and are surrounded by ellipsoidal plagioclase + quartz coronas. The garnet is concentrically zoned with spessartine decreasing from 9.7% to 1.7%, almandine increasing from 67.8% to 80.3%, and grossular decreasing from 15.7% to 8.1% (Fig. 3A). Coronas around garnet have been commonly interpreted as products of decompression; however, Stowell and Stein (2005) interpreted garnet zoning and corona textures as products of prograde growth of garnet and ferromagnesium depletion haloes, and not results of high-temperature decompression. Composition isopleths for garnet cores and rim thermobarometry suggest that garnet growth began at over 500 °C at 5–6.5 kbar and ended above 600 °C (Fig. 4B). The pseudosection predicts plagioclase compositions ranging from anorthite contents of 35% to 45% for temperatures of garnet growth between 550 and 700 °C at 6 kbar. Observed compositions of metamorphic plagioclase vary from 40% to 41% anorthite (Fig. 3A), matching predictions at temperatures of ~670 °C in the peak mineral assemblage field. Garnet modes on the *P-T* pseudosection (Fig. 4B) require at least a small pressure increase during garnet growth in the peak mineral assemblage field (≤1 kbar).

Nason Ridge

Nason Ridge is a prominent east-west ridge ~3 km north of the Mount Stuart batholith (Fig. 2B). It is almost exclusively underlain by rocks of the Nason Ridge Migmatitic Gneiss. These rocks comprise pelitic schist and gneiss intruded by sills and dikes of tonalite and trondhjemite. Several crosscutting generations of intrusive bodies display medium-grained foliated to coarse nonfoliated pegmatitic textures (Tabor et al., 1993, 2002; Zuluaga, 2004). The unit is inferred to have originated from intrusion of at least one generation of tonalite magma into the Chiwaukum Schist followed by local generation of small volumes of partial melt (Zuluaga, 2004). The age of metamorphism in these rocks has not been well constrained. As discussed below, *P-T* paths constructed for rocks along the Pacific Crest and Nason Ridge are compatible with metamorphic mineral growth during M_3.

Sample 02NC3b

This sample of pelitic gneiss from near the eastern terminus of Nason Ridge (Fig. 2B) contains a peak mineral assemblage of garnet + staurolite + biotite + kyanite + muscovite + plagioclase + quartz. Garnet porphyroblasts range up to 5 mm in diameter and have strong concentric zonation with spessartine decreasing from 12.1% to 0.5%, almandine increasing from 60.5% to 75.3%, and grossular decreasing from 22.8% to 7.2% (Fig. 3A). Zuluaga et al. (2005) present a *P-T* pseudosection, which allows interpretation of a *P-T* path. Garnet core composition isopleths and rim thermobarometry indicate that garnet growth began at ~540 °C at 6.8 kbar and ended ~655 °C at 6.7–8.2 kbar (Fig. 4C). The pseudosection predicts plagioclase compositions ranging from an anorthite content of 29% during garnet core growth to 28% during garnet rim growth. EPMA indicates that the cores of plagioclase grains have anorthite contents of 19%–20%, and rims are 27% anorthite (Fig. 3A). The discrepancy between low anorthite plagioclase core compositions and pseudosection predictions for plagioclase compositions during garnet growth likely reflects preservation of plagioclase compositions that predate garnet growth. The *P-T* path yields a maximum permissible pressure increase of 1.4 kbar.

Pacific Crest

The Pacific Crest is the north-south ridge along the western edge of the Nason terrane (Fig. 2B). This ridge is largely underlain by the Nason Ridge Migmatitic Gneiss, although Chiwaukum Schist can be found at the northernmost end of the ridge (Fig. 2B), and an ~5 km^2 body of tonalite orthogneiss (Pear Lake Orthogneiss) intruded along the southernmost segment (Fig. 2B). Detailed relationships between this orthogneiss body and intrusive sheets in the Nason Ridge Migmatitic Schist have not been published and the ages are not known; however, the contact between the units is generally sharp, and there is a strong shear fabric on the west margin of the Pear Lake Orthogneiss.

Sample 01NC52a

This sample of pelitic schist from the central part of the Pacific Crest near Skykomish Peak (Fig. 2B) contains a peak mineral assemblage of garnet + biotite + sillimanite + plagioclase + quartz. In addition, melt is inferred to have been present based on the ubiquitous presence of thin quartz + plagioclase lenses. Textural and compositional criteria suggest that only small volumes of melt were produced and that little or no melt migration occurred (Zuluaga, 2004). Garnet is concentrically zoned with spessartine decreasing from 11.6% to 1.4%, almandine increasing from 64.0% to 72.2%, and grossular decreasing from 16.3% to 5.6% (Fig. 3A). Zuluaga (2004) presents a *P-T* pseudosection that constrains the peak metamorphic conditions; however, garnet core compositional isopleths do not intersect. This could result from a change in the bulk chemical composition after initial metamorphic mineral growth, disequilibrium between initial garnet and other minerals in the rock, and/or elemental diffusion within garnet at peak temperatures. Regardless, no *P-T* estimate can be made for core garnet growth, and this precludes construction of a quantitative *P-T* path for this rock. However, mineral rim compositions are uniform in the rock, and rim *P-T* estimates of >663 °C at 6.1–7.7 kbar (Fig. 4D) are consistent with regional metamorphic conditions determined from samples nearby. The pseudosection predicts plagioclase compositions with an anorthite content of ~43% during garnet rim growth. These compositions are higher than those determined by EPMA (~36% anorthite; Fig. 3B).

Wenatchee Ridge

Wenatchee Ridge stretches from northwest to southeast through the central part of the Nason terrane. Wenatchee Ridge is partly underlain by rocks of the Nason Ridge Migmatitic Gneiss (central portion) and partly underlain by Chiwaukum Schist (eastern and western ends) (Figs. 2A and 2B). Ages for intrusive igneous rocks and garnet growth on Wenatchee Ridge are unknown; however, textural foliations and helicitic structures in porphyroblasts suggest that garnet grew during M_3 metamorphism.

Sample 01NC15b

This sample of pelitic gneiss from near the middle of Wenatchee Ridge (Fig. 2B) contains a peak mineral assemblage of garnet + biotite + kyanite + plagioclase + quartz. Garnet porphyroblasts up to 3 mm in diameter show strong concentric zoning with spessartine decreasing from 17% to 2%, almandine increasing from 60% to 73%, and grossular decreasing from 22% to 9% (Fig. 3A). Zuluaga (2004) presented a *P-T* pseudosection for this sample. Garnet core composition isopleths and rim thermobarometry indicate that garnet growth began at ~545 °C at 6.6 kbar and ended ~680 °C at 7.5–10.2 kbar (Fig. 4E). The pseudosection predicts plagioclase compositions ranging from an anorthite content of 34% during garnet core growth to 38% during garnet rim growth. EPMA indicates that the cores of plagioclase grains have anorthite contents of 24%, and rims are 38% anorthite (Fig. 3B). The discrepancy between low anorthite plagioclase core compositions and pseudosection predictions for plagioclase compositions during garnet growth is inferred to reflect preservation of plagioclase that predate garnet growth. The *P-T* path yields a maximum permissible pressure increase of 3.6 kbar—the largest possible increase for metamorphism calculated in this study.

Sample 98NC38

This sample of pelitic schist from near the eastern terminus of Wenatchee Ridge (Fig. 2B) contains a peak mineral assemblage of garnet + staurolite + biotite + kyanite + muscovite + plagioclase + quartz. Garnet porphyroblasts up to 2.5 mm in diameter show strong concentric zoning with spessartine decreasing from 11.5% to 0.5%, almandine increasing from 60.1% to 77.1%, and grossular decreasing from 22.9% to 6.8% (Tinkham, 2002). Some garnet grains have irregular shapes and matching zonation that may result from coalescence of nuclei and/or modification after growth (Fig. 3A). These grains contain numerous inclusions of biotite, muscovite, plagioclase, kyanite, quartz, rutile,

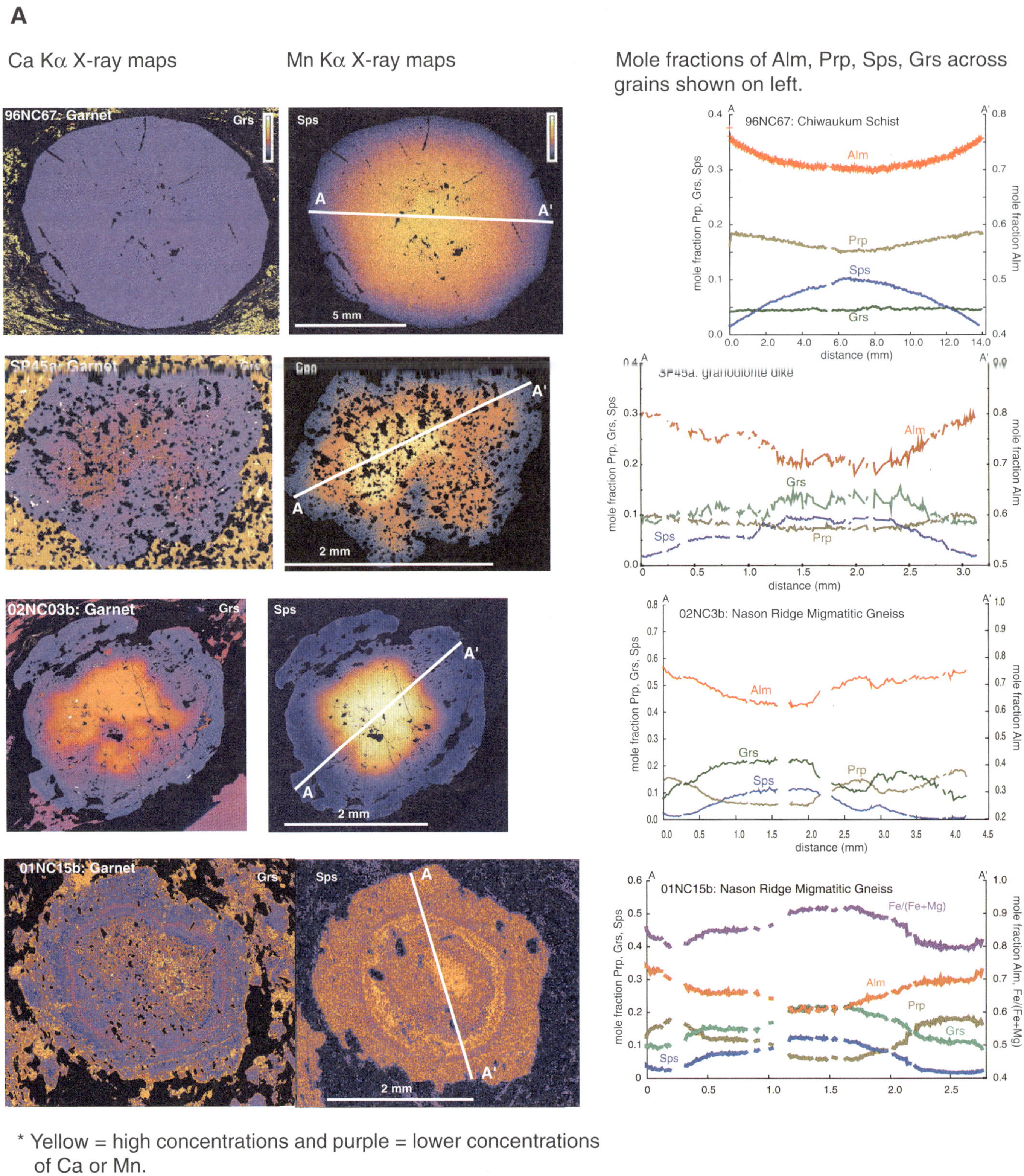

Figure 3 *(on this and following two pages)*. X-ray maps and mineral analyses taken across selected garnet and plagioclase grains from samples of Chiwaukum Schist and Nason Ridge Migmatitic Gneiss, Washington. (A) Garnet X-ray maps showing variation in Ca (Grs) and Mn (Sps). Yellow indicates high and purple indicates low concentrations. Mole fraction of garnet components is shown for lines from rim to rim across grains on left. (B) Plagioclase X-ray maps showing variation in Ca or Na. Light gray indicates high and dark grey indicates low concentrations. Mole fraction of anorthite is shown for lines from rim to rim across grains on right.

A

Ca Kα X-ray maps — Mn Kα X-ray maps — Mole fractions of Alm, Prp, Sps, Grs across grains shown on left.

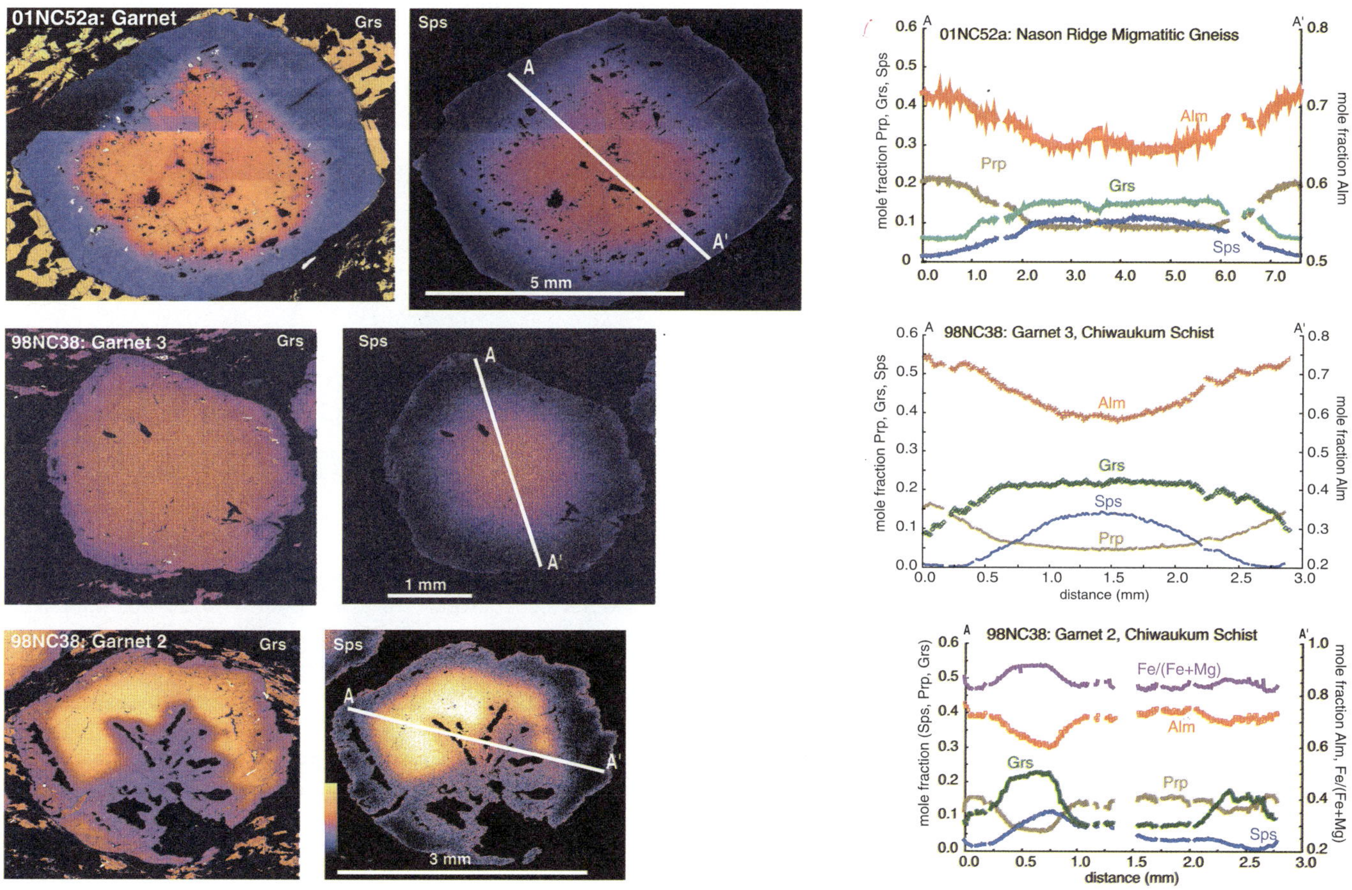

* Yellow = high concentrations and purple = lower concentrations of Ca or Mn.

B Mole fraction of An across grains shown on right. Ca or Na Kα X-ray maps.

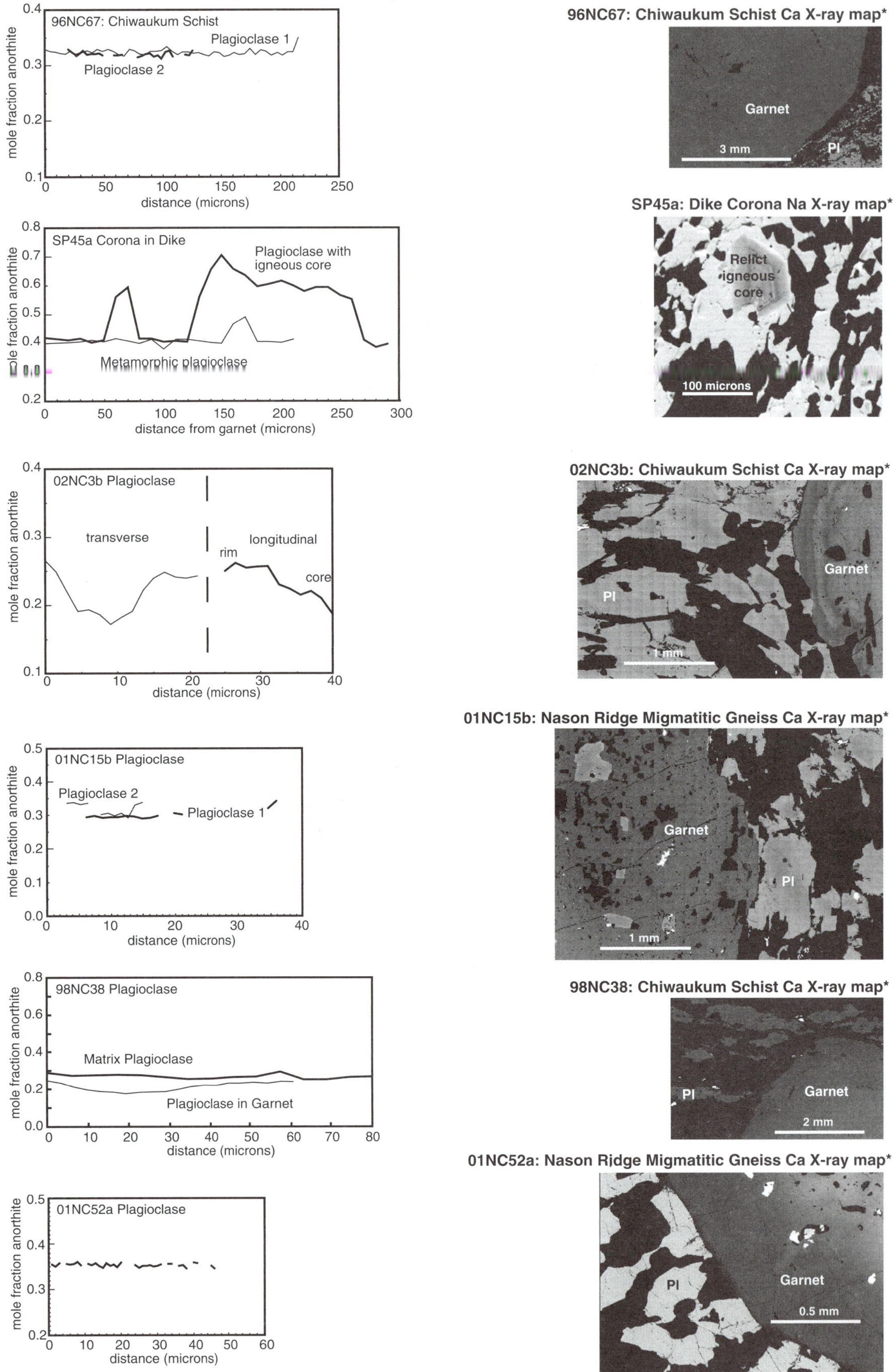

* Dark gray = low calcium concentrations and light gray = higher calcium concentrations.

and ilmenite (Fig. 5). Other garnet grains in the same thin section are subhedral with simple concentric zoning and are inferred to have developed during a single heating event (Tinkham, 2002). Tinkham (2002) and Miller et al. (2003) used these simply zoned garnet grains (Fig. 3A), the rock chemical composition, and phase equilibria to construct a *P-T* pseudosection and a garnet growth *P-T* path. Compositional isopleths for a garnet core and rim thermobarometry indicate that garnet growth began at ~535 °C at 6.8 kbar and ended ~600 °C at 5.5–8.8 kbar (Fig. 4F). EPMA indicates that the cores of plagioclase inclusions within garnet have anorthite contents of 19%, and rims are 24% anorthite (Fig. 3B). Low anorthite plagioclase core compositions are inferred to reflect preservation of plagioclase that predates garnet growth. Matrix plagioclase is nearly unzoned and is ~29% anorthite. The *P-T* path yields a maximum permissible pressure increase of 2 kbar.

Thermobarometry of mineral inclusions in garnet provide support for the *P-T* path determined from pseudosection phase equilibria. Biotite + muscovite + plagioclase and rutile + kyanite + ilmenite inclusions near the center of garnet (Fig. 5) indicate that garnet growth began at 530–540 ± 50 °C at 6.1–7.0 ± 1 kbar. Biotite inclusions record sequentially higher temperatures from core to rim, indicating that garnet growth continued during heating through 535, 540, 555, and 570 °C. Garnet and biotite rim thermometry indicates that garnet growth ended between 580 and 590 °C consistent with estimates based on the *P-T* stability of the peak mineral assemblage, presented above.

MAGNITUDES AND RATES OF CRUSTAL LOADING

Timing of M_3 Metamorphism

Garnet growth in the Icicle Canyon area between units of the Mount Stuart batholith (Fig. 2B) may have occurred prior to and during pluton emplacement. Low-precision garnet Sm-Nd ages (Magloughlin, 1993; Chace, 1998; Chace et al., 1998) for this region suggest a ca. 120 Ma metamorphic event (M_1; Stowell and Tinkham, 2003) prior to emplacement of the batholith. In contrast, high-precision garnet Sm-Nd ages for Chiwaukum Schist samples north of the Mount Stuart batholith (Fig. 2B) clearly indicate that much or all garnet growth postdates pluton emplacement at ca. 96 Ma (north) to ca. 91 Ma (south) (Matzel, 2004). Therefore, *P-T* paths presented here for the northern Mount Stuart batholith region can only elucidate the M_3 post–Mount Stuart history (after ca. 93 Ma) of the Nason terrane. Some of the garnet compositional zoning from Nason Ridge, the Pacific Crest, and Wenatchee Ridge is complex (Fig. 3A) and is compatible with multistage growth. Therefore, it is possible that the cores of some grains grew prior to M_3. However, *P-T* paths for Nason Ridge, Pacific Crest, and Wenatchee Ridge are inferred to represent M_3 metamorphism based on textural relationships and are assumed to be nearly synchronous with those for the northern Mount Stuart batholith region.

Magnitude of Loading During Metamorphism

The relatively small pressure increases determined for garnet growth *P-T* paths during M_3 metamorphism are of considerably lesser magnitude than those presented by Brown and Walker (1993), Brown et al. (1994), and Whitney et al. (1999). This may be attributed to (1) our choice of sampling localities, (2) comparison of M_3 loading (presented here) with M_2 plus M_3 loading (our interpretation of prior works), and/or (3) techniques. Many of the samples that were utilized in the earlier studies were taken from aureoles of ca. 90 Ma plutons. *P-T* paths presented here do not include samples from the northern Nason terrane (e.g., near the Eldorado pluton) studied by Brown et al. (1994). These authors inferred a pressure increase of 6 kbar during garnet growth in rocks northeast of the Entiat fault in the Chelan block. The earlier studies indicate pressure increases on the order of 4–5 kbar in rocks southwest of the Entiat fault in the Wenatchee block (see summary in Whitney et al., 1999). Garnet growth *P-T* paths for the Chiwaukum Schist along the northern edge of the Mount Stuart batholith (96NC67) and for the Nason Ridge Migmatitic Gneiss on Wenatchee Ridge (01NC15b) show the greatest permissible pressure increases. All but one (01NC15b) of the *P-T* paths indicate pressure increases that are considerably smaller than those determined in earlier studies. The *P-T* path for garnet growth in 96NC67 has a permissible pressure increase of ~1.5 kbar. Including the earlier pressures of M_2 andalusite stability in the *P-T* path enlarges the pressure increase to ~3 kbar of total M_2 plus M_3 loading. The most likely estimate of pressure increase, <3 kbar ignoring the uncertainties, is still considerably less than pressure increases reported in earlier studies. Although the permissible pressure increase of 3.6 kbar is large for 01NC15b, the preferred increase of ≤2.6 kbar (without uncertainties) and the *P-T* path for 98NC38 (nearby) suggest that pressure increases were less than 3.6 kbar. The larger pressure increases calculated in earlier studies used garnet core compositions combined with matrix mineral chemistry to determine early pressures. This differs considerably from the techniques reported here, which do not assume that any of the current matrix mineral chemistry was in equilibrium with garnet cores. Regardless, data presented here and earlier studies suggest that M_3 pressure increases recorded in garnet chemistry are most evident and perhaps of larger magnitude in the aureoles to ca. 90–96 Ma plutons. It is concluded that rocks outside aureoles of the plutons may not have been hot enough for mineral growth needed to record the early loading history.

Rates of M_3 Loading During Garnet Growth

Metamorphic and geochronologic data from the northern edge of the Mount Stuart batholith provide information on the rates of tectonic processes. Assuming the crystallization and garnet growth ages discussed above, minimum rates of loading directly after intrusion of the batholith must have been 0.2–0.4 kbar/m.y. (0.7–1.4 km/m.y.) using age uncertainties of 1 m.y. for

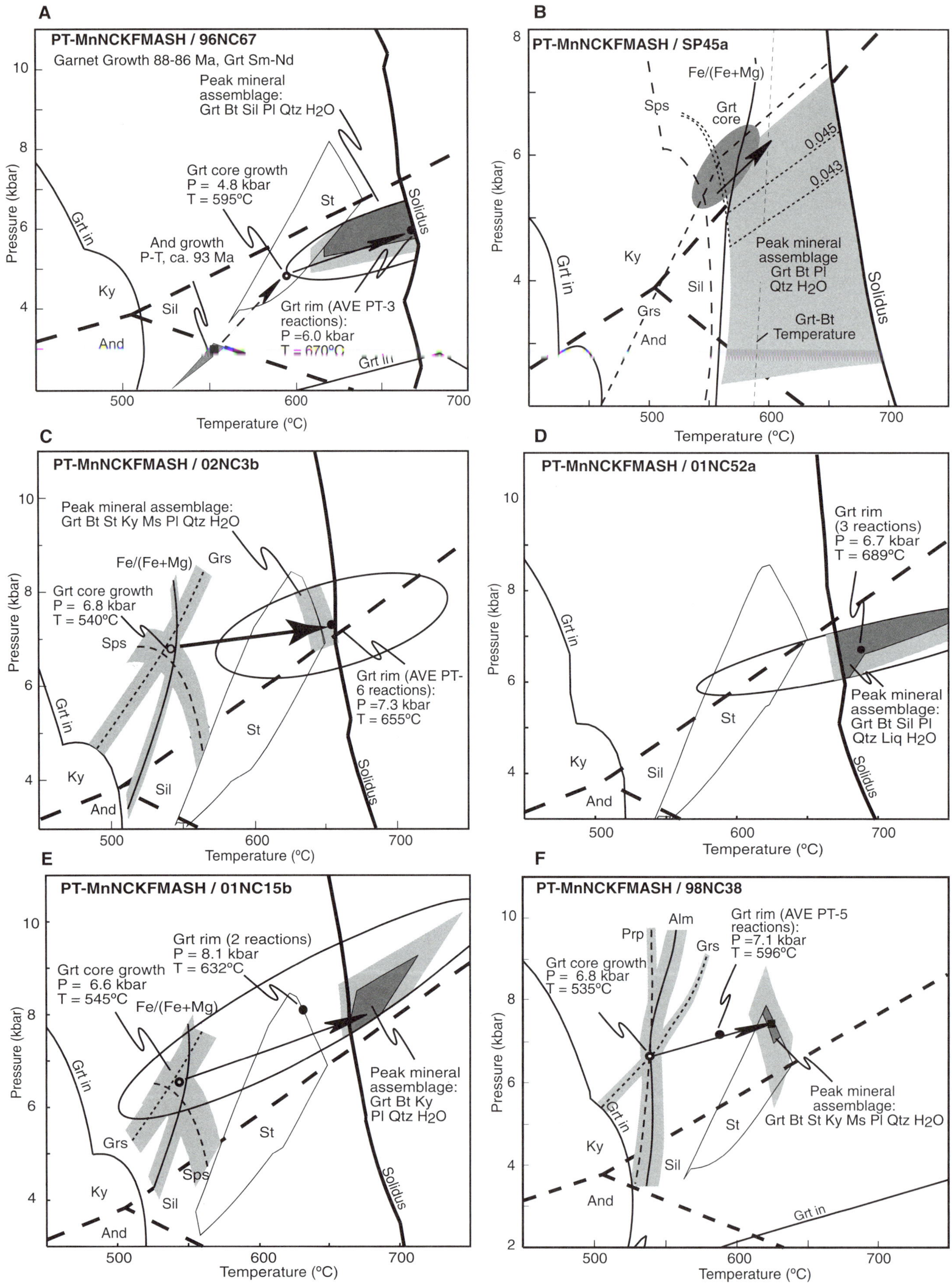
A
PT-MnNCKFMASH / 96NC67
Garnet Growth 88-86 Ma, Grt Sm-Nd
Peak mineral assemblage: Grt Bt Sil Pl Qtz H2O
Grt core growth P = 4.8 kbar T = 595ºC
And growth P-T, ca. 93 Ma
Grt rim (AVE PT-3 reactions): P =6.0 kbar T = 670ºC
Grt in
Ky
Sil
And
St
Solidus
Pressure (kbar)
Temperature (ºC)
B
PT-MnNCKFMASH / SP45a
Fe/(Fe+Mg)
Sps
Grt core
0.045
0.043
Peak mineral assemblage Grt Bt Pl Qtz H2O
Grt-Bt Temperature
Grt in
Ky
Sil
Grs
And
Solidus
Pressure (kbar)
Temperature (ºC)
C
PT-MnNCKFMASH / 02NC3b
Peak mineral assemblage: Grt Bt St Ky Ms Pl Qtz H2O
Fe/(Fe+Mg)
Grs
Grt core growth P = 6.8 kbar T = 540ºC
Sps
Grt rim (AVE PT-6 reactions): P =7.3 kbar T = 655ºC
Grt in
Ky
Sil
And
St
Solidus
Pressure (kbar)
Temperature (ºC)
D
PT-MnNCKFMASH / 01NC52a
Grt rim (3 reactions) P = 6.7 kbar T = 689ºC
Peak mineral assemblage: Grt Bt Sil Pl Qtz Liq H2O
Grt in
Ky
Sil
And
St
Solidus
Pressure (kbar)
Temperature (ºC)
E
PT-MnNCKFMASH / 01NC15b
Grt rim (2 reactions) P = 8.1 kbar T = 632ºC
Grt core growth P = 6.6 kbar T = 545ºC
Fe/(Fe+Mg)
Peak mineral assemblage: Grt Bt Ky Pl Qtz H2O
Grt in
Grs
Ky
Sps
Sil
And
St
Solidus
Pressure (kbar)
Temperature (ºC)
F
PT-MnNCKFMASH / 98NC38
Prp
Alm
Grs
Grt rim (AVE PT-5 reactions): P =7.1 kbar T = 596ºC
Grt core growth P = 6.8 kbar T = 535ºC
Peak mineral assemblage: Grt Bt St Ky Ms Pl Qtz H2O
Grt in
Ky
Sil
St
And
Grt in
Pressure (kbar)
Temperature (ºC)

Figure 4. *P-T* paths for garnet-bearing samples from the Chiwaukum Schist and Nason Ridge Migmatitic Gneiss. All paths were derived from a "complete" (see Appendix 3) *P-T* pseudosection and were calculated for $a_{water} = 1$. Rock compositions used for pseudosection construction are tabulated in Appendix Table A2. *P-T* Garnet growth segments are derived from intersections of garnet core compositional isopleths, and final garnet growth conditions are determined from rim mineral thermobarometry and peak mineral assemblage stability fields. Light gray areas around the peak mineral assemblages are the uncertainties calculated from THERMOCALC. Staurolite-bearing mineral assemblage fields and the aluminum silicate univariant reactions are included for reference in the pelitic rock pseudosections. (A) *P-T* path for 96NC67, Heather Lake. Total path is constrained by kyanite pseudomorphs after andalusite (M_2) found in nearby rocks ~0.5 km from the Mount Stuart batholith, garnet core composition, mineral rim thermobarometry, and the peak mineral assemblage. Modified from Stowell and Tinkham (2003) and Miller et al. (2003). (B) *P-T* path for SP45a, Lake Augusta. Path is constrained by garnet core composition, mineral rim thermobarometry, and the peak mineral assemblage. Dashed lines, labeled 0.043 and 0.045, are molar amounts of garnet predicted from THERMOCALC. (C) *P-T* path for 02NC3b, Nason Ridge. Path is constrained by garnet core composition, mineral rim thermobarometry, and the peak mineral assemblage. (D) *P-T* estimate for 01NC52a, Pacific Crest. Peak metamorphic conditions are constrained by mineral rim thermobarometry and the peak mineral assemblage that includes melt. (E) *P-T* path for 01NC15b, Wenatchee Ridge. Path is constrained by garnet core composition, mineral rim thermobarometry, and the peak mineral assemblage that includes melt. (F) *P-T* path for 98NC38, Wenatchee Ridge. Path is constrained by garnet core composition, mineral rim thermobarometry, and the peak mineral assemblage. And—andalusite; Bt—biotite; Grs—grossular; Grt—garnet; Ky—kyanite; Ms—muscovite; Pl—plagioclase; Prp—pyrope; Qtz—quartz; Sil—sillimanite; Sps—spessartine; St—staurolite.

initial garnet growth and 1.4 m.y. for pluton emplacement. Loading during garnet growth must have been between 0 and 2.1 km/m.y.

A TECTONIC MODEL

The timing and nature of metamorphism in the North Cascades provides important information about the accretion of terranes and subsequent tectonic activity in the magmatic arc. Many middle Cretaceous plutons that stitch the North Cascades terranes must have been emplaced prior to much or all of M_3 metamorphism based on pluton and metamorphic ages. Therefore, loading of the Nason terrane occurred subsequent to terrane accretion (cf. Walker and Brown, 1991; Brown and Walker, 1993), and details of the event provide information about intra-arc mid-crustal processes.

P-T paths constructed for garnet growth in rocks along the northeast margin of the Mount Stuart batholith are compatible with moderate pressure increases of ~1 kbar during M_3. However, additional pressure increases during metamorphism are required when the early M_2 growth of andalusite in these rocks is considered. Therefore, post–Mount Stuart emplacement loading of at least 2 kbar is inferred for these rocks. *P-T* paths constructed for garnet growth in rocks due north of Heather Lake along the Pacific Crest and northeast of the Mount Stuart batholith on Nason and Wenatchee ridges show little or no pressure increases during garnet growth. Maximum pressure increases permitted by the models presented here range from 0 to ~3.5 kbar for sample 01NC15b from Wenatchee Ridge. No pre-garnet mineral assemblages have been found in rocks from these areas, so little can be inferred about this early segment of the *P-T* path.

The relatively small pressure increases determined for garnet growth *P-T* paths during M_3 metamorphism are compatible with most if not all of the garnet growth occurring after loading to near-peak metamorphic pressures. Heating subsequent to loading—or a time lag between pressure increase and the onset of garnet growth—suggests thermal relaxation characteristic of thrust-loaded terranes (England and Thompson, 1984; Thompson and England, 1984). One-dimensional thermal modeling provides a test to determine whether loading and subsequent heating from thermal relaxation could occur in the <5 m.y. interval required by ages for the Mount Stuart batholith and garnet growth. Models utilizing the finite-difference techniques described in Stowell and Pike (2000) indicate that loading from a 6–9 km thrust sheet (2–3 kbar of loading) could result in heating of the footwall rocks by 50–100 °C in 1–2 m.y. after instantaneous emplacement of the hanging wall. If rocks in the footwall were already at temperatures above 500 °C, then the modeled temperature increase

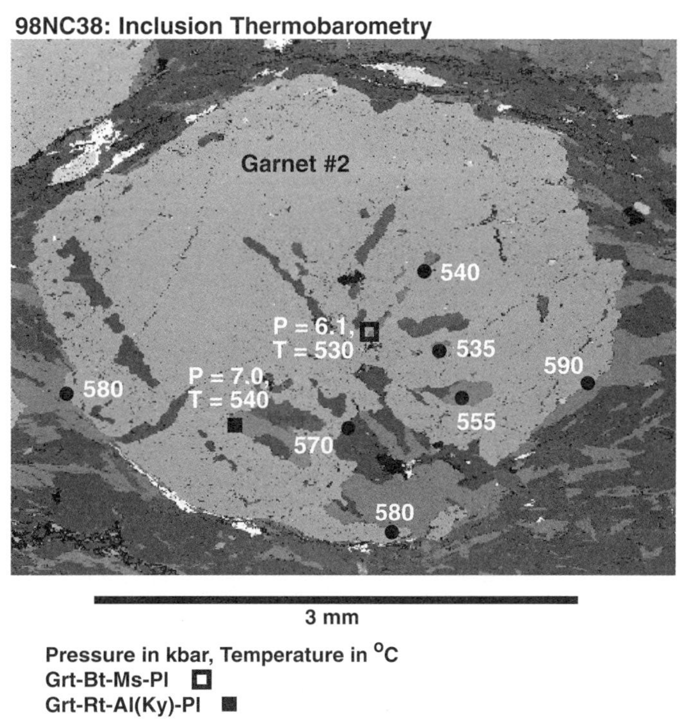

Figure 5. Backscattered electron image of garnet in sample 98NC38. Temperatures and pressures determined from thermobarometry are marked adjacent to the mineral inclusions used for the calculations. Al—aluminum silicate; Bt—biotite; Grt—garnet; Ky—kyanite; Ms—muscovite; Pl—plagioclase; Rt—rutile.

would be compatible with the M_3 garnet *P-T-t* path inferred here for Nason terrane rocks. Additional constraints on the possible cause for loading can be inferred from a thermal model for magma loading. In this model, a 4 km thick tabular body of tonalite with properties identical to those in Stowell and Pike (2000) is emplaced above the metamorphic rocks. The tonalite body is capped by 2 km of rock that initially has an equilibrium geothermal gradient typical of continental crust. Addition of heat from the horizontal pluton would cause rapid heating of the adjacent 5 km of metamorphic rocks below. Rocks 1 km and 2 km below the pluton would heat by up to 300 or >200 °C, respectively, in less than 1 m.y. So, in this case there would be essentially no time lag. Therefore, the ~2–4 m.y. age differences between M_3 garnet and known pluton ages are most compatible with thrust loading.

A useful tectonic model for the southern terminus of the Cascades core requires reconstruction of the Late Cretaceous to Paleogene crustal structure. Models rely on paleobarometric data that are tied to precise geochronology and structural relationships. Miller and Paterson (2001) presented a model for Paleogene crustal structure that placed high-pressure Swakane and Napeequa rocks northeast of the White River shear zone at the base of the North Cascades crust beneath the Wenatchee Ridge Gneiss. Miller et al. (2003) presented a revised model that takes into account recent identification of ca. 73 Ma detrital zircons in the Swakane Gneiss (Matzel et al., 2004). In this crustal section, no link is made between high-pressure metamorphism in rocks of the Swakane Gneiss and the medium-pressure metamorphism in rocks of the Wenatchee block southwest of the White River shear zone. The inferred sedimentary protolith for the Swakane Gneiss requires deposition of these rocks as sediment near the surface ca. 73 Ma, after M_3 metamorphism of the Chiwaukum Schist and Nason Ridge Migmatitic Gneiss. The schematic cross sections of the Wenatchee block presented in Figure 6 incorporate our paleobarometric data and all of the available structural data within the block. Near the structural base of the section, the Wenatchee Ridge Gneiss lies in the core of a late antiform near the highest pressure estimates within the overlying rocks (Fig. 6). Published cross sections constructed through these rocks have typically shown the Wenatchee Ridge Gneiss as the structurally lowermost unit in the Nason terrane (e.g., Tabor et al., 1987a; Miller and Paterson, 2001). Recent pressure estimates for pegmatite veins in this unit suggest an igneous crystallization pressure of >10 kbar (Magloughlin and Merkel, 2005), presumably ca. 93 Ma based on a U-Pb zircon age (J.P. Matzel, 2004, personal commun.). Pressure estimates for emplacement and metamorphism of the Wenatchee Ridge Gneiss are not available. However, pressure estimates presented here and the estimates for pegmatite emplacement are compatible with a large pressure difference between this unit and the overlying Nason Ridge Migmatitic Gneiss. This difference could result from fault juxtaposition or garnet growth in the Nason Ridge Migmatitic Gneiss after exhumation.

Structural and metamorphic data are compatible with the three tectonic models that are illustrated by cross sections in Figure 6. Near the base of the sections, a master thrust is shown as folded by late shortening (Fig. 6A) or as a late (reactivated?) structure (Figs. 6B and 6C). No data are available to constrain the timing of the most recent displacement on this fault; therefore, part or all of the 96–91 Ma Mount Stuart batholith may have intruded across this structure. Near the structural top of the sections, the Mount Stuart batholith is shown as thin sheets following the interpretation made by Paterson et al. (1994) for parts of the batholith. The first model (Fig. 6A) shows the Wenatchee Ridge Gneiss as a rootless antiform above a folded thrust. The thickness and geometry of this unit is unknown at depth; however, the Wenatchee Ridge Gneiss contains foliation and rotated porphyroclasts indicating intense deformation; therefore, Figure 6A shows this unit as a thin folded sheet containing a shear zone. In this shear model, the Wenatchee Ridge Gneiss could represent the lowermost Nason terrane above and/or within a detachment. The second model (Fig. 6B) shows the Wenatchee Ridge Gneiss within a horst block with ductile faults that separate the Nason Ridge Migmatitic Gneiss from rocks of Wenatchee Ridge Gneiss. This horst model implies distinct shear zones with opposite senses of shear on either side of the Wenatchee Ridge Gneiss, and ductile extension at high temperature. The third model (Fig. 6C) illustrates the Wenatchee Ridge Gneiss as a tabular igneous body that intruded the Chiwaukum Schist and was subsequently folded. This intrusive model interprets the Nason Ridge Migmatitic Gneiss as a contact zone of dikes and sills around the Wenatchee Ridge Gneiss, and high strain at the margins could be accounted for by rheological contrast during ductile folding. The cogenetic intrusive relationship between these two units implies a ca. 93 Ma, pre-M_3 age for at least some of the dikes and sills in the Nason Ridge Migmatitic Gneiss. The Napeequa Complex is shown at the base of the section following the paleobarometric and structural arguments invoked by Miller and Paterson (2001).

The models outlined above can be used to make predictions about deep crustal structure, ages, and small-scale structures around the Wenatchee Ridge Gneiss antiform. The rootless antiform and intrusive models show the Wenatchee Ridge Gneiss as a

Figure 6. Schematic cross sections from SW to NE across the Wenatchee block illustrating tectonic models for crustal structure. Folds, faults, and lithologic units are modified from Tabor et al. (1982, 1987a, 1993, 2002), Paterson et al. (1994), and Miller et al. (2000, 2003). Dashed lines indicate the approximate or inferred orientation of foliation. (A) Rootless antiform model. A thin layer of Wenatchee Ridge Gneiss, and overlying Nason Ridge Migmatitic Gneiss and Chiwaukum Schist, lie above a folded detachment. In this model, the thrust and adjacent rocks were folded during late-stage shortening. (B) Horst model. The Wenatchee Ridge Gneiss was uplifted along ductile faults with horst geometry. In this model, the Wenatchee Ridge Gneiss need not have any lithologic relation to the overlying Nason Ridge Migmatitic Gneiss. (C) Intrusive model. The Wenatchee Ridge Gneiss was intruded into the Chiwaukum Schist and then folded. Both units were folded with shearing along contacts. In this model, the Nason Ridge Migmatitic Gneiss represents a contact zone around the Wenatchee Ridge Gneiss.

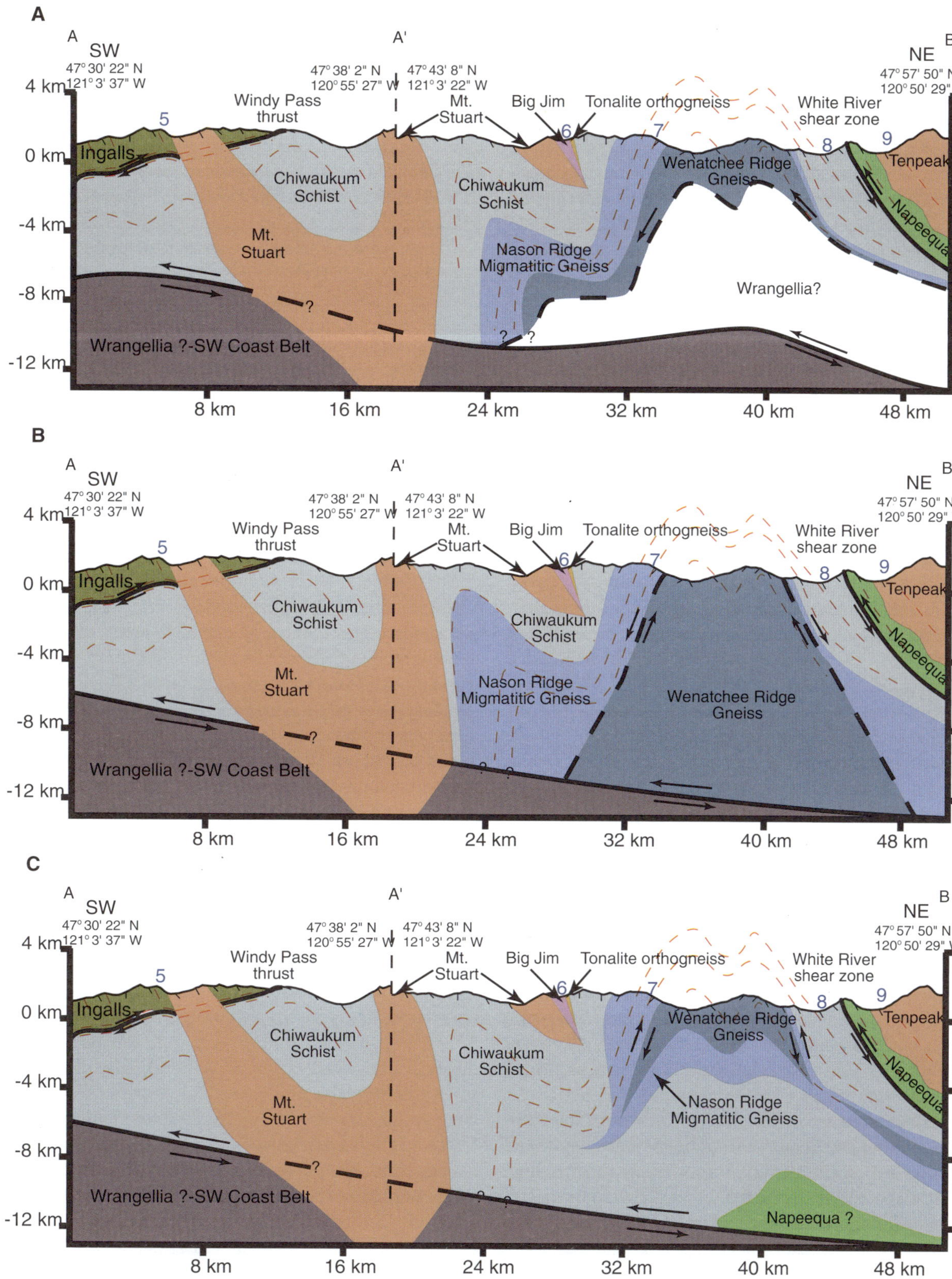

Vertical exaggeration = 1

5 = Pressure (kbar)

Moho depth ~ 35 km

tabular body based on its highly strained character and generally concordant contacts with the overlying Nason Ridge Migmatitic Gneiss. The intrusive model is compatible with Napeequa Complex at the base of the Nason terrane (Fig. 6A); however, the horst model, which is largely based on high-pressure mineralogy in the Wenatchee Ridge Gneiss, could include a thick sequence of Wenatchee Ridge Gneiss (Fig. 6B). The rootless antiform and horst models imply a change in shear sense across the antiform; however, the models predict opposite asymmetry on the limbs (Fig. 6). The horst model seems unlikely because high-temperature ductile extension implies early ca. 93 Ma extension that has not been documented in the North Cascades; therefore, the rootless antiform and intrusive models are preferred. Detailed structural analysis of the Wenatchee Ridge Gneiss and high-precision ages for intrusive rocks in the Nason Ridge Migmatitic Gneiss are needed to better constrain the tectonic models.

Assuming that the possible culmination in paleopressure within the Wenatchee Ridge Gneiss reflects the pressure at or about the time of intrusion at 93 Ma, then M_3 regional metamorphism is likely to be younger and may be interpreted independently. The Late Cretaceous M_3 pressure gradients across the Nason terrane can be interpreted to result from (1) rigid tilting of the crustal block, (2) tapering upper-crustal loading from northeast to southwest by magma or a thrust sheet, or (3) wedge thickening with variable shortening from northeast to southwest.

A tilted crustal block could only have horizontal, fossil pressure gradients up to the magnitude permitted by gravity and rock density: ~0.26 kbar/km for an average crustal density of 2700 kg/m^3. If the pressure gradients of ~0.15 kbar/km northeast of the batholith are a result of tilting alone, the crustal section must have been rotated more than 45° about a horizontal axis. No evidence readily supports this magnitude of rotation.

Loading during M_3 requires an overlying crustal block tapered from ~23 km thickness along the northeast margin of the Mount Stuart batholith to ~30 km thickness near the White River shear zone. In this model, loading resulted from a fault or faults that carried a thrust sheet onto the Wenatchee block (e.g., McGroder, 1991), magma that intruded above the current exposures (Brown and Walker, 1993), or a combination of the two. Regardless, the higher pressures observed in rocks to the northeast require significantly greater exhumation adjacent to the White River shear zone. In the case of thrust loading, this greater exhumation suggests a thrust sheet that moved from northeast to southwest that was subsequently preferentially eroded at the northeast end. The steep barometric gradient near the Mount Stuart batholith could result from a steep topographic slope in the overlying thrust wedge or differential shortening after loading. In the case of magma loading, the higher pressures to the northeast suggest greater volumes of magma and greater proximity to the source of magma. Sheeted intrusive rocks within the Nason Ridge Migmatitic Gneiss provide a possible root for now eroded plutons that loaded the Wenatchee block (Brown and Walker, 1993). However, the barometric gradient that, excepting possible high-pressure rocks in the Wenatchee Ridge Gneiss, continues across this unit with highest pressures to the northeast suggests that a more likely root zone for magma or a thrust would be northeast of the White River shear zone.

A shortened crustal wedge requires the same variation in exhumation noted for the loaded crustal block model above. However, in this model no overlying thrust is required. Instead, the fault or faults would lie beneath and/or within the current exposures in the Wenatchee block, and all of the loading would have resulted from thickening associated with internal shortening. The higher pressures to the northeast would require greater shortening and would be compatible with a wedge being transported from northeast to southwest. Internal deformation could be fold and/or fault related. This model is particularly difficult to evaluate due to a lack of unique stratigraphic markers in the Chiwaukum Schist, a unit for which quantitative shortening estimates are virtually absent, despite detailed structural work that has identified multiple generations of folding and significant shortening (Paterson et al., 1994; Miller et al., 2006). No faults or shear zones with well-constrained displacements have been identified between the Mount Stuart batholith and the White River shear zone; thus, by omission, fold-related shortening and thickening is currently the preferred strain accommodation mechanism at mid-crustal levels during loading. However, prefolding shortening by faulting cannot be dismissed. Potential structural evidence for coeval upper-crustal thrusting was eroded long ago, but as discussed above, a thrust-loading, thermal relaxation model is compatible with *P-T* data for peak metamorphism.

The observed proximity of rocks with *P-T* paths indicating large pressure increases to pluton aureoles (e.g., sample 96NC67 and Brown et al., 1994) suggests that rocks outside the aureoles were not hot enough for significant andalusite and garnet growth prior to M_3 loading. This is most compatible with rapid loading and subsequent heating. Because the high-pressure M_3 mineral assemblages are at least 5 m.y. younger than the Mount Stuart plutons, no link can be made between observed magma and metamorphism. Therefore, a tectonic loading model seems most plausible.

New pressure data for M_3 peak metamorphism integrated with M_3 pressure data from Evans and Davidson (1999) reveal a steep pressure gradient north of the Mount Stuart batholith and a moderate gradient across the Mount Stuart domain (Fig. 2B). The steep pressure gradient supports the Brown and Walker (1993) gradient of ~6 kbar north of the Mount Stuart batholith to ~8 kbar near the White River shear zone. However, the moderate gradient across the Mount Stuart domain differs appreciably from the steeper Brown and Walker (1993) gradient. The change in gradient between the batholith and the Nason terrane to the northeast may reflect an important relationship between structural competency and metamorphic gradient. Arc-normal contraction and consequent shortening was active during and after emplacement of the Mount Stuart batholith; strain is evident within both the batholith and host schist (Paterson et al., 1994; Paterson and Miller, 1998). However, as the batholith crystallized and became mechanically stronger, strain may have been preferentially accommodated in the weaker Chiwaukum Schist and Nason Ridge Migmatitic Gneiss (Miller et al., 1994; Paterson and Miller, 1998). This reasoning predicts that both shortening associated with M_3 loading and subsequent shortening, which may have effectively steepened the preserved baric gradient, should be more prominent northeast of the Mount Stuart domain. The baric gradients presented here fulfill this prediction and, when considered in conjunction with *P-T* data, support a model in which the middle crust experienced strain-partitioned, fold-related, tectonic thickening and shortening after and/or during thrust loading that caused M_3 regional metamorphism.

Nonetheless, alternative models for baric gradients and loading in the Nason terrane cannot be entirely dismissed. The models may be at least partly testable with additional data. For example, progressive movement of a southwest-directed thrust sheet above the Nason terrane should produce a progression of loading ages with the oldest metamorphism in the northeast, and magma loading models can be further evaluated with dates for intrusions in the Nason Ridge Migmatitic Gneiss.

APPENDIX 1. MINERAL COMPOSITIONS AND THERMOBAROMETRY

TABLE A1A. SUMMARY OF GARNET COMPOSITIONS (IN wt%) FOR METAMORPHIC ROCKS FROM THE NASON TERRANE, WASHINGTON

	96NC67		SP45a		01NC15b		01NC52a		02NC3b		98NC38	
Oxide	core	rim	core	rim	core[1] 5	rim 6	core 29	rim 9	core 16	rim 28	core	rim
SiO_2	36.77	37.25	36.42	36.35	36.71	37.31	37.13	37.66	36.80	37.03	36.19	37.49
Al_2O_3	21.52	21.23	20.93	21.06	21.20	21.66	21.74	22.04	21.44	21.78	21.32	21.65
FeO	31.21	33.93	31.37	35.03	26.45	31.74	28.71	32.24	27.58	33.47	27.06	33.21
MgO	3.79	4.64	1.79	2.32	1.33	4.32	2.30	5.44	1.38	4.24	1.22	3.41
MnO	4.50	0.72	3.86	0.75	5.59	0.98	5.30	0.71	4.96	0.55	6.25	1.26
CaO	1.55	1.64	4.17	2.78	8.06	3.59	5.33	2.33	7.76	2.91	7.83	3.15
Total	99.34	99.41	98.54	98.29	99.33	99.60	100.52	100.40	99.92	99.96	99.87	100.17
Cations based on 12 oxygens												
Si	2.97	2.99	2.98	2.98	2.96	2.97	2.96	2.96	2.96	2.95	2.93	2.99
Al	2.05	2.01	2.02	2.04	2.02	2.03	2.05	2.04	2.03	2.05	2.03	2.03
Fe	2.10	2.27	2.15	2.40	1.78	2.12	1.93	2.12	1.85	2.23	1.83	2.21
Mg	0.46	0.55	0.22	0.28	0.16	0.51	0.28	0.64	0.17	0.50	0.15	0.41
Mn	0.31	0.05	0.27	0.05	0.38	0.07	0.35	0.05	0.34	0.04	0.43	0.09
Ca	0.13	0.14	0.37	0.24	0.70	0.31	0.45	0.20	0.67	0.25	0.68	0.27
Total	8.01	8.01	8.01	7.99	8.00	8.01	8.01	8.01	8.01	8.02	8.04	7.99

[1]Number after core or rim designation refers to the number of analyses that were averaged.

TABLE A1B. SUMMARY OF PLAGIOCLASE COMPOSITIONS (IN wt%) FOR METAMORPHIC ROCKS FROM THE NASON TERRANE, WASHINGTON

Oxide	96NC67	SP45a	02NC3b	01NC52a	01NC15b	98NC38
SiO_2	59.67	55.64	60.78	58.38	58.72	62.06
Al_2O_3	25.89	27.65	25.00	26.65	26.14	25.26
FeO	0.10		0.00	0.00	0.00	0.00
CaO	6.53	8.28	5.40	7.34	6.65	5.15
Na_2O	7.72	6.73	8.39	7.44	7.73	8.68
K_2O	0.07	0.03	0.05	0.13	0.07	0.05
Total	99.98	98.33	99.62	99.95	99.32	101.20
Cations based on eight oxygens						
Si	2.66	2.53	2.70	2.61	2.63	2.72
Al	1.36	1.48	1.31	1.40	1.38	1.30
Fe	0.00		0.00	0.00	0.00	0.00
Ca	0.31	0.40	0.26	0.35	0.32	0.24
Na	0.67	0.59	0.72	0.64	0.67	0.74
K	0.00	0.00	0.00	0.01	0.00	0.00
Total	5.00	5.00	5.00	5.01	5.01	5.00

TABLE A1C. SUMMARY OF MICA AND STAUROLITE COMPOSITIONS (IN wt%) FOR METAMORPHIC ROCKS FROM THE NASON TERRANE, WASHINGTON

Oxide	96NC67 Bt	SP45a Bt	02NC3b Bt	02NC3b Ms	02NC3b St	01NC52a Bt	01NC15b Bt	98NC38 Bt
SiO_2	35.56	34.85	36.09	45.42	27.72	35.90	36.25	36.34
TiO_2	1.86	1.82	1.54	0.66	0.70	2.57	2.09	1.52
Al_2O_3	21.05	19.66	20.64	37.80	54.87	21.05	20.35	19.76
FeO	17.89	20.25	16.69	0.88	10.43	16.63	16.88	19.76
MgO	10.27	8.36	11.54	0.77	1.58	10.10	11.51	11.49
MnO	0.02	0.04	0.03	0.00	0.01	0.04	0.02	0.02
CaO	0.07	0.00	0.00	0.00	0.01	0.00	0.00	0.08
Na_2O	0.27	0.08	0.18	1.55	0.01	0.26	0.28	0.21
K_2O	8.28	9.49	8.93	8.76	0.00	8.95	8.56	7.58
ZnO	n.d.	n.d.	n.d.	n.d.	2.11	n.d.	n.d.	n.d.
Total	95.27	94.55	95.63	95.83	97.43	95.49	95.93	96.76
Cations based on 11 oxygens								
Si	2.66	2.75	2.68	2.98	7.67	2.67	2.68	2.69
Ti	0.11	0.11	0.09	0.03	0.15	0.14	0.12	0.09
Al	1.86	1.83	1.81	2.92	17.91	1.85	1.78	1.72
Fe	1.12	1.34	1.04	0.05	2.41	1.03	1.04	1.22
Mg	1.15	0.98	1.28	0.08	0.65	1.12	1.27	1.27
Mn	0.00	0.00	0.00	0.00	0.00	0.00	0.00	0.00
Ca	0.01	0.00	0.00	0.00	0.00	0.00	0.00	0.01
Na	0.04	0.01	0.03	0.20	0.00	0.04	0.04	0.03
K	0.79	0.48	0.85	0.73	0.00	0.85	0.81	0.72
Zn	n.d.	n.d.	n.d.	n.d.	0.43	n.d.	n.d.	n.d.
Total	7.72	7.48	7.76	6.99	29.23	7.71	7.74	7.74

Bt—biotite; Ms—muscovite; St—staurolite; n.d.—not determined.

TABLE A1D. SUMMARY OF AVERAGE *P-T* RESULTS FOR METAMORPHIC ROCKS FROM THE NASON TERRANE, WASHINGTON

Sample	Independent set of reactions for thermobarometry	Cor., SigFit*	*P*, *T* (kbar, °C)
96NC67	Grs + 2 Sil + Qtz = 3 An 3 East + Prp + 2 Grs + 6 Qtz = 6 An + 3 Phl 3 East + Alm + 2 Grs + 6 Qtz = 6 An + Ann + 2 Phl	0.760, 0.99	6.0 ± 1.0, 668 ± 59
SP45a	Ann + Prp = Alm + Phl	n.a.§	6†, 600 ± 50
02NC3b	2 Alm + Grs + 3 Cel = Phl + 2 Ann + 3 An + 6 Qtz 16 Grs + 48 Ky + 3 fSt + 12 Cel = 4 Alm + 48 An + 3 mSt + 12 Ms Alm + 6 Ilm + 3 Cel = Prp + 3 Ann + 3 Qtz + 6 Rt Ky + 9 Ilm + 3 Cel = Prp + 3 Ann + Qtz + 9 Rt 2 Prp + Grs + 2 Ann + 6 Rt = 2 Phl + 3 An + 3 Qtz + 6 Ilm 12 An + 12 Ilm + 3 mSt + 3 Cel = 5 Prp + 4 Grs + 12 Rt + 3 fSt + 3 Ms	0.549, 1.39	7.3 ± 1.1, 654 ± 81
01NC52a	Grs + 2 Sil + Qtz = 3 An Sil + 2 Qtz + 3 Ilm = Alm + 3 Rt Phl + 6 An + 3 Ilm = Prp + 2 Grs + Ann + 3 Sil + 3 Rt	0.846, 0.78	6.7 ± 1.0, 689 ± 117
01NC15b	Grs + 2 Ky + Qtz = 3 An Prp + Ann = Alm + Phl	0.931, n.a.	8.1 ± 2.3, 632 ± 132
98NC38	Grs + 2 Ky + Qtz = 3 An 3 Ilm + Ky + 2 Qtz = Alm + 3 Rt Qtz + 2 Ky + Phl = Ms + Prp 7 Qtz + 2 Ky + 3 East = 3 Ms + 2 Prp Qtz + 2 Ky + Ann = Ms + Alm	0.681, 0.99	7.5 ± 0.5, 585 ± 32

Note: Alm—almandine, An—anorthite, Ann—annite, Bt—biotite, Cel—celadonite, East—eastonite, fSt—iron staurolite, Grt—garnet, Grs—grossularite, Ilm—ilmenite, Ky—kyanite, Ms—muscovite, mSt—magnesium staurolite, Phl—phlogopite, Prp—pyrope, Qtz—quartz, Rt—rutile, Sil—sillimanite

*Correlation and significance of fit from THERMOCALC in average *P-T* mode.

†Assumed pressure. Activity models are the same as used for pseudosections and average *P-T* mode in THERMOCALC (see text).

§Not applicable.

APPENDIX 2. WHOLE-ROCK COMPOSITIONS FOR METAMORPHIC ROCKS FROM THE NASON TERRANE, WASHINGTON

TABLE A2. WHOLE-ROCK COMPOSITIONS USED IN PSEUDOSECTION CONSTRUCTION

	SiO_2	Al_2O_3	FeO	MgO	MnO	CaO	K_2O	Na_2O
96NC67	70.03	11.42	7.10	5.09	0.14	2.01	1.46	2.07
SP45a	70.77	11.42	5.06	2.41	0.10	4.91	1.09	4.25
02NC3b	70.84	10.42	6.05	4.43	0.10	2.81	1.48	3.24
01NC52a	68.00	12.17	3.36	5.22	0.10	3.36	1.32	3.01
01NC15b	74.87	8.66	5.46	4.30	0.08	2.85	1.21	2.58
98NC38	64.14	13.06	9.55	7.62	0.16	1.80	2.17	1.49

Notes: All values are in molar proportions. Analyses were normalized to 100% after removing TiO_2, P_2O_5, and H_2O.

APPENDIX 3: METHODS FOR CALCULATING *P-T* PATHS

Mineral Chemical Compositions

Quantitative mineral analyses and X-ray maps were collected with the JEOL 8600 electron probe microanalyzer at the University of Alabama using wavelength dispersion spectrometry. Major element analyses were collected with a 1–5 μm diameter beam at a current of 20 nA under a 15 kV accelerating potential. Trace element analyses were collected with a 1–5 μm diameter beam at a beam current of 50 nA under a 20 kV accelerating potential. Raw counts from characteristic X-ray peaks were converted to weight percent oxides by comparison to natural mineral and synthetic standards, using the CitZAF correction technique of Armstrong (1984). Count times ranged from 30 to 45 seconds. Operating conditions for collection of X-ray maps were 15 kV accelerating potential, 75–300 nA beam current, and a 1 μm beam. Count times ranged from 5 to 50 ms per pixel. Quantitative linescans across minerals were acquired to characterize chemical zoning. For garnet, X-ray maps were routinely acquired before linescans in order to determine where to collect linescans through the chemical core of garnet crystals, defined as the interior portion with the highest Mn content. X-ray maps and linescans were acquired from three or more of the largest garnets from each sample in order to identify the highest Mn concentrations, which are inferred to represent the initial garnet-growth core compositions. This strategy aids in finding the high-Mn cores, which are not always at the geometric center of grains. Near-rim garnet compositions with the highest Mg/Fe ratio were inferred to approximate those compositions present during the metamorphic peak.

Rock Chemical Compositions

X-ray fluorescence analyses were obtained at the University of Alabama, from fused disks. Bulk rock samples of ~30 g were used to obtain rock chemistry. Rock chips were ground on a diamond-embedded lap to remove surfaces that were cut by the rock saw or had signs of weathering, washed, and rinsed in acetone and 2 *M* HCl before jaw crushing and grinding to a powder in a steel ring-and-puck mill. Samples prepared and analyzed at the University of Alabama were mixed with several drops of saturated polyvinyl alcohol solution to enhance binding before pressing into a powder pellet. Pellets were analyzed on the Phillips PW2400 X-ray fluorescence spectrometer equipped with a Rh X-ray tube. Calibration was based on 15–20 certified rock standards per element.

Phase Equilibria Modeling

Pseudosections were constructed in the MnO-Na_2O-CaO-K_2O-FeO-MgO-Al_2O_3-SiO_2-H_2O (MnNCKFMASH) system using whole-rock compositions determined by X-ray fluorescence analysis, program THERMOCALC (v. 3.2), and the thermodynamic data set of Holland and Powell (1998) with the silicate melts model extension (Holland and Powell, 2001; White et al., 2001; th pdata files created 13 February 2002). The nine-component system was chosen for modeling because it is likely the minimum system necessary to quantitatively investigate and predict the composition of garnet in pelite (Vance and Mahar, 1998; Tinkham et al., 2001). Mineral equilibria were modeled with the mineral end members listed below. Activity models presented in Stowell and Tinkham (2003) were used for pseudosection construction. The *P-T* pseudosections were constructed assuming that the activity of water was equal to 1. If water activities were less than 1, the position of mode zero lines for hydrous phases would be shifted toward temperatures lower than those calculated.

P-T-t Path Construction

P-T pseudosections portray the stability fields of mineral assemblages and the compositions of solid solution phases at any *P-T* point for an effective bulk-rock composition. Therefore, these *P-T* pseudosections can be used to estimate *P-T* conditions for initial garnet growth by comparing sample and predicted garnet core compositions, following the procedures of Vance and Mahar (1998) and Stowell et al. (2001). MnNCKFMASH *P-T* pseudosections were calculated in order to estimate initial garnet growth *P-T* conditions. Garnet preferentially incorporates Mn during growth, and this fractionation combined with limited Mn availability in pelites results in crystals with Mn-rich cores. The net result for many unmodified crystals is a bell-shaped Mn growth zoning profile (Hollister, 1966). Therefore, the garnet core compositions are inferred to be those exhibiting the highest Mn content. Mn-rich garnet cores were first identified from X-ray maps, then quantitative analyses were collected along a line through the core. Five to ten analyses at the highest-Mn portion of the line were averaged for plotting compositional isopleths on *P-T* pseudosections. The range of pressure and temperature from the overlap of isopleths and their uncertainties, calculated from THERMOCALC, was used to determine precision for garnet core growth *P-T* estimates (e.g., Vance and Mahar, 1998). Realistic application of pseudosection modeling to construction of *P-T-t* paths for rocks requires estimating the composition of the equilibrium volume for reacting minerals at any point in time, a composition referred to as the effective bulk-rock composition. Zoned refractory phases, e.g., garnet, indicate that the effective bulk composition changed during the reaction history. However, Evans (2004) and Zuluaga et al. (2005) have shown that for many schist samples fractionation of less than 5% garnet is likely to produce little effect on a *P-T* pseudosection. Samples used here contain less than 5% garnet; therefore, whole-rock compositions obtained from X-ray fluorescence analysis were used to approximate the effective bulk-rock composition during initial garnet growth. Garnet near-rim *P-T* conditions were calculated using the multi-equilibrium thermobarometry approach (average *P-T*) described in Powell and Holland (1994) and program THERMOCALC. Garnet zoning for many samples shows small increases in Mn and Fe/(Fe + Mg) at the rim, suggesting minor diffusion re-equilibration (Fig. 3A). Therefore, near-rim garnet compositions, inside any chemical re-equilibration rind, were chosen to most closely represent compositions during the peak of metamorphism (Appendix 1 tables). This approximation may not provide peak compositions if a significant amount of garnet has been resorbed by net transfer reactions (Kohn and Spear, 2000). However, garnets discussed here show little or no Mn rise near the rim, and "retrograde" rims with lower Mg/(Mg + Fe) are less than 70 μm wide. Therefore, *P-T* estimates are likely to closely approximate peak compositions. Activity models used for thermobarometry are the same as those used in Stowell and Tinkham (2003). Mineral compositions, mineral reactions, and *P-T* results are in Appendix 1 tables. Precision for rim garnet pressure and temperature estimates was estimated from the intersection of independent equilibria calculated by THERMOCALC in average *P-T* mode (Powell and Holland, 1994). More precise estimates of peak metamorphic conditions were obtained from overlap of the uncertainty in the rim thermobarometry result and the pseudosection peak mineral assemblage field following the methods of Zuluaga et al. (2005). Inclusion thermobarometry for sample 98NC38 follows the methods of Ferry and Spear (1978), Hodges and Spear (1982), Hodges and Royden (1984), Hodges and Crowley (1985), and Bohlen et al. (1983).

Mineral End Members Considered in Phase Equilibria Modeling

- andalusite
- H_2O
- kyanite
- quartz
- sillimanite
- zoisite
- alkali feldspar
  - albite
  - sanidine
- biotite
  - annite
  - eastonite
  - Mn-biotite
  - ordered biotite
  - phlogopite
- chlorite
  - Al-free chlorite
  - amesite
  - clinochlore
  - daphnite
  - Mn-chlorite
- chloritoid
  - Fe-chloritoid
  - Mg-chloritoid
  - Mn-chloritoid
- cordierite
  - Fe-cordierite
  - hydrous cordierite
  - Mg-cordierite
  - Mn-cordierite
- garnet
  - almandine
  - grossular
  - pyrope
  - spessartine
- K-rich white mica
  - celadonite
  - Fe-celadonite
  - muscovite
  - paragonite
- Na-rich white mica
  - margarite
  - paragonite
- plagioclase
  - albite
  - anorthite
- staurolite
  - Fe-staurolite
  - Mg-staurolite
  - Mn-staurolite
- silicate melt
  - water
  - albite
  - anorthite
  - K-feldspar
  - quartz
  - sillimanite
  - forsterite
  - fayalite

ACKNOWLEDGMENTS

Helpful reviews from Lawford Anderson, Bernard Evans, and David Hirsch are greatly appreciated. The authors gratefully acknowledge assistance from S. Paterson. Research was supported by contributions from the Geological Society of America student research grant program (D.K.T., C.A.Z., and G.R.B.), Sigma Xi (D.K.T), the University of Alabama Department of Geological Sciences Hooks Fund (D.K.T., C.A.Z., G.R.B.), the University of Alabama Department of Geological Sciences Advisory Board (G.R.B.), the University of Alabama Philip E.

and Bunny LaMoreaux Foreign Geology Student Scholarship (C.A.Z.), and National Science Foundation grants EAR-0207777 (H.H.S.), EAR-9614758 (R.B.M.), and EAR-9980623 (R.B.M.). National Science Foundation grant EAR-9628232 provided funding for purchase of the X-ray fluorescence spectrometer used for bulk-rock analyses. Brent Miller provided valuable assistance in the isotope laboratory at the University of North Carolina, Chapel Hill. R. Chace, J. Hixon, K. Smith, C. Strickland, and K. Ouderkirk provided field assistance.

REFERENCES CITED

Armstrong, J.T., 1984, Quantitative analysis of silicate and oxide minerals: A reevaluation of ZAF corrections and proposal for new Bence-Albee coefficients, *in* Roming, A.D., Jr., and Goldstein, J.J., eds., Microbeam Analysis: San Francisco, California, San Francisco Press, p. 208.

Bohlen, S.R., Wall, V.J., and Boettcher, A.L., 1983, Experimental investigations and geological applications of equilibria in the system FeO-TiO_2-Al_2O_3-SiO_2-H_2O: American Mineralogist, v. 68, p. 1049–1058.

Brown, E.H., and Walker, N.W., 1993, A magma loading model for Barrovian metamorphism in the SE Coast Plutonic Complex, British Columbia and Washington: Geological Society of America Bulletin, v. 105, p. 479–500, doi: 10.1130/0016-7606(1993)105<0479:AMLMFB>2.3.CO;2.

Brown, E.H., Carey, J.A., Dougan, B.E., Dragovitch, J.D., Fluke, S.M., and McShane, D.P., 1994, Tectonic evolution of the Cascades crystalline core in the Cascade River area, Washington: Washington Division of Geology and Earth Resources Bulletin, v. 80, p. 93–113.

Bulman, G., 2005, Tectonic evolution of the Labyrinth Mountain area, Cascades crystalline core, Washington: Unraveling intertwined deformation, intrusion and metamorphism [M.S. thesis]: Tuscaloosa, Alabama, University of Alabama, 148 p.

Chace, R.J., 1998, Samarium-neodymium garnet geochronology and silicate mineral thermobarometry in the Nason and Chelan Mountains terranes of the North Cascades, Washington [M.S. thesis]: Tuscaloosa, Alabama, University of Alabama, 176 p.

Chace, R.J., Stowell, H.H., and Tinkham, D.K., 1998, The timing and conditions of high-pressure metamorphism in the Nason terrane, Cascades crystalline core, Washington: Geological Society of America Abstracts with Programs, v. 30, no. 5, p. 9.

England, P.C., and Thompson, A.B., 1984, Pressure-temperature-time paths of regional metamorphism, I: Heat transfer during the evolution of regions of thickened continental crust: Journal of Petrology, v. 25, p. 894–928.

Evans, B.W., and Berti, J.W., 1986, Revised metamorphic history for the Chiwaukum Schist, North Cascades, Washington: Geology, v. 14, p. 695–698, doi: 10.1130/0091-7613(1986)14<695:RMHFTC>2.0.CO;2.

Evans, B.W., and Davidson, G.F., 1999, Kinetic control of metamorphic imprint during synplutonic loading of batholiths: An example from Mount Stuart, Washington: Geology, v. 27, p. 415–418, doi: 10.1130/0091-7613(1999)027<0415:KCOMID>2.3.CO;2.

Evans, T.P., 2004, A method for calculating effective bulk composition modification due to crystal fractionation in garnet-bearing schist: Implications for isopleth thermobarometry: Journal of Metamorphic Geology, v. 22, p. 547–557, doi: 10.1111/j.1525-1314.2004.00532.x.

Ferry, J.M., and Spear, F.S., 1978, Experimental calibration of the partitioning of Fe and Mg between biotite and garnet: Contributions to Mineralogy and Petrology, v. 66, p. 113–117, doi: 10.1007/BF00372150.

Hodges, K.V., and Crowley, P.D., 1985, Error estimation and empirical geobarometry for pelitic systems: American Mineralogist, v. 70, p. 702–709.

Hodges, K.V., and Royden, L., 1984, Geologic thermobarometry of retrograded metamorphic rocks: An indication of the uplift trajectory of a portion of the northern Scandinavian Caledonides: Journal of Geophysical Research, v. 89, p. 7077–7090.

Hodges, K.V., and Spear, F.S., 1982, Geothermometry, geobarometry and the Al_2SiO_5 triple point at Mt. Moosilauke, New Hampshire: American Mineralogist, v. 67, p. 1118–1134.

Holdaway, M.J., 1971, Stability of andalusite and the aluminum silicate phase diagram: American Journal of Science, v. 271, p. 97–131.

Holland, T.J.B., and Powell, R., 1998, An internally consistent thermodynamic data set for phases of petrologic interest: Journal of Metamorphic Geology, v. 16, p. 309–343, doi: 10.1111/j.1525-1314.1998.00140.x.

Holland, T.J.B., and Powell, R., 2001, Calculation of phase relations involving haplogranitic melts using an internally consistent thermodynamic data set: Journal of Petrology, v. 42, p. 673–683, doi: 10.1093/petrology/42.4.673.

Hollister, L.S., 1966, Garnet zoning: An interpretation based on the Rayleigh fractionation model: Science, v. 154, p. 1647–1651, doi: 10.1126/science.154.3757.1647.

Kaneda, B.K., 1980, Contact metamorphism of the Chiwaukum Schist near Lake Edna, Chiwaukum Mountains, Washington [M.S. thesis]: Seattle, Washington, University of Washington, 150 p.

Kohn, M.J., and Spear, F., 2000, Retrograde net transfer reaction insurance for pressure-temperature estimates: Geology, v. 28, p. 1127–1130, doi: 10.1130/0091-7613(2000)28<1127:RNTRIF>2.0.CO;2.

Magloughlin, J.F., 1993, A Nason terrane trilogy; I, Nature and significance of pseudotachylyte; II, Summary of the structural and tectonic history; III, Major and trace element geochemistry and strontium and neodymium isotope geochemistry of the pseudosection and garnet Sm-Nd *P-T-t* paths, Chiwaukum Schist, amphibolite, and meta-tonalite gneiss of the Nason terrane [Ph.D. thesis]: Minneapolis, Minnesota, University of Minnesota, 325 p.

Magloughlin, J.F., and Merkel, I.S., 2005, Chemistry of multiple populations of magmatic epidote in tonalite pegmatite, North Cascade Mountains, Washington: Geological Society of America Abstracts with Programs, v. 37, no. 7, p. 308.

Matzel, J.P., 2004, Rates of tectonic and magmatic processes in the North Cascades continental magmatic arc [Ph.D. thesis]: Cambridge, Massachusetts, Massachusetts Institute of Technology, 249 p.

Matzel, J.P., Bowring, S.A., and Miller, R.B., 2004, Protolith age of the Swakane Gneiss, North Cascades, Washington: Evidence of rapid underthrusting of sediments beneath an arc: Tectonics, v. 23, no. 6, art. no. TC6009.

McGroder, M.F., 1991, Reconciliation of two-sided thrusting, burial metamorphism, and diachronous uplift in the Cascades of Washington and British Columbia: Geological Society of America Bulletin, v. 103, p. 189–209, doi: 10.1130/0016-7606(1991)103<0189:ROTSTB>2.3.CO;2.

Miller, R.B., 1985, The ophiolitic Ingalls Complex, north-central Cascade Mountains, Washington: Geological Society of America Bulletin, v. 96, p. 27–42, doi: 10.1130/0016-7606(1985)96<27:TOICNC>2.0.CO;2.

Miller, R.B., and Paterson, S.R., 2001, Influence of lithological heterogeneity, mechanical anisotropy, and magmatism on the rheology of an arc, North Cascades, Washington: Tectonophysics, v. 342, p. 351–370, doi: 10.1016/S0040-1951(01)00170-6.

Miller, R.B., Paterson, S.R., DeBari, S.M., and Whitney, D.L., 2000, North Cascades Cretaceous crustal section: Changing kinematics, rheology, metamorphism, pluton emplacement, and petrogenesis from 0 to 40 kilometers depth, *in* Woodsworth, G.J., Jackson, L.E., Jr., Nelson, J.L., and Ward, B.C., eds., Guidebook for geological field trips in southwestern British Columbia and northern Washington: Geological Society of America Field Guide, Cordilleran Section, Vancouver, April, p. 229–278.

Miller, R.B., Matzel, J.P., Paterson, S.R., and Stowell, H., 2003, Cretaceous to Paleogene Cascades Arc: Structure, metamorphism, and timescales of magmatism, burial, and exhumation of a crustal section, *in* Swanson, T.W., ed., Western Cordillera and adjacent areas: Geological Society of America Field Guide 4, p. 107–135.

Miller, R.B., Paterson, S.R., Lebit, H., Alsleben, H., and Luneburg, C., 2006, Significance of composite lineations in the mid to deep crust: A case study from the North Cascades, Washington: Journal of Structural Geology, v. 28, p. 302–322.

Monger, J.W.H., Price, R.A., and Templeman-Kluit, D.J., 1982, Tectonic accretion of the two major metamorphic and plutonic welts of the Canadian Cordillera: Geology, v. 10, p. 70–75, doi: 10.1130/0091-7613(1982)10<70:TAATOO>2.0.CO;2.

Paterson, S.R., and Miller, R.B., 1998, Magma emplacement during arc-perpendicular shortening: An example from the Cascades crystalline core, Washington: Tectonics, v. 17, no. 4, p. 571–586, doi: 10.1029/98TC01604.

Paterson, S.R., Miller, R.B., Anderson, J.L., Lund, S., Bendixen, J., Taylor, N., and Fink, T., 1994, Emplacement and evolution of the Mount Stuart batholith, *in* Swanson, D.A., and Haugerud, R.A., eds., Geologic field trips in the Pacific Northwest, 2: Annual Meeting of the Geological Society of America, 24–27 October, 1994: Department of Geological Sciences, Seattle, University of Washington, in conjunction with the Geological Society of America, p. 2F1–2F27.

Pattison, D.R.M., 2001, Instability of Al_2SiO_5 "triple point" assemblages in muscovite + biotite + quartz-bearing metapelites, with implications: American Mineralogist, v. 86, p. 1414–1422.

Plummer, C.C., 1980, Dynamothermal contact metamorphism superposed on regional metamorphism in the pelitic rocks of the Chiwaukum Mountains area, Washington Cascades: Geological Society of America Bulletin, v. 91, p. 386–388, doi: DOI: 10.1130/0016-7606(1980)91<386:DCMSOR>2.0.CO;2.

Powell, R., and Holland, T.J.B., 1988, An internally consistent data set with uncertainties and correlations, 3: Applications to geobarometry, worked examples and a computer program: Journal of Metamorphic Geology, v. 6, p. 173–204.

Powell, R., and Holland, T.J.B., 1994, Optimal geothermometry and geobarometry: American Mineralogist, v. 79, p. 120–133.

Powell, R., Holland, T., and Worley, B., 1998, Calculating phase diagrams involving solid solutions via non-linear equations, with examples using THERMOCALC: Journal of Metamorphic Geology, v. 16, p. 577–588, doi: 10.1111/j.1525-1314.1998.00157.x.

Rubin, C.M., Saleeby, J.B., Cowan, D.S., Brandon, M.T., and McGroder, M.F., 1990, Regionally extensive mid-Cretaceous west-vergent thrust system in the northwestern Cordillera: Implications for continent margin tectonism: Geology, v. 18, p. 276–280, doi: 10.1130/0091-7613(1990)018<0276:REMCWV>2.3.CO;2.

Sawyko, L., 1994, The geology and petrology of the Swakane Biotite Gneiss, North Cascades, Washington [M.S. thesis]: Seattle, Washington, University of Washington, 134 p.

Stowell, H.H., and Pike, M., 2000, A thermal model for cooling of foliated tonalite sill plutons, northern Coast Ranges SE Alaska, *in* Stowell, H.H., and McClelland, W.C., eds., Tectonics of the Coast Mountains, southeastern Alaska and British Columbia: Geological Society of America Special Paper 343, p. 183–192.

Stowell, H.H., and Stein, E., 2005, The significance of plagioclase-dominant coronas on garnet, Wenatchee block, northern Cascades, Washington, U.S.A.: Canadian Mineralogist, v. 42, p. 1191–1210.

Stowell, H.H., and Tinkham, D.K., 2003, Integration of phase equilibria modeling and garnet Sm-Nd chronology for construction of *P-T-t* paths: Examples from the Cordilleran Coast Plutonic Complex, USA, *in* Vance, D., et al., eds., Geochronology: Linking the isotopic record with petrology and textures: Geological Society [London] Special Publication 220, p. 119–145.

Stowell, H.H., Taylor, D.L., Tinkham, D.K., Goldberg, S.A., and Ouderkirk, K.A., 2001, Contact metamorphic *P-T-t* paths from Sm-Nd garnet ages, phase equilibria modeling, and thermobarometry, Garnet Ledge, southeastern Alaska: Journal of Metamorphic Geology, v. 19, p. 645–660.

Tabor, R.W., Waitt, R.B., Jr., Frizzell, V.A., Jr., Swanson, D.A., Byerly, G.R., and Bentley, R.D., 1982, Geologic map of the Wenatchee quadrangle, central Washington: U.S. Geological Survey Map I-1311, scale 1:100,000, 1 sheet, 26 p.

Tabor, R.W., Frizzell, V.A., Jr., Whetten, J.T., Waitt, R.B., Jr., Swanson, D.A., Byerly, G.R., Booth, D.B., Hetherington, M.J., and Zartman, R.E., 1987a, Geologic map of the Chelan 30′ by 60′ quadrangle, Washington: U.S. Geological Survey Map I-1661, scale 1:100,000, 1 sheet, 29 p.

Tabor, R.W., Zartman, R.E., and Frizzell, V.A., Jr., 1987b, Possible tectonostratigraphic terranes in the North Cascades crystalline core, Washington, *in* Schuster, J.E., ed., Selected papers on the geology of Washington: Washington Division of Geology and Earth Resources Bulletin, v. 77, p. 107–127.

Tabor, R.W., Frizzell, V.A., Jr., Booth, D.B., Waitt, R.B., Whetten, J.T., and Zartman, R.E., 1993, Geologic map of the Skykomish River 30- by 60-minute quadrangle, Washington: U.S. Geological Survey Map I-1963, scale 1:100,000, 1 sheet, 42 p.

Tabor, R.W., Booth, D.B., Vance, J.A., and Ford, A.B., 2002, Geologic map of the Sauk River 30- by 60-minute quadrangle, Washington: U.S. Geological Survey Map I-2592, 67 p.

Thompson, A.B., and England, P.C., 1984, Pressure-temperature-time paths of regional metamorphism, II: Their inference and interpretation using mineral assemblages in metamorphic rocks: Journal of Petrology, v. 25, p. 929–955.

Tinkham, D.K., 2002, MnNCKFMASH phase equilibria, garnet activity modeling, and garnet samarium-neodymium chronology: Applications to the Waterville Formation, Maine, and south-central Nason terrane, Washington [Ph.D. thesis]: Tuscaloosa, Alabama, University of Alabama, 176 p.

Tinkham, D.K., Zuluaga, C.A., and Stowell, H.H., 2001, Metapelite phase equilibria modeling in MnNCKFMASH: The effect of variable Al_2O_3 and MgO/(MgO + FeO) on mineral stability: Geological Materials Research, v. 3, no. 1, p. 1–42.

Valley, P.M., Whitney, D.L., Paterson, S.R., Miller, R.B., and Alsleben, H., 2003, Metamorphism of the deepest exposed arc rocks in the Cretaceous to Paleogene Cascades belt, Washington: Evidence for large-scale vertical motion in a continental arc: Journal of Metamorphic Geology, v. 21, p. 203–220, doi: 10.1046/j.1525-1314.2003.00437.x.

Vance, D., and Mahar, E., 1998, Pressure-temperature paths from *P-T* pseudosections and zoned garnets: Potential, limitations and examples from the Zanskar Himalaya, NW India: Contributions to Mineralogy and Petrology, v. 132, p. 225–245, doi: 10.1007/s004100050419.

Walker, N.W., and Brown, E.H., 1991, Is the southeast Coast Plutonic Complex the consequence of accretion of the Insular superterrane? Evidence from U-Pb zircon geochronometry in the northern Washington Cascades: Geology, v. 19, p. 714–717, doi: 10.1130/0091-7613(1991)019<0714:ITSCPC>2.3.CO;2.

White, R.W., Powell, R., and Holland, T.J.B., 2001, Calculation of partial melting equilibria in the system Na_2O-CaO-K_2O-FeO-MgO-Al_2O_3-SiO_2-H_2O (NCKFMASH): Journal of Metamorphic Geology, v. 19, p. 139–153, doi: 10.1046/j.0263-4929.2000.00303.x.

Whitney, D.L., and McGroder, M.F., 1989, Cretaceous crustal section through the proposed Insular-Intermontane superterrane suture: Geology, v. 17, p. 555–558, doi: 10.1130/0091-7613(1989)017<0555:CCSTTP>2.3.CO;2.

Whitney, D.L., Miller, R.B., and Paterson, S.R., 1999, *P-T-t* evidence for mechanisms of vertical tectonic motion in a contractional orogen, northwestern U.S. and Canadian Cordillera: Journal of Metamorphic Geology, v. 17, p. 75–90, doi: 10.1046/j.1525-1314.1999.00181.x.

Zuluaga, C.A., 2004, Thermodynamic modeling applied to metamorphic processes in migmatites: An example from the Nason Ridge Migmatitic Gneiss (Cascades core, Washington) [Ph.D. thesis]: Tuscaloosa, Alabama, University of Alabama, 176 p.

Zuluaga, C.A., Stowell, H.H., and Tinkham, D.K., 2005, The effect of zoned garnet on effective bulk rock chemical composition and metapelite pseudosection topology: Discussion and applications: American Mineralogist, v. 90, p. 1619–1628, doi: 10.2138/am.2005.1741.

Manuscript Accepted by the Society 1 February 2007

Printed in the USA

The Geological Society of America
Memoir 200
2007

Earth's first two billion years—The era of internally mobile crust

Warren B. Hamilton*
Department of Geophysics, Colorado School of Mines, Golden, Colorado, 80401, USA

ABSTRACT

The magmatic and tectonic processes of the pre–2.5 Ga hot, young Earth differed profoundly from those of the modern planet. The ancient rocks differ strikingly in individual and collective composition, occurrence, association, and structure from modern rocks. Widespread forcing of Archean geology into plate-tectonic frameworks reflects unwarranted faith in uniformitarianism and in inappropriate chemical discriminants, and disregard for the lack of features that characterize plate interactions. Archean crust records extreme and prolonged internal mobility and was far too weak and mobile to behave as rigid plates, required, by definition, for plate tectonics. None of the geologic indicators of subduction, arc magmatism, and continental sundering, separation, and convergence have been documented. No Archean oceanic crust or mantle has been recognized, and the only known basement to supracrustal rocks, including the thick basalts, high-Mg basalts, and ultramafic lavas that typify greenstone successions, consists of tonalite-trondhjemite-granodiorite (TTG) migmatites and gneisses. A thick global melabasaltic protocrust likely formed by ca. 4.45 Ga, and from it TTG suites were extracted by partial melting over the next 2 b.y. Delamination of the increasingly dense restitic protocrust enabled rise of lighter and hotter depleted mantle and hence more melting. The oldest known crustal materials are zircons, which scatter in age back to 4.4 Ga and are recycled in migmatites whose final crystallization was after 3.8 Ga, and in ancient sediments. Earth may have had a dense greenhouse atmosphere, not a hydrosphere, before 3.6 Ga, for the oldest proved supracrustal rocks are of that age, and older felsic crust may have been too hot to permit rise of dense melts. Rigid plates of lithosphere did not stabilize until a billion years after that and then were mostly small and local.

Dense lavas erupted atop mobile felsic crust after 3.6 Ga produced a density inversion that was partly righted by sinking of the volcanic rocks and rising of the subjacent TTG. In some places, the early dense rocks retained cohesion and sank as synclinal keels between rising domiform diapiric batholiths. In others, the early dense rocks sank deep into mobile TTG crust, and only later in Archean time was the felsic substrate strong enough to enable dome-and-keel style. The TTG substrate rose slowly, with variable amounts of partial melting to generate more-fractionated melts

*whamilto@mines.edu

Hamilton, W.B., 2007, Earth's first two billion years—The era of internally mobile crust, in Hatcher, R.D., Jr., Carlson, M.P., McBride, J.H., and Martínez Catalán, J.R., eds., 4-D Framework of Continental Crust: Geological Society of America Memoir 200, p. 233–296, doi: 10.1130/2007.1200(13). For permission to copy, contact editing@geosociety.org.

and with additions of new TTG from the underlying protocrust, for hundreds of millions of years. The mantle beneath preserved cratons generated ultramafic melts that required a temperature ~300°C hotter than modern asthenosphere ca. 3.5 Ga. Severe and prolonged lateral deformation was superimposed on large parts of some cratons during the era of volcanism and diapirism, obscuring dome-and-keel geology over broad tracts. Lower crust was at high temperature for prolonged periods and flowed pervasively, coupled discontinuously to the upper crust to produce lateral deformation therein.

Rifting, separation, rotation, and collision of internally more rigid lithosphere fragments began ca. 2.1 Ga, but may have been dominantly intracontinental deformation, quite distinct from modern plate tectonics. The products of this regime differ greatly from those of Phanerozoic plate tectonics, and reflect a transitional era of erratically stiffening lithosphere. An early-depleted upper mantle has been progressively re-enriched, by delamination and subduction of crustal materials, while new "juvenile" crust derived from it has become progressively more depleted, during Proterozoic and Phanerozoic time.

Keywords: Archean, Hadean, crustal evolution, mantle evolution, delamination, subduction.

INTRODUCTION

This essay is a study in alternatives. Most current interpretations of geodynamics and of evolution of Earth are forced to fit popular but dubious assumptions that Earth fractionated slowly and is still largely unfractionated, and that rocks of all ages must be explained with plate-tectonic processes combined with plumes rising from basal mantle. The data accord better with opposite interpretations. Earth largely fractionated very early in its history, plate tectonics did not begin operating until late Proterozoic time, and deep-mantle plumes do not operate now and did not affect the Archean Earth. This discussion progresses from, and importantly supersedes, earlier reports (Hamilton, 1998a, 1998b, 2002, 2003).

Our planet had a hot, violent beginning, and has evolved only slowly toward its present dynamic patterns. The geologic record of the young Earth differs profoundly from that of the modern one in crustal architecture and in rock types, assemblages, and structural and magmatic histories. Our goal should be to understand the evolving processes, but most geoscientists who study ancient complexes are imbued with dogmatic uniformitarianism. Many hundreds of published papers rationalize Archean geology in terms of plate-tectonic processes as much like those now operating as imagination allows. "Plate tectonics is the key to the past" even for Earth's most ancient rocks, wrote Windley (1993, p. 7). De Wit (1998, p. 215) asserted that "The geologic evidence in favour of plate tectonic processes operating in Late Archean time is solid; and that for Early Archean is more than compelling." The data "require a tectonic regime of lithospheric plates similar to the Phanerozoic Earth," wrote Cawood et al. (2006). Such statements are based largely on weak chemotectonic rationales and dismiss, with little or no evaluation, the lack of physical evidence and the possibility that the conditions that now enable plate tectonics did not exist in the hot young Earth. Venus and Mars lack plate tectonics, so the process is not inevitable throughout the history of a terrestrial planet.

Rocks older than ca. 2.5 Ga are exposed in least-modified form on about 35 large and small "cratons" (Bleeker, 2003), and also as large and small reworked complexes in, mostly, Paleoproterozoic orogenic belts. There is no proof that lithospherically distinct oceans and continents existed in Archean time.

This report emphasizes (as my previous papers did not) the extreme internal vertical and lateral mobility of Archean continental crust, which was incapable of behaving as the semi-rigid plates required, by definition, for plate tectonics. It proposes that thick global mafic crust formed very early in Earth history, that Archean felsic crust formed incrementally from this, and that delamination, and sinking through lighter depleted mantle, of parts of the increasingly dense restite of the mafic protocrust enabled rise of deeper lighter and hotter mantle and hence more melting of the remaining mafic lower crust. Although I know Archean geology firsthand only from extensive fieldtrips (in eight cratons), I have a half-century of experience in Phanerozoic plate-interaction geology and geophysics, including the most comprehensive synthesis yet made of active and late Phanerozoic convergent plate tectonics of a huge region (Hamilton, 1979).

Among plate-indicative features not known in Archean terrains is structural evidence, other than dikes, for continental rifting; stratigraphic evidence for development of stratal wedges on rifted margins; evidence for separation, rotation, and suturing of rigid continental fragments; evidence that sutures, magmatic arcs, or rigid plates existed; or presence of ophiolite, polymict mélanges, blueschists, or other indicators of disappearance of oceanic crust between landmasses. (Moyen et al., 2006, reported a high-pressure, low-temperature Mesoarchean meta-amphibolite within normal rocks, but their published data do not establish that the minerals used for thermobarometry in the complexly recrystallized and altered rock comprise an equilibrium assemblage.) I have seen hundreds of exposures of early Paleozoic to middle Tertiary subduction mélanges around the world, but not a suggestion of one in the Archean. Much of the Archean is exposed at upper-crustal levels where such features would be obvious if present. Nevertheless, only a small minority of geologists, among them Bateman et al. (2001), Bédard (2006), Bédard et al. (2003), Chardon et al. (1998, 2002), Choukroune et al. (1995), McCall (2003, 2005), Nemchin et al. (2006; they dealt only with pre-3.8 Ga rocks), and Stern (2005, 2007), have argued against the operation of plate tectonics in Archean time. Most Archean specialists regard the absence of geologic evidence as unimportant. Thus, Smithies et al. (2007, p. 51) dismissed discussion as quibbling about "an imperfect geological match with Phanerozoic subduction zone assemblages," and they attempted no explanation for the lack of physical indicators of subduction except to say that the major suture they themselves advocated is hidden.

I see no plate-tectonic interactions in the Archean record. Most plate-tectonic interpretations for the Archean are based on weak compositional analogies with modern rocks of known settings, or with imagined products of hypothetical plate-related settings that have no modern analogues. Although this is a valid approach in the search for explanations, its implicit predictions are not then tested against geologic data. Magmatic arcs are widely invoked in the Archean, yet the lack of evidence for the subduction systems to which modern analogy requires they be paired commonly is dismissed with appeals to cryptic or hidden sutures, or with designation as sutures of unremarkable shear zones. Widely disparate oceanic and continental settings are invoked for rocks interlayered within concordant greenstone sections, and juxtapositions are postulated by rootless megathrusts, usually younger-over-older and often multiple and closely spaced, for which there is no structural evidence. Areal variations in granitoid rocks beneath and between the greenstone packages are assigned to origins in separate places followed by amalgamation by unexplained plate-related processes. The cartoons drawn to illustrate such schemes (e.g., Kerrich and Polat, 2006) are unrelated to anything seen on the ground.

The world before 2.1 Ga was characterized by long-continued vertical and lateral infracrustal mobility incompatible with rigid-plate tectonics. Only felsic gneisses, and igneous rocks contained in them, are proved older than 3.6 Ga. Zircons from ancient gneisses have yielded igneous ages back to 4.2 Ga, and detrital zircons to 4.4 Ga have been found in derivative clastic sediments. Ages of igneous zircons in ancient gneisses commonly scatter over broad ranges, back to the local maximum age: the crustal material was repeatedly, if not continuously, near its solidus temperature for extremely long periods. No supracrustal rocks have been proved older than 3.6 Ga, from which I infer that not until that time were the ancient gneisses cool and dense enough to permit rise through them of mafic and ultramafic melts. (Claims made for supracrustal rocks as old as 3.8 Ga may be valid but are inadequately documented. The oldest proved mafic dikes cutting felsic gneisses are 3.5 Ga.) Archean mafic and ultramafic volcanic rocks commonly are assumed to be mostly ensimatic, but no ophiolites or other physical evidence for eruption on oceanic mantle have been found: only felsic gneisses have ever been seen as basement beneath supracrustal rocks in either outcrop or geophysics. If oceanic lithosphere existed during Archean time beyond surviving cratons, it should have left remnants within the Paleoproterozoic orogens that now separate many Archean cratons, but no Archean ophiolites have been found there either.

During Earth's second billion years, 3.5–2.5 Ga, first thin sedimentary rocks, then, typically, thick sections of mafic and ultramafic lavas followed by more varied igneous and sedimentary rocks, were deposited on top of the ancient gneisses, both blanketing them thermally and producing a density inversion. The inversion was partly righted by rise of domiform diapiric batholiths between sinking keels of supracrustal rocks. Variably pervasive lateral shearing accompanied the process in most upper-crustal terrains, further attesting to weakness of the crust. Early supracrustals largely sank into mobile gneisses and were swirled into them, whereas later supracrustals mostly stayed in the upper crust and retained their identities. The time of this change in style varied within cratons, as did the ending time of major diapiric rise, and then of even limited rise, of basement mobilized as batholiths. Cratonization ensued, at different times in different places, as lower continental crust lost its mobility. Archean specialists often miscite this lateral mobility as evidence for plate tectonics, whereas plate tectonics by definition requires quasi-rigid plates.

Starting ca. 2.1 Ga, masses of Archean crust retained coherence in some areas, pulled apart in others, rotated, and converged. The processes are widely assumed to include seafloor spreading and subduction, but may mostly record instead high mobility within mostly continental crust when that crust was still hot and weak. See, for example, McLaren et al. (2005) and Zhao et al. (2002). Much reworked Archean continental crust has been found within some Paleoproterozoic orogens. Modern-mode plate tectonics, with high-pressure, low-temperature metamorphism in its sutures, began only much later, in Neoproterozoic or very early Paleozoic time (Stern, 2005, 2007; Tsujimori et al., 2006).

Chemotectonics

Modern tectonic settings are characterized by suites of geologic and geophysical features. Plate-tectonic interpretations of Phanerozoic terrains can be based on analogies with these modern systems in petrology, rock associations, structure, stratigraphy, and so on. The plate-indicative features are not present in Archean terrains, so the many Archean geoscientists who believe that their observations must nevertheless be forced into plate-tectonic rationales assign plate-tectonic settings to igneous rocks on abstract geochemical criteria, either highly generalized or in the form of modern-setting discriminants such as those by Pearce and Cann (1973). When, as often is the case, the Archean analyses cannot be fitted into desired chemotectonic pigeonholes, hypothetical mixtures and derivatives of criteria are devised to postulate hybrid tectonic settings with no modern analogues. That the discriminants for modern rocks lack a valid statistical basis (Vermeesch, 2006), that they only poorly identify modern tectonic settings, and that they produce absurdities when applied to the Archean Earth are disregarded. Condie's (2005b, p. 33) admonition that "geochemical data can be used to help constrain tectonic settings, but it cannot be used alone to reconstruct ancient tectonic settings" goes unheeded, and elaborate tectonic syntheses are based on chemical rationales alone.

Trace-element rationales are favored and mostly (e.g., Hofmann, 1997) incorporate the petrologically disproved notion that melts are generated by spot melting at depth and are erupted on the surface with their initial trace-element signatures intact. This assumption is invalidated by the compositions of erupted rocks considered with phase petrology and thermodynamics. Almost all modern erupted melts are low-pressure equilibrates that cannot have the major- and minor-element compositions with which protomelts left their sites of first melting, complex evolution by polybaric crystallization and assimilation reactions instead being required (Dickson, 2006; Longhi, 2002; O'Hara and Herzberg, 2002; Presnall et al., 2002). Some changes are progressive whereas others are abrupt: e.g., equilibrium melt and solid compositions change sharply as the phase boundary is crossed, near a depth of 30 km, between the stability fields of spinel and plagioclase with olivine. Plagioclase and olivine cannot coexist deeper than this, yet widely different modern settings yield convergent basaltic melts, equilibrated at low pressure, that crystallize mostly plagioclase and clinopyroxene ± olivine. The simplistic rationales of chemotectonics thus ignore petrology and thermodynamics. Hundreds of speculative papers by chemists notwithstanding, tonalite melts cannot form in subducted slabs, rise through ultramafic mantle wedges, and intrude overlying crust. In Archean work, these same false assumptions are combined with weak chemical analogies to specify ancient tectonic settings even though the Archean assemblages lack the geologic features predicted by those assignments.

Chemotectonic analogies for Archean rocks most often are made from ratios of a few elements, mostly minor and trace, independent of bulk-rock compositions and considerations of occurrence and association. The compositions of modern igneous rocks of course do tend to vary with their tectonic settings. (One of my papers [Hamilton, 1963] may have been the first to apply major-element compositions and frequency distributions to identify oceanic island-arc materials now accreted to continents.) Although patterns are inconsistent, overlapping, and often ambiguous, idealized discriminants that work more often than not have been developed for modern rocks, and commonly are expressed on binary or ternary plots of selected minor and trace elements, or of their ratios. Archean rocks differ markedly from modern ones sharing the same broad rock names, so discriminants are selected that allow assignments accordant with the user's prejudices. Although the "discrimination diagrams seldom correctly classify samples from [modern] mid-ocean ridges, island arcs, and ocean islands with better than 60% accuracy" (Snow, 2006, p. 1), they are applied rigidly to small suites of Archean rocks. When individual modern provinces are considered, misassignments can be wildly in error. Thus, ratios of Nb, Y, and Zr often are used to define tectonic settings of Archean basalts, yet the archetypal ocean-island basalts (OIB) of Hawaii are scattered through the fields of oceanic plateau, oceanic island arc, OIB, and N-MORB ("normal" midocean-ridge basalt, discussed below) in Nb/Y versus Zr/Y space, and plot mostly in the field typical of oceanic plateaus in Zr/Nb versus Nb/Th space (Condie, 2005a, his Fig. 5).

Many Archean geologists make their plate-tectonic analogies from highly generalized lithologic assumptions, rather than from trace-element pigeonholes: basalt, high-Mg basalt, and ultramafic lava represent ocean floor (even though known to have been deposited on felsic crust in many places, even though they do not resemble modern oceanic rocks either individually or collectively, and even though no oceanic section of Archean crust and upper mantle has ever been found); calc-alkaline rocks formed in magmatic arcs (even though dissimilar petrologically to modern magmatic-arc rocks, not in belts, not paired to possible subduction systems, and lacking the predicted cross-strike compositional variations); and interbedded felsic and mafic rocks require rifting (not mantle-melt eruptions intercalated with eruptions from diapiric batholiths known to have been rising nearby). Brown et al. (2001) are among hundreds of authors whose rationales are based on these assumptions.

Archean igneous rocks—including tonalite and basalt, which respectively dominate crustal and supracrustal assemblages—are mostly quite different from modern rocks individually, and are altogether different in their occurrence and associations. The differences between most Archean basalts and the modern basalts with which tectonic analogies are favored often are greater than those between modern basalts in diverse tectonic settings, and the occurrences and assemblages of the ancient rocks do not resemble any of the modern ones. Predictions implicit in the tectonic-setting analogies of Archean with modern rocks are falsified when tested.

Granitic Rocks

Archean middle crust is dominated by tonalite, trondhjemite, and granodiorite (TTG) gneisses. These complexes are seen in many places to be basement unconformably beneath thick volcanic sections dominated by basalt and ultramafic lava. The diapiric and variably remobilized batholiths that rise into these supracrustal sections contain TTG gneisses and also much granodiorite, monzogranite, and granite recycled from TTG (e.g., Condie, 1981). Most Archean tonalite is much less magnesian and calcic, and more silicic, sodic, and potassic, than Phanerozoic tonalite, and differs also in minor elements, as in its generally much steeper rare-earth patterns (Condie, 1981;

Martin, 1987). Archean tonalite commonly has more quartz, more sodic plagioclase (mostly An<35, rather than 35–50), and less mafic material, and that mostly biotite rather than hornblende. Common Archean tonalite and trondhjemite have semiconstant K/Na with decreasing Ca, whereas only subordinate ones share the Phanerozoic characteristic of increasing K/Na, and the Archean rocks generally have higher contents of transition-group elements than their modern namesakes (Condie, 1981). Tonalite now forms typically in mature or reworked island arcs, where it is associated with abundant diorite, gabbro, and metavolcanic rocks, which by contrast are only minor associates with most Archean TTG. Tonalite is lacking in modern immature island arcs, which are basaltic, and commonly is only a minor component of arcs formed in continental crust, which are more potassic. Archean TTG occurs in vast terrains, not the narrow belts of Phanerozoic rocks. These striking dissimilarities to modern arc rocks are rationalized in terms of hypothetical subduction-related processes, unlike any now operating, by many authors favoring plate tectonics (e.g., Condie, 2005b; Polat and Münker, 2004; Smithies, 2000). Common Archean TTG does resemble chemically an uncommon subset (high-silica adakite) of silicic volcanic rocks present in some modern continental arcs and mature island arcs, and Martin et al. (2005) argued that this demonstrates subduction to have dominated Archean magmatism, a non sequitur. The ca. 2.70 Ga granitic rocks ubiquitous across most of the Superior craton, ~1500 × 2500 km, commonly are attributed to arc magmatism (e.g., Percival et al., 2001), although there is no modern analogue for arc magmatism on such a scale.

Silicic and Intermediate Volcanic Rocks

Archean supracrustal sections commonly include voluminous felsic extrusive and shallow-intrusive rocks, mostly high-silica dacite, rhyodacite, and rhyolite. Most of these are unlike most modern arc rocks, and have the same distinctive major- and minor-element characteristics as do Archean granitic rocks—low Mg and Ca, high Si and Na, steep rare earths, etc. (Condie, 1981). Andesites and mafic and calcic dacites, abundant in modern mature oceanic arcs, are uncommon in most of the Archean, and the andesites that do occur are lower in Al, and higher in Fe, FeO/Fe_2O_3, Mg, Ni, Cr, Co, and Zn, than modern andesites (Condie, 1981). As Archean felsic supracrustal rocks commonly have both zircon ages and distinctive compositions similar to those of the diapiric batholiths that rose into the sections, I see origins primarily by venting of those batholiths. Conventional plate-tectonic interpretations of Archean geology, by contrast, commonly assign the felsic rocks either to extensional settings or magmatic arcs.

Mafic and Ultramafic Volcanic Rocks

Basaltic and ultramafic lavas characterize many Archean supracrustal sections and commonly are assumed to be ensimatic ocean-floor rocks (usually, but not always, excepting the places where they are proved to lie depositionally upon continental crust) despite great dissimilarities to rocks now forming in such settings. No Archean ophiolites—mafic-crustal sections ending downward in depleted mantle rocks—are known. The ultramafic and high-Mg lavas commonly intercalated with the basalts have no modern analogues, seafloor or otherwise. Even the basalts on which seafloor analogies are primarily based are very different from modern seafloor rocks.

An unambiguous example of the inapplicability of modern-rock pigeonholes to tectonic settings of Archean igneous rocks is provided by the Neoarchean Fortescue lavas of the Pilbara craton in Western Australia, the chemistry of which is discussed here although the assemblage is otherwise described in the subsequent Pilbara section for its cratonization significance. The Fortescue rocks comprise a little-deformed kilometers-thick regional sheet, and their present area of exposure of 300 × 600 km is less, by an unknown amount, than their initial extent. The section is entirely ensialic, for Mesoarchean and early Neoarchean granite-and-greenstone terrain is widely exposed beneath it. The Fortescue rocks lack even small central volcanoes, and do not form narrow belts, yet application of popular chemotectonic criteria would wrongly assign most of the rocks to oceanic island arcs. Thorne and Trendall (2001) presented almost 400 major- and minor-element analyses of Fortescue samples, mostly mafic volcanic rocks, divided into many stratigraphic and areal subsets, in many tables and plots. The rocks fall mostly in the modern-analogue pigeonhole for basaltic andesites in silica-alkalies plots, whereas in iron-alumina-magnesia plots they are mostly high-Fe tholeiites. Many practitioners of Archean tectonics via spreadsheet base their assignments primarily on the relatively immobile high-field-strength elements—and most Fortescue mafic rocks plot in the field of modern oceanic-arc basalt in Ti/Zr/Y space (Fig. 1). Although the rocks obviously are continental flood lavas—I imply no particular heat source with that term—in none of the plots are more than stray samples put in the modern-analogue pigeonhole for within-plate basalts; and their major-element compositions do not resemble either olivine-alkalic or quartz-normative types of modern continental flood and rift basalts. Most of the Fortescue rocks are moderately enriched in light rare-earth elements relative to N-MORB, but are conspicuously depleted, despite their high SiO_2 contents, relative to modern quartz-normative continental basalts. Chemotectonicists encountering these rocks in deformed greenstone belts would misclassify most of them as oceanic-arc basalts. Indeed, Kojan and Hickman (1998) argued, on these chemical grounds, that two units they studied within the Fortescue section are "subduction related." Thorne and Trendall (2001) made no such obviously wrong assignment, but argued for a continental-rift origin although the only geologic evidence for rifting consists of dikes (which require extension but not crustal thinning), and the rocks are quite unlike any now forming in rifts. The Greenland pillow lavas assigned by Furnes et al. (2007) to a hypothetical ophiolite have the composition of Fortescue-type mafic andesite.

This continental flood-lava succession in many ways resembles a standard Archean greenstone association except that the

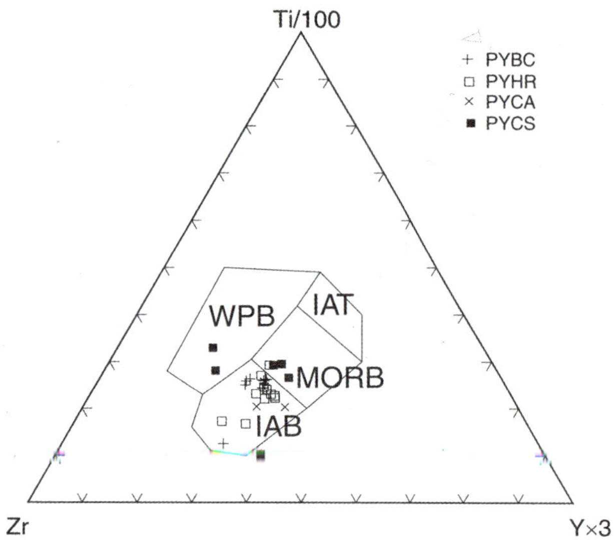

Figure 1. Chemotectonic misassignment of Archean mafic volcanic rocks. Ratios of the high-field-strength elements titanium, zirconium, and yttrium are used by many geochemists to assign tectonic settings to Archean rocks, applying discriminants like these by Pearce and Cann (1973; the midocean ridge basalt [MORB] field also contains island-arc rocks). This plot shows several (designated by four-letter codes) of many similar subsets of little-deformed Neoarchean regional-flood mafic lavas of the Fortescue Group, Pilbara craton, which unambiguously overlie older Archean continental crust, yet are here misclassified as calc-alkaline island-arc basalts (IAB) and MORB. Only two analyses here plot in the within-plate basalt (WPB) field. (IAT, island-arc tholeiite.) This plot is one of many similar Fortescue ones by Thorne and Trendall (2001, Figure 12.11E), who recognized the assemblage to be continental. Discriminants developed for modern tectonic regimes are not applicable to Archean rocks. Figure © by Geological Survey of Western Australia.

subjacent felsic crust had become stable enough to preclude more than incipient granite-and-greenstone dome-and-keel development. Conversely, the character of pre-batholithic supracrustal successions can be inferred from this arrested development. Had full development followed so that the Fortescue rock was now present as disrupted greenstone keels, most Archean investigators would wrongly interpret it to have been assembled by regional thrust sheets and plate amalgamations of island-arc rocks. Thus, among many Archean chemists who base tectonic assignments on the elements of Figure 1, Polat and Hofmann (2003, p. 197) stated confidently that nearby rocks in a Greenland greenstone assemblage formed in two unrelated oceanic island-arc sequences "juxtaposed as a consequence of Phanerozoic-style plate tectonic processes"—for which there is no structural or other geologic evidence.

"MORB" is an acronym for mid-ocean-ridge basalt, but commonly is used in geochemistry to apply restrictively to only the half of ridge basalts that fit a chemical definition thought to be consistent with derivation from "depleted upper mantle" as opposed to "primitive lower mantle" (conjectural plumes) and other sources. This filtered subset may be designated N-MORB, misleadingly meaning "normal MORB" (but more commonly referred to merely as "MORB"), only because it is selected to accord with that conjecture. The equally voluminous varied mid-ocean-ridge basalts that do not pass these filters, and that broadly overlap OIB and other types often assumed to be distinct on the basis of arbitrary pigeonholes, frequently are deemed products of variable hybridization by hypothetical plumes even if adjacent on the seafloor to N-MORB.

Condie (2005a, p. 491), although an advocate of Archean plate tectonics and plumes, emphasized the "near absence of Archean greenstone basalts similar to NMORB in composition." The dominant basalts of Archean greenstone assemblages are tholeiites, which commonly are assumed to be ocean-floor basalts even though they do not resemble any modern basalt suites in either their common compositions or in their usual association with komatiites (ultramafic lavas) and high-Mg basalts that have no analogues in modern oceanic rocks. Archean tholeiites commonly have higher Fe/Mg, and lower Al/(Fe+Mg), than either N-MORB or arc basalts, and lower Ti than N-MORB (Arndt et al., 1997; Cattell and Taylor, 1990; Condie, 1981, 1985, 2005a). On the trace-element discriminant plots favored by Condie (2005a, his Figs. 10–12), Nb/Y versus Zr/Y and Zr/Nb versus Nb/Th, the basalts he selected, with chemical criteria, as "non-arc" tholeiites from a number of Archean cratons scatter mostly in fields for modern "arc" and "oceanic plateau" basalts, so he speculated (overlooking the known ensialic setting of many of the rocks at issue) that they mostly formed in oceanic plateaus generated atop plumes. Ohta et al. (1996), certain that the distinctive, non-MORB–like Archean basalts must be ensimatic ocean-floor rocks, termed them "AMORB" ("Archean MORB").

Continuity of Sections

Where exposures, mapping, and zircon dating are all good, Archean supracrustal sections often are found to have subregional stratigraphy, incompatible with both the belt-like sources usually postulated where constraints are lacking, and with the hypothetical interthrusting of rocks formed in different tectonic settings often deduced by chemotectonics.

Mapping and petrology constrained by numerous zircon U-Pb dates show that the volcanic rocks of a well-studied 200 × 250 km Abitibi granite-and-greenstone region of the Superior craton have regional and semi-regional sheet stratigraphy and were neither erupted in narrow belts nor shuffled by bedding-parallel megathrusts (Ayer et al., 2002). Rock types that would be assigned widely diverse settings by chemotectonics are complexly intercalated stratigraphically throughout the thick section. Ayer et al. recognized seven mappable units of volcanic rocks within the age span 2.750–2.697 Ga, all dominantly mafic but all including thinly interbedded felsic and other rock types. Listing the lithologies in each unit, starting with the oldest, emphasizes their repeating character: (1) high Mg and high Fe basalts > komatiite + intermediate and felsic rocks; (2) mafic to felsic calc-alkalic rocks > tholeiite; (3) basalt, komatiite, and felsic rocks; (4) basalt, komatiite, and intermediate to felsic calc-alkalic rocks; (5) tholeiite, komatiite, and intermediate to felsic calc-alkalic rocks; (6) tholeiite > andesite + dacite; and (7), calc-alkalic basalt + andesite > tholeiitic rhyolite. The depositional base of the section is not exposed, being cut out by domiform granitoid masses—a setting that likely signifies remobilization of pre-greenstone basement. Although Stone and Stone (2000) documented derivation of high-Mg basalt by felsic-crustal contamination of ultramafic melt, the volcanic rocks have yielded no dated zircon xenocrysts older than the oldest supracrustal rocks, and Ayer et al. (2002) regarded the entire section as ensimatic. They postulated on chemical grounds that the section began as ocean floor and otherwise recorded rapidly alternating plume and oceanic-arc settings. Wyman (2003) and Wyman et al. (2002) recognized that trace elements in these diverse rocks do not fit modern chemotectonic pigeonholes, and invoked such hypothetical hybrids as arc-plume transition and extended reorganized island arc; and they proposed non-modern types of subduction to form the granitic rocks that intruded the supracrustal rocks at various times. The "arc-plume" assemblage of Ayer and Wyman and their associates, typified by lava-plain komatiite (ultramafic lava, an exclusively Archean and early Paleoproterozoic rock type at this composition) and high-Mg basalt, bears no resemblance to any modern arc or seafloor package in major-element composition, petrologic association, or occurrence—and heating a slab with a plume would destroy its negative buoyancy and preclude subduction. The schemes by Ayer and Wyman do not account for the thin repetitions of diverse rock types, for their areal distributions, nor for the stratigraphic demonstration that no sutures are present.

Thurston (2002) described many Superior craton examples of continuous stratigraphic sequences of supracrustal volcanic and sedimentary rocks, with or without unconformities, that included widely diverse rock types and yet clearly are autochthonous. A number of these sections are ensialic platform sequences that began with shallow-water sections of clastic, carbonate, and iron sediments and komatiite, deposited directly on TTG basement. Thurston nevertheless argued, with chemical rationales, that plate tectonics must have operated to form the original continental platforms by rifting, and later operated to amalgamate them with oceanic materials in between.

Parks et al. (2006) inferred 200 m.y. of regional stratigraphy to extend 400 km across strike in the western Superior craton, and thus to join tracts assigned to many different arc complexes by others. Supracrustal successions have been shown in other regions, where both mapping and dating are good and subsequent lateral disruption has been minor, to be concordant sections with subregional stratigraphy by Bleeker et al. (1999), Heather et al. (1995), Isachsen and Bowring (1994), Ketchum et al. (2004), Van Kranendonk et al. (2002, 2004a; the latter paper provides the most documentation for a large area), and many others. Individual units of course are often lenticular.

Misguided Assignments

The Archean chemotectonics game is played with few rules. Where detailed analogies fail because, as often is the case, the chemistry of ancient rocks does not fit the pigeonhole of a desired modern analogue, combinations of hypothetical plate settings are proposed and derivative melts are postulated to have mixed and fractionated as needed to yield desired products. If that is still not enough to account for whatever is observed, hypothetical plumes are added to melt and transport components. Thus, Hollings and Kerrich (2004) proposed that the misfits of mafic rocks from a small area in the Superior craton to their desired oceanic-arc analogy resulted because a depleted residue from melt extraction in a back-arc setting was carried under the arc and there subjected to partial melting to yield lavas unlike any modern rocks. Manikyamba et al. (2004) deduced from several trace-element ratios that basalts in the Dharwar craton, India, formed in an oceanic arc; but the Mg#, Cr, Co, and Ni are all too high for this analogy, so they added a plume to the hypothetical mantle wedge; and to explain the non-arclike behavior of Nb, Zr, and Hf, they also added a two-stage melting process.

Because geochemical constraints are so weak, different investigators may assign the same Archean complexes to different settings. Thus, a bimodal volcanic assemblage in the northwestern Pilbara craton, isolated by faults and cover, has been assigned, on chemotectonic grounds, to four mutually incompatible settings—forearc, backarc, oceanic rift, and oceanic arc—by different investigators (Smithies et al., 2005a). Smithies et al. (p. 221–222, 230) recognized that the rocks at issue did not much resemble any of these—but because "there is widespread acceptance that some form of plate tectonics operated throughout the Archaean," the rocks are "clearly[!] . . . an [oceanic] arc-related sequence." Their rationale (p. 221) for the dissimilarity to modern arc rocks enables them to include whatever is observed: "Distinct mantle sources are required and numerous hybrid magmas result from mixing of these sources or of primitive magmas." A rare type of, mostly, high-Mg rocks, boninites, occurs in some modern island arcs. Smithies et al. termed some of their Archean rocks boninites and cited their existence as evidence for subduction—but the Archean rocks are lower in Si, and higher in Al and heavy rare-earth elements (HREE), than modern boninites, a contrast they attributed to a plume. Believers (e.g., Tomlinson et al., 1999) in exotic plume models claim the ability to recognize melts from different parts of plumes that incorporated components from diverse plate-related complexes in the mantle. Kerrich et al. (1999) combined "heterogeneous multi-component" plumes with various hypothetical plate settings to explain whatever they found.

Alternatives

That there is no compositional or associational departure of Archean rocks from hoped-for modern analogues that cannot be rationalized with hypothetical modifying processes does not validate such speculations. Plate-tectonic settings are not indicated by Archean geology, so explanations compatible with the geology should be sought instead.

Ages and Histories

I use the nonstandardized terms Paleoarchean, Mesoarchean, and Neoarchean to refer to ages of, respectively, >3.5 Ga, 3.5–3.0 Ga, and 3.0–2.5 Ga, and early Paleoproterozoic to mean 2.5–2.0 Ga. These round-number divisions approximately enclose, respectively, many ancient gneisses, old granite-and-greenstone terrains, and most young granite-and-greenstone terrains, although the latter continued to form in some regions into the very early Paleoproterozoic. Sequences vary greatly from craton to craton and region to region. A common alternative usage adds the term "Hadean," and Hadean, Paleoarchean, Mesoarchean, Neoarchean, and Paleoproterozoic are separated at 3.6, 3.2, 2.8, and 2.5 Ga. I follow the convention that Ga and Ma refer to ages, or time ago, in gigayears and megayears, whereas b.y. and m.y. mean durations in the same units, billion years and million years.

Zircon Ages

The most reliable dates now available for Archean rocks are uranium-lead determinations on zircon, which crystallizes with uranium and thorium, but almost no lead, in its lattice. As ^{238}U and ^{235}U decay, with very different half-lives, to ^{206}Pb and ^{207}Pb, respectively, determination of these four isotopes permits calculation of apparent age by two independent U-Pb pairs. When the two ages agree, when they are concordant with the uranium/lead-evolution curve on a plot of $^{206}Pb/^{238}U$ versus $^{207}Pb/^{235}U$), the age of crystallization likely has been defined; where discordant, more complex explanations are needed. The most understandable data come from ion or laser-ablation microprobing of small parts of zircon crystals, which can be seen, with cathodoluminescence or scanning electron microscope back-scattering, to be cores, euhedral or discordant zones, overgrowths, or recrystallized areas. Euhedral prismatic cores and zones with relatively high Th/U ratios are probably igneous, whereas patches, rims, and anhedra with very low Th/U are probably solid-state metamorphic—but many sampled bits are ambiguous. More accurate measurements can be made on whole zircon grains or multiple grains, but such analyses smear components together and are misleading in rocks, widespread in the Archean, with multi-age zircons. Whole-grain determinations of only ^{207}Pb and ^{206}Pb have been widely used in, particularly, southern Africa but are unreliable because assumptions of concordance and single-age crystallization often are false. For simplicity, I round off ages in many citations, and otherwise cite dates without the analytical error bars assigned them by their authors.

A zircon determination dates, at best, crystallization of the minute volume analyzed. If all the zircon of apparent igneous character has about the same concordant age, as is often the case for homogeneous granites, then that age indeed likely defines crystallization of the rock. Where ages of apparently igneous spots scatter down concordia, as often is the case in migmatites and in granites with inherited crystals, interpretation is more difficult. Are we seeing repeated additions of new melt, or times of solution, remobilization and precipitation within a continuum at elevated temperature, or intermittent lead loss, or some of each?

The oldest zircon in a rock volume likely dates the most recent time that a melt undersaturated in zirconium was in contact with that zircon fraction. Zr solubility is lowest in cool, wet melts (Hanchar and Watson, 2003; Miller et al., 2003), so inheritance of zircon generally requires lack of contact with hot, dry melt. Ancient gneisses typically contain abundant biotite, so their final crystallization was from, or in equilibrium with, hydrous melts, but may long postdate the oldest igneous zircon cores present. Zr enters garnet, hornblende, and ilmenite as well as zircon, so secular variations in mineral assemblages in equilibrium with partial melts can cause resorption or precipitation of zircon. Water solubility in melt decreases with decreasing pressure, and solidus temperature increases with decreasing water, so only relatively hot and dry granitic melts commonly rise to shallow levels; resorption of old zircons in them is expected, and lack or scarcity of inherited ancient grains is not by itself evidence for lack of recycling from deeper gneisses.

Many Archean geologists state the age of a rock as that of its oldest abundant zircons, even though the major-mineral assemblage that now makes up the rock may be no older than the youngest igneous zircon. The common scarcity of zircons older than 3.9 Ga may be a function of fluctuations in Zr saturation rather than evidence for lack of preexisting felsic material. The oldest dates are not necessarily protolith ages, and the frequency distribution of zircon ages does not necessarily tell anything about volumes of continental crust that existed at or before the times of peaks or maxima. (Nutman, 2001, elaborated the contrary inference.) The extreme mixing and the prolonged or intermittent near-solidus and supra-solidus temperatures displayed by ancient gneisses require that volumes of material, on all scales, now adjacent often record different pressure-temperature-water histories.

ACCRETION AND FRACTIONATION OF EARTH

Popular explanations for both Archean and modern geodynamics incorporate the dubious assumption that Earth fractionated only slowly. This assumption dates from the 1950s, when Earth was thought to have accreted cold, then heated slowly by radioactivity and core separation; to still be largely unfractionated; and to retain much potential crustal material in the lower mantle whereas upper mantle had gradually lost much of its incompatible components to the crust. By the 1970s, slow accretion and delayed heating and separation of the core had mostly disappeared from conjectures, but the hypothesis of a mantle inverted in composition, unfractionated fertile mantle beneath depleted mantle, nevertheless was retained (e.g., DePaolo and Wasserburg, 1976). Speculation that Earth differentiated metal from silicates but that silicates remained unfractionated thereafter is now dogma for most geochemists (e.g., Hofmann, 1997) and geodynamicists, who refer confidently to "depleted upper mantle" and "primitive lower mantle." Rationales relating this hypothetical inverted-composition mantle to plumes and whole-mantle circulation are contrary to most of what has been learned about the inner solar system, and about mineral physics, in recent decades.

Chambers (2004) and Walter and Trønnes (2004) synthesized knowledge and concepts of meteoritics, isotopics, orbital simulations, and fractionation processes. A protoplanetary disk orbiting the Sun consisted overwhelmingly of gas, but in the vicinity of future Earth, about 0.5% of its mass consisted of solid grains of rock and metal with a high-temperature condensation history. Grains stuck together, and aggregated gravitationally and collisionally into masses up to about the size of Mars, within a million years of the beginning of condensation, ca. 4.57 Ga. The inner planets were largely aggregated within a few tens of millions of years by collision of these masses, and Earth fractionated as it was repeatedly or continuously partly or wholly molten. Accretion tailed off exponentially with time, and the volumetrically minor late additions were minimally involved in whole-planet fractionation. The Moon formed from an ejected molten mixture of mantle materials from Earth and a giant impactor before 4.50 Ga, and is highly depleted in both volatile elements and iron.

One important line of evidence for this scenario comes from the hafnium-tungsten isotopic system (review papers by Jacobsen, 2005, and Kleine et al., 2004a, 2004b). Lithophile ^{182}Hf decays to siderophile ^{182}W with a half life of 9 m.y., so ratios of Hf to W isotopes in rocks and metals constrain ages of differentiation of core from mantle. Metal cores of asteroidal masses were largely separated within ca. 5 m.y. of the beginning of condensation of the protoplanetary disk, presumably as a result of heating by short-lived radioisotopes ^{26}Al and ^{60}Fe. Earth's core was not, however, merely collected from differentiated planetoids, for separation of core and mantle continued for some tens of millions of years, although it was completed no later than 4.45 Ga. The elemental partitioning requires separation of metal from molten silicate: the early Earth was repeatedly melted, substantially or entirely, by large impacts. Dated rocks on the Moon reach ages of 4.45 or 4.50 Ga, and the oldest dated crustal zircons on Earth are almost 4.40 Ga, so apparently both Moon and Earth had fractionated crusts and most of their present sizes by those times. Subsequent bolides gardened the surfaces, and perhaps added much of Earth's volatile material, but they contributed only a tiny part of the total masses of Moon and Earth.

Samarium-Neodymium Model Ages

^{147}Sm-^{143}Nd model ages, also termed "mantle-separation ages" and "crustal residence ages," are widely used in Archean geology to designate felsic igneous rocks either as juvenile (generated directly from the mantle [which is often invoked although petrologically impossible] or with a brief intermediate residence in mafic rock for two-stage production of felsic melts) or as reworked (containing variable amounts of material from preexisting continental crust). Where model ages are approximately equal to magmatic ages, the rocks commonly are assigned oceanic-arc provenance.

^{147}Sm, halflife 1.06×10^{11} years, decays to ^{143}Nd (DePaolo, 1988). Sm and Nd are light rare earths, only two atomic numbers apart and similar in geochemical behavior. Both tend to go into melts rather than solids but are slightly fractionated from one another by partial melting or crystallization, Nd being enriched in, particularly, felsic melts. The bulk-mantle composition is assumed to have followed a secular depletion curve from an initially carbonaceous-chondritic Sm/Nd ratio of ~0.32. The Sm/Nd ratio (not their amounts) is near-chondritic in most Archean komatiites, high-Mg basalts, and basalts (e.g., Condie, 1981; Kerrich and Polat, 2006; Kerrich et al., 1999; Tomlinson et al., 1999). Strong decrease from chondritic Sm/Nd is shown primarily by more felsic fractionated or contaminated or secondary melts.

The model ages thus accord with the unconventional conclusion, reached here, that a thick global melabasaltic crust formed very early in Earth history and that this crust was the reservoir from which Archean TTG crust formed in turn. The model-age clocks of Archean TTG did not begin ticking until felsic melts were released from mafic progenitors.

If igneous rocks of a suite initially had varying amounts but constant ratios of Sm and Nd and have since remained closed systems, then plots of their ^{143}Nd/^{144}Nd versus ^{147}Sm/^{144}Nd yield isochrons that define that starting time. The qualifications often are not met, and real-world processes, including mixing of components, can result in pseudochrons. Thus, published whole-rock pseudochrons from the Archean Isua greenstone belt of southwest Greenland scatter from 4.0 to 2.4 Ga, and Furnes et al. (2007) selected two of these, of about 3.8 Ga each, to cite as "isochrones," using the age they deduced from other conjectures.

The widely used ε_{Nd} notation is the departure, in parts per 10,000, of measured ^{143}Nd/^{144}Nd from the ratio the sample would have if it had crystallized, at its known time of formation, with chondritic Sm/Nd. Zero ε_{Nd} thus represents the evolving ^{143}Nd/^{144}Nd of carbonaceous chondrite, and presumably of the bulk Earth, whereas a positive value indicates that the time-integrated history of the rock and its precursors has involved a higher Sm/Nd ratio. Figure 2B shows the secular change of ε_{Nd} in felsic igneous rocks. The mostly positive ε_{Nd} of Archean rocks requires very early mantle depletion, likely

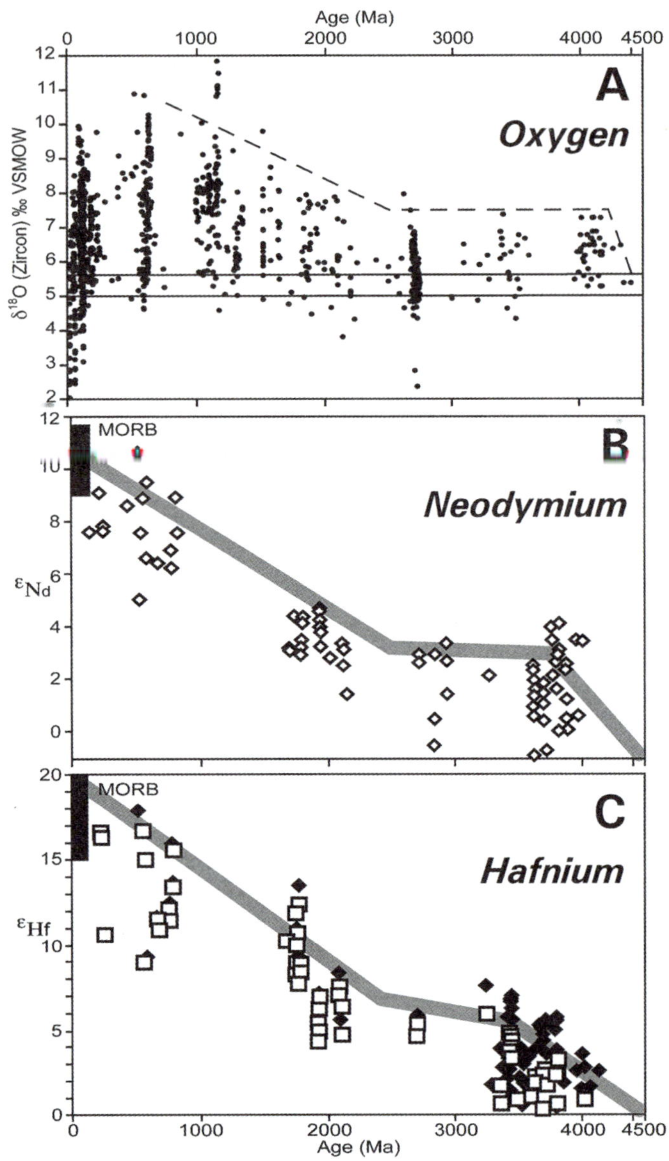

Figure 2. Isotopes in igneous rocks through time. Behavior of the upper limit of each parameter changes markedly at ca. 2.5 Ga. See text for discussion. (A) $\delta^{18}O$ in zircons in granitic rocks, (B) whole-rock ε_{Nd}, and (C) whole-rock ε_{Hf}, in igneous rocks. Adapted from Valley et al. (2005, Fig. 10; the oxygen data are their own, neodymium and hafnium are from other published reports, and the bounding lines are by Valley).

through generation of crust. A great change in the temporal pattern begins at about 2.5 Ga. The increasingly positive ε_{Nd} through post-Archean time is commonly assumed to indicate progressive growth of continents and complementary depletion of the mantle, but the data equally fit early separation of crust and its subsequent recycling back into the mantle (Armstrong, 1991; Vervoort et al., 1996). The latter option accords with upper-mantle evolution as determined from xenolith studies, whereas the former option is incompatible with knowledge of mantle heterogeneity.

Lutetium-Hafnium Systematics

The ^{176}Lu-^{176}Hf system is used, less widely, as another probe of crustal-separation ages, because the parental Lu tends to remain in ultramafic and mafic restites whereas the daughter Hf is enriched in felsic melts. The basic systematics and assumptions are similar to those of Sm-Nd modeling. Figure 2C shows the secular variation of ε_{Hf} (which is analogous to ε_{Nd}). The Lu-Hf system, like the Sm-Nd one, shows a striking change at ca. 2.5 Ga, and also is conventionally cited as evidence for progressive crystal growth and mantle depletion through time, but it too is equally compatible with early separation and subsequent recycling of crust.

Oxygen Isotopes

The striking change in Earth behavior at ca. 2.5 Ga is shown also by oxygen isotopes in zircons in granitoid rocks. Fractionation of oxygen isotopes varies with the inverse square of temperature and so variations are dominated by low-temperature processes (Valley et al., 2005). Geologic variations commonly are expressed as $\delta^{18}O$, the departure, in parts per thousand, of ^{18}O from its proportion to common ^{16}O in seawater. Continental and oceanic sedimentary processes mostly yield positive (heavier) values; mean modern seawater, by definition, is 0; and rainwater is negative. Low-temperature seawater alteration increases $\delta^{18}O$, whereas high-temperature alteration decreases it. Mantle-zircon $\delta^{18}O$ is mostly within the range 4.6 to 6.0.

Oxygen variations within zircon were related to U-Pb crystallization age by Valley et al. (2005, their Fig. 4; Fig. 2A of this paper) for 1200 samples of ages from 4.4 to 0 Ga. Archean zircon $\delta^{18}O$ is mostly below 7.0. Valley et al. regard values above 6 as requiring interaction with meteoric or hydrospheric water, whereas Nemchin et al. (2006) interpreted their data from 4.36–3.90 Ga zircons to indicate prolonged high-temperature history, in accord with conclusions reached here on other grounds. Perhaps oxygen isotopic partitioning at the extreme ranges of temperatures recorded by Archean crustal and upper-mantle magmatic rocks (much greater than ranges for modern rocks) account for the modest excess of $\delta^{18}O$ in Archean rocks, without involvement of the low-temperature water for which geologic evidence is otherwise lacking before 3.6 Ga. Or perhaps the very mobile hot Archean crust absorbed supercritical atmospheric water and mixed it downward.

Starting ca. 2.5 Ga, the common upper limit of zircon $\delta^{18}O$ increased progressively with time (Fig. 2A). Valley et al. (2005) inferred this change to mark the beginning of major cycling, by subduction, of surficially hydrated materials into the mantle, but as modern-style subduction is not required by geologic evidence during most of Proterozoic time, cycling primarily into the shallower source regions wherein Paleoproterozoic igneous rocks were generated likely accounts for the trend.

No Bolide Barrage at 3.9 Ga?

A great bombardment of large bolides, ca. 3.95 to 3.85 Ga, on the Moon was inferred from the scatter, between those limits but with a peak at ca. 3.90 Ga, of argon-argon dates of shock-melted glasses in impact breccias (Dalrymple and Ryder, 1993), and has been widely accepted. Such an event, if real, would necessarily have severely affected the Paleoarchean Earth. Haskin et al. (1998) showed, however, that all of the dated lunar glasses could have come came from the ejecta blanket of Imbrium Basin, the youngest large impact basin on the nearside of the Moon, and so may record only the Imbrium event, the spread in dates representing diffusion and analytical scatter. Stöffler et al. (2006) concur, whereas Norman et al. (2006) do not. Much smaller bolides have since continued to impact Moon and Earth, but there may have been no great terminal barrage. Lunar zircon dates, from granophyres and gabbros, define a frequency distribution decreasing exponentially with time from 4.3 to 3.9 Ga, with no sign of a lunar cataclysm (Meyer et al., 1996); I suggest that these rocks were fractionated in impact-melt lakes and record the exponential decline of accretion after the Moon had reached essentially its final size.

Anorthosite and Allied Fractionates May Record Impact-Melt Lakes

Although there is no support in the terrestrial geologic record for a great bolide bombardment ca. 3.9 Ga, there are many Archean igneous complexes that might have impact-melt origins. Ancient Archean gneisses often contain masses of calcic anorthosite and associated mafic and ultramafic fractionates that have been shredded and swirled into quartzofeldspathic gneisses as lenses and inclusions of all sizes from centimeters to kilometers. The oldest of these yet dated, 3.7 Ga, is in Western Australia (Kinny et al., 1988). In southern West Greenland, the Fiskenaesset complex, ca. 2.9 Ga (Ashwal et al., 1989), was shredded into older ductile gneiss as enclaves of anorthosite and allied rocks mixed throughout a large region (Friend et al., 2002; Myers, 1985; Nutman et al., 1996). An outcrop of a large enclave within Superior craton lower-crust gneisses is shown by Figure 3. Whatever the origin of the fractionates, these structural patterns show that the ancient gneisses had great mobility for prolonged periods.

Figure 3. Megacrystic Archean calcic anorthositic gabbro. Initial crystallization probably was at shallow depth, but it is now enclosed in granulite-facies lower-crustal gneisses. Strong stretching lineation, within subhorizontal foliation of variable flattening, is parallel to regional direction of elongation of plutons in decoupled upper-crustal granite-and-greenstone terrain. Pocketknife, left of center, gives scale. Shawmere Anorthosite, Kapuskasing uplift of Proterozoic age, Ontario.

Most Archean anorthosites contain very calcic plagioclase, typically ~An_{80}, and commonly are regarded as having fractionated from melts intruded into the gneisses from below, although each well-studied complex formed from melts representing diverse mixtures of crust and mantle rocks. An alternative origin of fractionation in melt pools (lopoliths), open to the Earth's surface, formed by impact melting of mixed crustal and mantle rocks, is possible. Layered calcic-plagioclase gabbros and leucogabbros, pyroxenites, and dunites are associated—all typical products of known fractionated magma lakes. The calcic plagioclase likely crystallized at shallow depths and is out of place in the deep-seated ductile gneisses into which it was churned. The An content of plagioclase decreases markedly with increasing pressure of crystallization from water-poor mafic aluminous melts. Near-liquidus plagioclase in two noritic compositions studied experimentally by Longhi et al. (1993) was ~An_{60-80} at 1 bar, but only ~An_{40-65} at 10 kbar. Capping felsic fractionates (peralkaline granophyre and rhyolite) and shock-fluidized breccias predicted by impact explanations have neither been recognized nor sought in Archean assemblages, so if indeed once present, they may have been enough modified during subsequent deformation and metamorphism at upper amphibolite to granulite conditions to be unexceptional parts of gneiss complexes.

An Archean fractionated magma pool of possible impact origin is exposed as the stratiform Stillwater complex, 2.7 Ga, in southwest Montana, which was little deformed until it was tilted and faulted in early Tertiary time. It crystallized from diverse melts of mixed crustal and mantle sources (Loferski et al., 1994), and many of its rocks have Nd and Pb isotopic features indistinguishable from those of nearby older granites and gneisses (Czamanske and Bohlen, 1990). The complex consists of a basal 2 km of cumulate harzburgite and orthopyroxenite, another 2 km mostly of norite, and an upper 3 km of anorthosite (typically ~An_{80}), anorthositic troctolite, norite, and gabbro. Higher parts are missing, but the contact-metamorphic rocks at the base of the complex formed at a depth of only ~10 km (Labotka, 1989), so the complex must have been extrusive. Page and Koski (1973) found masses of contact-metamorphosed breccia beneath much of the base of the complex, extending to at least 1.5 km from it, wherein angular clasts varied from microscopic to 5 m in diameter. They suggested that the breccias were Neoarchean tillites, but their photographs of hand specimens remind me of the shock-injected breccias I have seen beneath proved-impact Sudbury and Vredefort complexes. I suggest an impact origin for the Stillwater breccias and for the fractionated magma pool.

HEAT

Determining the proportion of Earth's heat loss that is due to current radioactivity is a model-driven exercise, for different assumptions permit calculation that either retained heat or current radioactivity accounts for most current heat flow. Either way, however, Earth's enormous heat content is largely retained from its early history and is lost only slowly. That the upper mantle has cooled several hundred degrees during the past 3 b.y. is indicated by the changing character of melts coming from it. Pollack (1997) discussed possible patterns of secular cooling.

Extreme early depletion of Archean high upper mantle is indicated by xenolith studies. The characteristic rock type brought up through Archean cratons is harzburgite, consisting of remarkably uniform olivine (Mg# mostly 92.0–93.6: Bernstein et al., 2007) and subordinate enstatite. The garnet, clinopyroxene, and somewhat less magnesian olivine in associated rocks apparently are products of metasomatism after that depletion (compare Bernstein et al., 2007, Griffin et al., 2004b, Malkovets et al., 2007, O'Reilly and Griffin, 2006, and Simon et al., 2003). Whether the depletion occurred by retention of magma-ocean olivine and upward loss of almost all crustal components or by depressurization melting and loss of almost all components save those of magnesian dunite, it was accomplished at extremely high temperature. The lost material must have been melabasaltic, or even enriched-komatiitic, in bulk composition—e.g., the extensive protocrust postulated here.

Radioactivity and Granite

Heat generated within Earth by radioactivity has decreased exponentially with time. The main heat-producing elements that remained live after a brief early period are potassium, uranium, and thorium, which occur in varied terrestrial rocks at relative abundances typically near K:U:Th ≈10,000:1:3.7. For these proportions, these elements generated ~4.5× more heat at 4.5 Ga, and 2.5× more at 3.0 Ga, than they do now (Van Schmus, 1995). All three elements are incompatible in most mantle minerals, and the small contents in mantle xenoliths occur primarily on grain boundaries, in microveins, and in late metasomatic minerals related to the kimberlitic or alkaline-basaltic melts that carried them to the surface (e.g., Rudnick et al., 1998). The three elements are concentrated in continental crust, within which their abundances increase markedly upward. Perhaps a third or half of Earth's total content of K, U, and Th now resides in continental crust (Rudnick and Fountain, 1995). If Paleoarchean continental crust was much more extensive than modern crust, then most of Earth's total supply of these elements would have been in that crust. The severe fractionation of heat-producing elements late in Archean time into high-crustal granites and their extrusive equivalents (e.g., Ridley, 1992), followed by erosion and, ultimately, subduction of much of the heat–producing elements, enriched the upper mantle in heat-producing elements available for subsequent re-refining into new continental crust.

Komatiite

Most Archean granite-and-greenstone terrains contain voluminous komatiite, and other ultramafic lavas and subordinate sills, crystallized from extremely low-viscosity melts with liquidus MgO contents up to at least 26% by weight (and probably 31%: Wilson, 2003), and hence at eruption temperatures at least as high as 1600°C (likely 1800°), 200 (or 400) °C higher than the rare most magnesian Phanerozoic lavas, and 300 (or 500) °C hotter than modern ambient asthenosphere (Abbott et al., 1994; Herzberg, 1999; Nisbet et al., 1993; Sproule et al., 2002). Ultramafic lavas are uncommon and small in early Paleoproterozoic successions, and are extremely rare, and are far less magnesian and hence lower in magmatic temperature, in younger sequences. That Archean komatiites were dominantly extrusive is demonstrated by their characteristics and field relationships (e.g., Dann, 2004; Groenewalt and Riganti, 2004; Viljoen et al., 1983). Komatiite lavas and associated dunites, peridotites, and other fractionates form composite flows recording rapid eruptions of volumes that reached tens of thousands of cubic kilometers, and wherein komatiite proper may mostly represent quietly crystallized overbank flows from major channels (Hill et al., 2001). Olivine commonly was the only liquidus phase over a broad cooling interval so that the distinctive spinifex texture of bladed extremely magnesian olivine (Fig. 4) could develop in the upper zones of ponded flows, although many ultramafic lavas lack this texture and are densely olivine-phyric throughout. That the dominant melt formed in the mantle during Archean time may have been komatiite, not basalt, is explored in a subsequent section.

A hypothetical uniformitarian Earth, wherein temperature and dynamics have always been about as now, has appeal for many geoscientists despite powerful contrary evidence. Among them, Parman et al. (1997) experimented with an odd-composition komatiitic melt and found that their laboratory clinopyroxene approximately matched the augite in the natural rock only when crystallized with >2 kbars of water pressure. The experimental runs severely mismatched the dominant mineral—highly magnesian olivine, MgO/FeO ~8—in the rock, producing instead olivine with MgO/FeO only ~2.5 to 4. Parman et al. dismissed this olivine result as somehow recording Mg-metasomatism, and argued from the augite alone in this sample that komatiites must be intrusive, at depths of at least 6 km, and that source-region temperatures need be no hotter than modern asthenosphere. Their claim has strengthened in repetition: "at least 3 wt% dissolved H_2O in the komatiite magma was required to reproduce the compositions of the igneous minerals [sic] preserved in the

Figure 4. Komatiite, ultramafic lava, was erupted at temperatures far hotter than modern asthenosphere. (A) Three of a thick stack of close-following overbank flows. Flows 1 (oldest), 2, and 3 are separated by, respectively, hammer and pocketknife. Middle flow: 2a, blocky flow top; 2b, massive spinifex (bladed olivine, interstitial clinopyroxene); 2c, downward-growing spinifex; 2d, skeletal olivine; 2e, cumulate peridotite; 2f, basal peridotite. Pike's Hill, Munro Township, Abitibi greenstone belt, Superior craton, eastern Ontario. (B) Thin komatiite flows, altered to talc-rich rocks. Outcrop ~10 m high. Ten km southeast of Kalgoorlie, Yilgarn craton, Western Australia. (C) Random spinifex. Blades of high-Mg olivine grew in ponded segment of lava. Warrawoona Group, Bamboo Creek mine, northeast Pilbara craton, Western Australia.

rock samples" (Grove and Parman, 2004, p. 179). Grove and Parman, unlike Parman et al., did acknowledge that komatiites are extrusive, but postulated that they erupted and crystallized with high water contents, despite both physical implausibility and lack of evidence. Or, they rationalized, if komatiite melts actually were hot, they were produced by transient plumes rising from uncommonly hot parts of the core-mantle boundary into mantle with ambient temperatures similar to modern ones (Grove and Parman, 2004, p. 181). Conjecture that komatiite requires such plumes is widely accepted among Archean specialists, and the obvious alternative (to me inescapable) that ambient Archean mantle was much hotter than the modern one is disregarded.

No Plumes or Whole-Mantle Circulation

Conventional geodynamics requires an unfractionated lower mantle, which could not plausibly have been maintained through the hot, violent events of Earth's first 50 or 100 million years even were it not incompatible with mineral-physics considerations (Anderson, 2007). Postulated whole-mantle circulation—including plumes that rise from deep to shallow mantle, and subducting slabs that sink through the lower mantle—is anchored to the obsolete assumption of an unfractionated lower mantle and also largely disregards mineral-physics information. Tomographic depictions of deeply subducting slabs appear to be artifacts of sampling biases consequent on the distribution of earthquakes and receivers. Actualistic plate tectonics is not incorporated in popular whole-mantle-circulation models, in terms of which almost nothing observed in either convergent or divergent plate behaviors and geometries can be explained; for example, hinges rolling back before advancing plates that bear undeformed forearc basins, a shrinking but fast-spreading Pacific whose ridges intersect trenches, and a slowly enlarging and nonsubducting Atlantic. Actual plate behavior shows plate tectonics to be driven by subduction, which is a result of cooling from the top of oceanic asthenosphere, and circulation systems likely are closed above the major boundary, at a depth of ~650 km, between upper and lower mantle (Hamilton, 2002, 2003, 2007).

Plumes are hypothetical features that have not been demonstrated to operate in the modern Earth (Anderson, 2006, 2007). They are nowhere needed, for ambient asthenosphere temperatures are high enough to account for the processes attributed to plumes, and "plume" lavas are no hotter than MORBs. All purported geophysical evidence cited for plumes is both ambiguous and suspect, whereas anti-plume geophysical evidence is strong. Plume conjectures are made ever more complex and unique to each example as predictions in both general and specific models are disproved. The notion that cylinders of hot material, fixed in deep-mantle space, stream from basal mantle to surface and there affect shallow tectonics and magmatism is disproved in the case of its type example, Hawaii, by the independent databases of paleomagnetism and of the geometry of plate motions recorded in seafloor magnetics and transform faults. Other Pacific "hotspots" once assumed to have had Hawaii-parallel behavior have since been proved to have complex, non-"track" behavior. The superabundant tiny volcanoes on Pacific seafloor mostly have OIB compositions where sampled. Purported tomographic and other evidence for whole-mantle circulation is, at best, ambiguous (Hamilton, 2002, 2003, 2007; papers by many authors in Foulger et al., 2005, and Foulger and Jurdy, 2007; many other relevant papers at www.mantleplumes.org). Intraplate magmatism presents problems not of heat sources, but of access to the surface of melts that could form anywhere; and, in some areas, of presence of low-melting-temperature materials within the upper mantle. Only highly selective use of data permits maintenance of the fixed-plume concept.

Most geodynamic modeling, both numerical and fishtank, merely presents visual aids for the assumption that whole-mantle convection and plumes operate, and is enabled to support this assumed conclusion by omitting or minimizing variations with temperature and pressure of physical properties that would preclude the desired result. Anderson (2002, 2006, 2007) and Hofmeister and Criss (2005) are among those who have presented powerful arguments, from such properties, against whole-mantle convection and plumes. The great decrease in thermal expansivity, and the enormous increase in viscosity, at lower-mantle pressure are alone enough to have forced irreversible chemical stratification, to preclude whole-mantle circulation, and to require that rapid circulation be limited to relatively shallow depths. The shallow concentration of the main radioactive heat-producing elements, and the great increase of radiative thermal conduction with increasing temperature (and hence with depth), require the same conclusion.

Popular geodynamic models of plumes and whole-mantle convection incorporate the assumption that the low seismic velocities that characterize several broad regions of the lowermost mantle are due to high temperatures, and hence that these are regions of positive buoyancy and potentially upwelling material. The relationships between velocities of S and P waves, mineral physics, and experimental petrology all show that these low basal-mantle velocities probably instead are products of high iron content and high density (Caracas and Cohen, 2005; Ishii and Tromp, 2004; Jacobsen et al., 2004; Mao et al., 2005; Trampert et al., 2004). The low-velocity material is an anchor, not a balloon; the favored source of plumes does not exist.

Appeals to mantle plumes nevertheless are almost ubiquitous in the Archean literature, and are presented as dogma needing no evaluation (e.g., Cawood et al., 2006; Kerrich and Polat, 2006; Smithies et al., 2005b; Tomlinson et al., 1999; Van Kranendonk et al., 2007). Because ultramafic lavas and lavas chemotectonically assigned to arcs often occur in the same sections, plumes frequently are postulated to have interacted with subducting slabs, even though the heating would destroy the negative buoyancy required for subduction. Proponents of Archean plumes who argue (as many have with me) that only with plumes can modern Hawaii, Iceland, Yellowstone, and magmatic temperatures be explained, and that therefore plumes should be invoked for the Archean also, are demonstrating their unawareness of the powerful geophysical and petrologic evidence in those regions that is incompatible with plume rationales.

Mafic Protocrust and Delamination

The obvious major material to extract as ancient crust is melabasalt, or even komatiite. Details of composition, thickness, and continuity could be specified only with chains of assumptions; but say global and on average 100 km thick, bearing feldspar and hornblende in its upper 35 km or so but dominated otherwise by broad-composition igneous garnet and clinopyroxene ± olivine (not the relatively low-temperature eclogitic variety of garnet-clinopyroxene rock), and likely graded in composition vertically. Kramers (2007) presented thermodynamic, petrologic, geochemical, and isotopic arguments for the very early formation and prolonged stability of such a protocrust, above near-liquidus peridotitic mantle, which accords with geologic deductions made here.

No such protocrust is now exposed at the surface except as enclaves in TTG gneisses, and if any survives directly beneath Archean felsic crust it must be thin and local. Archean cratons show by exposure, xenoliths, and seismic analysis that their crust now consists mostly of 30 or 40 km of felsic and intermediate rocks, dominantly tonalite and granodiorite, in addition to the discontinuous supracrustal greenstone belts and minor more mafic plutonic and granulitic rocks, and to the late mobilizates of more potassic granites. Seismic analysis shows that Archean crust commonly lacks the 10 km or more of gabbroic underplating that typifies Proterozoic and Phanerozoic crust (Durrheim and Mooney, 1994). Subduction of modern style did not operate, so presumably the vanished mafic crust broke away and sank after it became denser than the underlying depleted mantle (which consisted primarily of magnesian olivine and orthopyroxene) by crystallization to dense phases, notably garnet and clinopyroxene, and by upward loss of felsic components.

Refractory Archean continental lithospheric mantle has a composition appropriate for a residue after removal of voluminous komatiite and high-Mg basalt (Herzberg, 1999), but is much too voluminous to be complementary to the volume of komatiite and basalt erupted in supracrustal successions, and is far too refractory to be complementary to the surviving felsic crust. This mantle thus may be complementary primarily to vanished thick melabasaltic crust plus supracrustal rocks. Among those who have discussed aspects of these and related matters, in support of widely varying models, are Bédard (2006), Hamilton (2002, 2003), Pollack (1997), Ridley (1992), Ridley and Kramers (1990), and Sandiford and McLaren (2006). Sub-cratonic lithospheric mantle, as sampled by xenoliths in kimberlite, contains varying amounts of eclogite that is not in equilibrium with the dominant harzburgite. The eclogite typically has compositions appropriate for formation as a residue after removal of TTG from a basaltic protolith (Rollinson, 2006). Subduction is commonly inferred to have inserted eclogite into depleted mantle, whereas I infer sinking of delaminated protocrust. Refractory lithospheric mantle has also been enriched by post-Archean contributions from below, which I infer to have been released in part from protocrust that sank through the lithospheric harzburgite, and in part from slabs subducted into the transition zone after plate tectonics began.

Delamination may have been a major mechanism that supplied heat to Archean crust and enabled the magmatic and tectonic processes whose products are described in this report. As insulating depleted protocrust sank away, hot subjacent magnesian mantle rose to the base of the remaining lithosphere. Occasional delamination could deliver heat to the crust on time scales that might be mistaken for deep-seated cyclic processes of mantle circulation. Delamination, involving dense uppermost lithospheric mantle formed, for example, by cooling of underplated mafic igneous complexes, may now be a major process that adds heat and new melt, by asthenospheric upwelling and pressure-release melting, to the base of the remaining crust (Anderson, 2005, 2007; Elkins-Tanton, 2005; Lustrino, 2005), and may have been much more important in Archean time. Related conclusions were reached by Bédard (2006) and Percival and Pysklywec (2007).

Evolution of the Upper Mantle

The conventional assumption that the upper mantle has become progressively more depleted as continents enlarged is countered by powerful evidence that the early Earth was highly fractionated and that density inversions consequent on cooling have had the net result of progressively remixing fractionated material downward into an increasingly heterogeneous and re-enriched upper mantle.

Subcontinental lithospheric mantle has become less, not more, refractory with time (O'Reilly and Griffin, 2006). The average modal amount of clinopyroxene, the major potential source of basalt melt from peridotite, is ~2% in xenoliths of much Archean continental lithospheric mantle, 6% in Proterozoic, and 15% in Phanerozoic, and the respective contents of Al_2O_3 are ~1, 2.5, and 3.5 wt% (O'Reilly et al., 2001), although some xenolith-defined lithospheric mantle sections beneath Archean cratons are much richer in basaltic components than the typical low values (Griffin et al., 2004b). Sub-cratonic lithospheric mantle apparently was extremely depleted by generation of Archean protocrust and shallow magmatism, and then was re-enriched from above by stalled sinkers of delaminated densified protocrust, and from beneath by melts rising, from deeper in the upper mantle, from both delaminated and later-subducted material. The rising melts metasomatized the depleted lithospheric harzburgite, and some kimberlitic and other hybridized melts reached the surface. No materials need have sunk below the 660 km discontinuity. The sunken materials have re-enriched the upper mantle while greatly increasing its heterogeneity.

Magmatic processes related to plate tectonics—in particular, the generation of new oceanic crust and of the mantle component of arc magmatism—have operated in the direction to re-fractionate fusible components into new crust, but the net change has been to make the upper mantle more enriched, not more depleted, throughout Earth history since the very early fractionation of the planet. Further, felsic continental crust has become more mafic in bulk composition since Archean time. Archean felsic crust is more quartzose, sodic, and potassic, and less magnesian and calcic, than younger continental crust. Proterozoic and younger crust commonly has a thick mafic underplate, lacking beneath Archean crust, although it is argued here that a much thicker mafic Archean protocrust was removed by delamination after its felsic components were largely removed upward. Some of these matters were discussed by Hamilton (2002, 2003).

The observed trends fit a model wherein incompatible elements were concentrated early in crust much more voluminous than now and have since been progressively returned to the mantle by delamination and subduction, and wherein successive refining into new crust has been progressively less efficient in removing incompatible elements from the upper mantle. This is opposite to conventional explanations.

PETROGENESIS OF ARCHEAN IGNEOUS ROCKS

Archean magmatic rocks can be explained in terms of the conclusions reached in preceding sections, with no need for conjectural plate interactions that are contra-indicated by geologic evidence. Ambient upper mantle was then 200°–300°C hotter than now. Delamination and sinking of mafic protocrust that became increasingly dense as it both cooled and lost felsic components upward allowed this hot mantle to rise and to partially melt decompressively and to produce secondary melting in both sinking and remaining protocrust.

Felsic and Intermediate Rocks

The experimentally demonstrated ways to derive felsic melts are by fractionation of basaltic melt or by partial melting of mafic rocks (Clemens et al., 2006; Nakajima and Arima, 1998; Wyllie et al., 1993). Archean TTG may have been partly fractionated very early from developing thick mafic protocrust, and partly refined subsequently from protocrust by partial melting consequent upon delamination. Conversely, derivation from thick mafic protocrust would require delamination, for the melamafic rocks complementary to the felsic rocks are not now present. Densification of the restite by removal of felsic melts caused it to delaminate and sink through low-density depleted mantle (Nakajima and Arima, 1998; Wolf and Wyllie, 1993). The dominant rocks of Archean crust are tonalite, trondhjemite, and granodiorite, and the steeply fractionated and concave-upward rare-earth elements (REE) patterns characteristic of Archean TTG (but not of most modern rocks that share the same names) make probable derivation primarily by hornblende breakdown that left abundant garnet in the residue, which accords with derivation at a depth of >35 km from vanished mafic crust (Rollinson, 2006; he advocated subducted-slab sources). Detailed analytical and experimental data from the Barberton granite-and-greenstone terrain of South Africa (Clemens et al., 2006) accord with this scenario. Felsic crust existed by 4.4 Ga, for detrital zircons reaching almost that age are known in ancient sedimentary rocks in two areas, 400 km apart, in Western Australia (Crowley et al., 2005; Nemchin et al., 2006; Pidgeon and Nemchin, 2006; Wyche et al., 2004). Very early separation of a long-stable tonalitic magma layer within Archean continents, formed by fractionation atop a convecting and more or less global basaltic magma ocean above upper mantle, is thermally and mechanically plausible (Ridley and Kramers, 1990), but neodymium isotopes appear to require instead that separation of TTG from protocrust progressed throughout Archean time. As many Archean geologists have emphasized, the more potassic monzogranites and their kin that characterize late upper-crustal batholiths in many Archean cratons are explicable in terms of partial melting of older TTG that was, in turn, derived by partial melting of mafic rocks.

Modern subducting slabs sink because oceanic lithosphere forms by cooling from the top. The density inversion thus generated is righted by subduction, which provides the drive for plate tectonics (Hamilton, 2002, 2003, 2007). Prevalent speculations in Archean literature are based on vague notions contrary to information from modern arcs. For example, Smithies et al. (2003) proposed that Archean subduction was very rapid and flat: buoyant oceanic lithosphere passed directly beneath, and in contact with, upper-plate continental crust, and was so hot that it released, from depths of only 10–30 km, voluminous melts of tonalite directly into the overriding crust over broad horizontal spans. They did not explain how either the thin overriding pile of magmatic mush or the underflowing suprasolidus slab could comprise a rigid plate, or how suprasolidus oceanic crust could have negative buoyancy relative to denser mantle rocks. Such imaginative schemes, although common in Archean literature, do not represent plate tectonics.

Mafic and Ultramafic Volcanic Rocks

The dominant rocks of, particularly, the lower parts of Archean supracrustal assemblages are basalts, which commonly are intercalated with ultramafic lavas (komatiite and its kin, discussed previously) and with high-Mg basalts transitional between the types. The ultramafic lavas presumably were generated by high-fraction partial melting of hot mantle that rose as delaminated protocrust sank. Perhaps ultramafic melt was the dominant magma type generated in Archean mantle, and the high-Mg basalts and tholeiites, which are unlike modern mantle-melt basalts, were generated in substantial part by assimilation of both deep mafic protocrust and shallower TTG crust in ultramafic melts. The ultramafic melts had temperatures far above the liquidus temperatures of both mafic and felsic rocks and hence had great potential for assimilation. Thermal erosion of their substrates by ultramafic lavas is shown both by contact features and by widespread contamination of the lavas (Hill et al., 2001; Perring et al., 1996). There is a compositional spectrum in Archean lavas from ultramafic rocks through high-Mg basalts into the more common tholeiites, and many investigators have found that much Archean basalt and high-Mg basalt shows clear evidence for formation by assimilation of felsic material, such as TTG, in ultramafic melts. High-Mg basalts, and perhaps many other basalts, may have formed mostly by assimilation of continental crust in rising ultramafic melts (Cattell and Taylor, 1990; also Bateman et al., 2001, Sproule et al., 2002, and many others). Even the oldest mafic magmas may have assimilated large volumes of felsic crust (Arndt, 1999; Bickle et al., 1994; Green et al., 2000). Other basalts might represent secondary melts of protocrust.

Conventional literature commonly assigns Archean mafic and ultramafic sections to seafloor spreading. Other postulates are for rapid alternations of, say, plumes and subduction, or, where mapping and age constraints are poor, for shingling by invisible megathrusts of rocks formed in widely separated settings. As emphasized elsewhere in this report, many Archean sections of mafic and ultramafic rocks are proved to be ensialic, whereas none have been shown to be ensimatic.

Bimodal Assemblages

Batholiths rose from the lower crust as dense supracrustal rocks sank between them during protracted Archean periods, as described for specific examples subsequently. Felsic volcanic rocks are intercalated in most supracrustal assemblages, including those dominated by mafic and ultramafic rocks. Most batholithic and felsic-volcanic rocks share distinctive petrology, and commonly they have similar ages where well dated. Eruption from breaching batholiths provides a general explanation for the felsic supracrustal rocks. In terms of the model favored here, felsic, mafic, and ultramafic melts were all mobilized in response to delamination, but from different depths and materials.

Most Archean literature, however, assumes plate-tectonic settings, and assigns bimodal sections to either rifts or magmatic arcs. Such conjectures are enabled by unfamiliarity with modern magmatic assemblages. Modern magmatic arcs indeed can be bimodal on local scales, but they are unimodal on large scales. Many rifts indeed display bimodal volcanism, recording interlayering of mantle and crustal melts, but neither the mafic nor felsic components of Archean sections resemble those of modern rifts. Rift conjectures predict severe crustal thinning, which should be recorded by huge normal faults rotated to gentle dips and by thick down-rotated crustal blocks or clastic sections including scarp facies, which have not been found in Archean terrains even where subsequent deformation, metamorphism, and erosion have been minimal and such features should be obvious.

Dissimilarity to Modern Magmatic Arcs

Most papers on Archean geology published in the last 15 or 20 years (e.g., Cawood et al., 2006; Kerrich and Polat, 2006; Percival et al., 2006b; Van Kranendonk et al., 2007, and many hundreds of predecessors) have invoked subduction-produced magmatism and aggregation of island arcs and other plate-generated features, even though Phanerozoic complexes known to have formed in such fashions bear no similarity to Archean terrains. The analogies are based on the weak compositional rationales I critiqued previously, and are incompatible with the characteristics of actual arcs and their aggregates.

Intermediate and felsic rocks of Phanerozoic oceanic-island arcs are generated within dominantly mafic crust and are never themselves, as are Archean TTGs, the dominant crustal rocks. Observation of young arcs, both where active and where deeply eroded, indicates that rocks more evolved than basalt are generated in the crust (Hamilton, 1979, 1995). Only basalt, almost as primitive as MORB, erupts in young oceanic arcs. More evolved rocks become progressively more abundant as crustal dimensions of arcs increase by magmatic growth and plate aggregation. In the few exposed crustal sections through Phanerozoic arcs into the mantle—e.g., Talkeetna, a relatively primitive oceanic arc; Kohistan, an evolved oceanic arc with thick crust; and Ivrea, a continental arc—the basal crust consists exclusively of mafic and ultramafic igneous rocks and their granulitic equivalents, and includes 10 km or so of layered gabbro, apparently the underplated heat engine for partial melting of preexisting overlying rocks, from which the granitoid melts in turn separated and rose into the volcanic edifices. Xenolith studies indicate similar underplates elsewhere, as under the Neoproterozoic composite arcs of Saudi Arabia (Al-Mishwat and Nasir, 2004). Although intermediate and felsic magmatic-arc igneous rocks are commonly assumed to be produced primarily by partial melting of subducted slabs (e.g., Cawood et al., 2006), both geophysical (e.g., Hamilton, 1995) and basalt-petrologic data (e.g., Plank and Langmuir, 1988) indicate instead that the mantle component of arc melts is generated largely or entirely in the mantle wedge between slab and overriding plate—where olivine and plagioclase cannot coexist—and that andesites and tonalites form by recycling of mafic rocks within the crust.

For the complex ways in which oceanic materials aggregate by subduction, see my monograph (Hamilton, 1979) on the active and accreted arcs and other complexes of Indonesia, the Philippines, and Melanesia, a vast region still in its constructional phase.

Archean granite-and-greenstone terrains are typically eroded 10 km or so, and the Klamath Mountains of California and Oregon provide an appropriate region for comparison, for they too have been eroded mostly ~10 km so that what is seen here should be visible in Archean complexes if the latter were analogous. The Klamaths comprise a complex aggregate of arcs, ophiolites, mélanges, and other mostly oceanic materials, exposed for as much as 150 km across strike and 300 km along strike (Hamilton, 1978; Irwin, 1994; Snoke and Barnes, 2006; and references in each). Components range in age from late Proterozoic through Early Cretaceous. To the west lies the latest Jurassic to modern accretionary wedge of polymict mélange, broken formation, and blueschist, which dips eastward beneath the Klamath complex and extends westward from it 80–110 km, on- and offshore, to its toe at the active Oregon Trench. This accretionary wedge is also exposed in a window in the central Klamaths, which are floored at shallow depth by subducted materials. The Klamath Mountains are bounded on the east by post-Jurassic rifts. Farther east are the Cretaceous and Cenozoic magmatic arcs. The southern Klamaths are onlapped by the latest Jurassic through Cenozoic forearc basin of California. The northern Klamaths are onlapped by the Eocene and younger forearc basin of Oregon. Neither these bounding assemblages nor the Klamath Mountains proper contain Archean analogues.

The Klamath Mountains complexes were assembled by Late Jurassic time and are mostly well mapped and characterized although just what pairs to what, what the polarities of the various arcs were, how much of the assembly of components occurred against the continent, and how much took place somewhere in the ocean are partly disputed. There are two huge tectonic sheets of ophiolite, one 120 km long and 3–20 km wide, the other, including serpentinite-matrix mélange in continuity with it, 200 × 20–30 km, plus a 30 × 50 km sheet of arc ultramafic rocks. Island arcs are of assorted ages from Devonian to Late Jurassic, and occupy less area than do tracts of argillite-matrix polymict mélange and broken formation and serpentinite-matrix mélange which, with the great ophiolite sheets, record sutures. Blueschists are of Ordovician, Triassic, and Jurassic ages, and Cretaceous blueschist flanks much of the province on the west and is exposed beneath it in the medial window. The various island-arc remnants consist mostly of mafic and intermediate volcanic and volcaniclastic rocks. The subordinate arc plutonic rocks are mostly of Jurassic age and are dominantly mafic—gabbro, diorite, and mafic, calcic tonalite. None of this has any Archean analogue. Stocks and small batholiths in one area are of Jurassic leucotonalite, trondhjemite, and granodiorite, some of which superficially resemble similarly named Archean rocks, but these aggregate no more than a few percent of the province.

Although Archean terrains are often assumed to be aggregated arcs, there are no similarities between such terrains and Phanerozoic aggregated arcs. Archean specialists who nevertheless argue for modern-arc analogues betray their ignorance of modern arcs.

Paleozoic complexes between converged plates commonly are eroded more deeply than Archean and Klamath terrains and yet, where not subsequently plutonized, show the obvious stratigraphic, structural, and petrologic indicators of those plate interactions (e.g., Dewey, 2005). They also have no Archean analogues.

Were either the plutonic or extrusive Archean rocks at issue products of arcs, then they would be related systematically to sutures bearing evidence of subduction—and no suture of modern convergent-margin type has ever been documented in an

Archean terrain. The frequently made appeals to minor shear zones of local rock types as sutures are spurious. Many authors postulate that their predicted sutures are hidden or cryptic, which is unsatisfying as a universal explanation. Modern sutures can indeed be obscure within magmatically reworked midcrustal gneisses (the deep-crustal part of the western Idaho Mesozoic suture, between continental and oceanic complexes reworked by continental-arc magmatism, comes to mind), but in upper-crustal assemblages sutures commonly are spectacularly obvious, and Archean granite-and-greenstone terrains are upper crustal. I noted abundant evidence for many Phanerozoic sutures in Indonesia and surrounding regions and in the western United States (Hamilton, 1978, 1979, 1988). No such evidence has been found in Archean terrains.

ANCIENT GNEISSES AND LOWER-CRUST MOBILITY

The oldest rocks proved in Archean cratons are polycyclic migmatites dominated by TTG and including much pegmatite, amphibolite, and other rock types. These are the only rocks yet seen as basement beneath Archean supracrustal sequences, and they dominate middle and lower Archean crust. Ages of zircons reach 4.1 or 4.2 Ga in the gneisses of several cratons, and almost 4.4 Ga in detrital grains in Archean sandstones derived from these gneisses in two parts of the Yilgarn craton of Australia. Maximum ages of zircons reported by various investigators from ancient gneisses are 4.2 Ga in Western Australia and northwest Canada, 4.1 Ga in Labrador and Greenland, 4.0 Ga in Wyoming and Antarctica, and 3.9 to 3.6 Ga in many other cratons. Many more ancient finds are anticipated. The oldest proved supracrustal rocks, by contrast, are 3.6 Ga. (The claim that some supracrustal rocks reach 3.7–3.9 Ga is examined in the subsequent section on the Isua greenstone belt of Greenland.) The ancient ages are of zircons, not of crystallization of the final major-mineral assemblages of the rocks in which those zircons occur; the maximum age of those assemblages may be no more than 3.65 Ga (e.g., Whitehouse and Kamber, 2003). Whether the main mass of rock material has the age of its oldest preserved zircons, or is older still, or has mostly been added by younger magmatism is uncertain. Compositions of the ancient zircons show them to have crystallized from felsic, not mafic, melts (Barbara John, written commun., 2007).

Because zircons are retained better in cool, hydrous migmatites than in hotter, dryer melts, ancient zircons are found primarily in exposures of the middle and deep crust, where they are in complexes of varying heterogeneity and maximum age. Thus, the zircons in deep gneisses exposed at the surface by post-Archean processes in the north part of the North China craton yield ages to 3.8 Ga, whereas the southern part of the craton is upper-crustal granite-and-greenstone terrain that, although not extensively sampled, has yielded only Neoarchean zircons (Zheng et al., 2004). Lower-crustal felsic xenoliths erupted through that southern upper-crustal terrain contain zircons dated back to 3.6 Ga. Paleoarchean zircons have been found in the Slave craton (Bowring et al., 1990) and Superior craton (discussed in a subsequent section) primarily where Proterozoic deformation has resulted in exposure of middle- or lower-crustal gneisses.

That the felsic lower crust of Archean time was much hotter and more ductile for prolonged periods than is typical of the modern Earth is abundantly shown by geologic evidence. Protracted near-solidus temperatures, or repeated reheatings, are required by zircon age spectra. These indicators record processes operating over periods of hundreds of millions to a billion years, which heightens the contrast with the modern Earth.

Low effective viscosity is shown by extreme ductile flow. Ductile behavior, with variable remelting, also is demonstrated by the long-continuing rise of diapiric granites, with varying degrees of partial melting, into the upper crust in response to top-loading, and thermal blanketing, by dense volcanic and sedimentary sections. The common floating style of upper-crustal deformation required a weak and mobile lower crust. Upper crust was deformed by ductile shortening and extension, and by strike-slip faults, recording extreme deformation in many sectors, but this deformation commonly produced only minor vertical offsets. Fault blocks of Archean age, either normal or thrust, with vertical offsets large enough to markedly change crustal levels, are rare: the crust could not support high topographic relief.

Bailey (2006) showed that the higher temperature of Archean crust, indicated by both higher basal heat input and higher generation of internal heat, required profoundly different tectonic behavior than that of most modern continental crust, and provides a physical explanation for the empirical observations reported here. The low effective viscosity due to high Archean crustal temperature required decoupling between ductile lower crust and brittler upper crust (as is observed), which decreased through Archean and Proterozoic time. From his thermal extrapolations in time, Bailey (2006, p. 115) deduced that "during the early Archean . . . [crust was likely] hot enough to have its elevation limited to below sea level by continuous extensional free-boundary gravitational collapse. . . . Once above sea level, decoupled continental crust would have switched to extrusion collapse as exhibited today by the Tibetan Plateau . . . such extrusion collapse could have been endogenous throughout much or all of the Proterozoic, that is, generating mobile belts at continental boundaries which were not plate boundaries." Other important thermal modeling by Bodorkos and Sandiford (2006) emphasized the thermal and tectonic effects of blanketing of Archean crust by dense supracrustal successions.

Archean lower crust behaved, on geological time scales, almost as a fluid in hydrostatic equilibrium, and would have flowed out into ocean basins if such existed. Existence of hundreds of small, steep-sided microcontinents, as required by popular postulates of amalgamation of narrow arcs, is implausible. When upper-mantle temperature was markedly higher than now, maintenance of large hypsometric differences between continents and oceans would have been hindered. Complex structural and partial-melting reworking, including extensive interflowage, of initially widespread ancient gneisses instead accounts for observed relationships.

Continuous High Temperature or Episodic Intrusion?

Spot dates from zircons in ancient gneiss complexes, or in sandstones derived from them, commonly define spectra, or series of clusters, of nearly concordant ages. In many suites, ages scatter down or near concordia for hundreds of millions of years, or even for more than a billion years. Holden et al. (2006) identified 3000 detrital zircons from the northwest Yilgarn craton that had concordant spot U/Pb ages between 3.8 and 4.37 Ga. There is a population maximum at 4.1 Ga and a smaller maximum at 4.35 Ga, but no gaps in age distribution, so the data are "suggestive of a continuum of growth or recycling rather than episodic crustal development." The peaks and valleys in age distributions in much smaller samples of ancient Yilgarn zircons mismatch from sample to sample and area to area (Cavosie et al., 2004; Wyche et al., 2004) and likely also fit into regional continua. Nemchin et al. (2006, p. 230) deduced the 3.9–4.36 Ga zircons they studied to record "multiple episodic PT events and long term extreme temperature conditions." Amelin et al. (2004) found in their study of 4.2–3.8 Ga zircons that different grains had different initial $^{176}Hf/^{177}Hf$ ratios, consistent with "multi-episodic zircon growth rather than with ancient Pb loss." Crowley et al. (2005) found 4.2–3.8 Ga detrital grains to require substantial compositional and chronologic heterogeneity in protoliths.

Ages of zircons in young igneous rocks within ancient polycyclic migmatites, and in large and small discrete masses of younger gneisses, range down to Neoarchean and, in some cratons, Paleoproterozoic. In areally extensive studies in southern West Greenland, concordant ages of igneous zircons scatter from 4.1 to 2.6 Ga, with old and young limits varying both between nearby specimens and adjacent large areas (e.g., Nutman et al., 1996, 2004a). Some specimens yield concordant ages scattered down concordia for a billion years. The two specimens of Figure 4 of Nutman et al. (2004a) both yielded concordant-aged zircon spots ranging from 3.7 to 2.7 Ga. The Acasta Gneiss of the Slave craton shows similar patterns over lesser age spans. "No specific geological significance can be attributed to the multiple U-Pb ages" in most complex grains (Nemchin et al., 2006, p. 232), for the dates mostly are points in continua. Prolonged high-temperature histories of the lower crust and uppermost mantle are compatible with much other evidence; for example, with the eruptions of komatiite and high-Mg basalt in supracrustal assemblages, and with the severe upper-crust deformation of cratons-to-be.

Large tracts of lower-crustal crust rocks were repeatedly or continuously near their wet-solidus temperature for long periods. The proportion of felsic material in the present lower crust that was derived from the mantle very early, ca. 4.40 or 4.45 Ga, and remained near its solidus temperature throughout much of Archean time, and of younger felsic melt that came from since-delaminated mafic protocrust, is unclear. In any case, infracrustal partial melt formed repeatedly, and segregated locally or broadly to form discrete plutons. Vertical and horizontal motions and variable melt contents produced changing pressure-temperature effects and solid-liquid equilibria, and preservation of preexisting zircon was haphazard.

Hot Dry Melts, or Warm Wet Ones?

The melts recorded by the final times of crystallization of lower-crustal TTGs were mostly hydrous—their dominant mafic mineral commonly is biotite—but the oldest such rocks are <3.8 Ga. The melts in which 4.4–3.9 Ga zircons crystallized might have been hotter and dryer. Broad tracts of late Mesoarchean and Neoarchean orthopyroxene granitoids, recording relatively hot and dry middle- and upper-crustal conditions, as opposed to lower-crust granulites, also are known (e.g., Percival et al., 1992).

Internal Mobility of Lower Crust

Great and prolonged mobility of Archean lower crust is shown by its geologic features as well as by its zircon evidence for prolonged high temperature. For hundreds of millions of years, the crust flowed, churned, and sloshed slowly on lateral scales of scores or hundreds of kilometers and vertical scales of tens of kilometers. The entire crust was involved in the mixing until 3.5 Ga in some regions, and until as late as 2.7 Ga in others. The lower crust likely was continuously ductile, and may have been frequently above its solidus temperature and variably remelted, the higher-melt fractions tending to rise into the upper crust. As discussed later, the lower crust rose into the upper as diapiric batholiths, following surface loading by mafic-volcanic sequences, which in turn sank, at first as disrupted masses sinking deep, later as more coherent synclinal keels, still later only as gentle synforms. The upper crust floated on the mobile lower crust, which behaved effectively as a fluid at low strain rates.

Southwest Greenland

The Archean terrain of southern West Greenland exposes primarily orthogneisses of the middle and lower crust, and includes the best-exposed large tract of Paleoarchean gneisses known anywhere. Spectacular disruption and intermixing of rocks formed at different times, places, and depths are shown at all scales. Two of the many excellent field photographs by V.R. McGregor (e.g., 1973, 1993) are reproduced here as Figures 5 and 6. Paleoarchean gneisses, Mesoarchean and Neoarchean gneisses, granites, and dikes, supracrustal rocks younger than 3.6 Ga, and a large and initially shallow Neoarchean igneous layered complex are among the materials swirled together with very different time-depth histories. Disruption and mixing commonly are greater in the west, where exposed rocks were formed mostly at depths of 20–30 km, than in the shallower east.

A large part of the extreme ductile deformation of the rocks bearing 3.9–3.6 Ga zircons postdated abundant mafic dikes that

Figure 5. Extreme ductile mixing of lower-crust Paleoarchean gneiss and Mesoarchean additives, southern West Greenland. Amitsoq migmatite, which where dated contains mostly 3.8–3.6 Ga igneous zircons and subordinate younger ones, encloses dismembered amphibolitized Ameralik dikes (black; 3.5–3.2 Ga where dated elsewhere). Gray amphibolite and clinopyroxene-hornblende rock may include Mesoarchean basalt and komatiite that sank into ductile crust and were churned into it. Location ~10 km southeast of Nuuk. Photograph by Victor McGregor (McGregor, 1993, Fig. 12), © by Geological Survey of Denmark and Greenland.

are now contorted and disrupted within the gneisses and that, where dated by U-Pb analyses of zircon and baddelyite, are 3.5–3.2 Ga (Friend et al., 2002; Nutman et al., 2004b; White et al., 2000a, 2000b). Many of the southwestern dikes are now garnet amphibolite, so deformation there was at depths >20 km. The Fiskenaesset layered complex, age ca. 2.9 Ga (Ashwal et al., 1989; discussed previously in the context of the likely shallow origin of its anorthosite), has been shredded into gneisses with mostly older zircon ages over an exposed area of 90 km east-west (with parts farther east hidden beneath ice, and farther west beneath the sea), and 40 km, possibly 150 km, north-south (e.g., Allaart, 1982), and dispersed over a crustal depth range of ~30 km (e.g., Peck and Valley, 1996).

Figure 6. Mesoarchean dike sheath- and isocline-folded into ductile Paleoarchean Amitsoq migmatite, southern West Greenland. Folds in metamorphic rocks commonly record gradients and discontinuities in flow velocities, not bendings, which invalidates many studies that seek to discriminate and correlate generations of folds based on local geometry and superposition. Location ~20 km southeast of Nuuk. Photograph by Victor McGregor (McGregor, 1993, Fig. 14), © by Geological Survey of Denmark and Greenland.

Extremely deformed complexes that include gneisses with abundant 3.8–3.6 Ga zircons enclose known and possible supracrustal rocks with general structural concordance in many areas, and (despite the similar enclosure of young Fiskenaesset and dike rocks) these examples often are cited as requiring the supracrustal rocks to be older than the oldest numerous zircons in enclosing gneisses (e.g., Cates and Mojzsis, 2006; Friend et al., 2002; Mojzsis and Harrison, 2002; Nutman et al., 1996). McGregor (1975, 1993), who did far more fieldwork in these gneisses than anyone else, disagreed: Neoarchean gneisses commonly, but Paleoarchean gneisses never, are seen to intrude enclaves of supracrustal rocks, so the supracrustal rocks postdate the ancient gneisses and have been intercalated with them by profound ductile flow. Examine Figure 5 with this in mind. The mobility represents both ductile deformation, under upper amphibolite to granulite conditions, and fluid flow when partial melt was present. All concordant contacts with other rocks are likely to be tectonic, regardless of relative ages of protoliths. Some enclaves of supracrustal rocks long assumed to be older than most or all of the Paleoarchean and Mesoarchean protoliths of the gneisses are now dated as Neoarchean (Myers and Crowley, 2000; Nutman and Friend, 2007; Nutman et al., 2004a).

The oldest proved age of supracrustal rocks, in an enclave of highly metamorphosed iron formation and mafic and ultramafic volcanic(?) rocks, is 3.6 Ga (Manning et al., 2006). This is also the oldest proved age of supracrustal rocks anywhere.

Contortion at multi-kilometer scales is further shown throughout the region by the jumbled gneissic "terranes"—tracts of gneisses that vary irregularly in dominant lithologies, uniformity, and overlapping or offlapping of age spectra of zircons (e.g., Friend and Nutman, 2005; McGregor, 1993; Nutman and Friend, 2007; Nutman et al., 1996, 2004a). Igneous zircons within these various tracts range from 3.9 to 2.6 Ga but are dominantly 3.8–3.6 Ga in much of the strikingly polymict and polycyclic material (e.g., Cates and Mojzsis, 2006; Crowley, 2003; Crowley et al., 2002; Friend and Nutman, 2005; Mojzsis and Harrison, 2002 [who reported 4.1 Ga for a single zircon spot]; Nutman et al., 1996; Whitehouse and Kamber, 2003). Ages of metamorphic overgrowths on old grains scatter down to 2.6 or 2.5 Ga. Tectonic mixing was vertical as well as horizontal, providing opportunity for erratic solution or crystallization of zircon. Supracrustal rocks sank into the lower crust, and lower-crust granulites were mixed into the middle crust. Relative importance of new magmatism from the protocrust and of remobilization and remelting of ancient gneisses as causes of zircon-age spreads are unclear.

Most geologists working with these complexes assume that all TTG is produced by arc magmatism, and visualize the structure as consisting of grossly contorted megathrusts and sutures (Nutman and Friend, 2007, and scores of predecessors). As information accumulates, individual postulated sutures are proved nonexistent; thus, Hollis et al. (2005) found that one of the largest shear zones conjectured to be a suture did not separate the "terranes" attributed to it. The postulated structural coherence postulated is to me incompatible with the pervasively disruptive deformation. Although different vaguely delineated tracts indeed are typified by different ages of zircons, I see these as recording incomplete mixing of batches of old, variably remobilized, and new melts over a billion years.

Other Areas

Ancient gneisses elsewhere show similar mixing and hence mobility, though without spectacular exposures. The Narryer Gneiss of southwest Australia has yielded zircons as old as 4.2 Ga, and derivative Archean sedimentary rocks contain detrital zircons as old as 4.37 Ga. It appears to be a hash comparable to that of Greenland, and similarly contains much younger shredded

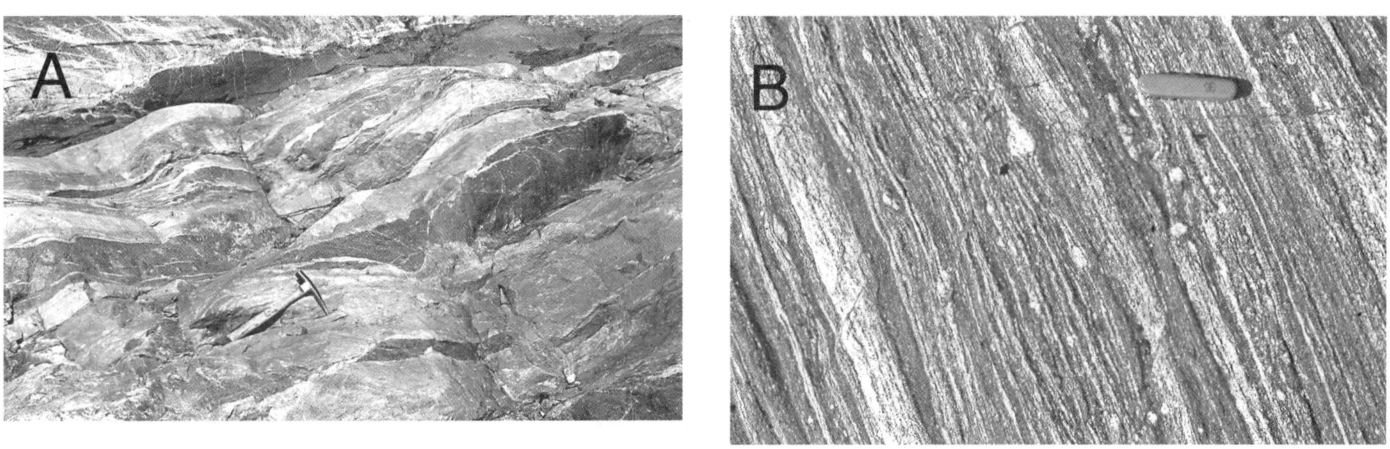

Figure 7. Paleoarchean and Mesoarchean midcrustal Acasta gneiss complex, basement to Neoarchean greenstone sections. These and nearby outcrops have yielded zircons with concordant U-Pb ages scattered from 4.2 to 2.9 Ga (e.g., Bowring et al., 1990; Iizuka et al., 2006) and record a billion years of crustal mobility, often with above-solidus temperatures. (A) Polycyclic migmatite of tonalite, hornblendite, pegmatite, and fragments of dark dikes. Hammer for scale. (B) Mylonitized migmatite. High-strain zones are common in metamorphic rocks and need not mark significant boundaries. Pocketknife for scale. Acasta Lake, west-central Slave craton, northwest Canada.

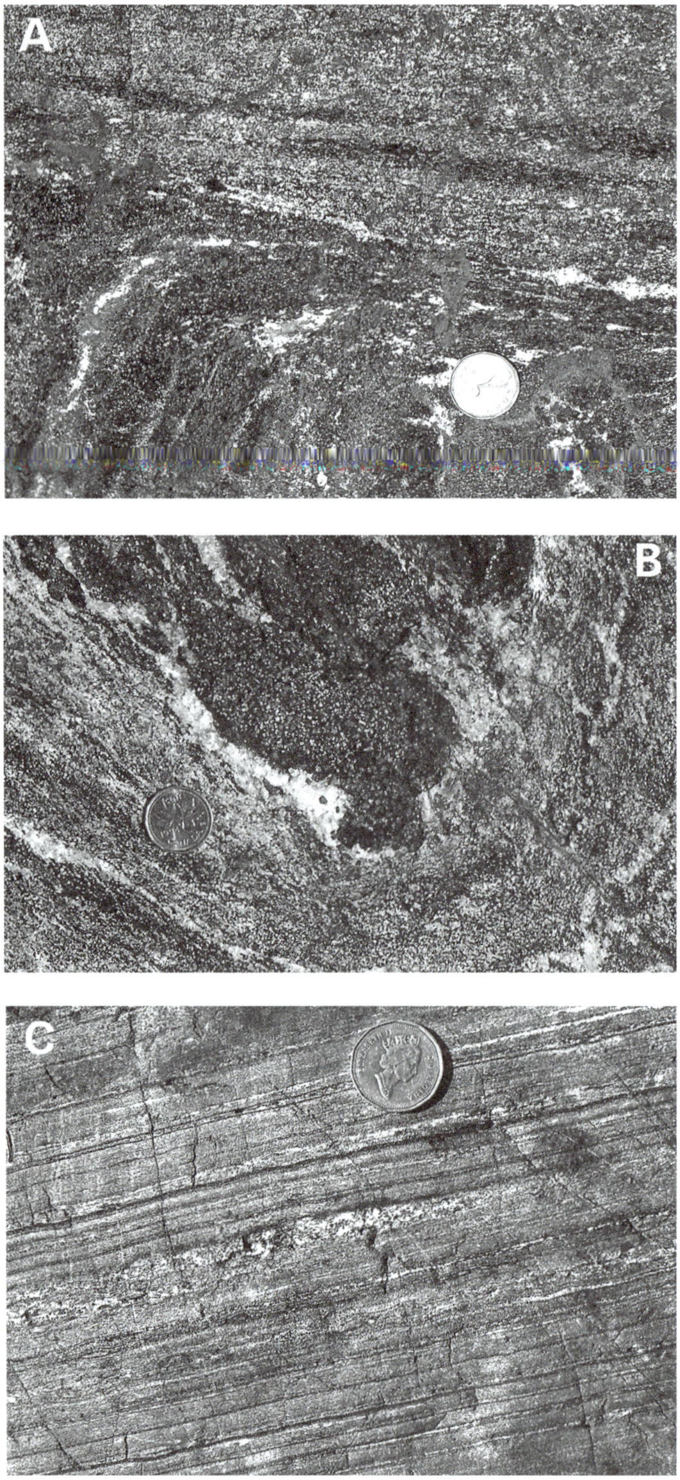

Figure 8. Archean lower-crust rocks, raised from beneath Neoarchean granite-and-greenstone upper crust in southeast part of Kapuskasing uplift, Superior craton, Ontario. (A) Top of enclave (black) of clinopyroxene-garnet-hornblende rock, possibly derived from subjacent mafic protocrust, enclosed in mylonitic tonalite gneiss (gray, top). Coin diameter ~2.5 cm. (B) similar enclave (black; light spots are garnets) in tonalitic gneiss. Coin diameter ~1.8 cm. (C) Granulite-facies mylonite ("plane gneiss," or "straight gneiss") of garnetiferous TTG. Coin diameter ~2.5 cm. Location ~10 km northeast of Chapleau.

calcic (low-pressure?) anorthosite and likely supracrustal rocks (Myers, 1988; Myers and Occhipinti, 2001). Deep-crustal migmatites exposed in the Vredefort Dome of the Kaapvaal craton contain supracrustal rocks (Hart et al., 2004). The Acasta Gneiss, which contains zircons to 4.2 Ga (Iizuka et al., 2006), of northwest Canada displays assemblages indicative of chaotic mixing (Fig. 7; Iizuka et al., 2007). Ductile TTG gneisses exposed in the Proterozoic Kapuskasing uplift in the Superior craton enclose large and small enclaves of clinopyroxene-garnet-hornblende granulite (Fig. 8)—the expected lithology of the initially global mafic protocrust from which the TTG was derived. Single- and batch-zircon ages of TTG and granulite are all about 2.65–2.85 Ga (Moser, 1994; Moser et al., 1996).

Archean lower crust is known in a number of cases to have remained hot long after cooling of overlying upper crust. Lower-crust xenoliths from the Slave craton have igneous zircons ~2.60 Ga, the age of overlying upper-crust granites, and also zircons as much as 100 m.y. younger (Davis et al., 2003). Lower-crust xenoliths from beneath the ~2.7 Ga Abitibi greenstone belt of the Superior craton have igneous zircon ages of ~2.8-2.6 Ga but granulite-facies metamorphic-zircon ages of 2.5-2.4 Ga (Moser and Heaman, 1997).

Mixing Process

Seismic-reflection profiles (e.g., Fig. 21) show Archean lower crust to be typified by broadly undulating patterns of flow in gneisses above subhorizontal Mohorovičić discontinuities. The flat basal discontinuities show the lower crust to have behaved effectively as a fluid, incapable of maintaining large surface loads as does the stronger crust of the modern Earth. The shredding displayed by lower-crust exposures requires that the flow patterns record great and pervasive differential motions. Changing patterns of flow around huge lozenges of temporarily lower-strain materials seem indicated.

"Continental Nuclei" and "Terranes" of Gneisses

Plate tectonics by definition requires internally rigid plates of lithosphere, and the preceding descriptions are incompatible with the existence of such plates. The internal mobility of the upper crust, as shown subsequently, also is incompatible with rigid plates. Nevertheless, many proponents of early-Earth plate tectonics postulate generation of the ancient gneisses in hundreds of different subduction systems, followed by amalgamation of these by plate convergences. This conjecture regarding early plate interactions is distinct from—although often combined with—that which proposes that the overlying concordant-section supracrustal rocks have themselves been interthrust by plate-tectonic processes, which have no modern analogues, from widely separated sites.

Because Archean magmatic rocks younger than ca. 2.72 Ga are similar across much of the Yilgarn craton whereas older Neoarchean granites show more diversity in zircon age spectra or Nd model ages, Cassidy and Champion (2004) proposed that cratonic nuclei formed in diverse locales by subduction processes and were amalgamated by plate-tectonic convergence prior to late unified evolution. Griffin et al. (2004a) saw different boundaries in the same region in their zircon ages and Hf isotopes. Friend and Nutman (2005) and Nutman and Friend (2007) postulated that tracts of Greenland TTG gneisses, discriminated partly by differently overlapping ages of zircons in sparse samples and partly by the presence or absence of polycyclic and heterolithic migmatites, in ~15,000 km^2 of southern West Greenland, represent six island arcs that were swept together, and complexly interthrust and interfolded on all scales, late in Neoarchean time. They appealed to thin zones of mylonitic gneisses, seen locally within vast tracts of deformed plutonic rocks, as possible sutures, and claimed (Friend and Nutman, 2005, p. 159) that this speculation provides "key evidence for the operation of some form of early Precambrian plate tectonics." Subsequently, Hollis et al. (2005) and Nutman and Friend (2007) recognized that some of these high-strain zones are within, not bounding, the hypothetical "terranes." Among many who have argued for amalgamation of old microcontinents of TTG are Percival et al. (2001, 2004a), in the Superior craton. There are no modern analogues for such hypothetical aggregates of TTG-dominated arcs, for no such arcs exist, individually or collectively.

The hypothetical nuclei usually are attributed to arc magmatism because they consist of TTG, although Kröner and Layer (1992) suggested that nuclei formed as felsic oceanic islands (for which also there are no modern analogues) atop small plumes (which are not proved to exist even in the modern Earth). Boundaries commonly are "cryptic" or "hidden" (e.g., Schmitz et al., 2004, for the Kaapvaal craton), or are placed arbitrarily between data spots, or are assigned to convenient shear zones, which commonly are too young to record the postulated ancient juxtapositions. Secondary conjectures invoke squashing or thrusting the TTG masses together as intervening high-density lithosphere disappeared, leaving none of the remnants such as abound in Phanerozoic sutures.

Nothing in the ancient assemblages resembles what is seen in known accretionary tracts in the modern Earth, nor has anything akin to a modern suture been found at any crustal level. Thin shear zones of local rock types in Archean terrains have been termed "mélanges" by some geologists but are utterly different. The Philippine Islands and the Klamath Mountains exemplify long-continuing amalgamation of island arcs, as noted previously, and share no apparent features with Archean terrains. Nor can Andean-arc batholiths, such as the Sierra Nevada and Idaho, be invoked as analogues, for they too are quite different.

Continuity of Felsic Crust

All cratons studied seismically have continuous felsic crust beneath greenstone belts. Many seismic studies of Archean crust show it to commonly lack the thick basal mafic layer that typifies younger continental crust, to be generally thin, and to have little relief on its Mohorovičić discontinuity (Moho). Kaapvaal crust has been studied in particularly high resolution (Nair et al., 2006; Niu and James, 2002). The Moho is sharp and flat, and even the basal crust is felsic, not mafic. Nair et al. recognized that this felsic crust likely was generated from deeper mafic, not ultramafic, material, and that the residual part of the mafic material must have sunk before the depleted-ultramafic present lithospheric mantle was stabilized. They postulated subduction-related mechanisms, but delamination of the residual part of a mafic protocrust would accomplish the result. I take the flat, sharp Moho (which is often seen in reflection studies of Archean crust elsewhere) as further evidence that lower crust was commonly too mobile to support large topographic loads during Archean time. The general lack of mafic basal crust that might record a source region for crustal TTG indicates that if such mafic crust once existed, it delaminated and sank long ago.

Many Archean upper-crustal batholiths include abundant gneiss raised from midcrustal depths (Fig. 9) and show obvious chemical, isotopic, and xenocrystic evidence for remobilization, with varying proportions of new melt, from old TTG. Many mafic and ultramafic lavas similarly display chemical and isotopic features, and even zircon xenocrysts, requiring that they assimilated old felsic material, and in many areas can be seen to lie unconformably upon TTG basement (e.g., Bleeker, 2002). Other tracts of granites and supracrustal rocks commonly are interpreted to be juvenile granites and ensimatic supracrustals unless they have yielded such geologic evidence, although as emphasized previously the Sm-Nd isotopic data cited for juvenile origins are satisfied by origins from ancient mafic protocrust rather than from the mantle. The general continuity of deeper felsic crust required by seismic and gravity studies, considered with the general continuity of dome-and-keel patterns across purported boundaries between tracts assumed to have had completely different crustal histories, indicates that broad continuity of felsic crust was established relatively early in the

Figure 9. Migmatitic gneiss raised in upper-crustal Mesoarchean diapiric batholith. Shallow Archean batholiths vary from mostly uniform young granitic rocks to mostly raised midcrustal materials: degrees of melting and mobilization of preexisting materials vary widely. Notebook is 21 cm long. Northern Shaw batholith, northeastern Pilbara craton.

history of each craton. The common little-varying areal density of diapiric batholiths, and their contacts primarily against the oldest strata locally preserved in synforms, are further evidence for continuity of felsic crust that rose into the overlying supracrustal rocks, and are evidence against unrelated juvenile origins for different batholiths.

Decoupling of Upper and Lower Crust

The very ductile lower crust of Archean cratons-to-be may have been generally decoupled from the upper crust. In the northeast Pilbara craton, seismic-reflection, aeromagnetic, and gravity surveys are all consistent with the extension of upper-crustal dome-and-keel structures down to a discontinuity at a depth of ~14 km, below which are subhorizontal or undulating gneisses (Wellman, 2000). Peschler et al. (2004) used an upward-continuation wavelet method of gravity modeling to deduce that Pilbara, east Yilgarn, and southeast Superior greenstone belts have similar maximum depth extents of ~10 km. Bailey (2006) recognized that decoupling was required by paleothermal considerations.

Lower-crust Archean gneisses have received relatively little structural study. Dips vary widely, from steep to undulating and subhorizontal but are dominantly gentle (e.g., Figs. 3, 5, 7, 10) . As seen at the scale of seismic-reflection profiles (Fig. 21), they mostly are undulating, consistent with deformation dominated by laminar flow. Upper crust shows dominantly steep structures, products of vertical dome-and-keel tectonics and of variably severe lateral deformation, as discussed subsequently. The scarcity of large vertical offsets (other than batholithic diapirism, and complementary sinking of supracrustal rocks, driven by density inversions) on upper-crustal structures shows that upper crust effectively floated on lower. Consideration of the laminar flow of the lower crust, maintenance of a flat Moho, great mixing within the lower crust, and the ubiquitous presence of felsic crust beneath the quasi-floating upper granite-and-greenstone crust together suggest that the discontinuous lateral deformation of the upper crust was a response to the more continuous flow in the lower crust. Among the few studies of lower-crust structure is that by Moser (1994) and Moser et al. (1996) in the Proterozoic Kapuskasing uplift in the Archean Superior craton, where sub-decoupling gneisses were flattened and transposed (e.g., Fig. 3) by combined simple and pure shear, and extended parallel to the elongation of domiform upper-crustal granites.

Flow of the lower crust presumably had a gravitational drive enabled by elevation differences that resulted from uneven distribution of advected and convected heat. Included were delamination and sinking of cool, dense material—but this was not rigid-plate tectonics.

Figure 10. Folded Neoarchean middle-crust amphibolitic and tonalitic migmatite and gneiss (peak on right; probably includes retrograded granulite) and monzogranite of Archean-Proterozoic boundary age (left distance). Mt. St. John, Teton Range, Wyoming craton. Identifications from Love et al. (1992).

Effect on Composition

Ponder Figures 5 and 8 and the effect of top-to-bottom crustal mixing on composition. Early TTG, felsic mobilizates rising from it, supracrustal rocks, intrusive rocks from diverse crustal and mantle sources, and the then-existing mafic protocrust were swirled together during half a billion or a billion years. The mixing minimized and delayed the transfer of radioactive elements to the top of the crust, thus maintaining high heat generation in the lower crust. To the extent that then-subjacent mafic crust was involved in the physical mixing, Sm/Nd would be maximized and the effective start of the Nd model-age clock delayed for not only the aggregate but for subsequent more felsic mobilizates from it.

ONLY FELSIC GNEISSES ARE KNOWN AS BASEMENT BENEATH SUPRACRUSTAL ROCKS

Archean supracrustal rocks, including mafic and ultramafic lavas, are ensialic wherever their initial bases have been seen. Seismic velocities show the lower crust beneath upper-crustal basalts and komatiites to be everywhere felsic and intermediate rocks, not mantle materials. Depositional bases of Archean supracrustal rocks have been found in many cratons, and invariably are with felsic gneisses and directly overlying sedimentary rocks, and never with mantle rocks such as are seen at the base of Phanerozoic oceanic crust and ophiolites. Basal strata on basement are typified by conglomerate and quartz-rich and feldspathic sandstones, often include banded iron formation and volcanic rocks, and are a few meters to a kilometer or more thick (Fig. 11; Bleeker, 2002; Blenkinsop et al., 1993; Böhm et al., 2003; Ketchum et al., 2004; Thurston, 2002; Wyche et al., 2004; and many others).

Archean supracrustal assemblages, deposited concordantly upon such basal strata, commonly contain pillow basalt, magnesian basalt, and komatiite. Some proponents of Archean plate tectonics (e.g., de Wit, 1998, and Furnes et al., 2007) refer to these as "ophiolitic" merely because of their mafic composition—but

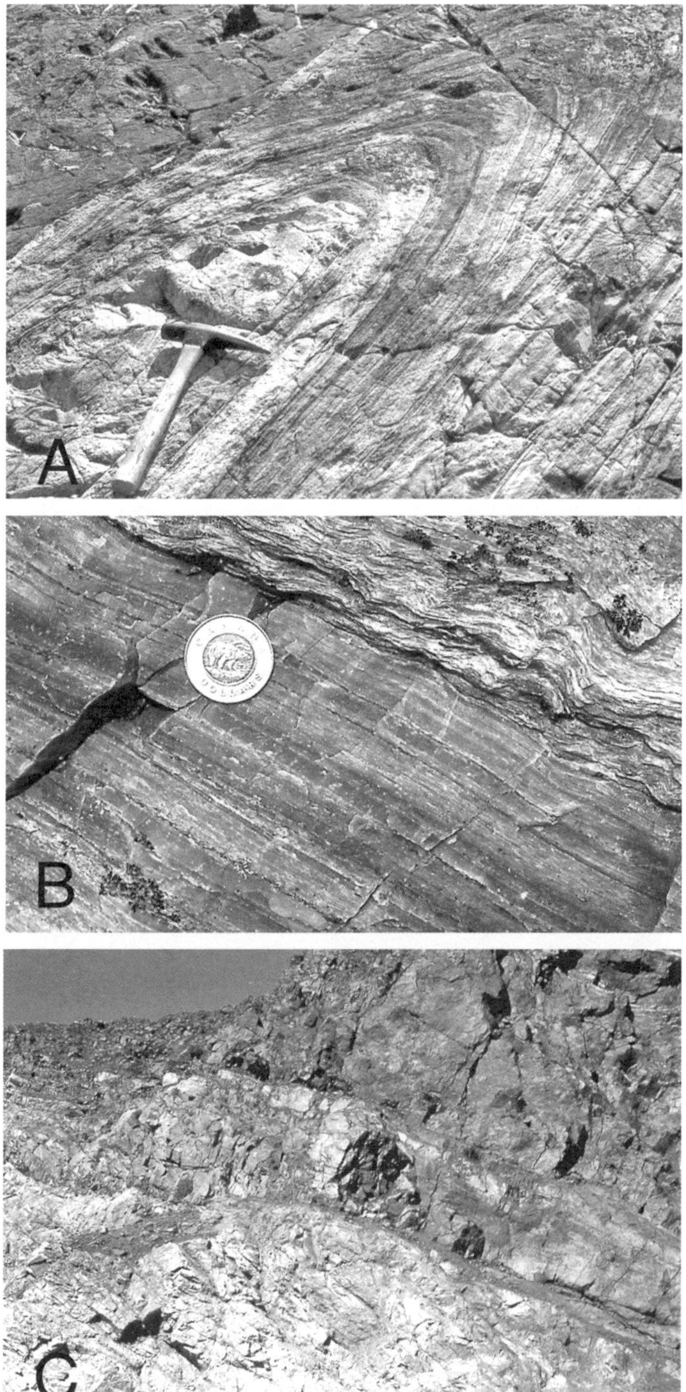

Figure 11. Basal sediments of Archean supracrustal successions. All greenstone belts whose depositional bases are known overlie sediments, such as these, deposited on felsic gneisses; no greenstone belt is proved ensimatic by field relationships. (A) Folded quartzite contains detrital zircons with igneous ages of 3.7–2.9 Ga (Isachsen and Bowring, 1994) and is cut by amphibolitized dike (upper left). It unconformably overlies ancient gneiss, and conformably underlies thick Neoarchean greenstone. Dwyer Lake, 30 km north of Yellowknife, Slave craton. Hammer for scale. (B) Fuchsitic quartzite and (upper right) metamorphosed feldspathic sandstone. This unit, the Lewis-Storey assemblage, contains detrital zircons with igneous ages of ca. 3.05–2.95 Ga, and underlies a Neoarchean greenstone succession (Percival et al., 2006a). Coin diameter 3 cm. East side of southern Lake Winnipeg, Superior craton, Manitoba. (C) Unconformity at base of Mesoarchean supracrustal succession. Gneiss (light, lower left) is overlain by several meters of quartzite (layer from upper left to lower right; age 3.4 or 3.3 Ga) and by thick iron formation (dark, upper right). Gneiss here has a zircon age of 3.44 Ga but this batholith has elsewhere yielded Paleoarchean zircon xenocrysts (Van Kranendonk et al., 2001). Batholith breached surface through ca. 3.5 Ga greenstone succession, and its further post-unconformity rise tilted the sediments. North margin of Muccan diapiric batholith, northeast Pilbara craton.

none of the rocks at issue compositionally resemble any of the components, even considered singly, of Phanerozoic ophiolites. Pillow basalts (of non-MORB composition) are common in the Archean, but otherwise not even vague analogues for any part of the typical oceanic succession—in order downward, pillow basalts, compositionally similar sheeted dikes, mafic plutonic rocks giving way to ultramafic ones, and, at the base, residual mantle—have ever been found in Archean assemblages. Bickle et al. (1994) demonstrated that a number of Archean mafic-and-ultramafic assemblages cannot be either ensimatic or ophiolitic. Kusky et al. (2001) claimed to have found an almost complete Neoarchean ophiolite in North China, but Zhai et al. (2002) reported that the purported sheeted dikes do not exist and that the gabbroic and ultramafic rocks are not Archean, the misidentified harzburgite tectonite being a minor rock type within a Mesozoic layered complex. Kusky (2002) acknowledged that much of his "stratigraphy" was based on Mesozoic gabbro and pyroxenite. Another misidentification of sheeted dikes by Kusky is illustrated in Figure 12. I noted elsewhere (Hamilton, 1998a), also on the basis of my own observations, that Kusky's (1991) purported subduction mélange in the Slave craton actually consists of thin shear zones of local rock types. Furnes et al. (2007) claimed to have found a small area of "cogenetic" ophiolitic sheeted dikes and pillow lavas in the Greenland Archean, but the dubiously characterized "sheeted dikes" [lava flows?] have the composition of pyroxenitic komatiite, and the pillow lavas are ferroandesitic; the latter rock type is proved, and the former suspected, to be ensialic on other cratons, and neither is oceanic-arc or MORB-like as Furnes et al. asserted. Even if the "dikes" are correctly identified, they cannot have fed the lavas, and they indicate local crustal extension but do not provide the "compelling structural evidence of . . . dike injection at a spreading ridge" wrongly claimed by Furnes et al. (2007, p. 1706).

Some geologists (e.g., Mueller et al., 2005) assume the clastic component of the thin basal sediments between felsic basement and volcanic sections to indicate continental rifting and, by distant extrapolation, plate tectonics. None of the actual characteristics of

Figure 12. Hornblende schist, a case of mistaken identity. Kusky (1991, his Fig. 5 and p. 824) claimed this outcrop to display sheeted diabase dikes formed at an oceanic spreading center. He depicted the structures dipping steeply left as 13 sheeted dikes, which "show preferential (70%) one-way chilling with most dikes indicating spreading center lay to the northwest" [left], and therefore to define part of an Archean ophiolite. In fact, the outcrop consists of uniform fine-grained hornblende schist, the parting planes wherewith Kusky delimited "dikes" are discontinuous joints along foliation, and there is no textural variation to suggest the presence of dikes or chilled margins. Further, this Neoarchean metavolcanic section is known to lie depositionally upon Mesoarchean tonalite-trondhjemite-granodiorite basement. Light spots are lichens. Location 8 on Kusky's Figure 2, 3 km south of south arm of Point Lake, Slave craton.

rifted margins—major normal faults, scarp-facies sediments, severe rotations of faults and strata, oceanward-thickening wedges of post-rift strata—have been demonstrated in these settings. I see instead a setting like that of the cratonic Upper Cambrian in the Grand Canyon region of the Colorado Plateau: low hills of basement rocks were flooded by shallow water, and locally derived sands and conglomerates gave way upward to quiet-water deposits.

Did Lithospherically Defined Oceans Exist?

Sediments and pillow basalts indicate that seawater existed from at least 3.6 Ga onward. Many Archean basalts are commonly assumed to be ensimatic, but nowhere have mantle rocks been seen beneath them either in outcrop or in seismic records. No scraps of Archean oceanic mantle have been found either in implausibly postulated Archean sutures or in Paleoproterozoic orogens between Archean cratons. Subaerial hills and low uplands are required to explain Archean clastic sediments after 3.5 Ga, but the surface of most preserved crust was below sea level, at ill-constrained water depths, while most supracrustal mafic and ultramafic rocks were erupted. Although the extreme mobility of Archean lower crust makes unlikely continents standing many kilometers above ocean floors, very large regions of water several kilometers deep must have existed if the volume of seawater was comparable to that now (Bailey, 2006; Bickle et al., 1994; Galer, 1991).

Archean Impact-Spherule Layers

One line of evidence cited against extensive felsic crust comes from the dozen or so thin layers in Mesoarchean, late Neoarchean, and early Paleoproterozoic stratigraphic sections that are known to contain abundant impact-melt spherules and microtektites recording bolide impacts, and which provide indirect evidence as to the character of Earth's crust as it then existed (Glikson, 2005; Glikson and Allen, 2004, and references in each). Some of the layers have relatively high contents of platinum-group elements that may record contributions from metallic bolides. Chlorite and other secondary mafic minerals, and sericite, commonly are conspicuous in the silicate fraction of the spherules. Impact melting and volatilization, and subsequent condensation, much decreased mobile-element concentrations, and severe hydration, and carbonatization or silicification, affected almost all spherules, so deducing target rocks from present compositions is difficult. Shocked quartz has been recognized in only one bed, but as quartz anneals at relatively low temperature the significance of this lack is unclear. Immobile element contents scatter widely, within and between units, but commonly are regarded as indicating derivation from dominantly mafic, partly intermediate or ultramafic, targets. Glikson and other investigators of these ejecta have concluded that the Archean Earth had much more crust of oceanic type than continental. The compositions plotted by Glikson (2005) do not, however, appear to be oceanic: I read his Figure 1A as suggestive of greenstone belt plus felsic crust, and his Figure 1B as TTG, perhaps with added basalt. The data cannot be fit to a dominantly ultramafic target, as would be required by major impacts on thin oceanic crust like that of the modern Earth.

GRANITE-AND-GREENSTONE TERRAINS: A BILLION YEARS OF PROGRESSIVE CONTINENTAL STABILIZATION

Archean cratons are exposed primarily at upper-crustal levels and are typified by granite-and-greenstone terrains. These display in many places a dome-and-keel tectonic style, whereby dense supracrustal rocks sank as synforms between rising batholiths. Some of these batholiths had abundant new melt and rose rapidly, and others were intermittently diapiric over hundreds of millions of years. The patterns of vertical tectonics are distorted or disrupted by lateral deformation that varies from slight through moderate (as, elongation of batholiths by synintrusive orthogonal shortening and extension) to severe pure and simple shear. Even the severe deformation, however, was seldom accompanied by major compressional or extensional faulting, for few abrupt large changes in crustal levels of exposure are known. The crust changed little in area because it effectively floated on lower crust too weak to support large topographic relief. Vertical and lateral styles grade together—granites rose and greenstones sank as the crust was being deformed laterally—but developments were prolonged and relative importances varied greatly in time and space. The pervasive internal mobility of the upper crust demonstrated by the lateral structures is commonly attributed to undefined plate-tectonic processes, but that very behavior, as well as the even greater long-continuing mobility of the lower crust, indicates that plate tectonics, which by definition requires quasi-rigid plates, was not operating. The rationales (e.g., Bodorkos and Sandiford, 2006, and Smithies et al., 2007) that lateral crustal deformation requires plate tectonics, and that Archean processes operating where dome-and-keel geology is well preserved were unrelated to those where there is lateral deformation, are, respectively, a non sequitur and a conjecture that is disputed here. Among the relatively few who have recognized diapiric rise and lateral deformation as long-continuing synchronous processes enabled by crustal mobility are Chardon et al. (2002), Gee et al. (1981), Lin (2005), and Parmenter et al. (2006).

The full granite-and-greenstone style, wherein supracrustal rocks generally remained in or on the upper crust whether in dome-and-keel or lateral-deformation mode, developed during a time-transgressive window that had begun to open by ca. 3.5 Ga, when felsic crust had cooled enough to permit dense mafic and ultramafic melts to rise through it. In some sectors, the crust was then stiff enough to support thick accumulations of the resulting still denser volcanic rocks crystallized on the surface, but in most regions the early volcanic rocks soon foundered into the deeper crust and became intermixed materials such as those noted elsewhere in this essay. Where the upper felsic crust had already cooled enough to provide substantial support, the density inversion was partly righted as coherent supracrustal rocks sank as synclinal keels between rising domiform diapiric batholiths. This degree of stiffening was not reached in most sectors until some time between 3.1 and 2.8 Ga, and the age of this transition was subuniform in some regions but erratic in others.

The window of full granite-and-greenstone development effectively closed ca. 2.8 Ga in the sectors where it opened earliest, but stayed open until 2.6 or 2.5 Ga where it opened last. Several cycles are recorded in some cratons. The final cycle in some cratons is recorded by full development of supracrustal rocks but only limited rise into them of diapiric batholiths, which continued to rise slowly in some sectors until ca. 2.1 Ga. Only then was the crust stabilized enough to form internally rigid cratons.

Archean supracrustal sequences begin with basal-unconformity sedimentary sections, commonly thin, overlying preexisting felsic basement, wherever depositional bases have been seen. The thick sections above these basal strata commonly are typified in their lower parts by plains of mafic and subordinate ultramafic submarine lavas, and in their upper parts by felsic volcanic rocks, which often form thick and thin intercalations in mafic rocks, in long-continuing single or multiple sequences. Many of these supracrustal successions have been shown, as noted previously, to be concordant sections with subregional stratigraphy. Conversely, speculation that the supracrustal rocks originated in narrow belts defined by plate-interaction magmatism or deformation is unsupported by geologic data. Rising diapiric batholiths fed concurrent felsic volcanism and shallow porphyries, produced domiform granites separated by sinking synclines of supracrustal rocks, and produced unconformities (e.g., Fig. 11C). Archean granitic assemblages, both as ancient gneisses and as young batholiths, often are assumed to require arc magmatism, but, as emphasized previously, they are not in primary linear complexes, and they

Figure 13. Coherent turbidites. Locally derived turbidites like these Neoarchean examples occur high in greenstone sections in many cratons. Although often referred to by proponents of Archean plate tectonics as "accretionary wedges" merely because they are turbidites, they nowhere show the mélange character required by that designation, and so provide evidence against, not for, operation of plate tectonics. (A) Subvertical graded beds have tops to left, and although contact-metamorphosed at amphibolite facies (sillimanite + andalusite + cordierite) are little deformed. Eastern Point Lake, Slave craton. (B) Folded turbidites, each bed consisting of light metasandstone grading up into dark metasiltstone, have tops to lower left. Metamorphosed at lower greenschist facies, but little internal deformation. North shore of Indin Lake, Slave craton. Hammer for scale.

differ in composition and lithologic associations from modern arc rocks. Thick sections of coherent sedimentary rocks, including voluminous turbidites such as those of Figure 13, blanketed older units in many cratons. That the late clastic sediments, including the turbidites, were eroded primarily from breaching batholiths is shown by their dominantly tonalitic and granitic composition (Condie, 1981). Though often misnamed "accretionary wedges" or "accretionary terrains" merely because they include turbidites (e.g., Hofmann et al., 2004), these late and widely distributed strata share no features of occurrence or disruption with such namesakes and so provide evidence that the strata are not related to plate convergence. In the few places where they comprise belts of relatively thick strata, as opposed to irregular veneers, the strata may have accumulated in ductile rifts formed as parts of the lateral deformation of the cratons-to-be.

The granite-and-greenstone, diapir-and-synform style, and its significance in righting density inversions, was first emphasized by Macgregor (1951). Among hundreds of papers expanding the thesis, describing the petrology and structure of individual and multiple examples of this style, are Bodorkos and Sandiford (2006), Chardon et al. (1998, 2002), Glikson (1979), Ridley (1992), and others cited in the following sections. Where minimally disrupted by lateral shear, greenstone belts are networks of synforms of supracrustal rocks crowded aside by, and sunk between, diapiric batholiths. The belts are defined by deformation of, mostly, semiconcordant volcanic successions, not by linear volcanic features. The batholiths contain both remobilized basement gneiss and new magmatic rocks, of which the latter might represent both more complete melting of felsic basement rocks and new partial melts of hydrated crustal mafic rocks. Metamorphism is mostly of contact type, the dominant strain in supracrustal rocks is batholith-side-up where lateral deformation is minimal, and both strain and metamorphism are low distant from batholiths.

Granite-and-greenstone and floating-tectonics styles of simultaneous vertical and lateral deformation over extremely long durations have no apparent analogues younger than early Paleoproterozoic. Individual systems commonly evolved over 100–500 million years, whereas modern magmatic arcs have maximum lifespans of a few tens of millions of years in one general location. Modern oceanic spreading systems have generated rocks in narrow bands only for short periods.

The following examples illustrate the range of granite-and-greenstone styles. Mostly Mesoarchean northeast Pilbara, and Neoarchean Zimbabwe, are of simple dome-and-keel style. Mesoarchean and Neoarchean northwest Pilbara and Kaapvaal, and Neoarchean northwest Superior, display both dome-and-keel and laterally disrupted styles. Neoarchean Yilgarn shows mostly the disrupted mode. South Pilbara and much of Kaapvaal show regionally continuous Neoarchean supracrustal sections, of greenstone-belt type, arrested in development because rise of diapiric batholiths into them was of only modest extent. Southern west Greenland shows widespread Paleoarchean to Neoarchean lower crust (discussed earlier), and also upper crust in partly disrupted dome-and-keel mode.

Pilbara Craton, Western Australia—Three Styles, Varying Ages

The Pilbara craton displays three variants of granite-and-greenstone upper-crustal Archean geology (Fig. 14). In the northeast is a Mesoarchean and early Neoarchean vertical-tectonics dome-and-keel complex (Fig. 15), the best-exposed and best-documented region of this type anywhere. In the northwest, a basically similar late Mesoarchean and early Neoarchean complex was disrupted by lateral deformation that in part was concurrent with the vertical tectonics. In the south is a late Neoarchean supracrustal accumulation that went only partway through granite-and-greenstone development. Thorne and Trendall (2001, pl. 1) presented a 1:500,000 geologic map of the entire craton compiled from 1:250,000 maps, which are available on DVD-ROM (Geological Survey of Western Australia, 2005a). New 1:100,000 geologic maps are available for parts of the region.

Northeast Pilbara—Mesoarchean Vertical-Tectonics Granite-and-Greenstone Terrain

The northeast part of the Pilbara craton superbly displays dome-and-keel geology, the minimally modified products of rise of diapiric felsic batholiths from the lower crust and complementary sinking of dense supracrustal rocks. Excellent reconnaissance mapping by Arthur Hickman and associates led to early recognition of this tectonic pattern (e.g., Hickman, 1984), which has been further documented by detailed fieldwork, supported by zircon U-Pb geochronology, by Hickman and many others since. Among recent papers summarizing structure, stratigraphy, petrology, geochronology, and structure in dome-and-keel and regional-stratigraphy terms are Bodorkos and Sandiford (2006), Hickman (2004), Hickman et al. (2001), Pawley et al. (2004), Sandiford et al. (2004), and Van Kranendonk et al. (2001, 2002, 2004a, 2004b, 2007). Figure 15 shows batholithic domes separated by keels of supracrustal rocks. Basement gneisses have not been identified in outcrop but much, and perhaps all, of the terrain must have had such a basement, for 3.7–3.5 Ga zircons occur as xenocrysts in younger granites and as detrital grains in Mesoarchean sedimentary rocks, and felsic contamination of 3.5 Ga basalts and komatiites is indicated by isotopic studies. In the eastern part of the terrain, the oldest supracrustal rocks are kilometers-thick 3.5–3.3 Ga basalt, high-Mg basalt, ultramafic lavas, and, scattered throughout the section, thin to thick units of felsic volcanic rocks, and cherts. A 3.4 Ga chert unit is continuously present throughout an area 200 km across, and this, plus many zircon ages of other mapped units, demonstrates stratal and bathymetric continuity and disproves subsequent amalgamation of initially separated tracts to form the supracrustal sequence. Late clastic strata in the greenstone sections formed ca. 2.9 Ga. Domiform granites rose, as the supracrustal rocks sank, from 3.4 to 2.7 Ga in the east—an enormous time span without modern

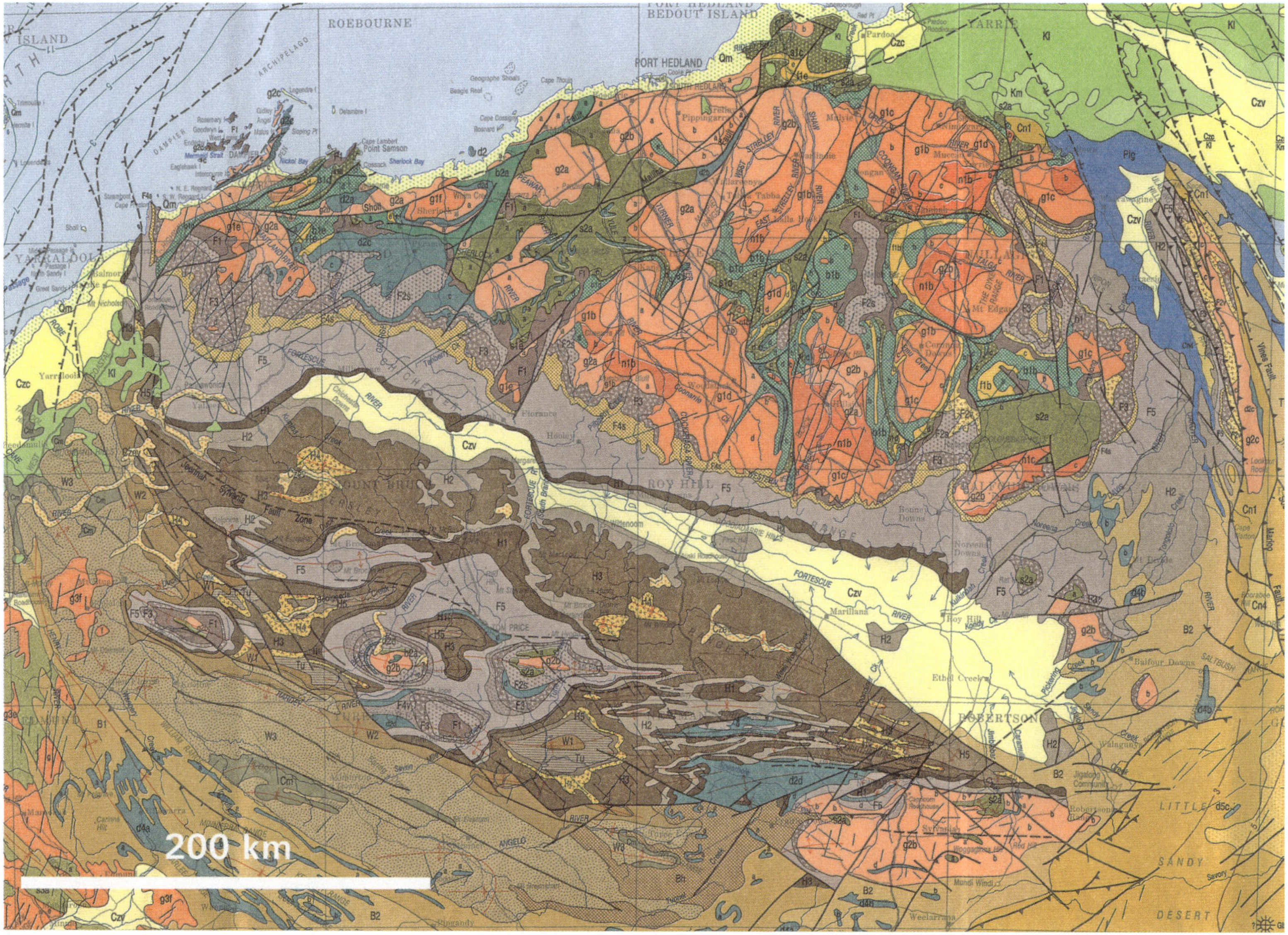

Figure 14. Geologic map of Pilbara craton, Western Australia, showing three styles of Archean granite-and-greenstone geology. Northeast area displays Mesoarchean and early Neoarchean dome-and-keel terrain: green, mostly mafic and ultramafic volcanic rocks; olive green, mostly sedimentary rocks; dark yellow, mostly felsic volcanic rocks; red and pink, granitic and gneissic rocks. Northwest area shows laterally disrupted granite-and-greenstone terrain, mostly late Mesoarchean and early Neoarchean, similar colors. Southern area, and remnants in north, preserve younger Archean sheet-greenstone terrain subjected to only modest rise of diapiric batholiths: violet, middle Neoarchean rocks, mostly mafic volcanics; chocolate brown, late Neoarchean and earliest Proterozoic sediments and subordinate volcanics. Limited continued rise of diapiric batholiths shown by synforms and scalloped patterns of middle Neoarchean volcanic rocks in north, and possibly by anticlines in south. Light brown, ocher, and enclosed colors, Proterozoic terrains bounding Pilbara craton; bright blue, Permian strata; light green, Cretaceous strata; pale yellow, Neogene materials. Area is approximately bounded by 20° and 24° S, and by 116° and 121.5° E. From Myers and Hocking, 1998, © by Geological Survey of Western Australia.

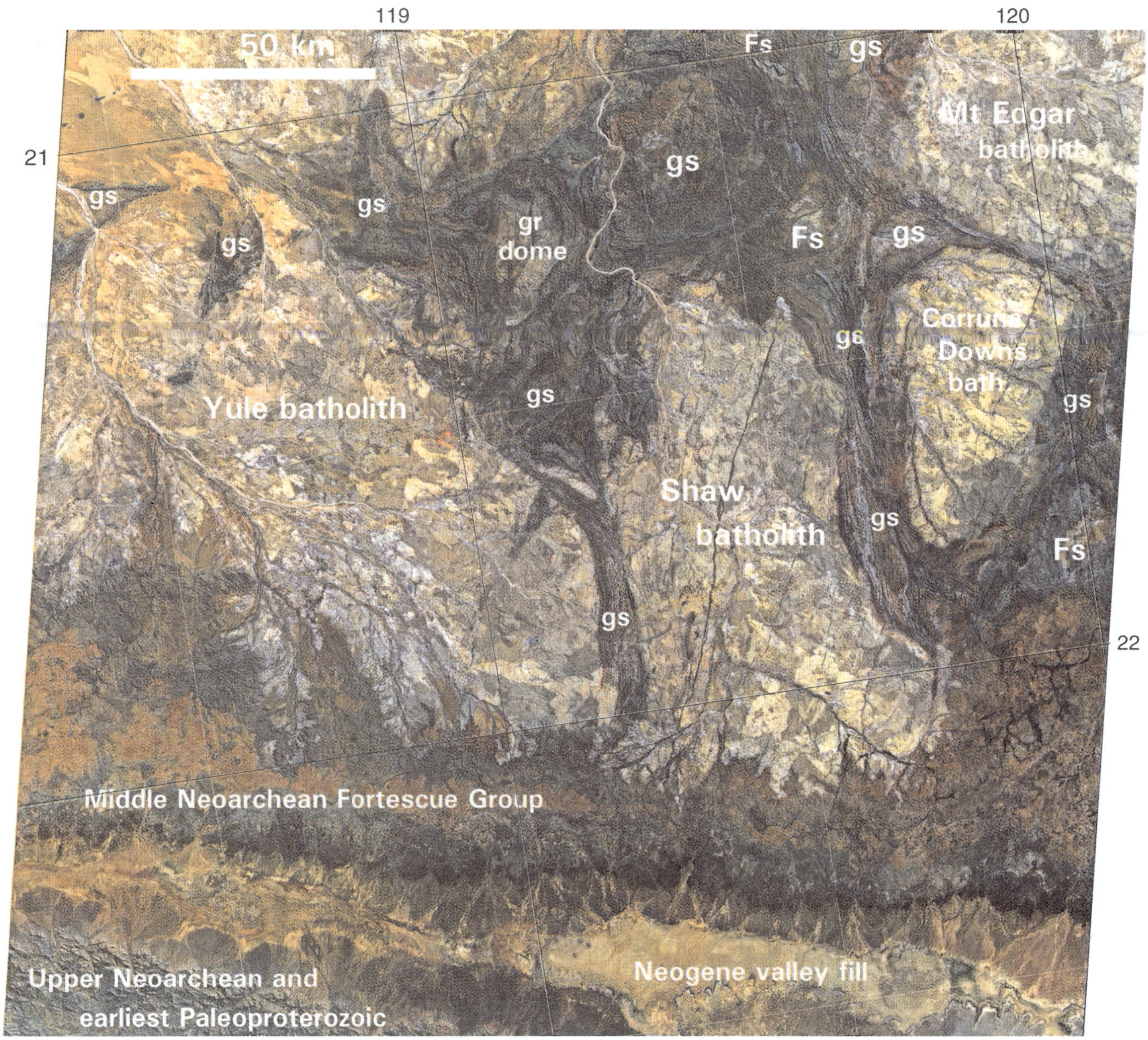

Figure 15. Landsat image of eastern Pilbara craton, showing two Archean tectonic styles. Older part of 3.5–2.9 Ga greenstone section (gs) formed with semi-regional sheet stratigraphy, and was deformed by diapiric batholiths (pale yellow; one small dome is marked gr) that rose progressively from 3.4 until after 2.7 Ga. Fortescue Group 2.8–2.7 Ga sedimentary and mafic-volcanic rocks were deposited across the older complex with regional sheet stratigraphy but were deformed into mostly gentle synclines (Fs) between diapiric batholiths that continued to rise. Fortescue Group represents a late greenstone succession that went only partway to granite-and-greenstone dome-and-keel stage. False-color image; blue assigned to spectral band 2 (visible green), green to 4 (near infrared), and red to 7 (mid-infrared). Image provided by CRA Exploration Pty., Ltd.

analogues—as shown both by ages of batholithic rocks (e.g., Fig. 16) and by decreasing deformation upward in the supracrustal section. The granites display abundant evidence for formation by remobilization, with varying degrees of solid-state deformation, remelting, and upward segregation, of mid-crustal TTG complexes. The composition of even the oldest recognized granitic rocks indicates probable origins by recycling of lower-crust TTG complexes, and successively younger granitoids record successively more thorough refining of older granitic rocks: the domiform granites were generated by mobilization, with variable remelting of its low-melting fraction, of ancient felsic basement (Bagas et al., 2002; Barley and Pickard, 1999; Champion and Smithies, 2000; Smithies et al., 2003, 2005b; Van Kranendonk et al., 2007). The rhyolites show parallel secular trends (Van Kranendonk et al., 2007). Farther west in the dome-and-keel terrain, mafic and ultramafic lavas are dominantly 3.3–3.0 Ga, also are intercalated with felsic volcanics, and are domed by granites with late-magmatic ages mostly 3.1–2.8 Ga. The abundant felsic eruptives in both older and younger sections show that granites were rising and breaching over a much longer time span than those defined by the known late batholithic zircons alone. Deformation (e.g., Fig. 17) and metamorphism commonly increase toward batholiths.

Figure 16. Two components of a diapiric batholith. Coarse granitic rock on right has igneous 3.47 Ga zircons; fine rock on left, 3.24 Ga (Martin Van Kranendonk, 1999, personal commun.). Many batholiths in granite-and-greenstone terrains rose intermittently for hundreds of millions of years. Coin diameter ~2 cm. North end of east lobe of Yule batholith, northeast Pilbara craton.

Figure 17. Contorted Mesoarchean chert (light gray), ferruginous chert (dark gray), and iron formation (black). Steep outcrop, pocketknife for scale. Severe batholith-side-up deformation is common near diapiric batholiths (here, several hundred meters to right of view). Coppin Gap, southwest part of Muccan 1:100,000 map sheet, Pilbara craton.

East Pilbara volcanic rocks are mostly between ca. 3.53 and 3.23 Ga, and there were long hiatuses centered on ca. 3.4 and 3.3 Ga. Van Kranendonk et al. (2007) attribute the episodicity to intermittent plumes from the deep mantle, whereas I infer intermittent delamination of subjacent mafic protocrust as a control on heat delivery.

Northwest Pilbara—Vertical and Horizontal Granite-and-Greenstone Tectonics

The much smaller far northwest part of the Pilbara craton displays a Mesoarchean and early Neoarchean granite-and-greenstone terrain disrupted by lateral motion (Fig. 14; Hickman, 2004; Hickman and Smithies, 2000; Hickman and Strong, 2003; Hickman et al., 2001; Krapez and Eisenlohr, 1998; Smithies et al., 2005a, 2007; Van Kranendonk et al., 2004b, 2007). A few granite domes and greenstone keels are well preserved, but this northwest terrain is cut by a major strike-slip fault overlapped by unfaulted 2.75 Ga rocks, and is broken by many lesser structures. Whereas the northeast Pilbara has an almost isotropic granite-and-greenstone network, northwest Pilbara tracts are markedly elongate northeastward and have arcuate map patterns suggestive of regional right-slip drag. Many rock occurrences are isolated by faults or by younger granites or sediments. Depth of erosion is generally deeper in the northwest Pilbara, where supracrustal rocks are metamorphosed mostly to upper greenschist or lower amphibolite facies, and are more deformed, than is typical in the northeast Pilbara, complicating analysis. Supracrustal rocks are within the usual range of greenstone types—mafic volcanic rocks, komatiite, quartzite, conglomerate, chert, banded iron formation, with felsic eruptives throughout many sections and abundant in some—and mostly range from ca. 3.3–3.0 Ga. There is a subregional chert, but no regional stratigraphy has otherwise been recognized. Granites range from ca. 3.3–2.9 Ga, with the younger ones being more evolved than the older and hence likely recording recycling, and the older themselves containing evidence for recycling from still older felsic crust (Hickman and Strong, 2003; Smithies and Champion, 1998). The assemblage is typical of dome-and-keel terrains, and not of some wholly different sort of setting, despite the interplay of horizontal and vertical tectonics.

Smithies et al. (2007, p. 60) term the strike-slip faulting "unequivocal evidence for major plate tectonic processes." To me, the faulting is a localized manifestation of the quasi-pervasive lateral deformation of most Archean upper crust—and it is not evidence for plate tectonics.

Paleoarchean and earliest Mesoarchean basement is unproved in outcrop but is recorded by one 3.7 Ga zircon xenocryst in younger granite, and by numerous 3.7–3.4 Ga detrital zircons in younger strata. Mapped relationships suggest to me that the oldest West Pilbara supracrustal rocks (the basal part of the Roebourne Group of Hickman and Strong, 2003) were deposited on felsic basement. These basal rocks comprise a thin platformal succession of basalt, komatiite, felsite, chert, iron formation, carbonate rocks, and clastic sediments, which lie quasi-concordantly on poorly exposed gneisses that have 3.25 Ga zircons in their one tested locality. This basal section is overlain by thick basalt and komatiite, with local chert—all suggestive of the basement and supracrustal successions known from many other Archean terrains. The section is repeated, in typical granite-and-greenstone style, in a tight syncline between the updomed gneisses and a younger batholith.

The northwest Pilbara is commonly assumed to consist of diverse products of plate-tectonic processes much like modern ones primarily because of the presence of lateral deformation and of chemotectonic rationales for the Whundo Group (e.g., Smithies et al., 2007, and Van Kranendonk et al., 2007). The Whundo is the assemblage, isolated by faults and younger cover, of ca. 3.1 Ga rocks noted earlier as given four mutually incompatible chemotectonic designations by different authors but now confidently assigned to an oceanic island arc (Smithies et al., 2007; Van Kranendonk et al., 2007). The compositions of the calc-alkaline mafic rocks, on which this assignment rests, are, however, too low in Al_2O_3 to have close analogues in modern arcs (cf. Smithies et al., 2007, their Table 1). Trace-element patterns are generally similar to those of northeast Pilbara rocks (compare Figs. 3 and 9 of Smithies et al., 2007), yet the northwest Pilbara patterns are rationalized in terms of subduction, and the northeast Pilbara patterns in terms of plumes, by Smithies et al. (2007).

Hypotheses of assembly from distant initial sites have also been based on sheared rocks presumed to record sutures. For example, a foliated subregional metachert was termed a mylonite by Hickman et al. (2001, their Figs. 27–29) and inferred to record a regional rootless thrust fault between unrelated assemblages, although the layer is parallel to bedding of units both above and below and the hypothetical thrust implausibly places younger on older strata. All foliated metamorphic rocks are in effect shear zones, and the photographs of the chert appear to show ordinary synmetamorphic foliation.

To me, the northwest Pilbara is merely a common variant on the granite-and-greenstone theme wherein mostly dense supracrustal rocks were deposited on, and sank into, mobile preexisting felsic crust, while simultaneously lateral deformation was under way. Batholiths rose from the laterally flowing lower crust, above which the floating upper crust was deforming less regularly. This interpretation is illustrated graphically by the Superior craton example, discussed subsequently.

South Pilbara—Non-Completed Neoarchean Granite-and-Greenstone Assemblage

Insight into the striking differences between Archean and modern tectonics and magmatism is provided by Neoarchean assemblages of the Pilbara craton, and of the Kaapvaal craton of South Africa, which record arrested development of granite-and-greenstone terrains. Thick wholly ensialic supracrustal sections of greenstone type were deformed only modestly by limited rise of remobilized-basement domes. In these Australian and African examples, subregional sheet stratigraphy, like that shown by the best-documented full-development granite-and-greenstone terrains, is obviously preserved. Typical granite-and-greenstone terrains formed in many other cratons during the same Neoarchean time interval, and I see these Australian and South African rocks as recording a granite-and-greenstone cycle whose dome-and-keel deformation went only partway to completion because the felsic crust had there become rigid enough to better support the dense overlying volcanic rocks erupted upon it.

The Mt. Bruce Supergroup, 2.8–2.2 Ga, unconformably overlies the Mesoarchean and early Neoarchean granite-and-greenstone terrains of the Pilbara craton. This thick section of middle and late Neoarchean and Paleoproterozoic volcanic and sedimentary rocks is preserved continuously in the Hamersley Basin on the southern part of the craton, through which older Archean granite-and-greenstone rocks are exposed in anticlines, and discontinuously on the northern parts of the craton (Figs. 14, 15). The Fortescue Group and most of the overlying Hamersley Group, each several kilometers thick, comprise the Archean part of the Mt. Bruce assemblage, and record deposition of mafic and felsic volcanic rocks, and clastic, carbonate, and iron sediments of remarkable regional continuity, from 2.8 to 2.5 Ga, over a preserved extent of ~300 × 600 km (Trendall et al., 2004).

The domiform Pilbara granites, which rose primarily in Mesoarchean and early Neoarchean time, continued to rise into the younger section, most obviously in the northeast Pilbara, where outliers are preserved in synclines, with gentle to moderate dips, between rejuvenated older batholiths (Figs. 15, 18). The incremental, protracted sinking of one northeast Pilbara syncline of Fortescue rocks, and complementary rise of the flanking domes, is recorded by eight unconformities separating rocks with

Figure 18. Old and young greenstone successions, northeast Pilbara craton. View south along steep Mesoarchean schist (foreground) to gently dipping Neoarchean sedimentary and mafic-volcanic rocks of Fortescue Group (distant ridge). Regional-sheet Fortescue rocks are deformed into syncline (gentle here, but dips to 40° beyond view) between three domiform batholiths (Muccan, Mt. Edgar, and Wawarragine) whose diapiric rise was primarily Mesoarchean but continued through Neoarchean time. Central part of Muccan map sheet.

upward-decreasing dips (Van Kranendonk, 2003). Symmetrical domes, some of them cored by exposed granite, are scattered through the central part of the main Hamersley Basin, and a large dome rises in the southeast (Fig. 14; Thorne and Trendall, 2001, plate 1A), and may record continuing rise of underlying Archean batholiths. Felsic volcanic rocks, mostly tuffs, intercalated in the middle and late Neoarchean section have still-rising batholiths as their likely source. One granite intrusive into the Fortescue volcanic rocks has the same 2.76 Ga age as an extensive rhyolite (Thorne and Trendall, 2001).

The Fortescue Group is dominated by mafic volcanic rocks, pillowed in large tracts in the south, with subordinate high-Mg basalt and felsic volcanic rocks and clastic strata, and spans ca. 2.80–2.71 Ga (Blake et al., 2004; Thorne and Trendall, 2001; Trendall et al., 2004). Like all Archean greenstone sections whose stratigraphic bases are exposed, the Fortescue section begins with sedimentary rocks, in this case fluvial and shallow-water sandstone, conglomerate, and shale, deposited in local lowlands. Most of the basal-clastic sections are only decimeters to tens of meters thick, but they reach almost 2 km. The upper parts of the thicker clastic sections contain much interbedded mafic lava, hyaloclastite, and tuff. Next upward are several km of volcanic rocks, mostly subaerial but partly shallow-subaqueous. Intercalated clastic sediments define by their facies and transport directions a broad interior upland flanked by lowlands (Thorne and Trendall, 2001). Succeeding thick sedimentary and volcanic units show a general trend toward southward transport into deepening water, although whether this represented a cratonic basin (as I presume) or proximity to a continental margin (which commonly is inferred) is not known. Uplands providing sediments extended beyond the present margins of the Pilbara craton at least to north and west (Thorne and Trendall, 2001). Burial metamorphism varied from low prehnite-pumpellyite facies to low greenschist facies, and records thermal gradients much higher than now (references in Thorne and Trendall, 2001).

Overlying this dominantly mafic-volcanic section are thick 2.6–2.2 Ga clastic sediments, carbonates, and iron formation, which have remarkably coherent and detailed regional stratigraphy and comprise the late Neoarchean Hamersley and early Paleoproterozoic Turee Creek Groups (Trendall et al., 2004). Included is a thick and regionally extensive 2.45 Ga rhyolite whose volume was at least 15,000 km^3 (Trendall, 1995). I presume that the unidentified granitic batholith that thus vented spectacularly during earliest Proterozoic time recorded the still-continuing rise of a domiform batholith in granite-and-greenstone mode.

The chemistry of the Fortescue volcanic rocks was discussed in the earlier section on chemotectonics. That these rocks are ensialic and occur in a regional sheet upon continental basement is obvious to all, and so they are not designated as the oceanic island arcs that the usual Archean spreadsheet tectonics would assign. Although they often are referred to as continental flood basalts, they are typically mafic ferroandesites, and they do not resemble either the quartz-normative or the alkaline olivine-basaltic rock types of modern continental flood-basalt provinces. Thorne and Trendall (2001) argued for a continental-rift origin, although the rocks also are quite unlike any now forming in that setting, and geologic indicators of significant crustal thinning are unknown. A Neoarchean swarm of mafic dikes in the northern Pilbara granite-and-greenstone terrain is of the same age as the main mafic section of the Fortescue and presumably records a major source region (Wingate, 1999). Small normal faults are reported, but not the major faults, with faults rotated to gentle dips and with strata, including scarp facies, rotated to steep dips, that characterize crustal thinning. The dikes are evidence for a hot substrate and for crustal mobility, but not for crustal thinning.

I see another Archean greenstone terrain that progressed through only early stages of doming by batholiths remobilized from the lower crust. A general crustal cooling trend is indicated, but continuing granite doming required a lower crust still much warmer than now, and abundant dikes attest to a continued floating style over a substrate hotter than now.

Northwest Superior Craton—Vertical Tectonics plus Lateral Mobility

The Superior craton of Canada is the largest preserved tract of Archean crust, ~1500 × 2500 km, and is rimmed by Proterozoic orogens. Vertical dome-and-keel tectonics is obvious in some sectors, and grades into other sectors in which horizontal tectonics appears dominant. Pervasive mobility of the crust in the northwest part of the craton is shown by shaded-relief magnetic anomalies (Fig. 19A) which, with the geologic and gravity maps (Figs. 19B and 19C), also illustrate effects of the more platelike

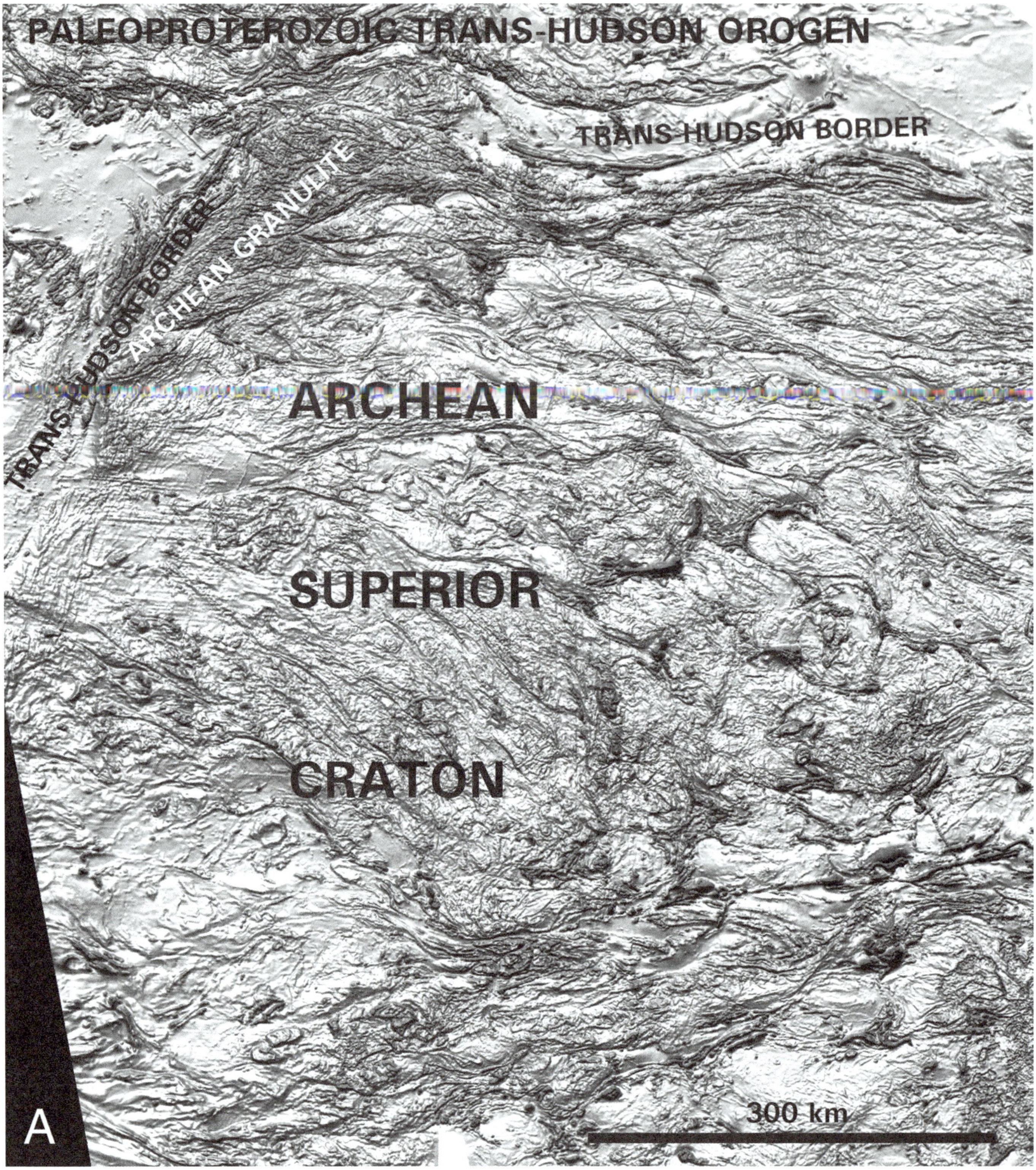

Paleoproterozoic truncations. The southwest and northeast corners of the map area are covered by thin Phanerozoic strata, but the magnetic and gravity maps reflect the underlying Precambrian geology. Exposures are poor in most of the region except along lakeshores, and access generally is difficult. Many zircon U-Pb ages, mostly Neoarchean but some Mesoarchean and a few Paleoarchean, constrain much of the geology. Recently compiled geologic maps of parts of the region (Bailes et al., 2003; Percival et al., 2002; Sanborn-Barrie et al., 2002, 2004; Stone et al., 2002, 2004; Stott et al., 2002) contain much textual, tabular, and interpretive-map information. These maps incorporate mapping at different times by many geologists with differing constraints

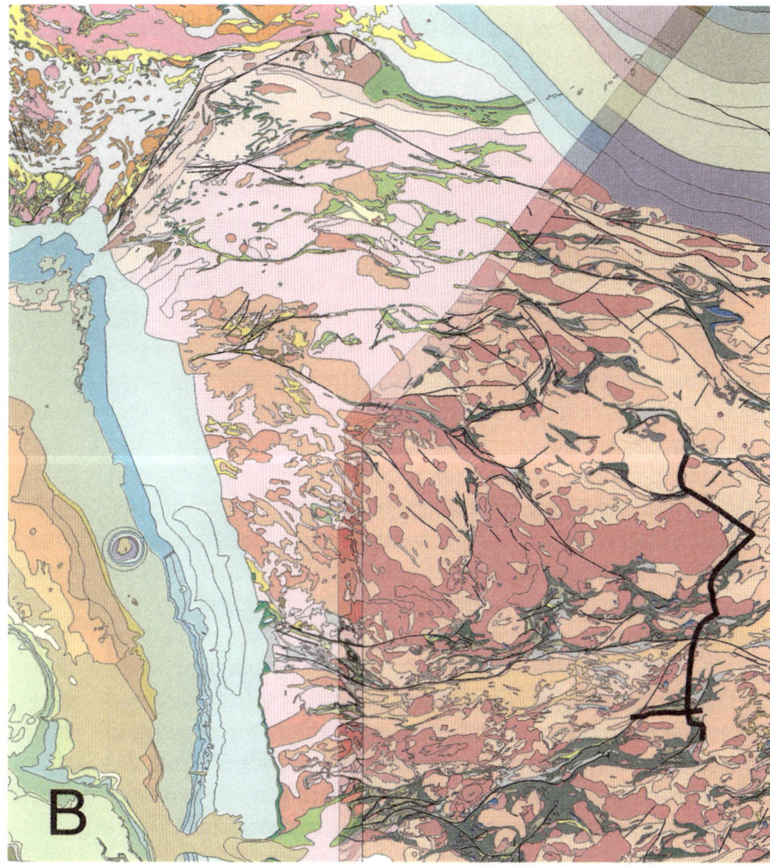

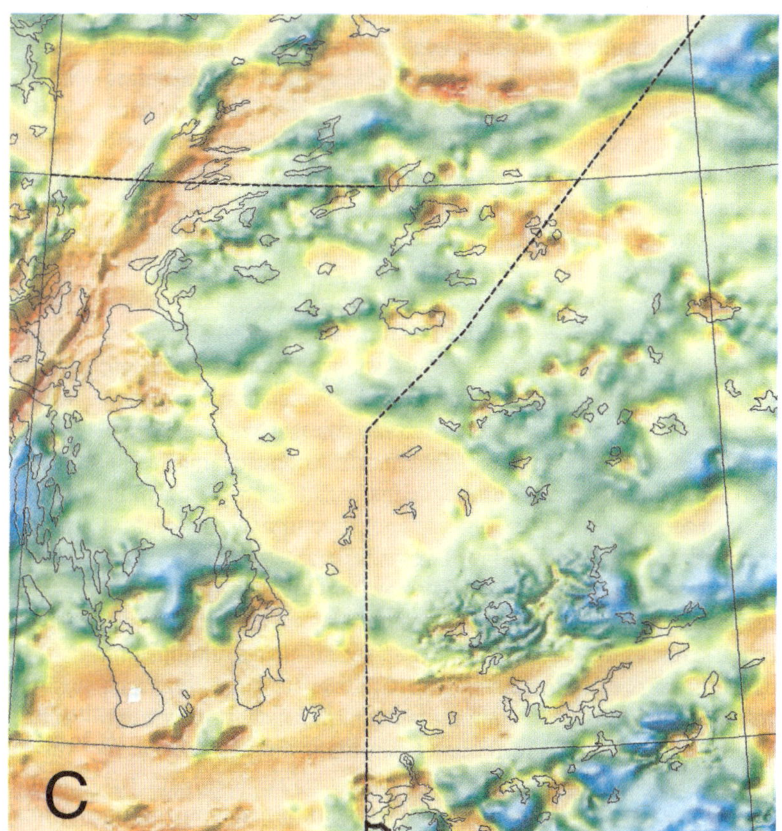

Figure 19. Magnetic, geologic, and gravity maps of northwestern Superior craton, Canada. Maps show approximately the same area, and have corners, clockwise from upper left, at ~56.6°N/100.7°W, 56.8°N/88.6°W, 49.0°N/89.5°W, and 49.0°N/97.3°W. Border between Manitoba (left) and Ontario is at right edge of color change in (B), and is dashed line in (C). (A) Shaded-relief total-field magnetic anomalies, from map prepared by Deborah Lemkow, Geological Survey of Canada. Archean craton is truncated by Paleoproterozoic rift and Trans-Hudson Orogen in northwest and north. (B) Geologic map, also provided by Lemkow. Thin, subhorizontal Phanerozoic strata are in pastel shades in northeast and southwest; geophysical maps show Precambrian geology through this cover. Archean rocks in Ontario: pale pink, granitic rocks; red, gneiss; olive green, mostly metavolcanic rocks; gray, and pale yellow, metasedimentary rocks. Archean rocks in Manitoba: pale pink, granitic rocks; pink, gneisses; green, mostly metavolcanic rocks. Dome-and-keel geology is conspicuous in east-central part of area, and in parts of southern third of area, but elsewhere is mostly disrupted. Two crossing heavy lines in southeast show locations of reflection profiles of Figure 21. (C) Bouguer gravity-anomaly map. Color spectrum goes from −15 mgals, dark brown, to −70 mgals, blue. Long gravity high trending north-northeast in northwest is due to lower-crust Archean rocks at Trans-Hudson border. From Miles et al. (2000).

and philosophies. Thus, Sanborn-Barrie et al. (2004) depicted 67 Archean supracrustal units that have, at best, only local stratigraphic significance, that overlap complexly in lithologies and known or assumed ages, and that defy regional synthesis.

Archean Tectonics

The dome-and-keel vertical tectonics typical of other Archean cratons was recognized by various early Superior investigators. Lin (2005) and Parmenter et al. (2006) presented detailed structural and geochronologic data to show that dome-and-keel vertical tectonism was concurrent with severe lateral deformation. Most geologists active during the past two decades have, however, sought plate-tectonic models and have invoked multiply oriented plate-tectonic accretion, cross-folding, and transpression rather than buoyancy effects, and have inferred distinct episodes of superimposed deformation and metamorphism. As intrusive and extrusive felsic rocks span long age intervals wherever dates are numerous, I presume that instead plutons were rising, venting, and deforming country rocks for prolonged periods, as in other granite-and-greenstone terrains. Lateral mobility of course is obvious in Figure 19A—but is that all there is, which oversimplifies the prevailing view, or did it operate as a variable disrupter of synchronous products of vertical tectonics driven by the usual Archean gravitational instabilities, as I perceive?

Almost all recent published interpretations (Lin, 2005, and Parmenter et al., 2006, are notable exceptions) are in terms of plate-tectonic interactions, with or without added plumes, and modern-style deformation, but vary widely because they are based primarily on ambiguous chemotectonics. Complexity of interpretations has increased as new data have confounded early explanations. Progressive southward tectonic accretion was inferred from early reconnaissance dating but has long since been disproved. Some models (e.g., Thurston, 2002) emphasize autochthonous volcanism on oceanic and continental platforms subsequently juxtaposed by plate convergence with other oceanic materials caught between. Other models postulate imbrication of rocks from widely differing settings by rootless megathrusts within the supracrustal piles. Most models advocate late welding and overprinting by batholith-generating magmatic arcs, for the younger granitic rocks commonly show the usual Archean incorporation of older materials (e.g., Whalen et al., 2004). The most complete synthesis in plate terms is that by Percival et al. (2006b). In a departure from modern plate models, Percival and Pysklywec (2007) proposed a form of delamination to provide heat for craton-wide melting. (My views and Percival's are converging in this regard.)

The region is conventionally split into subparallel belts (or subprovinces, domains, or terranes) assigned plate-tectonic significance (e.g., Card and Ciesielski, 1986; Davis et al., 2005; Percival et al., 2004a, 2004b, 2006b; Stone, 2005; Thurston et al., 1991; and the geologic maps cited above). The number and subdivisions of the proposed belts have undergone many changes, and many boundaries are arbitrary. Discrimination of belts is partly on the basis of dominant surface rock types—granite-and-greenstone, or mostly granitic rocks, or mostly metasedimentary rocks—and partly on the basis of chemotectonic rationales. Geologic evidence for deposition of supracrustal sections on older TTG basement is incorporated where recognized. Large tracts are assigned oceanic provenance but no rocks, including the widespread ultramafic lavas, have been proved ensimatic by geologic relationships. Nd model ages are widely used to discriminate juvenile granitoid rocks (model ages close to zircon ages) from reworked crust (model ages older; e.g., Tomlinson et al., 2004). I argued earlier that these "ages" record, at best, separation from ancient mafic protocrust, not from the mantle. Parks et al. (2006) showed that several of these belts probably contain the same general stratigraphic succession, and hence are not fundamental divisions.

Supracrustal rocks are dominantly of early and middle Neoarchean ages. Komatiite, high-Mg basalt, tholeiitic and calc-alkalic basalt, andesite, dacite, rhyolite, and assorted sedimentary rocks, including shallow-water stromatolitic limestones, are complexly intercalated through thick sections with broad age ranges. Each subregion shows magmatic activity over 200–600 million years, and middle Neoarchean magmatism is common to all of them—and both of these characteristics are incompatible with analogy to modern plate products. Except for lower-crustal granulite in the rifted northwest corner, metamorphic and granitic rocks now exposed crystallized mostly in the upper and middle crust, at depths of 5–16 km and only locally deeper (Easton, 2000; Easton and Berman, 2004; Stone et al., 2004). Supracrustal rocks are mostly metamorphosed at greenschist facies except within the contact aureoles, typically 2–3 km wide and at lower-amphibolite facies, of flanking batholiths. Popular interpretations are most often made in terms of suturing of diverse plutonic-arc TTG complexes together to form minicontinents to which accreted oceanic island arcs and seafloor, continental and oceanic plateaus, plume products, sedimentary piles, and other superimposed bits. Chemotectonic and plate-tectonic rationales are based as weakly in western Superior as elsewhere. For example, the supracrustals of the East Uchi area (Sanborn-Barrie et al., 2004), in the south-central part of the area of Figure 19, are assigned "mainly continental affinity" because of the Nd model ages of many units and their proximity to proved-basement tracts, but various intercalated parts of the section nevertheless are classed as ensimatic on chemical grounds. Intermittent plumes, superimposed on other plate-tectonic environments, are added to explain komatiites (Hollings et al., 1999; Tomlinson et al., 1998, 1999). Inferred sutures and thrusts are cryptic or hidden, or are assigned to nearby shear zones that might coincide with the anticipated boundaries. One such shear zone, repeatedly hypothesized to be a major suture, was shown by Culshaw et al. (2006) to be much too young to fit the conjecture. Other shear zones have proved to be noncoincident with hypothesized boundaries. Some conjectural major boundaries have disappeared with detailed work (e.g., Young and Helmstaedt, 2001). Van Staal (1998) showed that there was no shearing at stratigraphic boundaries that others had postulated to be sutures. Great older-over-younger or deeper-over-shallower thrust faults are predicted by suture rationales but have nowhere been demonstrated. Rifting and convergence features implicitly predicted to accompany postulated plate constructs have not been documented.

The basement that has been recognized in a few places beneath northwest Superior supracrustal rocks is felsic, and the usual thin, diverse strata intervene between it and the main greenstone sections. In three sectors of the West Uchi area, in the southwest part of the area shown in Figure 19, the oldest supracrustal rocks are seen to have been deposited on TTG basement (Bailes et al., 2003; Percival et al., 2006a). The basement gneiss, with dated zircons near 3.0 Ga, is overlain by slightly younger diverse sections of conglomerate, feldspathic and quartzose sandstone (which contain detrital zircon grains as old as 3.5 Ga), carbonates, iron formation, basalt, and komatiite (e.g., Fig. 11B). Supracrustal rocks as old as 3.3 Ga have been found in the southeast part of the area of Figure 19 but their basement is unknown (Sanborn-Barrie et al., 2002). Paleoarchean basement and supracrustal rocks are known in the far northwest part of the craton, discussed subsequently. Just south of the area shown in Figure 19 is another known example of

TTG basement, ca. 3.0 Ga, overlain by slightly younger sandstone, stromatolitic limestone, iron formation, komatiite, and basalt (Stone et al., 2002; Tomlinson et al., 1999).

The dominant western Superior trends are westward to northwestward toward an abrupt truncation at the Paleoproterozoic Trans-Hudson orogen. The region is a product of both vertical and lateral instability, as Lin (2005) and Parmenter et al. (2006) documented. That those authors were correct in writing that I overemphasized vertical tectonics in my previous papers on Archean geology is made obvious by the magnetic-anomaly map (Fig. 19A), part of which was included in their papers also (and in Percival et al., 2006b). The relatively shallow erosion level throughout most of the region precludes major change in surface area, so the deforming upper crust effectively floated on the lower crust. The magnetic map resembles an outcrop photograph of highly deformed metamorphic rocks. The pattern of northwest-striking shear has been recognized by all geologists working in the region in recent years, and commonly is attributed to plate interactions even though its quasipervasive character shows that no internally rigid plate existed. The shear zones generally are poorly exposed, but observations in many places show generally right-slip structural indicators, and often ductile-to-brittle transitions that, where dated, show that slip operated through much of the mid-Neoarchean period of regional magmatism. The shearing is not pervasive along strike, and I infer from map geometry that much syn-batholithic north-south shortening and east-west extension is recorded also—a mixture of bulk pure shear and simple shear. Tracts of pervasively sheared aspect give way along strike to elongate, but not monoclinic, dome-and-keel structures indicative of diapirism in a field of orthogonal shortening and extension.

A cluster of typical vertical-tectonics granitic domes and intervening tight synclinal keels of supracrustal rocks is obvious in the east-central part of Figures 19A and 19B, and includes five domes en echelon northeastward; more domes lie on strike to the southeast of the map area. Lin (2005, Fig. 2) and Thurston et al. (1991, their Figs. 5.2 and 5.23) presented geologic maps of a number of these classic dome-and-keel assemblies. Lin demonstrated their diapiric character with structural analysis, and Thurston noted that many of the greenstone belts are characterized by steep foliation and lineation, which accords with gravitational deformation as dominant. I infer from the northwestward elongation of the domes, and the apparent lack of throughgoing shears, that much of the diapiric rise was done in a regional strain field of southwestward shortening and northwestward extension. West-northwest of these obvious domes, shear zones take over, and most would-be domes and keels are variably shredded, although some are well preserved and show clear structural evidence for diapirism (Parmenter et al., 2006). Much of the lateral deformation was synchronous with diapirism (Lin, 2005; Parmenter et al., 2006), and all of it predated early Paleoproterozoic extension at the west margin of the craton. The deformation includes much north-south shortening, balanced by east-west elongation, as well as the right-slip translation emphasized by field geologists.

Other classic domiform batholiths (including the Trout Lake and Allison Lake batholiths of Sanborn-Barrie et al., 2004) that give way northwestward to shear structures, in this case in a mostly granitic tract, are shown in the south-central part of Figures 19A and 19B, in the Uchi belt, about 1/4 of the way north in the maps. The age, 2.74–2.69 Ga, of the dominant granitic rocks of the granite-and-greenstone terrain (Bailes et al., 2003; Sanborn-Barrie et al., 2004) is the same as that of those that dominate the mostly granitic tract (Berens River subprovince: Stone, 2000), but the latter granitic rocks are much more extensive at the surface. Petrobarometry shows erosion depths of the granite-and-greenstone terrain to be mostly <10 km (Easton, 2000), whereas erosion depth of the granite-dominated terrain is on average a little deeper, 10–15 km, except for the youngest plutons (Stone, 2000). Remnants of supracrustal rocks in both terrains go back to 3.0 Ga, but TTG basement has been recognized only in the granite-and-greenstone terrain and the south edge of the granite-dominated terrain. A major structural discontinuity, hidden beneath thin Phanerozoic strata but obvious on the magnetic map, trends northwestward through the "C" in "Craton" on Figure 19A, and here separates west-trending structures, to the southwest, from northwest-trending ones, to the northeast. In this short sector, the boundary looks superficially like a major strike-slip fault, but I infer from the geologic and magnetic patterns that instead it records a change in orientation of directions of shallow extension and shortening and thus is a discontinuity in plan-view flattening. Northwest of "C," trends curve to the west-northwest on both sides of the discontinuity, which splays into a concordant, and more pervasive, westward fabric in the southern terrain. In the other direction, east-southeastward from "Craton," the mostly granitic northern terrain includes the westernmost domes of normal granite-and-greenstone domes and keels in an array that curves eastward; and from there on elongations on both sides of the projected discontinuity are east-west.

Figure 20 shows three examples of vertically stretched rocks in dome-and-keel regions of the Superior craton east of the area shown in Figure 19—granite-side-up dip slip, with steep stretching lineations. Explanations in terms of superimposed compressional folds or thrust faults do not account for such structures.

Quasipervasive disruption by shear with dominantly west-northwestward strikes was mapped and described by Stone (2005) and Stone et al. (2004) in the northeast part of the area of Figure 19A (east and east-northeast of "Archean" in "Archean Superior craton"). Supracrustal rocks of early and middle Neoarchean age are intruded by voluminous plutons of the same ages; inherited zircons, to 3.6 Ga, have been found in several of the plutons. Young plutons define domes but older ones are elongate and typically have a "well developed mineral fabric with overall east-southeasterly trend and sub-vertical dip" (Stone, 2005, p. 74), and supracrustal rocks are mostly in narrow belts elongate in the regional direction. The supracrustal rocks commonly are assigned to diverse oceanic and continental plate-tectonic settings on chemotectonic bases. Shear zones show mostly ductile structures, of middle- to low-metamorphic grade, and slip

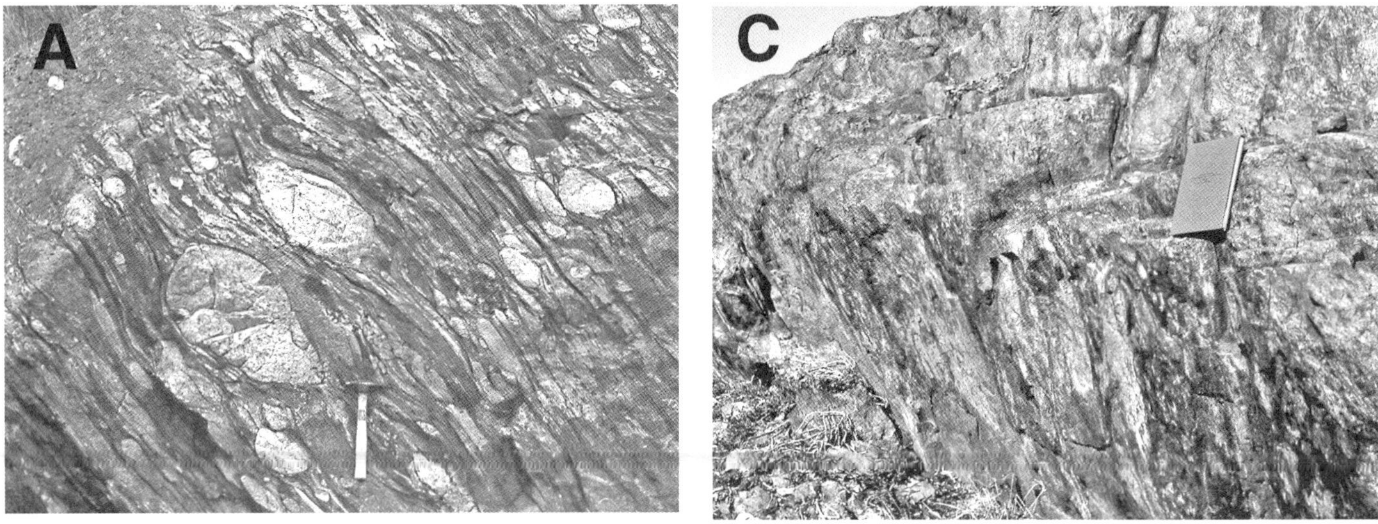

Figure 20. Dominant deformation in greenstone belts near diapiric batholiths is flattening parallel to contacts, downdip stretching, and batholith-side-up shear. Examples from Neoarchean of Superior craton, Canada. (A) Conglomerate derived from breached batholith; tonalite cobbles are flattened severely, pegmatite and potassic granite much less; Wawa belt, 5 km west of Wawa, Ontario. Hammer for scale. (B) Pillow basalt, stratigraphic top to right, moderate downdip stretching; outcrop 4 m high; Abitibi belt, 50 km east-northeast of Timmins, Ontario, on Highway 11. (C) Pillow basalt, pillows severely stretched into lenticular fragments, Abitibi belt, 3 km southeast of Noranda, Quebec. Notebook is 21 cm long.

indicators show right-slip with variable components of vertical offset. Depth of erosion tends to decrease with decreasing age of granitic rocks, and shows steps across some sectors of dominantly strike-slip faults.

A number of the west- to northwest-trending shear zones obvious on the magnetic map have yielded field evidence for right-slip motion. Northeast-trending left-slip structures also are present. The left-slip Miniss River fault (Bethune et al., 2006) is the conspicuous northeast-trending structure north of the center of the bar scale on Figure 19A. It has minor late semi-brittle right slip and ~40 km of earlier ductile-mylonitic left slip that was active at 2.681 Ga. The fault has an open-S configuration and has steep thrust offset at the southern restraining bend. A westward-widening belt of metamorphosed clastic sedimentary rocks (gray and pale yellow on the geologic map) deposited rapidly ca. 2.7 Ga, the western English River belt, is approximately bounded on the east by the Miniss River fault. The sediments postdate a large part of the regional magmatism, their TTG basement is exposed, and they are marked by a Bouguer gravity high rather than a low (Fig. 19C); they may have been deposited atop extensionally thinned basement. The metasediments record midcrustal erosion depths and probably are confined to the shallow, upper part of the preserved crust (Nitescu et al., 2006). The sediments unconformably overlap dominantly volcanic sections beyond the main belt and appear to be a thicker than usual variant of the late fills common in Archean terrains. Although these ensialic sediments have been termed an accretionary wedge (e.g., Bethune et al., 2006; Breaks, 1991) and inferred to represent a closed oceanic gap, no descriptions suggest the requisite polymict mélange to be present, although such mélange would be obvious where late synmetamorphic strain is low.

Lower Crust Tectonics

Western Superior lower crust is exposed in the far northwest, uplifted by Paleoproterozoic extension and convergence. Elsewhere, its character can be inferred from the undulating fabric of the deep crust on reflection profiles, and by analogy with the structural study of such crust in a Proterozoic uplift farther east in the Superior craton, by Moser and associates as cited earlier, where lower-crust gneisses were extended parallel to the upper-crust elongation of domes. The lateral deformation of the diapir-softened upper crust apparently is a passive response to pervasive flow of the mobile lower crust. The pervasive horizontal and vertical mobility, the lower crust feeding upper-crust diapirs even while flowing laterally, mostly decoupled, beneath it, is on this scale a uniquely Archean phenomenon, and is incompatible with the rigid-plate tectonics of popular inference.

Archean lower crust, the Pikwitonei granulite, exposed in the northwest part of the area of Figure 19, produces a conspicuous high-frequency magnetic pattern with an irregular short-wavelength west-trending fabric that perhaps records Archean lower-crust flow, and a Bouguer gravity high. (The very regular west-trending lineation in the west-central part of Figure 19A, south of the granulite, is a data-compilation artifact.) The dominant age of igneous crystallization of Archean granites in the rift-corner area is ca. 2.7–2.6 Ga. A granodiorite-leucotonalite probably of this age encloses abundant rafts of undated mafic and ultramafic lavas, and subordinate pelite, metamorphosed at granulite facies (Hartlaub et al., 2004), demonstrating vertical mixing in the crust. In one fault-bounded block in the northwest rift-corner area, gneissic granite, with a U-Pb zircon age of 3.2–3.1 Ga, intrudes feldspathic and quartzose sandstones and interbedded iron formation, tholeiitic basalt, komatiite, and basaltic andesite, metamorphosed and migmatized in the middle crust at uppermost amphibolite facies (Böhm et al., 2003)—likely a basal-supracrustal assemblage of the type common in Archean terrains. The sandstones were derived from felsic sources and contain abundant 3.9–3.2 Ga detrital zircons. The gneissic granite contains sparse relic zircons to 3.5 Ga, and its Nd model ages vary from 4.3 to 3.5 Ga, so likely it was reworked from basement to the sandstones. The supracrustal assemblage, which includes komatiite, is ensialic. Although Böhm et al. regarded this block as rafted in from some distant site because pre–3.5 Ga zircons are not known elsewhere in the Superior craton, I infer that this instead is another example of selective preservation of ancient zircons in deep-seated rocks, and that here, as elsewhere, early supracrustal rocks sank into the deeper crust rather than remaining near the surface to form granite-and-greenstone terrains.

Seismic Profiles

Crossing seismic-reflection profiles are shown by Figure 21. The steeply structured granite-and-greenstone upper crust is

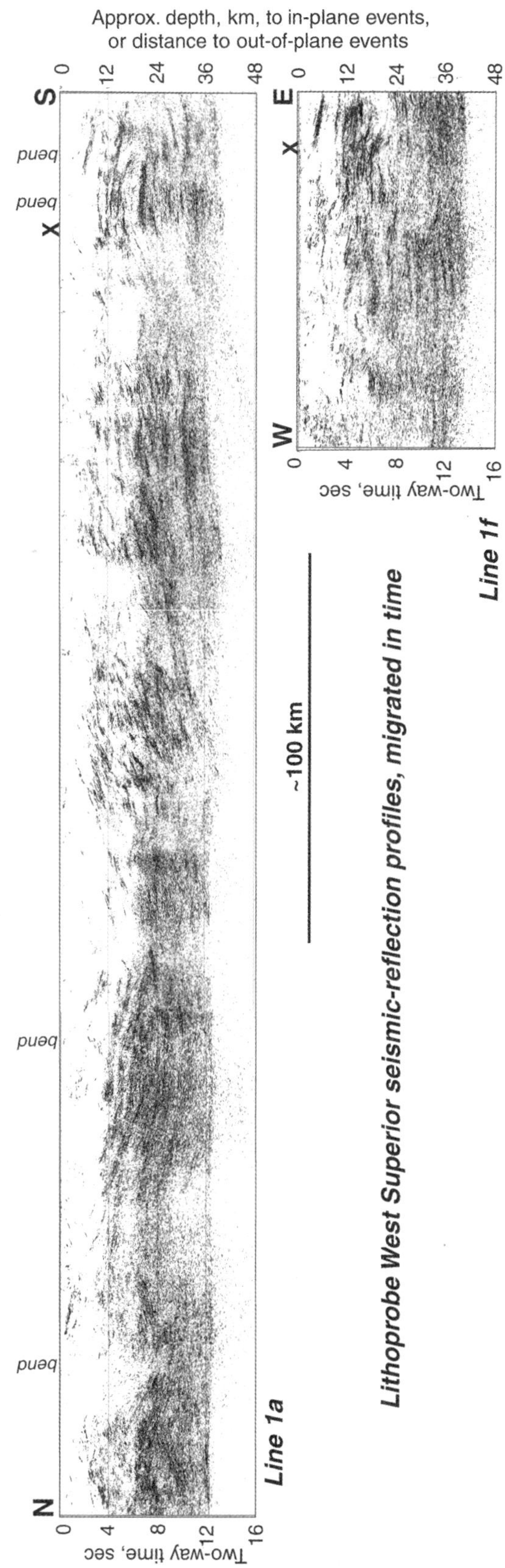

Figure 21. Vibroseis seismic-reflection profiles in Superior craton, locations shown on Figure 19B. Upper crust is semi-transparent; rest of crust consists of undulating gneisses; bottom of reflective crust is Mohorovičić discontinuity. Routes cross granite-and-greenstone structures of widely varying strikes and dips, and out-of-plane events must produce illusions within the profiles. Some investigators perceive thrust and extensional faults interlacing throughout the crust, others see subduction sutures, and I see a lower crust dominated by laminar flow. Plotted to be 1:1 for in-plane reflections at a velocity of 6 km/s. Major changes in traverse directions marked "bend." The profiles cross at (X), and the poor matches between the two lines there may be due to out-of-plane reflections. Profiles provided by the Lithoprobe project (http://www.litho.ucalgary.ca/atlas/atlas.html).

more or less transparent, whereas the undulating gneisses of the middle crust are highly reflective. The lower limit of abundant reflectors is the Moho, which is nearly horizontal in time. The traverses cross structures at all angles to their strikes and dips, and out-of-plane reflectors must account for many irregularities. As Hobbs et al. (2006, p. 490) emphasized, in such profiles "neither stacking nor migration can discriminate against out-of-plane energy and the 2-D stack represents the 3-D response of a broad swath centred on the profile." This problem is discussed further regarding the Yilgarn craton.

I infer from these profiles, considered with Archean surface geology, that upper and lower crust are decoupled. The lower crust consists of gneisses flattened pervasively with broadly undulating structures that have general flow-pattern continuity across the entire profile. The upper crust mostly appears acoustically transparent because its steep granite-and-greenstone structures are not imaged. Note that in the region of most coherent apparent upper-crust reflectors, above the scale bar, northward-inclined shallow structures appear to cross downward through gentle lower-crust structures, from which I infer that the apparently inclined reflectors are out-of-plane signals. The nearly flat reflection Moho is a product of the floating tectonic style and is evidence that the lower crust was too weak to support great topographic and structural loads. The lower-crust flow patterns permit explanation of the severe tectonic mixing shown by lower-crust outcrops.

The same line was interpreted, very differently, to show plate-tectonic aggregation by Percival et al. (2006b) and White et al. (2003), as was a coincident refraction line by Musacchio et al. (2004). The latter group interpreted two long crossing explosive-source seismic refraction and wide-angle reflection profiles, and Vp/Vs modeling, with the assumption that all recorded events are in the planes of the sections. They deduced the Moho to be a sharp velocity step, from Vp ~7.0 to >8.0 km/sec, gently undulating near 40 km, and the basal crust to be mostly intermediate in bulk composition but to be mafic in one subregion, which they inferred to be an underplated subducted slab of oceanic lithosphere. Their lines include one coincident with line 1a of Figure 21, another continuing the trend to the south, and a long crossing east-west line that includes coincidence with short line 1f of Figure 21. Their "minimum model" (their Fig. 8), derived with direct waves plus waves reflected from the Moho, shows monotonic downward increases in crustal and upper-mantle velocities. Their preferred model (their Fig. 5) incorporates highly ambiguous ray-tracing of what they interpreted as waves reflected from four undulating seismic discontinuities, at depths between ~50 and 120 km in the upper mantle, and has a velocity reversal high in the upper mantle in part of the north-south section. Musacchio et al. (2004) accepted this low-velocity zone as real and inferred it to overlie an Archean flat-subducted slab frozen in place. They attributed the lack of a matching velocity reversal in the east-west line, even where it crosses the north-south line, to strong velocity anisotropy, in different directions in overlying and underlying materials. They also inferred, from even more ambiguous evidence, a possible velocity reversal beneath the high-velocity zone, which also requires, in their terms, crossed anisotropies.

Percival et al. (2006b) and White et al. (2003) accepted the refraction interpretation of a frozen flat slab, assumed all apparent reflection alignments to record within-plane structures, and interpreted the north-south reflection profile of Figure 21, plus another line on trend to the south from it, in subduction terms. They perceived two subduction systems within the profile reproduced here as Figure 21, line 1a. The top of one, a "crustal suture," is the zone of reflections rising gently southward from near the reflection Moho at the center of the profile, then arcing through discontinuous horizontal events to a slight southward inclination at the south end of the line. The top of the other, the frozen slab of Musacchio et al., also can be perceived only within the crust, as discontinuous reflectors from 10 s, at the south end of the profile of Figure 21, to a non-imaged intersection with the Moho 70 km to the north. Although the low non-imaged north ends of both inferred subduction tops are in sectors where the reflection Moho is poorly defined, Percival et al. (2006b) and White et al. (2003) inferred that the Moho is cut, and slightly offset, by both. They interpreted the short reflectors with northward inclinations, above the scale bar and between ~2 or 3 and 6 or 7 s, as upper-crustal imbrication by subduction, and did not discuss the apparent crossing of deeper structure by these reflectors (which to me invalidates their within-plane inferences). They also suggested subduction-related explanations for several other parts of the profile.

The postulated crossed anisotropic velocities, to which the interpretations by Musacchio et al. (2004), Percival et al. (2006b), and White et al. (2003) are anchored, require unlikely coincidences. One layer must have its fast direction north-south and slow direction east-west, and the other layer the opposite orientations, both precisely as needed to cancel out the effect of an uncommonly low-velocity zone above an uncommonly high-velocity one. If instead ambiguities in the rays reflected complexly from hypothetical deep discontinuities, with which the structure was defined, were misinterpreted, the need for this explanation is obviated, and no slab is imaged. The interpretations by Musacchio et al., Percival et al., and White et al. require that the purported sutures cross the Moho and continue into the mantle. Such crossings are precluded where data are good and so are postulated to occur in sectors of the profile in which there are almost no data. The retention of a slab frozen in place for almost 3 Ga is incompatible with the inference by White et al. (2003) that the mantle into which the slab was subducted was so hot that the slab was extensively melted to release voluminous TTG: the dense restitic slab could not have remained so out of balance within extremely hot lower-density mantle, but would have sunk in it. The inferred subduction zones and upper-crust imbrication cannot be reconciled with the known surface geology or with modern analogues of the processes deduced.

A similar reflection profile 200 km to the west was presented by Calvert et al. (2004) and, with better processing, by Nitescu et al. (2006), and similarly shows a mostly transparent upper crust and an undulating fabric, less continuously imaged than that of Figure 21, in the lower two-thirds of the crust. Calvert et al. regarded incoherent, even locally crossing, reflections at 1–4 s in the upper crust of the Uchi granite-and-greenstone terrain to record great faults. I regard this interpretation as incompatible with the lack of large offsets of the surface geology, and presume the reflectors at issue to include out-of-plane features. Calvert et al. extrapolated their interpretations downward and drew complexly superimposed thrust and extensional faults through the entire crust, mostly through no-data parts of the display. These assumptions are not based on either data or analogy with such genuinely extended regions as the modern Basin and Range province, wherein such structures are lacking.

Paleoproterozoic Trans-Hudson Orogen

Like most Archean cratons, the Superior craton is outlined by Proterozoic orogenic belts, on the far sides of most of which are other Archean cratons, some of which are rotated to disparate trends. The east margin of the Trans-Hudson orogen sharply truncates Superior craton structures (Fig. 19). Thick Paleoproterozoic sedimentary rocks, and subordinate volcanic rocks including ultramafics, were deposited on the subsiding flank of the craton, and stratified rocks and Archean basement were severely modified by metamorphism, plutonism, and deformation (e.g., Böhm et al., 2003; Harris et al., 2000; Kraus and Williams, 1999; Lucas et al., 1996; Zwanzig and Böhm, 2002). Paleoproterozoic rifting of the north edge of the northwestern Superior craton was semi-concordant with the Archean structural grain (Fig. 19), and the sedimentary framework of the thick stratal wedge deposited on the thinned margin section is well preserved because subsequent deformation and metamorphism were much less severe than along the west edge (Peck et al., 2000). The northern trailing-edge stratigraphic section dips and faces north and is intruded by coeval mafic and ultramafic dikes and sills.

Archean rocks in the angle between the west and north Superior margins are exposed in part at lower-crust levels and in part are complexly deformed and retrograded. This tract includes the Pikwitonei granulite, which has a distinctive magnetic signature (Fig. 19A) and produces a Bouguer gravity high along the west margin of the Superior province (Fig. 19C). Depth of erosion into Archean granulites increases westward, oblique to the Archean strike (Weber, 1983), and rocks formed at depths of ~25 km are widely exposed (Percival, 1989). Granulite-facies crystallization is zircon-dated as 2.62 Ga, and its uplift and erosion, as well as subsequent deformation and retrograde metamorphism, were of Paleoproterozoic age (Bowerman et al., 2005). Extension began ca. 2.1 Ga, closure was completed ca. 1.8 Ga, and the broad and complex Trans-Hudson orogen was developed between these limits.

The Trans-Hudson orogen is commonly given plate-tectonic interpretations in terms of opening of a Paleoproterozoic ocean by seafloor spreading, development of island arcs, and closing of the ocean by subduction (e.g., Ansdell, 2005; Corrigan et al., 2005; White et al., 2005). The crust is thicker than Archean crust despite generally deeper erosion of the top. Plate rationales, as for Archean terrains, are based primarily on chemotectonics, and primarily on selected trace-element ratios because major-element analogies with modern assemblages are weak. Igneous rocks are bimodal, which is inappropriate for the purportedly ensimatic setting. The indicators of subduction in Phanerozoic orogens are lacking: there are no polymict mélanges, no high-pressure, low-temperature metamorphic rocks, no oceanic mantle rocks, no documented ophiolitic successions so far as I know. Strong reflection-seismic fabrics dip beneath both eastern (Superior) and western margins of the Trans-Hudson orogen (Corrigan et al., 2005; White et al., 1999, 2005), yet no Paleoproterozoic magmatic arcs were formed in the cratons to suggest that these fabrics are related to subduction. Reworked Archean basement rocks are known in many small and large windows eroded through Trans-Hudson orogen rocks (e.g., Annesley et al., 2005; Ashton et al., 2005; Bickford et al., 2005; Rayner et al., 2005; Zwanzig et al., 2006). Archean zircon xenocrysts are known in some Trans-Hudson orogen igneous rocks, and Nd and Pb isotopes suggest old-crust contamination in various sectors (e.g., Bickford et al., 2005). Compressional thickening of a great Paleoproterozoic sedimentary pile in the center of the Trans-Hudson orogen can quantitatively account for its dimensions, metamorphism, and anatectic plutonism (White, 2005).

The Superior craton was stabilized by early Proterozoic time, and behaved as an internally rigid plate bounded by mobile zones. But unambiguous evidence for seafloor spreading and subduction in those mobile zones is lacking. This was not plate tectonics as we see it operating in the modern Earth. Detailed critical analysis is overdue but will not be attempted here.

Yilgarn Craton, Western Australia—Vertical Tectonics plus Lateral Mobility

The Archean Yilgarn craton, ~750 × 900 km, of southwest Australia is also truncated on all sides by Proterozoic rifts closed by convergence, and further has rifts to modern oceans superimposed on the west and south. Exposures are mostly poor, but aeromagnetic maps add much information to that of geologic mapping. The first-vertical-derivative maps of reduced-to-pole data (Fig. 22; Chen et al., 2001a, Fig. 3; Whitaker, 2001, 2003) are particularly informative, and provide much detail regarding ductile and brittle shear zones. The east Yilgarn geologic map (Fig. 23) incorporates magnetic information with surface mapping. Whitaker and Bastrakova (2002) compiled a 1:500,000 map of the entire craton showing geologic units, in part characterized by their magnetic expression, and structures, including broad shear zones, inferred from the magnetic data. The 1:250,000 geologic maps of the craton are available on DVD-ROM (Geological Survey of Western Australia, 2005a), as are some 1:100,000 maps of the eastern part (Geological Survey of Western Australia, 2005b). Geologic summaries of sizeable areas include those by Brown et al. (2001), Cassidy and Champion (2004), Chen and Wyche

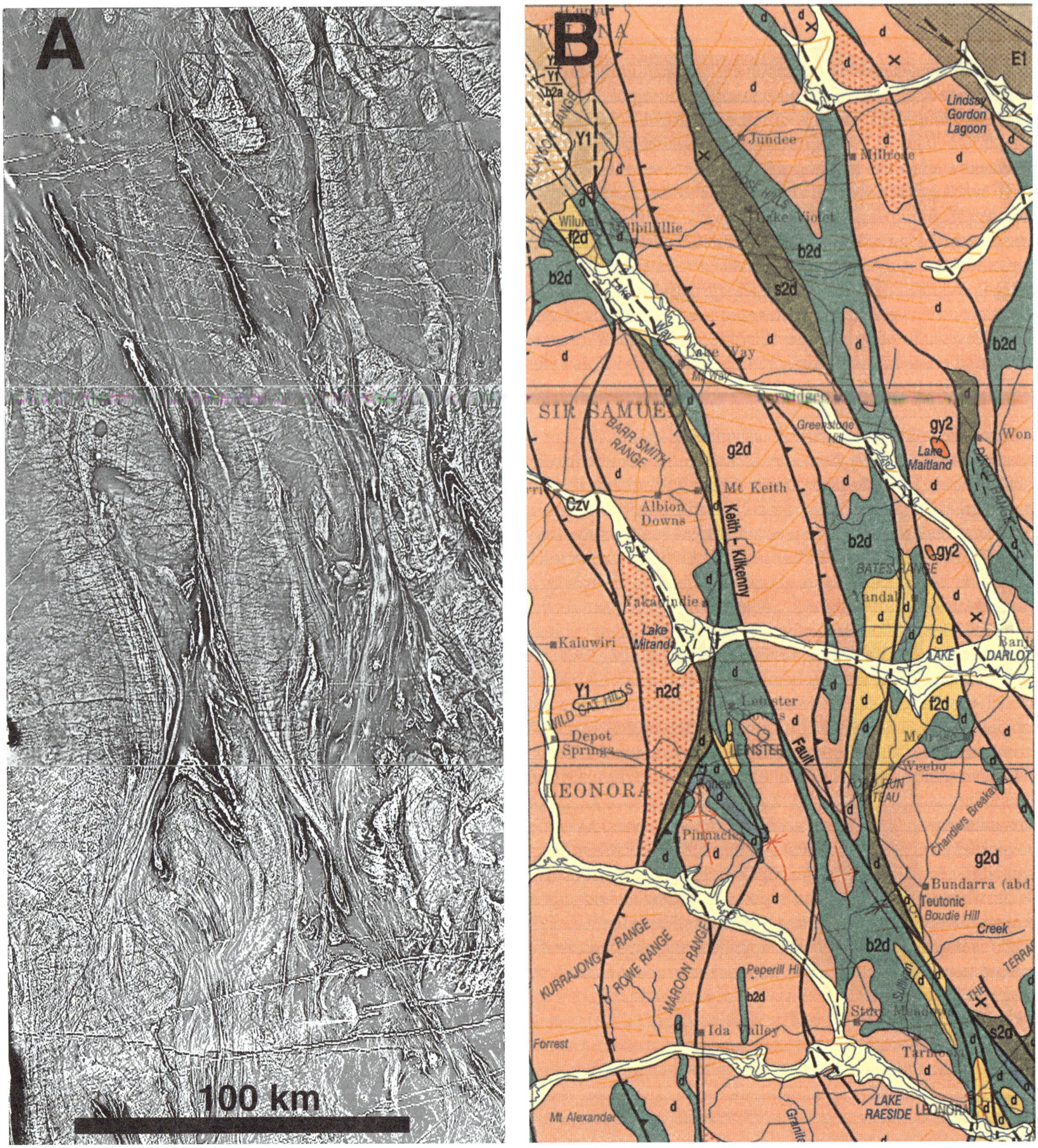

Figure 22. Magnetic and geologic maps of part of northeastern Yilgarn craton, Western Australia. Area extends from 120.0° to 121.5° E, and from 26.0° to 29.0° S. Granite and greenstone terrain, mostly of middle Neoarchean age, formed synchronously with shear zones, which trend generally northnorthwest, as batholiths rose diapirically from decoupled deeper crust. (A) First vertical derivative of reduced-to-pole magnetic field. Highly magnetic units are banded iron formation and magnetite-rich altered ultramafic lavas. Discrete faults appear to be less continuous and numerous than shown on the geologic map, and to have evolved within broad ductile shear zones (striped zones) in granitic rocks. Some late granite plutons (as, top center, and right center) may postdate ductile shear. Abundant Proterozoic dikes crosscut Archean units and have dominantly easterly trends. Flight-line spacing 400 m. Map © 2006 by Geoscience Australia. (B) Generalized geologic map. Plutonic rocks: pink, granitic rocks; spotted pink, gneissic granitic rocks. Supracrustal rocks: green, mostly mafic and ultramafic volcanic rocks; olive green, mostly sedimentary rocks; orange-yellow, mostly felsic volcanic rocks. Onlapping Paleoproterozoic strata (north corners), brown. Surficial deposits, pale yellow. From Myers and Hocking, 1998; map © by Geological Survey of Western Australia.

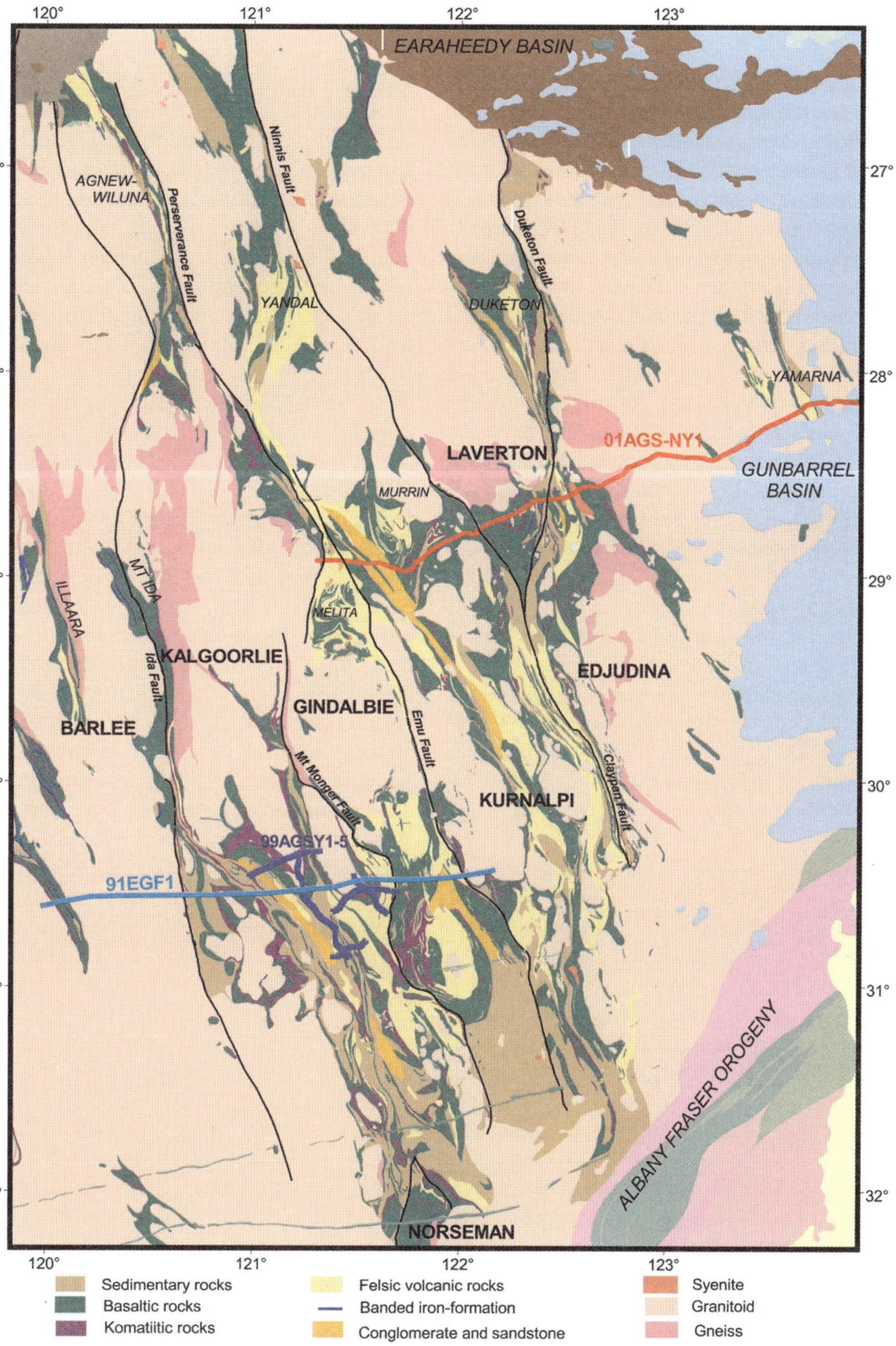

Figure 23. Geologic map of eastern Yilgarn craton, a Neoarchean granite-and-greenstone terrain wherein diapiric batholiths rose synchronously with lateral deformation. Explanation boxes are for Archean rocks only. Komatiitic and basaltic rocks occur mostly in lower parts of greenstone successions whereas felsic volcanics are mostly in higher parts; widespread intercalations of felsic in mafic rocks, and of mafic in felsic ones, are mostly too thin for discrimination here. Dominantly felsic sections typically overlie dominantly mafic ones, so broad distribution shows the tendency toward synformal structure of greenstone belts and domiform structure of granites, and hence the kinship with dome-and-keel geology. Sedimentary rocks, including conglomerate and sandstone, occur within the successions, at their bases, or unconformably overlying them, so as lumped here do not constrain structure. Light blue, dark blue, and red lines mark seismic-reflection profiles. Albany-Fraser orogen, Earaheedy Basin, and unlabeled brown unit in northwest corner are Proterozoic; pale blue strata in east are Paleozoic. Map reprinted from Groenewald et al. (2003, Fig. 2), and provided by P.B. Groenewald, Geological Survey of Western Australia. Most of area of Figure 22 is within the northwest quarter of this area.

(2001), Chen et al. (2001a, 2001b, 2003), Gee et al. (1981), Groenewald and Riganti (2004), Groenewald et al. (2000, 2003), Myers (1997), Myers and Occhipinti (2001), Passchier (1994), and Wilde (1990, 2001). Most of the craton is eroded only into upper-crustal granite-and-greenstone assemblages, although deeper-seated rocks are exposed in the far northwest (the region that includes the Paleoarchean gneisses discussed previously) and the southwest.

I see Yilgarn supracrustal rocks as having been deposited on felsic basement, which, variably remobilized and with new melts, rose into them as diapiric batholiths, while simultaneously the regional upper crust underwent lateral deformation. Rather similar interpretations have been made by Bodorkos and Sandiford (2006), Davis and Maidens (2003), and Gee et al. (1981). The contrary majority view accepts chemotectonic conjectures of plate-tectonic juxtapositions of continental nuclei and diverse oceanic assemblages, and great crustal faults drawn imaginatively on reflection profiles.

The broad granite-and-greenstone region is conventionally split into three large "terranes"—from west to east, Murchison, Southern Cross, and Eastern Goldfields—and many lesser tracts, widely presumed to have had separate histories and to have been juxtaposed by relatively late plate-tectonic sutures. Dissenters Passchier (1994) and Groenewald et al. (2003) recognized that there is no geologic evidence for plate tectonics or terrane amalgamation. Dissenter Whitaker (2001) emphasized that "aeromagnetic data . . . do not provide any support for craton-crossing faults. Such proposed faults either pass through felsic crust with no geophysical (or geological) evidence for their extent or continuity, or are coincident with shear zones across which the crust can be correlated." The craton-wide contemporaneous late magmatism and deformation of the Yilgarn craton has no analogue in plate processes, and there are no relics of oceanic crust or accretionary complexes where sutures have been postulated, nor any appropriate sedimentary basins.

The large- and small-scale juxtapositions postulated by plate-tectonics advocates are based not on structural evidence but on weak chemotectonic analogies and on the assumption that a tract so broad must have been amalgamated by plate processes. Thinly intercalated rocks are assigned to diverse plate settings that varied complexly in space and switched frequently to produce associations without modern analogues. Supracrustal sequences indeed are extensively faulted, but no megathrusts or great extensional faults have been demonstrated by mapped relationships. Myers and Hocking (1998; a bit of their map is shown in Fig. 22B) proposed that regional megathrust planes were precisely reactivated as regional normal faults, backslip offsetting earlier thrusting, but I see no evidence in detailed maps, nor any analogy to proved structures elsewhere, that renders this plausible. Gross stratigraphic sections are similar about the region, and some geologists have postulated that these similarities hide many juxtapositions of disparate bits that had parallel histories. Myers (1995, 1997) speculated that many subduction systems operated simultaneously to produce widely separated, but stratigraphically similar, buildups which then were sutured invisibly to produce the illusion of regional stratigraphy. Swager (1995, 1997) also appealed to parallel, simultaneous development: the "greenstone terranes represent stacked and collapsed basins." Swager's proposed sutures follow concordant lithologic contacts, across which there is no evidence for age reversals, that do not offset the units in any fashion suggestive of thrust faulting, and that abruptly end within continuous sections—so they cannot be sutures; and no scraps of possible lower oceanic crust or mantle are present along the hypothetical structures. As zircon U-Pb dates have become more numerous and constraining, such conjectures have become progressively less tenable. As in other cratons, none of the predictions implicit in the diverse plate conjectures are fulfilled.

Description

Yilgarn supracrustal rocks show the broad stratigraphic trends typical of Archean granite-and-greenstone terrains: thick lower sections dominated by tholeiitic and high-Mg basalts, plus subordinate komatiite, banded-iron formation, chert, clastic sediments, and felsic volcanic rocks, and upper sections dominated by felsic volcanic rocks and clastic sediments, with other types intercalated. Supracrustal rocks are dominantly of middle Neoarchean age in the east, but include early as well as middle Neoarchean in the west. In some central Yilgarn areas, older sedimentary rocks, ca. 3.0 Ga, are known beneath the mafic series, and the oldest strata in these basal sections are quartzose sandstones, which have yielded rare 4.4 Ga detrital zircons, and many ca. 3.8–3.1 Ga ones (Chen et al., 2003; Nelson, 1997; and 1997, 2000, and 2002 reports by D.R. Nelson in Geological Survey of Western Australia, 2005c). In the southwest Yilgarn craton, quartzite has yielded 3.7–3.2 Ga zircons (P.D. Kinny in Wilde, 1990). Basement to these early strata has not been recognized in the field, but as similar clastic rocks are known to overlie ancient felsic basement in other cratons, I presume that Paleoarchean basement was widely present in the Yilgarn but has been largely remobilized into younger batholiths. I expect that remobilized ancient gneisses and sunken Mesoarchean supracrustal rocks will be recognized when the broad tract of mid-crust gneisses and high-grade supracrustal rocks of the southwest Yilgarn is studied in detail. The lower parts of greenstone sections are dominated by tholeiitic basalt, high-Mg basalt, and komatiite, yet that they are at least partly ensialic is shown by the sediments that underlie them, by their chemical compositions and their occasional zircon xenocrysts, and by the intercalations in them of felsic volcanic rocks that likely record breaching of early-rising batholiths. The younger sections of the greenstone belts include similar mafic and ultramafic rock types, and also more varied mafic, intermediate, and voluminous felsic volcanic rocks, and contain abundant clastic and chemical sedimentary rocks, often very continuous.

None of these associations resemble the modern assemblages with which weak chemotectonic analogies are made. That, as in other cratons, batholiths rose over prolonged periods, and from a broad substrate rather than from linear sources, is indicated by the age range of the ubiquitous felsic volcanism. Rising of batholiths to the surface late in the era of density inversion is recorded by voluminous clastic sediments, from felsic sources, high in some stratigraphic sections, unconformable upon older supracrustal rocks.

A unit of komatiite and allied ultramafic lavas is preserved in the now-disconnected greenstone belts throughout a well-studied region in the eastern Yilgarn, 200 km wide across the strike of a number of greenstone belts and batholiths and 150 km long parallel to them (e.g., Hill et al., 2001; Walker and Blight, 1983; partly delineated in the south-central part of Fig. 23), and shows that region to have been a lava plain, not an amalgam of belts, when the thick lower greenstone succession was deposited. The ultramafic unit is dated in three places, and closely bracketed in several others, as 2.705 Ga, by ion-probe U-Pb zircon determinations (Nelson, 1997). The ultramafic unit mostly dips steeply and is 0.5–3 km thick. Thin, local interbeds of shale and basalt occur within the section, but the great composite sheet, with a volume of >10,000 km^3, was erupted within a brief period. Thick basalts and felsic volcanic rocks below and above the ultramafic unit, and the minor sedimentary rocks above it, are mostly within the narrow span of 2.71–2.66 Ga insofar as they are zircon-dated (Nelson, 1997). Much high-Mg basalt in the region may have formed by contamination of ultramafic melt by felsic crust (Hill et al., 2001).

As in the Superior craton, and unlike broad Phanerozoic orogenic terrains, the young granites exposed across most of the vast Yilgarn province were formed within a short period of time. Calcic granites, 2.72–2.66 Ga, late potassic granites, 2.66–2.63 Ga, and slightly older and coeval middle Neoarchean greenstone belts occur across much of the craton, whereas granitic rocks with diverse suites of old zircons, and varying Nd model ages, have more limited distributions (Cassidy and Champion, 2004). Detrital Paleoarchean zircons, in part discussed earlier, are known from ancient clastic sediments in various areas, but dated supracrustal rocks and late-igneous zircons in granitic rocks otherwise are latest Mesoarchean and Neoarchean. Cassidy and Champion postulated that cratonic nuclei, formed in diverse locales by subduction-related processes, had been amalgamated by plate-tectonic convergence prior to late unified volcanism, sedimentation, and plutonism. Griffin et al. (2004a) saw different boundaries in the same region in their zircon ages and Hf isotopes.

Yilgarn provides both a conflict between field data and conventional interpretation of Lu-Hf systematics and permissive evidence for a mafic protocrust. Griffin et al. (2004a) found zircons from Neoarchean granites of the Yilgarn Southern Cross "Province" to have high initial $^{176}Hf/^{177}Hf$, and concluded accordingly that these granites were juvenile and contained little if any material reworked from older crust, and that the area had no older felsic crust. Nearby, however, Wyche et al. (2004) found the oldest stratigraphic unit in two long greenstone belts, continuously underlying thick sections of basalt and high-Mg basalt, to be quartzite, the many analyzed zircons in which were 4.35–3.10 Ga, entirely older than the rest of the supracrustal succession and the Neoarchean granites. Although basement gneiss has not been recognized nearby, this is typical basal-greenstone ensialic geology, and I presume that the Neoarchean granites were derived from underlying mafic protocrust.

Metamorphism decreases away from granites, from low-pressure amphibolite facies at contacts to lower greenschist or

Figure 24. Undeformed Neoarchean pillow basalt. Lava flowed toward observer. Outcrop 2 m high, 8 km north of Norseman, Yilgarn craton.

prehnite-pumpellyite facies (e.g., Groenewald et al., 2000, Fig. 8). Little-deformed supracrustal rocks are widely preserved away from granites and shear zones (Fig. 24). As in dome-and-keel Archean terrains, batholiths did not come randomly into regionally deformed and metamorphosed tracts, but were active agents of deformation and metamorphism—although in the Yilgarn craton, as in the Superior craton, regional lateral deformation accompanied rise of batholiths. Granite petrology, Nd isotopes, and inherited zircons show some young batholiths to represent reworked older felsic crust. Other batholiths commonly are interpreted to be juvenile—but as in other Archean cratons, the data, even in conventional terms, may indicate separation from thick mafic protocrust, not mantle.

In the region of best exposure, Yilgarn supracrustal rocks are mostly strung out in irregular northward-trending belts (Fig. 23)—but this is also the region of most conspicuous shearing as shown both in outcrop and on magnetic maps, and of most known gold deposits, which are associated primarily with sheared greenstones. Within this relatively well-exposed region, there are no tidy domiform batholiths in networks of ovoidal greenstone keels, like the northeast Pilbara craton, parts of the Superior craton, and (discussed subsequently) the Zimbabwe craton, although partial examples can be recognized on Figure 23 and on detailed geologic maps. In most of the rest of the craton, exposures are too limited for confident generalizations. A broad tendency toward disrupted dome-and-keel patterns is indicated by the synclines that dominate many of the greenstone belts. (See many 1:250,000 geologic maps, including Barlee, Boorabbin, Edjudina, Glengarry, Hyden, Kurnalpi, Laverton, Ninghan, Wiluna, Yalgoo, and Youanmi sheets.) The trend toward synformal greenstone belts is shown even by the generalized rock distributions depicted in Figure 23, although this is lost in the still-more generalized compilation by

Myers and Hocking (1998; sample in Fig. 22B). Many granitic masses end in anticlinal noses beneath supracrustal rocks, in typical Archean dome-and-keel fashion (as Gee et al., 1981, recognized) but unlike patterns in subduction-related batholiths such as the Cretaceous Sierra Nevada of California.

Broad zones, mostly steep, of ductile shear, conspicuous on magnetic maps, anastomose northward through Yilgarn granite-and-greenstone terrain and, with more brittle and restricted faults that come and go within them, tend to define large lozenges of resistant granitic rocks (Fig. 22A; Whitaker, 2003). Chen et al. (2001a, 2001b, 2004) found a mixture of pure and simple shear in their central Yilgarn study areas; slip indicators on north-northwest–striking shear zones were mostly for left slip, whereas on north-northeastward sectors indicators were for right slip. As Chen et al. recognized in their papers, east-west shortening and north-south extension is indicated both by the shear zones and by longitudinal folding of supracrustal rocks, and broad structural patterns are controlled by the greater rigidity of granitic rocks than of supracrustals. Deformation was craton-wide but consistent strike-slip offsets have not been recognized. Discrete faults are given more continuity on geologic maps (Figs. 22B and 23) than appears to be permitted by detailed magnetic data (Fig. 22A; Whitaker, 2001, 2003). Ductile structures illustrated by photographs in the Chen et al. papers are what I expect to see in deformation at greenschist and lower amphibolite facies, at temperatures on the order of 350–600°C, so I presume shearing to have been generally synchronous with rise of batholiths. The youngest granites postdate all or much of the deformation (Fig. 22A).

Interpretation

The Yilgarn craton is a granite-and-greenstone terrain that was subjected to complex lateral flowage of decoupled upper crust as diapiric batholiths rose, with varying proportions of melt, from continuous felsic lower crust. Effectively floating upper crust underwent lateral deformation, without major-fault steps in crustal level and so with little change in surface area, while its felsic substrate was rising. The Yilgarn underwent lateral deformation synchronous with the usual Archean crustal inversion in response to density imbalance. Bodorkos and Sandiford (2006), Davis and Maidens (2003), and Gee et al. (1981) advocated similar models. The deformation is analogous to that of much of the Superior craton, which better displays the definitive domiform granites and supracrustal keels in many areas (Fig. 19).

The very different current consensus among most Yilgarn geologists regards the granite-and-greenstone assemblages as unrelated to dome-and-keel tectonics, and instead as products of episodes of Phanerozoic-type deformation, metamorphism, and plutonism. Up to eight regionally correlative episodes of diversely oriented contractional, extensional, and shear structural systems are invoked, within brief total time spans, to explain map patterns and conjectural plate-type juxtapositions (e.g., Brown et al., 2001; Chen et al., 2003; Henson et al., 2004). Density inversion of sinking greenstones, diapiric batholiths rising from a regional substrate, and rotations of structures in complex continua of deformation—the processes I see as dominant—are minimally considered.

Seismic-Reflection Profiles

Vibroseis reflection profiles in the eastern Yilgarn craton have been interpreted to show scores of great gentle thrust and normal faults interlacing through the entire crust and thus to require pervasive crustal shortening and extension due to plate-tectonic interactions (Drummond et al., 2000; Goleby et al., 2002, 2003, 2004, 2006; van der Velden et al., 2006). These interpretations are incorporated in most current Yilgarn syntheses. The profiles are located on Figure 23, but are not reproduced here. Their aspect is similar to that of the West Superior profiles of Figure 21 except that they are more chaotic in the time interval between ~4 and 8 s, which would be middle crust for in-plane reflections. The Yilgarn profiles show a largely transparent upper interval, ~2–5 s thick, with discontinuous gently to moderately inclined reflectors, above a main zone of packets of subhorizontal, gently inclined, and undulating reflectors that often intersect and occasionally cross through one another.

The Yilgarn interpreters treated all apparent events as within the planes of the sections, and, in my view, grossly overinterpreted the displays in drawing great crustal faults through them. The tracks cross granite-and-greenstone structures at all angles to widely varying strikes and dips of layered rocks, so out-of-section reflections must be important. "No amount of processing and manipulation of the data in 2D [profiles] can resolve these problems" of discriminating in-plane from out-of-plane reflections (Drummond et al., 2004, p. 223). Hobbs et al. (2006) illustrated some of the complex patterns that can be generated by out-of-plane reflections and mistaken for in-plane structure. Add filtering to generate signal coherence that may be illusory and post-stack migration to minimize telltale crossing events, and the constant-amplitude displays can be seriously misleading. In some profiled areas, thick dikes of Proterozoic gabbro are abundant. The line segment (Goleby et al., 2004, Fig. 4) with the most impressive bundles of gently inclined reflectors at two-way times appropriate for the middle crust (and on which many deep-fault interpretations elsewhere in the region are implicitly based) trends at a low angle through a sparse swarm of large subparallel dikes, conspicuous on magnetic maps, at distances and orientations from the traverse consistent with their being recorded as gently inclined reflectors. By contrast, the parts of the profiles that cross tracts of undiked granite—the west end of profile EGF1, and the part of NY1 between the Laverton complex and the Yamarna greenstone belts (Fig. 23)—show gentle and minimally complicated apparent mid-crust structure.

The great-fault interpretations are geometrically implausible. There are no surface structures with the predicted offsets, either normal or reverse. The conjectural structures are projected obliquely upward, through the almost data-free upper parts of the displays, to convenient mapped and possible shear zones on the surface; but those shear zones are mostly steep, not gentle, lack the requisite vertical offsets (they are strike-slip structures, not the crossing normal and thrust faults inferred by seismic interpreters), and were active at high temperatures when the lower crust could not have supported large discrete faults.

I see dominantly laminar flow in the middle and lower crustal parts of those profiles that are least likely contaminated by out-of-section events. Steep surface structures, including shear zones, apparently do not break the laminar patterns that begin at a depth of 6–10 km because the shallow, upper crust, with its sporadic deformation, was decoupled from the pervasively flowing deeper crust.

Yilgarn Deeper Crust

The western corners of the Yilgarn craton are bounded by Proterozoic orogens. The deeply eroded northwest corner includes the Paleoarchean gneiss complex discussed in an earlier section. The southwest corner exposes mostly Archean granitic and gneissic rocks (Gee et al., 1981; Wilde, 1990, 2001), and, as some geologists have reasoned (others have disagreed) may expose middle crust such as that beneath the granite-and-greenstone terrain of the rest of the craton. Orthopyroxene-bearing granites (low-pressure granulite facies) comprise one sizeable area, and supracrustal rocks are metamorphosed variously at low- to high-pressure amphibolite facies, low-pressure granulite facies, and, in one area, likely high-pressure granulite. Zircon U-Pb dates are sparse but are mostly ca. 2.7–2.6 Ga in both granitic and felsic-volcanic rocks, although a quartzite at the base of a greenstone section has yielded 3.7–3.2 Ga detrital grains, as noted previously.

Kaapvaal Craton, South Africa—Varying Ages, Varying Styles

The Kaapvaal Archean craton underlies much of South Africa, is surrounded by Proterozoic mobile belts, and is exposed discontinuously beneath Proterozoic and younger cover. The craton displays Paleoarchean basement gneiss and granite-and-greenstone terrains of diverse Mesoarchean and Neoarchean ages. Like the Pilbara craton, Kaapvaal contains thick late Neoarchean and early Paleoproterozoic volcanic and sedimentary rocks, older components of which overlap in age younger granite-and-greenstone terrains nearby, that represent greenstone-style sections whose deformation proceeded only partway to dome-and-keel development. Geologic and petrologic information is uneven. Eglington and Armstrong (2004) and Poujol et al. (2003) summarized the zircon geochronology of Archean and Paleoproterozoic rocks, but most of the dates represent Pb/Pb, or whole-grain U-Pb, analyses, which, as noted in the introduction, are ambiguous in rocks with complex histories.

Granite-and-Greenstone Terrains

The best-exposed and most-studied Kaapvaal granite-and-greenstone assemblage is the Mesoarchean, ensialic, dome-and-keel Barberton Greenstone Belt at the east edge of the craton. It comprises a complex, palmate, southwest-widening synclinorium, 100 km long and with a maximum width of 50 km, sunk between flanking granitic rocks in its narrow part, and fingering out southwestward into synclines between granitic domes and anticlines over those domes. Three of those southwestern domes are dominated by variably remobilized and intruded Paleoarchean basement gneiss, which has yielded 3.7–3.5 Ga zircons and is overlain by the basal units (Sandspruit and Theespruit Formations) of the supracrustal section (Kisters and Anhaeusser, 1995; cf. Dziggel et al., 2002). These basal strata consist of thick metasandstones and conglomerates derived from felsic migmatite, and subordinate intercalated mafic and felsic volcanic rocks. A kilometers-thick section (most of the Onverwacht Group) of mafic and ultramafic lavas, and subordinate felsic volcanic rocks, overlies the basal clastic sediments, and has an approximate age range of 3.5–3.4 Ga (Byerly et al., 1996; de Ronde and de Wit, 1994). Next higher are thick shales and turbidites (Fig Tree Group), with subordinate ultramafic and felsic lavas and banded-iron formation, ca. 3.4–3.3 Ga (Byerly et al., 1996). The lower turbidites were derived mostly from mafic volcanic rocks, and the upper from granitic rocks (Condie et al., 1970), showing progressive unroofing as batholiths domed and breached the section. Highest are thick quartzose sandstones and conglomerates (Moodies Group, 3.3–3.2 Ga) recording alluvial, fluvial, tidal, and shallow-marine settings (Fig. 25; Heubeck and Lowe, 1994). Domiform granites that rose into the supracrustal rocks have igneous ages from ca. 3.45 to 3.1 Ga, and in places as young as 2.7 Ga, and felsic volcanic rocks were erupted from them into and onto the deforming supracrustal strata. Metamorphism decreases inward from the granite contacts: 0–3 km of amphibolite facies, commonly less than 5 km of greenschist facies, and a broad interior region that mostly lacks both pervasive deformation and obvious metamorphism (Fig. 25; Saggerson and Turner, 1992).

Figure 25. Almost unmetamorphosed shallow-water Moodies sandstone, age 3.2 Ga, preserves undeformed ripple marks on steep, dark bedding surfaces. Metamorphism and deformation of Archean greenstone belts are due primarily to rise of domiform batholiths and sinking of dense volcanic rocks, and well-preserved material is common away from contacts. Outcrop ~2.5 m high, near center of Barberton greenstone synclinorium, ~5 km southeast of Barberton, South Africa.

Near-granite synmetamorphic deformation commonly consists of severe flattening and elongation, granite-up, greenstone-down (e.g., Jackson and Robertson, 1983, and Kisters et al., 2003). This is within the contact aureole and is typical synmetamorphic granite-and-greenstone deformation. (Kisters et al., seeking horizontal-tectonics explanations, attributed it to pre-granite deformation.) Amphibolite in the southwest locally contains garnet so depth of erosion reaches 18 or 20 km. Darracott (1975) inferred from gravity models that supracrustal rocks extend only 3–6 km downward into underlying granitic rocks. Complex systems of bedding-parallel, younger-over-older rootless megathrusts and cryptic sutures have been proposed to accommodate speculations regarding plate-tectonic settings of diverse igneous rocks (e.g., de Ronde and de Wit, 1994), but geologic relationships required by these conjectures have not been demonstrated.

Other dated Kaapvaal granite-and-greenstone assemblages are generally younger than Barberton. Sparse zircon U-Pb dates from them are primarily from granitic rocks and mostly are within the range 3.35–2.7 Ga (Eglington and Armstrong, 2004; Poujol et al., 2003). The still sparser dates from greenstone belts are mostly 3.2–2.9 Ga. Among these, the Murchison greenstone belt, ca. 3.0 Ga, displays characteristic granite-and-greenstone dome-and-keel geology (Minnitt and Anhaeusser, 1992; Poujol et al., 1996).

Deep Gneiss and Shallow Granite and Greenstone

The 2.0 Ga Vredefort impact structure exposes a steeply upturned concentric section through much of the Archean crust and exposes deep and shallow crust close together. An early Mesoarchean supracrustal package foundered into the deep crust, whereas a late Mesoarchean one remained on top and evolved in granite-and-greenstone mode (Hart et al., 1981). Gneisses with Mesoarchean zircons are cut by younger gneisses with Neoarchean zircons (Flowers et al., 2003), and enclose disrupted and highly metamorphosed relics of mafic, ultramafic, and sedimentary supracrustal rocks that contain detrital zircon with a U-Pb age ≥3.4 Ga; Re-Os isotopes indicate a likely age of ca. 3.5 Ga for ultramafic rocks; and metamorphic zircon is ca. 3.1 Ga (Hart et al., 1981, 2004). These sunken supracrustal rocks thus are approximately correlative with the Barberton section. Shallower Vredefort gneisses are overlain unconformably by a younger, but pre–3.0 Ga, greenstone assemblage of komatiite, basalt, and metasedimentary rocks, compositionally different from the older rocks enclosed in deep gneisses (Lana et al., 2003).

Incomplete Neoarchean Granite-and-Greenstone Assemblage

Unconformably overlying Kaapvaal granite-and-greenstone assemblages are kilometers-thick Neoarchean sections of relatively undeformed sedimentary and volcanic rocks preserved mostly in broad synformal basins. Initial continuities and relations of present basins of preservation to depositional systems are disputed, and tectonic settings are subjects of conflicting speculations. The section in the largest basin of preservation, Witwatersrand, begins with thin clastic sediments (Fig. 26A), followed by 1–2 km of basalt, komatiitic basalt,

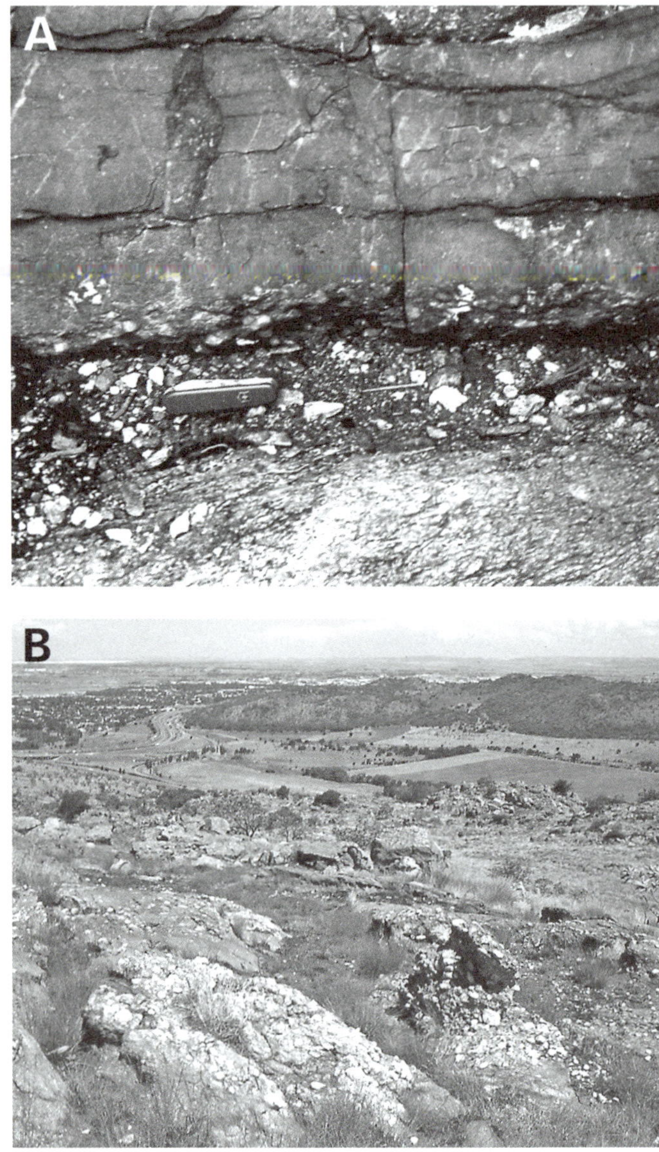

Figure 26. Neoarchean regional-sheet greenstone section of Witwatersrand Basin, Kaapvaal craton, went only partway to granite-and-greenstone structural development. (A) Basal unconformity. This section, like other greenstone sections, begins with clastic strata. Pocketknife is on rubble-covered zone, below which is older granite. Early Neoarchean quartzite above rubble has ~5 cm of quartz-pebble conglomerate at base. Central Rand. (B) At top of basal clastic section, Rand conglomerate (cobbles of quartzite, vein quartz, and chert) dips steeply, toward upper right, because of diapiric rise of unconformably underlying domiform granite. Later Neoarchean, ca. 2.7 Ga, Ventersdorp volcanic rocks in distance overlie conglomerate unconformably and dip gently. Valley is underlain by komatiitic basalt, and distant ridge is formed of basalt. View southeast in southern Johannesburg.

andesite, and felsic volcanic rocks, ca. 3.07 Ga (Cheney and Winter, 1995; Eriksson et al., 2001a; Robb and Meyer, 1995). I presume the felsic rocks to record venting of nearby batholiths. Above these strata are kilometers of sandstone, shale, minor basalt, and, high in the section, conglomerate. (As Kositcin and Krapež, 2004, noted, neither foreland basins nor post-rift passive-margin wedges, both often inferred to be the sites of deposition of the sediments, commonly contain basalts.) Next is several kilometers of mostly volcanic rocks (Fig. 26B)—basal komatiitic basalt, then basalt, then basalt with felsic intercalations—and minor sedimentary rocks, ca. 2.70 Ga (e.g., Crow and Condie, 1988). Other South African basins of preservation have thick sediments and minor volcanics up through the rest of the Archean, and then thick sections of Paleoproterozoic clastic, carbonate, and iron formations, plus local volcanic rocks, up to a young limit of ca. 2.0 Ga (e.g., Eriksson et al., 2001b).

Archean Kaapvaal felsic crust, like that of Pilbara, was remobilized, following blanketing by thick Neoarchean and early Paleoproterozoic volcanic and sedimentary rocks. Domiform diapiric batholiths, characteristic of granite-and-greenstone terrains, continued to rise. Doming is particularly obvious for the circular granitic Johannesburg Dome, which on its south side steeply tilted the Neoarchean strata of the north edge of the Witwatersrand basin (Fig. 26B). Further rise still later is shown by moderate tilting of Paleoproterozoic strata on the north side of the dome. Neoarchean rise and exposure of other granites are shown by igneous ages of the granites themselves, by ages of detrital zircons in Witwatersrand strata, and by felsic volcanism. The young age limit of detrital zircons in Witwatersrand strata becomes progressively younger upsection and lags the age of the strata that contain them but requires unroofing of syn-sedimentary granites (Kositcin and Krapež, 2004; Robb and Meyer, 1995). Slowly rising batholithic domes present discontinuously around the Witwatersrand basin account both for sedimentation patterns and much deformation (Brock and Pretorius, 1964). Venting of these remobilized domes presumably sourced the felsic volcanism in the basin.

The limited geochemical data from the Neoarchean volcanic rocks show them to be dominantly basaltic andesites, like those of the Fortescue assemblage in Australia.

Seismic-Reflection Profiles

A good seismic-reflection profile (Tinker et al., 2002, Fig. 3) shows the middle Neoarchean subsurface units to have sheet stratigraphy at the west edge of the Kaapvaal craton and to be overlapped by ca. 2.0 Ga units that thicken markedly westward toward a mobile belt beyond the profile. Both older and younger sections are broken by a gently west-dipping post–2.0 Ga fault that Tinker et al. regarded as normal but that cuts bedding at a low angle and has drag geometry appropriate for an east-directed thrust fault. I read these relationships as those expected of the onset of something more like plate tectonics ca. 2 Ga. The Archean continent was extensionally thinned; a stratal wedge was deposited on the thermally subsiding margin; subsequent convergence, by processes not yet defined, produced eastward thrusting of the stratal wedge onto the thinned Kaapvaal margin.

Regional-sheet stratigraphy, not belts, of late Neoarchean units is obvious also in other reflection profiles near the west margin of the craton (Tinker et al., 2002). Tinker et al. inferred much deformation of still-deeper, early-Neoarchean strata, but the very poor quality of the deep records provides no support for this conjecture. In still poorer records—only local discontinuous reflections in the top few kilometers, and little more than sparse artifacts deeper, all worse than ambiguous—from the central part of the craton, de Wit and Tinker (2004) conjectured Archean thrust and normal faults interlacing complexly through the entire crust. I see no basis for their speculation in the profiles, in geology known in outcrop, or in kinematic plausibility.

Zimbabwe Craton—Dome-and-Keel Tectonics, Ensialic Greenstones

The Zimbabwe (Rhodesian) craton is bounded on all sides by Proterozoic orogenic belts. Its median region, 250 × 500 km, exposes dome-and-keel granite-and-greenstone terrain that is strikingly obvious on the geological map of Zimbabwe (Geological Survey of Zimbabwe, 1994) and on detailed geologic map sheets (e.g., Baglow and van Beek, 1987). The batholiths of the granite-and-greenstone region typically have elliptical shapes, but a systematic orientation is not apparent. Severe lateral deformation, like that of Yilgarn and northwest Superior, does not exist. Macgregor's classic paper (1951), which recognized Archean geology as typified by diapiric "gregarious batholiths" and sinking synforms of dense volcanic rocks, was based on his long experience in this terrain. Many others, among them Jelsma et al. (1993) and Ramsay (1989), have since confirmed detailed aspects of this style in Zimbabwe.

Preserved greenstone belts are Neoarchean (2.90–2.65 Ga: Wilson et al., 1995), and most of them, likely all, are ensialic, despite widespread pillow basalt and komatiite. Felsic volcanic rocks intercalated in most of the major greenstone belts throughout the craton have yielded zircon xenocrysts 20 to 1000 m.y. older than the host volcanic rocks, proving the presence of ancient felsic lower crust (Horstwood et al., 1999; Wilson et al., 1995). Ancient gneiss, with zircons to 3.8 Ga, is known in outcrop in one part of the craton, where it is overlain by clastic sections, overlain in turn by thick mafic-volcanic sections (Blenkinsop et al., 1993; Jelsma et al., 1996). Inconclusive dating suggests highly metamorphosed enclaves of supracrustal rocks within the ancient gneiss complex to be ca. 3.5 Ga.

A spate of recent papers applied plate-tectonic speculations to Zimbabwe supracrustal rocks on the basis of vague lithologic analogies with modern rocks, or of local bedding-parallel high-strain zones within metamorphic rocks assumed to be, in the absence of dating, younger-over-older rootless, regional thrust faults.

Southwest Greenland—Early Supracrustals Foundered, Late Supracrustals Stayed High

The deeply eroded Archean terrain of southern West Greenland, a strip 120 km or so wide between coast and ice cap, is known primarily from reconnaissance mapping by many parties, patched together into 1:100,000 maps in the 1970s and 1980s, before reliable geochronology was available, and described in brief reports. Subsequent papers have been mostly geochronologic and geochemical studies of poorly characterized samples, interpreted nowadays with vague concepts of "terranes" amalgamated by plate-tectonic processes. Some new mapping has been done, but only a minority of studies integrate detailed field study and petrography with their chemistry and chronology, and major problems are unresolved.

I discussed earlier the extensive Paleoarchean gneisses, which contain tectonic enclaves of supracrustal rocks at least as old as 3.6 Ga. Younger TTG gneisses form both intermixtures and separate large masses with these ancient gneisses. The result is conventionally explained in terms of "terranes" assembled by hypothetical plate-tectonic processes, whereas I see felsic crust built incrementally by new materials, added from a deeper mafic protocrust, mixed into mobile older materials on diverse scales.

There also are many greenstone belts, exposed at generally deeper levels than are typical of Archean cratons and hence typically more metamorphosed. I concentrate here on some of the belts nearby to the south of the much-publicized Isua belt. Beech and Chadwick (1980) and Chadwick and Nutman (1979) recognized that some of the thicker sheets of supracrustal rocks lie in depositional contact on ancient gneiss. Beech and Chadwick found some of the thicker sections of supracrustal rocks to have a consistent stratigraphy: thin, discontinuous quartzite and other metasedimentary rocks lying directly on felsic basement, then intercalated mafic and ultramafic volcanic rocks, then mostly clastic rocks which are interbedded with what I presume, from their major-element chemistry, to be felsic volcanic rocks. These are classic ensialic greenstone successions. (Polat et al., 2007, nevertheless argued, with chemotectonics, that the volcanic rocks in one of those successions were ensimatic oceanic rocks.) Zircon U-Pb age determinations of ages of hundreds of detrital igneous grains in many of the sandstones show numerous detrital grains to be younger than 3.1 Ga in all samples, and felsic volcanic rocks in the successions also to be younger than 3.1 Ga (Hollis et al., 2005, 2006; Nutman and Friend, 2007; Nutman et al., 2004a, 2007).

The most obvious domes of typical Archean diapiric style yet mapped are in the area between 49.5° and 50.5° west long and 64.6° and 65.2° north lat. I see in published maps (Ivisârtoq and Isukasia: Chadwick and Coe, 1987, and Garde, 1988) four domes, 20 to 40 km long in the northeasterly elongation direction, outlined by elliptical belts of supracrustal rocks. Isua is the northern and most-studied of these domes. The others are the Ujaragssuit Nunât, Ivisârtoq, and Nunatarssuaq domes of Chadwick (1990), who assumed them to be upright folds of bedding-parallel rootless regional megathrusts, from unknown sources, that placed gneisses both above and below a very thin sheet of less-metamorphosed supracrustal rocks, even though the contacts he described are mostly unsheared. Each dome has an end hidden by ice or sea, and less than half of the Nunatarssuaq dome is well exposed. The supracrustal rocks are dominantly basaltic, with subordinate ultramafic and sedimentary units. As expected with the dome-and-keel analogy, deformation and metamorphism are highest close to the domiform batholiths, and pillow structures are well preserved in much of the lower-grade tracts, which are lower amphibolite facies. The three southern domes form a cluster that meet in a triple-junction synform of supracrustal rocks, as expected for a dome-and-keel analogy but not for a fold system.

Isua Dome and Greenstone Belt

The northern of the four domiform batholiths is that inside the elliptical Isua greenstone belt. This belt is widely assumed to contain the world's oldest supracrustal rocks, ca. 3.8–3.7 Ga. (Claims have been made for similar antiquity of supracrustal rocks in a small area in Labrador and another in Quebec, but published evidence is too cursory for evaluation; otherwise, the oldest supracrustal rocks proved anywhere, an enclave in lower-crust Greenland gneiss, are 3.6 Ga.) A large literature of geochemical and isotopic reports on Isua samples, and derivative interpretations of early-Earth evolution, builds on this assumption of uniquely old age. This extensive work has not included a quest for definitive dates and field relationships that would prove or disprove such an age, and no unambiguous evidence requires an age greater than ca. 3.0 Ga, which is similar to the maximum age, 3.1 Ga, of the many dated nearby greenstone successions. Resolution of this ambiguity is important for establishing milestones in crustal evolution.

The Isua belt, mostly 1–3 km wide, crops out as the rim of a nearly complete 12 × 20 km northeast-elongate ellipse (the tip is covered by inland ice), flanked inside and out by ancient gneisses and subordinate younger Archean granitic and gneissic rocks. Nutman (1986) compiled and summarized his own reconnaissance mapping and rock descriptions and those of others before him. Hanmer and Greene (2002), Myers (2001), and Rosing et al. (1996) made major revisions to Nutman's protolith identifications and structural interpretations. Both southwest and southeast parts of the rim were deformed primarily by dome-side-up shear (James, 1976), as expected for a dome-and-keel system. Both long limbs of the ring dip steeply southwestward but I presume it to have originated as the subvertical rim of a domiform batholith of typical granite-and-greenstone type, the stratified rocks having been deposited on felsic basement that was remobilized and rose through them. The assumption by most Isua investigators that the supracrustal rocks comprise a large xenolith in >3.6 Ga midcrustal igneous tonalite and that basement rocks are unknown is incompatible with the geometry, structure, and metamorphic petrology of the complex.

Parts of the surrounding polycyclic migmatites and gneisses have yielded many igneous zircon ages of 3.8–3.6 Ga, and metamorphic-zircon ages as young as 2.7 Ga, whereas other gneisses in the surrounding complex have igneous protolith ages as young as 2.7 Ga (Crowley, 2003; Crowley et al., 2002; Nutman et al., 1996, 1997, 2004a). The ancient gneisses were extremely deformed with supracrustal rocks where the latter are most metamorphosed—but did those gneisses intrude the supracrustal rocks (the conventional view), or do the gneisses instead comprise the basement on which the stratified rocks were deposited (my inference)?

Metamorphosed basalt and high-Mg basalt, including both pillow lavas and fragmental rocks where fabrics are preserved, dominate the supracrustal assemblage, but chert, banded iron formation, mafic andesite, and ultramafic rocks also are abundant. Minor metapelite and metasandstone are present, although their protoliths and distributions are poorly documented, and quartz-pebble conglomerate has been found locally. (The conglomerate contains detrital zircons that have not been dated.) Northeastern rocks were metamorphosed only at greenschist facies, and peak metamorphic conditions for the most-metamorphosed rocks elsewhere were ~500–600°C at a depth of 15 or 20 km (Boak et al., 1983; Rollinson, 2002). The latter metamorphic conditions are appropriate also for the retrograde metamorphism and deformation of the flanking and intercalated ancient gneisses, but even the highest-grade metamorphism of the supracrustal rocks is incompatible with the conventional interpretation that the supracrustal rocks predate the magmatic protoliths of the gneisses. The supracrustal rocks are nowhere migmatized, and their metamorphism does not accord with the conventional assumption that they were once deep in a huge tonalite magma chamber at 750°C.

The widely accepted 3.8–3.7 Ga age of the volcanic succession comes primarily from ion-microprobe U-Pb ages of igneous zircons in several specimens described only as "felsic metavolcanic rocks" (Nutman et al., 1997; metamorphic zircon ages scatter down to 2.7 Ga), and if that casual protolith identification is correct then so probably are the age assignments for those parts of the supracrustal succession. Myers (2001) and Rosing et al. (1996) refuted the volcanic designation, and characterized the dated rocks as structurally concordant and synmetamorphically deformed tonalite. Although Myers and Rosing et al. accepted these metatonalites as derived from intrusive dikes or sills and hence as providing a minimum age for the supracrustal rocks, the metatonalites may instead be isoclinal intercalations of basement rocks that provide a maximum age. Severely deformed metatonalite flanks much of the Isua belt and is intercalated with supracrustal rocks only where both are highly sheared and multiply deformed by isoclinal and sheath folds (e.g., Hanmer and Greene, 2002). Igneous zircons in several of these metatonalite intercalations were dated at ca. 3.8 Ga, with metamorphic-zircon ages down to 2.6 Ga, by Crowley (2003). Mylonitized tonalite intersheared and contorted with mafic schist in the extremely deformed margin of the Isua belt has an igneous-zircon age of 3.64 Ga but metamorphic-zircon ages of 2.9–2.6 Ga (White et al., 2000a, 2000b). Inclusions of similar (but not necessarily correlative) mylonite occur within less-deformed younger tonalite, near but outside the Isua belt (Hanmer and Greene, 2002), which has an igneous-crystallization age of 2.99 Ga (Hanmer et al., 2002). There are no dikes or sills of ancient tonalite within minimally deformed Isua rocks, wherein they should be were the conventional interpretation of supracrustal antiquity correct. I worked extensively in California with Proterozoic basement and Phanerozoic cover strata that were metamorphosed and extremely deformed together, with field relationships much like those attributed to intrusion at Isua. I tried to get to Isua to evaluate this option, but the logistics controller of the large Isua project would not support a questioning of the ancient age assignment.

A pegmatite that cuts supracrustal rocks but shares all or most of their deformation contains zircon with an igneous crystallization age of 2.95 Ga and a metamorphic age of 2.7 Ga (Hanmer et al., 2002). Mafic 3.5–3.2 Ga dikes postdate much of the gneissic foliation in the rocks (to me, basement) in the minimally deformed interior of the Isua granitic dome. One mafic dike that cuts the severely deformed margin of the dome and slightly penetrates the supracrustal rocks is presumed to be of similar age but is undated; otherwise, only young mafic dikes, ca. 2.8 Ga, which are much deformed with the supracrustal rocks, and undeformed dikes, ca. 2.2 Ga, are known to cut the supracrustals (White et al., 2000a, 2000b).

Supracrustal rocks are scattered throughout the 1000 km length of the Archean complex of southern West Greenland. Many clastic-sediment samples have been given maximum depositional ages by dates of detrital zircons, or minimum ages by dates of crosscutting dikes, and all of these ages are <3.1 Ga, despite the widespread occurrence of polymetamorphic gneisses that include Mesoarchean and Paleoarchean zircons (Hollis et al., 2005, 2006; Nutman et al., 2004a). Many of these young dated rocks are close to Isua, as noted previously. This 3.1 Ga limit accords with my reading of the data from Isua itself, which is near the center of the sample array. Well-dated supracrustal rocks of palinspastically nearby Labrador also are 3.1–2.9 Ga (James et al., 2002).

WHEN DID THE HYDROSPHERE CONDENSE?

The molten Earth, ca. 4.5 Ga, could not have had liquid water on its surface. A hydrosphere certainly existed by 3.6 Ga, the oldest proved age of waterlaid sedimentary and volcanic rocks, and it may have formed only at about that time. The preceding discussion emphasized the prolonged internal mobility, and hence high temperature, of Archean felsic crust. That crust was heated internally by radioactivity much more intense than now and was heated from the bottom by upper mantle much hotter than present asthenosphere. The oldest proved supracrustal rocks that foundered into the lower crust, 3.6 Ga in southern West Greenland, place a minimum age on cooling of felsic crust to a density higher than that of mafic melts. The oldest well-documented extrusive volcanic rocks that stayed near the surface in coherent masses, in Pilbara and Barberton, are 3.5 Ga, which may be the general age limit of upper crust stiff enough to hold such rocks, without foundering, on its surface. Should claims, discussed previously, for pre–3.6 Ga ages of supracrustal rocks (or of mafic dikes) be proved correct, they of course would demonstrate that the 3.6 Ga threshold I suggest is too young.

Earth's surface temperature is now controlled primarily by solar radiation, as modified by oceanic circulation, albedo, and atmospheric thermal blanketing including greenhouse effects. But what were the conditions in Paleoarchean time, when the Sun had lower luminosity than now but global heat flow was far higher? Could Earth's water and carbon dioxide then have been in a greenhouse atmosphere of several hundred bars? Quantitative evaluation (e.g., Kasting and Ackerman, 1986, and Pollack, 1997) is model-dependent.

OVERVIEW—THE FIRST BILLION YEARS

Condensation of the components of the solar system began ca. 4.57 Ga, and Earth had most of its present mass by 4.50 or 4.45 Ga. The late part of main accretion was violent and hot, total melting was probable, and core, lower mantle, and upper mantle fractionated irreversibly. The popular concept of an unfractionated lower mantle—the basis for much current geochemical and geodynamic speculation regarding both the ancient and modern Earth—is incompatible with mineral-physics information. A global melabasaltic protocrust, perhaps 100 km thick, formed by 4.4 Ga, leaving a highly refractory upper mantle. From that protocrust in turn came secondary felsic melts, which gradually formed a capping felsic crust. As the increasingly depleted mafic protocrust cooled, primarily to garnet-clinopyroxene rocks, parts of it delaminated and sank through the less-dense near-solidus upper mantle, which consisted mostly of olivine and orthopyroxene. This delamination allowed hot mantle to rise to the shallowing base of the protocrust, releasing ultramafic magmas from the mantle by decompression melting and inducing partial melting of the remaining protocrust to generate TTG. Incremental derivation of TTG started no later than 4.4 Ga and continued throughout Archean time, although much new felsic melt was generated by recycling TTG already in the felsic crust. The oldest felsic rocks—reasonably complete mineral assemblages dated as ca. 3.7 Ga by their youngest abundant igneous zircons—contain hydrous mafic minerals, and are extremely deformed. The composition of older inherited zircons, known to reach almost 4.4 Ga, fits the inference that earlier melts also were hydrous, although the water came from hornblende breakdown, without necessarily any inclusion of surface water. Derivation of TTG from the upper part, of hornblende-garnet-clinopyroxene mineralogy, of the mafic protocrust satisfies geologic, petrologic, and Sm-Nd constraints, despite widespread acceptance of the Sm-Nd data as indicating long-continued derivation of new felsic crust from the mantle.

The entire felsic crust may have behaved as a viscous fluid on geologic time scales before 3.5 Ga. Not only did rigid lithosphere plates not exist, but large hypsometric differences could not be maintained. Earth may have had a top-cooling hydrosphere only after 3.6 Ga, and the older felsic crust may have been kept warm at the top by a greenhouse atmosphere of several hundred bars of water and carbon dioxide.

THE SECOND BILLION YEARS

Liquid water covered most of Earth after 3.6 Ga, perhaps earlier, and the felsic crust cooled enough to permit transit of mafic and ultramafic melts to the surface. Continuing incremental delamination of mafic protocrust enabled further partial melting to generate new TTG that rose into the enlarging upper crust, and also enabled generation from rising mantle of ultramafic melts that became variably contaminated as they rose through protocrust and felsic crust. The compositions of ultramafic volcanic rocks require a mantle beneath the protocrust at least 300°C hotter than modern asthenospheric mantle at 3.5 Ga, and at least 200°C hotter at 2.5 Ga. These high temperatures recorded primarily retained Earth heat, and no heat transfer by deep-mantle plumes is needed. The felsic crust was kept hot by heat conducted and advected from beneath and by internal radioactivity much greater than now, and the lower felsic crust continued to flow and churn effectively as a liquid at geologic time scales, but now the upper crust was stiffening by top-down cooling and was variably decoupled in a slipping-clutch mode. The much lower temperatures of equilibration of mineral assemblages sampled by xenoliths reflect the geothermal gradients at later times of eruptions of kimberlites and alkaline volcanic rocks, fluxed by volatile-rich melts rising into the depleted mantle from crustal materials sunk by delamination and subduction.

Voluminous mafic and ultramafic melts erupted on the surface. Basaltic and high-Mg basaltic rocks may mostly represent contamination of mantle-sourced ultramafic melts by both mafic protocrust and new felsic crust. The thick volcanic rocks were denser than the felsic crust, and in most regions foundered and were mixed into the flowing subjacent gneisses. As early as 3.5 Ga in some regions, however, and at varying times as late as 2.8 Ga in others, the dense volcanic rocks accumulated more stably on the surface, and sank as coherent synformal keels between domiform batholiths that rose slowly from the substrate of old ductile gneisses augmented by new melts generated partly by remobilization of preexisting TTG and partly by new partial melts from the subjacent protocrust. Structural relief of up to 20 km was thus achieved. Early supracrustal rocks were primarily mafic and ultramafic. The batholiths breached to erupt increasing proportions of felsic volcanic rocks, and rose as uplands that became major sources of clastic sediments, which accumulated mostly in the subsiding synclines. Concentration of radionuclides high in the batholiths accelerated cooling of the crust, and the radionuclides were lost by erosion to sediments that eventually were cycled down into the upper mantle. Some diapiric batholiths rose intermittently for hundreds of millions of years, and others formed much more quickly.

There are many more age determinations of supracrustal rocks, and of final crystallization of granitic rocks, ca. 2.8–2.6

Ga than for any comparable earlier time span; but there are no global gaps. This concentration of determinations is commonly attributed to cyclical mantle circulation, but an alternative is that 2.8 Ga dates a global threshold of cooling, only after which was crust everywhere stiff enough to support thick sections of the dense supracrustal rocks such as previously had foundered.

Stabilization was gradual, with lessening dome-and-keel development continuing as long as several hundred million years after major deep-rooted structures had formed. The regionally subuniform spatial density and shallow crustal level of diapiric batholiths, and their contacts primarily against the oldest strata preserved in synforms, indicate derivation from regional lower crust, like salt domes from a salt layer, and not from belts or local sources. These associations have no modern analogues.

Superimposed on the vertical tectonics due to gravitational righting of density inversions was lateral deformation of the upper crust manifested in broad and narrow ductile shear zones, which developed, simultaneously with the rise of diapiric batholiths, with greatly varying intensity in various regions. The variable-strength granite-and-greenstone upper crust was largely decoupled from the pervasively flowing lower felsic crust but was deformed with it. The felsic crust was far too mobile to form internally rigid plates.

Archean igneous rocks lack close modern compositional analogues because they were generated by quite different processes. Archean upper-crustal behavior differed from that of the post–2.0 Ga Earth in its common floating tectonic style. Archean granite-and-greenstone terrains typically are eroded only ~5–15 km, far less than Proterozoic and Paleozoic orogenic terrains, because they lacked roots of thickened crust. The deep crust was too mobile to retain enough relief on the Mohorovičić discontinuity to support large highlands. Major thrust and normal faults of Archean age have been widely postulated but few mapped faults mark large steps in crustal level. Archean middle and lower crust is exposed primarily where raised by post-Archean uplift and then eroded deeply. This uplift occurred partly in intracratonic settings (e.g., the Proterozoic Kapuskasing uplift in the Superior craton), but mostly in Proterozoic and Phanerozic systems of rifting and convergence.

Most Archean specialists call upon plate-tectonic rifting and convergence to explain Archean geology, but semirigid plates of even upper continental crust existed only near the end of Archean time. Plate tectonics was not possible. The craton-wide character of much Archean magmatism and deformation precludes analogy with modern plate systems. Geochemical rationales are widely cited as requiring the presence of ocean-floor, oceanic plateau, and island-arc rocks, but no oceanic crust of Archean age has been shown to exist by either geologic or geophysical evidence, and the geochemical assignments are disproved wherever they can be tested. The only basement yet seen beneath any Archean volcanic rocks, including ultramafic ones, is older felsic crust, and where that is present the oldest stratiform rocks are clastic and chemical sediments beneath mafic and ultramafic volcanic rocks. Perhaps oceans developed subsequently where the least felsic crust developed by secondary melting of an initially global mafic protocrust that disappeared incrementally by delamination. No Archean ophiolites—sections of oceanic crust and uppermost mantle—have been found either within preserved cratons or in the Paleoproterozoic and younger orogenic belts between Archean cratons. The chemical analogies made with vaguely similar plate-related modern rocks are done without regard for the complete dissimilarities of modern plate-interaction complexes to anything defined by Archean structural, stratigraphic, petrologic, and regional associations, or even to detailed chemical comparisons. No Archean rifting of continental plates—cross-grain sundering and separation, with deposition of trailing-edge stratal wedges, nor structural evidence for major extensional faulting—has been shown. No direct evidence for subduction and collisons—oceanic debris caught in sutures, polymict mélange, magmatic arcs related to possible sutures, juxtaposed fragments of disjunct cratons—is known. All postulated sutures are cryptic, hidden, or assigned to local shear zones, and no predictions implicit in Archean plate conjectures have been validated.

The thick sections of minimally deformed volcanic and sedimentary rocks deposited during middle and late Neoarchean and early Paleoproterozoic time on the by then partly stabilized Kaapvaal and Pilbara cratons resemble the supracrustal successions of Mesoarchean and Neoarchean granite-and-greenstone terrains, and, by their arrested development, provide support for conclusions reached here about evolution of the many granite-and-greenstone terrains that progressed to more complete dome-and-keel architecture. These less-deformed sections also are ensialic and, like typical greenstone belts, begin with clastic strata, above which are thick sections dominated by mafic volcanic rocks, most of which would be misclassed as ensimatic by commonly used chemotectonic criteria. Thick mostly-sedimentary sections overlie these erupted mafic sections but include variable further amounts of mafic and felsic volcanic rocks. Stratigraphy is subregional, and varies modestly and irregularly by interlensing on scales of tens to hundreds of kilometers. Narrow belts, arcs, and rifts have not been demonstrated. Further, many of the subjacent batholiths continued to rise into those young sections, although not to the extent of fully developing steep synclinal greenstone keels like those formed earlier in the same regions.

These ensialic supracrustal successions—Witwatersrand-Ventersdorp and Fortescue-Hamersley—are of the same Neoarchean age as many typical granite-and-greenstone terrains, from which they differ primarily in the lesser rise of diapiric granites. The sections include all the greenstone belt lithologies—basalt, high-Mg basalt, komatiite, mafic andesite, felsic volcanic rocks, chert, iron formation, and clastic sediments. Semisolid granitic batholiths—in the case of northeast Pilbara at least, some of the same ones that earlier produced classic dome-and-keel geology—resumed their diapiric rise after thermal blanketing and weighting by the thick Neoarchean and early Paleoproterozoic sections, attesting to the continued weakness of the hot, felsic older Archean lower crust. Felsic volcanic rocks erupted into the young sequences indicate intermittent venting of batholiths and attest to their long-continuing rise. The successions evolved only partway into granite-and-greenstone terrains because their felsic basements had become too immobile to fully rise into the high, steep-sided batholiths of that mode.

The upper mantle was extremely depleted by 4.4 Ga, and has since been re-enriched by processes driven by top-down cooling, beginning with protocrust delamination.

THE LAST 2.5 BILLION YEARS

The early Paleoproterozoic, from 2.5 to 2.1 or 2.0 Ga, has left a major record in relatively few cratons. In the Kaapvaal and Pilbara cratons, Archean-style batholiths did their last, minor rising. Great dike swarms in Archean cratons attest to continuing high upper-mantle temperatures; the dikes show that by 2.0 Ga the subcontinental upper mantle had cooled to an ambient temperature only about 120°C above that of modern asthenosphere (Mayborn and Lesher, 2004). Then, ca. 2 Ga, a great burst of activity heralded the onset of orogenic belts that separated semi-rigid cratons of Archean rocks.

From ca. 2.1 to 1.8 or 1.6 Ga, Archean crust was extensionally thinned; sedimentary wedges were deposited on initially uplifted, then sagged, thinned zones; and extensive magmatism affected the orogenic belts (e.g., Fig. 19, and discussion of Trans-Hudson orogen; Hoffman, 1988; McLaren et al., 2005; Zhao et al., 2002). As with the Archean, modern-style plate-tectonic rationales commonly have been forced on the orogens—but as with the Archean, both the structural and stratigraphic assemblages and the individual rock types tend to be different from modern ones, and alternatives to plate tectonics should be sought. Tectonic assignments of igneous rocks are commonly made on chemical bases, and, as for Archean assignments, are forced on rocks that are quite different in composition and association from their purported modern analogues. Little evidence has been found to support the presumption that Paleoproterozoic oceans opened by sea-floor spreading and closed by subduction. A very few examples that may be dismembered crustal parts of Paleoproterozoic ophiolites—sediments, pillow basalts, sheeted dikes, gabbros and plagiogranites, cumulate ultramafic rocks—have been reported (Condie, 1992; Scott et al., 1992), although I know of no complete ophiolite, including tectonized harzburgite, of the type widely preserved in Phanerozoic orogens. The lack of high-pressure, low-temperature metamorphism, the lack, or scarcity, of ophiolites, and the lack, or scarcity, of direct analogues for many other modern plate-interaction products all indicate that operative processes were quite different from modern ones.

Some of the broad Paleoproterozoic convergence zones (e.g., Limpopo, between Kaapvaal and Zimbabwe cratons) are dominated by squashed and variably recycled Archean felsic crust, whereas many others (e.g., Trans-Hudson) have broad interior tracts of highly deformed sedimentary and igneous rocks. At least some of the latter type, including Trans-Hudson itself as discussed earlier, contain Archean substrates over broad tracts. McLaren et al. (2005) evaluated the characteristics of some Australian recycled-crust Proterozoic orogens, emphasized that they could not be fitted to conventional plate models, and related their features instead to thermal effects of varying distributions of crustal heat-producing elements: orogenic belts developed in the hotter regions, away from which continental plates retained integrity. Many Paleoproterozoic orogens may be primarily intraplate mobile zones. Some may have developed from Neoarchean and Paleoproterozoic cratonic sedimentary and volcanic rocks like those preserved in less altered form in the Witwatersrand and Fortescue sections discussed earlier. High-temperature Paleoproterozoic metamorphic rocks from mobile-zone interiors have been widely raised from depths of 30 km (e.g., Van Kranendonk, 1996), yet preserved crustal thicknesses still reach 50 km (e.g., Funck et al., 2000). Unlike the Archean floating style of deformation, mid-Paleoproterozoic crust was stiff enough to support high mountains and deep roots.

Not until very late Neoproterozoic or early Paleozoic time do more complete suites of features, including complete ophiolites and high-pressure, low-temperature metamorphism, indicative of subduction appear (Stern, 2005, 2007; Tsujimori et al., 2006). Only thereafter can modern-style plate tectonics be identified in the geologic record. Evaluations of all earlier rock assemblages should seek explanations for the differences from modern ones. Far too many reports mindlessly force inapplicable modern models on the data.

The geologic record shows the effects of progressive cooling and stiffening of the lithosphere throughout geologic time, and only for the last billion years does tectonic uniformitarianism approximate a reasonable concept. The widespread assumption that the early Earth was just like the modern one except for a dearth of trees and fish is disproved by data from every field that can be brought to bear on it.

ACKNOWLEDGMENTS

Discussions with scores of Archean-specialist geologists, geophysicists, and geochemists, often in the field, have greatly increased my understanding. Reviews of the manuscript by Wouter Bleeker and Martin Van Kranendonk, who disagree strongly with many of my conclusions, and of part of the paper by Carol Frost, resulted in major improvements. Illustrations provided by others are so credited in their captions.

REFERENCES CITED

Abbott, D., Burgess, L., Longhi, J., and Smith, W.H.F., 1994, An empirical thermal history of the Earth's upper mantle: Journal of Geophysical Research, v. 99, p. 13,835–13,850, doi: 10.1029/94JB00112.

Allaart, J.H., 1982, Geological Map of Greenland, Sheet 2, Frederikshåb Isblink-Søndre Strømfjord, 1:500,000: Geological Survey of Greenland.

Al-Mishwat, A.T., and Nasir, S.J., 2004, Composition of the lower crust of the Arabian Plate—A xenolith perspective: Lithos, v. 72, p. 45–72, doi: 10.1016/j.lithos.2003.08.003.

Amelin, Y., Davis, D., and Lee, D.-C., 2004, New Hf isotopic data from the Jack Hills zircons—Constraints on the Hadean crustal evolution: Eos (Transactions, American Geophysical Union), v. 85, no. 17, suppl., p. 470.

Anderson, D.L., 2002, The case for irreversible chemical stratification of the mantle: International Geology Review, v. 44, p. 97–116.

Anderson, D.L., 2005, Large igneous provinces, delamination, and fertile mantle: Elements, v. 1, p. 271–275.

Anderson, D.L., 2006, Speculations on the nature and cause of mantle heterogeneity: Tectonophysics, v. 416, p. 7–22, doi: 10.1016/j.tecto.2005.07.011.

Anderson, D.L., 2007, New theory of the Earth: New York, Cambridge University Press, 384 p.

Annesley, I.R., Madore, C., and Portella, P., 2005, Geology and thermotectonic evolution of the western margin of the Trans-Hudson Orogen—Evidence from the eastern sub-Athabasca basement, Saskatchewan: Canadian Journal of Earth Sciences, v. 42, p. 573–597, doi: 10.1139/e05-034.

Ansdell, K.M., 2005, Tectonic evolution of the Manitoba-Saskatchewan segment of the Paleoproterozoic Trans-Hudson Orogen, Canada: Canadian Journal of Earth Sciences, v. 42, p. 741–759, doi: 10.1139/e05-035.

Armstrong, R.L., 1991, The persistent myth of crustal growth: Australian Journal of Earth Sciences, v. 38, p. 613–630.

Arndt, N., 1999, Why was flood volcanism on submerged continental platforms so common in the Precambrian?: Precambrian Research, v. 97, p. 155–164, doi: 10.1016/S0301-9268(99)00030-3.

Arndt, N., Albarède, F., and Nisbet, E.G., 1997, Mafic and ultramafic magmatism, in De Wit, M.J., and Ashwal, L.D., eds., Greenstone belts: Oxford, Clarendon Press, p. 233–254.

Ashton, K.E., Lewry, J.F., Heaman, L.M., Hartlaub, R.P., Stauffer, M.R., and Tran, H.T., 2005, The Pelican thrust zone—Basal detachment between the Archean Sask Craton and Paleoproterozoic Flin Flon—Glennie complex, western Trans Hudson Orogen: Canadian Journal of Earth Sciences, v. 42, p. 685–706, doi: 10.1139/e04-035.

Ashwal, L.D., Jacobsen, S.B., Myers, J.S., Kalsbeek, F., and Goldstein, S.J., 1989, Sm-Nd age of the Fiskenaesset Anorthosite Complex, West Greenland: Earth and Planetary Science Letters, v. 91, p. 261–270, doi: 10.1016/0012-821X(89)90002-2.

Ayer, J., Amelin, Y., Corfu, F., Kamo, S., Ketchum, J., Kwok, K., and Trowell, N., 2002, Evolution of the southern Abitibi greenstone belt based on U-Pb geochronology—Autochtonous volcanic construction followed by plutonism, regional deformation and sedimentation: Precambrian Research, v. 115, p. 63–95, doi: 10.1016/S0301-9268(02)00006-2.

Bagas, L., Smithies, R.H., and Champion, D.C., 2002, Geochemistry of the Corunna Downs Granitoid Complex, East Pilbara Granite-Greenstone Terrane, Western Australia: Geological Survey of Western Australia Annual Review 2001–2002, p. 61–69.

Baglow, N., and van Beek, B.S., 1987, Filabusi [geologic map]: Geological Survey of Zimbabwe, scale 1:100,000.

Bailes, A.H., Percival, J.A., Corkery, M.T., McNicoll, V.J., Tomlinson, K.Y., Sasseville, C., Rogers, N., Whalen, J.B., and Stone, D., 2003, Geology and tectonostratigraphic assemblages, West Uchi map area, Manitoba and Ontario: Geological Survey of Canada Open File 1522, scale 1:250,000 with marginal notes.

Bailey, R.C., 2006, Crustal ductility and early continental tectonophysics: American Geophysical Union Geophysical Monograph 164, p. 115–130.

Barley, M.E., and Pickard, A.L., 1999, An extensive, crustally-derived, 3325 to 3310 Ma silicic volcanoplutonic suite in the eastern Pilbara Craton—Evidence from the Kelly Belt, McPhee Dome and Corunna Downs Batholith: Precambrian Research, v. 96, p. 41–62, doi: 10.1016/S0301-9268(99)00003-0.

Bateman, R., Costa, S., Swe, T., and Lambert, D., 2001, Archaean mafic magmatism in the Kalgoorlie area of the Yilgarn Craton, Western Australia—A geochemical and Nd isotopic study of the petrogenetic and tectonic evolution of a greenstone belt: Precambrian Research, v. 108, p. 75–112, doi: 10.1016/S0301-9268(00)00148-0.

Bédard, J.H., 2006, A catalytic delamination-driven model for coupled genesis of Archaean crust and sub-continental lithospheric mantle: Geochimica et Cosmochimica Acta, v. 70, p. 1188–1214, doi: 10.1016/j.gca.2005.11.008.

Bédard, J.H., Brouillette, P., Madore, L., and Berclaz, A., 2003, Archaean cratonization and deformation in the northern Superior Province, Canada—An evaluation of plate tectonic versus vertical tectonic models: Precambrian Research, v. 127, p. 61–87, doi: 10.1016/S0301-9268(03)00181-5.

Beech, E.M., and Chadwick, B., 1980, The Malene supracrustal gneisses of northwest Buksefjorden—Their origin and significance in the Archaean crustal evolution of southern West Greenland: Precambrian Research, v. 11, p. 329–355, doi: 10.1016/0301-9268(80)90071-6.

Bernstein, S., Kelemen, P.B., and Hanghøj, K., 2007, Consistent olivine Mg# in cratonic mantle reflects Archean mantle melting to the exhaustion of orthopyroxene: Geology, v. 35, p. 459–462, doi: 10.1130/G23336A.1.

Bethune, K.M., Helmstaedt, H.H., and McNicoll, V.J., 2006, Structural analysis of the Miniss River and related faults, western Superior Province—Post-collisional displacements initiated at terrane boundaries: Canadian Journal of Earth Sciences, v. 43, p. 1031–1054, doi: 10.1139/E06-017.

Bickford, M.E., Mock, T.D., Steinhart, W.E., III, Collerson, K.D., and Lewry, J.F., 2005, Origin of the Archean Sask craton and its extent within the Trans-Hudson orogen—Evidence from Pb and Nd isotopic compositions of basement rocks and post-orogenic intrusions: Canadian Journal of Earth Sciences, v. 42, p. 659–684, doi: 10.1139/e04-064.

Bickle, M.J., Nisbet, E.G., and Martin, A., 1994, Archean greenstone belts are not oceanic crust: Journal of Geology, v. 102, p. 121–138.

Bleeker, W., 2001, The late Archean record—a puzzle in ca. 35 pieces: Lithos, v. 71, p. 99-134.

Bleeker, W., 2002, Archaean tectonics—A review, with illustrations from the Slave craton: Geological Society [London] Special Publication 199, p. 151–181.

Bleeker, W., Ketchum, J.W.F., Jackson, V.A., and Villeneuve, M.E., 1999, The Central Slave Basement Complex, Part I, Its structural topology and autochthonous cover: Canadian Journal of Earth Sciences, v. 36, p. 1083–1109, doi: 10.1139/cjes-36-7-1083.

Blake, T.S., Buick, R., Brown, S.J.A., and Barley, M.E., 2004, Geochronology of a Late Archaean flood basalt province in the Pilbara Craton, Australia—Constraints on basin evolution, volcanic and sedimentary accumulation, and continental drift rates: Precambrian Research, v. 133, p. 143–173, doi: 10.1016/j.precamres.2004.03.012.

Blenkinsop, T.G., Fedo, C.M., Bickle, M.J., Eriksson, K.A., Martin, A., Nisbet, E.G., and Wilson, J.F., 1993, Ensialic origin for the Ngezi Group, Belingwe greenstone belt, Zimbabwe: Geology, v. 21, p. 1135–1138, doi: 10.1130/0091-7613(1993)021<1135:EOFTNG>2.3.CO;2.

Boak, J.L., Dymek, R.F., and Gromet, L.P., 1983, Petrology and rare earth element geochemistry of clastic metasedimentary rocks from the Isua supracrustal belt, West Greenland: Grønlands Geologiske Undersøgelse: Report, v. 112, p. 23–33.

Bodorkos, S., and Sandiford, M., 2006, Thermal and mechanical controls on the evolution of Archean crustal deformation—Examples from Western Australia: American Geophysical Union Geophysical Monograph 164, p. 131–147.

Böhm, C.L., Heaman, L.M., Stern, R.A., Corkery, M.T., and Creaser, R.A., 2003, Nature of Assean Lake ancient crust, Manitoba—A combined SHRIMP-ID-TIMS U-Pb geochronology and Sm-Nd isotope study: Precambrian Research, v. 126, p. 55–94, doi: 10.1016/S0301-9268(03)00127-X.

Bowerman, M.S., Heaman, L.M., Chacko, T., and Böhm, C.O., 2005, Constraining the assembly of the northwestern Superior Province—The polymetamorphic history of the Split Lake Block, Manitoba: Geological Association of Canada, Halifax 2005, Abstracts, p. 17.

Bowring, S.A., Housh, T.B., and Isachsen, C.E., 1990, The Acasta gneisses—Remnants of Earth's early crust, in Newsome, H., and Jones, J., eds., Origin of the Earth: Houston, Texas, Lunar and Planetary Institute, p. 319–343.

Breaks, F.W., 1991, English River Subprovince: Ontario Geological Survey Special 4, pt. 1, p. 239–277.

Brock, B.B., and Pretorius, D.A., 1964, Rand Basin sedimentation and tectonics, in Haughton, S.H., ed., The geology of some ore deposits in southern Africa: Johannesburg, Geological Society of South Africa, v. 1, p. 549–599.

Brown, S.J.A., Krapez, B., Beresford, S.W., Cassidy, K.F., Champion, D.C., Barley, M.E., and Cas, R.A.F., 2001, Archaean volcanic and sedimentary environments of the Eastern Goldfields Province, Western Australia—A field guide: Geological Survey of Western Australia Record 2001/13, 66 p.

Byerly, G.R., Kröner, A., Lowe, D.R., Todt, W., and Walsh, M.M., 1996, Prolonged magmatism and time constraints for sediment deposition in the early Archean Barberton greenstone belt—Evidence from the Upper Onverwacht and Fig Tree groups: Precambrian Research, v. 78, p. 125–138, doi: 10.1016/0301-9268(95)00073-9.

Calvert, A.J., Cruden, A.R., and Hynes, A., 2004, Seismic evidence for preservation of the Archean Uchi granite-greenstone belt by crustal-scale extension: Tectonophysics, v. 388, p. 135–143, doi: 10.1016/j.tecto.2004.07.043.

Caracas, R., and Cohen, R.E., 2005, Effect of chemistry on the stability and elasticity of the perovskite and post-perovskite phases in the $MgSiO_3$-$FeSiO_3$-Al_2O_3 system and implications for the lowermost mantle: Geophysical Research Letters, v. 32, no. 16, doi: 10.1029/2005GL023164, 4 p.

Card, K.D., and Ciesielski, A., 1986, DNAG#1—Subdivisions of the Superior Province of the Canadian Shield: Geoscience Canada, v. 13, p. 5–13.

Case, N.L., and Mojzsis, S.J., 2006, Chemical and isotopic evidence for widespread Eoarchean metasedimentary enclaves in southern West Greenland: Geochimica et Cosmochimica Acta (in press).

Cassidy, K.F., and Champion, D.C., 2004, Crustal evolution of the Yilgarn Craton from Nd isotopes and granite geochronology—Implications for metallogeny: University of Western Australia Centre for Global Metallogeny Publication 33, p. 317–320.

Cates, N.L., and Mojzsis, S.J., 2006, Chemical and isotopic evidence for widespread Eoarchean metasedimentary enclaves in southern West Greenland: Geochimica et Cosmochimica Acta, v. 70, no. 16, p. 4229–4257, doi: 10.1016/j.gca.2006.05.014.

Cattell, A.C., and Taylor, R.N., 1990, Archaean basic magmas, in Hall, R.J., and Hughes, D.J., eds., Early Precambrian basic magmatism: Glasgow, Blackie, p. 11–39.

Cavosie, A.J., Wilde, S.A., Liu, D., Weiblen, P.W., and Valley, J.W., 2004, Internal zoning and U-Th-Pb chemistry of Jack Hills detrital zircons—A mineral record of early Archean to Mesoproterozoic (4348–1576 Ma) magmatism: Precambrian Research, v. 135, p. 251–279, doi: 10.1016/j.precamres.2004.09.001.

Cawood, P.A., Kröner, A., and Pisarevsky, S., 2006, Precambrian plate tectonics—Criteria and evidence: GSA Today, v. 16, no. 7, p. 4–11, doi: 10.1130/GSAT01607.1.

Chadwick, B., 1990, The stratigraphy of a sheet of supracrustal rocks within high-grade orthogneiss and its bearing on late Archaean structure in southern West Greenland: Geological Society [London] Journal, v. 147, p. 639–652.

Chadwick, B., and Coe, K., 1987, Ivisârtoq: Geological Survey of Greenland, Geological Map 64 V.2 Nord, 1:100,000.

Chadwick, B., and Nutman, A.P., 1979, Archaean structural evolution in the northwest of the Buksefjorden region, southern West Greenland: Precambrian Research, v. 9, p. 199–226, doi: 10.1016/0301-9268(79)90003-2.

Chambers, J.E., 2004, Planetary accretion in the inner Solar System: Earth and Planetary Science Letters, v. 223, p. 241–252, doi: 10.1016/j.epsl.2004.04.031.

Champion, D.C., and Smithies, R.H., 2000, The geochemistry of the Yule Granitoid Complex, East Pilbara Granite-Greenstone Terrane; evidence for early felsic crust: Geological Survey of Western Australia Annual Review 1999–2000, p. 42–48.

Chardon, D., Choukroune, P., and Jayananda, M., 1998, Sinking of the Dharwar Basin (South India)—Implications for Archaean tectonics: Precambrian Research, v. 91, p. 15–39, doi: 10.1016/S0301-9268(98)00037-0.

Chardon, D., Peucat, J.-J., Jayananda, M., Choukroune, P., and Fanning, C.M., 2002, Archean granite-greenstone tectonics at Kolar (South India—Interplay of diapirism and bulk inhomogeneous contraction during juvenile magmatic accretion: Tectonics, v. 21, no. 3, paper 7, 17 p., doi: 10.1029/2001TC901032.

Chen, S.F., and Wyche, S., 2001, Archaean granite-greenstones of the central Yilgarn Craton, Western Australia—A field guide: Geological Survey of Western Australia Record 2001/14, 76 p.

Chen, S.F., Witt, W.K., and Liu, S., 2001a, Transpression and restraining jogs in the northeastern Yilgarn craton, Western Australia: Precambrian Research, v. 106, p. 309–328, doi: 10.1016/S0301-9268(00)00138-8.

Chen, S.F., Libby, J.W., Greenfield, J.E., Wyche, S., and Riganti, A., 2001b, Geometry and kinematics of large arcuate structures formed by impingement of rigid granitoids into greenstone belts during progressive shortening: Geology, v. 29, p. 283–286, doi: 10.1130/0091-7613(2001)029<0283:GAKOLA>2.0.CO;2.

Chen, S.F., Riganti, A., Wyche, S., Greenfield, J.E., and Nelson, D.R., 2003, Lithostratigraphy and tectonic evolution of contrasting greenstone successions in the central Yilgarn Craton, Western Australia: Precambrian Research, v. 127, p. 249–266, doi: 10.1016/S0301-9268(03)00190-6.

Chen, S.F., Libby, J.W., Wyche, S., and Riganti, A., 2004, Kinematic nature and origin of regional-scale ductile shear zones in the central Yilgarn Craton, Western Australia: Tectonophysics, v. 394, p. 139–154, doi: 10.1016/j.tecto.2004.08.001.

Cheney, E.S., and Winter, H.R., 1995, The late Archean to Mesoproterozoic major unconformity-bounded units in the Kaapvaal Province of southern Africa: Precambrian Research, v. 74, p. 203–223, doi: 10.1016/0301-9268(95)00011-S.

Choukroune, P., Bouhallier, H., and Arndt, N.T., 1995, Soft lithosphere during periods of Archaean crustal growth or crustal reworking, *in* Coward, M.P., and Ries, A.C., eds., Early Precambrian processes: Geological Society [London] Special Publication 95, p. 67–86.

Clemens, J.D., Yearron, L.M., and Stevens, G., 2006, Barberton (South Africa) TTG magmas—Geochemical and experimental constraints on source-rock petrology, pressure of formation and tectonic setting: Precambrian Research, v. 151, p. 53–78.

Condie, K.C., 1981, Archean greenstone belts: Amsterdam, Elsevier, 434 p.

Condie, K.C., 1985, Secular variation in the composition of basalts—An index to mantle evolution: Journal of Petrology, v. 26, p. 545–563.

Condie, K.C., 1992, The Proterozoic ophiolite problem, in Condie, K.C., ed., Proterozoic crustal evolution: Amsterdam, Elsevier, p. 55–95.

Condie, K.C., 2005a, High field strength element ratios in Archean basalts—A window to evolving sources of mantle plumes?: Lithos, v. 79, p. 491–504, doi: 10.1016/j.lithos.2004.09.014.

Condie, K.C., 2005b, TTGs and adakites—Are they both slab melts?: Lithos, v. 80, p. 33–44, doi: 10.1016/j.lithos.2003.11.001.

Condie, K.C., Macke, J.E., and Reimer, T.O., 1970, Petrology and geochemistry of early Precambrian graywackes from the Fig Tree Group, South Africa: Geological Society of America Bulletin, v. 81, p. 2759–2775, doi: 10.1130/0016-7606(1970)81[2759:PAGOEP]2.0.CO;2.

Corrigan, D., Hajnal, Z., Németh, B., and Lucas, S.B., 2005, Tectonic framework of a Paleoproterozoic arc-continent to continent-continent collisional zone, Trans-Hudson Orogen, from geological and seismic reflection studies: Canadian Journal of Earth Sciences, v. 42, p. 421–434, doi: 10.1139/e05-025.

Crow, C., and Condie, K.C., 1988, Geochemistry and origin of late Archean volcanics from the Ventersdorp Supergroup, South Africa: Precambrian Research, v. 42, p. 19–37, doi: 10.1016/0301-9268(88)90008-3.

Crowley, J.L., 2003, U-Pb geochronology of 3810–3630 Ma granitoid rocks south of the Isua greenstone belt, southern West Greenland: Precambrian Research, v. 126, p. 235–237, doi: 10.1016/S0301-9268(03)00097-4.

Crowley, J.L., Myers, J.S., and Dunning, G.R., 2002, Timing and nature of multiple 3700–3600 Ma tectonic events in intrusive rocks north of the Isua greenstone belt, southern West Greenland: Geological Society of America Bulletin, v. 114, p. 1311–1325, doi: 10.1130/0016-7606(2002)114<1311:TANOMM>2.0.CO;2.

Crowley, J.L., Myers, J.S., Sylvester, P.J., and Cox, R.A., 2005, Detrital zircons from the Jack Hills and Mount Narryer, Western Australia—Evidence for diverse >4.0 Ga source rocks: The Journal of Geology, v. 113, p. 239–263, doi: 10.1086/428804.

Culshaw, N., Purves, M., Reynolds, P., and Stott, G., 2006, Post-collisional upper crustal faulting and deep crustal flow in the eastern Wabigoon subprovince of the Superior Province, Ontario—Evidence from structural and ^{40}Ar-^{39}Ar data from the Humboldt Bay high strain zone: Precambrian Research, v. 145, p. 272–288, doi: 10.1016/j.precamres.2005.12.006.

Czamanske, G.K., and Bohlen, S.R., 1990, The Stillwater Complex and its anorthosites—An accident of magmatic underplating?: The American Mineralogist, v. 73, p. 37–45.

Dalrymple, G.B., and Ryder, G., 1993, $^{40}Ar/^{39}Ar$ Age spectra of Apollo 15 impact melt rocks by laser step-heating and their bearing on the history of lunar basin formation: Journal of Geophysical Research, v. 98, p. 13,085–13,095.

Dann, J., 2004, Spinifex-komatiite lava flows, 3.5 Ga Komati Formation, South Africa: Geological Society of America Abstracts with Programs, v. 36, no. 5, p. 247.

Darracott, B.W., 1975, The interpretation of the gravity-anomaly over the Barberton Mountain Land, South Africa: Geological Society of South Africa Transactions, v. 78, p. 123–128.

Davis, B.K., and Maidens, E., 2003, Archaean orogen-parallel extension—Evidence from the northern Eastern Goldfields Province, Yilgarn Craton: Precambrian Research, v. 127, p. 229–248, doi: 10.1016/S0301-9268(03)00189-X.

Davis, D.W., Amelin, Y., Nowell, G.M., and Parrish, R.R., 2005, Hf isotopes in zircon from the western Superior province, Canada—Implications for Archean crustal development and evolution of the depleted mantle reservoir: Precambrian Research, v. 140, p. 132–156, doi: 10.1016/j.precamres.2005.07.005.

Davis, W.J., Canil, D., MacKenzie, J.M., and Carbno, G.B., 2003, Petrology and U-Pb geochronology of lower crustal xenoliths and the development of a craton, Slave Province, Canada: Lithos, v. 71, p. 541–573, doi: 10.1016/S0024-4937(03)00130-0.

DePaolo, D.J., 1988, Neodymium isotope geochemistry, An introduction: Berlin, Springer-Verlag, 187 p.

DePaolo, D.J., and Wasserburg, G.J., 1976, Inferences about mantle sources and mantle structure from variations of 143Nd/144Nd: Geophysical Research Letters, v. 3, p. 743–746.

de Ronde, C.E.J., and de Wit, M.J., 1994, Tectonic history of the Barberton greenstone belt, South Africa—490 million years of Archean crustal evolution: Tectonics, v. 13, p. 983–1005, doi: 10.1029/94TC00353.

Dewey, J.F., 2005, Orogeny can be very short: Proceedings of the National Academy of Sciences of the United States of America, v. 102, p. 15,286–15,293, doi: 10.1073/pnas.0505516102.

de Wit, M.J., 1998, On Archean granites, greenstones, cratons and tectonics—Does the evidence demand a verdict?: Precambrian Research, v. 91, p. 181–226, doi: 10.1016/S0301-9268(98)00043-6.

de Wit, M.J., and Tinker, J., 2004, Crustal structures across the central Kaapvaal craton from deep-seismic reflection data: South African Journal of Geology, v. 107, p. 185–206, doi: 10.2113/107.1-2.185.

Dickson, F.W., 2006, Role of liquids in disequilibrium processes of earth and replacement in Papoose Flat pluton, California, *in* Rhoden, H.N., Steininger, R.C., and Vikre, P.G., eds., Window to the world: Geological Society of Nevada Symposium 2005, p. 161–178.

Drummond, B.J., Goleby, B.R., and Swager, C.P., 2000, Crustal signature of Late Archaean tectonic episodes in the Yilgarn craton, Western Australia—Evidence from deep seismic sounding: Tectonophysics, v. 329, p. 193–221, doi: 10.1016/S0040-1951(00)00196-7.

Drummond, B.J., Hobbs, R.W., and Goleby, B.R., 2004, The effects of out-of-plane seismic energy on reflections in crustal-scale 2D seismic sections: Tectonophysics, v. 388, p. 213–224, doi: 10.1016/j.tecto.2004.07.040.

Durrheim, R.J., and Mooney, W.D., 1994, Evolution of the Precambrian lithosphere: Seismological and geochemical constraints: Journal of Geophysical Research, v. 99, p. 15,359–15,374, doi: 10.1029/94JB00138.

Dziggel, A., Stevens, G., Poujol, M., Anhaeusser, C.R., and Armstrong, R.A., 2002, Metamorphism of the granite-greenstone terrane south of the Barberton greenstone belt, South Africa—An insight into the tectono-thermal evolution of the 'lower' portions of the Onverwacht Group: Precambrian Research, v. 114, p. 221–247, doi: 10.1016/S0301-9268(01)00225-X.

Easton, R.M., 2000, Metamorphism of the Canadian Shield, Ontario, Canada, 1. The Superior Province: Canadian Mineralogist, v. 38, p. 287–317.

Easton, R.M., and Berman, R.G., 2004, Tectonometamorphic map of Ontario, Canada, and parts of the United States of America: Geological Survey of Canada Open File 1810, scale 1:1,500,000.

Eglington, B.M., and Armstrong, R.A., 2004, The Kaapvaal Craton and adjacent orogens, southern Africa—A geochronological database and overview of the geological development of the craton: South African Journal of Geology, v. 107, p. 13–32, doi: 10.2113/107.1-2.13.

Elkins-Tanton, L.T., 2005, Continental magmatism caused by lithospheric delamination: Geological Society of America Special Paper 388, p. 449–461.

Eriksson, P.G., Martins-Neto, M.A., Nelson, D.R., Aspler, C.B., Chiarenzelli, J.R., Catuneanu, O., Sarkar, S., Altermann, W., and de Raufenbach, C.J., 2001a, An introduction to Precambrian basins—Their characteristics and genesis: Sedimentary Geology, v. 141–142, p. 1–35, doi: 10.1016/S0037-0738(01)00066-5.

Eriksson, P.G., Altermann, W., Catuneanu, O., van der Merwe, R., and Bumby, A.J., 2001b, Major influences on the evolution of the 2.67–2.1 Ga Transvaal basin, Kaapvaal craton: Sedimentary Geology, v. 141–142, p. 205–231, doi: 10.1016/S0037-0738(01)00075-6.

Flowers, R.M., Moser, D.E., and Hart, R.J., 2003, Evolution of the amphibolite-granulite facies transition exposed by the Vredefort impact structure, Kaapvaal craton, South Africa: The Journal of Geology, v. 111, p. 455–470, doi: 10.1086/375282.

Foulger, G.R., and Jurdy, D.M., eds., 2007, Plates, plumes, and planetary processes: Geological Society of America Special Paper 430 (in press).

Foulger, G.R., Natland, J.H., Presnall, D.C., and Anderson, D.L., editors, 2005, Plates, plumes, and paradigms: Geological Society of America Special Paper 388, 881 p.

Friend, C.R.L., and Nutman, A.P., 2005, New pieces to the Archaean terrane jigsaw puzzle in the Nuuk region, southern West Greenland—Steps in transforming a simple insight into a complex regional tectonothermal model: Geological Society [London] Journal, v. 162, p. 147–162, doi: 10.1144/0016-764903-161.

Friend, C.R.L., Bennett, V.C., and Nutman, A.P., 2002, Abyssal peridotites >3800 Ma from southern West Greenland—Field relationships, petrography, geochronology, whole-rock and mineral chemistry of dunite and harzburgite inclusions in the Itsaq Gneiss Complex: Contributions to Mineralogy and Petrology, v. 143, p. 71–92.

Funck, T., Louden, K.E., Wardle, R.J., Hall, J., Hobro, J.W., Salisbury, M.H., and Muzzatti, A.M., 2000, Three-dimensional structure of the Torngat orogen (NE Canada) from active seismic tomography: Journal of Geophysical Research, v. 105, p. 23,403–23,420, doi: 10.1029/2000JB900228.

Furnes, H., de Wit, M., Staudigel, H., Rosing, M., and Muehlenbachs, K., 2007, A vestige of Earth's oldest ophiolite: Science, v. 315, p. 1704–1707 plus electronic supplement, doi: 10.1126/science.1139170.

Galer, S.J.G., 1991, Interrelationships between continental freeboard, tectonics, and mantle temperature: Earth and Planetary Science Letters, v. 105, p. 214–218, doi: 10.1016/0012-821X(91)90132-2.

Garde, A.A., 1988, Isukasia: Geological Survey of Greenland, Geological Map 65 V.2 Syd, 1:100,000.

Gee, R.D., Baxter, J.L., Wilde, S.A., and Williams, I.R., 1981, Crustal development of the Archaean Yilgarn Block, Western Australia: Geological Society of Australia Special Publication, v. 7, p. 43–56.

Geological Survey of Western Australia, 2005a, Atlas of 1:250,000 Geological Series map images, Western Australia, January 2005 update: Geological Survey of Western Australia, DVD-ROM.

Geological Survey of Western Australia, 2005b, East Yilgarn 1:100,000 Geological Information Series, June 2005 update: Geological Survey of Western Australia, DVD-ROM.

Geological Survey of Western Australia, 2005c, Compilation of geochronology data, June 2005 update: Geological Survey of Western Australia, CD-ROM.

Geological Survey of Zimbabwe, 1994, Geological map of Zimbabwe, 1:1,000,000 [reprint of 1977 map].

Glikson, A.Y., 1979, Early Precambrian tonalite-trondhjemite sialic nuclei: Earth-Science Reviews, v. 15, p. 1–73, doi: 10.1016/0012-8252(79)90043-6.

Glikson, A.Y., 2005, Geochemical signatures of Archean to early Proterozoic Maria-scale oceanic impact basins: Geology, v. 33, p. 125–128, doi: 10.1130/G21034.1.

Glikson, A.Y., and Allen, C., 2004, Iridium anomalies and fractionated siderophile element patterns in impact ejecta, Brockman Iron Formation, Hamersley Basin, Western Australia—Evidence for a major asteroid impact on simatic crustal regions of the early Proterozoic earth: Earth and Planetary Science Letters, v. 220, p. 247–264, doi: 10.1016/S0012-821X(04)00062-7.

Goleby, B.R., Korsch, R.J., Fomin, T., Bell, B., Nicoll, M.G., Drummond, B.J., and Owen, A.J., 2002, Preliminary 3-D geological model of the Kalgoorlie region, Yilgarn craton, Western Australia, based on deep-reflection and potential-field data: Australian Journal of Earth Sciences, v. 49, p. 917–933, doi: 10.1046/j.1440-0952.2002.00967.x.

Goleby, B.R., Blewett, R.S., Groenewald, P.B., Cassidy, K.F., Champion, D.C., Korsch, R.J., Whitaker, A.J., Jones, L.E.A., Bell, B., and Carlson, G., 2003, Seismic interpretation of the northeastern Yilgarn craton seismic data: Geoscience Australia Record 2003/28, p. 85–125.

Goleby, B.R., Blewett, R.S., Korsch, R.J., Champion, D.C., Cassidy, K.F., Jones, L.E.A., Groenewald, P.B., and Henson, P., 2004, Deep seismic reflection profiling in the Archaean northeastern Yilgarn Craton, Western Australia—Implications for crustal architecture and mineral potential: Tectonophysics, v. 388, p. 119–133, doi: 10.1016/j.tecto.2004.04.032.

Goleby, B.R., Blewett, R.S., Fomin, T., Fishwick, S., Reading, A.M., Henson, P.A., Kennett, B.L.N., Champion, D.C., Jones, L., Drummond, B.J., and Nicoll, M., 2006, An integrated multi-scale 3D seismic model of the Archaean Yilgarn Craton, Australia: Tectonophysics, v. 420, p. 75–90, doi: 10.1016/j.tecto.2006.01.028

Green, M.G., Sylvester, P.G., and Buick, R., 2000, Growth and recycling of early Archaean continental crust—Geochemical evidence from the Coonterunah and Warrawoona Groups, Pilbara Craton, Australia: Tectonophysics, v. 322, p. 69–88, doi: 10.1016/S0040-1951(00)00058-5.

Griffin, W.L., Belousova, E.A., Shee, S.R., Pearson, N.J., and O'Reilly, S.Y., 2004a, Archean crustal evolution in the northern Yilgarn Craton—U-Pb and Hf-isotope evidence from detrital zircons: Precambrian Research, v. 131, p. 231–282, doi: 10.1016/j.precamres.2003.12.011.

Griffin, W.L., O'Reilly, S.Y., Doyle, B.J., Pearson, N.J., Coopersmith, H., Kivi, K., Malkovets, V., and Pokhilenko, N., 2004b, Lithosphere mapping beneath the North American plate: Lithos, v. 77, p. 873–922, doi: 10.1016/j.lithos.2004.03.034.

Groenewald, P.B., and Riganti, A., 2004, East Yilgarn 1:100 000 geological information series—An explanatory note: Geological Survey of Western Australia Report 95, 58 p.

Groenewald, P.B., Painter, M.G.M., Roberts, F.I., McCabe, M., and Fox, A., 2000, East Yilgarn geoscience database, 1:100 000 geology, Menzies to Norseman—An explanatory note: Geological Survey of Western Australia Report 78, 53 p.

Groenewald, P.B., Morris, P.A., and Champion, D.C., 2003, Geology of the Eastern Goldfields and an overview of tectonic models: Geoscience Australia Record 2003/28, p. 63–84.

Grove, T.L., and Parman, S.W., 2004, Thermal evolution of the Earth as recorded by komatiites: Earth and Planetary Science Letters, v. 219, p. 173–187, doi: 10.1016/S0012-821X(04)00002-0.

Hamilton, W.B., 1963, Metamorphism in the Riggins region, western Idaho: U.S. Geological Survey Professional Paper 436, 95 p. plus map.

Hamilton, W.B., 1978, Mesozoic tectonics of the western United States: Society of Economic Paleontologists and. Mineralogists, Pacific Section, Paleogeography Symposium 2, p. 33–70.

Hamilton, W.B., 1979, Tectonics of the Indonesian region: U.S. Geological Survey Professional Paper 1078, 345 p. plus 1:5,000,000 tectonic map.

Hamilton, W.B., 1988, Plate tectonics and island arcs: Geological Society of America Bulletin., v. 100, p. 1503–1527, doi: 10.1130/0016-7606(1988)100<1503:PTAIA>2.3.CO;2.

Hamilton, W.B., 1995, Subduction systems and magmatism: Geological Society [London] Special Publication 81, p. 3–28.

Hamilton, W.B., 1998a, Archean magmatism and tectonics were not products of plate tectonics: Precambrian Research, v. 91, p. 143–179, doi: 10.1016/S0301-9268(98)00042-4.

Hamilton, W.B., 1998b, Archean tectonics and magmatism: International Geology Review, v. 40, p. 1–39.

Hamilton, W.B., 2002, The closed upper-mantle circulation of plate tectonics: American Geophysical Union Geodynamics Series, v. 30, p. 359–409.

Hamilton, W.B., 2003, An alternative Earth: GSA Today, v. 13, no. 11, p. 4–12, doi: 10.1130/1052-5173(2003)013<0004:AAE>2.0.CO;2.

Hamilton, W.B., 2007, Driving mechanism and 3-D circulation of plate tectonics, *in* Sears, J.W., Harms, T.A., and Evenchick, C.A., eds., Whence the mountains? Inquiries into the evolution of orogenic systems: A volume in honor of Raymond A. Price: Geological Society of America Special Paper 433, doi: 10.1130/2007.2433(01) (in press).

Hanan, B.B., Blichert-Toft, J., Pyle, D.G., and Christie, D.M., 2004, Contrasting origins of the upper mantle revealed by hafnium and lead isotopes from the Southeast Indian Ridge: Nature, v. 432, p. 91–94, doi: 10.1038/nature03026.

Hanchar, J.M., and Watson, E.B., 2003, Zircon saturation thermometry: Mineralogical Society of America Reviews in Geochemistry, v. 53, p. 89–112, doi: 10.2113/0530089.

Hanmer, S., and Greene, D.C., 2002, A modern structural regime in the Paleoarchean (~3.64 Ga); Isua Greenstone Belt, southern West Greenland: Tectonophysics, v. 346, p. 201–222, doi: 10.1016/S0040-1951(02)00029-X.

Hanmer, S., Hamilton, M.A., and Crowley, J.L., 2002, Geochronological constraints on Paleoarchean thrust-nappe and Neoarchean accretionary tectonics in southern West Greenland: Tectonophysics, v. 350, p. 255–271, doi: 10.1016/S0040-1951(02)00120-8.

Hanmer, S., Sandeman, H.A., Davis, W.J., Aspler, L.B., Rainbird, R.H., Ryan, J.J., Relf, C., and Peterson, T.D., 2004, Geology and Neoarchean tectonic setting of the Central Hearne supracrustal belt, Western Churchill Province, Nunavit, Canada: Precambrian Research, v. 134, p. 63–83, doi: 10.1016/j.precamres.2004.04.005.

Harris, M.J., Symons, D.T.A., Blackburn, W.H., Peck, D.C., and Turek, A., 2000, Split Lake Block of the Superior boundary zone—Petrological, paleomagnetic and geobarometric data for the ca. 1818 Ma Fox Lake pluton (NTS 64A/1, /8): Manitoba Geological Survey Report of Activities 2000, p. 26–37.

Hart, R.J., Welke, H.J., and Nicolaysen, L.O., 1981, Geochronology of the deep profile through Archean basement at Vredefort, with implications for early crustal evolution: Journal of Geophysical Research, v. 86, p. 10,663–10,680.

Hart, R.J., McDonald, I., Tredoux, M., de Wit, M.J., Carlson, R.W., Andreoli, M., Moser, D.E., and Ashwal, L.D., 2004, New PGE and Re/Os-isotope data from lower crustal sections of the Vredefort Dome and a reinterpretation of its "crust on edge" profile: South African Journal of Geology, v. 107, p. 173–184, doi: 10.2113/107.1-2.173.

Hartlaub, R.P., Böhm, C.O., Kuiper, Y.D., Bowerman, M.S., and Heaman, L.M., 2004, Archean and Paleoproterozoic geology of the northwestern Split Lake Block, Superior Province, Manitoba (parts of NTS 54D4, 5, 6, and 64A1): Manitoba Geological Survey Report of Activities 2004, p. 187–194.

Haskin, L.A., Korotev, R.L., Rockow, K.M., and Joliff, B.L., 1998, The case for an Imbrium origin of the Apollo thorium-rich impact-melt breccias: Meteoritics & Planetary Science, v. 33, p. 959–975.

Heather, K.B., Shore, G.T., and van Breeman, O., 1995, The convoluted "layer cake"—An old recipe with new ingredients for the Swayze greenstone belt, southern Superior Province, Ontario: Geological Survey of Canada, Current Research 1995C, p. 1–10.

Henson, P.A., Blewett, R.S., Champion, D.C., Goleby, B.R., and Cassidy, K.F., 2004, A dynamic view of orogenesis and the development of the eastern Yilgarn Craton: Geoscience Australia: Record, 2004/09, p. 83–86.

Herzberg, C., 1999, Phase equilibrium constraints on the formation of cratonic mantle: Geochemical Society Special Publication 6, p. 241–257.

Heubeck, C., and Lowe, D.R., 1994, Depositional and tectonic setting of the Archean Moodies Group, Barberton Greenstone Belt, South Africa: Precambrian Research, v. 68, p. 257–290, doi: 10.1016/0301-9268(94)90033-7.

Hickman, A.H., 1984, Archaean diapirism in the Pilbara Block, Western Australia, *in* Kröner, A., and Greiling, R., eds., Precambrian tectonics illustrated: Stuttgart, Schweizerbart'sche Verlagsbuchhandlung, p. 113–127.

Hickman, A.H., 2004, Two contrasting granite-greenstone terranes in the Pilbara Craton, Australia—Evidence for vertical and horizontal tectonic regimes prior to 2900 Ma: Precambrian Research, v. 131, p. 153–172, doi: 10.1016/j.precamres.2003.12.009.

Hickman, A.H., and Smithies, R.H., 2000, Roebourne, sheet SF50–3, 2nd ed.: Geological Survey of Western Australia, 1:250,000 Geological Series [map].

Hickman, A.H., and Strong, C.A., 2003, Dampier-Barrow Island, Western Australia: Geological Survey of Western Australia, 1:250,000 Geological Series, Explanatory Notes, 75 p. plus geologic map.

Hickman, A.H., Smithies, R.H., Pike, G., Farrell, T.R., and Beintema, K.A., 2001, Evolution of the West Pilbara granite-greenstone terrane and Mallina Basin, Western Australia—A field guide: Geological Survey of Western Australia Record 2001–16, 65 p.

Hill, R.E.T., Barnes, S.J., and Dowling, S.E., 2001, Komatiites of the Norseman-Wiluna greenstone belt, Western Australia—A field guide: Geological Survey of Western Australia Record 2001/10, 71 p.

Hobbs, R.W., Drummond, B.J., and Goleby, B.R., 2006, The effects of three-dimensional structure on two-dimensional images of crustal seismic sections and on the interpretation of shear zone morphology: Geophysical Journal International, v. 164, p. 490–500, doi: 10.1111/j.1365-246X.2006.02814.x.

Hoffman, P.F., 1988, United plates of America, the birth of a craton—Early Proterozoic assembly and growth of Laurasia: Annual Review of Earth and Planetary Sciences, v. 16, p. 543–603, doi: 10.1146/annurev.ea.16.050188.002551.

Hofmann, A.F., 1997, Mantle geochemistry—The message from oceanic volcanism: Nature, v. 385, p. 219–229, doi: 10.1038/385219a0.

Hofmann, A., Dirks, P.H.G.M., and Jelsma, H.A., 2004, Clastic sedimentation in a late Archaean accretionary terrain, Midlands greenstone belt, Zimbabwe: Precambrian Research, v. 129, p. 47–69, doi: 10.1016/j.precamres.2003.09.017.

Hofmeister, A.M., and Criss, R.E., 2005, Heatflow and mantle convection in the triaxial Earth: Geological Society of America Special Paper 388, p. 289–302.

Holden, P., Ireland, T.R., Bruce, Z., and Harrison, T.M., 2006, Automated mining of detrital Hadean zircons from Jack Hills, Western Australia—Flash geochronology with SHRIMP II: 16th Annual V.M. Goldschmidt Conference, 27 August–1 September 2006, Melbourne, Australia, 1 p.

Hollings, P., and Kerrich, R., 2004, Geochemical systematics of tholeiites from the 2.86 Ga Pickle Crow Assemblage, northwestern Ontario—Arc basalts with positive and negative Nb-Hf anomalies: Precambrian Research, v. 134, p. 1–20, doi: 10.1016/j.precamres.2004.05.009.

Hollings, P., Wyman, D., and Kerrich, R., 1999, Komatiite-basalt-rhyolite volcanic associations in northern Superior Province greenstone belts—Significance of plume-arc interaction in the generation of the proto-continental Superior Province: Lithos, v. 46, p. 137–161, doi: 10.1016/S0024-4937(98)00058-9.

Hollis, J.A., Garde, A., Frei, D., and van Gool, J., 2005, Geochronology [of greenstone belts in the central Godthåbsfjord region, southern west Greenland]: Geological Survey of Denmark and Greenland Rapport 2005/42, p. 12–79.

Hollis, J.A., Frei, D., van Gool, J.A.M., Garde, A.A., and Persson, M., 2006, Using zircon geochronology to resolve the Archaean geology of southern

West Greenland: Geological Survey of Denmark and Greenland Bulletin 10, p. 49–52.

Horstwood, M.S.A., Nesbitt, R.W., Noble, S.R., and Wilson, J.F., 1999, U-Pb zircon evidence for an extensive early Archean craton in Zimbabwe—Reassessment of the timing of craton formation, stabilization, and growth: Geology, v. 27, p. 707–710, doi: 10.1130/0091-7613(1999)027<0707:UPZEFA>2.3.CO;2.

Iizuka, T., Horie, K., Komiya, T., Maruyama, S., Hirata, T., Hidaka, H., and Windley, B.F., 2006, 4.2 Ga zircon xenocryst in an Acasta gneiss from northwestern Canada—Evidence for early continental crust: Geology, v. 34, p. 245–248, doi: 10.1130/G22124.1.

Iizuka, T., Komiya, T., Ueno, Y., Katayama, I., Uchara, Y., Maruyama, S., Hirata, T., Johnson, S.P., and Dunkley, D.J., 2007, Geology and zircon chronology of the Acasta Gneiss Complex, northwestern Canada—New constraints on its tectonothermal history: Precambrian Research, v. 153, p. 179–208, doi: 10.1016/j.precamres.2006.11.017.

Irwin, W.P., 1994, Geologic map of the Klamath Mountains, California: U.S. Geological Survey Map I-2148, scale 1:500,000, 2 sheets.

Isachsen, C.E., and Bowring, S.A., 1994, Evolution of the Slave craton: Geology, v. 22, p. 917–920, doi: 10.1130/0091-7613(1994)022<0917:EOTSC>2.3.CO;2.

Ishii, M., and Tromp, J., 2004, Constraining large-scale mantle heterogeneity using mantle and inner-core sensitive normal modes: Physics of the Earth and Planetary Interiors, v. 146, p. 113–124, doi: 10.1016/j.pepi.2003.06.012.

Jackson, M.P.A., and Robertson, D.I., 1983, Regional implications of early-Precambrian strains in the Onverwacht Group adjacent to the Lochiel Granite, north-west Swaziland: Geological Society of South Africa Special Publication 9, p. 45–62.

Jacobsen, S.B., 2005, The Hf-W isotopic system and the origin of the Earth and Moon: Annual Review of Earth and Planetary Sciences, v. 33, p. 531–570, doi: 10.1146/annurev.earth.33.092203.122614.

Jacobsen, S.D., Spetzler, H., Reichmann, H.J., and Smyth, J.R., 2004, Shear waves in the diamond-anvil cell reveal pressure-induced instability in (Mg,Fe)O: Proceedings of the National Academy of Sciences of the United States of America, v. 101, p. 5867–5871, doi: 10.1073/pnas.0401564101.

James, D.T., Kamo, S., and Krogh, T., 2002, Evolution of 3.1 and 3.0 Ga volcanic belts and a new thermotectonic model for the Hopedale Block, North Atlantic craton (Canada): Canadian Journal of Earth Sciences, v. 39, p. 687–710, doi: 10.1139/e01-092.

James, P.R., 1976, Deformation of the Isua block, West Greenland—A remnant of the earliest stable continental crust: Canadian Journal of Earth Sciences, v. 13, p. 816–823.

Jelsma, H.A., van der Beek, P.A., and Vinyu, M.L., 1993, Tectonic evolution of the Bindura-Shamva greenstone belt (northern Zimbabwe)—Progressive deformation around diapiric batholiths: Journal of Structural Geology, v. 15, p. 163–176, doi: 10.1016/0191-8141(93)90093-P.

Jelsma, H.A., Vinyu, M.L., Valbracht, P.J., Davies, G.R., Wijbrans, J.R., and Verdurmen, E.A.T., 1996, Constraints on Archaean crustal evolution of the Zimbabwe craton—A U-Pb zircon, Sm-Nd, and Pb-Pb whole-rock isotope study: Contributions to Mineralogy and Petrology, v. 124, p. 55–70, doi: 10.1007/s004100050173.

Kasting, J.F., and Ackerman, T.P., 1986, Climatic consequences of very high carbon dioxide levels in the Earth's early atmosphere: Science, v. 234, p. 1383–1385, doi: 10.1126/science.11539665.

Kerrich, R., and Polat, A., 2006, Archean greenstone-tonalite duality—Thermochemical mantle convection models or plate tectonics in the early Earth global dynamics?: Tectonophysics, v. 415, p. 141–165, doi: 10.1016/j.tecto.2005.12.004.

Kerrich, R., Polat, A., Wyman, D., and Hollings, P., 1999, Trace element systematics of Mg- to Fe-tholeiitic basalt suites of the Superior Province—Implications for Archean mantle reservoirs and greenstone belt geophysics: Lithos, v. 46, p. 163–187, doi: 10.1016/S0024-4937(98)00059-0.

Ketchum, J.W.F., Bleeker, W., and Stern, R.A., 2004, Evolution of an Archean basement complex and its autochthonous cover, southern Slave Province, Canada: Precambrian Research, v. 135, p. 149–179, doi: 10.1016/j.precamres.2004.08.005.

Kinny, P.D., Williams, I.S., Froude, D.D., Ireland, T.H., and Compston, W., 1988, Early Archaean zircon ages from orthogneisses and anorthosites at Mount Narryer, Western Australia: Precambrian Research, v. 38, p. 325–341, doi: 10.1016/0301-9268(88)90031-9.

Kisters, A.F.M., and Anhaeusser, C.R., 1995, The structural significance of the Steynsdorp Pluton and anticline within the tectono-magmatic framework of the Barberton Mountain Land: South African Journal of Geology, v. 98, p. 43–51.

Kisters, A.F.M., Stevens, G., Dziggel, A., and Armstrong, R.A., 2003, Extensional detachment faulting and core-complex formation in the southern Barberton granite-greenstone terrain, South Africa—Evidence for a 3.2 Ga orogenic collapse: Precambrian Research, v. 127, p. 355–378, doi: 10.1016/j.precamres.2003.08.002.

Kleine, T., Mezger, K., Palme, H., and Münker, C., 2004a, The W isotope evolution of the bulk silicate Earth—Constraints on the timing and mechanisms of core formation and accretion: Earth and Planetary Science Letters, v. 228, p. 109–123, doi: 10.1016/j.epsl.2004.09.023.

Kleine, T., Mezger, K., Münker, C., Palme, H., and Bischoff, A., 2004b, ^{182}Hf-^{182}W isotope systematics of chondrites, eucrites, and martian meteorites—Chronology of core formation and early mantle differentiation in Vesta and Mars: Geochimica et Cosmochimica Acta, v. 68, p. 2935–2946.

Kojan, C.J., and Hickman, A.H., 1998, Late Archaean volcanism in the Kylena and Maddina Formations, Fortescue Group, west Pilbara: Geological Survey of Western Australia Annual Review, v. 1997–98, p. 43–53.

Kositcin, N., and Krapež, B., 2004, Relationship between detrital zircon age-spectra and the tectonic evolution of the Late Archaean Witwatersrand Basin, South Africa: Precambrian Research, v. 129, p. 141–168, doi: 10.1016/j.precamres.2003.10.011.

Kramers, J.D., 2007, Hierarchical Earth accretion and the Hadean Eon: Journal of the Geological Society, v. 164, p. 3–17, doi: 10.1144/0016-76492006-028.

Krapez, B., and Eisenlohr, B., 1998, Tectonic settings of archaean (3325–2775 Ma) crustal-supracrustal belts in the West Pilbara Block: Precambrian Research, v. 88, p. 173–205, doi: 10.1016/S0301-9268(97)00068-5.

Kraus, J., and Williams, P.F., 1999, Structural development of the Snow Lake Allochthon and its role in the evolution of the southeastern Trans-Hudson orogen in Manitoba, central Canada: Canadian Journal of Earth Sciences, v. 36, p. 1881–1889, doi: 10.1139/cjes-36-11-1881.

Kröner, A., and Layer, P.W., 1992, Crust formation and plate motion in the early Archean: Science, v. 256, p. 1405–1411, doi: 10.1126/science.256.5062.1405.

Kusky, T.M., 1991, Structural development of an Archean orogen, western Point Lake, Northwest Territories: Tectonics, v. 10, p. 820–841.

Kusky, T.M., 2002, Response [to Zhai et al., 2002]: Science Online, v. 295, 8 February 2002, 3 p.

Kusky, T.M., Li, J.-H., and Tucker, R.D., 2001, The Archean Dongwanzi ophiolite complex, North China craton—2.505-billion-year-old oceanic crust and mantle: Science, v. 292, p. 1142–1145, doi: 10.1126/science.1059426.

Labotka, T.C., 1989, Petrogenesis of metamorphic rocks beneath the Stillwater Complex—Assemblages and conditions of metamorphism: Montana Bureau of Mines and Geology Special Publication, v. 92, p. 70–76.

Lana, C., Gibson, R.L., Reimold, W.U., and Minnitt, R.C.A., 2003, Geology and geochemistry of a granite-greenstone association in the southeastern Vredefort dome, South Africa: South African Journal of Geology, v. 106, p. 291–314, doi: 10.2113/106.4.291.

Lin, S., 2005, Synchronous vertical and horizontal tectonism in the Neoarchean—Kinematic evidence from a synclinal keel in the northwestern Superior craton, Canada: Precambrian Research, v. 139, p. 181–194, doi: 10.1016/j.precamres.2005.07.001.

Loferski, P.J., Arculus, R.J., and Czamanske, G.K., 1994, Rare earth element evidence for the petrogenesis of the Banded Series of the Stillwater Complex, Montana, and its anorthosites: Journal of Petrology, v. 35, p. 1623–1649.

Longhi, J., 2002, Some phase equilibrium systematics of lherzolite melting, I: Geochemistry, Geophysics, Geosystems: G^3 [American Geophysical Union electronic journal], v. 3, no. 2, 33 p., doi: 10.1029/2001GC000204.

Longhi, J., Fram, M.S., Vander Auwera, J., and Montieth, J.N., 1993, Pressure effects, kinetics, and rheology of anorthositic and related magmas: American Mineralogist, v. 78, p. 1016–1030.

Love, J.D., Reed, J.C., Jr., and Christiansen, A.C., 1992, Geologic map of Grand Teton National Park, Teton County, Wyoming: U.S. Geological Survey Map I-2031, 1:62,500.

Lucas, S.B., Green, A., Hajnal, Z., White, D., Lewry, J., Ashton, K., Weber, W., and Clowes, R., 1993, Deep seismic profile across a Proterozoic collision zone—Surprises at depth: Nature, v. 363, p. 339–342, doi: 10.1038/363339a0.

Lucas, S.B., Stern, R.A., Syme, E.C., Reilly, B.A., and Thomas, D.J., 1996, Intra-oceanic tectonics and the development of continental crust—1.92–1.84 Ga evolution of the Flin Flon Belt, Canada: Geological Society of America Bulletin, v. 108, p. 602–629, doi: 10.1130/0016-7606(1996)108<0602:ITATDO>2.3.CO;2.

Lustrino, M., 2005, How the delamination and detachment of lower crust can influence basaltic magmatism: Earth-Science Reviews, v. 72, p. 21–38, doi: 10.1016/j.earscirev.2005.03.004.

Macgregor, A.M., 1951, Some milestones in the Precambrian of Southern Rhodesia: Geological Society of South Africa Proceedings, v. 54, p. 27–71.

Malkovets, V.G., Griffin, W.L., O'Reilly, S.Y., and Wood, B.J., 2007, Diamond, subcalcic garnet, and mantle metasomatism—Kimberlite sampling patterns define the link: Geology, v. 35, p. 339–342: doi: 10.1130/G23092A.1.

Manikyamba, C., Kerrich, R., Naqvi, S.M., and Mohan, M.R., 2004, Geochemical systematics of tholeiitic basalts from the 2.7 Ga Ramagiri-Hungund composite greenstone belt, Dharwar craton: Precambrian Research, v. 134, p. 21–39, doi: 10.1016/j.precamres.2004.05.010.

Manning, C.E., Mojzis, S.J., and Harrison, T.M., 2006, Geology, age and origin of supracrustal rocks at Akilia, West Greenland: American Journal of Science, v. 306, p. 303–366, doi: 10.2475/05.2006.02.

Mao, W.L., Meng, Y., Shen, G., Prakapenka, V.B., Campbell, A.J., Heinz, D.L., Shu, J., Caracas, R., Cohen, R.E., Fei, Y., Hemley, R.J., and Mao, H., 2005, Iron-rich silicates in the Earth's D" layer: Proceedings of the National Academy of Sciences of the United States of America, v. 102, p. 9751–9753, doi: 10.1073/pnas.0503737102.

Martin, H., 1987, Archaean and modern granitoids as indicators of changes in geodynamic processes: Revista Brasileira de Geociências, v. 17, p. 360–365.

Martin, H., Smithies, R.H., Rapp, R., Moyen, J.-F., and Champion, D., 2005, An overview of adakite, tonalite-trondhjemite-granodiorite (TTG), and sanukitoid—Relationships and some implications for crustal evolution: Lithos, v. 79, p. 1–24, doi: 10.1016/j.lithos.2004.04.048.

Mayborn, K.R., and Lesher, C.E., 2004, Paleoproterozoic mafic dike swarms of northeast Laurentia—Products of plumes or ambient mantle?: Earth and Planetary Science Letters, v. 225, p. 305–317, doi: 10.1016/j.epsl.2004.06.014.

McCall, G.J.H., 2003, A critique of the analogy between Archaean and Phanerozoic tectonics based on regional mapping of the Mesozoic-Cenozoic plate convergent zone in the Makran, Iran: Precambrian Research, v. 127, p. 5–17, doi: 10.1016/S0301-9268(03)00178-5.

McCall, G.J.H., 2005, How long has this been going on?: Geoscientist, v. 15, p. 11.

McGregor, V.R., 1973, The early Precambrian gneisses of the Godthåb district, West Greenland: Philosophical Transactions of the Royal Society of London, v. A-273, p. 343–358.

McGregor, V.R., 1975, A horizontal tectonic regime in the Archaean of Greenland and its implications for early crustal thickening—A reply: Precambrian Research, v. 2, p. 400–404, doi: 10.1016/0301-9268(75)90008-X.

McGregor, V.R., 1993, Qôqut 64 V.1 Syd, Geological map of Greenland 1:100 000, Descriptive Text: Grønlands Geologiske Undersøgelse, 40 p.

McLaren, S., Sandiford, M., and Powell, R., 2005, Contrasting styles of Proterozoic crustal evolution—A hot-plate tectonic model for Australian terranes: Geology, v. 33, p. 673–676, doi: 10.1130/G21544.1.

Meyer, C., Williams, I.S., and Compston, W., 1996, Uranium-lead ages for lunar zircons—Evidence for a prolonged period of granophyre formation from 4.32 to 3.88 Ga: Meteoritics & Planetary Science, v. 31, p. 370–387.

Miles, W.F., Roest, W.R., and Vo, M.P., 2000, Gravity anomaly map, Canada: Geological Survey of Canada Open File 3830a, scale 1:7,500,000.

Miller, C.F., McDowell, S.M., and Mapes, R.W., 2003, Hot and cold granites? Implications of zircon saturation temperatures and preservation of inheritance: Geology, v. 31, p. 529–532, doi: 10.1130/0091-7613(2003)031<0529:HACGIO>2.0.CO;2.

Minnitt, R.C.A., and Anhaeusser, C.R., 1992, Gravitational and diapiric structural history of the eastern part of the Archaean Murchison greenstone belt, South Africa: Journal of African Earth Sciences, v. 15, p. 429–440, doi: 10.1016/0899-5362(92)90026-9.

Mojzsis, S.J., and Harrison, T.M., 2002, Establishment of a 3.83 Ga magmatic age for the Akilia tonalite (southern West Greenland): Earth and Planetary Science Letters, v. 2002, p. 445–453.

Moser, D.E., 1994, The geology and structure of the mid-crustal Wawa gneiss domain—A key to understanding tectonic variation with depth and time in the late Archean Abitibi-Wawa orogen: Canadian Journal of Earth Sciences, v. 31, p. 1064–1080.

Moser, D.E., and Heaman, L.M., 1997, Proterozoic zircon growth in Archean lower crustal xenoliths, southern Superior craton—A consequence of Matachewan ocean opening: Contributions to Mineralogy and Petrology, v. 128, p. 164–175, doi: 10.1007/s004100050301.

Moser, D.E., Heaman, L.M., Krogh, T.E., and Hanes, J.A., 1996, Intracrustal extension of an Archean orogen revealed using single-grain U-Pb zircon geochronology: Tectonics, v. 15, p. 1093–1109, doi: 10.1029/96TC00130.

Moyen, J.-F., Stevens, G., and Kisters, A., 2006, Record of mid-Archean subduction from metamorphism in the Barberton terrain, South Africa: Nature, v. 442, p. 559–562, doi: 10.1038/nature04972.

Mueller, W.U., Corcoran, P.L., and Pickett, C., 2005, Mesoarchean continental breakup—Evolution and inferences from the >2.8 Ga Slave craton-cover succession, Canada: The Journal of Geology, v. 113, p. 23–45, doi: 10.1086/425967.

Musacchio, G., White, D.J., Asudeh, I., and Thomson, C.J., 2004, Lithospheric structure and composition of the Archean western Superior Province from seismic refraction/wide-angle reflection and gravity modeling: Journal of Geophysical Research, v. 109, no. B3, 28 p., doi: 10.1029/2003JB002427.

Myers, J.S., 1985, Stratigraphy and structure of the Fiskenaesset Complex, West Greenland: Grønlands Geologiske Undersøgelse Bulletin 150, 72 p.

Myers, J.S., 1988, Early Archaean Narryer Gneiss Complex, Yilgarn Craton, Western Australia: Precambrian Research, v. 38, p. 297–307, doi: 10.1016/0301-9268(88)90029-0.

Myers, J.S., 1995, The generation and assembly of an Archaean supercontinent—Evidence from the Yilgarn craton, Western Australia: Geological Society [London] Special Publication 95, p. 143–154.

Myers, J.S., 1997, Archaean geology of the Eastern Goldfields of Western Australia—Regional overview: Precambrian Research, v. 83, p. 1–10, doi: 10.1016/S0301-9268(97)00002-8.

Myers, J.S., 2001, Protoliths of the 3.8–3.7 Ga Isua greenstone belt: West Greenland: Precambrian Research, v. 105, p. 129–141, doi: 10.1016/S0301-9268(00)00108-X.

Myers, J.S., and Crowley, J.L., 2000, Vestiges of life in the oldest Greenland rocks? A review of early Archean geology in the Godthåbsfjord region, and reappraisal of field evidence for >3850 Ma life on Akilia: Precambrian Research, v. 103, p. 101–124, doi: 10.1016/S0301-9268(00)00089-9.

Myers, J.S., and Hocking, R.M., 1998, Geological map of Western Australia, scale 1:2,500,000: Geological Survey of Western Australia.

Myers, J.S., and Occhipinti, S.A., 2001, Narryer Terrane: Geological Survey of Western Australia Record 2001/8, p. 8–25.

Nair, S.K., Gao, S.S., Liu, K.H., and Silver, P.G., 2006, Southern African crustal evolution and composition—Constraints from receiver function studies: Journal of Geophysical Research, v. 111, no. B2, paper 304, 17 p., doi: 10.1029/2005JB003802.

Nakajima, K., and Arima, M., 1998, Melting experiments on hydrous low-K tholeiite—Implications for the genesis of tonalitic crust in the Izu-Bonin-Mariana arc: The Island Arc, v. 7, p. 359–373.

Nelson, D.R., 1997, Evolution of the Archaean granite-greenstone terranes of the Eastern Goldfields, Western Australia—SHRIMP U-Pb zircon constraints: Precambrian Geology, v. 83, p. 57–81, doi: 10.1016/S0301-9268(97)00005-3.

Nemchin, A.A., Pidgeon, R.T., and Whitehouse, M.J., 2006, Re-evaluation of the origin and evolution of >4.2 Ga zircons from the Jack Hills metasedimentary rocks: Earth and Planetary Science Letters, v. 244, p. 218–233, doi: 10.1016/j.epsl.2006.01.054.

Nisbet, E.G., Cheadle, M.J., Arndt, N.T., and Bickle, M.J., 1993, Constraining the potential temperature of the Archaean mantle—A review of the evidence from komatiites: Lithos, v. 30, p. 291–307, doi: 10.1016/0024-4937(93)90042-B.

Nitescu, B., Cruden, A.R., and Bailey, R.C., 2006, Integrated potential-field and seismic constraints on the structure of the Archean metasedimentary English River belt, Western Superior craton, Canada: Precambrian Research, v. 144, p. 261–277, doi: 10.1016/j.precamres.2005.11.008.

Niu, F., and James, D.E., 2002, Fine structure of the lowermost crust beneath the Kaapvaal craton and its implications for crustal formation and evolution: Earth and Planetary Science Letters, v. 200, v. 121–130.

Norman, M.D., Duncan, R.A., and Huard, J.J., 2006, Identifying impact events within the lunar cataclysm from ^{40}Ar–^{39}Ar ages and compositions of Apollo 16 impact melt rocks: Geochimica et Cosmochimica Acta, v. 70, p. 6032–6049, doi: 10.1016/j.gca.2006.05.021.

Nutman, A.P., 1986, The early Archean to Proterozoic history of the Isukasia area, southern West Greenland: Grønlands Geologiske Undersøgelse, Bulletin 154, 80 p.

Nutman, A.P., 2001, On the scarcity of >3900 Ma detrital zircons in ≥3500 Ma metasediments: Precambrian Research, v. 105, p. 93–114, doi: 10.1016/S0301-9268(00)00106-6.

Nutman, A.P., and Friend, C.R.L., 2007, Adjacent terranes with ca. 2715 and 2650 high-pressure metamorphic assemblages in the Nuuk region of the North Atlantic Craton, southern West Greenland—Complexities of Neoarchean collisional orogeny: Precambrian Research, v. 155, p. 159–203, doi: 10.1016/j.precamres.2006.12.009.

Nutman, A.P., McGregor, V.R., Friend, C.R.L., Bennett, V.C., and Kinny, P.D., 1996, The Itsaq Gneiss Complex of southern West Greenland; the world's most extensive record of early crustal evolution (3900–3600 Ma): Precambrian Research, v. 78, p. 1–39, doi: 10.1016/0301-9268(95)00066-6.

Nutman, A.P., Bennett, V.C., Friend, C.R.L., and Rosing, M.T., 1997, ~3710 and ≥3790 Ma volcanic sequences in the Isua (Greenland) supracrustal belt; structural and Nd isotope implications: Chemical Geology, v. 141, p. 271–287, doi: 10.1016/S0009-2541(97)00084-3.

Nutman, A.P., Friend, C.R.L., Barker, S.L., and McGregor, V.R., 2004a, Inventory and assessment of Palaeoarchaean gneiss terrains and detrital zircons in southern West Greenland: Precambrian Research, v. 135, p. 281–314, doi: 10.1016/j.precamres.2004.09.002.

Nutman, A.P., Friend, C.R.L., Bennett, V.C., and McGregor, V.R., 2004b, Dating of the Ameralik dyke swarms of the Nuuk district, southern West Greenland—Mafic intrusion events starting from c. 3510 Ma: Geological Society [London] Journal, v. 161, p. 421–436.

Nutman, A.P., Christiansen, O., and Friend, C.R.L., 2007, 2635 Ma amphibolite facies gold mineralization near a terrane boundary (suture?) on Storø, Nuuk region, southern West Greenland: Precambrian Research (in press).

O'Hara, M.J., and Herzberg, C., 2002, Interpretation of trace element and isotope features of basalts—Relevance of field relations, petrology, major element data, phase equilibria, and magma chamber modeling in basalt petrogenesis: Geochimica et Cosmochimica Acta, v. 66, p. 2167–2191, doi: 10.1016/S0016-7037(02)00852-9.

Ohta, H., Maruyama, S., Takahashi, E., Watanabe, Y., and Kato, Y., 1996, Field occurrence, geochemistry and petrogenesis of the Archean mid-oceanic ridge basalts (AMORBs) of the Cleaverville area, Pilbara Craton, Western Australia: Lithos, v. 37, p. 199–221, doi: 10.1016/0024-4937(95)00037-2.

O'Reilly, S.Y., and Griffin, W.L., 2006, Imaging global chemical and thermal heterogeneity in the subcontinental lithospheric mantle with garnets and xenoliths—Geophysical implications: Tectonophysics, v. 416, p. 289–309, doi: 10.1016/j.tecto.2005.11.014.

O'Reilly, S.Y., Griffin, W.L., Djomani, Y.H.P., and Morgan, P., 2001, Are lithospheres forever? Tracking changes in subcontinental lithospheric mantle through time: GSA Today, v. 11, no. 4, p. 4–10, doi: 10.1130/1052-5173(2001)011<0004:ALFTCI>2.0.CO;2.

Page, N.J., and Koski, R.A., 1973, A Precambrian diamictite below the base of the Stillwater Complex, southwestern Montana: U.S. Geological Survey Journal of Research, v. 1, p. 403–414.

Parks, J., Lin, S., Davis, D., and Corkery, T., 2006, New high-precision U-Pb ages for the Island Lake greenstone belt, northwestern Superior Province—Implications for regional stratigraphy and the extent of the North Caribou terrane: Canadian Journal of Earth Sciences, v. 43, p. 789–803, doi: 10.1139/E06-044.

Parman, S.W., Dann, J.C., Grove, T.L., and de Wit, M.J., 1997, Emplacement conditions of komatiite magmas from the 3.49 Ga Komati Formation, Barberton Greenstone Belt, South Africa: Earth and Planetary Science Letters, v. 150, p. 303–323, doi: 10.1016/S0012-821X(97)00104-0.

Parmenter, A.C., Lin, S., and Corkery, M.T., 2006, Structural evolution of the Cross Lake greenstone belt in the northwestern Superior Province, Manitoba—Implications for relationship between vertical and horizontal tectonism: Canadian Journal of Earth Sciences, v. 43, p. 767–787, doi: 10.1139/E06-006.

Passchier, C.W., 1994, Structural geology across a proposed Archaean terrane boundary in the eastern Yilgarn craton, Western Australia: Precambrian Research, v. 68, p. 43–64, doi: 10.1016/0301-9268(94)90064-7.

Pawley, M.J., Van Kranendonk, M.J., and Collins, W.J., 2004, Interplay between deformation and magmatism during doming of the Archaean Shaw Granitoid Complex, Pilbara Craton, Western Australia: Precambrian Research, v. 131, p. 213–230, doi: 10.1016/j.precamres.2003.12.010.

Pearce, J.A., and Cann, J.R., 1973, Tectonic setting of basic volcanic rocks determined using trace element analyses: Earth and Planetary Science Letters, v. 19, p. 290–300, doi: 10.1016/0012-821X(73)90129-5.

Peck, D.C., Potter, L., Desharnais, G., Scoates, R.F.J., Corkery, M.T., and Böhm, C.O., 2000, Geology of the western part of the Fox River Belt (parts of NTS 53M and 53N): Manitoba Geological Survey Report of Activities 2000, p. 38–41.

Peck, W.H., and Valley, J.W., 1996, The Fiskenaesset anorthosite complex—Stable isotope evidence for shallow emplacement into Archean oceanic crust: Geology, v. 24, p. 523–526, doi: 10.1130/0091-7613(1996)024<0523:TFACSI>2.3.CO;2.

Percival, J.A., 1989, Granulite terranes and the lower crust of the Superior Province: American Geophysical Union Geophysical Monograph 51, p. 301–310.

Percival, J.A., and Pysklywec, R.N., 2007, Are Archean lithospheric keels inverted?: Earth and Planetary Science Letters, v. 254, p. 393–403, doi: 10.1016/j.epsl.2006.11.047.

Percival, J.A., Mortensen, J.K., Stern, R.A., Card, K.D., and Bégin, N.J., 1992, Giant granulite terranes of northeastern Superior Province—The Ashuanipi complex and Minto block: Canadian Journal of Earth Sciences, v. 29, p. 2287–2308.

Percival, J.A., Stern, R.A., and Skulski, T., 2001, Crustal growth through successive arc magmatism—Reconnaissance U-Pb SHRIMP data from the northeastern Superior Province, Canada: Precambrian Research, v. 109, p. 203–238, doi: 10.1016/S0301-9268(01)00148-6.

Percival, J.A., Whalen, J.B., Tomlinson, K.Y., McNicoll, V., and Stott, G.M., 2002, Geology and tectonostratigraphic assemblages, north central Wabigoon Subprovince, Ontario: Geological Survey of Canada Open File 4270, scale 1:250,000 with marginal notes.

Percival, J.A., Bleeker, W., Cook, F.A., Rivers, T., Ross, G., and van Staal, C., 2004a, PanLITHOPROBE Workshop IV—Intra-orogen correlations and comparative orogenic anatomy: Geoscience Canada, v. 31, no. 1, p. 23–39.

Percival, J.A., McNicoll, V., Brown, J.L., and Whalen, J.B., 2004b, Convergent margin tectonics, central Wabigoon subprovince, Superior Province, Canada: Precambrian Research, v. 132, p. 213–244, doi: 10.1016/j.precamres.2003.12.016.

Percival, J.A., McNicoll, V., and Bailes, A.H., 2006a, Strike-slip juxtaposition of ca. 2.72 Ga juvenile arc and >2.98 Ga continent margin sequences and its implications for Archean terrane accretion, western Superior Province, Canada: Canadian Journal of Earth Sciences, v. 43, p. 895–927, doi: 10.1139/E06-039.

Percival, J.A., Sanborn-Barrie, M., Skulski, T., Stott, G.M., Helmstaedt, H., and White, D.J., 2006b, Tectonic evolution of the western Superior Province from NATMAP and Lithoprobe studies: Canadian Journal of Earth Sciences, v. 43, p. 1085–1117, doi: 10.1139/E06-062.

Perring, C.S., Barnes, S.J. and Hill, R.E.T., 1996, Geochemistry of komatiites from Forrestania, Southern Cross Province, Western Australia—Evidence for crustal contamination: Lithos, v. 37, p. 181–197, doi: 10.1016/0024-4937(95)00036-4.

Peschler, A.P., Benn, K., and Roest, W.R., 2004, Insights on Archean continental geodynamics from gravity modelling of gravity-greenstone terranes: Journal of Geodynamics, v. 38, p. 185–207, doi: 10.1016/j.jog.2004.06.005.

Pidgeon, R.T., and Nemchin, A.A., 2006, High abundance of early Archean grains and the age of detrital zircons in a sillimanite-bearing quartzite from Mt. Narryer, Western Australia: Precambrian Research, v. 150, p. 201–220.

Plank, T., and Langmuir, C.H., 1988, An evaluation of the global variations in the major element chemistry of arc basalts: Earth and Planetary Science Letters, v. 90, p. 349–370, doi: 10.1016/0012-821X(88)90135-5.

Polat, A., and Hofmann, A.W., 2003, Alteration and geochemical patterns in the 3.7–3.8 Ga Isua greenstone belt: West Greenland: Precambrian Research, v. 126, p. 197–218, doi: 10.1016/S0301-9268(03)00095-0.

Polat, A., and Münker, C., 2004, Hf-Nd isotope evidence for contemporaneous subduction processes in the source of late Archean arc lavas from the Superior Province, Canada: Chemical Geology, v. 213, p. 403–429, doi: 10.1016/j.chemgeo.2004.08.016.

Polat, A., Li, J., Fryer, B., Kusky, J., Gagnon, J., and Zhang, S., 2006, Geochemical characteristics of the Neoarchean (2800–2700 Ma) Taishan greenstone belt, North China Craton—Evidence for plume-craton interaction: Chemical Geology, v. 230, p. 60–87, doi: 10.1016/j.chemgeo.2005.11.012.

Polat, A., Appel, P.W.U., Frei, R., Pan, Y., Dilek, Y., Ordóñez-Calderón, J.C., Fryer, B., Hollis, J.A., and Raith, J.G., 2007, Field and geochemical characteristics of the Mesoarchean (~3075 Ma) Ivisaartoq greenstone belt, southern West Greenland—Evidence for seafloor hydrothermal alteration in supra-subduction oceanic crust: Gondwana Research, v. 11, p. 69–91.

Pollack, H.N., 1997, Thermal characteristics of the Archaean, *in* de Wit, M.J., and Ashwall, L.D., eds., Greenstone belts: Oxford, Oxford University Press, p. 223–232.

Poujol, M., Robb, L.J., Respaut, J.-P., and Anhaeusser, C.R., 1996, 3.07–2.97 Ga greenstone belt formation in the northeastern Kaapvaal craton—Implications for the origin of the Witwatersrand Basin: Economic Geology and the Bulletin of the Society of Economic Geologists, v. 91, p. 1455–1461.

Poujol, M., Robb, L.J., Anhaeusser, C.R., and Gericke, B., 2003, A review of the geochronological constraints on the evolution of the Kaapvaal Craton, South Africa: Precambrian Research, v. 127, p. 181–213, doi: 10.1016/S0301-9268(03)00187-6.

Presnall, D.C., Gudfinnson, G.H., and Walter, M.J., 2002, Generation of mid-ocean ridge basalts at pressures from 1 to 7 GPa: Geochimica et Cosmochimica Acta, v. 66, p. 2073–2090, doi: 10.1016/S0016-7037(02)00890-6.

Ramsay, J.G., 1989, Emplacement kinematics of a granite diapir—The Chindamora batholith, Zimbabwe: Journal of Structural Geology, v. 11, p. 191–209, doi: 10.1016/0191-8141(89)90043-6.

Rayner, N.M., Stern, R.A., and Bickford, M.E., 2005, Tectonic implications of new SHRIMP and TIMS U-Pb geochronology of rocks from the Sask Craton, Peter Lake Domain, and Hearne margin, Trans-Hudson Orogen, Saskatchewan: Canadian Journal of Earth Sciences, v. 42, p. 635–657.

Ridley, J.R., 1992, The thermal causes and effects of voluminous, late Archaean monzogranite plutonism: University of Western Australia Geology Department Publication 22, p. 275–285.

Ridley, J.R., and Kramers, J.D., 1990, The evolution and tectonic consequences of a tonalitic magma layer within Archean continents: Canadian Journal of Earth Sciences, v. 27, p. 219–228.

Robb, L.J., and Meyer, F.M., 1995, The Witwatersrand Basin, South Africa—Geological framework and mineralization processes: Ore Geology Reviews, v. 10, p. 67–94, doi: 10.1016/0169-1368(95)00011-9.

Rollinson, H., 2002, The metamorphic history of the Isua Greenstone Belt, West Greenland: Geological Society [London] Special Publication 199, p. 329–350.

Rollinson, H., 2006, Crustal generation in the Archean, *in* Brown, M., and Rushmer, T., eds., Evolution and differentiation of continental crust: New York, Cambridge University Press, p. 173-230.

Rosing, M.T., Rose, N.M., Bridgwater, D., and Thomsen, H.S., 1996, Earliest part of Earth's stratigraphic record—A reappraisal of the >3.7 Ga Isua (Greenland) supracrustal sequence: Geology, v. 24, p. 43–46, doi: 10.1130/0091-7613(1996)024<0043:EPOESS>2.3.CO;2.

Rudnick, R.L., and Fountain, D.M., 1995, Nature and composition of the continental crust—A lower crustal perspective: Reviews of Geophysics, v. 33, p. 267–309, doi: 10.1029/95RG01302.

Rudnick, R.L., McDonough, W.F., and O'Connell, R.J., 1998, Thermal structure, thickness and composition of continental lithosphere: Chemical Geology, v. 145, p. 395–411, doi: 10.1016/S0009-2541(97)00151-4.

Saggerson, E.P., and Turner, L.M., 1992, Metamorphic map of the Republics of South Africa, Transkei, Bophuthatswana, Venda and Ciskei and the Kingdoms of Lesotho and Swaziland: Geological Survey of South Africa, 1:1,000,000.

Sanborn-Barrie, M., Skulski, T., Percival, J.A., Whalen, J.B., Brown, J.L., and McNicoll, V., 2002, Geology and tectonostratigraphic assemblages, western Wabigoon Subprovince, Ontario: Geological Survey of Canada Open File 4255, scale 1:250,000 with marginal notes.

Sanborn-Barrie, M., Rogers, N., Skulski, T., Parker, J., McNicoll, V., and Devaney, J., 2004, Geology and tectonostratigraphic assemblages, east Uchi Subprovince, Red Lake and Birch-Uchi belts, Ontario: Geological Survey of Canada Open File 4256, scale 1:250,000 with marginal notes.

Sandiford, M., and McLaren, S., 2006, Thermo-mechanical controls on heat production distributions and the long-term evolution of the continents, *in* Brown, M., and Rushmer, T., eds., Evolution and differentiation of the continental crust: Cambridge, Cambridge University Press, p. 67–91.

Sandiford, M., Van Kranendonk, M.J., and Bodorkos, S., 2004, Conductive incubation and the origin of dome-and-keel structure in Archean granite-greenstone terrains—A model based on the eastern Pilbara Craton, Western Australia: Tectonics, v. 23, no. 10, paper 9, 19 p.

Schmitz, M.D., Bowring, S.A., de Wit, M.J., and Gartz, V., 2004, Subduction and terrane collision stabilize the western Kaapvaal craton tectosphere 2.9 billion years ago: Earth and Planetary Science Letters, v. 222, p. 363–376, doi: 10.1016/j.epsl.2004.03.036.

Scott, D.J., Helmstaedt, H., and Bickle, M.J., 1992, Purtuniq ophiolite, Cape Smith belt, northern Quebec, Canada—A reconstructed section of Early Proterozoic oceanic crust: Geology, v. 20, p. 173–176, doi: 10.1130/0091-7613(1992)020<0173:POCSBN>2.3.CO;2.

Simon, N.S.C., Irvine, G.J., Davies, G.R., Pearson, D.G., and Carlson, R.W., 2003, The origin of garnet and clinopyroxene in "depleted" Kaapvaal peridotites: Lithos, v. 71, p. 289–322, doi: 10.1016/S0024-4937(03)00118-X.

Smithies, R.H., 2000, The Archaean tonalite-trondhjemite-granodiorite (TTG) series is not an analogue for Cenozoic adakite: Earth and Planetary Science Letters, v. 182, p. 115–125, doi: 10.1016/S0012-821X(00)00236-3.

Smithies, R.H., and Champion, D.C., 1998, Secular compositional changes in Archaean granitoid rocks of the west Pilbara: Geological Survey of Western Australia Annual Review 1997–98, p. 71–76.

Smithies, R.H., Champion, D.C., and Cassidy, K.F., 2003, Formation of Earth's early Archaean continental crust: Precambrian Research, v. 127, p. 89–101, doi: 10.1016/S0301-9268(03)00182-7.

Smithies, R.H., Champion, D.C., Van Kranendonk, M.J., Howard, H.M., and Hickman, A.H., 2005a, Modern-style subduction processes in the Mesoarchean—Geochemical evidence from the 3.12 Ga Whundo intra-oceanic arc: Earth and Planetary Science Letters, v. 231–237.

Smithies, R.H., Van Kranendonk, M.J., and Champion, D.C., 2005b, It started with a plume—Early Archaean basaltic proto-continental crust: Earth and Planetary Science Letters, v. 238, p. 284–297, doi: 10.1016/j.epsl.2005.07.023.

Smithies, R.H., Van Kranendonk, M.J., and Champion, D.C., 2007, The Mesoarchean emergence of modern-style subduction: Gondwana Research, v. 11, p. 50–68, doi: 10.1016/j.gr.2006.02.001.

Snoke, A.W., and Barnes, C.G., 2006, The development of tectonic concepts for the Klamath Mountains province, California and Oregon, *in* Snoke, A.W., and Barnes, C.G., Geological Studies in the Klamath Mountains Province, California and Oregon: A Volume in Honor of William P. Irwin: Geological Society of America Special Paper 410, p. 1–29: doi: 10.1130/2006.2410(01).

Snow, C.A., 2006, A reevaluation of tectonic discrimination diagrams and a new probabilistic approach using large geochemical databases—Moving beyond binary and ternary plots: Journal of Geophysical Research, v. 111, no. B6, paper 6, 13 p., doi: 10.1029/2005JB003799.

Sproule, R.A., Lesher, C.M., Ayer, J.A., Thurston, P.C., and Herzberg, C.T., 2002, Spatial and temporal variations in the geochemistry of komatiites and komatiitic basalts in the Abitibi greenstone belt: Precambrian Research, v. 115, p. 153–186, doi: 10.1016/S0301-9268(02)00009-8.

Stern, R.J., 2005, Evidence from ophiolites, blueschists, and ultrahigh-pressure metamorphic terranes that the modern episode of subduction tectonics began in Neoproterozoic time: Geology, v. 33, p. 557–560, doi: 10.1130/G21365.1.

Stern, R.J., 2007, When and how did plate tectonics begin? Theoretical and empirical considerations: Chinese Science Bulletin, v. 52, p. 578–591.

Stöffler, D., Ryder, G., Ivanov, B.A., Artemieva, N.A., Cintala, M.J., and Grieve, R.A.F., 2006, Cratering history and lunar chronology: Reviews in Mineralogy and Geochemistry, v. 60, p. 519–596.

Stone, D., 2000, Temperature and pressure variations in suites of Archean felsic plutonic rocks, Berens River area, northwest Superior Province, Ontario, Canada: Canadian Mineralogist, v. 38, p. 455–470.

Stone, D., 2005, Geology of the northern Superior area, Ontario: Ontario Geological Survey, Open File Report 6140, 94 p.

Stone, D., Tomlinson, K.Y., Davis, D.W., Fralick, P., Hallé, J., Percival, J.A., and Pufahl, P., 2002, Geology and tectonostratigraphic assemblages, south-central Wabigoon Subprovince, Ontario: Geological Survey of Canada Open File 4284, scale 1:250,000 with marginal notes.

Stone, D., Corkery, M.T., Hallé, J., Ketchum, J., Lange, M., Skulski, T., and Whalen, J., 2004, Geology and tectonostratigraphic assemblages, eastern Sachigo Subprovince, Ontario and Manitoba: Geological Survey of Canada Open File 1582, scale 1:250,000 with marginal notes.

Stone, M.S., and Stone, W.E., 2000, A crustally contaminated komatiitic dyke-sill-lava complex, Abitibi greenstone belt, Ontario: Precambrian Research, v. 102, p. 21–46, doi: 10.1016/S0301-9268(99)00101-1.

Stott, G.M., Davis, D.W., Parker, J.R., Straub, K.J., and Tomlinson, K.Y., 2002, Geology and tectonostratigraphic assemblages, eastern Wabigoon Subprovince, Ontario: Geological Survey of Canada Open File 4285, scale 1:250,000 with marginal notes.

Swager, C.P., 1995, Geology of the greenstone terranes in the Kurnalpi-Edjudina region, southeastern Yilgarn Craton: Geological Survey of Western Australia, Report 47, 31 p.

Swager, C.P., 1997, Tectono-stratigraphy of late Archaean greenstone terranes in the southern Eastern Goldfields, Western Australia: Precambrian Geology, v. 83, p. 11–42, doi: 10.1016/S0301-9268(97)00003-X.

Thorne, A.M., and Trendall, A.F., 2001, Geology of the Fortescue Group, Pilbara Craton, Western Australia: Geological Survey of Western Australia Bulletin 144, 249 p.

Thurston, P.C., 2002, Autochthonous development of Superior Province greenstone belts?: Precambrian Research, v. 115, p. 11–36, doi: 10.1016/S0301-9268(02)00004-9.

Thurston, P.C., Osmani, L.A., and Stone, D., 1991, Northwestern Superior Province—Review and terrane analysis: Ontario Geological Survey Special Volume 4, Part 1, p. 81–142.

Tinker, J., de Wit, M., and Grotzinger, J., 2002, Seismic stratigraphic constraints on Neoarchean-Paleoproterozoic evolution of the western margin of the Kaapvaal Craton, South Africa: South African Journal of Geology, v. 105, p. 107–134, doi: 10.2113/105.2.107.

Tomlinson, K.Y., Stevenson, R.K., Hughes, D.J., Hall, R.P., Thurston, P.C., and Henry, P., 1998, The Red Lake greenstone belt, Superior Province—Evidence of plume-related magmatism at 3 Ga and evidence of an older enriched source: Precambrian Research, v. 89, p. 59–76, doi: 10.1016/S0301-9268(97)00078-8.

Tomlinson, K.Y., Hughes, D.J., Thurston, P.C., and Hall, R.P., 1999, Plume magmatism and crustal growth at 2.9 to 3.0 Ga in the Steep Rock and Lumby Lake area, western Superior Province: Lithos, v. 46, p. 103–136, doi: 10.1016/S0024-4937(98)00057-7.

Tomlinson, K.Y., Stott, G.M., Percival, J.A., and Stone, D., 2004, Basement terrane correlations and crystal recycling in the western Superior Province—Nd isotopic character of granitoid and felsic volcanic rocks in the Wabigoon subprovince, N. Ontario, Canada: Precambrian Research, v. 132, p. 245–274, doi: 10.1016/j.precamres.2003.12.017.

Trampert, J., Deschamps, F., Resovsky, J., and Yuen, D., 2004, Probabilistic tomography maps chemical heterogeneities throughout the lower mantle: Science, v. 306, p. 853–856, doi: 10.1126/science.1101996.

Trendall, A.F., 1995, The Woongarra Rhyolite—A giant lavalike felsic sheet in the Hamersley Basin of Western Australia: Geological Survey of Western Australia Report 42, 70 p.

Trendall, A.F., Compston, W., Nelson, D.R., De Laeter, J.R., and Bennett, V.C., 2004, SHRIMP zircon ages constraining the depositional chronology of the Hamersley Group, Western Australia: Australian Journal of Earth Sciences, v. 51, p. 621–644, doi: 10.1111/j.1400-0952.2004.01082.x.

Tsujimori, T., Sisson, V.B., Liou, J.G., Harlow, G.E., and Sorensen, S.S., 2006, Very-low-temperature record of the subduction process—A review of worldwide lawsonite eclogites: Lithos, v. 92, no. 3–4, p. 609–624.

Valley, J.W., Lackey, J.S., Cavosie, A.J., Clechenko, C.C., Spicuzza, M.J., Basel, M.A.S., Bindeman, I.N., Ferreira, V.P., Sial, A.N., King, E.M., Peck, W.H., Sinha, A.K., and Wei, C.S., 2005, 4.4 billion years of crustal maturation—Oxygen isotope ratios of magmatic zircons: Contributions to Mineralogy and Petrology, v. 150, p. 561–580, doi: 10.1007/s00410-005-0025-8.

van der Velden, A.J., Cook, F.A., Drummond, B.J., and Goleby, B.J., 2006, Reflections of the Neoarchean—A global perspective: American Geophysical Union Geophysical Monograph 164, p. 255–265.

Van Kranendonk, M.J., 1996, Tectonic evolution of the Paleoproterozoic Torngat orogen—Evidence from pressure-temperature-time-deformation paths in the North River map area: Labrador: Tectonics, v. 15, p. 843–869, doi: 10.1029/95TC03771.

Van Kranendonk, M.J., 2003, Stratigraphic and tectonic significance of eight local unconformities in the Fortescue Group, Pear Creek Centrocline, Pilbara Craton, Western Australia: Geological Survey of Western Australia Annual Review 2001–2002, p. 70–79.

Van Kranendonk, M.J., Hickman, A.H., Williams, I.R., and Nijman, W., 2001, Archaean geology of the East Pilbara granite-greenstone terrane, Western Australia—A field guide: Geological Survey of Western Australia Record 2001–9, 134 p.

Van Kranendonk, M.J., Hickman, A.H., Smithies, R.H., and Nelson, D.B., 2002, Geology and tectonic evolution of the Archean North Pilbara terrain, Pilbara Craton, Western Australia: Economic Geology and the Bulletin of the Society of Economic Geologists, v. 97, p. 695–732.

Van Kranendonk, M.J., Collins, W.J., Hickman, A., and Pawley, M.J., 2004a, Critical tests of vertical vs. horizontal tectonic models for the Archaean East Pilbara Granite-Greenstone terrane, Pilbara Craton, Western Australia: Precambrian Research, v. 131, p. 173–211, doi: 10.1016/j.precamres.2003.12.015.

Van Kranendonk, M.J., Smithies, R.H., Hickman, A.H., Bagas, L., Williams, I.R., and Farrell, T.R., 2004b, Event stratigraphy applied to 700 million years of Archaean crustal evolution, Pilbara Craton, Western Australia: Geological Survey of Western Australia Annual Review, v. 2003–2004, p. 49–61.

Van Kranendonk, M.J., Smithies, R.H., Hickman, A.H., and Champion, D.C., 2007, Review: Secular tectonic evolution of Archean continental crust: interplay between horizontal and vertical processes in the formation of the Pilbara Craton, Australia: Terra Nova, v. 19, p. 1–38, doi: 10.1111/j.1365-3121.2006.00723.x.

Van Schmus, W.R., 1995, Natural radioactivity of the crust and mantle: American Geophysical Union Reference Shelf, v. 1, p. 283–291.

van Staal, C.R., 1998, Some notes on the assemblage boundaries and internal structures of the Uchi-Confederation greenstone belt, northwestern Ontario: Geological Survey of Canada Current Research 1998-C, p. 107–114.

Vermeesch, P., 2006, Tectonic discrimination of basalts with classification trees: Geochimica et Cosmochimica Acta, v. 70, p. 1839–1848, doi: 10.1016/j.gca.2005.12.016.

Vervoort, J.D., Patchett, P.J., Gehrels, G.E., and Nutman, A.P., 1996, Constraints on early Earth differentiation from hafnium and neodymium isotopes: Nature, v. 379, p. 624–627, doi: 10.1038/379624a0.

Viljoen, M.J., Viljoen, R.P., Smith, H.S., and Erlank, A.J., 1983, Geological, textural, and geochemical features of komatiitic flows from the Komati Formation: Geological Society of South Africa Special Publication 9, p. 1–20.

Walker, I.W., and Blight, D.F., 1983, Barlee, Western Australia: Geological Survey of Western Australia 1:250,000 Geological Series, Explanatory Notes SH/50–8, 22 p. plus map.

Walter, M.J., and Trønnes, R.G., 2004, Early Earth differentiation: Earth and Planetary Science Letters, v. 235, p. 253–269.

Weber, W., 1983, The Pikwitonei granulite domain—A lower crustal level along the Churchill-Superior boundary in central Manitoba: Lunar and Planetary Institute Technical Report 83-03, p. 95–97.

Wellman, P., 2000, Upper crust of the Pilbara Craton, Australia—3D geometry of a granite/greenstone terrain: Precambrian Research, v. 104, p. 175–186, doi: 10.1016/S0301-9268(00)00092-9.

Whalen, J.B., Percival, J.A., McNicoll, V.J., and Longstaffe, F.J., 2004, Geochemical and isotopic (Nd-O) evidence bearing on the origin of late- to post-orogenic high-K granitoid rocks in the Western Superior Province—Implications for late Archean tectonomagmatic processes: Precambrian Research, v. 132, p. 303–326, doi: 10.1016/j.precamres.2003.11.007.

Whitaker, A.J., 2001, Components and structure of the Yilgarn Craton, as interpreted from aeromagnetic data: Geoscience Australia Record, no. 37, p. 536–538.

Whitaker, A.J., 2003, The geophysical characteristics of granites and shear zones in the Yilgarn Craton, and their implications for gold mineralization: Geoscience Australia Record 2003/14, p. 129–133.

Whitaker, A.J., and Bastrakova, I.V., 2002, Yilgarn Craton aeromagnetic interpretation: Geoscience Australia, scale 1:500,000.

White, D.J., 2005, High-temperature, low-pressure metamorphism in the Kisseynew domain, Trans-Hudson orogen—Crustal anatexis due to tectonic thickening?: Canadian Journal of Earth Sciences, v. 42, p. 707–721, doi: 10.1139/e04-087.

White, D.J., Jones, A.G., Lucas, S.B., and Hajnal, Z., 1999, Tectonic evolution of the Superior Boundary Zone from coincident seismic reflection and magnetotelluric profiles: Tectonics, v. 18, p. 430–451, doi: 10.1029/1999TC900002.

White, D.J., Musacchio, D., Helmstaedt, H.H., Harrap, R.M., Thurston, P.C., van der Velden, A., and Hall, K., 2003, Images of a lower-crustal oceanic slab—Direct evidence for tectonic accretion in the Archean western Superior province: Geology, v. 31, p. 997–1000, doi: 10.1130/G20014.1.

White, D.J., Thomas, M.D., Jones, A.G., Hope, J., Németh, B., and Hajnal, Z., 2005, Geophysical transect across a Paleoproterozoic continent-continent collision zone—The Trans-Hudson Orogen: Canadian Journal of Earth Sciences, v. 42, p. 385–402, doi: 10.1139/e05-002.

White, R.V., Crowley, J.L., and Myers, J.S., 2000a, Earth's oldest well-preserved mafic dyke swarms in the vicinity of the Isua greenstone belt, southern West Greenland: Geology of Greenland Survey Bulletin 186, p. 65–72.

White, R.V., Crowley, J.L., Parrish, R., and David, K., 2000b, Gate-crashing the billionth birthday party—Intrusion of dyke swarms in the Isua region: Eos (Transactions, American Geophysical Union), v. 81, no. 48, suppl., p. 1256.

Whitehouse, M.J., and Kamber, B.S., 2003, A rare earth element study of complex zircons from early Archean Amîtsoq gneisses, Godthåbstfjord, south-west Greenland: Precambrian Research, v. 126, p. 363–377, doi: 10.1016/S0301-9268(03)00105-0.

Wilde, S.A., 1990, Geology and crustal evolution of the southwestern Yilgarn Craton: Third International Archaean Symposium, Perth, Western Australia, Excursion Guidebook, p. 98–122.

Wilde, S.A., 2001, Jimperding and Chittering metamorphic belts, southwestern Yilgarn Craton, Western Australia—A field guide: Geological Survey of Western Australia Record 2001/12, 24 p.

Wilson, A.H., 2003, A new class of silica enriched, highly depleted komatiites in the southern Kaapvaal Craton, South Africa: Precambrian Research, v. 127, p. 125–141, doi: 10.1016/S0301-9268(03)00184-0.

Wilson, J.F., Nesbitt, R.W., and Fanning, C.M., Zircon geochronology of Archaean felsic sequences in the Zimbabwe craton—A revision of greenstone stratigraphy and a model for crustal growth: Geological Society [London] Special Publication 93, p. 109–126.

Windley, B.F., 1993, Uniformitarianism today—Plate tectonics is the key to the past: Geological Society [London] Journal, v. 150, p. 7–19.

Wingate, M.T.D., 1999, Ion microprobe baddeleyite and zircon ages for Late Archaean mafic dykes of the Pilbara Craton, Western Australia: Australian Journal of Earth Sciences, v. 46, p. 493–500, doi: 10.1046/j.1440-0952.1999.00726.x.

Wolf, M.B., and Wyllie, P.J., 1993, Garnet growth during amphibolite anatexis—Implications of a garnetiferous restite: The Journal of Geology, v. 101, p. 357–373.

Wyche, S., Nelson, D.R., and Riganti, A., 2004, 4350–3130 Ma detrital zircons in the Southern Cross granite-greenstone terrane, Western Australia—Implications for the early evolution of the Yilgarn Craton: Australian Journal of Earth Sciences, v. 51, p. 31–45, doi: 10.1046/j.1400-0952.2003.01042.x.

Wyllie, P.J., Wolf, M.B., and van der Laan, S.R., 1993, Conditions for formation of tonalites and trondhjemites—Magmatic sources and products, in de Wit, M.J., and Ashwal, L.D., eds., Greenstone belts: Oxford, Clarendon Press, p. 256–266.

Wyman, D.A., 2003, Upper mantle processes beneath the 2.7 Ga Abitibi belt, Canada—A trace element perspective: Precambrian Research, v. 127, p. 143–165, doi: 10.1016/S0301-9268(03)00185-2.

Wyman, D.A., Kerrich, R., and Polat, A., 2002, Assembly of Archean cratonic mantle lithosphere and crust—Plume-arc interaction in the Abitibi-Wawa subduction-accretion complex: Precambrian Research, v. 115, p. 37–62, doi: 10.1016/S0301-9268(02)00005-0.

Young, M., and Helmstaedt, H., 2001, Tectonic evolution of the northern Pickle Lake greenstone belt, northwestern Superior Province, Ontario: Geological Survey of Canada Current Research 2001-C20, 13 p.

Zhai, M., Zhao, G., and Zhang, Qi., 2002, Is the Dongwanzi complex an ophiolite?: Science Online, v. 295, 8 Feb. 2002, 1 p.

Zhao, G., Cawood, P.A., Wilde, S.A., and Sun, M., 2002, Review of global 2.1–1.8 Ga orogens—Implications for a pre-Rodinia supercontinent: Earth-Science Reviews, v. 59, p. 125–162, doi: 10.1016/S0012-8252(02)00073-9.

Zheng, J.-P., Griffin, W.L., O'Reilly, S.Y., Lu, F., Wang, C., Zhang, M., Wang, F., and Li, H., 2004, 3.6 Ga lower crust in central China—New evidence on the assembly of the North China craton: Geology, v. 32, p. 229–232, doi: 10.1130/G20133.1.

Zwanzig, H.V., and Böhm, C.O., 2002, Tectonostratigraphy, Sm-Nd isotope and U-Pb age data of the Thompson Nickel Belt and Kisseynew north and east margins (NTS 63J, 63P, 63Q, 64A, 64B), Manitoba: Manitoba Geological Survey Report of Activities 2002, p. 102–114.

Zwanzig, H.V., Murphy, L., Percival, J.A., Whalen, J.B., and Rayner, N., 2006, Thompson Nickel Belt-type units in the northeastern Kisseynew Domain, Manitoba (parts of NTS 63O): Manitoba Geological Survey Report of Activities 2006, p. 85–103.

Manuscript Accepted by the Society 1 February 2007

Printed in the USA

The Geological Society of America
Memoir 200
2007

Proterozoic accretionary belts in the Amazonian Craton

Umberto G. Cordani*
Wilson Teixeira*
Institute of Geosciences, University of São Paulo, Rua do Lago 562, Cidade Universitária, 05508-080 São Paulo, Brazil

ABSTRACT

Within the Amazonian Craton, Archean crust is restricted to the Carajás granite-greenstone terrain. The younger Maroni-Itacaiunas province, including supracrustal sequences and associated calc-alkaline granitoids, is linked with the Birimian system in West Africa, making up a large Paleoproterozoic cratonic nucleus. Beginning at ca. 2.0 Ga, accretionary belts formed along the southwestern margin of this nucleus, giving rise to the Ventuari-Tapajós (2000–1800 Ma), Rio Negro–Juruena (1780–1550 Ma), and Rondonian–San Ignacio (1500–1300 Ma) tectonic provinces. Continued soft-collision/accretion processes driven by subduction produced a very large "basement" in which granitoid rocks predominate, many of them with juvenile-like Nd isotopic signatures. Felsic volcanics are also widespread; however, there is no evidence of Archean basement inliers, and regions with high-grade metamorphics are restricted.

The Sunsas-Aguapeí (1250–1000 Ma) orogenic belt, at the southwestern end of the craton, was originated in an extensional environment, later deformed during the Grenvillian collision between Amazonia and Laurentia. Over the cratonic area, a widespread anorogenic granitic magmatism (1000–970 Ma) is a reflection of this orogeny over the stable foreland. After the termination of the Sunsas orogeny, continental fragmentation affected the eastern margin of the Amazonian Craton. The intra-oceanic Goiás magmatic arc, closely associated with the Transbrasiliano megasuture, is the evidence of a large oceanic domain that started its consumption between 900 and 800 Ma, giving rise to juvenile material represented by calc-alkaline orthogneisses. Later, these units were deformed during the Brasiliano orogeny (700–500 Ma), in the process of amalgamation of Gondwana.

Keywords: accretionary belts, Amazonian Craton, geochronology, Nd isotopes, tectonic evolution.

*ucordani@usp.br; wteixeir@usp.br

Cordani, U.G., and Teixeira, W., 2007, Proterozoic accretionary belts in the Amazonian Craton, *in* Hatcher, R.D., Jr., Carlson, M.P., McBride, J.H., and Martínez Catalán, J.R., eds., 4-D Framework of Continental Crust: Geological Society of America Memoir 200, p. 297–320, doi: 10.1130/2007.1200(14). For permission to copy, contact editing@geosociety.org.

INTRODUCTION

Classical models of orogeny involve either Andean-type belts or Wilson cycle episodes of ocean opening and closing, culminating in continent-continent collision (Dewey, 1969; Wilson, 1966). In both cases, a preexisting basement is involved in the petrogenetic processes related to the production of granitoid rocks, whose isotopic signature usually indicates mixing of juvenile mantle-derived and reworked crustal components in the parental magmas.

A somewhat different model of accretionary orogens arises when intra-oceanic magmatic arcs are formed within large oceanic domains, in areas of long-lived plate convergence and B-subduction, when both convergent plates consist of oceanic lithosphere. The relicts of such tectonic setting are domains that are typically as broad as they are long (e.g., the Altaids, the Arabian-Nubian Shield, or the Svecofennian orogen), and comprise a large proportion of juvenile mafic to silicic calc-alkaline igneous rocks, as well as their sedimentary products. The main tectonic processes include the formation and collision of island arcs, oceanic plateaus, and microcontinents. "Soft" collision of these arc terranes during the accretion process will not produce extensive crustal thickening, as is seen in Himalayan-type orogens. In the absence of high mountain ranges, exhumation will be minimized, resulting in the exposure of upper to middle crustal levels exhibiting low to middle metamorphic grades. These domains will preserve such plate-tectonic signatures as island-arc assemblages, oceanic plateaus, accretionary complexes, ophiolite suites, and fold-and-thrust belts. (For a complete summary, see Kröner et al., 1987.)

In this work we will address some relevant concepts related to accretionary belts, indicating their main structural and deformational features, as well as their petrogenetic and isotopic characteristics. Next, we will examine the extensive regions of accretionary belts with large proportions of mantle-derived magmatic arcs of Proterozoic age that occur at the southwestern portion of the Amazonian Craton, or marginal to its southeastern border. Their tectonic evolution will also be reassessed, by means of the available geochronological control, with some emphasis on robust U-Pb determinations in zircon crystals. Finally, the overall geodynamic significance of these large domains will be considered, as well as their bearing for the paleogeographic reconstruction of the Proterozoic supercontinents, Rodinia and Gondwana.

STRUCTURAL FEATURES OF ACCRETIONARY BELTS

Extensive areas can be formed by stacking and lateral accretion of arc complexes produced in a collection of successive subduction zones, in processes of "soft collision and accretion", characterized by the production of large amounts of granitoid magmas and associated felsic volcanics. These regions will not contain nappe complexes imbricating older continental basement, and high-grade collisional-type regional metamorphism (see for instance Kröner et al., 1987; Şengör and Natal'in, 1996; and Johnson, 2001, for the case of the Arabian-Nubian Shield). An "orogenic collage" is thereby produced, in which deformation, regional metamorphism, and crustal growth take place in a tectonic environment of ongoing plate convergence (Cordilleran, Pacific, Miyashiro-type, Turkic-type orogens). Deformational features of accretionary belts include extensional and compressive environments during steady-state convergence (arc/backarc versus accretionary prism), and they are normally overprinted by later compressive orogenic events (Kusky and Bradley, 1999).

At present, the Circum-Pacific system of orogenic zones is the paradigm for all kinds of accretionary belts. In the western Pacific, where the relative velocities of the convergent plates indicate a series of retreating orogens, island arcs grow oceanward, producing forearc accretion and backarc basin opening, for instance in the Japanese islands (Maruyama et al., 1997). In contrast, in the southeastern Pacific, the overriding plate advances toward the downgoing plate producing terrane accretion of previously rifted arc and microcontinental ribbons, and typical tectonic features on the overriding plate, such as retro-arc fold-and-thrust belts.

Accretionary orogens may also include some features of Himalayan-type collisional and Andean-type Cordilleran belts. In particular, the processes responsible for their amalgamation, cratonization, and incorporation into continental nuclei are not well understood, especially for pre-Mesozoic orogens. For instance, cratonization of orogenic belts may occur when there is a pause in lateral accretion. Renewed subduction may then result in a more conventional, Andean-type tectonic interface between an oceanic and a continental plate, marked by voluminous calc-alkaline magmatism as well as vertical crustal accretion at the margin of the now-cratonized block. In such cases volcano-sedimentary basins may cover large areas, subsequent to the respective orogenic pulses. Moreover, because of a long-term tectonomagmatic evolution, accretionary belts usually contain postaccretionary granitoid rocks, metamorphic rocks from various tectonic regimes, late-forming volcano-sedimentary basins, and anorogenic mafic to silicic bodies. However, the evidence for accretion in deeply exhumed crustal domains is difficult to recognize, since most of the high- and medium-level rocks are eroded away. In such cases, the "sea" of deformed amphibolite- to granulite-facies migmatitic gneisses remains as roots of the tectonic features summarized above.

PETROGENETIC AND ISOTOPIC CONSTRAINTS

Large portions of material from accretionary belts exhibit typical juvenile isotopic signatures, indicating essentially mantle-derived parental magmas. This indicates a significant growth of juvenile crust (Şengör and Natal'in, 1996), mainly within intra-oceanic island arcs, with subordinate proportions being added from accreted oceanic plateaus. The related rocks show nonradiogenic initial $^{87}Sr/^{86}Sr$ and positive $\varepsilon_{Nd(t)}$ isotopic parameters. Moreover, their Sm-Nd model ages are usually slightly older than their U-Pb or Rb-Sr whole-rock isochron ages indicative of the original crystallization process. In addition, in such tectonic units there is little evidence of basement inliers of continental character.

Besides the ubiquitous differentiated volcanic rocks, such as rhyodacites and andesites, that are common in island arcs, granitoid rocks are always relevant, and in most cases they are the most common constituents of accretionary belts. Most of them belong to the calc-alkaline series, and rock types consist of a variety of differentiated I-type granitoids, such as tonalites, quartz diorites, diorites, and granodiorites, but also including highly differentiated types such as the K-rich granitoids, typical end members of the alkaline and peralkaline series, as well as syenogranites and syenites. Regardless, granitoid rocks are the most representative materials of the continental crust and may be formed through many different petrogenetic processes, whose ultimate origin is the mantle.

Mantle–Continental Crust Differentiation

The amount of material formed in mantle-crust differentiation processes depends on the thermal regime of the planet, which has been losing heat throughout geologic time. Archean granitoids, formed at a time of high-heat regime, are typically associated with greenstone belts within granite-greenstone terrains. These granites are normally mantle-derived, as indicated by juvenile-like Nd isotopic signatures. On the other hand, purely juvenile material is not widespread in Middle to Late Proterozoic and Phanerozoic times. Crustal reworking is predominant in the orogenic belts of such ages, making it difficult to identify the proportion of mantle-derived material in the granitoid complexes. For instance, if we consider the whole of granitic-type rocks formed worldwide over the last 1600 m.y., those with crustal signatures (either those that assimilated significant crustal material or those formed directly by anatexis and crustal reworking) increasingly predominate with time over granitoids with juvenile signatures.

Although Sm-Nd fractionation may be associated with a variety of petrogenetic processes (e.g., Arndt and Goldstein, 1987), the principal REE (rare earth element) fractionation occurs when oceanic crust originates from asthenospheric melting. Because Nd will concentrate relative to Sm in the crust, a depleted mantle MORB (mid-oceanic-ridge basalt)–type component relative to CHUR (chondrite uniform reservoir) will dominate the asthenospheric sources.

Continental crustal material may originate either upon melting of the upper mantle to form mafic magmas, followed by differentiation, or through two-stage successive melting, or by melting plus fractionation to form silicic rocks (Ben Othman et al., 1984). Already within the continental crust, LREE (light rare earth element) enrichment from differentiation will decrease the Sm/Nd ratio and cause the slowing down of radiogenic Nd growth.

Over geologic time, the upper mantle has evolved in such a way that the asthenosphere has become depleted in LREE relative to the Bulk Earth, and its $\varepsilon_{Nd(t)}$ signature has become strongly positive relative to the chondrite uniform reservoir. Given this evolution, positive $\varepsilon_{Nd(t)}$ values in granitoid rocks indicate mantle-derived juvenile material (e.g., De Paolo, 1988). These rocks will also exhibit Sm-Nd model ages similar to their radiometric age, as well as low initial Sr isotopic ratios and Pb (μ_1) mantle-like indicators. As a consequence, the accretionary belts within the major continents are easily identified as areas where large amounts of juvenile material are present, as indicated by positive values of $\varepsilon_{Nd(t)}$. Examples of accretionary belts with dominantly granitic rocks with juvenile signature, encompassing the entire Proterozoic and the Paleozoic, are the Birimian of West Africa, the Svecofennian of the Baltic Shield, the Mazatzal-Yavapai of North America, the Arabian-Nubian Shield in northeastern Africa, and the Altaids of central Asia. Nevertheless, mantle-derived material may be difficult to characterize solely by isotopic geochemistry. Hence, the use of $\varepsilon_{Nd(t)}$ to trace juvenile processes is not completely unequivocal, because the possible mantle sources may not be typically depleted, such as pristine mantle from plumes, mantle affected by metasomatism, or lithospheric mantle enriched by earlier recycling of crustal material.

Cratonization of Accretionary Belts

Extraction of continental crust from mantle during periodic accretion and jamming of the active subduction zones will occur over time spans of a magnitude equal to that of the major orogenic cycles, on the order of 200 m.y. From one cycle to the next, cessation of each stage of subduction would be accompanied by cratonization of the previously accreted material, and possibly by a change in the regional stresses within the lithosphere, along and inboard of the active margin. When cratonization occurs, Andean-type, "Cordilleran" magmatic arcs may be formed, and their granitoid intrusions will be aligned and emplaced in coherent belts along and over the marginal zone of the newly cratonized continental block. In addition, volcano-sedimentary basins, including differentiated volcanic rocks such as rhyodacites and andesites, may also form within extensional tectonic zones during the final collapse subsequent to the orogenic pulses of the accretionary belt.

Posttectonic ("anorogenic") granitoids, formed within an intracratonic setting, may also be quite common in accretionary belts. Multiple origins are possible, but sources of material are predominantly mantle-derived. Voluminous anorogenic magmatism may originate in a multistage process, starting with underplating of massive basaltic magmas, followed by their partial melting, producing granitic liquids, and by successive steps of fractional crystallization (Jahn et al., 2000). Mixing models, with dominance of the mantle component over crustal material, are common in the genesis of the anorogenic granites. The starting point of such a process could be lithospheric extension, causing decompressional melting in the lower mantle lithosphere producing intraplate magma. According to Windley (1993), this process may be regarded as a very late consequence of an orogeny, and not a truly anorogenic process.

A special type of anorogenic magmatism produces rapakivi-type granites, formed by heat or magma transfer from the asthenosphere to the mantle lithosphere and to the base of the crust, leading to melting and assimilation (Rämö and Haapala, 1995). Ahäll et al. (2000), considering the tectonic setting of southern Sweden, suggested a process of "hybrid synorogenic response," considering the temporal correlation of the anorogenic magmatism with subduction processes occurring at the evolving active margin. In this way, the stepwise orogenic accretionary growth would be contemporary with inboard anorogenic magmatism.

From the above, an accretionary belt could be the result of complex juxtaposition of tectonic units, including a great deal of intra-oceanic material with positive $\varepsilon_{Nd(t)}$ signatures, but also containing in places Cordilleran-type granites, collisional-type belts, microcontinents, volcano-sedimentary basins, and posttectonic to anorogenic-type complexes.

HISTORY OF THE AMAZONIAN CRATON BEFORE 2.0 Ga

Figure 1, adapted from Cordani et al. (2000), shows the main geotectonic provinces of South America as of the late Neoproterozoic. For the large Amazonian Craton, in addition to the Roraima and Xingu-Iricoumé Archean nuclei, five major Proterozoic tectonic provinces are displayed (partly covered by Proterozoic and Phanerozoic sedimentary sequences): the Maroni-Itacaiunas, Ventuari-Tapajós, Rio Negro–Juruena, Rondonian–San Ignacio, and Sunsas-Aguapeí. The boundaries between these provinces, indicated in Figure 1, correspond to important structural features, as well as to changes in the geochronological pattern of both basement rocks and sedimentary covers. In most places such boundaries must be considered only tentative, and we are aware that more geological mapping, supported by geochronological studies, is needed for a better definition.

The Amazonian Craton, with an area of ~4,400,000 km^2, is one of the largest cratons of the world. In this work we will follow the mobilistic approach initially proposed by Cordani et al. (1979) and adopted and modified by Teixeira et al. (1989), Tassinari and Macambira (1999), Tassinari et al. (2000), and recently by Tassinari and Macambira (2004). These reviews, although taking into account the regional correlations of geologic units and major structures, have been based essentially on evaluation of regional radiometric age data, because the general geological knowledge of the region remains at reconnaissance scale. For many regions, the available outcrops are limited in number by the difficult access, intense alteration, and dense vegetation. As a consequence, the available petrological, geochemical, and structural data are insufficient to produce a detailed picture for most of the area.

Available radiometric age data are heterogeneously concentrated over an immense territory. The geochronological database consists of ~2000 dated samples, and most of the ages are produced by the Rb-Sr and K-Ar methods. U-Pb (SHRIMP [sensitive high-resolution ion microprobe] and isotopic dilution on zircon), Pb-Pb (zircon evaporation), Sm-Nd (whole rock), and $^{40}Ar/^{39}Ar$ mineral ages are also available but are restricted to a few areas. Some regions, such as the Carajás mineral province, French Guyana, and parts of Rondônia and Mato Grosso States of Brazil, are relatively better known, and for these there is much better control of the tectonic evolution, as well as for the succession of magmatic pulses and metamorphic events.

We are aware that different views concerning tectonic units, or different positions for the tentative boundaries between provinces, have been reported (Santos et al., 2000; Santos, 2003). These two articles include a number of references that report a great deal of geochronological work performed in the last few years, based on U-Pb zircon SHRIMP and TIMS (thermal ionization mass spectrometry) determinations, as well as Pb-Pb evaporation. Such analyses were carried out, altogether, in a few hundred samples distributed throughout large areas of the Amazonian Craton. These results are obviously more robust and precise than the previously available Rb-Sr and K-Ar age determinations. The general conformity of these new results with ages expected from a given geochronological province is evidence of the general robustness of the earlier geochronological results. Thus, we adhere to previous conclusions regarding geochronological provinces in this contribution.

The Amazonian Craton is here subdivided into two Archean nuclei and five Proterozoic tectonic provinces, showing internally coherent structural and age patterns (Fig. 1). Tables 1 and 2 summarize the main geologic and geochronologic characteristics of these provinces.

The primitive nuclei, the Xingu-Iricoumé and Roraima blocks, make up the Central Amazonian province (CA in Fig. 1) of Cordani et al. (2000). Nevertheless, genuine Archean continental crust is restricted to the relatively large Carajás granite-greenstone terrain of the Xingu-Iricoumé block (Fig. 2), whose rocks yielded radiometric ages between 2600 and 3200 Ma (Table 1). The Roraima block occurs in the northern part of the craton. Its basement rocks are covered by extensive and practically undeformed volcano-sedimentary sequences (Surumu, Iricoumé, etc., Fig. 3), interpreted by Tassinari et al. (2000) as stable foreland deposits marginal to the Paleoproterozoic Maroni-Itacaiunas province. These sequences yield well-constrained U-Pb zircon ages, always younger than 2000 Ma.

The clastic sedimentary rocks of the Roraima Formation and the roughly contemporary Urupi Formation (locations in Fig. 3) overlie in places the Iricoumé volcanics, which are well constrained in age to ca. 1880 Ma, dated by U-Pb age determinations in interbedded tuff layers (Santos et al., 2003). Moreover, the Roraima Formation is intruded by mafic sills and dikes of the voluminous Avanavero Suite (Gibbs and Barron, 1983, 1993). A tuff interbedded with undeformed sediments correlated with the Avanavero Suite yields a Rb-Sr whole-rock isochron age of ca. 1650 Ma (Priem et al., 1973). If these ages are significant, they imply a fairly long succession of distinct continental sedimentation events under relatively stable tectonic conditions within the Roraima block of the Central Amazonian province.

The Maroni-Itacaiunas province (MI in Fig. 1) is made up of mobile belts that surround the northern and northeastern border zones of the Central Amazonian province (eastern Venezuela, Guyana, Suriname, French Guyana, and parts of Amapá,

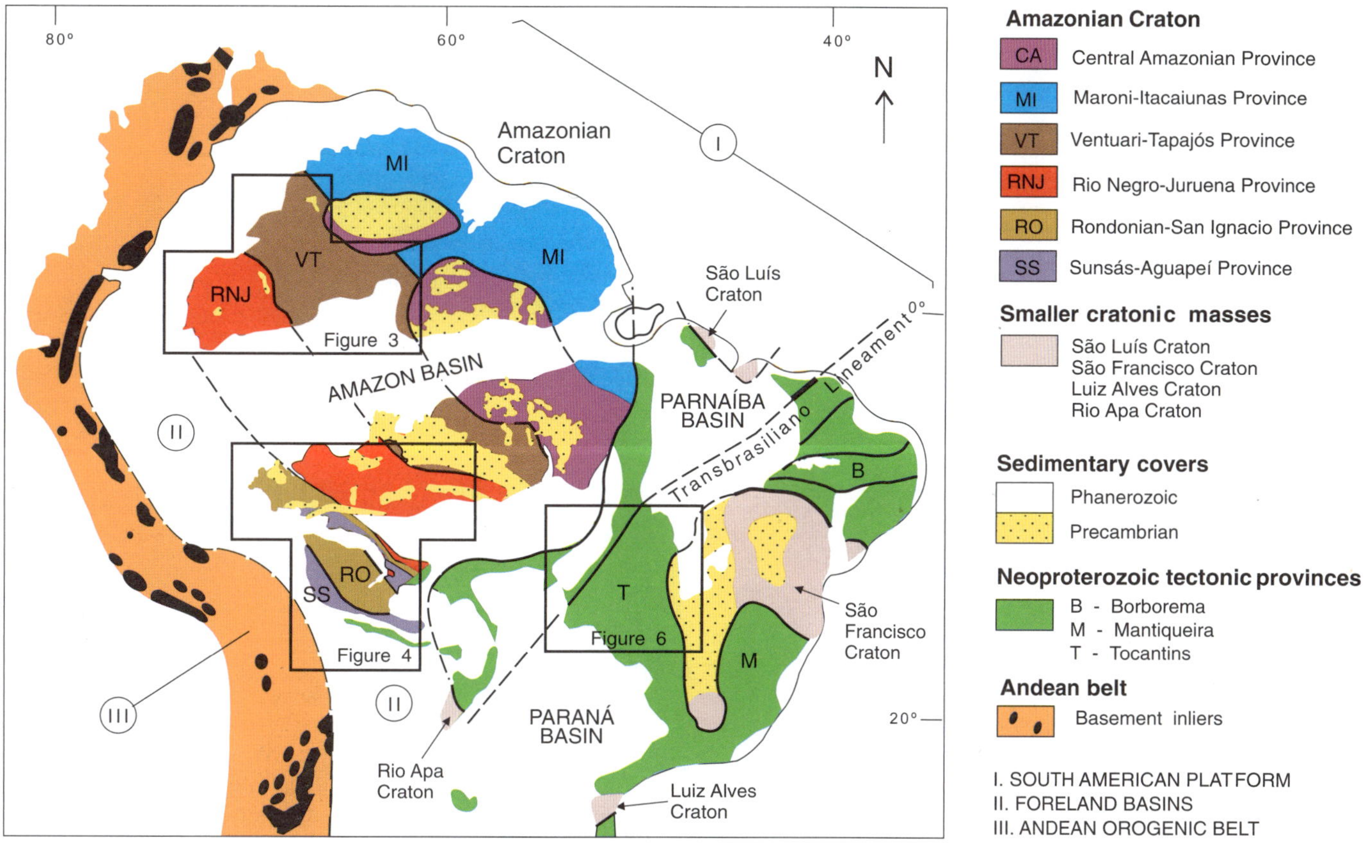

Figure 1. Main geotectonic provinces of South America as of the late Neoproterozoic, with emphasis on the Amazonian Craton and its tectonic provinces. The São Francisco Craton, the small São Luís and Luiz Alves cratonic fragments, as well as the Neoproterozoic Brasiliano/Pan-African orogens are shown in the eastern part of the figure. These belts are the Borborema province (B) in northeastern Brazil, the Tocantins province (T) in central Brazil, and the Mantiqueira province (M) in eastern and southeastern Brazil. The Andean belt is indicated in the figure, with its main Precambrian inliers. Adapted from Cordani et al., 2000. Locations of Figures 3, 4, and 6 are shown.

Roraima, Pará, and Amazonas States of Brazil). It consists of metavolcano-sedimentary sequences associated with juvenile calc-alkaline granitoid rocks (Table 1), with U-Pb zircon crystallization ages between 2050 and 2250 Ma, and positive $\varepsilon_{Nd(t)}$ values. It was affected by the widespread Transamazonian orogeny, initially defined by Hurley et al. (1967), which produced different generations of granitoids with ages down to ca. 1950 Ma. Large parts of this province comprise Paleoproterozoic greenstone belts consisting of mantle-derived material (Gibbs and Barron, 1993; Gruau et al., 1985; McReath and Faraco, 2006), while restricted parts are likely to be recycled older crust. However, a detailed overview of Maroni-Itacaiunas rocks in French Guyana, based on ~100 new Pb-Pb and Sm-Nd isotopic data (see Delor et al., 2003; Avelar et al., 2003, and references therein) confirms a complex multistage crustal growth for this province, with both juvenile Paleoproterozoic accretion (some $\varepsilon_{Nd(t)}$ values up to +3.4) and some recycled Meso- to Neoarchean materials. Regional cooling after the Transamazonian orogeny took place between 1900 and 1800 Ma, as indicated by the widely distributed K-Ar apparent ages of micas from the country rocks (Teixeira et al., 1989; Tassinari et al., 2000, and references therein).

High-grade metamorphic rocks are described from the Maroni-Itacaiunas province, some having originated under ultrahigh-temperature conditions (e.g., Bakhuis felsic granulites), indicating the importance of Paleoproterozoic collisional episodes (Tassinari et al., 2004, and references therein). These granulites extend southwards to the state of Roraima, Brazil (see Fig. 2), and include the Falawatra and Kanuku Complexes as part of the WSW-ENE–trending Central Guyana granulite belt. Their radiometric ages are 2050–2020 Ma and 1940–1850 Ma (Priem et al., 1978), and their Sm-Nd T_{DM} model ages are only generally older, closer to 2300 Ma (Ben Othman et al., 1984). These ages are clearly different from the typical ages of the Transamazonian orogeny, suggesting a slightly younger, Paleoproterozoic collisional event that may have taken place during a late stage of orogenic assembly. Moreover, the high-grade belt was intruded by anorogenic granitoids (Anorthosite-Mangerite-Charnokite-Granite [AMCG] Suite) at ca. 1560–1520 Ma (Gaudette et al.,

TABLE 1. MAIN GEOLOGIC AND GEOCHRONOLOGIC CHARACTERISTICS IN THE CENTRAL AMAZONIAN (CA) AND MARONI-ITACAIUNAS (MI) PROVINCES, NORTHEASTERN AMAZONIAN CRATON

General characteristics of the province	Main tectonometamorphic events	Regional geological and tectonic features	Anorogenic complexes
(CA; >2.6 Ga) Archean nuclei with cratonic covers Stable tectonic domain for the MI belt	Paleoproterozoic reactivation in many regions Shear zones with thermal activation related to the Nickerie/K'Mudku overprint (ca. 1000 Ma)	Xingu-Iricoumé and Roraima blocks: Archean and Paleoproterozoic supracrustal sequences (e.g., Carajás granite-greenstone terrain) Neoarchean and Paleoproterozoic granitoids Undeformed volcano-sedimentary covers (Uatumã, Iricoumé, Iriri, Rio Fresco) Continental sedimentary covers (Roraima and Urupi Formations)	Granitic intrusions (<1.9 Ga) Avanavero mafic sills and dikes (1.65 Ga)
(MI; 2.25–2.05 Ga) Paleoproterozoic mobile belts including magmatic arcs Transamazonian orogeny Regional cooling at 1.90 Ga Stable foreland for the accretionary belts	Regional Transamazonian metamorphism Juxtaposition of the Imataca terrane Central Guyana granulite belt Nickerie/K'Mudku overprint: regional shear zones, including reactivation of the Guri and Takutu faults	Extensive granite-greenstone terrain (e.g., Inini belt in French Guyana) Granulite complexes (Bakhuis, Falawatra, and Kanuku) Calc-alkaline granitoids Tectonomagmatic belts with Archean inliers (Amapá Block)	Granitoid intrusions (AMCG type; 1.56–1.52 Ga)

Note: See text for details. AMCG—anorthosite-mangerite-charnokite-granite.

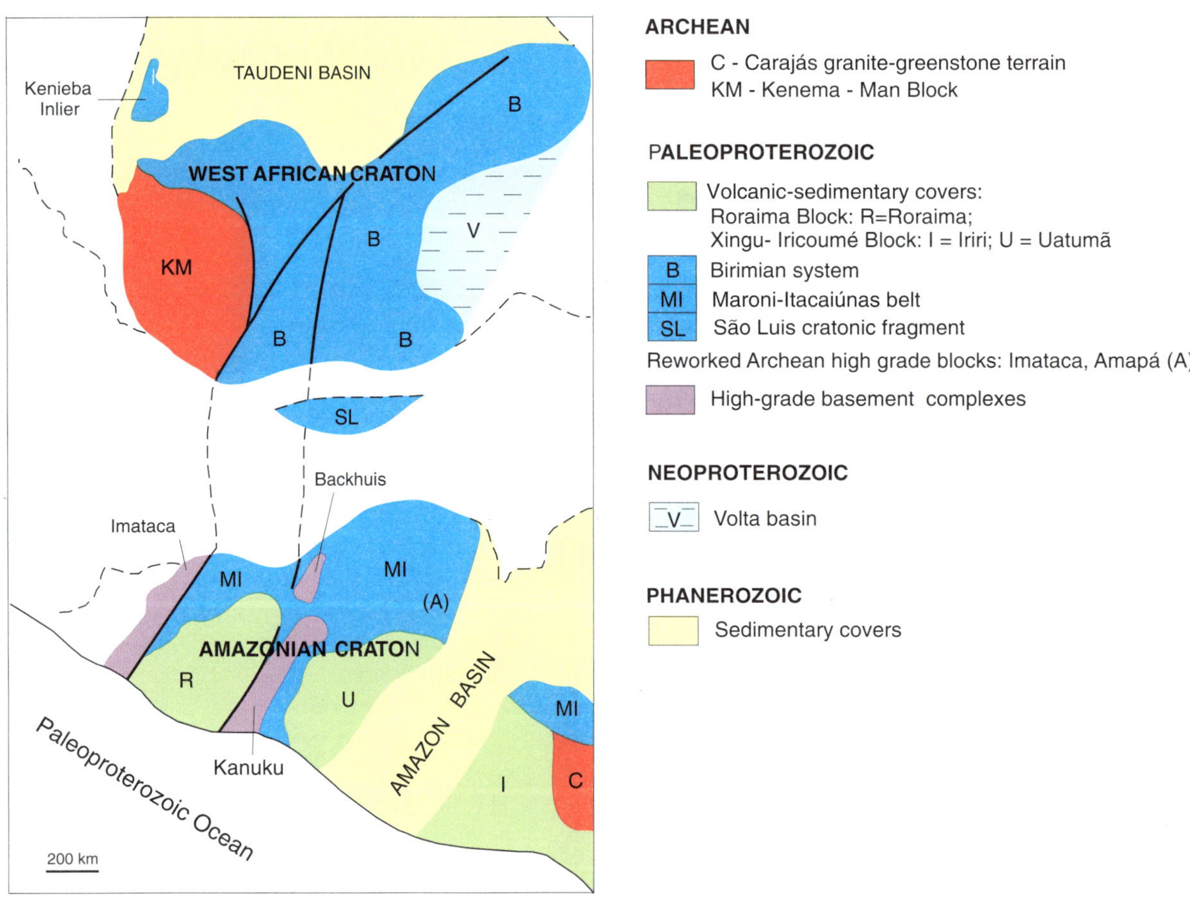

Figure 2. Reconstruction of the relative positions of the ancient nuclei of West Africa and northern South America in Paleoproterozoic times. Adapted from Nomade et al., 2003

TABLE 2. MAIN GEOLOGIC AND GEOCHRONOLOGIC CHARACTERISTICS IN THE PROTEROZOIC TECTONIC PROVINCES, SOUTHWESTERN AMAZONIAN CRATON

General characteristics of the province	Main tectonometamorphic events	Regional geological and tectonic features	Intrusive complexes
VT; 1.98–1.81 Ga			
Accretionary belts including roots of juvenile magmatic arcs Agglutination to Paleoproterozoic foreland Regional cooling at 1.80 Ga	Post-Transamazonian metamorphic episodes Tectonic reactivation related to stacking of the marginal arcs Rift structures associated with volcano-sedimentary covers	Calc-alkaline and sub-alkaline granite-gneissic complexes and granitoids (Parauari; 1.88 Ga) Low-grade supracrustal sequences (e.g., Jacareacanga Suite) Platform-type sedimentary rocks (Quasi Roraima) Foreland deposits (e.g., Beneficente basin) Felsic volcanism (Teles Pires; 1.76 Ga)	Posttectonic granites: Matupá and Maloquinha (1.87–1.86 Ga) Anorogenic rapakivi plutonism: El Parguaza, Surucucus, Auaris (1.56–1.55 Ga) Mafic and alkaline complexes
RNJ; 1.78–1.55 Ga			
Accretionary belts, with plutonic recurrent pulses Regional cooling at ca. 1.55–1.50 Ga	Medium-grade polycyclic metamorphism RSI metamorphic overprint near boundary, tectonic reactivations contemporary with the Sunsas orogeny Rift structures associated with volcano-sedimentary covers	Alto Jauru granite-greenstone terrain (1.79–1.75 Ga), coeval granitoid suites (e.g., São Romão and São Pedro; 1.78 Ga) Medium- to high-grade gneissic associations: tonalitic gneisses (1.75–1.73 Ga), granitic to charnockitic gneisses (1.60–1.53 Ga), Cachoeirinha calc-alkaline arc (1.59–1.52 Ga) Volcano-sedimentary sequences: Roosevelt (1.76–1.75 Ga), Mutum-Paraná (1.75 Ga), Comemoração (1.69 Ga), Caiabis and Dardanelos	Late-tectonic granitoids: Içana and Uaupés (1.54–1.51 Ga) Anorogenic bimodal plutonism: Serra da Providência Intrusive Suite (1.60–1.53 Ga), Rio Branco Suite (1.47–1.42 Ga) Rondonia Tin Province (Younger Granites of Rondonia): Oriente Novo (1.08 Ga), Santa Barbara (0.98 Ga), Santa Clara (1.08–1.07 Ga), São Domingos and Guapé (0.98–0.92 Ga)
RSI; 1.55–1.30 Ga			
Collisional orogeny involving a possible microcontinent combined with accretionary domains Voluminous syn- to posttectonic plutonism Regional cooling at ca. 1.32 Ga–1.02 Ga Foreland for the Sunsas orogeny	Regional metamorphism of the San Ignacio orogeny (1.35 Ga) Tectonic reactivation, deformation, and thermal overprint of the Sunsas orogeny: shear zones (Ji-Paraná), mylonitic belts, rift basins	Lomas Maneches Complex (1.69–1.66 Ga) and Chiquitania paragneisses Accretionary domains: Rio Crespo Intrusive Suite (1.50 Ga), Rio Alegre Complex (1.51–1.48 Ga), Santa Helena calc-alkaline batholith (1.45–1.42 Ga), Colorado Metamorphic Suite (1.36–1.30 Ga), Pensamiento granitoid complex (e.g., San Rafael granite; 1.36–1.30 Ga) Low-grade metasedimentary sequence (Tunui Group; 1.32 Ga) Pacaás Novos, Uopione, Aguapeí, and Nova Brasilândia rift basins (1.21–1.12 Ga) Nova Brasilândia medium- to high-grade belt (1.10 Ga)	Late- to posttectonic plutonism: Santo Antônio (1.41 Ga), Teotônio (1.39 Ga), Alto Candeias (1.34 Ga), São Lourenço–Caripunas (1.31 Ga) Anorogenic plutonism (Rondonia Tin Province): Maçangana (0.99–0.98 Ga) Mafic volcanism (1.00 Ga) Kimberlite plugs
SA; 1.25–1.00 Ga			
Collisional orogeny (Vibosi and Sunsas metamorphic belts) Final cratonization, regional cooling and exhumation: 1.00–0.92 Ga	Low-grade metamorphism Syntectonic granitoid intrusions Rift basins, aulacogen-type structures (Aguapeí; 1.16 Ga) Thrust and shear belts (e.g., Santa Catalina shear zone and Rio Negro Front)	Low- to medium-grade metamorphic sequences: Sunsas Belt Coeval syn- to late-tectonic granitoid plutons (1.10 Ga) Flat-lying sedimentary sequences (e.g., Huanchaca and Aguapeí)	Posttectonic plutonism: Taperas granite (1.07 Ga), São Domingos and Guapé (0.98–0.92 Ga) Rincón del Tigre basic complex (0.99 Ga)

KEY: VT—Venturi-Tapajós; RNJ—Rio Negro–Juruena; RSI—Rondonian–San Ignacio; SA—Sunsas-Aguapeí. See text for details.

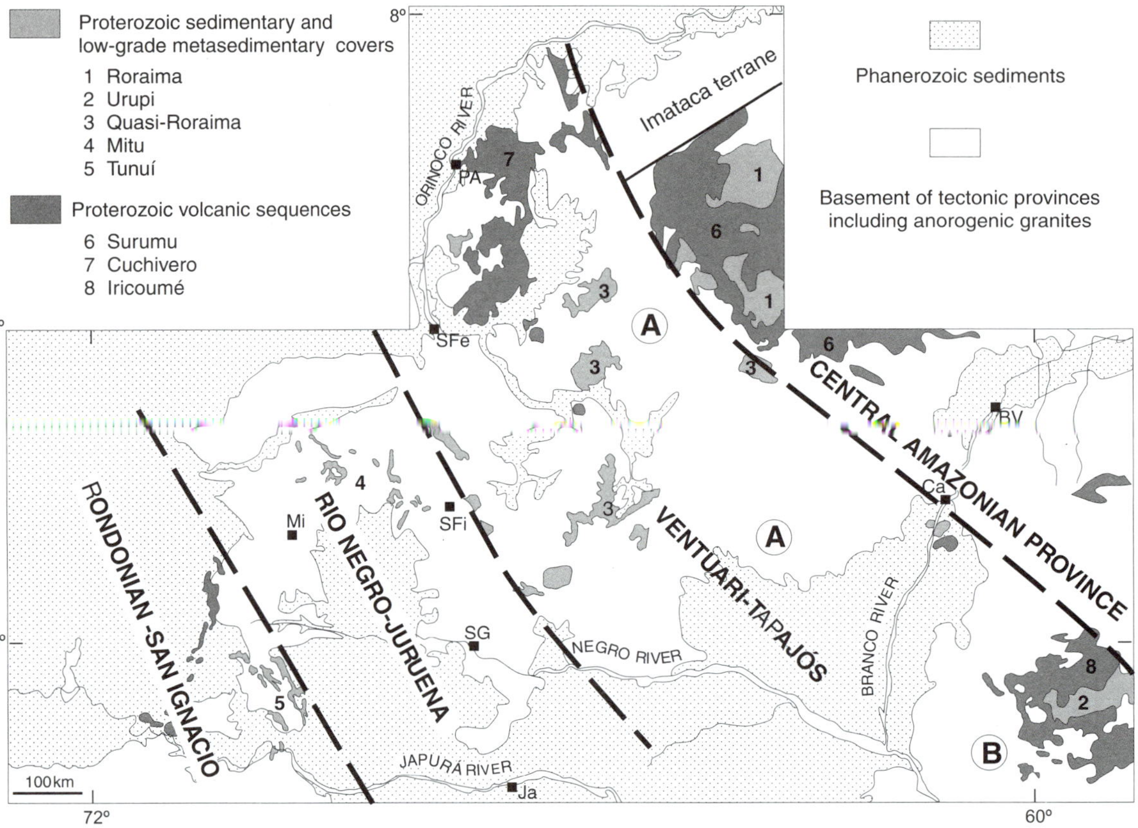

Figure 3. Tentative boundaries of the Proterozoic accretionary belts, north of the Amazon sedimentary basin. The principal volcanic and sedimentary sequences overlying the tectonic provinces are indicated in the figure. Urban settlements: BV—Boa Vista; Ca—Caracaraí; Ja—Japurá; Mi—Mitu; PA—Puerto Ayacucho; SFe—San Fernando de Atabapo; SFi—San Filipe; SG—São Gabriel da Cachoeira. (A) and (B) are specific geographic domains with predominance of basement rocks, discussed in the text.

1996; Reis et al., 2003) and was also affected by strong deformation associated with shearing and low-grade metamorphism at ca. 1100–1300 Ma (Gibbs and Barron, 1993). These geologic features are related to cataclastic belts and megashear zones that are widespread over large parts of the Guyana Shield and are designated as the Nickerie or K'Mudku event (Snelling and McConnell, 1969; Priem et al., 1971b).

Archean fragments are also present in this Paleoproterozoic province, such as the Imataca terrane at its western corner (Fig. 2), which may represent an allochthonous crustal fragment comprising Archean high-grade metamorphic rocks. Its rocks yielded SHRIMP U-Pb ages between 3200 and 2800 Ma. This terrane was probably juxtaposed against the province during the Transamazonian orogeny, along the Guri megafault zone, where $^{40}Ar/^{39}Ar$ ages of 1400–1200 Ma indicated a regional thermal episode of reactivation related to the K'Mudku event (Onstott et al., 1989). High-grade metamorphism (750–800 °C and 6–8 kbar) and crustal reworking at ca. 2000 Ma, in association with transpressive shearing and thrusting, was documented by Tassinari et al. (2004).

In addition, other inliers with Archean ages occur throughout the province, such as the Cupixi and Tartarugal Grande areas within Amapá State in Brazil (Tassinari and Macambira, 2004). More recent work by Rosa-Costa et al. (2006) greatly extended this Archean inlier to the entire southern portion of Amapá, an area more than 300 km long, and renamed it "Amapá block." These authors indicated the original Meso- to Neoarchean ages of the rocks, obtained through U-Pb zircon evaporation ages, and the long-lived regional tectonic evolution, including reworking and crustal accretion in the Paleoproterozoic, as well as the intrusion by several Transamazonian granites.

An interesting synthesis of the polycyclic history of the northwestern region of the Amazonian Craton has been made on the basis of U-Pb SHRIMP dating of 49 detrital zircon crystals from a sand sample taken from the Orinoco River, where it flows over the eastern edge of the Imataca block (Goldstein et al., 1997). Without taking into account the young Phanerozoic crystals coming from the Andes, the U-Pb apparent ages reflect individual magmatic or metamorphic events, and the resulting histogram shows discrete age groupings at 2800, 2100, and

1100–1200 Ma. The two older peaks are representative of the most important periods of crust formation within the Guyana Shield. The relatively large Grenvillian-like population of zircons may indicate derivation from basement inliers within the Colombian and Venezuelan Andes (Cordani et al., 2005).

In a broader context, the Maroni-Itacaiunas province correlates well with the Birimian system in West Africa (Vidal et al., 1996), affected by the Paleoproterozoic Eburnean orogeny. Taking this into account, the different sectors of the province seem to be a result of amalgamation of at least four large continental masses: the already mentioned Xingu-Iricoumé and Roraima blocks, the Kenema-Man block of the West African Craton, and the Imataca terrane. Nomade et al. (2003) presented the results of a paleomagnetic study performed on granitic and metavolcanic rocks of Paleoproterozoic age from French Guyana and the Ivory Coast. These authors conclude that the northern part of the Amazonian Craton and the southern part of the West African Craton belonged to the same continental mass by ca. 2000 Ma, but were separated before that time. Figure 2, adapted and modified from Nomade et al. (2003), is a reconstruction of the proposed "protocratonic mass" in existence after the Transamazonian orogeny. The large oceanic domain that must have occurred to the southwest of such continental mass is also indicated in that figure.

THE PROTEROZOIC ACCRETIONARY BELTS OF THE AMAZONIAN CRATON

The position of the Proterozoic tectonic provinces (Ventuari-Tapajós, Rio Negro–Juruena, Rondonian–San Ignacio, and Sunsas-Aguapeí) that cover the entire southwestern half of the Amazonian Craton is indicated in Figure 1. Figures 3 and 4, located to the north and to the south of the Amazon sedimentary basin, respectively, include a more detailed view of the Proterozoic tectonic provinces. The proposed boundary of the accretionary systems is indicated, as well as the volcano-sedimentary covers established upon them after cratonization. However, their "basement" is not subdivided, because we recognize that geological mapping and geological research still has not progressed to an adequate stage for a better characterization. For instance, although it is known that the entire region is formed mainly by granitic rocks, sensu lato, it is still not possible, at present, to make a clear distinction between "basement granitoids" and syn-, late-, and posttectonic, or anorogenic granitoid intrusions of different type. More robust and precise geochronological work is needed.

Legends are the same for Figures 3 and 4. Regrettably, most of the sites mentioned in the text, as well as the locations of many geologic bodies considered in this work, cannot be shown in these figures, and the interested readers are kindly invited to consult the indicated references. The main purpose of both figures is to show the general areas in which some control for the existence of juvenile material is available, employing Sm-Nd analyses carried out on isolated samples or groups of samples. As previously mentioned, the positive $\varepsilon_{Nd(t)}$ values, mostly derived from granitoids and felsic volcanic rocks, are interpreted to indicate mantle-derived material accreted to the crust by B-subduction of oceanic lithosphere. In this work, for the Proterozoic accretionary belts of the Amazonian Craton, we envisage and will describe the establishment of specific tectonic settings typical of oceanic domains, including the formation and stacking of intra-oceanic magmatic arcs.

Beginning at ca. 2.0 Ga, a series of successive magmatic arcs began to be accreted along the southwestern margin of the Central Amazonian province, producing the juvenile material of the Ventuari-Tapajós and the Rio Negro–Juruena tectonic provinces (Fig. 1). It is noteworthy that the relatively large continental masses colliding on the northern side during the so-called Transamazonian orogeny (producing the Maroni-Itacaiunas belt) were matched by the contemporaneous subduction of oceanic lithosphere and juvenile magmatic arcs along the opposite side of the Central Amazonian province. The continued and mainly intraoceanic soft-collision/accretion process produced a very large "basement" domain in which granites (sensu lato), gneisses, and migmatites predominate, at least 2700 km long and ~1000 km wide, which started with the formation of the magmatic arcs of the Ventuari-Tapajós province.

The Ventuari-Tapajós Province

Felsic volcanics and granitic rocks (sensu lato), formed essentially between 2000 and 1800 Ma (Table 2), are the main constituents of the Ventuari-Tapajós province. Most of them exhibit juvenile isotopic signatures, with positive $\varepsilon_{Nd(t)}$ values indicating their formation as magmatic arcs within oceanic domains. Moreover, these rocks have Sm-Nd model ages only slightly older than their U-Pb or Rb-Sr radiometric ages. There is no evidence of Archean basement inliers within the Ventuari-Tapajós province, and regions with high-grade metamorphics are absent or restricted. In our view, geologic evidence, coupled with the isotopic constraints, are consistent with a scenario of successive amalgamation of intra-oceanic magmatic arcs, essentially mantle-derived, of which the roots are now exhumed.

The main isotopic features of the Ventuari-Tapajós province can be divided according to three geographic domains (A and B in Fig. 3; C in Fig. 4. In Venezuela and northern Brazil (Fig. 3, A domain), calc-alkaline granite-gneiss complexes dominate, with U-Pb and Rb-Sr ages between 1980 and 1830 Ma (Tassinari et al., 1996; Wynn et al., 1993). Volcanic associations are also present. These rocks exhibit NW-SE structural trends, truncating the older NE-SW structures of the Imataca block in the area to the northeast of Puerto Ayacucho (Fig. 3). Both volcanic and plutonic rock types are mostly juvenile, yielding positive $\varepsilon_{Nd(t)}$ values between +0.70 and +3.05 (Sato and Tassinari, 1997; Tassinari et al., 1996).

Within this domain, a very large area is dominated by anorogenic granites with rapakivi textures, such as the El Parguaza, Surucucus, and Auaris plutons. In the regional geologic maps, the

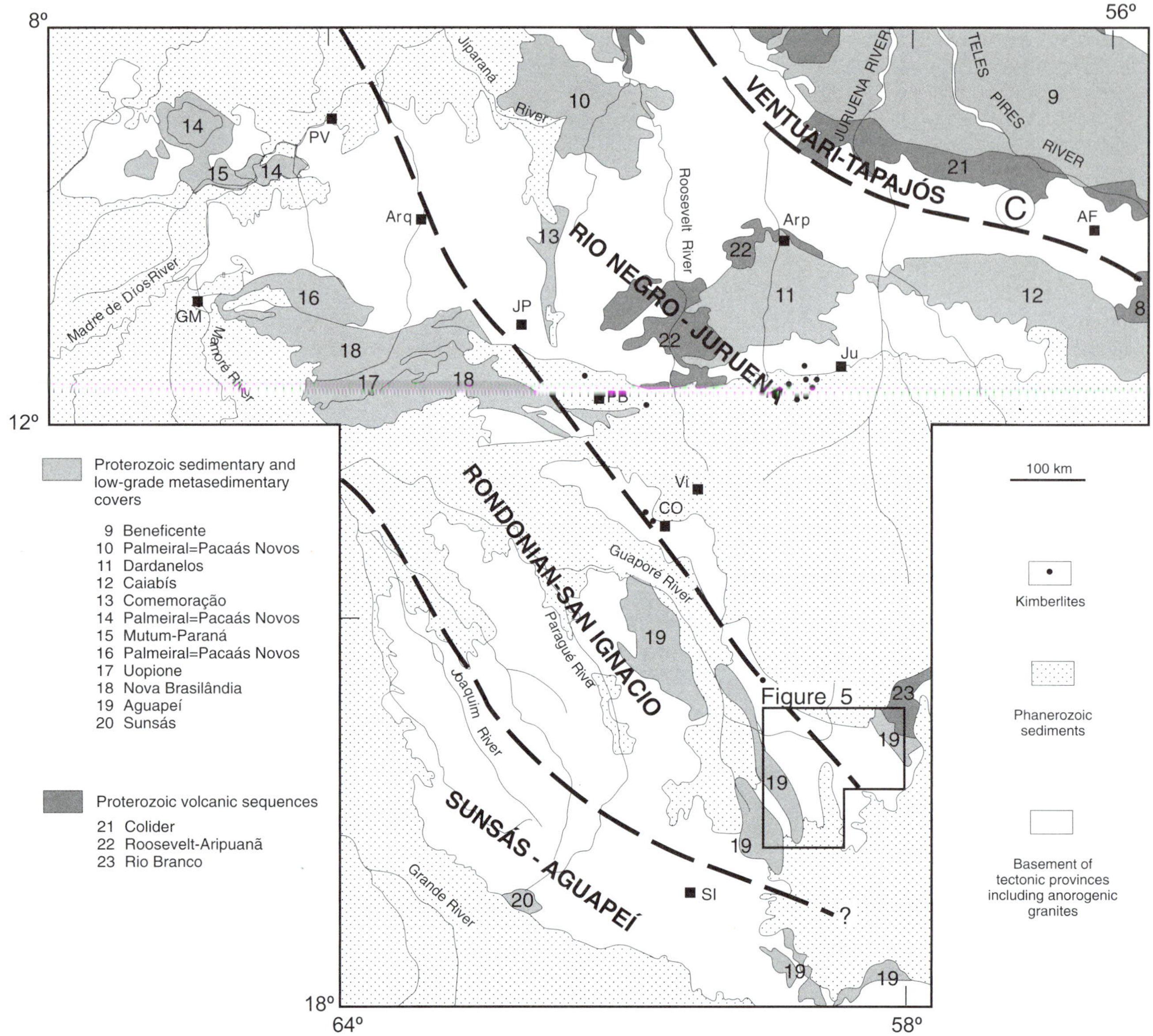

Figure 4. Tentative boundary of the Proterozoic accretionary belts, south of the Amazon sedimentary basin. The principal volcanic and sedimentary sequences overlying the tectonic provinces are indicated in the figure. Urban settlements: AF—Alta Floresta; Arp—Aripuanã; Arq—Ariquemes; CO—Colorado do Oeste; GM—Guajará Mirim; JP—Ji-Paraná; Ju—Juina; PB—Pimenta Bueno; PV—Porto Velho; SI—San Ignacio; Vi—Vilhena. Location of Figure 5 is shown. (C) Specific geographic domain with predominance of basement rocks, discussed in the text.

El Parguaza granite is a huge batholith for which a U-Pb zircon age of ca. 1550 Ma is reported (Tassinari et al., 1996). Santos (2003) reported a precise U-Pb SHRIMP age of 1551 ± 5 Ma for the Surucucus pluton. Several areas are marked by undeformed sedimentary cover rocks previously correlated with the Roraima Group (Fig. 3). Reis et al. (2003) reported a maximum age of ca. 1550 Ma for one of these rocks (U-Pb SHRIMP age of detrital zircon crystals), and included all of such sedimentary units as belonging to the Quasi Roraima Group.

The best control for age and geochemical features of the Ventuari-Tapajós province is found in the B and C domains (Figs. 3and 4). Distinct types of granitoids occur within domain B, located north of Manaus, yielding Pb-Pb and U-Pb SHRIMP zircon ages of ca. 1970 Ma, cut by late- to posttectonic plutons with U-Pb ages from 1890 to 1810 Ma (Santos, 2003). In the area between the Tapajós and Jamanxin Rivers, (domain C) occurs a succession of calc-alkaline to sub-alkaline rocks with $\varepsilon_{Nd(t)}$ values between +1.3 and +2.1 (Santos, 2003;

Santos et al., 2004). These rocks (Table 2) were formed within juvenile magmatic arcs between 1970 and 1880 Ma, as indicated by Pb-Pb and U-Pb SHRIMP ages. Posttectonic granites in this area, typified by the Maloquinha granite and the felsic volcanics of the Uatumã Group, yield ages of 1870–1860 Ma (Santos et al., 2004).

Similar Paleoproterozoic granites were studied by Lamarão et al. (2005) in the region of Riozinho do Anfrísio, to the north of the Ventuari-Tapajós province, well within the area covered by the felsic volcanics of the Iriri Formation, in the Xingu-Iricoumé block, Central Amazonian province (Table 1). They exhibit negative $\varepsilon_{Nd(t)}$ values between –0.7 and –5.2, and were interpreted by the authors, among other alternatives, as related to subduction in a Andean-type tectonic setting, indicating source magmas bearing an important juvenile component but contaminated by an older continental substratum. Moreover, the nondeformed Maloquinha-type granites in the same area contain inherited Archean-age zircons (Lamarão et al., 2002), indicating their at least partial crustal derivation, and reflecting heritage from ancient basement rocks belonging to the cratonic area of the Central Amazonian province.

In the southern part of the Ventuari-Tapajós province, near the town of Alta Floresta (Fig. 4), domain C contrasts with the A and B domains mainly by the occurrence of the greenschist-type supracrustal sequences of the Jacareacanga Suite. Santos (2003) reported some U-Pb ages from the adjacent Parauari granitoids intrusive into the country rocks and from detrital zircons. The ages were quite similar, within the 1900–2100 Ma interval. In the easternmost portion of the area (just outside Fig. 4), detrital zircon grains from a very low-grade metasedimentary sequence were dated by the U-Pb SHRIMP method, indicating a maximum depositional age of ca. 2080 Ma, as well as inheritance of Archean zircon grains up to 3100 Ma in age. In the same area, the 1872 ± 12 Ma intrusive Matupá granitic massif (Moura and Botelho, 2002) yielded a $\varepsilon_{Nd(t)}$ signature of about –3.0, suggesting reworking of basement and therefore a possible proximity to an ancient continental nucleus such as the Xingu-Iricoumé block.

Several large rifts and similar structures (not shown in Figures 3 and 4), associated with development of cratonic volcano-sedimentary basins and aulacogens, are found over the entire Ventuari-Tapajós province. Mafic and alkaline intrusions may also occur, related to these tectonic features, which are most likely formed by extensional tectonism, generally following the stacking of individual accretionary arcs.

After a period of ~200 m.y. during which active subduction produced the described succession of accretionary zones, we envisage temporary cessation of the process and the cratonization of the previously accreted material of the Ventuari-Tapajós province. This accretion at ca. 1800 Ma is delimited by the proposed boundary with the younger province, shown in Figures 3 and 4. In Figure 3, because of the general lack of structural and geochronological control, the boundary is located at the southernmost extension of the Quasi Roraima–type sedimentary cover.

The southern boundary is placed along what appears to be a major tectonic zone comprising a series of large overthrusts with WNW-ESE trends, located south of the town of Alta Floresta (Fig. 4). In that region, close to the Serra do Cachimbo, the Teles Pires Group, a practically unmetamorphosed volcano-sedimentary sequence bounded by extensional structures and located over the southernmost part of the Ventuari-Tapajós province, constrains the time of the cratonization stage. It comprises volcanic and plutonic rocks and underlies the very large Beneficente sedimentary basin, the age of which is bracketed between 1700 and 1400 Ma according to Tassinari et al. (2000).

Near Aripuanã, some shallow intrusives and extrusive felsic rocks of the Teles Pires Group exhibit gradational contacts. Zircon crystals from these rocks were dated by U-Pb SHRIMP (Neder et al., 2002), and yielded ages between 1762 and 1755 Ma. Their $\varepsilon_{Nd(t)}$ signatures showed a range between zero and –2. The Teles Pires–Beneficente volcano-sedimentary cover is located very close to the proposed boundary between the Ventuari-Tapajós and Rio Negro–Juruena provinces, and its geologic evolution is considered to reflect the main tectonic events of the latter province.

The Rio Negro–Juruena Province

A short time after the cratonization of the Ventuari-Tapajós province, subduction was resumed, initiating a second accretionary cycle, characterized by radiometric ages of granitoid rocks in the 1780–1550 Ma interval. The oldest ages are found in the region south of Alta Floresta, near the previously mentioned boundary with the province. In that region, some U-Pb SHRIMP zircon dates were obtained by Santos (2003) in granitic rocks of the São Romão and São Pedro Suites, which seem to characterize a magmatic arc of ca. 1780 Ma, formed adjacent to a previously cratonized region or cutting through the marginal zone within an Andean-type tectonic setting.

The Rio Negro–Juruena province is characterized by ages between 1780 and 1550 Ma, with a general younging of accretionary wedges from northeast to southwest (Tassinari et al., 1996). Like the Ventuari-Tapajós, the Rio Negro–Juruena province is composed predominantly of granite-gneiss and granitoid rocks, with positive to slightly negative $\varepsilon_{Nd(t)}$ signatures, roughly between +4.0 and –2.0, suggesting that juvenile accretionary events played a major role in their tectonic evolution. However, large parts of this province are poorly controlled by U-Pb geochronology and Nd isotope constraints.

Figure 5 shows the geologic setting of the Alto Jauru granite-greenstone terrain, at the southeastern end of the Rio Negro–Juruena province, where the geochronological pattern (see Table 2) is well constrained (Pinho et al., 1997; Geraldes et al., 2001, 2004b). The oldest rocks are the metavolcano-sedimentary sequences and the granite-gneissic rocks of the Alto Jauru Complex, yielding U-Pb zircon ages of 1750–1790 Ma and positive $\varepsilon_{Nd(t)}$ values between +2.2 and +2.6. In addition, toward the northwestern portion of Mato Grosso State, basement granitoid rocks yield whole-rock Rb-Sr and Pb-Pb ages of ca. 1700 Ma (Tassinari et al., 1996), with low initial Sr ratios and Pb μ_1 values of 8.1. The granitic rocks of the

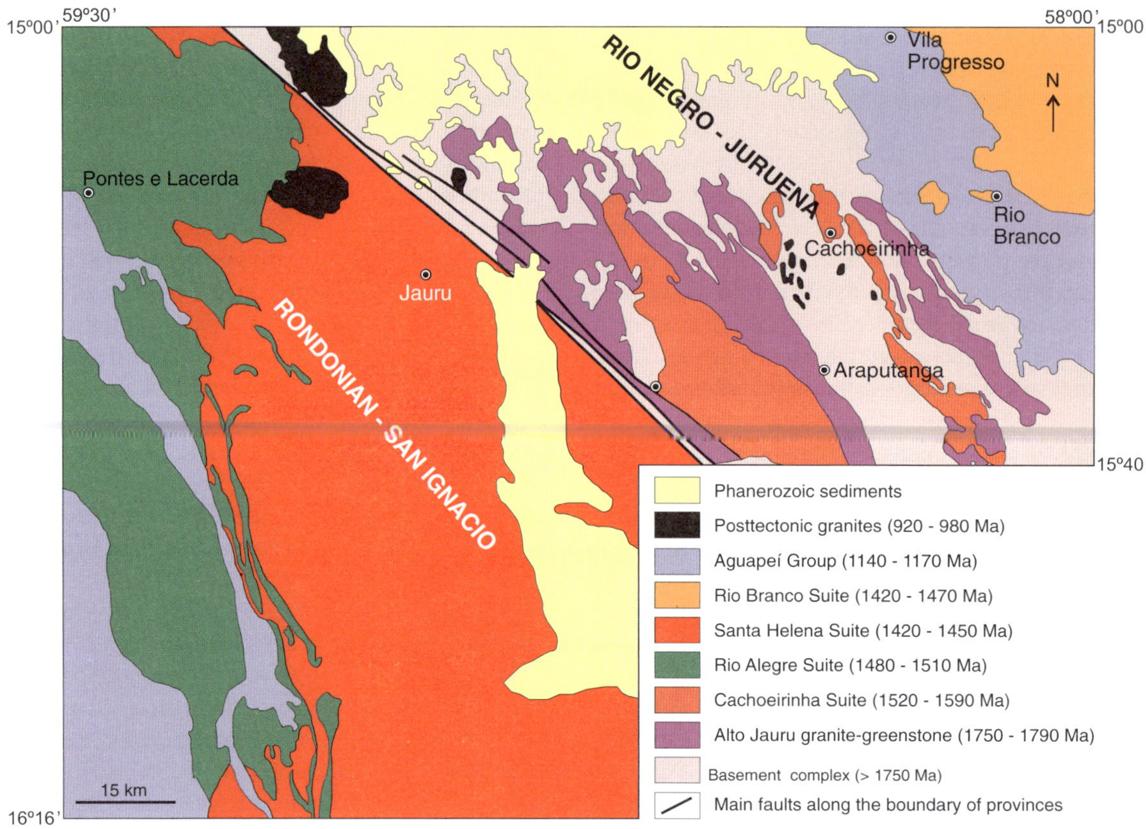

Figure 5. Simplified geological map of the Jauru region, western part of Mato Grosso State of Brazil, near the Bolivian border. Adapted from Geraldes et al., 2001.

Cachoeirinha magmatic arc are intrusive into the Alto Jauru granite-greenstone terrain, with U-Pb zircon ages between 1587 and 1522 Ma and $\varepsilon_{Nd(t)}$ values between –0.8 and +1.0. This geochronological framework can be roughly compared with that of the extreme northwest of the Rio Negro–Juruena province (border between Brazil and Venezuela, Fig. 3). There, conventional U-Pb and Rb-Sr ages between 1860 and 1810 Ma are available for calc-alkaline granitoids making up the basement complex (Tassinari et al., 1996; Gaudette and Olzewski, 1981), while U-Pb SHRIMP zircon ages around 1510–1540 Ma were obtained for younger granitic intrusions such as the Içana and Uaupés Suites (Santos et al., 2000).

In east-central Rondonia (Fig. 4), some U-Pb TIMS and SHRIMP zircon dates were obtained by Tassinari et al. (1996) and by Payolla et al. (2002) in the regional gneissic rocks. These results were grouped according to three distinct lithologic associations, with ages (1) ca. 1750 Ma, (2) between 1600 and 1530 Ma, and (3) ca. 1430 Ma. The oldest of these rocks, tonalitic gneisses and enderbitic granulites, are related to the development of the Rio Negro–Juruena province, exhibit calc-alkaline affinities, and their $\varepsilon_{Nd(t)}$ values range from –1.5 to +0.1, indicating a substantial contribution from juvenile magmatic source protoliths. The intermediate ones are more potassic and include the Serra da Providência Intrusive Suite, with its charnockitic and mangeritic granitoids, some of them with rapakivi texture. They occur in the southern part of the region near the border with the Rondonian–San Ignacio province, and exhibit $\varepsilon_{Nd(t)}$ values between –0.6 and +2.0, suggesting derivation from a mixture of predominantly juvenile sources and a slightly older crust. The youngest lithologies include fine-grained gneisses and granulites (Bettencourt et al., 1999a). Detrital zircons from a paragneiss within this region yielded U-Pb SHRIMP ages down to 1670 Ma, indicating that their provenance included sources from rocks of both the Ventuari-Tapajós and Rio Negro–Juruena provinces (Payolla et al., 2002).

As with the Ventuari-Tapajós province, volcano-sedimentary covers, practically undeformed and of a very low metamorphic grade, can also be found in the Rio Negro–Juruena province, filling rift-type or aulacogen-type basins (Figs. 3 and 4). Along the Juruena and Roosevelt Rivers, well into the province, felsic volcanics considered coeval with the Teles Pires Group were dated by Santos et al. (2004) to between 1773 and 1786 Ma (U-Pb

SHRIMP zircon ages), and showed $\varepsilon_{Nd(t)}$ values between –1.4 and +0.6. In the southern part of the province, the Roosevelt-Aripuanã volcano-sedimentary sequence occurs, from where a dacite yielded a U-Pb zircon age of 1740 Ma (Santos et al., 2000). In central Rondônia, the Comemoração volcano-sedimentary sequence crops out along the border of the Serra da Providência Intrusive Suite. Some zircon crystals extracted from acid-ash tuffs at the base of the sequence yielded a conventional U-Pb zircon age of 1690 Ma (concordia plot, upper intercept), indicating a maximum age for the sedimentation. Moreover, the low-grade Mutum-Paraná volcano-sedimentary formation, located close to Porto Velho, was dated using the U-Pb SHRIMP method in zircon grains from interbedded tuff layers, with an age of 1746 ± 4 Ma (Santos, 2003). Other sequences of probably similar tectonic origin, such as the Caiabis and Dardanelos, crop out discontinuously along more than 900 km within the Rio Negro–Juruena province. Where isotopic analytical data are available, the volcanic rocks exhibit positive $\varepsilon_{Nd(t)}$ values, and Sm-Nd T_{DM} model ages are quite close to their age of crystallization (Santos, 2003).

The southern boundary of the Rio Negro–Juruena with the Rondonian–San Ignacio province is very complex. It comprises several metamorphic belts, some of which reach granulite facies, several large shear zones, and recurrent plutonism (e.g., Bettencourt et al., 1999a; Payolla et al., 2002; Tohver et al., 2004, 2005b). Such features seem to be a good indication of the collisional nature of the boundary.

For the Ventuari-Tapajós and Rio Negro–Juruena provinces, the present authors envisage a practically continuous accretionary process involving oceanic domains, active during ~500 m.y. In this phase, crustal evolution combined predominantly juvenile processes with some reworking of crustal material formed in previous accretionary phases. In addition, the geochronological pattern of the "basement" granitoids seems to indicate a younging from northeast to southwest, suggesting the development of successive accretionary units, with a great deal of juvenile material formed in subduction environments. In our view, this large, continental region was formed as the result of two possible, successive Proterozoic mega-accretionary cycles, followed by cratonization, the earliest at 1980–1830 Ma and the younger at 1780–1550 Ma.

Many volcano-sedimentary rift-type structures of Mesoproterozoic age are found over both cratonized provinces. These basins may accompany numerous, circular intrusions of granitic composition, formed by anorogenic magmatism, indicating major processes of disruption of the already cratonized continental crust. A complete understanding of the tectonic and magmatic evolution of this very large and complex region cannot be reached without a much larger set of geochronological and isotopic data. In spite of this handicap, we suggest that many of the extensional structures located over the Ventuari-Tapajós province represent the reflection of the tectonic processes produced by the younger subduction regime outboard, especially those near the marginal zone. An example of basin formation linked with extensional tectonics is the large Beneficente Group at the Serra do Cachimbo. A similar tectonic context is visualized for the Rio Negro–Juruena province, where several aulacogenic basins and anorogenic structures may be related to the tectonic processes that took place as a reflection of the collisional episodes that established the younger Rondonian–San Ignacio, or even the Sunsas-Aguapeí provinces formed toward the southwest.

The Rondonian–San Ignacio Province

The Rondonian–San Ignacio province, characterized by the collisional-type orogeny bearing the same name, occurs in the southwestern part of the Amazonian Craton, and its tectonomagmatic episodes are attributed to the early to middle interval of the Mesoproterozoic, roughly between 1500 and 1300 Ma (Tassinari et al., 2004). This province includes large parts of the Precambrian shield of the Brazilian states of Rondônia and Mato Grosso, as well as a large area of the Santa Cruz province of Bolivia (Fig. 4). The total area exposed is at least 2000 km long and ~800 km wide.

The tentative boundary with the Rio Negro–Juruena province is not easy to trace, taking into account all the presently available geological evidence, as already presented (Figs. 3 and 4). It seems that the accretionary regime with soft-collision stacking of oceanic features, which was the rule for the Ventuari-Tapajós and part of the Rio Negro–Juruena provinces, was interrupted by a continental collision with a relatively large microcontinent, made up by a relatively thick continental crust. For the purpose of this paper, we may call it "Parecis microcontinent," speculating that it may occur below the sediments that fill up the very large Pimenta Bueno graben basin, which includes a thick Precambrian to Paleozoic sedimentary sequence that was at its turn covered by the Mesozoic Parecis Formation. Appropriate geophysical evidence, such as seismic data, necessary to confirm such possibility is scanty. However, the kimberlite provinces (Fig. 4) of Juina at the northern border of the Pimenta Bueno sedimentary basin, and of Colorado do Oeste at its southern border, may be an indirect indication of the existence of continental lithosphere with substantial thickness, below the Paleozoic and possibly Precambrian sediments (Tohver et al., 2004). These authors propose that this region represents the westward continuation of the Nova Brasilândia belt on the basis of large-scale aeromagnetic anomalies. A recent work using continental-size seismic tomography by Heintz et al. (2005), through surface wave inversion, seems to confirm the existence of a fairly large cratonic area with a lithospheric keel at least down to 200–250 km, under the Pimenta Bueno–Parecis basin, but extending to the neighboring areas, well into the Rio Negro–Juruena province.

Along the border region between the Rio Negro–Juruena and the Rondonian–San Ignacio provinces, in western Rondônia, additional evidence of a major collisional tectonometamorphic episode is given by large shear zones, such as the Ji-Paraná shear zone as described by Tohver et al. (2002). A few metamorphic belts are also indicated (Fig. 4), with variable metamorphic grade, up to granulite facies in the Nova Brasilândia belt (Tohver et al., 2004, 2005b). Some U-Pb SHRIMP measurements on zircon overgrowths from rocks of the Rio Negro–Juruena basement in east-central Rondônia yielded metamorphic ages around 1300–1350 Ma (Tassinari et al., 2000; Payolla et al., 2003; Bettencourt et al., 2006). These ages agree well with Sm-Nd whole-rock garnet ages of gneissic rocks (Payolla et al., 2002), diagnostic for the timing of collision-related, high-grade regional metamorphism of the Rondonian–San Ignacio orogeny. Moreover, a series of $^{40}Ar/^{39}Ar$ ages on amphibole and biotite from different rocks, between 1330 and 1300 Ma (Rizzotto et al., 2002; Tohver et al., 2005b, 2006b; Teixeira et al., 2006), provides additional evidence for this main Mesoproterozoic metamorphic-thermal event of the Rondonian–San Ignacio province.

The Rio Crespo Intrusive Suite, which comprises fine-grained granitic gneisses and charnockitic granulites, and occurs just south of Ariquemes (Fig. 4), is one of the oldest units of the Rondonian–San Ignacio province, with an age of 1492 ± 12 Ma obtained by Bettencourt et al. (2006), using the U-Pb SHRIMP method applied to the cores of several zircon crystals. These rocks exhibit slightly positive $\varepsilon_{Nd(t)}$ values of +0.6 to +1.2, suggesting that they represent a mixture of predominantly juvenile material and some older crust, possibly within a Andean-type continental margin. Bettencourt et al. (2006) obtained younger ages of ca. 1350 Ma for the overgrowth rims of the same zircons, evidence of the medium- to high-grade metamorphic imprint that affected the entire region during the collisional processes of Rondonian–San Ignacio orogeny.

Existence of an older continental substratum in the northern part of the Rondonian–San Ignacio province is supported by the polymetamorphic nature of some basement rocks, and by the several negative $\varepsilon_{Nd(t)}$ values of granitoid rocks found there, indicating some degree of crustal reworking. Litherland et al. (1986) identified some pre–San Ignacio basement regions, including the high-grade metamorphic rocks of the Lomas Maneches Complex in Bolivia, based on geological evidence as well as on a few very imprecise Rb-Sr whole-rock ages. More recently, Boger et al. (2005) presented more precise U-Pb SHRIMP zircon ages from two high-grade rocks of the Lomas Maneches Complex, which yielded ca. 1660 and ca. 1690 Ma.

The Rondonian–San Ignacio province also contains numerous anorogenic rapakivi granites that were emplaced over a long period of time. The ages of the following intrusive suites have been constrained by U-Pb geochronology (Bettencourt et al., 1999a; Payolla et al., 2002): Santo Antônio (1410 Ma), Teotônio (1390 Ma), Alto Candeias (1340 Ma), and São Lourenço–Caripunas (1310 Ma). These granites exhibit slightly positive $\varepsilon_{Nd(t)}$ values (Bettencourt et al., 1999b), indicating a high proportion of juvenile material in their magmas. In our view they may well be late or posttectonic-type intrusions related to inboard magmatism of different pulses of the Rondonian–San Ignacio orogeny.

Litherland et al. (1986) identified voluminous syn- to post-tectonic magmatism in the bulk of the Rondonian–San Ignacio province along its western margin. This magmatism is related to their San Ignacio orogeny, with Rb-Sr ages in the 1280–1380 Ma interval (Darbyshire, 2000). This region was named "Paragua Craton" because it was considered to be tectonically stable during the Meso- to Neoproterozoic deformation of the Sunsas belt (see below). The ubiquitous granitic rocks were all included in the "Pensamiento Granites," and their predominantly juvenile character was identified by Darbyshire (2000) on the basis of $\varepsilon_{Nd(t)}$ values between –0.9 and +3.9. Boger et al. (2005) shed some light on the evolution of the San Ignacio belt in Bolivia by means of some key U-Pb SHRIMP zircon ages. Firstly, U-Pb zircon ages of 1340–1320 Ma were obtained on the San Rafael granite, indicating emplacement synchronous with the San Ignacio orogeny, similar to those already mentioned, in Rondonia (Payolla et al., 2002; Rizzotto et al., 2002). Secondly, on detrital grains from paragneisses ages between 1690 and 1760 Ma were obtained. Moreover, R. Matos (2006, personal commun.) obtained U-Pb SHRIMP zircon ages from some rocks of the "Pensamiento Granites," with ages within the 1320–1400 Ma interval, coupled to positive to slightly negative $\varepsilon_{Nd(t)}$ values. This isotopic pattern strongly suggests that a relevant part of the "Paragua Craton" is made up by oceanic-type material, formed in magmatic arcs within an accretionary belt.

The southeasternmost part of Figure 4 may provide some better hints for the tectonic development of the Rondonian–San Ignacio province. In the Alto Jauru region (Fig. 5), much better access to the area, and more detailed studies by Geraldes et al. (2001, 2004a, 2004b), provide better understanding of the tectonic relations between the Rio Negro–Juruena province and the evolution of the Rondonian belt. First, a cratonized portion of the province is identified in the northeastern corner of Figure 5, which is covered by the undeformed Rio Branco Igneous Suite. The volcanic rocks of this suite are bimodal. They yield U-Pb zircon ages between 1470 and 1420 Ma and $\varepsilon_{Nd(t)}$ values between –1.0 and +1.9, indicating some assimilation of basement material (Geraldes et al., 2001). Extension of the Mesoproterozoic basement along the boundary of the Parecis sedimentary basin seems to be confirmed by a whole-rock Pb-Pb age of 1780 Ma (Tassinari et al., 1996) for a gneissic sample.

Two magmatic arcs of juvenile and possibly intra-oceanic character, related to the Rondonian–San Ignacio province, have been characterized in the same region (Fig. 5). The Rio Alegre Complex contains mafic-ultramafic bodies, BIFs (banded iron formations), chert, and granitoid plutonic rocks with U-Pb ages between 1510 and 1480 Ma and $\varepsilon_{Nd(t)}$ values between +2.5 and +4.7. The plutonic rocks of the Santa Helena calc-alkaline batholith exhibits U-Pb ages between 1450 and 1420 Ma and $\varepsilon_{Nd(t)}$ values between +2.6 and +4.0 (Geraldes et al., 2001, 2004b; Ruiz, 2005).

It is our view that the Rio Alegre and Santa Helena are juvenile crustal domains progressively built or amalgamated to the evolving continental margin. The U-Pb zircon ages, as well as the Sm-Nd constraints, indicate the onset of intermittent convergent-margin magmatism and soft accretion by stacking of outboard juvenile magmatic arcs during a continued subduction regime. Ancient basement has not been found here, high-grade metamorphics are not present, and some mafic rocks, as well as deep-ocean sediments, are described (Geraldes et al., 2001), indicating the presence of oceanic-type material. The Rio Alegre Complex is coeval with the already mentioned Pensamiento granitoids, and the entire region may well be regarded as assembled by several accretionary intra-oceanic magmatic arcs, revealing a large oceanic domain occurring in Bolivia at the end of Mesoproterozoic time.

In the northwestern corner of the Amazonian Craton, outcrops of Rondonian–San Ignacio province rocks are limited, being concealed beneath the Phanerozoic cover (Fig. 3). For the low-grade Tunui Group metasediments, $^{40}Ar/^{39}Ar$ muscovite ages indicate that the metamorphic imprint of the Rondonian–San Ignacio orogeny occurred at ca. 1320 Ma (Santos, 2003). In addition, some detrital zircon grains from the same unit were dated by U-Pb SHRIMP and yielded ages between 1875 and 1720 Ma (Santos, 2003), indicating provenance from rocks of the nearby Rio Negro–Juruena province.

The precise time of cratonization of the Rondonian–San Ignacio province is difficult to establish, because of the multiple tectonic reactivations related to the younger Sunsas-Aguapeí collisional belt (see below). The best approximation of the time of tectonic stabilization seems to be provided by the youngest K-Ar mica dates obtained from the Pensamiento Complex in Bolivia (Litherland et al., 1986), and by the $^{40}Ar/^{39}Ar$ muscovite and hornblende ages of Rizzotto et al. (2002) for the Colorado schists of Rondônia, which are slightly older than 1300 Ma. These authors used this interpretation to characterize their "Paragua Craton" as a region that was not affected by their Sunsas orogeny. Tohver et al. (2006a) have reviewed the available $^{40}Ar/^{39}Ar$ hornblende ages for the entire basement of Rondônia. The ages encompass the 1350–1550 Ma interval and seem to mime the duration of the Rondonian–San Ignacio orogeny. After cratonization, several rift-type structural basins located within the Rondonian–San Ignacio province were formed over the stabilized region, possibly as a result of collisional processes that established the younger Sunsas province to the southwest. The more important basins are the Pacaás Novos, Nova Brasilândia, and Uopione, to be dealt with later.

The Sunsas-Aguapeí Province

The collisional-type Sunsas orogeny occurred roughly between 1250 and 1000 Ma at the southwestern end of the Amazonian Craton. It is made of low- to medium-grade metamorphics and associated granitoid plutons (Litherland et al., 1986, 1989; Tassinari et al., 2000). According to Litherland et al. (1986), and later to Sadowski and Bettencourt (1996), the Sunsas belt was formed in an extensional environment, consisting of a passive margin sedimentary sequence (the Vibosi and Sunsas belts) that was subsequently deformed during a collisional event. This deformed sequence was intruded by syn- to late-tectonic granitoids, followed by the emplacement of posttectonic and later anorogenic plutons.

The evolution of this orogen involved early deposition of the Sunsas and Vibosi Groups, and their subsequent deformation and metamorphism (Litherland et al., 1986). These authors placed the boundary of the Sunsas belt along the southern boundary of Paragua Craton at the Rio Negro Front and Santa Catalina shear zone, where several mylonitic belts are observed (Fig. 4). Throughout the Paragua Craton, the unmetamorphosed and practically flat-lying sedimentary sequences of the Huanchaca Group (Litherland and Power, 1989), correlated with the Sunsas Group, overlie the Paragua basement rocks (Fig. 4). In Bolivia, the main phase of the Sunsas orogeny is constrained by the U-Pb SHRIMP zircon age of the undeformed posttectonic Taperas granite, dated by Boger et al. (2005) at 1076 ± 18 Ma.

Over the cratonic area, in Brazil, the Nova Brasilândia and Aguapeí belts were formed in rift-type structures that were affected by later transpression and crustal shortening. The history of deformation and metamorphism within these tectonic units seems to match in time the orogenic evolution of the Sunsas belt.

Within the Nova Brasilândia belt, U-Pb SHRIMP zircon ages of ca. 1120 Ma were produced by Santos et al. (2000). Moreover, Rizzotto et al. (2002) constrained the age of this unit more by U-Pb zircon geochronology, dating some detrital zircon crystals (1215 Ma), deformed granite intrusions (1113 Ma), basic intrusions (1110 Ma), and posttectonic granites (1005 Ma). Subsequent granulite-facies metamorphism took place at 1.09 Ga with cooling through 920 Ma, as recorded by the U-Pb dating of monazite and titanite, as well as $^{40}Ar/^{39}Ar$ ages from hornblende and biotite, respectively (Tohver et al., 2004). In addition, some mafic lava flows within the Pacaás Novos basin in Rondonia were dated by the K-Ar method between 1000 and 1200 Ma (Teixeira and Tassinari, 1984), and more recently a mafic sill yielded a $^{40}Ar/^{39}Ar$ age of 1198 ± 3 Ma (igneous biotite; Tohver et al., 2002). A maximum age for the sedimentary sequence making up the upper section in the basin was obtained by Santos (2003) at 1050 Ma, based on U-Pb determinations on detrital zircon grains.

The Aguapeí intracontinental rift, or aulacogen (Saes, 1999), is a structure more than 500 km long, with NW-SE trend, located in Mato Grosso State, Brazil, along the Bolivian border (Fig. 4). It is made up by clastic sediments, gently folded and affected by low-grade metamorphism. Santos (2003) and Santos et al. (2005) presented U-Pb SHRIMP ages from detrital zircon grains, the youngest of which was of 1161 ± 27 Ma, placing an upper limit on the age of deposition of the sediments.

The widespread anorogenic granitic magmatism of the Younger Granites of Rondônia, also known as the Rondônia Tin Province, which intruded large parts of the Rio Negro–Juruena province southeast of Porto Velho, can also be considered as a reflection of the Sunsas orogeny. This suite comprises alkali-granites, including rapakivi varieties, and associated mafic rocks. A few intrusive bodies were dated by several methods (Table 2), yielding ages between 970 and 1100 Ma (e.g., Priem et al., 1971a, 1989; Bettencourt et al., 1995, 1999a). Examples could be the ones dated by U-Pb SHRIMP method, such as the Santa Barbara Massif (1082–978 Ma; Leite et al., 2003), the Maçangana granite (990–980 Ma), and the Santa

Clara Intrusive Suite (1082–1074 Ma). The Younger Granites of Rondônia usually exhibit negative $\varepsilon_{Nd(t)}$ values (Bettencourt et al., 1999b), such as the Santa Barbara pluton, whose values are from −2.9 to −4.6, indicating crustal reworking (Sparrenberger et al., 2002). To the southeast, in Mato Grosso State, additional anorogenic granitic bodies are synchronous with the Younger Granites of Rondônia, such as the São Domingos and Guapé plutons, with Rb-Sr and U-Pb ages of ca. 980–920 Ma (Geraldes et al., 2001; Ruiz, 2005). The Rincón del Tigre basic complex, in Bolivia, dated at 990 Ma by the Rb-Sr method (Litherland et al., 1986) may also have taken part in the same tectonic history of the Sunsas orogeny. According to Bettencourt et al. (1999a), these intrusions were formed during the collisional stage, as inboard manifestations over the older and already cratonized Rio Negro–Juruena and Rondonian–San Ignacio provinces.

From the comprehensive 1000–950 Ma Rb-Sr and K-Ar dates reported by Litherland et al. (1986, 1989), the timing for the final cratonization, cooling, and exhumation of the Sunsas orogenic belt seems well constrained. A few $^{40}Ar/^{39}Ar$ mica ages in the 900–1000 Ma range, obtained by Ruiz (2005) in gneissic rocks on the basement of the Aguapeí aulacogen, close to the boundary between the Rondonian–San Ignacio and the Rio Negro–Juruena border, can also be attributed to the thermal reactivation followed by cooling as a response to the Sunsas orogeny.

The history of deformation and magmatism of the Sunsas-Aguapeí province matches relatively closely the tectonic events of the Laurentian Grenville province, prototype of the Grenvillian orogenic system. The main features are the high-grade metamorphism and three successive tectonometamorphic events (Rivers et al., 1989): the older at ca. 1190–1140 Ma (Elzevirian pulse), the intermediate and more widespread at 1080–1020 Ma (Ottawan), and the younger at 1000–980 Ma (Rigolet). The Elzevirian episode has been related to backarc closure and arc accretion, and the younger episodes have been related to continental collision (Rivers, 1997; Wasteneys et al., 1995). The Sunsas collisional belt is considered by most to be the South American counterpart of the Grenville belt, with a role in the agglutination of the supercontinent Rodinia (e.g., Hoffman, 1991; Sadowski and Bettencourt, 1996; Kröner and Cordani, 2003). However, high-grade metamorphic rocks, typical of the two younger pulses within the Grenville belt in Canada, are not present within the Sunsas belt, and therefore the details of the probable Grenville-Sunsas link are far from resolved.

A history similar to the Grenville belt is encountered in a few basement inliers with high-grade metamorphic rocks in the northern Andes (Cordani et al., 2005). The common tectonic events found in all of these Proterozoic rocks of South America, as well as in their correlative terranes in Mexico and the central Andes, strongly suggest that they are fragments of the once coherent Grenvillian collisional belt that participated in the agglutination of Rodinia. When the tectonic inliers from Colombia are merged with the correlative Mexican and central Andean domains, as well as with Laurentia and Amazonia, an extensive and continuous belt is formed, in the central region of Rodinia, as envisaged firstly by Hoffman (1991), and later by many others involved with global paleogeographic reconstructions.

ACCRETIONARY BELTS AT THE EASTERN MARGIN OF THE AMAZONIAN CRATON

The Sunsas belt is the youngest orogen in the Amazonian Craton. If we consider the southwestern half of the craton, the described continuous migration of the tectonic and magmatic processes from northeast to southwest must be considered as one of the major features of Earth's geodynamics for Proterozoic times. It must be linked to successive subduction processes during ~1.0 b.y., related to large-scale and systematic movement of the asthenosphere.

More or less synchronous with the termination of the tectonic events of the Sunsas belt is early Neoproterozoic continental breakup and fragmentation affecting the eastern margin of the craton, producing a passive margin and the observed truncation of the major structures. The intra-oceanic Goiás magmatic arc (Pimentel and Fuck, 1992; Pimentel et al., 2000) is evidence of the large oceanic basin that existed at 900–1000 Ma between the rifted margins of the Amazonian Craton and some other unidentified cratonic fragments. Candidates for a possible pre-Neoproterozoic correlation are the Baltica and São Francisco–Congo Cratons, the Trans-Saharan "metacraton," and the other smaller cratonic fragments in South America, such as the Paranapanema cratonic fragment (Mantovani et al., 2005). The Goiás magmatic arc is part of the Tocantins province of Central Brazil (Figs. 1 and 6), which, together with the Borborema province and the Trans-Saharan belt of Africa, were affected by a series of orogenic events that produced West Gondwana at the end of the Proterozoic (Kröner and Cordani, 2003).

Pimentel et al. (2000, and references therein) detailed the lithology, geochemistry, and the tectonic evolution of the geologic units belonging to the Tocantins province (Fig. 6). The main tectonic element in the region is a very large fault zone, a megasuture that was active throughout the Neoproterozoic, the Transbrasiliano Lineament, which crosses the entire region from northeast to southwest. The Araguaia and Brasilia marginal belts are identified at the borders of the Amazonian and São Francisco Cratons, respectively, evolving from passive margins with thick sedimentary piles into orogenic belts. Moreover, the Tocantins province comprises a mosaic of tectonic blocks, a series of juxtaposed terranes with variable lithologies, bounded by shear zones of different sizes. They make up a unit named Central Goiás Massif, considered a single heterogeneous cratonic fragment by Brito Neves and Cordani (1991).

The tectonic evolution of the Central Goiás Massif is very complex. Several ancient rock units are recognized in it, including the Archean nucleus with the Crixás, Goiás Velho, and other greenstone belts, some Paleoproterozoic medium- to high-grade complexes, and the three large layered mafic-ultramafic complexes: Canabrava, Niquelandia, and Barro Alto. These mafic-ultramafic complexes are very similar, and may have originated

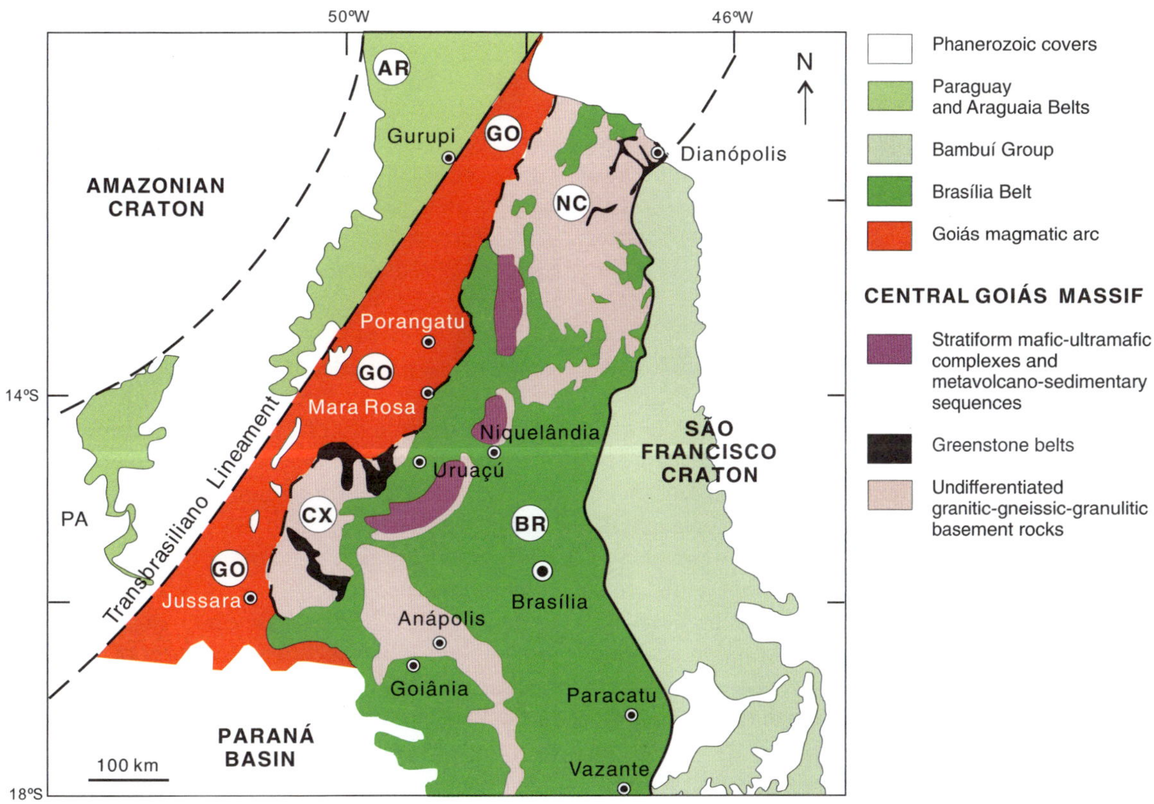

Figure 6. Tocantins province and its main tectonic domains. AR—Araguaia belt; PA—Paraguay belt; GO—Goiás magmatic arc; CX—Crixás granite-greenstone terrain; NC—Natividade-Cavalcante block; BR—Brasilia belt. Adapted from Cordani et al., 2000.

in the same large continental rift. They comprise an older lower layered series of gabbronorites, pyroxenites, and dunites, and a younger upper layered series in which gabbro-anorthosites predominate. The complexes are in contact with supracrustal sequences in their western part and were affected in their eastern part by high-grade metamorphism. Although many attempts were made (see Pimentel et al., 2000), the age of igneous crystallization of the mafic-ultramafic rocks is far from resolved. Age of metamorphism seems to be better constrained, ca. 750 Ma (Ferreira-Filho et al., 1998; Moraes et al., 2006). Pimentel et al. (2000) considered this high-grade metamorphic event to be related to the collision between the Central Goiás Massif and the São Francisco Craton. In addition, a high-grade metamorphic complex occurs in the southern part of the Central Goiás Massif (Fischel et al., 1998), where Sm-Nd garnet/whole-rock metamorphic ages are 630–610 Ma. These ages probably date the collision between the São Francisco Craton and the Paranapanema cratonic fragment, as suggested by Campos-Neto (2000), preceding other collisional events leading to the final amalgamation of the Gondwana supercontinent, accomplished some 100 m.y. later.

The Goiás magmatic arc is a major area of soft collision and accretion of the Amazonian Craton margin and contains Neoproterozoic granitoids, with the oldest of these clearly juvenile (Pimentel et al., 1999). It is a large and elongated region, closely associated with the Transbrasiliano megasuture, where calc-alkaline orthogneisses and associated rocks predominate. The principal rock types of this unit are plutonic, ranging in composition from tonalite to granodiorite. Several supracrustal belts occur in this region, with volcano-sedimentary sequences with mafic and ultramafic bodies, considered being ophiolite fragments (Pimentel et al., 1997). A deep seismic study carried out by Assumpção et al. (2004), using teleseismic P-wave tomography, showed that beneath the Goiás magmatic arc the crust is thinner and the lithospheric upper mantle has lower velocities compared to those of the adjacent cratons. This seems to characterize a differential uplift in the Neoproterozoic, accompanied by mafic intrusions in the lower crust, and eventually producing the observed gabbroic/granitic magmatism in the region of the Goiás magmatic arc. Many postorogenic granitic intrusions are also found in the region. The geochronological systematics of the Goiás magmatic arc indicates ages between 900 and 530 Ma. The juvenile signature of the granitoid rocks is demonstrated by their low initial Sr isotopic ratios, their positive $\varepsilon_{Nd(t)}$ values, and their Sm-Nd T_{DM} model ages between 0.9

and 1.2 Ga (Cordani and Sato, 1999; Pimentel et al., 2000). The interpretation of these authors is that the material represents the roots of a series of juvenile intra-oceanic island arcs produced by consumption of oceanic lithosphere. This implies the existence of a large oceanic domain, the Goiás Ocean, since at least 900 Ma, separating the Amazonian Craton from the São Francisco–Congo Craton and other smaller cratonic units. The disappearance of this ocean is one of the major processes bearing on the amalgamation of Gondwana.

THE POSITION OF THE AMAZONIAN CRATON WITHIN THE PROTEROZOIC SUPERCONTINENTS

Supercontinents are formed by the amalgamation of preexisting continental masses, with the concomitant disappearance of the intervening oceans. Reconstructions are based on the available paleomagnetic data that are normally affected by large uncertainties, especially regarding paleolongitudes. If it were possible to estimate the size of the oceans that separated the existing continental masses at a given time, the task of producing a reconstruction would be facilitated. Following the ideas put forward by Cordani et al. (2003), we maintain that intra-oceanic granitoid magmatic arcs, with juvenile isotopic signatures, indicate the existence of significantly large oceanic basins. In the case of the Amazonian Craton, this conclusion applies not only to the Ventuari-Tapajós and Rio Negro–Juruena provinces, but also to the Pensamiento–Rio Alegre territory in the southwestern corner and to the Goiás magmatic arc. In contrast, when granitoid material yields isotopic signatures indicating crustal reworking, consumption of small oceans is a more likely interpretation, indicating the relative proximity of the convergent continental masses. Such tectonic environments, which are common for the eastern part of the Brazilian Shield (see Cordani et al., 2000), are related to Wilson cycle processes of ocean opening and closing that usually lead to collisional belts.

Pre-Rodinia Times

It is not possible to indicate with confidence the possible position of the cratonic nuclei related to the Amazonian Craton before the onset of the Transamazonian-Eburnean orogeny, in the Paleoproterozoic. After this event, the northern part of the Guyana Shield was attached to West Africa. Between ca. 1900 and 1500 Ma, the series of intra-oceanic magmatic arcs found in the Ventuari-Tapajós and Rio Negro–Juruena tectonic provinces attests to the existence of a very large ocean. This means that in Paleo- and Mesoproterozoic times Amazonia was much smaller, and this must be taken into account in the reconstructions envisaged for those times.

The age and the character of the Paleo- to Mesoproterozoic belts of southwestern Amazonia permit correlation with the Svecofennian domain and the Transscandinavian igneous belt within Baltica, an idea first advanced by Almeida (1978), and also with the Midcontinent region of Laurentia. Pesonen et al. (2003) attempted to reconstruct the relative position of these three cratonic masses at ca. 1500 Ma. An additional fact indicating such a correlation is the global collisional event that produced Rodinia, described for the Grenville belt in Laurentia, the Sveconorvegian belt in Baltica, and the Sunsas and Nova Brasilândia belts in Amazonia.

The relation of Amazonia and Laurentia is a key issue for the reconstruction of the Mesoproterozoic. In our view, the juvenile rocks of Pensamiento–Rio Alegre, and related magmatic arcs, dated between 1400 and 1500 Ma, could well correspond to the accreted intra-oceanic material that attests to the existence of a large ocean that separated Amazonia from Laurentia.

From then on, different scenarios have been proposed by several authors, most of them envisaging a final Sunsas collision at 1000 Ma, as part of the global Grenvillian orogeny that formed Rodinia. In our view, the granitoid rocks forming the Paragua Craton of Litherland et al. (1986, 1989) and those of Mato Grosso and Rondonia States, south of the boundary with the Rio Negro–Juruena province (Fig. 4), were already united at the end of the Rondonian–San Ignacio orogeny, ca. 1320–1340 Ma. Therefore, we are at variance with the view of Tohver et al. (2005a, 2005b) and Boger et al. (2005), according to which the collision of the Paragua Craton with the Amazonian Craton took place at ca. 1.1 Ga, during the Sunsas orogeny.

We prefer one of the alternative models outlined by Boger et al. (2005) (Fig. 7), according to which the collision of the Paragua Block (our Pensamiento–Rio Alegre intra-oceanic accretionary terrane) with the southern part of the already cratonized Rio Negro–Juruena province originates the Rondonian–San Ignacio orogeny. In our interpretation, rifting of the Aguapeí and Nova Brasilândia Groups correspond to the final orogenic collapse within the Rondonian–San Ignacio province. The later activation of these could be a reflection of the successive collision that produced the Sunsas belt. Final suturing would be, more properly, between Amazonia and Laurentia, with the Sunsas as part of the Grenvillian belts.

As a somewhat different alternative, Sadowski and Bettencourt (1996) envisaged a large continent (a proto-Laurentia plus a proto-Amazonia?) that would be disrupted by the onset of two successive Wilson cycles. In this view, Laurentia and Amazonia were close together, as parts of the same continental mass, since at least 1.5 Ga. Their preferred evolutionary model includes formation of a reasonably large ocean floor during the first cycle (ca. 1.5 Ga) and only a small ocean during the second (ca. 1.25 Ga). Following this hypothesis, the onset of the second Wilson cycle may have produced the passive margin that later was inverted to produce the Sunsas belt, and also the initial rifting of the Aguapeí aulacogen, more or less along the weakness zone represented by the previous suture of the Rondonian–San Ignacio collision. The final closing of the second ocean would have produced a high-grade collisional belt on the Laurentian side and a series of transtensional features in Amazonia. Tohver et al. (2006a) produced recently a summary of the available robust paleomagnetic constraints for the Laurentia-Amazonia

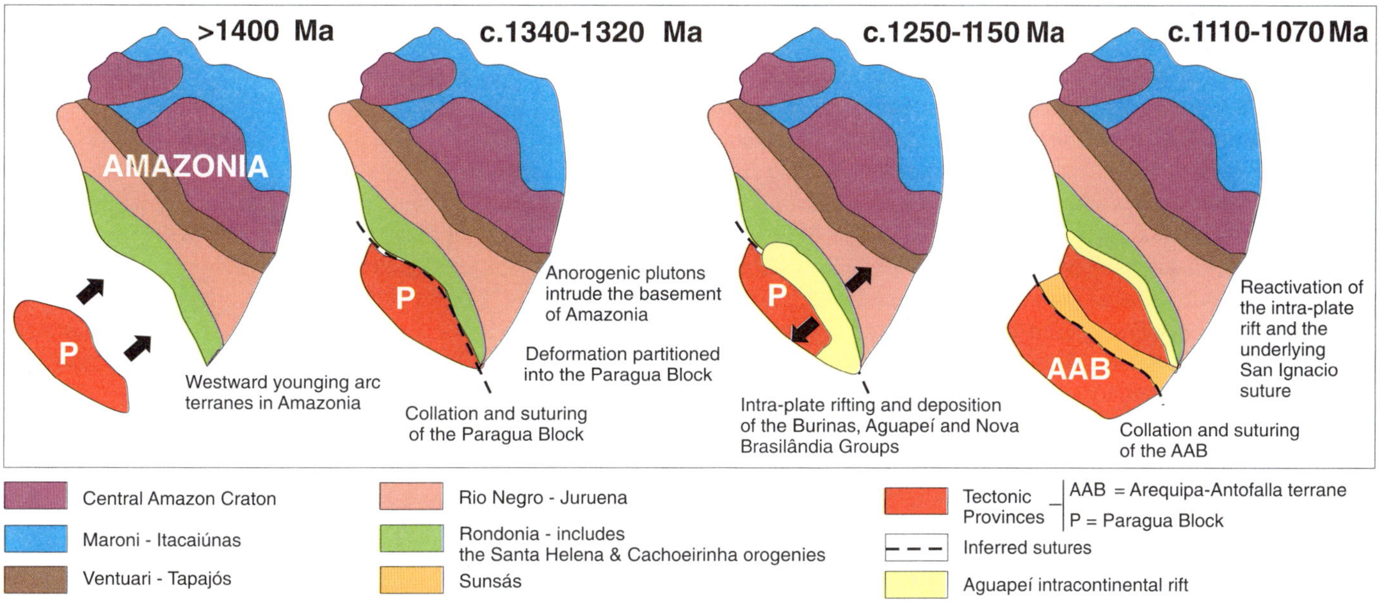

Figure 7. Cartoon of a possible scenario for the accretion of the Paragua block and the Arequipa-Antofalla terrane to the southwestern margin of the Amazonian Craton, after Boger et al., 2005.

link, where a common Rodinian paleogeography for both cratons was determined at least since 1200 Ma.

Amazonia in Rodinia

Rodinia was initially defined as a long-lived supercontinent that assembled all the continental fragments around Laurentia and remained stable from 1000 to 750 Ma (McMenamin and McMenamin, 1990; Hoffman, 1991). In most reconstructions (e.g., Hoffman, 1991; Sadowski and Bettencourt, 1996; Weil et al., 1998), Amazonia is placed against the eastern side of Laurentia by matching the Grenville and Sunsas belts. Tohver et al. (2002, 2004, 2005) suggested that a better matching for Amazonia would be against the Llano segment of Laurentia. Finally, other reconstructions, although keeping the general correlation for the ca. 1000 Ma mobile belts, place Laurentia and Amazonia in different positions (Starmer, 1996; Keppie and Ortega-Gutierrez, 1999) in order to accommodate other smaller cratonic fragments such as Oaxaquia of southern Mexico, Arequipa-Antofalla of western South America, and the Garzón Massif of Colombia (not shown in the figures).

Many of the basement units of the northern Andes, such as the Santa Marta, Santander, and Garzón Massifs in Colombia, yielded Grenvillian radiometric ages of ca. 1000 Ma (Tschanz et al., 1974; Cordani et al., 2005), a fact that suggested their correlation with the Sunsas belt. Kroonenberg (1982) suggested that Grenvillian basement may occur below the entire eastern Cordillera of Colombia. Restrepo-Pace et al. (1997) also suggested possible links with the southern terranes of the Arequipa Massif in Peru (Wasteneys et al., 1995) and the Precordillera in Argentina (Ramos, 1988). A review of the available evidence on the Andean basement fragments can be found in Ramos and Aleman (2000). A correlation of all Grenville-age terranes scattered in Mexico and their possible comparison with the Grenvillian-age belts of Laurentia, Amazonia, and Baltica was attempted by Keppie and Ortega-Gutierrez (1999), concluding for a close similarity between the tectonic evolution of all them. Following the model of Starmer (1996), these terranes may initially have formed at 1200 ± 100 Ma within a large magmatic arc that was generated by consumption of a Mesoproterozoic ocean located between Laurentia on one side and Amazonia plus Baltica on the other. These terranes subsequently collided during the Grenvillian orogeny.

Amazonia in Gondwana

According to Cordani et al. (2003) and Kröner and Cordani (2003), the geological, geochronological, and paleomagnetic database accumulated for South America and Africa in the last decade seems to demonstrate that most of these continental fragments were not part of Rodinia. Tohver et al. (2006a) reached a similar conclusion on the basis of their review of the available paleomagnetic data from Africa and South America. In the transition from Rodinia to Gondwana, it is necessary to envisage the disappearance of at least two very large oceans: the Goiás (or Brasiliano) Ocean in the West, and the Mozambique Ocean in the

East, this latter related to the collage of West and East Gondwana (Cordani et al., 2003). Figure 8 reconstructs the paleogeography for the time of amalgamation of West Gondwana at ca. 600 Ma, and shows that the Goiás Ocean extended into the Pharusian Ocean to the northeast (Brito-Neves et al., 1984; Trompette, 1994; Fetter et al., 2003; Caby, 1989, 2002). It may also have extended to the southwest, toward the Neoproterozoic to Cambrian Pampean belt, assuming that the granitoid rocks of that belt originated through subduction processes. Figure 8 also indicates that the megasuture of the Transbrasiliano Lineament plus the Hoggar 4°50′ Lineament, related to the closing of both ocean domains, crosses West Gondwana from northeast to southwest. Based on the close correlation between northeastern Brazil and west-central Africa, a large oceanic domain separating the West African Craton from the Borborema–Trans-Saharan cratonic fragment in the Neoproterozoic was already proposed by Castaing et al. (1994).

The oldest granitoids in the Goiás magmatic arc are 900–850 Ma (Pimentel et al., 1997), indicating that the oceanic lithosphere consumed in the production of the granitoid rocks must have been generated some 100 m.y. earlier. This seems to imply that, when the main nucleus of Rodinia formed around Laurentia, including Amazonia, a large ocean separated several continental masses and fragments from this nucleus, such as the São Francisco–Congo, Rio de La Plata, and Kalahari Cratons, plus the Borborema–Trans-Saharan provinces, the Central Goiás Massif, and the Paranapanema block (Cordani et al., 2003; Kröner and Cordani, 2003). This large oceanic domain would include the well-documented Goiás and Pharusian Oceans as well as the less well-defined Pampean Ocean.

The Pampean-Goiás-Pharusian oceanic lithosphere started its consumption in the earlier Neoproterozoic, by forming intra-oceanic magmatic arcs. Later these units were united in a series of soft collisions, in the process of amalgamation of Gondwana, along the Pampean, Paraguay, Araguaia, and Pharusian mobile belts. There is evidence for extensive granitoid magmatism in the period between 840 and 530 Ma, corresponding to the tectonomagmatic episodes of the Brasiliano/Pan-African orogenic cycle. The predominantly calc-alkaline chemistry of these magmatic rocks indicates subduction-related, active margin processes.

In conclusion, considering the possible relative positions of the ancient cratonic fragments now within South America, it is not difficult to envisage West Gondwana, still attached to Laurentia, at ca. 600 Ma (see for instance Cawood et al., 2001; Tohver et al., 2006a). When Laurentia rifted away, forming the Iapetus Ocean at ca. 580 Ma, many fragments of the Grenville belt were left behind, including the Blue Ridge province of Laurentia, which is derived from the Amazonian Craton (Tohver et al., 2006b). Along the western margin of Gondwana, these Grenvillian terranes, such as the Garzón Complex and the Dibulla, Bucaramanga, and Jojoncito gneisses of Colombia (see Cordani et al., 2005), as well as the Arequipa-Antofalla and Precordillera terranes (see Ramos, 2000), were later redistributed and accreted back to South America in Paleozoic times as part of the Hercynian-Alleghanian orogeny, in the process of amalgamation of Pangea.

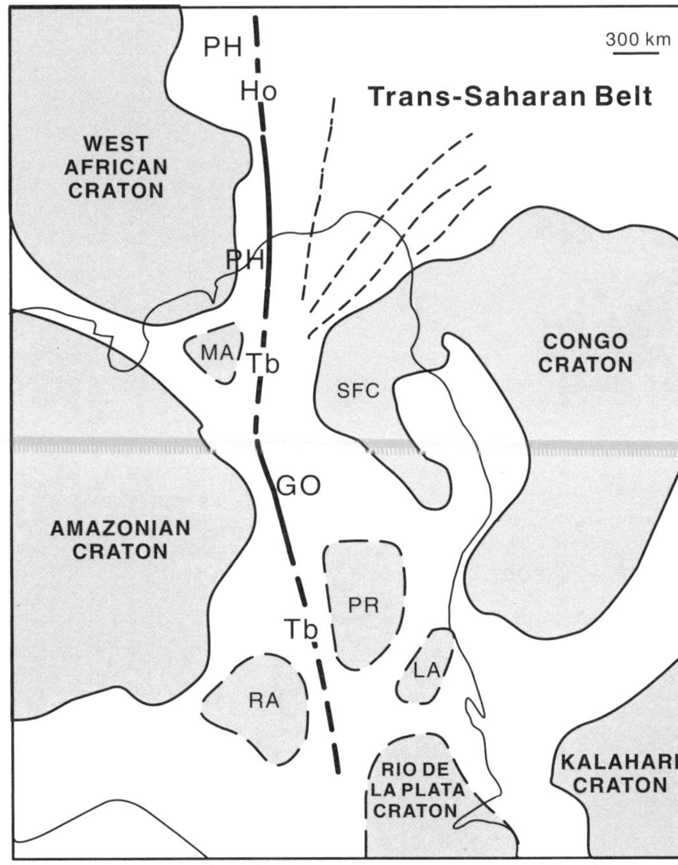

Figure 8. Relative position of cratons and mobile belts at ca. 600 Ma, during the final events of agglutination of West Gondwana. The Transbrasiliano-Hoggar 4°50′ megasuture is emphasized. For geographic reference, the present outline of South America is indicated. SFC—São Francisco Craton; PH—Pharusian belt; GO—Goiás magmatic arc; Tb—Transbrasiliano Lineament; Ho—Hoggar 4°50′ Lineament. Cratonic fragments: LA—Luiz Alves; MA—Maranhão; PR—Paranapanema; RA—Rio Apa.

ACKNOWLEDGMENTS

The authors thank colleagues Jorge Bettencourt and Colombo Tassinari for their comments and helpful suggestions, which improved an earlier version of the manuscript. They are also grateful to Eric Tohver for valuable discussions and especially for his careful and thoughtful review of the final version, which included an important amelioration of the English. The constructive reviews made by Bob Hatcher and Randy Van Schmus are also greatly appreciated. The authors also acknowledge the National Council of Scientific and Technological Development of Brazil (CNPq), for its continued support through grants 302851/2004-6 and 304300/2003-9.

REFERENCES CITED

Ahäll, K.I., Connelly, J.N., and Brewer, T.S., 2000, Episodic rapakivi magmatism due to distal orogenesis? Correlation of 1.69–1.50 Ga orogenic and inboard, "anorogenic"events in the Baltic Shield: Geology, v. 28, p. 823–826, doi: 10.1130/0091-7613(2000)28<823:ERMDTD>2.0.CO;2.

Almeida, F.F.M. de, 1978, A evolução dos cratons Amazônico e do São Francisco comparada com a de seus homólogos do Hemisfério Norte: Anais, Congresso Brasileiro de Geologia, Recife, v. 6, p. 2293–2407.

Arndt, N.T., and Goldstein, S.L., 1987, Use and abuse of crust formation ages: Geology, v. 15, p. 893–895, doi: 10.1130/0091-7613(1987)15<893:UAAOCA>2.0.CO;2.

Assumpção, M., An, M., Bianchi, M., Franca, G.S.L., Rocha, M., Barbosa, J.R., and Berrocal, J., 2004, Seismic studies of the Brasilia fold belt at the western border of the São Francisco Craton, central Brazil, using receiver function, surface-wave dispersion and teleseismic tomography: Tectonophysics, v. 388, p. 173–185, doi: 10.1016/j.tecto.2004.04.029.

Avelar, V.G., Lafon, J.-M., Delor, C., Guerrot, C., and Lahondere, D., 2003, Archean crustal remnants in the easternmost part of the French Guiana Shield: Pb-Pb and Sm/Nd geochronological evidence for Mesoarchean versus Neoarchean signatures: Geologie de la France, Special Guiana Shield: Bureau de recherches géologiques et minières–Société Géologique de France, no. 2–4, p. 83–89.

Ben Othman, D., Polvé, M., and Allegre, C.J., 1984, Nd-Sr isotopic composition of granulites and constraints on the evolution of the lower continental crust: Nature, v. 307, p. 510–515, doi: 10.1038/307510a0.

Bettencourt, J.S., Tosdal, R., Leite, W.B., Jr., and Payolla, B.L., 1995, Overview of the rapakivi granites of Rondonia Tin province (RTP), *in* Excursion Guide, Symposium on Rapakivi Granites and Related Rocks, Belém, Brazil, p. 5–48.

Bettencourt, J.S., Tosdal, R.M., Leite, W.B., Jr., and Payolla, B.L., 1999a, Mesoproterozoic rapakivi granites of Rondonia Tin province, southwestern border of the Amazonian Craton, Brazil, I: Reconnaissance U-Pb geochronology and regional implications: Precambrian Research, v. 95, p. 41–67, doi: 10.1016/S0301-9268(98)00126-0.

Bettencourt, J.S., Payolla, B.L., Leite, W.B., Jr., Tosdal, R.M., and Spiro, B., 1999b, Mesoproterozoic Rapakivi granites of Rondonia Tin province, SW border of Amazonian Craton, Brazil, III: Reconnaissance Nd, Sr, O, Pb isotopic geochemistry and regional implications, *in* Barbaran, B., ed., Fourth Hutton Symposium, Clermont-Ferrand, France: Abstracts, Documents du Bureau de recherches géologiques et minières, v. 290, 132 p.

Bettencourt, J.S., Payolla, B.L., Tosdal, R.M., Wooden, J.L., Leite, W.B., Jr., and Sparrenberger, I., 2006, SHRIMP-RG U-Pb zircon geochronology of gneiss from the Rio Crespo Intrusive Suite, SW Amazonian Craton, Rondonia, Brazil: New insight about protolith crystallization and metamorphic ages: Short Papers of the V SSAGI: Punta del Este, Uruguay, p. 49–52.

Boger, S.D., Raetz, M., Giles, D., Etchart, E., and Fanning, C.M., 2005, U-Pb age data from the Sunsas region of eastern Bolivia: Evidence for the allochthonous origin of the Paragua block: Precambrian Research, v. 139, p. 121–146, doi: 10.1016/j.precamres.2005.05.010.

Brito-Neves, B.B., and Cordani, U.G., 1991, Tectonic evolution of South America during the Late Proterozoic: Precambrian Research, v. 53, p. 23–40, doi: 10.1016/0301-9268(91)90004-T.

Brito-Neves, B.B., Fuck, R.A., Cordani, U.G., and Thomaz-Filho, A., 1984, Influence of basement structures on the evolution of the major sedimentary basins of Brazil: A case of tectonic heritage: Journal Geodynamics, v. 1, p. 495–510, doi: 10.1016/0264-3707(84)90021-8.

Caby, R., 1987, The Pan-African belt of West Africa from the Sahara Desert to the Gulf of Benin, *in* Schaer, J.P., and Rodgers, J., eds., The anatomy of mountain ranges: Princeton, New Jersey, Princeton University Press, p. 129–170.

Caby, R., 1989, Precambrian terranes of Benin-Nigeria and northeast Brazil and the Late Proterozoic South Atlantic fit, *in* Dallmeyer, R.D., ed., Terranes in the Circum-Atlantic Paleozoic orogens: Geological Society of America Special Paper 230, p. 145–158.

Caby, R., 2002, Neoproterozoic subductions and sutures in the Trans-Saharan belt (West Africa): 19th Colloquium on African Geology: El Jadida, Morocco, Abstracts Volume, p. 47–48.

Campos-Neto, M.C., 2000, Orogenic systems from southwestern Gondwana, *in* Cordani, U.G., Milani, E.J., Thomaz-Filho, A., and Campos, D.A., eds., Tectonic evolution of South America: Rio de Janeiro, Brazil, 31st International Geological Congress, p. 335–365.

Castaing, C., Feybesse, J.L., Thiéblemont, D., Triboulet, C., and Chèvremont, P., 1994, Paleo-geographical reconstructions of the Pan-African/Brasiliano orogen: Closure of an oceanic domain or intracontinental convergence between major blocks?: Precambrian Research, v. 69, p. 327–344, doi: 10.1016/0301-9268(94)90095-7.

Cawood, P.A., McCausland, P.J.A., and Dunning, G.R., 2001, Opening Iapetus: Constraints from the Laurentian margin in Newfoundland: Geological Society of America Bulletin, v. 113, p. 443–453, doi: 10.1130/0016-7606(2001)113<0443:OICFTL>2.0.CO;2.

Cordani, U.G., and Sato, K., 1999, Crustal evolution of the South American platform, based on Nd isotopic systematics on granitoid rocks: Episodes, v. 22, no. 3, p. 167–173.

Cordani, U.G., Tassinari, C.C.G., Teixeira, W., Basei, M.A.S., and Kawashita, K., 1979, Evolução tectônica da Amazônia com base nos dados geocronológicos: Actas, II Congresso Geologico Chileno, p. 137–148.

Cordani, U.G., Sato, K., Teixeira, W., Tassinari, C.C.G., and Basei, M.A.S., 2000, Crustal evolution of the South American platform, *in* Cordani, U.G., Milani, E.J., Thomaz-Filho, A., and Campos, D.A., eds., Tectonic evolution of South America: Rio de Janeiro, Brazil, 31st International Geological Congress, p. 19–40.

Cordani, U.G., D'Agrella-Filho, M.S., Brito-Neves, B.B., and Trindade, R.I.F., 2003, Tearing up Rodinia: The Neoproterozoic paleogeography of South American cratonic fragments: Terra Nova, v. 15, p. 350–359, doi: 10.1046/j.1365-3121.2003.00506.x.

Cordani, U.G., Cardona, A., Jimenez, D.M., Liu, D., and Nutman, A.P., 2005, Geochronology of Proterozoic basement inliers in the Colombian Andes: Tectonic history of remnants of a fragmented Grenville belt, *in* Vaughan, A.P.M., et al., eds., Terrane processes at the margins of Gondwana: Geological Society [London] Special Publication 246, p. 329–346.

Darbyshire, D.P.F., 2000, The Precambrian of eastern Bolívia—A Sm-Nd isotope study [abs.]: Rio de Janeiro, Brazil, 31st International Geological Congress, Abstract volume (CD-ROM).

Delor, C., Lahondère, D., Egal, E., Lafon, J.M., Cocherie, A., Guerrot, C., Rossi, P., Truffert, C., Thèveniaut, H., Phillips, D., and de Avelar, V.G., 2003, Transamazonian crustal growth and reworking as revealed by the 1:500,000 scale geological map of French Guyana (2nd edition), *in* Geologie de la France and surrounding areas, Special Guiana Shield: Bureau de recherches géologiques et minières–Société Géologique de France, no. 2–4,p. 5–57.

De Paolo, D.J., 1988, Age dependence of the composition of continental crust: Evidence from Nd isotopic composition in granitic rocks: Earth and Planetary Science Letters, v. 90, p. 263–271, doi: 10.1016/0012-821X(88)90130-6.

Dewey, J.F., 1969, Evolution of the Appalachian-Caledonian orogen: Nature, v. 222, p. 124–129, doi: 10.1038/222124a0.

Ferreira-Filho, C.F., Moraes, R., Fawcett, J.J., and Naldrett, A.J., 1998, Amphibolite to granulite progressive metamorphism in the Niquelandia Complex, central Brazil: Regional tectonic implications: Journal of South American Earth Sciences, v. 11, p. 35–50, doi: 10.1016/S0895-9811(97)00035-7.

Fetter, A.H., Santos, T.J.S., Van Schmus, W.R., Hackspacher, P.C., Brito-Neves, B.B., Arthaud, M.H., Nogueira-Neto, J.A., and Wernick, E., 2003, Evidence for Neoproterozoic continental arc magmatism in the Santa Quitéria batholith of Ceará State, NW Borborema province, NE Brazil: Implications for the assembly of West Gondwana: Gondwana Research, v. 6, no. 2, p. 265–273, doi: 10.1016/S1342-937X(05)70975-8.

Fischel, D.P., Pimentel, M.M., and Fuck, R.A., 1998, Idade do metamorfismo de alto grau no Complexo Anápolis-Itauçu, Goiás, determinada pelo método Sm-Nd: Revista Brasileira de Geociências, v. 28, p. 607–609.

Gaudette, H.E., and Olzewski, W.J., 1981, Geochronology of the basement rocks, Amazonas Territory, Venezuela, and the tectonic evolution of the Western Gúiana Shield: Geologie en Mijnbouw, v. 64, p. 131–144.

Gaudette, H.E., Olzewski, W.J., and Santos, J.O.S., 1996, Geochronology of Precambrian rocks from the northern part of Guiana Shield, state of Roraima, Brazil: Journal of South American Earth Sciences, v. 9, p. 183–195, doi: 10.1016/0895-9811(96)00005-3.

Geraldes, M.C., Van Schmus, W.R., Condie, K.C., Bell, S., Teixeira, W., and Babinski, M., 2001, Proterozoic geologic evolution of the SW part of the Amazonian Craton in Mato Grosso State, Brazil: Precambrian Research, v. 111, p. 91–128, doi: 10.1016/S0301-9268(01)00158-9.

Geraldes, M.C., Bettencourt, J.S., Teixeira, W., and Matos, R., 2004a, Geochemistry and isotopic constraints on the origin of the Mesoproterozoic Rio Branco "anorogenic" plutonic suite, SW of Amazonian Craton, Brazil: High heat flow and crustal extension behind the Santa Helena arc?: Journal of South American Earth Sciences, v. 16, p. 1–14.

Geraldes, M.C., Heilbron, M.C.P., and Teixeira, W., 2004b, Lithospheric versus asthenospheric source of the SW Amazonian Craton A-type granites: The role of the Paleo- and Mesoproterozoic accretionary belts for their coeval continental suites: Episodes, v. 27, no. 3, p. 1–5.

Gibbs, A.K., and Barron, C.N., 1983, The Guiana Shield revisited: Episodes, v. 2, p. 7–14.

Gibbs, A.K., and Barron, C.N., 1993, The geology of the Guyana Shield: Oxford Monographs on Geology and Geophysics, v. 22, 246 p.

Goldstein, S.L., Arndt, N.T., and Stallard, R.F., 1997, The history of a continent from U-Pb ages of zircons from Orinoco River sand and Sm-Nd isotopes in Orinoco basin river sediments: Chemical Geology, v. 139, p. 271–286, doi: 10.1016/S0009-2541(97)00039-9.

Gruau, G., Martin, H., Leveque, B., Capdevila, R., and Marot, A., 1985, Rb-Sr and Sm-Nd geochronology of lower Proterozoic granite-greenstone belts terrain in French Guiana: South America: Precambrian Research, v. 30, p. 63–80, doi: 10.1016/0301-9268(85)90029-4.

Heintz, M., Debayle, E., and Vauchez, A., 2005, Upper mantle structure of the South American continent and neighboring oceans from surface wave tomography: Tectonophysics, v. 406, p. 115–139, doi: 10.1016/j.tecto.2005.05.006.

Hoffman, P.F., 1991, Did the breakout of Laurentia turn Gondwana inside-out?: Science, v. 252, p. 1409–1412, doi: 10.1126/science.252.5011.1409.

Hurley, P.M., de Almeida, F.F.M., Melcher, G.C., Cordani, U.G., Rand, J.R., Kawashita, K., Vandoros, P., Pinson, W.H., and Fairbarn, H.W., 1967, Test of continental drift by comparison of radiometric ages: Science, v. 157, p. 495–500, doi: 10.1126/science.157.3788.495.

Jahn, B.M., Wu, F.Y., and Chen, B., 2000, Granitoids of Central Asian orogenic belt and continental growth in the Phanerozoic: Transactions of the Royal Society of Edinburgh, Earth Sciences, v. 91, p. 181–193.

Johnson, P.R., 2001, Oblique sinistral transpression in the Arabian Shield: The timing and kinematics of a Neoproterozoic suture zone: Precambrian Research, v. 107, p. 117–138, doi: 10.1016/S0301-9268(00)00157-1.

Keppie, J.D., and Ortega-Gutierrez, F.O., 1999, Middle American Precambrian basement: A missing piece of the reconstructed 1 Ga orogen, *in* Ramos, V.A., and Keppie, J.D., Laurentia-Gondwana connections before Pangea: Geological Society of America Special Paper 336, p. 199–210.

Kröner, A., and Cordani, U.G., 2003, African, southern Indian and South American cratons were not part of the Rodinia supercontinent: Evidence from field relationships and geochronology: Tectonophysics, v. 375, p. 325–352, doi: 10.1016/S0040-1951(03)00344-5.

Kröner, A., Greiling, R., Reischmann, T., Hussein, I.M., Stern, R.J., Dürr, S., Krüger, J., and Zimmer, M., 1987, Pan-African crustal evolution in the Nubian segment of northeast Africa, *in* Kröner, A., ed., Proterozoic Lithospheric Evolution: American Geophysical Union Geodynamics Series, v. 17, p. 235–257.

Kroonenberg, S.B., 1982, A Grenvillian granulite belt in the Colombian Andes and its relation to the Guiana Shield: Geologie en Mijnbouw, v. 61, p. 325–333.

Kusky, T.M., and Bradley, D.C., 1999, Kinematics of mélange fabrics: Examples and applications from the McHugh Complex, Kenai Peninsula, Alaska: Journal of Structural Geology, v. 21, p. 1773–1796, doi: 10.1016/S0191-8141(99)00105-4.

Lamarão, C.N., Dall'Agnol, R., Lafon, I.M., and Lima, E.F., 2002, Geology, geochemistry and Pb-Pb zircon geochronology of the Paleoproterozoic magmatism of Vila Riozinho, Tapajós gold province, Amazonian Craton, Brazil: Precambrian Research, v. 119, p. 189–223, doi: 10.1016/S0301-9268(02)00123-7.

Lamarão, C.N., Dall'Agnol, R., and Pimentel, M.M., 2005, Nd isotopic composition of Paleoproterozoic volcanic and granitoid rocks of Vila Riozinho: Implications for the crustal evolution of the Tapajós gold province, Amazon craton: Journal of South American Earth Sciences, v. 18, p. 277–292, doi: 10.1016/j.jsames.2004.11.005.

Leite, W.B., Jr., Bettencourt, J.S., and Payolla, B.L., 2003, Evidence for multiple sources inferred from Sr and Nd isotopic data from felsic rocks in the Santa Clara Intrusive Suite, Rondonia, Brazil: Short Papers of the IV SSAGI: Salvador, Brazil, p. 583–585.

Litherland, M., and Power, P.E.J., 1989, The geologic and geomorphologic evolution of Serrania Huanchaca, eastern Bolivia: The legendary "Lost World": Journal of South American Earth Sciences, v. 2, p. 1–17, doi: 10.1016/0895-9811(89)90023-0.

Litherland, M., Annels, R.N., Appleton, J.D., Berrange, J.P., Bloomfield, K., Darbyshire, D.P.F., Fletcher, C.J.N., Hawkins, M.P., Klinck, B.A., Mitchell, W.I., O'Connor, E.A., Pitfield, P.E.J., Power, G., and Webb, B.C., 1986, The geology and mineral resources of the Bolivian Precambrian shield: Overseas Memoir British Geological Survey, v. 9, 153 p.

Litherland, M., Annels, R.N., Hawkins, M.P., Klinck, B.A., O'Connor, E.A., Pitfield, P.E.J., Power, G., Darbyshire, D.P.F., Fletcher, C.N.J., Mitchell, W.I., and Webb, B.C., 1989, The Proterozoic of eastern Bolivia and its relationships to the Andean mobile belt: Precambrian Research, v. 43, p. 157–174, doi: 10.1016/0301-9268(89)90054-5.

Mantovani, M.S.M., Quintas, M.C.L., Shukowsky, W., and Brito-Neves, B.B., 2005, Delimitation of the Paranapanema Proterozoic block: A geophysical contribution: Episodes, v. 28, no. 1, p. 18–22.

Maruyama, S., Isozaki, Y., Kimura, G., and Terabayashi, M., 1997, Paleogeographic map of the Japanese islands: Plate tectonic synthesis from 750 Ma to the present: Island Arc, v. 6, p. 121–142, doi: 10.1111/j.1440-1738.1997.tb00043.x.

McMenamin, M.O., and McMenamin, D.L.S., 1990, The emergence of animals: The Cambrian breakthrough: New York, Columbia University Press, 217 p.

McReath, I., and Faraco, M.T.L., 2006, Paleoproterozoic greenstone-granite belts in northern Brazil and the former Guyana Shield–West African Craton province: Revista do Instituto de Geociências USP, v. 5, p. 49–63.

Moraes, R., Fuck, R.A., Pimentel, M.M., Gioia, S.M.C.L., Hollanda, M.H.B.M., and Armstrong, R., 2006, The bimodal rift-related Juscelândia volcanosedimentary sequence in central Brazil: Mesoproterozoic extension and Neoproterozoic metamorphism: Journal of South American Earth Sciences, v. 20, p. 287–301, doi: 10.1016/j.jsames.2005.07.016.

Moura, M.A., and Botelho, N.F., 2002, Petrologia do magmatismo associado `a mineralização do tipo Ouro pórfiro da Província aurífera Juruena Teles Pires (MT): Revista Brasileira de Geociências, v. 32, p. 377–386.

Neder, R.D., Leite, J.A.D., Figueiredo, B.R., and McNaughton, N.J., 2002, 1.76 Ga volcano-plutonism in the southwestern Amazonian Craton, Aripuanã-MT, Brazil: Tectono-stratigraphic implications from SHRIMP U-Pb zircon data and rock geochemistry: Precambrian Research, v. 119, p. 171–187, doi: 10.1016/S0301-9268(02)00122-5.

Nomade, S., Yan, C., Pouclet, A., Féraud, G., Théveniaut, H., Daouda, B.Y., Vidal, M., and Rigolet, C., 2003, The Guiana and the West African Shield Paleoproterozoic grouping: New paleomagnetic data for French Guiana and the Ivory Coast: Geophysical Journal, v. 154, p. 677–694, doi: 10.1046/j.1365-246X.2003.01972.x.

Onstott, T.C., Hall, C.M., and York, D., 1989, $^{40}Ar/^{39}Ar$ thermochronometry on the Imataca Complex, Venezuela: Precambrian Research, v. 42, p. 255–291, doi: 10.1016/0301-9268(89)90014-4.

Payolla, B.L., Bettencourt, J.S., Kozuch, M., Leite, W.B., Jr., Fetter, A.H., and Van Schmus, W.R., 2002, Geological evolution of the basement rocks in the east-central part of the Rondonia Tin province, SW Amazonian Craton, Brazil: U-Pb and Sm-Nd isotopic constraints: Precambrian Research, v. 119, p. 141–169, doi: 10.1016/S0301-9268(02)00121-3.

Payolla, B.L., Bettencourt, J.S., Tosdal, R.M., Wooden, J.L., and Leite, W.B., Jr., 2003, SHRIMP U-Pb zircon geochronology of high-grade paragneisses from NE Rondonia, SW Amazonian Craton, Brazil: Short Papers of the IV SSAGI: Salvador, Brazil, p. 248–251.

Pesonen, L.J., Elming, S.-Ä., Mertanen, S., Pisarewsky, S., D'Agrella-Filho, M.S., Meert, J.G., Schmidt, P.W., Abrahamsen, N., and Bylund, G., 2003, Paleomagnetic configuration of continents during the Proterozoic: Tectonophysics, v. 375, p. 289–324, doi: 10.1016/S0040-1951(03)00343-3.

Pimentel, M.M., and Fuck, R.A., 1992, Neoproterozoic crustal accretion in central Brazil: Geology, v. 20, p. 375–379, doi: 10.1130/0091-7613(1992)020<0375:NCAICB>2.3.CO;2.

Pimentel, M.M., Whitehouse, M.J., Viana, M.G., Fuck, R.A., and Machado, N., 1997, The Mara Rosa arc in the Tocantins province: Further evidence for Neoproterozoic crustal accretion in central Brazil: Precambrian Research, v. 81, p. 299–310, doi: 10.1016/S0301-9268(96)00039-3.

Pimentel, M.M., Fuck, R.A., and Botelho, N.F., 1999, Granites and the geodynamic history of the Neoproterozoic Brasilia belt: Lithos, v. 46, p. 463–483, doi: 10.1016/S0024-4937(98)00078-4.

Pimentel, M.M., Fuck, R.A., Jost, H., Filho, C.F.F., and de Araújo, S.M., 2000, The basement of the Brasilia fold belt and the Goiás magmatic arc, *in*

Cordani, U.G., Milani, E.J., Thomaz-Filho, A., and Campos, D.A., eds., Tectonic evolution of South America: Rio de Janeiro, Brazil, 31st International Geological Congress, p. 195–229.

Pinho, F.E.C., Fyfe, W.S., and Pinho, M.A.S.B., 1997, Early Proterozoic evolution of the Alto Jauru greenstone belt, southern Amazonian Craton, Brazil: International Geology Review, v. 39, p. 220–229.

Priem, H.N.A., Boelrijk, N.A.I.M., Hebeda, E.H., Verdumen, E.A.Th., and Bon, E.H., 1971a, Granitic complexes and associated Tin mineralization of Grenville age in Rondonia, western Brazil: Geological Society of America Bulletin, v. 82, p. 1095–1102, doi: 10.1130/0016-7606(1971)82[1095:GCAATM]2.0.CO;2.

Priem, H.N.A., Boelrijk, N.A.I.M., Hebeda, E.H., Verdumen, E.A.Th., and Verschure, R.H., 1971b, Isotopic ages of the Trans-Amazonian acidic magmatism and the Nickerie metamorphic episode in the Precambrian basement of Suriname, South America: Geological Society of America Bulletin, v. 82, p. 1667–1680, doi: 10.1130/0016-7606(1971)82[1667:IAOTTA]2.0.CO;2.

Priem, H.N.A., Boelrijk, N.A.I.M., Hebeda, E.H., Verdurmen, E.A.Th., and Verschure, R.H., 1973, Age of the Precambrian Roraima Formation in northeastern South America: Evidence from isotopic dating of Roraima pyroclastic rocks in Suriname: Geological Society of America Bulletin, v. 84, p. 1677–1684, doi: 10.1130/0016-7606(1973)84<1677:AOTPRF>2.0.CO;2.

Priem, H.N.A., Boelrijk, N.A.I.M., Hebeda, E.H., Kuyper, R.P., de Roever, E.W.F., Verdurmen, E.A.Th., Verschure, R.H., and Wilens, J.B., 1978, How old are the supposedly Archean charnockitic granulites in the Guyana Shield basement of western Suriname (South America)?: U.S. Geological Survey Open-File Report 78-701, p. 341–343.

Priem, H.N.A., Bon, E.H., Verdumen, E.A.Th., and Bettencourt, J.S., 1989, Rb-Sr chronology of Precambrian crustal evolution in Rondonia (western margin Brazilian Craton): Journal of South American Earth Sciences, v. 2, p. 163–170, doi: 10.1016/0895-9811(89)90044-8.

Rämö, O.T., and Haapala, I., 1995, One hundred years of rapakivi granite: Mineral Petrology, v. 52, p. 129–185, doi: 10.1007/BF01163243.

Ramos, V.A., 1988, Late Proterozoic–early Paleozoic of South America: A collisional history: Episodes, v. 11, no. 3, p. 168–174.

Ramos, V.A., 2000, The southern central Andes, *in* Cordani, U.G., Milani, E.J., Thomaz-Filho, A., and Campos, D.A., eds., Tectonic evolution of South America: Rio de Janeiro, Brazil, 31st International Geological Congress, p. 561–604.

Ramos, V.A., and Aleman, A., 2000, Tectonic evolution of the Andes, *in* Cordani, U.G., Milani, E.J., Thomaz-Filho, A., and Campos, D.A., eds., Tectonic evolution of South America: Rio de Janeiro, Brazil, 31st International Geological Congress, p. 635–685.

Reis, N.J., Fraga, L.M., Faria, M.S.G., and Almeida, M.E., 2003, Geologia do Estado de Roraima, Brasil, *in* Geologie de la France and surrounding areas, Special Guiana Shield: Bureau de recherches géologiques et minières–Société Géologique de France, no. 2–4, p. 121–134.

Restrepo-Pace, P.A., Ruiz, J., Gehrels, G., and Cosca, M., 1997, Geochronology and Nd isotopic data of Grenville-age rocks in the Colombian Andes: New constraints for Late Proterozoic–early Paleozoic paleocontinental reconstructions of the Americas: Earth and Planetary Science Letters, v. 150, p. 427–441, doi: 10.1016/S0012-821X(97)00091-5.

Rivers, T., 1997, Lithotectonic elements of the Grenville province: Review and tectonic implications: Precambrian Research, v. 86, p. 117–154, doi: 10.1016/S0301-9268(97)00038-7.

Rivers, T., Martignole, J., Gower, C.F., and Davidson, A., 1989, New tectonic divisions of the Grenville province, southeast Canada: Tectonics, v. 8, p. 63–84.

Rizzotto, G.J., Bettencourt, J.S., Teixeira, W., Pacca, I.G., D'Agrella-Filho, M.S., Vasconcelos, P.M., Basei, M.A.S., Onoe, A.T., and Passarelli, C.R., 2002, Geologia e geocronologia da Suite Metamorfica Colorado e suas encaixantes, SE de Rondônia: Implicações para a evolução mesoproterozóica do SW do Craton Amazônico: Revista do Instituto de Geociências USP: Série Científica, v. 2, p. 41–55.

Rosa-Costa, L.T., Lafon, J.M., and Delor, C., 2006, Zircon geochronology and Sm-Nd isotopic study: Further constraints for the Archean and Paleoproterozoic geodynamical evolution of the SE Guiana Shield, north of Amazonian Craton, Brazil: Gondwana Research, v. 10, no. 3-4, p. 277è300.

Ruiz, A.R., 2005, Evolução geológica do sudoeste do Craton Amazônico, região limítrofe Brasil–Bolivia–Mato Grosso [Ph.D. thesis]: State University of São Paulo, UNESP–Rio Claro, São Paulo, Brazil, 260 p.

Sadowski, G.R., and Bettencourt, J.S., 1996, Mesoproterozoic tectonic correlations between eastern Laurentia and the western border of the Amazonian Craton: Precambrian Research, v. 76, p. 213–227, doi: 10.1016/0301-9268(95)00026-7.

Saes, G.S., 1999, Evolução tectônica e paleogeográfica do Aulacógeno Aguapeí (1.2–1.0 Ga) e dos terrenos do seu embasamento na porção sul do Craton Amazônico [Ph.D. thesis]: Instituto de Geociências, University of São Paulo, 135 p.

Santos, J.O.S., 2003, Geotectônica dos Escudos das Guianas e Brasil Central, *in* Bizzi, L.A., Schobbenhaus, C., Vidotti, R.M., and Gonçalves, J.H., eds., Geologia, Tectônica e Recursos Minerais do Brasil: Companhia de Pesquisas de Recursos Minerais, Brasilia, Brazil, Cap. IV, parte II, p. 169–226.

Santos, J.O.S., Hartmann, L.A., Gaudette, H.E., Groves, D.I., McNaughton, N.J., and Fletcher, I.R., 2000, A new understanding of the provinces of the Amazon craton based integration of field mapping and U-Pb and Sm-Nd geochronology: Gondwana Research, v. 3, no. 4, p. 453–488, doi: 10.1016/S1342-937X(05)70755-3.

Santos, J.O.S., Potter, P.E., Reis, N.J., Hartmann, L.A., Flecher, I.R., and McNaughton, N.J., 2003, Age, source and regional stratigraphy of the Roraima Supergroup and Roraima-like outliers in northern South America based on U-Pb SHRIMP geochronology: Geological Society of America Bulletin, v. 115, p. 331–348, doi: 10.1130/0016-7606(2003)115<0331:ASARSO>2.0.CO;2.

Santos, J.O.S., Van Breemen, O.B., Groves, D.I., Hartmann, L.A., Almeida, M.E., McNaughton, N.J., and Fletcher, I.R., 2004, Timing and evolution of multiple Paleoproterozoic magmatic arcs in the Tapajós Domain, Amazon Craton: Constraints from SHRIMP and TIMS zircon, baddeleyite and titanite U-Pb geochronology: Precambrian Research, v. 131, p. 73–109, doi: 10.1016/j.precamres.2004.01.002.

Santos, J.O.S., McNaughton, N.J., Hartmann, L.A., Fletcher, I.R., and Matos, R., 2005, The age of the deposition of the Aguapeí Group, western Amazon Craton, based on U-Pb study on diagenetic xenotime and detrital zircon: Latin American Congress, Quito, Equador, Extended Abstracts, p. 1–4.

Sato, K., and Tassinari, C.C.G., 1997, Principais eventos de acresção continental do Craton Amazônico baseados em idades modelo Sm/Nd, calculadas em evolução de estágio único e estágio duplo, *in* Contribuições à Geologia da Amazônia, 5° Simpósio de Geologia de Amazônia: Sociedade Brasileria de Geologia, Núcleo Norte, Belém, p. 91–142.

Şengör, A.M.C., and Natal'in, B.A., 1996, Turkic-type orogeny and its role in the making of the continental crust: Annual Review of Earth and Planetary Sciences, v. 24, p. 263–337, doi: 10.1146/annurev.earth.24.1.263.

Snelling, N.J., and McConnell, R.B., 1969, The geochronology of Guyana: Geologie en Mijnbouw, v. 48, p. 201–213.

Sparrenberger, I., Bettencourt, J.S., Tosdal, R.M., and Wooden, J.L., 2002, Datações U-Pb convencional versus SHRIMP do Maciço Estanífero Santa Bárbara, Suite Granitos Últimos de Rondônia, Brasil: Revista do Instituto de Geociências USP: Série Científica, v. 2, p. 79–94.

Starmer, I.C., 1996, Accretion, rifting and collision in the North Atlantic supercontinent, 1700–950 Ma, *in* Brewer, T.S., ed., Precambrian crustal evolution in the North Atlantic region: Geological Society [London] Special Publication 112, p. 219–248.

Tassinari, C.C.G., and Macambira, M.J.B., 1999, Geochronological provinces of the Amazonian Craton: Episodes, v. 22, no. 3, p. 174–182.

Tassinari, C.C.G., and Macambira, M.J.B., 2004, A evolução tectônica do Craton Amazônico, *in* Mantesso-Neto, V., Bartorelli, A., Carneiro, C.D.R., and Brito-Neves, B.B., eds., Geologia do Continente Sul-Americano: São Paulo, Brazil, p. 471–485.

Tassinari, C.C.G., Cordani, U.G., Nutman, A.P., van Schmus, W.R., Bettencourt, J.S., and Taylor, P.N., 1996, Geochronological systematics on basement rocks from the Rio Negro–Juruena province (Amazonian Craton) and tectonic implications: International Geology Review, v. 38, p. 161–175.

Tassinari, C.C.G., Bettencourt, J.S., Geraldes, M.C., Macambira, M.J.B., and Lafon, J.M., 2000, The Amazon craton, *in* Cordani, U.G., Milani, E.J., Thomaz-Filho, A., and Campos, D.A., eds., Tectonic evolution of South America: Rio de Janeiro, Brazil, 31st International Geological Congress, p. 41–95.

Tassinari, C.C.G., Munhá, J.M.U., Teixeira, W., Palácios, T., Nutman, A.P., and Sosa, S.C., Santos, A.P., and Calado, B.O., 2004, The Imataca Complex, NW Amazonian Craton, Venezuela: Crustal evolution and integration of geochronological and petrological cooling histories: Episodes, v. 27, no. 1, p. 3–12.

Teixeira, W., and Tassinari, C.C.G., 1984, Caracterização geocronológica da província Rondoniana e suas implicações geotectônicas, Atas, 2° *in* Simpósio Amazônico: Sociedade Brasileira de Geologia Núcleo Norte–Departamento Nacional da Produção Mineral, Manaus, Brazil, P. 87–91.

Teixeira, W., Tassinari, C.C.G., Cordani, U.G., and Kawashita, K., 1989, A review of the geochronology of the Amazonian Craton: Tectonic implications: Precambrian Research, v. 42, p. 213–227, doi: 10.1016/0301-9268(89)90012-0.

Teixeira, W., Bettencourt, J.S., Girardi, V.A.V., Onoe, A.T., and Sato, K., 2006, Mesoproterozoic mantle heterogeneity in SW Amazonian Craton: $^{40}Ar/^{39}Ar$ and Nd-Sr evidence from mafic-felsic rocks, *in* Hanski, E., Mertanen, S., Rämö, T., and Vuollo, J., eds., Dyke Swarms—Time Markers of Crustal Evolution: Selected Papers, 5th International Dyke Conference, Rovaniemi, Finland, and 4th International Conference, Kawzulu-Natal, South Africa, p. 113–130.

Tohver, E., van der Pluijm, B.A., Van der Voo, R., Rizzotto, G., and Scandolara, J.E., 2002, Paleogeography of the Amazon craton at 1.2 Ga: Early Grenvillian collision with the Llano segment of Laurentia: Earth and Planetary Science Letters, v. 199, p. 185–200, doi: 10.1016/S0012-821X(02)00561-7.

Tohver, E., van der Pluijm, B.A., Mezger, K., Essene, E., Scandolara, J.E., and Rizzotto, G., 2004, Significance of the Nova Brasilândia metasedimentary belt in western Brazil: Refining the Mesoproterozoic boundary of the Amazon craton: Tectonics, v. 23, p. 1–20, doi: 10.1029/2003TC001563.

Tohver, E., van der Pluijm, B.A., Mezger, K., Scandolara, J.E., and Essene, E., 2005a, Two stage tectonic history of the SW Amazon craton in the late Mesoproterozoic: Identifying a cryptic suture zone: Precambrian Research, v. 137, p. 35–59, doi: 10.1016/j.precamres.2005.01.002.

Tohver, E., van der Pluijm, B.A., Scandolara, J.E., and Essene, E., 2005b, Late Mesoproterozoic deformation of SW Amazonia (Rondonia, Brazil): Geochronological and structural evidence for collision with southern Laurentia: Journal of Geology, v. 113, p. 309–323, doi: 10.1086/428807.

Tohver, E., D'Agrella-Filho, M.S., and Trindade, R.I.F., 2006a, Paleomagnetic record of Africa and South America for the 1200–500 Ma interval, and evaluation of Rodinia and Gondwana assemblies: Precambrian Research, v. 147, no. 3, p. 193–222.

Tohver, E., Teixeira, W., van der Pluijm, B.A., Geraldes, M.C., and Bettencourt, J.S., 2006b, A restored transect across the exhumed Grenville orogen of Laurentia and Amazonia, with implications for crustal architecture: Geology, v. 34, p. 669–672.

Trompette, R., 1994, Geology of western Gondwana (2000–500 Ma): Rotterdam, Balkema, 350 p.

Tschanz, C.M., Marvin, R.F., Cruz, J., Mehnert, H.H., and Cebula, G.T., 1974, Geologic evolution of the Sierra Nevada de Santa Marta, northeastern Colombia: Geological Society of America Bulletin, v. 85, p. 273–284, doi: 10.1130/0016-7606(1974)85<273:GEOTSN>2.0.CO;2.

Vidal, M., Delor, C., Pouclet, A., Simeon, Y., and Alric, G., 1996, Evolution geodynamique de l'Afrique de l'Ouest entre 2.2 Ga et 2 Ga: Le style "Archeen" des ceintures vertes et des ensembles sedimentaires birimiens du nord-est de la Cote d'Ivoire: Bulletin de la Société Géologique de France, v. 167, p. 307–319.

Wasteneys, H.A., Clark, A.H., Farrar, E., and Langridge, R.J., 1995, Grenvillian granulite-facies metamorphism in the Arequipa Massif, Peru: A Laurentia-Gondwana link: Earth and Planetary Science Letters, v. 132, p. 63–73, doi: 10.1016/0012-821X(95)00055-H.

Weil, A.B., Van der Voo, R., Mac Niocaill, C., and Meert, J.G., 1998, The Proterozoic supercontinent Rodinia: Paleomagnetically derived reconstructions for 1100 to 800 Ma: Earth and Planetary Science Letters, v. 154, p. 13–24, doi: 10.1016/S0012-821X(97)00127-1.

Wilson, J.T., 1966, Did the Atlantic close and then re-open?: Nature, v. 211, p. 676–681, doi: 10.1038/211676a0.

Windley, B.F., 1993, Proterozoic anorogenic magmatism and its orogenic connections: Geological Society [London] Journal, v. 150, p. 39–50.

Wynn, J.C., Sidder, G.B., Gray, F., Page, N., and Mendoza, S.V., 1993, Geology and mineral deposits of the Venezuelan Guyana Shield: U.S. Geological Survey Bulletin B-2124, p. A1–A7.

Manuscript Accepted by the Society 1 February 2007

Printed in the USA

The Geological Society of America
Memoir 200
2007

Precambrian accretionary history and Phanerozoic structures—A unified explanation for the tectonic architecture of the Nebraska region, USA

Marvin P. Carlson
Nebraska Geological Survey, University of Nebraska, Lincoln, Nebraska 68588-0517, USA

ABSTRACT

The Phanerozoic history in Nebraska and adjacent regions contains many patterns of structure and stratigraphy that can be directly related to the history of the Precambrian basement rocks of the area. A process is proposed that explains the southward growth of North America during the period 1.8–1.6 Ga. A series of families of accretionary events during the Proterozoic emplaced sutures that remained as fundamental basement weak zones. These zones were rejuvenated in response to a variety of continental stress events that occurred during the Phanerozoic. By combining the knowledge of basement history with the history of rejuvenation during the Phanerozoic, both the details of Proterozoic accretionary growth and an explanation for the patterns of Phanerozoic structure and stratigraphy is provided.

Keywords: Precambrian tectonics, Phanerozoic structures, accretionary terranes, midcontinent USA.

INTRODUCTION

The series of volumes of the *Decade of North American Geology* effectively summarized the geologic history of the continent. Separate volumes and selected authors described the general Phanerozoic history, while other volumes interpreted the Precambrian history. However, in these volumes there was not an effective effort to link these two periods of continental tectonics. The structures of the Phanerozoic, both current and historical, are fairly well known based on the limited outcrops and the thousands of deep well records available. There has not been, however, any comprehensive explanation for the varied trends and repetitive history of these Phanerozoic features. It has been shown that there is a history of rejuvenation for many of these structures, but it has been difficult to relate these structures to the varied patterns of continental stress fields present during the Phanerozoic. There is also no direct indication that there were causative vertical forces present across the region.

The Precambrian basement rocks in the Nebraska region are predominately accretionary arc material ranging in age from ca. 1.8 to 1.6 Ga—the Central Plains orogen of Sims and Peterman (1986) and Sims (1990) and a part of the Transcontinental Proterozoic province of Van Schmus et al. (1993). These publications describe these rocks as a broad granitic gneiss and schist terrane trending across the midcontinent and becoming younger in age southward as might be expected for an accretionary growth process. CD-ROM Working Group (2002, p. 9) noted "The Proterozoic orogen was rapidly assembled from oceanic terranes with no major continent-continent collisions in contrast to much of the Archean and Proterozoic lithosphere to the north." This

Carlson, M.P., 2007, Precambrian accretionary history and Phanerozoic structures—A unified explanation for the tectonic architecture of the Nebraska region, USA, *in* Hatcher, R.D., Jr., Carlson, M.P., McBride, J.H., and Martínez Catalán, J.R., eds., 4-D Framework of Continental Crust: Geological Society of America Memoir 200, p. 321–326, doi: 10.1130/2007.1200(15). For permission to copy, contact editing@geosociety.org.

Proterozoic orogen is bounded on the north by the Hudsonian continent. The 1.4 Ga Granite-Rhyolite province lies to the south. The purpose of this paper is to provide an interpretation for the accretionary process that created the Proterozoic basement of Nebraska and adjacent regions and demonstrate how the geometry of these crystalline rocks has controlled the pattern of tectonic rejuvenation during the Phanerozoic.

CURRENT STRUCTURES

Figure 1 illustrates the major structures now present in the region. The presence of these features has been well documented as well as their influence on depositional history over time. A pattern of tectonic rejuvenation during the Phanerozoic across the area was noted by Carlson (1999), wherein a series of paleogeologic maps illustrated both an obvious repetitive occurrence of structural highs and basins and the resulting impact on distribution, thickness, facies, sources, and erosion during the Phanerozoic. An understanding of these patterns is important in resource exploration. Regionally, there are structural grains trending northwest-southeast and northeast-southwest. The Rocky Mountains and the Nemaha uplift both have a general north-south alignment. It was suggested (Carlson, 2001) that most of these rejuvenated Phanerozoic structures in the Nebraska region are directly reflective of the patterns of the suturing of arc terranes that resulted in the southward growth of the North American continent during the period from ca. 1.8 to 1.6 Ga. A similar process was proposed by Karlstrom and Bowring (Van Schmus et al., 1993, p. 208): "Another implication of our assembly model is that lithospheric discontinuities, which may have initially represented terrane boundaries and deformation zones formed during assembly, persisted as major steeply dipping anisotropies after assembly and served to partition subsequent deformation, locus of plutonism, and later differential uplift." In Nebraska, the architecture of these accretionary terranes also had a direct effect on the location and trend of the 1.1 Ga Midcontinent rift system.

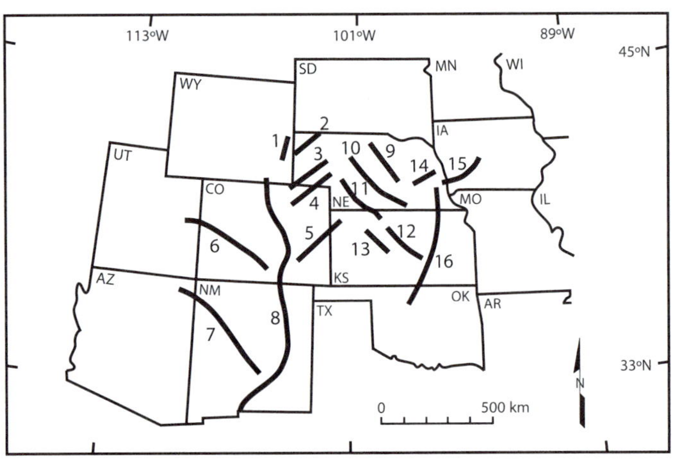

Figure 1. Important positive structures in the western Midcontinent region. 1—Hartville uplift; 2—Agate–White River fault; 3—Morrill County high; 4—Wattenberg high; 5—Los Animas arch; 6—Uncompahgre uplift; 7—Zuni-Pedernal uplift; 8—Rocky Mountains; 9—Ord arch; 10—North Platte arch; 11—Cambridge arch; 12—Central Kansas uplift; 13—Rush rib; 14—Eastern Nebraska high; 15—Redfield anticline; 16—Nemaha uplift.

NEBRASKA BASEMENT

Much of the Precambrian history of the central United States is contained within the Transcontinental Proterozoic province (Fig. 2) as summarized by Reed (1993). This broad pattern of Proterozoic history extends from Arizona to Wisconsin and includes most of the basement rocks underlying Nebraska as a part of the Central Plains orogen as defined by Sims and Peterman (1986). This orogen has been characterized as consisting mainly of schist and gneiss with an accretionary arc history. The metamorphic grade is mainly amphibolite facies but includes greenschist facies rocks that range in texture from granitoid to mylonitic. Scattered within the metamorphic terrane are isolated bodies of quartzite, volcanic rock, and anorthosite. The accretionary terranes were punctured by numerous plutons related both to the accretionary sutures and to the later (1.4 Ga) anorogenic granites coeval with the southern Granite-Rhyolite province. The wide distribution of the 1.4 Ga plutons suggests that the Granite-Rhyolite terrane was originally much more broadly distributed over the accretionary terranes, possibly as emplacement above a mantle upwelling (Hoffman, 1989).

Lidiak (1972) summarized the Precambrian history of Nebraska and generally described the distribution of basement rock types. Additional discussion was provided by Treves and Lowe (1984). Figure 3 is the current interpretation of the distribution of rock type at the Precambrian surface based on over 2200 drill holes that have penetrated into the Precambrian surface. Although few cores are available, fragmental thin sections

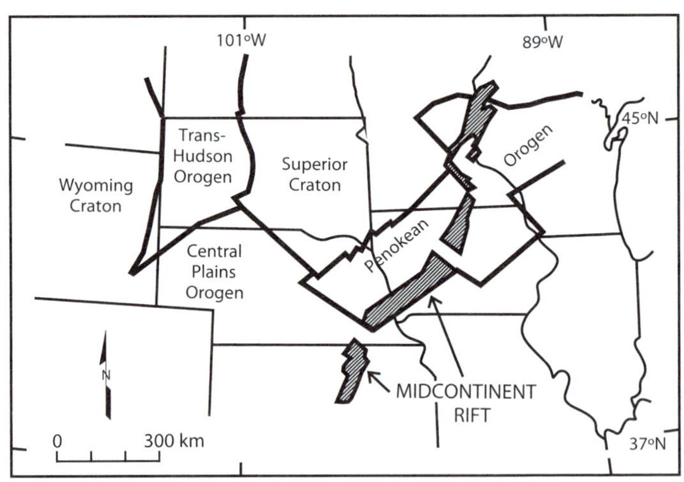

Figure 2. Major Precambrian provinces of the Nebraska region. Modified after Reed (1993).

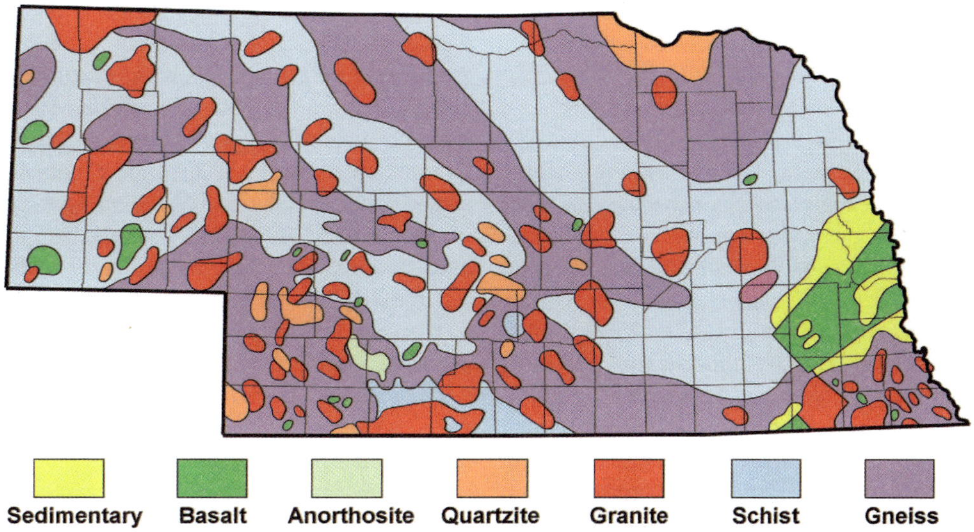

Figure 3. Interpretation of the distribution of rock type at the Precambrian surface. Based on petrologic descriptions by E.G. Lidiak and S.B. Treves; data on file at the Nebraska Geological Survey.

have been derived from the drill cuttings and described by E.G. Lidiak and S.B. Treves. These data are on open file and are being incorporated into a summary publication.

As illustrated on Figure 3, belts of schist and gneiss dominate the Precambrian surface in Nebraska. Van Schmus et al. (1993, p. 252) state that the petrography and bulk chemical composition of the orthogneisses are consistent with interpretation of them as metamorphosed remnants of accreted island arcs. They also conclude that the metasedimentary gneiss and schist are closely associated with the orthogneisses and may represent deposition within (1) a primary arc or (2) intra-arc basins formed by later extension.

Isotopic dates were obtained for a number of the granitic plutons and summarized by Van Schmus et al. (1993). The Precambrian of southeastern Nebraska contains a portion of the basalts and sediments characteristic of the Midcontinent rift system (1.1 Ga). The petrologic, geochemical, and paleomagnetic character of the rift in Nebraska was described by Marshall and Lidiak (1996). The basement surface in Nebraska is overlain by sediments ranging in age from Upper Cambrian to Cretaceous.

PROTEROZOIC HISTORY

Prior to ca. 1.83 Ga, the Wyoming and Superior provinces were separate continents accumulating accretionary material along their margins, notably the Penokean orogen. The collision of these two protocontinents ca. 1.83 Ga produced the intervening mixed terrane of the Trans-Hudson orogen. The combined units, the Hudsonian continent, became proto–North America. During the period from ca. 1.8 to 1.6 Ga, this continent grew southward by accretion of a series of arc terranes constituting the Central Plains orogen (Fig. 2). Each terrane is associated with a suture that created and has maintained a fundamental weakness in the crystalline basement. A simplistic concept of these accretionary processes is illustrated in Figure 4 as adapted from Hamilton (1988). This process was also described by Lucas et al. (1996) for the growth of North America by addition of the Flin Flon belt and the Penokean orogen. The rates of accretion described by Hamilton (1979) for Indonesia were at least 12 cm/yr. A similar velocity for the terranes proposed for this paper would allow migration toward the continent of ~500 km. The major portion of the terranes in Nebraska and adjoining regions consists of granitic gneiss and schist. The isolated quartzite bodies are interpreted as remnants of multicycle clastics, derived from the older continent and deposited in the forearc basins. The granitic bodies (older than the 1.4 Ga plutons) represent the inboard magmatic

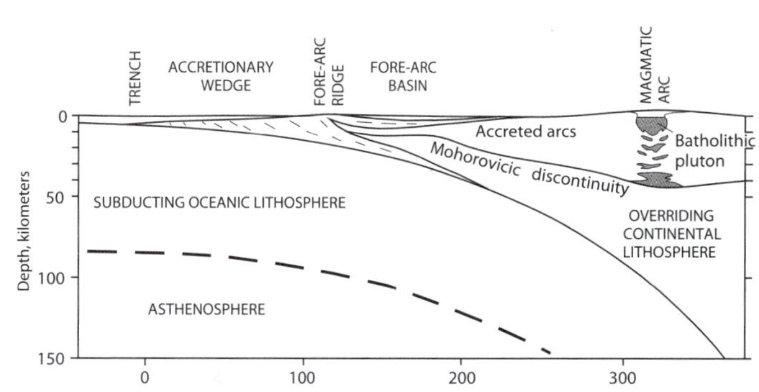

Figure 4. Section across a continental-margin subduction system. Forearc basins are sites of quartzite accumulation. Magmatic arcs are represented by plutons providing dates of suturing. Diagram modified after Hamilton (1988).

arcs with ages representing the time of accretion of each of the associated sutures.

SUTURES TO STRUCTURES

Figure 5 illustrates the location, trends, and ages proposed for the sutures of the defined accretionary terranes in the Nebraska region. This architecture was derived by combining the available data and interpretations for both the Precambrian and Phanerozoic history. Collectively, these data are not sufficient to arrive deductively at the proposed results. However, as stated by Sclater (2003, p. 137–138), "Neither Harry Hess nor Tuzo Wilson was testing a hypothesis. Rather, they were creating new concepts out of synthesis of poorly constrained observational information. They thought their concepts had validity because they explain the patterns they recognized in so many different sets of data." The interpretations presented in Figure 5 are derived from a similar inductive process and serve to explain the currently known tectonic history of the region. The implied correspondence of the current structures and the Precambrian patterns can be noted by comparing Figures 1, 5, and 6.

Dawes terrane. A subdivision of the Central Plains orogen, herein named "Dawes terrane" (Fig. 5), is represented by an assembly of arc-related rocks best illustrated in a series of wells in Dawes County, Nebraska. This terrane represents the earliest collisional event between the orogen and the Hudsonian craton at ca. 1.78 Ga as defined by a series of magmatic arc plutons. Dates (Van Schmus et al., 1993) defining the extent of the Dawes terrane are contained in a 1.77 Ga granite in Sioux County (northwestern Nebraska), a 1.78 Ga granite in Keya Paha County (north-central Nebraska), and a 1.78 Ga granite in Butler County (east-central Nebraska). This collisional event is also recorded in the eastern Wyoming Hartville uplift (Sims and Day, 1999) as a continent-arc collision (ca. 1.76 Ga). Collision along the Cheyenne belt produced comparable deformation across a 60 km wide zone in the Laramie Mountains at presumably the same time (ca. 1.76 Ga). The Dawes terrane underlies a portion of southeastern Wyoming, much of the panhandle of Nebraska, and a part of central Nebraska lying westward of the Superior craton and Penokean orogen. This basement suture was rejuvenated as the Agate–White River fault system (Fig. 1) in northwestern Nebraska during the Tertiary and as the Ord arch (rejuvenated several times during the Phanerozoic) across central Nebraska, and served as the southwestern termination of the Nebraska segment of the Midcontinent rift system in the Late Proterozoic.

Southern Iowa terrane. This terrane accreted at ca. 1.76 Ga, and the suture was rejuvenated as the southern boundary of the Nebraska segment of the Midcontinent rift system. The terrane suture was also active during the Phanerozoic as the Eastern Nebraska high and the Redfield anticline in western Iowa.

Frontier terrane. The next arc cycle contains a similar accumulation of lithotypes, herein termed "Frontier terrane" (Fig. 5), as represented in Frontier County, Nebraska. The Frontier collisional event, ca. 1.71 Ga, is recorded by a 1.71 Ga granite in Grant County (west-central Nebraska), reactivation along the Cheyenne belt in eastern Wyoming, and renewed activity in the Black Hills area (the Harney Peak batholith, 1.71 Ga; Dahl et al., 1999) and in the Hartville uplift (the granite of Haystack Range, 1.72 Ga; Sims and Day, 1999). This terrane underlies northeastern Colorado, southwestern and south-central Nebraska,

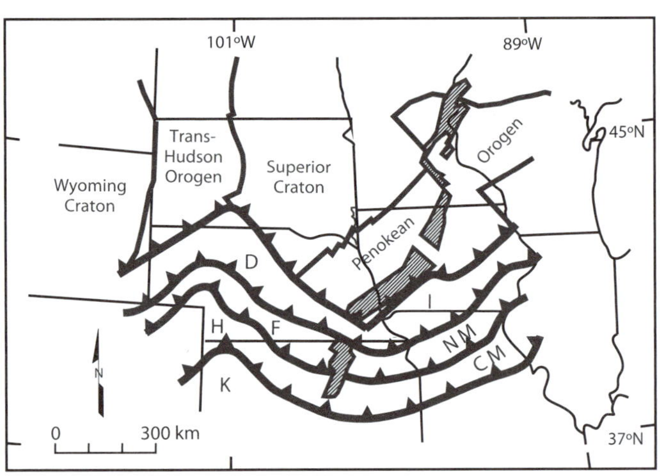

Figure 5. Proposed subdivisions and dates of suturing of the Precambrian accretionary terranes in the Nebraska region: D—Dawes (1.78 Ga); F—Frontier (1.71 Ga); H—Hitchcock (1.67 Ga); K—Kansas (1.61 Ga); I—Southern Iowa; NM—Northern Missouri; CM—Central Missouri. Shaded area is general extent of the Midcontinent rift system (ca. 1.1 Ga).

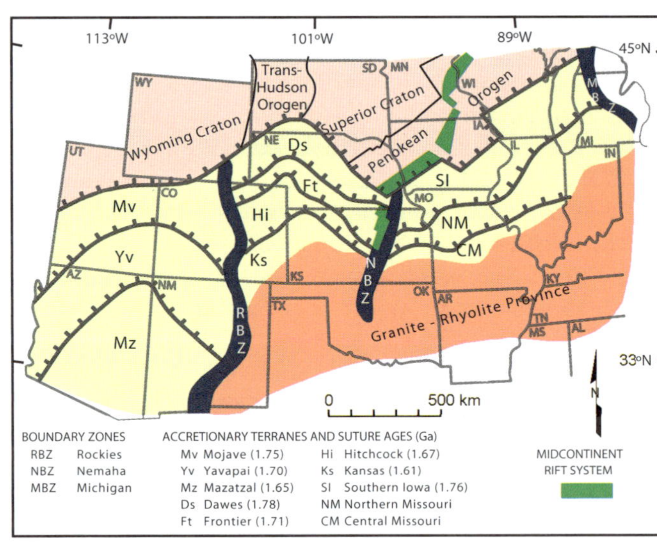

Figure 6. Proposed accretionary framework of the Precambrian provinces in the western Midcontinent region based on the interrelationships of basement tectonics and Phanerozoic structures.

and a portion of northern Kansas. Selverstone et al. (1999) suggested that in southwestern North America, the Mazatzal terrane subducted northward under the Colorado-Yavapai terrane at ca. 1.70 Ga and thus could be a time equivalent of the Frontier terrane suture in the Central Plains orogen. The area between the Dawes and Frontier sutures in southeastern Nebraska marks, and may be the cause of, the separation of segments of the Midcontinent rift system between Nebraska and Kansas. The 1.71 Ga Frontier suture was rejuvenated as the late Phanerozoic Morrill County high (Fig. 1) in western Nebraska and as the North Platte arch (active several times during the Phanerozoic), and formed the northern termination of the Kansas segment of the Midcontinent rift system (1.1 Ga).

Hitchcock terrane. This terrane is located in extreme southwestern Nebraska and adjacent Kansas and Colorado and is named "Hitchcock terrane" for the characteristic features present in Hitchcock County, Nebraska. This region was described by Van Schmus et al. (1993) as an arc-complex (volcanic-plutonic) suite (1.67–1.66 Ga) intruded by numerous plutons as young as 1.62 Ga. The Hitchcock terrane includes southwestern Nebraska, northeastern Colorado, and north-central Kansas. Rejuvenation of the Hitchcock suture (1.67 Ga) created the late Phanerozoic Wattenberg high and provided repetitive uplift along both the Cambridge arch in southwestern Nebraska and the Central Kansas uplift.

Kansas terrane. Another possible arc terrane of the Central Plains orogen is the Western Kansas terrane (Fig. 5). The age of this arc suture (1.61 Ga) is based primarily on the plutons of this age intruded into the older Hitchcock terrane. The Kansas terrane is bordered (and possibly interrupted by) the 1.30–1.40 Ga Granite-Rhyolite province that extends east-west across southern Kansas and Missouri. The Kansas terrane would be the youngest (1.61 Ga) arc-complex subdivision of the orogen. The Kansas suture (1.61 Ga) was rejuvenated as the Los Animas arch of eastern Colorado and the Rush rib of western Kansas, both active in the middle to late Phanerozoic.

Terranes of the eastern Central Plains orogen. Eastward from Nebraska and Kansas, along the southern margin of the Penokean orogen, the terranes designated as the Southern Iowa, Northern Missouri, and Central Missouri (Fig. 5) may be eastward age equivalents of the Nebraska units. It is probable that these eastern Central Plains orogen terranes are separate arc features that had angular convergent and subductive relationships with the Nebraska terranes. Times of the eastern arc-continent collisions would not necessarily be equivalent with the western Central Plains orogen terranes. A series of arc terranes in Wisconsin, similar to those in Nebraska in both lithotype and age, were described by Allen and Hinze (1992) and are an eastward continuation of the southward growth pattern of North America.

MIDCONTINENT RIFT SYSTEM

It is also proposed that the preexisting fabric of the older Precambrian controlled the geometry of the western limb of the Midcontinent rift system. The segments in Minnesota, Iowa, and Nebraska were emplaced along the sutures existing on both margins of the Penokean orogen (Fig. 2). The Kansas segment has a geometry that was controlled by the boundary zone discussed below. Reactivation of portions of the Midcontinent rift system has supplied many of the complicated depositional and structural patterns present in the Phanerozoic of eastern Nebraska and adjacent regions.

BOUNDARY ZONES

It should be noted that although most of the current structures in the area conform to the proposed sutures, there are two significant exceptions. Two generally north-south-trending features, the Rocky Mountains and the Nemaha uplift, share a very similar geometry and history. It is suggested that these trends represent boundary zones (Fig. 6) between several families of accretionary arcs that advanced individually from the southwest, south, and southeast onto the expanding North American continent during the period from 1.8 to 1.6 Ga. The antagonism between these families created these north-south zones of fundamental weakness in several basement trends susceptible to later rejuvenation as crosscontinental stress was applied. As noted by the CD-ROM Working Group (2002, p. 8), this portion of the Rocky Mountains south of the Cheyenne belt was much more susceptible to rejuvenation during the Phanerozoic. West of the Rocky Mountains boundary zone, features such as the Uncompahgre uplift and the Zuni-Pedernal uplift (Fig. 1) may also represent Phanerozoic rejuvenation of Precambrian sutures. The Rocky Mountains and the Nemaha uplift share a history of rejuvenation most notably during Late Mississippian to Pennsylvanian time.

SUMMARY AND CONCLUSIONS

A more comprehensive interpretation for the tectonic history of Nebraska and adjoining regions is proposed (Fig. 6) based on integrated research of both the process of Proterozoic accretionary growth of North America and the stratigraphic patterns in the Phanerozoic. An important consideration for the importance of the inherent basement sutures is that these trends may reactivate even though their orientation is not in direct concurrence with the impending continental stress. This broadens the possible interpretations for the causative stress patterns—for example, during Pennsylvanian time for such features as the Rocky Mountains and the Nemaha uplift. Even more important are the implications of these broad suture patterns as rejuvenated controls on sedimentation and local structure. Many decades of research on both the Precambrian and the Phanerozoic rocks in the midcontinent have provided only a general understanding of the geologic history. As pointed out by Oliver (2003, p. 163), "The other style of science, the deductive style, tends to limit us to what we are capable of imagining. We need not be so constrained. The inductive style of science improves our chances of learning things beyond our imagination. In a sense, the inductive style requires a bit more humility, but it offers the possibility of greater rewards." This inductive approach to the integration of Precambrian and Phanerozoic research has provided meaningful insights for a new approach to the tectonic history and geometry of the Nebraska region.

REFERENCES CITED

Allen, D.J., and Hinze, W.J., 1992, Wisconsin gravity minimum: Solution of a geologic and geophysical puzzle and implications for cratonic evolution: Geology, v. 20, p. 515–518, doi: 10.1130/0091-7613(1992)020<0515:WGMSOA>2.3.CO;2.

Carlson, M.P., 1999, Transcontinental arch—A pattern formed by rejuvenation of local features across central North America, *in* Marshak, S., et al., eds., Tectonics of continental interiors: Tectonophysics, v. 303, p. 225–235.

Carlson, M.P., 2001, Accretionary basement control on structure of central USA: Geological Society of America Abstracts with Programs, v. 33, no. 6, p. A-434.

CD-ROM Working Group (Continental Dynamics of the Rocky Mountains Working Group), 2002, Structure and evolution of the lithosphere beneath the Rocky Mountains: Initial results from the CD-ROM experiment: GSA Today, v. 12, no. 3, p. 4–10.

Dahl, P.S., Holm, D.K., Gardner, E.T., Hubacher, F.A., and Foland, K.A., 1999, New constraints on the timing of Early Proterozoic tectonism in the Black Hills (South Dakota), with implications for docking of the Wyoming province with Laurentia: Geological Society of America Bulletin, v. 111, p. 1335–1349, doi: 10.1130/0016-7606(1999)111<1335:NCOTTO>2.3.CO;2.

Hamilton, W.B., 1979, Tectonics of the Indonesian region: U.S. Geological Survey Professional Paper 1078, 345 p.

Hamilton, W.B., 1988, Plate tectonics and island arcs: Geological Society of America Bulletin, v. 100, p. 1503–1527, doi: 10.1130/0016-7606(1988)100<1503:PTAIA>2.3.CO;2.

Hoffman, P.F., 1989, Speculations on Laurentia's first gigayear (2.0 to 1.0 Ga): Geology, v. 17, p. 135–138, doi: 10.1130/0091-7613(1989)017<0135:SOLSFG>2.3.CO;2.

Lidiak, E.G., 1972, Precambrian rocks in the subsurface of Nebraska: Nebraska Geological Survey Bulletin 26, 41 p.

Lucas, S.B., Stern, R.A., Syme, E.C., Reilly, B.A., and Thomas, D.J., 1996, Intra-oceanic tectonics and the development of continental crust: 1.92–1.84 Ga evolution of the Flin Flon belt, Canada: Geological Society of America Bulletin, v. 108, p. 602–629, doi: 10.1130/0016-7606(1996)108<0602:ITATDO>2.3.CO;2.

Marshall, L.P., and Lidiak, E.G., 1996, Geochemistry and paleomagnetism of Keweenawan basalt in the subsurface of Nebraska: Precambrian Research, v. 76, p. 47–65, doi: 10.1016/0301-9268(95)00027-5.

Oliver, J., 2003, Earthquake seismology in the plate tectonics revolution, *in* Oreskes, N., ed., Plate tectonics: An insider's history of the modern theory of the Earth: Boulder, Colorado, Westview Press, p. 155–166.

Reed, J.C., 1993, Precambrian rocks of the conterminous United States, *in* Reed, J.C., et al., eds., Precambrian: Conterminous U.S.: Geological Society of America, Geology of North America, v. C-2, plate 1.

Sclater, J.B., 2003, Heat flow under the oceans, in Oreskes, N., ed., Plate tectonics: An insider's history of the modern theory of the Earth: Boulder, Colorado, Westview Press, p. 128–147.

Selverstone, J., Pun, A., and Condie, K.C., 1999, Xenolithic evidence for Proterozoic crustal evolution beneath the Colorado Plateau: Geological Society of America Bulletin, v. 111, p. 590–606, doi: 10.1130/0016-7606(1999)111<0590:XEFPCE>2.3.CO;2.

Sims, P.K., editor, 1990, Precambrian basement map of the northern midcontinent, U.S.A.: U.S. Geological Survey Miscellaneous Investigations Series Map I-1853-A, scale 1:1,000,000, 1 shet.

Sims, P.K., and Day, W.C., 1999, Geologic map of the Precambrian rocks of the Hartville uplift, southeastern Wyoming with a section on mineral deposits in the Hartville uplift by Terry Klein: U.S. Geological Survey Miscellaneous Investigations Series Map I-2661, scale 1:48,000, 1 sheet, 34 p.

Sims, P.K., and Peterman, Z.E., 1986, Early Proterozoic Central Plains orogen: A major buried structure in the north-central United States: Geology, v. 14, p. 488–491, doi: 10.1130/0091-7613(1986)14<488:EPCPOA>2.0.CO;2.

Treves, S.B., and Lowe, D.J., 1984, The Precambrian geology of Nebraska: Geological Association of Canada Program with Abstracts, v. 9, p. 112.

Van Schmus, W.R., Bickford, M.E., Anderson, J.L., Bender, E.E., Anderson, R.R., Bauer, P.W., Robertson, J.M., Bowring, S.A., Condie, K.C., Denison, R.E., Gilbert, M.C., Grambling, J.A., Mawer, C.K., Shearer, C.K., Hinze, W.J., Karlstrom, K.E., Kisvarsanyi, E.B., Lidiak, E.G., Reed, J.C., Jr., Sims, P.K., Tweto, O., Silver, L.T., Treves, S.B., Williams, M.L., and Wooden, J.L., 1993, Transcontinental Proterozoic provinces, *in* Reed, J.C., et al., eds., Precambrian: Conterminous U.S.: Geological Society of America, Geology of North America, v. C-2, p. 171–334.

Manuscript Accepted by the Society 1 February 2007

Printed in the USA

The Geological Society of America
Memoir 200
2007

Precambrian blocks and orogen boundaries in the north-central United States determined from gravity and magnetic data

Kevin Mickus
Department of Geosciences, Missouri State University, Springfield, Missouri 65897, USA

ABSTRACT

Newly compiled gravity and magnetic data within the north-central United States were used to investigate the crustal structure of the region's Precambrian geology. Gray-scale imaging, gravity stripping, enhanced magnetic gradients, and polynomial trend surfaces were used to delineate the potential field characteristics of each of the Precambrian orogens. This analysis expanded on the previous regional investigations by suggesting that several crustal blocks (or regions of different lithologies/crust thickness) may exist in each of the known orogens. In addition, the study found that the interpreted location of the orogen boundaries varied depending on the type of potential field analysis used. The Trans-Hudson orogen may extend farther south in Nebraska and the Penokean orogen may extend farther southwest into Nebraska than was previously thought. The reinterpretation of the blocks and orogen boundaries needs additional constraining geological and/or seismic evidence in order for the results of the gravity and magnetic analysis to be completely supported.

Keywords: Bouguer gravity, magnetic anomalies, Precambrian tectonics, orogenic belts.

INTRODUCTION

The Precambrian structure of the north-central section of the United States (Fig. 1) remains relatively unknown due to the lack of exposed lithologies in North and South Dakota, Nebraska, Iowa, and parts of Minnesota. The structure is known in general terms from scattered drill holes (Sims, 1990; Lidiak et al., 1993; Sims et al., 1993), outcrops in the Black Hills and the Superior Craton of Minnesota, potential field studies (e.g., Dutch, 1983; Green et al., 1985a, 1985b; Hinze, 1996), and scattered seismic studies (Gibbs et al., 1984; Hajnal et al., 1984; Morel-A-L'Huissier et al., 1987; Latham et al., 1988; Fischer et al., 2003). The deep crustal and upper mantle structure of the region is unknown due to the lack of deep seismic refraction studies in the Dakotas and Nebraska (Braile, 1989). A recent widely spaced (75–150 km in the Dakotas) broadband seismometer experiment (Fischer et al., 2003) showed that the crust thickened under the Williston basin and the Midcontinent rift.

These geologic and potential field studies have revealed several Precambrian tectonic provinces within the study area including the Superior Craton (2700 Ma), Penokean orogen (1830–1880 Ma), Trans-Hudson orogen (1800–1860 Ma), the Central Plains orogen (1630–1700 Ma), and the Midcontinent rift system (1000–1100 Ma). Figure 1 shows the commonly accepted boundaries of the above tectonic features (Sims et al., 1993; Van Schmus et al., 1993). Additional studies by Van Schmus et al. (1993, 1996) divided the Central Plains orogen into a northern and southern orogen based on age dates of drill hole samples.

Mickus, K., 2007, Precambrian blocks and orogen boundaries in the north-central United States determined from gravity and magnetic data, *in* Hatcher, R.D., Jr., Carlson, M.P., McBride, J.H., and Martínez Catalán, J.R., eds., 4-D Framework of Continental Crust: Geological Society of America Memoir 200, p. 327–340, doi: 10.1130/2007.1200(16). For permission to copy, contact editing@geosociety.org.

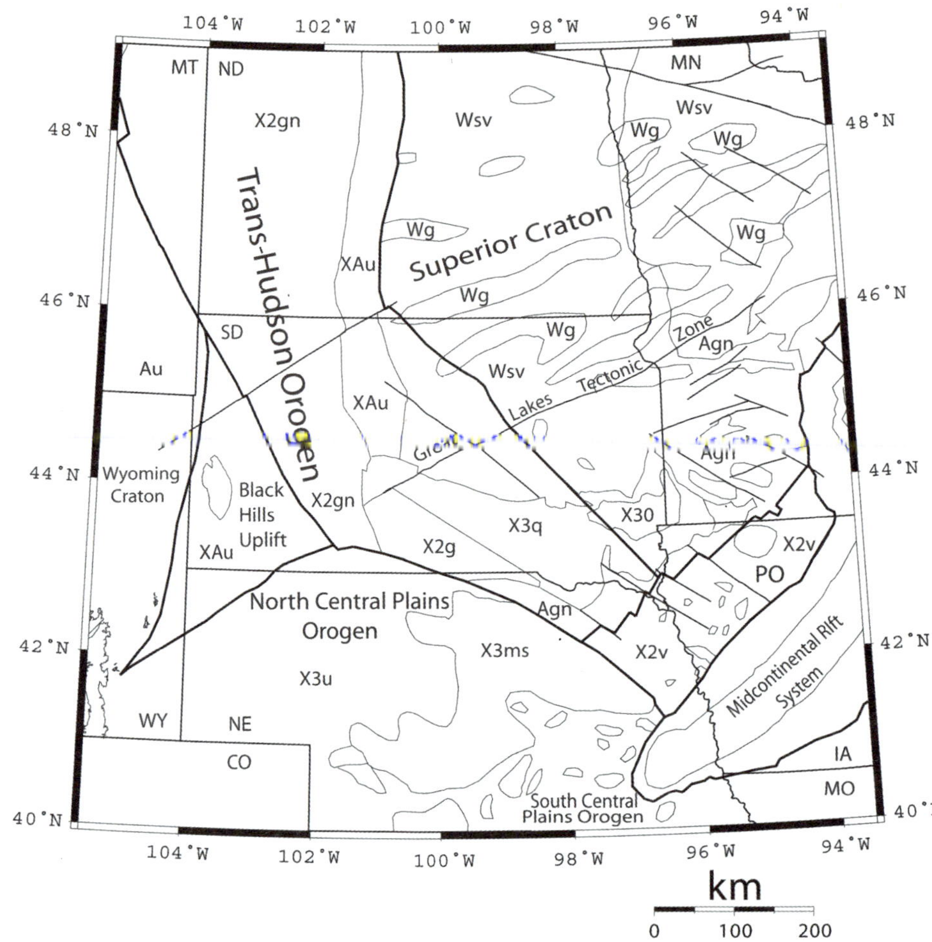

Figure 1. The study area with the major Proterozoic provinces and Proterozoic lithologic units (abbreviated) (Reed, 1993). The Proterozoic provinces (Van Schmus et al., 1993) are outlined by the thickest lines. The western boundary of the Trans-Hudson orogen is taken from Reed (1993). Thinner bolded lines represent major faults. PO—Penokean orogen; MT—Montana; WY—Wyoming; CO—Colorado; ND—North Dakota; SD—South Dakota; NE—Nebraska; MN—Minnesota; IA—Iowa; MO—Missouri. The Proterozoic lithologic units include: Wg—Archean granitic rocks; Wsv—Archean metasedimentary and metavolcanic rocks; Agn—Archean gneisses; Au—Archean rocks of the Wyoming Craton; XAu—Early Proterozoic gneiss and sedimentary rocks; X2v—Middle Proterozoic metavolcanic rocks; X2g—Middle Proterozoic granitic rocks; X3u—Middle Proterozoic rocks undivided; X30—Early Proterozoic quartzites; X3q—Early Proterozoic quartzites; X2gn—Early Proterozoic gneisses of the Hudson orogen; X3ms—Early Proterozoic rocks in the Central Plains orogen.

Carlson (2001) performed similar work in Nebraska to divide the North Central Plains orogen into a series of accreted terranes that decreased in age from north to south.

Potential field studies are particularly suited to examine the Precambrian geology of the north-central United States because most of the region is covered by Phanerozoic sediments of varied thickness, and lateral variations in basement physical properties can be detected by these techniques at the surface. Despite recent advances in the knowledge of the Precambrian geology of the region (Sims et al., 1993; Van Schmus et al., 1993; Carlson, 2001; McCombs et al., 2004), there are several questions about the location of boundaries of the provinces, existence of blocks, and general crustal thickness variations of the region.

This study examined new gravity and magnetic data compilations in order to suggest the boundaries of the above terranes and determine whether additional crustal blocks (defined as an area of similar geophysical character) exist within them (except for the Midcontinent rift system). Previous studies that used gravity and magnetic data (Dutch, 1983; Green et al., 1985a, 1985b; Klasner and King, 1986; Sims and Peterman, 1986; Thomas et al., 1987; Atekwana, 1996) delineated the major tectonic boundaries but did not use a wide variety of interpretational techniques that might delineate more subtle geophysical anomalies. Klasner and King (1986) used high-pass wavelength filtering and shaded relief gravity anomaly maps, Atekwana (1996) used vertical derivatives of gravity and magnetic data, and Thomas et al. (1987) used horizontal gravity gradients to interpret Precambrian features in portions of the study area. In this paper, polynomial trend surfaces and gravity stripping was applied to the gravity data while the enhanced horizontal gradients (EHG) technique was used on the magnetic data.

GENERAL GEOLOGICAL BACKGROUND

Figure 1 shows the major Precambrian rock units and tectonic features in the north-central United States (Reed, 1993; Van Schmus et al., 1993). The oldest rocks identified within the study area are associated with the Archean gneiss terrane within the Late Archean Superior Craton. Additional Archean rocks occur within the Wyoming Craton in the southeastern section of the study area (Fig. 1) but are not part of this study. Within the Superior Craton, these rock units are mostly covered by Pleistocene glacial deposits and have been described from scattered outcrops and potential field investigations (Chandler, 1985; Klasner and King, 1986; Atekwana, 1996). The gneissic terrane found in southern Minnesota may be a fragment of a protocontinent consisting of varying types of migmatitic gneisses (Agn, Fig. 1) (Morey et al., 1982). The Late Archean Superior Craton consists of two distinct regions: a granite-and-greenstone belt and an older high-grade gneiss terrane (Sims et al., 1993). The most common rock units consist of greenschist-facies volcanic and volcanogenic and metasedimentary rocks. In addition Late Archean plutons (Wg, Fig. 1), mostly tonalitic or granodioritic in composition, occur throughout the Superior Craton. The gneissic and granite-and-greenstone belt units are juxtaposed along the Great Lakes Tectonic Zone in South Dakota and Minnesota. Seismic and gravity/magnetic studies (Gibbs et al., 1984; Chandler et al., 1989) have indicated that the Great Lakes Tectonic Zone is a north-dipping collisional zone between the two lithological provinces.

The next tectonic event was the Penokean orogeny, which is defined by a zone of deformed and metamorphosed Early Proterozoic and Archean rocks along the southern margin of the Superior Craton (Sims et al., 1993). The Penokean orogeny illustrates a complete Wilson cycle with rifting, volcanic arcs, and continental convergence and suturing. The majority of rock units associated with the orogeny crop out northeast of the study area, but scattered drill holes in northwestern Iowa, northeastern Nebraska, and southeastern South Dakota (Fig. 1) indicate that the crust in these areas is also Early Proterozoic in age (Van Schmus et al., 1993) and may be related to the Penokean orogeny. These units may be part of the southern fold belt that is associated with the mapped folds and faults in Wisconsin (Sims et al., 1993). After the close of the Penokean orogeny, numerous quartz sandstones (later metamorphosed to quartzites [X3q and X30, Fig. 1]) were deposited in a tectonically stable environment in South Dakota, Minnesota, Nebraska, and Iowa.

Most of the study area consists of Middle and Early Proterozoic rocks that are associated with the Trans-Hudson orogen, Central Plains orogen, and the Midcontinent rift system. The location and nature of the Trans-Hudson orogen (Fig. 1) is based on scattered drill holes, trends of gravity and magnetic anomalies (Lidiak, 1971; Green et al., 1985a, 1985b; Klasner and King, 1986; Sims et al., 1993), and electrical conductivity studies (Alabi et al., 1975; Camfield and Gough, 1977; Jones et al., 1993). The boundaries of the Trans-Hudson orogen are mainly based on gravity and magnetic anomalies associated with tectonic domains in Canada (Green et al., 1985a) and extending these interpretations into the United States. Potential field studies since then (e.g., Dutch, 1983; Klasner and King, 1986) have generally concurred with the results of Green et al. (1985a). Additionally, Alabi et al. (1975) and Camfield and Gough (1977) proposed that a large-amplitude electrical conductivity anomaly, detected using low-resolution magnetometer arrays, along the eastern boundary of the Wyoming and Churchill Cratons corresponded to a Proterozoic plate boundary. More recent detailed magnetotelluric studies by Jones and Savage (1986) and Jones et al. (1993) confirm that this anomaly extends into Canada but lies to the east of the previous electrical anomaly and is discontinuous in nature. Gravity gradient studies by Thomas et al. (1987) also suggest that the conductivity anomaly is discontinuous and may not be entirely related to a plate or terrane boundary. Adding constraints to these geophysical interpretations are radiometrically age dated basement samples within the proposed Trans-Hudson orogen that fall within an Early Proterozoic age (Sims et al., 1993). Limited seismic data suggest that the orogen had a subduction zone dipping toward the Wyoming Craton (Baird et al., 1995).

The Central Plains orogen (North Central Plains and South Central Plains orogens) (Fig. 1) is a major Early Proterozoic orogen that spans most of Nebraska and parts of South Dakota, Colorado, and Wyoming in the study area (Lidiak, 1972; Sims and Peterman, 1986; Van Schmus et al., 1993, 1996). Using sparse basement samples, Van Schmus et al. (1993) divided the Central Plains orogen belt into a northern unit of primary crust of metaigneous and metasedimentary rocks that formed between 1700 Ma and 1800 Ma, with a southern unit (similar rock types) forming later (to 1600 Ma). Recently Carlson (2001) divided the Central Plains orogen, based on age dating of basement drill hole samples, into a series of progressively younger island arcs in Nebraska that were accreted to the Superior Craton from the south. In addition, post-tectonic granites are scattered throughout both orogens. The northern boundary of the orogen is provisionally defined by gravity anomalies (Klasner and King, 1986; Sims and Peterman, 1986). The boundary between the Central Plains orogen and the Wyoming Craton and the Penokean orogen is unclear because of the lack of definitive geophysical anomalies and drill hole information (Van Schmus et al., 1993).

The youngest Proterozoic rocks within the study area belong to the Middle Proterozoic Midcontinent rift system (Fig. 1). This feature is believed to be an intracontinental rift system consisting of thick deposits of bimodal volcanic rocks surrounded by extensional basins (Chandler et al., 1989; Anderson, 1992; Van Schmus, 1992). The most distinguishing feature of the rift system is the high-amplitude, linear gravity (to a lesser degree magnetic) anomaly that extends from northeastern Minnesota through Iowa into southeastern Nebraska within the study area.

POTENTIAL FIELD DATA AND PROCESSING

Gravity data were obtained from three sources: (1) United States Geological Survey (USGS [Kucks and Hill, 2002]), (2) National Geophysical Data Center, and (3) National Geospatial and Imaging Agency (formerly the Defense Mapping Agency). Each data set contained points that were not in either of the other two data sets, so the merged data that contains 11,714 points (Fig. 2) probably represents one of the most complete gravity data sets of the region currently available.

The merged gravity data were reduced using the 1967 International Gravity formula (Morelli, 1976). Free-air and Bouguer gravity corrections were made using sea level as a datum and 2670 kg/m^3 as a reduction density. The merged data were examined for possible spurious points (values two standard deviations greater or smaller than surrounding points) and these points were removed from any subsequent analysis. The merged Bouguer gravity anomaly data were gridded at a spacing of 2 min (~4.5 km) using the minimum curvature technique (Briggs, 1974) and contoured at a 10 mGal interval to produce a Bouguer gravity anomaly map (Fig. 3).

Total field aeromagnetic data were obtained from the United States Geological Survey (Bankey et al., 2002), which constructed a 1.0 km grid from more than 1000 individual surveys conducted over the past 50 yr (see Bankey et al., 2002, for all the processing details). The merged total-field aeromagnetic anomaly data were gridded at a spacing of 2 min (~4.5 km) and the dipolar effect of Earth's magnetic field was removed (reduction to the pole). The subsequent grid was contoured at a 200 gamma interval to produce a reduced-to-the-pole aeromagnetic intensity anomaly map (Fig. 4).

POTENTIAL FIELD DATA ANALYSIS

To effectively use the above data to study the nature of the tectonic regimes within the north-central United States, potential field anomalies resulting from individual features must be isolated. This is a difficult job because of the small density/magnetic susceptibility contrasts between most of the features. However, there are several techniques (e.g., wavelength filtering, polynomial trend surfaces, edge enhancements, isostatic residual anomalies) that can aid in accomplishing these tasks. All of these techniques were employed in this study and the techniques that were deemed useful are described below.

An examination of the Bouguer gravity anomaly map (Fig. 3) shows that average Bouguer gravity anomaly values decrease toward the west-southwest. This is due to two features: thickening of the crust (Braile, 1989) and higher elevations. There are additional anomalies due to the varying thickness of Phanerozoic sedimentary sequences especially within the Williston Basin of western North Dakota and eastern Montana that may obscure the gravity anomalies due to Precambrian features. The gravitational effects of these sediments can be removed by a technique called gravity stripping (Hammer, 1963), which produces a "deep-source" gravity anomaly map. This technique has been shown to be extremely useful in studying Precambrian geology by Adams and Keller (1996), who used the technique to delineate previously unknown Precambrian features beneath the Permian Basin in west Texas. .

To determine the deep-seated gravity anomaly, the thickness of the sedimentary units was digitized from available thickness maps (Burchett, 1984; Heck, 1988; Sims, 1990), which were gridded and contoured to produce a sedimentary thickness map (Fig. 5). The gravitational attraction resulting from the sediments was determined using a three-dimensional gravity modeling program (Montana et al., 1992), an elevation grid, the previously mentioned sedimentary thickness grid, and an average density contrast between the sedimentary rocks and the basement of −250 kg/m^3. This density was based on seismic refraction surveys across the northern section of the Williston basin (Hajnal et al., 1984), where the compressional wave velocities were converted to densities using an experimental density velocity relationship (Christiansen and Mooney, 1995). To obtain the deep-seated gravity anomaly map (Fig. 6), the gravity attraction of the sedimentary units was subtracted from the original Bouguer gravity anomaly grid. Slight errors in the final map exist because of density differences between the various regions but should not affect the final map patterns.

The deep-seated gravity anomaly map (Fig. 6) still contains the long-wavelength gravity anomaly. There are several methods to remove this long-wavelength anomaly, including wavelength filtering, isostatic gravity residuals, and polynomial trend surfaces. The best method is probably the isostatic residual gravity anomaly method as it is based on geologically derived parameters (Simpson et al., 1986). Within the study there is a lack of constraints (density of the load, density contrast across the root, compensation depth [Simpson et al., 1986]) needed to construct an isostatic residual gravity anomaly map, so one was not made. Thus, various orders of polynomial trend surfaces (Lance, 1982) were constructed to approximate the long-wavelength anomaly. Surfaces with orders between two and five basically produced the same anomaly pattern, so a third-order surface was used to approximate the long-wavelength anomaly. The third-order polynomial surface then was subtracted from the deep-seated gravity anomaly grid to produce a third-order polynomial residual gravity anomaly map (Fig. 7). Figure 7 shows the same general anomaly trends as the isostatic residual gravity anomaly map of Simpson et al. (1986). The difference between the two maps is mainly seen on the higher frequency anomalies shown in Figure 7. A similar long-wavelength anomaly is not seen in

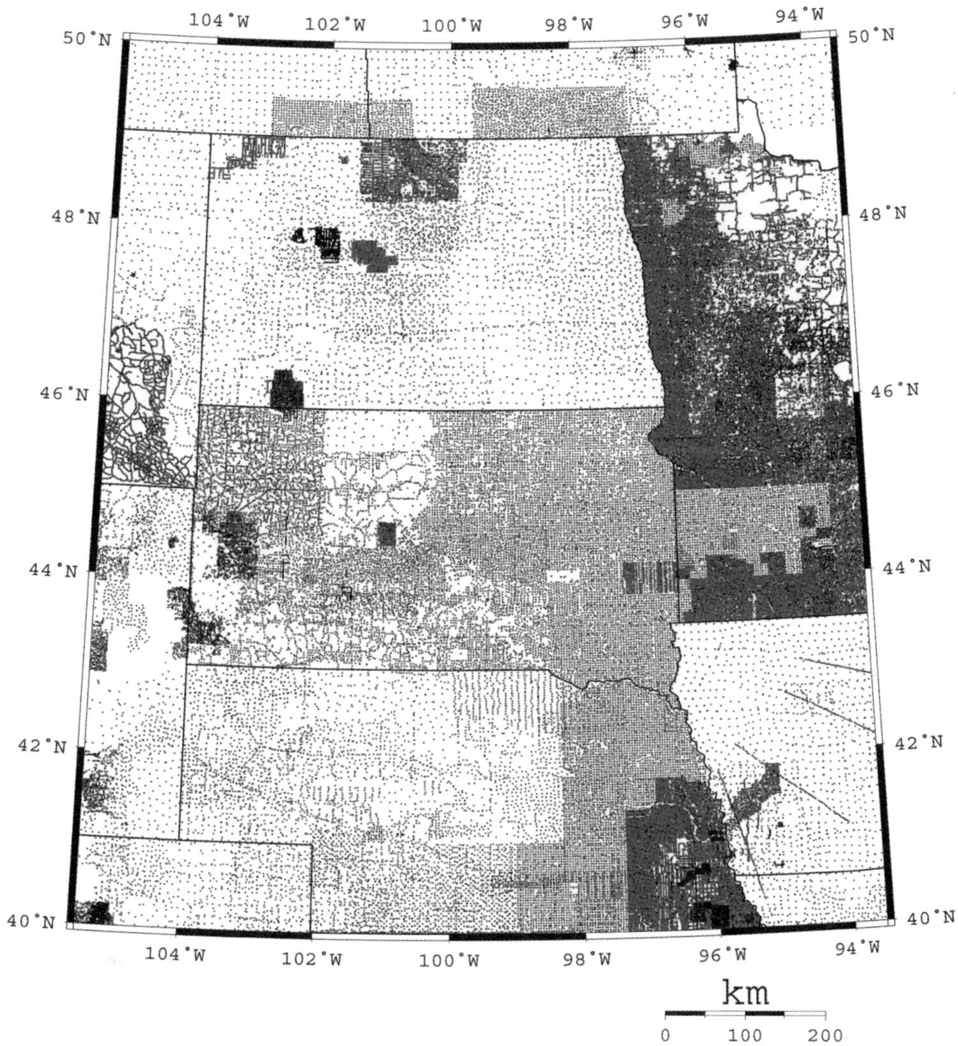

Figure 2. Location of gravity stations (+) within the study area. Also shown are the geologic features shown in Figure 1. These geologic features are shown on all the gravity and magnetic anomaly maps.

the magnetic anomaly data (Fig. 4), so a residual magnetic map was not constructed.

The reduced-to-the-pole magnetic anomaly fields were also analyzed by performing an edge enhancement using the EHG method (Fedi and Florio, 2001), as edge enhancement techniques can be used to locate the lateral boundaries of susceptibility contrasts and provide information on the location of geological units (Blakely and Simpson, 1986). The edges of magnetic susceptibility contrasts are shown with high positive amplitudes (lighter gray shades on Fig. 8). The use of edge enhancement techniques provides only a first-order geological interpretation of potential field data, and as Grauch and Cordell (1987) showed, that the results are often contaminated by nearby sources. Additionally, as Dehler and Lowe (1999) explained, the interpretation of the boundaries can provide reconnaissance information about general structural trends but a priori information (geological or seismic reflection/refraction data) is needed in order to obtain reliable results.

INTERPRETATION

The Bouguer gravity (Fig. 3) and reduced-to-the-pole magnetic intensity anomaly (Fig. 4) maps represent the combined effects of all the crustal and mantle density and/or magnetic susceptibility sources. By determining the horizontal location and depth of these sources, one can make inferences about the tectonic history of the north-central United States.

The most prominent gravity anomaly (Figs. 4, 6, and 7) is the gravity maximum associated with the thick layers of basalt and rhyolite of the Midcontinent rift system in Minnesota, Iowa,

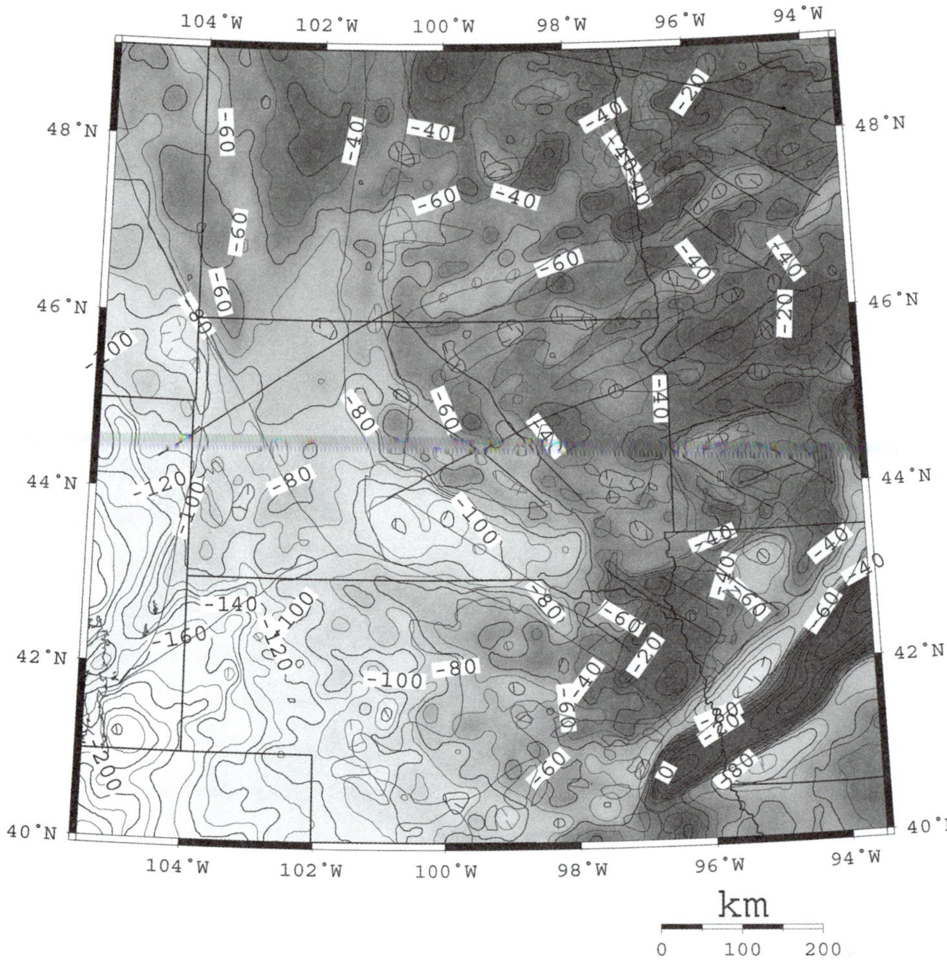

Figure 3. Bouguer gravity anomaly map of the study area. Contour interval is 10 mGal. On all the gravity and magnetic anomaly maps, darker gray scale represents relatively higher anomaly values.

and Nebraska. Flanking this positive anomaly are gravity minima associated with sediment-filled rift basins. The magnetic anomaly map also shows the Midcontinent rift as a series of positive anomalies with the magnetic anomalies being less distinctive.

The Superior Craton is characterized by a series of northeast-trending gravity and magnetic anomalies that shift to ~100° W in the Dakotas (Figs. 4 and 6). This relationship is more clearly seen on the transformed maps including the magnetic EHG (Fig. 8) and the polynomial residual gravity anomaly map (Fig. 7). These anomalies are an extension of the northeast-trending structures in the granite-and-greenstone terranes exposed in central Minnesota (Klasner and King, 1986). However, a closer examination of the maps reveals a more complicated relationship. Above 48° N, there is a weak west-northwest trend in the transformed gravity and magnetic anomaly maps. The amplitude of these anomalies (especially on the magnetic maps) is smaller than the northeast-trending anomalies. Thus, this fact and the change in the anomaly trend imply that there is either a change in rock types and/or a different structural grain. Superimposed on this west-northwest–trending anomaly is a weak north-south trend (anomaly 1 on Figures 4 and 6) that can also be seen on the transformed maps. All the known basement rocks are Archean in age, and clearly based on the gravity and magnetic anomalies, there are lithological and/or structural differences within the Archean rocks in the Dakotas.

An important boundary within the study area is the Great Lakes Tectonic Zone (Fig. 1), which is the boundary of greenstone-and-granite and gneissic terranes within the Superior Craton (Sims et al., 1993). Although exposed and imaged on seismic reflection lines in central Minnesota (Gibbs et al., 1984), its location southwest of its exposure is based on a linear trend of magnetic anomalies (Sims et al., 1993) that can be easily seen on the magnetic EHG map (Fig. 8), and gravity anomalies (Sims et al., 1993). However, the gravity evidence is not as strong, as there is little linearity in the gravity anomalies once the Great Lakes Tectonic Zone enters South Dakota (Fig. 6). Seismic reflection (Gibbs et al., 1984) and gravity studies (Southwick and Chandler, 1983) in central Minnesota have shown that the Great Lakes Tectonic Zone is a narrow zone with numerous

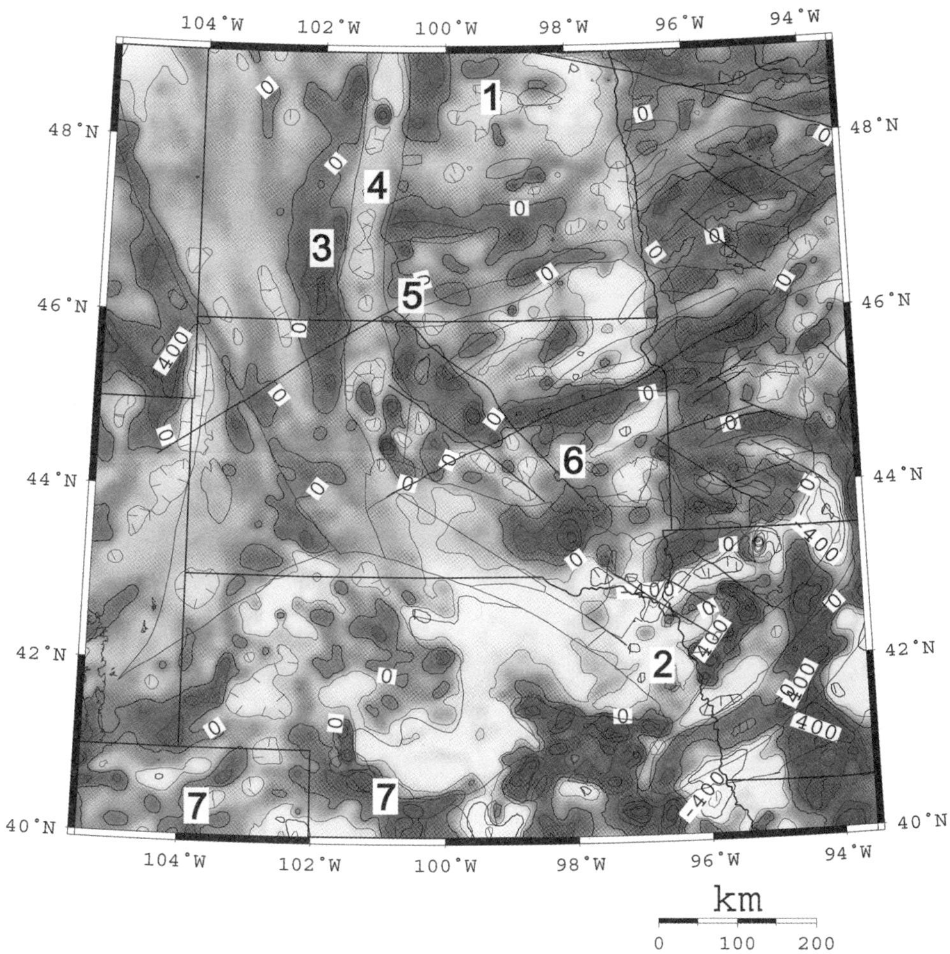

Figure 4. Reduced-to-the-pole magnetic intensity map. Contour interval is 200 gammas. On all the gravity and magnetic anomaly maps (Figs. 4, 6, 7, and 8), bolded numbers represent anomalies discussed in the text.

north-dipping reflectors that suggest that the structure is a major collisional zone within the Archean terranes. Extending the Great Lakes Tectonic Zone southwest into South Dakota is conjectural as there is a definite magnetic anomaly and only weak gravity evidence for it. Klasner and King (1986) used mainly magnetic anomalies to extend the structure to 100° W; however, based on the magnetic EHG map, the farthest west the Great Lakes Tectonic Zone should be extended is 99° W.

The gneissic terrane of the Superior Craton is known from drill holes southeast of the Great Lakes Tectonic Zone (Fig. 1). This region is characterized by relatively low-amplitude east-west– to northeast-trending gravity and magnetic anomalies (Figs. 4 and 6). The exception to this observation is a high-amplitude gravity anomaly (anomaly 2, Fig. 6) that trends northeast in Minnesota and east-west in South Dakota (Figs. 6 and 7). There is only a weak magnetic anomaly, implying that the source is dense but not magnetic (an iron-poor gabbroic body?). Additionally, the gneissic terrane contains three distinct potential field regions (anomalies 1, 2, and 3, Fig. 7). Anomaly 1 is separated from anomalies 2 and 3 by a high-amplitude gravity anomaly. This may imply that this high-amplitude anomaly is a boundary (southern extension of the Great Lakes Tectonic Zone?) between two distinct Archean-aged blocks. Magnetic minima can also be seen at the location of all the above gravity anomalies (Fig. 4). Anomaly 3 (Fig. 7) is centered over the quartzites and felsic volcanic rocks encountered in drill holes in South Dakota and is northeast of a large amplitude gravity minimum (anomaly 3, Fig. 6) and a magnetic quiet zone in southern South Dakota. This large amplitude anomaly is centered on the western extension of the quartzites and Middle Proterozoic granites and was interpreted by Klasner and King (1986) to be a Proterozoic batholith. To determine if any of the above anomalies (1, 2, or 3, Fig. 7) are due to a thickened crust, quartzites, batholiths, felsic volcanic rocks, or some combination of them, deep drill hole, and/or seismic studies are needed. However, based on the available potential field data, the gneissic terrane in southeastern Minnesota and

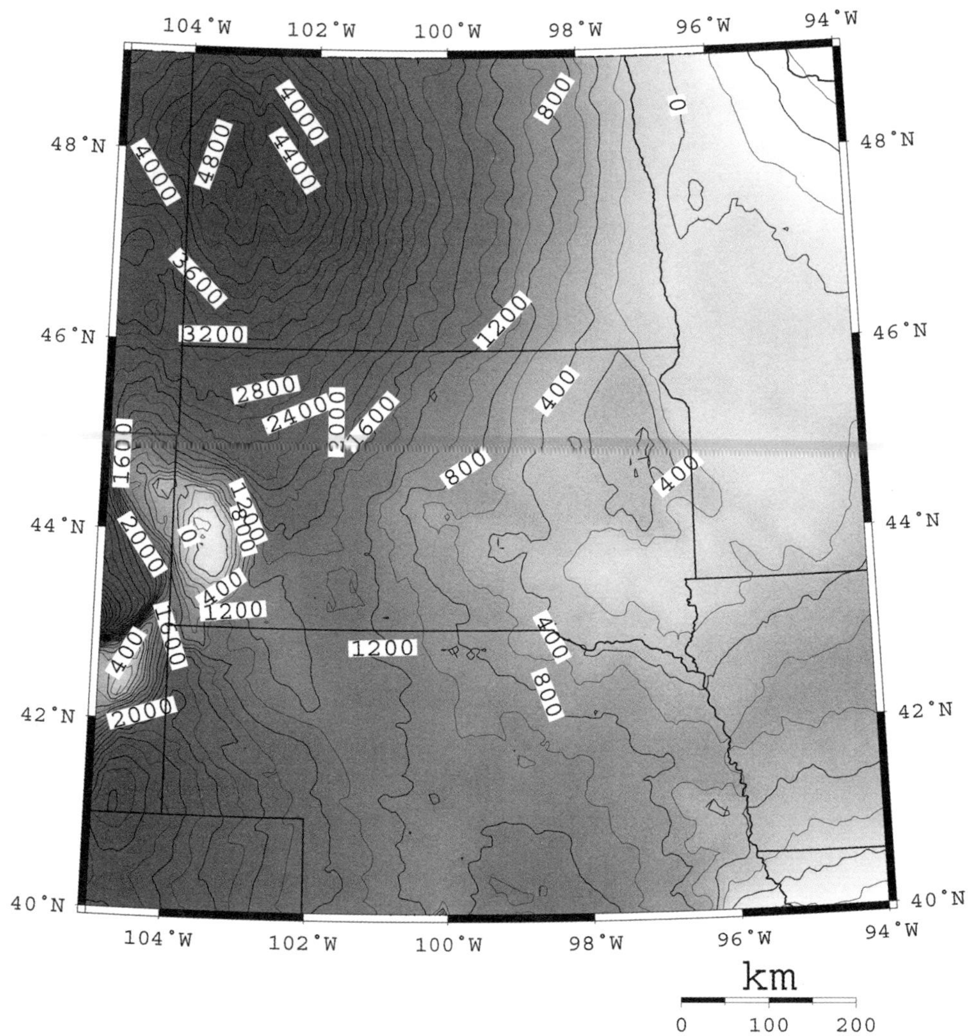

Figure 5. Thickness of the Phanerozoic sediments within the study area. Contour interval is 200 m.

eastern South Dakota may be divided into two distinct blocks (Fig. 9).

The Penokean orogen (Fig. 1) within the study area is poorly understood because of the lack of drill holes that penetrate the basement (Van Schmus et al., 1993). The gravity and magnetic anomalies can be broken into two main regions: (1) anomaly 4 (Fig. 7), which consists of a series of gravity maxima with a corresponding magnetic minima with a several short-wavelength magnetic maxima superimposed on it (anomaly 2, Fig. 4), and (2) anomaly 5 (Fig. 7), which consists of gravity and magnetic minima. The short-wavelength magnetic anomalies can be explained by the numerous magnetite-bearing granites that are known to exist in Iowa and Nebraska (Van Schmus et al., 1993). The origin of anomaly 4 (Fig. 7) is more difficult to explain as it is a longer-wavelength anomaly possibly related to a more regional geological feature. Anomaly 4 is located at the suspected northern boundary of the Penokean orogen and may represent a potential suture zone. Additionally, both the gravity and magnetic anomalies extend southwest of the poorly defined boundary with the Central Plains orogen (Fig. 1). Based on these anomalies, the Penokean orogen may extend farther southwest than previously thought.

The location of the Trans-Hudson orogen in the Dakotas is mainly based on the correlation of gravity and magnetic anomalies with similar anomalies occurring over outcrops in Canada (e.g., Green et al., 1985a; Thomas et al., 1987). The major anomalies within the orogen can be clearly seen on Figure 7 (anomalies 6 and 7) where the western region of the Trans-Hudson orogen is dominated by two gravity maxima. The magnetic anomalies are more variable with a magnetic quiet zone associated with anomaly 6, and a magnetic maximum (anomaly 3, Fig. 4) is associated with anomaly 7. The western boundary with the Wyoming Craton in Montana has been interpreted by Sims et al. (1993) to be associated with linear gravity and magnetic anomalies and

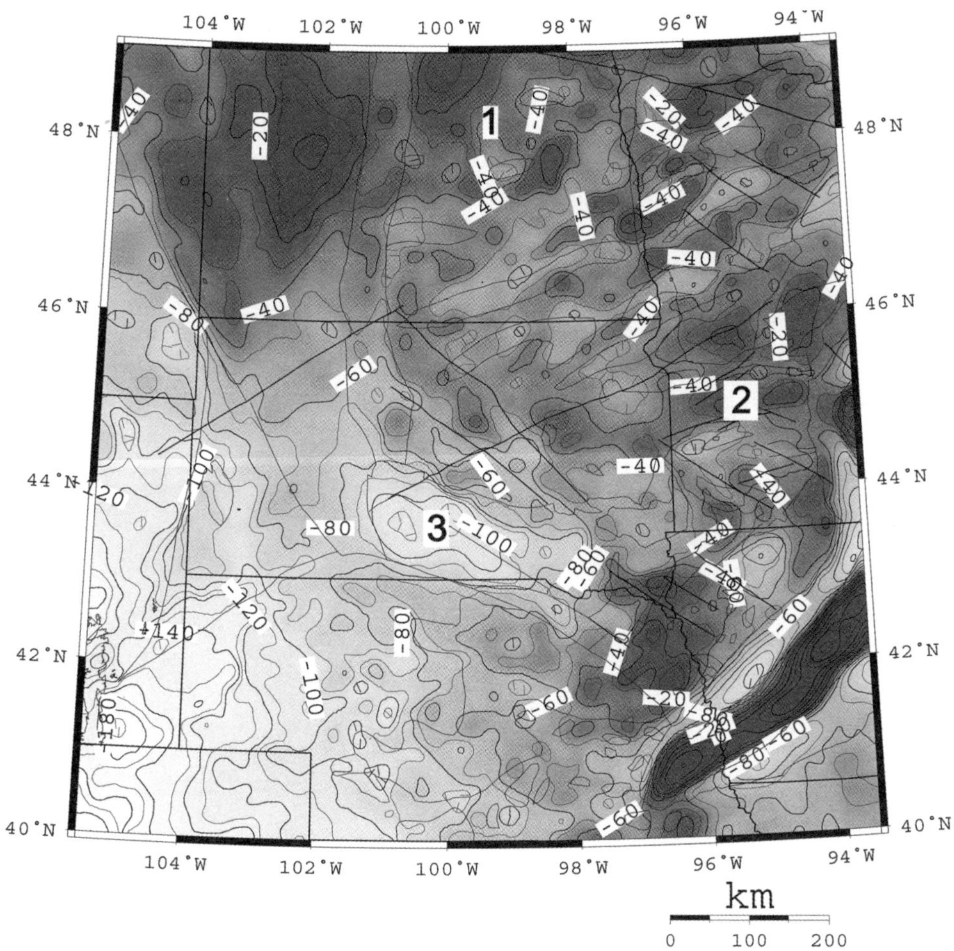

Figure 6. Deep-seated gravity anomaly map of the study area after removing the gravitational effects of the sediments shown in Figure 5. Contour interval is 10 mGal. Density contrast between the basement rock and the sediments used in the modeling was 250 kg/m^3.

can clearly be seen by a northwest-trending magnetic gradient (anomaly 4, Fig. 8). Green et al. (1985a) correlated this gradient with the Reindeer–South Indian Lakes Belt in northern Manitoba and interpreted the gravity anomalies (6 and 7, Fig. 7) as being caused by reworked oceanic material. The western boundary of the Trans-Hudson orogen is based on a maximum magnetic gradient (anomaly 1, Fig. 8) that trends to at least 43.5° N, where it veers to the east and continues into Nebraska (anomaly 2, Fig. 8) to at least 42.5° N.

Given the large amplitudes (40–50 mGal) and the large area they cover, anomalies 6 and 7 (Fig. 7) may be interpreted to represent to two distinct blocks within the Trans-Hudson orogen in the Dakotas. Part of the source of anomaly 6 has been attributed to basement faulting (up to 300 m offset) along the axis of the Cedar Creek Anticline (Sims et al., 1993). However the amplitude of the anomaly is 40–50 mGal and such an amplitude based on simple two-dimensional forward modeling cannot be caused by the suggested offset. A possible source of the gravity and magnetic maxima may be dense intracrustal features like those layers that have been imaged on seismic refraction profiles along the U.S.-Canada border in Canada (Morel-A-L'Huissier et al., 1987). Within North Dakota, a widely spaced broadband passive seismic array has shown that the crust is relatively thicker in western North Dakota than in the surrounding region but the low resolution of the survey did not detect intracrustal layers. However, the relatively broad potential field anomalies imply the existence of such layers.

The eastern boundary of the Trans-Hudson orogen coincides with the termination of the northeast-trending gravity and magnetic anomalies associated with the Superior Craton (Green et al., 1985a). The exact location of this termination is difficult to decipher, as Klasner and King (1986) showed that the northeast-trending anomalies terminated against a north-south–trending zone of circular gravity and magnetic maxima which changes direction to a southeast trend south of 46° N (anomaly 8, Fig. 7). However, in some locations (e.g., anomaly 5, Fig. 4), they terminate against a

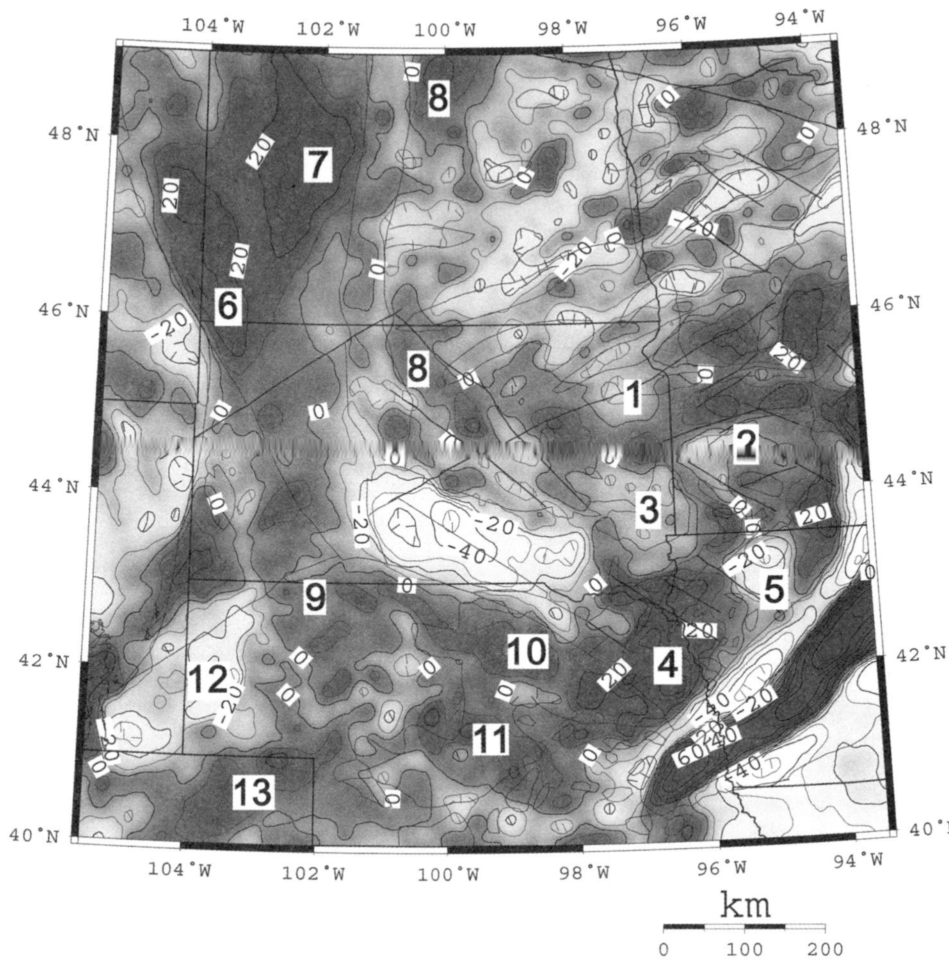

Figure 7. Third-order polynomial residual gravity anomaly map of the deep-seated gravity anomaly. Contour interval is 10 mGal.

north-south–trending zone of gravity and magnetic minima that extends into Canada (Green et al., 1985a). This uneven termination can be seen on the magnetic gradient map (Fig. 8). The lack of subsurface samples makes any interpretation of the boundary difficult. The few known basement samples have Early and Middle Proterozoic ages but these dates possibly represent overprinting of Archean rocks. Klasner and King (1986) considered this region a transition between Archean and Proterozoic terranes, and given the irregular nature of the potential field evidence, a distinct boundary could not be drawn. Based on the maps shown here, the northeast-trending anomalies can be clearly extended to 44° N (anomaly 6, Fig. 4), however south of this region, the potential field data do not provide evidence on the location of the boundary.

The southern end of the Trans-Hudson orogen has been interpreted to continue to southern South Dakota (Fig. 1 shows the interpreted boundary of the intersection of the Trans-Hudson and Central Plains orogens). This boundary is based on Bouguer gravity and magnetic anomalies and two drill holes in southern South Dakota that contain rocks younger than those associated with the Trans-Hudson orogen (Sims and Peterman, 1986). However, the northwest-trending anomalies associated with the western Trans-Hudson orogen continue into Nebraska (e.g., anomaly 2, Fig. 8), but these anomalies in Nebraska can also be associated with northwest-trending anomalies in central Missouri (Hinze, 1996) that have been related to the Central Plains orogen (Sims and Peterman, 1986). The complex nature of the anomalies in southern South Dakota and northern Nebraska can be explained by three possible scenarios: (1) the location of the Central Plains orogen is correct as seen on Figure 1 and the Trans-Hudson orogen extends beneath the Central Plains orogen, (2) the northern boundary of the Central Plains orogen is further south than previously known and the rocks encountered in southern South Dakota and northern Nebraska are isolated Early Proterozoic intrusions, or (3) the locations shown on Figure 1 are correct and the northwest-trending anomalies are associated with Central Plains orogen structures.

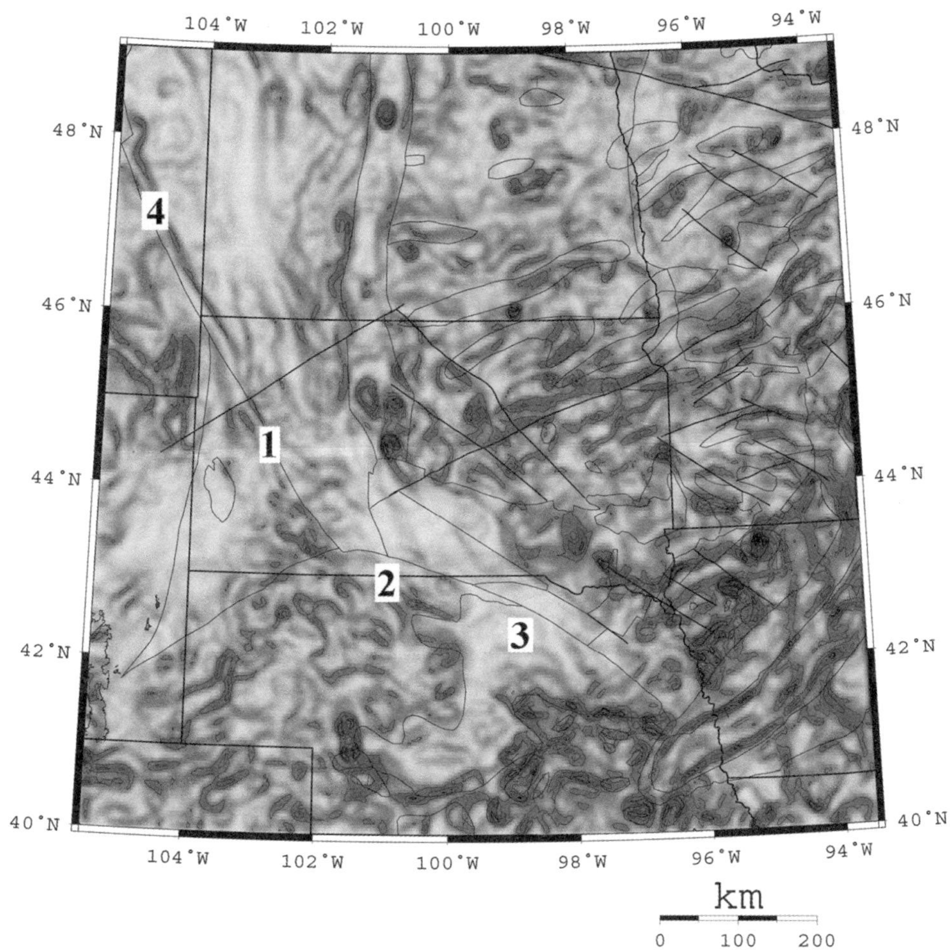

Figure 8. Enhanced horizontal magnetic gradient map. Contour interval is 20 gammas/km.

The last area of concern is the Central Plains orogen (Fig. 1). Sims and Peterman (1986) set the boundary (Fig. 1) based on curved gravity anomalies (Fig. 7); however, as discussed above, the location of this boundary may be different. Within the Northern Central Plains orogen, there are various trending gravity and magnetic anomalies that can be seen on all the anomaly maps. Middle Proterozoic metasedimentary rocks were mapped based on the subdued magnetic patterns (anomaly 3, Fig. 8). However, the same region on the polynomial residual gravity anomaly map contains two high-amplitude gravity maxima (anomalies 10 and 11, Fig. 7) that trend toward the northwest. These two anomalies are actually separated and roughly correspond to recently described accreted terranes with the Central Plains orogen based on the geochronology of basement samples (Carlson, 2001). However, these anomalies may reflect varying thickness of the metasediments and/or variations in the lithologies and/or crustal thickness beneath the metasediments. Anomaly 12 (Fig. 7) is a large-amplitude gravity minimum (and magnetic minimum and maximum) that may represent crustal thickening due to plate collision between the Wyoming Craton and the Central Plains orogen. This anomaly corresponds to the Dawes terrane (1.78 Ga) of Carlson (2001) and roughly extends to the gravity maximum (anomaly 11, Fig. 7). A gravity maximum (anomaly 13, Fig. 7) southeast of anomaly 12 is interpreted to be the corresponding gravity maximum that may occur to the ocean side in collisional orogens. This anomaly also corresponds to an accreted terrane (Frontier [1.71 Ga], Carlson, 2001) and can be extended to anomaly 11. However, the exact nature and correlation of these anomalies is difficult to interpret. Based on the residual gravity anomalies, anomalies 13 and 11 seem to connect and represent similar subsurface features. But the magnetic anomalies (anomaly 7, Fig. 4) suggest that the source of anomaly 13 may extend to the south of anomaly 11. Based on limited geochronology data (Carlson, 2001), the region of anomaly 13 may represent a third accretionary terrane (Hitchcock [1.67 Ga]).

Figure 8 shows a final interpretation of the gravity and magnetic anomalies as discussed above. Shown are the interpreted boundaries of the various major orogenic events and possible blocks within each orogeny. The exact location of each boundary is subjective due to the nature of interpreting gravity and magnetic data. Exact

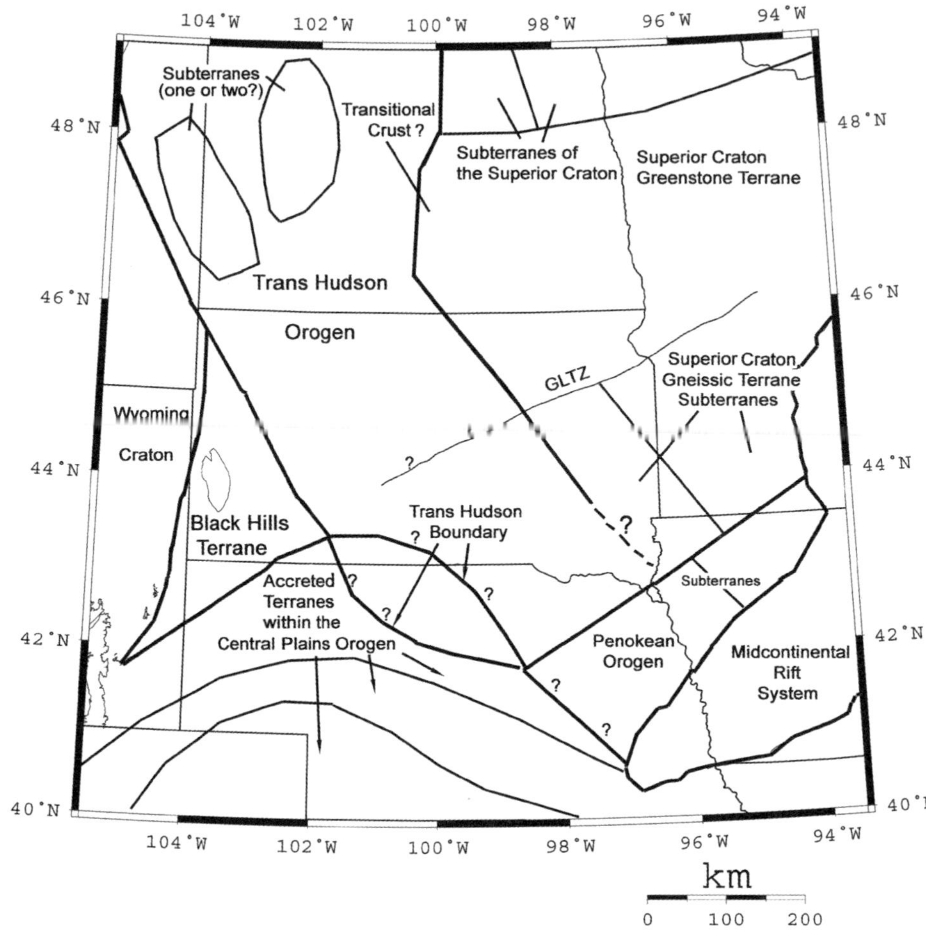

Figure 9. Tectonic features interpreted from the analysis of the gravity and magnetic data. Thickest lines represent the boundaries between the major orogens. Thinner lines represent blocks within each orogen. GLTZ—Great Lakes Tectonic Zone.

boundary locations will depend on more detailed basement sampling and seismic reflection/refraction studies.

CONCLUSIONS

The analysis of the gravity and magnetic fields within the north-central United States, using a variety of enhancement techniques, allows for the definition of orogen boundaries and the location of Precambrian blocks within these orogens. The above analysis has expanded on past potential field investigations by indicating that the region is more complicated than was previously known. Several possible blocks (or different lithologies/crust thickness) may exist within each of the previously described orogens including the Superior Craton (at least one and possibly three new blocks), the Penokean orogen (at least two different blocks and may extend toward the southwest), the Trans-Hudson orogen (may extend farther south and contain two different regions), and the Central Plains orogen in Nebraska (may be broken into two or three distinct regions).

ACKNOWLEDGMENTS

I wish to thank an anonymous reviewer and Marv Carlson for substantially improving this manuscript. I would also like to thank the editor of the volume, Robert Hatcher, for his patience during the editorial process.

REFERENCES CITED

Adams, D., and Keller, G., 1996, Precambrian basement geology of the Permian basin region of west Texas and eastern New Mexico: A geophysical perspective: American Association of Petroleum Geologists Bulletin, v. 80, p. 410–431.

Alabi, A., Camfield, P., and Gough, I., 1975, The North American Central Plains conductivity anomaly: Royal Astronomical Society Geophysical Journal, v. 43, p. 815–833.

Anderson, R., 1992, The midcontinent rift of Iowa [Ph.D. dissertation]: Ames, University of Iowa, 346 p.

Atekwana, E., 1996, Precambrian basement beneath the central Midcontinent United States as interpreted from potential field imagery, *in* van der Pluijm, B., and Catacosnios, P., eds., Basement and basins of eastern North America: Geological Society of America Special Paper 308, p. 33–44.

Baird, D., Knapp, J., Steer, D., Brown, L., and Nelson, K., 1995, Upper-mantle reflectivity beneath the Williston basin, phase-change Moho, and the origin of intracratonic basins: Geology, v. 23, p. 431–434, doi: 10.1130/0091-7613(1995)023<0431:UMRBTW>2.3.CO;2.

Bankey, V., Cuevas, A., Daniels, D., Finn, C., Hernandez, I., Hill, P., Kucks, P., Kucks, R., Miles, W., Pilkington, M., Roberts, C., Roest, W., Rystrom, V., Shearer, S., Snyder, S., Sweeney, R., Velez, J., Phillips, J., and Ravat, D., 2002, Digital data grids for the magnetic anomaly map of North America: United States Geological Survey Open-File Report 02-0414.

Blakely, R., and Simpson, R., 1986, Approximating edges of sources bodies from magnetic or gravity anomalies: Geophysics, v. 51, p. 1494–1498, doi: 10.1190/1.1442197.

Braile, L., 1989, Crustal structure of continental interior, *in* Pakiser, L., and Mooney, W., eds., Geophysical framework of the continental United States: Boulder, Colorado, Geological Society of America Memoir 172, p. 285–316.

Briggs, I.C., 1974, Machine contouring using minimum curvature: Geophysics, v. 39, p. 39–48, doi: 10.1190/1.1440410.

Burchett, R., 1984, Depth to Precambrian rocks in Nebraska: Nebraska Geological Survey Map BCT-33, scale 1:1,000,000.

Camfield, P., and Gough, I., 1977, A possible Proterozoic plate boundary in North America: Canadian Journal of Earth Sciences, v. 14, p. 1229–1238.

Carlson, M., 2001, Island arcs, accretionary terranes and Midcontinent structure; new understandings of the geologic architecture of the U.S. Midcontinent: Resource Notes-University of Nebraska–Lincoln, Conservation and Survey Division (1049-524X), v. 15, p. 15–18.

Chandler, V., 1985, Interpretation of Precambrian geology in Minnesota using low-altitude, high-resolution aeromagnetic data, *in* Hinze, W.J., ed., Utility of regional gravity and magnetic anomaly maps: Tulsa, Oklahoma, Society of Exploration Geophysicists, p. 375–391.

Chandler, V., McSwiggen, P., Morey, G., Hinze, W., and Anderson, R., 1989, Investigations of seismic reflection, gravity and magnetic data across Middle Proterozoic Mid-Continent Rift System, northwestern Wisconsin, eastern Minnesota and central Iowa: American Association of Petroleum Geologists Bulletin, v. 73, p. 261–275.

Christiansen, N., and Mooney, W., 1995, Seismic velocity and the composition of the continental crust: A global view: Journal of Geophysical Research, v. 100, p. 9751–9788.

Dehler, S., and Lowe, C., 1999, Determination of source depth and location from potential field data: Eos (Transactions, American Geophysical Union), v. 80, p. 551.

Dutch, S., 1983, Proterozoic structural provinces in the north-central United States: Geology, v. 11, p. 478–481, doi: 10.1130/0091-7613(1983)11<478:PSPITN>2.0.CO;2.

Fedi, M., and Florio, G., 2001, Detection of potential fields source boundaries by enhanced horizontal derivatives: Geophysical Prospecting, v. 49, p. 40–58, doi: 10.1046/j.1365-2478.2001.00235.x.

Fischer, K., Syracuse, E., Rondenay, S., Wysession, M., Salas, M., Doermann, L., and Welsh, M., 2003, Crustal structure and mantle anisotropy from Florida to Edmonton: Eos (Transactions, American Geophysical Union), v. 46, p. 84.

Gibbs, A., Payne, B., Setzer, T., Brown, L., Oliver, J., and Kaufman, S., 1984, Seismic reflection study of the Precambrian crust of central Minnesota: Geological Society of America Bulletin, v. 95, p. 280–296, doi: 10.1130/0016-7606(1984)95<280:SSOTPC>2.0.CO;2.

Grauch, V., and Cordell, L., 1987, Limitations of determining density or magnetic boundaries from the horizontal gradient of gravity or psuedogravity data: Geophysics, v. 52, p. 118–124, doi: 10.1190/1.1442236.

Green, A., Hajnal, Z., and Weber, W., 1985a, An evolutionary model of the western Churchill province and western margin of the Superior province in Canada and the north-central United States: Tectonophysics, v. 116, p. 281–322, doi: 10.1016/0040-1951(85)90212-4.

Green, A., Weber, W., and Hajnal, Z., 1985b, Evolution of Proterozoic terranes beneath the Williston basin: Geology, v. 13, p. 624–628, doi: 10.1130/0091-7613(1985)13<624:EOPTBT>2.0.CO;2.

Hajnal, Z., Fowler, C., Mereu, R., Kanasewich, E., Cumming, G., Green, A., and Mair, A., 1984, An initial analysis of the Earth's crust under the Williston basin: 1979 COCRUST experiment: Journal of Geophysical Research, v. 89, p. 9381–9400.

Hammer, S., 1963, Deep gravity interpretation by stripping: Geophysics, v. 28, p. 369–378, doi: 10.1190/1.1439186.

Heck, T., 1988, Precambrian structure map of North Dakota: North Dakota Geological Survey Miscellaneous Map 30, scale 1:1,000,000.

Hinze, W., 1996, The crust of the northern U.S. craton: A search for beginnings, *in* van der Pluijm, B., and Catacosnios, P., eds., Basement and basins of eastern North America: Geological Society of America Special Paper 308, p. 187–201.

Jones, A., and Savage, P., 1986, North American Central Plains conductivity anomaly goes east: Geophysical Research Letters, v. 13, p. 685–688.

Jones, A.G., Craven, J.A., McNeice, G.A., Ferguson, I.J., Boyce, T., Farquarson, C., and Ellis, R.G., 1993, The North American central plains conductivity anomaly within the Trans-Hudson orogen in northern Saskatchewan, Geology, v. 21, p. 1027–1030.

Klasner, J., and King, E., 1986, Precambrian basement geology of North and South Dakota: Canadian Journal of Earth Sciences, v. 23, p. 1083–1102.

Kucks, R., and Hill, P., 2002, South Dakota aeromagnetic and gravity data and maps: A website for distribution of data: United States Geological Survey Open-File Report 02-341.

Lance, J.O., 1982. Frequency domain analysis of least squares polynomial surfaces with application to gravity data in the Pedregosa Basin area [Ph.D. dissertation]: El Paso, University of Texas, 199 p.

Latham, T., Best, J., Chaimov, T., Oliver, J., Brown, L., and Kaufman, S., 1988, COCORP profiles form the Montana plains: The Archean cratonic crust and a lower crustal anomaly beneath the Williston basin: Geology, v. 16, p. 1073–1076, doi: 10.1130/0091-7613(1988)016<1073:CPFTMP>2.3.CO;2.

Lidiak, E., 1971, Buried Precambrian rocks of South Dakota: Geological Society of America Bulletin, v. 82, p. 1411–1422, doi: 10.1130/0016-7606(1971)82[1411:BPROSD]2.0.CO;2.

Lidiak, E., 1972, Precambrian rocks in the subsurface of Nebraska: Nebraska Geological Survey Bulletin 26, 41 p.

Lidiak, E., Bickford, M., and Kisvarsanyi, E., 1993, Proterozoic geology of the Eastern Midcontinent region, *in* Van Schmus, W.R., and Bickford, M.E., eds., and 23 others, Transcontinental Proterozoic provinces, *in* Reed, J.C., Bickford, M.E., Houston, R.S., Link, P.K., Rankin, D.W., Sims, P.K., and Van Schmus, W.R., eds., Precambrian: Conterminous U.S.: Boulder, Colorado, Geological Society of America, Geology of North America, v. C-2, p. 260–270.

McCombs, J., Dahl, P., and Hamilton, M., 2004, U-Pb ages of Neoarchean granitoids from the Black Hills, South Dakota, USA; implications for crustal evolution in the Archean Wyoming Province: Precambrian Research, v. 130, p. 161–184, doi: 10.1016/j.precamres.2003.11.004.

Montana, C., Mickus, K.L., and Peeples, W.J., 1992, Program to calculate the gravitational field and gravity gradient tensor due to right rectangular prisms: Computers & Geosciences, v. 18, p. 587–602, doi: 10.1016/0098-3004(92)90096-A.

Morel-A-L'Huissier, P., Green, A., and Pike, C., 1987, Crustal refraction surveys across the Trans-Hudson orogen/Williston basin of south central Canada: Journal of Geophysical Research, v. 92, p. 6403–6420.

Morelli, C., 1976, Modern standards for gravity surveys: Geophysics, v. 41, p. 1051, doi: 10.1190/1.1440661.

Morey, G., Sims, P., Cannon, W., Mudrey, M., and Southwick, D., 1982, Geologic map of the Lake Superior region; Minnesota, Wisconsin and northern Michigan: Minnesota Geological Survey State Map Series S-13, scale 1:1,000,000.

Reed, J., 1993, Map of Precambrian rocks of the conterminous United States and some adjacent parts of Canada, *in* Reed, J.C., Bickford, M.E., Houston, R.S., Link, P.K., Rankin, D.W., Sims, P.K., and Van Schmus, W.R., eds., Precambrian: Conterminous U.S.: Boulder, Colorado, Geological Society of America, Geology of North America, v. C-2.

Simpson, R., Jachens, R., and Blakey, R., 1986, A new isostatic residual gravity map of the conterminous United States with a discussion on the significance of isostatic residual anomalies: Journal of Geophysical Research, v. 91, p. 8348–8372.

Sims, P., 1990, Precambrian basement of the northern Midcontinent: United States Geological Survey Miscellaneous Investigation Map I-1903.

Sims, P., and Peterman, Z., 1986, Early Proterozoic Central Plains orogen: A major buried structure in the north-central United States: Geology, v. 14, p. 488–491, doi: 10.1130/0091-7613(1986)14<488:EPCPOA>2.0.CO;2.

Sims, P., Anderson, J.L., Bauer, R.L., Chandler, V.W., Hanson, G.N., Kalikoski, J., Morey, G.B., Mudrey, M.G., Jr., Ojakangas, R.W., Peterman, Z.E., Schultz, K.J., Shirey, S.B., Smith, E.I., Southwick, D.L., Van Schmus, W.R., and Weiblen, P.W., 1993, The Lake Superior region and Trans-Hudson orogen, *in* Reed, J.C., Bickford, M.E., Houston, R.S., Link, P.K., Rankin, D.W., Sims, P.K., and Van Schmus, W.R., eds., Precambrian: Conterminous U.S.: Boulder, Colorado, Geological Society of America, Geology of North America, v. C-2, p. 11–120.

Southwick, D., and Chandler, V., 1983, Subsurface investigations of the Great Lakes tectonic zone, west-central Minnesota: Geological Society of America Abstracts with Programs, v. 15, p. 692.

Thomas, M., Sharpton, V., and Grieve, R., 1987, Gravity patterns and Precambrian structure in the North American Central Plains: Geology, v. 15, p. 489–492, doi: 10.1130/0091-7613(1987)15<489:GPAPSI>2.0.CO;2.

Van Schmus, W., 1992, Tectonic setting of the Midcontinent rift system: Tectonophysics, v. 213, p. 1–15, doi: 10.1016/0040-1951(92)90247-4.

Van Schmus, W., Bickford, M., Sims, P., Anderson, R., Shearer, C., and Treves, S., 1993, Proterozoic geology of the western Midcontinent basement, *in* Reed, J.C., Bickford, M.E., Houston, R.S., Link, P.K., Rankin, D.W., Sims, P.K., and Van Schmus, W.R., eds., Precambrian: Conterminous U.S.: Boulder, Colorado, Geological Society of America, Geology of North America, v. C-2, p. 239–303.

Van Schmus, W., Bickford, M., and Turek, A., 1996, Proterozoic geology of the east-central Midcontinent basement, *in* van der Pluijm, B., and Catacosnios, P., eds., Basement and basins of eastern North America: Geological Society of America Special Paper 308, p. 7–32.

Manuscript Accepted by the Society 1 February 2007

Printed in the USA

The Geological Society of America
Memoir 200
2007

Variations in lithospheric structure across the margin of Baltica in Central Europe and the role of the Variscan and Carpathian orogenies

Marek Grad*
Institute of Geophysics, University of Warsaw, Pasteura 7, 02-093 Warsaw, Poland

Aleksander Guterch*
Institute of Geophysics, Polish Academy of Sciences, Ks. Janusza 64, 01-452 Warsaw, Poland

G. Randy Keller*
School of Geology and Geophysics, University of Oklahoma, Norman, Oklahoma 73071, USA

POLONAISE'97 and CELEBRATION 2000 Working Groups[†]

ABSTRACT

The European portion of the Eurasian plate formed as a result of a complex series of tectonic events that included the Caledonian, Variscan, and Carpathian orogenies. These orogenic events occurred along the western margin of the East European craton (a portion of the paleocontinent Baltica). In recognition of the complexity of the rifting that formed this margin in the Neoproterozoic-Cambrian and the subsequent tectonic events along it, the region adjacent to this margin has been called the Trans-European suture zone. In order to understand the processes at work during these tectonic events, a series of large integrated geophysical and geological investigations built around large seismic refraction-experiments (POLONAISE'97, CELEBRATION 2000, ALP 2002, and SUDETES 2003) were conducted between 1997 and 2003. In this study, we compare the results of the two longest seismic profiles and their tectonic implications. These studies showed that lithospheric structure of the East European craton directly inboard of its margin is relatively uniform, and the crust consists

*Grad: mgrad@mimuw.edu.pl; Guterch: aguterch@igf.edu.pl; Keller: grkeller@ou.edu.

[†]POLONAISE'97 and CELEBRATION 2000 Working Groups for profiles P4 and CEL05 included: E. Brückl (Austria); A.A. Belinsky (Belarus); I. Asudeh, R. Clowes, Z. Hajnal (Canada); P. Hrubcová, A. Špičák (Czech Republic); S.L. Jensen, H. Thybo (Denmark); K. Komminaho, U. Luosto, T. Tiira, J. Yliniemi (Finland); T. Fancsik, E. Hegedűs, A. Kovačs, R. Csabafi (Hungary); J. Jacyna, L. Korabliova, G. Motuza, V. Nasedkin (Lithuania); W. Czuba, E. Gaczyński, M. Grad, A. Guterch, T. Janik, M. Malinowski, A. Ślączka, P. Środa, M. Wilde-Piórko (Poland); S.L. Kostiuchenko (Russia); M. Bielik, J. Vozár (Slovakia); O. Selvi (Turkey); S. Harder, G.R. Keller, and K.C. Miller (USA).

Grad, M., Guterch, A., Keller, G.R., and POLONAISE'97 and CELEBRATION 2000 Working Groups, 2007, Variations in lithospheric structure across the margin of Baltica in Central Europe and the role of the Variscan and Carpathian orogenies, *in* Hatcher, R.D., Jr., Carlson, M.P., McBride, J.H., and Martínez Catalán, J.R., eds., 4-D Framework of Continental Crust: Geological Society of America Memoir 200, p. 341–356, doi: 10.1130/2007.1200(17). For permission to copy, contact editing@geosociety.org.

of three layers below the sedimentary cover, and it is ~45 km thick. The lithospheric mantle contains several reflectors, and its velocity is ~8.2 km/s.

In the Variscides (Paleozoic platform) of northern Poland along the Trans-European suture zone, the consolidated crust beneath the sedimentary cover consists of two distinct layers, and the total crustal thickness is 30–35 km. Within the core region of the Trans-European suture zone, the crust is ~40 km thick, but most of its upper half is the sedimentary fill of the late Paleozoic and Mesozoic Polish trough, which is underlain by a thick sequence of older continental-margin volcanic and sedimentary units. The lithospheric mantle structure is complex and indicative of a plate collision under the Trans-European suture zone region, and it has seismic velocities that are generally higher than those of the East European craton.

In the Carpathian region, the effects of younger collisions are present in addition to evidence of the Variscan orogeny. It is somewhat surprising that the crustal thickness beneath the Carpathians is 30–40 km, but it is not surprising that beneath the Pannonian Basin, an extensional feature, the crust is only 24–28 km thick. The thickest crust in this region (~50 km) occurs under the rifted margin of the East European craton. The sedimentary cover in this region varies greatly in age and thickness. It is 1–3 km thick beneath the East European craton, 10 km thick the beneath Polish Basin/Lublin trough region, ~18 km thick in the Carpathian foredeep, and 5–8 km thick beneath the Pannonian Basin. The velocity in the lithospheric mantle is 8.1–8.25 km/s beneath the East European craton, 8.2–8.4 km/s beneath the Variscides–Trans-European suture zone, and 7.8–8.0 km/s beneath Carpathian-Pannonian area. The crust and features in the lithospheric mantle appear to dip northward in this area.

In northern Poland, the rifted southwestern margin of the East European craton is abruptly bounded by the Trans-European suture zone and ultimately the Variscides over a region that is only ~100 km wide, while the collisional zone between the East European craton and the Carpathian-Pannonian area is ~300 km wide. In both areas, the lithospheric structure observed suggests that the Caledonian, Variscan, and Carpathian orogenies in this area were relatively "soft" collisions that left the East European craton passive margin largely intact. The structural model of the transition between the Pannonian Basin–Carpathians and the East European craton indicates northward "old" subduction under Baltica (Jurassic–Early Cretaceous). However, the thinning of the Pannonian lithosphere could be explained as the result of extension and high heat flow, with "young" southward subduction or slab rollback to the east, which took place in the Tertiary (Miocene).

Keywords: seismic refraction, lithospheric structure, crustal structure, Baltica, Trans-European suture zone, Carpathians, Alps, Pannonian Basin.

INTRODUCTION

The formation of the European portion of the Eurasian plate includes three geologically recent periods of mountain building due to accretion of terranes during the Caledonian and Variscan orogenies (Fig. 1A) and the collisional events of the Alpine-Carpathian orogeny (Fig. 1B). In addition, extension has affected the region on several occasions. First, the supercontinent Rodinia broke up near the Cambrian-Precambrian boundary and formed the rifted margin of southwest Baltica; second, extension was widespread after the Variscan orogeny, especially in the Polish–North German Basin, and third, the Eger rift formed during the Cenozoic. When viewed from a larger perspective, the Paleozoic tectonic evolution of Europe involved a series of orogenic pulses resulting from the collision and suturing of Baltica, Laurentia (the North American paleocontinent), Gondwana (Africa), intervening terranes like Avalonia and the Bohemian massif, and transported crustal blocks that formed the supercontinent Pangea (Figs. 1 and 2).

The Trans-European suture zone (Caledonides, Teisseyre-Tornquist zone area; Fig. 2) is a term that is used to refer to the broad zone of deformation that extends across Europe from the British Isles to the Black Sea region that formed as Europe was assembled from a patchwork of terranes during the late Paleozoic (Berthelsen, 1992a; Ziegler, 1990; Pharaoh, 1999). These terranes were accreted along the margin of Baltica (East European craton)

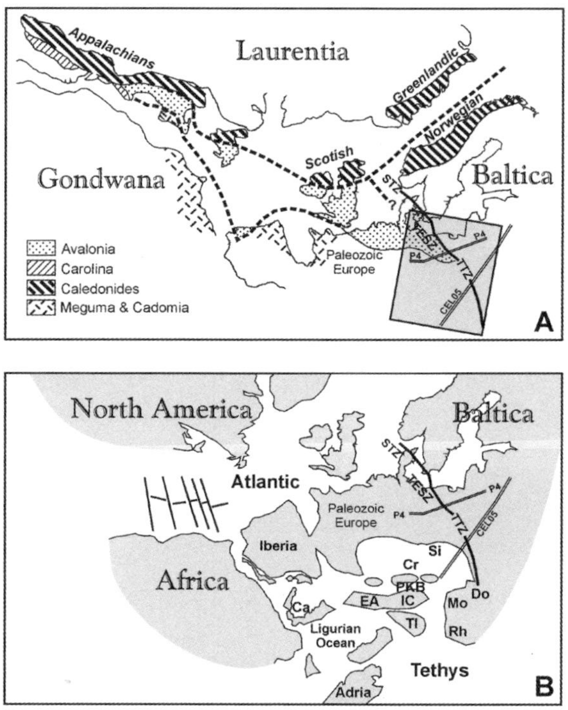

Figure 1. (A) Reconstruction of major tectonic units of the Appalachian-Caledonian orogeny (modified from MacNiocaill et al., 1998; Keller and Hatcher, 1999). STZ—Sorgenfrei-Tornquist zone, TESZ—Trans-European suture zone, TTZ—Teisseyre-Tornquist zone. Gray rectangle shows the area of recent seismic experiments in Central Europe (see Fig. 2) and the main profiles discussed in this paper (P4 and CEL05). (B) Paleogeography of the Western and Central Tethys during the Late Jurassic (152 Ma) (modified from Golonka et al., 2003a). Ca—Calabria; Cr—Czorsztyn ridge; Do—Dobrogea; EA—eastern Alps; IC—Inner Carpathians; Mo—Moesia plate; PKB—Pieniny Klippen belt basin; Rh—Rhodopes; Si—Silesian Basin; TI—Tisa plate.

that was formed during the breakup of Rodinia. The tectonic evolution of this region is in many ways analogous to the Appalachian-Ouachita orogeny in eastern North America (Keller and Hatcher, 1999; Golonka et al., 2003a, 2003b), and understanding this evolution is certainly of global importance to studies in terrane tectonics and continental evolution. The Trans-European suture zone is far more complex than a single suture, but in a broad sense it is the boundary between the accreted terranes and Baltica (Pożaryski, 1990; Franke, 1990; Berthelsen, 1992a, 1992b, 1998). The Bohemian massif is a large, complex terrane mostly located in the Czech Republic; its origin can be traced to northern Gondwana (Africa). In southern Poland, several structural blocks such as the Małopolska massif (MM in Fig. 2) are located adjacent to Baltica, and these were probably transported laterally along it similar to the Cenozoic movement of terranes along the western margin of North America (Keller and Hatcher, 1999; Dadlez et al., 2005).

The younger Carpathian Mountains and Pannonian Basin are the result of intricate Mesozoic-Cenozoic plate interactions in the Mediterranean region as the Tethys Ocean closed during convergence of Europe and Africa-Arabia (Fig. 1B). During the Cenozoic, complex interactions among small plates caused the Carpathian arc to evolve into its strongly arcuate shape. These plate interactions have been interpreted to involve subduction of oceanic areas, and they have produced considerable Neogene volcanism (Birkenmajer, 1986; Kováč et al., 1993; Kovačs et al., 2000; Szabó et al., 2004; Bielik et al., 2004). Back-arc extension was the dominant process that formed the Pannonian Basin, which contains up to 8 km of Neogene strata in its subbasins (e.g., Royden and Dövényi, 1988). This region is still tectonically active as evidenced by high seismicity of the southern Carpathians that extends to depths of ~200 km in the Vrancea region north of Bucharest.

NEW GENERATION OF SEISMIC EXPERIMENTS IN CENTRAL EUROPE

Beginning in 1997, a series of large seismic-refraction experiments covered Central Europe with an unprecedented network of seismic profiles with considerable three-dimensional (3-D) coverage (Fig. 2) and drew the attention of the international community to the crustal and lithospheric mantle structure and tectonics of the area. These experiments—POLONAISE'97, CELEBRATION 2000, ALP 2002, and SUDETES 2003—were only possible due to a massive international cooperative effort. Since the lithospheric structure in the region is very complex, the need for a 3-D approach was recognized early. The first of the new experiments was POLONAISE'97 (Guterch et al., 1998, 1999; Jensen et al., 1999; Środa and POLONAISE P3 Working Group, 1999; Janik et al., 2002), and it showed how much could be learned from even modest 3-D coverage (Środa et al., 2002; Czuba et al., 2002). A series of even larger experiments followed in rapid succession: CELEBRATION 2000, ALP 2002, and SUDETES 2003 (Guterch et al., 2001, 2003a, 2003b; Brückl et al., 2003; Malinowski et al., 2005; Janik et al., 2005; Grad et al., 2003b, 2006; Majdański et al., 2006). As a result of these experiments, the network of seismic-refraction profiles now extends along the Trans-European suture zone region of Poland, across the Bohemian massif, across Pannonian Basin, and through the western Carpathians and eastern Alps. The coverage extends along a swath that extends from northern Poland to the Adriatic Sea and the Dinarides (Fig. 2).

The scientific goals of this effort included attaining a better understanding of the formation of continental Europe in the Paleozoic and the Variscan and Caledonian orogenies, as well as the subsequent formation of the Alps-Carpathian mountain chain and the Pannonian Basin. Apart from producing two-dimensional (2-D) models along the many seismic profiles recorded, the ultimate goal was to construct a three-dimensional model of the lithospheric structure in the area and provide the

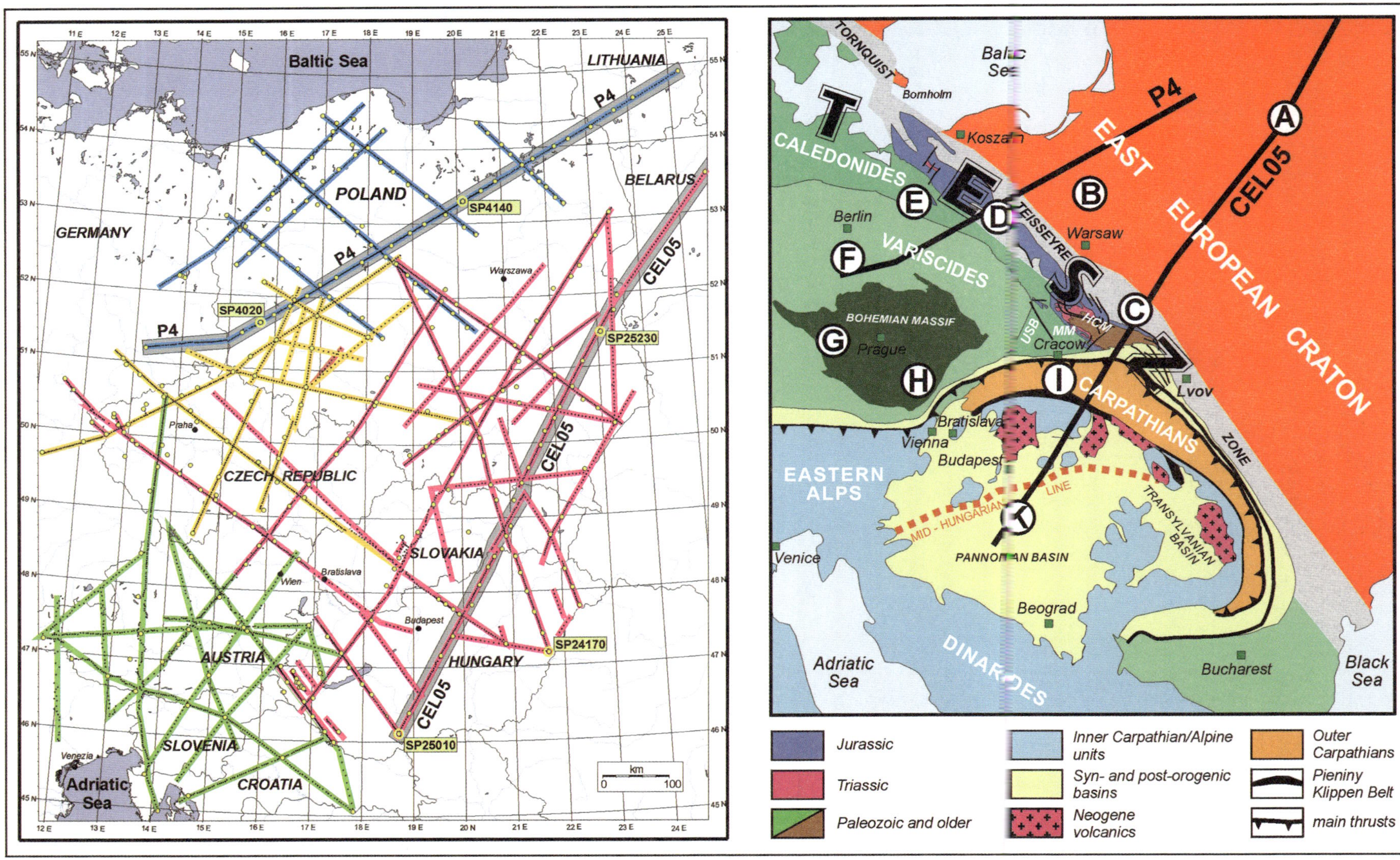

Figure 2. Location of seismic-refraction and wide-angle reflection profiles recorded in Central Europe during 1997–2003 (left). Profiles from experiments POLONAISE'97, CELEBRATION 2000, ALP 2002, and SUDETES 2003 are marked in blue, red, green, and yellow, respectively. The seismic profiles have a total length of ~20,000 km, with ~7000 seismic receiver positions (small black dots). Yellow dots show the locations of the 295 large shots. Gray lines show the positions of POLONAISE'97 profile P4 and CELEBRATION 2000 profile CEL05. Yellow circles show locations of shotpoints SP25010, SP4020, SP24170, SP25230, and SP4140 for which record sections are shown in Figures 3 and 4. On the right, locations of POLONAISE'97 profile P4 and CELEBRATION 2000 profile CEL05 are shown on the background of a simplified tectonic map of the Central Europe; the white circles (A–K) indicate locations of one-dimensional (1-D) models in Figure 7. HCM—Holy Cross Mountains; MM—Małopolska massif; TESZ—Trans-European suture zone; USB—Upper Silesia block.

structural background for other geophysical modeling (e.g., potential fields, heat flow). Finally, these 3-D models should be a starting point to evaluate and develop new geodynamic models for the tectonic evolution of the whole region.

The total length of all profiles made during POLONAISE'97, CELEBRATION 2000, ALP 2002, and SUDETES 2003 experiments is ~20,000 km. About 300 large explosive sources were employed, and seismic recordings were made at ~7000 sites (Fig. 2). Densely spaced shots and receivers along the network of profiles produced high-quality data extending to long offsets that resolve seismic models of the crust and lower lithosphere. Examples of seismic record sections from various tectonic domains are shown in Figures 3 and 4. Waves from the sedimentary cover are clearly visible in the area of the Pannonian Basin (SP24170 in Fig. 3) and the Trans-European suture zone (SP4140 in Fig. 4), indicating thick sediments there. Data from profiles on Baltica (East European craton) produce a wave field that is very regular (e.g., SP25230 in Fig. 4). The sedimentary cover is thin, and the diving wave traveling in the upper and middle crust (Pg) produces clear first arrivals with apparent velocities increasing smoothly from ~6 to 7 km/s. A strong reflected wave from the base of the crust (PmP) is observed in almost all of the record sections. In Baltica, a PmP wave is observed starting at offsets of ~110 km, and the intercept times of this arrival projected from the area of its critical point and the Pn waves are ~9–10 s. In the Pannonian Basin, the same wave is observed starting at offsets of 40–50 km, and corresponding intercept times for the PmP and Pn waves are only ~6 s. Such a big difference in the arrival times (reaching 3–6 s) for Pg, Pcrustal, PmP, and Pn phases (Fig. 4) reflects large variations in crustal and uppermost mantle structure across the area. These variations have been documented by careful modeling of these data, and models derived for the two longest profiles from the POLONAISE'97 and CELEBRATION 2000 experiments, namely profile P4 (Grad et al., 2003a) and profile CEL05 (Grad et al., 2006), are shown in Figures 5 and 6, respectively.

SEISMIC CHARACTERISTIC OF CENTRAL EUROPE CRUST

For the area spanned by our experiments, six principal types of crustal structure can be distinguished for major tectonic units: (1) Baltica in the northeast, including its margin; (2) Trans-European suture zone; (3) Variscides; (4) Bohemian massif; (5) Carpathians; and (6) Pannonian Basin (Fig. 2). The principal features of the upper, middle, and lower crust are shown in Figure 7 and discussed in turn below.

(1) The area of the East European craton (Baltica) is characterized by thin (1–2 km) sedimentary cover, and only along its margin does the cover attain a thickness of ~6 km. The crystalline crust of the East European craton has a three-layered structure characterized predominantly by the following velocities: 6.1–6.4 km/s in the upper layer, 6.5–6.8 km/s in the middle layer, and 6.9–7.2 km/s in the lower crust (Figs. 7A–7B). The total crustal thickness reaches a value of ~48 km and is slightly smaller in the East European craton margin (44 km in Fig. 7B). The Vp velocity beneath Moho is normal, 8.1–8.2 km/s. Further northeast into the East European craton, the EUROBRIDGE project discovered a three-layered crystalline crust, which is characteristic of the Svecofennian region to the east and southeast of the Sveconorwegian deformation front and Trans-Scandinavian Igneous Belt (Bogdanova, 1996; EUROBRIDGE Seismic Working Group, 1999; EUROBRIDGE'95 Seismic Working Group, 2001). To the northwest, in Baltica, a similar crustal structure was also observed in the BABEL A profile (BABEL Working Group, 1993), which was recorded in the Baltic Sea adjacent to Sweden. The relatively uniform velocity pattern and thicknesses of individual layers in the crystalline crust of Baltica point to its homogenization on a large scale (e.g., Grad and Luosto, 1987). We have been able to detect major inhomogeneities only in the upper crust, where high-velocity bodies have been delineated that correlate with the Kętrzyn and Suwałki anorthositic intrusions on profile P5 (Czuba et al., 2002) and on profile P4 (Fig. 5A; 600–670 km along profile) (Grad et al., 2003a). We are not arguing that the crust of Baltica is homogeneous in detail, but it does seem to have a consistent thickness and gross velocity structure over a large area, similar to that observed in other cratonal areas as the central United States (Braile et al., 1989).

(2) Along the Trans-European suture zone in Poland, at least three crustal blocks (terranes) can be identified based on seismic and geologic data: the Małopolska massif (in the southeast; Fig. 7C), Kuiavia (in central Poland; Fig. 7D), and Pomerania (to the northwest, Fig. 7E). As would be expected, the crust of the Trans-European suture zone region is complicated, particularly in the sedimentary layer that reaches a thickness of 9–12 km with velocity <5.4 km/s. The upper part of this layer (maximum 7–8 km) belongs to the Permian–Mesozoic sequence, and the lower part belongs to the earlier Paleozoic (?Devonian–Carboniferous) sequence, both of epicontinental origin. The consolidated crust, with relatively low velocities <6.0 km/s, is observed down to ~15–20 km depth. This part of the crust is commonly regarded as composed of a deformed and slightly metamorphosed early Paleozoic series of sedimentary and volcanic rocks, although boreholes have not penetrated them. Average velocities in middle and lower crust are 6.3–6.6 km/s and 6.8–7.2 km/s, respectively, and these values are similar to those in Baltica. The high-velocity lower crust, which has a high gradient of velocity and strong, ringing reflectivity (Fig. 7E), has a distinctly laminar seismic structure (Guterch et al., 1992; Jensen et al., 1999; Grad et al., 1999, 2002b). The postulated boundary between Kuiavia and Pomerania is rather sharp at particular levels of the crust (Janik et al., 2005). The attenuated lower and middle crust could belong to proximal terranes built of Baltican crust detached in the southeast and reaccreted to the Baltica due to the process of counterclockwise rotation of the Baltica paleocontinent during the Ordovician–early Silurian (Dadlez et al., 2005).

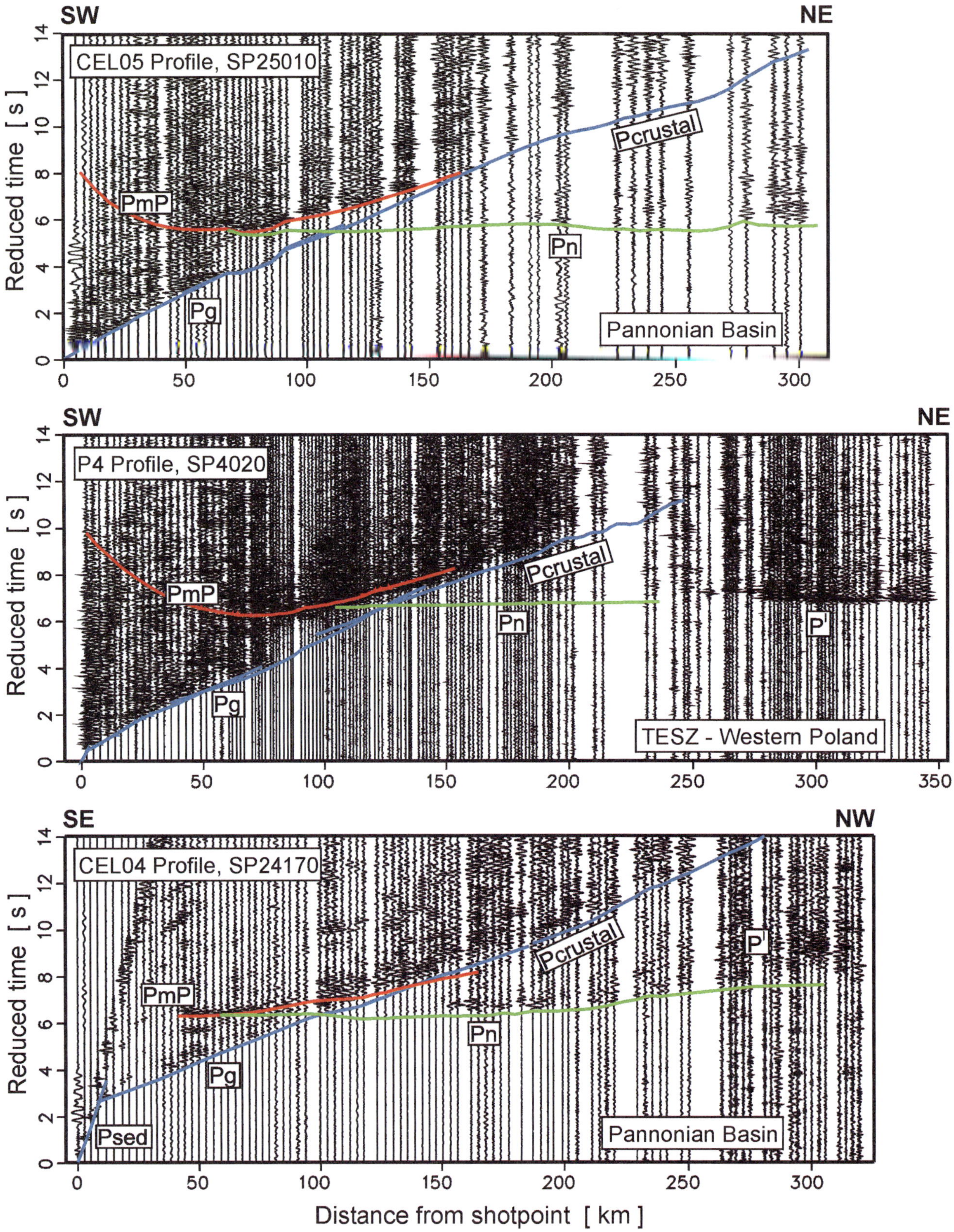

Figure 3. Examples of P-wave record sections from the Pannonian Basin and Trans-European suture zone (for shotpoint locations see Fig. 2). Pg, Pcrustal, PmP, and Pn—crustal and Moho phases; P^{I} and P^{II}—mantle lithosphere phases.

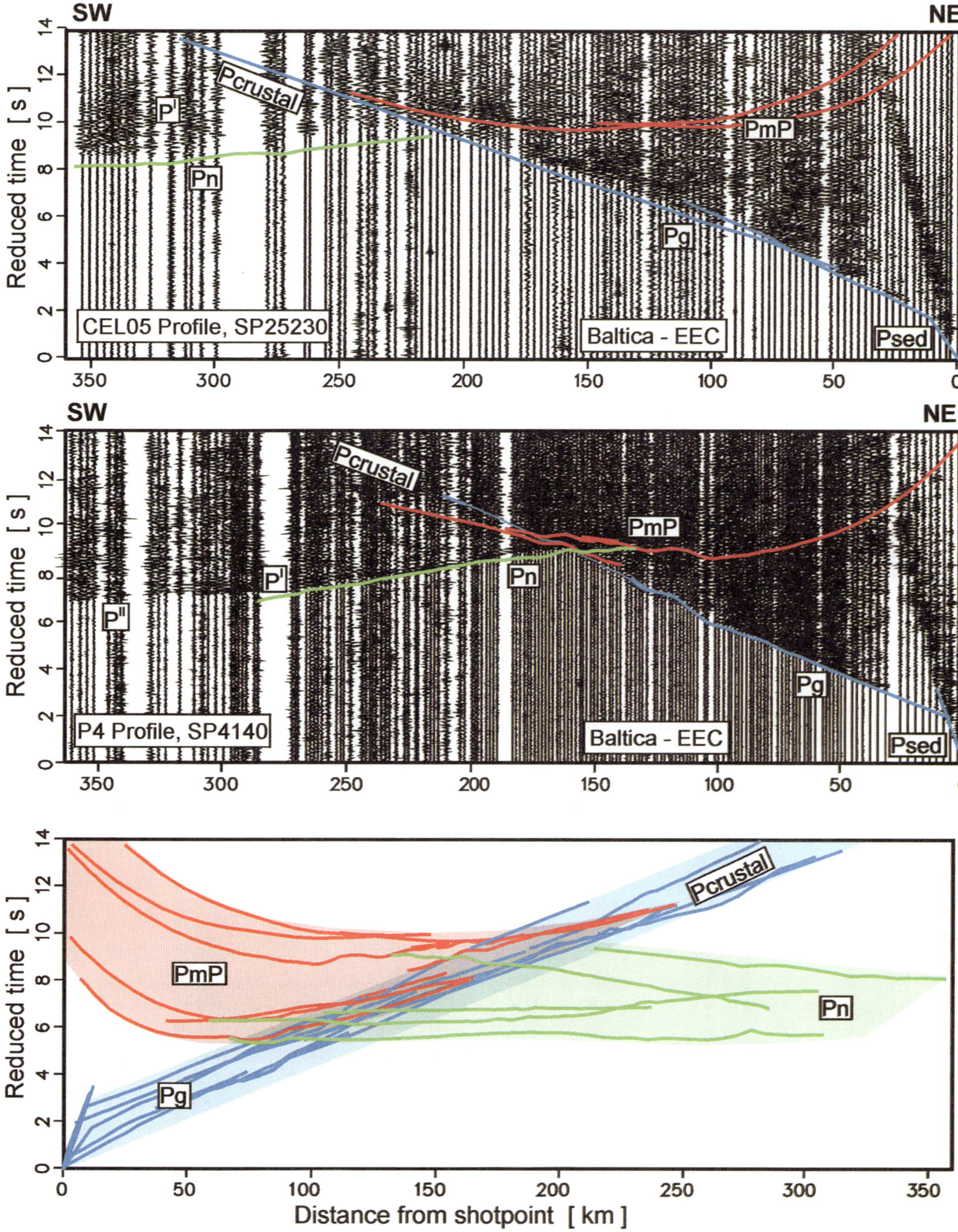

Figure 4. Examples of P-wave record sections for seismic profiles within the East European craton (EEC) (for shotpoint locations see Fig. 2; phases description as in Fig. 3). The lower figure is a comparison of traveltime Pg, Pcrustal, PmP, and Pn phases for the data shown in this figure and Figure 3. Note the large variation in the arrival times for each phase (reaching 3–6 s), which indicates large differences in the crustal and uppermost mantle structure across the region.

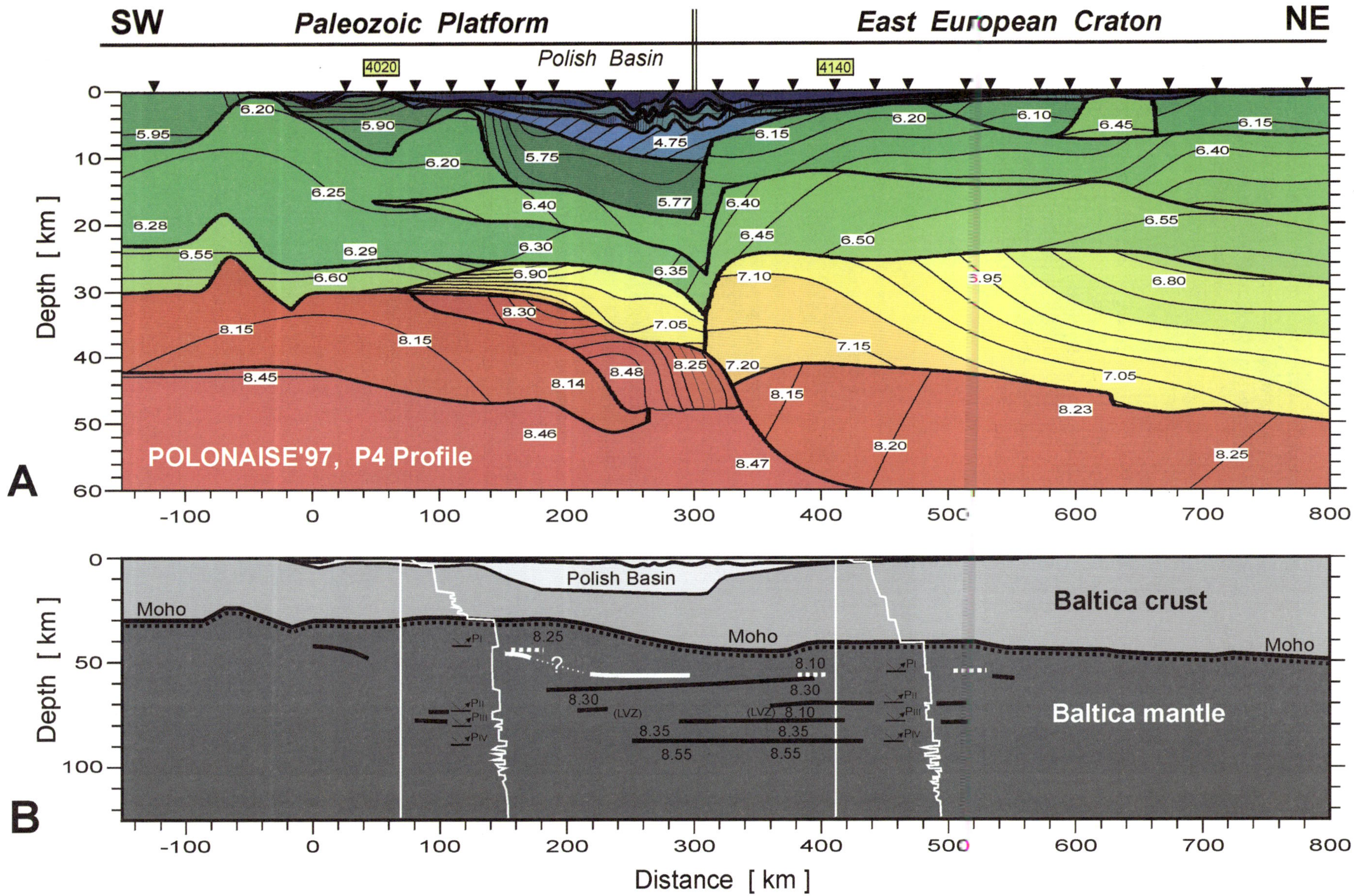

Figure 5. Two-dimensional P-wave velocity model for POLONAISE'97 profile P4 (A) derived by forward ray-tracing modeling using the SEIS83 package (Červený and Pšenčík, 1983). The profile P4 extends across the Trans-European suture zone region and the southwestern portion of the East European craton (Baltica). The thick solid lines are layer boundaries, and the thin lines are isovelocity contours (in km/s). Moho is crust-mantle boundary. Numbered triangles show shotpoints along the profile. Note thin Paleozoic Platform crust (~30 km), thick crust of the East European Craton (~45–50 km), and thick sedimentary basin in the Teisseyre-Tornquist zone–Polish Basin region in which P-wave velocities lower than 6 km/s extend to depths of ~20 km. Vertical exaggeration for the model is ~5.2 (modified from Grad et al., 2003a). (B) Simplified sketch of lithospheric structure along POLONAISE'97 profile P4 developed by forward ray-tracing. Lower lithospheric reflectors from beneath profile P4 (thick black lines), reflectors from the parallel profile P2 (thick white lines) and from profiles P1, P3, and P5 (dotted white lines) are shown (modified from Grad et al., 2002a).

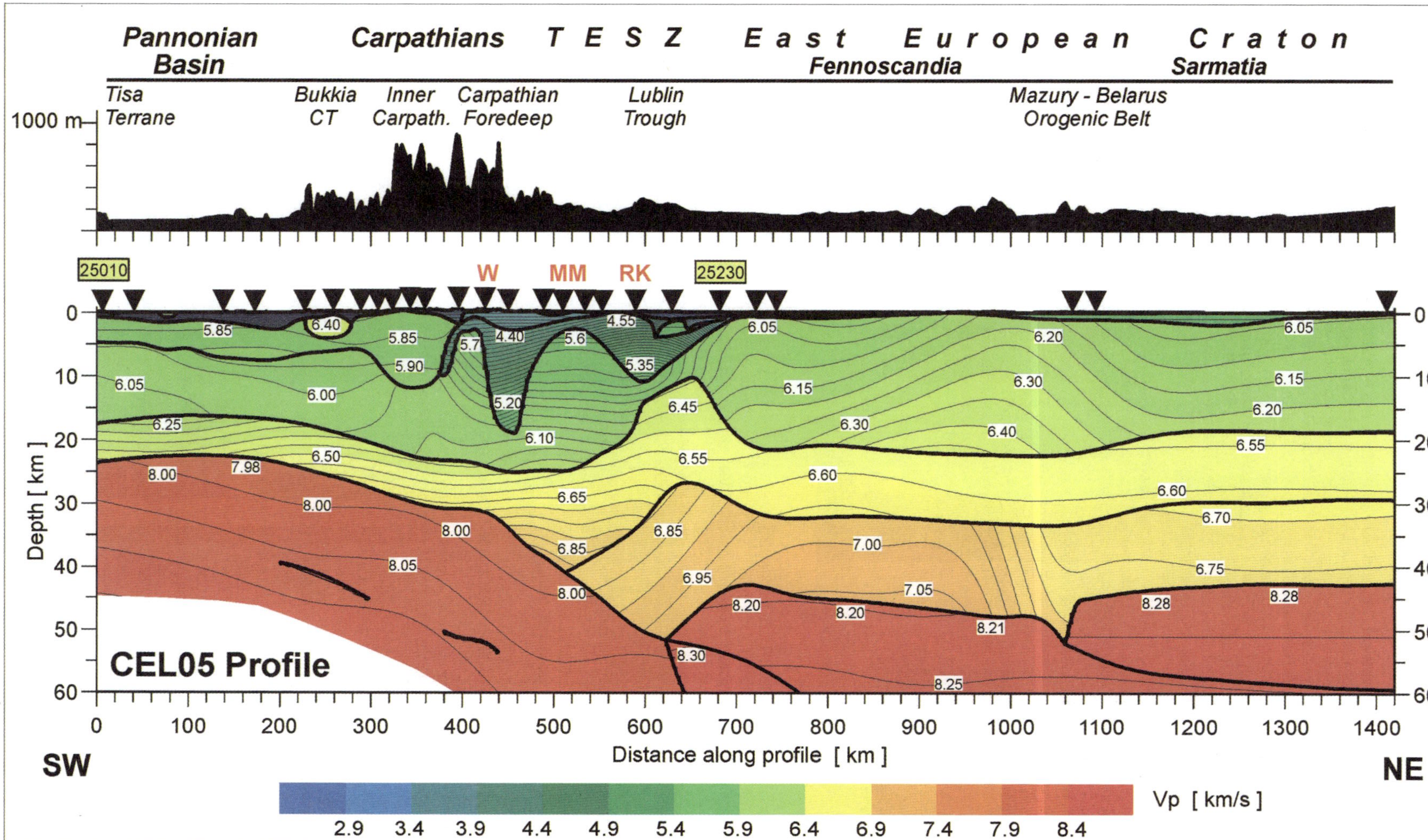

Figure 6. Two-dimensional P-wave velocity model for CELEBRATION 2000 profile CEL05 derived by forward ray-tracing modeling using the SEIS83 package (Červený and Pšenčík, 1983). The profile extends from the East European craton across the Trans-European suture zone and Carpathians to the Pannonian Basin. The thick solid lines are layer boundaries, and thin lines are isovelocity contours (in km/s); numbered triangles refer to shotpoints. In the upper part, topography along the profile is shown. Note very thin Pannonian crust (~25 km), thick crust of the East European craton (~45–50 km), and the thickest crust in the area investigated in the Trans-European suture zone (~52 km depth at distance 620 km). In the Carpathians and Trans-European suture zone, sedimentary basins with P-wave velocities lower than 6 km/s extend to 10–20 km depth. Bukkia CT—Bukkia composite terrane; MM—Małopolska massif; RK—Radom-Kraśnik uplift (at 580–600 km of profile); W—Wysowa (at ~420 km of profile). Vertical exaggeration for the model is ~6.8 (modified from Grad et al., 2006).

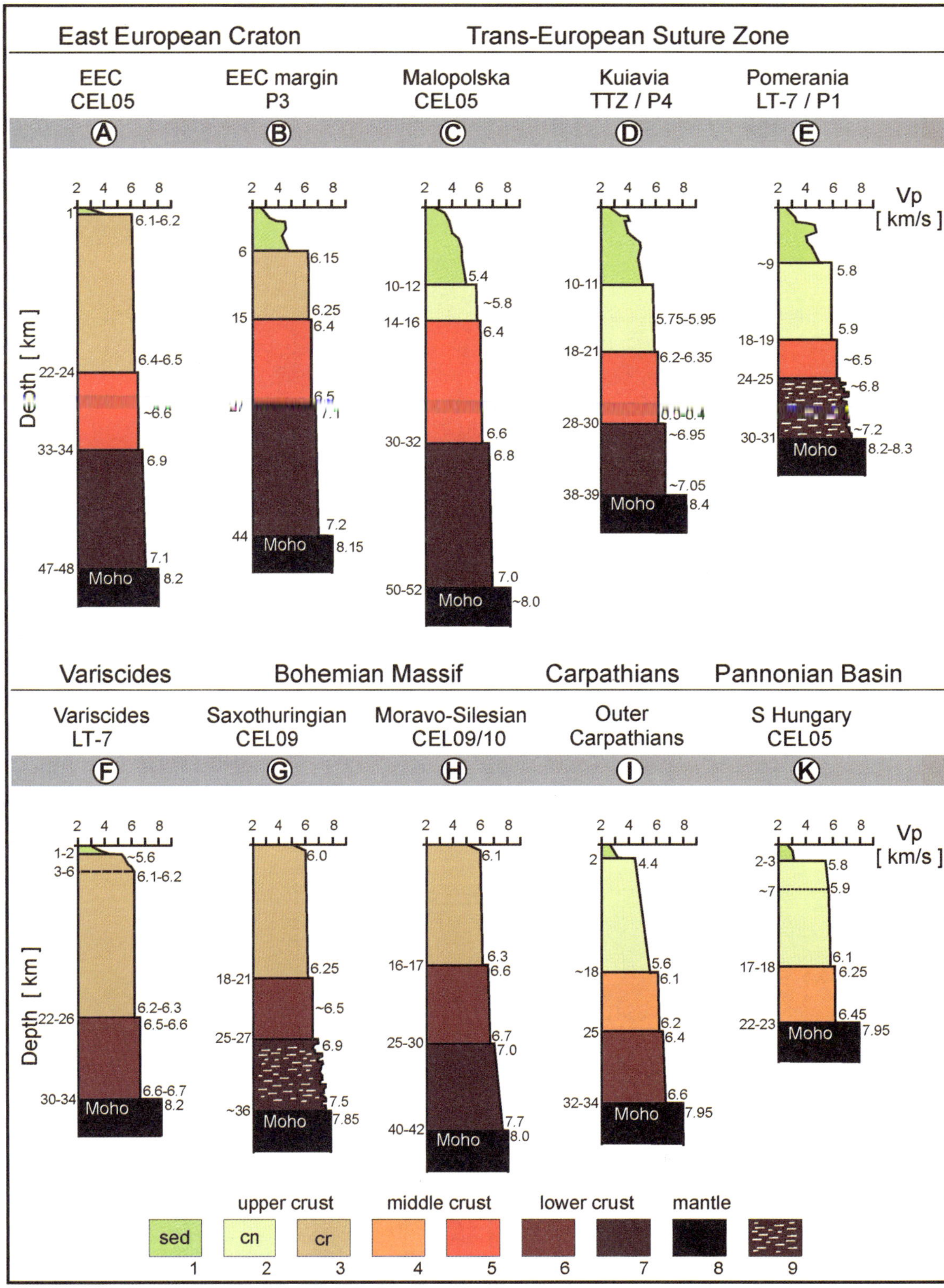

Figure 7. Characteristic crustal models for the East European craton, Trans-European suture zone, Variscides, Bohemian massif, Carpathians, and Pannonian Basin. For locations, see circles marked A–K in Figure 2. 1—sedimentary crust (sed, Vp < 5.7 km/s); 2—upper consolidated crust (cn, Vp = 5.7–6.1 km/s); 3—upper crystalline crust (cr, Vp = 6.1–6.35 km/s); 4—middle crystalline crust (Vp = 6.1–6.3 km/s); 5—middle crystalline crust of Trans-European suture zone and East European craton (Vp = 6.2–6.6 km/s); 6—lower crystalline Variscan crust (Vp = 6.5–6.7 km/s); 7—lower crust of Trans-European suture zone and East European craton (Vp = 6.8–7.3 km/s); 8—upper mantle (Vp = 8.05–8.4 km/s); 9—reflective zones in the lower crust.

(3) Variscan consolidated crust occurs below 1–2 km of sedimentary strata and is characterized by Vp = 5.6–6.3 km/s. The lower crustal layer is characterized by Vp = 6.5–6.65 km/s (Fig. 7F). These velocities are closely comparable with other Variscan areas of Central Europe (Wever et al., 1990; Grad et al., 2002b). It displays a two-layer structure with relatively low P-wave velocities continuing through the entire crust. The crust is much simpler than in Baltica and the Trans-European suture zone, due to the disappearance of the high-velocity lower crust.

(4) The Bohemian massif is a large, stable outcrop of Variscan rocks that lies mostly within the Czech Republic. The uppermost crust (2–4 km) is characterized by a high P-wave velocity gradient (~5.5–6.0 km/s). The upper and middle crust are characterized by velocities of ~6.0–6.3 and 6.5–6.7 km/s, respectively, with a generally small velocity gradient. The recorded Pg waves display an anisotropy of 1.5%–2% in the upper crust (Růžek et al., 2003; Vavryčuk et al., 2004). The crustal thickness of the Bohemian massif ranges in a wide interval from 29 to 40 km (e.g., Beránek and Dudek, 1972; Neunhöfer et al., 1981; Novotný et al., 1997; Wilde-Piórko et al., 2005; Hrubcová et al., 2005). Distinct lateral variations in the velocities at the crust-mantle transition correlate with different units of the Bohemian massif, namely the Saxothuringian, Moldanubian, and Moravo-Silesian blocks. The Moho in the Moldanubian unit is well defined as a first-order discontinuity with a large velocity contrast, while the lower crust in the Saxothuringian unit is highly reflective (Fig. 7G), and in the Moravo-Silurian unit, the crust-mantle transition is represented by a thick gradient zone (Fig. 7H) (Hrubcová et al., 2005).

(5) In the Carpathians, a complicated structure in the sedimentary cover was found, where rocks with velocities <5.6 km/s reach depths of ~20 km. Beneath this cover, a two-layer crystalline crust has velocities of 6.0–6.2 and 6.5–6.9 km/s, while further to northeast, a three-layer crust was found beneath the Trans-European suture zone and East European craton. The model presented in Figure 7I represents the Outer Carpathian crust with a thickness of 32–34 km. Across the Carpathians over a distance of ~300 km, the crustal thickness increases from ~25 to ~50 km, dipping to northeast (Fig. 6).

(6) The crustal structure of the Pannonian Basin is relatively simple. Below the 2–3-km-thick Neogene sediments (they have greater thicknesses in some places not crossed by our experiments), a layer of low-velocity rocks (5.8–6.1 km/s) with a very small velocity gradient occurs down to ~18 km depth (Fig. 7K). The lower crust has low velocities (Vp ~ 6.25–6.45 km/s). The uppermost mantle is characterized by velocities of 7.95–8.0 km/s, and the Moho lies here at a relatively shallow depth of only 22–23 km (Grad et al., 2006; Środa et al., 2006).

MANTLE LITHOSPHERE STRUCTURE

In all four experiments in Central Europe, upper mantle phases were well recorded in many of the record sections (see, for example, PI and PII phases from profiles P4, CEL04, and CEL05 for SP4020, SP24170, SP25230, and SP4140 in Figs. 3 and 4). Many reflected phases were recorded, but phases due to features in the mantle lithosphere were observed in the first arrivals only for large offsets (e.g., beneath the longest profiles P4 and CEL05 at 300–500 km, and more).

In all POLONAISE'97 profiles, the PI phase indicates a shallow mantle reflector at depths ~8–12 km deeper than the Moho (Grad et al., 2002a, 2003a; Jensen et al., 1999; Środa and POLONAISE P3 Working Group, 1999; Janik et al., 2002). Beneath profile P4, we found other reflectors at depths of ~70, 80, and 90 km (Fig. 5B) in the central part of the Trans-European suture zone region in the distance range of 180–420 km (Grad et al., 2002a). In general, the reflectivity of the uppermost mantle is stronger beneath the Paleozoic Platform and Trans-European suture zone than the East European craton. The deepest interpreted seismic reflector associated with a zone of high reflectivity may mark a change in upper mantle structure from an upper zone characterized by seismic scatterers of small vertical dimension to a lower zone with vertically larger seismic scatterers, possibly caused by inclusions of partial melt. Complex high and low velocities in the mantle lithosphere have been found also from other seismic experiments close to the Trans-European suture zone region, as well as within East European craton and the Baltic Shield (e.g., Guggisberg et al., 1991). The depth to the reflective zone correlates with the heat flow, such that it is shallower in "hot" areas than in "cold" areas, a feature also observed along the FENNOLORA seismic long-range profile.

For profile CEL05, waves reflected in the mantle lithosphere (PI and PII phases) were well recorded from many shotpoints. In the Pannonian-Carpathian area, the PI wave was recorded at offsets of 200–350 km (see, for example, PI wave for SP24170 in Fig. 3). This wave is usually much stronger than the Pn wave in the same distance range. The depth of corresponding reflector in the mantle lithosphere follows the shape of the Moho (~15 km deeper than Moho) and increases in depth from ~40 km at distance 200 km to ~55 km at distance 440 km, with a gap between 290 and 380 km along the profile (Fig. 6). In the area of East European craton, two phases (PI and PII) reflected in the mantle lithosphere were well observed, and the depths of corresponding reflectors were 75–80 and ~95 km (Fig. 8B). Some short reflecting elements were also found beneath Baltica at depths of ~55 and 60–70 km (Fig. 6). In the Trans-European suture zone area at distances of 520–650 km along profile CEL05, a very pronounced reflector was found at a depth of 60–70 km with a large, positive contrast of velocity (8.1 and 8.45 km/s for Pannonian and Baltica mantle, respectively).

TECTONIC MODELS

Based on the velocity structure observed along profiles P4 and CEL05 (Figs. 5 and 6), we propose general tectonic models for the regions traversed by these profiles and some integration of these results. Representations of possible tectonic interpretations for P4 and CEL05 are shown in Figure 8. From the northeasternmost extent of these profiles southwestward to the margin of the East

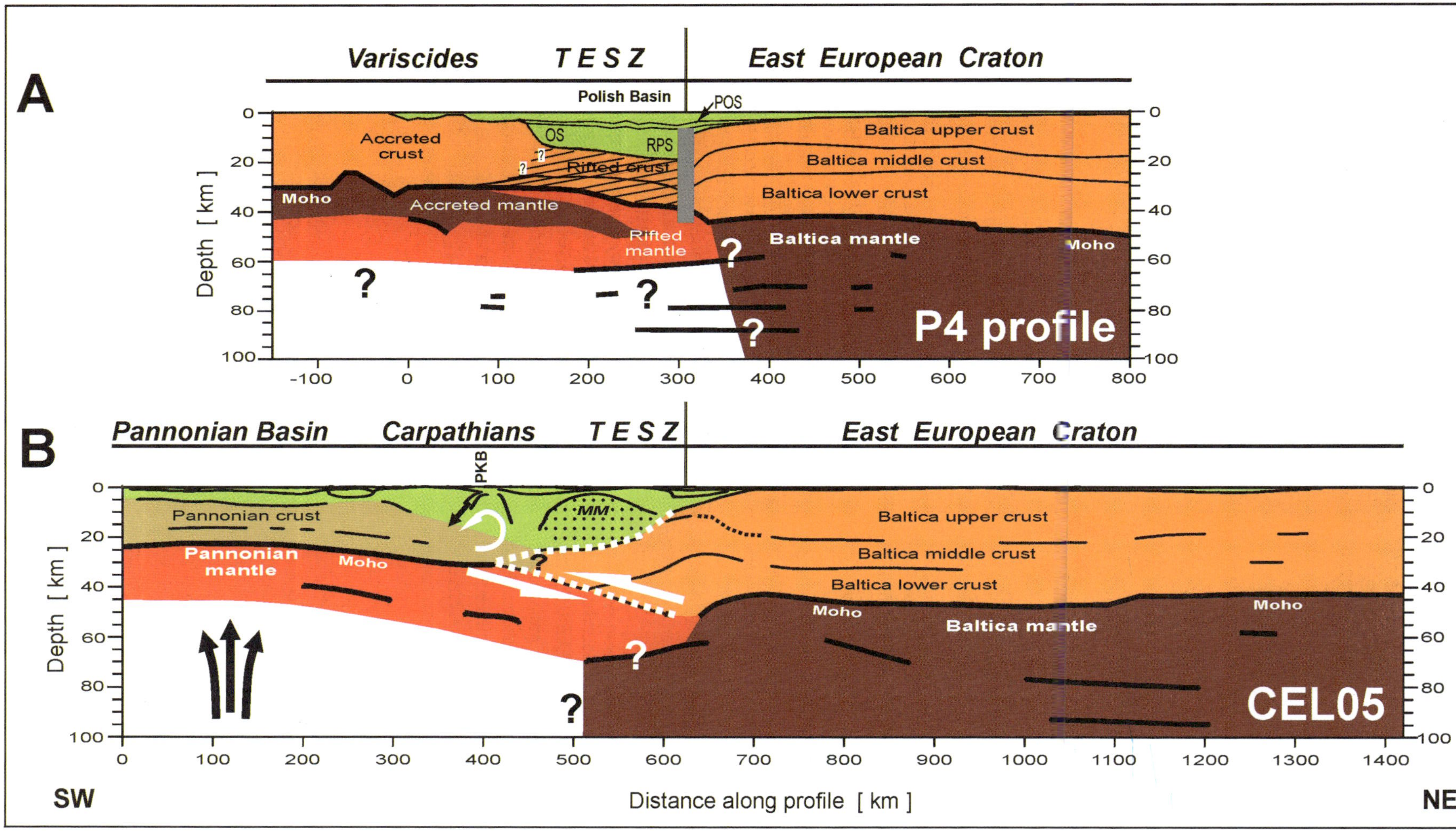

Figure 8. Tectonic interpretation of the transition from East European craton to Trans-European suture zone (TESZ) and Variscides along profile P4 (A) and from East European craton to Carpathians and Pannonian Basin along profile CEL05 (B). (A) Abrupt transtensional margin (gray bar) where the southwestern extent of Baltica lithosphere is minimal (modified from Grad et al., 2003a). POS—postorogenic sedimentary sequence; OS—orogenic sedimentary sequence; RPS—rift (pull-apart) sedimentary sequence. (B) Collisional model involving "old" northward subduction zone (white arrows) of the Pannonian lithosphere under East European craton in the Jurassic–Early Cretaceous (Książkiewicz, 1977) or by the end of the Jurassic, ca. 140 Ma (Golonka et al., 2003a), and thinning of the Pannonian lithosphere due to extension and high heat flow (black up-directed arrows), and "young" southward subduction (black down-directed arrow) or rollback (white twisted arrow) in the Tertiary (Miocene) in the Pieniny Klippen belt (PKB) (modified from Grad et al., 2006). MM—Małopolska massif.

European craton (Baltica), the simple and consistent structure of these models reflects the stable Phanerozoic history of the East European craton. However, considering the deformation associated with the Caledonian, Variscan, and Carpathian orogenies, the simplicity of the East European craton structure is puzzling. In contrast, the Appalachian-Ouachita orogeny caused tectonic inversion in the Southern Oklahoma aulacogen on the order of 10 km and extensive intraplate tectonism over a broad area (Keller and Stephenson, this volume).

The results from profile P4 and other POLONAISE'97 profiles have provided many new insights considering the structure and tectonic evolution of the Trans-European suture zone region in northwest Poland and the East European craton. However, there remain several possible explanations of how these tectonic events are reflected in the observed velocity structure (Grad et al., 2003a). In the model in Figure 8A, the low velocities that extend to 20 km represent the sedimentary and possibly volcanic strata that developed along the rifted margin of Baltica beginning in the Cambrian, as well as strata that were deposited after the Variscan deformation and are Permian and younger in age. These younger strata would classically be considered the fill of the Polish Basin. In addition, the orogenic sedimentary sequences (OS, Fig. 8A) in the southwestern portion of the Polish Basin could consist of accretionary wedge–foreland basin deposits. Since the model shows that the rifted passive margin of Baltica is largely intact, it appears that the Caledonian and Variscan orogenies in this area were relatively "soft" (possibly transpressive). A "soft" collision is also consistent with the lack of Paleozoic deformation within the adjacent portions of the East European craton.

Two main tectonic interpretations (Grad et al., 2003a) emerge, and the difference hinges on whether the apparent continuity of deep crustal layers beneath the 20-km-deep Polish Basin is real or just an artifact of the modeling technique that requires layers in the velocity to extend completely across the model space. A continuous deep structure beneath the basin implies that the rifting of Rodinia produced highly extended Baltican continental crust that continues to the southwest beneath the Polish Basin and beyond. Unlike the upper crustal structure that is clearly discontinuous, the layers that constitute the middle and lower crust do not change thickness or velocity dramatically along the model. This geometry can be interpreted as the result of a collision in which flake-type or crocodile-type collision (Meissner, 1989) occurred in which Baltica indented Avalonia, obducting its upper crust and underthrusting its lower crust–uppermost mantle. In this case, it is possible that Baltica-derived lower crust even extends more to the southwest, to a location such as the Elbe line, where previous studies show that high-velocity middle and lower crust terminate abruptly (Aichroth et al., 1992; Thybo, 2001). The implications of this model include extreme stretching of the lower crust while the upper crust was detached and removed by the rifting that formed the margin of Baltica.

An alternate explanation for the observed velocity structure is that it is the result of a more typical rifted continental-margin geometry (e.g., Holbrook et al., 1994) in which the extended and modified crust of Baltica does not extend a significant distance beyond the limits of the Polish Basin. Rifting can produce a variety of continental-margin geometries based on the amount of transtension present, and end members can be highly divergent (gradual thinning) or highly transtensional (abrupt) margins. These models then imply that the Paleozoic platform is solely underlain by the lithosphere of Avalonia-Variscan terranes, and they offer a simple explanation for the large difference in crustal structure across the Polish Basin.

A complex mantle structure was derived by forward ray-tracing modeling, and it is included in Figures 5B and 8A. This geometry is suggestive of a stage of northeast-dipping subduction and collision that would probably be related to the Variscan orogeny. The lack of Paleozoic magmatism along margin of Baltica does indicate that any period of subduction with this polarity would have been short-lived. Many analogies can be drawn between these models and tectonic models for the Ouachita orogenic belt in the south-central United States (Keller et al., 1989; Mickus and Keller, 1992; Keller and Hatcher, 1999). This orogenic belt formed along the highly irregular rifted margin of Laurentia that has been interpreted as being the result of transform faulting (Thomas, 1991). The overall geometry of the Ouachita margin is very similar to what is seen in Figure 8A, where the Sabine to the south would be analogous to the Bohemian massif with a deep basin lying between it and Laurentia.

The area in the vicinity of profile CEL05 has experienced a very different tectonic evolution that leads to another set of interpretations of the velocity model for this profile. Prior to the Alpine orogeny, the tectonic history of the region was probably very similar to that of the profile P4 area to the north and was dominated by the rifting of Baltica and Paleozoic sedimentation and deformation along its margin due to the accretion of Gondwanan terranes. However, the Alpine orogeny played a major role in the evolution of this region as manifested by the Carpathian Mountains. Finally, the Pannonian Basin is the result of complex plate interactions that are still active. The interpreted velocity model is shown in Figure 8B, and several elements of possible interpretations are not mutually exclusive (Grad et al., 2006). The northeast dip of the Moho and lower lithospheric reflectors are shown as being due to northward subduction under the East European craton. The geologic evidence for southward subduction is very strong, but an early (Jurassic–Early Cretaceous) phase of northward subduction has been suggested by several authors (Książkiewicz, 1977; Golonka et al. 2003a). Golonka et al. (2003a) produced a paleogeographic analysis of the Carpathian region in which northwest-northward subduction of the Meliata-Halstatt Ocean crust was completed by the end of the Jurassic, ca 140 Ma. However, strike-slip movements have played a major role in the Cenozoic evolution of the region (e.g., Csontos and Vörös, 2004), and they add an element of uncertainty to interpretations of older events. A model in which the Mesozoic-Cenozoic evolution is due to a collision similar to the Paleozoic collisions to the north is also possible. In that case, the lithosphere of Baltica acted as an indenter and caused northward obduction of the Carpathian-Pannonian crust, while the Pannonian mantle lithosphere underthrust Baltica, producing a crocodile structure. This model may well have occurred but cannot explain the entire history of the region. Finally, the apparent northward dip in the structure could be the result of thinning of the Pannonian lithosphere due to extension that has erased the effects of Miocene southward subduction of the rifted margin of Baltica. This model does not preclude the effects of Paleozoic and Mesozoic events producing some of the structure seen along the margin of Baltica.

CONCLUSIONS

The long profiles that were recorded during the POLONAISE'97 and CELEBRATION 2000 seismic experiments provide significant new data about the structure and tectonic evolution of Central Europe. As is always the case, it takes an integrated geological and geophysical analysis to interpret the velocity models that result from such experiments. In the case of profile P4 region, we cannot claim to resolve the debate about how far the thinned crust of the rifted margin of Baltica extends under Germany. However, profile P4 reaches the northwest margin of the Bohemian massif (Fig. 2), and the velocity models do not require that the remnants of Baltica extend far beyond the Polish Basin (Figs. 5 and 8A). In the case of profile CEL05, the north-dipping elements of the model in the vicinity of the margin of Baltica (Fig. 8B, 400–650 km) indicate a rather sharp termination of the lithosphere of Baltica. We suggest that this geometry is due mostly to Variscan and/or Mesozoic collision along an abrupt margin that formed during the breakup of Rodinia. To the south, it is hard to argue against Moho geometry being primarily due to late Cenozoic extension in the Pannonian Basin region. Reconciliation of the geologic and geophysical evidence for southward subduction during the Miocene with the deep portions of the seismic model is difficult unless one accepts that the younger extension in the Pannonian Basin area has erased the evidence or is even due in part to some delamination process.

REFERENCES CITED

Aichroth, B., Prodehl, C., and Thybo, H., 1992, Crustal structure along the central segment of the EGT from seismic refraction studies: Tectonophysics, v. 207, p. 43–64, doi: 10.1016/0040-1951(92)90471-H.

BABEL Working Group, 1993, Deep seismic reflection/refraction interpretation of crustal structure along BABEL profiles A and B in the southern Baltic Sea: Geophysical Journal International, v. 112, p. 325–343.

Beránek, B., and Dudek, A., 1972, The results of deep seismic sounding in Czechoslovakia: Zeitschrift für Geophysik, v. 38, p. 415–427.

Berthelsen, A., 1992a, Mobile Europe, *in* Blundell, D.J., Freeman, R., and Mueller, St., eds., A Continent Revealed—The European Geotraverse: Cambridge, Cambridge University Press, p. 11–32.

Berthelsen, A., 1992b, From Precambrian to Variscan Europe, *in* Blundell, D.J., Freeman, R., and Mueller, St., eds., A Continent Revealed—The European Geotraverse: Cambridge, Cambridge University Press, p. 153–164.

Berthelsen, A., 1998, The Tornquist zone northwest of the Carpathians: An intraplate-pseudosuture: Geologiska Föreningens i Stockholm Förhandlingar, v. 120, p. 223–230.

Bielik, M., Šefara, J., Sotak, J., Kováč, M., Bezák, V., and Plašienka, D., 2004, The Western Carpathians—Interaction of Hercynian and Alpine processes: Tectonophysics, v. 393, p. 63–86, doi: 10.1016/j.tecto.2004.07.044.

Birkenmajer, K., 1986, Stages of structural evolution of the Pieniny Klippen belt, Carpathians: Studia Geologica Polonica, v. 88, p. 7–32.

Bogdanova, S., 1996, Palaeoproterozoic accretion of Sarmatia and Fennoscandia, *in* Gee, D.G., and Zeyen, H.J., eds., EUROPROBE 1996—Lithosphere Dynamics: Origin and Evolution of Continents: Uppsala, The EUROPROBE Secretariat, Uppsala University, p. 81–89.

Braile, L.W., Hinze, W.J., von Frese, R.B.R., and Keller, G.R., 1989, Seismic properties of the crust and uppermost mantle of the conterminous United States and adjacent Canada, *in* Pakiser, L.C., and Mooney, W.D., eds., Geophysical Framework of the Continental United States: Geological Society of America Memoir 172, p. 655–680.

Brückl, E., Bodoky, T., Hegedűs, E., Hrubcová, P., Gosar, A., Grad, M., Guterch, A., Hajnal, Z., Keller, G.R., Špičák, A., Sumanovač, F., Thybo, H., Weber, F., and ALP 2002 Working Group, 2003, ALP 2002 seismic experiment: Studia Geophysica et Geodaetica, v. 47, p. 671–679, doi: 10.1023/A:1024780022139.

Červený, V., and Pšenčík, I., 1983, SEIS83—Numerical modeling of seismic wave fields in 2-D laterally varying layered structure by the ray method, *in* Engdahl, E.R., ed., Documentation of Earthquake Algorithm: Boulder, World Data Center A for Solid Earth Geophysics Report SE-35, p. 36–40.

Csontos, L., and Vörös, A., 2004, Mesozoic plate tectonic reconstruction of the Carpathian region: Palaeogeography, Palaeoclimatology, Palaeoecology, v. 210, p. 1–56, doi: 10.1016/j.palaeo.2004.02.033.

Czuba, W., Grad, M., Luosto, U., Motuza, G., Nasedkin, V., and POLONAISE P5 Working Group, 2002, Upper crustal seismic structure of the Mazury complex and Mazowsze massif within the East European craton in NE Poland: Tectonophysics, v. 360, p. 115–128, doi: 10.1016/S0040-1951(02)00352-9.

Dadlez, R., Grad, M., and Guterch, A., 2005, Crustal structure below the Polish Basin: Is it composed of proximal terranes derived from Baltica?: Tectonophysics, v. 411, p. 111–128, doi: 10.1016/j.tecto.2005.09.004.

EUROBRIDGE Seismic Working Group, 1999, Seismic velocity structure across the Fennoscandia-Sarmatia suture of the East European craton beneath the EUROBRIDGE profile through Lithuania and Belarus: Tectonophysics, v. 314, p. 193–217, doi: 10.1016/S0040-1951(99)00244-9.

EUROBRIDGE'95 Seismic Working Group, 2001, EUROBRIDGE'95: Deep seismic profiling within East European craton: Tectonophysics, v. 339, p. 153–175.

Franke, D., 1990, Der präpermische Untergrund der Mitteleuropäischen Senke Fakten und Hypothesen: Niedersächsische Akademie für Geowissenschaften Veröffentlichung, v. 4, p. 19–75.

Golonka, J., Krobicki, M., Oszczypko, N., Ślączka, A., and Słomka, T., 2003a, Geodynamic evolution and palaeogeography of the Polish Carpathians and adjacent areas during Neo-Cimmerian and preceding events (latest Triassic–earliest Cretaceous), *in* McCann, T., and Saintot, A., eds., Tracing Tectonic Deformation Using the Sedimentary Record: Geological Society [London] Special Publication 208, p. 138–158.

Golonka, J., Ślączka, A., and Picha, F., 2003b, Geodynamic evolution of the orogen: The west Carpathians and Ouachitas case study: Annales Societatis Geologorum Poloniae, v. 75, p. 145–167.

Grad, M., and Luosto, U., 1987, Seismic models of the crust of the Baltic Shield along the SVEKA profile in Finland: Annales Geophysicae, v. 5B, p. 639–650.

Grad, M., Janik, T., Yliniemi, J., Guterch, A., Luosto, U., Tiira, T., Komminaho, K., Środa, P., Höing, K., Makris, J., and Lund, C.-E., 1999, Crustal structure of the Mid-Polish trough beneath the Teisseyre-Tornquist zone seismic profile: Tectonophysics, v. 314, p. 145–160, doi: 10.1016/S0040-1951(99)00241-3.

Grad, M., Keller, G.R., Thybo, H., Guterch, A., and POLONAISE Working Group, 2002a, Lower lithospheric structure beneath the Trans-European suture zone from POLONAISE'97 seismic profiles: Tectonophysics, v. 360, p. 153–168, doi: 10.1016/S0040-1951(02)00350-5.

Grad, M., Guterch, A., and Mazur, S., 2002b, Seismic refraction evidence for crustal structure in the central part of the Trans-European suture zone in Poland, *in* Winchester, J.A., Pharaoh, T.C., and Verniers, J., eds., Palaeozoic Amalgamation of Central Europe: Geological Society [London] Special Publication 201, p. 295–309.

Grad, M., Jensen, S.L., Keller, G.R., Guterch, A., Thybo, H., Janik, T., Tiira, T., Yliniemi, J., Luosto, U., Motuza, G., Nasedkin, V., Czuba, W., Gaczyński, E., Środa, P., Miller, K.C., Wilde-Piórko, M., Komminaho, K., Jacyna, J., and Korabliova, L., 2003a, Crustal structure of the Trans-European suture zone region along POLONAISE'97 seismic profile P4: Journal of Geophysical Research, v. 108, no. B11, p. 2541, doi: 10.1029/2003JB002426.

Grad, M., Špičák, A., Keller, G.R., Guterch, A., Brož, M., Hegedűs, E., and SUDETES Working Group, 2003b, SUDETES 2003 seismic experiment: Studia Geophysica et Geodaetica, v. 47, p. 681–689, doi: 10.1023/A:1024732206210.

Grad, M., Guterch, A., Keller, G.R., Janik, T., Hegedűs, E., Vozár, J., Ślączka, A., Tiira, T., and Yliniemi, J., 2006, Lithospheric structure beneath trans-Carpathian transect from Precambrian platform to Pannonian Basin: CELEBRATION 2000 seismic profile CEL05: Journal of Geophysical Research, v. 111, p. B03301, doi: 10.1029/2005JB003647.

Guggisberg, B., Kaminski, W., and Prodehl, C., 1991, Crustal structure of the Fennoscandian Shield: A traveltime interpretation of the long-range FENNOLORA seismic refraction profile: Tectonophysics, v. 195, p. 105–137, doi: 10.1016/0040-1951(91)90208-A.

Guterch, A., Grad, M., Materzok, R., Perchuć, E., Janik, T., Gaczyński, E., Doan, T.T., Białek, T., Gadomski, D., Młynarski, S., and Toporkiewicz, S., 1992, Laminated structure of the lower crust in the fore-Sudetic region in Poland, derived from seismic data: Physics of the Earth and Planetary Interiors, v. 69, p. 217–223, doi: 10.1016/0031-9201(92)90138-L.

Guterch, A., Grad, M., Thybo, H., Keller, G.R., and Miller, K., 1998, Seismic experiment spreads across Poland: Eos (Transactions, American Geophysical Union), v. 79, no. 26, p. 302, 305.

Guterch, A., Grad, M., Thybo, H., Keller, G.R., and POLONAISE Working Group, 1999, POLONAISE'97—International seismic experiment between Precambrian and Variscan Europe in Poland: Tectonophysics, v. 314, p. 101–121, doi: 10.1016/S0040-1951(99)00239-5.

Guterch, A., Grad, M., Keller, G.R., and CELEBRATION 2000 Organizing Committee, 2001, Seismologists celebrate the new millennium with an experiment in Central Europe: Eos (Transactions, American Geophysical Union), v. 82, no. 45, p. 529, 534–535, doi: 10.1029/01EO00313.

Guterch, A., Grad, M., Špičák, A., Brückl, E., Hegedűs, E., Keller, G.R., Thybo, H., and CELEBRATION 2000, ALP 2002, SUDETES 2003 Working Groups, 2003a, An overview of recent seismic refraction experiments in Central Europe: Studia Geophysica et Geodaetica, v. 47, p. 651–657, doi: 10.1023/A:1024775921231.

Guterch, A., Grad, M., Keller, G.R., Posgay, K., Vozár, J., Špičák, A., Brückl, E., Hajnal, Z., Thybo, H., Selvi, O., and CELEBRATION 2000 Experiment Team, 2003b, CELEBRATION 2000 seismic experiment: Studia Geophysica et Geodaetica, v. 47, p. 659–669, doi: 10.1023/A:1024728005301.

Holbrook, W.S., Reiter, E.C., Purdy, G.M., Sawyer, D., Stoffa, P.L., Austin, J.A., Jr., Oh, J., and Makris, J., 1994, Deep structure of the U.S. Atlantic continental margin, offshore South Carolina, from coincident ocean bottom and multichannel seismic data: Journal of Geophysical Research, v. 99, no. B5, p. 9155–9178, doi: 10.1029/93JB01821.

Hrubcová, P., Środa, P., Špičák, A., Guterch, A., Grad, M., Keller, G.R., Brueckl, E., and Thybo, H., 2005, Crustal and uppermost mantle structure of the Bohemian massif based on CELEBRATION 2000 data: Journal of Geophysical Research, v. 110, p. B11305, doi: 10.1029/2004JB003080.

Janik, T., Yliniemi, J., Grad, M., Thybo, H., Tiira, T., and POLONAISE P2 Working Group, 2002, Crustal structure across the TESZ along POLONAISE'97 seismic profile P2 in NW Poland: Tectonophysics, v. 360, p. 129–152, doi: 10.1016/S0040-1951(02)00353-0.

Janik, T., Grad, M., Guterch, A., Dadlez, R., Yliniemi, J., Tiira, T., Gaczyński, E., Keller, G.R., and CELEBRATION 2000 Working Group, 2005, Lithospheric structure of the Trans-European suture zone along the TTZ-CEL03 seismic transect (from NW to SE Poland): Tectonophysics, v. 411, p. 129–156, doi: 10.1016/j.tecto.2005.09.005.

Jensen, S.L., Janik, T., Thybo, H., and POLONAISE Working Group, 1999, Seismic structure of the Palaeozoic platform along POLONAISE'97 profile P1 in southwestern Poland: Tectonophysics, v. 314, p. 123–143, doi: 10.1016/S0040-1951(99)00240-1.

Keller, G.R., and Hatcher, R.D., Jr., 1999, Some comparisons of the structure and evolution of the southern Appalachian-Ouachita orogen and portions of the Trans-European suture zone region: Tectonophysics, v. 314, p. 43–68, doi: 10.1016/S0040-1951(99)00236-X.

Keller, G.R., and Stephenson, R.A., 2007, The Southern Oklahoma and Dniepr-Donets aulacogens: A comparative analysis, *in* Hatcher, R.D., Jr., Carlson, M.P., McBride, J.H., and Martínez Catalán, J.R., eds., 4-D Framework of Continental Crust: Geological Society of America Memoir 200, doi: 10.1130/2007.1200(08).

Keller, G.R., Braile, L.W., McMechan, G.A., Thomas, W.A., Harder, S.H., Chang, W., and Jardine, W.G., 1989, The Paleozoic continent-ocean transition in the Ouachita Mountains imaged from PASSCAL wide-angle seismic reflection-refraction data: Geology, v. 17, p. 119–122, doi: 10.1130/0091-7613(1989)017<0119:PCOTIT>2.3.CO;2.

Kováč, M., Nagymarosy, A., Soták, J., and Šutovská, K., 1993, Late Tertiary paleogeographic evolution of the western Carpathians: Tectonophysics, v. 226, p. 401–415, doi: 10.1016/0040-1951(93)90129-8.

Kovačs, S., Haas, J., Csaszar, G., Szederkenyi, T., Buda, G., and Nagymaros, A., 2000, Tectonostratigraphic terranes in the pre-Neogene basement of the Hungarian part of the Pannonian area: Acta Geologica Hungarica, v. 43, p. 225–328.

Książkiewicz, M., 1977, Hypothesis of plate tectonics and the origin of the Carpathians: Annales de la Société Géologique de Pologne, v. 47, p. 329–353.

MacNiocaill, C., Smethurst, M.A., and Ryan, P.D., 1998, Oroclinal bending in the Caledonides of western Ireland: A mid-Paleozoic feature controlled by a pre-existing structural grain: Tectonophysics, v. 299, p. 31–47, doi: 10.1016/S0040-1951(98)00197-8.

Majdański, M., Grad, M., Guterch, A., and SUDETES 2003 Working Group, 2006, 2-D seismic tomographic and ray tracing modelling of the crustal structure across the Sudetes Mountains basing on SUDETES 2003 experiment data: Tectonophysics, v. 413, no. 3–4, p. 249–269, doi: 10.1016/j.tecto.2005.10.042.

Malinowski, M., Żelaźniewicz, A., Grad, M., Guterch, A., and Janik, T., 2005, Seismic and geological structure of the crust in the transition from Baltica to Palaeozoic Europe in SE Poland—CELEBRATION 2000 experiment, profile CEL02: Tectonophysics, v. 401, p. 55–77, doi: 10.1016/j.tecto.2005.03.011.

Meissner, R., 1989, Rupture, creep, lamellae and crocodiles: Happenings in the continental crust: Terra Nova, v. 1, p. 17–28.

Mickus, K.L., and Keller, G.R., 1992, Lithospheric structure of the south-central United States: Geology, v. 20, p. 335–338, doi: 10.1130/0091-7613(1992)020<0335:LSOTSC>2.3.CO;2.

Neunhöfer, H., Marillier, F., and Panza, G.F., 1981, Crust and upper mantle structure in the Bohemian massif from the dispersion of Rayleigh waves: Gerlands Beiträge Geophysik, v. 90, p. 514–520.

Novotný, O., Grad, M., Lund, C.-E., and Urban, L., 1997, Verification of the lithospheric structure along profile Uppsala–Prague using surface wave dispersion: Studia Geophysica et Geodaetica, v. 41, p. 15–28, doi: 10.1023/A:1023332522678.

Pharaoh, T.C., 1999, Palaeozoic terranes and their lithospheric boundaries within the Trans-European suture zone (TESZ): A review: Tectonophysics, v. 314, p. 17–41, doi: 10.1016/S0040-1951(99)00235-8.

Pożaryski, W., 1990, The Middle European Caledonides—Wrenching orogen composed of terranes: Przegląd Geologiczny, v. 38, p. 1–9.

Royden, L.H., and Dövényi, P.E., 1988, Variations in extensional styles at depth across the Pannonian Basin, *in* Royden, L.H., and Horváth, F., eds., The Pannonian Basin: A Study in Basin Evolution: American Association Petroleum Geologists Memoir 45, p. 235–255.

Růžek, B., Vavryčuk, V., Hrubcová, P., Zedník, J., and the CELEBRATION Working Group, 2003, Crustal anisotropy in the Bohemian massif, Czech Republic: Observations based on Central European lithospheric experiment based on refraction (CELEBRATION) 2000: Journal of Geophysical Research, v. 108, no. B8, 2392, doi: 10.1029./2002JB002242.

Środa, P., and POLONAISE P3 Working Group, 1999, P- and S-wave velocity model of the southwestern margin of the Precambrian East European craton; POLONAISE'97, profile P3: Tectonophysics, v. 314, p. 175–192, doi: 10.1016/S0040-1951(99)00243-7.

Środa, P., Czuba, W., Grad, M., Gaczyński, E., Guterch, A., and POLONAISE Working Group, 2002, Three-dimensional seismic modelling of crustal structure in the TESZ region based on POLONAISE'97 data: Tectonophysics, v. 360, p. 169–185, doi: 10.1016/S0040-1951(02)00351-7.

Środa, P., Czuba, W., Grad, M., Guterch, A., Tokarski, A.K., Janik, T., Rauch, M., Keller, G.R., Hegedűs, E., Vozár, J., and CELEBRATION 2000 Working Group, 2006, Crustal and upper mantle structure of the Western Carpathians from CELEBRATION 2000 profiles CEL01 and CEL04: Seismic models and geological implications: Geophysical Journal International, v. 167, p. 737–760; doi: 10.1111/j.1365-246X.2006.03104.x.

Szabó, C., Falus, Gy., Zajacz, Z., Kovács, I., and Bali, E., 2004, Composition and evolution of lithosphere beneath the Carpathian-Pannonian region: A review: Tectonophysics, v. 393, p. 119–137, doi: 10.1016/j.tecto.2004.07.031.

Thomas, W.A., 1991, The Appalachian-Ouachita rifted margin of southeastern North America: Geological Society of America Bulletin, v. 103, p. 415–431, doi: 10.1130/0016-7606(1991)103<0415:TAORMO>2.3.CO;2.

Thybo, H., 2001, Crustal structure along the EGT profile across the Tornquist Fan interpreted from seismic, gravity and magnetic data: Tectonophysics, v. 334, no. 3–4, p. 155–190, doi: 10.1016/S0040-1951(01)00055-5.

Vavryčuk, V., Hrubcová, P., Brož, M., Málek, J., and the A.L.P. 2002 Working Group, 2004, Azimuthal variation of Pg velocity in the Moldanubian, Czech Republic: Observations based on a multi-azimuthal common-shot experiment: Tectonophysics, v. 387, p. 189–203, doi: 10.1016/j.tecto.2004.06.015.

Wever, Th., Meissner, R., Sadowiak, P., and the DEKORP Group, 1990, Deep reflection seismic data along the central part of the European Geotraverse in Germany: A review: Tectonophysics, v. 176, p. 87–101, doi: 10.1016/0040-1951(90)90260-F.

Wilde-Piórko, M., Saul, J., and Grad, M., 2005, Differences in the crustal and uppermost mantle structure of the Bohemian massif from teleseismic receiver functions: Studia Geophysica et Geodaetica, v. 49, p. 85–107, doi: 10.1007/s11200-005-1627-3.

Ziegler, P.A., 1990, Geological Atlas of Western and Central Europe (2nd edition): Shell International Petroleum, Maatschappij B.V.: Bath, Geological Society Publishing House, 239 p., 56 enclosures.

Manuscript Accepted by the Society 1 February 2007

Printed in the USA

The Geological Society of America
Memoir 200
2007

The Uppermost Allochthon in the Scandinavian Caledonides: From a Laurentian ancestry through Taconian orogeny to Scandian crustal growth on Baltica

David Roberts*
Øystein Nordgulen
Victor Melezhik
Geological Survey of Norway, NO-7491 Trondheim, Norway

ABSTRACT

Thrust sheets of the Uppermost Allochthon in the Caledonides of Scandinavia are distinguished by lithological assemblages and magmatic units that are, in many ways, quite different from those in subjacent nappe complexes. Supracrustal successions are derived mainly from platformal, shelf-edge and basinal-slope environments and are characterized in particular by extensive developments of carbonate rock units that range in age from Late Riphean to Early Silurian. Metasedimentary iron ore formations are also present. Another prominent feature is the Ordovician, arc-type, granitoid plutons and batholiths that dominate the geology in certain parts of the allochthon. In addition to these lithological elements, the Uppermost Allochthon carries an Ordovician tectonothermal record and early Caledonian, NW-vergent thrust polarity that is unique in Norway. Taken together, these features are indicative of a history of development and crustal growth along the eastern margin of Laurentia, involving an outboard magmatic arc, or arcs, and Taconian accretionary orogenesis. This was followed by recycling and deposition in Late Ordovician to Early Silurian successor basins prior to Laurentia-Baltica collision and the onset of the Scandian orogeny. The Taconian thrust sheets were then detached from their Laurentian roots and incorporated into the Siluro-Devonian, Scandian orogenic wedge on the Baltoscandian margin of Baltica. Taking into account the widely reported sinistral megashear arising from the Scandian, oblique collision and plate rotation, the rock units that constitute the Uppermost Allochthon are likely to have originally been located closer to the northern Appalachian segment of the margin of Laurentia, in view of the strikingly similar lithostratigraphic, magmatic, and tectonothermal histories of these two, now widely separated terranes.

Keywords: Caledonides, Norway, Scandian, Taconian, Laurentia.

*David.Roberts@NGU.NO

Roberts, D., Nordgulen, Ø., and Melezhik, V., 2007, The Uppermost Allochthon in the Scandinavian Caledonides: From a Laurentian ancestry through Taconian orogeny to Scandian crustal growth on Baltica, *in* Hatcher, R.D., Jr., Carlson, M.P., McBride, J.H., and Martínez Catalán, J.R., eds., 4-D Framework of Continental Crust: Geological Society of America Memoir 200, p. 357–377, doi: 10.1130/2007.1200(18). For permission to copy, contact editing@geosociety.org.

INTRODUCTION

The Scandinavian Caledonides form part of a once extensive, but now fragmented, early Paleozoic orogenic belt that fringes the North Atlantic Ocean from the southeastern United States and northwest Africa via the British Isles to the Barents Sea. The Scandinavian segment, contiguous with East Greenland prior to the Cretaceous-Tertiary opening of the Atlantic, is characterized by its innumerable thrust sheets that carry stratigraphic and magmatic evidence of diverse and disparate origins. These thrust sheets, or nappes, were transported roughly from west to east onto the Baltoscandian margin of the paleocontinent Baltica in Siluro-Devonian time, during the Scandian orogeny (Gee, 1975).

It had long been known that in the foreland, along the Caledonian front in the east, thin sedimentary successions of Vendian-Cambrian age lie unconformably upon older Precambrian crystalline basement (Petersen, 1878), and these are succeeded, above a prominent sole thrust, by folded and imbricated, low-grade sedimentary rocks of similar character to those in the autochthon. Farther west, rocks of higher metamorphic grade are involved, in places with slices of basement rocks, and these relationships eventually convinced most nineteenth-century geologists, led by Törnebohm (1888, 1896), that only thrusting over vast distances could account for the juxtaposition, above sharp mylonitic contacts, of markedly different rock complexes.

Close to a century later, following extensive, detailed mapping and structural studies, geologists realized that many of these thrust sheets and nappes, in some cases already grouped together in "nappe complexes," could be combined into higher-rank "allochthons" based on a variety of lithological, magmatic, and geochemical criteria. Recognition of shelf and miogeoclinal successions, and more outboard oceanic- and arc-type terranes succeeded by even higher, exotic, thrust rock complexes, led to the establishment of a common tectonostratigraphy composed of Lower, Middle, Upper, and Uppermost Allochthons (Fig. 1) (Gee and Sturt, 1985; Roberts and Gee, 1985). Successions in the Lower and Middle Allochthons are clearly indigenous to the Baltoscandian continental margin, whereas the Upper Allochthon contains elements of the continent-ocean transition zone (the Seve Nappes; Andréasson, 1994) and the overlying, exotic, Iapetus Ocean terranes (the Köli Nappes). Faunas of mainly Laurentian affinity in one of the highest Köli Nappes clearly point to a foreign provenance (Bergström, 1979; Bruton and Bockelie, 1979, 1980; Spjeldnæs, 1985). In the Uppermost Allochthon, rock assemblages from a variety of environments, including extensive carbonate formations and voluminous granitoid plutons that do not occur in subjacent thrust sheets, led workers in the 1980s to suggest possible derivation of these exotic assemblages from either Laurentia or unknown microcontinents (Stephens and Gee, 1985; Roberts et al., 1985). These early ideas were taken further and elaborated upon (Grenne et al., 1999), and a Laurentian ancestry has been substantiated by chemostratigraphic, structural, and isotopic dating work during the last few years (e.g., Melezhik et al., 2001a, 2001b, 2003; Roberts et al., 2001, 2002; Yoshinobu et al., 2002).

In this contribution, we present a short description of the constituent units of the Uppermost Allochthon, concentrating on its more exotic elements and noting features of its tectonothermal history that are different from those recorded in the Baltoscandian and Iapetan allochthons. This, in turn, leads into a brief discussion of the accretionary development and crustal growth processes involved in the Uppermost Allochthon in particular, and within the Scandinavian Caledonide orogen as a whole. In addition to the emphasis on crustal growth and orogenic accretion, a main aim of this contribution has been to present a modern review of this highest allochthon and exotic terrane in the Scandinavian Caledonides.

REGIONAL EXTENT OF THE UPPERMOST ALLOCHTHON

The Uppermost Allochthon is almost entirely confined to north-central Norway in the counties of Troms, Nordland, and part of Nord-Trøndelag (Fig. 1), covering six degrees of latitude over a strike length of more than 700 km. It extends into Sweden in only one small area, southeast of Mo i Rana. The more-or-less continuous exposure of the Uppermost Allochthon is broken in just one part of northern Nordland, where the "Tysfjord window" of crystalline basement rocks forms a bridge that links to the Precambrian rocks of the Lofoten archipelago and thus disrupts the continuity of the overlying thrust sheets.

Since the time of the initial map compilation (Gee et al., 1985), additional detailed mapping, structural investigations, and geochemical studies have led to minor modifications to the general geographic extent of the Uppermost Allochthon. Later field studies have given rise to suggestions that outliers or fault-bounded units of typical Uppermost Allochthon complexes may also be present farther to the southwest in the coastal areas west of Trondheim, notably on the islands of Hitra (Nordgulen et al., 2002; Tucker et al., 2004) and Smøla (Fig. 1), where Arenig-Llanvirn faunas of North American aspect have been recorded (Bruton and Bockelie, 1979). In northern Norway, in the Ofotfjorden-Lyngen district, parts of the successions in the Uppermost Allochthon were relegated to the Upper Allochthon by Anderson (1989), Steltenpohl et al. (1990), and Anderson et al. (1992), but more recent regional mapping and compilations (Zwaan et al., 1998; K.B. Zwaan, 2005, personal commun.) still favor the original assignments of Gee et al. (1985).

Even farther north, in the county of Finnmark, recent studies in the highest turbidite formation of the tectonostratigraphic succession on Sørøya (Fig. 1) suggest that this unit is more likely to be Laurentian rather than Baltican and should preferably be assigned either to an exotic, high-level Köli Nappe or to the Uppermost Allochthon (Kirkland et al., 2005; Slagstad et al., 2006). Recent data also suggest that the Magerøy Nappe on the island of Magerøya (just off Fig. 1, northeast of Sørøya) may have been derived from the Laurentian margin, and could also

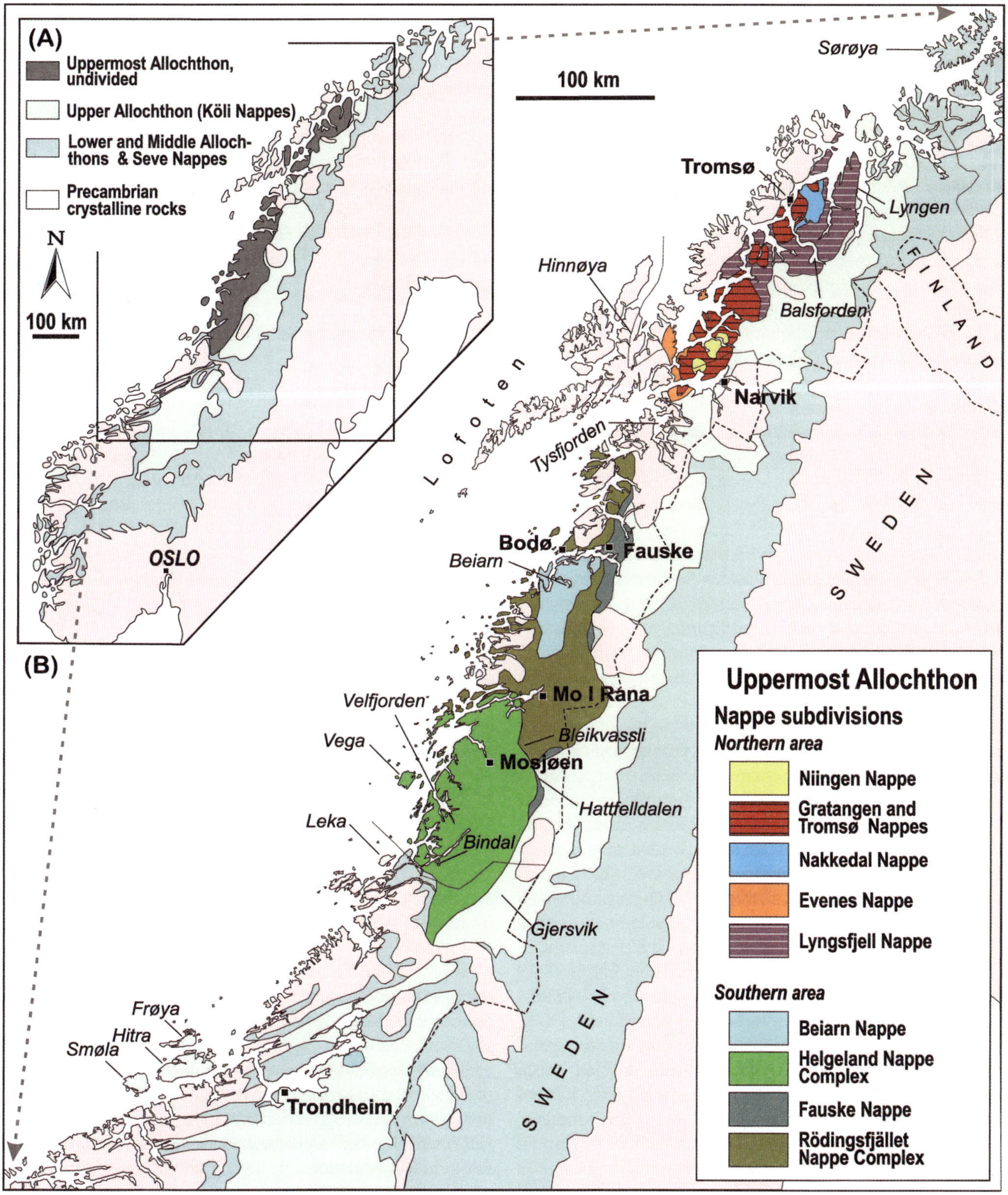

Figure 1. (A) Simplified map showing the principal nappes and nappe complexes that constitute the Uppermost Allochthon in the Scandinavian Caledonides. Names mentioned in the text are indicated. As described in the main text, parts of the "Smøla terrane" (Roberts, 1988), which extends northeast from the islands of Hitra and Smøla, may also belong to the Uppermost Allochthon. The same applies to the highest parts of the succession on Sørøya, in the far north (Kirkland et al., 2005; Slagstad et al., 2005). The inset map (B) shows the main allochthon subdivisions in the Caledonides of Norway and Sweden. The county of Troms extends approximately from Narvik-Hinnøya to ~30 km southwest of Sørøya; Nordland county extends from Narvik-Hinnøya to Bindal; and Nord-Trøndelag county extends from Bindal to ~30 km northeast of Trondheim.

possibly be regarded as an outlier of the Uppermost Allochthon (Kirkland et al., 2005; Corfu et al., 2006). Clearly, as new results and interpretations appear from different areas, further modifications to the allochthon hierarchy will be required.

TECTONOSTRATIGRAPHIC SUBDIVISIONS

Over the greater part of its exposure, south of the Tysfjord window, the Uppermost Allochthon is dominated by two tectonic mega-units, the Rödingsfjället Nappe Complex (Gavelin and Kulling, 1955) and the overlying Helgeland Nappe Complex (Ramberg, 1967) (Fig. 1). The basal contact of the allochthon, against the subjacent Köli Nappes of the Upper Allochthon, is characterized by a thick zone of mylonites and blastomylonites with some phyllonites, commonly imbricated into slices and lenses. Over large parts of its outcrop, the thrust-related mylonites are overprinted by late Scandian, ductile extensional structures (Osmundsen et al., 2003).

Each of these two nappe complexes is essentially a major, composite unit composed of several thrust sheets, which, in turn, are composed of diverse supracrustal rocks, basement gneisses, and a variety of intrusions. In the Helgeland Nappe Complex, no specific geographical names have been introduced for these separate thrust sheets, but their distinctive histories have been recognized in certain areas by the adoption of informal terms such as Lower, Middle, and Upper nappes (Thorsnes and Løseth, 1991; Yoshinobu et al., 2002).

One thrust sheet that was long thought to be a correlative of parts of the Helgeland Nappe Complex, but is geographically separated from it, north of Mo i Rana, is the Beiarn Nappe (Rutland and Nicholson, 1965) (Fig. 1). Later field work and regional map compilations have favored the view that the Beiarn Nappe is best considered to be a higher thrust unit within the Rödingsfjället Nappe Complex (Gustavson and Gjelle, 1991; Gustavson, 1996). Internal structural features of the Beiarn Nappe, however, are more akin to those in the Helgeland Nappe Complex (Stephens et al., 1985a) than the Rödingsfjället, such that the question of correlation is still open to discussion. A similar problem of thrust sheet affiliation, and one that highlights the rather artificial nature of allochthon subdivision, concerns the Fauske Nappe (Nicholson, 1974). On some map compilations, this has been interpreted as a Köli unit (Upper Allochthon) (e.g., Gjelle, 1988), but later work has placed the Fauske Nappe firmly at the base of the Rödingsfjället Nappe Complex (Gustavson, 1996; Roberts et al., 2002) (Fig. 1), in line with its original tectonostratigraphic position (Gee et al., 1985).

In the region north of the Tysfjord window, a variety of named and unnamed thrust sheets have been assigned to the Uppermost Allochthon (Binns, 1978; Andresen et al., 1985; Barker, 1986; Steltenpohl, 1987; Andresen and Steltenpohl, 1994; Zwaan et al., 1998; Melezhik et al., 2003), and specific correlations have been suggested (Steltenpohl et al., 1990). Based on modern regional map compilations, the subdivisions of Zwaan et al. (1998) for the northernmost areas (Lyngsfjell, Nakkedal, and Tromsø Nappes), and Barker (1986) and Melezhik et al. (2003) for the Ofotfjorden region (Bjerkvik, Evenes, Gratangen, and Niingen Nappes) are currently considered to be the most appropriate for these separate districts (Fig. 1). It should be noted, however, that the more recent compilations show that the Tromsø and Gratangen Nappes would appear to be one and the same thrust sheet.

LITHOSTRATIGRAPHIC SUCCESSIONS

General Features

Although there are differences in the lithological character of the many thrust sheets that constitute the Uppermost Allochthon, notably in those that are juxtaposed, there are also features that can be followed intermittently or, in some cases, almost continuously over distances of 300 km or more. In general terms, a distinction can be made between supracrustal successions, which are widespread, gneissic "basement" slices of more sporadic occurrence, and plutonic bodies that are most extensively developed in the south, in the Helgeland Nappe Complex (Barnes et al., 1992; Nordgulen et al., 1993). Depositional ages of the metasedimentary rocks are believed to range from Riphean to Early Silurian (Stephens et al., 1985a; Binns and Matthews, 1981; Bjørlykke and Olaussen, 1981; Melezhik et al., 2002a, 2002b).

The ubiquitous supracrustals are mostly composed of diverse, medium- to high-grade, psammitic, pelitic, and calcareous schists, calcite and dolomite marbles, and amphibolites of inferred Late Riphean to Cambrian age, with subordinate quartzites, polymict and monomict conglomerates, local diamictites, and laterally persistent stratabound iron formations (Gustavson, 1966; Nicholson and Rutland, 1969; Stephens et al., 1985a; Steltenpohl et al., 1990) (Fig. 2). Interestingly, true tillites, i.e., rocks of undisputed glaciogenic origin, have not been recorded in the Uppermost Allochthon. An appreciation of the diversity and detail of these thick supracrustal successions can be gained by studying many of the map-sheet compilations from this extensive region (e.g., Gustavson, 1981, 1996; Gustavson and Gjelle, 1991; Zwaan et al., 1998). Eclogites, eclogitized gneisses, rare ultramafite lenses, and diverse migmatized rocks have been described from the Tromsø Nappe (Krogh et al., 1990; Selbekk et al., 2000; Corfu et al., 2003), and dismembered ophiolites occur in a few places, both in northern (Boyd, 1983; Barker, 1986; Andresen and Steltenpohl, 1994; Furnes and Pedersen, 1995) and in southern areas (Prestvik, 1980; Thorsnes and Løseth, 1991; Heldal, 2001). The thick and laterally extensive marble units, as well as the iron formations, are special features of the Uppermost Allochthon and are described separately in the following.

Rocks considered to represent a crystalline basement to many of these successions occur in a few thrust sheets, notably in the Helgeland and Rödingsfjället Nappe Complexes. These take the form of various orthogneisses and some paragneisses, locally migmatized. It has been suggested that these units represent an original "basement-cover" situation (Riis and Ramberg, 1979), although no definite traces of unconformable contacts have yet

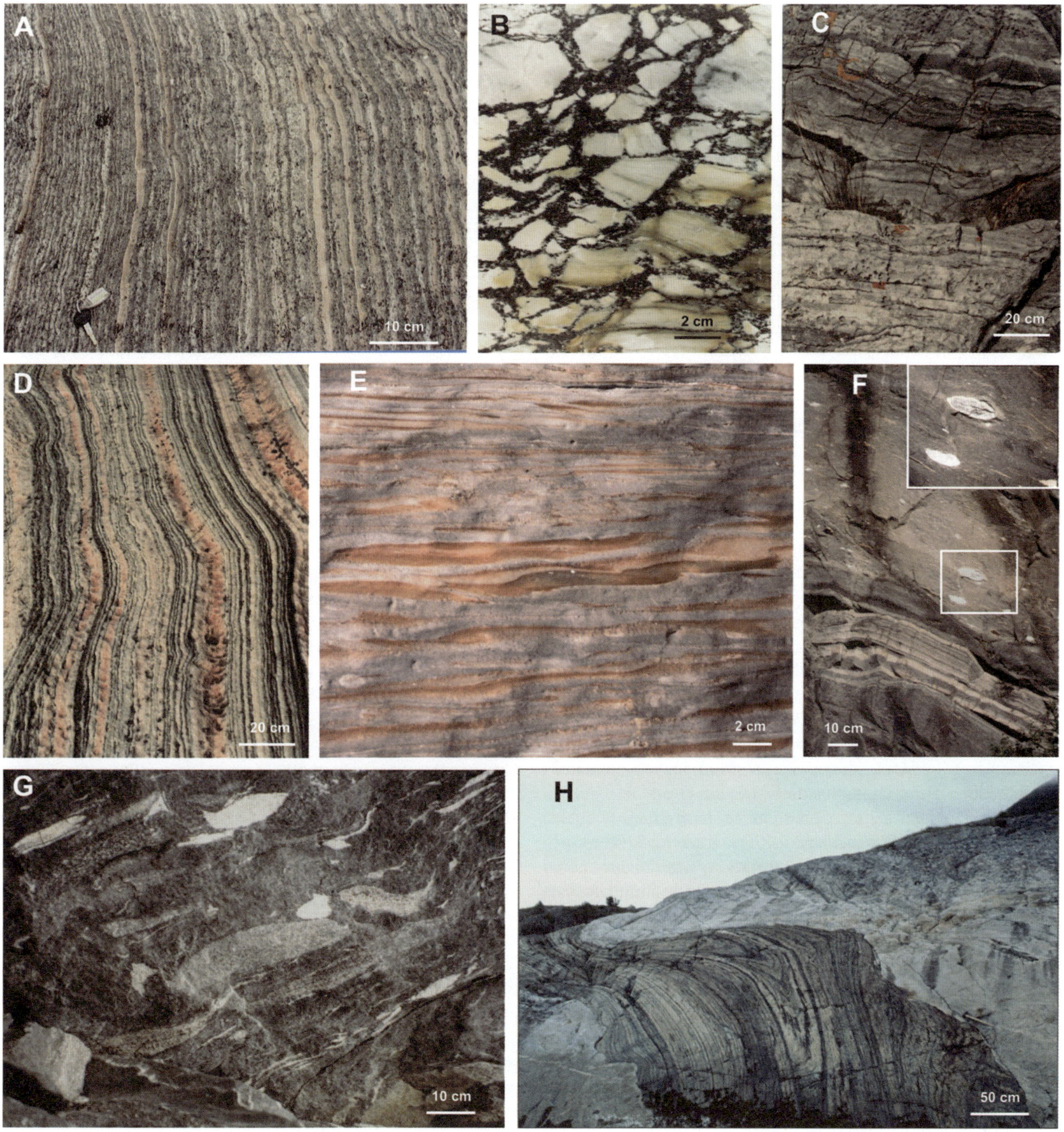

Figure 2. Representative lithologies from the Uppermost Allochthon. (A) Psammitic-pelitic couplets showing rhythmic bedding typical of a distal turbidite facies; from the Ofotfjorden area. (B) Well-preserved dolostone collapse breccia cemented by sphalerite (dark); from the Ofotfjorden area. (C) Dark-gray, laminated, calcite marble with $\delta^{13}C$ ranging between +4‰ and 6‰; basal part of the Late Ordovician–Early Silurian "tripartite marker unit," Ofotfjorden. (D) Variegated calcite marble with $\delta^{13}C$ ranging between −12‰ and −8‰; from the partly mylonitic, unfossiliferous, middle part of the "tripartite marker unit," Ofotfjorden. (E) Ripple cross-lamination typical of tidal-flat sediments, from the Ofotfjorden area. (F) Diamictite consisting of a dark, nonbedded, calcareous, graywacke matrix with scattered white dolostone fragments and tectonically flattened fragments of yellow limestones; enlarged area shows structural details; Evenes nappe, Ofotfjorden area. (G) Polymict, clast-supported conglomerate with clasts of gabbro and tonalite, and a "sandy" or "gravelly" matrix composed largely of comminuted gabbro; from the Ofotfjorden area. (H) Large metasedimentary xenolith within the western part of the Andalshatten pluton. An isoclinal F_2 fold (Nordgulen et al., 1993) is cut by the pluton.

been found (Heldal, 2001). Another form of "basement-cover" relationship is that of the generally low- to medium-grade, sedimentary successions that unconformably overlie deformed and eroded, fragmented ophiolites in the Lyngsfjell and Bjerkvik Nappes (Binns and Matthews, 1981; Boyd, 1983; Tull et al., 1985; Andresen and Steltenpohl, 1994; Zwaan et al., 1998), and the Helgeland Nappe Complex (Sturt et al., 1985; Thorsnes and Løseth, 1991; Heldal, 2001). Basal conglomerates to the successions are dominated by mafic clasts that are clearly derived from an ophiolite pseudostratigraphy (Fig. 2G). Conglomerates above this level are thoroughly polymict, including abundant clasts of psammites, schists, and some marbles from the older, Late Riphean to Cambrian cover sequences. Both the ophiolites and their sedimentary cover are discussed in a following section. A separate section is also devoted to the important plutonic component of the Uppermost Allochthon.

Carbonate Formations

Although variably metamorphosed carbonate formations occur within most of the major nappe complexes in the Scandinavian Caledonides, they are particularly abundant in the Uppermost Allochthon. In the Beiarn, Fauske, and Ofotfjorden areas (Fig. 1), the carbonate formations compose larger parts of the supracrustal complexes, whereas in other areas, they occur less commonly, either as laterally extensive, thick units or a series of lenses interbedded with various types of schist. Most of the carbonate formations are composed of calcite. Dolomite marbles are also present, although they are subordinate and occur as either thick coherent beds (i.e., the Fauske area) or thick lensoidal bodies embedded in calcite marble units (i.e., the Ofotfjorden area).

In general, the carbonate formations have been deformed and metamorphosed at medium to high grade and are mostly devoid of fossils. Therefore, until recently, very little was known about their true depositional ages. The only reported fossils, which enable us to date the sedimentation, are Llandovery-age corals, brachiopods, and gastropods (Binns and Matthews, 1981; Bjørlykke and Olaussen, 1981), which occur in lower-grade meta-limestone successions in the Balsfjorden area (Fig. 1). Recently, several specially targeted, isotope-chemostratigraphic studies have been carried out in an attempt to constrain the depositional ages of these carbonate formations. Within the limitations of isotope chemostratigraphy, particularly for Neo- and Mesoproterozoic time (Melezhik et al., 2001a), Late Ordovician–Early Silurian, Cambrian, and several different age groups of Neoproterozoic carbonate formations (Table 1) have been distinguished with a relatively high level of confidence. The isotope data also suggest that certain stratigraphic correlations and age assignments made in earlier literature are incorrect (see discussion in Melezhik et al., 2002a, 2002b), and also that several thick marble formations do not represent coherent stratigraphic units (Melezhik et al., 2002a). The isotopic data have revealed that in the Ofotfjorden area, the polymetamorphosed and polydeformed carbonate formations, of suggested Neoproterozoic, Cambrian, and Early Silurian age, were tectonically imbricated in nonstratigraphic order prior to final emplacement onto Baltica during the Scandian orogeny. This complex pattern of thrust-sheet stacking implies that an earlier, pre-Scandian, orogenic episode should be invoked to explain the tectonic juxtaposition of these assemblages, most likely the Middle to Late Ordovician Taconian event (Roberts et al., 2001, 2002; Melezhik et al., 2002a, 2002b).

For the most part, the carbonate formations lack primary depositional features, although in those areas where bedding and related sedimentary structures are preserved, various depositional settings have been reconstructed (Fig. 2E). In the Beiarn and Ofotfjorden area, Neoproterozoic, amphibolite-grade, calcite marbles represent one of the most exceptional examples of retention of original bedding and depositional chemistry (Melezhik et al., 2001b, 2005a). Trace-element changes in these marbles are less profound than those in Phanerozoic carbonates that have undergone only low-temperature diagenesis (e.g., Denison et al., 1994). The high Sr content and high preservational potential are indicative of an aragonitic lime mudstone precursor that could have easily escaped a diagenetic exchange. The Scandinavian Neoproterozoic marbles show close similarities in terms of their trace element and isotope geochemistry to the extensive Neoproterozoic shelf facies consisting of shales and massive limestones reported from East Greenland (Fairchild et al., 2000). Carbonate formations in both regions show evidence for primary aragonite mineralogy, high Sr contents, negligible Mn concentration, $^{87}Sr/^{86}Sr$ ratios around 0.70635, and similar carbon isotope values ranging between +4.3‰ and +7.5‰.

The dominant primary structure of the marbles of apparent Neoproterozoic age is rhythmic bedding, where the bedding cycles are of >10 cm thickness prior to tectonic thinning (Melezhik et al., 2001b), which is an important feature of nearly all pelagic carbonate sequences (Scholle et al., 1983). Numerous marble beds exhibit well-preserved graded bedding, and some can be clearly identified as turbidites, suggesting that the carbonate particles were transported by turbidity flows and were deposited within a pelagic system (Melezhik et al., 2001b). Thick units of siliciclastic rocks associated with the Neoproterozoic marbles may also show well-preserved, thin, rhythmic bedding characteristic of distal turbidites (Fig. 2A).

Marble formations of suggested Cambrian age provide evidence of diverse depositional settings. Those in the Ofotfjorden area do not exhibit any primary sedimentary features. However, their massive appearance and form of occurrence resemble buildups of reefal origin (Melezhik et al., 2003). In the Fauske area, a rather rare case of a clastic carbonate formation is locally developed and well exposed in a large quarry just 2 km north of the town. This is a 60-m-thick unit consisting of numerous 5-cm- to 3-m-thick beds, all of which show rapid lateral facies changes. A rapid vertical transition from basal carbonate debris (Fig. 3) through carbonate breccias to a calcarenite-graywacke turbidite lithofacies is accompanied by southeast-directed clast fining. Blocks in the basal facies are up to 6 × 4 m in size and represent fragments detached from the immediate substrate. Channelling

TABLE 1. APPARENT DEPOSITIONAL AGES OF MARBLE FORMATIONS FROM THE UPPERMOST ALLOCHTHON BASED ON Sr AND C ISOTOPE CHEMOSTRATIGRAPHY

Tectonostratigraphic affinity	$\delta^{13}C$ (‰)	$^{87}Sr/^{86}Sr$	Age (Ma)	Reference
Gratangen Nappe, Bogen Group				
Hekkelstrand Marble	+5	0.70615	710–700, 660	Melezhik et al. (2002a)
Fuglevann Marble	+6.5	0.70669	650–595	Melezhik et al. (2002a)
Liland Marble	+6.5	0.70655	650–595	Melezhik et al. (2003)
Evenes Nappe Complex				
Evenes Marble, Fm III	+5	0.70826	440–438	Melezhik et al. (2002b)
Evenes Marble, Fm II	+2.2	0.70878	550–500	Melezhik et al. (2002b)
Evenes Marble, Fm I	+5	0.70663	650–595	Melezhik et al. (2002b)
Evenestangen thrust sheet marble, dark marble	+6	0.70821	440–438	Melezhik et al. (2003)
Evenestangen thrust sheet marble, variegated marble	-8.3	0.70824	440–438	Melezhik et al. (2003)
Evenestangen thrust sheet marble, white marble	–5.9	0.70870	ca. 437	Melezhik et al. (2003)
Ramstad thrust sheet marble	+0.2	0.70920	515	Melezhik et al. (2003)
Steinsland thrust sheet marble	+6	0.70677	650–595	Melezhik et al. (2003)
Tangen sequence marble	+8	0.70720	620–610	Melezhik et al. (2002b)
Fauske Nappe, Rognan Group				
Fauske carbonate conglomerate	–0.2	0.70897	520	Melezhik et al. (2000a)
Helgeland Nappe Complex				
Hillstadfjellet carbonates	+1.5	0.70904	530–480	Sandøy (2003)
Saus carbonates	+6.0	0.70673	650–590	Sandøy (2003)
Beiarn Nappe Complex				
Sokumfjell Group marbles	+8.4	0.70642	635–600	Melezhik et al. (2001a)

and cross-bedding indicate southeast-directed paleocurrents. The depositional model involves a tectonically unstable, carbonate-shelf edge and southeast-directed transport of carbonate clasts, with finer carbonate debris accumulating on a basinal slope deepening to the southeast (Melezhik et al., 2000a) (Fig. 4), i.e., the opposite direction with respect to the Baltoscandian margin of Baltica, thus suggesting a foreign and inferred Laurentian paleogeographical ancestry (Roberts et al., 2001, 2002). Interestingly, the shelf-edge breccia facies (Figs. 3, 4A, and 4B) shows many similarities to less-deformed, Cambrian, bank-edge carbonate breccias in the Appalachians of the northeast United States (Rodgers, 1968) and Canada, e.g., the Cow Head breccia in Newfoundland (James and Stevens, 1986).

Carbonate formations of known or apparent Late Ordovician–Early Silurian age vary in lithofacies and metamorphic grade. In several places, these formations contain dolomite-hosted and dolomite collapse breccia–hosted, stratabound, ^{34}S-rich (+20‰ to +31‰), Zn (±Cu-Pb) occurrences (Fig. 2B) (Melezhik et al., 2000b). These have been interpreted as likely Mississippi Valley–type deposits (Bjørlykke and Olaussen, 1981; Grenne et al., 1999). Among diverse lithofacies, a "tripartite unit" has been described (Melezhik et al., 2002a, 2003). This is a succession composed of isotopically unusual (Melezhik et al., 2005b), dark gray (Fig. 2C), variegated (Fig. 2D), and white marbles that is developed discontinuously over a distance of 450 km from the Balsfjorden area (Bjørlykke and Olaussen, 1981) in the north via Ofotfjorden (Steltenpohl et al., 1990) to near Hattfjelldalen in the south (Fig. 1). Near Balsfjorden, the tripartite marble unit is a chemostratigraphic and lithostratigraphic equivalent of the lower-grade meta-limestone successions (Melezhik et al., 2002a). There, the dark gray and white carbonate rocks contain a varied Llandovery-age fauna assemblage (Binns and Matthews, 1981; Bjørlykke and Olaussen, 1981), whereas the strongly ^{13}C-depleted, thinly banded, and, in places, mylonitic, variegated unit is apparently nonfossiliferous. On a global basis, however, variegated carbonate rocks with comparable, extremely low $\delta^{13}C_{carb}$ values were deposited in the time interval 600–540 Ma (Melezhik et al., 2005b). Thus, although the precise depositional age of the banded, variegated marbles in Nordland and Troms is not known, a Vendian age is currently in favor. Whatever the case,

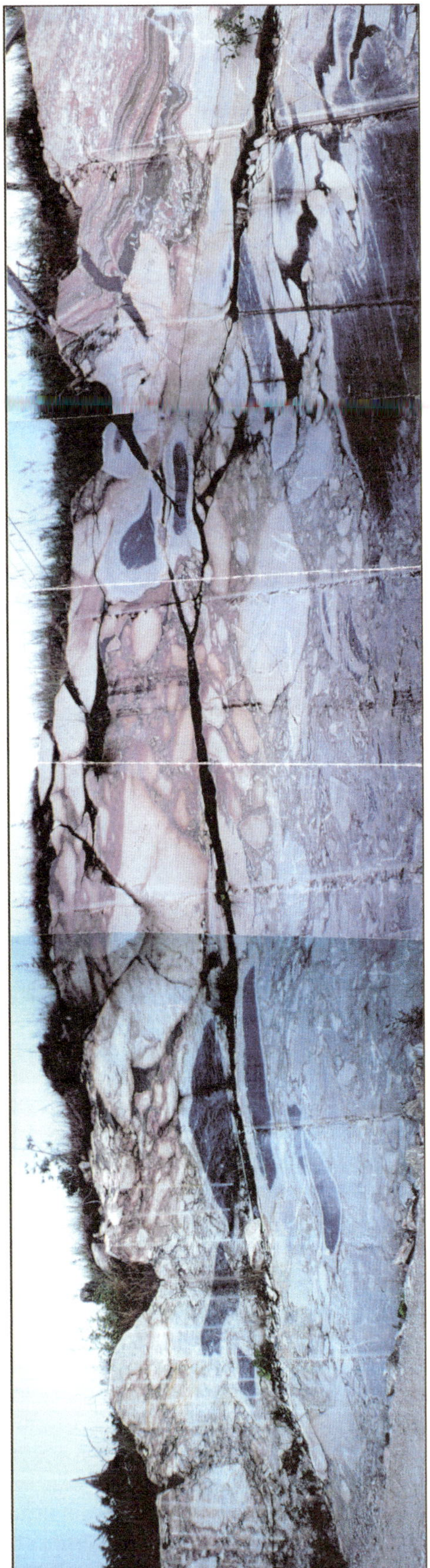

Figure 3. Part of the eastern wall of the NNW-SSE–trending "channel 2" in the Løvgavlen quarry near Fauske, in the Fauske Nappe. The greater part of the 4-m-high wall exposes a carbonate debris lithofacies with some blocks exceeding 2 m in size. Some of the darker, deformed blocks of "blue" calcite marble show prominent bleaching around their margins (see Melezhik et al., 2000a, for details). A disrupted mafic dike crosses the debris lithofacies from bottom left to top right. A carbonate conglomerate-breccia lithofacies overlies the blocky, debris material in the upper right of this panoramic photo. The apparent depositional age of this Fauske conglomerate formation is Middle Cambrian (Melezhik et al., 2000a).

the tripartite unit serves as a prominent chronostratigraphic marker horizon in the north-central Norwegian Caledonides.

In the far north in western Finnmark, a high-grade marble unit in a tectonostratigraphic succession on Sørøya has yielded isotope-chemostratigraphic data indicating a depositional age between 760 and 710 Ma (Slagstad et al., 2006). These carbonate rocks structurally underlie a turbidite formation that contains a tuffaceous psammite and two cross-cutting granite sheets that have yielded almost identical U-Pb zircon ages of ca. 438 Ma (Kirkland et al., 2005). Both the marble unit and the turbidite formation are now considered by Kirkland et al. to represent parts of either high-level Köli Nappes or the Uppermost Allochthon and to have been derived from the Laurentian margin.

Iron Formations

Metasedimentary oxide iron ores, in some places manganiferous, are numerous in the Uppermost Allochthon and are widespread over a distance of more than 500 km from the Mosjøen area of Nordland to the Tromsø district (Fig. 1). In the Mo i Rana area, where iron ores are most abundant, calcareous mica schists and marbles of the Dunderland Group are the principal host rocks (Bugge, 1948; Søvegjarto, 1977). Similarly, in other areas farther north, schists and marbles are hosts to the iron ores. Although the host-rock lithology is rather uniform, the ores vary in P and Mn content, which show stratigraphic and geographic controls. In the Mo i Rana ore district, the lower ore horizon is composed of magnetite, apatite, carbonate, and amphibole and contains 0.8–1.0 wt% P (Bugge, 1948; Søvegjarto, 1977), whereas the upper magnetite-hematite-quartz horizon has only 0.15–0.30 wt% P. Both ore types have low Mn contents, which contrast with Fe-spessartine ores from the Ofotfjorden area that contain up to 15 wt% Mn (Foslie, 1949).

The two ore horizons in the Dunderland Group show marked differences in gangue mineralogy, notably with a dominance of amphibole minerals in the lower, P-rich ores. These features and the presence of amphibolitic host rocks in the case of the lower horizon suggested to Grenne et al. (1999) the possibility of a volcanogenic source for these particular magnetite-rich ores. Bugge (1948) had originally concluded that both ore types were of sedimentary origin.

The isotope-chemostratigraphic approach applied to the dating of carbonate sedimentation, and thus the deposition of iron ores, suggests a Neoproterozoic age within the range 650–595 Ma (Melezhik et al., 2002b, 2003). This time interval overlaps with the onset of speculative "snowball Earth" conditions through the Neoproterozoic and the last appearance of banded iron formations on Earth after their long absence since 1800 Ma (Hoffman and Shrag, 2002). Their reappearance was assigned to the melting event of an ice-covered, stagnant ocean and transport of reduced Fe to oxic surface–shelf environments. However, there are several controversies involved with the snowball Earth hypothesis itself, and, consequently, this also applies to the genesis of the Neoproterozoic iron formations (Young, 2002).

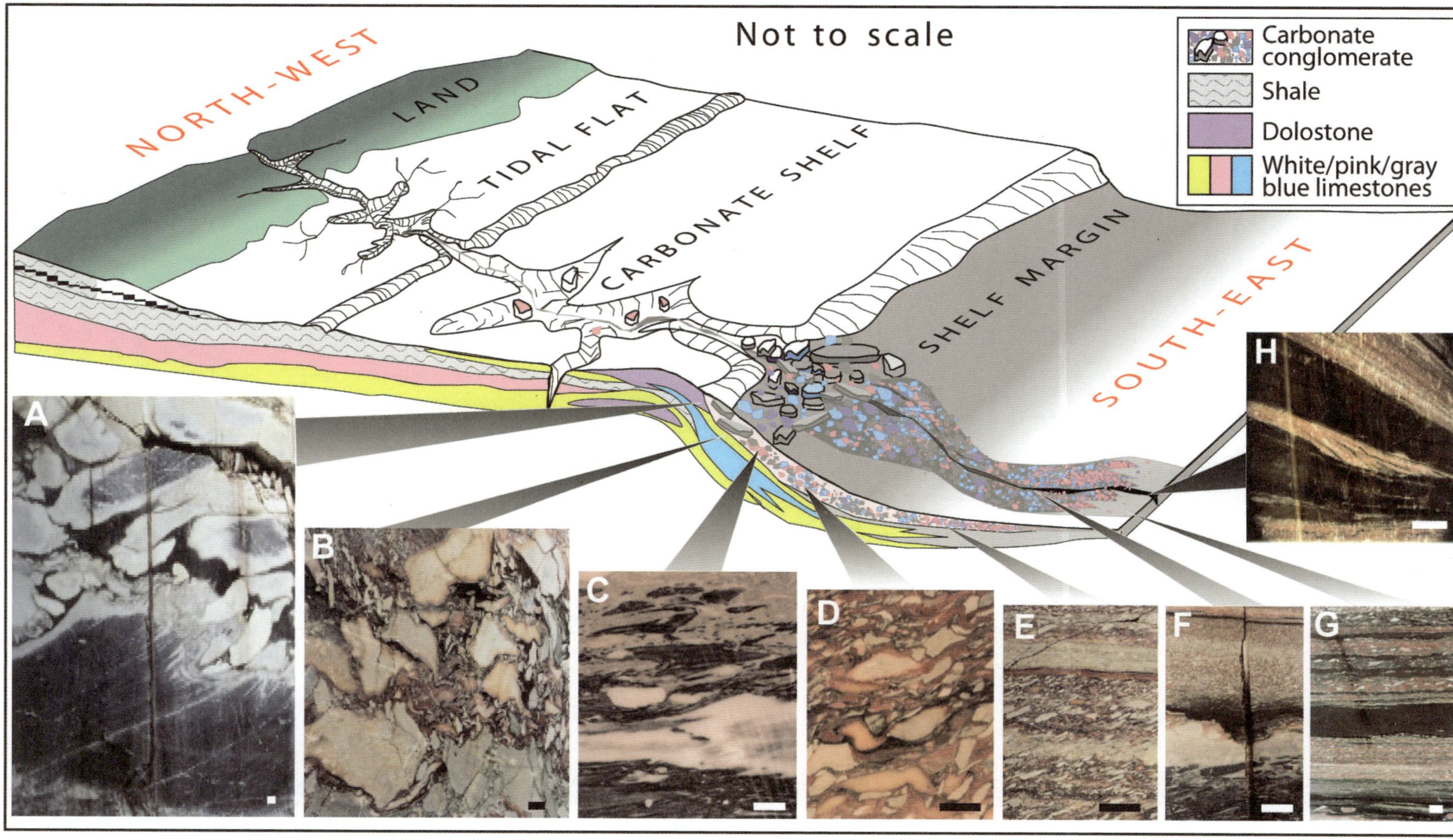

Figure 4. Depositional model depicting the different facies of the Fauske carbonate conglomerates in the Fauske Nappe (modified from Melezhik et al., 2000a). Paleocurrents, indicated by cross-lamination and channelling in interbedded calcareous graywacke and calcarenite, point to sediment transport to the southeast (present-day coordinates) on a basinal slope outboard of the unstable, carbonate-shelf edge. (A) Large, rounded and angular blocks of "blue" calcite marble bleached around their margins; carbonate debris, proximal mass-flow facies. (B) Chaotically deposited, unsorted blocks and angular fragments of pink and pale pink calcite marbles and white dolomite marbles in a dark calcareous schist matrix; carbonate debris, proximal mass-flow facies. (C) Large angular clasts of white and pale gray ("blue") marbles in a dark-gray calcarenite matrix; carbonate debris, distal mass-flow facies. (D) Carbonate breccia composed of angular fragments and subrounded clasts of pink calcite marble, white calcite, and white dolomite marble in a calcarenite matrix; proximal channel facies. (E) Originally horizontally bedded carbonate breccia showing a gradual upward decrease in fragment size accompanied by the gradual development of an internal planar lamination; proximal, submarine-channel facies. (F) Large-pebble conglomerates eroded and channelled, with a subsequent infill of the channel consisting of finely dispersed material (black and dark-gray silty graywacke); submarine-channel facies. (G) White, pale gray, and pink, small-pebble conglomerates interbedded with dark-gray graywackes; distal, turbidite facies. (H) Current-bedded, pink, carbonate gritstones and fine-pebble conglomerate filling erosional channel in laminated dark-gray graywacke; distal, submarine-channel facies. Scale bars: 20 cm long.

Ophiolites

In the far north, east of Tromsø, the Lyngen Magmatic Complex dominates the lower parts of the Lyngsfjell Nappe, and it constitutes the largest dismembered ophiolite in the Norwegian Caledonides (Boyd and Minsaas, 1984; Furnes et al., 1985). Gabbros, mafic volcanites, and scattered mafic dikes are the principal components. The complex is divided into western and eastern suites separated by a ductile, oceanic shear zone (Furnes and Pedersen, 1995). In a part of the volcanite section to the east, a coeval tonalite body has yielded a U-Pb zircon age of 469 ± 5 Ma (Oliver and Krogh, 1995). In brief, the geochemical signature of the gabbros of the western suite indicates that the parental magmas were tholeiites of back-arc basin origin. Eastern suite gabbros, on the other hand, are cumulates that are believed to have crystallized from ultradepleted, high-Ca boninitic magmas akin to those in forearcs (Furnes and Pedersen, 1995; Kvassnes et al., 2004). The Lyngen Magmatic Complex can be traced for more than 200 km to the southwest in a series of disconnected, tectonic lenses or thrust sheets of sheared mafic and ultramafic rocks (Boyd, 1983; Tull et al., 1985; Barker, 1986; Andresen and Steltenpohl, 1994).

A major unconformity separates the deformed, uplifted, and eroded Lyngen ophiolite from a fossil-bearing sedimentary succession (locally with a volcanic component) known as the Balsfjord Group, which is of latest Ordovician to Early Silurian age (Bjørlykke and Olaussen, 1981; Binns and Matthews, 1981; Minsaas and Sturt, 1985). This succession, with a basal mafic-pebble conglomerate, dominates the Lyngsfjell Nappe southwest of Lyngen (Minsaas and Sturt, 1985; Zwaan et al., 1998). A carbonate-evaporite tidal sequence that shows diverse sedimentary structures (Fig. 2E) is prominent within parts of the succession. The Lyngsfjell Nappe thins dramatically to the southwest, but it can be followed southward into the Narvik district as thin lensoid thrust sheets of sheared ophiolite and suprajacent mafic-clast conglomerate (Boyd, 1983; Steltenpohl et al., 1990; Andresen and Steltenpohl, 1994) (Fig. 2G). One such sheared ophiolite fragment, on Hinnøya, contains a tonalite gneiss that has provided a U-Pb zircon age of 479 ± 1 Ma (Northrup, 1997). In the Ofotfjorden district, carbonate rocks reappear in the thrust-dissected Evenes Nappe (or nappe complex), units of which have been correlated with those in the Balsfjord Group by Steltenpohl et al. (1990).

In southern areas of the Uppermost Allochthon, dismembered and fragmented ophiolites are important components in some of the thrust sheets of the Helgeland Nappe Complex (Thorsnes and Løseth, 1991; Nordgulen et al., 1993; Heldal, 2001). The Leka Ophiolite, which includes a spectacular exposure of harzburgite tectonite (Prestvik, 1980; Furnes et al., 1988) and which evolved in a suprasubduction-zone setting, is now considered to lie within the structurally lowermost part of the Helgeland Nappe Complex (Yoshinobu et al., 2002). A minor felsic sheet within the gabbro/sheeted dike unit on Leka has provided a U-Pb zircon age of 497 ± 2 Ma (Dunning and Pedersen, 1988). Fragmented ophiolites that occur farther north on several small islands (e.g., Bolvær complex; Heldal, 2001) are considered to lie at the same structural level as Leka, but there are also lensoid ophiolite remnants at higher levels in the tectonostratigraphy (Thorsnes and Løseth, 1991; Nordgulen et al., 1993). A common feature is that the ophiolite complexes were deformed and metamorphosed and overlain unconformably by mafic breccias and conglomerates at the base of low-grade, continental sedimentary successions of unknown but inferred Ordovician age (Gustavson, 1981, 1988; Sturt et al., 1985; Thorsnes and Løseth, 1991; Heldal, 2001). A minimum age for the thrusts beneath the ophiolite slices and their sedimentary cover has been provided by a U-Pb zircon date of 444 ± 11 Ma for the transecting Heilhornet Pluton (Nordgulen and Schouenborg, 1990). More recent studies in this coastal region have shown, however, that the various nappes and thrust sheets were imbricated during the time interval 477–447 Ma, prior to Late Ordovician exhumation (Yoshinobu et al., 2002). The low-grade cover successions in the Helgeland Nappe Complex, which probably predate 470 Ma (see the following), and their subjacent ophiolitic basements are therefore quite different in age from their counterparts in the Lyngen region in the far north.

PLUTONIC COMPLEXES

A variety of mainly calc-alkaline granitoid batholiths and intrusive complexes constitutes an important component of the Uppermost Allochthon. They were recognized early as a significant feature of the mountain belt, but prior to the 1980s, very few geochronological, geochemical, and structural studies had been undertaken. A major review of the igneous activity in the Scandinavian Caledonides as a whole is that of Stephens et al. (1985b). At that time, however, little information was available on the most prominent granitoid batholiths in Norway. Thus, attempts to reconstruct the igneous history of the mountain belt were based essentially on studies of the tectonostratigraphy and geochemistry of mafic volcanic rocks, whereas the granitoids were largely neglected (e.g., Stephens et al., 1985b).

Granitoids are particularly abundant within the Helgeland Nappe Complex, where the compositionally varied Bindal Batholith (Nordgulen et al., 1993) includes more than 50 individual plutons and composite intrusive complexes (Fig. 5). The batholith spans an age range from 477 to 430 Ma. An older group (477–468 Ma) of peraluminous to strongly peraluminous granites was intruded during a regional migmatization event that affected rocks in the western part of the nappe complex (Yoshinobu et al., 2002). These older plutons have isotopic compositions indicating mainly crustal sources (Birkeland et al., 1993). They include tourmaline granites and the major Vega pluton (Fig. 5), which is composed of heterogeneous anatectic granite.

A younger and texturally varied group of plutons was emplaced from 448 to 430 Ma (Nordgulen and Schouenborg, 1990; Nordgulen et al., 1993; Eide et al., 2002). They range in composition from olivine gabbro to leucogranite; however,

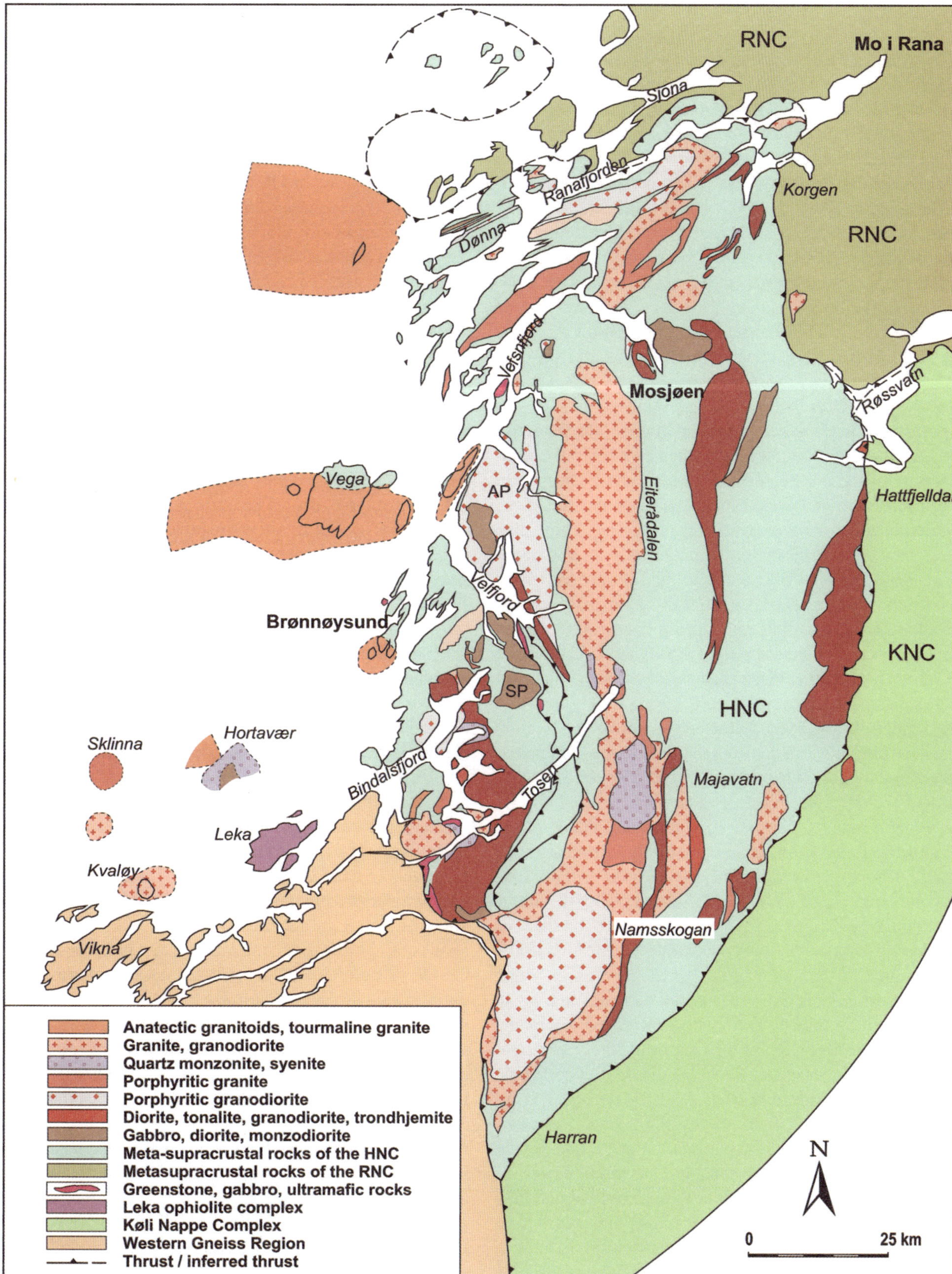

Figure 5. Geological map showing the distribution of granitoid rocks of the Bindal Batholith within the Helgeland Nappe Complex in north-central Norway. Subdivision of the Bindal Batholith is according to Nordgulen (1992). Abbreviations: HNC—Helgeland Nappe Complex; RNC—Rödingsfjället Nappe Complex; KNC—Köli Nappe Complex (part of the Upper Allochthon); AP—Andalshatten pluton; SP—Sausfjellet pluton.

tonalitic to granitic metaluminous rocks are clearly the most abundant. Most of the rocks are calc-alkaline, but there are significant occurrences of alkali-calcic gabbro to monzodiorite and syenite (Nordgulen, 1992; Barnes et al., 1992; Barnes and Prestvik, 2000). Detailed studies of some plutons show that emplacement was assisted by stoping and ductile flow in the wall-rock aureole (Barnes and Prestvik, 2000; Dumond et al., 2005). In some areas, hot, mafic plutons interacted with their pelitic and calcareous host rocks causing assimilation, mixing and mingling, and generation of hybridized rocks (Barnes et al., 2003, 2005; Dumond et al., 2005). Calc-alkaline plutons with U-Pb ages of 456 ± 2 and 458 ± 3 Ma have also been reported from the highest nappe of the Upper Allochthon in the Gjersvik area, directly beneath the Helgeland Nappe Complex (Roberts and Tucker, 1991; Meyer et al., 2003).

About 250 km farther southwest, the Smøla-Hitra Batholith (Gautneb and Roberts, 1989; Nordgulen et al., 1995, 2001) occupies large parts of the islands of Smøla, Hitra, and Frøya, and a narrow coastal zone on the adjacent mainland (Tucker et al., 2004), and it represents an intrusive complex that is probably correlative with that at Bindal. Apart from the crustally derived rocks of the Helgeland Nappe Complex, the chemistry and isotope composition of both batholiths (i.e., Bindal and Smøla-Hitra) are consistent with derivation from mixed crustal and mantle sources (Nordgulen and Sundvoll, 1992; Birkeland et al., 1993; Barnes et al., 2002). A common feature of the batholiths is their high Sr content (and Ba); many plutons may have Sr contents as high as 1000–1800 ppm (Nordgulen and Sundvoll, 1992; Barnes et al., 1992).

Granitoids also occur in the Rödingsfjället Nappe Complex, particularly in southern areas, although mainly as dike swarms that cross-cut the foliation in host rocks, which consist mainly of metasedimentary units interlayered with minor metavolcanic rocks (Ramberg, 1967; Bjerkgård et al., 1997). The Sr-isotope composition of marbles suggests Neoproterozoic ages (600–590 Ma) comparable to those of the Helgeland Nappe Complex (Bjerkgård et al., 1997). A Late Ordovician age for the dikes is suggested by a Rb-Sr whole-rock isochron (447 ± 7 Ma), i.e., similar to granites in the underlying Krutfjellet Nappe of the Upper Allochthon (Stephens et al., 1993). Little is known in terms of composition, but the generally evolved nature of the rocks suggests a mixed source for the magmas.

Large granitoid intrusions are abundant in the Beiarn Nappe (Fig. 1) between Mo i Rana and Bodø (e.g., Gjelle, 1980; Brattli and Tørudbakken, 1987; Solli, 1990). Porphyritic and equigranular granites and granodiorites are common in addition to subordinate gabbro, diorite, and tonalite. The plutons intrude medium-grade, foliated metasedimentary rocks of assumed Proterozoic to Cambrian age (Tørudbakken and Brattli, 1985). Age determinations on granitoids include Rb-Sr whole-rock dates of 495 ± 14 Ma for a quartz-monzonitic gneiss at Harefjell (Cribb, 1981), and 440 ± 30 Ma for the Høgtind Granite (Tørudbakken and Brattli, 1985). The only U-Pb dates so far reported are from a granite on an island ~5 km south of Bodø. These are on zircon and monazite and fall in the range 432–429 Ma (Bingen et al., 2002). Although compositional data and other precise U-Pb dates are lacking, the rocks appear to be analogous, in many ways, to the Bindal Batholith, and they probably were formed in a similar tectonic environment.

In the northernmost part of the Uppermost Allochthon, granitoids occur in the Ofotfjorden area within the stack of structural units (Gustavson, 1969; Steltenpohl et al., 2003), each of which contains early plutons that are texturally and chemically specific to each thrust sheet. The composition of these particular granitoids is relatively complex; most are S- and I-type, whereas early phase Snaufjell granitoids have A-type characteristics (Steltenpohl et al., 2003). The youngest granitoids have ages of ca. 430 Ma and cut nappe boundaries. This suggests that they are genetically related to processes that occurred during nappe stacking, which, according to Steltenpohl et al. (2003), was a product of the Scandian orogeny. The early granitoids, including the Snaufjell granite, remain undated, however, but Steltenpohl et al. (2003) suggested that they might record igneous activity related to the Taconian event.

In the far north, variably peraluminous to metaluminous granitic plutons, sheets, and dikes in the allochthons on Sørøya and Magerøya have yielded U-Pb zircon ages in the range 438–434 Ma (Kirkland et al., 2005; Corfu et al., 2006). These authors presented arguments favoring granite emplacement into successions draping the margin of Laurentia. The host rocks and intrusions were subsequently incorporated into either the Upper (Köli) or the Uppermost Allochthon during Scandian orogenesis.

The tectonic setting and geodynamic significance of the plutonic rocks that occur throughout the Uppermost Allochthon, and in some of the very highest Köli thrust sheets, have been considered on the basis of petrological and tectonostratigraphic evidence. The Early to Middle Ordovician peraluminous rocks of the Bindal Batholith originated during collision of microcontinental fragments and, possibly, concomitant ophiolite obduction in a peri-Laurentian setting (Yoshinobu et al., 2002). These conclusions are in general agreement with those derived from studies of ophiolites and arc sequences within the Upper Allochthon in western Norway (e.g., Pedersen and Dunning, 1997) and the Canadian Appalachians (e.g., Whitehead et al., 2000; Waldron and van Staal, 2001). The pattern of continued plutonism through the Ordovician and into the Silurian is also represented in various segments of the Appalachian belt (e.g., Kerr, 1997; Drummond et al., 1997; Whalen et al., 1997; Aleinikoff et al., 2002, Hollocher et al., 2002). The variable composition of the granitoids reflects derivation from multiple crustal and mantle sources, and it is, in general, related to a complex evolution of arc terranes, and eventual collisions between arcs, ribbon microcontinents, and the Laurentian margin during the final stages of the Taconian orogeny.

TECTONOTHERMAL RECORD

Until the last few years, comparatively few precise geochronological data existed on the rock complexes of the Uppermost

Allochthon. Together with a meager faunal record, restricted to one area in Troms, this hampered most attempts to interpret the likely Caledonian evolution of this major, composite thrust complex. Most workers considered the principal deformation and metamorphism, including thrust emplacement, to relate to the Scandian event (e.g., Nicholson, 1974; Gustavson, 1978). The discovery of fossils in a succession correlative with the Balsfjord Group (Olaussen, 1976; Bjørlykke and Olaussen, 1981), and the realization that clasts in a basal conglomerate above a major unconformity derived from the Lyngen ophiolite, led several workers to speculate about evidence for an earlier, pre-Scandian event (Minsaas and Sturt, 1985; Steltenpohl et al., 1990; Sturt and Roberts, 1991; Andresen and Steltenpohl, 1994). At that time, this early Caledonian event was considered to be Finnmarkian (Late Cambrian–Early Ordovician), based mainly on work in northern Norway but also on the belief that almost all Norwegian ophiolites were of Cambrian to earliest Ordovician age (Roberts and Gale, 1978; Furnes et al., 1985). Subsequently, publication of the minimum age (469 Ma) for the Lyngen ophiolite (Oliver and Krogh, 1995) opened up the probability that a younger Ordovician event was likely to be represented in the thrust sheets of this northern part of the Uppermost Allochthon. The main thrusting, however, is Scandian, with a peak metamorphism dated to 432 Ma (Northrup, 1997; Coker-Dewey et al., 2000).

Earlier, Rb-Sr isotopic investigations in areas farther south had, in fact, revealed loosely constrained evidence of Middle Ordovician tectonothermal activity. From the small area of Rödingsfjället Nappe Complex exposed in Sweden, Claesson (1979, p. 356) reported a Rb-Sr isochron age of 447 ± 7 Ma for a granitic dike that cut the regional schistosity and concluded that the penetrative deformation in the nappe "is probably Middle Ordovician in age." Comparable Rb-Sr studies in the Beiarn Nappe yielded a 440 ± 30 Ma age for a small granite pluton that transects both S2 schistosity and the mylonitic contact between internal thrust sheets (Tørudbakken and Brattli, 1985). These authors also concluded that the earlier D1 deformation in the Beiarn Nappe likely predated the inferred crystallization age of a quartz monzonite dated at 495 ± 14 Ma (Cribb, 1981).

In a study of a Zn-Pb deposit at Bleikvassli (Fig. 1), in a part of the Rödingsfjället Nappe Complex, Skauli et al. (1992) reported a Rb-Sr isochron age of 464 ± 22 Ma for a microcline gneiss, which they interpreted as a metamorphic age; and a nearby syn- to postmetamorphic granite pegmatite yielded a Rb-Sr date of 457 ± 5 Ma. Thus, metamorphism in this particular thrust sheet was believed to be of Middle Ordovician age. Rb-Sr studies in a part of the Helgeland Nappe Complex just west of Bleikvassli (Tørudbakken and Mickelson, 1986) indicated that the main D_2 tectonometamorphic event there preceded intrusion of a granite dike dated to 433 ± 11 Ma. A common thread in all these studies, however, was that a metamorphic overprint, and emplacement of the Uppermost Allochthon upon the Köli Nappes, occurred during the main, Scandian orogeny.

From this brief review, it is clear that there were strong indications of a pre-Scandian, inferred Middle Ordovician event in the Uppermost Allochthon well before the advent and application of more precise U-Pb and Ar-Ar dating methods. Dallmeyer and Andresen (1992) presented ^{40}Ar-^{39}Ar data from hornblende concentrates in the Nakkedal and Tromsø Nappes that pointed to tectonothermal activity at some stage during the Ordovician. Later, Selbekk et al. (2000) applied the U-Pb method to titanites from migmatite leucosomes and an associated anorthosite dike in the Nakkedal Nappe and produced ages of 456 ± 3 and 456 ± 4 Ma, respectively, which were interpreted as dating migmatization and dike intrusion. Eclogites in a passive-margin succession in the Tromsø Nappe were investigated by Corfu et al. (2003), who provided an isotope dilution–thermal ionization mass spectrometry (ID-TIMS) U-Pb zircon age of 452.1 ± 1.7 Ma, which was interpreted as dating the high-pressure metamorphic event. This was supported by 451–450 Ma ages of titanite from eclogite and rutile in neosome veins. Within the same nappe, a trondhjemitic orthogneiss yielded a U-Pb zircon age of 493 $^{+5}/_{-2}$ Ma, reflecting Late Cambrian magmatic activity (Corfu et al., 2003).

By far the most extensive U-Pb isotopic dating study of rocks in the Uppermost Allochthon is that from the southwestern parts of the Helgeland Nappe Complex (Nordgulen and Schouenborg, 1990; Nordgulen et al., 1993; Yoshinobu et al., 2002). The plutons and batholiths from this Velfjorden-Bindal region (Fig. 5) have already been described here. It is therefore sufficient to state that all of the thrust sheets in this coastal district are cut by the 447 Ma Andalshatten granodiorite. Thrust imbrication, moreover, is west-vergent, and superposed extensional structures, relating to exhumation, also predate 447 Ma (Yoshinobu et al., 2002). The oldest pluton affected by west-vergent thrusting is dated at 470 Ma, thus providing a minimum age for the low-grade successions that lie unconformably upon the ophiolites. In some thrust sheets, pulses of regional migmatization occurred during the time intervals 477–468 Ma and 448–445 Ma. Thus, most of Ordovician time between 477 and 445 Ma in this part of the Helgeland Nappe Complex was taken up by crustal growth processes—plutonism, migmatization, and contractional west-vergent deformation—and the emplacement of major plutons even continued up to 430 Ma, into the Early Silurian (Nordgulen et al., 1993). This was followed by further imbrication, metamorphism, and eastward thrusting during the Scandian orogeny. The significance of the earlier west-vergent structures is discussed later.

Structural Evidence

Structural information supporting the multiphase and poly-orogenic evolution of the Uppermost Allochthon has so far come from two main areas: (1) the Fauske district, in the basal part of the Rödingsfjället Nappe Complex, and (2) the Velfjord-Bindal region of the Helgeland Nappe Complex. In the Fauske Nappe, barren calcite marbles and carbonate conglomerates/breccias have been accorded an Early to Middle Cambrian (ca. 520 Ma) depositional age based on C and Sr isotope stratigraphy (Melezhik et al., 2000a). Sedimentological studies have revealed that deposition of the conglomerates occurred at the edge of an unstable carbonate shelf, with finer debris and turbiditic facies accumulating on a basinal slope deepening to the southeast (present-day coordinates) (Fig. 4). The earliest tectonic structures are NW-vergent, ramp-and-flat thrust faults and fault-bend folds (Fig. 6) with an associated S_1 schistosity (Roberts et al., 2001, 2002, 2004). These early structures and schistosity are overprinted by SE-verging folds and small-scale thrusts, and a penetrative, NW-dipping, S_2 cleavage. The shelf-to-basin geometry and paleocurrent data, coupled with the NW-verging, staircase thrusts and early folds, have been interpreted as recording initial deposition and subsequent Taconian deformation, respectively, along the eastern margin of Laurentia (Roberts et al., 2001). On the contrary, the diverse, later SE-verging structures and fabrics are interpreted as Scandian, relating to nappe emplacement onto Baltica. Although we have, as yet, no geochronology from the Fauske Nappe to confirm the age of the early, tectonic phase, the several indications of a Middle Ordovician tectonothermal event recorded elsewhere in the Rödingsfjället Nappe Complex do lend support to this conclusion.

In the Velfjorden-Bindal region, the west-vergent structures noted previously and described by Yoshinobu et al. (2002) are the earliest recorded tectonic structures in that part of the Helgeland Nappe Complex. Westward vergence is recognized in the entire stack of four, separate, internal thrust sheets, two of which include fragmented ophiolite complexes and their low-grade cover successions. This early deformation, which started at or just before ca. 470 Ma, and Ordovician magmatism have been interpreted by Yoshinobu et al. (2002) to be related to Taconian orogenesis close to the eastern margin of Laurentia. Subsequent thrust emplacement of the entire Helgeland Nappe Complex occurred during the Scandian orogeny, juxtaposing this allochthon against the Rödingsfjället thrust sheets and, in turn, above the Ordovician-Silurian successions of the Köli Nappes.

In the far north, in Troms, direct evidence for west-directed structural polarity at the mylonitic bases of thrust sheets carrying internal proof of Middle Ordovician metamorphism is lacking, but little detailed work has yet been done on these contacts. In the Ofotfjorden district, isotope-chemostratigraphical studies on many barren carbonate formations in the tectonically dissected Evenes and Gratangen Nappes have pointed to several likely depositional ages ranging from late Riphean to Silurian (Melezhik et al., 2002a, 2002b). While much of this imbrication is probably Scandian, it cannot be ruled out that juxtaposition of some Neoproterozoic and Cambrian units, as well as ophiolite obduction, might have occurred during the Taconian event along the eastern margin of Laurentia (Melezhik et al., 2002a). The minimum age of the Lyngen ophiolite, noted previously, also supports such an interpretation.

Figure 6. NW-verging thrust fault cutting up through multilayered, Fauske carbonate conglomerate, Løvgavlen quarry, Fauske; looking approximately northeast. A fault-bend fold has developed in the hanging wall, forward of the ramp, and the latter is cut by a thin, post-thrusting, mafic dike. This Cambrian conglomerate is part of the Fauske Nappe. The conglomerates and subjacent megabreccias are interpreted to have formed at the shelf edge and upper slope or rise along the eastern margin of Laurentia. The early, NW-verging thrusts and folds are considered to be Taconian. Later folds and cleavages verge southeast and fit into the local and regional pattern of Scandian deformation. For further details, see Roberts et al. (2001, 2002).

DISCUSSION

In this review, we have tried to show how the Uppermost Allochthon differs from the other major allochthons in the Caledonides of Scandinavia, principally in terms of its characteristic lithologies, which have disproportionately high abundances of platformal carbonate successions, granitoid batholithic complexes, and, in places, stratabound banded iron formations. In addition to these features, which are clearly unlike those that make up the thrust sheets that are indigenous to the Baltoscandian margin of Baltica, the Uppermost Allochthon carries a tectonothermal record and early Caledonian thrust polarity that is unique in Norway. The exotic assemblages alone led workers, over twenty years ago, to suggest that the suspect terranes represented by these high-level thrust sheets might have derived from the continental margin of eastern Laurentia or from an unknown microcontinent (Stephens and Gee, 1985, 1989; Roberts et al., 1985; Roberts, 1988). The nature of Caledonian metallogeny also favors such a notion (Grenne et al., 1999).

In more recent years, the revelations of the isotope-chemostratigraphic investigations on carbonate formations, structural studies in selected areas, and increasing geochronological evidence of a complex plutonic record and Ordovician, tectonothermal history support the earlier suggestions for a Laurentian provenance. This holds not only for Neoproterozoic to earliest Silurian sedimentation and volcanism, and the prolific Ordovician granitoid magmatism, but also for the pre-Scandian deformation, metamorphism, and migmatization. It has been suggested, in fact, that a large proportion, if not most, of this pre-Scandian, Ordovician deformation in the Uppermost Allochthon was actually an integral part of the Taconian orogenesis recorded along the Laurentian margin (Roberts et al., 2001, 2002; Yoshinobu et al., 2002; Roberts, 2003). Some of the highest Köli Nappes of the Upper Allochthon also provide evidence of this Ordovician event (Hall and Roberts, 1988; Stephens et al., 1993; Meyer et al., 2003). It cannot be ruled out, however, that a slightly earlier deformation, possibly equivalent to the Tremadoc-Arenig Penobscot arc and obduction event in the northern Appalachians (Colman-Sadd et al., 1992; van Staal et al., 1998), may be represented, e.g., in the Beiarn Nappe. The precise age of obduction of the Leka ophiolite is also unknown, maximum 497 Ma, with a younger constraint of 477 Ma. Whatever the case, subsequent emplacement of these Ordovician-deformed elements onto the oceanic Köli and subjacent Baltoscandian nappes first occurred during the Scandian orogeny at the time of collision between Baltica and Laurentia.

In the context of the theme of the 2004 International Basement Tectonics Association Conference, i.e., integrating our field and laboratory data toward a better understanding of the causative processes in crustal growth and its ultimate modification, we can examine briefly the overall development of the Uppermost Allochthon from the time of supracrustal accumulation in Neoproterozoic time and point to critical stages of crustal evolution. Continental crustal growth is known to occur by episodic accretion in diverse forms, such as contractional, accretionary orogeny, terrane amalgamation, and the addition of juvenile crust produced in arc systems. In the case of the Uppermost Allochthon of the Scandinavian Caledonides, crustal growth involved two major accretionary events, the earlier of which actually contributed to growth along the Laurentian continental margin. Later, during the Scandian orogeny, some of these very same, single or composite thrust sheets were transported onto the Baltoscandian margin, effectively contributing to crustal growth for a second time in their history.

Developments along the Laurentian margin in Late Neoproterozoic to early Paleozoic time have been described in many publications (e.g., Swett and Smit, 1972; Stanley and Ratcliffe, 1985; Williams, 1995; Smith et al., 2006; Higgins et al., 2004, and references therein). A passive, platformal margin with ensialic shelf basins hosted diverse siliciclastic successions and an extensive carbonate bank, with shelf-edge breccias passing oceanward into an expanding continental rise prism, outboard of which was the oceanic realm of Iapetus. Laurentia was persistently located at low latitudes during this late Riphean to Cambrian time interval, a position conducive to carbonate precipitation along its shelves. On the Baltican continent, on the other hand, which was positioned at high to intermediate southerly latitudes (Torsvik et al., 1996), carbonate formations are not so common, but tillites are well represented in the early to middle Vendian stratigraphic record (e.g., Reading and Walker, 1966; Banks et al., 1971; Edwards, 1984; Kumpulainen and Nystuen, 1985; Nystuen, 1985; Halverson et al., 2005). Initial oceanic contraction in Late Cambrian to Early Ordovician time partly coincided with the first stages of development of magmatic arcs. Sedimentation ultimately ceased in the platformal domain as the thrust sheets containing ophiolites, island-arc complexes, and slope-to-basin facies were successively incorporated, diachronously, and transported from east to west, in the developing Taconian orogenic wedge (Stanley and Ratcliffe, 1985; Williams, 1995). Following this major, crustal accretionary phase that culminated in Middle to early Late Ordovician time, uplift and erosion heralded a period of sedimentation and local volcanism in "successor basins," in Late Ordovician to Early Silurian time, representing recycling of the accretionary orogen. In certain terranes, however, magmatism continued into Early Silurian time. Several successor basins are known from the Appalachians (Robinson et al., 1998; Bream et al., 2004), and in Norway, in the Uppermost Allochthon, the Balsfjord Group succession, floored by debris from the Lyngen ophiolite, is considered to be one such example (Roberts et al., 2001). Some of the basinal successions overlying ophiolite fragments in the Helgeland Nappe Complex may possibly fall into the same category.

The collision between Baltica and Laurentia in Silurian-Devonian time that produced the Scandian orogeny and formation of the Caledonides in Scandinavia, Britain, and Ireland is known to have been oblique, as Baltica rotated counterclockwise and Laurentia drifted slowly southward (Torsvik et al., 1996; Cocks and Torsvik, 2002). This scissor-like collision inevitably led to a certain variation in the timing of peak metamorphism in the Scandinavian Caledonides, both transversely and laterally. A common thread, however, is one of vast magnitudes of thrust translation, in terms of hundreds of kilometers, particularly in the middle and higher nappe complexes (Gee and Sturt, 1985; Gayer et al., 1987). Subducted Baltican upper crust reached depths of 125 km at around 407 Ma (Terry et al., 2000) in a thrust sheet in part of northern west Norway. At the same time, Devonian sedimentation was proceeding at the surface in small extensional basins (Steel et al., 1985). Such recycling is also inferred to have occurred in an extensive foreland basin, now eroded, far to the east in Sweden (Samuelsson and Middleton, 1998; Larson et al., 1999; Huigen and Andriessen, 2004), although Hendriks and Redfield (2005) have argued for a measure of caution on this issue. During this period of Scandian contractional and accretionary orogeny, and crustal growth in the Baltican realm, the platformal successions, magmatic arc, and fragmented ophiolitic complexes were detached from their Laurentian roots and retransported onto the Baltoscandian margin—as the Uppermost Allochthon.

Although the subsequent late Scandian, extensional deformation, in the shape of many low-angle detachment faults (e.g., Hossack, 1984; Séranne, 1992; Braathen et al., 2000; Osmundsen et al., 2003, 2006), is strictly outside the topic of this contribution, the coeval link to ductile, sinistral shear along major vertical faults is an interesting facet of this story that requires elaboration from the point of view of terrane restoration. This Silurian-Devonian, left-lateral megashear has been described from many areas in East Greenland, Svalbard, Norway, the British Isles, and Ireland, and also from the northern Appalachians of New England and Newfoundland (Harland, 1971; Murphy, 1985; Grønlie and Roberts, 1989; Larsen and Bengaard, 1991; Soper et al., 1992; Strachan et al., 1992; Dewey and Strachan, 2003). East Greenland has no record of Taconian orogeny, and in its northernmost areas, carbonate sedimentation in fact persisted into Llandovery time (Higgins et al., 2004), just prior to Scandian deformation. The only record of Middle Ordovician arc magmatism is from an area in southern East Greenland (Hansen and Tembusch, 1978; Henriksen and Higgins, 1988). Thus, the present positions of Norway and Greenland in terms of Caledonide geology, arising from the opening of the North Atlantic, are somewhat anomalous. However, if we restore Baltica and the Scandinavian Caledonides southwestward to their likely position prior to the major, sinistral megashear, this would place Norway and the Uppermost Allochthon much closer to the northern Appalachians, near terranes that have similar magmatic and tectonothermal histories. The shelf-edge carbonate breccias of the Fauske Nappe would then have immediately recognizable counterparts in Newfoundland (Cow Head), Quebec, Vermont, and Maryland.

The magmatic arc granitoids described here have correlatives in the northern Appalachians that compare well in terms of both time of intrusion and geochemistry. In New England, for example, isotope geochronology has shown that arc-related plutons were emplaced during the time interval 480–442 Ma (Robinson et al., 1998). In detail, two linear arcs have been proposed, an earlier one (Shelburne Falls arc) above an ocean-facing subduction zone at 485–470 Ma, and a second, calc-alkaline arc (Bronson Hill arc) in late to post-Taconian time, 454–442 Ma (Karabinos et al., 1998). Clearly, this general situation has much in common with that recorded in the Helgeland Nappe Complex in Norway.

In East Greenland, Ordovician granodiorites and diorites occur only in the extreme south, in the Scoresby Sund region (Hansen and Tembusch, 1978; Higgins et al., 2004). It has been suggested that this occurrence may represent the northernmost tip of a Taconian arc (Roberts et al., 2004). The plutons are of I-type, calc-alkaline character and have been interpreted as indicative of landward subduction (Higgins et al., 2004).

SUMMARY

The highest nappe complexes in the Caledonides of Scandinavia, collectively termed the Uppermost Allochthon, are distinguished by their characteristic lithological assemblages and magmatic units that, in many ways, set them apart from the subjacent, Baltoscandian, platformal and continental-rise successions. Sandwiched between the two are the Iapetan, oceanic terranes of the Köli Nappes, which have Early Ordovician, North American faunas in one of the highest thrust sheets and sparse Baltican faunas at the lowest levels.

Principal features of the Uppermost Allochthon include an extensive outcrop of carbonate rocks (marbles, banded metalimestones, carbonate conglomerates, and local breccias) that can be followed for >600 km along strike, and the volumetrically significant granitoid plutons and batholiths that dominate the geology of some of the larger nappe complexes. Less prominent yet equally significant are the metamorphosed sedimentary iron formations that can also be traced throughout the length of the Uppermost Allochthon. All three of these main elements have no direct counterparts in the subjacent nappes and thrust sheets that derive from the Iapetus oceanic terranes and the Baltoscandian margin of Baltica. On the contrary, many of the features of the Uppermost Allochthon are highly reminiscent of those preserved in the thrust sheets and terranes of the northern Appalachians and East Greenland, derived from the Neoproterozoic to early Paleozoic continental margin of eastern Laurentia, and the magmatic arcs that developed just outboard of that margin. The tectonothermal record and evidence of early, NW- to W-vergent thrusting in the Uppermost Allochthon also point to the preservation of a Taconian history. Thus, taken as a whole, the Uppermost Allochthon appears to represent segments of the Laurentian margin and its conjoined arcs, and it carries a tectonothermal history of Taconian (and possible earlier Caledonian?) accretion onto that margin. Subsequently, during the Silurian-Devonian Scandian orogeny, the thrust sheets that now compose the Uppermost Allochthon were detached from their North American roots and incorporated into the Scandian orogenic wedge.

Crustal growth along the Laurentian margin during the early Paleozoic coincided with Taconian accretion and was augmented during this tectonomagmatic cycle by the emplacement of large volumes of granitoid plutonic rocks. Recycling of the accretionary orogen began in Late Ordovician time with accumulation in successor basins. This basin sedimentation and calc-alkaline magmatism continued into Early Silurian time. Scandian orogenesis followed, with the Taconian-deformed assemblages playing a major part in crustal growth on the Baltoscandian margin of Baltica.

Accepting the evidence for the major sinistral shear that has been widely reported to have accompanied and, in part, postdated the oblique, Scandian, continent-continent collision along the length of the Caledonian-Appalachian orogen, this would imply that the Scandinavian Caledonides lay at some unknown distance to the southwest, relative to their current position opposite Greenland. Similarly, the Appalachians would have been farther northeast, in relative terms, prior to the onset of left-lateral megashearing. It is therefore quite possible, given the similarities of their lithostratigraphic, magmatic, and tectonothermal histories, that the supracrustal rocks that now constitute the thrust sheets of the Uppermost Allochthon and the terranes of the northern Appalachians were much closer together before the final Scandian collision, nappe accretion, and initiation of major, sinistral, crustal movements along deep-seated, vertical faults.

ACKNOWLEDGMENTS

In completing this summary of our knowledge of the highest allochthon in the Scandinavian Caledonides, we are especially indebted to the Geological Survey of Norway for supporting our diverse research and mapping projects over the last two decades. Contributions from several research teams from American, British, and Norwegian universities have also provided significant advances in our understanding of these thrust complexes. The senior author is particularly grateful to Robert Hatcher Jr. for providing the financial support that enabled him to present this summary and overview at the International Basement Tectonics Association Conference in Oak Ridge. We are also grateful to our colleagues Peter Robinson and Arne Solli for their comments and suggestions on an early version of the manuscript. The final manuscript (formally accepted by the volume editors on 16 November 2005) has been improved following the constructive reviews of Robert Hatcher Jr. and Mark Steltenpohl. Translation of the Queen's English to the American version has been expertly performed by one of the Geological Society of America's desk editors.

REFERENCES CITED

Aleinikoff, J.N., Wright Horton, J., Jr., Drake, A.A., Jr., and Fanning, C.M., 2002, SHRIMP and conventional U-Pb ages of Ordovician granites and tonalites in the central Appalachian Piedmont: Implications for Paleozoic tectonic events: American Journal of Science, v. 302, p. 50–75, doi: 10.2475/ajs.302.1.50.

Anderson, M.W., 1989, Basement-cover evolution during Caledonian orogenesis, Troms, N. Norway, *in* Gayer, R.A., ed., The Caledonian Geology of Scandinavia: London, Graham and Trotman, p. 101–110.

Anderson, M.W., Barker, A.J., Bennett, D.J., and Dallmeyer, R.D., 1992, A tectonic model for Scandian terrane accretion in the northern Scandinavian Caledonides: Geological Society [London] Journal, v. 149, p. 1–15.

Andréasson, P.G., 1994, The Baltoscandian margin in Neoproterozoic–Early Palaeozoic time: Some constraints on terrane derivation and accretion in the Arctic Scandinavian Caledonides: Tectonophysics, v. 231, p. 1–32, doi: 10.1016/0040-1951(94)90118-X.

Andresen, A., and Steltenpohl, M.G., 1994, Evidence for ophiolite obduction, terrane accretion and polyorogenic evolution of the north Scandinavian Caledonides: Tectonophysics, v. 231, p. 59–70, doi: 10.1016/0040-1951(94)90121-X.

Andresen, A., Fareth, E., Bergh, S., Kristensen, S.E., and Krogh, E., 1985, Review of Caledonian lithotectonic units in Troms, north Norway, *in* Gee, D.G., and Sturt, B.A., eds., The Caledonide Orogen—Scandinavia and Related Areas: Chichester, John Wiley and Sons, p. 569–578.

Banks, N.L., Edwards, M.B., Geddes, W.P., Hobday, D.K., and Reading, H.G., 1971, Late Precambrian and Cambro-Ordovician sedimentation in East Finnmark: Norges Geologiske Undersøkelse, v. 269, p. 197–236.

Barker, A.J., 1986, The geology between Salangsdalen and Gratangenfjord, Troms, Norway: Norges Geologiske Undersøkelse Bulletin, v. 405, p. 41–56.

Barnes, C.G., and Prestvik, T., 2000, Conditions of pluton emplacement and anatexis in the Caledonian Bindal Batholith, north-central Norway: Norsk Geologisk Tidsskrift, v. 80, p. 259–274, doi: 10.1080/00291960051030581.

Barnes, C.G., Prestvik, T., Nordgulen, Ø., and Barnes, M.A., 1992, Geology of three dioritic plutons in Velfjord, Nordland: Norges Geologiske Undersøkelse Bulletin, v. 423, p. 41–54.

Barnes, C.G., Yoshinobu, A., Prestvik, T., Nordgulen, Ø., Karlsson, H., and Sundvoll, B., 2002, Mafic magma intraplating: Anatexis and hybridization in arc crust, Bindal Batholith, Norway: Journal of Petrology, v. 43, p. 2171–2190, doi: 10.1093/petrology/43.12.2171.

Barnes, C.G., Prestvik, T., Barnes, M.W., Anthony, E.Y., and Allen, C.M., 2003, Geology of a magma transfer zone: The Hortavær Igneous Complex, north-central Norway: Norwegian Journal of Geology, v. 83, p. 187–208.

Barnes, C.G., Prestvik, T., Sundvoll, B., and Surrat, D., 2005, Pervasive assimilation of carbonate and silicate rocks in the Hortavær igneous complex, north-central Norway: Lithos, v. 80, p. 179–199, doi: 10.1016/j.lithos.2003.11.002.

Bergström, S.M., 1979, Whiterockian (Ordovician) conodonts from the Hølonda Limestone of the Trondheim region, Norwegian Caledonides: Norsk Geologisk Tidsskrift, v. 59, p. 295–307.

Bingen, B., Nordgulen, Ø., and Solli, A., 2002, U-Pb geochronology of Paleozoic events in the Mid Scandinavian Caledonides, *in* Eide, E.A., coord., BATLAS—Mid Norway Plate Reconstruction Atlas with Global and Atlantic Perspectives: Trondheim, Geological Survey of Norway, p. 66–67.

Binns, R.E., 1978, Caledonian nappe correlation and orogenic history in Scandinavia north of lat. 67°N: Geological Society of America Bulletin, v. 89, p. 1475–1490, doi: 10.1130/0016-7606(1978)89<1475:CNCAOH>2.0.CO;2.

Binns, R.E., and Matthews, D.W., 1981, Stratigraphy and structure of the Ordovician-Silurian Balsfjord Supergroup, Troms, North Norway: Norges Geologiske Undersøkelse, v. 365, p. 39–54.

Birkeland, A., Nordgulen, Ø., Cumming, G.L., and Bjørlykke, A., 1993, Pb-Nd-Sr isotopic constraints on the origin of the Caledonian Bindal Batholith, central Norway: Lithos, v. 29, p. 257–271, doi: 10.1016/0024-4937(93)90020-D.

Bjerkgård, T., Larsen, R.B., and Marker, M., 1997, Regional setting of the Bleikvassli Zn-Pb deposit in Nordland, Norway [extended abs.]: Norges Geologiske Undersøkelse Bulletin, v. 433, p. 34–35.

Bjørlykke, A., and Olaussen, S., 1981, Silurian sediments, volcanics and mineral deposits in the Sagelvvatn area, Troms, North Norway: Norges Geologiske Undersøkelse, v. 365, p. 1–38.

Boyd, R., 1983, The Lillevik dyke complex, Narvik: Geochemistry and tectonic implications of a probable ophiolitic fragment in the Caledonides of the Ofoten region, North Norway: Norsk Geologisk Tidsskrift, v. 63, p. 39–54.

Boyd, R., and Minsaas, O., 1984, Lyngen 1634 III, Berggrunnsgeologisk kart, foreløpig utgave, M 1:50,000: Norges Geologiske Undersøkelse, Trondheim.

Braathen, A., Nordgulen, Ø., Osmundsen, P.T., Andersen, T.B., Solli, A., and Roberts, D., 2000, Devonian, orogen-parallel, opposed extension in the Central Norwegian Caledonides: Geology, v. 28, p. 615–618, doi: 10.1130/0091-7613(2000)28<615:DOOEIT>2.0.CO;2.

Brattli, B., and Tørudbakken, B., 1987, Arstaddal, beskrivelse til berggrunnsgeologisk kart 2028–4, M 1:50,000: Norges Geologiske Undersøkelse Skrifter, v. 81, 39 p.

Bream, B.R., Hatcher, R.D., Jr., Miller, C.F., and Fullargar, P.D., 2004, Detrital zircon ages and Nd isotopic data from the southern Appalachian crystalline core. GA-SC-NC-TN: New provenance constraints for Laurentian margin paragneisses, *in* Tollo, R.P., Corriveau, L., McLelland, J., and Bartholomew, M.J., eds., Proterozoic Evolution of the Grenville Orogen in North America: Geological Society of America Memoir 197, p. 450–475.

Bruton, D.L., and Bockelie, J.F., 1979, The Ordovician sedimentary sequence on Smøla, west central Norway: Norges Geologiske Undersøkelse, v. 348, p. 21–31.

Bruton, D.L., and Bockelie, J.F., 1980, Geology and paleontology of the Hølonda area, western Norway, a fragment of North America?, *in* Wones, D.R., ed., The Caledonides in the USA: Virginia Polytechnic Institute, Department of Geological Sciences Memoir 2, p. 41–47.

Bugge, J.A.W., 1948, Rana gruber; geologisk beskrivelse av jernmalmfeltene i Dunderlandsdalen: Norges Geologiske Undersøkelse, v. 171, p. 1–149.

Claesson, S., 1979, Pre-Silurian orogenic deformation in the north-central Scandinavian Caledonides: Geologiska Föreningen i Stockholm Förhandlingar, v. 101, p. 353–356.

Cocks, L.R.M., and Torsvik, T.H., 2002, Earth history 500–400 Ma: A faunal and palaeomagnetic review: Geological Society [London] Journal, v. 159, p. 631–644.

Coker-Dewey, J., Steltenpohl, M.G., and Andresen, A., 2000, Geology of western Ullsfjord, North Norway, with emphasis on the development of an inverted metamorphic gradient at the top of the Lyngen Nappe Complex: Norsk Geologisk Tidsskrift, v. 80, p. 111–128, doi: 10.1080/002919600750042609.

Colman-Sadd, S.P., Dunning, G.R., and Dec, T., 1992, Dunnage-Gander relationships and Ordovician orogeny in central Newfoundland: A sediment provenance and U/Pb study: American Journal of Science, v. 292, p. 317–355.

Corfu, F., Ravna, E.J.K., and Kullerud, K., 2003, A Late Ordovician U-Pb age for the Tromsø Nappe eclogites, Uppermost Allochthon of the Scandinavian Caledonides: Contributions to Mineralogy and Petrology, v. 145, p. 502–513, doi: 10.1007/s00410-003-0466-x.

Corfu, F., Torsvik, T.H., Andersen, T.B., Ashwal, L.D., Ramsay, D.M., and Roberts, R.J., 2006, Early Silurian mafic-ultramafic and granitic plutonism in contemporaneous flysch, Magerøy, northern Norway: U-Pb ages and regional significance: Geological Society [London] Journal, v. 163, p. 291–301, doi: 10.1144/0016-764905-014.

Cribb, S.J., 1981, Rb-Sr geochronological evidence suggesting a reinterpretation of part of the north Norwegian Caledonides: Norsk Geologisk Tidsskrift, v. 61, p. 97–110.

Dallmeyer, R.D., and Andresen, A., 1992, Polyphase tectonothermal evolution of Caledonian nappes in Troms, Norway: Evidence from $^{40}Ar/^{39}Ar$ mineral ages: Lithos, v. 29, p. 19–42, doi: 10.1016/0024-4937(92)90032-T.

Denison, R.E., Koepnick, R.B., Fletcher, A., Howell, M.W., and Callaway, W.S., 1994, Criteria for retention of original $^{87}Sr/^{86}Sr$ in ancient shelf limestone: Chemical Geology, v. 112, p. 131–143.

Dewey, J.F., and Strachan, R.A., 2003, Changing Silurian-Devonian relative plate motion in the Caledonides: Sinistral transpression to sinistral transtension: Geological Society [London] Journal, v. 160, p. 219–229.

Drummond, M.S., Neilson, M.J., Allison, D.T., and Tull, J.F., 1997, Igneous petrogenesis and tectonic setting of granitic rocks from the eastern Blue Ridge and Inner Piedmont, Alabama Appalachians, *in* Sihna, A.K., Whalen, J.B., and Hogan, J.P., eds., The Nature of Magmatism in the Appalachian Orogen: Geological Society of America Memoir 191, p. 147–164.

Dumond, G., Yoshinobu, A.S., and Barnes, C.G., 2005, Midcrustal emplacement of the Sausfjellet pluton, central Norway: Ductile flow, stoping, and in situ assimilation: Geological Society of America Bulletin, v. 117, p. 383–395, doi: 10.1130/B25464.1.

Dunning, G.R., and Pedersen, R.B., 1988, U/Pb ages of ophiolites and arc-related plutons of the Norwegian Caledonides: Implications for the development of Iapetus: Contributions to Mineralogy and Petrology, v. 98, p. 13–23, doi: 10.1007/BF00371904.

Edwards, M.B., 1984, Sedimentology of the Upper Proterozoic glacial record, Vestertana Group, Finnmark, North Norway: Norges Geologiske Undersøkelse Bulletin, v. 394, 76 p.

Eide, E.A., Osmundsen, P.T., Meyer, G.B., Kendrick, M.A., and Corfu, F., 2002, The Nesna shear zone, north-central Norway: An $^{40}Ar/^{39}Ar$ record of Early Devonian–Early Carboniferous ductile extension and unroofing: Norwegian Journal of Geology, v. 82, p. 317–339.

Fairchild, I.J., Spiro, B.F., Herrington, P.M., and Song, T., 2000, Controls on Sr and C isotope compositions of Neoproterozoic Sr-rich limestones of East Greenland and North China, *in* Grotzinger, J.P., and James, N.P., eds., Carbonate Sedimentation and Diagenesis in the Evolving Precambrian World: Society for Sedimentary Geology Special Publication 67, p. 297–313.

Foslie, S., 1949, Håfjellsmulden i Ofoten og dens sedimentære jern-manganmalmer: Norges Geologiske Undersøkelse, v. 174, p. 1–129.

Furnes, H., and Pedersen, R.B., 1995, The Lyngen magmatic complex: Geology and geochemistry: Geonytt, v. 22, p. 30.

Furnes, H., Ryan, P.D., Grenne, T., Roberts, D., Sturt, B.A., and Prestvik, T., 1985, Geological and geochemical classification of the ophiolitic fragments in the Scandinavian Caledonides, *in* Gee, D.G., and Sturt, B.A., eds., The Caledonide Orogen—Scandinavia and Related Areas: Chichester, John Wiley and Sons, p. 657–670.

Furnes, H., Pedersen, R.B., and Stillman, C.J., 1988, The Leka Ophiolite Complex, central Norwegian Caledonides: Field characteristics and geotectonic significance: Geological Society [London] Journal, v. 145, p. 401–412.

Gautneb, H., and Roberts, D., 1989, Geology and petrochemistry of the Smøla-Hitra Batholith, Central Norway: Norges Geologiske Undersøkelse, v. 416, p. 1–24.

Gavelin, S., and Kulling, O., 1955, Beskrivning til Berggrundskarta över Västerbottens län: Sveriges Geologiska Undersökning, Series Ca 37, 296 p.

Gayer, R.A., Rice, A.H.N., Roberts, D., Townsend, C., and Welbon, A., 1987, Restoration of the Caledonian Baltoscandian margin from balanced cross-sections: The problem of excess continental crust: Transactions of the Royal Society of Edinburgh, v. 78, p. 197–217.

Gee, D.G., 1975, A tectonic model for the central part of the Scandinavian Caledonides: American Journal of Science, v. 275A, p. 468–515.

Gee, D.G., and Sturt, B.A., eds., 1985, The Caledonide Orogen—Scandinavia and Related Areas: Chichester, John Wiley and Sons, 1266 p.

Gee, D.G., Kumpulainen, R., Roberts, D., Stephens, M.B., Thon, A., and Zachrisson, E., 1985, Scandinavian Caledonides. Tectonostratigraphic map, Scale 1:2 M, *in* Gee, D.G., and Sturt, B.A., eds., The Caledonide Orogen—Scandinavia and related areas: Chichester, John Wiley and Sons. (Printed by Swedmap, Huddinge, Sweden.)

Gjelle, S., 1980, Beiardalen, Berggrunnskart 2028–1, M 1:50,000: Norges Geologiske Undersøkelse, Trondheim.

Gjelle, S., 1988, Geologisk kart over Norge, Berggrunnskart Saltdal, M 1:250,000: Norges Geologiske Undersøkelse, Trondheim.

Grenne, T., Ihlen, P.M., and Vokes, F.M., 1999, Scandinavian Caledonide metallogeny in a plate tectonic perspective: Mineralium Deposita, v. 34, p. 422–471, doi: 10.1007/s001260050215.

Grønlie, A., and Roberts, D., 1989, Resurgent strike-slip duplex development along the Hitra-Snåsa and Verran faults, Møre-Trøndelag fault zone, central Norway: Journal of Structural Geology, v. 11, p. 295–305, doi: 10.1016/0191-8141(89)90069-2.

Gustavson, M., 1966, The Caledonian mountain chain of the southern Troms and Ofoten area. Part I: Basement rocks and Caledonian meta-sediments: Norges Geologiske Undersøkelse, v. 239, 162 p.

Gustavson, M., 1969, The Caledonian mountain chain of the southern Troms and Ofoten areas. Part II. Caledonian rocks of igneous origin: Norges Geologiske Undersøkelse, v. 261, 110 p.

Gustavson, M., 1978, Caledonides of North-Central Norway: Geological Survey of Canada Paper 78–13, p. 25–30.

Gustavson, M., 1981, Geologisk kart over Norge, Berggrunnskart Mosjøen, M 1:250,000: Norges Geologiske Undersøkelse, Trondheim.

Gustavson, M., 1988, Mosjøen. Beskrivelse til Berggrunnsgeologisk kart M: 1:250,000: Norges Geologiske Undersøkelse, Skrifter, v. 87, 42 p.

Gustavson, M., 1996, Geologisk kart over Norge. Berggrunnskart Sulitjelma, M 1:250,000: Norges Geologiske Undersøkelse, Trondheim.

Gustavson, M., and Gjelle, S., 1991, Geologisk kart over Norge. Berggrunnskart Mo i Rana, M 1:250,000: Norges Geologiske Undersøkelse, Trondheim.

Hall, L.M., and Roberts, D., 1988, Timing of Ordovician deformation in the Caledonian-Appalachian orogen, *in* Harris, A.L., and Fettes, D.J., eds., The Caledonian-Appalachian Orogen: Geological Society [London] Special Publication 38, p. 291–309.

Halverson, G.P., Hoffman, P.F., Schrag, D.P., Maloof, A.C., and Rice, A.H.N., 2005, Toward a Neoproterozoic composite carbon-isotope record: Geological Society of America Bulletin, v. 117, p. 1181–1207, doi: 10.1130/B25630.1

Hansen, B.T., and Tembusch, H., 1978, Rb-Sr isochron ages from east Milne Land, Scoresby Sund, East Greenland: Grønlands Geologiske Undersøgelse Rapport no. 95, p. 96–101.

Harland, W.B., 1971, Tectonic transpression in Caledonian Spitsbergen: Geological Magazine, v. 108, p. 27–42.

Heldal, T., 2001, Ordovician stratigraphy in the western Helgeland Nappe Complex in the Brønnøysund area, north-central Norway: Norges Geologiske Undersøkelse Bulletin, v. 438, p. 47–61.

Hendriks, B.W.H., and Redfield, T.F., 2005, Apatite fission track and (U-Th)/He data from Fennoscandia: An example of underestimation of fission track annealing in apatite: Earth and Planetary Science Letters, v. 236, p. 443–458, doi: 10.1016/j.epsl.2005.05.027.

Henriksen, N., and Higgins, A.K., 1988, Descriptive text to 1:100,000 sheets Rødefjord 70 Ø.3 N and Kap Leslie 70 Ø.2 N: Copenhagen, Grønlands Geologiske Undersøgelse, 34 p.

Higgins, A.K., Elvevold, S., Escher, J.C., Frederiksen, K.S., Gilotti, J.A., Henriksen, N., Jepsen, H.F., Jones, K.A., Kalsbeek, F., Kinny, P.D., Leslie, A.G., Smith, M.P., Thrane, K., and Watt, G.R., 2004, The foreland-propagating thrust architecture of the East Greenland Caledonides 72°–75°N: Geological Society [London] Journal, v. 161, p. 1009–1026.

Hoffman, P.F., and Shrag, D.P., 2002, The snowball Earth hypothesis: Terra Nova, v. 14, p. 129–155, doi: 10.1046/j.1365-3121.2002.00408.x.

Hollocher, K., Bull, J., and Robinson, P., 2002, Geochemistry of the metamorphosed Ordovician Taconian magmatic arc, Bronson Hill anticlinorium, western New England: Physics and Chemistry of the Earth, v. 27, p. 5–45.

Hossack, J.R., 1984, The geometry of listric growth faults in the Devonian basins of Sunnfjord, west Norway: Geological Society [London] Journal, v. 135, p. 705–711.

Huigen, Y., and Andriessen, P., 2004, Thermal effects of Caledonian foreland basin formation, based on fission-track analyses applied on basement rocks in central Sweden: Physics and Chemistry of the Earth, v. 29, p. 683–694.

James, N.P., and Stevens, R.H., 1986, Stratigraphy and correlation of the Cambro-Ordovician Cow Head Group, western Newfoundland: Geological Survey of Canada Bulletin, v. 366, 143 p.

Karabinos, P., Samson, S.D., Hepburn, J.C., and Stoll, H.M., 1998, Taconian orogeny in the New England Appalachians: Collision between Laurentia and the Shelburne Falls arc: Geology, v. 26, p. 215–218, doi: 10.1130/0091-7613(1998)026<0215:TOITNE>2.3.CO;2.

Kerr, A., 1997, Space-time-composition relationships among Appalachian-cycle plutonic suite in Newfoundland, *in* Sihna, A.K., Whalen, J.B., and Hogan, J.P., eds., The Nature of Magmatism in the Appalachian Orogen: Geological Society of America Memoir 191, p. 193–220.

Kirkland, C.L., Daly, J.S., and Whitehouse, M.J., 2005, Early Silurian magmatism and the Scandian evolution of the Kalak Nappe Complex, Finnmark, Arctic Norway: Geological Society [London] Journal, v. 162, p. 985–1003.

Krogh, E.J., Andresen, A., Bryhni, I., Broks, T.M., and Kristensen, S.E., 1990, Eclogites and polyphase P-T cycling in the Caledonian Uppermost Allochthon in Troms, northern Norway: Journal of Metamorphic Geology, v. 8, p. 289–309.

Kumpulainen, R., and Nystuen, J.P., 1985, Late Proterozoic basin evolution and sedimentation in the westernmost part of Baltoscandia, in Gee, D.G., and Sturt, B.A., eds., The Caledonide Orogen—Scandinavia and Related Areas: Chichester, John Wiley and Sons, p. 213–232.

Kvassnes, A.J.S., Strand, A.H., Moen-Eikeland, H., and Pedersen, R.B., 2004, The Lyngen Gabbro: The lower crust of an Ordovician incipient arc: Contributions to Mineralogy and Petrology, v. 148, p. 358–379, doi: 10.1007/s00410-004-0609-8.

Larsen, P.-H., and Bengaard, H.-J., 1991, Devonian basin initiation in East Greenland: A result of sinistral wrench faulting and Caledonian extensional collapse: Geological Society [London] Journal, v. 148, p. 355–368.

Larson, S.Å., Tullborg, E.-L., Cederbom, C., and Stiberg, J.-P., 1999, Sveconorwegian and Caledonian foreland basins in the Baltic Shield revealed by fission-track thermochronology: Terra Nova, v. 11, p. 210–215, doi: 10.1046/j.1365-3121.1999.00249.x.

Melezhik, V.A., Heldal, T., Roberts, D., Gorokhov, I.M., and Fallick, A.E., 2000a, Depositional environment and apparent age of the Fauske carbonate conglomerate, North Norwegian Caledonides: Norges Geologiske Undersøkelse Bulletin, v. 436, p. 147–168.

Melezhik, V.A., Lindahl, I., Pokrovsky, B., and Nilsson, L.P., 2000b, Sulphur source and genesis of polymetallic sulphide occurrences of the Ofoten district in the central-north Norwegian Caledonides: Evidence from sulphur isotopic studies: Mineralium Deposita, v. 35, p. 465–489, doi: 10.1007/s001260050256.

Melezhik, V.A., Gorokhov, I.M., Kuznetsov, A.B., and Fallick, A.E., 2001a, Chemostratigraphy of Neoproterozoic carbonates: Implications for 'blind dating': Terra Nova, v. 13, p. 1–11, doi: 10.1046/j.1365-3121.2001.00318.x.

Melezhik, V.A., Gorokhov, I.M., Fallick, A.E., and Gjelle, S., 2001b, Strontium and carbon isotope geochemistry applied to dating of carbonate sedimentation: An example from high-grade rocks of the Norwegian Caledonides: Precambrian Research, v. 108, p. 267–292, doi: 10.1016/S0301-9268(01)00135-8.

Melezhik, V.A., Roberts, D., Gorokhov, I.M., Fallick, A.E., Zwaan, K.B., Kuznetsov, A.B., and Pokrovsky, B.G., 2002a, Isotopic evidence for a complex Neoproterozoic to Silurian rock assemblage in the north-central Norwegian Caledonides: Precambrian Research, v. 114, p. 55–86, doi: 10.1016/S0301-9268(01)00218-2.

Melezhik, V.A., Gorokhov, I.M., Fallick, A.E., Roberts, D., Kuzhetsov, A.B., Zwann, K.B., and Pokrovsky, B.G., 2002b, Isotopic stratigraphy suggests Neoproterozoic ages and Laurentian ancestry for high-grade marbles from the north-central Norwegian Caledonides: Geological Magazine, v. 139, p. 375–393, doi: 10.1017/S0016756802006726.

Melezhik, V.A., Zwaan, B.K., Motuza, G., Roberts, D., Solli, A., Fallick, A.E., Gorokhov, I.M., and Kusnetzov, A.B., 2003, Isotope chemostratigraphy combined with a detailed mapping of high-grade marble sequences: Progress towards a new generation of geological maps in metamorphic terranes: Norwegian Journal of Geology, v. 83, p. 209–242.

Melezhik, V.A., Roberts, D., Fallick, A.E., Gorokhov, I.M., and Kuznetsov, A.B., 2005a, Geochemical preservation potential of high-grade calcite marble versus dolomite marble: Implication for isotope chemostratigraphy: Chemical Geology, v. 216, p. 203–224, doi: 10.1016/j.chemgeo.2004.11.020.

Melezhik, V.A., Fallick, A.E., and Pokrovsky, B.G., 2005b, Enigmatic nature of thick sedimentary carbonates depleted in ^{13}C beyond the canonical mantle value: The challenges to our understanding of the terrestrial carbon cycle: Precambrian Research, v. 137, p. 131–165, doi: 10.1016/j.precamres.2005.03.010.

Meyer, G.B., Grenne, T., and Pedersen, R.B., 2003, Age and tectonic setting of the Nesåa Batholith: Implications for Ordovician arc development in the Caledonides of central Norway: Geological Magazine, v. 140, p. 573–594, doi: 10.1017/S0016756803008069.

Minsaas, O., and Sturt, B.A., 1985, The Ordovician-Silurian clastic sequence overlying the Lyngen Gabbro Complex, and its environmental significance, *in* Gee, D.G., and Sturt, B.A., eds., The Caledonide Orogen—Scandinavia and Related Areas: Chichester, John Wiley and Sons, p. 379–393.

Murphy, F.C., 1985, Non-axial planar cleavage and Caledonian sinistral transpression in eastern Ireland: Geological Journal, v. 20, p. 257–279.

Nicholson, R., 1974, The Scandinavian Caledonides, in Nairn, N.E.M., and Stehli, F.G., eds., The Ocean Basins and Margins. 2. The North Atlantic: New York, Plenum Press, p. 161–203.

Nicholson, R., and Rutland, R.W.R., 1969, A section across the Norwegian Caledonides: Bodø to Sulitjelma: Norges Geologiske Undersøkelse, v. 260, 86 p.

Nordgulen, Ø., 1992, A Summary of the Petrography and Geochemistry of the Bindal Batholith: Trondheim: Norges Geologiske Undersøkelse Report 92-111, 103 p.

Nordgulen, Ø., and Schouenborg, B., 1990, The Caledonian Heilhornet pluton, north-central Norway: Geological setting, radiometric age and implications for the Scandinavian Caledonides: Geological Society [London] Journal, v. 147, p. 439–450.

Nordgulen, Ø., and Sundvoll, B., 1992, Strontium isotope composition of the Bindal Batholith, central Norwegian Caledonides: Norges Geologiske Undersøkelse Bulletin, v. 423, p. 19–39.

Nordgulen, Ø., Bickford, M., Nissen, A.L., and Wortman, G., 1993, U-Pb zircon ages from the Bindal Batholith, and the tectonic history of the Helgeland Nappe Complex, Scandinavian Caledonides: Geological Society [London] Journal, v. 150, p. 771–783.

Nordgulen, Ø., Solli, A., and Sundvoll, B., 1995, Caledonian granitoids in the Frøya-Froan area, central Norway: Norges Geologiske Undersøkelse Bulletin, v. 427, p. 48–51.

Nordgulen, Ø., Lindstrøm, M., Solli, A., Barnes, C.G., Sundvoll, B., and Karlsson, H.R., 2001, The Caledonian Smøla-Hitra Batholith, central Norway [abs.]: Geological Society of America Abstracts with Programs, Northeast Section, v. 33, no. 1, p. A32.

Nordgulen, Ø., Braathen, A., Corfu, F., Osmundsen, P.T., and Husmo, T., 2002, Polyphase kinematics and geochronology of the late-Caledonian Kollstraumen detachment, north-central Norway: Norwegian Journal of Geology, v. 82, p. 299–316.

Northrup, C.J., 1997, Timing, structural assembly, metamorphism, and cooling of Caledonian nappes in the Ofoten-Efjorden area, northern Norway: Tectonic insights from U-Pb and $^{40}Ar/^{39}Ar$ geochronology: The Journal of Geology, v. 105, p. 565–582.

Nystuen, J.P., 1985, Facies and preservation of glaciogenic sequences from the Varanger Ice Age in Scandinavia and other parts of the North Atlantic region: Palaeogeography, Palaeoclimatology, Palaeoecology, v. 51, p. 209–229, doi: 10.1016/0031-0182(85)90086-0.

Olaussen, S., 1976, Palaeozoic fossils from Troms, Norway: Norsk Geologisk Tidsskrift, v. 56, p. 457–459.

Oliver, G.J.H., and Krogh, T.E., 1995, U-Pb zircon age of 469 ± 5 Ma from a metatonalite from the Kjosen Unit of the Lyngen magmatic complex, northern Norway: Norges Geologiske Undersøkelse Bulletin, v. 428, p. 27–32.

Osmundsen, P.T., Braathen, A., Nordgulen, Ø., Roberts, D., Meyer, G.B., and Eide, E., 2003, The Devonian Nesna shear zone, north-central Norwegian Caledonides, and its regional implications: Geological Society [London] Journal, v. 160, p. 137–150.

Osmundsen, P.T., Eide, E.A., Haabesland, N.E., Roberts, D., Andersen, T.B., Kendrick, M., Bingen, B., Braathen, A., and Redfield, T.F., 2006, Kinematics of the Høybakken detachment zone and the Møre-Trøndelag fault complex, central Norway: Geological Society [London] Journal, v. 163, p. 303–318.

Pedersen, R.-B., and Dunning, G.R., 1997, Evolution of arc crust and relations between contrasting sources: U-Pb (age), Nd and Sr isotope systematics of the ophiolitic terrane of SW Norway: Contributions to Mineralogy and Petrology, v. 128, p. 1–15, doi: 10.1007/s004100050289.

Petersen, K., 1878, Det nordlige Sveriges og Norges geologi: Archiv for Mathematikk og Naturvidenskap: Christiania, v. 3, p. 1–38.

Prestvik, T., 1980, The Caledonian ophiolite complex of Leka, north-central Norway, *in* Panayiotou, A., ed., Ophiolites: Proceedings of an International Ophiolite Symposium, Cyprus 1979: Nicosia, Cyprus, Ministry of Agriculture and Natural Resources, Geological Survey Department, p. 555–566.

Ramberg, I., 1967, Kongsfjell-områdets geologi, en petrografisk og strukturell undersøkelse i Helgeland, Nord-Norge: Norges Geologiske Undersøkelse, v. 240, 152 p.

Reading, H.G., and Walker, R.G., 1966, Sedimentation of Eocambrian tillites and associated sediments in Finnmark, northern Norway: Palaeogeography, Palaeoclimatology, Palaeoecology, v. 2, p. 177–212, doi: 10.1016/0031-0182(66)90016-2.

Riis, F., and Ramberg, I.B., 1979, Nyere data fra Nordlands høymetamorfe dekkekompleks: Geolognytt, v. 13, p. 60–61.

Roberts, D., 1988, The terrane concept and the Scandinavian Caledonides: A synthesis: Norges Geologiske Undersøkelse Bulletin, v. 413, p. 93–99.

Roberts, D., 2003, The Scandinavian Caledonides: Event chronology, palaeogeographic settings and likely modern analogues: Tectonophysics, v. 365, p. 283–299, doi: 10.1016/S0040-1951(03)00026-X.

Roberts, D., and Gale, G.H., 1978, The Caledonian-Appalachian Iapetus Ocean, *in* Tarling, D.H., ed., Evolution of the Earth's Crust: New York, Academic Press, p. 255–342.

Roberts, D., and Gee, D.G., 1985, An introduction to the structure of the Scandinavian Caledonides, *in* Gee, D.G., and Sturt, B.A., eds., The Caledonide Orogen—Scandinavia and Related Areas: Chichester, John Wiley and Sons, p. 55–68.

Roberts, D., and Tucker, R.D., 1991, U-Pb zircon age of the Møklevatnet granodiorite, Gjersvik Nappe, central Norwegian Caledonides: Norges Geologiske Undersøkelse Bulletin, v. 421, p. 33–38.

Roberts, D., Sturt, B.A., and Furnes, H., 1985, Volcanite assemblages and environments in the Scandinavian Caledonides and the sequential development history of the mountain belt, *in* Gee, D.G., and Sturt, B.A., eds., The Caledonide Orogen—Scandinavia and Related Areas: Chichester, John Wiley and Sons, p. 919–930.

Roberts, D., Heldal, T., and Melezhik, V.A., 2001, Tectonic structural features of the Fauske conglomerates in the Løvgavlen quarry, Nordland, Norwegian Caledonides, and regional implications: Norsk Geologisk Tidsskrift, v. 81, p. 245–256.

Roberts, D., Melezhik, V.A., and Heldal, T., 2002, Carbonate formations and early NW-directed thrusting in the highest allochthons of the Norwegian Caledonides—Evidence of a Laurentian ancestry: Geological Society [London] Journal, v. 159, p. 1–5.

Roberts, D., Nordgulen, Ø., and Melezhik, V.A., 2004, The Uppermost Allochthon in the Scandinavian Caledonides: A terrane collage of Laurentian ancestry transported onto Baltica in Siluro-Devonian time [abs.], *in* Hatcher, R.D., Jr., Whisner, J.B., Thigpen, J.R., Whitmer, N.E., and Whisner, S.C., eds., 17th International Basement Tectonics Conference, June 2004, Oak Ridge, Tennessee, USA, p. 28–29.

Robinson, P., Tucker, R.D., Bradley, D., Berry, H.N., IV, and Osberg, P.H., 1998, Paleozoic orogens in New England, USA: Geologiska Föreningens i Stockholm Förhandlingar, v. 120, p. 119–148.

Rodgers, J., 1968, The eastern edge of the North American continent during the Cambrian and Early Ordovician, *in* Zen, E-An., White, W.S., Hadley, J.B., and Thompson, J.B., Jr., eds., Studies of Appalachian Geology—Northern and Maritime: New York, Wiley Interscience, p. 141–149.

Rutland, R.W.R., and Nicholson, R., 1965, Tectonics of the Caledonides of part of Nordland, Norway: Geological Society [London] Quarterly Journal, v. 121, p. 73–109.

Samuelsson, J., and Middleton, M.F., 1998, The Caledonian foreland basin in Scandinavia: Constrained by the thermal maturation of the Alum Shale: Geologiska Föreningens i Stockholm Förhandlingar v. 120, p. 307–314.

Sandøy, R., 2003, Geological Variations in Marble Deposits. The Geometry, Internal Structure and Geochemical Variations of the Industrial Mineral Marble Deposits in the Velfjord Area [Ph.D. thesis]: Trondheim, Norwegian University of Science and Technology, 534 p.

Scholle, P.A., Arthur, M.A., and Ekdale, A.A., 1983, Pelagic environments, *in* Scholle, P.A., Bebout, D.G., and Moore, C.H., eds., Carbonate Depositional Environments: Tulsa, Oklahoma, American Association of Petroleum Geologists, p. 619–691.

Selbekk, R.S., Skjerlie, K.P., and Pedersen, R.B., 2000, Generation of anorthositic magmas by H2O-fluxed anatexis of silica-undersaturated gabbro: An example from the north Norwegian Caledonides: Geological Magazine, v. 137, p. 609–621, doi: 10.1017/S0016756800004829.

Séranne, M., 1992, Late Palaeozoic kinematics of the Møre-Trøndelag fault zone and adjacent areas, central Norway: Norsk Geologisk Tidsskrift, v. 72, p. 141–158.

Skauli, H., Bjørlykke, A., and Thorpe, R.I., 1992, Lead-isotope study of the sulphide ore and alteration zone, Bleikvassli zinc-lead deposit, northern Norway: Mineralium Deposita, v. 27, p. 276–283.

Slagstad, T., Melezhik, V.A., Kirkland, C.L., Zwaan, K.B., Roberts, D., Gorokhov, I.M., and Fallick, A.E., 2005, Carbon isotope chemostratigraphy suggests revisions to the geological history of the West Finnmark Caledonides, northern Norway: Geological Society [London] Journal, v. 163, p. 277–289.

Smith, M.P., Rasmussen, J.A., Robertson, S., Higgins, A.K., and Leslie, A.G., 2006, Lower Palaeozoic stratigraphy of the East Greenland Caledonides: Geological Survey of Denmark and Greenland Bulletin v. 6, p. 5–28.

Solli, A., 1990, Saltstraumen, Berggrunnskart 2029-3, M 1:50,000: Norges Geologiske Undersøkelse, Trondheim.

Soper, N.J., Strachan, R.A., Holdsworth, R.E., Gayer, R.A., and Greiling, R.O., 1992, Sinistral transpression and the Silurian closure of Iapetus: Geological Society [London] Journal, v. 149, p. 871–880.

Søvegjarto, U., 1977, Kaledonske jernmalmer i Nordland, *in* Bjørlykke A., Lindahl, I., and Vokes, F.M., eds., Kaledoniske Malmforekomster: Trondheim, Bergverkenes Landssammenslutnings Industrigruppe's Tekniske Virksomheter, p. 95–105 (in Norwegian).

Spjeldnæs, N., 1985, Biostratigraphy of the Scandinavian Caledonides, *in* Gee, D.G., and Sturt, B.A., eds., The Caledonide Orogen—Scandinavia and Related Areas: Chichester, John Wiley and Sons, p. 317–329.

Stanley, R.S., and Ratcliffe, N.M., 1985, Tectonic synthesis of the Taconian orogeny in western New England: Geological Society of America Bulletin, v. 96, p. 1227–1250, doi: 10.1130/0016-7606(1985)96<1227:TSOTTO>2.0.CO;2.

Steel, R., Siedlecka, A., and Roberts, D., 1985, The Old Red Sandstone basins of Norway and their deformation: A review, *in* Gee, D.G. and Sturt, B.A., eds., The Caledonide Orogen—Scandinavia and Related Areas: Chichester, John Wiley and Sons, p. 293–315.

Steltenpohl, M.G., 1987, Tectonostratigraphy and tectonic evolution of the Skånland area, north Norway: Norges Geologiske Undersøkelse Bulletin, v. 409, p. 1–20.

Steltenpohl, M.G., Andresen, A., and Tull, J.M., 1990, Lithostratigraphic correlation of the Salangen (Ofoten) and Balsfjord (Troms) Groups: Evidence for post-Finnmarkian unconformity, north Norwegian Caledonides: Norges Geologiske Undersøkelse Bulletin, v. 418, p. 61–77.

Steltenpohl, M.G., Andresen, A., Lindstrøm, M., Gromet, P., and Steltenpohl, L.W., 2003, The role of felsic and mafic igneous rocks in deciphering the evolution of thrust-stacked terranes: An example from the north Norwegian Caledonides: American Journal of Science, v. 303, p. 149–185, doi: 10.2475/ajs.303.2.149.

Stephens, M.B., and Gee, D.G., 1985, A tectonic model for the evolution of the eugeoclinal terranes in the central Scandinavian Caledonides, *in* Gee, D.G., and Sturt, B.A., eds., The Caledonide Orogen—Scandinavia and Related Areas: Chichester, John Wiley and Sons, p. 953–970.

Stephens, M.B., and Gee, D.G., 1989, Terranes and polyphase accretionary history in the Scandinavian Caledonides, *in* Dallmeyer, R.D., ed., Terranes in the Circum-Atlantic Paelozoic Orogens: Geological Society of America Special Paper 230, p. 17–30.

Stephens, M.B., Gustavson, M., Ramberg, I.B., and Zachrisson, E., 1985a, The Caledonides of central-north Scandinavia—A tectonostratigraphic overview, *in* Gee, D.G., and Sturt, B.A., eds., The Caledonide Orogen—Scandinavia and Related Areas: Chichester, John Wiley and Sons, p. 135–162.

Stephens, M.B., Furnes, H., Robins, B., and Sturt, B.A., 1985b, Igneous activity within the Scandinavian Caledonides, *in* Gee, D.G., and Sturt, B.A., eds., The Caledonide Orogen—Scandinavia and Related Areas: Chichester, John Wiley & Sons, p. 623–652.

Stephens, M.B., Kullerud, K., and Claesson, S., 1993, Early Caledonian tectonothermal evolution in outboard terranes, central Scandinavian Caledonides: New constraints from U-Pb zircon dates: Geological Society [London] Journal, v. 150, p. 51–56.

Strachan, R.A., Holdsworth, R.E., Friderichson, J.D., and Jepson, H.F., 1992, Regional Caledonian structure within an oblique convergence zone, Dronning Louise Land, NE Greenland: Geological Society [London] Journal, v. 149, p. 359–371.

Sturt, B.A., and Roberts, D., 1991, Tectonostratigraphic relationships and obduction histories of Scandinavian ophiolitic terranes, *in* Peters, T., ed., Ophiolite Genesis and Evolution of the Ocean Lithosphere: Dordrecht, Kluwer Academic Publishers, p. 745–769.

Sturt, B.A., Andersen, T.B., and Furnes, H., 1985, The Skei Group, Leka: An unconformable clastic sequence overlying the Leka Ophiolite, *in* Gee, D.G. and Sturt, B.A., eds., The Caledonide Orogen—Scandinavia and Related Areas: Chichester, John Wiley and Sons, p. 395–405.

Swett, K., and Smit, D.E., 1972, Cambro-Ordovician shelf sedimentation of western Newfoundland, northwest Scotland and central East Greenland: Proceedings of the 24th International Geological Congress, Canada, v. 6, p. 33–41.

Terry, M.P., Robinson, P., Hamilton, M.A., and Jercinovic, M.J., 2000, Monazite geochronology of UHP and HP metamorphism, deformation and exhumation, Nordøyane, Western Gneiss region, Norway: The American Mineralogist, v. 85, p. 1651–1664.

Thorsnes, T., and Løseth, H., 1991, Tectonostratigraphy in the Velfjord-Tosen region, southwestern part of the Helgeland Nappe Complex, central Norwegian Caledonides: Norges Geologiske Undersøkelse Bulletin, v. 421, p. 1–18.

Törnebohm, A.E., 1888, Om fjällproblemet: Geologiska Föreningens i Stockholm Förhandlingar, v. 10, p. 328–336.

Törnebohm, A.E., 1896, Grunddragen av det centrala skandinaviens bergbyggnad: Kongelige Svenska Vetenskaps Akademiets Handlingar, v. 28, 212 p.

Torsvik, T.H., Smethurst, M., Meert, J., Van der Voo, R., McKerrow, W.S., Brazier, M., Sturt, B.A., and Walderhaug, H., 1996, Continental break-up and collision in the Neoproterozoic and Palaeozoic—A tale of Baltica and Laurentia: Earth-Science Reviews, v. 40, p. 229–258, doi: 10.1016/0012-8252(96)00008-6.

Tørudbakken, B.O., and Brattli, B., 1985, Ages of metamorphic and deformational events in the Beiarn Nappe Complex, Nordland: Norges Geologiske Undersøkelse Bulletin, v. 399, p. 27–39.

Tørudbakken, B.O., and Mickelson, M., 1986, A Rb-Sr age study from the Mosjøen unit, Helgeland Nappe Complex, and its bearing on the timing of tectonometamorphic events within the Uppermost Allochthon, central Scandinavian Caledonides, Norway: Norsk Geologisk Tidsskrift, v. 66, p. 263–270.

Tucker, R.D., Robinson, P., Solli, A., Gee, D.G., Thorsnes, T., Nordgulen, Ø., Krogh, T.E., and Bickford, M.E., 2004, Thrusting and extension in the Scandinavian hinterland, Norway: New U-Pb ages and tectonostratigraphic evidence: American Journal of Science, v. 304, p. 477–532, doi: 10.2475/ajs.304.6.477.

Tull, J.F., Bartley, J.M., Hodges, K.V., Andresen, A., Steltenpohl, M.G., and White, J.M., 1985, The Caledonides in the Ofoten region (68°–69°N), north Norway: Key aspects of tectonic evolution, *in* Gee, D.G., and Sturt, B.A., eds., The Caledonide Orogen—Scandinavia and Related Areas: Chichester, John Wiley and Sons, p. 553–568.

van Staal, C.R., Dewey, J.F., Mac Niocaill, C., and McKerrow, W.S., 1998, The Cambrian-Silurian Tectonic Evolution of the Northern Appalachians and British Caledonides: History of a Complex, West and Southwest Pacific-Type Segment of Iapetus: Geological Society [London] Special Publication 143, p. 199–242.

Waldron, J.W.F., and van Staal, C.R., 2001, Taconian orogeny and the accretion of the Dashwoods block: A peri-Laurentian microcontinent in the Iapetus Ocean: Geology, v. 29, p. 811–814, doi: 10.1130/0091-7613(2001)029<0811:TOATAO>2.0.CO;2.

Whalen, J.B., Jenner, G.A., Longstaffe, F.J., Gariépy, C., and Fryer, B.J., 1997, Implications of granitoid geochemical and isotopic (Nd, O, Pb) data from the Cambrian-Ordovician Notre Dame arc for the evolution of the Central Mobile belt, Newfoundland Appalachians, *in* Sihna, A.K., Whalen, J.B., and Hogan, J.P., eds., The Nature of Magmatism in the Appalachian Orogen: Geological Society of America Memoir 191, p. 367–395.

Whitehead, J., Dunning, G.R., and Spray, J.G., 2000, U-Pb geochronology and origin of granitoid rocks in the Thetford Mines ophiolite, Canadian Appalachians: Geological Society of America Bulletin, v. 112, p. 915–928, doi: 10.1130/0016-7606(2000)112<0915:UPGAOO>2.3.CO;2.

Williams, H., 1995, Summary and overview; Chapter 11, in Williams, H., ed., Geology of the Appalachian-Caledonian Orogen in Canada and Greenland: Geological Survey of Canada, Geology of Canada, v. 6, p. 843–890.

Yoshinobu, A., Barnes, C.G., Nordgulen, Ø., Prestvik, T., Fanning, M., and Pedersen, R.-B., 2002, Ordovician magmatism, deformation, and exhumation in the Caledonides of central Norway: An orphan of the Taconic orogeny: Geology, v. 30, p. 883–886, doi: 10.1130/0091-7613(2002)030<0883:OMDAEI>2.0.CO;2.

Young, G.M., 2002, Stratigraphic and tectonic settings of Proterozoic glaciogenic rocks and banded iron-formations; relevance to the snowball Earth debate: Journal of African Earth Sciences and the Middle East, v. 35, p. 451–466, doi: 10.1016/S0899-5362(02)00158-6.

Zwaan, K.B., Fareth, E., and Grogan, P.W., 1998, Geologisk kart over Norge, Berggrunnskart Tromsø, M 1:250,000: Norges Geologiske Undersøkelse, Trondheim.

Manuscript Accepted by the Society 1 February 2007

The Geological Society of America
Memoir 200
2007

Gondwana-derived terranes in the northern Hellenides

F. Himmerkus*
B. Anders
Institut für Geowissenschaften, Johannes Gutenberg-Universität, Graduiertenkolleg: Stoffbestand und Entwicklung von Kruste und Mantel, Becherweg 21, 55099 Mainz, Germany, and Max-Planck-Institut für Chemie, Abteilung Geochemie, 55099 Mainz, Germany

T. Reischmann
Max-Planck-Institut für Chemie, Abteilung Geochemie, 55099 Mainz, Germany

D. Kostopoulos
Faculty of Geology and GeoEnvironment, Department of Mineralogy and Petrology, National and Kapodistrian University of Athens, Panepistimioupoli Zographou, Athens 15784, Greece

ABSTRACT

The Hellenides constitute an integral part of the Alpine orogenic system in southeast Europe. Despite the recognition of several subparallel zones, which are interpreted as terranes separated by ophiolitic sutures (e.g., Pindos and Vardar sutures), the classical view of an orogen with a foreland fold-and-thrust belt, a central crystalline zone, and a rather undeformed hinterland is still under discussion. This paper concentrates on basement terranes of exotic provenance in two of the internal zones of the Hellenides that support the interpretation of the Hellenides as an accretionary orogen formed by amalgamation of crustal segments during the subduction of Tethyan oceanic basins. The oldest of these terranes, the Florina terrane in the Pelagonian zone, is composed of Neoproterozoic arc-related orthogneisses. Two other exotic terranes occur east of the Vardar zone within the Serbo-Macedonian Massif. The Pirgadikia terrane is a microterrane in the southern Chalkidiki Peninsula that consists of Pan-African mylonitic orthogneisses with volcanic arc–related trace-element geochemistry and Sr isotopic composition. The Vertiskos terrane occupies the northwestern part of the Serbo-Macedonian Massif and is primarily composed of coarse-grained, volcanic arc–related peraluminous orthogneisses of Silurian age. These terranes are exotic in relation to the other parts of the Hellenides. The provenance of the late Proterozoic Pan-African Florina and Pirgadikia terranes is assumed to be Gondwanan, whereas the Silurian Vertiskos terrane may have been part of the so-called Hun terrane, which formed at the northern active continental margin of Gondwana in the early Paleozoic.

Keywords: Geochronology, Internal Hellenides, Greece, Gondwana-derived terranes.

*Felix.Himmerkus@hdb.fzk.de

Himmerkus, F., Anders, B., Reischmann, T., and Kostopoulos, D., 2007, Gondwana-derived terranes in the northern Hellenides, *in* Hatcher, R.D., Jr., Carlson, M.P., McBride, J.H., and Martínez Catalán, J.R., eds., 4-D Framework of Continental Crust: Geological Society of America Memoir 200, p. 379–390, doi: 10.1130/2007.1200(19). For permission to copy, contact editing@geosociety.org.

INTRODUCTION

After the establishment of the paradigm of plate tectonics in the 1960s, the pace of orogenic processes came to be viewed in terms of the Wilson cycle (e.g., Miyashiro et al., 1982). In this scenario, an orogenic cycle was defined as the opening and closure of an ocean, starting with rifting, followed by seafloor spreading, and ending with subduction and, eventually, continental collision. However, this cycle does not take into account the preexisting geology of the plates involved or of movements parallel to the plate boundaries. The recognition of smaller fault-bounded crustal blocks, or terranes, led to the refinement of ideas on orogeny. The terrane concept was defined in the North American Cordillera, where similar crustal segments that occur along the strike of the orogen were recognized as terranes that had been accreted to North America by eastward subduction during the Mesozoic and later displaced by major right-lateral faults (e.g., Coney et al., 1980). A terrane is defined as a distinct fault-bounded crustal unit with a geological evolution that is significantly different from that of neighboring regions. In the case of a terrane made up of mainly sedimentary sequences, differences in facies and faunal diversity can be used to distinguish it from other terranes. In the case of crystalline basement units, geochemical and isotopic characteristics, crystallization ages, the age and grade of metamorphism and deformation, and paleomagnetic approaches can be used for terrane discrimination.

This paper describes the basement terranes in the northern Hellenides of Greece. Crystallization ages and the geochemical and isotopic characteristics of the basement rocks are used in conjunction with structural information to identify different terranes and their origin in the internal part of the Hellenides and, thereby, constrain the pre-Alpine to early Alpine history of the Hellenic orogen.

REGIONAL GEOLOGY

The Hellenides are an Alpine mountain chain in Greece, and they constitute the link between the Dinarides/Albanides to the northwest and the Taurides to the southeast in Asia Minor. This orogenic belt formed as a result of the closure of the different branches/oceans of the Tethys Ocean and the resultant accretion of the variably sized continental fragments that had been separated by these branches. The Hellenides form an elongate and strongly arcuate orogen, the present appearance of which is influenced by large-scale extension due to the retreat of the Hellenic subduction zone since the Miocene (Le Pichon and Angelier, 1979, 1981; Le Pichon et al., 1995; Gautier and Brun, 1994; Gautier et al., 1999; Jolivet, 2001; Jolivet et al., 2003; van Hinsbergen et al., 2005). This extension resulted in submergence of large parts of the central and internal units of the Hellenides in the Aegean Sea, and the exhumation of metamorphic core complexes and gneiss domes along flat-lying detachment faults (Lister et al., 1984; Brun and Sokoutis, 2004).

Over the past 25 yr, the Hellenides have become the focus of scientific attention mainly because of the well-preserved high-pressure and ultrahigh-pressure rocks in the metamorphic units of the Internal Hellenides (Altherr, 1979; Jacobshagen, 1986; Kostopoulos et al., 2000).

Scientific interest has also focused on neotectonics, which monitors the ongoing tectonic processes in the retreating Aegean subduction zone, as well as the extending continental crust in the Aegean Sea and along the North Anatolian fault zone (Le Pichon et al., 1995; Davies et al., 1997; Reilinger et al., 1997; Kahle et al., 2000; Jolivet, 2001; Royden and Papanikolaou, 2004). The pre-Alpine history of the Hellenides, however, has remained largely untouched. It is nevertheless an important field for research in order to establish a geological framework for the evolution of the Hellenides and for paleogeographic reconstructions of the Eastern Mediterranean. Our study contributes to this field of research.

The Hellenides trend approximately NNW-SSE and can be subdivided into several subparallel units or terranes separated by large fault zones. (Jacobshagen, 1986, and references therein; Papanikolaou, 1997). These units are grouped in two large zones: the External Hellenides, composed of mainly supracrustal rocks, and the Internal Hellenides, made up of basement units that include the Pelagonian zone, the Attico-Cycladic Massif (Dürr et al., 1978), the Vardar zone, the Serbo-Macedonian Massif, and the Rhodope Massif.

The westernmost zone of the Hellenides is the External Hellenide Platform (Fig. 1B), which includes the Paxon, Ionian, and Gavrovo-Tripolis zones and consists mainly of thick neritic Mesozoic carbonates and lesser clastic sediments that represent parts of the passive continental margin of southern Tethys (Apulian promontory of northern Gondwana).

This platform is separated from the Pelagonian zone to the east by remnants of the Pindos Ocean, which are composed of Mesozoic ophiolites and their accompanying pelagic sedimentary cover (Brun, 1956; Kostopoulos, 1988).

The Pelagonian zone is composed of crystalline basement (schists, orthogneisses, and paragneisses, amphibolites, and granitoids) in addition to Permian-Triassic volcano-sedimentary units, Mesozoic limestones, ophiolitic rocks (sensu latu), and abundant Tertiary to Quaternary sediments (e.g., Mountrakis, 1982, 1984; Kilias and Mountrakis, 1989; Jacobshagen, 1986, and references therein; Pe-Piper and Piper, 2002, and references therein). Geochronology of the basement rocks has revealed widespread Permian-Carboniferous magmatism with ages in the range of ca. 280–320 Ma (Yarwood and Aftalion, 1976; Schermer, 1990; Koroneos et al., 1993; Vavassis et al., 2000; Reischmann et al., 2001; Anders et al., 2003b, 2006). These ages are similar to crystallization ages found in the Attico-Cycladic Massif (Engel and Reischmann, 1998; Reischmann, 1998), which borders the Pelagonian zone to the south. Because of this similarity, the Pelagonian zone and the Attico-Cycladic Massif are interpreted as parts of a large Permian-Carboniferous magmatic arc (Reischmann et al., 2001). In the northwestern Pelagonian zone, pre–Permian-Carboniferous basement was inferred by Mountrakis (1984) on the basis of field observations. Recently, crystalline basement of Late Proterozoic age was identified by Anders et al. (2006) and interpreted as remnants of an exotic terrane.

East of the Pelagonian zone, there is the ophiolitic Axios-Vardar zone (Figs. 1B and 3), which is interpreted as a major suture zone on the basis of outcrops farther north in former Yugoslavia (Robertson and Karamata, 1994, and references therein; Dimitrijevic, 1974, 1997; Knezevic-Dordevic and Krstic, 1996). The Axios-Vardar zone has also been equated to the Izmir-Ankara suture zone farther east in Asia Minor (e.g., Stampfli et al., 2004).

The so-called Hellenic hinterland is made up of two discrete massifs east of the Vardar zone, namely the Serbo-Macedonian Massif and the Rhodope Massif. The Serbo-Macedonian Massif is an elongated polymetamorphic basement terrane that largely consists of orthogneisses (Kockel et al., 1977; Jacobshagen, 1986; Kilias et al., 1999). Metasedimentary rocks and amphibolites like those of the now-abandoned Circum-Rhodope belt (Kauffmann

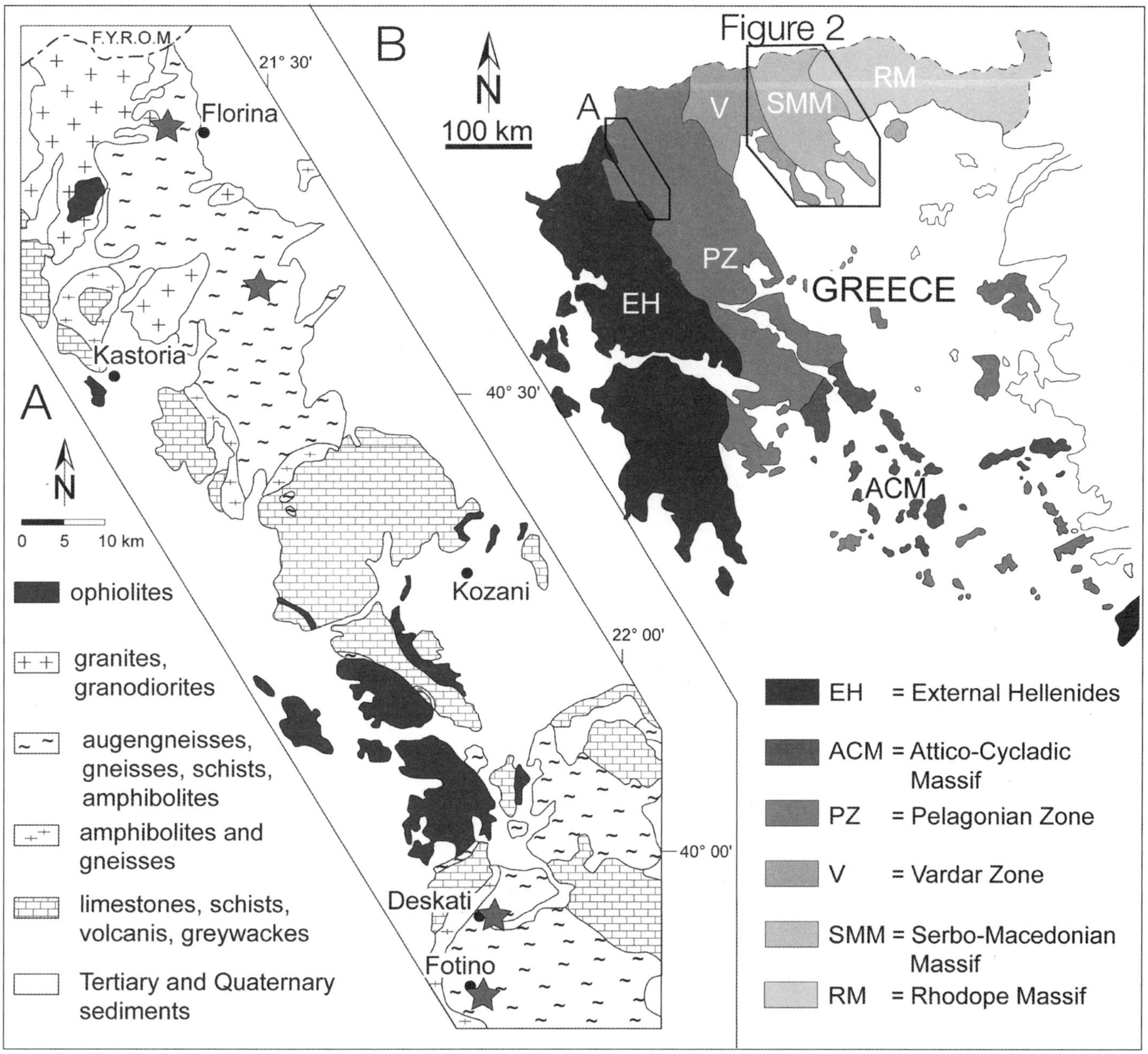

Figure 1. (A) Geological map of the northwestern part of the Greek Pelagonian zone (simplified from Institute of Geology and Mineral Exploration [1983] Geological Map of Greece). Occurrences of late Neoproterozoic rocks are indicated with stars. (B) Tectonostratigraphic overview of the Hellenides in Greece and the adjacent areas (modified after the IGME [1983] Geological Map of Greece). The External Hellenides mainly consist of metasediments, whereas the Internal Hellenides are made up of several subparallel zones of basement rocks.

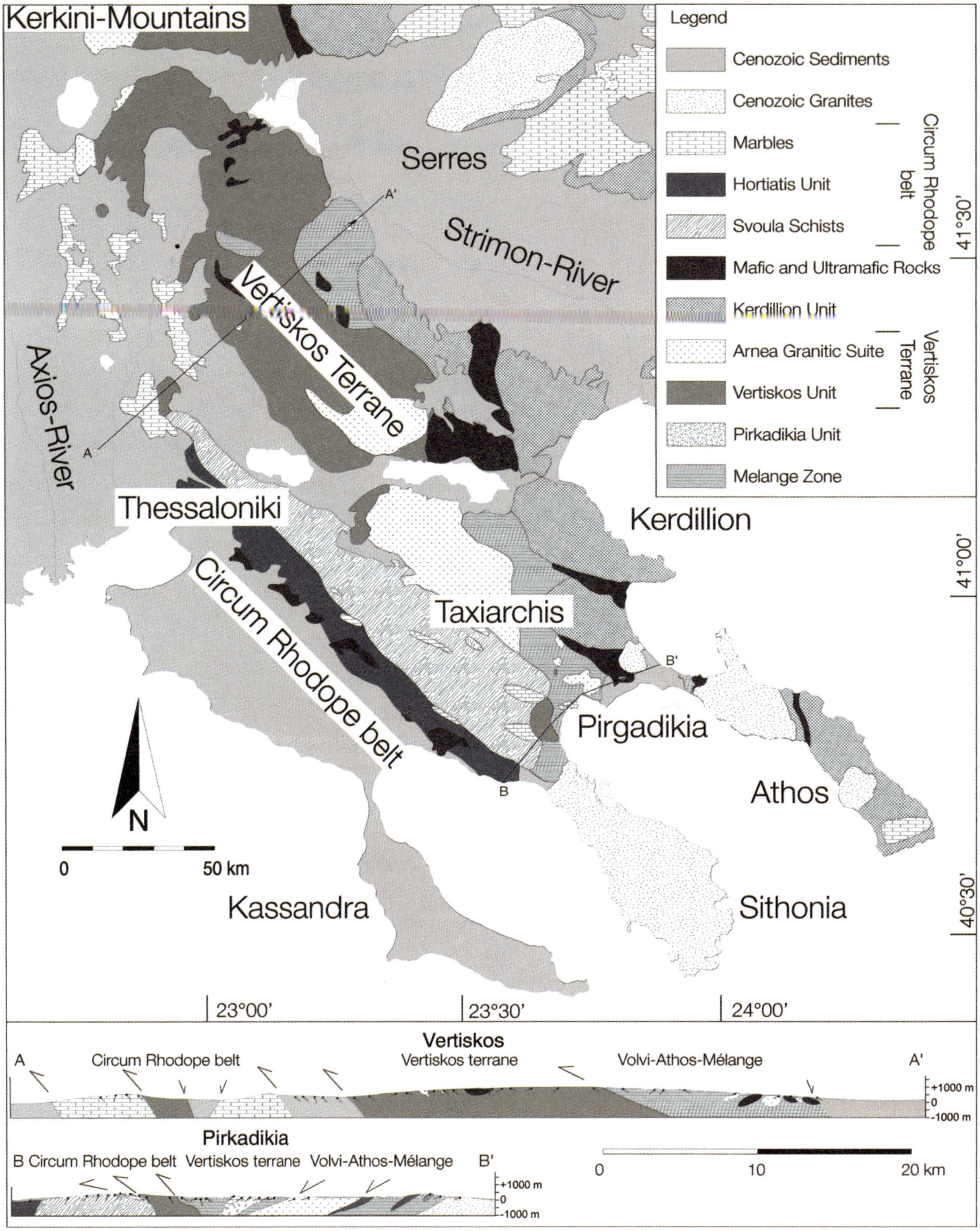

Figure 2. Geological-tectonic map of the Greek Serbo-Macedonian Massif with interpretative cross sections (simplified after Kockel et al., 1977). Scale of sections is five times that of map. The Circum-Rhodope belt (Kauffmann et al., 1976) is the eastern part of the mélange zone bordering the Serbo-Macedonian Massif and the adjacent Vardar zone.

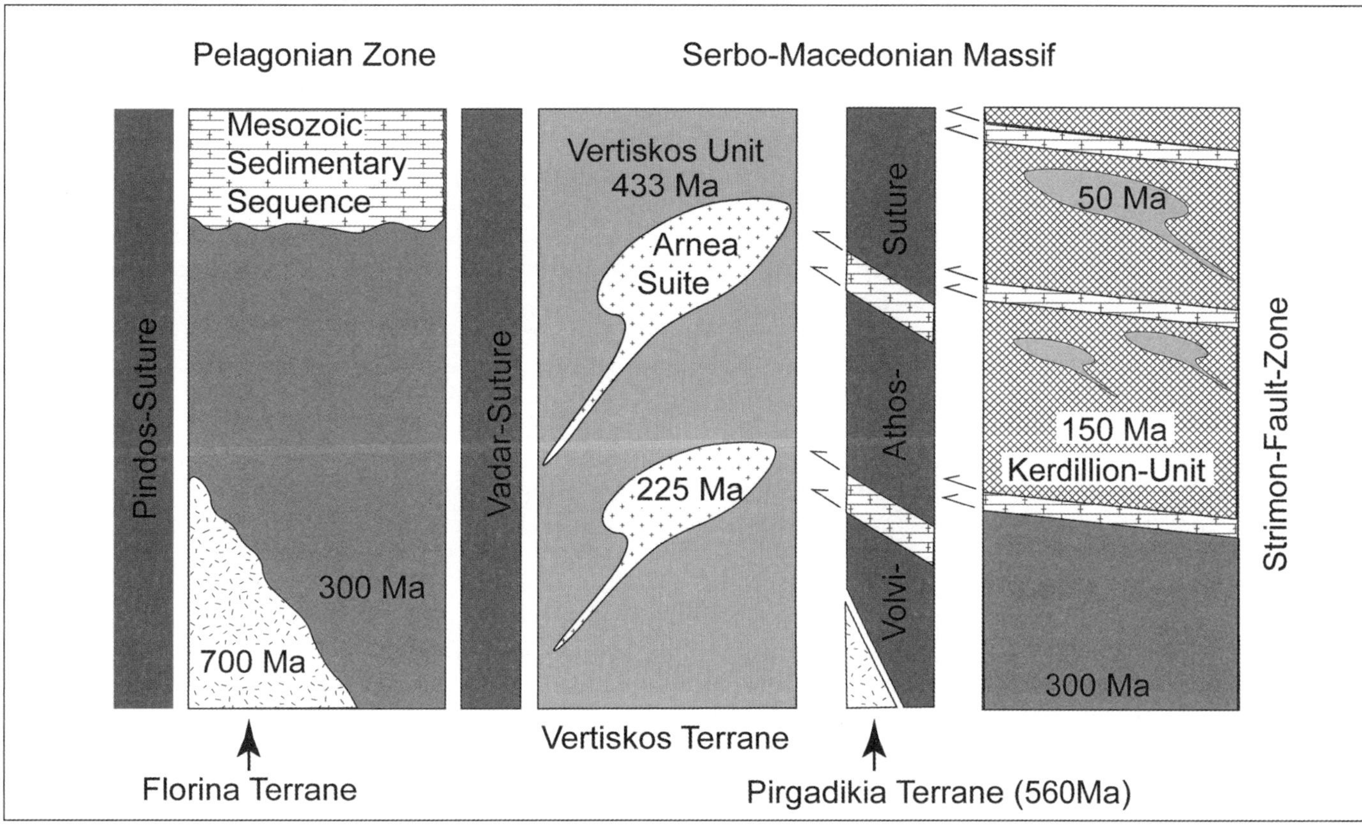

Figure 3. Tectonostratigraphic columns for units of the Pelagonian zone and the Serbo-Macedonian Massif in the Internal Hellenides. Basement units are separated by mélange zones containing ophiolitic material. The Kerdillion Unit is related to the adjacent Rhodope Massif east of the Strimon fault zone, which shows the same spectrum of ages. The Vertiskos terrane is composed of two lithological units (the Silurian Vertiskos Unit and the Triassic Arnea Unit), both of which are exotic to the Internal Hellenides, as is the case for the late Neoproterozoic terranes of Florina and Pirgadikia.

et al., 1976; Ricou et al., 1998) occur only in mélange zones bordering the basement complexes.

The Serbo-Macedonian Massif consists of four main basement units: the Pan-African Pirgadikia Unit, the Silurian Vertiskos Unit, the Triassic Arnea Unit, and the Upper Jurassic Kerdillion Unit. These basement units are separated by two major mélange zones (Fig. 2) that contain mafic and ultramafic rocks as well as marbles and clastic metasediments and slivers of the adjacent basement units. The northwestern part of the Serbo-Macedonian Massif is characterized by the association of Silurian augen gneisses of the Vertiskos Unit with Triassic leucogranites of the Arnea Unit. The Kerdillion Unit (Kockel et al., 1971) in the eastern part of the massif is lithologically similar to the adjacent Rhodope Massif and consists of banded biotite gneisses that are intruded by a large amount of leucocratic dikes and pegmatites. For this reason the gneisses have traditionally been regarded as migmatites. This unit is exposed in three large gneiss domes.

The Pirgadikia Unit, in the south of the Serbo-Macedonian Massif, occurs within the ophiolitic mélange zone that separates the Vertiskos Unit from the Vardar zone. This unit is composed of leucocratic mylonitic gneisses that originated in the late Neoproterozoic in a magmatic-arc setting.

The Rhodope Massif is the easternmost unit of the Hellenides and consists of an association of granitic gneisses, metapelites, eclogites, amphibolites, and very thick units of marbles (Burg et al., 1993, 1996; Ricou et al., 1998; Barr et al., 1999). It is situated east of the Strimon River Basin (Fig. 2) and is made up of a pile of nappes. Two principal units have been distinguished (Papanikolaou and Panagopoulos, 1981) as a lower tectonic unit composed of Permian-Carboniferous basement overlain by massive marbles of considerable thickness, and an upper tectonic unit containing Upper Jurassic gneisses, like those of the Kerdillion Unit, and Tertiary granites (Turpaud and Reischmann, 2003).

DESCRIPTION OF TERRANES IN THE INTERNAL HELLENIDES

Within the Internal Hellenides, three exotic terranes (Florina, Pirgadikia, and Vertiskos) have recently been identified. Since these areas include crystalline basement rocks, terrane identification has been accomplished using zircon geochronology and whole-rock isotope geochemistry combined with field work and petrographic studies.

The Florina Terrane

The westernmost terrane so far identified within the Internal Hellenides is the Florina terrane, situated in the northwestern part of the Greek Pelagonian zone. Recent geochronological studies of the crystalline basement have shown that whereas Permian-Carboniferous magmatism predominates in the Pelagonian zone, a well-defined group of orthogneisses and granites with Neoproterozoic (ca. 700 Ma) crystallization ages also exists in its northwestern part (Fig. 1) (Anders et al., 2003a, 2006). The Neoproterozoic basement rocks show a narrow age range between 699 Ma and 713 Ma (single-zircon U-Pb thermal ionization and secondary ion mass spectrometry [TIMS and SIMS] geochronology; Anders et al., 2006), and they are interpreted to have formed during the same magmatic event. The identification of pre–Permian-Carboniferous basement supports the theory of Mountrakis (1984), who proposed the existence of early Paleozoic or even older rocks in the Pelagonian zone on the basis of intrusive contacts and contact metamorphic relations between the Permian-Carboniferous Kastoria granite and the rocks that surround it. Mountrakis (1986) considered this pre–Permian-Carboniferous basement to be part of the Cimmerian continent, which rifted away from Gondwana in the Triassic. His proposal therefore implied the existence of far-traveled crustal units in the Pelagonian zone.

The Neoproterozoic basement rocks vary in color, grain size, degree of deformation, and development of feldspar "augen," and they are indistinguishable from the neighboring Permian-Carboniferous granites and gneisses. Based on crystallization ages, however, these rocks are clearly distinct from all other basement rocks of the Internal Hellenides. Consequently, Anders et al. (2006) proposed that these gneisses are remnants of a terrane. Geochemically, the Florina terrane basement rocks have the composition of felsic peraluminous granites with volcanic-arc/syn-collisional characteristics. Their ε_{Nd_i} values vary between +0.9 and −4.8, and Nd model ages (T_{DM}) range from 1.2 to 1.7 Ga, consistent with their formation in an active continental-margin setting (Anders et al., 2007).

The spatial distribution of the Neoproterozoic basement rocks is not well known. Their northernmost exposure lies west of Florina close to the Permian-Carboniferous Varnous pluton (Koroneos et al., 1993; Anders et al., 2007), while their southernmost outcrop is the Fotino granite south of Fotino village (Fig. 1A). The Neoproterozoic rocks are thus distributed across an area of ~20 × 100 km. Based on the chemical and mineralogical similarities, Anders et al. (2007) considered the Neoproterozoic rocks to constitute a single basement unit dissected during later tectonic events. Inherited components in zircon grains with ages of ca. 700 Ma were also identified in Pelagonian orthogneisses farther east (south of Veria) and southeast (close to Verdikoussa village) (localities not shown in Fig. 1A).

The Pirgadikia Terrane

This basement unit is a small terrane exposed in southeastern Chalkidiki north of the Sithonia Peninsula (Fig. 2). The rocks are mylonitic gneisses with two major occurrences at the shore near the village of Pirgadikia and around the village of Taxiarchis (Himmerkus et al., 2004a, 2006). Both outcrop areas are situated within a major mélange zone bordering the Vertiskos terrane to the southwest (Fig. 2). This mélange zone was formerly known as the Circum-Rhodope belt (Kauffmann et al., 1976; Kockel et al., 1977), and it is composed of low-grade metasedimentary rocks (Svoula Schist Formation), mafic and ultramafic rocks of the Hortiatis Unit (Thessaloniki and Sithonia ophiolites; e.g., Mussallam and Jung, 1986; Robertson, 2002), as well as gneisses from the adjacent basement terranes. The Mesozoic age of the metasediments incorporated in the succession and crystallization ages from the meta-igneous rocks (gneisses and amphibolites) indicate an Upper Jurassic age (Cimmerian) for the accretion.

The Pirgadikia gneisses form tectonic slivers floating within the metasedimentary rocks of the mélange zone. The individual basement exposures are less than one square kilometer in size. Similar gneissic rocks form other small exposures, but the number and size of these gneiss inliers remain unknown since good outcrops are rare. Nevertheless, the Pirgadikia gneisses have been identified over an area of ~10 × 30 km.

The gneisses near Taxiarchis are mylonitic quartzites, whereas those of Pirgadikia are leucocratic orthogneisses characterized by a strong lineation and grain-size reduction due to mylonitic deformation. The rocks on the Pirgadikia shore are in tectonic contact with massive marbles to the north and south. Two of the orthogneisses gave ages of 570 and 587 Ma (Himmerkus et al., 2006) by the single-zircon Pb-Pb evaporation method (Kober, 1986, 1987). The $^{87}Sr/^{86}Sr$ initial ratios calculated using the zircon ages are 0.70644 and 0.70734, respectively. The ε_{Nd_i} values are −7.59 and −6.24, respectively, which shows a strong contribution from preexisting crustal material in the source. The metaquartzite contains zircons with a uniform age of 555 Ma, indicating proximal sedimentation. The rocks of the Pirgadikia terrane differ from their country rocks in both rock type and metamorphic grade as well as in structural characteristics. The entire Serbo-Macedonian Massif and the western Rhodope Massif show a very consistent top-to-the-SW sense of shear (Burg et al., 1990, 1995; Kilias et al., 1999), which is also present in the metasedimentary and orthogneissic country rocks of the Pirgadikia terrane. By contrast, the mylonitic gneisses of Pirgadikia show top-to-the-E sense of shear. All contacts with the country rocks of the Vertiskos Unit and the mélange zone are tectonic and so do not permit correlations.

The basement unit of Pirgadikia is small but tectonically important since it relates the Greek part of the Serbo-Macedonian Massif to similar units of the Bulgarian Serbo-Macedonian Massif and the Bolu and Menderes Massifs in Turkey (see following).

The Vertiskos Terrane

The Vertiskos terrane occurs in the northwestern part of the Serbo-Macedonian Massif and is named after the Vertiskos Mountains northeast of Thessaloniki (Fig. 2). Kockel et al. (1977) defined the Vertiskos Unit as one of the principal units of the Serbo-Macedonian Massif based on its lithologies, which are mostly augen gneisses. These basement gneisses are intruded by a series of leucocratic granites termed the Arnea Granite Suite (AGS) after the main body, the Arnea Granite, east of Thessaloniki (Himmerkus et al., 2004b). The association of these two rock units defines the Vertiskos terrane, which is bordered by two large-scale shear zones interpreted to be mélange zones: the Vardar zone (including the Circum-Rhodope belt of Kauffmann et al., 1976) to the west, and the Athos-Volvi suture zone (TVG zone of Dixon and Dimitriadis, 1984) to the east (Himmerkus et al., 2005). The orthogneisses of the Vertiskos terrane form several lithological groups according to their mineralogy and texture, the most prominent of which consists of very coarse-grained augen gneisses with K-feldspars up to 10 cm in size. These rocks are mainly biotite gneisses with minor white mica, and they record a strong noncoaxial deformation. A second group is composed of fine-grained two-mica gneisses with small augen and a prominent C/S fabric. The third major group is composed of leucocratic muscovite gneisses that represent the most fractionated rocks of the succession. The rocks are peraluminous and classify as volcanic-arc granites in the discriminant diagrams of Pearce et al. (1984). The ages of the orthogneisses of the Vertiskos terrane are between 425.9 ± 4.2 Ma and 443.4 ± 5.5 Ma, with a mean of 432.2 Ma based on single-zircon Pb-Pb evaporation and conventional U-Pb dating of 22 samples (Himmerkus et al., 2002, 2003). Cathodoluminescence imaging of representative zircon grains shows clear magmatic zoning, suggesting that these are primary crystallization ages of the granitic precursor of the gneisses. The whole-rock $^{87}Sr/^{86}Sr$ initial ratio is 0.70924, indicating a significant contribution from preexisting continental crust in the source region of the granites. The individual samples also form a Rb-Sr error chron with an age of 430 ± 24 Ma (mean square of weighted deviates [MSWD] = 28) that is consistent with the zircon age. The ε_{Nd_i} values for two coarse-grained augen gneisses (–4.61 and –6.46) also suggest a strong contribution from old crustal material. These data show that the orthogneisses of the Vertiskos terrane originated in a Silurian magmatic-arc environment that was built on preexisting continental crust, a situation comparable to the modern Andean active continental margin.

PROVENANCE OF TERRANES IN THE INTERNAL HELLENIDES

The Precambrian and Silurian crystallization ages from the three terranes identified in the Internal Hellenides differ significantly from those of the surrounding basement units, which are Permian-Carboniferous or even Mesozoic (Yarwood and Aftalion, 1976; Engel and Reischmann, 1998; Reischmann, 1998; Vavassis et al., 2000; Reischmann et al., 2001; Turpaud and Reischmann, 2003; Anders, 2005). Two of these terranes (Vertiskos and Pirgadikia) form crustal blocks with distinct tectono-metamorphic histories separated by major faults that, in places, contain ophiolites (e.g., Mussallam and Jung, 1986; Dixon and Dimitriadis, 1984). They are therefore interpreted as exotic terranes accreted to the Hellenic orogen, a notion supported by their close proximity to the Vardar zone.

For the Florina terrane, there is no clear definition of its boundaries and contacts with the surrounding rocks. The crystallization ages of ca. 700 Ma, however, preclude a derivation from the neighboring rock units and, given their uniqueness in the Hellenides, support the interpretation of the Florina terrane as a far-traveled exotic block (Anders et al., 2007).

The fact that these three crustal units are exotic to their present location raises the question as to where they originated. The late Neoproterozoic crystallization ages in the Florina terrane of the Pelagonian zone suggest that it is derived from the northern margin of Gondwana, as are the Central European Variscides. An origin from Baltica cannot be entirely ruled out, but rocks with appropriate ages occur on the side of Baltica that was not close to Florina (e.g., Glasmacher et al., 1999). However, the question remains as to which part of the northern Gondwana margin represents the most likely provenance. Anders et al. (2007) pointed out that an affinity to East Avalonia is indicated by the Nd isotopic data and inherited zircon components, which indicate the presence of Proterozoic contributions (ca. 900–1330, 1500, and 1780 Ma). Geochronological and geochemical studies have identified several Avalonia-type crustal units in central Europe (e.g., Finger et al., 2000; Hegner and Kröner, 2000; Friedl et al., 2000). The Florina terrane can be compared with these Avalonia-type terranes.

The age and the chemical and isotopic characteristics of the Pirgadikia terrane are not unique in the Eastern Mediterranean. Similar Pan-African ages and isotopic characteristics of old recycled crust have been reported from basement units in the Serbo-Macedonian Massif of Bulgaria and in the Menderes Massif in western Turkey (Hetzel and Reischmann, 1996; Dannat and Reischmann, 1997; Loos and Reischmann, 1999). Similar ages were recently published for the Bolu Massif (Istanbul zone) in the western Pontides of northern Turkey (Ustaömer et al., 2005; Chen et al., 2002). The late Precambrian gneisses in the Serbo-Macedonian Massif of western Bulgaria form a domal structure known as the Osogovo-Lisets dome (Graf, 2001; Neubauer, 2002). Crystallization ages range from 545.1 ± 6.4 Ma to 568 ± 7.5 Ma (Graf et al., 1998). In the Menderes Massif, the predominant rock types are orthogneisses with a volcanic-arc signature. Their intrusion ages range from 520 to 570 Ma with a mean of 550 Ma (Hetzel and Reischmann, 1996; Loos and Reischmann, 1999). The Cadomian arc-related metagranites of the Bolu Massif and the Karadere basement have a similar age of 576.0 ± 6.0 Ma (Ustaömer et al., 2005) and are unconformably overlain by Lower Paleozoic clastic metasediments. In addition, large exposures of basement rocks with Pan-African ages are known from the Arabian-Nubian Shield (e.g., Reischmann, 2000) and other parts of North Africa.

The Vertiskos terrane, on the other hand, records a different history, and no basement rocks of similar lithology or age are known from the Hellenides or western Turkey. The terrane continues north into Bulgaria, where it is known as the Ograzhden block (Titorenkova et al., 2003), but disappears south of Blagoevgrad. Ordovician and Silurian ages, however, occur throughout the Alpine-Variscan chain (Stampfli et al., 1998; Neubauer, 2002; Von Raumer et al., 2003). The age and geotectonic environment of the Vertiskos augen gneisses suggest that this terrane may have been part of the so-called Hun superterrane defined by Stampfli and Borel (2002). This superterrane consisted of a series of basement and supracrustal rocks formed during Cambrian to Silurian times at the northern active continental margin of Gondwana. Here, an island arc was built on an archipelago of preexisting Cadomian crust that had rifted away from Gondwana in the earliest Cambrian during initial opening of the Rheic Ocean (Von Raumer et al., 2003). Subduction beneath Gondwana created peraluminous granitoids, which are typical of this setting. Parts of this superterrane occur in the Alpine and Variscan orogens from Western Europe to China. The basement units of the Pirgadikia and Vertiskos terranes, as well as the Pan-African rocks of the Osogovo-Lisets dome in the Bulgarian Serbo-Macedonian Massif, are consistent with such a setting, where they would represent Pan-African basement and Ordovician-Silurian volcanic-arc rocks formed by southward-directed subduction beneath Gondwana, respectively. Detrital zircon ages in metasediments of the mélange zones of the Serbo-Macedonian Massif along with inherited components in the magmatic zircons and Nd model ages of the basement units of the Serbo-Macedonian Massif indicate a strong contribution from Pan-African crust. However, Grenvillian ages of around 1 Ga are completely absent in the Vertiskos terrane (Fig. 4), though a few grains show ages of 1.5–2 Ga, and there is a well-defined inheritance of 2.4 Ga to 2.5 Ga. The absence of Grenvillian ages in the age spectrum of inherited components excludes western Gondwana and Baltica as sources of sediments and crustal material for the granites, since important Grenvillian age provinces exist in these cratons (e.g., Murphy et al., 2000; Linnemann et al., 2004).

PALEOGEOGRAPHIC RECONSTRUCTIONS

The study of the Alps and related young orogenic belts has produced a large data set from various parts of the Tethys Ocean, which included different basins that separating microcontinents or terranes. Paleogeographic reconstructions of the Mediterranean realm (Robertson and Dixon, 1984; Şengör et al., 1984;

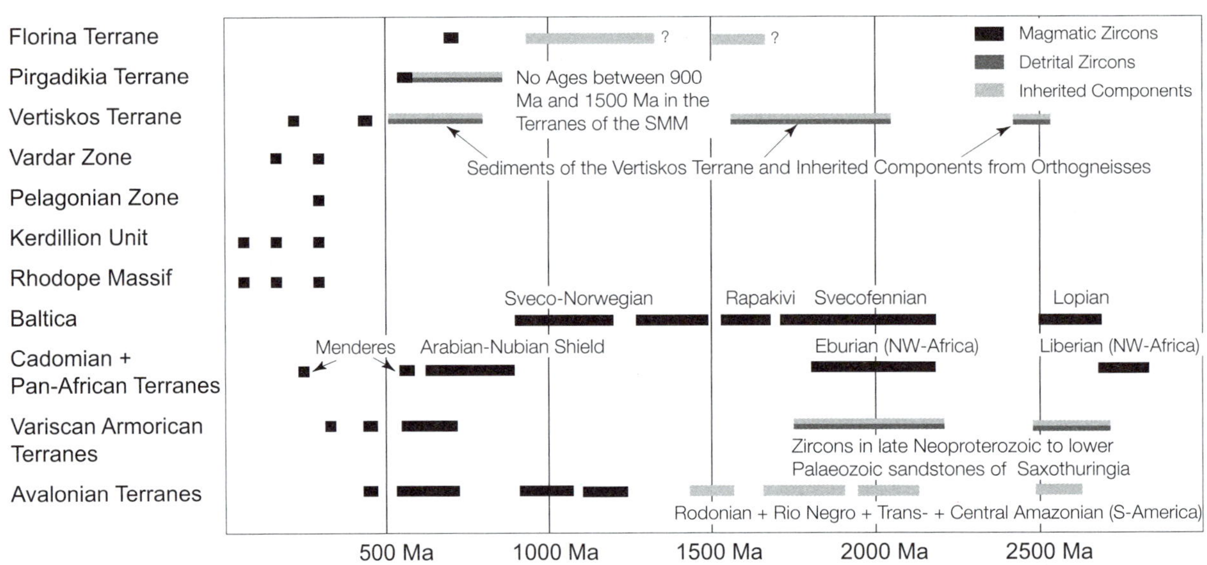

Figure 4. Compilation of age data for parts of the Alpine and Variscan chains in comparison with those for the shields of Baltica and northern Gondwana to constrain the terrane provenance. The age data for the Cadomian and Pan-African terranes are from Avigad et al. (2003) and Abdelsalam et al. (2002). The Variscan Avalonian terranes are from Linnemann et al. (2004), and the Avalonian terranes are from Nance and Murphy (1994, 1996), Nance et al. (2001), and Murphy et al. (2001). References for the Greek crystallization ages are in the text. The Pelagonian zone and the Florina terrane are separated for reasons of clarity. The terranes of the Serbo-Macedonian Massif (SMM) differ in their age spectra from the Florina terrane in the Mesoproterozoic. The Florina terrane resembles more the Avalonia-type terranes, whereas the Vertiskos and Pirgadikia terranes contain no Mesoproterozoic inheritance, which is considered to be a fingerprint of West African provenance in the peri-Gondwanan terranes (e.g., Linnemann and Romer, 2002). This lack of Grenvillian ages is encountered in the Cadomian terranes and excludes a provenance from Baltica and western Gondwana (see paleogeographic reconstruction, Fig. 5).

Stampfli and Borel, 2002) favor two large oceanic basins termed Paleotethys and Neotethys. Closure of the Paleotethys produced the Cimmerian orogeny in the Triassic or Late Jurassic (depending on the model), whereas closure of the Neotethys produced the Alpine orogeny in the Eocene. In Greece and the entire Balkan area, this simple subdivision is hard to apply, since there are several subparallel ophiolitic zones and evidence for several small ocean basins. This mosaic of microcontinents was accreted to the south European margin as a result of the closure of the ocean basins between them.

For the Paleozoic, we favor the reconstructions of Stampfli and Borel (2002) and Von Raumer et al. (2003), since these more readily account for the presence of exotic terranes (Fig. 5). According to these models, a piece of Pan-African crust rifted away from northern Gondwana in the early Paleozoic and became part of a magmatic arc that formed in the Ordovician and Silurian. We propose that the Vertiskos and the Pirgadikia terranes are part of this rifted continental fragment. We suggest that the Florina terrane can be correlated with the Avalonian-type terranes of southeastern Europe and tentatively place it next to the Bruno-Vistulikum terrane (Finger et al., 2000) in the reconstruction of Von Raumer et al. (2003) (Fig. 5).

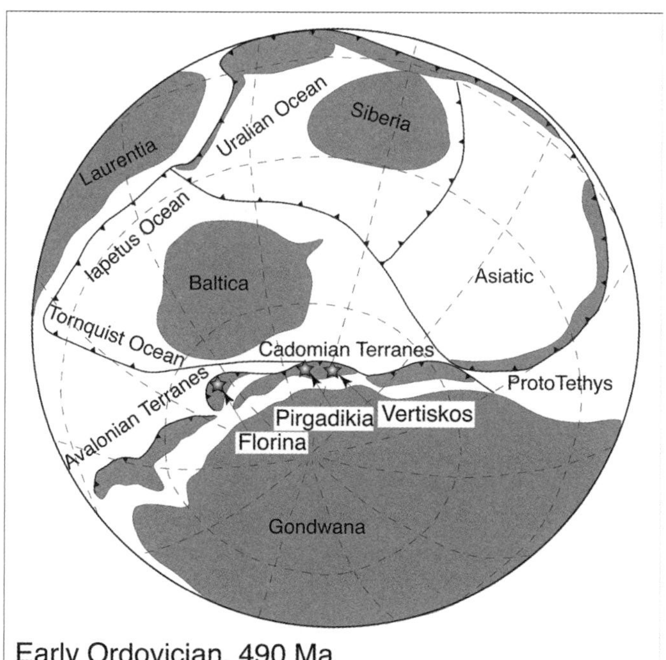

Figure 5. Paleogeographic reconstruction of the northern margin of Gondwana in early Ordovician times at ca. 490 Ma (modified after Von Raumer et al., 2003). The Florina terrane shows similarities to the Avalonian terranes of the central European Variscides. The two terranes in the Serbo-Macedonian Massif are placed in the vicinity of the Cadomian terranes because of the absence of Grenvillian-age inherited zircons.

CONCLUSIONS

The presence of Gondwana-derived terranes in different zones of the Internal Hellenides highlights the allochthonous character of parts of the Hellenic orogen. Distinct crustal units in the different zones were accreted to the European craton during closure of Tethyan oceanic basins. The Vardar zone is interpreted to represent a major suture, and the discovery of the exotic Vertiskos terrane in the neighboring basement units supports this view. The ubiquitous top-to-the-SW sense of shear in the Serbo-Macedonian Massif and the western Rhodope Massif and the transport direction of the western zones of the orogen (the External Hellenide Platform and the Sub-Pelagonian zone) indicate NE-directed accretion of various units in the late Paleozoic and Mesozoic. In the Tertiary, lateral movements parallel to the trend of the orogen and extension in the central Aegean overprinted the structure of amalgamated terranes. The identification of Gondwana-derived terranes shows the pre-Alpine to early Alpine history of the Hellenides to be far more complex than previously thought, since the main part of the Greek orogen came into being during this period. This new view of the Hellenic orogen can be compared with that of the central European Variscides, and it expands the region of Paleozoic to Mesozoic accretion orogeny and terrane amalgamation along the southern European margin to the southeast.

ACKNOWLEDGMENTS

B. Anders and F. Himmerkus would like to thank the Graduiertenkolleg “Stoffbestand und Entwicklung von Kruste und Mantels” at the Universität Mainz, the Max-Planck-Institut für Chemie, Abteilung Geochemie in Mainz, and the Deutsche Forschungsgemeinschaft for funding of the projects. Special thanks from F. Himmerkus are due to P. Zachariadis and G. Meinhold for useful discussions. The authors also thank Damian Nance and Anne Rassios for their helpful reviews.

REFERENCES CITED

Abdelsalam, M.G., Liègeois, J.-P., and Stern, R.J., 2002, The Saharan metacraton: Journal of African Earth Sciences, v. 34, no. 3–4, p. 119–136, doi: 10.1016/S0899-5362(02)00013-1.

Altherr, R., Schliestedt, M., Okrusch, M., Seidel, E., Kreuzer, H., Harre, W., Lenz, H., Wendt, I., and Wagner, G.A., 1979, Geochronology of high-pressure rocks on Sifnos (Cyclades, Greece): Contributions to Mineralogy and Petrology, v. 70, p. 245–255.

Anders, B., 2005, The Pre-Alpine Evolution of the Basement of the Pelagonian Zone and the Vardar Zone, Greece [Ph.D. thesis]: Mainz, Johannes Gutenberg-Universität, 153 p.

Anders, B., Reischmann, T., Poller, U., and Kostopoulos, D.K., 2003a, The oldest rocks in Greece: Geochronological evidence for remnants of a Precambrian basement within the central Hellenides: Goldschmidt Conference Abstracts, Geochimica et Cosmochimica Acta, 2003, v. 67, p. A18.

Anders, B., Reischmann, T., Poller, U., and Kostopoulos, D.K., 2003b, First zircon ages from south Pilion and Skiathos Island, Greece: Berichte der Deutschen Mineralogischen Gesellschaft: Beihefte zum European Journal of Mineralogy, v. 15, no. 1, p. 5.

Anders, B., Reischmann, T., Kostopoulos, D.K., and Poller, U., 2006, The oldest rocks of Greece: First evidence for a Precambrian terrane within the

Pelagonian zone: Geological Magazine, v. 143, p. 41–58.

Avigad, D., Kolodner, K., McWilliams, M., Persing, H., and Weissbrod, T., 2003, Origin of northern Gondwana Cambrian sandstone revealed by detrital zircon SHRIMP dating: Geology, v. 31, p. 227–230, doi: 10.1130/0091-7613(2003)031<0227:OONGCS>2.0.CO;2.

Barr, S.R., Temperley, S., and Tarney, J., 1999, Lateral growth of the continental crust through deep level subduction-accretion: A re-evaluation of central Greek Rhodope: Lithos, v. 46, p. 69–94, doi: 10.1016/S0024-4937(98)00055-3.

Brun, J.-P., and Sokoutis, D., 2004, North Aegean extension: From the Rhodope core complex to Neogene basins, *in* Chatzipetros, A.A., and Pavlides, S.B., eds., Proceedings of 5th International Symposium on Eastern Mediterranean Geology, Thessaloniki, Greece, 14–20 April, 2004, Ref: T1–26.

Brunn, J.H., 1956, Contribution à l'étude géologique du Pinde Septentrional et d'une partie de la Macédoine Occidental: Annales Géologiques des Pays Helléniques, v. 7, p. 1–358.

Burg, J.-P., Ivanov, Z., Ricou, L.-E., Dimov, D., and Klain, L., 1990, Implications of shear-sense criteria for the tectonic evolution of the central Rhodope Massif, southern Bulgaria: Geology, v. 18, p. 451–454, doi: 10.1130/0091-7613(1990)018<0451:IOSSCF>2.3.CO;2.

Burg, J.-P., Ricou, L.-E., Ivanov, Z., Godfriaux, I., Dimov, D., and Klain, L., 1993, Crustal-scale thrust complex in the Rhodope: Structure and kinematics: Bulletin of the Geological Society of Greece, v. 28, no. 1, p. 71–85.

Burg, J.-P., Godfriaux, I., and Ricou, L.-E., 1995, Extension of the Mesozoic Rhodope thrust units in the Vertiskos-Kerdillion Massifs (northern Greece): Comptes Rendus de la Académie des Sciences, Paris, v. 320, no. 9, p. 889–896.

Burg, J.-P., Ricou, L.-E., Ivanov, Z., Godfriaux, I., Dimov, D., and Klain, L., 1996, Syn-metamorphic nappe complex in the Rhodope Massif: Structure and kinematics: Terra Nova, v. 8, p. 6–15.

Chen, F., Siebel, W., Satir, M., Terzioglu, M.N., and Saka, K., 2002, Geochronology of the Karadere basement (NW Turkey) and implications for the geological evolution of the Istanbul zone: International Journal of Earth Sciences, v. 91, p. 469–481, doi: 10.1007/s00531-001-0239-6.

Coney, P.J., Jones, D.L., and Monger, J.W.H., 1980, Cordilleran suspect terranes: Nature, v. 288, p. 329–333, doi: 10.1038/288329a0.

Dannat, C., and Reischmann, T., 1997, A geochemical, isotopic and geochronological study of granitoid gneisses of the Menderes Massif: SW Turkey: Terra Nova, v. 9, p. 404.

Davies, R., England, P., Parsons, B., Billiris, H., Paradissis, D., and Veis, G., 1997, Geodetic strain of Greece in the interval 1892–1992: Journal of Geophysical Research, v. 102, p. 24,571–24,588, doi: 10.1029/97JB01644.

Dimitrijevic, M.D., 1974, Sur l'âge du métamorphisme et des plissements dans la masse Sérbo-Macédonienne: Bulletin de l'Association Géologique Carpatho-Balkanique, v. 21, p. 45–48.

Dimitrijevic, M.D., 1997, Geology of Yugoslavia: Special publication—Geological Institute, GEMINI, Belgrade, 187 p.

Dixon, J.E., and Dimitriadis, S., 1984, Metamorphosed ophiolitic rocks from the Serbo-Macedonian Massif, near Lake Volvi, north-east Greece, *in* Dixon, J.E, and Robertson, A.H.F., eds., The Geological Evolution of the Eastern Mediterranean: Geological Society [London] Special Publication 17, p. 603–618.

Dürr, S., Altherr, R., Keller, J., Okrusch, M., and Seidel, E., 1978, The median Aegean crystalline belt: Stratigraphy, structure, metamorphism, magmatism, *in* Closs, H., Roeder, D.H., and Schmidt, K., eds., Alps, Appenines, Hellenides: Stuttgart, Schweizerbart, International Union of Geological Sciences Report 38, p. 455–477.

Engel, M., and Reischmann, T., 1998, Single-zircon geochronology of the orthogneisses from Paros, Greece: Bulletin of the Geological Society of Greece, v. 32, no. 3, p. 91–99.

Finger, F., Handl, P., Pin, C., von Quadt, A., and Steyrer, H.P., 2000, The Brunovistulian: Avalonian Precambrian sequence at the eastern end of the Central European Variscides?, *in* Franke, W., Haak, V., Oncken, O., and Tanner, D., eds., Orogenic Processes: Quantification and Modelling in the Variscan Belt: Geological Society [London] Special Publication 179, p. 103–112.

Friedl, G., Finger, F., McNaughton, N.J., and Fletcher, I.R., 2000, Deducing the ancestry of terranes: SHRIMP evidence for South America derived Gondwana fragments in central Europe: Geology, v. 28, no. 11, p. 1035–1038, doi: 10.1130/0091-7613(2000)28<1035:DTAOTS>2.0.CO;2.

Gautier, P. and Brun, J.–P. 1994: Crustal-scale geometry and kinematics of late-orogenic extension in the central Aegean (Cyclades and Evia Island: Tectonophysics, v. 238, p. 399–424.

Gautier, P., Brun, J.-P., Moriceau, R., Sokoutis, D., Martinod, J., and Jolivet, L., 1999, Timing, kinematics and cause of Aegean extension: A scenario based on a comparison with simple analogue experiments: Tectonophysics, v. 315, p. 31–72, doi: 10.1016/S0040-1951(99)00281-4.

Glasmacher, U.A., Reynolds, P., Alekseyev, A.A., Puchkov, V.N., Taylor, K., Gorozhanin, V., and Walter, R., 1999, ^{40}Ar/^{39}Ar thermochronology west of the Main Uralian fault, southern Urals, Russia: Geologische Rundschau, v. 87, p. 515–525, doi: 10.1007/s005310050228.

Graf, J., 2001, Alpine Tectonics in Western Bulgaria: Cretaceous Compression of the Kraište Region and Cenozoic Exhumation of the Crystalline Osogovo-Lisec Complex [Ph.D. thesis]: Zürich, Switzerland, Eidgenössische Technische Hochschule, 185 p.

Graf, J., Bernoulli, D., Burg, J.-P., Ivanov, Z., and Von Quadt, A., 1998, Geochemistry and geochronology of igneous rocks of the central Serbo-Macedonian Massif (western Bulgaria), *in* Carpathian–Balkan Geological Association XVI Congress Abstracts: Vienna, Geological Survey of Austria, p. 191.

Hegner, E., and Kröner, A., 2000, Review of Nd isotopic data and xenocrystic and detrital zircon ages from the pre-Variscan basement in the eastern Bohemian Massif: Speculations on palinspastic reconstructions, *in* Franke, W., Haak, V., Oncken, O., and Tanner, D., eds., Orogenic Processes: Quantification and Modelling in the Variscan Belt: Geological Society [London] Special Publication 179, p. 113–129.

Hetzel, R., and Reischmann, T., 1996, Intrusion age of Pan-African augen gneisses in the southern Menderes massif and the age of cooling after Alpine ductile extensional deformation: Geologic Magazine, v. 133, p. 565–572.

Himmerkus, F., Reischmann, T., and Kostopoulos, D.K., 2002, First evidence for Silurian magmatism in the Serbo-Macedonian Massif, northern Greece: Goldschmidt Conference Abstract Volume: Geochimica et Cosmochimica Acta, v. 66, p. A330.

Himmerkus, F., Reischmann, T., and Kostopoulos, D.K., 2003, The Serbo-Macedonian Massif, the oldest crustal segment of the internal Hellenides, identified by zircon ages: Geophysical Research Abstracts, v. 5, p. 05671.

Himmerkus, F., Reischmann, T., and Kostopoulos, D.K., 2004a, The Pirgadikia Unit, the oldest crustal segment in the Serbo-Macedonian terrane assemblage, *in* Chatzipetros, A.A., and Pavlides, S.B., eds., Proceedings of 5th International Symposium on Eastern Mediterranean Geology Conference Proceedings, Thessaloniki, Greece, April 14–20, 2004, p. 84–85.

Himmerkus, F., Reischmann, T., and Kostopoulos, D.K., 2004b, Triassic rifting recorded in Gondwana-derived Tethyan terranes, Serbo-Macedonian Massif, northern Greece: Berichte der Deutschen Mineralogischen Gesellschaft: Beihefte zum European Journal of Mineralogy, v. 16, p. 57.

Himmerkus, F., Zachariadis, P., Reischmann, T., and Kostopoulos, D.K., 2005, The mafic complexes of the Athos-Volvi zone—A suture zone between the Serbo-Macedonian Massif and the Rhodope Massif?: Geophysical Research Abstracts, v. 7, p. 10,240.

Himmerkus, F., Reischmann, T., and Kostopoulos, D.K., 2006, Late Proterozoic and Silurian basement units within the Serbo-Macedonian Massif, northern Greece: The significance of terrane accretion in the Hellenides, *in* Mountrakis, D., and Robertson, A.H.F., eds., Tectonic Development of the Eastern Mediterranean Region: Geological Society [London] Special Publications 260, p. 35–50.

Institute of Geology and Mineral Exploration (IGME), 1983, Geological Map of Greece: Athens, Institute of Geology and Mineral Exploration, scale 1:500,000.

Jacobshagen, V., ed., 1986, Geologie von Griechenland: Berlin, Stuttgart, Gebrüder Borntraeger, 363 p.

Jolivet, L., 2001, A comparison of geodetic and finite strain pattern in the Aegean, geodynamic implications: Earth and Planetary Science Letters, v. 187, p. 95–104, doi: 10.1016/S0012-821X(01)00277-1.

Jolivet, L., Faccenna, C., Goffé, B., Burov, E., and Agard, P., 2003, Subduction tectonics and exhumation of high-pressure metamorphic rocks in the Mediterranean orogens: American Journal of Science, v. 303, p. 353–409, doi: 10.2475/ajs.303.5.353.

Kahle, H.G., Cocard, M., Peter, Y., Geiger, A., Reilinger, R., Barka, A., and Veis, G., 2000, GPS-derived strain rate fixed within the boundary zones of the Eurasian, African and Arabian plates: Journal of Geophysical Research, v. 105, p. 23,353–23,370, doi: 10.1029/2000JB900238.

Kauffmann, G., Kockel, F., and Mollat, H., 1976, Notes on the stratigraphic and palaeogeographic position of the Svoula Formation in the Innermost zone of the Hellenides (northern Greece): Bulletin de la Société Géologique de France, v. 18, p. 225–230.

Kilias, A., and Mountrakis, D., 1989, The Pelagonian Nappe: Tectonics, metamorphism and magmatism: Bulletin of the Geological Society of Greece, v. 23, no. 1, p. 29–46.

Kilias, A., Falalakis, G., and Mountrakis, D., 1999, Cretaceous-Tertiary structures and kinematics of the Serbo-Macedonian metamorphic rocks and their relation to the exhumation of the Hellenic hinterland (Macedonia, Greece): International Journal of Earth Sciences, v. 88, no. 3, p. 513–531, doi: 10.1007/s005310050282.

Knezevic-Dordevic, V., and Krstic, B., 1996, Terranes of Serbia: The formation of the Geologic framework of Serbia and the adjacent regions: Department of Petrology, Faculty of Mining and Geology, University of Belgrade & Committee for Geodynamics of the Serbian Academy of Sciences and Arts, Belgrade. 440 pp.

Kober, B., 1986, Whole grain evaporation for $^{207}Pb/^{206}Pb$-age investigations on single zircons using a double filament thermal ion source: Contributions to Mineralogy and Petrology, v. 93, p. 482–490, doi: 10.1007/BF00371718.

Kober, B., 1987, Single zircon evaporation combined with Pb+emitter-bedding for $^{207}Pb/^{206}Pb$-age investigations using thermal ion mass spectrometry, and implications to zirconology: Contributions to Mineralogy and Petrology, v. 96, p. 63–71, doi: 10.1007/BF00375526.

Kockel, F., Mollat, H., and Walther, H.W., 1971, Geologie des Serbo-Mazedonischen Massivs und seines Mesozoischen Rahmens (Nordgriechenland): Geologisches Jahrbuch, v. 89, p. 529–551.

Kockel, F., Mollat, H., and Walther, H.W., 1977, Erläuterungen zur Geologischen Karte der Chalkidiki und Angrenzender Gebiete 1:100,000 (Nord-Griechenland): Hannover, Bundesanstalt für Geowissenschaften und Rohstoffe, 119 p.

Koroneos, A., Christofides, G., del Moro, A., and Kilias, A., 1993, Rb-Sr geochronology and geochemical aspects of the Eastern Varnountas plutonite (NW Macedonia, Greece): Neues Jahrbuch für Mineralogie, Abhandlungen, v. 165, no. 3, p. 297–315.

Kostopoulos, D.K., 1988, Geochemistry, Petrogenesis and Tectonic Setting of the Pindos Ophiolite, NW Greece [Ph.D. thesis]: Newcastle-upon-Tyne, University of Newcastle-upon-Tyne, 528 p.

Kostopoulos, D.K., Ioannidis, N.M., and Sklavounos, S.A., 2000, A new occurrence of ultrahigh-pressure metamorphism, central Macedonia, northern Greece: Evidence from graphitized diamonds?: International Geology Review, v. 42, p. 545–554.

Le Pichon, X., and Angelier, J., 1979, The Hellenic arc and trench system: A key to the neotectonic evolution of the Eastern Mediterranean area: Tectonophysics, v. 60, p. 1–42, doi: 10.1016/0040-1951(79)90131-8.

Le Pichon, X., and Angelier, J., 1981, The Aegean Sea: Philosophical Transactions of the Royal Society of London, v. A300, p. 357–372.

Le Pichon, X., Chamot-Rooke, N., Lallement, S., Noomen, B., and Veis, G., 1995, Geodetic determination of the kinematics of central Greece with respect to Europe: Implications for Eastern Mediterranean tectonics: Journal of Geophysical Research, v. 100, p. 12,675–12,690, doi: 10.1029/95JB00317.

Linnemann, U., and Romer, R.L., 2002, The Cadomian orogeny in Saxo-Thuringia, Germany: Geochemical and Nd-Sr-Pb isotopic characterisation of marginal basins with constraints to geotectonic setting and provenance: Tectonophysics, v. 352, p. 33–64, doi: 10.1016/S0040-1951(02)00188-9.

Linnemann, U., McNaughton, N.J., Romer, R.L., Gehmlich, M., Drost, K., and Tonk, C., 2004, West African provenance for Saxo-Thuringia (Bohemian Massif): Did Armorica ever leave the pre-Pangean Gondwana?—U/Pb-SHRIMP zircon evidence and the Nd-isotopic record: International Journal of Earth Sciences, v. 93, p. 683–705, doi: 10.1007/s00531-004-0413-8.

Lister, G.S., Banga, G., and Feenstra, A., 1984, Metamorphic core complexes of Cordilleran type in the Cyclades, Aegean Sea, Greece: Geology, v. 12, p. 221–225, doi: 10.1130/0091-7613(1984)12<221:MCCOCT>2.0.CO;2.

Loos, S., and Reischmann, T., 1999, The evolution of the southern Menderes Massif in SW Turkey as revealed by zircon dating: Geological Society [London] Journal, v. 156, p. 1021–1030.

Miyashiro, A., Aki, K., and Şengör, A.M.C., 1982, Orogeny: Chichester, John Wiley and Sons, 242 p.

Mountrakis, D., 1982, Étude géologique des terrains métamorphiques de Macédoine occidentle (Grèce): Bulletin de la Société Géologique de France, v. 24, no. 4, p. 697–704.

Mountrakis, D., 1984, Structural evolution of the Pelagonian zone in northwestern Macedonia, Greece, *in* Dixon, J.E., and Robertson, A.H.F., eds., The Geological Evolution of the Eastern Mediterranean: Geological Society [London] Special Publication 17, p. 581–590.

Mountrakis, D., 1986, The Pelagonian zone in Greece: A polyphase-deformed fragment of the Cimmerian continent and its role in the geotectonic evolution of the Eastern Mediterranean: The Journal of Geology, v. 94, p. 335–347.

Murphy, J.B., Strachan, R.A., Nance, R.D., Parker, K.D., and Fowler, M.B., 2000, Proto-Avalonia: A 1.2–1.0 Ga tectonothermal event and constraints for the evolution of Rodinia: Geology, v. 28, p. 1071–1074, doi: 10.1130/0091-7613(2000)28<1071:PAGTEA>2.0.CO;2.

Murphy, J.B., Pisarevsky, S.A., Nance, R.D., and Keppie, J.D., 2001, Animated history of Avalonia in Neoproterozoic–early Palaeozoic, *in* Jessell, M.J., ed., General Contributions: Journal of the Virtual Explorer, v. 3, p. 45–58.

Mussallam, K., and Jung, D., 1986, Geologie und bau des Sithonia-ophioliths (Chalkidiki, NE-Griechenland: Anmerkungen zur bildung ozeanischer krusten): Geologische Rundschau, v. 75, no. 2, p. 383–409, doi: 10.1007/BF01820619.

Nance, R.D., and Murphy, J.B., 1994, Contrasting basement isotopic signatures and the palinspastic restoration of peripheral orogens: Example from the Neoproterozoic Avalonian-Cadomian belt: Geology, v. 22, p. 617–620, doi: 10.1130/0091-7613(1994)022<0617:CBISAT>2.3.CO;2.

Nance, R.D., and Murphy, J.B., 1996, Basement isotopic signatures and Neoproterozoic paleogeography of Avalonian-Cadomian and related terranes in the circum-North Atlantic, *in* Nance, R.D., and Thompson, M.D., eds., Avalonian and Related Peri-Gondwanan Terranes of the Circum-North Atlantic: Geological Society of America Special Paper 304, p. 333–346.

Nance, R.D., Murphy, J.B., and Keppie, J.D., 2002, A Cordilleran model for the evolution of Avalonia: Tectonophysics, v. 352, p. 11–31, doi: 10.1016/S0040-1951(02)00187-7.

Neubauer, F., 2002, Evolution of late Neoproterozoic to early Paleozoic tectonic elements in central and southeast European Alpine mountain belts: Review and synthesis: Tectonophysics, v. 352, p. 87–103, doi: 10.1016/S0040-1951(02)00190-7.

Papanikolaou, D., 1997, The tectonostratigraphic terranes of the Hellenides: Annales Géologiques des Pays Helléniques, v. 37, p. 495–514.

Papanikolaou, D., and Panagopoulos, A., 1981, On the structural style of southern Rhodope: Geologica Balcanica, v. 11, no. 3, p. 13–22.

Pearce, J.A., Harris, N.B.W., and Tindle, A.G., 1984, Trace element discrimination diagrams for the tectonic interpretation of granitic rocks: Journal of Petrology, v. 25, p. 956–983.

Pe-Piper, G., and Piper, D.J.W., 2002, The Igneous Rocks of Greece: The Anatomy of an Orogen: Berlin/Stuttgart, Gebrueder Borntraeger, 573 p.

Reilinger, R., McClusky, S., Oral, M., King, R., Toksoz, N., Barka, A., Kinik, I., Lenk, O., and Sanli, I., 1997, Global Positioning System measurements of present day crustal movements in the Arabia–Africa–Eurasia plate collision zone: Journal of Geophysical Research, v. 102, p. 9983–9999, doi: 10.1029/96JB03736.

Reischmann, T., 1998, Pre-Alpine origin of tectonic units from the metamorphic complex of Naxos, Greece, identified by single zircon Pb/Pb dating: Bulletin of the Geological Society of Greece, v. 32, no. 3, p. 101–111.

Reischmann, T., 2000, Ophiolites and island arcs in the late Proterozoic Nubian Shield: Ofioliti, v. 25, p. 1–13.

Reischmann, T., Kostopoulos, D.K., Loos, S., Anders, B., Avgerinas, A., and Sklavounos, S.A., 2001, Late Palaeozoic magmatism in the basement rocks southwest of Mt. Olympos, central Pelagonian zone, Greece: Remnants of a Permo-Carboniferous magmatic arc: Bulletin of the Geological Society of Greece, v. 34, no. 3, p. 985–993.

Ricou, L.-E., Burg, J.-P., Godfriaux, I., and Ivanov, Z., 1998, Rhodope and Vardar: The metamorphic and the olistostromic paired belts related to the Cretaceous subduction under Europe: Geodinamica Acta, v. 11, p. 285–309, doi: 10.1016/S0985-3111(99)80018-7.

Robertson, A.H.F., 2002, Overview of the genesis and emplacement of Mesozoic ophiolites in the Eastern Mediterranean Tethyan region: Lithos, v. 65, p. 1–67, doi: 10.1016/S0024-4937(02)00160-3.

Robertson, A.H.F., and Dixon, J.E., 1984, Introduction: Aspects of the geological evolution of the Eastern Mediterranean, *in* Dixon, J.E., and Robertson, A.H.F., eds., The Geological Evolution of the Eastern Mediterranean: Geological Society [London] Special Publication 17, p. 1–74.

Robertson, A.H.F., and Karamata, S., 1994, The role of subduction accretion processes in the tectonic evolution of the Mesozoic Tethys in Serbia: Tectonophysics, v. 234, no. 1–2, p. 73–94, doi: 10.1016/0040-1951(94)90205-4.

Royden, L.H., and Papanikolaou, D., 2004, Buoyancy of subducted lithosphere and the recent evolution of the Hellenic subduction system: Geological Society of America Abstracts with Programs, v. 36, no. 5, p. 482.

Schermer, E.R., 1990, Mechanisms of blueschist creation and preservation in an A-type subduction zone, Mount Olympus region, Greece: Geology, v. 18, p. 1130–1195, doi: 10.1130/0091-7613(1990)018<1130:MOBCAP>2.3.CO;2.

Şengör, A.M.C., Yilmaz, Y., and Sungurlu, O., 1984, Tectonics of the Mediterranean Cimmerides: Nature and evolution of the western termination of Palaeo-Tethys, *in* Dixon, J.E., and Robertson, A.H.F., eds., The Geological Evolution of the Eastern Mediterranean: Geological Society [London] Special Publication 17, p. 603–618.

Stampfli, G.M., and Borel, G.D., 2002, A plate tectonic model for the Paleozoic and Mesozoic constrained by dynamic plate boundaries and restored synthetic oceanic isochrones: Earth and Planetary Science Letters, v. 196, p. 17–33, doi: 10.1016/S0012-821X(01)00588-X.

Stampfli, G.M., Mosar, J., De Bono, A., and Vavasis, I., 1998, Late Palaeozoic, early Mesozoic plate tectonics of the western Tethys: Bulletin of the Geological Society of Greece, v. 23, p. 113–120.

Stampfli, G.M., Rosselet, F., and Bagheri, S., 2004, Tethyan oceans and sutures, *in* Chatzipetros, A.A., and Pavlides, S.B., eds., Proceedings of 5th International Symposium on Eastern Mediterranean Geology, Thessaloniki, Greece, April 14–20, 2004, Ref: T1–21.

Titorenkova, R., Macheva, L., Zidarov, N.V., Quadt, A., and Peytcheva, I., 2003, Metagranites from SW Bulgaria as a part of the Neoproterozoic to early Palaeozoic system in Europe: New insight from zircon typology, U-Pb isotope data and Hf-tracing: European Union of Geosciences XII: Geophysical Research Abstracts, v. 5, p. 08963.

Turpaud, P., and Reischmann, T., 2003, Zircon ages of granitic gneisses from the Rhodope (N. Greece), determination of basement age and evidences for a Cretaceous intrusive event: European Union of Geosciences XII: Geophysical Research Abstracts, v. 5, p. 04435.

Ustaömer, A.P., Mundil, R., and Renne, P.R., 2005, U/Pb and Pb/Pb zircon ages for arc-related intrusions of the Bolu Massif (W Pontides, NW Turkey): Evidence for Late Precambrian (Cadomian) age: Terra Nova, v. 17, p. 215–223, doi: 10.1111/j.1365-3121.2005.00594.x.

van Hinsbergen, D.J.J., Hafkenscheid, E., Spakman, W., Meulenkamp, E.J., and Wortel, R., 2005, Nappe stacking resulting from subduction of oceanic and continental lithosphere below Greece: Geology, v. 33, p. 325–328, doi: 10.1130/G20878.1.

Vavassis, I., De Bono, A., Stampfli, G., Giorgis, D., Valloton, A., and Amelin, Y., 2000, U-Pb and Ar-Ar geochronological data from the Pelagonian basement in Evia (Greece): Geodynamic implications for the evolution of Paleotethys: Schweizerische Mineralogische und Petrographische Mitteilungen, v. 80, p. 21–43.

Von Raumer, J.F., Stampfli, G.M., and Bussy, F., 2003, Gondwana-derived micro-continents—The constituents of the Variscan and Alpine collisional orogens: Tectonophysics, v. 365, p. 7–22, doi: 10.1016/S0040-1951(03)00015-5.

Yarwood, G.A., and Aftalion, M., 1976, Field relations and U-Pb geochronology of a granite from the Pelagonian zone of the Hellenides (High Pieria, Greece): Bulletin de la Société Géologique de France, v. 18, no. 2, p. 259–264.

Manuscript Accepted by the Society 1 February 2007

Printed in the USA

The Geological Society of America
Memoir 200
2007

Variscan thrust nappes, detachments, and strike-slip faults in the French Massif Central: Interpretation of the lineations

Philippe Matte
Laboratoire de Dynamique de la Lithosphere, UMR 5573, Centre National de la Recherche Scientifique, Université Montpellier 2, Montpellier 34095, France

ABSTRACT

The French Massif Central (average height 800 m) is the largest pre-Mesozoic, Variscan outcrop in France (80,000 km^2). It consists of lower Paleozoic metasediments, poorly dated mica schists, and high-grade gneisses (granitic orthogneisses, mafic, and ultramafic rocks). As part of the southern limb of the European Variscides, it is characterized by large south-vergent thrust nappes and recumbent folds. For a decade, the lineations in metamorphic rocks were systematically and sometimes incorrectly used as kinematic indicators of nappe transport. In the French Massif Central the tectonometamorphic history is long and complex (430–310 Ma), and the lineations are related to different events and different kinds of deformation, including thrusting, folding, detachments, and strike-slip faulting. We attempt to analyze these different lineations in various parts of the Massif Central and relate them to the successive stages of crustal thickening and unroofing of the Variscan crust. In the southern part of the French Massif Central, most of the lineations are transport lineations related to the emplacement of large thrust nappes (350–310 Ma). In the northern and western part, most of the NW-SE–trending lineations are not related to nappes transported to the NW but to detachments during the unroofing and thinning of the thickened crust during the same time span.

Keywords: Variscan, thrust nappes, lineations, detachments.

INTRODUCTION

The French Massif Central is one of the largest massifs of the southern European Variscan belt (Fig. 1), together with the Armorican and the Iberian massifs. All these massifs outcrop beneath the unconformably overlying continental latest Carboniferous and Permian-Triassic series and were not at all, or weakly, disturbed by Alpine events.

The French Massif Central consists mainly of high-grade metamorphic rocks (Fig. 2), and it was first considered to be an old (Precambrian-Caledonian) crystalline nucleus, the so-called "Noyau Arverne," surrounded to the south by younger and lower-grade lower Paleozoic sediments, the "schistes périphériques," which had been deformed and metamorphosed during the Mississippian (Roques, 1971). Demay (1942) was the first to describe large nappes in the metamorphic parts, and Gèze (1949) presented the first in the Paleozoic southern part. For the past 30 yr, progress in structural analysis and radiometric dating has provided a completely new image of this massif: most of the deformations and the high-grade metamorphism occurred during a long (100 m.y.) Variscan convergence between 420 and 320 Ma. The French Massif Central corresponds to a piece of strongly

Matte, P., 2007, Variscan thrust nappes, detachments, and strike-slip-faults in the French Massif Central: Interpretation of the lineations, *in* Hatcher, R.D., Jr., Carlson, M.P., McBride, J.H., and Martínez Catalán, J.R., eds., 4-D Framework of Continental Crust: Geological Society of America Memoir 200, p. 391–402, doi: 10.1130/2007.1200(20). For permission to copy, contact editing@geosociety.org.

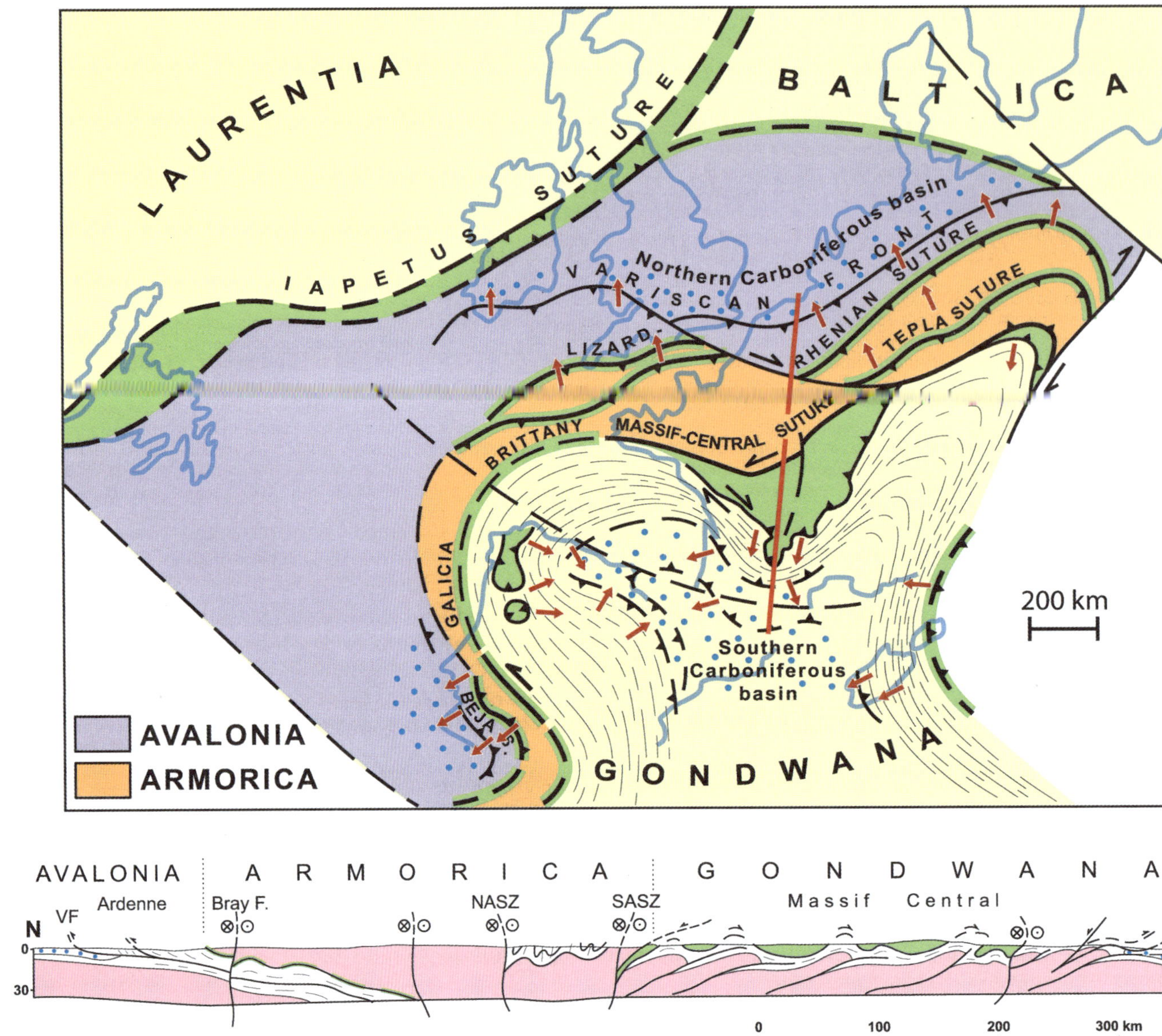

Figure 1. The west European Variscides and the main continental plates in Pennsylvanian times with a schematic section of the French Variscides. Red line—trace of the section. The arrows show the general vergence of folds and thrusts (adapted from Matte, 2001). NASZ—North Armorican Shear Zone; SASZ—South American Shear Zone; VF—Variscan Front.

thickened crust between 380 and 320 Ma. The structure and the southern vergence of the nappes were first defined by F. Arthaud (1969), in the southern part (Montagne Noire), and by Burg and Matte (1978) and Matte and Burg (1981), who described large nappes in the central metamorphic part of the French Massif Central. Burg et al. (1984) documented an inverted metamorphic zonation in this central area.

Since 1976, lineations and related shear-sense criteria have been analyzed to better define the nappe transport. This transport is clearly toward the south in the Cévennes and Montagne Noire and the southern low-grade part of the Massif Central (Arthaud, 1969; Mattauer and Etchecopar, 1977). In other areas, like the Velay anatectic dome, N-S–trending lineations are related to northward-directed detachments (Malavieille et al., 1990) and to a late Variscan general extension (Vanderhaeghe et al., 1999). In the western, high-grade part of the Massif Central (Limousin), most of the lineations trend NW-SE with top-to-the-NW shear sense. This shearing has been interpreted as related to synconvergence detachments (Mattauer et al., 1988) but also to northwestward thrusting (Burg et al., 1987; Friedrich et al., 1988; Roig et al., 1996; Roig and Faure, 2000). Other stretching lineations parallel to the trend of the belt in Montagne Noire, first interpreted

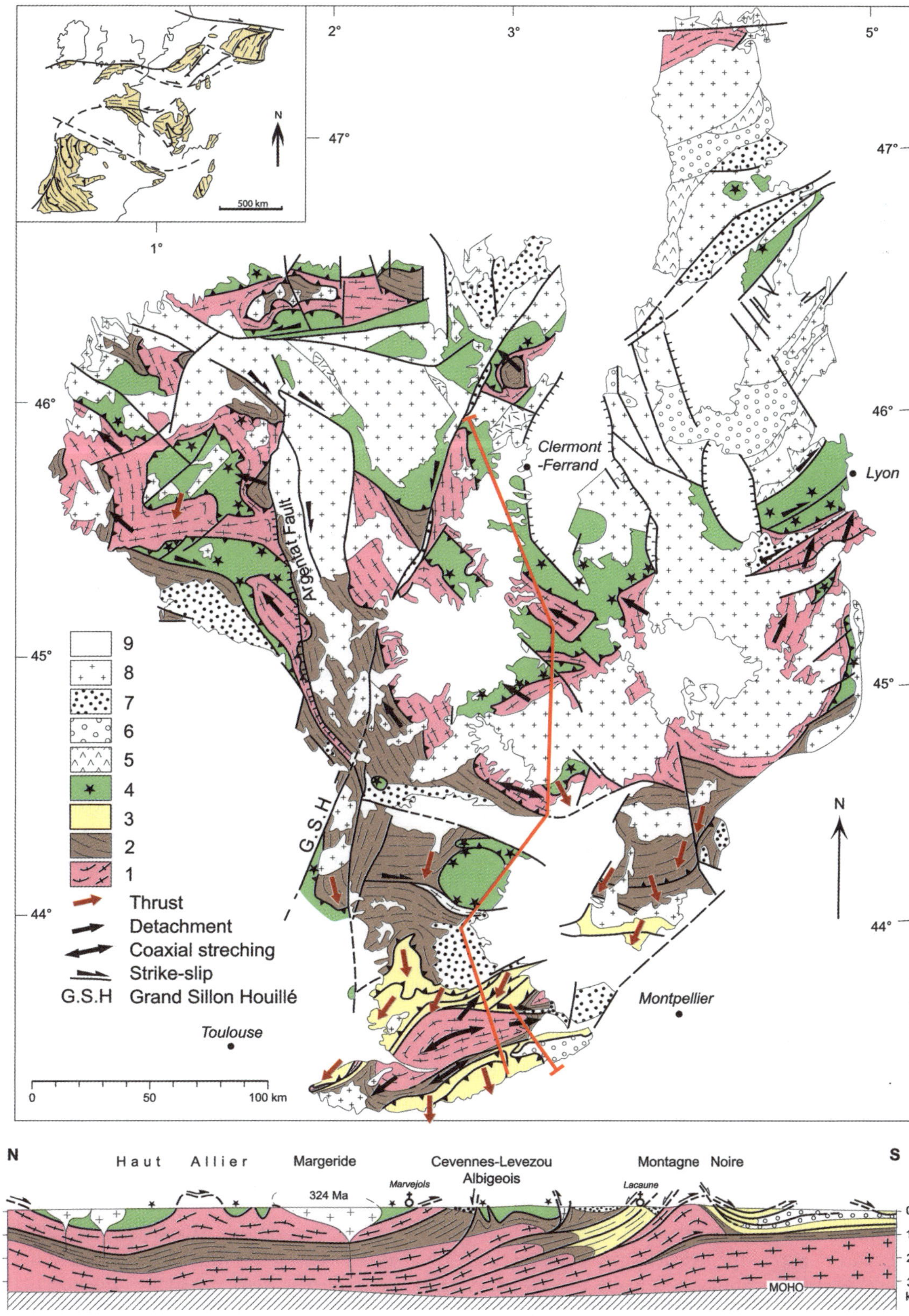

Figure 2. The main lithotectonic units of the French Massif Central. The arrows show the orientation of the lineations related to thrusts (red), detachments (black), and strike-slip-faults. 1—Gneissic high-grade unit, in part Proterozoic metasediments with Lower Paleozoic granitic sheets, and in part Upper Proterozoic and Lower Paleozoic undifferentiated. 2—Low- to high-grade essentially Lower Paleozoic metasediments. 3—Fossiliferous Lower Paleozoic. 4—Leptynite-Amphibolite Group (in part Lower Paleozoic oceanic lithosphere, in part thinned continental crust) with eclogites (stars). 5—Ophiolitic Upper Devonian (Brévennes series) and Upper Devonian–Lower Visean calc-alkaline volcanics (northern Massif Central). 6—Lower Carboniferous (Visean) turbidites in southern Massif Central and coal-bearing continental sediments and acid volcanics in northern Massif Central. 7—Upper coal-bearing Carboniferous and Permian. 8—Variscan granites. 9—Mesozoic-Cenozoic cover.

as related to large ductile wrenching (Nicolas et al., 1977), in fact result from stretching normal to the main shortening direction. In this paper, we discuss these different kinds of lineations, which formed between 360 and 310 Ma, and their relationships to thrusts, detachments, and strike-slip faults, in the frame of the Iberian-Variscan arc as defined by Matte (1968), Matte and Ribeiro (1975), Matte and Burg (1981), and Matte (1986).

THE MAIN LITHO-TECTONIC UNITS OF THE FRENCH MASSIF CENTRAL

From south to north, the main lithotectonic units of the French Massif Central, generally separated by main thrusts, are:

(1) The southern Montagne Noire, in the southernmost part, exhibits one of the best examples of large recumbent fold nappes (Fig. 3) described in detail by Arthaud (1969). These kilometer-scale synformal anticlines are made of low-grade fossiliferous sediments from lowermost Cambrian to Lower Carboniferous. The structures are evidenced by thick limestones in the Middle Cambrian and Devonian and thick quartzites in the Upper Cambrian and Lower Ordovician. The Lower Carboniferous consists in a thick flysch with large Silurian to Namurian olistolites in the most proximal part (Engel et al., 1982).

(2) The Montagne Noire axial zone consists of a high-grade metamorphic dome (Thompson and Bard, 1982), which is an antiform of foliation made of paragneisses and orthogneisses derived from pre-Variscan Ordovician granites (Roger et al., 2004), intrusive into a sedimentary pelitic series of probable Upper Proterozoic age.

(3) The northern Montagne Noire, separated from the axial zone by a large dextral detachment, is made of low-grade lower Paleozoic metasedimentary units from Lower Cambrian to Silurian, separated by southwestward-directed thrusts. The bottom of the southernmost unit is made of Upper Proterozoic sediments intruded by Cambrian granites transformed into orthogneisses.

(4) The Cévennes-Lot-Albigeois schists and mica schists are a monotonous unfossiliferous quartz-pelitic thick series (4–5 km), with some basaltic rocks. The metasediments are probably Lower Paleozoic, and, in part, Upper Proterozoic in age, where the mica schists were intruded by 540 Ma diorites (Pin and Lancelot, 1982), and granites of similar ages. A very strong deformation is revealed by a pervasive composite slaty-strain-slip cleavage to foliation dipping gently to the north and a prominent stretching lineation trending roughly N-S. Metamorphism increases from greenschist up to the kyanite and the sillimanite isograd and appears to be inverted below a major thrust of the so-called Leptynite-Amphibolite Group (Burg et al., 1984). The Cévennes-Albigeois pelitic series probably represents distal facies of the Cambrian-Ordovician sediments on the thinned passive margin of Gondwana.

(5) The Leptynite-Amphibolite Group is a composite complex unit (2000–5000 m). This unit lies within the sillimanite isograd, even the southernmost klippes (Fig. 2), which are directly thrust on top of biotite mica schists. The main characteristic is that it contains many boudins of high-pressure mafic and acid rocks: granulites, eclogites. A coesite-bearing eclogite was recently found in the Lyon area (Lardeaux et al., 2001), and garnet peridotites (meta-lherzolites) are known in various parts (west and east) of the French Massif Central (Lasnier, 1971; Gardien et al., 1988; Bonnot and Piboule, 1980).

The lower part consists of alternating felsic and mafic metaigneous rocks interbedded with metasediments. The mafic rocks are amphibolites with boudins of more or less retrogressed coronitic gabbros, eclogites, and basic granulites. The felsic rocks are white to pink gneisses (leptynites) of volcanic and plutonic origin. Conventional U/Pb ages ca. 480 Ma (upper intercept) have been found on a leptynite and a coronitic gabbro with well-preserved magmatic textures in the Marvéjols area (Pin, 1979) and on a norite transformed into eclogite in the Decazeville small klippe (Burg and Matte, 1977; Paquette et al., 1995). The high-pressure metamorphism has been dated ca. 420–430 Ma in various parts of the Leptynite-Amphibolite Group by U/Pb on zircons (Ducrot et al., 1983; Pin and Lancelot, 1982), and by U/Pb on zircons and Nd/Sm (Paquette et al., 1995). That is the earliest metamorphic Variscan event probably related to oceanic subduction.

The upper part is made of anatectic, locally cordierite–bearing gneisses with various relics of high-pressure rocks: acid and mafic granulites, eclogites, and more or less serpentinized peridotites, which, in places, form swarms with hand- to hectometer-scale lenses, with harzburgites and garnet peridotites still recognizable. After the high-pressure–ultrahigh-pressure (HP-UHP) metamorphism preserved in boudins, the Leptynite-Amphibolite Group underwent medium- to low-pressure, high-temperature (HT) metamorphism dated by $^{39}Ar/^{40}Ar$ around 380 Ma (Costa, 1990, 1992).

The geodynamic setting of the Leptynite-Amphibolite Group is not well understood. It was considered to be a part of a dismembered ophiolite by Dubuisson et al. (1988) on the basis of the presence of harzburgites and as a back-arc and/or arc crust by Bodinier et al. (1988) from geochemistry. However, by taking into account the high proportion of metasediments and felsic gneisses and Sr-Nd isotopes and trace elements, Pin (1989) and Pin and Marini (1993) favored a thinned continental crustal origin.

The upper part of the Leptynite-Amphibolite Group, with serpentinized peridotite swarms and various relics of HP rocks in a metagraywacke matrix, could represent an oceanic accretionary prism in a subduction zone (Lardeaux et al., 2001).

(6) The Brévennes-Morvan series is a complex pile of sediments and volcanic rocks from Late Devonian to Early Carboniferous. This epizonal pile lies on top of the meso-catazonal rocks of the Lyonnais unit, which is the eastern continuation of the Leptynite-Amphibolite Group, and in which are found coesite-bearing eclogites (Lardeaux et al., 2001). This boundary is complex and variable: the southern boundary in the Brévennes valley between the epizonal mafic volcanic Brévennes series and the high-grade underlying rocks is a southward thrust with a dextral wrench component (Feybesse et al., 1988). The volcanic series

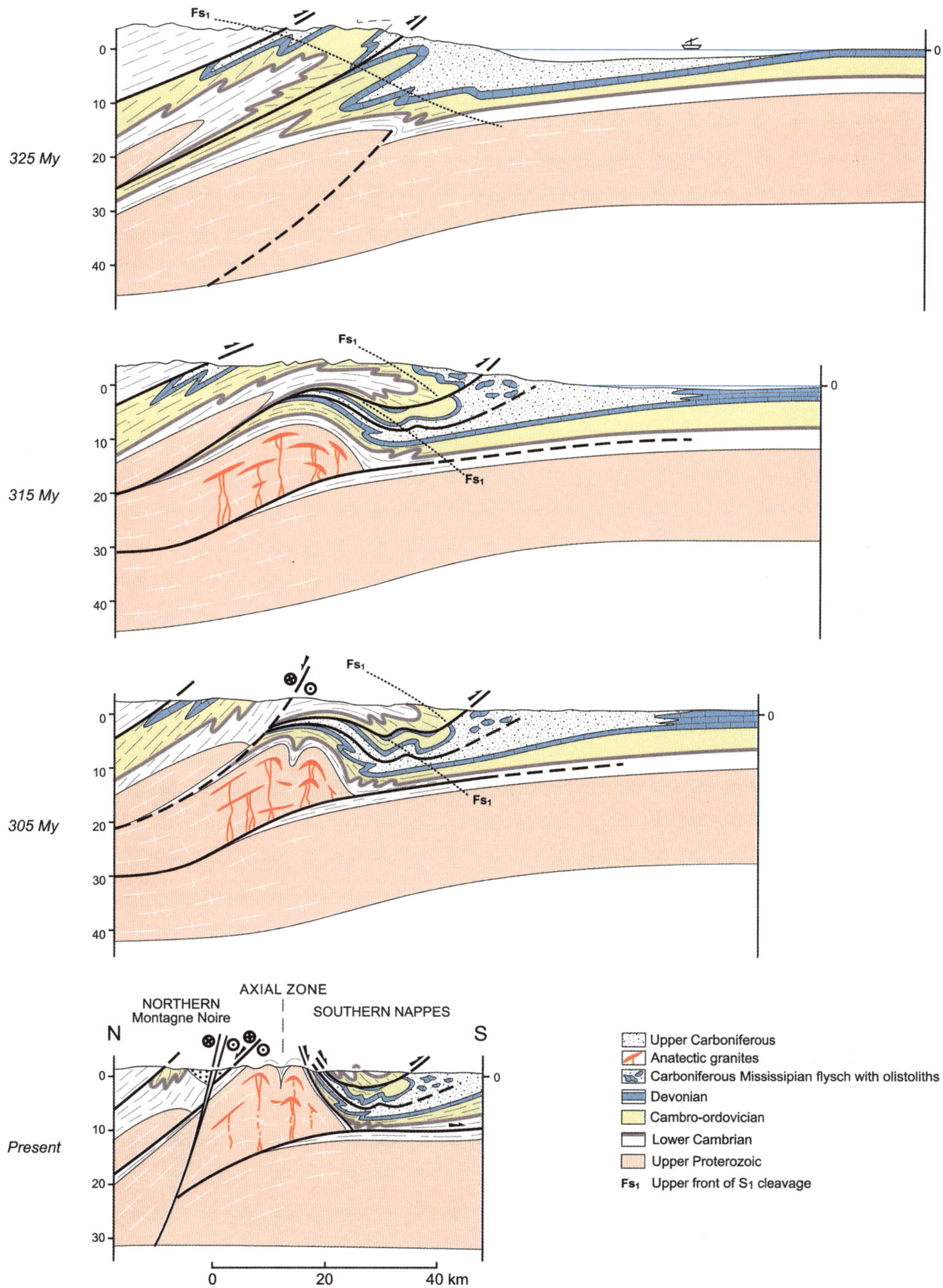

Figure 3. Tentative reconstruction of the emplacement of the Montagne Noire nappes (southern edge of the French Massif Central) during the Pennsylvanian.

is strongly shortened with folds and stretched pillow lavas. In another place to the north, epizonal schistose mafic rocks of probable early Visean age overlie with a clear stratigraphic unconformity onto high-grade sillimanite-bearing mica schists. In other places (Tarare), the epizonal mafic series of probable Tournaisian age are on top of the high-grade gneiss with a contact dipping gently to the NW with top-to-the-NW shear indicators. The Brévennes-Morvan series is in turn unconformably overlain by the weakly deformed Upper Visean "Tufs anthracifères," which is made up of alternating dacitic tuffs, conglomerates, pelites, sandstones, and coal-bearing beds.

The Brévennes Upper Devonian bimodal association of spilite-keratophyre with gabbros, serpentinites, and mid-ocean-ridge basalt (MORB)–type basalts was considered by Pin and Paquette (1997) to be the result of subduction-related rift volcanism. More to the north (Morvan, Beaujolais), the Upper Devonian calc-alkaline volcanic mafic rocks and the diorites show a clear mantelic origin and were interpreted by Pin and Paquette (2002) as related to a southward subduction.

ATTITUDE AND MEANING OF THE LINEATIONS

Lineations on Gently Dipping Foliations

In the Montagne Noire, stretching lineations are not developed in the southern nappes where buckling and flattening dominate. It is only in the deepest part of the nappe pile, in the metamorphic axial zone and in the metasediments to the south, that E-W– to NE-SW–trending stretching lineations develop parallel to the trend of recumbent folds (Fig. 4). They are particularly prominent in the Upper Devonian red goniatite-rich limestones. The strong coaxial stretching is the result of two superposed cleavages (Mattauer et al., 1996) and of pressure-solution deformation comparable to that of the Purgatory Conglomerate in Rhode Island (Mosher, 1987).

In the metamorphic axial zone, the same kind of coaxial stretching lineations trending ENE-WSW are well expressed, especially in the Ordovician orthogneisses (Roger et al., 2004). They are not related to a large ductile strike-slip fault as proposed by Nicolas et al. (1977), but they developed normal to the post-folial NNW-SSE shortening that formed the metamorphic dome. The deformation becomes noncoaxial at the periclinal ends of the dome, with opposite detachments toward ENE and WSW (Matte et al., 1998). This deformation occurred between 327 and 305 Ma when anatectic granites and pegmatites, dated on monazites and zircons (Matte et al., 1998; Roger, 2006, personal commun.), were emplaced.

In the northern epizonal part of the Montagne Noire, where slaty cleavage is present in the whole Lower Paleozoic series, stretching lineations trend NNE at high angle to the recumbent F1 folds, which trend roughly NW-SE with a south to southwestward vergence. The 30°–60° NW-dipping thrusts bear the same kind of lineations, but here they have a clear top-to-the-SW shear sense and thus have a sinistral strike-slip displacement component.

In the Cévennes-Albigeois pelitic series, lineations have a relatively homogeneous pattern. They roughly trend N-S, varying from

Figure 4. NE-SW–trending fold overturned toward the southeast in the Devonian limestones of the Montagne Noire (Moulin de Graïs) with axial-planar cleavage and stretching lineation parallel to the intersection lineation. Photo of Aude Gebelin.

NW to NE (Mattauer and Etchecopar, 1977; Arnaud and Burg, 1993,), and may be interpreted as transport lineations indicating a top-to-the-S shear sense related to the main phase of southward thrusting. The pattern becomes more complex to the north, near the contact with the Leptynite-Amphibolite unit: south of Marvéjols, coaxial lineations trending E-W to WNW-ESE are present in granitic orthogneisses and Cambrian metadiorites that intrude into the pelitic series. These lineations, with equivocal shear sense, seem to result from the superposition of two flat foliations: post-F1 folding with axial-planar strain-slip foliation, and F2 folds, which are probably related to the late stage of thrusting and amphibolite-facies metamorphism. The boundary between the pelitic series and the overlying Leptynite-Amphibolite Group is sharp, and synmetamorphic mylonites developed in some places with N-S–trending lineations and southward shear indicators (Faure et al., 1979). The $^{39}Ar/^{40}Ar$ ages determined by Costa (1989) on biotites and muscovites of the mica schists, metadiorites, and mylonites (350–340 Ma) give an upper limit (exhumation) for the thrusting.

In the more internal and metamorphic (sillimanite grade) parts of the Massif Central, where the pelitic series and granitic orthogneisses outcrop in windows beneath the Leptynite-Amphibolite Group, NW-SE–trending lineations correspond to the axes of isoclinal folds with no clear shear sense.

In the Leptynite-Amphibolite Group, stretching lineations are rare. In the upper part, a strong anatexis has masked earlier structures including foliations. In the lower part, flattening generally dominates with widespread chocolate-tablet boudinage. Only near the basal thrust are lineations and shear criteria visible. The best examples outcrop north of Marvéjols, along the National Road 9, which shows a good section through the Leptynite-Amphibolite Group The flat foliation, which gently dips north, bears a NW-SE–trending lineation normal to the necks of large 10-m-scale boudins of amphibolites. These boudins, which, in some places, show relics of equigranular eclogites or mafic granulites in the core, are asymmetric with a clear top-to-the-SE shear sense (Fig. 5). A U/Pb age by Pin (1989) on zircons in a pegmatitic neck of a boudin (344 +12/–28 Ma, upper intercept) is close to the $^{39}Ar/^{40}Ar$ ages found below in the parautochthonous series.

Lineations Related to Ductile Strike-Slip Faults

These lineations are nearly horizontal on vertical or steeply dipping mylonitic foliations. Dextral and sinistral strike-slip faults that deform two-mica granites are well developed essentially in the northwestern part of the French Massif Central (Gebelin, 2004) and are the continuation of the large dextral South Armorican shear zone. The granites were in part emplaced during this strike-slip deformation ca. 315–320 Ma (Gebelin et al., 2007). Some westward detachments, like the Argentat fault (Mattauer et al., 1988) (Fig. 2), may be related to the South Armorican–French Massif Central dextral shear zone, like the detachment related to the Karakoram dextral fault in western Tibet (Murphy et al., 2002).

Figure 5. Asymmetric amphibolite boudin in the Leptynite-Amphibolite Group, 2 km to the north of Marvejols on National Road 9, showing a top-to-the-SE shearing. Photo of Philippe Turpaud.

DISCUSSION

In the French Massif Central, different kinds of lineations were formed during different Variscan tectonometamorphic events, which took place between 430 and 310 Ma. There is no lineation clearly related to the early subduction and thrusting events (430–390 Ma). Apart from the lineations related to strike-slip mylonites along the steeply dipping strike-slip faults, most of the lineations occur on flat-lying foliations. Among the lineations resulting from noncoaxial deformation, some are clearly related to the southward-directed emplacement of the French Massif Central nappes, particularly near and beneath the main thrust between the Leptynite-Amphibolite Group and the underlying pelitic and gneissic Cévennes series. Others that trend E-W to NW-SE, with top-to-the-NW shear sense, mainly in western French Massif Central, have been interpreted in different ways. Brun and Burg (1982) and Burg et al. (1987) proposed that these lineations resulted from a "corner effect" along the Iberian-Armorican arc, which deflected the N-S lineations toward the west. Roig and Faure (2000) proposed that this top-to-the-NW shearing corresponded to a northwestward thrusting event that occurred at 350 Ma, sandwiched between two southward thrust events dated 380 Ma and 320 Ma. This latter hypothesis is hard to reconcile with the general vergence of thrusts and recumbent folds toward the inner, concave side of the Iberian-Armorican arc (Figs. 1 and 6), and an abrupt change of the orogenic polarity seems unlikely in such a thrust belt. Burg et al. (1994) proposed that E-W– to NW-SE–trending stretching lineations, near-parallel to the trend of the Variscan belt in southern Brittany and Massif Central, resulted from a synthickening extension from 325 to 310 Ma. I agree and think that the top-to-the-NW shearing represents synconvergence detachments that formed during progressive south to southeastward thrusting. These detachments probably began at the beginning of the Carboniferous and helped to exhume the Leptynite-Amphibolite unit, with its high-pressure relics; this agrees with the metamorphism related to this shearing that shows increasing temperature and decreasing pressure. In the western French Massif Central (Limousin), northwestward shearing occurred at 350 Ma, younger than Barrovian metamorphism (Bellot, 2001). Exhumation continued during the Pennsylvanian (320–310 Ma) in the eastern part of the French Massif Central with northeastward shearing, and to the west along the Iberian-Armorican arc as shown in southern Brittany by Gapais et al. (1993). In the Iberian part of the arc, synconvergence detachments also formed early in Mississippian times during the emplacement of the allochthonous units (Martínez Catalán et al., 2002a) and later in more external zones during the eastward progradation of the thrusts (Fig. 6).

CONCLUSION

The French Massif Central, characterized by large high-grade metamorphic nappes and recumbent folds is part of the so-called Iberian-Armorican arc. From the internal meso-catazonal zones near the Galicia–Brittany–French Massif Central suture to the external epizonal parts, the vergence is toward the inner part of the arc, where deformation reaches the southern Carboniferous flysch basin. In both western Iberia and the French Massif Central, these structures developed during a complex convergence history, from the early high- to ultrahigh-pressure stage (ca. 400 Ma) to the latest thrusts in the external Carboniferous basins (310–300 Ma), with a clear forelandward migration of the deformation and metamorphism through time. Both areas show a similar major thrust that transported a high-grade allochthon with early Variscan HP to UHP relics to the east and the south. This allochthon is better defined in western Iberia (Martínez Catalán et al., 2002b), where there is a clear ophiolitic unit. Though the structure and the tectonothermal history of the French Massif Central are less well known, due to the lack of a deep seismic-reflection profile and to the scarcity of geochronological data, we propose the following evolution (Fig. 7):

In Late Ordovician times (ca. 450 Ma), a probably small Galicia–South Brittany–French Massif Central ocean closed by subduction to the northwest (Matte, 2001).

During the Silurian (420 Ma), the thinned passive continental margin of Gondwana was subducted to great depth together with the overlying accretionary prism, and both underwent high- to ultrahigh-pressure metamorphism.

Upward buoyancy-driven rising of the thinned continental slice dragging the subducted sediments and parts of the oceanic crust and mantle (Chemenda et al., 1997) occurred during the Early to Middle Devonian.

Continued northward subduction of continental crust during the Devonian caused progressive southward thrusting and further exhumation of the Leptynite-Amphibolite Group by northwestward detachments, with probable breakoff of the northward-subducting lithosphere and beginning of delamination.

During the Late Devonian and early Carboniferous, strong rifting and calc-alkaline volcanism with marine series developed in the northeastern part of the French Massif Central (Brévennes-Morvan series) after complete exhumation and strong erosion of the Leptynite-Amphibolite Group rocks. Erosion of the internal parts of the belt is recorded in the external Carboniferous flysch basin by the presence of gneissic pebbles and metamorphic minerals. The Brévenne-Morvan volcanism can hardly have been related to a continuous northward subduction as proposed by Lardeaux et al. (2001), but it could have been the result of slab rollback or breakoff. Another possibility is that this volcanism was related to a southward subduction of the Rheic Ocean (Pin and Paquette, 1997), as proposed in Figure 7. At the same time (360–350 Ma), the latest stages of southward-directed thrusting of the Leptynite-Amphibolite Group occurred, along with continuous northwestward detachment.

During the Early Carboniferous (340–330 Ma), the Brévennes-Morvan series were in turn strongly folded with out-of-sequence southward thrusts and epizonal slaty-cleavage before the deposit of unconformable Late Visean volcanics and coal-bearing series (Tufs anthracifères). At the same time,

Figure 6. Ages (estimated from stratigraphy and U-Pb and Ar/Ar dating) and sense of thrusting, detachment, and strike-slip faulting in the Iberian-Armorican arc. Green—allochthonous complexes in western Iberia, southern Brittany, and French Massif Central (FMC). Red arrows—thrusting, blue arrows—detachments.

southward-directed thrusting continued in the southernmost flank of the French Massif Central with the emplacement of the Montagne Noire fold nappes (Fig. 3) during the Namurian (320–315 Ma).

Many questions still remain in this history: how did the extent of the Leptynite-Amphibolite Group nappe reach 300 km? Did the Permian-Carboniferous late extension drastically increase the size of the nappe as proposed by Burg et al. (1994)? Also, the timing of exhumation of so large a nappe is still not very well constrained, and the compatibility and timing among thrusts, detachments, and strike-slip faults are incompletely understood.

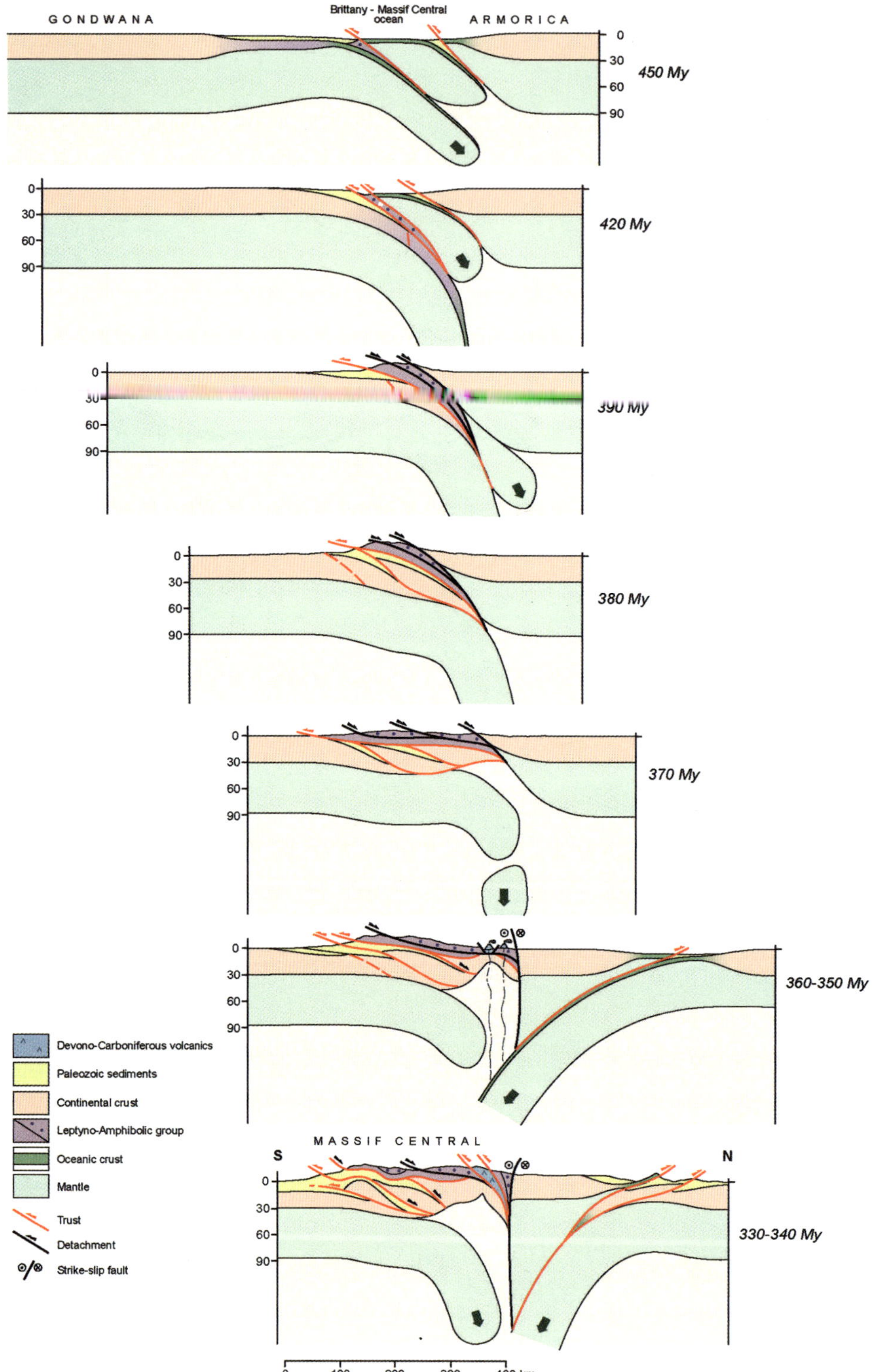

Figure 7. Schematic cartoon showing the subduction history of the French Massif Central. The Leptynite-Amphibolite Group complex with high-pressure to ultrahigh-pressure relics (400–430 Ma and Barrovian metamorphism, 390–360 Ma) was probably exposed before Upper Devonian–Lower Carboniferous times in the northern part of the French Massif Central and intruded by subduction-related Devonian-Carboniferous volcanism. The vertical scale is exaggerated.

ACKNOWLEDGMENTS

I thank Bob Hatcher, who invited me to the International Basement Tectonic Conference in Oak Ridge, Tennessee, June 2004; Anne Delplanque, for the final version of the figures; Jean-Philippe Devaux, who helped me with the English; and the two reviewers, J.P. Burg and Bob Hatcher, for their constructive remarks.

REFERENCES CITED

Arnaud, F., and Burg, J.P., 1993, Microstructures des mylonites schisteuses: Cartographie des chevauchements varisques dans les Cévennes et détermination de leur cinématique: Comptes Rendus de l'Académie des Sciences de Paris, v. 317, p. 1441–1447.

Arthaud, F., 1969, Etude tectonique et microtectonique comparée de deux domaines Hercyniens: Les nappes de la Montagne Noire (France) et l'anticlinorium de l'Iglesiente (Sardaigne): Montpellier, Publications de l'Université des Sciences et Techniques du Languedoc (USTELA), Série Géologie Structurale, 1970, 175 p.

Bellot, J.-Ph., 2001, La Structure de la Croûte Varisque du Sud-Limousin (Massif Central Français) [these]: Montpellier, Université Montpellier 2, 320 p.

Bodinier, J.L., Burg, J.P., Leyreloup, A., and Vidal, H., 1988, Reliques d'un bassin d'arrière-arc subducté puis obducté dans la région de Marvéjols (Massif Central): Bulletin de la Société Géologique de France, v. 8, p. 115–123.

Bonnot, H., and Piboule, M., 1980, Mise en évidence d'une dualité d'origine des ultrabasites à grenat du Limousin et recherche de la signification des péridotites d'origine mantellique dans le Massif Central Français: Comptes Rendus de l'Académie des Sciences de Paris, v. 291, p. 129–132.

Brun, J.-P., and Burg, J.P., 1982, Combined thrusting and wrenching in the Ibero-Armorican arc: A corner effect during continental collision: Earth and Planetary Science Letters, v. 61, p. 319–332, doi: 10.1016/0012-821X(82)90063-2.

Burg, J.-P., and Matte, Ph., 1977, La klippe de la Bessenoits (Decazeville, Aveyron), un nouvel argument en faveur de l'existence d'un chevauchement majeur vers le sud dans le Massif Central: Comptes Rendus Sommaires de la Société Géologique de France, v. 6, p. 325–329.

Burg, J.-P., and Matte, Ph., 1978, A cross-section through the Massif Central and the scope of its Variscan geodynamic evolution: Zeitschrift Deutschland. Geologische Gesselschaft, v. 129, p. 429–460.

Burg, J.-P., Leyreloup, A., Marchand, J., and Matte, Ph., 1984, Inverted metamorphic zonation and large scale thrusting in the Variscan belt: An example in the French Massif Central, *in* Hutton, D.H.M., and Sanderson, P.J., eds., Variscan Tectonics of the North Atlantic Region: Geological Society [London] Special Publication 14, p. 47–61.

Burg, J.-P., Balé, P., Brun, J.P., and Girardeau, J., 1987, Stretching lineation and transport direction in the Ibero-Armorican arc during the Siluro-Devonian collision: Geodinamica Acta, v. 1, p. 71–87.

Burg, J.-P., Van den Driessche, J., and Brun, J.P., 1994, Syn- to post-thickening extension in the Variscan belt of Western Europe: Modes and structural consequences: Géologie de la France, v. 3, p. 33–51.

Chemenda, A., Matte, Ph., and Sokolov, V., 1997, Model of Paleozoic obduction and exhumation of high pressure/low temperature rocks in southern Urals: Tectonophysics, v. 276, p. 217–227, doi: 10.1016/S0040-1951(97)00057-7.

Costa, S., 1989, Age radiométrique $^{40}Ar/^{39}Ar$ du metamorphisme des séries du Lot et du charriage du Groupe Leptyno-Amphibolique de Marvéjols (M.C.F.): Comptes Rendus de l'Académie des Sciences de Paris, v. 309, p. 561–567.

Costa, S., 1990, De la Collision Continentale à l'Extension Tardi-Orogénique: 100 millions d'Années d'Histoire Varisque dans le Massif Central Français: Une Étude Chronologique par la Méthode $^{40}Ar/^{39}Ar$ [thèse]: Montpellier, l'Université Montpellier 2, 441 p.

Costa, S., 1992, East-west diachronism of the collisional age in French Massif Central: Implications for the European Variscan orogen: Geodinamica Acta, v. 5, p. 51–68.

Demay, A., 1942, Microtectonique et tectonique profonde: Cristallisations et injections magmatiques syntectoniques: Memoire Explicatif de la Carte Géologique dé France, 260 p.

Dubuisson, G., Hirn, A., Girardeau, J., Mercier, J.-C.C., and Veinante, J.L., 1988, Multiple Variscan nappes in Limousin, western Massif Central, France: Geophysical constraints to the geological model and geodynamic implications: Tectonophysics, v. 147, p. 19–31, doi: 10.1016/0040-1951(88)90145-X.

Ducrot, J., Lancelot, J.R., and Marchand, J., 1983, Datation U-Pb sur zircons de l'éclogite de la Borie (Haut Allier, France) et consequences sur l'évolution ante-hercynienne de l'Europe occidentale: Earth and Planetary Science Letters, v. 62, p. 385–394.

Engel, W., Feist, R., and Franke, W., 1982, Le Carbonifère anté-stéphanien de la Montagne Noire: Rapports entre mise en place des nappes et sédimentation: Bulletin du Bureau de Recherches Géologique et Minières, Orléans, v. 4, p. 341–389.

Faure, M., Pin, Chr., and Mailhé, D., 1979, Les roches mylonitiques associées au charriage du Groupe Leptyno-Amphibolique de Marvéjols, (Lozère, Massif Central Français): Comptes Rendus de l'Académie des Sciences de Paris, v. 288, p. 1267–1270.

Feybesse, J.L., Lardeaux, J.M., Johan, V., Tegyev, M., Dufour, E., Lemiere, B., and Delfour, J., 1988, La série de la Brévenne (Massif Central Français): Une unité Dévonienne charriée sur le complexe métamorphique des monts du Lyonnais à la fin de la collision Varisque: Comptes Rendus de l'Académie des Sciences de Paris, v. 307, p. 991–996.

Friedrich, M., Marignac, Chr., and Floc'h, J.P., 1988, Sur l'existence de trois chevauchements ductiles "Himalayens" successifs à vergence NW en Limousin: Comptes Rendus de l'Académie des Sciences de Paris, v. 306, p. 663–669.

Gapais, D., Lagarde, J.-L., Le Corre, C., Audren, C., Jegouzo, P., Casas Sainz, A., and Van Den Driessche, J., 1993, La zone de cisaillement de Quiberon: Témoin d'extension de la chaîne Varisque en Bretagne méridionale au Carbonifère: Comptes Rendus de l'Académie des Sciences de Paris, v. 316, p. 1123–1129.

Gardien, V., Lardeaux, J.-M., and Miseri, M., 1988, Les péridotites des Monts du Lyonnais (M.C.F.): Témoins privilégiés d'une subduction de lithosphère Paléozoïque: Comptes Rendus de l'Académie des Sciences de Paris, v. 307, p. 1967–1972.

Gebelin, A., 2004, Déformation et mise en place des granites (360–300 Ma) dans un segment de la Chaîne Varisque (Plateau de Millevaches, Massif Central) [these]: Montpellier, Université Montpellier 2, 235 p.

Gebelin A., Brunel, M., Monié, P., Faure, M., and Arnaud, N., 2007, Transpressional tectonics and Carboniferous magmatism in the Limousin (Massif Central, France): Structural and $^{40}Ar/^{39}Ar$ investigations: Tectonics, v. 26, no. 2, TC 2008.

Gèze, B., 1949, Etude géologique de la Montagne Noire et des Cévennes méridionales: Mémoires de la Société Géologique de France, no. 62, p. 1–215.

Lardeaux, J.-M., Ledru, P., Daniel, I., and Duchene, S., 2001, The Variscan French Massif Central—A new addition to the ultra-high pressure metamorphic "club": Exhumation processes and geodynamic consequences: Tectonophysics, v. 332, p. 143–167, doi: 10.1016/S0040-1951(00)00253-5.

Lasnier, B., 1971, Les péridotites et pyroxénolites à grenat du Bois des Feuilles (Monts du Lyonnais, France): Contributions to Mineralogy and Petrology, v. 34, p. 29–42, doi: 10.1007/BF00376029.

Malavieille, J., Guilhot, S., Costa, S., Lardeaux, J.-M., and Gardien, V., 1990, Collapse of a thickened Variscan crust in the French Massif Central: Mont du Pilat extensional shear zone and St. Etienne Carboniferous basin: Tectonophysics, v. 177, p. 139–149, doi: 10.1016/0040-1951(90)90278-G.

Martínez Catalán, J.R., Hatcher, R.D., Jr., Arenas, R., and Diaz Garcia, F., eds., 2002a, Variscan-Appalachian Dynamics: The Building of the Late Paleozoic Basement: Geological Society of America Special Paper 364, 305 p.

Martínez Catalán, J.R., Diaz Garcia, F., Arenas, R., Abati, J., Castinieira, P., Gonzalez Cuadra, P., Gomez Barreiro, J., and Rubio Pascal, F.J., 2002b, Thrust and detachment systems in the Ordenes complex (northwestern Spain): Implications for the Variscan-Appalachian geodynamics, *in* Martínez Catalán, J.R., Hatcher, R.D., Jr., Arenas, R., and Diaz Garcia, F., eds., Variscan-Appalachian Dynamics: The Building of the Late Paleozoic Basement: Geological Society of America Special Paper 364, p. 163–182.

Mattauer, M., and Etchecopar, A., 1977, Arguments en faveur de chevauchements de type Himalayen dans la chaîne Hercynienne du Massif Central Français: Colloque International du Centre National de la Recherche Scientifique, v. 268, p. 261–267, Paris 1976.

Mattauer, M., Brunel, M., and Matte, P., 1988, Failles ductiles et grands chevauchements: Une nouvelle analogie entre l'Himalaya et la chaîne hercynienne du massif central Français: Comptes Rendus de l'Académie des Sciences de

Paris, v. 306, p. 671–676.

Mattauer, M., Laurent, Ph., and Matte, Ph., 1996, Plissement Hercynien synschisteux post-nappe et étirement subhorizontal dans le versant sud de la Montagne Noire (Sud du Massif Central, France): Comptes Rendus de l'Académie des Sciences de Paris, v. 322, p. 309–315.

Matte, Ph., 1968, La structure de la virgation Hercynienne de Galice (Espagne): Travaux du Laboratoire de Géologie de la Faculté des Sciences de Grenoble: Géologie Alpine, v. 44, 128 p.

Matte, Ph., 1986, Tectonics and plate tectonics model for the Variscan belt of Western Europe: Tectonophysics, v. 126, p. 329–374, doi: 10.1016/0040-1951(86)90237-4.

Matte, Ph., 2001, The Variscan collage and orogeny (480–290 Ma) and the tectonic definition of the Armorica microplate: A review: Terra Nova, v. 13, p. 122–128, doi: 10.1046/j.1365-3121.2001.00327.x.

Matte, Ph., and Burg, J.-P., 1981, Sutures, thrusts and nappes in the Variscan arc of Western Europe: Plate-tectonic implications, *in* McClay, K.R., and Price, N.J., eds., Thrust and Nappe Tectonics: Geological Society [London] Special Publication 9, p. 353–358.

Matte, Ph., and Ribeiro, A., 1975, Forme et orientation de l'ellipsoïde de déformation dans la virgation Hercynienne de Galice. Relations avec le plissement et hypothèses sur la genèse de l'arc Ibéro-Armoricain: Comptes Rendus de l'Académie des Sciences de Paris, v. 280, p. 2825–2828.

Matte, P., Lancelot, J., and Mattauer, M., 1998, La zone axiale hercynienne de la Montagne Noire n'est pas un "metamorphic core complex" extensif mais un anticlinal post-nappe à cœur anatectique: Geodinamica Acta, v. 11, p. 13–22, doi: 10.1016/S0985-3111(98)80025-9.

Mosher, S., 1987, Pressure-solution deformation of the Purgatory Conglomerate, Rhode Island (U.S.A.): Quantification of volume change, real strains and sedimentary shape factor: Journal of Structural Geology, v. 9, p. 221–232, doi: 10.1016/0191-8141(87)90027-7.

Murphy, M.A., Yin, An, Knapp, P., Harrison, T. M., Menning, C. E., Ryerson, F.J., Lin, Dong, and Guo, Jingi, 2002, Structural evolution of the Gurla Mandhata detachment system, southwest Tibet: Implications for the eastward extent of the Karakoram fault system: Geological Society of America Bulletin, v. 114, p. 428–447.

Nicolas, A., Bouchez, J.L., Blaise, J.L., and Poirier, J.P., 1977, Geological aspects of deformation in continental shear-zones: Tectonophysics, v. 42, p. 55–73, doi: 10.1016/0040-1951(77)90017-8.

Paquette, J.-L., Monchoux, P., and Couturier, M., 1995, Geochemical and isotopic study of a norite-eclogite transition in the European Variscan belt: Implications for U-Pb zircon systematics in metabasic rocks: Geochimica et Cosmochimica Acta, v. 59, p. 1611–1622.

Pin, C., 1979, Age à 482 Ma des roches orthodérivées du Groupe Leptyno-Amphibolique de Marvejols (Lozère Massif Central Français), déterminé par la méthode U-Pb sur zircons: Comptes Rendus de l'Académie des Sciences de Paris, v. 28, p. 291–293.

Pin, C., 1989, Essai sur la Chronologie et l'Évolution Géodynamique de la Chaîne Hercynienne d'Europe [thèse d'Etat]: Clermont-Ferrand, Université de Clermont-Ferrand, 455 p.

Pin, Chr., and Lancelot, J.R., 1982, U-Pb dating of an early bimodal Paleozoic magmatism in the French Massif Central and of its further metamorphic evolution: Contributions to Mineralogy and Petrology, v. 79, p. 1–12, doi: 10.1007/BF00376956.

Pin, Chr., and Marini, F., 1993, Early Ordovician continental break-up in Variscan Europe: Nd-Sr isotope and trace element evidence from bimodal igneous associations of the southern Massif Central, France: Lithos, v. 29, p. 177–196, doi: 10.1016/0024-4937(93)90016-6.

Pin, C., and Paquette, J.-L., 1997, A mantle-derived bimodal suite in the Hercynian belt: Nd isotope and trace element evidence for a subduction related rift origin of the Late Devonian Brévenne metavolcanics, Massif Central (France): Contributions to Mineralogy and Petrology, v. 129, p. 222–238, doi: 10.1007/s004100050334.

Pin, C., and Paquette, J.-L., 2002, Sr-Nd isotope and trace element evidence for a late Devonian active margin in northern Massif Central (France): Geodinamica Acta, v. 15, p. 63–77.

Roger, F., Respaut, J.P., Brunel, M., Matte, Ph., and Paquette, J.-L., 2004, Première datation U-Pb des orthogneiss oeillés de la zone axiale de la (Sud du Massif Central): Montagne Noire nouveaux témoins du magmatisme varisque Ordovicien dans la chaîne varisque: Comptes Rendus de Géoscience, v. 336, p. 19–28.

Roig, J.-Y., and Faure, M., 2000, La tectonique cisaillante polyphasée du Sud Limousin (Massif Central Français) et son interprétation dans un modèle d'évolution polycyclique de la chaîne Hercynienne: Bulletin de la Société Géologique de France, v. 171, p. 295–307, doi: 10.2113/171.3.295.

Roig, J.-Y., Faure, M., and Ledru, P., 1996, Polyphase wrench tectonics in the southern French Massif Central. Kinematic inferences for pre- and synkinematic granitoids: Geologische Rundschau, v. 85, p. 138–153, doi: 10.1007/s005310050063.

Roques, M., 1971, Structure géologique du Massif Central, *in* Symposium J. Jung: Géologie, Géomorphologie et Structure du Massif Central Français (Plein Air-Service edit): Clermont-Ferrand, p. 17–32.

Thompson, P.H., and Bard, J.-P., 1982, Isograds and mineral assemblages in the Eastern axial zone, Montagne Noire (France). Implications for temperature gradients and P/T history: Canadian Journal of Earth Sciences, v. 19, p. 129–143.

Vanderhaeghe, O., Burg, J.P., and Teyssier, C., 1999, Exhumation of migmatites in two collapsed orogens: Canadian Cordillera and French Variscides, *in* Ring, U., Brandon, M.T., Gordon, S., Willett, and Sean, D., eds., Exhumation Processes: Normal Faulting, Ductile Flow and Erosion: Geological Society [London] Special Publication 154, p. 181–204.

Manuscript Accepted by the Society 1 February 2007

Printed in the USA

The Geological Society of America
Memoir 200
2007

Space and time in the tectonic evolution of the northwestern Iberian Massif: Implications for the Variscan belt

José R. Martínez Catalán
Departamento de Geología, Universidad de Salamanca, 37008 Salamanca, Spain

Ricardo Arenas
Departamento de Petrología y Geoquímica, Universidad Complutense, 28040 Madrid, Spain

Florentino Díaz García
Departamento de Geología, Universidad de Oviedo, 33005 Oviedo, Spain

Pablo González Cuadra
Área de Geodinámica Externa, Universidad de León, 24007 León, Spain

Juan Gómez-Barreiro
Jacobo Abati
Pedro Castiñeiras
Javier Fernández-Suárez
Sonia Sánchez Martínez
Pilar Andonaegui
Departamento de Petrología y Geoquímica, Universidad Complutense, 28040 Madrid, Spain

Emilio González Clavijo
Alejandro Díez Montes
Instituto Geológico y Minero de España, Azafranal 48, 37001 Salamanca, Spain

Francisco J. Rubio Pascual
Instituto Geológico y Minero de España, Ríos Rosas 23, 28003 Madrid, Spain

Beatriz Valle Aguado
Departamento de Geociências, Universidade de Aveiro, 3810-193 Aveiro, Portugal

ABSTRACT

Recent advances in geochemical studies of igneous rocks, isotopic age data for magmatism and metamorphism, quantitative pressure-temperature *(P-T)* estimates of metamorphic evolution, and structural geology in the northwestern Iberian Massif are integrated into a synthesis of the tectonic evolution that places the autochthonous

Martínez Catalán, J.R., Arenas, R., Díaz García, F., González Cuadra, P., Gómez-Barreiro, J., Abati, J., Castiñeiras, P., Fernández-Suárez, J., Sánchez Martínez, S., Andonaegui, P., González Clavijo, E., Díez Montes, A., Rubio Pascual, F.J., and Valle Aguado, B., 2007, Space and time in the tectonic evolution of the northwestern Iberian Massif: Implications for the Variscan belt, *in* Hatcher, R.D., Jr., Carlson, M.P., McBride, J.H., and Martínez Catalán, J.R., eds., 4-D Framework of Continental Crust: Geological Society of America Memoir 200, p. 403–423, doi: 10.1130/2007.1200(21). For permission to copy, contact editing@geosociety.org.

and allochthonous terranes in the framework of Paleozoic plate tectonics. Because northwestern Iberia is free from strike-slip faults of continental scale, it is retrodeformable and preserves valuable information about the orthogonal component of convergence of Gondwana with Laurentia and/or Baltica, and the opening and closure of the Rheic Ocean.

The evolution deduced for northwest Iberia is extended to the rest of the Variscan belt in an attempt to develop a three-dimensional interpretation that assigns great importance to the transcurrent components of convergence. Dominant Carboniferous dextral transpression following large Devonian and Early Carboniferous thrusting and recumbent folding is invoked to explain the complexity of the belt without requiring a large number of peri-Gondwanan terranes, and its ophiolites and high-pressure allochthonous units are related to a single oceanic closure.

Palinspastic reconstruction of the Variscan massifs and zones cannot be achieved without restoration of terrane transport along the colliding plate margins. A schematic reconstruction is proposed that involves postcollisional strike-slip displacement of ~3000 km between Laurussia and Gondwana during the Carboniferous.

Keywords: Variscan belt, exotic terranes, accretionary history, strike-slip tectonics, Iberian Massif.

INTRODUCTION

A paleogeographic continental reconstruction for the late Paleozoic (Fig. 1) shows that three important Paleozoic belts, the northern Appalachians, the British and Scandinavian Caledonides, and the North German–Polish Caledonides, meet relatively close to the NW corner of the Iberian Massif. These belts mark the collision of three continental masses, Laurentia, Baltica, and Avalonia, which closed the Iapetus and Tornquist Oceans (which had opened during the Late Proterozoic and early Paleozoic). Avalonia likely formed close to Gondwana, because of its fauna (Cocks and Fortey, 1988), and it could also be considered a part of the Pan-African assemblage. It is viewed as a microcontinent or terrane assemblage detached from Gondwana in the Late Cambrian–Early Ordovician. It drifted away, creating the Rheic Ocean, and its subsequent closure during the Devonian gave rise to the Variscan belt in central and western Europe and northern Africa and to the Alleghanian orogen in North America (Hatcher, 1989, 2002; Winchester et al., 2002; van Staal et al., 1998).

Oceanic closure, which ultimately led to the formation of Pangea, occurred in several steps and gave rise to three groups of orogenic episodes, which are themselves diachronous. The first convergence-related events took place between the Late Cambrian and the Middle Ordovician. They have been described in the Scandinavian and British Caledonides (Finnmarkian and Grampian, respectively) and in the Appalachians (Taconic and Penobscottian), and they were related to arc-continent collisions in both sides of the Iapetus realm (Kelling et al., 1985; Hossack and Cooper, 1986; Stephens and Gee, 1989; van Staal, 2005; van Staal et al., 1998, this volume).

The second group of events took place between the Early Silurian and Late Devonian and was induced by the closure of Iapetus and the collision of Laurentia with Baltica (Scandian) in the north, and Laurentia with Avalonia in the south (Soper, 1988; Rey et al., 1997). Details of the latter include the accretion of Ganderia—an arc formed at the Avalonian side of Iapetus—to Laurentia in the Early Silurian (Salinic), the collision of Avalonia in the Late Silurian–Early Devonian (Acadian), and subsequent dextral transpression during the late Early Devonian–Early Carboniferous (Neoacadian) (van Staal et al., 1998; Winchester et al., 2002; van Staal, 2005). Deformation and metamorphism occurred at the same time in exotic terranes that were later incorporated to the Variscan belt. They have been referred to as Ligerian (Cogné, 1977; Faure et al., 1997) and Eo-Variscan or early Variscan events, and they have been described in the Bohemian Massif (Franke, 2000; Franke and Zelazniewicz, 2002), the French Massif Central (Santallier et al., 1994), the Armorican Massif (Ballèvre et al., 1994), the Alps (von Raumer and Neubauer, 1993, 1994), and northwestern Iberia (Gómez Barreiro et al., 2006, 2007; Fernández-Suárez et al., this volume).

The third group of events spanned the Carboniferous and Early Permian, and it gave rise to intense deformation in Europe and Africa (Variscan) and the Appalachians (Alleghanian). These events resulted from the closure of the Rheic and Theic Oceans and collision of Gondwana with the previously formed Laurussia continent (Laurentia, Baltica, and Avalonia; Lefort, 1989).

The Variscan-Appalachian belt is linear but sinuous, with several oroclinal bends. In Europe (Fig. 2), the belt runs between two arcuate structures, the Bohemian Massif (Franke and Zelazniewicz, 2002) and the Iberian-Armorican arc (Bard et al., 1971; Ribeiro et al., 1995). An additional feature of the Variscan-Appalachian belt is the extent of transcurrent movements that have been active during most, if not all of the orogeny, and that have resulted in the terrane dispersion that hinders paleogeographic restoration (Gates et al., 1986; Hatcher, 1989, 2002; Martínez Catalán, 1990; Shelley and Bossière, 2000, 2002).

The northwestern corner of the Iberian Massif includes the Spanish regions of Galicia and the Cantabrian Mountains, as well

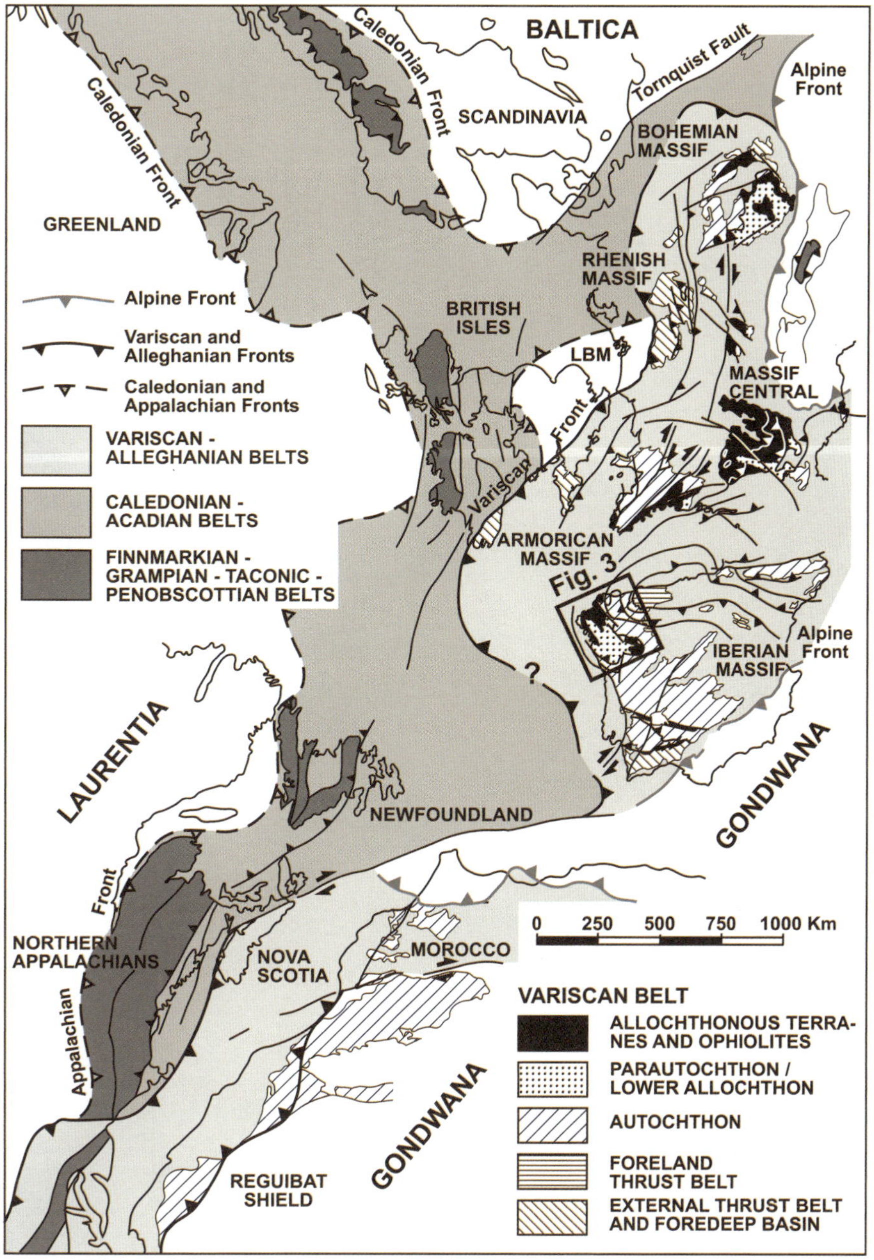

Figure 1. Sketch showing the position of Iberia in relation to the Appalachian, Caledonian, and Variscan belts at the end of Variscan convergence (modified from Martínez Catalán et al., 2002). LBM—London-Brabant Massif.

as northern Portugal, and it is located at the hinge zone of the Iberian-Armorican arc (Figs. 2 and 3). Coherent with its setting inside the Variscan mobile belt, the region preserves relics of one of the oceanic realms that once separated the early Paleozoic continents and recorded large amounts of orogenic shortening during their amalgamation. Strike-slip faults and shear zones exist but are not of continental scale, which implies that the Galician–northern Portugal section, including the Cantabrian Mountains, is retrodeformable and provides information about the orthogonal components of Gondwana-Laurussia plate convergence.

The northwestern Iberian basement consists of plutonic and metamorphic rocks with grades ranging from very low to catazonal, and a clear separation can be established between autochthonous and allochthonous terranes. The autochthon consists of a thick metasedimentary sequence deposited in northern Gondwana during the Late Proterozoic and Paleozoic, whereas the allochthon consists of the remnants of a huge and structurally complex nappe pile preserved in the core of late Variscan synforms. Both are separated by a thrust sheet, several kilometers thick, consisting of metasediments and volcanics derived

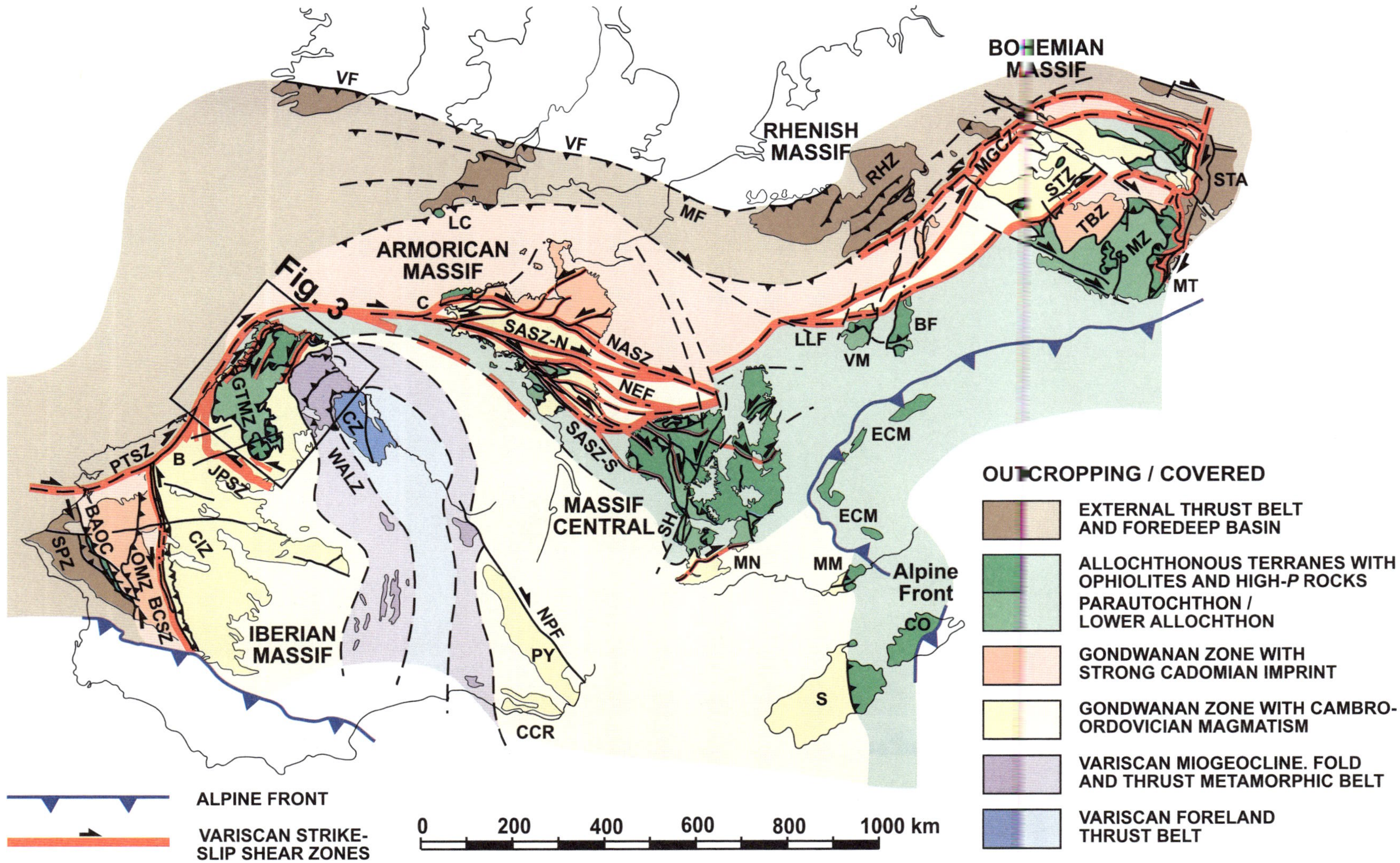

Figure 2. Subdivision of the Variscan belt showing the allochthonous terranes and the main transcurrent shear zones. Abbreviations: B—Buçaco; BAOC—Beja-Acebuches ophiolitic complex; BCSZ—Badajoz-Córdoba shear zone; BF—Black Forest; C—Crozon; CCR—Catalonia Coast Ranges; CIZ—Central Iberian zone; CO—Corsica; CZ—Cantabrian zone; ECM—External crystalline massifs of the Alps; GTMZ—Galicia-Trás-os-Montes zone; JPSZ—Juzbado-Penalva shear zone; LC—Lizard Complex; LLF—Layale-Lubine fault; MF—Midi fault; MGCZ—Mid-German crystalline zone; MM—Maures Massif; MN—Montagne Noire; MT—Moldanubian thrust; MZ—Moldanubian zone; NASZ—North Armorican shear zone; NEF—Nort-sur-Erdre fault; NPF—North Pyrenean fault; OMZ—Ossa-Morena zone; PTSZ—Porto-Tomar shear zone; PY—Pyrenees; RHZ—Rhenian-Hercynian zone; S—Sardinia; SASZ—South Armorican shear zone (N and S—northern and southern branches); SH—Sillon Houillier; SPZ—South Portuguese zone; STA—Silesian terrane assemblage; STZ—Saxo-Thuringian zone; TBZ—Teplá-Barrandian zone; VF—Variscan front; VM—Vosges Massif; WALZ—West Asturian-Leonese zone.

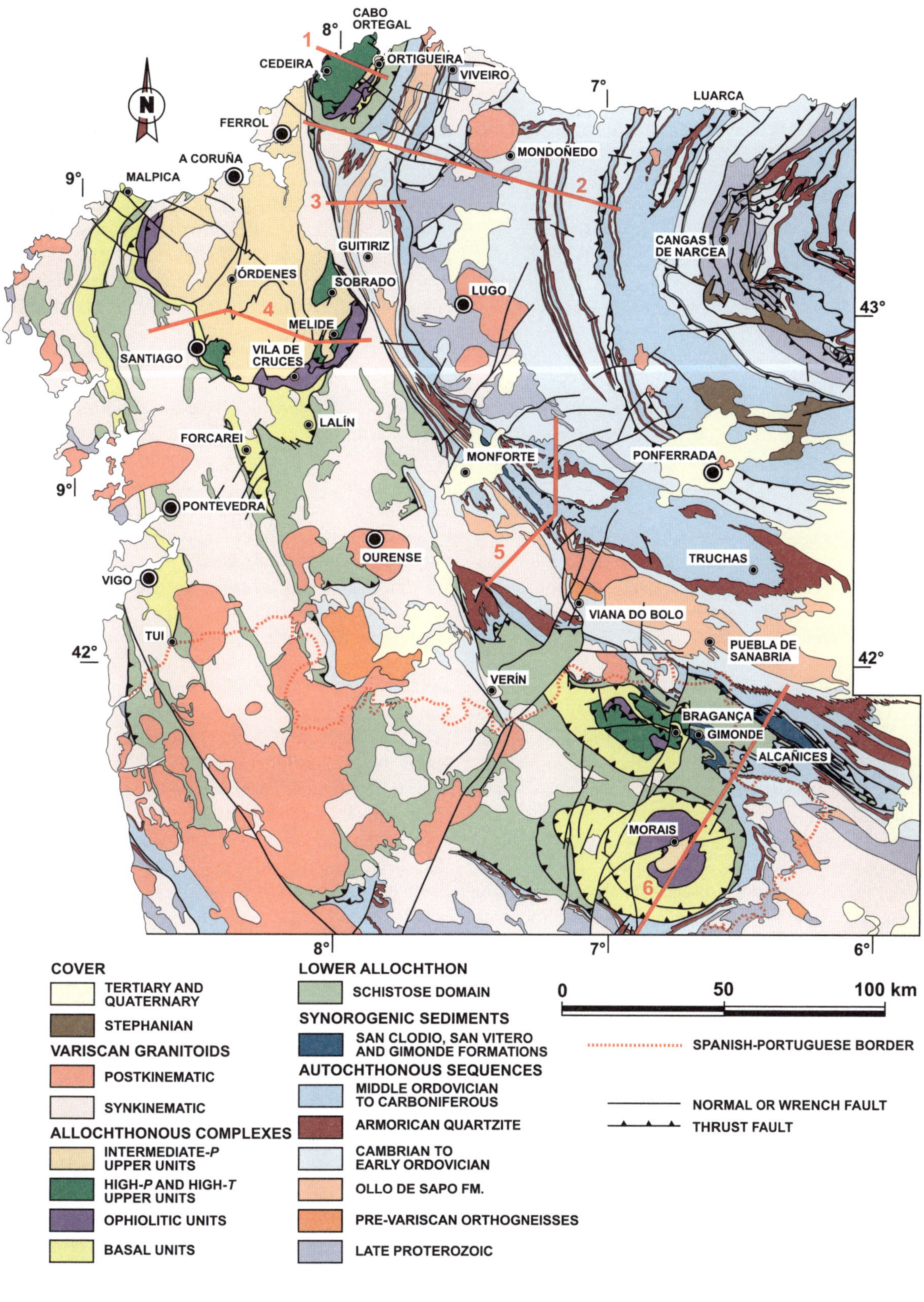

Figure 3. Geological sketch map of northwestern Iberia, showing the allochthonous complexes and their units. For location, see Figures 1 and 2. The locations of cross sections in Figure 4 are indicated.

from the outer margin of Gondwana (Farias et al., 1987), often described as a parautochthon (Ribeiro et al., 1990). However, because stratigraphic continuity with the autochthon is broken (Valverde-Vaquero et al., 2005), it will be referred to here as the lower allochthon. The allochthonous terranes, together with the lower allochthon, are included in the so-called Galicia-Trás-os-Montes zone (Farias et al., 1987).

TECTONIC SETTING OF ALLOCHTHONOUS TERRANES

There are three allochthonous complexes in Galicia (Cabo Ortegal, Órdenes, and Malpica-Tui), and two in northern Portugal (Bragança and Morais). They consist of a pile of allochthonous units characterized by unique lithologic associations and tectonometamorphic evolution. These units are separated from each other by faults, either thrusts or extensional detachments (Figs. 3 and 4). Three groups of allochthonous units can be recognized from bottom to top in ascending structural order: basal, ophiolitic, and upper units.

The basal units form a rather continuous thrust sheet consisting of schists and paragneisses alternating with felsic and mafic igneous rocks, of which granitic and peralkaline orthogneisses have yielded Rb-Sr and U-Pb ages of 490–470 Ma (Van Calsteren et al., 1979; García Garzón et al., 1981; Santos Zalduegui et al., 1995). The bimodal, partially alkaline magmatism reflects Ordovician rifting (Ribeiro and Floor, 1987; Pin et al., 1992). Since the basal units are not separated from the lower allochthon by ophiolites, it is assumed that they were part of Gondwana. Because the ophiolitic units overlie them, they are viewed as fragments of the most external edge of the Gondwanan continental margin. The Early Ordovician magmatism, partly peralkaline, probably resulted from the drift of a broken-away peri-Gondwanan terrane.

The ophiolitic units crop out discontinuously surrounding the upper units (Fig. 3) and form part of a formerly continuous and strongly imbricated nappe stack with at least two different types of ophiolite (see Arenas et al., this volume; Sánchez Martínez et al., this volume). Ophiolitic units occupying a relatively higher structural position represent the basal section of an ophiolite sequence that contains serpentinized harzburgitic ultramafic rocks, pegmatitic gabbros, and diabase dikes. Their geochemistry indicates a suprasubduction character, whereas zircons from leucogabbros yield a concordant U-Pb age of 395 Ma (Díaz García et al., 1999; Pin et al., 2002), providing evidence for oceanic crust generation, and consumption, in Early Devonian time. The structurally lower ophiolitic units consist of greenschist-facies volcanic and plutonic mafic rocks (greenstones) and metapelites, with rare felsic orthogneisses, serpentinites, and cherts—all strongly deformed.

The upper units occupy the core of the allochthonous complexes (Fig. 4) and have been subdivided according to their metamorphic evolution into high-pressure (*P*) and high-temperature (*T*) upper units, below, and intermediate-*P* upper units, above. Both groups consist of terrigenous metasediments, orthogneisses, and metabasites, with the additional presence of ultramafic rocks in the high-*P* and high-*T* upper units. The metabasites include metagabbros, eclogites, high-*P* and high-*T* mafic granulites, and amphibolites. The gabbros and orthogneisses have yielded U-Pb ages around 500 Ma, whereas detrital zircons in the metasediments indicate a maximum depositional age of 480 Ma for the uppermost, greenschist-facies metagraywackes (Fernández-Suárez et al., 2003), and 507 Ma for the structurally lower, high-*P* and high-*T* paragneisses (Schäfer et al., 1993).

Most of the mafic rocks are metagabbros with tholeiitic compositions. Their geochemical signature has been compared to mid-ocean-ridge basalt (MORB) (Gil Ibarguchi et al., 1990) and related to continental rifting in the case of the high-*P* and high-*T* upper units (Galán and Marcos, 1997), whereas the intermediate-*P* upper units have arc-tholeiitic affinities (Andonaegui et al., 2002; Castiñeiras, 2003). The additional presence of intermediate plutonic rocks, such as diorites and tonalites, in the intermediate-*P* upper units, reinforces the interpretation that these rocks were generated in an arc setting. Furthermore, geochemical studies of the ultramafic rocks of Cabo Ortegal are consistent with this hypothesis (Santos et al., 2002).

In spite of the arc affinities shown by some of them, the upper units seem to represent a terrane that had drifted away from Gondwana. The oldest ages obtained from upper intercepts and inherited zircons from the upper units (between 2.7 and 1.8 Ga; Kuijper, 1980; Peucat et al., 1990; Dallmeyer and Tucker, 1993; Schäfer et al., 1993) and the basal units (1.8 Ga; Santos Zalduegui et al., 1995) are similar to those found in the orthogneisses of the autochthon (Lancelot et al., 1985; Gebauer, 1993), and they are also similar to those of the West African craton. These ages point to a common Gondwanan basement for the upper and basal units and for the Iberian autochthon. Moreover, detrital zircon ages have been determined for graywackes from low-grade metasediments of an intermediate-*P* upper unit in the Órdenes Complex, which yielded three age populations of 2.5–2.4 Ga, 2.1–1.9 Ga, and 610–480 Ma (Fernández-Suárez et al., 2003), which also record the major events in the West African craton of northern Gondwana.

On the other hand, Late Cambrian to Early Ordovician magmatism is widespread, not only in the allochthonous units, but also in the autochthon, and is a little older in the upper units (ca. 500 Ma; Dallmeyer and Tucker, 1993; Abati et al., 1999) than in the basal units or the autochthon (490–470 Ma; Van Calsteren et al., 1979; García Garzón et al., 1981; Vialette et al., 1987; Santos Zalduegui et al., 1995; Gebauer, 1993; Valverde Vaquero and Dunning, 2000). Magmatism has calc-alkaline and arc affinities in the autochthon (Ortega et al., 1996), in some of the allochthonous upper units, and in many granitoids of the allochthonous basal units, but some granites and mafic rocks of the latter are alkaline to peralkaline (Floor, 1966; Pin et al., 1992).

To reconcile the tectonic stability registered by the Early Ordovician passive-margin sediments of the autochthon (Pérez-Estaún et al., 1991), the rift-related magmatism of the basal units,

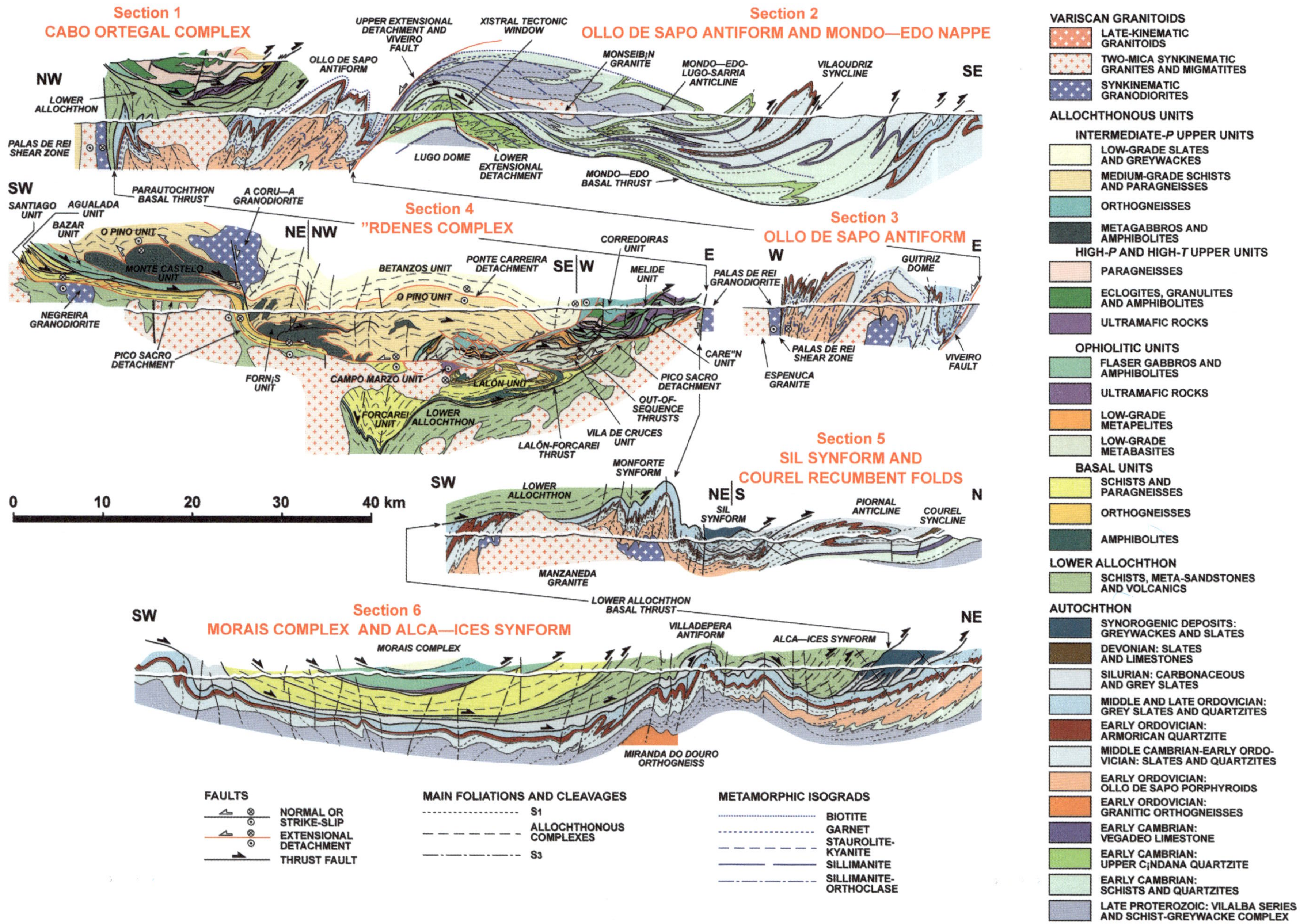

Figure 4. Geological sections across central and eastern Galicia and northern Portugal showing the main Variscan structures. Isograds of Variscan regional metamorphism have been drawn in sections 1, 2, and 3 to show their relationships to folding. See Figure 3 for locations. Sections are based on the following contributions: 1—Marcos et al. (1984) and Arenas (1988); 2—Bastida et al. (1982) and Martínez Catalán et al. (2003); 3—González Lodeiro et al. (1981); 4—Martínez Catalán et al. (2002); 5—Martínez Catalán et al. (2004); 6—Ribeiro (1974) and González Clavijo and Martínez Catalán (2002).

the widely accepted early Paleozoic terrane dispersion in the peri-Gondwanan realm, and the calc-alkaline and arc affinities in the upper units and the autochthon, Valverde Vaquero and Dunning (2000) suggested that the rifting was located in a back-arc setting behind a subduction zone. This hypothesis is supported in reconstructions of peri-Gondwanan terranes by Stampfli et al. (2002), Winchester et al. (2002), von Raumer et al. (2003), and van Staal et al. (1998). Rollback of the subducting slab may have pulled Avalonia apart from Gondwana, and also the terrane partially preserved in the Galician upper allochthonous units. This terrane was possibly part of a discontinuous continental ribbon in the eastern continuation of Avalonia (Gómez Barreiro et al., 2007), and it recorded active-margin arc-related magmatism during separation (Fig. 5).

OROGENIC EVOLUTION: THE CROSS-SECTION VIEW

Early Variscan Accretionary History

In the upper allochthonous units, there is an increase in metamorphic grade from top to bottom, showing a transition from the uppermost epizonal units to mesozonal and catazonal units below. U-Pb dating of metamorphism has yielded two main age populations (Schäfer et al., 1993; Santos Zalduegui et al., 1996; Abati et al., 1999; Ordóñez Casado et al., 2001; Fernández-Suárez et al., 2002a, this volume). The older, dated around 500–490 Ma on monazite, whole zircon grains, and magmatic domains in zircon, is coeval with widespread magmatism and probably reflects high-*T* and low- to intermediate-*P* metamorphism in the magmatic arc. The younger age population, dated between 410 and 390 Ma in monazite, zircon, and metamorphic overgrowths of zircon, is linked to initial Variscan convergence and high-*P* and high-*T* metamorphism. In fact, start of Variscan convergence should be older, because 410 Ma would be the age of decompression melting associated with the onset of exhumation (Fernández-Suárez et al., this volume). Actually, $^{40}Ar/^{39}Ar$ data, which give a ca. 425 Ma age for retrogressive amphibolite-facies foliation in high-*P* and high-*T* units, suggest an even older age for the high-*P* and high-*T* metamorphism (Dallmeyer et al., 1997; Gómez Barreiro, 2004; Gómez Barreiro et al., 2006).

This compressional event of Silurian to Early Devonian age produced a thick metamorphic pile, and the deep parts register pressures of 1.8 GPa or higher (Gil Ibarguchi et al., 1990; Mendia Aranguren, 2000). Thickening of the upper units and the subduction of some of them probably reflect their underthrusting following accretion to a large continental mass, either Baltica or Laurentia (Fig. 5). Accretion was followed by retrograde amphibolite-facies metamorphism in the lower parts of the accretionary wedge at 390–375 Ma (Dallmeyer et al., 1991, 1997; Valverde Vaquero and Fernández, 1996) related to the beginning of exhumation. The fact that units with differences of more than 1.2 GPa in peak pressure occur presently in a sheet less than 10 km thick indicates that the original pile has been largely attenuated. Actually, the different upper units are separated from each other by extensional faults (Figs. 4 and 5), interpreted as detachments developed in different stages of their stacking and emplacement (Martínez Catalán et al., 2002).

The ophiolitic units were stacked in several slices (Díaz García et al., 1999) and reached metamorphic conditions ranging between 625 °C and 680 °C and 1.1 and 1.2 GPa. Amphibolite-facies prograde metamorphism, dated at 390–380 Ma (Dallmeyer and Gil Ibarguchi, 1990; Dallmeyer et al., 1991, 1997), was coeval with retrograde metamorphism in the upper units, thus suggesting that underthrusting and imbrication of oceanic lithosphere caused exhumation of the overlying upper units (Fig. 5).

The basal units record a high-*P* regional metamorphic event not found in the lower allochthon or in the autochthon. Peak pressures reached 1.5–1.7 GPa in a west-directed subduction zone (in present coordinates), as deduced from the pressure gradient along both limbs of a huge recumbent anticline (Fig. 4, section 4; Arenas et al., 1995; Martínez Catalán et al., 1996). Subduction may have started ca. 380 Ma and ended ca. 365 Ma (Van Calsteren et al., 1979; Santos Zalduegui et al., 1995; Rodríguez et al., 2003).

Variscan Collisional Deformation: Thrust and Nappe Tectonics

The structural evolution of the autochthon and lower allochthon is relatively simple, and the structures are related to three main compressional events that developed during convergence following collision (Pérez-Estaún et al., 1991). The first event (D_1) produced recumbent folds with east vergence and axial planar cleavage (S_1), which is the oldest penetrative fabric recognized. Large D_1 folds occur in the Mondoñedo nappe (Fig. 4, section 2): the Mondoñedo-Lugo-Sarria anticline and the Vilaoudriz syncline (Matte, 1968; Bastida et al., 1986). Their common overturned limb reaches 15–30 km due to late horizontal shearing related partly to thrusting and partly to late orogenic extension. To the west, recumbent folds are common in low-grade areas, but they are smaller, and their reverse limbs rarely attain 5 km.

The youngest deposits preserved in the core of one of the recumbent synclines are Early Devonian (Fig. 4, section 5), and $^{40}Ar/^{39}Ar$ dating of regional S_1 cleavage yielded ages of 359 Ma close to the allochthonous complexes and 336 Ma to the east, far away from them and adjacent to the Cantabrian zone (Dallmeyer et al., 1997). When compared with deformation ages in the allochthonous units, S_1 in the autochthon developed immediately after the greenschist-facies foliation in the structurally lower ophiolites (363–367 Ma; Dallmeyer et al., 1997) and the end of subduction-related metamorphism in the basal units (365 Ma; Rodríguez et al., 2003). It seems that once continental subduction of the outermost edge of Gondwana became locked, shortening began in inner parts of its continental platform, giving rise first to recumbent folds (D_1), and then to large thrust sheets, described as the second deformational event (D_2).

Four large thrusts developed in the internal zones of northwestern Iberia (Figs. 4 and 5). The Lalín-Forcarei thrust carried

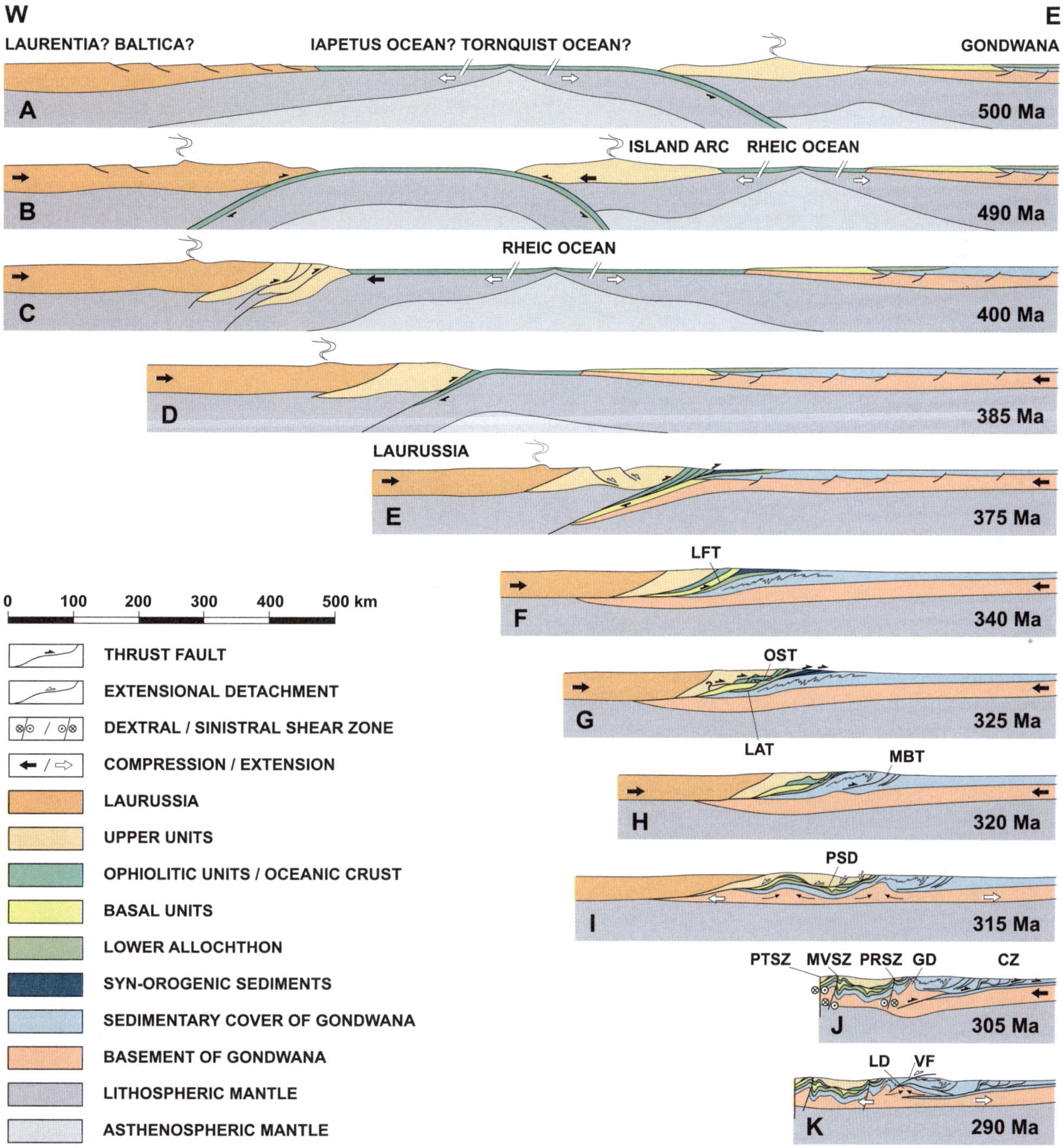

Figure 5. Proposed stages in the tectonic evolution of northwest Iberia. A: Individualization of a peri-Gondwanan terrane by slab rollback, and arc-related magmatic activity during the Late Cambrian to Early Ordovician. B: Drifting of the peri-Gondwanan terrane away from Gondwana and spreading of the Rheic Ocean. C: Building of an accretionary wedge by underthrusting and imbrication of the peri-Gondwanan terrane during the Silurian to Early Devonian. D: Closure of the Rheic Ocean and imbrication of oceanic slices. E: Subduction of outer edge of the Gondwanan continental margin in the Middle to Late Devonian. F: Thrusting of allochthonous units over the lower allochthon during the Early Carboniferous. G: Development of out-of-sequence thrusts in the Early to Middle Carboniferous. H: Thrusting in more external parts of the belt during the Middle Carboniferous. I: Collapse and extension of Gondwanan thickened crust with formation of extensional detachments and domes. J: Late upright folding, strike-slip faulting, and thin-skinned tectonics in the foreland during the Late Carboniferous. K: Extensional collapse migrates to the east, probably in response to crustal thickening induced by underthrusting of basement to the west during thin-skinned shortening in the foreland. Note that W-E coordinates correspond to present. Past coordinates might have varied from NE-SW to NW-SE during the time interval covered by the cartoon, according to the paleoposition of continental masses in the reconstruction by Winchester et al. (2002). Abbreviations: CZ—Cantabrian zone; GD—Guitiriz dome; LAT—lower allochthon thrust; LD—Lugo dome; LFT—Lalín-Forcarei thrust; MBT—Mondoñedo basal thrust; MVSZ—Malpica-Vigo shear zone; OST—out-of-sequence thrusts; PRSZ—Palas de Rei shear zone; PSD—Pico Sacro detachment; PTSZ—Porto-Tomar shear zone; VF—Viveiro fault.

the basal units over the lower allochthon after 346 Ma, which is the age of migmatization in allochthonous paragneisses (U-Pb in monazite and rutile; Abati and Dunning, 2002). The S_2 cleavage in the underlying lower allochthon, which developed during emplacement of the Lalín-Forcarei thrust, was dated at 340 Ma ($^{40}Ar/^{39}Ar$; Dallmeyer et al., 1997). The upper and ophiolitic allochthonous units subsequently moved over the basal units and the lower allochthon, becoming strongly imbricated along the sole thrust. As they had been previously stacked and internally imbricated in sequence, the new thrust system developed out of sequence (Martínez Catalán et al., 2002). The age of the out-of-sequence thrusts is constrained between 340 Ma, age of the Lalín-Forcarei thrust, and 323 ± 11 Ma, age of the Palas de Rei granodiorite (Bellido et al., 1992), or 317 ± 15 Ma, age of the Espenuca granite (Ortega Cuesta, 1998), both of which postdate nappe emplacement and predate the Pico Sacro extensional detachment. A 325 Ma $^{40}Ar/^{39}Ar$ age obtained by Dallmeyer et al. (1997) in an ultramylonite from a high-*P* and high-*T* upper unit may be a representative age for the out-of-sequence thrusts.

The isotopic data are consistent with the age and structural relationships of synorogenic flysch deposits that crop out close to the eastern boundary of the Bragança Complex in Portugal, in the Alcañices synform in Zamora, and in a narrow synform in central Galicia (Fig. 3; Fig. 4, sections 5 and 6). They consist of low-grade slates, graywackes, and conglomerates with plant debris and metamorphic pebbles (Riemer, 1966; Matte, 1968; Martínez García, 1972; Pérez-Estaún, 1974; Ribeiro and Ribeiro, 1974) cropping out in imbricates inside and in front of the lower allochthon thrust sheet (González Clavijo and Martínez Catalán, 2002; Martínez Catalán et al., 2004). The synorogenic deposits are turbiditic and have been dated as Late Devonian (Frasnian) in Portugal, using palynomorphs (Pereira et al., 1999), and as early Namurian using the age of the youngest detrital zircon in the Sil synform in central Galicia (Martínez Catalán et al., 2004). Here, zircon age populations are more compatible with those of the allochthonous terranes than with the autochthon because they lack the 1.1–1 Ga Mesoproterozoic population, which is well represented in the autochthonous succession (Fernández-Suárez et al., 2000b, 2002b; Martínez Catalán et al., 2004). This suggests that the synorogenic turbidites were deposited in a trough that developed in front of the allochthonous terranes during their emplacement by thrusting (González Clavijo and Martínez Catalán, 2002; Martínez Catalán et al., 2004). Their age is older near the Bragança Complex than in central Galicia, which reflects the advance of the allochthonous sheet.

The emplacement of the lower allochthon, carrying the allochthonous terranes piggyback, took place along a nearly horizontal detachment, the lower allochthon thrust. It has an apparent displacement of nearly 200 km and is a wonderful example of thin-skinned tectonics in the hinterland of a collisional belt (Fig. 4, sections 5 and 6). Silurian carbonaceous slates were the weak layer that accommodated the detachment, and they became strongly phyllonitized (Farias et al., 1987; Farias Arquer, 1990; González Clavijo and Martínez Catalán, 2002; Marcos and Llana Fúnez, 2002). The low amplitude of the previous (D_1) recumbent folds left the stratigraphic sequence nearly undisturbed, allowing the thrust surface to utilize the graphite-rich Silurian slates. Fan-like imbricates at the lower allochthon thrust front are well preserved in the Alcañices synform (González Clavijo and Martínez Catalán, 2002). In the Sil synform, the younger detrital zircon (dated 324 ± 7 Ma) constrains the age of the lower allochthon thrust as late Visean–early Namurian (Martínez Catalán et al., 2004). Its motion was nearly synchronous with the later out-of-sequence thrusts, or somewhat younger.

The fourth large fault, the Mondoñedo basal thrust, developed further east, carrying the large recumbent folds previously formed and possibly enlarging their amplitude by ductile flow concentrated at its basal shear zone (Bastida et al., 1986; Aller and Bastida, 1993; Martínez Catalán et al., 2003). The precise age of the Mondoñedo basal thrust is unknown, but two thrust faults east of the Mondoñedo basal thrust were dated by Dallmeyer et al. (1997) at ca. 320 Ma.

Several other minor thrust faults developed to the east of the Mondoñedo basal thrust, and thrusting became very important in the Cantabrian zone (Figs. 2 and 5), a thin-skinned foreland thrust belt where these structures are the main ones responsible for orogenic shortening (Pérez-Estaún et al., 1988).

Variscan Orogenic Collapse and Late Variscan Folding

Closely following the emplacement of the allochthonous terranes and thrust imbrication of the autochthon, relatively deep parts of the crust underwent a temperature increase associated in part with decompression. A pervasive subhorizontal tectonic foliation developed in the middle and lower parts of the autochthonous section, from the biotite zone down to the deepest accessible parts of the crust, cropping out in the core of late-orogenic extensional domes. The foliation is a crenulation cleavage in the upper parts, but it passes quickly to a schistosity and a gneissose banding downward, in the sillimanite–K-feldspar zone.

Extension is demonstrated by the thinning and even disappearance of some of the previously metamorphic zones at several map-scale shear zones, equivalent to ductile detachments, and by the metamorphic evolution on both sides of them: isobaric heating at their hanging wall and isothermal decompression at their footwall (Escuder Viruete et al., 1994; Díez Balda et al., 1995; Arenas and Martínez Catalán, 2003; Martínez Catalán et al., 2004). Kinematic criteria demonstrate a noncoaxial component of deformation with a sense of shear that varies from one detachment to other and indicate extension normal, oblique, and parallel to the orogenic trend.

Extension occurred under the allochthonous terranes, and also to the east, in the Ollo de Sapo antiform and in the Mondoñedo nappe. For instance, the Pico Sacro detachment developed between the Órdenes Complex and the migmatite and granite assemblage below (Fig. 4, section 4; Fig. 5). The Guitiriz dome is also an extensional structure, which, as many others, evolved into a dome (Fig. 4, section 3; Fig. 5).

The Lugo dome (Fig. 4, section 2) developed in the internal parts of the Mondoñedo nappe, which has footwall units that crop out in two tectonic windows related to doming. There, internal extension and two extensional ductile detachments stretched the nappe and its relative autochthon and are responsible for most of the stretching undergone by the reverse limb of the two largest recumbent folds (Fig. 5). One of the extensional detachments affected the footwall unit, whereas the other strongly attenuated the thrust sheet and evolved into a brittle structure, the Viveiro fault, which cuts across the whole Mondoñedo nappe and its footwall unit (Arenas and Martínez Catalán, 2003; Martínez Catalán et al., 2003). The difference in peak pressure between both sides of the upper extensional detachment and the Viveiro fault has been estimated to be 0.4–0.5 GPa (Reche et al., 1998), roughly equivalent to a subtraction of 15–19 km by the shear zone and the fault.

One of the main characteristics of the geological map of northwestern Iberia is the alternation of domes and basins (Martínez et al., 1988), which in many cases implies crustal-scale boudinage enhanced by deep crustal flow. Heat accumulation due to crustal thickening and some advection of mantle-derived rocks (Galán et al., 1996) caused partial melting, lowering the viscosity of the middle and lower crust, and facilitated viscous flow that accommodated extension of the whole crust, probably in response to gravitational forces. High-grade autochthonous rocks—the youngest parageneses of which are high-T and low-P—crop out in the domes, accompanied by abundant Variscan granitoids, whereas the basins are occupied by low-grade autochthonous metasediments and, in five cases, by the remnants of the allochthonous terranes preserved as klippen (Fig. 3).

Late upright folds are related to the third compressional event (D_3) and are associated with a crenulation cleavage (S_3). D_3 macrostructures interfere with D_1 recumbent folds and are easy to identify because they fold the regional metamorphic isograds. Large D_3 folds commonly nucleated in previously developed domes and basins, and they vary from open to tight. Variations in flattening are due to heterogeneous strain associated with subvertical, transcurrent ductile shear zones (Iglesias Ponce de León and Choukroune, 1980).

The main phase of gravitational collapse and extension occurred between 320 and 310 Ma, which are the ages of many synkinematic granitoids (Fernández-Suárez et al., 2000a). This part of collapse is considered intra-orogenic for two reasons. One is that the extensional domes and basins were overprinted by upright folds and transcurrent ductile shear zones, as in the case of the Guitiriz dome (Fig. 4, section 3; Fig. 5). The other is that extension in the internal zone was followed by shortening in the Cantabrian zone (Figs. 2 and 5), a thin-skinned foreland thrust belt that developed to the east between 312 and 300 Ma (Pérez-Estaún et al., 1988).

Upright D_3 folds have been dated by synkinematic granitoids at 314 ± 6 Ma (Capdevila and Vialette, 1970; Ries, 1979), whereas strike-slip shear zones closely related to their development moved between 315 and 305 Ma (Regêncio Macedo, 1988; Valle Aguado et al., 2005).

However, the Lugo dome developed later, as demonstrated by $^{40}Ar/^{39}Ar$ cooling ages around 300 Ma (Dallmeyer et al., 1997), and also because the continuation to the south of the Viveiro fault, which bounds its western flank, cuts and deforms a late-kinematic granodiorite massif with a Rb-Sr age of 286 ± 6 Ma (Román-Berdiel et al., 1995; Ortega et al., 2000), which belongs to the same series as others dated by U-Pb at ca. 295 Ma (Fernández-Suárez et al., 2000a). Late development of the Lugo dome was probably a consequence of migration of the extension to the external zones of the orogen with time (Fig. 5), in the same way that compressional episodes D_1 and D_2 had done before, as shown by the diachronous character of their associated cleavages (Dallmeyer et al., 1997).

The amount of extension undergone by the orogenic crust is difficult to estimate, but it seems to be very important given the abundance of extensional detachments and the high strains associated with them and the accompanying regional fabrics. Consequently, the apparently huge displacement shown by the main Variscan thrusts in the allochthonous terranes, including ~200 km for the lower allochthon thrust, may to a large extent be a consequence of late orogenic extension and would have been originally much less (Fig. 5).

GLOBAL VIEW: THE THIRD AND FOURTH DIMENSIONS

The Pieces of the Puzzle

Iberian geology can be correlated with that of central Europe by comparing its stratigraphic, metamorphic, and magmatic features with those of the different zones of the European massifs. The zoning of the Variscan belt was first established in central Europe by Kossmat (1927), and in Iberia by Lotze (1945), and correlations are being continuously updated as more information becomes available (Bard et al., 1971; Julivert et al., 1972; Tollman, 1982; Franke, 1989; Martínez Catalán, 1990; Matte, 2002).

In Figure 2, a correlation has been attempted using a few simple criteria. The autochthonous Central Iberian zone can be compared with the central domain of the Armorican Massif in France based on continuity across the Iberian-Armorican arc and strong stratigraphic similarities (Robardet et al., 1990; Young, 1990). The Ossa-Morena zone of southern Iberia is usually correlated with the northern domain of the Armorican Massif based on the presence of a strong Cadomian imprint and also on stratigraphic grounds (Cogné, 1974; Eguíluz et al., 1984, 2000; Chantraine et al., 1994). It is important to note that the sedimentary and faunal records in Iberia indicate that these zones were part of the northern Gondwanan shelf, distal in the case of the Ossa-Morena zone and proximal in the case of the Central Iberian zone (Robardet and Gutiérrez-Marco, 2004).

The correlation can be continued to the Bohemian Massif based on the presence of Cadomian crust in the Mid-German crystalline zone and the Saxo-Thuringian zone, and also based on the presence of a Cambrian-Ordovician rift sequence in the latter, which shows a stronger similarity to the Central Iberian zone than to the Ossa-Morena zone (Franke, 1989, 2000; Linnemann and Romer, 2002; Linnemann et al., 2003; Robardet and Gutiérrez-Marco, 2004). The Teplá-Barrandian zone has also evident affinities with the autochthonous terranes of northern Gondwana (Franke, 2000), and it includes some of the best-preserved Cadomian basement in Europe. This fact and the Paleozoic succession and faunal similarities (Gutiérrez-Marco et al., 1999, 2001) suggest a connection with the Iberian Ossa-Morena zone, or with a zone transitional between the Ossa-Morena zone and the Central Iberian zone.

Exotic terranes with Paleozoic ophiolites, remnants of Cambrian-Ordovician volcanic arcs, and early Variscan high-*P* metamorphism similar in age and evolution to the northwestern Iberian terranes exist along the whole length of the Variscan belt (Fig. 2). They occur in the southern domain of the Armorican Massif (Hanmer, 1977; Marchand, 1981; Balé and Brun, 1986; Ballèvre et al., 1994), in the French Massif Central (Burg and Matte, 1978; Girardeau et al., 1994; Ledru et al., 1994a, 1994b), the Vosges and Black Forest massifs (Wimmenauer and Lim, 1988; Eisbacher et al., 1989; Franke, 1989, 2000), the Saxo-Thuringian and Moldanubian zones of the Bohemian Massif (Tollman, 1982; Behr et al., 1982, 1984; Franke, 1989, 2000; Crowley et al., 2002), and the Polish Sudetes (O'Brien et al., 1997; Kröner and Hegner, 1998; Timmermann et al., 2000; Aleksandrowski and Mazur, 2002; Floyd et al., 2002). Furthermore, the exotic terranes seem to continue to the south in the External crystalline massifs of the Alps and the Maures Massif in southern France, Corsica, and northern Sardinia (Bourrouilh et al., 1980; Frisch et al., 1984, 1987; Becker et al., 1987; Ménot et al., 1988; Vauchez and Bufalo, 1988; von Raumer and Neubauer, 1993, 1994; von Raumer et al., 2002).

Finally, the Rhenian-Hercynian zone wraps around the other zones and can be traced from the Bohemian Massif to the southern British Isles and to southern Iberia, where it is represented by the South Portuguese zone (Oliveira et al., 1979). The Rhenian-Hercynian zone is an external thrust belt and also a foredeep basin that developed during the Middle Devonian and Carboniferous, possibly on Avalonian crust adjacent to the developing Variscan mountain belt.

The correlation shown in Figure 2 by itself does not provide a straightforward interpretation of the history of terrane evolution, convergence, and collision. One of the most important problems to be solved is the paleoposition and origin of the allochthonous terranes. Considering the paleogeographic continental reconstruction for the late Paleozoic (Fig. 1), the ophiolites of northwest Iberia and similar units in central Europe seem to witness an oceanic realm between Gondwana, represented by the autochthon, and Avalonia, represented by the London-Brabant Massif and the eastern Appalachians (Rast and Skehan, 1983; Williams and Hatcher, 1983).

According to current reconstructions (Scotese, 2001; Winchester et al., 2002), that ocean would have been the Rheic. However, it is unrealistic to postulate that the ca. 500 Ma arc-type magmatism preserved in the upper allochthonous units occurred inside the Rheic Ocean when it was beginning to open. Actually, arc development at that time was widespread in the Iapetus Ocean, on the northern side of Avalonia (van Staal et al., 1998; Winchester et al., 2002; van Staal, 2005). Therefore, correlation of the upper allochthonous units with, for instance, an arc occurring outboard of the Iapetus margin of Avalonia is more reasonable.

These facts can be reconciled if the upper units are remnants of a peri-Gondwanan continental block that was drifting at the same time as Avalonia (Gómez Barreiro et al., 2007) or that was detached from its Iapetus margin. These units would have registered active-margin magmatism and, later, would have docked to Laurussia, facing the Rheic Ocean, with Gondwana at the opposite margin, and without any intervening Avalonian terrane (Fig. 5). This possibility would imply that Avalonia did not form a continuous ribbon between Gondwana and Laurussia, at least, not for its easternmost part. Moreover, it is possible that the unstable ocean that drove the slab rollback that detached the upper allochthonous units from Gondwana was not the Iapetus, but the Tornquist Ocean (Fig. 5).

A major problem in interpreting the Variscan belt is how many peri-Gondwanan terranes were involved and how many oceans developed among them. Some interpretations suggest that nearly every ophiolitic unit represents a suture, so that several microcontinents, arcs, and oceans were involved (Matte, 1986, 1991, 2002; Franke, 1989, 2000; Franke and Zelazniewicz, 2002). However, different ophiolites and associated allochthonous units occur in terranes separated from each other by strike-slip shear zones, which suggests that different possible sutures could in fact be the same, repeated by wrench tectonics. The correlation among the Variscan exotic terranes was explored by Martínez Catalán (1990), who concluded that all of them could be remnants of a single gigantic, tongue-shaped allochthonous sheet and that a single ocean might account for all terranes of oceanic affinity. Although possible on purely geometrical grounds, the tongue shape of the allochthon seems mechanically unreasonable and can be replaced by a rather continuous strip along the northern Gondwana platform if the transpressional character of the orogen is considered.

Strike-Slip Tectonics

One of the clues for any interpretation of the Variscan belt resides in the relationship between northwest Iberia and the Armorican Massif, on both sides of the Iberian-Armorican arc. There is a close stratigraphic similarity between the Central Iberian zone and the central Armorican Massif, which has Ordovician and Devonian sections that are identical in Buçaco (western Portugal) and Crozon (western Armorican Massif; Henry et al., 1974; Paris and Robardet, 1977; Robardet et al., 1990; Young, 1990; Paris, 1998), precluding the possibility that both were separated by an oceanic domain. However, the Central Iberian zone and the central domain of the Armorican Massif lie on different sides of the allochthonous terranes (Fig. 2), suggesting that they were separated by a suture, the Massif Central suture of Matte (1991).

This apparent contradiction may be solved from an Iberian perspective. In the southern Armorican Massif, the allochthonous terranes occur adjacent to the southern branch of the South Armorican shear zone, which is considered to overprint the root there (Ballèvre et al., 1994). However, in Iberia, the allochthonous terranes overlie the Central Iberian zone, and their root zone lies outside, to the north, west, or south of the Central Iberian zone. It seems reasonable that the Armorican terranes are also allochthonous and do not root in the southern domain of the massif, in spite of the fact that they seem to root there because they have been overprinted and masked by subvertical shear zones. Therefore, the suture would be rootless in both domains, and the present terrane distribution may be a consequence of wrench tectonics.

Shelley and Bossière (2000, 2002) developed the hypothesis that the terrane collage was essentially due to dextral transpression induced by sliding of Laurentia along the northern margin of Gondwana. Their interpretation relies largely on continental reconstructions by Dalziel et al. (1994) and Dalziel (1997), and in well-established evidence for pervasive Devonian-Carboniferous dextral shearing in the Variscan-Appalachian belt (Gates et al., 1986; Rolet et al., 1994; van Staal and De Roo, 1995; Franke and Zelazniewicz, 2002; Hatcher, 2002).

Shelley and Bossière were right in stressing the importance of strike-slip motion, but paid little attention to the orthogonal component of convergence. However, the importance of orthogonal components is suggested by the subduction and subsequent exhumation of high-*P* allochthonous units and by the large displacement of allochthonous terranes in northwest Iberia and in the Bohemian Massif. Figure 6 is a simplified attempt to incorporate both orthogonal and transcurrent components into a model for the development of the Variscides in a way that is geometrically feasible. It avoids the need to invoke "extra" peri-Gondwanan terranes and intervening oceans.

Our foundations are the tectonic evolution of northwestern Iberia, the evidence that its section is retrodeformable, and the fact that the allochthonous terranes in the Galicia-Trás-os-Montes zone contain a rootless suture.

Both the Galicia-Trás-os-Montes zone and the Central Iberian zone are truncated to the west by the Porto-Tomar dextral shear zone (Ribeiro et al., 1980), which continues into the South Armorican shear zone (Fig. 2). The Galicia-Trás-os-Montes zone might root at a possible cryptic suture at the boundary between the Central Iberian zone and the Ossa-Morena zone, or farther south, between the latter and the South Portuguese zone, where a true suture exists.

The first possibility was explored by Simancas et al. (2002), and it is supported by the presence of a unit comparable to the basal units of the Galicia-Trás-os-Montes zone in the Badajoz-Córdoba sinistral shear zone, which represents the Central Iberian zone–Ossa-Morena zone boundary. However, no ophiolites have been found here, and it is not clear whether these units actually root there or are a narrow klippe pinched at the Badajoz-Córdoba sinistral shear zone. Furthermore, the existence of an oceanic domain between the Central Iberian zone and the Ossa-Morena zone in the Paleozoic is not favored by faunal studies (Robardet, 2002, 2003; Robardet and Gutiérrez-Marco, 2004).

Conversely, ophiolites occur in the Beja-Acebuches ophiolitic complex, at the Ossa-Morena zone–South Portuguese zone boundary (Crespo-Blanc, 1991; Fonseca and Ribeiro, 1993; Quesada et al., 1994; Figueiras et al., 2002). The age of these ophiolites is unknown, but their position in the Variscan belt is similar to those of Lizard in south Cornwall (Fig. 2), dated at 390–400 Ma (U-Pb; Clark et al., 1998; Nutman et al., 2001), so that the Beja-Acebuches ophiolitic complex may be coeval with and perhaps linked to the upper ophiolites of the Galicia-Trás-os-Montes zone, dated at 395 Ma (see also Sánchez Martínez et al., this volume).

Vergences are opposite in the Beja-Acebuches ophiolitic complex suture and the Galicia-Trás-os-Montes zone allochthonous terranes: while emplacement of the Galicia-Trás-os-Montes zone has an eastward component in present coordinates, the recumbent folds and thrusts in the South Portuguese zone, Beja-Acebuches ophiolitic complex, and Ossa-Morena zone show a southwest-directed motion (Silva et al., 1990; Crespo Blanc, 1991; Onézime et al., 2002; Expósito et al., 2002, 2003; Simancas, 2004). This does not necessarily imply that both sutures represent two different oceans, as the opposite vergences may indicate a change in subduction polarity along the plate boundary of a single ocean.

Faunal evidence indicates that the Central Iberian zone and the Ossa-Morena zone were never separated from each other by an ocean during the Paleozoic (Robardet, 2002, 2003; Robardet and Gutiérrez-Marco, 2004), and we assume that both the Beja-Acebuches ophiolitic complex and the Galicia-Trás-os-Montes zone sutures represent the closure of the Rheic Ocean by collision between Gondwana and Laurussia. However, the evolution of each suture was very different. To the east, closure of the Rheic Ocean built an accretionary wedge in the active Laurussia margin, which was later emplaced as a gigantic thrust over the northern Gondwana continental platform (Figs. 5 and 6).

To the west, south-directed subduction of oceanic lithosphere created the Rhenian-Hercynian zone (Kossmat, 1927), where an accretionary wedge developed locally during the Early to Middle Devonian (Silva et al., 1990; Eden and Andrews, 1990; Onézime et al., 2002). Afterward, terrigenous sediments and volcanics were deposited in a foredeep basin during the Late Devonian to Middle Carboniferous (Oliveira, 1990), and they deformed closely following deposition, forming a thin-skinned thrust belt (Franke, 2000; Oncken et al., 2000; Onézime et al., 2002). A transform fault might have separated the two parts of the Laurussia-Gondwana plate boundary with opposite subduction polarities.

We begin our cartoon (Fig. 6) at the Early Carboniferous, when the Rheic Ocean had been closed, the northern Gondwana

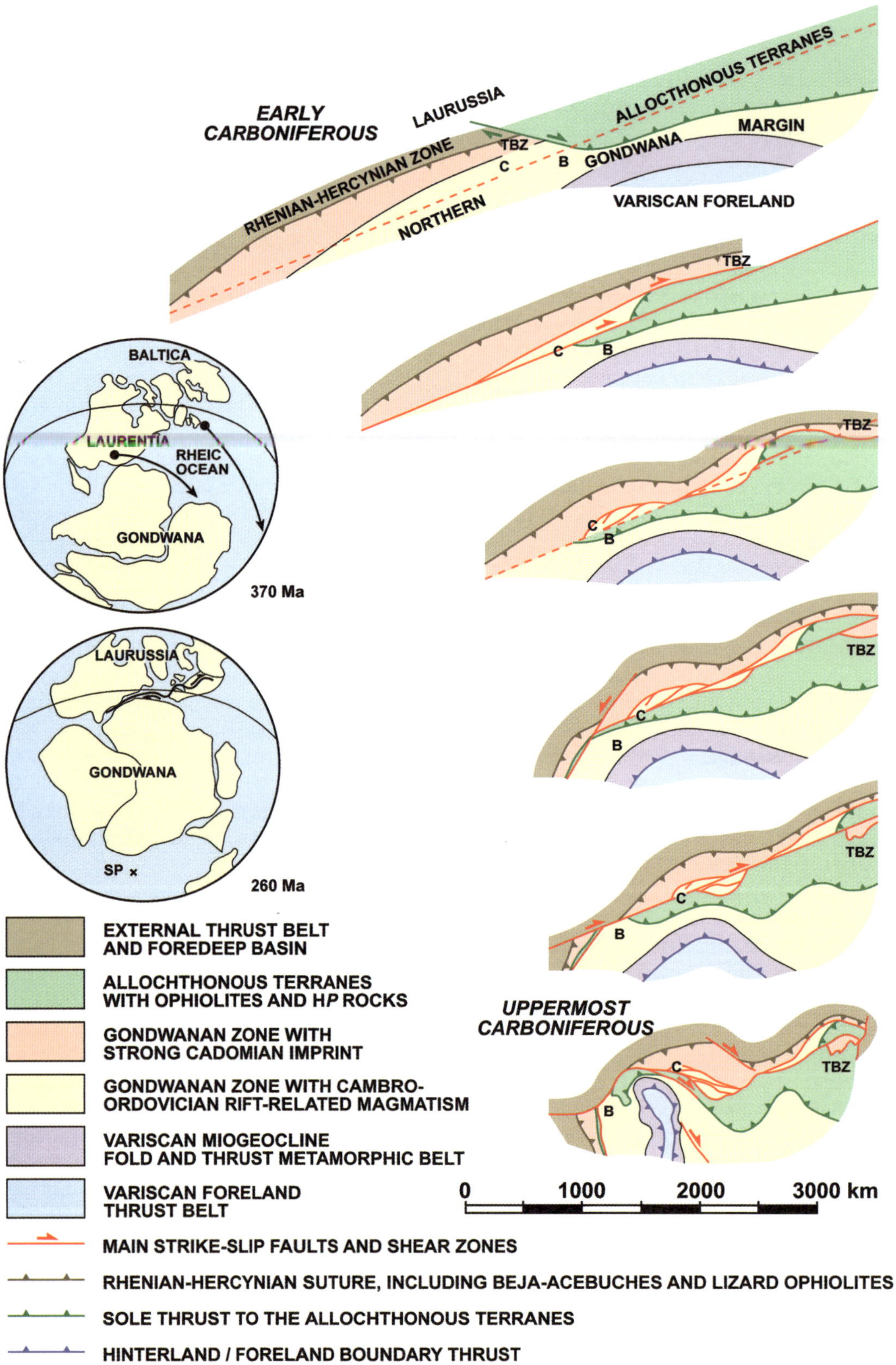

Figure 6. Map view of the proposed evolution of the Variscan belt during the Early and Middle Carboniferous, explaining the different allochthonous terranes as remnants of a huge nappe stack thrust onto the northern platform of Gondwana. Compare with Figure 2 to locate the main massifs and zones and their evolution during strike-slip motion. Note in particular the original neighborhood of Buçaco (B), Crozon (C), and the Teplá-Barrandian zone (TBZ), characterized by similar Paleozoic successions and closely related faunal assemblages. The inset shows the clockwise rotation of Laurentia and Baltica, which provided the convergent component of transpression, and the Carboniferous dextral shearing along the northern margin of Gondwana (after Shelley and Bossière, 2002).

platform had recorded the early deformation events related to collision, and the allochthonous terranes, including ophiolites and previously subducted rocks, had been obducted onto it. The complexities of this stage were described in a previous section and are sketched in Figure 5. The Rhenian-Hercynian basin had developed already and was the site of synorogenic sedimentation and deformation. This is the starting point of our model in the third dimension, which explains how the allochthonous terranes and the Gondwanan autochthon became subsequently dismembered by strike-slip shear zones and faults that produced ~3000 km of dextral displacement, and how the Rhenian-Hercynian zone came to be placed to the north of the allochthonous terranes, duplicating the Rheic suture and producing the double vergence characteristic of the Variscan belt.

Probably, wrench components did not appear then for the first time. In fact, dextral transpression has been identified in the Appalachians since the Early Devonian (van Staal and De Roo, 1995), and Late Devonian–Early Carboniferous dextral displacements have been suggested for the central European Variscides by Franke (2000) and for the Armorican Massif by Rolet (1994). It is possible that oblique convergence was active from the beginning of the Variscan cycle, when the Rheic Ocean began to close, and it may have been responsible for the orogen-parallel stretching lineations so common in many allochthonous units (Burg, 1981; Quinquis and Choukroune, 1981; Vauchez and Bufalo, 1988; Ribeiro et al., 1990; Llana-Fúnez, 2002).

The Iberian-Armorican arc has been interpreted as an oroclinal bend that developed from rigid-plastic indentation during collision (Matte, 1986). The arc is rather tight when considering its inner zones, which have a paleogeographic significance (Lotze, 1945; Julivert et al., 1972, 1980). Conversely, it is more open when affecting the main dextral wrench system (Porto-Tomar dextral shear zone–South Armorican shear zone). Even admitting the existence of a primary open arcuate structure, as indicated by paleomagnetism (Perroud and Bonhommet, 1981; Perroud, 1982; Hirt et al., 1992), Figure 2 suggests that oroclinal bending was coeval with late orogenic stages, and also with the concomitant wrenching. This may explain the sinistral motion of shear zones, which limited the Ossa-Morena zone in the southern branch of the arc, and the formation of fan imbricates and strike-slip duplexes such as those apparently present in the Armorican Massif, which could have formed by the master fault cutting across oroclinal bends (Fig. 6).

The timing of the different steps of the proposed sequence remains imprecise, but the strike-slip activity is bracketed between 345 and 300 Ma, the interval of motion of dextral shear zones in the Armorican Massif (Diot et al., 1983; Peucat et al., 1984; Rolet, 1994), which is similar to ages in Iberia, including 342 Ma amphibolite-facies deformation in the Beja-Acebuches ophiolitic complex ($^{40}Ar/^{39}Ar$ age; Dallmeyer et al., 1993). It is coeval with or predates sinistral movement, dated at 315–305 Ma for the Porto-Tomar and Juzbado-Penalva shear zones (Rb-Sr ages, Regêncio Macedo, 1988; U-Pb ages, Valle Aguado et al., 2005).

When comparing cross-section and map-view evolution, it is clear that orthogonal shortening and strike slip were active during most of the Carboniferous. This may explain repeated thrusting in northwest Iberia (Fig. 5) and perhaps northwest-directed thrusting of the allochthonous terranes in the Bohemian Massif (Franke, 1989, 2000; Collins et al., 2000). Thrust tectonics reflect the orthogonal component of transpression and tend to hide or delete the traces of previous along-strike components (Johnston, 2001). Therefore, well-preserved strike-slip structures are those formed after thrust and nappe tectonics have ceased in a region. Both orogenic mechanisms reflect a partition of deformation during oblique convergence and may act at the same time, and northwest Iberia is a clear example of such behavior: the Porto-Tomar and Juzbado-Penalva shear zones (Fig. 2) moved between 315 and 305 Ma (Regêncio Macedo, 1988; Valle Aguado et al., 2005), when thin-skinned tectonics were active in the Cantabrian zone (Fig. 5).

CONCLUSIONS

The northwestern Iberian Massif offers a clue to understanding the evolution of the Variscan belt, mainly because strike-slip structures are subordinate inside the section, which makes it retrodeformable. A key aspect of the section is the presence of exotic terranes that form a huge and complex allochthonous sheet emplaced upon the sequences deposited on the passive margin of northern Gondwana. The geochemistry of igneous rocks and isotopic age data show that the exotic terranes include parts of a peri-Gondwanan terrane that evolved as a Late Cambrian–Early Ordovician island arc, suprasubduction ophiolites, and pieces of the outermost edge of the Gondwanan continental margin. A precise matching between the ages of metamorphic fabrics and development of large structures has been attempted and the structures have been interpreted in the context of a two-stage evolutionary model consisting of an early Variscan accretionary stage, related to the closure of the Rheic Ocean, followed by a Variscan collisional stage.

Correlation of Iberian allochthonous terranes with those in central European massifs, and also of strike-slip structures along the Variscan belt, have resulted in the elaboration of a transpressional tectonic model for the Variscides that avoids the multiplication of microcontinents and narrow oceanic domains. The model explains the whole belt in terms of the closure of a single ocean, the Rheic, and the subsequent oblique collision between Gondwana and Laurussia. Overprint of a gigantic allochthonous sheet by strike-slip shear zones and faults with a total displacement of ~3000 km may account for the apparent multiplicity of sutures and peri-Gondwanan terranes.

Given the importance of dextral transpression in the Variscan-Appalachian belt, the paleoposition of Iberia along the Gondwanan margin might have been more to the east than suggested by late Paleozoic reconstructions. It is possible that northwest Iberia faced the Tornquist Ocean during the early Paleozoic, that early Variscan deformation was a consequence of Gondwana-Baltica convergence, and that the early Variscan accretionary prism preserved in the allochthonous terranes developed in the southern margin of Baltica.

ACKNOWLEDGMENTS

This contribution has been funded by the Spanish government agency Dirección General de Investigación, through projects BTE2001-0963-C02 and CGL2004-04306-CO2/BTE. The paper has benefited from constructive reviews by R.D. Hatcher Jr., D.T. Secor, and C.R. van Staal, whose criticisms and suggestions are kindly acknowledged.

REFERENCES CITED

Abati, J., and Dunning, G.R., 2002, Edad U-Pb en monacitas y rutilos de los paragneises de la Unidad de Agualada (Complejo de Ordenes, NW del Macizo Ibérico): Geogaceta, v. 32, p. 95–98.

Abati, J., Dunning, G.R., Arenas, R., Díaz García, F., González Cuadra, P., Martínez Catalán, J.R., and Andonaegui, P., 1999, Early Ordovician orogenic event in Galicia (NW Spain): Evidence from U-Pb ages in the uppermost unit of the Ordenes complex: Earth and Planetary Science Letters, v. 165, p. 213–228, doi: 10.1016/S0012-821X(98)00268-4.

Aleksandrowski, P., and Mazur, S., 2002, Collage tectonics in the northeasternmost part of the Variscan belt: The Sudetes, Bohemian Massif, *in* Winchester, J.A., Pharaoh, T.C., and Verniers, J., eds., Palaeozoic Amalgamation of Central Europe: Geological Society [London] Special Publication 201, p. 237–277.

Aller, J., and Bastida, F., 1993, Anatomy of the Mondoñedo Nappe basal shear zone (NW Spain): Journal of Structural Geology, v. 15, p. 1405–1419, doi: 10.1016/0191-8141(93)90002-R.

Andonaegui, P., González del Tánago, J., Arenas, R., Abati, J., Martínez Catalán, J.R., Peinado, M., and Díaz García, F., 2002, Tectonic setting of the Monte Castelo gabbro (Ordenes Complex, northwestern Iberian Massif): Evidence for an arc-related terrane in the hanging wall to the Variscan suture, *in* Martínez Catalán, J.R., Hatcher, R.D., Jr., Arenas, R., and Díaz García, F., eds., Variscan-Appalachian Dynamics: The Building of the Late Paleozoic Basement: Geological Society of America Special Paper 364, p. 37–56.

Arenas, R., 1988, Evolución petrológica y geoquímica de la unidad alóctona inferior del complejo metamórfico básico-ultrabásico de Cabo Ortegal (Unidad de Moeche) y del Silúrico paraautóctono, Cadena Hercínica Ibérica (NW de España): A Coruña, Spain, Laboratorio Geológico de Lage, Corpus Geologicum Gallaeciae, v. 4, 543 p.

Arenas, R., and Martínez Catalán, J.R., 2003, Low-*P* metamorphism following a Barrovian-type evolution: Complex tectonic controls for a common transition, as deduced in the Mondoñedo thrust sheet (NW Iberian Massif): Tectonophysics, v. 365, p. 143–164, doi: 10.1016/S0040-1951(03)00020-9.

Arenas, R., Rubio Pascual, F.J., Díaz García, F., and Martínez Catalán, J.R., 1995, High-pressure micro-inclusions and development of an inverted metamorphic gradient in the Santiago schists (Ordenes Complex, NW Iberian Massif, Spain): Evidence of subduction and syn-collisional decompression: Journal of Metamorphic Geology, v. 13, p. 141–164.

Balé, P., and Brun, J.P., 1986, Les complexes métamorphiques du Léon (NW Bretagne): Un segment du domain éo-Hercynien sud Armoricain translaté au Dévonien: Bulletin de la Société Géologique de France, v. 8, p. 471–477.

Ballèvre, M., Marchand, J., Godard, G., Goujou, J.C., and Wyns, R., 1994, Eo-Hercynian events in the Armorican Massif, structure and metamorphism: The Variscan orogeny in the Armorican Massif, *in* Keppie, J.D., ed., Pre-Mesozoic Geology in France and Related Areas: Berlin, Germany, Springer-Verlag, p. 183–194.

Bard, J.P., Capdevila, R., and Matte, Ph., 1971, Structure de la chaîne Hercynienne de la Meseta Ibérique: Comparaison avec les segments voisins, *in* Histoire Structurale du Golfe de Gascogne: Paris, France, Publications de l'Institut Français du Pétrole, Technip, Symposium CNEXO I, v. 4, p. 1–68.

Bastida, F., Marcos, A., Marquínez, J., Martínez Catalán, J.R., Pérez-Estaún, A., and Pulgar, J.A., 1982, Mapa Geológico Nacional, Hoja 1, La Coruña: Madrid, Spain, Instituto Geológico y Minero de España, escala 1:200,000.

Bastida, F., Martínez Catalán, J.R., and Pulgar, J.A., 1986, Structural, metamorphic and magmatic history of the Mondoñedo nappe (Hercynian belt, NW Spain): Journal of Structural Geology, v. 8, p. 415–430, doi: 10.1016/0191-8141(86)90060-X.

Becker, L.P., Frank, W., Höck, V., Kleinschmidt, G., Neubauer, F., Sassi, F.P., and Schramm, J.M., 1987, Outlines of the pre-Alpine metamorphic events in the Austrian Alps, *in* Flügel, H.V., Sassi, F.P., and Grecula, P., eds., Pre-Variscan and Variscan Events in the Alpine-Mediterranean Mountain Belts: Mineralia Slovaca, Monography: Bratislava, Alfa, p. 69–106.

Behr, H.J., Engel, W., and Franke, W., 1982, Variscan wildflysch and nappe tectonics in the Saxothuringian zone (northeast Bavaria, West Germany): American Journal of Science, v. 282, p. 1438–1470.

Behr, H.J., Engel, W., Franke, W., Giese, P., and Weber, K., 1984, The Variscan belt in Central Europe: Main structures, geodynamic implications, open questions: Tectonophysics, v. 109, p. 15–40, doi: 10.1016/0040-1951(84)90168-9.

Bellido, F., Brandle, J.L., Lasala, M., and Reyes, J., 1992, Consideraciones petrológicas y cronológicas sobre las rocas graníticas Hercínicas de Galicia: Cuadernos do Laboratorio Xeolóxico de Laxe, v. 17, p. 241–261.

Bourrouilh, R., Cocozza, T., Demange, M., Durand-Delga, M., Gueirard, S., Guitard, G., Julivert, M., Martínez, F.J., Massa, D., Mirouse, R., and Orsini, J.B., 1980, Essai sur l'évolution paléogéographique, structurale et métamorphique du Paléozoïque du Sud de la France et de l'Ouest de la Méditerrané, *in* 26 Congrés Géologique International, Colloque C6, Géologie de l'Europe: Mémoires du Bureau des Recherches Géologiques et Minières, v. 108, p. 159–188.

Burg, J.P., 1981, Tectonique tangentielle Hercynienne en Vendée littorale: Signification del linéations d'étirement E-W dans les porphyroïdes à foliation horizontale: Comptes Rendus de l'Académie des Sciences Paris, v. 293, p. 849–854.

Burg, J.P., and Matte, Ph., 1978, A cross section through the French Massif Central and the scope of its Variscan geodynamic evolution: Zeitschrift der Deutschen Geologischen Gesellschaft, v. 129, p. 429–460.

Capdevila, R., and Vialette, Y., 1970, Estimation radiométrique de l'âge de la deuxième phase tectonique Hercynienne en Galice Moyenne (nord-ouest de l'Espagne): Comptes Rendus de l'Académie des Sciences Paris, v. 270, p. 2527–2530.

Castiñeiras, P., 2003, Origen y evolución tectonotermal de las unidades de O Pino y Cariño (Complejos Alóctonos de Galicia) [Ph.D. thesis]: Madrid, Spain, Universidad Complutense, 281 p.

Chantraine, J., Auvray, B., Brun, J.P., Chauvel, J.J., and Rabu, D., 1994, The Cadomian orogeny in the Armorican Massif: Introduction, *in* Keppie, J.D., ed., Pre-Mesozoic Geology in France and Related Areas: Berlin, Germany, Springer-Verlag, p. 75–80.

Clark, A.H., Scott, D.J., Sandeman, H.A., Bromley, A.V., and Farrar, E., 1998, Siegenian generation of the Lizard ophiolite: U-Pb zircon age data for plagiogranite, Porthkerris, Cornwall: Geological Society [London] Journal, v. 155, p. 595–598.

Cocks, L.R.M., and Fortey, R.A., 1988, Lower Palaeozoic facies and faunas around Gondwana, *in* Audley-Charles, M.G., and Hallam, A., eds., Gondwana and the Tethys: Geological Society [London] Special Publication 37, p. 183–200.

Cogné, J., 1974, Le Massif Armoricain, *in* Debelmas, J., ed., Géologie de la France: Paris, Doin, v. 1, p. 105–161.

Cogné, J., 1977, La chaîne Hercynienne ouest-Européenne corespond-elle à un orògene par collision? Propositions pour une interprétation géodynamique globale, *in* Géologie de l'Himalaya: Paris, France, Colloque International du Centre National de la Recherche Scientifique, v. 268, p. 111–129.

Collins, A.S., Kryza, R., and Zalasiewicz, J., 2000, Macrofabric fingerprints of Late Devonian–Early Carboniferous subduction in the Polish Variscides, the Kaczawa complex, Sudetes: Geological Society [London] Journal, v. 157, p. 283–288.

Crespo Blanc, A., 1991, Evolución geotectónica del contacto entre la Zona de Ossa-Morena y la Zona Surportuguesa en las sierras de Aracena y Aroche (Macizo Ibérico meridional): Un contacto mayor de la Cadena Hercínica Europea: Granada, Spain, Facultad de Ciencias, Granada, 328 p.

Crowley, Q.G., Floyd, P.A., Stedrá, V., Winchester, J.A., Kachlík, V., and Holland, J.G., 2002, The Mariánské-Lázne Complex, NW Bohemian Massif: Development and destruction of an early Palaeozoic seaway, *in* Winchester, J.A., Pharaoh, T.C., and Verniers, eds., Palaeozoic Amalgamation of Central Europe: Geological Society [London] Special Publication 201, p. 177–195.

Dallmeyer, R.D., and Gil Ibarguchi, J.I., 1990, Age of amphibolitic metamorphism in the ophiolitic unit of the Morais allochthon (Portugal): Implications for early Hercynian orogenesis in the Iberian Massif: Geological Society [London] Journal, v. 147, p. 873–878.

Dallmeyer, R.D., and Tucker, R.D., 1993, U-Pb zircon age for the Lagoa augen gneiss, Morais Complex, Portugal: Tectonic implications: Geological Society [London] Journal, v. 150, p. 405–410.

Dallmeyer, R.D., Ribeiro, A., and Marques, F., 1991, Polyphase Variscan emplacement of exotic terranes (Morais and Bragança Massifs) onto Iberian successions: Evidence from $^{40}Ar/^{39}Ar$ mineral ages: Lithos, v. 27, p. 133–144, doi: 10.1016/0024-4937(91)90025-G.

Dallmeyer, R.D., Fonseca, P., Quesada, C., and Ribeiro, A., 1993, $^{40}Ar/^{39}Ar$ mineral age constraints for the tectonothermal evolution of a Variscan suture in southwest Iberia: Tectonophysics, v. 222, p. 177–194, doi: 10.1016/0040-1951(93)90048-O.

Dallmeyer, R.D., Martínez Catalán, J.R., Arenas, R., Gil Ibarguchi, J.I., Gutiérrez Alonso, G., Farias, P., Aller, J., and Bastida, F., 1997, Diachronous Variscan tectonothermal activity in the NW Iberian Massif: Evidence from $^{40}Ar/^{39}Ar$ dating of regional fabrics: Tectonophysics, v. 277, p. 307–337, doi: 10.1016/S0040-1951(97)00035-8.

Dalziel, I.W.D., 1997, Neoproterozoic-Paleozoic geography and tectonics: Review, hypothesis and environmental speculation: Geological Society of America Bulletin, v. 109, p. 16–42, doi: 10.1130/0016-7606(1997)109<0016:ONPGAT>2.3.CO;2.

Dalziel, I.W.D., Dalla Salda, L.H., and Gahagan, L.M., 1994, Paleozoic Laurentian-Gondwana interaction and the origin of the Appalachian-Andean mountain system: Geological Society of America Bulletin, v. 106, p. 243–253, doi: 10.1130/0016-7606(1994)106<0243:PLGIAT>2.3.CO;2.

Díaz García, F., Arenas, R., Martínez Catalán, J.R., González del Tánago, J., and Dunning, G., 1999, Tectonic evolution of the Careón ophiolite (northwest Spain): A remnant of oceanic lithosphere in the Variscan belt: The Journal of Geology, v. 107, p. 587–605, doi: 10.1086/314368.

Díez Balda, M.A., Martínez Catalán, J.R., and Ayarza, P., 1995, Syn-collisional extensional collapse parallel to the orogenic trend in a domain of steep tectonics: The Salamanca detachment zone (Central Iberian zone, Spain): Journal of Structural Geology, v. 17, p. 163–182, doi: 10.1016/0191-8141(94)E0042-W.

Diot, H., Bouchez, J.L., and Blaise, J., 1983, La bande ductile Lanvaux-Les-Ponts-de-Cé entre Redon et Angers. Une composante de cisaillement Hercynien du Massiff Armoricain (France): Bulletin de la Société Géologique de France, v. 7, p. 155–167.

Eden, C.P., and Andrews, J.R., 1990, Middle to Upper Devonian mélanges in SW Spain and their relationship to the Meneage Formation in South Cornwall: Proceedings of the Ussher Society, v. 7, p. 217–222.

Eguíluz, L., Fernández-Carrasco, J., Garrote, A., Liñán, E., and Quesada, C., 1984, Sucesiones estratigráficas del Anticlinorio de Olivenza-Monesterio en la transversal Montemolín- Arroyomolinos: Cuadernos do Laboratorio Xeolóxico de Laxe, v. 8, p. 117–123.

Eguíluz, L., Gil Ibarguchi, J.I., Abalos, B., and Apraiz, A., 2000, Superposed Hercynian and Cadomian orogenic cycles in the Ossa-Morena zone and related areas of the Iberian Massif: Geological Society of America Bulletin, v. 112, p. 1398–1413, doi: 10.1130/0016-7606(2000)112<1398:SHACOC>2.0.CO;2.

Eisbacher, G.H., Lüschen, E., and Wickert, F., 1989, Crustal-scale thrusting and extension in the Hercynian Schwarzwald and Vosges, Central Europe: Tectonics, v. 8, p. 1–21.

Escuder Viruete, J., Arenas, R., and Martínez Catalán, J.R., 1994, Tectonothermal evolution associated with Variscan crustal extension in the Tormes Gneiss Dome (NW Salamanca, Iberian Massif, Spain): Tectonophysics, v. 238, p. 117–138, doi: 10.1016/0040-1951(94)90052-3.

Expósito, I., Simancas, J.F., González Lodeiro, F., Azor, A., and Martínez Poyatos, D.J., 2002, Estructura de la mitad septentrional de la Zona de Ossa-Morena: Deformación en el bloque inferior de un cabalgamiento cortical de evolución compleja: Revista de la Sociedad Geológica de España, v. 15, p. 3–14.

Expósito, I., Simancas, J.F., González Lodeiro, F., Bea, F., Montero, P., and Salman, K., 2003, Metamorphic and deformational imprint of Cambrian–Lower Ordovician rifting in the Ossa-Morena zone (Iberian Massif, Spain): Journal of Structural Geology, v. 25, p. 2077–2087, doi: 10.1016/S0191-8141(03)00075-0.

Farias Arquer, P., 1990, La geología de la región del Sinforme de Verín (Cordillera Herciniana, NW de España): A Coruña, Spain, Serie Nova Terra, Ediciós do Castro, Sada, v. 2, 201 p.

Farias, P., Gallastegui, G., González-Lodeiro, F., Marquínez, J., Martín Parra, L.M., Martínez Catalán, J.R., de Pablo Maciá, J.G., and Rodríguez Fernández, L.R., 1987, Aportaciones al conocimiento de la litoestratigrafía y estructura de Galicia Central: Memórias da Faculdade de Ciências, Universidade do Porto, v. 1, p. 411–431.

Faure, M., Leloix, C., and Roig, J.Y., 1997, L'évolution polyciclique de la chaîne Hercynienne: Bulletin de la Société Géologique de France, v. 168, p. 695–705.

Fernández-Suárez, J., Dunning, G.R., Jenner, G.A., and Gutiérrez-Alonso, G., 2000a, Variscan collisional magmatism and deformation in NW Iberia: Constraints from U-Pb geochronology of granitoids: Geological Society [London] Journal, v. 157, p. 565–576.

Fernández-Suárez, J., Gutiérrez-Alonso, G., Jenner, G.A., and Tubrett, M.N., 2000b, New ideas on the Proterozoic–early Paleozoic evolution of NW Iberia: Insights from U-Pb detrital zircon ages: Precambrian Research, v. 102, p. 185–206, doi: 10.1016/S0301-9268(00)00065-6.

Fernández-Suárez, J., Corfu, F., Arenas, R., Marcos, A., Martínez Catalán, J.R., Díaz García, F., Abati, J., and Fernández, F.J., 2002a, U-Pb evidence for a polyorogenic evolution of the HP-HT units of the NW Iberian Massif: Contributions to Mineralogy and Petrology, v. 143, p. 236–253, doi: 10.1007/s00410-001-0337-2.

Fernández-Suárez, J., Gutiérrez Alonso, G., and Jeffries, T.E., 2002b, The importance of along-margin terrane transport in northern Gondwana: Insights from detrital zircon parentage in Neoproterozoic rocks from Iberia and Brittany: Earth and Planetary Science Letters, v. 204, p. 75–88, doi: 10.1016/S0012-821X(02)00963-9.

Fernández-Suárez, J., Díaz García, F., Jeffries, T.E., Arenas, R., and Abati, J., 2003, Constraints on the provenance of the uppermost allochthonous terrane of the NW Iberian Massif: Inferences from detrital zircon U-Pb ages: Terra Nova, v. 15, p. 138–144, doi: 10.1046/j.1365-3121.2003.00479.x.

Figueiras, J., Mateus, A., Gonçalves, M.-A., Waerenborgh, J., and Fonseca, P., 2002, Geodynamic evolution of the South Variscan Iberian suture as recorded by mineral transformations: Geodinamica Acta, v. 15, p. 45–61, doi: 10.1016/S0985-3111(01)01078-6.

Floor, P., 1966, Petrology of an aegirine-riebeckite gneiss-bearing part of the Hesperian Massif: The Galiñeiro and surrounding areas, Vigo, Spain: Leidse Geologische Mededelingen, v. 36, p. 1–203.

Floyd, P.A., Kryza, R., Crowley, Q.G., Winchester, J.A., and Abdel Wahed, M., 2002, Sleza Ophiolite: Geochemical features and relationship to lower Paleozoic rift magmatism in the Bohemian Massif, *in* Winchester, J.A., Pharaoh, T.C., and Verniers, J., eds., Palaeozoic Amalgamation of Central Europe: Geological Society [London] Special Publication 201, p. 197–215.

Fonseca, P., and Ribeiro, A., 1993, Tectonics of the Beja-Acebuches ophiolite: A major structure in the Iberian Variscan foldbelt: Geologische Rundschau, v. 82, p. 440–447, doi: 10.1007/BF00212408.

Franke, W., 1989, Tectonostratigraphic units in the Variscan belt of central Europe, *in* Dallmeyer, R.D., ed., Terranes in the Circum-Atlantic Paleozoic Orogens: Geological Society of America Special Paper 230, p. 67–90.

Franke, W., 2000, The mid-European segment of the Variscides: Tectonostratigraphic units, terrane boundaries and plate tectonic evolution, *in* Franke, W., Haak, V., Oncken, O., and Tanner, D., eds., Orogenic Processes: Quantification and Modelling in the Variscan Belt: Geological Society [London] Special Publication 179, p. 35–61.

Franke, W., and Zelazniewicz, A., 2002, Structure and evolution of the Bohemian Arc, *in* Winchester, J.A., Pharaoh, T.C., and Verniers, J., eds., Palaeozoic Amalgamation of Central Europe: Geological Society [London] Special Publication 201, p. 279–293.

Frisch, W., Neubauer, F., and Satir, M., 1984, Concepts of the evolution of the Austroalpine basement complex (eastern Alps) during the Caledonian-Variscan cycle: Geologische Rundschau, v. 73, p. 47–68, doi: 10.1007/BF01820360.

Frisch, W., Neubauer, F., Bröcker, M., Brückmann, W., and Haiss, N., 1987, Interpretation of geochemical data from the Caledonian basement within the Austroalpine basement complex, *in* Flügel, H.V., Sassi, F.P., and Grecula, P., eds., Pre-Variscan and Variscan Events in the Alpine-Mediterranean Mountain Belts: Mineralia Slovaca, Monography: Bratislava, Alfa, p. 209–226.

Galán, G., and Marcos, A., 1997, Geochemical evolution of high-pressure mafic granulites from the Bacariza Formation (Cabo Ortegal complex, NW Spain): An example of a heterogeneous lower crust: Geologische Rundschau, v. 86, p. 539–555, doi: 10.1007/s005310050162.

Galán, G., Pin, C., and Duthou, J.L., 1996, Sr-Nd isotopic record of multi-stage interactions between mantle-derived magmas and crustal components in a collision context—The ultramafic-granitoid association from Vivero (Hercynian belt, NW Spain): Chemical Geology, v. 131, p. 67–91, doi: 10.1016/0009-2541(96)00027-7.

García Garzón, J., de Pablo Maciá, J.G., and Llamas Borrajo, J.F., 1981, Edades absolutas obtenidas mediante el método Rb/Sr en dos cuerpos de ortoneises en Galicia Occidental: Boletín Geológico y Minero, v. 92, p. 463–466.

Gates, A.E., Simpson, C., and Glover, L., III, 1986, Appalachian Carboniferous dextral strike-slip faults: An example from Brookneal, Virginia: Tectonics, v. 5, p. 119–133.

Gebauer, D., 1993, Intra-grain zircon dating within the Iberian Massif: Ollo de Sapo augengneisses, bimodal gneisses from the Massif de Guilleries (Girona), graywacke of the Tentudía Group (Serie Negra, SW Spain) and the HP/HT-rock association at Cabo Ortegal (Galicia): Évora, Portugal, Comunicações da XII Reunião de Geologia do Oeste Peninsular, Universidade de Évora, v. 2, p. 41–46.

Gil Ibarguchi, J.I., Mendia, M., Girardeau, J., and Peucat, J.J., 1990, Petrology of eclogites and clinopyroxene garnet metabasites from the Cabo Ortegal Complex (northwestern Spain): Lithos, v. 25, p. 133–162, doi: 10.1016/0024-4937(90)90011-O.

Girardeau, J., Dubuisson, G., and Mercier, J.C., 1994, Igneous activity, Caledono-Hercynian magmatism in the French Massif Central. Early Paleozoic mantle formations. The Limousin ophiolite complexes: Evidence for oceanic lithosphere, *in* Keppie, J.D., ed., Pre-Mesozoic Geology in France and Related Areas: Berlin, Germany, Springer-Verlag, p. 349–354.

Gómez Barreiro, J., 2004, La Unidad de Fornás: Evolución Tectonometamórfica del SO del Complejo de Órdenes [Ph.D. thesis]: Salamanca, Spain, Universidad de Salamanca, 294 p.

Gómez Barreiro, J., Wijbrans, J.R., Castiñeiras, P., Martínez Catalán, J.R., Arenas, R., Díaz García, F., and Abati, J., 2006, ^{40}Ar/^{39}Ar laserprobe dating of mylonitic fabrics in a polyorogenic terrane of NW Iberia: Geological Society [London] Journal, v. 163, p. 61–73, doi: 10.1144/0016-764905-012.

Gómez Barreiro, J.G., Martínez Catalán, J.R., Arenas, R., Castiñeiras, P., Abati, J., Díaz García, F., and Wijbrans, J.R., 2007, Tectonic evolution of the upper allochthon of the Órdenes Complex (northwestern Iberian Massif): Structural constraints to a polyorogenic peri-Gondwanan terrane, *in* Linnemann, U., Nance, R.D., Kraft, P., and Zulauf, G., eds., The Evolution of the Rheic Ocean: From Avalonian-Cadomian Active Margin to Alleghenian-Variscan Collision: Geological Society of America Special Paper 423, p.315–332, doi: 10.1130/2007.2423(15).

González Clavijo, E., and Martínez Catalán, J.R., 2002, Stratigraphic record of preorogenic to synorogenic sedimentation, and tectonic evolution of imbricate units in the Alcañices synform (northwestern Iberian Massif), *in* Martínez Catalán, J.R., Hatcher, R.D., Jr., Arenas, R., and Díaz García, F., eds., Variscan-Appalachian Dynamics: The Building of the Late Paleozoic Basement: Geological Society of America Special Paper 364, p. 17–35.

González Lodeiro, F., Hernández Urroz, J., Klein, E., Martínez Catalán, J.R., and de Pablo Maciá, J.G., 1981, Mapa Geológico Nacional, Hoja 8, Lugo: Madrid, Spain, Instituto Geológico y Minero de España, escala 1:200,000.

Gutiérrez Marco, J.C., Rábano, I., Sarmiento, G.N., Aceñolaza, G.F., San José, M.A., Pieren, A.P., Herranz, P., Couto, H.M., and Piçarra, J.M., 1999, Faunal dynamics between Iberia and Bohemia during the Oretanian and Dobrotivian (late Middle–earliest Upper Ordovician), and biogeographic relations with Avalonia and Baltica: Acta Universitatis Carolinae-Geologica, v. 43, p. 487–490.

Gutiérrez Marco, J.C., Sarmiento, G.N., Robardet, M., Rábano, I., and Vanek, J., 2001, Upper Silurian fossils of Bohemian type from NW Spain and their palaeogeographical significance: Journal of the Czech Geological Society, v. 46, p. 247–258.

Hanmer, S.K., 1977, Age and tectonic implications of the Baie d'Audierne basic-ultrabasic complex: Nature, v. 270, p. 336–338, doi: 10.1038/270336a0.

Hatcher, R.D., Jr., 1989, Tectonic synthesis of the U.S. Appalachians, *in* Hatcher, R.D., Jr., Thomas, W.A., and Viele, G.W., eds., The Appalachian-Ouachita Orogen in the United States: Boulder, Colorado, Geological Society of America, Geology of North America, v. F-2, p. 511–535.

Hatcher, R.D., Jr., 2002, Alleghanian (Appalachian) orogeny, a product of zipper tectonics: Rotational, transpressive continent–continent collision and closing of ancient oceans along irregular margins, *in* Martínez Catalán, J.R., Hatcher, R.D., Jr., Arenas, R., and Díaz García, F., eds., Variscan-Appalachian Dynamics: The Building of the Late Paleozoic Basement: Geological Society of America Special Paper 364, p. 199–208.

Henry, J.L., Nion, J., Paris, F., and Thadeu, D., 1974, Chitinozoaires, Ostracodes et Trilobites de l'Ordovicien du Portugal (Serra de Buçaco) et du Massif Armoricain: Essai de comparación et signification paléogéographique: Comunicaçoes dos Serviços Geologicos de Portugal, v. 57, p. 303–345.

Hirt, A.M., Lowrie, W., Julivert, M., and Arboleya, M.L., 1992, Paleomagnetic results in support of a model for the origin of the Asturian arc: Tectonophysics, v. 213, p. 321–339, doi: 10.1016/0040-1951(92)90461-E.

Hossack, J.R., and Cooper, M.A., 1986, Collision tectonics in the Scandinavian Caledonides, *in* Coward, M.P., and Ries, A.C., eds., Collision Tectonics: Geological Society [London] Special Publication 19, p. 287–304.

Iglesias Ponce de León, M., and Choukroune, P., 1980, Shear zones in the Iberian arc: Journal of Structural Geology, v. 2, p. 63–68, doi: 10.1016/0191-8141(80)90035-8.

Johnston, S.T., 2001, The Great Alaskan terrane wreck: Reconciliation of paleomagnetic and geological data in the northern Cordillera: Earth and Planetary Science Letters, v. 193, p. 259–272, doi: 10.1016/S0012-821X(01)00516-7.

Julivert, M., Fontboté, J.M., Ribeiro, A., and Conde, L., 1972, Mapa Tectónico de la Península Ibérica y Baleares: Madrid, Spain, Instituto Geológico y Minero de España, escala 1: 1,000,000.

Julivert, M., Fontboté, J.M., Ribeiro, A., and Conde, L., 1980, Mapa Tectónico de la Península Ibérica y Baleares, Memoria: Madrid, Spain, Instituto Geológico y Minero de España, 113 p.

Kelling, G., Phillips, W.E.A., Harris, A.L., and Howells, M.F., 1985, The Caledonides in the British Isles: A review and appraisal, *in* Gee, D.G., and Sturt, B.A., eds., The Caledonide Orogen—Scandinavia and Related Areas: New York, John Wiley & Sons, p. 1125–1146.

Kossmat, F., 1927, Gliederung des varistischen gebirgsbaues: Abhandlungen des Sächsischen Geologischen Landesamtes, v. 1, 39 p.

Kröner, A., and Hegner, E., 1998, Geochemistry, single zircon ages and Sm-Nd systematics of granitoid rocks from the Góry Sowie (Owl Mts.), Polish West Sudetes: Evidence for early Paleozoic arc-related plutonism: Geological Society [London] Journal, v. 155, p. 711–724.

Kuijper, R.P., 1980, Precambrian U-Pb zircon ages from western Galicia (NW Spain): Earth-Science Reviews, v. 16, p. 313–316, doi: 10.1016/0012-8252(80)90051-3.

Lancelot, J.R., Allegret, A., and Iglesias-Ponce de León, M., 1985, Outline of Upper Precambrian and Lower Paleozoic evolution of the Iberian Peninsula according to U-Pb dating of zircons: Earth and Planetary Science Letters, v. 74, p. 325–337, doi: 10.1016/S0012-821X(85)80004-2.

Ledru, P., Autran, A., and Santallier, D., 1994a, Lithostratigraphy of Variscan terranes in the French Massif Central: A basis for paleogeographical reconstruction, *in* Keppie, J.D., ed., Pre-Mesozoic Geology in France and Related Areas: Berlin, Germany, Springer-Verlag, p. 276–288.

Ledru, P., Costa, S., and Echtler, H., 1994b, Structure, *in* Keppie, J.D., ed., Pre-Mesozoic Geology in France and Related Areas: Berlin, Germany, Springer-Verlag, p. 305–323.

Lefort, J.P., 1989, Basement Correlation across the North Atlantic: Berlin, Germany, Springer-Verlag, 148 p.

Linnemann, U., and Romer, R.L., 2002, The Cadomian orogeny in Saxo-Thuringia, Germany: Geochemical and Nd-Sr-Pb isotopic characterization of marginal basins with constraints to geotectonic setting and provenance: Tectonophysics, v. 352, p. 33–64, doi: 10.1016/S0040-1951(02)00188-9.

Linnemann, U., Elicki, O., and Gaitzsch, B., 2003, Die Stratigraphie das Saxothuringikums: Geologica Saxonica, v. 48/49, p. 29–70.

Llana-Fúnez, S., 2002, Quartz *c*-axis texture mapping of a Variscan regional foliation (Malpica-Tui Unit, NW Spain): Journal of Structural Geology, v. 24, p. 1299–1312, doi: 10.1016/S0191-8141(01)00138-9.

Lotze, F., 1945, Zur Gliederung der Varisziden der Iberischen Meseta: Geotektonische Forschungen, v. 6, p. 78–92.

Marchand, J., 1981, Ecaillage d'un "mélange tectonique" profond: Le complexe cristallophyllien de Champtoceaux (Bretagne méridionale): Comptes Rendus de l'Académie des Sciences, Paris, v. 293, p. 223–228.

Marcos, A., and Llana Fúnez, S., 2002, Estratigrafía y estructura de la lámina tectónica del para-autóctono y de su autóctono en el área de Chantada (Galicia, NO de España): Trabajos de Geología, Universidad de Oviedo, v. 23, p. 53–72.

Marcos, A., Marquínez, J., Pérez-Estaún, A., Pulgar, J.A., and Bastida, F., 1984, Nuevas aportaciones al conocimiento de la evolución tectonometamórfica del Complejo de Cabo Ortegal (NW de España): Cuadernos do Laboratorio Xeolóxico de Laxe, v. 7, p. 125–137.

Martínez, F.J., Julivert, M., Sebastian, A., Arboleya, M.L., and Gil Ibarguchi, J.I., 1988, Structural and thermal evolution of the high-grade areas in the northwestern parts of the Iberian Massif: American Journal of Science, v. 288, p. 969–996.

Martínez Catalán, J.R., 1990, A non-cylindrical model for the northwest Iberian allochthonous terranes and their equivalents in the Hercynian belt of Western Europe: Tectonophysics, v. 179, p. 253–272, doi: 10.1016/0040-1951(90)90293-H.

Martínez Catalán, J.R., Arenas, R., Díaz García, F., Rubio Pascual, F.J., Abati, J., and Marquínez, J., 1996, Variscan exhumation of a subducted Paleozoic continental margin: The basal units of the Ordenes Complex, Galicia: NW Spain: Tectonics, v. 15, p. 106–121, doi: 10.1029/95TC02617.

Martínez Catalán, J.R., Díaz García, F., Arenas, R., Abati, J., Castiñeiras, P., González Cuadra, P., Gómez Barreiro, J., and Rubio Pascual, F., 2002, Thrust and detachment systems in the Ordenes Complex (northwestern Spain): Implications for the Variscan-Appalachian geodynamics, *in* Martínez Catalán, J.R., Hatcher, R.D., Jr., Arenas, R., and Díaz García, F., eds., Variscan-Appalachian Dynamics: The Building of the Late Paleozoic Basement: Geological Society of America Special Paper 364, p. 163–182.

Martínez Catalán, J.R., Arenas, R., and Díez Balda, M.A., 2003, Large extensional structures developed during emplacement of a crystalline thrust sheet: The Mondoñedo nappe (NW Spain): Journal of Structural Geology, v. 25, p. 1815–1839, doi: 10.1016/S0191-8141(03)00038-5.

Martínez Catalán, J.R., Fernández-Suárez, J., Jenner, G.A., Belousova, E., and Díez Montes, A., 2004, Provenance constraints from detrital zircon U-Pb ages in the NW Iberian Massif: Implications for Paleozoic plate configuration and Variscan evolution: Geological Society [London] Journal, v. 161, p. 461–473.

Martínez García, E., 1972, El Silúrico de San Vitero (Zamora): Comparación con las series vecinas e importancia orogénica: Acta Geológica Hispánica, v. 7, p. 104–108.

Matte, Ph., 1968, La structure de la virgation Hercynienne de Galice (Espagne): Revue de Géologie Alpine, v. 44, p. 157–280.

Matte, Ph., 1986, Tectonics and plate tectonics model for the Variscan belt of Europe: Tectonophysics, v. 126, p. 329–374, doi: 10.1016/0040-1951(86)90237-4.

Matte, Ph., 1991, Accretionary history and crustal evolution of the Variscan belt in Western Europe: Tectonophysics, v. 196, p. 309–337, doi: 10.1016/0040-1951(91)90328-P.

Matte, Ph., 2002, Variscides, between the Appalachians and the Urals: Similarities and differences between Paleozoic subduction and collision belts, *in* Martínez Catalán, J.R., Hatcher, R.D., Jr., Arenas, R., and Díaz García, F., eds., Variscan-Appalachian Dynamics: The Building of the Late Paleozoic Basement: Geological Society of America Special Paper 364, p. 237–248.

Mendia Aranguren, M.S., 2000, Petrología de la Unidad Eclogítica del Complejo de Cabo Ortegal (NW de España): A Coruña, Spain, Laboratorio Xeolóxico de Laxe: Nova Terra, v. 16, p. 424.

Ménot, R.P., Peucat, J.J., Scarenzi, D., and Piboule, M., 1988, 496 My age of plagiogranites in the Chamrousse ophiolite complex (external crystalline massifs in the French Alps): Evidence of a Lower Paleozoic oceanization: Earth and Planetary Science Letters, v. 88, p. 82–92, doi: 10.1016/0012-821X(88)90048-9.

Nutman, A.P., Green, D.H., Cook, C.A., Styles, M.T., and Holdsworth, R.E., 2001, SHRIMP U-Pb zircon dating of the exhumation of the Lizard Peridotite and its emplacement over crustal rocks: Constraints for tectonic models: Geological Society [London] Journal, v. 158, p. 809–820.

O'Brien, P.J., Corner, A., Jaeckel, P., Hegner, E., Zelazniewicz, A., and Kryza, R., 1997, Petrological and isotopic studies on Paleozoic high pressure granulites, Góry Sowie Mts. Polish Sudetes: Journal of Petrology, v. 38, v. 433–456.

Oliveira, J.T., 1990, South Portuguese zone: Stratigraphy and synsedimentary tectonism, *in* Dallmeyer, R.D., and Martínez García, E., eds., Pre-Mesozoic Geology of Iberia: Berlin, Springer-Verlag, p. 334–347.

Oliveira, J.T., Horn, M., and Paproth, E., 1979, Preliminary note on the stratigraphy of the Baixo Alentejo Flysch Group, Carboniferous of Portugal, and on the paleogeographic development compared to corresponding units in northwest Germany: Comunicaçoes dos Serviços Geologicos de Portugal, v. 65, p. 151–168.

Oncken, O., Plesch, A., Weber, J., Ricken, W., and Schrader, S., 2000, Passive margin detachment arc-continent collision (Central European Variscides), *in* Franke, W., Haak, V., Oncken, O., and Tanner, D., eds., Orogenic Processes: Quantification and Modelling in the Variscan Belt: Geological Society [London] Special Publication 179, p. 199–216.

Onézime, J., Charvet, J., Faure, M., Chauvet, A., and Panis, D., 2002, Structural evolution of the southernmost segment of the West European Variscides: The South Portuguese zone (SW Iberia): Journal of Structural Geology, v. 24, p. 451–468, doi: 10.1016/S0191-8141(01)00079-7.

Ordóñez Casado, B., Gebauer, D., Schäfer, H.J., Gil Ibarguchi, J.I., and Peucat, J.J., 2001, A single Devonian subduction event for the HP/HT metamorphism of the Cabo Ortegal complex within the Iberian Massif: Tectonophysics, v. 332, p. 359–385, doi: 10.1016/S0040-1951(00)00210-9.

Ortega Cuesta, L.A., 1998, Estudio petrogenético del granito sincinemático de dos micas de A Espenuca (A Coruña): A Coruña, Spain, Serie Nova Terra, Ediciós do Castro, Sada, v. 14, 377 p.

Ortega, L.A., Carracedo, M., Larrea, F.J., and Gil Ibarguchi, J.I., 1996, Geochemistry and tectonic environment of volcanosedimentary rocks from the Ollo de Sapo Formation (Iberian Massif, Spain), *in* Demaiffe, D., ed., Petrology and Geochemistry of Magmatic Suites of Rocks in the Continental and Oceanic Crusts: Brussels, Belgium, Université Libre de Bruxelles, p. 277–290.

Ortega, L.A., Aranguren, A., Menéndez, M., and Gil Ibarguchi, I., 2000, Petrogénesis, age and emplacement of the late-Hercynian Veiga granite (Ollo de Sapo antiform, northwestern Spain): Cuadernos do Laboratorio Xeolóxico de Laxe, v. 25, p. 265–268.

Paris, F., 1998, Early Palaeozoic palaeobiogeography of northern Gondwana regions: Acta Universitatis Carolinae-Geologica, v. 42, p. 473–483.

Paris, F., and Robardet, M., 1977, Paléogéographie et relations Ibéro-Armoricaines au Paléozoïque anté-Carbonifére: Bulletin de la Société Géologique de France, v. 19, p. 1121–1126.

Pereira, Z., Meireles, C., and Pereira, E., 1999, Upper Devonian palynomorphs of the NE sector of Trás-os-Montes (Central Iberian zone): Badajoz, Spain, XV Reunión de Geología del Oeste Peninsular (International Meeting on Cadomian Orogens, Extended Abstracts), p. 201–206.

Pérez-Estaún, A., 1974, Aportaciones al conocimiento del Carbonífero de San Clodio (Prov. de Lugo): Breviora Geologica Asturica, v. 18, p. 3–8.

Pérez-Estaún, A., Bastida, F., Alonso, J.L., Marquínez, J., Aller, J., Alvarez-Marrón, J., Marcos, A., and Pulgar, J.A., 1988, A thin-skinned tectonics model for an arcuate fold and thrust belt: The Cantabrian zone (Variscan Ibero-Armorican arc): Tectonics, v. 7, p. 517–537.

Pérez-Estaún, A., Martínez Catalán, J.R., and Bastida, F., 1991, Crustal thickening and deformation sequence in the footwall to the suture of the Variscan belt of northwest Spain: Tectonophysics, v. 191, p. 243–253, doi: 10.1016/0040-1951(91)90060-6.

Perroud, H., 1982, Contribution à l'étude paléomagnétique de l'arc Ibéro-Armoricain: Bulletin de la Societé Géologique et Minéralogique de Bretagne, v. 14, p. 1–114.

Perroud, H., and Bonhommet, N., 1981, Paleomagnetism of the Ibero-Armorican arc and the Hercynian orogeny in Western Europe: Nature, v. 292, p. 445–448, doi: 10.1038/292445a0.

Peucat, J.J., Auvray, B., Hirbec, Y., and Calvez, J.Y., 1984, Granites et cisaillements Hercyniens dans le Nord du Massif Armoricain: Géochronologie Rb-Sr: Bulletin de la Société Géologique de France, v. 7, p. 1365–1373.

Peucat, J.J., Bernard-Griffiths, J., Gil Ibarguchi, J.I., Dallmeyer, R.D., Menot, R.P., Cornichet, J., and Iglesias Ponce de León, M., 1990, Geochemical and geochronological cross section of the deep Variscan crust: The Cabo Ortegal high-pressure nappe (northwestern Spain): Tectonophysics, v. 177, p. 263–292, doi: 10.1016/0040-1951(90)90285-G.

Pin, C., Ortega Cuesta, L.A., and Gil Ibarguchi, J.I., 1992, Mantle-derived, early Paleozoic A-type metagranitoids from the NW Iberian Massif: Nd isotope and trace-element constraints: Bulletin de la Société Géologique de France, v. 163, p. 483–494.

Pin, C., Paquette, J.L., Santos Zalduegui, J.F., and Gil Ibarguchi, J.I., 2002, Early Devonian supra-subduction zone ophiolite related to incipient collisional processes in the western Variscan belt: The Sierra de Careón unit, Ordenes Complex, Galicia, *in* Martínez Catalán, J.R., Hatcher, R.D., Jr., Arenas, R., and Díaz García, F., eds., Variscan-Appalachian Dynamics: The Building of the Late Paleozoic Basement: Geological Society of America Special Paper 364, p. 57–72.

Quesada, C., Fonseca, P.E., Munhá, J., Oliveira, J.T., and Ribeiro, A., 1994, The Beja-Acebuches ophiolite (southern Iberia Variscan fold belt): Geological characterization and geodynamic significance: Boletín Geológico y Minero, v. 105, p. 3–49.

Quinquis, H., and Choukroune, P., 1981, Les schistes bleus de l'Ile de Groix dans la Chaîne Hercynienne: Implications cinématiques: Bulletin de la Société Géologique de France, v. 7, p. 409–418.

Rast, N., and Skehan, J.W., 1983, The evolution of the Avalonian plate: Tectonophysics, v. 100, p. 257–286, doi: 10.1016/0040-1951(83)90191-9.

Reche, J., Martínez, F.J., Arboleya, M.L., Dietsch, C., and Briggs, W.D., 1998, Evolution of a kyanite-bearing belt within a HT-LP orogen: The case of NW Variscan Iberia: Journal of Metamorphic Geology, v. 16, p. 379–394, doi: 10.1111/j.1525-1314.1998.00142.x.

Regêncio Macedo, C.A., 1988, Granitóides, Complexo Xisto-Grauváquico e Ordovícico na Região Entre Trancoso e Pinhel (Portugal Central) [Ph.D. thesis]: Coimbra, Portugal, Universidade de Coimbra, 430 p.

Rey, P., Burg, J.P., and Casey, M., 1997, The Scandinavian Caledonides and their relationship to the Variscan belt, *in* Burg, J.P., and Ford, M., eds., Orogeny through Time: Geological Society [London] Special Publication 121, p. 179–200.

Ribeiro, A., 1974, Contribution a l'étude tectonique de Trás-os-Montes Oriental: Memorias dos Serviços Geológicos de Portugal, v. 24, 179 p.

Ribeiro, A., Pereira, E., and Severo, L., 1980, Análise da deformaçao da zona de cisalhamento Porto-Tomar na transversal de Oliveira de Aeméis: Comunicaçoes dos Serviços Geológicos de Portugal, v. 66, p. 3–9.

Ribeiro, A., Pereira, E., and Dias, R., 1990, Central Iberian zone: Allochthonous sequences. Structure in the northwest of the Iberian Peninsula, *in* Dallmeyer, R.D., and Martínez García, E., eds., Pre-Mesozoic Geology of Iberia: Berlin, Germany, Springer-Verlag, p. 220–236.

Ribeiro, A., Dias, R., and Brandao Silva, J., 1995, Genesis of the Ibero-Armorican arc: Geodinamica Acta, v. 8, p. 173–184.

Ribeiro, M.L., and Floor, P., 1987, Magmatismo peralcalino no Maciço Hespérico: Sua distribução e significado geodinámico, *in* Bea, F., Carnicero, A., Gonzalo, J.C., López-Plaza, M., and Rodríguez Alonso, M.D., eds., Geología de los Granitoides y Rocas Asociadas del Macizo Hespérico: Madrid, Spain, Rueda, p. 211–221.

Ribeiro, M.L., and Ribeiro, A., 1974, Signification paléogéographique et tectonique de la présence de galets de roches metamorphiques dans un flysch d'âge Dévonien supérieur du Tras-os-Montes oriental (nord-est du Portugal): Comptes Rendus de l'Académie des Sciences, Paris, v. 278, p. 3161–3163.

Riemer, W., 1966, Datos para el conocimiento de la estratigrafía de Galicia: Notas y Comunicaciones del Instituto Geológico y Minero de España, v. 81, p. 7–20.

Ries, A.C., 1979, Variscan metamorphism and K-Ar dates in the Variscan fold belt of S Brittany and NW Spain: Geological Society [London] Journal, v. 136, p. 89–103.

Robardet, M., 2002, Alternative approach to the Variscan belt of southwestern Europe: Preorogenic paleobiogeographical constraints, *in* Martínez Catalán, J.R., Hatcher, R.D., Jr., Arenas, R., and Díaz García, F., eds., Variscan-Appalachian Dynamics: The Building of the Late Paleozoic Basement: Geological Society of America Special Paper 364, p. 1–15.

Robardet, M., 2003, The Armorica 'microplate': Fact or fiction? Critical review of the concept and contradictory paleobiogeographical data: Palaeogeography, Palaeoclimatology, Palaeoecology, v. 195, p. 125–148, doi: 10.1016/S0031-0182(03)00305-5.

Robardet, M., and Gutiérrez-Marco, J.C., 2004, The Ordovician, Silurian and Devonian sedimentary rocks of the Ossa-Morena zone (SW Iberian Peninsula, Spain): Journal of Iberian Geology, v. 30, p. 73–92.

Robardet, M., Paris, F., and Racheboeuf, P.R., 1990, Paleogeographic evolution of southwestern Europe during early Paleozoic times, *in* McKerrow, W.S., and Scotese, C.R., eds., Paleozoic Palaeogeography and Biogeography: Geological Society [London] Memoir 12, p. 411–419.

Rodríguez, J., Cosca, M.A., Gil Ibarguchi, J.I., and Dallmeyer, R.D., 2003, Strain partitioning and preservation of 40Ar/39Ar ages during Variscan exhumation of a subducted crust (Malpica-Tui complex, NW Spain): Lithos, v. 70, p. 111–139, doi: 10.1016/S0024-4937(03)00095-1.

Rolet, J., 1994, Carboniferous tectonics and pull-apart basin development: The Variscan orogeny in the Armorican Massif, structure and metamorphism, *in* Keppie, J.D., ed., Pre-Mesozoic Geology in France and Related Areas: Berlin, Germany, Springer-Verlag, p. 215–219.

Rolet, J., Gresselin, F., Jegouzo, P., Ledru, P., and Wyns, R., 1994, Intracontinental Hercynian events in the Armorican Massif: The Variscan orogeny in the Armorican Massif, structure and metamorphism, *in* Keppie, J.D., ed., Pre-Mesozoic Geology in France and Related Areas: Berlin, Germany, Springer-Verlag, p. 195–219.

Román-Berdiel, T., Pueyo-Morer, E.L., and Casas-Sáiz, A.M., 1995, Granite emplacement during contemporary shortening and normal faulting: Structural and magnetic study of the Veiga Massif (NW Spain): Journal of Structural Geology, v. 17, p. 1689–1706, doi: 10.1016/0191-8141(95)00062-I.

Santallier, D., Lardeaux, J.M., Marchand, J., and Marignac, Ch., 1994, Metamorphism: The Massif Central, *in* Keppie, J.D., ed., Pre-Mesozoic Geology in France and Related Areas: Berlin, Germany, Springer-Verlag, p. 324–340.

Santos Zalduegui, J.F., Schärer, U., and Gil Ibarguchi, J.I., 1995, Isotope constraints on the age and origin of magmatism and metamorphism in the Malpica-Tuy allochthon, Galicia, NW-Spain: Chemical Geology, v. 121, p. 91–103, doi: 10.1016/0009-2541(94)00123-P.

Santos Zalduegui, J.F., Schärer, U., Gil Ibarguchi, J.I., and Girardeau, J., 1996, Origin and evolution of the Paleozoic Cabo Ortegal ultramafic-mafic complex (NW Spain): U-Pb, Rb-Sr and Pb-Pb isotope data: Chemical Geology, v. 129, p. 281–304, doi: 10.1016/0009-2541(95)00144-1.

Santos, J.F., Schärer, U., Gil Ibarguchi, J.I., and Girardeau, J., 2002, Genesis of pyroxenite-rich peridotite at Cabo Ortegal (NW Spain): Geochemical and Pb-Sr-Nd isotope data: Journal of Petrology, v. 43, p. 17–43, doi: 10.1093/petrology/43.1.17.

Schäfer, H.J., Gebauer, D., Gil Ibarguchi, J.I., and Peucat, J.J., 1993, Ion-microprobe U-Pb zircon dating on the HP/HT Cabo Ortegal Complex (Galicia, NW Spain): Preliminary results: Terra Abstracts, v. 5, no. 4, p. 22.

Scotese, C.R., 2001, Atlas of Earth History: Arlington, Texas, PALEOMAP Project, 52 p.

Shelley, D., and Bossière, G., 2000, A new model for the Hercynian orogen of Gondwanan France and Iberia: Journal of Structural Geology, v. 22, p. 757–776, doi: 10.1016/S0191-8141(00)00007-9.

Shelley, D., and Bossière, G., 2002, Megadisplacements and the Hercynian orogen of Gondwanan France and Iberia, *in* Martínez Catalán, J.R., Hatcher, R.D., Jr., Arenas, R., and Díaz García, F., eds., Variscan-Appalachian Dynamics: The Building of the Late Paleozoic Basement: Geological Society of America Special Paper 364, p. 209–222.

Silva, J.B., Oliveira, J.T., and Ribeiro, A., 1990, South Portuguese zone: Structural outline, *in* Dallmeyer, R.D., and Martínez García, E., eds., Pre-Mesozoic Geology of Iberia: Berlin, Germany, Springer-Verlag, p. 348–362.

Simancas, F., 2004, Zona Sudportuguesa, estructura, *in* Vera, J.A., ed., Geología de España: Madrid, Spain, Sociedad Geológica de España–Instituto Geológico y Minero de España, p. 211–214.

Simancas, F., González Lodeiro, F., Expósito Ramos, I., Azor, A., and Martínez Poyatos, D., 2002, Opposite subduction polarities connected by transform faults in the Iberian Massif and western European Variscides, *in* Martínez Catalán, J.R., Hatcher, R.D., Jr., Arenas, R., and Díaz García, F., eds., Variscan-Appalachian Dynamics: The Building of the Late Paleozoic Basement: Geological Society of America Special Paper 364, p. 249–258.

Soper, N.J., 1988, Timing and geometry of collision, terrane accretion and sinistral strike-slip events in the British Caledonides, *in* Harris, A.L., and Fettes, D.J., eds., The Caledonian-Appalachian Orogen: Geological Society [London] Special Publication 38, p. 481–492.

Stampfli, G.M., von Raumer, J.F., and Borel, G.D., 2002, Paleozoic evolution of pre-Variscan terranes: From Gondwana to the Variscan collision, *in* Martínez Catalán, J.R., Hatcher, R.D., Jr., Arenas, R., and Díaz García, F., eds., Variscan-Appalachian Dynamics: The Building of the Late Paleozoic Basement: Geological Society of America Special Paper 364, p. 263–280.

Stephens, B., and Gee, D.G., 1989, Terranes and polyphase accretionary history in the Scandinavian Caledonides, *in* Dallmeyer, R.D., ed., Terranes in the Circum-Atlantic Paleozoic Orogens: Geological Society of America Special Paper 230, p. 17–30.

Timmermann, H., Parrish, R.R., Noble, S.R., and Kryza, R., 2000, New U-Pb monazite and zircon data from the Sudetes Mountains: Evidence for a single-cycle Variscan orogeny: Geological Society [London] Journal, v. 157, p. 265–268.

Tollman, A., 1982, Großräumiger variszischer Deckenbau im Moldanubikum und neue Gedanken zum Variszikum Europas: Geotektonische Forschungen, v. 64, p. 1–91.

Valle Aguado, B., Azevedo, M.R., Schaltegger, U., Martínez Catalán, J.R., and Nolan, J., 2005, U-Pb zircon and monazite geochronology of Variscan magmatism related to syn-convergence extension in central northern Portugal: Lithos, v. 82, p. 169–184, doi: 10.1016/j.lithos.2004.12.012.

Valverde Vaquero, P., and Dunning, G.R., 2000, New U-Pb ages for Early Ordovician magmatism in Central Spain: Geological Society [London] Journal, v. 157, p. 15–26.

Valverde Vaquero, P., and Fernández, F.J., 1996, Edad de enfriamiento U/Pb en rutilos del Gneis de Chímparra (Cabo Ortegal, NO de España): Geogaceta, v. 20, p. 475–478.

Valverde-Vaquero, P., Marcos, A., Farias, P., and Gallastegui, G., 2005, U-Pb dating of Ordovician felsic volcanism in the schistose domain of the Galicia-Trás-os-Montes zone near Cabo Ortegal (NW Spain): Geológica Acta, v. 3, p. 27–37.

Van Calsteren, P.W.C., Boelrijk, N.A.I.M., Hebeda, E.H., Priem, H.N.A., Den Tex, E., Verdurmen, E.A.T.H., and Verschure, R.H., 1979, Isotopic dating of older elements (including the Cabo Ortegal mafic-ultramafic complex) in the Hercynian orogen of NW Spain: Manifestations of a presumed early Paleozoic mantle-plume: Chemical Geology, v. 24, p. 35–56, doi: 10.1016/0009-2541(79)90011-1.

van Staal, C.R., 2005, The northern Appalachians, *in* Selley, R.C., Cocks, L.R.M., and Plimer, I.R., eds., Encyclopedia of Geology: Oxford, Elsevier, p. 81–91.

van Staal, C.R., and De Roo, J.A., 1995, Mid-Paleozoic tectonic evolution of the Appalachian Central Mobile Belt in northern New Brunswick, Canada: Collision, extensional collapse and dextral transpression, *in* Hibbard, J.P., van Staal, C.R., and Cawood, P.A., eds., Current Perspectives in the Appalachian-Caledonian Orogen: Geological Association of Canada Special Paper 41, p. 367–389.

van Staal, C.R., Dewey, J.F., MacNiocaill, C., and McKerrow, W.S., 1998, The Cambrian-Silurian tectonic evolution of the northern Appalachians and British Caledonides: History of a complex, west and southwest Pacific-type segment of Iapetus, *in* Blundell, D.J., and Scott, A.C., eds., Lyell: The Past Is the Key to the Present: Geological Society [London] Special Publication 143, p. 199–242.

Vauchez, A., and Bufalo, M., 1988, Charriage crustal, anatexie et decrochements ductiles dans les Maures orientales (Var, France) au cours de l'orogenese Varisque: Geologische Rundschau, v. 77, p. 45–62, doi: 10.1007/BF01848675.

Vialette, Y., Casquet, C., Fúster, J.M., Ibarrola, E., Navidad, M., Peinado, M., and Villaseca, C., 1987, Geochronological study of orthogneisses from the Sierra de Guadarrama (Spanish Central System): Neues Jahrbuch für Mineralogie-Monatshefte, v. 10, p. 465–479.

von Raumer, J.F., and Neubauer, F., 1993, Late Precambrian and Paleozoic evolution of the Alpine basement—An overview, *in* von Raumer, J.F., and Neubauer, F., eds., Pre-Mesozoic Geology in the Alps: Berlin, Germany, Springer-Verlag, p. 625–639.

von Raumer, J.F., and Neubauer, F., 1994, The Paleozoic evolution of the Alps: Schweizerische Mineralogische und Petrographische Mitteilungen, v. 74, p. 459–467.

von Raumer, J.F., Stampfli, G.M., Borel, G., and Bussy, F., 2002, Organization of pre-Variscan basement areas at the north-Gondwanan margin: International Journal of Earth Sciences, v. 91, p. 35–52, doi: 10.1007/s005310100200.

von Raumer, J.F., Stampfli, G.M., and Bussy, F., 2003, Gondwana-derived microcontinents—The constituents of the Variscan and Alpine collisional orogens: Tectonophysics, v. 365, p. 7–22, doi: 10.1016/S0040-1951(03)00015-5.

Williams, H., and Hatcher, R.D., Jr., 1983, Appalachian suspect terranes, *in* Hatcher, R.D., Jr., Williams, H., and Zietz, I., eds., The Tectonics and Geophysics of Mountain Chains: Geological Society of America Memoir 158, p. 33–53.

Wimmenauer, W., and Lim, S.K., 1988, L'association leptyno-amphibolique de la Forêt-Noire (R.F.A.): Bulletin de la Société Géologique de France, v. 8, p. 35–41.

Winchester, J.A., Pharaoh, T.C., and Verniers, J., 2002, Palaeozoic amalgamation of Central Europe: An introduction and synthesis of new results from recent geological and geophysical investigations, *in* Winchester, J.A., Pharaoh, T.C., and Verniers, J., eds., Palaeozoic Amalgamation of Central Europe: Geological Society [London] Special Publication 201, p. 1–18.

Young, T.P., 1990, Ordovician sedimentary facies and faunas of southwest Europe: Palaeogeographic and tectonic implications, *in* McKerrow, W.S., and Scotese, C.R., eds., Palaeozoic Palaeogeography and Biogeography: Geological Society [London] Memoir 12, p. 421–430.

Manuscript Accepted by the Society 1 February 2007

Printed in the USA

The Geological Society of America
Memoir 200
2007

Paleozoic ophiolites in the Variscan suture of Galicia (northwest Spain): Distribution, characteristics, and meaning

Ricardo Arenas*
Departamento de Petrología y Geoquímica, Universidad Complutense, 28040 Madrid, Spain

José R. Martínez Catalán
Departamento de Geología, Universidad de Salamanca, 37008 Salamanca, Spain

Sonia Sánchez Martínez
Departamento de Petrología y Geoquímica, Universidad Complutense, 28040 Madrid, Spain

Florentino Díaz García
Departamento de Geología, Universidad de Oviedo, 33005 Oviedo, Spain

Jacobo Abati
Javier Fernández-Suárez
Pilar Andonaegui
Departamento de Petrología y Geoquímica, Universidad Complutense, 28040 Madrid, Spain

Juan Gómez-Barreiro
Departamento de Geología, Universidad de Salamanca, 37008 Salamanca, Spain

ABSTRACT

Allochthonous ophiolitic units in the northwestern Iberian Massif are remnants of peri-Gondwanan Paleozoic oceans sandwiched among other exotic terranes of continental and volcanic-arc derivation. All these terranes define an intricate suture zone that marks the convergence and collision between Laurussia and Gondwana. The suture is defined by three different ophiolitic ensembles: upper ophiolitic units, lower ophiolitic units, and the Somozas mélange. The lower ophiolitic units were derived from an alternation of basalts and sediments intruded by gabbros and scarce granitoids, and they formed during the opening of a marginal basin, the Galician ocean, during Late Cambrian to Early Ordovician time. This ocean was created as a back arc by the severance of a volcanic arc that had developed at the northern margin of Gondwana and formed part of the Rheic oceanic realm. The upper ophiolitic units formed during the Early Devonian from intraoceanic subduction in the early Paleozoic lithosphere of the Rheic Ocean. These suprasubduction ophiolites were formed

*arenas@geo.ucm.es

Arenas, R., Martínez Catalán, J.R., Sánchez Martínez, S., Díaz García, F., Abati, J., Fernández-Suárez, J., Andonaegui, P., and Gómez-Barreiro, J., 2007, Paleozoic ophiolites in the Variscan suture of Galicia (northwest Spain): Distribution, characteristics, and meaning, *in* Hatcher, R.D., Jr., Carlson, M.P., McBride, J.H., and Martínez Catalán, J.R., eds., 4-D Framework of Continental Crust: Geological Society of America Memoir 200, p. 425–444, doi: 10.1130/2007.1200(22). For permission to copy, contact editing@geosociety.org.

just before the ocean closed, preceding the collision between Gondwana and Laurussia. The Somozas mélange appears in an anomalous position at the base of the Cabo Ortegal Complex. The ophiolites involved in this tectonic mélange represent an imbricate of highly dismembered oceanic lithosphere, slivers of subducted outer edge of the Gondwanan continental margin, and Paleozoic metasediments of the northern Gondwanan platform. The ophiolites might either record the development of a different peri-Gondwanan oceanic domain, or they might be equivalent to any of the other ophiolitic ensembles, and their anomalous structural position is simply a consequence of complex thrusting.

Keywords: ophiolites, Rheic Ocean, Variscan belt.

INTRODUCTION

The Variscan belt is the European part of the large orogen that developed during the late Paleozoic as a result of the progressive collision between Laurussia and Gondwana. It is probably the longest mountain belt that developed during the last stages of the formation of Pangea, and it is currently traceable from North America across northern Africa and southwestern Europe to the Bohemian Massif in Poland and the Czech Republic in Central Europe. Parts of this orogen can also be seen in southeastern Europe, where they appear in the basement of Alpine belts (Von Raumer and Neubauer, 1993; Matte, 1995, 2001). A large exposure of the Variscan basement forms the Iberian Massif, in the western Iberian Peninsula. In Galicia and Asturias, northwest of the Iberian Massif, a long and continuous section of the Variscan belt, nearly 400 km wide, developed in Devonian and Carboniferous times. Deformation progressed from west to east (Dallmeyer et al., 1997), and the western parts represent the more internal domain, which includes the remnants of a complex suture zone that contains several ophiolitic units (Arenas et al., 1986). The ophiolites form part of the so-called allochthonous complexes of Galicia-Trás-os-Montes; they separate the main colliding elements and represent the remnants of the Paleozoic oceanic domains that closed during the long convergence between Laurentia-Baltica and Gondwana and their final collision.

The Galicia-Trás-os-Montes zone structurally overlies the autochthonous sequences of the Central Iberian zone. According to its original description (Farias et al., 1987; Arenas et al., 1988), the Galicia-Trás-os-Montes zone consists of two different and superimposed domains (Fig. 1): a lower assemblage represented by the schistose domain of Galicia-Trás-os-Montes, and an upper assemblage formed by the allochthonous complexes of Galicia-Trás-os-Montes.

The schistose domain is made up of Paleozoic metasediments, mostly schists and sandstones, with interbedded felsic volcanic rocks. It is limited by a basal thrust and shows an imbricate internal structure (Ribeiro, 1974; Pereira et al., 1989, 2000). Even though the character of the Paleozoic sedimentary succession in the schistose domain differs markedly from the related autochthon of the Central Iberian zone, it is still possible to make stratigraphic correlations between both domains because they record a similar Variscan tectonothermal evolution. For these reasons, the schistose domain is also known as the parautochthon (Ribeiro et al., 1990). Considering that ophiolites do not appear between the Central Iberian zone and the Galicia-Trás-os-Montes zone, there is no doubt that the schistose domain formed part of the same paleogeographic realm as the Central Iberian zone: the northern continental margin of Gondwana (Arenas et al., 1986; Martínez Catalán et al., 1999). It was displaced much less than the allochthonous complexes, but the magnitude of the actual translation of the schistose domain over the Central Iberian zone is unknown, and no palinspastic reconstruction of its position on the Gondwanan paleomargin has been attempted.

The allochthonous complexes occupy the uppermost structural position in the NW Iberian Massif (Fig. 1). They were probably thrust over the schistose domain during an early deformational stage, and, later on, the already assembled Galicia-Trás-os-Montes zone overrode the Central Iberian zone. The allochthonous complexes consist of a stack of largely displaced allochthonous units, which were part of a gigantic nappe pile that was assembled during the first stages of the Variscan collision (Ries and Shackleton, 1971). The allochthonous complexes are presently preserved as megaklippen in the cores of open synforms, and they are considered to be remnants of this pile. Their denomination is based on the distinctive tectonothermal evolution recorded in each of the allochthonous units, as well as on the very large displacements that preceded their final emplacement. The allochthonous units include ophiolitic assemblages and terranes with continental and island-arc affinities (Arenas et al., 1986; Martínez Catalán et al., 1997, 1999). The ophiolites are remnants of oceanic lithosphere that was almost completely removed in pre-Variscan and Variscan times, whereas the other units are remnants of colliding

→

Figure 1. Schematic geologic map and cross-section of the allochthonous complexes of Galicia, showing the distribution of the main allochthonous units and their structural relationships. Six different ophiolitic units can be distinguished in the Cabo Ortegal and Órdenes Complexes. These ophiolites are located below the upper units, and, with the exception of the Somozas mélange, they overlie the basal units. IP—intermediate pressure; HP-HT—high pressure, high temperature; GTMZ—Galicia-Trás-os-Montes zone; CIZ—Central Iberian zone.

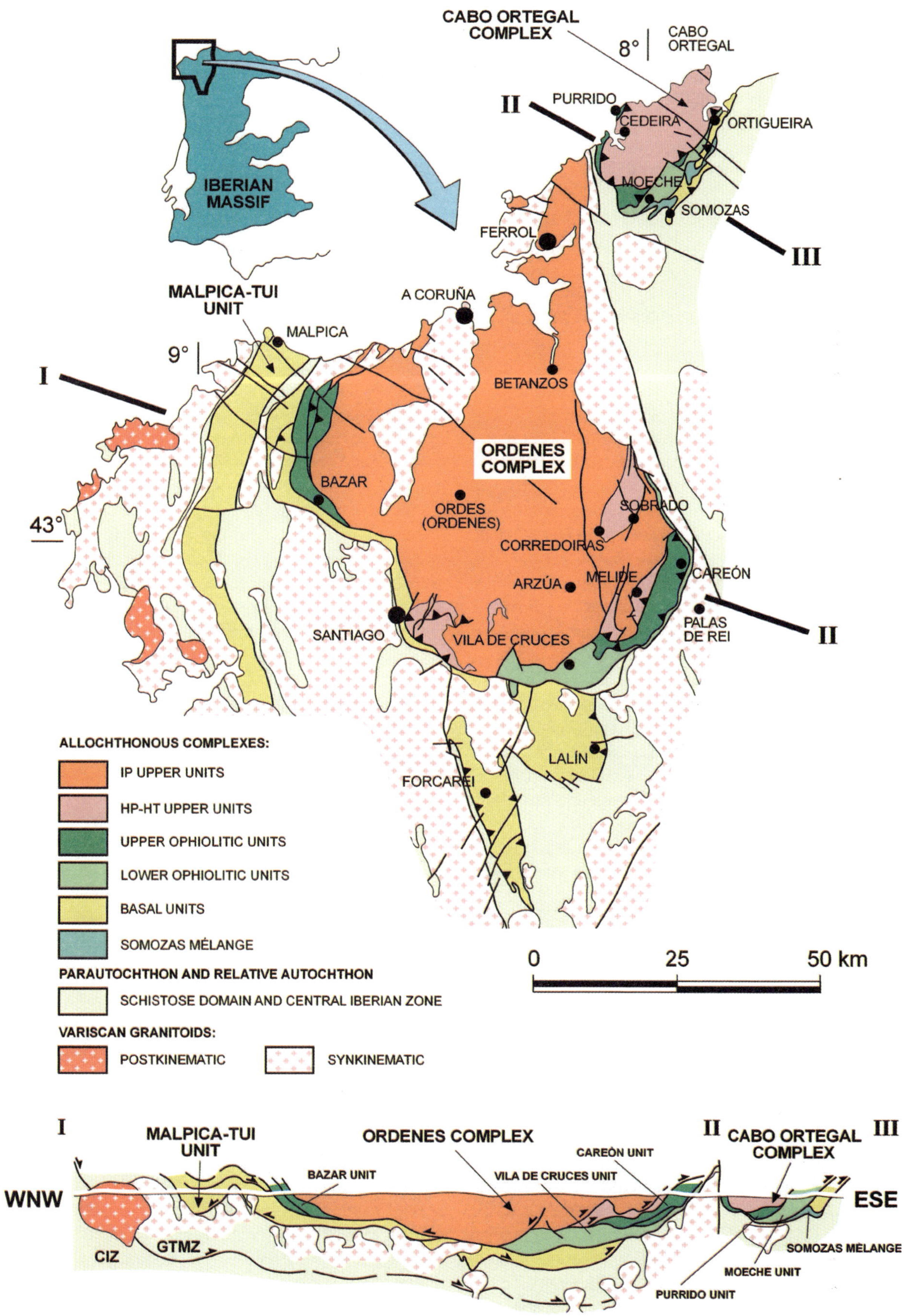

CABO ORTEGAL COMPLEX
8°
CABO ORTEGAL
PURRIDO
CEDEIRA
ORTIGUEIRA
MOECHE
SOMOZAS
FERROL
IBERIAN MASSIF
MALPICA-TUI UNIT
A CORUÑA
MALPICA
9°
BETANZOS
ORDENES COMPLEX
BAZAR
ORDES (ORDENES)
SOBRADO
CORREDOIRAS
CAREÓN
ARZÚA
MELIDE
SANTIAGO
VILA DE CRUCES
PALAS DE REI
43°
LALÍN
FORCAREI
I
II
III
ALLOCHTHONOUS COMPLEXES:
IP UPPER UNITS
HP-HT UPPER UNITS
UPPER OPHIOLITIC UNITS
LOWER OPHIOLITIC UNITS
BASAL UNITS
SOMOZAS MÉLANGE
PARAUTOCHTHON AND RELATIVE AUTOCHTHON
SCHISTOSE DOMAIN AND CENTRAL IBERIAN ZONE
VARISCAN GRANITOIDS:
POSTKINEMATIC
SYNKINEMATIC
0
25
50 km
MALPICA-TUI UNIT
ORDENES COMPLEX
CABO ORTEGAL COMPLEX
CAREÓN UNIT
BAZAR UNIT
VILA DE CRUCES UNIT
WNW
ESE
CIZ
GTMZ
SOMOZAS MÉLANGE
MOECHE UNIT
PURRIDO UNIT

continental margins and other exotic terranes located between them. Therefore, the allochthonous complexes include a rootless Variscan suture, which has provided key data for reconstructions of the evolution of the Paleozoic peri-Gondwanan realm.

THE ALLOCHTHONOUS COMPLEXES OF GALICIA

Five allochthonous complexes exist in the northwestern Iberian Massif. Three of them are located in Galicia, northwest Spain (Cabo Ortegal, Órdenes, and Malpica-Tui; Fig. 1), and two are in the Portuguese region of Trás-os-Montes (Bragança and Morais). Ophiolites occur in all but Malpica-Tui, where only the basal allochthonous units have been identified. Hence, it cannot be considered as a true complex, although it is truly allochthonous, and is often named the Malpica-Tui unit. The following description introduces the Galician complexes.

The Órdenes Complex is the largest (135 × 75 km) in the Galicia-Trás-os-Montes zone, and it contains representatives of nearly all the allochthonous units recognized in the NW Iberian Massif. Detailed descriptions of these units and interpretations of their meaning have been published previously (Díaz García, 1990; Abati, 2002; Castiñeiras, 2003; Martínez Catalán et al., 2002), and a synthesis of the structure and evolution of the entire complex has been presented by Arenas et al. (2000). Figure 1 shows its general structure, which is an open synform, and the distribution and tectonostratigraphic arrangement of the principal units. A more detailed view of the internal structure of the complex is presented in Martínez Catalán et al. (this volume). Different units are separated from each other by several generations of thrusts and extensional detachments, and a thrust separates the basal units from the parautochthon in the southern part of the complex, whereas late extensional detachments and strike-slip shear zones delimit its eastern and western parts.

The Cabo Ortegal Complex, located to the north (Fig. 1), is characterized by a spectacular coastal section. It has an ellipsoidal present-day shape and a small size (30 × 20 km). The more outstanding aspects about the petrology and structure of this complex have been described by several authors (Vogel, 1967; Arenas, 1988; Azcárraga, 2000; Mendia, 2000; Castiñeiras, 2003; Puelles, 2003), but some additional compilations have been presented recently (Ábalos et al., 2000; Gil Ibarguchi et al., 2000; Marcos et al., 2000, 2002). The complex is separated from the schistose domain by a large thrust fault (Fig. 1), which is a relatively young structure that belongs to the out-of-sequence thrust generation described by Martínez Catalán et al. (2002). This basal thrust shows crosscutting relationships with older generations of thrusts that separate several allochthonous units in the complex. Some of the older thrusts also overprint a first generation of extensional detachments with an important metamorphic jump (Castiñeiras, 2003). In spite of its size, the Cabo Ortegal Complex includes representatives of all the units described in all of the allochthonous complexes (Fig. 1), although some of them are poorly represented when compared with their equivalents in the Órdenes Complex.

The Malpica-Tui unit is a narrow synform structure (150 × 10 km) located in western Galicia (Fig. 1). Descriptions of its main characteristics have been presented by Floor (1966), Arps (1970), Llana-Fúnez (2001), Llana-Fúnez and Marcos (2002), and Rodríguez (2003). The unit consists of three lithological assemblages that are probably separated by thrusts (Gil Ibarguchi and Ortega Gironés, 1985), but the basal thrust fault is rarely preserved (Llana-Fúnez, 2001) because its present boundaries are mostly extensional and strike-slip shear zones and other faults.

The units that form the allochthonous complexes can be grouped into three main lithologic assemblages, which appear in a regular stacking order. From bottom to top, they consist of the basal, ophiolitic, and upper units (Fig. 1). The basal and upper units have lithologic signatures characteristic of both continental crust and island arcs, including forearc and back-arc domains, both in an arc setting. The ophiolitic units consist of lithologic units characteristic of oceanic domains, including marginal basins and transitional zones between continents and oceans. Moreover, the basal and ophiolitic units exhibit a relatively simple tectonothermal evolution of early Variscan to Variscan events, but the upper units are polymetamorphic assemblages where both Variscan and pre-Variscan tectonothermal events have been described. In the following sections, their main characteristics will be described, with emphasis placed on ophiolitic units, the primary objective of this contribution. The origin and evolution of all units will be discussed in the context of early Paleozoic peri-Gondwanan evolution and later Laurussia-Gondwana collision.

BASAL UNITS

These units consist of metasediments alternating with granitic orthogneisses and metabasites, which attain a maximum thickness of 3000–4000 m in the Malpica-Tui unit (Llana-Fúnez, 2001). The metasediments include metagraywackes, pelitic and semipelitic schists with garnet and albite-oligoclase porphyroblasts, paragneisses and migmatitic paragneisses, as well as scarce chert, black slate, calc-silicate, and quartzite lenses. It is still possible to recognize turbiditic facies in the low-grade successions (Llana-Fúnez, 2001).

The deformation is heterogeneous; it is possible to find weakly deformed metaigneous rocks surrounded by strongly deformed to mylonitic metasedimentary rocks. In little deformed turbiditic layers of the Malpica-Tui unit, Fombella Blanco (1984) described palynomorphs that suggested an Early to Middle Ordovician age. The metabasic rocks include garnet-bearing and garnet-free amphibolites, amphibolites with albite porphyroblasts, and eclogites with a variety of textures. Orthogneisses include mylonitic felsic gneisses and biotite gneisses of calc-alkaline affinity, and also alkaline and peralkaline gneisses (Floor, 1966; Montero, 1993; Rodríguez, 2003). Rb-Sr isotopic dating of some gneisses has yielded protolith ages ranging from 460 to 480 Ma (Priem et al., 1966; Van Calsteren et al., 1979; García Garzón et al., 1981), whereas a more precise age of ca.

480 Ma (U-Pb on zircon by thermal ionization mass spectrometry [TIMS]) has been obtained from a metagranodiorite (Santos Zalduegui, 1995). The protoliths of the peralkaline gneisses were probably the youngest because they never include mafic rocks, which are viewed as dikes and are very common in the other types of orthogneisses (Floor, 1966; Gil Ibarguchi and Ortega Gironés, 1985). Apparently, initial magmatism probably had a bimodal felsic-mafic character and was followed by another peralkaline magmatic event. According to Pin et al. (1992), alkaline and peralkaline gneisses have geochemical signatures typical of A-type granitoids, whereas the mafic rocks have chemical compositions of alkaline basalts. It has been suggested that Ordovician magmatism was active in the basal units during the first stages of continental rifting, although available data do not rule out a more complex evolution for this magmatism and hence for the origin of the basal units.

The basal units were affected first by a high-pressure (*P*) metamorphic event related to subduction at the beginning of Variscan deformation. The timing of this metamorphism has been dated by Rodríguez et al. (2003) at 365–370 Ma ($^{40}Ar/^{39}Ar$) using white micas from eclogites, although the same authors place the beginning of subduction and the coeval eclogite metamorphic event somewhat earlier, 380–385 Ma. According to the intensity of the high-*P* Variscan metamorphism, the basal units can be divided into two different assemblages: a lower unit containing low- to intermediate-temperature (*T*) metamorphism, and an upper unit affected by medium- to high-*T* metamorphism (Arenas et al., 1995, 1997; Martínez Catalán et al., 1996). Nevertheless it is clear that all these units were part of a relatively continuous and little disturbed slab of continental high-*P* rocks (Fig. 1).

It is possible to recognize the metamorphic gradient associated with the first high-*P* event in this slab in the Órdenes Complex: pressure and temperature increase from east to west (present coordinates). This has been interpreted by Martínez Catalán et al. (1996) to indicate a westward polarity for early Variscan subduction. This subduction developed below a mantle wedge, which was probably responsible for the thermal increase in the basal units during uplift. Heating was caused by convective heat transport from the overlying mantle, and it gave rise to an inverted metamorphic gradient (Arenas et al., 1995; Martínez Catalán et al., 1996). Data on the tectonothermal evolution of the basal units have also been presented by Gil Ibarguchi and Ortega Gironés (1985), Arenas (1988), Gil Ibarguchi (1995), Arenas et al. (1997), Rubio Pascual et al. (2002), Rodríguez (2003), and Alcock et al. (2005).

The deformational history of the basal units after the high-*P* event is complex and has been described by Martínez Catalán et al. (1996, 1999). Exhumation occurred during continental collision, and was mainly caused by thrusting, aided by the coeval extensional detachments related to extensional collapse of the overlying part of the orogenic edifice. The evolutionary history is similar to that proposed by Platt (1986) for the dynamics of orogenic wedges.

OPHIOLITIC UNITS

Five ophiolitic units are recognized in Galicia, three in the Órdenes Complex (Vila de Cruces, Careón, and Bazar units; Fig. 1), and two in the Cabo Ortegal Complex (Moeche and Purrido units; Fig. 1). In addition, remnants of an extremely fragmented ophiolite are involved in the Somozas mélange, a complex ophiolitic and sedimentary mélange imbricated with the basal units of the Cabo Ortegal Complex (Fig. 1). The ophiolite involved in this mélange exhibits a characteristic lithological composition and a unique structural position. It includes most of the lithologies described in common ophiolites, and they are usually not penetratively deformed.

Based on their similar lithological composition and structural position beneath the upper units of the allochthonous complexes, the Bazar, Careón, and Purrido units are considered to be equivalent, and they are generally described as the upper ophiolitic units (Fig. 1). The Moeche and Vila de Cruces units underlie the upper ophiolitic units, and, based on their equivalent lithologies, similar structural position, and common tectonothermal evolution, both are correlated and described as the lower ophiolitic units (Fig. 1). The Somozas mélange contains a unique and distinctive lithological assemblage and will be described separately.

Upper Ophiolitic Units

The Careón unit is the best-preserved lithological ensemble of this group (Figs. 2 and 3). According to Díaz García et al. (1999a), it consists of a stack of three different thrust sheets, the Orosa, Careón, and Vilouriz slices (Fig. 3), which together represent an early Variscan imbricate. The Careón slice reveals a more varied lithologic association, with 500 m of serpentinized ultramafic rocks and some 600 m of gabbro. This assemblage may contain a mid-Paleozoic crust-mantle transition of oceanic lithosphere. It also contains stocks of gabbro pegmatite, diabase dikes, some wehrlite sills, and uncommon leucogabbro and leucotonalite (Fig. 3). Díaz García et al. (1999a) obtained an age of ca. 395 Ma (U-Pb on zircon by TIMS) in a sample of quartz-bearing leucogabbro from the top of the Careón slice (Fig. 3), which they interpreted to date gabbro protolith crystallization and the age of this ophiolite. This age has been confirmed by Pin et al. (2002; U-Pb on zircon by TIMS) with another gabbro sample in the Careón slice.

The amphibolite-facies regional foliation, dated at ca. 376 Ma ($^{40}Ar/^{39}Ar$ in hornblende; Dallmeyer et al., 1997), suggests that the accretion of the ophiolitic slices took place during the Middle Devonian, although ages as old as 390 Ma (Early Devonian) have been obtained by the same method in similar units of the Cabo Ortegal and Portuguese Complexes (Peucat et al., 1990; Dallmeyer and Gil Ibarguchi, 1990; Dallmeyer et al., 1991). Pervasive shearing affected the ophiolite during accretion, transforming the mafic rocks into greenschist-facies rocks and garnet-free or garnet-bearing amphibolites characterized by an inverted metamorphic gradient that increases upward toward the

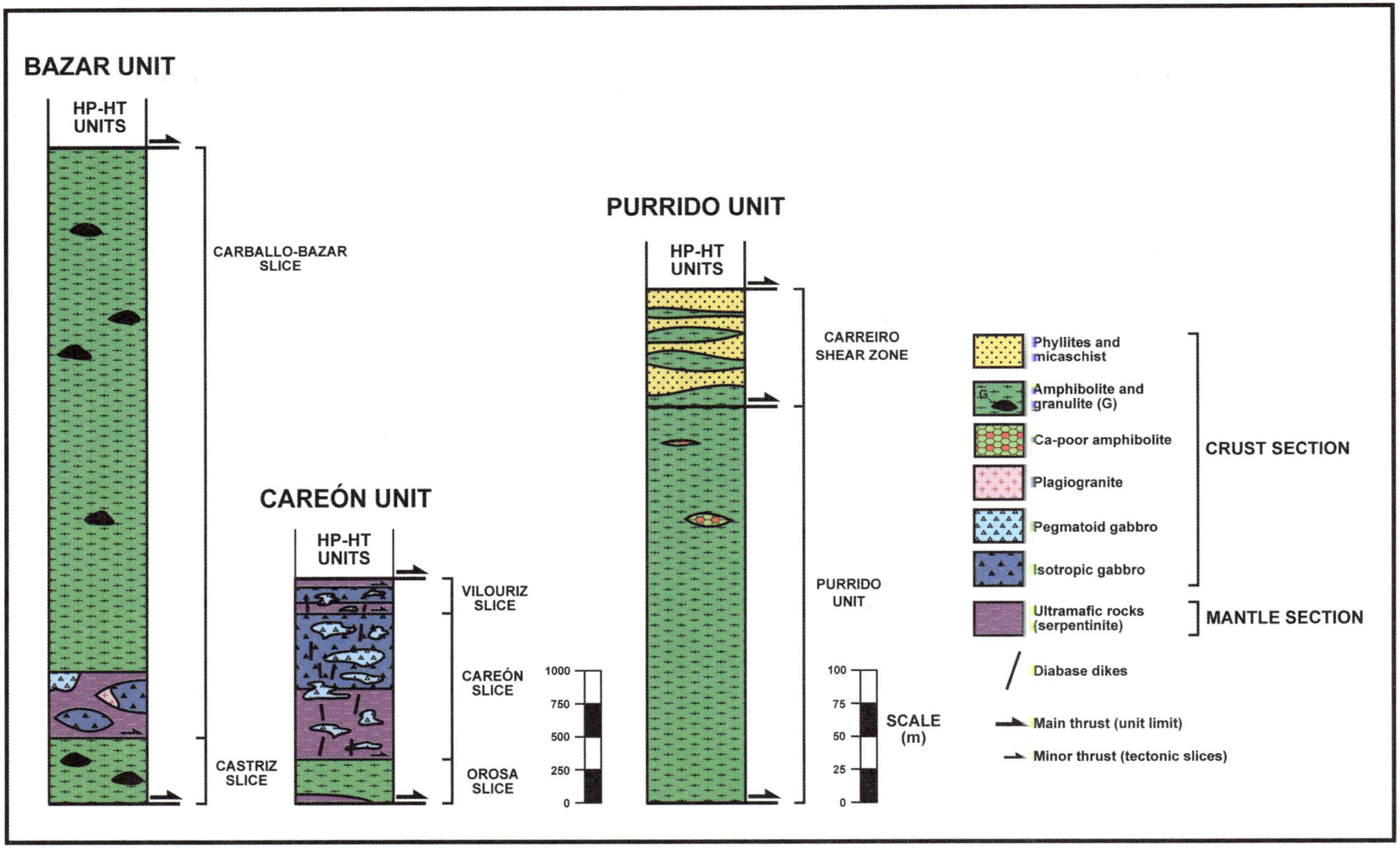

Figure 2. Schematic columns showing the lithological constitution of the upper ophiolitic units. The Bazar and Careón units occur in the Órdenes Complex, while the Purrido unit forms part of the Cabo Ortegal Complex. The columns also show the internal structure of these ophiolitic units, characterized by the presence of several tectonic slices without large internal recumbent folds. Note the strong contrast in thickness among the units. HP-HT—high pressure, high temperature.

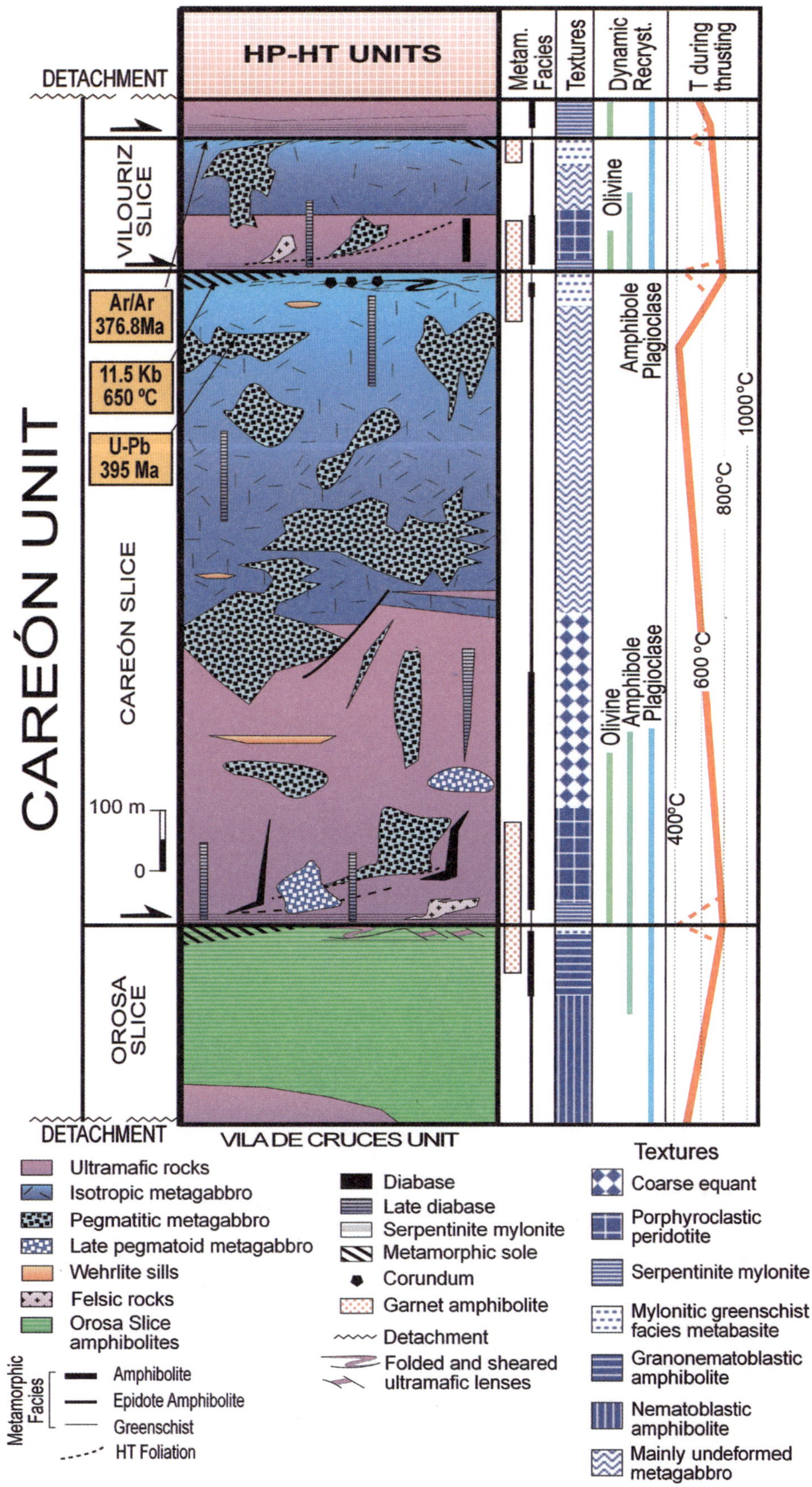

Figure 3. Detailed section of the Careón ophiolite (Díaz García et al., 1999a), composed of three tectonic slices that repeat an original oceanic section. The Careón slice shows the most complete remnant of the oceanic lithosphere. HP-HT—high pressure, high temperature.

ultramafic rocks of the overlying tectonic slice (Fig. 3). *P-T* conditions of 650 °C and 11.5 kbar have been calculated from the mineral assemblages of the garnet-bearing amphibolites (Díaz García et al., 1999a), and they provide an indication of the tectonic setting where both accretion and imbrication of the ophiolite slices took place. Moreover, spectacular centimeter-size red corundum porphyroblasts grew at some thrust contacts between the mafic and ultramafic rocks (Fig. 3). These contacts are considered metamorphic soles developed during the imbrication of hot oceanic lithosphere.

According to Díaz García et al. (1999a), the Careón unit represents one of the last portions of oceanic lithosphere formed in the peri-Gondwanan realm some 395 m.y. ago, during Late Silurian to Early Devonian times. This young oceanic lithosphere was subducted promptly below an advancing eo-Variscan accretionary wedge, preserving at least a part of its original thermal

Figure 4. Field aspects of the upper (A–C) and lower (D–F) ophiolitic units in the Cabo Ortegal Complex. (A–C) Metagabbroic amphibolite and garnet amphibolite (B) from the Purrido unit. (D–F) Greenschist from the Moeche unit; the photograph in E shows one of the recumbent folds that characterizes the internal structure of this unit.

structure. Accretion and imbrication of hot ophiolitic slices can explain the steep metamorphic gradients characteristic of this unit, which have their best expression in the development of the metamorphic soles. According to Pin et al. (2002), the geochemistry of the mafic rocks of the Careón unit is characteristic of a suprasubduction zone setting. Consequently, the oceanic lithosphere that it represents was not generated in a mid-oceanic ridge but probably in the initial stages of development of a marginal basin, or in a forearc during intraoceanic subduction.

The Purrido unit is a 300–400-m-thick sheet consisting of common amphibolite and garnet-bearing amphibolite of gabbroic origin (Vogel, 1967; Azcárraga, 2000; Figs. 2 and 4). Its lithology and thickness are similar to those of the Orosa slice in the Careón unit (Figs. 2 and 3). Its internal structure is rather monotonous, and internal folding or imbrications have not been described. A hornblende concentrate from the regional amphibolite-facies foliation yielded a $^{40}Ar/^{39}Ar$ age of 391.3 ± 6.6 Ma (Peucat et al., 1990), which is also viewed as dating the accretion of oceanic slices. Furthermore, the geochemical characteristics of the Purrido metabasites also point to a suprasubduction zone setting (Sánchez Martínez, 2003), supporting the proposed correlation between the Purrido and Careón units.

The Bazar unit is located in the western part of the Órdenes Complex (Figs. 1 and 2). Though its internal structure is poorly known, its thickness, up to 5000 m, and the presence of several thrust faults suggests that it is an imbricate. It is made up of monotonous metagabbroic amphibolite, with a relatively undeformed layer of gabbro, leucogabbro, pegmatoid gabbro, pyroxenite, and ultramafic rocks located toward the lower part of the unit (Díaz García, 1990; Abati, 2002; Fig. 2). The unit was affected by an early high-*T* metamorphism, which formed mafic granulites transitional between the low- and medium-*P* types. The origin and meaning of this high-*T* metamorphic event are not clear, but it can be tentatively interpreted as related to subduction of very young oceanic lithosphere, culminating with subduction of a mid-ocean ridge.

Lower Ophiolitic Units

In the Órdenes Complex, the Vila de Cruces unit (Fig. 5) is a thrust sheet that may be 3500 m thick; it consists of variably deformed greenschist, metapelitic mica schists, and schists with porphyroblasts of albite and garnet, subordinate black chert, felsic orthogneiss, metagabbro, and serpentinite (Martínez Catalán et al., 2002). Another 700 m should be added below, if the ultramafic rocks of the Campo Marzo slice, separated from Vila de Cruces by a late extensional detachment, are included in this ophiolite (Martínez Catalán et al., 1996).

The pelitic schists occur as intercalations in the greenschist-facies mafic rocks and are repeated by recumbent folds in the lower part of the unit (Vila de Cruces slice) and by thrust faults in its upper part (Sampaio slice; Fig. 5). The metapelites included in the Vila de Cruces slice are low-*T* mica schists with occasional chloritoid and albite porphyroblasts. In the Sampaio slice, the metapelitic rocks appear as medium-*T* schists with frequent porphyroblasts of garnet and albite. In spite of the strong deformation and retrogression recorded by these schists, the earliest mineral assemblage containing garnet does not include biotite, which suggests high-*P* and low- to intermediate-*T* conditions during the first metamorphic event. This event preceded or was contemporaneous with development of the first generation of kilometer-scale recumbent folds. A new generation of smaller overturned folds deformed the unit, developing the regional axial-planar greenschist-facies foliation. Finally, the unit was deformed by E-SE–vergent thrusts, which created the present imbricate structure (Martínez Catalán et al., 2002).

The regional foliation has been dated at 363–367 Ma ($^{40}Ar/^{39}Ar$ in muscovite concentrates from pelitic schist; Dallmeyer et al., 1997), which also establishes a minimum age for the first high-*P* metamorphic event and for the first deformation affecting the unit. Moreover, the isotopic dating of an orthogneiss of the Vila de Cruces unit yielded and age of ca. 500 Ma (U-Pb on zircon by TIMS; Arenas et al., 2005), which establishes a minimum age limit for generation of that lower ophiolitic unit.

In the Cabo Ortegal Complex, the 500-m-thick Moeche unit was described by Arenas (1988) as a layer of ultramylonitic greenschist. Although more strongly deformed, this unit is lithologically similar to the Vila de Cruces unit, and it occurs at an equivalent structural position (Figs. 4 and 5). The Moeche unit can be correlated with the lower part of its equivalent in the Órdenes Complex, the Vila de Cruces slice, because it does not contain schist with albite and garnet porphyroblasts. The tectonothermal evolution of the Moeche unit is also similar to that of the Vila de Cruces ophiolite, and the regional foliation has been dated here at ca. 364 Ma ($^{40}Ar/^{39}Ar$, whole rock in a mica schist; Dallmeyer et al., 1997).

In the lower ophiolitic units, Variscan deformation was so strong that it is generally impossible to recognize the nature of the protoliths. However, some scarce metabasites have textures suggesting a gabbroic origin. In the Vila de Cruces orthogneisses, porphyroclasts wrapped by the mylonitic foliation reveal fine-grained granular textures indicative of a history that postdates hypabyssal granitoids. Most of the greenschist, by far the most abundant lithology, was probably derived from volcanic rocks (basalts), an interpretation supported by the geochemical composition of the mafic rocks, absence of plutonic textures, and the presence of interbedded schist and chert (Martínez Catalán et al., 2002). Moreover, an origin as weathered volcanic rocks may explain the strong mylonitization characteristic of many greenschists, as these rocks may show a weaker rheological behavior against deformation.

An important fact related to the origin of the lower ophiolitic units is the common occurrence of metasedimentary rocks not found in the upper ophiolitic units. This suggests an origin near an emerged domain. Furthermore, the lower ophiolites were obducted over the basal units, and some of their lithologies, like the albite-bearing schists and the orthogneisses, are rather similar. Therefore, and considering that the Vila de Cruces unit has a Late

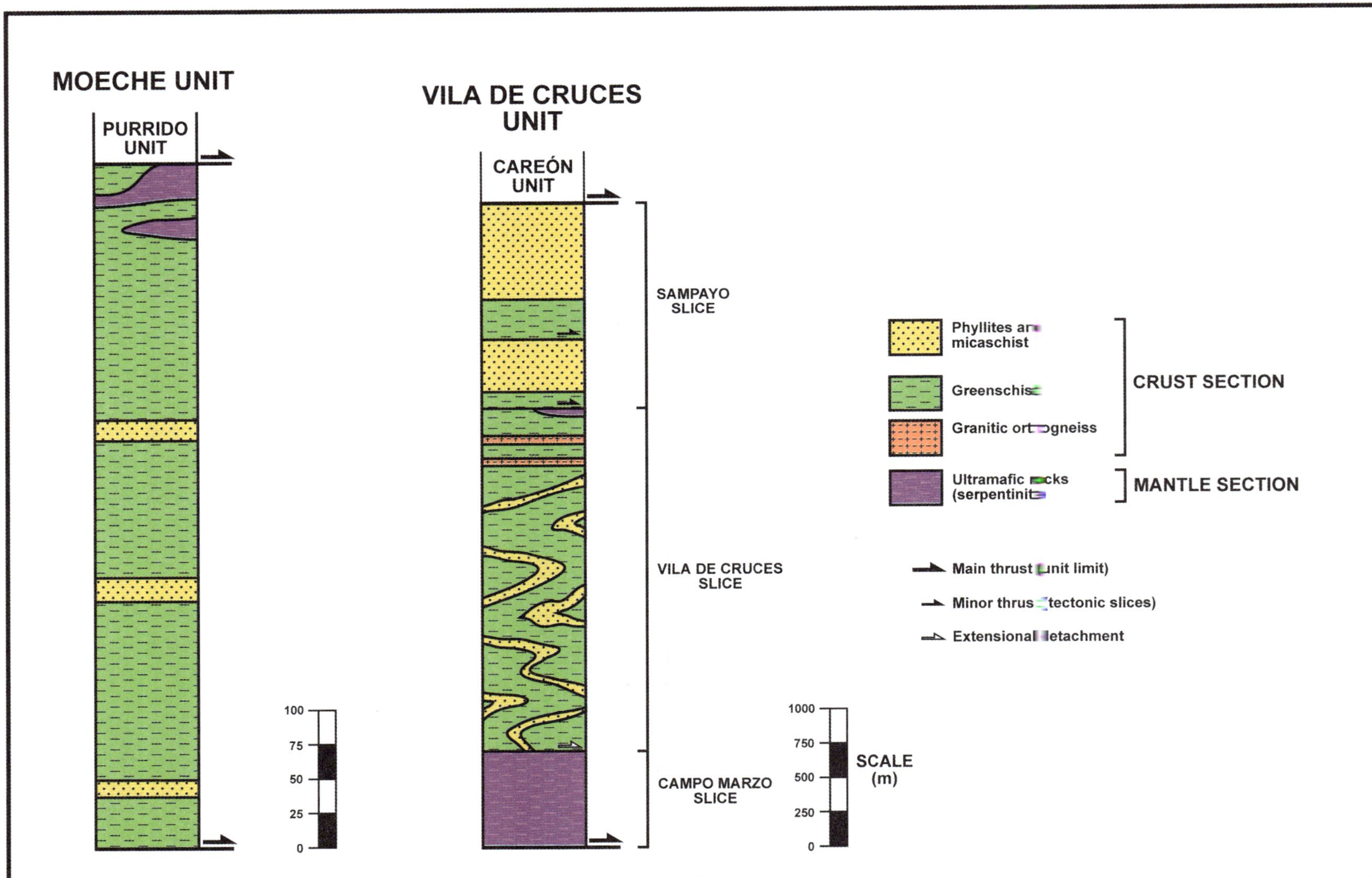

Figure 5. Schematic columns showing the lithologic makeup of the lower ophiolitic units. The Moeche unit (Cabo Ortegal Complex) does not show internal imbrication or significant folding. On the contrary, the Vila de Cruces unit contains several tectonic slices, and the central one is characterized by internal recumbent folding. Note the difference in thickness between both ophiolitic units.

Cambrian age, it is very likely that the lower ophiolitic units were generated near the Gondwanan continental margin, represented by the basal units in the Galicia-Trás-os-Montes allochthonous complexes. In this context, these ophiolites would probably be related to the early opening of an oceanic basin in the peri-Gondwanan realm and, more specifically, to the first stages of development of a marginal basin above a subduction zone. This oceanic domain is here named the Galician ocean. Though it is not clear whether or not this basin represents the first stages in the opening of the Rheic Ocean, a relationship with this ocean can be postulated. New geochemical data support this interpretation (Sánchez Martínez, 2003; Sánchez Martínez et al., this volume).

Somozas Mélange

The lithologic characteristics and structural position of the Somozas mélange were described by Arenas (1988), who also pointed out the dismembered character of the unit. Marcos et al. (2002) suggested that the internal structure of this unit is typical of a tectonic mélange. It is a 500-m-thick unit where meter- to hectometer-sized ellipsoidal tectonic blocks can be recognized (Fig. 6). The tectonic blocks include ophiolitic lithologies, metasedimentary rocks, and high-*T* orthogneisses and amphibolites, all wrapped by a low-*T*, highly sheared matrix of serpentinite or phyllite. In detail, the internal structure is rather complex, and

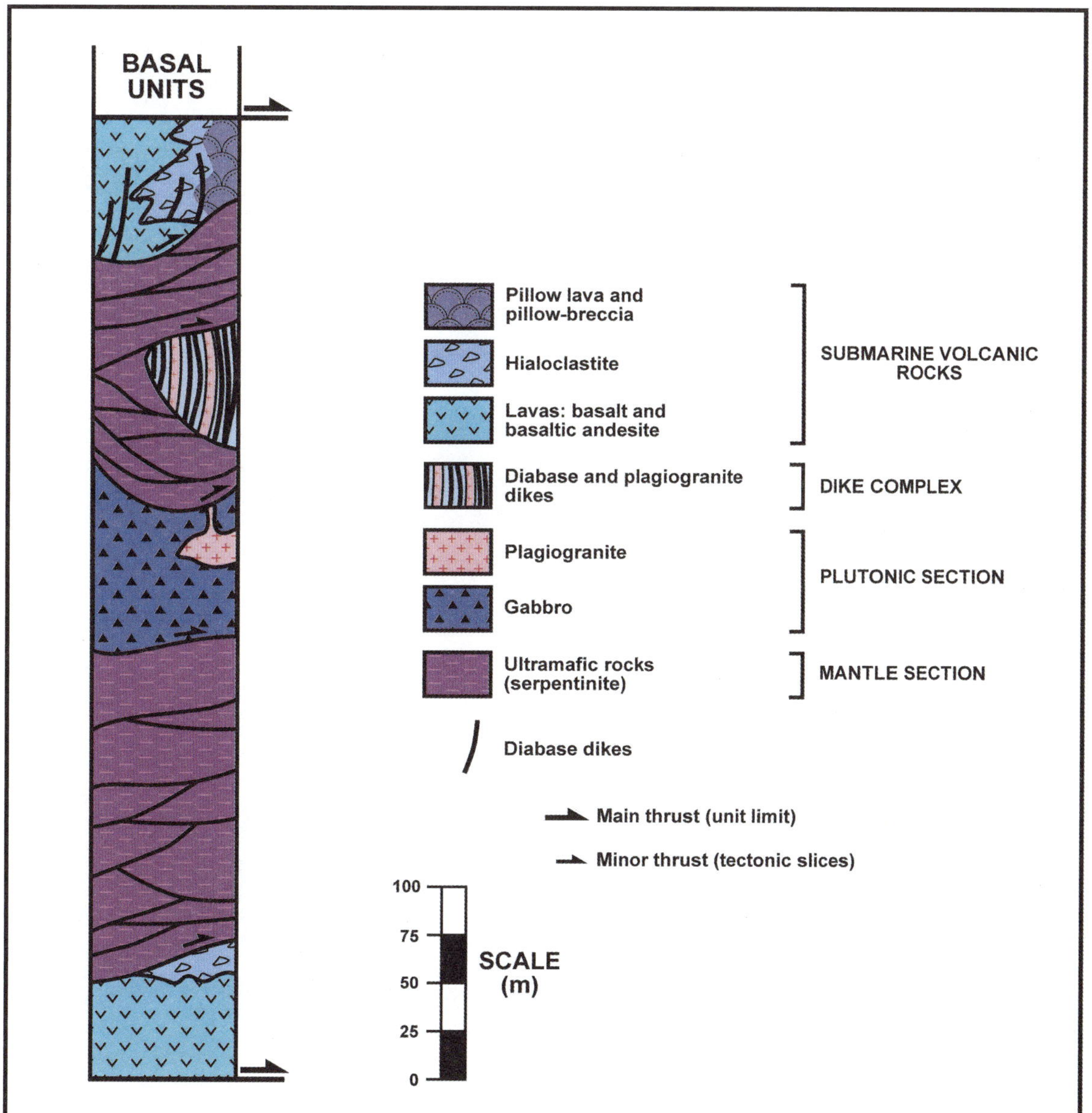

Figure 6. Schematic section showing the lithologic characteristics and internal structure of the ophiolitic mélange that forms part of the Somozas mélange (Cabo Ortegal Complex).

Figure 7. Field aspects in the Somozas mélange (Cabo Ortegal Complex). (A) Pillow breccia preserving large pillows. (B) Pillow breccia. (C) Sheared pillow breccia with large individual pillows. (D) Close-packed pillows. (E) Massive submarine volcanic rocks intruded by diabase dikes. (F) Imbricate of serpentinite (lower part of the quarry) and sheared conglomerate (upper part). (G) Imbricate of serpentinite and marble.

two different superimposed mélange units can be distinguished. The lower part of the mélange is mostly a sedimentary mélange, where few blocks of ophiolitic and high-*T* rocks are distributed between sheared brownish phyllite and bluish phyllonite, with occasional sheared conglomerate, metagraywacke, and rhyolite beds. The upper part consists of typical ophiolitic mélange with few sedimentary blocks (phyllite, phyllonite, conglomerate, and marble). Some scarce high-*T* blocks also appear in this upper part. The section in Figure 6 is typical of the upper ophiolitic mélange. Deformation in the different tectonic blocks is very heterogeneous, with highly sheared tectonic blocks and almost undeformed ophiolitic lithologies mixed even at outcrop scale (Fig. 7). Conversely, the serpentinitic or metapelitic matrix surrounding the blocks is always strongly sheared. Some of the blocks contain other fabrics and mineral associations that predate their involvement in the mélange, and different blocks reveal different metamorphic conditions. For example, the metasedimentary rocks are always affected by low- to very low-*T* metamorphism, but the felsic orthogneisses and some amphibolites preserve medium- to high-*T* mineral assemblages. Most of the blocks of the ophiolitic lithologies contain mineral assemblages of low- to very low-*T*, but some of them exhibit higher-grade mineral assemblages that can include kyanite and garnet.

Ophiolitic lithologies consist of massive metavolcanic rocks, pillow-lava, submarine breccia and pillow-breccia, hyaloclastite, diabase, gabbro and microgabbro, plagiogranite, and

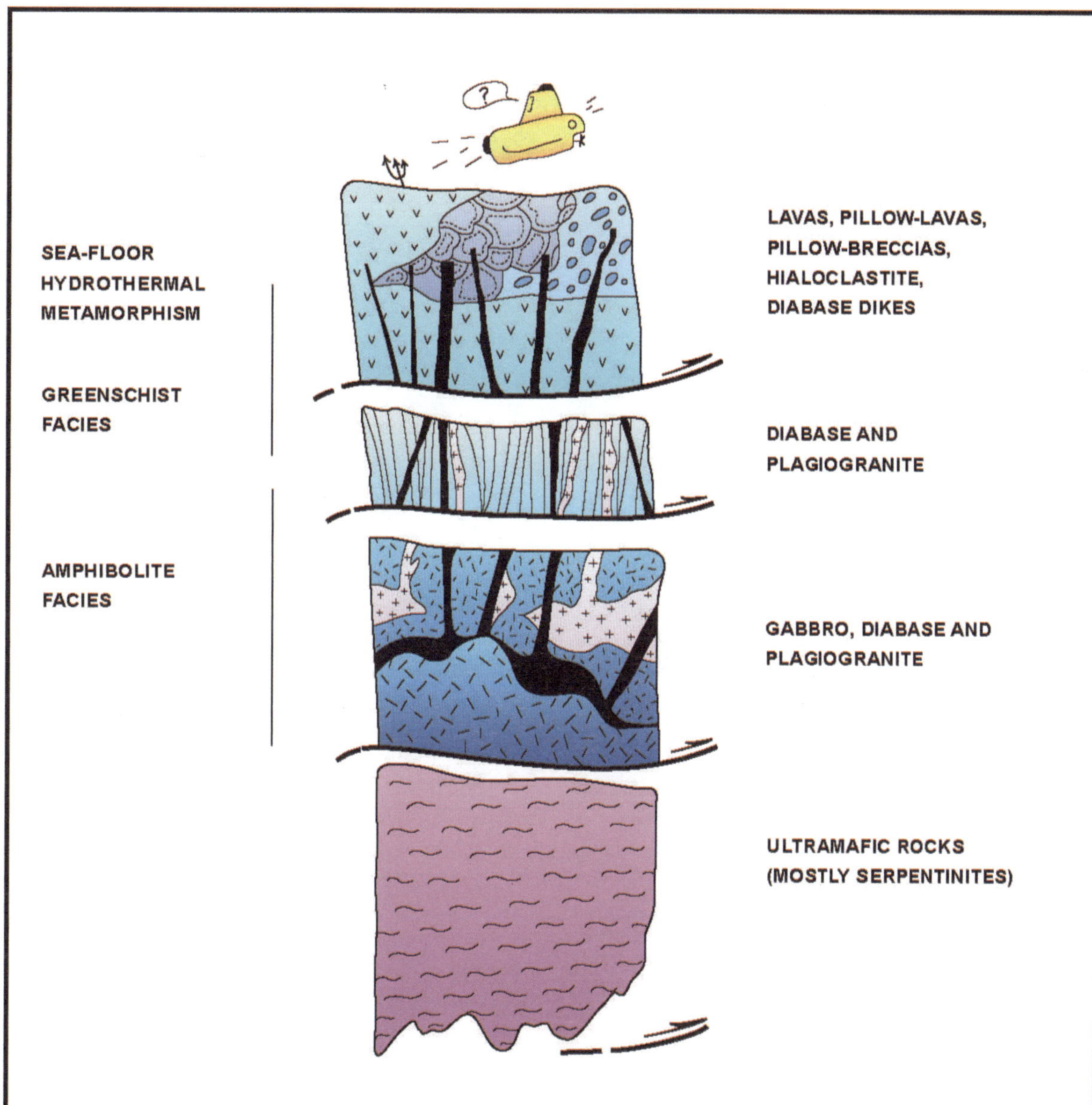

Figure 8. Idealized reconstruction of the ophiolite that forms part of the Somozas mélange. This ophiolite is currently fragmented and involved in a composite tectonic mélange.

highly serpentinized spinel-bearing ultramafic rocks (Arenas and Peinado, 1981; Arenas, 1988; Fig. 7). The serpentinites are the more abundant lithology in the ophiolitic mélange. Figure 8 depicts an idealized reconstruction of ophiolite involved in the Somozas mélange (Arenas, 1988). This column (Fig. 8) contains the first description of one ophiolite involved in the Variscan suture of the northwest Iberian Massif. Variation of the seafloor hydrothermal metamorphism was taken into account in building this ideal section. Metamorphic grade increases from greenschist facies in the upper part of the ophiolite to low-*T* amphibolite facies in the gabbro layer (hornblende-plagioclase-clinozoisite-quartz-ilmenite). Despite that only hectometer-size original sections of this ophiolite are preserved, it is possible to observe all the different parts of a typical oceanic lithosphere in the field, including large blocks of submarine volcanic rocks with varied lithologies, parts of a sheeted dike complex, and parts of a gabbro-ultramafic complex. According to Arenas (1988), most of the mafic rocks in this ophiolite have characteristic mid-ocean-ridge basalt (MORB) chemical compositions, but volcanic rocks in some tectonic blocks have calc-alkaline basaltic andesite compositions.

So far, no geochronologic data of the lithologies in the Somozas mélange are available, and the age and provenance of the high-*T* orthogneiss and amphibolite blocks remain unsolved. Some tectonic blocks of marble contain fragments of corals, crinoids, and foraminifers, suggesting an age no older than the Middle Ordovician (Van der Meer Mohr, 1975). Moreover, the rhyolites involved in the lower part of the mélange are almost identical to those of the upper part of the Silurian series that underlies the Somozas mélange. Hence, a Silurian age seems reasonable for the metasedimentary series involved in the mélange.

The Somozas mélange represents tectonic mixing of ophiolite and sediments that probably occurred during the obduction of the ophiolite over the Gondwanan continental margin. The mélange was formed after accretion of both the ophiolites and part of the sediments below the advancing orogenic wedge. The subduction reached only a moderate depth because high-*P* metamorphism is lacking. This interpretation is consistent with the presence in many tectonic blocks of tectonic fabrics and mineral assemblages that record events earlier than their incorporation in the mélange. Nevertheless, the age and nature of the ocean basin where the ophiolites originated are unknown, and the existence of igneous rocks with different geochemical affinities makes this interpretation even more difficult. In any case, considering the local character of this ophiolitic mélange in the allochthonous complexes of Galicia-Trás-os-Montes (Fig. 1), the possibility that these ophiolites mark the opening of a relatively restricted basin must be considered.

UPPER UNITS

The upper units consist of a very thick allochthonous lithologic pile (Fig. 1) that records the oldest tectonothermal events and represents the oldest terranes accreted to the eo-Variscan orogenic wedge. The upper and lower ophiolitic units were sequentially accreted below them, followed by the basal units.

Considering the characteristics of their tectonothermal evolution, the upper units are divided in two different assemblages: high-*P* and high-*T* upper units (HP-HT upper units) below, and intermediate-*P* upper units above (Díaz García et al., 1999b; Martínez Catalán et al., 1999; Arenas et al., 2000; Fig. 1). These two lithologic assemblages are always separated from each other by large extensional detachments (Martínez Catalán and Arenas, 1992; Díaz García et al., 1999b; Martínez Catalán et al., 2002; Castiñeiras, 2003). In spite of this tectonic separation, both assemblages are considered parts of the same far-traveled allochthonous terrane (Arenas et al., 1986; Martínez Catalán et al., 1997), based on their lithologic similarity and coeval tectonothermal evolution. During accretion, the lower parts of this terrane reached great depths, developing high-*P* and high-*T* metamorphism, whereas the intermediate and upper parts evolved under an intermediate-*P* gradient.

The more salient feature in the tectonothermal evolution of the upper units is their polymetamorphic nature. Abati et al. (1999) and Fernández-Suárez et al. (2002) have shown that the medium- and high-*T* units underwent a first metamorphic event at the Cambrian-Ordovician boundary (500–480 Ma; U-Pb on zircon and monazite by TIMS in paragneisses), in addition to a younger, Early Devonian event (400–380 Ma; U-Pb in monazite, titanite, and rutile by TIMS in paragneisses). In this context, an accurate interpretation of the chronology of the high-*P* and high-*T* event is difficult, and it is only possible by precise analysis of zircons (U-Pb with the secondary ion mass spectrometry (SIMS) method). The most recent data suggest that the Early Devonian event is responsible for the high-*P* and high-*T* event (see Fernández-Suárez et al., this volume). Hence, this metamorphism is viewed as early Variscan and related to the early development of the orogenic wedge (Ordóñez Casado et al., 2001). Interpretation of the older Cambrian-Ordovician metamorphic event is much more uncertain, though a relationship with volcanic-arc activity has been suggested (Abati et al., 1999, 2002).

High-Pressure and High-Temperature Upper Units

The HP-HT upper units consist of a varied lithological succession that includes migmatitic paragneisses, eclogite, high-*P* mafic granulite, gabbro and coronitic metagabbro, metaperidotite, metapyroxenite, and scarce orthogneisses, as well as their retrogression products, including several types of amphibolite, greenschist, chlorite schist, and serpentinite (Vogel, 1967; Gil Ibarguchi et al., 1990; Arenas, 1991; Mendia, 2000; Arenas and Martínez Catalán, 2002).

Complex internal structure hinders precise reconstruction of the original lithologic succession, but in the Cabo Ortegal Complex, the sequence consists of, from bottom to top, a thick layer of mantle-like ultramafic rocks, mafic rocks of probable gabbroic origin, and a thick pile of sediments intruded by mafic rocks (gabbro and diabase). Considering that the presence of a thick terrigenous sedimentary succession is not characteristic of common oceanic domains, Arenas et al. (1986) suggested that an island-arc environment, either a back arc, a fore-arc or an intra-arc, seems to be the more probable dynamic setting for the origin of the upper units. The isotopic ages obtained for the protoliths of some high-*P* mafic and ultramafic rocks (ca. 500 Ma; Ordóñez Casado et al., 2001; Santos et al., 2002; Fernández-Suárez et al., this volume) place the age of this arc at the Cambrian-Ordovician boundary.

Intermediate-Pressure Upper Units

These units consist of a thick pile of terrigenous metasedimentary rocks intruded by large gabbro and granitic bodies, as well as many smaller igneous bodies with variable composition (diorite, tonalite, and some more felsic lithologies). These units crop out widely in the central Órdenes Complex (Fig. 1) and are poorly represented in the Cabo Ortegal Complex (Castiñeiras, 2003). Geochronologic data from the largest igneous bodies (U-Pb on zircon by TIMS; Abati et al., 1999) have yielded 499 ± 2 Ma in the Monte Castelo Gabbro, in the western Órdenes Complex, and 500 ± 2 Ma in the Corredoiras Orthogneiss, in the eastern part of the same complex. These data are similar to other protolith ages in the HP-HT upper units and confirm the coeval nature of felsic-mafic magmatism in the upper units. Moreover, Fernández-Suárez et al. (2003) obtained a maximum age of sedimentation of ca. 480 Ma for turbidite deposits of the uppermost part of these units (U-Pb laser-ablation inductively coupled plasma–mass spectrometry [ICP-MS] on detrital zircons). It is clear from these data that rock formation in the intermediate-*P*

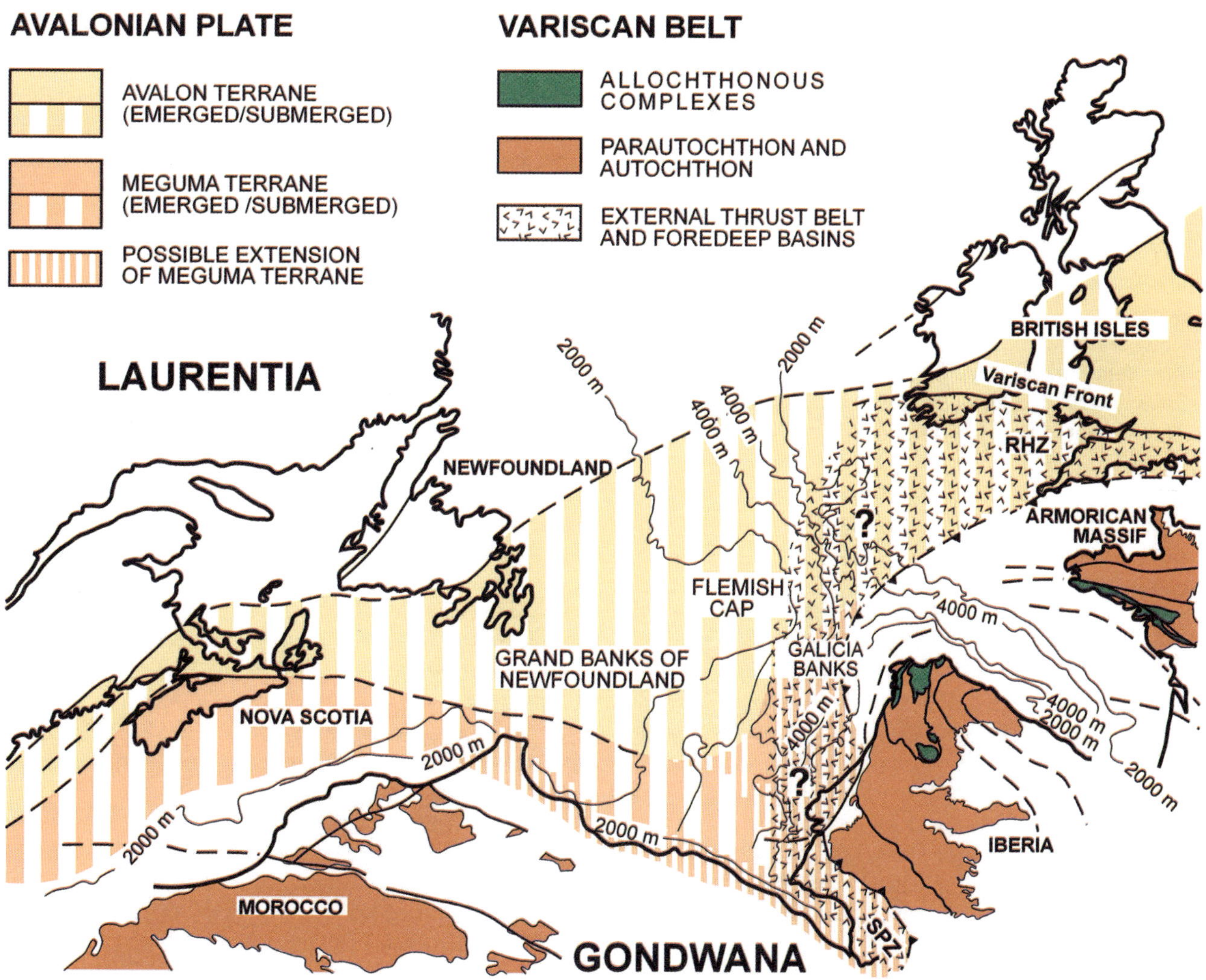

Figure 9. Reconstruction of Pangea in the vicinity of Iberia (based on Lefort, 1989; Skehan and Rast, 1995; Martínez Catalán et al., 1997). The figure shows the present position of the Avalon terrane and the allochthonous complexes of northwest Iberia. RHZ—Rhenohercynian zone; SPZ—South Portuguese zone.

upper units mostly occurred between the Late Cambrian and the Early Ordovician, although present data cannot exclude the presence of older lithologies in the metasedimentary series surrounding the large igneous massifs.

The uppermost parts of the intermediate-*P* units in the Órdenes Complex consist of 2000–3000 m of turbidites with a probable Ordovician age (Fernández-Suárez et al., 2003). The lithologies are flysch-like and consist of alternating graywackes and pelites, with scarce conglomerate. Coarse-grained graywacke and conglomerate contain abundant igneous rock, slate, chert, and bipyramidal quartz fragments. The abundance of igneous rock fragments and the immature character of the graywackes, together with the geochemical characteristics of some metaigneous rocks of other intermediate-*P* upper units (Andonaegui et al., 2002; Castiñeiras, 2003), suggest an island-arc setting for the assemblage. The lack of volcanic successions indicates that the upper parts of the arc edifice have not been preserved, whereas the turbidites probably represent the filling of a forearc or backarc basin. Such an island arc would have been generated in the peri-Gondwanan realm, according to the age populations yielded by the detrital zircons from the turbidites (Fernández-Suárez et al., 2003).

Considering the tectonothermal evolution of the intermediate-*P* upper units, their lowest parts exhibit metamorphic conditions of intermediate-*P* granulite facies (Abati, 2002; Abati et al., 2003), while the middle parts were recrystallized in the amphibolite facies, and the uppermost parts only reached the greenschist facies. According to their structural and metamorphic characteristics, the upper intermediate-*P* and HP-HT units define a section of an orogenic wedge with up to 10 km of present thickness, where the metamorphic grade increases gradually from top to bottom. However, the currently observed section is only a remnant of a thick pile, which, according to the *P-T* conditions deduced for the HP-HT mineral assemblages, attained a thickness of 70–80 km.

TERRANES AND ACCRETIONARY HISTORY

The allochthonous complexes of the Galicia-Trás-os-Montes zone can be grouped in several terranes, some of which underwent very large displacements before accretion to the Variscan orogenic wedge and so are considered far-traveled terranes. The origin and meaning of these terranes as well as their evolution during the Paleozoic have been the object of recent contributions (Martínez Catalán et al., 1997; Díaz García et al., 1999a; Martínez Catalán et al., 1999; Andonaegui et al., 2002; Castiñeiras, 2003; Sánchez Martínez, 2003), which have improved the understanding of the Paleozoic evolution of the peri-Gondwanan realm where the Iberian Massif is located (Figs. 9 and 10).

As described in the previous section, the upper units of the allochthonous complexes (both HP-HT and intermediate-*P* upper units) seem to form part of a single allochthonous terrane, accreted to Laurussia during an eo-Variscan event (ca. 420–400 Ma; Fig. 10C), the lower parts of which were affected by high-*P* and high-*T* metamorphism. The accretion to Laurussia occurred after the complete closure of an intervening ocean, which could have been the Iapetus (Scotese, 2001; Stampfli and Borel, 2002) or the Tornquist Ocean, depending on the unconstrained paleoposition of Iberia at that time. The ophiolites related to that suture could have overlain the upper units of the allochthonous complexes (Fig. 10C) but are not represented in the northwest Iberian Massif. They could be presently located in the submerged continental margin of Galicia, in the Grand Banks of Newfoundland, and/or in the poorly exposed and poorly known Tornquist suture, depending on the correlation between the upper units of the Galicia-Trás-os-Montes zone and the allochthonous terranes of North America and Central Europe (Figs. 9 and 10). It seems that the first eo-Variscan tectonothermal event recorded in the upper units of Galicia occurred ca. 420–430 Ma, and hence, it is somewhat younger than the estimated time of accretion of Avalon to Laurussia, dated at 460–430 Ma (Cawood et al., 1995; Murphy et al., 1995).

The geochemical characteristics of the metaigneous rocks in the upper units suggest generation in an island-arc setting (Andonaegui et al., 2002; Santos et al., 2002; Castiñeiras, 2003). The immature and turbiditic character of the low-*T* metasediments that occupy the uppermost parts of these units also favors an island-arc setting. Hence, the upper units probably represent different parts of an island arc that evolved near the Cambrian-Ordovician boundary in a peri-Gondwanan realm (Abati et al., 1999; Santos et al., 2002; Fernández-Suárez et al., 2003). The high-*T* metamorphic event that affected the upper units at 500–480 Ma (Abati et al., 1999; Fernández-Suárez et al., 2002) was probably related to the evolution of this arc.

The more problematic aspects of the evolution of this island arc are its precise location in the peri-Gondwanan realm and whether or not it formed part of the Avalon microcontinent. Taking into account the different terranes preserved in northwest Iberia, this island arc was separated from Gondwana by an oceanic domain represented by the lower ophiolitic units. The name Galician ocean is used in this contribution, but it can equally be considered as a part of the large Rheic Ocean, which probably developed behind an island arc (see the paleomap of Scotese [2001] for the Avalonian domain at 480 Ma).

In Galicia, the metabasic rocks of the lower ophiolites have compositions suggesting an origin in a marginal basin (Sánchez Martínez et al., this volume), and some of the ophiolites include thick metasedimentary formations alternating with the mafic rocks. These facts suggest that most of the mafic rocks in these ophiolites were derived from volcanic rocks. In this context, the island arc represented by the upper units might have originated as a volcanic arc associated with the northern continental margin of Gondwana (Fig. 10A). The drift of the arc from the continental margin caused the opening of a marginal basin (Fig. 10B), which, according to the available continental reconstructions for the Cambrian-Ordovician boundary, should have been connected with the early opening of the Rheic Ocean. The age populations of detrital zircons from turbidites support

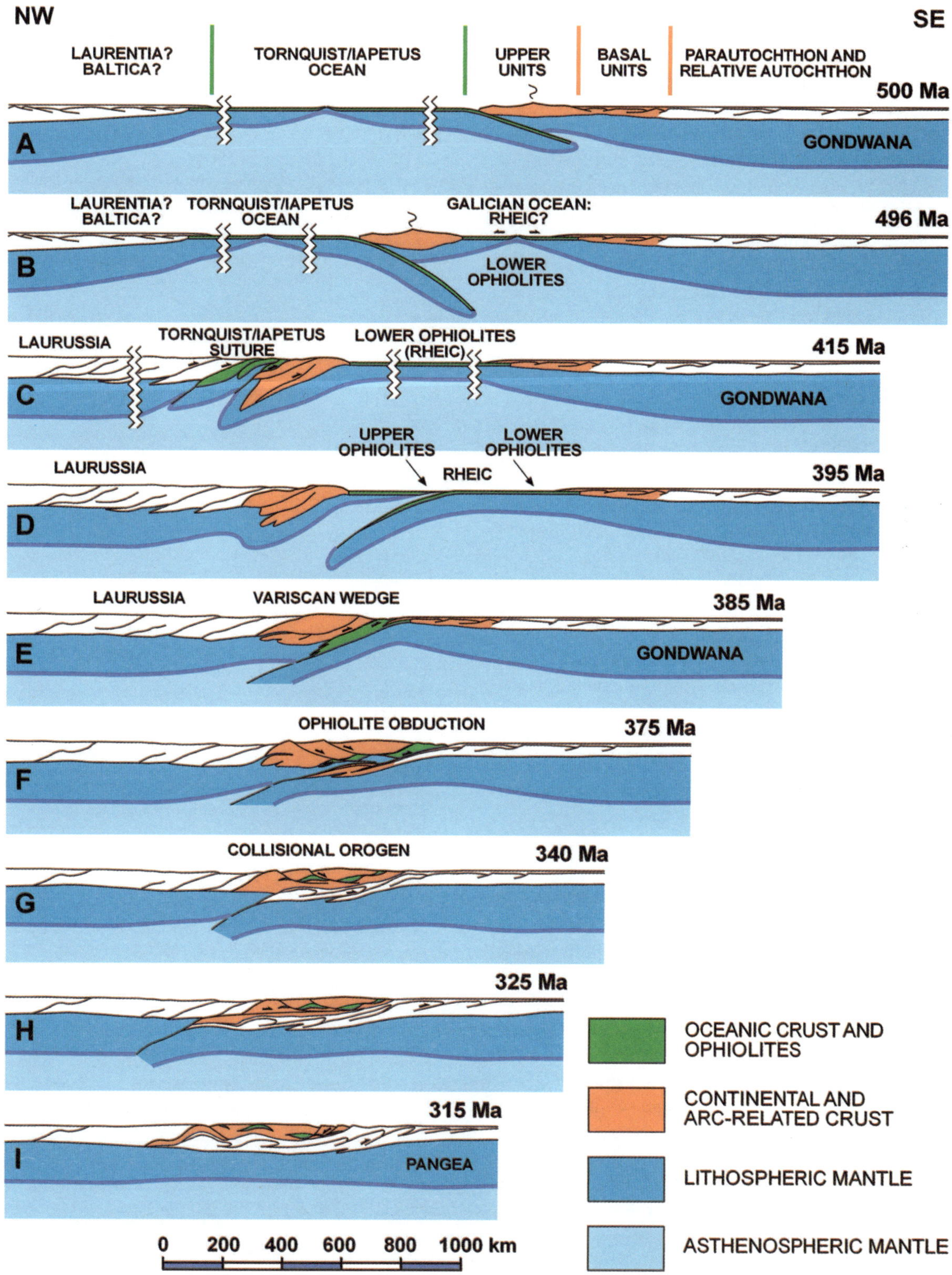

Figure 10. Proposed Paleozoic evolution of the terranes included in the allochthonous complexes of northwestern Iberian Massif. The reconstruction shows the tectonic setting of terrane generation in the peri-Gondwanan realm during Cambrian and Ordovician times, the accretion of these terranes to Laurussia, and the final collision between Laurussia and Gondwana. Plate configuration during Cambrian and Ordovician times is largely based in Scotese (2001). The name Galician ocean defines the oceanic domain originated by the opening of a marginal basin in the peri-Gondwanan realm. This Galician ocean could represent either a part of the larger Rheic Ocean, or the first stages of the opening of this ocean in its eastern branch (see Scotese, 2001).

this interpretation (Fernández-Suárez et al., 2003). This model is similar to that proposed for the rapid opening of the Sea of Japan in middle Miocene time, which caused a very rapid drift of a volcanic arc from the continental margin of Eurasia toward the Pacific Ocean. In the present island arc, represented by the Japanese Islands, only the metamorphic belt located to the west, the Hida belt, preserves remnants of the Eurasian basement (Dallmeyer and Takasu, 1992).

Closing of the Galician ocean probably began in Late Silurian–Early Devonian times and seems to have occurred through intraoceanic subduction (Fig. 10D). The upper ophiolitic units, which are dated at ca. 395 Ma (Díaz García et al. 1999a) and that reveal chemical compositions characteristic of suprasubduction zone ophiolites (Pin et al., 2002; Sánchez Martínez et al., this volume), were formed at this time (Fig. 10D). These suprasubduction zone ophiolites were soon subducted and accreted below the upper units (ca. 391–376 Ma). Later on, the accreted remnants of these ophiolites, together with some remnants of the early opening of the Galician ocean, located closer to the Gondwanan continental margin, were accreted to the orogenic wedge and preserved as the lower ophiolitic units (Fig. 10E).

There is general agreement on the interpretation of the basal units as the outer edge of the Gondwanan continental margin (Arenas et al., 1986; Martínez Catalán et al., 1996). This relatively thin continental crust was subducted during the first stages of the collision, when the Variscan deformation reached Gondwana (ca. 380–370 Ma; Rodríguez et al., 2003; Fig. 10F). This margin was the site of voluminous but poorly studied 460–480 Ma magmatism during the Early to Middle Ordovician. The meaning of this magmatism is unclear, although it is generally assigned to rifting on the Gondwanan margin. However, this crustal extension is somewhat younger than the drifting away of the upper units from Gondwana, and it could reflect the local opening of a small ocean basin separating the basal units from mainland Gondwana. The Somozas mélange, structurally below the basal units of the Cabo Ortegal Complex that include an ophiolitic association, makes this hypothesis feasible, although confirmation will have to wait until new isotopic ages and a detailed geochemical data are available.

No ophiolites underlie the schistose domain, which is in thrust contact above the metasedimentary series of the Central Iberian zone. So, no suture is present between the autochthon and parautochthon, and no significant differences in the metamorphic evolution have been found between both assemblages. Consequently, the parautochthon cannot be considered an exotic terrane. Rather, it may be a tectonic unit that belonged to the Gondwanan continental margin and was incorporated below the allochthonous complexes during emplacement.

ACKNOWLEDGMENTS

The authors thank Harry Y. McSween for revision of the first version of the manuscript. Important editorial help and enriching correction by Robert Hatcher Jr. is kindly acknowledged. This research was funded by the Spanish Dirección General de Investigación (Ministerio de Ciencia y Tecnología and Ministerio de Educación y Ciencia) and includes results from projects BTE2001-0963-CO2 and CGL2004-04306-CO2/BTE.

REFERENCES CITED

Ábalos, B., Azcárraga, J., Gil Ibarguchi, J.I., Mendia, M., and Puelles, P., 2000, Mapa Geológico del Complejo de Cabo Ortegal (NO de España): Instituto Universitario de Xeoloxía Isidro Parga Pondal, Universidade da Coruña, Coruña, Spain, Servicio de Publicacións, 60 p.

Abati, J., 2002, Petrología metamórfica y geocronología de la unidad culminante del Complejo de Órdenes en la región de Carballo (Galicia, NW del Macizo Ibérico): Nova Terra, v. 20, 269 p.

Abati, J., Dunning, G.R., Arenas, R., Díaz García, F., González Cuadra, P., Martínez Catalán, J.R., and Andonaegui, P., 1999, Early Ordovician orogenic event in Galicia (NW Spain): Evidence from U-Pb ages in the uppermost unit of the Órdenes Complex: Earth and Planetary Science Letters, v. 165, p. 213–228, doi: 10.1016/S0012-821X(98)00268-4.

Abati, J., Arenas, R., Martínez Catalán, J.R., and Díaz García, F., 2003, Anticlockwise *P-T* path of granulites from the Monte Castelo Gabbro (Órdenes Complex, NW Spain): Journal of Petrology, v. 44, p. 305–327, doi: 10.1093/petrology/44.2.305.

Alcock, J., Arenas, R., and Martínez Catalán, J.R., 2005, Shear stress in subducting continental margin from high-pressure, moderate-temperature metamorphism in the Órdenes Complex, Galicia: NW Spain: Tectonophysics, v. 397, p. 181–194, doi: 10.1016/j.tecto.2004.11.009.

Andonaegui, P., González del Tánago, J., Arenas, R., Abati, J., Martínez Catalán, J.R., Peinado, M., and Díaz García, F., 2002, Tectonic setting of the Monte Castelo Gabbro (Órdenes Complex, NW Iberian Massif): Evidence for an arc-related terrane in the hanging wall to the Variscan suture, *in* Martínez Catalán, J.R., Hatcher, R.D., Jr., Arenas, R., and Díaz García, F., eds., Variscan-Appalachian Dynamics: The Building of the Late Paleozoic Basement: Geological Society of America Special Paper 364, p. 37–56.

Arenas, R., 1988, Evolución petrológica y geoquímica de la unidad alóctona inferior del complejo metamórfico básico-ultrabásico de Cabo Ortegal (Unidad de Moeche) y del Silúrico parautóctono, Cadena Hercínica Ibérica (NW de España): Corpus Geologicum Gallaecia, v. IV, 543 p.

Arenas, R., 1991, Opposite *P,T,t* paths of Hercynian metamorphism between the upper units of the Cabo Ortegal Complex and their substratum (northwest of the Iberian Massif): Tectonophysics, v. 191, p. 347–364, doi: 10.1016/0040-1951(91)90067-3.

Arenas, R., and Martínez Catalán, J.R., 2002, Prograde development of corona textures in metagabbros of the Sobrado unit (Órdenes Complex, northwestern Iberian Massif), *in* Martínez Catalán, J.R., Hatcher, R.D., Jr., Arenas, R., and Díaz García, F., eds., Variscan-Appalachian Dynamics: The Building of the Late Paleozoic Basement: Geological Society of America Special Paper 364, p. 73–88.

Arenas, R., and Peinado, M., 1981, Presencia de pillow-lavas en las metavolcanitas submarinas de las proximidades de Espasante, Cabo Ortegal, NW de España: Cuadernos de Geología Ibérica, v. 7, p. 105–119.

Arenas, R., Gil Ibarguchi, J.I., González Lodeiro, F., Klein, E., Martínez Catalán, J.R., Ortega Gironés, E., Pablo Maciá, J.G., and Peinado, M., 1986, Tectonostratigraphic units in the complexes with mafic and related rocks of the NW of the Iberian Massif: Hercynica, v. II, p. 87–110.

Arenas, R., Farias, P., Gallastegui, G., Gil Ibarguchi, J.I., González Lodeiro, F., Klein, E., Marquínez, J., Martín Parra, L.M., Martínez Catalán, J.R., Ortega, E., de Pablo Maciá, J.G., Peinado, M., and Rodríguez Fernández, L.R., 1988, Características geológicas y significado de los dominios que componen la Zona de Galicia-Trás-os-Montes: Actas del II Congreso Geológico de España, Simposios, p. 75–84.

Arenas, R., Rubio Pascual, F.J., Díaz García, F., and Martínez Catalán, J.R., 1995, High-pressure micro-inclusions and development of an inverted metamorphic gradient in the Santiago schists (Ordenes Complex, NW Iberian Massif, Spain): Evidence of subduction and syncollisional decompression: Journal of Metamorphic Geology, v. 13, p. 141–164.

Arenas, R., Abati, J., Martínez Catalán, J.R., Díaz García, F., and Rubio Pascual, F.J., 1997, *P-T* evolution of eclogites from the Agualada unit (Órdenes Complex, NW Iberian Massif, Spain): Implications for crustal subduction: Lithos, v. 40, p. 221–242, doi: 10.1016/S0024-4937(97)00029-7.

Arenas, R., Díaz García, F., Martínez Catalán, J.R., Abati, J., González Cuadra, P., Andonaegui, P., González del Tánago, J., Rubio Pascual, F., Castiñeiras, P., and Gómez Barreiro, J., 2000, Structure and evolution of the Ordenes Complex: A Coruña, Spain, Basement Tectonics 15, Pre-Conference Field Trip, 160 p.

Arenas, R., Martínez Catalán, J.R., Sánchez Martínez, S., Fernández-Suárez, J., Andonaegui, P., Pearce, J.A., and Corfu, F., 2005, The Vila de Cruces Ophiolite: A remnant of the early Rheic Ocean in the Variscan suture of Galicia (NW Iberian Massif) [abs.], *in* IGCP 497: The Rheic Ocean—Its Origin Evolution and Correlatives: Portsmouth, University of Portsmouth, p. 19–20.

Arps, C.E.S., 1970, Petrology of a part of the western Galicia basement between the Río Jallas and the Ría de Arosa (NW Spain) with emphasis on zircon investigations: Leidse Geologische Mededelingen, v. 46, p. 57–155.

Azcárraga, J., 2000, Evolución tectónica y metamórfica de los mantos inferiores de grado alto y alta presión del complejo de Cabo Ortegal: Nova Terra, v. 17, 346 p.

Castiñeiras, P., 2003, Origen y evolución tectonotermal de las unidades de O Pino y Cariño (Complejos Alóctonos de Galicia) [Ph.D. thesis]: Madrid, Spain, Universidad Complutense, 281 p.

Cawood, P.A., Van Gool, J.A.M., and Dunning, G.R., 1995, Collisional tectonics along the Laurentian margin of the Newfoundland Appalachians, *in* Hibbard, J.P., van Staal, C.R., and Cawood, P.A., eds., Current Perspectives in the Appalachian-Caledonian Orogen: Geological Association of Canada Special Paper 41, p. 283–301.

Dallmeyer, R.D., and Gil Ibarguchi, J.I., 1990, Age of amphibolitic metamorphism in the ophiolitic unit of the Morais allochthon (Portugal): Implications for early Hercynian orogenesis in the Iberian Massif: Geological Society [London] Journal, v. 147, p. 873–878.

Dallmeyer, R.D., and Takasu, A., 1992, Plate tectonic setting of the Japanese Islands, *in* Dallmeyer, R.D., ed., Terrane Accretion in Southwest Japan: Shimane, Japan, A Geotraverse Excursion, Chapter I, p. 3–5.

Dallmeyer, R.D., Ribeiro, A., and Marques, F., 1991, Polyphase Variscan emplacement of exotic terranes (Morais and Bragança Massifs) onto Iberian successions: Evidence from $^{40}Ar/^{39}Ar$ mineral ages: Lithos, v. 27, p. 133–144, doi: 10.1016/0024-4937(91)90025-G.

Dallmeyer, R.D., Martínez Catalán, J.R., Arenas, R., Gil Ibarguchi, J.I., Gutiérrez-Alonso, G., Farias, P., Aller, J., and Bastida, F., 1997, Diachronous Variscan tectonothermal activity in the NW Iberian Massif: Evidence from $^{40}Ar/^{39}Ar$ dating of regional fabrics: Tectonophysics, v. 277, p. 307–337, doi: 10.1016/S0040-1951(97)00035-8.

Díaz García, F., 1990, La geología del sector occidental del Complejo de Órdenes (Cordillera Hercínica, NW de España): Nova Terra, v. 3, 269 p.

Díaz García, F., Arenas, R., Martínez Catalán, J.R., González del Tánago, J., and Dunning, G.R., 1999a, Tectonic evolution of the Careón Ophiolite (northwest Spain): A remnant of oceanic lithosphere in the Variscan belt: The Journal of Geology, v. 107, p. 587–605, doi: 10.1086/314368.

Díaz García, F., Martínez Catalán, J.R., Arenas, R., and González Cuadra, P., 1999b, Structural and kinematic analysis of the Corredoiras detachment: Evidence for early Variscan synconvergent extension in the Ordenes Complex, NW Spain: International Journal of Earth Sciences, v. 88, p. 337–351, doi: 10.1007/s005310050269.

Farias, P., Gallastegui, G., González Lodeiro, F., Marquínez, J., Martín-Parra, L.M., Martínez Catalán, J.R., de Pablo Maciá, J.G., and Rodríguez Fernández, L.R., 1987, Aportaciones al conocimiento de la litoestratigrafía y estructura de Galicia Central: Memorias Museo e Laboratorio Mineralógico e Geológico, Faculdade de Ciências, Universidade de Porto, v. 1, p. 411–431.

Fernández-Suárez, J., Corfu, F., Arenas, R., Marcos, A., Martínez Catalán, J.R., Díaz García, F., Abati, J., and Fernández, F.J., 2002, U-Pb evidence for a polyorogenic evolution of the HP-HT units of the NW Iberian Massif: Contributions to Mineralogy and Petrology, v. 143, p. 236–253, doi: 10.1007/s00410-001-0337-2.

Fernández-Suárez, J., Díaz García, F., Jeffries, T.E., Arenas, R., and Abati, J., 2003, Constraints on the provenance of the uppermost allochthonous terrane of the NW Iberian Massif: Inferences from detrital zircon U-Pb ages: Terra Nova, v. 15, p. 138–144, doi: 10.1046/j.1365-3121.2003.00479.x.

Fernández-Suárez, J., Arenas, R., Abati, J., Martínez Catalán, J.R., Whitehouse, M.J., and Jeffries, T.E., 2007, this volume, U-Pb chronometry of polymetamorphic high-pressure, *in* Hatcher, R.D., Jr., Carlson, M.P., McBride, J.H., and Martínez Catalán, J.R., eds., 4-D Framework of Continental Crust: Geological Society of America Memoir 200, doi: 10.1130/2007.1200(24).

Floor, P., 1966, Petrology of an aegirine-ribeckite gneiss-bearing part of the Hesperian Massif: The Galiñeiro and surrounding areas, Vigo, Spain: Leidse Geologische Mededelingen, v. 36, p. 1–204.

Fombella Blanco, M., 1984, Age palynologique du Blastomilonitic Graben, Zone Occidentale de la Galice: Revue de Micropaleontologie, v. 27, p. 113–117.

García Garzón, J., de Pablo Maciá, J.G., and Llamas Borrajo, J.F., 1981, Edades absolutas obtenidas mediante el método Rb-Sr de dos cuerpos de ortogneises en Galicia Occidental: Boletín Geológico y Minero, v. 92, p. 443–455.

Gil Ibarguchi, J.I., 1995, Petrology of jadeite-metagranite and associated orthogneiss from the Malpica-Tui allochthon (northwest Spain): European Journal of Mineralogy, v. 7, p. 403–415.

Gil Ibarguchi, J.I., and Ortega Gironés, E., 1985, Petrology, structure and geotectonic implications of glaucophane-bearing eclogites and related rocks from the Malpica-Tui (MT) unit, Galicia, northwest Spain: Chemical Geology, v. 50, p. 145–162, doi: 10.1016/0009-2541(85)90117-2.

Gil Ibarguchi, J.I., Mendia, M., Girardeau, J., and Peucat, J.J., 1990, Petrology of eclogites and clinopyroxene-garnet metabasites from the Cabo Ortegal Complex (northwestern Spain): Lithos, v. 25, p. 133–162, doi: 10.1016/0024-4937(90)90011-O.

Gil Ibarguchi, J.I., Ábalos, B., Azcárraga, J., Mendia, M., and Puelles, P., 2000, A petrological and structural excursion through the high-grade/high-pressure allochthonous units of the Cabo Ortegal complex (NW Spain): A Coruña, Spain, Basement Tectonics 15, Mid-Conference Field Trip, 59 p.

Lefort, J.P., 1989, Basement Correlation across the North Atlantic: Berlin, Springer, 148 p.

Llana-Fúnez, S., 2001, La estructura de la Unidad de Malpica-Tui (Cordillera Varisca en Iberia): Madrid, Instituto Geológico y Minero de España, Serie Tesis Doctorales, v. 1, 295 p.

Llana-Fúnez, S., and Marcos, A., 2002, Structural record during exhumation and emplacement of high-pressure–low-to intermediate-temperature rocks in the Malpica-Tui unit (Variscan belt of Iberia), *in* Martínez Catalán, J.R., Hatcher, R.D., Jr., Arenas, R., and Díaz García, F., eds., Variscan-Appalachian Dynamics: The Building of the Late Paleozoic Basement: Geological Society of America Special Paper 364, p. 125–142.

Marcos, A., Fernández Rodríguez, J.J., and Llana-Fúnez, S., 2000, Structure of the Cabo Ortegal Nappe: A Coruña, Spain, Basement Tectonics 15, Mid-Conference Field Trip, 15 p.

Marcos, A., Farias, P., Galán, G., Fernández, J.J., and Llana-Fúnez, S., 2002, Tectonic framework of the Cabo Ortegal Complex: A slab of lower crust exhumed in the Variscan orogen (northwestern Iberia Peninsula), *in* Martínez Catalán, J.R., Hatcher, R.D., Jr., Arenas, R., and Díaz García, F., eds., Variscan-Appalachian Dynamics: The Building of the Late Paleozoic Basement: Geological Society of America Special Paper 364, p. 143–162.

Martínez Catalán, J.R., and Arenas, R., 1992, Deformación extensional de las unidades alóctonas superiores de la parte oriental del Complejo de Órdenes (Galicia): Geogaceta, v. 11, p. 108–111.

Martínez Catalán, J.R., Arenas, R., Díaz García, F., Rubio Pascual, F.J., Abati, J., and Marquínez, J., 1996, Variscan exhumation of a subducted Paleozoic continental margin: The basal units of the Órdenes Complex, Galicia: NW Spain: Tectonics, v. 15, p. 106–121, doi: 10.1029/95TC02617.

Martínez Catalán, J.R., Arenas, R., Díaz García, F., and Abati, J., 1997, Variscan accretionary complex of northwest Iberia: Terrane correlation and succession of tectonothermal events: Geology, v. 25, p. 1103–1106, doi: 10.1130/0091-7613(1997)025<1103:VACONI>2.3.CO;2.

Martínez Catalán, J.R., Arenas, R., Díaz García, F., and Abati, J., 1999, Allochthonous units in the Variscan belt of NW Iberia: Terranes and accretionary history, *in* Sinha, A.K., ed., Basement Tectonics, Volume 13: Amsterdam, Kluwer Academic Publishers, p. 65–84.

Martínez Catalán, J.R., Díaz García, F., Arenas, R., Abati, J., Castiñeiras, P., González Cuadra, P., Gómez Barreiro, J., and Rubio Pascual, F.J., 2002, Thrusts and detachment systems in the Órdenes Complex (northwestern Spain): Implications for the Variscan-Appalachian geodynamics, *in* Martínez Catalán, J.R., Hatcher, R.D., Jr., Arenas, R., and Díaz García, F., eds., Variscan-Appalachian Dynamics: The Building of the Late Paleozoic Basement: Geological Society of America Special Paper 364, p. 163–182.

Martínez Catalán, J.R., Arenas, R., Díaz García, F., Gómez-Barreiro, J., González Cuadra, P., Abati, J., Castiñeiras, P., Fernández-Suárez, J., Sánchez Martínez, S., Andonaegui, P., González Clavijo, E., Díez Montes, A., Rubio Pascual, F.J., and Valle Aguado, B., 2007, Space and time in the tectonic evolution of the northwestern Iberian Massif: Implications for the comprehension of the Variscan belt, *in* Hatcher, R.D., Jr., Carlson, M.P., McBride, J.H., and Martínez Catalán, J.R., eds., 4-D Framework of Continental Crust: Geological Society of America Memoir 200, doi: 10.1130/2007.1200(21).

Matte, Ph., 1995, Southern Uralides and Variscides: Comparison of their anatomies and evolutions: Geologie en Mijnbouw, v. 74, p. 151–166.

Matte, Ph., 2001, The Variscan collage and orogeny (480–290 Ma) and the tectonic definition of the Armorica microplate: A review: Terra Nova, v. 13, p. 122–128, doi: 10.1046/j.1365-3121.2001.00327.x.

Mendia, M., 2000, Petrología de la unidad eclogítica del Complejo de Cabo Ortegal (NW de España): Nova Terra, v. 16, 424 p.

Montero, P., 1993, Geoquímica y petrogénesis del Complejo Peralcalino de la Sierra del Galiñeiro (Pontevedra, España) [Ph.D. thesis]: Oviedo, Spain, Universidad de Oviedo, 318 p.

Murphy, J.B., Nance, R.D., Keppie, J.D., Dostal, J., and Coursens, B.L., 1995, Odyssey of West Avalonia: Isotopic constraints for Late Proterozoic. III: Early Silurian paleogeography, *in* Hibbard, J.P., van Staal, C.R., and Cawood, P.A., eds., Current Perspectives in the Appalachian-Caledonian Orogen: Geological Association of Canada Special Paper 41, p. 227–237.

Ordóñez Casado, B., Gebauer, D., Schäfer, H.J., Gil Ibarguchi, J.I., and Peucat, J.J., 2001, A single Devonian subduction event for the HP/HT metamorphism of the Cabo Ortegal Complex within the Iberian Massif: Tectonophysics, v. 332, p. 359–385, doi: 10.1016/S0040-1951(00)00210-9.

Pereira, E., Ribeiro, A., de Carvalho, G., Noronha, F., Ferreira, N., and Monteiro, J.M., 1989, Carta Geológica de Portugal, Folha 1: Lisboa, Serviços Geológicos de Portugal, escala 1:200,000.

Pereira, E., Ribeiro, A., Marques, F., Munhá, J., Meireles, C., Ribeiro, M.A., Pereira, D, Noronha, F., and Ferreira, N., 2000, Carta Geológica de Portugal, Folha 2: Lisboa, Serviços Geológicos de Portugal, escala 1:200,000.

Peucat, J.J., Bernard-Griffiths, J., Gil Ibarguchi, J.I., Dallmeyer, R.D., Menot, R.P., Cornichet, J., and Iglesias Ponce de León, M., 1990, Geochemical and geochronological cross section of the deep Variscan crust: The Cabo Ortegal high-pressure nappe (northwestern Spain): Tectonophysics, v. 177, p. 263–292, doi: 10.1016/0040-1951(90)90285-G.

Pin, C., Ortega Cuesta, L.A., and Gil Ibarguchi, J.I., 1992, Mantle-derived, early Paleozoic A-type metagranitoids from the NW Iberian Massif: Nd isotope and trace-element constraints: Bulletin de la Société Géologique de France, v. 163, p. 483–494.

Pin, C., Paquette, J.L., Santos Zalduegui, J.F., and Gil Ibarguchi, J.I., 2002, Early Devonian suprasubduction-zone ophiolite related to incipient collisional processes in the western Variscan belt: The Sierra de Careón unit, Órdenes Complex, Galicia, *in* Martínez Catalán, J.R., Hatcher, R.D., Jr., Arenas, R., and Díaz García, F., eds., Variscan-Appalachian Dynamics: The Building of the Late Paleozoic Basement: Geological Society of America Special Paper 364, p. 57–71.

Platt, J.P., 1986, Dynamics of orogenic wedges and the uplift of high-pressure metamorphic rocks: Geological Society of America Bulletin, v. 97, p. 1037–1053, doi: 10.1130/0016-7606(1986)97<1037:DOOWAT>2.0.CO;2.

Priem, H.N.A., Boelrijk, N.A.I.M., Verschure, R.H., Hebeda, E.H., and Floor, P., 1966, Isotopic evidence for Upper-Cambrian or Lower-Ordovician granite emplacement in the Vigo area, north-western Spain: Geologie en Mijnbouw, v. 45, p. 36–40.

Puelles, P., 2003, Deformación, metamorfismo y exhumación de las granulitas de alta presión de La Bacariza (Complejo de Cabo Ortegal, NO España), [Ph.D. thesis]: Leioa, Spain, Universidad del País Vasco, 372 p.

Ribeiro, A., 1974, Contribution a l'Etude Tectonique de Trás-os-Montes Oriental: Memorias dos Serviços Geológicos de Portugal 24, 179 p.

Ribeiro, A., Pereira, E., and Dias, R., 1990, Central-Iberian zone: Allochthonous sequences. Structure in the northwest of the Iberian Peninsula, *in* Dallmeyer, R.D., and Martínez García, E., eds., Pre-Mesozoic Geology of Iberia: Berlin, Springer-Verlag, p. 220–236.

Ries, A.C., and Shackleton, R.M., 1971, Catazonal complexes of north-west Spain and north Portugal, remnants of a Hercynian thrust plate: Nature Physical Sciences, v. 234, p. 65–68.

Rodríguez, J., 2003, Recristalización y deformación de litologías supracorticales sometidas a metamorfismo de alta presión (complejo de Malpica-Tuy, NO del Macizo Ibérico) [Ph.D. thesis]: Leioa, Spain, Universidad del País Vasco, 410 p.

Rodríguez, J., Cosca, M.A., Gil Ibarguchi, J.I., and Dallmeyer, R.D., 2003, Strain partitioning and preservation of $^{40}Ar/^{39}Ar$ ages during Variscan exhumation of a subducted crust (Malpica-Tui complex, NW Spain): Lithos, v. 70, p. 111–139, doi: 10.1016/S0024-4937(03)00095-1.

Rubio Pascual, F.J., Arenas, R., Díaz García, F., Martínez Catalán, J.R., and Abati, J., 2002, Contrasting high-pressure metabasites from the Santiago unit (Órdenes Complex, northwestern Iberian Massif, Spain), *in* Martínez Catalán, J.R., Hatcher, R.D., Jr., Arenas, R., and Díaz García, F., eds., Variscan-Appalachian Dynamics: The Building of the Late Paleozoic Basement: Geological Society of America Special Paper 364, p. 105–124.

Sánchez Martínez, S., 2003, Geoquímica de las unidades ofiolíticas de Purrido y Moeche (Complejo de Cabo Ortegal, NW del Macizo Ibérico) [D.E.A.]: Madrid, Spain, Universidad Complutense, 206 p.

Sánchez Martínez, S., Arenas, R., Andonaegui, P., Martínez Catalán, J.R., and Pearce, J.A., 2007, Geochemistry of two associated ophiolites from the Cabo Ortegal Complex (Variscan belt of NW Spain), *in* Hatcher, R.D., Jr., Carlson, M.P., McBride, J.H., and Martínez Catalán, J.R., eds., 4-D Framework of Continental Crust: Geological Society of America Memoir 200, doi: 10.1130/2007.1200(23).

Santos, J.F., Schärer, U., Gil Ibarguchi, J.I., and Girardeau, J., 2002, Genesis of pyroxenite-rich peridotite at Cabo Ortegal (NW Spain): Geochemical and Pb-Sr-Nd isotope data: Journal of Petrology, v. 43, p. 17–43, doi: 10.1093/petrology/43.1.17.

Santos Zalduegui, J.F., 1995, Geocronología y geoquímica isotópica de diferentes unidades de los complejos alóctonos de Cabo Ortegal y Malpica-Tuy (NO de España): Nova Terra, v. 11, 178 p.

Scotese, C.R., 2001, Atlas of Earth History: Arlington, Texas, PALEOMAP Project, 52 p.

Skehan, J.V., and Rast, N., 1995, Late Proterozoic to Cambrian evolution of the Boston Avalon terrane, *in* Hibbard, J.P., van Staal, C.R., and Cawood, P.A., eds., Current Perspectives in the Appalachian-Caledonian Orogen: Geological Association of Canada Special Paper 41, p. 207–225.

Stampfli, G.M., and Borel, G.D., 2002, A plate tectonic model for the Paleozoic and Mesozoic constrained by dynamic plate boundaries and restored synthetic oceanic isochrons: Earth and Planetary Science Letters, v. 196, p. 17–33, doi: 10.1016/S0012-821X(01)00588-X.

Van Calsteren, P.W.C., Boelrijk, N.A.I.M., Hebeda, E.H., Priem, H.N.A., Den Tex, E., Verdurmen, E.A.T.H., and Verschure, R.H., 1979, Isotopic dating of older elements (including the Cabo Ortegal mafic-ultramafic complex) in the Hercynian orogen of NW Spain: Manifestations of a presumed early Paleozoic mantle plume: Chemical Geology, v. 24, p. 35–56, doi: 10.1016/0009-2541(79)90011-1.

Van der Meer Mohr, C.G., 1975, The Palaeozoic strata near Moeche in Galicia: NW Spain: Leidse Geologische Mededelingen, v. 49, p. 32–37.

Vogel, D.E., 1967, Petrology of an eclogite- and pyrigarnite-bearing polymetamorphic rock complex at Cabo Ortegal: NW Spain: Leidse Geologische Mededelingen, v. 40, p. 121–213.

Von Raumer, J.F., and Neubauer, F., 1993, Late Precambrian and Paleozoic evolution of the Alpine basement—An overview, *in* Von Raumer, J.F., and Neubauer, F., eds., Pre-Mesozoic Geology in the Alps: Berlin, Springer-Verlag, p. 625–639.

Manuscript Accepted by the Society 1 February 2007

Printed in the USA

The Geological Society of America
Memoir 200
2007

Geochemistry of two associated ophiolites from the Cabo Ortegal Complex (Variscan belt of NW Spain)

Sonia Sánchez Martínez*
Ricardo Arenas
Pilar Andonaegui
Departamento de Petrología y Geoquímica, Universidad Complutense, 28040 Madrid, Spain

José R. Martínez Catalán
Departamento de Geología, Universidad de Salamanca, 37008 Salamanca, Spain

Julian A. Pearce
School of Earth, Ocean and Planetary Sciences, Cardiff University, CF10 3YE Cardiff, UK

ABSTRACT

Two Paleozoic ophiolites of different age mark the Variscan suture in the Cabo Ortegal Complex (northwestern Iberian Massif). The Moeche and Purrido allochthonous ophiolites are structurally located between an exotic terrane of continental affinity (basal units), below, and another terrane with arc affinity (upper units), above. The mafic rocks of the Moeche unit are greenschist with a composition typical of tholeiitic basalt. Their rare earth element (REE) concentrations range from 12 to 31 times the chondritic abundances, with almost flat chondrite-normalized REE patterns ($[La_N/Yb_N]_{avg}$ = 0.83), and they have no significant Eu anomalies. According to their contents in some immobile trace elements, the geochemical composition of these metabasalts seems to be transitional between normal mid-ocean-ridge basalt (N-MORB) and island-arc tholeiites. Their abundances in the most immobile trace elements with the highest discriminating power (Th, Nb, Ce, Zr, Ti, and Y), normalized to an average tholeiitic N-MORB, depict quite flat patterns close to the average N-MORB concentrations, although they are slightly enriched. A small negative Nb anomaly points to an origin linked to a subduction zone.

The Purrido unit consists of amphibolites with compositions also equivalent to tholeiitic basalt. REE compositions range from 6 to 22 times chondritic abundance, with essentially flat chondrite-normalized REE patterns ($[La_N/Yb_N]_{avg}$ = 0.99) and very small positive Eu anomalies ($[Eu/Eu^*]_{avg}$ = 1.08). The abundance of some trace

*Current address: Departamento de Petrología y Geoquímica, Facultad de Geología, Universidad Complutense, C./José Antonio Novais, nº 2, 28040 Madrid, Spain; s.sanchez@geo.ucm.es.

Sánchez Martínez, S., Arenas, R., Andonaegui, P., Martínez Catalán, J.R., and Pearce, J.A., 2007, Geochemistry of two associated ophiolites from the Cabo Ortegal Complex (Variscan belt of NW Spain), *in* Hatcher, R.D., Jr., Carlson, M.P., McBride, J.H., and Martínez Catalán, J.R., eds., 4-D Framework of Continental Crust: Geological Society of America Memoir 200, p. 445–467, doi: 10.1130/2007.1200(23). For permission to copy, contact editing@geosociety.org.

elements in these metabasites shows an affinity with island-arc tholeiites, but, unlike the greenschist of the Moeche unit, their N-MORB–normalized trace-element patterns are slightly depleted in Nb, Zr, Ti, and Y, with a somewhat steeper negative Nb anomaly. These features are found in basalts generated in some back-arc basins.

Both ophiolites seem to represent oceanic lithosphere that developed in a suprasubduction zone context. The Moeche unit was probably formed during the opening of a back-arc basin at the Cambrian-Ordovician boundary, whereas the Purrido unit was generated later, during the Early Devonian, as a result of the intraoceanic subduction of the older oceanic lithosphere represented by the Moeche unit. The two ophiolites are representative of the Variscan suture in northwest Iberia, and they preserve a significant part of the history of the oceanic realms that closed during the final amalgamation of Pangea.

Keywords: ophiolites, geochemistry, suprasubduction zone, NW Iberia.

INTRODUCTION

The ophiolite concept has undergone an important evolution through time concerning to its origin and meaning (Dilek, 2003; Moores, 2003). At first, ophiolites were interpreted as remnants of oceanic crust generated at mid-ocean ridges and, therefore, related to divergent plate margins (Anonymous, 1972). Later, Boudier and Nicolas (1985) proposed a classification of ophiolitic types based on their structure and the nature of their mantle section. According to these authors, it is possible to distinguish between harzburgite ophiolite type (HOT), interpreted as having originated at fast spreading ridges, and lherzolite ophiolite type (LOT), supposedly linked to slow spreading ridges. Subsequently, the geochemical fingerprinting of ophiolite complexes has revealed that most of the ophiolites emplaced over continental margins or island arcs display subduction zone affinities. Consequently, the term suprasubduction zone ophiolite (Pearce et al., 1984) was introduced to designate ophiolites that are products of limited seafloor spreading above a subduction zone. Thus, the origin of suprasubduction zone ophiolites is connected to convergent plate margins, compared to mid-ocean-ridge basalt (MORB) ophiolites, which are generated at divergent plate margins (Hawkins, 2003; Pearce, 2003). The latter are hardly ever preserved in orogens because their thermal state favors complete subduction.

The European Variscan belt includes a complex suture zone that can be followed from the southwestern Iberian Peninsula, through the Iberian-Armorican arc and the French Massif Central, to the Bohemian Massif (Franke, 1989, 2000; Matte, 1991, 2001; Martínez Catalán et al., 1997, this volume). In this suture, several variably fragmented ophiolite assemblages of oceanic affinity have been described. Where detailed geological sections exist, for example, the northwest Iberian Massif (Martínez Catalán et al., 1996; Díaz García et al., 1999; Martínez Catalán et al., 2002), ophiolites are structurally located below an upper exotic terrane of arc affinity that has been affected by a high-pressure (P) and high-temperature (T) eo-Variscan metamorphism (upper units of Cabo Ortegal Complex in northwest Spain, or the equivalent Góry Sowie Massif in the Polish Sudetes), and above a basal terrane that is considered the most external part of the Gondwanan margin. This early to middle Paleozoic passive margin was subducted and underwent high-P and low- to medium-T Variscan metamorphism (Malpica-Tui unit in northwest Spain; Champtoceaux Complex in the Armorican Massif). In the Bohemian Massif, only the Ślęża ophiolite, dated at the Late Silurian–Early Devonian boundary, has a typically oceanic lithological sequence (Pin et al., 1988; Oliver et al., 1993; Dubińska et al., 2004). Other units usually described as ophiolites in the Bohemian Massif, such as the Mariánské Lázně ophiolite (Kastl and Tonika, 1984; Crowley et al., 2002; Timmermann et al., 2004), have lithologies and a tectonothermal evolution that are more compatible with the arc-related upper exotic terrane of northwestern Iberia. In Cornwall, the Lizard Complex includes an ophiolitic assemblage dated at 380–400 Ma (Clark et al., 1998; Nutman et al., 2001) that is correlative with the Ślęża ophiolite. In the northwest Iberian Massif (Galicia and Trás-os-Montes), there is a varied group of ophiolites of different age and origin (protoliths at ca. 500 Ma and ca. 395 Ma; see Arenas et al., 2007b, this volume). Some of them are comparable to the ophiolites from the Sudetes and Cornwall. Finally, in the southwestern Iberian Massif, the Beja-Acebuches ophiolite resembles the Early Devonian ophiolites, although its protoliths are still undated (Fonseca and Ribeiro, 1993; Crespo-Blanc and Orozco, 1991; Quesada et al., 1994; Castro et al., 1996).

The correlation between different ophiolitic units of the European Variscan belt is essential in order to accurately delimit the suture or sutures, as well as to understand the nature and extent of the terranes that accreted during the orogeny; furthermore, precise chronology, geochemical features, and tectonic setting of these units will permit their correlation with ophiolitic types generated at divergent (MORB) or convergent (suprasubduction zone) settings and will delineate the Paleozoic evolution in the peri-Gondwanan realm of Europe.

This contribution describes the geochemical features of two ophiolitic units that form part of the Cabo Ortegal Complex (Moeche and Purrido ophiolitic units). These two ophiolites, which have very different ages, have equivalents in other allochthonous complexes of northwest Iberia, which always display a systematic association. This work is part of a broader attempt to systematize the knowledge about the geochemistry and U-Pb and $^{40}Ar/^{39}Ar$ geochronology of the ophiolites from the northwestern Iberian Massif (Dallmeyer et al., 1997; Díaz García et al., 1999; Arenas et al., 2007a, 2007b [this volume]).

GEOLOGICAL SETTING

The remnants of a rootless Variscan suture are preserved in the so-called allochthonous Galicia-Trás-os-Montes complexes, located at the most internal part of the Variscan belt in the northwestern Iberian Massif. These complexes represent mega-klippen of a huge nappe pile of allochthonous units, which are characterized by their different internal structure and a distinctive tectonothermal evolution (Ries and Shackleton, 1971; Arenas et al., 1986; Martínez Catalán et al., 2007, this volume). Overthrusting took place during Devonian and Carboniferous times in the earliest stages of the collision between Gondwana and Laurussia (Martínez Catalán et al., 1997). One of the colliding terranes, which are the upper units, reveals a more complex history because it is polymetamorphic and was accreted to Laurussia in pre-Variscan times (Gómez-Barreiro et al., 2006). The Cabo Ortegal Complex, one of the five allochthonous complexes in northwest Iberia (Cabo Ortegal, Órdenes, and Malpica-Tui in Galicia, Spain, and Bragança and Morais in Trás-os-Montes, Portugal), is located along the northern coastal section. Despite its relatively reduced dimensions, it contains representatives of each allochthonous unit described in northwest Iberia. The main lithologic and structural features of the units that make up the Cabo Ortegal Complex were described first by Vogel (1967), Engels (1972), and Arenas (1988). Other works that more recently have described geological aspects of the complex are those of Santos Zalduegui (1995), Fernández (1997), Azcárraga (2000), Mendia (2000), Castiñeiras (2003), and Puelles (2003).

The Cabo Ortegal Complex (Fig. 1) includes two main ophiolites, the Moeche and Purrido units, which represent the main Variscan suture of the orogen. The overlying terrane (high-P and high-T upper units: La Capelada, Cedeira, and Peña Escrita units) is composed of a diverse polymetamorphic assemblage of gneisses and mafic and ultrabasic rocks affected by high-P and high-T early Variscan metamorphism (Gil Ibarguchi et al., 1990; Arenas, 1991; Ordóñez Casado et al., 2001; Fernández-Suárez et al., 2002; Marcos et al., 2002). This is a terrane of Gondwanan affinity (Fernández-Suárez et al., 2003) that is interpreted to be a volcanic arc detached from the continent during Late Cambrian–Early Ordovician times (Arenas et al., 1986; Abati et al., 1999, 2003; Andonaegui et al., 2002; Castiñeiras, 2003). Underlying the ophiolitic units, the so-called basal units are represented by the Espasante unit, which consists of orthogneiss, schist, and metabasite of continental affinity that have experienced high-P and low- to medium-T Variscan metamorphism. The basal units have been interpreted as the most external part of the Gondwanan margin, which subducted at the beginning of the Variscan deformation (Arenas et al., 1995, 1997; Martínez Catalán et al., 1996). In the Cabo Ortegal Complex, the basal units appear to be imbricated within the Somozas mélange, a tectonic mélange with ophiolitic and sedimentary lithologies, which crops out at the contact between the allochthonous complex and its relative autochthon (Fig. 1).

The Moeche and Purrido ophiolitic units are structurally stacked and separated by a thrust fault (Fig. 1), although their lithologic composition and ages are different. Analogous units also appear in the allochthonous complexes of northwest Iberia, and they always occupy the same structural position and show similar lithological features and tectonothermal evolution (see Arenas et al., 2007, this volume). These two units represent a pair of related ophiolites with a genetic link that will be discussed in the following sections.

MOECHE OPHIOLITIC UNIT

The structurally lower Moeche ophiolite crops out east of the Cabo Ortegal Complex, as a 500-m-thick, NE-SW–striking unit thrust between the Purrido and Espasante units, which are located at the top and bottom, respectively (Fig. 1). Its contacts with these units are tectonic and originally should correspond to thrusts generated during the initial stacking of the nappe pile.

This unit consists of mylonitic greenschists with thin interlayers of metasedimentary rocks that presently display greenschist-facies mineral paragenesis (Arenas, 1988). The greenschists are fine-grained rocks with nematoblastic texture and a strong regional foliation that probably has a decompressive origin. The most common types (Fig. 2A) have a paragenesis consisting of: chlorite + actinolite + epidote-clinozoisite + quartz + albite + opaque minerals (ilmenite-pyrite) + sphene ± white mica ± stilpnomelane. Amphibole-free types dominated by chlorite with heterogeneous mineralogical and chemical composition have been also identified (heterogeneous greenschists, Fig. 2B). They probably were derived from igneous rocks hydrothermally altered prior to the regional metamorphism. Apart from these two types of metabasites, coarser-grained greenschists of recognizable blastogranular texture (Fig. 2C), probably derived from diabase or fine-grained gabbroic types, are also locally present. Ore deposits of copper and minor amounts of gold and silver, which have been interpreted as volcanogenic massive sulfide deposits similar to those found in the Cyprus ophiolites (Cyprus-type; Fernández Pompa et al., 1976; Castroviejo et al., 2004), are associated with the greenschists. The metasedimentary rocks of this unit are phyllite and fine-grained mica schists composed of: quartz + albite + chlorite + stilpnomelane + white mica + ilmenite ± rutile ± sphene (Fig. 2D). The mica-rich metasediments have a pelitic origin, whereas the quartz- and plagioclase-rich ones seem to derive from graywackes. An increase in grain size and degree of recrystallization of the metasedimentary rocks from north to south suggests a slight increase in the intensity of metamorphism in this direction. Based on these features, the Moeche ophiolite

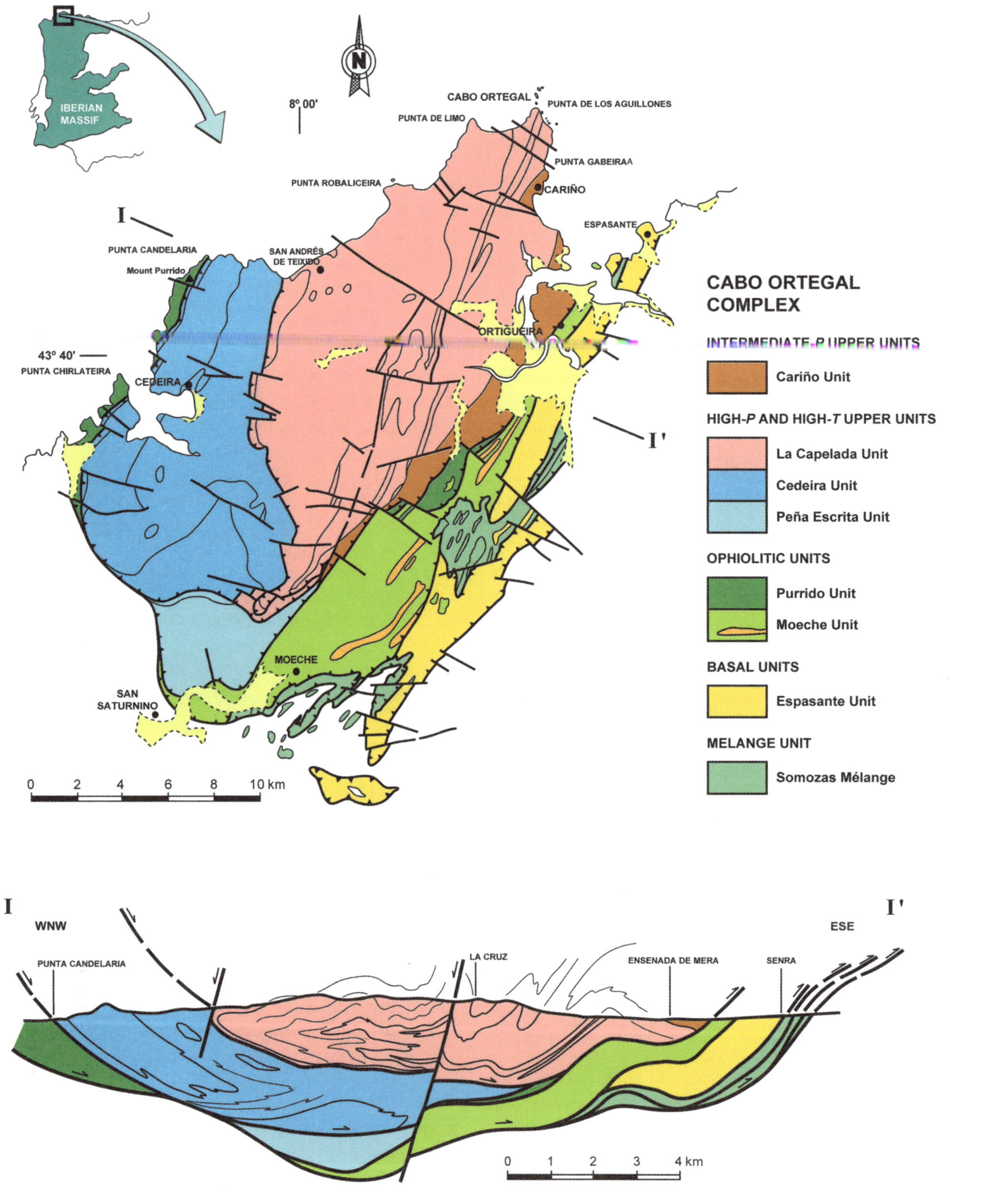

Figure 1. Geologic map and cross section of the Cabo Ortegal Complex (northwestern Iberian Massif), showing the position of the Moeche and Purrido ophiolites as well as the other allochthonous units in this complex (modified from Vogel, 1967; Marcos et al., 1984; Arenas et al., 1986).

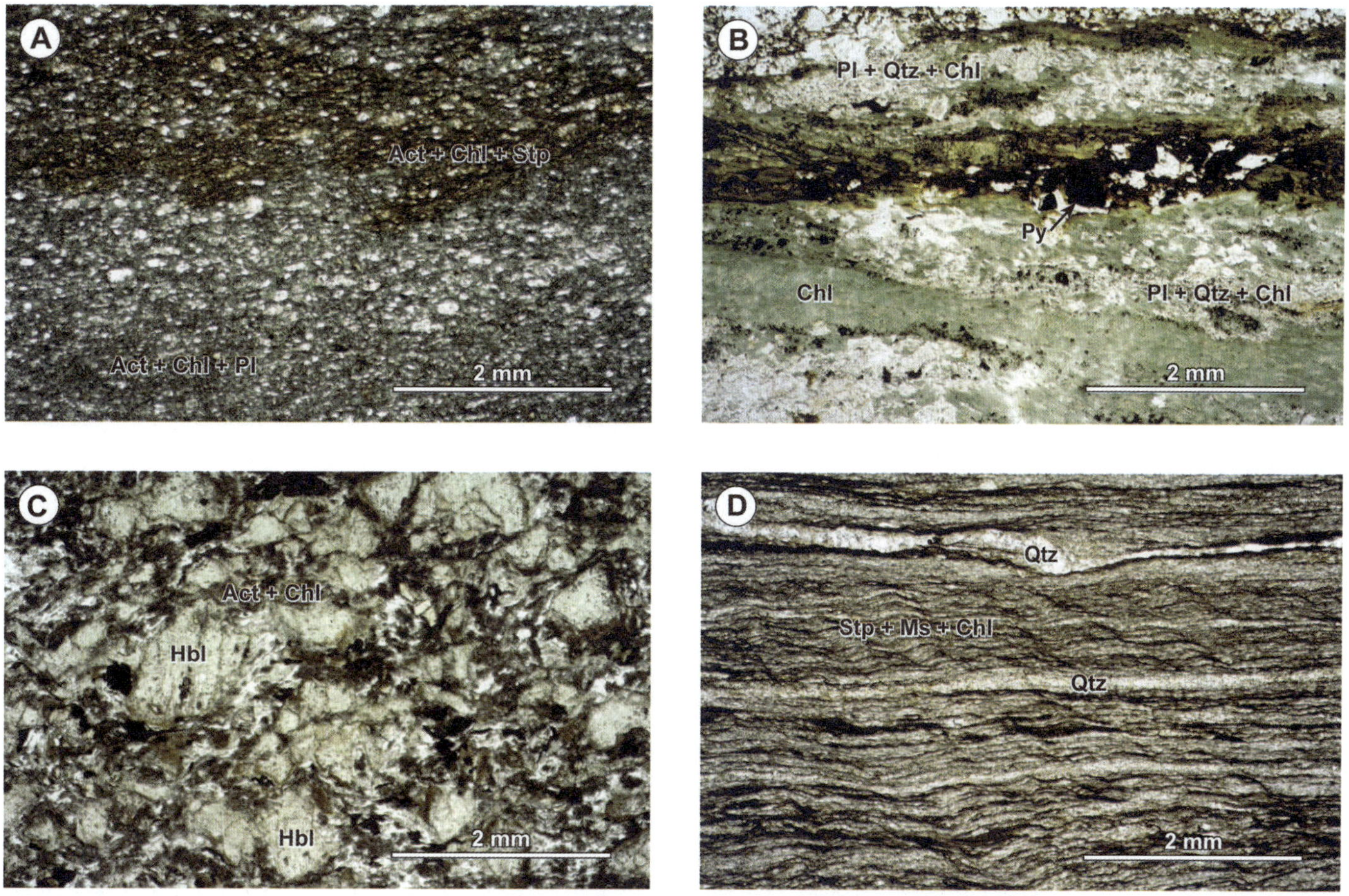

Figure 2. Thin sections of the different petrographic types that can be identified in the Moeche Unit. (A) Texture of the common mylonitic greenschists; these are fine-grained rocks with micronematoblastic texture, mainly composed of actinolite (Act), chlorite (Chl), plagioclase (Pl), and epidote. (B) Amphibole-free greenschist of the so-called heterogeneous greenschist group; this sample shows the characteristic compositional layering of this type of lithology, with a layer containing preferential concentration of sulfides. (C) Greenschist with blastogranular texture probably derived from a dolerite or gabbro protolith. (D) Phyllite with lepidoblastic texture and a strong mylonitic foliation.

is interpreted as having derived from a sequence of volcanic rocks alternating with terrigenous sediments intruded by some gabbroic stocks.

In the Órdenes Complex, located to the west, the Vila de Cruces unit is considered analogous to the Moeche unit because it has an identical structural position and the same lithologic assemblage (see Arenas et al., 2007b, this volume), and, in addition, it has a mappable orthogneiss intruding the mafic schists. The protolith of the Vila de Cruces unit has been dated at ca. 500 Ma (U-Pb in zircon; Arenas et al., 2007a), which is considered a reference age for generation of both the Vila de Cruces and Moeche ophiolitic units. Furthermore, an early high-*P* and low-*T* metamorphic event has been also recognized in the Vila de Cruces unit (Martínez Catalán et al., 2002; Arenas et al., submitted), although it seems to have been obliterated in the Moeche unit. The regional foliation in both units postdates the development of this high-*P* event and has a decompressive nature. It is associated with the development of a second generation of recumbent folds, and it has been dated at 364.4 ± 0.7 Ma in the Moeche unit (whole-rock $^{40}Ar/^{39}Ar$ in a phyllite interbedded between mafic schists; Dallmeyer et al., 1997).

PURRIDO OPHIOLITIC UNIT

Magnificent outcrops of the structurally upper ophiolite occur along the cliffs of the western coastline of the Cabo Ortegal Complex, where the Purrido unit appears as a 300-m-thick, NNE-SSW–striking band underlying the high-*P* and high-*T* Cedeira unit (Fig. 1). Both units are separated by the Carreiro shear zone (Vogel, 1967; Azcárraga, 2000), which is a thrust fault that represents the suture between an upper arc-affinity exotic terrane and an oceanic domain. To the east, the Purrido unit becomes dismembered, but its structural position overlying the Moeche unit can be established.

This ophiolitic unit has a very homogeneous lithologic composition, with well-foliated massive nematoblastic amphibolites, occasionally with garnet-bearing types, and without significant compositional layering (Vogel, 1967). These features suggest that the rocks resulted from deformation and metamorphism of massive gabbro. Some meter-size lensoid alternations of Ca-poor amphibolite have been found and are recognizable in outcrops by their high garnet content. The Ca-poor amphibolites probably derive from hydrothermal alteration of basic rocks that occurred prior to regional metamorphism.

The petrographic features of the common metabasites of the Purrido unit have already been described by Vogel (1967) and Engels (1972). These are medium-grained amphibolites with a well-defined nematoblastic planar fabric developed during prograde metamorphism associated with the underthrusting of the unit. Their mineralogical composition is very uniform and consists of: hornblende + plagioclase + epidote-clinozoisite + quartz + sphene ± rutile ± opaque minerals (ilmenite) ± chlorite (Fig. 3A). Occasionally, they include garnet-rich bands (Fig. 3B). They are texturally similar to the garnet-free types, although their grain size is somewhat coarser, and they are slightly more granoblastic. The scarce Ca-poor amphibolites lack ferromagnesian amphiboles or aluminous phases typical of this type of lithology (Spear, 1993), but they are characterized by development of very abundant large garnet porphyroblasts (1.5 cm) and by a low CaO chemical composition (Fig. 3C; see also the geochemistry section). Finally, strongly retrogressed, sheared micronematoblastic to granonematoblastic mylonitic amphibolite (Fig. 3D) is found at the bottom of the unit. The mineralogical composition of these rocks consists of actinolite + hornblende + plagioclase + epidote-clinozoisite + quartz ± opaque minerals (ilmenite) ± sphene, and it reflects a retrogressive assemblage.

According to its lithological features and structural position, the Purrido unit is considered to be correlative with the Careón unit, which crops out in the western part of the Órdenes Complex (Díaz García et al., 1999; Arenas et al., 2007b, this volume). Specifically, the amphibolites and garnet-bearing amphibolites of the Purrido unit are similar to those of the lower sheet of the Careón

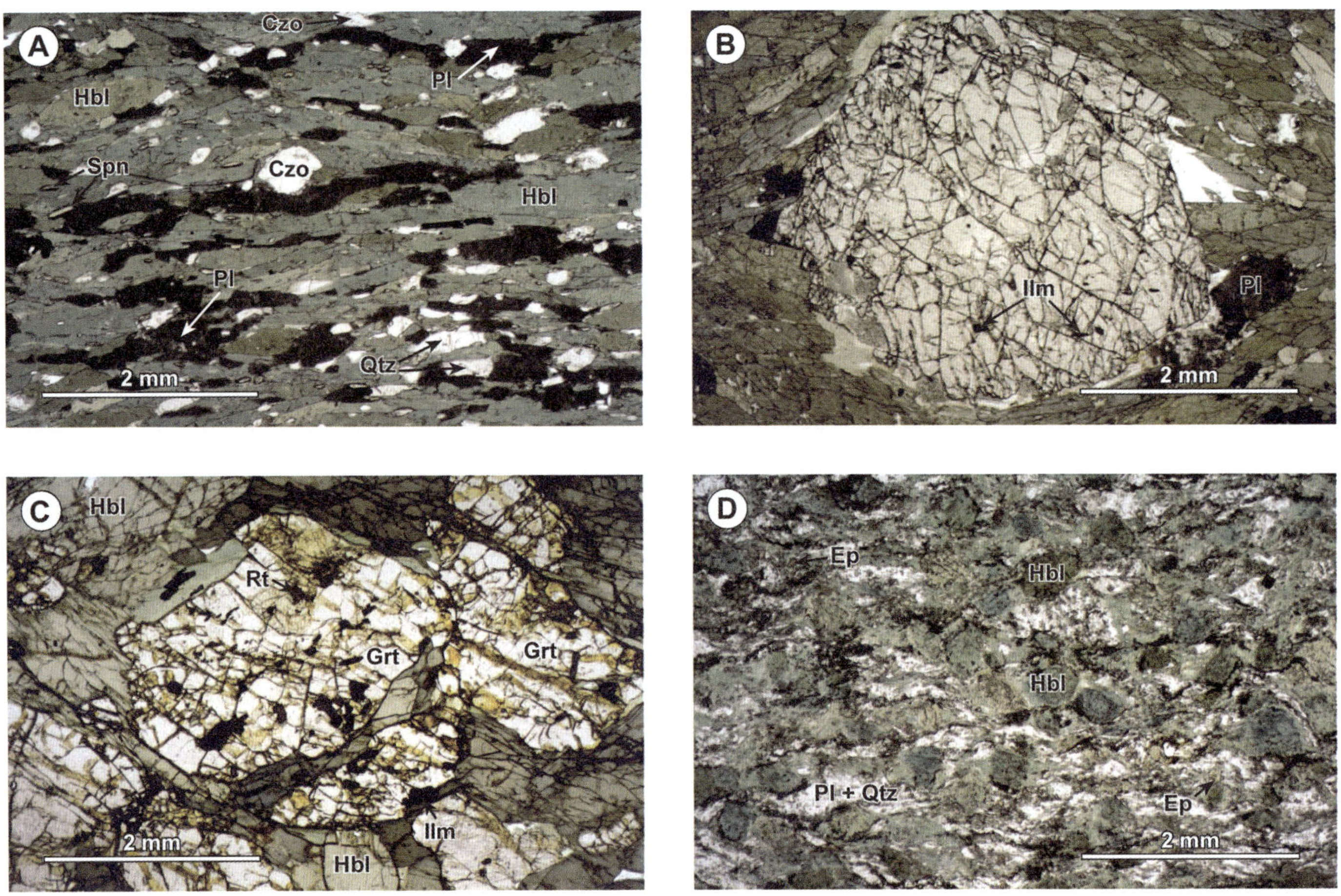

Figure 3. Thin sections of the different petrographic types identified in the Purrido unit. (A) Texture of a common amphibolite, the predominant lithology in this ophiolite; this typically nematoblastic and well-foliated rock is composed of hornblende (Hbl), plagioclase (Pl) (usually saussuritized), clinozoisite (Czo), quartz (Qtz), and ilmenite (Ilm), and sphene (Spn) as accessory minerals. (B) Garnet-bearing amphibolite; a large porphyroblast of synkinematic garnet (Grt) appears wrapped by the amphibolite matrix. (C) Texture of a Ca-poor amphibolite characterized by the abundance of large garnet porphyroblasts. Rt—rutile. (D) Sheared retrogressed amphibolite from the base of the Purrido unit. Ep—epidote. There, the less-deformed samples developed a granonematoblastic texture where relict hornblende-type amphiboles are surrounded by actinolite aggregates.

unit (Orosa sheet), and hence both are considered to have followed an analogous tectonothermal evolution. Thermobarometric data obtained in the Careón unit show that the garnet-bearing amphibolites equilibrated at 640–680 °C and 11.1–11.9 kbar (Díaz García et al., 1999). Amphiboles of the prograde nematoblastic fabric of the Purrido unit have been dated at 391.3 ± 6.6 Ma ($^{40}Ar/^{39}Ar$ in a hornblende concentrate from a garnet-bearing amphibolite; Peucat et al., 1990). No isotopic dating of the gabbroic protoliths of the Purrido amphibolites has been attempted, but metagabbros of the equivalent Careón unit have been dated at 395 ± 2 Ma (U-Pb in zircon from metagabbros; Díaz García et al., 1999; Pin et al., 2002).

GEOCHEMISTRY

Sample Locations and Analytical Methods

A set of 31 samples was collected for geochemical characterization of both ophiolitic units, 13 from the Moeche unit and 18 from the Purrido unit. The location of these samples is shown in Figure 4. Sampling was focused on picking the most representative lithologies of each unit and selecting the freshest samples that had no visible alteration. Sample preparation was carried out in the sample preparation laboratory of the Departamento de Petrología y Geoquímica of the Universidad Complutense in Madrid. This process consisted of: (1) washing and drying samples; (2) removing the weathered portions; (3) crushing the rocks in a jaw crusher and disk mill; and (4) quartering and final powdering with an agate mill to prevent metallic element contamination. Analyses were performed by Activation Laboratories Ltd. (Actlabs), Canada. The digestion procedure used by these laboratories was lithium metaborate/tetraborate fusion, and the analytical technique used to measure the elemental concentrations was inductively coupled plasma–mass spectrometry (ICP-MS). The results are given in Tables 1–4. Moreover, the data quality was determined by replicate analyses (acid digestion, ICP-MS) of two samples of each unit at Cardiff University laboratories.

General Geochemical Features

Nine of the 13 samples from the Moeche unit correspond to common mafic greenschist, and four correspond to the amphibole-free types (heterogeneous greenschist). All but one of the samples (with SiO_2 < 45%) have compositions typical of basic rocks (SiO_2 = 46.54%–51.00%). The compositional ranges of the rest of the major elements for the common greenschists vary: 13.77%–18.28% Al_2O_3, 11.27%–14.06% Fe_2O_3(T), 0.162%–0.203% MnO, 5.14%–8.45% MgO, 5.14%–9.71% CaO, 2.68%–4.95% Na_2O, 0.01%–0.39% K_2O, 1.651%–2.147% TiO_2, and 0.10%–0.27% P_2O_5 (Table 1). The heterogeneous greenschists are distinguished by notably lower CaO contents (1.21%–6.00%), which results in the lack of calcic amphiboles in their mineralogy; they also have slightly lower Na_2O (0.39%–4.98%) values and higher Fe_2O_3(T) and TiO_2 values (15.38%–20.80% and 2.13%–2.59%, respectively). Their MgO contents can be lower or higher depending on the sample (Table 2). Normative compositions (CIPW norms; Fe_2O_3 calculated according to the method of Irvine and Baragar, 1971) of most of the common greenschist samples are silica-saturated, with variable percentages in normative olivine and hypersthene (0.18%–13.11% and 1.86%–28.63%, respectively). Only two of the samples are silica-oversaturated, with normative quartz and hypersthene values of 0.19%–2.45% and 24.94%–26.93%, respectively. The heterogeneous greenschists generally exhibit a silica-oversaturated normative composition, with normative quartz, hypersthene, and corundum; only one of them has a normative composition similar to those of the common greenschists. The percentage of normative hypersthene (24.48%–55.71%) is quite high in these samples—a feature that, together with the presence of corundum, suggests a modified chemical composition, at least with respect to major elements.

Thirteen of the 18 samples of metabasites from the Purrido unit were collected from the most common lithology, garnet-free amphibolite. Four samples are retrogressed and sheared amphibolites located at the bottom of the unit, and the remaining sample is a Ca-poor hypergarnetiferous-type amphibolite. All of them, except the Ca-poor amphibolite (SiO_2 = 40.43%), exhibit a narrow range of variation in SiO_2 (47.83%–51.94%), which permits classifying them as basic rocks. The common Purrido amphibolites have a narrower compositional range in the rest of the major elements than the Moeche unit samples; their ranges are: 13.73%–15.72% Al_2O_3, 8.64%–10.45% Fe_2O_3(T), 0.117%–0.204% MnO, 8.39%–9.23% MgO, 10.13%–12.72% CaO, 2.01%–2.93% Na_2O, 0.01%–0.14% K_2O, 0.846%–1.171% TiO_2, and 0.08%–0.12% P_2O_5 (Table 3). The retrogressed amphibolites at the bottom of the unit exhibit higher Fe_2O_3(T) (11.07%–12.99%), Na_2O (3.00%–3.66%), TiO_2 (1.60%–2.11%), and P_2O_5 (0.14%–0.26%) and a little lower Al_2O_3 (13.52%–13.94%) and CaO (8.19%–8.55%) (Table 4) compositions. The Ca-poor amphibolite is characterized by enrichment in Al_2O_3 and Fe_2O_3(T) and depletion in Na_2O and, especially, CaO, which becomes half the composition of the common types (Table 3). Most samples of common and retrogressed amphibolites are silica-saturated, containing normative olivine and hypersthene (3.58%–10.99% and 8.43%–15.65%, respectively). Only three turned out to be silica-oversaturated, with variable percentages in normative quartz and hypersthene (0.69%–3.32% and 18.51%–23.13%). The Ca-poor amphibolite exhibits a silica-saturated normative composition with normative olivine, hypersthene, and corundum. The high percentage of hypersthene and the presence of corundum are typical of a MgO- and Al_2O_3-rich rock of modified chemical composition.

Alteration Effects and Classification

The effects of deformation and metamorphism, which could modify their chemical composition, should be considered in order to classify the metabasites (Cann, 1970; Humphris and Thompson, 1978; Menzies and Seyfried, 1979). The most significant

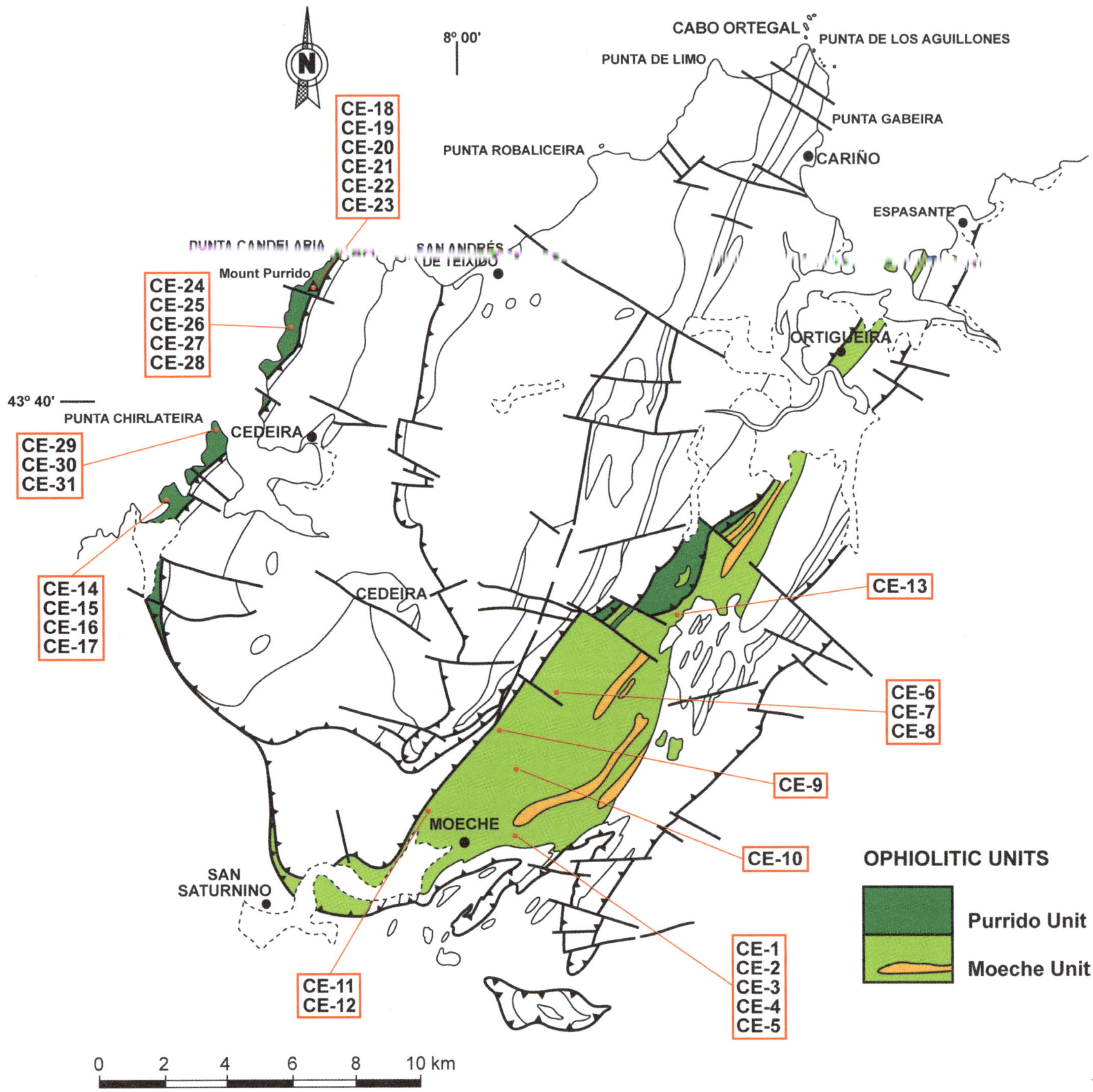

Figure 4. Location of the metabasites sampled for whole-rock chemical analysis. CE-1 to CE-13: Moeche unit samples. CE-14 to CE-31: Purrido unit samples.

TABLE 1. WHOLE-ROCK MAJOR AND TRACE ELEMENT DATA OF COMMON MOECHE UNIT GREENSCHISTS

Sample	CE-1	CE-3	CE-5	CE-7	CE-9	CE-10	CE-11	CE-12	CE-13
SiO_2	51.00	46.60	50.11	46.54	48.50	48.42	48.61	49.43	48.53
Al_2O_3	14.76	18.28	15.05	16.45	13.77	15.12	13.88	13.79	14.84
Fe_2O_3	11.27	12.35	11.65	14.06	11.88	12.44	12.54	12.09	12.98
MnO	0.195	0.193	0.165	0.162	0.203	0.184	0.184	0.206	0.179
MgO	6.41	5.14	6.07	5.31	7.93	7.57	8.45	8.29	8.01
CaO	6.59	7.74	8.07	8.18	9.71	5.36	7.56	8.68	6.08
Na_2O	4.95	3.57	4.37	4.09	3.12	4.04	3.16	2.68	3.39
K_2O	0.13	0.39	0.26	0.09	< 0.01	0.13	0.09	0.05	0.33
TiO_2	1.653	1.651	1.662	2.147	1.811	1.692	1.919	1.784	2.063
P_2O_5	0.12	0.10	0.12	0.23	0.18	0.17	0.27	0.22	0.21
LOI	2.70	4.01	2.56	2.76	2.31	3.70	3.54	2.95	3.72
TOTAL	99.78	100.02	100.09	100.02	99.42	98.83	100.20	100.17	100.33
Sc	44	47	44	44	42	41	39	39	41
V	327	288	321	304	330	319	300	317	333
Cr	350	379	334	210	177	89	136	230	176
Co	67	52	40	32	32	29	33	36	35
Ni	143	90	105	65	36	37	40	51	57
Cu	65	168	79	82	64	159	54	63	52
Zn	106	117	111	156	117	251	95	104	104
Ga	15	17	14	20	18	19	19	19	19
Rb	3	8	6	< 1	< 1	1	< 1	1	7
Sr	186	195	271	208	153	101	140	171	103
Y	37.7	49.7	36.0	53.9	42.7	53.4	54.4	45.9	48.3
Zr	93	101	87	133	129	135	195	151	157
Nb	6.7	1.8	4.3	2.5	2.8	2.9	3.7	3.4	3.1
Cs	0.4	0.5	0.6	0.3	<0.1	<0.1	0.2	0.3	0.3
Ba	28	84	100	17	12	21	17	15	36
Hf	2.6	3.3	2.4	4.4	3.9	4.0	5.6	4.4	4.6
Ta	0.10	0.06	0.09	0.12	0.13	0.12	0.20	0.18	0.16
Pb	<5	<5	<5	<5	<5	<5	<5	<5	<5
Th	0.18	0.21	0.13	0.26	0.24	0.31	0.50	0.34	0.31
U	0.10	0.10	0.11	0.24	0.10	0.58	0.18	0.13	0.17
La	4.20	9.85	4.36	5.78	4.93	8.15	7.46	5.94	6.07
Ce	10.3	13.9	10.5	16.7	16.0	23.5	24.0	18.9	19.0
Pr	1.94	3.52	1.88	2.77	2.49	3.57	3.58	2.87	2.95
Nd	11.0	17.7	10.8	16.0	13.7	19.1	19.2	15.4	16.2
Sm	3.83	5.57	3.69	5.54	4.56	5.85	6.10	4.95	5.25
Eu	1.49	2.16	1.54	2.20	1.75	2.20	2.12	1.82	1.97
Gd	5.43	7.67	5.14	7.65	6.15	7.83	7.96	6.57	7.17
Tb	1.04	1.43	0.99	1.48	1.18	1.47	1.53	1.26	1.38
Dy	6.56	9.46	6.28	9.75	7.87	9.74	10.1	8.37	9.11
Ho	1.43	1.93	1.38	2.04	1.66	2.02	2.10	1.76	1.88
Er	4.16	5.62	4.09	6.17	5.06	6.06	6.29	5.31	5.68
Tm	0.607	0.803	0.596	0.899	0.736	0.840	0.942	0.757	0.839
Yb	3.59	4.82	3.50	5.40	4.53	5.19	5.87	4.81	5.26
Lu	0.524	0.680	0.502	0.802	0.701	0.741	0.865	0.709	0.768
Σ REE	56.101	85.113	55.248	83.181	71.317	96.261	98.117	79.426	83.527
Eu/Eu*	1.004	1.016	1.087	1.039	1.016	0.999	0.935	0.981	0.987
$(La/Sm)_N$	0.677	1.091	0.729	0.644	0.667	0.860	0.755	0.740	0.713
$(Gd/Yb)_N$	1.206	1.268	1.171	1.129	1.082	1.203	1.081	1.089	1.087
$(La/Yb)_N$	0.782	1.367	0.833	0.716	0.728	1.050	0.850	0.826	0.772

Note: Oxides are in wt%.Trace and rare earth elements are in parts per million (ppm). Element concentrations expressed with the < sign are below detection limit. LOI—loss on ignition.

TABLE 2. WHOLE-ROCK MAJOR AND TRACE ELEMENT DATA OF HETEROGENEOUS MOECHE UNIT GREENSCHISTS

Sample	CE-2	CE-4	CE-6	CE-8
SiO_2	47.22	50.33	47.81	42.44
Al_2O_3	16.76	16.17	14.05	13.91
Fe_2O_3	15.38	15.74	16.02	20.80
MnO	0.213	0.175	0.257	0.297
MgO	4.11	4.76	10.44	11.16
CaO	6.10	1.21	1.36	2.00
Na_2O	4.63	4.98	1.52	0.39
K_2O	0.07	0.17	<0.01	0.04
TiO_2	2.590	2.505	2.129	2.476
P_2O_5	0.17	0.17	0.19	0.24
LOI	2.85	3.60	6.22	6.54
TOTAL	100.09	99.81	100.01	100.20
Sc	48	48	41	43
V	406	388	368	401
Cr	245	213	193	98
Co	47	49	28	46
Ni	128	57	51	44
Cu	59	47	360	347
Zn	156	143	591	410
Ga	25	21	20	20
Rb	<1	<1	<1	<1
Sr	113	63	22	3
Y	52.2	42.4	96.7	54.0
Zr	170	162	145	158
Nb	5.8	3.1	2.8	2.9
Cs	<0.1	<0.1	0.2	<0.1
Ba	19	23	6	<3
Hf	3.7	5.1	4.5	5.0
Ta	0.16	0.17	0.08	0.15
Pb	<5	<5	<5	10
Th	0.22	0.33	0.26	0.25
U	0.17	0.15	0.11	0.10
La	7.17	5.75	7.60	4.93
Ce	18.0	18.2	18.4	16.9
Pr	2.97	3.12	3.77	2.76
Nd	16.8	17.2	21.3	16.0
Sm	5.55	5.65	7.46	5.65
Eu	2.15	1.98	2.96	2.10
Gd	7.73	6.94	11.6	7.75
Tb	1.49	1.38	2.33	1.55
Dy	9.31	9.33	15.2	10.2
Ho	2.03	1.91	3.15	2.09
Er	6.03	5.95	9.75	6.28
Tm	0.891	0.915	1.37	0.936
Yb	5.29	5.77	7.82	5.71
Lu	0.785	0.845	1.12	0.868
Σ REE	86.196	84.940	113.830	83.724
Eu/Eu*	1.009	0.972	0.978	0.976
$(La/Sm)_N$	0.797	0.628	0.629	0.538
$(Gd/Yb)_N$	1.165	0.959	1.182	1.082
$(La/Yb)_N$	0.906	0.666	0.650	0.577

Note: Oxides are in wt%. Trace and rare earth elements are in ppm. Element concentrations expressed with the < sign are below detection limit. LOI—loss on ignition.

TABLE 3. WHOLE-ROCK MAJOR AND TRACE ELEMENT DATA OF PURRIDO UNIT AMPHIBOLITES

Sample	CE-18	CE-19	CE-20	CE-21	CE-22	CE-23	CE-24
SiO_2	48.54	48.33	48.47	47.83	48.31	48.38	49.33
Al_2O_3	15.49	15.63	15.72	15.67	14.94	15.63	15.53
Fe_2O_3	8.64	10.16	10.45	10.33	10.01	9.34	9.29
MnO	0.160	0.155	0.155	0.140	0.144	0.149	0.117
MgO	8.73	8.86	8.69	8.40	8.96	9.23	8.49
CaO	12.17	10.62	11.22	11.93	12.23	11.60	12.64
Na_2O	2.40	2.93	2.76	2.49	2.10	2.55	2.01
K_2O	0.11	<0.01	0.11	0.08	0.11	0.14	0.03
TiO_2	0.999	0.978	0.960	1.002	0.905	0.846	1.027
P_2O_5	0.09	0.11	0.09	0.14	0.10	0.10	0.12
LOI	1.82	1.46	1.70	1.81	1.67	1.54	1.71
TOTAL	99.15	99.24	100.33	99.82	99.48	99.51	100.29
Sc	37	34	38	37	38	35	37
V	246	281	251	287	285	248	260
Cr	321	392	301	247	391	423	327
Co	31	46	39	40	45	43	37
Ni	85	172	110	99	136	166	90
Cu	33	61	51	23	85	79	20
Zn	<30	71	54	44	48	51	<30
Ga	15	16	16	17	15	15	16
Rb	<1	<1	1	<1	<1	<1	1
Sr	143	206	134	249	181	174	183
Y	24.1	24.5	24.6	28.6	22.4	20.0	24.9
Zr	55	47	44	66	34	33	67
Nb	1.7	2.8	1.3	2.6	2.3	2.4	1.9
Cs	<0.1	<0.1	<0.1	<0.1	<0.1	<0.1	<0.1
Ba	15	24	17	20	18	19	13
Hf	1.8	1.6	1.7	2.1	1.2	1.1	2.0
Ta	0.06	0.08	0.03	0.12	0.05	0.08	0.08
Pb	<5	<5	<5	<5	<5	7	<5
Th	0.25	0.37	0.15	0.59	0.25	0.26	0.22
U	0.08	0.14	0.05	0.19	0.10	0.10	0.08
La	4.05	6.39	2.57	7.39	4.73	4.18	4.48
Ce	11.2	14.0	7.36	18.8	11.0	9.48	12.9
Pr	1.56	1.87	1.11	2.42	1.51	1.30	1.80
Nd	8.18	9.05	6.23	11.3	7.46	6.54	9.29
Sm	2.49	2.47	2.22	3.01	2.14	1.91	2.77
Eu	1.02	0.984	0.955	1.15	0.895	0.808	1.09
Gd	3.40	3.45	3.30	3.85	3.06	2.78	3.56
Tb	0.67	0.63	0.67	0.75	0.59	0.51	0.70
Dy	4.47	4.16	4.47	5.12	3.83	3.32	4.52
Ho	0.93	0.94	0.95	1.09	0.86	0.75	0.98
Er	2.84	2.83	2.89	3.39	2.62	2.31	2.95
Tm	0.428	0.430	0.426	0.505	0.389	0.342	0.437
Yb	2.62	2.51	2.65	3.32	2.35	2.07	2.75
Lu	0.397	0.369	0.413	0.496	0.347	0.309	0.403
Σ REE	44.255	50.083	36.214	62.591	41.781	36.609	48.630
Eu/Eu*	1.078	1.036	1.085	1.038	1.075	1.078	1.067
$(La/Sm)_N$	1.004	1.596	0.714	1.515	1.364	1.350	0.998
$(Gd/Yb)_N$	1.034	1.096	0.993	0.924	1.038	1.071	1.032
$(La/Yb)_N$	1.034	1.702	0.649	1.488	1.346	1.350	1.089

(*continued*)

TABLE 3. (*continued*)

Sample	CE-25	CE-26	CE-27	CE-29	CE-30	CE-31	CE-28[#]
SiO_2	48.14	49.20	49.09	49.18	51.94	48.96	40.43
Al_2O_3	15.37	15.16	15.15	14.84	13.73	15.42	18.05
Fe_2O_3	9.98	10.18	9.83	10.40	10.23	9.66	21.12
MnO	0.135	0.153	0.140	0.181	0.204	0.150	0.181
MgO	9.15	8.80	9.03	8.39	7.84	8.73	8.90
CaO	11.76	11.51	12.72	11.90	10.13	11.75	6.19
Na_2O	2.44	2.56	2.10	2.54	2.81	2.46	1.74
K_2O	0.05	0.03	0.09	0.02	0.05	0.12	0.07
TiO_2	1.079	1.171	0.892	1.027	1.117	1.114	1.028
P_2O_5	0.10	0.10	0.08	0.10	0.11	0.11	0.06
LOI	1.58	1.50	1.21	1.44	1.42	1.75	1.11
TOTAL	99.78	100.36	100.33	100.02	99.58	100.22	98.88
Sc	37	40	40	40	40	39	45
V	272	261	246	267	275	241	296
Cr	397	366	388	243	235	333	279
Co	46	37	39	34	31	36	94
Ni	138	88	89	62	52	84	66
Cu	22	54	24	24	25	43	104
Zn	30	39	<30	72	58	50	49
Ga	15	16	15	16	16	16	25
Rb	1	<1	<1	<1	<1	<1	1
Sr	123	126	126	144	171	132	50
Y	24.9	28.5	23.5	25.4	28.3	26.0	33.4
Zr	41	63	44	60	61	64	75
Nb	2.0	1.5	1.1	1.5	1.5	1.5	1.4
Cs	<0.1	<0.1	0.1	<0.1	<0.1	<0.1	0.1
Ba	9	11	9	12	10	11	11
Hf	1.4	2.2	1.5	1.9	2.0	2.1	2.4
Ta	0.06	0.04	0.02	0.05	0.06	0.33	0.04
Pb	<5	<5	<5	<5	<5	<5	<5
Th	0.11	0.13	0.10	0.16	0.16	0.12	0.25
U	0.05	0.05	0.03	0.06	0.08	0.05	0.17
La	2.70	2.89	2.14	3.40	3.59	2.94	4.17
Ce	7.43	8.95	6.67	10.2	10.5	9.36	13.5
Pr	1.18	1.40	1.03	1.49	1.60	1.40	2.11
Nd	6.51	7.83	5.89	7.97	8.84	7.90	11.8
Sm	2.25	2.79	2.13	2.58	2.96	2.66	3.92
Eu	0.921	1.17	0.935	1.02	1.25	1.12	0.824
Gd	3.23	3.99	3.09	3.46	3.97	3.64	5.12
Tb	0.63	0.77	0.62	0.68	0.77	0.73	0.94
Dy	4.06	5.21	4.25	4.52	5.06	4.84	6.05
Ho	0.93	1.11	0.91	0.98	1.08	1.00	1.28
Er	2.85	3.41	2.77	3.00	3.25	3.03	3.87
Tm	0.418	0.490	0.409	0.442	0.465	0.457	0.569
Yb	2.55	3.00	2.53	2.72	2.94	2.77	3.61
Lu	0.374	0.460	0.393	0.415	0.456	0.422	0.555
Σ REE	36.033	43.470	33.767	42.877	46.731	42.269	58.318
Eu/Eu*	1.050	1.078	1.120	1.049	1.121	1.106	0.565
$(La/Sm)_N$	0.740	0.639	0.620	0.813	0.748	0.682	0.656
$(Gd/Yb)_N$	1.010	1.060	0.974	1.014	1.076	1.047	1.131
$(La/Yb)_N$	0.708	0.644	0.566	0.836	0.817	0.710	0.772

Note: Oxides are in wt%. Trace and rare earth elements are in ppm. Element concentrations expressed with the < sign are below detection limit. LOI—loss on ignition.

[#]Calcium-poor amphibolite.

TABLE 4. WHOLE-ROCK MAJOR AND TRACE ELEMENT DATA OF SHEARED RETROGRESSED PURRIDO UNIT AMPHIBOLITES

Sample	CE-14	CE-15	CE-16	CE-17
SiO_2	48.47	48.40	48.52	48.53
Al_2O_3	13.72	13.52	13.94	13.88
Fe_2O_3	12.99	12.46	11.91	11.07
MnO	0.216	0.191	0.209	0.149
MgO	7.80	7.84	8.01	8.87
CaO	8.55	8.52	8.19	8.54
Na_2O	3.00	3.52	3.66	3.48
K_2O	0.04	0.09	0.09	0.18
TiO_2	1.986	2.106	1.868	1.601
P_2O_5	0.18	0.21	0.26	0.14
LOI	2.93	2.62	2.65	2.96
TOTAL	99.88	99.48	99.31	99.40
Sc	42	43	42	42
V	364	366	340	317
Cr	164	130	232	274
Co	33	33	32	42
Ni	87	97	131	97
Cu	55	41	30	68
Zn	102	81	124	69
Ga	20	20	18	18
Rb	2	<1	<1	<1
Sr	138	130	125	149
Y	48.7	50.6	39.8	38.3
Zr	155	181	109	110
Nb	4.4	4.0	3.3	3.1
Cs	<0.1	<0.1	<0.1	<0.1
Ba	6	9	12	11
Hf	3.5	3.8	2.7	2.8
Ta	0.14	0.13	0.10	0.11
Pb	<5	<5	<5	<5
Th	0.22	0.26	0.14	0.17
U	0.11	0.13	0.08	0.08
La	5.12	5.50	3.77	4.10
Ce	15.2	16.2	11.4	12.0
Pr	2.49	2.64	1.90	1.97
Nd	14.3	15.0	11.2	11.3
Sm	4.90	5.04	3.85	3.81
Eu	1.84	1.87	1.53	1.48
Gd	6.95	7.13	5.60	5.39
Tb	1.30	1.35	1.10	1.03
Dy	8.29	8.58	6.91	6.46
Ho	1.84	1.93	1.52	1.42
Er	5.56	5.83	4.51	4.35
Tm	0.826	0.876	0.664	0.661
Yb	4.90	5.09	3.90	3.87
Lu	0.708	0.752	0.592	0.562
Σ REE	74.224	77.788	58.446	58.403
Eu/Eu*	0.969	0.959	1.013	1.004
$(La/Sm)_N$	0.645	0.673	0.604	0.664
$(Gd/Yb)_N$	1.131	1.117	1.145	1.110
$(La/Yb)_N$	0.699	0.723	0.646	0.708

Note: Oxides are in wt%.Trace and rare earth elements are in ppm. Element concentrations expressed with the < sign are below detection limit. LOI—loss on ignition.

example corresponds to some greenschists of the Moeche unit that exhibit evidence of chemical variation. This is also the case for the retrogressed amphibolites at the base of the Purrido unit and the Ca-poor amphibolites in the same unit. In these lithologies, the elements with more mobile behavior might not presently reflect the protolith composition, and thus great care must be taken to choose the most suitable classification diagrams for these kinds of rocks. It is important to avoid classifications based only on major element (especially Na_2O and K_2O) or large ion lithophile element (LILE) diagrams.

For instance, the TAS (total alkali-silica) classification diagram of volcanic rocks (Le Maitre et al., 1989) must be used with extreme caution. According to this diagram, the most representative metabasites of Moeche and Purrido units can be classified as basalts (Figs. 5A and 5C). The Nb/Y-Zr/TiO_2 diagram is more reliable for classifying metabasites because it is based on immobile elements (Winchester and Floyd, 1977; Pearce, 1996). Accordingly, the greenschists of the Moeche unit plot as basalt and, to a lesser extent, subalkaline basaltic andesite (Fig. 5B). Purrido unit amphibolites plot in a narrower compositional range in the subalkaline basalt field (Fig. 5D).

Rare Earth Elements

The total rare earth element (REE) compositions for greenschists of the Moeche unit range between 55.2 and 98.1 ppm (Table 1), with concentrations between 12 and 31 times chondritic abundances (value of Nakamura, 1974). Their chondrite-normalized REE patterns (Fig. 6A) vary from flat to slightly light (L) REE-enriched ($[La/Yb]_N$ = 0.72–1.37), without significant Eu anomalies (Eu/Eu* = 0.94–1.09; calculated according to Taylor and McLennan, 1985), which are similar to those displayed by normal (N) MORB derived from depleted asthenospheric mantle. The total REE content of heterogeneous greenschist samples ranges from

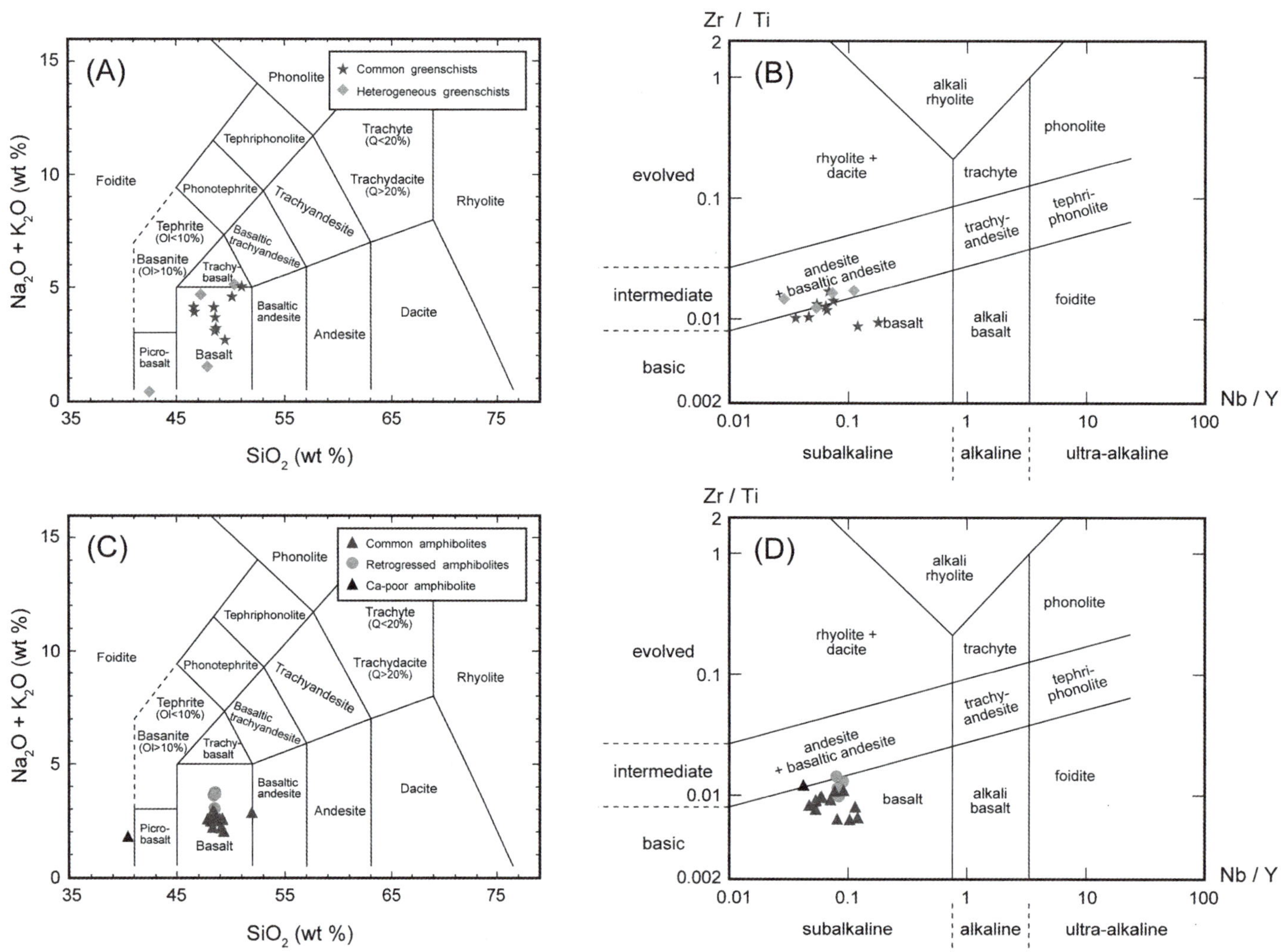

Figure 5. Chemical classification diagrams for the Moeche (A–B) and Purrido (C–D) ophiolites. (A) and (C) Total alkalis versus SiO_2 diagram (Le Maitre et al., 1989). (B) and (D) Nb/Y-Zr/TiO_2 diagram of Winchester and Floyd (1977) modified by Pearce (1996). Ol—olivine, Q—quartz.

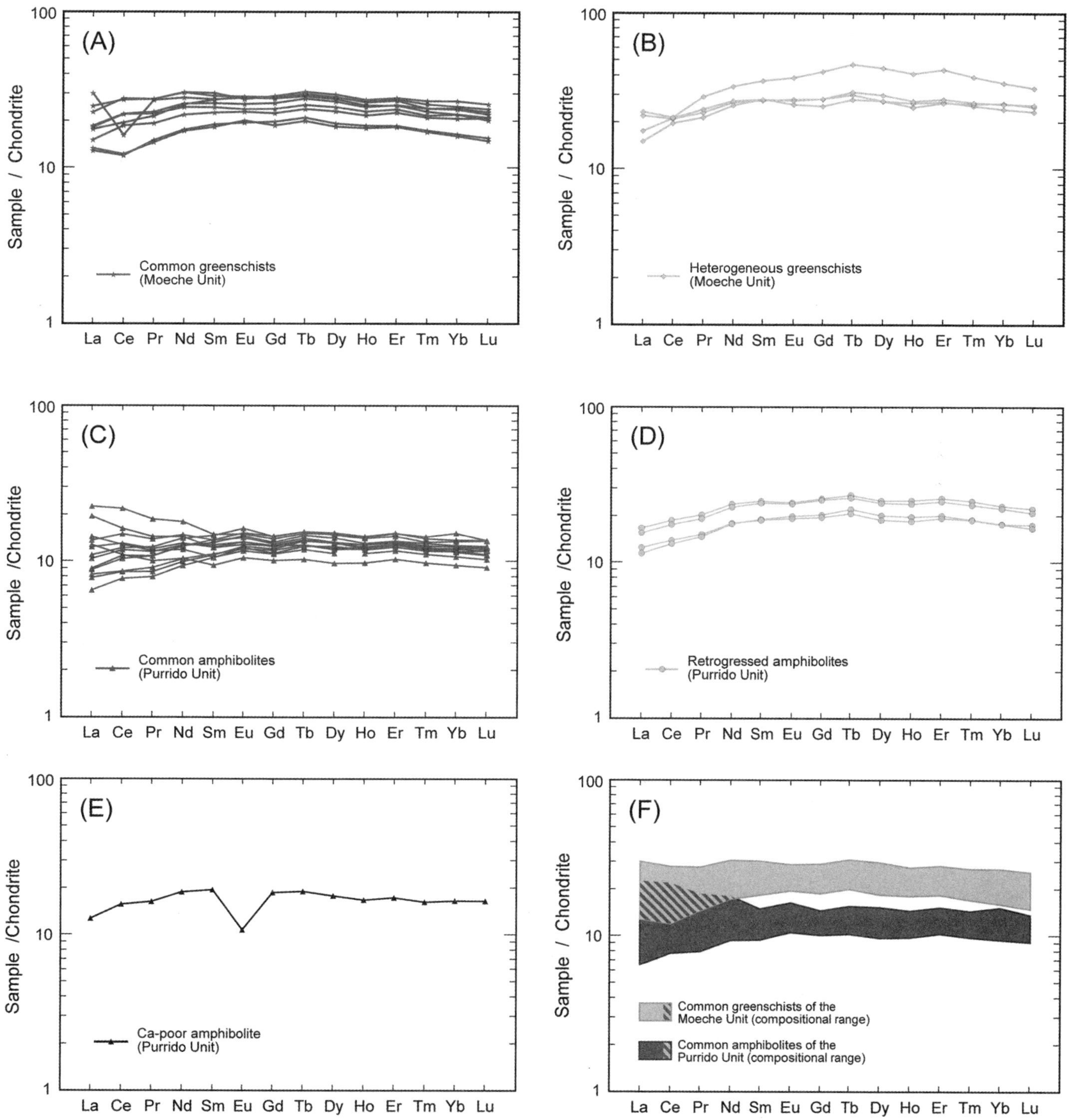

Figure 6. Chondrite-normalized rare earth element (REE) plots corresponding to Moeche (A–B) and Purrido (C–E) units samples and the compositional range of the most representative metabasites of each unit (F). Normalizing values are from Nakamura (1974).

83.7 to 113.8 ppm (Table 2), indicating more enrichment than the common greenschists; their REE concentrations range from 15 to 47 times chondritic abundance. They display generally quite flat chondrite-normalized REE patterns (Fig. 6B) or slightly LREE-depleted patterns compared to heavy (H) REE ($[La/Yb]_N$ = 0.58–0.91), without significant Eu anomalies (Eu/Eu* = 0.97–1.01).

Common amphibolites of the Purrido unit are slightly more REE-depleted than the metabasites of the Moeche unit (Fig. 6F), with total REE values ranging from 33.8 to 62.6 ppm (Table 3). They exhibit concentrations between 6 and 22 times chondritic abundance, with almost flat HREE patterns ($[Gd/Yb]_N$ = 0.92–1.13), and slightly enriched or depleted LREE ($[La/Sm]_N$ = 0.62–1.60) (Fig. 6C); they also have slightly positive Eu anomalies (Eu/Eu* = 1.04–1.12). The REE total content of the retrogressed amphibolites at the base of the Purrido unit (58.4–77.8 ppm, (Table 4)) is slightly higher than in the common amphibolites. Their concentrations range from 11 to 27 times chondritic abundance, with almost flat chondrite-normalized REE patterns (Fig. 6D), although they are slightly LREE-depleted ($[La/Yb]_N$ = 0.65–0.72) and do not have Eu anomalies (Eu/Eu* = 0.96–1.01). Ca-poor amphibolite exhibits concentrations between 11 and 19 times chondritic abundance. Its REE plot (Fig. 6E) displays a flat HREE ($[Gd/Yb]_N$ = 1.13) pattern and has a slope indicating a slight depletion in LREE related to HREE ($[La/Sm]_N$ = 0.66). It has a large negative Eu anomaly (Eu/Eu* = 0.56), which clearly sets apart the Ca-poor amphibolite from the other mafic

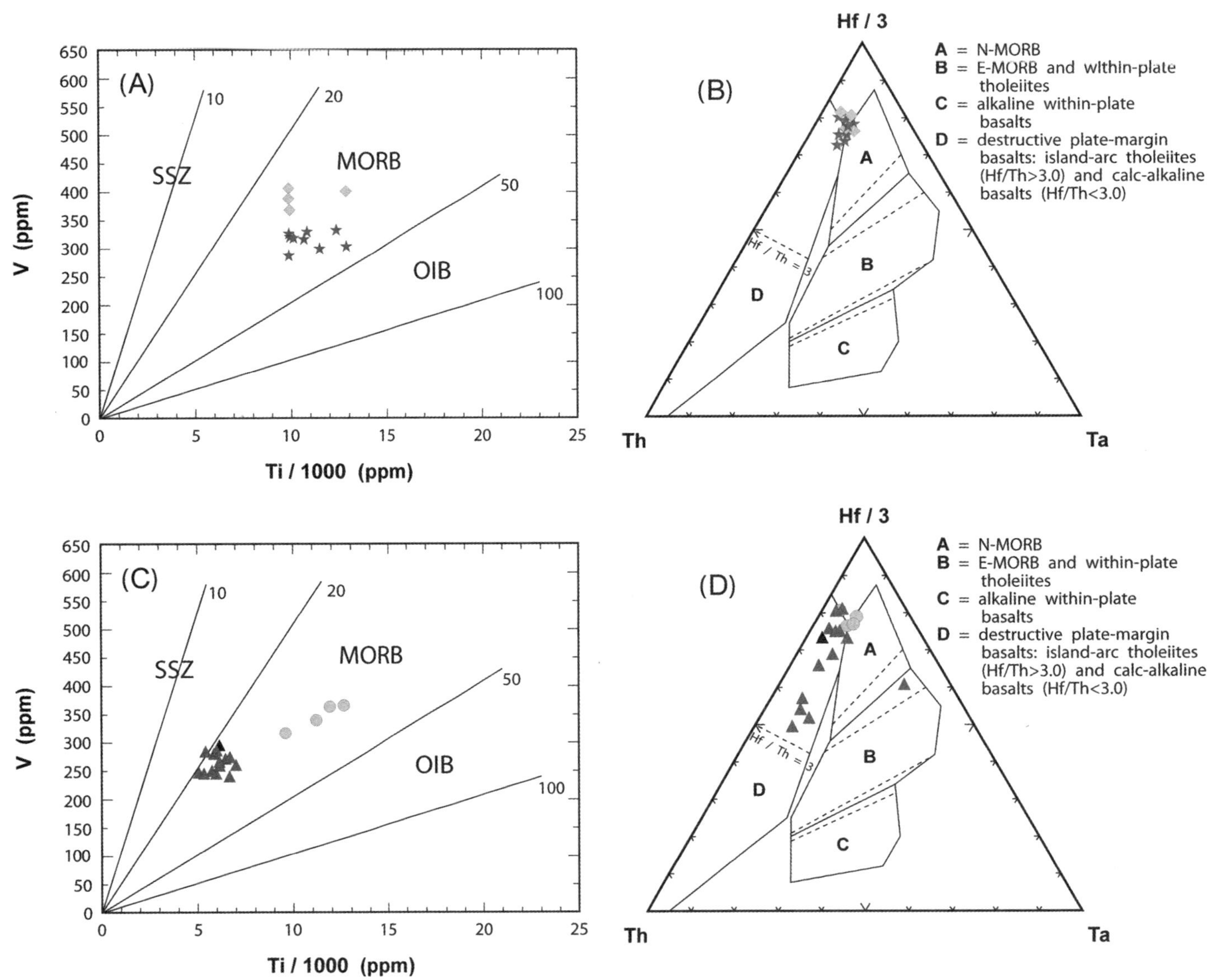

Figure 7. Trace-element tectonic discrimination diagrams of the Moeche (A–B) and Purrido (C–D) metabasites. (A) and (C) Ti-V diagram (after Shervais, 1982), (B) and (D) Th-Hf-Ta diagram (Wood, 1980). Symbols are as in Figure 5. SSZ—suprasubduction zone, MORB—mid-ocean-ridge basalt (E—enriched, N—normal), OIB—ocean-island basalt.

lithologies of the unit. Eu is compatible with plagioclase, and the decrease of this calcic phase in the Ca-poor amphibolites could justify the development of the strong anomaly they exhibit relative to this element.

Trace Elements and Tectonic Setting of Generation

Geochemical fingerprinting of ophiolite assemblages is one of the main tools to identify their original tectonic setting, but this fingerprinting only turns out to be efficient in altered or metamorphosed rocks if it is performed using immobile trace elements (Cann, 1970). With this aim, samples of the most representative lithologies are plotted in discrimination diagrams based on these kinds of elements, and correlations are established with lithologies of well-known tectonic setting, such as N-MORB, within-plate, or other basaltic types.

Shervais's diagram (1982), one of the most commonly used to distinguish between different kinds of ophiolites, is based on plotting two immobile elements; one incompatible (Ti) and another compatible (V), which indicate the degree of melting as well as the existence of a subduction component. Subduction-related basalts have both higher Ti and V, which allows them to be distinguished from those generated in other tectonic settings. A plot of the Moeche unit greenschist samples on the Ti-V diagram (Fig. 7A) indicates that all plot in the MORB field because their Ti/V ratio is higher than 20 (Ti/V = 30.3–42.3). The same

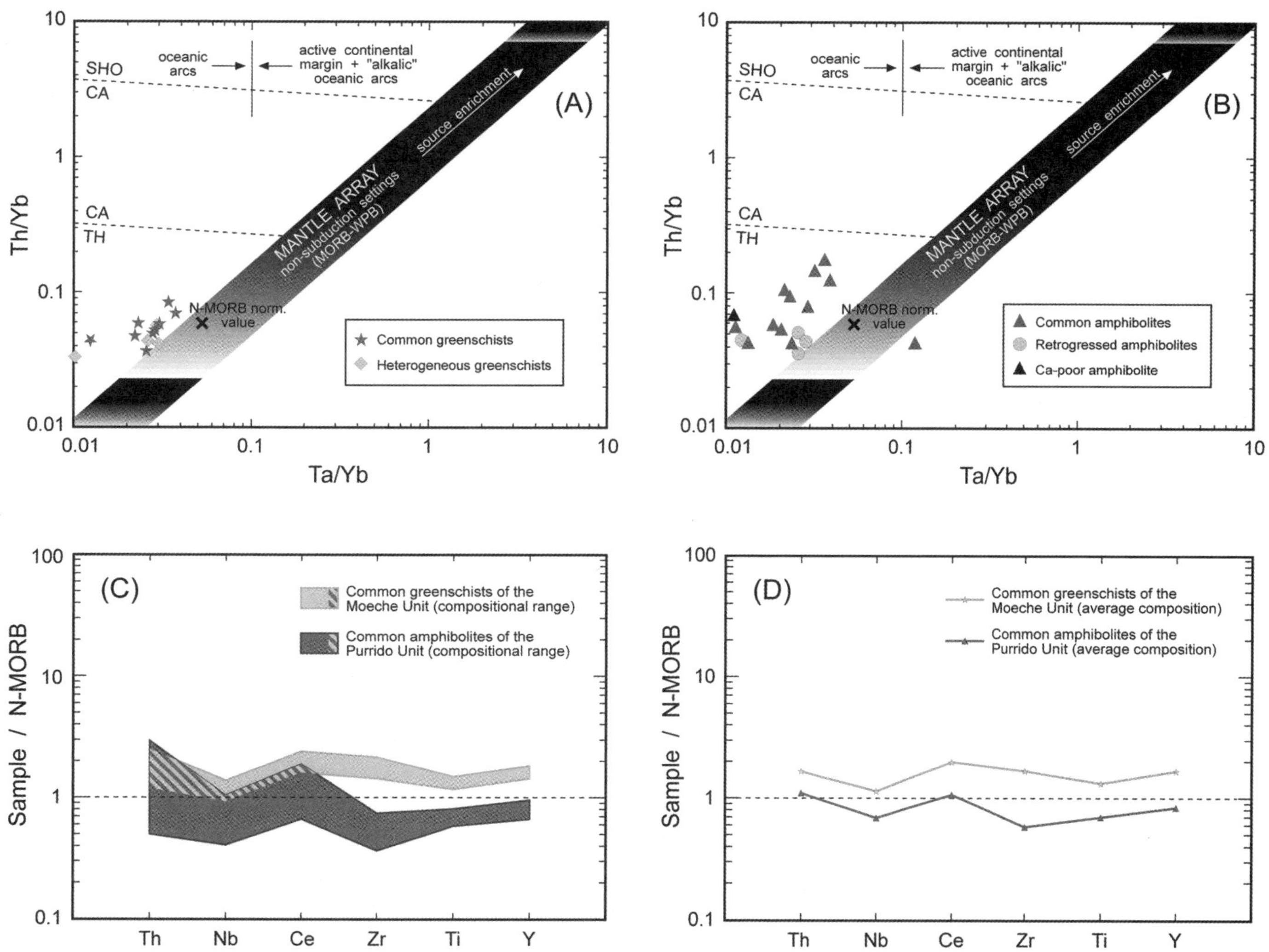

Figure 8. Incompatible element ratio plot of Ta/Yb-Th/Yb (after Pearce, 1983) for Moeche (A) and Purrido (B) samples. (C) and (D) Normal mid-ocean-ridge basalt (N-MORB)–normalized trace-element patterns of the most representative metabasites of Moeche and Purrido ophiolitic units, corresponding to the compositional range for the samples and their average compositions, respectively. Normalizing values are from Pearce (1996). SHO—shoshonite, CA—calc-alkaline, TH—tholeiite, WPB—within-plate basalts.

happens with the Purrido unit amphibolites (Fig. 7C), although their Ti/V ratio is closer to that of subduction-related basalt, particularly the common amphibolites (Ti/V = 19.0–27.7).

Another especially useful diagram to recognize subduction-related lavas (Wood, 1980) is based on three high field strength elements (HFSE), Th, Hf, and Ta. Th is an element with immobile behavior in subduction zones, so the lavas generated in this tectonic setting are enriched in Th. Furthermore, Th is also immobile during alteration and metamorphism, up to near melting temperatures (Johnson and Plank, 1999). Thus, subduction-related lavas are enriched in Th compared to Ta and Hf. Plots of the Moeche samples in this diagram (Fig. 7B) suggest that they have a transitional composition between island-arc tholeiite and N-MORB, whereas the Purrido samples exhibit a clearer arc affinity (Fig. 7D).

The Th/Yb-Ta/Yb diagram (Pearce, 1982) (Figs. 8A and 8B) permits estimation of the relative contributions of subduction and mantle components in the generation of basaltic magmas, which also provide information about tectonic setting. In this diagram, Yb is used as a normalizing factor because it reduces variations related to partial melting and fractional crystallization that the other two elements have undergone. On the other hand, Yb is an element unaffected by other enrichment processes, such as those related to dehydration of oceanic crust at subduction zones. In principle, the processes that involve only the subcontinental mantle equally impact Th and Ta concentration, and the basic rocks generated in this way are projected along the mantle array. Nevertheless, the subduction processes entail a preferential Th enrichment that can be even stronger if crustal contamination exists. Moeche unit greenschist samples plot close to the mantle array in the Th/Yb-Ta/Yb diagram (Fig. 8A), but their Ta/Yb ratios (Ta/Yb_{avg} = 0.026) are slightly lower than those of typical N-MORB, so that they are slightly displaced toward the island-arc tholeiite field. The Purrido amphibolites exhibit higher Th/Yb ratios than the Moeche metabasites (Th/Yb_{avg} = 0.073 in Purrido, compared to 0.052 in Moeche), which are different from average N-MORB ratios, and they also exhibit Ta/Yb ratios lower than N-MORB, which remove them from the mantle array and locate them mostly within the island-arc tholeiite field (Fig. 8B). On the other hand, the Th/Yb-Ta/Yb diagram suggests that the mantle source for the basic rocks of both units is similar to that of N-MORB, although some influence of subduction zone activity is evident, slight for the Moeche unit and more obvious for the Purrido unit.

Finally, trace-element abundance diagrams normalized to the average composition of a rock of known origin permit more accurate determination of the tectonic setting of an igneous suite. The standard diagrams of this kind (Pearce, 1982; Sun and McDonough, 1989; Pearce and Parkinson, 1993) are not the most suitable to study variably metamorphosed rocks because most of their trace elements exhibit mobile behavior under these conditions, especially the LILEs (Sr, K, Rb, Ba). An effective trace-element abundance diagram for metamorphic rocks can be obtained by plotting a few elements with the most immobile behavior during alteration and metamorphism and with the highest discriminating power. Pearce (1996) devised a diagram that fulfills these conditions, based only on six trace HFSE: Th, Nb, Ce, Zr, Ti, and Y. These elements appear arranged according to their degree of incompatibility with garnet lherzolite, increasing from right to left (Pearce, 1982), and the normalizing factor used is the average N-MORB composition in these elements (Pearce, 1983, 1996). The abundance diagrams for the Moeche and Purrido metabasites were plotted using only samples with chemical composition unaffected by alteration, so they more accurately reflect the chemistry of the original igneous protolith (Figs. 8C and 8D). Both the compositional range of the samples from each unit (Fig. 8C) and the average composition of the Moeche and Purrido common metabasites (Fig. 8D) are represented.

The compositional range for all trace elements in Figure 8C is quite narrow for the Moeche metabasites, and it is slightly enriched compared to the average N-MORB composition. The N-MORB–normalized trace-element pattern for the average composition of the Moeche greenschists is close to the average N-MORB concentration, although it is slightly enriched in most of the trace elements (Fig. 8D). Most significant is the presence of a small negative Nb anomaly, which indicates that the origin of this ophiolite is to some extent related to subduction zone activity. Regarding the Purrido metabasites, their trace-element compositional range is somewhat wider, although it becomes narrower from Th to Y. Th, Nb, and Ce concentrations are similar to those of N-MORB, whereas Zr, Ti, and Y are variably depleted. Unlike the Moeche unit metabasites, the N-MORB–normalized trace-element pattern corresponding to the average Purrido amphibolite composition is slightly depleted in Nb, Zr, Ti, and Y relative to N-MORB, although it is still quite flat and close to unit (Fig. 8D). In this case, there is also a somewhat steeper negative Nb anomaly, again suggesting a subduction-related origin.

DISCUSSION

Different kinds of data can provide significant information that can clarify the tectonic setting of ophiolite generation. These include lithologic composition and pseudostratigraphy (the relationships among its different parts or members), the presence and nature of the accompanying sediments, the location of the ophiolite, its structural relationships with the surrounding terranes, and the nature and origin of these terranes. However, determining the origin of the ophiolitic assemblages obducted in the most internal part of orogens may be an especially arduous task because of the difficulties created by a generally complex tectonothermal evolution and the uncertain nature of associated exotic terranes. In these cases, accurate geochemical characterization of the ophiolite components is essential in order to determine their tectonic origin.

Two fundamental kinds of ophiolites can be distinguished based on their geochemical characteristics: MORB-type ophiolites, the origins of which are related to divergent plate margins, and suprasubduction zone ophiolites, which are related to

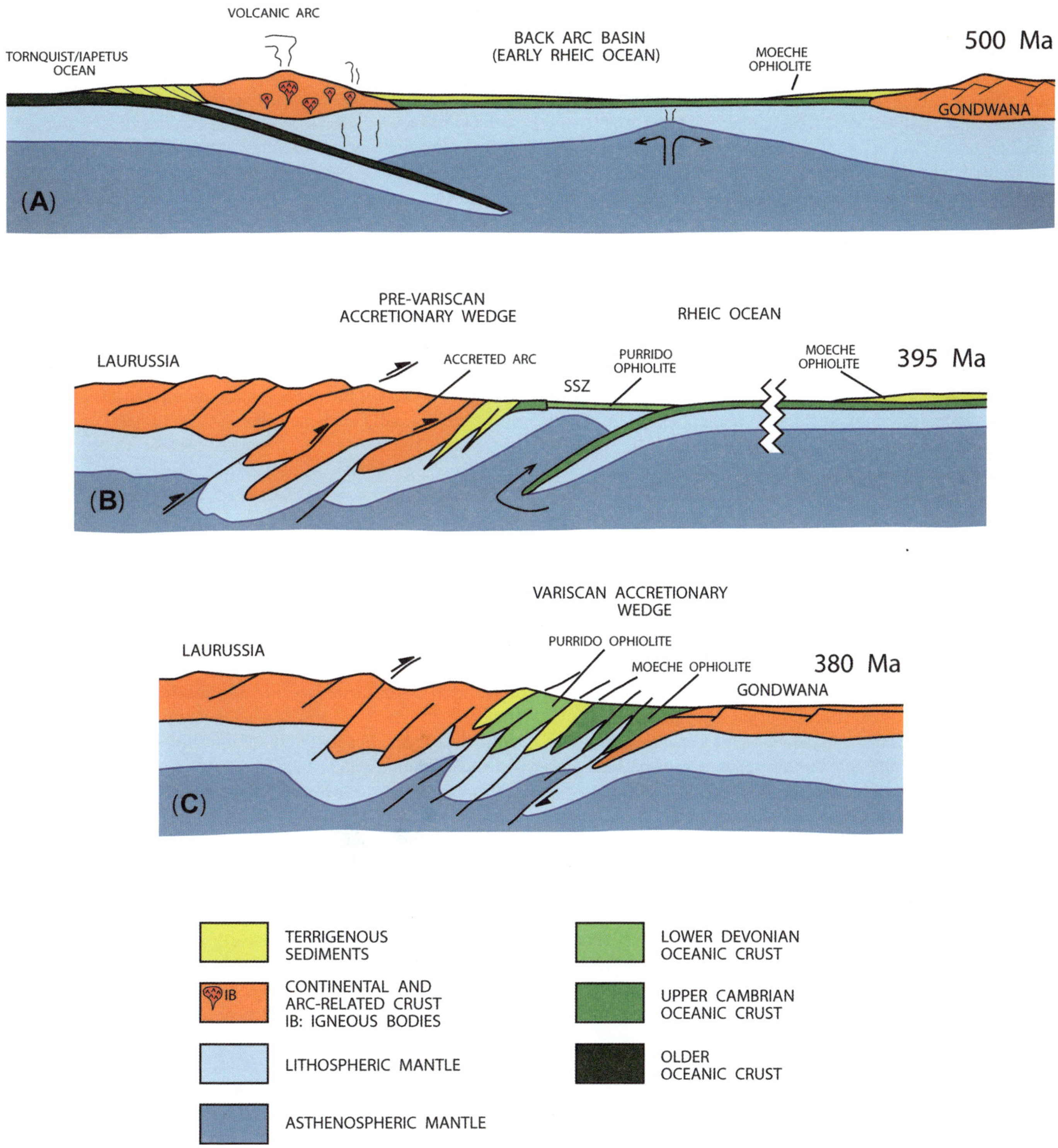

Figure 9. Cartoon showing the main steps in the tectonic evolution and later accretion of the Moeche and Purrido ophiolites to the Variscan orogenic wedge. Additional explanations are in the text. SSZ—suprasubduction zone.

convergent plate margins (Pearce et al., 1984; Hawkins, 2003; Pearce, 2003). Nevertheless, sometimes, the tectonic regime where new crust is generated may be quite complex, giving rise to magmas with geochemical features transitional between the two principal types of ophiolites previously mentioned (Pearce, 1996). This could be the case of the two ophiolitic units studied in this contribution whose compositions do not fit the ideal characteristics of either MORB or suprasubduction zone types. However, when lithologic composition, geochemical features, and relationships with the surrounding exotic terranes of northwest Iberia are considered, a clear tectonic setting for their generation can be proposed.

The Moeche unit, which essentially consists of mafic mylonitic greenschist with phyllite and mica schist interlayers, probably derives from a sequence of basaltic rocks alternating with pelitic and graywacke sedimentary layers. Although the metabasites are strongly deformed, the geochemical signature of the igneous protoliths is still clearly recognizable, so that they can be classified as tholeiitic basalts. The discrimination diagrams based on immobile trace elements (Wood, 1980; Shervais, 1982; Pearce, 1983) do not shed much light on the precise tectonic setting of this ophiolite, since they clearly show a rather transitional chemistry between N-MORBs and island-arc tholeiites. However, the N-MORB–normalized trace-element patterns (Pearce, 1996) are more illuminating, and they suggest that the metabasites of the Moeche unit are comparable to normal oceanic basalts (N-MORB), although minor trace-element enrichment and a small negative Nb anomaly points out that their generation was influenced to some extent by a subduction zone.

The Purrido unit entirely consists of amphibolites derived from a gabbro protolith. It does not include metasedimentary rocks, except for the ones that belong to the Carreiro shear zone, located at the top of the unit, and the relationships of these with the ophiolite components are presently uncertain. The major- and trace-elements content of the Purrido amphibolites also reveal that they are equivalent to tholeiitic basalts. In this instance, the discrimination diagrams are somewhat more definitive for the tectonic setting of the ophiolite, and they suggest a clearer arc affinity. Their N-MORB–normalized trace-element pattern is quite close to that of oceanic basalt, but, unlike the Moeche unit metabasites, it is slightly depleted in some elements (Nb, Zr, Ti, Y) and also exhibits a small negative Nb anomaly, both of which are found in basalt generated in some back-arc basins in suprasubduction zone settings (Pearce, 1996).

The lithology, geochemistry, and age of both ophiolites can be integrated with the regional context of northwest Iberia exotic terranes to formulate a model that links them genetically and provides an explanation for their spatial association (Fig. 9). The Moeche ophiolite was probably generated during the opening of a back-arc basin in the peri-Gondwanan realm during the Late Cambrian (ca. 500 Ma; Fig. 9A). The opening of this marginal basin led to the pulling apart of the terrane represented by the upper units of the allochthonous complexes of the northwest Iberian Massif with arc affinities (Arenas et al., 1986; Abati et al., 1999; Castiñeiras, 2003). No magmatic activity occurred subsequent to ca. 500 Ma in the upper units. This is likely the age of the Moeche ophiolite, which suggests fast opening of the marginal basin that interrupted the volcanic activity in the drifting terrane. Taking into account the timing considered reasonable for opening of this back-arc basin in the peri-Gondwanan realm, as well as the projected location of the terranes that presently constitute southern Europe at that time, the basin probably evolved into the Rheic Ocean (Scotese, 2001; Stampfli and Borel, 2002). The presence of abundant clastic sediments in the Moeche unit raises the likelihood of a nearby emergent source area, and hence this ophiolite could be related to the initial stages of the opening of the ocean basin (Fig. 9A).

Later, during Late Silurian–Early Devonian times, intraoceanic subduction of the dense cold lithosphere represented by the Moeche unit started (Fig. 9B). According to Molnar and Atwater (1978), subduction of dense cold oceanic lithosphere favors the retreat of the upper plate and promotes an extensional context that finally leads to the generation of new oceanic lithosphere. We suggest that the Purrido ophiolite was generated by this process, which allows us to explain its structural position in the orogenic belt, along with its age (ca. 395 Ma) and geochemistry, relative to suprasubduction zone ophiolites (Fig. 9B). The new ocean lithosphere, younger and lighter than that represented by the Moeche unit, would have been rapidly accreted to the orogenic wedge, explaining the closeness in time between the protoliths and metamorphic ages of the Purrido and Careón units (Díaz García et al., 1999; Peucat et al., 1990). On the other hand, the lack of sediments in the Purrido unit may indicate the ephemeral character of intraoceanic subduction, which did not generate an island arc and was soon stopped by the advancing orogenic wedge attached to the Laurussian margin (Fig. 9C).

According to our data, the Purrido ophiolite preserves information of the late history of the Rheic Ocean (see also Díaz García et al., 1999; Pin et al., 2002). If this unit was the equivalent to other ophiolites considered the most characteristic of the European Variscan belt (e.g., Ślęża ophiolite in the Bohemian Massif, Lizard ophiolite in the Lizard Complex, and Beja-Acebuches and Careón ophiolites in the southwestern and the northwestern Iberian Massif, respectively) as previously suggested, these oceanic units associated with the Variscan suture might only have witnessed the most recent evolutionary stages of the extinct Rheic Ocean basin. Moreover, as is common in other orogenic systems, many ophiolite units preserved in Europe mostly reflect rather late suprasubduction histories, and rarely the old, cold oceanic lithosphere of the closed ocean. However, this old oceanic lithosphere would probably be represented by the Moeche unit, which may hold the key data for the earliest evolution of the Rheic Ocean. Units equivalent to the Moeche ophiolite have not been described in the European Variscan belt until now. Even though this fact reflects the expected poor preservation due to its ability to disappear easier by subduction, the information supplied by the northwest Iberian Massif illustrates the importance of studying ophiolites made of very contrasted lithological assemblages. These ophiolites can be complex and include highly diachronous sections, but they may hold the keys to the long-term evolution of large oceans.

ACKNOWLEDGMENTS

The authors sincerely acknowledge the review of the first version of the manuscript by Thomas B. Hanley. We also kindly thank Robert Hatcher Jr. for the editorial help and careful review. This research has been funded by the Spanish projects BTE2001-0963-CO2 and CGL2004-0463-CO2/BTE (Ministerio de Educación y Ciencia, Dirección General de Investigación).

REFERENCES CITED

Abati, J., Dunning, G.R., Arenas, R., Díaz García, F., González Cuadra, P., Martínez Catalán, J.R., and Andonaegui, P., 1999, Early Ordovician orogenic event in Galicia (NW Spain): Evidence from U-Pb ages in the uppermost unit of the Ordenes Complex: Earth and Planetary Science Letters, v. 165, p. 213–228, doi: 10.1016/S0012-821X(98)00268-4.

Abati, J., Arenas, R., Martínez Catalán, J.R., and Díaz García, F., 2003, Anticlockwise *P-T* path of granulites from the Monte Castelo Gabbro (Órdenes Complex, NW Spain): Journal of Petrology, v. 44, p. 305–327, doi: 10.1093/petrology/44.2.305.

Andonaegui, P., González del Tánago, J., Arenas, R., Abati, J., Martínez Catalán, J.R., Peinado, M., and Díaz García, F., 2002, Tectonic setting of the Monte Castelo Gabbro (Órdenes Complex, NW Iberian Massif): Evidence for an arc-related terrane in the hanging wall to the Variscan suture, *in* Martínez Catalán, J.R., Hatcher, R.D., Jr., Arenas, R., and Díaz García, F., eds., Variscan-Appalachians Dynamics: The Building of the Late Paleozoic Basement: Geological Society of America Special Paper 364, p. 37–56.

Anonymous, 1972, Penrose field conference on ophiolites: Geotimes, v. 17, p. 24–25.

Arenas, R., 1988, Evolución petrológica y geoquímica de la unidad alóctona inferior del complejo metamórfico básico-ultrabásico de Cabo Ortegal (Unidad de Moeche) y del Silúrico parautóctono, Cadena Hercínica Ibérica (NW de España): Corpus Geologicum Gallaecia, v. IV, 543 p.

Arenas, R., 1991, Opposite *P,T,t* paths of Hercynian metamorphism between the upper units of the Cabo Ortegal Complex and their substratum (northwest of the Iberian Massif): Tectonophysics, v. 191, p. 347–364, doi: 10.1016/0040-1951(91)90067-3.

Arenas, R., Gil Ibarguchi, J.I., González Lodeiro, F., Klein, E., Martínez Catalán, J.R., Ortega Gironés, E., Pablo Maciá, J.G., and Peinado, M., 1986, Tectonostratigraphic units in the complexes with mafic and related rocks of the NW of the Iberian Massif: Hercynica, v. II, p. 87–110.

Arenas, R., Rubio Pascual, F.J., Díaz García, F., and Martínez Catalán, J.R., 1995, High-pressure micro-inclusions and development of an inverted metamorphic gradient in the Santiago schists (Órdenes Complex, NW Iberian Massif, Spain): Evidence of subduction and syncollisional decompression: Journal of Metamorphic Geology, v. 13, p. 141–164.

Arenas, R., Abati, J., Martínez Catalán, J.R., Díaz García, F., and Rubio Pascual, F.J., 1997, *P-T* evolution of eclogites from the Agualada unit (Órdenes Complex, NW Iberian Massif, Spain): Implications for crustal subduction: Lithos, v. 40, p. 221–242, doi: 10.1016/S0024-4937(97)00029-7.

Arenas, R., Martínez Catalán, J.R., Sánchez Martínez, S., García, F., Abati, J., Fernández- Arenas, R., Martínez Catalán, J.R., Sánchez Martínez, S., Fernández-Suárez, J., Andonaegui, P., and Corfu, F., 2007a, The Vila de Cruces Ophiolite: A remnant of the early Rheic Ocean in the Variscan suture of Galicia (NW Iberian Massif): Journal of Geology, v. 115, p. 129–148.

Arenas, R., Martínez Catalán, J.R., Sánchez Martínez, S., García, F., Abati, J., Fernández-Arenas, R., Martínez Catalán, J.R., Sánchez Martínez, S., Fernández-Suárez, J., Andonaegui, P., and Gómez-Barreiro, J., 2007b, this volume, Paleozoic ophiolites in the Variscan suture of Galicia (northwest Spain): Distribution, characteristics, and meaning, *in* Hatcher, R.D., Jr., Carlson, M.P., McBride, J.H., and Martínez Catalán, J.R., eds., 4-D Framework of Continental Crust: Geological Society of America Memoir 200, doi: 10.1130/2007.1200(22).

Azcárraga, J., 2000, Evolución tectónica y metamórfica de los mantos inferiores de grado alto y alta presión del complejo de Cabo Ortegal: A Coruña, Nova Terra, v. 17, 346 p.

Boudier, F., and Nicolas, A., 1985, Harzburgite and lherzolite subtypes in ophiolitic and oceanic environments: Earth and Planetary Science Letters, v. 76, p. 84–92, doi: 10.1016/0012-821X(85)90150-5.

Cann, J.R., 1970, Rb, Sr, Y, Zr and Nb in some ocean-floor basaltic rocks: Earth and Planetary Science Letters, v. 10, p. 7–11, doi: 10.1016/0012-821X(70)90058-0.

Castiñeiras, P., 2003, Origen y Evolución Tectonotermal de las Unidades de O Pino y Cariño (Complejos Alóctonos de Galicia) [Ph.D. thesis]: Madrid, Spain, Universidad Complutense, 281 p.

Castro, A., Fernández, C., De la Rosa, J.D., Moreno Ventas, I., and Rogers, G., 1996, Significance of MORB-derived amphibolites from the Aracena metamorphic belt: Southwest Spain: Journal of Petrology, v. 37, p. 235–260, doi: 10.1093/petrology/37.2.235.

Castroviejo, R., Armstrong, E., Lago, A., Martínez Simón, J.M., and Argüelles, A., 2004, Geología de las mineralizaciones de sulfuros masivos en los cloritoesquistos de Moeche (Complejo de Cabo Ortegal, La Coruña): Boletín Geológico y Minero, v. 115, p. 4–32.

Clark, A.H., Scott, D.J., Sandeman, H.A., Bromley, A.V., and Farrar, E., 1998, Siegenian generation of the Lizard ophiolite: U-Pb zircon age data for plagiogranite, Porthkerries, Cornwall: Journal of the Geological Society of London, v. 155, p. 595–598.

Crespo-Blanc, A., and Orozco, M., 1991, The boundary between the Ossa-Morena and South Portuguese zones (southern Iberian Massif): A major suture in the European Hercynian Chain: Geologische Rundschau, v. 80, p. 691–702, doi: 10.1007/BF01803695.

Crowley, Q.G., Floyd, P.A., Štědrá, V., Winchester, J., Kachlík, V., and Holland, G., 2002, The Mariánské Lázně Complex (northwestern) Bohemian Massif: Development and destruction of an early Paleozoic Saxothuringian seaway: Geological Society of London Special Publication 201, p. 157–177.

Dallmeyer, R.D., Martínez Catalán, J.R., Arenas, R., Gil Ibarguchi, J.I., Gutiérrez Alonso, G., Farias, P., Bastida, F., and Aller, J., 1997, Diachronous Variscan tectonothermal activity in the NW Iberian Massif: Evidence from $^{40}Ar/^{39}Ar$ dating of regional fabrics: Tectonophysics, v. 277, p. 307–337, doi: 10.1016/S0040-1951(97)00035-8.

Díaz García, F., Arenas, R., Martínez Catalán, J.R., González del Tánago, J., and Dunning, G.R., 1999, Tectonic evolution of the Careón ophiolite (northwest Spain): A remnant of oceanic lithosphere in the Variscan belt: The Journal of Geology, v. 107, p. 587–605, doi: 10.1086/314368.

Dilek, Y., 2003, Ophiolite concept and its evolution, *in* Dilek, Y., and Newcomb, S., eds., Ophiolite Concept and the Evolution of Geological Thought: Geological Society of America Special Paper 373, p. 1–16.

Dubińska, E., Bylina, P., Kozlowski, A., Dörr, W., and Nejbert, K., 2004, U-Pb dating of serpentinization: Hydrothermal zircon from a metasomatic rodingite shell (Sudetic ophiolite, SW Poland): Chemical Geology, v. 203, p. 183–203, doi: 10.1016/j.chemgeo.2003.10.005.

Engels, J.P., 1972, The catazonal polymetamorphic rocks of Cabo Ortegal (NW Spain): A structural and petrofabric study: Leidse Geologische Mededelingen, v. 48, p. 83–133.

Fernández, F.J., 1997, Estructuras Desarrolladas en Gneises Bajo Condiciones de Alta P y T (Gneises de Chimparra, Cabo Ortegal, A Coruña, Galicia, España): A Coruña, Nova Terra, v. 13, 249 p.

Fernández-Pompa, F., Montserín, V., and Peinado Moreno, M., 1976, Mapa Geológico de España, Hoja no. 7 (Cedeira): Madrid, Instituto Geológico y Minero de España, escala 1:50,000.

Fernández-Suárez, J., Corfu, F., Arenas, R., Marcos, A., Martínez Catalán, J.R., Díaz García, F., Abati, J., and Fernández, F.J., 2002, U-Pb evidence for a polyorogenic evolution of the HP-HT units of the NW Iberian Massif: Contributions to Mineralogy and Petrology, v. 143, p. 236–253, doi: 10.1007/s00410-001-0337-2.

Fernández-Suárez, J., Díaz García, F., Jeffries, T.E., Arenas, R., and Abati, J., 2003, Constraints on the provenance of the uppermost allochthonous terrane of the NW Iberian Massif: Inferences from detrital zircon U-Pb ages: Terra Nova, v. 15, p. 138–144, doi: 10.1046/j.1365-3121.2003.00479.x.

Fonseca, P., and Ribeiro, A., 1993, Tectonics of the Beja-Acebuches ophiolite: A major suture in the Iberian Variscan fold-belt: Geologische Rundschau, v. 82, p. 440–447, doi: 10.1007/BF00212408.

Franke, W., 1989, Tectonostratigraphic units in the Variscan belt of central Europe, *in* Dallmeyer, R.D., ed., Terranes in the Circum-Atlantic Paleozoic Orogens: Geological Society of America Special Paper 230, p. 67–90.

Franke, W., 2000, The mid-European segment of the Variscides: Tectono-stratigraphic units, terranes boundaries and plate tectonic evolution, *in* Franke, W., Haak, V., Oncken, O., and Tanner, D., eds., Orogenic Processes: Quantification and Modelling in the Variscan Belt: Geological Society [London] Special Publication 179, p. 35–61.

Gil Ibarguchi, J.I., Mendia, M., Girardeau, J., and Peucat, J.J., 1990, Petrology of eclogites and clinopyroxene garnet metabasites from the Cabo Ortegal Complex (northwestern Spain): Lithos, v. 25, p. 133–162, doi: 10.1016/0024-4937(90)90011-O.

Gómez-Barreiro, J., Wijbrans, J.R., Castiñeiras, P., Martínez Catalán, J.R., Arenas, R., Díaz García, F., and Abati, J., 2006, $^{40}Ar/^{39}Ar$ laserprobe dating of mylonitic fabrics in a polyorogenic terrane of the NW Iberia: Geological Society [London] Journal, v. 163, p. 61–73, doi: 10.1144/0016-764905-012.

Hawkins, J.W., 2003, Geology of supra-subduction zones—Implications for the origin of ophiolites, *in* Dilek, Y., and Newcomb, S., eds., Ophiolite Concept and the Evolution of Geological Thought: Geological Society of America Special Paper 373, p. 227–268.

Humphris, S.E., and Thompson, G., 1978, Trace element mobility during hydrothermal alteration of oceanic basalts: Geochimica et Cosmochimica Acta, v. 42, p. 127–136, doi: 10.1016/0016-7037(78)90222-3.

Irvine, T.N., and Baragar, W.R.A., 1971, A guide to the chemical classification of the common volcanic rocks: Canadian Journal of Earth Sciences, v. 8, p. 523–548.

Johnson, M.C., and Plank, T., 1999, Dehydration and melting experiments constrain the fate of subducted sediments: Geochemistry, Geophysics, Geosystems, v. 1, 1999GC000014.

Kastl, E., and Tonika, J., 1984, The Mariánské Lázně Metaophiolite Complex (west Bohemia): Krystalinikum, v. 17, p. 59–76.

Le Maitre, R.W., Bateman, P., Dudek, A., Keller, J., Lameyre Le Bas, M.J., Sabine, P.A., Schmid, R., Sorensen, H., Streckeisen, A., Wooley, A.R., and Zanettin, B., 1989, A Classification of Igneous Rocks and Glossary of Terms: Oxford, Blackwell Scientific Publications, 193 p.

Marcos, A., Marquínez, J., Pérez-Estaún, A., Pulgar, J.A., and Bastida, F., 1984, Nuevas aportaciones al conocimiento de la evolución tectonometamórfica del Complejo de Cabo Ortegal (NW de España): Cuadernos del Laboratorio Xeolóxico de Laxe, v. 7, p. 125–137.

Marcos, A., Farias, A., Galán, G., Fernández, F.J., and Llana-Fúnez, S., 2002, Tectonic framework of the Cabo Ortegal Complex: A slab of lower crust exhumed in the Variscan orogen (northwestern Iberian Peninsula), *in* Martínez Catalán, J.R., Hatcher, R.D., Jr., Arenas, R., and Díaz García, F., eds., Variscan-Appalachians Dynamics: The Building of the Late Paleozoic Basement: Geological Society of America Special Paper 364, p. 143–162.

Martínez Catalán, J.R., Arenas, R., Díaz García, F., Rubio Pascual, F.J., Abati, J., and Marquínez, J., 1996, Variscan exhumation of a subducted Paleozoic continental margin: The basal units of the Órdenes Complex, Galicia: NW Spain: Tectonics, v. 15, p. 106–121, doi: 10.1029/95TC02617.

Martínez Catalán, J.R., Arenas, R., Díaz García, F., and Abati, J., 1997, Variscan accretionary complex of northwest Iberia: Terrane correlation and succession of tectonothermal events: Geology, v. 25, p. 1103–1106, doi: 10.1130/0091-7613(1997)025<1103:VACONI>2.3.CO;2.

Martínez Catalán, J.R., Díaz García, F., Arenas, R., Abati, J., Castiñeiras, P., González Cuadra, P., Gómez Barreiro, J., and Rubio Pascual, F., 2002, Thrusts and detachment systems in the Órdenes Complex (northwestern Spain): Implications for the Variscan-Appalachian geodynamics, *in* Martínez Catalán, J.R., Hatcher, R.D., Jr., Arenas, R., and Díaz García, F., eds., Variscan-Appalachians Dynamics: The Building of the Late Paleozoic Basement: Geological Society of America Special Paper 364, p. 163–182.

Martínez Catalán, J.R., Arenas, R., Díaz García, F., Gómez-Barreiro, J., González Cuadra, P., Abati, J., Castiñeiras, P., Fernández-Suárez, J., Sánchez Martínez, S., Andonaegui, P., González Clavijo, E., Díez Montes, A., Rubio Pascual, F.J., and Valle Aguado, B., 2007, Space and time in the tectonic evolution of the northwestern Iberian Massif: Implications for the comprehension of the Variscan belt, *in* Hatcher, R.D., Jr., Carlson, M.P., McBride, J.H., and Martínez Catalán, J.R., eds., 4-D Framework of Continental Crust: Geological Society of America Memoir 200, doi: 10.1130/2007.1200(21).

Matte, Ph., 1991, Accretionary history and crustal evolution of the Variscan belt in Western Europe: Tectonophysics, v. 196, p. 309–337, doi: 10.1016/0040-1951(91)90328-P.

Matte, Ph., 2001, The Variscan collage and orogeny (480–290 Ma) and the tectonic definition of the Armorican microplate: A review: Terra Nova, v. 13, p. 122–128, doi: 10.1046/j.1365-3121.2001.00327.x.

Mendia, M., 2000, Petrología de la unidad eclogítica del Complejo de Cabo Ortegal (NW de España): A Coruña, Nova Terra, v. 16, 424 p.

Menzies, M.A., and Seyfried, W., 1979, Basalt-seawater interaction: Trace element and isotopic variation in experimentally altered basalts: Earth and Planetary Science Letters, v. 44, p. 463–472, doi: 10.1016/0012-821X(79)90084-0.

Molnar, P., and Atwater, T., 1978, Interarc spreading and Cordilleran tectonics as alternates related to the age of subducted oceanic lithosphere: Earth and Planetary Science Letters, v. 41, p. 330–340, doi: 10.1016/0012-821X(78)90187-5.

Moores, E.M., 2003, A personal history of the ophiolite concept, *in* Dilek, Y., and Newcomb, S., eds., Ophiolite Concept and the Evolution of Geological Thought: Geological Society of America Special Paper 373, p. 17–29.

Nakamura, N., 1974, Determination of REE, Ba, Fe, Mg, Na and K in carbonaceous and ordinary chondrites: Geochimica et Cosmochimica Acta, v. 38, p. 757–775, doi: 10.1016/0016-7037(74)90149-5.

Nutman, A.P., Green, D.H., Cook, C.A., Styles, M.T., and Holdsworth, R.E., 2001, SHRIMP U-Pb zircon dating of the exhumation of the Lizard Peridotite and its emplacement over crustal rocks: Constraints for tectonic models: Journal of the Geological Society of London, v. 158, p. 809–820.

Oliver, G.J.H., Corfu, F., and Krough, T.E., 1993, U-Pb ages from SW Poland: Evidence for a Caledonian suture zone between Baltica and Gondwana: Journal of the Geological Society of London, v. 150, p. 355–369.

Ordóñez Casado, B., Gebauer, D., Schäfer, H.J., Gil Ibarguchi, J.I., and Peucat, J.J., 2001, A single Devonian subduction event for the HP/HT metamorphism of the Cabo Ortegal Complex within the Iberian Massif: Tectonophysics, v. 332, p. 359–385, doi: 10.1016/S0040-1951(00)00210-9.

Pearce, J.A., 1982, Trace element characteristics of lavas from destructive plate boundaries, *in* Thorpe, R.S., ed. Andesites: Orogenic Andesites and Related Rocks: Chichester, Wiley, p. 525–548.

Pearce, J.A., 1983, Role of the sub-continental lithosphere in magma genesis at active continental margins, *in* Hawkesworth, C.J., and Norry, M.J., eds., Continental Basalts and Mantle Xenoliths: Nantwich, Shiva, p. 230–249.

Pearce, J.A., 1996, A users' guide to basalt discrimination diagrams, *in* Wyman, D.A., ed., Trace Element Geochemistry of Volcanic Rocks: Applications for Massive Sulphide Exploration: Geological Association of Canada, Short Course Notes, v. 12, p. 79–113.

Pearce, J.A., 2003, Supra-subduction zone ophiolites: The search for modern analogues, *in* Dilek, Y., and Newcomb, S., eds., Ophiolite Concept and the Evolution of Geological Thought: Geological Society of America Special Paper 373, p. 269–293.

Pearce, J.A., and Parkinson, I.J., 1993, Trace element models for mantle melting; application to volcanic arc petrogenesis, *in* Prichard, H.M., Alabaster, T., Harris, N.B.W., and Neary, C.R., eds., Magmatic Processes and Plate Tectonics: Geological Society of London Special Publication 76, p. 373–403.

Pearce, J.A., Lippard, S.J., and Roberts, S., 1984, Characteristics and tectonic significance of supra-subduction zone ophiolites, *in* Kokelaar, B.P., and Howells, M.F., eds., Marginal Basin Geology: Geological Society [London] Special Publication 16, p. 77–94.

Peucat, J.J., Bernard-Griffiths, J., Gil Ibarguchi, J.I., Dallmeyer, R.D., Menot, R.P., Cornichet, J., and Iglesias Ponce de León, M., 1990, Geochemical and geochronological cross section of the deep Variscan crust: The Cabo Ortegal high-pressure nappe (northwestern Spain): Tectonophysics, v. 177, p. 263–292, doi: 10.1016/0040-1951(90)90285-G.

Pin, C., Majerowicz, A., and Wojciechowska, I., 1988, Upper Paleozoic oceanic crust in the Polish Sudetes: Nd-Sr isotope and trace element evidence: Lithos, v. 21, p. 195–209, doi: 10.1016/0024-4937(88)90009-6.

Pin, C., Paquette, J.L., Santos Zalduegui, J.F., and Gil Ibarguchi, J.I., 2002, Early Devonian suprasubduction-zone ophiolite related to incipient collisional processes in the western Variscan belt: The Sierra de Careón unit, Órdenes Complex, Galicia, *in* Martínez Catalán, J.R., Hatcher, R.D., Jr., Arenas, R., and Díaz García, F., eds., Variscan-Appalachians Dynamics: The Building of the Late Paleozoic Basement: Geological Society of America Special Paper 364, p. 57–71.

Puelles, P., 2003, Deformación, Metamorfismo y Exhumación de las Granulitas de Alta Presión de La Bacariza (Complejo de Cabo Ortegal, NO España) [Ph.D. thesis]: Leioa, Spain, Universidad del País Vasco, 372 p.

Quesada, C., Fonseca, P.E., Munhá, J., Oliveira, J.T., and Ribeiro, A., 1994, The Beja-Acebuches ophiolite (southern Iberia Variscan fold belt): Geological characterization and geodynamic significance: Boletin Geológico y Minero, v. 105, p. 3–49.

Ries, A., and Shackleton, R.M., 1971, Catazonal complexes of north-western Spain and north Portugal; remnants of a Hercynian thrust plate: Natural and Physical Science, v. 234, p. 65–69.

Santos Zalduegui, J.F., 1995, Geocronología y geoquímica isotópica de diferentes unidades de los complejos alóctonos de Cabo Ortegal y Malpica-Tuy (NO de España): A Coruña, Nova Terra, v. 11, 178 p.

Scotese, C.R., 2001, Atlas of Earth History: Arlington, Texas, PALEOMAP Project, 52 p.

Shervais, J.W., 1982, Ti-V plots and the petrogenesis of modern and ophiolitic lavas: Earth and Planetary Science Letters, v. 59, p. 101–118, doi: 10.1016/0012-821X(82)90120-0.

Spear, F.S., 1993, Metamorphic phase equilibria and pressure-temperature-time paths: Mineralogical Society of America Monograph 1, 799 p.

Stampfli, G.M., and Borel, G.D., 2002, A plate tectonic model for the Paleozoic and Mesozoic constrained by dynamic plate boundaries and restored synthetic oceanic isochrons: Earth and Planetary Science Letters, v. 196, p. 17–33, doi: 10.1016/S0012-821X(01)00588-X.

Sun, S.-S., and McDonough, W.F., 1989, Chemical and isotopic systematics of oceanic basalts: Implications for mantle composition and processes, *in* Saunders, A.D., and Norry, M.J., eds., Magmatism in the Ocean Basins: Geological Society [London] Special Publication 42, p. 313–345.

Taylor, S.R., and McLennan, S.M., 1985, The Continental Crust: Its Composition and Evolution: Oxford, Blackwell, 328 p.

Timmermann, H., Štědrá, V., Gerdes, A., Noble, S.R., Parrish, R.R., and Dörr, W., 2004, The problem of dating high-pressure metamorphism: A U-Pb isotope and geochemical study on eclogites and related rocks of the Mariánské Lázně Complex, Czech Republic: Journal of Petrology, v. 45, p. 1311–1338, doi: 10.1093/petrology/egh020.

Vogel, D.E., 1967, Petrology of an eclogite- and pyrigarnite-bearing polymetamorphic rock complex at Cabo Ortegal: NW Spain: Leidse Geologische Mededelingen, v. 40, p. 121–213.

Winchester, J.A., and Floyd, P.A., 1977, Geochemical discrimination of different magma series and their differentiation products using immobile elements: Chemical Geology, v. 20, p. 325–343, doi: 10.1016/0009-2541(77)90057-2.

Wood, D.A., 1980, The application of a Th-Hf-Ta diagram to problems of tectonomagmatic classification and to establishing the nature of crustal contamination of basaltic lavas of the British Tertiary Volcanic Province: Earth and Planetary Science Letters, v. 50, p. 11–30, doi: 10.1016/0012-821X(80)90116-8.

Manuscript Accepted by the Society 1 February 2007

Printed in the USA

The Geological Society of America
Memoir 200
2007

U-Pb chronometry of polymetamorphic high-pressure granulites: An example from the allochthonous terranes of the NW Iberian Variscan belt

J. Fernández-Suárez*
R. Arenas
J. Abati
Departamento de Petrología y Geoquímica, Universidad Complutense, 28040 Madrid, Spain

J.R. Martínez Catalán
Departamento de Geología, Universidad de Salamanca, 33708 Salamanca, Spain

M.J. Whitehouse
NORDSIM Facility, Swedish Museum of Natural History, SE-104 05 Stockholm, Sweden

T.E. Jeffries
Department of Mineralogy, The Natural History Museum, London, SW7 5BD, UK

ABSTRACT

Secondary ion mass spectrometry (SIMS) U-Pb geochronology of zircons from high-pressure granulite units within the allochthonous complexes of the northwestern Iberian Variscan belt illustrates the complexity of dating high-pressure events using the U-Pb chronometer. Zircons from four rocks belonging to the high-pressure (*P*), high-temperature (*T*) units were dated by SIMS. A gabbro body with preserved igneous texture within the Órdenes Complex high-*P* granulites yielded a crystallization age of ca. 515 Ma. A high-*P* mafic granulite within the same unit contained zircons that had U-Pb ages clustered around 387 Ma, and it preserved no record of events between the crystallization of precursor gabbros and the granulite event (sensu lato). A mafic high-*P* granulite from an equivalent structural unit in the Cabo Ortegal Complex contained abundant zircons that recorded crystallization of igneous protolith at 490–520 Ma and crystallization of new zircon from a melt phase starting at ca. 404 Ma, and zircons with $^{206}Pb/^{238}U$ ages between ca. 480 and 430 Ma that are interpreted to reflect partial Pb loss during the granulite-facies event. Bright luminescent cores in zircons from a leucosome pod within the outcrop area of the latter granulite

*jfsuarez@geo.ucm.es

Fernández-Suárez, J., Arenas, R., Abati, J., Martínez Catalán, J.R., Whitehouse, M.J., and Jeffries, T.E., 2007, U-Pb chronometry of polymetamorphic high-pressure granulites: An example from the allochthonous terranes of the NW Iberian Variscan belt, *in* Hatcher, R.D., Jr., Carlson, M.P., McBride, J.H., and Martínez Catalán, J.R., eds., 4-D Framework of Continental Crust: Geological Society of America Memoir 200, p. 469–488, doi: 10.1130/2007.1200(24). For permission to copy, contact editing@geosociety.org.

yielded a mean age of 397 Ma, whereas dark U-rich rims in zircons from this leucosome yielded a mean age of 390 Ma. These data, in conjunction with previous isotope dilution–thermal ionization mass spectrometry (ID-TIMS; zircon, monazite, titanite) U-Pb ages and $^{40}Ar/^{39}Ar$ dating of fabrics from rocks within the same units, point to a Silurian age for the high-*P* metamorphism. Timing of the peak pressure cannot be further constrained with available data because, in this case, we infer that the U-Pb system in zircon recorded only a segment of the retrograde path. We suggest that the high-*P* metamorphic event is related to the accretion and understacking of these units to the margin of Baltica or Laurentia following the closure of the Iapetus or Tornquist Oceans and concomitant opening of the Rheic Ocean.

Keywords: U-Pb dating, zircon, granulite, Variscan belt, NW Iberia.

INTRODUCTION

Obtaining reliable time constraints on high-grade metamorphic events using the U-Pb chronometer depends heavily on our ability to interpret textural features of zircon and to decipher the U-Pb isotopic information encoded in this mineral. Paradoxically, the natural features of zircon play both in favor and against the successful completion of such an endeavor. On the one hand, the U-Pb system in zircon is able to retain age information from more than one geological event. On the other hand, the structural, textural, chemical, and isotopic complexity of zircon in high-grade metamorphic rocks sometimes makes the interpretation of the age information rather difficult. The occurrence of zircon features and zircon processes, such as inheritance, partial to complete resetting, growth of new zircon, recrystallization, dissolution and reprecipitation, episodic Pb loss, continuous high-temperature Pb loss, alteration, etc., can preclude a straightforward interpretation of isotopic age data.

High-spatial-resolution ion microprobe U-Pb geochronology offers the substantial advantage of analysis at 10–30 μm, which makes it possible to avoid mixed zircon domains. Unfortunately, this high spatial resolution comes at the expense of lower precision in the measured U-Pb and $^{207}Pb/^{206}Pb$ ratios; in the case of Paleozoic or younger zircons, this limitation compromises the assessment of natural discordance (e.g., Pb loss or inheritance of only slightly older zircons; cf. McFarlane et al., 2005).

High-pressure granulites are key rocks in deciphering the tectonothermal processes that take place in the lower crust and lithospheric mantle and thus are fundamental to understanding collisional orogeny. Dating the peak of high-pressure granulite metamorphism is a challenging task regardless of the isotopic chronometer used, and this also applies to the commonly used U-Pb system (e.g., Root et al., 2004; Timmermann et al., 2004). One of the key issues in interpreting U-Pb data from high-pressure (*P*) and high-temperature (*T*) rocks arises from the fact that zircon formation (from a melt phase, via solid state reactions or fluid assisted processes) can take place at peak metamorphic conditions, but it can also take place in the prograde path and/or in the retrograde or exhumation path (e.g., Roberts and Finger, 1997; Fernández-Suárez et al., 2002; Whitehouse and Platt, 2003). This scenario is further complicated when the protolith has undergone more than one episode of high-grade metamorphism. A particularly difficult situation is one in which two events have taken place within a short time interval, which impinge on the limited precision of ion microprobe U-Pb data. Such an example is found in the high-*P* granulites within the allochthonous complexes of northwest Iberia (Martínez Catalán et al., 1997, 1999, this volume), where previous geochronological studies have spurred a long-standing discussion on the age of the high-pressure event. Some authors have proposed a Middle Devonian age (ca. 380–390 Ma; e.g., Ordoñez Casado et al., 2001), whereas others have favored a Late Cambrian–Early Ordovician age (ca. 480–495 Ma; Fernández-Suárez et al., 2002). In an attempt to clarify this disagreement, we have undertaken an ion microprobe U-Pb study of zircon from rocks belonging to these high-*P* granulite units. Constraints for at least the minimum age of the high-*P* and high-*T* event are pivotal to understanding the early Variscan (eo-Variscan) geodynamic evolution of northwest Iberia.

GEOLOGICAL SETTING

The allochthonous complexes of northwest Iberia (Figs. 1, 2, and 3), the main geological features and geodynamic significance of which are described in detail by Martínez Catalán et al. (1999, this volume), consist of units stacked upon a relative autochthon considered to have been the continental margin of northern Gondwana. These complexes, exposed as structural synforms, are formed by units with different tectonothermal evolution and are interpreted to be dismembered fragments of a Variscan and eo-Variscan accretionary complex (Martínez Catalán et al., 1997, 1999, this volume). The allochthonous complexes have been divided into three units (Arenas et al., 1995) based on their structural position and tectonometamorphic evolution: basal, intermediate (ophiolitic), and upper units.

The basal units consist of schists, paragneisses, and Early Ordovician felsic and mafic igneous rocks that underwent subduction and high-*P* metamorphism at 380–365 Ma during the closure of an oceanic realm (Arenas et al., 1995; Martinez Catalán et al., 1996). They are considered to represent the outer margin of northern Gondwana involved in the Variscan collision.

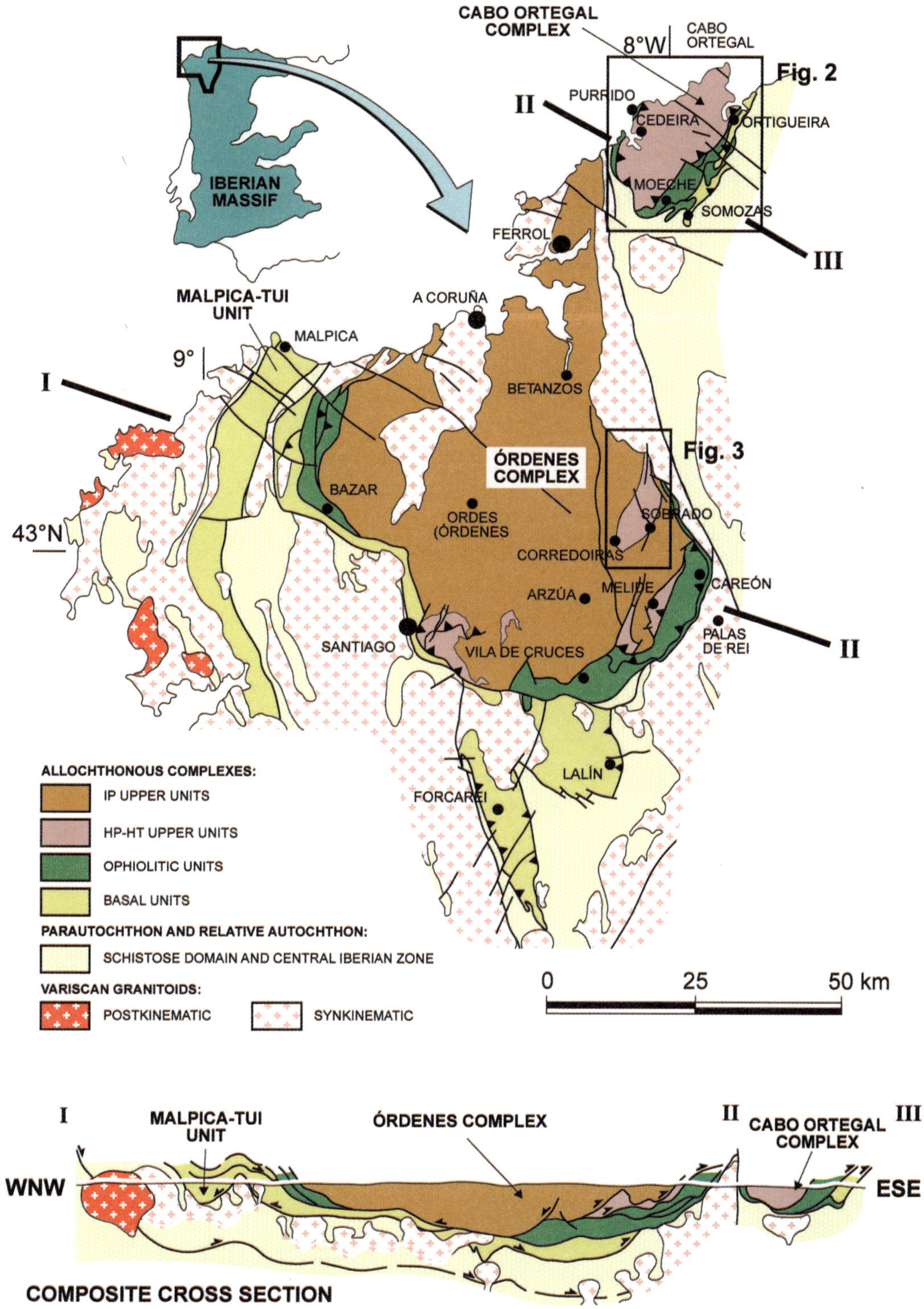

Figure 1. Simplified geological map and cross section of northwestern Spain showing the Órdenes and Cabo Ortegal allochthonous complexes. Boxes indicate location of Figures 2 and 3. IP—intermediate pressure; HP-HT—high-pressure, high-temperature.

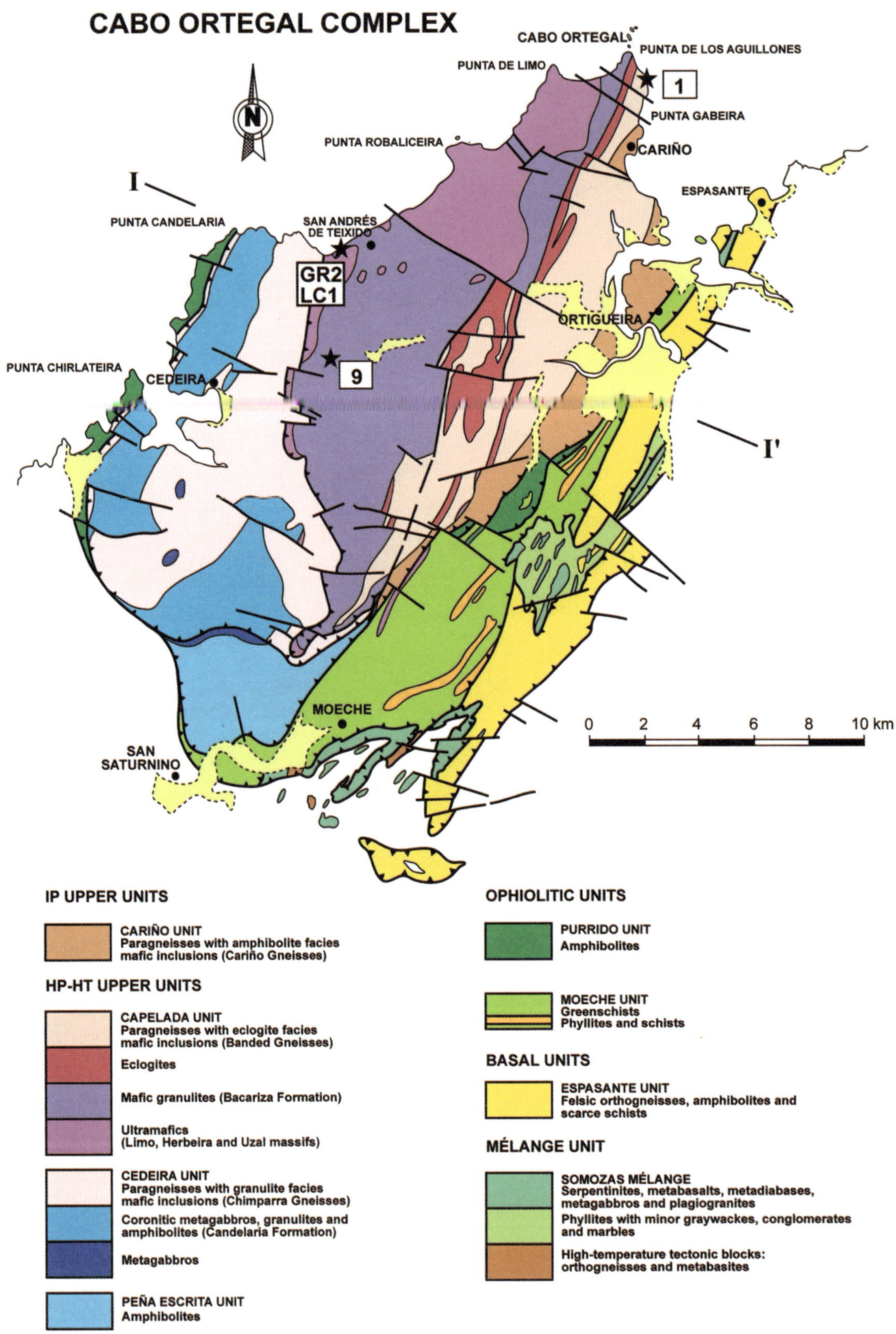

Figure 2. Geological map of the Cabo Ortegal Complex showing the main structures, units, and location of samples GR2 and LC1. Also shown are the locations of leucosome 9 and paragneiss 1 studied by Fernández-Suárez et al. (2002) and mentioned in the text. IP—intermediate pressure; HP-HT—high-pressure, high-temperature.

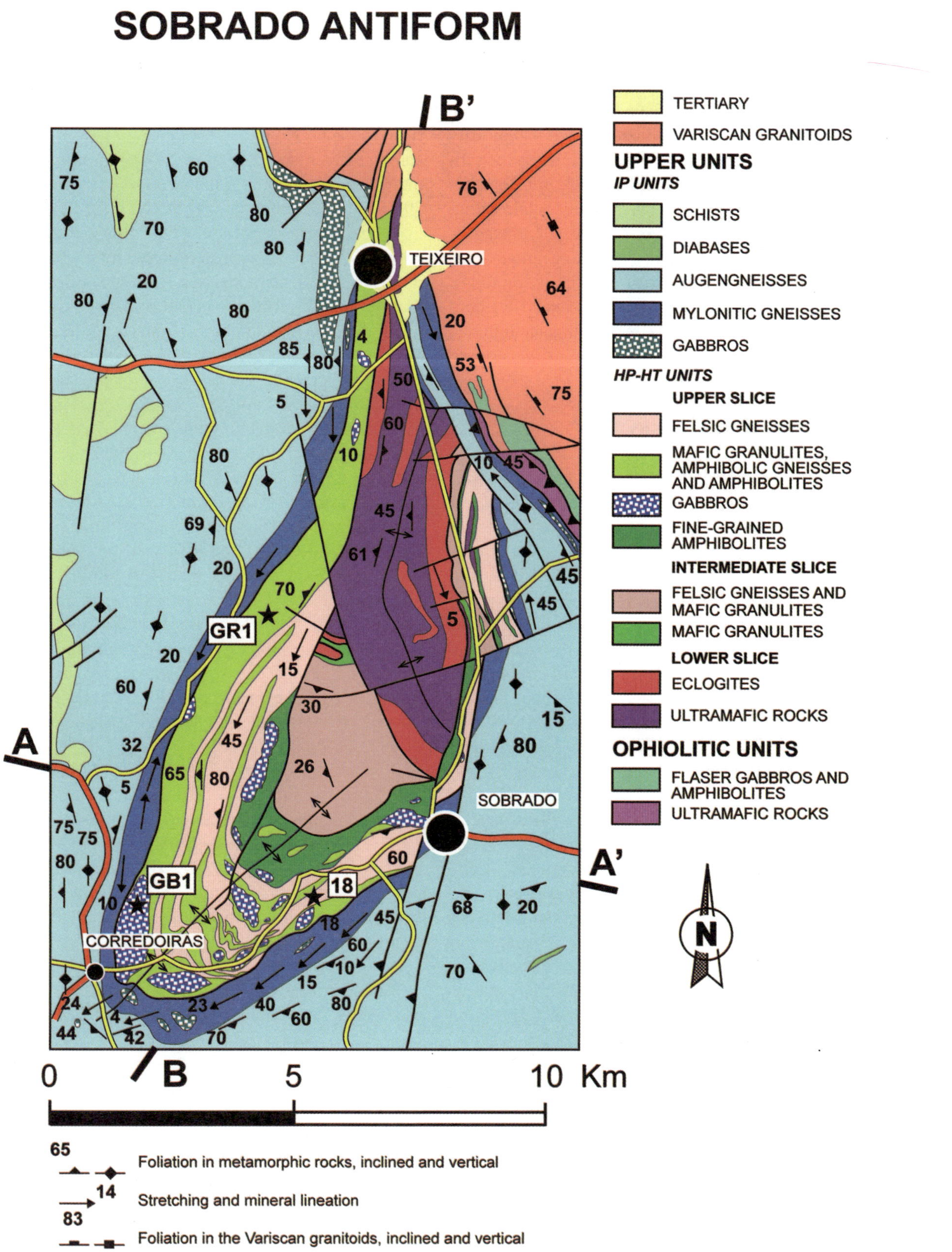

Figure 3. Geological map of the eastern part of the Órdenes Complex showing the location of samples GB1 and GR1. Also shown is the location of paragneisses with ages ca. 485–495 Ma concordant monazite studied by Fernández-Suárez et al. (2002) and discussed in the text (sample 18). IP—intermediate pressure; HP-HT—high-pressure, high-temperature.

The ophiolitic units (Díaz García et al., 1999; Arenas et al., this volume) structurally overlie the basal units and represent an Early Devonian oceanic crust and older oceanic lithosphere that developed during the opening of the Rheic Ocean.

The upper units constitute an outboard terrane tectonically emplaced upon the ophiolitic units. These upper units can be subdivided into two subtypes: (1) high-*P* and high-*T* units, and (2) intermediate-*P* units. The high-*P* and high-*T* types contain eclogite- to granulite-facies paragneisses, ultramafic rocks, metagabbros, and orthogneisses of continental and arc affinities. The units of the intermediate-*P* type are thrust structurally on top of the high-*P* and high-*T* units along a wide shear zone. They consist of a sequence of low- to high-grade metasediments intruded by tholeiitic metagabbros and granodioritic to tonalitic orthogneisses dated at ca. 500 Ma (Abati et al., 1999).

GEOCHRONOLOGICAL BACKGROUND

Available geochronologic data previous to this study cover a variety of rocks from the allochthonous complexes in northwest Iberia, and they have been obtained by different methods (U-Pb isotope dilution–thermal ionization mass spectrometry [ID-TIMS] and secondary ion mass spectrometry [SIMS], Sm-Nd, Rb-Sr, and Ar-Ar). A compilation of published geochronologic data on these units can be found in Martínez Catalán et al. (1999) and Fernández-Suárez et al. (2002). In the high-*P* and high-*T* units (upper unit), ID-TIMS multigrain zircon U-Pb data on eclogites and mafic granulites from the Cabo Ortegal and Órdenes Complexes yielded ages in the range 480–490 Ma (Peucat et al., 1990). The authors suggested that this age dates a high-grade metamorphic event, which possibly occurred very shortly after the formation of the magmatic protoliths. Sensitive high-resolution ion microprobe (SHRIMP) U-Pb data on presumed magmatic domains of zircons from the high-*P* mafic granulites (Schäfer et al., 1993; Ordóñez Casado et al., 2001) yielded ages in the range 480–500 Ma, which is considered by the authors to represent the crystallization age of the igneous protoliths. Ages from cathodoluminescence (CL) distinct rims in the same zircons are not reported. U-Pb ages of zircons from eclogites are between 510 and 470 Ma, and their rims range in age between 380 and 390 Ma. Ordóñez Casado et al. (2001) contended that the older ages date protolith crystallization, whereas the Devonian ages represent new zircon crystallization under high-*P* conditions, and they proposed that a single subduction event at ca. 380–390 Ma caused the high-*P* and high-*T* metamorphism recorded by these units. Within ultramafic bodies, the U-Pb ages (both ID-TIMS and SHRIMP) reported for pyroxenite veins and pegmatoids range between ca. 385 and 395 Ma (Schäfer et al., 1993; Santos Zalduegui et al., 1996; Ordóñez Casado, 1998).

There are fewer ages on paragneisses from the high-*P* and high-*T* units. ID-TIMS U-Pb data for Cabo Ortegal paragneisses (Peucat et al., 1990) yielded lower-intercept ages of 417 +3/–2 Ma and 422 ± 4 Ma. Kuijper (1980) studied a sample of migmatitic paragneisses from the Órdenes Complex and obtained a lower-intercept zircon age of 476 ± 12 Ma and a subconcordant monazite fraction with a $^{207}Pb/^{206}Pb$ age of 487 Ma. SHRIMP data on zircons from the eclogite-facies Chimparra paragneiss (Cabo Ortegal Complex; Ordóñez Casado, 1998) yielded ages spreading from 515 Ma up to 2.14 Ga. Analyses on zircon rims, interpreted as metamorphic overgrowths, yielded ages around 390 Ma. Fernández-Suárez et al. (2002) reported ID-TIMS monazite ages from paragneisses of the high-*P*, high-*T* units between 495 and 485 Ma. They also reported monazite ages in the range 385–395 Ma in paragneisses from areas that have undergone intense Variscan tectonothermal reworking during the emplacement of the high-*P* and high-*T* units onto the underlying ophiolitic units. The same authors reported a subconcordant multigrain fraction of acicular zircon in an eclogite-facies paragneiss of the Cabo Ortegal Complex (sample 1 in Fig. 2) with an age of 476 +8/–2 Ma, which is within error of the lower-intercept age defined by highly discordant multigrain zircon fractions in the same paragneiss (472 ± 27 Ma).

The presence of concordant ca. 486 Ma zircons from a leucosome within a mafic granulite (sample 9 in Fig. 2) led Fernández-Suárez et al. (2002) to suggest that the high-*P* granulite metamorphism could be older than 486 Ma. The same leucosome also contained low-U zircons with U-Pb ages of ca. 430–440 Ma, but $^{207}Pb/^{206}Pb$ ages were imprecise and were interpreted as possible mixtures of ca. 480–490 cores and younger ca. 380–390 overgrowths on the basis that there was no independent evidence for a geological event at ca. 430–440 Ma. In the same leucosome, two pale-green titanite fractions yielded a concordia age of 387 ± 3 Ma.

Within the intermediate-*P* units, U-Pb (ID-TIMS) zircon data from gabbros and granitoids indicate crystallization ages of ca. 495–500 Ma (Dallmeyer and Tucker, 1993; Abati et al., 1999). Within the same units, Abati et al. (1999) obtained concordant monazite ages in granulite-facies paragneisses in the range 493–498 Ma, but these authors found no isotopic evidence for younger events in the samples investigated, with the exception of two rutile fractions that yielded $^{206}Pb/^{238}U$ ages of ca. 380–390 Ma.

In summary, previous U-Pb geochronological studies have established the existence of a magmatic event (intrusion of gabbros and granitoids) at ca. 500 Ma, a high-grade metamorphic event at 480–495 Ma, and a later metamorphic event for which most ages reported so far cluster around 380–390 Ma. It is undisputed that there was (at least) one episode of high-pressure granulite facies metamorphism that transformed mafic igneous rocks into high-*P* mafic granulites and metasedimentary rocks into high-*P* paragneisses. The issue yet to be resolved is: does either group of ages (495–480 or 390–380 Ma) date the peak high-*P* granulite event?

The present work reopens the issue of the age of the high-*P* granulite metamorphism in the allochthonous complexes of northwest Iberia through a SIMS study of zircons from gabbros, granulites, and leucosomes within the high-*P* and high-*T* units.

SAMPLE DESCRIPTION

The investigated sample locations are shown in Figures 2 and 3. Sample GB1 is a gabbro with preserved igneous texture within a high-*P* granulite in the Órdenes Complex. The gabbro is composed of plagioclase + clinopyroxene + orthopyroxene ± olivine + titanite + ilmenite (with minor secondary biotite and hornblende). Sample GR1 is a mafic high-*P* granulite that is interpreted to represent an equivalent of the former gabbro transformed into a granulite. The mineralogy of GR1 is garnet + plagioclase + clinopyroxene + quartz + rutile (with minor hornblende and biotite). Sample GR2 is a mafic high-*P* granulite from the Cabo Ortegal Complex where no vestiges of the igneous protolith appear to have been preserved. The mineralogy of GR2 is garnet + plagioclase + clinopyroxene + quartz + rutile (minor hornblende). At the outcrop scale, both granulites (GR1 and GR2) show evidence of localized melting. Sample LC1 is a leucosome pod within the GR2 granulite exposure. The mineralogy of LC1 is plagioclase (~80%) + quartz + hornblende + rutile + titanite + ilmenite (retrograde veins with epidote, clinozoisite, and titanite).

ANALYTICAL METHODS: SIMS U-Pb GEOCHRONOLOGY

Samples were disaggregated by a jaw crusher and pulverized with a disc mill. Zircons were separated at the Universidad Complutense of Madrid by heavy fraction enrichment on a Wilfley™ table followed by density separation using di-iodomethane (CH_2I_2) and magnetic separation in a Frantz™ isodynamic separator. Zircons were selected from the least magnetic fraction and handpicked in alcohol under a binocular microscope. Zircon grains representative of the main sizes and morphological types present in each sample were selected for SIMS analyses. Grains were set in synthetic resin mounts along with several chips of the standard zircon 91500 (Wiedenbeck et al., 1995) and polished to expose equatorial sections.

U-Pb isotope analyses were performed by ion microprobe (CAMECA IMS 1270) at the NORDSIM facility, Stockholm, Sweden. Analytical procedures and protocol, data reduction, and age calculation followed those described in Whitehouse et al. (1999). The location of analyses within the zircon grains was guided by cathodoluminescence (CL) images that revealed the internal structure of each of the analyzed grains. The nominal beam size for all analyses was 30 μm. The results of U-Th-Pb analyses in the four samples studied are given in Table 1.

Given the low precision of SIMS $^{207}Pb/^{206}Pb$ measurements in Paleozoic zircons, the possible effects of Pb loss cannot be realistically evaluated, and therefore intercept ages derived from discordant analyses may be unreliable. Corrections based on measured ^{204}Pb are equally unreliable in most cases given the low common Pb content (and hence poor precision in the measured $^{206}Pb/^{204}Pb$ ratio) of most zircon domains unaffected by metamictization, fractures, alteration, etc., such as those selected for analyses. Therefore, given that the use of CL-guided analyses ensures that mixed domains are not included in one spot, we used the ^{207}Pb-corrected $^{206}Pb/^{238}U$ ages (Table 1) except where otherwise stated. In this case, we used the present-day Stacey and Kramers (1975) common Pb composition (see Whitehouse et al., 1999), but the use of a slightly different model Pb composition would not affect the results beyond the analytical errors reported, and therefore it is not a significant issue in the context of the data as used herein.

ZIRCON FEATURES AND U-Pb RESULTS

Gabbro sample GB1 contained few zircons with sector and oscillatory zoning (Fig. 4), relatively high U (400–1200 ppm) and Th (380–580 ppm) concentrations, and Th/U ratios between ~0.4 and 0.7, which yielded a concordia age (Ludwig, 1998) of 515 ± 7 Ma (Fig. 5). The weighted average of the ^{207}Pb-corrected $^{206}Pb/^{238}U$ ages is 517.6 ± 5 Ma, which is within the error of the concordia age calculated from uncorrected ratios. This age is interpreted to date the crystallization of the gabbro.

Zircons from granulite GR1 are equant, stubby, football-like multifaceted crystals with mainly polygonal or sector CL zoning, and they show no core-rim features (Fig. 6). The Th/U ratios range from ~0.1 to 0.5, with a median value of 0.28, and the U and Th concentrations are about ten times lower than those in zircons from the alleged protolith gabbro GB1. Seventeen analyses were performed on grain centers and edges of these zircons, and they yielded a tight cluster of data points with a mean age of 387 ± 5 Ma (Fig. 7).

Granulite GR2 contained very abundant zircon with a remarkable variety of external morphology, size, and internal CL features (Fig. 8). Thirty analyses were performed on 18 zircons representing the range of internal features of zircon as revealed by CL imaging. The results (Table 1; Fig. 9) can be summarized as follows: (1) zircon cores and whole zircons with oscillatory zoning or CL banding with diffuse limits yielded ages between ca. 490 and 520 Ma. We interpret these ages to represent the igneous crystallization age of the protolith. (2) Cores featuring blurred or weak CL patterns (but original zoning still visible in cases) had $^{206}Pb/^{238}U$ ages that ranged from ca. 480 to 430 Ma. We interpret these as disturbed and partially reset igneous protolith zircons. (3) Zircon overgrowths with either oscillatory or no zoning had variable Th/U ratios (<0.05–0.5) and ages between ca. 410 and 385 Ma. We interpret these as new zircon growth over ovoid or irregular cores that had previously been partially resorbed (Fig. 8).

The leucosome sample LC1 contained abundant zircons. These were mostly stubby, subhedral to euhedral prisms that typically displayed bright CL ovoid cores and dark rims, both with weak oscillatory or sector zoning (Fig. 10). Analyses performed on cores yielded a mean age of 397 ± 4 Ma (range 404–392 Ma; Fig. 11), and analyses on rims yielded a younger mean age of 391 ± 4 Ma (Fig. 12). Th/U ratios were <0.07 in rims but ranged from 0.05 to 0.6 in cores. Ages of cores and rims in LC1 were

TABLE 1. ION MICROPROBE U-Th-Pb DATA FOR SAMPLES GB1, GR1, GR2, AND LC1

Sample Analysis no.	Comments	U (ppm)	Pb (ppm)	Th/U[†]	$^{206}Pb/^{204}Pb$ (measured)	Isotopic ratios $^{206}Pb/^{238}U$	±σ (%)	$^{207}Pb/^{235}U$	±σ (%)	$^{207}Pb/^{206}Pb$	±σ (%)	Age (Ma) $^{206}Pb/^{238}U$	±σ	$^{207}Pb/^{235}U$	±σ	$^{207}Pb/^{206}Pb$	±σ	$^{206}Pb/^{238}U_{corr}$[§]	±σ
GB1																			
n1147-4-20a	Dark CL center	636	67	0.68	35936	0.0830	1.38	0.6571	1.52	0.0574	0.64	514	7	513	6	508	14	514	7
n1147-4-20b	Bright CL border	405	41	0.51	32549	0.0838	1.34	0.6554	1.69	0.0567	1.03	519	7	512	7	480	23	520	7
n1147-4-2a	Center sector z.	519	55	0.74	22639	0.0832	1.40	0.6578	1.61	0.0574	0.80	515	7	513	7	505	17	515	7
n1147-4-2b	Border oscil. z.	442	44	0.39	23489	0.0836	1.34	0.6711	1.53	0.0582	0.75	517	7	521	6	539	16	517	7
n1147-4-3a	Dark CL center	1193	122	0.53	53956	0.0847	1.33	0.6615	1.41	0.0567	0.48	524	7	516	6	478	11	525	7
n1147-4-7a	Fine oscil. z.	601	62	0.65	50922	0.0824	1.32	0.6526	1.47	0.0574	0.65	511	6	510	6	508	14	511	7
n1147-4-9a	Bright CL sector	355	36	0.44	20000	0.0846	1.47	0.6842	1.77	0.0586	0.99	524	7	529	7	554	22	523	8
GR1																			
n1148-7-10a	Center polyg z.	35	3	0.27	7116	0.0629	1.95	0.4233	5.49	0.0488	5.13	393	7	358	17	141	116	396	8
n1148-7-11a	Center polyg z.	249	17	0.21	11490	0.0626	1.87	0.4536	2.47	0.0525	1.62	391	7	380	8	309	36	392	7
n1148-7-18b	Dark CL border	192	13	0.18	6702	0.0625	1.89	0.4818	2.37	0.0559	1.43	391	7	399	8	448	31	390	7
n1148-7-21a	Blurred CL center	43	3	0.11	8735	0.0624	1.97	0.4979	3.55	0.0578	2.95	390	7	410	12	524	63	389	8
n1148-7-23a	Bright CL center	48	3	0.08	2071	0.0609	1.92	0.4656	4.48	0.0554	4.04	381	7	388	15	429	88	381	7
n1148-7-28a	Sector z. center	40	3	0.32	1235	0.0618	1.92	0.4870	3.65	0.0571	3.10	387	7	403	12	496	67	386	7
n1148-7-31a	Center	27	2	0.46	1207	0.0602	1.95	0.4713	3.95	0.0568	3.44	377	7	392	13	483	74	376	7
n1148-7-38a	Bright CL center	142	10	0.37	5430	0.0618	1.86	0.4599	2.48	0.0540	1.64	387	7	384	8	370	36	387	7
n1148-7-39a	Center	37	3	0.29	2172	0.0630	1.92	0.4990	3.64	0.0575	3.09	394	7	411	12	510	67	392	7
n1148-7-40a	Center	41	3	0.29	1314	0.0622	1.93	0.4674	3.63	0.0545	3.07	389	7	389	12	392	67	389	7
n1148-7-42a	Polyg z. center	190	13	0.18	4123	0.0614	1.88	0.4349	3.01	0.0514	2.34	384	7	367	9	257	53	386	7
n1148-7-49a	Center	39	3	0.26	6417	0.0596	1.93	0.4660	3.43	0.0567	2.83	373	7	388	11	481	61	372	7
n1148-7-49b	Border	31	2	0.27	2163	0.0598	2.01	0.4469	3.85	0.0542	3.29	374	7	375	12	381	72	374	7
n1148-7-5a	Center	166	12	0.22	4793	0.0636	1.88	0.4810	2.38	0.0548	1.47	398	7	399	8	406	33	397	7
n1148-7-6a	Bright CL center	38	3	0.94	2739	0.0624	1.88	0.4817	3.70	0.0560	3.18	390	7	399	12	452	69	389	7
n1148-7-8a	Sector z. border	30	2	0.24	1239	0.0610	1.87	0.4850	4.75	0.0576	4.37	382	7	402	16	516	93	380	7
n1148-7-9a	Sector z. border	103	7	0.23	7105	0.0633	1.85	0.4885	2.60	0.0559	1.83	396	7	404	9	450	40	395	7

(Continued)

TABLE 1. ION MICROPROBE U-Th-Pb DATA FOR SAMPLES GB1, GR1, GR2, AND LC1 (*Continued*)

Sample Analysis no.	Comments	U (ppm)	Pb (ppm)	Th/U†	$^{206}Pb/^{204}Pb$ (measured)	Isotopic ratios $^{206}Pb/^{238}U$	±σ (%)	$^{207}Pb/^{235}U$	±σ (%)	$^{207}Pb/^{206}Pb$	±σ (%)	Age (Ma) $^{206}Pb/^{238}U$	±σ	$^{207}Pb/^{235}U$	±σ	$^{207}Pb/^{206}Pb$	±σ	$^{206}Pb/^{238}U_{corr}$§	±σ
GR2																			
n1149-1-10a	Blurred CL core	58	5	0.81	2092	0.0703	1.27	0.5297	5.27	0.0547	5.11	438	5	432	19	399	111	438	6
n1149-1-10b	Overgrowth oscil	59	4	0.31	1455	0.0624	1.36	0.4695	6.39	0.0546	6.24	390	5	391	21	395	134	390	5
n1149-1-11a	Dark CL center	124	12	0.51	6273	0.0793	1.28	0.6517	1.97	0.0596	1.49	492	6	510	8	589	32	490	6
n1149-1-11b	Oscil z. border	95	9	0.32	6909	0.0809	1.31	0.6515	2.01	0.0584	1.52	502	6	509	8	544	33	501	6
n1149-1-12a	Dark CL center	106	7	0.10	3882	0.0645	1.26	0.5085	2.11	0.0572	1.69	403	5	417	7	499	37	402	5
n1149-1-12b	Bright CL rim	41	3	0.41	809	0.0626	1.39	0.4536	8.93	0.0525	8.82	392	5	380	29	309	189	392	6
n1149-1-14a	Dark CL center	166	15	0.52	2912	0.0737	1.33	0.5789	2.82	0.0569	2.49	459	6	464	11	489	54	458	6
n1149-1-15a	Sector z. core	48	4	0.56	498	0.0666	1.44	0.4380	10.16	0.0477	10.06	416	6	369	32	84	223	419	6
n1149-1-15b	Dark rim	105	7	0.06	2896	0.0632	1.26	0.5069	2.15	0.0582	1.74	395	5	416	7	537	38	393	5
n1149-1-15c	Dark rim	141	9	0.05	1233	0.0634	1.31	0.4400	4.26	0.0504	4.05	396	5	370	13	211	91	398	5
n1149-1-20a	Oscil z. center	107	8	0.34	2882	0.0650	1.31	0.5354	2.09	0.0598	1.64	406	5	435	7	595	35	403	5
n1149-1-27a	Dark CL core	132	10	0.45	1929	0.0620	1.35	0.4389	3.87	0.0514	3.63	388	5	370	12	257	81	389	5
n1149-1-27b	Bright rim	20	1	0.01	772	0.0616	1.40	0.4206	13.36	0.0495	13.29	385	5	357	41	172	284	388	5
n1149-1-29a	Blurred CL center	82	8	0.53	1597	0.0792	1.40	0.5749	4.66	0.0526	4.44	492	7	461	17	312	98	494	7
n1149-1-33a	Blurred CL center	103	8	0.46	1848	0.0691	1.55	0.5067	4.04	0.0532	3.73	431	6	416	14	336	82	432	7
n1149-1-33b	Rim oscil. z.	118	8	0.05	2645	0.0650	1.26	0.4712	3.12	0.0526	2.86	406	5	392	10	312	64	407	5
n1149-1-35a	Structureless CL	65	5	0.15	2321	0.0660	1.31	0.5331	2.42	0.0586	2.04	412	5	434	9	551	44	410	5
n1149-1-38a	Faint oscil. z. core	117	11	0.53	3505	0.0767	1.29	0.5796	2.53	0.0548	2.17	476	6	464	9	405	48	477	6
n1149-1-38b	Faint oscil. rim	50	4	0.06	1525	0.0660	1.26	0.5559	2.59	0.0611	2.26	412	5	449	9	643	48	409	5
n1149-1-39a	Dark CL core	13	1	0.03	332	0.0604	1.47	0.3248	30.89	0.0390	30.86	378	5	286	80	-416	659	385	6
n1149-1-42a	Oscil z. center	125	12	0.32	2000	0.0839	1.31	0.6254	2.95	0.0541	2.64	519	7	493	12	375	58	522	7
n1149-1-47a	Core	109	10	0.42	1201	0.0814	1.46	0.5821	4.35	0.0519	4.09	504	7	466	16	280	91	508	7
n1149-1-47b	Structureless rim	46	3	0.32	482	0.0622	1.26	0.3172	15.31	0.0370	15.26	389	5	280	38	-556	368	397	5
n1149-1-53a	Faint oscil. z. center	72	6	0.53	1291	0.0712	1.29	0.5194	4.37	0.0529	4.18	444	6	425	15	324	92	445	6
n1149-1-57a	Dark core	193	14	0.12	3159	0.0654	1.34	0.4785	2.46	0.0531	2.06	408	5	397	8	332	46	409	5
n1149-1-58a	Sector z. center	95	8	0.71	2235	0.0649	1.34	0.5092	3.25	0.0569	2.96	405	5	418	11	488	64	404	5
n1149-1-58b	Dark border	157	15	1.35	4159	0.0657	1.30	0.4715	2.45	0.0520	2.08	410	5	392	8	287	47	412	5
n1149-1-8a	Faint oscil. border	135	13	0.63	2688	0.0770	1.39	0.5881	3.09	0.0554	2.75	478	6	470	12	429	60	479	7

(Continued)

TABLE 1. ION MICROPROBE U-Th-Pb DATA FOR SAMPLES GB1, GR1, GR2, AND LC1 (*Continued*)

Sample Analysis no.	Comments	U (ppm)	Pb (ppm)	Th/U[†]	$^{206}Pb/^{204}Pb$ (measured)	Isotopic ratios $^{206}Pb/^{238}U$	±σ (%)	$^{207}Pb/^{235}U$	±σ (%)	$^{207}Pb/^{206}Pb$	±σ (%)	Age (Ma) $^{206}Pb/^{238}U$	±σ	$^{207}Pb/^{235}U$	±σ	$^{207}Pb/^{206}Pb$	±σ	$^{206}Pb/^{238}U_{corr}$[§]	±σ
LC1																			
n1147-3-10a	Core	141	10	0.29	7984	0.0639	1.30	0.4879	2.02	0.0553	1.54	400	5	403	7	426	34	399	5
n1147-3-14a	Core	70	5	0.15	3594	0.0645	1.54	0.4777	2.80	0.0537	2.34	403	6	396	9	357	52	404	6
n1147-3-14b	Rim	541	36	0.01	33037	0.0633	1.32	0.4704	1.54	0.0539	0.80	396	5	391	5	366	18	396	5
n1147-3-15a	Core	662	44	0.00	42808	0.0632	1.34	0.4758	1.52	0.0546	0.71	395	5	395	5	395	16	395	5
n1147-3-16a	Core	550	37	0.01	39310	0.0631	1.33	0.4737	1.60	0.0545	0.89	394	5	394	5	391	20	394	5
n1147-3-17a	Core	94	7	0.67	2225	0.0627	1.33	0.4570	3.77	0.0529	3.53	392	5	382	12	325	78	393	5
n1147-3-17b	Rim	503	33	0.01	29087	0.0614	1.36	0.4642	1.61	0.0548	0.85	384	5	387	5	404	19	384	5
n1147-3-19a	Core	170	13	0.31	6561	0.0633	1.44	0.4879	2.03	0.0559	1.43	396	6	403	7	448	31	395	6
n1147-3-19b	Rim	652	43	0.01	38716	0.0617	1.31	0.4613	1.59	0.0542	0.90	386	5	385	5	379	20	386	5
n1147-3-22a	Core	72	5	0.07	3901	0.0645	1.32	0.5024	2.51	0.0565	2.13	403	5	413	9	471	47	402	5
n1147-3-22b	Rim	620	41	0.01	44820	0.0631	1.32	0.4735	1.52	0.0545	0.75	394	5	394	5	390	17	394	5
n1147-3-2a	Core	117	9	0.37	3240	0.0645	1.37	0.5054	2.32	0.0569	1.87	403	5	415	8	486	41	402	5
n1147-3-2b	Rim	474	31	0.00	25258	0.0628	1.42	0.4724	1.66	0.0545	0.87	393	5	393	5	393	19	393	5
n1147-3-3a	Core	126	9	0.27	6213	0.0632	1.30	0.4824	2.07	0.0554	1.60	395	5	400	7	426	35	395	5
n1147-3-3b	Rim	437	29	0.00	38906	0.0626	1.36	0.4751	1.62	0.0551	0.89	391	5	395	5	416	20	391	5
n1147-3-9a	Core	652	43	0.01	24325	0.0627	1.31	0.4700	1.50	0.0544	0.73	392	5	391	5	387	16	392	5

Note: U/Pb ratios include a propagated component representing the scatter and the standard data; z—zoned, oscil—oscillatory; polyg.—polygonal, CL—cathodoluminescence.

[†]Calculated from measured ThO intensity.

[§]^{207}Pb-corrected age. Calculated from uncorrected ratios assuming 0 Ma Stacey and Kramers (1975) average terrestrial Pb composition.

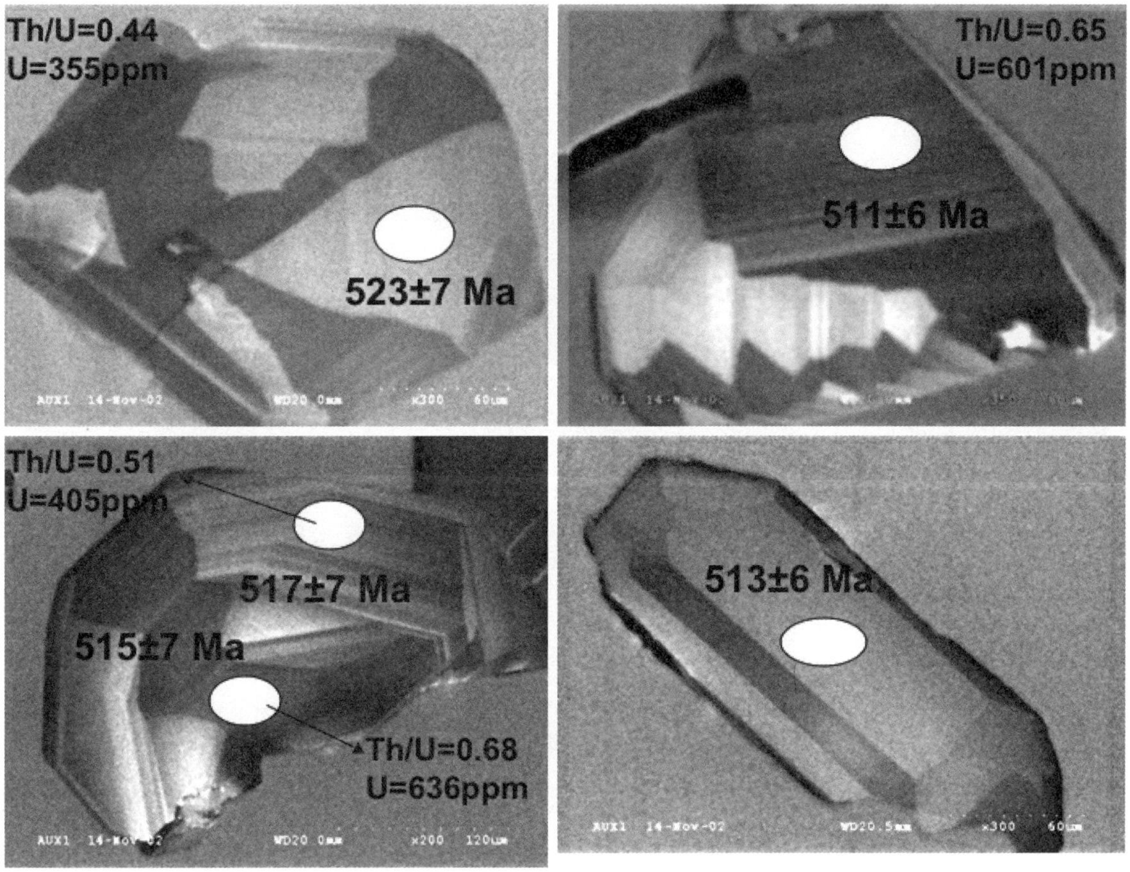

Figure 4. Cathodoluminescence images of zircons from gabbro GB1. Ages and 1σ errors of spot analyses, Th/U ratios, and U concentration are shown in some cases.

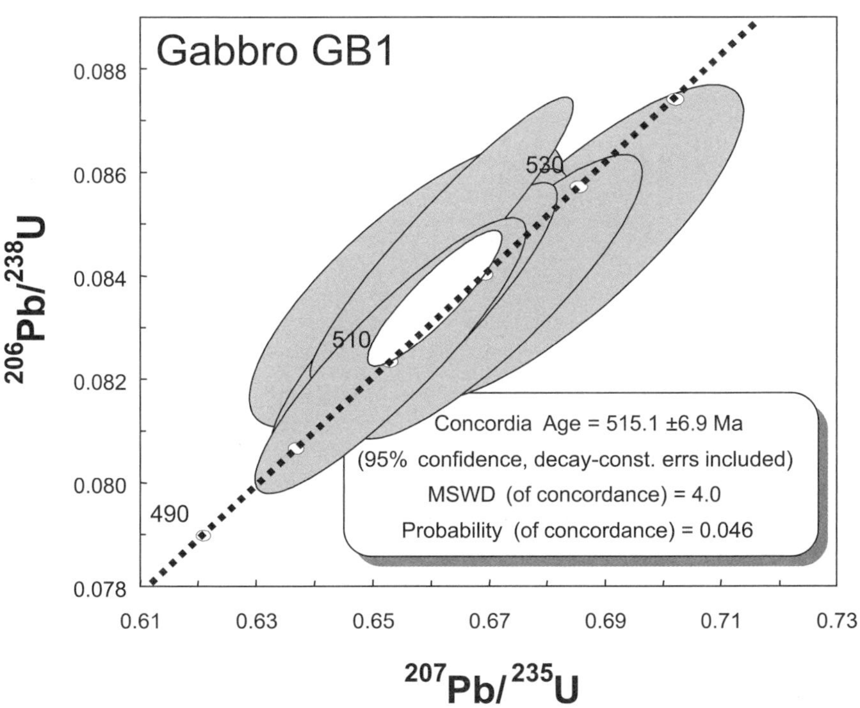

Figure 5. Concordia plot of U-Pb analyses from gabbro GB1. White ellipse represents concordia age and error (Ludwig, 1998) calculated with Isoplot 3.00 (Ludwig, 2003). Gray ellipses represent 2σ uncertainties. MSWD—mean square of weighted deviates.

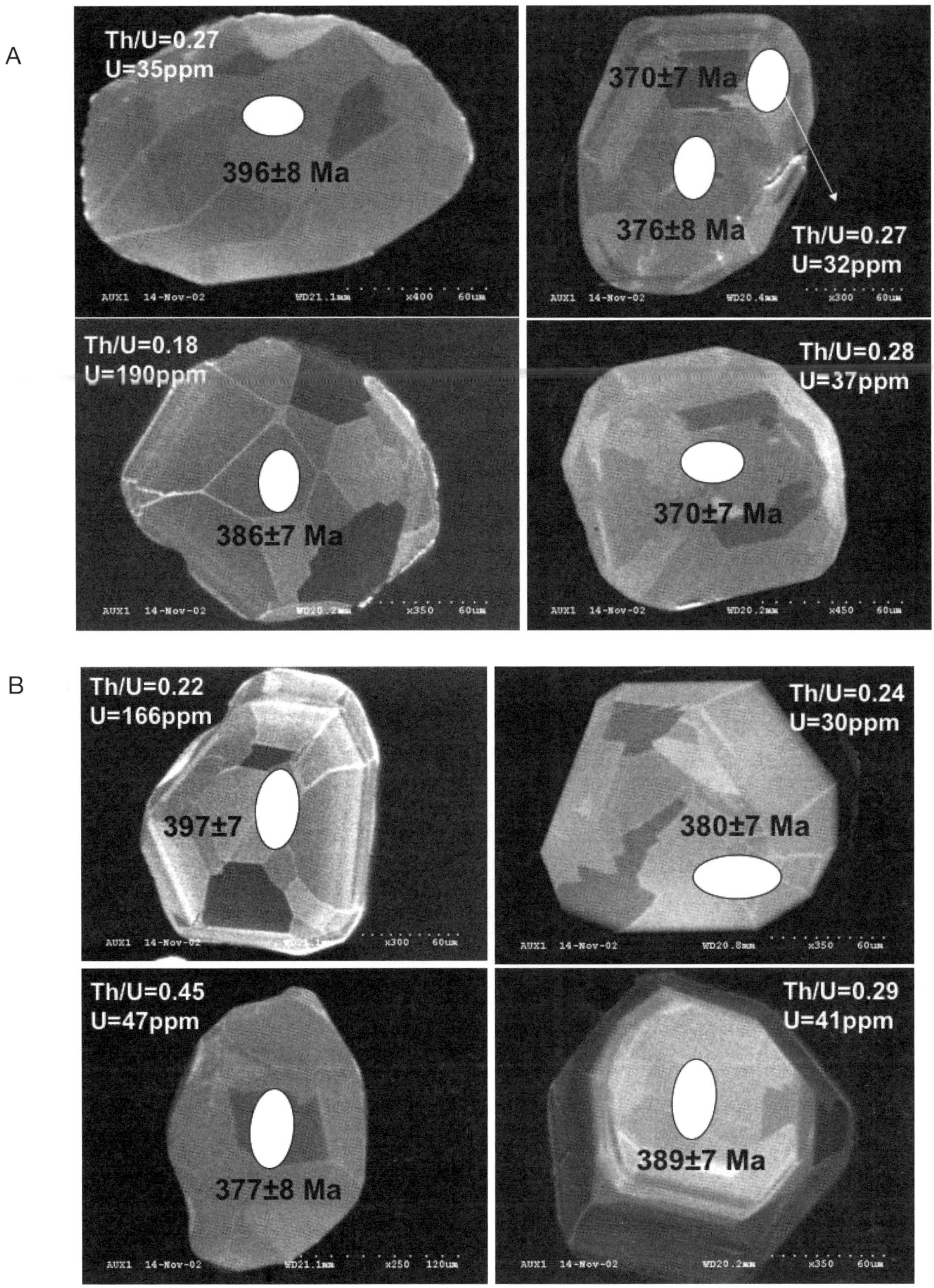

Figure 6. Cathodoluminescence images of zircons from granulite GR1. Note the lack of core-rim features. See text for details.

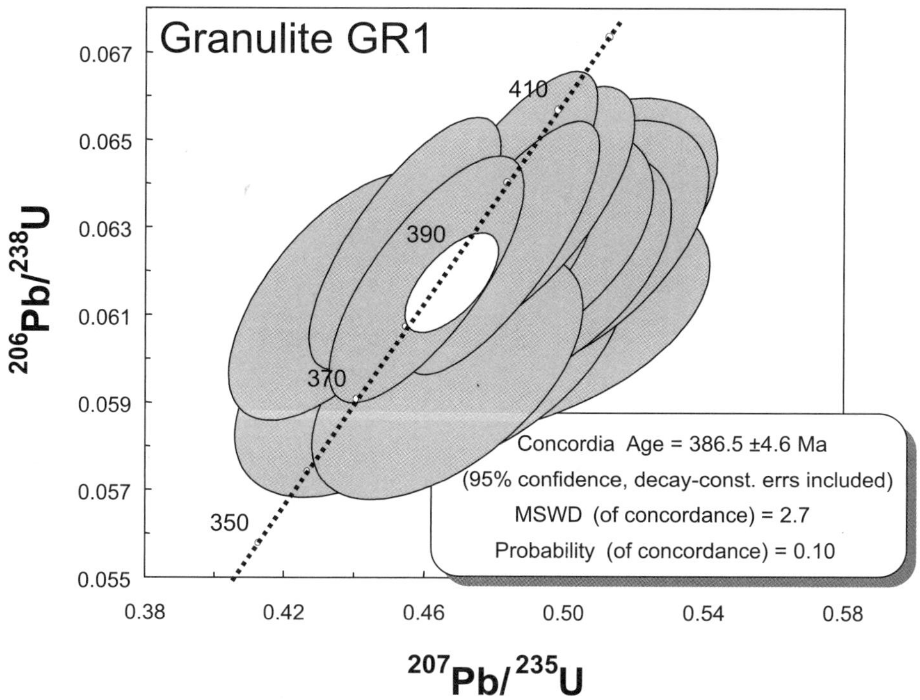

Figure 7. Concordia plot of U-Pb analyses from granulite GR1. White ellipse represents concordia age and error (Ludwig, 1998) calculated with Isoplot 3.00 (Ludwig, 2003). Gray ellipses represent 2σ uncertainties. MSWD—mean square of weighted deviates.

analytically indistinguishable from ages in zircon type (3) of the parent granulite GR2 (same outcrop as LC1), which strengthens the interpretation that zircon type (3) in GR2 represents new zircon crystallization from a melt phase.

DISCUSSION: AGE CONSTRAINTS ON HIGH-*P* GRANULITE METAMORPHISM

Zircons from granulite GR1 yielded mostly concordant ages between 395 and 370 Ma (weighted mean of $^{206}Pb/^{238}U$ ages = 387 ± 5 Ma, mean square of weighted deviates [MSWD] = 1.1, $N = 17$). The following features are of interest: (1) these zircons had no preserved cores of the ca. 515 Ma protolith zircons (as GB1), and no intermediate (reset) ages between 515 Ma and ca. 395–370 Ma were found, and (2) ages of centers and borders of crystals were analytically indistinguishable (consistent with the absence of core-rim features in CL images). Given these two features, two interpretations are possible: (1) zircons from granulite GR1 might represent complete or almost complete resetting of the protolith zircons during granulite metamorphism (highly unlikely)—this would imply the assumption that the protolith gabbro of this particular sample is equivalent in zircon features to sample GB1; or (2) the actual protolith of GR1 is not equivalent to gabbro GB1, and it contained very little or no igneous zircon, and therefore all or most zircon in GR1 is of metamorphic origin (as suggested by CL features). Assuming that these are metamorphic zircons, their mean age (387 ± 5 Ma) represents a minimum age constraint for the high-*P* granulite event in this unit.

The history of granulite GR2 started with crystallization of the igneous protolith at ca. 490–520 Ma (coeval with the age of gabbro GB1), but, as described already, the zircon features are more complex in this rock. Some analyses of cores in granulite GR2 yielded concordant ages (i.e., 2σ error ellipses intercept the concordia curve) between ca. 480 and 420 Ma (Fig. 9), but the imprecise $^{207}Pb/^{206}Pb$ ages, the large errors, and the slope of the concordia curve in this region preclude the possibility of ascertaining whether some of these analyses represent (undisturbed) crystallization ages or simply fall on a discordia mixing line. Given the spread of measured $^{206}Pb/^{238}U$ ratios, we consider that these ages are likely to represent protolith (igneous) zircons that have lost part of their Pb during the high-*P* and high-*T* event. This is shown graphically in Figure 9B, where the $^{206}Pb/^{238}U$ ages have been statistically treated with the TuffZirc algorithm of Ludwig (2003)—the ages cascade from the protolith age and reach a coherent plateau at an age of 404 ± 6 Ma, which can be considered a minimum statistically robust age for the onset of new zircon growth in GR2. It should be noted, however, that our data set is limited, and the existence of older overgrowths cannot be ruled out.

Based on this, we infer that the peak pressure of the high-*P* and high-*T* granulite event took place before the onset of new zircon crystallization, which started at ca. 404 Ma or earlier. Since neither the *P-T* trajectory nor the exhumation rate of these granulites is tightly constrained, it is not possible at present to estimate the time elapsed between the peak of high-*P* granulite metamorphism and decompression melting.

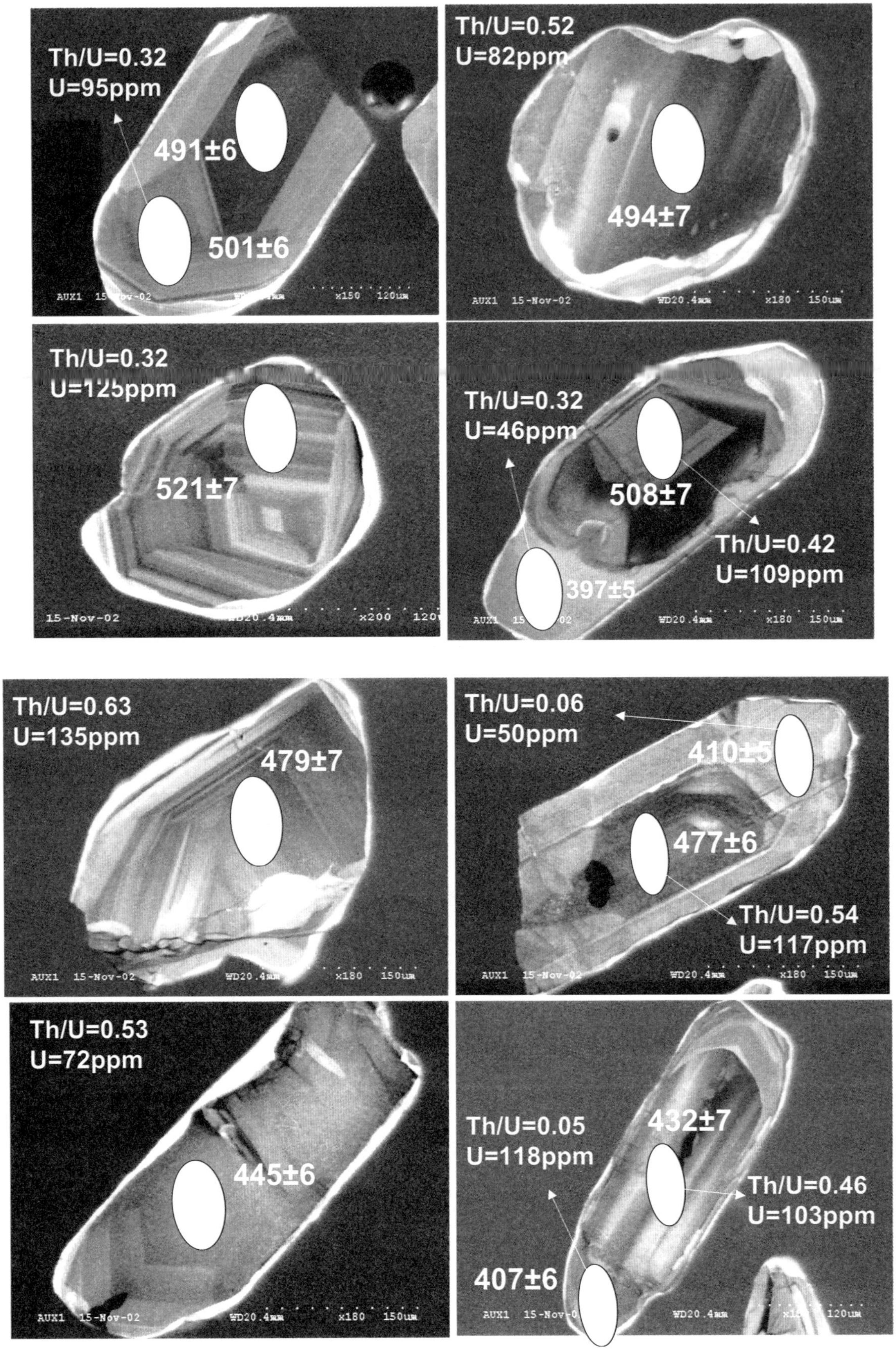

Figure 8 (*on this and following page*). Cathodoluminescence images of zircons from granulite GR2. See text for details.

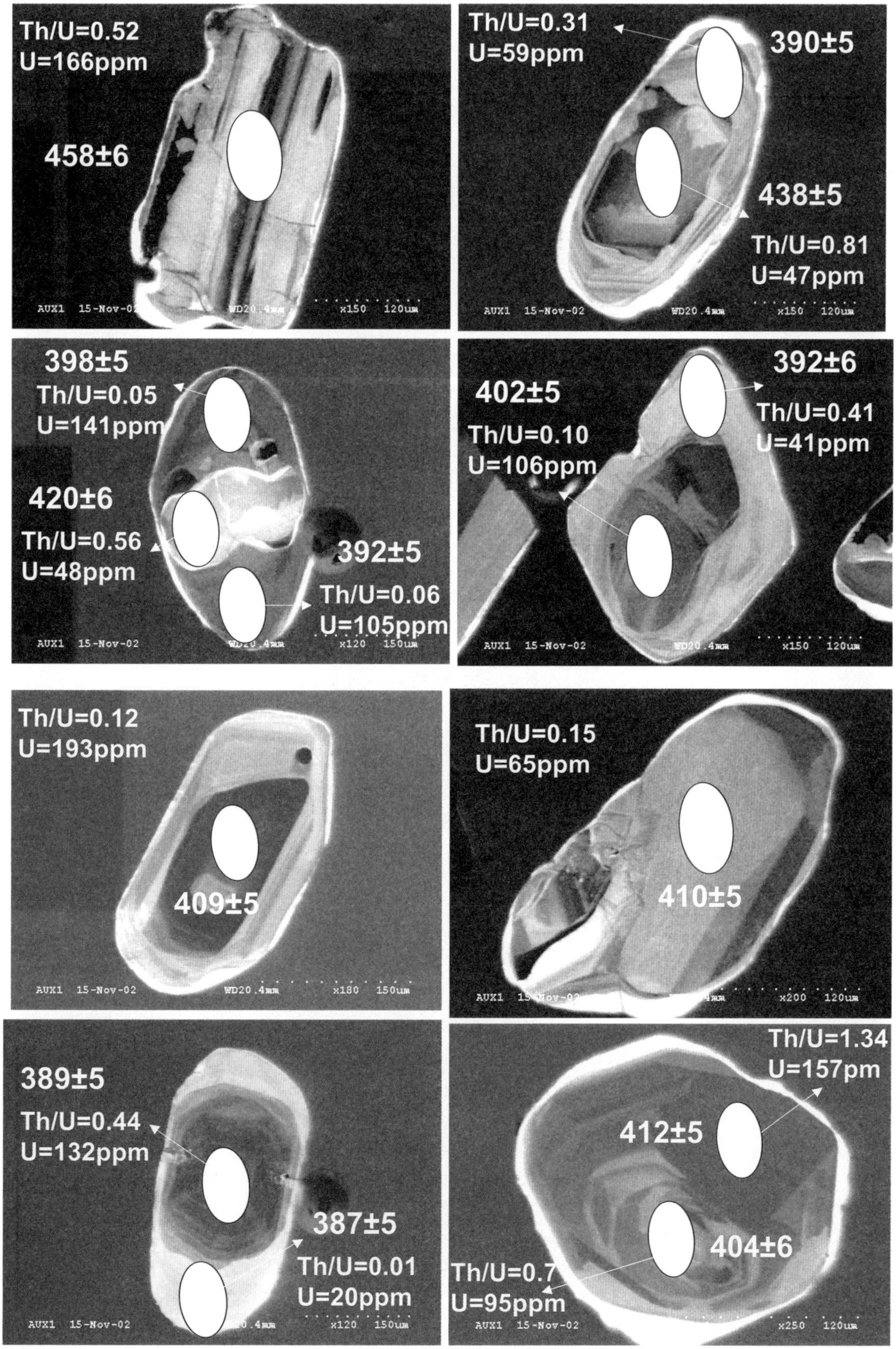

Figure 8. (*continued*)

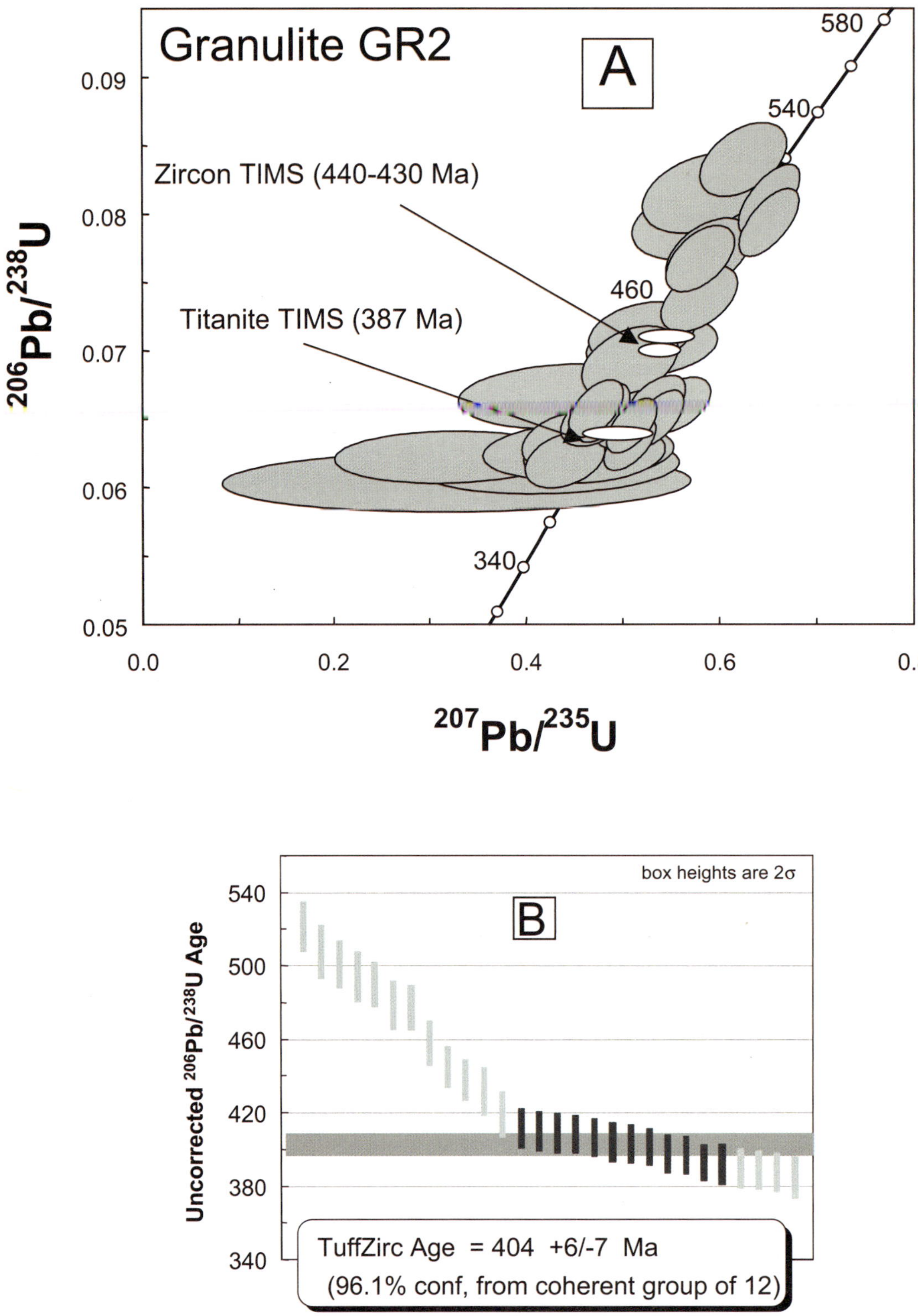

Figure 9.(A) Concordia plot of all U-Pb analyses from granulite GR2 (gray ellipses, 2σ uncertainties). The white ellipses represent isotope dilution–thermal ionization mass spectrometry (ID-TIMS) zircon and titanite U-Pb analyses reported by Fernández-Suárez et al. (2002) in a leucosome from the same unit (location is given in Fig. 1, sample 9). (B) Median age value of U-Pb ages (uncorrected $^{206}Pb/^{238}U$ ages) from granulite GR2 obtained with the TuffZirc algorithm (Ludwig, 2003). A statistically coherent group of ages, the median value of which is given in the graph, is interpreted as an indication that 404 Ma is the minimum age for the onset of new zircon growth in the granulite. The "cascading" gray bars are interpreted to represent analyses of partially reset protolith zircons with variable amounts of Pb loss during high-*P* and high-*T* granulite metamorphism.

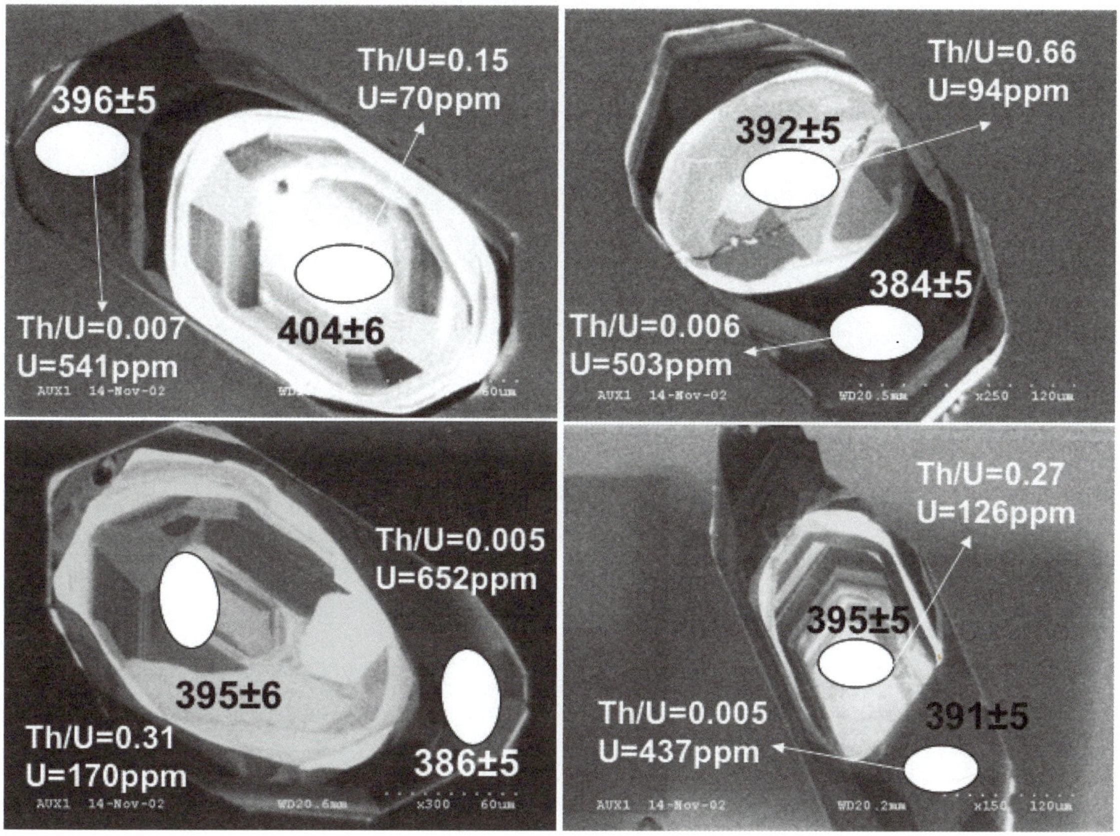

Figure 10. Cathodoluminescence images of zircons from leucosome LC1. See text for details.

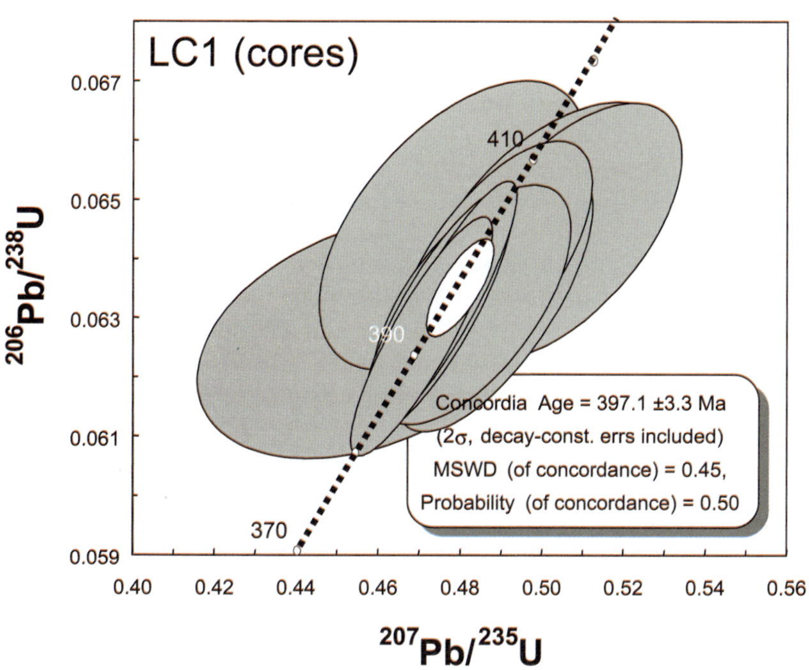

Figure 11. Concordia plot of U-Pb analyses from bright cathodoluminescence (CL) zircon cores in leucosome LC1. White ellipse represents concordia age and error calculated with Isoplot 3.00 (Ludwig, 2003). Gray ellipses represent 2σ uncertainties. MSWD—mean square of weighted deviates.

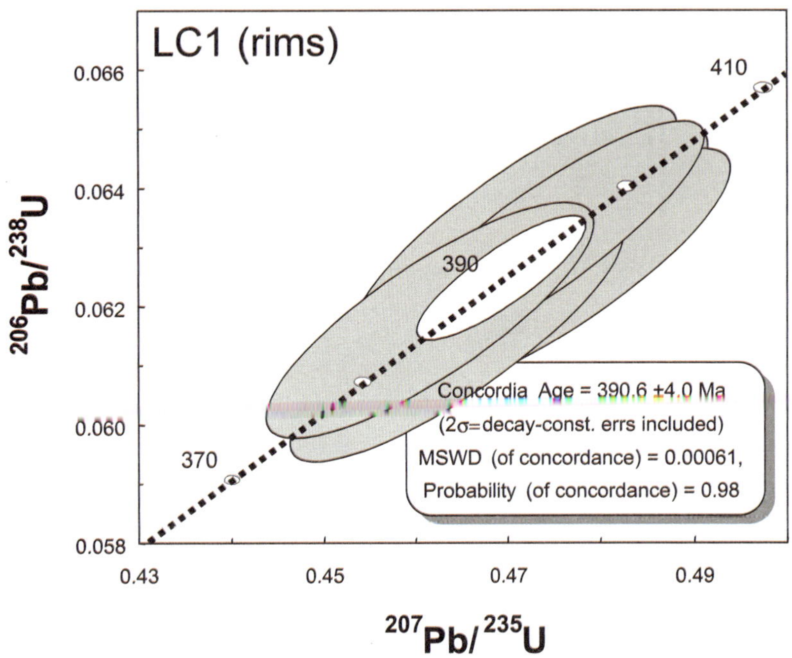

Figure 12. Concordia plot of U-Pb analyses from dark, high-U cathodoluminescence (CL) zircon rims in leucosome LC1. White ellipse represents concordia age and error calculated with Isoplot 3.00 (Ludwig, 2003). Gray ellipses represent 2σ uncertainties. MSWD—mean square of weighted deviates.

Fernández-Suárez et al. (2002) reported two ID-TIMS analyses of abraded multigrain low-U (<13ppm) equant multifaceted zircon in a leucosome sample from the same unit as GR2 (sample 9 in Fig. 2) with $^{206}Pb/^{238}U$ ages at 443 and 429 Ma (Fig. 9). The imprecise $^{207}Pb/^{206}Pb$ ages of 372 and 450 Ma, respectively, and the multigrain nature of the fractions preclude a straightforward interpretation of these analyses in conjunction with the new SIMS data. However, in the light of the data presented here, it is conceivable to view those low-U zircons as related to the high-*P* granulite metamorphism, the peak of which must have predated the oldest zircons of leucosome LC1 and the type (3) zircon in sample GR2. Fernández-Suárez et al. (2002) also reported three overlapping concordant ages obtained on abraded fractions of stubby zircon prisms with $^{207}Pb/^{206}Pb$ ages of 486, 492, and 494 Ma in leucosome 9 (Fig. 2). These ages are at variance with ion microprobe ages obtained on zircons from leucosome LC1 and suggest that the 486–494 Ma ages could represent igneous protolith zircons entrained in the leucosome melt.

It is evident from this that an interpretation based only on data from samples GB1 and GR1 (Órdenes Complex) would point to a ca. 390–380 Ma age for the high-*P* granulite event. However, results from sample GR2 and LC1 (Cabo Ortegal Complex) suggest that melting started at 404 Ma or earlier, and therefore the age of the high-*P* granulite event must be older than 404 Ma and younger than 490–520 Ma (age of protolith gabbros; Abati et al., 1999). Currently accepted geological and petrologic constraints (e.g., Martínez Catalán et al., 1997, 1999) argue against the possibility that the units containing the GR1 and GR2 samples have had a significantly different tectonothermal evolution, but this possibility cannot be entirely dismissed in view of the data presented herein.

The contention of this study, that the pressure peak of granulite metamorphism must be older than ca. 404 Ma, is in agreement with amphibole $^{40}Ar/^{39}Ar$ ages as old as 400–420 Ma for retrogressive amphibolite-facies foliation in high-*P* and high-*T* units (Dallmeyer et al., 1997; Gómez Barreiro et al., 2006). The $^{40}Ar/^{39}Ar$ data are an indication that these units were being exhumed prior to 400 Ma, and therefore the U-Pb ages of zircon reported herein cannot be said to date the peak pressure of the granulite event but a segment of the exhumation path. The relatively large errors of both U-Pb (this study) and $^{40}Ar/^{39}Ar$ (Dallmeyer et al., 1997; Gómez Barreiro et al., 2006) ages preclude any speculation on the relative chronology recorded by the U-Pb and $^{40}Ar/^{39}Ar$ systems in these units; even when taken at face value, they suggest that zircon could have been forming from a melt or fluid phase as the rock went through the amphibole closure temperature for the $^{40}Ar/^{39}Ar$ system. Further to this, the titanite U-Pb ages reported by Fernández-Suárez et al. (2002) at 387 Ma are within error of the youngest ages of zircon in granulite GR2 and leucosome LC1. This is significant because low-Al titanite is on the low-*P* side of metamorphic reactions involving Ti phases in mafic granulites (e.g., Bohlen and Liotta, 1986), which strengthens the notion that zircon grew in the retrograde path, and therefore none of the U-Pb ages reported so far (including this study) dates the peak pressure of the high-*P* and high-*T* event, but, instead, they date a section of the retrograde path.

Paragneisses associated with these mafic high-*P* and high-*T* granulites (e.g., sample 18 in Fig. 3) preserve monazites that yield concordant ages at 495–485 Ma (Fernández-Suárez et al., 2002), and these ages attest to the existence of a Late Cambrian–Early Ordovician metamorphic event shortly following (and partially overlapping) the crystallization of the igneous protoliths (Abati et al., 1999). If we assume, as discussed already, that the high-*P* and high-*T* granulite metamorphism is younger than this high-*T* event, then caution should be used in the interpretation of metamorphic monazite ages because the U-Pb system may remain unperturbed at high-*P* granulite-facies conditions. This is consistent with recent experimental data that suggest that closure temperature in monazite may exceed that of zircon (Cherniak et al., 2004). The fact that the ages for this early high-*T* metamorphism overlap with the crystallization ages of gabbros and granitoids suggests that this event could be related to a crustal-scale thermal increase that took place in an arc environment, as suggested by Abati et al. (1999) and Martínez Catalán et al. (this volume).

FURTHER CONSIDERATIONS

Although the *P-T* paths of these rocks have been established (Arenas, 1991; Arenas and Martínez Catalán, 2002), it is not constrained whether or not initial decompression occurred with increasing temperature (curvature of the *P-T* loop), which could in turn be indicative of how close to peak pressure the decompression melting started. There is no independent evidence for the exhumation rate of these units, so it is not possible to estimate where in the gap between 495 and 480 and 420–400 Ma the peak pressure is most likely to have been.

The examples shown by GR1 and GR2 are also an indication that what we see in zircon from high-*P* granulite rocks depends on the abundance (presence) and setting of zircon (and or zirconium) in the protolith. Rocks such as GR2 with abundant "protolith" zircon can be expected to have a wider variety of textural, chemical, and isotopic features of zircon after a high-grade metamorphic event and concomitant changes in *P-T*–driven mineral reactions, fluid activity, presence of melt, etc. Therefore, rocks such as GR2 where "premetamorphic" zirconium is mostly tied up in zircon are likely to record a more complete and complex history of metamorphic event(s) in their U-Pb system.

Rocks like GR1, with very little or no initial zircon, and in which much of the premetamorphic zirconium is likely to have been tied up in major phases, can record an apparently different history with slightly different (and possibly misleadingly simple) chronology. In the case reported here, the U-Pb system in zircons from GR1 seems not to have recorded any events between the crystallization of the gabbros and 390–380 Ma and would give a simplistic view that the high-*P* metamorphic event took place at ca. 390–380 Ma. In linking petrogenetic processes with zircon formation in rocks where most of the initial zirconium is tied up in essential minerals, a question arises: which reactions free zirconium and make it available for zircon crystallization during granulite metamorphism: Prograde reactions? Retrograde reactions? Or both?

GEODYNAMIC SIGNIFICANCE

In the context of the geodynamic evolution of the allochthonous complexes of northwest Iberia, the data presented herein support the hypothesis of Martínez Catalán et al. (this volume) whereby the high-*P* granulite event probably reflects the understacking of these units following accretion to a large continental mass, either Baltica or Laurentia (Fig. 5 in Martínez Catalán et al., this volume). Accretion took place in eo-Variscan times and is therefore unrelated to the Variscan collisional orogenesis (sensu stricto). This accretion was heralded by the closure of the Iapetus or Tornquist Oceans and concomitant opening of the Rheic Ocean (Murphy et al., 2006). The time frame proposed here for the high-*P* event is compatible with the age of Acadian events, and therefore it is possible that the high-*P* and high-*T* units of the allochthonous complexes of northwest Iberia were part of the Laurussian margin of the Rheic Ocean until they were thrust upon the autochthonous terranes of northwest Iberia (representing the Gondwanan margin of the Rheic) in the course of the Variscan-Alleghanian collision.

ACKNOWLEDGMENTS

Fernández-Suárez wishes to acknowledge a Training and Mobility of Researchers (TMR) grant (HIGH_LAT program) to access the NORDSIM facility. Kjell Bilstrom is thanked for his kindness during Fernández-Suárez's stay in Stockholm. Robert Hatcher, Jr., is thanked for organizing the BTA 17 meeting and making it a success. The NORDSIM facility is financed and operated under an agreement between the joint Nordic Research Councils (NOS-N), the Geological Survey of Finland, and the Swedish Museum of Natural History. This is NORDSIM contribution #135. John Aleinikoff and Brent Miller are kindly acknowledged for insightful reviews of the manuscript.

REFERENCES CITED

Abati, J., Dunning, G.R., Arenas, R., Díaz García, F., González Cuadra, P., Martínez Catalán, J.R., and Andonaegui, P., 1999, Early Ordovician orogenic event in Galicia (NW Spain): Evidence from U-Pb ages in the uppermost unit of the Órdenes Complex: Earth and Planetary Science Letters, v. 165, p. 213–228, doi: 10.1016/S0012-821X(98)00268-4.

Arenas, R., 1991, Opposite *P,T,t* paths of Hercynian metamorphism between the upper units of the Cabo Ortegal Complex and their substratum (northwest of the Iberian Massif): Tectonophysics, v. 191, p. 347–364, doi: 10.1016/0040-1951(91)90067-3.

Arenas, R., and Martínez Catalán, J.R., 2002, Prograde development of corona textures in metagabbros of the Sobrado unit (Órdenes Complex, northwestern Iberian Massif), *in* Martínez Catalán, J.R., Hatcher, R.D., Jr., Arenas, R., and Díaz García, F., eds., Variscan-Appalachian Dynamics: The Building of the Late Paleozoic Basement: Geological Society of America Special Paper 364, p. 73–88.

Arenas, R., Rubio Pascual, F.J., Díaz García, F., and Martínez Catalán, J.R., 1995, High-pressure micro-inclusions and development of an inverted metamorphic gradient in the Santiago schists (Órdenes Complex, NW Iberian Massif, Spain): Evidence of subduction and syn-collisional decompression: Journal of Metamorphic Geology, v. 13, p. 141–164.

Arenas, R., Martínez Catalán, J.R., Sánchez Martínez, S., García, F., Abati,

J., Fernández-Suárez, J., Andonaegui, P., and Gómez-Barreiro, J., 2007, this volume, Paleozoic ophiolites in the Variscan suture of Galicia (northwest Spain): Distribution, characteristics, and meaning, *in* Hatcher, R.D., Jr., Carlson, M.P., McBride, J.H., and Martínez Catalán, J.R., eds., 4-D Framework of Continental Crust: Geological Society of America Memoir 200, doi: 10.1130/2007.1200(22).

Bohlen, S.R., and Liotta, J.J., 1986, A barometer for garnet amphibolites and garnet granulites: Journal of Petrology, v. 27, p. 1025–1034.

Cherniak, D.J., Watson, E.B., Grove, M., and Harrison, T.M., 2004, Pb diffusion in monazite: A combined RBS/SIMS study: Geochimica et Cosmochimica Acta, v. 68, p. 829–840, doi: 10.1016/j.gca.2003.07.012.

Dallmeyer, R.D., and Tucker, R.D., 1993, U-Pb zircon age for the Lagoa augen gneiss, Morais Complex, Portugal: Tectonic implications: Journal of the Geological Society of London, v. 150, p. 405–410.

Dallmeyer, R.D., Martínez Catalán, J.R., Arenas, R., Gil Ibarguchi, J.I., Gutiérrez Alonso, G., Farias, P., Aller, J., and Bastida, F., 1997, Diachronous Variscan tectonothermal activity in the NW Iberian Massif: Evidence from $^{40}Ar/^{39}Ar$ dating of regional fabrics: Tectonophysics, v. 277, p. 307–337, doi: 10.1016/S0040-1951(97)00035-8.

Díaz García, F., Arenas, R., Martínez Catalán, J.R., González del Tánago, J., and Dunning, G., 1999, Tectonic evolution of the Careón ophiolite (northwest Spain): A remnant of oceanic lithosphere in the Variscan belt: The Journal of Geology, v. 107, p. 587–605, doi: 10.1086/314368.

Fernández Suárez, J., Corfu, F., Arenas, R., Marcos, A., Martínez Catalán, J.R., Díaz García, F., Abati, J., and Fernández, F.J., 2002, U-Pb evidence for a polyorogenic evolution of the HP-HT units of the NW Iberian Massif: Contributions to Mineralogy and Petrology, v. 143, p. 236–253, doi: 10.1007/s00410-001-0337-2.

Gómez Barreiro, J., Wijbrans, J.R., Castiñeiras, P., Martínez Catalán, J.R., Arenas, R., Díaz García, F., and Abati, J., 2006, $^{40}Ar/^{39}Ar$ laserprobe dating of mylonitic fabrics in a polyorogenic terrane of NW Iberia: Geological Society [London] Journal, v. 163, p. 61–73.

Kuijper, R.P., 1980, Precambrian U-Pb zircon ages from western Galicia (NW Spain): Earth-Science Reviews, v. 16, p. 313–316, doi: 10.1016/0012-8252(80)90051-3.

Ludwig, K.R., 1998, On the treatment of concordant uranium-lead ages: Geochimica et Cosmochimica Acta, v. 62, p. 665–676, doi: 10.1016/S0016-7037(98)00059-3.

Ludwig, K.R., 2003, Isoplot 3.00. A Geochronological toolkit for Microsoft Excel: University of California–Berkeley Geochronology Center Special Publication No. 4, 70 p.

Martínez Catalán, J.R., Arenas, R., Díaz García, F., Rubio Pascual, F.J., Abati, J., and Marquínez, J., 1996, Variscan exhumation of a subducted Paleozoic continental margin: The basal units of the Órdenes Complex, Galicia: NW Spain: Tectonics, v. 15, p. 106–121, doi: 10.1029/95TC02617.

Martínez Catalán, J.R., Arenas, R., Díaz García, F., and Abati, J., 1997, The Variscan accretionary complex of NW Iberia: Involved terranes and succession of tectonothermal events: Geology, v. 25, p. 1103–1106, doi: 10.1130/0091-7613(1997)025<1103:VACONI>2.3.CO;2.

Martínez Catalán, J.R., Arenas, R., Díaz García, F., and Abati, J., 1999, Allochthonous units in the Variscan belt of NW Iberia: Basement Tectonics, v. 13, p. 65–84.

Martínez Catalán, J.R., Arenas, R., Díaz García, F., Gómez-Barreiro, J., González Cuadra, P., Abati, J., Castiñeiras, P., Fernández-Suárez, J., Sánchez Martínez, S., Andonaegui, P., González Clavijo, E., Díez Montes, A., Rubio Pascual, F.J., and Valle Aguado, B., 2007, this volume, Space and time in the tectonic evolution of the northwestern Iberian Massif: Implications for the comprehension of the Variscan belt, *in* Hatcher, R.D., Jr., Carlson, M.P., McBride, J.H., and Martínez Catalán, J.R., eds., 4-D Framework of Continental Crust: Geological Society of America Memoir 200, doi: 10.1130/2007.1200(21).

McFarlane, C.R.M., Connelly, J.N., and Carlson, W.D., 2005, Intracrystalline redistribution of Pb during high-temperature contact metamorphism: Chemical Geology, v. 217, p. 1–20.

Murphy, J.B., Gutiérrez Alonso, G., Nance, R.D., Fernández-Suárez, J., Keppie, J.D., Quesada, C., Strachan, R.A., and Dostal, J., 2006, Origin of the Rheic Ocean: Rifting along a Neoproterozoic suture?: Geology, v. 34, p. 325–328, doi: 10.1130/G22068.1.

Ordóñez Casado, B., 1998, Geochronological studies of the pre-Mesozoic basement of the Iberian Massif: The Ossa Morena zone and the allochthonous complexes within the Central Iberian zone [Ph.D. dissertation]: Zurich, Swiss Federal Institute of Technology, 235 p.

Ordóñez Casado, B., Gebauer, D., Schäfer, H.J., Gil Ibarguchi, J.I., and Peucat, J.J., 2001, A single Devonian subduction event for the HP/HT metamorphism of the Cabo Ortegal Complex within the Iberian Massif: Tectonophysics, v. 332, p. 359–385, doi: 10.1016/S0040-1951(00)00210-9.

Peucat, J.J., Bernard-Griffiths, J., Gil Ibarguchi, J.I., Dallmeyer, R.D., Menot, R.P., Cornichet, J., and Iglesias Ponce de León, M., 1990, Geochemical and geochronological cross section of the deep Variscan crust: The Cabo Ortegal high-pressure nappe (northwestern Spain): Tectonophysics, v. 177, p. 263–292, doi: 10.1016/0040-1951(90)90285-G.

Roberts, M.P., and Finger, F., 1997, Do U-Pb zircon ages from granulites reflect peak metamorphic conditions?: Geology, v. 25, p. 319–322, doi: 10.1130/0091-7613(1997)025<0319:DUPZAF>2.3.CO;2.

Root, D.B., Hacker, B.R., Mattinson, J.M., and Wooden, J.L., 2004, Zircon geochronology and ca. 400 Ma exhumation of Norwegian ultrahigh-pressure rocks: An ion microprobe and chemical abrasion study: Earth and Planetary Science Letters, v. 228, p. 325–341, doi: 10.1016/j.epsl.2004.10.019.

Santos Zalduegui, J.F., Schärer, U., Gil Ibarguchi, J.I., and Girardeau, J., 1996, Origin and evolution of the Paleozoic Cabo Ortegal ultramafic-mafic complex (NW Spain): U-Pb, Rb-Sr and Pb-Pb isotope data: Chemical Geology, v. 129, p. 281–304, doi: 10.1016/0009-2541(95)00144-1.

Schäfer, H.J., Gebauer, D., Gil Ibarguchi, J.I., and Peucat, J.J., 1993, Ion-microprobe U-Pb zircon dating on the HP/HT Cabo Ortegal Complex (Galicia, NW Spain): Preliminary results: Terra Abstracts, v. 5, no. 4, p. 22.

Stacey, J.S., and Kramers, J.D., 1975, Approximation of terrestrial lead isotope evolution by a two-stage model: Earth and Planetary Science Letters, v. 26, p. 207–221, doi: 10.1016/0012-821X(75)90088-6.

Timmermann, H., Stedra, V., Gerdes, A., Noble, S.R., Parrish, R.R., and Dörr, W., 2004, The problem of dating high-pressure metamorphism: A U-Pb isotope and geochemical study on eclogites and related rocks of the Marianske Lazne Complex, Czech Republic: Journal of Petrology, v. 45, p. 1311–1338, doi: 10.1093/petrology/egh020.

Wiedenbeck, M., Allé, P., Corfu, F., Griffin, W.L., Meier, M., Orbeli, F., von Quadt, A., Roddick, J.C., and Spiegel, W., 1995, Three natural zircon standards for U-Th-Pb, Lu-Hf, trace element and REE analyses: Geostandards Newsletter, v. 19, p. 1–23.

Whitehouse, M.J., and Platt, J.P., 2003, Dating high-grade metamorphism—Constraints from rare-earth elements in zircon and garnet: Contributions to Mineralogy and Petrology, v. 145, p. 61–74.

Whitehouse, M.J., Kamber, B., and Moorbath, S., 1999, Age and significance of U-Th-Pb zircon data from early Archaean rocks of west Greenland—A reassessment based on combined ion microprobe and imaging studies: Chemical Geology, v. 160, p. 201–224, doi: 10.1016/S0009-2541(99)00066-2.

Manuscript Accepted by the Society 1 February 2007

Printed in the USA

Geological Society of America
Memoir 200
2007

Flexural to broken foreland basin evolution as a result of Variscan collisional events in northwestern Spain

M. Keller*
Institut für Geologie und Mineralogie, Universität Erlangen, Schlossgarten5, D-91054 Erlangen, Germany

H. Bahlburg
Geologisch-Paläontologisches Institut, Universität Münster, Corrensstraße 24, D-48149 Münster, Germany

C.D. Reuther
Geologisch-Paläontologisches Institut und Museum, Universität Hamburg, Bundesstrasse 55, 20146 Hamburg, Germany

A. Weh
Institut für Geologie und Mineralogie, Universität Erlangen, Schlossgarten5, D-91054 Erlangen, Germany

ABSTRACT

The Variscan Cantabrian Zone of northern Spain records the transition from passive-margin sedimentation to foreland basin deposition during the Late Devonian. Initiation of the foreland basin system is reflected in the evolution of a hinge line that separated the foredeep from the adjacent peripheral bulge. Bulge erosion provided siliciclastic detritus that was deposited in the evolving foredeep. After cessation of sedimentary input, the foredeep achieved starved conditions during the earliest Tournaisian. These events were a long-distance effect of initial subduction of the Iberian continental margin prior to collision between northwestern Gondwana and North America. During the Namurian, propagation of the deformation front and the rise of the orogen led to increasing input of detritus that began to fill the foredeep.

During the late Namurian and the early Westphalian, deformation started to affect the Cantabrian Zone. The Palencian nappes were emplaced onto the Valsurvio Dome, after which thick wedge-top deposits formed (the Curavacas Conglomerate). When the Esla nappe and the Valsurvio Dome became involved in deformation and approached the rigid lithosphere of the Cantabrian Block, anticlinal stacks began to form, and the lithosphere broke. Two major faults developed, the León fault and the Ruesga fault, which dissected the lithosphere to accommodate further compression. The resulting topography led to the gravitationally induced gliding of the Palencian nappes into the Pisuerga Carrión unit, where sedimentation and deformation continued under broken

*keller@geol.uni-erlangen.de

Keller, M., Bahlburg, H., Reuther, C.D., and Weh, A., 2007, Flexural to broken foreland basin evolution as a result of Variscan collisional events in northwestern Spain, *in* Hatcher, R.D., Jr., Carlson, M.P., McBride, J.H., and Martínez Catalán, J.R., eds., 4-D Framework of Continental Crust: Geological Society of America Memoir 200, p. 489–510, doi: 10.1130/2007.1200(25). For permission to copy, contact editing@geosociety.org.

foreland basin conditions. The Cantabrian Mountains preserve a foreland basin system that in its early stage was formed under flexural deformation of the lithosphere induced by subduction. When deformation affected the autochthonous continental lithosphere, the size of the flexurally induced basins changed from a single, large-scale depocenter to smaller-scale, interconnected basins. In its late stage of development, the Cantabrian foreland basin system records the transition to broken foreland conditions as the thrust front approached the rigid Cantabrian Block, a basement feature that has influenced the tectonostratigraphic evolution of the Cantabrian Zone since Silurian times. Incipient oroclinal bending during the main stage of thrusting might have been an additional factor triggering the failure of the lithosphere.

Keywords: Cantabrian Mountains, Devonian, Carboniferous, broken foreland basin, León fault, Ruesga fault, Variscan orogeny.

INTRODUCTION

Foreland basins systems (DeCelles and Giles, 1996) are depositional centers that form in response to compressional plate-tectonic events and the resulting deformation of the continental lithosphere. The foreland basin system discussed in this paper developed during the late Variscan formation of Pangea, which included the collision between Gondwana and Laurussia. During this collision, several smaller-scale continental fragments were trapped between the colliding landmasses, and each of them has a distinct deformational history (Franke, 2000; Stampfli et al., 2002). In many cases, neither the exact timing nor the interactions between the terranes during accretion are clear (Franke et al., 2000; Robardet, 2000, 2003; McKerrow et al., 2000).

The foreland basin system in the Cantabrian Mountains (Figs. 1 and 2) developed during the Late Devonian (Keller, 2000) and persisted until the close of the Variscan orogeny during the Stephanian. In the Cantabrian Mountains, recognition of foreland basin initiation is based on the geometry and stratigraphy of the Upper Devonian deposits and on a basinwide unconformity that is interpreted to represent the erosion of the peripheral bulge (Keller, 2000). The first episode of foreland basin sedimentation

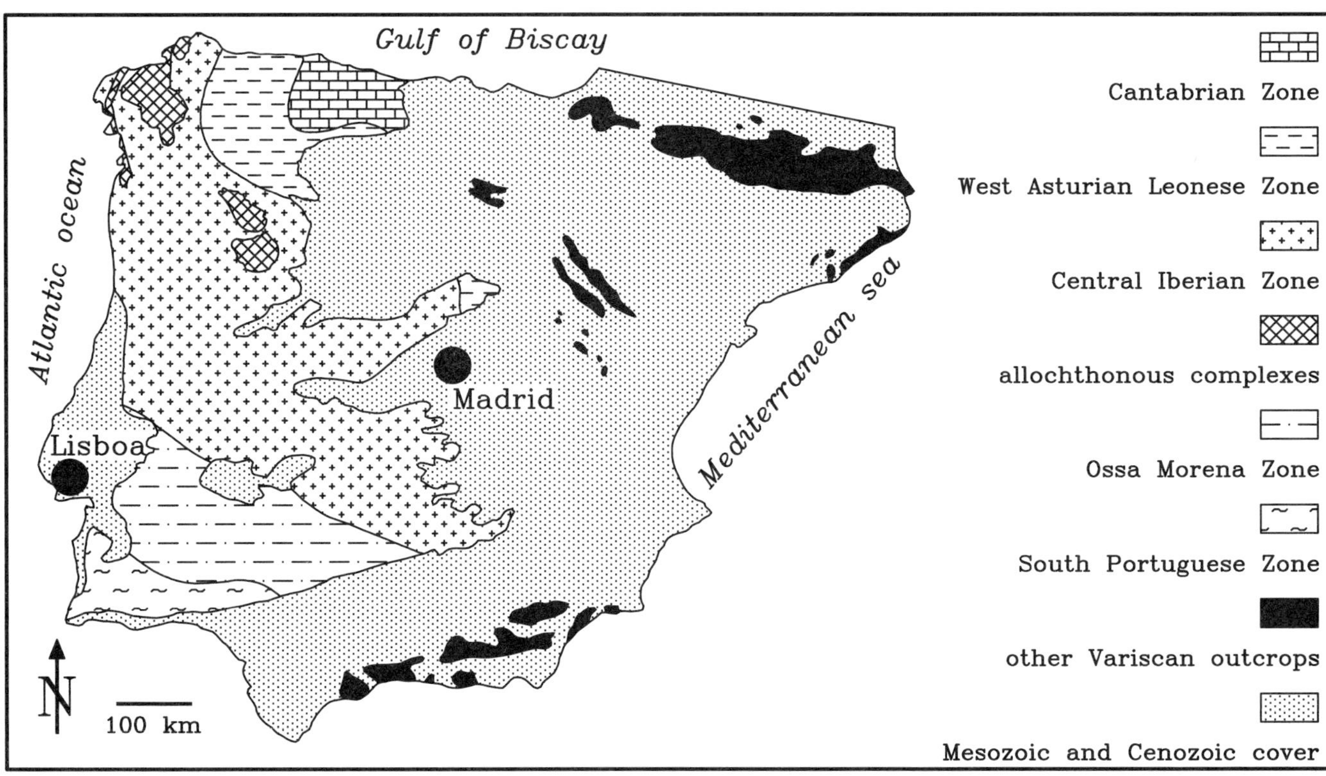

Figure 1. Subdivision of the Iberian Massif into zones following Lotze (1945) and Julivert (1971)

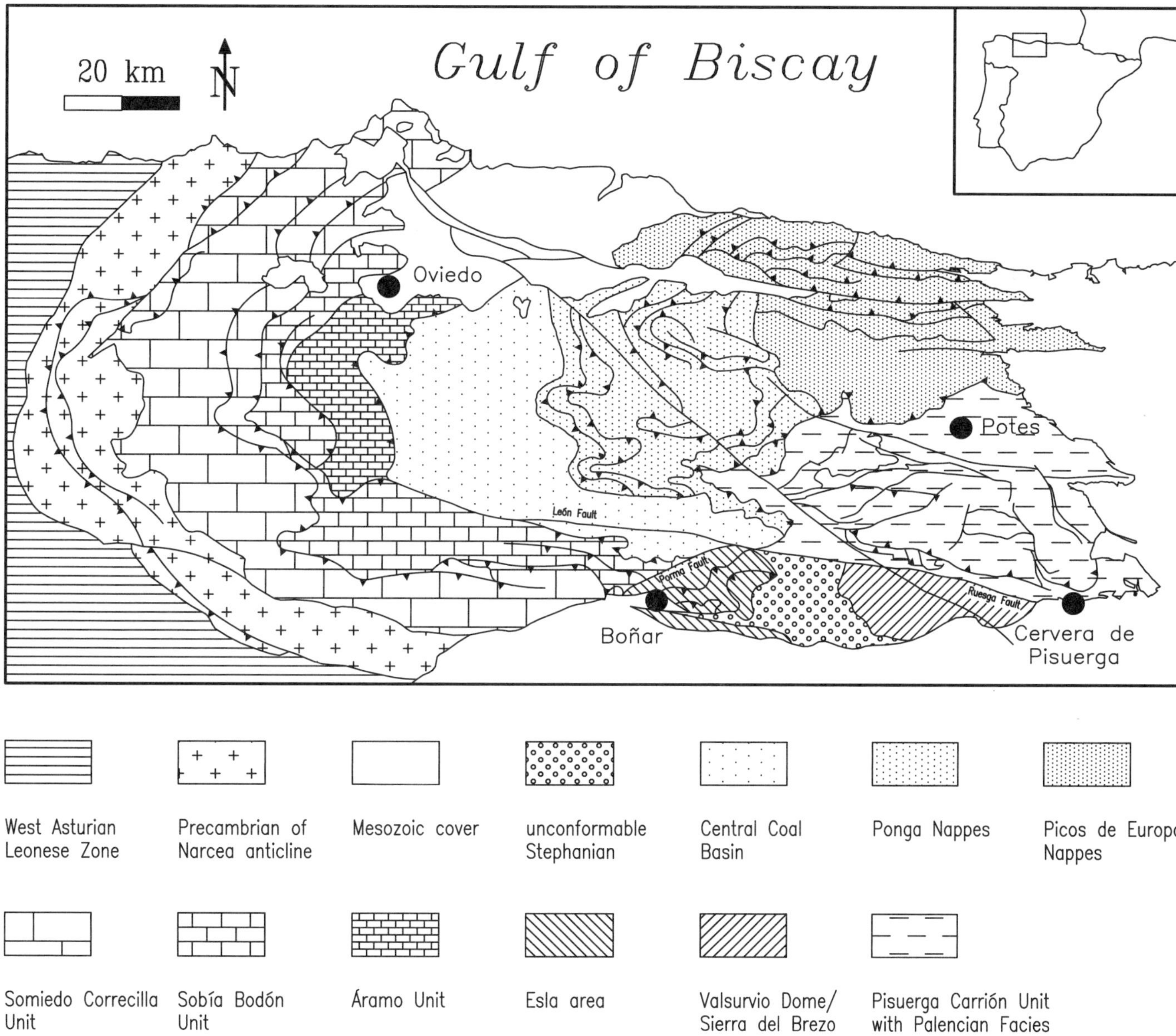

Figure 2. Structural and paleogeographic components of the Cantabrian Zone. Paleogeographically, the Central Coal Basin, the Ponga Nappe unit, and the Picos de Europa nappes are part of the Cantabrian Block (Radig, 1962). The Somiedo Correcilla unit, the Sobian Bodón unit, and the Áramo unit are the Fold-and Nappe Province (modified from Julivert, 1971). Not all occurrences of unconformable Stephanian are shown.

continued for several millions of years and is characterized by pelagic, condensed successions.

During the Namurian and Westphalian, a depocenter developed in the southeastern Cantabrian Zone that accumulated thick successions of synorogenic detritus. This basin is known as the Pisuerga Carrión Unit (Figs. 2 and 3). Its basin fill traditionally has been interpreted to represent syntectonic sedimentation in a foreland basin, the "flysch basin" of the Cantabrian Mountains sensu Julivert (1978), Reuther (1977), and Maas (1974). In recent models (Rodríguez Fernández, 1993, 1994; Rodríguez Fernández and Heredia, 1987), unconformity-bounded sedimentary units are attributed to the advance of discrete nappe piles, and a succession of individual foreland basins has been reconstructed.

The intentions of this paper are (1) to document the Late Devonian through Pennsylvanian foreland basin evolution of the Cantabrian Mountains as represented by the sedimentary basin fill; (2) to document that there was a single, long-lived foreland basin, which reflects flexural bending of the lithosphere in its early evolution and the abrupt change to broken foreland conditions during its late stage; and (3) to propose a new model for the

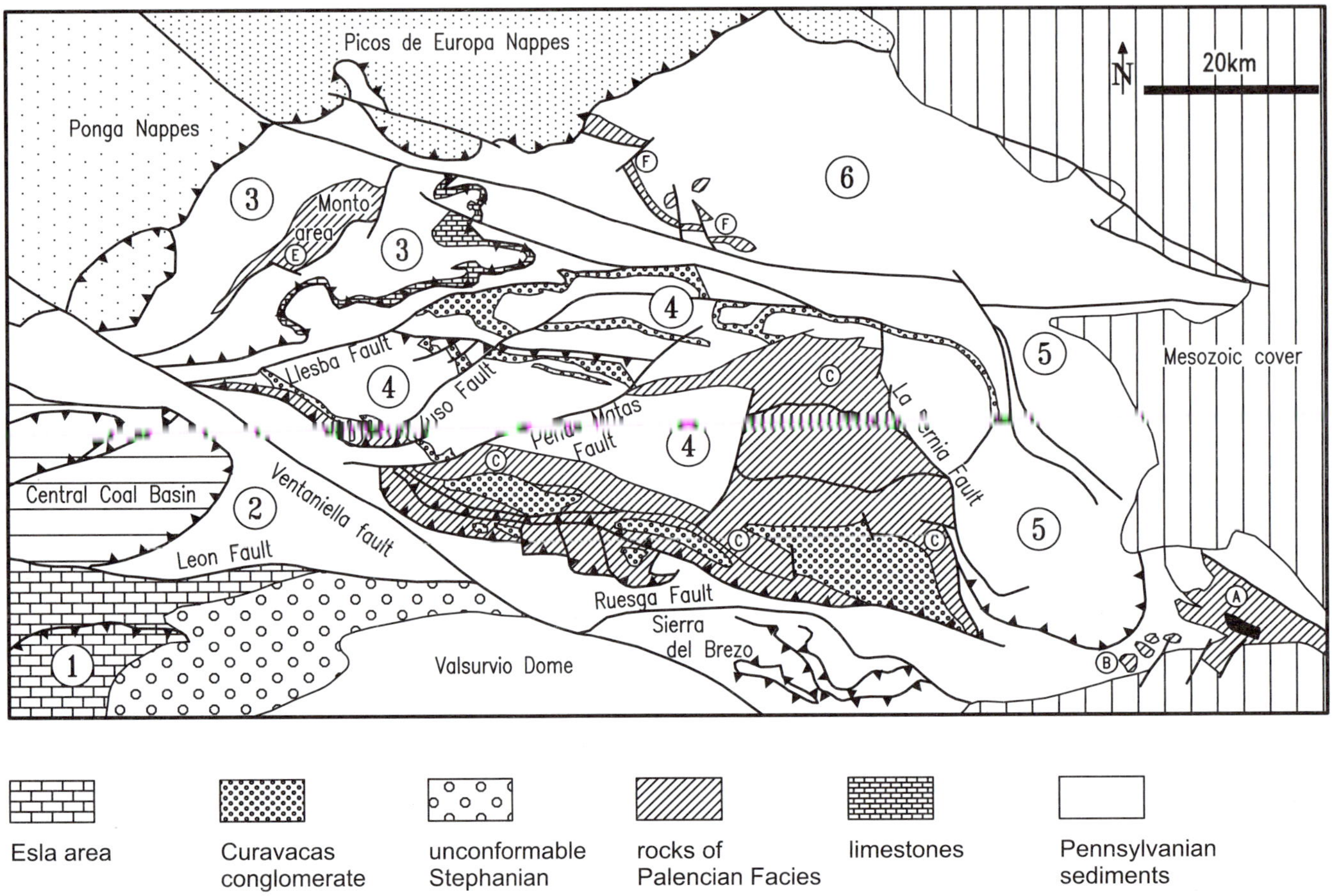

Figure 3. Simplified geological map of the Pisuerga Carrión unit (modified from Alonso and Rodríguez Fernández, 1983) and major fault systems. Areas 1–6 are those of Figure 4B. Individual occurrences of Palencian Facies are (A) Barruelo area with San Julián klippe, (B) Muda nappes, (C) Arauz or Alto Carrión unit, (D) Sierra de Hormas, (E) Gildar-Montó area, and (F) Liébana area. Black is Revilla nappe.

interaction between sedimentation and tectonics in the Pisuerga Carrión unit.

Our interpretation of the Carboniferous sedimentation in the Cantabrian Zone is based on different models that were developed for individual depocenters to explain sedimentary patterns during the Variscan orogeny. Together with our own observations, these models are combined in a single comprehensive model that tries to duplicate the events that created a foreland basin system during the Late Devonian and that are responsible for a complex architecture of basin-fill sediments.

REGIONAL SETTING

The Northern Iberian Massif

East-west cross sections through the Variscan rocks of northwestern Spain (Julivert, 1971, 1979, 1981; Marcos and Pulgar, 1982; Fernández Viejo et al., 2000; Gallastegui et al., 1997; Pérez-Estaún et al., 1988, 1994) show the Central Iberian Zone (Julivert, 1971) in the core of the Iberian Massif, where Variscan deformation started during the Devonian (Dallmeyer et al., 1997; Martínez Catalán et al., 1997). During the Late Devonian, west-directed subduction of the northwestern Iberian margin beneath an accretionary wedge preceded the main collisional event that formed the Iberian Variscan orogen (Rodríguez et al., 2003). During the Mississippian, several allochthonous complexes including oceanic crust were obducted onto the autochthonous Central Iberian Zone, where they are preserved as klippen in northwestern Spain and northern Portugal (Fig. 1; Arenas et al., 2000).

The Central Iberian Zone (Julivert, 1971) is the most internal zone of the Iberian Massif. It is composed of Upper Proterozoic terrigenous sediments and volcanic rocks (today mainly gneisses) overlain by Lower Ordovician marine metasediments. The youngest pre-orogenic rocks are Silurian and Lower Devonian platform deposits. The first deformation affected the Central Iberian Zone at ca. 359 Ma (Dallmeyer et al., 1997), shortly after the Devonian-Carboniferous boundary.

The West Asturian Leonese Zone (Fig. 1) shows many characteristics of thick-skinned tectonics (Martínez Catalán et al., 1990). Stratigraphically, it is composed of a very thick succession of Cambrian through Silurian sedimentary rocks (Julivert, 1983; Pérez-Estaún et al., 1990) that locally accumulated to almost 10 km of strata. In contrast, Lower Devonian rocks are very thin and are only present in a few localities (Drot and Matte, 1967; Truyols Massoni, 1986). Emsian through Westphalian sediments are nowhere preserved. Stephanian molasse deposits unconformably overlie the Lower Paleozoic rocks (Pérez-Estaún et al., 1990). Carls (1988) claimed that many of the intervening deposits had been stripped off and transported toward the Cantabrian Zone as the Palencian nappes.

During the Mississippian, at around 336 Ma (Dallmeyer et al., 1997), deformation began to affect strata close to the eastern edge of the West Asturian Leonese Zone. During the early Namurian, the boundary between the West Asturian Leonese zone and the Cantabrian Zone was incorporated in the deformation. The accompanying metamorphism is estimated to have taken place post–early Emsian and pre–late Westphalian (Suárez et al., 1990).

THE CANTABRIAN ZONE

By Westphalian B time, thrusting and deformation had migrated farther east, and the westernmost parts of the Cantabrian Zone (the Somiedo Correcilla unit; Fig. 2) experienced their first deformation (Pérez-Estaún et al., 1988; Dallmeyer et al., 1997). Today, the Cantabrian Zone is the northeastern thin-skinned foreland fold-and-thrust belt of the external Iberian autochthon (Quesada, 1992). There is still considerable discussion about the arcuate pattern (Iberian-Armorican or Cantabrian-Asturian Arc) of the Variscan rocks in northwestern Spain (see Aller et al., 2002): whether it is a primary feature or of secondary origin and whether the curvature was initiated by oroclinal bending or by strike-slip deformation (Van der Voo et al., 1997; Hirt et al., 1992; Parés et al., 1994; Steward, 1995; Weil et al. 2000, 2001; Shelley and Bossiere, 2000). The most controversial issues are the age of the presumed bending and the intensity of rotation that affected the Cantabrian Zone between the Westphalian and the Permian. The most radical reconstruction was proposed by Weil et al. (2000, 2001) and Kollmeier et al. (2000), who claimed that the original extension of the shelf was essentially north-south and that all of the curvature is of secondary origin.

Tectonosedimentary Subdivision of the Cantabrian Zone

The Cantabrian Zone is subdivided into several tectonostratigraphic regions (Fig. 2; Julivert, 1971) or units (Pérez-Estaún et al., 1988).

1. The Fold-and-Nappe Province is composed of three major thrust complexes, the Somiedo Correcilla unit, the Sobia Bodón unit, and the Áramo unit (Fig. 2). East of the Porma fault (Fig. 2), the Esla nappe, the Valsurvio "Dome," and the Sierra del Brezo are the southeastern extensions of the Somiedo Correcilla unit. Still farther east, a small fragment of the Fold-and-Nappe Province is preserved in the Revilla nappe (Fig. 3).

To the south and west, Precambrian sediments separate the Fold-and-Nappe Province from the West Asturian Leonese Zone (Fig. 2) and act as a rheologic boundary between the zones (Gutiérrez Alonso, 1992, 1996). Along the southern branch of the Cantabrian Zone, the León fault (Fig. 2) and the Ruesga fault (Fig. 3) separate the Fold-and-Nappe Province from the Central Coal Basin and the Pisuerga Carrión unit. There is ongoing debate about the role of these faults during sedimentation, their age, and their origin (Marcos, 1968a, 1968b, 1979; Kullmann and Schönenberg, 1978; Wagner et al., 1984; Espina et al., 1996; Weh, 2006). The Fold-and-Nappe Province (Fig. 2) has a continuous succession of Silurian and Devonian rocks along its western and southern margin. These sediments were deposited on a shelf in a passive-margin setting. Toward the northeast, there is a systematic change toward proximal depositional environments and an increase in number and magnitude of unconformities and hiatuses.

A profound change in depositional style took place during the Frasnian, when the shelf became dissected along a hinge line into a southern basinal area with rapid subsidence and a tilted northern area of uplift and erosion. The corresponding unconformity is mantled by uppermost Devonian rocks of the Ermita Formation and will further on be referred to as "pre- Ermita unconformity." These tectonically induced changes reflect the initiation of the Cantabrian foreland basin. The area of uplift and erosion corresponds to the peripheral bulge, the area south of the hinge line to the foredeep. This hinge line, called "Intra-Asturo Leonesian faciesline" by Raven (1983), is recognized from the Asturian coast to the Esla area. From there, the hinge can be traced to the Valsurvio Dome and the Sierra del Brezo into a major shear zone, which today separates these two units.

Soon after the formation of the pre-Ermita unconformity, and close to the Devonian-Carboniferous boundary, most of the Cantabrian Zone became deeply submerged, and a very thin and strongly condensed succession represents the entire Mississippian. This succession is composed of Tournaisian black shales with phosphorite nodules (Vegamian Formation), and Visean nodular limestones, red siliceous shales, and abundant debris-flow deposits (Genicera or Alba Formation). Together with the sandstones and crinoidal limestones above the pre-Ermita unconformity, this succession will further on be called the Mississippian condensed succession.

2. The Pisuerga Carrión unit is the relative autochthon to all the surrounding nappe piles (Julivert, 1978; Aller et al., 2002; Rodríguez Fernández, 1993, 1994; Rodríguez Fernández and Heredia, 1987). Successive emplacement of the different nappes led to the deposition of discrete clastic wedges in front of the individual thrust piles and the formation of multiple unconformities between the individual wedges (Fig. 4B; Alonso and Rodríguez Fernández, 1983; Rodríguez Fernández and Heredia, 1987). The age of the oldest sediments in the autochthon is not well constrained. Heredia et al. (1990) described Tournaisian to Namurian deposits, which in their interpretation form the pre-orogenic succession. Recent field work suggests that these deposits are

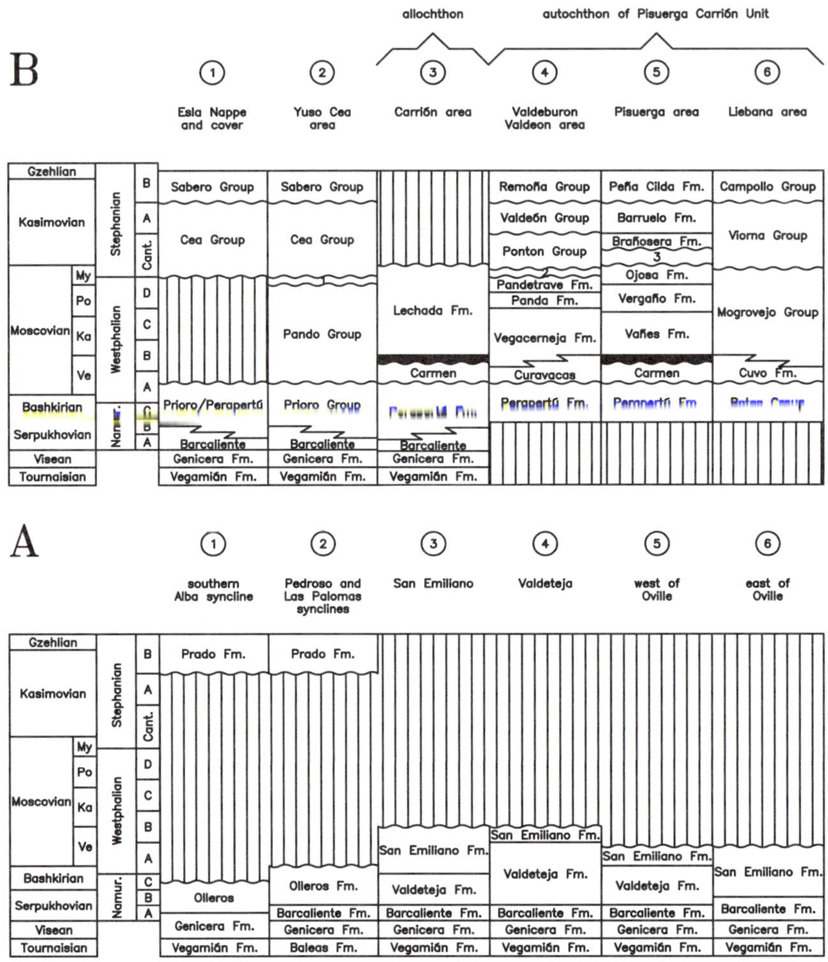

Figure 4. Namurian and Westphalian lithostratigraphic units of the Pisuerga Carrión unit and southern Fold-and-Nappe Province (modified from Rodríguez Fernández and Heredia, 1987; Heredia et al., 1990). For sections 1–6 in A, see Figure 5; for sections 1–6 in B, see Figure 3. Ve—Vereiskian; Ka—Kashirskian; Po—Podolskian; My—Myachkovskian; 1—Conjas Formation; 2—Maraña Formation; 3—Verdeña and San Salvador Formations; black—Curavacas Conglomerate

contained within allochthonous complexes and that the oldest preserved autochthonous successions (Perapertú Formation and equivalents) are of late Namurian age (Fig. 4B). The siliciclastic strata of the Pisuerga Carrión unit are generally poorly dated, and their spatial relationships are not clear, mainly because differentiation of strata within the allochthon and the autochthon is controversial (Fig. 4B).

3. The Palencian nappes (Fig. 3) are contained within the Pisuerga Carrión unit as allochthonous, apparently gravitationally induced structures (Wagner, 1971; Ambrose, 1974; Frankenfeld, 1983, 1984; Marquínez and Marcos, 1984; Rodríguez Fernández, 1993, 1994; Rodríguez Fernández and Heredia, 1987). Middle and Upper Devonian sediments are represented by pelagic facies not known from elsewhere within the Cantabrian Zone. These sediments have been called the "Palencian Facies" (Brouwer, 1964). The allochthonous units comprise strata from the Silurian through the Mississippian. They are present within the Alto Carrión unit (Fig. 3; Rodríguez Fernández, 1994), the Gildar-Montó area (Marquínez and Marcos, 1984; Frankenfeld, 1983), the San Julian "klippen" and Mudá area (Wagner, 1971), and the Liébana area, where the Palencian Facies was redeposited as large olistoliths (Maas, 1974). Rocks of the Palencian Facies are unconformably overlain by a thick succession of the Curavacas Conglomerate (Kanis, 1956; Brouwer and Van Ginkel, 1964; Colmenero et al., 1988; Fig. 4B). According to our field work, this angular unconformity is only present within the allochthonous units. It was attributed to the Palencian folding phase (Wagner 1959, 1965) of early Westphalian age and used to confine the emplacement of the Palencian nappes into the Pisuerga Carrión flysch basin to the late Namurian or early Westphalian.

Wagner (1971), Ambrose (1974), and Wagner and Winkler Prins (2000) claimed that the Palencian Facies was transported from the north toward the south either from within the Pisuerga Carrión unit (Ambrose, 1974) or from outside (Wagner, 1971;

Wagner and Winkler Prins, 2000). This would place the Palencian nappes into a position now occupied by the Picos de Europa nappes (Fig. 2) or still farther north. Frankenfeld (1983, 1984) argued in favor of a southern provenance for the Palencian Facies within the West Asturian Leonese Zone. Structural and sedimentologic evidence supports this view (Marquínez and Marcos, 1984; Henn and Jahnke, 1984; Carls, 1988; Rodríguez Fernández and Heredia, 1987; Rodríguez Fernández, 1993, 1994).

4. The Ponga nappe unit (Fig. 2) was thrust eastward onto the Pisuerga Carrión unit between late Westphalian D and Stephanian A time. It consists of a large number of individual thrust sheets, which show strong interference patterns of the various deformations due to their position close to the core of the Asturian Arc.

5. The Picos de Europa unit (Fig. 2) was the last major thrust unit to be emplaced. There is general agreement that thrusting occurred during the Stephanian B and that transport was essentially south directed (Rodríguez Fernández and Heredia, 1990; Aller et al., 2002). The corresponding deformation is recognized in the Ponga nappe unit, the Pisuerga Carrión unit, and the Valsurvio Dome.

Stratigraphically, the Ponga nappes, the Picos de Europa nappes, and the Central Coal Basin (Fig. 2) are characterized by the absence of Silurian and most of the Devonian strata. Very thin, uppermost Devonian conglomerates, sandstones, and crinoidal limestones mantle the pre- Ermita unconformity, which in this area spans the Middle and Late Ordovician, the Silurian, and the Devonian. This major hiatus led to the postulation of the Cantabrian Block (Fig. 2; Radig, 1962) or the Asturian geanticline (Van Adrichem Boogaert, 1967) as a structural high with no sedimentation or, alternatively, erosion during Silurian-Devonian times. The Cantabrian Block was the source area for most of the siliciclastic rocks deposited on the Asturo Leonese shelf. After deposition of the Mississippian condensed succession, the Central Coal Basin and the Ponga nappe unit received abundant and thick successions of synorogenic detritus from the Westphalian onward. In contrast, the Picos de Europa province has an almost exclusively calcareous succession that spans most of the Carboniferous.

FORELAND BASIN RESTORATION

Relative Position of Tectonic Units Prior to Thrusting

The paleogeography of the convex side of the Cantabrian Zone prior to thrusting and the organization of the foreland basin system are reflected in the stratigraphy and sedimentology of the Silurian through Namurian strata within the major thrust sheets (Fig. 4).

Alba Syncline (Figure 5)

Located south of the hinge line, the Silurian-Devonian stratigraphy shows the most complete record on the shelf and the most distal depositional environments. During the Late Devonian, several individual prograding clastic successions record a northern source area (Van Loevezijn, 1986; Raven, 1983). After deposition of the Mississippian condensed succession, siliciclastic input from a southern source is interpreted to reflect the

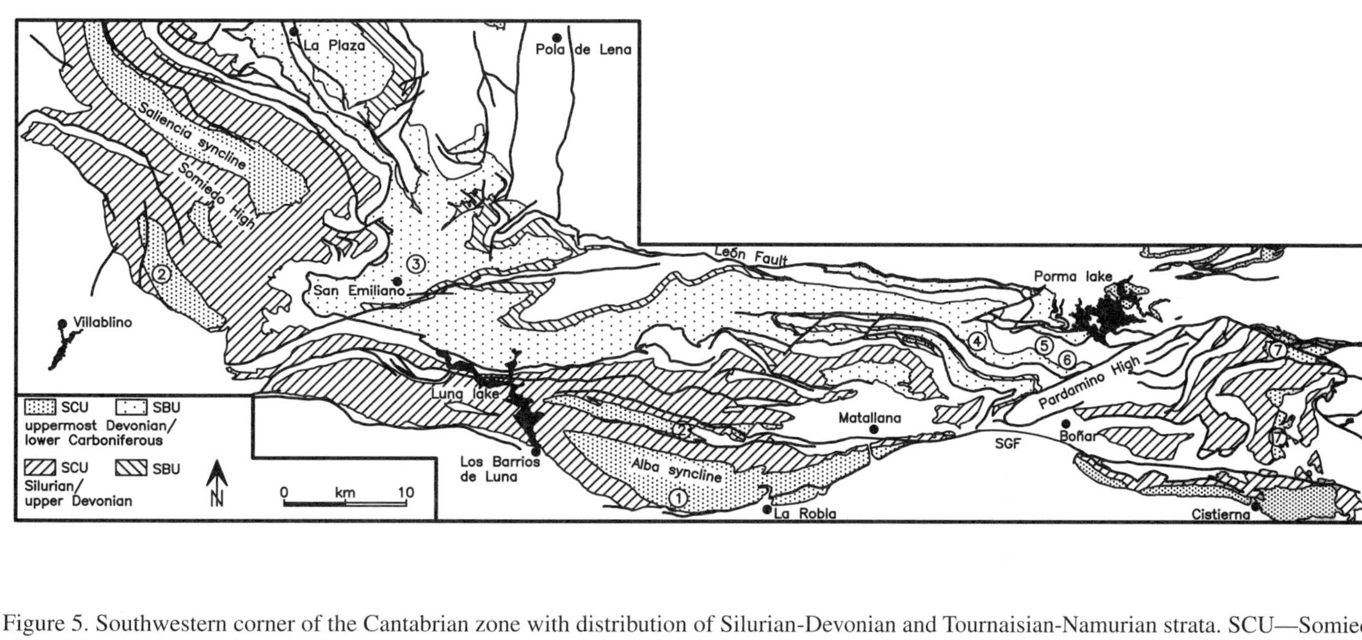

Figure 5. Southwestern corner of the Cantabrian zone with distribution of Silurian-Devonian and Tournaisian-Namurian strata. SCU—Somiedo Correcilla unit; SBU—Sobia Bodón unit; SGF—Sabero Gordón fault. Sections: 1—southern Alba syncline; 2—Pedroso syncline; 3—San Emiliano; 4—Valdeteja; 5—Oville west; 6—Oville east; 7—autochthon in Esla area.

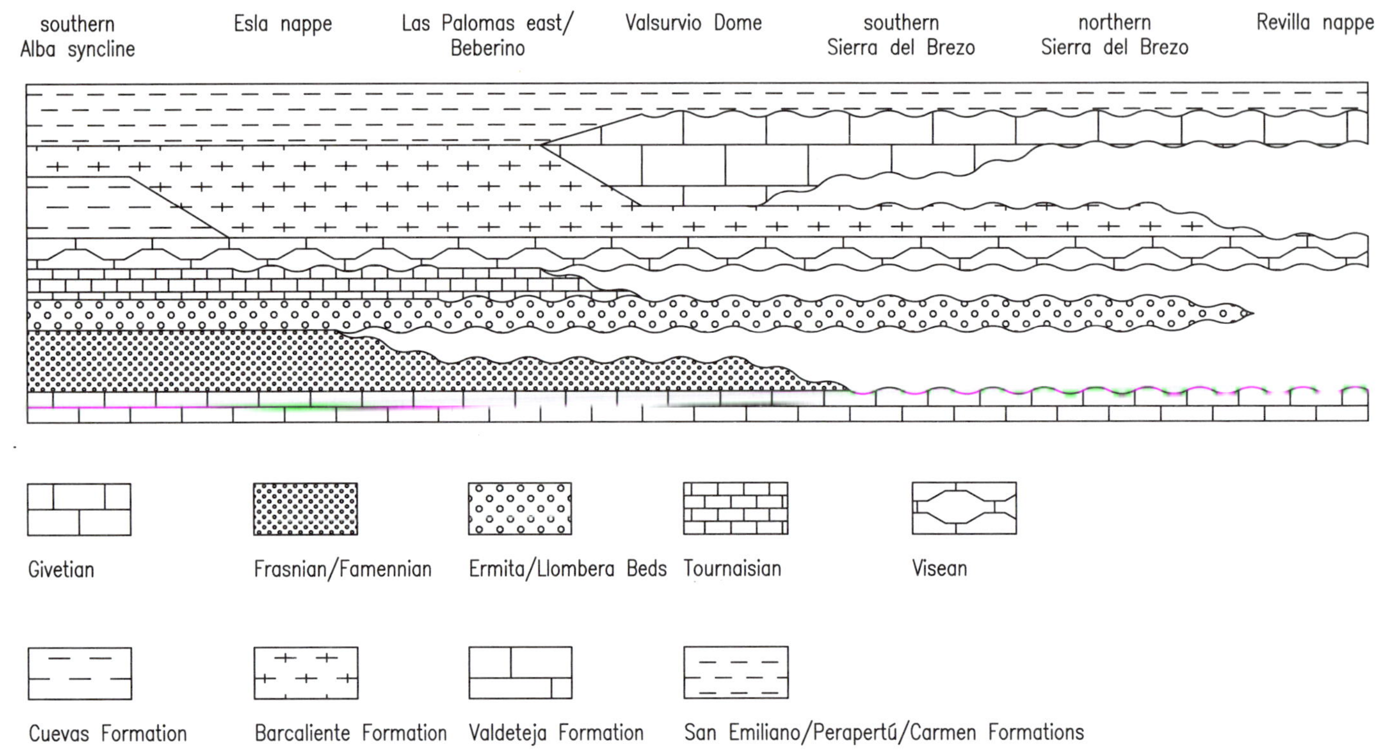

Figure 6. Paleogeographic reconstruction of structural units based on Givetian through Namurian lithostratigraphy.

approach of the orogenic front (Reuther, 1977; Julivert, 1978). This evolution of the Carboniferous succession is unique to the Alba syncline (Fig. 6).

Eastern Las Palomas Syncline, Pedroso Syncline, Esla Autochthon (Figures 5 and 6)

In the eastern Las Palomas syncline and in the northern Pedroso syncline, a similar succession is observed during the Carboniferous. However, the absence of most of the Upper Devonian rocks (Fig. 6) indicates that these sections were located north of the hinge line. Likewise, the autochthon of the Esla nappe is characterized by the absence of Upper Devonian deposits and the pre-Ermita erosion cut into Givetian limestones.

Valsurvio Dome (Figure 6)

The oldest strata in the Valsurvio Dome are of late Emsian age. Despite its greenschist facies metamorphic overprint, all stratigraphic equivalents of the Lower and Middle Devonian succession in the Esla area are recognized. The Upper Devonian is represented by a very thick succession of coarse-grained siliciclastic deposits. The facies of these pre-Ermita sediments is similar to that of the Somiedo area (Las Palomas syncline and adjacent areas). This very thick Upper Devonian succession of the Valsurvio Dome is interpreted to have been deposited south of the Upper Devonian hinge line. Above the Mississippian condensed succession, Namurian turbiditic limestones (Barcaliente Formation) are observed. However, there is a marked loss in thickness across the Valsurvio Dome from south to north beneath the overlying Valdeteja carbonate platform (Reuther, 1977; Wagner and Winkler Prins, 2000). Unconformable siliciclastic turbidites with intercalated quartz pebble conglomerates of Late Namurian and Westphalian A age ("upper Namurian turbidites": Carmen Formation or Cervera Formation) represent the youngest pre-Stephanian rocks preserved in the Valsurvio Dome.

Sierra del Brezo (Figures 3 and 6)

Pre–Upper Devonian strata are equivalent to those of the Valsurvio Dome. In the Sierra del Brezo, however, the Ermita unconformably rests on the Givetian limestones with a well-developed karst surface (Raven, 1983; Wagner and Winkler Prins, 2000). Recognition of the pre-Ermita unconformity in the Sierra del Brezo points to an original position north of the Upper Devonian hinge line. The overall Carboniferous sedimentation resembles that of the Valsurvio Dome, but the cut-out of strata beneath the Valdeteja carbonates is much more pronounced (Wagner and Winkler Prins, 2000).

Revilla Nappe (Figure 3)

The oldest strata exposed are of Givetian age and are unconformably overlain by the Visean nodular limestones (Fig. 6; Wagner et al., 1984). Another erosional unconformity separates the Visean rocks from the Valdeteja Formation; no Barcaliente Formation is preserved. Yet another erosional unconformity brings the upper Namurian turbidites, the youngest strata preserved in this nappe, on top of the Valdeteja carbonates (Fig. 6).

Sobia Bodón Unit (Figure 3)

In the Sobia Bodón unit, the proximity to the Cantabrian Block during Silurian–Devonian time is manifested by abundant erosional unconformities and proximal facies in both carbonate and siliciclastic systems (Van den Bosch, 1969; Evers, 1967; Rupke, 1965; Van Loevezijn, 1986; Buggisch et al., 1982; Keller, 1988; 1997). In vast areas, the pre-Ermita unconformity is cut into Lower–Middle Devonian limestones. At the eastern end of the Sobia Bodón unit (sections 5 and 6 in Fig. 4A), the Ermita locally rests on Ordovician rocks (Fig. 4). Above the Mississippian condensed succession, there is the Barcaliente Formation and then the Valdeteja carbonate platform (Fig. 4A). This platform interfingers with and is finally destroyed by the turbidites of the San Emiliano Formation. The interfingering, the successive onlap onto the platform, and its final destruction are evident in the eastern Sobia Bodón unit (near the village of Oville; Frankenfeld, 1976). Locally, no carbonate platform could develop, and siliciclastic turbidites rest on the Barcaliente Formation.

Palencian Nappes

If the Palencian Facies are restored into a position just south or west of the convex boundary of the Cantabrian Zone (Fig. 7), as originally proposed by Frankenfeld (1983, 1984), then Silurian to Lower Emsian strata have a very close match in sections in the southern Alba syncline and the Las Palomas syncline of the Fold-and-Nappe Province. The differences between the Palencian Facies and the distal shelf facies in their restored position are much smaller than the facies changes that are observed if the shelf strata are traced from the distal sections to sections proximal to the Cantabrian Block. After a drowning event during the early Emsian (Keller, 1997), the shelf returned to carbonate-platform environments during the late Emsian, whereas the Palencian Facies remained in pelagic environments. Why the two areas that formerly formed a single shelf became separated during the late Emsian remains unclear.

Regional Correlation

Distribution of sedimentary successions and the position of unconformities within the restored paleogeographic units indicate that a simple north-south or east-west arrangement of transects is an oversimplification of the likely paleogeography. During the Silurian through Devonian, the Cantabrian Block acted as the dominant sediment source. Paleocurrent measurements in the Upper Devonian siliciclastic successions indicate a sediment transport away from the Cantabrian Block toward the south and west (Raven, 1983; Van Loevezijn, 1986). Proximal-distal trends in the older formations support these polarities (e.g., Buggisch et al., 1982; Suarez de Centi, 1988; García Ramos, 1978; Keller, 1988, 1997). It has also been demonstrated that the boundary between the shelf and the Cantabrian Block was not a linear feature but that there was a succession of promontories and reentrants, which exerted major control on the sedimentary patterns and erosional unconformities. Two well-documented examples are the Pardamino High (Fig. 5; Rupke, 1965; Evers, 1967) and the Somiedo High (Fig. 5; Raven, 1983).

This polarity around the Asturian Arc disappeared with the deposition of the Mississippian condensed succession, and, subsequently, a different orientation of the sedimentary surfaces developed. This is first evident during the deposition of the Visean nodular limestones, which unconformably rest on uppermost Devonian siliciclastic rocks of the Ermita Formation in the Valsurvio Dome area and farther east. In the Revilla nappe, finally, the nodular limestones are found on top of Givetian sedimentary rocks (Fig. 6).

The thickness of the Barcaliente Formation also decreases toward the east as it is successively cut out beneath the overlying Valdeteja Formation (Fig. 6). The Barcaliente Formation is not preserved in the Revilla nappe. Similarly, from the Valsurvio Dome toward the east, the erosional contact beneath the upper Namurian turbidites (Wagner, 1971) is always marked by paleokarst wherever these rest on older limestones. In contrast, in the Sobia Bodón unit, both the Barcaliente and the Valdeteja Formations are very thick, and no major unconformities are reported from there. During the Namurian and Lower Westphalian, a west-east gradient developed from a rapidly subsiding area close the advancing thrust sheets in the west toward an area in the east in which several subaerial unconformities testify to repeated uplift (Reuther, 1977; Wagner and Winkler Prins, 2000).

The Flexural Basin

Along the (present-day) western margin of Iberia, subduction of the Iberian autochthon under an accretionary wedge to the west or northwest started around 385–380 Ma (Rodríguez et al., 2003) and had ceased between 374 and 365 Ma (Martínez Catalán et al., 2002), subsequently giving way to collisional events during the Mississippian (Martínez Catalán et al., 2004a). Concomitantly, a foreland basin system developed that extended from the Central Iberian Zone across the West Asturian Leonese Zone into the Cantabrian Zone (Keller, 2000). Polarities across the subduction/collision zone, with west-directed subduction of the Gondwana margin and a foreland basin system to the east, indicate the evolution of a peripheral foreland basin (sensu Dickinson, 1974; Miall, 1995).

Distal remnants of the foredeep are preserved in the Palencian nappes, which at that time were part of the West Asturian Leonese Zone, and along the southern margin of the Somiedo Correcilla unit. The flysch-like deposits of San Clodio at the

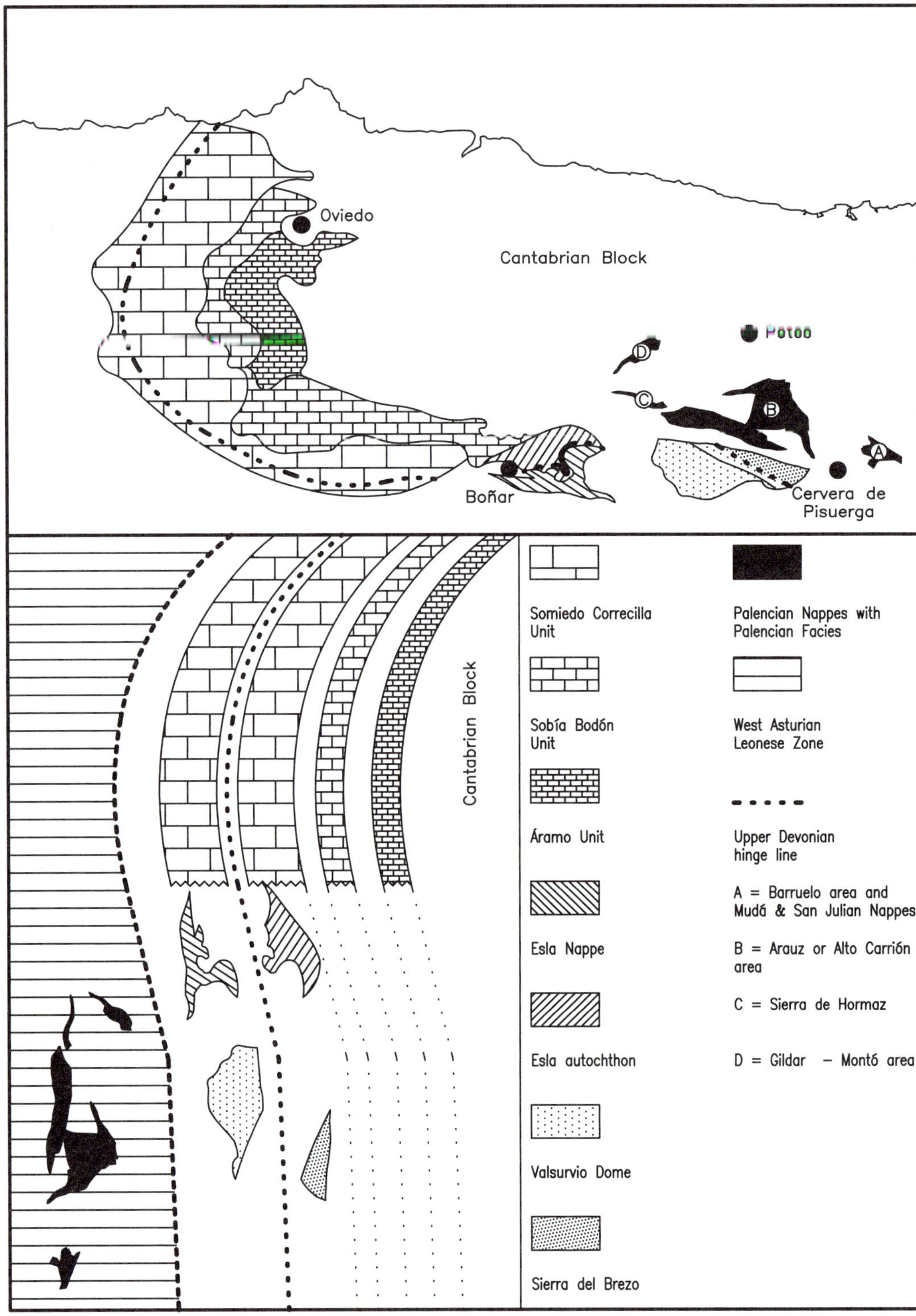

Figure 7. Reconstruction of the major tectonosedimentary units of the southern branch of the Cantabrian Zone to their prethrusting, pre-oroclinal bending position. The reconstruction is based on the sedimentary record shown in Figure 6 and the relative position of the units with respect to the Upper Devonian hinge line.

western margin of the West Asturian Leonese Zone (Riemer, 1963; Pérez-Estaún, 1974) have been attributed to this episode (Frankenfeld, 1984; Eichmüller and Seibert, 1984; Ábalos et al., 2002). However, as these deposits recently have been dated as early Namurian (Martínez Catalán et al., 2004b), they are younger than the events discussed here, so that there is no record of this episode in the northern Iberian massif outside of the Cantabrian Zone.

A hinge line separated the distal foredeep from the peripheral bulge, which was subject to erosion almost immediately after its rise (Keller, 2000). Current directions in the Upper Devonian rocks (Raven, 1983; Van Loevezijn, 1986) indicate that sediment export was toward the foredeep, perpendicular to the basin axis, from the uplifted area in the north and east into the newly formed basin. In the uplifted area north of the hinge line, the pre-Ermita unconformity developed because of forebulge erosion. Several prograding wedges are recognized, and one of them extended into the Palencian Facies area, where it is preserved as the Murcia Quartzite. The unconformity is not recognized in the Palencian Facies, which was deposited farther outboard and under deeper bathymetrical conditions.

There are no indications of Late Devonian activity along the Sabero Gordón fault or the León fault (Fig. 5) that would have accommodated the differential movements between uplift and subsidence, although synsedimentary activity along these faults was repeatedly postulated in the older literature (e.g., de Sitter, 1962; Kullmann and Schönenberg, 1978; Reijers, 1974; Krans, 1982; Raven, 1983; Van Loevezijn, 1986). Instead, it was the Late Devonian hinge line (Fig. 7) that acted as a flexural node of the newly established foreland basin. With the initiation of the foreland basin system during the Late Devonian, the Cantabrian Block was transformed from a sediment shedding hinterland into the foreland of the evolving Variscan orogen of northern Spain, and the former shelf was transformed into the distal foredeep. With the migration of the peripheral bulge toward the foreland, the entire region became successively submerged, and starved sedimentation prevailed during the Mississippian. A diachronous base of the lithostratigraphic units documents migration of the forebulge and the foredeep during Tournaisian and Visean times (Frankenfeld, 1984; Van Adrichem Boogaert, 1967). These pelagic environments extended from the West Asturian Leonese Zone, where they had persisted since the Middle Devonian, across the entire Cantabrian Zone. Sedimentation is characterized by strongly condensed successions of thin crinoidal grainstones, pelagic carbonates, red and green shales, siliceous shales, and black shales with phosphorite nodules.

During the Tournaisian, the sedimentary basin was dissected into a succession of ridges and troughs (Raven, 1983). On the ridges, shallow-water carbonates (Baleas/Cándamo Limestones) were deposited; in the troughs, black shales with phosphorite nodules were deposited (Vegamián Formation). This interval has been interpreted as a "levelling stage" by Kullmann and Schönenberg (1975) or as a stage of tectonic quiescence by Reuther (1977), Heredia et al. (1990), and Rodríguez Fernández (1993, 1994). Marcos and Pulgar (1982) regarded the uppermost Devonian–Mississippian succession as transitional between stable shelf sedimentation and synorogenic deposition.

Stockmal et al. (1986, 1992) showed that under the assumption of a relatively old and thus rigid continental crust, deep-water, starved conditions may prevail in a foreland basin for several tens of millions of years. This is especially true if the corresponding margin was a passive margin, which is basically able to accommodate thick overthrust loads without creating substantial subaerial relief (Stockmal et al., 1986; Stockmal and Beaumont, 1987). Northwestern Iberia is interpreted to be an integral part of the northern Gondwana passive margin (e.g., Martínez Catalán et al., 1997, 2004a): as thrusting propagated toward the foreland relatively slowly (~5 km/m.y.; Dallmeyer et al., 1997), several millions of years had to pass before the thrust sheets of the Central Iberian Zone and the West Asturian Leonese Zone finally developed a topographic expression above sea level. Until that time, the peripheral foreland basin remained in an underfilled stage (Flemings and Jordan, 1989) and the Mississippian condensed succession was deposited. Underfilled basins are often characterized by a submarine topographic depression caused by thrust loads, where the basin axis is oriented parallel to the thrust front. Consequently, away from the margins of the basin, sediment transport is more or less parallel to the basin axis. This general pattern is recognizable in the Cantabrian Zone (Sanchez de la Torre et al., 1983; Raven, 1983; Van Loevezijn, 1986) up to the basal Namurian.

Rheologically, this stage of foreland basin evolution as interpreted for the Cantabrian Zone can be explained by flexural bending of the lithosphere under elastic or viscoelastic conditions (see Jordan [1981] and Beaumont [1981] for details and differences in the models). From a tectonosedimentary viewpoint, a foreland basin and a forebulge of the northern Iberian foreland basin system (DeCelles and Giles, 1996) are identified. Wedge-top deposits are not preserved, whereas the potential back-bulge area on the opposite side of the Cantabrian Block is unknown.

In the Mississippian condensed succession, many contacts between the stratigraphic units are unconformable, and slumping and sliding are common phenomena (Keller, 2000). Toward the end of the Visean and during the basal Namurian, breccia beds or debris-flow deposits become increasingly abundant. A well-documented example is the Brezo Breccia of Reuther (1977), which is a widespread deposit in the Valsurvio Dome–Sierra del Brezo. It might have a correlative in a debris-flow horizon close to the top of the Genicera Formation in the Alba syncline. Another well-known example is the Porma Breccia (Reuther, 1977) of Arnsbergian (Namurian A) age, which is found in almost all sections of the Barcaliente Formation.

Compartmentalization of the Foreland Basin

In the Cantabrian Zone, the Namurian through Westphalian A is the time of major lateral changes of facies (Figs. 8 and 9). Along the southern limb of the Alba syncline, there is a thick

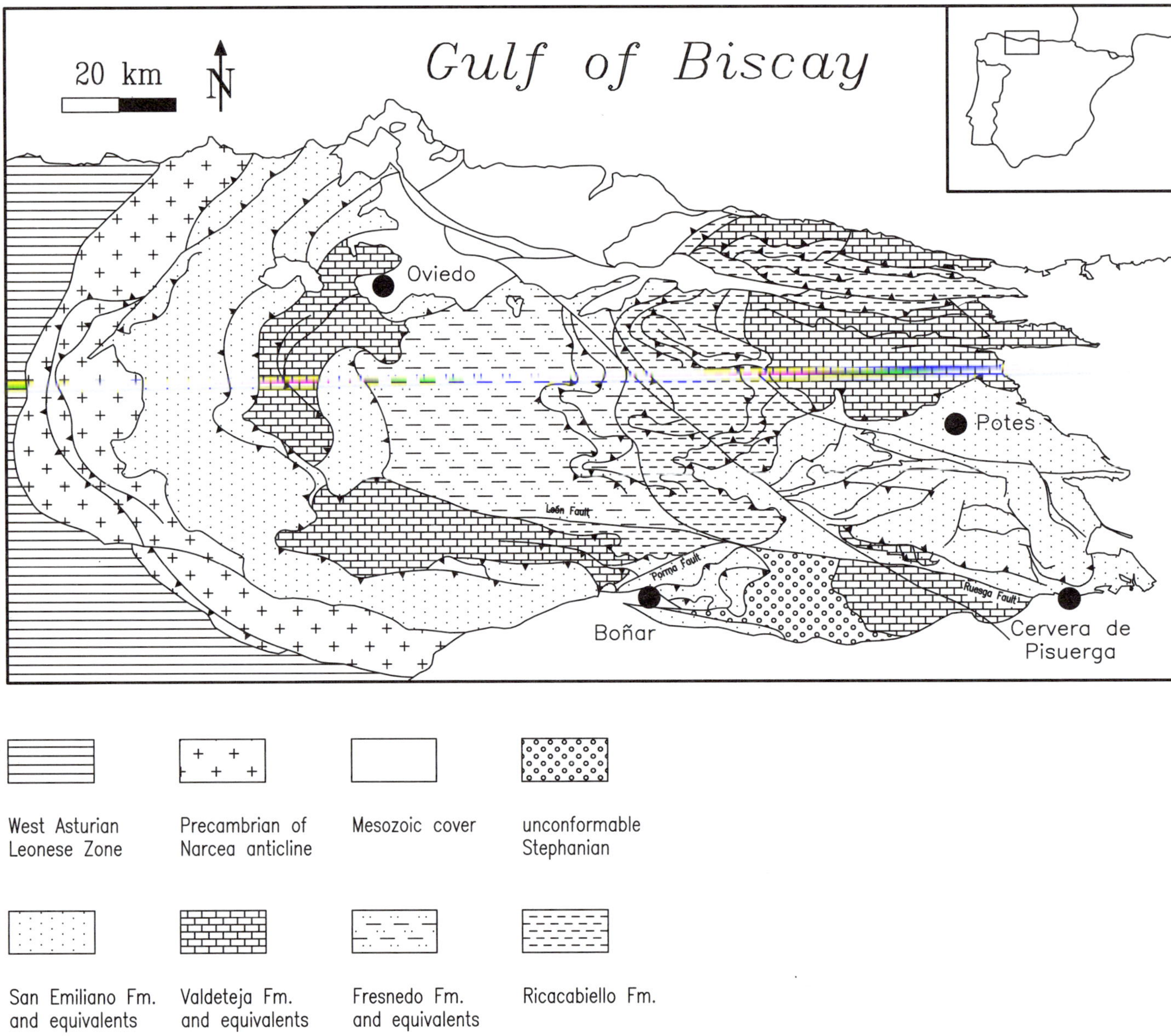

Figure 8. Sediment distribution in the Cantabrian Zone during the Namurian and early Westphalian (modified from Eichmüller and Seibert, 1984; Fernández, 1993). San Emiliano Formation and equivalents represent foredeep deposits; Valdeteja limestones represent the forebulge. Fresnedo and Ricacabiello Formations were deposited in back-bulge area.

succession of siliciclastic sediments (Cuevas Formation) composed of turbiditic, coarsening- and thickening-upward cycles (Boschma and Van Staalduinen, 1968; Wagner et al., 1971). Internally, several higher-order successions with similar themes are recognized (Moser, 2001). Thick conglomerate horizons in the upper part of the succession are composed of siliceous shales that yielded a Late Devonian–Mississippian radiolarian fauna. Petrographic analysis also shows the input from a magmatic-metamorphic hinterland (Moser, 2001). These siliciclastic sediments were shed from the south and west during the Namurian A through Namurian C (?) (Reuther, 1977; Julivert, 1978; Sanchez de la Torre et al., 1983). Lithologically, many of the clasts can be attributed to the Upper Devonian of the Palencian Facies and to the Mississippian condensed succession (Moser, 2001) where siliceous shales are abundant. Within this succession, there are a few tongues of clastic carbonates that are regarded as southern extensions of a carbonate turbidite system (Barcaliente Formation; Wagner et al., 1971). These turbidites apparently were shed from the north and east (Hemleben and Reuther, 1980). In this direction, the depositional environments of the Barcaliente

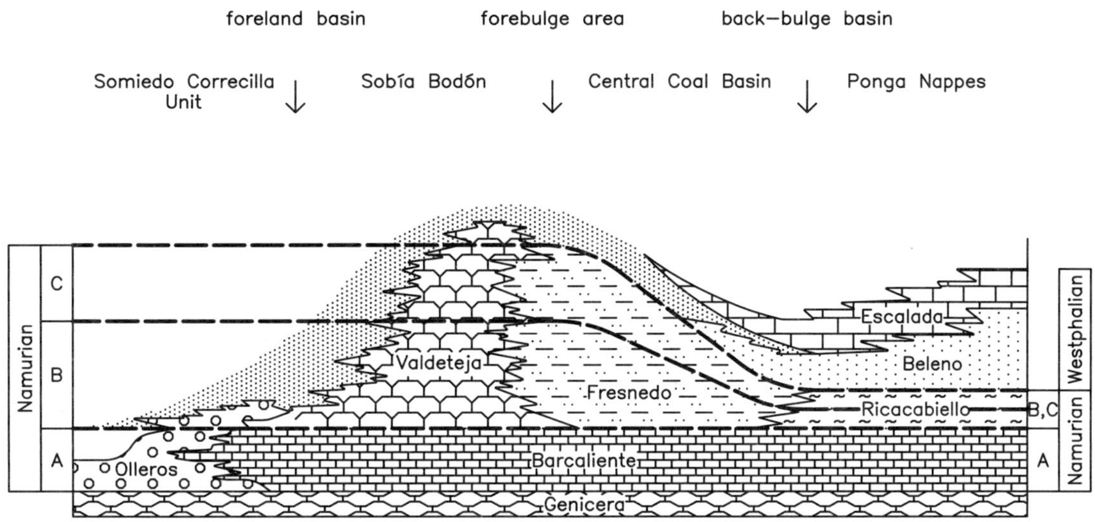

Figure 9. Cross section through southwestern Cantabrian Zone during Namurian and Early Westphalian times showing the foreland basin system (modified from Martínez Chacón et al., 1985). Stippled area—San Emiliano Formation. Note unusual starved succession (Ricacabiello Formation) in back-bulge area.

Formation shallow (González Lastra, 1978), and these shallow-marine environments apparently provided the carbonate debris for the calcareous turbidites. With the progradation of the siliciclastic system toward the foreland, the carbonate turbidites became buried (Fig. 9).

After deposition of the Mississippian condensed succession, subsidence in the Somiedo Correcilla unit must have increased considerably to accommodate the thick siliciclastic detritus provided by the rising orogen and to create a relief for the shedding of turbiditic carbonates from the foreland side of the basin. Increasing subsidence in the Somiedo Correcilla unit is contrasted by relative uplift in the Sobia Bodón unit, where the carbonate platform of the Valdeteja Formation was established above the carbonate turbidites (Fig. 9). Some of the prograding siliciclastic sediments bypassed this platform and were deposited in an area otherwise characterized by starved basin conditions. Typical sediments of these starved areas in the form of red and brown siliceous shales and marlstones with manganese nodules are found in the Ponga nappes (Ricacabiello Formation; Sjerp, 1967; Bahamonde and Colmenero, 1993) and the Central Coal Basin (Fernández, 1993; Salvador, 1993).

A characteristic feature of the carbonate platforms is their marked diachronism, with strata older in the south and west and younger to the north and east (Eichmüller, 1985; Eichmüller and Seibert, 1984). Restoration of the Palencian nappes to a southern and western position provides an elegant solution for a long-lived controversy: Is there a succession of individual carbonate platforms of different age or do the scattered outcrops of carbonates represent one long-lived depositional system that interfingers with the carbonate turbidites? The contradicting interpretations are based on carbonate-platform deposits along the southern margin of the Palencian nappes (within the Espigüete and Santa Lucia thrust sheets), which were dated as Namurian A (Frankenfeld, 1983) and thus considered coeval to the Barcaliente Formation (Barba et al., 1990; Rodríguez Fernández et al., 1985). Similarly, the limestones in the Valsurvio Dome and the Sierra del Brezo are older than the Valdeteja Formation but resemble their lithology and facies (Wagner, 1971; Wagner and Winkler Prins, 2000), and the carbonate turbidite system is very thin.

The reconstruction presented here offers a solution to these contradictory interpretations of the carbonate platforms. With the approach of the orogenic front and the siliciclastic turbidites from the west, subsidence increased in the Somiedo Correcilla unit, generating a depocenter for the detritus. This subsidence was compensated by a flexural bulge in the Sobia Bodón unit, where thick carbonate platforms developed. With the migration of subsidence toward the foreland, the site of carbonate-platform construction on the corresponding bulge also had to migrate to escape burial by detritus. This migration caused a pronounced diachronism of the carbonate-platform deposits (Eichmüller, 1985; Eichmüller and Seibert, 1984). As the Palencian nappes were located closer to the orogenic front at that time, the carbonates preserved there are older than those of the shelf. Local uplift by bulge migration is reflected in the Valsurvio Dome, the Sierra del Brezo, and the Revilla nappe, where large parts of the carbonate turbidite system are missing and where there are laterally extensive breccias at the base of the Valdeteja Formation (Brezo Breccia of Reuther, 1977). We interpret this erosion to reflect bulge movement and erosion.

A major event of Namurian–Westphalian A time is the onset of autochthonous synorogenic sedimentation in the Pisuerga Carrión unit during the Bashkirian. In the Liébana area (Figs. 3 and

4B), the Cosgaya and Vejo Formations (combined in the Potes Group; Fig. 4B) are a succession of conglomerates, breccias, sandstones, and shales of turbiditic origin. Lithostratigraphically and chronostratigraphically, they are equivalents of the upper Namurian turbidites in the southern Pisuerga Carrión unit and in the Palencian nappes (Fig. 4B). During the Visean, thrusting had prograded toward the eastern parts of the West Asturian Leonese Zone (Dallmeyer et al., 1997) and it, together with the increasing morphologic relief, is held to be responsible for the formation and deposition of breccias well into the Namurian. It is also responsible for the subdivision into local troughs and highs.

The rising orogen began to shed siliciclastic detritus into the foredeep from approximately the Visean-Namurian boundary. In the Alba syncline, this is documented in the prograding siliciclastic turbidites overlying the Visean nodular limestones. Depositional environments in the Las Palomas syncline and corresponding settings show northerly derived carbonate turbidites on top of the Mississippian condensed succession and only thereafter the onset of siliciclastic turbiditic sedimentation (Fig. 6).

This sedimentary evolution was caused by increased subsidence close to the orogenic front, which enabled the accommodation of thick siliciclastic successions. From the foreland side of the basin, carbonate debris was shed and deposited probably in a ramp-like setting. This morphology was enhanced by uplift so that the depositional environment was brought to depths favorable for carbonate production (Valdeteja Formation and equivalents). The paleogeography (Fig. 8), with an orogen-derived siliciclastic belt, a carbonate belt migrating toward the foreland, and a basin behind the reef belt, is here interpreted to reflect the formation of a subsequent peripheral bulge with a foredeep and a back-bulge area (Fig. 9). Rodríguez Fernández (1993), Eichmüller (1985), Bahamonde and Colmenero (1993), and Fernández (1993) have presented similar interpretations of this Namurian-Westphalian paleogeography. With the migration of the basin, the bulge, too, was forced to migrate, and so was the carbonate factory. As a result, the carbonate systems display a marked diachronism, where the oldest sediments of Namurian A age are in the Palencian nappes and the Valsurvio Dome–Sierra del Brezo–Revilla nappes.

Several pathways for the siliciclastic detritus dissected the belt of carbonate platforms (Fig. 8). In the west, siliciclastic detritus that bypassed the platform in the Sobia Bodón unit was trapped in the Central Coal Basin (Salvador, 1993). In other areas of the Sobia Bodón unit, the carbonate barrier effectively protected the back-bulge area from major siliciclastic deposition. As a result, the Ponga Nappe area became a starved basin during deposition of the Ricacabiello Formation (Sjerp, 1967; Maas, 1974; Bahamonde and Colmenero, 1993). Farther to the southeast, the barrier was more permeable, and sediment began to spill over into the Pisuerga Carrión basin and finally buried the carbonate belt there. From this time onward (late Namurian), the Pisuerga Carrión unit is recognized as a sedimentary basin, which began to accommodate thick successions of synorogenic detritus.

The debris shed from the rising mountain chain became increasingly coarse grained. It is found over vast areas of the Valsurvio Dome, the Sierra del Brezo, and the Pisuerga Carrión unit. It is also present in the Palencian nappes. Conglomerate horizons in these successions are often almost exclusively composed of metamorphosed quartz arenites (Kanis, 1956; Aller et al., 1985; Colmenero et al., 1988). A likely source for these conglomerates is the thrust sheets containing the Lower Paleozoic section of the West Asturian Leonese Zone (Aller et al., 1985), which is dominated by quartz arenites. These successions had already experienced deformation and metamorphism by Namurian time (Dallmeyer et al., 1997).

In the Namurian–basal Westphalian foreland basin system (Fig. 9), we identify a foredeep, a migrating forebulge area with a reef belt, and a back-bulge area. In contrast to the Upper Devonian peripheral bulge, the Namurian-Westphalian forebulge formed on a relatively local scale and was a narrow but elongated feature. Its rise reflects the approach of the orogenic front to the boundary between the West Asturian Leonese Zone and Cantabrian Zone during the Namurian (Dallmeyer et al., 1997). The bulk of the detritus was delivered to the Pisuerga Carrión area. Its spatial relations to the Ponga nappe area, where unusually condensed sediments were deposited, are not clear. The starved nature of the sediments in some areas of the back-bulge area testifies to maximum protection against siliciclastic detritus by the carbonate belt but also to minimal sediment input from the foreland side of the basin. This might have been the result of a high eustatic sea level during the Mississippian (Ross and Ross, 1988) and the absence of a hinterland due to peneplanation of the preceding bulge.

In comparison with the preceding episode, the evolution of the foreland basin system during the Namurian records a dramatic change in size of the tectonosedimentary units. Two thrust sheets, the Trones and the La Espina thrust sheets, mark the tectonic boundary between the West Asturian Leonese Zone and the Cantabrian Zone (Gutiérrez Alonso, 1992, 1996). Syntectonic movements along the Trones thrust have been dated at ca. 321 Ma (Dallmeyer et al., 1997), which corresponds to the early Namurian. We correlate the onset of accelerating siliciclastic turbidite deposition to the emplacement and morphologic rise of the easternmost thrust units of the West Asturian Leonese Zone. The actual foredeep fill (San Emiliano Formation; Fernández, 1993) is relatively well known, as it was constrained to the area between the rising thrust sheets and the peripheral bulge with the evolving carbonate platforms. Back-bulge deposits are found in the Ponga nappes, the Central Coal Basin, and the Pisuerga Carrión unit. The evolution and the propagation of the sedimentary environments and their strong diachronism are well documented (Eichmüller, 1985; Bahamonde and Colmenero, 1993; Fernández, 1993).

Fill of the foredeep during the Namurian B through Westphalian A includes the transition from the underfilled to the overfilled stage sensu Flemings and Jordan (1990), at least in the Sobia Bodón unit and the Áramo unit (Fernández, 1993), where depositional environments evolved from deep-marine turbidite systems during the late Namurian to fluvial-deltaic deposits during the early Westphalian, which finally overwhelmed the carbonate platforms. As thrusting propagated toward the foreland, the Palencian Facies were incorporated into deformation, and this area was transformed into the Palencian nappes (Fig. 10B). They were transported onto the Valsurvio Dome–Sierra del Brezo area, where the Pisuerga Carrión unit was the adjacent foredeep. Soon after the emplacement, the rising orogen began to shed increasing amounts of quartzite-clast conglomerates (the Curavacas Conglomerate), which were unconformably deposited across the deformed and thrusted Palencian Facies. This unconformity represents the Palencian folding phase of Wagner (1960) and Wagner and Bowman (1983). The conglomerate succession represents a proximal fan-delta deposit and is many hundred meters thick (Colmenero et al., 1988). It is here interpreted to represent wedge-top deposits (sensu DeCelles and Giles, 1996) that originated close to the rising mountain chain (Colmenero et al., 1988). The fluvial-deltaic sediments laterally grade into and interfinger with shales and turbiditic sandstones of the Lechada Formation (Fig. 4). Some of the boulders and pebbles that constitute the Curavacas Formation were transported beyond the limit of the wedge top and deposited in the Pisuerga Carrión basin (Fig. 10B). There, they are associated with turbiditic sequences.

The presence of these tongues of conglomerate with the (assumed) same age and lithology as the Curavacas Conglomerate led Wagner (1960), Rodríguez Fernández and Heredia (1987), and Rodríguez Fernández (1994) to the assume that the Palencian folding phase affected the entire Pisuerga Carrión unit and that the Palencian nappes must had been emplaced by Westphalian B time (the age of the basal Curavacas Formation). However, we are not aware of any deformation similar in style and intensity to the type-Palencian deformation that affected the Pisuerga Carrión unit prior to the deposition of the Curavacas Conglomerate tongues. In contrast, maps and cross sections (Maas, 1974; Savage, 1977; Rodríguez Fernández et al., 1994) show a more or less uninterrupted Namurian-Westphalian succession, an aspect also mentioned by Alonso and Rodríguez Fernández (1983). The erosional contact observed in some localities at the base of the conglomerate units is a purely depositional feature. Sedimentation continued with thick turbiditic successions. Locally, however, some isolated carbonate platforms were able to persist.

The Broken Foreland

Breaking of the foreland crust has been documented from a variety of orogens (see summary in Jordan, 1995), notably from the Laramide orogen of the western United States (Dickinson et al., 1988; McQueen and Beaumont, 1989) and from the Andes of South America (Jordan and Almendinger, 1986; Fielding and Jordan, 1988). The characteristics of a broken foreland are major deep-seated faults dipping at an angle of 30° to 45°. These faults bound small rigid blocks or intermediate blocks with an additional flexural component at the block margins. The faults may evolve from mid-crustal or even deeper levels. Finally, differential movements and tilting of the blocks may create a synoptic topographic relief of several kilometers (Jordan, 1995; McQueen and Beaumont, 1989).

During the Westphalian B, the Somiedo Correcilla unit and the Esla nappe were incorporated into thrusting (Pérez-Estaún et al., 1988). Emplacement of the Esla nappe and the Valsurvio Dome–Sierra del Brezo seems to have been coeval (Rodríguez Fernández, 1993), so that the onset of thrust deformation there can also be attributed to the Westphalian B. However, Pérez-Estaún et al. (1988) pointed out the differences between deformational styles in the Somiedo Correcilla unit and the Esla nappe. Whereas the Somiedo nappe shows little imbrication or duplex formation, the Esla nappe is characterized by the formation of three superimposed thrust sheets and three duplexes that together accommodated almost 90 km of horizontal displacement. The León fault, across which there is an abrupt change in deformational style, limits this anticlinal stack (Alonso, 1987; Pérez-Estaún et al., 1988) to the north. Similarly, the Valsurvio Dome and the Sierra del Brezo were subject to important horizontal shortening after emplacement of and together with the Palencian nappes (Fig. 10C).

The dynamic metamorphism observed in the Valsurvio Dome during its first deformation (Koopmans, 1962; Weh, 2006) may be attributed to the combined thickness of the Palencian Facies plus the overlying conglomerates which had been thrust upon the Valsurvio Dome. Here, the Ruesga fault is the tectonic boundary to the adjacent Pisuerga Carrión unit. Another anticlinal stack was described by Marcos (1968a) from the Sobia Bodón unit close to the León fault, where the Cueto Negro anticlinal stack implies severe crustal shortening compensated by vertical piling of thrust sheets.

The association of anticlinal stacks and major fault zones, the León fault and the Ruesga fault, is here interpreted to indicate a genetic link between their structures. As the three areas migrated toward the north-northeast (Arboleya, 1981), they finally approached the zone in which, during the Silurian and Devonian, the Cantabrian Block was bounded by the adjacent subsiding shelf. In addition, close to this transition, the Upper Devonian hinge line had developed.

In the foreland basin system of the Cantabrian Zone, the approach of the thrust front toward the rigid Cantabrian Block involved high in-plane stress and the formation of several anticlinal stacks in the Sobia Bodón unit, the Esla area, and the Valsurvio Dome area. It is likely that during the intended thrusting of the Valsurvio Dome and the Esla nappe, with their thick and competent Upper Devonian succession, toward the rigid Cantabrian Block, the zone of crustal weakness between the Cantabrian Block and the shelf was transformed into deep-seated faults: the foreland broke (Fig. 10C) along the León and Ruesga faults. Shortening was accomplished not by thrusting alone but by folding and uplift along these newly formed faults, which gave rise to internal deformation in all areas. In the Esla area, much of the shortening was transformed into the formation of an anticlinal stack (Alonso, 1987; Pérez-Estaún et al., 1988) where the stratigraphic succession was at least triplicated. Thrusting also occurred in the Valsurvio Dome,

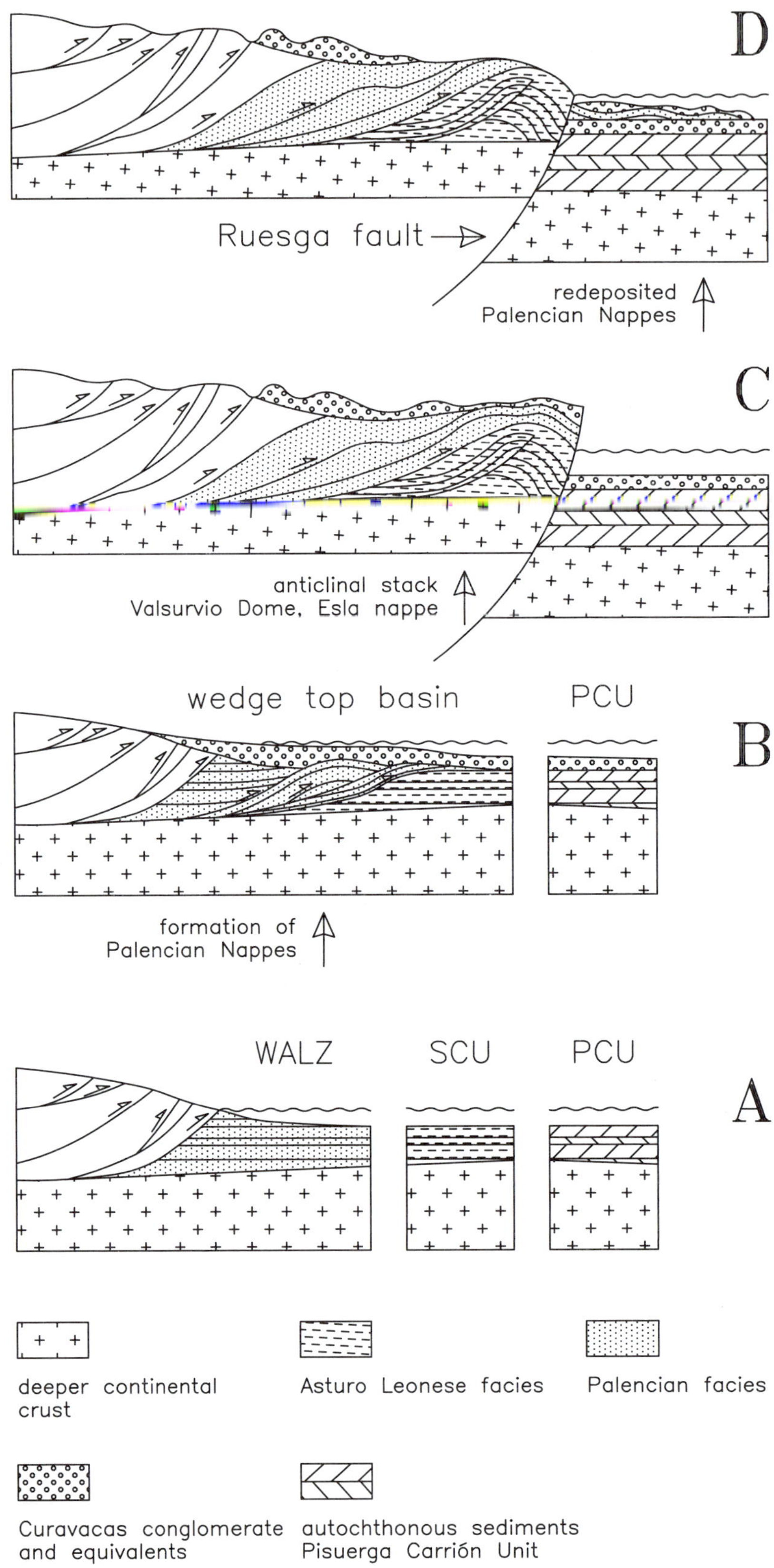

Figure 10. Evolutionary model of the foreland basin system of the Cantabrian Zone. (A) Prior to the onset of orogeny in West Asturian Leonese Zone (WALZ) and Cantabrian Zone (Somiedo Correcilla unit [SCU]). The Palencian Facies or Palencian Nappes are restored to a position in the WALZ. (B) Namurian–early Westphalian: Emplacement of the Palencian Nappes onto the Valsurvio Dome area and subsequent deposition of the wedge-top sediments (Curavacas conglomerate). Note that similar sediments are transported to the Pisuerga Carrión unit (PCU). (C) Late Westphalian: Forward-propagating deformation leads to evolution of anticlinal stacks and breaking of the foreland along deep-seated faults. (D) Palencian Nappes gravitationally glide into the Pisuerga Carrión unit.

and there, almost all stratigraphic contacts were converted into thrust planes (Bittner, 1992; Weh, 2006).

Whereas the southern fault block began to rise, the foreland basin deepened dramatically. The growing topographic relief together with the buckling of the lithosphere in the Valsurvio Dome area during the Westphalian seem to have triggered the gravitational gliding of the Palencian nappes into the foredeep of the Pisuerga Carrión unit with the Curavacas Conglomerate "riding piggyback" (Fig. 10D). In this scenario, the wedge-top deposits together with parts of the wedge itself (the Palencian nappes) were overriding their distal equivalents of the Pisuerga Carrión unit. This probably happened during the late Westphalian, since some olistoliths within the Lechada Formation of the Palencian nappes yield Westphalian C ages (Alonso and Rodríguez Fernández, 1983).

During the late Westphalian, the Pisuerga Carrión unit was strongly compartmentalized by a mosaic of deep-seated faults so that small carbonate platforms developed on the tilted and uplifted edges of blocks. They shed abundant debris into the adjacent foredeep. Several of the faults, e.g., the Peñas Matas and Polentinos faults, served as pathways for ascending magmas. Gabbroic and gabbrodioritic sills are reported from the Ruesga and León faults, and the main granitic stocks in the Pisuerga Carrión unit are also confined to this system of faults (Corretgé and Suárez, 1990). The subsidence history of the Central Coal Basin shows a dramatic increase from the Mississippian into the late Westphalian (Salvador, 1993). Sedimentation rates increased from 1 cm/1000 yr during the Mississippian to 10 or 15 cm/1000 yr during the Bashkirian to 40 cm/1000 yr during the Westphalian A and B, and finally to almost 150 cm/1000 yr during the Upper Westphalian (Salvador, 1993). We interpret this increase not only to be the result of flexural bending by emplacement of the Somiedo Correcilla and the Sobia Bodón nappes but also of mechanical failure of the foreland.

A deep-seated origin of the Ruesga fault is an explanation for the concentration of small magmatic intrusions and the high heat fluxes and metamorphism observed in small Stephanian basins (Raven and van der Pluijm, 1986; Frings, 2002) along their traces. Uplift of the southern block and stripping off of the Palencian nappes exposed relatively lower crustal levels in the Valsurvio Dome. This might be an explanation for the higher degree of syndeformational metamorphism exposed in this area when compared to other regions. It also explains the strong thickening of the crust to the north of the fault, which led to underplating and abundant magmatism in the SE Cantabrian Zone (Fernández-Suárez et al., 2000; Valverde Vaquero et al., 1999).

The presumed age of both faults is in good agreement with the model presented here. The León fault is interpreted to have formed during the Westphalian, shortly after emplacement of the Sobia Bodón unit but prior to the Stephanian (Marcos, 1968a, 1968b, 1979). The Ruesga fault is attributed to the Westphalian A or B (Wagner and Winkler Prins, 2000). Both faults were apparently reactivated as strike-slip or tear faults during the Upper Westphalian emplacement of the Ponga nappes onto the Pisuerga Carrión unit (Marcos, 1979; Heredia, 1998; Wagner et al., 1984). The complex history of both faults, including several reactivations, continued well into the Mesozoic and Cenozoic (Alonso et al., 1996; Pulgar et al., 1999) and underlines their structural importance.

During the Westphalian D, a basin reorganization is recognized in the Pisuerga Carrión unit and adjacent areas. This reorganization is known as the "Leonian phase" or "Leonian movements" (Wagner, 1959; Wagner and Martínez García, 1974). The main characteristic of this Leonian phase in the Pisuerga Carrión unit is the dominance of faulting. First, reverse faults developed, which were rapidly followed by normal faults. Folding and thrusting were not involved in this deformation (Wagner et al., 1984; Wagner and Martínez García, 1974; Heredia et al., 1990; Weh, 2006). Concomitantly, steep valleys filled with coarse continental debris formed. Laterally, they grade into marine deltaic complexes. Locally, carbonate platforms were able to survive. Other local basins were filled with turbiditic sandstones and debris-flow deposits. Many of the basin-fill interpretations invoke tilting and differential subsidence to explain the rapid lateral changes of facies and thickness on either side of the Leonian unconformity, e.g., Wagner et al. (1984), Rodríguez Fernández (1994), Van de Graaff (1971a, 1971b, 1971c) and Wagner and Varker (1971). We conclude from our data and observations in the Pisuerga Carrión unit that there seems to be continuity in the overall tectonosedimentary evolution from the higher Westphalian into the Cantabrian. Local unconformities (Leonian unconformity) are present and probably testify to tilting of individual blocks or to increasing activity along normal faults.

The entire evolution as interpreted here indicates that breaking of the foreland lithosphere occurred on a regional scale and that the León and Ruesga faults are only two of the more prominent representatives. Other faults dissected the Pisuerga Carrión unit, and their activation-reactivation led to the tilting and the formation of tilt-block basins. Following Rodríguez Fernández (1994), Rodríguez Fernández and Heredia (1987), and Weh (2006), the mainly west-east–directed deformation during this time may have been related to the formation of the Central Coal Basin and the emplacement of the Ponga nappes unit. Clastic wedges associated with their emplacement are confined to relatively narrow depocenters. This gives further indication of a strong compartmentalization of the basin.

The next major tectonic event was a south-directed compression, which led to a strong overprint of the former structures. There is general agreement that the emplacement of the Picos de Europa nappe pile onto the Pisuerga Carrión unit induced this event. The sedimentary record of this event is relatively sparse in the southern Pisuerga Carrión unit, so that a detailed analysis of the relationship between sedimentation and tectonics is not possible. The subsequent late stages of foreland basin deformation are recorded mainly in the Picos de Europa unit. Wagner and Martínez García (1974), Wagner et al. (1984), and Martínez García (1990) have described these final events of the Variscan orogeny in the Cantabrian Zone of northern Spain in detail, and the reader is referred to these publications for further information.

One problem that we could not address in this paper is the effect of oroclinal bending on the evolution of the broken foreland basin system. Neither its timing nor the prebending paleogeography is clear. However, Hirt et al. (1992) and Steward (1995) emphasized that bending began relatively early during the compressional history. If so, this early bending must have created additional space problems in the core of the Asturian Arc (the Pisuerga Carrión unit). These problems might have contributed to the breaking of the lithosphere.

CONCLUSIONS

The Cantabrian Zone of northwestern Spain records the transition from a passive-margin setting into the peripheral foreland basin system of the Variscan orogen during the late Paleozoic. Evolution of the foreland basin system started during the Late Devonian with the formation of a foredeep and the adjacent peripheral bulge. These large-scale features developed as a response to subduction prior to collision along the western margin of Iberia. After erosion of the bulge and its subsequent submergence, the entire basin system migrated toward the foreland. This is documented in the overall diachronism, facies, and areal distribution displayed by the Upper Devonian and Mississippian lithostratigraphic units. Erosional debris from the rising orogen reached the Cantabrian Zone during the Namurian.

In contrast to previous models, we propose a succession of events that started with the emplacement of the Palencian nappes onto the Valsurvio Dome and the Sierra del Brezo without the postulated immediate gravitational transport of these nappes into the Pisuerga Carrión flysch basin. In our model, and in agreement with the general style of deformation in the Cantabrian Zone (Pérez-Estaún et al., 1988; Pérez-Estaún and Bastida, 1990), deformation followed a simple forward-propagating style. After emplacement of the Palencian nappes onto the Valsurvio Dome, erosion of the rearward thrust sheets with rocks of the West Asturian Leonese Zone provided large amounts of detritus. This debris, mainly composed of quartzites, was deposited in a wedge-top setting on the Palencian nappes but also in the adjacent foredeep.

During the Westphalian B, thrusting began to affect the Esla area and the Valsurvio Dome area. Their proximity to the rigid Cantabrian Block prevented a simple forward propagation of thrusting. Instead, the lithosphere began to buckle and finally broke. A system of fractures originated that was rooted in deeper crustal levels. The León fault and the Ruesga fault are two of the most important faults of this system. The concomitant uplift along these faults provided the topographic relief necessary for the gravitationally induced emplacement of the Palencian nappes in the Pisuerga Carrión unit, including part of the wedge-top system. Further on, broken foreland conditions controlled sedimentation in the Pisuerga Carrión unit. With the successive emplacement of the Central Coal Basin, the Ponga nappes unit, and the Picos de Europa unit, many of the faults became reactivated, so that today they exhibit complex superposition of structural indicators.

The Variscan Cantabrian foreland basin system is a long-lived system that originated under flexural bending of the lithosphere during the Late Devonian. It was transformed into a broken foreland basin during the Westphalian and achieved its final geometry through oroclinal bending, which might have had additional influence on the breaking of the crust underlying the basin, and through the formation of large-scale faults.

ACKNOWLEDGMENTS

We thank J.R. Martínez Catalán (Salamanca) for inspiring discussions about the orogenic events and their timing in northwestern Spain. D. Barbeau (Columbia), J.R. Martínez Catalán (Salamanca), and an anonymous referee made constructive reviews on an initial version of the manuscript. These comments and suggestions were greatly appreciated and helped to improve the paper considerably. Field work was supported by the Deutsche Forschungsgemeinschaft through various grants, which are gratefully acknowledged.

REFERENCES CITED

Ábalos, B., Carreras, J., Druguet, E., Escuder Viruete, J., Gómez Pugnaire, M.T., Lorenzo Alvarez, S.L., Quesada, C., Rodríguez Fernández, L.R., and Gil-Ibarguchi, J.I., 2002, Variscan and pre-Variscan tectonics, *in* Gibbons, W., and Moreno, T. eds., The Geology of Spain: Geological Society [London], p. 155–184.

Aller, J., Alonso, J.L., and Pérez-Estaún, A., 1985, Prodedencia de los Conglomerados Cantabrienses del Grupo Cea en la región del Esla (Zona Cantábrica, valles del Esla y Cea): Boletín Geológico y Minero, v. 46, p. 581–591.

Aller, J., Bastida, F., and Rodríguez Fernández, L.R., 2002, Cantabrian Zone: General geological features, *in* García López, S., and Bastida, F., eds., Palaeozoic Conodonts from Northern Spain: Madrid, Instituto Geológico y Minero de España, Cuadernos del Museo Geominero, p. 3–33.

Alonso, J.L., 1987, Sequences of thrust and emplacement in the superimposed duplexes of the Esla Nappe region (Cantabrian Zone, NW Spain): Journal of Structural Geology, v. 9, p. 969–983, doi: 10.1016/0191-8141(87)90005-8.

Alonso, J.L., and Rodríguez Fernández, L.R., 1983, Las discordancias Carboníferas en la región del Pisuerga-Carrión (Cordillera Cantábrica, NO de España): Madrid, Congreso Internacional de Estratigrafía y Geología del Carbonífero, p. 533–540.

Alonso, J.L., Pulgar, J.A., Garcá Ramos, J.C., and Barba, P., 1996, Tertiary basins and alpine tectonics in the Cantabrian Mountains (NW Spain), *in* Friend, P.F. and Dabrio, C.J., eds., Tertiary Basins of Spain: Cambridge, Cambridge University Press, p. 214–227.

Ambrose, T., 1974, The Stratigraphy and Structure of the Pre-Carboniferous Rocks of North-West of Cervera del Pisuerga, Cantabrian Mountains, Spain [Ph.D. thesis]: Sheffield, University of Sheffield, 200 p.

Arboleya, M.L., 1981, La estructura del manto del Esla (Cordillera Cantábrica, León): Boletín Geológico y Minero, v. 42, p. 19–40.

Arenas, R., Diaz García, F., Martínez Catalán, J.R., Abati, J., Gonzalez Cuadra, P., Andonaegui, P., Gonzalez del Tanago, J., Rubio Pascual, F., Castiñeiras, P., and Gomez Barreiro, J., 2000, Structure and evolution of the Ordenes complex: Basement Tectonics, 15. Pre-Conference field trip, A Coruña, 160 p.

Bahamonde, J.R., and Colmenero, J.R., 1993, Análisis estratigráfico del Carbonífero Medio y Superior del Manto del Ponga (Zona Cantábrica): Trabajos de Geologío, v. 19, p. 155–193.

Barba, P., Fernández, L.P., Rodríguez Fernández, L.R., and Colmenero, J.R., 1990, Memoria del Mapa Geológico de España (Riaño): Madrid, Instituto Geológica y Minero de España, escala 1:50,000, 182 p.

Beaumont, C., 1981, Foreland basins: Geophysical Journal of the Royal Astronomical Society, v. 65, p. 291–329.

Bittner, M., 1992, Stratigraphie und Metamorphose im Balsurbio Dom (Kantabrisches Gebirge, Nordspanien) [Diploma thesis]: Erlangen, Universität Erlangen, 102 p.

Boschma, D., and Van Staalduinen, C.J., 1968, Mappable units of the Carboniferous of northwestern Spain: Leidse Geologische Mededelingen, v. 43, p. 221–232.

Brouwer, A., 1964, Deux facies dans le Dévonien des montagnes Cantabriques méridionales: Breviora Geológica Astúrica, v. 8, p. 3–10.

Brouwer, A., and Van Ginkel, A.C., 1964, La sucession Carbonifère dans la partie méridionale des Montagnes Cantabriques: Paris, V Congrés du Stratigraphie et Géologie du Carboniföre, p. 307–319.

Buggisch, W., Meiburg, P., and Schumann, D., 1982, Facies, paleogeography and intra-Devonian gaps of the Asturo-Leonese Basin (Cantabrian Mountains, Spain): Neues Jahrbuch für Geologie und Palaontologie, Abhandlungen, v. 163, p. 212–230.

Carls, P., 1988, The Devonian of Celtiberia (Spain) and Devonian paleogeography of SW Europe, *in* McMillan, N.J., Embry, A.F., and Glass, D.J., eds., Devonian of the World: Canadian Society of Petroleum Geologists Memoir 14, p. 421–466.

Colmenero, J.R., Agueida, J.A., Fernández, L.P., Salvador, C.I., Bahamonde, J.R., and Barba, P., 1988, Fan-deltas systems related to the Carboniferous evolution of the Cantabrian Zone, northwestern Spain, *in* Nemec, W., and Steel, R., eds., Fan-Deltas: Sedimentology and Tectonic Settings: London, Blackie & Son, p. 267–285.

Corretgé, L.G., and Suárez, O., 1990, Igneous rocks, *in* Dallmeyer, R.D., and Martínez García, E., eds., Pre-Mesozoic Geology of Iberia: Berlin, Springer-Verlag, p. 72–79.

Dallmeyer, R.D., Martínez Catalán, J.R., Arenas, R., Gil Ibarguchi, J.I., Gutiérrez Alonso, G., Farias, P., Bastida, F., and Aller, J., 1997, Diachronous Variscan tectonothermal activity in the NW Iberian Massif: Evidence from ^{40}Ar/^{39}Ar dating of regional fabrics: Tectonophysics, v. 277, p. 307–337, doi: 10.1016/S0040-1951(97)00035-8.

DeCelles, P.G., and Giles, K.A., 1996, Foreland basin systems: Basin Research, v. 8, p. 105–123, doi: 10.1046/j.1365-2117.1996.01491.x.

de Sitter, L.U., 1962, The structure of the southern slope of the Cantabrian Mountains: Explanation of a geological map with sections, scale 1:100,000: Leidse Geologische Mededelingen, v. 26, p. 225–264.

Dickinson, W.R., 1974, Plate tectonics and sedimentation, *in* Dickinson, W.R., ed., Tectonics and Sedimentation: Society of Economic Paleontologists and Mineralogists Special Publication 22, p. 1–27.

Dickinson, W.R., Klute, M.A., Hayes, M.J., Janecke, S.U., McKittrick, M.A., and Olivares, M.D., 1988, Paleogeographic and paleotectonic setting of Laramide sedimentary basins in the Central Rocky Mountain region: Geological Society of America Bulletin, v. 100, p. 1023–1039, doi: 10.1130/0016-7606(1988)100<1023:PAPSOL>2.3.CO;2.

Drot, J., and Matte, P., 1967, Sobre la presencia de capas del Devóniano en el límite de Galicia y León (NW de España): Notas y Communicaciones Instituto Geológico y Minero de España, v. 93, p. 87–92.

Eichmüller, K., 1985, Die Valdeteja Formation: Aufbau und Geschichte einer oberkarbonischen Karbonatplattform (Kantabrisches Gebirge, Nordspanien): Facies, v. 13, p. 45–154, doi: 10.1007/BF02536901.

Eichmüller, K., and Seibert, P., 1984, Faziesentwicklung zwischen Tournai und Westfal D im Kantabrischen Gebirge (NW-Spanien): Zeitschrift der Deutschen Geologischen Gesellschaft, v. 135, p. 163–191.

Espina, R.G., Alonso, J.L., and Pulgar, J.A., 1996, Growth and propagation of buckle folds determined from syn-tectonic sediments (the Ubierna fold belt, Cantabrian Mountains, N Spain): Journal of Structural Geology, v. 18, p. 431–441, doi: 10.1016/0191-8141(95)00103-K.

Evers, H.J., 1967, Geology of the Leonides between the Bernesga and Porma Rivers, Cantabrian Mountains: NW Spain: Leidse Geologische Mededelingen, v. 41, p. 83–151.

Fernández, L.P., 1993, La Formación San Emiliano (Carbonífero de la Zona Cantábrica, NO de España): Estratigrafía y extensión lateral. Algunas implicaciones paleogeográficas: Trabajos de Geología, v. 19, p. 97–122.

Fernández Suárez, J., Dunning, G.R., Jenner, G.A., and Gutierrez Alonso, G., 2000, Variscan collisional magmatism and deformation in NW Iberia: Constraints from U-Pb geochronology of granitoids: Geological Society [London] Journal, v. 157, p. 565–576.

Fernández Viejo, G., Gallart, J., Pulgar, J.A., Córdoba, D., and Danobeitia, J.J., 2000, Seismic signature of Variscan and Alpine tectonics in NW Iberia: Crustal structure of the Cantabrian Mountains and Duero Basin: Journal of Geophysical Research, v. 105, p. 3001–3018, doi: 10.1029/1999JB900321.

Fielding, E.J., and Jordan, T., 1988, Active deformation at the boundary between the Precordillera and Sierras Pampeanas, Argentina, and comparison with ancient Rocky Mountain deformation, *in* Schmidt, C.J. and Perry, W.J., Jr., eds., Interaction of the Rocky Mountain Foreland and the Cordilleran thrust belt: Geological Society of America Memoir 171, p. 143–163.

Flemings, P.B., and Jordan, T.E., 1989, Synthetic stratigraphic model of foreland basin development: Journal of Geophysical Research, v. 94, p. 3851–3866.

Flemings, P.B., and Jordan, T.E., 1990, Stratigraphic modelling of foreland basins: Interpreting thrust deformation and lithosphere rheology: Geology, v. 18, p. 430–434, doi: 10.1130/0091-7613(1990)018<0430:SMOFBI>2.3.CO;2.

Franke, W., 2000, The mid-European segment of the Variscides: Tectonostratigraphic units, terrane boundaries and plate tectonic evolution, *in* Franke, W., Haak, V., Oncken, O., and Tanner, D., eds., Orogenic Processes: Quantification and Modelling in the Variscan Belt: Geological Society [London] Special Publication 179, p. 35–61.

Franke, W., Haak, V., Oncken, O., and Tanner, D., 2000, Orogenic Processes: Quantification and Modelling in the Variscan belt, *in* Franke, W., Haak, V., Oncken, O., and Tanner, D., eds., Orogenic Processes: Quantification and Modelling in the Variscan Belt: Geological Society [London] Special Publication 179, p. 1–3.

Frankenfeld, H., 1976, Verzahnung einer Flachwasserkalk-Fazies mit einer Beckenfazies im Namur/Westfal bei Oville in der Provinz León im Kantabrischen Gebirge, Spanien [Diploma thesis]: Tübingen, Universität Tübingen, 101 p.

Frankenfeld, H., 1983, El manto de Montó Arauz: Interpretación estructural de la región del Pisuerga-Carrión (Zona Cantábrica, NW de España): Trabajos de Geología, v. 13, p. 37–47.

Frankenfeld, H., 1984, Deckenbewegungen im Namur des östlichen Kantabrischen Gebirges und die damit verbundenen Vorstellungen über die Paläogeographie der Namur- und prä-Namur-Schichten: Zeitschrift der Deutschen Geologischen Gesellschaft, v. 135, p. 223–241.

Frings, K., 2002, Paläotemperatur-Anomalien in spät-Variskischen Kohlebecken am Beispiel des Ciñera-Matallana Beckens, Kantabrisches Gebirge, 9: Spanien [Ph.D. thesis]: Heidelberg, Universität Heidelberg, 134 p.

Gallastegui, J., Pulgar, J.A., and Alvarez Marrón, J., 1997, 2-D seismic modelling of the Variscan foreland thrust and fold belt crust in NW Spain from Escin-1 deep seismic reflection data: Tectonophysics, v. 269, p. 21–32, doi: 10.1016/S0040-1951(96)00166-7.

García Ramos, J.C., 1978, Estudio e interpretación de las principales facies sedimentarias comprendidas en las formaciones Naranco y Huergas (Devónico Medio) en la Cordillera Cantábrica: Trabajos de Geología, v. 10, p. 195–247.

González Lastra, J., 1978, Facies Salinas en la Caliza de Montaña (Cordillera Cantábrica): Trabajos de Geología, v. 10, p. 249–265.

Gutiérrez Alonso, G., 1992, El antiforme del Narcea y su relación con los mantos occidentales de la Zona Cantábrica [Ph.D. thesis]: Oviedo, Universidad de Oviedo, 320 p.

Gutiérrez Alonso, G., 1996, Strain partitioning in the footwall of the Somiedo Nappe: Structural evolution of the Narcea tectonic window, NW Spain: Journal of Structural Geology, v. 18, p. 1217–1229, doi: 10.1016/S0191-8141(96)00034-X.

Hemleben, C., and Reuther, C., 1980, Allodapic limestones of the Barcaliente Formation (Namurian A) between Luna and Cea Rivers (southern Cantabrian Mountains, Spain): Neues Jahrbuch für Geologie und Paläontologie, Abhandlungen, v. 159, p. 225–255.

Henn, A., and Jahnke, H., 1984, Die palentinische Faziesentwicklung im Devon des Kantabrischen Gebirges: Zeitschrift der Deutschen Geologischen Gesellschaft, v. 135, p. 131–147.

Heredia, N., 1998, Los cabalgamientos del sector suroriental de las unidades del Ponga y de la Cuenca Carbonífera Central (Zona Cantábrica, NO de España): Trabajos de Geología, v. 20, p. 53–127.

Heredia, N., Rodríguez Fernández, L.R., and Wagner, R.H., 1990, Part II: Carboniferous of the Cantabrian Zone, *in* Dallmeyer, R.D., and Martínez García, E., eds., Pre-Mesozoic Geology of Iberia: New York, Springer, p. 34–38.

Hirt, A.M., Lowrie, W., Julivert, M., and Arboleya, M.L., 1992, Paleomagnetic results in support of a model for the origin of the Asturian Arc: Tectonophysics, v. 213, p. 321–339, doi: 10.1016/0040-1951(92)90461-E.

Jordan, T.E., 1981, Thrust loads and foreland basin evolution, Cretaceous, western United States: American Association of Petroleum Geologists Bulletin, v. 65, p. 2506–2520.

Jordan, T.E., 1995, Retroarc foreland and related basins, *in* Busby, C., and Ingersoll, R.V., eds., Tectonics of Sedimentary Basins: Cambridge, Blackwell Science, p. 331–362.

Jordan, T.E., and Almendinger, R.W., 1986, The Sierras Pampeanas of Argentina: A modern analogue of Rocky Mountain foreland deformation: American Journal of Science, v. 286, p. 737–764.

Julivert, M., 1971, Décollement tectonics in the Hercynian Cordillera of northwest Spain: American Journal of Science, v. 270, p. 1–29.

Julivert, M., 1978, Hercynian orogeny and Carboniferous paleogeography in northwestern Spain: A model of deformation-sedimentation relationship: Zeitschrift der Deutschen Geologischen Gesellschaft, v. 129, p. 565–592.

Julivert, M., 1979, A cross-section through the northern part of the Iberian Massif: Its position within the Hercynian fold belt: Krystallinikum, v. 14, p. 51–67.

Julivert, M., 1981, A cross section through the northern part of the Iberian massif: Geologie en Mijnbouw, v. 60, p. 107–128.

Julivert, M., 1983, La estructura de la Zona Asturoccidental Leonesa, Libro Jubilar J.M. Ríos: Geología de España, v. 1, p. 381–408.

Kanis, J., 1956, Geology of the eastern zone of the Sierra del Brezo (Palencia, Spain): Leidse Geologische Mededelingen, v. 21, p. 377–445.

Keller, M., 1988, Die La Vid Gruppe-Fazies, Paläogeographie und synsedimentäre Tektonik im Unterdevon des Kantabrischen Gebirges (NW-Spanien): Erlanger Geologische Abhandlungen, v. 115, p. 77–154.

Keller, M., 1997, Evolution and sequence stratigraphy of an Early Devonian carbonate ramp, Cantabrian Mountains, northern Spain: Journal of Sedimentary Research, section B, Stratigraphy and Global Studies, v. 67, p. 638–652.

Keller, M., 2000, La cuenca de antepaís Cantábrica, ¿un fenómeno Devónico o Carbonífero?: Geotémas, v. 1, p. 285–288.

Kollmeier, J.M., van der Pluijm, B.A., and Van der Voo, R., 2000, Analysis of Variscan dynamics: Early bending of the Cantabria-Asturias Arc: Earth and Planetary Science Letters, v. 181, p. 203–216, doi: 10.1016/S0012-821X(00)00203-X.

Koopmans, B.N., 1962, The sedimentary and structural history of the Valsurvio Dome, Cantabrian Mountains, Spain: Leidse Geologische Mededelingen, v. 26, p. 121–232.

Krans, T.F., 1982, Block-movements and sedimentation in the Upper Silurian and Lower Devonian of the Cantabrian Mountains (NW Spain): Neues Jahrbuch für Geologie und Paläontologie, Abhandlungen, v. 163, p. 163–172.

Kullmann, J., and Schönenberg, R., 1975, Geodynamische und palökologische Entwicklung im Kantabrischen Variszikum (Nordspanien): Neues Jahrbuch für Geologie und Paläontologie, Monatshefte, v. 1975, p. 151–166.

Kullmann, J., and Schönenberg, R., 1978, Facies differentiation caused by wrench deformation along a deep-seated fault system (León line, Cantabrian Mountains, north Spain): Tectonophysics, v. 48, p. 15–22, doi: 10.1016/0040-1951(78)90082-3.

Lotze, F., 1945, Zur Gliederung der Varisziden in der Iberischen Meseta: Geotektonische Forschungen, v. 6, p. 78–92.

Maas, K., 1974, The geology of the Liébana, Cantabrian Mountains. Deposition and deformation in a flysch area: Leidse Geologische Mededelingen, v. 49, p. 379–465.

Marcos, A., 1968a, La tectónica de la unidad de Sobia-Bodón: Trabajos de Geología, v. 2, p. 59–87.

Marcos, A., 1968b, Nota sobre el significado de la 'León line': Breviora Geológica Astúrica, v. 12, p. 1–15.

Marcos, A., 1979, Facies differentiation caused by wrench deformation along a deep-seated fault system (León line, Cantabrian Mountains, north Spain): Discussion: Tectonophysics, v. 60, p. 303–308, doi: 10.1016/0040-1951(79)90168-9.

Marcos, A., and Pulgar, J.A., 1982, An approach to the tectonostratigraphic evolution of the Cantabrian foreland thrust and fold belt, Hercynian Cordillera of NW Spain: Neues Jahrbuch für Geologie und Palaontologie. Abhandlungen, v. 163, p. 256–260.

Marquínez, J., and Marcos, A., 1984, La estructúra de la Unidad de Gildar-Montó (Cordillera Cantábrica): Trabajos de Geología, v. 14, p. 53–64.

Martínez Catalán, J.R., Pérez-Estaún, A., Bastida, F., Pulgar, J.A., and Marcos, A., 1990, Part III West-Asturian Leonese Zone: Structure, *in* Dallmeyer, R.D., and Martínez García, E., eds., Pre-Mesozoic Geology of Iberia: Berlin, Springer, p. 103–114.

Martínez Catalán, J.R., Arenas, R., Díaz García, F., and Abati, J., 1997, Variscan accretionary complex of northwest Iberia: Terrane correlation and succession of tectonothermal events: Geology, v. 25, p. 1103–1106, doi: 10.1130/0091-7613(1997)025<1103:VACONI>2.3.CO;2.

Martínez Catalán, J.R., Díaz García, F., Arenas, R., Abati, J., Castiñeiras, P., González Cuadra, P., Gómez Barreiro, J., and Rubio Pascual, F.J., 2002, Thrusts and detachment systems in the Ordenes Complex (northwestern Spain): Implications for the Variscan-Appalachian geodynamics, *in* Martínez Catalán, J.R., Hatcher, R.D., Jr., Arenas, R., and Díaz García, F., eds., Variscan-Appalachian Dynamics: The Building of the Late Paleozoic Basement: Geological Society of America Special Paper 364, p. 163–182.

Martínez Catalán, J.R., Arenas, R., Abati, J., Andonaegui, P., Castiñeiras, P., Díaz García, F., Díez Montes, A., Fernández Suárez, J., Gómez Barreiro, J., González Clavijo, E., González Cuadra, P., and Rubio, P., 2004a, Space and time in the tectonic evolution of the Variscan belt of NW Spain, *in* 17th International Basement Tectonics Association Conference: Oak Ridge, Tennessee, p. 67.

Martínez Catalán, J.R., Fernández-Suárez, J., Jenner, G.A., Belousova, E., and Díez Montes, A., 2004b, Provenance constraints from detrital zircon U-Pb ages in the NW Iberian Massif: implications for Palaeozoic plate configuration and Variscan evolution: Geological Society [London] Journal, v. 161, p. 463–476.

Martínez Chacón, M.L., Menéndez Álvarez, J.R., Sánchez de Posada, L.C., and Truyols, J., 1985, Aportaciones al conocimiento de la Formación Ricacabiello (Carbonífero de la Zona Cantábrica, N de España) y su contenido paleontológico: Trabajos de Geología, v. 15, p. 53–65.

Martínez García, E., 1990, Part II Cantabrian and Palencian zones: Stephanian and Permian basins, *in* Dallmeyer, R. D., and Martínez García, E., eds., Pre-Mesozoic Geology of Iberia: Berlin, Springer, p. 39–54.

McKerrow, W.S., MacNiocaill, C., Ahlberg, P.E., Clayton, G., Cleal, C.J., and Eagar, R.M.C., 2000, The late Palaeozoic relations between Gondwana and Laurussia, *in* Franke, W., Haak, V., Oncken, O., and Tanner, D., eds., Orogenic Processes: Quantification and Modelling in the Variscan Belt: Geological Society [London] Special Publication 179, p. 9–20.

McQueen, H.W.S., and Beaumont, C., 1989, Mechanical models of tilted block basins: in Price, R.A., Hutchinson, R.W., and Macqueen, R.W., eds., Origin and Evolution of Sedimentary Basins and Their Energy and Mineral Resources, American Geophysical Union Geophysical Monograph 48, p. 65–71.

Miall, A.D., 1995, Collision-related foreland basins, *in* Busby, C., and Ingersoll, R.V., eds., Tectonics of Sedimentary Basins: Cambridge, Blackwell Science, p. 393–424.

Moser, T., 2001, Sedimentologie und Stratigraphie des Namurs am Südrand der Albasynklinale mit einer Kartierung der südwestlichen Muldenflanke bei Cuevas, südwestliches Kantabrisches Gebirge, NW-Spanien [Diploma thesis]: Erlangen, Universität Erlangen, 149 p.

Parés, J.P., Van der Voo, R., Stamatakos, J., and Pérez-Estaún, A., 1994, Remagnetizations and postfolding oroclinal rotations in the Cantabrian/Asturian Arc, northern Spain: Tectonics, v. 13, p. 1461–1471, doi: 10.1029/94TC01871.

Pérez-Estaún, A., 1974, Aportaciones al conocimiento del Carbonífero de San Clodio (Prov. de Lugo): Breviora Geológica Astúrica, v. 18, p. 3–8.

Pérez-Estaún, A., and Bastida, F., 1990, Part II. Cantabrian and Palencian zones: Structure, *in* Dallmeyer, R.D., and Martínez García, E., eds., Pre-Mesozoic Geology of Spain: Berlin, Springer, p. 55–69.

Pérez-Estaún, A., Bastida, F., Alonso, J.L., Marquínez, J., Aller, J., Alvarez Marrón, M.A., and Pulgar, J.A., 1988, A thin-skinned tectonics model for an arcuate fold and thrust belt: The Cantabrian Zone (Variscan Armorican Arc): Tectonics, v. 7, p. 517–537.

Pérez Estaún, A., Bastida, F., Martínez Catalán, J.R., Gutierrez Marco, J.C., Marcos, A., and Pulgar, J.A., 1990, Structure, *in* Dallmeyer, R.D., and Martínez García, E., eds., Pre-Mesozoic Geology of Iberia: Berlin, Springer-Verlag, p. 92–102.

Pérez-Estaún, A., Pulgar, J.A., Banda, E., Alvarez Marrón, J., and Team ENR, 1994, Crustal structure of the external Variscides in northwestern Spain from deep seismic reflection profiling: Tectonophysics, v. 232, p. 91–118, doi: 10.1016/0040-1951(94)90078-7.

Pulgar, J.A., Alonso, J.L., Espina, R.G., and Marín, J.A., 1999, La deformación alpina en el basamento varísco de la Zona Cantábrica: Trabajos de Geología, v. 21, p. 283–294.

Quesada, C., 1992, Evolución tectónica del Macizo Ibérico, *in* Gutierrez Marco, J.C., Saavedra, J., and Rábano, I., eds., Paleozóico Inferior de Ibéro-América: Mérida, Publicaciones Universidad Extremadura, p. 173–190.

Radig, F., 1962, Ordovizium und Silur und die Frage prävariskischer Faltungen in Nordspanien: Geologische Rundschau, v. 52, p. 346–357, doi: 10.1007/BF01840085.

Raven, J.G.M., 1983, Conodont biostratigraphy and depositional history of the Middle Devonian to Lower Carboniferous in the Cantabrian Zone (Cantabrian Mountains, Spain): Leidse Geologische Mededelingen, v. 52, p. 265–339.

Raven, J.G.M., and van der Pluijm, B.A., 1986, Metamorphic fluids and transtension in the Cantabrian Mountains of northern Spain: An application of the conodont colour alteration index: Geological Magazine, v. 123, p. 673–681.

Reijers, T.J.A., 1974, Hinge movements influencing deposition during the Upper Devonian in the Esla area of the Cantabrian Mountains, Spain: Geologie en Mijnbouw, v. 53, p. 13–21.

Reuther, C.D., 1977, Das Namur im südlichen Kantabrischen Gebirge (Nordspanien)—Krustenbewegungen und Faziesdifferenzierung im Übergang Geosynklinale-Orogen: Clausthaler Geologische Abhandlungen, v. 28, 122 p.

Riemer, W., 1963, Die Entwicklung des Paläozoikums in der südlichen Provinz Lugo (Spanien): Neues Jahrbuch für Geologie und Paläontologie, Abhandlungen, v. 117, p. 273–285.

Robardet, M., 2000, Alternative approach to the Variscan belt in southwestern Europe: Preorogenic palaeobiogeographical constraints, *in* Martínez Catalán, J.R., Hatcher, R., Jr., Arenas, R., and Díaz García, F., eds., Variscan-Appalachian Dynamics: The Building of the Late Paleozoic Basement: Geological Society of America Special Paper 364, p. 1–15.

Robardet, M., 2003, The Armorica "microplate": Fact or fiction? Critical review of the concept and contradictory palaeobiogeographical data: Palaeogeography, Palaeoclimatology, Palaeoecology, v. 195, p. 125–148, doi: 10.1016/S0031-0182(03)00305-5.

Rodríguez, J., Cosca, M.A., Gil-Ibarguchi, J.I., and Dallmeyer, R.D., 2003, Strain partitioning and preservation of $^{40}Ar/^{39}Ar$ ages during Variscan exhumation of a subducted crust (Malpica-Tui complex, NW Spain): Lithos, v. 70, p. 111–139, doi: 10.1016/S0024-4937(03)00095-1.

Rodríguez Fernández, L.R., 1993, Tectonosedimentary evolution of a Carboniferous foreland basin related with arcuated fold-thrust belt. The example of NW Iberian Variscan belt, 12: Buenos Aires, International Congress of the Carboniferous-Permian Stratigraphy and Geology, p. 435–445.

Rodríguez Fernández, L.R., 1994, La estratigrafía del Paleozóico y la estructura de la región de Fuentes Carrionas y areas adyacentes (Cordillera Herciniana, N.O. de España): Laboratório Xeolóxico de Laxe, Serie Terra Nova, v. 9, p. 240.

Rodríguez Fernández, L.R., and Heredia, N., 1987, La estratigrafía del Carbonífero y la estructura de la unidad del Pisuerga Carrión, NO de España: Cuadernos do Laboratorio Xeológico de Laxe, v. 12, p. 207–229.

Rodríguez Fernández, L.R., and Heredia, N., 1990, Part II. Cantabrian and Palencian zones: Structure Palencian zone, *in* Dallmeyer, R.D., and Martínez García, E., eds., Pre-Mesozoic Geology of Iberia: Berlin, Springer, p. 69–71.

Rodríguez Fernández, L.R., Heredia, N., Lobato, L., and Velando, F., 1985, Memoria del Mapa Geológico de España, 1:50,000, no. 106 (Camporredondo de Alba): Madrid, Instituto Geológico y Minero de España, 98 p.

Rodríguez Fernández, L.R., Barba, P., and Heredia, N., 1994, Memoria del Mapa Geológico de España, 1:50,000, no. 81 (Potes): Madrid, Instituto Geológico y Minero de España, 128 p.

Ross, C.A., and Ross, J.R.P., 1988, Late Paleozoic transgressive-regressive deposition, *in* Wilgus, C.K., Hastings, B.S., Kendall, C.G.St.C., Posamentier, H.W., Ross, C.A., and Van Wagoner, J.C., eds., Sea-Level Changes: An Integrated Approach: Society of Economic Paleontologists and Mineralogists Special Publication 42, p. 227–247.

Rupke, J., 1965, Geology of the Leonides between the Bernesga and Porma Rivers, Cantabrian Mountains: NW Spain: Leidse Geologische Mededelingen, v. 32, p. 1–74.

Salvador, C.I., 1993, La sedimentación durante el Westfaliense en una cuenca de antepaís (Cuenca Carbonífera Central de Asturias, N de España): Trabajos de Geología, v. 19, p. 195–254.

Sánchez de la Torre, L., Agueda Villar, J.A., Colmenero Navarro, J.R., García Ramos, J.C., and González Lastra, J., 1983, Evolución sedimentária y paleogeográfica del Carbonífero en la Cordillera Cantábrica: Carbonífero y Pérmico de España: Madrid, p. 133–150.

Savage, J. F., 1977, Geological maps of the southern Cantabrian Mountains, sheets Carrión and Yuso: Leidse Geologische Mededelingen, v. 50.

Shelley, D., and Bossiere, G., 2000, A new model for the Hercynian orogen of Gondwana in France and Iberia: Journal of Structural Geology, v. 22, p. 757–776, doi: 10.1016/S0191-8141(00)00007-9.

Sjerp, N., 1967, The geology of the San Isidro–Porma area (Cantabrian Mountains, Spain): Leidse Geologische Mededelingen, v. 39, p. 55–128.

Stampfli, G.M., von Raumer, J.F., and Borel, G.D., 2002, Paleozoic evolution of pre-Variscan terranes: From Gondwana to the Variscan collision, *in* Martínez Catalán, J.R., Hatcher, R.D., Jr., Arenas, R., and Díaz García, F., eds., Variscan-Appalachian Dynamics: The Building of the Late Paleozoic Basement: Geological Society of America Special Paper 364, p. 263–280.

Steward, S.A., 1995, Paleomagnetic analysis of fold kinematics and implications for geological models of the Cantabrian/Asturian Arc, north Spain: Journal of Geophysical Research, v. 100, p. 20,079–20,094, doi: 10.1029/95JB01482.

Stockmal, G.S., and Beaumont, C., 1987, Geodynamic models of convergent margin tectonics: The southern Canadian Cordillera and the Swiss Alps, *in* Beaumont, C., and Tankard, A.J., eds., Sedimentary Basins and Basin-Forming Mechanisms: Canadian Society of Petroleum Geologists Memoir 12, p. 393–411.

Stockmal, G.S., Beaumont, C., and Boutilier, R., 1986, Geodynamic models of convergent margin tectonics, transition from rifted margin to overthrust belt and consequences for foreland-basin development: American Association of Petroleum Geologists Bulletin, v. 70, p. 181–190.

Stockmal, G.S., Cant, D.J., and Bell, J.S., 1992, Relationship of the stratigraphy of the western Canada foreland basin to Cordilleran tectonics: Insights from geodynamic models, *in* Leckie, D.A., and McQueen, R.W., eds., Foreland Basins and Fold Belts: American Association of Petroleum Geologists Memoir 55, p. 107–124.

Suárez, O., Corretgé, L.G., and Galan, G., 1990, Part III. West Asturian-Leonese Zone: Distribution and characteristics of Hercynian metamorphism, *in* Dallmeyer, R.D., and Martínez García, E., eds., Pre-Mesozoic Geology of Iberia: Berlin, Springer-Verlag, p. 129–133.

Suarez de Centi, C.A., 1988, Estratigrafía y sedimentología de la unidad FM. Furada/San Pedro (Silúrico superior–Devónico inferior), Zona Cantábrica, cuenca Astur-Leonesa: Madrid, Instituto Geológico y Minera España, Informe, v. 7, 530 p.

Truyols-Massoni, M., 1986, *Nowakia acuaria* (Praguiense, Devónico inferior) de la Zona Asturoccidental-Leonesa (NO de España): Breviora Geológica Astúrica, v. 27, p. 12–16.

Valverde Vaquero, P., Cuesta Fernández, A., Gallastegui, G., Suárez, O., Corretge, L.G., and Dunning, G.R., 1999, U-Pb dating of late Variscan magmatism in the Cantabrian Zone (northern Spain): Journal of Conference Abstracts, v. 4, p. 101.

Van Adrichem Boogaert, H.A., 1967, Devonian and Lower Carboniferous conodonts of the Cantabrian Mountains (Spain) and their stratigraphic application: Leidse Geologische Mededelingen, v. 39, p. 129–192.

Van de Graaff, W.J.E., 1971a, Facies distribution and basin configuration in the Pisuerga area before the Leonian phase: Trabajos de Geología, v. 3, p. 161–177.

Van de Graaff, W.J.E., 1971b, The Piedrasluengas Limestone, a possible model of limestone facies distribution in the Carboniferous of the Cantabrian Mountains: Trabajos de Geología, v. 3, p. 151–159.

Van de Graaff, W.J.E., 1971c, Three Upper Carboniferous, limestone-rich, high destructive, delta systems with submarine fan deposits, Cantabrian Mountains, Spain: Leidse Geologische Mededelingen, v. 46, p. 157–235.

Van den Bosch, J.W., 1969, Geology of the Luna-Sil region, Cantabrian Mountains (NW Spain): Leidse Geologische Mededelingen, v. 44, p. 137–225.

Van der Voo, R., Stamatakos, J.A., and Parés, J.M., 1997, Kinematic constrains on thrust belt curvature from syndeformational magnetizations in the Lago del Valle syncline in the Cantabrian Arc: Journal of Geophysical Research, v. 102, p. 10105–10120.

Van Loevezijn, G.B.S., 1986, Stratigraphy and facies of the Nocedo, Fueyo and Ermita Formations (Upper Devonian to lowermost Carboniferous) in León, N Spain: Scripta Geologica, v. 81, p. 1–116.

Wagner, R.H., 1959, Flora fósil y estratigrafía del Carbonífero en España NW y Portugal N: Estudios Geológicos, v. 15, p. 398–420.

Wagner, R.H., 1960, Middle Westphalian floras from northern Palencia (Spain) in relation with the Curavacas phase of folding: Estudios Geológicos, v. 16, p. 55–92.

Wagner, R.H., 1965, Palaeobotanical dating of Upper Carboniferous folding phases in NW Spain: Memorias Instituto Geológico y Minero de Espana, v. 66, p. 1–169.

Wagner, R.H., 1971, Carboniferous nappe structures in northwestern Palencia (Spain): Trabajos de Geología, v. 4, p. 431–459.

Wagner, R.H., and Bowman, M.B.U., 1983, The position of the Bashkirian-Moscovian boundary in West European chronostratigraphy: Newsletters on Stratigraphy, v. 12, p. 132–161.

Wagner, R.H., and Martínez García, E., 1974, The relation between geosynclinal folding phases and foreland movements in northwest Spain: Studia Geológica, v. 7, p. 131–1158.

Wagner, R.H., and Varker, W.J., 1971, The distribution and development of post-Leonian strata (upper Westphalian D, Cantabrian, Stephanian A) in northern Palencia, Spain: Trabajos de Geología, v. 4, p. 533–601.

Wagner, R.H., and Winkler Prins, C.F., 2000, Carboniferous stratigraphy of the Sierra del Brezo in northern Palencia: Evidence of major uplifts: Trabajos de Geología, v. 20, p. 385–402.

Wagner, R.H., Winkler Prins, C.F., and Riding, R.E., 1971, Lithostratigraphic units of the lower part of the Carboniferous in northern León, Spain: Trabajos de Geología, v. 4, p. 603–663.

Wagner, R.H., Carballeira, J., Ambrose, T., and Martínez García, E., 1984, Memoria del Mapa Geológico de España, 1:50,000, no. 107 (Barruelo de Santullan): Madrid, Instituto Geológico y Minero de España, 113 p.

Weh, A., 2006, Polythermale Alterationen in der südöstlichen Ecke der Kantabrischen Zone in Hinsicht auf Sedimentologie, Strukturgeologie und Metamorphose: Norderstedt, Books on Demand, 273 p.

Weil, A.B., Van der Voo, R., van der Pluijm, B.A., and Parés, J.M., 2000, The formation of an orocline by multiphase deformation: A paleomagnetic investigation of the Cantabria-Asturias Arc (northern Spain): Journal of Structural Geology, v. 22, p. 735–756, doi: 10.1016/S0191-8141(99)00188-1.

Weil, A.B., Van der Voo, R., and van der Pluijm, B.A., 2001, Oroclinal bending and evidence against the Pangea megashear; the Cantabria-Asturias Arc (northern Spain): Geology, v. 29, p. 991–994, doi: 10.1130/0091-7613(2001)029<0991:OBAEAT>2.0.CO;2.

Manuscript Accepted by the Society 1 February 2007

Printed in the USA

The Geological Society of America
Memoir 200
2007

The Notre Dame arc and the Taconic orogeny in Newfoundland

C.R. van Staal*
Geological Survey of Canada, NRCan, 601 Booth Street, Ottawa, Ontario K1A 0E8, Canada, and Department of Earth Sciences, University of Ottawa, and Ottawa-Carleton Geoscience Centre, 140 Louis Pasteur, Ottawa, Ontario K1N 6N5, Canada

J.B. Whalen
V.J. McNicoll
S. Pehrsson
Geological Survey of Canada, NRCan, 601 Booth Street, Ottawa, Ontario K1A 0E8, Canada

C.J. Lissenberg*
Department of Earth Sciences, University of Ottawa, and Ottawa-Carleton Geoscience Centre, 140 Louis Pasteur, Ottawa, Ontario K1N 6N5, Canada

A. Zagorevski
Geological Survey of Canada, NRCan, 601 Booth Street, Ottawa, Ontario K1A 0E8, Canada, and Department of Earth Sciences, University of Ottawa, and Ottawa-Carleton Geoscience Centre, 140 Louis Pasteur, Ottawa, Ontario K1N 6N5, Canada

O. van Breemen
Geological Survey of Canada, NRCan, 601 Booth Street, Ottawa, Ontario K1A 0E8, Canada

G.A. Jenner
Department of Earth Sciences, Memorial University of Newfoundland, St. John's, Newfoundland A1B 3X5, Canada

ABSTRACT

The Taconic orogeny in Newfoundland consisted of three accretionary events (Taconic 1, 2, and 3). Taconic 1 is represented by ca. 495 Ma, west-directed obduction of the infant-arc Lushs Bight oceanic tract (510–501 Ma) onto the peri-Laurentian Dashwoods microcontinent. Subduction is inferred to have initiated at a spreading center abandoned during an inboard ridge jump responsible for separation of Dashwoods from Laurentia and opening of the Humber seaway. Clogging of the subduction zone by Dashwoods forced subduction to step back into the Humber seaway. Inception of the new subduction zone led to formation of the ca. 490 Ma Baie Verte oceanic tract.

*Present addresses: van Staal—Geological Survey of Canada, 625 Robson Street, Vancouver, British Columbia V6B 5J3, Canada; Lissenberg—Equipe de Géosciences Marines, Institute de Physique du Globe de Paris, 4 Place Jussieu, tour 14, 75252 Paris Cedex 05, France.

van Staal, C.R., Whalen, J.B., McNicoll, V.J., Pehrsson, S., Lissenberg, C.J., Zagorevski, A., van Breemen, O., and Jenner, G.A., 2007, The Notre Dame arc and the Taconic orogeny in Newfoundland, *in* Hatcher, R.D., Jr., Carlson, M.P., McBride, J.H., and Martínez Catalán, J.R., eds., 4-D Framework of Continental Crust: Geological Society of America Memoir 200, p. 511–552, doi: 10.1130/2007.1200(26). For permission to copy, contact editing@geosociety.org.

Closure of the Humber seaway formed the Notre Dame arc (489–477 Ma) built on Dashwoods and the coeval Snooks Arm arc built on the Baie Verte oceanic tract. Seaway closure led to collision (Taconic 2) between the arcs and Laurentia, which caused significant shortening of the Notre Dame arc. After a magmatic gap of 7–10 m.y., the Notre Dame arc records a voluminous flare-up of predominantly tonalite magmatism (464–459 Ma) during the waning stages of Taconic 2. Magmatism overlaps with deformation and includes both arc and non-arc-like tonalite. This flare-up was related to break-off of the oceanic lithosphere of the downgoing slab. The rapidly upwelling asthenosphere that replaced the broken-off slab induced melting in the subarc mantle and arc infrastructure.

Taconic 3 is represented by 455–450 Ma accretion of a peri-Laurentian arc that had formed after the ca. 480 Ma initiation of west-directed subduction in the Iapetus Ocean outboard of the Dashwoods microcontinent.

Keywords: Taconic, arc collision, slab break-off, tonalite flare-up.

INTRODUCTION

Structures in ancient mountain belts are generally grouped and classified into orogenies on the basis of time rather than on their relationships to kinematically and spatially distinct tectonic events formed in response to plate-convergence processes (Şengör, 1990). This legacy of the pre–plate tectonics era commonly works on a regional scale in most orogens, but pitfalls become noticeable when it is applied globally to coeval but kinematically and spatially unrelated orogenic systems. This problem is compounded where unrelated, but coeval orogenic systems have been juxtaposed during subsequent plate convergence in accretionary orogens. For example, several recent, but unrelated arc-continent collisions in the SW Pacific may well be juxtaposed in the near geological future due to the northerly motion of Australia (van Staal et al., 1998).

The Taconic orogeny of the northern Appalachians has been and still is a contentious mountain-building event because there is much disagreement on its time span, effects, and cause. Historically, this orogen has been restricted to only include Middle to Late Ordovician tectonism (see Pavlides et al., 1968, and references therein). Rodgers (1971), on the other hand, proposed that the Taconic orogeny included all Early to Late Ordovician structural events recognized in the Appalachians. However, fossils and paleomagnetic data have shown that several kinematically unrelated Early to Late Ordovician collision-related events (Penobscot, Taconic and Famatinian) took place roughly contemporaneously on opposite sides of the Iapetus Ocean and within the Iapetus Ocean itself (van Staal, 1994; van Staal et al., 1998). Other workers have questioned the intensity and scale of crustal deformation during the Taconic orogeny. For example, rocks classically described as being Taconic, such as the metamorphic tectonites of the Laurentian margin of western Newfoundland, were shown to have formed at least in part, if not completely, during the Early Silurian Salinic orogeny (Cawood et al., 1994).

The object of this paper is to discuss the effects, causes, kinematics, and time span of the Late Cambrian–Late Ordovician Taconic orogeny in the northern Appalachians, using the excellent exposures and well-studied structural relationships of western and central Newfoundland as a template. Particularly, we will discuss the tectonic and time constraints provided by the structural history and petrological characteristics of plutonic rocks of the Notre Dame arc. We will present evidence that the Taconic orogeny in Newfoundland consisted of three distinct phases of Late Cambrian to Late Ordovician (495–450 Ma) infant-arc and arc accretion in the peri-Laurentian realm. The second phase was the most intense due to Early to Middle Ordovician collision between Laurentia and a magmatic arc built upon a ribbon microcontinent. We will present new petrological and geochronological constraints on a very voluminous, late syncollision magmatic flare-up, previously interpreted as products of normal arc activity. Significantly, we interpret this flare-up to have formed as a result of slab break-off beneath an arc infrastructure that was significantly thickened due to shortening and tectonic underplating. These events can be recognized, at least in part, along the length of the northern Appalachians.

A THREE-STAGE MODEL FOR THE TACONIC OROGENY

The Taconic Orogeny: Previous Geological Constraints

The Taconic orogeny was originally defined in the Hudson Valley of New York, where there is an angular unconformity between Ordovician and Silurian rocks. This is also the site of a major far-traveled, rootless allochthon (Taconic allochthon) that emplaced Cambrian-Ordovician slope/rise sedimentary rocks above rocks that formed part of the Laurentian carbonate platform (Rowley and Kidd, 1981; Zen, 1972). The rocks affected by the Taconic orogeny form a relatively narrow belt that can be traced continuously from New England into Newfoundland (Taconide zone of Zen, 1972). The Taconic zone is restricted to the western part of the northern Appalachians, and its eastern boundary approximately coincides with the Red Indian Line in

Newfoundland (Williams et al., 1988) and its extension in New England (van Staal, 2005). The Red Indian Line separates Lower Ordovician and older rocks that formed close to Laurentia from those that formed near Gondwana (van Staal et al., 1998), i.e., the Taconic orogeny principally involved rocks that formed in the periphery of Laurentia (van Staal, 2005). At present, the cause of the Taconic orogeny in New England is generally ascribed to the collision of an outboard Ordovician magmatic arc with the Laurentian margin, following easterly subduction (present coordinates) of the intervening oceanic tract (Stanley and Ratcliffe, 1985). However, others working mainly in Quebec and Newfoundland have related the emplacement of the west-directed Ordovician allochthons primarily to obduction of suprasubduction-zone ophiolite(s) (e.g., Cawood and Suhr, 1992; Dewey and Shackleton, 1984) or to a combination of forearc ophiolite obduction followed by arc-continent collision (e.g., Hibbard, 1983; Casey and Dewey, 1984; Pinet and Tremblay, 1995). Arrival of the Laurentian margin at the trench, and the inception of ophiolite obduction and/or arc-continent collision, is relatively well constrained by chromite-bearing, foreland-basin clastics deposited above Laurentian slope/rise sediments preserved in far-traveled allochthons in Newfoundland (ca. 475 Ma, middle Arenig; Knight et al., 1991; Stevens, 1970; Waldron et al., 1998) and Quebec reentrants (Hiscott, 1978). An Early to Middle Ordovician obduction and collision is consistent with Middle Ordovician (470–460 Ma) radiometric age constraints on Taconic deformation and metamorphism of deeply buried Laurentian margin rocks from both Quebec and New England (Castonguay et al., 2001; Laird et al., 1993; Whitehead et al., 1996). However, emplacement of the Taconic allochthon and some spatially associated metamorphic thrust sheets in New England took place later during the Caradoc and/or early Ashgill (e.g., Zen, 1972; Rowley and Kidd, 1981; Bradley, 1989; Ratcliffe et al., 1999), indicating that deformation and metamorphism had to be younger in the west and were at least time-transgressive or unrelated.

Based on radiometric age constraints on deformation, metamorphism, and foreland-basin evolution, Karabinos et al. (1998) and van Staal et al. (1998) argued that the younger, Late Ordovician part of the Bronson Hill arc (ca. 454–442 Ma; Tucker and Robinson, 1990) is too young to have formed during east-directed closure of the oceanic basin that separated the Taconic arc from Laurentia. It is a well-known observation in modern orogenic belts (e.g., Taiwan; Huang et al., 2000) that arc magmatism ceases shortly after collision, probably as a result of the disappearance of the fertile subarc mantle wedge during shallowing of the dip angle (flattening) of the downgoing plate following the entrance of buoyant crust into the trench (van Hunen et al., 2002; English et al., 2003). Instead, Karabinos et al. (1998) and van Staal et al. (1998) proposed that the Early Ordovician phases of the Shelburne Falls and Notre Dame arcs in New England and in Newfoundland, respectively, formed as a result of Taconic convergence. They suggested that these arcs were built on peri-Laurentian ribbon microcontinent(s) that had rifted off Laurentia during the Early Cambrian (Waldron and van Staal, 2001), following an earlier Late Neoproterozoic opening of Iapetus (Cawood et al., 2001). Waldron and van Staal (2001) proposed that the return of the ribbon microcontinent(s) was the primary cause of the Taconic orogeny. However, this model is incomplete, because it does not explain Early Ordovician development of both oceanic and continental peri-Laurentian arcs in Newfoundland (Bédard et al., 2000), a large flare-up of Middle Ordovician (464–459 Ma), dominantly arc-like magmatism during the waning stages of arc-continent collision (van Staal et al., 2004; van Staal, 2005), and the continuation of orogenesis into the Late Ordovician.

As will be presented herein, there is ample evidence for multiple Cambrian to Late Ordovician accretionary events in the peri-Laurentian realm. These events all took place in a relatively short time period and, in part, may have overlapped due to diachroneity along the length of the northern Appalachians. To avoid confusion and incorrect correlation between temporally close, but spatially and/or kinematically unrelated events, particularly in areas in the northern Appalachians where the Taconic rocks are poorly exposed and/or severely overprinted by younger orogenic events such as in parts of New England, we have adapted a broad definition for the Taconic orogeny, such that it includes all Cambrian–Late Ordovician tectonic events that involve accretion of peri-Laurentian rocks on or near the Laurentian margin. Separation of the discrete Taconic-related accretionary events may not always be possible in every part of the Appalachian orogen, in part because they can be locally absent. However, whether accreted rock units or terranes have a peri-Laurentian provenance or not can generally be determined (van Staal et al., 1998; van Staal, 2005; Gerbi et al., 2006). This broad definition of the Taconic orogeny is therefore readily applicable along the entire length of the Appalachians.

Formation and Accretion of Peri-Laurentian Oceanic Terranes in Newfoundland Prior to the Notre Dame Arc—Laurentian-Margin Collision

The rocks added to Laurentian margin during the Cambrian and Ordovician in Newfoundland, and hence the ones most likely to preserve clues to the cause(s) of the Taconic orogeny as defined here, principally occur in the Notre Dame subzone of Newfoundland's Dunnage zone (Williams et al., 1988). The rocks in this subzone preserve a complex and protracted record of formation of suprasubduction-zone oceanic crust, their interaction with Laurentian continental crust, and overprint by continental-arc magmatism (Swinden et al., 1997; Whalen et al., 1997a; van Staal et al., 1998).

The Notre Dame subzone is sandwiched between the Baie Verte–Brompton Line to the west and the Red Indian Line to the east (Fig. 1). Both structures are associated with ophiolites, mélanges, and highly deformed rocks (Williams and St. Julien, 1982; Williams et al., 1988), which is consistent with the idea that they represent important sutures. The ophiolitic rocks of the Humber Arm and Hare Bay allochthons (Fig. 1), which occur as

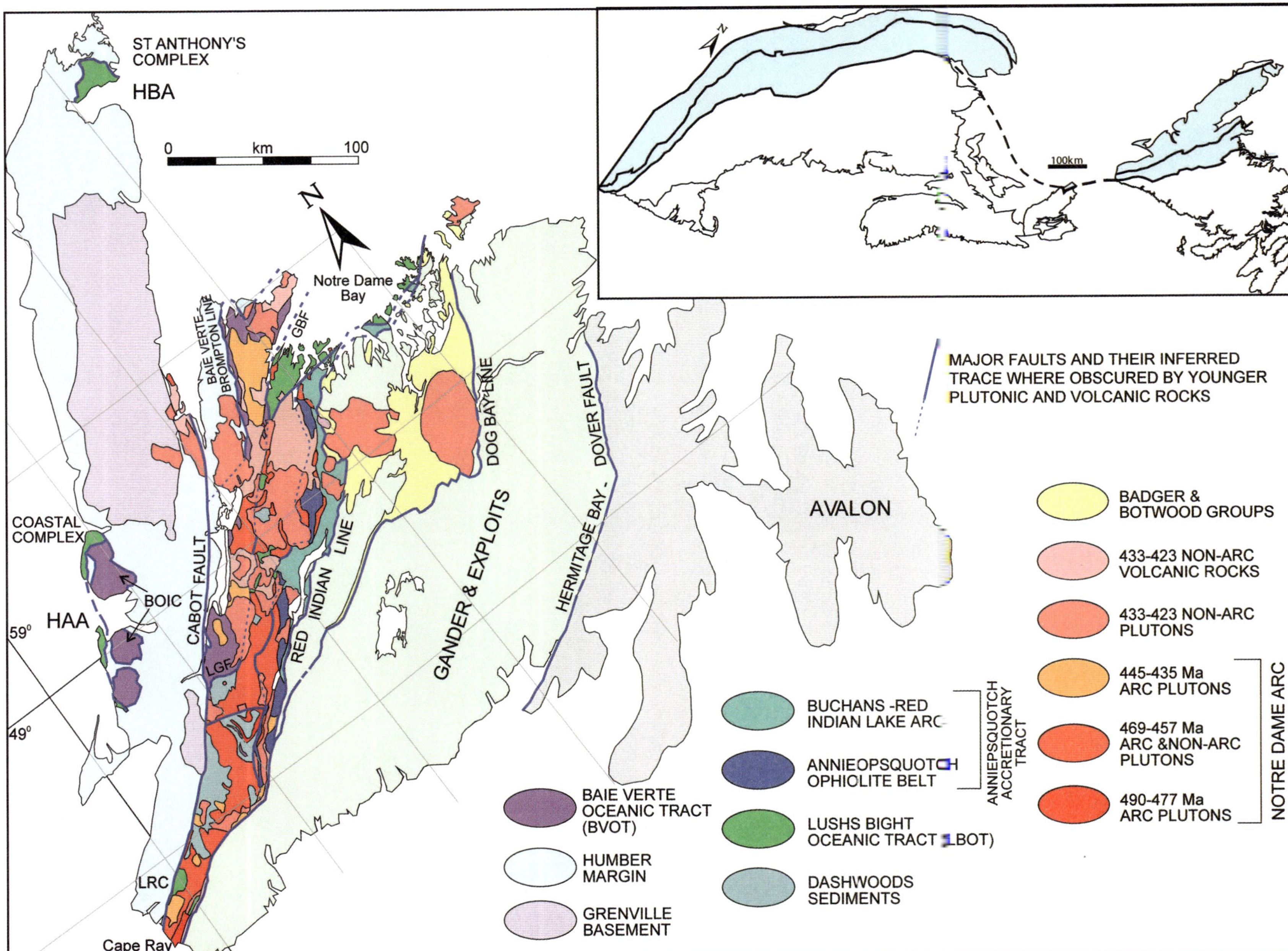

Figure 1. Tectonostratigraphic zones of Newfoundland with detailed distribution of the various ophiolitic, sedimentary, and plutonic rocks in the Notre Dame subzone and related adjacent regions. Inset shows the Taconide zone of the northern Appalachians. BOIC—Bay of Islands complex; GBF—Green Bay fault; HAA—Humber Arm allochthon; HBA—Hare Bay allochthon; LGF—Little Grand Lake fault; LRC—Long Range mafic-ultramafic complex.

klippen above rocks of Laurentia's margin, are generally regarded as structural outliers of oceanic rocks that originally formed part of the Notre Dame subzone (Williams, 1975). Fragments of three unrelated oceanic terranes are preserved in the Notre Dame subzone (Fig. 1): the Lushs Bight and Baie Verte oceanic tracts and the Annieopsquotch accretionary tract (van Staal et al., 1998). The Lushs Bight and Baie Verte oceanic tracts largely correspond to the Notre Dame Bay area and western ophiolite belt of Swinden et al. (1997), respectively. The Annieopsquotch accretionary tract mainly consists of the Annieopsquotch (eastern) ophiolite and the Buchans and Roberts Arm belts (Lissenberg et al., 2005a, 2005b; Zagorevski et al., 2006). These terranes, which range in age from Middle Cambrian to Middle Ordovician (510–460 Ma), are bounded by faults and can be distinguished based on their age and tectonic evolution. They are linked together by an extensive swath of Ordovician to Early Silurian plutons of the Notre Dame arc (see following) that crosscut and stitch their faulted boundaries and divide the oceanic terranes into a complex outcrop pattern of isolated bodies of oceanic rocks surrounded by younger plutonic rocks.

The oldest known oceanic rocks occur in the Lushs Bight oceanic tract, which principally includes the Lushs Bight, lower part of the Western Arm, Moreton's Harbour, and Sleepy Cove Groups (Swinden et al., 1997). These rocks are superbly exposed along the indented coast of Notre Dame Bay in north-central Newfoundland (Fig. 1). Radiometric ages of gabbroic and trondhjemitic intrusions, and numerous crosscutting mafic dikes, indicate that the bulk of the volcanic rocks of the Lushs Bight oceanic tract have a Middle to Upper Cambrian age (510–501 Ma; Elliot et al., 1991; Szybinski, 1995). The association of pillow basalts, sheeted dikes, gabbro, and rare ultramafic rocks (Kean et al., 1995; Williams and Payne, 1975) indicates an ophiolitic affinity. Large intrusive bodies of juvenile Middle Cambrian trondhjemite and diorite (Fryer et al., 1992; Jenner et al., 1991) and the abundance of boninite and primitive island-arc tholeiite (Swinden, 1996) suggest that this tract represents a fragment of suprasubduction-zone oceanic crust. Based on its petrological makeup and a lack of older arc rocks, the Lushs Bight oceanic tract has been interpreted to be an infant-arc terrane formed during subduction initiation at ca. 510 Ma (van Staal et al., 1998). Paleomagnetic studies of the Moreton's Harbour Group have yielded a low southerly latitude of ~10°S and suggest that the Lushs Bight oceanic tract formed near the Laurentian margin, which was situated at a similar latitude at that time (Johnson et al., 1991). This hypothesis depends in part on the assumption that the undated Moreton's Harbour Group is not significantly younger than the remainder of the Lushs Bight oceanic tract. However, formation near the Laurentian margin rather than an intraoceanic setting is in keeping with the low ε_{Nd} values (0 to +2.8) of the Lushs Bight oceanic tract, which suggest the presence of older crust recycled via subduction of continent-derived sediment (Swinden et al., 1997). Age constraints on early shear zones and related structures, combined with extensive isotopic evidence of crustal contamination in the crosscutting Upper Cambrian–Lower Ordovician mafic dikes and plutons, suggest that the Lushs Bight oceanic tract was deformed and emplaced onto Precambrian continental crust between 500 and 493 Ma (Szybinski, 1995; Swinden et al., 1997). Similar constraints are provided by large isolated ophiolitic enclaves, such as the ophiolitic Long Range mafic-ultramafic complex (Figs. 1, 2, and 3) in the southern Notre Dame subzone (Dashwoods subzone of Williams, 1995[1]). The ophiolitic Long Range mafic-ultramafic complex (mainly mantle and lower-crustal rocks), a probable correlative of the Lushs Bight oceanic tract, and the underlying Mischief mélange are stitched by the ca. 488 Ma Cape Ray granodiorite (Hall and van Staal, 1999), which contains inherited Precambrian zircons (Dubé et al., 1996) and has a ε_{Nd} value of –4.5 and T_{DM} (depleted mantle model age) of 1.3 Ga (Whalen et al., 1997b), suggesting that it interacted with Precambrian continental crust at depth. These relationships suggest that the Lushs Bight oceanic tract formed a once-continuous blanket of obducted suprasubduction-zone oceanic lithosphere across the Notre Dame subzone. Although Precambrian continental basement is nowhere exposed in the Notre Dame subzone, its existence at depth is consistently indicated by inherited zircons and isotopic data extracted from the extensive swath of Early to Middle Ordovician tonalite and granodiorite plutons of the Notre Dame arc (Whalen et al., 1997a; see following). These data support earlier inferences that the unexposed Precambrian basement, which is referred to as the Dashwoods block (Waldron and van Staal, 2001), but herein abbreviated to Dashwoods, has a Laurentian provenance (see also following). Dashwoods has been interpreted as a microcontinent (Waldron and van Staal, 2001) rather than Laurentia's leading edge (cf. van Staal et al., 1998) because this Precambrian basement was subjected to obduction by the Lushs Bight oceanic tract 20–15 m.y. before the first evidence of tectonic loading of the Laurentian margin to the west (Fig. 3). Such an interpretation is further supported by major time discrepancies between the paleomagnetism-based Neoproterozoic (ca. 570 Ma) timing of Iapetus Ocean opening and a well-defined Early Cambrian (ca. 530 Ma) rift-drift transition preserved in the Laurentian margin, where the younger Early Cambrian rifting event was related to departure of the Dashwoods block (Cawood et al., 2001). Separation of a narrow microcontinent probably happened as a result of an inboard ridge jump (Müller et al., 2001), which, if correct, demands that a spreading ridge system was present near the Laurentian margin during the Neoproterozoic-Cambrian boundary (Fig. 4). The alternative, a ridge jump from a spreading center positioned in the center of a 4000–5000-km-wide ocean (Torsvik et al., 1996; Cawood et al., 2001) is highly unlikely. An inboard ridge jump leaves an abandoned spreading center and related transform faults immediately outboard of the newly created microcontinent (Fig. 4), which are

[1]We adhere to the original definition of the Notre Dame subzone of Williams et al. (1988) because ophiolitic rocks and Early to Middle Ordovician plutons cross the boundary between Dashwoods and Notre Dame subzones without a change in composition or style of deformation and metamorphism (see also following discussion).

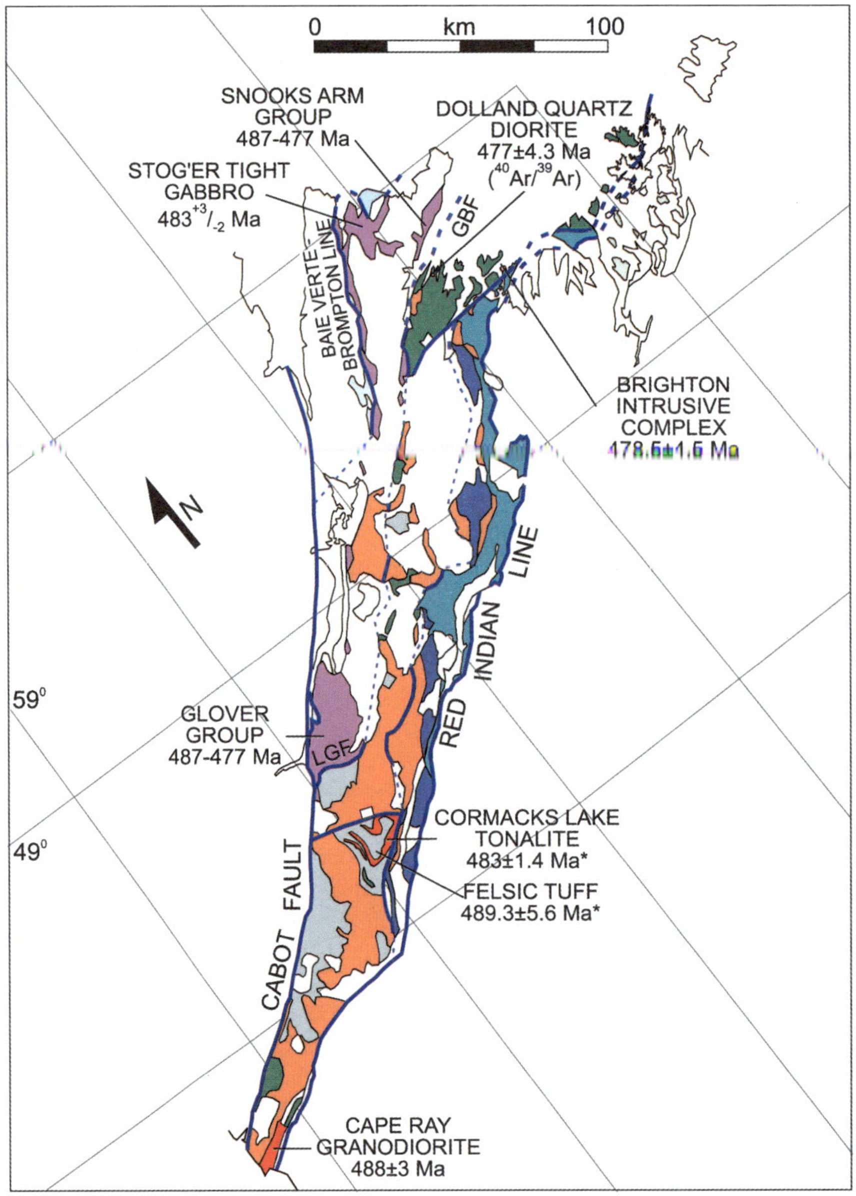

Figure 2. Distribution of the Ordovician magmatic rocks of the first phase of the Notre Dame arc and coeval oceanic Snooks Arm arc. Ages determined in this paper are marked by an asterisk. Other ages were compiled from Szybinski (1995), Dubé et al. (1996), and Ramezani et al. (2000). Color scheme, abbreviations, and legend are same as in Figure 1.

favorable sites for subduction initiation during subsequent ocean closure (Hall et al., 2003). We postulate that the subduction initiation responsible for generation of the Lushs Bight oceanic tract used such an abandoned ridge or associated transform fault.

Early to Middle Cambrian subduction initiation near the Laurentian margin is also consistent with the age, deformation history, and makeup of the Cambrian Coastal complex (Karson and Dewey, 1978), which forms part of the Humber Arm allochthon, (Fig. 1; Cawood and Suhr, 1992), and the ophiolitic St. Anthony's complex of the Hare Bay allochthon in northern Newfoundland (Fig. 1). The latter has an Upper Cambrian metamorphic sole (ca. 495 Ma; Jamieson, 1988; G. Dunning, 2004, personal commun.), which also suggests that subduction and detachment of relatively young oceanic lithosphere (Wakabayashi and Dilek, 2003) had already initiated during the Late Cambrian near Laurentia. It is unclear at this stage whether Dashwoods formed a continuous ribbon continent in front of the Laurentian margin along the length of the northern Appalachians or a series of isolated, small ribbon microcontinents separated by oceanic gaps. In contrast to the Lushs Bight oceanic tract, magmatic

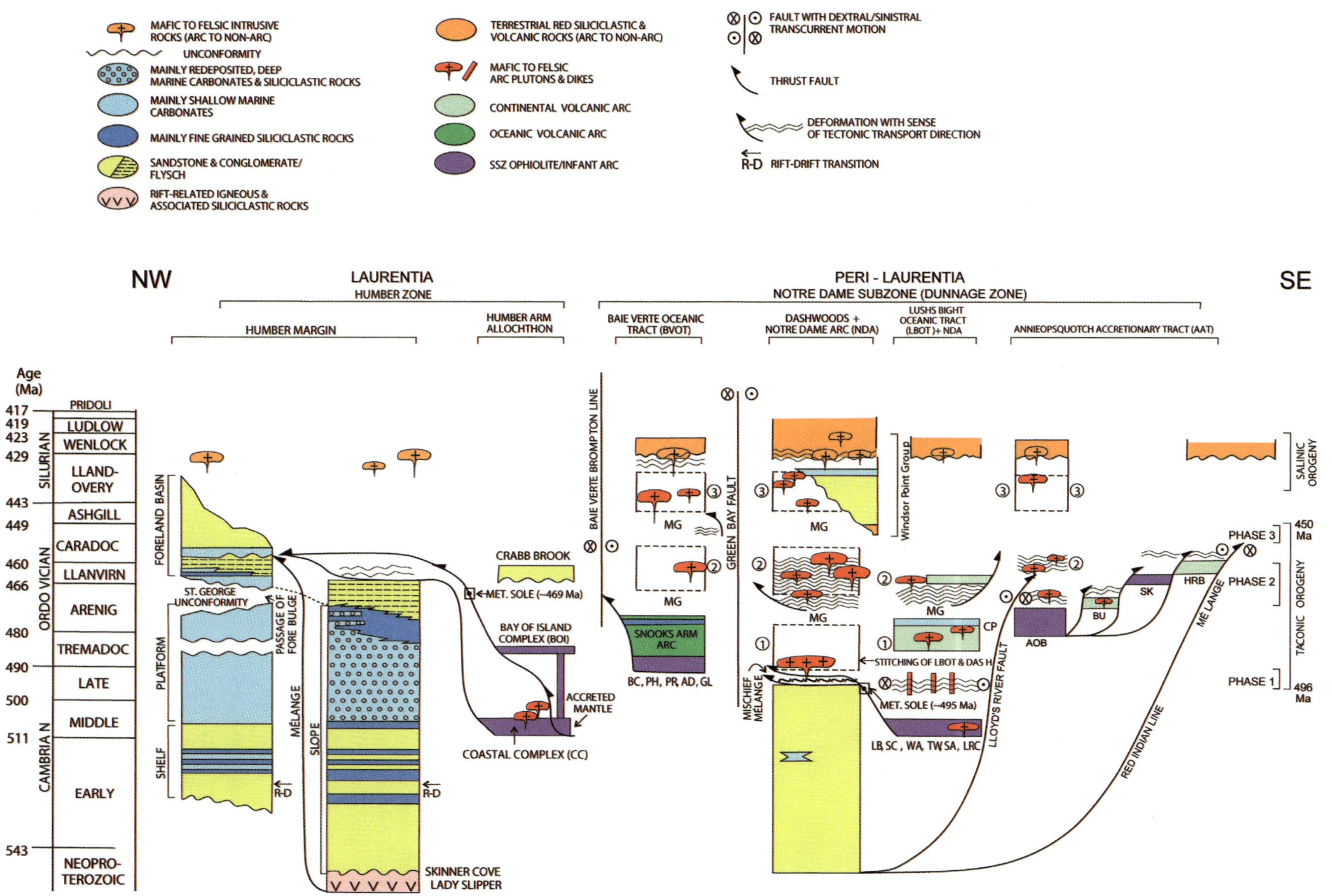

Figure 3. Summary of the stratigraphic and tectonic evolution of the Laurentian margin and Notre Dame subzone during the Ordovician–Early Silurian. AD—Advocate complex; AOB—Annieopsquotch ophiolite belt; BC—Betts Cove complex; BU—Buchans and Robert Arm Groups; GL—Grand Lake complex; HRB—Harbour Round Formation; LB—Lushs Bight Group; LRC—Long Range mafic-ultramafic complex; MG—magmatic gap; PH—Pacquet Harbour Group; PR—Point Rousse complex; SA—St. Anthony complex; SC—Sleepy Cove Group; SK—Skidder Formation; TW—Twillingate Trondhjemite; WA—Western Arm Group.

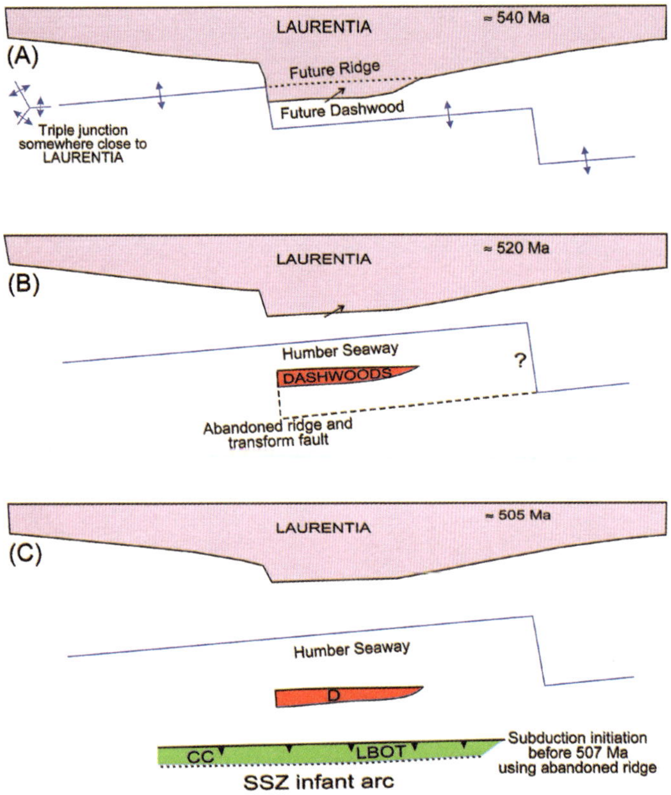

Figure 4. Hypothetical formation of Dashwoods (D) by an inboard ridge jump near a promontory in the Laurentian margin. This process may have separated a series of small ribbon microcontinents by intersection of the new spreading center(s) with several promontories along strike or one large ribbon. Spreading in the Humber seaway was slow because fossils and paleomagnetic data suggest that Dashwoods remained near Laurentia. CC—Coastal complex; LBOT—Lushs Bight oceanic tract.

evidence that the Coastal complex was ever emplaced over Dashwoods prior to its obduction onto the Laurentian margin is lacking. Although this is consistent with the latter paleogeography, the former paleogeography cannot be ruled out at present.

The Late Cambrian (500–493 Ma) obduction of the Lushs Bight oceanic tract onto the Dashwoods microcontinent (Fig. 5) is the first phase of the Taconic orogeny (Taconic 1) as defined herein (Fig. 3).

The Baie Verte oceanic tract, which occurs to the west of the Lushs Bight oceanic tract (Figs. 1 and 2), is largely preserved as a series of ophiolitic slivers and enclaves in a narrow fault-bounded belt, wedged between the Baie Verte Brompton Line and the Green Bay fault in the north (Hibbard, 1983). To the south, it terminates against the Little Grand Lake fault (Figs. 1 and 2), which bends to the north at its eastern extremity and probably was connected to the ancestor of the Green Bay fault, prior to intrusion of the Silurian Topsails Igneous Suite (Whalen et al., 1987, 1996). The Baie Verte oceanic tract includes the Betts Cove, Pointe Rousse, Advocate, and Grand Lake ophiolite complexes, and related volcanic rocks such as the Pacquet Harbour Group (Hibbard, 1983; Swinden et al., 1997). U-Pb zircon dating of the Betts Cove and Grand Lake ophiolites, which occur on opposite sides of the Baie Verte oceanic tract, has yielded two nearly identical ages of 489 $^{+3}/_{-2}$ and 490 ± 4 Ma, respectively (Cawood et al., 1996; Dunning and Krogh, 1985), which fall on the Cambrian-Ordovician boundary (Fig. 3). Although compositionally similar, including an abundance of boninite (Bédard et al., 1998), the Baie Verte oceanic tract ophiolites are significantly younger (20–15 m.y.) than the oceanic rocks of the Lushs Bight oceanic tract (Fig. 3). They either formed during the final stages of obduction of the Lushs Bight oceanic tract, which is an unlikely setting for boninite genesis, or they postdate it and are tectonically related to inception of a new subduction zone in the Humber seaway (Fig. 5; Waldron and van Staal, 2001). Furthermore, the Baie Verte oceanic tract became basement to an oceanic-arc–back-arc complex, e.g., the Snooks Arm and Glover Groups (Bédard et al., 2000; Szybinski et al., 1995), during the Tremadoc and early Arenig (487–476 Ma; Williams, 1992; Ramezani et al., 2000), herein referred to as the Snooks Arm arc (Figs. 3 and 6). The Snooks Arm arc developed contemporaneously with the first phase of the Notre Dame arc in Dashwoods. The simplest model to explain these relationships is that the Baie Verte oceanic tract formed when subduction stepped back into the Humber seaway, which separated the Dashwoods microcontinent from the Humber margin at ca. 490 Ma, shortly after obduction of the Lushs Bight oceanic tract (Figs. 5 and 6). Clogging of the original subduction zone due to entrance of the buoyant Dashwoods microcontinent probably forced the stepping back (Mueller and Phillips, 1991) into the Humber seaway. Such a mechanism provides a simple explanation for formation of the first phase of the continental Notre Dame arc in Dashwoods and the coeval oceanic Snooks Arm arc above the Baie Verte oceanic tract (Fig. 5). The existence of a coeval arc system built both on oceanic (Snooks Arm arc) and continental (Notre Dame arc) substrates can be explained as follows: (1) the Snooks Arm arc was originally located along strike of the Notre Dame arc (Fig. 6), analogous to the alignment of the oceanic Banda and continental Sunda arcs of the Indonesian island arc, before their present juxtaposition due to arc-parallel translation; (2) the arc magmatic axis moved locally westward onto forearc oceanic crust due to slab rollback; or (3) there was a combination of both.

Constraints on Humber Seaway Closure and Laurentian Margin–Dashwoods Collision

The ca. 484 Ma suprasubduction-zone ophiolitic Bay of Island complex (Fig. 1) provides additional constraints concerning the tectonic evolution of the Humber seaway (Jenner et al., 1991; Kurth et al., 1998). Relationships in the Lewis Hills indicate that formation of the Bay of Island complex resulted from rifting of Lower to Middle Cambrian infant-arc crust of the Coastal complex (Fig. 1; Kurth et al., 1998; Suhr and Edwards, 2000; Suhr and Cawood, 2001) following transfer of the Coastal

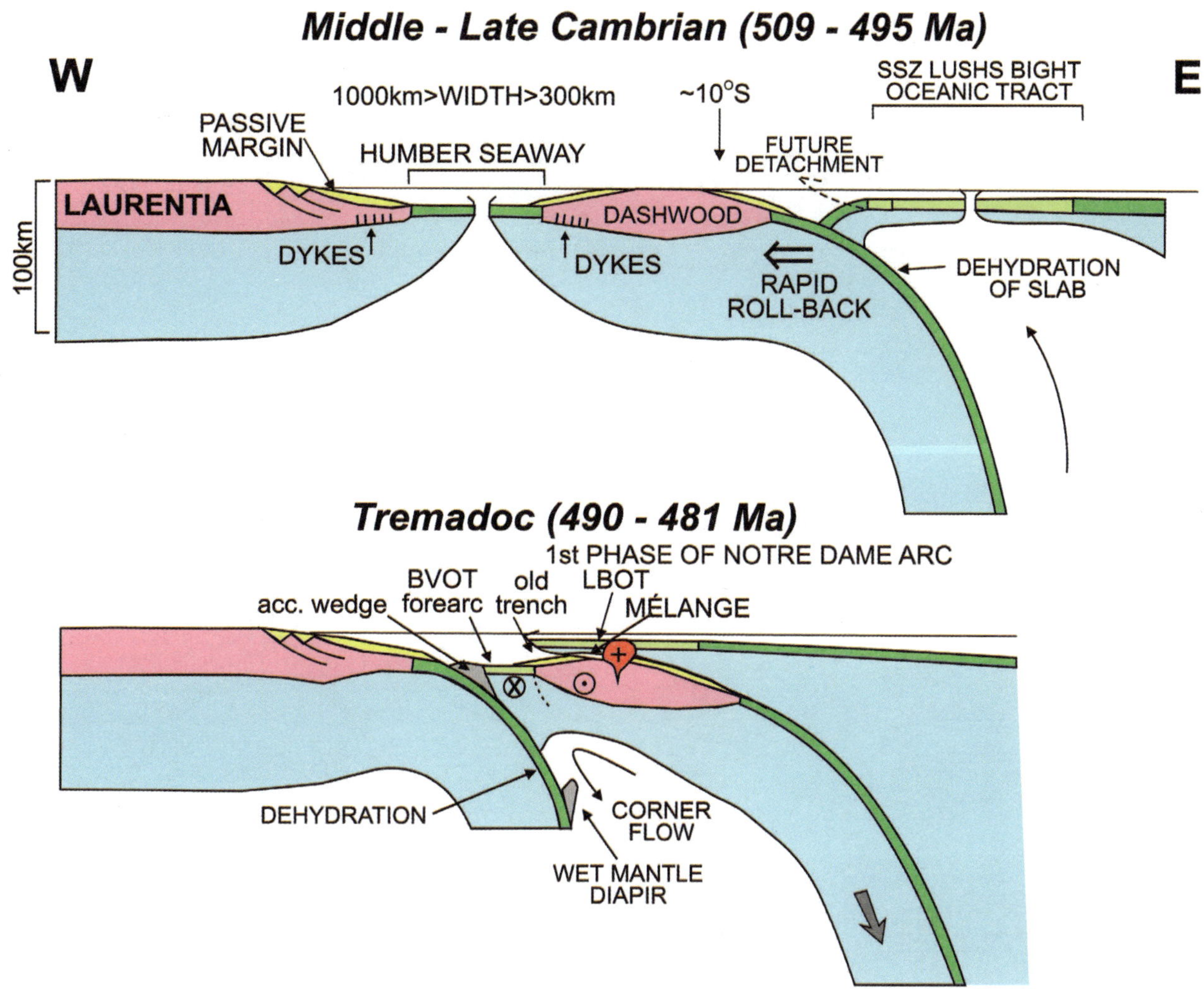

Figure 5. Diagram with Cambrian–Early Ordovician tectonic evolution of Dashwoods and the Humber margin. Subduction initiation is assumed to have used zones of lithospheric weakness such as the abandoned ridge or transform fault immediately outboard of Dashwoods. BVOT—Baie Verte oceanic tract; LBOT—Lushs Bight oceanic tract

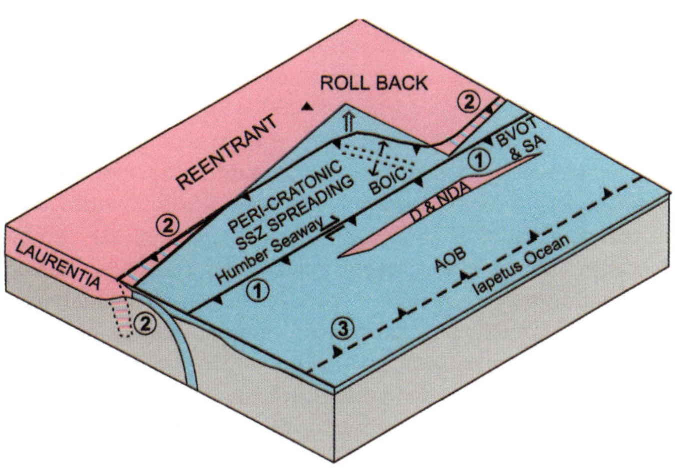

Figure 6. (1) Early Ordovician (Tremadoc) subduction of the Humber seaway with formation of the oceanic Snooks Arm (SA) and Notre Dame arcs (NDA) on the Baie Verte oceanic tract (BVOT) and composite Dashwoods (D), respectively. Dashwoods is inferred to have come from, in part, the latitude of northern Newfoundland or Labrador (see section Zircon Inheritance and Provenance of the Notre Dame Arc), consistent with dextral oblique convergence (Brem et al., 2003). Collision of arc system with promontories (2) slows down convergence in Humber seaway. Closure of seaway mainly continues by rollback of the subduction zone into oceanic lithosphere trapped in reentrants, causing formation of the Bay of Islands complex (BOIC) by pericollisional spreading. This process may have been accompanied and/or aided by dextral transtension. Progressive reduction in convergence within the Humber seaway eventually created a new west-dipping subduction zone (3) at ca. 480 Ma, which was responsible for formation of the infant arc Annieopsquotch ophiolite belt (AOB).

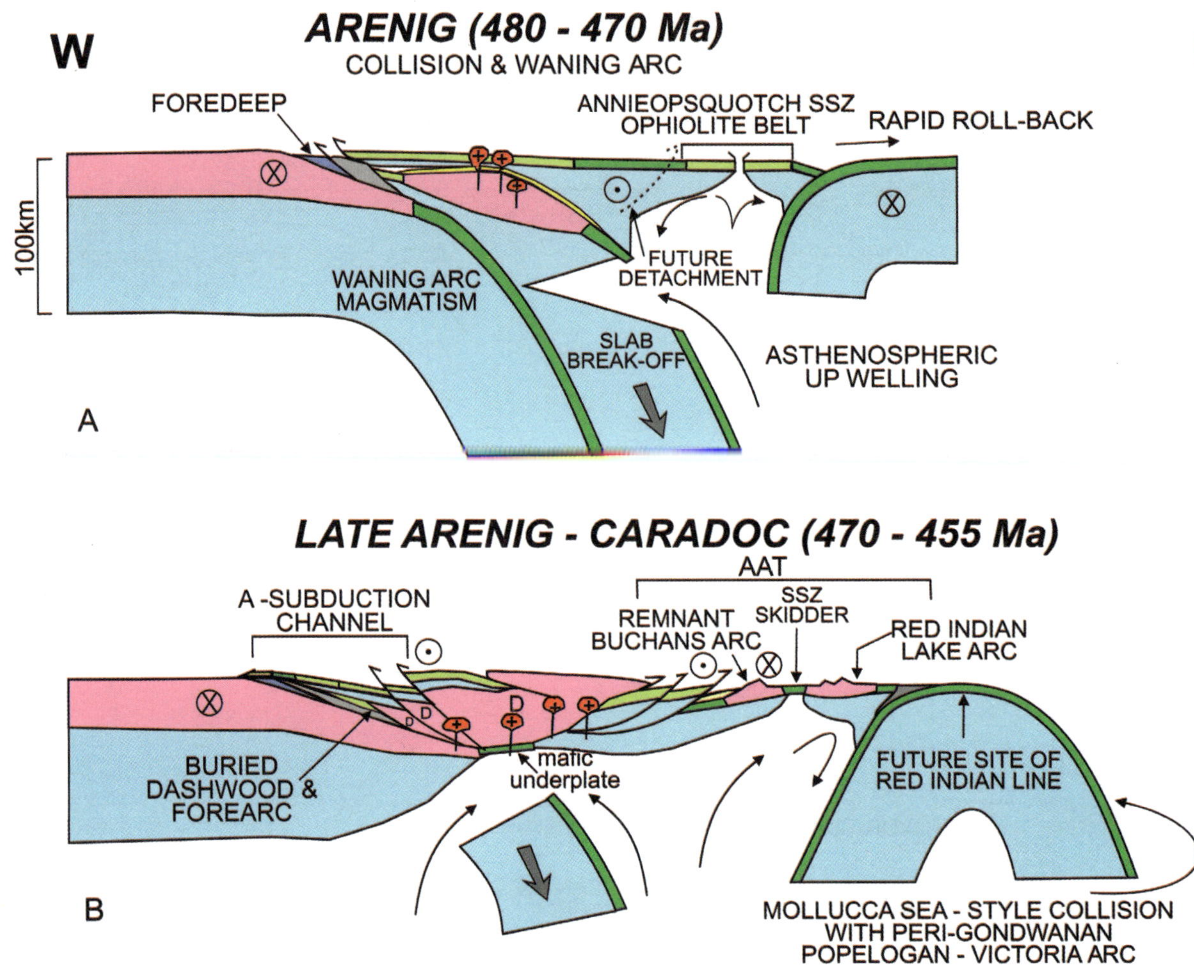

Figure 7. (A) Diagram showing the Middle Ordovician collision (Taconic 2) of Dashwoods with the Laurentian margin. Slowing down of convergence due to the start of collision near promontories (Fig. 6) initiated west-directed subduction outboard of Dashwoods. (B) Collisional thickening of the Notre Dame arc. East-directed thrusting, first related to Taconic 2 and later to Taconic 3, caused accretion of elements of the Annieopsquotch accretionary tract to Dashwoods. Break-off of the Laurentian margin slab allowed upwelling of the asthenosphere and melting of the lower part of the Notre Dame arc's infrastructure, including a mafic underplate.

complex into a forearc position with respect to the subduction zone responsible for closing the Humber seaway (Figs. 5 and 6). The Coastal complex, although a correlative of the Lushs Bight oceanic tract, was magmatically inactive during the Early Ordovician when spreading responsible for formation of the Bay of Island complex took place. Hence, the Bay of Island complex cannot have formed in a back-arc basin sensu stricto (cf. Suhr and Edwards, 2000). More likely, it formed by pericollisional spreading (Harris, 1992) due to rollback of the downgoing slab into a second-order reentrant in the Laurentian margin following impingement of the trench and forearc with nearby promontories (Fig. 6) and/or forearc transtension (Cawood and Suhr, 1992; Dewey, 2002; van Staal, 2005).

The metamorphic sole of the Bay of Island complex has a hornblende $^{40}Ar/^{39}Ar$ age of 469 ± 5 Ma (Dallmeyer and Williams, 1975), which is surprisingly young when compared with the ophiolite's U-Pb zircon age of ca. 484 Ma (cf. Hacker et al., 1996; Wakabayashi and Dilek, 2000). Nevertheless, assuming that no undetectable excess argon is present (cf. Dallmeyer and Williams, 1975), it suggests that the ophiolite was obducted onto the Laurentian margin during the Middle Ordovician, probably after the inception of tectonic loading of the Laurentian margin, which has a minimum age of ca. 475 Ma. Instead, already-obducted segments of the Coastal complex and/or Baie Verte oceanic tract near promontories probably were the source of the ophiolitic detritus in the foreland basin (Fig. 6). The relatively young detachment and emplacement of the Bay of Island complex, with respect to inception of Laurentian-margin tectonic loading, signal nearly complete Humber seaway closure at the latitude of Newfoundland (Fig. 6). Hence, this

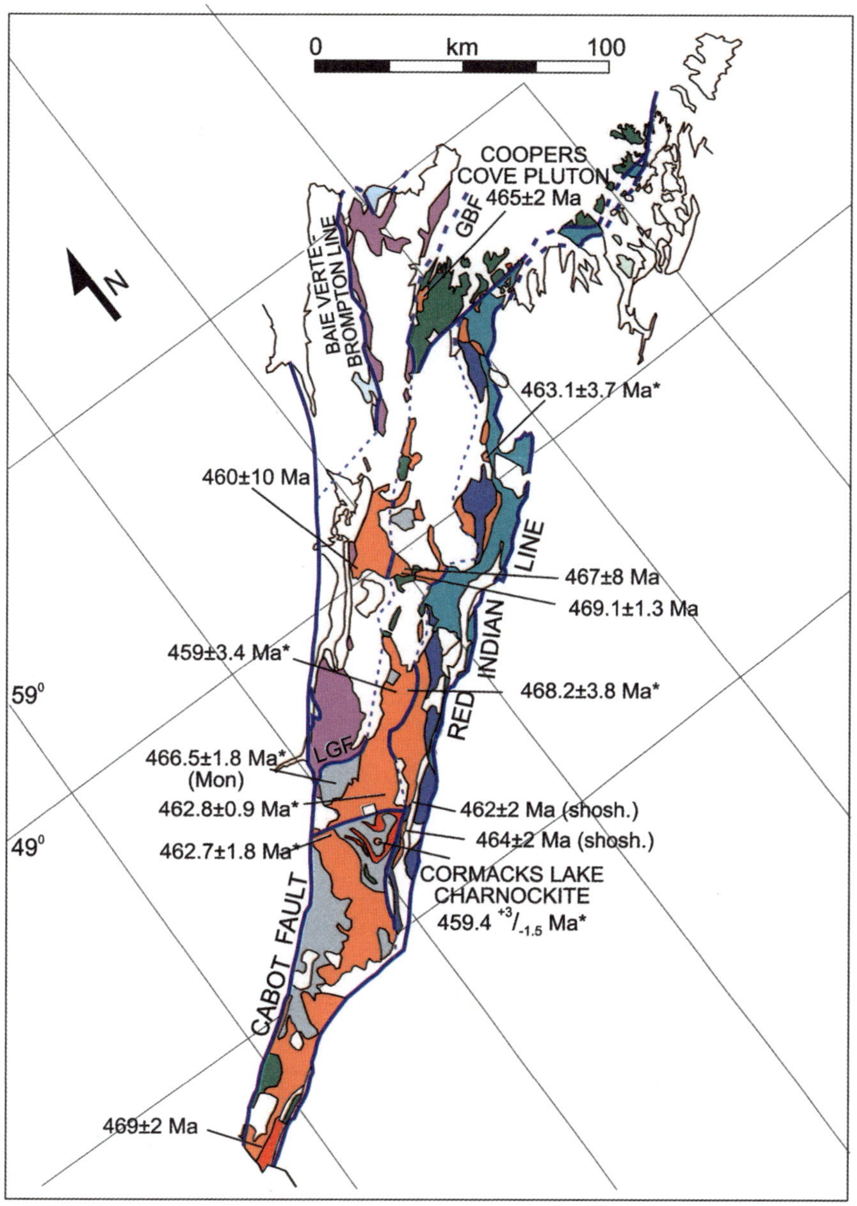

Figure 8. Distribution of the Ordovician plutons of the second phase of the Notre Dame arc with available U-Pb ages (469–459 Ma). Ages determined in this paper are marked by an asterisk. Also shown is the monazite (mon) age near Island Pond. Note that the second phase is much more voluminous than the first phase. Shosh.—shoshonitic members of the second phase of the Notre Dame arc (Lissenberg et al., 2005b). Color scheme, abbreviations, and legend are same as in Figure 1.

implies full-scale collision between the Laurentian margin and Dashwoods during the Middle Ordovician. The tectonic events related to Humber seaway ophiolite emplacement and Laurentia-Dashwoods collision (herein grouped as Taconic 2) had a major impact on the Notre Dame arc and its Dashwoods host rocks in general (see following discussion).

Arc-continent collisions generally start at promontories in the continental margin (cf. Cawood and Suhr, 1992). This commonly stops or significantly slows down convergence between the two approaching elements. We postulate that a significant decrease in convergence rate due to collision of Dashwoods with Laurentian-margin promontories transferred convergence from the Humber seaway to the Iapetus Ocean, which lay outboard (east) of Dashwoods. This process may have produced a west-dipping subduction zone in Iapetus immediately outboard of Dashwoods at ca. 480 Ma (Figs. 6 and 7A). Evidence for this, the infant-arc–like Annieopsquotch ophiolite belt, is preserved in the highest thrust sheets of the Annieopsquotch accretionary tract (Lissenberg et al., 2005a, 2005b; Lissenberg and van Staal, 2006). Middle Ordovician arc-continent collision in Newfoundland is consistent with the rapid accretion

TABLE 1. U-Pb SENSITIVE HIGH-RESOLUTION ION MICROPROBE ANALYTICAL DATA

Spot name	U (ppm)	Th (ppm)	Th/U	Pb* (ppm)	^{204}Pb (ppb)	^{204}Pb/^{206}Pb	± ^{204}Pb/^{206}Pb	f(206)^{204†}	^{208}Pb/^{206}Pb	± ^{208}Pb/^{206}Pb	^{207}Pb/^{235}U	± ^{207}Pb/^{235}U
PQB-SP-01007 (z7138): Felsic tuffaceous gneiss, Cormacks Lake complex (UTM zone 21, 433868E-5344937N)												
7138-24.1*	367	2	0.007	23	0	0.000019	0.000071	0.0003	0.0055	0.0027	0.5270	0.0132
7138-16.1*	56	24	0.435	4	1	0.000292	0.000325	0.0051	0.1234	0.0130	0.5524	0.0505
7138-12.1*	104	55	0.546	8	1	0.000132	0.000192	0.0023	0.1792	0.0081	0.5409	0.0308
7138-1.1*	241	101	0.434	17	2	0.000124	0.000062	0.0021	0.1387	0.0032	0.5213	0.0127
7138-9.1*	1145	1406	1.269	100	1	0.000015	0.000013	0.0003	0.4063	0.0019	0.5348	0.0066
7138-11.1*	240	225	0.971	19	3	0.000213	0.000097	0.0037	0.3062	0.0051	0.5271	0.0177
7138-23.1*	414	2	0.006	26	4	0.000165	0.000052	0.0029	-0.0008	0.0020	0.5255	0.0109
7138-5.1*	220	11	0.051	14	1	0.000044	0.000092	0.0008	0.0212	0.0036	0.5655	0.0177
7138-20.1	151	86	0.584	12	2	0.000180	0.000176	0.0031	0.1913	0.0082	0.6116	0.0317
7138-15.1	272	217	0.824	24	2	0.000123	0.000081	0.0021	0.2622	0.0047	0.6217	0.0177
7138-21.1	167	75	0.467	13	4	0.000323	0.000113	0.0056	0.1361	0.0062	0.6013	0.0226
7138-4.1	172	85	0.512	14	4	0.000336	0.000098	0.0058	0.1609	0.0059	0.6113	0.0203
7138-10.1	101	[illegible]	[illegible]	[illegible]	[illegible]	0.000574	0.000140	[illegible]	[illegible]	0.0062	0.5707	0.0294
7138-18.1	129	51	0.409	11	2	0.000197	0.000178	0.0034	0.1346	0.0072	0.6520	0.0361
7138-3.1	201	129	0.664	18	1	0.000059	0.000054	0.0010	0.2182	0.0035	0.6475	0.0143
7138-2.1	285	184	0.667	25	6	0.000295	0.000088	0.0051	0.2068	0.0041	0.6563	0.0187
WXNF-48 (z2821): Hornblende granodiorite, Hungry Mountain complex (UTM zone 21, 540646E-5434846N)												
2821-84.1	463	501	1.119	40	3	0.000120	0.000202	0.0021	0.3460	0.0130	0.5739	0.0339
2821-92.1	251	217	0.894	21	4	0.000251	0.000214	0.0044	0.2885	0.0120	0.5619	0.0384
2821-91.1	795	1080	1.404	75	3	0.000055	0.000037	0.0010	0.4259	0.0060	0.5595	0.0114
2821-99.1	713	479	0.694	57	25	0.000555	0.000109	0.0096	0.2027	0.0059	0.5425	0.0206
2821-8.1	138	83	0.623	11	6	0.000680	0.000727	0.0118	0.1990	0.0322	0.5279	0.1184
2821-86.1	131	40	0.318	10	0	0.000007	0.000203	0.0001	0.1154	0.0105	0.6578	0.0391
2821-48.1	410	403	1.015	36	3	0.000108	0.000083	0.0019	0.3207	0.0083	0.5454	0.0178
2821-94.1	658	615	0.965	57	6	0.000151	0.000126	0.0026	0.2920	0.0087	0.5702	0.0307
2821-96.1	326	390	1.236	30	2	0.000100	0.000207	0.0017	0.4009	0.0138	0.5848	0.0366
2821-100.1	863	1143	1.368	82	2	0.000031	0.000148	0.0005	0.4217	0.0092	0.5824	0.0262
2821-36.1	1080	1569	1.501	106	7	0.000094	0.000036	0.0016	0.4569	0.0082	0.5649	0.0106
2821-55.1	447	285	0.660	36	1	0.000045	0.000135	0.0008	0.1963	0.0087	0.5940	0.0251
2821-88.2	405	403	1.028	36	3	0.000114	0.000073	0.0020	0.3142	0.0069	0.5939	0.0191
2821-66.1	667	772	1.196	61	10	0.000236	0.000144	0.0041	0.3504	0.0085	0.5717	0.0264
2821-61.1	1162	2111	1.877	125	8	0.000099	0.000056	0.0017	0.5771	0.0067	0.5871	0.0188
2821-69.1**	1113	110	0.102	157	22	0.000154	0.000057	0.0027	0.0274	0.0024	1.4464	0.0269
2821-4.1**	355	52	0.151	143	3	0.000029	0.000020	0.0005	0.0367	0.0013	9.1842	0.1480
2821-54.1**	81	134	1.716	52	3	0.000081	0.000166	0.0014	0.4738	0.0099	11.8008	0.4100
WXNF-175 (z3405): Granodiorite, Cariboo Lakes gneiss complex (UTM zone 21, 475287E-5394211N)												
3405-6.1	273	127	0.482	20	4	0.000227	0.000141	0.0039	0.1526	0.0078	0.5300	0.0243
3405-1.3	111	68	0.629	9	2	0.000356	0.000489	0.0062	0.2183	0.0208	0.5716	0.0779
3405-43.2	223	139	0.645	17	1	0.000078	0.000108	0.0014	0.1996	0.0086	0.5461	0.0214
3405-41.1	175	84	0.497	13	4	0.000339	0.000156	0.0059	0.1617	0.0090	0.5240	0.0281
3405-21.2	181	88	0.501	14	3	0.000262	0.000171	0.0045	0.1697	0.0089	0.5760	0.0351
3405-10.2	214	124	0.598	17	3	0.000231	0.000102	0.0040	0.1848	0.0071	0.5475	0.0241
3405-13.1	279	168	0.622	22	3	0.000190	0.000158	0.0033	0.1998	0.0139	0.5659	0.0319
3405-11.1	332	243	0.757	27	5	0.000247	0.000088	0.0043	0.2422	0.0087	0.5825	0.0187
3405-17.1	247	137	0.573	19	5	0.000351	0.000194	0.0061	0.1623	0.0110	0.5468	0.0331
3405-9.2	186	102	0.568	15	0	0.000010	0.000010	0.0002	0.1855	0.0061	0.5837	0.0133
3405-84.1	112	42	0.387	8	5	0.000685	0.000373	0.0119	0.1236	0.0179	0.5482	0.0631
3405-1.2	154	89	0.597	12	0	0.000010	0.000010	0.0002	0.1900	0.0092	0.5926	0.0214
3405-77.1	83	46	0.580	7	2	0.000400	0.000528	0.0069	0.1963	0.0253	0.5900	0.0893
3405-7.2	395	271	0.709	32	1	0.000025	0.000099	0.0004	0.2148	0.0064	0.6196	0.0216
3405-99.1	240	133	0.575	19	5	0.000333	0.000166	0.0058	0.1851	0.0106	0.5527	0.0299
3405-90.1	194	87	0.463	15	3	0.000235	0.000196	0.0041	0.1460	0.0212	0.5819	0.0397
3405-64.1	127	26	0.211	9	9	0.001124	0.000479	0.0195	0.0551	0.0194	0.5348	0.0820
3405-52.1	518	483	0.962	45	10	0.000295	0.000095	0.0051	0.2937	0.0088	0.5678	0.0218
3405-2.1	187	185	1.026	17	3	0.000226	0.000216	0.0039	0.3150	0.0138	0.5743	0.0475
3405-83.1	135	62	0.476	11	6	0.000716	0.000678	0.0124	0.1449	0.0280	0.5740	0.1164
3405-76.1	112	44	0.404	9	4	0.000537	0.000468	0.0093	0.1121	0.0198	0.5173	0.0852
3405-20.2**	46	21	0.467	5	2	0.000425	0.000542	0.0074	0.2414	0.0255	0.9407	0.1260
3405-85.1**	411	217	0.546	48	89	0.002253	0.000188	0.0390	0.1647	0.0087	1.5636	0.0606

(continued)

TABLE 1. (*continued*)

Spot name	U (ppm)	Th (ppm)	Th/U	Pb* (ppm)	^{204}Pb (ppb)	^{204}Pb/^{206}Pb	± ^{204}Pb/^{206}Pb	f(206)^{204†}	^{208}Pb/^{206}Pb	± ^{208}Pb/^{206}Pb	^{207}Pb/^{235}U	± ^{207}Pb/^{235}U
WXNF-175 (z3405): Granodiorite, Cariboo Lakes gneiss complex (UTM zone 21, 475287E-5394211N)												
3405-88.1**	219	81	0.383	30	5	0.000228	0.000117	0.0040	0.1928	0.0087	1.3938	0.0507
3405-86.1**	79	45	0.591	11	6	0.000698	0.000402	0.0121	0.1686	0.0169	1.2740	0.1250
3405-20.1**	46	48	1.082	8	4	0.000670	0.000535	0.0116	0.3660	0.0265	1.2830	0.1788
3405-75.1**	306	69	0.232	46	2	0.000048	0.000066	0.0008	0.0830	0.0043	1.5925	0.0501
3405-100.1**	117	53	0.474	19	8	0.000478	0.000184	0.0083	0.1402	0.0095	1.5486	0.0762
3405-51.1**	55	26	0.483	10	2	0.000235	0.000349	0.0041	0.1666	0.0161	1.7078	0.1323
3405-36.1**	152	53	0.360	26	4	0.000198	0.000092	0.0034	0.1078	0.0068	1.7084	0.0567
3405-50.1**	152	80	0.545	28	3	0.000131	0.000171	0.0023	0.1604	0.0143	1.8200	0.0763
3405-61.1**	86	3	0.031	14	3	0.000241	0.000142	0.0042	-0.0028	0.0163	1.8731	0.0793
3405-91.1**	157	43	0.280	29	3	0.000114	0.000106	0.0020	0.0859	0.0082	2.0146	0.0725
3405-91.1**	157	43	0.280	29	3	0.000114	0.000106	0.0020	0.0859	0.0082	2.0146	0.0725
3405-97.1**	231	85	0.381	44	5	0.000130	0.000084	0.0023	0.1205	0.0054	2.0480	0.0584
3405-32.1**	243	102	0.435	48	4	0.000106	0.000066	0.0018	0.1266	0.0045	2.2428	0.0551
3405-65.1**	100	45	0.469	21	2	0.000129	0.000160	0.0022	0.1375	0.0090	2.3253	0.1157
3405-29.1**	123	67	0.564	26	10	0.000465	0.000190	0.0081	0.1543	0.0093	2.0885	0.0994
3405-89.1**	296	124	0.434	68	10	0.000195	0.000050	0.0034	0.1622	0.0038	4.2808	0.0850
3405-43.1**	315	136	0.445	68	11	0.000190	0.000085	0.0033	0.1134	0.0047	2.4142	0.0677
3405-96.1**	238	53	0.231	50	0	0.000010	0.000010	0.0002	0.0724	0.0028	2.8180	0.0531
3405-44.1**	398	228	0.591	96	35	0.000467	0.000060	0.0081	0.1904	0.0040	2.8723	0.0694
3405-21.1**	278	55	0.205	66	4	0.000074	0.000041	0.0013	0.0561	0.0031	4.7479	0.1002
3405-74.1**	281	95	0.351	67	4	0.000063	0.000061	0.0011	0.1046	0.0038	2.8879	0.0574
3405-11.2**	241	190	0.815	68	9	0.000188	0.000047	0.0033	0.2492	0.0045	3.0979	0.0554
3405-1.1**	818	283	0.358	209	8	0.000044	0.000062	0.0008	0.1064	0.0034	3.2050	0.0700
3405-9.1**	158	55	0.357	41	0	0.000010	0.000010	0.0002	0.1053	0.0037	3.3005	0.0791
3405-37.1**	194	118	0.627	56	12	0.000280	0.000109	0.0049	0.1774	0.0054	3.8862	0.1016
3405-34.1**	221	529	2.478	93	7	0.000136	0.000047	0.0024	0.7204	0.0072	3.6676	0.0894
3405-10.1**	171	181	1.094	59	3	0.000072	0.000038	0.0012	0.3266	0.0076	3.9670	0.0746
3405-31.1**	114	81	0.738	37	0	0.000010	0.000010	0.0002	0.2420	0.0073	4.2629	0.0949
3405-5.1**	187	101	0.562	61	6	0.000117	0.000109	0.0020	0.1551	0.0058	5.1154	0.1440
3405-72.1**	254	85	0.347	83	11	0.000169	0.000050	0.0029	0.0874	0.0042	5.8850	0.1800
3405-41.1**	220	129	0.607	91	7	0.000109	0.000035	0.0019	0.1476	0.0031	9.4070	0.1592
3405-47.1**	109	25	0.234	43	12	0.000340	0.000087	0.0059	0.0585	0.0042	9.2429	0.2409
3405-46.1**	122	39	0.329	51	6	0.000143	0.000060	0.0025	0.0707	0.0045	10.2070	0.1738
3405-87.1**	78	33	0.444	39	3	0.000116	0.000059	0.0020	0.1087	0.0039	13.6425	0.2223
3405-45.1**	41	26	0.654	24	16	0.000924	0.000289	0.0160	0.1873	0.0141	13.5449	0.4837

(*continued*)

TABLE 1. (*continued*)

Spot name	^{206}Pb/^{238}U	± ^{206}Pb/^{238}U	Corr. coef.	^{207}Pb/^{206}Pb	± ^{207}Pb/^{206}Pb	Ages (Ma)§ ^{206}Pb/^{238}U	± ^{206}Pb/^{238}U	^{207}Pb/^{206}Pb	± ^{207}Pb/^{206}Pb
PQB-SP-01007 (z7138): Felsic tuffaceous gneiss, Cormacks Lake complex (UTM zone 21, 433868E-5344937N)									
7138-24.1*	0.0674	0.0008	0.568	0.0567	0.0012	421	5	479	47
7138-16.1*	0.0677	0.0010	0.288	0.0592	0.0052	422	6	574	204
7138-12.1*	0.0678	0.0009	0.349	0.0579	0.0031	423	5	525	123
7138-1.1*	0.0690	0.0008	0.550	0.0548	0.0011	430	5	404	47
7138-9.1*	0.0691	0.0007	0.894	0.0561	0.0003	431	4	458	12
7138-11.1*	0.0692	0.0009	0.505	0.0553	0.0016	431	6	422	67
7138-23.1*	0.0694	0.0008	0.617	0.0550	0.0009	432	5	410	37
7138-5.1*	0.0707	0.0009	0.531	0.0580	0.0016	440	6	531	60
7138-20.1	0.0768	0.0009	0.349	0.0578	0.0028	477	6	522	111
7138-15.1	0.0778	0.0011	0.581	0.0579	0.0014	483	6	528	52
7138-21.1	0.0785	0.0010	0.463	0.0555	0.0019	487	6	434	76
7138-4.1	0.0787	0.0010	0.483	0.0564	0.0017	488	6	466	66
7138-13.1	0.0795	0.0011	0.392	0.0520	0.0025	493	7	287	113
7138-18.1	0.0795	0.0011	0.372	0.0595	0.0031	493	7	584	117
7138-3.1	0.0796	0.0010	0.650	0.0590	0.0010	494	6	567	37
7138-2.1	0.0804	0.0009	0.486	0.0592	0.0015	499	5	574	55

(*continued*)

TABLE 1. (*continued*)

Spot name	$^{206}Pb/^{238}U$	$\pm^{206}Pb/^{238}U$	Corr. coef.	$^{207}Pb/^{206}Pb$	$\pm^{207}Pb/^{206}Pb$	Ages (Ma)[§] $^{206}Pb/^{238}U$	$\pm^{206}Pb/^{238}U$	$^{207}Pb/^{206}Pb$	$\pm^{207}Pb/^{206}Pb$
WXNF-48 (z2821): Hornblende granodiorite, Hungry Mountain complex (UTM zone 21, 540646E-5434846N)									
2821-84.1	0.0724	0.0009	0.331	0.0575	0.0032	450	5	512	129
2821-92.1	0.0732	0.0014	0.390	0.0556	0.0035	456	8	438	148
2821-91.1	0.0735	0.0008	0.646	0.0552	0.0009	457	5	421	35
2821-99.1	0.0735	0.0009	0.436	0.0535	0.0018	457	5	350	80
2821-8.1	0.0736	0.0016	0.219	0.0520	0.0115	458	9	286	439
2821-86.1	0.0737	0.0013	0.412	0.0647	0.0035	459	8	764	119
2821-48.1	0.0742	0.0009	0.467	0.0533	0.0016	461	5	342	67
2821-94.1	0.0742	0.0010	0.361	0.0557	0.0028	462	6	441	117
2821-96.1	0.0743	0.0009	0.314	0.0571	0.0034	462	5	496	138
2821-100.1	0.0749	0.0009	0.370	0.0564	0.0024	466	5	468	96
2821-30.1	[illegible]	[illegible]	[illegible]	[illegible]	0.0008	468	5	389	32
2821-55.1	0.0754	0.0009	0.398	0.0571	0.0022	469	5	496	88
2821-88.2	0.0755	0.0012	0.582	0.0571	0.0015	469	7	493	59
2821-66.1	0.0760	0.0010	0.393	0.0546	0.0023	472	6	394	99
2821-61.1	0.0762	0.0009	0.477	0.0559	0.0016	474	5	447	64
2821-69.1**	0.1498	0.0016	0.668	0.0701	0.0010	900	9	930	29
2821-4.1**	0.3861	0.0047	0.824	0.1725	0.0016	2105	22	2582	15
2821-54.1**	0.4528	0.0093	0.682	0.1890	0.0048	2408	41	2734	43
WXNF-175 (z3405): Granodiorite, Cariboo Lakes gneiss complex (UTM zone 21, 475287E-5394211N)									
3405-6.1	0.0705	0.0009	0.389	0.0545	0.0023	439	5	391	99
3405-1.3	0.0710	0.0013	0.260	0.0584	0.0078	442	8	546	320
3405-43.2	0.0713	0.0010	0.473	0.0556	0.0019	444	6	435	80
3405-41.1	0.0714	0.0011	0.410	0.0532	0.0026	445	7	338	116
3405-21.2	0.0716	0.0011	0.361	0.0584	0.0033	446	6	543	130
3405-10.2	0.0726	0.0011	0.450	0.0547	0.0022	452	6	400	91
3405-13.1	0.0726	0.0009	0.349	0.0565	0.0030	452	6	474	122
3405-11.1	0.0734	0.0009	0.497	0.0576	0.0016	457	5	513	63
3405-17.1	0.0734	0.0009	0.318	0.0540	0.0031	457	5	372	136
3405-9.2	0.0738	0.0009	0.650	0.0574	0.0010	459	6	507	39
3405-84.1	0.0726	0.0012	0.266	0.0548	0.0061	452	7	404	272
3405-1.2	0.0726	0.0013	0.586	0.0592	0.0018	452	8	573	66
3405-77.1	0.0730	0.0015	0.261	0.0587	0.0086	454	9	554	359
3405-7.2	0.0738	0.0013	0.589	0.0609	0.0017	459	8	634	62
3405-99.1	0.0741	0.0009	0.346	0.0541	0.0028	461	5	376	119
3405-90.1	0.0744	0.0012	0.361	0.0568	0.0036	462	7	482	148
3405-64.1	0.0749	0.0015	0.249	0.0518	0.0078	466	9	277	311
3405-52.1	0.0750	0.0010	0.471	0.0549	0.0019	466	6	409	78
3405-2.1	0.0761	0.0015	0.354	0.0547	0.0043	473	9	400	185
3405-83.1	0.0764	0.0019	0.245	0.0545	0.0108	475	11	391	392
3405-76.1	0.0768	0.0016	0.250	0.0488	0.0079	477	10	140	339
3405-20.2**	0.1010	0.0025	0.305	0.0675	0.0087	620	15	854	293
3405-85.1**	0.1080	0.0021	0.592	0.1050	0.0033	661	12	1714	59
3405-88.1**	0.1231	0.0023	0.611	0.0822	0.0024	748	13	1249	58
3405-86.1**	0.1358	0.0023	0.295	0.0681	0.0064	821	13	871	209
3405-20.1**	0.1414	0.0031	0.280	0.0658	0.0089	852	18	800	311
3405-75.1**	0.1490	0.0027	0.673	0.0775	0.0018	895	15	1135	47
3405-100.1**	0.1587	0.0024	0.417	0.0708	0.0032	950	13	951	95
3405-51.1**	0.1620	0.0027	0.336	0.0764	0.0056	968	15	1107	154
3405-36.1**	0.1674	0.0029	0.615	0.0740	0.0020	998	16	1042	54
3405-50.1**	0.1741	0.0023	0.426	0.0758	0.0029	1035	13	1090	79
3405-61.1**	0.1794	0.0034	0.553	0.0757	0.0027	1063	19	1088	73
3405-91.1**	0.1818	0.0032	0.589	0.0804	0.0024	1077	17	1206	59
3405-91.1**	0.1818	0.0032	0.589	0.0804	0.0024	1077	17	1206	59
3405-97.1**	0.1838	0.0031	0.689	0.0808	0.0017	1088	17	1217	42
3405-32.1**	0.1896	0.0023	0.600	0.0858	0.0017	1119	13	1333	39
3405-65.1**	0.1981	0.0053	0.630	0.0851	0.0033	1165	28	1319	77
3405-29.1**	0.2010	0.0032	0.443	0.0754	0.0032	1181	17	1078	89
3405-89.1**	0.2035	0.0032	0.861	0.1525	0.0016	1194	17	2375	17
3405-43.1**	0.2091	0.0025	0.529	0.0837	0.0020	1224	13	1287	47

(*continued*)

TABLE 1. (*continued*)

Spot name	$^{206}Pb/^{238}U$	$\pm^{206}Pb/^{238}U$	Corr. coef.	$^{207}Pb/^{206}Pb$	$\pm^{207}Pb/^{206}Pb$	Ages (Ma)[§] $^{206}Pb/^{238}U$	$\pm^{206}Pb/^{238}U$	$^{207}Pb/^{206}Pb$	$\pm^{207}Pb/^{206}Pb$
WXNF-175 (z3405): Granodiorite, Cariboo Lakes gneiss complex (UTM zone 21, 475287E-5394211N)									
3405-96.1**	0.2096	0.0026	0.752	0.0975	0.0012	1227	14	1577	24
3405-44.1**	0.2176	0.0043	0.877	0.0957	0.0011	1269	23	1542	22
3405-21.1**	0.2272	0.0037	0.845	0.1516	0.0017	1320	20	2363	20
3405-74.1**	0.2314	0.0029	0.711	0.0905	0.0013	1342	15	1437	27
3405-11.2**	0.2433	0.0031	0.787	0.0924	0.0010	1404	16	1475	21
3405-1.1**	0.2479	0.0040	0.812	0.0938	0.0012	1428	21	1503	24
3405-9.1**	0.2538	0.0041	0.752	0.0943	0.0015	1458	21	1514	30
3405-37.1**	0.2634	0.0044	0.723	0.1070	0.0020	1507	22	1749	34
3405-34.1**	0.2684	0.0043	0.745	0.0991	0.0016	1533	22	1607	31
3405-10.1**	0.2812	0.0037	0.780	0.1023	0.0012	1598	19	1666	22
3405-31.1**	0.2826	0.0044	0.772	0.1094	0.0016	1604	22	1790	26
3405-5.1**	0.2970	0.0058	0.773	0.1249	0.0023	1677	29	2027	32
3405-72.1**	0.3112	0.0063	0.744	0.1372	0.0028	1747	31	2191	36
3405-41.1**	0.3597	0.0054	0.936	0.1897	0.0011	1981	26	2740	10
3405-47.1**	0.3682	0.0070	0.806	0.1821	0.0028	2021	33	2672	26
3405-46.1**	0.3885	0.0052	0.849	0.1906	0.0017	2116	24	2747	15
3405-87.1**	0.4381	0.0058	0.875	0.2259	0.0018	2342	26	3023	13
3405-45.1**	0.4950	0.0103	0.674	0.1985	0.0053	2592	44	2814	44

Note: See Stern (1997): Uncertainties are reported at 1σ (absolute) and are calculated by numerical propagation of all known sources of error.
*Metamorphic zircon analyses.
**Inherited zircon analyses.
[†]f(206)[204] refers to mole fraction of total ^{206}Pb that is due to common Pb, calculated using the ^{204}Pb method; common Pb composition used is the surface blank
[§]204-corrected ages (Stern, 1997).

of the Annieopsquotch ophiolite belt to Dashwoods by retroarc underthrusting at ca. 470 Ma, ~10 m.y. after its formation (van der Velden et al., 2004; Lissenberg et al., 2005b). Upper-plate compression as a result of Dashwoods–Humber margin collision probably caused the bivergent structures (Fig. 7B).

Accretion of all Annieopsquotch accretionary tract peri-Laurentian elements to Dashwoods, and by implication Laurentia, was accomplished between 455 and 450 Ma. Accretion terminated in a Moluccan Sea–style arc-arc collision between the youngest Annieopsquotch accretionary tract arc phase (Red Indian Lake arc; Zagorevski et al., 2006) and the peri-Gondwanan Popelogan-Victoria arc, the suture of which is the Red Indian Line (Figs. 1 and 7B; van Staal et al., 1998). This arc-arc collision thus terminated the Taconic orogeny because all peri-Laurentian elements had been accreted to Laurentia (van Staal, 2005). This terminal phase is herein referred to as Taconic 3.

THE NOTRE DAME ARC

Introduction

The Notre Dame subzone is mainly underlain by a narrow linear belt of Ordovician to Silurian plutonic and volcanic rocks (Fig. 1) with an average width of ~50 km (Figs. 1 and 2). None of its Ordovician arc plutons crosses the bounding Baie Verte Brompton and Red Indian Lines. Hence, the Ordovician part of the Notre Dame arc has the typical geometry of a linear magmatic-volcanic arc system bounded by sutures on both sides. Systematic petrological and geochronological studies of Ordovician to Silurian intrusions of the Notre Dame subzone of Newfoundland revealed the presence of three major magmatic pulses of predominantly immature arc-like magmatism (Figs. 1, 2, and 8): (1) 489–477 Ma, (2) 469–459 Ma, and (3) 445–435 Ma, separated by two magmatic gaps. These magmatic gaps (Fig. 3) are significant because high-quality U-Pb ages exist for virtually all magmatic suites and plutons recognized by us during mapping of the Notre Dame subzone. Close correspondence between these magmatic gaps and the respective ages of Humber margin–Dashwoods and Dashwoods–Red Indian Lake arc–Popelogan-Victoria arc collisions, respectively, which have been constrained by other means, suggests a causal relationship (Fig. 3). Furthermore, the syntectonic intrusion of part of the Notre Dame arc's second-phase plutonic rocks (Whalen et al., 1987, see following) at ambient metamorphic conditions corresponding to amphibolite and/or granulite facies suggests a linkage between this magmatism and active Middle Ordovician tectonism. Hence, the spatial and temporal distribution, and petrogenesis of Notre Dame arc magmatism, as well as the associated Ordovician deformation and metamorphism, have a major bearing on understanding the tectonic processes responsible for the Taconic orogeny.

The results of new geological, structural, geochemical, isotopic (Nd), and geochronological (U-Pb) studies of the magmatic rocks of the Notre Dame arc are discussed later. For the sake

TABLE 2: U-Pb THERMAL IONIZATION MASS SPECTROMETRY ANALYTICAL DATA

Fract.[1]	Description[2]	Wt. (ug)	U (ppm)	Pb[3] (ppm)	$^{206}Pb/^{204}Pb^4$	Pb[5] (pg)	Isotopic Ratios[6] $^{208}Pb/^{206}Pb$	$^{207}Pb/^{235}U$	±1SE Abs	$^{206}Pb/^{238}U$	±1SE Abs	Corr.[7] Coeff.	$^{207}Pb/^{206}Pb$	±1SE Abs	Ages (Ma)[8] $^{206}Pb/^{238}U$	±2SE	$^{207}Pb/^{235}U$	±2SE	$^{207}Pb/^{206}Pb$	±2SE	% Disc
PQB-SP-01017 (z7139): Tonalitic orthogneiss, Cormacks Lake Complex (UTM zone 21, 433820E-5344832N)																					
Z1-2A (Z; 2)	Co,Clr,Eu,St,fln,Dia	19	401	30	15125	2	0.09	0.60393	0.00075	0.07714	0.00008	0.931	0.05678	0.00003	479.0	0.9	479.7	1.0	483.1	2.1	0.9
Z1-2B (Z; 4)	Co,Clr,Eu,St,rln,Dia	19	303	23	10352	3	0.08	0.60773	0.00141	0.07761	0.00017	0.960	0.05679	0.00004	481.9	2.0	482.1	1.8	483.4	2.9	0.3
Z2 (Z; 12)	Co,Clr,Eu,St,rln,Dia	16	312	23	1451	16	0.07	0.59126	0.00148	0.07555	0.00010	0.660	0.05676	0.00011	469.5	1.2	471.7	1.9	482.3	8.4	2.8
Z2-1 (Z; 17)	Co,Clr,Eu,St,rln,Dia	18	380	29	5990	6	0.09	0.60854	0.00087	0.07773	0.00009	0.897	0.05678	0.00004	482.6	1.1	482.6	1.1	483.0	2.8	0.1
M3A (M; 1)	Y,Clr,An,Eq,M5	2	10176	3882	14903	6	5.28	0.53128	0.00066	0.06949	0.00007	0.920	0.05545	0.00003	433.1	0.8	432.7	0.9	430.4	2.3	−0.7
M6A (M; 4)	pY,Clr,Eu,Eq,M5	2	1448	1135	724	18	11.93	0.53160	0.00206	0.06963	0.00010	0.594	0.05537	0.00018	433.9	1.2	432.9	2.7	427.2	14.4	−1.6
VK-87-NF2 (z1131): Charnockite, Cormack's Lake Complex (UTM zone 21, 432041E-5351903N)																					
Z1 (Z)	Co,Clr,Eu,Pr,fln,NM1	72	353	27	6443	18	0.16	0.55972	0.00064	0.07242	0.00007	0.958	0.05605	0.00002	450.7	0.9	451.3	0.8	454.5	1.5	0.9
Z2 (Z)	Co,Clr,Eu,Pr,fln,NM1	109	765	58	16860	23	0.16	0.56054	0.00061	0.07251	0.00007	0.966	0.05606	0.00002	451.3	0.8	451.9	0.8	454.9	1.3	0.8
Z3 (Z)	Co,Clr,Eu,Pr,fln,NM1	79	413	31	8806	17	0.14	0.56933	0.00057	0.07345	0.00007	0.939	0.05621	0.00002	456.9	0.8	457.6	0.7	460.8	1.6	0.9
A1 (Z; 8)	Co,Clr,Eu,Pr,NM1	14	308	24	5023	4	0.15	0.56958	0.00097	0.07352	0.00010	0.898	0.05619	0.00004	457.3	1.1	457.7	1.3	459.9	3.5	0.6
B1 (Z; 5)	Co,Clr,Eu,Pr,NM1	16	216	16	2884	5	0.15	0.56767	0.00117	0.07328	0.00012	0.773	0.05618	0.00007	455.9	1.4	456.5	1.5	459.7	5.8	0.9
C1 (Z; 5)	Co,Clr, Eu,St,NM1	15	96	7	674	10	0.18	0.56660	0.00176	0.07304	0.00010	0.703	0.05626	0.00013	454.4	1.3	455.8	2.3	462.7	10.3	1.9
VL-02-21A (z7506): Tonalite, Southwest Brook Complex (UTM zone 21, 444364E-5364922N)																					
Z1A (Z; 12)	Co,Clr,Eu,Pr,fln,Dia	33	15	1	546	2	0.23	0.57956	0.00359	0.07440	0.00017	0.506	0.05650	0.00031	462.6	2.1	464.2	4.6	472.1	23.8	2.1
Z1B (Z; 20)	Co,Clr,Eu,Pr,fln,Dia	36	134	11	3437	6	0.23	0.57616	0.00106	0.07453	0.00011	0.850	0.05607	0.00005	463.4	1.4	462.0	1.4	455.1	4.3	−1.9
Z2A (Z; 13)	Co,Clr,Eu,El,fln,Dia	26	112	9	3462	3	0.23	0.57807	0.00110	0.07440	0.00014	0.479	0.05635	0.00011	462.6	1.6	463.2	1.4	466.2	8.4	0.8
Z3A (Z; 12)	Co,Clr,Eu,St,rln,Dia	32	53	4	248	34	0.22	0.57548	0.00468	0.07431	0.00019	0.625	0.05617	0.00039	462.1	2.2	461.6	6.0	459.0	30.1	−0.7
T1 (T; 22)	Br,Clr,An,Frag,fln,M1	89	147	10	361	163	0.06	0.52764	0.00344	0.06860	0.00012	0.641	0.05579	0.00031	427.7	1.4	430.2	4.6	443.9	24.6	3.8
T2 (T; 25)	Br,Clr,An,Frag,fln,M1	114	89	6	264	178	0.12	0.52778	0.00395	0.06903	0.00015	0.674	0.05546	0.00035	430.3	1.8	430.3	5.3	430.6	27.6	0.1
T3 (T; 30)	pBr,Clr,An,Frag,fln,M1	69	36	3	128	97	0.16	0.52015	0.00814	0.06864	0.00028	0.674	0.05496	0.00073	428.0	3.4	425.3	10.9	410.5	58.1	−4.4
T4 (T; 30)	pBr,Clr,An,Frag,fln,M1	77	48	3	159	113	0.15	0.52823	0.00743	0.06902	0.00024	0.653	0.05551	0.00067	430.3	2.9	430.6	9.9	432.7	53.0	0.6
PQB-SP-02-79B (z7516): Quartz diorite, Southwest Brook Complex (UTM zone 21, 421693E- 5357980N)																					
Z1A (Z; 24)	Co,Clr,El,Eu,fln,Dia	22	299	24	6310	4	0.22	0.57079	0.00079	0.07367	0.00008	0.868	0.05619	0.00004	458.2	1.0	458.5	1.0	459.9	3.1	0.4
Z1B (Z; 31)	Co,Clr,El,Eu,fln,Dia	10	341	28	4006	4	0.23	0.56803	0.00078	0.07326	0.00007	0.865	0.05623	0.00004	455.8	0.9	456.7	1.0	461.6	3.2	1.3
Z1C (Z; 11)	Co,Clr,El,Eu,rln,Dia	15	270	22	3501	5	0.22	0.57159	0.00075	0.07367	0.00007	0.861	0.05627	0.00004	458.2	0.8	459.0	1.0	463.1	3.1	1.1
Z2 (Z; 23)	Co,Clr,El,Eu,rln,Dia	33	374	30	12338	3	0.20	0.57415	0.00081	0.07394	0.00009	0.914	0.05632	0.00003	459.8	1.1	460.7	1.0	465.0	2.6	1.2
Z3 (Z; 9)	Co,Clr,St,Eu,Dia	33	194	24	8671	5	0.23	1.27565	0.00204	0.10987	0.0002	0.927	0.08421	0.00005	672.0	1.9	834.9	1.8	1297.4	2.3	50.7
Z4 (Z; 30)	Co,Clr,El,Eu,rln,Dia	24	290	24	3952	8	0.22	0.57300	0.00077	0.07387	0.00009	0.736	0.05626	0.00005	459.4	1.1	460.0	1.0	462.5	4.2	0.7
T2 (T; 25)	pBr,Clr,An,fln,Frag,M75	159	362	35	289	913	0.55	0.52654	0.00347	0.06938	0.00014	0.702	0.05504	0.00030	432.4	1.7	429.5	4.6	413.9	23.8	−4.6
T3 (T; 24)	Br,Clr,An,fln,Frag,M75	279	428	39	230	2427	0.49	0.52760	0.00462	0.06919	0.00017	0.687	0.05530	0.00040	431.3	2.1	430.2	6.1	424.5	32.0	−1.7
WXNF01-230 (z7425): Paragneiss, Southwest Brook Complex (UTM zone 21, 435240E-5372920N)																					
M1 (M; 1)	Y,Clr,Sh,Eq,M5	6	1632	553	9773	4	4.16	0.58328	0.00085	0.07502	0.00009	0.919	0.05639	0.00003	466.3	1.1	466.6	1.1	467.7	2.6	0.3
M2 (M; 1)	Y,Clr,Sh,Eq,M5	5	589	631	1729	8	15.37	0.58323	0.00201	0.07512	0.00024	0.910	0.05631	0.00008	467.0	2.9	466.5	2.6	464.5	6.3	−0.6
M4 (M; 1)	pY,Clr,Sh,Eq,M5	1	1340	1762	1563	4	19.15	0.58055	0.00139	0.07508	0.00012	0.750	0.05608	0.00009	466.7	1.5	464.8	1.8	455.7	7.0	−2.5
R1 (R; >250)	Y,Clr,Eu,M5	342	73	5	158	772	0.04	0.53116	0.00669	0.06899	0.00025	0.705	0.05584	0.00058	430.0	3.0	432.6	8.9	446.1	45.4	3.7
R2 (R; >250)	Y,Clr,Eu,M5	421	67	4	167	817	0.01	0.53197	0.00699	0.06912	0.00023	0.671	0.05582	0.00062	430.8	2.8	433.1	9.3	445.3	48.8	3.4
WXNF-94 (z3058): Hornblende gabbro, Hungry Mountain Complex (UTM zone 21, 504763E-5415398N)																					
A (Z; 70)	Co,Clr,An,Frag,NM1	517	184	15	23270	19	0.17	0.58709	0.00063	0.07555	0.00007	0.966	0.05636	0.00002	469.5	0.9	469.0	0.8	466.7	1.3	−0.6
B (Z; 100)	Co,Clr,An,Frag,NM1	422	171	13	15075	22	0.16	0.58664	0.00062	0.07543	0.00007	0.958	0.05640	0.00002	468.8	0.8	468.7	0.8	468.3	1.4	−0.1
C (Z; 120)	Co,Clr,An,Frag,NM1	185	378	30	26124	13	0.17	0.58757	0.00062	0.07551	0.00007	0.944	0.05644	0.00002	469.2	0.8	469.3	0.8	469.7	1.5	0.1
WXNF-63 (z2823): Granodiorite, Cariboo Lakes Complex (UTM zone 21, 475700E-5388900N)																					
A (Z; 20)	Co,Clr,Eu,St,fln, rFr,Dia	44	166	14	1230	28	0.21	0.58299	0.00118	0.07504	0.00010	0.666	0.05635	0.00009	466.4	1.2	466.4	1.5	466.2	6.7	−0.1
B (Z; 12)	Co,Clr,Eu,St,fln, rFr,Dia	39	236	19	1100	40	0.21	0.57839	0.00122	0.07441	0.00008	0.547	0.05637	0.00010	462.7	0.9	463.4	1.6	467.0	7.8	1.0
C (Z; 20)	Co,Clr,Eu,St,fln, rFr,Dia	75	185	15	1092	60	0.22	0.57790	0.00105	0.07424	0.00007	0.701	0.05646	0.00008	461.7	0.8	463.1	1.3	470.3	5.9	1.9

Notes:

[1] Z—zircon; M—monazite; T—titanite ; R—rutile. All zircon fractions are abraded following the method of Krogh (1982). Number in brackets refers to the number of grains in the analysis.

[2] Fraction descriptions: Co—colorless, Br—brown, pBr—pale brown, Y—yellow, pY—pale yellow, Clr—clear, fln—few Inclusions, rln—rare inclusions, fFr—few fractures, rFr—rare fractures, Eu—euhedral, An—anhedral, Sh—subhedral, Pr—prismatic, El—elongate, Eq—equant, St—stubby prism, Fr—fragments, Dia—diamagnetic, NM1—nonmag @1.8A 1°SS, M1—magnetic @ 1.0 A 10°SS, M75—magnetic @ 0.75A 10°SS, M5—magnetic @ 0.5A 10°SS.

[3] Radiogenic Pb.

[4] Measured ratio, corrected for spike and fractionation.

[5] Total common Pb in analysis corrected for fractionation and spike.

[6] Corrected for blank Pb and U and common Pb, errors quoted are 1σ absolute; Pb blank isotopic composition is based on the analysis of procedural blanks; corrections for common Pb were made using Stacey-Kramers compositions.

[7] Correlation coefficient.

[8] Corrected for blank and common Pb, errors quoted are 2σ in Ma.

TABLE 3. SUMMARY OF U-Pb GEOCHRONOLOGY AND GEOCHEMISTRY FROM THE NOTRE DAME ARC SAMPLES

Sample	Sample number	Age (Ma)	Method	$\varepsilon_{Nd}(T)$ on dated sample	$\varepsilon_{Nd}(T)$ on other samples from unit	Comment
Cormacks Lake complex						
Felsic tuffaceous gneiss	PQB-SP-01007 (z7138)	489.3 ± 5.6	Concordia age, U-Pb SHRIMP (zircon)	1.9		Metamorphic zircon at ca. 430 Ma; metamorphic monazite and rutile at ca. 432 Ma
Tonalitic orthogneiss	PQB-SP-01017 (z7139)	483.1 ± 1.4	Upper-intercept age, U-Pb TIMS (zircon)	2.2	0.8, 1.5	Metamorphic monazite at ca. 433 Ma
Charnockite	VK-87-NF2 (z1131)	459.4 +3.0/–1.5	Upper-intercept age, U-Pb TIMS (zircon)		4.3, 6.0	Metamorphic titanite and allanite at ca. 430 Ma
Southwest Brook Complex						
Tonalite	VL-02-21A (z7506)	462.8 ± 0.9	Concordia age, U-Pb TIMS (zircon)	–6.6		Metamorphic titanite at ca. 429 Ma
Quartz diorite	PQB-SP-02-79B (z7516)	462.7 ± 1.8	Upper-intercept age, U-Pb TIMS (zircon)	–1.7		Metamorphic titanite at ca. 432 Ma
Paragneiss	WXNF01-230 (z7425)	466.5 ± 1.8	Concordia age, U-Pb TIMS (monazite)	–10.4		Metamorphic rutile at ca. 430 Ma
Hungry Mountain and Cariboo Lakes complexes						
Hornblende gabbro	WXNF-94 (z3058)	469.1 ± 1.3	Concordia age, U-Pb TIMS (zircon)	–9.7		Published Ar-Ar (hornblende) age of 469.7 ± 4.2 Ma
Hornblende tonalite	WXNF-48 (z2821)	463.1 ± 3.7	Concordia age, U-Pb SHRIMP (zircon)	–6.1		Inheritance of ca. 930 Ma, discordant Archean analyses
Granodiorite	WXNF-63 (z2823)	468.2 ± 3.8	Upper-intercept age, U-Pb TIMS (zircon)	–6.8		
Granodiorite	WXNF-175 (z3405)	459.0 ± 3.4	Concordia age, U-Pb SHRIMP (zircon)	–9.3		Abundant inheritance ranging from ca. 600, 850–1600, discordant Archean analyses

Note: SHRIMP—sensitive high-resolution ion microprobe; TIMS—thermal ionization mass spectrometry.

of brevity, the new U-Pb thermal ionization mass spectrometry (TIMS) and sensitive high-resolution ion microprobe (SHRIMP) ages as well the geochemical and Nd-isotopic data are concisely summarized within the text. The U-Pb results are presented in concordia diagrams in Figure 9. The U-Pb SHRIMP and TIMS analytical data, along with sample locations, are presented in Tables 1 and 2, respectively. Details of geochemical and isotopic techniques are presented in Appendix 1. More details concerning the geochemical and Nd-isotopic data, U-Pb samples, and age interpretations are presented in Appendix 2 and summarized in Table 3. Whole-rock geochemical and Nd isotopic data are presented in Tables 4 and 5.

Geological and Petrological Overview

The Early to Middle Ordovician high-grade metamorphic plutonic rocks of the Notre Dame arc are coeval with dominantly greenschist or subgreenschist facies calc-alkaline volcanic rocks in the Cutwell Group and the upper part of the Western Arm Group in the northern part of the Notre Dame subzone (Fig. 1). These volcanic rocks represent an ensialic arc suprastructure to the obducted Cambrian oceanic rocks of the Lushs Bight oceanic tract (Szybinski, 1995). A series of relatively small Ordovician diorite to granodiorite plutons (Figs. 1, 2, and 8), which were emplaced into volcanic rocks of the Lushs Bight oceanic tract in Notre Dame Bay, links this volcanic-dominated part of the Notre Dame arc to the arc-plutonic domain farther to the south. This regional distribution of igneous rocks suggests that the Notre Dame subzone preserves an oblique north-south section through a continental magmatic-volcanic arc system, with progressively deeper rocks exposed in the south.

Large bodies and screens of variably migmatized sillimanite-bearing paragneiss and schist are relatively common in the southern Notre Dame subzone. These metasedimentary rocks in part lithologically resemble rocks of the Neoproterozoic to Lower Ordovician Fleur de Lys Supergroup of the Laurentian margin (e.g., Currie et al., 1992). They probably were deposited on Dashwoods when it rifted and moved away from Laurentia during the Early Cambrian. Their relative abundance in this part of the Notre Dame subzone led to introduction of the Dashwoods subzone (Williams, 1995). However, as several large bodies of similar paragneiss and schist occur also in the northern part of the Notre Dame subzone (e.g., Whalen et al., 1987; Whalen, 1993), we consider the subdivisions into Dashwoods and Notre Dame subzones to be mainly related to level of denudation, and thus they are only of geographic, not tectonic, significance. In contrast, the Baie Verte oceanic tract is tectonically different from the remainder of the Notre Dame subzone, as evidenced by: (1) the Early Ordovician Snooks Arm suprastructure, which represents an oceanic rather than a continental arc cover; (2) metamorphism, which is generally below the amphibolite facies; and (3) the rare occurrence or absence of Middle Ordovician felsic arc plutons. Its oldest known continentally contaminated arc plutons are the ca. 440 Ma Burlington and Glover Island granodiorites (Cawood

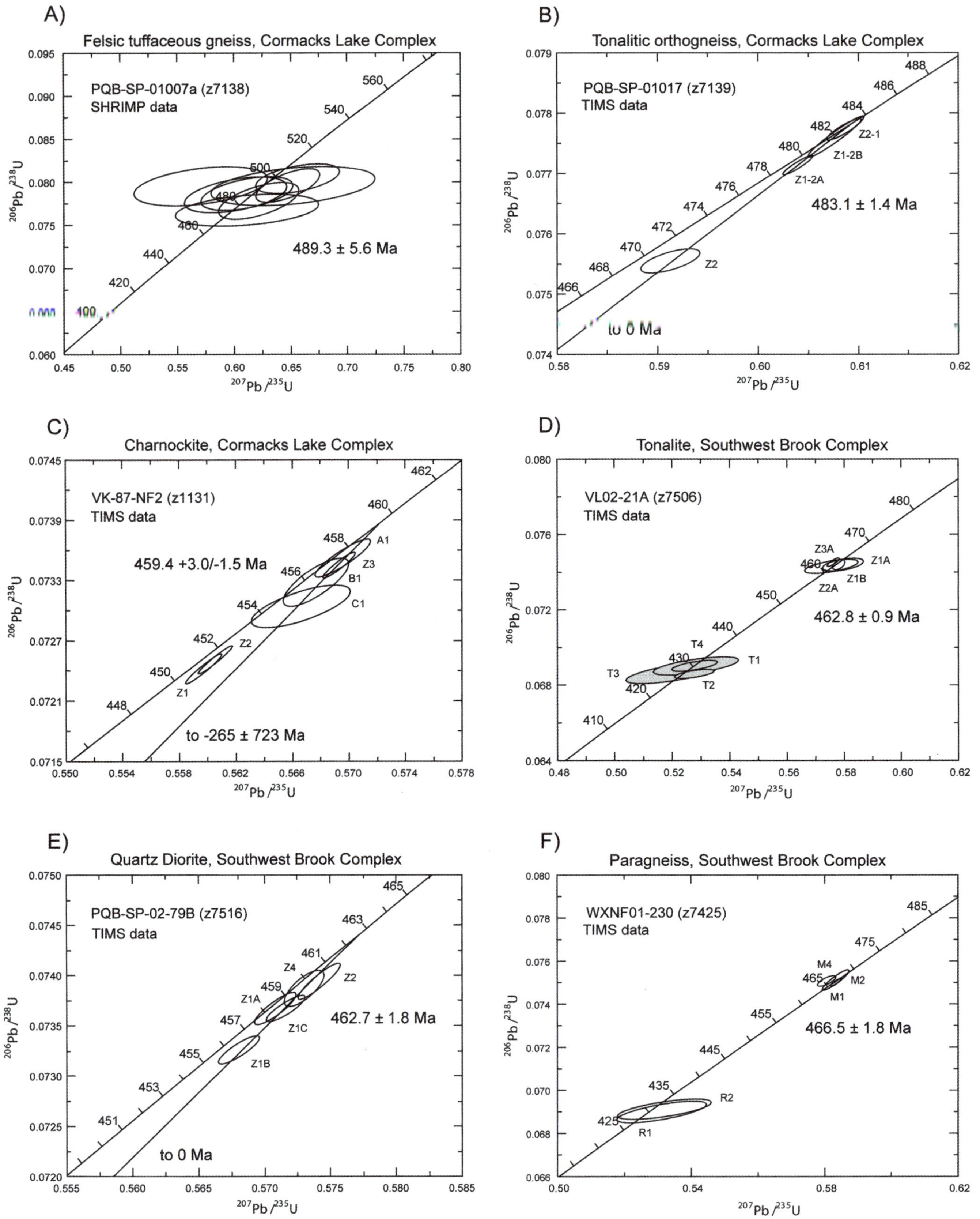
A)
Felsic tuffaceous gneiss, Cormacks Lake Complex
PQB-SP-01007a (z7138)
SHRIMP data
489.3 ± 5.6 Ma
^{206}Pb/^{238}U
^{207}Pb/^{235}U
B)
Tonalitic orthogneiss, Cormacks Lake Complex
PQB-SP-01017 (z7139)
TIMS data
483.1 ± 1.4 Ma
Z2-1
Z1-2B
Z1-2A
Z2
to 0 Ma
C)
Charnockite, Cormacks Lake Complex
VK-87-NF2 (z1131)
TIMS data
459.4 +3.0/-1.5 Ma
A1
Z3
B1
C1
Z2
Z1
to -265 ± 723 Ma
D)
Tonalite, Southwest Brook Complex
VL02-21A (z7506)
TIMS data
462.8 ± 0.9 Ma
Z3A
Z1A
Z1B
Z2A
T4
T1
T3
T2
E)
Quartz Diorite, Southwest Brook Complex
PQB-SP-02-79B (z7516)
TIMS data
462.7 ± 1.8 Ma
Z4
Z2
Z1A
Z1C
Z1B
to 0 Ma
F)
Paragneiss, Southwest Brook Complex
WXNF01-230 (z7425)
TIMS data
466.5 ± 1.8 Ma
M4
M2
M1
R2
R1

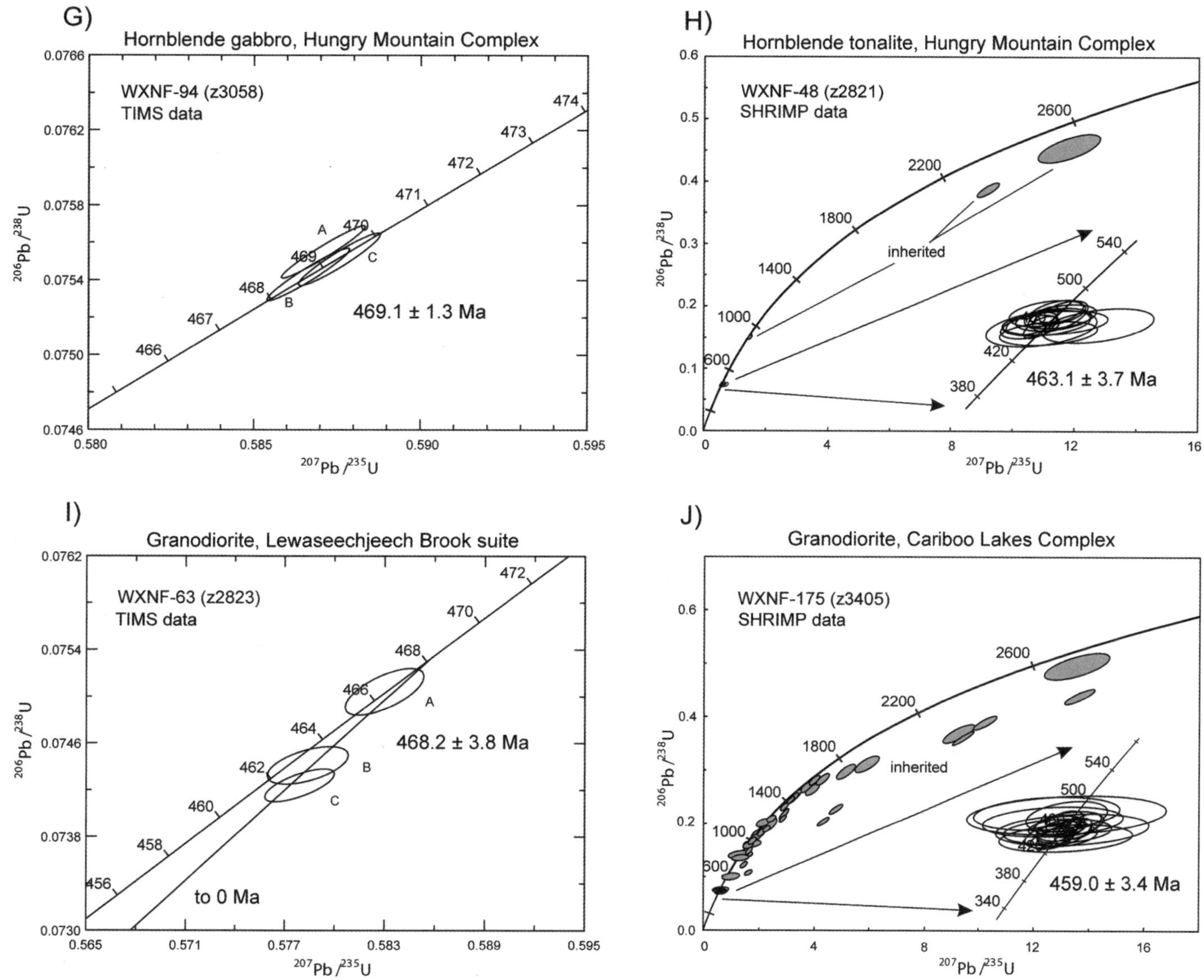

Figure 9 (on this and previous page). 12 U-Pb concordia diagrams of sensitive high-resolution ion microprobe (SHRIMP) and thermal ionization mass spectrometry (TIMS) analyses with errors plotted at 2σ.

and van Gool, 1998), which belong to the third and last phase of the Notre Dame arc and occur throughout the Notre Dame subzone (Fig. 3; van Staal, 2005; Whalen et al., 2006). The fault between the Baie Verte oceanic tract and the remainder of the Notre Dame subzone (Figs. 1 and 2), therefore, separates two fundamentally different Ordovician terranes. Unfortunately, the nature of this fault is largely obscured by stitching Lower Silurian plutons of the Topsails suite (Fig. 2; Whalen et al., 1996). The east-west–trending Little Grand Lake fault, its southern termination (Fig. 1), is a north-directed thrust, which was active during the Late Ordovician (Brem et al., 2003). Hence, part of the Baie Verte oceanic tract and Snooks Arm arc was thrust beneath the Notre Dame arc (Fig. 7B) after the first phase of the Notre Dame arc but before the Early Silurian magmatism.

Ordovician Notre Dame arc plutonic rocks have been described previously by Whalen et al. (1997a), from which the following has been summarized. Tonalite to granodiorite represent the predominant compositions. However, hornblende-bearing gabbro, diorite, monzonite, and rare granite also occur. The felsic plutonic rocks have low-, medium-, and high-K compositions and, based on Al_2O_3 content, include both low- and high-Al tonalites (Barker, 1979). In keeping with their generally hornblende-bearing mineralogy, almost all samples are metaluminous to only slightly peraluminous (Al index <1.1). These characteristics, along with common igneous microgranitoid enclaves and their $\delta^{18}O$ (Vienna standard mean ocean water [VSMOW]) values, which lie mainly within the range of mantle-derived granitoid rocks (i.e., +6‰ to +8‰; Taylor, 1968, 1978), indicate that Notre

TABLE 4. REPRESENTATIVE WHOLE-ROCK ANALYSES FROM NOTRE DAME ARC ROCKS

Sample number	Complex/map unit	Rock type	Major elements (wt%)												
			SiO_2	TiO_2	Al_2O_3	Fe_2O_3	FeO	MnO	MgO	CaO	Na_2O	K_2O	P_2O_5	LOI	Total
PQB-01-SP-007B	Cormack Lake C	Amphibolite	54.2	1.39	17.70	1.95	9.40	0.19	6.53	1.54	5.80	0.09	0.13	0.60	99.52
PQB-02-S144	Cormack Lake C	Layered amphibolite	45.3	0.38	12.41	15.31	–	0.90	8.45	16.24	0.94	0.38	0.03	0.55	100.87
PQB-01-SP-005	Cormack Lake C	Tonalitic orthogneiss	76.4	0.16	12.38	–	2.12	0.04	1.12	0.80	4.65	1.31	0.03	0.60	99.64
PQB-01-SP-017	Cormack Lake C	Tonalitic orthogneiss	75.1	0.18	12.90	1.38	2.00	0.07	0.42	2.74	4.00	0.87	0.03	0.60	100.29
WX01NF-193	Cormack Lake C	Tuffaceous gneiss	75.8	0.43	11.80	1.10	2.30	0.32	1.64	0.30	2.90	0.51	0.02	0.70	97.82
PQB-01-SP-015	Cormack Lake C	Tuffaceous gneiss	85.2	0.11	10.70	0.25	0.40	0.00	0.10	0.04	0.40	1.45	0.01	0.60	99.26
VL-92-LH-11	Cape Ray C	K-feldspar porphyritic granodiorite	68.5	0.62	13.70	1.84	1.40	0.06	1.33	3.44	2.80	4.19	0.35	0.80	99.03
VL-92-LH-68	Cape Ray C	K-feldspar porphyritic granodiorite	58.5	1.29	14.80	2.88	4.70	0.13	4.40	6.15	3.30	2.51	0.23	0.80	99.69
VL-92-84	Stag Hill gneiss	Granodioritic orthogneiss	78.8	0.06	11.40	0.38	0.20	0.14	0.11	0.19	3.90	4.05	0.01	0.40	99.64
PQB-01-SP-036	Cormack Lake C	Charnockite	69.1	0.19	15.44	0.00	2.64	0.07	0.08	0.90	5.16	5.26	0.00	0.50	99.32
WX92NF-53	Southwest Brook C	Low-Th tonalite	67.3	0.40	14.45	2.87	2.70	0.12	1.65	6.22	3.07	0.18	0.10	0.47	99.50
WX01NF-184	Southwest Brook C	Low-Th tonalite	69.4	0.88	12.20	2.60	2.40	0.07	1.26	5.06	2.80	0.31	0.17	0.20	97.35
WX01NF-187	Southwest Brook C	Low-Th tonalite	74.4	0.76	10.90	1.40	1.60	0.05	0.97	4.14	2.75	0.23	0.12	0.50	97.87
WX01NF-188	Southwest Brook C	Low-Th tonalite	67.6	0.88	13.10	2.30	2.80	0.07	1.47	5.66	2.90	0.22	0.22	0.30	97.52
WX92NF-54	Southwest Brook C	Mafic enclave in NF53	46.8	0.41	16.20	3.98	7.40	0.22	8.47	12.30	1.32	0.78	0.04	1.58	99.45
VL01J-213	Southwest Brook C	Mafic enclave in NF184	47.3	2.15	15.17	13.66	–	0.23	7.40	10.29	2.84	0.40	0.23	0.54	100.25
VL01J-215	Southwest Brook C	Mafic enclave in NF187	48.3	1.59	15.16	11.51	–	0.29	8.58	9.76	3.26	0.70	0.17	1.07	100.41
VL01J-216	Southwest Brook C	Mafic enclave in NF188	46.3	2.51	14.08	15.21	–	0.36	7.59	10.52	2.65	0.37	0.28	0.70	100.52
WX01NF-209	Southwest Brook C	Comagmatic mafic dike	48.1	0.76	16.10	2.90	7.80	0.23	6.84	10.51	2.50	1.16	0.07	0.50	97.47
WX01NF-211	Southwest Brook C	Comagmatic mafic dike	50.9	1.17	17.00	3.60	7.30	0.17	3.64	8.98	3.00	0.30	0.29	0.90	97.25
WX01NF-212	Southwest Brook C	Comagmatic mafic dike	46.5	1.80	15.30	4.70	7.40	0.19	7.46	10.94	2.40	0.27	0.05	0.30	97.31
WX92NF-99	Hungry Mountain C	Comagmatic mafic dike	47.9	1.88	16.11	3.46	7.60	0.23	7.38	10.56	2.63	0.22	0.16	1.08	99.27
WX92NF-100	Hungry Mountain C	Comagmatic mafic dike	43.4	1.44	18.35	8.67	7.00	0.23	5.54	11.05	2.02	0.24	0.18	0.68	98.74
WX92NF-97	Hungry Mountain C	Comagmatic mafic dike	57.4	0.71	15.65	4.21	4.45	0.26	3.69	6.61	3.47	0.89	0.09	1.40	98.79
WX02NF-254	Southwest Brook C	Blue quartz-eye tonalite	63.9	0.53	15.80	1.77	3.80	0.11	2.42	6.12	3.20	1.26	0.18	0.40	99.49
PQB-SP-02-079b	Southwest Brook C	Quartz diorite	60.9	0.75	15.50	1.87	3.80	0.10	3.25	5.47	3.80	3.21	0.37	0.50	99.52

(*continued*)

TABLE 4. REPRESENTATIVE WHOLE-ROCK ANALYSES FROM NOTRE DAME ARC ROCKS *(continued)*

Sample number	Trace elements (ppm)																		
	Cr	Ni	Co	Sc	V	Cu	Pb	Zn	Sn	Rb	Cs	Ba	Sr	Ga	Ta	Nb	Hf	Zr	Y	Th
PQB-01-SP-007B	<10	<1	55	44	234	<1	2	75	1.6	1	0.0	67	85	21	0.6	2.1	1.80	59	27	1.20
PQB-02-S144	339	94	75	34	233	<1	2	81	0.6	3	0.1	73	301	11	0.2	0.4	0.56	19	11	0.57
PQB-01-SP-005	<10	2	3	9	17	3	1	48	2.0	22	0.4	208	51	13	0.2	3.0	4.40	110	63	6.20
PQB-01-SP-017	<10	<1	89	14	16	79	3	45	1.0	20	0.3	222	111	13	1.8	3.8	2.80	98	37	3.40
WX01NF-193	<10	<1	<1	27	<5	38	8	148	3.3	8	0.6	980	59	17	0.1	0.7	2.60	71	53	0.40
PQB-01-SP-015	<10	<1	50	10	<5	11	127	63	8.1	18	0.1	5708	37	12	0.9	1.2	2.80	80	51	1.60
VL-92-LH-11	26	12	17	7	61	<1	20	51	2.4	128	2.0	1950	1010	18	1.6	23.0	9.70	415	19	35.00
VL-92-LH-68	126	59	36	22	170	31	13	78	2.0	81	2.2	472	422	18	0.9	12.0	4.50	175	36	9.50
VL-92-84	<10	<1	<1	6	5	30	31	7	1.4	200	1.6	51	23	21	5.4	62.0	7.30	82	36	22.00
PQB-01-SP-036	<10	1	<1	8	<5	2	1	54	–	60	–	1191	94	31	2.1	94.6	19.40	831	62	11.40
WX92NF-53	15	0	<1	16	93	14	<1	31	–	4	–	21	154	14	–	1.0	0.03	6	4	0.17
WX01NF-184	13	<1	8	25	71	37	12	61	1.9	9	0.4	170	375	13	0.2	4.8	12.00	590	24	0.49
WX01NF-187	15	<1	7	19	46	45	9	46	3.7	4	0.1	185	389	11	0.2	4.6	12.00	495	21	0.50
WX01NF-188	13	<1	7	29	88	47	13	72	3.9	3	0.1	190	394	15	0.2	4.4	15.00	600	35	0.34
WX92NF-54	134	44	<1	69	319	71	<1	75	–	31	–	135	132	14	–	0.6	0.44	12	20	0.06
VL01J-213	172	73	45	44	340	128	60	184	1.2	5	0.1	95	341	19	0.4	5.8	3.28	134	31	0.43
VL01J-215	259	113	45	39	260	41	12	132	1.1	16	0.2	166	408	16	0.3	5.2	2.59	105	25	0.43
VL01J-216	103	53	48	48	351	126	7	196	1.5	3	0.1	75	262	18	0.4	6.4	3.01	121	36	0.38
WX01NF-209	60	18	39	48	299	85	7	87	0.8	24	0.2	78	266	16	0.2	3.3	1.20	33	20	0.63
WX01NF-211	17	<1	22	41	158	192	4	86	0.8	3	0.1	240	366	19	0.1	2.2	0.40	11	18	0.26
WX01NF-212	196	100	48	39	308	62	3	87	1.0	6	0.4	160	426	17	0.1	1.1	0.78	26	15	0.06
WX92NF-99	114	95	<1	49	316	58	5	76	–	6	–	88	277	21	–	3.7	2.74	104	27	0.41
WX92NF-100	<10	<1	<1	71	486	70	<1	78	–	7	–	44	294	22	–	2.3	0.68	17	22	0.19
WX92NF-97	24	8	<1	45	274	77	8	79	–	25	–	441	299	17	–	4.4	1.68	48	35	0.51
WX02NF-254	<10	<1	45	20	134	13	8	59	1.0	42	0.6	791	377	15	0.9	4.0	3.40	141	22	1.60
PQB-SP-02-079b	68	35	47	15	131	14	19	86	2.6	100	1.7	1128	600	20	1.7	16.0	6.70	247	20	16.00

(continued)

TABLE 4. REPRESENTATIVE WHOLE-ROCK ANALYSES FROM NOTRE DAME ARC ROCKS (*continued*)

Sample number	Trace elements (ppm)														
	U	La	Ce	Pr	Nd	Sm	Eu	Gd	Tb	Dy	Ho	Er	Tm	Yb	Lu
PQB-01-SP-007B	0.75	7.3	19.0	2.7	12.0	3.5	1.30	4.30	0.68	4.50	0.97	2.90	0.42	2.90	0.47
PQB-02-S144	0.72	1.8	4.2	0.6	2.9	0.9	0.38	1.20	0.23	1.60	0.35	1.10	0.17	1.20	0.19
PQB-01-SP-005	2.00	19.5	43.4	5.1	25.1	6.3		7.23	1.32	7.93	2.12	6.83	1.10	7.42	1.23
PQB-01-SP-017	1.10	8.8	20.0	2.4	11.0	3.0	0.68	3.90	0.73	5.00	1.20	3.60	0.61	4.40	0.72
WX01NF-193	0.29	3.7	9.2	1.4	7.6	2.9	1.50	5.30	1.10	7.60	1.90	5.50	0.88	6.30	1.00
PQB-01-SP-015	1.10	5.7	13.0	1.8	8.5	3.0	0.48	4.90	1.00	7.30	1.80	5.40	0.92	5.80	0.93
VL-92-LH-11	2.90	86.0	190.0	22.0	78.0	12.0	2.10	6.40	0.82	3.60	0.62	1.50	0.23	1.40	0.22
VL-92-LH-68	2.90	36.0	78.0	9.3	36.0	7.2	1.50	6.60	1.00	5.80	1.20	3.30	0.52	3.30	0.53
VL-92-84	11.00	3.7	27.0	1.8	6.9	3.5	0.08	4.00	0.82	5.50	1.00	2.70	0.42	2.80	0.39
PQB-01-SP-036	2.60	99.0	265.7	20.4	84.2	18.3	3.23	13.88	2.40	11.80	2.52	6.3[illegible]	0.91	5.49	0.81
WX92NF-53	–	1.6	3.7	0.5	2.4	0.6	0.48	0.69	0.09	0.66	0.17	0.5[illegible]	0.07	0.52	0.07
WX01NF-184	0.49	16.0	33.0	4.5	22.0	5.0	1.60	5.10	0.73	4.20	0.88	2.3[illegible]	0.33	2.20	0.35
WX01NF-187	0.37	13.5	28.5	4.0	18.5	4.3	1.50	4.30	0.62	3.55	0.73	1.95	0.30	1.90	0.31
WX01NF-188	0.25	16.0	35.0	5.1	26.0	6.6	1.90	6.90	1.10	6.30	1.30	3.40	0.49	3.30	0.49
WX92NF-54	–	1.0	3.6	0.7	4.4	1.7	0.50	2.61	0.46	3.32	0.80	2.33	0.35	2.26	0.31
VL01J-213	0.18	7.3	20.7	3.3	16.5	4.7	1.78	5.38	0.87	5.50	1.15	3.23	0.48	2.94	0.42
VL01J-215	0.20	9.3	24.3	3.5	15.8	4.0	1.71	4.40	0.74	4.48	0.93	2.53	0.38	2.40	0.36
VL01J-216	0.20	8.0	23.0	3.7	17.7	5.3	1.92	6.10	0.99	6.24	1.33	3.64	0.53	3.30	0.48
WX01NF-209	0.23	11.0	26.0	3.3	14.0	3.0	1.10	3.00	0.50	3.10	0.68	1.80	0.29	2.00	0.31
WX01NF-211	0.13	12.0	26.0	3.4	16.0	3.5	1.20	3.60	0.53	3.20	0.65	1.60	0.24	1.50	0.24
WX01NF-212	0.08	3.6	9.2	1.4	7.6	2.3	1.30	2.80	0.45	2.70	0.56	1.50	0.21	1.40	0.20
WX92NF-99	–	6.6	15.9	2.5	12.8	3.6	1.38	4.43	0.73	4.66	0.97	2.75	0.38	2.44	0.36
WX92NF-100	–	3.8	10.9	1.7	8.4	2.7	0.94	3.63	0.56	4.19	0.83	2.30	0.32	2.02	0.28
WX92NF-97	–	9.6	25.2	3.7	17.3	4.5	1.32	4.82	0.78	5.14	1.14	3.35	0.51	3.44	0.52
WX02NF-254	0.60	23.0	45.0	5.5	23.0	4.7	1.10	4.20	0.64	3.60	0.75	2.00	0.30	1.80	0.30
PQB-SP-02-079b	4.90	51.0	103.0	12.0	46.0	7.2	1.60	4.90	0.65	3.30	0.64	1.70	0.28	1.90	0.30

Note: "–" indicates not determined. C—complex.

TABLE 5. WHOLE-ROCK Sm-Nd ISOTOPIC DATA FOR NOTRE DAME ARC ROCKS IN WESTERN NEWFOUNDLAND

Complex (C)/map unit	Rock type	Sample no.	SiO_2 (wt%)	Age (Ma)	ε_{Nd} (T)	T_{DM}* (Ga)	$^{143}Nd/^{144}Nd$ (m)	±2SD	$^{147}Sm/^{144}Nd$ (m)	Nd (ppm)	Sm (ppm)
Caribou Lake C	Granodioritic orthogneiss	WX01NF-175†	71.4	459	−9.30	1.45	0.511828	6	0.0857	12.63	1.79
Island Pond paragneiss	Biotite-rich paragneiss	WX01NF-230	66.7	467	−10.37	1.79	0.511848	5	0.1121	38.21	7.09
Southwest Brook C	Tonalite	WX02NF-254	63.9	463	−6.57	1.81	0.512118	7	0.1361	21.14	4.76
Southwest Brook C	Low-Th tonalite	WX92NF-53†	72.1	470	4.77	NC	0.512793	8	0.1677	2.14	0.59
Southwest Brook C	Low-Th tonalite	WX92NF-75†	67.4	470	−6.75	1.57	0.512048	41	0.1173	14.95	2.90
Southwest Brook C	Low-Th tonalite	WX92NF-86†	67.1	470	−6.04	1.87	0.512160	17	0.1419	12.20	2.86
Southwest Brook C	Low-Th tonalite	WX92NF-114†	65.1	470	−8.36	1.68	0.511961	17	0.1158	12.54	2.40
Southwest Brook C	Low-Th tonalite	WX01NF-188	67.6	470	−6.98	NC	0.512159	6	0.1572	21.67	5.64
Southwest Brook C	Mafic raft in NF-53	WX92NF-54	46.8	470	4.54	NC	0.513087	34	0.2671	4.83	2.14
Southwest Brook C	Comagmatic mafic dike	WX01NF-211	50.9	470	−4.72	1.76	0.512234	6	0.1439	13.72	3.27
Cormacks Lake C	Amphibolitic metavolcanic	PQB01SP-007B	54.2	489	4.20	NC	0.512747	6	0.1637	11.11	3.01
Cormacks Lake C	Felsic tuffaceous gneiss	PQB01SP-015	85.2	489	1.95	NC	0.512632	5	0.1637	7.79	2.11
Cormacks Lake C	Felsic tuffaceous gneiss	WX01NF-193	75.8	489	3.38	NC	0.512682	8	0.1564	6.34	1.64
Cormacks Lake C	Tonalitic orthogneiss	PQB01SP-017	75.1	483	2.21	NC	0.512657	17	0.1669	10.52	2.91
Cormacks Lake C	Charnockite	PQB01SP-036	69.1	459	4.29	0.69	0.512624	6	0.1189	76.90	15.13
Cormacks Lake C	Quartz diorite	PQB02SP-079B	60.9	463	−6.86	1.39	0.511987	3	0.0979	42.82	6.94
Stag Hill gneiss	Granodioritic orthogneiss	VL92-84	78.8	462	−7.20	NC	0.512602	17	0.3066	6.20	3.14
		VL-92-84 (duplicate)		462	−7.54	NC	0.512565	20	0.3002	6.57	3.26

*Depleted mantle ages (T_{DM}) are based on model of DePaolo (1981); values were not calculated (NC) for samples with $^{147}Sm/^{144}Nd$ >0.15.
†Whole-rock major- and trace-element analyses for these samples were published in Whalen et al. (1997a).

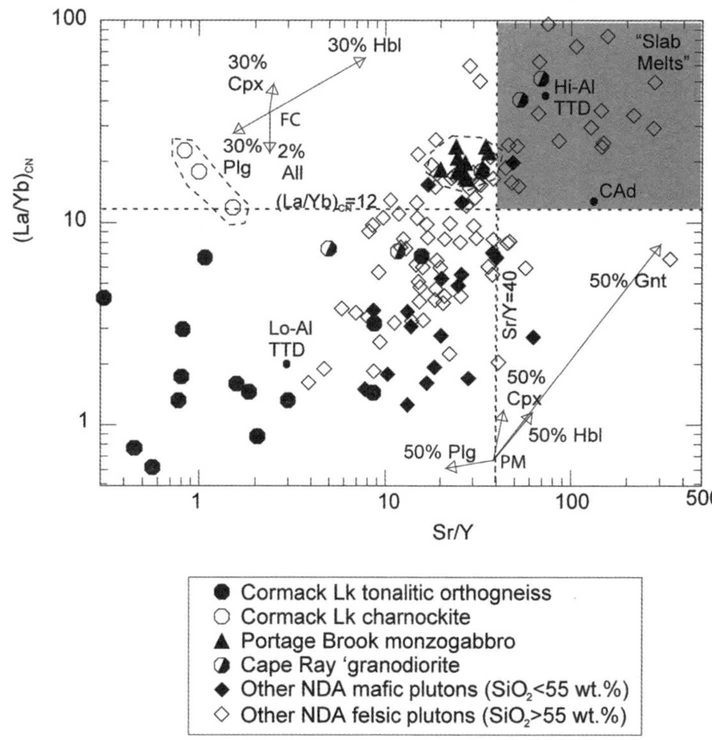

Figure 10. Plot of $(La/Yb)_{CN}$ versus Sr/Y for Notre Dame arc rocks with symbols legend. Shown are various compositional criteria suggested by Drummond and Defant (1990) and Defant and Drummond (1990) to be characteristic of "slab-derived" melts, or, more generally, melts of mafic protoliths leaving residual garnet but not plagioclase. Also shown are Rayleigh fractionation (FC) and equilibrium partial melting (PM) trends for crystallization of new or residual minerals; weight % values are as indicated. Mineral abbreviations: plagioclase (Plg), clinopyroxene (Cpx), hornblende (Hbl), garnet (Gnt), and allanite (All). Partition coefficients of Martin (1987) (Plg, Gnt, All) and Nutman et al. (1999) (Hbl, Cpx) were employed in the calculations. Also shown with a solid dot symbol are average compositions from Drummond et al. (1996) for Cenozoic adakite (N = 140) (CAd), Archean high-Al tonalite-trondhjemitic-dacite (N = 174) (Hi-Al TTD), and low-Al tonalite-trondhjemite-dacite plutons (N = 121) (Lo-Al TTD).

Dame arc plutons were mainly derived from nonweathered infracrustal protoliths. However, their ε_{Nd}(T) values, ranging between +2.6 and –13.5, require significant contributions from old crustal and/or subcontinental lithospheric material. Pb-isotopic data are most compatible with a dominantly lower-crustal origin for this old component. Associated gabbros indicate that they received contributions from a mantle-derived component.

A subordinate proportion of the Notre Dame arc plutonic rocks also displays geochemical characteristics typical of Phanerozoic adakites and high-Al trondhjemite-tonalite-dacite (TTD) plutonic suites, commonly interpreted as slab melts formed at sites where young (<20 m.y.), hot oceanic crust was being subducted (Drummond et al., 1996). Their elevated La/Yb and Sr/Y ratios (Fig. 10) have been interpreted as a fingerprint of partial melting of a mid-ocean-ridge basalt (MORB)–like protolith (i.e., the downgoing slab) under relatively high-pressure and high-temperature conditions, leaving a plagioclase-poor, garnet-rich residue (e.g., Rushmer, 1993; Rapp and Watson, 1995). However, since most Notre Dame arc samples within this group only display some geochemical features typical of slab-derived melts and none of the Notre Dame arc plutons displays MORB-like Nd-isotopic signatures, a more complex petrogenetic model is required than simple slab melting alone. Instead, Whalen et al. (1997a) interpreted the Notre Dame arc plutons as the products of continental arc magmatism that contained contributions from both mantle- and slab-derived magmas modified through assimilation-fractional-crystallization processes in the lower crust, and also partial melts of solidified, previously hybridized, lower-crustal rocks.

Structures, Textures, Metamorphism, and Geochronology of the Lower to Middle Ordovician Notre Dame Arc Plutons

The Ordovician felsic Notre Dame arc plutons commonly contain abundant mafic enclaves, which, based on frequently well-preserved commingling textures, have been interpreted as evidence of coexisting felsic and mafic magmas (Whalen et al., 1997a). Where the felsic host is moderately to strongly deformed, the mafic enclaves (Fig. 11C) are generally transformed into lenticular lenses of amphibolite that parallel the main fabric (Fig. 12F). Enclaves are locally so abundant that strong deformation and metamorphism have transformed the rocks into banded gneisses (Fig. 11C). Pressure shadows, feldspar tails, and ubiquitous dynamic recrystallization of quartz and feldspar indicate that the L-S fabric, even where it is moderately to weakly developed, was predominantly formed during solid-state deformation under amphibolite facies conditions. Nevertheless, since some plutons preserve evidence for synemplacement deformation (Lissenberg et al., 2005b; Lissenberg and van Staal, 2006), such deformation may have been more widespread than is evident at present.

Structural relationships critical to delineating the timing of Ordovician deformation and metamorphism in the Notre Dame subzone were investigated in a few key areas where age relationships between various deformed plutonic rock phases are locally well preserved. Most of our combined structural and geochronological investigations were carried out between Cape Ray, at the southern end of the Notre Dame subzone, and Little Grand Lake, in its center (Figs. 1 and 2), but they were augmented by investigations in the Hungry Mountain–Topsails area in the northern part of the Notre Dame subzone (Whalen et al., 1987; this study).

Structural analysis and regional correlation of structures in polyphase deformed plutonic rocks are notoriously difficult, particularly in their interiors, where regional structural markers are generally absent and deformation is commonly highly heterogeneous. Therefore, our studies were focused on areas where metaplutonic rocks were interleaved or associated with metasedimentary rocks. The metasedimentary rocks commonly have a compositional layering on centimeter to meter scale, which at least in part, probably represents bedding (Pehrsson et al., 2003a). Therefore, overprinting relationships within metasedimentary rocks provide a template for correlation and classification of structures in adjacent plutonic rocks and also provide insights to metamorphic conditions accompanying deformation.

First Phase of Notre Dame Arc Magmatism

Tonalitic Orthogneiss of the Cormacks Lake Complex

The Cormacks Lake tonalite is a relatively narrow, generally strongly deformed sheet-like body, which is commonly transformed into orthogneiss. It was probably emplaced into the surrounding metasedimentary and metavolcanic rocks of the Cormacks Lake complex as a sill (Fig. 2; Herd and Dunning, 1979; van Staal et al., 2005). Intrusive relationships are locally preserved on outcrop scale, but low-angle truncations of marker units against the orthogneiss, such as a thin, but regionally extensive metapillow basalt layer, probably also represent relict intrusive contacts (Pehrsson et al., 2003a). The tonalite sill and surrounding paragneiss are folded into a complex mushroom-like geometry, which is probably an interference pattern between F_3 and F_4 folds (Pehrsson et al., 2003a). The main fabric, a steeply dipping S_2 transposition foliation, is axial planar to steeply inclined or upright, shallowly to steeply plunging, tight to isoclinal F_2 folds (Figs. 11B, 12C, and 12D). F_2 folds are consistently folded by younger upright F_3 and F_4 structures, both on meso- and macroscales. An older foliation (S_1) is evident in F_2 fold hinges in both the para- and orthogneiss. The U-Pb crystallization ages of 489.3 ± 5.6 Ma and 483.1 ± 1.4 Ma obtained for a felsic tuffaceous schist (PQB-SP-01007) and tonalitic orthogneiss (PQB-SP-01017), respectively, provide maximum ages for the oldest deformation (Figs. 9A, 9B; Appendix 2; Tables 1 and 2). Metamorphism is pervasive throughout the Cormacks Lake complex and locally has attained relatively high-pressure granulite facies conditions as evidenced by remnants of garnet and clinopyroxene in mafic rocks (Herd and Dunning, 1979; Owen and Greenough, 2000; Pehrsson et al., 2003a). Minerals associated with peak metamorphism, such as sillimanite and biotite, are either folded by intrafolial F_2 microfolds (i.e., they formed part of S_1; Fig. 11D), lie in S_2 or occur in S_2-parallel leucosomes. All are folded and deformed by F_3, and, hence, we interpret peak metamorphic mineral growth to have taken place before, during, and possibly slightly after F_2, but before F_3. In addition, S_2 is weakly developed, and S_1 is absent in a distinct high-K charnockite intrusion, suggesting that this pluton intruded late during F_2 under granulite facies metamorphic conditions (Currie et al., 1992). Combined, these relationships suggest that peak metamorphic conditions were reached before and/or during F_2. The charnockite intrusion (VK-87-NF2) yielded a U-Pb age of $459.4\ ^{+3.0}/_{-1.5}$ Ma (Fig. 9C; Appendix 2; Table 2) and forms part of the terminal stage of the second phase of the Notre Dame arc (see following).

Cape Ray Igneous Complex

Although the Cape Ray igneous complex occurs in the southernmost part of the Notre Dame subzone, it is linked by a

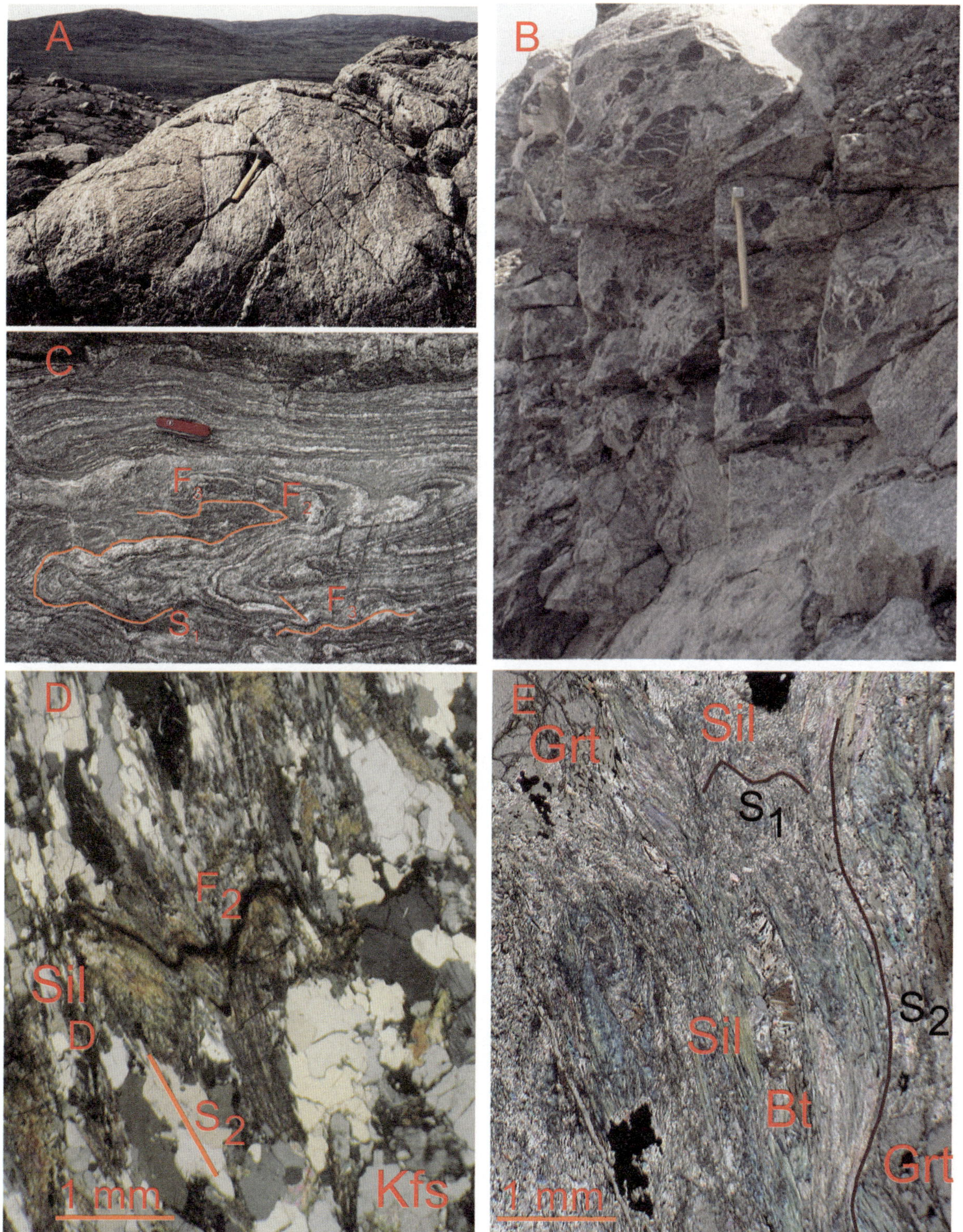

Figure 11. (A) Tonalite and granodiorite veins of the Cape Ray igneous complex intruding into mafic and ultramafic rocks of the ophiolitic Long Rang mafic-ultramafic complex (LRC). Note both host and guest are strongly foliated. (B) Southwest Brook complex tonalite with abundant enclaves of net-veined amphibolite. Tonalite is moderately foliated (compared to same rock in Fig. 12F). (C) F_2 intrafolial folds in strongly banded orthogneiss made up of tonalite and granodiorite and thin lenticular mafic layers mainly consisting of biotite and/or hornblende. F_2 structures are refolded by F_3 on the same outcrop. S_3 is only well-developed in biotite-rich layers. (D) F_2 microfolds of S_1 sillimanite (Sil), quartz, and K-feldspar (Kfs) paragneiss of Cormacks Lake complex. (E) S_2 transposition foliation defined by sillimanite and biotite (Bt) in garnet-bearing paraschist adjacent to and interlayered with plutonic rocks of Cape Ray igneous complex. Sillimanite and biotite are tightly kinked and crenulated by F_2 microfolds within S_2 and locally still outline S_1. S_2 is deflected by the garnet porphyroblasts.

Figure 12. (A) Pink Cape Ray granodiorite (488 ± 3 Ma) with F_2 folds in well-developed S_1. A new axial plane foliation (S_2) defined by biotite is present. (B) Strongly foliated (S_1) metagabbro of the ophiolitic Long Range mafic-ultramafic complex. S_1 is folded by F_2 and F_3 and has the same ductile deformation history as the rocks of the Cape Ray igneous complex. Note that tonalite veins are folded by F_2. (C) Cormacks Lake orthogneiss (483.1 ± 1.4 Ma) is intruded by Southwest Brook quartz diorite with bluish gray quartz grains. PQB-SP-02-79B (462.7 ± 1.8 Ma) was collected on this outcrop. (D) Gedrite-bearing paragneiss interlayered with Cormacks Lake orthogneiss. S_1 is folded into tight F_2 structures that are overprinted by F_3 and F_4 folds. (E) Southwest Brook tonalite (same as in Figs. 9C and 10F) with a strongly developed composite S_1/S_2 foliation. Samples VL-02-21A (463 ± 1 Ma) and VL-02-21B (431 ± 2 Ma) were collected here. (F) Southwest Brook tonalite with mafic enclaves flattened parallel to S_1 that are folded by F_2 folds, forming a composite S_1/S_2 transposition foliation. S_1/S_2 is refolded by open to tight nearly symmetrical F_3 kinks, which are accompanied by a weakly developed axial plane foliation (S_3). Tonalite is same body as in Figures 11B and 12E.

large continuous plutonic body of tonalite and granodiorite with the Cormacks Lake and Southwest Brook complexes (Figs. 1, 2, and 8). The oldest phase, the Cape Ray granodiorite (Fig. 12A), varies in composition from feldspar porphyritic granodiorite to monzonite, and it intruded the ophiolitic Long Range mafic-ultramafic complex (Fig. 11A) and garnet-sillimanite paragneiss (Fig. 11E). It yielded a U-Pb zircon age of 488 ± 3 Ma, and it is cut by equigranular tonalite with an age of 469 ± 2 Ma (Dubé et al., 1996). In many places, these two phases are intimately mixed, prohibiting mapping of their boundaries accurately. Both phases are pretectonic and contain a strong S_2 transposition foliation formed by tight to isoclinal folding of S_1 (Hall and van Staal, 1999), which is continuous with S_2 in the adjacent, migmatitic paragneiss and the strongly deformed ophiolitic Long Range mafic-ultramafic complex (Fig. 12B). This tonalite was in turn intruded by a younger tonalite (Stag Hill orthogneiss) that cuts S_1 but generally contains S_2. Locally, in the hinges of F_2 folds, Stag Hill dikes are oriented parallel to S_2, suggesting that they intruded during F_2. All three of these phases are unconformably overlain by the Upper Ordovician–Silurian Windsor Point Group, which contains foliated cobbles of these felsic plutons in its basal conglomerate. Interlayered ignimbritic rhyolite yielded a U-Pb zircon age of 453 $^{+5}/_{-4}$ Ma (Dubé et al., 1996), suggesting that F_2 and earlier structures formed between 470 and 455 Ma and were accompanied and/or followed by Notre Dame arc uplift and exhumation.

Sillimanite and biotite in paragneiss screens locally define S_1 in the hinges of F_2 microfolds, but they also grew along S_2. Garnet locally grew across S_2 but is also wrapped by it (Fig. 11E), suggesting that overall relationships between deformation and metamorphism are the same as in the Cormacks Lake and Southwest Brook complexes (see following).

Second Phase of Notre Dame Arc Magmatism

Southwest Brook Complex

Metasedimentary and metaplutonic rocks of the Cormacks Lake complex were intruded by voluminous tonalite and quartz diorite of the Southwest Brook complex, which surrounds the Cormacks Lake complex (Fig. 8). Our mapping (van Staal et al., 2005) has revealed several distinct phases in the Southwest Brook complex. The oldest felsic phase, a white hornblende– and biotite-tonalite, has been dated at 462.8 ± 0.9 Ma (VL02–21A; Figure 9D; Appendix 2; Table 2). The tonalite contains abundant amphibolitic mafic enclaves largely derived from consanguineous dioritic and gabbroic magmas (Figs. 11C). Flattened enclaves are locally folded into tight to isoclinal folds and have an axial planar foliation (Fig. 12F) that is equated with S_2 in rocks of the adjacent Cormacks Lake complex. Hence, folds in the flattened enclaves are F_2 structures, and the flattened enclaves preserve an older structure (S_1). S_2 is folded by open to tight F_3 folds accompanied by a new S_3 axial-planar foliation that is co-planar with S_3 in the Cormacks Lake complex. Pink granitic dikes, which yielded a U-Pb zircon age of 431 ± 2 Ma (Whalen et al., 2006) cut across the S_2 foliation (Fig. 12E). Quartz diorite, which is generally less deformed than the tonalite (Figs. 12F), cuts the Cormacks Lake orthogneiss where the latter has a well-developed S_2 transposition foliation (Fig. 12C). Both the orthogneiss and quartz diorite were deformed by F_3, such that a weak to moderate S_3 foliation developed in the quartz diorite. S_3 is axial planar to F_3 folds in the S_2 transposition foliation present in the orthogneiss. From these relationships, we tentatively conclude that the tonalite intruded before or during F_2, and the quartz diorite intruded after or late during F_2 but before F_3. Early Silurian mafic and felsic dikes that have intruded the Ordovician rocks (Fig. 12E) are folded by F_3. In the Cormacks Lake complex, these Silurian dikes show a pronounced, amphibolite facies Silurian metamorphic overprint (Fig. 9; Appendix 2; Table 2), which is also present in the host rocks and immediately surrounding tonalite of the Southwest Brook complex (Table 2) but rarely elsewhere. These ages indicate that F_3 and associated high-grade metamorphism formed during the Silurian Salinic orogeny. The Cormacks Lake complex rocks probably remained buried at deep levels within the arc's infrastructure until the end of the Early Silurian third phase of the Notre Dame arc (Fig. 3), and then they were rapidly exhumed to high crustal levels (Pehrsson et al., 2003b). This late Early Silurian metamorphic overprint (433–429 Ma; Appendix 2) largely, but not completely, masked geochronological evidence for the preceding Taconic metamorphism (Pehrsson et al., 2003b) in this area.

The Southwest Brook complex is surrounded by psammitic to pelitic, kyanite- or sillimanite-bearing paragneiss and schist, suggesting a Barrovian-style metamorphism. Sillimanite needles, together with biotite, define S_1 and S_2 and are interpreted to record peak metamorphic conditions reached before and/or during F_2. Metamorphic monazite in a sample of biotite-rich paragneiss (WXNF01-230), which occurs ~20 km to the north of the Cormacks Lake complex, has been dated at 466.5 ± 1.8 Ma (Fig. 9F; Appendix 2; Table 2) and is interpreted as the age of peak regional metamorphism in the central part of the Notre Dame subzone.

Hungry Mountain and Cariboo Lakes Complexes

The Hungry Mountain–Cariboo Lakes complexes underlie most of the north-central part of the Notre Dame subzone (Figs. 2 and 8). Their Ordovician plutonic rocks are connected and consanguineous with those farther to the south in the Southwest Brook complex (Whalen et al., 1997a). Hornblende tonalite of the Hungry Mountain complex has yielded a U-Pb zircon age of 467 ± 8 Ma (Whalen et al., 1987), whereas hornblende tonalite (WXNF-48) and hornblende gabbro (WXNF-94) from this complex have U-Pb crystallization ages of 469.1 ± 1.3 Ma and 463.1 ± 3.7 Ma, respectively (Figs. 9G and 9H; Appendix 2; Tables 1 and 2). Farther south, halfway between the Southwest Brook and Hungry Mountain complexes, granodiorite and tonalite of the Cariboo Lakes complex (WXNF-175) and Lewaseechjeech Brook suite (WXNF-63) have been dated at 459.0 ± 3.4 Ma and 468.2 ± 3.8 Ma, respectively (Figs. 9I and 9J; Appendix 2; Tables 1 and 2). These data confirm that the very voluminous second-stage Notre Dame arc plutonism was coeval throughout the Notre Dame subzone. Although no detailed structural investigations were carried out in this part of the Notre Dame subzone, reconnaissance studies did not indicate any significant differences in structural history compared to elsewhere. Generally, these rocks are heterogeneously deformed, ranging from weakly foliated rocks to very strongly foliated, amphibolite facies mylonites with intrafolial folds near major shear zones, such as the southeast-directed (retroarc) Hungry Mountain thrust (e.g., Calon and Green, 1987). This transposition foliation, which is also the dominant structure in screens of interlayered sillimanite paragneiss, predates crosscutting late Lower Silurian dikes (ca. 429 Ma) of the unmetamorphosed Topsails intrusive suite (Whalen et al., 1987). Locally strongly foliated and boudinaged tonalite and amphibolite dikes are cut by openly folded, unfoliated tonalite veins at high angles, suggesting that deformation overlapped with Notre Dame arc magmatism. The presence of inverted metamorphism in the footwall of the Hungry Mountain thrust and correlative faults along strike (Thurlow, 1981; Lissenberg and van Staal, 2006) supports syntectonic magmatism because it requires that the Hungry Mountain complex was emplaced while it was still hot. Further support is provided by syntectonic intrusion of a ca. 468 Ma granodiorite pluton (Lissenberg et al., 2005b) in an along-strike segment of the Hungry Mountain thrust. Hence, the main foliation in the Hungry Mountain and Caribou Lakes complexes is correlated with the regional S_2 transposition foliation present elsewhere in the Notre Dame arc.

Structural Conclusions

Our observations suggest that the first phase of Notre Dame arc magmatism predated regional deformation and metamorphism, while the second Notre Dame arc magmatic phase largely overlapped with F_2, and possibly also S_1, in most parts of the Notre Dame subzone. If the latter is correct, F_2 and S_1 may have formed during a regionally continuous deformation, referred to herein as D_{1-2}. S_1 probably had shallow to moderate dips prior to tight to isoclinal F_2 upright folding and formation of a steeply dipping S_2 transposition foliation. Considering the structural and metamorphic evidence for deep burial of volcanic and sedimentary rocks formed during the first phase of the Notre Dame arc (e.g., Cormacks Lake complex) before and during F_2 (e.g., Fig. 11D), S_1 probably formed as a result of thrusting and/or recumbent folding (Fig. 7B). With the exception of the northernmost part of the Notre Dame subzone, D_{1-2} was accompanied by a Barrovian-style amphibolite to granulite facies metamorphism. These high-grade conditions were maintained, at least locally, until after F_2. Thus, emplacement of the second phase of the Notre Dame arc was preceded and/or accompanied by significant structural thickening and metamorphism in most parts of the Notre Dame subzone.

Discussion and Implications of the Geochronology

The crystallization ages of the pretectonic Cormacks Lake complex felsic tuffaceous gneiss and orthogneiss, 489.3 ± 5.6 Ma and 483.1 ± 1.4 Ma, respectively, indicate that they formed during the first magmatic phase of the Notre Dame arc (490–477 Ma). These ages also provide a maximum age for S_1 and F_2.

The ages of the Southwest Brook tonalite (462.8 ± 0.9 Ma), quartz diorite (462.7 ± 1.8 Ma), and Cormacks Lake complex charnockite (459.4 $^{+3}/_{-1.5}$ Ma) suggest that in this part of the Notre Dame arc, F_2 mainly took place between 464 and 458 Ma (Middle Ordovician). The age of 466 ± 1 Ma, obtained from metamorphic monazite in the biotite-rich paragneiss, is interpreted to be the age of peak regional metamorphism in the central part of the Notre Dame subzone. On the basis of microstructures, this generally occurred before or early during F_2. Hence, D_{1-2} in the Cormacks Lake and Southwest Brook complexes had started by at least 467 Ma. $^{39}Ar/^{40}Ar$ and U-Pb dating of metamorphic fabrics and stitching plutons, respectively, has been carried out in the boundary zone between the Notre Dame arc and the adjacent Annieopsquotch accretionary tract (Lissenberg et al., 2005b). This work indicates that Notre Dame arc deformation, involving southeast-directed retroarc thrusting above the Annieopsquotch ophiolite belt (Fig. 7B), was active by at least 468 Ma and probably had started at ca. 470 Ma. Although individual deformation events may have been slightly time-transgressive from one place to another, a 470–458 Ma age for D_{1-2} is consistent with age constraints both to the south (Cape Ray) and to the north (Hungry Mountain–Cariboo lakes) over a strike length of more than 200 km (Figs. 1 and 2). Thus, D_{1-2} regionally represents a major shortening and arc-thickening event, consistent with collision between Dashwoods and the Laurentian margin at this time.

There are abundant U-Pb data from accessory minerals (zircon, monazite, titanite, allanite, and rutile) from samples of the Cormacks Lake complex and surrounding rocks (see previous) that are Early Silurian in age (433–429 Ma) and are interpreted to be metamorphic in origin (Appendix 2). These Silurian metamorphic data are related to tectonic processes accompanying Salinic accretion of Ganderia to Laurentia (van Staal, 1994, 2005; Pehrsson et al., 2003b; Valverde-Vaquero et al., 2006; Whalen et al., 2006).

Zircon Inheritance and Provenance of the Notre Dame Arc

With the exception of Cormacks Lake complex rocks, members of the first and second phases of Notre Dame arc magmatism commonly contain significant populations of Proterozoic inherited zircons. Based on an extensive petrological, isotopic, and geochemical data set, Whalen et al. (1997a) suggested that second-phase Notre Dame arc plutons formed by mixing of subduction-derived magmas and an old crustal component present in the arc infrastructure. Considering the predominance of intermediate hornblende-bearing magmatic rocks in the Notre Dame arc, the mantle-like O-isotopic signatures, and the scarcity of granites sensu stricto, it is unlikely that this old crustal component was derived from assimilation of underthrusted and accreted forearc sediments by arc magmas that invaded them following arc-trench migration as a result of slab rollback. Hence, zircon inheritance and tracer isotopic data indicate the presence of continental basement within the Notre Dame arc infrastructure (Dashwoods microcontinent; Waldron and van Staal, 2001). In general, the additional new data we have collected support this conclusion. For example, paragneiss WXNF-01-230 yielded a $\varepsilon_{Nd}(T)$ value of –10.2, indicating a major contribution of an old crustal source, probably Laurentian basement. Furthermore, SHRIMP studies of inherited zircons in WXNF-175 and WXNF-48 yielded generally typical Laurentian zircon populations at ca. 600 Ma and in the 850–1600 Ma age range (Figs. 9H and 9J), which corroborate earlier TIMS studies by Whalen et al. (1987), Dunning et al. (1989), and Dubé et al. (1996). The presence of Early Mesoproterozoic (1300–1650 Ma) inherited ages is significant because it suggests a linkage to the Pinware terrane in Labrador (Tucker and Gower, 1994) or the Long Range Inlier in northern Newfoundland (Heaman et al., 2002). Thus, xenocrystic Notre Dame arc zircons provide information concerning the provenance of the Dashwoods microcontinent.

New Geochemical and Isotopic Evidence Concerning the Origin of Notre Dame Arc Magmatism

First Phase of Notre Dame Arc Magmatism

The Cormacks Lake complex mafic volcanic rocks (ca. 489 Ma; Fig. 9A; Appendix 2) exhibit compositions typical of an island-arc setting, including primitive mantle–normalized patterns that very closely resemble those of island-arc tholeiite and oceanic-arc gabbro (Fig. 13). Petrochemical characteristics of the ca. 483 Ma Cormacks Lake tonalitic orthogneiss (Fig. 9B), including the possibly consanguineous tuffaceous felsic metavolcanic rocks (Appendix 2), resemble felsic oceanic-arc plutonic rocks and average low-Al tonalite (Figs. 13B and 13C). They represent compositional equivalents to the type II rhyolites of Lesher et al. (1986). Such compositions are thought to form by low degrees of partial melting of mafic protoliths at intermediate to deep levels (~5–10 kbar; Wolf and Wyllie, 1994; Rapp and Watson, 1995) or by shallow-level fractionation from low-K mafic magmas (Barker and Arth, 1976; Drummond and Defant, 1990). Based on their low Sr/Y and La/Yb ratios (Defant and Drummond, 1990), crystal-liquid equilibria in both processes involved little or no garnet and instead were dominated by plagioclase, pyroxene, and possibly amphibole (Fig. 10).

The $\varepsilon_{Nd}(T)$ values for the ca. 483 Ma Cormacks Lake tonalitic orthogneiss and ca. 489 Ma meta-tuffaceous samples (four analyses) range between +0.8 and +2.2, and a ca. 489 Ma metabasalt sample yielded a $\varepsilon_{Nd}(T)$ value of +4.2 (Table 5). These Nd isotopic values are less positive than modern intraoceanic ensimatic arcs and are more typical of island arcs built on thin continental crust (Swinden et al., 1997). Although no inherited zircons have yet been detected in first-phase Cormacks Lake complex igneous rocks, subsequent Southwest Brook complex plutonic rocks that surround and cut these Cormacks Lake complex rocks contain abundant inherited zircon (see previous section and Appendix 2.2) and exhibit old crustal $\varepsilon_{Nd}(T)$ signatures (17 analyses, ranging from –1.3 to –13.5, average –6.8; Table 5 and data of Whalen et al., 1997a). In the southernmost Notre Dame subzone, inherited Proterozoic zircons are present in the ca. 488 Ma Cape Ray granodiorite (Dubé et al., 1996). This first-phase intrusion exhibits a ε_{Nd} (T) value of –4.5, indicating interaction with old continental crust. Subsequent second-phase ca. 469 Ma Cape Ray tonalite is also negative ($\varepsilon_{Nd}[T]$ = –9.2) (Whalen et al., 1997b), as is the proximal syn-F_2 (ca. 462 Ma) Stag Hill orthogneiss ($\varepsilon_{Nd}[T]$ = –7.2; Table 5). In the northernmost part of the Notre Dame subzone (Fig. 2), the ca. 479 Ma calc-alkaline Brighton gabbro, which intruded into the Lushs Bight oceanic tract (Szybinski, 1995), is also negative ($\varepsilon_{Nd}[T]$ = –1.9; Swinden et al., 1997). Although overall these data support an ensialic setting, the marked contrast in Nd-isotopic signatures between first- and second-phase Notre Dame arc magmatism of the Cormacks Lake complex indicates that the presence of, or degree of, assimilation of old continental crust into first-phase Notre Dame arc magmas apparently varied along the length of the arc.

However, formation on thin continental crust during the early stages of Notre Dame arc construction is at odds with the elevated $(La/Yb)_{CN}$ and Sr/Y values of some Cape Ray orthoclase porphyritic monzonite to granodiorite samples, which resemble high-Al trondhjemite-tonalite-dacite (TTD) plutonic suites (cf. Drummond and Defant, 1990) that were generated at lower crust or mantle depths in the presence of residual garnet (Fig. 10). On the other hand, the Cape Ray granodiorite bears no resemblance to TTD suites in other characteristics, such as its K_2O contents, which are equivalent to shoshonitic to high-K suites, and negative ε_{Nd} value. In these features, and their $(La/Yb)_{CN}$ and Sr/Y values, it resembles Grenvillian orthogneisses (Fig. 13) that occur in inliers in Newfoundland's Laurentian margin (Fig. 1; Whalen et al., 1997a). For this reason, we interpret the Cape Ray granodiorite to have formed from high-percentage crustal partial melts that had recycled geochemical and isotopic characteristics of the Grenvillian basement. This interpretation does not require deep crustal melting and thus permits the presence of a similar thin crustal substrate along the length of the Notre Dame arc within Newfoundland.

Second Phase of Notre Dame Arc Magmatism

First-order features of the second phase of Notre Dame arc magmatism include its high-volume domination by tonalite (Fig. 8) and generation during a relatively short flare-up (464–459 Ma). This voluminous tonalite flare-up occurred during the waning stages of a significant phase of arc shortening and thickening inferred to be induced by Dashwoods–Laurentian margin collision. Although arc magmatism may continue briefly after the start of a collision, which may explain rare pre–flare-up ca. 469 Ma tonalite intrusions (Fig. 8), it generally shuts off rapidly due to shallowing of the underthrusted continental slab because this

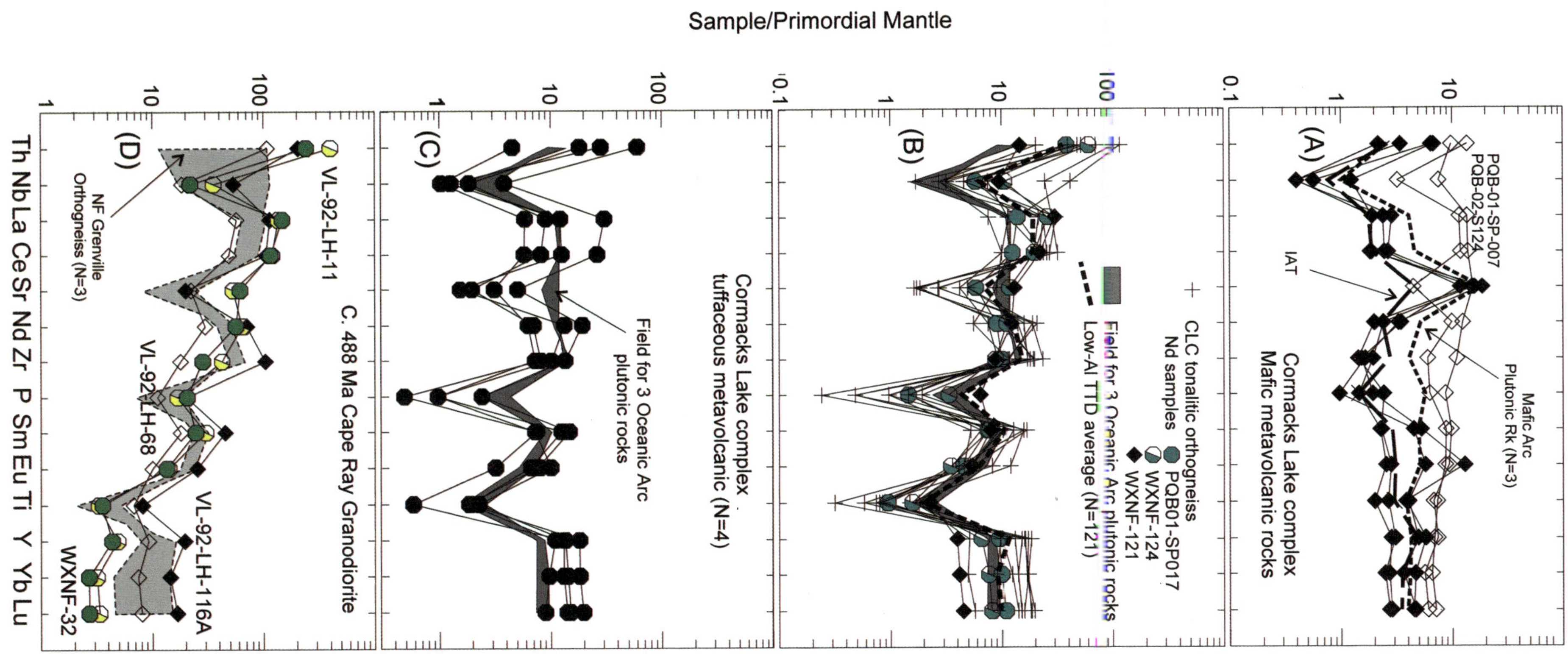

Figure 13. Primordial mantle–normalized extended-element plots for: (A) Cormacks Lake complex mafic metavolcanic rocks (N = 5) along with the average for three mafic oceanic-arc plutonic rocks (data of Whalen, 1985) and an island-arc tholeiite (IAT) from Rautenschlein et al. (1985). (B) Cormacks Lake complex tonalitic orthogneiss samples (N = 13) along with the compositional field for three felsic oceanic-arc plutonic rocks (data of Whalen, 1985) and average low-Al tonalite from Drummond et al. (1996). (C) Cormacks Lake complex (N = 4) tuffaceous metavolcanic samples along with the felsic oceanic-arc plutonic field shown in B. (E) Cape Ray ca. 488 Ma feldspar porphyritic phase (N = 4) along with the compositional field for three Newfoundland Grenville orthogneiss samples (data of Whalen et al., 1997a). Normalizing values are those of Sun and McDonough (1989)

progressively destroys the fertile mantle wedge, either by displacement, quenching, or a combination of both. The flare-up observed here also cannot be related in any way to west-directed subduction of Iapetus beneath Dashwoods. At this time, the subduction zone that produced the (ca. 464–460 Ma) Red Indian Lake volcanic arc was situated outboard (east) and was separated from Dashwoods by a wide oceanic back-arc basin (Fig. 7B; Zagorevski et al., 2006). Hence, it is unlikely that the second-phase flare-up represents normal arc magmatism. This inference is consistent with the results of our new geochemical and isotopic investigations of rocks of the Southwest Brook, Cormacks Lake, and Hungry Mountain complexes, which revealed that the second-phase Notre Dame arc magmatism, besides being dominated by arc-like tonalite, also includes contemporaneous non-arc-like charnockite, gabbro, and tonalite. Mafic and felsic plutonic rocks, herein termed non-arc-like, have Nb/Th values > 7 (Fig. 15) and exhibit either nonsymmetrical negative Nb anomalies or no negative Nb anomalies on primitive mantle–normalized plots (Figs. 14A and 14B).

The ca. 459 Ma non-arc-like Cormacks Lake complex charnockite pluton resembles Silurian A-type granites in western Newfoundland (Whalen et al., 1996) (Fig. 14A) in its high field strength element (HFSE) and heavy rare earth element (HREE) enrichment (Fig. 14A) and juvenile character (ε_{Nd}[T] values of +4.3 and +6.1). However, in contrast to the A-type granites, the Cormacks Lake complex charnockite lacks a negative Nb anomaly. Compositionally similar non-arc-like granulitic granodiorite to tonalite (so-called pincinite) xenoliths in Pliocene basaltic tuffs have been described from the Pannonian Basin in the western Carpathians (Huraiova et al., 2005) (Fig. 14A). The pincinites are interpreted as lower-crustal melts related to Tertiary basaltic intrusions produced during slab break-off (Girbacea and Frisch, 1998). Other than their negative Sr, P, Eu, and Ti anomalies, charnockite and pincinite patterns parallel but tend to be more incompatible element–enriched than ocean-island basalts (OIB). This and their juvenile character are compatible with derivation via crystal fractionation (plagioclase, apatite, and ilmenite) from an enriched mantle–derived (OIB-type) magma (Sun and McDonough, 1989).

Non-arc-like tonalite, characterized by relatively flat normalized patterns and Zr enrichment (Fig. 14B), constitutes ~20% of the Southwest Brook complex. Its mafic enclaves and amphibolitic dikes are generally also non-arc-like (Figs. 14C and 14D). Five non-arc-like tonalite samples yielded ε_{Nd}(T) values of –8.8, –7.0, –6.8, –6.1, and +4.8, where the sole positive value was from the most normalized element–depleted (Zr = 6 ppm) sample (Table 5). The mafic enclave contained in this depleted positive ε_{Nd}(T) sample also yielded a similar ε_{Nd}(T) value of +4.5. At four sites where mafic xenoliths hosted by this tonalite were sampled, both xenoliths and tonalite exhibited almost identical primitive mantle–normalized patterns and abundances (Fig. 14C), except the more enriched tonalite samples showed marked positive Zr anomalies (Zr = 495–600 ppm). The enclaves probably represent dismembered remnants of deep-level comagmatic mafic dikes, which were transported upward within their tonalite host. Higher-level comagmatic amphibolitic dikes are locally preserved in the tonalite bodies. Five out of six of these mafic dikes sampled in three Southwest Brook and two Hungry Mountain complex sites were non-arc-like (Figs. 14D, 14E, and 15). The one analyzed non-arc-like mafic dike sample yielded a ε_{Nd}(T) value of –4.7 (Table 5).

Modern arc-related igneous rocks almost invariably exhibit symmetrical negative Nb anomalies with respect to Th and La, which are nonconservative elements added to the mantle wedge as a result of fluid fluxing (cf. McDonald et al., 2000). This relationship even persists in immature arcs above intraoceanic subduction zones starved of sediments, suggesting that subducted oceanic crust by itself carries enough Th component to supply the magma source region through fluid-fluxing of the overlying mantle wedge. Thus, the mafic xenolith and comagmatic mafic dike data document that second-phase Notre Dame arc-like tonalitic magmatism coexisted with non-arc-like, possibly mainly asthenospheric mantle-derived, mafic magmatism (see following). The petrogenetic difficulty of forming high Nb/Th tonalites via subduction-related processes within arcs was examined in considerable detail, including quantitative modeling, by Whalen et al. (2003). This earlier work, plus the presence of fairly similar composition felsic rocks (see field for the 3 Ga high Nb/Th tonalites in Fig. 14B) within second-phase Notre Dame arc magmatism, along with their association with non-arc-like mafic rocks, argues against a typical arc-type setting for the second-phase magmatic flare-up. However, the high Nb/Th compositions still present a petrogenetic problem, not only regarding their high Nb/Th values, but also their ε_{Nd}(T) range (+4.8 to –8.8) and the positive Zr anomalies in tonalites. Since these rocks are not alkaline, high Zr (>500 ppm) in felsic compositions cannot be attributed to simple fractionation or partial-melting processes but rather requires the presence of excess (cumulate?) zircon. Based on the observations that the dated Southwest Brook complex tonalitic rocks contain abundant old xenocrystic zircon (see previous section) and the fact that high Nb/Th tonalite containing excess zircon also exhibits negative ε_{Nd}(T) values, we suggest that high Nb/Th tonalite petrogenesis involved selective input from a variably zircon-enriched, Th-poor, and/or Nb-undepleted, isotopically evolved component. Based on Pb-Nd-isotopic evidence presented in Whalen et al. (1996), this component was likely Notre Dame arc lower-continental crust. Heat for partial melting of this protolith could have been supplied by both asthenospheric and enriched mantle–derived mafic magmas, now represented by the tonalite-associated high Nb/Th mafic xenoliths and dikes and fractionated Cormacks Lake complex charnockite, respectively. It may not be coincidental that these high Nb/Th rocks occur proximal to or within the Cormacks Lake complex, the deepest exposed structural level within the Notre Dame arc, where they may not have risen far from their sources. If they had risen to higher crustal levels, assimilation of typical low Nb/Th (2.3–1.3; Rudnick and Fountain, 1995) middle to upper crust during their rise would likely have reduced their high Nb/Th signatures.

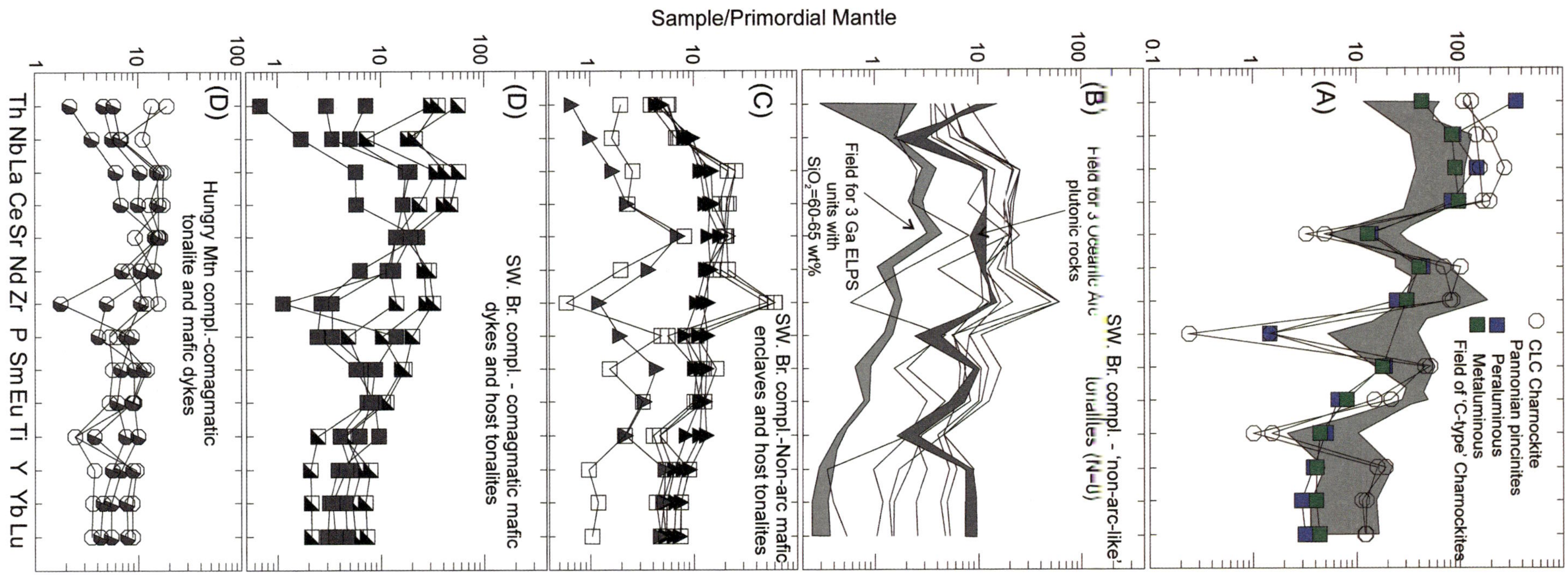

Figure 14. Primordial mantle–normalized extended-element plots for: (A) Cormacks Lake complex (CLC) charnockite samples along with the compositional field for eight C-type charnockite samples (data of Kilpatrick and Ellis, 1992) and two Pannonian pincinite xenolith glass samples from Huraiova et al. (2005). (B) Southwest Brook complex (SWBC) "non-arc-like" tonalites along with the felsic oceanic-arc plutonic field shown in Figure 13B and the field for high Nb/Th 3 Ga English Lake complex tonalites from Figure 6d of Whalen et al. (2003). (C) Southwest Brook complex non-arc-like mafic enclaves and host tonalite; the two lowest-abundance samples are from the same site and other similar pairs for three different sites. (D) Southwest Brook complex comagmatic mafic dikes (N = 3) and host tonalite. (E) Hungry Mountain complex comagmatic mafic dikes and host tonalite; two mafic dike samples are from the same site. Normalizing values are those of Sun and McDonough (1989).

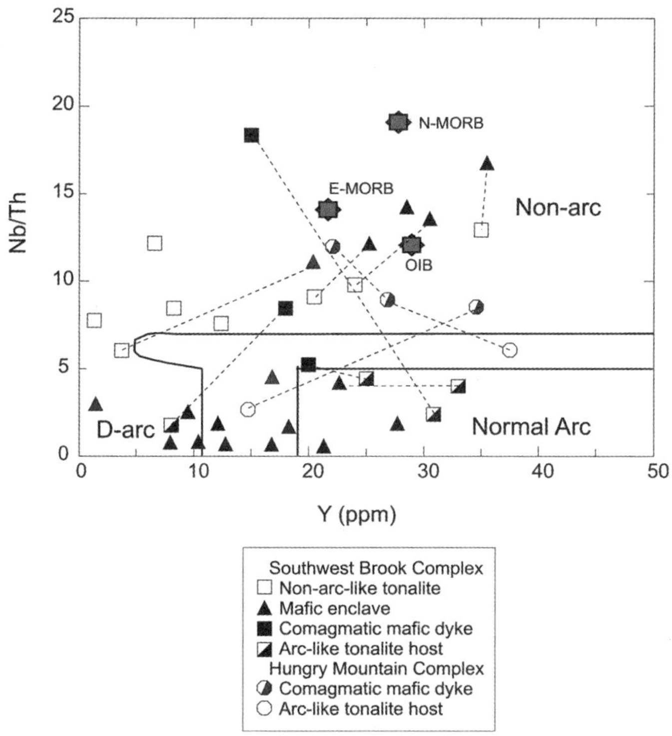

Figure 15. Plot of selected Notre Dame arc felsic and mafic plutonic rocks on Nb/Th-Y diagram (after Swinden et al., 1989). In this plot, mafic enclave or dike and host tonalite samples collected from the same site are joined by dashed lines. Abbreviations: N-, E-MORB and OIB—normal and enriched mid-ocean-ridge and ocean-island basalt; D-arc—depleted arc. Representative non-oceanic-arc basalt compositions are from Sun and McDonough (1989).

The ca. 462 Ma Portage Lake monzogabbro, which exhibits absarokite to shoshonite compositions, was emplaced in the wide, ductile east-vergent thrust zone that separates Dashwoods from the Annieopsquotch accretionary tract (Lissenberg and van Staal, 2006), immediately east of the Cormacks Lake complex (Fig. 8). This pluton has been interpreted as deriving from Dashwoods subduction-metasomatized subcontinental lithospheric mantle (Lissenberg et al., 2005b). In a normal arc setting, such high-K magmatism generally occurs to the rear of contemporaneous normal frontal arc magmatism. Its spatial proximity to, and temporal overlap with, both Southwest Brook complex arc-like and non-arc-like magmatism and Cormacks Lake complex charnockite emplacement further argues against second-stage Notre Dame arc magmatism occurring within a typical arc-type setting.

A New Tectonic Model for Second-Phase Notre Dame Arc Magmatism

There are four possible tectonic scenarios in which to generate TTD suite melts with "slab-melt-like" characteristics (high-Al rocks with elevated Sr/Y and La/Yb ratios). These include formation by partial melting of: (1) Grenvillian TTD-type basement rocks; (2) young and hot subducted oceanic slabs (e.g., Defant and Drummond, 1990); (3) dense lower subarc crust that had delaminated or was dripping into the underlying asthenosphere (e.g., Kay and Kay, 1993; Xu et al., 2002); and (4) hydrous mafic lower-arc crust and/or a basaltic underplate at the base of a thick arc infrastructure (>40 km; Atherton and Petford, 1993; Petford and Atherton, 1996; Saleeby et al., 2003).

The possibility that the high Sr/Y and La/Yb signatures for the bulk of our TTD-like plutonic rocks, like negative $\varepsilon_{Nd}(T)$ values, represent recycled characteristics from melted Grenvillian TTD-type basement rocks, a model we invoked to explain the petrogenesis of the first-phase Cape Ray granodiorite (see previous), is not considered to be a viable option for the second-phase tonalites. Experimental studies (e.g., Rutter and Wyllie, 1988; Skjerlie et al., 1996) have demonstrated that fluid-absent partial melting of tonalitic compositions under lower-to middle-crustal conditions generates liquids that, with increasing pressure and temperature, vary from granitic to tonalitic in composition, with the later requiring very high degrees of melting. Hence, the paucity of granites, sensu stricto, and granodiorites containing liquidus K-feldspar within second-phase Notre Dame arc magmatism is a strong argument against this model.

Melting of subducted oceanic crust is also rejected because the underthrusted slab mainly is composed of continental rather than oceanic crust during the waning stages of an arc-continent collision. In any case, even if this argument is dismissed, the subducted oceanic crust of the Humber seaway mainly formed during Early Cambrian seafloor spreading (Waldron and van Staal, 2001) and, hence, was probably too old and cold to generate slab melts.

Considering the tectonic constraints, partial melting of sinking or dripping delaminated arc crust (Kay and Kay, 1993; Xu et al., 2002) is not a realistic mechanism either. Constraints include: (1) the tonalite flare-up took place very shortly (<10 m.y.) after the start of full-scale collision, leaving little time to develop a gravitationally unstable root upon cooling; (2) generated felsic melts would be expected to inherit the arc signatures of their source; and (3) the lack of dispersion to high Mg#'s at low silica levels in second-phase Notre Dame arc plutonic rocks argues against any interaction with the surrounding mantle during ascent. These arguments leave garnet-stable partial melting of hydrous mafic crust present in the Notre Dame arc infrastructure, including potentially subcreted basalt, either as a result of tectonic (Saleeby et al., 2003) or magmatic (Petford and Atherton, 1996) underplating, as the most viable mechanism. However, an extra heat source is needed to melt the mafic rocks in a collisional (A-subduction) setting. Underthrusting of a cold continental slab cools the overlying arc lithosphere and destroys the mantle wedge (Billen and Gurnis, 2001; van Hunen et al., 2002). Estimates of minimum crustal thickness required to generate slab melt–like characteristics range from 27 km (Tate and Johnson, 2000) to >40 km (Petford and Atherton, 1996). Considering the collisional setting, our preferred model is that the tonalite flare-up was produced by slab break-off–induced partial melting of both thickened subarc mafic crust and structurally underplated MORB-like oceanic crust (Fig. 7B) derived from the downgoing slab (Humber seaway). In a slab break-off setting, the rapidly upwelling hot asthenosphere that replaces the broken-off slab can significantly increase the temperature in the overlying, thickened Notre Dame arc infrastructure and trigger melting (van de Zedde and Wortel, 2001). Whalen et al. (1997a) argued that Notre Dame arc magmas received input from various crustal- and mantle-derived components, including Precambrian mafic lower crust, as evidenced by inherited zircon and old Nd- and Pb- isotopic signatures. Melting of such a complex mixture of old and juvenile subarc crust could create the observed non-arc-like and arc-like tonalite melts. The tonalite associated non-arc-like mafic xenoliths and comagmatic mafic dikes described previously may represent remnants of mafic magmatism that fluxed lower-crustal melting. Mafic magmas were likely derived from ascending and partially melting asthenosphere shortly after slab break-off. During mantle upwelling, OIB-like mafic magmas may also have been generated, from which the Cormacks Lake complex charnockite magma could have formed, either directly by crystal fractionation, or indirectly by remelting of an OIB-like underplate. Similar deep-crustal melt xenoliths (termed pincinites) have been identified from the western Carpathians (Fig. 14A), where they are interpreted as basaltic intrusion-related crustal melts produced during slab break-off (Girbacea and Frisch, 1998). Immediately to the east (Fig. 7B), mafic magma-driven partial melting of preserved subarc lithospheric mantle could have generated the rare and aerially restricted shoshonitic plutonic rocks (Lissenberg et al., 2005b) that form but one component of this magmatically diverse flare-up.

In such a tectonic scenario, only locally preserved remnants of subarc mantle would be expected due to significant thinning and/or delamination of the lithospheric mantle caused by mechanical removal (tectonic erosion) during underthrusting of the Laurentian margin (shallow A-subduction). Evidence for uplift of the Notre Dame arc induced by this collision and accompanying thinning/delamination of its lithospheric mantle is provided by upper Caradoc (ca. 453 Ma) conglomerates and subaerial ignimbrites that unconformably overlie the Ordovician plutonic rocks of the Notre Dame arc near Cape Ray (Dubé et al., 1996). Another attraction of this model is that slab break-off would create space in the subarc mantle that could accommodate west-directed subduction of oceanic lithosphere beneath composite Laurentia, which would have occurred after the Late Ordovician arc-arc collision. This post-Taconic, Salinic-related subduction beneath Laurentia caused the third, Late Ordovician–Early Silurian phase (445–433 Ma) of the Notre Dame arc (van Staal et al., 1998; Whalen et al., 2006).

SUMMARY AND REGIONAL SIGNIFICANCE OF THE TACONIC EVENTS

The Taconic orogeny in Newfoundland consisted of three phases of accretion of oceanic and continental arc crust in the peri-Laurentian realm. The first phase (Taconic 1) took place during the Late Cambrian (500–493 Ma) and is represented by accretion of infant-arc crust of the Lushs Bight oceanic tract to the peri-Laurentian Dashwoods microcontinent. The Cambrian (505 ± 2 Ma) metamorphic sole to the ophiolitic Belvidere complex in northern Vermont (Laird et al., 1993) indicates that intraoceanic detachment and subduction had also started there during the Cambrian. The Belvidere complex and the ca. 504 Ma Mount Orford ophiolite in southern Quebec (Huot et al., 2002) are probably correlatives of the Lushs Bight oceanic tract, which suggests that Taconic 1 may have been an orogen-scale tectonic event rather than being confined to Newfoundland. The regional significance of this event awaits positive identification of Dashwoods-like crust beneath much of the Connecticut Valley–Gaspé trough in New England and southern Quebec.

Taconic 2 represents the accretion of the Notre Dame arc together with its Dashwoods infrastructure, and the Snooks Arm arc with its Baie Verte oceanic tract ophiolitic basement, to Laurentia. This collision was under way in the Arenig (ca. 475 Ma) in Newfoundland and was terminated at the end of the Middle Ordovician (464–459 Ma) by break-off of the oceanic slab (Humber seaway) attached to the Laurentian margin. Shallow underthrusting of the Humber margin beneath Dashwoods led to emplacement of foreland-directed thrust sheets in the Laurentian margin and strong tectonic shortening of the plutonic rocks of the Notre Dame arc. The latter deformation involved bivergent thrusting and folding, which was accompanied by amphibolite to granulite facies metamorphism. Slab break-off at the end of collision triggered partial melting of the arc's lower mafic crust, including underplated Humber seaway oceanic crust. Intrusion of the resulting plutons overlapped with the waning stages of collision-related deformation. Deformation continued for a short period after slab break-off either because the negative buoyancy of the unruptured part of the downgoing slab elsewhere was still driving convergence, i.e., tearing of the slab was diachronous along the length of the arc (e.g., Sperner et al., 2001), and/or it was related to compressive stresses generated in the upper plate after slab detachment (Price and Audley-Charles, 1987; Buiter et al., 2002). Middle Ordovician deformation and metamorphism related to Taconic 2 have been recognized everywhere in the northern Appalachians and represent the most important phase of the Taconic orogeny. Locally, however, the evidence of Taconic orogenesis is masked or severely overprinted by younger tectonism (cf. Cawood et al., 1994). The time span and characteristics of the main phase of the Grampian orogeny of the British Caledonides (Friedrich et al., 1999) indicate that it directly correlates with Taconic 2. Accretion of the continental and oceanic arcs to Laurentia was followed by Late Ordovician (455–450 Ma) accretion of peri-Laurentian arc rocks that were present outboard of Dashwoods to composite Laurentia (Zagorevski et al., 2006). This Taconic event (Taconic 3) was terminated by soft collision of the peri-Gondwanan Popelogan-Victoria arc with composite Laurentia (van Staal et al., 1998) and is reflected in late Caradoc–Ashgill uplift and deposition of Notre Dame arc detritus in a Late Ordovician–Early Silurian marine forearc basin (Badger Group in Fig. 1), which formed immediately east of the Red Indian Line (van Staal, 1994, 2005; McNicoll et al., 2001) during Salinic convergence between Laurentia and Ganderian elements. Accretion related to Taconic 3 led to protracted loading of the Laurentian margin into the Late Ordovician (Fig. 3). The Late Ordovician emplacement of the Taconic allochthon in the Hudson Valley of New York may have been related to this phase. Alternatively, its emplacement could have been due to final closure of a reentrant in the Humber seaway as depicted by stage 2 in Figure 6. The latter process was probably responsible for the Caradoc obduction (457–454 Ma) of the Mont Albert ophiolite in Gaspé (Pincivy et al., 2003).

APPENDIX 1. ANALYTICAL TECHNIQUES

Whole-rock major elements plus FeO and F (Table 5) were analyzed at the Geological Survey of Canada (GSC) as outlined in Whalen et al. (1997a). Trace-element data were obtained at the GSC using a combination of inductively coupled plasma–emission spectrometry (ICP-ES) and mass spectrometry (ICP-MS).

Sm-Nd-isotopic ratios were measured using the Nu Plasma multicollector ICP-MS at the GSC. $^{143}Nd/^{144}Nd$ isotopic ratios are reported relative to the value of 0.51186 for the LaJolla Nd standard (Table 5). Nine spiked and unspiked analyses of BCR-1 yielded a weighted average value of 0.512636 ± 0.000009 for $^{143}Nd/^{144}Nd$. The ε ^{143}Nd value was calculated for the igneous age relative to the accepted chondritic uniform reservoir using $^{143}Nd/^{144}Nd$ = 0.512636 and $^{147}Sm/^{144}Nd$ = 0.1966. Depleted-mantle model ages (T_{DM}) were calculated according to the model of DePaolo (1988).

SHRIMP II analyses were conducted at the GSC using analytical procedures described by Stern (1997), with standards and U-Pb calibration methods following Stern and Amelin (2003). Zircons from the samples were cast in 2.5-cm-diameter epoxy mounts (GSC mount #257 for sample PQB-SP-01017 [z7139] and GSC mount #278 for samples WXNF-48 [z2821] and WXNF-175 [z3405]) along with fragments of the GSC laboratory standard zircon (z6266, with $^{206}Pb/^{238}U$ age = 559 Ma).

The mid-sections of the zircons were exposed using 9, 6, and 1 μm diamond compound, and the internal features of the zircons were characterized with backscatter electrons (BSE) and cathodoluminescence (CL) utilizing a Cambridge Instruments scanning electron microscope (SEM). Mount surfaces were evaporatively coated with 10 nm of high-purity Au. Analyses were conducted using a $^{16}O^-$ primary beam, projected onto the zircons at 10 kV. The sputtered area used for analysis was ~25 μm in diameter with a beam current of ~13 nA for PQB-SP-01007a (z7138) and 5 nA for WXNF-48 (z2821) and WXNF-175 (z3405). The count rates of ten isotopes of Zr^+, U^+, Th^+, and Pb^+ in zircon were sequentially measured over six scans with a single electron multiplier and a pulse counting system with dead time of 35 ns. Offline data processing was accomplished using customized in-house software. The 1σ external errors of $^{206}Pb/^{238}U$ ratios reported in Table 1 incorporate a ±1.0% error for PQB-SP-01017 (z7139) on mount #257 and a ±1.5% error for samples WXNF-48 (z2821) and WXNF-175 (z3405) on mount #278 in calibrating the standard zircon (see Stern and Amelin, 2003). No fractionation correction was applied to the Pb-isotope data; common Pb correction utilized the measured $^{204}Pb/^{206}Pb$ and compositions modeled after Cumming and Richards (1975). Isoplot v. 2.49 (Ludwig, 2001) was used to generate concordia plots; data were plotted with errors at the 2σ level (Fig. 12).

U-Pb TIMS analytical methods utilized in this study are outlined in Parrish et al. (1987) and Davis et al. (1997). Heavy mineral concentrates were prepared using standard crushing, grinding, Wilfley™ table, and heavy liquid techniques. Mineral separates were sorted by magnetic susceptibility using a Frantz™ isodynamic separator. Multigrain zircon fractions analyzed were very strongly air abraded following the method of Krogh (1982). Multigrain fractions of titanite were also lightly air abraded. Treatment of analytical errors followed Roddick et al. (1987), with errors on the ages reported at the 2σ level (Table 2). Procedural blank values for this study had ranges of 0.1–0.2 pg U and 2–10 pg Pb for zircon analyses, 1–2 pg U and 7–11 pg Pb for titanite analyses, and 0.1 pg U and 1–2 pg Pb for monazite analyses. U-Pb TIMS concordia diagrams are presented in Figure 12.

A concordia age (Ludwig, 1998) was calculated for some of the samples presented in this paper. The concordia age incorporates errors on the decay constants and includes both an evaluation of concordance and an evaluation of equivalence of the data (how well the data fit the assumption that they are repeated measurements of the same point). The calculated concordia ages and errors quoted in the text are at 2σ with decay constant errors included.

APPENDIX 2. DETAILED GEOCHRONOLOGY SAMPLE DESCRIPTIONS

U-Pb SHRIMP and TIMS analyses for the geochronology samples described here are presented in Tables 1 and 2, respectively, and are plotted in concordia diagrams with errors at the 2σ level (Fig. 9). Further details on the U-Pb analytical techniques are presented in Appendix 1. Geochronology sample locations are indicated in Tables 1 and 2. New U-Pb ages and $\varepsilon_{Nd}(T)$ values for these samples/units are provided in Table 3 along with the identification of the samples for which geochemistry and isotopic data are available in Whalen et al. (1997a, 1997b). Geochemical and Nd analyses are presented in Tables 4 and 5 for those samples with data that were not previously published in Whalen et al. (1997a, 1997b) (see Table 3). For dated units, discussion of geochemical and Nd isotopic data has been integrated in the following with their geochronology descriptions. For some other units, a geochemical-isotopic section has been added at the end of this appendix

Cormacks Lake Complex

Felsic Tuffaceous Gneiss (PQB-SP-01007a; z7138)

A sample of bedded and graded gedrite-bearing, fine-grained gneiss was collected on the southwestern shore of Cormacks

Lake (Fig. 2). The gneiss mainly consists of quartz, feldspar, biotite, and gedrite. Primary structures, combined with the relative abundance of gedrite, suggest that its protolith was altered felsic tuffaceous sediment, either a tuff or a tuffite. Except for variable large ion lithophile element (LILE) concentrations, four samples from this unit exhibit extended element–normalized patterns that closely resemble those of Cormacks Lake complex tonalitic orthogneiss, low-Al tonalite, and type II rhyolites of Lesher et al. (1986) (Figs. 13B and 13C). If, as inferred from field observations, these rocks are of metavolcanic origin, they were likely derived from volcanism of this composition. Also consistent with such an origin are: (1) similarly positive $\varepsilon_{Nd}(T)$ values (+ 1.9) to Cormacks Lake complex tonalite (+0.8 to +2.2), and (2) low Cr (<1–12 ppm) and Ni (<1–2 ppm), identical to the tonalite but much lower than quartz-rich graywackes (Cr = 51–63; Ni = 19–20; Taylor and McLennan, 1985).

The sample contains a small amount of zircon, predominantly composed of stubby prismatic grains, and therefore U-Pb analysis was performed on SHRIMP. The SHRIMP data yielded two age clusters. The older population of data was collected from oscillatory-zoned euhedral zircon and is interpreted to be magmatic in origin. A concordia age for this population is calculated to be 489.3 ± 5.6 Ma (mean square of weighted deviates [MSWD] of concordance and equivalence = 1.5; n = 8) (Fig. 9A; Table 1). A younger population of data was obtained from zircon rims and grains with faint or no zoning. This younger population had a concordia age of 429 ± 4 Ma (MSWD of concordance and equivalence = 1.4; n = 8) and is interpreted to be metamorphic in origin (Table 1, not plotted). Metamorphic monazite and rutile with ages of ca. 432 Ma were also analyzed from this rock. The date of 489.3 ± 5.6 Ma is interpreted to be the crystallization age of this rock.

Tonalitic Orthogneiss (PQB-SP-01017; z7139)

Tonalitic orthogneiss was sampled northeast of Cormacks Lake. Its dominant mafic mineral is biotite and to a lesser extent hornblende. Minor clinopyroxene is partially to completely altered to hornblende. Based on major-element plutonic rock classification plots, most Cormacks Lake complex orthogneiss samples are of tonalite to trondhjemite composition, and, based on their Al_2O_3 content at 70% SiO_2, which is less than 15 wt%, plus other geochemical criteria such as low $(La/Yb)_{CN}$ values (Fig. 10), they belong to a low-Al tonalite suite (Barker, 1979). The orthogneiss samples show noticeable variability in K_2O content; in the K_2O-SiO_2 diagram of LeMaitre (1989) (not shown), eight out of eleven analyzed samples plot in or close to the low-K field, two are high-K, and WXNF-121 is high-medium-K. This K_2O variation likely is due to alkali-element mobility during deformation and metamorphism, an interpretation that is supported by the orthogneiss sample Al-index variation ranging from 0.9 to 1.2. Orthogneiss primitive mantle–normalized patterns (Fig. 13B) are also relatively flat, and most samples exhibit well-developed negative Nb, Sr, P, and Ti anomalies. Of note is the significant enrichment of two high-K samples in LILE relative to the other samples and also in Nb relative to Th and La, such that they lack significant negative Nb anomalies. This feature may reflect significant degrees of fractionation, rather than alteration. The $\varepsilon_{Nd}(T)$ values for the two low La/Yb samples are positive (+2.2 and +1.5). In addition, a third Nd sample collected from within this map unit, although geochemically distinct from most other samples, also exhibits a positive $\varepsilon_{Nd}(T)$ value of +0.8. As noted previously, orthogneiss-normalized patterns closely resemble felsic end members of oceanic-arc plutonic complexes and average low-Al tonalite (Fig. 13B).

The dated sample of foliated, coarse-grained, white to pale-pink weathering orthogneiss contains abundant small stubby euhedral prismatic zircon grains (Table 2). A linear regression, which has a forced lower intercept at the origin and includes all of the zircon fractions analyzed, has an upper intercept of 483.1 ± 1.4 Ma (MSWD = 0.03) (Fig. 9B). This date of 483.1 ± 1.4 Ma is interpreted to be the crystallization age of the rock. The orthogneiss also contains monazite, which has been dated at 433 ± 1 Ma and is interpreted to be metamorphic in origin (Table 2, not plotted).

Charnockite (VK-87-NF2; z1131)

Weakly foliated partially recrystallized and altered charnockite was sampled north of Cormacks Lake (Fig. 8). Its principal mineralogy consists of string mesoperthite, hornblende (partially to completely replacing augite), and biotite. Iddingsite pseudomorphs possibly after fayalite are also present. Orthopyroxene is absent in the analyzed sample but is present in nearby rocks of this unit (Currie et al., 1992). This plutonic rock is strongly enriched in LILE and HFSE, with pronounced negative Sr, P, and Ti anomalies (Fig. 14A), which together with its juvenile isotopic composition are distinctive features shared with the Silurian A-type granites in the Notre Dame subzone (Whalen et al., 1996). However, it lacks the pronounced negative Nb anomalies of Silurian A-type granites.

The charnockite contains abundant stubby to prismatic euhedral zircon with abundant fluid inclusions. A linear regression including fractions A1, B1, C1, and Z3 has a lower intercept near the origin (–265 ± 723 Ma) and an upper intercept of 459.4 $^{+3.0}/_{-1.5}$ Ma (MSWD = 0.2) (Table 2; Fig. 9C). Zircon fractions Z1 and Z2, which are composed of a large number of grains, are interpreted to have undergone Pb loss and were not included in the linear regression. The date of 459.4 + $^{3}/_{-1.5}$ Ma is interpreted to be the crystallization age of the charnockite. Titanite and allanite from this sample have been dated at ca. 430 Ma and are interpreted to be metamorphic in origin (not plotted).

Southwest Brook Complex

Tonalite (VL-02-21A; WX02NF-254; z7506)

A sample of well-foliated and lineated pinkish-white hornblende and biotite-bearing tonalite (Fig. 12E) was collected for U-Pb geochronology. The tonalite is cut by a pink monzogranite dike, which yielded an age of 431 ± 2 Ma (Whalen et al., 2006)

The Southwest Brook tonalite contains abundant high-quality zircons, which have a fairly low U content (15–135 ppm) and range in morphology from elongate to stubby prismatic grains (Table 2). Four multigrain zircon fractions were analyzed from this sample (Fig. 9D). A concordia age, using all four zircon analyses, was calculated to be 462.8 ± 0.9 Ma (MSWD of concordance and equivalence = 1.7), which is interpreted to be the crystallization age of the tonalite. The tonalite also contains abundant titanite, which has been dated at 429 ± 2 Ma and is interpreted to be metamorphic in origin (Table 2; Fig. 9D). This intermediate plutonic rock (SiO_2 = 61 wt%) exhibits a strongly fractionated REE pattern ($[La/Yb]_{CN}$ = 18), an extended element–normalized pattern with arc-type LILE enrichment and Nb depletion, and an old crustal $\varepsilon_{Nd}(T)$ value of –6.6.

Quartz Diorite (PQB-SP-02-79B; z7516)

Weakly foliated, white hornblende-bearing quartz diorite was collected along the shoreline of Cross Pond where it cuts intrafolially folded and tectonically banded Cormacks Lake tonalitic orthogneiss, described previously (Fig. 12C). This moderately felsic (SiO_2 = 64 wt%) plutonic rock exhibits a $(La/Yb)_{CN}$ value of 18, an extended element normalized–pattern with arc-type LILE enrichment and Nb depletion, and an old crustal $\varepsilon_{Nd}(T)$ value of –6.9.

The quartz diorite contains abundant euhedral zircon; six fractions were analyzed from the rock. Fraction Z3, which was composed of stubby prismatic zircon, is discordant and contains an older inherited component, which is probably of Proterozoic age (Table 2, not plotted). The other zircon fractions were composed of delicate elongate crystals, and a linear regression including all of these analyses has an upper intercept of 462.7 ± 1.8 Ma and a forced lower intercept at the origin (MSWD = 1.8) (Fig. 9E). This date of 462.7 ± 1.8 Ma is interpreted to be the crystallization age of the quartz diorite. The quartz diorite also contains abundant anhedral fragments of titanite that have been dated 432 ± 2 Ma and are interpreted to be metamorphic in origin (Table 2, not plotted).

Paragneiss (WXNF01-230)

A regionally extensive unit of biotite-rich gneiss, interlayered with sillimanite or kyanite schist, largely follows the western boundary of the southern part of the Notre Dame subzone (Fig. 1). This unit was sampled immediately east of Island Pond (Fig. 8), at least 2 km north of its contact with tonalite of the Southwest Brook complex and ~20 km to the north of the Cormacks Lake complex, with its pronounced high-temperature Silurian metamorphic overprint (Pehrsson et al., 2003b). Four analyzed Island Pond paragneiss samples exhibit moderately fractionated REE patterns with $(La/Yb)_{CN}$ = 7–13 and consistent LILE enrichment (not shown). As would be predicted for a flysch-type sedimentary rock, their sample patterns closely resemble the upper crust compositional estimate of Taylor and McLennan (1985), except that they lack a negative Ti anomaly. The dated Island Pond paragneiss sample exhibits a ε_{Nd}(466 Ma) value of –10.2, indicating sedimentary input from old continental crustal sources. The paragneiss contains abundant monazite, which was analyzed in order to constrain the age of regional metamorphism. As the metamorphic grade in the paragneiss does not increase toward its contact with the tonalite, here or elsewhere, monazite growth is unlikely to be due to contact metamorphism.

Three single subhedral yellow grains of monazite were analyzed from the paragneiss (Table 2). All three monazite analyses overlap each other and intersect concordia (Fig. 9F). A concordia age was calculated to be 466.5 ± 1.8 Ma (MSWD of concordance and equivalence = 2.3), and this is interpreted as the age of regional metamorphism. The paragneiss also contains abundant rutile, which has been dated at 430 ± 2 Ma.

Hungry Mountain and Cariboo Lakes Complexes

Hornblende Gabbro (WXNF-94; z3058)

A pegmatitic hornblende gabbro to diorite was sampled in a quarry southeast of Hinds Lake (Whalen et al., 1997a). Equivalents of this gabbro contain numerous mafic to ultramafic xenoliths that have been interpreted as remnants of the ca. 480 Ma (Dunning and Krogh, 1985) Annieopsquotch ophiolite belt (Whalen et al., 1997a). This gabbro (SiO2 = 49 wt%) exhibits a $(La/Yb)_{CN}$ value of 2.8, an extended element–normalized pattern with arc-type LILE enrichment and Nb depletion, and an old crustal $\varepsilon_{Nd}(T)$ value of –7.8.

The coarse-grained gabbro contains abundant fragments of high-quality zircon. Three analyzed fractions overlap each other and concordia (Fig. 9G; Table 2). A concordia age, utilizing the three analyses, was calculated to be 469.1 ± 1.3 Ma (MSWD of concordance and equivalence = 2.6), which is interpreted to be the crystallization age of the gabbro.

Hornblende Tonalite (WXNF-48; z2821)

A sample of medium-grained, equigranular tonalite was collected in Kings Brook where this tonalite and associated granodiorite intruded lithologically similar hornblende gabbro to dated sample WXNF-94. This felsic (SiO_2 = 66 wt%) plutonic rock exhibits a $(La/Yb)_{CN}$ value of 3.4, an extended element–normalized pattern with arc-type LILE enrichment and Nb depletion, and an old crustal $\varepsilon_{Nd}(T)$ value of –6.1.

The tonalite contains abundant euhedral zircon, ranging in morphology from stubby prismatic to elongate. Multigrain fractions previously analyzed by TIMS (not shown) produced discordant analyses, revealing the presence of inherited Pb components. In situ analyses of zircons from this sample were carried out on SHRIMP. Backscattered electron and cathodoluminescence imaging of the grains revealed that most of the zircons have fine oscillatory zoning, suggesting a magmatic origin. Some of the grains have small (10–30 μm) cores, which are most likely the cause of the inheritance in the multigrain TIMS analyses. A cluster of SHRIMP analyses (n = 15) overlaps each other and concordia (Fig. 9H; Table 1). A concordia age was calculated to be 463.1 ± 3.7 Ma (MSWD of concordance and equivalence = 1.4), which is interpreted to be the crystallization age of the tonalite. Analysis of small cores provided inherited ages of ca. 930

Ma and discordant Archean analyses with ages of ca. 2.6–2.7 Ga (Fig. 9H; Table 1).

Granodiorite (WXNF-63; z2823)

A sample of moderately foliated, medium-grained, rare pink K-feldspar porphyritic hornblende-biotite granodiorite was collected from the Lewaseechjeech Brook suite adjacent to the stream that drains Cariboo Lakes. Although contacts were not observed, on the basis of contrasting degrees of deformation, this granodiorite was considered to intrude units comprising the Cariboo Lake complex, such as dated orthogneiss sample WXNF-175 (see following). This intermediate (SiO_2 = 61 wt%) plutonic rock exhibits a $(La/Yb)_{CN}$ value of 8.4, an extended element–normalized pattern with arc-type LILE enrichment and Nb depletion, and an old crustal $\varepsilon_{Nd}(T)$ value of –6.8.

The granodiorite contains abundant stubby prismatic zircon with fluid inclusions (Table 2). Three fractions were analyzed from this rock. A linear regression of these analyses with a lower intercept anchored at the origin has an upper intercept of 468.2 ± 3.8 Ma (MSWD = 0.48), which is interpreted to be the crystallization age of the granodiorite (Fig. 9I).

Granodiorite (WXNF-175; z3405)

Foliated, white, medium- to coarse-grained, biotite-hornblende granodiorite orthogneiss was sampled from the Cariboo Lakes gneiss complex. On an outcrop scale, this unit exhibits compositional banding consisting of variations in mafic to felsic mineral concentrations. This compositional banding is folded, and the unit contains dismembered mafic dikes that exhibit arc tholeiite compositions (C.J. Lissenberg, unpublished data). Additional units comprising the Cariboo Lakes gneiss complex are migmatitic paragneiss and biotite-muscovite monzonitic orthogneiss (Whalen et al., 1997a). This felsic (SiO_2 = 71 wt%) plutonic rock exhibits a $(La/Yb)_{CN}$ value of 14.8, an extended element–normalized pattern with arc-type LILE enrichment and Nb depletion, and an old crustal $\varepsilon_{Nd}(T)$ value of –9.3.

The sample contains abundant euhedral zircon; however, previous attempts to date this sample by TIMS resulted in discordant analyses and revealed evidence of abundant inheritance in this rock. Therefore, in situ analyses of zircons from this sample were carried out on SHRIMP. Backscattered electron and cathodoluminescence imaging of the zircons revealed the presence of abundant cores ranging in size from 10 to 60 μm. Large euhedral rims on the cores all have fine oscillatory zoning, suggesting a magmatic origin. Analysis of these magmatic rims resulted in a cluster of SHRIMP analyses (n = 16) overlapping each other and concordia (Fig. 9J; Table 1). A concordia age was calculated to be 459.0 ± 3.4 Ma (MSWD of concordance and equivalence = 1.0), which is interpreted to be the crystallization age of the rock. Taking into account analytical errors, this granodiorite orthogneiss is a minimum of 2 m.y. younger than dated sample WXNF-63 (468.2 ± 3.8 Ma; see previous), which was thought to be the younger unit. As the contacts between them were not observed, it is possible that these units are structurally juxtaposed. Analysis of the cores provides a range of inherited ages in this rock: ca. 600 Ma, 850–1600 Ma, and discordant Archean analyses (Fig. 9J; Table 1).

Additional Geochemical Descriptions

Cormacks Lake Complex Amphibolites

Six Cormacks Lake complex amphibolite samples, interpreted as metamorphosed mafic volcanic rocks, were analyzed. Three samples (PQB-02–124, 140, and 144) exhibit highly flattened pillow textures, and the other three variably foliated amphibolites (PQB-01-SP007, 8A, and 8B) were interlayered with the dated 489 gedrite metavolcanic gneiss. Although these highly strained rocks do not show any evidence for pillow structures like samples 124–140, their association and interlayering with the layered gedrite gneiss and the presence locally of dolomite interbeds are consistent with a volcanic as opposed to an intrusive origin. On Th-Zr, (Hf)-Nb (Wood, 1980) and La-Y-Nb (Cabanis and Lecolle, 1989) tectonomagmatic discrimination diagrams (not shown), amphibolite samples plot in the calc-alkaline basalt and island-arc tholeiite fields, respectively. Four amphibolite sample primitive mantle–normalized patterns very closely resemble those of island-arc tholeiite and gabbro phases of an oceanic-arc plutonic complex (Fig. 13A). The other two sample patterns (PQB-01-SP007 and 02-S124) lack pronounced negative Sr anomalies and exhibit less well-developed negative Nb anomalies. Although the juvenile $\varepsilon_{Nd}(T)$ value of +4.2 obtained from sample PQB-01-SP007 is compatible with formation in a primitive arc environment, it does not rule out input from older crustal material, since the $\varepsilon_{Nd}(T)$ value of (DePaolo, 1981) depleted mantle is +7.1.

Cape Ray Igneous Complex

Whalen et al. (1997b) presented geochemical and Nd isotopic data from a ca. 488 Ma Cape Ray K-feldspar porphyritic monzonite to granodiorite sample (WXNF-32). This sample exhibits a $\varepsilon_{Nd}(T)$ value of –4.5 and a $(La/Yb)_{CN}$ value of 52. As this sample was collected from near the coast, rather than proximal to the U-Pb zircon sample site of Dubé et al. (1996), three additional archived feldspar porphyritic samples from closer to the dated exposure were geochemically analyzed. The most felsic (SiO_2 = 69 wt%) of these (VL-92-LH011) has a $(La/Yb)_{CN}$ value of 41 and exhibits an almost identical normalized pattern to monzonite sample WXNF-32 (Fig. 13D). Although the other two samples have lower $(La/Yb)_{CN}$ values of 7, they closely resemble other Cape Ray granodiorite samples in there overall elemental abundances and patterns (except HREE).

ACKNOWLEDGMENTS

This paper is a contribution to the 2000–2004 Red Indian Line Target Geoscience Initiative Project. Cees van Staal is grateful to Jean Bédard, Arjan Brem, Ken Currie, Lindsay Hall, and Hank Williams for numerous field trips and discussions on the geology of the Notre Dame subzone and/or petrology of the tonalites. Tracy Rushmer is thanked for discussions on amphibolite melting and generation of slab-melt–like rocks and companionship during the course of this research. Jim Hibbard, Neil Rogers, and John Percival edited an early version of the manuscript and, together with journal reviewers Peter Cawood and John Dewey, made thoughtful comments and suggestions that improved the manuscript. McNicoll thanks Julie Peressini, Diane Bellerive, and Carole Lafontaine for their assistance in the Geological Survey Geochronology laboratory. All field assistants are thanked for their enthusiastic help in the field, at times under abominable Newfie conditions. This is Geological Survey of Canada contribution 2005208.

REFERENCES CITED

Atherton, M., and Petford, N., 1993, Generation of sodium-rich magmas from newly underplated basaltic crust: Nature, v. 362, p. 144–146, doi: 10.1038/362144a0.

Barker, F., 1979, Trondhjemite: Definition, environment and hypotheses of origin, *in* Barker, F., ed., Trondhjemites, Dacites and Related Rocks: New York, Elsevier, Developments in Petrology Series, v. 6, p. 1–12.

Barker, F., and Arth, J.G., 1976, Generation of trondhjemite-tonalitic liquids and Archean bimodal trondhjemite-basalt suites: Geology, v. 4, p. 596–600, doi: 10.1130/0091-7613(1976)4<596:GOTLAA>2.0.CO;2.

Bédard, J., Lauziere, K., Tremblay, A., and Sangster, A., 1998, Evidence for forearc seafloor spreading from the Betts Cove ophiolite, Newfoundland: Oceanic crust of boninitic affinity: Tectonophysics, v. 284, p. 233–245, doi: 10.1016/S0040-1951(97)00182-0.

Bédard, J.H., Lauziere, K., Tremblay, A., Sangster, A., Douma, S.L., and Dec, T., 2000, Betts Cove Ophiolite and its Cover Rocks, Newfoundland: Geological Survey of Canada Bulletin 550, 76 p.

Billen, M.I., and Gurnis, M., 2001, A low viscosity wedge in subduction zones: Earth and Planetary Science Letters, v. 193, no. 1–2, p. 227–236, doi: 10.1016/S0012-821X(01)00482-4.

Bradley, D.C., 1989, Taconic plate kinematics as revealed by foredeep stratigraphy, Appalachian orogen: Tectonics, v. 8, p. 1037–1049.

Brem, A. G., Lin, S., and van Staal, C. R., 2003, Structural relationships south of Grand Lake Newfoundland, *in* Current Research 2002: Newfoundland Department of Mines and Energy, Geological Survey Report 2002-1, p. 1–13.

Buiter, S.J.H., Govers, R., and Wortel, M.J.R., 2002, Two-dimensional simulations of surface deformation caused by slab detachment: Tectonophysics, v. 354, p. 195–210, doi: 10.1016/S0040-1951(02)00336-0.

Cabanis, B., and Lecolle, M., 1989, The La/10-Y/15-Nb/8 diagram: A tool for distinguishing volcanic series and discovering crustal mixing and/or contamination: Comptes Rendus de L'Academie des Sciences, Serie 2, v. 309, no. 20, p. 2023–2029.

Calon, T.J., and Green, F.K., 1987, Preliminary results of a detailed structural analysis of the Buchans Mine area, *in* Kirkham, R.V., ed., Buchans Geology Newfoundland: Geological Survey of Canada Paper 86–24, p. 273–288.

Casey, J.F., and Dewey, J.F., 1984, Initiation of subduction zones along transform and accreting plate boundaries, triple-junction evolution, and forearc spreading centres—implications for ophiolitic geology and obduction, *in* Gass, I.G., Lippard, S.J., and Shelton, A.W. eds., Ophiolites and Oceanic Lithosphere: Geological Society of London Special Publication 13, p. 131–144.

Castonguay, S., Ruffet, G., Tremblay, A., and Feraud, G., 2001, Tectonometamorphic evolution of the southern Quebec Appalachians; $^{40}Ar/^{39}Ar$ evidence for Middle Ordovician crustal thickening and Silurian–Early Devonian exhumation of the internal Humber zone: Geological Society of America Bulletin, v. 113, p. 144–160, doi: 10.1130/0016-7606(2001)113<0144:TEOTSQ>2.0.CO;2.

Cawood, P.A., and Suhr, G., 1992, Generation and obduction of ophiolites; constraints for the Bay of Islands complex, western Newfoundland: Tectonics, v. 11, p. 884–897.

Cawood, P.A., and van Gool, J.A.M., 1998, Geology of the Corner Brook–Glover Island Region, Newfoundland: Geological Survey of Canada Bulletin 427, 96 p.

Cawood, P.A., Dunning, G.A., Lux, D., and van Gool, J.A.M., 1994, Timing of peak metamorphism and deformation along the Appalachian margin of Laurentia in Newfoundland: Silurian, not Ordovician: Geology, v. 22, p. 399–402, doi: 10.1130/0091-7613(1994)022<0399:TOPMAD>2.3.CO;2.

Cawood, P.A., van Gool, J.A.M., and Dunning, G.R., 1996, Geological development of eastern Humber and western Dunnage zones; Corner Brook–Glover Island region, Newfoundland: Canadian Journal of Earth Sciences, v. 33, p. 182–198.

Cawood, P.A., McCausland, P.J.A., and Dunning, G.R., 2001, Opening Iapetus: Constraints from the Laurentian margin in Newfoundland: Geological Society of America Bulletin, v. 113, p. 443–453, doi: 10.1130/0016-7606(2001)113<0443:OICFTL>2.0.CO;2.

Cumming, G.L., and Richards, J.R., 1975, Ore lead in a continuously changing Earth: Earth and Planetary Science Letters, v. 28, p. 155–171, doi: 10.1016/0012-821X(75)90223-X.

Currie, K.L., van Breemen, O., Hunt, P.A., and van Berkel, J.T., 1992, Age of high-grade gneisses south of Grand Lake, Newfoundland: Atlantic Geology, v. 28, p. 153–161.

Dallmeyer, R.D., and Williams, H., 1975, $^{40}Ar/^{39}Ar$ ages from the Bay of Islands metamorphic aureole: Their bearing on the timing of Ordovician ophiolite obduction: Canadian Journal of Earth Sciences, v. 12, p. 1685–1690.

Davis, W.J., McNicoll, V.J., Bellerive, D.R., Santowski, K., and Scott, D.J., 1997, Modified chemical procedures for the extraction and purification of uranium from titanite, allanite and rutile in the Geochronology Laboratory, Geological Survey of Canada: Radiogenic Age and Isotopic Studies Report 10: Geological Survey Canada Current Research, v. 1997-F, p. 33–35.

Defant, M.J., and Drummond, M., 1990, Derivation of some modern arc magmas by melting of young subducted lithosphere: Nature, v. 347, p. 662–665, doi: 10.1038/347662a0.

DePaolo, D.J., 1981, Trace element and isotopic effects of combined wallrock assimilation and fractional crystallization: Earth and Planetary Letters, v. 52, p. 189–202.

DePaolo, D. J., 1988. Neodymium Isotope geochemistry: An introduction: Springer, Berlin, Heidelberg, New York, 187 p.

Dewey, J.F., 2002, Transtension in arcs and orogens: International Geology Reviews, v. 44, p. 402–439.

Dewey, J.F., and Shackleton, R.M., 1984, A model for the evolution of the Grampian tract in the early Caledonides and Appalachians: Nature, v. 312, no. 5990, p. 115–121, doi: 10.1038/312115a0.

Drummond, M.S., and Defant, M.J., 1990, A model for trondhjemite-tonalite-dacite genesis and crustal growth via slab melting: Archean to modern comparisons: Journal of Geophysical Research, v. 95, p. 21,503–21,521.

Drummond, M.S., Defant, M.J., and Kepezhinskas, P.K., 1996, Petrogenesis of slab-derived trondhjemite-tonalite-dacite/adakite magmas: Transactions of the Royal Society of Edinburgh, Earth Sciences, v. 87, p. 205–215.

Dubé, B., Dunning, G.R., Lauziere, K., and Roddick, J.C., 1996, New insights into the Appalachian orogen from geology and geochronology along the Cape Ray fault zone, southwest Newfoundland: Geological Society of America Bulletin, v. 108, p. 101–116, doi: 10.1130/0016-7606(1996)108<0101:NIITAO>2.3.CO;2.

Dunning, G.R., and Krogh, T.E., 1985, Geochronology of ophiolites of the Newfoundland Appalachians: Canadian Journal of Earth Sciences, v. 22, p. 1659–1670.

Dunning, G.R., Wilton, D.H.C., and Herd, R.K., 1989, Geology, geochemistry and geochronology of a Taconic batholith, southwestern Newfoundland: Transactions of the Royal Society of Edinburgh, Earth Sciences, v. 80, p. 159–168.

Elliott, C.G., Dunning, G.R., and Williams, P.F., 1991, New U/Pb zircon age constraints on the timing of deformation in north-central Newfoundland and implications for early Paleozoic Appalachian orogenesis: Geological Society of America Bulletin, v. 103, p. 125–135, doi: 10.1130/0016-7606(1991)103<0125:NUPZAC>2.3.CO;2.

English, J.M., Johnston, S.T., and Wang, K., 2003, Thermal modelling of the Laramide orogeny: Testing the flat-slab subduction hypothesis: Earth and Planetary Science Letters, v. 214, p. 619–632, doi: 10.1016/S0012-821X(03)00399-6.

Friedrich, A.M., Bowring, S.A., Martin, M.W., and Hodges, K.V., 1999, Short-lived continental magmatic arc at Connemara, western Irish Caledonides: Implications for the age of the Grampian orogeny: Geology, v. 27, p. 27–30, doi: 10.1130/0091-7613(1999)027<0027:SLCMAA>2.3.CO;2.

Fryer, B.J., Kerr, A., Jenner, G.A., and Longstaffe, F.J., 1992, Probing the crust with plutons: Regional isotopic geochemistry of granitoid intrusions across insular Newfoundland: Current Research: Government of Newfoundland and Labrador Geological Survey Report 92–1, p. 119–139.

Gerbi, C.C., Johnson, S.E., and Aleinikoff, J.N., 2006, Origin and orogenic role of the Chain Lakes Massif, Maine and Quebec: Canadian Journal of Earth Sciences, v. 43, p. 339–366, doi: 10.1139/E05-112.

Girbacea, R., and Frisch, W., 1998, Slab in the wrong place: Lower lithospheric mantle delamination in the last stage of the Eastern Carpathian subduction retreat: Geology, v. 26, p. 611–614, doi: 10.1130/0091-7613(1998)026<0611:SITWPL>2.3.CO;2.

Hacker, B.R., Mosenfelder, J.L., and Gnos, E., 1996, Rapid emplacement of the Oman ophiolite: Thermal and geochronologic constraints: Tectonics, v. 15, p. 1230–1247, doi: 10.1029/96TC01973.

Hall, C., Gurnis, M., Sdrolias, M., Lavier, L.L., and Müller, R.D., 2003, Catastrophic initiation of subduction following forced convergence across fracture zones: Earth and Planetary Science Letters, v. 212, p. 15–30, doi: 10.1016/S0012-821X(03)00242-5.

Hall, L., and van Staal, C.R., 1999, Geology of Southern End of Long Range Mountains (Dashwoods Subzone), Newfoundland: Geological Survey of Canada Open-File 3727, scale 1:50,000.

Harris, R.A., 1992, Peri-collisional extension and the formation of Oman type ophiolites in the Banda arc and Brooks Range, *in* Parson, L.M., Murton, B.J., and Browning, P., eds., Ophiolites and their Modern Oceanic Analogues: Geological Society, London, Special Publication 60, p. 301–325.

Heaman, L.M., Erdmer, P., and Owen, J.V., 2002, U-Pb geochronologic constraints on the crustal evolution of the Long Range Inlier, Newfoundland: Canadian Journal of Earth Sciences, v. 39, p. 845–865, doi: 10.1139/e02-015.

Herd, R.K., and Dunning, G.R., 1979, Geology of Puddle Pond map-area, southwestern Newfoundland: Geological survey of Canada, Paper 79-1A, p. 305–310.

Hibbard, J., 1983, Geology of the Baie Verte Peninsula, Newfoundland: Mineral Development Division Department of Mines and Energy, Government of Newfoundland and Labrador Memoir 2, 279 p.

Hiscott, R.N., 1978, Provenance of Ordovician deep-water sandstones, Tourelle Formation, Quebec, and implications for initiation of the Taconic orogeny: Canadian Journal of Earth Sciences, v. 15, p. 1579–1597.

Huang, C.Y., Yuan, P.B., Lin, C.W., Wang, T.K., and Chang, C.P., 2000, Geodynamic processes of Taiwan arc-continent collision and comparison with analogs in Timor, Papua New Guinea, Urals and Corsica: Tectonophysics, v. 325, p. 1–21, doi: 10.1016/S0040-1951(00)00128-1.

Huot, F., Hebert, R., and Turcotte, B., 2002, A multistage magmatic history for the genesis of the Orford ophiolite (Quebec, Canada): A study of the Mont Chagnon Massif: Canadian Journal of Earth Sciences, v. 39, p. 1201–1217, doi: 10.1139/e02-030.

Huraiova, M., Dubessy, J., Konecny, P., Simon, K., Kral, J., Zielinski, G., Lipka, J., and Hurai, V., 2005, Glassy orthopyroxene granodiorites of the Pannonian Basin: Tracers of ultra-high temperature deep-crustal anatexis triggered by Tertiary basaltic volcanism: Contributions to Mineralogy and Petrology, v. 148, p. 615–633, doi: 10.1007/s00410-004-0625-8.

Jamieson, R.A., 1988, Metamorphic P-T-t data from western Newfoundland and Cape Breton—Implications for Taconian and Acadian tectonics [abs.]: Geological Society of Canada, Program with Abstracts, v. 13, p. A60.

Jenner, G.A., Dunning, G.R., Malpas, J., Brown, M., and Brace, T., 1991, Bay of Islands and Little Port complexes, revisited: Age, geochemical and isotopic evidence confirm suprasubduction-zone origin: Canadian Journal of Earth Sciences, v. 28, p. 1635–1652.

Johnson, R.J.E., van der Pluijm, B.A., and van der Voo, R., 1991, Paleomagnetism of the Moreton's Harbour Group, northeastern Newfoundland Appalachians: Evidence for an Early Ordovician island arc near the Laurentian margin of Iapetus: Journal of Geophysical Research, ser. B, Solid Earth and Planets, v. 96, p. 11,689–11,701.

Karabinos, P., Samson, S.D., Hepburn, J.C., and Stoll, H.M., 1998, Taconian orogeny in the New England Appalachians: Collision between Laurentia and the Shelburne Falls arc: Geology, v. 26, p. 215–218, doi: 10.1130/0091-7613(1998)026<0215:TOITNE>2.3.CO;2.

Karson, J., and Dewey, J.F., 1978, Coastal complex, western Newfoundland: An Early Ordovician oceanic fracture zone: Geological Society of America Bulletin, v. 89, p. 1037–1049, doi: 10.1130/0016-7606(1978)89<1037:CCWNAE>2.0.CO;2.

Kay, R.W., and Kay, S.M., 1993, Delamination and delamination magmatism: Tectonophysics, v. 219, p. 177–189, doi: 10.1016/0040-1951(93)90295-U.

Kean, B.F., Evans, D.T.W., and Jenner, G.A., 1995, Geology and Mineralization of the Lushs Bight Group: Department of Mines and Energy Mineral Development Division, Government of Newfoundland and Labrador Report 95–2, 204 p.

Kilpatrick, J.A., and Ellis, D.J., 1992, C-type magmas: Igneous charnockites and their extrusive equivalents: Transactions of the Royal Society Edinburgh, Earth Sciences, v. 83, p. 155–164.

Knight, I., James, N.P., and Lane, T.E., 1991, The Ordovician St. George unconformity, northern Appalachians: The relationship of plate convergence at the St. Lawrence Promontory to the Sauk/Tippecanoe Sequence boundary: Geological Society of America Bulletin, v. 103, p. 1200–1225, doi: 10.1130/0016-7606(1991)103<1200:TOSGUN>2.3.CO;2.

Krogh, T. E., 1982, Improved accuracy of U-Pb ages by creation of more concordant systems using an air abrasion technique: Geochimica Cosmochimica Acta, v. 46, p. 637–649.

Kurth, M., Sassen, A., Suhr, G., and Mezger, K., 1998, Precise ages and isotopic constraints for the Lewis Hills (Bay of Islands ophiolite): Preservation of an arc-spreading ridge intersection: Geology, v. 26, p. 1127–1130, doi: 10.1130/0091-7613(1998)026<1127:PAAICF>2.3.CO;2.

Laird, J., Trzcienski, W.E., and Bothner, W.A., 1993, High-pressure, Taconian, and subsequent polymetamorphism of southern Quebec and northern Vermont, *in* Cheney, J.T., and Hepburn, J.C., eds., Field Trip Guidebook for the Northeastern United States: Boulder, Colorado, Geological Society of America, p. 1–23.

Le Maitre, R. W., 1989. A classification of igneous rocks and glossary of terms: London, Blackwell, 193 p.

Lesher, C.M., Goodwin, A.M., Campbell, I.H., and Gorton, M.P., 1986, Trace-element geochemistry of ore-associated and barren, felsic metavolcanic rocks in the Superior Province: Canada: Canadian Journal of Earth Sciences, v. 23, p. 222–237.

Lissenberg, C.J., and van Staal, C.R., 2006, Feedback between deformation and magmatism in the Lloyd's River fault zone, central Newfoundland: An example of episodic fault reactivation in an accretionary setting: Tectonics, v. 25, TC4004, 18 p.

Lissenberg, C.J., van Staal, C.R., Bédard, J.H., and Zagorevski, A., 2005a, Geochemical constraints on the origin of the Annieopsquotch ophiolite belt, southwest Newfoundland: Geological Society of America Bulletin, v. 117, p. 1413–1426, doi: 10.1130/B25731.1.

Lissenberg, C.J., Zagorevski, A., McNicoll, V.J., van Staal, C.R., and Whalen, J.B., 2005b, Assembly of the Annieopsquotch accretionary tract, southwest Newfoundland: Age and geodynamic constraints from syn-kinematic intrusions: The Journal of Geology, v. 113, p. 553–570, doi: 10.1086/431909.

Ludwig, K.R., 1998, On the treatment of concordant uranium-lead ages: Geochimica et Cosmochimica Acta, v. 62, p. 665–676, doi: 10.1016/S0016-7037(98)00059-3.

Ludwig, K.R., 2001, User's Manual for Isoplot/Ex rev. 2.49: A Geochronological Toolkit for Microsoft Excel: Berkeley Geochronology Center Special Publication 1a, 55 p.

Martin, H., 1987, Genesis of Archean trondhjemites, tonalites, and granodiorites from eastern Finland: Major and trace element geochemistry: Journal of Petrology, v. 28, p. 921–953.

MacDonald R., Hawkesworth, C.J., and Heath, E., 2000, The lesser Antilles volcanic chain: a study in arc magmatism: Earth Science Reviews, v. 49, p. 1–76.

McNicoll, V., van Staal, C.R., and Waldron, J.W.F., 2001, Accretionary history of the northern Appalachians: SHRIMP study of Ordovician-Silurian syntectonic sediments in the Canadian Appalachians, *in* Geological Association of Canada/Mineralogical Association of Canada Annual Meeting Abstracts: St. John's, Newfoundland, p. 100.

Mueller, S., and Phillips, R.J., 1991, On the initiation of subduction: Journal of Geophysical Research, ser. B, Solid Earth and Planets, v. 96, no. 2, p. 651–665.

Müller, R.D., Gaina, C., Roest, W.R., and Hansen, D.L., 2001, A recipe for microcontinent formation: Geology, v. 29, p. 203–206, doi: 10.1130/0091-7613(2001)029<0203:ARFMF>2.0.CO;2.

Nutman, A.P., Bennett, C., Friend, C.R.L., and Norman, M.D., 1999, Meta-igneous (non-gneissic) tonalites and quartz-diorites from an extensive ca. 3800 Ma terrain south of the Isua supracrustal belt, southern West Greenland: Constraints on early crust formation: Contributions to Mineralogy and Petrology, v. 137, p. 364–388, doi: 10.1007/s004100050556.

Owen, J.V., and Greenough, J.D., 2000, Petrology of the Cormacks Lake complex, Newfoundland; decompressional reaction relations in cordierite + orthoamphibole-bearing gneisses and associated rocks: Mineralogical Magazine, v. 64, p. 711–724, doi: 10.1180/002646100549562.

Parrish, R.R., Roddick, J.C., Loveridge, W.D., and Sullivan, R.W., 1987, Uranium-lead analytical techniques at the Geochronology Laboratory, Geological Survey of Canada: Radiogenic age and isotope studies, Report 1: Geological Survey Canada Paper 87–2, p. 3–7.

Pavlides, L., Boucot, A.J., and Skidmore, W.B., 1968, Stratigraphic evidence for the Taconic orogeny in the northern Appalachians, *in* Zen, E-an, White, W.S., Hardley, J.B., and Thompson, J.B., eds., Studies of Appalachian Geology: New York, Northern and Maritime Interscience Publishers, p. 61–82.

Pehrsson, S., van Staal, C.R., Herd, R.K., and McNicoll, V., 2003a, The Cormacks Lake complex, Dashwoods subzone: A window into the deeper levels of the Notre Dame arc, in Current Research 2002: Newfoundland Department of Mines and Energy, Geological Survey Report 2002–1, p. 115–125.

Pehrsson, S., van Staal, C.R., McNicoll, V., and Whalen, J., 2003b, Tectonic evolution of the Dashwoods microcontinent and its accretion to the Laurentian margin: Geological Society of America, Abstracts with Programs, v. 35, no. 3, p. 98.

Petford, N., and Atherton, M., 1996, Na-rich partial melts from newly underplated basaltic crust: The Cordillera Blanca batholith, Peru: Journal of Petrology, v. 37, p. 1491–1521, doi: 10.1093/petrology/37.6.1491.

Pincivy, A., Malo, M., Ruffet, G., Tremblay, A., and Sacks, P.E., 2003, Regional metamorphism of the Appalachian Humber zone of Gaspé Peninsula; ^{40}Ar/^{39}Ar evidence for crustal thickening during the Taconian orogeny: Canadian Journal of Earth Sciences, v. 40, p. 301–315, doi: 10.1139/e02-076.

Pinet, N., and Tremblay, A., 1995, Tectonic evolution of the Quebec-Maine Appalachians: From oceanic spreading to obduction and collision in the northern Appalachians: American Journal of Science, v. 295, p. 173–200.

Price, N.J., and Audley-Charles, M.G., 1987, Tectonic collision processes after plate rupture: Tectonophysics, v. 140, p. 121–129, doi: 10.1016/0040-1951(87)90224-1.

Ramezani, J., Dunning, G.R., and Wilson, M.R., 2000, Geologic setting, geochemistry of alteration, and U-Pb age of hydrothermal zircon from the Silurian Stog'er Tight gold prospect, Newfoundland Appalachians, Canada: Exploration and Mining Geology, v. 9, p. 171–188, doi: 10.2113/0090171.

Rapp, R.P., and Watson, E.B., 1995, Dehydration melting of metabasalt at 8–32 kbar: Implications for continental growth and crust-mantle recycling: Journal of Petrology, v. 36, p. 891–931.

Ratcliffe, N.M., Harris, A.G., and Walsh, G.J., 1999, Tectonic and regional metamorphic implications of the discovery of Middle Ordovician conodonts in cover rocks east of Green Mountain massif, Vermont: Canadian Journal of Earth Sciences, v. 36, p. 371–382, doi: 10.1139/cjes-36-3-371.

Rautenschlein, M., Jenner, G.A., Hertogen, J., Hofmann, A.W., Kerrich, R., Schmincke, H.U., and White, W.M., 1985, Isotopic and trace element composition of volcanic glasses from the Akaki Canyon, Cyprus: Implications for the origin of the Troodos Ophiolite: Earth and Planetary Science Letters, v. 75, p. 369–383, doi: 10.1016/0012-821X(85)90180-3.

Roddick, J.C., Loveridge, W.D., and Parrish, R.R., 1987, Precise U-Pb dating of zircon at the sub-nanogram Pb level: Chemical Geology, v. 66, p. 111–121.

Rodgers, J., 1971, The Taconic orogeny: Geological Society of America Bulletin, v. 82, p. 1141–1177, doi: 10.1130/0016-7606(1971)82[1141:TTO]2.0.CO;2.

Rowley, D.B., and Kidd, W.S.F., 1981, Stratigraphic relationships and detrital composition of the medial Ordovician flysch of western New England: Implications for the tectonic evolution of the Taconic orogeny: The Journal of Geology, v. 89, p. 199–218.

Rudnick, R.L., and Fountain, D.M., 1995, Nature and composition of the continental crust: A lower crustal perspective: Reviews of Geophysics, v. 33, p. 267–309, doi: 10.1029/95RG01302.

Rushmer, T., 1993, Experimental high-pressure granulites: Some applications to natural mafic xenoliths suites and Archean granulite terranes: Geology, v. 21, p. 411–414, doi: 10.1130/0091-7613(1993)021<0411:EHPGSA>2.3.CO;2.

Rutter, J.M., and Wyllie, P., 1988, Melting of vapour-absent tonalite at 10 kbar to simulate dehydration-melting in the deep crust: Nature, v. 331, p. 159–160, doi: 10.1038/331159a0.

Saleeby, J.B., Ducea, M.N., and Clemens, K.D., 2003, Production and loss of high-density batholithic root, southern Sierra Nevada, California: Tectonics, v. 22, p. 3-1–3-24.

Şengör, A.M.C., 1990, Plate tectonics and orogenic research after 25 years: A Tethyan perspective: Earth-Science Reviews, v. 27, no. 1–2, p. 1–201, doi: 10.1016/0012-8252(90)90002-D.

Skjerlie, K.P., and Johnson, A.D., 1996, Vapour-absent melting from 10 to 20 kbar of crustal rocks that contain multiple hydrous phases: Implications for anatexis in the deep to very deep continental crust and active continental margins: Journal of Petrology, v. 37, p. 661–691, doi: 10.1093/petrology/37.3.661.

Sperner, B., Lorenz, F., Bonjer, K., Hettel, S., and Wenzel, F., 2001, Slab break-off—Abrupt cut or gradual detachment? New insights from the Vrancea region (SE Carpathians, Romania): Terra Nova, v. 13, p. 172–179, doi: 10.1046/j.1365-3121.2001.00335.x.

Stanley, R.S., and Ratcliffe, N.M., 1985, Tectonic synthesis of the Taconian orogeny in western New England: Geological Society of America Bulletin, v. 96, p. 1227–1250, doi: 10.1130/0016-7606(1985)96<1227:TSOTTO>2.0.CO;2.

Stern, R.A., 1997, The GSC sensitive high resolution ion microprobe (SHRIMP): Analytical techniques of zircon U-Pb-Th age determinations and performance evaluation: Radiogenic age and isotope studies, Report 10: Geological Survey Canada Current Research, v. 1997-F, p. 1–31.

Stern, R.A., and Amelin, Y., 2003, Assessment of errors in SIMS zircon U-Pb geochronology using a natural zircon standard and NIST SRM 610 glass: Chemical Geology, v. 197, p. 111–146, doi: 10.1016/S0009-2541(02)00320-0.

Stevens, R.K., 1970, Cambro-Ordovician flysch sedimentation and tectonics in west Newfoundland and their possible bearing on a proto–Atlantic Ocean, *in* Lajoie, P., ed., Flysch Sedimentology in North America: Geological Association of Canada Special Paper 7, p. 165–177.

Suhr, G., and Cawood, P.A., 2001, Southeastern Lewis Hills (Bay of Islands ophiolite): Geology of a deeply eroded, inside corner, ridge transform intersection: Geological Society of America Bulletin, v. 113, p. 1025–1038, doi: 10.1130/0016-7606(2001)113<1025:SLHBOI>2.0.CO;2.

Suhr, G., and Edwards, S.J., 2000, Contrasting mantle sequences exposed in the Lewis Hills massif: Evidence for the early, arc-related history of the Bay of Islands ophiolite, *in* Dilek, Y., Moores, E.M., Elthon, D., and Nicolas, A., eds., Ophiolites and Oceanic Crust: New Insights from Field Studies and the Ocean Drilling Program: Geological Society of America Special Paper 349, p. 433–442.

Sun, S.S., and McDonough, W.F., 1989, Chemical and isotopic systematics of oceanic basalt: Implications for mantle composition and processes, *in* Saunders, A.D., and Norry, M.J., eds., Magmatism in the Ocean Basins: Geological Society of London Special Publication 42, p. 313–345.

Swinden, H.S., 1996, Geochemistry of volcanic rocks in the Moreton's Harbour–Twillingate area, Notre Dame Bay: Current Research: Government of Newfoundland and Labrador Geological Survey Report 96–1, p. 207–226.

Swinden, H.S., Jenner, G.A., Kean, B.F., and Evans, D.T.W., 1989, Volcanic rock geochemistry as a guide for massive sulphide exploration in central Newfoundland: Current Research: Newfoundland Department of Energy and Mines, Geological Survey Report 89–1, p. 201–219.

Swinden, H.S., Jenner, G.A., and Szybinski, Z.A., 1997, Magmatic and tectonic evolution of the Cambrian-Ordovician Laurentian margin of Iapetus: Geochemical and isotopic constraints from the Notre Dame subzone, Newfoundland, *in* Sinha, K., Whalen, J.B., and Hogan, J.P., eds., The Nature of Magmatism in the Appalachian Orogen: Geological Society of America Memoir 191, p. 337–365.

Szybinski, Z.A., 1995, Paleotectonic and structural setting of the western Notre Dame Bay area, Newfoundland Appalachians [PhD thesis]: St. John's, Newfoundland, Memorial University of Newfoundland, 690 p.

Szybinski, Z.A., House, S., and Jenner, G.A., 1995, Stratigraphy and structure of the Glover Group, Grand Lake–Little Grand Lake area, Newfoundland: Current Research: Geological Survey of Canada, v. 1995-E, p. 245–251.

Tate, M.C., and Johnson, S.E., 2000, Subvolcanic and deep-crustal tonalite genesis beneath the Mexican Peninsular Ranges: The Journal of Geology, v. 108, p. 721–728, doi: 10.1086/317948.

Taylor, H.P., Jr., 1968, The oxygen isotope geochemistry of igneous rocks: Contributions to Mineralogy and Petrology, v. 19, p. 1–71, doi: 10.1007/BF00371729.

Taylor, H.P., Jr., 1978, Oxygen- and hydrogen-isotope studies of plutonic rocks: Earth and Planetary Science Letters, v. 38, p. 177–210, doi: 10.1016/0012-821X(78)90131-0.

Taylor, S.R., and McLennan, S.M., 1985, The Continental Crust: Its Composition and Evolution: Oxford, Blackwell, 312 p.

Thurlow, J.G., 1981, The Buchans Group: Its stratigraphic and structural setting, *in* Swanson, E.A., Strong, D.F., and Thurlow, J.G., eds., The Buchans Orebodies: Fifty Years of Geology and Mining: Geological Association of Canada Special Paper 22, p. 79–89.

Torsvik, T.H., Smethurst, M.A., Meert, J.G., van der Voo, R., McKerrow, W.S., Brasier, M.D., Sturt, B.A., and Walderhaug, H.J., 1996, Continental break-up and collision *in* the Neoproterozoic and Paleozoic—A tale of Baltica and Laurentia: Earth-Science Reviews, v. 40, p. 229–258, doi: 10.1016/0012-8252(96)00008-6.

Tucker, R.D., and Gower, C.F., 1994, A U-Pb geochronological framework for the Pinware terrane, Grenville province, southeast Labrador: The Journal of Geology, v. 102, p. 67–78.

Tucker, R. D. and Robinson, P., 1990. Age and setting of the Bronson Hill magmatic arc: A re-evaluation based on U-Pb zircon ages in southern New England: Geological Society of America Bulletin, v. 102, p. 1404–1419.

Valverde-Vaquero, P., van Staal, C.R., McNicoll, V., and Dunning, G., 2006, Middle Ordovician magmatism and metamorphism along the Gander margin in Central Newfoundland: Geological Society [London] Journal, v. 163, p. 347–362.

van der Velden, A.J., van Staal, C.R., and Cook, F.A., 2004, Crustal structure, fossil subduction and the tectonic evolution of the Newfoundland Appalachians: Evidence from a reprocessed seismic reflection survey: Geological Society of America Bulletin, v. 116, p. 1485–1498, doi: 10.1130/B25518.1.

van de Zedde, D.M.A., and Wortel, M.J.R., 2001, Shallow slab detachment as a transient source of heat at midlithospheric depths: Tectonics, v. 20, p. 868–882, doi: 10.1029/2001TC900018.

van Hunen, J., van der Berg, A.P., and Vlaar, N.J., 2002, On the role of subducting oceanic plateaus in the development of shallow flat subduction: Tectonophysics, v. 352, p. 317–333, doi: 10.1016/S0040-1951(02)00263-9.

van Staal, C.R., 1994, Brunswick subduction complex in the Canadian Appalachians: Record of the Late Ordovician to Late Silurian collision between Laurentia and the Gander margin of Avalon: Tectonics, v. 13, p. 946–962, doi: 10.1029/93TC03604.

van Staal, C.R., 2005, The northern Appalachians, *in* Selley, R.C., Cocks, L.R.M., and Plimer, I.R., eds., Encyclopedia of Geology: Oxford, Elsevier, p. 81–91.

van Staal, C.R., Dewey, J.F., Mac Niocaill, C., and McKerrow, W.S., 1998, The Cambrian-Silurian tectonic evolution of the northern Appalachians and British Caledonides: History of a complex, west and southwest Pacific-type segment of Iapetus, *in* Blundell, D., and Scott, A.C., eds., Lyell: The Past is the Key to the Present: Geological Society [London] Special Publication 143, p. 199–242.

van Staal, C.R., Lissenberg, C.J., Zagorevski, A., McNicoll, V., Whalen, J., Bédard, J., and Pehrsson, S., 2004, A new look at the tectonic processes involved in the Ordovician Taconic orogeny in the northern Appalachians: Evidence for multiple accretion of (infant) arc terranes: Geological Society of America Abstracts with Programs, v. 36, no. 5, p. 481.

van Staal, C.R., Lissenberg, J., Pehrsson, S., Zagorevski, A., Valverde-Vaquero, P., Herd, R.K., McNicoll, V., and Whalen, J., 2005, Geology, Puddle Pond, Newfoundland (NTS 12-A/05): Geological Survey of Canada Open-File OF1664, scale 1:50,000, 1 sheet.

Wakabayashi, J., and Dilek, Y., 2000, Spatial and temporal relationships between ophiolites and their metamorphic soles: A test of models of forearc ophiolite genesis, *in* Dilek, Y., Moores, E.M., Elthon, D., and Nicolas, A., eds. Ophiolites and Oceanic Crust: New Insights from Field Studies and the Ocean Drilling Program: Geological Society of America Special Paper 349, p. 53–64.

Wakabayashi, J.W., and Dilek, Y., 2003, What constitutes 'emplacement' of an ophiolite?: Mechanisms and relationship to subduction initiation and formation of metamorphic soles, *in* Dilek, Y., and Robinson, P.T., eds., Ophiolites in Earth History: Geological Society [London] Special Publication 218, p. 427–447.

Waldron, J.W.F., and van Staal, C.R., 2001, Taconian orogeny and the accretion of the Dashwoods block: A peri-Laurentian microcontinent in the Iapetus Ocean: Geology, v. 29, p. 811–814, doi: 10.1130/0091-7613(2001)029<0811:TOATAO>2.0.CO;2.

Waldron, J.W.F., Anderson, S.D., Cawood, P.A., Goodwin, L.B., Hall, J., Jamieson, R.A., Palmer, S.E., Stockmal, G.S., and Williams, P.F., 1998, Evolution of the Appalachian Laurentian margin: Lithoprobe results in western Newfoundland: Canadian Journal of Earth Sciences, v. 35, p. 1271–1287, doi: 10.1139/cjes-35-11-1271.

Whalen, J.B., 1985, Geochemistry of an island-arc plutonic suite: The Uasilau–Yau Yau intrusive complex, New Britain, P.N.G.: Journal of Petrology, v. 26, p. 603–632.

Whalen, J. B., 1993. Geology of the Star Lake sheet, Newfoundland (NTS 12A/11)-1:50.000 colour map with descriptive notes: Geological Survey of Canada, Open file 2736.

Whalen, J. B., Currie, K.L., and van Breemen, O., 1987, Episodic Ordovician-Silurian plutonism in the Topsails terrane, western Newfoundland: Transactions of the Royal Society of Edinburgh, Earth Sciences, v. 78, p. 17–28.

Whalen, J.B., Jenner, G.A., Longstaffe, F.J., Robert, F., and Gariépy, C., 1996, Geochemical and isotopic (O, Nd, Pb and Sr) constraints on A-type granite petrogenesis based on the Topsails igneous suite, Newfoundland Appalachians: Journal of Petrology, v. 37, p. 1463–1489, doi: 10.1093/petrology/37.6.1463.

Whalen, J.B., Jenner, G.A., Longstaffe, F.J., Gariépy, C., and Fryer, B.J., 1997a, Implications of granitoid geochemical and isotopic (Nd, O, Pb) data from the Cambrian-Ordovician Notre Dame arc for the evolution of the Central Mobile belt, Newfoundland Appalachians, *in* Sinha, K., Whalen, J.B., and Hogan, J.P., eds., The Nature of Magmatism in the Appalachian Orogen: Geological Society of America Memoir 191, p. 367–395.

Whalen, J.B., van Staal, C.R., Longstaffe, F.J., Gariepy, C., and Jenner, G.A., 1997b, Insights into tectonostratigraphic zone identification in southwestern Newfoundland based on isotopic (Nd, O, Pb) and geochemical data: Atlantic Geology, v. 33, p. 231–241.

Whalen, J.B., Percival, J.A., McNicoll, V.J., and Longstaffe, F.J., 2003, Intra-oceanic production of continental crust in a Th-depleted ca. 3.0 Ga arc complex, western Superior Province, Canada: Contributions to Mineralogy and Petrology, v. 146, p. 78–99, doi: 10.1007/s00410-003-0484-8.

Whalen, J.B., McNicoll, V.J., van Staal, C.R., Lissenberg, C.J., Longstaffe, F.J., Jenner, G.A., and van Breeman, O., 2006, Spatial-temporal and geochemical characteristics of Silurian collision-zone magmatism, Newfoundland Appalachians: An example of a rapidly evolving magmatic system related to slab break-off: Lithos, v. 89, p. 377–404, doi: 10.1016/j.lithos.2005.12.011.

Whitehead, J., Reynolds, P.H., and Spray, J.G., 1996, 40Ar/39Ar age constraints on Taconian and Acadian events in the Quebec Appalachians: Geology, v. 24, p. 359–362, doi: 10.1130/0091-7613(1996)024<0359:AAACOT>2.3.CO;2.

Williams, H., 1975, Structural succession, nomenclature, and interpretation of transported rocks in western Newfoundland: Canadian Journal of Earth Sciences, v. 12, p. 1874–1894.

Williams, H., 1995, Geology of the Appalachian-Caledonian orogen in Canada and Greenland: Geological Survey of Canada, Geology of Canada, v. 6 (also Geological Society of America, Geology of North America, v. F-1), 921 p.

Williams, H., and Payne, J.G., 1975, The Twillingate granite and nearby volcanic groups: An island arc complex in northeast Newfoundland: Canadian Journal of Earth Sciences, v. 12, p. 982–995.

Williams, H., and St. Julien, P., 1982, The Baie Verte–Brompton Line: Early Paleozoic continent-ocean interface in the Canadian Appalachians, *in* St. Julien, P., and Beland, J., eds., Major Structural Zones and Faults of the Northern Appalachians: Geological Association of Canada Special Paper 24, p. 177–208.

Williams, H., Colman-Sadd, S.P., and Swinden, H.S., 1988, Tectonic-stratigraphic subdivisions of central Newfoundland, in Current research: Part B. Eastern and Atlantic Canada: Geological Survey of Canada Paper 88–1B, p. 91–98.

Williams, S.H., 1992, Lower Ordovician (Arenig-Llanvirn) graptolites from the Notre Dame subzone, central Newfoundland: Canadian Journal of Earth Sciences, v. 29, p. 1717–1733.

Wolf, M.B., and Wyllie, P.J., 1994, Dehydration-melting of amphibolite at 10 kbar: The effects of temperature and time: Contributions to Mineralogy and Petrology, v. 115, p. 369–383, doi: 10.1007/BF00320972.

Wood, D.A., 1980, The application of a Th-Hf-Ta diagram to problems of tectonomagmatic classification and to establishing the nature of crustal contamination of basaltic lavas of the British Tertiary volcanic province: Earth and Planetary Science Letters, v. 50, p. 11–30.

Xu, J.-F., Shinjo, R., Defant, M., Wang, Q., and Rapp, R.P., 2002, Origin of adakitic intrusive rocks in the Ningzhen area of east China: Partial melting of delaminated lower continental crust?: Geology, v. 30, p. 1111–1114, doi: 10.1130/0091-7613(2002)030<1111:OOMAIR>2.0.CO;2.

Zagorevski, A., Rogers, N., van Staal, C.R., McNicoll, V.J., Lissenberg, C.J., and Valverde-Vaquero, P., 2006, Lower to Middle Ordovician evolution of peri-Laurentian arc and back-arc complexes in the Iapetus: Constraints from the Annieopsquotch accretionary tract, central Newfoundland: Geological Society of America Bulletin, v. 118, p. 324–342, doi: 10.1130/B25775.1.

Zen, E.A., 1972, The Taconide Zone and the Taconide Orogeny in the Western Part of the Northern Appalachian Orogen: The Geological Society of America Special Paper 135, 72 p.

Manuscript Accepted by the Society 1 February 2007

The Geological Society of America
Memoir 200
2007

The Cat Square terrane: Possible Siluro-Devonian remnant ocean basin in the Inner Piedmont, southern Appalachians, USA

Arthur J. Merschat
Robert D. Hatcher Jr.
Department of Earth and Planetary Sciences and Science Alliance Center of Excellence, University of Tennessee, Knoxville, Tennessee 37996-1410, USA

ABSTRACT

The Inner Piedmont is a large, composite, sillimanite-grade terrane that extends from near the Virginia–North Carolina border to central Alabama and consists of the eastern Tugaloo and Cat Square terranes. It is bound to the west by the Brevard fault zone and to the east by the central Piedmont suture. It is the core of the Neoacadian (360–350 Ma) orogen in the southern Appalachians and records Late Devonian–Mississippian closure and high-grade metamorphism (sillimanite I and II) of Siluro-Devonian sediments deposited in the remnant Rheic ocean basin.

The Cat Square terrane is bounded by the younger-over-older Brindle Creek fault to the west and the central Piedmont suture to the east. It consists of a unique sequence of Siluro-Devonian metapsammite and pelitic schist that was intruded by Devonian anatectic granitoids (Toluca Granite, ~378 Ma, and Walker Top Granite, ~366 or ~407 Ma). Rare mafic and ultramafic rocks occur in the eastern Cat Square terrane. Minimum sediment thickness is estimated at 4 km (13,000 ft). Detrital zircons indicate that Cat Square terrane rocks have a maximum age of ~430 Ma, with both Laurentian (2.8, 1.8, 1.4, 1.1 Ga) and peri-Gondwanan (600, 500 Ma) affinities. Deposition on oceanic crust explains the existence of several mafic and ultramafic bodies and the absence of continental basement in the Cat Square terrane. The Cat Square terrane petrotectonic assemblage represents a Siluro-Devonian remnant ocean basin between Laurentia and the approaching Carolina superterrane. Metapsammite and pelitic schist may represent turbidites shed from approaching tectonic highlands on both flanks of the closing ocean. Palinspastic restoration of the Inner Piedmont constrains the location of the Cat Square basin to the Pennsylvania embayment, and links the mid-Devonian to Mississippian deformation in the Neoacadian core to the SW-migrating pulses of the diachronous Acadian-Neoacadian clastic wedge. Location and SW migration of the clastic wedge in concert with structural patterns in the Inner Piedmont support a transpressive NW-directed collision of the Carolina superterrane with the New York promontory and zippering the basin shut from NE to SW.

Keywords: Appalachians, Inner Piedmont, remnant ocean basin, Cat Square terrane, transpression, Neoacadian, Acadian clastic wedge.

Merschat, A.J., and Hatcher, R.D., Jr., 2007, The Cat Square terrane: Possible Siluro-Devonian remnant ocean basin in the Inner Piedmont, southern Appalachians, USA, *in* Hatcher, R.D., Jr., Carlson, M.P., McBride, J.H., and Martínez Catalán, J.R., eds., 4-D Framework of Continental Crust: Geological Society of America Memoir 200, p. 553–565, doi: 10.1130/2007.1200(27).

INTRODUCTION

Accretionary margins are the sites of greatest addition of new crust and are therefore important in understanding the four dimensional (4-D) evolution of continental crust. These are sites of accelerated rates and processes, especially during continent-continent and continent-arc collisions. Ingersoll et al. (2003) concluded that few sedimentary systems match the rates and volumes of sediment shed during continent-continent collision (e.g., India and Asia). Sediment shed during collision is transported through major river systems and deposited in deltas and in remnant ocean basins as part of huge submarine fans that dwarf all other depositional systems on Earth (Graham et al., 1975; Ingersoll et al., 1995, 2003). Ingersoll et al. (1995) defined a remnant ocean basin as a shrinking ocean basin flanked by at least one convergent margin and filled with recycled sediment shed from adjacent uplifted terranes (Fig. 1). Remnant ocean basins are destined to become deformed accretionary wedges or are further recycled and eventually preserved in terrigenous clastic wedges (Ingersoll et al., 1995, 2003). Preservation of remnant ocean basins is poor, and their recognition is difficult because of complex deformation in suture zones (Ingersoll and Busby, 1995). Recognition of remnant ocean basins, however, is important to understanding the spatial and temporal development of orogens, and processes related to formation of new continental crust. Petrotectonic assemblages are key to identifying deformed remnant ocean basins and unraveling their tectonic evolution.

The Paleozoic Appalachians are a long-lived accretionary-collisional orogen that contains numerous accreted terranes and a complete Wilson cycle (Hatcher, 1978; Williams and Hatcher, 1983; Abbott and Raymond, 1984; Hatcher, 1987, 1989, 2002a; Rankin et al., 1989). Previously, little evidence supported a middle to late Paleozoic orogeny in the southern Appalachians (Osberg et al., 1989), but recent geochronologic data have yielded overwhelming support for a Neoacadian, 360–350 Ma, tectonothermal event in the southern Appalachians (Dennis and Wright, 1997; Carrigan et al., 2001; Bream, 2002, 2003). One of the most intriguing discoveries is the Cat Square terrane, a Siluro-Devonian terrane in the eastern Inner Piedmont of mixed-affinity metasedimentary rocks metamorphosed to upper-amphibolite facies by latest Devonian time (Merschat et al., 2005a). This paper attempts to summarize existing geochronology, geochemistry, map relationships, and petrotectonic assemblage to ascertain the pre-Neoacadian tectonic setting of the Cat Square terrane and the early kinematics of Late Devonian to Mississippian collision of the Carolina superterrane with Laurentia and previously accreted terranes. Based on these data, we speculate that the Cat Square terrane developed as a mid-Paleozoic remnant ocean basin.

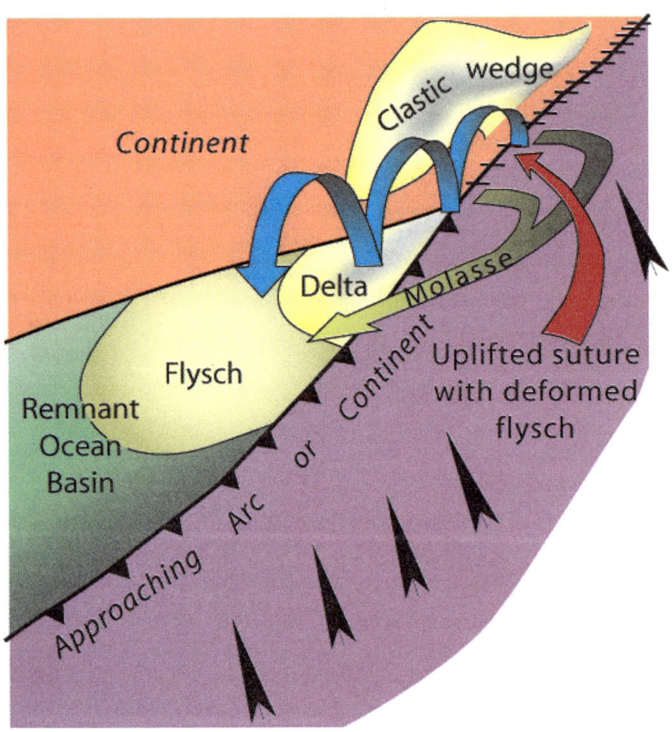

Figure 1. Conceptual diagram of a remnant ocean basin (after Graham et al., 1975) illustrating the relationships between sequential closure of a remnant ocean basin, longitudinal sediment dispersal between uplifted terranes, and sediment recycling indicated by blue and yellow arrows. Black arrows indicate direction of approaching arc or continent.

TECTONIC SETTING

The Inner Piedmont extends from central Alabama to near the Virginia–North Carolina border (Fig. 2). It is bounded to the west by the Brevard fault zone and to the east by the central Piedmont suture. Gently dipping structures and an extensive sillimanite and higher-grade core distinguish the Inner Piedmont from surrounding terranes and belts. The Inner Piedmont forms part of the southern Appalachian Neoacadian metamorphic core, a product of Late Devonian–Mississippian oblique collision of the Carolina superterrane with Laurentia (Hatcher, 2002b; Merschat et al., 2005a; Hatcher and Merschat, 2006).

The polydeformed Inner Piedmont is characterized by a gently dipping stack of crystalline Type F thrust sheets (fold nappes; Hatcher and Hooper, 1992) containing various para- and orthogneisses, a curved subhorizontal mineral lineation pattern, and a broad core of migmatitic, sillimanite and higher-grade rocks. The Inner Piedmont thrust stack consists of small-to-large Type F thrust sheets overlain and dwarfed by the much larger Brindle Creek thrust sheet. These thrust sheets nucleated during dextral transpression and record a net displacement of several hundred kilometers, of which 200–400 km are associated with dextral orogen-parallel (SW-directed) displacement along the mid-Paleozoic Brevard fault zone (Merschat et al., 2005a). Metamorphic grade remains sillimanite or higher across the Inner Piedmont core and drops to kyanite and garnet grade along the western flank and kyanite along parts of the eastern flank (Hatcher and Merschat, 2006, their Fig. 10). Paragneisses are pervasively migmatitic, and reached sillimanite I and II grade metamorphism

A

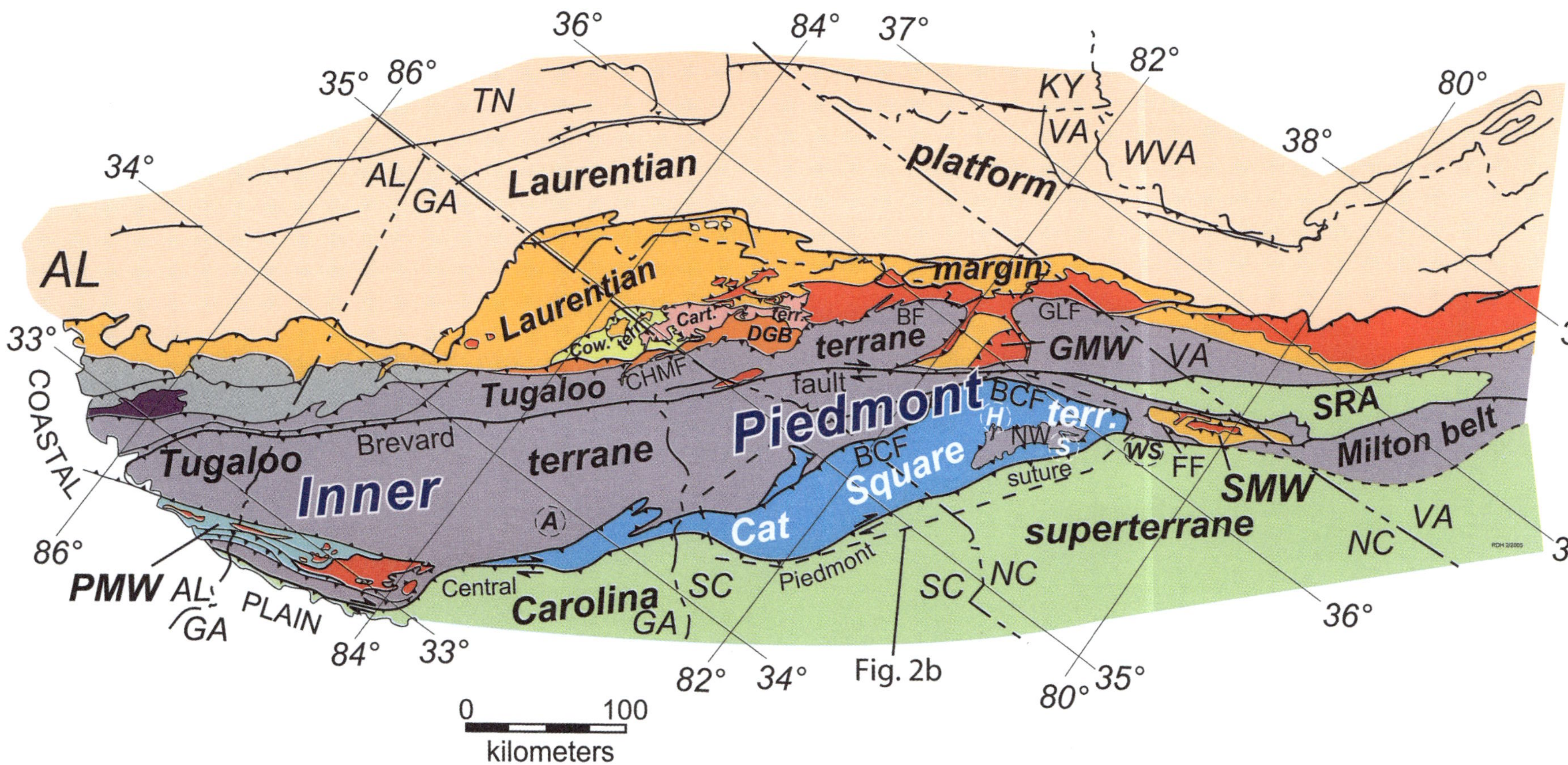

Figure 2. (A) Simplified tectonic map of the southern Appalachians showing the location of the Inner Piedmont, and the Tugaloo (lavender) and Cat Square (blue-green) terranes. The Laurentian platform contains the Alleghanian foreland fold-thrust belt driven in front of the Blue Ridge–Piedmont megathrust sheet (the Laurentian margin, gold with red inliers of Grenvillian, not necessarily Laurentian Grenville, and older inherited crust), and all components to the east. All terranes west of the Cat Square terrane have a Laurentian source (Bream et al., 2004). Cat Square terrane contains Siluro-Devonian metasedimentary rocks with a mixed Laurentian and peri-Gondwanan source. Carolina superterrane is peri-Gondwanan and exotic to Laurentia. BCF—Brindle Creek fault; BF—Burnsville fault; Cart. terr.—Cartoogechaye terrane; CHMF—Chattahoochee–Holland Mountain fault; Cow. terr.—Cowrock terrane; DGB—Dahlonega gold belt; FF—Forbush fault; GLF—Gossan Lead fault; GMW—Grandfather Mountain window; NW—Newton window; SRA—Smith River allochthon, an outlier of Carolina superterrane rocks; SMW—Sauratown Mountains window; PMW—Pine Mountain window. Light gray—probable western Tugaloo terrane rocks in Alabama and Georgia; purple—Ordovician Elkahatchee Quartz Diorite. A—Athens; H—Hickory; S—Statesville; WS—Winston–Salem.

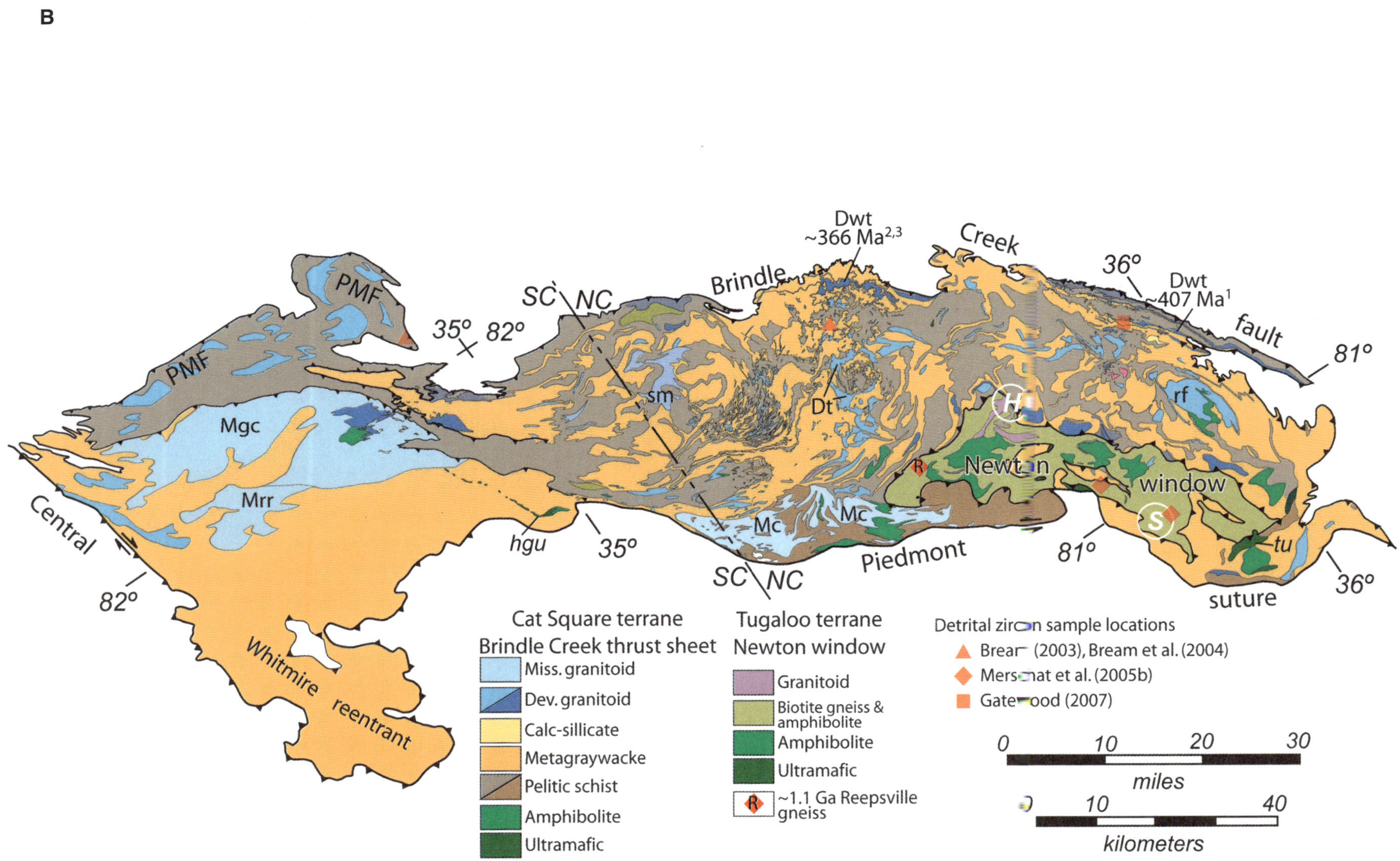

Figure 2(B) Simplified geologic map of the Cat Square terrane from its northern extent to near the Georgia–South Carolina border (modified from a compilation by Merschat and Hatcher, 2003). Dt—Toluca Granite; Dwt—Walker Top Granite with U-Pb ages from Gatewood (2007)[1], Giorgis et al. (2002)[2], and Ma es (2002)[3]; hgu—Hammett Grove meta-igneous mafic-ultramafic body; Mc—Cherryville Granite; Mgc—Gray Court granite; Mrr—Reedy River granite; PMF—Paris Mountain thrust sheet, a remnant of the original Brindle Creek thrust sheet that broke through as an out-of-sequence fault; R—~1050 Ma Reepsville gneiss; rf—Rocky Face granite; sm—Sandy Mush granite; tu—Turnersberg ultramafic.

(800–850 °C and 5–6 kbar; Mirante and Patino Douce, 2000; Merschat, 2003). Numerous studies have demonstrated that the Inner Piedmont is polydeformed, and the dominant structural elements are S_2, L_2, and F_2 (e.g., Hopson and Hatcher, 1988; Hatcher, 1993, 2001, 2002b; Merschat et al., 2005a). The gently dipping S_2 foliation distinguishes the Inner Piedmont from other terranes, but recent studies (e.g., Davis, 1993a, 1993b; Hatcher, 2001; Merschat et al., 2005a) have focused on the curved L_2 pattern and its coaxial relationship with F_2, a pattern initially recognized by Goldsmith (1981). Migmatitic layering parallel to S_2 and the dominant fabrics, S_2 and L_2, commonly defined by high-grade minerals, hornblende, sillimanite, and kyanite, suggest that deformation occurred at or near peak metamorphic conditions (Davis, 1993a, 1993b; Bier et al., 2002; Merschat et al., 2005a). U-Pb ages of monazite and zircon metamorphic rims suggest Neoacadian (360–350 Ma) timing of metamorphism (Dennis and Wright, 1997; Carrigan et al., 2001; Bream 2003).

The composite Inner Piedmont consists of the eastern Tugaloo and overlying Cat Square terranes (Bream et al., 2001, 2004; Bream, 2002, 2003; Merschat et al., 2005a). The Brindle Creek fault separates the Laurentian-affinity, eastern Tugaloo terrane rocks of the western Inner Piedmont from mixed Laurentian- and Gondwanan-affinity rocks of the eastern Inner Piedmont–Cat Square terrane (Bream, 2002, 2003; Bream et al., 2004) (Fig. 2). The eastern Tugaloo terrane consists of Neoproterozoic to Cambrian(?) Tallulah Falls Formation metagraywacke and pelite overlain by Middle Ordovician amphibolite, quartzite, marble, and felsic metatuff of the Poor Mountain Formation. Various Middle Ordovician to Silurian granitoids intrude the eastern Tugaloo terrane.

CAT SQUARE TERRANE

The Brindle Creek fault extends from south of Athens, Georgia, to northwest of Winston-Salem, North Carolina, and is truncated at both ends by the central Piedmont suture (Fig. 2). The Brindle Creek thrust sheet is the largest Inner Piedmont thrust sheet, with an exposed area of 13,765 km^2. It juxtaposes younger Cat Square terrane rocks against older eastern Tugaloo terrane rocks. Bream et al. (2001, 2004) first suggested that the Brindle Creek fault is a terrane boundary to explain contrasting zircon suites. Bream (2002, 2003) and Bream et al. (2001, 2004) recognized four detrital zircon suites from the Cat Square terrane: (1) dominance of Laurentian, 1.4, 1.2, and 1.1 Ga; (2) abundant Paleozoic 480 and 430 Ma; (3) lesser Neoproterozoic peri-Gondwanan 600 and 500 Ma; and (4) a small component of Archean 2.8 and 2.5 Ga zircons. The Cat Square terrane consists of pelitic schist, metapsammite, calc-silicate, minor amphibolite, and ultramafic rocks intruded by mostly anatectic Devonian and Mississippian granitoids (Bier et al., 2002; Merschat and Kalbas, 2002; Merschat et al., 2005a) (Fig. 2B). Primary sedimentary structures were obliterated by multiple transposition and high metamorphic grade. Schist and metapsammite are interlayered from outcrop to map-scale, but they form separable mappable units throughout the terrane.

Mafic and ultramafic rocks are not common in the Cat Square terrane, but reconnaissance and detailed geologic mapping documents several mafic and lesser ultramafic bodies (e.g., Overstreet et al., 1963; Privett, 1984; Mittwede et al., 1987; Goldsmith et al., 1988; Merschat, 2003; Wilson, 2006). Amphibolites are commonly layered, consisting of alternating dark layers of hornblende, plagioclase, and other minor minerals separated by thin layers of quartz and feldspar. Ultramafic rocks are foliated and altered to chlorite, scapolite, tremolite, serpentine, talc schists with rare olivine, and relict pyroxene (Privett, 1984; Mittwede et al., 1987; Goldsmith et al., 1988). These bodies are enveloped in both metapsammite and pelitic schist of the Cat Square terrane. Size of the bodies varies significantly, but the largest mafic and ultramafic bodies occur in the eastern part of the Cat Square terrane near the central Piedmont suture (Fig. 2B). Several large mafic and ultramafic bodies also occur in the Newton window (Fig. 2B) near the central Piedmont suture, but recent geochronology suggests that this antiformal structure is a window through the Brindle Creek thrust sheet exposing the eastern Tugaloo terrane (Merschat et al., 2005b). Wilson (2006) reported trace- and major-element geochemical data of Cat Square terrane amphibolites from the Brushy Mountains are representative of oceanic crust with enriched mid-ocean-ridge basalt (E-MORB) and back-arc affinities. The lenticular Hammett Grove meta-igneous suite in the eastern Cat Square terrane in northwestern South Carolina (Fig. 2B) consists of steatized and serpentinized ultramafic rocks, metapyroxenite, metagabbro, and amphibolite (Mittwede et al., 1987). Based on rock assemblage and chemical composition, Mittwede et al. (1987) concluded that the Hammet Grove meta-igneous suite represents a deformed ophiolite emplaced in front of the Carolina superterrane.

Numerous peraluminous Devonian and Mississippian granitoids intrude the Cat Square terrane. Devonian intrusions include the medium- to coarse-grained, weakly to well foliated ~378 Ma Toluca Granite (Mapes, 2002) and disconnected bodies of texturally similar megacrystic granites of the Walker Top plutonic suite, yielding ages of ~407 Ma (Brushy Mountains; Gatewood, 2007) and ~366 Ma (South Mountains; Giorgis et al., 2002; Mapes, 2002). Chemistry and isotopic signature of the Toluca and Walker Top Granites indicate that they are anatectic and likely derived from melting Cat Square terrane metapelite and metapsammite protoliths (Giorgis et al., 2002; Mapes, 2002). The ~355 Ma Cherryville Granite, ~357 Ma Gray Court, and ~325 Ma Reedy River granitoids occur along the eastern flank of the Cat Square terrane and have a similar origin (Mapes, 2002). Concordant contacts, lack of contact aureoles, and metamorphic assemblages of country rocks suggest that these are catazonal plutons. Assuming that these Devonian and Mississippian granitoids are anatectic melts and did not add new material to the terrane, a minimum thickness of the Cat Square terrane can be estimated at ~4 km (~13,000 ft) from various small-scale cross sections (Goldsmith et al., 1988; Nelson et al., 1998; Hatcher and Merschat, 2006).

SILURO-DEVONIAN CAT SQUARE REMNANT OCEAN BASIN

Despite the Cat Square terrane being highly deformed and metamorphosed, map relationships combined with geochronology and geochemistry permit a limited reconstruction of the Cat Square basin. Detrital zircons indicate that sediment was derived from both Laurentia and a peri-Gondwanan terrane (Bream et al., 2004). The closest and likeliest peri-Gondwanan source is the Carolina superterrane. Electron-microprobe monazite ages of ~530 Ma link the Smith River allochthon to the Carolina superterrane (Hibbard et al., 2003) (Fig. 2). The present structural position of the Smith River allochthon suggests that the Carolina superterrane overrode the Inner Piedmont.

Timing of deposition is bracketed by the youngest detrital zircons, ~430 Ma, and ages of the oldest anatectic Devonian intrusions. Deposition ended prior to the ~407 Ma formation of anatectic Walker Top Granite (Brushy Mountains), when the rocks would have been buried 15–20 km. Intrusion of the subduction-related Concord and Salisbury plutonic suites into the Carolina superterrane (McSween et al., 1984; McSween and Harvey, 1997; Esawi, 2004), and the oldest metamorphic rims (~400 Ma) observed on Cat Square terrane detrital zircons (Bream, 2002, 2003) support eastward subduction of material beneath the Carolina superterrane during this same interval (430–400 Ma). The entire volume (minimum ~55,059 km^3) of the Cat Square terrane would have been deposited over a 23 m.y. time span (a minimum rate of 0.17 mm/yr).

An enigmatic aspect of the Inner Piedmont is the lack of continental basement on which Inner Piedmont rocks were deposited. The 1143 ± 33 Ma Forbush Gneiss (Carrigan et al., 2003) exposed in the cores of breached antiforms along the Forbush fault in the southwestern Sauratown Mountains window (McConnell, 1990; Horton and McConnell, 1991), and the 1050 ± 18 Ma Reepsville gneiss exposed in the Newton window (Merschat et al., 2005b) are the only reported continental basement rocks in the Inner Piedmont (Fig. 2). Both occur exclusively in the eastern Tugaloo terrane, so they are not basement to the Cat Square terrane. An argument that the older eastern Tugaloo terrane is basement to the Cat Square terrane can be entertained, but the restriction of Devonian magmatism to the Cat Square terrane and truncations of Devonian granitoids against the Brindle Creek fault (e.g., Gatewood et al., 2004; Wilson et al., 2005; Wilson, 2006; Gatewood, 2007) preclude juxtaposition with the eastern Tugaloo–western Inner Piedmont until after ~407 Ma. At Vale, North Carolina, just east of Cat Square crossroads, the type locality of the Cat Square terrane (Hatcher, 2002b), a reported charnockite (termed Cat Square charnockite by Kish [1997]) was initially thought to be Grenville basement. Kish (1997), however, reported a U-Pb age of 358–347 Ma and chemistry indicating that the charnockite has a metamorphic origin. Regardless of whether or not the eastern Tugaloo terrane was deposited on continental basement, no evidence supports deposition of Cat Square terrane rocks on continental basement.

We propose that the Cat Square terrane was deposited on oceanic crust, possibly represented by the few large amphibolite bodies throughout the Cat Square terrane. Closely associated altered ultramafic rocks may represent oceanic mantle. Geochemical studies of eastern Inner Piedmont mafic and ultramafic rocks are limited (Mittwede et al., 1987; Wilson, 2006) but generally support a simatic origin. These bodies may constitute pieces of middle to late Paleozoic(?) ophiolites on which the Siluro-Devonian Cat Square terrane metasedimentary rocks were deposited. Emplacement of the mafic and ultramafic bodies likely occurred in an accretionary wedge that developed in front of the Carolina superterrane as the basin closed. The mafic and ultramafic bodies were subsequently modified by high-temperature Neoacadian deformation.

The Devonian to Mississippian Acadian-Neoacadian clastic wedge extends over 1000 km from New England to Alabama and is 400–700 km wide, prograding into the Illinois basin (Dennison, 1985; Ettensohn, 1985, 2004). The Acadian-Neoacadian clastic wedge consists of an aggregate of delta-alluvial complexes that are a mosaic of open-marine, marginal-marine, and terrestrial facies (Dennison, 1985; Ettensohn, 1985, 2004). Several large and well-documented delta complexes of the Acadian-Neoacadian clastic wedge include the Catskill, Bedford-Berea, Price-Pocono-Grainger, and Borden (Fig. 3). The typical unconformity-bound lithologic sequence recognized throughout the basin begins with transgressive, shallow-marine limestone, overlain by deep-water black shale, followed by deep-water turbidites, and finally marginal-marine sediments with various red beds (Ettensohn, 2004). The black basinal shales mark periods of subsidence after crustal loading and are used to track basin subsidence (Ettensohn, 1985). Common to all complexes is a progressive westward and southwestward facies migration indicative of an eastern source that shifted southwestward over time. Dennison (1985) recognized seven to eight lobes of the Catskill delta complex beginning in New York and becoming progressively younger to the southwest. Ettensohn (1985, 2004) proposed four tectophases, or spatially and temporally distinct pulses of sedimentation, during the Acadian-Neoacadian orogeny. The first tectophase was focused in the St. Lawrence promontory in northern New England during the Early and Middle Devonian, correlating temporally and spatially with the traditional Acadian orogeny (410–380 Ma; Osberg et al., 1989). The second (Middle Devonian) and third (Middle to Late Devonian) tectophases resulted from collisions with the New York and Virginia promontories, respectively. Combined, these two tectophases deposited over 3000 m of sediment in the New York promontory and over 2400 m in the Virginia Promontory, and are represented by the Catskill, Bedford-Berea, Price-Pocono, Borden, and Grainger–Ft. Payne delta complexes (Ettensohn, 1985, 2004). In the southern Appalachian foreland, these tectophases are represented by the Devonian-Mississippian Chattanooga Shale–Ft. Payne–Grainger Formation unconformably overlying folded and faulted Ordovician and lower Silurian sequences (Hatcher et al., 2005; Hatcher et al., this volume, their Fig. 13). Additional clastics deposited in the southern Appalachian Talladega belt at this time are interpreted as related to the

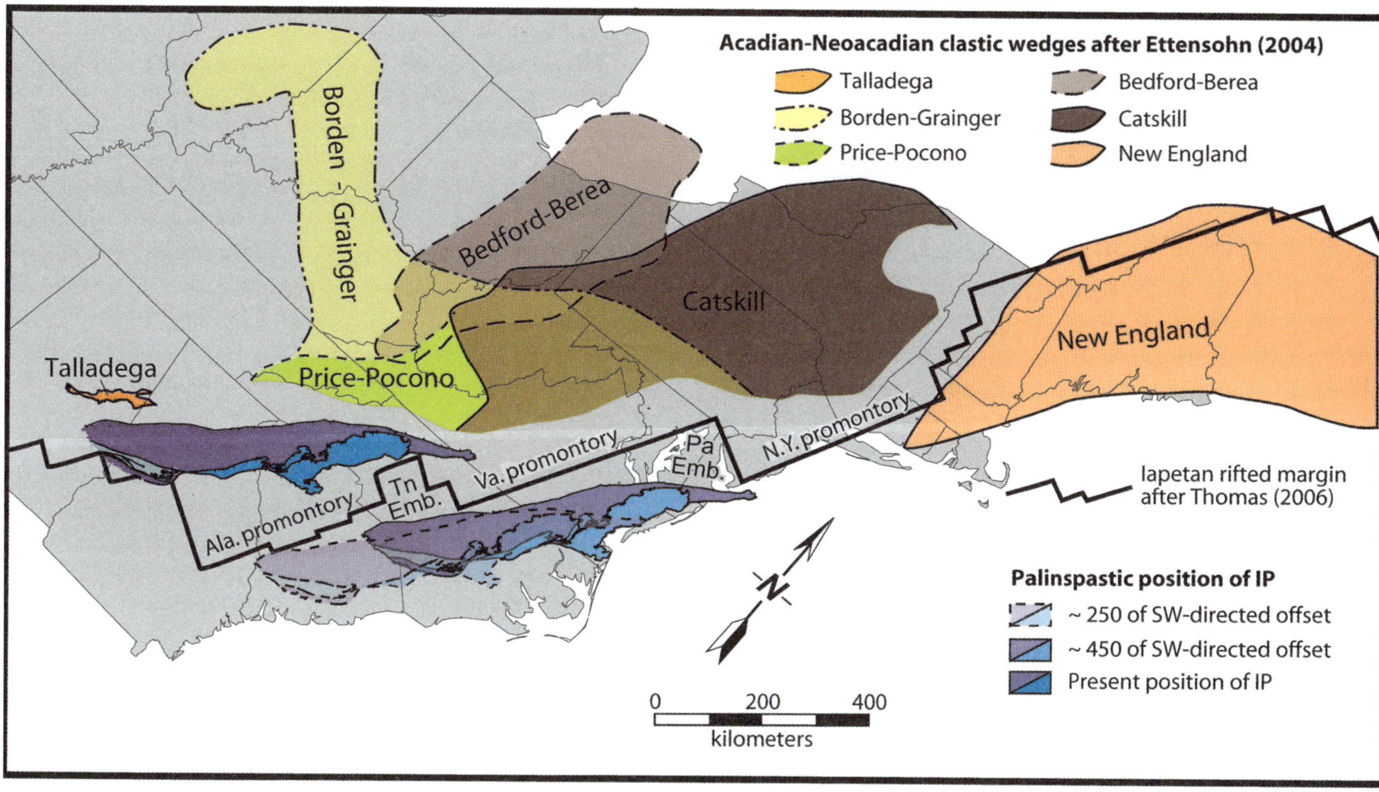

Figure 3. Possible palinspastic location of the Inner Piedmont and distribution of major delta complexes of the Acadian-Neoacadian clastic wedge. Locations and shape of clastic wedges are after Ettensohn (2004), and Iapetan rifted margin is from Thomas (2006). Emb.—embayment.

traditional Acadian clastic wedge (Tull, 2002). The final tectonic phase persisted from Mississippian to early Pennsylvanian time, during which it migrated southward from its initiation at the Virginia promontory (Ettensohn, 1985, 2004). Tectophases two through four temporally overlap Neoacadian deformation in the Inner Piedmont, but significant lateral offset of the Inner Piedmont from the thickest accumulations in the central Appalachians makes correlations difficult (Fig. 3). Collectively, southwestward migration of the Acadian-Neoacadian clastic wedge supports an oblique, transpressive, zippered collision of the Avalon-Carolina arc with Laurentia during the Devonian (Dennison, 1985; Ettensohn, 1985, 2004). Collision progressed from NE to SW: occurring first in the St. Lawrence promontory, then moving southwestward to the New York and Virginia promontories (Dennison, 1985; Ettensohn, 2004).

Palinspastic restoration of the Inner Piedmont links Devonian-Mississippian high-temperature deformation in the crystalline core with the SW-migrating Acadian-Neoacadian clastic wedge in the foreland. The Brevard fault zone has a complex history that records at least two phases of dextral strike-slip activity during the Neoacadian (~360 Ma) and early Alleghanian (~325 Ma; Hatcher, 2001). Merschat et al. (2005a) suggested that the high-temperature mid-Paleozoic Brevard fault zone (Neoacadian) has at least 200 km of SW-directed dextral displacement, and possibly 400 km or greater (Dennis, 2005). SW-directed displacement associated with the early Alleghanian Brevard fault zone has been estimated to be at least 50 km by Hatcher (1989), but may be considerably more when compared with estimates for similar Alleghanian dextral faults (e.g., Bailey et al., 2004). Removal of the SW-directed displacement during Neoacadian and early Alleghanian reactivation of the Brevard fault zone restores the Inner Piedmont to a position in the central Appalachians during the Early Devonian (Fig. 3). This palinspastic position provides critical insight into Appalachian tectonics. First, the location of the Cat Square basin was in the Pennsylvania embayment, between the New York and Virginia promontories, and possibly extended into the Tennessee embayment. Uplifted crust from colliding promontories shed sediment into both the peripheral foreland basin and Cat Square basin. Secondly, this position links mid-Paleozoic deformation in the Inner Piedmont with the prograding Acadian-Neoacadian clastic wedge and its southwestward migration. Structural patterns in the Inner Piedmont and the SW-migrating clastic wedge are consistent with oblique transpressive collision, and sequential NE to SW closure of the basin.

A corollary to our analysis is that the western Tugaloo terrane also had to be transported southwestward by a distance similar to that of the eastern Tugaloo and Cat Square terranes. Otherwise, the same stratigraphic sequence (Tallulah Falls–Ashe

Formation) would likely not occur in proximity across the Brevard fault zone (Hatcher, 2002b). Middle Devonian, amphibolite facies, dextral strike-slip kinematics have been documented along the Burnsville fault, bounding the western Tugaloo terrane in northwestern North Carolina (Adams et al., 1995; Trupe et al., 2003) and supporting SW-directed transport of the western Tugaloo during the Neoacadian.

Our reconstruction of the Siluro-Devonian Cat Square basin shares several similarities with the remnant ocean-basin models of Graham et al. (1975) and Ingersoll et al. (1995, 2003), including: (1) a shrinking ocean basin; (2) a short life span; (3) a sedimentary petrotectonic assemblage; (4) mixed crustal affinity; and (5) oblique collision and sequential NE-to-SW closure of the basin. The Cat Square basin was likely the last remnant of the Rheic ocean separating Laurentia and the approaching Carolina superterrane (Fig. 4). The remnant ocean basin existed from, at most, 430 to 407 Ma, and its closure marked the Late Devonian subduction of the Rheic ocean and accretion of the Carolina superterrane to Laurentia. The <25 m.y. duration of the Cat Square basin is consistent with the life span of other modern and ancient remnant ocean basins (Ingersoll and Busby, 1995). The Cat Square terrane is the deformed equivalent of the flysch facies and fragments of an ophiolitic suite in an idealized sequence of a closed remnant ocean basin (Ingersoll et al., 1995) (Fig. 5). Interlayered pelitic schist and metapsammite are lithologies consistent with sediment derived from uplifted basement terranes (Laurentia) and magmatic arcs (Carolina superterrane) (Ingersoll et al., 1995). Graham et al. (1975) recognized that sequential closure of remnant ocean basins typically causes longitudinal sediment dispersal parallel to the suture and the axis of the basin. Recent models explaining the complex structural patterns in the Inner Piedmont have proposed oblique, NW-directed convergence of the Carolina superterrane (e.g., Merschat et al., 2005a). These are supported by the palinspastic position of the Inner Piedmont and the SW-migrating clastic wedge. Oblique collision would produce a diachronous suture that closed from north to south, like a zipper. Sediment would have been dispersed longitudinally southward into the Siluro-Devonian Cat Square basin and would sample both terranes, resulting in mixed-affinity rocks (Bream et al., 2004). We suggest that the Cat Square terrane represents a Siluro-Devonian remnant ocean basin that separated Laurentia from the Carolina superterrane, and records the Devonian closing of the Rheic ocean and subduction beneath the Carolina superterrane (Fig. 4).

TECTONIC SYNTHESIS

The following tectonic scenario describes the likely evolution and tectonic ramifications of the Cat Square terrane from remnant ocean basin with provenance from approaching terranes to its current position as part of the Inner Piedmont. Deposition of the Cat Square terrane occurred in a remnant ocean basin between Laurentia and the approaching Carolina superterrane from 430 to 410 Ma. Palinspastic restoration of the Inner Piedmont northward to the central Appalachians suggests that the location of the Cat Square remnant ocean basin was in the Pennsylvania embayment between the New York and Virginia promontories. Eastward subduction of the basin beneath the Carolina superterrane produced Early Devonian magmatism in the Carolina superterrane (Figs. 4A and 4B). The combination of SW-prograding Acadian-Neoacadian clastic wedge and structural patterns in the Inner Piedmont supports an oblique, NW-directed transpressive collision initiated in the St. Lawrence promontory and then New York and Virginia promontories. The Cat Square remnant ocean basin was zippered closed from NE to SW and subducted beneath the Carolina superterrane along with part of the Tugaloo terrane, and incorporated into the Neoacadian orogen by ~380 Ma (Fig. 4C). Metamorphic conditions associated with anatectic formation of the ~378 Ma Toluca Granite require burial of the Cat Square terrane to 15–20 km and temperatures of 800 °C, consistent with subduction beneath the Carolina superterrane. Suturing of the Carolina superterrane to Laurentian elements likely produced flysch and molasse that were shed into the central Appalachian foreland basin by ~380 Ma and migrated southwestward into the southern Appalachian foreland basin by ~340 Ma. Peak metamorphic conditions (sillimanite I and II) coeval with peak deformation, D_2 fabrics, and structures affected the Cat Square terrane as they became part of the Inner Piedmont by 360–350 Ma. Curved high-temperature NW-rotating to SW-directed flow resulting from oblique convergence of the Carolina superterrane transported Inner Piedmont thrust sheets west and eventually southwest several hundred kilometers along the mid-Paleozoic Brevard fault zone (Figs. 4C and 4D). Hatcher and Merschat (2006) have suggested that the high-temperature curved ductile flow in the Inner Piedmont is a tectonically forced orogenic channel.

CONCLUSIONS

1. The Cat Square terrane consists of pelitic schist, metapsammite, and lesser amphibolite and ultramafic rocks. These rocks were intruded by anatectic Devonian-Mississippian granitoids.

2. The Cat Square terrane was deposited in a Siluro-Devonian remnant ocean basin between eastern Laurentian terranes and the approaching peri-Gondwanan Carolina superterrane.

3. Palinspastic restoration of the Inner Piedmont delimits the location of the Cat Square basin to the Pennsylvania embayment between the New York and Virginia promontories and links uplift as a product of 380–350 Ma deformation in the Inner Piedmont with the Acadian clastic wedge.

4. Southwest migration of the Acadian-Neoacadian clastic wedge and structural patterns in the Inner Piedmont are compatible with oblique NW-directed collision of the Carolina superterrane with the Laurentian margin. Closure of the Cat Square basin zippered shut from NE to SW.

5. The Cat Square terrane is the deformed equivalent of a remnant ocean basin. Rare amphibolite and ultramafic bodies may represent a deformed middle to late Paleozoic ophiolite suite consisting of metamorphosed mantle, and oceanic crust overlain

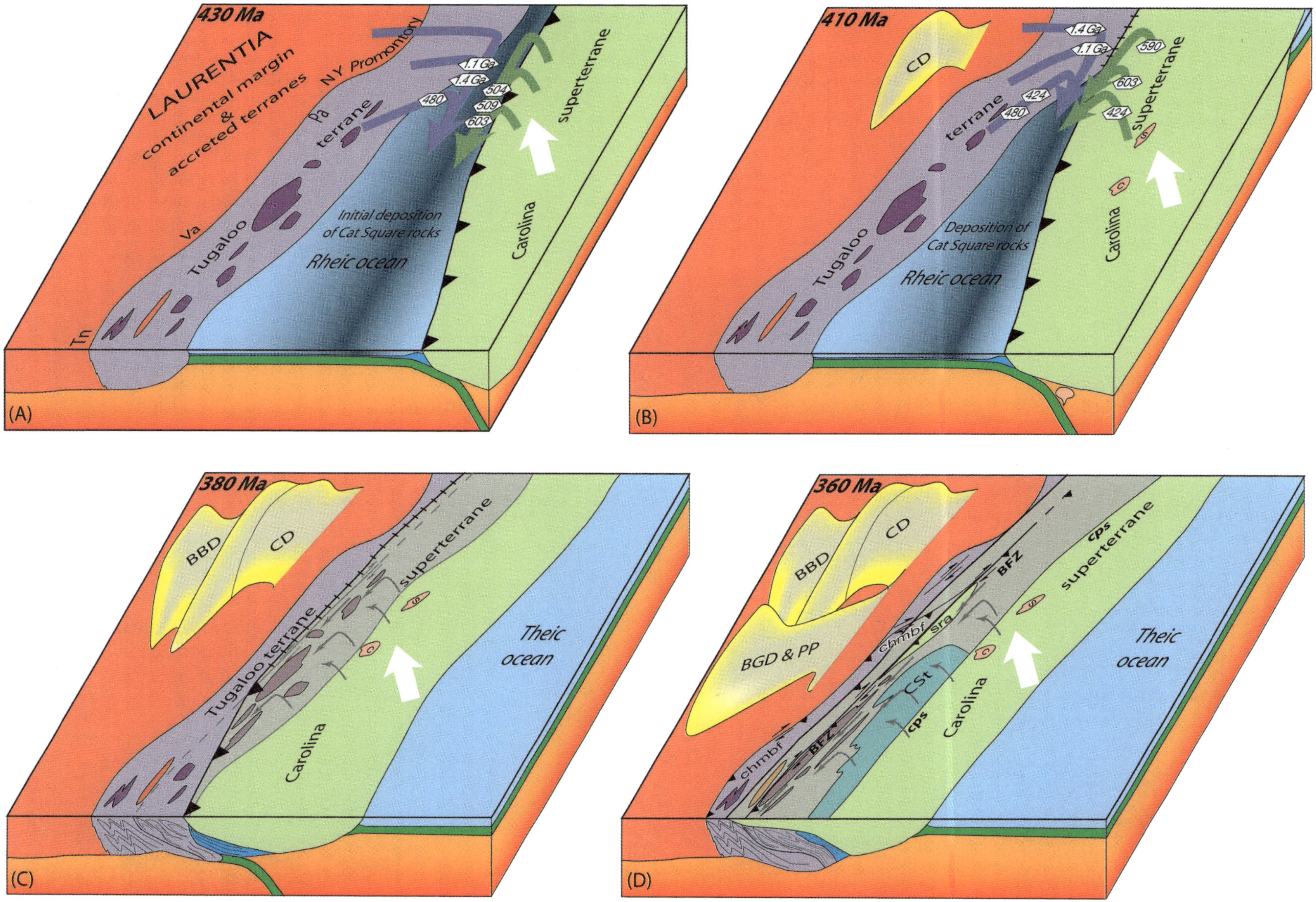

Figure 4. Three-dimensional block diagrams representing the sequential tectonic history of the Cat Square terrane. (A) Silurian configuration of Laurentia, Tugaloo terrane with mid-Ordovician plutons (dark purple), and Carolina superterrane separated by Rheic ocean. White arrows indicate motion of the Carolina superterrane. (B) Continued closure of the Rheic ocean and deposition of Cat Square terrane rocks during the Early Devonian. Magmatism in the Carolina superterrane, indicated by the Concord and Salisbury plutonic suites, is the product of subduction of the basin beneath the Carolina superterrane. (C) Closure of the Rheic ocean and suturing of Carolina superterrane by ~380 Ma. Gray lines indicate location of Brevard fault zone and ductile flow of Inner Piedmont material beneath the Carolina superterrane. (D) Neoacadian configuration resulting from dextral transpression. BBD—Bedford-Berea delta; BFZ—Brevard fault zone; BGD—Borden-Grainger delta; c—Concord Plutonic suite; CD—Catskill delta; chmbf—Chattahoochee–Holland Mountain–Burnsville fault; cps—central Piedmont suture; CSt—Cat Square terrane; PP—Price-Pocono delta; s—Salisbury Plutonic suite; sra—Smith River allochthon; Emb.—embayment.

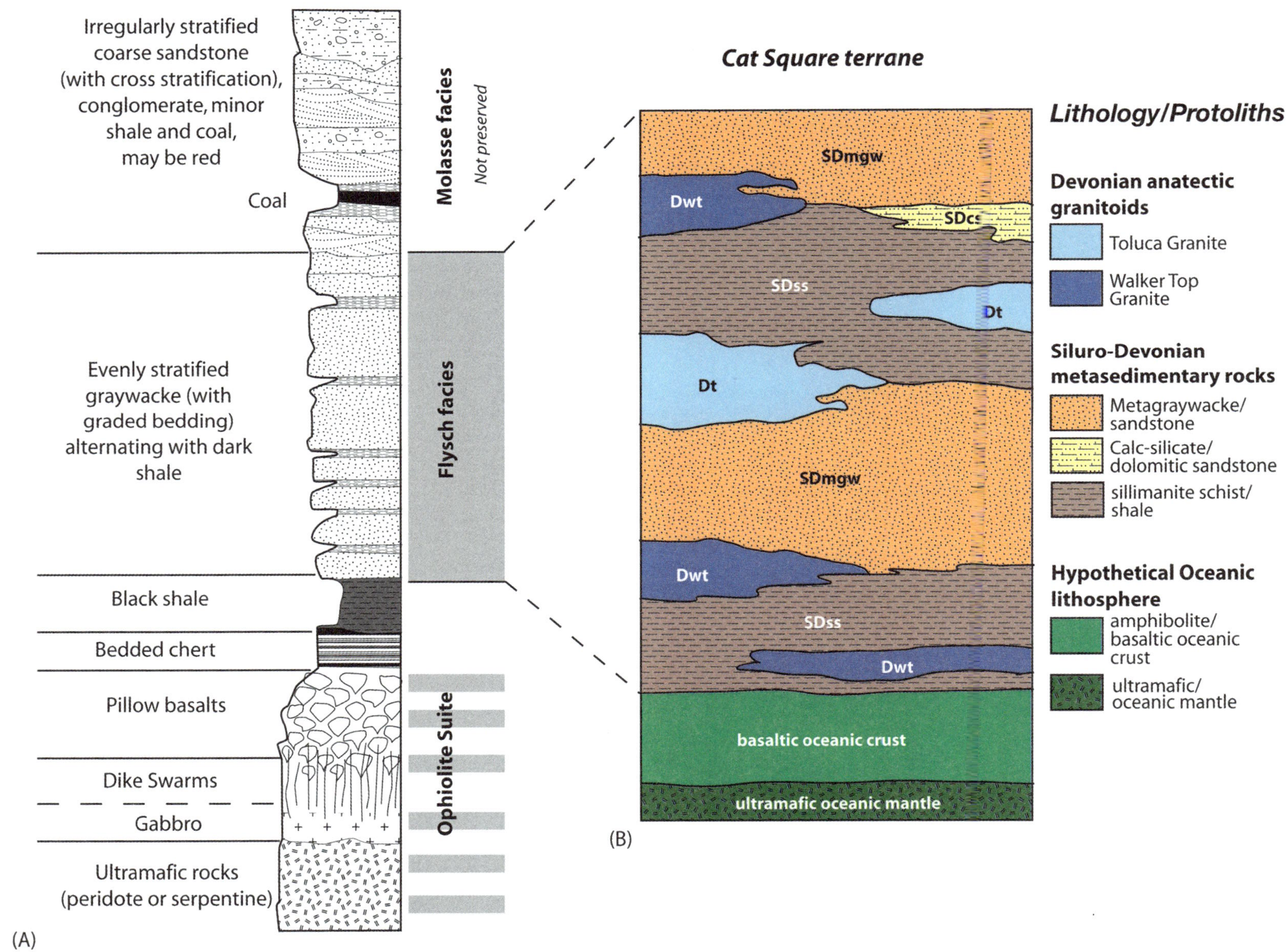

Figure 5. (A) Idealized stratigraphic column of a remnant ocean basin (modified from Ingersoll et al., 1995). Gray shading indicates parts of the idealized stratigraphic section likely preserved in the Cat Square terrane. (B) Schematic stratigraphic column highlighting the similarities of the petrotectonic assemblage in the Cat Square terrane and the idealized stratigraphic column of a remnant ocean basin in A. Rocks of the Cat Square terrane represent the deformed flysch facies and ophiolite suite, respectively. An equivalent proximal molasse facies is not recognized and may never have existed, because the rocks were subducted to mid-crustal levels before deposition of the molasse facies. Anatectic Walker Top and Toluca Granites are included because they were derived from metasedimentary rocks of the Cat Square terrane and constrain the minimum age of deposition. Dt—Toluca Granite; Dwt—Walker Top Granite; SDcs—calc-silicate; SDmgw—metapsammite; SDss—sillimanite schist.

by interlayered pelitic and psammitic rocks of the Brindle Creek thrust sheet. Interlayered schist and metapsammite sequences likely represent metamorphosed turbidites (flysch facies) shed from the evolving suture zone between Laurentia and the approaching Carolina superterrane.

6. The tectonic history of the Cat Square terrane provides insight into the brief life span of an ancient remnant ocean basin. The basin existed for less than 25 m.y. before being subducted to mid-crustal depths, metamorphosed to upper amphibolite facies (sillimanite I and II assemblages), and transported ~400 km southwestward from the central to southern Appalachians by 350 Ma during the Neoacadian orogeny. Evolution of the Cat Square terrane from basin to metamorphic core occurred in less than 80 m.y.

ACKNOWLEDGMENTS

U.S. National Science Foundation grants GA-1409, GA-20321, EAR-810852, EAR-8206949, EAR-8417894, and EAR-9004604 to Hatcher supported much of the field work in the Carolinas and northeast Georgia Inner Piedmont by Hatcher and graduate students during the 1970s, 1980s, 1990s, and early 2000s. EAR-9814800 to Hatcher supported early detrital zircon sensitive high-resolution ion microprobe geochronology. Additional support during the late 1960s and early 1970s was provided to Hatcher by the South Carolina Geological Survey, Henry S. Johnson, Jr., and Norman K. Olson, state geologists. Graduate student field work in the late 1990s and 2000s has been supported by the EDMAP component of the National Cooperative Geologic Mapping Program administered by the U.S. Geological Survey. The University of Tennessee Science Alliance Center of Excellence also has provided considerable support for Hatcher and students since 1986. Reviews by William R. Dickinson and Raymond V. Ingersoll significantly improved the manuscript. We remain culpable, however, for any errors of fact or interpretation.

REFERENCES CITED

Abbott, R.N., Jr., and Raymond, L.A., 1984, The Ashe metamorphic suite, northwest North Carolina: Metamorphism and observations on geologic history: American Journal of Science, v. 284, p. 350–375.

Adams, M.G., Stewart, K.G., Trupe, C.H., and Willard, R.A., 1995, Tectonic significance of high-pressure rocks and dextral strike-slip faulting along the Taconic suture, *in* Hibbard, J.P., van Staal, C.R., and Cawood, P.A., eds., Current Perspectives in the Appalachian-Caledonian Orogen: Geological Association of Canada Special Paper 41, p. 21–42.

Bailey, C.M., Francis, B.E., and Fahrney, E.E., 2004, Strain and vorticity analysis of transpressional high-strain zones from the Virginia Piedmont, USA, *in* Alsop, G.I., Holdsworth, R.E., McCaffrey, K.J.W., and Hand, M., eds., Flow Processes in Faults and Shear Zones: Geological Society [London] Special Publication 224, p. 249–264.

Bier, S.E., Bream, B.R., and Giorgis, S.D., 2002, Inner Piedmont stratigraphy, metamorphism, and deformation in the Marion–South Mountains area, North Carolina, *in* Hatcher, R.D., Jr., and Bream, B.R., eds., Inner Piedmont Geology in the South Mountains–Blue Ridge Foothills and the Southwestern Brushy Mountains, Central-Western North Carolina: Knoxville, Tennessee, North Carolina Geological Survey, Carolina Geological Society Guidebook, p. 65–100.

Bream, B.R., 2002, The southern Appalachian Inner Piedmont: New perspectives based on recent detailed geologic mapping, Nd isotopic evidence, and zircon geochronology, *in* Hatcher, R.D., Jr., and Bream, B.R., eds., Inner Piedmont Geology in the South Mountains–Blue Ridge Foothills and the Southwestern Brushy Mountains, Central-Western North Carolina: Knoxville, Tennessee, North Carolina Geological Survey, Carolina Geological Society Guidebook, p. 45–63.

Bream, B.R., 2003, Tectonic Implications of Geochronology and Geochemistry of Para- and Orthogneisses from the Southern Appalachian Crystalline Core [Ph.D. dissertation]: Knoxville, University of Tennessee, 296 p.

Bream, B.R., Hatcher, R.D., Jr., Miller, C.F., and Fullagar, P.D., 2001, Geochemistry and provenance of Inner Piedmont paragneisses, NC and SC: Evidence for an internal terrane boundary?: Geological Society of America Abstracts with Programs, v. 33, no. 2, p. 65.

Bream, B.R., Hatcher, R.D., Jr., Miller, C.F., and Fullagar, P.D., 2004, Detrital zircon ages and Nd isotopic data from the southern Appalachian crystalline core, GA-SC-NC-TN: New provenance constraints for Laurentian margin paragneisses, *in* Tollo, R.P., Corriveau, L., McLelland, J., and Bartholomew, M.J., eds., Proterozoic Evolution of the Grenville Orogen in North America: Geological Society of America Memoir 197, p. 459–475.

Carrigan, C.W., Bream, B.R., Miller, C.F., and Hatcher, R.D., Jr., 2001, Ion microprobe analyses of zircon rims from the eastern Blue Ridge and Inner Piedmont, NC-SC-GA: Implications for the timing of Paleozoic metamorphism in the southern Appalachians: Geological Society of America Abstracts with Programs, v. 33, no. 2, p. 7.

Carrigan, C.W., Miller, C.F., Fullagar, P.D., Bream, B.R., Hatcher, R.D., Jr., and Coath, C.D., 2003, Ion microprobe age and geochemistry of southern Appalachian basement, with implications for Proterozoic and Paleozoic reconstructions: Precambrian Research, v. 120, p. 1–36, doi: 10.1016/S0301-9268(02)00113-4.

Davis, T.L., 1993a, Lithostratigraphy, Structure, and Metamorphism of a Crystalline Thrust Terrane, Western Inner Piedmont, North Carolina [Ph.D. dissertation]: Knoxville, University of Tennessee, 245 p.

Davis, T.L., 1993b, Geology of the Columbus Promontory, western Piedmont, North Carolina, southern Appalachians, *in* Hatcher, R.D., Jr., and Davis, T.L., eds., Studies of Inner Piedmont Geology with a Focus on the Columbus Promontory: Knoxville, Tennessee, North Carolina Geological Survey, Carolina Geological Society Guidebook, p. 17–43.

Dennis, A.J., 2005, Middle Paleozoic clastic wedge provenance, dextral collision, and the southern Appalachian Inner Piedmont: The Spechty Kopf and Rockwell Formations of the central Appalachians as tectonic markers: Geological Society of America Abstracts with Programs, v. 37, no. 2, p. 5.

Dennis, A.J., and Wright, J.C., 1997, Middle and late Paleozoic monazite U-Pb ages, Inner Piedmont, South Carolina: Geological Society of America Abstracts with Programs, v. 29, no. 3, p. 12.

Dennison, J.M., 1985, Catskill delta shallow marine strata, *in* Woodrow, D.L., and Sevon, W.D., eds., The Catskill Delta: Geological Society of America Special Paper 201, p. 91–106.

Esawi, E.K., 2004, Evidence from the Farmington pluton for Early Devonian subduction-related magmatism in the Carolina zone of central North Carolina: Journal of Geodynamics, v. 37, p. 531–548, doi: 10.1016/j.jog.2004.02.019.

Ettensohn, F.R., 1985, The Catskill delta complex and the Acadian orogeny: A model, *in* Woodrow, D.L., and Sevon, W.D., eds., The Catskill Delta: Geological Society of America Special Paper 201, p. 39–49.

Ettensohn, F.R., 2004, Modeling the nature and development of major Paleozoic clastic wedges in the Appalachian basin, USA: Journal of Geodynamics, v. 37, p. 657–681, doi: 10.1016/j.jog.2004.02.009.

Gatewood, M.P., 2007, Tectonics of the northeastern Inner Piedmont, northwestern North Carolina, from detailed geologic mapping, geochronologic, geochemical, and petrologic studies with macro- and mesostructural analyses of ductile fault zones [M.S thesis]: Knoxville, University of Tennessee, 279 p.

Gatewood, M.P., Merschat, A.J., Wilson, C.G., and Hatcher, R.D., Jr., 2004, Inner Piedmont timing relationships and Neoacadian crustal flow, Brushy Mountains, NC: Results from detailed geologic mapping, structural and petrological analyses: Geological Society of America Abstracts with Programs, v. 36, no. 5, p. 504.

Giorgis, S.D., Mapes, R.W., and Bream, B.R., 2002, The Walker Top Granite: Acadian granitoid or eastern Inner Piedmont basement?, *in* Hatcher, R.D., Jr., and Bream, B.R., eds., Inner Piedmont Geology in the South Mountains–Blue Ridge Foothills and the Southwestern Brushy Mountains, Central-Western North Carolina: Knoxville, Tennessee, North Carolina Geological Survey, Carolina Geological Society Guidebook, p. 33–44.

Goldsmith, R., 1981, Structural patterns in the Inner Piedmont of the Charlotte and Winston-Salem 2-degree quadrangles, North Carolina and South Carolina, *in* Horton, J.W., Jr., Butler, J.R., and Milton, D.M., eds., Geological Investigations of the Kings Mountain Belt and Adjacent Areas in the Carolinas: Columbia, South Carolina Geological Survey, Carolina Geological Society Field Trip Guidebook, p. 19–27.

Goldsmith, R., Milton, D.J., and Horton, J.W., Jr., 1988, Geologic map of the Charlotte 1-degree × 2-degree quadrangle, North Carolina and South Carolina: U.S. Geological Survey Map I-1251-E, scale 1:250,000.

Graham, S.A., Dickinson, W.R., and Ingersoll, R.V., 1975, Himalayan-Bengal model for flysch dispersal in the Appalachian-Ouachita system: Geological Society of America Bulletin, v. 86, p. 273–286, doi: 10.1130/0016-7606(1975)86<273:HMFFDI>2.0.CO;2.

Hatcher, R.D., Jr., 1978, Tectonics of the western Piedmont and Blue Ridge, southern Appalachians: Review and speculations: American Journal of Science, v. 278, p. 276–304.

Hatcher, R.D., Jr., 1987, Tectonics of the southern and central Appalachian internides: Annual Review of Earth and Planetary Sciences, v. 15, p. 337–362, doi: 10.1146/annurev.ea.15.050187.002001.

Hatcher, R.D., Jr., 1989, Tectonic synthesis of the U.S. Appalachians, Chapter 14, *in* Hatcher, R.D., Jr., Thomas, W.A., and Viele, G.W., eds., The Appalachian Ouachita Orogen in the United States: Boulder, Colorado, Geological Society of America, Geology of North America, v. F-2, p. 511–535.

Hatcher, R.D., Jr., 1993, Perspective on the tectonics of the Inner Piedmont, southern Appalachians, *in* Hatcher, R.D., Jr., and Davis, T.L., eds., Studies of Inner Piedmont Geology with a Focus on the Columbus Promontory: Knoxville, Tennessee, North Carolina Geological Survey, Carolina Geological Society Guidebook, p. 17–43.

Hatcher, R.D., Jr., 2001, Rheological partitioning during multiple reactivation of the Palaeozoic Brevard fault zone, southern Appalachians, USA, *in* Holdsworth, R.E., Strachan, R.A., Magloughlin, J.F., and Knipe, R.J., eds., The Nature and Tectonic Significance of Fault Zone Weakening: Geological Society [London] Special Publication 186, p. 255–269.

Hatcher, R.D., Jr., 2002a, Alleghanian (Appalachian) orogeny, a product of zipper tectonics: Rotational transpressive continent-continent collision and closing of ancient oceans along irregular margins, *in* Martínez Catalán, J.R., Hatcher, R.D., Jr., Arenas, R., and Díaz García, F., eds., Variscan-Appalachian Dynamics: The Building of the Late Paleozoic Basement: Geological Society of America Special Paper 364, p. 199–208.

Hatcher, R.D., Jr., 2002b, An Inner Piedmont primer, *in* Hatcher, R.D., Jr., and Bream, B.R., eds., Inner Piedmont Geology in the South Mountains–Blue Ridge Foothills and the Southwestern Brushy Mountains, Central-Western North Carolina: Knoxville, Tennessee, North Carolina Geological Survey, Carolina Geological Society Guidebook, p. 1–18.

Hatcher, R.D., Jr., and Hooper, R.J., 1992, Evolution of crystalline thrust sheets in the internal parts of mountain chains, *in* McClay, K.R., ed., Thrust Tectonics: London, Chapman and Hall, p. 217–234.

Hatcher, R.D., Jr., and Merschat, A.J., 2006, The Appalachian Inner Piedmont: An exhumed strike-parallel, tectonically forced orogenic channel, *in* Law, R.D., Searle, M., and Godin, L., eds., Channel Flow, Ductile Extrusion and Exhumation of Lower-Mid Crust in Continental Collision Zones: Geological Society of London Special Publication 268, p. 517–540.

Hatcher, R.D., Jr., Bream, B.R., Merschat, A.J., Mapes, R.W., and Miller, C.F., 2005, Evidence for the (Neo-)Acadian orogeny in the southern Appalachians: Geological Society of America Abstracts with Programs, v. 37, no. 2, p. 5.

Hibbard, J.P., Tracy, R.J., and Henika, W.S., 2003, Smith River allochthon: A southern Appalachian peri-Gondwanan terrane emplaced directly on Laurentia?: Geology, v. 31, p. 215–218, doi: 10.1130/0091-7613(2003)031<0215:SRAASA>2.0.CO;2.

Hopson, J.L., and Hatcher, R.D., Jr., 1988, Structural and stratigraphic setting of the Alto allochthon, NE Georgia: Geological Society of America Bulletin, v. 100, p. 339–350, doi: 10.1130/0016-7606(1988)100<0339:SASSOT>2.3.CO;2.

Horton, J.W., and McConnell, K.I., 1991, The western Piedmont, *in* Horton, J.W., and Zullo, V.A., eds., The Geology of the Carolinas: Knoxville, Tennessee, University of Tennessee Press, p. 36–58.

Ingersoll, R.V., and Busby, C.J., 1995, Tectonics of sedimentary basins, *in* Busby, C.J., and Ingersoll, R.V., eds., Tectonics of Sedimentary Basins: Cambridge, Massachusetts, Blackwell Science, p. 1–52.

Ingersoll, R.V., Graham, S.A., and Dickinson, W.R., 1995, Remnant ocean basins, *in* Busby, C.J., and Ingersoll, R.V., eds., Tectonics of Sedimentary Basins: Cambridge, Massachusetts, Blackwell Science, p. 363–392.

Ingersoll, R.V., Dickinson, W.R., and Graham, S.A., 2003, Remnant-ocean submarine fans: Largest sedimentary systems on Earth, *in* Chan, M.A., and Archer, A.W., eds., Extreme Depositional Environments: Mega End Members in Geologic Time: Geological Society of America Special Paper 370, p. 191–208.

Kish, S.A., 1997, The Cat Square charnockite—A Paleozoic charnockite in the Inner Piedmont of North Carolina: Geological Society of America Abstracts with Programs, v. 29, no. 3, p. 28.

Mapes, R.W., 2002, Geochemistry and Geochronology of Mid-Paleozoic Granitic Plutonism in the Southern Appalachian Piedmont Terrane, North Carolina–South Carolina–Georgia [M.S. thesis]: Nashville, Vanderbilt University, 150 p.

McConnell, K.I., 1990, Geology and Geochronology of the Sauratown Mountains Anticlinorium, Northwestern North Carolina [Ph.D. dissertation]: Columbia, South Carolina, University of South Carolina, 232 p.

McSween, H.Y., Jr., and Harvey, R.P., 1997, Concord Plutonic Suite: Pre-Acadian gabbro-syenite intrusion in the southern Appalachians, *in* Sinha, A.K., Whalen, J.B., and Hogan, J.P., eds., The Nature of Magmatism in the Appalachian Orogen: Geological Society of America Memoir 197, p. 221–234.

McSween, H.Y., Jr., Sando, T.W., Clark, S.R., Harden, J.T., and Strange, E.A., 1984, The gabbro-metagabbro association of the southern Appalachian Piedmont: American Journal of Science, v. 284, p. 437–461.

Merschat, A.J., 2003, Inner Piedmont Tectonics in the Southwestern Brushy Mountains, North Carolina: Field and Laboratory Data Revealing 3-D Crustal Flow and Sillimanite I and II Metamorphism [M.S. thesis]: Knoxville, University of Tennessee, 198 p.

Merschat, A.J., and Hatcher, R.D., Jr., 2003, Digital geologic map of the northern Inner Piedmont, NC-SC-GA: Geological Society of America Abstracts with Programs, v. 35, no. 1, p. 59–60.

Merschat, A.J., and Kalbas, J.L., 2002, Geology of the southwestern Brushy Mountains, North Carolina Inner Piedmont: A summary and synthesis of recent studies, *in* Hatcher, R.D., Jr., and Bream, B.R., eds., Inner Piedmont Geology in the South Mountains–Blue Ridge Foothills and the Southwestern Brushy Mountains, Central-Western North Carolina: Knoxville, Tennessee, North Carolina Geological Survey, Carolina Geological Society Guidebook, p. 101–126.

Merschat, A.J., Hatcher, R.D., Jr., and Davis, T.L., 2005a, The northern Inner Piedmont, southern Appalachians, USA: Kinematics of transpression and SW-directed mid-crustal flow: Journal of Structural Geology, v. 27, p. 1252–1281, doi: 10.1016/j.jsg.2004.08.005.

Merschat, A.J., Gatewood, M.P., Fisher, C.M., Miller, C.F., Hatcher, R.D., Jr., Wooden, J.L., and Stahr, D.W., III, 2005b, The Newton antiform, NC: A previously unrecognized window through the allochthonous Inner Piedmont thrust stack: Geological Society of America Abstracts with Programs, v. 37, no. 7, p. 20.

Mirante, D.C., and Patino-Douce, A.E., 2000, Melting and migmatization in the southern Appalachian Inner Piedmont of northeast Georgia; the Athens gneiss: Geological Society of America Abstracts with Programs, v. 33, no. 7, p. 297.

Mittwede, S.K., Ødegård, M., and Sharp, W.E., 1987, Major chemical characteristics of the Hammet Grove meta-igneous suite, northwestern South Carolina: Southeastern Geology, v. 28, p. 49–63.

Nelson, A.E., Horton, J.W., and Clarke, J.W., 1998, Geologic Map of the Greenville 1° × 2° Quadrangle, Georgia, South Carolina, and North Carolina: U.S. Geological Survey Map I-2175, scale 1:250,000.

Osberg, P.H., Tull, J.F., Robinson, P., Hon, R., and Butler, J.R., 1989, The Acadian orogen, Chapter 4, *in* Hatcher, R.D., Jr., Thomas, W.A., and Viele, G.W., eds., The Appalachian-Ouachita Orogen in the United States: Boulder, Colorado, Geological Society of America, Geology of North America, v. F-2, p. 179–232.

Overstreet, W.C., Yates, R.G., and Griffits, W.R., 1963, Geology of the Shelby Quadrangle, North Carolina: U.S. Geological Survey Map I-384, scale 1:62,500.

Privett, D.R., 1984, The Turnersberg intrusive: Petrogenesis of a metamorphosed alpine ultramafite in the eastern Inner Piedmont Iredell County, North Carolina: Southeastern Geology, v. 25, p. 55–60.

Rankin, D.W., Drake, A.A., Jr., Glover, L., III, Goldsmith, R., Hall, L.M., Murray, D.P., Ratcliff, N.M., Read, J.F., Secor, D.T., Jr., and Stanley, R.S., 1989, Pre-orogenic terranes, *in* Hatcher, R.D., Jr., Thomas, W.A., and Viele, G.W., eds., The Appalachian Ouachita Orogen in the United States: Boulder, Colorado, Geological Society of America, Geology of North America, v. F-2, p. 7–100.

Thomas, W.A., 2006, Tectonic inheritance at a continental margin: GSA Today, v. 16, no. 2, p. 4–11, doi: 10.1130/1052-5173(2006)016[4:TIAACM]2.0.CO;2.

Trupe, C.H., Stewart, K.G., Adams, M.G., Waters, C.L., Miller, B.V., and Hewitt, L.K., 2003, The Burnsville fault: Evidence for the timing and kinematics of southern Appalachian Acadian dextral transform tectonics: Geological Society of America Bulletin, v. 115, p. 1365–1376, doi: 10.1130/B25256.1.

Tull, J.F., 2002, Southeastern margin of the middle Paleozoic shelf, southwesternmost Appalachians: Regional stability bracketed by Acadian and Alleghanian tectonism: Geological Society of America Bulletin, v. 114, p. 643–655, doi: 10.1130/0016-7606(2002)114<0643:SMOTMP>2.0.CO;2.

Williams, H., and Hatcher, R.D., Jr., 1983, Appalachian suspect terranes, *in* Hatcher, R.D., Jr., Williams, H., and Zietz, I., eds., Contributions to the Tectonics and Geophysics of Mountain Chains: Geological Society of America Memoir 158, p. 33–53.

Wilson, C.G., 2006, Origin and Tectonic Evolution of the Southern Appalachian Neoacadian Crystalline Core: Evidence from the Geology of the Gilreath 7.5-Minute Quadrangle, North Carolina [M.S. thesis]: Knoxville, University of Tennessee, 219 p.

Wilson, C.G., Merschat, A.J., Gatewood, M.P., and Hatcher, R.D., Jr., 2005, Metamorphism, kinematics, and crustal flow of the southern Appalachian orogenic core evidence from detailed geologic mapping in the Brushy Mountains, NC: Geological Society of America Abstracts with Programs, v. 37, no. 2, p. 40.

Manuscript Accepted by the Society 1 February 2007

Printed in the USA

The Geological Society of America
Memoir 200
2007

Rifted-margin architecture, cover stratigraphy, and structure of basement culminations, frontal Appalachian Blue Ridge, Georgia, USA

James F. Tull*
Department of Geological Sciences, Florida State University, Tallahassee, Florida 32306, USA

ABSTRACT

The three southernmost external Appalachian Grenville basement massifs occur in the Georgia western Blue Ridge, where they core map-scale, inclined-to-recumbent, west-vergent isoclinal anticlinoria formed during peak greenschist-facies Paleozoic metamorphism. The massifs are in thrust contact with Cambrian rocks of the adjacent foreland thrust belt. The basal cover unit of each massif, the Pinelog Formation, is likely correlative with the Late Proterozoic Snowbird Group of the Ocoee Supergroup. This sequence was deposited in fluvial-alluvial to shallow-water deltaic to intertidal environments adjacent to sharp basement uplifts, most probably along extensional fault scarps during initial Late Proterozoic rifting of Laurentia. A second phase of more extensive continental rifting and subsidence followed with deposition of the Great Smoky Group, a mostly deep-water turbiditic sequence. During the second rift cycle, a large (>2000 km^2) block, which included much of the Georgia western Blue Ridge and contained the deposits of the earlier rift basin, was tilted westward with the underlying basement along a southeast-dipping extensional fault system (stratigraphic offset >1 km) flanking the western margin of the Blue Ridge. Great Smoky units lie unconformably above this block and progressively cut deeper into the underlying cover sequence and then into the basement toward the east. This northwest-border fault system was bounded to the southwest by a large continental transfer (transform) fault, which marked the southern limit of the Ocoee basin, and across which the polarity of faulting along the rifted margin was reversed. Internal thrusts associated with Alleghanian continental collision, e.g., the frontal Blue Ridge thrust, were likely rooted below the basement massifs because the massifs had been previously detached from autochthonous basement by the two earlier cycles of rift-related faulting.

Keywords: two-stage rifting, Appalachian, basement massifs, Blue Ridge, Ocoee supergroup.

*tull@gly.fsu.edu

Tull, J.F., 2007, Rifted-margin architecture, cover stratigraphy, and structure of basement culminations, frontal Appalachian Blue Ridge, Georgia, *in* Hatcher, R.D., Jr., Carlson, M.P., McBride, J.H., and Martínez Catalán, J.R., eds., 4-D Framework of Continental Crust: Geological Society of America Memoir 200, p. 567–594, doi: 10.1130/2007.1200(28). For permission to copy, contact editing@geosociety.org.

INTRODUCTION

Numerous allochthonous Mesoproterozoic (Grenvillian, ca. 1.1 Ga) basement massifs occur in the Appalachian orogen from Newfoundland to Georgia in a discontinuous band in structural juxtaposition with Paleozoic sedimentary rocks immediately outboard of the southeastern flank of the Appalachian foreland. In the southern Appalachians, these external basement massifs occur as anticlinorial culminations within the western Blue Ridge, the frontal metamorphic allochthon (Fig. 1). The Blue Ridge massifs were detached with their metasedimentary cover rocks and thrust large distances northwestward over adjacent rocks of the foreland fold-and-thrust belt. Formation of the anticlinoria and the thrust sheets containing the massifs occurred during Paleozoic deformation, and the geometry and kinematic evolution of this deformation have been relatively well described and studied (e.g., Bartholomew, 1992). The positions and structural settings of these Mesoproterozoic blocks during latest Proterozoic to earliest Paleozoic Iapetan rifting and drifting are issues that are more obscure, however. Stratigraphic data are an important part of the database from which inferences can be made. On the basis of stratigraphic, structural, and/or isotopic evidence, several geologists have suggested that latest Proterozoic to earliest Paleozoic rifting along southeastern Laurentia either involved a very long protracted episode (possibly as much as ~140 m.y.) or more than one stage of rifting (Badger and Sinha, 1988; Bartholomew, 1992; Li and Tull, 1998). Different segments of the rifted margin may have different rift histories that record different rift stages (Bartholomew and Tollo, 2004). Multistage rifting can be expected to involve significant modification of basins formed during an earlier stage, as well as the production of complex basement and cover relationships.

As a preamble to the opening of the Iapetus Ocean, the Ocoee Supergroup, composed predominantly of clastic sediments, was formed in a large intracratonic (intra-Rodinian) rift basin(s) along the future southeast margin of Laurentia at the current latitude of the Tennessee structural salient (Fig. 1), attaining a thickness >10 km (King et al., 1958; Rankin, 1975). These rocks are now contained within the western Blue Ridge allochthon, and they extend for ~350 km along structural strike from north Georgia at least into northwesternmost North Carolina, and across strike for ~100 km. Li and Tull (1998) proposed that the Ocoee is a composite sequence that records two stages of rifting. Because most of the margins of this regionally extensive rift sequence are no longer exposed, their geometries must be inferred. Metamorphism, polydeformation, and thrust dismemberment of the basin and its margins make these inferences difficult.

The three southernmost external basement massifs in the orogen (Corbin, Salem Church, and Fort Mountain; Fig. 2) occur in north Georgia. These small massifs are isolated from the other massifs and extend discontinuously over a strike length of ~75 km. The structural and stratigraphic settings of these Mesoproterozoic rocks and their immediate cover sequences during Iapetan rifting are the focus of this study. It will be shown here that these three crystalline basement massifs share a very similar cover stratigraphy and structure. Each occurs within the core of a series of west-verging, steeply inclined to recumbent, doubly plunging, refolded isoclinal anticlines, immediately above the frontal western Blue Ridge thrust. Each of these basement-cored anticlines was thrust above a footwall composed of Lower Cambrian units in the immediately adjacent Appalachian foreland, and each massif is directly overlain by two unconformably bounded,

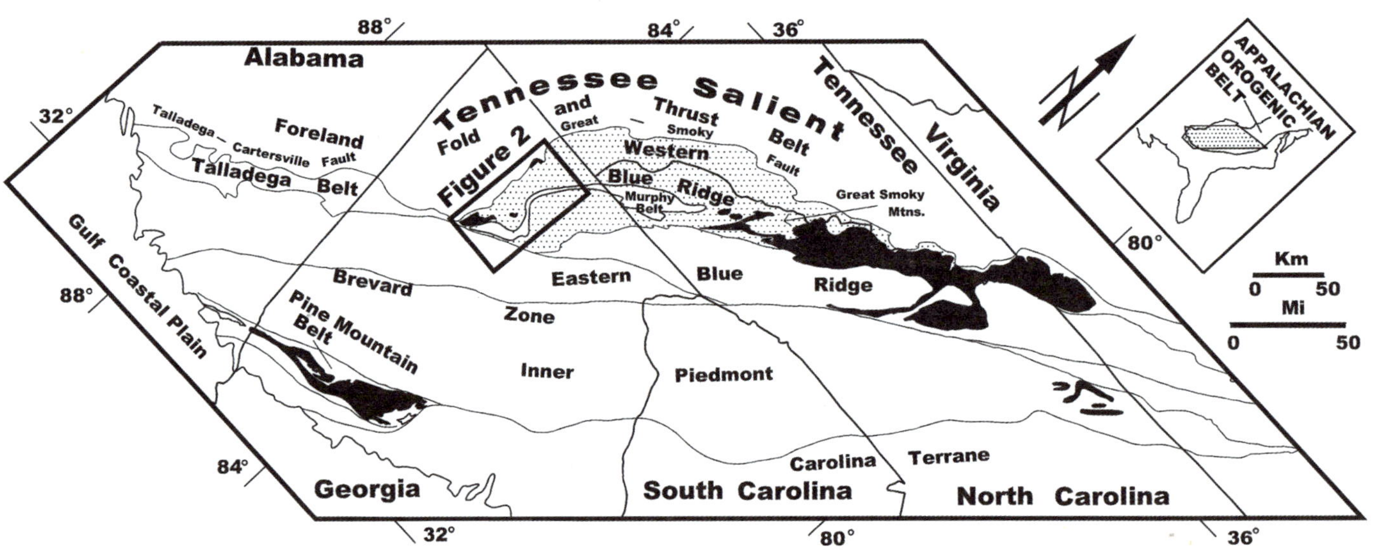

Figure 1. Map of the southern Appalachian orogen showing the major tectonic subdivisions, distributions of basement massifs (black) and the Ocoee Supergroup (stipple), and location of Figure 2.

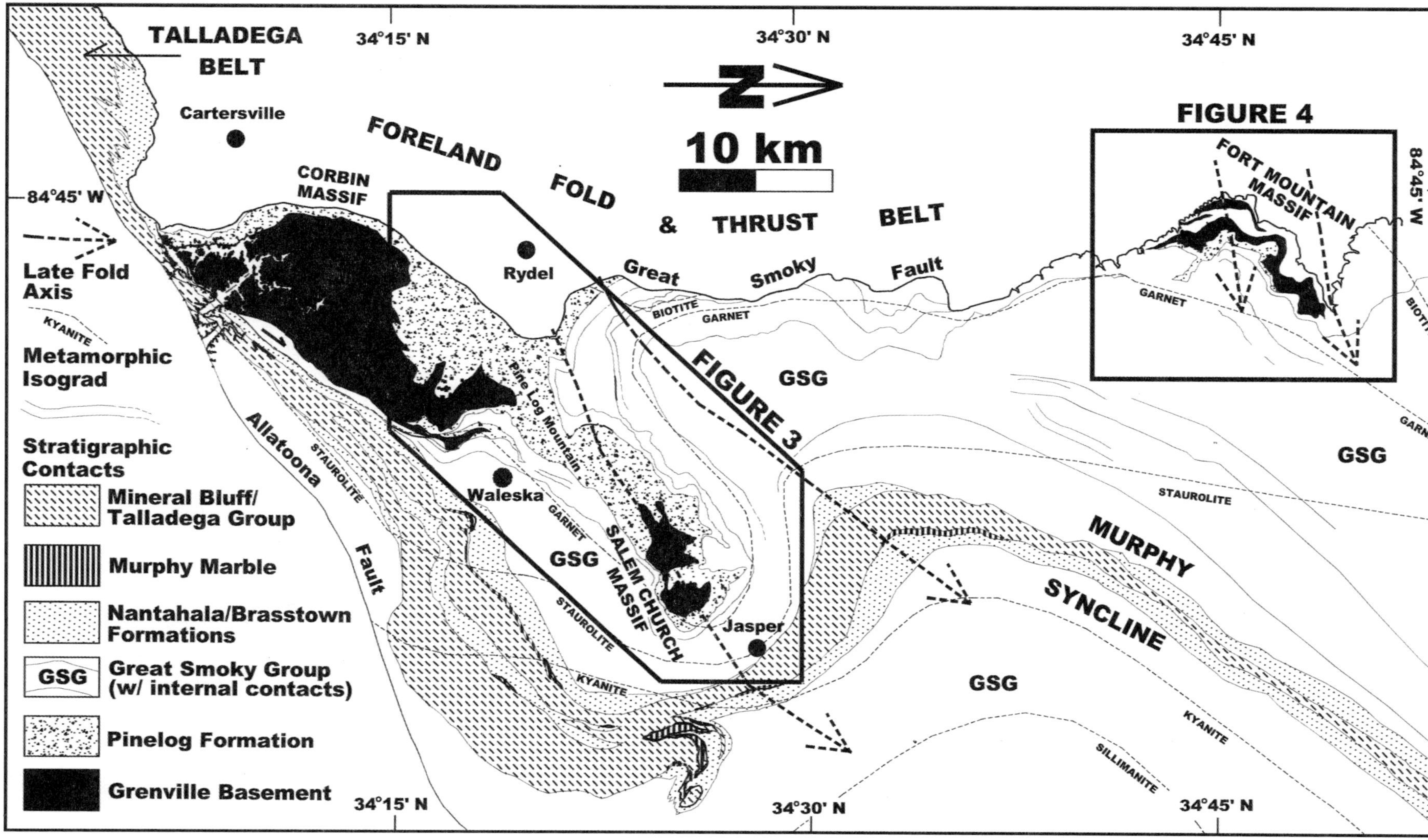

Figure 2. Generalized geologic map of the western Blue Ridge in Georgia showing the locations of the three southernmost external basement massifs (solid) and Figures 3 and 4.

greenschist-facies cover sequences of the Late Proterozoic Ocoee Supergroup, which represent two distinct cycles of rift-basin formation during early developmental stages of the Iapetus Ocean.

GEORGIA BLUE RIDGE GRENVILLE BASEMENT MASSIFS

C.W. Hayes (unpublished manuscript cited in Hull et al., 1919), a geologic pioneer in the southern Appalachians, first reported that masses of metaplutonic rocks along the west flank of the Blue Ridge in northwest Georgia predate surrounding metasedimentary sequences, and thus they form what was once the basement of the Blue Ridge stratified metasedimentary rocks. La Forge (1919, p. 40) also argued that these metaplutonic rocks had "furnished much of the material for the overlying metamorphosed sediments." Three decades later, in their study of the Fort Mountain area, Furcron and Teague (1947) recognized that heterogeneous metaplutonic rocks occur nonconformably beneath rocks of the Late Proterozoic Ocoee "series" (now Supergroup). Thus, during the first half of the last century, the insightful recognition by these men of the presence of basement rocks in northwest Georgia that predate the Appalachian orogenic cycle helped to establish an important foundation for understanding of Appalachian tectonics. Much later, their ideas related to the stratigraphic position of the Corbin (1.1 and 1.5 Ga), Salem Church, and Fort Mountain (1.03 Ga) massifs were confirmed isotopically (U-Pb zircon) (Odom et al., 1973; Russell, 1976; Heatherington et al., 1996).

The two southern basement massifs (Corbin and Salem Church massifs) are found in adjacent culminations, whereas the third (Fort Mountain) is isolated along the thrust front some 40 km to the north in three adjacent parallel basement lenses (Fig. 2). The Corbin massif is by far the largest and best studied (Kesler, 1950; Martin, 1974; Costello, 1978; McConnell and Costello, 1980; Gargi, 1985; Li, 1996; Kath and Crawford, 1996; Higgins et al., 1996). Coarse-grained granitic metaplutonic rocks of calc-alkaline affinity constitute most of the basement complex in all three of the massifs and include the Corbin Metagranite (Gneiss) (Costello, 1978; Gargi, 1985; Higgins et al., 1996), the Salem Church Granite (Bayley, 1928), and the Fort Mountain Gneiss (Furcron and Teague, 1947). The Corbin Metagranite ranges from megacrystic granodiorite through quartz monzonite to granite (Martin, 1974; Li, 1996). The Fort Mountain Gneiss overlaps the Corbin in composition somewhat, but is dominated by more siliceous, "quartz-rich" granitoids. More than one granitic phase appears to be present in each of the massifs. Metamorphic alteration of original granitoids is highly variable, from slight sericitization around the rims of igneous feldspars, to more pervasive modification to porphyroclastic augen gneiss and cataclastic schist, including extensive alteration of feldspars to white mica, and locally, calcite. No metamorphic fabric, however, has been recognized within the metaplutonic units that predates that within the cover rocks.

Field mapping indicates that the metaplutonic rocks contain numerous xenoliths and roof pendants(?), the largest of which is 4 km long and 800 m wide, consisting of supracrustal metasedimentary(?) and meta-igneous rocks (Kesler, 1950; McConnell and Costello, 1980, 1984; Gargi, 1985; Kath and Crawford, 1996). In the Corbin massif, these rocks are divided into the Rowland Springs and Red Top Mountain Formations, both of which were intruded by the Corbin Metagranite (Higgins et al., 1996). The former, of diorite/andesite composition, contains mineral assemblages that indicate it was metamorphosed to pyroxene granulite facies (Gargi, 1985; Kath and Crawford et al., 1996). The Red Top Mountain Formation, locally intrusive into the Rowland Springs Formation, is represented by small bodies of mafic meta-igneous rocks (Higgins et al., 1996). The central and eastern lenses of the Fort Mountain Gneiss contain distinctive bodies of magnesium-rich mafic and ultramafic schist, up to several hundred meters long, described by Furcron and Teague (1947) as the Cohutta Schist. Most of these bodies contain lenses and pods of talc that were commercially mined in the area for over half a century. The origin of these bodies is enigmatic but they probably represent low-grade metamorphic alteration of mafic and ultramafic (trace element contents: Cr = 2292 ppm, Ni = 819 ppm) xenoliths or intrusions within the Fort Mountain Gneiss (Needham, 1972).

Higgins et al. (1996) argued that the granulite facies event documented in the Rowland Springs Formation predated intrusion of the Mesoproterozoic Corbin Metagranite. Heatherington et al. (1996), however, interpreted ion microprobe U-Pb zircon ages to suggest that the main Corbin phase is no younger than 1106 ± 13 Ma, approximately coincident with the main (Ottawan) phase of the Grenville orogeny. They interpreted younger zircon rim ages (at 1013–1061 Ma) as possibly indicative of later granulite-facies metamorphism and potentially increased thermal gradients associated with Neoproterozoic rifting. It appears, however, that no granulite-facies mineral assemblages, if they ever existed, are preserved in any of the region's most dominant phases of metaplutonic rocks proper (Gargi, 1985). The metaplutonic rocks of each of the three massifs contain low-grade metamorphic mineral assemblages and fabrics compatible with those in the surrounding Late Proterozoic to Paleozoic cover rocks. These assemblages were therefore most likely established during Paleozoic orogeny. In many zones in the basement metaplutonic rocks, this Paleozoic deformation was so intense that the protoliths were converted to mylonites, low-grade schists, and phyllonites (Gargi, 1985).

COVER SEQUENCES

The stratigraphic sequences overlying each of the three northwest Georgia basement massifs are strikingly similar, both lithologically and in terms of lithostratigraphic sequence, and they display similar geometric relationships to the basement and to one another around each massif. Knowledge of the age and tectonic setting of these cover sequences is critical to understanding the tectonic evolution of the basement massifs, but for decades, much debate has focused upon these issues. Therefore, a detailed discussion of the age and correlation of the cover sequences is necessary.

Basement and Cover Contact

The basement and cover contact separates basement country rocks, which underwent high-grade Middle(?) Proterozoic metamorphism and were subsequently enveloped by deep-seated Mesoproterozoic plutonic rocks, from stratigraphically overlying stratified rocks that underwent only low-grade Paleozoic metamorphism. Most geologists agree that this contact represents a locally highly sheared nonconformity (Hayes, unpublished manuscript cited in Hull et al., 1919; La Forge, 1919; Furcron and Teague, 1947; McConnell and Costello, 1984; Higgins et al., 1996; Li, 1996; Li and Tull, 1998). The composition of the immediately overlying stratified sequence indicates direct derivation via erosion of the basement massifs. For example, metaconglomerates and metawackes contain abundant pebbles and grains of angular microcline and distinctive subrounded to rounded blue quartz, both characteristic of the underlying Corbin Metagranite. Cobble-bearing conglomerates contain lithic clasts of Corbin-like granite (McConnell and Costello, 1984; Higgins et al., 1996; Li, 1996), and cross-beds immediately above the contact consistently indicate younging toward the overlying cover sequence. Some paleotopography may have existed at the base of the cover sequence, but transposition along this contact makes any paleotopography difficult to recognize.

Pinelog Formation

Description

The stratigraphically lowest cover unit in the region is the Pinelog Formation (Pinelog conglomerate of Hayes, 1895, 1901; unpublished manuscript cited in Hull et al., 1919). This unit lies directly on the basement rocks and consists of a thick sequence of metawacke/metasandstone and metaconglomerate, and subordinate metapelite. Heterogeneous strain, polyfolding, and erosion beneath an overlying unconformity make thickness estimates of the unit only approximate. The thickness of the unit is ~1100 m near Pine Log Mountain, but it thins northeastward to ~400 m on the northwest side, and ~360 m on the southeast side of the Salem Church culmination (Fig. 3). At Fort Mountain, the Pinelog has a maximum thickness of only ~250 m. The unit is locally absent on the southeast side of the Corbin culmination and is absent around the western and central basement lenses and on the west flank of the eastern basement lens at Fort Mountain (Figs. 3 and 4). These thickness variations are interpreted to be the result of erosion beneath an overlying unconformity (Li and Tull, 1998) (see following).

Because of the presence of this unconformity, the original maximum thickness of the Pinelog Formation is unknown. Previous workers did not differentiate the Pinelog as a separate unit at Fort Mountain, mapping it as undifferentiated Ocoee Series (Furcron and Teague, 1947) or Great Smoky Group (Ocoee Supergroup) (McConnell and Costello, 1984). However, its stratigraphic position and lithologic character unequivocally link the Fort Mountain rocks to the Pinelog in the type area 50 km to the south. Around the Corbin and Salem Church massifs, where the unit reaches its maximum thickness, Li and Tull (1998) divided the Pinelog into three mappable subunits (Fig. 3), with maximum estimated thicknesses of the lower, middle, and upper subunits of ~600 m, ~100 m, and ~400 m, respectively.

The resistant lower Pinelog subunit is responsible for some of the highest elevations in the Georgia western Blue Ridge (900 m elevation and a relief of >600 m at Fort Mountain; 730 m elevation and a relief of >400 m at Pine Log Mountain). This subunit is dominated by poorly to moderately sorted, coarse-grained, mostly matrix-supported quartzofeldspathic rocks, including metaquartzwacke, feldspathic and arkosic metawacke, subgraywacke, pebbly metasandstone, and granule and pebble metaconglomerate, with minor metapelite and metadiamictite. As Hayes (unpublished manuscript cited in Hull et al., 1919) first recognized, the composition of much of the lower Pinelog Formation indicates direct derivation via erosion of the basement massifs without significant transport. Some of the rocks range so closely in mineralogical composition to the basement rocks that it is difficult to differentiate them from basement where the contact zone is highly strained. As noted by LaForge (1919), many of these rocks probably represent gruss (residual or sedentary arkose) derived from in situ disintegration of the basement granite. Lower Pinelog metaconglomerates, metasandstones, and metawackes contain abundant angular to subangular feldspar and distinctive subrounded blue quartz. The feldspar is dominantly K-feldspar (microcline, perthite), but several samples also include appreciable plagioclase, with K-feldspar/plagioclase ratios ≅1. Conglomerates locally contain lithic clasts of granite and gray phyllite. Accessory minerals include calcite, chlorite, epidote, clinozoisite, pyrite, hematite, limonite, and trace amounts of zircon, sphene, apatite, graphite, ilmenite and magnetite. Tourmaline is notably very rare to absent as an accessory phase. Figure 5B compares the mineralogy of the lower Pinelog at Fort Mountain with that of the lower and middle Pinelog around the two basement massifs to the south. Siliceous marble occurs within the lower Pinelog Formation in at least two localities (McCallie, 1907; Kesler, 1950; Higgins et al. 1988, 1989; Li, 1996).

The middle Pinelog subunit (not preserved at Fort Mountain) occurs gradationally above the lower subunit around the Corbin and Salem Church culminations, and consists predominantly of thinly bedded, grain-supported, locally trough cross-bedded quartzite, ranging from muddy sandstone/subarkose to quartz arenite (Fig. 5B), with local very thin interbeds of gray sericite phyllite. Bed thickness ranges from a few to ~20 cm. A distinctive part of the unit contains thinly bedded, very mature, fine-grained metaquartz arenite several tens of meters thick. This subunit is generally less feldspathic and contains somewhat less pelitic material than the lower subunit (Fig. 5B). The quartzite

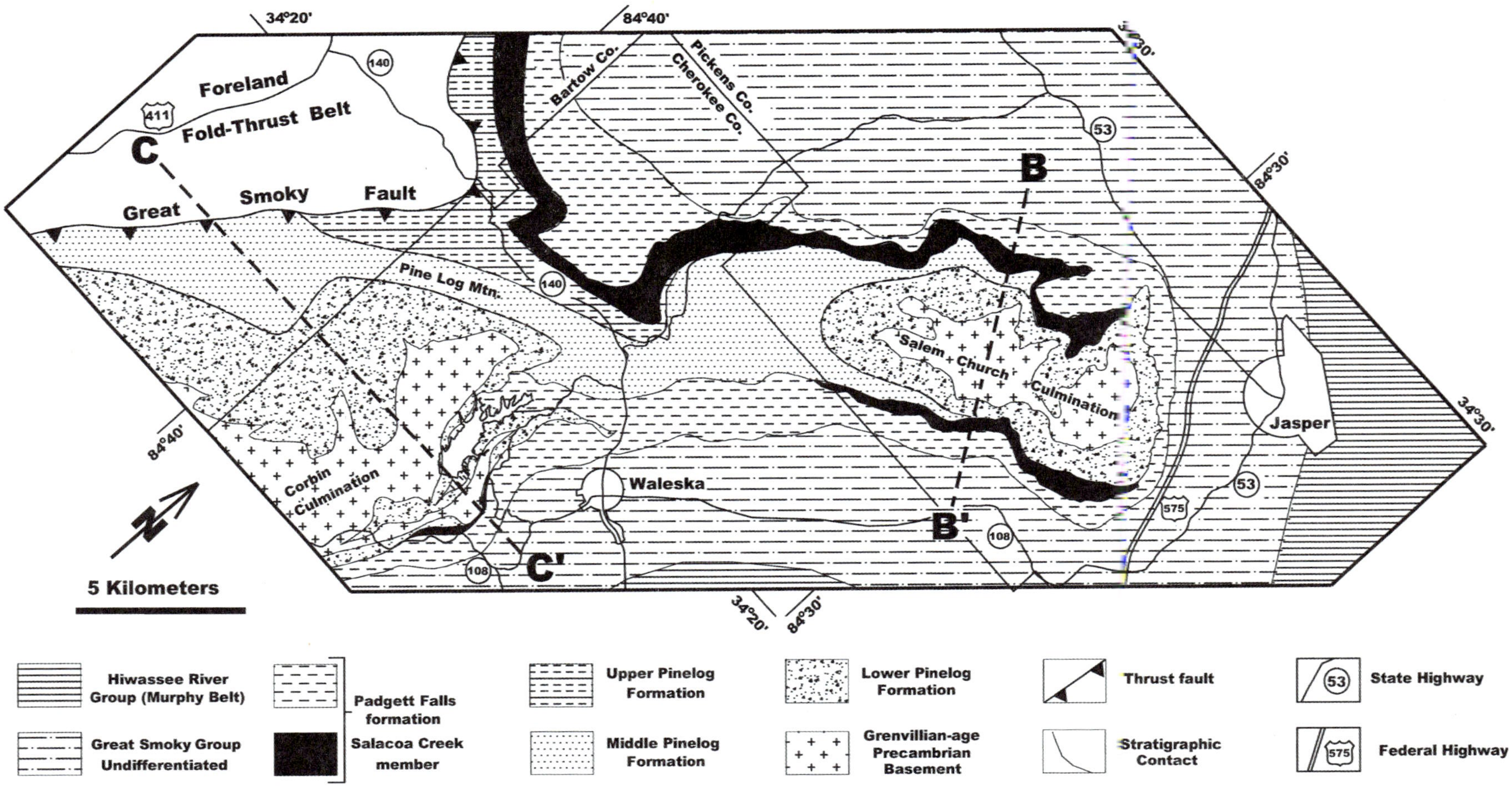

Figure 3. Geologic map of the Corbin and Salem Church basement culminations (modified from Li and Tull, 1998). Locations of cross section B–B' and C–C' in Figure 7 are shown.

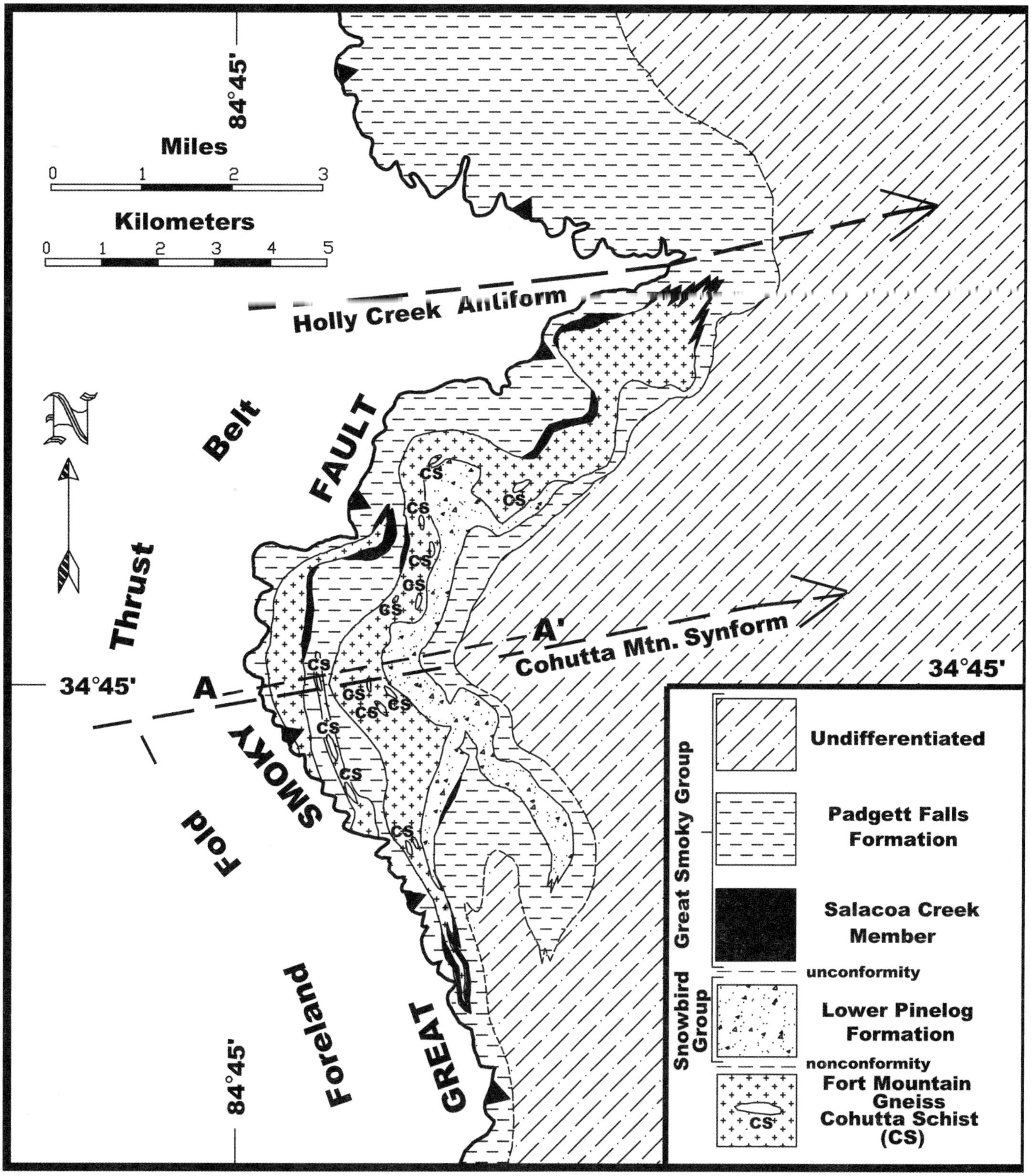

Figure 4. Geologic map of the Fort Mountain basement culmination. Location of cross section A–A' in Figure 7 is shown.

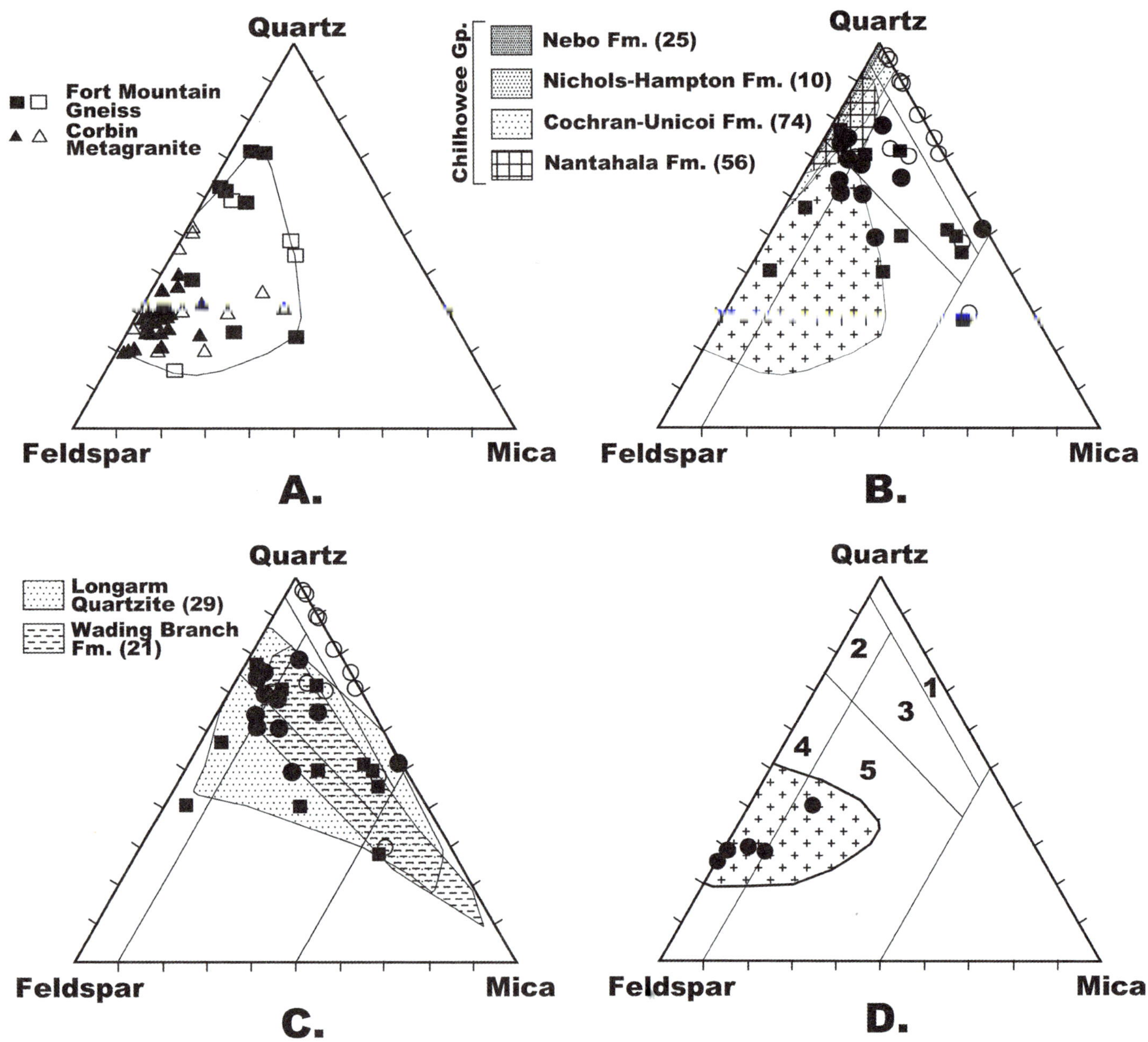

Figure 5. (A) Quartz-feldspar-mica (Q-F-M) ternary of Corbin Metagranite (triangles) and Fort Mountain Gneiss (squares). Data sources: closed triangles—Martin (1974; 23 samples), open triangles—Li (1996; 13 samples), closed squares—current study (8 samples), open squares—McConnell and Costello (1984; 4 samples). (B) Q-F-M ternary of Pinelog Formation metawackes and metasandstones showing fields of basement (plus symbols) (from Fig. 5A), and Chilhowee Group sandstones (from Whisonant, 1967), and Nantahala Formation sandstones (from Aylor, 1994). Sample totals are in parentheses. Lower Pinelog from Corbin and Salem Church massifs are denoted by solid squares (Li and Tull, 1998; 11 samples); lower Pinelog from Fort Mountain massif are denoted by solid circles (12 samples); middle Pinelog from Corbin and Salem Church massifs are denoted by open circles (Li and Tull, 1998; 12 samples). Classification fields are numbered and named in D. (C) Q-F-M ternary of Pinelog metawackes and metasandstones (symbols and samples same as in B) showing fields of the Snowbird Group's Wading Branch Formation (dashed) (21 samples) and Longarm Quartzite (stippled) (29 samples) (from Hadley and Goldsmith, 1963). (D) Q-F-M ternary of samples of granite pebbles in the Salacoa Creek Member diamictite from localities around the Corbin and Salem Church massifs. Field of Corbin Metagranite is denoted by plus symbols. Classification of fields in B and C: 1—muddy sandstone, 2—subarkose, 3—muddy subarkose and feldspathic siltstone, 4—arkose, 5—muddy arkose and feldspathic siltstone.

is highly recrystallized, but relict primary grains indicate that its original grain size was fine- to medium-grained sand. The percent of fine-grained muscovite increases gradually toward the top of this subunit, and northwest of Pinelog Mountain (Fig. 3), the quartzite grades into quartz mica phyllite/schist and mica phyllite/schist of the upper Pinelog Formation.

The upper Pinelog subunit is preserved only along the northwest flank of the Corbin culmination where it gradationally overlies the middle Pinelog Formation (Fig. 3). This subunit is up to ~400 m thick and consists predominantly of quartz-sericite phyllite/schist and mica phyllite/schist, locally containing coarse clasts of quartz and feldspar interlayered with minor, thinly bedded micaceous metasandstone/metawacke and pebbly metasandstone.

Depositional Setting

Poor to moderate roundness of clasts, poor sorting, relative abundance of feldspar, plus locally preserved graded beds, trough cross-beds, and other stream-channel features (consistently facing away from the basement) suggest that rocks of the lower Pinelog were deposited in a fluvial-alluvial environment. Collectively, these features suggest sharp adjacent uplifts, creating high relief, and rapid mechanical weathering of granitic basement, most probably along extensional fault scarps. Although some zones of the middle Pinelog are also matrix-rich and feldspathic, Li and Tull (1998) suggested that, based upon higher degrees of roundness and sorting and more mature composition and texture, sediments of the middle subunit underwent significant transport and/or reworking in a high-energy environment, and thus were probably deposited in fluvial-alluvial to shallow-water deltaic to intertidal environments. Feldspar clasts become less abundant and finer-grained up-section. Cross-beds show both unidirectional and bidirectional geometry. Trace fossils in this part of the sequence near Salem Church suggest intergrading of fluvial and shoreface facies (Martin and Crawford, 1996).

Because of the predominantly fine size of primary grains and the abundance of pelitic material, Li and Tull (1998) interpreted the upper Pinelog to be the result of mainly suspended load, and they suggested that it was deposited more distally from its source than the lower two subunits. It was most likely deposited in a prodelta to delta-front environment.

In summary, the characteristics of the Pinelog Formation siliciclastic rocks suggest that this sequence represents a transgressive sequence that fines upward. It was initially deposited in a terrestrial alluvial fan-type environment well supplied by a nearby granitic source, and it is overlain by units that suggest a transition into a shallow-marine environment. These relationships were interpreted by Li and Tull (1998) to indicate that the Pinelog Formation consists of sediments formed within a continental rift basin, initially as fluvial-alluvial fan deposits adjacent to faulted and sharply uplifted Laurentian basement, followed by stabilization and marine incursion.

Correlation

Accurate stratigraphic assignment of the metasedimentary cover sequences that mantle the basement massifs in northwest Georgia is an essential step in understanding the early tectonic development of this segment of the Laurentian margin. However, correlations by various geologists of the cover rocks in contact with basement have focused on two sequences of fundamentally different tectonic setting and age: (1) the rift-associated Late Proterozoic Ocoee Supergroup (Hayes, unpublished manuscript cited in Hull et al., 1919; Hadley, 1970; McConnell and Costello, 1980, 1984; Costello et al., 1982; McConnell and Abrams, 1984; Costello, 1988, 1993; Higgins et al., 1988; Li, 1996; Li and Tull, 1998; Tull and Holm, 2005) versus (2) the drift-related Lower Cambrian Chilhowee Group (La Forge in Hull et al., 1919; Kesler, 1950; Cressler, 1970; Sears, 1988; Higgins et al., 1989, 1996). Until recently, however, correlations of cover units have been based only upon very general lithologic characteristics. No detailed comparisons with sequences 1 and 2 have been made, in part because of a previous lack of detailed petrologic descriptions of the cover units, and because no internal stratigraphy of the cover units has been established.

The Ocoee Supergroup, which is composed predominantly of clastic sediments, is interpreted to have formed along the southeast margin of Laurentia within a large intracratonic rift basin (King et al., 1958; Rankin, 1975) (Fig. 1). Over four decades ago, geologists with the U.S. Geological Survey conducted comprehensive studies of this sequence in the Great Smoky Mountains of Tennessee and North Carolina (Fig. 1) and succeeded in subdividing it into three groups: the basal Snowbird Group (~4 km thick), the overlying Great Smoky Group (~7 km thick), and an upper Walden Creek Group (~2.6 km thick) (King et al., 1958). Difficulties were noted, however, when correlation of these units across tectonic boundaries was attempted (King et al., 1958; Rast and Kohles, 1986). Recently, for example, Unrug et al. (2000) argued that some formations of the Walden Creek Group contain Paleozoic fossils indicative of a Silurian or younger age. However, most geologists still argue on various stratigraphic or isotopic grounds for a Late Proterozoic age for the Snowbird and Great Smoky Groups (e.g., Aleinikoff et al., 2004).

In contrast, the Lower Cambrian Chilhowee Group, which stratigraphically overlies the Ocoee, is a continental to marine shelf clastic sequence that marks the tectonic transition from continental rifting and seafloor spreading to stabilization of the Laurentian margin as a passive margin during widening of the Iapetus Ocean (Rankin, 1993; Simpson and Eriksson, 1989). This unit records the initial postrift thermotectonic subsidence, and it extends along the full strike length of the ancient Laurentian margin, as well as for significant distances across that margin. Northeast and northwest of the Ocoee basin, it oversteps the older Ocoee rift-related sequences and associated bounding faults and lies directly on the basement (King, 1970). To the southeast, in the center of the western Blue Ridge, the Murphy belt occupies the core of the Murphy synclinorium, a first-generation, northwest-overturned, doubly plunging, large-amplitude and large-wavelength (>5 km) isoclinal fold that extends for >170 km (Fig. 1). For many years, geologists have believed that units stratigraphically above the Ocoee sequence in the core of the Murphy synclinorium (Fig. 2) were Paleozoic, and this was confirmed by the discovery of a Paleozoic fossil at the base of the Mineral Bluff Group near the top of the Murphy stratigraphy (Tull et al., 1993). On the limbs of this synclinorium, the Great Smoky Group is gradationally overlain by the Nantahala Formation (Bayley, 1928; Hurst, 1955; Fairley, 1965, 1973; Mohr, 1973; McConnell and Costello, 1980, 1984; McConnell and Abrams, 1984; Costello, 1988, 1993; Aylor, 1994; Groszos, 1996) (Fig. 2). The Nantahala and the stratigraphically overlying Brasstown Formation likely represent the distal, outer margin, shallow-marine equivalents of the Chilhowee Group (Power and Forrest, 1973; Aylor, 1994). These units are overlain by the Murphy Marble, a unit that most geologists since Keith (1907) interpret as the lowermost part of the Cambrian-Ordovician peritidal-shelf carbonate bank. These correlations indicate that during the passive drifting stage of Iapetus, the drift-facies sequence now found in the Murphy belt prograded southeastward across a broad section of crust previously significantly thinned and extended during Neoproterozoic rifting. Therefore, at the paleoposition of the Murphy belt, the Chilhowee Group must have been separated from the Grenville basement by several kilometers of rift-facies Ocoee Group sediments. Thus, any major rift-boundary fault(s) associated with the northwest flank of the Ocoee basin must lie northwest of the Murphy belt.

Next, the two possible alternative correlations of the cover sequence directly mantling the Georgia basement massifs (Pinelog = Chilhowee Group versus Pinelog = Ocoee Supergroup) are contrasted relative to several key issues:

1. If the Chilhowee and Pinelog are equivalent, then the thick Ocoee Supergroup rift-facies sequence is absent above basement over an extensive area along the west flank of the southern Blue Ridge, making the Georgia basement massifs the only locations along a >200-km-long segment south of the Great Smoky Mountains of the Tennessee structural salient (Fig. 1) where the Ocoee is absent between the Chilhowee and the basement.

2. If the Pinelog and the Nantahala-Brasstown Formations in the Murphy belt are equivalent to the Chilhowee Group, then the thick Ocoee sequence beneath the Murphy belt must pinch out between that belt and the basement massifs, implying a significant palinspastic distance between the former and the latter, and probably initial separation between them by a large-displacement (several km), down-to-the-east, rift-boundary fault (e.g., Li and Tull, 1998, Fig. 9 therein). An Ocoee-equivalent Pinelog, however, does not require a down-to-the-east rift-boundary fault separating the basement massifs from the Ocoee stratigraphy beneath the Murphy belt, which would allow for continuity of the Ocoee between the two entities.

3. Because it faces upward (eastward) into the structurally overlying, upright Ocoee Supergroup (Great Smoky Group), a Chilhowee-equivalent Pinelog must be in contact with the overlying sequence along a major fault. Thus, geologists (e.g., Higgins, et al., 1996) who correlate the Pinelog and Chilhowee interpret the Pinelog to be bounded above by a postmetamorphic fault, although the geographic location, lateral continuity, stratigraphic position, and sense of slip of such a proposed fault (e.g., Lost Town Creek fault) vary among the different authors proposing one (McConnell and Costello, 1980; Costello et al., 1982; McConnell and Abrams, 1984; Costello, 1988, 1993). The proposed fault neither offsets nor telescopes metamorphic isograds. Abundant evidence exists for a late-stage strike-slip fault(s) along the southeast flank of the Corbin massif (McConnell and Costello, 1980; Costello et al., 1982; Higgins, et al., 1996; Tull and Holm, 2005) (Fig. 2), but extension of this fault northeastward, where it is interpreted as a major thrust fault at or near the upper boundary of the Pinelog (Higgins et al., 1996; Costello, 1988,1993) is a point of contention (e.g., Li and Tull, 1998, p. 851). Importantly, previous workers did not map a fault at Fort Mountain between the Pinelog and the overlying Great Smoky Group; they interpreted the cover rocks as a continuous sequence (Furcron and Teague, 1947; McConnell and Costello, 1984). With an Ocoee-equivalent Pinelog, there is no stratigraphic requirement that the upright Pinelog be separated by a major fault from the structurally overlying and also upright Ocoee Supergroup to the east (northwest limb of Murphy syncline).

4. A Chilhowee-equivalent Pinelog should contain an internal stratigraphy similar to that of the type Chilhowee Group immediately to the north, or, alternatively, similar to the presumed more-distal Chilhowee-equivalent Nantahala-Brasstown Formations immediately to the east. The lower Pinelog Formation should thus be correlative with the basal Chilhowee Formation (Cochran-Unicoi), or with the Nantahala Formation in the Murphy belt. Some similarities exist with the Cochran-Unicoi, which contains locally graded, poorly sorted coarse conglomerates with mudstone fragments and quartzofeldspathic lithic clasts. The Cochran-Unicoi conglomerates are interbedded with massive to thinly laminated, fine- to very coarse, poorly sorted, arkosic sandstone and greenish siltstone and mudstone (Mack, 1980). In Tennessee, the minimum thickness of the Cochran is 220–470 m (Whisonant, 1967). The upper part (<160 m thick) contains sandstones that have a low feldspar content, comparable perhaps to the middle Pinelog Formation (~100 m thick). The Cochran is overlain by the Nichols Shale (Hampton Formation in Virginia and northeast Tennessee), which is ~125 m thick and consists of gray to black, micaceous mudstone interbedded with minor siltstone and very fine-grained sandstone (Mack, 1980). The Nichols could correlate with the upper Pinelog Formation, which consists of gray sericite phyllite interbedded with thin-bedded metasandstone. If these correlations are valid, however, then Chilhowee units above the Nichols must be absent from the basement massifs. This includes the upper formations of the group (Wilson Ridge and Weisner Formations), which consists of interbedded orthoquartzites, feldspathic quartz arenites, and conglomerates, with an aggregate thickness of ~575 m.

The Nantahala Formation conformably overlies the Great Smoky Group (Ocoee Supergroup) along the flanks of the Murphy synclinorium (Fig. 2). In a reference section west of Murphy, North Carolina, it is 1550 m thick, but it thins to 310 m in the southern Murphy belt in Georgia (Aylor, 1994). Most of the Nantahala is a dark laminated and banded, carbonaceous metasiltstone, but this lithology is interbedded with metasandstone (Tusquitee Quartzite of Keith, 1907) concentrated in the upper half or near the top of the formation. Metasiltstone strongly dominates the lower half. In ten measured sections along its outcrop length, Aylor (1994) found that metasandstone averaged only 22% of the unit. The overlying Brasstown Formation, predominantly a dark-gray interlaminated metasiltstone-metasandstone sequence, ranges in thickness between 215 m in the southwest and 2396 m in the northeast (Aylor, 1994). Sandstones are fine-grained and rarely >1 cm thick. Thus, there is some internal stratigraphic similarity between the Pinelog and the lower half of the Chilhowee Group, but the upper three-quarters of the Chilhowee must be missing from the Pinelog. There is very little lithologic or internal stratigraphic similarity between the Pinelog and the presumed Chilhowee-equivalent clastic sequence in the Murphy belt.

If the Pinelog is equivalent to part of the Ocoee, however, then it should contain an internal stratigraphy similar to that found in other regions containing the basal Ocoee unit (Snowbird Group). In the Great Smoky Mountains, the Snowbird Group nonconformably overlies the basement complex (Hadley and Goldsmith, 1963). Although the unit is locally as thick as ~4 km, individual formations thin significantly toward the southeast in the Great Smoky Mountains (Montes and Hatcher, 1999). Hadley and Goldsmith (1963) divided the group into five formations: the basal Wading Branch Formation, Longarm Quartzite, Roaring Fork Sandstone, Pigeon Siltstone, and Metcalf Phyllite. In the Greenbrier fault's hanging wall, only the first three occur (Hadley and Goldsmith, 1963). The Wading Branch (30–450 m thick) is a heterogeneous sequence of sandy argillite, siltstone, wacke, and poorly sorted, pebbly, feldspathic sandstone with quartz pebble conglomerate lenses, nonconformably overlying basement (Hadley and Goldsmith, 1963). Sandstones are locally graded, contain chips and slabs of metapelite, and have a high matrix mica content (Fig. 5C). The formation has been interpreted as reworked alluvial deposits in discontinuous depressions overlying a highly weathered Grenville basement (Hadley, 1970). In gradational contact above the Wading Branch is the Longarm Quartzite (15–1500 m thick), which is dominated by medium- to coarse-grained, moderately well-sorted, cross-bedded, feldspathic quartzite (subarkose to arkose) (Hadley and Goldsmith, 1963). Longarm primary structures are indicative of a transgressive fluvial-deltaic setting (Hadley, 1970; De Windt, 1975). Intertonguing with the Longarm is the overlying Roaring Fork Sandstone, a sequence of interbedded medium- to fine-grained, well-sorted sandstone, argillite, and siltstone locally >2500 m thick (King, 1964). Gradationally above the Roaring Fork is the Pigeon Siltstone, a very thick, uniform sequence of meta-argillite. Although King (1964) interpreted the Pigeon and overlying Metcalf Phyllite to be lateral equivalents, De Windt (1975) considered the Roaring Fork and Pigeon succession to have been deposited as delta-front, interdeltaic, and prodeltaic sediments, with the overlying Metcalf representing more basinal deposits. A number of important stratigraphic similarities thus exist between the Pinelog Formation and the Snowbird Group. For example, the lower Pinelog is interpreted to have formed in an alluvial to fluvial-deltaic setting (Li and Tull, 1998) similar to the Snowbirds' Wading Branch Formation and Longarm Quartzite (Hadley and Goldsmith, 1963; De Windt, 1975). The middle Pinelog also has lithologic similarities to the Longarm; both are interpreted to represent transgressive fluvial-alluvial to shallow-water deltaic to intertidal sequences, including littoral sands (De Windt, 1975; Li, 1996; Martin and Crawford, 1996; Li and Tull, 1998). The upper Pinelog has lithologic characteristics resembling the Roaring Fork Sandstone. Uppermost Snowbird units possibly equivalent to the Pigeon Siltstone and Metcalf Phyllite appear to be absent above the Pinelog, but the latter unit is interpreted to be incomplete due to erosion beneath an overlying unconformity (see following).

5. A Chilhowee-equivalent Pinelog should be petrologically similar to the former or its equivalents in the Murphy belt. Close petrographic examination demonstrates, however, that this is not the case. Cochran sandstones have an average matrix content of 15%–20% (Whisonant, 1967), but other Chilhowee Group sandstones are overwhelmingly grain supported (Whisonant, 1967). Likewise, 85% of Nantahala-Brasstown sandstones have less than 10% matrix (Aylor, 1994). In contrast, the matrix fraction (grains smaller than fine sand <0.125 mm) for most of the lower Pinelog is volumetrically significant (33 samples, average 37%; mean 34%), consisting predominantly of recrystallized quartz and mica (muscovite and biotite). Grain-supported textures predominate in the middle Pinelog, but these sandstones, like those in the lower subunit, are also petrologically distinct from those in the Chilhowee (Fig. 5B). In contrast to the relatively low K-feldspar/plagioclase ratios of many of the coarse-grained rocks of the Pinelog, K-feldspar in the Chilhowee is without exception overwhelmingly the dominant feldspar (Whisonant, 1967). Another notable difference between the Pinelog and Chilhowee sandstones occurs in the heavy mineral suite. Tourmaline occurs only as a rare trace mineral in the Pinelog (Li, 1996), whereas it is a ubiquitous accessory phase in the Chilhowee Group (Whisonant, 1967), and it is common in the Murphy belt's Nantahala and Brasstown Formations (Aylor, 1994; Groszos, 1996).

If the Pinelog is equivalent to part of the Ocoee Supergroup, however, then these units should be petrologically similar. An important contrast between quartzite and arkose within the Chilhowee Group versus wacke in the Ocoee Supergroup has long been noted as a distinguishing characteristic between these two sequences (King, 1949; Fairley, 1966). Keith (1904) noted that the composition of the Snowbird (subarkosic to arkosic metasandstone/metawacke), like that of the Pinelog, was indicative of direct derivation from the underlying basement. Both the lower Pinelog and the Wading Branch Formation commonly display graded bedding and contain extensive wacke and other high matrix content coarse-grained rocks. Like the Longarm, the lower Pinelog contains medium- to coarse-grained subarkosic to arkosic metasandstone and muddy subarkosic to arkosic metasandstone. Blue quartz is a distinctive component of most Snowbird metasandstones as well as those of the Pinelog. On a quartz-feldspar-mica (Q-F-M) diagram (Fig. 5C), lower Pinelog samples plot in or near the field of samples from the Wading Branch, as well as almost exclusively within the field of the Longarm. Conversely, the Pinelog, Wading Branch, and Longarm are distinct from the fields of the Chilhowee Group (Figs. 5B and 5C). Unlike Chilhowee sandstones, the Pinelog and the Snowbird are essentially devoid of detrital tourmaline (Hadley and Goldsmith, 1963). In summary, the Pinelog Formation is petrographically distinct from the Chilhowee Group and its equivalents, but it has a number of important petrographic similarities to the Snowbird Group.

6. If the Pinelog and Chilhowee are equivalent, then they should contain similar current direction indicators. Whisonant (1967) presented abundant current bedding data for Chilhowee sandstone units (Cochran-Unicoi, Nebo, Hesse-Erwin, and Hampton). These data were acquired in eastern Tennessee, a few tens of kilometers north of the current study area. They indicate that paleocurrent directions were predominantly south-southeast in the Cochran-Unicoi, southeast in the Nebo, and east-southeast in the Hampton and Hesse-Erwin (Figs. 6A–6D). In northeastern Tennessee and southwestern Virginia, Cochran-Unicoi paleocurrent directions are east-southeast and east-northeast (Fig. 6I) (Brown, 1970; Cudzil, 1985). In the Murphy belt, in the Nantahala Formation, paleocurrent directions are southeastward, whereas in the Brasstown Formation, they are northeastward (Aylor, 1994) (Figs. 6E and 6F). In contrast to Chilhowee sandstones in the southeastern foreland, and their presumed equivalents in the Blue Ridge (Nantahala and Brasstown), cross-beds in the Pinelog indicate paleocurrent directions of west-southwest and southwest (Fig. 6G), which differ significantly from all of the Chilhowee units. A second significant difference between Chilhowee cross-beds and those in the Pinelog is the foreset bed dips—those in the Cochran-Unicoi average between 8° and 11° (Whisonant, 1967; Skelly, 1987), whereas those in the Pinelog are much steeper, averaging ~23° (Fig. 6H). Current bedding data demonstrate therefore that the paleocurrent directions in the Pinelog were toward the southwest, which is significantly different from those in the Chilhowee Group (Fig. 6I). Published current bedding data are not abundant from the Snowbird Group. In contrast to the Chilhowee, however, they are to the west and northwest in the lower part of the Snowbird (Longarm Quartzite) (Hadley and Goldsmith, 1963), and predominantly west-southwest in the upper part of the unit (Pigeon Siltstone) (Hamilton, 1961). The west and southwest paleocurrents are similar to those of the Pinelog.

7. If the Pinelog and Chilhowee are equivalent, then units stratigraphically above the Pinelog should be correlative with the Cambrian-Ordovician carbonate bank sequence. In the southeastern foreland, the Chilhowee is overlain by the Shady Dolomite, and in the Murphy belt, the Brasstown is overlain by the Murphy Marble. Rare impure carbonate occurs within the lower Pinelog Formation (McCallie, 1907; Kesler, 1950; Li and Tull, 1998), but no carbonate-dominated unit occurs above the formation. Li and Tull (1998) interpreted the siliciclastic unit (Padgett Falls Formation) above the Pinelog to overlie the Pinelog unconformably, and they assigned this unit to the base of the Great Smoky Group (see following).

The forgoing analysis demonstrates that the Pinelog Formation has a much closer stratigraphic and petrologic correlation with the Late Proterozoic Snowbird Group of the Ocoee Supergroup than with the Cambrian Chilhowee Group (Table 1).

Padgett Falls Formation

Description

Directly overlying the Pinelog Formation, there is a thick sequence of carbonaceous sericite slate and phyllite containing minor meta-arkose/metawacke referred to as the Padgett Falls Formation (Costello, 1988, 1993; Li and Tull, 1998). Much of this unit is thinly laminated, but much also shows little compositional layering. Locally, along the southeast side of the Corbin massif south of Waleska, and over the western two-thirds of the Fort Mountain massif (Figs. 3 and 4), the Pinelog is absent between basement rocks and the Padgett Falls Formation, and the latter rests directly against the basement rocks. These relationships were astutely noted over a century ago by Hayes (p. 401, unpublished manuscript cited in Hull et al., 1919), who wrote: "Whenever the granite is not bordered by coarse conglomerate or quartzite [Pinelog Formation] it is in contact with black graphitic slates [Padgett Falls Formation], which generally overlie the coarser sediments." One distinctive lithofacies found discontinuously at the base of the Padgett Falls in both the Fort Mountain area and the basement massifs to the south (Figs. 3 and 4) is a pebble-to-boulder–bearing polymictic metadiamictite and metagraywacke/metawacke. Regionally, this metadiamictite facies, referred to as the Salacoa Creek Member of the Padgett Falls Formation by Li and Tull (1998), is in contact with various stratigraphic levels of the Pinelog Formation, as well as with the basement gneiss at a number of localities (Figs. 3 and 4). The matrix of the diamictite is a sandy carbonaceous and locally calcareous muscovite slate or phyllite/metawacke typical of the Padgett Falls Formation, into which this lithofacies grades laterally and vertically. Pebble- to boulder-size lithic clasts in the diamictite are overwhelmingly granite, mineralogically similar to the basement units (Fig. 5D), but they also include minor arkosic sandstone and metapelite (probably derived from the Pinelog Formation) and rare mafic igneous clasts. The granite clasts are as large as 5 m in diameter. The igneous clasts in this lithologically unique facies are not nearly as well foliated as the adjacent basement gneisses. Some are well rounded and essentially unstrained, suggesting that strain was heavily partitioned into the matrix of the host graphitic phyllite/metawacke and that the clasts acted as semirigid

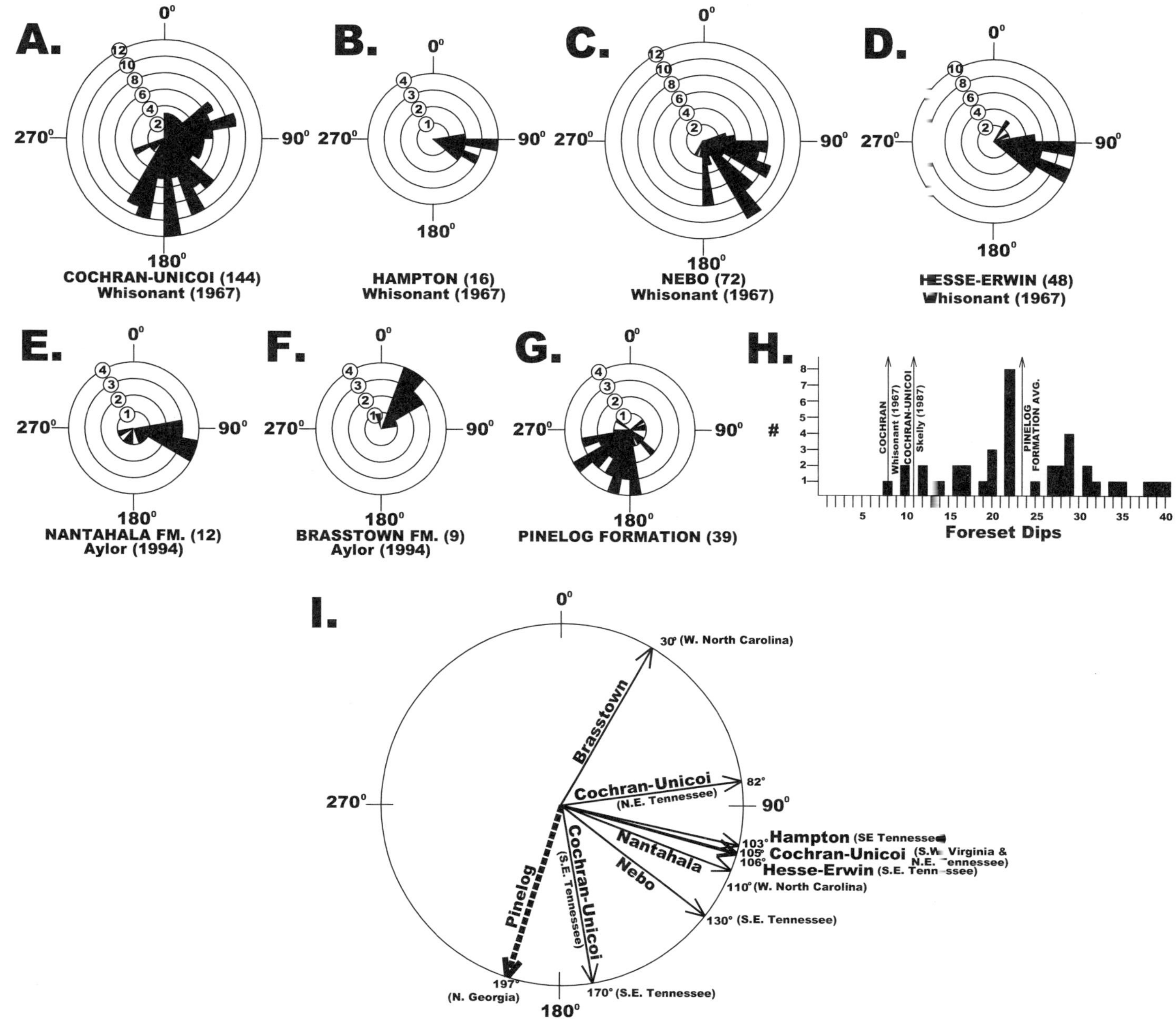

Figure 6. Cross-bed paleocurrent directions from the Chilhowee Group (A–D) and Nantahala (E), Brasstown (F), and Pinelog Formations (G). Numbers in circles are number of measurements. (H) Histogram of Pinelog cross-bed foreset dips, with averages from the Cochran-Unicoi Formation (Whisonant, 1967; Skelly, 1987) for comparison. (I) Vector means of cross-bed paleocurrent directions (A–G above) from the Chilhowee Group and its equivalents (Nantahala and Brasstown Formations) compared to the Pinelog Formation.

TABLE 1. CORRELATION OPTIONS: PINELOG = CHILHOWEE GROUP VS. PINELOG = OCOEE SUPERGROUP

Pinelog = Chilhowee Group	Pinelog = Ocoee Supergroup
A. Thick Ocoee Supergroup rift-facies sequence absent over an extensive area of basement along the W flank of Tennessee salient.	A. Ocoee Supergroup present between Chilhowee Group and basement in S Tennessee salient.
B. Ocoee Supergroup must pinch out W of Murphy belt and E of basement massifs, likely requiring separation by down-to-the-east rift-boundary fault.	B. Continuity of Ocoee Supergroup between basement massifs and Murphy belt. No requirement of down-to-the-east rift-boundary fault in between.
C. Because it faces upward into the overlying, upright Ocoee Supergroup to the E, the Pinelog must be in contact with the Ocoee along a major fault.	C. No stratigraphic requirement that the upright Pinelog be separated by a major fault from the structurally overlying, upright Ocoee Supergroup.
D. Pinelog internal stratigraphy should be similar to the Chilhowee Group, or to Chilhowee-equivalent Nantahala/Brasstown formations, but these stratigraphies are dissimilar.	D. Pinelog internal stratigraphy is most similar to that of the basal Ocoee's Snowbird Group.
E. Pinelog shows significant petrologic differences from both the Chilhowee Group and the Nantahala and Brasstown formations, including much higher matrix and feldspar content and a different heavy mineral suite.	E. Pinelog shows significant petrologic similarities with units of the Snowbird Group.
F. Paleocurrent directions in the Chilhowee and Nantahala/Brasstown formations. are southeastward to northeastward, whereas those in the Pinelog are southwestward.	F. Pinelog paleocurrent directions compare favorably with those in the Snowbird Group.
G. Chilhowee Group and Nantahala/Brasstown formations, are overlain by carbonate units: Shady Dolomite and Murphy Marble.	G. No carbonate-dominated unit occurs above the Pinelog Formation.

bodies during peak greenschist-facies metamorphic strain. The metawacke is poorly bedded and contains angular to subrounded very coarse sand to granules of blue quartz, microcline, and plagioclase, and minor granitic rock fragments, detrital muscovite, zircon, and tourmaline in a matrix of fine muscovite and quartz, and locally extensive calcite and minor biotite. The distribution of the metadiamictite is probably much greater than shown in Figures 3 and 4, but it is difficult to differentiate from "normal" Padgett Falls phyllite except where it is well exposed along streams. The Padgett Falls is overlain by units of the Great Smoky Group.

Depositional Setting

The protolith of the graphitic sericite phyllite facies of the Padgett Falls Formation was a carbonaceous pelite, locally intermixed with rhythmically repeated lenses of graded, locally cross-bedded, matrix-rich metasandstone and granule metaconglomerate and minor, graphitic, thinly bedded carbonate rocks. The matrix-rich metasandstone facies contains rip-up clasts of graphitic phyllite as well as sorted ungraded clasts of blue quartz and feldspar throughout entire beds. This suggests resedimentation of gravel and mud and transport by gravity or debris flow, with rapid deposition in a relatively deep-water environment (Palmes, 1998). The more abundant interbedded graphitic sericite phyllite facies, which contains abundant pyrite, suggests slower depositional rates in a highly reducing anaerobic, probably relatively deep-water environment (Li, 1996; Palmes, 1998). The Salacoa Creek Member metadiamictite originated as a pebbly to bouldery and sandy mudstone. Clast size and immaturity indicate only relatively minor transport for much of the detritus prior to being deposited in a basin bordered by a significant basement-cover uplift. The most likely depositional setting of the metadiamictite facies was as lobes of a proximal submarine fan system, in which muddy sediments were transported by debris or turbidity flow immediately offshore of basement-cover highlands.

Li and Tull (1998) interpreted the discordance with underlying units at the base of the Padgett Falls to be the result of a regional unconformity that cut through the Pinelog into the basement. They based this conclusion on the following observations: (1) Primary structures (cross-beds and graded beds) indicate that the Pinelog faces toward the Padgett Falls Formation, whereas the latter faces away from the Pinelog. (2) As noted already, the base of the Padgett Falls is discordant to the internal stratigraphy of the Pinelog, regionally cutting through at least ~1100 m of the Pinelog to lie against the basement units. (3) Across much of the region, the base of the Padgett Falls Formation consists of metadiamictite and metawacke of the Salacoa Creek Member. This lithofacies lies in contact with various stratigraphic levels of the Pinelog and directly overlies basement as well (Figs. 3 and 4). The along- (>75 km) and across- (tens of km) strike regional distribution of the diamictite implies that it represents an aerially extensive lithofacies. (4) The Salacoa Creek Member contains an assortment of clasts (granite, granitic gneiss, mafic igneous clasts, quartzite, metasandstone, metapelite, marble, feldspar, and blue quartz) similar to lithologies within the Pinelog Formation and the underlying basement (Fig. 5D), suggesting significant uplift and erosion of underlying units prior to deposition of the diamictite. (5) The contact between the Pinelog and the Padgett Falls Formations marks an abrupt and significant change in depositional environment, from the alluvial, fluvial/deltaic, or shallow-marine setting of the lower, middle, and upper Pinelog Formation, to the relatively deep-water (submarine fan[s]?) Padgett Falls Formation deposited by debris or turbidity flow in a region of much greater local relief.

Correlation

Following Hayes' (unpublished manuscript cited by Hull et al., 1919) initial correlations with sequences that he had mapped in the southeastern Tennessee Blue Ridge (Hayes, 1895), several previous workers (Furcron and Teague, 1947; McConnell and Costello, 1980, 1984; Costello et al., 1982; McConnell and Abrams, 1984) have correlated much of the graphitic slate/phyllite (referred to here as the Padgett Falls Formation) with the Wilhite Slate, or Wilhite Formation, of the Walden Creek Group. As noted above, the Walden Creek Group is believed to occur stratigraphically above the Great Smoky Group (Hurst and Shlee, 1962; Costello and Hatcher, 1986, 1991), and therefore correlation of the Padgett Falls with the Walden Creek Group requires a fault between the Padgett Falls and the overlying upright Great Smoky Group. McConnell and Costello (1980, 1984) and Costello et al. (1982) proposed such a fault along the southeast flank of the Corbin Metagranite and extended it northeastward as the Lost Town Creek fault or fault zone along the southeastern (upper) boundary of the graphitic phyllite unit, separating their Wilhite Formation from the overlying Great Smoky Group. However, still farther northeast, McConnell and Costello (1980) interpreted the Wilhite (Padgett Falls)–Great Smoky contact as stratigraphic, even though it was upright. McConnell and Abrams (1984) also mapped a fault that they called the Lost Town Creek fault, but they placed it at the base of the graphitic phyllite unit, immediately east of the Corbin basement. Costello (1988, 1993) later correlated both the Pinelog Formation and the lower part of what is referred to here as the Padgett Falls Formation with the Walden Creek Group. However, recognizing that the top of the graphitic unit graded into units at the base of the Great Smoky Group, Costello separated the upper part from the Walden Creek Group, placing it in the hanging wall of the Lost Town Creek fault and renaming it the Padgett Falls Formation, which he placed at the base of the Great Smoky Group. If any or all of the graphitic unit is indeed correlative with the Walden Creek Group, then a fault separating it from the upright Great Smoky is clearly required.

Important key features relating to the "Wilhite correlation" issue are found at Fort Mountain. Here, Furcron and Teague (1947) mapped the "Wilhite Slate" only between the eastern and western belts of basement gneiss, and they mapped all boundaries of this unit as thrust faults. However, they separated very similar black slates structurally above (east of) and below (west of) the belts of basement from the Wilhite Slate, mapping them instead as Ocoee "series" rocks. They interpreted their Ocoee "series" black slates structurally above (east of) the easternmost band of basement at Fort Mountain to lie unconformably above basement, and those below (west of) this basement band to be completely bounded by thrust faults. The Salacoa Creek Member diamictite was apparently originally recognized at Fort Mountain by Hayes as the "Cohutta conglomerate" in his unpublished report (unpublished manuscript cited by Hull et al., 1919), and granite boulder–bearing "conglomerates," assigned in the present study to the Salacoa Creek Member, were described by Furcron and Teague (1947) from the northern terminus of the eastern basement lens. These authors assigned these conglomerates to the Ocoee "series" and interpreted them as representing fanglomerates or channel conglomerates derived from erosion of the adjacent basement. McConnell and Costello (1984) later differentiated these conglomeratic rocks at the northern terminus of the eastern basement lens as their Parr Branch Formation, correlating them with the Pinelog Formation to the south at Pine Log Mountain bordering the Corbin massif (Fig. 3). At the latter locality, McConnell and Costello (1984) also recognized the Salacoa Creek diamictites but included them within the Pinelog Formation, assigning them to the Snowbird Group and interpreting them to be overlain conformably by the Wilhite Formation. Differing with Furcron and Teague (1947), however, McConnell and Costello (1984) interpreted the "lithic conglomerate" at Fort Mountain as resting unconformably against the basement, rather than being in fault contact, although at the localities that they examined, the unconformity is overturned, with basement structurally above the diamictite. Although McConnell and Costello (1984) described the "lithic conglomerate" (their Parr Branch Formation) as being bounded on either side by carbonaceous phyllite, they separated this unit from the other carbonaceous phyllites of the area, which they mapped as Wilhite Formation, interpreting the Wilhite Formation to be stratigraphically above the Parr Branch. Thus, both at the Corbin and Fort Mountain massifs, McConnell and Costello (1984) interpreted what is mapped in the current interpretation as the Salacoa Creek diamictite member, to be *stratigraphically* overlain by the graphitic pelite unit (their Wilhite Formation, the Padgett Falls Formation of the current study). At the former locality, however, the conglomerates occur at the top of the Pinelog Formation, whereas at Fort Mountain, they occur directly against the Precambrian basement. Following Furcron and Teague (1947), McConnell and Costello (1984) mapped the graphitic sequence between the bands of basement as Wilhite Formation, and the graphitic slate/phyllite to the west of the westernmost basement lens as Great Smoky Group. Differing with Furcron and Teague (1947), however, McConnell and Costello (1984) mapped both what is correlated here with the Pinelog Formation east of the easternmost lens of basement gneiss, as well as the carbonaceous phyllite sequence above it, as being in thrust contact above basement gneiss rather than being in nonconformable contact.

All previous workers recognized nonconformities between basement and cover rocks at Fort Mountain: Furcron and Teague (1947) at the structural top of the eastern band of basement with what is mapped here as both Pinelog and Padgett Falls Formations, and McConnell and Costello (1984) at the structural base and northeastern tip of this basement band with what is correlated here with the Salacoa Creek Member of the Padgett Falls Formation. My interpretation concurs with both of these earlier interpretations, but it extends these interpretations (admittedly with much additional mapping data) to include all basement-cover contacts in the area (see following).

Correlation of the type Wilhite Formation in southeastern Tennessee with the graphitic pelites described here around the three Georgia basement massifs has previously been argued to be in error on strictly stratigraphic grounds (Fairley, 1966; Li and Tull, 1998; Palmes, 1998). The report of middle Paleozoic fossils from the type Wilhite Formation (Unrug et al., 2000) is consistent with the sequence mantling the basement massifs not being correlative with the Wilhite Formation. Following Costello (1988, 1993) and Li and Tull (1998), I refer to this unit as the Padgett Falls Formation and place it at the base of the Great Smoky Group in stratigraphic continuity below the northwest (upright) limb of the Murphy syncline.

In the Great Smoky Mountains of North Carolina and Tennessee (Fig. 1), the Greenbrier fault places the Snowbird Group and overlying Great Smoky Group over a much thicker section of the Snowbird and the overlying Rich Butt Sandstone, correlative with the lower Great Smoky Group (Hadley and Goldsmith, 1963). East of this fault, the basal Great Smoky unit is the Elkmont Sandstone, which, very similar to the Padgett Falls Formation, is predominantly a black, massive, monotonous metasiltstone that contains beds of conglomeratic metasandstone (Montes and Hatcher, 1999). West of the fault, the Rich Butt consists of dark-gray argillite interbedded with lenticular sandstone containing rip-up clasts of dark metasiltstone, also similar to the Padgett Falls Formation. No diamictites similar to those of the Salacoa Creek Member, however, are known from the base of the Great Smoky or Rich Butt in their type sections. Although extension of the Greenbrier fault southwestward from the Great Smoky Mountains is not well defined, the area of the current study likely occurs in that fault's hanging wall (Li and Tull, 1998).

In the eastern Great Smoky Mountains southeast of the Greenbrier fault in the Dellwood Quadrangle, Hadley and Goldsmith (1963) showed that the basal Great Smoky Group crosses northwestward from Precambrian basement across the Snowbird Group's Longarm Quartzite onto the Roaring Fork Sandstone. Northwest of the fault, where the Snowbird Group is ten times thicker than it is to the southeast, these authors documented that the Rich Butt Sandstone, believed to be equivalent to the basal Great Smoky Group, steps across the Roaring Fork Sandstone, to lie near the top of the Pigeon Siltstone. In a reconstruction, Hadley and Goldsmith (1963) estimated the palinspastic distance to be ~60 km from where the Great Smoky Group lies directly above basement on the southeast to where the Rich Butt lies above the Pigeon Siltstone on the northwest. This implies a discordance of a few degrees between Snowbird units and the base of the Great Smoky. Thus, the geometric relationships in the Great Smoky Mountains, in which the base of the Great Smoky Group (and its likely equivalent Rich Butt Sandstone) cuts up-section toward the northwest from basement to near the top of the Snowbird Group, are very similar to those observed at the base of the Padgett Falls Formation around the Corbin and Salem Church areas in Georgia (Li and Tull, 1998). However, Hadley and Goldsmith (1963) interpreted the Great Smoky Group to be conformable above the Snowbird, in spite of the regional truncation of underlying Snowbird units. In a dramatically different interpretation of these geometric relationships, Montes and Hatcher (1999) argued that the base of the Great Smoky Group in the southeastern Great Smoky Mountains is the Greenbrier fault, and the entire Snowbird Group is in the footwall of that fault.

Cover Sequence Correlation Summary

Because of distinctive similarities in lithology, primary structures, stratigraphic sequence, and inferred depositional environments with cover sequences of basement massifs elsewhere in the Blue Ridge, and because of their stratigraphic continuity with the overlying sequence in the Murphy belt to the east, the cover rocks surrounding the Georgia basement massifs are believed to constitute parts of the Snowbird and Great Smoky groups of the Ocoee Supergroup (see previous). The basement gneisses, as well as the nonconformably overlying Snowbird-equivalent Pinelog Formation, are both interpreted to be overlain along a low-angle unconformity by the Padgett Falls Formation. The latter formation is conformably overlain by the rest of the Great Smoky Group. Current bedding in the Padgett Falls and overlying units of the Great Smoky consistently suggest younging toward structurally overlying units on the west (upright) limb of the Murphy syncline. Immediately above the Padgett Falls, the Great Smoky Group consists of sericite phyllite that contains thick lenses of massive, poorly sorted, feldspathic, matrix-supported metasandstone and metagraywacke, and conglomerate. The phyllites are locally carbonaceous, but carbonaceous pelites are much less abundant than in the underlying Padgett Falls Formation. This section of the Great Smoky corresponds to units that have been differentiated between the Fort Mountain and Salem Church massifs as the Carters Dam and Harris Branch Formations by Palmes (1998). Based upon primary depositional structures and distribution of lithofacies, Palmes (1998) interpreted the depositional environments of these units to be one in which relatively coarse-grained turbiditic lenses were emplaced within a basin that was also accumulating anoxic basin-floor pelitic deposits. On the basis of lithologic similarity and stratigraphic position, Palmes (1998) correlated the Carters Dam Formation with the Copper Hill Formation (Hurst, 1955), and the Harris Branch Formation with the Wehutty Formation (Hernon, 1967). The Copper Hill and Wehutty Formations occur within the stratigraphically lower part of the Great Smoky Group. Sedimentary structures (scoured surfaces, graded beds, and channels), lithofacies (including both oligomictic and polymictic metaconglomerates), and composition of Great Smoky Group units overlying the Padgett Falls Formation suggest a complex relatively deep-marine basin with intertonguing of multiple wedges of coarse clastic rocks with more basinal facies.

BASEMENT AND COVER DEFORMATION

Structure of the Corbin and Salem Church Culminations

The Corbin basement culmination is interpreted to be an inclined, tight to isoclinal, overturned, doubly plunging F_1 anticlinorium (McConnell and Abrams, 1984; McConnell and Costello, 1980, 1984; Li, 1996; Li and Tull, 1998) (Fig. 7C). On both limbs of this structure, bedding and schistosity within the cover rocks dip moderately eastward or southeastward. Both map- and outcrop-scale F_1 folds deform the bedding within the cover rocks and produce a penetrative greenschist-facies axial-planar foliation subparallel to bedding, except in the hinge areas of folds. Within the basement, this fabric is locally represented by a mylonitic or phyllonitic foliation (see following).

In contrast to the Corbin massif, around the Salem Church culmination, cover rock bedding and schistosity dip gently outward in all directions, leading previous workers to interpret the structure of this massif as a simple northeast-southwest elongated dome, referred to as the Salem Church anticline (Bayley, 1928; McConnell and Costello, 1980, 1984; Costello, 1988, 1993). However, based on the distribution of cover units and the inferred facing directions, Li and Tull (1998) demonstrated that the dome arches an earlier regional, basement-cored, recumbent F_1 anticline with an amplitude of up to 7 km, equivalent in scale to the anticlinorium cored by the Corbin massif (Fig. 7B). Basement rocks are stratigraphically and structurally overlain by upright cover rocks on the southeast side of the culmination, but in the center and along the northwest side, basement occurs structurally above overturned cover rocks. The overturned cover sequence has been traced continuously along strike to the northeast and southwest to where the cover sequence occurs in an upright position above the Salem Church basement (Li and Tull, 1998). The two massifs (Corbin and Salem Church), separated horizontally at the surface by 11 km of cover rocks, are interpreted to be a single heterogeneous basement massif (McConnell and Abrams, 1984; McConnell and Costello, 1980, 1984; Li, 1996). The F_1 anticlines coring the two massifs may actually be the same structure, where the intervening cover rocks occur within an axial depression. Li and Tull (1998) considered this regional-scale isoclinal anticlinorium cored by the two massifs to be a companion fold of the regional-scale isoclinal Murphy synclinorium to the southeast, with the two major northwest-vergent synmetamorphic Blue Ridge structures sharing a common limb, which was locally faulted along the southeast flank of the Corbin massif (Fig. 2). The entire western Blue Ridge stratigraphic column (>10 km), including the Grenville basement, is encompassed within the structural relief of these two structures.

The early F_1 structures of the two southern massifs have been redeformed by at least three subsequent fold phases, including two nearly orthogonal map-scale sets (F_3 and F_4) associated with open, upright folding and broad doming. Superposition of these latter map-scale fold phases on each other and the earlier regional F_1 structures produced Ramsay types 1 and 3 map-scale interference patterns (Ramsay, 1967), which cause the Salem Church culmination, representing a higher structural level than the Corbin, to appear as a dome (Li and Tull, 1998) (Fig. 7B).

Structure of the Fort Mountain Culmination

Geologists since Furcron and Teague (1947) have interpreted the overall structure of the Fort Mountain area to be the result of thrust imbrication of basement and its low-metamorphic-grade metasedimentary cover, although, as mentioned already, Furcron and Teague (1947) mapped the eastern boundary of the easternmost basement lens with both the Pinelog and Padgett Falls Formations as an unconformity. McConnell and Costello (1984) followed Furcron and Teague's interpretation of thrust faults with two exceptions. They mapped the eastern boundary of the easternmost basement lens as a thrust fault rather than an unconformity, but they interpreted the northwest boundary of this basement lens as an overturned unconformity and traced this unconformity around the northern tip of basement in an isoclinal fold closure. Importantly, McConnell and Costello (1984) interpreted the thrust contact above the uppermost (eastern) basement lens to terminate into stratigraphic contacts between their Wilhite Formation and the overlying Great Smoky Group, both to the immediate north and south of the basement lens, implying very little displacement on this proposed fault. Several lines of evidence cited next suggest that the sinuous, narrow, linear basement culminations are the result of refolded, map-scale isoclinal folds rather than thrust imbrication (Fig. 7A).

Petrologic and Stratigraphic Evidence for Large-Scale Isoclinal Folds

Both petrologic and stratigraphic data suggest that the low-grade metasedimentary units adjacent to the Fort Mountain basement occur nonconformably above it. (1) Arkosic rocks from both the Pinelog and Padgett Falls Formations contain abundant blue quartz, K-feldspar, and plagioclase, indicating direct derivation via erosion of the Fort Mountain basement terrane. (2) Numerous cross-beds as well as graded beds in the Pinelog Formation consistently face away from the basement. (3) Lenses of metadiamictite in the basal Padgett Falls Formation (Salacoa Creek Member) sheath the lenses of basement (Fig. 4) and contain granule- to boulder-sized clasts of granite and gneissic granite petrographically similar to the Fort Mountain Gneiss. All of these relationships indicate that the cover rocks rest nonconformably against basement.

Structural Evidence for Large-Scale Isoclinal Folds

Wherever basement-cover contacts are observed, they are sharp and overprinted by metamorphism. No fault-zone materials were observed at the contacts. At the along-strike truncations of basement, basement and cover rocks are isoclinally interdigitated and commonly appear interlayered at outcrop scale. Synmetamorphic cleavage and mineral and stretching lineations (generally plunging downdip to the southeast; Fig. 8, stereonet C) in

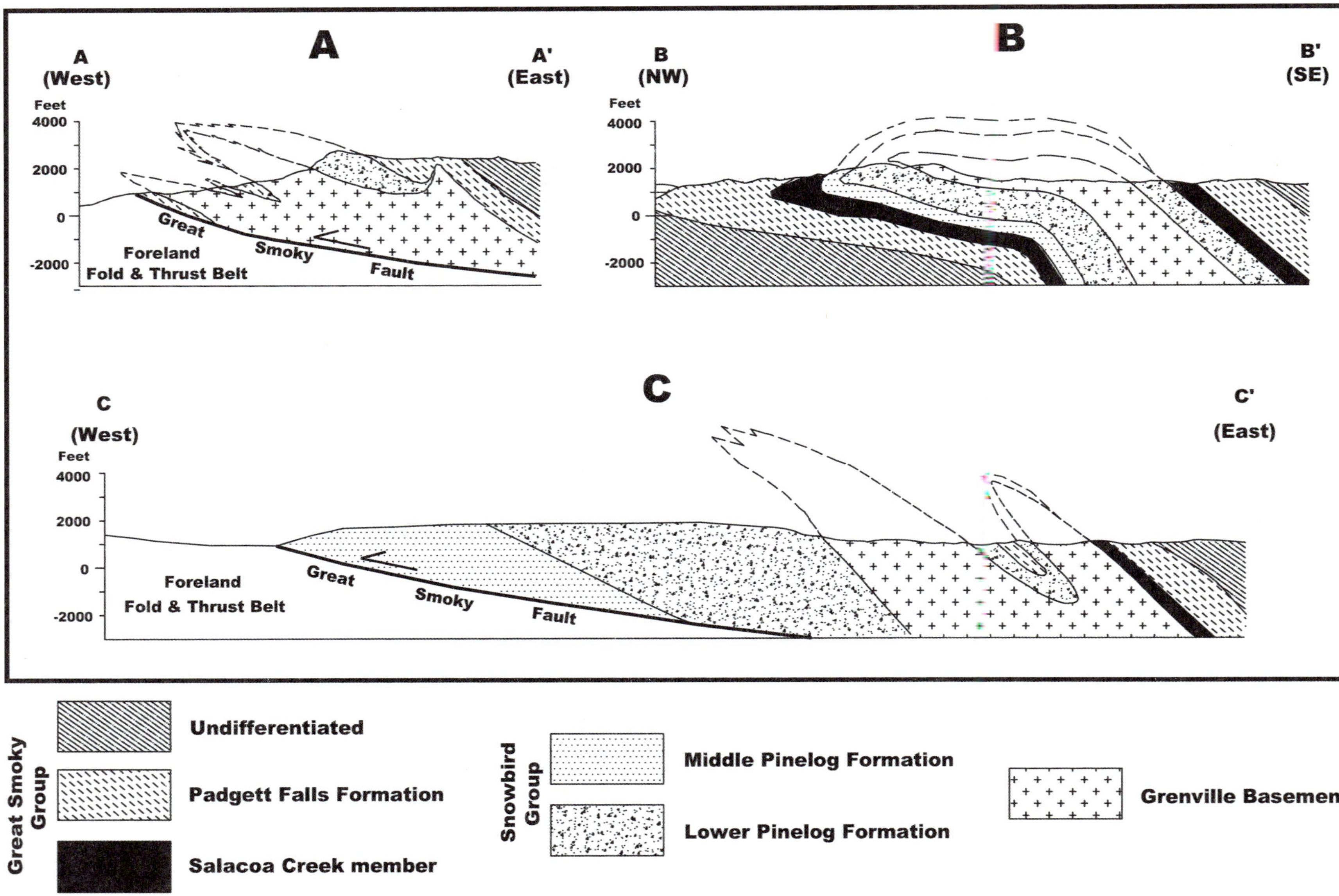

Figure 7. Cross sections through the three Georgia western Blue Ridge basement massifs: A–A' Fort Mountain (Fig. 4), B–B' Salem Church, and C–C' Corbin (Fig. 3). No vertical exaggeration.

both basement and cover rocks are concordant and parallel. No discordance of fabrics is observed across these contacts, and the metamorphic grade of basement and cover is the same. At several localities (at both the northern and southern tip of the eastern belt of basement and at the northern end of the western belt of basement) where the basement lenses terminate along strike, field mapping demonstrates that the basement is continuously sheathed in granite-gneissic granite-pebble to boulder diamictite, and the diamictites can be traced completely around the terminations of the map-scale basement lenses from the upright limb to the overturned limb (Fig. 4). At these localities, synmetamorphic cleavage can be traced at a high angle to the contact from basement into cover rocks (Fig. 8), suggesting that this is an axial-planar cleavage. Except for the isolated lens of Pinelog Formation on the east flank of the eastern basement lens, the Padgett Falls Formation completely surrounds the basement (Fig. 4).

All of these relationships support the interpretation that the basement lenses in the Fort Mountain area occur as the cores of three refolded, shallowly inclined, doubly plunging, isoclinal anticlines (Fig. 7A). These three basement-cored anticlines are the Fort Mountain, the Georgia Mine, and the Rock Creek anticlines, respectively (Fig. 8). The largest, the Fort Mountain anticline, is digitated into a series of smaller, map-scale folds to the south (Fig. 8). The Georgia Mine and Rock Creek anticlines are cut in map-view by the Great Smoky fault (Fig. 8).

Postmetamorphic Folding Events

The Great Smoky fault bounds the basement-cored fold nappes immediately to the west in the Fort Mountain area, and this thrust is believed by most workers to be latest Paleozoic in age because, in several places, its trace cuts Carboniferous rocks. In the Fort Mountain area, the Great Smoky fault and rocks of the Blue Ridge allochthon have been deformed by at least two postemplacement, map-scale fold sets, which thus must also be late Paleozoic (Fig. 8). These fold sets have a similar geometry, kinematic sequence, and orientation to the two map-scale fold sets that refold the Corbin and Salem Church basement-cored anticlinoria to the south (see Li and Tull, 1998, Fig. 11 therein). At Fort Mountain, as at the two southern massifs, the earlier of the two fold sets (F_3) forms southeast-trending, open, upright, shallowly southeast-plunging folds (Fig. 8, fold 5, stereonet A). The latest of these (F_4) forms an east-northeast–trending, open (interlimb angle of ~157°), upright, shallowly east-northeast–plunging cross antiform-synform pair, which produces an indentation in the thrust front of ~7 km (Fig. 8, folds 6 and 7, stereonet B). The F_1 basement-cored anticlines are restricted to the hinge and limbs of the Cohutta Mountain synform and are not found at the surface north of the axial trace of the Holly Creek antiform (Fig. 8).

Characteristics of Basement Deformation

Peak deformation associated with formation of the basement-cored anticlines took place under lower-greenschist-facies conditions. Mineral assemblages in immediately overlying pelitic cover rocks contain white mica, chlorite, and biotite. Garnet is absent in these cover rocks and was not observed in basement rocks in this study, although Furcron and Teague (1947) mentioned it as an accessory phase in their study of the Fort Mountain Gneiss. Mafic and ultramafic rocks (Cohutta Schist) at Fort Mountain enclosed in the basement metagranite consist of metamorphic assemblages containing chlorite, talc, serpentine, and dolomite. At Fort Mountain, peak metamorphic temperatures were therefore in the range of ~400 °C (Bucher and Frey, 1994).

The geometry of the basement-cored folds (isoclinal, large amplitude, and large wavelength) indicates that the basement rocks within all three massifs experienced a significant bulk ductile deformation accommodated by strain-softening mechanisms. Strain localization in the metagranites, however, was extreme, and deformation was very heterogeneous. The highly vegetated and scree-covered basement slopes at Fort Mountain have prevented a systematic study of the spatial distribution of the various basement strain facies, but Gargi (1985) made some effort in this regard with respect to the Corbin massif exposed along the shores of Lake Allatoona. For the most part, however, only a qualitative assessment has been made based upon samples and fabrics obtained from isolated exposures. At Fort Mountain, the predominant volume of the basement metagranite exposed within this anticlinorium is composed of relatively low-strain orthogneiss in which the igneous hypidiomorphic granular texture is mostly preserved. In less-deformed parts of the basement, quartz exhibits ubiquitous undulatory extinction, serrated grains, and dynamic recrystallization of subgrains, but feldspars are only slightly sheared and broken. White mica–chlorite retrogression of feldspar grain boundaries is essentially absent. Zones in which basement strain is somewhat higher were apparently associated with higher water activity, and they exhibit growth of fine-grained white mica and chlorite around the peripheries and within fractures of brittlely deformed (broken, sheared, and bent) feldspar phenocrysts, accompanied by increased undulose extinction of quartz. An anastomosing white mica, and locally biotite, chlorite, stilpnomelane, and epidote planar fabric is developed around and locally through feldspar porphyroclasts. Zones of still higher strain exhibit an overall grain-size reduction, more extensive retrogression of feldspar to white mica, ribboning of quartz grains and production of peripheral quartz subgrains, and an increase in the matrix/porphyroclast ratio. In these zones, a mylonitic schistosity is defined by quartz blades with undulatory extinction and narrow (millimeter-scale) well-developed white mica–biotite shear bands. Feldspars occur as cloudy augen porphyroclasts with undulatory extinction, wrapped by the shear band schistosity. Perthite is locally recrystallized to quartz-feldspar aggregates. A relatively sharp transition occurs between the low-strain metagranite and the mylonitic orthogneiss. Deformation of feldspars in both basement and cover rocks of each massif is mostly brittle, whereas in most cases, quartz exhibits abundant signs of intracrystalline plastic deformation. Ductility of quartz begins at about ~350 °C, whereas dynamic recrystallization of feldspar begins at ~500 °C, a temperature interval consistent with

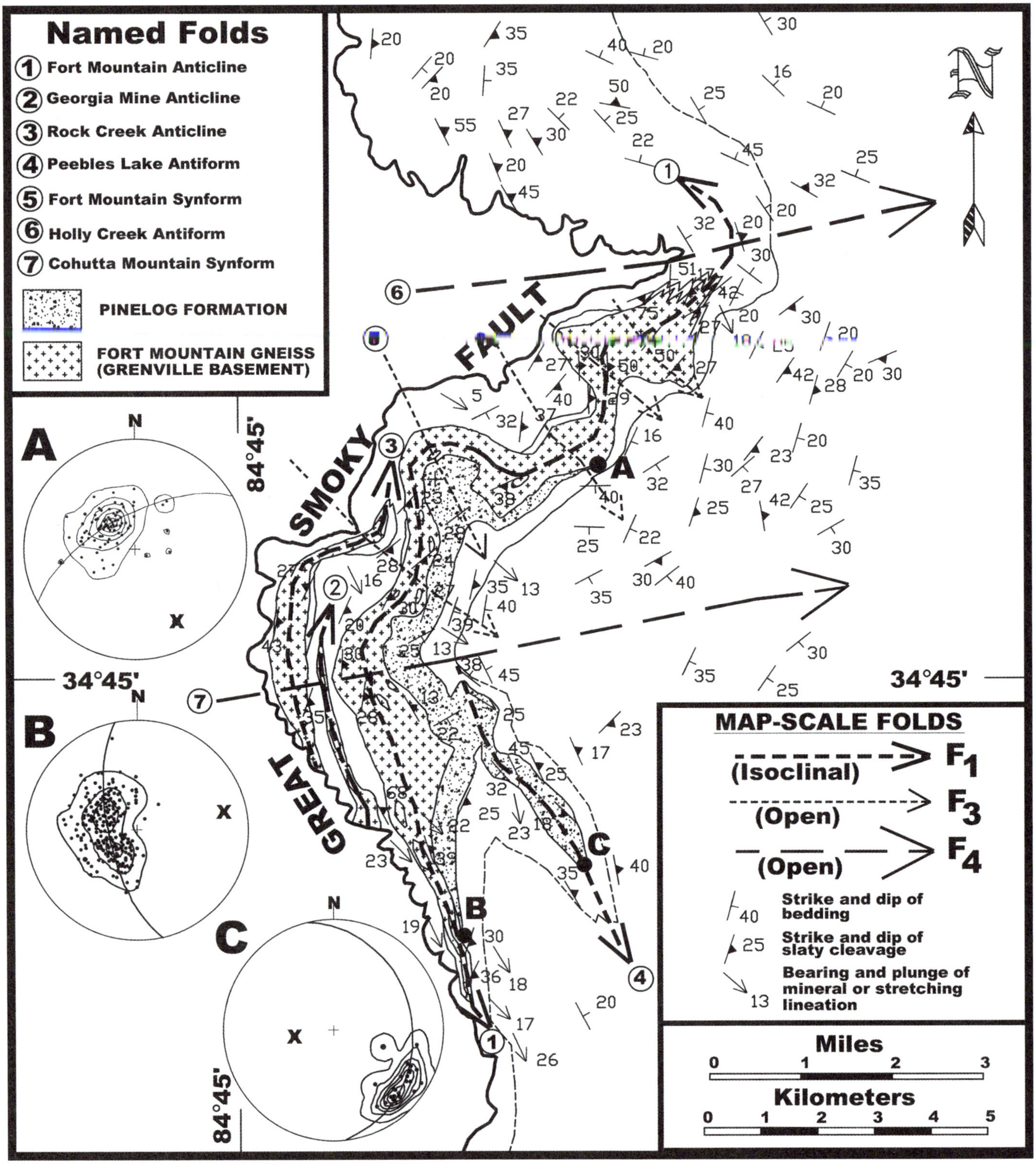

Figure 8. Structure map of the Fort Mountain massif showing trends and geometries of major structures. Stereonets represent lower-hemispheric equal-area projections of selective structural data. (A) Slaty cleavage from Fort Mountain synform (fold 5): contour interval = 1, 5, 10, 15, 20, 25% (97 data points), fold axis = 25°S29°E. (B) Slaty cleavage from Holly Creek antiform (fold 6): contour interval = 1, 5, 10% (289 data points), fold axis = 22°N79°E. (C) Synmetamorphic mineral and stretching lineations from whole map area: contour interval = 1%, and then 5% intervals (24 data points).

the metamorphic grade of the area. Basement deformation at Fort Mountain therefore took place below the brittle-ductile transition within feldspars. A similar progressive development of superposed mylonitic fabrics was described by Gargi (1985) in the Corbin massif. In all three massifs, Paleozoic deformation temperatures were at the low- to mid-greenschist facies, estimated at ~350–400 °C by Gargi (1985) for the Corbin massif.

Boyer and Mitra (1988) and Gapais (1989) noted that most prograde deformational histories of granites, such as expected during reworking of pretectonic basement like that found within the three northwest Georgia basement massifs, commonly result in heterogeneous strain patterns associated with discrete shear zones. At temperatures below the epidote amphibolite facies, i.e., below ~500 °C, highly feldspathic (i.e., granitic) rocks exhibit discrete deformation zones separating relatively low-strain material with an abundance of relict igneous grains. According to Gapais (1989), discrete shear zones anastomosing around low-strain domains should be representative of pretectonic basement. The Fort Mountain basement rocks contain a higher percentage (generally >60%; Fig. 5A) of the weak (ductile) grain fraction (quartz) than found in most granitic rocks (~25%). Therefore, an effective ductile state of the Fort Mountain basement would be expected to have been reached at somewhat lower temperatures than that seen in other basement-cored nappes with higher feldspar contents, such as the Aar and Gotthard Massifs of the western Alps, which were deformed at 450–500 °C and 500–550 °C, respectively (Gapais, 1989). Basement rocks within the Corbin and Salem Church massifs also generally contain a higher percentage of feldspar than those at Fort Mountain (Fig. 5A), but these ductile, map-scale, basement-cored isoclinal folds (Li and Tull, 1998) formed at temperatures similar to those accompanying the ductile deformation at Fort Mountain.

Age of Deformation

Correlations of cover units bounding the three Georgia Blue Ridge massifs suggest that they are part of the Late Proterozoic Snowbird and Great Smoky Groups of the Ocoee Supergroup. The Great Smoky fault, believed by most workers to be latest Paleozoic, bounds the basement-cored fold nappes of all three massifs immediately to the west. The frontal Blue Ridge fault and the basement massifs carried in the allochthon were deformed by at least two postemplacement fold sets, which must also be late Paleozoic (see previous).

Ductile deformation associated with the F_1 basement-cored anticlines was synchronous with regional metamorphism. The age of peak western Blue Ridge metamorphism is not well resolved and has been under debate for many years. Most studies suggest that it is early to middle Paleozoic (Glover, et al., 1983; Kish, 1990, 1991; Connelly and Dallmeyer, 1993; Goldberg and Dallmeyer, 1997; Moecher et al., 2004). As noted already, Unrug et al. (2000) reported that part of the Walden Creek Group in Tennessee contains Paleozoic fossils indicative of a Silurian or younger age. Russell (1976) and Hatcher and Odom (1980) reported a Rb-Sr isochron of 368 ± 9 Ma ($^{87}Sr/^{86}Sr = 0.722$) for mylonitic gneiss from the Fort Mountain Gneiss, and they interpreted this age to represent the time of Sr isotopic homogenization associated with ductile deformation. These data suggest that peak metamorphism and formation of the large-scale isoclinal F_1 folds that deformed the basement massifs occurred during the Late Devonian.

RIFT-RELATED FAULTING AND RIFTED-MARGIN CONFIGURATION

Even though the Ocoee Supergroup is considered by most geologists to have formed within a Late Proterozoic rift basin (Rodgers, 1972; Rankin, 1975), thus far, extensive mapping efforts have failed to identify any rift-related faults, although they have been inferred on stratigraphic grounds based on restored sections (Boyer and Mitra, 1988; Li and Tull, 1998). In a few instances, growth faults within the Laurentian rift stratigraphy have also been inferred from contrasts in stratigraphic thicknesses of individual units (Wehr and Glover, 1985; Schwab, 1986; Montes and Hatcher, 1999). Additionally, Late Proterozoic extensional displacements on subsurface faults beneath the Blue Ridge allochthon have been inferred from seismic-reflection data (Harris et al., 1981).

Analysis of the distribution of units that appear in subcrop beneath the pre–Padgett Falls unconformity reveals a similar sense of truncation of the Pinelog Formation and the underlying basement across all three basement massifs. Li and Tull (1998) demonstrated that this unconformity bevels regionally southeastward across the Pinelog Formation into the basement around the Corbin and Salem Church massifs. Palinspastic restoration of the Padgett Falls subcrop surface around these two massifs, using restored stratigraphic cutoff lines as reference guides, indicates that, at the time of initiation of Great Smoky deposition (Padgett Falls Formation), the Pinelog Formation dipped northwestward at a low angle beneath the pre–Padgett Falls unconformity (see Li and Tull, 1998, Fig. 16 therein). Across this area, the Pinelog thickens northwestward to ~1100 m. Li and Tull (1998) speculated that the Pinelog around these massifs was rotated downward toward the northwest by a pre– or syn–Padgett Falls Formation east- or southeast-dipping normal fault system with a minimum stratigraphic separation of 1 km. Escarpments along this fault system could have supplied much of the coarse basement and cover-derived detritus to the accumulating turbiditic Great Smoky sediments. This fault (and associated faults) resulted from a second cycle of rifting related to formation of the Great Smoky Group, and it modified the geometry of the earlier rift-related Snowbird (Pinelog) basin. The immediately underlying basement rocks therefore probably experienced two distinct episodes of Late Proterozoic rift faulting and were detached from autochthonous basement during these episodes, making them susceptible to incorporation into the Blue Ridge allochthon during later Alleghanian thrusting. At the latitude of the Corbin and Salem Church massifs, the Pinelog Formation extends unfaulted above the basement westward to the Alleghanian-age Great Smoky thrust. Here, the frontal thrust has detached the Blue Ridge allochthon structurally above the proposed second-phase rift fault, such that all of the basement and cover rocks in the Corbin and Salem Church areas occur within the hanging wall of the second-phase rift-related fault (Li and Tull, 1998). Relationships at Fort Mountain, however, indicate that north of Corbin and Salem Church, the Great Smoky thrust may have cut deeper into the rift-related architecture, because a second-phase rift-related fault is interpreted to occur within the Blue Ridge allochthon there.

The Pinelog Formation is restricted to the upright (eastern) limb of the Fort Mountain anticline (Figs. 4 and 8), where it extends along strike to the north-northeast for ~7.5 km. At both its northern and southern extremities (points A and B, Fig. 8), the Pinelog pinches out in map-view between the basement and the base of the Padgett Falls Formation. Between these two points, the Pinelog reaches a maximum thickness of ~250 m. The intersection of the unconformity at the base of the Padgett Falls with the Pinelog–Fort Mountain Gneiss contact is interpreted as a stratigraphic cutoff line. Where such a cutoff line intersects the ground surface obliquely, the intersection is a stratigraphic cutoff point. Points A and B on Figure 8 are therefore two such stratigraphic cutoff points on the line of intersection between the pre–Padgett Falls unconformity and the Pinelog–Fort Mountain Gneiss contact. This line must project in the subsurface between points A and B and must extend down the dip of the base of the Padgett Falls at least to the east of the section of Pinelog that surfaces in the core of the Peebles Lake anticline (Fig. 8, point C). Figure 9 illustrates the outcrop and subcrop projection of the Pinelog Formation at Fort Mountain based upon the assumed approximated subsurface projection of cutoff line A–B. Between Fort Mountain and Cohutta Mountain, where the Pinelog reaches its maximum thickness in the area, the Pinelog dips southeastward at ~25°, whereas the overlying Padgett Falls dips southeastward at ~36°. Cross sections X–X', Y–Y', and Z–Z' (Fig. 9) illustrate the approximated downdip truncation of the Pinelog along the base of the Padgett Falls Formation between the Pinelog outcrop and projected cutoff line A–B. Along the west (overturned) limb of the Fort Mountain anticline, the Padgett Falls Formation rests unconformably in an overturned position against the basement gneiss, and no Pinelog strata are found between the two units (Figs. 4 and 10A, point O). This is in spite of the fact that the Pinelog thickens to the west on the east limb of the anticline (Figs. 9 and 10A). Therefore, the Pinelog disappears between the Padgett Falls and the basement somewhere updip (to the west) of the east limb in the hinge area of the Fort Mountain anticline (Figs. 10A and 10B). Rotation of the base of the Padgett Falls Formation to an upright horizontal position on the two limbs of the Fort Mountain anticline (Fig. 10B) indicates that prior to deposition of the Padgett Falls, the Pinelog was tilted westward ~11°. The only way that *westward* truncation of the Pinelog within the hinge and west limb of the anticline could have resulted from erosion along the pre–Padgett Falls unconformity would be for the Pinelog to have been nested within an open pre–Padgett Falls syncline (Fig. 10C), with the west limb eroded from view in the hinge area of the Fort Mountain anticline. However, in map-view of the sinuous Fort Mountain anticline, no trace exists of the west limb of such a hypothetical pre–Padgett Falls syncline.

A more probable explanation for the westward truncation of the Pinelog Formation at Fort Mountain is that it was cut out to the west of present outcrops by a pre–Padgett Falls southeast-dipping normal fault, here termed the "Fort Mountain fault," that now occurs somewhere within the hinge of the Fort Mountain anticline (Fig. 10D). Along most of its trace across the present erosion surface, such a fault would place basement against basement, and because it is a premetamorphic fault, it would have been subsequently recrystallized and isoclinally folded and would be very difficult to identify given the nature of the exposures on the highly vegetated and scree-covered mountain slopes. If a significant mechanical contrast was maintained across this fault during later Paleozoic ductile deformation, the presence and/or position of the Fort Mountain anticline may have been controlled by the anisotropy created by this southeast-dipping fault. On the northwest slope of Fort Mountain, a small (displacement ~10 cm) premetamorphic normal fault, which may be a subsidiary fracture paralleling a basement fault like the one suggested previously, was observed within Pinelog Formation metasandstone with an orientation of N75°E-64°SE.

If the Ocoee sequence was deposited during two distinct cycles of continental rifting, as suggested by Li and Tull (1998), then the previously described Fort Mountain fault was associated with the second cycle. The earlier cycle probably involved only limited extension and formation of the Snowbird Group (Pinelog Formation), which is composed of mostly highly feldspathic alluvial, fluvial, and relatively shallow-marine clastic rocks. As continental extension became more extensive, the second rift cycle followed as the Great Smoky Group was deposited. This sequence is characterized by deeper-water, mostly turbiditic rocks. Both of these rift cycles were ensialic, and unlike rift basins that formed to the northeast (north of latitude 36°N) and southeast, which formed the thinner (<3 km thick) Mount Rogers and Grandfather Mountain Formations, Ocoee basin extension was not accompanied by volcanism or other igneous activity.

It has been proposed that the southwestern margin of the Ocoee basin was a northwest-striking continental transfer (transform) fault (Cartersville transfer fault), across which the polarity of the rifted Laurentian margin was reversed (Thomas, 1993; Tull and Holm, 2005). The Ocoee basin(s) within the Tennessee salient developed across an oceanward-dipping, lower-plate (distal) margin, whereas to the southwest, the Talladega belt along the Alabama recess (Figs. 1 and 2) developed above a continent-dipping, upper-plate (proximal) margin (Thomas, 1993).

A generalized pre–Padgett Falls Formation subcrop map (Fig. 11) of a large area of the Georgia western Blue Ridge shows the interpreted distribution of Pinelog Formation subunits and basement as they existed beneath the pre–Padgett Falls unconformity prior to deposition of the Padgett Falls Formation. The curvature of units in the subcrop map, which mirrors the "S" curve in the Murphy belt, is mainly an artifact

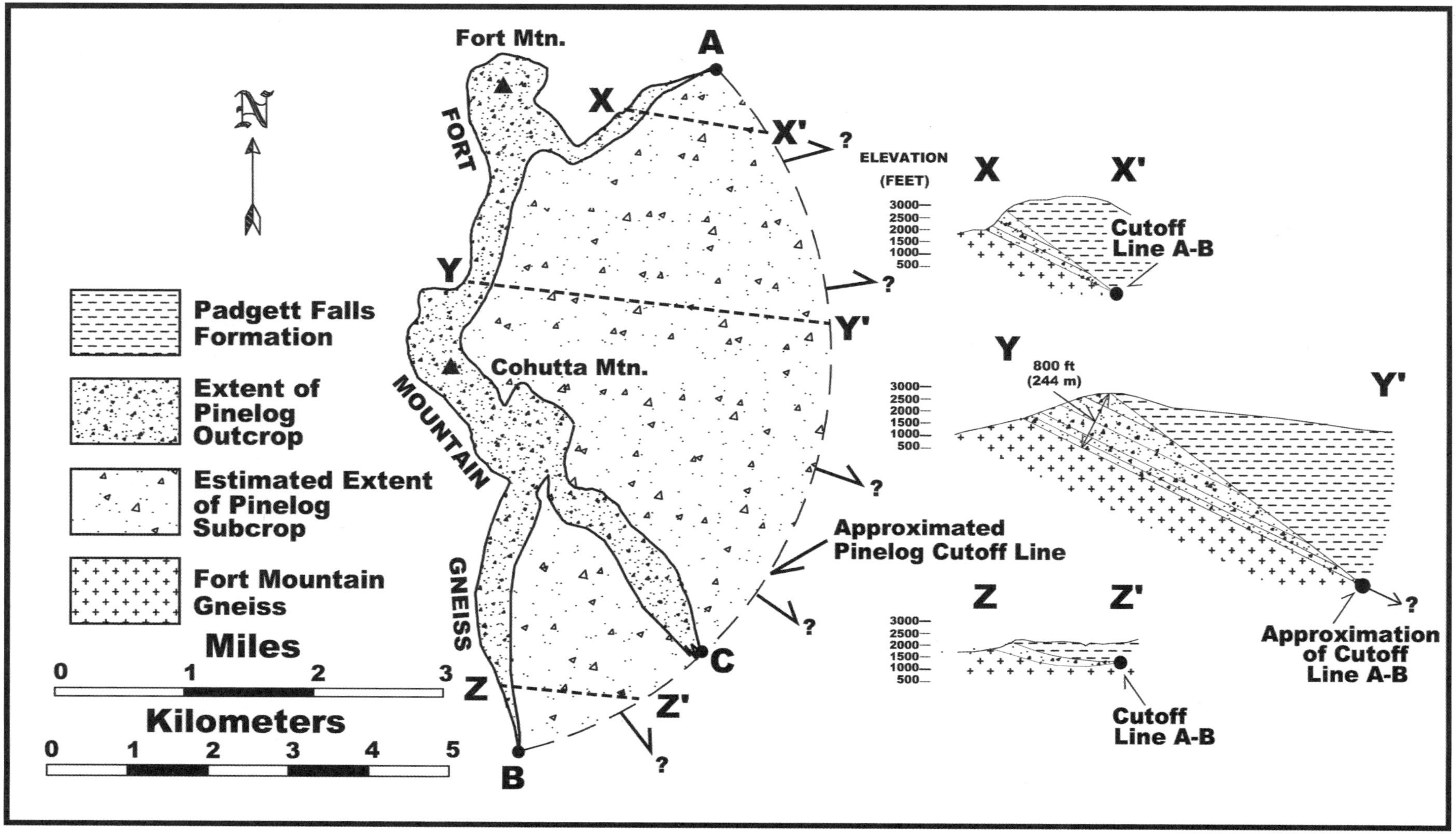

Figure 9. Map showing outcrop (from Fig. 8) and approximated subcrop distribution of Pinelog Formation at Fort Mountain. A and B are stratigraphic cutoff points of the Pinelog Formation. See text for explanation. No vertical exaggeration in cross sections.

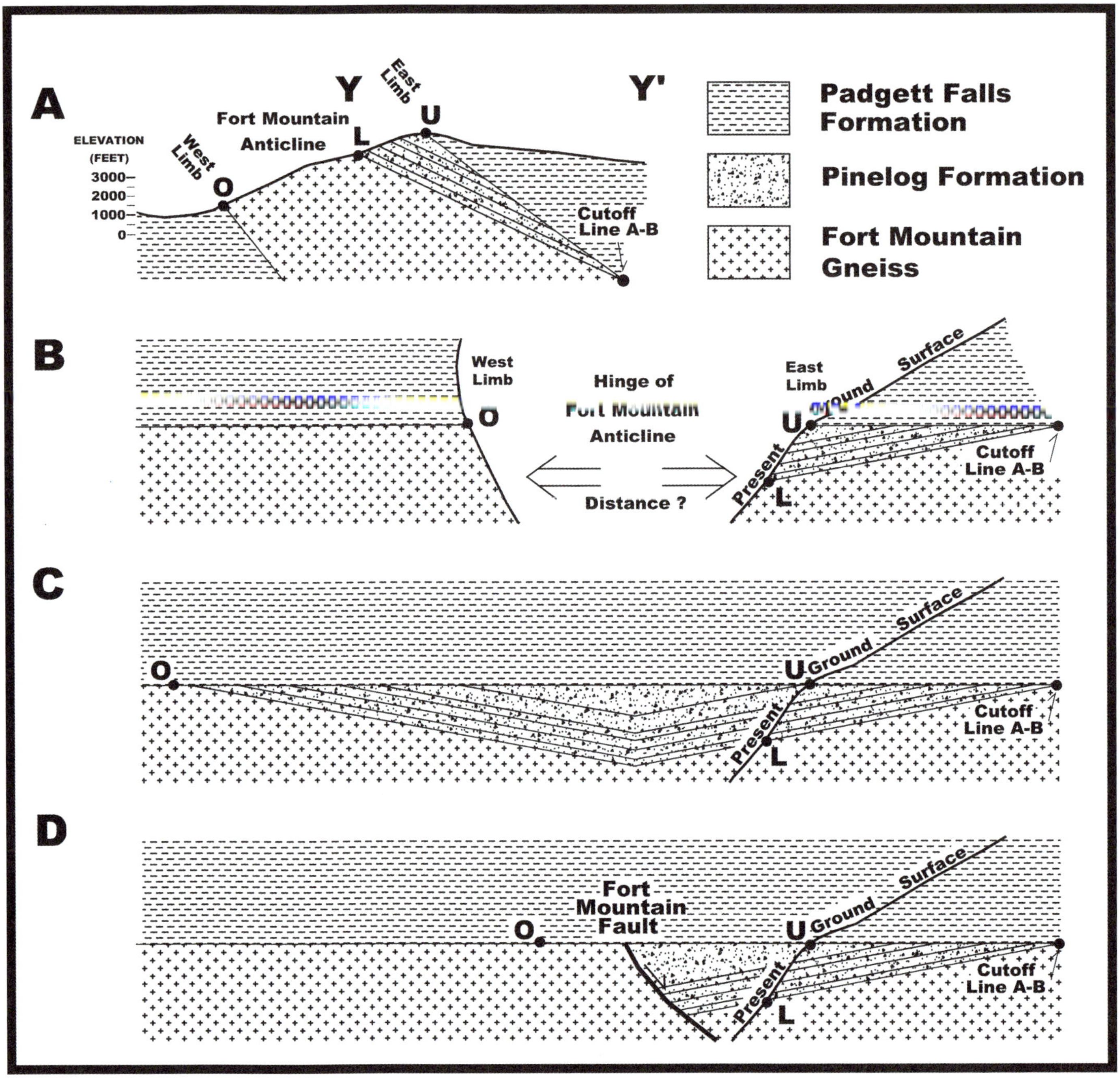

Figure 10. (A) Cross section through the Fort Mountain anticline of the extension of line of section Y–Y' in Figure 9 into the overturned (west) limb. Points U, L, and O represent the upper and lower contacts of the Pinelog Formation and the overturned basement-cover contact, respectively, in the line of section. (B) Cross section showing the limbs of the Fort Mountain anticline in A unfolded. (C) Explanation for the westward truncation of the Pinelog Formation as a result of folding. (D) Explanation for the westward truncation of the Pinelog Formation as a result of faulting. No vertical exaggeration. See text for explanation.

resulting from the inflection caused by map-scale Alleghanian folds. The progressive northward intersections of the cutoff lines with the projected trace of the Fort Mountain fault indicate that the pre–Padgett Falls surface cut deeper into the rifted block in that direction. The map indicates that the three Georgia external basement massifs form part of an aerially extensive (>2000 km^2), contiguous, rifted continental block of basement and cover that existed in a northwest-tilted position immediately prior to deposition of the Great Smoky Group. The Great Smoky Group is characterized by diverse directions of sediment transport and significant lateral variability in thickness and composition, suggesting an assortment of internal basement uplifts (Rast and Kohles, 1986). These uplifts in the Georgia Blue Ridge are potentially positioned within areas designated as "highlands" (Fig. 11). The

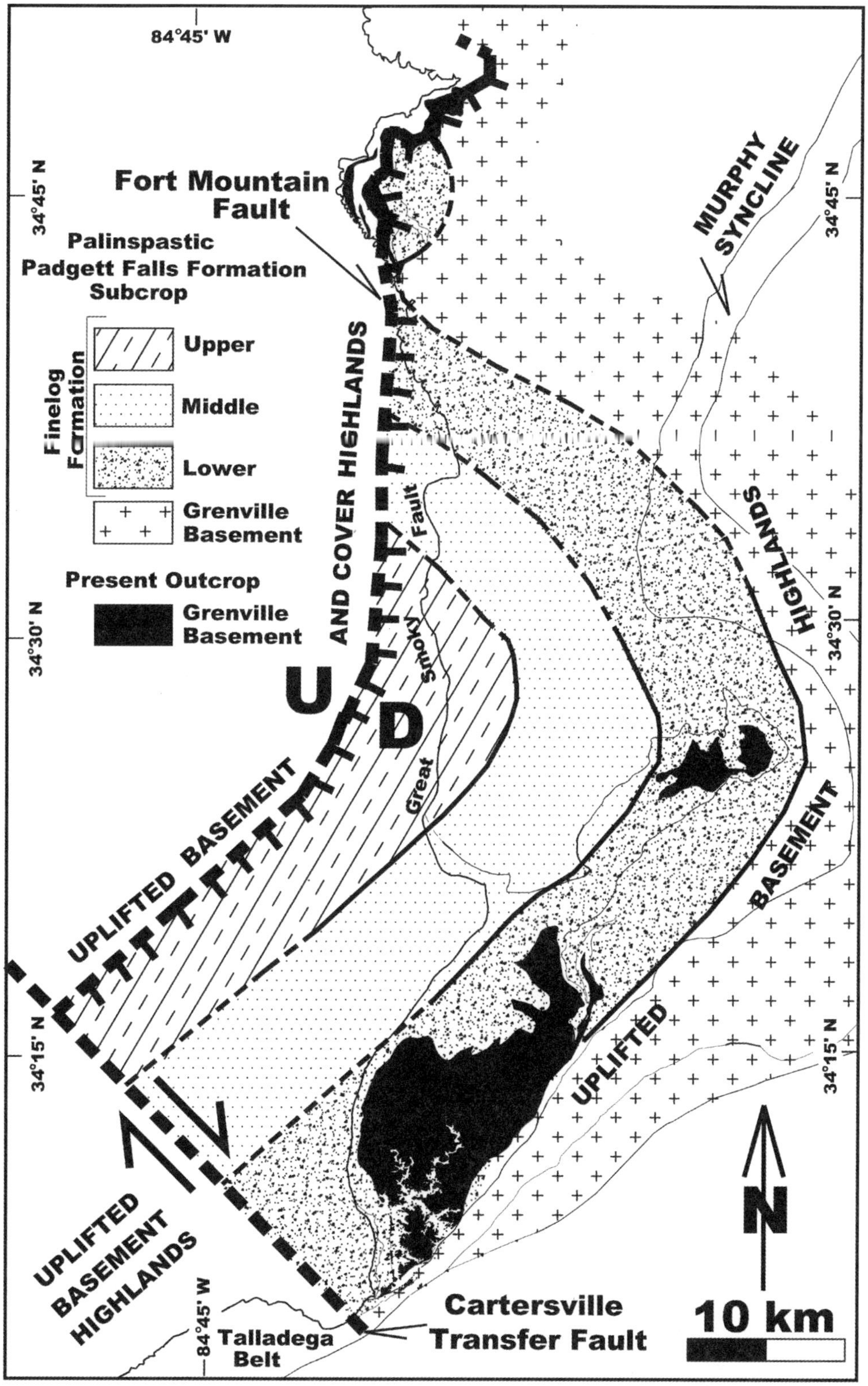

Figure 11. Generalized subcrop map beneath the pre–Padgett Falls (Great Smoky Group) unconformity superimposed upon the geologic map illustrated in Figure 2 showing the palinspastic positions (internal to the Blue Ridge allochthon) of stratigraphic cutoff lines and major faults associated with Late Proterozoic rifting and formation of the Great Smoky Group basin. Solid lines separating units represent stratigraphic cutoff lines beneath the retrodeformed pre–Padgett Falls surface in the Corbin and Salem Church areas (after Li and Tull, 1998), as well as the projected Pinelog–basement cutoff line at Fort Mountain (Fig. 9), and dashed lines are the inferred traces of these lines. For reference, the Pinelog–basement nonconformity cutoff line is fixed to its current position along the east flank of the massifs, whereas the other two cutoff lines are placed in their relative retrodeformed positions to the west (Li and Tull, 1998). See text for explanation.

Fort Mountain fault and the Cartersville transfer fault (Fig. 11) are examples of the complex internal structural framework within the lower-plate rifted continental margin now represented by the Tennessee salient.

SUMMARY

The three external Grenville basement massifs in the Appalachian Blue Ridge of Georgia form the cores of isoclinal inclined-to-recumbent doubly plunging anticlinoria immediately southeast of (above) the foreland fold-and-thrust belt. The massifs are composed predominantly of coarse granitic Mesoproterozoic plutonic rocks that have intruded older mafic and ultramafic units, some of which have been subjected to granulite facies metamorphism. The Mesoproterozoic and older basement rocks underwent Paleozoic greenschist-facies metamorphism and accompanying ductile deformation. This retrogressive event imparted variably developed mylonitic fabrics on the basement rocks. Two unconformably bounded, greenschist-facies cover sequences of the Late Proterozoic Ocoee Supergroup directly overlie the basement complexes, representing two distinct cycles of continental rift basin formation during early stages of the evolution of the Iapetus Ocean. Detailed mapping indicates that the younger of these two sequences, the Great Smoky Group, cuts deeper into the underlying cover sequence and into the basement toward the southeast across an area of several thousand square kilometers. Northwestward tilting of the underlying rift sequence with the basement is believed to be the result of downward rotation of this large block toward the northwest along a large displacement normal fault (Fort Mountain fault). Along most of the southernmost Blue Ridge, this rift-related fault was located palinspastically west of current exposures of the Blue Ridge allochthon, but stratigraphic relationships at Fort Mountain indicate that this fault lies within the eastern basement lens of that massif, i.e., within the Blue Ridge thrust sheet. Southwest of the Tennessee salient, near the southern boundary of the Corbin massif, the Ocoee basin was bounded by a northwest-striking continental transfer (transform) fault (Thomas, 1993; Tull and Holm, 2005). To the southwest, the polarity of the rifted margin of Laurentia was reversed, and no Late Proterozoic continental rift basins like the Ocoee basin are known beneath the Lower Cambrian drift-facies succession.

ACKNOWLEDGMENTS

I am fortunate to have worked with several geologists whose studies have contributed to understanding the geology surrounding the northwest Georgia basement massifs. These include Li Li, Stephen Palmes, Steve Petrushak, and Bill Waite. Reviews by Robert D. Hatcher Jr. and Melvin J. Bartholomew are acknowledged for improving the quality of the manuscript. Support from the U.S. Geological Survey's National Cooperative Geologic Mapping Program, EDMAP Component, was invaluable in several of these studies.

REFERENCES CITED

Aleinikoff, J.N., Southworth, C.S., Yacob, E.Y., and Fanning, C.M., 2004, Isotopic constraints on the time of thee Thunderhead Sandstone, Ocoee Supergroup: U-Pb geochronology of monazite and xenotime overgrowths: Geological Society of America Bulletin, v. 36, no. 5, p. 98.

Aylor, J.G., Jr., 1994, Paleozoic History of the Nantahala and Brasstown Formations of the Basal Murphy Belt, South Westernmost North Carolina and North Westernmost Georgia [Ph.D. dissertation]: Tallahassee, Florida, Florida State University, 421 p.

Badger, R.L., and Sinha, A.K., 1988, Age and Sr isotopic signature of the Catoctin volcanic province: Implications for subcrustal mantle evolution: Geology, v. 16, no. 8, p. 692–695, doi: 10.1130/0091-7613(1988)016<0692:AASISO>2.3.CO;2.

Bartholomew, M.J., 1992, Structural characteristics of the Late Proterozoic (post-Grenville) continental margin of the Laurentian craton, *in* Bartholimew, M.J., Hyndman, D.W., Mogk, D.W., and Mason, R., eds., Basement Technics 10, Characterization and Comparison of Ancient and Mesozoic Continental Margins: Dordrecht, Netherlands, Kluwer Academic Publishers, p. 443–467.

Bartholomew, M.J., and Tollo, R.P., 2004, Northern ancestry for the Goochland terrane as a displaced fragment of Laurentia: Geology, v. 32, no. 8, p. 669–672, doi: 10.1130/G20520.1.

Bayley, W.S., 1928, Geology of the Tate Quadrangle, Georgia: Georgia Geological Survey Bulletin 43, 170 p.

Boyer, S.E., and Mitra, G., 1988, Relations between deformation of crystalline basement and sedimentary cover at the basement/cover transition zone of the Appalachian Blue Ridge Province, *in* Mitra, G., and Wojtal, S., eds., Geometries and Mechanisms of Thrusting, with Special Reference to the Appalachians: Geological Society of America Special Paper 222, p. 119–136.

Brown, W.R., 1970, Investigations of the sedimentary record in the Piedmont and Blue Ridge of Virginia, *in* Fisher, G.W., Pettijohn, F.J., Reed, J.C., Jr., and Weaver, K.N., eds., Study of Appalachian Geology—Central and Southern: New York, Interscience Publishers, p. 335–349.

Bucher, K., and Frey, M., 1994, Petrogenesis of Metamorphic Rocks: Berlin, Springer-Verlag, 318 p.

Connelly, J.B., and Dallmeyer, R.D., 1993, Polymetamorphic evolution of the western Blue Ridge: Evidence from $^{40}Ar/^{39}Ar$ whole-rock slate/phyllite and muscovite ages: American Journal of Science, v. 293, p. 322–359.

Costello, J.O., 1978, Shear Zones in the Corbin Gneiss of Georgia: Georgia Geologic Survey Bulletin, v. 93, p. 32–37.

Costello, J.O., 1988, Structural controls on southern Murphy syncline geometry: Atlanta, Georgia, Georgia Geological Society, Guidebook, v. 8, p. 7–19.

Costello, J.O., 1993, Studies in Appalachian Foreland to Hinterland Transition Zone Geology, Georgia and Tennessee [Ph.D. dissertation]: Columbia, South Carolina, University of South Carolina, 144 p.

Costello, J.O., and Hatcher, R.D., Jr., 1986, Contact relationships between the Walden Creek Group and Great Smoky Group in Ocoee Gorge, Tennessee: Implications on the regional extent of the Greenbrier fault: Geological Society of America Abstracts with Programs, v. 18, no. 3, p. 216.

Costello, J.O., and Hatcher, R.D., Jr., 1991, Problems of stratigraphic correlation between the Great Smoky, Snowbird, and Walden Creek Groups between the Great Smoky Mountains National Park, central-east Tennessee and Ocoee Gorge, southeastern Tennessee, *in* Kish, S.A., ed., Study of Precambrian and Paleozoic Stratigraphy in the Western Blue Ridge: Durham, North Carolina, USA, Carolina Geological Society, Guidebook, p. 13–25.

Costello, J.O., McConnell, K.I., and Power, W.R., 1982, Geology of late Precambrian and early Paleozoic rocks in and near the Cartersville district, Georgia: Atlanta, Georgia Geological Society, Guidebook, 17th Annual Field Trip, 42 p.

Cressler, C.W., 1970, Geology and Groundwater Resources of Floyd and Polk Counties, Georgia: Georgia Geological Survey Information Circular 39, 95 p.

Cudzil, M.R., 1985, Fluvial, Tidal and Storm Sedimentation in the Chilhowee Group (Lower Cambrian), Northeastern Tennessee [M.S. thesis]: Knoxville, University of Tennessee, 164 p.

De Windt, J.T., 1975, Geology of the Great Smoky Mountains, Tennessee and North Carolina, with road log for field excursion, Knoxville–Clingmans

Dome–Maryville: The Compass of Sigma Gamma Epsilon, v. 52, no. 4, p. 73–129.

Fairley, W.M., 1965, The Murphy Syncline in the Tate Quadrangle, Georgia: Georgia Geological Survey Bulletin 75, 71 p.

Fairley, W.M., 1966, The Cartersville fault at Cartersville Georgia, *in* Bentley, R.D., Fairley, W.M., Fields, H.H., Power, W.R., and Smith, J.M., eds., The Cartersville Fault Problem: Georgia Geological Survey Geological Guidebook Series, v. 4, p. 28–33.

Fairley, W.M., 1973, Correlations of stratigraphic belts of the northwest Georgia Piedmont and Blue Ridge: American Journal of Science, v. 273, p. 686–697.

Furcron, A.S., and Teague, K.H., 1947, Talc Deposits of Murray County, Georgia: Georgia Geological Survey Bulletin 75, 71 p.

Gapais, D., 1989, Shear structures within deformed granites: Mechanical and thermal indicators: Geology, v. 17, p. 1144–1147, doi: 10.1130/0091-7613(1989)017<1144:SSWDGM>2.3.CO;2.

Gargi, S.P., 1985, Geometric and Kinematic Analysis, and Geochemical Study of the Corbin Gneiss Complex and its Associated Sheared Rocks in the Blue Ridge of Northwest Georgia [Ph.D. dissertation]: Oxford, Ohio, Miami University, 250 p.

Glover, L., III, Speer, J.A., Russell, G.S., and Farrar, S.S., 1983, Ages of regional metamorphism and ductile deformation in the central and southern Appalachians: Lithos, v. 16, p. 223–245, doi: 10.1016/0024-4937(83)90026-9.

Goldberg, S.A., and Dallmeyer, R.D., 1997, Chronology of Paleozoic metamorphism and deformation in the Blue Ridge thrust complex, North Carolina and Tennessee: American Journal of Science, v. 297, p. 488–526.

Groszos, M.S., 1996, The Geology of the Murphy Belt in North Central Georgia [Ph.D. dissertation]: Tallahassee, Florida, Florida State University, 366 p.

Hadley, J.B., 1970, The Ocoee Series and its possible correlatives, *in* Fisher, G.W., Pettijohn, F.J., Reed, J.C., Jr., and Weaver, K.N., eds., Study of Appalachian Geology—Central and Southern: New York, Interscience Publishers, p. 247–259.

Hadley, J.B., and Goldsmith, R., 1963, Geology of the Eastern Great Smoky Mountains, North Carolina and Tennessee: U.S. Geological Survey Professional Paper 349-B, 118 p.

Hamilton, W.B., 1961, Geology of the Richardson Cove and Jones Cove Quadrangles, Tennessee: U.S. Geological Survey Professional Paper 349-A, p. A1–A55.

Harris, L.D., Harris, A.G., DeWitt, W., Jr., and Bayer, K.C., 1981, Evaluation of southeastern overthrust belt beneath the Blue Ridge–Piedmont thrust: American Association of Petroleum Geologists Bulletin, v. 65, p. 2497–2505.

Hatcher, R.D., Jr., and Odom, A.L., 1980, Timing of thrusting in the southern Appalachians, USA: Model for orogeny?: Geological Society [London] Journal, v. 137, p. 321–327.

Hayes, C.W., 1895, Description of Cleveland Sheet: U.S. Geological Survey Geologic Atlas, folio 20.

Heatherington, A.L., Mueller, P.A., Smith, M.S., and Allen, P.N., 1996, The Corbin Gneiss: Evidence for Grenvillian magmatism and older continental basement in the southernmost Blue Ridge: Southeastern Geology, v. 36, no. 1, p. 15–25.

Hernon, R.M., 1967, The Wehutty Formation of North Carolina: U.S. Geological Survey Bulletin 1274A, p. A49–A52.

Higgins, M.W., Atkins, R.L., Crawford, T.J., Crawford, R.F., III, Brooks, R., and Cook, R.B., 1988, The structure, tectonostratigraphy, and evolution of the southernmost part of the Appalachian orogen: U.S. Geological Professional Paper 1475, 173 p.

Higgins, M.W., Crawford, T.J., and Crawford, R.F., III, 1989, Windows in the crystalline terrane of Georgia, southwestern North Carolina, and southeastern Tennessee and the geologic framework of the southern Appalachians, *in* Fritz, W.J., ed., Excursions in Georgia Geology: Georgia Geological Society–Southeast Geological Society of America, Field Trip Guidebook, v. 9, no. 1, p. 239–262.

Higgins, M.W., Crawford, R.F., Crawford, T.J., III, Costello, J.O., Offield, T.O., and Kath, R., 1996, Geology of the Cartersville District and the Cartersville fault problem—A progress report, *in* Kath, R.L. ed., The Cartersville Fault Problem: Revisited: Atlanta, Georgia, Georgia Geological Society, 31st Annual Field Trip Guidebook, p. 9–53.

Hull, J.P.D., La Forge, L., and Crane, W.R., 1919, Report on the Manganese Deposits of Georgia: Georgia Geologic Survey Bulletin 35, 295 p.

Hurst, V.J., 1955, Stratigraphy, Structure, and Mineral Resources of the Mineral Bluff Quadrangle, Georgia: Georgia Geological Survey Bulletin 63, 137 p.

Hurst, V.J., and Shlee, J.S., 1962, Field Excursion, Ocoee Metasediments, North-Central Georgia–Southern Tennessee: Georgia Geological Survey Guidebook, no. 3, 28 p.

Kath, R.L., and Crawford, T.J., 1996, Mineralogy, geothermometry, and geobarometry of the Rowland Springs Formation, Allatoona Complex, Georgia, *in* Kath, R.L. ed., The Cartersville Fault Problem: Revisited: Atlanta, Georgia, Georgia Geological Society, 31st Annual Field Trip Guidebook, p. 63–79.

Keith, A., 1904, Geologic Atlas of the United States: Asheville Folio, Tennessee–North Carolina: U.S. Geological Survey Folio 116, 10 p.

Keith, A., 1907, Description of the Nantahala Quadrangle, North Carolina and Tennessee: U.S. Geological Survey Geologic Atlas, Folio 143, 11 p.

Kesler, T.L., 1950, Geology and Mineral Deposits of the Cartersville District, Georgia: U.S. Geological Survey Professional Paper 224, 97 p.

King, P.B., 1949, The base of the Cambrian in the southern Appalachians, Part 2: American Journal of Science, v. 247, p. 622–645.

King, P.B., 1964, Geology of the Central Great Smoky Mountains, Tennessee: U.S. Geological Survey Professional Paper 349-C, 148 p.

King, P.B., 1970, The Precambrian of the United States of America: Southeastern United States, *in* Rankama, K., ed., The Geologic Systems: The Precambrian, Volume 4: New York, John Wiley and Sons, Interscience, p. 1–71.

King, P.B., Hadley, J.B., Neuman, R.B., and Hamilton, W., 1958, Stratigraphy of the Ocoee Series, Great Smoky Mountains, Tennessee and North Carolina: Geological Society of America Bulletin, v. 69, p. 947–966, doi: 10.1130/0016-7606(1958)69[947:SOOSGS]2.0.CO;2.

Kish, S.A., 1990, Timing of middle Paleozoic (Acadian) metamorphism in the southern Appalachians- K-Ar studies in the Talladega belt, Alabama: Geology, v. 18, p. 650–653, doi: 10.1130/0091-7613(1990)018<0650:TOMPAM>2.3.CO;2.

Kish, S.A., 1991, Potassium-argon dating in the western Blue Ridge of North Carolina and Tennessee, *in* Kish, S.A., ed., Studies of Precambrian to Paleozoic Stratigraphy in the Western Blue Ridge: Durham, North Carolina, USA, Carolina Geological Society, Field Trip Guidebook, p. 69–78.

LaForge, L., 1919, The Cartersville district, *in* Hull, J.P.D., LaForge, L., and Crane, W.R., eds., Report on the Manganese Deposits of Georgia: Georgia Geological Survey Bulletin 35, p. 25–67.

Li, L., 1996, The Stratigraphy and Structure of the Corbin and Salem Church Basement Culminations in the Southern Appalachian Blue Ridge, Northwest Georgia [Ph.D. dissertation]: Tallahassee, Florida, Florida State University, 210 p.

Li, L., and Tull, J.F., 1998, Cover stratigraphy and structure of the southernmost external basement massifs in the southern Appalachian Blue Ridge: Evidence for two-stage Late Proterozoic rifting: American Journal of Science, v. 298, p. 829–867.

Mack, G.H., 1980, Stratigraphy and depositional environments of the Chilhowee Group (Cambrian) in Georgia and Alabama: American Journal of Science, v. 280, p. 497–517.

Martin, A.J., and Crawford, R.F., 1996, Trace fossils in the Pinelog Formation, Piedmont–Blue Ridge, Georgia, and their application to sedimentology, stratigraphy, and structural geology, *in* Kath, R.L. ed., The Cartersville Fault Problem: Revisited: Atlanta, Georgia Geological Society, 31st Annual Field Trip Guidebook, p. 89–96.

Martin, B.F., Jr., 1974, The Petrology of the Corbin Gneiss [M.S. thesis]: Athens, University of Georgia, 113 p.

McCallie, S.W., 1907, Marbles of Georgia: Georgia Geological Survey Bulletin 1, 126 p.

McConnell, K.I., and Abrams, C.E., 1984, Geology of the Greater Atlanta Region: Georgia Geologic Survey Bulletin 96, 89 p.

McConnell, K.I., and Costello, J.O., 1980, Guide to geology along a traverse through the Blue Ridge and Piedmont provinces of north Georgia, *in* Frey, R.W., ed., Excursions in Southeastern Geology: Geological Society of America, Field Trip Guidebook, v. 1, no. 12, p. 241–258.

McConnell, K.I., and Costello, J.O., 1984, Basement–cover rock relationships along the western edge of the Blue Ridge thrust sheet in Georgia, *in* Bartholomew, M.J., ed., The Grenville Event in the Appalachians and Related Topics: Geological Society of America Special Paper 194, p. 263–280.

Moecher, D.P., Samson, S.D., and Miller, C.F., 2004, Precise time and conditions of peak Taconian granulite facies metamorphism in the southern Appalachian orogen, U.S.A., with implications for zircon behavior during crustal melting events: The Journal of Geology, v. 112, no. 3, p. 289–304, doi: 10.1086/382760.

Mohr, D.W., 1973, Stratigraphy and structure of part of the Great Smoky and Murphy Belt Groups, western North Carolina: American Journal of Science, v. 273-A, p. 41–71.

Montes, C., and Hatcher, R.D., Jr., 1999, Documenting Late Proterozoic rifting in the Ocoee basin, western Blue Ridge, North Carolina: Southeastern Geology, v. 39, no. 1, p. 37–50.

Needham, R.E., 1972, The Geology of the Murray County, Georgia, Talc District [M.S. thesis]: Athens, University of Georgia, 107 p.

Odom, A.L., Kish, S.A., and Leggo, P.J., 1973, Extension of "Grenville basement" to the southern extremity of the Appalachians: U-Pb ages of zircons: Geological Society of America Abstracts with Programs, v. 5, no. 5, p. 425.

Palmes, S.A., 1998, Stratigraphy, Structure, and Metamorphism of the Western Blue Ridge, North Georgia [Ph.D. dissertation]: Tallahassee, Florida, Florida State University, 205 p.

Power, W.R., and Forrest, J.T., 1973, Stratigraphy and paleogeography in the Murphy Marble belt: American Journal of Science, v. 273, p. 698–711.

Ramsay, J.G., 1967, Folding and Fracturing of Rocks: New York, McGraw-Hill, 568 p.

Rankin, D.W., 1975, The continental margin of eastern North America in the southern Appalachians: The opening and closing of the proto–Atlantic Ocean: American Journal of Science, v. 275-A, p. 298–336.

Rankin, D.W., 1993, Proterozoic rocks east and southeast of the Grenville front, *in* Reed, J.C., Jr., et al., eds., Precambrian: Conterminous United States: Boulder, Colorado, Geological Society of America, Geology of North America, v. C-2, p. 335–338.

Rast, N., and Kohles, K.M., 1986, The origin of the Ocoee Supergroup: American Journal of Science, v. 286, p. 593–616.

Rodgers, J., 1972, Latest Precambrian (post-Grenville) rocks of the Appalachian region: American Journal of Science, v. 272, p. 507–530.

Russell, G.S., 1976, Rb-Sr evidence from cataclastic rocks for Devonian faulting in the southern Appalachians: Geological Society of America Abstracts with Programs, v. 8, no. 6, p. 1081.

Schwab, F.L., 1986, Latest Precambrian–earliest Paleozoic sedimentation, Appalachian Blue Ridge and adjacent areas: Review and speculation, *in* McDowell, R.C., and Glover, L., III, eds., Studies in Appalachian Geology: Virginia Tech Department of Geological Sciences, Memoir 3, p. 115–137.

Sears, J.W., 1988, Origin and palinspastic significance of a fault duplex near Cartersville, Georgia, *in* Mitra, G., and Wojtal, S., eds., Geometries and Mechanisms of Thrusting, with Special Reference to the Appalachians: Geological Society of America Special Paper 222, p. 179–184.

Simpson, E.L., and Eriksson, K.A., 1989, Sedimentology of the Unicoi Formation in southern and central Virginia: Evidence for late Proterozoic to Early Cambrian rift-to-passive margin transition: Geological Society of America Bulletin, v. 101, p. 42–54, doi: 10.1130/0016-7606(1989)101<0042:SOTUFI>2.3.CO;2.

Skelly, R., 1987, Sedimentology of Fluvial-to-Marine Transition: The Chilhowee Group (Lower Cambrian), Southeast Tennessee [M.S. thesis]: Knoxville, University of Tennessee, 211 p.

Thomas, W.A., 1993, Low-angle geometry of the late Precambrian–Cambrian Appalachian-Ouachita rifted margin of southeastern North America: Geology, v. 21, p. 921–924, doi: 10.1130/0091-7613(1993)021<0921:LADGOT>2.3.CO;2.

Tull, J.F., and Holm, C.S., 2005, Structural evolution of a major Appalachian salient-recess junction: Consequences of oblique collisional convergence across a continental margin transform fault: Geological Society of America Bulletin, v. 117, p. 482–499, doi: 10.1130/B25578.1.

Tull, J.F., Ausich, W.I., Groszos, M.S., and Thompson, T.W., 1993, Appalachian Blue Ridge cover sequence ranges at least into the Ordovician: Geology, v. 21, p. 215–218, doi: 10.1130/0091-7613(1993)021<0215:ABRCSR>2.3.CO;2.

Unrug, R., Ausich, W.I., Bednarczyk, J., Cuffey, R.J., Mamet, B.J., Palmes, S.L., and Unrug, S., 2000, Paleozoic age of the Walden Creek Group, Ocoee Supergroup, in the western Blue Ridge, southern Appalachians: Implications for evolution of the Appalachian margin of Laurentia: Geological Society of America Bulletin, v. 112, no. 7, p. 982–996, doi: 10.1130/0016-7606(2000)112<0982:PAOTWC>2.3.CO;2.

Wehr, F., and Glover, L., III, 1985, Stratigraphy and tectonics of the Virginia–North Carolina Blue Ridge: Evolution of a later Proterozoic–early Paleozoic hinge zone: Geological Society of America Bulletin, v. 96, p. 285–295, doi: 10.1130/0016-7606(1985)96<285:SATOTV>2.0.CO;2.

Whisonant, R.C., 1967, Stratigraphy and Petrology of the Basal Cambrian Chilhowee Group in Central-Eastern and Southeastern Tennessee [Ph.D. dissertation]: Tallahassee, Florida, Florida State University, 237 p.

Manuscript Accepted by the Society 1 February 2007

Printed in the USA

The Geological Society of America
Memoir 200
2007

Tectonic map of the southern and central Appalachians: A tale of three orogens and a complete Wilson cycle

Robert D. Hatcher Jr.
Department of Earth and Planetary Sciences and Science Alliance Center of Excellence, University of Tennessee, Knoxville, Tennessee 37996-1410, USA

Brendan R. Bream
Department of Earth and Environmental Sciences, Vanderbilt University, Nashville, Tennessee 37235, USA

Arthur J. Merschat
Department of Earth and Planetary Sciences and Science Alliance Center of Excellence, University of Tennessee, Knoxville, Tennessee 37996-1410, USA

If you do not like my model, build your own.
—John Rodgers

ABSTRACT

A new tectonic map of the southern and central Appalachians incorporates modern field and structural-stratigraphic, geochronologic (mostly sensitive high-resolution ion microprobe–reverse geometry, SHRIMP–RG, and Sm-Nd), geochemical, and geophysical data to identify crustal boundaries and blocks. Major tectonic units include the ~735 Ma Laurentian failed rift, ~565 Ma rifted margin sedimentary-volcanic assemblage deposited on Grenvillian and pre-Grenvillian crust, the Laurentian platform, and a series of distal Laurentian terranes (Cowrock, Cartoogechaye, Tugaloo-Chopawamsic-Potomac) accreted to Laurentia during the Taconian (Ordovician) or Neoacadian (Late Devonian–early Mississippian) orogenies. The Dahlonega gold belt consists of more proximal metasandstone and pelitic schist; it also contains Ordovician arc volcanic rocks, and a mixed detrital zircon suite of Laurentian and Gondwanan, or Penokean, affinity. The newly recognized Cat Square terrane contains Laurentian, Avalonian, and 430 Ma detrital zircons, and is considered a remnant ocean basin that closed during Acadian-Neoacadian accretion of the Carolina superterrane. The Pine Mountain terrane (southernmost exposed Appalachians) consists of Grenvillian basement and a cover sequence bearing Gondwanan or Penokean detrital zircons. The Carolina superterrane contains numerous peri-Gondwanan terranes that

Hatcher, R.D., Jr., Bream, B.R., and Merschat, A.J., 2007, Tectonic map of the southern and central Appalachians: A tale of three orogens and a complete Wilson cycle, *in* Hatcher, R.D., Jr., Carlson, M.P., McBride, J.H., and Martínez Catalán, J.R., eds., 4-D Framework of Continental Crust: Geological Society of America Memoir 200, p. 595–632, doi: 10.1130/2007.1200(29). For permission to copy, contact editing@geosociety.org.

were deformed, metamorphosed, and amalgamated prior to 530 Ma, then accreted to Laurentia during the Neoacadian along the central Piedmont suture. The Raleigh-Goochland terrane contains blocks of Laurentian basement and cover that moved SW (dextrally) out of the collision zone to the north as the Theic ocean closed north to south during the early Alleghanian orogeny. This event also produced the Kiokee-Raleigh belt high-grade metamorphic core in the eastern Piedmont, and includes faults of the Pine Mountain window. The latter is framed by Alleghanian thrust and dextral faults formed at different crustal depths (times?). Subsurface components of the southern and central Appalachians are recognized in potential field and limited drill data. The Carolina superterrane extends beneath the Coastal Plain—possibly eastward to the East Coast magnetic anomaly. South of the Carolina superterrane and north of the Wiggins suture is the Brunswick (Charleston) terrane, another peri-Gondwanan terrane. The east–west Alleghanian Wiggins suture with the Suwannee terrane is recognizable to the south beneath Georgia and Alabama in potential field data, truncating all Appalachian structures and older crustal blocks west of the Appalachians. South of the suture, African basement and cover lie in the eastern Florida subsurface, while to the west are other Gondwanan or peri-Gondwanan components that may have originally connected with Yucatan.

Keywords: Southern and Central Appalachians, tectonic map, Paleozoic orogen, Wilson cycle

INTRODUCTION

Our purpose here is to summarize the tectonic framework of the southern and central Appalachians in a modern context (Fig. 1; Plate 1 [on the CD-ROM accompanying this volume, or in the GSA Data Repository[1]]). This will be accomplished through a new tectonic map that has been compiled at a scale of 1:1,000,000 during the past three years. This map contains up-to-date data related to terrane boundaries, their geometry, and the nature of tectonostratigraphic terranes based on current knowledge of stratigraphic sequences and modern geochronologic and isotopic data from basement, metasedimentary rocks, and plutons. Previous tectonic maps of this region include the U.S. Geological Survey–American Association of Petroleum Geologists' (1961) tectonic map of North America, King's tectonic map of North America (1969) and the southern and central Appalachians (1970), Rodgers' (1970) map of the Appalachians, Williams' (1978) tectonic-lithofacies map of the Appalachians, Rankin et al.'s (1989) tectonic map of the early Paleozoic tectonic elements of the U.S. Appalachians, Horton et al.'s (1989) tectonic map of the southern and central Appalachians, Hatcher et al.'s (1990) tectonic map of the U.S. Appalachians, Muehlberger's (1996) tectonic map of North America, and the recently published tectonic map of the Appalachians (Hibbard et al., 2006a). All of these maps chronicle an evolution of the state of knowledge, particularly with respect to the internal parts of the Appalachian orogen. In contrast to geologic maps, tectonic maps are by definition interpretive, both of the nature of boundaries, and in terms of the nature of tectonostratigraphic units: here terranes of Laurentian affinity, suspect terranes, and exotic terranes. Several terranes defined by Williams and Hatcher (1982, 1983) as suspect, e.g., the Piedmont (now Cowrock, Cartoogechaye, Dahlonega gold belt, Tugaloo, Cat Square, Milton, Chopawamsic, Potomac, Baltimore, Westminster, Philadelphia, and Hamburg terranes), are with one exception (Cat Square) now known to be partly of Laurentian provenance, and partially of Theic oceanic origin (Faill, 1997a). A previously unrecognized terrane, Cat Square (Hatcher, 2002a; Bream, 2003; Bream et al., 2004), has been defined based on Laurentian, peri-Gondwanan, and 430 Ma detrital zircons, and dominance of Devonian and younger plutons. Other subsurface terranes are also delineated on the new tectonic map based primarily on aeromagnetic, gravity, and limited drilling and geochronologic data (Fig. 1B). Our tectonic map [Fig. 1A(1), Plate 1] is biased toward the best current data to minimize the amount of speculation involved in defining tectonic units and boundaries; nevertheless, tectonic maps become outdated in a decade or so with the appearance of new data. There is no reason to suspect this iteration will prove any different. It should be noted, however, that all of the major boundaries in King's 1969 tectonic map remain unchanged in Muelhberger's 1996 map. The principal difference is that, today, we have identified many more important boundaries, which can be represented at the scales of these maps.

The reference frame used in this discussion is the present geography. We recognize, however, that the equator trended north-south (present geographic frame) bisecting North America during the early and middle Paleozoic (e.g., Hatcher, 1974). Laurentia

[1]GSA Data Repository item 2007239, Plate 1, is available at www.geosociety.org/pubs/ft2007.htm, or on request from editing@geosociety.org, Documents Secretary, GSA, P.O. Box 9140, Boulder, CO 80301, USA.

rotated counterclockwise during the middle to late Paleozoic bringing the continent into its present orientation, but still at equatorial latitude. Most subsequent movement of Laurentia during the Mesozoic and Cenozoic has been northward to its present location north of the equator (van der Voo, 1988).

TECTONIC FRAMEWORK

The Appalachians were created following the rifting of supercontinent Rodinia during three or more Paleozoic orogenies: the Ordovician-Silurian Taconian, the Devonian Acadian and Devonian-Mississippian Neoacadian, and the Pennsylvanian-Permian Alleghanian orogenies (Fig. 2). All orogenies were diachronous along the length of the chain. The character of the orogen changes roughly at the latitude of New York City from one constructed largely from west to east in the north to an orogen built from the inside out to the south (Hatcher and Odom, 1980), with the exception of the Acadian thrust belt, which extends up the Hudson River Valley to just south of Albany, New York (Marshak, 1986). This along-strike change in character is related to changes in mechanical stratigraphy, relationships to the shape of the original continental margin (e.g., Thomas, 2006), accretion of arcs and microcontinents (e.g., van Staal et al., 2007, this volume), and the shape of the Gondwanan indenter that collided with Laurentia at the end of the Paleozoic. The internal parts of the chain to the southeast initially consisted of deformed rifted margin sedimentary rocks and recycled Grenville basement. These give way eastward to accreted terranes of distal Laurentian metasedimentary oceanic and arc affinity material (Tugaloo, Milton, Chopawamsic, Potomac, and Baltimore terranes), then a sequence of mid-Paleozoic metasedimentary rocks (Cat Square terrane), and a truly exotic peri-Gondwanan (Panafrican), composite Carolina (-Avalon from New England northward) superterrane (Carolinia of Hibbard et al., 2006a) of Neoproterozoic to Ordovician(?) arc volcanic and volcaniclastic rocks [Figs. 1A(1) and 2B, Plate 1]. The Brunswick (Charleston) terrane occurs in the subsurface of South Carolina and Georgia, and from limited drill data, appears to consist of another peri-Gondwanan component (Mueller et al., 2005). The subsurface Suwannee terrane lies south of an east-west–trending suture zone in southern Georgia and Alabama, and contains Gondwanan-affinity basement and cover sedimentary rocks to the south (Pojeta et al., 1976; Mueller et al., 1994). The Appalachian collage records a complete Wilson cycle of the opening and closing of a series of Paleozoic oceans starting with rifting of supercontinent Rodinia and ending with the assembly of supercontinent Pangea (e.g., Thomas, 2006).

DATA THAT HELP GENERATE TECTONIC MAPS

New data and ideas are the principal drivers that alter our knowledge of the crust and demand construction of new tectonic maps. These new data include: (1) state-of-the-art geochronologic data from instruments not in widespread use two decades ago (sensitive high-resolution ion microprobe [SHRIMP]; laser ablation inductively coupled plasma mass spectrometer, [LA ICP–MS]); (2) isotopic data formerly not available (e.g., Pb isotope and Sm-Nd); (3) high quality chemical analyses of (meta-) volcanic and plutonic rocks amenable to tectonic discriminant analysis; (4) larger areas of detailed geologic mapping that delineate previously unrecognized major boundaries and new tectonostratigraphic terranes; (5) better understanding of the tectonic links between the internal and external parts of the southern and central Appalachians; and (6) modern digital compilations of potential field (aeromagnetic, gravity) data of the exposed and subsurface Appalachians.

Pre-Appalachian Basement—Setting the Stage for the Appalachian Wilson Cycle

Supercontinent Rodinia formed 1.2–1.0 Ga by amalgamation of most existing continental landmasses in several possible configurations (e.g., Hoffman, 1991; Karlstrom et al., 2001; Sears and Price, 2003). The Rodinia-forming event that affected eastern Laurentia is the Grenvillian (Ottawan 1.15–1.0 Ga) orogeny, which was preceded by the 1.24–1.18 Ga Elzeverian orogeny (Tollo et al., 2004a). Blue Ridge and Piedmont basement massifs record the Grenville orogeny, but southern and central Appalachian Grenvillian rocks may have a South American—not Laurentian—origin, based on Pb isotopic data (Loewy et al., 2003; Tohver et al., 2004) (Fig. 3). This conclusion requires that a previously unidentified suture exist beneath the platform sedimentary sequence under the Alleghanian Blue Ridge–Piedmont megathrust sheet. The known likely candidate for this suture is the New York–Alabama magnetic-gravity lineament (King and Zietz, 1978) (Fig. 1, Plate 1).

Southern Appalachian pre-Appalachian basement rocks ages are remarkably consistent around 1.15–1.0 Ga (Heatherington et al., 1996; Carrigan et al., 2003; Bream et al., 2004) (Fig. 4). Most southern Appalachian Grenvillian rocks are granitic orthogneisses that were emplaced between 1.15 and 1.0 Ga (Carrigan et al., 2003). The 1.8 Ga Mars Hill terrane in northwestern North Carolina and adjacent Tennessee (Gulley, 1985; Raymond and Johnson, 1994) is a block of pre-Grenville basement that was incorporated into the Grenville orogen; it contains the oldest rocks in the Appalachians, and their age has been reconfirmed using the modern SHRIMP dating technique (Ownby et al., 2004) (Figs. 1 and 4, Plate 1). It contains para- and orthogneisses that were metamorphosed and intruded during the Grenville orogeny (Gulley, 1985; Carrigan et al., 2003; Ownby et al., 2004) and a unique geochemical signature (Heatherington et al., 2007). The 1.19 Ga date on the Reepsville gneiss (Merschat et al., 2005b; A.J. Merschat, personal commun.) is the first known occurrence of Grenville basement in the internal parts of the Inner Piedmont. Basement rocks in central and northern Virginia are petrologically more diverse (Tollo et al., 2004c), including a suite of meta-igneous rocks (charnockite, metagranite, monzonite, etc.) that was intruded during three pulses of magmatic activity: 1.16–1.14 Ga, ~1.12 Ga, and 1.08–1.05 Ga (Tollo et al., 2004b).

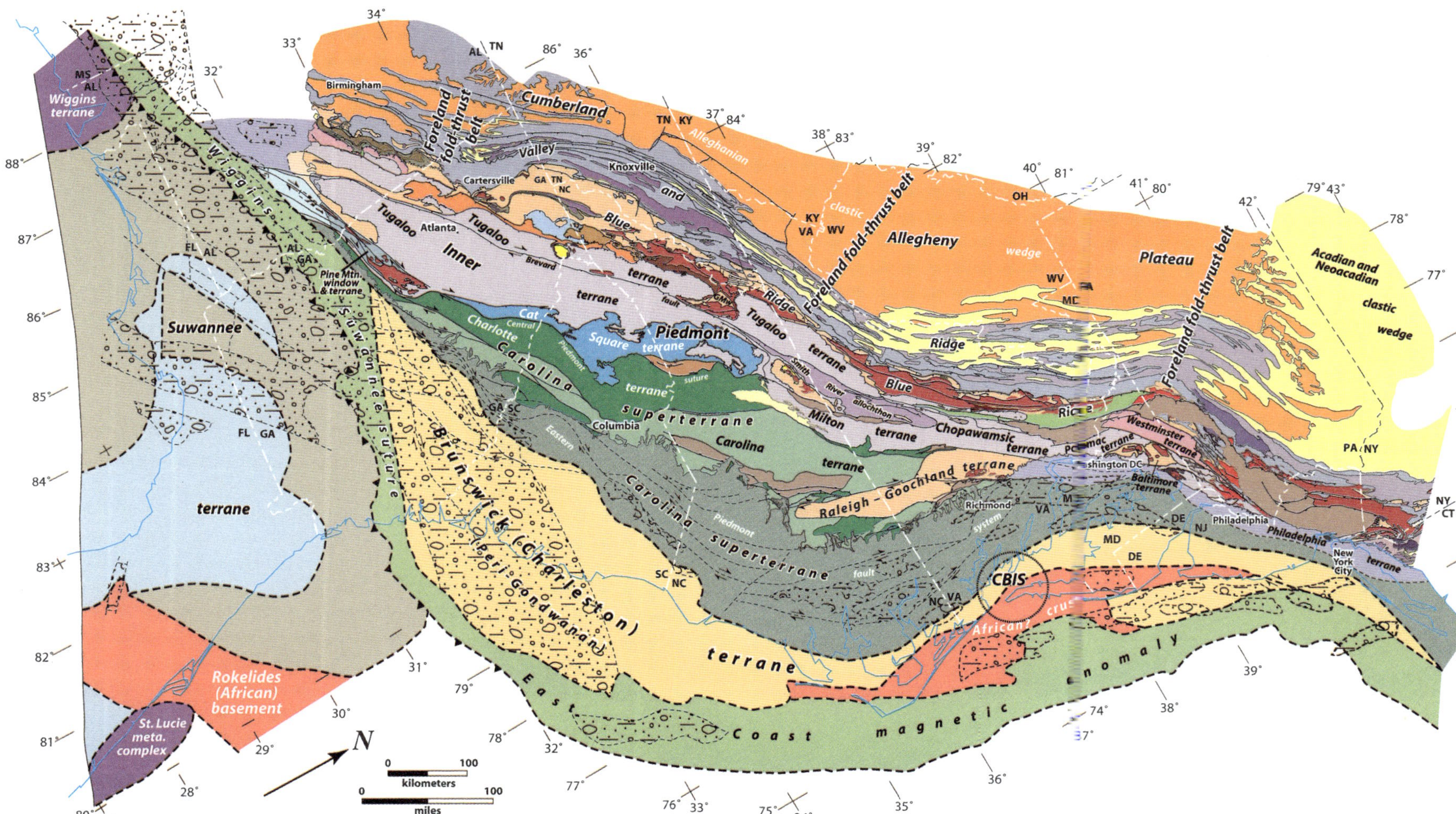

Figure 1(A)(1). Tectonic map simplified from Plate 1, showing the distribution of major tectonic units in the southern and central Appalachians, as well as the location of the foreland fold-thrust belt, major boundaries, and the names of several structures. Explanation of tectonic units not identified on the map: Medium blue from Alabama to New York–Laurentian platform. Dark purple in the Valley and Ridge from Georgia to Virginia–Middle Ordovician clastic wedge. Lighter purple in the Valley and Ridge from Tennessee to New York–Martinsburg clastic wedge (mostly Upper Ordovician). Dark purple in the SE Pennsylvania Valley and Ridge—Hamburg complex. Tan in the Blue Ridge, Alabama Inner Piedmont, and Goochland terrane—rifted margin succession. Pink in the Alabama Blue Ridge—Wedowee Group. Brown in the western Alabama Blue Ridge—Talladega belt. Dark Purple and brown in the Georgia and North Carolina Blue Ridge—Murphy syncline sequence. Red—Grenvilian basement. Maroon-brown in the North Carolina and Tennessee Blue Ridge—Mars Hill terrane. Light blue in the Georgia and North Carolina Blue Ridge—Cowrock terrane. Burnt orange NE of the Cowrock terrane in North Carolina and Georgia—Cartoogechaye terrane. Bright orange in the Georgia and North Carolina Blue Ridge—Dahlonega gold belt. Light green in Virginia to Maryland Blue Ridge—Catoctin Greenstone. GMW—Grandfather Mountain window. Light yellow in NE Georgia along the NW South Carolina state line—Tallulah Falls Quartzite. Brown—Mesozoic (Triassic–Jurassic) basins. Subsurface Mesozoic basins indicated by a conglomerate pattern. Orange unit west of the Carolina superterrane along the South Carolina–North Carolina border—Kings Mountain belt. Light tan unit NE of the Kings Mountain belt in North Carolina—Churchland stitching pluton. CBIS—Chesapeake Bay impact structure.

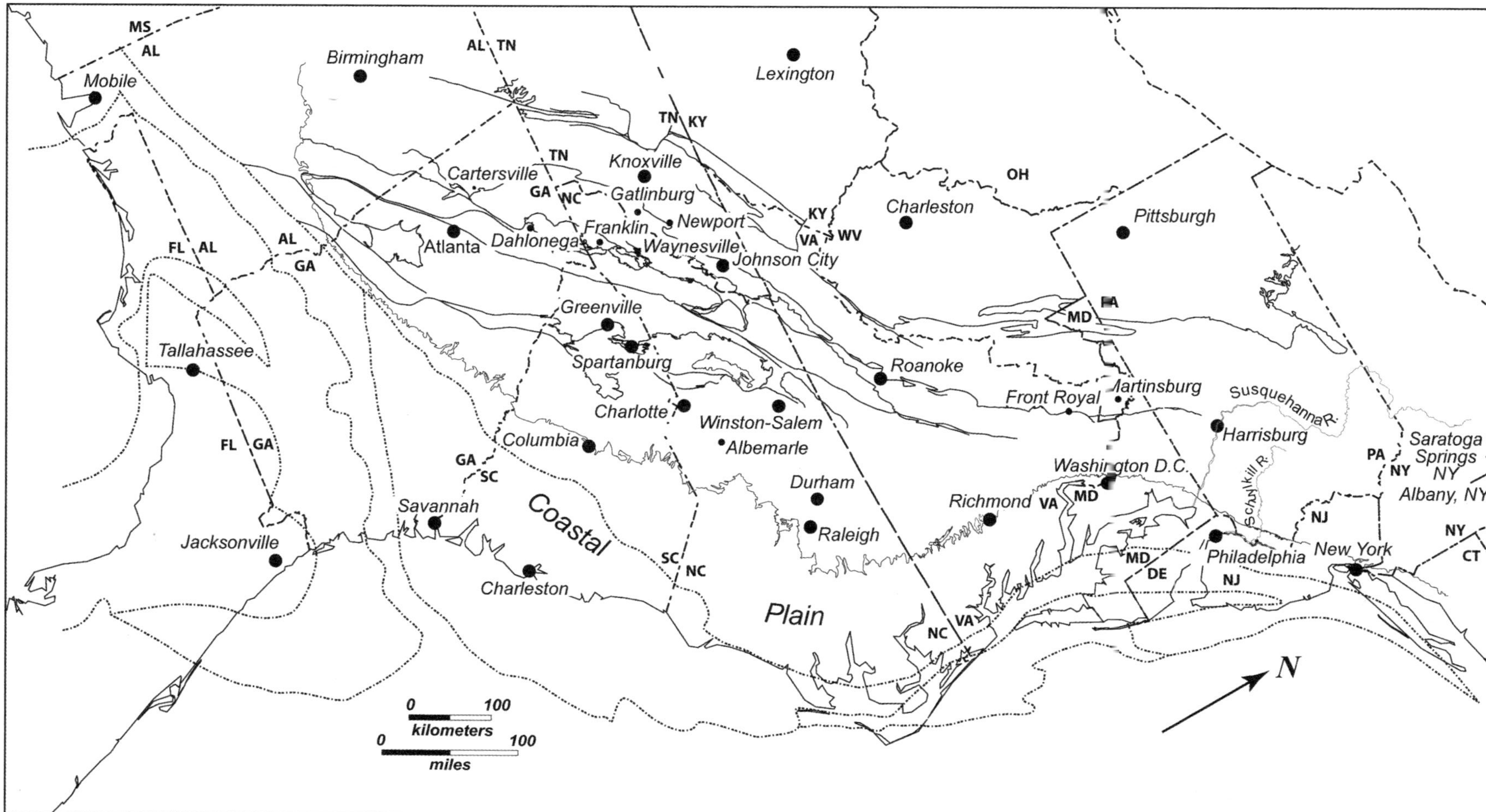

Figure 1(A)(2). Map showing state and major tectonic boundaries in the southern and central Appalachians, geographic localities menti-ned in the paper, and a few cities not mentioned in the text. Exposed tectonic boundaries are shown as solid lines; subsurface tectonic boundaries are dotted; and state lines are shown as do–dash lines. The light, sinuous boundary is the edge of the Gulf and Atlantic Coastal Plain.

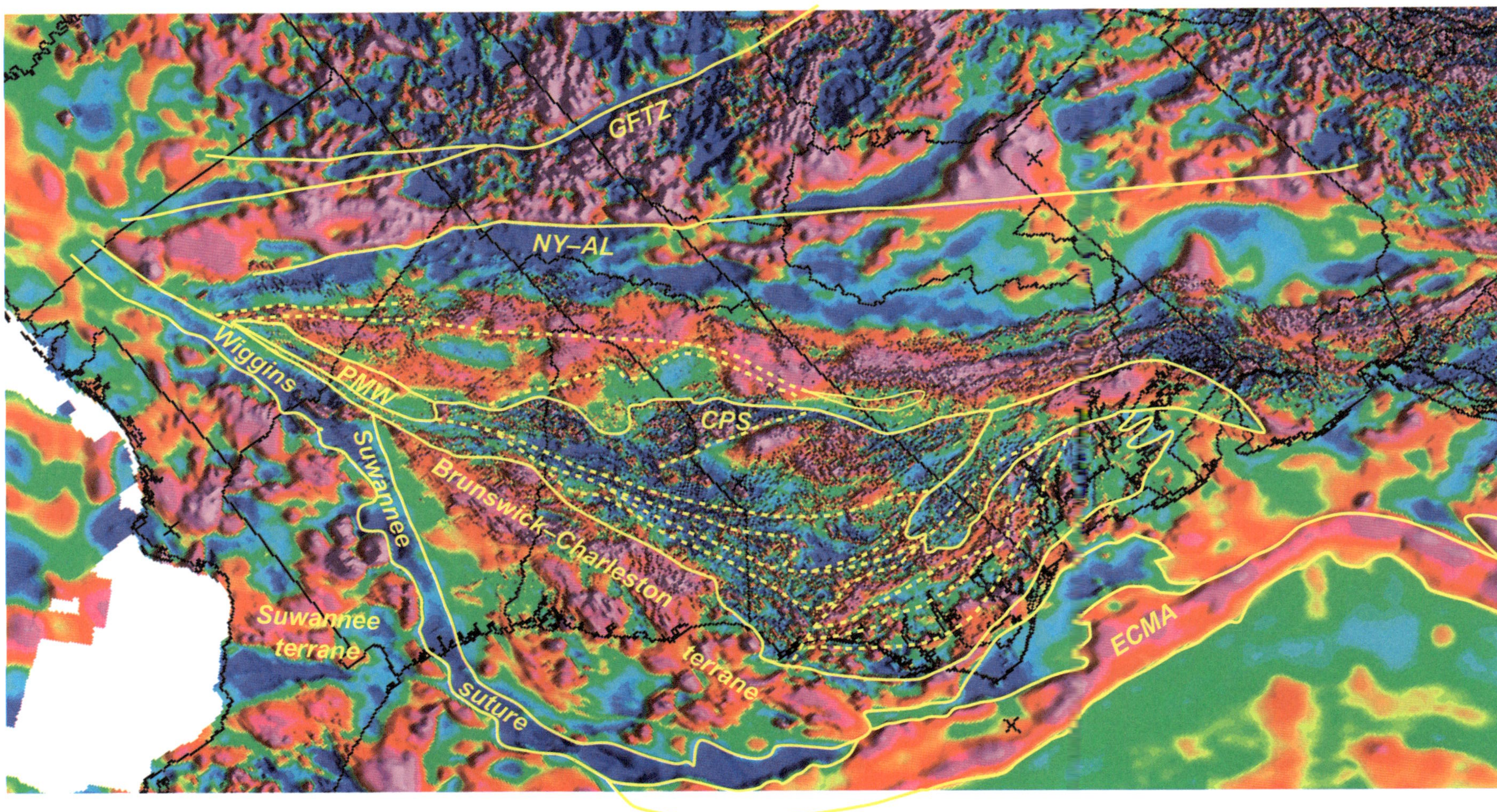

Figure 1(B). Aeromagnetic map of the southern and central Appalachians from "Magnetic map of North America," Geological Survey of Canada, U.S. Geological Survey, Universidad Autónoma de Mexico. Red colors indicate magnetic highs; blue and purple colors indicate magnetic lows. Contours are in nanoteslas (nT). Solid lines—major tectonic boundaries; Dashed lines—minor boundaries; GFTZ—Grenville front tectonic zone; NY-AL—New York–Alabama lineament; PMW—Pine Mountain window; CPS—Central Piedmont suture; ECMA—East Coast magnetic anomaly.

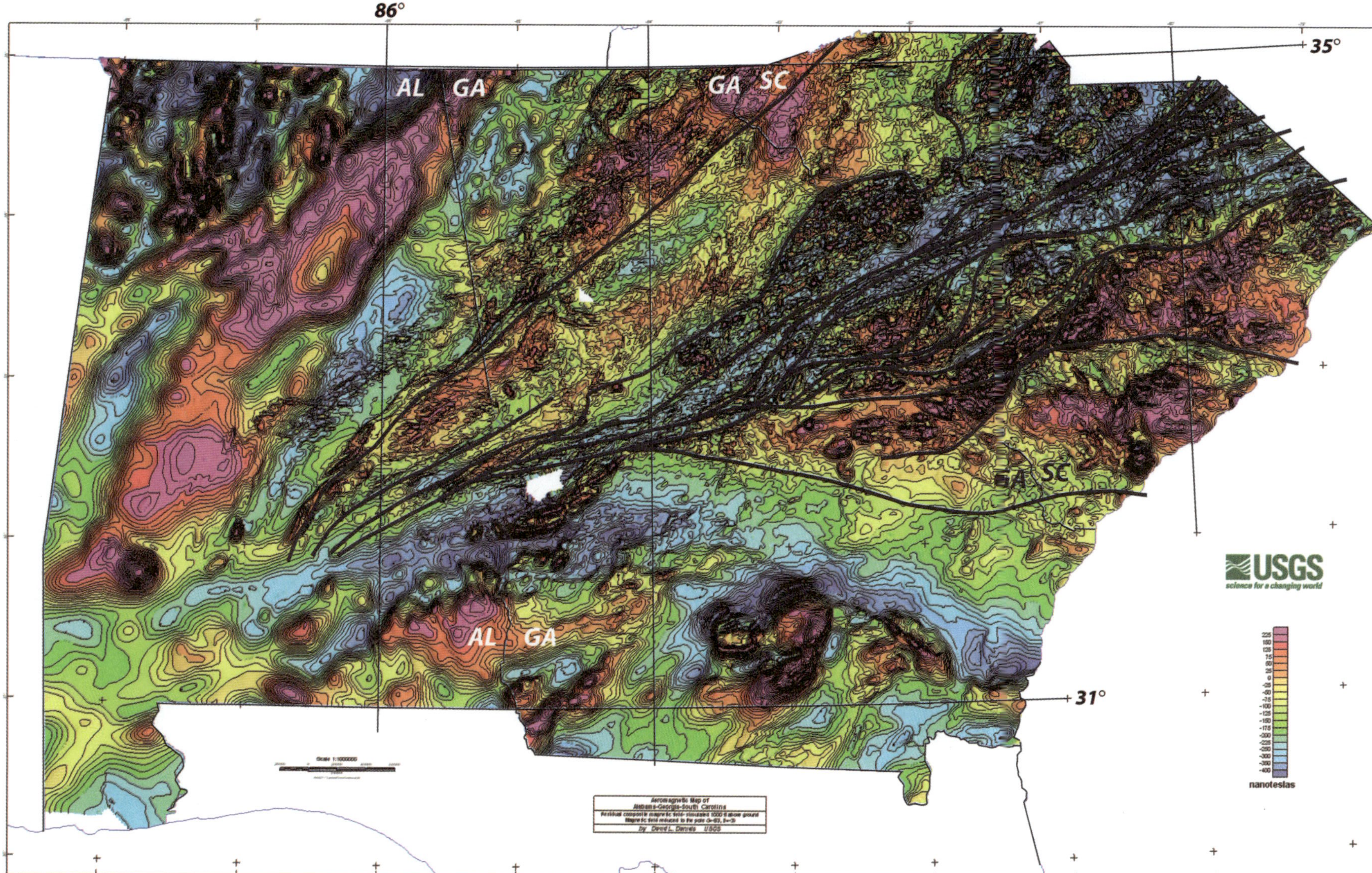

Figure 1(C). Composite aeromagnetic map of Alabama, Georgia, and South Carolina. Heavy lines indicate tectonic boundaries interpreted (by Hatcher) from the aeromagnetic data, and form the basis for much of the interpretation of subsurface crustal structure in these states. See Plate 1 for identification of these boundaries and tectonic units. Aeromagnetic data compiled by (and courtesy of) D.L. Daniels, U.S. Geological Survey.

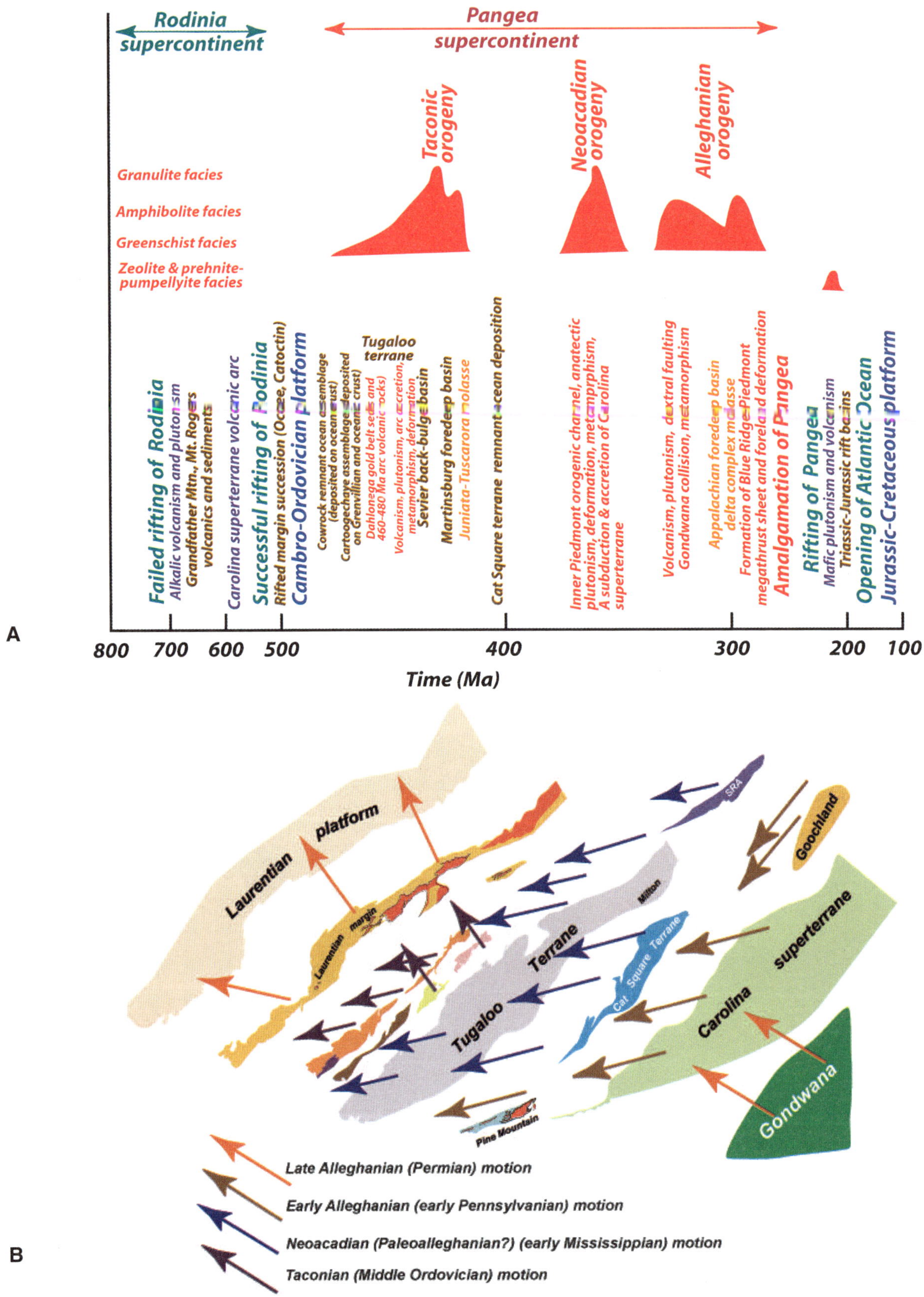

Figure 2. (A) Events distributed through time in the Appalachian Wilson cycle. High points on the red curves denote orogenies and indicate locations of maxima in time and maximum pressure-temperature conditions. Compressional events and related activities are red; arc and rifted margin volcanism is purple; clastic wedges are brown and orange; carbonate platforms are blue; rifting events are green. Note that the horizontal scale is expanded between 500 and 200 Ma. (B) Southern Appalachian tectonostratigraphic terranes retrodeformed. Arrows indicate broad kinematics of assembly and are color coded according to time of accretion, using present-day knowledge. SRA—Smith River allochthon.

Detrital Zircons

SHRIMP and LA ICP–MS detrital zircon ages have probably had as great an impact on southern and central Appalachians tectonics as any other data set (Fig. 5). These data (e.g., Bream, 2003; Bream et al., 2004) have overwhelmingly confirmed that the Tugaloo-Milton-Potomac terrane is a single tectonic unit, had a distal Laurentian provenance, and once-and-for-all that the Brevard fault zone is not a suture. This requires that the contrasts in structural style across the Brevard fault zone be explained otherwise, e.g., as a SW-directed orogenic channel in the Inner Piedmont (Merschat et al., 2005a; Hatcher and Merschat, 2006).

Detrital zircon SHRIMP geochronology is also responsible for recognition of the Cat Square terrane in the Inner Piedmont (Giorgis, 1999; Bream et al., 2001; Hatcher, 2002a). The bounding faults had been delineated, but high-grade rocks east of the Brindle Creek fault lithologically similar to those in the western Inner Piedmont contain a zircon suite consisting of 430 Ma zircons, along with both Laurentian and peri-Gondwanan zircons (Bream, 2003) (Figs. 5 and 6). Trujillo and Sinha (2004) obtained 434-431 Ma ages from high Mg-number mafic complexes in the Virginia Piedmont. They concluded that these and similar complexes in southeastern Pennsylvania formed during an extensional event related to the end of the Taconic orogeny, but they may be related to the early Silurian "Salinic" orogeny that is documented in the northern Appalachians (e.g., van Staal et al., 2004).

Detrital zircon geochronology has given rise to some paradoxical situations in the Dahlonega gold belt and Pine Mountain window. Presumed Gondwanan detrital zircons (2.2–1.9 Ga) were discovered in one sample from each terrane, but these zircons were not present in several additional samples. More likely, these zircons were derived from the Laurentian Penokean orogen (2.2–1.865 Ga; Sims et al., 1993), which contains granitoids in this age range.

Thomas et al. (2004) recognized that detrital zircons in Taconian clastic wedges do not contain synorogenic zircons, and that the first Ordovician zircons appear in the Catskill sequence. Similarly, they recognized that clastic wedges derived from the Acadian (and Neoacadian) clastic wedges do not contain any middle Paleozoic zircons. Likewise, only the youngest parts of the Alleghanian clastic wedges contain late Paleozoic zircons (Thomas et al., 2004; Becker et al., 2005). Perhaps the relationship: *clastic wedges in*

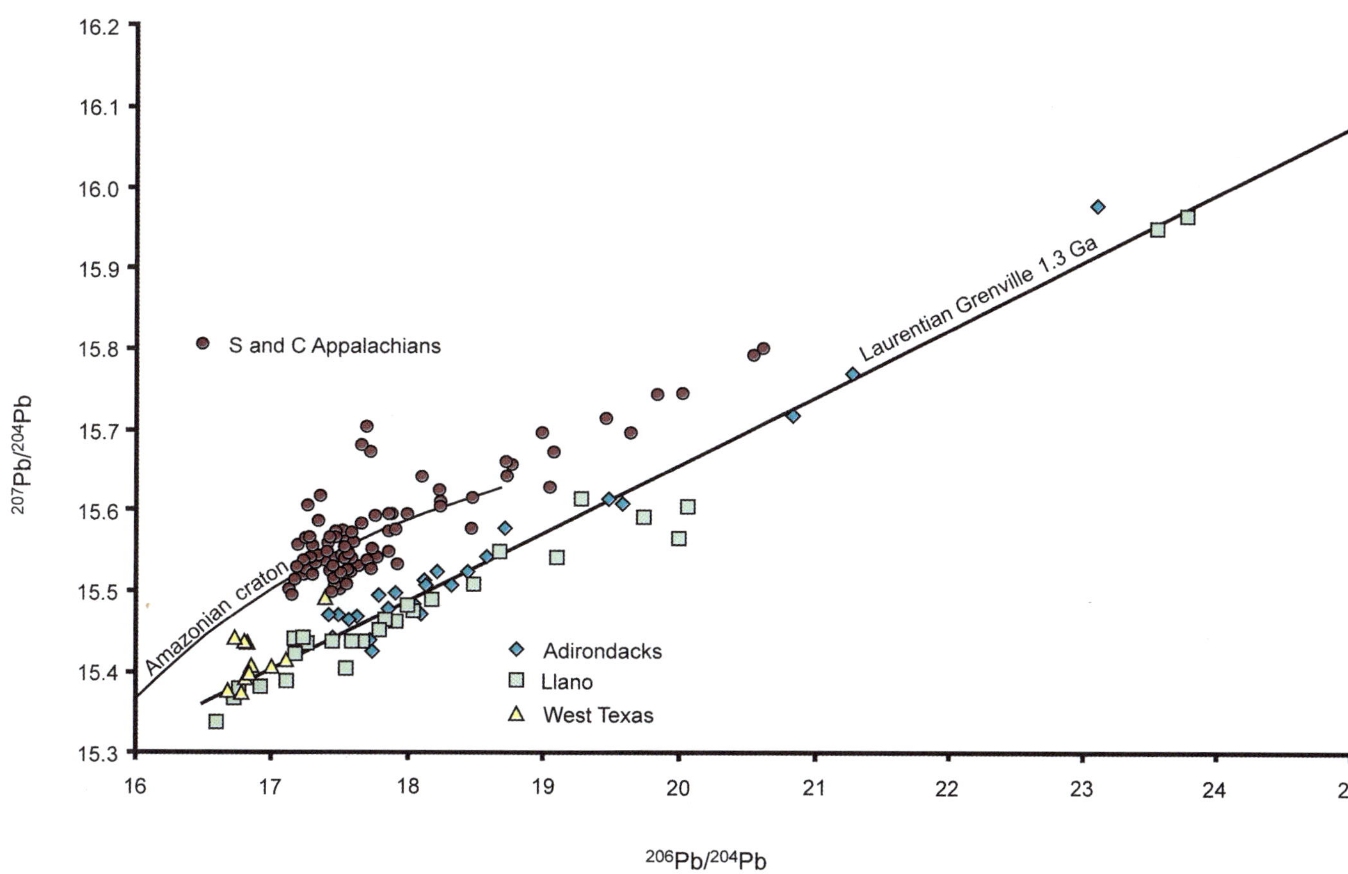

Figure 3. Comparison of Pb isotopic data from the southern and central Appalachians Grenvillian basement rocks with similar data from known Laurentian sources (from Sinha and McLelland, 1999; Loewy et al., 2003).

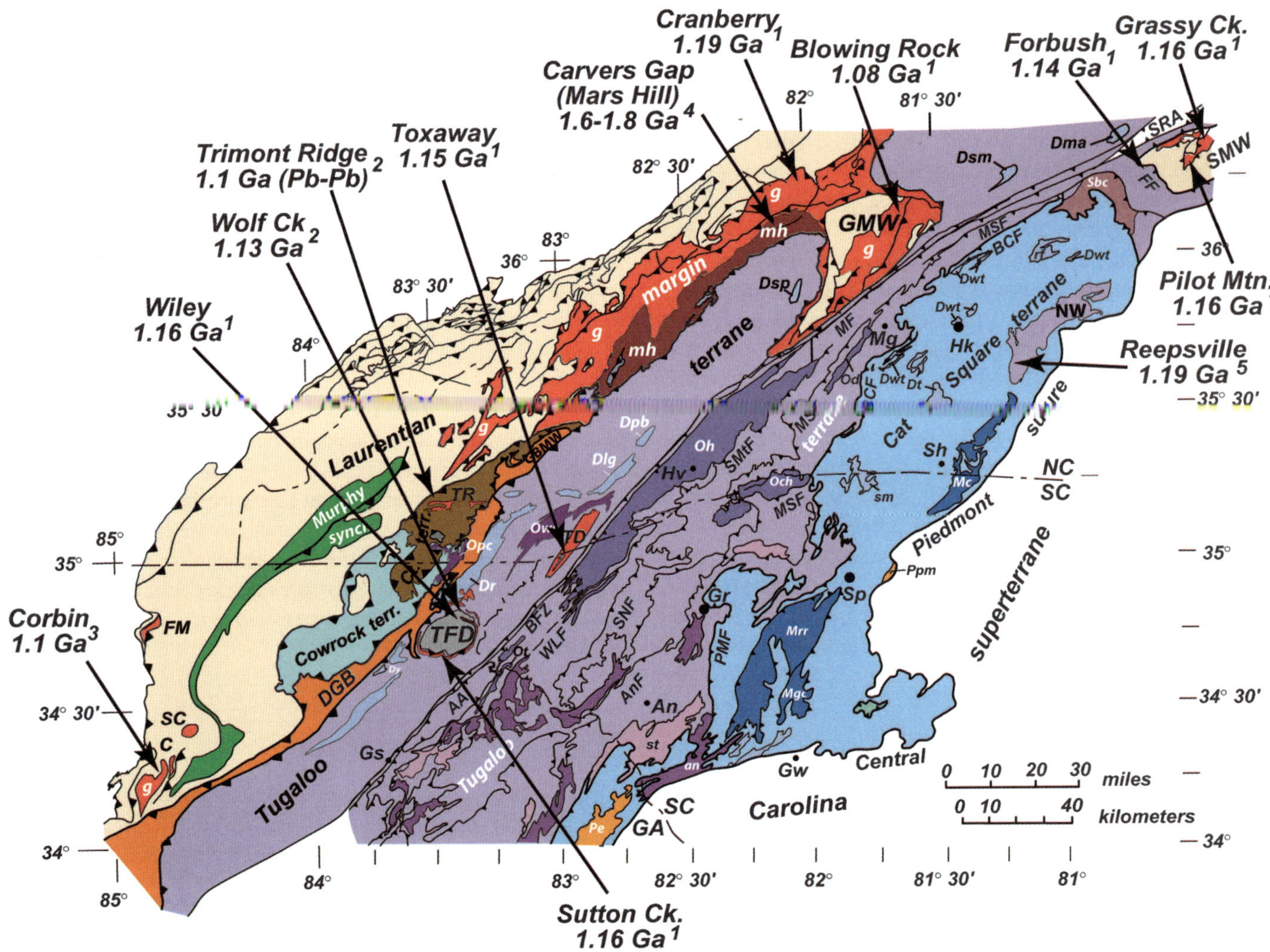

Figure 4. Southern and central Appalachian basement rocks and ages determined by sensitive high-resolution ion microprobe (SHRIMP) and ion microprobe zircon ages, unless otherwise noted. Ages of basement rocks are representative of Grenville and granite-rhyolite provinces, except for the Mars Hill terrane ages. Sources: 1—Carrigan et al. (2003). 2—Hatcher et al. (2004). 3—Heatherington et al. (1996). 4—Ownby et al. (2004). 5—Merschat et al. (2005b). G—mostly Grenvillian orthogneisses; Mh—Mars Hill terrane; DGB—Dahlonega gold belt; GMW—Grandfather Mountain window; GBMW—Great Balsam Mountains window; NW—Newton window; FM—Fort Mountain massif; SC—Salem Church massif; C—Corbin massif; TR—Trimont Ridge massif; TFD—Tallulah Falls dome; TD—Toxaway dome; Ct. terr.—Cartoogechaye terrane; AA—Alto allochthon. Faults: AnF—Anderson; BCF—Brindle Creek; BFZ—Brevard; FF—Forbush; MF—Marion; MSF—Mill Spring; PMF—Paris Mountain; RF—Ridgeway; SNF—Six Mile; WLF—Walhalla. Named Ordovician (purple) and Ordovician(?) (lighter purple) granitoids: Och—Caesars Crossroads. Devonian and Devonian(?) plutons (light blue): Drf—Rocky Face; Dt—Toluca; Dwt—Walker Top; sm—Sandy Mush. Mississippian granitoids: Mc—Cherryville; Mgc—Gray Court; Mr—Rabun; Mrr—Reedy River; Mw—Walnut Creek; My—Yonah (Mapes, 2002). Pennsylvanian granitoids: Pe—Elberton; Ppm—Pacolet Mills (Carolina superterrane). Towns: Gr—Greenville; Gs—Gainesville; Gw—Greenwood; Hk—Hickory; Hv—Hendersonville; Mg—Morganton; Sh—Shelby.

an orogen contain sediments from the previous orogeny, not the one that produced the clastic wedge, should be referred to as "Thomas' Law."

Plutons and Inheritance

Another important breakthrough with modern age dating technology is the ability to accurately date parts of individual zircons (Fig. 7). The zircons in Paleozoic plutons in the western Tugaloo terrane are all composite, revealing several generations of oscillatory-zoned igneous growth commonly surrounding a pre-Paleozoic xenocrystic core; this composite assemblage is frequently surrounded by a metamorphic rim (Figs. 7 and 8). Ironically, few plutons east of the Brevard fault dated using modern technology contain inherited components (Bream, 2003) (Fig. 7). The obvious interpretation would be (again) that the Brevard fault is a suture, but that interpretation is negated by incontrovertible detrital zircon and field evidence. More likely, the plutons east of the Brevard fault were either: (1) intruded through a thinner crust; (2) ascended rapidly to where they crystallized in the middle crust; higher temperature, and/or contained more fluids; or (3) formed from Zr-deficient magma so that all older zircons were resorbed.

Sm-Nd Data

Sm-Nd isotopic data have been collected from several southern Appalachian terranes (Fig. 9). Data from the Tugaloo terrane (western Inner Piedmont, eastern Blue Ridge), western Blue Ridge (Laurentian margin) metasandstones, Blue Ridge basement, and Milton terrane rocks clearly have 1.6–1.2 Ga depleted mantle model ages, separate from those of the more juvenile Carolina superterrane rocks (Bream et al., 2004, and references therein). Mars Hill terrane rocks, as expected, reveal an even more primitive source (Carrigan et al., 2003; Heatherington et al., 2007) (Fig. 9).

Timing of Metamorphism

The ability to date parts of zircons and monazite has also permitted us to more definitively address the questions related to timing of metamorphism in the southern and central Appalachians (Fig. 2A). Unzoned metamorphic rims thick enough to date frequently occur on detrital zircons and less so on zircons in plutons (Fig. 8). Metamorphic zircons and chemical ages of monazite have clearly demonstrated that the western Blue Ridge reached upper-amphibolite- to granulite-facies conditions during the Ordovician (Moecher and Miller, 2000; Moecher et al., 2004). SHRIMP ages on zircon rims (Carrigan et al., 2001; Bream, 2003; Merschat et al., 2006) reveal that a 360-350 Ma Neoacadian event and an early Alleghanian 325 Ma event affected the eastern Blue Ridge and Inner Piedmont. TIMS (thermal ionization mass spectrometer) ages on monazite (Dennis and Wright, 1997) document similar events in the Inner Piedmont.

More conventional TIMS ages on Carolina superterrane plutons have yielded a pre–530 Ma amphibolite-facies metamorphic event in the Charlotte terrane (Barker et al., 1998) that likely produced greenschist-facies metamorphism in the lower-grade rocks farther east (Secor and Snoke, 2004). Alleghanian 320–300 Ma amphibolite-facies metamorphism affected an elongate antiform extending from east of the Pine Mountain window in Georgia through South Carolina, the Kiokee belt, which becomes the Goochland terrane in North Carolina and Virginia [Fig. 1A (1), Plate 1] (Evans and Johnson, 1993; Spears et al., 2004).

Geophysical and Drill Data—Keys to Subsurface Terranes

Ideally, knowledge of Appalachian terranes buried beneath the Atlantic and Gulf Coastal Plains would be gained via numerous drill holes. Unfortunately, the density of holes drilled into pre-Mesozoic basement is very low, and many of these holes had specific targets, e.g., plutons, in mind. The nature of crustal materials, and particularly terrane boundaries, beneath the Coastal Plain, however, is best revealed with high-quality aeromagnetic data that have been collected by state geological surveys and the U.S. Geological Survey for the past four decades, and are now digitized (Figs. 1B and 1C). These data provide some information about the crust down to the Curie temperature, ~35 km in the eastern United States. Fortunately, the Coastal Plain cover is relatively thin from the Fall Line to the Outer Continental Shelf (e.g., Colquhoun et al., 1992; Hatcher et al., 1994), so the eastern and southern limits of many exposed terranes can be determined from aeromagnetic data [Figs. 1B, 1C, and Plate 1]. In addition, there are several terranes, e.g., Suwannee, Brunswick/Charleston, and Wiggins suture zone that have been recognized only with aeromagnetic data. The few drill holes that penetrate these terranes do, however, provide useful information on rock types and limited material for age dating. For example, there may be only one drill hole into the Brunswick/Charleston terrane, and a TIMS zircon age date from some of the cuttings yielded an Avalonian (-Carolinian)-type age (604 ± 1 Ma; Mueller et al., 2005).

More importantly, aeromagnetic data provide useful areal limits on the shape and continuity of these buried terranes facilitating a partial solution to the puzzle of the buried and exposed Appalachians. For example, the generally high magnetic signature of the buried Brunswick/Charleston terrane appears to narrow and be traceable northeastward just west of the Suwannee terrane, which likewise traces northward along the western flank of the East Coast Magnetic Anomaly (Fig. 1B). These data also reveal that this terrane is broken by dextral and sinistral faults of the eastern Piedmont fault system (Plate 1). The curved mineral lineation pattern in the Inner Piedmont depicted by red arrows in the Carolinas (Plate 1) is mimicked by a curved pattern of aeromagnetic anomalies in Georgia and Alabama (shown in blue lines in Plate 1). The magnetic pattern in Georgia and Alabama becomes asymptotic to a very linear magnetic anomaly that trends northeast across Georgia, and joins the mapped trend of the Brindle Creek fault in South Carolina (Fig. 1C, Plate 1).

The distribution of the Brunswick/Charleston terrane to the northeast [Fig. 1A(1), Plate 1] is based totally on speculatively projecting the magnetic signature of the terrane from the southwest. Similarly, the area shown as African(?) crust immediately west of the East Coast magnetic anomaly from North Carolina to Maryland is inferred from the geophysical signature of this block, and its identification is pure speculation.

MAJOR BOUNDARIES AND STRUCTURES

Western Limit of Appalachian Deformation

The Appalachian orogen would appear to have western limits, such as the Pine Mountain fault and Burning Springs anticline in the foreland fold-thrust belt (Plate 1), which would be recognizable on a tectonic map (Fig. 1, Plate 1). Macroscopic features are readily delimited on these maps, but more far-field effects have been recognized that extend well into regions that appear

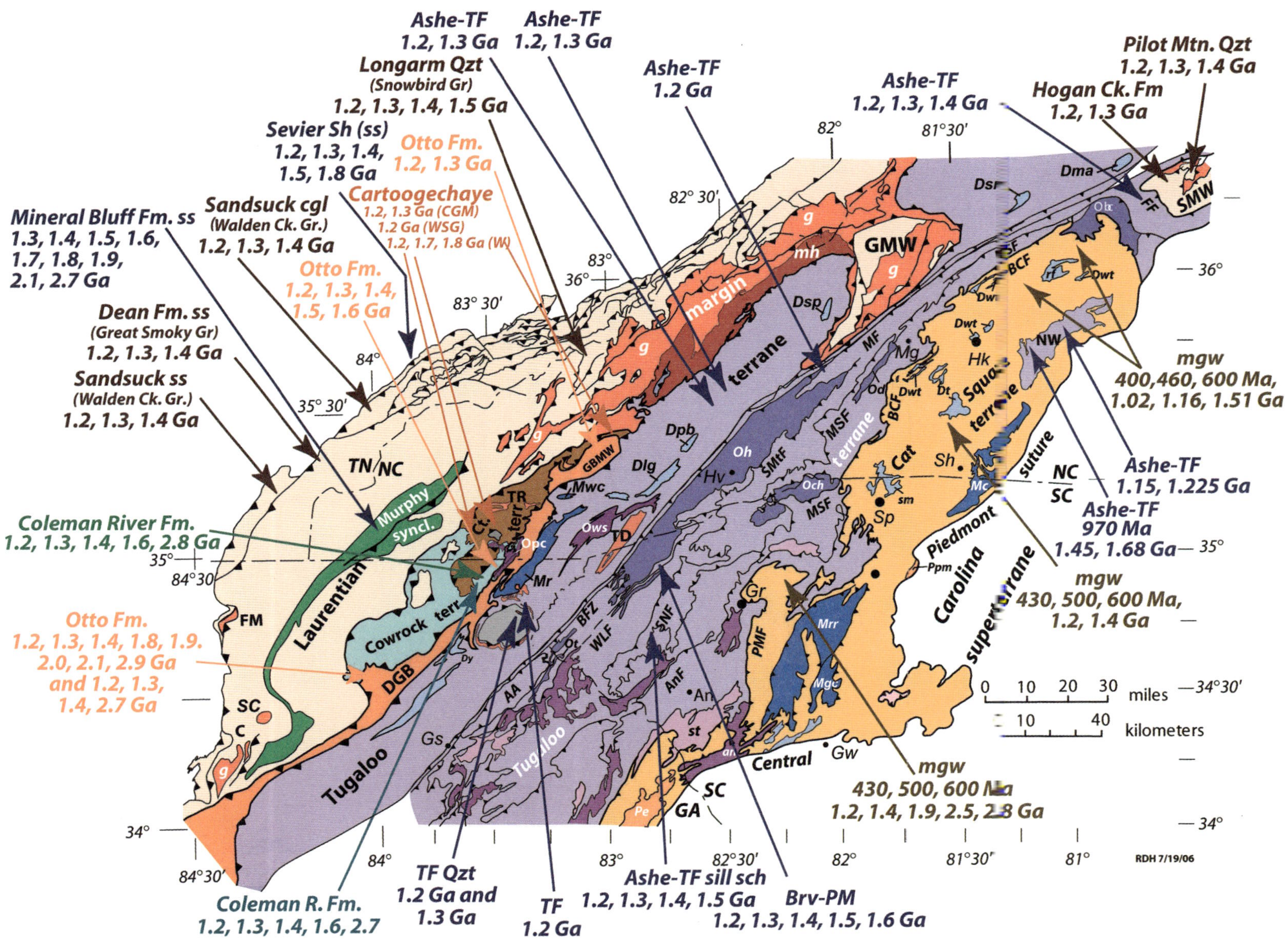

Figure 5. (A) Detrital zircon ages from the southern Appalachians west of the Carolina superterrane (from Bream, 2003; Merschat et al., 2005b; and Gatewood, 2007). All samples contain, and most are dominated by, 1.1–1.0 Ga zircons, so only the additional significant age populations are shown. Sensitive high-resolution ion microprobe, reverse geometry (SHRIMP–RG) and ion microprobe ages of detrital zircons from mostly sandstones and metasandstones from the Valley and Ridge, Laurentian margin, and Tugaloo terrane clearly indicate a Laurentian provenance from Grenville and granite-rhyolite province sources in the Laurentian basement to the west. The Dahlonega gold belt contains possible Gondwanan zircons, in addition to a dominant Laurentian zircon population. The Cat Square terrane contains Archean, Penokean(?), granite-rhyolite province, Grenvillian, Avalonian (Carolina), and Silurian zircons, requiring mid- to late Paleozoic docking of the Carolina superterrane. See Figure 4 for explanation of abbreviations.

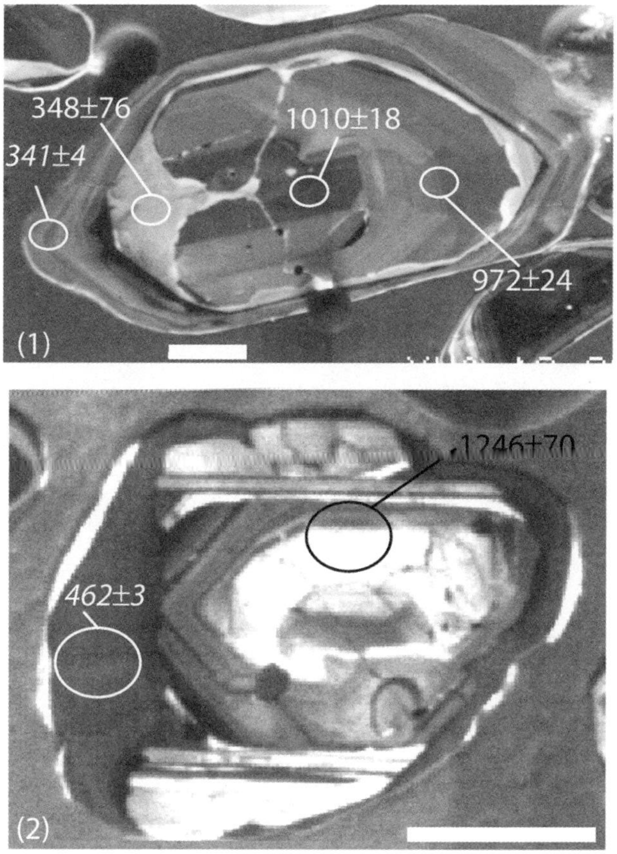

Figure 5 (B) Cathodoluminescence images of zircons analyzed with the SHRIMP. Ages are $^{207}Pb/^{206}Pb$ and $^{206}Pb/^{238}U$. All scale bars are 100 µm. (1) Large magmatic zircon with sector zoning truncated by Paleozoic metamorphic rims from the informally named Reepsville gneiss (Merschat et al., 2005b), a Middle Proterozoic orthogneiss exposed in the Newton window, Inner Piedmont south of Hickory, North Carolina. (2) Xenocrystic detrital zircon with Paleozoic metamorphic rim from the Dahlonega gold belt, Otto Formation, exposed in the Great Balsam Mountains window near Waynesville, North Carolina.

undeformed. These features take the form of meso- or microfabric elements, such as layer-parallel shortening-related cleavage or deformation-related calcite twinning (e.g., Geiser and Engelder, 1983; Craddock and van der Pluijm, 1989), map-scale structures that may be buried and truncated beneath unconformities (e.g., Hatcher et al., 2005a, and see "Mid-Paleozoic Sedimentation and Neoacadian Docking" section), or regional basins and arches west of the orogen that may be related to the response of loading of the continental margin with obducted arcs or ophiolites (e.g., Quinlan and Beaumont, 1984). The western limit of Appalachian deformation is therefore not everywhere a line that can readily be drawn on a map, but may represent a series of structures formed during more than one orogeny that may be expressed in different forms.

New York–Alabama Lineament

The subsurface New York–Alabama lineament was first recognized by King and Zietz (1978) as a prominent boundary in regional aeromagnetic data underlying part of the western Appalachians. It also is recognizable in regional gravity data (Society of Exploration Geophysicists, 1982) in the eastern U.S., but it has no surface expression. Seismic-reflection lines that cross the lineament (e.g., McBride et al., 2007 this volume, their Fig. 8) neither image the feature nor reveal any displacement of the contact between the Precambrian basement and the overlying Cambrian cover. This feature is therefore confined to the pre-Paleozoic rocks and probably is a major tectonic boundary that formed during the Grenville orogeny. It may represent the suture within the Grenville orogen between Laurentian rocks to the west and Gondwanan rocks to the east (Hatcher et al., 2004).

Blue Ridge Thrust

The term Blue Ridge thrust is used collectively to denote the frontal thrust that separates the western Blue Ridge geologic province from the Valley and Ridge province. One or more faults make up the western boundary from Alabama to near Front Royal, Virginia [Fig. 1A(2)], and are known by different names. In Alabama, the name Talladega–Emerson fault gives way to the Cartersville fault in western to central Georgia (Plate 1). North of Cartersville, the fault has been called Cartersville-Great Smoky fault (Rodgers, 1970), but this is a misnomer. The Great Smoky fault is largely eroded south of northernmost Georgia and the frontal fault is properly the Alaculsy Valley-Miller Cove fault (Costello, 1984). The Great Smoky fault, which carried largely unmetamorphosed Chilhowee and uppermost Walden Creek Group rocks, terminates against the Alaculsy Valley–Miller Cove fault immediately south of the Georgia-Tennessee line (Salisbury, 1961). The Miller Cove fault actually transports previously penetratively deformed and metamorphosed rocks of the Walden Creek and older rock units, and truncates the Great Smoky fault as an out-of sequence basement fault (transporting Paleozoic basement; Hatcher, 1999). The Great Smoky fault is thus the easternmost Valley and Ridge thrust sheet, with a detachment in the weak, subchlorite-grade shales of the uppermost Walden Creek Group; it ramps down into a still-lower detachment along the ductile-brittle transition in Paleozoic basement to the southeast. The same thing happens immediately northeast of Gatlinburg, Tennessee, where the Miller Cove fault becomes the frontal fault. A block of Chilhowee Group and underlying upper Walden Creek Group rocks again forms the frontal block immediately south of Newport, Tennessee, and the fault that forms its base has been called the English Mountain fault (Rodgers, 1953; King et al., 1958; Hamilton, 1961). Again, this is the Great Smoky fault with the same bedding fault characteristics as in the main Chilhowee Mountain block to the south.

In northeastern Tennessee and southwestern Virginia, the Great Smoky fault consists of the Holston-Iron Mountain fault

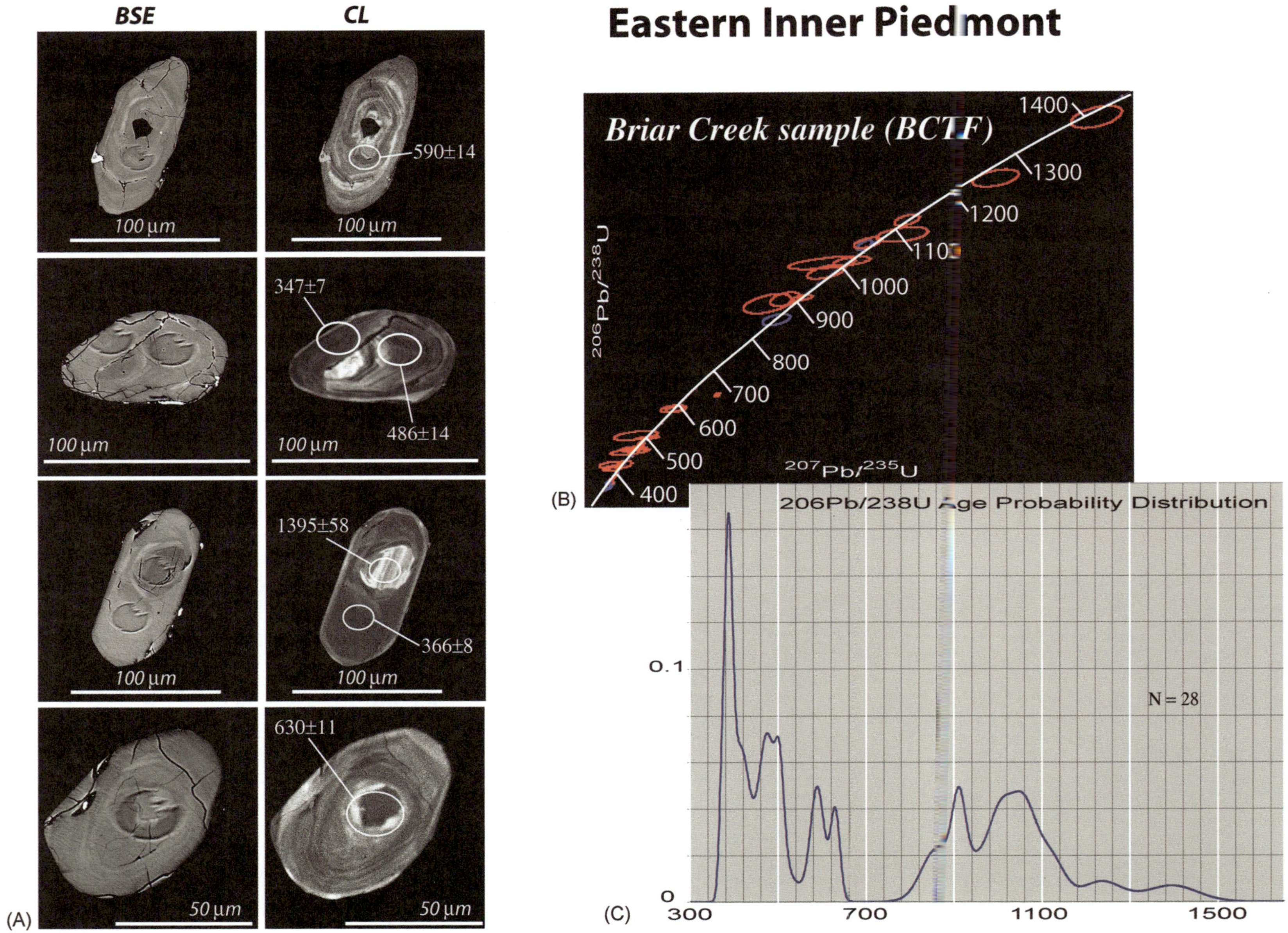

Figure 6. Example of Cat Square terrane detrital zircons (from Bream, 2003). (A) Back-scattered electron and cathodoluminescence images with spot ages of zircons from one sample from the Cat Square terrane near Briar Creek, South Mountains, North Carolina. (B) Tera-Wasserberg plot of data showing the range of ages of zircons in the sample. (C) Summed probability plot of 28 zircons analyzed in the Briar Creek sample. Note the dominance of 430 Ma zircons and significant populations of 500 and 600 Ma (peri-Gondwanan) and Grenvillian zircons.

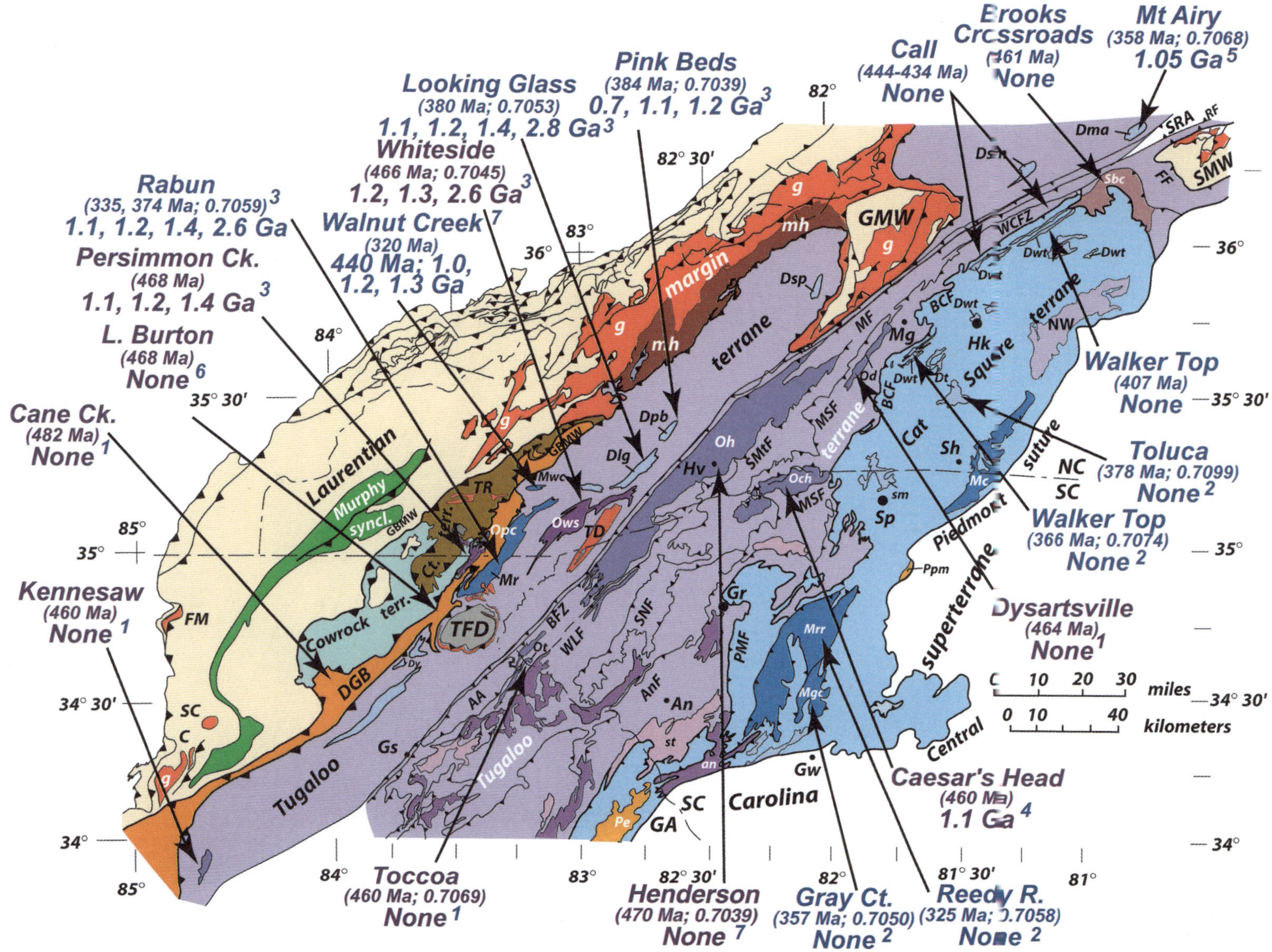

Figure 7. Sensitive high-resolution ion microprobe (SHRIMP) and ion microprobe Paleozoic pluton ages, $^{87}Sr/^{86}Sr$ initial ratios, and inheritance in the western half of the southern Appalachians. Ordovician plutons dominate the Tugaloo terrane, with a suite of Devonian plutons in the western part of the terrane. Several Carboniferous plutons occur in the Tugaloo terrane in Georgia (from older age data). Cat Square terrane plutons are dominantly Acadian to Neoacadian (380, 360–350 Ma), and younger. The Devonian-early Mississippian plutons are anatectic. Remarkably little inheritance in Inner Piedmont plutons contrasts with abundant and varied inheritance in the western Tugaloo terrane (west of the Brevard fault). This break in inheritance at the Brevard fault zone likely represents contrasting crustal types on either side of the fault at the times these plutons were intruded, because the detrital zircon suite across the fault contains no break. Sources of information: 1—Bream (2003); 2—Mapes (2002); 3—Miller et al. (2000) and Mapes et al. (2002); 4—Ranson et al. (1999); 5—A. K. Sinha (personal commun.); 6—Thomas (2001); 7—Stahr et al. (2005); 8—Vinson (1999); 9—Gatewood (2007). See Figure 4 for explanation of abbreviations.

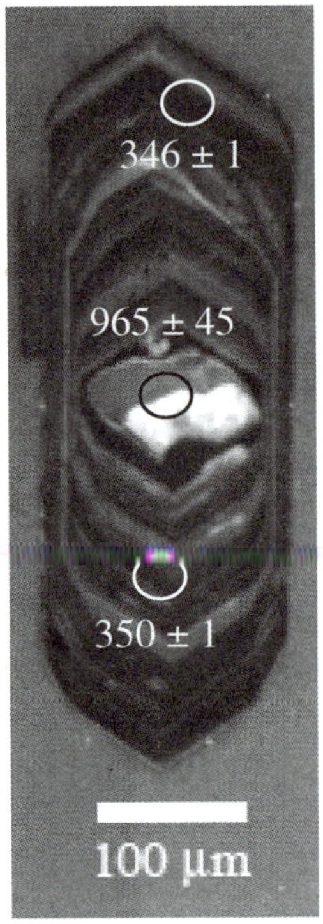

Figure 8. Igneous zircon with an inherited core and sector-zoned overgrowths from the Rabun Granodiorite, southwestern North Carolina, showing locations of spot ages and inheritance. Core age is a $^{207}Pb/^{206}Pb$ age; zoned portions are $^{238}U/^{206}Pb$ ages. Image courtesy of Donald W. Stahr III.

system. In its hanging wall are small pieces of Grenvillian basement overlain by Chilhowee Group sedimentary rocks, then a Cambrian to Cambro-Ordovician section that is as young as the Knox Group, which again strongly suggests that the frontal Blue Ridge thrust is the easternmost Valley and Ridge thrust. To the southeast, the Holston–Iron Mountain fault system is truncated by the Stone Mountain fault, which primarily carries Grenville basement. Northward into Virginia the frontal fault is called the Blue Ridge thrust (Rodgers, 1970), and it transports basement and Chilhowee Group rocks to just south of Roanoke, Virginia, where the Blue Ridge becomes a distinct basement-cored anticline, flanked by the Neoproterozoic Catoctin volcanic and sedimentary sequence, and overlain by Chilhowee Group rocks. Harris (1979) compared this segment to the Pine Mountain block on the west side of the Valley and Ridge. Wickham (1972) concluded that the Blue Ridge fault ends near Front Royal, Virginia, but Harris (1979) concluded that the thrust simply went blind at that point and continues in the subsurface into Maryland. This is consistent with other reconstructions (e.g., Kulander and Dean, 1986) and attempts to balance the Valley and Ridge-Blue Ridge relationships farther north, even into Pennsylvania (Geiser, *in* Hatcher et al., 1989).

Hayesville–Soque River Fault

The Cartoogechaye and Cowrock terranes in the central Blue Ridge of North Carolina are bounded by the Hayesville-Soque

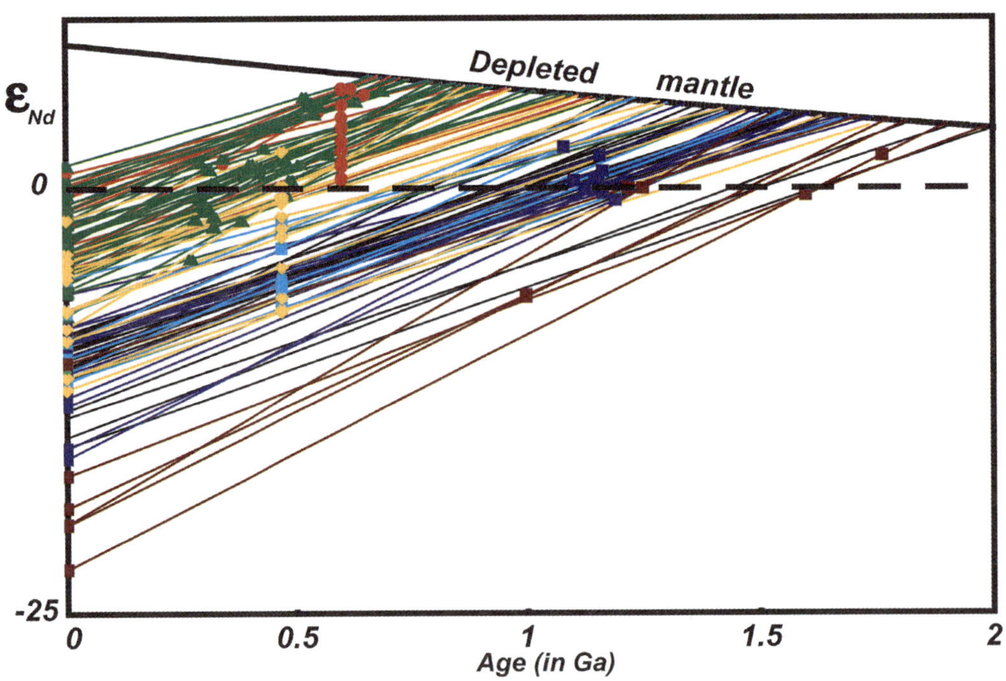

Figure 9. Sm-Nd data from the southern Appalachians. Red—Carolina superterrane (Samson et al., 1995); green—Carolina superterrane (Fullagar et al., 1997); yellow—Chopawamsic terrane (Coler et al., 2000); black—Blue Ridge and Inner Piedmont sedimentary and metasedimentary rocks (Bream et al., 2004); light blue—Milton terrane (Coler et al., 2000); dark blue—eastern and western Blue Ridge basement rocks (Carrigan et al., 2003; Bream et al., 2004); Brown—Mars Hill terrane (Ownby et al., 2004).

River fault. The extent of this fault system is not so great as that of some of the other major boundaries, but it separates mafic-ultramafic-poor western Blue Ridge rocks from rocks of the central Blue Ridge that contain moderate to abundant amounts of mafic and ultramafic rocks. There also is a contrast in metasandstone compositions across these faults from two-mica, two-feldspar metasandstones in the western Blue Ridge to dominance of one-mica (biotite) one-feldspar (plagioclase) metasandstones in the Cowrock and Cartoogechaye terranes (Hatcher, 1978, 1987, 1989, and see "Central Blue Ridge Terranes" below), as well as in the Tugaloo terrane farther east.

Chattahoochee–Holland Mountain–Burnsville Fault

The western boundary of the Tugaloo terrane is the Chattahoochee-Holland Mountain fault in the eastern Blue Ridge [Fig. 1A (1), Plate 1], which truncates the 335 Ma Rabun Granodiorite (Stahr et al., 2005; Miller et al., 2006). This fault transported migmatitic kyanite- to sillimanite-grade Tugaloo terrane rocks southwestward over staurolite-kyanite-sillimanite–grade rocks of the Dahlonega gold belt in southern North Carolina and northern Georgia, and was tightly folded after emplacement. The Chattahoochee–Holland Mountain must therefore be an Alleghanian fault. Miller et al. (2006) depicted the Burnsville fault cutting previously accreted Ashe rocks and subsequently being transported southwestward with those and other rocks during the Alleghanian orogeny (Trupe et al., 2003; Stahr et al., 2005; Miller et al., 2006).

Brevard Fault Zone

The Brevard is a very linear northeast-trending fault zone that is traceable some 750 km from Alabama to Virginia in the crystalline southern Appalachians (Fig. 1A, Plate 1). It ranges from 1 to 3 km wide, dips gently to moderately toward the southeast, and appears listric in seismic-reflection profiles (e.g., Costain et al., 1989). This fault zone has been the subject of controversy for many years because it separates two major geological realms and has a complex history. It also is easily recognized and studied in the field, on topographic maps, and in satellite imagery. The attributes of the Brevard fault zone on which there is agreement include its extent from Alabama to Virginia; its linearity; a polyphase history of ductile and brittle motion resulting in retrogression of earlier prograde rock sequences; listric flattening of the fault zone toward the southeast in seismic-reflection profiles (Costain et al., 1989); a mappable lithostratigraphy for at least 200 km (see Hatcher, 2001, his Figures 2 and 10); stratigraphically controlled faults that propagated into strongly anisotropic (Christensen and Szymanski, 1988) graphitic and micaceous rock; and exotic horses of platform carbonate (Cambro–Ordovician Knox Group) rocks in the fault zone along the northwestern brittle boundary with the Blue Ridge (called the Rosman fault by Horton and Butler, 1986) and a horse of basement(?) rocks on its southeastern flank (Hatcher, 2001; Hatcher et al., in press). The stratigraphy described here into which the fault zone was localized ends in northeastern Atlanta (M.W. Higgins, 2006, personal commun.). Kath and Crawford (2006), however, pointed out that the Brevard fault zone changes character in Atlanta from that present to the northeast to a predominantly high- temperature fault zone that overprints a series of ortho- and paragneisses, and pelitic schist. This fault zone has also been overprinted by brittle deformation, which produced only small-displacement faults. Bentley and Neathery (1970) noted a change in behavior in Alabama, where the fault changed to a more northerly strike and propagated into another predominantly siliciclastic sequence, the Jacksons Gap Group (Osborne et al., 1988).

The Brevard is a complex fault zone that had an earlier (Late Devonian–early Mississippian Neoacadian) high-temperature history of oblique to strike-slip motion and a later (Alleghanian, Pennsylvanian–Early Permian) history of both strike-slip (high-temperature and greenschist-facies retrogressive) and later brittle dip-slip motion. The last brittle, dip-slip event produced a narrow fault zone (Rosman fault) along its northwest boundary that functioned as a major out-of-sequence thrust within the Blue Ridge–Piedmont megathrust sheet (Hatcher, 1971a, 1989, 2001; Hatcher and Hooper, 1992). Hatcher (1978, 1989, 2001) summarized the reasons that the Brevard fault zone cannot be a suture, principally because the same physical stratigraphic sequence of paragneisses (Tallulah Falls–Ashe Formation; Hatcher, 1971b) occurs on both sides of it. In addition, we now know that this sequence (on both sides of the Brevard fault zone) contains detrital zircons with the same Laurentian provenance (Bream, 2003; Bream et al., 2004), and is unconformably overlain in the Inner Piedmont by the Middle Ordovician Poor Mountain Formation arc to MORB volcanic and sedimentary rocks (Hatcher, 2002a; Bream, 2003).

The Brevard fault zone had an early movement history that was coeval with formation of many of the penetrative structures (foliations, mineral lineations) in the eastern Blue Ridge and Inner Piedmont. This history produced at least three sets of superposed ductile folds (two sets are not coaxial), and at least two of these events transposed the rocks sufficiently to produce regionally penetrative foliations (see Hopson and Hatcher, 1988). Both Neoacadian and early Alleghanian high-temperature foliations are present in the eastern to central Blue Ridge. The events that produced these foliations have been recognized from U-Pb ages of foliated plutons, such as the 467 Ma Whiteside Granite, the 468 Ma Persimmon Creek tonalite (McDowell et al., 2002), the 335 Ma Rabun pluton in the eastern Blue Ridge (Miller et al., 2000; Stahr et al., 2005; Miller et al., 2006) and the 464 Ma Dysartsville granodiorite (Bream et al., 2000; Bream, 2003) in the Inner Piedmont. Relict Taconian or early Neoacadian(?) foliations present in boudins and intrafolial folds are wrapped by the dominant (Neoacadian) foliation in the easternmost Blue Ridge and Inner Piedmont. These earliest structures have been transposed by the Alleghanian overprint in the Brevard fault zone, and are visible only as relict high-grade assemblages preserved as aligned inclusion trails in microlithons and unretrograded garnets (see photos in Roper and Dunn 1973, their Figures 2, 4, and 8).

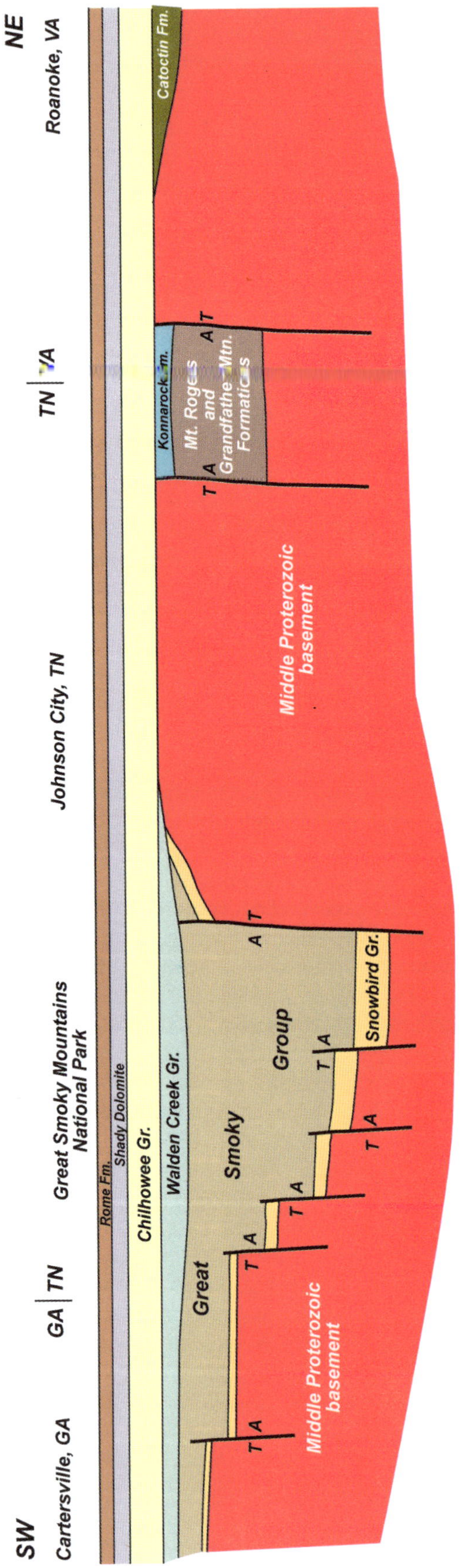

Figure 10. Along-strike changes in the relative thicknesses of the rifted-margin assemblages from Georgia to Virginia. Note the abrupt termination of the basin to the northeast, gradual termination to the southwest, and the appearance of the Catoctin volcanic-sedimentary sequence to the northeast in Virginia. Relative thicknesses are based on King et al. (1958) and King and Ferguson (1960).

Evidence for the early (Neoacadian?) history of the Brevard fault zone consists primarily of ties to the high-temperature structures formed under middle- to upper-amphibolite-facies conditions in the Inner Piedmont. Brevard phyllonite (part of the Chauga River Formation) occurs southeast of the Brevard fault zone in South Carolina and southern North Carolina in a lower-grade (garnet) part of the Inner Piedmont called the Chauga belt (Hatcher, 1972). There it consists of a lower- to middle-amphibolite-facies assemblage of metapelite, and metapsammite, and does not contain the strong Alleghanian retrograde overprint characteristic of the Alleghanian Brevard fault zone. Moreover, the quartz slip system in the rocks of this part of the Inner Piedmont is typical of high-temperature ductile deformation processes (Davis, 1993). Davis et al. (1991) and Davis (1993) first suggested that this lineation tracks the ductile flow of the plastic Type-F (fold-related) thrust sheets (Hatcher and Hooper, 1992) that dominate here. The primordial Brevard fault zone in this model served as a buttress to deflect westward-moving thrust sheets in the central part of the Inner Piedmont toward the southwest along the western margin (see Hatcher, 2001, his Fig. 3). The buttressing effect is indicated by the strong coaxial northeast–southwest mineral lineation within and immediately southeast of the Brevard fault zone and less strong alignment farther east. The net result would have been dextral flow with respect to the Blue Ridge to the northwest, and sinistral flow with respect to the more internal (central and eastern) parts of the Inner Piedmont (Hatcher, 2001; Merschat et al., 2005a). This phenomenon is very similar to that described by Jones et al. (1997) and Dewey et al. (1998) as lateral extrusion (or escape) in transpression zones. In the present example, the Brevard fault zone would have buttressed the obliquely converging Inner Piedmont thrust sheets and caused them to extrude (escape) toward the southwest. At least two of their three boundary conditions, wrench simple shear and pure shear in XY, can be identified in the mesofabrics of the Brevard fault zone (Hatcher, 2001; Merschat et al., 2005a).

Brindle Creek Fault and Terrane Boundary

Giorgis (1999) defined the Brindle Creek fault as a low-angle fault and extensive mylonite zone traceable across parts of the South Mountains, west-central North Carolina. Although not indicated as a fault, this boundary was originally recognized as a lithologic boundary and traced regionally throughout the Charlotte 1° × 2° quadrangle by Goldsmith et al. (1988). As mapped by Goldsmith et al. (1988), we now know that this same boundary separates the western Inner Piedmont of recognizable Tallulah Falls and Poor Mountain Formation stratigraphy, and Ordovician granitoid plutons (Dysartsville–Caesars Head–Toccoa suite and Henderson Gneiss), from the eastern two-thirds of the Inner Piedmont composed predominantly of sillimanite schist and biotite gneiss, and Acadian-Neoacadian granitoid plutons (Cherryville, Walker Top, and Toluca orthogneisses), with some amphibolite. The fault has since been traced in detail northeastward throughout the Brushy Mountains. It is recognized not only by truncation of both hanging-wall and footwall units, but also by mylonite along the contact up to one km thick (Merschat et al., 2005a; Wilson, 2006; Gatewood, 2007). Dennis (2007) has suggested that the contact is an unconformity, but these and other data provide incontrovertible evidence that the Brindle Creek is a fault with >100 km of displacement.

A major change in recognizable physical stratigraphy and ages of plutons occurs at the Brindle Creek fault. The characteristic Tallulah Falls Formation stratigraphy is no longer recognizable, and no Poor Mountain Formation rocks have been recognized east of the Brindle Creek fault. Instead, the thick sillimanite schist appears to be overlain by a biotite gneiss (metagraywacke) unit. No Ordovician plutons have been recognized to date east of the fault, and none of the Acadian-Neoacadian plutons occur west of the fault in North Carolina, except the Pink Beds, Looking Glass, Mount Airy, and Stone Mountain (North Carolina) plutons in the eastern Blue Ridge. A megacrystic granitoid was originally mapped by Goldsmith et al. (1988) and correlated with the Ordovician Henderson Gneiss. The Walker Top Granite (Giorgis et al., 2002) has since yielded SHRIMP crystallization ages of ~366 Ma (Mapes, 2002) and 407 Ma (Gatewood, 2007), suggesting it is a series of Acadian to Neoacadian plutons, which occur only in the Brindle Creek thrust sheet. One of these, the Toluca Granite, has yielded an ~378 Ma Acadian SHRIMP crystallization age (Mapes, 2002).

The northern and southern extent of the Brindle Creek fault is only speculative at present (Fig. 1B), but this fault is the most fundamental boundary in the Inner Piedmont (Bream et al., 2001; Merschat et al., 2005a). A sinuous low-angle boundary separating biotite-gneiss-biotite-amphibolite gneiss (bag) to the southeast from Paleozoic (Pzg) granitoid to the northwest in the Winston Salem 1° × 2° sheet (Rankin et al., 1972; Espenshade et al., 1975) could trace the northeast continuation of the Brindle Creek thrust. Recent detailed geologic mapping in the Brushy Mountains by Wilson (2006) and Gatewood (2007) has traced the fault to east of Moravian Falls, North Carolina. This boundary continues to the northeast to the vicinity of the Sauratown Mountains window, turns abruptly east, then southeast, and is truncated by the western border fault of the Davie County Mesozoic basin (Espenshade et al., 1975). Farther southwest, it appears to also have a more due-southerly trend, causing its outcrop trace to track toward the eastern parts of the Inner Piedmont into South Carolina. Such a boundary was not recognized by Nelson et al. (1998) in the Greenville 1° × 2° quadrangle, so its trace most likely occurs farther east. The Paris Mountain thrust sheet in the Greenville 1° × 2° sheet is a truncated remnant of the Brindle Creek thrust sheet. If the boundary identified above in the Winston-Salem 1° × 2° sheet (Espenshade et al., 1975) represents the Brindle Creek thrust sheet, and the Brindle Creek thrust sheet indeed continues into South Carolina between Greenville and Spartanburg (Nelson et al., 1998), the southern end of the fault may terminate against the central Piedmont suture, possibly as far southwestward as central Georgia [Fig. 1A(1)], or against another Inner Piedmont thrust (Plate 1).

The Brindle Creek fault separates assemblages containing major differences in: (1) stratigraphic sequence; (2) ages of detrital zircons (1.1 Ga to the west from small populations of 2.7, 1.8, 1.4 Ga plus a major population of 1.1 Ga plus small populations of 600, 500, and 430 Ma old zircons; Bream et al., 2004); (3) relatively few mafic and ultramafic rocks; and (4) age and character of granitoid plutonism. The rocks between the Brindle Creek fault and the central Piedmont suture comprise a previously unrecognized suspect terrane, the Cat Square terrane, which was accreted during the Neoacadian event. If so, the 1.1 Ga detrital zircons in this terrane could have been derived either from the Laurentian margin or from a different 1.1 Ga source, e.g., a Gondwanan or peri-Gondwanan terrane, like the 600 and 500 Ma zircons. Nevertheless, the older zircon population remains difficult to explain.

The maximum upper bound for the depositional age of the paragneisses in the Brindle Creek thrust sheet thus may be 405–390 Ma (Gatewood, 2007). If so, this would prohibit early Paleozoic docking of the Carolina superterrane (Hibbard, 2000), but would still permit docking to begin at ~405 Ma or later. The Siluro-Devonian plutonic suite in the western Carolina superterrane (McSween et al., 1991) likely formed during the initial stages of B subduction of Theic ocean crust beneath Carolina superterrane and Avalon to the north (Hatcher and Merschat, 2006). Docking of Carolina superterrane shortly before ~405 Ma and partial A-subduction of Tugaloo and Cat Square terranes beneath the Carolina superterrane could provide the burial conditions needed to reach sillimanite-grade metamorphic conditions by ~407 Ma (Gatewood, 2007) and form the 380 (Toluca), 407 and 366 (Walker Top), and 350 Ma (Cherryville of Kish, 1983) plutonic suites. Renewed heating of the already warm Inner Piedmont could have reset the younger 320 Ma TIMS ages reported by Dennis and Wright (1997), reset the eastern Inner Piedmont, and emplaced the nearby suite of 330–315 Ma plutons (e.g., the 317 Ma High Shoals Granite and the 325 Ma Elberton pluton; Horton et al., 1987; Goldsmith et al., 1988; Nelson et al., 1998). This would be the first shared thermal event between the Inner Piedmont and Carolina terrane (Dennis et al., 2000), indicating probable Neoacadian to early Alleghanian accretion of the Carolina superterrane. Regardless of whether the Cat Square terrane was accreted during the Neoacadian or the early Alleghanian, the frequently invoked "Newfoundland" model of Ordovician accretion of Carolina superterrane, and other terranes to the west (Williams, 1979; Hibbard, 2000; Hibbard et al., 2002) must be considered invalid for the southern and central Appalachians.

Central Piedmont Suture

The central Piedmont suture separates the Carolina superterrane from all Laurentian and other accreted elements to the west (Fig. 1A, Plate 1). This boundary has a complex history: an earlier

Neoacadian transpressional history followed by early Alleghanian to late Alleghanian reactivation in dextral strike slip in the Carolinas and Georgia (e.g., McConnell, 1990; Hooper et al., 1997; West, 1998), with dip-slip segments, e.g., that in the Whitmire reentrant in central South Carolina (West, 1998). Segments in Virginia and north-central North Carolina (Espenshade et al., 1975; Goldsmith et al., 1988; Evans and Johnson, 1993) were suitably oriented to also be reactivated as Mesozoic rift faults (Fig. 1, Plate 1).

Several structures truncate against the Central Piedmont suture, implying that the central Piedmont suture is a late Neoacadian structure as several have proposed (e.g., Dennis et al., 2000), or that the suture was reactivated as a thrust during the Alleghanian, making it no longer a suture, as proposed by Hibbard (2000) and Hibbard et al. (2002). The latter hypothesis is difficult to understand and reconcile because, once a suture joins two terranes, reactivation, low-angle crosscutting by a later thrust, or other modification does not make the boundary any less of a suture. U-Pb age dating of fault rocks that frame the Pine Mountain window (Box Ankle and Ocmulgee faults; Hooper et al., 1997) yields consistent Alleghanian ages (Student and Sinha, 1992). Neither of these faults exhibits evidence of reactivation around the Pine Mountain window. The Ocmulgee fault is doubtlessly the southwest continuation of the central Piedmont suture (Hooper and Hatcher, 1990). If the time of accretion of the Carolina superterrane is early Alleghanian, what is the source of the 600–430 Ma zircons discussed later and the age of the metasedimentary rocks that contain them? The trace of the central Piedmont suture does undergo an abrupt change in strike in central South Carolina, where it produces a reentrant in the fault that West (1998) called the Whitmire reentrant (see Plate 1 and discussion). The outcrop pattern around the Whitmire reentrant (West, 1998), and aeromagnetic data (Fig. 1C), indicate the central Piedmont suture has a gentle dip here. Moreover, the aeromagnetic pattern of the frontal part of the Carolina superterrane southwest of the Whitmire reentrant in northeasatern Georgia, and northeastward from the reentrant across South Carolina, and into North Carolina (Fig. 1C) indicates all of this segment may be part of a local thrust sheet here called the Abbeville thrust sheet (Plate 1). The aeromagnetic data also reveal a partially linear anomaly from the southwest end of this proposed thrust sheet in Georgia that traces through part of the Charlotte terrane in South Carolina, into North Carolina where it tracks along the northwest side of the High Shoals pluton (Plate 1), and then crosses the Kings Mountain belt to rejoin the exposed Central Piedmont suture (shown by a dotted line in Plate 1, and also labeled Central Piedmont suture). This line is also traced by a series of gravity anomalies in gravity data filtered to remove the long-wavelength components (D.L. Daniels personal communication, 2007). We interpret these linear anomalies as the subsurface northeastern extent of the Central Piedmont suture, possibly defining the locus of a crustal ramp over which the thrust component of the Central Piedmont suture rode locally. If this speculation is correct, a corollary to this is that the Kings Mountain shear zone may be a dextral tear fault at the northeastern end of the Abbeville thrust sheet (Plate 1). This interpretation permits the Abbeville thrust sheet to form at the same time as the collision of Carolina superterrrane with the various Laurentian components to the west, and takes away the need for long-distance transport of the Carolina superterrane (Hibbard et al., 1998; Dennis, 2004). The maximum displacement needed on the thrust component is 55–65 km—a small fraction of the >400 km of transport suggested for the Carolina superterrane (Hatcher and Merschat, 2006; Merschat and Hatcher, this volume).

Eastern Piedmont Fault System and Alleghanian Metamorphic Core

The eastern Piedmont fault system is an extensive array of mostly dextral strike-slip faults that bound the Alleghanian metamorphic core of the southern Appalachians (e.g., Hatcher et al., 1977; Snoke and Mosher *in* Hatcher et al., 1989) [Fig. 1A(1)]. Although this fault system was originally recognized based in aeromagnetic data (Hatcher et al., 1977), modern aeromagnetic maps (Figs. 1B and 1C) reveal that it is much more extensive beneath the Coastal Plain than was originally thought (Fig. 1, Plate 1).

The eastern Piedmont fault system contains the Alleghanian metamorphic core that includes faults that frame the Pine Mountain window in Georgia and Alabama, the Kiokee belt in Georgia and South Carolina, and the Raleigh belt, including the Goochland terrane, in North Carolina and Virginia (Farrar, 1985; Secor et al., 1986; Hooper et al., 1997) [Fig. 1A(1), Plate 1]. This assemblage developed in Cambrian and older greenschist-facies rocks in South Carolina and eastern Georgia and formed a steep gradient to amphibolite-facies assemblages in the core of the Kiokee belt across the Modoc and other faults (Secor et al., 1986). The core of the Goochland terrane similarly contains assemblages as high as granulite facies (Farrar, 1985).

Pine Mountain Window

The Pine Mountain window and terrane in west-central Georgia and eastern Alabama is a complex window exposing 1.1–1.0 Ga Grenvillian basement (Heatherington et al., 2006) and a thin cover sequence of early Paleozoic(?) schist, quartzite-dolomite, and a higher schist [Fig. 1A(1)]. The window is framed by the dextral Towaliga, and Goat Rock and Dean Creek, faults to the north and south respectively, but is closed at the east end by the Box Ankle thrust (Hooper et al., 1997; McBride et al., 2005). All of these faults moved during the Alleghanian (Steltenpohl et al., 1992; Student and Sinha, 1992), but the Box Ankle fault was emplaced at sillimanite-grade pressure and temperature conditions, whereas the Towaliga and Goat Rock faults moved during garnet grade conditions, and the Dean Creek fault moved under chlorite-grade conditions (the same as the Modoc farther east). Detrital zircons from the cover-sequence Hollis Quartzite inside the window yield the usual dominance of 1.1 Ga ages, but one sample contained a significant component of 2.2–1.9 Ga zircons (Steltenpohl et al., 2004) suggesting either Gondwanan provenance for this sequence and basement, or a source in the Laurentian Penokean orogen. Additional samples from the Hollis Quartzite, however, do not contain the 2.2–1.9 Ga suite (Mueller et al., 2005). If this *is* a rifted block of Gondwanan basement, the emplacement kinematics would be very complex, because of the late Paleozoic ages of all of the faults framing the window. This basement block and its cover have traditionally been considered Laurentian and the cover sequence is thought to be metamorphosed platform rocks, and Steltenpohl et al. (2004), Mueller et al. (2005), Steltenpohl (2005), and others continue to interpret it as having a Laurentian (Penokean?) origin. It remains a suspect terrane.

Neoproterozoic to Ordovician Iapetan Margin Development and Destruction

Rifting of Rodinia

The supercontinent Rodinia formed 1.2–1 Ga as a product of the worldwide Grenvillian orogeny (Hoffman, 1988, 1991), remained intact for ~300 m.y., and then rifted apart ~750 Ma in both western and northeastern North America (Cawood et al., 2001; Karlstrom et al., 2001). Failed rifting began ~735 Ma in the southern and central Appalachians (Aleinikoff et al., 1995). This failed rifting produced the alkalic A-type Crossnore plutonic-volcanic suite (Tollo et al., 2004c), the Grandfather Mountain and Mount Rogers Formations marine, nonmarine, and glaciogenic sedimentary and bimodal volcanic rocks in North Carolina, Tennessee, and Virginia (Rankin, 1993), and the Robertson River igneous complex in central Virginia (Tollo et al., 2004c). The successful rifting that followed created basins that deposited the Ocoee Supergroup in Tennessee, North Carolina, and Georgia, and possibly more distal facies on old continental and early Paleozoic ocean crust (Lynchburg Formation, Ashe-Tallulah Falls Formation and equivalents farther north and south) as new ocean crust was created offshore. Tull (2007, this volume) has suggested that rifting was a two-phase process.

Successful rifting of Rodinia along the southern-central Appalachian margin began ~565 Ma (Aleinikoff et al., 1995), forming the Catoctin volcanic-sedimentary assemblage in Virginia, although initial rifting along this margin probably began as early as 850 Ma, based on SHRIMP ages of authigenic overgrowths of xenotime and monazite on zircon (Aleinikoff et al., 2006). Neither the Catoctin nor the Mt. Rogers–Grandfather Mountain nor the Ocoee Supergroup assemblages is physically connected, because initial rifting created an irregular margin in southeastern Laurentia leaving basement highs that separated rift basins (Thomas, 2006) (Fig. 10). Ocoee Supergroup rocks occupy the western Blue Ridge terrane from near Cartersville, Georgia to south of Johnson City, Tennessee, and from the frontal Blue Ridge eastward to the Allatoona and Hayesville-Soque River faults.

Deposition of the Ocoee Supergroup locally reached thicknesses exceeding 15 km, with most of the thickness contained in the Great Smoky Group (King et al., 1958). The Ocoee Supergroup contains three groups: (1) the Snowbird Group that rests nonconformably on Grenvillian basement; (2) the Great Smoky Group that has been observed in direct contact with basement rocks in relatively few places (e.g., the Fort Mountain area in northwest Georgia, Tull, 2007, this volume) and is faulted onto the Snowbird Group rocks; and (3) the Walden Creek Group, which conformably overlies both the Great Smoky and Snowbird Groups, revealing an extraordinarily complex relationship between the components of the Ocoee Supergroup. The Walden Creek Group unconformably overlies the lower Snowbird Group near the northeast end of the Ocoee basin, and it overlies basement farther northeast beyond the pinchout of the Snowbird Group. Remarkably, for a unit that formed along a rifted continental margin, it contains no volcanic rocks, and only a few scattered gabbro and diorite bodies occur in the entire outcrop belt (Hurst, 1955; Hadley and Goldsmith, 1963; King, 1964; Tull et al., 1998).

The Snowbird Group contains an immature basal sandstone unit overlain by much cleaner quartz arenite to arkosic sandstone, then a pelitic unit. The Great Smoky Group consists of rapidly deposited deep-water immature graywacke and pelite, some of which was deposited in an anoxic environment, minor amounts of cleaner sedimentary rocks, and polymictic vein quartz-dominated conglomerate that also contains quartzite, sandstone, granitoid, black shale rip-ups, and variably textured limestone and dolostone clasts. The Walden Creek Group consists of banded chlorite-sericite slate (Wilhite Slate) that also contains shale, clean sandstone, and limestone. The overlying Sandsuck Formation consists of gray shale, limestone, sandstone, and conglomerate. The Sandsuck conformably underlies the Chilhowee Group. The Sandsuck Formation beneath the Chilhowee has yielded soft-bodied wormlike fossils (in sandstone) (Broadhead et al., 1991), suggesting a Tommotian (earliest Cambrian) age. Repetski et al. (2006), however, reported mid-Paleozoic conodonts from two localities in the Sandsuck(?), in fault blocks separated from the rest of the sequence, while Southworth and Aleinikoff (2007) reported 923 Ma overgrowths of xenotime and monazite on detrital zircons from the Sandsuck type locality. The date on the authigenic zircon overgrowths is difficult to reconcile with well-established field relationships between the Sandsuck and adjacent units, along with the 850-750 Ma timing of rifting of Rodinia along the southeastern Laurentian margin (e.g., Gray and Zeitler, 1997; Cawood and Nemchin, 2001; Cawood et al., 2001). While the Ocoee Supergroup thins gradually southward to near Cartersville, Georgia (Tull and Holm, 2005; Tull, this volume), its thickness rapidly decreases northward from its maximum of 15 km in the Great Smoky Mountains National Park to zero southwest of Johnson City, Tennessee (Fig. 10).

Overlying the Great Smoky Group in the central-western Blue Ridge of southwestern North Carolina and northern Georgia are the rocks of the Murphy syncline [Fig. 1A(1), Plate 1]. This sequence consists of a banded slate with interbedded quartz arenite that likely is correlative with the lower Walden Creek Group (Wilhite Slate) across the Ducktown anticline

(Hurst, 1955; Hurst and Schlee, 1962). This unit is overlain by metasiltstone, then a calcareous schist and a marble unit. Overlying this sequence is a pelitic schist that contains some sandstone. This unit was interpreted by Tull and Groszos (1988) as unconformably overlying the remainder of the sequence. The marble beneath yielded an echinoid stem, suggesting the underlying sequence could be as young as Ordovician to mid-Paleozoic (Tull et al., 1993). This could be the Middle Ordovician, post-Knox, unconformity and the clastic unit overlying it, the Middle Ordovician clastic wedge. Tull and Groszos (1990), however, preferred to interpret this younger unit as being deposited in a middle Paleozoic successor basin, but timing of metamorphism in this part of the Blue Ridge (Ordovician, Moecher et al., 2004) makes this interpretation less tenable.

In central Virginia, the Ocoee-equivalent ~565 Ma Catoctin Formation consists of basalt and rhyolite with some interlayered clastic sedimentary rocks that are locally underlain by an older suite of immature rift-related sedimentary rocks (Meechum River Formation, Snowbird Group equivalent?). East of the Blue Ridge anticlinorium in northern North Carolina and Virginia is a sequence of distal, deeper-water clastic sedimentary and rift-related volcanic rocks, the Ashe–Alligator Back–Lynchburg sequence. More-distal eastward and southeastward equivalents of these metasedimentary and metavolcanic rocks (Ashe–Tallulah Falls–Sandy Springs) have been recognized in the eastern Blue Ridge and Inner Piedmont (Tugaloo terrane) from the Carolinas to Alabama. All have a Laurentian source (Bream, 2003; Bream et al., 2004). They are overlain by Cambrian-Ordovician(?) metasedimentary rocks (Candler, Chauga River), and by Middle Ordovician(?) mid-ocean-ridge basalt (MORB)-arc volcanic rocks (Slippery Creek-Poor Mountain-Ropes Creek). Farther east in the Virginia Piedmont is a sequence of ~470 Ma felsic-mafic volcanic rocks (Chopawamsic) that were intruded by ~455 Ma plutons, then overlain by fossiliferous Late Ordovician (Ashgillian) fine-grained clastic rocks comprising the Chopawamsic terrane (Spears et al., 2004). The Baltimore mafic complex and associated James Run Formation (Faill, 1997a) along strike in Maryland are probably equivalent units.

Farther southwest in western Georgia and Alabama sequences equivalent to the Ashe-Tallulah Falls occur in similar structural positions in the Inner Piedmont and eastern Blue Ridge. These units, called Emuckfaw and Wedowee Groups (e.g., Neathery and Reynolds, 1973; Tull, 1978; Osborne et al., 1988), are similar lithologically to the Ashe-Tallulah Falls. Rocks in the Alabama and western Georgia western Blue Ridge overlie the last remnants of the rifted margin succession (Kahatchee Mountain Group in the Talladega belt), but they contain Early Cambrian Archeocyathids (Jumbo Dolomite, lower Sylacauga Marble Group), Ordovician conodonts (upper Sylacauga Marble Group), Silurian-Devonian conodonts and Devonian macrofossils in the upper Talladega Group, and Mississippian plant fossils in the overlying Erin Slate (Tull et al., 1988; Gastaldo, 1995). Correlation problems exist, however, between the brittlely deformed, fossiliferous Devonian section (Jemison Chert) near the Coastal Plain overlap in Alabama, and the penetratively deformed rocks correlated with these rocks immediately to the east. It is possible that the more easterly rocks belong to the Neoproterozoic to early Paleozoic rifted margin succession and not to the mid-Paleozoic sequence, as suggested in Plate 1. The Mississippian Erin Slate probably unconformably overlies (or is faulted against) all of these rocks regardless of how they might be correlated, and the unconformity may be the equivalent of the pre-Chattanooga Shale unconformity in the foreland.

Rift-to-Drift Transition and Platform Development

The Lower Cambrian Chilhowee Group was deposited on the Laurentian margin of Iapetus during the rift-to-drift transition, and consists of upward-maturing alternating sandstone and shale units. The lowest unit consists of immature sandstone and graywacke with some basalt in northeastern Tennessee and southwestern Virginia, but no volcanic rocks are present in central to southeastern Tennessee and Georgia. The upper part of the Chilhowee Group was probably deposited along the open ocean margin, because it becomes calcareous shale (Helenmode Member, King and Ferguson, 1960), then grades upward into the first successful carbonate bank assemblage (Shady Dolomite). Both the Chilhowee Group and the overlying Shady (-Tomstown) Dolomite are continuous throughout the Appalachian margin, but they thin markedly onto the Alabama and New York promontories (e.g., Adams et al., 1926; Rodgers, 1953, 1970).

Once the first carbonate bank was established, the southeastern Laurentian margin continued to develop, with pulses of progressively finer clastic sediment derived from the eroding Grenville mountains. This is clearly reflected in Cambrian facies, which are more carbonate-rich to the east and more clastic-rich to the west (e.g., Rodgers, 1953; Rankin et al., 1989). From Late Cambrian to earliest Middle Ordovician time, carbonate deposition transgressed across the entire margin into the central interior of North America, extending westward as far as present-day New Mexico, and northward into southern and eastern Canada (Sloss, 1963, 1988). This formed the first extensive carbonate platform across interior Laurentia.

Detrital zircon suites from the rifted-margin succession (Ocoee Supergroup, Ashe Formation, and equivalents) across the southern Appalachian Blue Ridge and western Inner Piedmont (Tugaloo terrane) are all dominated by ~1.1 Ga zircons, with smaller components of 1.5–1.3 Ga zircons (Bream et al., 2004) (Fig. 5). Dahlonega gold belt detrital zircons, while also dominated by 1.4–1.1 Ga components, contain 2.9–2.7 Ga, 2.2–1.9 Ga, and 700–650 Ma zircons. These rocks have traditionally been considered only a more distal facies of the Ocoee Supergroup, but 2.2–1.9 Ga detrital zircons suggest that they may be more exotic, although they could have been derived from the Laurentian Penokean orogen. In addition, the Dahlonega gold belt contains 460 and 480 Ma volcanic arc complexes (Hopson, 1989; Thomas, 2001; Settles, 2002; Bream, 2003; McClellan et al., 2007).

Taconian Events

Following this great incursion of shallow-water marine carbonate deposition into interior North America, the entire carbonate platform, except an area in northern Virginia, Maryland, and Pennsylvania, was eustatically uplifted, producing a karst surface, but carbonate deposition resumed during the Middle Ordovician (Llanvirn) (Ross et al., 1982). Shortly afterward, the eastern margin began to rapidly subside as Middle Ordovician volcanic arcs and accompanying sediments were obducted onto the eastern Laurentian margin. A foredeep formed in front of the arriving thrust sheets, and to the west of that, a forebulge (primordial Cincinnati arch). This foredeep basin accumulated >3 km of clastic sediment in the southern Appalachians forming the Sevier clastic wedge that graded westward into the thin platform limestone sequence. K-bentonite beds (late Llanvirn) from continuing volcanic activity to the east (e.g., the abundant Mid–Ordovician granitoids in the eastern Tugaloo and Milton terranes) were deposited throughout the southern and central Appalachians and on the adjacent platform to the west (Bergström et al., 2004). This clastic wedge is continuous throughout the orogen from Alabama to Newfoundland.

The younger (Llanvirn-Caradocian) Martinsburg clastic wedge formed a thin successor basin atop the Sevier wedge in the southern Appalachians, but accumulated some 5 km of deep-water turbiditic clastics in Pennsylvania, probably in a true foredeep (Faill, 1997a). In northeastern Tennessee, the Martinsburg unconformably overlies the Bays Formation (upper Llanvirn–Black River; Rodgers, 1953). Deposition in the Martinsburg basin, however, began in the Llanvirn, because the Deike and Millbrig K–bentonites are present in the Martinsburg sequence at the type locality near Martinsburg, West Virginia (Faill, 1997a). The Martinsburg assemblage grades upward into the shallower-water Ashgillian Juniata red bed succession, and then into the Llandoverian Clinch–Tuscarora quartz arenite and Shawangunk conglomerate.

The Martinsburg basin, however, was more than a successor basin in southeastern Pennsylvania. It was preceded by deposition of the dark shale and limestone of the Myerstown Formation (Black River and older), followed by deposition of the Martinsburg, which began here at ~460 Ma. The Myerstown is locally overlain by the Cocalico Formation of variegated shale and sandstone; Stose (1946), Lash and Drake (1984), and Drake et al. (1989) considered the Cocalico a candidate for another Taconic allochthon like those of New York and Vermont because of its lithologic assemblage, and the fact that it contains Middle Ordovician and older components. Wise and Ganis (2006), however, suggested that the Cocalico may represent autochthonous Martinsburg. Between the Susquehanna and Schuylkill Rivers, the Martinsburg contains allochthons emplaced during the Middle Ordovician, the isolated Jonestown volcanics (perhaps similar to the Starks Knob volcanics northeast of Saratoga Springs, New York), and olistoliths of older units (Wise and Ganis, 2006). Wise and Ganis (2006) preferred to refer to this segment of the Martinsburg basin as the "Hamburg complex" (Plate 1), and, based on new paleontological data (e.g., Ganis et al., 2001) do not consider any of the contacts between the units here, such as the Yellow Breeches thrust, to be pre-Alleghanian faults. They would thus abandon the Taconian klippe hypothesis for several of these units previously thought to belong to the Hamburg klippe in southeastern Pennsylvania.

While the source of the clastic sedimentary assemblage in the Sevier and Martinsburg foredeep basins was clearly to the east, detrital zircon suites contain no Paleozoic components (Bream et al., 2001; Thomas et al., 2004). All are dominated by zircons derived from Grenvillian and older Laurentian sources, suggesting provenance from the exhumed and cannibalized platform, and rifted-margin sedimentary and rift-facies volcanic rocks. Clasts in Llanvirn polymictic conglomerates in the Sevier basin sample platform carbonate and the entire section down to the basement, so none are derived from sources in the Taconian orogen (Kellberg and Grant, 1956).

The Taconic orogeny is clearly diachronous in the southern and central Appalachians; earlier pulses began in the late Arenig in the southern Appalachians; and the younger event—the type Taconic orogeny—beginning sometime in the Llandeilo in both the southern and central Appalachians, with deposition continuing into the Silurian in the central Appalachians (Silurian clastic deposition extends southward into Tennessee). Corresponding suites of plutons (Aleinikoff et al., 2002; Bream, 2003; Steltenpohl, 2005), penetrative deformation, and metamorphism to conditions as high as granulite facies (Force, 1976; Hatcher and Butler, 1979; Absher and McSween, 1985; Eckert et al., 1989; Moecher et al., 2004) were a product of subduction, obduction, and arc accretion. The 470–455 Ma plutonic-volcanic suite (Hillabee, Ropes Creek, Poor Mountain, Chopawamsic, Baltimore) in the southern and central Appalachians has an oceanic MORB to arc affinity (Tull and Stow, 1980; Spell and Norrell, 1990; Coler et al., 2000; Hatcher, 2002a, his Figure 7; Settles, 2002; McClellan et al., 2007). At this time the Neoproterozoic to Cambrian rifted margin assemblage was A-subducted beneath the arc, penetratively deformed, and metamorphosed to lower mid-crustal metamorphic assemblages.

During the Taconic events, the Laurentian margin underwent A-subduction and several oceanic and arc assemblages were obducted onto the margin with accompanying emplacement of the Cowrock, Cartoogechaye, and Tugaloo terranes in the southern Appalachians, and the Chopawamsic, Potomac, and Baltimore terranes in the central Appalachians [Fig. 1A(1)]. Grenvillian basement was remobilized brittlely in the core of the recumbent Musconnectong nappe in New Jersey and eastern Pennsylvania (Drake, 1970). This basement was also remobilized to the northeast (Hudson–New Jersey Highlands) and southwest, forming external massifs in the core of the Blue Ridge from Virginia to Georgia. Although not yet accreted to Laurentia, internal massifs in the eastern Virginia Goochland terrane, the Tugaloo terrane west of the Brevard fault zone in the Carolinas and northeastern Georgia, and possibly in the Pine Mountain window in Georgia and Alabama, were remobilized at this time. Basement remobilization in the Pine Mountain window probably occurred later (see following).

A suite of thrust sheets carrying deep-water sedimentary rocks was emplaced onto the New England and Canadian Maritimes margin during the Ordovician, locally carrying oceanic crust and mantle in ophiolite sheets (Québec and Newfoundland) (Drake et al., 1989). The Hamburg complex in southeastern Pennsylvania is an assemblage of smaller allochthons and olistoliths (Wise and Ganis, 2006). No Taconic allochthons (or candidates) remain intact farther south. Geochemical data from both mafic and ultramafic rocks in the southern Appalachians, however, suggest that these rocks had an oceanic-to-arc origin and many were likely emplaced as ophiolites (Hatcher et al., 1984; Misra and McSween, 1984; Misra and Conte, 1991; Ryan et al., 2005; Swanson et al., 2005). Other premetamorphic thrusts formed at this time from the Pennsylvania Piedmont southward. The Martic line (Wise, 1970) is probably a premetamorphic fault with a complex history that carried the Westminster and Potomac terranes onto the Laurentian margin (Wise and Ganis, 2006). These two terranes also contain ophiolite fragments (Faill, 1997a). The premetamorphic Greenbrier fault thrusted Great Smoky Group rocks over Snowbird Group and basement rocks at this time (Hadley and Goldsmith, 1963). Amalgamation of old and new crust generated during Taconian events formed a large accreted mass along the southeastern Laurentian margin. This accreted mass consists of the Laurentian margin, and terranes farther north (Baltimore, Potomac, Westminster) thrust onto it, the central Blue Ridge terranes and Dahlonega gold belt discussed below.

Central Blue Ridge Terranes

Although Hadley and Goldsmith (1963) and Hadley and Nelson (1971) recognized a major boundary in the central Blue Ridge, the Hayesville fault was first recognized as a premetamorphic fault separating Great Smoky Group rocks of the western Blue Ridge from high-grade central Blue Ridge gneiss, schist, and amphibolite (Hatcher, 1978). Rock units on both sides of the fault are truncated (e.g., Eckert, 1984). Part of the basis for recognition of this fault, however, is the occurrence of abundant mafic and ultramafic rocks to the east (Hatcher, 1978), whereas to the west there are few or no igneous rocks (e.g., Hurst, 1955; Hadley and Goldsmith, 1963). Moreover, metasandstones to the east are dominated by quartz-biotite-plagioclase, whereas those to the west are two-mica, two-feldspar metasandstones. Two separate tectonostratigraphic terranes have been recognized in the Hayesville thrust sheet. The southernmost is the Cowrock terrane—located mostly in Georgia—truncated to the north and east by the Chunky Gal Mountain and Shope Fork faults (Hatcher et al., 2004) [Figs. 1A(1) and 11]. The Cowrock terrane sequence—the Coweeta Group (Hatcher, 1988)—consists of metasandstone, schist, and some mafic and ultramafic rocks in the Coleman River Formation, particularly around the Shooting Creek and Brasstown Bald windows (ophiolite?) (Hartley, 1973), along with additional mafic rocks in Georgia near the southwest end of the terrane (Nelson and Gillon, 1985). The Coleman River Formation of metasandstone and pelitic schist is overlain by the lithologically more diverse Ridgepole Mountain Formation, which consists of biotite schist, metasandstone, metagraywacke, and quartzite (Hatcher, 1979, 1988). Coleman River metasandstone contains Grenvillian and older detrital zircons (Bream et al., 2004), indicating these rocks are not basement, but are more likely a more distal Laurentian margin sequence. The Coweeta Group was intruded by the 466 Ma Persimmon Creek Gneiss in the southeastern part of the terrane (Hatcher, 1979; McDowell et al., 2002; Miller et al., 2006). The eastern boundary of the Cowrock terrane is the Soque River fault, which truncates the Shope Fork fault, and continues northward into North Carolina, where it forms the western boundary of the Dahlonega gold belt (DGB Plate 1) and the eastern flank of both the Cowrock and Cartoogechaye terranes (Fig. 11). Settles (2002) demonstrated that the Hayesville and Soque River faults join around the southwest end of the Cowrock terrane just west of Dahlonega, Georgia, where they truncate the fault commonly mapped as the Allatoona fault [Fig. 1A(1)]. J.F. Tull (2006, written commun.), however, reported that the Allatoona fault farther southwest formed very late and truncates Mississippian rocks in its footwall in Alabama. So, the fault that Settles (2002) mapped as the Allatoona fault is likely not the same fault as that mapped in Alabama.

The Cartoogechaye terrane lies to the northeast of and atop the Cowrock terrane. It consists of metasandstone, pelitic schist, and an assemblage of mafic and ultramafic rocks, along with some Grenville basement (Hatcher et al., 2004). These were metamorphosed to assemblages as high as granulite facies during the Ordovician (Force, 1976; Hatcher and Butler, 1979; Absher and McSween, 1985; Eckert et al., 1989; Moecher and Miller, 2000; Moecher et al., 2004). Metasandstone at Chunky Gal Mountain and Winding Stair Gap west of Franklin, North Carolina, and another metasandstone from near Waynesville, North Carolina, yielded Grenvillian and older detrital zircons (Miller et al., 2000; Merschat et al., 2006), indicating that this terrane cannot consist predominantly of Grenville basement. The Trimont Ridge gneiss does, however, appear to be a basement unit (Hatcher et al., 2004), along with other undated felsic gneiss bodies in this terrane. The Cartoogechaye terrane is traceable northward from the Georgia-North Carolina line to the region just east of the Great Smoky Mountains National Park. Basement rocks in the core of the Blue Ridge anticlinorium to the northeast plunge southwest beneath the Hayesville fault [Fig. 1A(1), Plate 1].

The lithologic assemblages in the Cowrock and Cartoogechaye terranes suggest an oceanic association. The mafic and ultramafic rocks here have been demonstrated to have an oceanic origin (Hatcher et al., 1984; Misra and McSween, 1984; McSween and Hatcher, 1985; Ryan et al., 2005; Swanson et al., 2005), but the exact origin of the metasedimentary assemblage has never been resolved. The association of mafic and ultramafic rocks with metasandstone and pelitic schist, along with some quartzite, suggests the possibility that these rocks were ocean crust and mantle overlain by ocean floor sediments. The association, particularly the presence of small amounts of quartzite

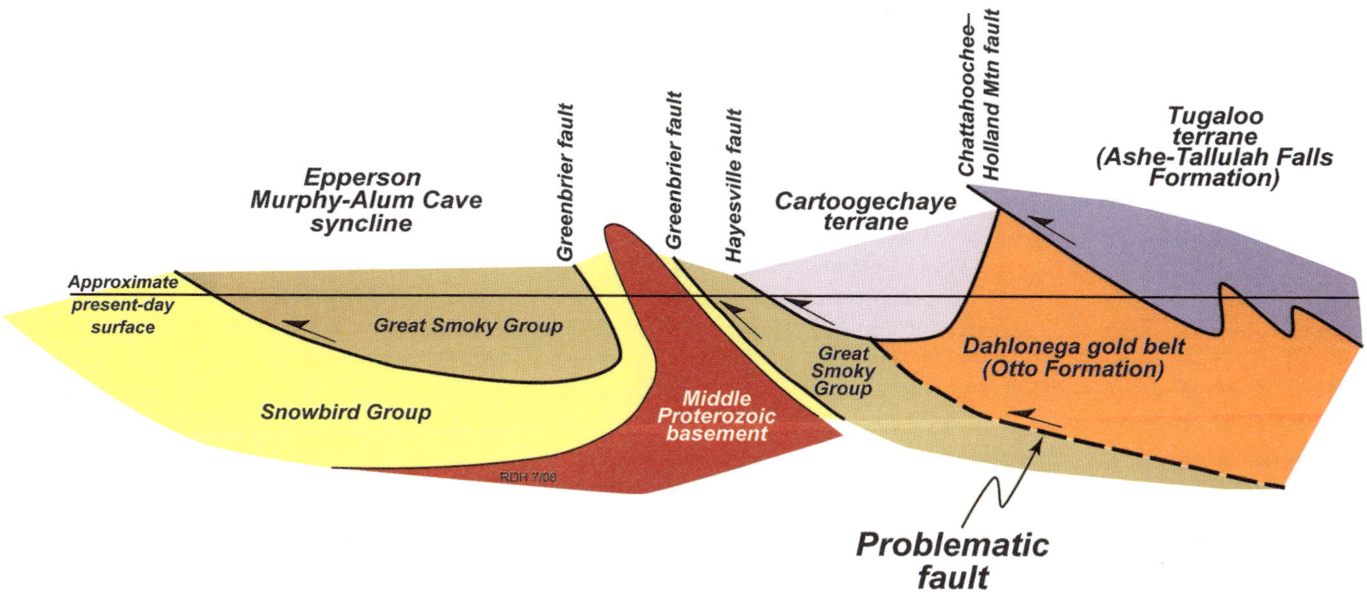

Figure 11. Schematic cross section through part of the Blue Ridge at approximately at 36°N latitude showing the relationships between major tectonic units and the need for a uniquely located additional fault if the Dahlonega gold belt rocks are *not* a more easterly facies of the Great Smoky Group.

(chert and felsic tuff protoliths?), also suggests that some of these rocks could have been part of an accretionary wedge, but no geochemical data exist to test this hypothesis.

Dahlonega Gold Belt

The Dahlonega gold belt is a narrow, linear exposure beneath the Soque River and Chattahoochee faults. It is composed predominantly of metasandstone and schist (Otto Formation) that also contains several mafic and mafic-ultramafic complexes and traceable mafic units (German, 1985; Hopson, 1989). Volumetrically, metavolcanic rocks comprise no more than 10% of the rocks of the belt, despite the fact that the belt has been characterized by some (e.g., Hibbard et al., 2006a) as a metamorphosed magmatic arc. It extends from western Georgia to Waynesville, North Carolina, where it is exposed between the Cowrock and Cartoogechaye terranes to the west and the Tugaloo terrane to the east [Fig. 1A(1)]. The western boundary is the Soque River fault, and the eastern boundary is the Chattahoochee fault. Immediately north of Franklin, North Carolina, the Chattahoochee fault overrides the Soque River fault, so the Dahlonega gold belt is covered, but it reemerges in the Great Balsam Mountains window to the northeast (Hatcher et al., 2005b) (Fig. 11, Plate 1). Toward the southwest, and possibly into Alabama, the Dahlonega gold belt is bounded to the west by the Allatoona fault (Osborne et al., 1988).

Detrital zircons from Dahlonega gold belt metasandstones reflect a North American source in four samples, but one sample contains Gondwanan(?) (2.0–1.8 Ga) zircons, in addition to Grenvillian and older North American detrital zircons (Bream, 2003; A.J. Merschat, personal commun.). The Gondwanan ages have not been reproduced in the other samples. Zircons in the one sample containing the possible Gondwanan suite, however, could have been derived from the 2.0–1.8 Ga Laurentian Penokean orogen. Dahlonega gold belt rocks were likely metamorphosed during the Taconian event, with ~460 Ma metamorphic zircon rims on detrital zircons (Merschat et al., 2006).

Another perplexing problem is the relationships between the Dahlonega gold belt and the Cowrock and Cartoogechaye terranes, along with its relationships to the Laurentian margin (Ocoee Supergroup) rocks to the west. Hadley (1970) considered the Dahlonega gold belt rocks to be correlative with the Ocoee Supergroup, no doubt because of the similarities between Dahlonega gold belt metasandstone and schist to those of the Great Smoky Group. Both metasandstones are two-mica, two-feldspar sandstones that resemble those of the more fine-grained parts of the Great Smoky Group, like the Ammons Formation (Mohr, 1975; Ausburn et al., 1998), but they could be distal facies of any part of the Great Smoky Group. One difficulty in making this correlation, however, lies in the presence of the possible Gondwanan detrital zircons (2.0, 1.8 Ga) in one sample (Fig. 5) (Bream, 2003). At the northeastern end of the Dahlonega gold belt, southwest of Waynesville, North Carolina, is a very narrow belt of Cartoogechaye terrane rocks to the west that separate Dahlonega gold belt from western Blue Ridge rocks. If the Dahlonega gold belt rocks (Otto Formation, Hatcher, 1988) are a more eastern facies of the Great Smoky Group, there would be no problem to connect these to western Blue Ridge Ocoee Supergroup rocks beneath the Cartoogechaye terrane. If the Dahlonega gold belt rocks are not correlative with the Great Smoky Group, a hidden fault must be present that remains beneath the Cartoogechaye terrane throughout the extent of the Dahlonega gold belt (Fig. 11), even beneath the narrow segment of the Cartoogechaye terrane farther north [Fig. 1A(1), Plate 1]. A candidate fault exists at the southwest end of the Cowrock terrane west of Dahlonega, Georgia, where the Allatoona fault passes beneath the southwestern extent of the Hayesville–Soque River fault (Settles, 2002). It is difficult to perceive, however, that this fault continues beneath the entire length of the Cowrock and Cartoogechaye terranes and fortuitously does not emerge even where the Cartoogechaye terrane reaches its narrowest width [Fig. 1A(1), Plate 1]. The simplest solution would be to consider the Dahlonega gold belt rocks a more outboard facies of the Great Smoky Group that contain small amounts of arc and MORB volcanic rocks.

Tugaloo-Milton-Potomac Terrane

East of the Chattahoochee–Holland Mountain fault lies the Tugaloo terrane, which consists of Ashe–Tallulah Falls Formation rocks (and equivalents), and small amounts of Grenville basement in the Toxaway and Tallulah Falls domes [Figs. 1A(1)]. In addition, several large granitic plutons occur here that were intruded during either the middle or late Paleozoic, long after intrusion of the Middle Ordovician Whiteside pluton (Miller et al., 2000) [Fig. 1A(1)]. Most of these plutons consist of granodiorite, with some tonalite (Miller et al., 1998; McDowell et al., 2002; Stahr et al., 2005). In addition, numerous small ultramafic bodies and several mafic-ultramafic complexes occur in the Tugaloo terrane. The Tugaloo terrane extends from the Chattahoochee-Holland Mountain fault eastward across the Brevard fault zone to the Brindle Creek fault, which separates the eastern Inner Piedmont Cat Square terrane from western Inner Piedmont Tugaloo terrane rocks (Hatcher, 2002a) [Fig. 1A(1)].

The Tallulah Falls–Ashe Formation consists of a lower unit of metasandstone, pelitic schist, and interlayered amphibolite, which is overlain by an aluminous, white mica-dominated schist unit that contains either kyanite or sillimanite, abundant garnet, and minor amphibolite. The aluminous schist unit is overlain by a metasandstone and pelitic schist unit that contains small amounts of amphibolite—or none (Hatcher, 1971b, 1973). This sequence is readily mappable in detail throughout the eastern Blue Ridge and western Inner Piedmont of northeast Georgia and western North Carolina. These subdivisions have been recognized wherever detailed mapping has been conducted in the Carolinas and northeastern Georgia.

A series of more northerly terranes roughly in the same present tectonic position, and composed of a variety of clastic rocks similar to the assemblages farther south were deposited in different parts of Faill's (1997a) Cambrian and Ordovician "Octararo sea," an arm of the Theic ocean. These terranes, called Potomac, Philadelphia, Hamburg complex (e.g., Wise and Ganis, 2006), Westminster, etc., were variously accreted to Laurentia in the Taconian and possibly later events (Faill, 1997a). The Hamburg complex, the westernmost of these terranes, was emplaced as a series of blocks and allochthons into the Martinsburg clastic wedge (Drake et al., 1989; Faill, 1997a; Wise and Ganis, 2006). This complex is the southernmost vestige of the Taconic klippen that are so prominent in the Appalachians from New England northward (Rodgers, 1970), but the Hamburg complex is not easily subdivided into distinct thrust sheets (G.R. Ganis, 2006, 2007, personal commun.).

MID-PALEOZOIC SEDIMENTATION AND NEOACADIAN DOCKING OF THE CAROLINA SUPERTERRANE—CAT SQUARE TERRANE

Evidence for the classic 410–380 Ma Acadian orogeny in the southern and central Appalachian internides is mostly lacking. A suite of 382–360 Ma granitoid plutons (Pink Beds; Looking Glass; Mt. Airy; Stone Mountain, North Carolina; and Almond granite bodies in Alabama) intruded the eastern Blue Ridge of the Carolinas and Georgia (Higgins et al., 1997; Miller et al., 1998, 2000; Mapes, 2002), and a suite of 407–366 Ma anatectic granitoids (Toluca) exists in the Cat Square terrane (Giorgis et al., 2002; Mapes, 2002; Gatewood, 2007).

The diachronous Catskill delta spread Middle and Upper Devonian sediment from New England into New York and Pennsylvania but was time transgressive, younging until it pinches to the south (Ettensohn, 1985, 2004). Late Silurian-Early Devonian clastic sedimentary rocks are thought to comprise the Talladega Group in Alabama (Tull, 2002). Deposition of the Upper Devonian–early Mississippian Chattanooga-Brallier-Grainger (and equivalents) black shale and siliciclastic rocks unconformably on folded and faulted Silurian and older rocks on the platform (Fig. 12) may be far-field indicators of the Acadian-Neoacadian orogeny in the southern and central Appalachians. Carbonate rocks that overlie the basal Mississippian clastic sequence become more clastic-rich toward the mountain chain (Hatcher et al., 2005a) (Fig. 13) and provide clear evidence of the separation of the Acadian-Neoacadian clastic wedge from the later Alleghanian wedge.

Deposition continued until post–430 Ma in the still-open remnant of the Theic ocean that lay between southeastern Laurentia and the approaching Carolina superterrane. Here a sequence of immature deep-water sandstones (subgraywacke to graywacke) and pelites that today consist of aluminous schist and biotite gneiss were probably deposited on oceanic(?) crust (Merschat and Hatcher, this volume). Detrital zircons include the usual prominent 1.4–1.1 Ga Laurentian suite, but with the addition of major components of 600, 500, and 430 Ma zircons, along with small populations of 2.2–1.9 and 2.9–2.7 Ga zircons (Bream, 2003). The remnant ocean basin in which these sediments were being deposited closed incrementally north-to-south, zipper fashion, as the more westerly terranes were subducted beneath the Carolina superterrane. The diverse detrital zircon population suggests that sediment was derived from both Laurentia and Carolina (Bream et al., 2004; Merschat and Hatcher, 2007 this volume). Presence of younger zircons precludes docking of the Carolina superterrane before the Late Devonian or early Mississippian (Merschat

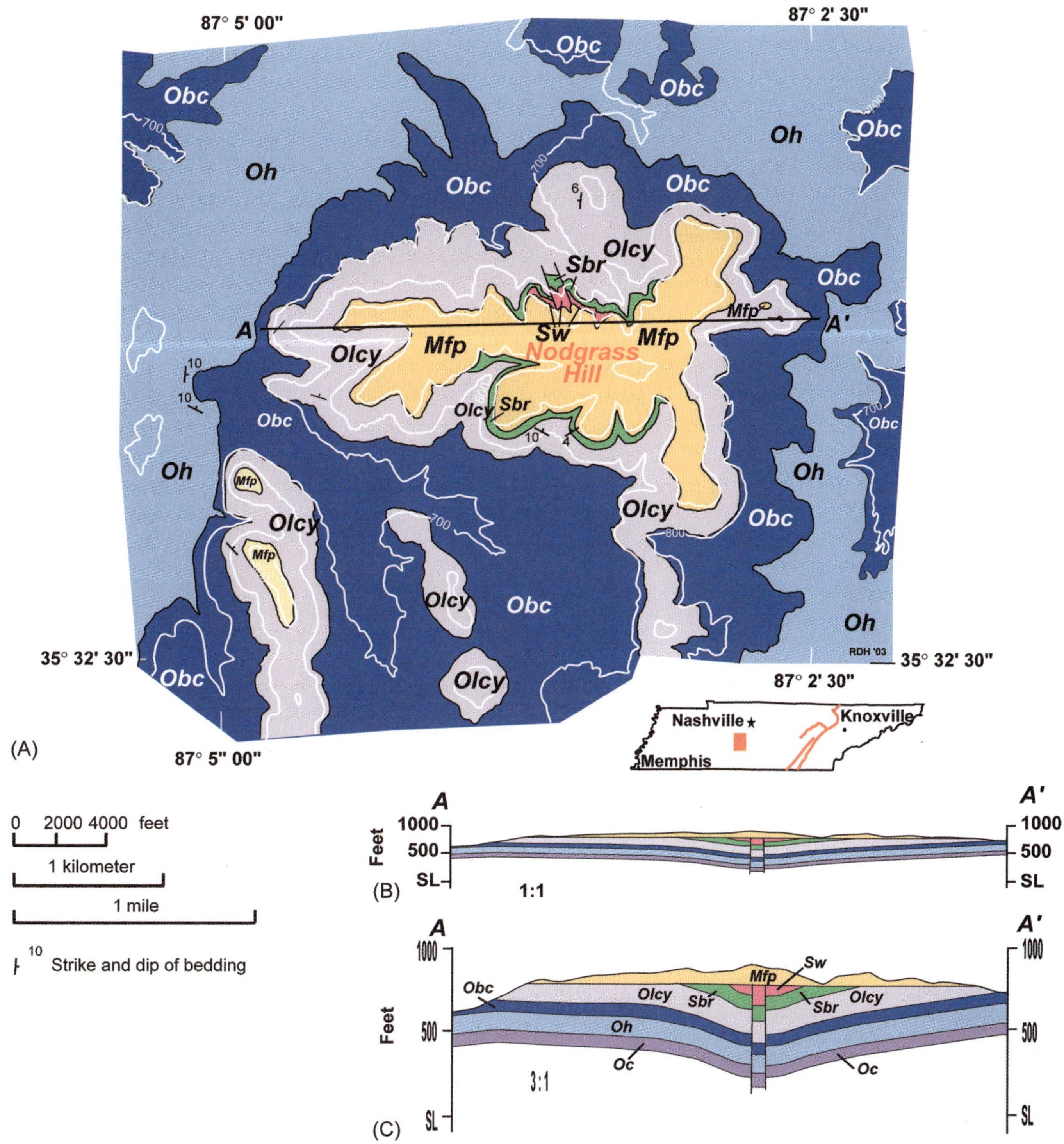

Figure 12. (A) Part of the geologic map of the Columbia, Tennessee, 7.5-min quadrangle (Wilson et al., 1964; geology mapped by R.D. Hatcher in 1962) showing the truncating relationships between the pre-Chattanooga rocks and Chattanooga Shale–Ft. Payne Formation (Mfp) on Nodgrass Hill and surrounding area. A faulted syncline of Silurian and Ordovician rocks is truncated by the unconformity. (B–C) Cross section A–A′ with no vertical exaggeration (B) and with a 3:1 vertical exaggeration (C). Oc—Carters Limestone; Oh—Hermitage Formation; Obc—Bigby–Cannon Limestone; Olcy—Catheys and Leipers Formations; Sbr—Brassfield Limestone; Sw—Wayne Group limestone and calcareous siltstone; SL—sea level. Topographic contours in (A) are white; contours in feet; contour interval is 100 ft. The red rectangle in the index map is the location of the Columbia quadrangle.

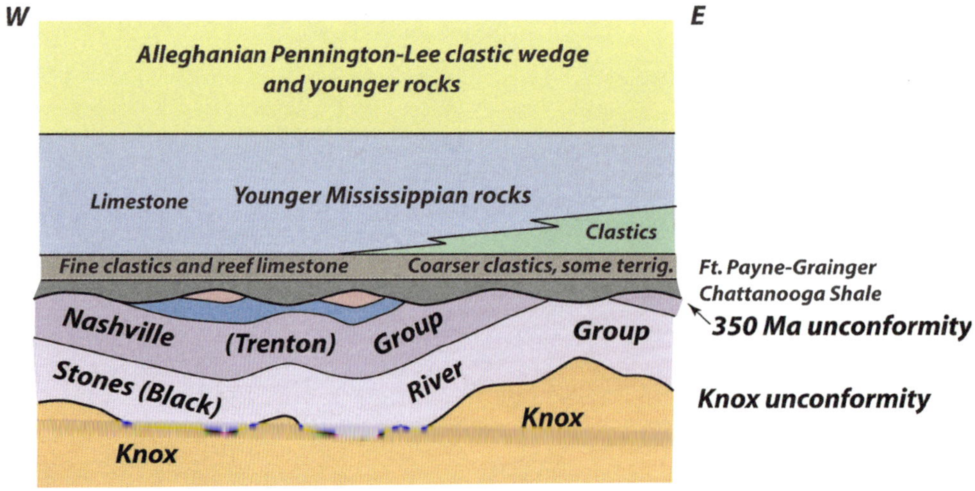

Figure 13. Schematic relationships between deformed Ordovician and Silurian rocks beneath the 360 Ma pre–Chattanooga unconformity in the southern Appalachians and the Neoacadian clastic wedge above the Chattanooga Shale and the overlying Alleghanian clastic wedge. The Chattanooga is the first evidence of Neoacadian tectonic activity at this latitude (~Knoxville, Tennessee, 36°N) on the foreland. Mississippian rock units above the Chattanooga also reflect the orogeny in the internal parts of the mountain chain by becoming more clastic-rich to the east.

et al., 2005a; Hatcher and Merschat, 2006), despite arguments to the contrary from paleomagnetic (Vick et al., 1987) and geologic (Hibbard, 2000) data. Additional paleomagnetic data from low-grade, slightly deformed, Carolina slate belt rocks in the Albermarle, North Carolina, area (Noel et al., 1988) are inconclusive about the earlier docking of Carolina.

Detrital zircon ages as young as 430 Ma (Bream et al., 2004) require that the Cat Square terrane, which lies immediately east of the Tugaloo terrane and west of Carolina [Fig. 1A(1)], be accreted during the middle to late Paleozoic. Anatectic granites formed at ~407, 380, and 366 Ma (Thomas, 2001; Gatewood, 2007). Metamorphic rims on zircons not only in the Cat Square terrane, but also throughout the Tugaloo terrane, along with deformed 366 Ma plutons in the Cat Square terrane, indicate peak metamorphism occurred from 360 to 350 Ma (Carrigan et al., 2001; Thomas, 2001; Bream, 2003). A 320 Ma high-temperature-medium-pressure event is also recorded in zircon rims (Bream, 2003; Merschat et al., 2005a). Monazite U-Pb TIMS ages yielded similar 360 and 320 Ma ages (Dennis and Wright, 1997). Metamorphism to upper-amphibolite- and granulite- facies conditions probably corresponds to mid-Paleozoic docking of the Carolina superterrane, during which the previously assembled superterrane plus the Cat Square terrane were A-subducted beneath Carolina superterrane, where they achieved the burial depths necessary to produce wholesale migmatization in the Inner Piedmont and eastern Blue Ridge (Merschat et al., 2005a; Hatcher and Merschat, 2006). Components of the Carolina superterrane suprastructure that formerly rested atop the Inner Piedmont have now been eroded [Fig. 1A(1), Plate 1].

The Inner Piedmont, which now consists of the Cat Square and the eastern Tugaloo terranes, records a unique pattern of crustal flow that probably contains the key to the docking kinematics of the Carolina superterrane (Hatcher, 2001; Merschat et al., 2005a; Hatcher and Merschat, 2006). This flow pattern, from southeast to northwest, consists of a northwest-directed suite of thrust sheets that to the west became southwest-directed as they were buttressed against the eastern Blue Ridge rocks along the Neoacadian Brevard fault zone (Hatcher, 2001, his Figure 3; Merschat et al., 2005a, their Fig. 2). The docking process was probably thus zippered—oblique transpressional—beginning in the north and closing southwestward. As the Carolina superterrane subducted the components previously joined to Laurentia, the Inner Piedmont may have decoupled at mid-crustal level to form an orogenic channel through which the partially melted plastic mass escaped and extruded southwestward (Hatcher and Merschat, 2006). Ironically, the Avalon terrane in New England was A-subducted westward beneath central New England about the same time, or during the early Alleghanian (Phinney, 1986; Robinson et al., 1998).

While the detrital zircon suites in the Tugaloo terrane have quite uniform ages and probably reflect a Laurentian source (Bream et al., 2004), both Ordovician and Devonian plutons in the western Tugaloo terrane contain a wide range of inherited components (Fig. 7), from dominant 1.1 Ga to Neoarchean zircons. In contrast, almost none of the Ordovician (~466 Ma) plutons in the eastern Tugaloo terrane and Devonian 406–380 Ma plutons in the Cat Square terrane contain any inheritance. Magmas in the western Tugaloo terrane must have passed through an appreciable thickness of old crust, whereas magmas crystallizing the plutons in the Inner Piedmont east of the Brevard fault zone, which were generated in more oceanic to arc environments, did not pass through older continental crust en route to crystallization. Alternatively, inherited zircons could have been consistently resorbed in Zr–undersaturated magmas throughout the region between the Brevard fault zone and the central Piedmont suture. This may be a more likely explanation, because all metasandstones in the western Tugaloo terrane contain old zircons.

Inner Piedmont and eastern Blue Ridge rocks cooled sufficiently to block the 360–350 Ma metamorphic event. The oldest direct thermal connection between the Carolina superterrane suprastructure and the Cat Square and Tugaloo terrane infrastructure consists of the 408–399 Ma Devonian mafic-alkalic Concord plutonic suite, which occurs in the Charlotte terrane, and the felsic 403–378 Ma Salisbury plutonic suite (McSween et al., 1984, 1991; Miller et al., 2000; Samson and Secor, 2000). $^{87}Sr/^{86}Sr$ initial ratios for these plutons average 0.703, supporting a lower-crustal or oceanic origin (Butler and Fullagar, 1978). In addition to the plutons, a 360 Ma thermal overprint is also recorded in $^{40}Ar/^{39}Ar$ plateau ages in the western Carolina superterrane (Dallmeyer et al., 1986), probably coupled to the high-grade 360–350 Ma metamorphic event, which occurred at greater depth.

PERI-GONDWANAN CAROLINA SUPERTERRANE

Coeval with the rifting of Rodinia was the development of a number of peri-Gondwanan terranes in the intervening oceans separating Laurentia from Gondwana. All of these terranes developed proximal to Gondwana, formed composite superterranes, and were accreted to the Laurentian assemblage during the early to mid-Paleozoic (e.g., Rast and Skehan, 1983; van Staal et al., this volume). The peri-Gondwanan component located in the southern and central Appalachians consists of the Carolina superterrane (or Carolinia) [Fig. 1A(1)]. These terranes also contain the basic components of most of the coeval Panafrican terranes: dominance of 625–550 Ma volcanic arcs and associated volcanogenic sediments, subarc plutonic complexes, and difficult identification of basement (except in the Mozambique belt; e.g., Kröner et al., this volume). Basement from which the arc materials in the Carolina superterrane consisted of evolved continental crust, based on inherited zircons at 1.0, 2.1, and 2.5 Ga, and Sm-Nd whole-rock isotopic data (Ingle et al., 2003).

The Neoproterozoic component of the Carolina superterrane consists of a mafic to felsic volcanic arc assemblage and associated volcaniclastic sedimentary rocks. A number of plutons in the range of 618–550 Ma were generated and intruded into the underpinning of the arc complex (Hibbard et al., 2002; Ingle et al., 2003; Bradley et al., 2006). This assemblage was metamorphosed during the Cambrian because 530 Ma plutons cut upper-amphibolite- to greenschist-facies metamorphic rocks (Dennis and Wright, 1997; Barker et al., 1998). Neoproterozoic Ediacaran metazoan fossils have been identified on bedding planes in low-grade felsic volcanic rocks near Durham, North Carolina (Cloud et al., 1976), and in the volcano-sedimentary sequence in the Albermarle area farther south in North Carolina (Hibbard et al., 2006c; Weaver et al., 2006).

Unconformably overlying the arc complex is a sequence of Cambrian and possible Ordovician clastic sedimentary rocks. A Middle Cambrian Acado-Baltic (*Paradoxides*) fauna occurs in these sedimentary rocks in South Carolina (Secor et al., 1983), and Ordovician conodonts have also been reported in another part of the sequence in North Carolina (Koeppen et al., 1995), but Hibbard et al. (2006b, 2006c) reported Ediacaran fossils from at least one of the same localities, and their age is corroborated by U-Pb radiometric ages.

The high-grade western Carolina superterrane contains a 360–350 Ma metamorphic overprint and numerous younger (Devonian to Carboniferous) granitoids and gabbros (McSween et al., 1991; Hibbard et al., 2002) probably related to the middle to late Paleozoic docking of the Carolina terrane (see above). The Smith River allochthon [Fig. 1A(1), Plate 1] has been described as a peri-Gondwanan terrane and outlier of the Carolina superterrane based on chemical ages of monazite (Hibbard et al, 2003), but detrital zircon data (A.J. Merschat, 2007, personal comm.) do not support this conclusion. The greenschist- and lower-grade central and eastern Carolina superterrane consists mostly of volcanic and volcaniclastic rocks interrupted by the Alleghanian Kiokee belt metamorphic core. Southeast of this metamorphic core are more low-grade volcanic and volcaniclastic rocks that have been named different terranes by Hibbard et al. (2002), although the Carolina superterrane was subdivided differently by Horton et al. (1989).

LATE MISSISSIPPIAN TO PERMIAN ALLEGHANIAN ZIPPERED COLLISION WITH GONDWANA AND THE AMALGAMATION OF PANGEA

Collision of Gondwana with crust amalgamated during the early to middle Paleozoic orogenies actually began the final chapter in the construction of the Appalachian Mountain Chain (Fig. 14). This collision was oblique, transpressive, rotational, and involved north-to-south collision zipper-like closing of Gondwana with the previously assembled Laurentian-peri–Gondwanan collage (Hatcher, 2002b). Initial collision produced a suite of largely S-type granites (Sinha and Zietz, 1982). Additionally, the Brevard fault zone was reactivated dextrally under chlorite-grade conditions, indicating considerable unroofing since 335 Ma, and additional dextral faults—the Hylas and other components of the eastern Piedmont fault system (Roanoke Rapids, Nutbush Creek, Modoc-Dean Creek, Augusta, Towaliga, Goat Rock, Bartletts Ferry) (Hatcher et al., 1977)—formed as crustal blocks attempted to escape from the collision zone by moving southwestward from the collision zone, the vicinity of present-day northern Virginia-Maryland-Pennsylvania-New Jersey. In addition to the 325 Ma plutonic suite, there is a younger suite of shallow 300-265 Ma plutons largely confined to the Carolina superterrane, with the exception of several of these plutons in the Georgia Inner Piedmont. These plutons were likewise generated during various stages of collision. A fault-bounded metamorphic belt called the Pine Mountain-Kiokee-Raleigh-Goochland terrane formed in the eastern Carolina superterrane, and it extends from Georgia to

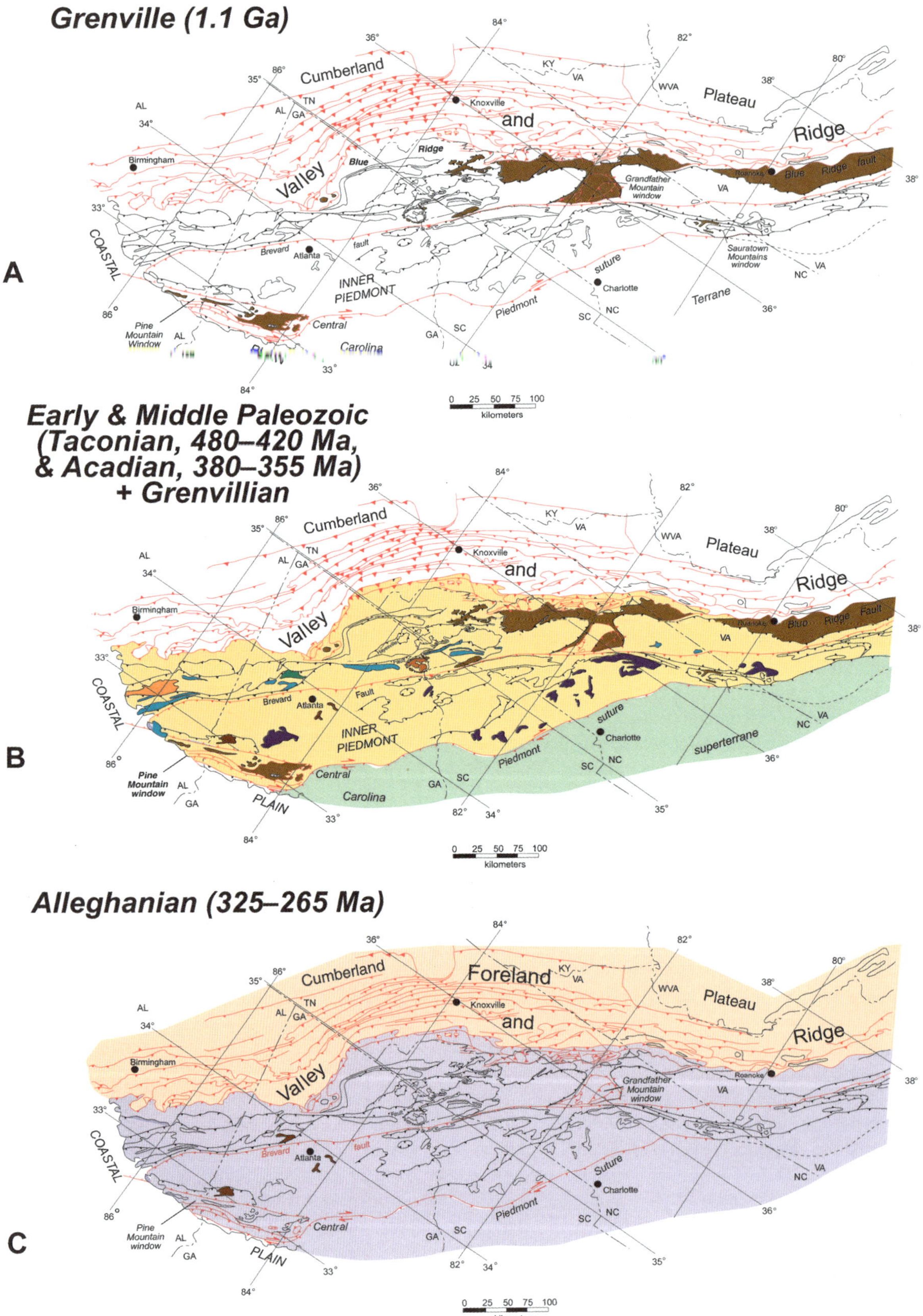

Figure 14. Evolution of Appalachian "basements" (modified from Hatcher, 1999). (A) Grenvillian basement. (B) Taconian and Neoacadian basement. (C) Alleghanian crust.

the eastern Virginia Piedmont [Fig. 1A(1)]. It was also generated during the early and middle stages of collision and likely derived from the present–day latitude of New Jersey (Bartholomew and Tollo, 2004). Low-grade Carolina superterrane rocks were metamorphosed to middle- and upper-amphibolite-facies assemblages. In South Carolina and Georgia, the northwest flank of this metamorphic belt consists of the dextral Modoc fault, whereas the down-to-the-south Augusta fault comprises the southeastern flank (Maher et al., 1994). The Modoc fault has been traced southwestward to the southeastern flank of the Pine Mountain window in Georgia and Alabama where it joins the Dean Creek fault (Hooper et al., 1997; McBride et al., 2005).

The collision process, with its accompanying faulting and other indications of compression, caused the mountain chain interior to be uplifted beginning in the late Mississippian (325–320 Ma). The uplift process was complex and is recorded by current indicators, variable sediment thickness, and decreased maturity of sediments that demonstrate the orogen was unroofed piecemeal from Alabama to Pennsylvania (Thomas *in* Hatcher et al., 1989; Faill, 1997b, 1998; Thomas et al., 2004; Becker et al., 2005). At this time, a near-sea-level delta formed on the western flank of the chain on the Laurentian margin, and the first sediments that contain Paleozoic detrital zircons were deposited from Alabama to Pennsylvania (Thomas et al., 2004; Becker et al., 2005). The transition from marine to nonmarine deltaic depositional conditions took place from the late Mississippian into the early Pennsylvanian, and abundant land plants formed the coal and associated sedimentary rocks of the Appalachian basin (Thomas, *in* Hatcher et al., 1989).

The final stage of the Alleghanian scenario involved head-on (west-to-east, present coordinates) collision of southeastern Laurentia with Gondwana (Fig. 2B). The major product of head-on collision is the Blue Ridge-Piedmont megathrust sheet that transported crust amalgamated during all previous Paleozoic events at least 350 km onto the North American platform (Fig. 14). The Brevard fault remained a suitably oriented weak fault and was again reactivated, this time as a brittle out-of-sequence thrust in the interior of the megathrust sheet (Hatcher, 1971a, 2001). The late Mississippian to Permian clastic wedge was ultimately deformed as the Blue Ridge Piedmont megathrust sheet pushed into the continent deforming the platform into a foreland fold-thrust belt [Fig. 1A(1), Plate 1], and deforming the delta complex that had formed to the west of the rising mountain chain. While final collision was head-on and the megathrust sheet moved westward, uplift of the interior of the chain was nonuniform. The oldest parts of the delta complex are in the southern part of the orogen, possibly related to uplift and erosion of the earlier Ouachita orogen (Becker et al., 2005). This part of the foredeep was superceded to the north by a younger delta complex that fed southwestward along the trough of the foredeep (Thomas, *in* Hatcher et al., 1989). Contrasts in late Mississippian-Pennsylvanian sequences in the southern and central Appalachians also indicate the Alleghanian event began earlier in the south (Faill, 1998). Map-scale overprinting relationships at the transition between the southern and central Appalachians suggest that deformation was also earlier in the southern Appalachians (Dean et al., 1988; Evans and Johnson, 1993).

The collision ended ~265 Ma, forming supercontinent Pangea, and completing the Paleozoic Wilson Cycle of opening and closing of oceans. The subsurface Suwannee terrane south of the almost east-west-trending suture in southern Georgia and Alabama consists of Gondwanan basement and cover Ordovician to Devonian sedimentary rocks of West African affinity (Pojeta et al., 1976; Mueller et al., 1994). These rocks were left on the Laurentian side in the wake of the breakup of Pangea. The southern and central Appalachians constitute a microcosm in the assembly of all existing continents along the huge suture zone that extended from southeastern Laurentia through northwest Africa, western Europe, and the Ural Mountains in Russia.

BREAKUP OF PANGEA

The breakup of Pangea began during the Late Triassic, some 50 m.y. following amalgamation, beginning the initial stages of opening of the present Atlantic Ocean. The assembly of Pangea took place over a span of at least 100 m.y., but, paradoxically, its breakup occurred over a much shorter time span. This contrasts with the 200 m.y. or more time frame for assembling Rodinia and the 300 m.y. that it took for it to break apart. While lithospheric mantle processes were responsible for breakup of both supercontinents, either different processes triggered breakup or the same process operated at different rates.

CONCLUSIONS

1. A new tectonic map has been compiled for the exposed and subsurface southern and central Appalachians. It is intended to illustrate the tectonic history of this part of the orogen, which has recorded part or all of two complete Wilson cycles separating three Paleozoic orogenies.

2. Compilation was aided by large amounts of detailed geologic mapping; modern geochronologic, isotopic, and felsic and mafic igneous rocks geochemical data; and compilations of aeromagnetic and gravity data.

3. Major faults separating tectonic and tectonostratigraphic units are shown, with an attempt to indicate the best state of knowledge of time(s) of movement. Provenance of tectonostratigraphic terranes, based on the best available data (commonly SHRIMP analyses of detital zircons), and ages of plutons are also depicted on the map.

ACKNOWLEDGMENTS

Research support was provided by National Science Foundation Grants GA-1409, GA-20326, EAR-7615564, EAR-7826316, EAR-7911802, EAR-810852, EAR-8206949, EAR-8305832, EAR-8417894, EAR-8816343, EAR-9004604 and EAR-9814800 (to Hatcher), several grants to Hatcher from the EDMAP program managed by the U.S. Geological Survey (Grants 1434HQ97AG01720, 98HQAG2026, 99HQAG0026, 00HQAG0035, 01HQPA0004, 03HQAG0021, 04HQAG0026, 05HQAG0004, 06HQPA0004), and a U.S. Department of Energy, Fossil Energy Program grant (DE-FC26-02NT15341) to Hatcher. The tectonic map [Fig. 1A(1); Plate 1] is an extensive revision of the tectonic map by Hatcher et al. (1990). A.K. Sinha graciously shared a digital map of plutons in Virginia compiled from Evans and Johnson (1993), and M.W. Higgins and R.A. Crawford shared critical unpublished maps of the Georgia Piedmont. Constructive reviews of the paper and tectonic map by R.T. Faill and J.F. Tull resulted in considerable improvement of both. Discussions of central Appalachians geology with D.U. Wise and G.R. Gates also enhanced Hatcher's geologic knowledge about that region. The authors, however, remain culpable for all errors of fact or interpretation.

REFERENCES CITED

Absher, B.S., and McSween, H.Y., 1985, Granulites at Winding Stair Gap, North Carolina: The thermal axis of Paleozoic metamorphism in the southern Appalachians: Geological Society of America Bulletin, v. 96, p. 588–599, doi: 10.1130/0016-7606(1985)96<588:GAWSGN>2.0.CO;2.

Adams, G.I., Butts, C., Stephenson, L.W., and Cooke, C.W., eds., 1926, Geology of Alabama: Alabama Geological Survey Special Report 14, 312 p.

Aleinikoff, J.N., Zartman, R.E., Walter, M., Rankin, D.W., Lyttle, P.T., and Burton, W.C., 1995, U-Pb ages of metarhyolites of the Catoctin and Mt. Rogers Formations, central and southern Appalachians: Evidence of two pulses of Iapetan rifting: American Journal of Science, v. 295, p. 428–454.

Aleinikoff, J.N., Horton, J.W., Jr., Drake, A.A., Jr., and Fanning, C.M., 2002, SHRIMP and conventional U-Pb ages of Ordovician granites and tonalites in the Appalachian Piedmont; implications for Paleozoic tectonic events: American Journal of Science, v. 302, p. 50–75, doi: 10.2475/ajs.302.1.50.

Aleinikoff, J.N., Southworth, S., and Fanning, C.M., 2006, Geochronologic constraints on the timing of deposition of the Ocoee Supergroup: SHRIMP U-Pb dating of xenotime and monazite overgrowths on detrital minerals: Geological Society of America Abstracts with Programs, v. 38, no. 3, p. 65.

Ausburn, M.P., Hatcher, R.D., Jr., Bryan, J.G., Godfrey, S.E., and Acker, L.L., 1998, Geologic map and mineral resources summary of the Topton quadrangle, North Carolina: North Carolina Geological Survey, GM 159-NW, scale 1/24,000.

Barker, C.A., Secor, D.T., Pray, J.R., and Wright, J.E., 1998, Age and deformation of the Longtown metagranite, South Carolina Piedmont—A possible constraint on the origin of the Carolina terrane: The Journal of Geology, v. 106, p. 713–725.

Bartholomew, M.J., and Tollo, R.P., 2004, Northern ancestry for the Goochland terrane as a displaced fragment of Laurentia: Geology, v. 32, no. 8, p. 669–672.

Becker, T.P., Thomas, W.A., Samson, S.D., and Gehrels, G.E., 2005, Detrital zircon evidence of Laurentian crustal dominance in the lower Pennsylvanian deposits of the Alleghanian clastic wedge in eastern North America: Sedimentary Geology, v. 182, p. 59–86, doi: 10.1016/j.sedgeo.2005.07.014.

Bentley, R.D., and Neathery, T.N., 1970, Geology of the Brevard Zone and Related Rocks of the Inner Piedmont of Alabama: Tuscaloosa, Alabama Geological Society, Eighth Annual Field Trip Guidebook, 119 p.

Bergström, S.M., Huff, W.D., Saltzman, M.R., Kolata, D.R., and Leslie, S.A., 2004, The greatest volcanic ash falls in the Phanerozoic: Trans-Atlantic relations of the Ordovician Millbrig and Kinnekulle K-bentonites: The Sedimentary Record, v. 2, no. 4, p. 4–8.

Bradley, P., Gay, K., and Clark, T., 2006, An overview of new geologic mapping of the Chapel Hill, Hillsborough and Efland 7.5-minute quadrangles, Orange and Durham Counties, Carolina terrane, North Carolina: Raleigh, North Carolina Geological Survey, Carolina Geological Society Field Trip guidebook, p. 1–16.

Bream, B.R., 2003, Tectonic Implications of Geochronology and Geochemistry of Para- and Orthogneisses from the Southern Appalachian Crystalline Core [Ph.D. dissertation]: Knoxville, University of Tennessee, 296 p.

Bream, B.R., Hatcher, R.D., Jr., Miller, C.F., and Fullagar, P.D., 2000, Paragneiss geochemistry and preliminary ion microprobe geochronology of zircons from the southern Appalachian crystalline core: Geological Society of America Abstracts with Programs, v. 32, no. 7, p. A-31.

Bream, B.R., Hatcher, R.D., Jr., Miller, C.F., and Fullagar, P.D., 2001, Geochemistry and provenance of Inner Piedmont paragneisses, NC and SC: Evidence for an internal terrane boundary?: Geological Society of America Abstracts with Programs, v. 33, no. 2, p. A-65.

Bream, B.R., Hatcher, R.D., Jr., Miller, C.F., and Fullagar, P.D., 2004, Detrital zircon ages and Nd isotopic data from the southern Appalachian crystalline core, GA-SC-NC-TN: New provenance constraints for part of the Laurentian margin, *in* Tollo, R.P., Corriveau, L., McLelland, J., and Bartholomew, M.J., eds., Proterozoic Evolution of the Grenville Orogen in North America: Geological Society of America Memoir 197, p. 459–475.

Broadhead, T.W., Hatcher, R.D., Jr., and Costello, J.O., 1991, Tectonic and stratigraphic implications of mid–Paleozoic(?) fossils from the Late Proterozoic(?) Walden Creek Group rocks in the Foothills Belt, southeastern Tennessee, *in* Kish, S.A., ed., Studies of Precambrian and Paleozoic Stratigraphy in the Western Blue Ridge: Raleigh, Carolina Geological Society Guidebook, p. 39–44.

Butler, J.R., and Fullagar, P.D., 1978, Petrochemical and geochronological studies in the southern Appalachians: III. Leucocratic adamellites of the Charlotte belt near Salisbury, North Carolina: Geological Society of America Bulletin, v. 89, p. 460–466, doi: 10.1130/0016-7606(1978)89<460:PAGSOP>2.0CO;2.

Carrigan, C.W., Bream, B.R., Miller, C.F., and Hatcher, R.D., Jr., 2001, Ion microprobe analyses of zircon rims from the eastern Blue Ridge and Inner Piedmont, NC-SC-GA: Implications for the timing of Paleozoic metamorphism in the southern Appalachians: Geological Society of America Abstracts with Programs, v. 33, no. 2, p. A-7.

Carrigan, C.W., Miller, C.F., Fullagar, P.D., Hatcher, R.D., Jr., Bream, B.R., and Coath, C.D., 2003, Ion microprobe age and geochemistry of southern Appalachian basement, with implications for Proterozoic and Paleozoic reconstructions: Precambrian Research, v. 120, p. 1–36, doi: 10.1016/S0301-9268(02)00113-4.

Cawood, P.A., and Nemchin, A.A., 2001, Paleogeographic development of the east Laurentian margin: Constraints from U-Pb dating of detrital zircons in the Newfoundland Appalachians: Geological Society of America Bulletin, v. 113, p. 1234–1246.

Cawood, P.A., McCausland, P.J.A., and Dunning, G.R., 2001, Opening Iapetus: Constraints from the Laurentian margin in Newfoundland: Geological Society of America Bulletin, v. 113, p. 443–453, doi: 10.1130/0016-7606(2001)113<0443:OICFTL>2.0CO;2.

Christensen, N.I., and Szymanski, D.L., 1988, Origin of reflections from the Brevard fault zone: Journal of Geophysical Research, v. 93, p. 1087–1102.

Cloud, P.E., Wright, J., and Glover, L., III, 1976, Traces of animal life from 620–million–year old rocks in North Carolina: American Scientist, v. 64, p. 396–406.

Coler, D.G., Wortman, G.L., Samson, S.D., Hibbard, J.P., and Stern, R., 2000, U-Pb, Nd isotopic, and geochemical evidence for the correlation of the Chopawamsic and Milton terranes, Piedmont zone, southern Appalachian orogen: The Journal of Geology, v. 108, p. 363–380, doi: 10.1086/314411.

Colquhoun, D.J., Popenoe, P., Dillon, W.P., Valentine, P.C., Huddlestun, P.F., Poag, C.W., Hatcher, R.D., Jr., and Arthur, M.A., 1992, Coastal–shelf stratigraphy and structure along COSUNA II, southeastern United States Atlantic Coast Corridor, South Carolina and Georgia: Tulsa, Oklahoma, American Association of Petroleum Geologists.

Costain, J.K., Hatcher, R.D., Jr., Çoruh, C., Pratt, T.L., Taylor, S.R., Litehiser, J.J., and Zietz, I., 1989, Geophysical characteristics of the Appalachian crust, Chapter 8, *in* Hatcher, R.D., Jr., Thomas, W.A., and Viele, G.W., eds., The Appalachian-Ouachita Orogen in the United States: Boulder, Colorado, Geological Society of America, The Geology of North America, v. F-2, p. 385–416.

Costello, J.O., 1984, Relationships between the Cartersville Fault and Great Smoky Fault in the Southern Appalachians: A Reinterpretation [M.S. thesis]: Columbia, University of South Carolina, 75 p.

Craddock, J.P., and van der Pluijm, B.A., 1989, Late Paleozoic deformation of the cratonic carbonate cover of eastern North America: Geology, v. 17, p. 416–419, doi: 10.1130/0091-7613(1989)017<0416:LPDOTC>2.3CO;2.

Dallmeyer, R.D., Wright, J.E., Secor, D.T., Jr., and Snoke, A.W., 1986, Character of the Alleghanian orogeny in the southern Appalachians: Part II. Geochronological constraints on the tectonothermal evolution of the eastern Piedmont in South Carolina: Geological Society of America Bulletin, v. 97, p. 1329–1344, doi: 10.1130/0016-7606(1986)97<1329:COTAOI>2.0.CO;2.

Davis, T.L., 1993, Lithostratigraphy, Structure, and Metamorphism of a Crystalline Thrust Terrane, Western Inner Piedmont, North Carolina, [Ph.D. dissertation]: Knoxville, Tennessee, University of Tennessee, 245 p.

Davis, T.L., Hatcher, R.D., Jr., Liu, A., and Tabor, J.R., 1991, Southern Appalachian western Inner Piedmont: A progressive crustal-scale shear zone: Geological Society of America Abstracts with Programs, v. 23, no. 5, p. A138.

Dean, S.L., Kulander, B.R., and Skinner, J.M., 1988, Structural chronology of the Alleghanian orogeny in southeastern West Virginia: Geological Society of America Bulletin, v. 100, p. 299–310, doi: 10.1130/0016-7606(1988)100<0299:SCOTAO>2.3.CO;2.

Dennis, A.J., 2004, Trans Appalachians internides geotraverse: Day 4, Overview of recent developments in the geology of the central Piedmont of South Carolina, *in* Merschat, A. J., and Hatcher, R. D., eds., Jr., 2004, Trans Appalachians internides geotraverse: Oak Ridge, Tennessee, 17th International Basement Tectonics Conference Field Trip Guide, p. 67–84.

Dennis, A.J., 2007, Cat Square basin, Catskill clastic wedge: Silurian-Devonian orogenic events in the central Appalachians and the crystalline southern Appalachians, *in* Sears, J.W., Harms, T.A., and Evenchick, C.A., eds., Whence the mountains? Inquiries into the evolution of orogenic systems: A volume in honor of Raymond A. Price: Boulder, Colorado, Geological Society of America Special Paper 433, doi: 10.1130/2007.2433(15) (in press).

Dennis, A.J., and Wright, J.E., 1997, The Carolina terrane in northwestern South Carolina, U.S.A.: Late Precambrian-Cambrian deformation and metamorphism in a peri-Gondwanan oceanic arc: Tectonics, v. 16, p. 460–473, doi: 10.1029/97TC00449.

Dennis, A.J., Secor, D.T., and Shervais, J.W., 2000, Fault-bounding eclogite-bearing gneisses, Newberry, South Carolina: Geological Society of America Abstracts with Programs, v. 32, no. 2, p. A–14.

Dewey, J.F., Holdsworth, R.E., and Strachan, R.A., 1998, Transpression and transtension zones, *in* Holdsworth, R.E., Strachan, R.A., and Dewey, J.F., eds., Continental Transpressional and Transtensional Tectonics: Geological Society [London] Special Publication 135, p. 1–14.

Drake, A.A., 1970, Structural geology of the Reading Prong, *in* Fisher, G.W., Pettijohn, F.J., Reed, J.C., and Weaver, K.N., eds., Studies of Appalachian geology: Central and southern: New York, Wiley-Interscience, p. 271–294.

Drake, A.A., Jr., Sinha, A.K., Laird, J., and Guy, R.E., 1989, The Taconic orogen, *in* Hatcher, R.D., Jr., Thomas, W.A., and Viele, G.W., eds., The Appalachian–Ouachita Orogen in the United States: Boulder, Colorado, Geological Society of America, The Geology of North America, v. F–2, p. 101–177.

Eckert, J.O., Jr., 1984, Stratigraphy, Structure, and Metamorphism of the East Half of the Wayah Bald Quadrangle, North Carolina: Evidence for Paleozoic Granulite Facies Metamorphism in the Southern Appalachians [M.S. thesis]: Columbia, South Carolina, University of South Carolina, 411 p.

Eckert, J.O., Hatcher, R.D., Jr., and Mohr, D.W., 1989, The Wayah granulite–facies metamorphic core, southwestern North Carolina: High–grade culmination of Taconic metamorphism in the southern Blue Ridge: Geological Society of America Bulletin, v. 101, p. 1434–1447, doi: 10.1130/0016-7806(1989)101<1434:TWGFMC>2.3.CO;2.

Espenshade, G.H., Rankin, D.W., Shaw, K.W., and Neuman, R.B., 1975, Geologic Map of the East Half of the Winston-Salem Quadrangle, North Carolina-Virginia: U.S. Geological Survey Miscellaneous Investigations Map I-709-B, scale 1:250,000.

Ettensohn, F.R., 1985, The Catskill delta complex and the Acadian orogeny: A model, *in* Woodrow, D.L., and Sevon, W.D., eds., The Catskill delta: Geological Society of America Special Paper 201, p. 39–49.

Ettensohn, F.R., 2004, Modeling the nature and development of major Paleozoic clastic wedges in the Appalachian basin, USA: Journal of Geodynamics, v. 37, p. 657–681, doi: 10.1016/j.jog.2004.02.009.

Evans, N.H., and Johnson, S.S., 1993, Geologic map of Virginia: Virginia Division of Mineral Resources, scale 1:500,000.

Faill, R.T., 1997a, A geologic history of the north-central Appalachians: Part 1. Orogenesis from the Mesoproterozoic through the Taconic orogeny: American Journal of Science, v. 297, p. 551–569.

Faill, R.T., 1997b, A geologic history of the north-central Appalachians: Part 2: The Appalachian basin from the Silurian through the Carboniferous: American Journal of Science, v. 297, p. 729–761.

Faill, R.T., 1998, A geologic history of the north-central Appalachians: Part 3: The Alleghany orogeny: American Journal of Science, v. 298, p. 131–179.

Farrar, S.S., 1985, Tectonic evolution of the easternmost Piedmont, North Carolina: Geological Society of America Bulletin, v. 96, p. 362–380, doi: 10.1130/0016-7606(1985)96<362:TEOTEP>2.0.CO;2.

Force, E.R., 1976, Metamorphic Source Rocks of Titanium Placer Deposits—a Geochemical Cycle: U.S. Geological Survey Professional Paper 959–B, 13 p.

Fullagar, P.D., Goldberg, S.A., and Butler, J.R., 1997, Nd and Sr isotopic characterization of crystalline rocks from the southern Appalachian Piedmont and Blue Ridge, North and South Carolina, *in* Sinha, A.K., Whalen, J.B., and Hogan, J.P., eds., The Nature of Magmatism in the Appalachian Orogen: Geological Society of America Memoir 191, p. 165–179.

Ganis, G.R., Williams, S.H., and Repetski, J.R., 2001, New biostratigaphic evidence from the eastern part of the Hamburg klippe, Pennsylvania, and its significance for interpreting the depositional and tectonic history of the klippe: Geological Society of America Bulletin, v. 113, p. 109–128, doi: 10.1130/0016-7606(2001)113<0109:NBIFTW>2.0.CO;2.

Gastaldo, R.A., 1995, New occurrence of *Periastron reticulatum* Unger emend. Beck, an enigmatic Mississippian fossil plant: Journal of Paleontology, v. 69, p. 388–392.

Gatewood, M.P., 2007, Tectonics of the Northeastern Inner Piedmont, Northwestern North Carolina, from Detailed Geologic Mapping, Geochronologic, Geochemical, and Petrologic Studies with Macro- and Mesostructural Analyses of Ductile Fault Zones [M.S. thesis]: Knoxville, University of Tennessee, 279 p.

Geiser, P., and Engelder, T., 1983, The distribution of layer-parallel shortening fabrics in the Appalachian foreland of New York and Pennsylvania: Evidence for two non-coaxial phases of the Alleghanian orogeny, *in* Hatcher, R.D., Jr., Williams, H., and Zietz, I., eds., Contributions to the Tectonics and Geophysics of Mountain Chains: Geological Society of America Memoir 158, p. 161–176.

German, J.M., 1985, The Geology of the Northeastern Portion of the Dahlonega Gold Belt: Georgia Geologic Survey Bulletin 100, 41 p.

Giorgis, S.D., 1999, Inner Piedmont Geology of the Northwestern South Mountains near Morganton, North Carolina [M.S. thesis]: Knoxville, University of Tennessee, 191 p.

Giorgis, S.D., Mapes, R.W., and Bream, B.R., 2002, The Walker Top Granite: Acadian granitoid or eastern Inner Piedmont basement? *in* Hatcher, R.D., Jr., and Bream, B.R., eds., Inner Piedmont Geology in the South Mountains-Blue Ridge Foothills and the Southwestern Brushy Mountains, central-western North Carolina: Knoxville, Tennessee, Carolina Geological Society Guidebook, p. 33–43.

Goldsmith, R., Milton, D.J., and Horton, J.W., Jr., 1988, Geologic Map of the Charlotte 1° x 2° Quadrangle, North Carolina and South Carolina: U.S. Geological Survey Map I-1251-E, scale 1:250.000.

Gray, M.B., and Zeitler, P.K., 1997, Comparison of clastic wedge provenance in the Appalachian foreland using U/Pb ages of detrital zircons: Tectonics, v. 16, p. 151–160.

Gulley, G.L., Jr., 1985, A Proterozoic granulite facies terrane on Roan Mountain, western Blue Ridge belt, North Carolina–Tennessee: Geological Society of America Bulletin, v. 96, p. 1428–1439, doi: 10.1130/0016-7606(1985)96<1428:APGTOR>2.0CO;2.

Hadley, J.B., 1970, The Ocoee series and its possible correlatives, *in* Fisher, G.W., Pettijohn, F.J., Reed, J.C., Jr., and Weaver, K.N., eds., Studies of Appalachian Geology, Central and Southern: New York, Interscience Publishers, p. 247–259.

Hadley, J.B., and Goldsmith, R., 1963, Geology of the eastern Great Smoky Mountains, North Carolina and Tennessee: U.S. Geological Survey Professional Paper 349–B, 118 p.

Hadley, J.B., and Nelson, A.E., 1971, Geologic map of the Knoxville quadrangle, North Carolina, Tennessee, and South Carolina: U.S. Geological Survey Miscellaneous Geologic Investigations Map I-654, scale 1:250,000.

Hamilton, W.B., 1961, Geology of the Richardson Cove and Jones Cove quadrangles, Tennessee: U.S. Geological Survey Professional Paper 349-A, 55 p.

Harris, L.D., 1979, Similarities between the thick-skinned Blue Ridge anticlinorium and the thin-skinned Powell Valley anticline: Geological Society of America Bulletin, v. 90, p. 525–539, doi: 10.1130/0016-7606(1979)90<525:SBTTBR>2.0.CO;2.

Hartley, M.E., III, 1973, Ultramafic and related rocks in the vicinity of Lake Chatuge: Geological Survey of Georgia Bulletin 85, 61 p.

Hatcher, R.D., Jr., 1971a, Structural, petrologic, and stratigraphic evidence favoring a thrust solution to the Brevard problem: American Journal of Science, v. 270, p. 177–202.

Hatcher, R.D., Jr., 1971b, Geology of Rabun and Habersham Counties, Georgia: A reconnaissance study: Georgia Department of Mines, Mining, and Geology Bulletin 83, 48 p.

Hatcher, R.D., Jr., 1972, Developmental model for the southern Appalachians: Geological Society of America Bulletin, v. 83, p. 2735–2760, doi: 10.1130/0016-7606(1972)83[2735:DMFTSA]2.0.CO;2.

Hatcher, R.D., Jr., 1973, Basement versus cover rocks in the Blue Ridge of northeast Georgia, northwestern South Carolina and adjacent North Carolina: American Journal of Science, v. 273, p. 671–685.

Hatcher, R.D., Jr., 1974, North American Paleozoic foldbelts and their deformational histories: A plate tectonics anomaly?: American Journal of Science, v. 274, p. 135–147.

Hatcher, R.D., Jr., 1978, Tectonics of the western Piedmont and Blue Ridge: Review and speculation: American Journal of Science, v. 278, p. 276–304.

Hatcher, R.D., Jr., 1979, The Coweeta Group and Coweeta syncline: Major features of the North Carolina-Georgia Blue Ridge: Southeastern Geology, v. 21, p. 17–29.

Hatcher, R.D., Jr., 1987, Tectonics of the southern and central Appalachian internides: Annual Review of Earth and Planetary Sciences, v. 15, p. 337–362.

Hatcher, R.D., Jr., 1988, Bedrock geology and regional geologic setting of Coweeta Hydrologic Laboratory in the eastern Blue Ridge, with some discussion of Quaternary deposits and structural controls of topography, *in* Swank, W.T., and Crossley, D.A., Jr., eds., Coweeta Symposium Volume 66: New York, Springer-Verlag, p. 81–92.

Hatcher, R.D., Jr., 1989, Tectonic synthesis of the U.S. Appalachians, Chapter 14, *in* Hatcher, R.D., Jr., Thomas, W.A., and Viele, G.W., eds., The Appalachian-Ouachita Orogen in the United States: Boulder, Colorado, Geological Society of America, The Geology of North America, v. F-2, p. 511–535.

Hatcher, R.D., Jr., 1999, Crust-forming processes, *in* Sinha, A.K., ed., Basement tectonics 13: Dordrecht, Netherlands, Kluwer Academic Publishers, p. 99–118.

Hatcher, R.D., Jr., 2001, Rheological partitioning during multiple reactivation of the Paleozoic Brevard fault zone, Southern Appalachians, USA, *in* Holdsworth, R.E., Strachan, R.A., Macloughlin, J.F., and Knipe, R.J., eds., The Nature and Significance of Fault Zone Weakening: Geological Society [London] Special Publication 186, p. 255–269.

Hatcher, R.D., Jr., 2002a, An Inner Piedmont primer, *in* Hatcher, R.D., Jr., and Bream, B.R., eds., Inner Piedmont Geology in the South Mountains and the Southwestern Brushy Mountains, central–western North Carolina: Knoxville, Tennessee, North Carolina Geological Survey, Carolina Geological Society Guidebook, p. 1–18.

Hatcher, R.D., Jr., 2002b, The Alleghanian (Appalachian) orogeny, a product of zipper tectonics: Rotational transpressive continent–continent collision and closing of ancient oceans along irregular margins, *in* Martínez Catalán, J.R., Hatcher, R.D., Jr., Arenas, R., and García, F.D., eds., Variscan-Appalachian Dynamics: The Building of the Late Paleozoic Basement: Geological Society of America Special Paper 394, p. 199–208.

Hatcher, R.D., Jr., and Butler, J.R., 1979, Guidebook for southern Appalachian field trip in the Carolinas, Tennessee and northeastern Georgia: International Geological Correlation Program Project 27, Chapel Hill, University of North Carolina, 117 p.

Hatcher, R.D., Jr., and Hooper, R.J., 1992, Evolution of crystalline thrust sheets in the internal parts of mountain chains, *in* McClay, K.R., ed., Thrust Tectonics: London, Chapman and Hall, p. 217–234.

Hatcher, R.D., Jr., and Merschat, A.J., 2006, The Appalachian Inner Piedmont: An exhumed strike-parallel, tectonically forced orogenic channel, *in* Law, R.D., Searle, M., and Godin, L., eds., Channel Flow, Ductile Extrusion, and Exhumation of Lower-Mid Crust in Continental Collision Zones: Geological Society [London] Special Publication 268, p. 517–540.

Hatcher, R.D., Jr., and Odom, A.L., 1980, Timing of thrusting in the southern Appalachians, USA: Model for orogeny?: Journal of the Geological Society: v. 137, p. 321–327.

Hatcher, R.D., Jr., Howell, D.E., and Talwani, P., 1977, Eastern Piedmont fault system: Speculations on its extent: Geology, v. 5, p. 636–640, doi: 10.1130/0091-7613(1977)5<636:EPFSSO>2.0.CO;2.

Hatcher, R.D., Jr., Hooper, R.J., Petty, S.M., and Willis, J.D., 1984, Structure and chemical petrology of three southern Appalachian mafic-ultramafic complexes and their bearing upon the tectonics of emplacement and origin of Appalachian ultramafic bodies: American Journal of Science, v. 284, p. 484–506.

Hatcher, R.D., Jr., Thomas, W.A., Geiser, P.A., Snoke, A.W., Mosher, S., and Wiltschko, D.V., 1989, Alleghanian orogen, Chapter 5, *in* Hatcher, R.D., Jr., Thomas, W.A., and Viele, G.W., eds., The Appalachian–Ouachita Orogen in the United States: Boulder, Colorado, Geological Society of America, The Geology of North America, v. F-2, p. 233–318.

Hatcher, R.D., Jr., Osberg, P.H., Robinson, P., and Thomas, W.A., 1990, Tectonic map of the U.S. Appalachians, Plate 1 *in* Hatcher, R.D., Jr., Thomas, W.A., and Viele, G.W., eds., The Appalachian–Ouachita Orogen in the United States: Boulder, Colorado, Geological Society of America, The Geology of North America, v. F-2, scale 1/2,500,000.

Hatcher, R.D., Jr., Colquhoun, D.J., Secor, D.T., Jr., Cook, F.A., Dillon, W.P., Klitgord, K., Merschat, C.E., Wiener, L.S., Milici, R.C., Nelson, A.E., Sheridan, R.E., and Snoke, A.W., 1994, Transect E5—Cumberland Plateau to Blake Plateau: Geological Society of America, Decade of North American Geology Centennial Transect E5.

Hatcher, R.D., Jr., Bream, B.R., Miller, C.L., Eckert, J.O., Jr., Fullagar, P.D., and Carrigan, C.W., 2004, Paleozoic structure of southern Appalachian Blue Ridge Grenvillian internal basement massifs, *in* Tollo, R.P., Corriveau, L., McLelland, J., and Bartholomew, M J., eds., Proterozoic evolution of the Grenville orogen in North America: Boulder, Colorado, Geological Society of America Memoir 197, p. 525–547.

Hatcher, R.D., Jr., Bream, B.R., Merschat, A.J., Mapes, R.W., and Miller, C.F., 2005a, Evidence for the (neo-) Acadian orogeny in the southern Appalachians: Geological Society of America Abstracts with Programs, v. 37, no. 2, p. 5.

Hatcher, R.D., Jr., Merschat, A.J., and Thigpen, J.R., 2005b, Blue Ridge primer, *in* Hatcher, R.D., Jr., and Merschat, A.J., eds., Blue Ridge geology geotraverse east of the Great Smoky Mountains National Park, Western North Carolina: Knoxville, Tennessee, Carolina Geological Society Annual Field Trip Guide, p. 1–24.

Hatcher, R.D., Jr., Acker, L.L., and Liu, A., in press, Geology of the Whetstone, Holly Springs, Tugaloo Lake, and Rainy Mountain quadrangles, South Carolina, Georgia: South Carolina Geological Survey, scale 1/24,000.

Heatherington, A.L., Mueller, P.A., Smith, M.S., and Nutman, A.P., 1996, The Corbin Gneiss: Evidence for Grenville magmatism and older continental basement in the southernmost Blue Ridge: Southeastern Geology, v. 36, p. 15–25.

Heatherington, A.L., Mueller, P.A., Kamenov, G., Steltenpohl, M.G., Hanley, T.B., and Wooden, J.L., 2006, Pine Mountain terrane: Geochemistry and geochronology of basement gneisses and xenoliths: Geological Society of America Abstracts with Programs, v. 38, no. 3, p. 63.

Heatherington, A.L., Mueller, P.A., Kamenov, G., and Cherukumilli, K., 2007, Hafnium isotopic evidence for Paleoproterozoic crust in the Mars Hill terrane and comparison with other terranes: Geological Society of America Abstracts with Programs, v. 39, no. 2, p. 79.

Hibbard, J.P., 2000, Docking Carolina: Mid-Paleozoic accretion in the southern Appalachians: Geology, v. 28, p. 127–130, doi: 10.1130/0091-7613(2000)28<127:DCMAIT>2.0.CO;2.

Hibbard, J.P., Shell, G.S., Bradley, P.J., Samson, S.D., and Wortman, G.L., 1998, The Hyco shear zone in North Carolina and southern Virginia: Implications for the Piedmont zone–Carolina zone boundary in the southern Appalachians: American Journal of Science, v. 298, p. 85–107.

Hibbard, J.P., Stoddard, E.F., Secor, D.T., Jr., and Dennis, A.J., 2002, The Carolina zone: Overview of Neoproterozoic to early Paleozoic peri-Gondwanan terranes along the eastern flank of the southern Appalachians: Earth-Science Reviews, v. 57, p. 299–339, doi: 10.1016/S0012-8252(01)00079-4.

Hibbard, J.P., Tracy, R.J., and Henika, W.S., 2003, Smith River allochthon: A southern Appalachian peri-Gondwanan terrane emplaced directly on Laurentia?: Geology, v. 31, p. 215–218, doi: 10.1130/0091-7613(2003)031<0215:SRAASA>2.0.CO;2.

Hibbard, J.P., van Staal, C.R., Rankin, D.W., and Williams, H., 2006a, Lithotectonic map of the Appalachian orogen, Canada–United States of America: Geological Survey of Canada, Map 2096A, scale 1:1,500,000.

Hibbard, J.P., McMenamin, M., Pollock, J., Weaver, P.G., Tacker, R.C., Miller, B.V., Samson, S., and Secor, D., 2006b, Significance of a new Ediacaran fossil find in the Carolina terrane of North Carolina: Geological Society of America Abstracts with Programs, v. 38, no. 2, p. 91.

Hibbard, J.P., McMenamin, M., Pollock, J., Weaver, P.G., Tacker, R.C., Miller, B.V., Samson, S., and Secor, D., 2006c, Significance of a new Ediacaran fossil find and U-Pb zircon ages from the Albermarle Group, Carolina terrane of North Carolina, *in* Bradley, P.J., and Clark, T.W., eds., The Geology of the Chapel Hill, Hillsborough, and Efland 7.5-Minute Quadrangles, Orange and Durham Counties, Carolina Terrane, North Carolina: Raleigh, North Carolina Geological Survey, Carolina Geological Society Field Trip Guidebook, p. 29–33.

Higgins, M.W., Arth, J.G., Wooden, J.L., Crawford, T.J., Stern, T.W., and Crawford, R.E., 1997, Age and origin of the Austell Gneiss, western Georgia Piedmont-Blue Ridge, and its bearing on the ages of orogenic events in the southern Appalachians, *in* Sinha, A.K., Whalen, J.B., and Hogan, J.P., eds., The Nature of Magmatism in the Appalachian Orogen: Boulder, Colorado, Geological Society of America Memoir 191, p. 181–192.

Hoffman, P.F., 1988, United plates of America: The birth of a craton—Early Proterozoic assembly and growth of Laurentia: Annual Review of Earth and Planetary Sciences, v. 16, p. 542–603, doi: 10.1146/annurev.ea.16.050188.002551.

Hoffman, P.F., 1991, Did the breakout of Laurentia turn Gondwanaland inside-out?: Science, v. 252, p. 1409–1412, doi: 10.1126/science.252.5011.1409.

Hooper, R.J., and Hatcher, R.D., Jr., 1990, The Ocmulgee fault: The Piedmont–Avalon terrane boundary in central Georgia: Geology, v. 18, p. 708–711, doi: 10.1130/0091-7613(1990)018<0708:OFTPAT>2.3.CO;2.

Hooper, R.J., Hatcher, R.D., Jr., Troyer, P.K., Dawson, R.J., and Redmond, C.G., 1997, The character of the Avalon terrane and its boundary with the Inner Piedmont in central Georgia, *in* Glover, L., III, and Gates, A.E., eds., Central and Southern Appalachian Sutures: Results from the EDGE Project and Related Studies: Geological Society of America Special Paper 314, p. 1–14.

Hopson, J.L., 1989, Structure, stratigraphy, and petrogenesis of the Lake Burton mafic-ultramafic complex, *in* Fritz, W.J., Hatcher, R.D., Jr., and Hopson, J.L., eds, Geology of the Eastern Blue Ridge of Northeast Georgia and the Adjacent Carolinas: Georgia Geological Society Guidebook, v. 9, no. 3, p. 93–111.

Hopson, J.L., and Hatcher, R.D., Jr., 1988, Structural and stratigraphic setting of the Alto allochthon, northeast Georgia: Geological Society of America Bulletin, v. 100, p. 339–350, doi: 10.1130/0016-7606(1988)100<0339:SASSOT>2.3.CO;2.

Horton, J.W., Jr., and Butler, J.R., 1986, The Brevard fault zone at Rosman, Transylvania County, North Carolina, *in* Neathery, T. L., ed., Southeastern Section of the Geological Society of America Centennial Field Guide, Boulder, Colorado, Geology of North America, v. 6, p. 251–256.

Horton, J.W., Jr., Sutter, J.F., Stern, T.W., and Milton, D.J., 1987, Alleghanian deformation, metamorphism, and granite emplacement in the central Piedmont of the southern Appalachians: American Journal of Science, v. 287, p. 635–660.

Horton, J.W., Jr., Drake, A.A., Jr., and Rankin, D.W., 1989, Tectonostratigraphic terranes and their Paleozoic boundaries in the central and southern Appalachians, *in* Dallmeyer, R.D., ed., Terranes in the Circum-Atlantic Paleozoic Orogens: Geological Society of America Special Paper 230, p. 213–245.

Hurst, V.J., 1955, Stratigraphy, Structure, and Mineral Resources of the Mineral Bluff Quadrangle, Georgia: Atlanta, Georgia Geological Survey Bulletin 63, 137 p.

Hurst, V.J., and Schlee, J.S., 1962, Ocoee metasediments, north-central Georgia, southeastern Tennessee: Geological Society of America Southeastern Section Guidebook no. 3: Atlanta, Georgia Department of Mines, Mining, and Geology, 28 p.

Ingle, S., Mueller, P.A., Heatherington, A.L., and Kozuch, M., 2003, Isotopic evidence for the magmatic and tectonic histories of the Carolina terrane: Implications for stratigraphy and terrane accretion: Tectonophysics, v. 371, p. 187–211, doi: 10.1016/S0040-1951(03)00228-2.

Jones, R.R., Holdsworth, R.E., and Bailey, W., 1997, Lateral extrusion in transpression zones: the importance of boundary conditions: Journal of Structural Geology, v. 19, p. 1201–1217, doi: 10.1016/S0191-8141(97)00034-5.

Karlstrom, K.E., Åhäll, K.-I., Harlan, S.S., Williams, M.L., McLelland, J., and Geissman, J.W., 2001, Long-lived (1.8–1.0 Ga) convergent orogen in southern Laurentia, its extensions to Australia and Baltica, and implications for refining Rodinia: Precambrian Research, v. 111, p. 5–30, doi: 10.1016/S0301-9268(01)00154-1.

Kath, R.L., and Crawford, T.J., 2006, New insight into the Brevard zone in Georgia: Results from detailed surface and subsurface geologic mapping of three large tunnels in the Metro-Atlanta area of Georgia: Geological Society of America Abstracts with Programs, v. 38, no. 3, p. 79.

Kellberg, J.M., and Grant, L.F., 1956, Coarse conglomerates of the Middle Ordovician in the southern Appalachian Valley: Geological Society of America Bulletin, v. 67, p. 697–716.

King, E.R., and Zietz, I., 1978, The New York–Alabama lineament: Geophysical evidence for a major crustal boundary in the basement beneath the Appalachian basin: Geology, v. 6, p. 312–318, doi: 10.1130/0091-7613(1978)6<312:TNYLGE>2.0.CO;2.

King, P.B., 1964, Geology of the Central Great Smoky Mountains, Tennessee: U.S. Geological Survey Professional Paper 349–C, 148 p.

King, P.B., 1969, Tectonic Map of North America: U.S. Geological Survey, scale 1:5,000,000.

King, P.B., 1970, Epilogue, *in* Fisher, G.W., Pettijohn, F.J., Reed, J.C., Jr., and Weaver, K.N., eds., Studies of Appalachian Geology: Central and Southern: New York, Wiley-Interscience, p. 437–439.

King, P.B., and Ferguson, H.W., 1960, Geology of Northeasternmost Tennessee: U.S. Geological Survey Professional Paper 311, 136 p.

King, P.B., Hadley, J.B., Neuman, R.B., and Hamilton, W.B., 1958, Stratigraphy of the Ocoee series, Great Smoky Mountains, Tennessee and North Carolina: Geological Society of America Bulletin, v. 69, p. 947–966, doi: 10.1130/0016-7606(1958)69[947:SOOSGS]2.0.CO;2.

Kish, S.A., 1983, A Geochronological Study of Deformation and Metamorphism in the Blue Ridge and Piedmont of the Carolinas [Ph.D. dissertation]: Chapel Hill, University of North Carolina, 220 p.

Koeppen, R.P., Repetski, J.E., and Weary, D.J., 1995, Microfossil assemblages indicate Ordovician or Late Cambrian age for Tillery Formation and mudstone member of Cid Formation, Carolina Slate belt, North Carolina: Geological Society of America Abstracts with Programs, v. 27, no. 6, p. 397.

Kröner, A., Windley, B.F., Badarch, G., Tomurtogoo, O., Heger, E., Jaahn, B.M., Gruschka, S., Khain, E.V., Demoux, A., and Wingate, M.T.D., 2007, this volume, Accretionary growth and crust-formation in the central Asian orogenic belt and comparison with the Arabian-Nubian Shield, *in* Hatcher, R.D., Jr., Carlson, M.P., McBride, J.H., and Martínez Catalán, J.R., The 4-D Framework of Continental Crust: Geological Society of America Memoir 200, doi: 10.1130/2007.1200(11).

Kulander, B.R., and Dean, S.L., 1986, Structure and tectonics of central and southern Appalachian Valley and Ridge and Plateau provinces, West Virginia and Virginia: American Association of Petroleum Geologists Bulletin, v. 70, p. 1674–1684.

Lash, G.G., and Drake, A.A., Jr., 1984, The Richmond and Greenwich Slices of the Hamburg Klippe in Eastern Pennsylvania—Stratigraphy, Sedimentology, and Plate Tectonic Implications: U.S. Geological Survey Professional Paper 1312, 40 p.

Loewy, S.L., Connelly, J.N., Dalziel, I.W.D., and Gower, C.F., 2003, Eastern Laurentia in Rodinia: constraints from whole-rock Pb and U/Pb geochronology: Tectonophysics, v. 375, p. 169–197, doi: 10.1016/S0040-1951(03)00338-X.

Maher, H.D., Jr., Dallmeyer, R.D., Secor, D.T., Jr., and Sacks, P.E., 1994, ^{40}Ar/^{39}Ar constraints on chronology of Augusta fault zone movement and late Alleghanian extension, southern Appalachian Piedmont, South Carolina and Georgia: American Journal of Science, v. 294, p. 428–448.

Mapes, R.W., 2002, Geochemistry and geochronology of mid-Paleozoic granitic plutonism in the southern Appalachian Piedmont terrane, North Carolina-South Carolina-Georgia [M.S. thesis]: Nashville, Vanderbilt University, 150 p.

Mapes, R.W., Maybin, A.H., Miller, C.F., Fullagar, P.D., and Bream, B.R., 2002, Geochronology and geochemistry of mid-Paleozoic granitic magmatism, central and eastern Inner Piedmont, North Carolina and South Carolina: Geological Society of America Abstracts with Programs, v. 34, no. 2, p. 92.

Marshak, S., 1986, Structure and tectonics of the Hudson Valley fold-thrust belt eastern New York State: Geological Society of America Bulletin, v. 97, p. 354–368, doi: 10.1130/0016-7606(1986)97<354:SATOTH>2.0.CO/2.

McBride, J.H., Hatcher, R.D., Jr., and Stephenson, W.J., 2005, Integrating seismic reflection and geological data and interpretations across an internal basement massif: The southern Appalachian Pine Mountain window, USA: Geological Society of America Bulletin, v. 117, p. 669–686, doi: 10.1130/B25313.1.

McBride, J.H., Pugin, A.J.M., and Hatcher, R.D., Jr., 2007, this volume, Scale independence of décollement thrusting, *in* Hatcher, R.D., Jr., Carlson, M.P., McBride, J.H., and Martínez Catalán, J.R., eds., The 4-D Framework of Continental Crust: Geological Society of America Memoir 200, doi: 10.1130/2007:1200(07).

McClellan, E.A., Steltenpohl, M.G., Thomas, C., and Miller, C.F., 2007, Isotopic age constraints and metamorphic history of the Talladega belt: New evidence for timing of arc magmatism and terrane emplacement along the southern Laurentian margin: The Journal of Geology, v.115, p. 541–561.

McConnell, K.I., 1990, Geology and Geochronology of the Sauratown Mountains Anticlinorium, Northwestern North Carolina [Ph.D. thesis]: Columbia, University of South Carolina, 232 p.

McDowell, S.M., Miller, C.F., Fullagar, P.D., Bream, B.R., and Mapes, R.W., 2002, The Persimmon Creek Gneiss, eastern Blue Ridge, North Carolina-Georgia: Evidence for the missing Taconic arc?: Southeastern Geology, v. 41, p. 103–117.

McSween, H.Y., and Hatcher, R.D., Jr., 1985, Ophiolites (?) of the southern Appalachian Blue Ridge, *in* Woodward, N.B., ed., Field Trips in the Southern Appalachians: University of Tennessee Department of Geological Sciences Studies in Geology 9, p. 144–170.

McSween, H.Y., Jr., Sando, T.W., Clark, S.R., Harden, J.T., and Strange, E.A., 1984, The gabbro-metagabbro association of the southern Appalachian Piedmont: American Journal of Science, v. 284, p. 437–461.

McSween, H.Y., Speer, J.A., and Fullagar, P.D., 1991, Plutonic rocks, *in* Horton, J.W., Jr., and Zullo, V.A., eds., The Geology of the Carolinas—Carolina Geological Society 50th Anniversary Volume: Knoxville, Tennessee, The University of Tennessee Press, p. 109–126.

Merschat, A.J., and Hatcher, R.D., Jr., 2007, this volume, The Cat Square terrane: Possible Siluro-Devonian remnant ocean basin in the Inner Piedmont, southern Appalachians, *in* Hatcher, R.D., Jr., Carlson, M.P., McBride, J.H., and Martínez Catalán, J. R., eds., The 4-D Framework of Continental Crust: Geological Society of America Memoir 200, doi: 10.1130/2007.1200(27).

Merschat, A.J., Hatcher, R.D., Jr., and Davis, T.L., 2005a, The northern Inner Piedmont, USA: kinematics of transpression and SW-directed mid-crustal flow: Journal of Structural Geology, v. 27, p. 1252–1281, doi: 10.1016/j.jsg.2004.08.005.

Merschat, A.J., Gatewood, M.P., Fisher, C.M., Miller, C.F., Hatcher, R.D., Jr., Wooden, J.L., and Stahr, D.W., III, 2005b, The Newton antiform, North Carolina: A previously unrecognized window through an allochthonous Inner Piedmont thrust stack: Geological Society of America Abstracts with Programs, v. 37, no. 7, p. 20.

Merschat, A.J., Hatcher, R.D., Jr., Miller, C.F., and Wooden, J.L., 2006, Provenance of central and western Blue Ridge paragneisses, southwestern North Carolina: Geological Society of America Abstracts with Programs, v. 38, no. 3, p. 63.

Miller, B.V., Fetter, A.H., and Stewart, K.G., 2006, Plutonism in three orogenic pulses, eastern Blue Ridge province, southern Appalachians: Geological Society of America Bulletin, v. 118, p. 171–184, doi: 10.1130/B25580.1.

Miller, C.F., Hatcher, R.D., Jr., Harrison, T.M., Coath, C.D., and Gorisch, E.B., 1998, Cryptic crustal events elucidated through zone imaging and ion microprobe studies of zircon, southern Appalachian Blue Ridge, North Carolina–Georgia: Geology, v. 26, p. 419–422, doi: 10.1130/0091-7613(1998)026<0419:CCEETZ>2.3.CO;2.

Miller, C.F., Hatcher, R.D., Jr., Ayers, J.C., Coath, C.D., and Harrison, T.M., 2000, Age and zircon inheritance of eastern Blue Ridge plutons, southwestern North Carolina and northeastern Georgia, with implications for magma history and evolution of the southern Appalachian orogen: American Journal of Science, v. 300, p. 142–172, doi: 10.2475/ajs.300.2.142.

Misra, K.C., and Conte, J.A., 1991, Amphibolites of the Ashe and Alligator Back Formations, North Carolina: Samples of Late Proterozoic–early Paleozoic oceanic crust: Geological Society of America Bulletin, v. 103, p. 737–750, doi: 10.1130/0016-7606(1991)103<0737:AOTAAA>2.3.CO;2.

Misra, K.C., and McSween, H.Y., Jr., 1984, Mafic rocks of the southern Appalachians: A review: American Journal of Science, v. 284, p. 294–318.

Moecher, D.P., and Miller, C.F., 2000, Precise age for peak granulite facies metamorphism and melting in the eastern Blue Ridge from SHRIMP U-Pb analysis of zircon: Geological Society of America Abstracts with Programs, v. 32, no. 2, p. 63.

Moecher, D.P., Samson, S.D., and Miller, C.F., 2004, Precise time and conditions of peak Taconian granulite facies metamorphism in the southern Appalachian orogen, USA, with implications for zircon behavior during crustal melting events: The Journal of Geology, v. 112, p. 289–304, doi: 10.1086/382760.

Mohr, D.W., 1975, Geologic Map and Mineral Resources of the Noland Creek Quadrangle, North Carolina: North Carolina Geological Survey GM-158 NE, scale 1:24,000.

Muehlberger, W.R., 1996, Tectonic Map of North America: American Association of Petroleum Geologists, scale 1:5,000,000.

Mueller, P., Heatherington, A., Wooden, J.L., Shuster, R.D., Nutman, A.P., and Williams, I.S., 1994, Precambrian zircons from the Florida basement: A Gondwanan connection: Geology, v. 22, p. 119–122, doi: 10.1130/0091-7613(1994)022<0119:PZFTFB>2.3.CO;2.

Mueller, P.A., Heatherington, A.L., Wooden, J.L., Steltenpohl, M.G., and Hanley, T.B., 2005, Age and provenance of Precambrian crust in the Southernmost Appalachians, *in* Steltenpohl, M.G., ed., Southernmost Appalachian Terranes, Alabama and Georgia: Southeastern Section, Geological Society of America Field Trip Guide: Tuscaloosa, Alabama Geological Society, p. 98–114.

Neathery, T.L., and Reynolds, J.W., 1973, Stratigraphy and metamorphism of the Wedowee Group: A reconnaissance: American Journal of Science, v. 273, p. 723–741.

Nelson, A.E., and Gillon, K.A., 1985, Stratigraphic Nomenclature in the Richard Russell and Helen Thrust Sheets, Georgia and North Carolina: U.S. Geological Survey Bulletin 1605-A, p. A59–A62.

Nelson, A.E., Horton, J.W., Jr., and Clarke, J.W., 1998, Geologic map of the Greenville 1° x 2° Quadrangle, South Carolina, Georgia, and North Carolina: U.S. Geological Survey Map I-2175, scale 1:250,000.

Noel, J.R., Spariosu, D.J., and Dallmeyer, R.D., 1988, Paleomagnetism and $^{40}Ar/^{39}Ar$ ages from the Carolina slate belt, Albemarle, North Carolina: Implications for terrane amalgamation with North America: Geology, v. 16, p. 64–68, doi: 10.1130/0091-7613(1988)016<0064:PAAAAF>2.3.CO;2.

Osborne, W.E., Szabo, M.W., Neathery, T.L., and Copeland, C.W., Jr., compilers, 1988, Geologic Map of Alabama, Northeast Sheet: Geological Survey of Alabama Special Map 220, scale 1:250,000.

Ownby, S.E., Miller, C.F., Berquist, P.J., Carrigan, C.W., Wooden, J.L., and Fullagar, P.D., 2004, U-Pb geochronology and geochemistry of a portion of the Mars Hill terrane, North Carolina-Tennessee: Constraints on origin, history, and tectonic assembly, *in* Tollo, R.P., Corriveau, L., McLelland, J., and Bartholomew, M.J., eds., Proterozoic Evolution of the Grenville Orogen in North America: Boulder, Colorado, Geological Society of America Memoir 197, p. 609–632.

Phinney, R.A., 1986, A seismic cross section of the New England Appalachians: The orogen exposed, *in* Barazangi, M., and Brown, L.D., eds., Proceedings of International Symposium on the Continental Crust: Results from Reflection Seismology: American Geophysical Union Geodynamics Series Monograph 14, p. 157–172.

Pojeta, J., Jr., Kriz, J., and Berdan, J.M., 1976, Silurian–Devonian pelecypods and Paleozoic stratigraphy of subsurface rocks in Florida and Georgia and related Silurian pelecypods from Bolivia and Turkey: U.S. Geological Survey Professional Paper 879, p. 1–32.

Quinlan, G.M., and Beaumont, C., 1984, Appalachian thrusting, lithospheric flexure, and the Paleozoic stratigraphy of the eastern interior of North America: Canadian Journal of Earth Sciences, v. 21, p. 973–996.

Rankin, D.W., 1993, The Volcanogenic Mount Rogers Formation and the Overlying Glaciogenic Konnarock Formation—Two Late Proterozoic Units in Southwestern Virginia: U.S. Geological Survey Bulletin 2029, 26 p.

Rankin, D.W., Espenshade, G.J., and Neuman, R.B., 1972, Geologic Map of the West Half of the Winston-Salem Quadrangle, North Carolina, Virginia, and Tennessee: U.S. Geological Survey Map I-907-A, scale 1:250,000.

Rankin, D.W., Drake, A.A., Jr., Glover, L., III, Goldsmith, R., and Hall, L., M., Murray, D.P., Ratcliffe, N.M., Read, J.F., Secor, D.T., Jr., and Stanley, R.S., 1989, Pre-orogenic terranes, Chapter 2, *in* Hatcher, R.D., Jr., Thomas, W.A., and Viele, G.W., eds., The Appalachian Ouachita Orogen in the United States: Boulder, Colorado, Geological Society of America, The Geology of North America, v. F-2, p. 7–100.

Ranson, W.A., Williams, I.S., and Garihan, J.M., 1999, SHRIMP zircon U-Pb ages of granitoids from the Inner Piedmont of South Carolina: Evidence for Ordovician magmatism involving mid to Late Proterozoic crust: Geological Society of America Abstracts with Programs, v. 31, no. 7, p. A-167.

Rast, N., and Skehan, J.W., 1983, The evolution of the Avalonian plate: Tectonophysics, v. 100, p. 257–286, doi: 10.1016/0040-1951(83)90191-9.

Raymond, L.A., and Johnson, P.A., 1994, The Mars Hill terrane: An enigmatic southern Appalachian terrane: Geological Society of America Abstracts with Programs, v. 26, no. 4, p. 59.

Repetski, J.W., Hatcher, R.D., Jr., Southworth, C.S., and Thigpen, J.R., 2006, Conodonts from the Walden Creek Group, southeastern Tennessee: Geological Society of America Abstracts with Programs, v. 38, no. 3, p. 66.

Robinson, P., Tucker, R.D., Bradley, D., Berry, H.N., IV, and Osberg, P.H., 1998, Paleozoic orogens in New England, USA: Geologiska Föreningens Stockholm Forhandlingar, v. 120, p. 119–148.

Rodgers, J., 1953, Geologic Map of East Tennessee with Explanatory Text, Part II: Tennessee Division of Geology Bulletin 58, 168 p.

Rodgers, J., 1970, The Tectonics of the Appalachians: New York, Wiley-Interscience, 271 p.

Roper, P.J., and Dunn, D.E., 1973, Superposed deformation and polymetamorphism, Brevard zone, South Carolina: Geological Society of America Bulletin, v. 84, p. 3373–3386, doi: 10.1130/0016-7606(1973)84<3373:SDAPBZ>2.0.CO;2.

Ross, R.J., Jr., et al., 1982, The Ordovician System in the United States: Correlation chart and explanatory notes: International Union of Geological Sciences Publication 12, 73 p. and 3 oversize charts.

Ryan, J.G., Yurkovich, S., Peterson, V., Burr, J., and Kruse, S., 2005, Geology and petrogenesis of mafic and ultramafic rocks of the Willets-Addie area, central Blue Ridge, North Carolina, *in* Hatcher, R.D., Jr., and Merschat, A.J., eds., Blue Ridge Geology Geotraverse East of the Great Smoky National Park, Western North Carolina: Knoxville, Tennessee, North Carolina Geological Survey, Carolina Geological Society Annual Field Trip Guidebook, p. 91–98.

Salisbury, J.W., 1961, Geology and Mineral Resources of the Northwest Quarter of the Cohutta Mountain Quadrangle: Georgia Geological Survey Bulletin 11, 61 p.

Samson, S.D., and Secor, D., 2000, New U-Pb geochronological evidence for a Silurian magmatic event in central South Carolina: Geological Society of America Abstracts with Programs, v. 32, no. 2, p. A71.

Samson, S.D., Hibbard, J.P., and Wortman, G.L., 1995, Nd isotopic evidence for juvenile crust in the Carolina terrane, southern Appalachians: Contributions to Mineralogy and Petrology, v. 121, p. 171–184, doi: 10.1007/s004100050097.

Sears, J.W., and Price, R.A., 2003, Tightening the Siberian connection to western Laurentia: Geological Society of America Bulletin, v. 115, p. 943–953, doi: 10.1130/B25229.1.

Secor, D.T., and Snoke, A.W., 2004, Trans-Appalachians Internides Geotraverse: Day 5, Carolina exotic terrane and Alleghanian metamorphic core, eastern Piedmont, South Carolina, *in* Merschat, A.J., and Hatcher, R.D., Jr., eds., Trans Appalachians Internides Geotraverse: Knoxville, University of Tennessee, Oak Ridge, Tennessee, 17th International Basement Tectonics Conference Field Trip Guide, p. 85–111.

Secor, D.T., Jr., Samson, S.L., Snoke, A.W., and Palmer, A.R., 1983, Confirmation of the Carolina slate belt as an exotic terrane: Science, v. 221, p. 649–651, doi: 10.1126/science.221.4611.649.

Secor, D.T., Jr., Snoke, A.W., Bramlett, K.W., Costello, O.P., and Kimbrell, O.P., 1986, Character of the Alleghanian orogeny in the southern Appalachians: Part I. Alleghanian deformation in the eastern Piedmont of South Carolina: Geological Society of America Bulletin, v. 97, p. 1319–1328, doi: 10.1130/0016-7606(1986)97<1319:COTAOI>2.0.CO;2.

Settles, D.J., 2002, Defining the Hayesville-Soque River and Allatoona Faults and an Ordovician Arc Assemblage within the Central Blue Ridge Northwest of Dahlonega, Georgia [M.S. thesis]: Knoxville, University of Tennessee, 148 p.

Sims, P.K., Anderson, J.L., Bauer, R.L., Chandler, V.W., Hanson, G.N., Kallikoski, J., Morey, G.B., Mudrey, M.G., Jr., Ojakangas, R.W., Peterman, Z.E., Schulz, K.J., Shirey, S.B., Smith, E.I., and Southwick, D.L., Van Schmus, W.R., and Weiblen, P.W., 1993, Lake Superior region and Trans–Hudson orogen, *in* Reed, J.C., Jr., Bickford, M.E., Houston, R.S., Link, P.K., Rankin, D.W., Sims, P.K., and Van Schmus, W.R., eds., Precambrian: Conterminous U.S.: Boulder, Colorado, Geological Society of America, The Geology of North America, v. C-2, p. 11–120.

Sinha, A.K., and McLelland, J., 1999, Lead isotope mapping of crustal reservoirs within the Grenville superterrane: II. Adirondack massif, New York, *in* Sinha, A.K., ed., Basement Tectonics 13: Dordrecht, Netherlands, Kluwer Academic Publishers, p. 297–312.

Sinha, A.K., and Zietz, I., 1982, Geophysical and geochemical evidence for a Hercynian magmatic arc, Maryland to Georgia: Geology, v. 10, p. 593–596, doi: 10.1130/0091-7613(1982)10<593:GAGEFA>2.0.CO;2.

Sloss, L.L., 1963, Sequences in the cratonic interior of North America: Geological Society of America Bulletin, v. 74, p. 93–114, doi: 10.1130/0016-7606(1963)74[93:SITCIO]2.0.CO;2.

Sloss, L.L., 1988, Tectonic evolution of the craton in Phanerozoic time, *in* Sloss, L.L., ed., Sedimentary cover—North American craton, U.S.: Boulder, Colorado, Geological Society of America, The Geology of North America, v. D-2, p. 25–51.

Society of Exploration Geophysicists, 1982, Gravity anomaly map of the United States: Tulsa, Oklahoma, Society of Exploration Geophysicists, scale 1:2,500,000.

Southworth, S., and Aleinikoff, J.N., 2007, Time versus tectonic significance: The disconformity at the base of the Chillhowee Group and the Neoproterozoic-Cambrian boundary: Geological Society of America Abstracts with Programs, v. 39, no. 2, p. 9.

Spears, D.B., Owens, B.E., and Bailey, C.M., 2004, The Goochland–Chopawamsic terrane boundary, Central Virginia Piedmont, *in* Southworth, S., and Burton, W., eds., Geology of the National Capitol region—Field trip guidebook: U.S. Geological Survey Circular 1264, p. 223–245.

Spell, T.L., and Norrell, G.T., 1990, The Ropes Creek assemblage: Petrology, geochemistry, and tectonic setting of an ophiolitic thrust sheet in the southern Appalachians: American Journal of Science, v. 290, p. 811–842.

Stahr, D.W., Miller, C.F., Hatcher, R.D., Jr., Wooden, J., and Fisher, C.M., 2005, Evidence for high-temperature ductile Alleghanian deformation in the eastern Blue Ridge: Implications of new structural, petrologic, and geochronologic data from southwestern North Carolina: Geological Society of America Abstracts with Programs, v. 37, no. 7, p. 72.

Steltenpohl, M.G., ed., 2005, Southernmost Appalachian Terranes, Alabama and Georgia, Southeastern Section Geological Society of America Field Trip Guide: Boulder, Colorado, Alabama Geological Society, 162 p.

Steltenpohl, M.G., Goldberg, S.A., Hanley, T.B., and Kunk, M.J., 1992, Alleghanian development of the Goat Rock fault zone, southernmost Appalachians: Temporal compatibility with the master décollement: Geology, v. 20, p. 845–848, doi: 10.1130/0091-7613(1992)020<0845:ADOTGR>2.3.CO;2.

Steltenpohl, M.G., Heatherington, A., Mueller, P., and Wooden, J.L., 2004, Pre-Appalachian tectonic evolution of the Pine Mountain window in the southernmost Appalachians, Alabama and Georgia, *in* Tollo, R.P., Corriveau, L., McLelland, J., and Bartholomew, M.J., eds., Proterozoic Evolution of the Grenville Orogen in North America: Boulder, Colorado, Geological Society of America Memoir 197, p. 633–646.

Stose, G.W., 1946, The Taconic sequence in Pennsylvania: American Journal of Science, v. 244, p. 665–696.

Student, J.J., and Sinha, A.K., 1992, Carboniferous U-Pb ages of zircons from the Box Ankle and Ocmulgee faults, central Georgia: Implications for accretionary models: Geological Society of America Abstracts with Programs, v. 24, no. 2, p. 69.

Swanson, S.E., Raymond, L.A., Warner, R.D., Ryan, J.G., Yurkovich, S.P., and Peterson, V.L., 2005, Petrotectonics of mafic and ultramafic rocks in Blue Ridge terranes of western North Carolina and northern Georgia, *in* Hatcher, R.D., Jr., and Merschat, A.J., eds., Blue Ridge Geology Geotraverse East of the Great Smoky Mountains National Park, Western North Carolina: Knoxville, Tennessee, North Carolina Geological Survey, Carolina Geological Society Annual Field Trip Guidebook, p. 73–90.

Thomas, C.W., 2001, Origins of Mafic-Ultramafic Complexes of the Eastern Blue Ridge Province, Southern Appalachians: Geochronological and Geochemical Constraints [M.S. thesis]: Nashville, Tennessee, Vanderbilt University, 154 p.

Thomas, W.A., 2006, Tectonic inheritance at a continental margin: GSA Today, v. 16, no. 2, p. 4–11, doi: 10.1130/1052-5173(2006)016[4:TIAACM]2.0.CO;2.

Thomas, W.A., Becker, T.P., Samson, S.D., and Hamilton, M.A., 2004, Detrital zircon evidence of a recycled orogenic foreland provenance for Alleghanian clastic-wedge sandstones: The Journal of Geology, v. 112, p. 23–37, doi: 10.1086/379690.

Tohver, E., Bettencourt, J.S., Tosdal, R., Mezger, K., Leite, W.B., and Payolla, B.L., 2004, Terrane transfer during the Grenville orogeny: tracing the Amazonian ancestry of southern Appalachian basement through Pb and Nd isotopes: Earth and Planetary Science Letters, v. 228, p. 161–176, doi: 10.1016/j.epsl.2004.09.029.

Tollo, R.P., Corriveau, L., McClelland, J.M., and Bartholomew, M.J., 2004a, Introduction, *in* Tollo, R.P., Corriveau, L., McLelland, J., and Bartholomew, M J., eds., Proterozoic Evolution of the Grenville Orogen in North America: Boulder, Colorado, Geological Society of America Memoir 197, p. 1–18.

Tollo, R.P., Aleinikoff, J.N., Borduas, E.A., Hackley, P.C., and Fanning, C.M., 2004b, Petrologic and geochronologic evolution of the Grenville orogen, northern Blue Ridge province, Virginia, *in* Tollo, R.P., Corriveau, L., McLelland, J., and Bartholomew, M.J., eds., Proterozoic Evolution of the Grenville Orogen in North America: Boulder, Colorado, Geological Society of America Memoir 197, p. 647–677.

Tollo, R.P., Aleinikoff, J.N., Bartholomew, M.J., and Rankin, D.W., 2004c, Neoproterozoic A-type granitoids of the central and southern Appalachians: intraplate magmatism associated with episodic rifting of the Rodinian supercontinent: Precambrian Research, v. 128, p. 3–38, doi: 10.1016/j.precamres.2003.08.007.

Trujillo, J., and Sinha, A.K., 2004, Extensional tectonics in collisional orogens: A case study of early Silurian extension in the central Appalachians associated with the Taconic orogeny: Geological Society of America Abstracts with Programs, v. 36, no. 2, p. 75.

Trupe, C.H., Stewart, K.G., Adams, M.G., Waters, C.L., Miller, B.V., and Hewitt, L.K., 2003, The Burnsville fault: Evidence for the timing and kinematics of southern Appalachian Acadian dextral transform tectonics: Geological Society of America Bulletin, v. 115, p. 1365–1376, doi: 10.1130/B25256.1.

Tull, J.F., 1978, Structural development of the Alabama Piedmont northwest of the Brevard zone: American Journal of Science, v. 278, p. 442–460.

Tull, J.F., 2002, Southeastern margin of the middle Paleozoic shelf, southwesternmost Appalachians: Regional stability bracketed by Acadian and Alleghanian tectonism: Geological Society of America Bulletin, v. 114, p. 643–655, doi: 10.1130/0016-7606(2002)114<0643:SMOTMP>2.0.CO;2.

Tull, J.F., 2007, this volume, Rift-margin architecture, cover stratigraphy and structure of basement culminations, frontal Appalachian Blue Ridge, Georgia, *in* Hatcher, R.D., Jr., Carlson, M.P., McBride, J.H., and Martínez-Catalán, J.R., eds., The 4-D Framework of Continental Crust: Geological Society of America Memoir 200, doi: 10.1130/2007.1200(28).

Tull, J.F., and Groszos, M.S., 1988, Murphy belt: Stratigraphic complexities and regional correlations, *in* Fritz, W.J., and La Tour, T.E., eds., Geology of the Murphy belt and related rocks, Georgia and North Carolina: Georgia Geological Society Guidebook, v. 8, p. 35–74.

Tull, J.F., and Groszos, M.S., 1990, Nested Paleozoic "successor" basins in the southern Appalachian Blue Ridge: Geology, v. 18, p. 1046–1049, doi: 10.1130/0091-7613(1990)018<1046:NPSBIT>2.3.CO;2.

Tull, J.F., and Holm, C.S., 2005, Structural evolution of a major Appalachian salient-recess junction: Consequences of oblique collisional convergence across a continental margin transform fault: Geological Society of America Bulletin, v. 117, p. 482–499, doi: 10.1130/B25578.1.

Tull, J.F., and Stow, S.H., 1980, The Hillabee Greenstone: A mafic volcanic complex in the Appalachian Piedmont of Alabama: Geological Society of America Bulletin, v. 91, p. 27–36, doi: 10: 10.1130/0016-7606(1980)91<27:THGAMV>2.0.CO;2.

Tull, J.F., Harris, A.J., Repetski, J.E., Mckinney, F.K., Garrett, C.B., and Bearce, D.N., 1988, New paleontologic evidence constraining the age and paleotectonic setting of the Talladega slate belt, southern Appalachians: Geological Society of America Bulletin, v. 100, p. 1291–1299, doi: 10.1130/0016-7606(1988)100<1291:NPECTA>2.3.CO;2.

Tull, J.F., Ausich, W.I., Groszos, M.S., and Thompson, T.W., 1993, Appalachian Blue Ridge cover sequence ranges at least into the Ordovician: Geology, v. 21, p. 215–218, doi: 10.1130/0091-7613(1993)021<0215:ABRCSR>2.3.CO;2.

Tull, J.F., Kish, S.A., Groszos, M.S., and Campbell, S.K., 1998, Laurentian magmatism of the southern Blue Ridge: Post-Iapetan rifting: Geological Society of America Bulletin, v. 110, p. 1281–1303, doi: 10.1130/0016-7606(1998)110<1281:LMOTSA>2.3.CO;2.

U.S. Geological Survey and American Association of Petroleum Geologists, 1961, Tectonic map of the United States: Tulsa, Oklahoma, U.S. Geological Survey and American Association of Petroleum Geologists, scale 1:2,500,000.

van der Voo, R., 1988, Paleozoic paleogeography of North America, Gondwana, and intervening displaced terranes: Comparisons of paleomagnetism with paleoclimatology and biogeographical patterns: Geological Society of America Bulletin, v. 100, p. 311–324, doi: 10.1130/0016-7606(1988)100<0311:PPONAG>2.3.CO;2.

van Staal, C.R., McNiocall, V., Valverde-Vaquero, P., Barr, S.M., Fyffe, L.R., and Reusch, D.N., 2004, Ganderia, Avalonia, and the Salinic and Acadian orogenies: Geological Society of America Abstracts with Programs, v. 36, no. 2, p. 128–129.

van Staal, C.R., Whalen, J.B., McNicoll, V., Pehrsson, S., Lissenberg, C.J., Zagorevski, A., van Breemen, O., and Jenner, G.A., 2007 this volume, The Notre Dame arc and the Taconic orogeny in Newfoundland, *in* Hatcher, R.D., Jr., Carlson, M.P., McBride, J.H., and Martínez-Catalán, J. R., eds., The 4-D Framework of Continental Crust: Geological Society of America Memoir 200, doi: 10.1130/2007.1200(26).

Vick, H.K., Channell, J.E.T., and Opdyke, N.D., 1987, Ordovician docking of the Carolina slate belt: Paleomagnetic data: Tectonics, v. 6, p. 573–584.

Vinson, S., 1999, Ion Probe Geochronology of Granitoid Gneisses of the Inner Piedmont, North Carolina and South Carolina [M.S. thesis]: Nashville, Tennessee, Vanderbilt University, 84 p.

Weaver, P.G., Tacker, R.C., McMenamin, M.A.S., and Webb, R.A., 2006, Ediacaran body fossils of South-Central North Carolina: Preliminary report, *in* Bradley, P.J., and Clark, T.W., eds., The Geology of the Chapel Hill, Hillsborough, and Efland 7.5-Minute Quadrangles, Orange and Durham Counties, Carolina Terrane, North Carolina: Raleigh, North Carolina Geological Survey, Carolina Geological Society Field Trip Guidebook, p. 34–41.

West, T.E., Jr., 1998, Structural analysis of the Carolina-Inner Piedmont terrane boundary: Implications for the age and kinematics of the central Piedmont suture, a terrane boundary that records Paleozoic Laurentia–Gondwana interactions: Tectonics, v. 17, p. 379–394, doi: 10.1029/98TC01081.

Wickham, J.S., 1972, Structural history of a portion of the Blue Ridge, northern Virginia: Geological Society of America Bulletin, v. 83, p. 723–760, doi: 10.1130/0016-7606(1972)83[723:SHOAPO]2.0.CO;2.

Williams, H., 1978, Tectonic-Lithofacies Map of the Appalachian Orogen: Memorial University of Newfoundland, map 1, scale 1:1,000,000.

Williams, H., 1979, The Appalachian orogen in Canada: Canadian Journal of Earth Sciences, v. 16, p. 792–807.

Williams, H., and Hatcher, R.D., Jr., 1982, Suspect terranes and accretionary history of the Appalachian orogen: Geology, v. 10, p. 530–536, doi: 10.1130/0091-7613(1982)10<530:STAAHO>2.0.CO;2.

Williams, H., and Hatcher, R.D., Jr., 1983, Appalachian suspect terranes, *in* Hatcher, R.D., Jr., Williams, H., and Zietz, I., eds., Contributions to the Tectonics and Geophysics of Mountain Chains: Geological Society of America Memoir 158, p. 33–53.

Wilson, C.G., 2006, Origin and Tectonic Evolution of the Southern Appalachian Neoacadian Crystalline Core: Evidence from the Geology of the Gilreath 7.5-Minute Quadrangle, North Carolina [M.S. thesis]: Knoxville, University of Tennessee, 219 p.

Wilson, C.W., Jr., Kirby, J.P., and Miller, R.A., 1964, Geologic Map of the Columbia Quadrangle, Tennessee: Tennessee Division of Geology, GM 57-SE, scale 1:24,000.

Wise, D.U., 1970, Multiple deformation, geosynclinal transitions and the Martic problem in Pennsylvania, *in* Fisher, G.W., Pettijohn, F.J., Reed, J.C., Jr. and Weaver, K.N., eds., Studies of Appalachian Geology: Central and Southern: New York, Wiley-Interscience, p. 317–334.

Wise, D.U., and Ganis, G.R., 2006, Taconian nappe structures (mostly) of Lancaster and Dauphin Counties, Pennsylvania, *in* de Wet, A., ed., Four Field Trips in Central Pennsylvania: Lancaster, Pennsylvania, Geological Society of America Northeastern Section Field Trip Guidebook, p. 27–65.

Manuscript Accepted by the Society 1 February 2007

Printed in the U.S.A.

Index